# R.C.C. DESIGNS
## (REINFORCED CONCRETE STRUCTURE)

# REINFORCED CONCRETE STRUCTURE
## (R.C.C. Designs)

(As Per IS : 456–2000)

*By*

### Dr. B.C. PUNMIA
*Formerly,*
Professor and Head, Deptt. of Civil Engineering, &
Dean, Faculty of Engineering
M.B.M. Engineering College,
**Jodhpur**

### ELEVENTH EDITION

(THOROUGHLY REVISED AND ENLARGED INCORPORATING
WORKING STRESS AND LIMIT STATE DESIGN METHODS)

# LAXMI PUBLICATIONS (P) LTD
(An ISO 9001:2015 Company)

BENGALURU • CHENNAI • GUWAHATI • HYDERABAD • JALANDHAR
KOCHI • KOLKATA • LUCKNOW • MUMBAI • RANCHI
NEW DELHI

# REINFORCED CONCRETE STRUCTURE (R.C.C. DESIGNS)

Copyright © with Author & Publishers

All rights reserved including those of translation into other languages. In accordance with the Copyright (Amendment) Act, 2012, no part of this publication may be reproduced, stored in a retrieval system, translated into any other language or transmitted in any form or by any means, electronic, mechanical, photocopying, recording or otherwise. Any such act or scanning, uploading, and or electronic sharing of any part of this book without the permission of the publisher constitutes unlawful piracy and theft of the copyright holder's intellectual property. If you would like to use material from the book (other than for review purposes), prior written permission must be obtained from the publishers.

Printed and bound in India
*First Edition* : 1975
*Eleventh Edition* : 2022, *Edition* : 2023
ISBN : 978-81-318-0942-6

**Limits of Liability/Disclaimer of Warranty:** The publisher and the author make no representation or warranties with respect to the accuracy or completeness of the contents of this work and specifically disclaim all warranties. The advice, strategies, and activities contained herein may not be suitable for every situation. In performing activities adult supervision must be sought. Likewise, common sense and care are essential to the conduct of any and all activities, whether described in this book or otherwise. Neither the publisher nor the author shall be liable or assumes any responsibility for any injuries or damages arising herefrom. The fact that an organization or Website if referred to in this work as a citation and/or a potential source of further information does not mean that the author or the publisher endorses the information the organization or Website may provide or recommendations it may make. Further, readers must be aware that the Internet Websites listed in this work may have changed or disappeared between when this work was written and when it is read.

All trademarks, logos or any other mark such as Vibgyor, USP, Amanda, Golden Bells, Firewall Media, Mercury, Trinity, Laxmi appearing in this work are trademarks and intellectual property owned by or licensed to Laxmi Publications, its subsidiaries or affiliates. Notwithstanding this disclaimer, all other names and marks mentioned in this work are the trade names, trademarks or service marks of their respective owners.

**Revised by :**
**DR. K. JAGANNADHA RAO**
Professor, Deptt. of Civil Engineering and
Joint Director, Student Affairs and Progression
Chaitanya Bharathi Institute of Technology (A)
Gandipet, Hyderabad-500075

*Typeset at* : Excellent Graphics
*Printed at* : Ajit Printers, Delhi
C—00235/022/06

**Branches**

| | | |
|---|---|---|
| © | Bengaluru | 080-26 75 69 30 |
| © | Chennai | 044-24 34 47 26 |
| © | Guwahati | 0361-254 36 69 |
| © | Hyderabad | 040-27 55 53 83 |
| © | Jalandhar | 0181-222 12 72 |
| © | Kochi | 0484-405 13 03 |
| © | Kolkata | 033-40 04 77 79 |
| © | Lucknow | 0522-430 36 13 |
| © | Ranchi | 0651-224 24 64 |

PUBLISHED IN INDIA BY

**Laxmi Publications (P) Ltd.**
(An ISO 9001:2015 Company)
113, GOLDEN HOUSE, GURUDWARA ROAD, DARYAGANJ,
NEW DELHI - 110002, INDIA
*Telephone* : 91-11-4353 2500, 4353 2501
www.laxmipublications.com    info@laxmipublications.com

# Preface

Reinforced Concrete occupies a leading position in modern construction. Proper reinforced concrete construction depends upon thorough understanding of the action of the structure and upon the knowledge of characteristics and limitations of those materials that are used in its construction. The personal element — the care with which work is excecuted in the field — is also of major importance in concrete construction. The textbook, in its two volumes, presents modern methods of design for both ordinary and prestressed concrete structures. Volume I has 19 chapters and deals with the design of more common and elementary structures. Volume II incorporates the design principles of advanced structures, including prestressed concrete structures.

In the present volume, the subject matter has been introduced gradually, beginning with properties and characteristics of the materials used in the reinforced concrete. Chapters 2, 3, 4 and 5 deal with the basic design principles of various types of beams and slabs while chapter 6 gives their detailed designs. In chapter 7, the designs of various types of stairs, including cinema balcony, are given. Chapters 8, 9, 10 and 11 present the detailed design procedures for various types of roof slabs. Axially loaded columns are given in chapter 12 while problems of combined direct and bending stresses are discussed in chapter 13. Chapters 14 and 15 deal with reinforced concrete foundations and reinforced concrete retaining walls respectively. Practically all the common types of foundations and retaining walls have been included. Reinforced concrete domes are given in chapter 16. The designs of water tanks have been divided into two parts: (*i*) tanks resting on ground surface, and (*ii*) underground and overhead tanks. The first part has been given in chapter 17. The underground and overhead tanks have been given in Volume II of the book. Chapter 18 describes the formwork of various concrete structures. The tests for cement and concrete have been described in detail in chapter 19.

Each chapter begins with clear statements of pertinent definitions, design principles and theories, and the design procedures. The basic principles are supplemented with numerous design examples and illustrations, along with the detailed drawings. The book is designed for students of civil engineering discipline. The designing and construction engineers will also find it a useful guide.

Various designs in this volume are based on the latest Indian Standards. The tables or curves giving permissible stresses and/or design coeffcients, as well as the basic rules for design, taken from the various Indian Standards are gratefully acknowledged. In spite of every care taken to check the numerical work, some errors may remain, and I shall be obliged for any intimation of these the readers may discover.

*Jodhpur*
*1-4-75*

**B.C. PUNMIA**

# Preface to the Eleventh Edition

'Design of Reinforced Concrete Structures' authored by Dr B.C. Punmia is a useful comprehensive textbook for undergraduate students of Civil Engineering and post-graduate students of Structural Engineering as well. This book includes chapters on the design of simple elements of a building and advanced structures, based on working stress and limit state design philosophies, besides theory on materials. Further, the book consists of clear diagrams, ample illustrations, and unsolved questions for better understanding and practice. This book proves to be a valuable possession for civil engineers preparing for competitive examinations like GATE, IES and other public service examinations.

This book has been thoroughly revised as per the latest amendments of IS 456 and other relevant BIS codes. Solutions to numerous questions appeared in the question papers of various universities and boards of Technical Education of the country have been included in the current edition of the book.

As the detailed drawing is an essential requirement for construction at the site, this book provides the necessary guidance for the preparation of precise structural drawings with an emphasis on clarity of drawings, dimensions, adoption of scale and use of symbols. Another attraction of this book is the colourful drawings that distinguish the reinforcement from the matrix and help the students understand the subject better.

I am confident that this book will be of immense use to students of civil engineering and architecture disciplines, academicians, practising professionals and field engineers of the construction industry.

**K. JAGANNADHA RAO**

# Contents

| S. No. | Chapters | Page No. |
|---|---|---|

1. **INTRODUCTION** ............................................................................................................. 1–28
    - 1.1. Cement Concrete ................................................................................................. 1
    - 1.2. Classification and Composition of Cement ......................................................... 1
    - 1.3. Specifications for Portland Cement ..................................................................... 6
    - 1.4. Aggregates ........................................................................................................... 7
    - 1.5. Water ................................................................................................................... 10
    - 1.6. Measurement of Materials .................................................................................. 10
    - 1.7. Water Cement Ratio ............................................................................................ 11
    - 1.8. Properties and Tests on Concrete ....................................................................... 12
    - 1.9. Durability of Concrete ......................................................................................... 12
    - 1.10. Workability of Concrete ...................................................................................... 12
    - 1.11. Methods of Proportioning Concrete Mixes ......................................................... 13
    - 1.12. Grades of Concrete and Characteristic Strength (IS : 456–2000) ................... 17
    - 1.13. Concrete Mix Proportioning (IS : 456–2000) .................................................... 18
    - 1.14. Formwork (IS : 456–2000) ................................................................................. 20
    - 1.15. Transporting, Placing, Compaction and Curing (IS : 456–2000) ..................... 21
    - 1.16. Permissible Stresses in Concrete (IS : 456–2000) ............................................ 23
    - 1.17. Steel Reinforcement ............................................................................................ 25

## PART I : WORKING STRESS METHOD

2. **THEORY OF REINFORCED BEAMS AND SLABS** ............................................................ 30–66
    - 2.1. Introduction ......................................................................................................... 30
    - 2.2. Modular Ratio ...................................................................................................... 30
    - 2.3. Singly Reinforced Beam ...................................................................................... 31
    - 2.4. Neutral Axis of Beam Section ............................................................................. 33
    - 2.5. Moment of Resistance ......................................................................................... 35
    - 2.6. Balanced, Under-Reinforced and Over-Reinforced Section .............................. 37
    - 2.7. Types of Problems in Singly-Reinforced Beams ................................................ 38
    - 2.8. Variation of $M_r$ with p ........................................................................................ 49
    - 2.9. Beam of Triangular Section ................................................................................ 54
    - 2.10. Beam of Trapezoidal Section .............................................................................. 56
    - 2.11. Slab Spanning in One Direction ......................................................................... 63

3. **SHEAR AND BOND** ......................................................................................................... 67–100
    - 3.1. Shear Stress in R.C. Beams ................................................................................ 67
    - 3.2. Effects of Shear : Diagonal Tension ................................................................... 69
    - 3.3. Regions of Cracks in Beams ............................................................................... 69

(viii)

|     |                                                                                              |      |
| --- | -------------------------------------------------------------------------------------------- | ---- |
| 3.4.  | Mechanisms of Shear Transfer in R.C. Beam Without Shear Reinforcement                      | 70   |
| 3.5.  | Shear Span                                                                                  | 71   |
| 3.6.  | Modes of Shear Failure                                                                      | 71   |
| 3.7.  | Factors Affecting Shear Resistance of a R.C. Member                                         | 73   |
| 3.8.  | Reinforcement for Diagonal Tension                                                          | 73   |
| 3.9.  | Types of Shear Reinforcement                                                                | 74   |
| 3.10. | Vertical Stirrups                                                                           | 75   |
| 3.11. | Inclined Bars                                                                               | 76   |
| 3.12. | Lattice Girder Effect                                                                       | 77   |
| 3.13. | Indian Standard Code Recommendations (*IS : 456–2000*)                                      | 78   |
| 3.14. | Critical Section for Design Shear : *IS : 456–2000*                                         | 82   |
| 3.15. | Enhanced Shear Strength of Sections Close to Supports (*IS : 456–2000*)                     | 88   |
| 3.16. | Bond, Anchorage and Development Length                                                      | 89   |
| 3.17. | Flexural Bond Stress                                                                        | 90   |
| 3.18. | Anchorage Bond Stress : Development Length                                                  | 91   |
| 3.19. | Hooks and Bends                                                                             | 91   |
| 3.20. | IS Code on Bond and Anchorage Requirements (*IS : 456–2000*)                                | 92   |
| 3.21. | Checking Development Lengths of Tension Bars                                                | 94   |
| 3.22. | Development Length Requirements at Simple Supports                                          | 94   |
| 3.23. | Development Length at Point of Inflexion                                                    | 96   |
| 3.24. | Conditions for Curtailment of Reinforcement                                                 | 97   |
| 3.25. | Reinforcement Splicing                                                                      | 99   |

## 4. TORSION ......................................................................................................... 101–106

|     |                                                                                              |      |
| --- | -------------------------------------------------------------------------------------------- | ---- |
| 4.1. | Introduction                                                                                | 101  |
| 4.2. | Torsional Resistance : Elastic Behaviour                                                    | 101  |
| 4.3. | Indian Standard Recommendations on Design for Torsion (*IS : 456–2000*)                     | 102  |

## 5. DOUBLY REINFORCED BEAMS .................................................................. 107–120

|     |                                                                                              |      |
| --- | -------------------------------------------------------------------------------------------- | ---- |
| 5.1. | Introduction                                                                                | 107  |
| 5.2. | Location of Neutral Axis                                                                    | 108  |
| 5.3. | Moment of Resistance                                                                        | 109  |
| 5.4. | Steel Beam Theory                                                                           | 110  |
| 5.5. | Types of Problems in Doubly Reinforced Beams                                                | 110  |
| 5.6. | Shear Stress, Bond Stress and Development Length                                            | 117  |

## 6. DESIGN OF T AND L-BEAMS ....................................................................... 121–146

|     |                                                                                              |      |
| --- | -------------------------------------------------------------------------------------------- | ---- |
| 6.1.  | Introduction                                                                               | 121  |
| 6.2.  | Dimensions of a T-beam                                                                     | 122  |
| 6.3.  | Position of Neutral Axis                                                                   | 123  |
| 6.4.  | Lever Arm and Moment of Resistance                                                         | 124  |
| 6.5.  | Moment of Resistance Taking Compression in RIB into Account                                | 125  |
| 6.6.  | Depth of Balanced Section of T-beam                                                        | 126  |
| 6.7.  | Economical Depth of T-beam                                                                 | 127  |
| 6.8.  | Shear, Bond and Development Length                                                         | 128  |
| 6.9.  | Types of Problems in T-beam                                                                | 129  |
| 6.10. | Doubly Reinforced T-beams                                                                  | 137  |
| 6.11. | L-beam                                                                                     | 142  |

## 7. DESIGN OF BEAMS AND SLABS .................................................................. 147–215
   7.1. Design of Beams ........................................................................................... 147
   7.2. Design of Lintel Beams ................................................................................. 159
   7.3. Design of Cantilever ..................................................................................... 165
   7.4. Design of One-way Slab ............................................................................... 167
   7.5. Design of Cantilever Chajja .......................................................................... 170
   7.6. Design of Lintel with Sunshade .................................................................... 172
   7.7. Design of Continuous Slab ........................................................................... 174
   7.8. Design of Doubly Reinforced Beam ............................................................. 179
   7.9. Design of T-beam .......................................................................................... 182
   7.10. Design of Cycle Stand Shade ....................................................................... 186
   7.11. Design of T-beam Roof ................................................................................. 190
   7.12. Design of Inverted T-beam Roof ................................................................... 194
   7.13. Design of Overhanging T-beam Roof ........................................................... 196
   7.14. Design of Cantilever Canopy ........................................................................ 201
   7.15. Design of L-beam : Design for Torsion ........................................................ 206

## 8. DESIGN OF STAIR CASES ............................................................................. 216–237
   8.1. Introduction ................................................................................................... 216
   8.2. General Notes on Design of Stairs ............................................................... 217
   8.3. Design of Stairs Spanning Horizontally ....................................................... 219
   8.4. Design of Dog-legged Stair ........................................................................... 220
   8.5. Design of Stairs with Quarter Space Landing ............................................. 221
   8.6. Design of Open Newel Stair with Quarter Space Landing ......................... 223
   8.7. Design of Staircase with Central Stringer Beam ......................................... 226
   8.8. Design of Cinema Balcony ............................................................................ 229

## 9. REINFORCED BRICK AND HOLLOW TILE ROOFS ..................................... 238–243
   9.1. Reinforced Brick Work .................................................................................. 238
   9.2. Design of Reinforced Bricks Slab ................................................................. 239
   9.3. Hollow Tile Roof ............................................................................................ 241
   9.4. Design of Hollow Tile Roof ........................................................................... 241

## 10. TWO-WAY SLABS ............................................................................................ 244–268
   10.1. Introduction ................................................................................................... 244
   10.2. Slab Simply Supported on the Four Edges, with Corners not Held Down and Carrying U.D.L. ................. 244
   10.3. Slab Simply Supported on the Four Edges with Corners Held Down and Carrying U.D.L. ....................... 248
   10.4. Slab with Edges Fixed *or* Continuous and Carrying U.D.L. ...................... 252
   10.5. Other Cases of Slabs ..................................................................................... 255
   10.6. Indian Standard Code Method (*IS : 456–2000*) ......................................... 261

## 11. CIRCULAR SLABS ........................................................................................... 269–283
   11.1. Introduction ................................................................................................... 269
   11.2. Slab Freely Supported at Edges and Carrying U.D.L. ................................. 270
   11.3. Slabs Fixed at Edges and Carrying U.D.L. .................................................. 270
   11.4. Slab Simply Supported at the Edges with Load *w* Uniformly Distributed Along the Circumference of a Concentric Circle ..................................................... 271
   11.5. Slab Simply Supported at Edges, with U.D.L. Inside a Concentric Circle ................. 272
   11.6. Slab Simply Supported at Edges, with a Central Hole and Carrying U.D.L. .............. 272

- 11.7. Slab Simply Supported at the Edges with a Central Hole and Carrying w Distributed along the Circumference of a Concentric Circle .................................................................. 273

## 12. FLAT SLABS .................................................................................................................. 284–302
- 12.1. Introduction ............................................................................................................ 284
- 12.2. Components of Flat Slab Construction .................................................................. 284
- 12.3. Indian Code Recommendations (IS : 456–2000) .................................................... 285
- 12.4. Direct Design Method ........................................................................................... 286
- 12.5. Equivalent Frame Method .................................................................................... 289
- 12.6. Shear in Flat Slab .................................................................................................. 291
- 12.7. Slab Reinforcement ............................................................................................... 292
- 12.8. Openings in Flat Slab ............................................................................................ 292

## 13. DESIGN OF AXIALLY LOADED COLUMNS ........................................................... 303–319
- 13.1. Introduction ............................................................................................................ 303
- 13.2. Short and Long (or Slender) Columns .................................................................. 304
- 13.3. Types of Columns ................................................................................................. 304
- 13.4. Load Carrying Capacity of Short Columns .......................................................... 305
- 13.5. Indian Standard Recommendations (IS : 456–2000) ............................................. 306
- 13.6. Design Procedure .................................................................................................. 311

## 14. COLUMNS WITH AXIAL LOAD AND MOMENTS ................................................ 320–349
- 14.1. Introduction ............................................................................................................ 320
- 14.2. Case 1: Compressive Load at Eccentricity Smaller than D/4 ............................... 320
- 14.3. Bending about Two Axes ...................................................................................... 322
- 14.4. Design of Columns Subjected to Combined Bending and Direct Stresses (IS: 456–2000) ........ 322
- 14.5. Case 2: Compressive Load at Large Eccentricity (e > 1.5 D) ............................... 331
- 14.6. Case 3: Compressive Load at Moderate Eccentricity [D/4 < e < 3 D/2] .............. 333
- 14.7. Circular Section Subjected to Eccentric Load ....................................................... 338
- 14.8. Case 4: Tensile Load at Small Eccentricity .......................................................... 342
- 14.9. Case 5: Tensile Load at Large Eccentricity .......................................................... 342
- 14.10. Case 6: Tensile Load at Moderate Eccentricity .................................................... 344
- 14.11. Sections of Irregular Shape ................................................................................... 346
- 14.12. Design Examples ................................................................................................... 346

## 15. DESIGN OF ISOLATED FOOTINGS ......................................................................... 350–385
- 15.1. Introduction ............................................................................................................ 350
- 15.2. Pressure Distribution Beneath Footings ................................................................ 351
- 15.3. Bearing Capacity of Soil and Settlement of Footings ........................................... 351
- 15.4. Indian Standard Code Recommendations for Design of Footings (IS: 456–2000) ..... 354
- 15.5. Design of Continuous Footings ............................................................................. 356
- 15.6. Isolated Footing of Uniform Depth ....................................................................... 358
- 15.7. Isolated Sloped Footing ......................................................................................... 361
- 15.8. Isolated Stepped Footing ....................................................................................... 363
- 15.9. Isolated Footing for Circular Columns .................................................................. 364
- 15.10. Isolated Footing Subjected to Eccentric Load ...................................................... 379

## 16. COMBINED FOOTINGS .............................................................................................. 386–431
- 16.1. Introduction ............................................................................................................ 386

|  |  |  |
|---|---|---|
| 16.2. | Combined Rectangular Footing | 386 |
| 16.3. | Combined Trapezoidal Footing | 389 |
| 16.4. | Strap Footing | 413 |
| 16.5. | Raft Footing | 420 |

## 17. PILE FOUNDATIONS ................................................................................. 432–442

|  |  |  |
|---|---|---|
| 17.1. | Types of Piles | 432 |
| 17.2. | Pile Driving | 433 |
| 17.3. | Load Carrying Capacity of Piles | 433 |
| 17.4. | Group Action in Pile | 435 |
| 17.5. | Structural Design of R.C. Pile | 436 |
| 17.6. | Design of Pile Cap | 437 |

## 18. DESIGNS OF RETAINING WALLS .............................................................. 443–487

|  |  |  |
|---|---|---|
| 18.1. | Introduction | 443 |
| 18.2. | Types of Retaining Walls | 443 |
| 18.3. | Active Earth Pressure: Rankine's Theory | 444 |
| 18.4. | Passive Earth Pressure | 446 |
| 18.5. | Stability of Cantilever Retaining Wall | 447 |
| 18.6. | Design Principles of Cantilever Retaining Wall | 449 |
| 18.7. | Design of Cantilever Retaining Wall with Horizontal Backfill | 451 |
| 18.8. | Design of Cantilever Retaining Wall with Horizontal Backfill and Traffic Load | 457 |
| 18.9. | Design of Cantilever Retaining Wall with Sloping Backfill | 464 |
| 18.10. | Design of Counterfort Retaining Wall | 470 |
| 18.11. | Back Anchoring of Retaining Wall | 472 |

# PART II : WATER TANKS

## 19. DOMES ...................................................................................................... 490–500

|  |  |  |
|---|---|---|
| 19.1. | Introduction | 490 |
| 19.2. | Nature of Stresses in Spherical Domes | 490 |
| 19.3. | Analysis of Spherical Domes | 491 |
| 19.4. | Stresses Due to Wind Load | 494 |
| 19.5. | Design of R.C. Domes | 494 |
| 19.6. | Conical Domes | 498 |

## 20. BEAMS CURVED IN PLAN ........................................................................ 501–531

|  |  |  |
|---|---|---|
| 20.1. | Introduction : Torsional Moments in Beams | 501 |
| 20.2. | Circular Beam Supported Symmetrically | 502 |
| 20.3. | Semicircular Beam Simply Supported on Three Equally Spaced Columns | 505 |
| 20.4. | Curved Beam Simply Supported at Ends and Continuous over Two Equally Spaced Intermediate Supports | 509 |
| 20.5. | Curved Beam Fixed at Ends | 510 |
| 20.6. | Semi-Circular Beam with Slab | 514 |
| 20.7. | Torsion Factor | 516 |
| 20.8. | Stresses Due to Torsion in Concrete Beams | 518 |
| 20.9. | Reinforcement Due to Torsion | 520 |
| 20.10. | Indian Standard Code for Design for Torsion (IS : 456–2000) | 523 |

## 21. WATER TANKS-I : SIMPLE CASES .................................................................. 532–545
- 21.1. Introduction .................................................................................................... 532
- 21.2. General Design Requirements According to Indian Standard Code of Practice (IS : 3370 Part II, 1965) ........................................................................ 533
- 21.3. Joints in Water Tanks .................................................................................... 535
- 21.4. Circular Tank with Flexible Joint Between Floor and Wall ........................ 536
- 21.5. Circular Tank with Rigid Joint Between Floor and Wall ............................. 539
- 21.6. IS Code Method for Circular Tanks ............................................................. 543

## 22. WATER TANKS-II : CIRCULAR AND INTZE TANKS ..................................... 546–626
- 22.1. Circular Tank with Rigid Joint between Floor and Wall ............................. 546
- 22.2. IS Code Methods and Other Methods for Cylindrical Tanks ..................... 551
- 22.3. Design of Flat Base Slab for Elevated Circular Tanks ................................. 568
- 22.4. Circular Tank with Domed Bottom and Roof .............................................. 575
- 22.5. Intze Tank ....................................................................................................... 587
- 22.6. Effects of Continuity ..................................................................................... 589
- 22.7. Design of Tank Supporting Towers .............................................................. 590
- 22.8. Design of Foundations .................................................................................. 596

## 23. WATER TANKS-III : RECTANGULAR TANKS ............................................... 627–642
- 23.1. Introduction .................................................................................................... 627
- 23.2. Approximate Method ..................................................................................... 627
- 23.3. Exact Method ................................................................................................. 632

## 24. WATER TANKS-IV : UNDERGROUND TANKS ............................................. 643–654
- 24.1. Introduction .................................................................................................... 643
- 24.2. Earth Pressure on Tank Walls ........................................................................ 643
- 24.3. Uplift Pressure on the Floor of the Tank ...................................................... 645
- 24.4. Design of Rectangular Tank .......................................................................... 645

# PART III : MISCELLANEOUS STRUCTURES

## 25. REINFORCED CONCRETE PIPES ................................................................... 656–666
- 25.1. Loads on Pipes ............................................................................................... 656
- 25.2. Stresses Due to Hydrostatic Pressure ........................................................... 657
- 25.3. Stresses Due to Self Weight .......................................................................... 657
- 25.4. Stresses Due to Weight of Water Inside ....................................................... 659
- 25.5. Stresses Due to Earthfill over Haunches ...................................................... 659
- 25.6. Stresses Due to Uniformly Distributed Load on Top .................................. 660
- 25.7. Stresses Due to Uniform Pressure from Sides ............................................. 660
- 25.8. Stresses Due to Triangularly Distributed Load ............................................ 660
- 25.9. Stresses Due to Point Load on Crown .......................................................... 661
- 25.10. Stresses Due to Over-Burden and External Loads ...................................... 661

## 26. BUNKERS AND SILOS ...................................................................................... 667–687
- 26.1. Introduction .................................................................................................... 667
- 26.2. Janssen's Theory ............................................................................................ 667
- 26.3. Airy's Theory ................................................................................................. 669
- 26.4. Bunkers ........................................................................................................... 672

(xiii)

|  |  |  |
|---|---|---|
| 26.5. | Hooper Bottom | 673 |
| 26.6. | Indian Standard on Design of Bins (*IS: 4995–1968*) | 675 |

## 27. CHIMNEYS .................................................................................................. 688–702
| 27.1. | Introduction | 688 |
|---|---|---|
| 27.2. | Wind Pressure | 688 |
| 27.3. | Stresses in Chimney Shaft Due to Self-Weight and Wind Loads | 689 |
| 27.4. | Stresses in Horizontal Reinforcement Due to Force Shear | 691 |
| 27.5. | Stresses Due to Temperature Difference | 692 |
| 27.6. | Combined Effect of Self Load, Wind and Temperature | 693 |
| 27.7. | Temperature Stresses in Horizontal Reinforcement | 695 |
| 27.8. | Design of R.C. Chimney | 696 |

## 28. PORTAL FRAMES ........................................................................................ 703–712
| 28.1. | Introduction | 703 |
|---|---|---|
| 28.2. | Analysis of Portal Frames | 703 |
| 28.3. | Design of Rectangular Portal Frame with Vertical Loads | 704 |
| 28.4. | Design of Hinge at the Base | 711 |

## 29. BUILDING FRAMES .................................................................................... 713–736
| 29.1. | Introduction | 713 |
|---|---|---|
| 29.2. | Substitute Frames | 713 |
| 29.3. | Analysis for Vertical Loads | 715 |
| 29.4. | Methods of Computing B.M. | 716 |
| 29.5. | Analysis of Frames Subjected to Horizontal Forces | 722 |
| 29.6. | Portal Method | 722 |
| 29.7. | Cantilever Method | 724 |
| 29.8. | Factor Method | 728 |

# PART IV : CONCRETE BRIDGES

## 30. AQUEDUCTS *AND* BOX CULVERTS ......................................................... 738–750
| 30.1. | Aqueducts and Syphon Aqueducts | 738 |
|---|---|---|
| 30.2. | Design of an Aqueduct | 738 |
| 30.3. | Box Culvert | 741 |
| 30.4. | Design of Box Culvert | 742 |

## 31. CONCRETE BRIDGES ................................................................................. 751–824
| 31.1. | Introduction: Various Types of Bridges | 751 |
|---|---|---|
| 31.2. | Selection of Type of Bridge and Economic Span Length | 752 |
| 31.3. | Types of Loads, Forces and Stresses | 753 |
| 31.4. | Live Load | 753 |
| 31.5. | Impact Effect | 757 |
| 31.6. | Wind Load | 758 |
| 31.7. | Longitudinal Forces | 759 |
| 31.8. | Lateral Loads | 759 |
| 31.9. | Centrifugal Force | 759 |
| 31.10. | Width of Roadway and Footway | 760 |
| 31.11. | General Design Requirements | 760 |

- 31.12. Solid Slab Bridges ... 762
- 31.13. Deck Girder Bridges ... 771
- 31.14. B.M. in Slab Supported on Four Edges ... 772
- 31.15. Distribution of Live Loads on Longitudinal Beams ... 779
- 31.16. Method of Distribution Coefficients ... 779
- 31.17. Courbon's Method ... 790
- 31.18. Design of a T-beam Bridge ... 792

# PART V : LIMIT STATE DESIGN

## 32. DESIGN CONCEPTS ... 826–829
- 32.1. Methods of Design ... 826
- 32.2. Safety and Serviceability Requirements (IS : 456–2000) ... 827
- 32.3. Characteristic and Design Values and Partial Safety Factors ... 827

## 33. SINGLY REINFORCED SECTIONS ... 830–843
- 33.1. Limit State of Collapse in Flexure ... 830
- 33.2. Stress Strain Relationship for Concrete ... 831
- 33.3. Stress-Strain Relationship for Steel ... 831
- 33.4. Stress Block Parameters ... 833
- 33.5. Design Stress Block Parameters (IS : 456–2000) ... 834
- 33.6. Singly Reinforced Rectangular Beams ... 835
- 33.7. Procedure for Finding Moment of Resistance ... 837
- 33.8. Design of Rectangular Beam Section ... 839

## 34. DOUBLY REINFORCED SECTIONS ... 844–853
- 34.1. Necessity ... 844
- 34.2. Stress Block and Neutral Axis (NA) ... 844
- 34.3. Types of Problems ... 846
- 34.4. Determination of Moment of Resistance ... 846
- 34.5. Design of a Doubly Reinforced Section ... 847

## 35. T AND L-BEAMS ... 854–881
- 35.1. Introduction ... 854
- 35.2. T-beam ... 855
- 35.3. L-beam ... 856
- 35.4. Stress Block and Neutral Axis ... 857
- 35.5. Moment of Resistance when $x_u \le D_f$ : (As per IS: 456–2000 Annexure-G) ... 857
- 35.6. Moment of Resistance when N.A. Falls in the web ... 858
- 35.7. IS Code Procedure for Finding Moment of Resistance (IS: 456–2000; Annexure-G) ... 861
- 35.8. Types of Problems ... 863
- 35.9. Design of T-beam ... 863
- 35.10. Analysis of Doubly Reinforced T-beams ... 870
- 35.11. Design of Doubly Reinforced T-beam ... 873

## 36. SHEAR, BOND AND TORSION IN BEAMS ... 882–891
- 36.1. Limit State of Collapse : Shear ... 882
- 36.2. Development Length ... 887

| | | |
|---|---|---|
| 36.3. | Development Length Requirements at Simple Supports | 888 |
| 36.4. | Torsion: Limit State of Collapse | 889 |

## 37. DESIGN OF RCC BEAMS AND SLABS ............................................................. 892–909

| | | |
|---|---|---|
| 37.1. | Design of RCC Beams | 892 |
| 37.2. | Design of Cantilever | 897 |
| 37.3. | Design of Doubly Reinforced Beam | 899 |
| 37.4. | Design of One Way Slab | 902 |
| 37.5. | Design of One Way Continuous Slab | 904 |
| 37.6. | Design of T-beam Roof | 909 |

## 38. AXIALLY LOADED COLUMNS ........................................................................... 910–919

| | | |
|---|---|---|
| 38.1. | Introduction | 910 |
| 38.2. | Limit State of Collapse : Compression (*As per Clause 39 of IS : 456–2000*) | 910 |
| 38.3. | Short Columns | 911 |
| 38.4. | Short Axially Loaded Members in Axial Compression | 911 |
| 38.5. | Short Axially Loaded Column with Minimum Eccentricity | 912 |
| 38.6. | Design Charts (*Sp* 16 Design Charts 24 to 26) | 913 |
| 38.7. | Compression Members with Helical Reinforcement | 913 |
| 38.8. | Design Specifications (*IS : 456–2000*) | 915 |

## 39. COLUMNS WITH UNIAXIAL AND BIAXIAL BENDING ....................................... 920–937

| | | |
|---|---|---|
| 39.1. | Introduction | 920 |
| 39.2. | Combined Axial Load and Uniaxial Bending | 920 |
| 39.3. | Construction of Interaction Curves for Column Design | 922 |
| 39.4. | Short Columns Subjected to Axial Load and Biaxial Bending | 934 |

## 40. DESIGN OF STAIR CASES ................................................................................ 938–945

| | | |
|---|---|---|
| 40.1. | General Notes on Design of Stairs | 938 |
| 40.2. | Design of Stairs Spanning Horizontally | 940 |
| 40.3. | Design of Dog-Legged Stair | 941 |
| 40.4. | Design of Stair with Quarter Space Landing | 943 |

## 41. TWO WAY SLABS ............................................................................................... 946–962

| | | |
|---|---|---|
| 41.1. | Introduction | 946 |
| 41.2. | Simply Supported Slab with Corners Free to Lift (*IS Code Method*) | 946 |
| 41.3. | Restrained Slabs (*IS Code Method*) | 949 |

## 42. CIRCULAR SLABS .............................................................................................. 963–968

| | | |
|---|---|---|
| 42.1. | Introduction | 963 |
| 42.2. | Slab Freely Supported at Edges and Carrying U.D.L. | 963 |
| 42.3. | Slabs Fixed at Edges and Carrying U.D.L. | 964 |
| 42.4. | Slab Simply Supported at the Edges with Load *W* Uniformly Distributed Along the Circumference of a Concentric Circle | 964 |
| 42.5. | Slab Simply Supported at Edges, with U.D.L. Inside a Concentric Circle | 964 |

## 43. YIELD LINE THEORY AND DESIGN OF SLABS ................................................ 969–986

| | | |
|---|---|---|
| 43.1. | Introduction | 969 |
| 43.2. | Yield Line Patterns | 969 |
| 43.3. | Moment Capacity Along an Yield Line | 971 |

|     |       |                                                                                          |
| --- | ----- | ---------------------------------------------------------------------------------------- |
|     | 43.4. | Ultimate Load on Slabs.................................................................. 972 |
|     | 43.5. | Analysis by Virtual Work Method...................................................... 972 |
|     | 43.6. | Analysis by Equilibrium Method....................................................... 979 |

## 44. FOUNDATIONS................................................................................. 987–994

    44.1. Indian Standard Code Recommendations for Design of Footings (*IS : 456–2000*)........ 987
    44.2. Isolated Footing of Uniform Depth........................................................ 989
    44.3. Isolated Sloped Footing...................................................................... 992

## PART VI : PRESTRESSED CONCRETE AND MISCELLANEOUS TOPICS

## 45. PRESTRESSED CONCRETE................................................................. 996–1079

    45.1. Introduction..................................................................................... 996
    45.2. Basic Concepts................................................................................. 997
    45.3. Classification and Types of Prestressing................................................ 1005
    45.4. Prestressing Systems : End Anchorages................................................ 1006
    45.5. Losses of Prestress........................................................................... 1013
    45.6. Computation of Elongation of Tendons................................................ 1018
    45.7. Properties of Materials...................................................................... 1024
    45.8. Merits and Demerits of Prestressed Concrete........................................ 1027
    45.9. Basic Assumptions............................................................................ 1028
    45.10. Analysis of Beams for Flexure........................................................... 1028
    45.11. Kern Distances and Efficiency of Section............................................. 1031
    45.12. Design of Sections for Flexure : Magnel's Method................................ 1034
    45.13. Rectangular Section......................................................................... 1039
    45.14. I-Section........................................................................................ 1041
    45.15. Alternative Design Procedure........................................................... 1050
    45.16. Shear and Diagonal Tension.............................................................. 1053
    45.17. Stresses at Anchorage...................................................................... 1055
    45.18. Indian Standard Code Recommendations (*IS : 1343–1980*).................... 1058
    45.19. Procedure for Limit State Design....................................................... 1069

## 46. SHRINKAGE AND CREEP.................................................................. 1080–1089

    46.1. Introduction.................................................................................... 1080
    46.2. Shrinkage of Concrete...................................................................... 1080
    46.3. Shrinkage Stresses in Symmetrically Reinforced Sections....................... 1082
    46.4. Shrinkage Stresses in Singly Reinforced Beams.................................... 1083
    46.5. Instantaneous and Repeated Loading on Concrete................................ 1084
    46.6. Sustained Loading : Creep................................................................ 1085
    46.7. Factors Affecting Creep.................................................................... 1086
    46.8. Effect of Creep on $E_c$ and $m$........................................................ 1087
    46.9. Effect of Shrinkage and Creep in Columns........................................... 1087
    46.10. Effect of Shrinkage and Creep in Beams............................................. 1088

## 47. FORMWORK..................................................................................... 1090–1099

    47.1. Introduction.................................................................................... 1090
    47.2. Indian Standard on Formwork (*IS : 456–2000*).................................... 1091

|  |  |  |
|---|---|---|
| 47.3. | Loads on Formwork | 1092 |
| 47.4. | Shuttering for Columns | 1093 |
| 47.5. | Shuttering for Beam and Slab Floor | 1095 |

## 48. TESTS ON CEMENT *AND* CEMENTS CONCRETE ............ 1100–1109

|  |  |  |
|---|---|---|
| 48.1. | Introduction | 1100 |
| 48.2. | Test on Fineness of Cement | 1100 |
| 48.3. | Test for Consistency of Cement Paste | 1102 |
| 48.4. | Test for Determination of Initial/Final Setting Times | 1103 |
| 48.5. | Test for Soundness of Cement | 1104 |
| 48.6. | Test for Determination of Compressive Strength of Cement | 1104 |
| 48.7. | Test for Tensile Strength of Cement | 1105 |
| 48.8. | Test for Workability | 1106 |
| 48.9. | Test for Compressive Strength | 1107 |
| 48.10. | Test for Flexural Strength | 1108 |

**APPENDIX–A**    B.M. and S.F. Coefficients .................. 1110

**APPENDIX–B**    Properties of Materials and Concrete .................. 1114

**APPENDIX–C**    Reinforcement .................. 1116

**APPENDIX–D**    Loadings .................. 1117

**INDEX** .................. 1121–1127

# Symbols

$A$ = Total area of section.
$A_b$ = Equivalent area of helical reinforcement.
$A_c$ = Area of compressive steel.
$A_e$ = Equivalent area of section.
$A_k$ = Area of concrete core.
$A_m$ = Area of steel *or* iron core.
$A_{sc}$ = Area of longitudinal reinforcement (comp.).
$A_{st}$ = Area of steel (tensile).
$A_l$ = Area of longitudinal torsional reinforcement.
$A_{sv}$ = Total cross-sectional area of stirrup legs *or* bent up bars within distance $s_v$.
$A_w$ = Area of web reinforcement.
$A_\Phi$ = Area of cross-section of one bar.
$a$ = Lever arm.
$a_c$ = Area of concrete.
$B$ = Flange width of T-beam.
$b$ = Width.
$b_r$ = Width of rib.
$C$ = Compressive force.
$c$ = Compressive stress in concrete.
$c'$ = Stress in concrete surrounding compressive steel.
$c_s$ = Permissible tensile stress in concrete.
$c_1$ = Compressive stress at the junction of flange and web.
$D$ = Depth.
$d$ = Effective depth.
$d_c$ = Cover to compressive steel.
$d_s$ = Depth of slab.
$d_t$ = Cover to tensile steel.
$e$ = Eccentricity.
 = Compressive steel depth factor (= $d_c/d$).
$F$ = Shear force.
$F_d$ = Design load.

(*xviii*)

| | | |
|---|---|---|
| $F_r$ | = | Radial shear force. |
| $f$ | = | Stress (in general). |
| $f_{ck}$ | = | Characteristic compressive stress. |
| $f_y$ | = | Characteristic strength of steel. |
| $H$ | = | Height. |
| $I$ | = | Moment of inertia. |
| $I_e$ | = | Equivalent moment of inertia of section. |
| $j$ | = | Lever arm factor. |
| $K_a$ | = | Coefficient of active earth pressure. |
| $K_p$ | = | Coefficient of passive earth pressure. |
| $k$ | = | Neutral axis depth factor ($n/d$). |
| $L$ | = | Length. |
| $L_d$ | = | Development length. |
| $l$ | = | Effective length of column ; Length ; Bond length. |
| $M$ | = | Bending moment ; moment. |
| $M_r$ | = | Moment of resistance ; Radial bending moment. |
| $M_t$ | = | Torsional moment. |
| $M_\theta$ | = | Circumferential bending moment. |
| $m$ | = | Modular ratio. |
| $n$ | = | Depth of neutral axis. |
| $n_c$ | = | Depth of critical neutral axis. |
| $\Sigma 0$ | = | Sum of perimeter of bars. |
| $P_a$ | = | Active earth pressure. |
| $P_p$ | = | Passive earth pressure. |
| $P_u$ | = | Ultimate Load. |
| $p$ | = | Percentage steel. |
| $p'$ | = | Reinforcement ratio ($A_{st}/bd$). |
| $p_a$ | = | Active earth pressure intensity. |
| $p_e$ | = | Net upward soil pressure. |
| $p_p$ | = | Passive earth pressure intensity. |
| $Q$ | = | Shear resistance. |
| $q$ | = | Shear stress (due to bending). |
| $q'$ | = | Shear stress due to torsion. |
| $R$ | = | Radius ; Resistance factor $\left(=\dfrac{1}{2}cjk\right)$. |
| $r$ | = | Radius ; cost ratio of steel and concrete ; $L/B$ ratio. |
| $s$ | = | Spacing of bars ; standard deviation. |

$s_a$ = Average bond stress.

$s_b$ = Local bond stress.

$T$ = Tensile force ; Thickness of wall ; Torsional moment.

$T_u$ = Torsional moment (limit state design).

$t$ = Tensile stress in steel.

$t_{c'}$ = Compressive stress in compressive steel.

$V_u$ = Shear force due to design load (limit state design).

$V_{us}$ = Strength of shear reinforcement (limit state design).

$W$ = Point load ; Total load.

$X$ = Co-ordinate.

$x_u$ = Depth of neutral axis (limit state design).

$Z$ = Distance : Co-ordinate.

$Y, y$ = Co-ordinate.

$Z_B, Z_L$ = Bending moment coefficients.

$\alpha$ = Inclination ; coefficient.

$\beta$ = Surcharge angle.

$\gamma$ = Unit weight of soil.

$\gamma'$ = Submerged unit weight of soil.

$\gamma_f$ = Partial safety factor appropriate to the loading.

$\gamma_m$ = Partial safety factor appropriate to the material.

$\sigma_{cc}$ = Permissible stress in concrete (direct comp).

$\sigma_{cc'}$ = Direct compressive stress in concrete.

$\sigma_{cbc}$ = Permissible compressive stress in concrete due to bending.

$\sigma_{cu}$ = Ultimate compressive stress in concrete cubes.

$\sigma_{sc}$ = Permissible compressive stress in bars.

$\sigma_{sh}$ = Permissible stress in helical reinforcement.

$\sigma_{sp}$ = Permissible punching shear stress.

$\sigma_{st}$ = Permissible tensile stress in reinforcement.

$\sigma_{sy}$ = Yield point compressive stress in steel.

$\mu$ = Coefficient of friction.

$\Phi$ = Diameter of bar ; angle of internal friction.

$\phi$ = angle.

$\tau_{bd}$ = Design bond stress.

$\tau_c$ = Shear stress in concrete.

$\tau_{cmax}$ = Max. shear stress in concrete with shear reinforcement.

$\tau_v$ = Nominal shear stress.

# INTRODUCTION

## 1.1. CEMENT CONCRETE

Cement concrete is a product obtained by mixing of cement, sand, aggregates and water in predetermined proportions. When these materials are mixed, they form a plastic mass which can be poured in suitable moulds, called *forms,* and set into hard solid mass called concrete. The chemical reaction of cement and water, in the mix, is relatively slow and requires time and favourable temperature for its completion. This time, known as *setting time* may be divided into three phases. The first phase, designated as time of *initial setting*, requires from 30 minutes to 60 minutes for completion. During this phase, the mixed concrete decrease its plasticity and develops pronounced resistance to flow. The second phase, known as *final setting*, may vary between 5 to 6 hours after the mixing operation. During this phase, concrete appears to be relatively soft solid without surface hardness. The third phase consists of *progressive hardening* and *increase in strength*. The process is rapid in the initial stage, until about one month after mixing, at which time the concrete almost attains the major portion of its potential hardness and strength.

Depending on the quality and proportions of the ingredients used in the mix, the properties of concrete vary. Concrete has enough strength in compression, but has little strength in tension. Due to this, concrete is weak in bending, shear and torsion. Hence the use of *plain concrete,* is limited to applications where compressive strength is the principal requirement and where tensile stresses are either totally absent or are extremely low. However, to use cement concrete for common structures such as beams, slabs, retaining structures etc, *steel bars* may be placed at tensile zones of the structure. The steel bars, known as *steel reinforcement*, embedded in the concrete, takes the tensile stresses. The concrete so obtained is termed as *reinforced cement concrete*, commonly abbreviated as R.C.C.

## 1.2. CLASSIFICATION AND COMPOSITION OF CEMENT

### 1. *Classification*

Cement may be classified into five groups : (1) Portland Cement, (2) High Alumina Cement, (3) Super Sulphated Cement, (4) Natural Cements and (5) Special Cements, with the following subdivisions :

1. *Portland Cements are further subdivided as:*
    (a) Ordinary Portland Cement (OPC)
    (b) Rapid Hardening Cement
    (c) Extra Rapid Hardening Cement
    (d) Low Heat Portland Cement

    (e) Portland Blast furnace Slag Cement     (f) Portland-Puzzolana Cement (PPC)
    (g) Sulphate Resisting Portland Cement     (h) White Portland Cement
    (i) Coloured Portland Cement
2. *High Alumina Cement*
3. *Super Sulphated Cement*
4. *Natural Cements*
5. *Special Cements*
    (a) Masonry Cement     (b) Trief Cement
    (c) Expansive Cement     (d) Oil Well Cement

## Recommendations of IS 456 : 2000

The cement used for *manufacture of concrete* shall be any of the following:
(a) 33 Grade Ordinary Portland Cement Conforming to IS 269
(b) 43 Grade Ordinary Portland Cement Conforming to IS 8112
(c) 53 Grade Ordinary Portland Cement Conforming to IS 12269.
(d) Rapid Hardening Portland Cement Conforming to IS 8041
(e) Portland Slag Cement Conforming to IS 455.
(f) Portland Pozzolana Cement (Fly ash based), conforming to IS 1489 (Part 1)
(g) Portland Pozzolana Cement (Calcined clay based), conforming to IS 1489 (Part 2)
(h) Hydrophobic Cement Conforming to IS 8043
(i) Low Heat Portland Cement Conforming to IS 12600
(j) Sulphate Resisting Portland Cement Conforming to IS 12330

Other combinations of Portland cement with mineral admixtures conforming to relevant Indian Standards laid down may also be used in the manufacture of concrete provided that there are satisfactory data on their suitability, such as performance test on concrete containing them. Low heat Portland cement conforming to IS 12600 shall be used with adequate precautions with regard to the removal of form work etc. High alumina cement conforming to IS 6452 or super sulphated cement conforming to IS 6909 may be used only under special circumstances with the prior approval of the engineer-in-charge.

### 2. Composition of Portland Cement

The principal raw materials used in the manufacture of cement are :
(a) Argillaceous or silicates of alumina in the form of clays and shales.
(b) Calcareous or calcium carbonate, in the form of lime stone, chalk and marl which is a mixture of clay and calcium carbonate.

The ingradients are mixed in the proportion of about two parts of calcareous material to one part of argillaceous material and then crushed and ground in ball mills in a dry state or mixed in a wet state. The dry powder or the wet slurry is then burnt in a rotary kiln at a temperature between 1400° to 1500°C. The clinker obtained from the kiln is first cooled and then passed on to ball mills where gypsum is added and it is ground to the requisite fineness.

The chief chemical constituents of Portland cement are as follows :

| | |
|---|---|
| Lime (CaO) | 60 to 67% |
| Silica ($SiO_2$) | 17 to 25% |
| Alumina ($Al_2O_3$) | 3 to 8% |
| Iron Oxide ($Fe_2O_3$) | 0.5 to 6% |
| Magnesia (MgO) | 0.1 to 4% |
| Sulphur Trioxide ($SO_3$) | 1 to 3% |
| Soda and/or Potash ($Na_2O+K_2O$) | 0.5 to 1.3% |

The above constituents forming the raw materials undergo chemical reactions during burning and fusion, and combine to form the following compounds (*called Bogue compounds*) in the finished products :

| Compound | Abbreviated designation |
|---|---|
| 1. Tricalcium silicate ($3CaO.SiO_2$) | $C_3S$ |
| 2. Dicalcium silicate ($2CaO.SiO_2$) | $C_2S$ |
| 3. Tricalcium aluminate ($3CaO.Al_2O_3$) | $C_3A$ |
| 4. Tetracalcium alumino-ferrite ($4CaO.Al_2O_3.Fe_2O_3$) | $C_4AF$ |

The proportions of the above four compounds vary in the various Portland cements. Tricalcium silicate and dicalcium silicates contribute most to the eventual strength. Initial setting of Portland cement is due to the tricalcium aluminate. Tricalcium silicate hydrates quickly and contributes more to the early strength. The contribution of dicalcium silicate takes place after 7 days and may continue for upto 1 year. Tricalcium aluminate hydrates quickly, generates much heat and makes only a small contribution to the strength within the first 24 hours. Tetracalcium alumino-ferrite is comparatively inactive. All the four compounds generate heat when mixed with water, the aluminate generating the maximum heat and the dicalcium silicate generating the minimum. Due to this, tricalcium aluminate is responsible for most of the undesirable properties of concrete. Cement having less $C_3A$ will have higher ultimate strength, less generation of heat and less cracking. Table 1.1 gives the composition and percentage of the four compounds for normal, rapid hardening and low heat portland cement.

TABLE 1.1. Composition and Compound Content of Portland Cement (After Lea)

|  | *Normal* | *Rapid hardening* | *Low heat* |
|---|---|---|---|
| **(a) Composition: Percent** | | | |
| Lime | 63.1 | 64.5 | 60 |
| Silica | 20.6 | 20.7 | 22.5 |
| Alumina | 6.3 | 5.2 | 5.2 |
| Iron Oxide | 3.6 | 2.9 | 4.6 |
| **(b) Compound: Percent** | | | |
| $C_3S$ | 40 | 50 | 25 |
| $C_2S$ | 30 | 21 | 45 |
| $C_3A$ | 11 | 9 | 6 |
| $C_4AF$ | 12 | 9 | 14 |

**3. Ordinary Portland Cement (IS : 269) :** The properties of various types of portland cements differ because of relative proportions of the four compounds and the fineness to which the cement clinker is ground. The *Ordinary Portland Cement* is the basic Portland cement and is manufactured in larger quantities than all the others. It is mostly suited for use in general concrete construction where there is no exposure to sulphates.

**4. Rapid Hardening Portland Cement (IS : 269) :** This cement is also known as *high- early strength cement*. It is similar to ordinary Portland cement except that it is ground finer, possesses more $C_3S$ and less $C_2S$ than the ordinary Portland Cement. The magnitude of the increase in strength is gauged from the fact that the strength developed at the age of 3 days is about the same as 7-day strength of ordinary Portland cement with the same water-cement- ratio. The main advantage of a rapid hardening cement is that shuttering can be removed much earlier, thus saving considerable time and expenses. In the concrete products in industry, moulds can be released quicker. Rapid hardening cement is also used for road work where it is necessary to open the road traffic with the minimum delay.

**5. *Extra Rapid Hardening Cement* :** Extra rapid hardening cement is obtained by intergrinding calcium chloride with rapid Hardening Portland cement. The normal addition of $CaCl_2$ is 2% (of the commercial 70% $CaCl_2$) by weight of the rapid hardening cement. The addition of $CaCl_2$ also imparts quick setting properties. Hence this cement should be placed and fully compacted within 30 minutes of mixing.

**6. *Low Heat Portland Cement (IS : 269)* :** When concrete is poured in any structure, an increase in temperature occurs and a certain amount of heat is generated. This is due to the chemical reaction that takes place while the cement is setting and hardening. Low heat Portland cement is used in massive constructions like abutments, retaining walls, dams, etc., where the rate at which the heat can be lost at the surface is lower than at which the heat is generated. The heat generated in ordinary cement at the end of 3 days may be of the order of 80 calories per gram of cement, while in low heat cement it is 50 calories per gram. It has low percentage of $C_3A$ and relatively more $C_2S$ and less $C_3S$ than ordinary Portland cement. This is achieved by restricting the amount of calcium and increasing the silicates present in the raw materials of manufacture. Therefore, it has low rate of gain of strength, but the ultimate strength is practically the same.

**7. *Portland Blast Furnace Cement (IS : 455)* :** This cement is made by intergrinding Portland cement clinker and blast furnace slag, the proportion of the slag being not less than 25% or more than 65% by weight of cement, as prescribed by IS : 455. The slag should be granulated blast furnace slag of high lime content, which is produced by rapid quenching of molten slag obtained during the manufacture of pig iron in a blast furnace. It is usual for the Portland cement clinker to be ground with a slag, a small percentage of gypsum being added to regulate the setting time. The blending of the Portland cement clinker with the slag, by no means detracts from any desired property of cement. Indeed, it confers upon it some additional advantage. This is because the granulated slag itself possesses latent hydraulic properties which are tremendously activated when the slag is crystalised and integrated with Portland cement clinker. In general, blast furnace cement will be found to gain strength more slowly than the ordinary Portland cement. It has less heat of hydration than ordinary Portland cement. In view of its low heat evolution, it can be used in mass concrete structures such as dams, retaining walls, foundations and bridge abutments.

**8. *Portland Pozzolana Cement (IS : 1489)* :** Portland Pozzolana cement is manufactured either by intergrinding Portland cement clinker and pozzolana or by intimately and uniformly blending Portland cement and fine pozzolana. While intergrinding presents no difficulty, *blending* tends to result in a non-uniform product and Indian Standard is specific in specifying that the latter method should be confined to factories and other such works where intimate blending can be ensured through mechanical means. As per Indian Standard, the proportion of pozzolana may vary from 10 to 25% by weight of cement. The pozzolana used in the manufacture of Portland Pozzolana cement in India is, at present, burnt clay or shale, or fly ash. Although pozzolanas have no cementing value themselves, they have the property of combining with free lime to produce a stable lime pozzolana compound which has definite cementitious properties. This cement has higher resistance to chemical agencies and to attack by sea water, because of absence of free lime. Portland Pozzolana cement also has a lower heat of evolution. Portland Pozzolana cement have a lower rate of development of strength than ordinary Portland cement. However, when the pozzolana is selected with care and is refined and ground with Portland cement clinker under controlled conditions, the compressive strengths reached by Portland pozzolana cement are comparabale with those reached by ordinary Portland cement. This can be seen from the following table which compares the strength at different ages of Portland Pozzolana cement and Ordinary Portland cement manufactured at the cement works of the Associated Cement Companies (ACC) Ltd., India :

|             | Compressive Strength, $N/mm^2$ | |
| --- | --- | --- |
| Age in days | Portland Pozzolana Cement | Ordinary Portland Cement |
| 3 | 19.6 – 21.6 | 18.6 – 22.6 |
| 7 | 25.5 – 32.4 | 26.5 – 31.4 |
| 28 | 36.3 – 47.1 | 35.3 – 51.0 |

**9. *Sulphate Resisting Cement* :** In sulphate resisting cement, the quantity of tricalcium aluminate is strictly limited. They are normally ground finer than Portland cement. The action of sulphates is to form sulpho-aluminates which have expansive properties and so cause disintegration of the concrete. Sulphate resistisng cement should be allowed to harden as long as possible to allow a resistant skin to be formed through carbonation by the action of atmospheric carbon dioxide.

**10. *White and Coloured Cements*** : The greyish colour of Portland cements is due to the presence of iron oxide. White Portland cement is manufactured in such a way that the percentage of iron oxide is limited to less than 1%. To achieve this, superior raw materials, such as chalk and lime stone having low percentage of iron, and white clay (China clay) are used. Sodium aluminium fluoride (cryolite) is added to act as flux in the absence of iron oxide. Oil fuel is used in place of pulverised coal in the kilning process in order to avoid contamination by coal ash. Coloured Portland cements are usually obtained by adding strong pigments, upto 10% to the ordinary or white cement, during grinding of clinker. The essential requirements of a good pigment are that it should be permanent and should be chemically inert when mixed with cement.

**11. *High Alumina Cement*** : High Alumina cement, also known as *aluminous cement foundu* is manufactured in entirely different way from that of Portland cements. The raw materials used for its manufacture are chalk and bauxite which is a special clay of extremely high alumina content. The manufacture of this type of cement is more expensive than the Portland cements, though it has many advantages over other types of cements. High alumina cement is characterised by its dark colour, high early strength, high heat of hydration and resistance to chemical attack. It thus produces concrete of far greater strength and in considerably less time even than Rapid-Hardening Portland cement, allowing earlier removal of the formwork. Its rapid hardening properties arise from the presence of calcium aluminate, chiefly monocalcium aluminate ($Al_2O_3 . CaO$), as the predominant compound in place of calcium silicates of Portland cement and after setting and hardening there is no free hydrated lime as in the case of Portland cement. However, great care should be taken in the use of high alumina cement, and it must not be mixed with any other type of cement, since the heat given off on setting is greater than with other cement.

**12. *Super Sulphated Cement* (IS : 6909)** : Super sulphated cement is made from well granulated blastfurnace slag (80 to 85%), calcium sulphate (10 to 15%) and Portland cement (1 to 2%) and is ground finer than the Portland cement. One of its most important properties is its low total heat of hydration. It is, therefore, very suitable for construction of dams and mass concreting work. Concrete made from super sulphated cement may expand if cured in water and may shrink if the concrete is cured in air. Another big advantage of super sulphated cement is its comparatively high resistance to chemical attack.

**13. *Natural cement*** : Natural cements are those cements which are manufactured from naturally occurring *cement rocks* which have compositions similar to the artificial mix of argillaceous and calcareous materials from which Portland cement is manufactured. However, the natural cement rocks are burned at somewhat lower temperatures than those used for the production of Portland cement clinker. The properties of such cement depend upon the composition of the natural cement rock.

**14. *Masonry Cement*** : For a long time, lime mixed with sand was used for mortar for laying brick work. However, in order to increase the strength and rapidity of gaining strength it became common to mix Portland cement with the lime. The usual proportions of cement : lime : sand may range from 1 : 1 : 6 for heavy loads to 1 : 3 : 12 for light loads. Cement sand mortars are too harsh, while lime makes the mortar easier to work. In order to avoid the necessity for mixing cement and lime, masonry cements have recently been introduced. According to Wuerpel, most successful masonry cement are composed of Portland cement clinker, lime stone, gypsum and air-entraining agent. These constituents are ground to an even greater fineness than that of high early strength Portland cement. The plasticity and workability of masonry cements are imparted by the lime stone and air-entraining agents. The ease of working masonry cements and their water retentive properties help to increase their adhesion to bricks or other building units and this is further assisted by the fact their shrinkage is fairly low.

**15. *Trief Cement*** : Trief cement is practically the same as blastfurnace cement except that the blastfurnace slag is ground wet and separately from the cement. Wet grinding results in a fine product, with a specific surface of at least 3000 $cm^2/g$. Due to this, the slow rate of gain of strength normally associated with blastfurnace cement is avoided and strength from early ages equal to those of ordinary Portland cement are obtained. This cement has smaller shrinkage and a smaller heat of evolution while setting than ordinary Portland cement.

**16. *Expansive Cement* :** Expansive cement expands while hardening. Ordinary concrete shrinks while hardening, resulting in shrinkage cracks. This can be avoided by mixing expansive cement with the normal cements in the concrete, which will neither shrink nor expand. Another useful application of expansive cement is in repair work where the opened up joints can be filled with this cement so that after expansion a tight joint is obtained. Expansive cements have been used in France for underpinning and for the repair of bomb damaged arch bridges.

**17. *Oil Well Cements* :** In the drilling of oil wells, cement is used to fill the space between the steel lining tube and the wall of the well, and to grout up porous strata and to prevent water or gas from gaining access to oil-bearing strata. The cement used may be subject to very high pressure, and the temperature may rise to 400°F. Cement used must be capable of being pumped for up to about 3 hours. It must also harden quickly after setting. These properties can be achieved by (a) adjusting the composition of the cement and (b) by adding retarders to ordinary Portland cement. In case (a), the proportion of $Fe_2O_3$ is adjusted so that it is above that required to combine with all the $Al_2O_3$ to form tetra calcium alumino-ferrite $4CaO.Al_2O_3.Fe_2O_3$. The proportion of tricalcium aluminate $3CaO.Al_2O_3$ formed is therefore very small and the setting time is accordingly increased. Setting times of upto 4 hours at a temperature of 200°F and 6 hours at a temperature of 70°F can be obtained with a Portland cement containing no tricalcium aluminate. By the use of retarders setting time of upto $6\frac{1}{2}$ hours at temperatures of upto 220°F can be obtained.

## 1.3. SPECIFICATIONS FOR PORTLAND CEMENT

For the quality control of Portland cement used for plain and reinforced concrete, the Indian Standard Institution has recommended the following specifications and tests : (1) Chemical composition (2) fineness (3) soundness (4) setting time (5) compressive strength, and (6) heat of hydration. *The detailed procedures for the above tests are given in a separate chapter at the end of the book.*

*Comparison of properties of various type of cements* : Table 1.2 gives the *physical requirements* of various common types of cements, as specified by the Indian Standards. Similarly, Table 1.3 gives the chemical requirements for various types of cements.

**TABLE 1.2.** Physical Requirements (Indian standard specification for cements)

| | Type of Cement | Ordinary Portland | Rapid Hardening Portland | Low heat Portland | Portland blast Furnace Slag | Portland Pozzolana | High alumina | Super Sulphated |
|---|---|---|---|---|---|---|---|---|
| | *Indian Standard* | *269-1967* | *269-1967* | *269-1967* | *455-1967* | *1489-1967* | *6452-1972* | *6909-1973* |
| **1. Fineness** | Residue by Wt. on 90-micron IS sieve Min. specific surface (Blain's air permeability) $cm^2/gm$. | ≯ 10% 2550 | ≯ 5% 3253 | — 3200 | ≯ 10% 2250 | ≯ 5% 3000 | — 2250 | — 4000 |
| **2. Soundness** | Explansion[1] (Lechatelier) Autoclave test[2] | ≯ 10 mm ≯ 0.8% | ≯ 10 mm ≯ 0.8% | ≯ 10 mm ≯ 0.8% | ≯ 10 mm ≯ 0.8% | ≯ 10 mm ≯ 0.8% | ≯ 5 mm — | ≯ 5 mm — |
| **3. Setting time** | Initial Final | ≮ 30 min ≯ 600 | ≮ 30 min ≯ 600 | ≮ 60 min ≯ 600 | ≮ 30 min ≯ 600 | ≮ 30 min ≯ 600 | ≮ 30 min ≯ 600 | ≮ 30 min ≯ 600 |
| **4. Min. compressive strength[4] ($N/mm^2$)** | 24 hours[3] 72 hours[3] 168 hours[3] 762 hours[3] | — 16 22 — | 16 27.5 — — | — 10 16 35 Heat of hydration to be not more than 65 cal/gm at 7 days and 75 cal/gm at 28 days | — 16 22 — | — 32 31 Average drying shrinkage of mortars to be not more than 0.15% | 30 35 — — | — 15 22 30 |

**Notes:**
1. In the event of the cements failing to comply with this requirement, the test is repeated with another portion of the same sample after aeration for 7 days, when the expansion for the first five types shall not be more than 5 mm.
2. When specified by purchaser and for all cements having a magnesia contents more than 3 per cent.
3. There should be a progressive increase in strength from the strength at 24, 72 or 168 hours.
4. Alternatively, see amendment 1 of IS : 269–1967, amendment 2 of IS : 455–1967 and amendment 2 of IS : 1489–1967 for revised compressive strength.

**TABLE 1.3.** Chemical Requirements (Indian Standard Specifications for cements)

| Type of Cement | Ordinary and rapid hardening Portland | Low heat Portland | Portland blast furnace Slag | Portland pozzolana | High alumina | Super sulphated |
|---|---|---|---|---|---|---|
| *Indian Standard* | **269-1967** | **269-1967** | **455-1967** | **1489-1972** | **6452-1972** | **6909-1973** |
| 1. Insoluble residue | 2.0% | 2.0% | $(x' + 0.025) + 2.0$ | $x + 2.0 \dfrac{(100 - x)}{100}$ | – | 4% |
| 2. Megnesia (Mgo) | 6% | 6% | 8% | 6% | – | 10% |
| 3. Sulphur as sulphuric anhydride($SO_2$) as sulphide (S) | 2.75% – | 2.75% – | 3% 1.5% | 2.75 – | – – | 6% 1.5% |
| 4. Manganic oxide ($Mn_2O_3$) | – | – | 2% | – | – | – |
| 5. Loss on ignition | 4% | 4% | 4% | 5% | – | – |

where : $x$ is the declared percentage of pozzalana in the given cement.

$x'$ is the declared percentage of granulated blast furnace slag in the cement.

## 1.4. AGGREGATES

*Aggregate* is a general term applied to those inert or chemically inactive materials which, when bonded together by cement, form concrete. Most of the aggregates used are naturally occurring aggregates such as crushed rock and sand. Artificial and processed aggregates may be broken brick or crushed air-cooled blastfurnace slag. Light weight aggregates, such as pumice, furnace clinker, coke, breeze, sawdust, foamed slag, expanded slates, expanded vermiculite etc. are used for the production of concrete of low density.

*Classification* : Aggregates may be divided into two groups : (*a*) coarse aggregates and (*b*) fine aggregate. Aggregates less than 4.75 mm are known as *fine aggregates* while those more than 4.75 mm in size are known as *coarse aggregate*. For large and important works it has become usual to separate the coarse aggregate also into two or more sizes, and these fractions are kept separate until the proper quantity of each has been weighed out for a batch of concrete.

*Quality of aggregate* : Natural aggregate used for concrete construction is required to comply with the norms laid down in IS : 383–1970 'Specification for coarse and fine aggregates from natural sources for concrete'.

Important characteristics of aggregates are : (1) strength (2) size (3) particle shape (4) surface texture (5) grading (6) impermeability (7) cleanliness (8) chemical inertness (9) physical and chemical stability at high temperatures (10) coefficient of thermal expansion, and (11) cost.

Aggregate should be chemically inert, strong, hard, durable, of limited porosity, free from adherent coatings clay lumps, coal, and coal residues and should contain no organic or other admixture that may cause corrosion of the reinforcement, or impair the strength or durability of the concrete. The limits of the content of deleterious materials in aggregate are given in Table 1.4.

**TABLE 1.4.** Limits of Content of Deleterious Materials* (IS : 383–1970)

| Deleterious substances | Fine Aggregate | | Coarse Aggregates | |
|---|---|---|---|---|
| | uncrushed | crushed | uncrushed | crushed |
| Coal and lignite | 1.00 | 2.90 | 1.00 | 1.00 |
| Clay lumps | 1.00 | 1.00 | 1.00 | 1.00 |
| Soft fragments | – | – | 3.00 | – |
| Materials finer than 75 μ IS sieve | 3.00 | 15.00 | 3.00 | 3.00 |
| Shale | 1.50 | – | – | – |
| Total of percentage of all deleterious materials** | 1.00 | 2.00 | 5.00 | 5.00 |

\* Percentage by weight of aggregate.  \*\* Mica is excluded

*Coarse aggregate* : The material retained on 4.75 mm sieve is termed as coarse aggregate. Crushed stone and natural gravel are the common materials used as coarse aggregate for concrete. Natural gravels can be quarried from pits where they have been deposited by alluvial or glacial action, and are normally composed of flint, quartz, schist and igneous rocks. Coarse aggregates are obtained by crushing various types of granites (such as syenites, dolerites, diorites, quartzites etc.), schist, gneiss, crystalline hard lime stone and good quality sand stones. When very high strength concrete is required, a very fine-grained granite is perhaps the best aggregate. Coarse grained rocks make harsh concrete, and need high proportion of sand and high water-cement ratio to get reasonable degree of workability. Harder types of sand stones, having fine grained texture, are suitable as coarse aggregate, but softer varieties should be used with caution. Concrete made with sand stone aggregate give trouble due to cracking, because of high degree of shrinkage. Similarly hard and close-grained crystalline lime stones are very suitable for aggregate, is cheap, but should be used only in plain concrete. Blast furnace slag, coal ashes, coke-breeze etc. may also be used as aggregate to obtain light weight and insulating concrete of low strength.

*Fine aggregate* : The material smaller than 4.75 mm size is called fine aggregate. Natural sands are generally used as fine aggregate. Sand may be obtained from pits, river, lake or sea shore. When obtained from pits, it should be washed to free it from clay and silt. Sea shore sand may contain chlorides which may cause efflorescence, and may cause corrosion of reinforcement. Hence it should be thoroughly washed before use. Similarly, if river sand contains impurities such as mud etc. it should be washed before use. Angular grained sand produces good and strong concrete, because it has good interlocking property, while round grained particles of sand do not afford such interlocking.

*Grading of aggregates* : Gradation of the aggregates is almost as important as its quality is. The grading of the aggregates has marked effect on the workability, uniformity, and finishing qualities of concrete. The grading of coarse aggregate may be varied through wider limits than that of sand without appreciably affecting the workability of concrete.

*Fineness Modulus* : The *fineness modulus* of an aggregate is an index number which is roughly proportional to the average size of the particles in the aggregate. The coarse the aggregate, the higher the fineness modulus. The fineness modulus is obtained by adding the percentage of the weight of material retained on the following IS sieves and dividing it by 100 : 80 mm, 40 mm, 20 mm, 10 mm, 4.75 mm, 2.36 mm, 1.18 mm, 600-micron, 300-micron and 150-micron (total 10 sieves).

TABLE 1.5. Determination of Fineness Modulus

| IS Sieve | Coarse aggregate (10 kg) | | | Fine aggregate (1 kg) | | |
|---|---|---|---|---|---|---|
| | Weight retained (kg) | Total wt. retained (kg) | % Weight retained | Weight retained (kg) | Total wt. retained (kg) | % Weight retained |
| 80 mm | 0.0 | 0.0 | 0.0 | 0.0 | 0.0 | 0.0 |
| 40 mm | 0.0 | 0.0 | 0.0 | 0.0 | 0.0 | 0.0 |
| 20 mm | 3.5 | 3.5 | 35.0 | 0.0 | 0.0 | 0.0 |
| 10 mm | 3.0 | 6.5 | 65.0 | 0.0 | 0.0 | 0.0 |
| 4.75 mm | 2.8 | 9.3 | 93.0 | 0.0 | 0.0 | 0.0 |
| 2.36 mm | 0.7 | 10.0 | 100.0 | 0.1 | 0.1 | 10.0 |
| 1.18 mm | 0.0 | 10.0 | 100.0 | 0.25 | 0.35 | 35.0 |
| 600 micron | 0.0 | 10.0 | 100.0 | 0.35 | 0.70 | 70.0 |
| 300 micron | 0.0 | 10.0 | 100.0 | 0.20 | 0.90 | 90.0 |
| 150 micron | 0.0 | 10.0 | 100.0 | 0.10 | 1.00 | 100.0 |
| | Sum : 693.0 | | | Sum : 305.0 | | |
| Fineness modulus | $\frac{693.0}{100} = 6.93$ | | | $\frac{305.0}{100} = 3.05$ | | |

Table 1.5 illustrates the method of determining fineness modulus of both coarse and fine aggregates. It has been found that certain values of fineness moduli for the fine and coarse aggregates give good workability, with a minimum quantity of cement. The limits of fineness moduli are given in Table 1.6.

TABLE 1.6. Limits of Fineness Moduli

| Maximum size of aggregate | | Fineness modulus | |
|---|---|---|---|
| | | max. | min. |
| (a) Fine aggregate | | 2.0 | 3.5 |
| (b) Coarse aggregate | (i) 20 mm | 6.0 | 6.9 |
| | (ii) 40 mm | 6.9 | 7.5 |
| | (iii) 75 mm | 7.5 | 8.0 |
| | (iv) 150 mm | 8.0 | 8.5 |
| (c) Mixed aggregate | (i) 20 mm | 4.7 | 5.1 |
| | (ii) 25 mm | 5.0 | 5.5 |
| | (iii) 32 mm | 5.2 | 5.7 |
| | (iv) 40 mm | 5.4 | 5.9 |
| | (v) 75 mm | 5.8 | 6.3 |
| | (vi) 150 mm | 6.5 | 7.0 |

## 1.5. WATER

Water acts as lubricant for the fine and coarse aggregates and acts chemically with cement to form the binding paste for the aggregate and reinforcement. Water is also used for curing the concrete after it has been cast into the forms.

Water used for mixing and curing shall be clean and free from injurious amount of oils, acids, alkalis, salts, sugar, organic materials or other substances that may be deleterious to concrete or steel. Potable water is generally considered satisfactory for mixing concrete. As a guide, the following concentrations represent the maximum permissible values :

(a) To neutralize 100 ml sample of water, using phenolphthalein as an indicator, it should not require more than 5 ml of 0.02 normal NaOH.

(b) To neutralize 100 ml sample of water, using mixed indicator, it should not require more than 25 ml of 0.02 normal $H_2SO_4$.

(c) The permissible limits of solids shall be as given in Table 1.7.

**TABLE 1.7.** Permissible Limits for Solids in Water (IS : 456–2000)

| S. N. | Type of solid | Permissible limit (max.) |
|---|---|---|
| (i) | Organic | 200 mg/l |
| (ii) | Inorganic | 300 mg/l |
| (iii) | Sulphates (as $SO_3$) | 400 mg/l |
| (iv) | Chlorides (as Cl) | 2000 mg/l for concrete not containing embeded steel and 500 mg/l for reinforced concrete work |
| (v) | Suspended matter | 2000 mg/l |

***pH value*** : The pH value of water shall not be less than 6.

***Water for curing*** : Water found satisfactory for mixing is also suitable for curing concrete. However, water used for curing should not produce any objectionable stain or unsightly deposit on the concrete surface. The presence of tannic acid or iron compounds is objectionable.

***Sea water*** : Mixing and curing with sea water is not recommended because of presence of harmful salts in sea water. Under unavoidable circumstances, sea water may be used for mixing or curing in plain concrete with no embedded steel after having given due consideration to possible disadvantages and precautions including use of appropriate cement system.

## 1.6. MEASUREMENT OF MATERIALS

The materials used for preparation of concrete are (i) cement (ii) fine aggregate (iii) coarse aggragate and (iv) Water. Their accurate measurement before mixing is very important so that the required quantities in the proportion of the concrete mix are obtained.

**1. *Cement*** : It is preferable to measure cement in terms of its weight, and not in terms of volume. The volume of cement changes with the conditions of measurement. In our country, cement is supplied in bags, each bag weighing 50 kg. Under normal conditions, the volume of cement in the bag is considered equivalent to 34.5 litres. However, if the same cement is shovelled, the bag may measure upto 42 litres. Before mixing, therefore, cement is measured in terms of weight.

**2. *Fine aggregates*** : Fine aggregate (*i.e.* sand) may be measured by weight for accurate works, and by volume for ordinary works. However, when dry sand absorbs water from atmosphere, or when water is mixed to it artificially, its volume increases. This increase in volume due to moisture in sand is known as 'bulking of sand'. Water particles lubricate the sand particles, causing surface tension, and due to this, particles are pulled apart. Thus increase in volume results. This increase in volume depends on the gradation of sands, but may be taken to be maximum at a moisture content of about 4% by weight of dry sand. Further increase in moisture results in decrease in the percent increase of volume, as shown in Fig. 1.1. The bulking increases with fineness, and may be about 25% by volume.

Due to this, if sand is measured by volume, bulking should be properly accounted for.

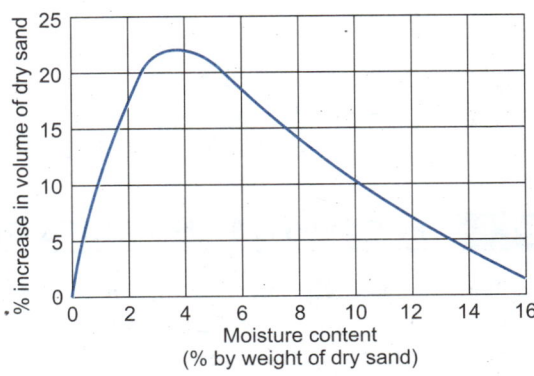

Fig. 1.1.   Bulking Curve for Sand

**3. Coarse aggregate :** There is no problem of bulking in coarse aggregate, and hence it may be measured either by volume or by weight. However, the weight of a given volume of aggregate is influenced by the size of the measuring box. Hence for accurate and large scale works, measurement should be done by weight. The unit weight of coarse aggregate in loose and dry state is found exactly in the same manner as for fine aggregate, except that a bigger container is used. Since the size of container has effect on the determination, Indian Standards specify the following container for carrying out the tests :

1. Maximum size of aggregate 5 mm to 40 mm : 15 litre capacity cylinder of 25 cm diameter.
2. Maximum size of aggregate over 40 mm : 40 litre capacity cylinder of 35 cm diameter.

**4. Water :** Water is normally measured by volume, and specified as so many litres per bag of cement. For a given quantity of water to be mixed in concrete, adjustment should be made for the amount of water present in sand and aggregate. The amount of the water present in the aggregate, due to hygroscopic action etc. should be subtracted from the total required quantity of water. However, if the aggregate is dry, and found to absorb water, extra water should be *added* to account for this. The percentage absorption should be determined first.

## 1.7. WATER CEMENT RATIO

Water-cement ratio is the ratio of volume of water mixed in concrete to volume of cement used. The strength and workability of concrete depend to a great extent on the amount of water used. For a given proportion of the materials, there is an amount of water which gives the greatest strength. Amount of water less than this optimum water decreases the strength and about 10 percent less may be insufficient to ensure complete setting of cement. Similarly, more than the optimum water increases the workability but decreases the strength. An increase in 10% above the optimum may decrease the strength approximately by 15%, while an increase in 50% may decrease the strength to one half. The use of an excessive amount of water not only produces low strength but increases shrinking and decreases density and durability.

According to Abram's water-cement ratio law, for any given conditions of test the strength of workable concrete mix is dependent only on the water-cement ratio. Lesser the water-cement ratio greater will be its strength. From Abram's law, it follows that provided the concrete is fully compacted, the strength is not affected by aggregate shape, type or surface texture, or the aggregate grading, the workability and the richness of the mix.

According to Powers, cement does not combine chemically with more than half the quantity of water in the mix. Cement requires about 1/5 to 1/4 of its weight of water to become completely hydrated. This suggests that if water cement ratio is less than 0.4 to 0.5, complete hydration will not be secured. Some practical values of water cement ratio for structural reinforced concrete are about 0.45 for 1 : 1 : 2 concrete,

0.50 for $1 : 1\frac{1}{2} : 3$ concrete and 0.55 to 0.60 for $1 : 2 : 4$ concrete. However, concrete vibrated by efficient mechanical vibrators require less water cement ratio, and hence have more strength. Sometimes, plasticising agents may be mixed to increase the workability of the mix. For such concrete, water-cement ratio is reduced, resulting in an increase in the strength.

## 1.8. PROPERTIES AND TESTS ON CONCRETE

The important properties of concrete, which govern the design of a concrete mix are (*i*) strength (*ii*) durability (*iii*) workability and (*iv*) economy. The aim of proportioning a concrete mix will be to find the economic proportions of cement, coarse aggregate, fine aggregate and water so as to get a mix of given strength, proper workability and durability. The various tests conducted on concrete are discussed in detail in a separate chapter 'Tests on cement and concrete'.

## 1.9. DURABILITY OF CONCRETE

Durability is the property of concrete in resisting its disintegration and decay. The concrete should be durable with regard to various weathering conditions such as atmospheric gases, moisture changes, temperature variations. Disintegration and decay of concrete may be due to the following reasons :

(1) Use of unsound cement, which due to some delayed chemical reactions, undergo volume changes after the concrete has hardened.

(2) Use of less durable aggregate, which may either react with cement, or may be reacted upon by atmospheric gases.

(3) Entry of harmful gases and salts through excessive pores and voids present in unsound concrete, causing its disintegration.

(4) Freezing and thawing of water sucked through the cracks or crevices, by capillary action causing its disintegration.

(5) Expansion and contraction resulting from temperature changes or alternate wetting and drying.

As stated earlier, water required for chemical reaction is about 25% of the weight of cement. Hence excess water present in concrete later evaporates, leaving voids and pores. These pores or voids are later responsible for decay of concrete. Hence for durable concrete, water cement ratio should be as small as possible. A well compacted concrete has less voids and pores and has more durability. The entrainment of air in concrete has been found to increase very considerably the resistance of concrete to freezing and thawing. The improvement in this respect is due to relief, occassioned by the minute dispersed air bubbles which act as expansion chambers, of stresses and pressures, caused by temperature and moisture changes and by expansion of the moisture contained in concrete on freezing. Vinsol resin is sometimes mixed with concrete to have the property of entrapping innumerable minute air bubbles in concrete. In order to prevent Vinsol resin reacting chemically with the cement, and to make it soluble in water, it is first neutralised by the addition of sodium hydroxide which converts it into a soap. The quantity of resin required for such purpose is extremely small ranging from 0.005 to 0.05 of 1 percent of the weight of cement.

## 1.10. WORKABILITY OF CONCRETE

It is difficult to properly define and measure the 'workability' of concrete, despite its being the most important property. In its simplest form, the term 'workability' may be defined as the ease with which concrete can be mixed, handled, transported, placed in position and compacted. According to Indian Standard (IS : 1199), *workability of concrete is that property of concrete which determines the amount of internal work necessary to produce full compaction.* The greatest single factor affecting the workability is the amount of water in the mix. A workable concrete does not show any bleeding or segregation. Bleeding of concrete takes place when excess of water in the mix comes up at the surface, causing small pores through the mass of concrete. Segregation is caused when coarse aggregate separate out from the finer materials, resulting in large voids, less durability and less strength.

*Several tests which have been developed to measure the workability of concrete are:*

(1) Slump test (2) Compacting factor test (3) Vee-Bee test and (4) Vibro-workability test. These have been described in detail separately.

*Slump test* is the simplest and commonly used test. In this test, concrete is compacted in a vessel of the shape of the frustum of a cone and open at both the ends. Concrete is compacted with the help of standard tamping rod, in four equal layers. Immediately after the vessel is filled, it is raised vertically, without giving any jerks etc. The concrete in the vessel becomes free and therefore slump. The vertical settlement, measured in mm, is termed as slump. The following table gives a rough guide of workability of concrete, in terms of slump for various types of work :

| Type of work | Slump (mm) |
|---|---|
| 1. Concrete for road work | 20 to 30 |
| 2. Ordinary R.C.C. work for beams and slabs etc. | 50 to 100 |
| 3. Columns, retaining walls and thin vertical sections | 75 to 150 |
| 4. Vibrated concrete | 12 to 25 |
| 5. Mass concrete | 25 to 50 |

*The compaction factor test* measures the workability of concrete in terms of internal energy required to compact the concrete fully. In this test concrete is compacted in a lower cylindrical mould by making it to fall through two vertically placed hoppers. The weight of concrete in mould is determined. The theoretical weight of materials, required to fill the mould without air voids is also calculated from the knowledge of the proportions of the mix. The compacting factor is then calculated by dividing the observed weight of concrete in the mould by the theoretical weight. A concrete of low workability is represented by a compaction factor of about 0.85, of medium workability for a compaction factor of 0.92 and of good workability for a compaction factor of 0.95.

The various factors which influence the workability of concrete are (1) water in the mix (2) maximum size of particles (3) ratio of coarse and fine aggregates (4) particles interference (5) particle interlocking and (6) admixtures. Out of these, water in the mix is greatest single factor affecting the workability. Addition of water increases workability. The larger the maximum aggregate size and coarser the grading the smaller is the amount of water required for a given workability. In general, the grading requiring the least amount of water for a given workability will be that which gives the smallest surface area for a given amount of aggregate. A smooth rounded aggregate requires less water for a given workability than the irregular shaped aggregate. For a given aggregate-cement ratio, if the quantity of coarse aggregate is increased, the total surface area is reduced and hence more water would be available for lubrication, for a constant water cement ratio, resulting in increase in workability.

## 1.11. METHODS OF PROPORTIONING CONCRETE MIXES

### 1. ARBITRARY METHOD

This method is adopted only for works of small magnitude or of moderate importance. The combined aggregate should be dense and should have least voids. For this, the quantity of fine aggregates should be sufficient to fill the voids of coarse aggregate. The ratio of coarse aggregate to fine is found to lie between $1\frac{1}{2}$ to $2\frac{1}{2}$, for a dense mix of aggregates. However, a common practice is to take the quantities of fine and coarse aggregate in the proportion of 1 : 2, and hence to express the quantities of cement, sand and coarse aggregate in the proportions of $1 : n : 2n$ by volume. The ratios of 1 : 1 : 2 and 1 : 1.2 : 2.4 are considered suitable for very high strength concrete, the ratios $1 : 1\frac{1}{2} : 3$ and $1 : 2 : 4$ are used for normal reinforced concrete work and ratios $1 : 3 : 6$ and $1 : 4 : 8$ are used for foundations and mass concrete work. The amount of water to be used in the above mixes is decided on the basis of workability of the mix. The workability depends upon the type of work and the method of compaction. This method is widely used for all works of small magnitude.

## 2. MINIMUM VOIDS METHOD

In this method, the voids of coarse aggregate and fine aggregate are determined separately. The quantity of sand used should be such that it completely fills the voids of the coarse aggregate. Similarly, the quantity of cement used should be such that it fills the voids of sand, so that a dense mix having minimum voids is obtained. However, in actual practice, the quantity of sand used in the mix is kept 10% more than the voids in the coarse aggregate and the quantity of cement is taken 15% more than the voids in the sand. To the mix of cement, sand and coarse aggregate so obtained, sufficient water is added to make the mix workable. However, this method does not give satisfactory result because the presence of cement, sand and water separates the constituents of the coarse aggregate, thereby increasing its voids determined previously in absence of sand and cement. Similarly, the voids of sand are increased due to the addition of cement and water. Hence we do not always get a dense concrete. At the same time, the grading of aggregates has not been done so as to require least amount of water (and hence least w/c ratio) resulting in higher strength.

## 3. MAXIMUM DENSITY METHOD

This method of minimum voids was later improved by Fuller, to get a grading of materials to get maximum density. Based on wide scale experiments, he gave the following expression for the grading of materials :

$$P = 100 \left(\frac{d}{D}\right)^{\frac{1}{2}} \qquad \ldots(1.1)$$

where  $D$ = maximum size of aggregate

$P$ = percentage by weight, of material finer than diameter $d$.

The coarse and fine aggregates should be fully graded according to the above rule. For example, let the maximum size of coarse aggregate be 20 mm and the maximum size of fine aggregate be 4 mm, the percentage of material finer than 4 mm is given by

$$P = 100 \left(\frac{4}{20}\right)^{\frac{1}{2}} = 44.7\%$$

*i.e.* 44.7 kg of the aggregate, including the weight of cement, are to be mixed with 55.3 kg of coarse aggregate. The quantity of various intermediate sizes should also correspond to this formula.

Let us prepare the mix having a ratio of cement to aggregates (fine + coarse) as 1 : 6 by weight.

∴  Quantity of cement in 100 kg. of mix = $\frac{100}{7}$ = 14.3 kg

∴  Quantity of sand = 44.7 − 14.3 = 30.4 kg.

Hence the ratio of cement, sand and coarse aggregate by weight will be 14.3 : 30.4 : 55.3. Let us assume the unit weights of cement, fine aggregate and coarse aggregate as 1440, 1750 and 1600 kg per cubic metre respectively. Then the ratio of the three constituents, by volume will be

$$\frac{14.3}{1440} : \frac{30.4}{1750} : \frac{55.3}{1600}$$

or $\qquad 1 : \left(\frac{30.4}{1750} \times \frac{1440}{14.30}\right) : \left(\frac{55.3}{1600} \times \frac{1440}{14.30}\right)$

or $\qquad 1 : 1.75 : 3.48$

or $\qquad 1 : 1\frac{3}{4} : 3\frac{1}{2}$ (nominally)

After having decided the proportions of various materials, sufficient quantity of water is added to make the mix workable. Table 1.8 gives the grading of mixed aggregate for 40 mm and 20 mm maximum size of aggregate. The method is not so popular since grading cannot be accurately achieved in field, and there is no control over the strength.

**TABLE 1.8.** Grading of Mixed Aggregate

| Max. Size of coarse aggregate | Percentage passing the I.S size | | | | | | | | |
|---|---|---|---|---|---|---|---|---|---|
| | 40 mm | 20 mm | 10 mm | 4 mm | 2 mm | 1 mm | 500 micron | 250 micron | 125 micron |
| 40 mm | 100 | 71 | 50 | 32 | 22 | 16 | 11 | 8 | 6 |
| 20 mm | – | 100 | 71 | 44 | 32 | 22 | 16 | 11 | 8 |

## 4. FINENESS MODULUS AND WATER CEMENT RATIO METHOD

**(a) Fineness modulus :** It has been observed that strength of mix is dependent wholly on the water cement ratio while the grading of the particles is important from workability and economic point of view. The grading of particles by Fullers formula, to get maximum density, is difficult and sometimes uneconomical to achieve in practice. Fineness modulus method essentially is a substitute for Fuller's maximum density method, aimed at standardisation of the grading of aggregates. The term *fineness modulus*, suggested by Abram, is a numerical index of fineness of both fine as well as coarse aggregates.

Certain values of fineness modulus for mixed aggregates are found to give the best result. Let $p$ be the desired fineness modulus for a mix of fine and coarse aggregates. If $p_1$ and $p_2$ are the fineness moduli of fine and coarse aggregates respectively, than the proportion $R$ of finer aggregate to the combined aggregate, by weight is given by :

$$R = \frac{p_2 - p}{p - p_1} \times 100 \qquad \ldots(1.2)$$

For example of Table 1.5 if the desired fineness modulus of the combined aggregate is 5.3, we have

$$R = \frac{6.93 - 5.3}{5.3 - 3.05} \times 100 = 72.5\%$$

**(b) Abram's water-cement ratio law :** Abram's water-cement ratio (w/c ratio) law states that for any given conditions of test, the strength of workable concrete mix is dependent only on the water-cement ratio. It follows from this law that provided the concrete is fully compacted, the strength is not affected by aggregate shape, type or surface texture, or the aggregate grading, the workability and the richness of mix. We know that workability of mix (defined as 'that property of the concrete which determines the amount of useful internal work necessary to produce full compaction') is dependent on the amount of water in the mix. But amount of water in the mix, corresponding to a given strength, is governed by the water-cement ratio law. Hence the only way to increase the quantity of water to increase the workability of the mix is to increase the amount of cement also. However, as the grading of aggregate does not affect the strength of the concrete directly, the object must be to choose the grading to give the best workability with lowest water content. The grading of the particles should, therefore, be such that in the fully compacted state the total surface area of particles of aggregates as well as voids in them are the least. This means that larger the maximum aggregate size and the coarser the grading, the smaller is the amount of water required for a given workability. However, beyond a certain limit, the further increase in the maximum size and coarseness of grading results in harsh and undersanded mixtures causing honeycombing, thus requiring more cement for smoothness. On the other hand, increase in the proportion of fine aggregate (*i.e.* sand) gives smooth mix but requires more cement and hence results, in uneconomical mix. Between these two limits lies the grading, which can be either determined by Fuller's maximum optimum density method or the fineness modulus method.

According to Abram's law, the strength of mix increases with the decrease in the water cement ratio. In terms of crushing strength after 7 days curing, the law can be expressed as follows :

$$p_7 = \frac{984}{7^x} \qquad \ldots(1.3)$$

where $p_7$ = cylinder crushing strength, in kg/cm², after 28 days curing ;

$x$ = water cement ratio by volume.

In the above expression, the constant 984 and 7 may vary slightly with the quality of aggregates and cement, method of curing and method of testing.

Expressed in terms of strength after 28 days curing, the law can be written as:

$$p_{28} = \frac{984}{4^x} \qquad \ldots(1.4)$$

Fig. 1.2 shows the relations between the crushing strength and water cement ratio (by weight) for various periods of curing. Fig 1.3 gives the relationship between 28 days compressive strength(cube) of concrete mixes with different water-cement ratios by weight and the 7 days compressive strength of cement (IS : 456–1964). Both Figs. 1.2 and 1.3 may be used as a guide for the selection of proper water cement ratio for a mix of given strength.

**(c) Procedure for design of mix** : The procedure for the design of mix can be summarised as follows:

1. For the requirements of strength of the mix, choose suitable water cement ratio from Figs. 1.2 and 1.3.

2. Determine the maximum size of the aggregate available. Also determine the fineness modulus of both coarse and fine aggregates.

3. Determine the grading of the aggregates by Fuller's maximum density method. If this does not correspond to the grading of available materials, try to improve that grading to make it similar to that obtained by Fuller's method. If this is difficult to achieve, design the grading by fineness modulus method. For that, choose suitable fineness modulus of combined aggregate and determine the ratio of fine aggregates to coarse aggregates by *Eq 1.2*.

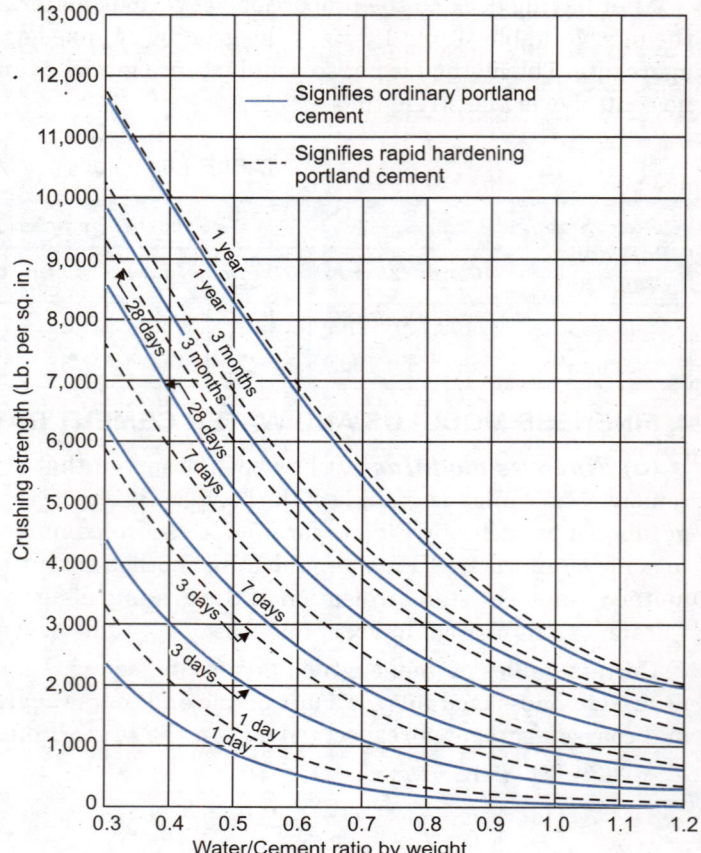

**Fig. 1.2.** Relation between Crushing Strength and Water-Cement Ratio for Fully Compacted Concrete.

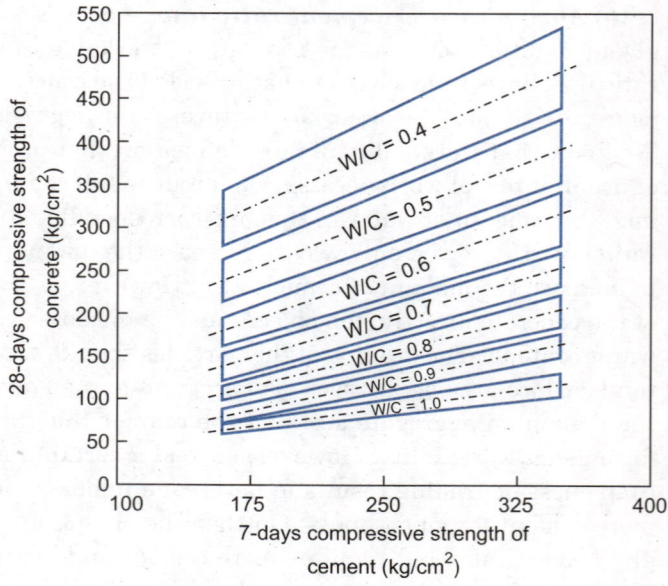

**Fig. 1.3.** Relationship between Compressive Strength of Concrete Mixes with Different Water-Cement Ratio and the 7 day Compressive Strength of Cement

4. Determine the workability of the mix in terms of slump or compaction factor, required for the work.

5. Fix the ratio of cement to that of combined aggregate, mix quantity of water corresponding to the water cement ratio determined in step 1, and determine the workability of mix. Change the cement-aggregate ratio by trial to get the desired workability.

6. Determine the actual proportions of cement, fine aggregates, coarse aggregate and water, from the above steps, so that economical concrete of desired strength and workability is obtained.

## 1.12. GRADES OF CONCRETE AND CHARACTERISTIC STRENGTH (IS : 456–2000)

Indian standard IS 456–2000 specifies *fifteen* grades of concrete (Table 1.9), designated as M 10, M 15, M 20, M 25, M 30, M 35, M 40, M 45, M 50, M 55, M 60, M 65, M 70, M 75 and M 80. In the designation of concrete mix, letter M refers as the mix and the number to the specified *characteristic strength* ($f_{ck}$) of 150 mm cube at 28 days, expressed in N/mm². The *characteristic strength* is defined as the strength of the material below which not more than 5% of the test results are expected to fall.

*IS : 456–2000 further recommends that minimum grade of concrete shall be not less than M 20 in reinforced concrete work.*

For concrete of compressive strength greater than M 55, design parameters given in the standard may not be applicable and values may be obtained from specialized literatures and experimental results.

TABLE 1.9. Grades of Concrete (IS : 456–2000)

| Group | Grade Designation | Specified characteristic compressive strength of 150 mm cube at 28 days (N/mm²) |
|---|---|---|
| *Ordinary concrete* | M 10 | 10 |
| | M 15 | 15 |
| | M 20 | 20 |
| *Standard concrete* | M 25 | 25 |
| | M 30 | 30 |
| | M 35 | 35 |
| | M 40 | 40 |
| | M 45 | 45 |
| | M 50 | 50 |
| | M 55 | 55 |
| *High strength concrete* | M 60 | 60 |
| | M 65 | 65 |
| | M 70 | 70 |
| | M 75 | 75 |
| | M 80 | 80 |

**Increase in strength with age :** There is normally a gain of strength beyond 28 days. The quantum of increase depends upon the grade and type of cement, curing and environmental conditions etc. The design should be based on 28 days characteristic strength of concrete unless there is evidence to justify a higher strength for a particular structure due to age. For concrete of grade M 30 and above, the ratio of increase of compressive strength with age shall be based on actual investigations.

**Tensile strength of concrete :** The flexural and splitting tensile strength shall be obtained as described in IS 516 and IS 5816 respectively. When the designer wishes to use an estimate of tensile strength from the compressive strength, the following formula may be used :

Flexural strength, $f_{cr} = 0.7\sqrt{f_{ck}}$ N/mm² ...(1.5)

where $f_{ck}$ is the characteristic cube compressive strength of concrete in N/mm².

**Elastic deformation :** The modulus of elasticity is primarily influenced by the elastic properties of the aggregate and to a lesser extent by the conditions of curing and age of concrete, the mix proportions and the type of cement. The modulus of elasticity is normally related to the compressive strength of concrete. The modulus of elasticity of concrete can be assumed as follows :

$$E_c = 5000\sqrt{f_{ck}} \quad\quad ...(1.6)$$

where $E_c$ is the short term modulus of elasticity in N/mm².

Actual measured values may differ by ± 20 percent from the values obtained from the above expression.

**Shrinkage :** The shrinkage of concrete depends upon the constituents of concrete, size of the member and environmental conditions. For a given humidity and temperature, the shrinkage of concrete is most influenced by the total amount of water present in the concrete at the time of mixing and, to a lesser extent, by the cement content.

In the absence of test data, the approximate value of shrinkage strain for design may be taken as 0.0003. For more information see IS 1343.

**Creep of concrete :** Creep of concrete depends in addition to the factors listed above under shrinkage, on the stress in the concrete, age at loading and the duration of loading. As long as the stress in concrete does not exceed one third of its characteristic compressive strength, creep may be assumed to be proportional to the stress.

**TABLE 1.10.** Values of Creep Coefficient

| Age at loading | Creep coefficient |
| --- | --- |
| 7 days | 2.2 |
| 28 days | 1.6 |
| 1 year | 1.1 |

In the absence of experimental data and detail information on the effect of the variables, the ultimate creep strain may be estimated from the following values (Table 1.10) of creep coefficient (that is ultimate creep strain divided by elastic strain at the age of the loading); for long span structure, it is advisable to determine actual creep strain, likely to take place. Note that the ultimate creep strain estimated as described above does not include the elastic strain.

**Thermal coefficient :** The coefficient of thermal expansion depends on nature of cement, the aggregate, the cement content, the relative humidity and the size of section. The value of coefficient of thermal expansion for concrete with different aggregates may be taken from Table 1.11.

**TABLE 1.11.** Values of Coefficient of Thermal Expansion for Concrete

| Type of aggregate | Coefficient of thermal expansion for concrete/°C |
| --- | --- |
| 1. Quartzite | 1.2 to 1.3 × 10⁻⁵ |
| 2. Sandstone | 0.9 to 1.2 × 10⁻⁵ |
| 3. Granite | 0.7 to 0.95 × 10⁻⁵ |
| 4. Basalt | 0.8 to 0.95 × 10⁻⁵ |
| 5. Limestone | 0.6 to 0.9 × 10⁻⁵ |

## 1.13. CONCRETE MIX PROPORTIONING (IS : 456–2000)

**1. Mix Proportion :** The mix proportions shall be selected to ensure the workability of the fresh concrete and when concrete is hardened, it shall have the required strength, durability and surface finish.

**1.1** The determination of the proportions of cement, aggregates and water to attain the required strengths shall be made as follows:

(a) By designing the concrete mix; such concrete shall be called 'Design mix concrete', or

(b) By adopting nominal concrete mix: such concrete shall be called 'Nominal mix concrete'.

Design mix concrete is preferred to nominal mix. If design mix concrete cannot be used for any reason on the work for grade of M 20 or lower, nominal mixes may be used with the permission of engineer-in-charge.

## 1.2. Information Required

*In specifying a particular grade of concrete, the following information shall be included:*

(a) Type of mix, that is, design mix concrete or nominal mix concrete:
(b) Grade designation;
(c) Type of cement;
(d) Maximum nominal size of aggregate;
(e) Minimum cement content (for design mix concrete);
(f) Maximum water-cement ratio;
(g) Workability;
(h) Mix proportion (for nominal mix concrete);
(i) Exposure conditions
(j) Maximum temperature of concrete at the time of placing;
(k) Method of placing; and
(l) Degree of supervision.

*In appropriate circumstances, the following additional information may be specified:*

(a) Type of aggregate,
(b) Maximum cement content, and
(c) Whether an admixture shall or shall not be used and the type of admixture and the condition of use.

## 2. Design Mix Concrete

**2.1.** To guarant quality of concrete used in the construction, we have to carry out the mix design.

**2.2.** The mix shall be designed to produce the grade of concrete having the required workability and a characteristic strength not less than appropriate values given in Table 1.9. The target mean strength of concrete mix should be equal to the characteristic strength plus 1.65 times the standard deviation.

**2.3.** Mix design done earlier not prior to one year may be considered adequate for later work provided there is no change in source and the quality of the materials.

## 2.4. Standard Deviation

The standard deviation for each grade of concrete shall be calculated, separately.

*2.4.1 Standard deviation based on test strength of sample*

(a) *Number of test results of samples* : The total number of test strength of samples required to constitute an acceptable record for calculation of standard deviation shall be not less than 30. Attempts should be made to obtain the 30 samples, as early as possible, when a mix is used for the first time.

(b) *In case of significant changes in concrete* : When significant changes are made in the production of concrete batches (for example changes in the materials used, mix design, equipment or technical control), the standard deviation value shall be separately calculated for such batches of concrete.

(c) *Standard deviation to be brought up to data* : The calculation of the standard deviation shall be brought up to data after every change of mix design.

*2.4.2. Assumed standard deviation*

Where sufficient test results for a particular grade of concrete are not available, the value of standard deviation given in Table 1.12 may be assumed for design of mix in the first instance. As soon as the results of sample are available, actual calculated standard deviation shall be used and the mix designed property.

**TABLE 1.12.** Assumed Standard Deviation

| Grade of concrete | Assumed Standard Deviation (N/mm²) |
|---|---|
| M 10 | 3.5 |
| M 15 | |
| M 20 | 4.0 |
| M 25 | |

However, when adequate past records for a similar grade exist and justify to the designer a value of standard deviation different from that shown in Table 1.12, it shall be permissible to use that value.

**Note:** The above values correspond to the site control having proper storage of cement; weight batching of all materials; controlled addition of water; regular checking of all materials, aggregate gradings and moisture content; and periodical checking of workability and strength. Where there is deviation from the above, the values given in the above table shall be increased by 1 N/mm$^2$.

**3. Nominal Mix Concrete :** Nominal mix concrete may be used for concrete of M 20 or lower. The proportions of materials for nominal mix concrete shall be in accordance with Table 1.13.

**TABLE 1.13.** Proportions for Nominal Mix Concrete (*Clauses 3 and 3.1*)

| Grade of Concrete | Total Quantity of Dry Aggregates by Mass per 50 kg of Cement; to be Taken as the Sum of the Individual Masses of Fine and Coarse Aggregates; kg; Max | Proportion of Fine Aggregate to Coarse Aggregate (by Mass) | Quantity of Water per 50 kg of Cement; Max litres |
|---|---|---|---|
| (1) | (2) | (3) | (4) |
| M 5 | 800 | Generally 1 : 2 but subject to an upper limit of 1 : 1$\frac{1}{2}$ and a lower limit of 1 : 2$\frac{1}{2}$ | 60 |
| M 7.5 | 625 | | 45 |
| M 10 | 480 | | 34 |
| M 15 | 330 | | 32 |
| M 20 | 250 | | 30 |

**Note:** The proportion of the fine to coarse aggregates should be adjusted from upper limit to lower limit progressively as the grading of fine aggregates becomes finer and the maximum size of coarse aggregate becomes larger. Graded coarse aggregate shall be used.

**Example:** For an average grading of fine aggregate, the proportions shall be 1 : 1$\frac{1}{2}$, 1 : 2 and 1 : 2$\frac{1}{2}$ for maximum size of aggregates 10 mm, 20 mm and 40 mm respectively.

**3.1** The cement content of the mix specified in Table 1.13 for any nominal mix shall be proportionately increased if the quantity of water in a mix has to be increased to overcome the difficulties of placement and compaction, so that the water-cement ratio as specified is not exceeded.

## 1.14. FORMWORK (IS : 456–2000)

**1. General :** The formwork shall be designed and constructed so as to remain sufficiently rigid during placing and compaction of concrete, and shall be such as to prevent loss of slurry from the concrete. For further details regarding design, detailing, etc. reference may be made to IS 14687.

**2. Cleaning and Treatment of Formwork :** All rubbish, particularly, clippings, shavings and sawdust shall be removed from the interior of the forms before the concrete is placed. The face of formwork in contact with the concrete shall be cleaned and treated with form release agent. Release agents should be applied so as to provide a thin uniform coating to the forms without coating the reinforcement.

**3. Stripping Time :** Forms shall not be released until the concrete has achieved a strength of at least twice the stress to which the concrete may be subjected at the time of removal of formwork. The strength referred to shall be that of concrete using the same cement and aggregates and admixture, if any with the same proportions and cured under conditions of temperature and moisture similar to those existing on the work.

**3.1.** While the above criteria of strength shall be the guiding factor for removal of formwork, in normal circumstances where ambient temperature does not fall below 15°C and where ordinary Portland cement is used and adequate curing is done, following striking period (Table 1.14) may deem to satisfy the guideline given in para 3.

**TABLE 1.14.** Stripping Time

| Type of Formwork | Minimum Period before Striking Formwork |
|---|---|
| (a) Vertical formwork to columns; wallsbeams | 16-24 hours |
| (b) Soffit formwork to slabs (Props to be refixed immediately after removal of formwork) | 3 days |
| (c) Soffit formwork to beams (Props to be refixed immediately after removal of formwork) | 7 days |
| (d) Props to slabs:<br>(1) Spanning up to 4.5 m<br>(2) Spanning over 4.5 m | 7 days<br>14 days |
| (e) Props to beams and arches:<br>(1) Spanning up to 6 m<br>(2) Spanning over 6 m | 14 days<br>21 days |

**3.2.** The number of props left under, their sizes and disposition shall be such as to be able to safely carry the full dead load of the slab, beam or arch as the case may be together with any live load likely to occur during curing or further construction.

**3.3.** Where the shape of the element it such that the formwork has re-entrant angles, the formwork shall be removed as soon as possible after the concrete has set, to avoid shrinkage cracking occurring due to the restraint imposed.

## 1.15. TRANSPORTING, PLACING, COMPACTION AND CURING (IS : 456–2000)

### 1. Transporting and Handling

After mixing, concrete shall be transported to the formwork as rapidly as possible by methods which will prevent the segregation or loss of any of the ingredients or ingress of foreign matter or water and maintaining the required workability.

**1.1.** During hot or cold weather, concrete shall be transported in deep containers. Other suitable methods to reduce the loss of water by evaporation in hot weather and heat loss in cold weather may also be adopted.

**2. Placing :** The concrete shall be deposited as nearly as practicable in its final position to avoid rehandling. The concrete shall be placed and compacted before initial setting of concrete commences and should not be subsequently disturbed. Methods of placing should be such as to preclude segregation. Care should be taken to avoid displacement of reinforcement or movement of formwork. As a general guidance, the maximum permissible free fall of concrete may be taken as 1.5 m.

**3. Compaction :** Concrete should be thoroughly compacted and fully worked around the reinforcement, around embedded fixtures and into corners of the formwork.

**3.1** Concrete shall be compacted using mechanical vibrators complying with IS 2505, IS 2506, IS 2514 and IS 4656. Over vibration and under vibration of concrete are harmful and should be avoided. Vibration of very wet mixes should also be avoided.

Whenever vibration has to be applied externally, the design of formwork and the disposition of vibrators should receive special consideration to ensure efficient compaction and to avoid surface blemishes.

**4. Construction Joints :** Joints are a common source of weakness and therefore, it is desirable to avoid them. If this is not possible, their number shall be minimized. Concreting shall be carried out continuously up to construction joints, the position and arrangement of which shall be indicated by the designer. Construction joints should comply with IS 11817.

Construction joints shall be placed at accessible locations to permit cleaning out of laitance, cement slurry and unsound concrete, in order to create rough/uneven surface. It is recommended to clean out laitance and cement slurry by using wire brush on the surface of joint immediately after initial setting of concrete and to clean out the same immediately thereafter. The prepared surface should be in a clean saturated surface dry condition when fresh concrete is placed, against it.

In the case of construction joints at locations where the previous pour has been cast against shuttering, the recommended method of obtaining a rough surface for the previously poured concrete is to expose the aggregate with a high pressure water jet or any other appropriate means.

Fresh concrete should be thoroughly vibrated near construction joints so that mortar from the new concrete flows between large aggregates and develop proper bond with old concrete. Where high shear resistance is required at the construction joints, shear keys may be provided. Sprayed curing membranes and release agents should be thoroughly removed from joints surfaces.

**5. Curing :** Curing is the process of preventing the loss of moisture from the concrete whilst maintaining a satisfactory temperature regime. The prevention of moisture loss from the concrete is particularly important if the water-cement ratio is low, if the cement has a high rate of strength development, if the concrete contains granulated blast furnace slag or pulverised fuel ash. The curing regime should also prevent the development of high temperature gradients within the concrete.

The rate of strength development at early ages of concrete made with supersulphated cement is significantly reduced at lower temperatures. Supersulphated cement concrete is seriously affected by inadequate curing and the surface has to be kept moist for at least seven days.

**5.1. *Moist Curing* :** Exposed surfaces of concrete shall be kept continuously in a damp or wet condition by ponding or by covering with a layer of sacking, canvas, hessian or similar materials and kept constantly wet for at least seven days from the date of placing concrete in case of ordinary Portland Cement and at least 10 days where mineral admixture or blended cement are used. The period of curing shall not be less than 10 days for concrete exposed to dry and hot weather conditions. In the case of concrete where mineral admixtures or blended cements are used, it is recommended that above minimum periods may be extended to 14 days.

**5.2. *Membrane Curing* :** Approved curing compounds may be used in lieu of moist curing with the permission of the engineer-in-charge. Such compounds shall be applied to all exposed surfaces of the concrete as soon as possible after the concrete has set. Impermeable membranes such as polyethylene sheeting covering closely the concrete surface may also be used to provide effective barrier against evaporation.

**5.3.** For the concrete containing Portland pozzolana cement, Portland slag cement or mineral admixture, period of curing may be increased.

**6. Supervision :** It is difficult to alter concrete once placed. Hence, constant and strict supervision of all the items of the construction is necessary during the progress of the work, including the proportioning and mixing of the concrete. Supervision is also of extreme importance to check the reinforcement and its placing before being covered.

**6.1.** Before any important operation, such as concreting or stripping of the formwork is started, adequate notice shall be given to the construction supervisor.

## 1.16. PERMISSIBLE STRESSES IN CONCRETE (IS : 456–2000)

### 1. DIRECT TENSILE STRESS

The direct tensile stress $f_t$ is calculated from the expression

$$f_t = \frac{F_t}{A_c + m\,A_{st}}$$

where
- $F_t$ = Total tension on member minus pre-tension in steel, if any, before concreting
- $A_c$ = Cross-sectional area of concrete, excluding any finishing material and reinforcing steel
- $m$ = Modular ratio, and
- $A_{st}$ = Cross-sectional area of reinforcing steel in tension.

For members in direct tension, when full tension is taken by the reinforcement alone, the tensile stress ($f_t$) shall be not greater than the values given below (Table 1.15):

**TABLE 1.15.** Permissible Direct Tensile Stress

| Grade of concrete | M 10 | M 15 | M 20 | M 25 | M 30 | M 35 | M 40 |
|---|---|---|---|---|---|---|---|
| Tensile stress (N/mm²) | 1.2 | 2.0 | 2.8 | 3.2 | 3.6 | 4.0 | 4.4 |

### 2. COMPRESSIVE STRESS AND BOND STRESS

The permissible compressive stresses and average bond stresses shall be as given in Table 1.16.

**TABLE 1.16.** Permissible Stresses in Concrete (IS : 456–2000)

| Grade of concrete | Permissible stress in compression (N/mm²) | | Permissible stress in Bond (Average) for plain bars in tention (N/mm²) |
| --- | --- | --- | --- |
| | Bending ($\sigma_{cbc}$) | Direct ($\sigma_{cc}$) | $\tau_{bd}$ |
| M 10 | 3.0 | 2.5 | – |
| M 15 | 5.0 | 4.0 | 0.6 |
| M 20 | 7.0 | 5.0 | 0.8 |
| M 25 | 8.5 | 6.0 | 0.9 |
| M 30 | 10.0 | 8.0 | 1.0 |
| M 35 | 11.5 | 9.0 | 1.1 |
| M 40 | 13.0 | 10.0 | 1.2 |
| M 45 | 14.5 | 11.0 | 1.3 |
| M 50 | 16.0 | 12.0 | 1.4 |

The bond stress given in Table 1.16 shall be increased by 25 percent for bars in compression. In the case of deformed bars conforming to IS : 1786, the bond stress given in Table 1.16 may be increased by 60 percent.

### 3. SHEAR STRESS

The permissible shear stress in concrete in beams without shear reinforcement shall be as given in Table 1.17.

**TABLE 1.17.** Permissible Shear Stress in Concrete (*IS : 456–2000*)

| $\dfrac{100 A_s}{bd}$ | Permissible shear stress in concrete $\tau_c$, N/mm² for grades of concrete | | | | | |
|---|---|---|---|---|---|---|
| | M 15 | M 20 | M 25 | M 30 | M 35 | M 40 and above |
| ≤ 0.15 | 0.18 | 0.18 | 0.19 | 0.20 | 0.20 | 0.20 |
| 0.25 | 0.22 | 0.22 | 0.23 | 0.23 | 0.23 | 0.23 |
| 0.50 | 0.29 | 0.30 | 0.31 | 0.31 | 0.31 | 0.32 |
| 0.75 | 0.34 | 0.35 | 0.36 | 0.37 | 0.37 | 0.38 |
| 1.00 | 0.37 | 0.39 | 0.40 | 0.41 | 0.42 | 0.42 |
| 1.25 | 0.40 | 0.42 | 0.44 | 0.45 | 0.45 | 0.46 |
| 1.50 | 0.42 | 0.45 | 0.46 | 0.48 | 0.49 | 0.49 |
| 1.75 | 0.44 | 0.47 | 0.49 | 0.50 | 0.52 | 0.52 |
| 2.00 | 0.44 | 0.49 | 0.51 | 0.53 | 0.54 | 0.55 |
| 2.25 | 0.44 | 0.51 | 0.53 | 0.55 | 0.56 | 0.57 |
| 2.50 | 0.44 | 0.51 | 0.55 | 0.57 | 0.58 | 0.60 |
| 2.75 | 0.44 | 0.51 | 0.56 | 0.58 | 0.60 | 0.62 |
| 3.00 and above | 0.44 | 0.51 | 0.57 | 0.60 | 0.62 | 0.63 |

**Note:** $A_s$ is that area of longitudinal tension reinforcement which continues at least one effective depth beyond the section being considered except at supports where the full area of tension reinforcement may be used.

## 4. MODULAR RATIO

The modular ratio $m$ has the value $\dfrac{280}{3\,\sigma_{cbc}}$ where $\sigma_{cbc}$ is the permissible compressive stress due to bending in concrete in N/mm² as specified in Table 1.16. The above expression for $m$ partially takes into account long term effects such as creep. Based on the above expression, the values of $m$ (rounded off to the nearest integral value) for various grades of concrete will be as given in Table 1.18.

**Note:** The expression given for $m$ partially takes into account long term effects such as creep. Therefore, this $m$ is not the same as the modular ratio derived based on the values of $E_c$ given in *Eq. 1.6*.

**TABLE 1.18.** Modular Ratio

| Grade of concrete | M 10 | M 15 | M 20 | M 25 | M 30 | M 35 | M 40 |
|---|---|---|---|---|---|---|---|
| Modular ratio (m) | 31 (31.11) | 19 (18.67) | 13 (13.33) | 11 (10.98) | 9 (9.33) | 8 (8.11) | 7 (7.18) |

**Note:** The figures in the bracket are the exact values.

## 5. INCREASE IN PERMISSIBLE STRESSES

Where stresses due to wind (or earthquake) and temperature effects are combined with those due to dead, live and impact load, the stresses specified in Table 1.15, 1.16, 1.17, may be exceeded upto a limit of $33\tfrac{1}{3}$ percent. Wind and seismic forces need not be considered as acting simultaneously.

## 1.17. STEEL REINFORCEMENT

Steel reinforcement used in reinforced concrete may be of the following types:
(a) 1. Mild steel bars conforming to IS : 432 (Part I)–1966.
   2. Hot rolled mild steel deformed bars conforming to IS : 1139–1966.
(b) 1. Medium tensile steel conforming to IS : 432 (Part I)–1966.
   2. Hot rolled medium tensile steel deformed bars conforming to IS : 1139–1966.
(c) 1. Hot rolled high yield strength deformed bars (HYSD bars) conforming to IS : 1139–1966.
   2. Cold-worked steel high strength deformed bars conforming to IS : 1786–1979 (Grade Fe 415 and Grade Fe 500).
(d) 1. Hard drawn steel wire fabric conforming to IS : 1566–1967.
   2. Rolled steel made from structural steel conforming to IS : 226–1975.

The most important characteristic of a reinforcing bar is its *stress strain curve* and the important property is the *yield stress* or *0.2% proof stress*, as the case may be. The permissible stresses in steel reinforcement as per IS : 456–2000 are given in Table 1.19, in which columns 2, 3 and 4 correspond to types (a), (b) and (c) respectively, mentioned above. The modulus of elasticity $E$ for these steels may be taken as $2 \times 10^5$ N/$mm^2$.

Type (a) bars have yield strength of 250 N/mm². Hence it is sometimes known as Fe 250 steel, having characteristic strength of 250 N/mm².

Type (c) bars are high yield strength bars, also known as, HYSD bars. These may be *hot rolled* high yield strength bars or *cold worked* steel high strength deformed bars. The latter are also known as CTD bars (cold twisted deformed bars) or *Tor Steel*, and are available in two grades : (i) Fe 415 (or Tor 40) and (ii) Fe 500 (Tor 50) having 0.2% proof stress as 415 N/mm² and 500 N/mm² respectively.

The hot rolled mild steel bars, in common use for nearly six decades, has low yield stress but has significant *yield plateau* having plastic strain nearly 10 times the limiting strain at yield point. Due to such an excessive yielding, there is considerable readjustment of stress in concrete so as to maintain equilibrium of internal stresses, resulting in excessive cracking and deformations in the structural member. The idealised stress strain curve for mild steel bars, is shown in curve 1 of Fig. 1.5.

The *yield plateau* can be eliminated by straining the bar beyond the yield plateau either by twisting or *stretching* and then unloading it as shown

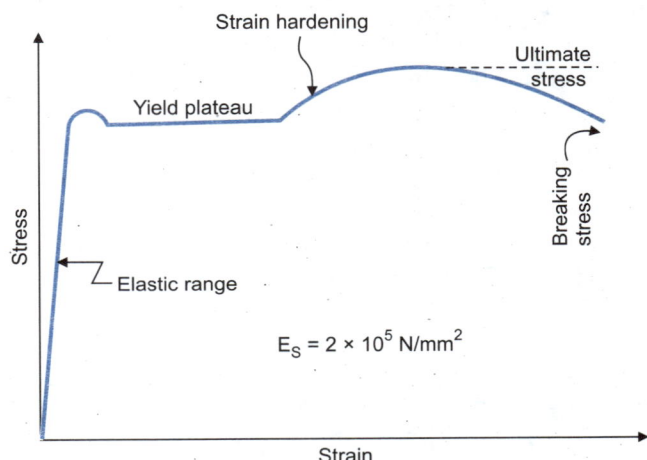

Fig. 1.4. Stress Strain Curve for Mild Steel

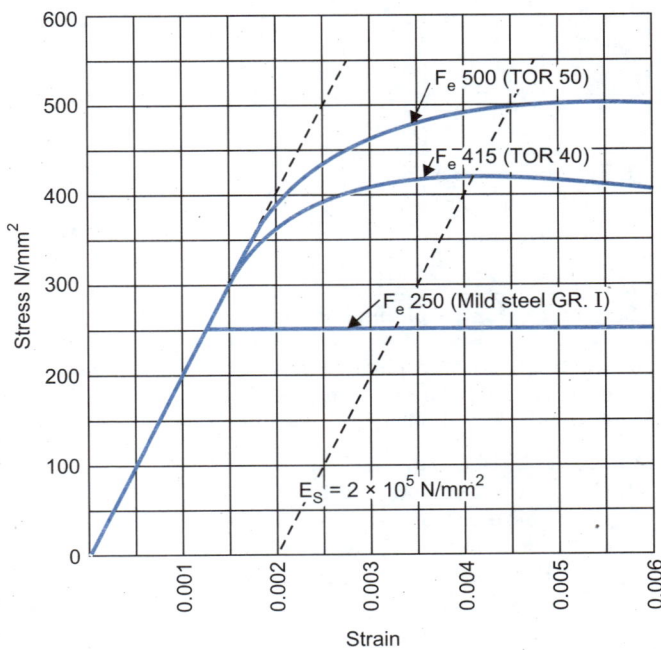

Fig. 1.5. Stress Strain Curve for CTD Bars

in Fig. 1.6. This process is known as cold working, and the *Tor-Steel reinforcement bars*, available in India are of this type. By this process, higher yield stress can be attained with much smaller strains. A *twisted bar* has considerable increased yield stress, about 50% to 100% more than the ordinary mild steel bars. Their use can permit higher working stress and hence considerable saving in quantity of steel can be achieved. Bond between concrete and steel can be improved by use of *deformed bars*.

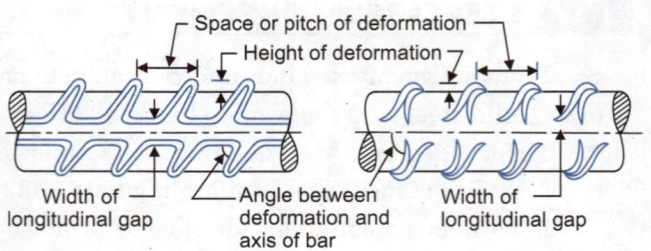

**Fig. 1.6.** Deformed Bars

A deformed bar is a bar of steel provided with lugs or ribs deformation on surface of the bar to minimise the slippage of the bar in concrete, as shown in Fig. 1.6.

The process of cold workings has the disadvantage that useful ductility of mild steel is reduced. It is to be noted that if the cold worked bar is reheated, the original yield plateau is recovered. Since cold-worked bars lack a well-defined yield plateau, yield stress is specified at 0.2% proof stress.

In reinforced concrete, a long time trend is evident towards the use of high strength materials, both steel and concrete. In big housing projects, *High yield strength deformed bars* (HYSD bars) are in common use, having yield stress (0.2 percent proof stress) equal to 230 to 415 N/$mm^2$ and permissible stress equal to 230 N/$mm^2$ for grade Fe 415. For grade Fe 500 HYSD bars, the yield stress is as high as 500 N/$mm^2$ while the permissible tensile stress is equal to 275 N/$mm^2$. *Modern high strength steels* of Fe 415 and Fe 500 grades require the simultaneous use of high strength concrete of M 20 grade and higher. Table 1.19 gives the values of permissible stresses in various types of steel reinforcement.

**TABLE 1.19.** Permissible Stresses in Steel Reinforcement

| Type of stress in steel reinforcement | Permissible stresses | | |
|---|---|---|---|
| | (a) Mild steel bars conforming to Grade I of IS : 432 (Part I) | (b) Medium tensile steel conforming to IS : 432 (Part I) | (c) High yield strength deformed bars (HYSD bars) conforming to IS : 1786 (Grade Fe 415) |
| (1) | (2) | (3) | (4) |
| 1. Tension ($\sigma_{st}$ or $\sigma_{sv}$) <br> (a) up to and including 20 mm <br> (b) over 20 mm | 140 N/$mm^2$ <br> 130 N/$mm^2$ | Half the guaranteed yield stress subject to a maximum of 190 N/$mm^2$ | 230 N/$mm^2$ <br> 230 N/$mm^2$ |
| 2. Compression in column bars ($\sigma_{sc}$) | 130 N/$mm^2$ | 130 N/$mm^2$ | 190 N/$mm^2$ |
| 3. Compression in bars in a beam or a slab where the compressive resistance of the concrete is taken into account. | The calculated compressive stress in the surrounding concrete multiplied by 1.5 times the modular ratio or $\sigma_{sc}$ whichever is lower. | | |
| 4. Compression in bars in a beam or a slab where compressive resistance of the concrete is taken into account. <br> (a) Upto and including 20 mm. <br> (b) Over 20 mm. | 140 N/$mm^2$ <br> 130 N/$mm^2$ | Half the guaranteed yield stress subject to a maximum of 190 N/$mm^2$ | 190 N/$mm^2$ <br> 190 N/$mm^2$ |

**Note 1.** For high yield strength deformed bars of Grade Fe 500, the permissible stress in direct tension and flexural tension shall be 0.55 $f_y$. The permissible stresses for shear and compression reinforcement shall be as for Grade Fe 415.

**Note 2.** For welded wire fabric conforming to IS : 1566, the permissible value in tension is 230 N/mm².

**Note 3.** For the purpose of this Code, the yield stress of steels for which there is no clearly defined yield point should be taken to be 0.2 percent proof stress.

**Note 4.** When mild steel conforming to Grade II of IS : 432 (Part I) is used the permissible stresses shall be 90 percent of permissible stress in column 2, or if the design details have already been worked out on the basis of mild steel conforming to Grade I of IS : 432 (Part I), the area of reinforcement shall be increased by 10 percent of that required for Grade I steel.

**Per cent Elongation.** Minimum requirement of *% elongation* should be as follows :

TABLE 1.20

|  | M.S. bars Grade I | Medium tensile steel bars | Medium tensile steel bars | Medium steel deformed bars | HYSD and CTD bars |
|---|---|---|---|---|---|
| Dia < 10 mm | 20 | 17 | 23 | 20 | 14.5 |
| Dia ≥ 10 mm | 23 | 20 | 23 | 20 | 14.5 |

**Characteristic Strength of Steel Reinforcement :** The *characteristic strength* ($f_y$) means that value of strength below which not more than 5% of the test results are expected to fall. Until the relevant Indian Standard specifications for reinforcing steel are modified to include the concept of characteristic strength, the characteristic value shall be assumed as the minimum yield/0.2 percent proof stress specified in the relevant Indian Standard Codes. The values of characteristic strength for three common types of reinforcements are given in Table 1.21.

TABLE 1.21.  Characteristic Strength of Steel Reinforcement

| Type of reinforcement | Yield stress or 0.2% proof stress (N/mm²) | Characteristic strength $f_y$ (N/mm²) | Permisible tensile strength $\sigma_{st}$ (N/mm²) |
|---|---|---|---|
| 1. Mild steel bars conforming to Grade I of IS : 432 (Part I) or deformed m.s. bars of I.S. 1139 | 250 (average) | 250 | 140 (upto 20 mm dia.) 130 (over 20 mm dia.) |
| 2. High yield strength deformed bars conforming to IS : 1109 or Grade Fe 415 of IS : 1786–1979 | 415 | 415 | 230 |
| 3. High yield strength deformed bars of Grade Fe 500 of IS : 1786–1979 | 500 | 500 | 275 |

## PROBLEMS

1. Discuss in brief various types of cements used in cement concretes.
2. Discuss salient specifications for Portland cement used for cement concrete.
3. Write a note on 'aggregates' used for cement concrete.
4. Explain the following: (*i*) Determination of fineness modulus (*ii*) Grading aggregates (*iii*) Quality of water for cement concrete (*iv*) Durability of concrete (*v*) Workability of Concrete (*vi*) Water cement Ratio.
5. Discuss the salient properties of various grades of concrete as per IS : 456–2000. What are the corresponding nominal mixes?
6. Explain in brief various types of reinforcement used in R.C.C.

# PART – I
# WORKING STRESS METHOD

2. THEORY OF REINFORCED BEAMS AND SLABS
3. SHEAR AND BOND
4. TORSION
5. DOUBLY REINFORCED BEAMS
6. DESIGN OF T AND L-BEAMS
7. DESIGN OF BEAMS AND SLABS
8. DESIGN OF STAIR CASES
9. REINFORCED BRICK AND HOLLOW TILE ROOFS
10. TWO-WAY SLABS
11. CIRCULAR SLABS
12. FLAT SLABS
13. DESIGN OF AXIALLY LOADED COLUMNS
14. COLUMNS WITH AXIAL LOAD AND MOMENTS
15. DESIGN OF ISOLATED FOOTINGS
16. COMBINED FOOTINGS
17. PILE FOUNDATIONS
18. DESIGN OF RETAINING WALLS

# 2
# THEORY OF REINFORCED BEAMS AND SLABS

## 2.1. INTRODUCTION

Plain cement concrete has high compressive strength, but its tensile strength is low. Normally, the tensile strength of concrete is about 10 to 15% of its compressive strength. Hence, if a beam is made of plain cement concrete, it has a very low load carrying capacity since its low tensile strength limits its overall strength. It is, therefore, *reinforced* by placing steel bars in the tensile zone of the concrete beam so that the compressive bending stress is carried by concrete and tensile bending stress is carried entirely by steel reinforcing bars.

Fig. 2.1(*a*) shows a simply supported reinforced concrete beam subjected to transverse loads, bending it downwards. The reinforcement, consisting of steel bars, is placed at a suitable depth below the *neutral axis*. Similarly, Fig. 2.1(*b*) shows a cantilever bending downwards. Since the tensile zone is above the neutral axis in this case, the steel bars are provided at some suitable height above the neutral axis. In both the cases, steel reinforcement is provided in the tensile zone only; such beams are known as singly reinforced beams. However, if reinforcement is provided in compressive zone also, to carry the compressive stresses, it is known as a *doubly reinforced section* (see Chapter 4).

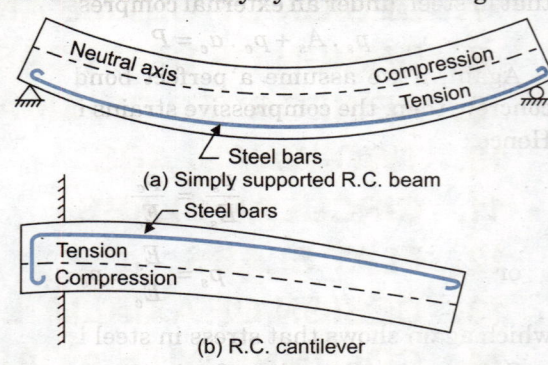

Fig. 2.1. Reinforced Concrete Beam.

The concrete, reinforced by steel bars, is called *reinforced cement concrete* and is abbreviated as R.C.C. Reinforced cement concrete is thus a rational union of concrete and steel combined to act jointly. This joint action of steel and concrete in a reinforced concrete section is dependent on the following important factors :

(*i*) Bond between concrete and steel bars.

(*ii*) Absence of corrosion of steel bars embedded in the concrete.

(*iii*) Practically equal thermal expansion of both concrete and steel.

## 2.2. MODULAR RATIO

Fig. 2.2 shows a composite section consisting of two bars, one of concrete and other of steel, well bonded together with the help of end plugs, and subjected to a compressive load $P$.

Let $E_s$ = Modulus of elasticity of steel and

$E_c$ = Modulus of elasticity of concrete.

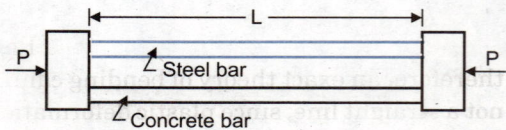

Fig. 2.2. Composite Section.

Let the area of cross-section of each bar be equal to $A$. Let $P_s$ be the load carried by steel and $P_c$ be the load carried by concrete. Hence from statics:

$$P_s + P_c = P \qquad \ldots(1)$$

Also, from compatibility,

$$\Delta_s = \Delta_c = \Delta$$

where $\Delta$ is the shortening of each bar under the external load.

Hence
$$\frac{P_s L}{A E_s} = \frac{P_c L}{A E_c} \quad \text{or} \quad \frac{P_s}{P_c} = \frac{E_s}{E_c} \qquad \ldots(2)$$

From (1) and (2) we get

$$P_s = \frac{P \cdot E_s}{E_c + E_s} = \frac{mP}{1+m} \qquad \ldots(2.1)$$

and
$$P_c = \frac{P \cdot E_c}{E_c + E_s} = \frac{P}{1+m} \qquad \ldots[2.1(a)]$$

where $m = E_s/E_c$ = modular ratio.

From *Eqs. 2.1* and *2.1(a)*, it is seen that the load carried by steel is $m$ times the load carried by concrete.

### Equivalent Areas of Composite Section

Fig. 2.3 shows a reinforced concrete column having area of cross-section $A_s$ of steel bars and $a_c$ of that of concrete such that $A = A_s + a_c$. If $p_c$ is the stress set up in concrete and $p_s$ that in steel, under an external compressive load $P$, we have,

$$p_s \cdot A_s + p_c \cdot a_c = P \qquad \ldots(1)$$

Again, if we assume a perfect bond between steel and concrete then, the compressive strains in both will be equal. Hence :

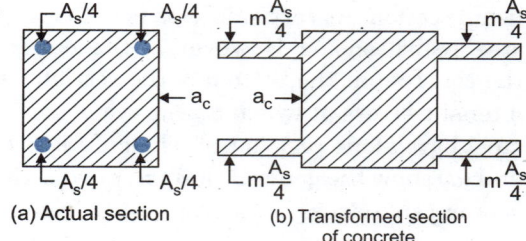

(a) Actual section  (b) Transformed section of concrete

Fig. 2.3

$$\frac{p_s}{E_s} = \frac{p_c}{E_c}$$

or
$$p_s = \frac{E_s}{E_c} \cdot p_c = m \cdot p_c \qquad \ldots(2)$$

which again shows that stress in steel is equal to $m$ times the stress in concrete.

Substituting the value of $p_s$ in (1), we get $A_s \cdot m \cdot p_c + a_c \cdot p_c = P$

$$p_c = \frac{P}{a_c + m \cdot A_s} = \frac{P}{A_e} \qquad \ldots(2.2)$$

In *Eq. 2.2*, the expression $(a_c + m A_s)$ is called the *equivalent area of the section in terms of concrete*. In other words, the area of steel $A_s$ can be replaced by an equivalent area of concrete equal to $m A_s$ as shown in Fig. 2.2(a).

Again
$$A_e = a_c + mA_s = (A - A_s) + m A_s = A + (m-1) A_s \qquad \ldots(2.3)$$

where $A$ is the total cross-sectional area of the original column section.

## 2.3. SINGLY REINFORCED BEAM

(*a*) **Fundamental Assumptions.** Reinforced concrete beams are non-homogeneous in nature and, therefore, an exact theory of bending cannot be developed. The stress-strain curve for reinforced concrete is not a straight line, since plastic deformations accompany the elastic strains. It thus follows that Hooke's law applies only in a narrow range of stress. There are two fundamental theories on which analysis of reinforced concrete can be done : (*i*) *straight-line theory* or the *elastic theory*, and (*ii*) *ultimate load theory* or *limit state* *theory*. In the straight line theory the ultimate compressive strength of concrete and the yield point stress of steel are divided by the same or different appropriate safety factors to determine the allowable stresses that may be permitted in the two materials under working conditions. In the ultimate load theory, the resistance of the beam to pure bending is determined either by the ultimate strength of concrete or by the yield point stress of steel, and then this resistance can be divided by proper safety factor to determine the bending resistance that can be relied upon under working conditions. We shall first take up the straight line theory or the elastic theory. The limit state design has been discussed separately in detail.

The application of elastic theory to the beams is based on the following assumptions:

1. At any cross-section, plane sections before bending remain plane after bending. This means that unit strain above and below the neutral axis are proportional to the distance from the neutral axis.

2. The concrete and steel reinforcement are well bonded. This means that the tensile strain in concrete surrounding the steel is equal to the tensile strain in the steel reinforcement.

3. All tensile stresses are taken by the reinforcement alone and none by the concrete. This means that while calculating the moment of resistance of the beam, the contribution of the concrete in the tensile zone is to be completely neglected.

4. Modulus of elasticity of concrete is constant at all stresses and is not a function of duration of stress. Young's modulus of concrete is generally found by secant method from the stress-strain curve.

5. Steel reinforcement is free from initial stresses when it is embedded in concrete.

Fig. 2.4(*a*) shows a singly reinforced concrete beam section. At very small loads both concrete and steel resist tension. However, as the load increases, the concrete at the bottom of the section will reach a tensile stress at which cracks occur, and thus the steel bars alone resists the tension. Figs. 2.4(*b*), (*c*) and (*d*) show the general character of distribution of compressive stress in concrete as the load increases. Since the stress-strain diagram for concrete [Fig. 2.4(*f*)] is not linear, the stresses in the concrete do not increase in direct proportion to increase in strain. Therefore, the stress diagram is slightly curved. The straight line theory assumes that the stress distribution in the concrete at working stresses of approximately half the ultimate compressive strength of concrete ($\sigma_{cu}$) is like that of Fig. 2.4(*c*) but the slight curvature in the compression diagram is neglected.

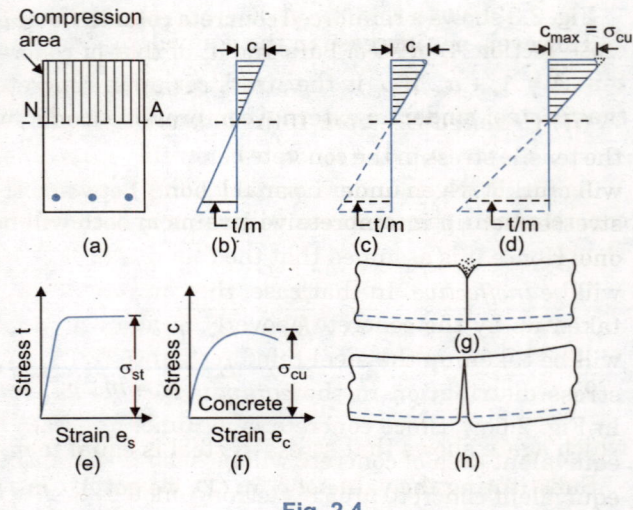

Fig. 2.4

When the loads increase further so that maximum compressive stress occurs at the top of the section, the concrete flows plastically. The corresponding stress diagram in concrete is shown in Fig. 2.4(*d*), where $\sigma_{cu}$ represents the ultimate compressive strength of concrete. At such a stage, the beam may continue to deform without offering proportionately increased resistance, and a small increase in load may cause complete failure of beam, without sufficient warning. Such compression failure is indicated in Fig. 2.4(*g*). However, if the tensile stress in steel reaches its ultimate value first, the beam would "pop open" [Fig. 2.4(*h*)] and the beam will fail in tension or by local crushing of concrete at top of the concrete. However, upto the elastic limit of steel, the reinforcing bars resist tension in proportion to their strain.

### (*b*) Equivalent Area

(*i*) *Uncracked Section.* Fig. 2.5(*a*) shows a R.C. beam section having $A_{st}$ = area of steel reinforcement, $a_c$ = area of concrete and $A$ = total area of cross-section. If the load on the beam is *small* so that the tensile stress set-up in the concrete below the neutral axis is smaller than the permissible, the concrete area below the N.A. will not crack. In that case, $e_s = e_c$ at the centre of the reinforcement.

$$\therefore \qquad \frac{t}{E_s} = \frac{c'}{E_c}$$

or $\qquad t = \dfrac{E_s}{E_c} c' = m \cdot c' \qquad$ ...(2.4)

or $\qquad c' = \dfrac{t}{m} \qquad$ ...[2.4(a)]

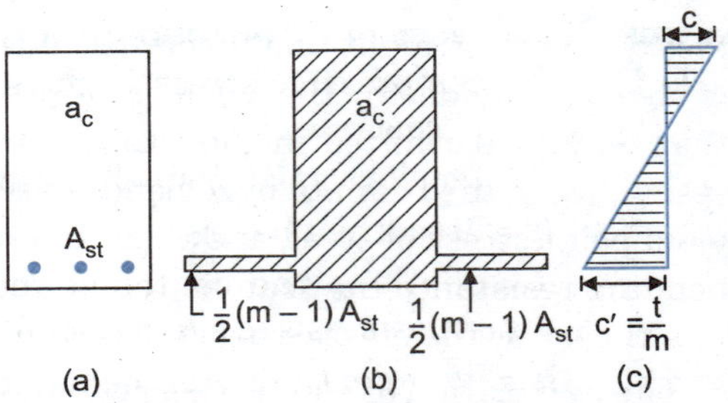

where $e_s$ = strain in steel

$e_c$ = strain in concrete surrounding steel

$t$ = stress in steel

$c'$ = stress in concrete surrounding steel, assuming that concrete is not cracked.

$c$ = compressive stress in concrete at the extreme fibre.

Fig. 2.5. Equivalent Area : Uncracked Section.

Eq. 2.4 suggests that stress in steel is $m$ times the stress in concrete, or conversely, the stress in concrete is $1/m$ times the stress in steel. Fig. 2.5(c) shows the stress distribution in the beam section consisting of an equivalent concrete section of Fig. 2.5(b). If $a_c'$ is the additional area of concrete equivalent to steel area, we have

Load carried by steel = Load carried by equivalent concrete

or $\qquad A_{st} \cdot t = a_c' \, c' \quad$ or $\quad a_c' = \dfrac{t}{c'} \cdot A_{st} = m \, A_{st}$ (since $t/c' = m$) $\qquad$ ...(2.5)

However, net increase in the equivalent concrete area (uncracked) is

$$a_{ce} = a_c' - A_{st} = m \, A_{st} - A_{st} = (m-1) \, A_{st} \qquad \text{...[2.5(a)]}$$

**(ii) Cracked Section.** If the load on the beam is increased, the tensile stress in the concrete below the N.A. will increase and will crack it. Even under normal loading conditions, the tensile stresses set up in concrete will be more than the permissible one. Hence it is assumed that the concrete area in tension zone will be *ineffective*. In that case, the compressive stress will be taken up by the concrete above N.A. and the tensile stresses will be taken up the steel reinforcement. Fig. 2.6(c) shows the stress distribution in the *equivalent concrete section* shown in Fig. 2.6(b). Since concrete is assumed to have cracked, the equivalent area of concrete will be such that the load carried by steel is the same as the load carried by an equivalent concrete area $a_{ce}$ (strong enough to carry that tensile load),

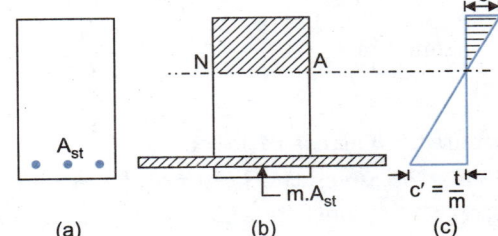

Fig. 2.6. Equivalent area: Cracked Section.

*i.e.,* $\qquad A_{st} \cdot t = a_{ce} \cdot c' \quad$ or $\quad a_{ce} = \dfrac{t}{c'} A_{st} = m \, A_{st} \qquad$ ...(2.6)

where $c'$ is the tensile stress in the *equivalent concrete area*.

The difference between the stress diagrams of Fig. 2.5(c) and 2.6(c) must be clearly borne in mind. In Fig. 2.5(c), the tensile stress distribution is all over the portion of the concrete below the N.A., while in Fig. 2.6(c), the tensile stress $c'$ (= $t/m$) is in the equivalent concrete area concentrated at the level of steel reinforcement.

## 2.4. NEUTRAL AXIS OF BEAM SECTION

In order to determine the position of the neutral axis in a reinforced concrete beam section, we shall consider two cases: (*i*) when the stresses developed in the concrete section are known and (*ii*) when the dimensions of the section are known but stresses are not known. Fig. 2.7(a) shows the reinforced concrete beam section having steel reinforcement placed in the tension zone. Such a section is known as *singly reinforced section*.

## Case (i) Stresses in Concrete and Steel given

Let  $c$ = compressive stress in the extreme fibre of concrete

$t$ = tensile stress in steel reinforcement

$b$ = breadth of beam

$d$ = depth to the centre of reinforcement (known as the *effective depth*)

$n = kd$ = depth of N.A. below the top of the beam

$k$ = neutral axis depth factor = $n/d$

$A_{st}$ = area of tensile reinforcement,

Since there is no resultant force across the section,

Total Compression = Total Tension.

Neglecting the tensile stress in concrete, we get $\frac{1}{2} c \cdot b \cdot kd = t \cdot A_{st}$

or
$$\frac{A_{st}}{b \cdot d} = k \cdot \frac{c}{2t} \qquad \ldots(2.7)$$

If $t$ is the stress in steel, the stress $c'$ in the equivalent concrete area will be equal to $t/m$ as proved in Eq. 2.4. Hence from the diagram [Fig. 2.7(c)], we get

$$\frac{c}{k \cdot d} = \frac{t}{m} \cdot \frac{1}{d - kd} = \frac{t}{md(1-k)}$$

$$\frac{c}{k} = \frac{t}{m(1-k)} \quad \text{or} \quad mc - mck = kt$$

From which
$$k = \frac{mc}{mc + t} = \frac{1}{1 + \frac{1}{m}\frac{t}{c}} = \frac{1}{1 + \frac{r}{m}} \qquad \ldots(2.8)$$

where $r$ = stress ratio = $t/c$

Eq. 2.8 gives the location of the neutral axis when the stresses in concrete and the steel are known. If the stress ratio $r$ (= $t/c$) and the modular ratio $m$ are known, N.A. can be located from Eq. 2.8 Fig. 2.8 shows the graphical representation of Eq. 2.8 for various values of stress ratio and modular ratio.

Let, $\sigma_{st}$ = Permissible tensile stress in reinforcement (*i.e.*, permissible value of $t$)

$\sigma_{cbc}$ = Permissible compressive stress in concrete due to bending (*i.e.*, permissible value of $c$)

$k_c$ = Critical neutral axis depth factor

$n_c$ = Depth of critical neutral axis, corresponding to permissible values of stresses in steel and concrete.

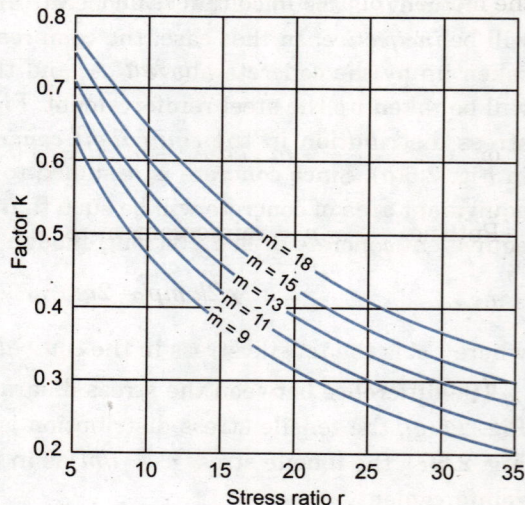

Fig. 2.8. Determination of $k$

Then,
$$k_c = \frac{m \sigma_{cbc}}{m \sigma_{cbc} + \sigma_{st}} \qquad \ldots[2.8(a)]$$

$$n_c = k_c \cdot d = \frac{m \sigma_{cbc}}{m \sigma_{cbc} + \sigma_{st}} \cdot d = \frac{1}{1 + \frac{r_c}{m}} d \qquad \ldots[2.8(b)]$$

where $r_c$ = stress ratio = $\sigma_{st}/\sigma_{cbc}$

The section in which the stresses in concrete and steel reach their corresponding permissible values simultaneously is called a *balanced section* (See § 2.6).

**Note.** As per IS : 456 – 2000, the modular ratio is given by

$$m = \frac{280}{3\,\sigma_{cbc}} \quad \text{or} \quad m\,.\,\sigma_{cbc} = 280/3 = 93.33$$

Hence *Eq. 2.8* reduces to the following form

$$k_c = \frac{m\,\sigma_{cbc}}{m\,\sigma_{cbc} + \sigma_{st}} = \frac{93.33}{93.33 + \sigma_{st}} = \frac{1}{1 + 0.0107\,\sigma_{st}} \qquad \ldots[2.8(c)]$$

This suggests that value of $k_c$ does not depend upon the grade of the concrete. It depends only on the permissible stress in reinforcement See Table 2.1.

The area of steel for the section is given by *Eq. 2.7*. If $p$ is the *percentage steel*, we have

$$p = \frac{A_{st}}{b\,.\,d} \times 100\,; \quad \text{But} \quad \frac{A_{st}}{b\,.\,d} = \frac{k\,.\,c}{2t} \quad \text{(from } Eq.\ 2.7\text{)}$$

$$\therefore \quad p = \frac{k\,.\,c}{2t} \times 100 = 50\,\frac{c}{t}\,.\,\frac{mc}{mc+t} = \frac{50\,mc^2}{t(mc+t)} \qquad \ldots(2.9)$$

Expressed in terms of the permissible values of the stresses, the above equation may be written as :

$$p_c = \frac{50\,m\,\sigma_{cbc}^2}{\sigma_{st}\,(m\,\sigma_{cbc} + \sigma_{st})} \qquad \ldots[2.9(a)]$$

### Case (*ii*) Dimensions of the Beam Known

Let us now take the case when the dimensions of the section, including the reinforcement are known, but the stresses in concrete and steel are *not known*. Refer Fig. 2.7. The neutral axis of a homogeneous beam is the horizontal line that passes through the centre of gravity of the cross-section. Hence equating the moment of area in compression to the moment of the *equivalent area* in tension about the N.A., we get,

$$b\,.\,kd\,.\,\frac{kd}{2} = mA_{st}\,(d - kd)$$

or $\quad k^2\,.\,\dfrac{bd^2}{2} + m\,.\,kd\,A_{st} - m\,.\,dA_{st} = 0 \quad$ or $\quad k^2 + 2k\,.\,m\,.\,\dfrac{A_{st}}{bd} - 2m\,\dfrac{A_{st}}{bd} = 0$

Putting $\dfrac{A_{st}}{bd} = p' = $ reinforcement ratio, we get

$$k^2 + 2k\,mp' - 2mp' = 0,$$

From which,
$$k = \frac{-2mp' \pm \sqrt{4m^2 p'^2 + 8mp'}}{2}$$

$$= \sqrt{2mp' + (mp')^2} - mp'. \qquad \ldots(2.10)$$

Thus if $p'$ is known (*i.e.*, if the dimensions $b$ and $d$, and the area of steel $A_{st}$ are known) the neutral axis depth factor $k$ can be calculated from *Eq. 2.10* and the N.A. can be located.

## 2.5. MOMENT OF RESISTANCE

Since there is no resultant force across the section, the total compressive force acting at the centre of gravity of the compressive forces is equal to the total tensile force acting at the centre of gravity of the steel reinforcement. Hence the moment of resistance of the reinforced concrete beam section is equal to the moment of the couple consisting of the compressive force and the tensile force. The total compressive force $C = \frac{1}{2} c.k.b.d$ $\left(= \frac{1}{2} \sigma_{cbc}.k_c d.b\right)$ and acts at a distance of $kd/3$ from the top of the section. Similarly, the total tensile force $T = t.A_{st} (= \sigma_{st} A_{st})$ and acts at a distance $d$ from the top of the section. If $jd$ is the lever arm (i.e., the distance between the compressive force and the tensile force), we have

Lever arm $\qquad a = jd = d - \dfrac{kd}{3} = d\left(1 - \dfrac{k}{3}\right)$

or $\qquad j = 1 - \dfrac{k}{3}$ ...(2.11)

*Eq. 2.11 gives the lever arm factor $j$ in terms of the neutral axis depth factor $k$.* Now, moment of resistance $M_r$ is given by

$$M_r = \text{force (either compressive or tensile)} \times \text{lever arm}$$

∴ $\qquad M_r = \dfrac{1}{2} c.kd.b(jd) = \left(\dfrac{1}{2}cjk\right)bd^2$ ...[2.12(a)]

or $\qquad M_r = Rbd^2$ ...(2.12)

where $\qquad R = \dfrac{1}{2} cjk \quad \text{or} \quad R_c = \dfrac{1}{2} \sigma_{cbc}.j k_c$ ...(2.13)

Table 2.1 gives the values of design constants $k_c$, $j$, $R_c$, and $p_c$ for balanced design.

**TABLE 2.1.** Values of Design Constants.

| Grade of concrete → | | M 15* | M 20 | M 25 | M 30 | M 35 | M 40 |
|---|---|---|---|---|---|---|---|
| Modular ratio $m$ → | | 18.67 | 13.33 | 10.98 | 9.33 | 8.11 | 7.18 |
| $\sigma_{cbc}$ N/mm² → | | 5.0 | 7.0 | 8.5 | 10.0 | 11.5 | 13.0 |
| $m\sigma_{cbc}$ | | 93.33 | 93.33 | 93.33 | 93.33 | 93.33 | 93.33 |
| (a) $\sigma_{st} = 140$ N/mm² (Fe 250) | $k_c$ | 0.400 | 0.400 | 0.400 | 0.400 | 0.400 | 0.400 |
| | $j$ | 0.867 | 0.867 | 0.867 | 0.867 | 0.867 | 0.867 |
| | $R_c$ | 0.867 | 1.214 | 1.474 | 1.734 | 1.994 | 2.254 |
| | $p_c$ (%) | 0.714 | 1.000 | 1.214 | 1.429 | 1.643 | 1.857 |
| (b) $\sigma_{st} = 190$ N/mm² | $k_c$ | 0.329 | 0.329 | 0.329 | 0.329 | 0.329 | 0.329 |
| | $j$ | 0.890 | 0.890 | 0.890 | 0.890 | 0.890 | 0.890 |
| | $R_c$ | 0.732 | 1.025 | 1.244 | 1.464 | 1.684 | 1.903 |
| | $p_c$ (%) | 0.433 | 0.606 | 0.736 | 0.866 | 0.997 | 1.127 |
| (c) $\sigma_{st} = 230$ N/mm² (Fe 415) | $k_c$ | 0.289 | 0.289 | 0.289 | 0.289 | 0.289 | 0.289 |
| | $j$ | 0.904 | 0.904 | 0.904 | 0.904 | 0.904 | 0.904 |
| | $R_c$ | 0.653 | 0.914 | 1.110 | 1.306 | 1.502 | 1.698 |
| | $p_c$ (%) | 0.314 | 0.440 | 0.534 | 0.628 | 0.722 | 0.816 |
| (d) $\sigma_{st} = 275$ N/mm² (Fe 500) | $k_c$ | 0.253 | 0.253 | 0.253 | 0.253 | 0.253 | 0.253 |
| | $j$ | 0.916 | 0.916 | 0.916 | 0.916 | 0.916 | 0.916 |
| | $R_c$ | 0.579 | 0.811 | 0.985 | 1.159 | 1.332 | 1.506 |
| | $p_c$ (%) | 0.230 | 0.322 | 0.391 | 0.460 | 0.530 | 0.599 |

\* As per IS 456 : 2000, the minimum grade of concrete shall be not less than M 20 in R.C.C. work.

From *Eq. 2.12*, the dimensions of the beam section can be determined to develop a given moment of resistance. *It should be noted that for a given type of concrete and steel reinforcement, factor $R_c$ is a constant and does not depend upon the beam dimensions.*

Also, $M_r = t \cdot A_{st} \cdot jd = \sigma_{st} \cdot A_{st} \cdot jd$ ...(2.14)

Hence, $A_{st} = \dfrac{M_r}{t \cdot jd} = \dfrac{M}{t \cdot jd} = \dfrac{M}{\sigma_{st} \cdot jd}$ ...(2.15)

where $M$ is the external bending moment and $\sigma_{st}$ is permissible stress in steel reinforcement.

*Eq. 2.15* is used for determining the cross-sectional area of steel reinforcement for given moment of resistance and effective depth.

If the section of a beam, including the steel reinforcement, is completely known, its moment of resistance is calculated both *Eqs. 2.12(a)* and *2.14*, using maximum permissible values of stresses $c(=\sigma_{cbc})$ and $t(=\sigma_{st})$. Thus the values of $M_r$ will be obtained and if the two values do not agree, it will be concluded that both concrete and steel do not develop their allowable stresses simultaneously. The moment of resistance of the beam will then be equal to the lesser of the two values.

## 2.6. BALANCED, UNDER-REINFORCED AND OVER-REINFORCED SECTION

(*a*) **Balanced Section.** In a beam section, if the area of steel reinforcement $A_{st}$ is of such magnitude that the permissible stresses $c(=\sigma_{cbc})$ and $t(=\sigma_{st})$ in concrete and steel respectively, are developed simultaneously, the section is known as the balanced section, critical section or *economical section*. For such a section (Fig. 2.9), the neutral axis factor $k_c$ is determined by *Eq. 2.8(b)*, using $c(=\sigma_{cbc})$ and $t(=\sigma_{st})$ as the maximum permissible stress in the two materials and depth $n_c$ of the critical N.A. is taken equal to $k_c$ times $d$. For such a balanced section, the moment of resistance obtained from the compressive force will be equal to the moment of resistance obtained from the tensile force.

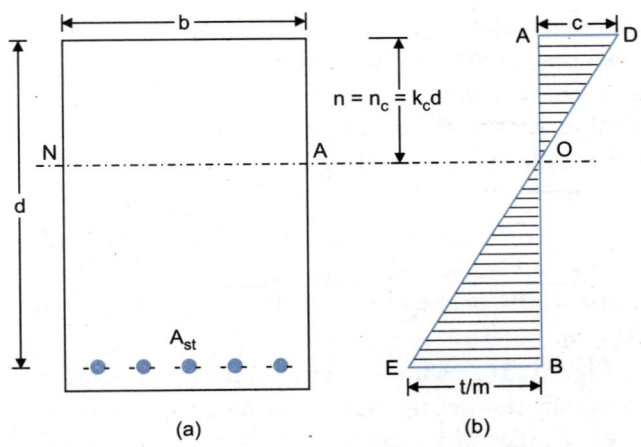

Fig. 2.9. Balanced Section.

The percent area of reinforcement for a balanced section is obtained from *Eq. 2.9(a)*:

$$p_c = \dfrac{1}{2} k \cdot \dfrac{c}{t} \times 100 = \dfrac{1}{2} k_c \cdot \dfrac{\sigma_{cbc}}{\sigma_{st}} \times 100$$

or $p = \dfrac{50 \, k_c}{r_c}$, where $r_c$ is the stress ratio *i.e.*, $\dfrac{\sigma_{st}}{\sigma_{cbc}}$

For example, if $c = \sigma_{cbc} = 5$ N/mm$^2$, $t = \sigma_{st} = 140$ N/mm$^2$ and $m = 19$, we have

$$k_c = \dfrac{1}{1 + \dfrac{1}{m}\dfrac{t}{c}} = \dfrac{1}{1 + \dfrac{140}{19 \times 5}} = 0.404$$

and $p_c = \dfrac{1}{2} \times 0.404 \times \dfrac{5}{140} \times 100 = 0.72\%$

Thus for a reinforced concrete having the above properties if the reinforcement provided is 0.72% of the cross-section ($b \times d$), a balanced section will be obtained. The N.A. of a balanced section is called the *critical neutral axis*.

(*b*) **Under-Reinforced Section.** An under-reinforced section is the one in which the percentage steel provided is *less* than that given in *Eq. 2.9* and therefore full strength of concrete in compression is not developed. The actual neutral axis of such a section will fall *above* the critical neutral axis of a balanced section.

Fig. 2.10(a) shows the under-reinforced section in which the actual N.A. is above the critical N.A. Let $c = \sigma_{cbc}$ be the permissible stress in concrete and $t = \sigma_{st}$ be the permissible stress in steel. If full permissible compressive stress $c = \sigma_{cbc} = AD_1$ is permitted to be developed in concrete, the corresponding stress $t_1$ in steel will be such that $BE_1 = t_1/m$. Evidently, since $t_1 > t$, full compressive stress $c$ cannot develop in concrete. Instead, a compressive stress $c_1 = AD$ is developed of such a magnitude that $BE = t/m = \sigma_{st}/m$ is obtained. Thus, in a under-reinforced concrete, the concrete is not fully stressed to its permissible value when stress in steel reaches its maximum value of $t = \sigma_{st}$. The moment of resistance of an under-reinforced section is, therefore, computed on the basis of the tensile force in steel :

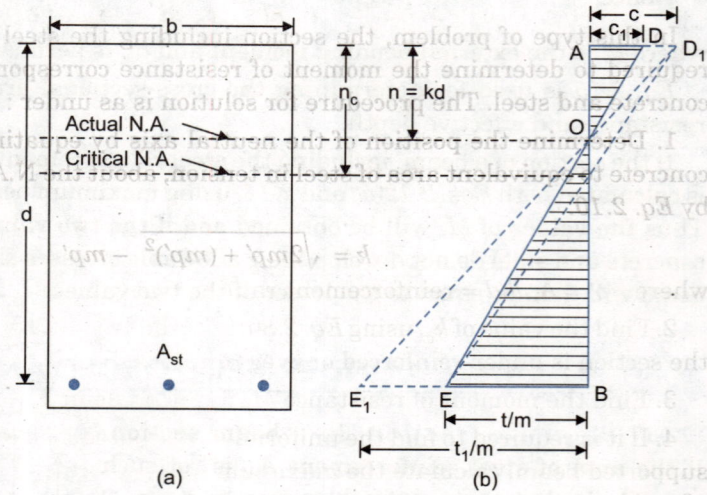

Fig. 2.10. Under-reinforced Section.

$$M_r = t \, A_{st} \cdot jd = \sigma_{st} \cdot A_{st} \cdot jd \qquad \text{(From Eq. 2.14)}$$

**(c) Over-Reinforced Section.** In an over-reinforced section, the reinforcement provided is more than critical one and therefore the actual N.A. of such a section falls below the critical N.A. of a balanced section.

Fig. 2.11(a) shows a over-reinforced section in which the actual N.A. falls below the critical one. If stress in steel is permitted to be equal to maximum permissible value $t = (\sigma_{st})$ the corresponding stress in concrete will be equal to $AD_1 = c_1$ which is greater than permissible value $c = (\sigma_{cbc})$ in concrete. Hence the stress distribution will be governed by line $DOE$ (and not $D_1 O E_1$), giving rise to a stress $t_1$ in steel which is lesser than $t$. Thus, in a over-reinforced section, steel reinforcement is not fully stressed to its permissible value and the moment of resistance is determined on the basis of compressive force developed in concrete :

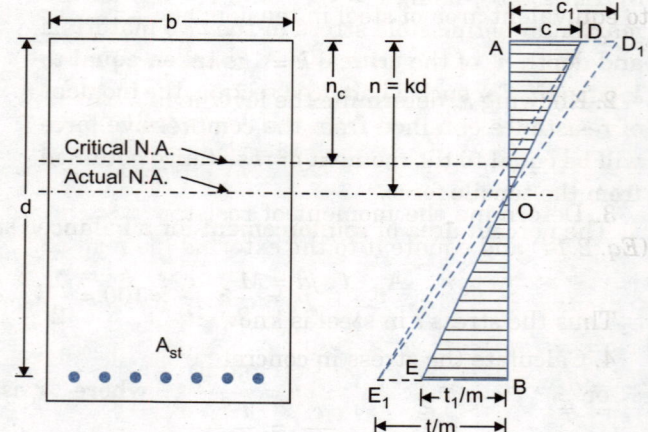

Fig. 2.11. Over-reinforced Section.

$$M_r = \tfrac{1}{2} c \cdot kd \cdot b \cdot jd = \tfrac{1}{2} c \cdot j \cdot k \cdot bd^2 = \tfrac{1}{2} \sigma_{cbc} \cdot j \cdot k \cdot bd^2 \qquad \ldots[2.12(a)]$$

or

$$M_r = \tfrac{1}{2} c \cdot n \cdot \left(d - \tfrac{n}{3}\right) b = \tfrac{1}{2} \sigma_{cbc} \cdot n \left(d - \tfrac{n}{3}\right) b \qquad \ldots[2.12(b)]$$

where $n$ = depth of the neutral axis.

## 2.7. TYPES OF PROBLEMS IN SINGLY-REINFORCED BEAMS

There may be three types of problems in singly-reinforced concrete beams :

Type I : Determination of the moment of resistance of the given section.

Type II : Determination of stresses in the given section subjected to a given bending moment.
Type III : Design of section to resist a given bending moment.

### Type I : Determination of Moment of Resistance

In this type of problem, the section including the steel reinforcement is completely defined and it is required to determine the moment of resistance corresponding to a given set of permissible stresses in concrete and steel. The procedure for solution is as under :

1. Determine the position of the neutral axis by equating the moment of the compressive area of the concrete to equivalent area of steel in tension, about the N.A. Thus, the neutral axis depth factor $k$ is given by *Eq. 2.10*.

$$k = \sqrt{2mp' + (mp')^2} - mp' \qquad \qquad \ldots(2.10)$$

where $p' = A_{st}/bd$ = reinforcement ratio.

2. Find the value of $k_c$, using *Eq. 2.8(a)*, for the critical N.A. of the balanced section and determine whether the section is under-reinforced or over-reinforced.

3. Find the moment of resistance $M_r$ by either using *Eq. 2.12(a)* or *Eq. 2.14*, whichever may be the case.

4. If it is required to find the uniformly distributed load or point or any other type of the load on a simple-supported beam, calculate the maximum B.M. in terms of the unknown load and equate it to the moment of resistance determined in step 3.

### Type II : Determination of Stresses in the Section

In this type of problem, the dimensions of the section, including the reinforcement, are completely given and it is required to find the stresses developed in steel and concrete, when subjected to a given bending moment. The solution is done in the following steps:

1. Determine the position of neutral axis by equating the moment of the compressive area of the concrete to equivalent area of steel in tension, about the N.A. Thus, the N.A. depth factor $k$ is given by

$$k = \sqrt{2mp' + (mp')^2} - mp' \quad \text{where} \quad p' = A_{st}/bd \qquad \ldots(2.10)$$

2. Knowing $k$, determine the lever arm factor $j$ from Fig. 2.11.

$$j = 1 - \frac{k}{3}.$$

3. Determine the moment of resistance of the section in terms of tensile force in the reinforcement (*Eq. 2.14*) and equate it to the external moment $M$. Thus

$$A_{st} \cdot t \cdot jd = M \qquad \ldots(2.16)$$

Thus the stress $t$ in steel is known.

4. Calculate the stress in concrete from the relation (Fig. 2.7)

$$\frac{c}{kd} = \frac{t}{m} \frac{1}{d - kd} \quad \text{or} \quad c = \frac{kt}{m(1-k)} = \frac{nt}{m(d-n)} \qquad \ldots(2.17)$$

(where $n = kd$ = depth of the neutral axis).

### Type III : Design of the Section

This is more common type of problem in which it is required to determine the dimensions of the section, including the reinforcement, to develop a given moment of resistance corresponding to a given set of permissible stresses in concrete and steel. The design is done in the following steps :

1. Determine the position of the critical neutral axis of the balanced section from *Eq. 2.8(a)* or *Eq. 2.8(b)*

$$k_c = \frac{1}{1 + \dfrac{1}{m}\dfrac{t}{c}} = \frac{1}{1 + \dfrac{1}{m}\dfrac{\sigma_{st}}{\sigma_{cbc}}} = \frac{1}{1 + \dfrac{r_c}{m}}, \quad \text{where} \quad r_c = \text{stress ratio}$$

Also determine 
$$j = 1 - \frac{k_c}{3}.$$

2. Determine the factor $R_c = \frac{1}{2} c j k = \frac{1}{2} \sigma_{cbc} j \cdot k_c$

3. Assume a suitable $b/d$ ratio for the beam section and calculate its moment of resistance from Eq. 2.12 : $M_r = R_c \cdot bd^2$.

4. Equate this to be maximum bending moment to which the beam section is subjected. To compute the maximum bending moment, the approximate self-weight of the beam of any suitable assumed section should also be taken into account. Thus the dimensions $b$ and $d$ are known.

5. Determine the area of steel reinforcement from Eq. 2.15 :

$$A_{st} = \frac{M}{t \cdot jd} = \frac{M}{\sigma_{st} \cdot j d}$$

*The detailed design features of a singly reinforced beam are given in Chapter 7.*

## Analysis Problems:

**Example 2.1.** *A reinforced concrete beam 250 mm wide 475 mm over all depth is reinforced with 3 bars of 16 mm dia and effective cover 50 mm. Use M20 grade concrete and Fe 415 steel. Find the depth of neutral axis.*

**Solution:** Given data:

$$b = 250 \text{ mm}$$
$$D = 475 \text{ mm}$$
$$d = 475 - 50 = 425 \text{ mm}$$
$$d_c = 50 \text{ mm}$$

**Materials:** M20 grade concrete Fe 415 steel.

$$A_{st} = 3 \times \frac{\pi}{4} \times 16^2 = 603.18 \text{ mm}^2.$$

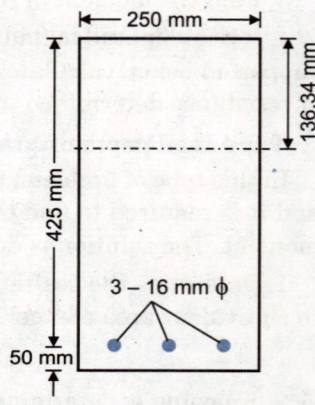

Fig. 2.12

**Step 1:** Permissible stress in concrete

$$\sigma_{cbc} = 7 \text{ N/mm}^2 \quad \left( \begin{array}{l} \text{As per table no. 21 and 22 of} \\ \text{IS 456 – 2000} \end{array} \right)$$
$$\sigma_{st} = 230 \text{ N/mm}^2$$

**Step 2:** Modular ratio $(m) = \dfrac{280}{3 \times \sigma_{cbc}} = \dfrac{280}{3 \times 7} = 13.33$

**Step 3:** Depth of neutral axis

$$c = T$$

$$\frac{b n_a^2}{2} = m \cdot A_{st} (d - n_a)$$

$$\frac{1}{2} \times 250 \times n_a^2 = 13.13 \times 603.18 \, (425 - n_a)$$

$$n_a^2 + 64.33 \, n_a - 27360 = 0$$

Quadratic equation $ax^2 + bx + c = 0$ $\quad \therefore x = \left\{ \dfrac{-b \pm \sqrt{b^2 - 4ac}}{2a} \right\}$

then $\quad n_a = \dfrac{-64.33 + \sqrt{(64.33)^2 + (4 \times 1 \times 27360)}}{(2 \times 1)}$

$$n_a = 136.34 \text{ mm}.$$

# THEORY OF REINFORCED BEAMS AND SLABS

**Example 2.2.** *A reinforced concrete beam of 300 mm width overall depth is 550 mm is reinforced with 4 bars of 20 mm ϕ and effective over 50 mm. Using M20 concrete and Fe 415 steel. Estimate moment of resistance of section.*

**Solution:** Given data:

$$b = 300 \text{ mm}$$
$$D = 550 \text{ mm}$$
$$d = 500 \text{ mm}$$
$$d_c = 50 \text{ mm}$$
$$A_{st} = 4 \times \frac{\pi}{4} \times 20^2 = 1256.66 \text{ mm}^2.$$

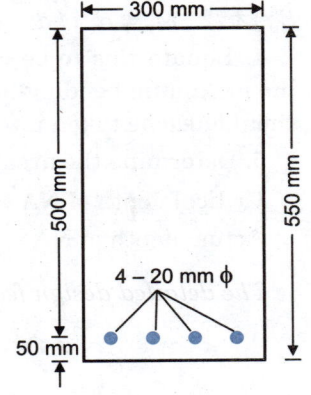

Fig. 2.13

**Step 1:**
$$\sigma_{cbc} = 7 \text{ N/mm}^2$$
$$\sigma_{st} = 230 \text{ N/mm}^2$$

**Step 2:** Modular ratio $= \dfrac{280}{3 \times 7} = 13.33$; Take m = 13

**Step 3:** Critical neutral axis
$$K_c = 0.289$$
$$n_c = K_c \times d = 0.289 \times 500 = 144.5 \text{ mm}$$

Actual depth of N.A.

$$\frac{1}{2} \times b \times n_a^2 = m \cdot A_{st}\left(d - \frac{n_a}{3}\right)$$

$$150\, n_a^2 = 8.168 \times 10^6 - 5445.5\, n_a.$$

$$n_a^2 + 36.3\, n_a - 54453.3 = 0$$

$$n_a = 215.9 \text{ mm}$$

$$n_a > n_c$$

Hence the section is over reinforced

$$\therefore \quad MR = \frac{1}{2} \times \sigma_{cbc} \times b \times n_a \times \left(d - \frac{n_a}{3}\right)$$

$$= \frac{1}{2} \times 7 \times 300 \times 215.9 \left(500 - \frac{215.9}{3}\right)$$

$$= 97 \times 10^6 \text{ kN/m}$$

**Example 2.3.** *A reinforced R.C.C beam 230 mm width 500 mm effective depth, 4–16 mm ϕ reinforcement. Find the depth of neutral axis, moment of resistance of section materials used are M20 grade concrete and Fe 415 steel.*

**Solution:** Given data:

$$b = 230 \text{ mm}$$
$$d = 500 \text{ mm}$$
$$A_{st} = 4 - 16 \text{ mm } \phi$$
$$= 4 \times \frac{\pi}{4} \times 16^2 = 804.57 \text{ mm}^2.$$

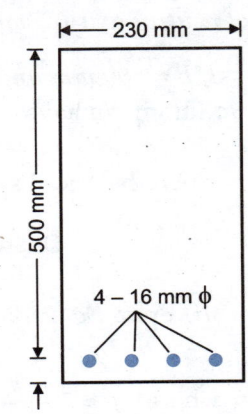

Fig. 2.14

**Materials:**
M20 grade concrete = $\sigma_{cbc} = 7 \text{ N/mm}^2$
Fe 415 steel. $\sigma_{st} = 230 \text{ N/mm}^2$

**Step 1:** Modular ratio

$$m = \frac{280}{3 \times 7} = 13.33, \text{ Take m = 13}$$

## 42 REINFORCED CONCRETE STRUCTURE

**Step 2:** Critical neutral axis

$$\text{Neutral axis factor } (K_c) = \frac{1}{1 + \dfrac{\sigma_{st}}{m \cdot \sigma_{cbc}}}$$

$$= \frac{1}{1 + \dfrac{230}{13 \times 7}}$$

$$n_c = 0.283$$

Critical depth of NA $= K_c \times d = 0.283 \times 500 = 141.5$ mm

Actual depth of NA

$$c = T$$

$$\frac{1}{2} \times b \times n_a^2 = m \cdot A_{st}\left(d - \frac{n_a}{3}\right)$$

$$\frac{1}{2} \times 230 \times n_a^2 = 13 \times 804.57 \left(500 - \frac{n_a}{3}\right)$$

$$115 n_a^2 = -3486.5 n_a + 5.23 \times 10^6$$

$$n_a^2 + 30.3 n_a - 45.5 \times 10^3 = 0.$$

Quadratic equation $ax^2 + bx + c = 0$

$$x = \frac{-b \pm \sqrt{b^2 - 4ac}}{2a}$$

$$\therefore \quad n_a = \frac{-30.3 + \sqrt{(30.3)^2 + 4 \times 45.5 \times 10^3}}{2 \times 1}$$

$$n_a = 198.7 \text{ mm}$$

$n_a > n_c$ {Hence the section is over reinforced}

$$\therefore \quad MR = \frac{1}{2} \times \sigma_{cbc} \times b \times n_a \left(d - \frac{n_a}{3}\right)$$

$$= \frac{1}{2} \times 7 \times 230 \times 198.7 \left(500 - \frac{198.7}{3}\right)$$

$$= 69.38 \times 10^6 \text{ kN/m}$$

**Example 2.4.** *For a balanced rectangular section (b × d) of a singly reinforced beam, determine (i) depth of neutral axis (ii) moment of resistance and (iii) percentage of steel using M 20 concrete and Fe 415 steel.*

*If b = 200 mm and d = 300 mm, determine the numerical values of n, $M_r$ and $A_{st}$.*

**Solution:** We have $c = \sigma_{cbc} = 7$ N/mm$^2$, $m = 13$ and $t = \sigma_{st} = 230$ N/mm$^2$

(i) For balanced section, $k_c = \dfrac{mc}{mc + t} = \dfrac{13 \times 7}{13 \times 7 + 230} = 0.283$

$\therefore$ Depth of N.A. $= n_c = k_c d = \mathbf{0.283\ d}$

(ii) From Eq. 2.12(a), $M_r = \dfrac{1}{2} cjk \cdot bd^2 = R_c \cdot bd^2$

where $j_c = 1 - \dfrac{k_c}{3} = 1 - \dfrac{0.283}{3} = 0.906$

$$\therefore \quad R_c = \frac{1}{2} cj_c k_c = \frac{1}{2} \times 7 \times 0.906 \times 0.283 = 0.897$$

$$\therefore \quad M_r = R_c \, bd^2 = \mathbf{0.897\, bd^2} \qquad \text{(where } M_r \text{ is in N-mm units)}$$

(*iii*) From *Eq. 2.9* percentage of steel is given by

$$p_c = \frac{50\, mc^2}{t(mc + t)} = \frac{50 \times 13\,(7)^2}{230(13 \times 7 + 230)} = 0.43\%$$

Alternatively, 
$$p_c = \frac{k_c \cdot c}{2t} \times 100 = \frac{0.283 \times 7}{2 \times 230} \times 100 = 0.43\%$$

For the given numerical values,

$$n_c = 0.283 \times 300 = \mathbf{84.9\ mm}$$
$$M_r = 0.897\,(200)\,(300)^2 = 16.146 \times 10^6 \text{ N-mm} = \mathbf{16.146\ kN\text{-}m}$$

$$A_{st} = p_c \cdot \frac{bd}{100} = \frac{0.43 \times 200 \times 300}{100} = \mathbf{258\ mm^2.}$$

**Example 2.5.** *Determine the moment of resistance of a singly reinforced beam 160 mm wide and 300 mm deep to the centre of reinforcement, if the stresses in steel and concrete are not to exceed 140 N/mm² and 5 N/mm². The reinforcement consists of 4 bars of 16 mm diameter. Take m = 18. If the above beam is used over an effective span of 5 m, find the maximum load the beam can carry, inclusive of its own weight.*

**Solution:** 
$$A_{st} = 4 \times \frac{\pi}{4}\,(16)^2 = 804\ mm^2$$

$$c = \sigma_{cbc} = 5\ \text{N/mm}^2\ ; \quad t = \sigma_{st} = 140\ \text{N/mm}^2\ ; \quad m = 18.$$

Equating the moment of area in compression to the moment of the equivalent area in tension, about the N.A., we get

$$b \times n \times \frac{n}{2} = m\, A_{st}\,(d - n)$$

or
$$\frac{160 \times n^2}{2} = 18 \times 804\,(300 - n)$$

or $\quad n^2 + 181\,n - 54287 = 0$

$$\Rightarrow \quad n = \frac{-181 \pm \sqrt{(181)^2 - 4 \times 1 \times (-54287)}}{2 \times 1}$$

From which $n = 159.5$ mm

If $n_c$ is the depth of critical neutral axis, we have, from *Eq. 2.8*

$$n_c = k_c \cdot d = \frac{18 \times 5}{(18 \times 5) + 140} \times 300 = 117.4\ mm$$

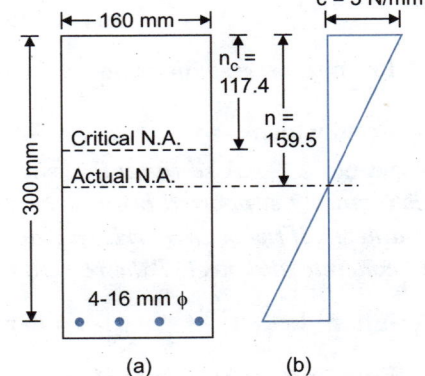

Fig. 2.15

Since the depth of actual neutral axis is more than the critical one, the section is over-reinforced. Thus, concrete reaches its maximum stress earlier to steel. Hence the moment of resistance is found on the basis of compressive force developed in concrete.

$$\text{Lever arm} = d - \frac{n}{3} = 300 - \frac{159.5}{3} = 246.8\ mm$$

$$\therefore \quad M_r = \frac{1}{2}c \cdot n \cdot b \cdot \left(d - \frac{n}{3}\right)$$

$$= \frac{1}{2} \times 5 \times 159.5 \times 160\,(246.8) \times 10^{-6} = 15.75\ kN\text{-}m$$

Let $w$ = uniformly distributed load in kN/m inclusive of the self-weight of the beam.

$$\text{Maximum B.M.} = \frac{wL^2}{8} = \frac{w \times 25}{8} \text{ kN-m}$$

Equating this to the moment of resistance of the beam we get

$$w = \frac{15.75 \times 8}{25} = \mathbf{5.04 \text{ kN/m}}.$$

**Example 2.6.** *The cross-section of a simply supported reinforced beam is 200 mm wide and 300 mm deep to the centre of the reinforcement which consists of 3 bars of 16 mm dia. Determine from the first principles the depth of N.A. and the maximum stress in concrete when steel is stressed to 120 N/mm². Take m = 19.*

**Solution:** Here, the dimensions are fully known

$$A_{st} = 3 \times \frac{\pi}{4}(16)^2 = 603.2 \text{ mm}^2.$$

Let the depth of N.A. be $n$. Equating the moment of the compressive area to the moment of equivalent area of steel, about N.A., we get

$$b \cdot n \cdot \frac{n}{2} = m A_{st}(d-n)$$

or $\qquad 100 n^2 = 19 \times 603.2 (300 - n)$

or $\qquad n^2 + 114.6 n - 34382 = 0$

From which $n = \mathbf{136.8 \text{ mm}}$

Fig. 2.16(b) shows the stress diagram.

Stress in steel = $t = 120$ N/mm² (Given)

$\therefore \qquad \dfrac{t}{m} = \dfrac{120}{19} = 6.316$

From the stress diagram $\dfrac{c}{n} = \dfrac{t/m}{d-n}$ or $\dfrac{c}{136.8} = \dfrac{6.316}{300 - 136.8} = 0.0387$

From which $\qquad c = \mathbf{5.29 \text{ N/mm}^2}$

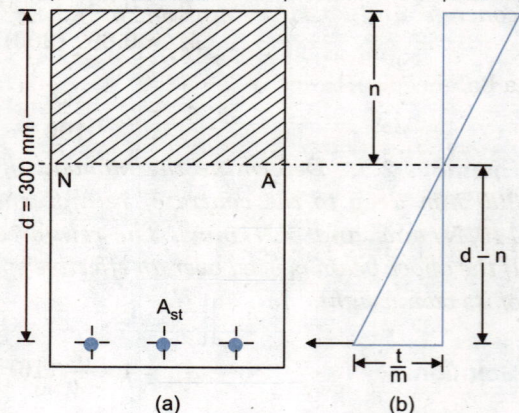

Fig. 2.16

**Example 2.7.** *A rectangular, singly reinforced beam, 300 mm wide and 500 mm effective depth is used as a simply supported beam over an effective span of 6 m. The reinforcement consists of 4 bars of 20 mm diameter. If the beam carries a load of 12 kN/m, inclusive of the self-weight, determine the stresses developed in concrete and steel. Take m = 19.*

**Solution:** Let $\qquad n$ = depth of N.A. ; $A_{st} = 4 \times \dfrac{\pi}{4}(20)^2 = 1256.6 \text{ mm}^2$.

Equating the moments of two areas about N.A. we get

$$b \times n \times \frac{n}{2} = m A_{st}(d-n) \quad \text{or} \quad 300 \times \frac{n^2}{2} = 19 \times 1256.6 (500 - n)$$

or $\qquad n^2 + 159.2 n - 79587 = 0 \quad$ From which $n = 213.5$ mm

Lever arm, $\qquad a = d - \dfrac{n}{3} = 500 - \dfrac{213.5}{3} = 428.8$ mm

$$\text{Maximum B.M.} = \frac{wl^2}{8} = \frac{12(6)^2}{8} = 54 \text{ kN-m} = 54 \times 10^6 \text{ kN-mm}$$

Let $c$ be the compressive stress in concrete

$\therefore \qquad M_r = \dfrac{1}{2} c \cdot n \cdot b \times a = \dfrac{1}{2} c \times 300 \times 213.5 \times 428.8 = 13.732 \times 10^6 \, c$ N-mm

Equating $M_r$ to the external B.M. we get

$$13.732 \times 10^6 \, c = 54 \times 10^6 \quad \text{or} \quad c = \frac{54}{13.732} = \mathbf{3.93 \text{ N/mm}^2}$$

If $t$ is the corresponding stress in steel, we get

$$t = \frac{mc}{n}(d-n) = \frac{19 \times 3.93}{213.5}(500 - 213.5) = \mathbf{100.2 \text{ N/mm}^2}$$

*Alternatively,* $\quad M_r = A_{st} \cdot t \, a = 1256.6 \times t \times 428.8$

$\therefore \quad 1256.6 \times 428.8 \, t = 54 \times 10^6 \quad \text{or} \quad t = \dfrac{54 \times 10^6}{1256.6 \times 428.8} = 100.2 \text{ N/mm}^2.$

**Example 2.8.** *Design a reinforced concrete beam subjected to a bending moment of 20 kN-m. Use M 20 concrete, and Fe 415 reinforcement. Keep the width of the beam equal to half the effective depth.*

**Solution:** For M 20 concrete, $c = \sigma_{cbc} = 7 \text{ N/mm}^2$ and $m = 13.33$. For Fe 415 steel, $\sigma_{st} = 230 \text{ N/mm}^2$. For a balanced section,

$$k_c = \frac{m \, \sigma_{cbc}}{m \, \sigma_{cbc} + \sigma_{st}} = \frac{13.33 \times 7}{13.33 \times 7 + 230} = 0.289;$$

$$j_c = 1 - \frac{k_c}{3} = 1 - \frac{0.289}{3} = 0.904$$

$$R_c = \frac{1}{2} c \cdot j_c \cdot k_c = \frac{1}{2} \times 7 \times 0.904 \times 0.289 = 0.914$$

Now $\quad M_r = R_c \cdot bd^2 = \dfrac{R_c \, d^3}{2} = \dfrac{0.914}{2} d^3 = 0.457 \, d^3 \text{ N-mm}$

Given $\quad$ B.M. $= M = 20 \text{ kN-m} = 20 \times 10^6 \text{ N-mm}$

$\therefore \quad 0.457 \, d^3 = 20 \times 10^6$

or $\quad d = \left(\dfrac{20 \times 10^6}{0.457}\right)^{\frac{1}{3}} \approx \mathbf{352.4 \text{ mm}}$

$\quad b = \dfrac{1}{2} d = \mathbf{176.2 \text{ mm}}.$

Area of steel is given by $A_{st} = \dfrac{M}{\sigma_{st} \, j_c \, d} = \dfrac{20 \times 10^6}{230 \times 0.904 \times 352.4} = \mathbf{273 \text{ mm}^2}$

Thus provide a beam of 176.2 mm width and 352.4 mm effective depth, with area of tensile reinforcement equal to 273 mm².

**Example 2.9.** *A reinforced concrete beam 200 mm × 400 mm effective depth is used over an effective span of 5 m. It is subjected to a uniformly distributed load of 5 kN/m inclusive of its own weight. Find the necessary steel reinforcement at the centre of the span. Take allowable stresses in steel and concrete as 130 N/mm² and 4 N/mm² respectively, and m = 16.*

**Solution:** Given: $\quad c = \sigma_{cbc} = 4 \, ; t = \sigma_{st} = 130 \text{ and } m = 16$

$$M = \frac{wl^2}{8} = \frac{5 \times 25}{8} \text{ kN-m} = 15.625 \text{ kN-m} = 15.625 \times 10^6 \text{ N-mm}$$

$\quad M_r = R \cdot bd^2 = 15.625 \times 10^6 \quad \text{or} \quad R = \dfrac{15.625 \times 10^6}{200(400)^2} = 0.488$

Hence $\quad M_r = 0.488 \, bd^2$ ...(i)

For the balanced section, $k_c = \dfrac{mc}{mc+t} = \dfrac{16 \times 4}{16 \times 4 + 130} = 0.33$

$$j = 1 - \dfrac{k_c}{3} = 1 - 0.11 = 0.89$$

$\therefore \quad R_c = \dfrac{1}{2} c j k_c = \dfrac{1}{2} \times 4 \times 0.89 \times 0.33 = 0.587$

$\therefore \quad M_r = 0.587\, bd^2$ ...(ii)

The moment of resistance of the given beam has to be less than the moment of resistance of the critical section. Hence, the steel reinforcement will be corresponding to an under-reinforced beam. For such a section, the stress in steel will reach the maximum value of 130 N/mm².

Let $n$ = depth of N.A. of the actual section. Then the corresponding stress in concrete is given by

$$\dfrac{c}{t/m} = \dfrac{n}{d-n} \quad \text{or} \quad c = \dfrac{t}{m} \cdot \dfrac{n}{d-n} = \dfrac{130}{16} \times \dfrac{n}{400-n} \quad \text{...(iii)}$$

Now $\quad M_r = \dfrac{1}{2} c \cdot n \cdot b \left( d - \dfrac{n}{3} \right)$

$\therefore \quad 15.625 \times 10^6 = \dfrac{1}{2} \left\{ \dfrac{130}{16} \times \dfrac{n}{400-n} \right\} n \times 200 \left( 400 - \dfrac{n}{3} \right)$

or $\quad n^2 \dfrac{(1200 - n)}{400 - n} = 57700$

This is a cubic equation in $n$. Solving it by trial and error, we get, $n = 122$ mm.

Hence from (iii) $\quad c = \dfrac{130}{16} \times \dfrac{122}{400 - 122} = 3.566$ N/mm²

Now, Total compression = Total tension, $\therefore \dfrac{1}{2} cnb = A_{st} \cdot t$

or $\quad A_{st} = \dfrac{1}{2} \dfrac{cnb}{t} = \dfrac{1}{2} \dfrac{3.566 \times 122 \times 200}{130} = \mathbf{334.6\ mm^2}$

**Check.** For the actual section,

$$k = \dfrac{mc}{mc+t} = \dfrac{16 \times 3.566}{16 \times 3.566 + 130} = 0.305$$

$$j = 1 - \dfrac{0.305}{3} = 0.8983$$

$$R = \dfrac{1}{2} c j k = \dfrac{1}{2} \times 3.566 \times 0.8983 \times 0.305 = 0.4885$$

$M_r = 0.4855 \times 200 \times (400)^2 \simeq 15.63 \times 10^6$ N-mm

Also, $\quad M_r = A_{st}\, t\, j\, d = 334.6 \times 130 \times 0.8983 \times 400 \simeq 15.63 \times 10^6$ N-mm

Hence O.K.

**Example 2.10.** *A reinforced concrete beam 200 mm × 400 mm effective depth is used over an effective span of 5 m. It is subjected to a uniformly distributed load of 7 kN/m inclusive of its own weight. Find the necessary steel reinforcement at the centre of the span. Take allowable stresses in steel and concrete as 130 N/mm² and 4 N/mm² respectively and m = 16.*

**Solution:** This example is exactly the same as the previous example except that $w$ has been increased to 7 kN/m.

$$c = \sigma_{cbc} = 4 \text{ N/mm}^2 \,; \quad t = \sigma_{st} = 130 \text{ N/mm}^2 \quad \text{and} \quad m = 16.$$

$$M = \frac{wL^2}{8} = \frac{7 \times 25}{8} = 21.875 \text{ kN-m} = 21.875 \times 10^6 \text{ N-mm}$$

$$\therefore \quad R = \frac{M_r}{bd^2} = \frac{M}{bd^2} = \frac{21.875 \times 10^6}{200 \times (400)^2} = 0.6836$$

Hence, $\quad M_r = 0.6836 \, bd^2$ ...(i)

For the balanced section

$$k_c = \frac{mc}{mc + t} = \frac{16 \times 4}{16 \times 4 + 130} = 0.33 \, ;$$

$$j = 1 - \frac{0.33}{3} = 1 - 0.11 = 0.89.$$

$$R_c = \frac{1}{2} c \, j_c \, k_c = \frac{1}{2} \times 4 \times 0.89 \times 0.33 = 0.5874 \, ;$$

$$M_r = 0.5874 \, bd^2 \qquad \qquad \text{...(ii)}$$

Since the moment of resistance of the given beam is greater than the moment of resistance of the critical section, the steel reinforcement will be corresponding to an over-reinforced beam. For such a section, the stress in concrete will reach the maximum value of 4 N/mm².

Now, $\quad M_r = \frac{1}{2} c \, n \, b \left( d - \frac{n}{3} \right) \quad$ or $\quad 21.875 \times 10^6 = \frac{1}{2} \times 4 \, n \times 200 \left( 400 - \frac{n}{3} \right)$

$\therefore \quad n^2 + 1200 \, n + 164100 = 0$. From which, $n = 157.5$ mm

Hence, $\quad n = 157.5$ mm

(**Note.** For critical section, $n_c = k \, d = 0.33 \times 400 = 132$ mm). The stress in steel is given by

$$t = \left( \frac{d - n}{n} \right) m \, c = \frac{400 - 157.5}{157.5} \times 16 \times 4 = 98.5 \text{ N/mm}^2$$

Now, $\qquad$ total tension = total compression

$\therefore \qquad t \cdot A_{st} = \frac{1}{2} cn \cdot b$

or $\qquad A_{st} = \frac{1}{2} \frac{c \, n \, b}{t} = \frac{1}{2} \frac{4 \times 157.5 \times 200}{98.5} = \mathbf{639.6 \text{ mm}^2}$

**Check.** $\qquad M = A_{st} \cdot t \, j \, d$

$$= A_{st} \cdot t \left( d - \frac{n}{3} \right) = 639.6 \times 98.5 \left[ 400 - \frac{157.5}{3} \right] = 21.89 \times 10^6 \text{ N-mm}$$

**Note.** In the above example of an over-reinforced section, the stress in steel is only 98.5 N/mm² which is very much less than the permissible value of 130 N/mm². Thus the strength of steel is not fully utilized. Such a design is undesirable and uneconomical. In circumstances where the dimensions of the beam are limited, and where the section has to develop greater moment of resistance than that of the balanced section, it is always desirable to design the section as *doubly reinforced* (see Chapter 4). In doubly reinforced section, steel reinforcement is placed in the compression zone also, which increases its moment of resistance.

*Hence a doubly reinforced section is always preferred over an over-reinforced section.*

**Example 2.11. (a)** *The moment of resistance of a rectangular reinforced concrete beam, of breadth b cm and effective depth d cm is $0.9 \, bd^2$. If the stresses in the outside fibre of the concrete and in the steel do not exceed 5 N/mm² and 140 N/mm² respectively, and the modular ratio equals 18, determine the ratio of depth of the neutral axis from the outside compression fibre to the effective depth of the beam and the ratio of area of tension steel to the effective area of the beam. The beam is reinforced for tension only.*

**Solution:** Given : $c = \sigma_{cbc} = 5$ ; $t = \sigma_{st} = 140$ and $m = 18$. Let the ratio of depth of the neutral axis to the effective depth be $k$.

Now $$R = \frac{1}{2} c j k = 0.9 \quad \text{(given)}$$

∴ $$\frac{1}{2} c \left(1 - \frac{k}{3}\right) k = 0.9 \quad \text{or} \quad \left(1 - \frac{k}{3}\right) k = \frac{1.8}{c} = \frac{1.8}{5} = 0.36$$

or $k^2 - 3k + 1.08 = 0$. From which, $k = \mathbf{0.43}$.

For the balanced section, $k_c = \dfrac{mc}{mc + t} = \dfrac{18 \times 5}{18 \times 5 + 140} = 0.392$

Since the actual N.A. is below the critical N.A., the section is *over-reinforced*. Hence the stress in concrete reaches a maximum value of 5 N/mm². The corresponding value of stress in steel is given by

$$t = \frac{mcd(1-k)}{kd} = \frac{mc(1-k)}{k} = \frac{18 \times 5(1 - 0.43)}{0.43} = 119 \text{ N/mm}^2$$

Now $$p' = \frac{A_{st}}{bd} = \frac{M}{tjd} \cdot \frac{1}{bd} = \frac{0.9 \, bd^2}{tjbd^2} = \frac{0.9}{t \cdot j}$$

Here $j = 1 - \dfrac{k}{3} = 1 - \dfrac{0.43}{3} = 0.857$ ; $t = 119$ N/mm²

∴ $$p' = \frac{0.9}{t j} = \frac{0.9}{119 \times 0.857} = 0.00882$$

or $p = 100 \, p' = \mathbf{0.882\%}$

**Note.** For a balanced section $p_c = \dfrac{50 \, k_c \, \sigma_{cbc}}{\sigma_{st}} = 50 \times \dfrac{0.392 \times 5}{140} = 0.7\%$

**Example 2.11. (b)** *Solve example 2.11 (a) if M 20 grade concrete and Fe 415 grade steel reinforcement are used for the beam.*

**Solution:** Given $c = \sigma_{cbc} = 7$ N/mm² ;

$m \triangleq 13$ ; $\sigma_{st} = 230$ N/mm² ; $R = \dfrac{1}{2} cjk = 0.9$ (given)

∴ $$\frac{1}{2} c \left(1 - \frac{k}{3}\right) k = 0.9 \quad \text{or} \quad \left(1 - \frac{k}{3}\right) k = \frac{1.8}{c} = \frac{1.8}{7} = 0.2571$$

∴ $k^2 - 3k + 0.7714 = 0$ From which $k = \mathbf{0.284}$

For the balanced section, $k_c = \dfrac{mc}{mc + t} = \dfrac{13 \times 7}{13 \times 7 + 230} = 0.2825 \approx 0.284$

Thus the beam is nearly balanced. Hence the stresses in concrete and steel reach their permissible values simultaneously.

For a balanced section, $p_c = 50 \dfrac{k_c \, c}{t} = 50 \times \dfrac{0.284 \times 7}{230} = \mathbf{0.432\%}$

**Check.** $p_c = \dfrac{A_{st}}{bd} \times 100 = \dfrac{m}{t \, jd} \times \dfrac{1}{bd} \times 100 = \dfrac{0.9 \, bd^2}{t \, j \, bd^2} \times 100 = \dfrac{90}{t \, j}$

Here $j_c = 1 - \dfrac{k_c}{3} = 1 - \dfrac{0.284}{3} = 0.9053$ ∴ $p_c = \dfrac{90}{230 \times 0.9053} = 0.432\%$

**Example 2.11. (c)** *Solve example 2.11 (a) if M 15 grade concrete and Fe 415 grade steel reinforcement is used for beam.*

**Solution:** Given $c = \sigma_{cbc} = 5$ N/mm$^2$; $m = 19$ and $t = \sigma_{st} = 230$ N/mm$^2$; $R = \frac{1}{2} cjk = 0.9$

$\therefore \quad \frac{1}{2}c\left(1 - \frac{k}{3}\right)k = 0.9 \quad$ or $\quad \left(1 - \frac{k}{3}\right)k = \frac{1.8}{c} = \frac{1.8}{5} = 0.36$

$\therefore \quad k^2 - 3k + 1.08 = 0$. From which $k = \mathbf{0.43}$

For the balanced section, $k_c = \dfrac{m\,\sigma_{cbc}}{m\,\sigma_{cbc} + \sigma_{st}} = \dfrac{19 \times 5}{19 \times 5 + 230} = 0.2923$

Since the actual N.A. is below the critical N.A., the section is *over-reinforced*. Hence the stress in concrete reaches a maximum value of 5 N/mm$^2$. The corresponding value of stress in steel is given by

$$t = \frac{mcd(1-k)}{kd} = \frac{mc(1-k)}{k} = \frac{19 \times 5(1-0.43)}{0.43} = 125.93 \text{ N/mm}^2$$

Now $\quad p' = \dfrac{A_{st}}{bd} = \dfrac{M}{tjd} \cdot \dfrac{1}{bd} = \dfrac{0.9\,bd^2}{tjbd^2} = \dfrac{0.9}{t \cdot j}$

where $j = 1 - \dfrac{k}{3} = 1 - \dfrac{0.43}{3} = 0.857$ and $t = 125.93$ N/mm$^2$

$\therefore \quad p' = \dfrac{0.9}{125.93 \times 0.857} = 0.00834 \quad$ or $\quad p = 100\,p' = \mathbf{0.834\%}$

**Note 1.** For a balanced section, $p_c = \dfrac{50\,k_c\sigma_{cbc}}{\sigma_{st}} = 50 \times \dfrac{0.2923 \times 5}{230} = 0.318\%$

**2.** As per IS 456 : 2000, the minimum grade of concrete shall be not less than M 20 in RCC work.

## 2.8. VARIATION OF $M_r$ WITH p

We have seen that a *balanced section* is the one in which the stresses in steel and concrete reach the permissible values simultaneously. The percentage steel for a balanced section is given by *Eq. 2.9* :

$$p = p_c = \frac{k_c \cdot \sigma_{cbc}}{2\,\sigma_{st}} \times 100 \qquad \ldots[2.9\,(b)]$$

The moment of resistance of a balanced section is computed from *Eq. 2.12*. If, however, the section is under-reinforced, $M_r$ is computed on the basis of tensile force in steel (*Eq. 2.14*). If the section is over-reinforced $M_r$ is determined on the basis of compressive force developed in concrete (*Eq. 2.12a*). For a beam with given dimensions (*i.e., b, d, p'*) the neutral axis depth factor $k$ is computed from *Eq. 2.10*. In general,

$$M_r = R \cdot bd^2$$

Hence $M_r$ depends upon $R$, and the variation of $M_r$ is directly proportional to variation of $R$.

$\therefore \quad R = \dfrac{M_r}{bd^2} \text{ N/mm}^2$

(where $M_r$ is in N-mm, $b$ in mm and $d$ in mm)

For illustration purposes, let us study the variation of $R$ (or $M_r$) with $p$ for M 20 concrete using (*a*) Mild steel bars (Fe 250) having $\sigma_{st} = 140$ N/mm$^2$ and (*b*) HYSD bars (Fe 415) having $\sigma_{st} = 230$ N/mm$^2$.

(*a*) **Fe 250 :** $\qquad \sigma_{st} = 140$ N/mm$^2$

$$k_c = \frac{m\,\sigma_{cbc}}{m\,\sigma_{cbc} + \sigma_{st}} = \frac{1}{1 + 0.0107\,\sigma_{st}} = \frac{1}{1 + 0.0107 \times 140} = 0.400$$

$$j = 1 - \frac{0.400}{3} = 0.867$$

$$p_c = \frac{k_c \cdot \sigma_{cbc}}{2\,\sigma_{st}} \times 100 = \frac{0.4 \times 7}{2 \times 140} \times 100 = 1.00$$

$$R_c = \frac{1}{2}\sigma_{cbc}.j\,k_c = \frac{1}{2} \times 7 \times 0.867 \times 0.4 = 1.213 \text{ N/mm}^2$$

For any other value of $k$ is given by *Eq. 2.10*

$$k = \sqrt{2mp' + (mp')^2} - mp' \qquad \text{where } p' = \frac{p}{100} = \frac{A_{st}}{bd} \text{ . and } m = 13$$

For $p < p_c$, $\qquad M_r = \sigma_{st}.A_{st}.jd = \sigma_{st}(p'.bd)jd. = (\sigma_{st}.p'j)\,bd^2 = Rb\,d^2$
where $R = \sigma_{st}.p'.j$.

For example, where $\qquad p = 0.4, \quad p' = 0.004$

$$k = \sqrt{(2 \times 13 \times 0.004) + (13 \times 0.004)^2} - 13 \times 0.004 = 0.275$$

$$j = 1 - \frac{k}{3} = 0.908; \qquad\qquad R = \sigma_{st}.p'.j = 140 \times 0.004 \times 0.908 = 0.508$$

Similarly, $\qquad p > p_c, \quad M_r = \frac{1}{2}\sigma_{cbc}.j.k.bd^2 = R\,bd^2 \quad$ where $\quad R = \frac{1}{2}\sigma_{cbc}.j.k$

For example, when $\qquad p = 1.2, \quad p' = 0.012$

$$k = \sqrt{(2 \times 13 \times 0.012) + (13 \times 0.012)^2} - 13 \times 0.012 = 0.424$$

$$j = 1 - \frac{k}{3} = 0.859$$

$$R = \frac{1}{2}\sigma_{cbc}.j.k = \frac{1}{2} \times 7 \times 0.866 \times 0.403 = 1.275$$

Values of $k, j\left(=1-\dfrac{k}{3}\right)$ and $R$ for various values of $p$ are tabulated below:

| p | 0.4 | 0.6 | 0.8 | 1.0 | 1.2 |
|---|---|---|---|---|---|
| p' | 0.004 | 0.006 | 0.008 | 0.01 | 0.012 |
| k | 0.275 | 0.325 | 0.364 | 0.396 | 0.424 |
| j | 0.908 | 0.892 | 0.879 | 0.868 | 0.859 |
| R | 0.508 | 0.749 | 0.984 | 1.215 | 1.275 |

**(b) Fe 415:** $\qquad \sigma_{st} = 230 \text{ N/mm}^2$

$$k_c = \frac{1}{1 + 0.0107 \times 230} = 0.289 \; ; \quad j = 1 - \frac{k_c}{3} = 0.904$$

$$p_c = \frac{0.289 \times 7}{2 \times 230} \times 100 = 0.440 \; ; \quad R_c = \frac{1}{2} \times 7 \times 0.904 \times 0.289 = 0.914$$

The values of $k, j$ and $R$ for various values of $p$ are tabulated below:

| p | 0.1 | 0.2 | 0.3 | 0.440 | 0.5 | 0.6 |
|---|---|---|---|---|---|---|
| k | 0.150 | 0.206 | 0.246 | 0.289 | 0.304 | 0.328 |
| j | 0.950 | 0.931 | 0.918 | 0.904 | 0.898 | 0.891 |
| R | 0.218 | 0.428 | 0.634 | 0.914 | 0.956 | 1.023 |

The variation of $R\left(=\dfrac{M_r}{bd^2}\right)$ with $p$, for both the types of bars are shown in Fig. 2.17. It is seen that upto $p = p_c$ the moment of resistance increases almost *linearly*. Provision of reinforcement in excess of $p_c$ does not lead to proportionate increase in $M_r$ and is not economical.

As per IS : 456-2000, the minimum reinforcement is given by

$$p' = \dfrac{A_{st}}{bd} = \dfrac{0.85}{f_y} \quad \text{(see Chapter 7)}$$

or $\quad p = \dfrac{A_{st}}{bd} \times 100 = \dfrac{85}{f_y}$ percent

For mild steel bars, having
$f_y = 250$ N/mm$^2$,
$p = 85/250 = 0.34\%$

For HYSD bars having
$f_y = 415$ N/mm$^2$
$p = 85/415 = 0.205\%$

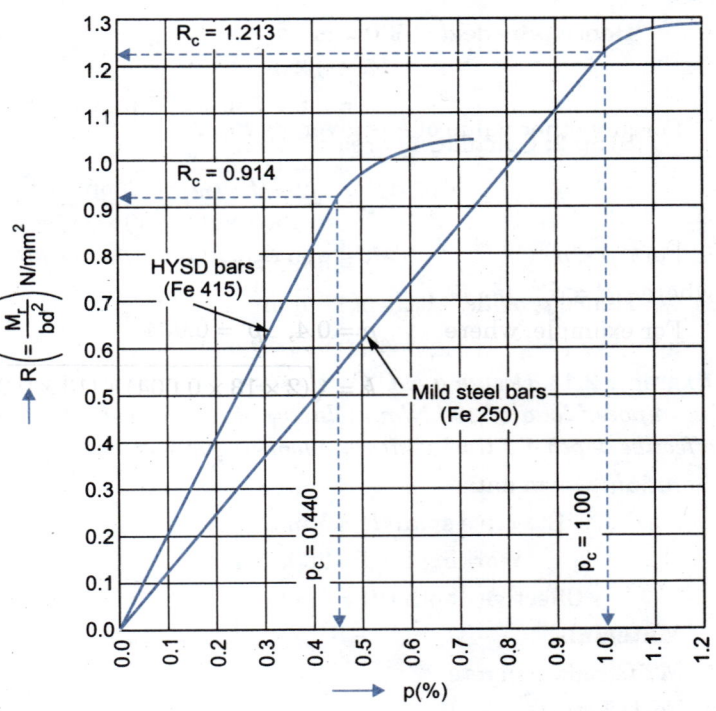

Fig. 2.17. Variation of $R$ and $M_r$ with $p$.

## Design Problems:

**Example 2.12.** *Design a rectangular R.C.C. beam to resist a bending moment of 65 kN-m use M20 grade concrete and Fe 415 steel assume B = 0.5 D.*

**Solution:** Given data:
  Bending moment = 65 kN-m

**Materials:**
M20 grade concrete $\sigma_{cbc} = 7$ N/mm$^2$
  Fe 415 steel = 230 N/mm$^2$

**Step 1:** Design constants

$$m = \dfrac{280}{3 \times \sigma_{cbc}} = \dfrac{280}{3 \times 7} = 13.33$$

$$K_c = \dfrac{1}{1 + \dfrac{\sigma_{st}}{m \cdot \sigma_{cbc}}} = 0.289$$

$$J = 1 - \dfrac{K_c}{3} = \left(1 - \dfrac{0.289}{3}\right) = 0.904$$

$$Q_c = \dfrac{1}{2} \times K \times J \times \sigma_{cbc} = \dfrac{1}{2} \times 0.289 \times 0.904 \times 7$$
$$= 0.914$$

Depth required
$$MR = Q\,bd^2$$
$$65 \times 10^6 = 0.914 \times 0.5\,d \times d^2$$
$$65 \times 10^6 = 0.914 \times 0.5 \times d^3$$
$$d = 522 \text{ mm}$$

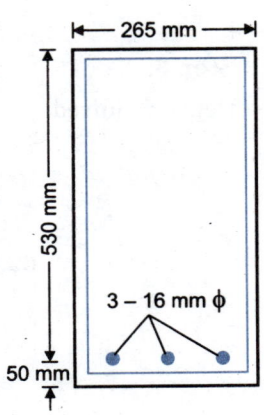

Fig. 2.18

Hence provide $d$ = 530 mm
$$b = 0.5\, d$$
$$b = 0.5 \times 530 = 265 \text{ mm}.$$

**Step 2:** Calculating area of steel.
$$A_{st} = \frac{MR}{\sigma_{st} \times J \times d} = \frac{65 \times 10^6}{230 \times 0.904 \times 530}$$
$$= 589.85 \text{ m}$$

Hence provide 3 bars of 16 mm
$$A_{st} = 603.19 \text{ mm}^2.$$

**Example 2.13.** *Design a R.C.C. rectangular beam simply supported over an effective span of 5 m to support on imposed load of 20 kN/m inclusive of its self weight adopt M20 grade concrete and Fe 415 steel provide effective depth 1.5 time width use working stress method.*

**Solution:** Given data:

Effective span ($l$) = 5 m
Working load = 20 kN/m
Effective depth ($d$) = 1.5 $b$

**Materials:**
M20 grade concrete
Fe 415 steel

**Step 1:** Permissible stresses
M20 grade concrete ($\sigma_{cbc}$) = 7 N/mm²
Fe 415 steel    $\sigma_{st}$ = 230 N/mm²

**Step 2:** Calculating Design Constant.
$$m = \frac{280}{3 \times 7} = 13.33$$

$$K = \frac{1}{1 + \dfrac{\sigma_{st}}{m \cdot \sigma_{cbc}}} = \frac{1}{1 + \dfrac{230}{13.33 \times 7}} = 0.289$$

$$J = \left(1 - \frac{K}{3}\right) = \left(1 - \frac{0.289}{3}\right) = 0.904$$

$$Q = \frac{1}{2} \times K \times J \times \sigma_{cbc} = \frac{1}{2} \times 0.289 \times 0.904 \times 7$$
$$= 0.914$$

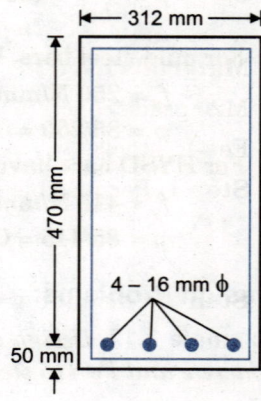

Fig. 2.19

**Step 3:**

Depth required
$$MR = Qbd^2$$
$$MR = \text{max. B.M.} = \frac{wl^2}{8} = \frac{20 \times 5^2}{8} = 62.5 \text{ kN-m}.$$
$$62.5 \times 10^6 = 0.914 \times b \times (1.5\, b)^2 \qquad \{\because \; d = 1.5\, b\}$$
$$b = \sqrt[3]{\frac{62.5 \times 10^6}{0.914 \times (1.5)^2}},\; b = 312.06 \text{ mm}$$
$$\therefore \quad d = 1.5 \times b = 1.5 \times 312.06$$
$$= 468 \text{ mm (say 470 mm)} = (470 + d_c)$$
$$D = 470 + 50 = 520 \text{ mm}$$

**Step 4:** Area of steel $(A_{st}) = \dfrac{MR}{\sigma_{st} \times J \times d}$

$$= \dfrac{62.5 \times 10^6}{230 \times 0.904 \times 470} = 639.56$$

Assume 16 mm ϕ bar of 4 bars

∴ Area of steel $A_{st}$ = 804.25 mm².

**Example 2.14.** *Design a simply supported beam over an effective span of 6.5 m to support the imposed load of 28 kN/m inclusive of its self weight use M20 grade concrete Fe 415 steel provide effective depth 1.5 breadth design in {working stress method}.*

**Solution:** Given data.

Length $(l)$ = 6.5 m = 6.5 × 10³ mm
Load $(w)$ = 28 kN/m
Depth $(d)$ = 1.5 b

**Materials used :**

M20 grade concrete

Fe 415 steel

**Step 1:** Permissible stresses

$$\sigma_{cbc} = 7 \text{ N/mm}^2$$
$$\sigma_{st} = 230 \text{ N/mm}^2$$

**Step 2:** Calculating constant

$$m = \dfrac{280}{3 \times \sigma_{cbc}} = \dfrac{280}{3 \times 7} = 13.33$$

$$K = \dfrac{1}{1 + \dfrac{280}{13.33 \times 7}} = 0.289$$

$$J = 1 - \dfrac{K}{3} = 1 - \dfrac{0.289}{3} = 0.904$$

$$M = \dfrac{1}{2} \times 0.289 \times 0.904 \times 7 = 0.914$$

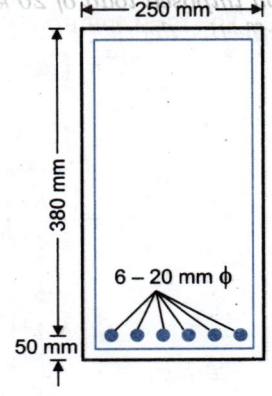

Fig. 2.20

**Step 3:** Moment of Resistance = $Q \, bd^2$

$$MR = \dfrac{wl^2}{8} = \dfrac{28 \times 6500^2}{8}$$

$$= 147.88 \times 10^6 \text{ N/mm}$$

$147.88 \times 10^6 = 0.914 \times b \times (1.5 \, b)^2$

$147.88 \times 10^6 = 0.914 \times 2.25 \, b^3$

$b$ = 238.79 mm (say 250 mm)

$d$ = 1.5 × (250) = 375 (say 380 mm)

$D = d_e + d_c$ = 380 + 50 = 430 mm

$$A_{st} = \dfrac{MR}{\sigma_{st} \times J \times d} = \dfrac{147.88 \times 10^6}{230 \times 0.904 \times 380}$$

$$= 1871.67 \text{ mm}^2$$

Say 6 bars of 20 mm ϕ

$$A_{st} = 1884.95 \text{ mm}^2$$

## 2.9. BEAM OF TRIANGULAR SECTION

### (a) Balanced Section.

Fig. 2.21(a) shows a triangular section having effective depth $d$ and width $b$ at the level of the reinforcement. Let the N.A. be at a depth $n$ below the apex. At depth $x$ below the apex, the width $b_x$ and stress $c_x$ are given by

$$b_x = \frac{b}{d} x \; ; \quad c_x = \frac{kd - x}{kd} c$$

Hence the compressive force $dC_x$ of elementary strip of thickness $dx$ is given by

$$dC_x = b_x \cdot c_x \cdot d_x$$

The total compressive force is given by

$$C = \int dC_x = \int_0^{kd} b_x \cdot c_x \cdot dx$$

$$= \int_0^{kd} \frac{b}{d} x \cdot \frac{kd - x}{kd} c \, dx$$

$$= \frac{cb}{k d^2} \int_0^{kd} x \cdot (kd - x) dx$$

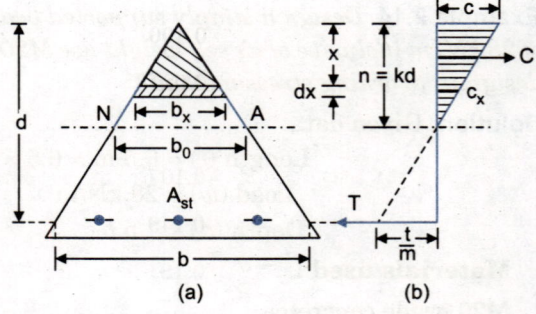

Fig. 2.21

$$= \frac{cb}{k d^2} \left[ kd \cdot \frac{x^2}{2} - \frac{x^3}{3} \right]_0^{kd} = \frac{k^2 bd}{6} c \qquad \ldots(2.18)$$

Taking $c = \sigma_{cbc}$ and $k = k_c$ for a balanced design

$$C = \frac{k_c^2 bd}{6} \cdot \sigma_{cbc} \qquad \ldots[2.18(a)]$$

where $k_c = \dfrac{m \sigma_{cbc}}{m \sigma_{cbc} + \sigma_{st}} = \dfrac{1}{1 + 0.0107 \sigma_{st}}$ ...[2.18(b)]

Total tensile force $T = A_{st} \cdot \sigma_{st}$. Equating the two, for equilibrium,

$$A_{st} \cdot \sigma_{st} = \frac{k_c^2 \cdot bd}{6} \cdot \sigma_{cbc} \quad \text{or} \quad A_{st} = \frac{k_c^2 \cdot bd}{6} \cdot \frac{\sigma_{cbc}}{\sigma_{st}} \qquad \ldots(2.19)$$

*Eq. 2.19* will be useful in determining reinforcement for a *balanced section*.

In order to find the moment of resistance of the section, take the moment of compressive force about the centre of tensile steel.

$$\therefore \quad M_r = \int_0^{kd} (d - x) \, dC_x = \int_0^{kd} (d - x) \, b_x \cdot c_x \, dx$$

$$= \int_0^{kd} (d - x) \frac{b}{d} x \cdot \frac{kd - x}{kd} \cdot c \, dx = \frac{bc}{k d^2} \int_0^{kd} x \cdot (kd - x)(d - x) \, dx$$

$$= \frac{bc}{kd^2} \left[ kd^2 \cdot \frac{x^2}{2} - d(k+1) \cdot \frac{x^3}{3} + \frac{x^4}{4} \right]_0^{kd}$$

$$M_r = \frac{bc}{kd^2} \left[ \frac{kd^2}{2} \cdot (kd)^2 - \frac{d(k+1)}{3} (kd)^3 + \frac{1}{4} (kd)^4 \right]$$

or $\quad M_r = \dfrac{c \cdot k^2 (2-k)}{12} bd^2 \qquad \ldots(2.20)$

Taking $c = \sigma_{cbc}$ and $k = k_c$ for a balanced section,

$$M_r = \frac{k_c^2 (2 - k_c)}{12} \sigma_{cbc} \cdot bd^2 = Q bd^2 \qquad \ldots(2.21)$$

where $\quad Q = \dfrac{k_c^2 (2 - k_c)}{12} \cdot \sigma_{cbc} = K \cdot \sigma_{cbc}$ ...[2.21(a)]

and $\quad K = \dfrac{1}{12} k_c^2 (2 - k_c)$ ...[2.21(b)]

and $k_c$ is given by *Eq. 2.8* (c).

The values of $K$ and $Q$ are given in the table below.

| $\sigma_{st}$ (N/mm$^2$) | 140 (Fe 250) | 190 | 230 (Fe 415) | 275 (Fe 500) |
|---|---|---|---|---|
| $k = k_c$ | 0.400 | 0.330 | 0.289 | 0.254 |
| $K$ | 0.0213 | 0.0151 | 0.0119 | 0.0094 |
| **Q for** | | | | |
| $\sigma_{cbc} = 5$ (M 15) | 0.107 | 0.076 | 0.060 | 0.047 |
| $\sigma_{cbc} = 7$ (M 20) | 0.149 | 0.106 | 0.083 | 0.066 |
| $\sigma_{cbc} = 8.5$ (M 25) | 0.181 | 0.129 | 0.101 | 0.080 |
| $\sigma_{cbc} = 10.0$ (M 30) | 0.213 | 0.151 | 0.119 | 0.094 |
| $\sigma_{cbc} = 11.5$ (M 35) | 0.245 | 0.174 | 0.137 | 0.108 |

**(b) Determination of $M_r$ and stresses.** If the dimensions of the section (*i.e.* $b$, $d$ and $A_{st}$) are given, the $M_r$, and the actual stresses can be determined by first locating the N.A. This can be done by equating the moment of the compressive area about N.A. to the moment of the equivalent tensile area about N.A. Thus, from Fig. 2.21, width $b_0$ at the N.A. is given by

$$b_0 = \dfrac{b}{d} \cdot kd = bk = \dfrac{bn}{d}$$

∴ Moment of compressive area about N.A.

$$= \left(\dfrac{1}{2} b_0 \cdot n\right) \dfrac{n}{3} = \dfrac{1}{2} \times \dfrac{bn}{d} \cdot n \cdot \dfrac{n}{3} = \dfrac{1}{6} \dfrac{b}{d} n^3.$$

Moment of equivalent tensile area about N.A. $= m A_{st} (d - n)$

Equating the two, we get, $\dfrac{1}{6} \dfrac{b}{d} n^3 = m A_{st} (d - n)$

or $\quad n^3 + \dfrac{6 md A_{st}}{b} n - \dfrac{6 md^2 A_{st}}{b} = 0$

This is a cubic equation, the solution of which (by trial) would yield the value of $n$.

The depth of critical N.A. is given by $n_c = \dfrac{m \sigma_{cbc}}{m \sigma_{cbc} + \sigma_{st}} \cdot d = k_c \cdot d$

(i) If $n = n_c$, the section is *balanced*, and the $M_r$ is obtained by *Eq. 2.21*.

(ii) If $n < n_c$, the section is *under-reinforced*, and the stress in steel will reach its maximum value first. Thus $t = \sigma_{st}$. Hence the corresponding stress $c$ in concrete is given, from the stress diagram (Fig. 2.21 b), by

$$\dfrac{c}{n} = \dfrac{\sigma_{st} / m}{d - n}; \quad \text{or} \quad c = \dfrac{\sigma_{st}}{m} \cdot \dfrac{n}{d - n} \qquad ...(2.17)$$

Knowing $c$, the moment of resistance of the section is found from *Eq. 2.20*.

Thus $\quad M_r = c \cdot \dfrac{k^2 (2 - k)}{12} bd^2, \quad$ (where $k = \dfrac{n}{d}$) ...(2.20)

**56** REINFORCED CONCRETE STRUCTURE

(iii) If $n > n_c$, the section is *over-reinforced* and the stress in concrete will reach its maximum value first. Thus $c = \sigma_{cbc}$. The corresponding stress in steel is given by

$$\frac{\sigma_{cbc}}{n} = \frac{t/m}{d-n} \quad \text{or} \quad t = \sigma_{cbc} \cdot \frac{m(d-n)}{n} \qquad \ldots[2.17(a)]$$

Also, $M_r$ is given by *Eq. 2.20* by substituting $c = \sigma_{cbc}$

Thus, $$M_r = \sigma_{cbc} \cdot \frac{k^2(2-k)}{12} bd^2 \quad \left(\text{where } k = \frac{n}{d}\right)$$

## 2.10. BEAM OF TRAPEZOIDAL SECTION

**(a) Balanced Section.** The moment of resistance for a trapezoidal section can be found by taking the sum of $M_r$ for middle rectangular and $M_r$ for triangular sections.

Thus, $$M_r = M_{r1} + M_{r2}$$

where $M_{r1}$ = moment of resistance of triangular portion

$$= \frac{ck^2(2-k)}{12}(b-b_1)d^2 \qquad \ldots(2.20)$$

and $M_{r2}$ = moment of resistance of rectangular portion

$$= \frac{1}{2}\sigma_{cbc} \cdot n\left(d - \frac{n}{3}\right)b \qquad \ldots[2.12(b)]$$

by substituting $n = kd$ and $\sigma_{cbc} = c$

$$M_{r2} = \frac{1}{2}c\left(1 - \frac{k}{3}\right)k\, b_1\, d^2$$

$$\therefore \quad M_r = \left[\frac{ck^2(2-k)}{12}(b-b_1) + \frac{1}{2}c\left(1 - \frac{k}{3}\right)k\, b_1\right]d^2 \qquad \ldots(2.22)$$

For a balanced section, $c = \sigma_{cbc}$ and $k = k_c$

$$M_r = \left[\sigma_{cbc}\frac{k_c^2(2-k_c)}{12}(b-b_1) + \frac{1}{2}\sigma_{cbc}\left(1 - \frac{k_c}{3}\right)k_c\, b_1\right]d^2$$

or $$M_r = [Q(b-b_1) + Rb_1]d^2 \qquad \ldots[2.22(a)]$$

where $$Q = \frac{k_c^2(2-k_c)}{12} \cdot \sigma_{cbc} \qquad \ldots[2.21(a)]$$

$$R = \frac{1}{2}\sigma_{cbc}\cdot\left(1 - \frac{k_c}{3}\right)\cdot k_c \quad \text{and} \quad k_c = \frac{m\,\sigma_{cbc}}{m\,\sigma_{cbc} + \sigma_{st}} = \frac{1}{1 + 0.0107\,\sigma_{st}}$$

The total compressive force is given by
$$C = C_1 + C_2$$

where $C_1$ = compressive force for triangular portion = $\dfrac{k^2}{6}c(b-b_1)d$ ...(2.18)

and $C_2$ = compressive force for rectangular portion = $\dfrac{1}{2}cb_1kd$

$$\therefore \quad C = \frac{k^2}{6}c(b-b_1)d + \frac{1}{2}cb_1kd = \frac{kd}{2}\left[\frac{k(b-b_1)}{3} + b_1\right]c \qquad \ldots(2.23)$$

For a balanced section, $c = \sigma_{cbc}$ and $k = k_c$

$$\therefore \quad C = \frac{k_c d}{2}\left[b_1 + \frac{k_c(b-b_1)}{3}\right]\sigma_{cbc} \quad \ldots[2.23(a)]$$

Total tensile force, for a balanced section is $T = A_{st} \cdot \sigma_{st}$
Equating the two, we get

$$A_{st} = \frac{k_c \cdot d}{2}\left[b_1 + \frac{k_c}{3}(b-b_1)\right]\frac{\sigma_{cbc}}{\sigma_{st}} \quad \ldots(2.24)$$

**(b) Determination of $M_r$ and stresses.** If the dimensions of the section are given, the $M_r$ and actual stresses can be determined by first locating N.A. This can be done by equating the moment of compressive area about N.A. to the moment of equivalent tensile area about N.A.

Thus, from Fig 2.22, moment of compressive area about N.A. $= \dfrac{1}{6}\dfrac{b-b_1}{d} \cdot n^3 + \dfrac{b_1 n^2}{2}$

Moment of equivalent tensile area about N.A. $= m \cdot A_{st}(d-n)$

Equating the two, we get, $\dfrac{1}{6}\dfrac{b-b_1}{d} \cdot n^3 + \dfrac{b_1 n^2}{2} - m \cdot A_{st}(d-n) = 0$

This is cubic equation, the solution of which (by trial) would yield the value of $n$.
Hence $\quad k = n/d$.

The depth of critical N.A. is given by, $n_c = \dfrac{m\sigma_{cbc}}{m\sigma_{cbc} + \sigma_{st}} \cdot d = k_c d$.

(i) If $n = n_c$, the section is *balanced*, and the $M_r$ is obtained by *Eq. 2.22 (a)*.

(ii) If $n < n_c$, the section is *under-reinforced* and stress in steel will reach its maximum value first. Thus $t = \sigma_{st}$. The corresponding stress $c$ in concrete is given by

$$c = \frac{\sigma_{st}}{m} \cdot \frac{n}{d-n} \quad \text{(from stress diagram)}$$

Knowing $c$, $M_r$ is found from *Eq. 2.22*.

(iii) If $n > n_c$, the section is *over-reinforced*, and the stress in concrete will reach its maximum value first. Thus $c = \sigma_{cbc}$. The corresponding stress $t$ in steel is given by

$$t = \sigma_{cbc} \cdot \frac{m(d-n)}{n} \quad \text{(from stress diagram).}$$

and $M_r$ is given by *Eq. 2.22*, by taking $c = \sigma_{cbc}$.

Thus, $$M_r = \left[\frac{k^2(2-k)}{12}(b-b_1) + \frac{1}{2}\left(1 - \frac{k}{3}\right)kb_1\right]\sigma_{cbc} \cdot d^2 \quad \left(\text{where } k = \frac{n}{d}\right)$$

**(c) Trapezoidal section with $b_1 > b$**

Since the area of concrete in tension zone is not useful, it is more economical to keep $b_1$ greater than $b$. All the *equations* (i.e., 2.22, 2.23, 2.24) will be valid, except that the term containing $(b - b_1)$ factor will be negative. Alternatively these equations can be *rearranged* in the following form:

$$M_r = \left[\frac{1}{2}c\left(1 - \frac{k}{3}\right)kb_1 - \frac{ck^2(2-k)}{12}(b_1 - b)\right]d^2 \quad \ldots(2.25)$$

$$M_r = [Rb_1 - Q(b_1 - b)]d^2 \quad \ldots[2.25(a)]$$

$$C = \frac{kd}{2}\left[b_1 - \frac{k}{3}(b_1 - b)\right]c \quad \ldots(2.26)$$

$$C = \frac{k_c d}{2}\left[b_1 - \frac{k_c}{3}(b_1 - b)\right]\sigma_{cbc} \quad \ldots[2.26(a)]$$

$$A_{st} = \frac{k_c d}{2}\left[b_1 - \frac{k_c}{3}(b_1 - b)\right]\frac{\sigma_{cbc}}{\sigma_{st}} \quad \ldots(2.27)$$

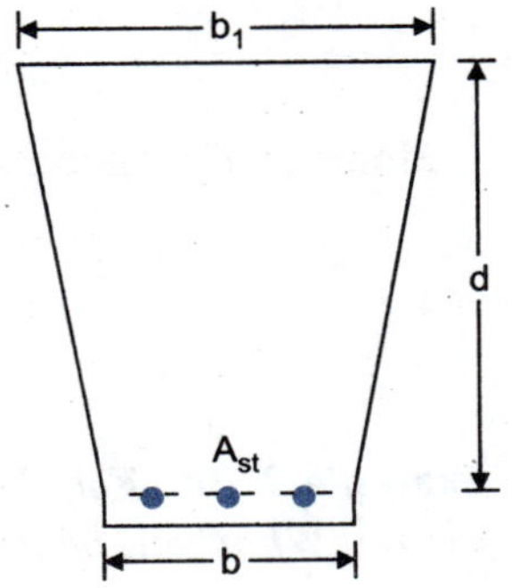

Fig. 2.23

(**d**) **Inverted triangular section** (Fig. 2.24). Such a section can be analysed or designed by taking $b \approx 0$ at the level of reinforcement, and using the equations of case (c) of the trapezoidal section. The corresponding equations can be rearranged in the following form.

$$M_r = \left[\frac{1}{2} c \left(1 - \frac{k}{3}\right) k - \frac{ck^2 (2-k)}{12}\right] b_1 d^2 \qquad \ldots(2.28)$$

$$M_r = [R - Q] b_1 d^2 \qquad \ldots(2.29)$$

$$C = \frac{kd(3-k)}{6} b_1 \, c = \frac{k_c \cdot d(3-k_c)}{6} b_1 \cdot \sigma_{cbc} \qquad \ldots(2.30)$$

$$A_{st} = \frac{k_c \cdot d(3-k_c)}{6} b_1 \cdot \frac{\sigma_{cbc}}{\sigma_{st}} \qquad \ldots(2.31)$$

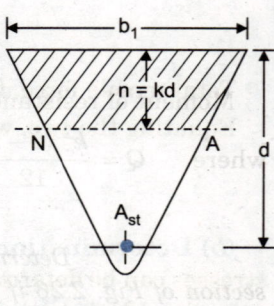

Fig. 2.24

**Example 2.15.** *A triangular section, 600 mm deep and 300 mm wide is reinforced as shown in Fig. 2.25. Taking $\sigma_{cbc} = 5 \, N/mm^2$, $\sigma_{st} = 140 \, N/mm^2$ and $m = 19$, find the moment of resistance of the beam section.*

**Solution:** Given : $\quad b = 300$ mm and $d = 600$ mm

$$A_{st} = 4 \frac{\pi}{4} (12)^2 = 452.4 \text{ mm}^2$$

Let the actual N.A. be at a depth $n$ below the apex.

Width at N.A. $= \frac{b}{d} \cdot n$.

Moment of compressive area about N.A. $= \frac{1}{2} \frac{bn}{d} \cdot n \cdot \frac{n}{3} = \frac{bn^3}{6d}$

Moment of equivalent tensile area about N.A. $= m A_{st} (d - n)$.

Equating the two, $\dfrac{bn^3}{6d} = m A_{st} (d - n)$

Substituting the numerical values, we get

$$\frac{300 \, n^3}{6 \times 600} = 19 \times 452.4 \, (600 - n)$$

or $\quad n^3 + 103147 \, n - 61888320 = 0$

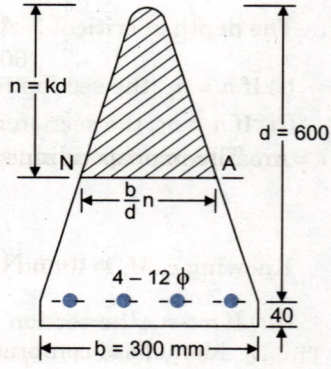

Fig. 2.25

Solving this by trial and error, we get :

$$n = 311 \text{ mm}; \, k = n/d = 0.518$$

$$k_c = \frac{m \, \sigma_{cbc}}{m \, \sigma_{cbc} + \sigma_{st}} = \frac{19 \times 5}{19 \times 5 + 140} = 0.404$$

$$n_c = 0.404 \times 600 = 242.6 \text{ mm}$$

Since $n > n_c$, the section is over-reinforced. Hence the stress in concrete reaches its maximum value first. Therefore, $c = \sigma_{cbc} = 5 \text{ N/mm}^2$

$$t = \sigma_{cbc} \cdot \frac{m(d-n)}{n} = \frac{5 \times 19 \, (600 - 311)}{311} = \mathbf{88.3 \text{ N/mm}^2}$$

Moment of resistance is given by Eq. 2.20 in which $c = \sigma_{cbc} = 5 \text{ N/mm}^2$

$$M_r = \sigma_{cbc} \cdot \frac{k^2 (2-k)}{12} bd^2 = 5 \times \frac{(0.518)^2 (2 - 0.518)}{12} \times 300 \, (600)^2$$

$$= 17.89 \times 10^6 \text{ N-mm} = \mathbf{17.89 \text{ kN-m.}}$$

**Example 2.16.** *For the beam section of Example 2.15, find the reinforcement corresponding to balanced section. What will be its moment of resistance ?*

**Solution:** For balanced section : $k_c = 0.404$; $n_c = 0.404 \times 600 = 242.6$ mm.

The reinforcement corresponding to balanced section is given by Eq. 2.19

$$A_{st} = k_c^2 \cdot \frac{bd}{6} \cdot \frac{\sigma_{cbc}}{\sigma_{st}} = (0.404)^2 \times \frac{300 \times 600}{6} \times \frac{5}{140} = \mathbf{174.9 \text{ mm}^2}.$$

Moment of resistance is given by *Eq. 2.21* : $M_r = Q bd^2$

where $\quad Q = \dfrac{k_c^2 (2-k)}{12} \cdot \sigma_{cbc} = \dfrac{(0.404)^2 (2-0.404)}{12} \times 5 = 0.1085$

$\therefore \qquad M_r = 0.1085 \times 300 (600)^2 = 11.72 \times 10^6 \text{ N-mm} = \mathbf{11.72 \text{ kN-m}}.$

**Example 2.17.** *Determine the moment of resistance of the beam section of Fig. 2.26 if the beam is inverted and reinforcement is provided at the apex. Also, determine the reinforcement and moment of resistance corresponding to the balanced section.*

**Solution:** Assume width at the c.g. of reinforcement as zero. Hence $b = 0$

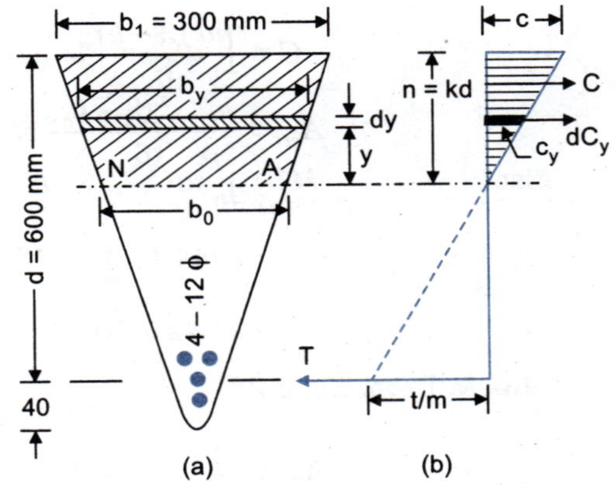

$$A_{st} = 4 \times \frac{\pi}{4} (12)^2 = 452.4 \text{ mm}^2.$$

The width $b_0$ at the N.A. is given by the relation

$$b_0 = \frac{d-n}{d} \cdot b_1 = \frac{600-n}{600} \times 300$$

$$= \frac{600-n}{2}$$

**Fig. 2.26**

Area of compression zone $= \dfrac{b_0 + b_1}{2} \times n = \dfrac{1}{2}\left[\dfrac{600-n}{2} + 300\right] n$

$\bar{x}$ from N.A. $= \dfrac{2b_1 + b_0}{b_1 + b_0} \times \dfrac{n}{3} = \dfrac{600 + \dfrac{600-n}{2}}{300 + \dfrac{600-n}{2}} \times \dfrac{n}{3}$

$\therefore$ Moment of compression area about N.A.

$$= \frac{1}{2}\left[\frac{600-n}{2} + 300\right] n \times \frac{\frac{600-n}{2} + 600}{\frac{600-n}{2} + 300} \times \frac{n}{3} = \frac{n^2}{6}\left[600 + \frac{600-n}{2}\right] \qquad \ldots(1)$$

Moment of equivalent tensile area about N.A. $= m A_{st} (d-n) = 19 \times 452.4 (600-n)$ ...(2)

Equating the two, we get : $\dfrac{n^2}{6}\left[600 + \dfrac{600-n}{2}\right] = 19 \times 452.4 (600-n)$

or $\qquad n^3 - 1800 n^2 - 103147 n + 61888320 = 0$

Solving this by trial, we get : $n = 165$ mm;

Hence $k = 165/600 = 0.275$. For the balanced section, $k_c = 0.404$; $n_c = 0.404 \times 600 = 242.4$ mm.

Since $n > n_c$ the section is *under-reinforced*, and the stress in steel will reach its max. permissible value first. Hence $t = \sigma_{st} = 140$ N/mm$^2$

$\therefore \qquad c = \dfrac{\sigma_{st}}{m} \cdot \dfrac{n}{d-n} = \dfrac{140}{19} \times \dfrac{165}{600-165} = 2.795$ N/mm$^2$.

In order to find the moment of resistance, consider a strip of thickness $dy$ at a distance of $y$ from N.A. Width $b_y$ is given by : $\dfrac{b_y}{b_1} = \dfrac{d-n+y}{d}$

$\therefore \qquad b_y = \dfrac{d-n+y}{d} b_1 = \dfrac{(435+y)}{2}$

**60  REINFORCED CONCRETE STRUCTURE**

The stress $c_y$ at the level of strip is given by : $\dfrac{c_y}{c} = \dfrac{y}{n}$

or $\qquad c_y = \dfrac{y}{n} \cdot c = \dfrac{2.795}{165} y$

$\therefore \qquad dC_y = b_y \cdot c_y \, dy = \dfrac{435 + y}{2} \times \dfrac{2.795}{165} y \, dy = 0.00847 \, (435 + y) y \, dy$

$\therefore \qquad C = \int_0^n dC_y = \int_0^{n=165} 0.00847 \, (435 + y) \, y \, dy = 0.00847 \left[ \dfrac{435 \, y^2}{2} + \dfrac{y^3}{3} \right]_0^{n=165}$

$\qquad\qquad = 62838 \text{ N}$

Also $\qquad M_r = \int_0^n (dC_y)(d - n + y) = \int_0^{165} 0.00847 \, (435 + y) \, y \, [600 - 165 + y] \, dy$

$\qquad\qquad = 0.00847 \int_0^{165} (435 + y)^2 \, y \, dy = 0.00847 \left[ 189225 \dfrac{y^2}{2} + \dfrac{y^4}{4} + 870 \dfrac{y^3}{3} \right]_0^{165}$

$\qquad\qquad = 34.42 \times 10^6 \text{ N-mm} = \mathbf{34.42 \text{ kN-m}}$

**Check.** From *Eq. 2.28*,

$$M_r = \left[ \dfrac{1}{2}\left(1 - \dfrac{k}{3}\right) k - \dfrac{k^2 \, (2 - k)}{12} \right] c b_1 d^2$$

$$= \left[ \dfrac{1}{2}\left(1 - \dfrac{0.275}{3}\right) \times 0.275 - \dfrac{(0.275)^2 \, (2 - 0.275)}{12} \right] 2.795 \times 300 \, (600)^2$$

$$= 34.42 \times 10^6 \text{ N-mm.}$$

This is about *three times more* than the one of the example 2.15, with the same dimensions but placement reversed.

### For the Balanced Section

$\qquad\qquad k_c = 0.404 ; \; n_c = 0.404 \times 600 = 242.4 \text{ mm}$

$\qquad\qquad c_y = \dfrac{y}{n_c} \cdot \sigma_{cbc} = \dfrac{5}{242.4} \cdot y = 0.02062 \, y$

$\qquad\qquad b_y = \dfrac{d - n_c + y}{d} b_1 = \dfrac{600 - 242.4 + y}{600} \times 300 = \dfrac{357.6 + y}{2}$

$\therefore \qquad dC_y = \dfrac{357.6 + y}{2} \times 0.02062 \, y \, dy = 0.01031 \, (357.6 + y) \, y \, dy$

$\qquad\qquad C = \int_0^{n_c} dC_y = \int_0^{242.4} 0.01031 \, (357.6 + y) \, y \, dy$

$\qquad\qquad = 0.01031 \left[ \dfrac{357.6 \, y^2}{2} + \dfrac{y^3}{3} \right]_0^{242.4} = 157264 \text{ N.}$

Equating this to total tensile area, we get : $A_{st} = \dfrac{C}{\sigma_{st}} = \dfrac{157264}{140} = \mathbf{1123.3 \text{ mm}^2}.$

**Check** : From *Eq. 2.31*

$$A_{st} = \dfrac{k_c \, d \, (3 - k_c)}{6} b_1 \dfrac{\sigma_{cbc}}{\sigma_{st}}$$

$$= \dfrac{0.404 \times 600 \, (3 - 0.404)}{6} \times 300 \times \dfrac{5}{140} = 1123.7 \text{ mm}^2$$

Also
$$M_r = \int_0^{n_c} (d\,C_y)(d - n_c + y)$$
$$= \int_0^{242.4} 0.01031\,(357.6 + y)\,y\,(600 - 242.4 + y)\,dy$$
$$= 0.01031 \int_0^{242.4} (357.6 + y)^2\,y\,dy$$
$$= 0.01031 \left[\frac{127878 y^2}{2} + \frac{y^4}{4} + \frac{715.2 y^3}{3}\right]_0^{242.4}$$
$$= 82.64 \times 10^6 \text{ N-mm} = \mathbf{82.64\ kN\text{-}m}$$

**Check** : From *Eq. 2.29* : $M_r = (R - Q)\,b_1 d^2$

where  $R = \dfrac{1}{2} \times 5\left(1 - \dfrac{0.404}{3}\right) \times 0.404 = 0.874$

$Q = \dfrac{k_c^2\,(2 - k_c)}{12}\,\sigma_{cbc} = \dfrac{(0.404)^2\,(2 - 0.404)}{12} \times 5 = 0.1085$

∴  $M_r = (0.874 - 0.1085) \times 300\,(600)^2 = 82.67 \times 10^6$ N-mm

**Example 2.18.** *A simply supported beam of trapezoidal shape is shown in Fig. 2.27(a). Taking $\sigma_{cbc} = 5\,N/mm^2$, $\sigma_{st} = 140\,N/mm^2$ and $m = 18$, determine the moment of resistance of the beam section and the actual stresses in the section.*

**Solution:** Let $n$ be the depth of the N.A. Width $ab = 500 - \dfrac{200}{600}\,n = 500 - \dfrac{n}{3}$

Width $b$ at reinforcement level $= 500 - \dfrac{200}{600} \times 560 = 313.3$ mm.

The trapezium $abdc$ of compression zone can be divided into rectangle $abhg$ and two triangles $acg$ and $bhd$. Width $cg = hd = \dfrac{1}{6}\,n$.

Hence moment of the compressive area about N.A. is

$= n\left(500 - \dfrac{n}{3}\right)\dfrac{n}{2} + \dfrac{1}{2}\,n \times \dfrac{1}{3}\,n \times \dfrac{2}{3}\,n$

$= 250\,n^2 - \dfrac{n^3}{6} + \dfrac{n^3}{9} = 250\,n^2 - \dfrac{n^3}{18}$  ...(i)

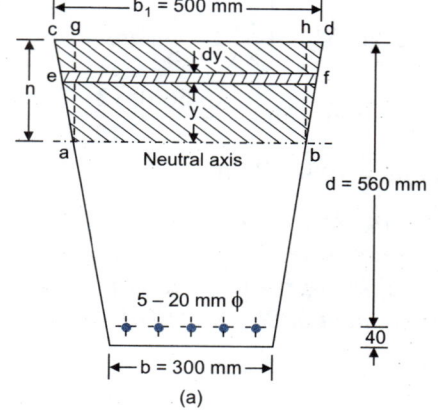
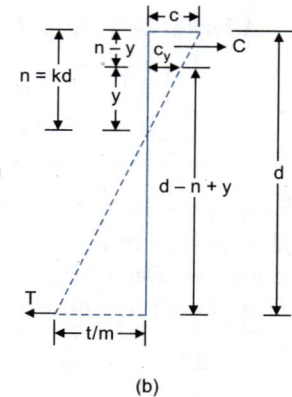

**Fig. 2.27**

$A_{st} = 5 \times \dfrac{\pi}{4}(20)^2 = 1570.8$ mm$^2$

∴  Moment of equivalent tensile area, about N.A. $= 18 \times 1570.8\,(560 - n)$  ...(ii)

Equating the two area moments, we get  $250\,n^2 - \dfrac{n^3}{18} = 18 \times 1570.8\,(560 - n)$

or  $n^3 - 4500\,n^2 - 508940\,n + 285 \times 10^6 = 0$

Solving it by trial and error, we get $n = 205$ mm; $k = \dfrac{n}{d} = \dfrac{205}{560} = 0.366$

$k_c = \dfrac{m\,\sigma_{cbc}}{m\,\sigma_{cbc} + \sigma_{st}} = \dfrac{18 \times 5}{18 \times 5 + 140} = 0.391$;

$n_c = 0.391 \times 560 = 219$ mm.

**62** REINFORCED CONCRETE STRUCTURE

Since the actual N.A. is *above* the critical N.A., the section is *under-reinforced*, and the stress in steel will reach the maximum value earlier. Hence $t = \sigma_{st} =$ **140 N/mm²**.

Corresponding stress in concrete is given by
$$c = \frac{t}{m} \times \frac{n}{d-n} = \frac{140}{18} \times \frac{205}{560-205} = 4.49 \text{ N/mm}^2$$

In order to find the moment of resistance of the beam, consider an elementary strip of area $b_y \cdot dy$ at a distance $y$ from the N.A.

Width of strip $\quad ef = b_y = 500 - \frac{n-y}{3} = 500 - \frac{205-y}{3}$

Stress $c_y$ at $\quad ef = c \cdot \frac{y}{n} = \frac{4.49}{205} y = 0.0219 \, y$

$\therefore \quad dC_y = b_y \cdot dy \cdot c_y = \left(500 - \frac{205-y}{3}\right)(0.0219 \, y) \, dy$

$= (1295 + y) \times 0.0073 \, y \, dy$

Moment of the force about the centre of steel area is
$$d M_y = b_y \cdot dy \cdot c_y (d - n + y) = 0.0073 \, (1295 + y) \, y \, dy \, (560 - 205 + y)$$

$$M_r = \int_0^n 0.0073 \, (1295 + y)(355 + y) \, y \, dy$$

$$= 0.0073 \int_0^{205} (y^3 + 1650 \, y^2 + 459725 \, y) \, dy$$

$$= 0.0073 \left[\frac{y^4}{4} + \frac{1650 \, y^3}{3} + \frac{459725 \, y^2}{2}\right]_0^{205}$$

$$= 108.33 \times 10^6 \text{ N-mm} = \textbf{108.33 kN-m}$$

*Alternatively*, from Eq. 2.25, $M_r = \left[\frac{1}{2}\left(1 - \frac{k}{3}\right) kb_1 - \frac{k^2(2-k)}{12}(b_1 - b)\right] cd^2$

$= \left[\frac{1}{2}\left\{1 - \frac{0.366}{3}\right\} \times 0.366 \times 500 - \frac{(0.366)^2 (2 - 0.366)}{12}(500 - 313.3)\right] \times 4.49 \, (560)^2$

$= 108.3 \times 10^6$ N-mm = **108.3 kN-m**.

**Example 2.19.** *A singly reinforced concrete beam 50 cm wide and 60 cm deep has a 10 cm wide and 10 cm deep groove in the compression side, as shown in Fig. 2.28. The reinforcement consists of 4 bars of 20 mm diameter. Determine (i) position of the neutral axis (ii) maximum compressive and tensile stresses when it is subjected to a bending moment of 80 kN-m. Take m = 15.*

**Solution:** Area of steel $= 4 \times \frac{\pi}{4} (20)^2$
$= 1256.6$ mm²

Let the depth of the N.A. be $n$.

$\therefore$ Moment of tensile area about N.A.
$= 15 \times 1256.6 \, (560 - n) \quad \quad ...(i)$

Moment of compressive area about N.A.
$= 400 \, n \times \frac{n}{2} + 100 \, (n - 100) \frac{(n - 100)}{2}$

$= 200 \, n^2 + 50 \, (n^2 - 200 \, n + 10000)$

$= 250 \, n^2 - 10000 \, n + 500000 \quad \quad ...(ii)$

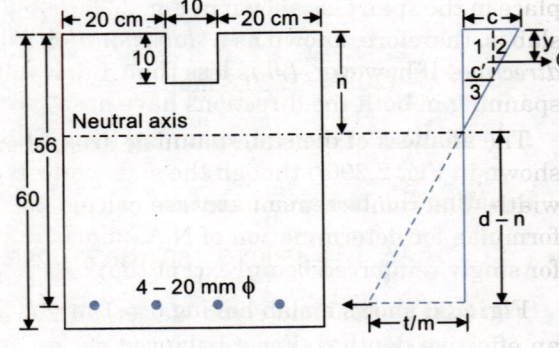

Fig. 2.28

Equating the two we get : $250 \, n^2 - 10000 \, n + 500000 = 15 \times 1256.6 \, (560 - n)$
or $\quad n^2 + 35.4 \, n - 42022 = 0$. From which $n = 188$ mm

Stress $c'$ in concrete at the bottom level of groove is
$$c' = c\frac{n-100}{n} = c\frac{188-100}{188} = 0.468\,c$$

Let $\bar{y}$ be the distance of C.G. of total compressive force, from top fibre. Let us divide the total compressive force into three parts as marked in the stress diagram.

Force $C_1 = 400 \times 100\, c' = 40000 \times 0.468\, c = 18720\, c$, acting at 50 mm from top.

Force $C_2 = 400 \times 100 \times \tfrac{1}{2}(c - c') = 20000 \times 0.532\, c = 10640\, c$, acting at $\dfrac{100}{3}$ m from top

Force $C_3 = 500(188 - 100)\tfrac{1}{2}\, c' = 22000 \times 0.468\, c = 10296\, c$

acting at $100 + \dfrac{1}{3}(188 - 100) = 129$ mm from the top.

Hence taking the moments of individual forces about the top we get

$$\bar{y} = \frac{(18720\, c \times 50) + (10640\, c \times 33.3) + (10296\, c \times 129)}{18720\, c + 10640\, c + 10296\, c} = 66 \text{ mm}.$$

Lever arm $= d - \bar{y} = 560 - 66 = 494$ mm.

Now $\quad M = A_{st}\cdot t \times \text{Lever arm} = A_{st}\cdot t\, a$

$\therefore \quad t = \dfrac{M}{a\cdot A_{st}} = \dfrac{80 \times 10^6}{494 \times 1256.6} = 128.9$ N/mm$^2$

From the stress diagram, $\dfrac{c}{t/m} = \dfrac{n}{d-n}$

$\therefore \quad c = \dfrac{t}{m}\cdot\dfrac{n}{d-n} = \dfrac{128.9}{15} \times \dfrac{188}{560-188} = 4.34$ N/mm$^2$.

## 2.11. SLAB SPANNING IN ONE DIRECTION

Fig. 2.29(a) shows a room of size $l \times L$, in which size $L$ is much greater than $l$. The roof of the room consists of a reinforced concrete slab. When uniformly distributed load is applied to the slab, the bending generally takes place about two axes, parallel to direction of $L$ and $l$. However, if the ratio is greater than 1.5, 84% of the total load is carried by the shorter span ($l$) and 16% by the longer span ($L$). Hence if the ratio $L/l$ is the greater than 1.5, it is usual to consider the bending of the slab to take place in the span $l$, as shown in Fig. 2.29(b). Such a slab is, therefore, known as a *slab spanning in one direction*. If however, $L/l$ is less than 1.5, bending takes place in both the directions. The analysis of slabs spanning in both the directions have been given in Chapter 10.

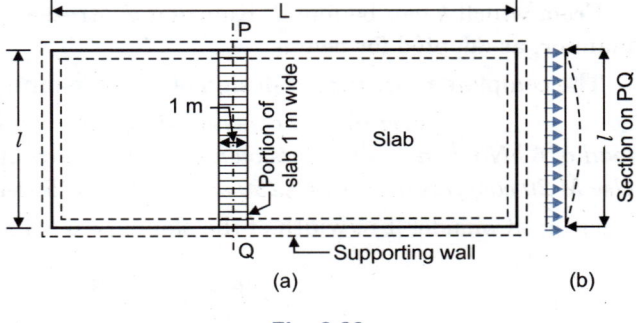

Fig. 2.29

The analysis of the slab spanning in one direction is done by assuming it to be a beam of 1 m width, as shown in Fig. 2.29(a) though the slab width is continuous and is not composed of individual beams of 1 m width. The reinforcement etc. are calculated for 1 m width and the bars are disturbed accordingly. The formulae for determination of N.A., moment of resistance and steel reinforcement are the same as found for singly reinforced beam, except that $b$ is taken to be equal to 1000 mm.

Fig. 2.30 shows a slab having $b = 1$ m $= 1000$ mm and an effective depth $d$. For a balanced design the depth of critical N.A. is given by

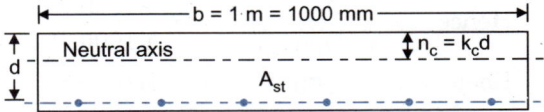

Fig. 2.30

$$k_c = \frac{m\,\sigma_{cbc}}{m\,\sigma_{cbc} + \sigma_{st}}\,;\quad j_c = 1 - \frac{k_c}{3}\,;\quad R_c = \frac{1}{2}\sigma_{cbc}\cdot j_c \cdot k_c \text{ and } M_r = R_c\,bd^2$$

Equating this to the external B.M. $M$, we get

$$d = \sqrt{\frac{M}{bR_c}} = \sqrt{\frac{M}{1000\,R_c}} \qquad \ldots(2.32)$$

Thus the effective depth $d$ of the slab can be known. The area of steel per 1000 mm width is given by:

$$A_{st} = \frac{M}{\sigma_{st}\cdot j_c \cdot d} \qquad \ldots(2.33)$$

Let    $\Phi$ = diameter of the bar chosen (mm)

$A_\Phi$ = area of one bar (mm²);

$s$ = spacing of bars (mm), c/c

Then No. bars = $\dfrac{1000}{s}$ and total area = $\dfrac{1000}{s}\cdot A_\Phi$

Equating this to the total area ($A_{st}$) required, $A_{st} = \dfrac{1000}{s} A_\Phi$

or 

$$s = \frac{1000\,A_\Phi}{A_{st}} \qquad \ldots(2.34)$$

For a slab of given dimensions $d$ and $A_{st}$, the N.A. can be located by equating the moments of equivalent areas about the N.A.

Thus, if $kd$ is the depth of N.A., we get: $b\cdot kd\,\dfrac{kd}{2} = m\,A_{st}\,(d - kd)$

or 

$$k^2 + 2km\,\frac{A_{st}}{bd} - 2m\,\frac{A_{st}}{bd} = 0$$

or 

$$k^2 + 2km\,\frac{A_{st}}{1000\,d} - 2m\,\frac{A_{st}}{1000\,d} = 0 \qquad \ldots(2.35)$$

From which $k$ can be found. Knowing $k$, stresses in the section can be determined exactly in the same manner, as adopted for beams.

The complete procedure of design of slabs spanning in one direction is given in Chapter 8.

**Example 2.20.** *A reinforced concrete slab has an effective span of 4 m, and carries a uniformly disturbed load of 6 kN/m² inclusive of its own weight. Determine (i) effective depth of the slab (ii) steel reinforcement. Use M 20 concrete and (a) Fe 250 steel, (b) Fe 415 steel.*

**Solution:** Consider 1 m width of the slab. $w = 6 \times 1 \times 1 = 6$ kN/m length of slab

$$M = \frac{wL^2}{8} = \frac{6 \times 4 \times 4}{8}\ \text{kN-m} = 12\ \text{kN-m} = 12 \times 10^6\ \text{N-mm}$$

**(a) Mild steel (Fe 250) bars**

Given $\sigma_{cbc} = 5$ N/mm²; $\sigma_{st} = 140$ N/mm² and $m = 13$

∴ 

$$k_c = \frac{13 \times 7}{13 \times 7 + 140} = 0.394\,;\ j_c = 1 - \frac{k_c}{3} = 0.869$$

$$R_c = \frac{1}{2}\sigma_{cbc}\cdot j_c \cdot k_c = \frac{1}{2} \times 7 \times 0.869 \times 0.394 = 1.198$$

Hence 

$$d = \sqrt{\frac{M}{1000\,R_c}} = \sqrt{\frac{12000000}{1000 \times 1.198}} = \mathbf{100\ mm}$$

Keep    $d = 100$ mm

$$A_{st} = \frac{M}{\sigma_{st}\,j_c\,d} = \frac{12000000}{140 \times 0.869 \times 100} = \mathbf{986.36\ mm^2}$$

Choosing 12 mm dia. bars,

$$A_\Phi = \frac{\pi}{4}(12)^2 = 113.1 \text{ mm}^2$$

$$\therefore\quad s = \frac{1000\, A_\Phi}{A_{st}} = \frac{1000 \times 113.1}{986.36} = 114.66 \text{ mm}$$

Hence provide 12 mm dia. bars @ 110 mm c/c.

**(b) Fe 415 Steel :** $\sigma_{cbc} = c = 7 \text{ N/mm}^2$, $m = 13$ and $\sigma_{st} = 230 \text{ N/mm}^2$

$$\therefore\quad k_c = \frac{13 \times 7}{13 \times 7 + 230} = 0.283\ ;\ j_c = 1 - \frac{0.283}{3} = 0.905$$

$$R_c = \frac{1}{2}\sigma_{cbc}\, j_c\, k_c = \frac{1}{2} \times 7 \times 0.905 \times 0.283 = 0.896$$

$$d = \sqrt{\frac{M}{1000\, R_c}} = \sqrt{\frac{12 \times 10^6}{1000 \times 0.896}} = 115.7 \text{ mm}$$

Keep $\quad d = 115$ m

$$A_{st} = \frac{M}{\sigma_{st}\, j_c\, d} = \frac{12 \times 10^6}{230 \times 0.905 \times 115} = 501.3 \text{ mm}^2$$

Choosing 10 mm dia. bars,

$$A_\Phi = \frac{\pi}{4}(10)^2 = 78.54 \text{ mm}^2$$

$$\therefore\quad \text{Spacing } s = \frac{1000\, A_\Phi}{A_{st}} = \frac{1000 \times 78.54}{501.3} = 156.67 \text{ mm}$$

Hence provide 10 mm dia. bars at 150 mm c/c. It is seen that by using HYSD bars, though the depth of slab is increased, there is considerable saving in steel reinforcement.

## PROBLEMS

1. (a) Discuss in brief basic assumptions of straight line theory.
   (b) Drive expressions for the position of neutral axis and moment of resistance of balanced rectangular section.
   (c) How do you find the moment of resistance of a beam section?
2. Working from the first principles, determine the moment of resistance and percentage of steel in a singly reinforced rectangular section ($b \times d$) for a balanced design, if the stresses in concrete and steel are not to exceed 5 N/mm² and 140 N/mm² respectively. Take $m = 18$.
3. Design a R.C. beam to carry a load of 6 kN/m inclusive of its own weight on an effective span of 6 m. Keep the breath to be 2/3rd of effective depth. The permissible stresses in concrete and steel are not to exceed 5 N/mm² and 140 N/mm² respectively. Take $m = 18$.
4. The cross-section of singly reinforced beam is 320 mm wide and 450 mm deep to the centre of the reinforcement. The reinforcement consists of 4 bars of 16 mm. Determine the moment of resistance of the beam section. The permissible stresses in concrete and steel are 5 N/mm² and 140 N/mm² respectively. Take $m = 18$.
5. Determine the moment of resistance, per meter width of a R.C. slab having the following details:
   Overall thickness: 100 mm
   Reinforcement: 12 Φ @ 100 mm c/c
   Cover to the centre of reinforcement: 20 mm
   Take the permissible stresses in concrete and steel as 6 N/mm² and 140 N/mm² respectively, and $m = 15$.
6. A R.C.C. beam 300 mm wide and 600 mm deep has 4 bars of 20 mm diameter as tension reinforcement, the centre of the bars being 50 mm from the bottom of the beam. Determine the uniformly distributed load (exclusive

of its own weight) the beam can carry over an effective simply supported span of 6 m. Take the permissible stresses in concrete and steel as 5.2 N/mm² and 126 N/mm² respectively and $m = 18$.

7. A R.C. beam has a section 400 mm wide and 600 mm deep. Find the necessary reinforcement to carry a load of 10 kN/m in addition to its own weight, over an effective span of 6 m. The beam has an effective cover of 50 mm to the centre of reinforcement. The permissible stresses in concrete and steel may be taken as 5 N/mm² and 140 N/mm² respectively. Take $m = 18$.

8. Solve problem 7 if the beam has size 400 mm × 450 mm.

9. The moment of resistance of a rectangular reinforced concrete beam, of breadth $b$ (mm) and effective depth $d$ (mm) is $0.7\, bd^2$ N-mm. If the stresses in the outside fibre of concrete and in the steel do not exceed 4.2 N/mm² and 140 N/mm² respectively, and the modular ratio equals 18, determine the ratio of the neutral axis from the outside compression fibres to the effective depth of the beam and the ratio of the area of tension steel to the effective area of the beam.

10. A reinforced concrete beam has section 300 mm wide and 600 mm deep. The reinforcement consists of mild steel bars of 25 mm diameter with a cover of 50 mm to the centre of reinforcement. If it is subjected to a bending moment of 120 kN-m, determine the stresses developed in steel and concrete. Take $m = 15$.

11. Determine the moment of resistance of the beam section shown in Fig. 2.31. Take $\sigma_{cbc} = 5$ N/mm² and $m = 18$.

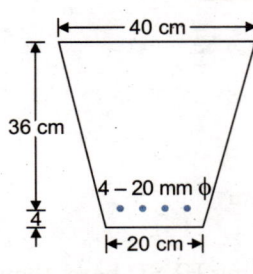

**Fig. 2.31**

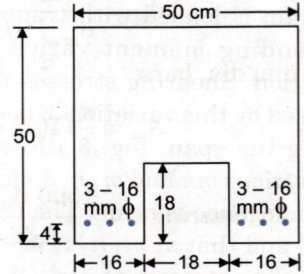

**Fig. 2.32**

12. Determine the moment of resistance of the section shown in Fig. 2.32. Take $\sigma_{cbc} = 5$ N/mm² and $\sigma_{st} = 140$ N/mm² and $m = 18$.

13. An RC cantilever beam of 2 m long is a rectanguler section 240 × 400 mm effective. Tensile reinforcement consists of four 20 mm $\phi$ bars. What maximum udl may be allowed without exceeding stress of 7 N/mm² and 190 N/mm² in concrete and steel respectively? $m = 13.33$ and weight of concrete is 25 kN/m³.

14. An RCC beam 320 × 800 mm overall size is reinforced with four bars of 32 mm $\phi$ bars. The centre of steel is 30 mm from the bottom of the beam. The beam is subjected to a bending moment of 250 kN-m. Find stresses developed in steel and concrete ($m = 13.33$).

## ANSWERS

2. $0.85\, bd^2$; 0.698%
3. $b = 240$ mm ; $d = 360$ mm ; $A_{st} = 616$ mm²
4. 44.9 kN-m
5. $8.5 \times 10^6$ N-mm
6. 12.38 kN/m
7. 1040 mm²
8. 1970 mm²
9. $k = 0.38$ ; $p = 0.65\%$
10. 6.57 N/mm² ; 106 N/mm²
11. 45.5 kN-m
12. 68.5 kN-m
13. Udl ($w$) = 22.895 kN/m
14. $\sigma_{st} = 118.3$ N/mm², $\sigma_{cbc} = 5.388$ N/mm²

# CHAPTER 3
# SHEAR AND BOND

## 3.1. SHEAR STRESS IN R.C. BEAMS

When a beam is loaded with transverse loads, the bending moment varies from section to section. Shearing stresses in the beam are caused by this variation of bending moment along the span. Fig. 3.1(*b*) shows two vertical sections $mn$ and $m_1 n_1$ distant $dx$ apart, of a homogeneous beam. The B.M. at $mn$ is $M$ (say) and that at $m_1 n_1$ is $M + \delta M$. Hence the bending stresses at section $m_1 n_1$ are greater than those at section $mn$. Thus, at the *same fibre*, there are unequal bending stresses at two cross-sections distant $dx$ apart. This inequality of stresses produces a tendency in each fibre to slide over adjacent fibre in a horizontal plane, causing horizontal shear stress which is accompanied by the complementary shear stress in the vertical direction.

This should be resisted by shear reinforcement provided in the form of vertical stirrups (*or*) bent up bars along with stirrups.

In the case of homogeneous beams, the shear stress $q$ at any plane is given by

$$q = \frac{V}{I \cdot z}(A \bar{y});$$

where  $V$ = shear force at the section
  $I$ = moment of inertia of the beam section
  $z$ = width of the section
  $A\bar{y}$ = moment of area above the section, about N.A.

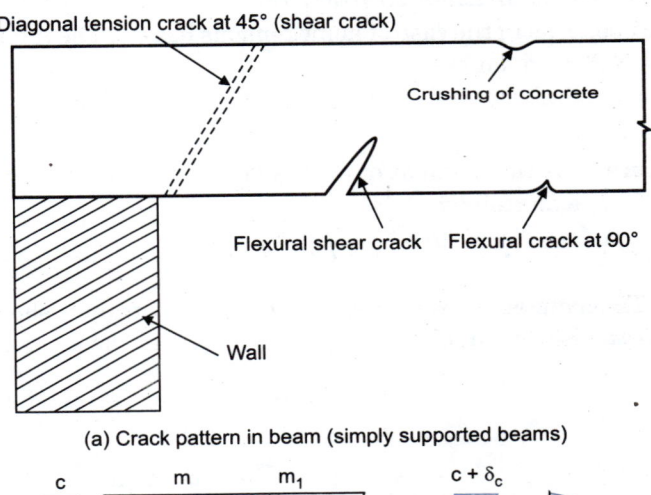

(a) Crack pattern in beam (simply supported beams)

(b) Homogeneous beam

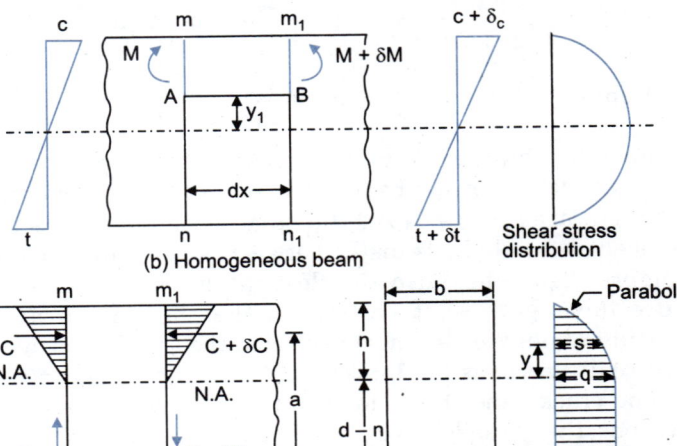

(c) Reinforced concrete beam

Fig. 3.1. Shear Stresses in Beams

Fig. 3.1(c) shows two sections $mn$ and $m_1 n_1$ distant $dx$ apart, of a R.C. beam. Let the bending moment at $mn$ be $M$ and that at $m_1 n_1$ be $M + \delta M$. Due to this variation of bending moment in a length $\delta x$, the compressive forces in concrete at these two sections will be $C$ and $C + \delta C$, and the tensile forces in steel will be $T$ and $T + \delta T$ as shown. Let us assume that the shear force $V$ remains constant in the length $dx$.

Let us assume that concrete does not take any tension. Consider any plane between N.A. and centre of steel reinforcement. Let $q$ be the intensity of shear at the plane. Hence total horizontal shear $= q \cdot b \cdot \delta x$ at the layer. Also, the total horizontal force that tends to slide this layer past the adjacent one is equal to $(T + \delta T) - (T) = \delta T$. Hence equating the two, we get

$$q \cdot b \cdot \delta x = \delta T \quad \text{or} \quad q = \frac{\delta T}{b \cdot \delta x} \qquad \ldots(i)$$

In order to get the value of $q$, it is first essential to determine the value of $\delta T$.

For equilibrium $\Sigma M = 0$,

$$\therefore \quad (T + \delta T) jd = Tjd + V \cdot \delta x \quad \text{or} \quad \delta T = \frac{V \cdot \delta x}{jd} \qquad \ldots(ii)$$

Substituting in (i), we get $\quad q = \dfrac{V}{bjd} \qquad \ldots(3.1)$

This value of $q$ is evidently the same for any layer between the N.A. and the centre of steel. Thus the shear stress distribution below the N.A. is rectangular. The shear stress distribution above the N.A. is parabolic, as in the case of homogeneous beams. The intensity of shear stress $s$ at any layer distant $y$ above the N.A. is given by

$$s = \frac{V}{Ib} \int_y^n b \cdot y \cdot dy = \frac{V}{2I}(n^2 - y^2) \qquad \ldots(3.2)$$

where $I$ is the moment of inertia of the R.C. beam section about the N.A. and can be found the following expression:

$$I = \frac{1}{3} b \cdot n^3 + m A_{st} (d - n)^2 \qquad \ldots(3.3)$$

The complete shear stress distribution diagram is shown in Fig. 3.1(b). The intensity of maximum shear ($q$) can also be found by equating the external shear force to the shearing resistance of the beam, as under:

$$V = \text{(Area of shear stress diagram)} \times b = \left\{ \left(\frac{2}{3} qn\right) + q(d - n) \right\} b$$

$$= \left(\frac{2}{3} qn + qd - qn\right) b = qb\left(d - \frac{n}{3}\right) = q \cdot bjd$$

Hence $\quad q = \dfrac{V}{b \cdot jd} \qquad \ldots(3.1)$

Eq. 3.1, obtained from the stress based approach, gives the value of maximum shear stress $q$ at N.A. and for all layers below N.A. in a cracked reinforced beam. Though in pure flexure, we neglect the concrete area below N.A., the same concrete between the cracks is needed for shear transfer between steel reinforcement and the compression zone. Also the flexural cracks (assumed vertical in pure flexure) are not vertical in the shear-flexural zone under consideration. We have also neglected the shear deformations, which are really responsible for the mobilization of aggregate interlock at the cracks and dowel forces in the reinforcement. Due to these reasons, the *stress-based approach* followed above in the derivation of Eq. 3.1 may not represent the real behaviour of the beam in shear and flexure. Hence Eq. 3.1, representing conventional shear stress formula $q = V/bjd$, has now been abandoned. Indian standard Code (IS : 456–2000) recommends to use *nominal shear stress* given by the expression.

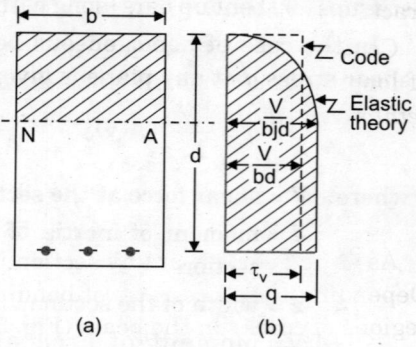

Fig. 3.2

$$\tau_v = \frac{V}{bd} \qquad \ldots(3.4)$$

where  $b$ = breadth of beam in 'mm'
  $d$ = effective depth of beam in 'mm'

where  $\tau_v$ = nominal shear stress in N/mm² (or) MPa and $V$ = shear force at the section, due to design loads.

Thus in Fig. 3.2(b), the maximum shear stress $q$ (= $V/bjd$) obtained from elastic theory, is greater than the *nominal shear stress* (or average shear stress) $\tau_v$ suggested by the code.

*The guidelines in the design for shear are:*
1. Shear stress in concrete is considered
2. Minimum stirrups spacing is defined
3. Bent up bars are allowed for half the total shear
4. Check for shear stress in slab is introduced in the new code.

## 3.2. EFFECTS OF SHEAR : DIAGONAL TENSION

We know that the bending stress and the shearing stress vary across the cross-section of R.C. beam. Below the N.A., the bending stress $(f)$ is tensile while the shear stress $q$ is constant. Fig. 3.3(b) shows a small element, taken from the portion below the N.A. This element is subjected to a longitudinal tensile stress $f$ and horizontal shear stress $q$ alongwith the vertical complementary shearing stress $q$.

$$p = \frac{f}{2} \pm \sqrt{\left(\frac{f}{2}\right)^2 + q^2} \qquad \ldots(3.5)$$

and the inclination of the principal plane with horizontal is given by

$$\tan 2\theta = -\frac{2q}{f} \qquad \ldots[3.5(a)]$$

The major principal stress $p_1$ is tensile and is given by

$$p_1 = \frac{f}{2} + \sqrt{\left(\frac{f}{2}\right)^2 + q^2} \qquad \ldots[3.5(b)]$$

At the supports, where bending stress $f$ is practically zero, the value of the principal (tensile) stress is equal to the shear stress $q$, and it is inclined at 45° to the horizontal, *i.e.* it acts diagonally. Hence it is known as *diagonal tension*. At the centre of the beam, where shearing stresses are practically negligible, the principal tensile stress is equal to the longitudinal tensile stress $f$, its direction will be horizontal and the cracks will be vertical.

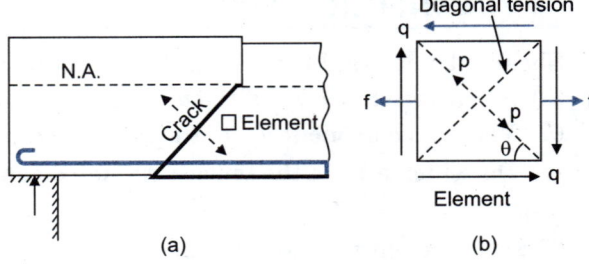

Fig. 3.3.  Diagonal Tension

Concrete is quite strong in shear, but the diagonal tension, which is caused by the combined action of longitudinal tension and the transverse shearing stresses, is to be resisted by the provision of *shear reinforcement or diagonal (or inclined tension) reinforcement or web reinforcement.*

## 3.3. REGIONS OF CRACKS IN BEAMS

As stated earlier, at any section in a beam, there exists both bending moment $(M)$ as well as shear $(V)$. Depending upon the ratio of bending moment $(M)$ to shear $(V)$ at different sections, there may be three regions of cracks in the beam (Fig. 3.4) :

(a) Region I    : Region of *flexure cracks*
(b) Region II   : Region of *flexure-shear cracks*
(c) Region III  : Region of *web-shear cracks* or *diagonal tension cracks*

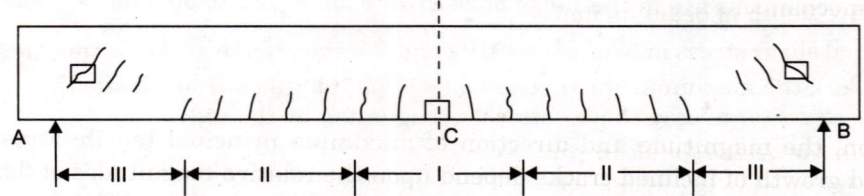

Fig. 3.4. Different Regions of Cracks in Beam

(a) **Region I : Region of flexural cracks**

This region normally occurs adjacent to mid-span where B.M. is *large* and shear force is either zero or very small. The principal planes are perpendicular to the beam axis. when the principal tensile stress reaches the *tensile strength* of concrete (which is quite low) *tensile cracks develop vertically*. These cracks are known as *flexural cracks* resulting primarily due to flexure.

(b) **Region II : Regions of flexure-shear cracks**

These regions are near the quarter span, to both the sides, where B.M. is considerable and at the same time shear force is significant. The cracks in this region are initiated at the tension face, travel vertically (due to flexure) and gradually tend to develop in the inclined direction towards the N.A., as the shear stress goes on increasing towards the N.A. Since the cracks develop under the combined action of bending moment and shear, these cracks are known as flexure-shear cracks.

(c) **Region III : Regions of diagonal tension cracks**

These regions are adjacent to each support of a beam, where shear force is predominant. Since shear stress is maximum at the N.A., inclined cracks start developing at the N.A. along the diagonal of an element subjected to the action of pure shear. Hence these cracks are known as *diagonal tension cracks* or *web-shear cracks*.

## 3.4. MECHANISMS OF SHEAR TRANSFER IN R.C. BEAM WITHOUT SHEAR REINFORCEMENT

Shear is transferred between two adjacent planes in a R.C. beam by the following mechanisms :
(a) Shear resistance $V_{cz}$ of the *uncracked* portion of concrete
(b) Vertical component $V_{ay}$ of the *interface shear* or *aggregate interlock force* $V_a$ and
(c) Dowel force $V_d$ in the tension reinforcement, due to dowel action
Thus,  $V = V_{cz} + V_{ay} + V_d$  ...(3.6)

The relative contribution of each of the above three mechanisms depend upon the stage of loading and extent of cracking. In the initial stage, before the flexural cracking starts, the entire shear is resisted by the shear resistance of the concrete (*i.e.* $V = V_{cz}$). As the flexural cracking starts, interface shear comes into action, resulting in the redistribution of stresses. Further extension of flexural crack results in sharing the shear by the dowel force $V_d$ of the tension

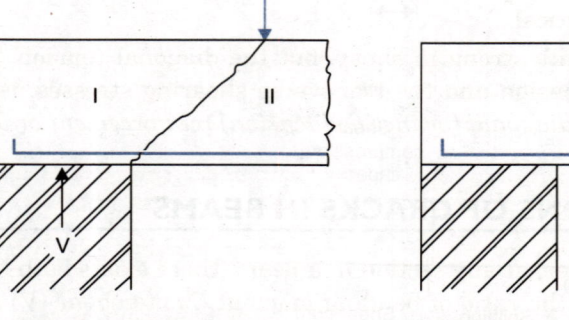

(a) Diagonal tension crack     (b) Flexural shear crack

Fig. 3.5. Mechanism of Shear Transfer

reinforcement. Thus, at the final stage of collapse, the shear is borne by all the three mechanism, expressed by *Eq. 3.6*.

For beams of rectangular section, without shear reinforcement, the proportion of shear transferred by the above three mechanisms are in the range of 20 to 40% for $V_{cz}$, 30 to 50% for $V_{ay}$ and 15 to 25% by $V_d$.

## 3.5. SHEAR SPAN

At any section, the magnitude and direction of maximum principal tensile stress, and hence the development and growth of inclined cracks depend upon the relative magnitudes of flexural stress $f$ and the shear stress $\tau$. Since $f$ and $\tau$ at any section are *proportional* to $M/bd^2$ and $V/bd$ respectively, we have

$$\frac{f}{\tau} = \frac{A_1\ M/bd^2}{A_2\ V/bd} = A_3 \left[\frac{M}{Vd}\right] \qquad \ldots(3.7)$$

where $A_1, A_2, A_3$ are the constants of proportionality.

For a beam subjected to concentrated load, the ratio $M/V$ at the initial section subjected to *maximum V* is expressed by a distance $a_v$ called *shear span* between the support and the load (Fig. 3.6). The distance $a_v$ ($= M/V$) represent the dominance of shear over flexure. *Many tests on the shear have established that diagonal tension is a function of M/V and hence on the shear span $a_v$.* It also depends on the effective depth $d$, as clear from the following expressions :

$$\frac{f}{\tau} \propto \frac{a_v}{d} \qquad \ldots(3.8)$$

Thus, the diagonal tension failure (*i.e.*, shear failure) is a function of ratio $a_v/d$. It is also interesting to note that the dimension $a_v/d$ (or $M/V.d$) provides a measure of the relative magnitudes of the flexural stress and shear stress, and hence enables the prediction of the mode of failure in the beam in flexural shear.

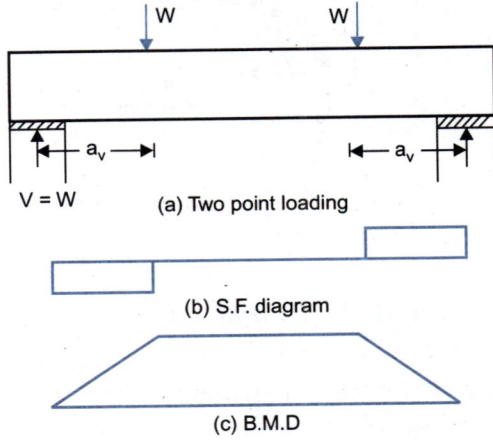

Fig. 3.6. Shear Span

## 3.6. MODES OF SHEAR FAILURE

The shear failure of a R.C. beam, without shear reinforcement is governed by $a_v/d$ ratio. Depending upon the value of $a_v/d$, ratio, a beam may experience following types of shear failure.

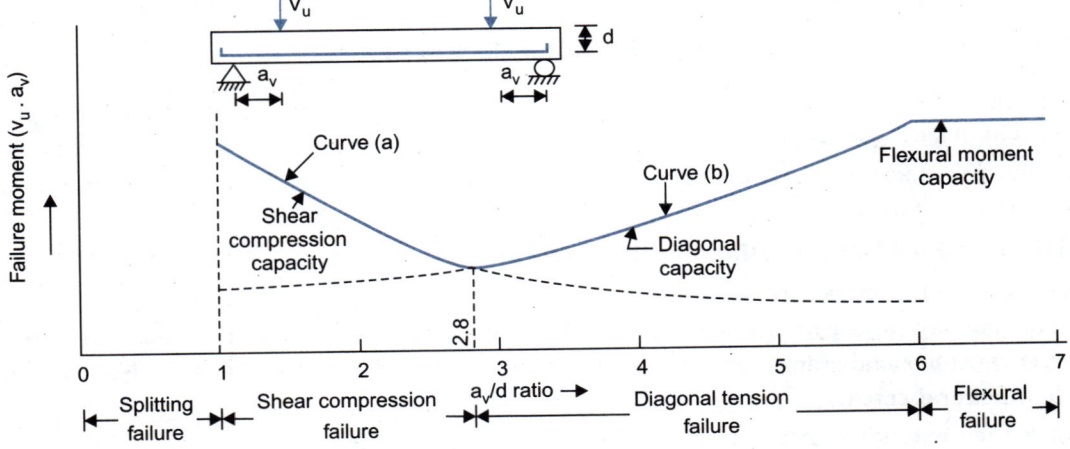

Fig. 3.7. Effect of $a_v/d$ on Shear Strength of R.C. Beam

1. Case I     :  $a_v/d < 1$         :  Splitting or compression failure (deep beams)
2. Case II    :  $1 < a_v/d < 2.5$   :  Shear compression or shear tension failure
3. Case III   :  $2.5 < a_v/d < 6$   :  Diagonal tension failure
4. Case IV    :  $a_v/d > 6$         :  Flexure failure

Fig. 3.7 shows the plot of failure moment ($V_u \cdot a_v$) versus $a_v/d$ ratio.

### Case I : $a_v/d < 1$ (deep beams) : Splitting or compression failure

This case corresponds to a deep beam without shear reinforcement, where the inclined cracking transforms the beam into a *tied arch* [Fig. 3.8 a]. The load is carried by (i) direct compression in the concrete between the load and reaction point by crushing of concrete and by (ii) tension in the longitudinal steel by yielding or fracture or anchorage failure or bearing failure.

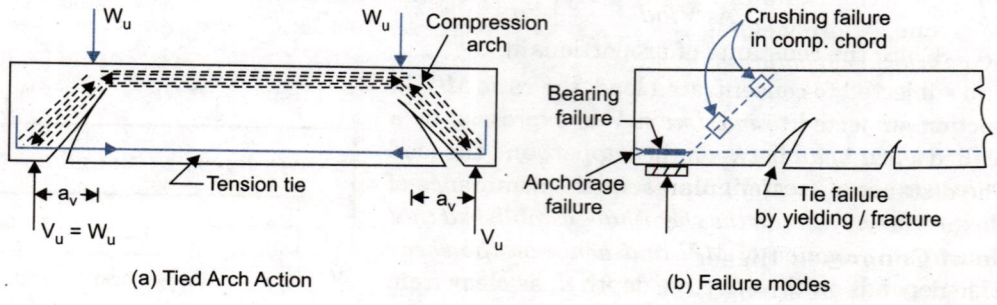

**Fig. 3.8.** Case I : $a_v/d < 1$ (Deep Beams)

### Case II : $1 < a_v/d < 2.5$ : Shear compression/tension failure

This case is common in short beam with $a_v/d$ ratio between 1 to 2.5, where failure is initiated by an inclined crack – more commonly a *flexural shear crack*. Fig. 3.9 (a) shows the *shear compression failure* due to vertical compressive stresses developed in the vicinity of the load.

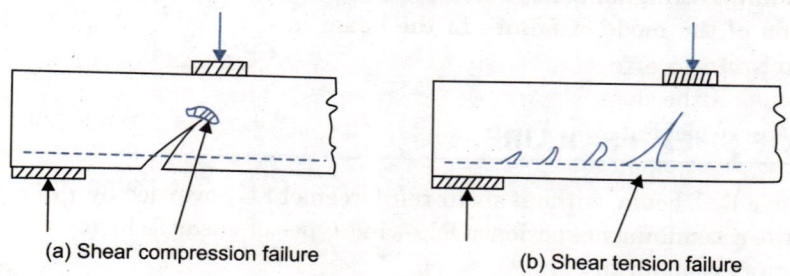

**Fig. 3.9.** Case II : $1 < a_v/d < 2.5$

Similarly, the vertical compressive stress over the reaction limits the bond splitting and diagonal cracking along the steel. The crack extends towards the tension reinforcement and then propagates along the reinforcement (Fig. 3.9b) resulting in the failure of the beam by *anchorage failure* of tension reinforcement called shear tension failure.

### Case III : $2.5 a_v/d < 6$ : Diagonal tension failure

Diagonal tension failure occurs when the shear span to the effective depth ratio is in the range of 2.5 to 6. *Normal beams* have $a_v/d$ ratio in excess of 2.5. Such beams may fail either in shear or in flexure. In the beginning, several flexural cracks (nearly vertical) are formed and then the diagonal flexural crack starts from the last flexural crack.

The crack then extends in compression zone in the inclined direction, first steeply and then at a flat slope. Shortly before reaching the critical failure, the bottom inclined crack gets widened. If shear (web) reinforcement is not provided, the cracks extends rapidly to the top of the beam and failure occurs *rapidly*. Addition of shear reinforcement enhances the shear strength considerably.

### Case IV : $a_v/d > 6$ : Flexural failure

Flexural failure is encountered when $a_v/d$ ratio is greater than 6. Two cases may be encountered :

(*i*) under-reinforced beam and

(*ii*) over reinforced beam.

In the case of under reinforced beam, tension reinforcement is less than the limiting one, due to which failure is initiated by yielding of tension reinforcement, leading to the ultimate failure due to crushing to concrete in compression zone. Such a *ductile failure* is known as *flexural tension failure* which is quite slow giving enough warning. In the over-reinforced sections, failure occurs due to crushing of concrete in compression zone, before yielding of tension reinforcement. Such a failure, known as *flexural compression-failure*, is quite sudden.

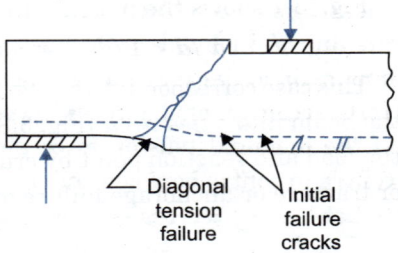

Fig. 3.10.   Case III : $2.5\, a_v/d < 6$

## 3.7. FACTORS AFFECTING SHEAR RESISTANCE OF A.R.C. MEMBER

The *shear resistance* of rectangular beams, *without shear reinforcement*, depends upon the following factors.

**1. Grade of Concrete :** Higher grade of concrete has higher characteristic strength which in turn results in (*i*) higher tensile strength (*ii*) greater dowel shear resistance, (*iii*) greater aggregate interlock capacity and (*iv*) greater concrete strength in compression zone. Hence *shear resistance increases with the increase in the grade of concrete.*

**2. Percentage** *and* **grade of longitudinal tensile reinforcement :** The increase in percentage ($p_t$) of longitudinal tensile reinforcement results in the increase in dowel shear ($V_d$). Due to this reason, the design Codes make the shear strength ($\tau_c$) of concrete a function of $p_t$ and grade of concrete (See Table 7.1). However, higher grade of steel results in lesser shear resistance of a R.C. beam because the percentage of steel ($p_t$) corresponding to a higher grade steel is less than that required for a low grade steel, say mild steel.

**3. Ratio of shear span to effective depth** (*i.e. $a_v/d$ ratio*) **:** As discussed in the previous article, for $a_v/d$ ratio between 6 and 2.5, the shear capacity, being governed by inclined crack resistance, decreases with decrease in $a_v/d$ ratio (curve b of Fig. 3.7). However, for a value of $a_v/d$ less than 2.5, the shear capacity, being dependent on *shear-compression or shear-bond capacity*, increases rapidly. The minimum shear capacity is at $a_v/d$ ratio around 2.5.

**4. Compressive force :** Presence of axial compressive force results in increase of shear capacity. The effect of axial compression on the design shear strength has been taken into account by I.S. Code by increasing the design shear strength by a modification factor δ (see Fig. 7.11).

**5. Compressive reinforcement :** The shear resistance is found to increase with the increase in the percentage of compressive steel ($p_c$).

**6. Axial tensile force :** Axial tensile force reduces marginally the shear resistance of concrete, as is evident from *Eq. 7.12*.

**7. Shear reinforcement :** The shear resistance of a R.C. beam increases with the increase in *shear reinforcement ratio*. This is due to two reasons (*i*) concrete gets confined between stirrup spacing, and (*ii*) the shear/web reinforcement itself provides shear resistance, $V_s$.

## 3.8. REINFORCEMENT FOR DIAGONAL TENSION

As stated above, proper reinforcement must be provided to resist the diagonal tension. Fig. 3.11 shows a portion of the beam near its support, reinforced with inclined bars 1, 2, 3 etc. spaced at distance s apart.

Let the web reinforcement, be inclined at an angle $\alpha$ with the horizontal and the direction of diagonal tension be at 45° with the horizontal. Near the support, it is assumed that the value of the diagonal tension is equal to the maximum shearing stress $q$. The beam is likely to crack along the line $ab$ inclined at 45°; hence total tension in one inclined bar (say bar 2) must be such that its vertical component balances the vertical component of the diagonal tension force along the length $ef$ centre to centre between the bars.

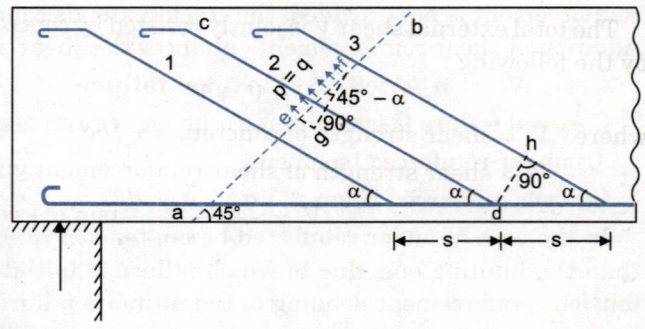

Fig. 3.11. Web Reinforcement

Let $A_{sv}$ = cross-sectional area of bar 2 = web reinforcement
$\sigma_{sv}$ = permissible tensile stress in shear reinforcement

Vertical component to the force in the bar = $\sigma_{sv} \cdot A_{sv} \sin \alpha$
Vertical component of the diagonal tension in the length $ef = q \times b \,(ef) \sin 45°$
Equating the two, we get : $\sigma_{sv} \cdot A_{sv} \sin \alpha = q \cdot b \,(ef) \sin 45°$ ...(i)

In order to get the value of the length $ef$, draw $fg$ perpendicular to bar 2; also draw $eg$ perpendicular to $fg$.
Evidently, $\qquad fg = hd = s \sin \alpha$
$\therefore \qquad ef = fg \cdot \sec efg = fg \cdot \sec (45° - \alpha)$
or $\qquad ef = s \sin \alpha \cdot \sec (45° - \alpha)$ ...(ii)

Substituting the value of $ef$ in (i), we get
$$\sigma_{sv} \cdot A_{sv} \sin \alpha = (q \cdot b \cdot \sin 45°) \times s \sin \alpha \sec (45° - \alpha)$$
$$s = \frac{\sigma_{sv} A_{sv}}{q \cdot b \cdot \sin 45°} \cos (45° - \alpha) = \frac{\sigma_{sv} A_{sv}}{q \cdot b} (\sin \alpha + \cos \alpha) \qquad \text{...[3.9(a)]}$$

Replacing $q$ by $\tau_v$ (as envisaged by the Code), we have
$$s = \frac{\sigma_{sv} A_{sv}}{\tau_v \cdot b} (\sin \alpha + \cos \alpha) \quad \text{where } \tau_v = \frac{V}{bd} \qquad \text{...(3.9)}$$

$\therefore \qquad s = \dfrac{\sigma_{sv} A_{sv} \cdot d}{V} (\sin \alpha + \cos \alpha)$ ...(3.10)

Eq. 3.10 gives the spacing of bars inclined at $\alpha$ with the horizontal.
If the web reinforcement is vertical, i.e. if $\alpha = 90°$ we get
$$s = \frac{\sigma_{sv} A_{sv} \cdot d}{V} = \frac{\sigma_{sv} A_{sv}}{\tau_v \cdot b} \qquad \text{...(3.11)}$$

If the bars are inclined at 45°, the spacing is given by
$$s = \frac{\sigma_{sv} A_{sv} \cdot d}{V} \sqrt{2} = \frac{\sigma_{sv} A_{sv}}{\tau_v \cdot b} \sqrt{2} \qquad \text{...(3.12)}$$

## 3.9. TYPES OF SHEAR REINFORCEMENT

Shear reinforcement will be necessary if the nominal shear stress exceeds the allowable shear stress $\tau_c$. The Code recommendations are discussed is § 3.13.

However, in general, shear reinforcement may be provided in the following three forms:
 (i) Shear reinforcement in the form of vertical bars, known as *stirrups*.
 (ii) Shear reinforcement in the form of inclined bars.
 (iii) Shear reinforcement in the form of combination of stirrups and inclined bars.

The total external shear $V$ is jointly resisted by concrete as well as shear reinforcement, and is represented by the following :

$$V = V_c + V_s$$

where  $V_c$ = shear strength of concrete = $\tau_c \cdot bd$
$V_s$ = shear strength of shear reinforcement given by rearranging *Eq. 3.10* in the following form :

$$V_s = \frac{\sigma_{sv} \cdot A_{sv} \cdot d}{s}(\sin \alpha + \cos \alpha) \qquad ...(3.13)$$

## 3.10. VERTICAL STIRRUPS

Shear reinforcement in the form of vertical *stirrups* consists of 5 mm to 16 mm diameter steel bars bent round the tensile reinforcement where it is anchored to 6 to 12 mm dia. anchor bars or holding bars. Depending upon the magnitude of the shear stress to be resisted, a stirrup may be one legged, two legged, four legged or multi-legged, as shown in Fig. 3.12.

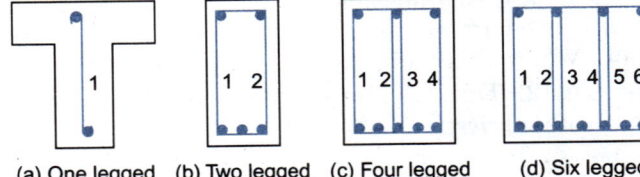

(a) One legged  (b) Two legged  (c) Four legged  (d) Six legged

Fig. 3.12. Forms of Vertical Stirrups

The spacing of vertical stirrups will depend upon the net shear force to be resisted at the section, determined from the expression:

$$F = V - V_c = V_s$$

Evidently, the net shear force $F$ should not exceed the shear strength $V_s$ of stirrup.

Hence from *Eq. 3.11*, replacing $V$ by $F$, we get

$$s = \frac{\sigma_{sv} \cdot A_{sv} \cdot d}{F} \qquad ...(3.14)$$

Alternatively, when vertical stirrups are used in conjunction with other forms of shear reinforcement the strength of vertical stirrups, spaced at $s$ is given by

$$V_s = \frac{\sigma_{sv} \cdot A_{sv} \cdot d}{s} \qquad ...[3.11(a)]$$

where  $A_{sv}$ = cross-sectional area of stirrups
= (Area of cross-section of stirrups bar) × number of legs = $A_\phi \cdot n$
$s$ = centre to centre spacing of stirrups.

*Eq. 3.14* can also be *alternatively* obtained from Fig 3.13. Let us assume that in absence of shear reinforcement, the beam fails in diagonal tension, the inclination of the tension crack being at 45° to the axis of the beam and extended up to a horizontal distance equal to $(d - d_c) \approx d$

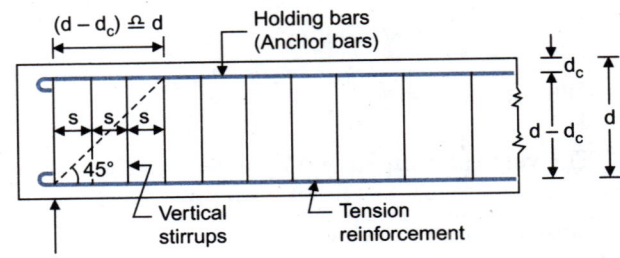

Fig. 3.13. Spacing of Vertical Stirrups

∴ No. of stirrups resisting shear force = $d/s$
∴ [Shear force $F$ resisted by stirrups] = [force resisted by each stirrup] × [No. of stirrups]

or $$F = \sigma_{sv} \cdot A_{sv} \times \frac{d}{s}$$

or $$s = \frac{\sigma_{sv} \cdot A_{sv} \cdot d}{F} = \frac{\sigma_{sv} \cdot A_{sv} \cdot d}{V - V_c} \qquad ...(3.14)$$

where  $F$ = net S.F. to be resisted by stirrups = $V - V_c$

**Spacing diagram for vertical stirrups.** The spacing of the stirrups at any section along the length of the beam can be determined from *Eq. 3.9*, since the net shear force $F$ at that section can be easily determined. This will give variable spacings all along the length of the beam. These variable spacings can be *averaged* out by the construction of *spacing diagram*.

**Procedure.** (1) Determine net shear force at various sections distant $x_1, x_2, x_3$ etc. from the support and calculate the spacings at these sections from *Eq. 3.14*.

(2) Plot the spacing diagram with the calculated spacing as ordinate (Fig. 3.14).

(3) Calculate the spacing of the first stirrup, on the basis of net S.F. at the support. Set the first stirrup at half the distance of this spacing, from the support.

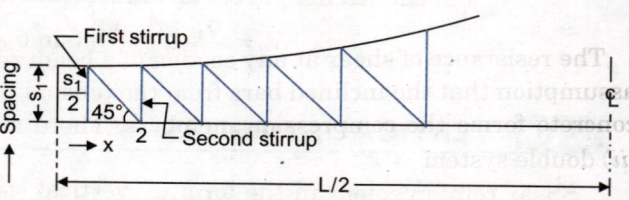

Fig. 3.14. Spacing Diagram.

(4) Where this ordinate at (1) intersects the spacing diagram, set a line at 45°, meeting the span line in (2). This will give the position of the second stirrup. Repeat this procedure to get the position of other stirrups.

## 3.11. INCLINED BARS

Some of the main reinforcing bars, provided to resist the tensile bending stress, can be bent up near the supports where they are no longer required since bending moment is very much reduced near the support. Such bent up bars resist the diagonal tension. To maintain symmetry about the vertical axis of the section, bars are generally bent up in *pairs*. These bars are bent usually at 45°.

Fig. 3.15 shows a number of pairs of inclined bars, spaced equally $s_v$ apart, and inclined at $\alpha$ with the axis of the beam. Let $\beta$ be the inclination of the crack, usually assumed equal to 45° at the support where S.F. is maximum. The number of pairs of inclined bars, crossing such a crack (Fig. 3.15) is given by

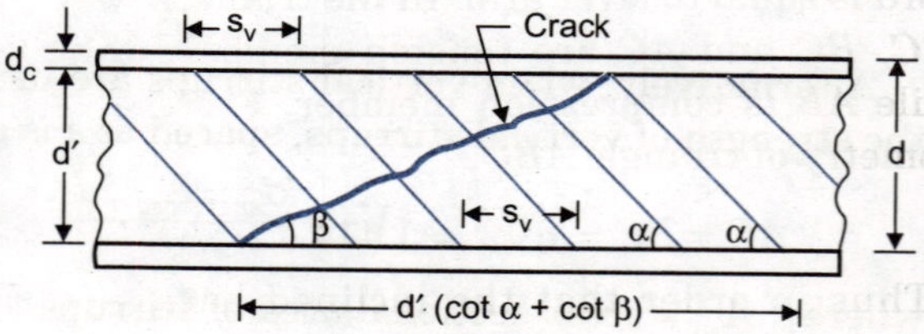

Fig. 3.15. Inclined Bars

$$n = \frac{(d - d_c)(\cot \alpha + \cot \beta)}{s_v}, \quad \text{where} \quad d_c = \text{cover from top}$$

Taking $d - d_c = d'$,
$$n = \frac{d'(\cot \alpha + \cot \beta)}{s_v} \qquad \ldots(3.15)$$

The vertical component ($V_s$) of the tensile force on inclined bars crossing the diagonal crack is given by
$$V_s = n \cdot A_{sv} \cdot \sigma_{sv} \cdot \sin \alpha \qquad \ldots(3.16)$$

Substituting the value of $n$,
$$V_s = A_{sv} \cdot \sigma_{sv} \cdot (\cos \alpha + \sin \alpha \cot \beta) \frac{d'}{s_v} \qquad \ldots(3.17)$$

If $n = 1$ (*i.e.* only one pair of inclined bars), $s_v = d'(\cot \alpha + \cot \beta)$  ...[3.18(a)]

$\therefore \quad V_s = A_{sv} \cdot \sigma_{sv} \cdot \sin \alpha$  ...[3.18(b)]

Taking $\alpha = 45°$, $V_s = 0.707 \cdot A_{sv} \cdot \sigma_{sv}$  ...(3.18)

If $n = 2$ (*i.e.* two pairs of inclined bars), $s_v = \frac{1}{2} d'(\cot \alpha + \cot \beta)$  ...(3.19)

$\therefore \quad V_s = 2 A_{sv} \cdot \sigma_{sv} \cdot \sin \alpha$  ...(3.20)

The above expressions are also valid in the case of *inclined stirrups*, in which case $A_{sv}$ is the total area of the all the legs of each inclined stirrups. The above analysis helps in understanding the *truss analogy* or *lattice girder effect*.

## 3.12. LATTICE GIRDER EFFECT

The resistance of shear at any section of a beam reinforced with inclined bars may be calculated on the assumption that the inclined bars from the tension members of a system of the lattice girder in which the concrete forms the compression members. There are two systems of lattice girders : (*i*) single system (*ii*) double system

(*i*) **Single System** (*Fig. 3.16*) : In this system, the bars are bent at 45° and the imaginary compression members (known as *struts*) AB, CD etc. are assumed to be at $67\frac{1}{2}°$ with horizontal. Member AC is an imaginary horizontal portion of the inclined bar acting as *tension chord* and is assumed to have the same cross-sectional area as of inclined bars. The member BD is an imaginary *compression chord*. It is assumed that the distance between the top compression chord and bottom tension chord is equal to lever arm. In the triangle ABC, BC and AC are tension members while AB is compression member. From geometry of triangle ABC,

$$AC = BC = a\sqrt{2} = 1.414\, a$$

Thus in order that the inclined bars are to be effective, the bars nearest to the support are bent at 45° at a distance of 1.414a from the support. Since the triangle ABC is isosceles, the total force in the inclined bar BC is equal to the total force in the horizontal bar. Thus, the force in the inclined bar is $A_{sv}\, \sigma_{sv}$ where $A_{sv}$ is the area of cross-section of the bent up bars. The vertical component of this force to resist the shear force is equal to:

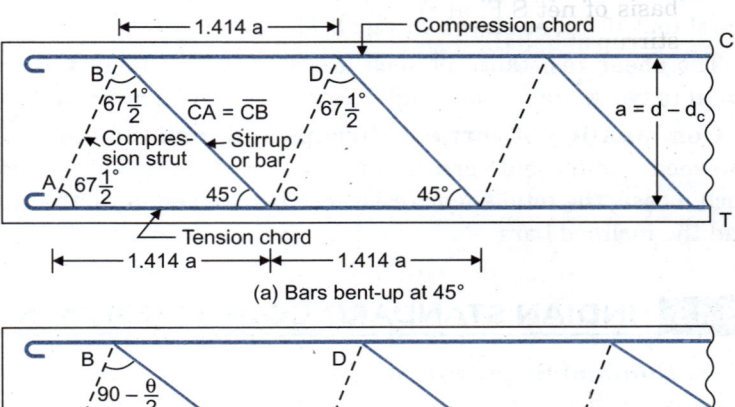
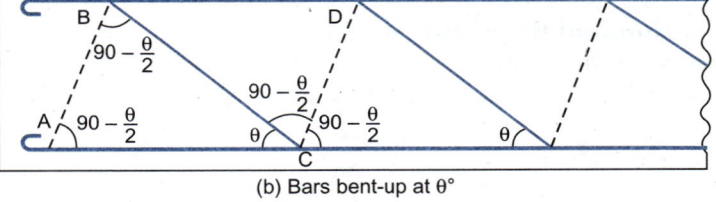
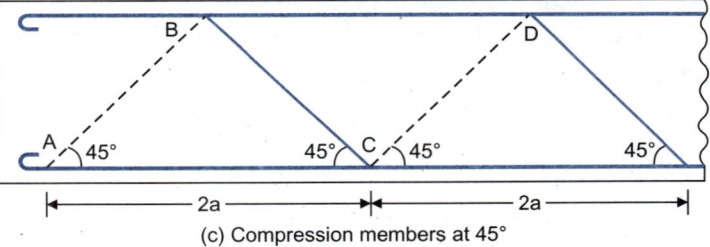

Fig. 3.16. Single System of Lattice Girder.

$$V_s = A_{sv} \cdot \sigma_{sv} \cdot \sin 45° = 0.707\, A_{sv} \cdot \sigma_{sv}. \qquad \ldots(3.18)$$

*The inclined bar is effective for a length equal to lever arm on either side measured along the N.A.* If, however, the bars are bent up at an angle θ, the compression member is assumed to be inclined at 90° − θ/2 with the horizontal as shown in Fig. 3.16(*b*).

Sometimes, the imaginary compression members are assumed to be inclined at 45° as shown in Fig. 3.16(*c*). In this case, the lengths AB and BC will be equal, and the spacing will be equal to 2a. The total force in the bar BC can be calculated from the force triangle ABC are under :

$$\frac{\text{Force in } BC}{\text{Force in } AC} = \frac{\sin 45°}{\sin 90°}$$

$$\therefore \qquad \text{Force in } BC = \frac{A_{sv} \cdot \sigma_{sv} \cdot \sin 45°}{\sin 90°} = A_{sv} \frac{\sigma_{sv}}{\sqrt{2}}$$

This shows that the permissible stresses in $BC$ is reduced to $\dfrac{\sigma_{sv}}{\sqrt{2}}$. This ensures that the stress in the horizontal portion of the inclined bars is equal to the permissible value of $\sigma_{sv}$.

Hence the vertical S.F. resisted by $BC = V_s = A_{sv} \cdot \dfrac{\sigma_{sv}}{\sqrt{2}} \cdot \sin 45°$ or $V_s = 0.707 \cdot A_{sv} \cdot \dfrac{\sigma_{sv}}{\sqrt{2}}$ ...(3.21)

(*ii*) **Double System** (*Fig. 3.17*) : In this system, the bars are bent up at spacing equal to half the spacing of the single system. The compression members are assumed to be inclined at $67\frac{1°}{2}$, but the spacing between them is assumed to be equal to $1.414\,a$, while the spacing between the inclined bars is kept equal to $0.707\,a$, as shown in Fig. 3.17.

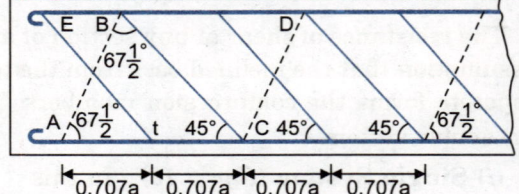

Fig. 3.17. Double System

The shear resistance of such a system will evidently be equal to twice that of the single system *i.e.*, it will be equal to : $2 \times 0.707\,A_{sv} \cdot \sigma_{sv}$.

**Combination of vertical stirrups and inclined bars.** If the bent up bars are not sufficient to resist the shear, additional vertical stirrups are also provided to act in combination with the inclined bars. In such a case, the total shear resistance is taken equal to the sum of the resistances of the vertical stirrups and the inclined bars.

## 3.13. INDIAN STANDARD CODE RECOMMENDATIONS (IS : 456 – 2000)

**1. Nominal Shear Stress :** Indian Standard Code (IS : 456 – 2000) uses a concept of *nominal shear stress* ($\tau_v$) rather than *shear stress* (*q*) given by *Eq. 3.1*. The nominal shear stress in beams or slabs of uniform depth shall be calculated by the following equations:

$$\tau_v = \dfrac{V}{bd} \qquad ...(3.4)$$

where $V$ = shear force due to design loads

$b$ = breadth of the member, which for flanged section shall be taken as the breadth of web ($b_r$)

and $d$ = effective depth

In the case of beams of varying depth, the equation shall be modified as :

$$\tau_v = \dfrac{V \pm \dfrac{M_v}{d} \tan \beta}{b \cdot d} \qquad ...[3.4(a)]$$

where $M_v$ = factored bending moment at the section

$Z_v$ = nominal shear stress

where $\tau_v$, $V$, $b$ and $d$ have the same meanings as above and $\beta$ is the angle between the top and the bottom edges of the beam.

The negative sign in the formula applies when the bending moment $M$ increases numerically in the same direction as the effective depth increases, and the positive sign when the moment decreases numerically in this direction.

**2. Design Shear Strength of Concrete**

**(*a*) Beams without Shear Reinforcement :** The permissible shear stress ($\tau_c$) in concrete in beams without shear reinforcement is given in Table 3.1. The permissible shear stress depends upon two factors : grade of concrete and percentage reinforcement ($100\,A_s/bd$).

This value depends mainly on grade of concrete and area of steel provided for resisting bending moment as per table no. '19' of IS: 456–2000 and Table '61' of SP-16 gives the design shear strength of concrete in beams as a function of grade of concrete and percentage of tension steel.

# SHEAR AND BOND

**TABLE 3.1.** Permissible Shear Stress ($\tau_c$) in Concrete (IS : 456–2000)

| $\dfrac{100 A_s}{bd}$ | Permissible shear stress in concrete ($\tau_C$) N/mm² for grade of concrete | | | | | |
|---|---|---|---|---|---|---|
| | M 15 | M 20 | M 25 | M 30 | M 35 | M 40 |
| ≤ 0.15 | 0.18 | 0.18 | 0.19 | 0.20 | 0.20 | 0.20 |
| 0.25 | 0.22 | 0.22 | 0.23 | 0.23 | 0.23 | 0.23 |
| 0.50 | 0.29 | 0.30 | 0.31 | 0.31 | 0.31 | 0.32 |
| 0.75 | 0.34 | 0.35 | 0.36 | 0.37 | 0.37 | 0.38 |
| 1.00 | 0.37 | 0.39 | 0.40 | 0.41 | 0.42 | 0.42 |
| 1.25 | 0.40 | 0.42 | 0.44 | 0.45 | 0.45 | 0.46 |
| 1.50 | 0.42 | 0.45 | 0.46 | 0.48 | 0.49 | 0.49 |
| 1.75 | 0.44 | 0.47 | 0.49 | 0.50 | 0.52 | 0.52 |
| 2.00 | 0.44 | 0.49 | 0.51 | 0.53 | 0.54 | 0.55 |
| 2.25 | 0.44 | 0.51 | 0.53 | 0.55 | 0.56 | 0.57 |
| 2.50 | 0.44 | 0.51 | 0.55 | 0.57 | 0.58 | 0.60 |
| 2.75 | 0.44 | 0.51 | 0.56 | 0.58 | 0.60 | 0.62 |
| 3.00 and above | 0.44 | 0.51 | 0.57 | 0.60 | 0.62 | 0.63 |

**(b) Solid slabs :** For solid slab, the permissible shear stress in concrete shall be $k \cdot \tau_c$ where $k$ has the value given in Table 3.2.

**TABLE 3.2.** Factor $k$

| Overall depth of slab (mm) | 300 or more | 275 | 250 | 225 | 200 | 175 | 150 or less |
|---|---|---|---|---|---|---|---|
| $k$ | 1.00 | 1.05 | 1.10 | 1.15 | 1.20 | 1.25 | 1.30 |

The above requirement does not apply to flat slabs.

**Note.** $A_s$ is that area of longitudinal tensile reinforcement which continues at least one effective depth beyond the section being considered except at supports where the full area of tension reinforcement may be used.

**(c) Shear strength of members under axial compression**

For members subjected to axial compression, $P$, the permissible shear stress in concrete $\tau_c$, given in Table 3.1, shall be multiplied by the following factor:

$$\delta = 1 + \frac{5P}{A_g f_{ck}} \text{ (but not exceeding 1.5)} \quad \ldots(3.22)$$

where $P$ = axial compressive force in N
$A_g$ = gross area of concrete section in mm²
and $f_{ck}$ = characteristic compressive strength of concrete

**(d) Beams with Shear Reinforcement :** When shear reinforcement is provided the nominal shear stress $\tau_v$ in beam shall not exceed $\tau_{c.max}$ given in Table 3.3.

To avoid compression failure of the section in shear, the nominal shear stress $\tau_v$ should not exceed the maximum shear stress in concrete $\tau_{c.max}$ values given in table '20' of IS : 456–2000.

**TABLE 3.3.** Maximum Shear Stress $\tau_{c.max}$ N/mm² (IS : 456–2000)

| Concrete grade | M 15 | M 20 | M 25 | M 30 | M 35 | M 40 onwards. |
|---|---|---|---|---|---|---|
| $\tau_{cmax}$ (N/mm²) | 2.5 | 2.8 | 3.1 | 3.5 | 3.7 | 4.0 |

For slabs, $\tau_v$ shall not exceed half the value of given in Table 3.3.

### (3) Minimum Shear Reinforcement

The minimum quantity of shear reinforcement that should be provided for all beams expect those of minor importance like lintel is given as per clause 26.5.1 of IS: 456–2000.

When $\tau_v$ is less than $\tau_c$ given in Table 3.1, minimum shear reinforcement shall be provided as explained below :

Minimum shear reinforcement in the form of stirrups shall be provided such that

$$\frac{A_{sv}}{b \cdot s_v} \geq \frac{0.4}{0.87 f_y} \qquad \text{...(3.23)}$$

where $A_{sv}$ = total cross-sectional area of stirrup legs effective in shear.

$s_v$ = stirrup spacing along the length of the member

$b$ = breadth of beam or breadth of the web of flanged beam and,

$f_y$ = characteristic strength of the stirrup reinforcement, in N/mm², which shall not be taken greater than 415 N/mm². The values of $f_y$ for three types of steel reinforcement are given in Table 1.21.

However, in members of minor structural importance such as lintels or where the maximum shear stress calculated is less than half the permissible value, this provision need not be complied with.

The term *characteristic strength* means that value of the strength below which not more than 5 percent of the test results are expected to fall. *Until the relevant Indian Standard specification* for reinforcing steel are modified to include the concept of characteristic strength, the characteristic value shall be assumed as the minimum yield/0.2 percent proof stress specified in the relevant Indian Standard specifications.

*Eq. 3.23* can be rearranged in the following form to give the maximum spacing of stirrups:

$$s_v \leq \frac{0.87 f_y A_{sv}}{0.4 b} \leq \frac{2.175 A_{sv} \cdot f_y}{b} \qquad \text{...[3.23(a)]}$$

*Eq. 3.23* may also be written in the form

$$\frac{A_{sv} \cdot f_y \cdot d}{s_v} = 0.46 \, bd \qquad \text{...[3.23(b)]}$$

In the above expression, the left hand term $\frac{A_{sv} \cdot f_y \cdot d}{s_v}$ is equal to the shear resistance $V_s$ of the vertical stirrups while the right hand term $0.46 \, bd$ represents $\tau_c \cdot bd$ (*i.e.* shear resistance $V_c$ for concrete). Thus *Eq. 3.23* ensures provision of reinforcement for a net shear stress of 0.46 N/mm². The *minimum shear reinforcement* for two types of steel bars for $b = 300$ mm works out as under :

(a) **Mild Steel Bars** : Taking $f_y = 250$ N/mm²

$$s_v = \frac{2.175 \, A_{sv} \times 250}{300} = 1.8125 \, A_{sv} \text{ mm}$$

Taking 8 mm dia bars, $A_{sv} = 2 \times \frac{\pi}{4} (8)^2 = 100.5 \text{ mm}^2$

Hence $s_v = 1.8125 \times 100.5 = 182$ mm.

(b) **HYSD Bars** : Taking $f_y = 415$ N/mm²

$$s_v = \frac{2.175 \, A_{sv} \times 415}{300} = 3.009 \, A_{sv} \text{ mm}$$

For 8 mm dia. bars, $s_v = 3.009 \times 100.5 = 302$ mm.

### (4) Design of Shear Reinforcement

Shear reinforcement has to be provided against diagonal tensile stresses caused by the shear force. The longitudinal bars do not prevent the diagonal tension failure.

The inclined shear crack strats at the bottom near the support and extends towards compression zone.

When $\tau_v$ exceeds $\tau_c$ given in Table 3.1, shear reinforcement shall be provided in any of the following forms :

(i) Vertical stirrups

(ii) Bent-up bars along with stirrups and

(iii) Inclined stirrups.

Where bent up bars are provided, their contribution towards shear resistance shall not be more than half that of the total shear reinforcement. Shear reinforcement shall be provided to carry a shear equal to $V_s = V - \tau_c \cdot bd$. The strength of shear reinforcement $V_s$ shall be calculated as below :

(i) **For vertical stirrups**

$$V_s = \frac{\sigma_{sv} \cdot A_{sv} \cdot d}{s_v} \qquad \ldots(3.24)$$

(ii) **For inclined stirrups or a series of bars bent-up at different cross-section :**

$$V_s = \frac{\sigma_{sv} \cdot A_{sv} \cdot d}{s_v}(\sin\alpha + \cos\alpha) \qquad \ldots(3.25)$$

(iii) **For single bar or single group of parallel bars, all bent up the same cross-section**

$$V_s = \sigma_{sv} \cdot A_{sv} \cdot \sin\alpha \qquad \ldots(3.26)$$

where  $A_{sv}$ = total cross-sectional area of stirrups legs or bent up bars within a distance $s_v$.

$s_v$ = spacing of the stirrups or bent up bars along the length of the member

$\tau_c$ = design shear strength of the concrete.

$b$ = breadth of the member which for flanged beams, shall be taken as breadth of the web $b_w$.

$\sigma_{sv}$ = permissible tensile stress in shear reinforcement which shall not be taken greater than 230 N/mm$^2$.

$\alpha$ = angle between the inclined stirrups or bent up bar and the axis of the member, not less than 45° and

$d$ = effective depth.

**Note:** where more than one type of shear reinforcement is used to reinforce the same portion of the beam, the total shear resistance shall be computed as the sum of the resistance of the various types separately. The area of stirrups shall not be less than the minimum specified in para 3 above.

**(5) Maximum Spacing of Shear Reinforcement**

Spacing of vertical stirrups should not exceed 0.75 $d$ (or) 300 mm whichever is less as per clause 26.5.1.5 of IS : 456–2000 and the diameter should not be less than 6 mm.

The maximum spacing of shear reinforcement measured along the axis of the member shall not exceed 0.75 $d$ for vertical stirrups and $d$ for inclined stirrups at 45°, where $d$ is the effective depth of the section under consideration. In no case shall the spacing exceed 300 mm.

**(6) Summary**

(i) **If** $\tau_v \leq \frac{1}{2}\tau_c$ or $V \leq \frac{1}{2}V_c$ : No shear reinforcement is really needed.

(ii) **If** $\frac{1}{2}V_c \leq V \leq V_c$ : Provide nominal reinforcement given by *Eq. 3.23*.

However, the maximum spacing is restricted to 0.75 $d$ or 300 mm, whichever is less.

(iii) **If** $V_c \leq V \leq V_{c.max}$ Design the transverse reinforcement for the net S.F.

$$F = (V - V_c).$$

If $\Sigma V_s$ is the sum of the resistances of various types of shear reinforcements (*Eq. 3.24, 3.25, 3.26*) then $\Sigma V_s$ should be equal to or greater than $(V - V_c)$.

However, the maximum spacing ($s_v$) is restricted to 0.75 $d$ or 300 mm, whichever is less. Also the contribution of bent-up bars should not be more than half of the total shear reinforcement and the area of stirrups should not be less than the one provided by *Eq. 3.23*.

**(iv) If $V \geq V_{c.max}$**, redesign the web of the section such that $V_{c.max}$ becomes equal to or more than external shear $V$.

## 3.14. CRITICAL SECTION FOR DESIGN SHEAR : *IS : 456–2000*

A support to a beam may offer either a *compressive reaction* or a *tensile reaction*. Fig. 3.18 (*a*) shows some typical situations, such as in the case of a beam resting on wall (Fig. 3.18 *a-i*) or on column (Fig. 3.18 *a-ii*) and 3.18 (*a-iii*), where the support offers a compressive reaction. The critical section for such a support is at a distance $d$ from the face of the support because the diagonal crack nearest to the support starts from the face of the support and runs upwards at 45° (approx) towards the load. Similarly Figs. 3.18 *b* (*i*), (*ii*), (*iii*) show situations where the support offers a tensile reaction; the critical section for such a situation is at the face of the support. Also, in a case where there is a concentrated load between the face of the support and the distance $d$ (Fig. 3.18 *b iv*), the critical section is at the face of the support.

**Note :** The above provisions are applicable for beams generally carrying uniformly distributed load or where the principal load is located farther than $2d$ from the face of the support.

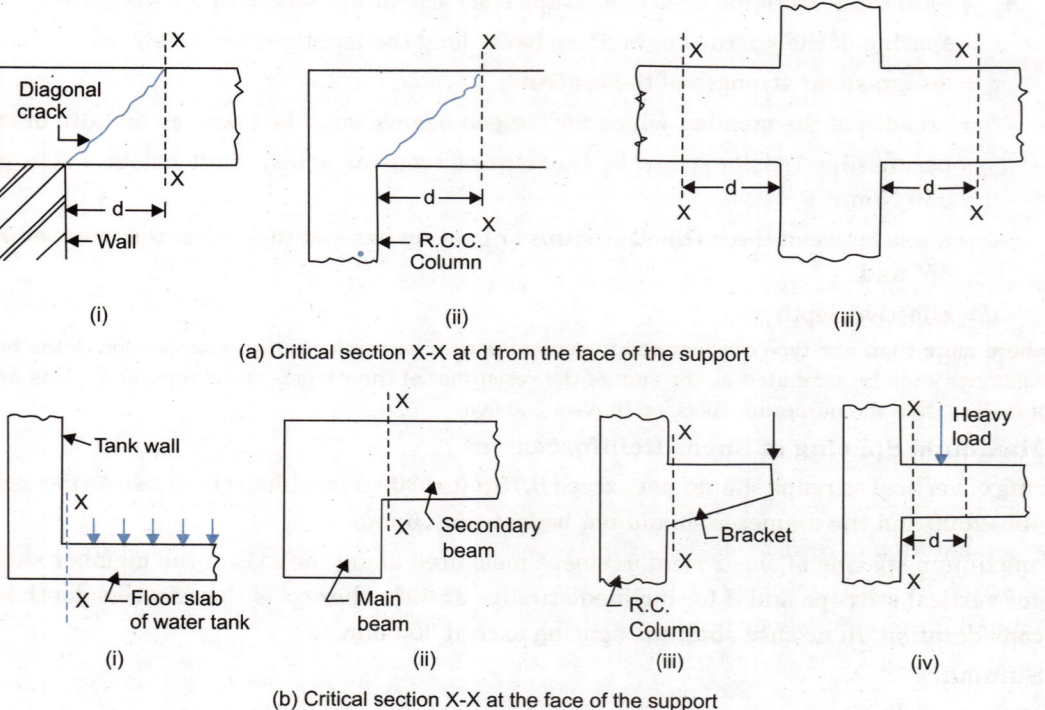

Fig. 3.18. Critical Sections for Shear X-X at the Face of the Support

**Example 3.1.** *A reinforced concrete beam 250 mm wide and 400 mm effective depth is subjected to a shear force of 95 kN at the supports. The tensile reinforcement at the support is 0.5 percent. Find the spacing of 12 mm diameter 2-legged stirrups to resist the shearing stress at supports, for M 15 concrete. Use (a) mild steel reinforcement (b) HYSD bars. Also, design the minimum shear reinforcement at the mid-span of the beam.*

**Solution:** Given $p = 100 A_{st}/bd = 0.5\%$

Hence from Table 3.1, for M 15 concrete, we get, $\tau_c = 0.29$ N/mm$^2$

Also from Table 3.3, maximum shear stress, $\tau_{c.max} = 1.6$ N/mm$^2$

$$\tau_v = \frac{V}{bd} = \frac{95 \times 1000}{250 \times 400} = 0.95 \text{ N/mm}^2$$

This is greater than $\tau_c$ but less than $\tau_{c,max}$.

The shear reinforcement is provide to carry a shear $V_s$ equal to $V - \tau_c \cdot bd$.

$$\therefore \quad V_s = V - \tau_c bd = 95000 - 0.29 \times 250 \times 400 = 66000 \text{ N}$$

**(a) Mild steel bars** (Fe 250) $\sigma_{sv} = 140$ N/mm$^2$

The spacing of stirrups at supports is given by *Eq. 3.24*, rearranged as follows :

$$s_v = \frac{\sigma_{sv} \, d \, A_{sv}}{V_s} = \frac{140 \times 400}{66000} A_{sv} = 0.848 \, A_{sv} \text{ mm.}$$

Using 12 mm Φ bars two legged stirrups, $A_{sv} = 2 \frac{\pi}{4} (12)^2 = 226.2$ mm$^2$

$$\therefore \quad s_v = 0.848 \times 226.2 = 191.9 \text{ mm}$$

Hence provide stirrups at spacing of 190 mm $c/c$ at the supports.

For nominal (minimum) shear reinforcement at mid span, the spacing at 12 mm Φ 2-lgd stirrups is given by *Eq. 3.23 (a)*.

$$s_v = \frac{2.175 \, A_{sv} \, f_y}{b} = \frac{2.175 \times 226.2 \times 250}{250} = 492 \text{ mm.}$$

However, max. spacing = $0.75\,d$ or 300 mm which ever is less = $0.75 \times 400 = 300$ mm.

Thus, use of 12 mm dia stirrups will be uneconomical. Using 10 mm dia bars for stirrups,

$$A_{sv} = 2 \times \frac{\pi}{4} (10)^2 = 157.1 \text{ mm}^2$$

$$\therefore \quad s_v = \frac{2.175 \, A_{sv} \cdot f_y}{b} = \frac{2.175 \times 157 \times 250}{300} = 341 \text{ mm.}$$

Subject to a maximum of $0.75\,d = 0.75 \times 400 = 300$ mm. Hence use 10 mm dia 2-lgd. stirrups at 300 mm spacing at the location where nominal reinforcement is required.

**(b) HYSD bars**

$$s_v = \frac{\sigma_{sv} \cdot d \cdot A_{sv}}{V_s} = \frac{230 \times 400}{66000} A_{sv} = 1.3939 \, A_{sv} \text{ mm.}$$

Using 12 mm Φ bars two legged stirrups, $A_{sv} = 2 \times \frac{\pi}{4} (12)^2 = 226.2$ mm$^2$

$$\therefore \quad s_v = 1.3939 \times 226.2 = 315.3 \text{ mm}$$

subject to a maximum of $0.75\,d = 0.75 \times 400 = 300$ mm.

Hence provide the stirrups @ 300 mm c/c.

For nominal (minimum) shear reinforcement, let us use 8 mm dia bars.

$$A_{sv} = 2 \times \frac{\pi}{4} (8)^2 = 100.53 \text{ mm}^2. \text{ The spacing is given by } Eq. 3.23 \,(a)$$

$$s_v = \frac{2.175 \, A_{sv} \cdot f_y}{b} = \frac{2.175 \times 100.53 \times 415}{250} = 362 \text{ mm.}$$

subject to a maximum spacing equal to $0.75d = 300$ mm

Hence provide 8 mm dia. 2-lgd stirrups at 300 mm spacing at the location where nominal reinforcement is required.

**Example 3.2.** *A reinforced concrete beam has 300 mm width and 500 mm effective depth. The shear reinforcement consists of 8 mm diameter 4-lgd stirrups spaced at 100 mm centre to centre at the supports. If the beam is subjected to a shear force of 120 kN at the ends, calculate the maximum shear stress developed in the shear reinforcement. The beam has 0.5% reinforcement at the ends.*

**Solution:** From Table 3.1, for $(A_s/bd) \times 100 = 0.5$, $\tau_c = 0.29$ N/mm$^2$

$$V_c = \tau_c \cdot bd = 0.29 \times 300 \times 500 = 43500 \text{ N};$$

$$V_s = V - V_c = 120000 - 43500 = 76500 \text{ N}$$

Now, for four legged stirrups, $A_{sv} = 4 \times \dfrac{\pi}{4}(8)^2 = 201.1$ mm$^2$.

Let $t_{sv}$ = stress developed in the stirrups.

Then, $V_s = \dfrac{A_{sv} \cdot t_{sv} \cdot d}{s_v}$ or $t_{sv} = \dfrac{V_s \cdot s_v}{A_{sv} \cdot d} = \dfrac{76500 \times 100}{201.1 \times 500} =$ **76 N/mm$^2$**.

**Example 3.3.** *A simply supported beam, 300 mm wide and 500 mm effective depth carries a uniformly distributed load of 50 kN/m including its own weight, over an effective span of 4 metres. Design the shear stirrups in the form of vertical stirrups. Use M 15 concrete. Take $\sigma_{st} = \sigma_{sv} = 140$ N/mm$^2$ and $f_y = 250$ N/mm$^2$. Assume that the beam contains 0.75% reinforcement throughout the length.*

**Solution:** The nominal shear stress in the beam is given by

$$\tau_v = \dfrac{V}{bd} = \dfrac{V}{300 \times 500} = \dfrac{V}{150000} \text{ N/mm}^2$$

At the ends, $V = \dfrac{wL}{2} = \dfrac{50000 \times 4}{2} = 100000$ N

$$\therefore \quad \tau_v = \dfrac{100000}{150000} = 0.67 \text{ N/mm}^2$$

From Table 3.1, for $\dfrac{100 A_s}{bd} = 0.75\%$, permissible shear stress $\tau_c = 0.34$ N/mm$^2$. Since the nominal shear stress is more than this, shear reinforcement is necessary at the end section. Using 10 mm Φ 2-lgd vertical stirrups, $A_{sv} = 2 \dfrac{\pi}{4}(10)^2 = 157$ mm$^2$

Shear resistance to be provided by the stirrups at end section is

$$V_s = V - \tau_c bd = 100000 - 0.34 \times 300 \times 500 = 49000 \text{ N}$$

$$\therefore \quad s_v = \dfrac{\sigma_{sv} \cdot A_{sv} \cdot d}{V_s} = \dfrac{140 \times 157 \times 500}{49000} = 224.3 \text{ mm} \approx 220 \text{ mm.}$$

Maximum spacing of stirrups is given by

$$s_v = \dfrac{2.175 A_{sv} \cdot f_y}{b} = \dfrac{2.175 \times 157 \times 250}{300} = 284 \approx 280 \text{ mm.} \quad \ldots(1)$$

In no case should the spacing exceed $0.75\,d$ (= $0.75 \times 500$) or 300 mm which ever is less.

Thus, the spacing is to be varied from 220 mm at end section to 280 mm at a section distant $x$ (say) from the mid span. Let us locate this section, where shear is equal to $V_x$ where

$$V_x = \dfrac{V}{2}(x) = \dfrac{100000}{2} x = 50000\, x$$

$$\therefore \quad V_s = V_x - \tau_c bd = 50000\, x - 0.34 \times 300 \times 500 = 50000\, x - 51000$$

$$\therefore \quad s_v = \dfrac{\sigma_{sv} \cdot A_{sv} \cdot d}{V_s} = \dfrac{140 \times 157 \times 500}{50000\, x - 51000} \quad \ldots(2)$$

Equating (1) and (2) we get $\dfrac{140 \times 157 \times 500}{50000 x - 51000} = 280$

From which $x \approx 1.8$ m. Thus vary the spacing from 220 mm c/c at supports to 280 mm c/c at 1.8 m from the mid span. For the remaining length, provide the stirrups at 280 mm c/c.

**Example 3.4.** *Redesign the shear stirrups of the beam of example 3.3, using HYSD bars, taking* $\sigma_{st} = \sigma_{sv} = 230 \, N/mm^2$ *and* $f_y = 415 \, N/mm^2$.

**Solution:** $V = \dfrac{wL}{2} = \dfrac{50000 \times 4}{2} = 100000 \, N$ ; $\tau_v = \dfrac{V}{bd} = \dfrac{100,000}{300 \times 500} = 0.67 \, N/mm^2$

For $100 \, A_s/bd = 0.75\%$, we get $\tau_c = 0.34 \, N/mm^2$ from Table 3.1. Hence shear reinforcement is necessary at the ends. Using 8 mm dia. 2-lgd vertical stirrups,

$$A_{sv} = 2 \cdot \frac{\pi}{4}(8)^2 = 100.53 \, mm^2$$

Now $\quad V_s = V - \tau_c \, bd = 100000 - 0.34 \times 300 \times 500 = 49000 \, N$

$\therefore \quad s_v = \dfrac{\sigma_{sv} \cdot A_{sv} \cdot d}{V_s} = \dfrac{230 \times 100.53 \times 500}{49000} = 236 \, mm \, (= 230 \, mm, \text{ say})$

Maximum spacing of stirrups is given by

$$s_v = \dfrac{2.175 \, A_{sv} \cdot f_y}{V_s} = \dfrac{2.175 \times 100.53 \times 415}{300} = 302 \, mm \approx 300 \, mm \qquad \ldots(1)$$

In no case should the spacing exceed $0.75 \, d \, (= 0.75 \times 500)$ or 300 m, whichever is less.

Hence the spacing is to be varied from 230 mm at end section to 300 mm at a section distant $x$ m from the mid span where minimum shear reinforcement, given by eq. (1) is to be provided.

At that section, $\quad V_x = \dfrac{V}{2}(x) = \dfrac{100000}{2} x = 50000 x$

$\therefore \quad V_s = V_x - \tau_c \, bd = 50000 \, x - 0.34 \times 300 \times 500 = 50000 \, x - 51000$

$\therefore \quad s_v = \dfrac{\sigma_{sv} \cdot A_{sv} \cdot d}{V_s} = \dfrac{230 \times 100.53 \times 500}{50000 \, x - 51000} \qquad \ldots(2)$

Equating (1) and (2), we get

$$\dfrac{230 \times 100.53 \times 500}{50000 \, x - 51000} = 300 \quad \text{From which we get } x = 1.79 \, m.$$

Thus vary the spacing from 230 mm c/c at supports to 300 mm c/c at 1.79 m from mid-span. For the remaining length, provide the stirrups at 340 mm c/c.

**Example 3.5.** *A singly reinforced concrete beam has an effective depth of 600 mm and width of 300 mm and is reinforced with 8 bars of 20 mm diameter at the centre of the span. From the requirements of bending stresses, it is known that four of these bars can be safely bent up at a distance of 1.6 m from the support. The maximum S.F. at the ends is 160 kN and the distance from the either support in which shear reinforcement is needed is 1.3 m. Explain how you would use the bent up bars as diagonal tension reinforcement. Also design the stirrup reinforcement required. Take* $\sigma_{sv} = 140 \, N/mm^2$. *Use M 15 concrete.*

**Solution:** For M 15 concrete, $j = 0.867$ with mild steel bars

$\therefore \quad$ Lever arm $\quad a = jd = 0.867 \times 600 = 520 \, mm$

Let us bend two bars at a distance $= 1.414 \, a = 1.414 \times 520 = 735 \, mm$ from the support.

Shear resistance of the two bars bent at 45° is

$$V_{s1} = A_{sv} \cdot \sigma_{sv} \sin 45° = 0.707 \cdot A_{sv} \cdot \sigma_{sv}$$

$$= 0.707 \left\{ 2 \times \frac{\pi}{4}(20)^2 \right\} \times 140 = 62159 \, N.$$

However, maximum contribution towards shear resistance shall not be more than half the total shear reinforcement.

Also, $\quad \dfrac{100 \, A_s}{bd} = \dfrac{100 \times 4 \left(\frac{\pi}{4}\right)(20)^2}{300 \times 600} = 0.7\%.$ $\quad$ Hence from Table 3.1, $\tau_c = 0.33 \, N/mm^2$

∴  $V_c = \tau_c \cdot bd = 0.33 \times 300 \times 600 = 59400$ N

and  $V_s = V - V_c = 160000 - 59400$ N $= 100600$ N

Max. Contribution $V_{s1}$ due to bent up bars $= \dfrac{1}{2} V_s = \dfrac{1}{2} \times 100600 = 50300$ N.

∴  $V_{s2}$ from stirrups $= 50300$ N.

Using 8 mm Φ two-legged stirrups, $A_{sv} = 2 \dfrac{\pi}{4} (8)^2 = 100.5$ mm$^2$

$$s_v = \dfrac{\sigma_{sv} \cdot A_{sv} \cdot d}{V_{s2}} = \dfrac{140 \times 100.5 \times 600}{50300} = 167.8 \text{ mm}.$$

This spacing should not be more than $s_v$ corresponding to minimum reinforcement requirement.

i.e.  $s_v \leq \dfrac{2.175 \cdot A_{sv} \cdot f_y}{b} = \dfrac{2.175 \times 100.5 \times 250}{300} = 182$ mm.

Also, spacing should not be more than $0.75\, d$ ($= 0.75 \times 600 = 450$ mm) nor more than 300 mm. Hence provide 2-lgd 8 mm dia. stirrups @ 160 mm c/c.

Let us bend the other two bars at a distance of $2 \times 1.414\, a = 2 \times 735 = 1470$ mm from the support. This can safely be done since from B.M. point of view, we can bend them even at a distance of 1600 mm from the support. Effective length of these bars resisting shear $= 2 \times 1.414\, a = 1470$ mm. This is more than the length of 1300 mm in which shear reinforcement is required. Hence no extra shear reinforcement is required. However, provide minimum shear reinforcement, in the form of 8 mm Φ 2-lgd stirrups @ 182 mm ($\approx 180$ mm, say) c/c for the remaining length. The arrangement is shown in Fig. 3.19.

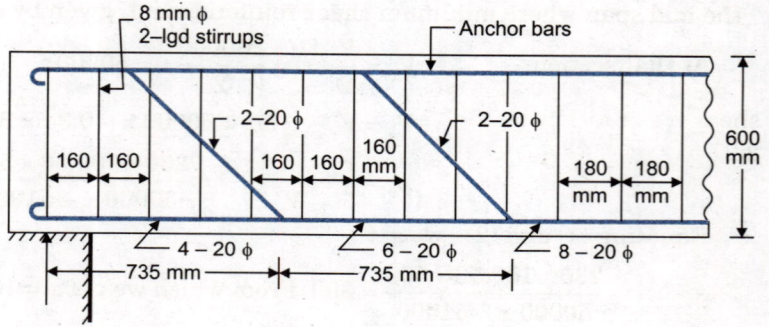

Fig. 3.19

**Example 3.6.** *A simply supported beam, 300 mm wide and 600 mm effective depth carries a uniformly distributed load of 70 kN/m including its own weight, over an effective span of 6 m. The reinforcement consists of 5 bars of 25 mm diameter. Out of these, two bars can be safely bent at 1 m distance from the support. Design suitable shear reinforcement for the beam. Use M 15 concrete Take $\sigma_{st} = \sigma_{sv} = 140$ N/mm$^2$ and $f_y = 250$ N/mm$^2$.*

**Solution:** Max. S.F.  $V = \dfrac{wL}{2} = \dfrac{70000 \times 6}{2} = 210000$ N

Nominal shear stress  $= \dfrac{V}{bd} = \dfrac{210000}{300 \times 600} = 1.17$ N/mm$^2$

Area of reinforcement  $= (5 - 2) \dfrac{\pi}{4} (25)^2 = 1473$ mm$^2$; $p = \dfrac{1473}{300 \times 600} \times 100 \approx 0.82\%$

Hence from Table 3.1, for $P = 0.82\%$ and M 15 concrete, we get, $\tau_c = 0.35$ N/mm$^2$

From Table 3.3, $\tau_{cmax} = 1.6$ N/mm$^2$.

Since the nominal shear stress is less than this, the beam is safe.

For M 15 concrete, let $j = 0.87$. Lever arm $jd = 0.87 \times 600 = 522$ mm.

Let us bend 2 bars at 45° at a distance $= 1.414 \times jd = 1.414 \times 522 \approx 730$ mm from the support.

Area of cross-section of 2 bent up bars $= 2 \dfrac{\pi}{4} (25)^2 = 982$ mm$^2$

Shear resistance of concrete = $V_c = \tau_c \, bd = 0.35 \times 300 \times 600 = 63000$ N
Shear resistance to be provided by shear reinforcement = $210000 - 63000 = 147000$ N
Shear resistance of two bent up bars = $\sigma_{sv} \cdot A_{sv} \cdot \sin \alpha = 140 \times 982 \times 0.707 \approx 97198$ N

However the maximum shear resistance to be assigned to the bent up bars should not exceed half the resistance to be provided by shear reinforcement, *i.e.*, it should not exceed $\frac{1}{2} \times 147000 = 73500$ N. Hence, the remaining resistance to be provided by the vertical stirrups = 73500 N

Thus $\qquad V_s = 73500$ N

Hence, the spacing of vertical stirrups is given by

$$s_v = \frac{\sigma_{sv} \cdot A_{sv} \cdot d}{V_s} = \frac{140 \times A_{sv} \times 600}{73500} = 1.143 \, A_{sv}.$$

Using 10 mm $\Phi$ 2 lgd stirrups, having $A_{sv} = 2 \frac{\pi}{4}(10)^2 = 157$ mm$^2$

$$s_v = 1.143 \times 157 \approx 180 \text{ mm}$$

The maximum spacing of stirrups is given by

$$s_v = \frac{2.175 \cdot A_{sv} \cdot f_y}{b} = \frac{2.175 \times A_{sv} \times 250}{300} = 1.8125 \, A_{sv}$$

$$= 1.8125 \times 157 = 284 \text{ mm} \approx 280 \text{ mm}$$

At distance greater than 730 mm (0.73 m) from the supports, bent up bars are not available and the shear resistance is available only from the stirrups as well as from concrete. At 0.73 m from supports (or 2.27 m from mid span), the S.F. is given by

$$V = \frac{210000}{3} \times 2.27 = 158900 \text{ N} \, ; \quad A_{st} = 5 \frac{\pi}{4}(25)^2 = 2454 \text{ mm}^2$$

$$\therefore \qquad p = \frac{2454}{300 \times 600} \times 100 \approx 1.36\%.$$

Hence from Table 3.1, $\tau_c = 0.41$ N/mm$^2$.

$\therefore$ Shear resistance of concrete = $\tau_c \, bd = 0.41 \times 300 \times 600 = 73800$ N

Hence shear resistance to be provided by stirrups is given by

$$V_s = V - \tau_c \, bd = 158900 - 73800 = 85100 \text{ N}$$

Thus, the spacing of 10 mm $\Phi$ 2-lgd stirrups is

$$s_v = \frac{\sigma_{sv} \cdot A_{sv} \cdot d}{V_s} = \frac{140 \times 157 \times 600}{85100} = 155 \text{ mm} \approx 150 \text{ mm (say)}.$$

To locate the section where maximum spacing of 280 mm is provided, proceed as follows: Let the distance of the section be $x$ m from mid-span.

$$V = \frac{210000}{3} \times x = 70000 \, x \text{ N}$$

Shear resistance of concrete = $\tau_c \, bd = 0.41 \times 300 \times 600 = 73800$ N

$\therefore \qquad V_s = 70000 \, x - 73800$

$$s_v = \frac{\sigma_{sv} \cdot A_{sv} \cdot d}{V_s} = \frac{140 \times 157 \times 600}{70000 \, x - 73800} \text{ mm}.$$

Equating this to 280 mm we get, $\dfrac{140 \times 157 \times 600}{70000 \, x - 73800} = 280$ mm.

which gives $x = 1.73$ m from mid-span (or 1.27 m from supports).

Thus, the shear reinforcement may be summarised as follows after slight adjustment in the spacings:
(a) *At supports* : 10 mm Φ 2-*lgd* stirrups @ 180 mm c/c plus 2 bent up bars from the main reinforcement.
(b) *At 0.73 m from supports to 127 m from supports* : vary the spacing of the 10 mm Φ 2-*lgd* stirrups from 150 m c/c to 280 m c/c .
(c) *Beyond 1.27 m from supports* : 10 mm Φ 2-lgd stirrups @ 280 mm c/c. Provide 2-12 mm Φ holding bars at top. Fig. 3.20 shows the shear reinforcement.

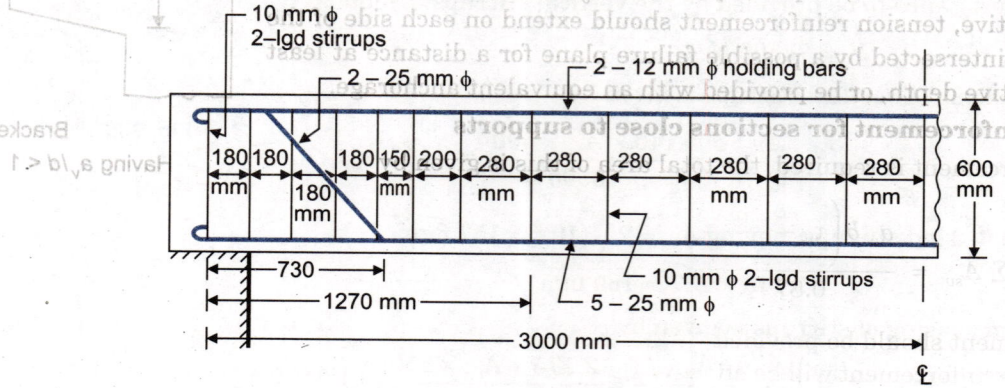

Fig. 3.20

## 3.15. ENHANCED SHEAR STRENGTH OF SECTIONS CLOSE TO SUPPORTS (IS : 456–2000)

### 1. General

Shear failure at sections of beams and cantilevers *without* shear reinforcement will normally occur on plane inclined at an angle 30° to the horizontal, such as the failure plane *AC* shown dotted in Fig. 3.21 (*a*). If however, the failure plane is *forced* to be inclined more steeply than this, because the section *X-X* considered in Fig. 3.21 (*a*) is close to the support, or for other reasons, the *shear force required to produce failure is increased*. This occurs when the ratio $a_v/d$ is less than 2. As is evidenced by extensive test results on R.C. beams without shear reinforcement, when $a_v/d$ is less than 2, the shear capacity increases parabolically while for $a_v/d$ ratio more than 2, the shear capacity almost remains constant (Fig. 3.21 b).

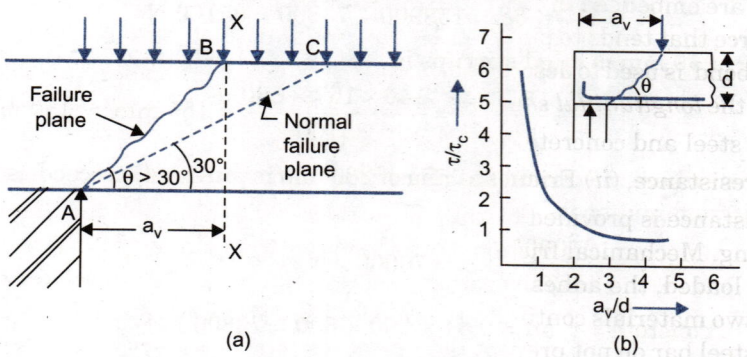

Fig. 3.21. Enhancement of Shear Strength

### 2. Enhancement of shear strength

As per IS 456 : 2000, the *enhancement* of shear strength may be taken into account in the design of sections near a support by increasing design shear strength ($\tau_c$) of concrete to $2d\,\tau_c/a_v$ provided that the design stress at the face of support remains less than the value $\tau_{c.max}$ given in Table. 3.3.

Thus, enhanced design shear strength $\tau_{ce} = \tau_c \cdot \dfrac{2d}{a_v}$ (But not greater than $\tau_{c.max}$) ...(3.27)

Account may be taken of the enhancement in any situation where the section considered is closer to the face of the support or concentrated load than twice the effective depth, $d$. Such an enhancement may be particularly useful for the design of corbels (or brackets), or pile caps where $a_v/d$ ratio may be less than unity. To be effective, tension reinforcement should extend on each side of the point where it is intersected by a possible failure plane for a distance at least equal to the effective depth, or be provided with an equivalent anchorage.

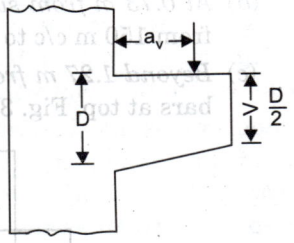

Fig. 3.22. Bracket Having $a_v/d < 1$

**3. Shear reinforcement for sections close to supports**

If shear reinforcement is required, the total area of this is given by

$$\sum A_{sv} = \dfrac{a_v\, b \left( \tau_v - \dfrac{2 d\, \tau_c}{a_v} \right)}{0.87 f_y} \geq \dfrac{0.4\, a_v\, b}{0.87 f_y} \qquad ...(3.28)$$

This reinforcement should be provided within the middle three quarters of $a_v$; where $a_v$ is less than $d$, horizontal shear reinforcement will be effective than vertical. In the above treatment $a_v$ is either the *shear span* or the distance of concentrated load from the face of the support.

**4. Simplified approach for enhanced shear strength near supports**

The approach given is para 1, 2 and 3 above may be used for all beams. However, for beams carrying generally uniform load or where the principal load is located farther than $2d$ from the face of the support, the shear stress may be calculated at a section a distance $d$ from the face of the support. The value of $\tau_c$ is calculated in accordance with Table 3.1 and appropriate shear reinforcement is provided at sections closer to the support, no further check for shear at such section is required.

## 3.16. BOND, ANCHORAGE AND DEVELOPMENT LENGTH

One of the most important assumptions in the behaviour of reinforced concrete structures is that there is proper bond between concrete and reinforcing bars. This means they have to act together without any slip as due to bond only the force will be transferred to the steel *from the surrounding concrete* and *to concrete from steel*. When steel bars are embedded in concrete, the concrete, after setting, *adheres* to the surface of the bars and thus resist any force that tends to pull or push this rod. The intensity of this adhesive force is called *bond stress*. Thus the term 'bond' is used to describe the means by which slip between steel and concrete is prevented. The *bond stresses* are the *longitudinal shearing stresses* acting on the surface between the steel and concrete.

The bond between steel and concrete comprise of three resistances :

(*i*) Pure adhesive resistance, (*ii*) Frictional resistance, and (*iii*) Mechanical resistance,

Pure adhesive resistance is provided by the relatively weak adhesion of the 'chemical gum' produced by concrete during setting. Mechanical friction is provided by the shrinkage of the concrete thereby gripping the steel bars. When loaded, the adhesive resistance is taken first. But even after the adhesion is broken, friction between the two materials continues to provide a considerable bond resistance. The hooks and other means of *anchoring* steel bar do not prevent the initial slipping ; but they delay the collapse of structure as the member acts as a tied arch with the uncracked concrete representing the arch and the anchored bars as the tie rods. The frictional resistance is low for a smooth bar surface, such as that of cold-rolled steel, and is higher for rough surfaces. Hence it is essential that the bars must be clean and free from dirt, grease, scale and loose rust so that pure adhesive resistance and frictional resistance can be developed fully. Anything that destroys the ability of the concrete to grip the steel may prove to be serious because it will prevent the stress in the latter from being fully developed.

## 90 REINFORCED CONCRETE STRUCTURE

The *mechanical resistance* is provided by *deformed bars* only, and not by the plain bars. The deformed bars have lugs, or corrugations, and give higher bond resistance by providing an interlock between steel and concrete. In the case of deformed bars, adhesion and friction become minor elements, and the bond strength is primarily dependent on the bearing (compressive) strength of the concrete against the lugs or corrugations. The plain bars slip easily and cause localised serious cracking while the deformed bars tend to distribute the deformation in the concrete in a series of closely spaced hair cracks.

Bond stress may be visualised as *interface shear*. The bond stress in reinforced *concrete members* arises due to two distinct situations (*a*) from the change in tensile force carried by the bar, along its length, due to *change in the bending moment* along the length of the member, and (*b*) from the *anchorage of bar* in case of tension or compression. The first type is known as *flexural bond stress* ($\tau_{bf}$) and the second type is known as *anchorage bond stress* ($\tau_{ba}$ or $\tau_{bd}$)

### 3.17. FLEXURAL BOND STRESS

Fig. 3.23 shows two sections *ab* and *cd*, spaced *dx* apart of a reinforced concrete beam. Let *M* be the bending moment at section *ab* and *M + dM* be the bending moment at section *cd*. Similarly let *T* and *T + dT* be the tensile forces developed in steel reinforcement at *ab* and *cd* respectively.

Now $\qquad M = T \cdot jd \qquad$ and $\qquad M + dM = (T + dT)\, jd$

Hence $\qquad dM = dT \cdot jd \qquad$ or $\qquad dT = \dfrac{dM}{jd}$ ...(i)

The difference in the tensile force (*dT*) in steel bars between sections is resisted by bond stress developed in the length *dx*.

Let $\quad \tau_{bf}$ = bond stress developed in concrete around the steel reinforcement.

$\Sigma 0$ = sum of the perimeters of the steel bars resisting tension.

Total surface area of bars = $dx\, \Sigma 0$.

Hence, $\qquad \tau_{bf} [dx\, \Sigma 0] = dT$ ...(ii)

Substituting the value of *dT* in (i), we get, $\tau_{bf}(dx\,\Sigma 0) = dM/jd$

or $\qquad \tau_{bf} = \dfrac{dM}{dx} \cdot \dfrac{1}{jd\, \Sigma 0}$

But $\qquad \dfrac{dM}{dx}$ = shear force = *V*

Hence $\qquad \tau_{bf} = \dfrac{V}{jd\, \Sigma 0} = \dfrac{V}{a\, \Sigma 0}$ ...(3.29)

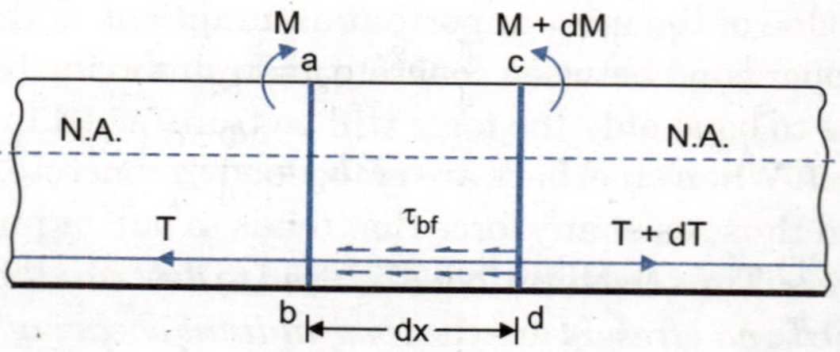

The stress $\tau_{bf}$ at a particular section is called *local bond stress*.

Fig. 3.23

### General expression for bond stress

Eq 3.29 is applicable only when all the reinforcing bars placed are equidistant from the neutral axis. However, there are cases, such as the case of a beam of circular section, in which it is not possible to place the bars equidistant from the N.A. Also, *Eq. 3.29* is not applicable for finding bond stress in the reinforcement provided in the compression zone of the beam. In such cases, the expression derived below can be used to find bond stress in the individual bars. Let a reinforcing bar of diameter $\Phi$ be placed at a distance *y* from the N.A. Let *dM* be the change in bending moment in a length *dx*.

∴ Change in the stress in the bar due to change in the bending moment *dM* is equal to $m \cdot \dfrac{dM}{I} \cdot y$,

where *I* the moment of inertia of the equivalent section and *m* is the modular ratio.

∴ Force causing slip = stress × area = $\dfrac{\pi}{4} \Phi^2 \cdot m \dfrac{dM}{I} \cdot y$ ...(i)

Bond force resisting the slip = $\tau_{bf} \cdot \pi \Phi \cdot dx$. ...(ii)

Equating the two, we get, $\tau_{bf} \cdot \pi \Phi \cdot dx = \dfrac{\pi}{4} \Phi^2 \cdot m \dfrac{dM}{I} y$

or $\qquad \tau_{bf} = \dfrac{dM}{dx} \dfrac{my}{I} \cdot \dfrac{\Phi}{4}$  But $\dfrac{dM}{dx}$ = shear force = V

$\therefore \qquad \tau_{bf} = \dfrac{V}{4I} my \cdot \Phi$ ...[3.29 (a)]

## 3.18. ANCHORAGE BOND STRESS : DEVELOPMENT LENGTH

*Eqs. 3.29 and 3.29 (a) give the localised flexural bond stress at a particular section. However, bond stresses are developed over a specified length and any attempt to visualise bond on the basis of stresses at a particular cross-section is bound to be erroneous. In the case of bond, it has been recently concluded that the concept of anchorage bond is more meaningful than concept of flexural bond stresses of localised nature. Current design methods disregard such local bond stresses, even though they may result in localized slip between steel and concrete adjacent to the cracks. Thus, the concepts of development length (in the case of flexure) and anchorage length (in case of tension or compression) have gained prominence.*

Fig. 3.24 shows a steel bar embedded in concrete, and subjected to a tensile force T. Due to this force, there will be a tendency of the bar to slip out and this tendency is resisted by the bond stress developed over the perimeter of the bar, along its length of embedment. Let us assume that average, uniform bond stress is developed along the length. The required length, necessary to develop full resisting force, is called *anchorage length* in case of axial tension (or compression) and *development length* in the case of flexural tension, and is designated by symbol $L_d$.

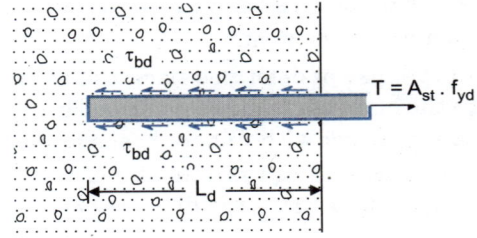

Fig. 3.24

Hence if $\Phi$ is the nominal diameter of the bar, we have

$$\dfrac{\pi}{4} \Phi^2 \cdot \sigma_{st} = \tau_{bd} \cdot \pi \Phi \cdot L_d$$

or $\qquad L_d = \dfrac{\Phi \cdot \sigma_{st}}{4 \tau_{bd}}$ ...(3.30)

The permissible bond stress $\tau_{bd}$ is specified in the Code (Table 3.4) for various grades of concrete.

*Eq. 3.30 shows that a bar must extend a length $L_d$ beyond any section at which it is required to develop its full strength so that sufficient bond resistance is mobilised.*

For M 15 concrete, $\tau_{bd} = 0.6$ N/mm$^2$ while for mild steel bars, $\sigma_{st} = 140$ N/mm$^2$. Hence from *Eq. 3.30*,

$$L_d = \dfrac{140 \Phi}{4 \times 0.6} = 58.3 \Phi.$$

The *development length* for other grades of concrete, and for mild steel and HYSD bars, is given in Table 3.5.

## 3.19. HOOKS AND BENDS

Quite often, space available at the end of beam is limited to accommodate the full development length $L_d$. In that case, hooks or bends are provided. The anchorage value ($L_e$) of hooks or bend is taken equal to 4 φ for every 45° bend and this value is accounted as contribution to the development length $L_d$.

Fig. 3.25 (*ai*) shows a semicircular hook, fully dimensioned, with respect to a factor K. The value of K is taken as 2 in the case of mild steel conforming to IS : 432–1966, (specifications for Mild-Steel and

**92** REINFORCED CONCRETE STRUCTURE

Medium Tensile Steel bars and Hard-Drawn steel wires for concrete reinforcement) or IS : 1139–1959, (specifications for 'Hot rolled mild steel and medium tensile steel deformed bars for concrete reinforcement'). The hook with $K = 2$ is shown in Fig. 3.15.(aii) with equivalent horizontal length of the hook. For the case of Medium Tensile Steel conforming to IS : 432–1966 or IS : 1139–1959, $K$ is taken as 3. In the case of cold worked steel conforming to IS : 1986–1961, (specifications for cold twisted steel bars for concrete reinforcement), $K$ is taken as 4. In the case of bars above 25 mm, however, it is desirable to increase the value of $K$ to 3, 4 and 6 respectively.

Fig. 3.25 (b) shows a right angled bend, with dimensions in terms of K, the value of which may be taken as 2 for ordinary mild steel for diameters below 25 mm and 3 for diameters above 25 mm.

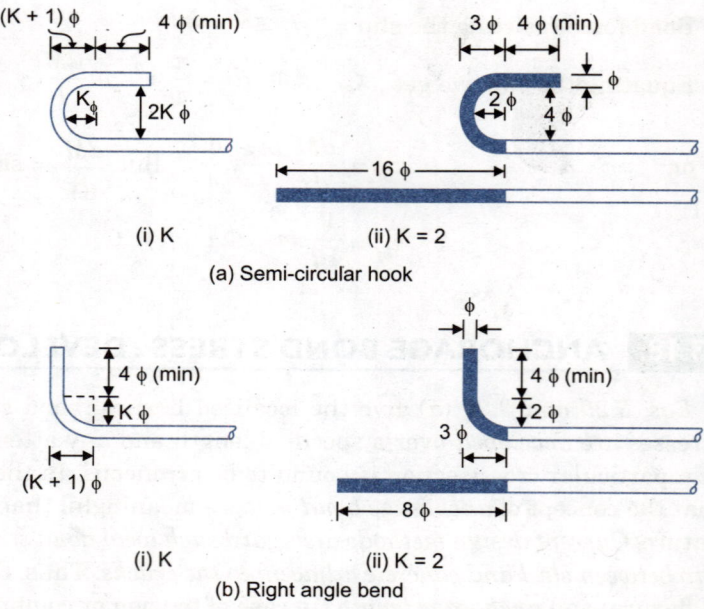

**Fig. 3.25.** Dimension and Equivalent Bond Length of Bends and Hooks.

In the case of deformed bars, the value of bond stress for various grades of concrete is greater by 60% than the plain bars. Hence deformed bars may be used without hooks, provided anchorage requirements are adequately met with. Where hooks are formed in deformed bars, the internal radius of the bend should be at least three times the diameter of the bar. The length of straight bar beyond the end of the curve should be at least four times the diameter of the bars.

## 3.20. IS CODE ON BOND AND ANCHORAGE REQUIREMENTS (IS : 456–2000)

### (a) Permissible Stress in Bond

The permissible stress in bond (average) for plain bars in tension are given in table 1.29. However, the values are reproduced in Table 3.4 for ready reference.

The values of design bond stress prescribed and given in table 3.4 for plain round bars in tension as per Clause 26.2.1.1 in IS: 456–2000.

**TABLE 3.4.** Permissible Bond Stress $\tau_{bd}$ (IS : 456–2000) in Plain Bars in Tension.

| Grade of concrete | M 20 | M 25 | M 30 | M 35 | M 40 and above |
|---|---|---|---|---|---|
| Permissible stress in bond $\tau_{bd}$ (N/mm²) | 1.2 | 1.4 | 1.5 | 1.7 | 1.9 |

**Note 1.** *The bond stress given shall be increased by 25 percent for bars in compression.*

**Note 2.** *In the case of deformed bars conforming to IS : 1139–1966 and IS : 1786–1979, the bond stress given above may be increased by 60%.*

### (b) Development of Stress in Reinforcement

The calculated tension or compression in any bar at any section shall be developed on each side of the section by the appropriate development length of end anchorage or by a combination thereof.

**Development length of bars.** The development length $L_d$ is given by

$$L_d = \frac{\Phi \cdot \sigma_s}{4\tau_{bd}} = k_d \, \varphi \qquad \qquad \text{...(by 3.30)}$$

where  $\Phi$ = nominal diameter of the bar
  $\sigma_s$ = stress in bar at the section considered at design load, and
  $\tau_{bd}$ = design bond stress given in Table 3.4
  $k_d$ = development length factor $\sigma_s/4\tau_{bd}$

The values of $L_d$ for plain mild steel bars ($\sigma_s = \sigma_{st} = 140$ N/mm²) and high yield strength deformed bars (HYSD bars having $\sigma_s = \sigma_{st} = 230$ N/mm²), for various grades of concrete are summarised in Table 3.5.

**TABLE 3.5.** Development Length in Tension

| Grade of Concrete | Plain m.s. bars | | HYSD bars | |
|---|---|---|---|---|
| | $\tau_{bd}$ (N/mm²) | $k_d = L_d/\Phi$ | $\tau_{bd}$ (N/mm²) | $k_d = L_d/\Phi$ |
| M 15 | 0.6 | 58 | 0.96 | 60 |
| M 20 | 0.8 | 44 | 1.28 | 45 |
| M 25 | 0.9 | 39 | 1.44 | 40 |
| M 30 | 1.0 | 35 | 1.60 | 36 |
| M 35 | 1.1 | 32 | 1.76 | 33 |
| M 40 | 1.2 | 29 | 1.92 | 30 |
| M 45 | 1.3 | 27 | 2.08 | 28 |
| M 50 | 1.4 | 25 | 2.24 | 26 |

When the bars are in compression, the bond stress given in Table 3.5 shall be increased by 25%. Hence the development length in compression is given by

$$L_{dc} = \frac{\Phi \, \sigma_{sc}}{5 \, \tau_{bd}} \qquad \ldots[3.30(a)]$$

**Note 1.** *The development length includes anchorage value of hooks in tension reinforcement.*

**Note 2.** *For bars of sections other than circular, the development length should be sufficient to develop the stress in the bar by bond.*

**Note 3.** *When the actual reinforcement provided is more than that theoretically required so that actual stress ($\sigma_s$) is steel is less than full resistance, the development length required may be reduced by the relation:*

**Reduced development length**  $L_{dr} = L_d \left( \dfrac{A_{st} \text{ required}}{A_{st} \text{ provided}} \right) \qquad \ldots(3.31)$

This principle is used in the design of footing and other shear bending members where bond is critical. By providing more steel, the bond requirements are satisfied.

**Bars bundled in contact.** The development length of each bar or bundled bars shall be that for the individual bar, increased by 10 percent for the two bars in contact, 20 percent for three bars in contact and 33 percent for four bars in contact.

### (c) Anchoring Reinforcing Bars

*(i) Anchoring Bars in Tension.* Deformed bars may be used without end anchorages provided development length requirement is satisfied. Hooks should normally be provided for plain bars in tension. The anchorage value of bend shall be taken as 4 times the diameter of the bar for each 45° bend subject to a maximum of 16 times the diameter of the bar. The anchorage value of a standard U-type hook shall be equal to 16 times the diameter of the bar.

*(ii) Anchoring Bars in Compression.* The anchorage length of straight bar in compression shall be equal to the development length of bars in compression. The projected length of hooks, bends and straight lengths beyond bends if provided for a bar in compression, shall be considered for development length.

### (iii) Anchoring Shear Reinforcement :

**Inclined bars.** The development length shall be as for bars in tension ; this length shall be measured as under : (1) in tension zone from the end of the sloping or inclined portion of the bar, and (2) in the compression zone, form mid depth of the beam.

**Stirrups.** Not with standing any of the provisions of this standard, in case of secondary reinforcement, such as stirrups and transverse ties, complete development lengths and anchorage shall be deemed to have been provided when the bar is bent through an angle of at least 90° round a bar of at least its own diameter and is continued beyond the end of the curve for a length of at least eight diameters, or when the bar is bent through an angle of 135° and is continued beyond the end of curve for a length of atleast six bar diameters or when the bar is bent through an angle of 180° and is continued beyond the end of the curve for a length at least four bar diameters.

## 3.21. CHECKING DEVELOPMENT LENGTHS OF TENSION BARS

The flexural bond is maximum at the section where the shear force is maximum. As stated earlier, the computed stress ($\sigma_s$) in a reinforcing bar, at every section must be *developed* on both the sides of the section. This is done by providing development length $L_d$ (computed from *Eq. 3.30.* corresponding to $\sigma_s$) to both the sides of the section. Such a development length is usually available at mid-span location where positive (or sagging) B.M. is maximum for simply supported beams. Similarly, such a development is usually available at the intermediate support of a continuous beam where negative (or hogging) B.M. is maximum. Hence no special checking may be necessary in such locations. *However special checking for development length is essential* at the following :

1. At simple supports
2. At cantilever supports
3. In flexural members that have relatively short spans
4. At points of contraflexure
5. At lap splices
6. At point of bar cutoff
7. For stirrups and transverse ties.

We shall consider some of the above cases in the forthcoming sections.

## 3.22. DEVELOPMENT LENGTH REQUIREMENTS AT SIMPLE SUPPORTS

The Code stipulates that at the simple supports (and at the point of inflection), the positive moment tension reinforcement shall be *limited* to a diameter such that

$$L_d \leq \frac{M_1}{V} + L_0 \qquad \ldots[3.31\,(a)]$$

where  $L_d$ = development length computed for $\sigma_{st}$ from *Eq. 3.30*
      $M_1$ = moment of resistance of the section, assuming all reinforcement at the section to be stressed to $\sigma_{st}$.
      $V$ = shear force at the section due to design loads.
      $L_0$ = sum of anchorage beyond the centre of support and the equivalent anchorage value of any hook or mechanical anchorage at the simple support (at the point of inflections, $L_0$ is limited to $d$ or 12 Φ which ever is greater).

The Code further recommends that the value of $M_1/V$ in *Eq. 3.31 (a)*. may be increased by 30 percent when the ends of the reinforcement are confined by a compressive reaction. This condition of 'confinement' of reinforcing bars may not be available at all the types of simple supports.

Four types of simple supports are shown in Fig. 3.26. In Fig. 3.26(a), the beam is simply supported on a wall which offers a compressive reaction which confines the ends of reinforcement. Hence a factor of 1.3 will be applicable with the term $M_1/V$. Similarly in Fig. 3.26(b), the slab is simply supported on a wall and hence factor 1.3 will be applicable. However, in Fig. 3.26(c) and (d), though a simple support is available, the reaction does not confine the ends of the reinforcement and hence the factor 1.3 will not be applicable with $M_1/V$ term. Similarly, for the case of slab connected to a beam [Fig. 3.26(e)] or for the case of a secondary beam connected to a main beam [3.26(f)] tensile reaction is induced and hence factor 1.3 will not be available. Thus, at simple supports, where the compressive reaction confines the ends of reinforcing bars, we have

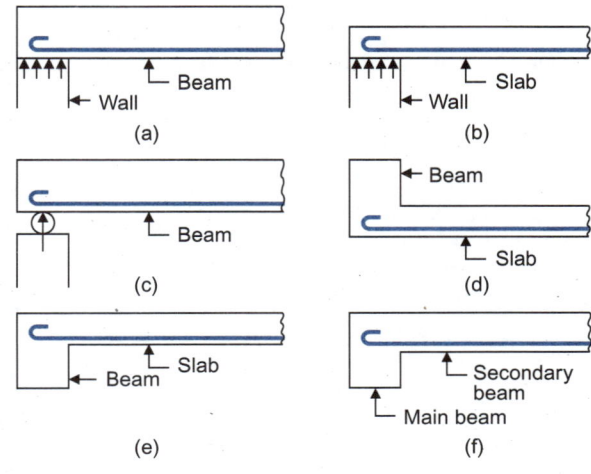

Fig. 3.26

$$L_d \leq 1.3 \frac{M_1}{V} + L_0 \qquad \qquad \ldots[3.31\,(b)]$$

In *Eq. 3.31(a)* or *3.31(b)*, $M_1$ is the moment of resistance of bars available at the support.

**Computation of length $L_0$.** For the computation of $L_0$, the support width should be known. Fig. 3.27 shows a beam with end support, in which $x'$ is the side cover and $x_0$ is the distance of the *beginning* of the hook from the centre line of the support. Now $L_0$ = sum of anchorage beyond the centre of the support and the equivalent anchorage value.

If no hook is provided, as in case of deformed bars, $L_0 = \frac{l_s}{2} - x'$. However, if a hook or bend is provided, $L_0$ is computed as under.

**(a) Provision of standard hook at the end** [Fig. 3.27 (a)]

The dark portion shows the hook which has an anchorage value of $16\varphi$ (IS 456 : 2000), for all types of steel. The distance of the beginning of the hook from its apex of the semi-circle is equal to $(K + 1)\varphi$. For mild steel bars, $K = 2$ and for HYSD bars, $K = 4$. Hence this distance is equal to $3\varphi$ for mild steel bars and $5\varphi$ for HYSD bars. Let $l_s$ be the width of the support.

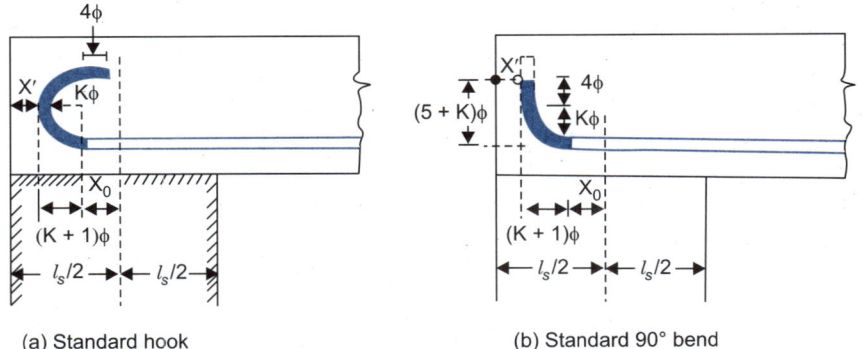

(a) Standard hook        (b) Standard 90° bend

**Fig. 3.27.** Computation of $L_0$

Then $\qquad L_0 = x_0 + 16\,\varphi$ where $x_0 = \frac{l_s}{2} - x' - (K+1)\,\varphi$

$\therefore \qquad L_0 = \left(\frac{l_s}{2} - x' - (K+1)\,\varphi\right) + 16\,\varphi = \frac{l_s}{2} - x' + (15 - K)\,\varphi \qquad \ldots(3.32)$

Taking $\qquad K = 2$ for mild steel bars, $L_0 = \frac{l_s}{2} - x' + 13\,\varphi \qquad \ldots[3.32\,(a)]$

Taking $\qquad K = 4$ for HYSD bars, $\qquad L_0 = \frac{l_s}{2} - x' + 11\,\varphi \qquad \ldots[3.32\,(b)]$

### (b) Provision of 90° standard bend [Fig. 3.27 (b)]

The dark portion shows the 90° bend which has an anchorage value of 8φ (IS : 456–2000), for all types of steel. Here also, the distance of beginning of the hook from its apex of the semi-circle is equal to $(K + 1)\varphi$.

Then $\qquad L_0 = x_0 + 8\varphi \qquad$ where $x_0 = \dfrac{l_s}{2} - x' - (K + 1)\varphi$

$\therefore \qquad L_0 = \left(\dfrac{l_s}{2} - x' - (K + 1)\varphi\right) + 8\varphi = \dfrac{l_s}{2} - x' + (7 - K)\varphi \qquad \ldots(3.33)$

Taking $\qquad K = 2$ for mild steel bars, $L_0 = \dfrac{l_s}{2} - x' + 5\varphi \qquad \ldots[3.33\,(a)]$

Taking $\qquad K = 4$ for HYSD bars, $\qquad L_0 = \dfrac{l_s}{2} - x' + 3\varphi \qquad \ldots[3.33\,(b)]$

**Remedies to get development length**

If the check for the satisfaction of *Eq. 3.31* or *3.31 (a)* is not obtained, following remedial measures may be adopted to satisfy the check :

1. Reduce the diameter φ of the bar, thereby reducing the value of $L_d$, keeping the area of steel at the section unchanged. This is the standard procedure envisaged in the Code, *i.e.* reducing the value of $L_d$ by *limiting* the diameter of the bar to such a value that *Eq. 3.31* or *3.31 (a)* is satisfied.

2. Increasing the value of $L_0$ by providing *extra length* of the bend over and above the standard value $(5 + K)\varphi$ shown by dotted lines in Fig. 3.27 (b).

3. By increasing the number of bars (there by increasing $A_{st1}$) to be taken into the support. This method is uneconomical.

4. By providing adequate mechanical anchorage.

We shall discuss the first remedial method in the following section.

**Limiting bar diameter to satisfy** *Eq. 3.31* **or** *3.31 (a)*

IS 456 : 2000 stipulates that at simple supports having compressive reaction, the positive moment reinforcement shall be limited to a diameter such that

$$L_d \leq \dfrac{1.3\, M_1}{V} + L_0 \qquad \ldots[3.31\,(b)].$$

Substituting $\qquad L_d = \dfrac{\sigma_s\, \varphi}{4\, \tau_{bd}}$, in the above condition, we get

$$\dfrac{\sigma_s \cdot \varphi}{4\, \tau_{bd}} \leq 1.3 \dfrac{M_1}{V} + L_0 \qquad \text{or} \qquad \varphi \leq \dfrac{4\, \tau_{bd}}{\sigma_s}\left(1.3 \dfrac{M_1}{V} + L_0\right) \qquad \ldots(3.34)$$

Similarly for simple supports not offering compressive reaction,

$$\varphi \leq \dfrac{4\, \tau_{bd}}{\sigma_s}\left(\dfrac{M_1}{V} + L_0\right) \qquad \ldots(3.35)$$

## 3.23. DEVELOPMENT LENGTH AT POINT OF INFLEXION

Fig. 3.28 shows the conditions at a point of inflection (P.I.). As already indicated earlier, the Code states that the following condition be satisfied

$$\left(\dfrac{M_1}{V} + L_0\right) \geq L_d \qquad \ldots\text{(by 3.31)}$$

where $L_0$ should not be greater than $d$ or $12\,\varphi$ whichever is greater, and $V$ is the shear force at the point of inflexion.

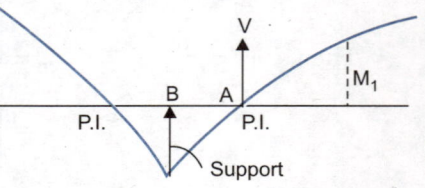

**Fig. 3.28.** Development Length at a Point of Inflection

# SHEAR AND BOND

## 3.24. CONDITIONS FOR CURTAILMENT OF REINFORCEMENT

In most of the cases, the B.M. varies appreciable along the span of the beam. From the point of view of economy, the moment of resistance of the beam should be reduced along the span according to the variation of B.M. This is effectively achieved by reducing the area of reinforcement, *i.e.* by curtailing the reinforcement provided for maximum B.M. In general, all steel, whether in tension or in compression should extend $d$ or 12 φ (which ever is greater) beyond the theoretical point of cutoff (TPC).

**1. Rules for continuation of positive moment reinforcement in the span region of supports.** (IS : 456–2000)

(*a*) At least one third the positive moment reinforcement in *simple members* and one-fourth the positive moment reinforcement in *continuous members* shall extend along the same face of the member into the support to a length equal to $L_d/3$.

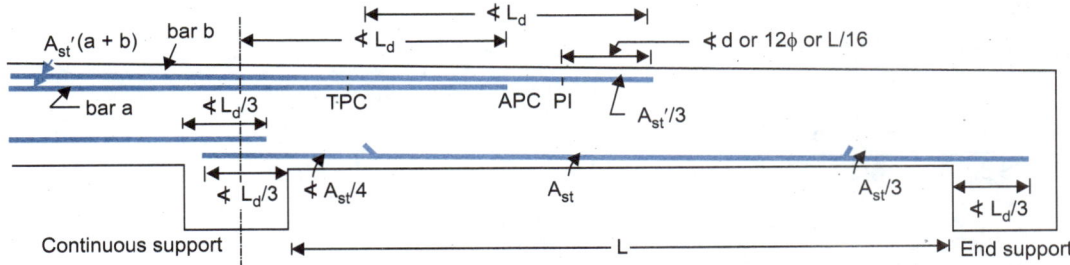

Fig. 3.29. Rules for Continuation of Bars into and from Supports

(*b*) When a flexural member is part of the primary lateral load resisting system, the positive reinforcement required to be extended into the support as described in (*a*) shall be anchored to develop its design stress in tension at the face of the support.

(*c*) At simple supports *Eq. 8.31* (or *8.31 a*) should be satisfied.

**2. Rules for continuation of negative moment reinforcement beyond the point of inflexion (P.I.)**

At least one-third of the total reinforcement provided for negative moment at the support (*i.e.* $A_{st}'/3$) shall extend beyond the point of inflexion (P.I.) for a distance not less than the effective depth ($d$) or 12 φ or one-sixteenth of the clear span which ever is greater.

**3. Rules for continuation of a curtailed bar beyond theoretical point of the cutoff**

As per IS Code, every bar that is to be curtailed shall continue for a distance equal to the effective depth of the member or 12 φ, whichever is greater, beyond the theoretical point of cutoff, except at the simple support or end of cantilever. Thus the distance between the *theoretical point of cutoff* (TPC) and the *actual of cutoff* (APC) should be greater than $d$ or 12 φ. Also, the distance of the actual point of cutoff (APC) from the point of max. B.M. should be at least $L_d$.

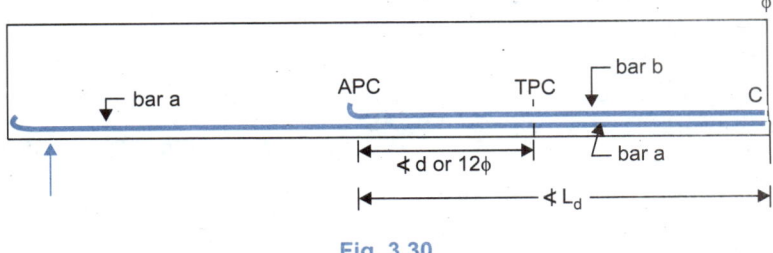

Fig. 3.30

**4. Rules for maintaining shear capacity at region of cutoff of flexural reinforcement in tension zone**

Curtainment of flexural tension reinforcement results in the loss of shear strength in the region of cutoff, and hence it is necessary to make provision to guard against such loss. Flexural reinforcement shall not be terminated in a tension zone unless any one of the following condition is satisfied:

(a) The shear at the cutoff point does not exceed two thirds that permitted, including the shear strength of web reinforcement. In other would the *shear capacity* shall be atleast 1.5 times the applied shear at the point of curtailment. Thus,

$$V \geqslant \frac{2}{3}(V_c + V_s) \quad \text{or} \quad V_c + V_s \geq 1.5\, V \qquad \ldots(3.36)$$

where   $V_c$ = shear capacity of concrete, based on continuing reinforcement only
   $V_s$ = shear capacity of shear reinforcement
   $V$ = applied shear at the point of curtailment

(b) Stirrup area in excess of that required for shear and torsion is provided along each terminated bar over a distance, from cutoff point equal to three-fourth the effective depth of the member. Excess area of shear reinforcement is given by :

Excess $\qquad A_{sv} \geq \dfrac{0.4\, b\, s_v}{f_y} \qquad \ldots(3.37)$

where $\quad s_v \geqslant \dfrac{d}{8\beta_b} \geqslant \dfrac{0.87\, f_y\, A_{sv}}{0.4\, b} \qquad \ldots(3.38)$

$$\beta_b = \frac{\text{area of bars cutoff at the section}}{\text{total area of bars at the section}} \qquad \ldots(3.39)$$

(c) For 36 mm or smaller bars, the continuing bars provide double the area required for flexure at the cutoff point and the shear does not exceed three-fourth that permitted.

Thus, $\qquad M_r \geq 2\, M \quad \text{and} \quad V_c + V_s \geq 1.33\, V \qquad \ldots(3.40)$

where   $M_r$ = moment of resistance of remaining (or continued) bars
   $M$ = B.M. at cut off point   and   $V$ = S.F. at cutoff point

### 5. Scheme of Curtailment of Tension Reinforcement in Simply Supported Beams

Some bars of tensile reinforcement can be curtailed if not required from flexure point of view. The distance $x$, measured from the centre of the span, at which $N_x$ number of bars (out of that $N_c$ bars) can be curtailed, is given by *Eq. 3.41*.

$$x = \frac{L}{2}\sqrt{\frac{N_x}{N_c}} \qquad \ldots(3.41)$$

Alternatively, a simple graphical method can be used for such curtailment. Fig. 3.31 shows the bending moment diagram for half the span of the beam, in which $N_c$ (= 4, say) number of bars are required at the mid span. Each bar has area equal to $A_{st}/N_c$, where $A_{st}$ is the total area of tensile reinforcement. The stress in the bars is close to the permissible value ($\sigma_{st}$) at the mid span, but decreases with the B.M. towards the point $B_1$ at which one of the bars can be curtailed and the remaining bars are stressed to

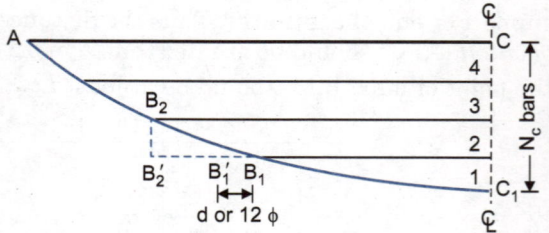

**Fig. 3.31.** Curtailment of Bars

the permissible value. However, one of the bars can be terminated the point $B_1$ only. If the length $C_1 B_1$ is equal to or greater than the development length $L_d$ for that bar. Similarly, the second bar can be curtailed at point $B_2$ if $B_1 B'_2$ is equal to or greater than $L_d$ for the second bar. The Code specifies that the reinforcement shall extend the point at which it is no longer required to resist flexure, for a distance equal to effective depth $d$ or 12 Φ whichever is greater. Thus, for the first bar, if $B_1 C_1 > L_d$, it should extend, beyond $B_1$, to a point $B'_1$ such that $B_1 B'_1$ is equal to $d$ 12 Φ whichever is more. Another important recommendation of the Code is that in simple beams, one-third of the positive reinforcement should be continued to the supports, and the diameter should be so chosen that *Eq. 3.31* is satisfied.

## 3.25. REINFORCEMENT SPLICING

Splices are provided when the length of the bar available is less than that required. Splicing of a reinforcing bars becomes essential in the field due to either requirements of construction or due to non-availability of bars of desired length. The purpose of *splicing* is to transfer effectively the axial force from the terminating bar to the commencing (continuing) bar with the same line of action at the junction (Fig. 3.32 a). The following are the recommendations of IS code (IS 456 : 2000).

(a) Lap splices shall not be used for bars larger than 36 mm ; for larger diameters, bars may be welded. In cases where welding is not practicable, lapping of bars larger than 36 mm may be permitted, in which case additional spirals should be provided around the lapped bars (Fig. 3.32 d).

(b) Lap splices shall be considered as staggered if the centre to centre distance of the splices is not less than 1.3 times the lap length calculated as described in (c).

(c) The lap length including anchorage value of hooks for bars in *flexural tension* shall be $L_d$ or 30 φ whichever is greater and for *direct tension* shall $2 L_d$ or 30 φ whichever is greater. The *straight length*, ($L'$) of the lap shall not be less than 15 φ or 200 mm (Fig. 3.32). The following provision shall also apply :

(1) top of a section as cast and the minimum cover is less than twice the diameter of the lapped bar, the lapped length shall be increased by a factor of 1.4.

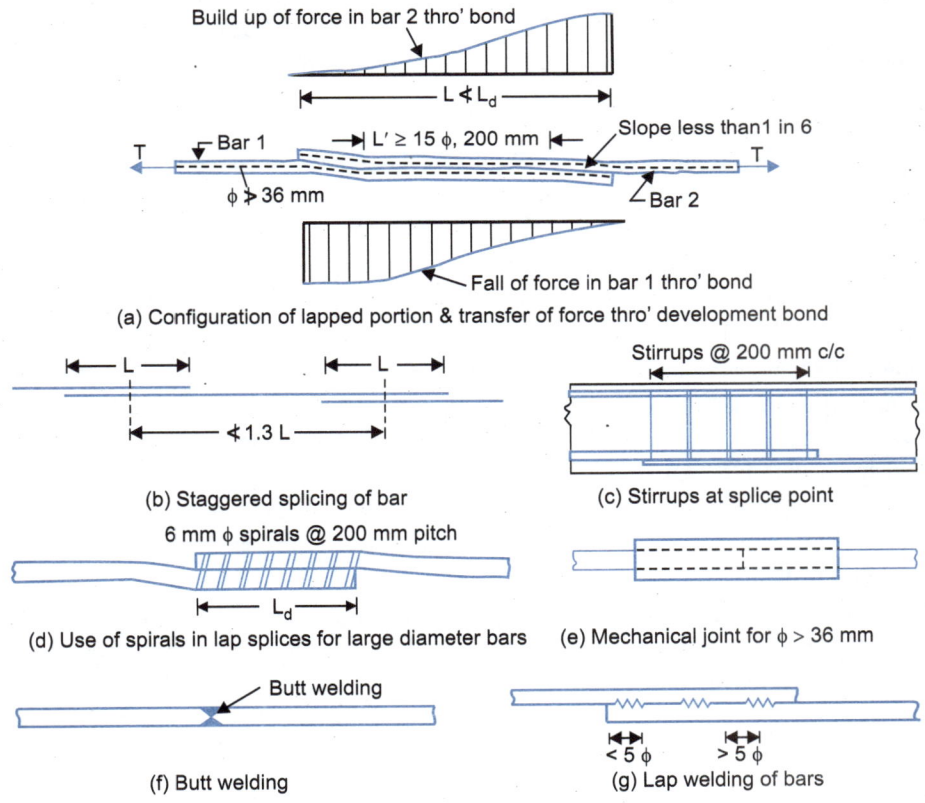

Fig. 3.32. Reinforcement Splicing

(2) corner of a section and the minimum cover to either face is less than twice the diameter of the lapped bar or where the clear distance between adjacent laps is less than 75 mm or 6 times the diameter of bar, whichever is greater, the lap length should be increased by a factor of 1.4. Where both condition (1) and (2) apply, the lap length should be increased by a factor of 2.0.

**Note:** *Splices in tension members shall be enclosed in spirals made of bars not less than 6 mm diameter with pitch not more than 100 mm.*

(d) The lap length in compression shall be equal to the development length in compression, but not less than 24 φ.

(e) When bars of two different diameter are to be spliced, the lap length shall be calculated on the basic of diameter of the smaller bar.

(f) When splicing of welded wire fabric is to be carried out, lap splices of wires shall be made so that overlap measured between the extreme cross-wires shall be not less than the spacing of cross wires plus 100 mm.

(g) In case of bundled bars, lapped splices of bundled bars shall be made by splicing one bar at a time; such individual splices within a bundle shall be staggered.

**Strength of welds :** The following values may be used where the strength of the weld has been proved by tests to be at least as great as than of the parent bar.

**(a) Splices in compression :** For welded splices and mechanical connection, 100 percent of the design strength of joined bars.

**(b) Splices in tension :** (1) 80 percent of the design strength of welded bars (100 percent if welding is strictly supervised and if at any cross-section of the member not more than 20 percent of the tensile reinforcement is welded).

**(c)** 100 percent of design strength of mechanical connection.

**End-bearing splices :** End-bearing splices shall be used only for bars in compression. The ends of the bars shall be square cut and concentric bearing ensured by suitable devices.

## PROBLEMS

1. A R.C. beam supports a total uniformly distributed load of 80 kN inclusive of its own weight and has 16 cm width and 32 cm effective depth. Determine (i) maximum shear stress (ii) spacing of 10 mm diameter two lagged stirrups near the support. Use M 20 concrete and Fe 415 steel.

2. A beam has an effective depth of 40 cm. At a particular section, it has 8 mm Φ 2 lgd stirrups provided at 20 cm centre to centre. Calculate the shear resistance of the stirrups. Take $\sigma_{sv}$ = 140 mm² and $j$ = 0.87.

3. A hook made from 12 mm diameter rod is embedded in a concrete roof, for a distance of 100 mm. Calculate the maximum load that can be suspended from the hook. Take permissible bond stress = 0.6 N/mm².

4. A R.C. beam supports a total uniformly distributed load of 160 kN. The beam has 20 cm width and 36 cm effective depth. Determine the spacing of 10 mm diameter two legged vertical stirrups, in addition to two main reinforcing bars of 16 mm bent up at 45° to resist stear. Take $\sigma_{cbc}$ = 5 N/mm², $\sigma_{st}$ = 140 N/mm², and $m$ = 18.

5. An RCC beam 250 × 550 mm effective carries a load of 25 kN/m (including self weight) over span of 6 m. If the tensile steel is 1%, find nominal shear stress in the beam and shear strength of concrete. Assume M20 grade concrete and Fe 415 steel.

6. An RCC slab 100 mm thick carries an Udl of 5.1 kN/m² over a span of 3 m, reinforcement consists of 10 mm φ bars @ 150 mm/c/c. The slab is supported on a brick wall 230 mm thick. Find the development length required from the face of support. Take M25 grade and Fe 415 steel cover 15 mm.

# CHAPTER 4

# TORSION

## 4.1. INTRODUCTION

Torsion when encountered in reinforced concrete members, usually occurs in combination with flexure and transverse shear.

Torsion may be induced in a reinforced concrete in various ways during the process of load transfer in a structural system.

Axial loads, flexure, shear and torsion are the basic loading situations for which independent theories have been developed, and the more complicated interactive loading situations have been analysed as combinations of these basic effects. A common example of torsional loading is that of *peripheral beams* in each floor of a multistoreyed building, in which the slab is cast monolithic with the beam giving rise to L-beam configuration (see chapters 6 and 7). Another example of torsional loading is that of a ring beam provided at the bottom of an elevated circular water tank. Such a ring beam is subjected to bending moment, shear force and torsional moment. The beams supporting cantilevered canopy slabs are also subjected to significant torsional loading. Other prominent examples of torsional loadings are edge beams of concrete shell roofs, and helicoidal staircases.

Fig. 4.1(a) shows the loading on the peripheral beam received from the slab. Due to monolithic beam-slab construction, the end beam has a flange to one side, giving rise to L-beam configuration. A portion of the slab, of length $B$, acts as flange of the L-beam. Evidently, such an L-beam is not symmetrically loaded. Due to load on the flange, an eccentric load $W$ is transferred to the beam. This eccentric load may be considered as a central load $W$ plus a moment $M_t$ as shown in Fig. 4.1(b). The central load $W$ causes bending moment in the beam, which is jointly resisted by the rectangular portion of the beam as well as the flange (of width $B$). The torsional moment $M_t$ causes torsion in the beam which is resisted by the rectangular portion alone and the flange does not contribute to any torsional moment of resistance.

Fig. 4.1. Torsional Loading in L-Beams.

## 4.2. TORSIONAL RESISTANCE : ELASTIC BEHAVIOUR

When a shaft of circular section is subjected to torsional moment (or torque) $M_t$, the maximum shearing stress induced in the material is given by the classical formula:

$$\frac{M_t}{J} = \frac{N\theta}{l} = \frac{f_s}{R} \qquad \ldots(4.1)$$

where  $M_t$ = applied torque;  $\qquad$ $J$ = polar moment of inertia.
$\qquad$ $N$ = shear modulus; $\qquad$ $l$ = length of the shaft.
$\qquad$ $f_s$ = maximum shear stress at the extreme fibres.
$\qquad$ $R$ = radius of circular shaft.

However, circular sections are very rarely used in reinforced concrete construction. When a beam of rectangular and other non-circular section is subjected to torsion, considerable warping of the cross-section takes place, and plane sections do not remain plane.

For a beam of rectangular section subjected to torsional moment $M_t$, the maximum torsional shear stress occurs at the middle of the *long side*, and its magnitude is given by

$$\tau_l = \tau_{max} = \frac{M_t}{\alpha b^2 D} = \frac{M_t}{Z_t} \qquad \ldots(4.2)$$

The torsional shear stress at the middle of the *short side* of the rectangular section is given by:

$$\tau_s = \eta \, \tau_{max}. \qquad \ldots(4.3)$$

where  $D$ = depth or long side of the section.
$\qquad$ $b$ = short side of the section.
$\qquad$ $Z_t$ = polar or torsional section modulus
$\qquad$ $\alpha, \eta$ = coefficients depending upon $D/b$ ratio (Table 4.1)

### TABLE 4.1

| $D/b$ | 1.0 | 1.5 | 2.0 | 3.0 | 6.0 | 10.0 | $\infty$ |
|---|---|---|---|---|---|---|---|
| $\alpha$ | 0.208 | 0.231 | 0.246 | 0.267 | 0.299 | 0.313 | 0.333 |
| $\eta$ | 1.00 | 0.859 | 0.795 | 0.753 | 0.743 | 0.742 | 0.742 |

For the design purposes, the *torsional section modulus* for rectangular sections may be taken as:

$$Z_t = \alpha b^2 D \qquad \ldots[4.4\,(a)]$$

or $\qquad Z_t \approx \dfrac{b^2 D}{3 + \dfrac{2b}{D}} \qquad \ldots[4.4\,(b)]$

For the T and L-sections, Cowan gives the following expressions for maximum torsional shear stress:

$$\tau_{max} = \frac{3 M_t \cdot t_i}{\Sigma b_i t_i^3} \qquad \ldots(4.5)$$

where  $t_i$ is $t_1$ or $t_2$.

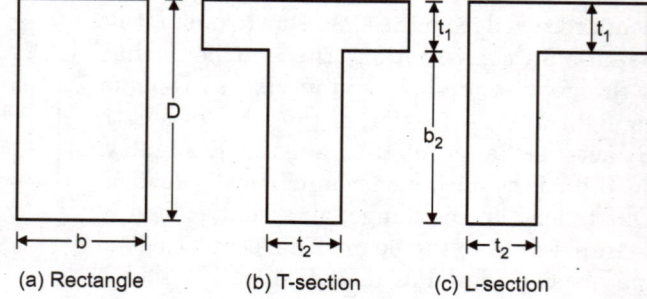

(a) Rectangle  (b) T-section  (c) L-section

Fig. 4.2

## 4.3. INDIAN STANDARD RECOMMENDATIONS ON DESIGN FOR TORSION
(IS : 456–2000)

When a beam (either rectangular or flanged) is subjected to torsional moment, shear stresses are developed in it. Shear stresses resulting from torsional moment can be accurately assessed by membrane analogy or by any other recognised theory. However, such an analysis is beyond the scope of the present book.

The Indian Standard (IS : 456–2000) recommendations given in para 2 and 3 below are applicable to beams of solid rectangular cross-section. However, these clauses may also be applied to flanged beams, by substituting $b_w$ for $b$, in which case they are generally conservative; therefore, specialist literature may be referred to.

The code of IS: 456–2000 (Clause No. 41.4.2) recommends a simplified skew bending based formulation for the design of longitudinal reinforcement to resist torsion combined with flexure in beam with rectangle Section.

### 1. Critical Section

Section located less than a distance $d$ from the face of the support may be designed for the same torsion as computed at the distance $d$ where $d$ is the effective depth.

### 2. Shear and Torsion

***Equivalent shear*** : Equivalent shear $V_e$ shall be calculated from the formula:

$$V_e = V + 1.6 \frac{T}{b} \qquad \ldots(4.6)$$

where   $V_e$ = equivalent shear ;        $V$ = S.F.
        $T$ = torsional moment, and      $b$ = breadth.

*The equivalent nominal shear stress* ($\tau_{ve}$) *shall be calculated from the expression:*

$$\tau_{ve} = \frac{V_e}{bd} \quad \text{(where } b = b_w \text{ for flanged sections).}$$

(*i*) The value of $\tau_{ve}$ shall not exceed that values of $\tau_{c\,max}$ given in Table 3.3. If it exceeds, the section should be redesigned by increasing the concrete area.

(*ii*) If the equivalent nominal shear stress $\tau_{ve}$ does not exceed $\tau_c$ given in Table 3.1, minimum shear reinforcement shall be provided in the form of stirrups, such that

$$\frac{A_{sv}}{b \cdot s_v} \geq \frac{0.4}{0.87 f_y} \qquad \ldots(4.7)$$

The above equation can be rearranged in the following form to give the maximum spacing of stirrups:

$$s_v \leq \frac{2.175 \, A_{sv} \, f_y}{b} \qquad \ldots[4.7\,(a)]$$

where   $A_{sv}$ is the total cross-sectional area of stirrups legs effective in shear,
         $s_v$ is the stirrup spacing
and     $f_y$ is the characteristic strength of the stirrup reinforcement, in N/mm², which shall not be taken greater than 415 N/mm².

(*iii*) If $\tau_{ve}$ exceeds $\tau_c$ given in Table 3.1, both longitudinal and transverse reinforcement shall be provided in accordance with para 3 below.

### 3. Reinforcement of Member Subjected to Torsion

Reinforcement in torsion, when required, shall consist of longitudinal and transverse reinforcement.

#### (a) Longitudinal Reinforcement

The longitudinal reinforcement shall be designed to resist an equivalent bending moment $M_{e1}$ given by:

$$M_{e1} = M + M_T \qquad \text{where} \quad M = B.M. \text{ at the cross-section, and}$$

$$M_T = T \frac{(1 + \frac{D}{b})}{1.7} \qquad \ldots(4.8)$$

where $T$ is the torsional moment, $D$ is the overall depth of the beam and $b$ is the breadth of the beam.

If the numerical value of $M_T$ as defined above exceeds the numerical value of $M$, longitudinal reinforcement shall be provided on the flexural compression face, such that beam can also withstand an equivalent moment $M_{e2}$ given by

$$M_{e2} = M_T - M$$

the moment $M_{e2}$ being taken as acting in the opposite sense to the moment $M$.

**(b) Transverse Reinforcement**

The transverse reinforcement as per Clause 41.4.3 of IS: 456 – 2000 for design of transverse stirrups are considered.

Two legged closed hoops enclosing the corner longitudinal bars shall have an area of cross-section $A_{sv}$ given by:

$$A_{sv} = \frac{T \cdot s_v}{b_1 d_1 \sigma_{sv}} + \frac{V \cdot s_v}{2.5 d_1 \sigma_{sv}} \qquad ...(4.9)$$

However, the total transverse reinforcement shall not be less than $\dfrac{(\tau_{ve} - \tau_c) b \cdot s_v}{\sigma_{sv}}$

where  $T$ = torsional moment
  $V$ = shear force
  $s_v$ = spacing of the stirrup reinforcement
  $b_1$ = centre to centre distance between corner bars in the direction of width.
  $d_1$ = centre to centre distance between corner bars in the direction of depth.
  $b$ = breadth of member
  $\sigma_{sv}$ = permissible tensile stress in shear reinforcement
  $\tau_{ve}$ = equivalent shear stress
  $\tau_c$ = shear strength of the concrete as specified in Table 3.1.

**Note 1. *Distribution of torsional reinforcement :***

As per Clause 26.5.1.7 (a) specifies maximum limit to the spacing ($s_v$) of stirrups provided as torsional reinforcement. When a member is designed for torsion, torsion reinforcement shall be provided as below.

(a) The transverse reinforcement for torsion shall be rectangular closed stirrups placed perpendicular to the axis of the member. The spacing of the stirrups shall not exceed the least of $x_1$, $\dfrac{x_1 + y_1}{4}$ and 300 mm, where $x_1$ and $y_1$ are respectively the short and long dimensions of the stirrup.

(b) Longitudinal reinforcement shall be placed as close as is practicable to the corners of the cross-section and in all cases, there shall be at least one longitudinal bar in each corner of the ties. When the cross-sectional dimension of the member exceeds 450 mm, additional longitudinal bars shall be provided to satisfy the requirements of minimum reinforcement and spacing given in Note 2.

**Note 2. *Side face reinforcement :***

Where the depth of the web in a beam exceeds 750 mm, the side face reinforcement shall be provided along the two faces. The total area of such reinforcement shall be not less than 0.1 percent of the web area and shall be distributed equally on two faces at a spacing not exceeding 300 mm or web thickness whichever is less.

## Design

**Example 4.1.** *A rectangular beam 400 mm wide is subjected to the following at a section: (i) B.M. of 45 kN-m (ii) S.F. of 30 kN, and (iii) Torsional moment of 20 kN-m. Design the section. Take the following values*:

(i) *Permissible stress in steel in tension and shear : 140 N/mm²*

(ii) *Permissible bending compressive stress in concrete : 5 N/mm²*

(iii) *m = 18*

**Solution:**

**(1) Design constants :** For the given set of stresses ($c = \sigma_{cbc}$ = 5 N/mm², $t = \sigma_{st}$ = 140 N/mm² and $m$ = 18) we have $k_c$ = 0.39, $j_c$ = 0.87 and $R_c$ = 0.85.

**(2) Design of section for B.M. :** $b = 400$ mm ; $M = 45$ kN-m $= 45 \times 10^6$ N-mm

$$\therefore \quad d = \sqrt{\frac{M}{R_c \cdot b}} = \sqrt{\frac{45 \times 10^6}{0.85 \times 400}} = 363.80 \text{ mm.}$$

Keep total $D = 420$ mm. Using 16 mm $\Phi$ bars, 12 mm $\varphi$ rings and a nominal cover of 30 mm, available $d = 420 - 30 - 12 - 16/2 = 370$ mm.

**(3) Equivalent shear:** $V_e = V + 1.6 \times \dfrac{T}{b} = 30000 + 1.6 \times \dfrac{20 \times 10^6}{400} = 110000$ N

$$\tau_{ve} = \frac{V_e}{bd} = \frac{110000}{400 \times 370} = 0.743 \text{ N/mm}^2$$

For M 15 concrete, $\tau_c$ for 0.5% reinforcement is 0.29 N/mm² and $\tau_c$ for 1.0% reinforcement is 0.37 N/mm² (Table 3.1). Though the percentage reinforcement is not known, it is certain that permissible shear stress $\tau_c$ will be less than $\tau_{ve}$. Hence shear reinforcement is necessary.

Also, from Table 3.3, $\tau_{c\,max}$ for M 15 concrete is 1.6 N/mm². Thus $\tau_{ve}$ is less than $\tau_{c\,max}$. Hence the dimensions of the beam are acceptable.

**(4) Longitudinal reinforcement:** Since $\tau_{ve}$ exceeds $\tau_c$, both longitudinal and transverse reinforcement are necessary. The longitudinal reinforcement is designed to resist an equivalent bending moment $M_{e1}$, given by

$$M_{e1} = M + M_T \qquad \text{where} \quad M = \text{B.M.} = 45 \times 10^6 \text{ N-mm}$$

$$M_T = T\,\frac{(1 + D/b)}{1.7} = 20 \times 10^6 \left( \frac{1 + 420/400}{1.7} \right) = 24.11 \times 10^6 \text{ N-mm}$$

$$\therefore \quad M_{e1} = 45 \times 10^6 + 24.11 \times 10^6 = 69.11 \times 10^6 \text{ N-mm}$$

$$\therefore \quad A_{st} = \frac{M_{e1}}{\sigma_{st} \cdot jd} = \frac{69.11 \times 10^6}{140 \times 0.87 \times 370} = 1533 \text{ mm}^2.$$

Using 16 mm $\Phi$ bars, $A_\Phi = \dfrac{\pi}{4}(16)^2 = 201$ mm² $\therefore$ No. of bars $= \dfrac{1533}{201} = 7.629 \cong 8$.

**(5) Transverse reinforcement :** Transverse reinforcement is provided in the form of vertical stirrups. Provide two corners bars of 16 mm $\Phi$, at the top face, at a nominal cover of 30 mm, (moderate exposure) to hold the stirrups, as shown in Fig. 4.3.

$\therefore \quad b_1 =$ centre to centre distance between corner bars in the direction of width

$$= 400 - 2\left(30 + 12 + \frac{16}{2}\right) = 300 \text{ mm}$$

$d_1 =$ centre to centre distance between corner bars in the direction of depth

$$= 420 - 2\left(30 + 12 + \frac{16}{2}\right) = 320 \text{ mm}$$

The area of cross-section $A_{sv}$ of the stirrups is given by:

$$A_{sv} = \frac{T\,s_v}{b_1\,d_1\,\sigma_{sv}} + \frac{V \cdot s_v}{2.5\,d_1\,\sigma_{sv}}.$$

Using 12 mm $\Phi$ 2-lgd stirrups, $A_{sv} = 2\,\dfrac{\pi}{4}(12)^2 = 226$ mm²

$$\therefore \quad 226 = \left[ \frac{20 \times 10^6}{300 \times 320 \times 140} + \frac{30000}{2.5 \times 320 \times 140} \right] s_v.$$

From which $s_v = 149.18$ mm $\cong 140$ mm

However, the spacing should not exceed the least of $x_1$, $\dfrac{x_1 + y_1}{4}$ and 300 mm, where, $x_1$ = short dimension of stirrup

$= 300 + 16 + 12 = 328$ mm

$y_1 = 300 + 16 + 12 = 328$

$\therefore \quad \dfrac{x_1 + y_1}{4} = \dfrac{328 + 328}{4} = 164$ mm.

Thus $s_v = 140$ mm is permissible.

Hence, provide 12 mm Φ two-legged stirrups @ 140 mm centre to centre. The section of the beam is shown in Fig. 4.3.

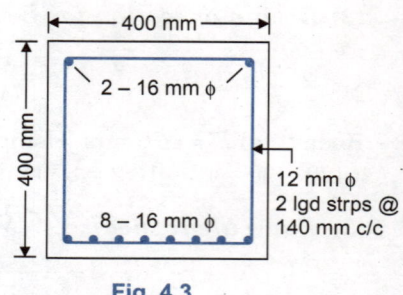

Fig. 4.3

## PROBLEMS

1. Discuss IS Code provisions for design for torsion.
2. A rectangular beam 300 mm wide is subjected to the following at a given section
   (i) B.M. : 30 kN-m
   (ii) S.F. : 20 kN
   (iii) Torsional moment 40 kN-m.

   Design the section. Use (a) M 15 concrete and mild steel reinforcement.
   (b) M 20 concrete and Fe 415 steel, and (c) M 15 concrete and Fe 415 steel.
3. Design the torsional reinforcement in a rectangular beam section 380 mm wide and 760 mm deep subjected to an ultimate twisting moment 140 kN-m considered with an ultimate (logging) bending moment of 280 kN-m and an ultimate shear force of 110 kN. Assume M25 grade Fe 415 steel and mild exposure conditions.

# CHAPTER 5: DOUBLY REINFORCED BEAMS

## 5.1. INTRODUCTION

When the dimensions of a beam are restricted by architectural (or) structural considerations the sections will have insufficient Area of concrete to provide for compressive stresses. In such cases, steel is placed in compression and tension zones and such beams are called doubly reinforced beam sections. We have seen that the balanced section of a beam is the most economical section from the requirement of steel point of view. If the area of tensile steel reinforcement is doubled, the moment of resistance of the beam is increased only by about 22%. The moment of resistance of a balanced section is equal to $R_c\, bd^2$. If a beam of specified dimensions ($b \times d$) is required to resist a moment much greater than $R_c\, bd^2$, there are two alternatives: (*i*) to use an over-reinforced section, or (*ii*) to use doubly reinforced section. An over-reinforced section is always uneconomical since the increase in the moment of resistance is not in proportion to the increase in the area of tensile reinforcement since the concrete, having reached maximum allowable stress, cannot take more additional load without adding compression steel. The other alternative is to provide reinforcement in the compression side of the beam and thus to increase the moment of resistance of the beam beyond the value $R_c\, bd^2$ for a singly reinforced balanced section. Sections reinforced with steel in compression and tension are known as *doubly reinforced sections*.

Apart from the requirement of developing more resistance than $R_c\, bd^2$, doubly reinforced sections are used in the following circumstances:

(1) When the members are subjected to alternate external loads and the bending moment in the sections reverses, such as in concrete piles, etc.

(2) When the members are subjected to loading eccentric to either side of the axis, such as in columns subjected to wind loads.

(3) When the members are subjected to accidental lateral loads, shock or impact.

The steel reinforcement provided in the compression zone is subjected to compressive stress. However, concrete undergoes creep strains due to continued compressive stress, with the result that the strain in concrete goes on increasing with time. This increases compressive strain in steel also in addition to creep strain in compressive steel. *Thus the total compressive strain in compressive steel will be much greater than the strain in surrounding concrete due to flexure alone.* According to IS : 456–2000, the compressive stress in steel should be calculated by multiplying the stress in the surrounding concrete by $1.5\, m$. In other words, *the modified modular ratio ($m_c$) to be used for compression steel is 1.5 times the modular ratio normally used (i.e. $m_c = 1.5\, m$).* The stress in compression steel so found should not exceed the permissible value ($\sigma_{sc}$) specified in Table 1.32.

## 5.2. LOCATION OF NEUTRAL AXIS

Fig. 5.1 shows a doubly reinforced section. The theory of doubly reinforced section is based on the same assumptions as for singly reinforced section. The neutral axis of a doubly reinforced section can be found by finding the centre of gravity of the combined section consisting of concrete in compression only and steel in compression and tension. Let,

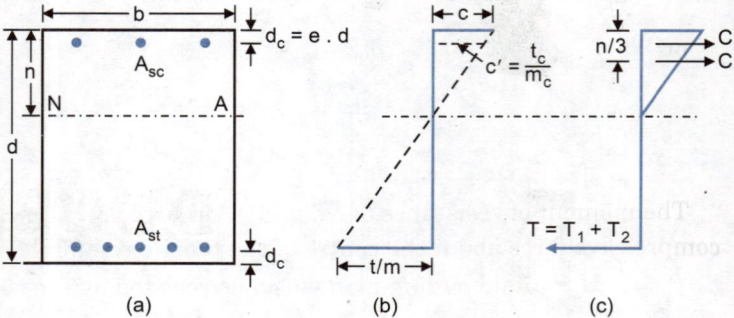

Fig. 5.1. Doubly Reinforced Section.

$b$ = breadth of the beam.
$d$ = effective depth of the beam
$d_c$ = depth of the centre of the compressive steel = $e.d$
$e$ = compressive steel depth factor = $d_c/d$
$c$ = maximum stress in concrete
$t$ = maximum stress in steel
$c'$ = stress in concrete surrounding compressive steel
$t_c$ = stress in compressive steel = $m_c \cdot c' = 1.5\, m \cdot c'$
$A_{st}$ = area of tensile steel
$A_{sc}$ = area of compressive steel
$m_c$ = modular ratio for compression zone = $1.5\, m$

From Fig. 5.1 (b) we get : $\dfrac{c}{c'} = \dfrac{kd}{kd - ed} = \dfrac{k}{k - e}$ or $c' = \dfrac{k-e}{k} c$

Hence stress in steel $\quad t_c = m_c c' = m_c \dfrac{k-e}{k} c = 1.5\, m \dfrac{k-e}{k} \cdot c$

If maximum stresses in concrete and tensile steel are known, the N.A. can be located from stress diagram [Fig. 5.1(b)] exactly in the same manner as in singly reinforced section:

$$\dfrac{c}{t/m} = \dfrac{n}{d-n} = \dfrac{kd}{d-kd} = \dfrac{k}{1-k} \quad \text{or} \quad k = \dfrac{mc}{mc+t} = \dfrac{m\,\sigma_{cbc}}{m\,\sigma_{cbc} + \sigma_{st}}$$

where $\sigma_{cbc}$ and $\sigma_{st}$ are the permissible stresses in concrete and steel respectively. Thus the presence of steel in compression zone does not affect the depth of N.A.

Again neglecting the concrete in the tensile zone and equating the moment of the compressive area about N.A. to the moment of the tensile area about N.A. we get

$$\dfrac{bn^2}{2} + m_c \cdot A_{sc}(n-d_c) - A_{sc}(n-d_c) = m\, A_{st}(d-n)$$

or $\qquad \dfrac{bn^2}{2} + (m_c - 1) A_{sc}(n-d_c) = m\, A_{st}(d-n) \qquad$ ...[5.1(a)]

or $\qquad \dfrac{bn^2}{2} + (1.5m - 1) A_{sc}(n-d_c) = m\, A_{sc}(d-n) \qquad$ ...(5.1)

where $(m_c - 1) A_{sc}$ or $(1.5\, m - 1) A_{sc}$ is the concrete area equivalent to compressive steel.

In the above equation, all the terms except $n$ is known. Hence the N.A. can be located. The above equation can also be expressed in terms of the neutral axis depth factor $k$ and compressive steel depth factor $e$ as follows:

$$\frac{b}{2} k^2 d^2 + (m_c - 1) A_{sc} (k - e) d = m A_{st} (1 - k) d$$

or
$$\frac{b k^2 d}{2} + (m_c - 1) A_{sc} (k - e) = m A_{st} (1 - k) \qquad \ldots(5.2)$$

## 5.3. MOMENT OF RESISTANCE

The moment of resistance of a doubly reinforced section can be determined by taking the moment of the compressive force about the centre of the tensile reinforcement. Let,

$M$ = total bending moment to be resisted by the section
  = moment of resistance of the section = $M_1 + M_2$
$M_1$ = moment that can be developed by the balanced section, without compressive reinforcement
  = moment of the compressive force in concrete about the centre of tensile reinforcement.
$M_2$ = moment, in excess of $M_1$, that is to be provided by the compression steel
  = moment of the compressive force in steel, about the centre of the tensile reinforcement.
$A_{st}$ = total tensile reinforcement = $A_{st1} + A_{st2}$.
$A_{st1}$ = tensile steel for the balanced section, corresponding to the moment $M_1$.
$A_{st2}$ = additional tensile steel necessary to develop the remaining moment $M_2$

Refer Fig. 5.1. Taking the moments of the compressive forces about the centre of the tensile steel, we get

$$M = b n \frac{c}{2}\left(d - \frac{n}{3}\right) + (m_c - 1) A_{sc} \cdot c' (d - d_c) \qquad \ldots(5.3)$$

(where $c$ may be equal to or less than $\sigma_{cbc}$. See note 1).

Expressed in terms of $k$ and $e$,

we get
$$M = b \, kd \, \frac{c}{2}\left(d - \frac{kd}{3}\right) + (m_c - 1) A_{sc} \cdot c'(d - ed)$$

or
$$M = \frac{1}{2} jck \cdot bd^2 + (m_c - 1) A_{sc} \cdot c' d (1 - e) \qquad \ldots[5.3\,(a)]$$

or
$$M = \underset{\downarrow}{R_c} b \, d^2 + \underset{\downarrow}{c' d} (m_c - 1) A_{sc} (1 - e)$$

or
$$M = \quad M_1 \ + \ M_2 \qquad \ldots[5.3\,(b)]$$

Now
$$A_{st1} = \frac{M_1}{t \times \text{lever arm}} = \frac{M_1}{t \, j \, d} = \frac{M_1}{\sigma_{st} \cdot jd} \qquad \ldots(5.4)$$

and
$$A_{st2} = \frac{M_2}{t \times \text{lever arm}} = \frac{M_2}{t(d - d_c)} \qquad \ldots(5.5)$$

(where $t$ may be equal to or less than $\sigma_{st}$. See Note 3 below)

The additional tensile area $A_{st2}$ is actually required to balance the compressive steel. Hence $A_{st2}$ can also alternatively be determined by equating the moment of the equivalent concrete area of the compressive steel to the moment of equivalent concrete area of the tensile steel $A_{st2}$ about N.A. Thus,

$$(m_c - 1) A_{sc} (n - d_c) = m A_{st2} (d - n)$$

or
$$A_{st2} = A_{sc} \frac{m_c - 1}{m} \left(\frac{n - d_c}{d - n}\right) = A_{sc} \frac{m_c - 1}{m} \left(\frac{k - e}{1 - k}\right) \qquad \ldots(5.6)$$

The total area $A_{st}$ of tensile steel will be equal to $A_{st1} + A_{st2}$.

**Note 1.** The value of $c$ in *Eq 5.3* will be equal to the permissible value of stress $\sigma_{cbc}$ in concrete if the total area of steel $A_{st}$ is so provided that it balances both the compressive force in concrete as well as compressive force in steel. However, there are *two cases* in which the value of $c$ in *Eq. 5.3* may be less than the working or permissible stress.

*Case (i)* : In this case the tensile steel may be insufficient even to balance the compressive concrete, with the result that neutral axis will be above the critical one and the stress in steel will reach its permissible value first.

*Case (ii)* : In this case the compressive steel $A_{sc}$ may be more than the amount required to balance the additional tensile reinforcement $A_{st2}$ (*Eq 5.6*) with the result that the N.A. will again be above the critical N.A. and the stress in tensile steel will reach its maximum permissible value first. In *both these cases*, the value of $c$ has to be first determined from the known value of $t$ from the following:

$$c = \frac{t}{m}\frac{n}{d-n} = \frac{\sigma_{st}}{m}\cdot\frac{n}{d-n}$$

Except in these two cases, the value of $c$ to be substituted in *Eq. 5.3* will be equal to its permissible value $\sigma_{cbc}$.

**Note 2.** The stress in concrete ($c'$) surrounding the compression steel can also be found from the following relation easily derived from the stress diagram [Fig. 5.1(b)]:

$$\frac{c'}{c} = \frac{n-d_c}{n} \text{ or } c' = c\frac{n-d_c}{n} = c\frac{k-e}{k} \qquad \ldots(5.7)$$

In this equation, the value of $c$ should be determined as explained in note 1.

**Note 3.** If the additional tensile reinforcement $A_{st2}$ required to balance the compressive reinforcement cannot be provided in the same tier or layer as that of $A_{st1}$, the value of stress $t$, to be used in *Eq. 5.5*, will be slightly less than the permissible value $\sigma_{st}$ and will have to be first determined from the stress diagram.

## 5.4. STEEL BEAM THEORY

We have seen that the total moment of resistance of a doubly reinforced beam consists of two terms: the moment of resistance $M_1$ of the compression concrete and the corresponding tensile steel ($A_{st1}$) to balance it, and the moment of resistance $M_2$ of the compression steel ($A_{sc}$) and the remaining tensile steel ($A_{st2}$) to balance it. In the steel beam theory, the concrete is completely neglected and the moment of resistance is taken equal to the moment of steel couple, taking the permissible value of stresses in compressive steel equal to permissible value in tensile steel, $\sigma_{st}$.

$$M = A_{st}\cdot t\,(d-d_c) = A_{st}\cdot\sigma_{st}\,(d-d_c) \qquad \ldots(5.8)$$

Thus, in the steel beam theory, the concrete serves only as a web and the compressive and tensile steels as the flanges of an imaginary steel joist having equal area of both the flanges. However, there seems to be no justification in neglecting the compressive area of concrete and permitting the stress in compressive steel equal to stress in tensile steel, as the stress in concrete would pass the limit of the safe working stress. However, the method can be used to find the approximate value of the moment of resistance of the section specially when the area of compression steel is equal to or more than the area of the tensile steel and when these areas are within 2 to 3 percent of the cross-sectional area.

## 5.5. TYPES OF PROBLEMS IN DOUBLY REINFORCED BEAMS

There are three types of problems in doubly reinforced beams. The methods of solving these problems are practically similar to those of singly reinforced beams.

### Type 1. Determination of Moment of Resistance

In this type of problem, the dimensions of the section, including the area of tensile and compression reinforcement is completely known and it is required to determine the moment of resistance of the beam, corresponding to a given set of permissible stresses in concrete and steel. The procedure for solution is as under:

## DOUBLY REINFORCED BEAMS

1. Determine the position of the neutral axis by equating the moments of area of compressive concrete and the area of compressive steel expressed in equivalent concrete to the moment of the area of the concrete equivalent to the area of steel in tension, about the N.A.. Thus, from *Eq. 5.1*,

$$\frac{bn^2}{2} + (m_c - 1) A_{sc} (n - d_c) = mA_{st} (d - n)$$

or

$$\frac{bk^2 d}{2} + (m_c - 1) A_{sc} (k - e) = mA_{st} (1 - k)$$

From which either $n$ or $k$ can be determined.

2. Find the position of the critical neutral axis from the relation

$$n_c = \frac{mc}{mc + t} \cdot d = \frac{m\sigma_{cbc}}{m\sigma_{cbc} + \sigma_{st}} \cdot d$$

3. If the actual N.A. falls below the critical neutral axis, the stress $c$ in concrete reaches its maximum permissible value $\sigma_{cbc}$ first. If, however, the actual N.A. falls above the critical neutral axis, the stress $t$ in tensile steel reaches its maximum permissible value $\sigma_{st}$ first. From this value of $t$, calculate the value of $c$ from the stress diagram:

$$\frac{c}{n} = \frac{t/m}{d - n} \quad \text{or} \quad c = \frac{nt}{m(d - n)} = \frac{kt}{m(1 - k)}$$

Having known the value of $c$ in either case, find the value of $c'$ from the stress diagram:

*Eq. 5.7*
$$c' = c \frac{n - d_c}{n} = c \frac{k - e}{k}$$

4. Having known $c$ and $c'$, find the moment of resistance of the beam by taking the moment of all the forces about the tensile steel. Thus from *Eq. 5.3*.

$$M = bn \frac{c}{2} \left( d - \frac{n}{3} \right) + (m_c - 1) A_{sc} \cdot c' (d - d_c)$$

or

$$M = \frac{1}{2} cjkbd^2 + (m_c - 1) A_{sc} \cdot c' d (1 - e)$$

### Type II. Determination of Stresses in the Section

In this type of problem, the dimensions of the section are completely known, including the tensile and compressive reinforcement, and it is required to find the stresses developed in concrete, tensile reinforcement and compressive reinforcement, when subjected to a given bending moment. The solution is done in the following steps:

(1) Determine the position of the neutral axis as explained in previous case [*i.e.*, from *Eq. 5.1* or *5.1 (a)*].

(2) Determine the value of stress $c'$ in the concrete surrounding the compressive reinforcement, in terms of maximum stress $c$ in concrete, from the stress diagram. (*Eq. 5.7*)

$$c' = \frac{c(n - d_c)}{n} = c \frac{k - e}{k}$$

(3) Determine the moment of resistance of the beam by taking the moments of forces about the centre of the tensile steel, and equate it to the external moment $M$. Thus :

$$M = bn \cdot \frac{c}{2} \left( d - \frac{n}{3} \right) + (m_c - 1) A_{sc} \cdot c' (d - d_c)$$

or

$$M = bn \cdot \frac{c}{2} \left( d - \frac{n}{3} \right) + (m_c - 1) A_{sc} \cdot \left\{ \frac{c}{n} (n - d_c) \right\} (d - d_c)$$

This equation contains only one unknown *i. e.*, stress $c$ in concrete, which can be determined.

(4) Determine the stress $t$ in tensile steel from following relation:

$$t = m c \left( \frac{d - n}{n} \right)$$

Similarly, determine the stress $t_c$ in the compressive steel from *Eq. 5.7*:

$$t_c = m_c c' = \frac{1.5\,m \cdot c}{n}(n - d_c) \qquad \ldots(5.9)$$

### Type III. Design of the Section

This is more common type of problem in which it is required to determine the dimensions of the section, to develop a *given* moment of resistance corresponding to a given set of permissible stresses in concrete and steel. Generally, the size of the beam ($b$ and $d$) is known and it is required to determine the area of tensile and compressive reinforcement. The design is done in the following steps.

(1) Determine the position of the *critical neutral axis* of the section for the given set of stress:

$$n_c = \frac{mc}{mc + t}\,d = \frac{m\sigma_{cbc}}{m\sigma_{cbc} + \sigma_{st}} \cdot d$$

(2) Determine the area of tensile reinforcement $A_{st1}$ and the moment of resistance $M_1$ corresponding to the singly reinforced balanced section: from *Eq. 5.4*

$$M_1 = R_c\,bd^2 = \frac{1}{2}\,cjk \cdot bd^2 = bn_c\,\frac{c}{2}\left(d - \frac{n_c}{3}\right)$$

and

$$A_{st1} = \frac{M_1}{tjd} = \frac{M_1}{\sigma_{st} \cdot jd}$$

where $c = \sigma_{cbc}$ and $t = \sigma_{st}$.

**Note.** If the external moment $M$ is greater than $M_1$, the beam is designed as doubly reinforced beam. This test should always be applied before step 3.

(3) Determine the additional area of tensile reinforcement $A_{st2}$ for the remaining bending moment $M_2$ from *Eq. 5.5*.

$$A_{st2} = \frac{M_2}{t(d - d_c)} = \frac{M - M_1}{t(d - d_c)}$$

where $t = \sigma_{st}$ if $A_{st2}$ and $A_{st1}$ are provided at the same level.

Thus, the total area of tensile reinforcement will be equal to: $A_{st} = A_{st1} + A_{st2}$

(3) Determine the area of compressive steel $A_{sc}$ by equating the moments of equivalent areas of compressive steel ($A_{sc}$) and additional tensile area ($A_{st2}$). Thus from *Eq. 5.6*:

$$A_{sc} = \frac{m(d - n)}{(m_c - 1)(n - d_c)}\,A_{st2} = \frac{m(1 - k)}{(m_c - 1)(k - e)}\,A_{st2}$$

**Note.** If $A_{sc}$ comes out to be equal to or more than $A_{st}$, the beam should redesigned using steel beam theory.

**Example 5.1.** *A beam section, 300 mm wide and 560 mm deep is reinforced with 4 bars of 25 mm diameter in the tensile zone and 4 bars of 12 mm diameter in the compression zone. The cover to the centre of both the reinforcement is 40 mm. Determine the moment of resistance of the section, if M 20 concrete and HYSD bars are used.*

**Solution:** $\sigma_{cbc} = 7$ N/mm$^2$ : $\sigma_{st} = 230$ N/mm$^2$ ; $m = 13.33$ ; $m_c = 13.33 \times 1.5 \cong 20$

$$A_{sc} = 4 \times \frac{\pi}{4}(12)^2 = 452.4 \text{ mm}^2$$

and

$$A_{st} = 4 \times \frac{\pi}{4}(25)^2 = 1963.5 \text{ mm}^2$$

To determine the position of neutral axis

$$\frac{bn^2}{2} + (mc^{-1})A_{sc}(n - d_c) = m\,A_{st}(d - n)$$

$$\frac{300}{2}n^2 + (20 - 1) \times 452.4\,(n - 40) = 13.33 \times 1963.5\,(520 - n)$$

or $\qquad n^2 + 231.8\,n - 93027 = 0$

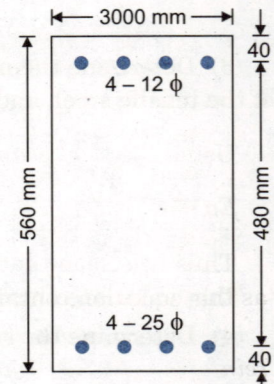

Fig. 5.2

From which $n = 210.4$ mm. The critical N.A. is given by

$$n_c = \frac{m\sigma_{cbc}}{m\sigma_{cbc} + \sigma_{st}} d = \frac{13.33 \times 7}{13.33 \times 7 + 230} \times 520 = 150.1 \text{ mm}$$

Since the actual N.A. is below the critical N.A., the stress in concrete will reach its maximum permissible value first.

$$\therefore \quad c = \sigma_{cbc} = 7 \text{ N/mm}^2$$

$$c' = c \frac{n - d_c}{n} = 7 \times \frac{210.4 - 40}{210.4} = 5.669 \text{ N/mm}^2$$

$$M_r = 300 \times 210.4 \times \frac{7}{2}\left(520 - \frac{210.4}{3}\right) + (20 - 1) \times 452.4 \times 5.669\,(520 - 40)$$

$$= 122.77 \times 10^6 \text{ N-mm}$$

$$= \mathbf{122.77 \text{ kN-m}}$$

**Note** $t_c = m_c \, c' = 20 \times 5.669 = 113.34$ N/mm² which is less than the permissible value $\sigma_{sc}$ (= 190 N/mm²).

**Example 5.2.** *A doubly reinforced rectangular beam of breadth b and effective depth d has tensile reinforcement equal to 1.2 percent and compressive reinforcement also 1.2 percent. The compressive reinforcement is at 0.1d below the top. Determine the moment of resistance of the section if the permissible stresses in concrete and steel are 5 N/mm² and 140 N/mm² respectively, and m = 19. What will be the moment of resistance by the steel beam theory?*

**Solution:** Given : $\sigma_{cbc} = 5$ N/mm² ; $\sigma_{st} = 140$ N/mm² ; $m = 19$

$$\frac{bn^2}{2} + (1.5 \times 19 - 1)\frac{1.2}{100} bd\,(n - 0.1\,d) = 19 \times \frac{1.2}{100} bd\,(d - n)$$

or $\qquad n^2 + 1.116\,n\,d - 0.522\,d^2 = 0$ which gives $n = 0.355\,d$.

$$n_c = \frac{m\sigma_{cbc}}{m\sigma_{cbc} + \sigma_{st}} d = \frac{19 \times 5}{19 \times 5 + 140} d = 0.404\,d$$

The actual N.A. is above the critical one ; hence the stress in tensile steel reaches its maximum value of 140 N/mm² before the stress in concrete can reach the value of 5 N/mm². Corresponding to $t = 140$ N/mm², $c$ and $c'$ can be found as under :

$$c = \frac{t}{m} \cdot \frac{n}{d-n} = \frac{140}{19} \times \frac{0.355\,d}{0.645\,d} = 4.06 \text{ N/mm}^2$$

$$c' = \frac{t}{m} \cdot \frac{n - d_c}{d-n} = \frac{140}{19}\left(\frac{0.355 - 0.1}{0.645}\right) = 2.91 \text{ N/mm}^2.$$

$$M_r = b \times 0.355\,d \times \frac{4.06}{2}\left(d - \frac{0.355\,d}{3}\right) + (1.5 \times 19 - 1) \times \frac{1.2\,bd}{100} \times 2.91\,(d - 0.1\,d)$$

$$= 0.653\,bd^2 + 0.864\,bd^2 = \mathbf{1.517\,bd^2 \text{ N-mm}} \quad \text{(where } b \text{ and } d \text{ are in mm)}$$

Using the *steel beam theory*

Eq. 5.3 $\qquad M_r = A_{st} \cdot \sigma_{st}\,(d - d_c) = \dfrac{1.2\,bd}{100} \times 140\,(d - 0.1\,d) = \mathbf{1.512\,bd^2 \text{ N-mm}}$

Thus it is seen that the moment of the resistance found by the approximate theory is nearly the same as that calculated by the exact theory, the difference being 0.3%.

**Example 5.3.** *A doubly reinforced concrete beam is 400 mm wide and 600 mm deep to the centre of tensile reinforcement. The compression reinforcement consists of 4 bars of 16 mm diameter, and is placed with its centre at a depth of 40 mm from the top. The tensile reinforcement consists of 4 bars of 20 mm diameter. The* section is subjected to a bending moment of 100 kN-m. Determine the stresses in concrete and steel. Take m = 16.

**Solution:** Area of steel in compression,

$$A_{sc} = 4 \times \frac{\pi}{4}(16)^2 = 805 \text{ mm}^2$$

Area of steel in tension, $A_{st} = 4 \times \frac{\pi}{4}(20)^2 = 1257 \text{ mm}^2$

Taking moment of equivalent areas, about N.A.

$$\frac{bn^2}{2} + (m_c - 1) A_{sc} (n - d_c) = m A_{st} (d - n)$$

or $\dfrac{400}{2} n^2 + (1.5 \times 16 - 1) 805 (n - 40) = 16 \times 1257 (600 - n)$

or $n^2 + 193.35 n - 64039 = 0$

From which n = 174.3 mm

$$c' = \frac{n - d_c}{n} c = \frac{174.3 - 40}{174.3} c = 0.77 c$$

Now, $M_r = b n \cdot \dfrac{c}{2}\left(d - \dfrac{n}{3}\right) + (m_c - 1) A_{sc} \cdot c' (d - d_c)$

$$100 \times 10^6 = 400 \times 174.3 \; \frac{c}{2}\left(600 - \frac{174.3}{3}\right) + (1.5 \times 16 - 1) 805 \times 0.77c (600 - 40)$$

From which **c = 3.73 N/mm²**; $c' = 0.77 c = 0.77 \times 3.73 = 2.87$ N/mm²

The stress in compressive steel = $1.5 mc' = 1.5 \times 16 \times 2.87 =$ **68.9 N/mm²**

Stress in tensile steel is $t = \dfrac{cm(d-n)}{n} = \dfrac{3.73 \times 16 (600 - 174.3)}{174.3} =$ **145.7 N/mm²**

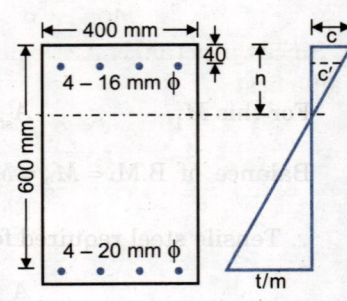

Fig. 5.3

**Example 5.4.** *A doubly reinforced rectangular beam is 240 mm wide and 500 mm deep. If the limiting stresses in concrete and steel are 5 N/mm² and 230 N/mm² respectively determine the steel areas for bending moment of 80 kN-m. Assume that steel is buried on both faces with its centre 40 mm from either face. Take m = 19.*

**Solution:** Given $c = \sigma_{cbc} = 5$ N/mm² ; $t = \sigma_{st} = 230$ N/mm²

$$n = n_c = \frac{mc}{mc + t} d = \frac{19 \times 5}{19 \times 5 + 230} \times 460 = 134.5 \text{ mm.}$$

$$c' = \frac{n - d_c}{n} \cdot c = \frac{134.5 - 40}{134.5} \times 5 = 3.513 \text{ N/mm}^2.$$

$$M_r = b \cdot n \cdot \frac{c}{2}\left(d - \frac{n}{3}\right) + (m_c - 1) A_{sc} \cdot c' (d - d_c)$$

or $80 \times 10^6 = 240 \times 134.5 \times \dfrac{5}{2}\left(460 - \dfrac{134.5}{3}\right) + (1.5 \times 19 - 1) A_{sc} \times 3.513 (460 - 40)$

From which $A_{sc} =$ **1146 mm²**.

Now, total compression = total tension

∴ $bn \dfrac{c}{2} + (m_c - 1) A_{sc} \cdot c' = A_{st} \cdot t$

or $240 \times 134.5 \times \dfrac{5}{2} + (1.5 \times 19 - 1) \times 1146 \times 3.513 = A_{st} \times 230$

From which $A_{st} =$ **832 mm²**

**Alternative solution.** $n = n_c = 134.5$ mm, as found above.

Lever arm $a = d - \dfrac{n}{3} = 460 - \dfrac{134.5}{3} = 415.2$ mm.

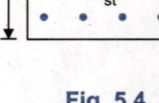

Fig. 5.4

Now $M_1$ for singly reinforced section

$$= b.n\frac{c}{2}\left(d - \frac{n}{3}\right) = 240 \times 134.5 \times \frac{5}{2}\left(460 - \frac{134.5}{3}\right)$$

$$= 33.5 \times 10^6 \text{ N-mm} = 33.5 \text{ kN-m}$$

For this $M_1$, $\quad A_{st1} = \dfrac{M_1}{t \times a} = \dfrac{33.5 \times 10^6}{230 \times 415.2} = 350.8 \text{ mm}^2.$

Balance of B.M. = $M_2 = M - M_1 = (80 - 33.5) \, 10^6 = 46.5 \times 10^6$ N-mm.

∴ Tensile steel required for this B.M. is given by

$$A_{st2} = \frac{M_2}{t(d - d_c)} = \frac{46.5 \times 10^6}{230(460 - 40)} = 481.4 \text{ mm}^2.$$

Compression steel to balance this is given by *Eq. 5.6*

$$A_{sc} = \frac{m(d - n)}{(m_c - 1)(n - d_c)} \times A_{st2} = \frac{19(460 - 134.5)}{(1.5 \times 19 - 1)(134.5 - 40)} \times 481.4 = \mathbf{1145.6 \text{ mm}^2}.$$

Total $\quad A_{st} = A_{st1} + A_{st2} = 350.8 + 481.4 = \mathbf{832.2 \text{ mm}^2}.$

**Note.** It is found here that $A_{sc} > A_{st}$. Such a design is uneconomical because the HYSD bars ($A_{sc}$) provided in compression zone are highly under-stressed. For such a circumstance, it is better to re-design the beam, using steel beam theory, by providing equal areas of steel in tension and compression zones. Thus,

$$A_{sc} = A_{st} = \frac{M}{\sigma_{st}(d - d_c)} = \frac{80 \times 10^6}{230(460 - 40)} = 828 \text{ mm}^2.$$

Thus, total steel area = 828 + 828 = 1656 mm², against a total area of 1145.6 + 832.2 = 1977.8 mm² found by balanced design. The reader may now find the actual stresses developed in concrete, compression steel and tensile steel, with the above provision of equal steel areas.

**Example 5.5.** *A reinforced concrete section is subjected to a reversal of bending moment of equal magnitude of 120 kN-m in either direction. Design the section if the permissible stresses in concrete and steel are 6 N/mm² and 140 N/mm² respectively, and m = 15. Assume b = 0.6 d and effective cover to steel equal to 0.1 d.*

**Solution:** Given $c = \sigma_{cbc} = 5$ N/mm² and $t = \sigma_{st} = 140$ N/mm².

Since the moment can act in either direction (*i.e.*, hogging as well as sagging), we will have to provide equal reinforcement at the top as well as bottom. Hence $A_{sc} = A_{st} = A_s$ (say).

Now $\quad n = n_c = \dfrac{mc}{mc + t} d = \dfrac{15 \times 6}{15 \times 6 + 140} d = 0.391 d$

Equating the moment of equivalent areas about the N.A., we get

$$\frac{bn^2}{2} + (m_c - 1) A_{sc}(n - d_c) = m A_{st}(d - n)$$

∴ $\quad \dfrac{0.6 \, d \, (0.391 d)^2}{2} + (1.5 \times 15 - 1) A_s (0.391 - 0.1) d = 15 A_s (d - 0.391 d)$

or $\quad 0.046 \, d^2 + 6.26 \, A_s = 9.14 \, A_s$

∴ $\quad A_s = A_{sc} = A_{st} = 0.016 \, d^2 \text{ (mm}^2\text{)}$ ...(1)

Now $\quad M_r = bn\dfrac{c}{2}\left(d - \dfrac{n}{3}\right) + (m_c - 1) A_{sc} \cdot c'(d - d_c)$

where $\quad c' = \dfrac{n - d_c}{n} c = \dfrac{0.391 - 0.1}{0.391} \times 6 = 4.46$ N/mm²

$A_{sc} = 0.016 \, d^2$ mm² ; $M_r = M = 120 \times 10^6$ N-mm

∴ $120 \times 10^6 = 0.6\,d \times 0.391 d \times \dfrac{6}{2}\left(d - \dfrac{0.391}{3}d\right) + (1.5 \times 15 - 1) \times 0.016\,d^2 \times 4.46\,(d - 0.1\,d)$

or $120 \times 10^6 = 0.612\,d^3 + 1.381\,d^3 = 1.993\,d^3$

From which $\qquad d \approx \mathbf{392\ mm}$. Hence, $b = 0.6 \times 392 = \mathbf{235\ mm}$

$\qquad A_{sc} \approx A_{st} = 0.016\,(392)^2 = \mathbf{2460\ mm^2}$ ; Cover to steel $= 0.1\,d \approx \mathbf{40\ mm}$.

**Example 5.6.** *Determine the moment of resistance of a doubly reinforced beam of trapezoidal section shown in Fig. 5.5. Take $\sigma_{cbc} = 5\ N/mm^2$, $\sigma_{st} = 140\ N/mm^2$ and $m = 18$.*

**Solution:** Let $n$ be the depth of N.A. To get the position of N.A., take the moments of the equivalent areas about the N.A. The concrete area consisting of trapezium $abdc$ can be divided into the rectangle $abhg$ and two triangles $acg$ and $bhd$ [Fig. 5.5(b)].

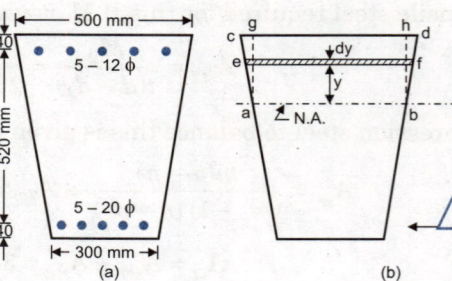

Fig. 5.5

Width $\qquad ab = 500 - \dfrac{200}{600}n = 500 - \dfrac{n}{3}$

Width $\qquad cg = hd = \dfrac{1}{6}n$

$\qquad A_{sc} = 5 \times \dfrac{\pi}{4}(12)^2 = 565.5\ mm^2$

$\qquad A_{st} = 5 \times \dfrac{\pi}{4}(20)^2 = 1570.8\ mm^2$

Equating the moment of equivalent areas about the N.A., we get

$n\left[500 - \dfrac{n}{3}\right]\dfrac{n}{2} + \dfrac{1}{2}n \times \dfrac{n}{3} \times \dfrac{2n}{3} + (m_c - 1)A_{sc}\,[n - d_c] = m \times A_{st}\,(d - n)$

$\qquad 250\,n^2 - \dfrac{n^3}{18} + (1.5 \times 18 - 1) \times 565.5\,[n - 40] = 18 \times 1570.8 \times (560 - n)$

or $\qquad n^3 - 4500\,n^2 - 773593\,n + 2.956 \times 10^8 = 0$

Solving this by trial and error, we get, $n = 187$ mm. For the given stresses, the position of critical N.A. is given by : $n_c = \dfrac{m\,\sigma_{cbc}}{m\,\sigma_{cbc} + \sigma_{st}}\,d = \dfrac{18 \times 5}{18 \times 5 + 140} \times 560 = 219$ mm.

Since the actual N.A. is above the critical N.A. the stress in tensile steel will reach its maximum permissible value first. Thus $t = \sigma_{st} = 140\ N/mm^2$.

The corresponding stress in concrete is given by

$\qquad c = \dfrac{t}{m}\dfrac{n}{d - n} = \dfrac{140}{18} \times \dfrac{187}{560 - 187} = 3.9\ N/mm^2$

The stress in concrete surrounding compressive steel is given by

$\qquad c' = \dfrac{n - d_c}{n} \times c = \dfrac{187 - 40}{187} \times 3.9 = 3.07\ N/mm^2$

The moment of resistance, $M_r$ of the section is found by taking the moment of the compressive forces about the centre of tensile steel.

Let $\qquad M_{r1}$ = moment of compressive force in concrete, about the centre of tensile steel,

$\qquad M_{r2}$ = moment of compressive force in steel, about the centre of tensile steel,

so that $\ M_r = M_{r1} + M_{r2}$

To find $M_{r1}$, consider an elementary strip of thickness $dy$ at a distance $y$ from the N.A. [Fig. 5.5(b)] and let the stress at the level be $c_y$.

$$\text{Breadth of strip } ef = b_y = 500 - \frac{500-300}{600}[n-y] = 500 - \frac{1}{3}[187-y]$$

$$\text{Stress } c_y = c \times \frac{y}{n} = \frac{3.9}{187}y = 0.0209\, y$$

Force in elementary strip $= b_y \cdot dy \cdot c_y$

Moment of this force about the centre of steel area is given by

$$dM_{r1} = b_y \cdot dy \cdot c_y (d-n+y)$$

$$\therefore \quad M_{r1} = \int_0^n b_y \cdot c_y (d-n+y)\, dy$$

$$= \int_0^{187} \left\{500 - \frac{1}{3}(187-y)\right\} 0.0209\, y\, (560-187+y)\, dy$$

$$= \frac{0.0209}{3}\int_0^{187}(y^3 + 1686\, y^2 + 489749\, y)\, dy$$

$$= \frac{0.0209}{3}\left[\frac{y^4}{4} + \frac{1686}{3}y^3 + \frac{489749}{2}y^2\right]_0^{187}$$

$$= 87.39 \times 10^6 \text{ N-mm.}$$

Moment of compressive forces in the compression reinforcement is given by

$$M_{r2} = (m_c - 1)A_{sc} \cdot c'(d - d_c)$$
$$= (1.5 \times 18 - 1)\, 565.5 \times 3.07\,(560 - 40) = 23.47 \times 10^6 \text{ N-mm.}$$

Hence total moment of resistance is given by

$$M_r = M_{r1} + M_{r2} = 87.39 \times 10^6 + 23.47 \times 10^6 \text{ N-mm}$$

$$= 110.86 \times 10^6 \text{ N-mm} = \mathbf{110.86 \text{ kN-m.}}$$

## 5.6. SHEAR STRESS, BOND STRESS AND DEVELOPMENT LENGTH

### (a) Maximum Shear Stress

Fig. 5.6 (b) shows the shear stress distribution in a doubly reinforced beam. The distribution of the shear stress at any plane, distant $y$ above the N.A. is given by $q = \frac{VA\bar{y}}{bI}$. There will be sudden increase in the shear stress below the level of the compression steel.

This is so because the quantity $A\bar{y}$ suddenly increases for all planes below the level compression steel.

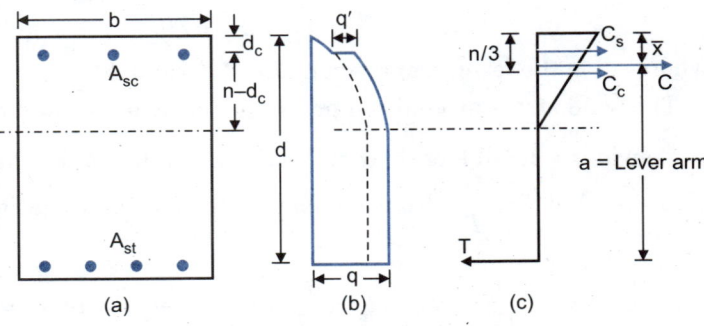

Fig. 5.6. Shear Stress Distribution.

Let $q'$ be the increase in the shear stress just below the level of the compression steel. Then

$$q' = \frac{V}{bI} \times (\text{increase in } A\bar{y}) = \frac{V}{bI} \times (m_c - 1)A_{sc}(n - d_c) \qquad \dots(5.10)$$

where $I$ is the moment of inertia of the equivalent section.

For all planes below the N.A., the shear stress will be maximum and of a constant magnitude given by

$$q = \frac{V}{ba} \qquad \ldots(5.11)$$

where  $a$ = Lever arm = distance between tensile force and the centre of the compressive force.

Fig. 5.6 (c) shows the tensile and compressive forces.

Let  $C_s$ = compressive force in steel, acting at $d_c$ below the top of the beam.

$C_c$ = compressive force in concrete, acting at $n/3$ below the top of the beam.

$C$ = resultant compressive force acting at $\bar{x}$ below the top of the beam.

Now  $C_s = (m_c - 1) A_{sc} \cdot \dfrac{n - d_c}{n} c$ ; $C_c = b \times n \times c/2$

$$\therefore \quad \bar{x} = \frac{C_s \times d_c + C_c \cdot \dfrac{n}{3}}{C_s + C_c} = \frac{(m_c - 1) A_{sc} \cdot \dfrac{n - d_c}{n} \cdot c \cdot d_c + bn \dfrac{c}{2} \times \dfrac{n}{3}}{(m_c - 1) A_{sc} \cdot \dfrac{n - d_c}{n} \cdot c + \dfrac{bnc}{2}}$$

$$\therefore \quad a = d - \bar{x} = d - \frac{\dfrac{bn^2}{6} + (m_c - 1) A_{sc} \cdot \dfrac{n - d_c}{n} d_c}{\dfrac{bn}{2} + (m_c - 1) A_{sc} \cdot \dfrac{n - d_c}{n}} \qquad \ldots(5.12)$$

*Alternatively*, lever arm can also be found from the following relation if $M$ and $A_{st}$ are known. Thus $M = t \cdot A_{st} \cdot a$ from which $a = \dfrac{M}{t \, A_{st}} = \dfrac{M}{\sigma_{st} \cdot A_{st}}$  ...(5.13)

### (b) Nominal Shear Stress

IS Code (IS : 456–2000) recommends the use of *nominal stress* in the beam, given by

$$\tau_v = \frac{V}{bd} \qquad \ldots(5.14)$$

### (c) Flexural Bond Stress

In the case of doubly reinforced beam, flexural bond stress is determined both in the tensile steel as well as the compressive steel, though the bond stress around the compressive steel will be of smaller magnitude. The bond stress around tensile steel may be found by *Eq. 3.28*.

$$\tau_{bf} = \frac{V}{j \, d \, \Sigma 0} = \frac{V}{a \Sigma 0} \qquad \ldots(5.15)$$

where  $a$ is the lever arm given by *Eq. 5.13* or *5.14*.

The bond stress around compression steel can be found from expression derived below.

Consider a small length $\delta x$ of the beam and let $\delta M$ be the change in the bending moment in the length. Knowing that $p = \dfrac{M}{I} y$ from elementary theory, the variation in the stress $\delta t_c$ in compression steel is given by

$$\delta t_c = \frac{\delta M}{I} (n - d_c) m_c \text{ where } m_c = 1.5 \, m$$

$\therefore$ Variation in force ($\delta C_s$) in compression steel, in a length $\delta x$ is given by

$$\delta C_s = \frac{\delta M}{I} (n - d_c) m_c \cdot A_{sc}$$

Let  $\tau'_{bf}$ = bond stress around compression steel.

$\Sigma 0'$ = sum of the perimeters of the steel bars of the compression reinforcement.

Then $\tau_{bf}' \delta x \times \Sigma 0' = \dfrac{\delta M}{I} (n - d_c) m_c . A_{sc}$

or $\tau_{bf}' = \dfrac{\delta M}{\delta x} . \dfrac{n - d_c}{I} . \dfrac{m_c . A_{sc}}{\Sigma 0'} = \dfrac{F}{I \Sigma 0'} (n - d_c) m_c A_{sc}$ ...(5.16)

where $I = \dfrac{bn^3}{3} + (m_c - 1) A_{sc} (n - d_c)^2 + m . A_{st} (d - n)^2$.

### (d) Development Length

*Eqs. 5.12* and *5.13* give the localized bond stress at a *particular section*. However, bond stresses are *developed* over a specified length and any attempt to visualise bond on the basis of stress at a particular cross-section is bound to be erroneous. In the case of bond, it has been recently concluded that the concept of *anchorage bond* is more meaningful than the concept of flexural bond stress of localised nature. *Current design methods disregard such local bond stresses even though they may result in localised slip between steel and concrete adjacent to the cracks.* Thus the concept of *development length* has gained prominence.

As per IS : 456 – 2000, the development length for tensile reinforcement is given by

$$(L_d)_{tension} = \dfrac{\Phi \sigma_s}{4 \tau_{bd}} \qquad \text{...(5.17)}$$

where $\sigma_s$ = stress in bar at the section considered at the design load.

For bars in compression, the permissible bond stress is increased by 25%. Hence

$$(L_d)_{comp.} = \dfrac{\Phi . \sigma_{sc}}{5 \tau_{bd}} \qquad \text{...(5.18)}$$

where $\sigma_{sc}$ = *compressive stress* in the bars, at the section considered, at the design load.

## PROBLEMS

1. A doubly reinforced rectangular beam is 300 mm wide and 450 mm deep and is subjected to a bending moment of 90 kN-m. If the limiting stresses in concrete and steel are 5 N/mm² and 140 N/mm², determine the steel areas. Assume that steel is buried on both faces, with its centre 40 mm from either face. Take $m = 18$.

2. A rectangular beam of breadth $b$ and effective depth $d$ is reinforced with 1.2% tensile and 1.2% compressive steel. Determine the moment of resistance of the section (a) by accurate method and (b) by steel beam theory, neglecting concrete. Take compression reinforcement to be placed at 0.1 effective depth below top. Take the permissible stress in steel and concrete as 140 N/mm² and 5.2 N/mm² respectively, and $m = 18$.

3. A beam 250 mm × 500 mm in section is reinforced with 3 bars of 14 mm diameter at top and 5 bars of 20 mm diameter at the bottom, each at an effective cover of 40 mm. Determine the moment of resistance of the beam section. Take the permissible stress in steel and concrete as 126 N/mm² and 5.2 N/mm² respectively, and $m = 18$.

4. A doubly reinforced concrete beam is 250 mm wide and 500 mm deep from the compression edge to the centre of tensile steel. The area of compression and tensile steel are both 1300 mm² each. The centre of compression steel is 50 mm from the compression edge. If the beam is subjected to a total bending moment of 70 kN-m, determine the stress in concrete, and tension and compression steel. Take $m = 18$.

5. Determine the moment of resistance of a doubly reinforced beam of trapezoidal section shown in Fig. 5.7. What will be the moment of resistance using steel beam theory? Take permissible stress in concrete and steel as 5 N/mm² and 140 N/mm² and $m = 18$.

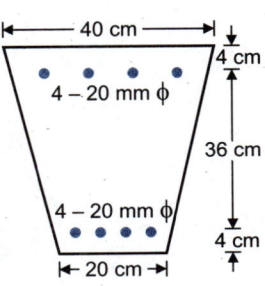

Fig. 5.7

6. A pile 300 mm × 300 mm is reinforced with 4 bars of 20 mm diameter with concrete cover of 40 mm over the centre of steel. When being hauled up to the frame, it is subjected to a bending moment of 25 kN-m. Determine the maximum stresses set up in concrete and steel. Take $m = 18$.

7. An RCC beam 230 × 400 mm overall is reinforced with four 22 mm $\phi$ bars with centres 30 mm from the bottom edge and three 20 mm $\phi$ bars with centres 25 mm from top edge. Find the neutral axis of the beam if ($m = 18.66$).

## ANSWERS

1. $A_{st} = 1770$ mm$^2$; $A_{sc} = 1300$ mm$^2$
2. (a) $1.594\ bd^2$ N-mm  (b) $1.598\ bd^2$ N-mm
3. 73.5 kN-m.
4. $c = 3.56$ N/mm$^2$, $t = 123.8$ N/mm$^2$, $t_c = 68.3$ N/mm$^2$
5. (a) 56.7 kN-m  (b) 56.2 kN-m
6. $c = 3.19$ N/mm$^2$; $t = 88.9$ N/mm$^2$
7. 200 mm

# CHAPTER 6

# DESIGN OF T AND L-BEAMS

## 6.1. INTRODUCTION

In RCC construction, beams and slabs are generally cast at one time. Hence the construction becomes monolithic.

The slab portion is called 'Flange' and the beam 'Web'. Such a joint action beam of flange and web looks like a letter 'T' and hence it is known as 'T-beam'.

We have seen that the concrete below the N.A. does not resist any bending moment but simply serves to embed the tensile steel. Also, the portion of the concrete just above the N.A. carries only very little compressive force since the intensity of compressive stress there is of a very small magnitude. This suggests that the section of the beam should be such that it has greater width at the top (compression side) in comparison to the width below the neutral axis. Such a section is known as a T-section as shown in Fig. 6.1 [$a(ii)$].

In actual practice, a beam of T-section, known as T-beam is more common than the rectangular section. Fig. 6.1 ($b$) shows the plan of a big size room. The roof system of the room may be of two types:

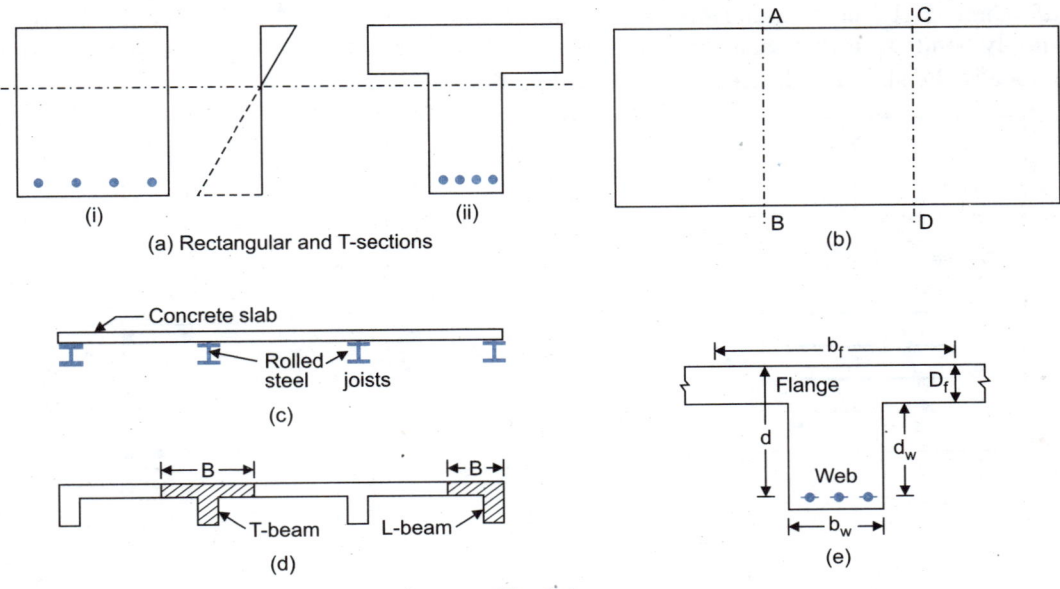

Fig. 6.1

(i) a number of beams of rolled steel section with a concrete roof slab, both acting independently, as shown in Fig. 6.1 (c) or

(ii) a number of R.C. beams with a roof slab acting monolithically [Fig. 6.1 (d)]. The second system is more common because in this system a portion of the slab of width B acts monolithically with the rectangular section of the beam, thus giving rise to T-beam. In fact, the slab forms the compression flange of the T-beam, while the comparatively narrow rectangular section forms the web of the T-beam.

In such a system, the roof slab is designed to bend over the span BD and the reinforcement for positive and negative bending moment is provided in the direction at right angles to the direction AB of the beam. The beam, on the other hand, takes uniformly distributed load from the slab, and bends over the span AB. The slab, which forms the upper part of the T-beam, is stressed laterally in compression, but this does not reduce its capacity to carry longitudinal compression as a part of the beam.

## 6.2. DIMENSIONS OF A T-BEAM

Fig. 6.1 (e) shows the following four important dimensions of a T-beam section:

(i) Breadth of the flange ($b_f$)  (ii) Thickness of the flange ($D_f$)
(iii) Breadth of the rib ($b_w$) and  (iv) Depth of the rib ($d_w$).

**(1) Breadth of the Flange ($b_f$):** The breadth of the flange of the T-beam is that portion of the slab which acts monolithic with the beam and which resists the compressive stresses.

According to Indian standard code (IS : 456-2000), a slab which is assumed to act as a flange of a T-beam (or of a L-beam) shall satisfy the following:

(a) The slab be cast integrally with the web, or the web and the slab shall be effectively bonded together in any other manner, and

(b) If the main reinforcement of the slab is parallel to the beam, transverse reinforcement shall be provided as shown in Fig. 6.2. Such reinforcement shall not be less than 60% of the main reinforcement at the mid-span of the slab.

It is to be noted that in normal circumstances, slab spans across T-beams, whereby the main reinforcement in the slab (i.e. flange of T-beam) run at right angles to the main span of the beam. This main reinforcement helps in bonding the slab and beam together, thus ensuring monolithic action between the slab and the beam. However, some times, the slab spans parallel to the span of the main beam, with the result that the main reinforcement ($A_{st}$) of the slab runs parallel to the main beam. The slab (i.e. flange) and the beam are not adequately bonded. In this case, therefore, the code recommends to provide transverse reinforcement at the top face of the slab (i.e. flange) as shown in Fig. 6.2.

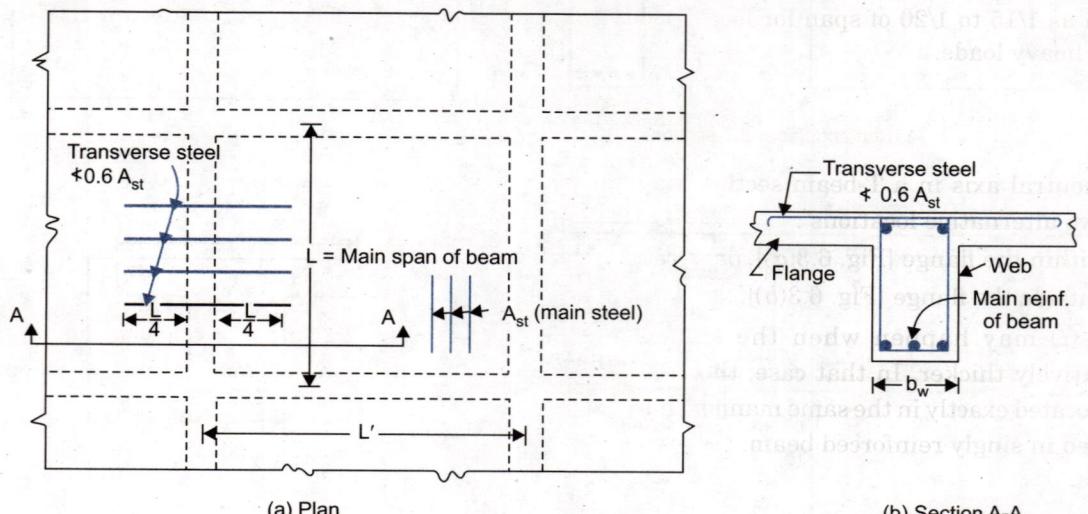

**Fig. 6.2.** Provision of Transverse Steel for Flange Action

In absence of more accurate determinations, the effective width $b_f$ of the flange may be taken as the least of the following (IS 456 : 2000):

(a) $\qquad b_f = \dfrac{l_0}{6} + b_w + 6 D_f$

(b) $\qquad b_f = b_w + \dfrac{1}{2}$ (sum of clear distances to the adjacent beams on either side)

(c) for isolated beams, the flange shall be obtained as below but in no case greater than actual width:

$$b_f = \dfrac{l_0}{\left(\dfrac{l_0}{b}\right) + 4} + b_w$$

where  $b_f$ = effective width of flange
 $l_0$ = distance between points of zero moments in the beam
 $b_w$ = breadth of web
 $D_f$ = thickness of flange and
 $b$ = actual width of flange.

**Note :** For continuous beams and frames, $l_0$ may be assumed as 0.7 times the effective span.

**2. Thickness of the Flange ($D_f$)** : The thickness of the flange is taken equal to the total thickness of the slab, including cover. The slab spans in a direction transverse to the span of the beam, and, therefore, the thickness of the slab is determined on the basis of its bending in transverse direction. As far as design of the T-beam is concerned, $D_f$ is the fixed dimension known from the considerations of bending etc. of the slab.

**3. Breadth of the Rib *or* Web ($b_w$)** : The breadth of the rib is equal to width of the portion of the beam in the tensile zone. This width should be sufficient to accommodate the tensile reinforcement with proper spacing between the bars. The width of the beam should also be sufficient to provide lateral stability ; from this point of view, it should atleast be equal to one-third of the height (depth) of the rib. From architectural point of view, the breadth of the rib should be equal to the width of the column supporting the beam.

**4. Depth of Rib *or* Web ($d_w$)** : The depth of rib is the vertical distance between the bottom of the flange and the centre of the tensile reinforcement, and is dependent on the effective depth ($d$) of the beam. The effective depth ($d$) of the T-beam is the distance between the top of the flange and the centre of the tensile reinforcement. Thus $d = d_w + D_f$. The assumed overall depth of the T-beam may be taken as 1/12 to 1/15 of the span when it is simply supported at the ends. When it is continuous, the assumed overall depth may be taken as 1/15 to 1/20 of span for light loads, 1/12 to 1/15 of span for medium loads and 1/10 to 1/12 of span for heavy loads.

## 6.3  POSITION OF NEUTRAL AXIS

The neutral axis in a T-beam section may fall in two alternative locations :

(i) within the flange [Fig. 6.3(a)], or

(ii) outside the flange [Fig. 6.3(b)].

Case (i) may happen when the slab is comparatively thicker. In that case, the N.A. may be located exactly in the same manner that is adopted in singly reinforced beam:

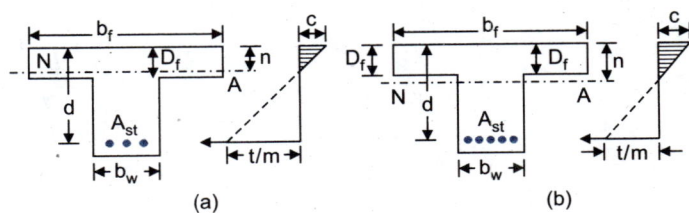

Fig. 6.3. Location of Neutral Axis.

$$\dfrac{b_f\, n^2}{2} = m\, A_{st}\, (d - n) \qquad \ldots(6.1)$$

If, however, the N.A. falls exactly at the junction of the flange and rib, the above equation reduces to:

$$\frac{b_f D_f^2}{2} = m A_{st} (d - D_f) \qquad \ldots(6.2)$$

In the second position, which is the most common one, the N.A. falls outside the flange. In that case, N.A. can be located by equating the moments of compressive and tensile areas about the N.A.:

$$b_f \cdot D_f \left(n - \frac{D_f}{2}\right) + \frac{b_w (n - D_f)^2}{2} = m A_{st} (d - n) \qquad \ldots(6.3)$$

In the above equation, the term $\dfrac{b_w (n - D_f)^2}{2}$ is usually very small and it is customary to reject this term. In that case, the above expression reduces to:

$$b_f D_f \left(n - \frac{D_f}{2}\right) = m A_{st} (d - n) \qquad \ldots(6.4)$$

Thus, if the dimensions of the T-beam section are completely known, the N.A. can be located either by *Eq. 6.1* or by *Eq. 6.4*. If by application of *Eq. 6.1*, $n$ comes out to be more than $D_f$, calculations should be repeated by using *Eq. 6.4*. Similarly, if *Eq. 6.4* is used first and $n$ comes out to be less than $D_f$ calculations should be repeated by using *Eq. 6.1*.

If, however, the maximum stresses in concrete ($c = \sigma_{cbc}$) and steel ($t = \sigma_{st}$) are *known*, the N.A. is located by using the stress diagram:

$$\frac{n}{d - n} = \frac{c}{t/m} \qquad \ldots(6.5)$$

## 6.4. LEVER ARM AND MOMENT OF RESISTANCE

(*a*) **Lever Arm:** Fig. 6.4 (*b*) shows the stress diagram for a T-beam section in which the N.A. falls outside the flange, which is more general case.

Let $c$ be the stress in the concrete at the top (extreme fibre) and $c_1$ be the stress in the concrete at the junction of flange with the rib. Since the compression area lying between the junction and the N.A. is very small, the compressive force of this area is *neglected*. Therefore, the shaded area in Fig. 6.4(*b*) shows the net compressive stress diagram, the centroid of which is at a depth $\bar{y}$ (say) below the top. The value of $\bar{y}$ can be found by dividing the trapezoidal stress diagram into two triangles by a diagonal (shown dotted) and taking moments at the top:

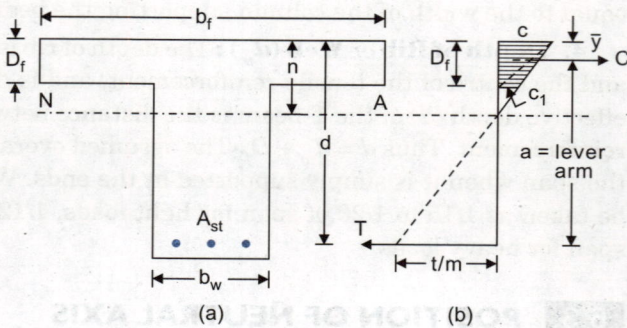

Fig. 6.4. Lever Arm.

$$\bar{y} = \frac{\left[c \cdot \dfrac{D_f}{2} \cdot \dfrac{D_f}{3}\right] + \left[c_1 \cdot \dfrac{D_f}{2} \cdot \dfrac{2}{3} D_f\right]}{\left[c \cdot \dfrac{D_f}{2}\right] + \left[c_1 \dfrac{D_f}{2}\right]} = \left[\frac{c + 2c_1}{c + c_1}\right] \frac{D_f}{3} \qquad \ldots(6.6)$$

But from stress diagram, $\dfrac{c_1}{n - D_f} = \dfrac{c}{n}$ or $c_1 = \left(\dfrac{n - D_f}{n}\right) c \qquad \ldots(6.7)$

Substituting this value of $c_1$ in *Eq. 6.6*, we get

$$\bar{y} = \frac{1 + 2\left\{\dfrac{n - D_f}{n}\right\}}{1 + \left\{\dfrac{n - D_f}{n}\right\}} \times \frac{D_f}{3} = \frac{3n - 2D_f}{2n - D_f} \times \frac{D_f}{3} \qquad \ldots[6.7(a)]$$

Lever arm
$$a = d - \bar{y} = d - \frac{D_f}{3}\left\{\frac{3n - 2D_f}{2n - D_f}\right\}$$

$$= d - \frac{D_f}{3}\left\{\frac{3}{2} - \frac{D_f}{2(2n - D_f)}\right\}$$

or
$$a = d - \frac{D_f}{2} + \frac{D_f^2}{6(2n - D_f)} \qquad \ldots(6.8)$$

If the actual N.A. is above the critical N.A., the section is under-reinforced, the stress ($t$) in steel reaches its maximum permissible value $\sigma_{st}$ first and the stress $c$ in concrete should be found from *Eq. 6.5* first, before substituting *Eq. 6.6*. If, however, the actual N.A. falls at or below the critical N.A., the section is over-reinforced and the stress ($c$) in concrete reaches its maximum permissible value ($\sigma_{cbc}$) first.

**(b) Moment of Resistance:** The moment of resistance of the T-beam is found by multiplying the total compression and the lever arm.

If the N.A. lies in the flange, $M_r = b_f \cdot n \cdot \dfrac{c}{2}\left(d - \dfrac{n}{3}\right)$ ...[6.9]

If the N.A. lies in the web, the moment of resistance is given by

$$M_r = b_f \cdot D_f \cdot \frac{c + c_1}{2} \times a \quad \text{(Neglecting the compression of the rib)} \qquad \ldots[6.9(a)]$$

Hence, $\quad M_r = \dfrac{b_f \cdot D_f \cdot c}{2}\left[1 + \dfrac{n - D_f}{n}\right]\left[d - \dfrac{D_f}{2} + \dfrac{D_f^2}{6(2n - D_f)}\right] \qquad \ldots[6.9(b)]$

Also, $\quad M_r = t \cdot A_{st} \cdot a = \sigma_{st} \cdot A_{st} \cdot a \qquad \ldots(6.10)$

## 6.5. MOMENT OF RESISTANCE TAKING COMPRESSION IN RIB INTO ACCOUNT

When the compression of the rib is taken into account, the neutral axis can be located by:

$$b_f \cdot D_f \cdot \left(n - \frac{D_f}{2}\right) + \frac{b_w}{2}(n - D_f)^2 = m A_{st}(d - n) \qquad \ldots(6.11)$$

The moment of resistance of the beam can be found by two methods:

**Method 1.** In this method, the distance $\bar{y}$ of the C.G. of compressive force is found first and then lever arm is determined. The moment of resistance is then found by multiplying the total compressive force by the lever arm.

Thus, taking moments at the top of the beam:

$$\bar{y} = \frac{b_f \cdot D_f \cdot \dfrac{c + c_1}{2} \times \dfrac{D_f}{3}\left[\dfrac{2c_1 + c}{c_1 + c}\right] + b_w(n - D_f) \times \dfrac{c_1}{2}\left[D_f + \dfrac{n - D_f}{3}\right]}{b_f \cdot D_f \cdot \dfrac{c + c_1}{2} + b_w(n - D_f)\dfrac{c_1}{2}} \qquad \ldots(6.12)$$

where $\quad c_1 = \dfrac{n - D_f}{n} \cdot c$

$\therefore\quad$ Lever arm $\quad a = d - \bar{y}$

Hence, $M_r = \left\{ b_f \cdot D_f \cdot \dfrac{c + c_1}{2} + b_w (n - D_f) \dfrac{c_1}{2} \right\} a$

...(6.13)

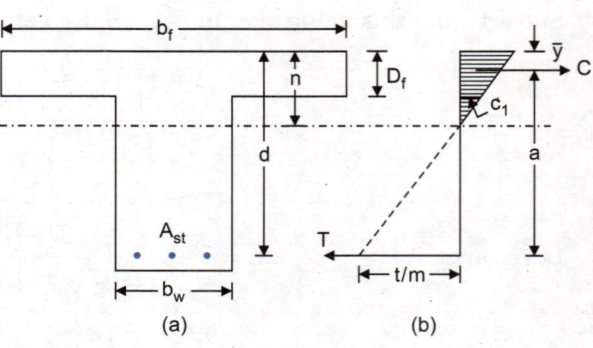

Fig. 6.5

**Method 2.** The moment of resistance of the beam is the sum of the moment of resistance due to force in flange and the moment of resistance due to force in rib. Thus,

$$M_r = M_{r1} + M_{r2}$$

where $M_{r1}$ = moment of resistance of the force in flange

or $\quad M_{r1} = \dfrac{b_f \cdot D_f \cdot c}{2} \left[ 1 + \dfrac{n - D_f}{n} \right] \left[ d - \dfrac{D_f}{2} + \dfrac{D_f^2}{6(2n - D_f)} \right]$ ...(6.14)

and $\quad M_{r2}$ = moment of resistance of the force in rib.

$= b_w (n - D_f) \times \dfrac{c_1}{2} \left[ d - n + \dfrac{2}{3}(n - D_f) \right]$

$= \dfrac{b_w}{2} (n - D_f) \left\{ \dfrac{n - D_f}{n} c \right\} \left[ d - \dfrac{n}{3} - \dfrac{2}{3} D_f \right]$

$= \dfrac{c \cdot b_w}{2n} (n - D_f)^2 \left[ d - \dfrac{n + 2D_f}{3} \right]$ ...(6.15)

## 6.6. DEPTH OF BALANCED SECTION OF T-BEAM

Let us find the depth of balanced section of T-beam, corresponding to external moment $M$. The critical neutral axis depth factor $k$ is given by:

$$k = \dfrac{mc}{mc + t} = \dfrac{m\sigma_{cbc}}{m\sigma_{cbc} + \sigma_{st}} \quad ...(i)$$

The depth of balanced section of T-beam is determined on the assumption that the compression of the rib is negligible and can be ignored. Hence the compressive force (C) of beam is equal to the compressive force in flange, given by

$$C = b_f \cdot D_f \cdot \dfrac{1}{2} (c + c_1)$$

or $\quad C = b_f \cdot D_f \cdot \dfrac{1}{2} \left[ c + \dfrac{n - D_f}{n} c \right] = \dfrac{b_f \cdot c \cdot D_f}{2} \left[ \dfrac{2n - D_f}{n} \right]$

$= \dfrac{b_f \cdot c \cdot D_f}{2} \left[ \dfrac{2kd - D_f}{kd} \right]$ ...(ii)

Lever arm $= a = d - \dfrac{D_f}{2} + \dfrac{D_f^2}{6(2n - D_f)}$ ...(From Eq. 6.8) ...(iii)

$M = M_r$ = compressive force $C$ × lever arm $a$

or $\quad M = \dfrac{b_f \cdot c \cdot D_f}{2} \left[ \dfrac{2kd - D_f}{kd} \right] \left[ d - \dfrac{D_f}{2} + \dfrac{D_f^2}{6(2kd - D_f)} \right]$

or $\quad M = \dfrac{b_f \cdot c \cdot D_f}{12 kd} [12 k d^2 - 6 d \cdot D_f - 6 k d D_f + 3 D_f^2 + D_f^2]$

DESIGN OF T AND L-BEAMS **127**

or
$$M = \frac{b_f \cdot c \cdot D_f}{12\,kd}[12\,kd^2 - 6dD_f(1+k) + 4D_f^2]$$

or
$$M = b_f \cdot c \cdot D_f\left[d - \frac{D_f(1+k)}{2k} + \frac{D_f^2}{3kd}\right] \qquad ...(6.16)$$

In the above expression, all quantities except $d$ are known. Hence $d$ can be calculated from the solution of the above quadratic equation.

*Alternatively,* $d$ can be computed by *assuming* the lever arm $a$ equal to $0.9\,d$, though its actual value may vary from $0.88\,d$ to $0.95\,d$. Then:

$$M = M_r = C \times a = \frac{b_f \cdot c \cdot D_f}{2}\left[\frac{2kd - D_f}{k \cdot d}\right] \times 0.9\,d \qquad ...[6.17(a)]$$

$$M = 0.45\,b_f \cdot c \cdot D_f\left[\frac{2kd - D_f}{k}\right] \qquad ...(6.17)$$

This is a linear equation in terms of the unknown $d$.

## 6.7. ECONOMICAL DEPTH OF T-BEAM

The balanced depth found above is not economical. This is because of the fact that in T-beam large compression area is provided by the flange and consequently balanced depth is comparatively reduced and area of steel is increased. The cost of steel is several times more than the cost of concrete and thus overall cost of T-beam, having balanced depth, is increased. To find an expression for the economical depth of beam, we will have to take into account the *cost-ratio* of steel and concrete. Let,

$R_s$ = cost of steel, per cubic mm = $r \cdot R_c$
$R_c$ = cost of concrete, per cubic mm
$r$ = cost ratio = $\dfrac{\text{cost of steel}}{\text{cost of concrete}}$ per unit volume = $\dfrac{R_s}{R_c}$
$d_c$ = depth of cover, measured below the centre of the tensile steel.

The volume of concrete provided by the flange is *fixed*, and is, therefore, not included in the cost calculations.

Volume of concrete, per mm length, in the web = $b_w(d - D_f + d_c)$

∴ Cost of concrete in the rib, per mm length = $b_w(d - D_f + d_c)\,R_c$

Similarly, cost of steel, per mm length of beam = $A_{st}\,R_s = r \cdot R_c \cdot A_{st}$

But $\qquad A_{st} = \dfrac{M}{tjd}$, where $t = \sigma_{st}$ = permissible stress in steel

∴ Total cost ($Q$) of the beam, per mm length is given by

$$Q = R_c \cdot b_w(d - D_f + d_c) + r \cdot R_c \cdot \frac{M}{tjd} = R_c \cdot \left[b_w(d - D_f + d_c) + \frac{r \cdot M}{tjd}\right]$$

In the above expression, $b_w$, $D_f$ and $d_c$ are fixed dimensions while depth $d$ is variable. For the cost to be *minimum*,

$$\frac{dQ}{dd} = 0 = R_c\left[b_w - \frac{rM}{tjd^2}\right]$$

∴ $\qquad d^2 = \dfrac{r \cdot M}{tjb_w}, \qquad$ where $t = \sigma_{st}$

or
$$d = \sqrt{\frac{r \cdot M}{tj\, b_w}} = \sqrt{\frac{r \cdot M}{\sigma_{st}\, j\, b_w}} \qquad \ldots(6.18)$$

Taking $t = \sigma_{st} = 140$ N/mm² and $j = 0.9$, the above expression reduces to

or
$$d = \sqrt{\frac{r \cdot M}{126\, b_w}} \qquad \ldots[6.18(a)]$$

Similarly, taking $\sigma_{st} = 230$ N/mm² and $j = 0.9$, we have

$$d = \sqrt{\frac{r \cdot M}{207\, b_w}} \qquad \ldots[6.18(b)]$$

In the above expressions, the cost of shuttering has not been taken into account. For deeper beams, the cost of shuttering may be appreciable. In that case the depth of beam found from *Eq. 6.18* may be further reduced by 10 to 15% to account for the additional cost of shuttering.

The width of web ($b_w$) is kept a certain minimum, so as to accommodate tensile steel. Sometimes, steel may be provided in two or three layers, but this reduces the lever arm. Suitable value of $b_w$ is also selected on the consideration of shear.

## 6.8. SHEAR, BOND AND DEVELOPMENT LENGTH

### (a) Shear Stress Distribution

The distribution of shear stress in T-beam can be found exactly in the same way as for a rectangular beam. Fig. 6.6 (b) shows the shear stress distribution across the depth.

The maximum shear ($q$) is induced in all fibres between the N.A. and the centre of tensile reinforcement and its magnitude is given by *Eq. 3.1*:

$$q = \frac{V}{b_w \cdot a} \qquad \ldots(6.19)$$

where $a$ = lever arm

$$= d - \frac{D_f}{2} + \frac{D_f^2}{6(2n - D_f)}$$

The shear stress ($s$) at any layer $y$ above the N.A. is given by *Eq. 3.2*.

$$s = \frac{V}{2I}(n^2 - y^2). \qquad \ldots(6.20)$$

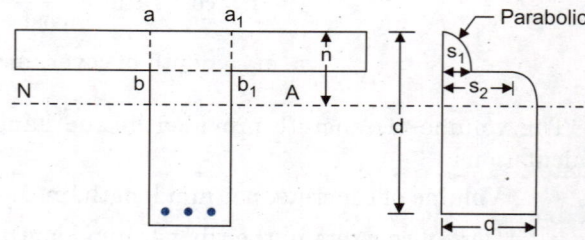

Fig. 6.6. Shear Stress Distribution.

where $I$ is the moment of inertia of the T-beam section, about the N.A.

The ratio of stress $s_1$ in the flange to stress $s_2$ in the rib, at the junction is evidently given by

$$\frac{s_1}{s_2} = \frac{b_w}{b_f}$$

The complete shear stress distribution across the section is shown in Fig. 6.6 (b).

### (b) Nominal Shear Stress

Indian Standard Code (IS : 456–2000) recommends the use of nominal shear stress ($\tau_v$) in design given by the expression,

$$\tau_v = \frac{V}{b_w \cdot d} \qquad \ldots(6.21)$$

# DESIGN OF T AND L-BEAMS

### (c) Flexural Bond and Development Length

The flexural bond stress around main reinforcement is given by *Eq. 3.29*:

$$\tau_{bf} = \frac{V}{a \Sigma 0}$$

However, in the case of bond, it has been recently concluded that the concept of *anchorage bond* is more meaningful than the concept of flexural bond stresses of localised nature. Current design methods disregard such local bond stresses, and recommend the use of development length given by

$$L_d = \frac{\Phi \cdot \sigma_s}{4 \tau_{bd}}$$

## 6.9. TYPES OF PROBLEMS IN T-BEAM

There may be three types of problems in T-beam:

**Type I.** Determination of moment of resistance of the given section.
**Type II.** Determination of stresses in the given section subjected to a given bending moment.
**Type III.** Design of section to resist a given bending moment.

### TYPE I. Determination of Moment of Resistance

In this type of problem, the section including the steel reinforcement is completely defined and it is required to determine the moment of resistance corresponding to a given set of permissible stresses in concrete and steel. The procedure for solution is as under:

1. Determine the position of the neutral axis by equating the moment of the compressive area of concrete to the area of concrete equivalent to the area of steel in tension. Assuming that the N.A. lies in the flange, we get from *Eq. 6.1*:

$$\frac{b_f \cdot n^2}{2} = m \cdot A_{st}(d-n)$$

If $n$ comes out to be greater that $D_f$, N.A. falls in the rib and *Eq. 6.4* is used:

$$b_f \cdot D_f \left[ n - \frac{D_f}{2} \right] = m \cdot A_{st}(d-n). \text{ Thus } n \text{ is known.}$$

2. Determine the position of critical neutral axis, on the basis of permissible stresses and $m$, using *Eq. 6.5*:

$$\frac{n_c}{d-n_c} = \frac{c}{t/m} = \frac{\sigma_{cbc}}{\sigma_{st}/m}$$

or

$$n_c = k_c \cdot d = \frac{d}{1 + t/mc} = \frac{d}{1 + \sigma_{st}/m \cdot \sigma_{cbc}}$$

3. If the actual neutral axis is *above* the critical N.A., the section is *under-reinforced* and steel reaches maximum stress ($\sigma_{st}$) first. The actual stress $c$ in concrete is calculated from the relation

$$c = \frac{t}{m} \cdot \frac{n}{d-n} \qquad \text{(where } t = \sigma_{st} = \text{permissible stress in steel)}$$

If the N.A. lies in the flange, the moment of resistance is given by

$$M_r = b_f \cdot n \cdot \frac{c}{2}\left(d - \frac{n}{3}\right)$$

If the N.A. lies in the web, the moment of resistance is given by

$$M_r = \frac{b_f \cdot D_f \cdot c}{2}\left[1 + \frac{n - D_f}{n}\right]\left[d - \frac{D_f}{2} + \frac{D_f^2}{6(2n - D_f)}\right]$$

*Alternatively*, the moment of resistance may be calculated from *Eq. 6.10:*

$$M_r = t \cdot \sigma_{st} \cdot a = t A_{st}\left[d - \frac{D_f}{2} + \frac{D_f^2}{6(2n - D_f)}\right]$$

4. If the actual neutral axis is below the critical N.A., the section is *over-reinforced*, and concrete reaches maximum stress ($\sigma_{cbc}$) first.

If the N.A. falls in the flange, the moment of resistance is given by *Eq. 6.9 (a)*.

$$M_f = b_f \cdot n \cdot \frac{c}{2}\left[d - \frac{n}{3}\right] \quad \text{(where } c = \sigma_{cbc} = \text{permissible stress in concrete)}$$

If the N.A. falls in the web, the moment of resistance is given by *Eq. 6.9*.

$$M_r = \frac{b_f \cdot D_f \cdot c}{2}\left[1 + \frac{n - D_f}{n}\right]\left[d - \frac{D_f}{2} + \frac{D_f^2}{6(2n - D_f)}\right]$$

5. If it is further required to find the uniformly distributed load or point load or other type of load on a simply supported beam, calculate the maximum B.M. in terms of the unknown load and equate it to the moment of resistance determined in step 3 or 4.

## TYPE II. Determination of Stresses in the Section

In this type of problem, the dimensions of the section, including the reinforcement, are completely known and it is required to find the stresses developed in steel and concrete, when subjected to a given bending moment. The solution is done in the following steps:

1. Determine the position of the neutral axis by equating the moments of equivalent areas about the N.A. Assuming that the N.A. lies in the flange, we get from *Eq. 6.1*.

$$\frac{b_f \cdot n^2}{2} = m A_{st}(d - n).$$

If $n$ comes out to be greater than $D_f$, N.A. falls in the rib and *Eq. 6.4* is used,

$$b_f \cdot D_f\left(n - \frac{D_f}{2}\right) = m \cdot A_{st}(d - n). \text{ Thus } n \text{ is known.}$$

2. Equate the moment of resistance of the beam to the external moment $M$, and determine the compressive stress in concrete.

Thus, if the N.A. falls in the flange, we have

$$M = M_r = b_f \cdot n \cdot \frac{c}{2}\left(d - \frac{n}{3}\right) \text{ from which } c \text{ can be calculated.}$$

If the N.A. falls in the rib, we have from *Eq. 6.9*,

$$M = M_r = \frac{b_f \cdot D_f \cdot c}{2}\left[1 + \frac{n - D_f}{n}\right]\left[d - \frac{D_f}{2} + \frac{D_f^2}{6(2n - D_f)}\right]$$

Thus maximum compressive stress $c$ in concrete is known.

3. Calculate the tensile stress in steel by the stress relationship: $t = mc\,\dfrac{d - n}{n}$.

## TYPE III. Design of the Section

This is more common type of problem in which it is required to determine the dimensions of the rib including tensile reinforcement to develop a given moment of resistance corresponding to a given set of permissible stresses in concrete ($c = \sigma_{cbc}$) and steel ($t = \sigma_{st}$). The dimensions of the flange (*i.e.* $b_f$ and $D_f$) are treated as known.

1. Assume some suitable value of width of rib based on criteria discussed in § 6.2.
2. Determine the effective width $b_f$ of the flange, as discussed in § 6.2.

3. Determine the effective depth of the beam on three considerations:
   (i) balanced section  (ii) economy
   (iii) shear stress.

For balanced section, the effective depth is given by *Eq. 6.16*

$$M = b_f \cdot c \cdot D_f \left[ d - \frac{D_f(1+k)}{2k} + \frac{D_f^2}{3kd} \right]$$

*Alternatively*, taking lever arm to be equal to $0.9d$, and use *Eq. 6.17*.

$$M = 0.45 \, b_f \cdot c \cdot D_f \left[ \frac{2kd - D_f}{k} \right]$$

For economical depth, use *Eq. 6.18*: $d = \sqrt{\dfrac{rM}{t \, j \, b_w}}$

For shear considerations, use *Eq. 6.21*: $\tau_v = \dfrac{V}{b_w \cdot d}$  or  $d = \dfrac{V}{b_w \cdot \tau_v}$

where $\tau_v$ may be taken equal to $\tau_{c.max}$ given in Table 3.3.
The *maximum* value of $d$ obtained from the above three criteria is taken as the effective depth of the beam.

Take the approximate lever arm $a = d - \dfrac{D_f}{2}$, and determine the area of tensile reinforcement

$$A_{st} = \frac{M}{t \times a} = \frac{M}{\sigma_{st} \times a}.$$

Arrange suitably the steel reinforcement.
4. Determine the actual N.A. and check the stresses in concrete and steel.
5. Check the section for shear and provide shear reinforcement if necessary.
6. Check for bond.

The complete design procedure for T-beam is given in Chapter 7.

**Example 6.1.** *An isolated T-beam, simply supported over a span of 6 m has following dimensions: Width of flange 750 mm, thickness of flange 125 mm; over all depth of 400 mm; width of web 260 mm; effective cover to tensile reinforcement, 40 mm. The beam is reinforced with 4 bars of 20 mm dia. Determine the moment of resistance of the beam if (a) mild steel bars are used (b) Fe 415 steel bars are used. Take $\sigma_{cbc} = 5 \, N/mm^2$ and $m = 19$ in each case.*

**Solution:** The section of the T-beam, is shown in Fig. 6.7
  Assume that N.A. lies in flange.

  Area of steel $= A_{st} = 4 \times \dfrac{\pi}{4}(20)^2 = 1256.6 \, mm^2$

For an isolated T- beam, effective flange width

$$b_f = \frac{l_0}{\frac{l_0}{b} + 4} + b_w = \frac{6000}{\frac{6000}{750} + 4} + 260 = 760 \, mm$$

Available  $b = 750$ mm.
∴         $b_f = b = 750$ mm

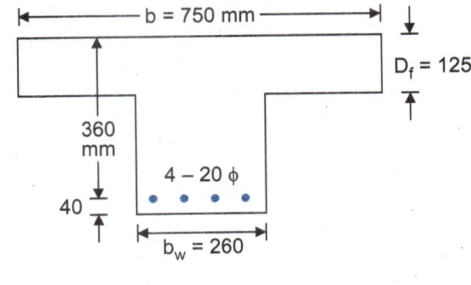

Fig. 6.7

Equating the moments of equivalent areas about the N.A., we get

$$750 \times \frac{n^2}{2} = 19 \times 1256.6 \, (360 - n)$$

or         $n^2 = 22921 - 63.67 \, n.$     From which,   $n = 122.9$ mm
This is less than $D_f (= 125$ mm$)$. Hence the N.A. just lies inside the flange.

## 132 REINFORCED CONCRETE STRUCTURE

(a) **Mild steel reinforcement.**

$$n_c = \frac{d}{1 + \frac{\sigma_{st}}{m\sigma_{cbc}}} = \frac{360}{1 + \frac{140}{19 \times 5}} = 145.5 \text{ mm}$$

Since the actual N.A. falls *above* the critical N.A., the section is *under-reinforced* and the stress in steel reaches its maximum value first. Thus, $t = \sigma_{st} = 140$ N/mm². The corresponding stress in concrete is given by

$$c = \frac{t}{m} \cdot \frac{n}{d-n} = \frac{140}{19} \times \frac{122.9}{360-122.9} = 3.82 \text{ N/mm}^2$$

The moment of resistance of the beam is given by

$$M_r = b_f \cdot n \cdot \frac{c}{2}\left(d - \frac{n}{3}\right) = 750 \times 122.9 \times \frac{3.82}{2}\left(360 - \frac{122.9}{3}\right)$$
$$= 56.17 \times 10^6 \text{ N-mm} = \mathbf{56.17 \text{ kN-m.}}$$

(b) **Fe 415 Steel.**

$$n_c = \frac{d}{1 + \frac{\sigma_{st}}{m\sigma_{cbc}}} = \frac{360}{1 + \frac{230}{19 \times 5}} = 105.2 \text{ mm}$$

Since the actual N.A. falls *below* the critical N.A., the section is *over-reinforced* and the stress in concrete reaches its maximum value first. Thus $c = \sigma_{cbc} = 5$ N/mm². The corresponding stress in steel is given by

$$t = \frac{m \cdot c(d-n)}{n} = \frac{19 \times 5(360 - 122.9)}{122.9} = 183.3 \text{ N/mm}^2$$

The moment of resistance of the beams is given by

$$M_r = b_f \cdot n \cdot \frac{c}{2}\left(d - \frac{n}{3}\right) = 750 \times 122.9 \times \frac{5}{2}\left(360 - \frac{122.9}{3}\right)$$
$$= 73.52 \times 10^6 \text{ N-mm} = \mathbf{73.52 \text{ kN-m.}}$$

**Example 6.2.** *An isolated T-beam has a flange width of 1000 mm, flange thickness of 80 mm and effective depth of 400 mm. The rib is 240 mm wide and reinforced with 5 bars of 20 mm diameter. Determine the moment of resistance of the section, if the permissible stresses in concrete and steel are 5.2 N/mm² and 140 N/mm² and m = 18. Neglect the compressive force in the web. The beam is simply supported over a span of 4 m.*

**Solution:** The effective width

$$b_f = \frac{l_0}{\left(\frac{l_0}{b}\right) + 4} + b_w$$

$$= \frac{4000}{\left(\frac{4000}{1000}\right) + 4} + 240 = 740 \text{ mm}$$

Thus the effective width of flange is very much less than the actual flange width.

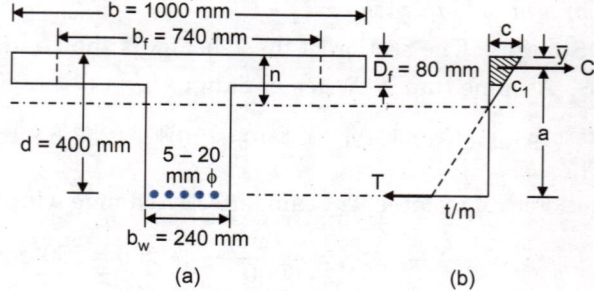

Fig. 6.8

Area of steel $\quad A_{st} = 5\frac{\pi}{4}(20)^2 = 1570 \text{ mm}^2$

Equating the moments of equivalent areas about N.A. and assuming that N.A. falls in the flange, we get

$$b_f \cdot \frac{n^2}{2} = m \cdot A_{st}(d - n) \quad \text{or} \quad 740 \times \frac{n^2}{2} = 18 \times 1570(400 - n)$$

or $\quad n^2 + 76.4n - 30550 = 0$. From which $n = 141$ mm $> D_f$. Hence N.A. falls in the web.

Again, equating the moments of equivalent areas, and neglecting the compression of the web, we get from *Eq. 6.4:*

or $\quad 740 \times 80 \left(n - \dfrac{80}{2}\right) = 18 \times 1570 (400 - n)$

or $\quad 1.477 n = 230.9.$ From which $n = 156.4$ mm

Taking $\sigma_{cbc} = 5.2$ N/mm², $\sigma_{st} = 140$ N/mm² and $m = 18$,

$$n_c = \dfrac{d}{1 + \dfrac{\sigma_{st}}{m\sigma_{cbc}}} = \dfrac{400}{1 + \dfrac{140}{18 \times 5.2}} = 160.3 \text{ mm}$$

Since the actual N.A. falls above the critical N.A., the stress in steel reaches the maximum value first. Hence $t = \sigma_{st} = 140$ N/mm². The corresponding stress in concrete at the outer fibre is given by

$$c = \dfrac{t}{m} \dfrac{n}{d-n} = \dfrac{140}{18} \times \dfrac{156.4}{400 - 156.4} = 4.99 \text{ N/mm}^2$$

$$c_1 = \dfrac{n - D_f}{n} c = \dfrac{156.4 - 80}{156.4} \times 4.99 = 2.44 \text{ N/mm}^2$$

Hence $\quad \bar{y} = \left[\dfrac{c + 2c_1}{c + c_1}\right] \dfrac{D_f}{3} = \left[\dfrac{4.99 + 4.88}{4.99 + 2.44}\right] \dfrac{80}{3} = 35.4$ mm.

Lever arm $\quad a = d - \bar{y} = 400 - 35.4 = 364.6$ mm

$$M_r = b_f d_f \dfrac{c + c_1}{2} \times a = 740 \times 80 \dfrac{4.99 + 2.44}{2} \times 364.6$$

$= 80.1 \times 10^6$ N-mm $= $ **80.1 kN-m**

Alternatively, $\quad M_r = \sigma_{st} A_{st} \cdot a = 140 \times 1570 \times 364.6 = 80.1 \times 10^6$ N-mm $= 80.1$ kN-m

**Example 6.3.** *Solve Example 6.2 taking web compression into account.*

**Solution:** The N.A. will lie in the web at a distance $n$ below the top of the flange. Equating the moments of equivalent areas about the N.A., we have from *Eq. 6.3*.

$740 \times 80(n - 40) + 120 (n - 80)^2 = 18 \times 1570 (400 - n)$

or $\quad n^2 + 568.8n - 107533 = 0$

From which $\quad n = 149.7$ mm.

As before $\quad n_c = 160.3$ mm.

The actual N.A. falls above the critical one and hence the section is under-reinforced. The stress in steel reaches its maximum value first. Thus, $t = \sigma_{st} = 140$ N/mm². The corresponding value of stress in concrete at the outer fibre is given by

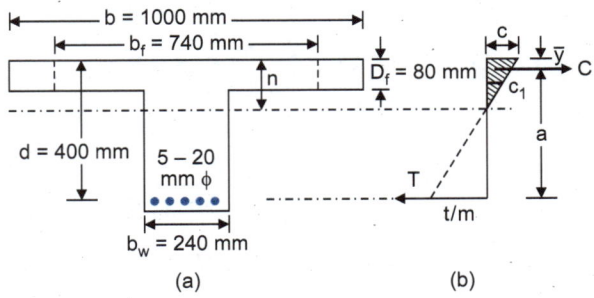

Fig. 6.9

$$c = \dfrac{t}{m} \cdot \dfrac{n}{d - n}$$

$$= \dfrac{140}{18} \times \dfrac{149.7}{400 - 149.7} = 4.65 \text{ N/mm}^2$$

$$c_1 = \dfrac{n - D_f}{n} \cdot c = \dfrac{149.7 - 80}{149.7} \times 4.65 = 2.16 \text{ N/mm}^2$$

$$\bar{y} = \dfrac{b_f \cdot D_f \cdot \dfrac{c + c_1}{2} \times \dfrac{D_f}{3}\left\{\dfrac{2c_1 + c}{c_1 + c}\right\} + b_w (n - D_f) \times \dfrac{c_1}{2}\left\{D_f + \dfrac{n - D_f}{3}\right\}}{b_f \cdot D_f \cdot \dfrac{c + c_1}{2} + b_w (n - D_f) \dfrac{c_1}{2}}$$

$$= \left[740 \times 80 \left\{\frac{4.65 + 2.16}{2}\right\} \times \frac{80}{3} \left\{\frac{4.32 + 4.65}{2.16 + 4.65}\right\} + 240\{149.7 - 80\} \times \frac{2.16}{2} \left\{80 + \frac{149.7 - 80}{3}\right\}\right]$$

$$\div \left[740 \times 80 \left\{\frac{4.65 + 2.16}{2}\right\} + 240\{149.7 - 80\} \frac{2.16}{2}\right] = 40.8 \text{ mm}$$

$\therefore$ L.A. $= a = d - \bar{y} = 400 - 40.8 = 359.2$ mm

$$M_r = \left\{b_f \cdot D_f \frac{c + c_1}{2} + b_w (n - D_f)\frac{c_1}{2}\right\} a$$

$$= \left\{740 \times 80 \left(\frac{4.65 + 2.16}{2}\right) + 240 (149.7 - 80)\frac{2.16}{2}\right\} 359.2$$

$$= 79 \times 10^6 \text{ N-mm} = 79 \text{ kN-m}$$

Alternatively, $M_r = \sigma_{st} \cdot A_{st} \cdot a = 140 \times 1570 \times 359.2 = 79 \times 10^6$ N-mm $= 79$ kN-m.

**Example 6.4.** *Given the T-beam of the dimensions of Ex. 6.2 except that the reinforcement consists of 4 bars of 25 mm diameter. Determine the maximum uniformly distributed load, inclusive of its self-weight, that the beam can carry over an effective span of 4 m. Neglect the compression of the rib.*

**Solution:** $A_{st} = 4 \frac{\pi}{4} (25)^2 = 1963$ mm$^2$.

Let the N.A. be in the web at a distance $n$ below the top of the flange. Equating the moments of the equivalent areas about the N.A., we get from *Eq. 6.4*.

$$740 \times 80 \left(n - \frac{80}{2}\right) = 18 \times 1963 (400 - n)$$

from which $n = 174.6$ mm. But $n_c = 160.3$ mm as before.

Since the actual N.A. falls below the critical N.A., the section is over-reinforced and hence the stress in concrete reaches its maximum value of $c = \sigma_{cbc} = 5.2$ N/mm$^2$ first.

$\therefore \quad c_1 = \frac{n - D_f}{n} c = \frac{174.6 - 80}{174.6} \times 5.2 = 2.82$ N/mm$^2$

and $\quad t = mc \frac{d - n}{n} = 18 \times 5.2 \frac{400 - 174.6}{174.6} = 120.83$ N/mm$^2$

$$\bar{y} = \frac{c + 2c_1}{c + c_1} \times \frac{D_f}{3} = \frac{5.2 + 5.64}{5.2 + 2.82} \times \frac{80}{3} = 36 \text{ mm}$$

L.A. $\quad a = d - \bar{y} = 400 - 36 = 364$ mm

$$M_r = b_r \cdot D_f \cdot \frac{c + c_1}{2} \times a = 740 \times 80 \frac{5.2 + 2.82}{2} \times 364$$

$$= 86.4 \times 10^6 \text{ N-mm} = 86.4 \text{ kN-m}$$

(Alternatively $M_r = t \cdot A_{st} \cdot a = 120.83 \times 1963 \times 364 = 86.4 \times 10^6$ N-mm)

Let $w$ kN/m be the safe uniformly distributed load, inclusive of the safe weight.

$\therefore \quad M_{max} = \frac{wL^2}{8} = \frac{w \times 16}{2} = 2w$ kN-m.

Equating this to the moment of resistance of the beam.

$2w = 86.4 \quad$ or $\quad w = \mathbf{43.2}$ **kN/m.**

**Example 6.5.** *The flange of a T-beam is 120 mm thick and 1200 mm wide. The web of the beam is 300 mm wide and depth upto the centre of tensile reinforcement is 700 mm. The reinforcement consists of 5 bars of 25 mm diameter arranged in one row. If the beam section is subjected to a B.M. of 200 kN-m, determine the stresses developed in concrete and steel. Take m = 18. The beam is simply supported over a span of 6 m.*

**Solution:** $b_f = \dfrac{6000}{\dfrac{6000}{1200}+4} + 300 = 967$ mm

This is less than the actual width ($b$) of the flange.

$$A_{st} = 5 \times \dfrac{\pi}{4}(25)^2 \approx 2460 \text{ mm}^2$$

Let the depth of N.A. be $n$. Equating the moments of equivalent areas above N.A. and neglecting the compression of web, we get from *Eq. 6.4*.

$$967 \times 120(n-60) = 18 \times 2460(700-n)$$

From which $n = 236.8$ mm.

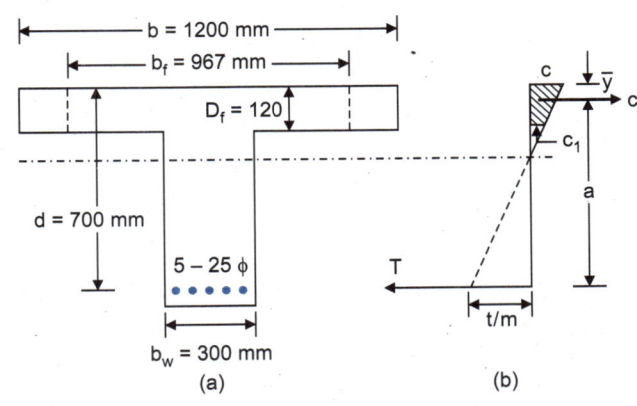

Fig. 6.10

Let $c$ be the maximum compressive stress in concrete at the top of the flange and $c_1$ be the compressive stress at the bottom of the flange.

$$c_1 = \dfrac{n-D_f}{n} \cdot c = \dfrac{236.8-120}{236.8} \times c = 0.493\,c$$

∴ $\bar{y} = \left[\dfrac{c+2c_1}{c+c_1}\right]\dfrac{D_f}{3} = \left[\dfrac{1+0.986}{1+0.493}\right] \times \dfrac{120}{3} = 53.2$ mm.

L.A. $a = D - \bar{y} = 700 - 53.2 = 646.8$ mm.

∴ $M_r = b_f \cdot D_f \cdot \dfrac{c+c_1}{2} \times a = 967 \times 120\,\dfrac{1+0.493}{2}\,c \times 646.8 = 56.03 \times 10^6\,c$

External B.M. $= 200$ kN-m $= 200 \times 10^6$ N-mm. Equating the two, we get

$$56.03\,c = 200 \quad \text{or} \quad c = \mathbf{3.57 \text{ N/mm}^2}$$

$$t = mc\,\dfrac{d-n}{n} = 18 \times 3.57\,\dfrac{700-236.8}{236.8} = \mathbf{125.68 \text{ N/mm}^2}$$

Alternatively, $t = \dfrac{M}{A_{st} \times a} = \dfrac{200 \times 10^6}{2460 \times 646.8} = 125.68$ N/mm².

**Example 6.6.** *A reinforced concrete T-beam has a flange 1200 mm wide and 100 mm thick. The effective depth of the beam is 400 mm and width of web is 250 mm. Determine the moment of resistance of the beam and the area of steel in tension, if $m = 19$, and the permissible stresses in concrete and steel are $5 \text{ N/mm}^2$ and $140 \text{ N/mm}^2$ respectively. Assume that the beam is simply supported over a span of 6 m.*

**Solution:** $b_f = \dfrac{l_0}{\dfrac{l_0}{b}+4} + b_w = \dfrac{6000}{\dfrac{6000}{1200}+4} + 250 \approx 917$ mm.

The area of tensile steel is to be found for balanced section, for which the

$$n_c = \dfrac{d}{1+\dfrac{\sigma_{st}}{m\sigma_{cbc}}} = \dfrac{400}{1+\dfrac{140}{19 \times 5}} \approx 162 \text{ mm}$$

Equating the moments of equivalent areas about the N.A. and neglecting the compressive area of the web, we get from *Eq. 6.4*

$$917 \times 1000\left(162 - \dfrac{100}{2}\right) = 19 \times A_{st}(400-162)$$

From which $A_{st} = \mathbf{2271 \text{ mm}^2}$.

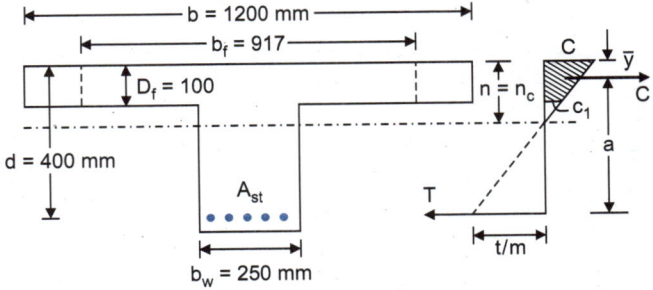

Fig. 6.11

$$M_r = b_f \cdot D_f \left(\frac{c+c_1}{2}\right) \times \text{lever arm}$$

Now
$$c_1 = \frac{n_c - D_f}{n_c} \cdot c = \frac{162 - 100}{162} \times 5 = 1.914 \text{ N/mm}^2$$

$$\bar{y} = \left(\frac{c + 2c_1}{c + c_1}\right)\frac{D_f}{3} = \left(\frac{5 + 2 \times 1.914}{5 + 1.914}\right)\frac{100}{3} = 42.6 \text{ mm.}$$

∴ $\quad a = d - \bar{y} = 400 - 42.6 = 357.4 \text{ mm}$

∴ $\quad M_r = 917 \times 100 \left(\dfrac{5 + 1.914}{2}\right) \times 357.4 \approx 113.4 \times 10^6 \text{ N-mm}$

$\quad\quad\quad = \mathbf{113.4 \text{ kN-m}}$

*Alternatively*, $\quad M_r = \sigma_{st} \cdot A_{st} \, a = 140 \times 2271 \times 357.4 \approx 113.6 \times 10^6 \text{ N-mm}$

**Example 6.7.** *A T-beam roof consists of 100 mm thick reinforced concrete slab cast monolithic with 300 mm beams spaced 4 metres centre to centre. The super-imposed load over the slab is 3.9 kN/m², and the effective span of the beam is 5 metres. Determine the depth and area of steel in one of the intermediate beams. Take permissible stresses in concrete and steel as 5 N/mm² and 140 N/mm² respectively and m = 18.*

**Solution:** Given $D_f$ = 100 mm and $b_w$ = 300 mm.

$$b_f = \frac{l_0}{6} + b_w + 6D_f = \frac{5000}{6} + 300 + 6 \times 100 = 1733 \text{ mm}$$

This is less than $b_w$ plus half the sum of clear distances to the adjacent beams on either side.

The effective depth $d$ can be determined on the basis of three criteria: (*i*) B.M. or maximum allowable compressive stress in concrete (*ii*) shear considerations and (*iii*) economy.

For the balanced section, taking $c = \sigma_{cbc} = 5$ N/mm² and $t = \sigma_{st} = 140$ N/mm²,

$$k = k_c = \frac{mc}{mc + t} = \frac{18 \times 5}{18 \times 5 + 140} = 0.391$$

Assume $a = 0.9d$ for preliminary calculations. Hence from *Eq. 6.17 (a)*.

$$M_r = b_f \cdot c \cdot \frac{D_f}{2} \left[\frac{2kd - D_f}{kd}\right] \times a, \quad \text{where } a = 0.9d$$

∴ $\quad M_r = 1733 \times 5 \times 50 \left[\dfrac{2 \times 0.391\, d - 100}{0.391\, d}\right] \times 0.9\, d$

or $\quad M_r = 779850 \, [0.782\, d - 100] \text{ N-mm} \approx 0.78 \, [0.782\, d - 100] \text{ kN-m}$

(*Alternatively*, we can use *Eq. 6.16* in which it is not necessary to assume $a$).
Unit weight of concrete = 25000 N/m³

Now, load of slab per sq. m = $\dfrac{(25000) \times 100}{1000}$ = 2500 N

Super-imposed load per sq. m. = 3900 N

$\quad\quad\quad\quad\quad\quad\quad\quad$ Total = 6400 N

∴ Load per metre run of beam = 6400 × 4 = 25600 N

Weight of web, per metre run = 2400 N

$\quad\quad\quad\quad\quad\quad$ Total = 28000 N = 28 kN

B.M. $\quad\quad M = \dfrac{wL^2}{8} = \dfrac{28 \times 25}{8} = 87.5 \text{ kN-m}$

Equating this to the moment of resistance, we get : $0.78 [0.782 d - 100] = 87.5$

From which $d \approx 272$ mm.    S.F. $= V = \dfrac{wl}{2} = \dfrac{28 \times 5}{2} = 70$ kN

From shear considerations, $\tau_v = \dfrac{V}{bd} = \dfrac{V}{b_w \cdot d}$

Taking $\tau_v = \tau_{c.max} = 1.6$ N/mm$^2$, we get: $d_{min} = \dfrac{V}{b_w \cdot \tau_v} = \dfrac{70000}{300 \times 1.6} \approx 146$ mm.

Economical depth is given by *Eq. 6.18*

$$d = \sqrt{\dfrac{r \cdot M}{\sigma_{st} \cdot jb_w}}, \text{ where } M = 87.5 \times 10^6 \text{ N-mm}$$

Taking $r = 60$, $\quad d = \sqrt{\dfrac{60 \times 87.5 \times 10^6}{140 \times 0.9 \times 300}} = 372$ mm

Adopt $d = 370$ mm.   Lever arm $a = 0.9 \times 370 = 333$ mm

Area of steel, $\quad A_{st} = \dfrac{M}{\sigma_{st} \times a} = \dfrac{87.5 \times 10^6}{140 \times 333} = 1876.9$ mm$^2$

Using 5-22 mm Φ bars, actual area provided is

$A_{st} = 5 \times \dfrac{\pi}{4}(22)^2 = 1900$ mm$^2$ > $1876.9$ mm$^2$

Thus the section is completely known (Fig. 6.12). However, we have to determine the actual stresses induced in the section. To find the position of the N.A., we have from *Eq. 6.4*

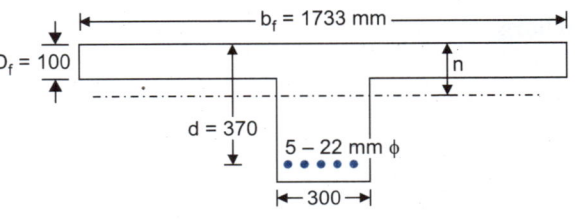

Fig. 6.12

$1733 \times 100 (n - 50) = 18 \times 1900 (370 - n)$   From which $n = 102.7$ mm

$$\bar{y} = \dfrac{3n - 2D_f}{2n - D_f} \times \dfrac{D_f}{3} = \dfrac{3 \times 102.7 - 200}{2 \times 102.7 - 100} \times \dfrac{100}{3} = 34.2 \text{ mm}$$

$a = d - \bar{y} = 370 - 34.2 = 335.8$ mm.     (against the value of 333 mm assumed earlier).

$M_r = b_f \cdot D_f \dfrac{c + c_1}{2} \cdot a = b_f D_f \dfrac{c}{2}\left[1 + \dfrac{n - D_f}{n}\right] \times a = b_f \cdot D_f \dfrac{c}{2} \cdot \dfrac{2n - D_f}{n} \times a$

$= 1733 \times 100 \times \dfrac{c}{2} \cdot \dfrac{2 \times 102.7 - 100}{102.7} \times 335.8 = 29.86 \times 10^6 \, c$ N-mm $= 29.86 \, c$ kN-m

Equating this to the external B.M. of 87.5 kN-m

$29.86 c = 87.5 \qquad$ or $\qquad c = \mathbf{2.93 \text{ N/mm}^2}$

$t = mc \dfrac{d - n}{n} = 18 \times 2.93 \dfrac{370 - 102.7}{102.7} = \mathbf{137.3 \text{ N/mm}^2}$.

## 6.10. DOUBLY REINFORCED T-BEAMS

Sometimes, the dimensions of the web are restricted from architectural or other considerations. If the beam is expected to carry more bending moment than it can take as a balanced section, compression reinforcement is provided to increase its moment of resistance. Such a beam is known as doubly reinforced T-beam. Let:

$A_{sc}$ = area of compressive reinforcement

$d_c$ = cover for the compressive reinforcement

$c'$ = compressive stress in concrete surrounding compressive steel.

The design of such a beam is done in the following steps:

1. Find the moment of resistance $M_1$ of the T-beam as the balanced section.

$$M_1 = b_f \cdot D_f \cdot \frac{c}{2}\left[1 + \frac{kd - D_f}{kd}\right]\left[d - \frac{D_f}{2} + \frac{D_f^2}{6(2kd - D_f)}\right] \quad ...(6.22)$$

2. Calculate the moment of resistance ($M_2$) to be provided by the compression steel.
$$M_2 = M - M_1, \quad \text{where } M = \text{total B.M.}$$

But
$$M_2 = (m_c - 1) A_{sc} \times c' (d - d_c) \quad \text{where } c' = \frac{kd - d_c}{kd} \cdot c \text{ and } m_c = 1.5\, m$$

$\therefore$
$$M_2 = (m_c - 1) A_{sc} \times c \times \frac{kd - d_c}{kd} (d - d_c) \quad ...(6.23)$$

Thus, $A_{sc}$ is known from this equation.

3. Find the tensile reinforcement as under:
$$A_{st1} = \frac{M_1}{\sigma_{st} \times a} \quad (\text{where } t = \sigma_{st}) \quad ...(6.24)$$

where $a = d - \dfrac{D_f}{2} + \dfrac{D_f^2}{6(2kd - D_f)}$

$$A_{st2} = \frac{M_2}{\sigma_{st}(d - d_c)} \quad ...(6.25)$$

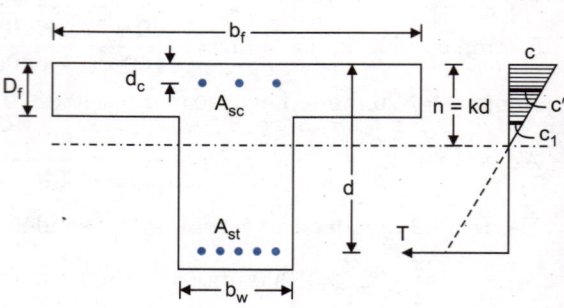

Fig. 6.13

Total area of steel
$$A_{st} = A_{st1} + A_{st2} \quad ...(6.26)$$

For a doubly reinforced T-beam of given dimensions, the neutral axis can be located by equating the moments of equivalent areas about the neutral axis. Thus:

$$b_f \cdot D_f \left(n - \frac{D_f}{2}\right) + (m_c - 1) A_{sc}(n - d_c) = m A_{st}(d - n) \quad ...(6.27)$$

(Assuming the N.A. to lie in the web)

The moment of resistance of such a beam, neglecting the compressive force in the rib, is given by

$$M_r = b_f \cdot D_f \left(\frac{c + c_1}{2}\right) \times a + (m_c - 1) A_{sc} \times c \times \frac{n - d_c}{n} (d - d_c) \quad ...(6.28)$$

Taking $a$ equal to $d - \dfrac{D_f}{2} + \dfrac{D_f^2}{6(2n - D_f)}$, the above expression reduces to:

$$M_r = b_f \cdot D_f \cdot c \left[\frac{2n - D_f}{2n}\right]\left[d - \frac{D_f}{2} + \frac{D_f^2}{6(2n - D_f)}\right] + (m_c - 1) A_{sc} \cdot \frac{c(n - d_c)}{n}(d - d_c) \quad ...(6.29)$$

where $c = \sigma_{cbc}$ = permissible compressive stress in concrete.

**Example 6.8.** *An isolated T-beam carries a uniformly distributed load of 40 kN/m run, inclusive of its own weight, over an effective span of 6 m. The beam has the following dimensions: width of flange: 800 mm; thickness of flange: 100 mm; effective depth of the beam : 480 mm and width of rib 300 mm. Determine the necessary areas of tensile and compressive reinforcement. Take $\sigma_{cbc}$ = 5 N mm², $\sigma_{st}$ = 140 N/mm² and m = 18.*

**Solution:** Given $\sigma_{cbc}$ = 5 N/mm²; $t = \sigma_{st}$ = 140 N/mm².

$$b_f = \frac{l_0}{\left(\dfrac{l_0}{b}\right) + 4} + b_w = \frac{6000}{\left(\dfrac{6000}{800}\right) + 4} + 300 = 822 \text{ mm}.$$

But this cannot be more than $b$ (=800 mm).
Hence $b_f$ = 800 mm;

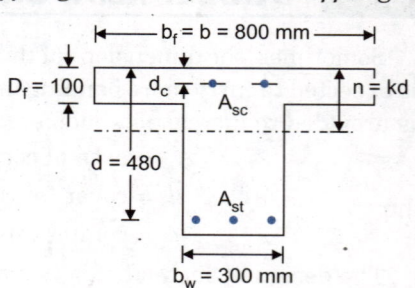

Fig. 6.14

$$M = \frac{40 \times 36}{8} = 180 \text{ kN-m} = 180 \times 10^6 \text{ N-mm}.$$

For balanced section,

$$n = kd = \frac{mc}{mc+t}d = \frac{18 \times 5}{18 \times 5 + 140} \times 480 = 187.8 \text{ mm}.$$

The moment of resistance of the balanced section is

$$M_1 = b_f \cdot D_f \frac{c}{2}\left[\frac{2kd - D_f}{kd}\right]\left[d - \frac{D_f}{2} + \frac{D_f^2}{6(2kd - D_f)}\right]$$

$$= 800 \times 100 \times \frac{5}{2}\left[\frac{2 \times 187.8 - 100}{187.8}\right] \times \left[480 - 50 + \frac{100 \times 100}{6(2 \times 187.8 - 100)}\right]$$

$$= 293504 \times [436] = 128 \times 10^6 \text{ N-mm; where } 436 = a = \text{L.A.}$$

$$A_{st1} = \frac{M_1}{t \times a} = \frac{128 \times 10^6}{140 \times 436} = 2097 \text{ mm}^2$$

$$M_2 = M - M_1 = (180 - 128)10^6 = 52 \times 10^6 \text{ N-mm}$$

∴ $\quad 52 \times 10^6 = (m_c - 1) A_{sc} \times c \left[\dfrac{kd - d_c}{kd}\right](d - d_c)$, Assuming $d_c = 40$ mm,

$$A_{sc} = \frac{(52 \times 10^6) \times (187.8)}{(1.5 \times 18 - 1) \times 5 (187.8 - 40)(480 - 40)} = 1155 \text{ mm}^2, \text{ and}$$

$$A_{st2} = \frac{M_2}{t(d - d_c)} = \frac{52 \times 10^6}{140(480 - 40)} = 844 \text{ mm}^2$$

∴ Total tensile steel $A_{st} = 2097 + 844 = \textbf{2941}$ mm$^2$ and Comp. steel $A_{sc} = \textbf{1155 mm}^2$.

**Example 6.9.** *Fig 6.15 shows the section of a doubly reinforced T-beam section. The tensile reinforcement consist of 5 bars of 24 mm diameter while the compressive reinforcement consists of 3 bars of 20 mm diameter. Taking $\sigma_{cbc} = 5$ N/mm$^2$, $\sigma_{st} = 140$ N/mm$^2$ and m = 18, determine the moment of resistance of the section (a) neglecting compression of web (b) taking compression of web into account. The beam is used over an effective span of 5 m.*

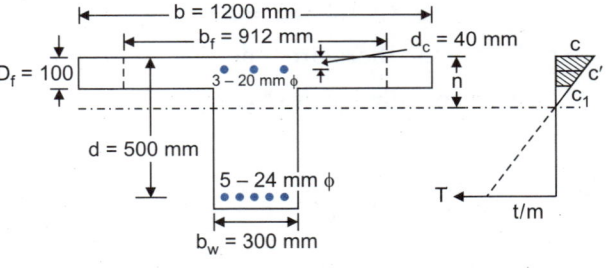

Fig. 6.15

**Solution:** The effective width of flange is given by

$$b_f = \frac{l_0}{(l_0/b) + 4} + b_w = \frac{5000}{(5000/1200) + 4} + 300 = 912 \text{ mm}$$

**(a) Neglecting Web Compression:**

Let the N.A. lie in the web at a distance $n$ below the top of the flange. Taking the moment of equivalent areas about N.A. we get

$$b_f D_f\left(n - \frac{D_f}{2}\right) + (m_c - 1) A_{sc}(n - d_c) = m A_{st}(d - n)$$

where $A_{sc} = 3\dfrac{\pi}{4}(20)^2 = 942.5$ mm$^2$ and $A_{st} = 5\dfrac{\pi}{4}(24)^2 = 2262$ mm$^2$

∴ $\quad 912 \times 100(n - 50) + (1.5 \times 18 - 1)$

$$942.5(n - 40) = 18 \times 2262(500 - n)$$

or $\quad\quad\quad\quad\quad\quad\quad n - 50 + 0.269\, n - 10.78 = 223 - 0.446\, n$

**140** REINFORCED CONCRETE STRUCTURE

Given expression is correct
from which $n = 165.5$ mm. Depth of critical N.A. is given by

$$n_c = \frac{m\sigma_{cbc}}{m\sigma_{cbc} + \sigma_{st}} d = \frac{18 \times 5}{18 \times 5 + 140} \times 500 = 196 \text{ mm}$$

Since the actual N.A. falls above the critical N.A., the section is under-reinforced and the stress in steel will reach its maximum value first. Hence $t = \sigma_{st} = 140$ N/mm².

$$\therefore \quad c = \frac{t}{m} \cdot \frac{n}{d-n} = \frac{140}{18} \times \frac{165.5}{500 - 165.5} = 3.848 \text{ N/mm}^2$$

The moment of resistance is given by

$$M_r = b_f \cdot D_f \frac{c}{2}\left[1 + \frac{n - D_f}{n}\right]\left[d - \frac{D_f}{2} + \frac{D_f^2}{6(2n - D_f)}\right] + (m_c - 1) A_{sc} \cdot c \times \frac{n - d_c}{n} (d - d_c)$$

$$= 912 \times 100 \times \frac{3.848}{2}\left[1 + \frac{165.5 - 100}{165.5}\right] \times \left[500 - 50 + \frac{(100)^2}{6(2 \times 165.5 - 100)}\right]$$

$$+ (1.5 \times 18 - 1) \times 945 \times 3.848 \; \frac{165.5 - 40}{165.5} \; (500 - 40)$$

$$= 111.98 \times 10^6 + 32.98 \times 10^6 = 144.96 \times 10^6 \text{ N-mm} = \mathbf{144.96 \text{ kN-m}}$$

**(b)** *Taking Compressive Force of Web into Account*

$$b_f \cdot D_f \left(n - \frac{D_f}{2}\right) + \frac{b_w (n - D_f)^2}{2} + (m_c - 1) A_{sc} (n - d_c) = m A_{st}(d - n)$$

or $\quad 912 \times 100 (n - 50) + 150(n - 100)^2 + (1.5 \times 18 - 1)945(n - 40) = 18 \times 2260 (500 - n)$

or $\quad 608n - 30400 + n^2 + 10000 - 200n + 164n - 6552 = 135600 - 271n$

or $\quad n^2 + 843 n - 162552 = 0$. From which $n = 161.8$ mm. As found earlier, $n_c = 196$ mm.
The section is, therefore, under-reinforced, and $t = \sigma_{st} = 140$ N/mm².
The corresponding stresses $c$, $c'$ and $c_1$ are as follows:

$$c = \frac{t}{m} \cdot \frac{n}{d - n} = \frac{140}{18} \times \frac{161.8}{500 - 161.8} = 3.72 \text{ N/mm}^2.$$

$$c' = \frac{t}{m} \cdot \frac{n - d_c}{d - n} = \frac{140}{18} \times \frac{161.8 - 40}{500 - 161.8} = 2.80 \text{ N/mm}^2.$$

$$c_1 = \frac{t}{m} \cdot \frac{n - D_f}{d - n} = \frac{140}{18} \times \frac{161.8 - 100}{500 - 161.8} = 1.42 \text{ N/mm}^2.$$

The moment of resistance of the beam will be equal to the sum of the moment of resistance due to (*i*) compressive force of flange (*ii*) compressive force of rib, and (*iii*) compressive force of steel.

or $\quad M_r = M_{r1} + M_{r2} + M_{r3}$

Now $\quad M_{r1} = $ M.R. due to compression in flange $= b_f \cdot D_f \left(\dfrac{c + c_1}{2}\right)(d - \bar{y})$

where $\quad \bar{y} = \left(\dfrac{c + 2c_1}{c + c_1}\right)\dfrac{D_f}{3} = \left(\dfrac{3.72 + 2.84}{3.72 + 1.42}\right)\dfrac{100}{3} = 42.54$ mm

$\therefore \quad M_{r1} = 912 \times 100 \left(\dfrac{3.72 + 1.42}{2}\right)(500 - 42.54) = 107.22 \times 10^6$ N-mm

$M_{r2}$ = M.R. due to compressive force in rib = $b_w (n - D_f) \dfrac{c_1}{2} \left[ d - D_f - \dfrac{1}{3}(n - D_f) \right]$

$= 300(161.8 - 100) \dfrac{1.42}{2} \left[ 500 - 100 - \dfrac{1}{3}(161.8 - 100) \right] = 4.99 \times 10^6$ N-mm

$M_{r3}$ = M.R. due to compressive force in steel = $(m_c - 1) A_{sc} \cdot c' (d - d_c)$
$= (1.5 \times 18 - 1) \, 945 \times 2.80 \, (500 - 40) = 31.65 \times 10^6$ N-mm

∴ Total $M = (107.22 + 4.99 + 31.65) \, 10^6 = 143.86 \times 10^6$ N-mm = **143.86 kN-m**.

**Example 6.10.** *A T-beam has the following dimensions: width of flange 1000 mm, thickness of flange 120 mm, width of web 250 mm; effective depth 500 mm. The beam is reinforced with 2400 sq. mm of steel in tension and 1200 sq. mm of steel in compression. Effective cover of compressive steel = 40 mm. Determine the following:*

 (i) *depth of N.A.*
 (ii) *stress in compressive and tensile steel, if maximum stress in concrete at the top edge is $4 \, N/mm^2$.*
 (iii) *moment of resistance of the section.*
 (iv) *lever arm.*

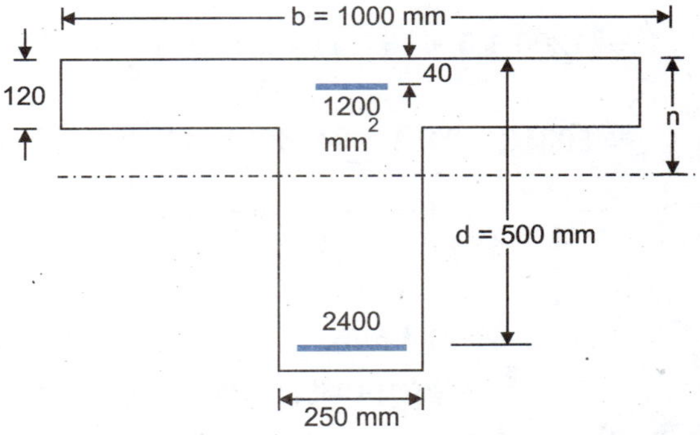

*Take m = 16 and assume that the beam is simply supported over a span of 12.5. m.*

**Solution:** For an isolated T-beam,

$$b_f = \dfrac{l_0}{(l_0 / b) + 4} + b_w = \dfrac{12500}{(12500/1000) + 4} + 250 = 1007.6 \text{ mm} > b.$$

Hence $b_f = b = 1000$ mm. Let the depth of N.A. be $n$ below the top of flange. Taking moments of the equivalent areas about the N.A., we get

$$b_f \cdot D_f \left( n - \dfrac{D_f}{2} \right) + \dfrac{b_w(n - D_f)^2}{2} + (m_c - 1) A_{sc} (n - d_c)$$
$$= m A_{st} (d - n)$$

∴ $1000 \times 120 \left( n - \dfrac{120}{2} \right) + \dfrac{250(n - 120)^2}{2} + (1.5 \times 16 - 1) \times 1200 \, (n - 40)$
$= 16 \times 2400 \, (500 - n)$

or $120000 n - 7200000 + 125 n^2 + 1800000 - 30000 n + 27600 n - 110400$
$= 19200000 - 38400 n$

or $n^2 + 1248 n - 205632 = 0.$

From which $n = $ **147 mm.**

Maximum compressive stress in concrete = $4 \, N/mm^2$ (given)

$t = mc \dfrac{d - n}{n} = 16 \times 4 \times \dfrac{500 - 147}{147}$
$= $ **153.7 N/mm²**

$c' = c \dfrac{n - d_c}{n} = 4 \times \dfrac{147 - 40}{147} = $ **2.91 N/mm²**

∴ Stress $t_c = m_c \cdot c' = 1.5 \times 16 \times 2.91 = $ **69.8 N/mm²**

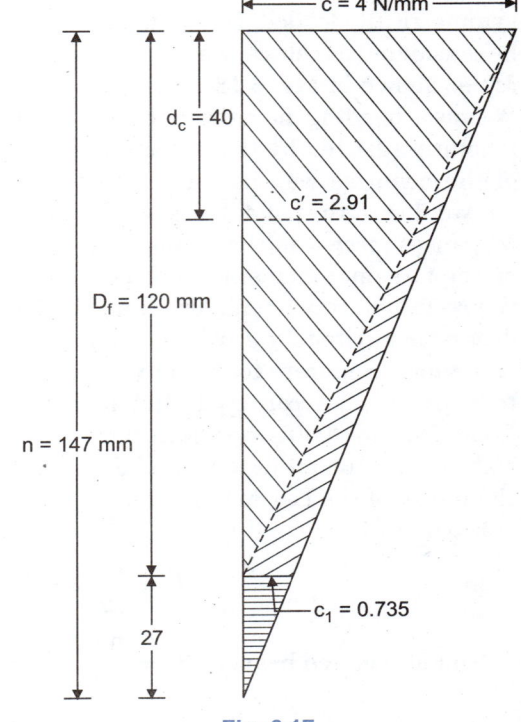

Fig. 6.16

Fig. 6.17

Stress $\quad c_1 = c \dfrac{n - D_f}{n} = 4 \times \dfrac{147 - 120}{147} = 0.735$ N/mm².

In order to find the *lever arm*, let us find distance $\bar{y}$ of the centroid of the compressive force. Fig. 6.17 shows the compressive stress diagram in which the total compressive force has been split into four components: (*i*) compressive force in steel  (*ii*) compressive force in rib  (*iii*) compressive force in flange, due to upper triangle and  (*iv*) compressive force in flange, due to lower triangle.

Total compressive force

$$C = A_{sc}(m_c - 1)c' + \dfrac{1}{2} c_1 (n - D_f) b_w + b_f \cdot D_f \cdot \dfrac{c}{2} + b_f \cdot D_f \cdot \dfrac{c_1}{2}$$

$$= [1200(1.5 \times 16 - 1) \times 2.91] + \left[\dfrac{1}{2} \times 0.735 (147 - 120) 250\right] + \left[1000 \times 120 \times \dfrac{4}{2}\right] + \left[1000 \times 120 \times \dfrac{0.735}{2}\right]$$

$$= 80316 + 2481 + 240000 + 44100 = 366897 \text{ N}$$

$$\bar{y} = \dfrac{(80316 \times 40) + 2481\left[120 + \dfrac{27}{3}\right] + \left[240000 \times \dfrac{120}{3}\right] + \left[44100 \times \dfrac{120 \times 2}{3}\right]}{366897}$$

or $\quad \bar{y} = 5.4$ mm. $\quad \therefore \quad$ Lever arm $a = d - \bar{y} = 500 - 45.4 = 454.6$ mm

$\therefore \quad M_r$ = compressive force × lever arm = $366897 \times 454.6 \approx 167 \times 10^6$ N-mm = **167 kN-m**

**Check** $\quad M_r = A_{st} \cdot \sigma_{st} \cdot a = 2400 \times 153.7 \times 454.6 \approx 167 \times 10^6$ N-mm

## 6.11. L-BEAM

In the case of T-beam, the rectangular beam has flanges on both the sides and hence load is transferred to it in vertical plane passing through the middle of its width. However, in the case of monolithic beam-slab construction, the end beams have flanges to one side only, as shown in Fig. 6.18(*a*). Such a beam having the shape of letter 'L' is known as L-beam.

Fig. 6.18 (*b*) shows the loading on the beam received from the slab. Evidently, the beam is not symmetrically loaded. Due to this, an eccentric load $W$ is transferred to the beam. This eccentric load may be considered as central load $W$ plus a moment $M_t$, as shown in Fig. 6.18(*c*). The central load $W$ causes bending moment in the beam, which is jointly resisted by the rectangular portion of the beam as well as the flange (of width $B$), similar to that of a T-beam. The moment $M_t$ causes *torsion* in the beam, and is known as the torsional moment which is resisted by the rectangular portion alone, and the flange does not contribute to any torsional moment of resistance. Separate torsional resistance has to be provided. According to Indian Standard Code (IS: 456-2000), the width $B$ (or $b_f$) of the slab, acting monolithic with the beam, forming the flange of the L-beam is taken as the least of the following:

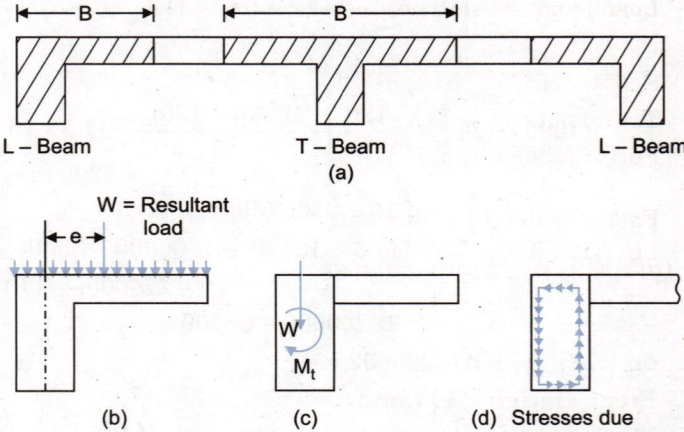

Fig. 6.18. L-Beam.

(*a*) $\quad b_f = \dfrac{l_0}{12} + b_w + 3 D_f$

(*b*) For isolated beam, $\quad b_f = \dfrac{0.5 \, l_0}{\dfrac{l_0}{b} + 4} + b_w$

# DESIGN OF T AND L-BEAMS

(but in no case greater than the actual width of flange) where $l_0$ is the distance between points of zero moments in the beam. For continuous beams and frames, $l_0$ may be assumed as 0.7 times the effective span.

The location of the N.A. and the calculations of the moment of resistance of L-beam are done exactly in the same way as that for T-beam, and all the equations developed for T-beam are also applicable for L-beam. The design for bending moment is also done exactly in the way as that for T-beam. However, additional reinforcement, as discussed in chapter 4, has to be provided to resist shear induced by torsional moment.

The design procedure for L-beam is given chapter 7.

**Example 6.11.** *A Tee-beam floor consists of 150 mm thick R.C.C slab monolithic with 300 mm wide beams. The beams are spaced at 3.5 m centre to centre and their effective span is 6 m. If the super imposed load on the slab is 5 kN/m², design an intermediate tee-beam. Use $M_{20}$ mix and Fe 415 steel.*

**Solution:** Given data
Span $\quad l = 6$ m
Spacing of beams $\quad = 3.5$ m c/c
Depth $\quad D_f = 150$ mm
$\quad b_w = 300$ mm
$\quad f_{ck} = 20$ N/mm²
$\quad f_y = 415$ N/mm²

(*i*) **Depth of beam:**
Selecting the depth in range of $l/12$ to $l/15$ based on stiffness.
$$d = \frac{6000}{15} = 400 \text{ mm.}$$
Adopt, $\quad d = 400$ mm
$\quad D = 450$ mm.

(*ii*) **Loads:**
Dead load of the slab $\quad = 0.15 \times 25 = 3.75$ kN/m²
Live load from the slab $\quad = 5$ kN/m²
Total load $\quad = 8.75$ kN/m².
Load from the slab per meter run of the beam = Load on slab × c/c distance beams
$\quad = 8.75 \times 3.5 = 30.63$ kN/m.
Self weight of the beam $\quad = 0.3 \times 0.3 \times 1 \times 25 = 2.25$ kN/m.
Total load $\quad = 32.88$ kN/m.
Factored load($w_u$) $\quad = 1.5 \times 32.88 = 49.32$ kN/m

Factored Bending Moment: $\quad \dfrac{M_u l^2}{8} = \dfrac{49.32 \times 6^2}{8} = 221.94$ kN-m.

(*iii*) **Effective width of flange:**
$$b_f = \frac{l_0}{6} + b_w + 6D_f$$
$$= \frac{6000}{6} + 300 + 6 \times 150 = 2200 \text{ mm, limited to 3500 mm.}$$
Hence, $\quad b_f = 2200$ mm.

(*iv*) Assuming $x_u$ is with in flange
$$x_u = \frac{0.87 f_y A_{st}}{0.36 f_{ck} b_f}$$
$$= \frac{0.87 \times 415 \times A_{st}}{0.36 \times 20 \times 2200} = 0.0228 A_{st}.$$

(v) Reinforcement:
$$M_u = 0.87 f_y A_{st} (d - 0.42 x_u)$$
$$221.94 \times 10^6 = 0.87 \times 415 \times A_{st} (400 - 0.42 \times 0.0228 A_{st})$$
$$A_{st}^2 - 41776 A_{st} + 64.2 \times 10^6 = 0.$$

Solving the Quadratic Equation
$$A_{st} = 1597.9 \text{ mm}^2$$
$$x_u = 0.0228 A_{st} = 0.228 \times 1597.9$$
$$= 36.4 \text{ mm} < D_f$$

Hence, our assumption is correct
Minimum area of tension steel.
$$A_{st\ min} = \frac{0.85\ b_w d}{f_y}$$
$$= \frac{0.85 \times 300 \times 400}{415}$$
$$= 245.8 \text{ mm}^2 < A_{st} \text{ provided.}$$
$$A_{st\ max} = 0.04\ b_w d$$
$$= 0.04 \times 300 \times 450$$
$$= 5400 \text{ mm}^2 > A_{st}$$

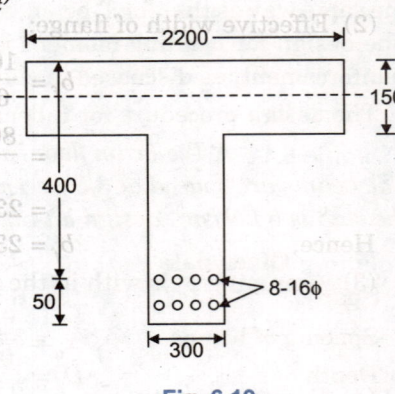

Fig. 6.19

So, provide 8–16 mm φ bars.
$$A_{st} = 1608.5 \text{ mm}^2.$$

**Example 6.12.** *Calculate the area of reinforcement required for a T-beam with the following data:*

*span = 8 m, Ends simply supported*
*spacing of beams = 3 m,*
*Super imored load = 3 kN/m²*
*Thickness of slab = 130 mm*
*Weight of wall on beam = 15 kN/m.*
*Assume width of web 230 mm, total depth 680 mm, use $M_{20}$ and $Fe_{250}$.*

**Solution: Given data**
$$\text{Span, } l = 8 \text{ m}$$
$$D_f = 130 \text{ mm}$$
$$D = 680 \text{ mm}$$
$$b_w = 230 \text{ mm}$$
$$f_{ck} = 20 \text{ N/mm}^2$$
$$f_y = 250 \text{ N/mm}^2.$$

(1) Loads:
Dead load of the slab = 0.13 × 25 = 3.25 kN/m.
Live load from the slab = 3kN/m².
Total load from slab = 6.25 kN/m².
Load from the slab = Load on slab × c/c distance between beams
= 6.25 × 3 = 18.75 kN/m,
Self weight of the beam = 0.55 × 0.23 × 1 × 25 = 3.16 kN/m.
Weight of wall on the beam = 15 kN/m.
Total load = 36.91 kN/m.
Factored load $W_u$ = 1.5 × 36.91
= 55.37 kN/m.

Factored Bending Moment

$$M_u = \frac{W_u l^2}{8} = \frac{55.37 \times 8^2}{8} = 442.96 \text{ kN-m}$$

(2) Effective width of flange:

$$b_f = \frac{l_0}{6} + b_w + 6 D_f$$

$$= \frac{8000}{6} + 230 + 6 \times 130$$

$$= 2343 \text{ mm, limiting to 3000 mm.}$$

Hence, $b_f = 2343$ mm.

(3) Assuming $x_u$ is with in the flange

$$x_u = \frac{0.87 f_y A_{st}}{0.36 f_{ck} b_f} = \frac{0.87 \times 250 \times A_{st}}{0.36 \times 20 \times 2343} = 0.0129 A_{st}.$$

(4) Reinforcement:

$$M_u = 0.87 f_y A_{st} (d - 0.42 x_u)$$
$$442.96 \times 10^6 = 0.87 \times 250 \times A_{st} (650 - 0.42 \times 0.0129 A_{st})$$
$$A_{st}^2 - 120012.7 A_{st} + 376 \times 10^6 = 0.$$

Solving Quadratic Equation

$$A_{st} = 3219.4 \text{ mm}^2$$
$$x_u = 0.0129 A_{st}$$
$$= 0.0219 \times 3219.4$$
$$= 41.5 \text{ mm} < D_f$$

So, assumption is correct.

Minimum area of tension steel

$$A_{st \, min} = \frac{0.85 \, b_w d}{f_y}$$

$$= \frac{0.85 \times 230 \times 650}{250} = 508.3 \text{ mm}^2 < A_{st}$$

Maximum area of tension steel

$$A_{st \, max} = 0.04 \, b_w D$$
$$= 0.04 \times 230 \times 680$$
$$= 6256 \text{ mm}^2 > A_{st}$$

So, provide 11–20 mm $\phi$ bars

$$A_{st} = 3455.8 \text{ mm}^2.$$

## PROBLEMS

1. Determine the moment of resistance of T-beam section having the following dimensions: width of flange 600 mm; thickness of slab 80 mm; tensile steel-area 3200 mm² centred at 500 mm from the top; width of web 250 mm. Take the permissible value of stresses in concrete and steel as 4.2 N/mm² and 126 mm² and $m = 18$. Neglect the compression of the rib. The beam is used over an effective span of 4 m.

2. Solve problem 1, taking into account compressive force in the rib.

3. A T-beam has flange width of 800 mm, flange thickness 80 mm and an effective depth of 500 mm. The reinforcement consist of 5 bars of 20 mm diameter and the width of rib is 300 mm. Determine the maximum uniformly distributed load that the beam can carry over an effective span of 5 m. Neglect compression of rib. Take permissible stress in concrete and steel as 5 N/mm² and 40 N/mm² and $m = 18$.

4. A T-beam has a flange of 1000 mm width and 80 mm thickness. The effective depth is 300 mm and the width of rib is 200 mm. Taking $\sigma_{cbc}$ = 5 N/mm$^2$, $\sigma_{st}$ = 140 N/mm$^2$ and $m$ = 18, determine (a) area of steel in tension and (b) moment of resistance of the beam if it is used over an effective span of 3 m.

5. A R.C. T-beam has the following dimensions: width of flange 1300 mm; thickness of flange 100 mm; effective depth 560 mm; width of rib 300 mm; reinforcement: 6 bars of 25 mm diameter. Determine the stresses developed in the concrete and steel, when the beam is subjected to a bending moment of 180 kN-m. The beam has an effective span of 5 m.

6. A T-beam roof consists 100 mm thick reinforced concrete slab cast monolithic with 250 mm wide beams spaced 3 metres centre to centre, the effective span of each beam being 4 metres. Design one of the intermediate beams for a super imposed load of 3 kN/m$^2$ over the slab.

Take $\sigma_{cbc}$ = 5 N/mm$^2$, $\sigma_{st}$ = 140 N/mm$^2$ and $m$ = 18. The ratio of cost of steel to that of concrete may be taken as 50.

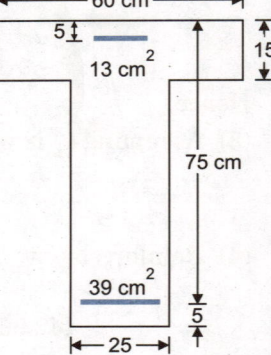

7. The sectional dimensions of a T-beam are shown in Fig. 6.20. The beam is reinforced width 3900 sq. mm of steel in tension and 1300 sq. mm in compression. Taking $c$ = 7 N/mm$^2$ and $m$ = 15, calculate from the first principles:

(a) The depth of N.A. (b) The stress in the soffit of the top slab in terms of the top edge stress of $c$ = 7 N/mm$^2$ (c) Compressive stress in the top steel (d) Moment of resistance of the section (e) Lever arm.

Assume that the beam is simply supported over a span of 5 m.

Fig. 6.20

8. The flange of an L-beam is 800 × 130 mm. The breadth of rib is 300 mm and effective depth is 560 mm. Find the area of tensile steel required at a section where the bending moment is 350 kN-m. Assume $\sigma_{cbc}$ = 5 N/mm$^2$, $\sigma_{st}$ = 140 N/mm$^2$, $m$ = 18.66.

9. An RC 'T' beam 10 metres long is reinforced with five 30 mm$\phi$ bars on tension side. The flange is 1500 ×120 mm and web is 300 × 370 mm effective. Total load on beam including self weight is 20 kN/m. Determine the stresses in steel and concrete at the centre of span. ($m$ = 13.33)

10. Design a T-beam spanning 5.8 mtrs, supporting a one way slab thickness 150 mm, and subjected to a live load of 3.5 kN/m$^2$ and a dead load (due to Floor Finish, Partition etc) of 1.2 kN/m$^2$, in addition to it's self weight. Assume Fe 415, steel and $M_{25}$ concrete and the centre to centre of beams as 4.2 mtr.

## ANSWERS

1. 80 kN-m
2. 101 kN-m
3. 32.47 kN-m
4. (a) 1185 mm$^2$ (b) 44.25 kN-m
5. $c$ = 4.79 N/mm$^2$; $t$ = 127.4 N/mm$^2$
7. (a) 282 mm (b) 3.28 N/mm$^2$ (c) 174.3 N/mm$^2$ (d) 460 kN-m (e) 677 mm.
8. $A_{st}$ = 4961.8 mm$^2$
9. $\sigma_{cbc}$ = 5.227 N/mm$^2$
   $\sigma_{st}$ = 159.46 N/mm$^2$

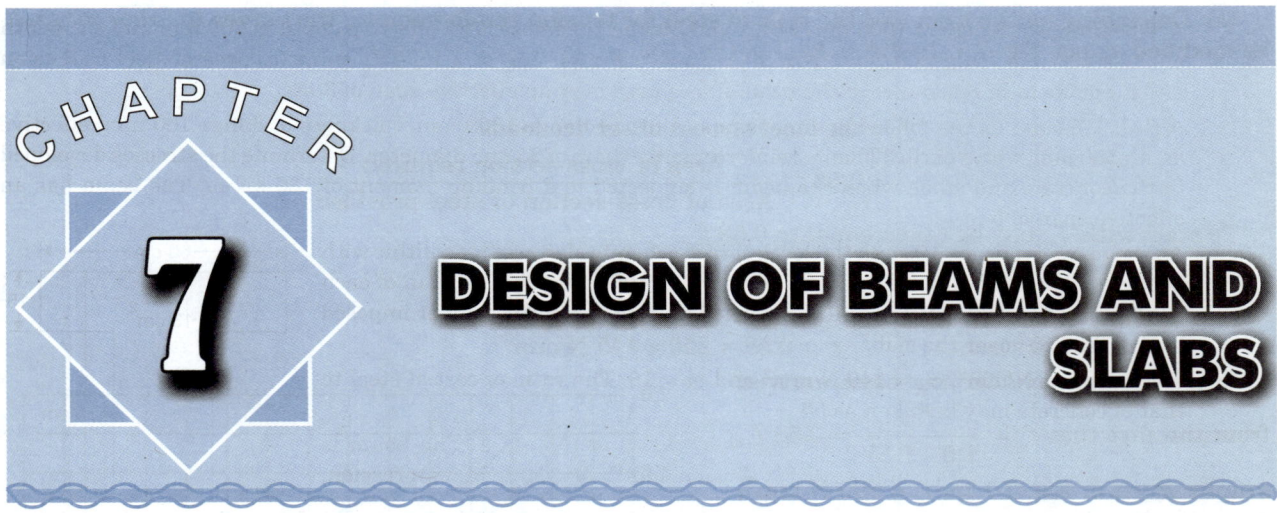

# DESIGN OF BEAMS AND SLABS

## 7.1. DESIGN OF BEAMS

### (A) BASIC RULES FOR DESIGN (IS : 456–2000)

We have already discussed the theory of singly reinforced beams. While designing the beams, following important rules must be kept in mind:

1. **Effective Span.** Unless otherwise specified, the effective span of a member shall be as follows:

   (a) *Simply supported beam or slab.* The effective span of a member that is not built integrally with its supports shall be taken as clear span plus the effective depth of slab or beam or centre to centre of supports, whichever is less.

   (b) *Continuous beam or slab.* In case of continuous beam or slab, if the width of the support is less than 1/12 of the clear span, the effective span shall be as in (a) above. If the supports are wider than 1/12 of the clear span or 600 mm whichever is less, the effective span shall be taken as under:

   (i) For the end span with one end fixed and the other continuous or for intermediate spans, the effective span shall be clear span between supports, and

   (ii) For the end span with one end free and the other continuous, the effective span shall be equal to the clear span plus half the effective depth of the beam or slab or the clear span plus half the width of the discontinuous support, whichever is less.

   (iii) In the case of spans with roller or rocker bearings, the effective span shall always be the distance between the centres of bearings.

   (c) *Cantilever.* The effective length of cantilever shall be taken as its length to the face of the support plus half the effective depth except where if forms the end of a continuous beam where the length to the centre of the support shall be taken.

   (d) *Frames.* In the analysis of a continuous frame, centre to centre distance shall be used.

2. **Control of Deflection.** The deflection of a structure or part there of shall not adversely affect the appearance or efficiency of the structure or finishes or partitions. For beams and slabs, the vertical deflection limits may generally be assumed to be satisfied provided that the span to depth ratios are not greater than the values obtained as below:

   (a) Basic values of span to effective depth ratios for span upto 10 m:

   | | |
   |---|---|
   | Cantilever | 7 |
   | Simply supported | 20 |
   | Continuous | 26 |

   (b) For spans above 10 m, the values in (a) may be multiplied by 10/span in metres, except for cantilever in which case deflection calculations should be made.

(c) Depending on the area and the type of steel for tension reinforcement, the values in (a) or (b) shall be modified as per Fig. 7.1. In Fig. 7.1.

$$f_s = \text{the steel streess of service loads}$$
$$= 0.58 f_y \frac{\text{Area of cross-section required}}{\text{Area of cross-section of steel provided}}$$

To start with, $f_s$ may be taken equal to $0.58 f_y$
Thus, for Fe 500 steel, $\quad f_s = 0.58 \times 500 = 290 \text{ N/mm}^2$
For Fe 415 steel, $\quad f_s = 0.58 \times 415 \approx 240 \text{ N/mm}^2$
For Fe 250 steel, $\quad f_s = 0.58 \times 250 \approx 145 \text{ N/mm}^2$

Since the factor 0.58 has been arrived from the fact that $f_s = \dfrac{f_y}{1.5 \times 1.15} = 0.58 f_y$, graphs given in Fig. 7.1 for $f_s = 290 \text{ N/mm}^2$, $f_s = 240 \text{ N/mm}^2$ and $f_s = 145 \text{ N/mm}^2$ can be directly used for steel of grade Fe 500, Fe 415 and Fe 250 respectively *corresponding to the percentage of steel required*. Initially, $p_t\%$ is not known. Hence assume an initial values of $p_t$, and find the modification factor $m_{f_1}$. Since it is desirable to use *under-reinforced* section, the value of $p_t$ should be assumed to be equal to 40 to 50% of $p_{t \, . \, lim.}$ for mild steel bars and equal to about 30% of $p_{lim.}$ for HYSD bars.

Finally, when the beam has been designed, the required $p_t(\%)$ is known. Also, $A_{st}$ required and $A_{st}$ actually provided are also known. Hence one can compute $f_s$ as:

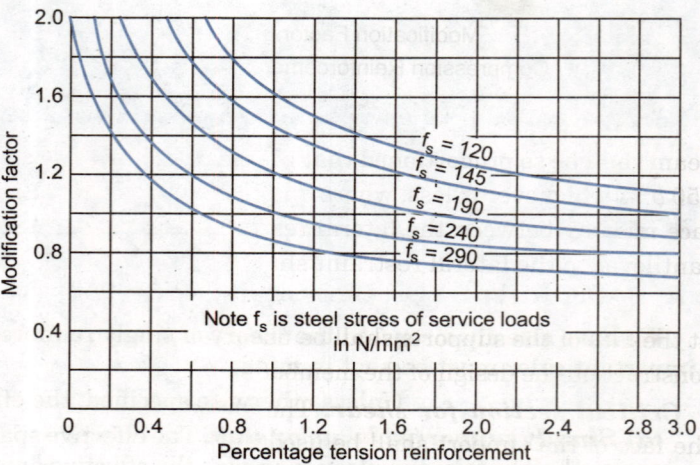

$$f_s = 0.58 f_y \frac{\text{Area of cross-section of steel required}}{\text{Area of cross-section of steel provided}}$$

**Fig. 7.1.** Modification Factor ($m_{f1}$) for Tension Reinforcement.

$$f_s = 0.58 f_y \frac{\text{Area of cross-section required}}{\text{Area of cross-section provided}} = 0.58 f_y \cdot r_a$$

Since the ratio $r_a$ is always less than 1, the value of $f_s$, to be taken to determine $m_{f_1}$, will be less than $0.58 f_y$.
For example, let $\quad f_y = 415 \quad$ and let $\quad r_a = 0.9$
Then $\quad f_s = 0.58 f_y \cdot r_a = 0.58 \times 415 \times 0.9 \approx 217 \text{ N/mm}^2$

Hence the final value of modification factor has to be found corresponding to an *imaginary curve* of $f_s = 217 \text{ N/mm}^2$ value, which falls between $f_s = 240$ and $f_s = 190$ values.

(d) Depending on the area of compression reinforcement, the value of span to depth ratio be further modified as per modification factor $m_{f2}$ given in Fig. 7.2.

(e) For flanged beams, the value of (a) or (b) be modified as per Fig. 7.3 and the reinforcement percentage for use in Fig. 7.1 and 7.2 should be based on area of section of equal to $b_f \, d$.

**Note**: Where deflection are required to be calculated, the method given in Appendix C of the Code (IS : 456-2000) may be followed.

**Solid Slabs:** The above provisions for beams also apply for slabs.

**Note 1.** For slabs spanning in two directions, the shorter of the two spans should be used for calculating the span to effective depth ratios.

**Note 2.** For two-way slabs of small spans (up to 3.5 m), with mild steel reinforcement, the span to overall depth ratio given below may generally be assumed to satisfy vertical deflection limits for loading class upto 3 kN/m².

Simply supported slabs: 35 $\hspace{4cm}$ Continuous slabs: 40

For high strength deformed bars of grade Fe 415, the values given above should be multiplied by 0.8.

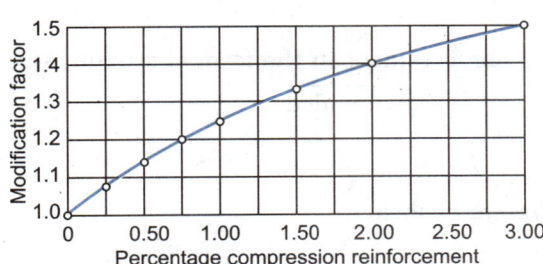

Fig. 7.2. Modification Factor $m_{f2}$ for Compression Reinforcement.

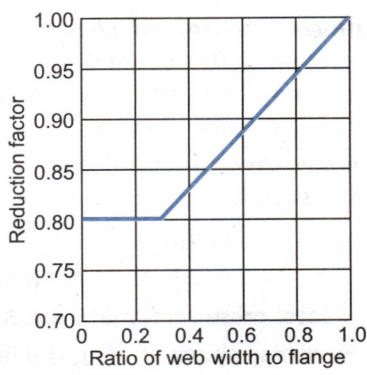

Fig. 7.3. Reduction Factor for Flanged Beams.

3. **Slenderness Limits for Beams to Ensure Lateral Stability.** A simply supported or continuous beam shall be so proportioned that the clear distance between the lateral restraints does not exceed $60\,b$ or $250\,b^2/d$ whichever is less, where $d$ is the effective depth of the beam and $b$ the breadth of the compression face midway between the lateral restraints. For a cantilever, the clear distance from the free end of the cantilever to the lateral restraint shall not exceed $25\,b$ or $100\,b^2/d$ whichever is less.

4. **Critical Sections for Moments and Shear.** For monolithic construction, the moments computed at the face of the supports shall be used in the design of the member at those sections. For non-monolithic construction the design of the member shall be done keeping in view para 1.

*Critical section for shear.* The shear computed at the face of the support shall be used in the design of the member at that section except as in case mentioned below.

When the reaction in the direction of the applied shear introduces compression into the end region of the member, sections located at the distance less than $d$ from the face of the support may be designed for the same shear as that computed at distance $d$ (see Fig. 7.4.) (Also, see § 3.14 and Fig. 3.18).

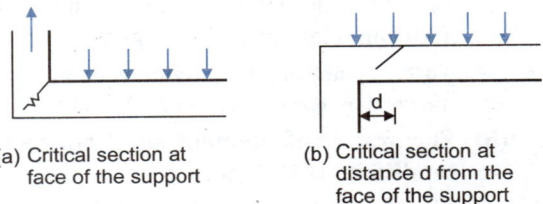

(a) Critical section at face of the support

(b) Critical section at distance d from the face of the support

Fig. 7.4. Critical Sections for Shear.

5. **Reinforcement in Beams**

(a) **Tension reinforcement**

(i) *Minimum reinforcement.* The minimum area of the tension reinforcement shall not be less than that given by the following:

$$\frac{A_{st}}{bd} = \frac{0.85}{f_y}$$ where $f_y$ is the characteristic strength of reinforcement in N/mm².

(ii) *Maximum reinforcement.* The maximum area of tension reinforcement shall not exceed $0.04\,bD$.

(b) **Compression reinforcement.** The maximum area of compression reinforcement shall not exceed $0.04\,bD$. Compression reinforcement in beams shall be enclosed by stirrups for effective lateral restraint.

(c) **Side face reinforcement.** Where the depth of the web in a beam exceeds 750 mm, side face reinforcement shall be provided along the two faces. The total area of such reinforcement shall be not less than 0.1 percent of the web area and shall be distributed equally on two faces at a spacing not exceeding 300 mm or web thickness whichever is less.

(d) **Transverse reinforcement in beams for shear and torsion.** The transverse reinforcement in beams shall be taken around the outer-most tension and compression bars. In T-beams and L-beams, such reinforcement shall pass around longitudinal bars located close to the outer face of the flange.

(e) **Maximum spacing of shear reinforcement.** The maximum spacing of shear reinforcement measured along the axis of the member shall not exceed $0.75\,d$ for vertical stirrups and $d$ for inclined stirrups at 45°, where $d$ is the effective depth of the sections under consideration. In no case shall the spacing exceed 300 mm.

(f) **Minimum shear reinforcement.** Minimum shear reinforcement in the form of stirrups shall be provided such that

$$\frac{A_{sv}}{b \cdot s_v} \geq \frac{0.4}{0.87\, f_y}$$

where  $A_{sv}$ = total cross-sectional area of stirrup legs effective in shear
  $s_v$ = stirrup spacing along the length of member
  $b$ = breadth of the beam or breadth of the web of flanged beam, and
  $f_y$ = characteristic strength of the stirrup reinforcement in N/mm² which shall not be taken greater than 415 N/mm².

Based on the above

$$s_v \leq \frac{0.87\, f_y\, A_{sv}}{0.4\, b} \leq \frac{2.175\, f_y\, A_{sv}}{b}$$

(g) **Maximum diameter.** The diameter of the reinforcing bars shall not exceed one eighth of the total thickness of the slab.

6. **Curtailment of Tension Reinforcement in Flexural Member:** See also § 3.24.

(a) For curtailment, reinforcement shall extend beyond the point at which it is no longer required to resist flexure for a distance equal to the effective depth of the member or 12 times the bar diameter, whichever is greater except at simple support or end of cantilever. In addition, clauses (b) to (f) given below shall be satisfied.

*Note.* A point at which reinforcement is no longer required to resist flexure is where the resistance moment of the section, considering only the continuing bars, is equal the design moment.

(b) Flexural reinforcement shall not be terminated in a tension zone unless any one of the following conditions is satisfied:

(i) The shear at the cut-off point does not exceed two thirds that permitted, including the shear strength of web reinforcement provided.

(ii) Stirrup area in excess of the required for shear and torsion is provided along each terminated bar over a distance from the cut-off point equal to three fourths the effective depth of the member. The excess stirrup area shall be not less than $0.4\,bs/f_y$, where $b$ is the width of the beam, $s$ is the spacing and $f_y$ is the characteristic strength of reinforcement in N/mm². The resulting spacing shall not exceed $d/8\,\beta_0$ where $\beta_0$ is the ratio of the area of bars cut-off to the total area of bars at the section, and $d$ is the effective depth.

(iii) For 36 mm and smaller bars, the continuing bars provide double the area required for flexure at the cut-off point and the shear does not exceed three-fourth that permitted.

(c) **Positive moment reinforcement.**

(i) At least one-third the positive moment reinforcement in simple members and one-fourth the positive moment reinforcement in continuous members shall extend along the same face of the member into the support, to a length equal to $L_d/3$.

(ii) When a flexural member is part of the primary lateral load resisting system, the positive reinforcement required to be extended into the support as described in (i) above shall be anchored to develop its design stress in tension at the face of the support.

7. **Development Length Requirements at Supports**

At simple supports and at point of inflection, positive moment tension reinforcement shall be limited to a diameter such that $L_d$ computed for $\sigma_{st}$ (by the relation $L_d = \Phi\,\sigma_{st}/4\,\tau_{bd}$) does not exceed $\left\{\dfrac{M_1}{V} + L_0\right\}$, where   $M_1$ = moment of resistance of the section assuming all reinforcement at the section to be stressed to $\sigma_{st}$.

   $V$ = S.F. at the section due to design loads

   $L_0$ = Sum of the anchorage beyond the centre of the support and the equivalent anchorage value of any look or mechanical anchorage at simple support; and at a point of inflection, $L_0$ is limited to the effective depth of the member or 12 $\Phi$ whichever is greater, and

   $\Phi$ = diameter of bars.

The value of $M_1/V$ in the above expression may be increased by 30 percent when the ends of the reinforcement are confined by a compression reaction. (See Fig. 3.26).

**Computation of $L_0$.** For the computation of $L_0$, the support width should be known. Fig. 7.5 shows a beam with end support, in which $x'$ is the side cover and $x_0$ is the distance of the *beginning* of the hook or bend from the centre line of the support.

If no hook is provided, as in case of deformed bars, $L_0 = \dfrac{l_s}{2} - x'$. However, if a hook or bend is provided, $L_0$ is computed as under.

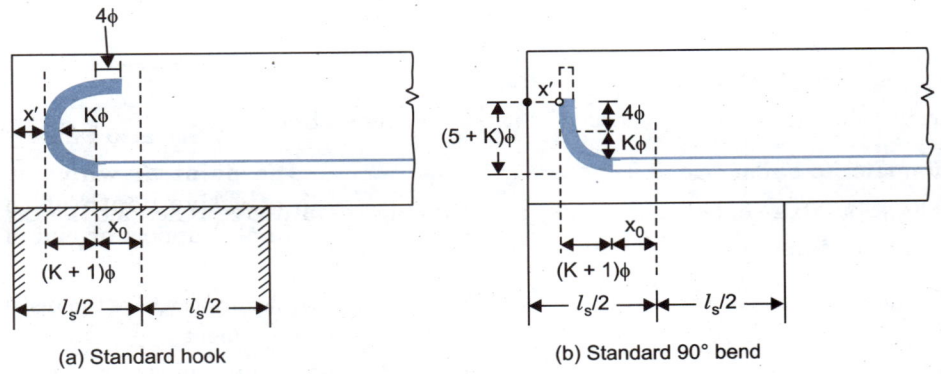

Fig. 7.5.  Computation of $L_0$

### (a) Provision of standard hook at the end [Fig. 7.5 (a)]

The dark portion shows the hook which has an anchorage value of 16 $\varphi$ (IS 456 : 2000), for all types of steel. The distance of the beginning of the hook from its apex of the semi-circle is equal to $(K + 1)\varphi$. For mild steel bars, $K = 2$ and for HYSD bars, $K = 4$. Hence this distance is equal to $3\varphi$ for mild steel bars and $5\varphi$ for HYSD bars. Let $l_s$ be the width of the support.

Then $\qquad L_0 = x_0 + 16\,\varphi \qquad$ where $\qquad x_0 = \dfrac{l_s}{2} - x' - (K+1)\,\varphi$

$\therefore \qquad L_0 = \left(\dfrac{l_s}{2} - x' - (K+1)\,\varphi\right) + 16\,\varphi = \dfrac{l_s}{2} - x' + (15 - K)\,\varphi \qquad$ ...(3.32)

Taking $\qquad K = 2$ for mild steel bars, $\quad L_0 = \dfrac{l_s}{2} - x' + 13\,\varphi \qquad$ ...[3.32 (a)]

Taking $\qquad K = 4$ for HYSD bars, $\quad L_0 = \dfrac{l_s}{2} - x' + 11\,\varphi \qquad$ ...[3.32 (b)]

### (b) Provision of 90° standard bend (Fig. 7.5 b)

The dark portion shows the 90° bend which has an anchorage value of $8\varphi$ (IS : 456: 2000), for all types of steel. Here also, the distance of beginning of the hook from its apex of the semi-circle is equal to $(K + 1)\varphi$.

Then $\qquad L_0 = x_0 + 8\,\varphi \quad$ where $\quad x_0 = \dfrac{l_s}{2} - x' - (K+1)\,\varphi$

$$L_0 = \left(\frac{l_s}{2} - x' - (K+1)\varphi\right) + 8\varphi = \frac{l_s}{2} - x' + (7-K)\varphi \qquad ...(3.33)$$

Taking $\qquad K = 2$ for mild steel bars, $\qquad L_0 = \frac{l_s}{2} - x' + 5\varphi \qquad ...[3.33\,(a)]$

Taking $\qquad K = 4$ for HYSD bars, $\qquad L_0 = \frac{l_s}{2} - x' + 3\varphi \qquad ...[3.33\,(b)]$

### Remedies to get development length

If the check for the satisfaction of *Eq. 3.31 or 3.31 (a)* is not obtained, following remedial measures may be adopted to satisfy the check:

1. Reduce the diameter $\varphi$ of the bar, thereby reducing the value of $L_d$, keeping the area of steel at the section unchanged. This is the standard procedure envisaged in the Code, *i.e.* reducing the value of $L_d$ by *limiting* the diameter of the bar to such a value that *Eq. 3.31 or 3.31 (a)* is satisfied.

2. Increasing the value $L_0$ by providing *extra length* of the bend over and above the standard value $(5 + K)\varphi$ shown by dotted lines in Fig. 7.5 (b).

3. By increasing the number of bars (thereby increasing $A_{st1}$) to be taken into the support. This method is uneconomical.

4. By providing adequate mechanical anchorage.

We shall discuss the first remedial method in the following section.

### Limiting bar diameter to satisfy *Eq. 3.31 or 3.31 (a)*

IS 456 : 2000 stipulates that at simple supports having compressive reaction, the positive moment reinforcement shall be limited to a diameter such that

$$L_d \leq \frac{1.3\,M_1}{V} + L_0 \qquad ...[3.31\,(a)]$$

Substituting $\qquad L_d = \frac{\sigma_s\,\varphi}{4\,\tau_{bd}}$, in the above condition, we get

$$\frac{\sigma_s \cdot \varphi}{4\,\tau_{bd}} \leq 1.3\frac{M_1}{V} + L_0 \quad \text{or} \quad \varphi \leq \frac{4\,\tau_{bd}}{\sigma_s}\left(1.3\frac{M_1}{V} + L_0\right) \qquad ...(3.34)$$

Similarly, for simple supports not offering compressive reaction,

$$\varphi \leq \frac{4\,\tau_{bd}}{\sigma_s}\left(\frac{M_1}{V} + L_0\right) \qquad ...(3.35)$$

**8. Spacing of Reinforcement.** For the purpose of the clause, the diameter of a round bar shall be its nominal diameter and in the case of bars which are not round, or in case of deformed bars or crimped bars; the diameter shall be taken as the diameter of a circle giving an equivalent effective area. Where spacing limitations and minimum concrete cover (see clause 9) are based on bar diameter, a group of bars bundled in contact shall be treated as a single bar of diameter derived from the total equivalent area.

#### (a) Minimum distance between individual bars

The following shall apply for spacing for bars:

(*i*) The horizontal distance between two parallel main reinforcing bars shall usually be not less than the greatest of the following:

(1) The diameter of the bar if the diameters are equal

(2) the diameter of the larger bar if the diameters are unequal, and

(3) 5 mm more than the nominal maximum size of coarse aggregate.

**Note.** This does not preclude the use of larger size of aggregates beyond the congested reinforcement in the same member; the size of aggregates may be reduced around congested reinforcement to comply with this provision.

(*ii*) Greater horizontal distance than the minimum specified in (*i*) should be provided wherever possible. However when needle vibrators are used the horizontal distance between bars of a group may be reduced to two-thirds the nominal maximum size of the coarse aggregate, provided that sufficient space is left between groups of bars to enable the vibrator to be immersed.

(*iii*) Where there are two or more rows of bars, the bars shall be vertically in line and the minimum vertical distance between the bars shall be 15 mm, two-thirds the nominal maximum size of aggregate or the maximum size of bar whichever is the greatest.

(**b**) *Maximum distance between bars in tension:* Unless the calculation of crack widths shows that a greater spacing is acceptable, the following rules shall be applied to flexural members in normal internal or external conditions of exposure. The horizontal distance between parallel reinforcement bars, or groups, near the tension face of a beam shall not be greater than the value of in Table 7.1 depending on the amount of redistribution carried out in analysis and the characteristic strength of the reinforcement.

TABLE 7.1. Clear Distance between Bars

| $f_y$ | Percentage redistribution to or from section considered | | | | |
|---|---|---|---|---|---|
| | − 30 | − 15 | 0 | +15 | +30 |
| | Clear distance between bars | | | | |
| N/mm² | mm | mm | mm | mm | mm |
| 250 | 215 | 260 | 360 | 300 | 300 |
| 415 | 125 | 155 | 180 | 210 | 235 |
| 500 | 105 | 130 | 150 | 175 | 135 |

**Note.** The spacings given in the Table 7.1 are not applicable to members subjected to particularly aggressive environments unless in the calculation of the moment of resistance, $f_y$ has been limited to 300 N/mm² in limit state design and $\sigma_{st}$ limited to 165 N/mm² in working stress design.

### 9. Nominal Cover to Reinforcement

(**a**) *Nominal cover:* Nominal cover is the design depth of concrete cover to all steel reinforcements, including links. It is the dimension used in design and indicated in the drawings. *It shall not be less than the diameter of the bar.*

(**b**) *Nominal cover to meet durability requirement:* Minimum values for the nominal cover of normal weight aggregate concrete which should be provided to all reinforcement, including links depending on the condition of exposure shall be as given in Table 7.2.

TABLE 7.2. Nominal Cover to Meet Durability Requirements

| Exposure | Nominal concrete cover in mm (Not less than) |
|---|---|
| 1. Mild | 20 |
| 2. Moderate | 30 |
| 3. Severe | 45 |
| 4. Very severe | 50 |
| 5. Extreme | 75 |

**Note 1.** For main reinforcement upto 12 mm dia bar for mild exposure, the nominal cover may be reduced by 5 mm.

2. For exposure conditions 'severe' and 'very severe', reduction of 5 mm may be made, where concrete grade is M 35 and above.

(**c**) *Nominal cover to meet specified period of fire resistance:* Minimum values of nominal cover of normal weight aggregate concrete to be provided to all reinforcement including links to meet specified period of the fire resistance shall be as given in table 7.3.

**TABLE 7.3.** Nominal Cover to Meet Specified Period of Fire Resistance

| Fire resistance (h) | Nominal cover (mm) | | | | | | |
|---|---|---|---|---|---|---|---|
| | Beams | | Slabs | | Ribs | | Columns |
| | Simply supported | Continuous | Simply supported | Continuous | Simply supported | Continues | |
| 0.5 | 20 | 20 | 20 | 20 | 20 | 20 | 40 |
| 1 | 20 | 20 | 20 | 20 | 20 | 20 | 40 |
| 1.5 | 20 | 20 | 25 | 20 | 35 | 20 | 40 |
| 2 | 40 | 30 | 35 | 25 | 45 | 35 | 40 |
| 3 | 60 | 40 | 45 | 35 | 55 | 45 | 40 |
| 4 | 70 | 50 | 55 | 45 | 65 | 55 | 40 |

**Note:** Cases that lie below the bold lines require attention to the additional measures necessary to reduce the risks of spalling.

(d) *Tolerance for cover:* Unless specified otherwise, actual concrete cover should not deviate from the required nominal cover by $^{+10}_{\phantom{+}0}$ mm.

### 10. Anchoring Reinforcing Bars
(a) *Anchoring bars in tension*

(i) Deformed bars may be used without end anchorages provided development length requirement is satisfied. Hooks should normally be provided for plain bars in tension.

(ii) *Bends and hooks.* The anchorage value of bend shall be taken as 4 times the diameter of the bar for each 45° bend subject to a maximum of 16 times and diameter of the bar. The anchorage value of a standard U-type hook shall be equal to 16 times the diameters of the bar.

(b) *Anchoring bar-in-compression:* The anchoring length of straight bar in compression shall be equal to the development length of bar in compression. The projected length of hooks, bends, straight length beyond bends if provided for a bar in compression shall be considered for development length.

### 11. Moment and Shear Coefficients for Continuous Beams and Slabs

(a) Unless more exact estimates are made, for beams and slabs of uniform cross-section which support substantially uniformly distributed loads over three or more spans which do not differ by more than 15 percent of the longest, the bending moments and shear forces used in design may be obtained by using coefficients given in Table 7.4 and Table 7.5 respectively. For moment at supports where two unequal spans meet or in case where the spans are not equally loaded, the average of the two values for the negative moment at the support may be taken for design.

(b) Where a member is built into a masonry wall which develops only partial restraint, the member shall be designed to resist a negative moment at the face of the support of $Wl/24$ where $W$ is the total design load and $l$ is the effective span, or such other restraining moment as may be shown to be applicable. For such a condition, shear coefficient given in Table 7.5 at the end support may be increased by 0.05.

**TABLE 7.4.** Bending Moment Coefficients

| Type of load | Span moment | | Support moment | |
|---|---|---|---|---|
| | Near middle of end span | At middle of interior span | At support next to the end support | At other interior supports |
| 1. Dead load and imposed load (fixed) | $+\dfrac{1}{12}$ | $+\dfrac{1}{16}$ | $-\dfrac{1}{10}$ | $-\dfrac{1}{12}$ |
| 2. Imposed load (not fixed) | $+\dfrac{1}{10}$ | $+\dfrac{1}{12}$ | $-\dfrac{1}{9}$ | $-\dfrac{1}{9}$ |

**Note.** For obtaining the B.M., the coefficient shall be multiplied by the total design load and effective span.

TABLE 7.5. Shear Force Coefficients

| Type of load | At end support | At support next to the end support | | At all other Interior supports |
| --- | --- | --- | --- | --- |
| | | Outer side | Inner side | |
| 1. Dead load and imposed load (fixed) | 0.4 | 0.6 | 0.55 | 0.5 |
| 2. Imposed load (not fixed) | 0.45 | 0.6 | 0.6 | 0.6 |

**Note.** For obtaining the S.F., the coefficient shall be multiplied by the total design load.

### 12. Unit Weight of Concrete

Indian Standard Code (IS : 456–2000) specifies that *unless more accurate calculations are warranted*, the unit weights of plain concrete and reinforced concrete made with standard gravel or crushed natural stone aggregate may be taken as 24000 N/m³ (24 kN/m³) and 25000 N/m³ (25 kN/m³) respectively.

## (B) DESIGN PROCEDURE

The design of a singly reinforced beam is done in the following steps:

1. **Calculation of constants.** For the given set of stresses, determine $k_c$, $j_c$ and $R_c$.
2. **Calculation of bending moment.** Assume suitable values of overall depth and breadth of the beam, and determine the effective span. Calculate the self weight of the beam, and add it to the given superimposed load to get the total U.D.L. per metre length of the beam. Calculate the maximum bending moment in the beam.
3. **Design of the section.** Calculate the effective depth of the beam by the expression:

$$d = \sqrt{\frac{M}{R_c \cdot b}}$$

Find out the total depth of the beam by adding suitable cover to the above calculated effective depth. If this is nearly equal to the value of depth assumed in (1) it is alright. If the difference is more, recalculate the self-weight on the basis of the revised depth of the beam. Recalculate the depth on the basis of the new bending moment.

4. **Steel reinforcement.** Calculate the area of steel by the formula

$$A_{st} = \frac{M}{\sigma_{st} \cdot j_c \, d},$$ where $\sigma_{st}$ = permissible tensile stress in steel.

Choose suitable diameter and number of bars. Arrange them suitably in the width.

**Curtailment of Reinforcement.** Curtail the reinforcement at other locations where it is not longer required from the bending moment point of view.

Let $M$ = bending moment at the centre of the span.

$M_x$ = bending moment at other section distance $x$ from the centre

$N$ = numbers of bars at the centre of the span

$N_x$ = number of bars curtailed at the section distance $x$ from the centre

$N - N_x$ = number of bars at the given section distant $x$ from the centre

Since $A_{st}$ is proportional to the bending moment, we have

$$\therefore \quad \frac{N - N_x}{N} = \frac{M_x}{M} \quad \text{or} \quad 1 - \frac{N_x}{N} = \frac{M_x}{M}$$

or

$$\frac{N_x}{N} = 1 - \frac{M_x}{M} = \frac{M - M_x}{M} \qquad \ldots(i)$$

If the beam carries U.D.L., $M = \dfrac{wL^2}{8}$ and $M_x = \left(\dfrac{wL^2}{8} - \dfrac{wx^2}{2}\right)$

$$\therefore \qquad M - M_x = \frac{wx^2}{2} \quad \text{or} \quad \frac{M - M_x}{M} = \frac{wx^2}{2} \bigg/ \frac{wL^2}{8} = \frac{4x^2}{L^2} \qquad \ldots(ii)$$

Equating (i) and (ii), $\qquad \dfrac{N_x}{N} = \dfrac{4x^2}{L^2}$

or $\qquad x = \dfrac{L}{2}\sqrt{\dfrac{N_x}{N}} \qquad \ldots(7.1)$

Thus, the above expression gives the distance $x$ from the centre where $N_x$ bars can be curtailed out of total bars $N$. However, the bars have to develop full tension at the centre of span, and hence the distance $x$ should be equal to or greater than the development length $L_d$ of the bar which is to be curtailed. The Code also specifies that the reinforcement shall extend the point at which it is no longer required to resist flexure, for distance equal to effective depth $d$ or 12 Φ whichever is greater. Hence the bar should be curtailed at a distance $x + d$ (or $x + 12Φ$) from the centre; and this distance should be greater than $L_d$. See § 3.24 and Fig. 3.31 also for graphical method.

Another important recommendation of the Code is that in simple beams, one-third of the positive reinforcement should be continued to the supports, and the diameter should be so chosen that *Eq. 3.31* or *3.31 (a)* is satisfied.

However, if the bar is to be bent up, it may be safely bent up at a distance $x$ given by *Eq. 7.1*.

**5. Shear reinforcement.** Calculate the maximum shear force in the beam. Find the nominal shear stress in the beam by the expression $\tau_v = V/bd$. Also, determine (or estimate) design stress $\tau_c$ using Table 3.1.

(i) If $\tau_v \le \dfrac{1}{2}\tau_c$ or $V < \dfrac{1}{2}V_c$ (where $V_c = \tau_c \cdot bd$), no shear reinforcement is really necessary.

(ii) If $\dfrac{1}{2}V_c \le V \le V_c$, provide nominal shear reinforcement given by:

$$\frac{A_{sv}}{b \cdot s_v} \ge \frac{0.4}{0.87 f_y}$$

Choosing bar diameter, $A_{sv}$ is known, and hence $s_v$ can be computed from the above expression. However, the maximum spacing is restricted to $0.75\,d$ or 300 mm, whichever is less.

(iii) If $V_c \le V \le V_{c(max.)}$ [where $V_{c(max.)} \cdot bd$]

Design the transverse reinforcement for the net S.F. $= V - V_c$.

If $\Sigma V_s$ is the sum of the resistances of various types of shear reinforcement (*Eq. 3.24, 3.25* and *3.26*) then $\Sigma V_s$ should be equal to or greater than $(V - V_c)$. However, the contribution of bent up bars should not be more than half, the area of stirrups should not be less than the one given by *Eq. 3.23*, and maximum $s_v$ should be restricted to $0.75\,d$ (or 300 mm).

(iv) If $V \ge V_{c(max.)}$, redesign the web of the section such that $V_{c.max.}$ becomes equal to or more than external shear $V$.

**6. Check for development length at the end.** Check for development length of discussed in para $a(6)$, i.e. $L_d \le \dfrac{M_1}{V} + L_0$. See Fig. 7.5 for computation of $L_0$.

**Design: Example 7.1.** *A reinforced concrete beam is supported on two balls 750 mm thick, spaced at clear distance of 6 metres. The beam carries a superimposed load of 9.8 kN/m. Using M 20 concrete, design the beam. Take the permissible tensile and shear stress in steel as 230 N/mm² for HYSD bars.*

**Solution:**

**1. Calculation of Design Constants.** For M 20 concrete, $c = \sigma_{cbc} = 7$ N/mm² and $m = 13.33$. Also, $t = \sigma_{st} = 230$ N/mm².

$$\therefore \quad k_c = \frac{mc}{mc+t} = \frac{13.33 \times 7}{13.33 \times 7 + 230} = 0.289;$$

$$j_c = 1 - k_c/3 = 1 - 0.289/3 = 0.904$$

$$R_c = \frac{1}{2} c \, j_c \, k_c = \frac{1}{2} \times 7 \times 0.904 \times 0.289 = 0.914$$

**2. Calculation of B.M.**

Let the effective depth of beam ≈ 1/10 span ≈ 6000/10 = 600 mm

Assume total depth of beam = 600 mm (say), for computation of dead weight of beam. Let the width of the beam = $\frac{1}{2} d$ = 300 mm (say).

Self load of beam per metre run = (0.6 × 0.3 × 1) 25000 = 4500 N.
External load = 9800 N/m.

$\therefore$ Total load per metre run = 9800 + 4500 = 14300 N

Effective span = $L = l + d$ = 6 + 0.6 = 6.6 m

(This is smaller than the centre to centre distance of 6.75 m between the supports).

$$M = \frac{wL^2}{8} = \frac{14300(6.6)^2}{8} = 77870 \text{ N-m} = 77.87 \times 10^6 \text{ N-mm}$$

**3. Design of section.**

$$d = \sqrt{\frac{M}{R_c b}} = \sqrt{\frac{77.87 \times 10^6}{0.914 \times 300}} = 533 \text{ mm}$$

Let us take $d$ = 545 mm and $D$ = 580 mm.
Revised self load of beam = 0.3 × 0.58 × 25000 = 4350 N/m

$\therefore$ $w$ = 9800 + 4350 = 14150 N/m.

Effective span = 6 + 0.545 = 6.545 m

$$M = \frac{14150 (6.545)^2}{8} \approx 75770 \text{ N-m} = 75.77 \times 10^6 \text{ N-mm}$$

$$d = \sqrt{\frac{75.77 \times 10^6}{0.914 \times 300}} = 526 \text{ mm}$$

Assuming that 16 mm Φ bars will be used, with 8 mm dia. links and a nominal cover of 25 mm, $D$ = 526 + 25 + 8 + 16/2 = 567 mm.

Keeping $D$ = 570 mm and providing a nominal cover of 25 mm, and using 8 mm φ links, available effective depth = 570 – 25 – 8 – 16/2 = 529. Hence OK.

**4. Steel reinforcement**

$$A_{st} = \frac{M}{\sigma_{st} \cdot j_c d} = \frac{75.77 \times 10^6}{230 \times 0.904 \times 529} = 688.9 \text{ mm}^2$$

$A_\Phi$ for 16 mm Φ bars = $\frac{\pi}{4} (16)^2 = 201.06 \text{ mm}^2.$  $\therefore$ No. of bars = 688.9/201.06 = 3.4

Hence provide 4 bars, having $A_{st}$ = 4 × 201.06 = 804.24 mm². Keeping 25 mm nominal side cover clear spacing between bars will be = $\frac{1}{3}$(300 – 25 × 2 – 2 × 8 – 4 × 16) = 56.7 mm, which is much more than the diameter of the bar.

Min. reinforcement is given by:

$$\frac{A_{st}}{bd} = \frac{0.85}{f_y}$$

Taking $f_y$ = 415 N/mm², $A_{st} = \frac{0.85 (300 \times 529)}{415} = 325 \text{ mm}^2$

Since the actual $A_{st}$ provided is much more than this, the design is OK.

**5. Check for deflection.** $l/d = 6600/529 \approx 12.5 <$ basic $L/d$ ratio of 20. Hence OK.

**Note:** Since $d$ on the basis of deflection is normally much less than the one provided on the basis of flexure, modification factor have not been applied.

**6. Check for shear and design of shear reinforcement.**

The reaction at the wall supports will be uniformly distributed over the full width. Hence the shear force will be maximum at the edge of the support.

Max. $\quad V = \dfrac{wl}{2} = \dfrac{14150 \times 6}{2} = 42450$ N

and $\quad \tau_v = \dfrac{V}{bd} = \dfrac{42450}{300 \times 529} = 0.267$ N/mm$^2$.

Assuming that out of 4 bars of main reinforcement, 2 bars will be bent up near the support, and hence only 2 bars will be available.

$\therefore \quad \dfrac{100 A_s}{bd} = \dfrac{100}{300 \times 529} \times 402.12 = 0.25\%$

Hence from Table 3.1, permissible shear ($\tau_c$) for M 20 concrete, for 0.25% steel = 0.22 N/mm$^2$, which is less than the nominal shear stress. Hence shear reinforcement is required.

$$V_c = \tau_c \, bd = 0.22 \times 300 \times 529 = 34914 \text{ N}$$

$\therefore \quad V_s = V - V_c = 42450 - 34914 = 7536$ N

Using 8 φ 2 lgd stirrups, $\quad A_{sv} = 2 \dfrac{\pi}{4}(8)^2 = 100.5$

$\therefore \quad s_v = \dfrac{\sigma_{sv} A_{sv} d}{V_s} = \dfrac{230 \times 100.5 \times 529}{7536} = 1622$ mm

However, minimum shear reinforcement is governed by the expression:

$$s_v = \dfrac{2.175 \, A_{sv} \cdot f_y}{b} = \dfrac{2.175 \times 100.5 \times 415}{300} = 302.4 \text{ mm}.$$

Subject to max. of $0.75 \, d$ or 300 mm which ever is less. Hence provide the stirrups 300 mm c/c. Provide 2 – 10 mm Φ holding bars at top.

**7. Check for development length at supports.**

The code stipulates that at the simple supports, where the reinforcement is confined by a compressive reaction, the diameter of the reinforcement be such that $1.3 \dfrac{M_1}{V} + L_0 \geq L_d$

Assuming that 2 bars are bent up and 2 bars are available at the supports,

$$A_{st} = 2 \dfrac{\pi}{4}(16)^2 = 402.12 \text{ mm}^2$$

$M_1$ = moment of resistance of the section, assuming all reinforcement stressed to $\sigma_{st}$
   = $230 \times 402.12 \times 0.904 \times 529 = 44.23 \times 10^6$ N-mm

$V = 42450$ N

$L_0$ = sum of anchorage value of hooks.

Let us provide a support equal to the width of the wall, i.e. 600 mm. Let the clear side cover $x' = 40$ mm. For a 90° bend having anchorage value of 8Φ, we have

$$L_0 = \left(\dfrac{l_s}{2} - x' + 3\varphi\right) = \dfrac{600}{2} - 40 + 3 \times 16 = 308 \text{ mm}$$

$$1.3 \dfrac{M_1}{V} + L_0 = 1.3 \times \dfrac{44.23 \times 10^6}{42450} + 308 = 1354 + 308 = 1662 \text{ mm}$$

Development length $\quad L_d = \dfrac{\Phi \, \sigma_{st}}{4 \, \tau_{bd}} = \dfrac{16 \times 230}{4 \times (1.6 \times 0.8)} = 719$ mm

Alternatively, $L_d \approx 45\,\Phi$ (Table 3.5) $= 45 \times 16 = 720$ mm.

Thus, $\left(1.3\dfrac{M_1}{V} + L_0\right) > L_d$. Hence Code requirement are satisfied.

**8. Details of reinforcement.** Due the partial fixity that may be caused at the supports, some reinforcement is always provided at the top of the beam near the ends. Let us bend two bars, one at a distance $x_1$ from the support and the other at a distance $x_2$ from the support.

B.M. at $\quad x_1 = \dfrac{wL}{2}x_1 - \dfrac{w\,x_1^2}{2}$

This should be $\dfrac{3}{4}$ of the maximum B.M.

$\therefore \quad \dfrac{wL}{2}x_1 - \dfrac{wx_1^2}{2} = \dfrac{3}{4} \cdot \dfrac{wL^2}{8}$

This gives $x_1 = 0.25\,L = 0.25 \times 6.545 = 1.64$ m.

However, bend one bar at a distance of 1.60 m from the support.

Similarly, for the second bar,

$\dfrac{wLx_2}{2} - \dfrac{wx_2^2}{2} = \dfrac{1}{2}\dfrac{wL^2}{8}$

This gives $x_2 = 0.14L \approx L/7 \approx 0.14 \times 6.545 = 0.92$ m. However, bend the second bar at a distance of 0.9 m from the face of the support. The remaining 2 bars shall be taken straight into the support. Fig. 7.6 shows the longitudinal section and the cross-section of the beam.

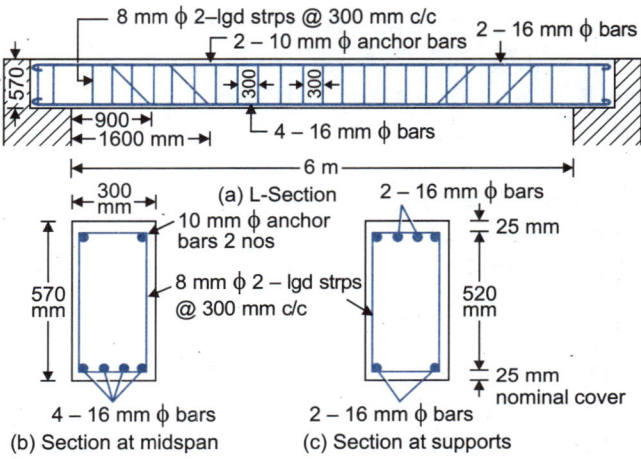

Fig. 7.6

## 7.2. DESIGN OF LINTEL BEAMS

Lintels are provided over the openings of doors, windows, almirahs, etc. Generally, they support the load of the wall over it, and sometimes also the live load are transferred by the sub-roof of the room. Following five cases may arise from point of view of distribution of load over the lintels:

(1) When the length of wall on each side is more than half the effective span of the lintel.
(2) When the length of wall to one side is less than half the effective span.
(3) When the length of wall on both sides is less than half the effective span.
(4) When there are openings over the lintel.
(5) When there is load carrying slab over the lintel.

**Case 1. *When the length of wall on each side is more than half the effective span of the lintel.***

This is most general case. Because of arch-action in the masonry, all the load of the wall above the lintel is not transferred to the lintel. It is assumed that the load transferred is in the form of equilateral triangle, and the load on the lintel is equal to the weight of the masonry in the triangular portion, as shown in Fig. 7.7(*a*). If, however, the height of the wall above the lintel is insufficient (*i.e.* if the apex of the triangle falls above the top of the wall), whole of the rectangular load above the lintel is taken to act on the lintel, as shown in Fig. 7.7(*b*). Let:

$H$ = height of wall above the lintel.
$l$ = actual opening.
$L$ = effective length of the opening.
$h$ = effective height of masonry, above the lintel.

Thus, $\quad h = L \sin 60° = L \cdot \dfrac{\sqrt{3}}{2} = \dfrac{L}{2}\sqrt{3}$.

∴ Area of triangle $= \frac{1}{2} \times L \times \frac{L}{2}\sqrt{3} = \frac{L^2}{4}\sqrt{3}$.

**Case 2. When the length of the wall to one side is less than half the effective span.**

Fig 7.7 (c) shows the situation where the length of wall to one side is less than half the effective span, but the length to the other side is more than half the effective span. In that case, the load transferred to the lintel will be equal to the weight of masonry contained in the rectangle of height $h$ equal to the effective span.

**Case 3. When the length of the walls to each side is less than half the effective span.**

This is shown in Fig. 7.7 (d). The load acting on the lintel will be equal to the weight of the masonry contained in the rectangle of height $h$ equal to the full height $H$ of the wall.

**Case 4. When there are openings on the lintel.**

If there are openings, due to the provision of ventilators etc. and if these openings are intersected by the 60° lines, the loading will be calculated by allowing dispersion lines at 60° from the top edges of the openings, as shown in Fig. 7.7 (e). The total load on the lintel will be equal to the weight of the masonry contained in the shaded area.

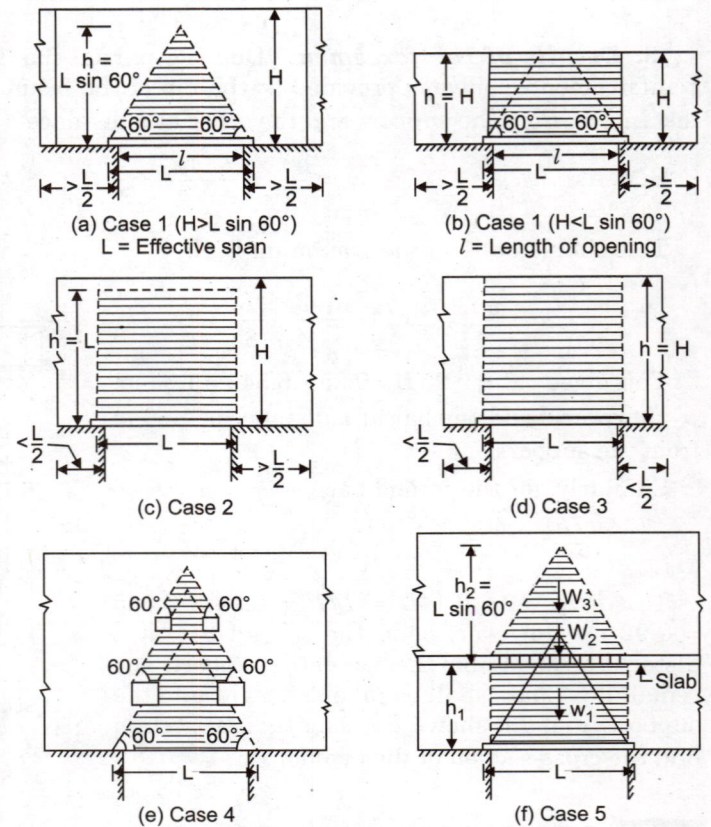

**Fig. 7.7.** Various Cases of Loading on Lintels.

**Case 5. When there is load carrying slab falling within the dispersion triangle** [*Fig. 7.7 (f)*].

If the roof slab is provided at a level well above the apex of the dispersion triangle, the uniformly distributed load carried by the slab is *not* transferred to the lintel. If, however, the slab intersects the dispersion triangle, three types of loads are transferred to the lintel:

(*i*) Load $W_1$ due the weight of the masonry contained in the rectangle of height $h_1$ equal to the height of the slab above the lintel.

(*ii*) Load $W_2$ carried by the slab, in a length $L$.

(*iii*) Load $W_3$ due to the weight of masonry contained in the equilateral triangle above the slab.

The design of the lintel is similar to that R.C. beam discussed in § 7.1. The width of lintel is normally kept equal to width of the wall.

The max. bending moment $M_1$ due to the triangular load is taken as $M_1 = WL/6$
where $W$ = total weight of masonry contained in the equilateral triangle. The max. bending moment $M_2$ due the uniformly distributed self-weight $w$ of the lintel is: $M_2 = wL^2/8$

∴ Total $M = \dfrac{WL}{6} + \dfrac{wL^2}{8}$

The maximum shear will act at the edge of the opening and is given by

$F = \dfrac{W}{2} + \dfrac{wl}{2}$

**Example 7.2.** *Design a R.C. lintel is to be provided over an opening of 2 m wide. The height of masonry above the lintel is 3 m. The opening is centrally located in a long wall, 300 mm thick. The unit weight of masonry may be taken as 1.9 kN/m³. Use M 20 concrete and Fe 415 steel.*

# DESIGN OF BEAMS AND SLABS

**Solution:**

**1. Calculation of constants:**

$$k_c = \frac{mc}{mc + t} = \frac{13.33 \times 7}{13.33 \times 7 + 230} = 0.289;$$

$$j_c = 1 - k_c/3 = 0.904$$

$$R_c = \frac{1}{2} c j_c k_c = \frac{1}{2} \times 7 \times 0.904 \times 0.289 = 0.914$$

**2. Calculation of bending moment**

Let the total depth of lintel = 24 cm; Width of lintel = 30 cm.
Self weight of lintel, per m length = $0.24 \times 0.30 \times 1 \times 25000 = 1800$ N/m
Effective span will be the minimum of the following:
(i) Centre to centre of bearing: Providing a bearing 20 cm,
$$L = 2 + 0.2 = 2.2 \text{ m}$$
(ii) Clear span + effective depth: Assuming effective cover of 3 cm, $d = 24 - 3 = 21$ cm.
∴ $L = 2 + 0.21$ m $= 2.21$ m. Hence adopt $L = 2.2$ m.
Height of equilateral triangle, assuming 60° dispersion

$$= L \sin 60° = \frac{L}{2}\sqrt{3} = \frac{2.2}{2} \times \sqrt{3} = 1.9 \text{ m}$$

This is less than the height of the wall above the lintel. Hence load on the lintel will be equal to the weight of the masonry contained in the triangular portion.

∴ $$W = \left(\frac{1}{2} \times 2.2 \times 1.9\right)(0.3 \times 19000) = 11913 \text{ N}$$

B.M. $$M = \frac{WL}{6} + \frac{wL^2}{8} = \left[\frac{11913 \times 2.20}{6} + \frac{1800 \times (2.20)^2}{8}\right] = 5457 \text{ N-m}$$
$$= 5457000 \text{ N-mm}$$

**3. Design of section**

∴ $$d = \sqrt{\frac{M}{R_c b}} = \sqrt{\frac{5457000}{0.914 \times 300}} = 141 \text{ mm}$$

Adopt total depth $D = 180$ mm. Using 10 mm dia main bars, 8 mm φ stirrups and 20 mm nominal cover, available $d = 180 - 20 - 8 - 10/2 = 147$ mm. The total depth is much less than the assumed value. However, recalculation of the load is not necessary, because self-weight of lintel is comparatively smaller than the super-imposed load.

**4. Reinforcement:** $$A_{st} = \frac{M}{\sigma_{st} j_c d} = \frac{5457000}{230 \times 0.904 \times 147} = 178.6 \text{ mm}^2$$

Provide 10 mm diameter bars. No. of bars required $= \frac{178.6}{\pi/4(10)^2} = 2.27$

Hence provide 3 Nos. of 10 mm Φ bars.

Total area provided $= 3 \times \frac{\pi}{4}(10)^2 = 235.6$ mm². Minimum reinforcement is given by

$$\frac{A_{st}}{bd} = \frac{0.85}{f_y} \text{ or } A_{st} = \frac{0.85}{415} \times (300 \times 147) = 90.3 \text{ mm}^2$$

Since the actual steel provided (= 235.6 mm²) is more than this, the design is O.K. Bend one of the bars at a distance of $L/7 = 2200/7 \approx 300$ mm from the face of each support. Keep a nominal cover = 20 mm.

**5. Check for shear and design of shear reinforcement**

$$V = \frac{W}{2} + \frac{wl}{2} = \frac{11913}{2} + \frac{1800 \times 2}{2} = 7757 \text{ N}.$$

∴ $$\tau_v = \frac{V}{bd} = \frac{7757}{300 \times 147} = 0.176 \text{ N/mm}^2$$

At supports, $\dfrac{100 A_s}{bd} = \dfrac{100}{300 \times 147}\left(\dfrac{2}{3} \times 235.6\right) = 0.35\%$

From Table 3.1, for M 20 concrete, and for 0.35% steel, $\tau_c = 0.25$ N/mm². Hence only nominal shear reinforcement is required, given by the relation:

$$\dfrac{A_{sv}}{b \cdot s_v} \geq \dfrac{0.4}{0.87 f_y}.$$

Using 8 Φ – two legged stirrups,

$$A_{sv} = 2 \times \dfrac{\pi}{4}(8)^2 = 100.5 \text{ mm}^2$$

∴ $s_v = \dfrac{2.175 A_{sv}}{b} \times f_y = \dfrac{2.175 \times 100.5 \times 415}{300} = 302.4$ mm,

Max. spacing = $0.75 \times 147 = 110$ mm. Hence provide these @ 110 mm c/c throughout. Provide 2 – 8Φ anchor bars (holding bars) at top, throughout the length.

### 6. Check for development length at supports

At the supports, $1.3 \dfrac{M_1}{V} + L_0 \geq L_d$ ;

$A_{st} = \dfrac{2}{3} \times 235.6 = 157.1$ mm²

$M_1 = A_{st} \cdot j_c d \cdot \sigma_{st}$
$\phantom{M_1} = 157.1 \times 230 \times 0.904 \times 147$
$\phantom{M_1} = 4.8 \times 10^6$ N-mm

$V = 7757$ N. ;

$L_d = \dfrac{\Phi \sigma_{st}}{4 \tau_{bd}} = \dfrac{10 \times 230}{4 \times 1.6 \times 0.8} = 450$ mm.

Taking bars straight in the support, without any hook or bend, with $x' = 25$ mm

$$L_0 = L_s/2 - x' = 200/2 - 25 = 75 \text{ mm}$$

∴ $1.3 \dfrac{M_1}{V} + L_0 = 1.3 \times \dfrac{4.80 \times 10^6}{7757} + 75 = 804 + 75 = 879$ mm

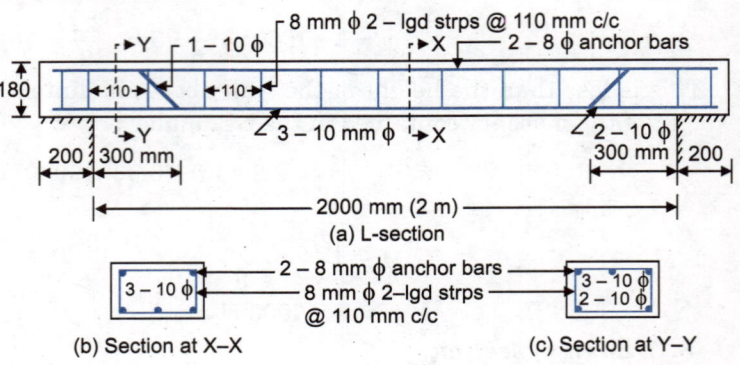

Fig. 7.8

Hence Code requirements are satisfied.

### 7. Details of reinforcement: Fig. 7.8.

**Example 7.3.** *Fig. 7.9(a) shows the cross-section of a wall of room, 5 m wide and 18 m long from inside. Design the lintel for the window having an opening of 2 m. The reinforced concrete slab, 16 cm thick, may be assumed to transfer half the load to the wall shown. Use the following data for design:*

(i) *Weight of lime concrete terracing*
$= 19$ kN/m³

(ii) *Weight of masonry*
$= 19.2$ kN/m³

(iii) *Weight of reinforced concrete*
$= 24$ kN/m³

(iv) *Super-imposed load on the slab*
$= 1.6$ kN/m²

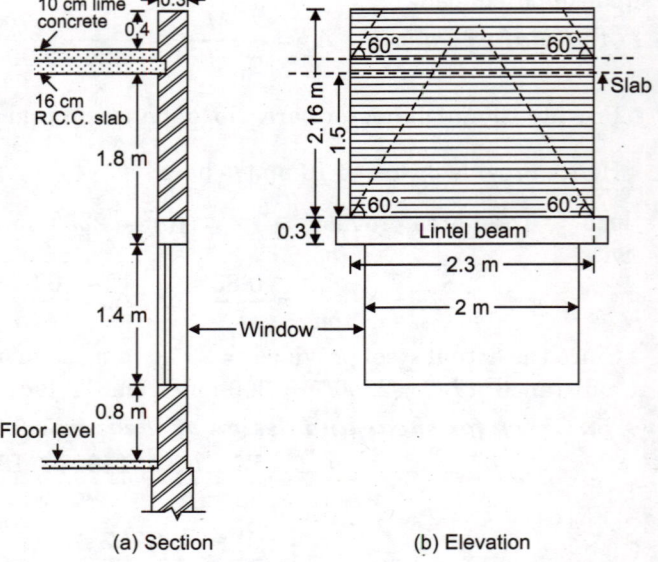

Fig. 7.9

Use M 20 concrete and Fe 415 steel. The window may be assumed to be centrally located along the length of the room.

**Solution:**

**1. Loading and bending moment.** Let us assume the overall depth of the lintel to be 300 mm, and a bearing of 300 mm on either side. Hence effective span of lintel will be equal to 2 + 0.3 = 2.3 m. Assuming the dispersion to be along 60° lines, the height of load triangle, measured above the top of the lintel = 2.3 sin 60° = $\dfrac{2.3 \times \sqrt{3}}{2}$ ≈ 2 m, which is more than height (1.5 m) of the slab above the lintel. Hence the load of the slab will also be transferred to lintel.

Fig. 7.9 (b) shows the elevation, showing all the heights. If we construct equilateral triangle above the top of the slab, its apex will fall very much above the top of the parapet wall. Hence the weight of the whole wall above the lintel will be transferred to the lintel. Thus, the loads per metre length of the lintel will consist of the following:

(i) *Weight of the wall:* Weight per metre run = 0.3 × 2.16 × 19000 = 12300 N

(ii) *Load transferred by the slab:* Consider 1 m strip of the slab of 5 m span.

| | | |
|---|---|---|
| Live load on this strip | = $\dfrac{1}{2}(5 \times 1) \times 1600$ | = 4000 N |
| Dead weight of 16 cm thick slab | = $\dfrac{1}{2}\left(5 \times 1 \times \dfrac{16}{100}\right) 24000$ | = 9600 N |
| Dead weight of lime concrete | = $\dfrac{1}{2}\left(5 \times 1 \times \dfrac{10}{100}\right) 19000$ | = 4750 N |
| ∴ Total load per metre run of the lintel | | = 18350 N |

(iii) *Self-weight of lintel:* Weight per metre run = (0.3 × 0.3 × 1) 24000 = 2160 N

∴ Total $w$ = 12300 + 18350 + 2160 = 32810 N

Max. bending moment $M = \dfrac{wL^2}{8} = \dfrac{32810 \times (2.3)^2}{8}$ = 2.17 × 10⁴ N-m = 21.7 × 10⁶ N-mm

**2. Calculation of constants**

$$k_c = \dfrac{mc}{mc + t} = \dfrac{13.33 \times 7}{13.33 \times 7 + 230} = 0.289; \quad j_c = 1 - k_c/3 = 0.904$$

$$R_c = \dfrac{1}{2} c j_c k_c = \dfrac{1}{2} \times 7 \times 0.904 \times 0.289 = 0.914$$

**3. Design of section**

Effective depth $\quad d = \sqrt{\dfrac{M}{R_c \cdot b}} = \sqrt{\dfrac{21.7 \times 10^6}{0.914 \times 300}}$ = 281.3 mm

Provide $D$ = 320 mm. Using 12 mm Φ main bars and 8 mm diameter stirrups and keeping a nominal cover of 20 mm, available effective depth = 320 – 8 – 12/2 – 20 = 286 mm. This will slightly increase the dead load, but since the dead load of lintel is very small in comparison to the other loads, revision of calculations with new dimensions is not necessary.

**4. Steel reinforcement:**

$$A_{st} = \dfrac{M}{\sigma_{st} \cdot j_c d} = \dfrac{21.7 \times 10^6}{230 \times 0.904 \times 286} = 365 \text{ mm}^2$$

No. of 12 mm Φ bars $= \dfrac{365}{\dfrac{\pi}{4}(12)^2}$ ≈ 3.3. Hence provide 4 nos. 12 mm Φ bars.

Total $A_{st} = 4 \times \dfrac{\pi}{4}(12)^2 = 452.4$ mm². Out of these, bend 2 bars up a distance of $L/7$ = 2300/7 ≈ 300 mm from the edge of each support.

### 5. Check for shear and design for shear reinforcement

Max. S.F. $\quad V = \dfrac{w \cdot l}{2} = \dfrac{32810 \times 2}{2} = 32810$ N  (At the edge of the support)

$\therefore \quad \tau_v = \dfrac{V}{bd} = \dfrac{32810}{300 \times 286} = 0.382$ N/mm²

At support, $A_{st} = \dfrac{1}{2} \times 452.4 = 226.2$ mm², $\quad \therefore \dfrac{100 A_s}{bd} = \dfrac{100 \times 226.2}{300 \times 286} = 0.26\%$

Hence from Table 3.1. $\tau_c \approx 0.225$ N/mm². Since $\tau_v$ is more than $\tau_c$, shear reinforcement is necessary.
$V_c = \tau_c \cdot bd = 0.225 \times 300 \times 286 = 19305$ N.

$\therefore \quad V_s = V - V_c = 32810 - 19305 = 13505$ N

Shear resistance $V_{s1}$ of bent up bars is

$V_{s1} = \sigma_{sv} \cdot A_{sv} \sin \alpha = 0.707 \, \sigma_{sv} \cdot A_{sv} = 0.707 \times 230 \times 226.2 = 36782$ N

But $V_{s1}$ assigned to inclined bars cannot be more than $\dfrac{1}{2} V_s = \dfrac{1}{2} \times 13505 = 6752.5$ N.

Hence vertical stirrups are to be provided for a shear $V_{s2} = 6752.5$ N only.

Using 8 mm Φ 2-lgd stirrups, $A_{sv} = 2 \dfrac{\pi}{4} (8)^2 = 100.5$ mm²

Hence, $\quad s_v = \dfrac{\sigma_{sv} \cdot A_{sv} \cdot d}{V_{s2}} = \dfrac{230 \times 100.5 \times 286}{6752.5} = 979$ mm. This is too large.

However, max. spacing corresponding to nominal shear stirrups is given by

$s_v \leq \dfrac{2.175 \, A_{sv} \cdot f_y}{b} = \dfrac{2.175 \times 100.5 \times 415}{300} \leq 302$ mm,

subject to lesser than least of (0.75 $d$ or 300 mm).

$s_{v\,max.} = 0.75 \times 286 = 214.5$ mm.

Hence provide 8 mm Φ 2-lgd stirrups @ 200 mm c/c. Provide 2-8 mm Φ holding bars at top.

### 6. Check for development length at the end:
At the support, the Code requirements are that

$1.3 \dfrac{M_1}{V} + L_0 \geq L_d; \quad A_{st} = \dfrac{1}{2} \times 452.4 = 226.2$ mm²

$M_1 = A_{st} \cdot \sigma_{st} \cdot j_c \, d$
$= 226.2 \times 230 \times 0.904 \times 286$
$= 13.45 \times 10^6$ N-mm

$V = 32810$ N

Providing straight bars without hook or bend

$L_0 = \dfrac{L_s}{2} - x'.$

$L_s = 300$ mm (say);
$x'$ = side cover = 25 mm
$L_0 = 300/2 - 25 = 125$ mm

$\therefore \quad 1.3 \dfrac{M_1}{V} + L_0$

$= 1.3 \times \dfrac{13.45 \times 10^6}{32810} + 125 = 533 + 125 = 658$ mm

$L_d \approx 45 \, \Phi = 540$ mm. Hence the Code requirement are satisfied.

### 7. Details of reinforcements.
Shown in Fig. 7.10.

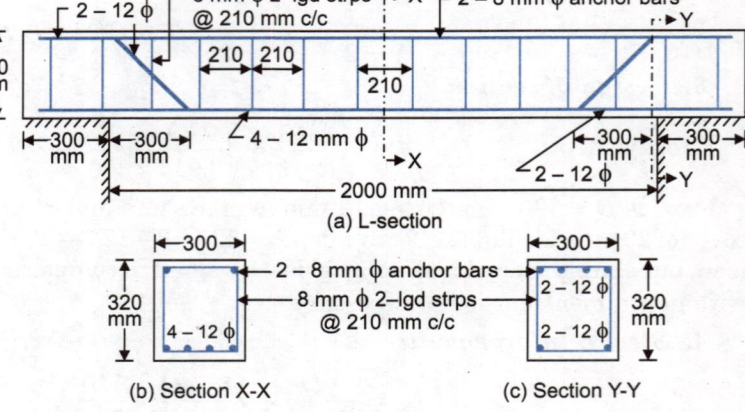

Fig. 7.10

## 7.3. DESIGN OF CANTILEVER

When a cantilever beam or slab is subjected to downward loads, it bends downwards with its convexity upwards. Due to this, the upper fibres of the section are subjected to tensile stress and the lower fibres are subjected compressive stresses. Hence the main reinforcement is provided at the *upper face*, and not at the lower face. The bending moment varies from a zero value at the free section to maximum value at the fixed end (support). Hence the depth of section and the steel reinforcement is varied as per requirement of bending moment. In order that the whole cantilever does not overturn, it should be properly fixed in the wall. The reinforcement should be embedded in the support for a distance of $L_d$ from the edge of the support.

When curtailment of bar is done, it should be kept in mind that the embedment of the bar is not less than $L_d$ from the point it is needed the most, nor less than 12 diameters (or $d$) from the point it can be spared from the bending moment point of view.

**Effective length:** The effective length of a cantilever shall be taken as it length of the face of the support plus half the effective depth except where if forms the end of a continuous beam where the length to the centre of support shall be taken.

**Example 7.4.** *A cantilever (beam) projects 2.5 m beyond the fixed end, and carries a superimposed load of 12 kN per metre run. Design the cantilever using M 20 concrete and Fe 415 steel. Take width of support = 400 mm.*

**Solution: 1. *Calculation of bending moment:***

From deflection point of view, $d = L/7 = 2500/7 \simeq 360$ mm. Let the section of the cantilever be 250 mm wide and 500 mm deep at the fixed end, and 250 mm wide and 200 mm deep at the free end.

$$\therefore \quad \text{Weight of beam} = \frac{1}{2}(0.5 + 0.2) \times 0.25 \times 2.5 \times 25000 = 5469 \text{ N}$$

Acting at $\dfrac{0.5 + 2 \times 0.2}{0.5 + 0.2} \times \dfrac{2.5}{3} = 1.07$ m from the fixed end.

$$M_{max.} = \left[(5469 \times 1.07) + \frac{12000 \times 2.5^2}{2}\right] = 4.3352 \times 10^4 \text{ N-m}$$

$$= 43.352 \times 10^6 \text{ kN-m}$$

S.F. at the edge of the support = $5469 + 12000 \times 2.5 = 35469$ N

**2. *Design constants:*** For M 20 concrete – Fe 415 steel.

$$k_c = \frac{mc}{mc + t} = \frac{13.33 \times 7}{13.33 \times 7 + 230} = 0.289;$$

$$j_c = 1 - k_c/3 = 0.904$$

$$R_c = \frac{1}{2} c j_c \, k_c = \frac{1}{2} \times 7 \times 0.904 \times 0.289 = 0.914$$

**3. *Design of section***

Effective depth $\quad d = \sqrt{\dfrac{M}{R_c \cdot b}} = \sqrt{\dfrac{43.352 \times 10^6}{0.914 \times 250}} = 436$ mm

Keep $D = 490$ mm. Using 16 mm $\Phi$ bars, and 25 mm nominal cover, and 8 mm $\varphi$ rings, available effective depth = $490 – 25 – 8 – 16/2 = 449$ mm. Hence OK. Keep total depth at free end = 200 mm.

**4. *Steel reinforcement***

$$A_{st} = \frac{M}{\sigma_{st} \cdot j_c \, d} = \frac{43.352 \times 10^6}{230 \times 0.904 \times 449} = 464.4 \text{ mm}^2.$$

No. of 16 mm $\Phi$ bars = $\dfrac{464.4}{\dfrac{\pi}{4}(16)^2} = 2.31$

Hence provide 3 Nos. 16 mm $\Phi$ bars giving total $A_{st} = 603.2$ mm$^2$

Since the B.M. decreases to zero at the end, let us curtail few bars. Let 1 bar be *curtailed* at a distance $x$ from the free end. Assuming the B.M.D. to be parabolic, the B.M. at this section may be approximately taken to be equal to

$$\left(\frac{x}{2.5}\right)^2 \times 43.352 \times 10^6 = 6.936 \times 10^6\, x^2 \text{ N-mm. Area of 2-16 }\Phi \text{ bars} = 402 \text{ mm}^2.$$

$$\therefore \qquad 402 = \frac{6.936 \times 10^6\, x^2}{230 \times 0.934\, d_x}$$

(where $d_x$ = effective depth at that section)  ...(i)

Total depth at the section = $\left[200 + \dfrac{490 - 200}{2.5} x\right]$

$$\therefore \quad d_x = \left[200 + \frac{290}{2.5} x\right] - [8 + 8 + 25]$$

$$d_x = 159 + 116\, x \qquad \qquad \qquad \qquad \qquad \qquad \ldots(ii)$$

Substituting in (i), $\quad 402 = \dfrac{6.936 \times 10^6\, x^2}{230 \times 0.904 \times (159 + 116\, x)}$, From which $x \approx 2.3$ m

Minimum embedment requirement beyond this = $12\Phi = 12 \times 16 = 192$ mm, or equal to $d_x = 159 + 116 \times 2.3 = 425$ mm whichever is more.

∴ Two bars may be curtailed at $2.5 - 2.3 + 0.425 = 0.625$ m from the edge of the support. This distance should be greater than $L_d \approx 45\Phi = 45 \times 16 = 720$ mm = 0.72 m. Hence the bars can be curtailed at a distance of 0.72 m from the support.

**5. *Check for shear and design of shear reinforcement***

$$V = 35469 \text{ N}; \quad M = 43.352 \times 10^6 \text{ N-mm}$$

$$\tau_v = \frac{V - \dfrac{M \tan \beta}{d}}{bd}, \quad \text{where} \quad \tan \beta = \frac{490 - 200}{2500} = 0.116$$

$$\therefore \quad \tau_v = \frac{35469 - \dfrac{43.352 \times 10^6}{449} \times 0.116}{250 \times 449} = 0.216 \text{ N/mm}^2$$

For M 20 grade concrete and $\dfrac{100\, A_s}{bd} = \dfrac{100 \times 603.2}{250 \times 449} = 0.54$, we have $\tau_c = 0.31$

Since $\tau_v > \tau_c$, provide nominal shear reinforcement given by

$$s_v = \frac{2.175\, A_{sv}\, f_y}{b}$$

$$= \frac{2.175\, A_{sv}\, 415}{250} = 3.61\, A_{sv}.$$

Using 8 mm φ 2 lgd strips,

$$s_v = 3.61 \times 100.6 = 363 \text{ mm}.$$

However $s_v$ should be lesser than least of $0.75\, d$ (= $0.75 \times 449 = 336$) or 300 mm.

Hence provide 8 φ 2 – lgd stirrups @ 300 mm c/c at supports and reduce this spacing gradually to $0.75 (200 - 25 - 8 - 8) \approx 120$ mm at the ends.

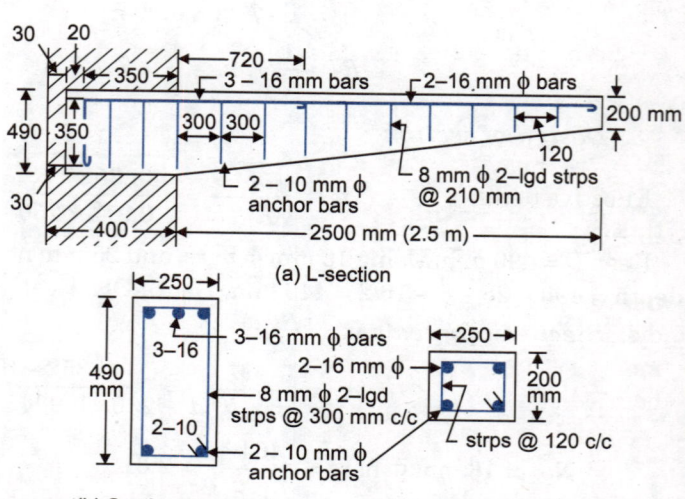

(a) L-section

(b) Section at support    (c) Section at the end

**Fig. 7.11**

**6. *Embedment of reinforcement in the support***

In order develop full tensile strength at the face of the support, each of the three bars must be embedded into the support by a length equal to $L_d = 45\,\varphi = 45 \times 16 = 720$ mm. This could be best achieved by providing one bend of 90° where anchorage value of the bend is $8\,\varphi = 8 \times 16 = 128$ mm. Thus, total anchorage length achieved = 350 + 350 + 128 = 828 mm.

**7. *Details of reinforcement:*** As shown in Fig. 7.11.

## 7.4. DESIGN OF ONE-WAY SLAB

### (a) Longitudinal and Transverse Reinforcements

The analysis of slab spanning in one direction is done by assuming it to be a beam of 1 m width. The reinforcement etc. are calculated for 1 m width and bars are distributed accordingly.

Thus,
$$d = \sqrt{\frac{M}{1000\,R_c}}, \quad \text{where } M \text{ is in N-mm}$$

$$A_{st} = \frac{M}{\sigma_{st} \cdot j_c \cdot d}, \quad \text{and} \quad \text{spacing } s = \frac{1000\,A_\Phi}{A_{st}} \text{ mm}$$

In addition to the main reinforcement (called longitudinal reinforcement), *transverse reinforcement* (also called distribution reinforcement) is also provided in a direction at right angles to the span of the slab. The transverse reinforcement is provided to serve the following purposes:

(*i*) It distributes the effects of point load on the slab more evenly and uniformly.

(*ii*) It distributes the shrinkage and temperature cracks more evenly.

(*iii*) It keeps the main reinforcement in position.

### (b) Basic Rules for Design (IS : 456–2000)

**1. *Effective span.*** The basic rules for the effective span of a slab are the *same* as given by beams (clause *a*-1 of § 7.1). Thus, in the case of freely supported slab, the effective span is taken equal to the distance between centre to centre of supports, or the clear distance between the supports plus the effective depth of the slab whichever is less.

**2. *Control of deflection.*** The same rules apply, as given for beams, in clause *a*-2 of § 7.1.

Thus, the *basic* values of span of effective depth ratios, for span upto 10 m are as follows:

| | |
|---|---|
| Cantilever | 7 |
| Simply supported | 20 |
| Continuous | 26 |

**3. *Minimum reinforcement.*** The mild steel reinforcement in either direction in slabs shall not be less than 0.15% of the total cross-sectional area. However, this value can be reduced to 0.12% when high strength deformed bars or welded wire fabric are used.

**4. *Maximum diameter.*** The diameter of the reinforcing bars shall not exceed one eighth of the total thickness of the slab.

**5. *Spacing of bars.*** The horizontal distance between parallel main reinforcement bars shall not be more than three times the effective depth of a solid slab or 300 mm, whichever is smaller. The horizontal distance between parallel reinforcement bars provided against shrinkage and temperature shall not be more than five times the effective depth of the slab or 450 mm whichever is smaller. The horizontal distance between two parallel main reinforcing bars shall usually be not less than the greatest of the following:

(*i*) The diameter of the bar if the diameters are equal.

(*ii*) The diameter of the larger bar if the diameters are unequal, and

(*iii*) 5 mm more than the nominal maximum size of coarse aggregate.

**6. *Cover to reinforcement.*** The same rules apply, as given in clause *A*-8 of § 7.1.

**7. Curtailment of tension reinforcement.** The same rules, as given for beams in clause A-6 of § 7.1 apply for slabs also. Thus, for curtailment, reinforcement shall extend beyond the point at which it is no longer required to resist flexure, for a distance equal to effective depth or 12 times the bars diameter whichever is greater, except at supports or end of cantilever.

At least one-third the positive moment reinforcement in simple members and one-fourth of the positive moment reinforcement in continuous members shall extend along the same face of the member into the support, to a length equal $L_d/3$.

At simple supports, and at point of inflexion, positive moment tension reinforcement shall be limited to such a diameter that $\left(\dfrac{M_1}{V} + L_0\right) \geq L_d$. The value of $M_1/V$ in the expression may be increased by 30% when the ends of the reinforcement are confined by a compressive reaction.

At least one-third of the total reinforcement provided for negative moment at the support shall extend beyond the point of inflexion, for a distance not less than the effective depth of the member or 12 Φ or one-sixteenth of the clear span whichever is greater.

### (c) Bar Bending Scheme

As per IS 456 : 2000, atleast one third of the maximum positive reinforcement should extend along the same face of the slab into the support, to a length equal to $L_d/3$. Some positive reinforcement should also be bent up, near support, to take up negative B.M. which may develop due to partial fixity. Unfortunately, IS 456 : 2000 does not give any clear cut guidance about the bar bending and cut-off details, as given in CP 110. Due to this, different practices about bar bending is followed. One *practice* is to bend alternate bars up to a distance of 0.15 L (or L/7) from the centre of the supports, so that bars available at the top face, for atleast a

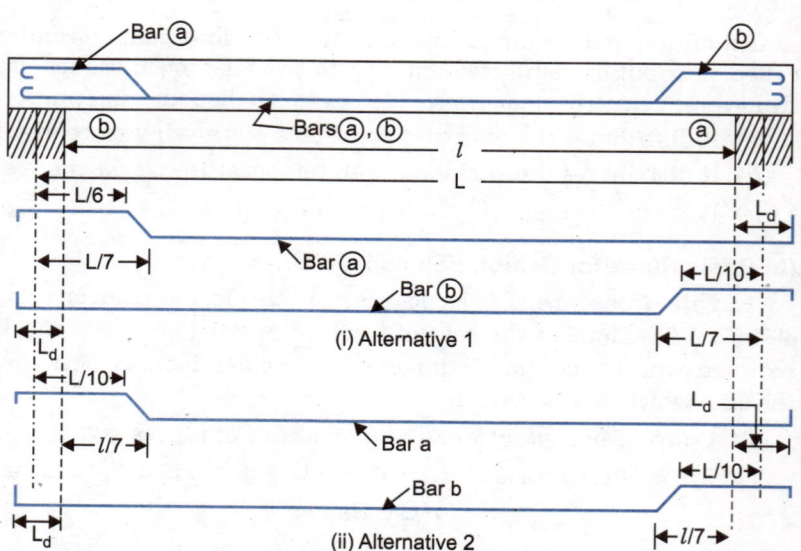

Fig. 7.12. Bar Bending Scheme in One Way Slab

length equal to 0.1 L (or L/10) from the centre of the support. The other *popular practice* is to bend alternate bars up at a distance of 0.15 l (or l/7) from the *face of the support*, where B.M. reduces to less than half its maximum values; this will ensure in majority of cases, that the bars available at the upper face is more than L/10 from the centre of support.

**Design: Example 7.5.** *Design a R.C. slab for a room having inside dimensions 3 m × 6 m. The thickness of the supporting wall is 300 mm. The slab carries 100 mm thick lime concrete at its top, the unit weight of which may be taken as 19000 N/m³. The live load on the slab may be taken as 2500 N/m². Assume the slab to the simply supported at the ends. Use M 20 concrete and Fe 415 steel.*

**Solution:**

**1. Calculation of constants**

For M 20 concrete $\quad c = \sigma_{cbc} = 7$ N/mm²; Also, $\sigma = \sigma_{st} = 230$ N/mm² and $m = 13.33$.

∴ $\quad k_c = 0.289$; $\quad j_c = 0.904$ and $R_c = 0.914$ as in the previous example.

## 2. B.M. and S.F.

The slab will evidently bend over a clear span of 3 m. Let the overall depth of slab be 125 mm, and the approximate effective depth be $125 - 20 - 4 \approxeq 100$ mm. Hence effective span $= 3 + 0.1 = 3.1$ m, which is evidently less than the centre to centre distance between the supports.

(i) Load due to self weight of slab $= (1 \times 1 \times 0.125 \times 25000)$ $= 3125$ N
(ii) Load due to lime concrete finish $= 1 \times 1 \times 0.1 \times 19000$ $= 1900$ N
(iii) Superimposed live load $= 1 \times 1 \times 2500$ $= 2500$ N

Total U.D.L. ($w$), per metre run of slab $= 7525$ N

$$\therefore \quad M = \frac{wL^2}{8} = \frac{7525(3.10)^2}{8} = 9039 \text{ N-m/m width}$$

$$V = \frac{wl}{2} = \frac{7525 \times 3}{2} = 11288 \text{ N at the face of the support.}$$

## 3. Design of section:
$M = 9039$ N-m $= 9.039 \times 10^6$ N-mm per m or 1000 mm width.

$$d = \sqrt{\frac{M}{bR_c}} = \sqrt{\frac{9.039 \times 10^6}{1000 \times 0.914}} \approx 100 \text{ mm}$$

From stiffness (or deflection) point of view, span/effective depth ratio = 20
For a balanced design, percentage reinforcement

$$= \frac{k\,\sigma_{cbc}}{2\,\sigma_{st}} \times 100 = \frac{0.289 \times 7}{2 \times 230} \times 100 = 0.44\%.$$

For Fe 415 steel, $f_s = 0.58$, $f_y = 0.58 \times 415 \approxeq 240$ mm². Hence from Fig. 7.1, modification factor $\approx 1.3$.

$$\therefore \quad \frac{\text{span}}{d} = 20 \times 1.3 = 26. \quad \text{Hence} \quad d = \frac{\text{span}}{26} = \frac{3100}{26} = 119 \text{ mm}.$$

Keep $D = 135$ mm. Using a nominal cover of 15 mm for mild exposure, and using 8 mm φ bars, $d = 135 - 15 - 8/2 = 116$ mm.

## 4. Main reinforcement

$$A_{st} = \frac{M}{\sigma_{st} \cdot j_c \cdot d} = \frac{9.039 \times 10^6}{230 \times 0.904 \times 116} = 375 \text{ mm}^2$$

This is more than minimum reinforcement $= \dfrac{0.12 \times 135 \times 1000}{100} = 162$ mm².

Using 8 mm Φ bars, area of each bar, $A_\Phi = \dfrac{\pi}{4}(8)^2 = 50.27$ mm²

$$\therefore \quad \text{Pitch } s = \frac{1000\,A_\Phi}{A_{st}} = \frac{1000 \times 50.27}{375} = 134 \text{ mm} \approx 130 \text{ mm}.$$

Actual $\quad A_{st} = \dfrac{1000 \times 50.27}{130} = 386.37$ mm².

$$\% \text{ reinforcement} = \frac{386.37}{135 \times 1000} \times 100 = 0.286$$

It is to be noted that the above pitch of 130 mm is less than 3 times effective depth of the slab. Bend every alternate bar up at the support, at a distance of $L/5 = 3.10/5 = 0.6$ m from the edge of the support. $A_{st}$ at the support $= \dfrac{1}{2} \times 386.37 = 193.19$ mm² or 0.143% which is more than the minimum of 0.12%.

## 5. Check for development length and shear at the support.

The Code stipulates that at simple supports, the diameter of the reinforcement be such that $1.3 \dfrac{M_1}{V} + L_0 \geq L_d$, where $M_1 = 230 \times 193.19 \times 0.904 \times 116 = 4.659 \times 10^6$ N-mm.

$$V = 11288 \text{ N}$$

Let us assume that a nominal cover of 25 mm is provided at the side (end) and the slab has a support of 300 mm width.

Providing 90° bend, $L_0 = \dfrac{l_s}{2} - x' + 3\Phi = \dfrac{300}{2} - 25 + 3 \times 10 = 155$ mm

$$\therefore \quad 1.3\dfrac{M_1}{V} + L_0 = \dfrac{1.3 \times 4.659 \times 10^6}{11288} + 155 = 536 + 155 = 691 \text{ mm}$$

$$L_d = \dfrac{\Phi \sigma_{st}}{4 \tau_{bd}} = \dfrac{8 \times 230}{4 \times 0.8 \times 1.6} = 360 \text{ mm}$$

(Alternatively, $L_d \approx 45\Phi = 360$ mm)

$$\therefore \quad 1.3\dfrac{M_1}{V} + L_0 \geq L_d$$

Hence anchorage requirements are satisfied.

**Note:** The reinforcement should extend by a length equal to $L_d/3 = 368/3 = 120$ mm beyond the face of the support. This suggests that the width of the support should not be less than $125 + 25 = 150$ mm. In the present case, the support width is 300 mm.

$$\tau_v = \dfrac{V}{bd} = \dfrac{11288}{1000 \times 116} \approx 0.097 \text{ N/mm}^2.$$

This is much less than the permissible value of $\tau_c = 1.3 \times 0.18 = 0.234$ N/mm² for M 20 concrete for $p' = 0.143$. Hence safe.

**6. Distribution reinforcement:**

$$A_{sd} = \dfrac{0.12 \, bD}{100} = \dfrac{0.12 \times 1000 \times 135}{100} = 162 \text{ mm}^2$$

Using 8 mm $\Phi$ bars,

Pitch $s = \dfrac{1000 A_\Phi}{A_{sd}} = \dfrac{1000 \times 50.27}{162} = 310$ mm.

This is less than five times the effective depth of the slab and is also less than 450 mm. Hence provide 6 mm $\Phi$ bars @ 300 mm c/c. Near the edge of the support, the distribution reinforcement may be provided both at top as well as the bottom. The details of reinforcement etc. are shown in Fig. 7.13.

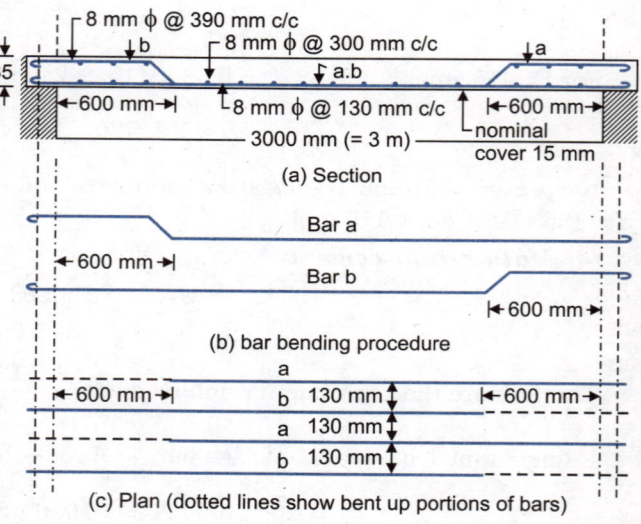

Fig. 7.13

## 7.5. DESIGN OF CANTILEVER CHAJJA

A cantilever slab bends downwards, with the result that tension is developed at the upper face. Hence reinforcement is also provided at the upper face. The span of the slab is taken equal to the actual length of overhang plus half the effective depth. If the width of the cantilever is long, 1 metre width of the cantilever is taken for the design purposes. However, if the width of the cantilever is short, whole of the width may be taken as the width of the slab for the design purposes.

**Design: Example 7.6.** *Design a cantilever slab having an overhang of 1.25 m. Take live load intensity of 1000 N/m² on the cantilever. Use M 20 concrete and HYSD bars. Assume weight of finishing at the top as 800 N/m².*

**Solution:** *(i) Design constants:* For $\sigma_{cbc} = 7$ N/mm², $\sigma_{st} = 230$ N/mm² and $m \approx 13.33$, we have, $k_c = 0.289$; $j_c = 0.904$ and $R_c = 0.914$.

(ii) **B.M. and S.F.:** Assume the cantilever to be of average total thickness of 100 mm.

Dead weight, per m² = $0.1 \times 1 \times 1 \times 25000 = 2500$ N

Dead weight of finish per m² = 800 N ; Live load per m² = 1000 N

Total $w$ = 4300 N

$$\therefore \quad M = \frac{wL^2}{2} = \frac{4300(1.25)^2}{2} = 3359 \text{ N-m} = 3.359 \times 10^6 \text{ N-mm}$$

$$V_{max.} = wL = 4300 \times 1.25 = 5375 \text{ N}$$

(iii) **Design of section:**

$$d = \sqrt{\frac{M}{R_c b}} = \sqrt{\frac{3.359 \times 10^6}{0.914 \times 1000}} = 61 \text{ mm}$$

From stiffness (i.e. deflection) point of view, $L/d = 7$ for a cantilever where $L = l + d/2$ = 1250 + 60 = 1310 mm (say). For M 20 – Fe 415 combination $p_{t,\,lim} = 0.44\%$. Hence modification factor for HYSD bars ≃ 1.3 mm. Hence $d = L/1.3 \times 7 = 1310/1.3 \times 7$ ≃ 144 mm. However since this is a structure of minor importance keep $D$ = 150 mm at the supports. Keeping nominal cover of = 20 mm, and using 8 mm φ bars, $d$ = 150 – 20 – 8/2 = 126 mm. Reduce $D$ = 100 mm at free end.

(iv) **Reinforcement:** $A_{st} = \dfrac{M}{\sigma_{st}\,j_c d} = \dfrac{3.359 \times 10^6}{230 \times 0.904 \times 126} = 128 \text{ mm}^2.$

Choosing 8 mm Φ bars, $A_\Phi = \dfrac{\pi}{4}(8)^2 = 50.3 \text{ mm}^2.$

Pitch $s = \dfrac{1000\,A_\Phi}{A_{st}} = \dfrac{1000 \times 50.3}{128} = 392 \text{ mm}$

Max. permissible spacing = $3d$ or 300 mm which ever is smaller

Hence provide 8 mm Φ bars @ 300 mm c/c. Actual $A_{st} = \dfrac{1000 \times 50.3}{300} = 167 \text{ mm}^2.$

(v) **Embedment of Reinforcement in the support:** In order to develop full tensile strength at the face of the support, each bar should be embedded into the support by a length equal to $L_d = 45\,\varphi = 45 \times 8$ = 360 mm. This could be best achieved by providing one bend of 90° where anchorage value of this bend = $8\,\varphi = 8 \times 8 = 64$ mm. Thus, total anchorage value achieved = (300 – 20) + 64 + (150 – 2 × 20 – 4) ≃ 450 mm > $L_d$.

(vi) **Check for shear:** Neglecting the taper and taking an average $d$ = 110 mm

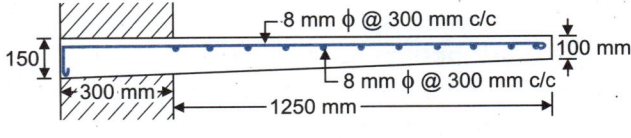

**Fig. 7.14**

$$\tau_v = \frac{V}{bd} = \frac{5375}{1000 \times 110} = 0.049 \text{ N/mm}^2$$

This is much less than the permissible value of $\tau_c = 1.3 \times 0.18 = 0.234 \text{ N/mm}^2$ for M 20 concrete for

$$p' = \frac{100\,A_s}{bd} = \frac{100 \times 169}{1000 \times 110} = 0.15\%. \quad \text{Hence safe.}$$

(vii) **Distribution reinforcement:** Taking average $D$ = 125 mm

$$A_{st} = \frac{0.12\,bD}{100} = \frac{0.12 \times 1000\,D}{100} = 1.2\,D \text{ mm}^2 = 1.2 \times 125 = 150 \text{ mm}^2$$

Using 8 mm Φ bars, each having $A_\Phi = 50.27 \text{ mm}^2$, pitch $s = \dfrac{1000\,A_\Phi}{A_{sd}} = \dfrac{1000 \times 50.27}{150} = 335$ mm.

However, provide these @ 300 mm c/c.

The section of the cantilever slab is shown in Fig. 7.14.

# 172 REINFORCED CONCRETE STRUCTURE

## 7.6. DESIGN OF LINTEL WITH SUNSHADE

**Design: Example 7.7.** *A lintel, covering a verandah opening of 3 m, carries a sunshade of 1 m span, monolithic with it. The bottom of the sunshade and the lintel are at the same level, as shown in Fig. 7.15. The verandah roof of 150 mm total thickness carries a live load of 1500 N/m². Design the lintel and the sunshade. Use M 20 concrete and Fe 415 steel. Take thickness of waterproof as 100 mm with unit weight of 20 kN/m³.*

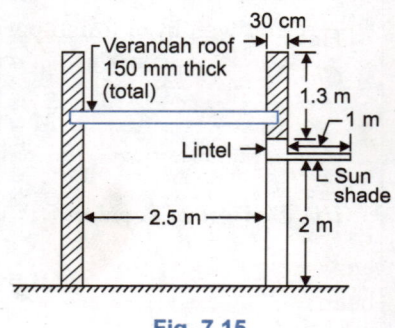

Fig. 7.15

**Solution: Design constants:**

Given
$$\sigma_{cbc} = 7 \text{ N/mm}^2;$$
$$\sigma_{st} = 230 \text{ N/mm}^2 ; m \approx 13.33$$
$$k = \frac{13.33 \times 7}{13.33 \times 7 + 230} = 0.289;$$
$$j_c = 1 - \frac{k_c}{3} = 0.904$$
$$R_c = \frac{1}{2} \sigma_{cbc} j_c k_c = \frac{1}{2} \times 7 \times 0.904 \times 0.289 = 0.914.$$

**1. Design of sunshade:** Let us design the sunshade for half the live load that is carried by the verandah roof. Let the total thickness of the sunshade be 100 mm (average). Consider one metre length of the sunshade. Live load, per m run = 1000 N.

$$\text{Dead load, per m run} = \frac{1 \times 1 \times 100}{1000} \times 25000 = 2500 \text{ N}.$$

Weight of finishing/waterproofing = 500 N (say)

∴
$$w = 1000 + 2500 + 500 = 4000 \text{ N/m}$$
$$M = \frac{wL^2}{2} = \frac{4000 \times 1 \times 1}{2} = 2000 \text{ N-m} = 2000 \times 10^3 \text{ N-mm}$$

∴
$$d = \sqrt{\frac{M}{R_c b}} = \sqrt{\frac{2000 \times 10^3}{0.914 \times 1000}} = 47 \text{ mm}$$

From stiffness point of view, assuming a modification factor of 1.3 as in the previous example, $L/d = 7 \times 1.3 = 9.1$. Taking effective length $L = 1000 + 150/2 = 1075$ mm $d = 1075/9.1 = 118$ mm. Provide $D = 140$ mm at the supports and 100 mm at ends. Hence at support, providing 15 mm nominal cover, $d = 140 - 15 - 8/2 = 121$ mm.

$$A_{st} = \frac{M}{\sigma_{st} j_c d} = \frac{2000 \times 10^3}{230 \times 0.904 \times 121} = 80 \text{ mm}^2.$$

$$\text{Minimum steel} = \frac{0.12 \times 1000 \times 120}{100} = 144 \text{ mm}^2. \text{ Choosing 8 mm } \Phi \text{ mm bars,}$$

$$\text{Pitch } s = \frac{1000 A_\Phi}{A_{st}} = \frac{1000 \times 50.3}{144} \approx 349 \text{ mm}$$

Max. spacing = $3d = 3 \times 120 = 360$ mm or 300 mm which ever is smaller. Hence provide 8 mm $\Phi$ bars @ 300 mm c/c. Anchor the bars inside the lintel as shown.

**Transverse reinforcement**

$$A_{sd} = \frac{0.12}{100} \times bD = \frac{0.12}{100} \times 1000 \, D = 1.2 \, D = 1.2 \times 120 = 144 \text{ mm}^2$$

($D$ = average total depth = 120 mm),

$$s = \frac{1000 \times 50.3}{144} \approx 349. \text{ Max. } s = 5d \text{ or } 450 \text{ mm whichever is small.}$$

Hence provide 8 mm Φ distribution bars @ 345 mm c/c.

*Check for shear and embedment length at the support*

$$V = wL = 4000 \text{ N}; \quad \text{Neglecting the taper in the slab,}$$

$$\tau_v = \frac{V}{bd} = \frac{4000}{1000 \times 121} = 0.03 \text{ N/mm}^2. \quad \text{Hence safe.}$$

In order to develop full tensile strength at the face of the support (i.e. lintel beam), each bar should be embedded in the lintel beam by a length = $L_d$ = 45 × 8 = 360 mm. This is best achieved by providing one bend of 90° and extending it by about 100 mm into the beam. The amount of anchorage length thus provided = (300 − 30) + 8 × 8 + 100 = 434 mm > $L_d$.

**2. Design of Lintel beam:** Let the width of Lintel beam = 300 mm. Let, total depth of Lintel = 1/10 span ≈ 400 mm (say). Consider one metre length of the lintel.

(*i*) Dead load of verandah roof  = $\frac{1}{2}$ (2.5 × 1 × 0.15 × 25000) ≈ 4688 N

(*ii*) Dead load of water proofing  = $\frac{1}{2}$ (2.5 × 1 × 0.1 × 20000)  = 2500 N

(*iii*) Live load from verandah roof  = $\frac{1}{2}$ (2.5 × 1 × 1500)  = 1875 N

(*iv*) Dead weight of masonry  = 0.3 × 1.3 × 1 × 20000  = 7800 N

(*v*) Self weight of lintel  = 0.3 × 0.4 × 1 × 25000  = 3000 N

(*vi*) Dead weight of sunshade  = 1 × 0.120 × 1 × 25000  = 3000 N

(*vii*) Dead weight of finishing on sunshade  = 500 × 1 × 1  = 500 N

Total of (*i*) to (*vii*): Load per m run  = 23363 N  ≈ 23500 N/m.

**Note:** The live load on the sunshade is not likely to be there simultaneously with live load on varandah roof and hence has not been included. Similarly, torsional moment due to load on the sunshade will be extremely small and hence neglected.

Assume a bearing of 400 mm on either end. Effective span = $L$ = 3 + 0.4 = 3.4 m.

$$M = \frac{wL^2}{8} = \frac{23500(3.4)^2}{8} = 33958 \text{ N-m} = 33.958 \times 10^6 \text{ N-mm}$$

$$\therefore \quad d = \sqrt{\frac{M}{R_c b}} = \sqrt{\frac{33.958 \times 10^6}{0.914 \times 300}} = 352 \text{ mm}.$$

From stiffness point of view $L/d$ = 20. Using a modification factor of 1.3 for M 20 –Fe 415 combination $d$ = 3400/(20 × 1.3) = 131 mm only. Hence keep $D$ = 400 mm. Using 16 mm φ main bars, 8 mm φ rings and 25 mm nominal cover, available $d$ = 400 − 25 − 8 − 16/2 = 359 mm.

$$A_{st} = \frac{M}{\sigma_{st} j_c d} = \frac{33.958 \times 10^6}{230 \times 0.904 \times 359} = 455 \text{ mm}^2$$

No. of 16 mm Φ bars = $\dfrac{455}{\frac{\pi}{4}(16)^2}$ = 2.3

No. of 12 mm Φ bars = $\dfrac{455}{\frac{\pi}{4}(12)^2}$ ≈ 4.

Hence provide 4-12 Φ bars. Out of these, bend 2 bars up, at a distance of $L/7$ = 3.4/7 ≈ 0.5 m from the edge of each support.

Remaining $A_s = 2 \times \dfrac{\pi}{4}(12)^2 = 226.2$ mm$^2$.

**Check for shear and development length**

$$V = \dfrac{wl}{2} = \dfrac{23500 \times 3}{2} = 35250 \text{ N};$$

$$\tau_v = \dfrac{V}{bd} = \dfrac{35250}{300 \times 359} = 0.327 \text{ N/mm}^2,$$

$$\dfrac{100 A_s}{bd} = \dfrac{100 \times 226.2}{300 \times 359} = 0.21\%.$$

Hence $\tau_c \approx 0.328$ N/mm$^2$ and shear reinforcement is not necessary. However provide only nominal shear stirrups, determined from the expression:

$$s_v \leq \dfrac{2.175 A_{sv} f_y}{b} \leq \dfrac{2.175 \times 415}{300} A_{sv} \leq 3.009 A_{sv}$$

Using 8 mm $\Phi$ 2-lgd stirrups, having $A_{sv} = 2 \times \dfrac{\pi}{4}(8)^2 = 100.5$ mm$^2$,

$s \leq 3.009 \times 100.5 = 302$ mm $\approx 300$ mm.
Max. spacing = $0.75 d = 0.75 \times 359$
= 269. Hence provide these @ 250 mm c/c.
Provide 2-10 $\varphi$ anchor bars at top.

Also, at the support, the Code requirements are: $1.3 \dfrac{M_1}{V} + L_0 \geq L_d$

Here
$M_1 = A_{st} \cdot \sigma_{st} \cdot j_c d$
$= 226.2 \times 230 \times 0.904 \times 359$
$= 16.88 \times 10^6$ N-mm
$V = 35250$ N  $L_s = 400$ mm ;
$x'$ = side cover = 30 mm (say)
Providing 90° bend at the ends,

$$L_0 = \dfrac{L_s}{2} - x' + 3\Phi$$

$$= \dfrac{400}{2} - 30 + 3 \times 12 = 206 \text{ mm}.$$

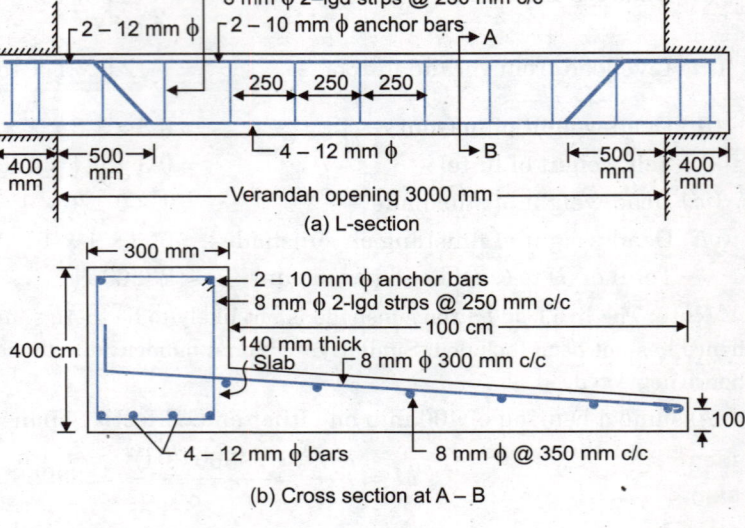

Fig. 7.16

$$1.3 \dfrac{M_1}{V} + L_0 = 1.3 \times \dfrac{16.88 \times 10^6}{35250} + 206 = 622 + 206 = 828 \text{ mm}$$

$L_d = 45 \Phi = 45 \times 12 = 540$ mm. Hence Code requirements are satisfied.

The details of reinforcement etc. are shown in Fig. 7.16.

## 7.7. DESIGN OF CONTINUOUS SLAB

When a slab is continuous over several spans, negatives (*i.e.* hogging) bending moment is induced over the intermediate supports and hence reinforcement has to be provided at the top of the slab portion over the intermediate supports. These intermediate supports may either be monolithic with the slab, or may be of some different material such as rolled steel joist etc. Following are some of the *basic* rules for design, as per IS : 456-2000.

1. *Effective span.* In the case of continuous slab, if the width of the support is less than 1/12 of the clear span, the effective span shall be the distance between centres of supports, or the clear distance between supports plus the effective depth of the slab, whichever is smaller. If the supports are wider than 1/12 the clear span or 600 mm, whichever is less, the effective span shall be taken as under:

   (a) For end span with one end fixed and the other continuous, or for intermediate spans, the effective span shall be the clear span between supports; and

   (b) For end span with one end free and the other continuous, the effective span shall be equal to the clear span plus half the effective depth of the slab or the clear span plus half the width of the continuous support, whichever is less.

   **Note:** In the case of spans with roller or rocker bearings, the effective span shall always be the distance between the centres of bearings.

2. *Limiting stiffness.* The ratio of span to effective depth should not exceed 26. Modification factor for tension reinforcement will be extra as per Fig. 7.1.

3. *Moment and shear coefficients.* Same rules as for continuous beams, given in clause A-10 of § 7.1. Thus, the B.M. values are indicated on Fig. 7.17.

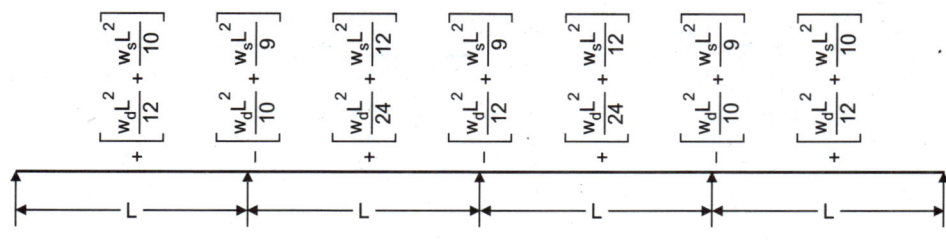

Fig. 7.17

4. *Reinforcement and its arrangement.* The rules for main reinforcement and transverse/distribution reinforcement are the same as discussed for simply supported one way slab. However, the Code (IS 456 : 2000) makes the following recommendations for continuation/curtailment of reinforcement.

   **(a) Positive moment reinforcement:** At least *one third* the positive moment reinforcement in simple members and one fourth the positive moment reinforcement in continuous members shall extend along the same face of the member into the member, to a length equal to $L_d/3$.

   **(b) Negative moment reinforcement:** At least one third of the total reinforcement provided for negative moment at the support shall extend beyond the point of inflexion for a distance not less than the effective depth of the member or 12 φ or one sixteenth of the clear span whichever is greater.

   Except for the above two recommendations, IS 456 : 2000 is silent in regard to simplified rules for curtailment of reinforcement is one way continuous slab. In absence of any such recommendation for IS Code, various practices are followed, and one such practice is given in Fig. 7.18.

   However, CP 110-1972 has given some simplified rules, as given in Fig. 7.19 (a). Two more forms of reinforcement detailing are shown in Fig. 7.19 (b) and (c) respectively. It may be noted that detailing type II shown in Fig. 7.19 (b) is similar to the one shown in Fig. 7.18, except that the bend up point is at a distance of 0.25 L from the centre of support (in place of L/5) and the negative reinforcement extends to a distance of 0.3 L to the either side of the support (in place of L/4). In detailing type III, no bar is bent up, and the positive reinforcement and the negative reinforcement are provided separately at appropriate faces. The curtailment of positive reinforcement in type III detailing is as per IS Code recommendations.

**176** REINFORCED CONCRETE STRUCTURE

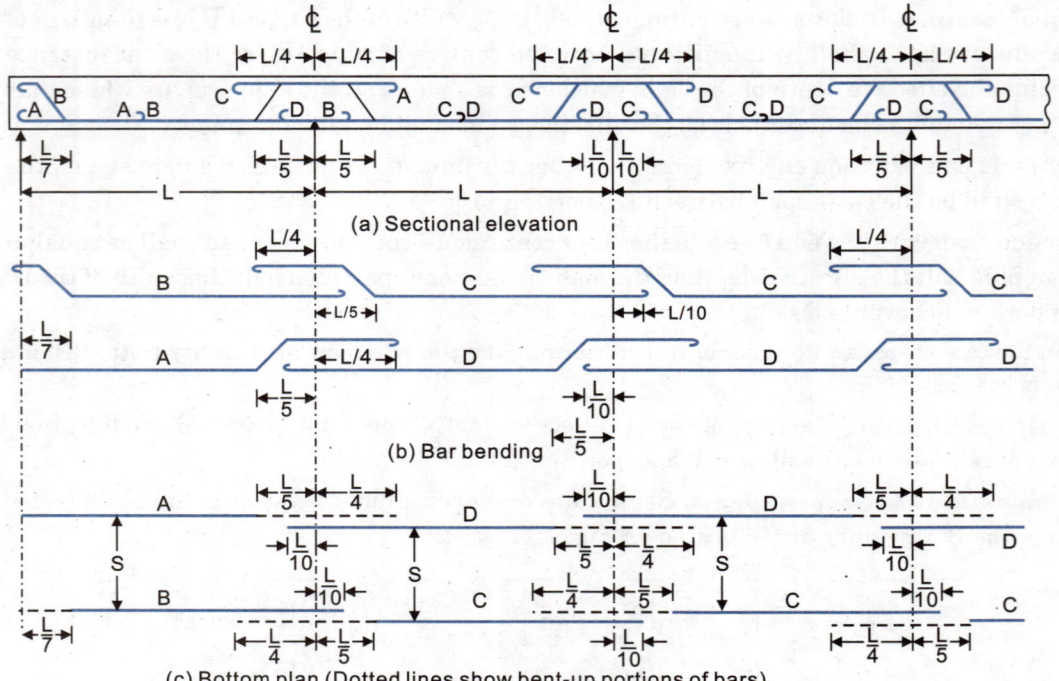

Fig. 7.18

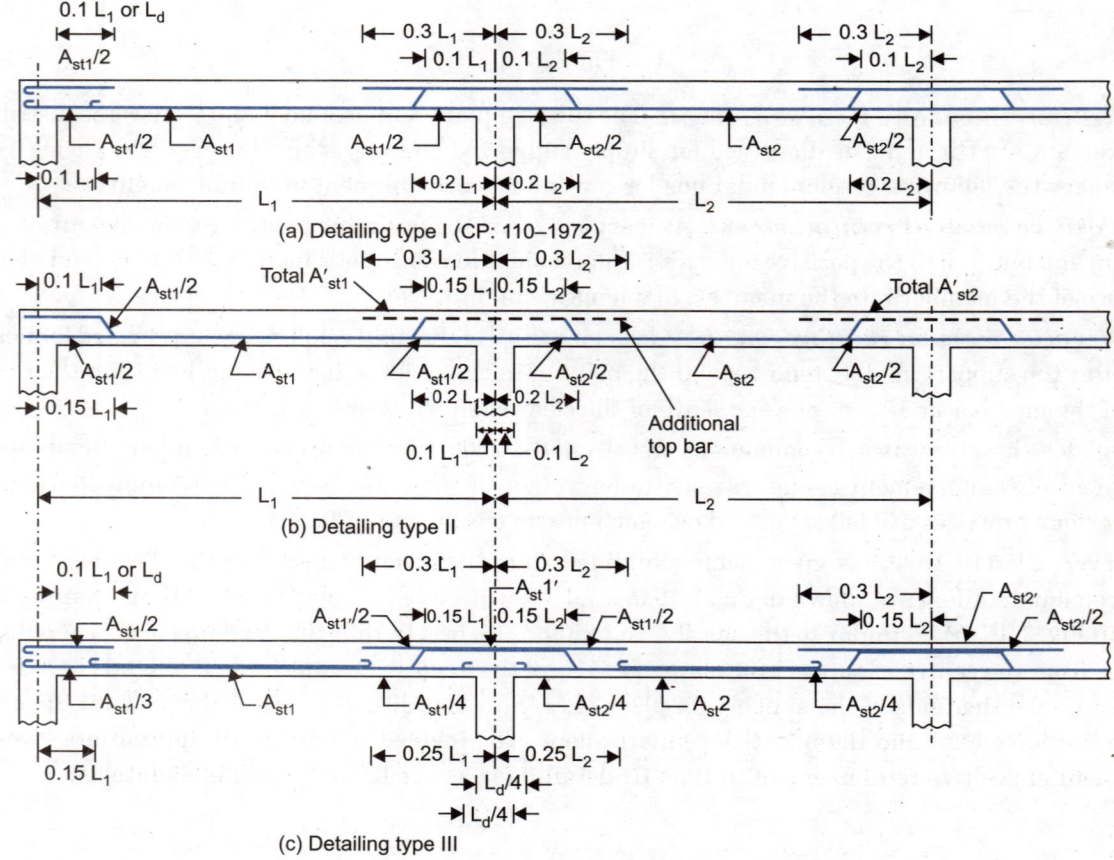

Fig. 7.19. Alternative forms of Reinforcement Detailing

**Design: Example 7.8.** *Design a continuous R.C. slab for a hall 6.5 m wide and 13.5 long. The slab is supported on R.C.C. beams, each 200 mm wide. The end bearing of the slab on the walls is 120 mm. The super-imposed load is 2500 N/m². The floor carries a finishing weighing 600 N/m². Use M 20 concrete and Fe 415 steel reinforcement.*

**Solution:** **1. Design constants**

We have $\sigma_{cbc} = 7$ N/mm², $m \approx 13.33$; $\sigma_{st} = 230$ N/mm². Hence,

$$k_c = \frac{m\,\sigma_{cbc}}{m\,\sigma_{cbc} + \sigma_{st}} = \frac{13.33 \times 7}{13.33 \times 7 + 230} = 0.289$$

$$j_c = 1 - k_c/3 = 0.904\,;\qquad R_c = \frac{1}{2} \times 7 \times 0.904 \times 0.289 = 0.914$$

**2. Effective spans.** Let us arrange the spans as shown in Fig. 7.20. Let the effective depth of slab = 100 mm. The width of the intermediate supports is less than 1/12 of the clear span. Hence the effective span will be as under:

(i) For end spans, effective span:

$L$ = centre to centre of supports = $3.25 + \dfrac{0.20}{2} = 3.35$ m

or $L$ = clear distance + $d$ = 3.15 + 0.10 = 3.25 m, whichever is less. Hence $L$ = 3.25 m.

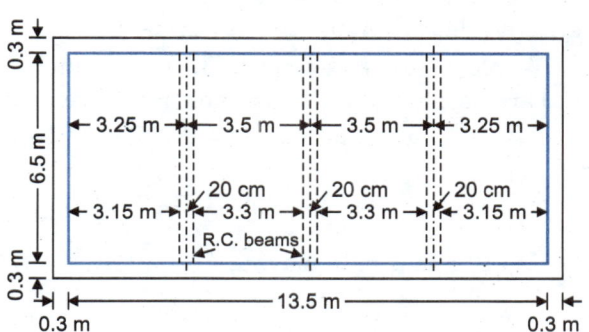

(ii) For intermediate spans: $L$ = centre to centre of supports = 3.5 m or $L$ = 3.3 + 0.1 = 3.4 m, whichever is less. Hence $L$ = 3.4 m. Thus, in this case, the lengths of effective span are not equal. However, the difference is very much less than 15% of the longer spans; hence they may be treated as being equal for the purposes of using the bending moment coefficients given in the previous article.

**Fig. 7.20.** Positioning of Beam in the Hall.

**3. Calculation of bending moment and shear force**

Assume overall depth of slab = 120 mm. Consider 1 m width of slab.

(i) Self wight of slab = 0.12 × 1 × 1 × 25000 = 3000 N
(ii) Weight of finishing = 600 N

∴ $w_d$ = 3600 N/m

Super-imposed load = $w_s$ = 2500 N/m run.

For end spans:

$$M \text{ (near the centre)} = \frac{w_d L^2}{12} + \frac{w_s L^2}{10} = \frac{3600\,(3.25)^2}{12} + \frac{2500\,(3.25)^2}{10}$$

$$= 5809 \text{ N-m} = 5.8 \times 10^6 \text{ N-mm.} \qquad \ldots(i)$$

$$M \text{ (at support next to end support)} = -\left(\frac{w_d L^2}{10} + \frac{w_s L^2}{9}\right)$$

$$= -\left[\frac{3600\,(3.25)^2}{10} + \frac{2500\,(3.25)^2}{9}\right]$$

$$= -6737 \text{ N-m} = -6.737 \times 10^6 \text{ N-mm} \qquad \ldots(ii)$$

S.F. at end support = $0.4\,w_d \cdot L + 0.45\,w_s \cdot L = (0.4 \times 3600 + 0.45 \times 2500)\,3.25 = 8336$ N

S.F. (at the exterior face of the support next to end support)

$= 0.6\,w_d \cdot L + 0.6\,w_s \cdot L = (0.6 \times 3600 + 0.6 \times 2500)\,3.25 = 11895$ N

*For intermediate spans:*

$$M \text{ (at the middle)} = \frac{w_d L^2}{24} + \frac{w_s L^2}{12} = \frac{3600 (3.4)^2}{24} + \frac{2500 (3.4)^2}{12}$$

$$= 4142 \text{ N-m} = 4.142 \times 10^6 \text{ N-mm}. \qquad \ldots(iii)$$

$$M \text{ (at interior support)} = -\left(\frac{w_d L^2}{12} + \frac{w_s L^2}{9}\right) = -\left[\frac{3600 (3.4)^2}{12} + \frac{2500 (3.4)^2}{9}\right]$$

$$= -6679 \text{ N-m} = 6.679 \times 10^6 \text{ N-mm} \qquad \ldots(iv)$$

$F$ (interior faces) $= 0.5 \, w_d \cdot L + 0.6 \, w_s \cdot L = (0.5 \times 3600 + 0.6 \times 2500) \, 3.4 = 11220$ N

**4. *Design of section.*** The effective depth will be calculated on the basis of maximum bending moment out of four values calculated above.

$$M_{max.} = 6.737 \times 10^6 \text{ N-mm. Hence, } d = \sqrt{\frac{6.737 \times 10^6}{0.914 \times 1000}} = 86 \text{ mm}$$

Keep overall depth = 110 mm, and a nominal cover equal to 15 mm. Using 10 mm Φ bars, available effective depth = 110 – 15 – 5 = 90 mm.

**5. *Main reinforcement.*** The maximum bending moments in the end spans (*ii*) and intermediate span (*iv*) are nearly equal. Hence same reinforcement will be provided in all spans.

$M_{max.}$ for design $= 6.737 \times 10^6$ N-mm

$$\therefore \quad A_{st} = \frac{M}{\sigma_{st} \, j_c \, d} = \frac{6.737 \times 10^6}{230 \times 0.904 \times 90} = 360 \text{ mm}^2.$$

Use 10 mm Φ bars, having $A_\Phi = \frac{\pi}{4}(10)^2 = 78.54$ mm²

$$\therefore \quad \text{Pitch } s = \frac{1000 \, A_\Phi}{A_{sd}} = \frac{1000 \times 78.54}{360} = 218 \text{ mm}$$

Hence keep pitch = 210 mm. Alternate bars from the bottom can be bent up near supports to bear the negative bending moment.

Actual $A_{st}$ provided per metre width $= \dfrac{1000 \times 78.54}{210} = 374$ mm²

**6. *Distribution reinforcement***

$$A_{st} = 0.12\% = \frac{0.12}{100} \times 1000 \, D = 1.2 \, D = 1.2 \times 110 = 132$$

Using 8 mm Φ bars, pitch $s = \dfrac{1000 \, A_\Phi}{A_{sd}} = \dfrac{1000 \times 50.27}{132} = 380$ mm

Hence provide pitch at 380 mm c/c, which is less than five times the effective depth of the slab.

**7. *Check for development length at support***

The Code stipulates that the diameter of the bars should be such that $1.3 \dfrac{M_1}{V} + L_0 \geq L_d$ at the end support.

$$M_1 = A_{st} \, \sigma_{st} \, j_c \, d = \left(\frac{1}{2} \times 374\right)(230 \times 0.904 \times 90) = 3.499 \times 10^6 \text{ N-mm}$$

$V = 8336$ N

$L_s = 200$ mm; $x' = 25$ mm (say).

$L_0 = \dfrac{L_s}{2} - x' = \dfrac{200}{2} - 25 = 75$ mm. (Without providing any hook or bend)

$1.3 \dfrac{M_1}{V} + L_0 = 1.3 \times \dfrac{3.499 \times 10^6}{8336} + 75 = 545 + 75 = 620$ mm.

$L_d \approx 45 \, \Phi = 45 \times 10 = 450$ mm. Thus Code requirements are satisfied.

**8. Check for shear.** Maximum shear = 11895

$$\tau_v = \frac{V}{bd} = \frac{11895}{1000 \times 90} = 0.132 \text{ N/mm}^2 \;;\; \frac{100 A_s}{bd} = \frac{100 \times 187}{1000 \times 90} = 0.208\%$$

Hence from Table 3.1 and 3.2, $\tau_c = 0.20 \times 1.3 = 0.26$ N/mm² for M 20 concrete. Thus the slab is safe in shear. The details of reinforcement etc. are shown in Fig. 7.21.

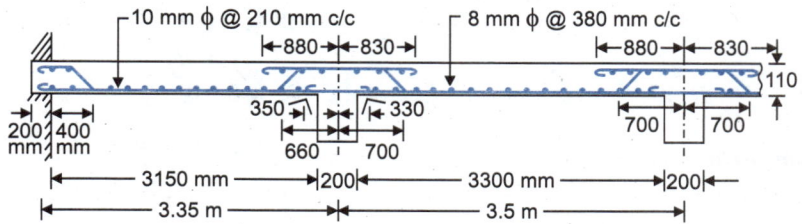

**Fig. 7.21.** Detailing of Beam-Longitudinal Section

## 7.8. DESIGN OF DOUBLY REINFORCED BEAM

The theory of doubly reinforced beam has already been discussed in Chapter 5. In addition to the basic rules for design for singly reinforced beams, the following two important points are noteworthy. (IS : 456–2000).

**1. Percentage of compression reinforcement:** The maximum area of compression reinforcement shall not exceed 0.04 *bD*. Compression reinforcement in beams shall be enclosed by stirrups for effective lateral restraint.

**2. Anchoring bars in compression:** The anchoring length of straight bar in compression shall be equal to the development length of bar in compression. The projected length of hooks, bends and straight length beyond bends if provided for a bar in compression shall be considered for development length.

### Design procedure

1. Determine the position of the critical neutral axis of the section of the given set of stresses:
$n_c = \dfrac{m \, \sigma_{cbc}}{m \, \sigma_{cbc} + \sigma_{st}} \cdot d$, where $d$ is the known effective depth of the beam.

2. Determine the area of tensile reinforcement $A_{st1}$ and the moment of resistance $M_1$ corresponding to the singly reinforced balanced section:

$$M_1 = R_c b d^2 = \frac{1}{2} \sigma_{cbc} j_c k_c \cdot bd^2 = bn_c \frac{\sigma_{cbc}}{2}\left(d - \frac{n_c}{3}\right) \quad \text{and} \quad A_{st} = \frac{M_1}{\sigma_{st} \cdot j_c d}.$$

3. Determine the additional area of tensile reinforcement $A_{st2}$ for the remaining bending moment $M_2$:

$$A_{st2} = \frac{M_2}{\sigma_{st}(d - d_c)} = \frac{M - M_1}{\sigma_{st}(d - d_c)}$$

Thus, the total area of tensile reinforcement will be equal to $A_{st} = A_{st1} + A_{st2}$.

4. Determine the area of compressive steel $A_{sc}$ by equating moments of equivalent areas of compressive steel ($A_{sc}$) and additional tensile area ($A_{st2}$). Thus

$$A_{sc} = \frac{m(d - n_c)}{(m_c - 1)(n_c - d_c)} \cdot A_{st2} = \frac{m(1 - k_c)}{(m_c - 1)(k_c - e)} \cdot A_{st2}$$

where $m_c = 1.5 \, m$ and $e - d_c/d$

5. From the point of view of economy, curtail the tensile and compressive reinforcement at suitable sections.

6. Check the sections for shear stress and development length. Design shear reinforcement if the beam is unsafe for shear. Otherwise provide nominal shear reinforcement.

**Design: Example 7.9.** *Design a doubly reinforced beam to carry a super-imposed load of 60 kN per meter run. The overall depth and width of the beam are restricted to 840 mm and 300 mm respectively. The beam has a clear span of 5 m and a bearing of 50 cm on each end. Use M 20 concrete and Fe 415 steel.*

**Solution:**

**1. Design constants:** $c = \sigma_{cbc} = 7$ N/mm$^2$ ; $t = \sigma_{st} = 230$ N/mm$^2$ ; $m = 13.33$

$$k_c = \frac{m \cdot c}{mc + t} = \frac{13.3 \times 7}{13.3 \times 7 + 230} = 0.289 \; ; j_c = 1 - k_c/3 = 0.904$$

$$R_c = \frac{1}{2} c \, j_c k_c = \frac{1}{2} \times 7 \times 0.904 \times 0.289 = 0.914.$$

**2. Bending moment and shear force:** Effective span $= 5 + 0.5 = 5.5$ m

Wt. of beam per m run = $0.3 \times 0.84 \times 25000$ = 6300 N
Superimposed load per m run = 60000 N
Total U.D.L. per m run = 66300 N

$$\therefore \quad M = \frac{wL^2}{8} = \frac{66300 \times (5.5)^2}{8} = 250700 \text{ N-m} = 250.7 \times 10^6 \text{ N-mm}$$

$$F = \frac{wl}{2} = \frac{66300 \times 5}{2} = 165750 \text{ N}.$$

**3. Moment of resistance $M_1$ and reinforcement $A_{st1}$.** Let us assume that the centre of the tensile reinforcement will be at 65 mm above the bottom fibre.

$\therefore \qquad d = 840 - 65 = 775$ mm ; $n_c = k_c d = 0.289 \times 775 = 224$ mm.

For a singly reinforced balanced section,

$$M_1 = R_c b d^2 = 0.914 \times 300 (775)^2 = 164.69 \times 10^6 \text{ N-mm}$$

Area of tensile reinforcement is given by

$$A_{st1} = \frac{M_1}{\sigma_{st} \, jd} = \frac{164.69 \times 10^6}{230 \times 0.904 \times 775} = 1022 \text{ mm}^2.$$

**4. Moment of resistance $M_2$ and reinforcement $A_{st2}$**

$$M_2 = M - M_1 = (250.7 - 164.69)10^6 = 86.01 \times 10^6 \text{ N-mm}.$$

This remaining B.M. has to be resisted by a couple provided by the tensile and compressive reinforcements. Let the centre of compressive reinforcement be placed at 45 mm below the top fibre, *i.e.* $d_c = 45$ mm.

$$\therefore \quad A_{st2} = \frac{M_2}{\sigma_{st}(d - d_c)} = \frac{86.01 \times 10^6}{230(775 - 45)} = 513 \text{ mm}^2$$

$\therefore \qquad$ Total $A_{st} = A_{st1} + A_{st2} = 1022 + 513 = 1535$ mm$^2$.

**5. Compressive reinforcement $A_{sc}$**

$$A_{sc} = \frac{m(d - n_c)}{(m_c - 1)(n_c - d_c)} A_{st2}, \text{ where } n_c = 0.289 \times 775 = 224$$

$$= \frac{13.33 (775 - 224)}{(1.5 \times 13.33 - 1)(224 - 45)} \times 513 = 1108 \text{ mm}^2$$

**6. Reinforcing bars**

$A_{st} = 1535$ mm$^2$. Using 16 mm Φ bars, having $A_\phi = 201.06$ mm$^2$

$\therefore \qquad$ No. of 16 mm φ bars = 1535/201.06 = 7.63

Hence provide 8-16 mm Φ bars. Place 5 bars in the bottom tier and 3 bars in the top tier, keeping a clear distance of 25 mm between the two tiers. Keep a nominal cover of 25 mm. Use 25 mm Φ spacers bars @ 1 m c/c.

$A_{sc} = 1108$ mm$^2$. No. of 20 mm Φ bars = 1108/314 = 3.8. Hence provide 4-20 Φ bars at the top, in one tier, keeping a nominal cover of 25 mm.

DESIGN OF BEAMS AND SLABS **181**

**7. *Curtailment of reinforcement:*** The bending at any point distant $x$ metres from the centre of the span is given by

$$M_x = \left(\frac{wL^2}{8} - \frac{wx^2}{2}\right)1000 = \left(M - \frac{wx^2}{2} \times 1000\right)$$

(where the moments $M_x$ and $M$ are in N-mm units).

At the point where compressive reinforcement is not required, the bending moment should be equal to $M_1$

$\therefore \qquad M_1 = M - \frac{wx^2}{2} \times 1000$

$\therefore \qquad x = \sqrt{\frac{2(M_1 - M)}{1000\,w}} = \sqrt{\frac{2M_2}{1000\,w}} = \sqrt{\frac{2 \times 86.01 \times 10^6}{1000 \times 66300}} \approx 1.61 \text{ m}$

Hence at $x = 1.61$ m from the centre, compressive reinforcement is no longer required and it may, therefore, be curtailed. However, curtail only 2 bars and continue 2 bars upto the support, to hold the shear stirrups. At this section, the bending moment is $M_1$ only. Hence tensile reinforcement required = $A_{st1}$ = 1022 mm², which will need only 5 bars. Hence curtail 3 bars of the upper tier at this point and continue rest of the bars to the support.

**8. *Shear reinforcement:*** Near the support, where the S.F. is maximum, the section is singly reinforced (since the two compressive reinforcing bars serve as holding bars of the shear stirrups).

Available effective depth $d = 840 - 25 - 8 - 16/2 = 799$ mm

$$\tau_v = \frac{V}{bd} = \frac{165750}{300 \times 799} = 0.69 \text{ N/mm}^2$$

Available $\qquad A_s = 5 \times 201.06 = 1005.3 \text{ mm}^2$ ;

$\therefore \qquad \frac{100\,A_s}{bd} = \frac{100 \times 1005.3}{300 \times 799} = 0.42\%$

Hence from table 3.1, $\tau_c \approx 0.27$ N/mm². Hence shear reinforcement is necessary.

$V_c = \tau_c \cdot bd = 0.27 \times 300 \times 799 = 64719$ N ;
$V_s = V - V_c = 165750 - 64719 = 101031$ N.

The shear stirrups are governed by

$$s_v = \frac{\sigma_{sv} \cdot d \cdot A_{sv}}{V_s} = \frac{230 \times 799}{101031} A_{sv} = 1.819\,A_{sv}.$$

Using 8 mm Φ 2-lgd stirrups,

$$A_{sv} = 2\frac{\pi}{4}(8)^2 = 100.53 \text{ mm}^2.$$

Hence, $\qquad s_v = 1.819 \times 100.53 = 183$ mm.

Max. spacing is given by

$$s_v = \frac{2.175\,fy}{b} \cdot A_{sv} = \frac{2.175 \times 415}{300} A_{sv}$$

$$= 3.009\,A_{sv} = 3.009 \times 100.53 = 302 \text{ mm}.$$

Hence provide 8 mm Φ 2-lgd stirrups @ 180 mm c/c near the support. This spacing can be increased to 300 mm toward the mid span of the beam.

**9. *Check for development length at supports***

At the supports $1.3\dfrac{M_1}{V} + L_0 \geq L_d$.

Here $\quad M_1 = \sigma_{st} \cdot A_{st} \cdot j_c d = (230 \times 5 \times 201.06 \times 0.904 \times 799) = 162.83 \times 10^6$ N-mm

$V = 165750$ N ; $\quad L_s = 500$ mm ; $x' = 30$ mm (say).

Providing bars straight, is without any hook or bend.

$$L_0 = \frac{L_s}{2} - x' = \frac{500}{2} - 30 = 220 \text{ mm} ; \quad L_d \approx 45 \, \Phi = 45 \times 16 = 720 \text{ mm}$$

$$\therefore \quad 1.3 \frac{M_1}{V} + L_0 = 1.3 \times \frac{162.83 \times 10^6}{165750} + 220 = 1277 + 220 = 1497 \text{ mm}$$

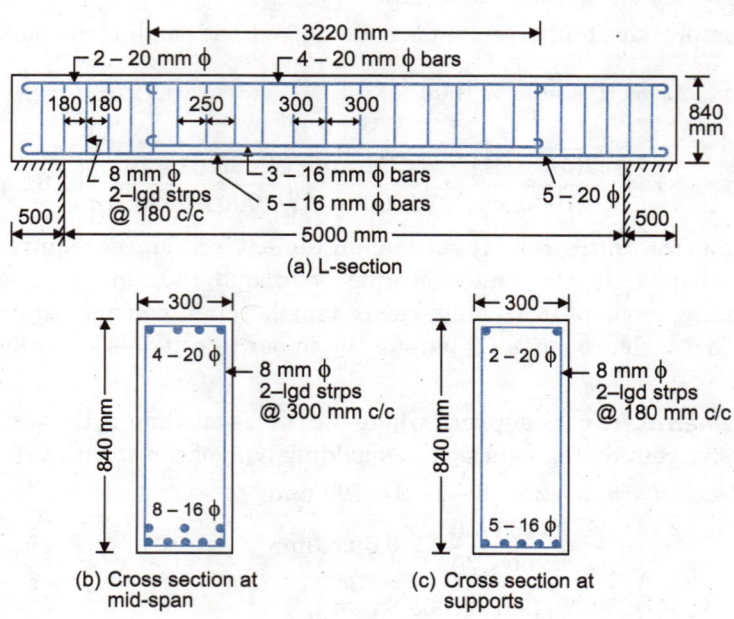

Fig. 7.22. Detailing of Beam

Hence Code requirements are satisfied. The details of reinforcement etc. are shown in Fig. 7.22.

## 7.9. DESIGN OF T-BEAM

The theory of T-beam has been discussed in detail in Chapter 6. The rules for selection of width of flange, breadth of rib and depth of rib have been in given in that chapter. The general notes given in § 7.1 for rectangular beam also apply for this beam. The design of T-beam is done in the following steps:

1. Assume some suitable value of width of web based on criteria discussed § 6.2.
2. Determine the effective width $b_f$ of the flange as discussed in § 6.2.
3. Assume effective span of the beam and calculate the bending moment and S.F.
4. Determine the effective depth of the beam on any one of the three considerations:
    (*i*) balanced section           (*ii*) economy
   (*iii*) shear stress.

For balanced section, the effective depth is given by *Eq. 6.16*:

$$M = b_f \cdot \sigma_{cbc} \cdot D_f \left[ d - \frac{D_f(1 + k_c)}{2k_c} + \frac{D_f^2}{3k_c \, d} \right]$$

Alternatively, taking lever arm to be equal to 0.9d, we have from *Eq. 6.17*,

$$M = 0.45 \, b_f \cdot \sigma_{cbc} \cdot D_f \left[ \frac{2k_c \, d - D_f}{k_c} \right]$$

For economical depth, use *Eq. 6.18*, $d = \sqrt{\dfrac{r \, M}{\sigma_{st} \, j_c \, b_w}}$

For shear consideration, use *Eq. 6.21*, $d_{min} = \dfrac{V}{b_w \cdot \tau_v} = \dfrac{V}{b_w \cdot \tau_{c.max}}$

where $\tau_v$ may be taken equal to $\tau_{c.cmax}$ given in Table 3.3.

The *maximum* value of $d$ obtained from the above three criteria is taken as the effective depth of the beam.

5. Take the approximate lever arm $a = d - \dfrac{D_f}{2}$ and determine the area of tensile reinforcement:

$$A_{st} = \dfrac{M}{\sigma_{st} \cdot a}. \quad \text{Arrange suitably the steel reinforcement.}$$

6. Determine the actual N.A. and check the stresses in concrete and steel.

7. Check the section for shear and provide shear reinforcement, if necessary. Design of shear reinforcement is done in the same manner as for rectangular beams.

8. Check for development length.

**Design: Example 7.10.** *The floor of a hall measures 16 m × 6 m to the faces of the supporting walls. The floor consists of three beams spaced at 4 m centre to centre, and the slab thickness is 120 mm. The floor carries a uniformly distributed load of 5 kN/m², inclusive of the floor finishes. Design the intermediate beam. Use M 20 concrete and Fe 415 steel. The ratio of cost of steel to cost of concrete may be taken equal to 60. The support width may be assumed equal to 500 mm.*

**Solution:**

1. **Calculation of constants:** $\sigma_{cbc} = 7$ N/mm² ; $\sigma_{st} = 230$ N/mm² ; $m = 13.33$

$$k_c = \dfrac{m\,\sigma_{cbc}}{m\,\sigma_{cbc} + \sigma_{st}} = \dfrac{13.33 \times 7}{13.33 \times 7 + 230} = 0.289 \,; \quad j_c = 1 - k_c/3 = 0.904$$

$$R_c = \dfrac{1}{2} \times 7 \times 0.904 \times 0.289 = 0.914$$

2. **Determination of $b_w$ and $b_f$.** The width of web should be sufficient to accommodate the tensile reinforcement. Assume that the reinforcement consists of 5 bars of 25 mm diameter. Keeping the clear distance between bars equal to 25 mm, and clear cover at the end to be 25 mm, the total width with 8 mm φ rings is $b_w = (5 \times 25) + (4 \times 25) + (2 \times 25) + 2 \times 8 = 291$ mm. Keep $b_w = 300$ mm.

The width of flange $b_f$ is taken the least of the following:

(a) $\quad b_f = \dfrac{l_0}{6} + b_w + 6 D_f$, where $l_0 = 6 + 0.5 = 6.5 = 6500$ mm

∴ $\quad b_f = \dfrac{6500}{6} + 300 + 6 \times 120 \approx 2100$ mm

(b) $\quad b_f = b_w + \dfrac{1}{2}$ (sum of clear distances to adjacent beams on either side)

$= 300 + \dfrac{1}{2} [2\{4000 - 300\}] = 4000$ mm. Hence adopt $b_f = 2100$ mm.

3. **Bending moment and S.F.**

Effective span = 6 + 0.5 = 6.5 m.

| | | |
|---|---|---|
| Dead load of slab per m² = 0.12 × 1 × 1 × 25000 | = | 3000 N |
| Super-imposed load per m² | = | 5000 N |
| Total load | = | 8000 N/m² |
| ∴ Load per metre run of beam = 8000 × 4 | = | 32000 N/m |
| Weight of web per metre run | = | 3500 N (say) |
| Total load $w$ | = | 35500 N/m |

$$\therefore \qquad M = \frac{wL^2}{8} = \frac{35500\,(6.5)^2}{8} = 187484 \text{ N-m} \approx 187.5 \times 10^6 \text{ N-mm}$$

$$\text{S.F.} \quad V = \frac{wl}{2} = \frac{35500 \times 6}{2} = 106500 \text{ N}$$

### 4. Effective depth of beam

For balanced section, we have from Eq. 6.17

$$M = 0.45\, b_f \cdot \sigma_{cbc} \cdot D_f \left( \frac{2k_c\, d - D_f}{k_c} \right). \text{ Hence,}$$

$$187.5 \times 10^6 = 0.45 \times 2100 \times 7 \times 120 \left[ \frac{2 \times 0.289\, d - 120}{0.289} \right], \text{ from which } d = 326 \text{ mm}$$

$$d_{min.} = \frac{V}{b_w \cdot \tau_{c.\,max.}}.$$

For M 20 concrete, $\tau_{c.\,max.} = 1.8 \text{ N/mm}^2$ (Table 3.3)

$$\therefore \qquad d_{min.} = \frac{106500}{300 \times 1.8} = 197 \text{ mm}$$

From economic consideration,

$$d = \sqrt{\frac{r \cdot M}{\sigma_{st} \cdot j_c \cdot b_w}} = \sqrt{\frac{60 \times 187.5 \times 10^6}{230 \times 0.914 \times 300}} \approx 425 \text{ mm}$$

Hence keep $D = 500$ mm. Using nominal cover of 25 mm, and using 25 mm φ bars and 8 mm φ rings, available $d = 500 - 25 - 8 - 25/2 \simeq 454$ m.

### 5. Area of steel

$$\text{Approx. lever arm } a = d - \frac{D_f}{2} = 454 - \frac{120}{2} = 394 \text{ mm.}$$

(Alternatively, $a = 0.9d = 0.9 \times 454 = 409$ mm). Adopt $a = 394$ mm for preliminary calculations.

$$A_{st} = \frac{M}{\sigma_{st} \cdot a} = \frac{187.5 \times 10^6}{230 \times 394} = 2069 \text{ mm}^2.$$

Choosing 25 mm Φ bars, No. of bars $= \dfrac{2069}{490.9} = 4.21$

Hence provide 5 Nos. of 25 mm Φ bars. Actual $A_{st}$ provided $= 5 \times 490.9 = 2454 \text{ mm}^2$.

### 6. Determination of actual N.A. and stresses in steel and concrete

Assuming that the N.A. falls below the flange, we have

$$b_f \cdot D_f \left( n - \frac{D_f}{2} \right) = m A_{st}(d-n) \text{ or } 2100 \times 120 \left( n - \frac{120}{2} \right) = 13.33 \times 2454\,(454-n), \text{ from which } n = 105.3 \text{ mm}.$$

Hence N.A. falls in the flange.

$$\therefore \qquad \frac{b_f\, n^2}{2} = m A_{st}(d-n) \quad \text{or} \quad \frac{2100\, n^2}{2} = 13.33 \times 2454\,(454 - n)$$

From which $n = 104.4$ mm

$$\bar{y} = \frac{3n - 2D_f}{2n - D_f} \times \frac{D_f}{3} = \frac{3 \times 104.4 - 2 \times 120}{2 \times 104.4 - 120} \times \frac{120}{3} = 32.7 \text{ mm}$$

$\therefore$ Lever arm $a = d - \bar{y} = 454 - 32.7 \approx 421$ mm (against an assumed value of 394 mm)

$$\therefore \qquad \text{Stress in steel} = \frac{M}{A_{st} \cdot a} = \frac{187.5 \times 10^6}{2454 \times 421} = 181.5 \text{ N/mm}^2$$

Corresponding stress in concrete is given by
$$c = \frac{t}{m} \cdot \frac{n}{d-n} = \frac{181.5}{13.33} \times \frac{104.4}{454 - 104.4} = 4.07 \text{ N/mm}^2.$$

*Alternatively,* $\quad M_r = b_f n \dfrac{c}{2}\left(d - \dfrac{n}{3}\right)$

or $\quad c = \dfrac{2 M_r}{b_f n \left(d - \dfrac{n}{3}\right)} = \dfrac{2 \times 187.5 \times 10^6}{2100 \times 104.4 \left(454 - \dfrac{104.4}{3}\right)} = 4.08 \text{ N/mm}^2$

**7. Shear reinforcement:** $\tau_v = \dfrac{V}{b_w \cdot d} = \dfrac{106500}{300 \times 454} = 0.78 \text{ N/mm}^2$

At supports, assuming to continue only 3 bars up to supports
$$\frac{100 A_s}{b_w d} = \frac{100 \times 3 \times 490.9}{300 \times 454} \approx 1.1\%.$$

Hence from Table 3.1, for M 20 concrete,
$\tau_c \approx 0.40 \text{ N/mm}^2$. Hence shear reinforcement is necessary.
$V_c = \tau_c \cdot bd = 0.4 \times 300 \times 454 = 54480 \text{ N}$
$\therefore \quad V_s = V - V_c = 106500 - 54480 = 52020 \text{ N}.$

Let us bend 2 bars up at 45°, at a distance $1.414\, a = 1.414 \times 421 \approx 590$ mm from the edge of support. Shear resistance of the 2-bent up bars is:
$$V_{s1} = 0.707 A_{sv} \cdot \sigma_{sv} = 0.707 [2 \times 490.9] \times 230 = 159650 \text{ N}.$$

However, max. resistance assigned to bent up bars $= \dfrac{1}{2} V_s = \dfrac{1}{2} \times 52020 = 26010 \text{ N}.$

Hence provide vertical stirrups for the remaining 26010 N shear $(= V_{s2})$.

Using 8 mm Φ-2-lgd. stirrups, $A_{sv} = 2\dfrac{\pi}{4}(8)^2 = 100.5 \text{ mm}^2$

$\therefore \quad s_v = \dfrac{\sigma_{sv} \cdot A_{sv} \, d}{V_{s2}} = \dfrac{230 \times 100.5 \times 454}{26010} = 403 \text{ mm}.$

The maximum spacing corresponding to nominal stirrups is given by
$$s_v \leq \frac{2.175 A_{sv} \cdot f_y}{b} = \frac{2.175 \times 100.5 \times 415}{300} = 302 \text{ mm}$$

Also, the spacing should not be more than $0.75 d = 0.75 \times 454 = 340$ mm or more than 300 mm. Hence provide 8 mm Φ 2 lgd-stirrups @ 300 mm c/c throughout.

**8. Check for development length at supports**

The Code envisages that the diameter of bars should be so selected that following relation is satisfied at the supports:
$$1.3 \frac{M_1}{V} + L_0 \geq L_d.$$

Here, $M_1 = 3 \times 490.9 \times 230 \times 421$
$= 1.426 \times 10^8 \text{ N-mm}$
$V = 106500 \text{ N}$
$L_s = 500 \text{ mm}$
$x' = 30 \text{ mm (say)}$

Without any hook or bend
$L_0 = L_s/2 - x'$
$= \dfrac{500}{2} - 30 = 220 \text{ mm}$

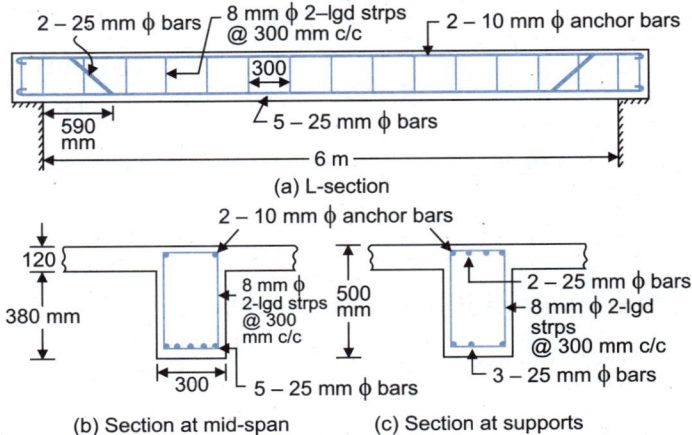

Fig. 7.23. Detailing of Beam

$$1.3\frac{M_1}{V} + L_0 = 1.3 \times \frac{1.426 \times 10^8}{106500} + 220 = 1740 + 220 = 1960 \text{ mm}$$

$$L_d \approx 45\,\Phi = 45 \times 25 = 1125 \text{ mm.}$$

Hence the Code requirements are satisfied. The details of reinforcements etc. are shown in Fig. 7.23.

## 7.10. DESIGN OF CYCLE STAND SHADE

**Design: Example 7.11.** *A cycle stand shade consists of a R.C. slab which cantilevers 3 metres on each side of central R.C. beam and is monolithic with it. The R.C. beam is simply supported on columns 400 mm wide, at the ends, over a clear span of 6 metres. Design the shade for a superimposed load of 2000 N/m². Use M 20 concrete and Fe 415 steel.*

**Solution:** For M 20 concrete we have $c = \sigma_{cbc} = 7$ N/mm² and $m = 13.33$. Also, $t = \sigma_{st} = 230$ N/mm².

∴  $k_c = 0.289; j_c = 0.904$   and   $R_c = 0.914$

### (A) Design of Cantilevering Slab

For stiffness, the ratio of span of effective depth of cantilever = 7. For a balanced design, the percentage reinforcement for M 20 concrete = 0.44%. Assuming $p_t = 0.25$, we get from Fig. 7.1 modification factor, using Fe 415 steel ≈ 1.6.  ∴  span/d = 7 × 1.6 = 11.2.

∴ Effective depth = $\frac{\text{span}}{11.2} = \frac{3000}{11.2} = 268$ mm. Keep the total depth of 290 mm and reduce it to 100 mm at the ends. Mean thickness of slab = $(290 + 100)\frac{1}{2} = 195$ mm.

Keeping 15 mm nominal cover and using 10 mm φ bars, available $d = 290 - 15 - 10/2 = 270$ mm at supports and $d = 100 - 15 - 10/2 = 80$ mm at the free end.

Dead load of slab per sq. metre = $0.195 \times 1 \times 1 \times 25000 = 4875$ N.

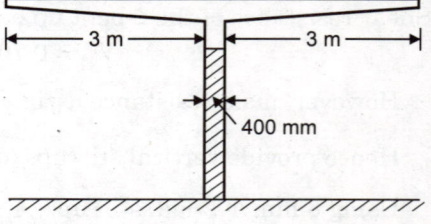

Fig. 7.24. Balanced Cantilever Slab and Beam

Superimposed load = 2000 N/m²

∴  Total load/m² = 2000 + 4875 = 6875 N/m²

∴ $$M = \frac{wL^2}{2} = \frac{6875 \times 3 \times 3}{2} = 30938 \text{ N-m} = 30.938 \times 10^6 \text{ N-mm/m}$$

$$V = wL = 6875 \times 3 = 20625 \text{ N/m width.}$$

Considering $b = 1$ m = 1000 mm,

$$d = \sqrt{\frac{M}{R_c b}} = \sqrt{\frac{30.938 \times 10^6}{0.914 \times 1000}} = 184 \text{ mm}$$

$$A_{st} = \frac{M}{t j_c d} = \frac{30.938 \times 10^6}{230 \times 0.904 \times 270} = 550.7 \text{ mm}^2$$

For 10 mm Φ,   $A_\Phi = \frac{\pi}{4}(10)^2 = 78.54$ mm².

∴   Pitch $s = \frac{1000\,A_\Phi}{A_{st}} = \frac{1000 \times 78.54}{550.7} = 142.6$ mm.

Hence provide 10 mm Φ bars @ 140 mm c/c.

Actual area of steel provided = $\frac{1000\,A_\Phi}{s} = \frac{1000 \times 78.54}{140} = 561$ mm².

$$p_t = \frac{561}{1000 \times 270}\,100 = 0.21\%$$

Let us curtail alternate bars at half the cantilever length.

$$\text{B.M. at the point} = \frac{wx^2}{2} = \frac{6875\,(1.5)^2}{2} \times 1000 = 7.73 \times 10^6 \text{ N-mm}$$

$$\text{Effective depth of slab} = \frac{270 + 80}{2} = 175 \text{ mm}$$

$$\therefore \quad A_{st} \text{ required} = \frac{7.73 \times 10^6}{230 \times 0.904 \times 175} = 212.5 \text{ mm}^2$$

$$\text{Actual } A_{st} \text{ available} = \frac{1}{2} \times 561 = 280.5 \text{ mm}^2. \quad \text{Hence O.K.}$$

**Check for shear.** $\quad \tau_v = \dfrac{V}{bd} = \dfrac{20625}{1000 \times 270} \approx 0.08$ N/mm$^2$ which is quite small.

**Distribution reinforcement**

$$A_{sd} = \frac{0.12\,bD}{100} = \frac{0.12 \times 1000 \times 290}{100} = 348 \text{ mm}^2.$$

Using 8 mm Φ bars, $\quad A_\Phi = 50.3$ mm$^2$

$$\therefore \quad \text{Pitch } s = \frac{1000\,A_\Phi}{A_{sd}} = \frac{1000 \times 50.3}{348} = 144 \text{ mm}.$$

Hence provide 8 mm Φ bars @ 140 mm c/c.

At half the cantilever length, average depth = $(290 + 100)\dfrac{1}{2} = 195$ mm.

$$\therefore \quad A_{sd} = \frac{0.12\,bD}{100} = \frac{0.12 \times 1000}{100}\,D = 1.2\,D = 1.2 \times 195 = 234 \text{ mm}^2$$

$$\text{Pitch } s = \frac{1000 \times 50.3}{234} \approx 210 \text{ mm c/c}.$$

## (B) Design of Beam

The beam will be designed as T-beam with tapered flanges. However we can approximately assume the tapered flanges to be equivalent to flanges of uniform average thickness. Width of web = $(b_w)$ = width of column = 400 mm.

$$\text{Average depth of flange, } D_f = \frac{290 + 100}{2} = 195 \text{ mm} = 0.195 \text{ m}$$

Total width of flange $\quad = b = 6.4$ m $= 6400$ mm.
Let the effective span beam $\quad = l_0 \approx 6 + 0.4 = 6.4$ m
For an isolated T-beam, the width of the flange is

$$b_f = \frac{l_0}{\left(\dfrac{l_0}{b}\right) + 4} + b_w = \frac{6400}{\left(\dfrac{6400}{6400}\right) + 4} + 400 = 1680 \text{ mm}.$$

Assume the overall depth of beam = 400 mm, for the purpose of calculating the dead load.
(i) Dead load of slab    = $(0.195 \times 6.0 \times 1)25000$   = 29250 N
(ii) Dead load of beam = $0.4 \times 0.4 \times 1 \times 25000$     =  4000 N
(iii) Superimposed load = $2000\,(3 + 3 + 0.4) \times 1$   = 12800 N

$$\text{Total } w = 46050 \text{ N}$$

$$M = \frac{wl_0^2}{8} = \frac{46050\,(6.4)^2}{8} \times 1000 = 235.77 \times 10^6 \text{ N-mm}.$$

### Effective depth of the beam:
For a balanced section, we have $M = 0.45 \, b_f c \cdot D_f \left( \dfrac{2k_c \, d - D_f}{k_c} \right)$

or $235.77 \times 10^6 = 0.45 \times 1680 \times 7 \times 195 \left[ \dfrac{2 \times 0.289 \, d - 195}{0.289} \right]$, from which $d = 451$ mm.

Keep $D = 500$ mm. Using 30 mm bars, 8 mm rings and 130 mm nominal cover,
$$d = 500 - 30 - 8 - 15 = 447 \text{ mm.}$$

Revised dead load of beam $= 0.4 \times 0.50 \times 1 \times 25000 = 5000$

$\therefore$ Total $w = 29200 + 5000 + 12800 = 47000$ N/m

$$M = \dfrac{47000 \, (6.4)^2}{8} \times 1000 = 240.64 \times 10^6 \text{ N-mm}$$

Approximate L.A. $= a = d - \dfrac{D_f}{2} = 447 - \dfrac{195}{2} \approx 350$ mm

$$A_{st} = \dfrac{M}{\sigma_{st} \times a} = \dfrac{240.64 \times 10^6}{230 \times 350} = 2989.3 \text{ mm}^2$$

For 30 mm $\Phi$ bars $A_\Phi = 706.86$ mm$^2$. $\therefore$ No. of 30 mm $\Phi$ bars $= 2989.3/706.80 \approx 4.3$.

Provide 4 – 30 mm $\varphi$ plus 1 – 20 mm $\varphi$ bars, actual $A_{st} = 706.86 \times 4 + 314 = 3141$ mm$^2$

### Determination of N.A. and actual stresses:
Assuming that the N.A. and falls in the flange, we have

$$\dfrac{1680}{2} (n)^2 = 13.33 \times 3141 \, (477 - n)$$

or $\quad n^2 + 49.84 \, n - 22280 = 0$.

From which $n \approx 126.4$ mm

$\therefore \quad$ L.A. $= d - \dfrac{n}{3} = 447 - \dfrac{126.4}{3} \stackrel{\sim}{=} 404.8$ mm

Average stress in steel $= \dfrac{M}{A_{st} \times a} = \dfrac{240.64 \times 10^6}{3141 \times 404.8} = 189.3$ N/mm$^2$

Corresponding stress in concrete is given by

$$c = \dfrac{t}{m} \times \dfrac{n}{d-n} = \dfrac{189.3}{13.33} \times \dfrac{126.4}{447 - 126.4} = 5.6 \text{ N/mm}^2 \text{ (Safe)}$$

### Shear reinforcement:
$$V = \dfrac{wl}{2} = \dfrac{47000 \times 6.0}{2} = 141000 \text{ N} \; ; \; \tau_v = \dfrac{141000}{400 \times 447} = 0.79 \text{ N/mm}^2.$$

Assuming that at the ends, two bars will be bent up, available $\dfrac{100 \, A_s}{bd} = \dfrac{100 \times 1727.3}{400 \times 447} = 0.97\%$. Hence from Table 3.1, for M 20 concrete, we get $\tau_c \stackrel{\sim}{=} 038$ N/mm$^2$. Since $\tau_v > \tau_c$ shear reinforcement is necessary. Also from Table 3.3, for M 20 concrete, $\tau_{c\,.\,max} = 1.8$ N/mm$^2$. Since $\tau_v$ is less than this, the section is O.K.

Shear reinforcement will be provided both in the from of inclined bars as well as in the form of vertical stirrups. Out of 4–30 mm $\varphi$ bars, let us bend 2 bars at a distance of $\sqrt{2} \, a = \sqrt{2} \times 338.3 \stackrel{\sim}{=} 480$ mm from the edge of the support.

Area of cross-section of 2 bent-up bars $= 2 \, \dfrac{\pi}{4} (30)^2 \stackrel{\sim}{=} 1413.7$ mm$^2$

$\therefore$ Shear resistance of two bent up bars $= V_1$

$$= \sigma_{sv} \cdot A_{sv} \cdot \sin \alpha = 230 \times 1413.7 \times 0.707 \approx 229880 \text{ N.}$$

This is more than the S.F. itself. Though shear resistance of concrete ($V_2 = \tau_c \cdot bd$) is additionally available, we will have to provide nominal shear stirrups. Using 8 mm $\Phi$ – 2 lgd stirrups, $A_{sv} = 2\frac{\pi}{4}(8)^2 = 100.53$ mm$^2$.

The maximum spacing of stirrups is given by,

$$s_v = \frac{2.175\,A_{sv}\,f_y}{b} = \frac{2.175 \times 100.53 \times 415}{400} = 226 \text{ mm.}$$

Hence provide 8 mm $\varphi$ 2-lgd stirrups @ 200 mm c/c upto 960 mm from the edge of the support. At 960 mm (= 0.96 m) from the edge of the support,

$$\text{Shear force} = \frac{141000}{3}(3 - 0.96) = 95880 \text{ N.}$$

Shear resistance of concrete = $\tau_c \cdot b_w \cdot d = 0.38 \times 400 \times 447 = 67944$ N.

Hence shear to be resisted by the stirrups is given by

$$V_s = V - \tau_c \cdot b_w \cdot d = 95880 - 67944 = 27936$$

The spacing of 8 mm $\Phi$-2 lgd stirrups is

$$s_v = \frac{\sigma_{sv} \cdot A_{sv} \cdot d}{V_s} = \frac{230 \times 100.53 \times 447}{27936} = 370 \text{ mm.}$$

But maximum spacing cannot exceed 226 mm. Hence provide 8 mm $\varphi$ 2-lgd stirrups @ 220 mm c/c throughout the length of the beam. Provide 3-12 mm $\varphi$ holding bars at the top.

### Check for development length at the end:

The Code requires that $1.3\,\dfrac{M_1}{V} + L_0 \geq L_d$. $A_{st}$ available at end = 1727.3 mm$^2$

$\therefore\quad M_1 = 1727.3 \times 230 \times 0.904 \times 447 = 160.54 \times 10^6$ N-mm; $\quad V = 141000$ N.

Let us assume the width of support equal to 400 mm and a side cover $x' = 30$ mm, and providing 90° bend at the ends,

$$\therefore\quad L_0 = \frac{l_s}{2} - x' + 3\Phi = \frac{400}{2} - 30 + (3 \times 30) = 260 \text{ mm.}$$

$$L_d = \frac{\Phi \cdot \sigma_{st}}{4\tau_{bd}} = \frac{\Phi \times 230}{4 \times 0.8 \times 1.6} = 45 \times \Phi = 45 \times 30 = 1350.$$

Now $\quad 1.3\,\dfrac{M_1}{V} + L_0 = 1.3 \times \dfrac{160.54 \times 10^6}{141000} + 260 = 1480 + 260 = 1740$ mm.

Thus $\left(\dfrac{M_1}{V} + L_0\right) > L_d$. Hence the anchorage requirements are satisfied.

The details of reinforcement etc. are shown in Fig. 7.25.

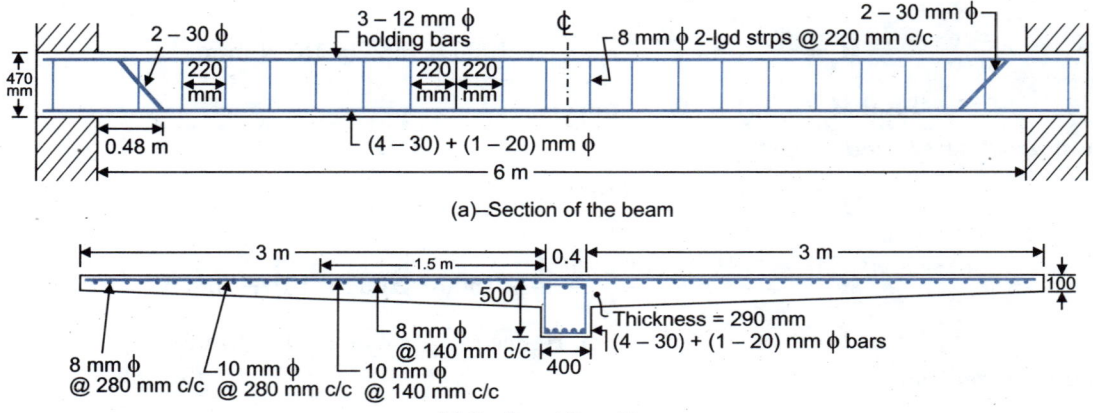

(a)–Section of the beam

(b) Section at the mid-span

**Fig. 7.25.** Detailing of Beam

## 7.11. DESIGN OF T-BEAM ROOF

**Design: Example 7.12.** *A hall measures 10 m × 6 m from inside, and walls 400 mm thick. Design a suitable R.C. T-beam roof to carry a super-imposed load of 2000 N/m². Use M 20 concrete and Fe 415 steel.*

**Solution:** Let us keep end spans of slab slightly less than the middle span so that approximately equal maximum B.M. is induced in all the three spans. Assuming the T-beams to be 300 mm wide, the following spans of the slab are selected:

(i) End spans = 3 m (clear) (ii) Intermediate span = 3.4 m (clear).

The clear span of each T-beam will be 6 m, while the R.C slab monolithic with the beam will be continuous over the three spans. For M 20 concrete, $\sigma_{cbc} = c = 7$ N/mm² and $m \approx 13.33$. Taking $\sigma_{st} = t = 230$ N/mm², we have $k_c = 0.289$; $j_c = 0.904$ and $R_c = 0.914$.

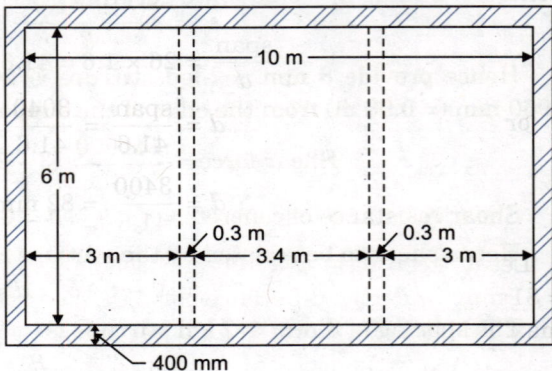

Fig. 7.26. Position of Beams in the Hall

**(A) Design of continuous slab**

Let us assume total thickness of slab = 100 mm and effective thickness $\triangleq$ 80 mm.

**1. Effective spans.** The width of the supports is more than $\frac{1}{12}$ the clear span.

Effective end span = clear spans + $\frac{1}{2}$ (effective thickness of slab) = 3 + 0.04 = 3.04 m.

Effective middle span = clear distance between supports = 3.4 m.

**2. Maximum B.M.**

$$\text{Dead load/m}^2 = 0.1 \times 1 \times 25000 = 2500 = w_d$$
$$\text{Superimposed load} = 2000 \text{ N/m}^2 = w_s$$

*For end span*: Near the middle of end span

$$M \text{ due to dead load} = \frac{w_d L^2}{12} = \frac{2500 (3.04)^2}{12} \times 1000 \approx 1.925 \times 10^6 \text{ N-mm.}$$

$$M \text{ due to superimposed load} = \frac{w_s L^2}{10} = \frac{2000 (3.04)^2}{10} \times 1000 = 1.848 \times 10^6 \text{ N-mm.}$$

Total $M = 3.773 \times 10^6$ N-mm (+)

*For intermediate span* (at the middle)

$$M_d = \frac{w_d L^2}{24} = \frac{2500 (3.4)^2}{24} \times 1000 = 1.204 \times 10^6 \text{ N-mm}$$

$$M_s = \frac{w_s L^2}{12} = \frac{2000 (3.4)^2}{12} \times 1000 = 1.927 \times 10^6 \text{ N-mm}$$

Total $M = 3.131 \times 10^6$ N-mm (+)

*At the support next to end support*

$$M_d = \frac{w_d L^2}{10} = \frac{2500 (3.04)^2}{10} \times 1000 = 2.310 \times 10^6 \text{ N-mm}$$

$$M_s = \frac{w_s L^2}{9} = \frac{2000 (3.04)^2}{9} \times 1000 = 2.054 \times 10^6 \text{ N-mm}$$

Total $M = 4.364 \times 10^6$ N-mm. Hence $M_{max} = 4.364 \times 10^6$ N-mm.

**3. Design of section:**

$$d = \sqrt{\frac{M_{max}}{R_c\, b}} = \sqrt{\frac{4.364 \times 10^6}{0.914 \times 1000}} \approx 70 \text{ mm,}$$

Let us check this depth from stiffness (deflection) point of view. Span/effective depth ratio = 26 for continuos slab. For balanced section, % reinforcement = $\dfrac{kc}{2t} \times 100 = \dfrac{0.284 \times 7}{2 \times 230} \times 100 = 0.44\%$ for M 20 concrete. Taking $p_t = 0.25\%$, we get from Fig. 7.1, modification factor $\approx 1.6$.

$$\therefore \quad \dfrac{\text{span}}{d} = 26 \times 1.6 = 41.6$$

or
$$d = \dfrac{\text{span}}{41.6} = \dfrac{3040}{41.6} = 73 \text{ mm for end spans and}$$

$$d = \dfrac{3400}{41.6} = 82 \text{ mm for intermediate span.}$$

Let us keep total $D = 100$ mm. Using 8 mm $\Phi$ bars, and a nominal cover of 15 mm, $d = 100 - 4 - 15 = 81$ mm. Since the difference between the various bending moments is not large, reinforcement found for max. B.M. will be provided for other bending moments also.

$$A_{st} = \dfrac{M}{\sigma_{st} \cdot jd} = \dfrac{4.364 \times 10^6}{230 \times 0.904 \times 81} = 259 \text{ cm}^2$$

For 8 mm $\Phi$ bars, spacing
$$s = \dfrac{1000 \, A_\Phi}{A_{st}} = \dfrac{1000 \times 50.3}{259} = 194 \text{ mm}$$

Hence provide 8 mm $\Phi$ bars @ 190 mm c/c at the middle of each span. Bend up alternative bars from the bottom at a distance of $L/5 = 3040/5 \approx 600$ mm from the centre of intermediate support and at $L/7 \approx 450$ mm from the end support. Thus reinforcement available at each intermediate support will be 8 mm $\Phi$ bars @ 190 mm c/c.

$$\text{Actual } A_{st} \text{ provided} = \dfrac{100 \times 50.3}{190} = 264.7 \text{ cm}^2 \,;\, p_t = \dfrac{264.7}{1000 \times 100} \times 100 = 0.26\%$$

### Distribution reinforcement

$$A_{sd} = \dfrac{0.12 \times 1000 \times 100}{100} = 120 \text{ mm}^2$$

Using 8 mm $\Phi$ bras, $\quad s = \dfrac{1000 \times 50.3}{120} = 419$ mm. Hence provide these at 400 mm c/c.

### 4. Check for shear

$$V = 0.4 \, w_d \cdot L + 0.45 \, w_s \cdot L \text{ at the outer support}$$
$$= (0.4 \times 2500 + 0.45 \times 2000) \, 3.04 = 5776 \text{ N}$$

$$\therefore \quad \tau_v = \dfrac{V}{bd} = \dfrac{5776}{1000 \times 81} = 0.07 \text{ N/mm}^2.$$

This is much less than the permissible value of 0.18 N/mm$^2$ even for minimum reinforcement of 0.15% (see Table 3.1). Hence safe.

### 5. Check for development length at supports

The Code envisages that the diameter of the reinforcing bars should be such that the following equation is satisfied at the end supports:

$$1.3 \, \dfrac{M_1}{V} + L_0 \geq L_d$$

where $M_1 = \sigma_{st} \cdot A_{st} \cdot jc \, d \,;\quad V = 5776$ N

$$A_{st} \text{ at ends} = \dfrac{1000 \times 50.3}{380} = 132.4 \text{ mm}^2$$

$$\therefore \quad M_1 = 132.4 \times 230 \times 0.904 \times 81 = 2.229 \times 10^6 \text{ N-mm}$$

Let us provide a side cover of 30 mm and assume a support width of 400 mm. Taking bars straight without providing any hook or bend,

$$L_0 = \frac{l_s}{2} - x' = 400/2 - 30 = 170 \text{ mm.} \quad L_d \approx 45\Phi = 45 \times 8 = 360 \text{ mm.}$$

Now $\quad 1.3 \dfrac{M_1}{V} + L_0 = 1.3 \times \dfrac{2.229 \times 10^6}{5776} + 170 = 502 + 170 = 672 \text{ mm.}$

$\therefore \quad 1.3 \dfrac{M_1}{V} + L_0 > L_d$. Hence the Code requirements are satisfied.

### (B) Design of T-beams

Let width of web = $b_w$ = 300 mm.

Effective span of beam ($l_0$) = c/c of bearing = 6.4 m.

$$b_f = \frac{l_0}{6} + b_w + 6D_f = \frac{6400}{6} + 300 + 6 \times 100 = 1966 \text{ mm}$$

**1. Loading per metre run**

Load from slab $= (w_d + w_s)\left[\dfrac{3}{2} + 0.3 + \dfrac{3.4}{2}\right] = (2500 + 2000)(3.5) \approx 15750$ N/m.

Weight of web (assuming depth of rib = 300 mm) = $0.3 \times 0.3 \times 25000 \approx 2250$ N/m.

$\therefore \quad$ Total $w$ = 15750 + 2250 = 18000 N/m.

**2. Design of section**

$$M = \frac{wL^2}{8} = \frac{18000\,(6.4)^2}{8} \times 1000 = 92.16 \times 10^6 \text{ N-mm}$$

For balanced section, we have from *Eq. 5.17*.

$$M = 0.45\, b_f\, \sigma_{cbc} \cdot D_f \left(\frac{2k_c\,d - D_f}{k_c}\right). \text{ Hence,}$$

$$92.16 \times 10^6 = 0.45 \times 1966 \times 7 \times 100 \left(\frac{2 \times 0.289\,d - 100}{0.289}\right).$$

From which $d \approx 247$ mm $\approx 250$ mm (say).

**3. Area of steel**

$$a \approx d - \frac{D_f}{2} = 250 - \frac{100}{2} = 200 \text{ mm. Alternatively, } a \approx 0.9\,d$$
$$= 0.9 \times 250 = 225 \text{ mm}$$

Adopt $a$ = 210 mm for preliminary calculations.

$$A_{st} = \frac{M}{\sigma_{st} \times a} = \frac{92.16 \times 10^6}{230 \times 210} = 1908 \text{ mm}^2.$$

Using 25 mm $\Phi$ bars, $A_\Phi = 490.87$ mm$^2$

$\therefore \quad$ No. of bars = 1908/490.87 = 3.89. Provide 4-25 mm $\Phi$ bars.

Providing $D$ = 300 m, keeping 25 mm nominal cover and using 8 mm $\varphi$ rings available $d = 300 - 25 - 8 - 25/2 = 254.5$ mm. Actual $A_{st} = 4 \times 490.87 = 1963.48$ mm$^2$.

Bend 2 bars up at $L/7 \approx 1$ m from support, to take up negative B.M.

**4. Determination of actual N.A. and stresses**

Assuming the N.A. to fall below the flange, we have:

$$b_f \cdot D_f \left(n - \frac{D_f}{2}\right) = m\,A_{st}\,(d - n)$$

$\therefore \quad 1966 \times 100\,(n - 50) = 13.33 \times 1963.48\,(254.5 - n)$

From which $n = 74$ mm. Thus, the N.A. falls in the flange

$$\therefore \quad \frac{b_f \cdot n^2}{2} = mA_{st}(d-n)$$

or $\quad \dfrac{1966}{2} n^2 = 13.33 \times 1963.48 (254.5 - n)$ which gives $n \simeq 70$ mm

$$\therefore \quad a = d - \frac{n}{3} = 254.5 - \frac{70}{3} = 231$$

$\therefore$ Average stress in steel $= \dfrac{M}{A_{st} \times a} = \dfrac{92.16 \times 10^6}{1963.48 \times 231} = 203$ N/mm²

Corresponding stress in concrete is given by,

$$c = \frac{t}{m} \cdot \frac{n}{d-n} = \frac{203}{13.33} \times \frac{70}{254.5 - 70} = 5.78 \text{ N/mm}^2$$

**5. Check for shear**

$$V = \frac{wl}{2} = \frac{18000 \times 6}{2} = 54000 \text{ N};$$

$$\tau_v = \frac{V}{b_w \cdot d} = \frac{54000}{300 \times 254.5} = 0.7 \text{ N/mm}^2$$

Assuming that 2 out of 4 bars will be bent up, $\dfrac{100 A_s}{bd}$ at the support $= \dfrac{100}{300 \times 254.5}\left(\dfrac{1}{2} \times 1963.48\right)$
= 1.29%. Hence from Table 3.1 permissible $\tau_c = 0.425$ N/mm². Thus, shear reinforcement is required. Shear resistance of concrete is, $V_c = \tau_c \, bd = 0.425 \times 300 \times 254.5 = 32449$ N.

Let us bend 2 bars, out of 4 bars of 25 mm Φ at a distance $= \sqrt{2}\, a = \sqrt{2} \times 231 \approx 325$ mm from the support.

$$A_{sv} = 2\, \frac{\pi}{4}(25)^2 = 981.7 \text{ mm}^2.$$

Hence shear resistance of bent up bar is $V_s = \sigma_{sv} A_{sv} \sin \alpha = 230 \times 981.7 \times 0.707 = 159634$ N. Subject to a max. of $V/2 = 54000/2 = 27000$ N. Hence only nominal stirrups will be required, the spacing of which is given by

$$s_v = \frac{2.175\, A_{sv} \cdot f_y}{b}. \text{ Using 8 mm Φ 2-lgd stirrups, having}$$

$$A_{sv} = 100.53 \text{ mm}^2$$

$$s_v = \frac{2.175 \times 100.53 \times 415}{300} = 302 \text{ mm}.$$

However, provide these @ 300 mm c/c throughout the length of the beam. Provide 2-12 mm Φ holding bars at the top.

**6. Check for development length at support**

At the support, the following equation should be satisfied: $1.3 \dfrac{M_1}{V} + L_0 \geq L_d$

At the end only two bars are available, out of total 4 bars. Hence,

$$A_{st} = 2 \times 490.87 = 981.7 \text{ mm}^2$$

$$M_1 = 981.7 \times 230 \times 231 = 52.16 \times 10^6 \text{ N-mm}; \quad V = 54000 \text{ N}$$

$l_s$ = length of support available = 400 mm. Let the side cover

$x' = 30$ mm.

Assuming no hook or bend provided at the end,

$$L_0 = \frac{l_s}{2} - x' = \frac{400}{2} - 30 = 170 \text{ mm}.$$

$$L_d \approx 45\varphi = 45 \times 25 = 1125 \text{ mm}$$

Now 
$$1.3\frac{M_1}{V} + L_0 = 1.3 \times \frac{52.16 \times 10^6}{54000} + 170 = 1255 + 170 = 1425 \text{ mm}.$$

∴ $1.3\frac{M_1}{V} + L_0 > L_d$. Hence the Code requirements are satisfied. The details of reinforcement etc. are shown in Fig. 7.27.]

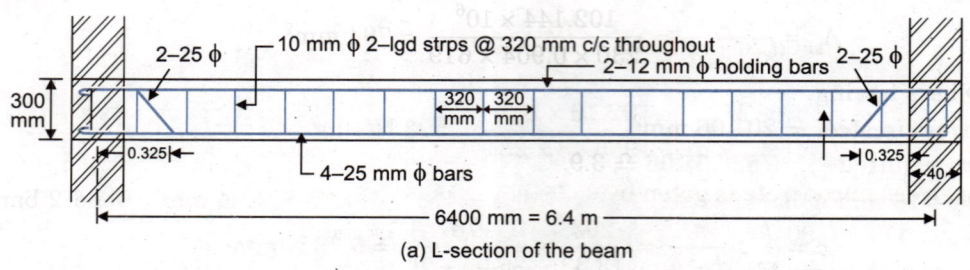

(a) L-section of the beam

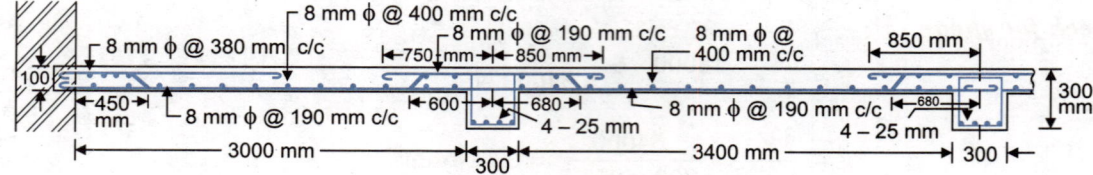

**Fig. 7.27.** Detailing of Reinforcement

## 7.12. DESIGN OF INVERTED T-BEAM ROOF

We have seen that if the size of a room is large, we may use monolithic beam-slab construction, forming the T-beam roof. However, in some cases, the provision of the beam (*i.e.* rib) below the slab may be undesirable from architectural or other point of view. In that case, the beam is provided *above* the slab, forming what is known as the inverted T-beam system. In such a case, the ceiling composed of the roof slab is plane and the beam projects above the top of the slab. The slab is thus provided at the 'tension side' of the beam and is therefore, not helpful in resisting the tensile stresses with the beam. The beam, therefore, acts as a simple rectangular beam, which may either be singly reinforced or doubly reinforced. The depth of this beam will be equal to the depth measured between top of the beam and the bottom of the slab. Though the slab is cast monolithic with the beam it is tied to the beam by providing suitable vertical ties or stirrups. The design of the slab, however, is done in the same manner as for an ordinary T-beam roof.

**Design: Example 7.13.** *A hall measures 10 m × 6 m from inside and has walls 400 mm thick. Design a suitable inverted T-beam roof to carry a super-imposed load of 2000 N/m². Use M 20 concrete and Fe 415 steel.*

**Solution:** This example is the same as example 7.12 except that the system is inverted. Hence the beam will act as a rectangular beam. However, since the slab is continuous over the two beams, its design will be the same as done in example 7.12. We will use the same set of stresses.

### Design of rectangular beam

**(i) Loading per metre run**

Total thickness of slab = 100 mm. Let the width of the beam = 300 mm

∴ Load from slab $= (2500 + 2000)\left(\frac{3}{2} + 0.3 + \frac{3.4}{2}\right) = 15750$ N

Weight of rib (assuming $d_w$ = 560 mm) = 0.3 × 0.56 × 25000 = 4200

∴ Total load/m run = 15750 + 4200 = 19950 N

**(ii) Design of section.** Effective span L = c/c of bearing = 6.4 m.

∴ $$M = \frac{wL^2}{8} = \frac{19950\,(6.4)^2}{8} \approx 102144 \text{ N-m} = 102.144 \times 10^6 \text{ N-mm}$$

$$\therefore \quad d = \sqrt{\frac{M}{R_c d}} = \sqrt{\frac{102.144 \times 10^6}{0.914 \times 300}} = 610 \text{ mm}$$

Keep total depth = 660 mm. Using 16 mm Φ bars, 8 mm φ rings and a nominal cover of 25 mm, available $d = 660 - 25 - 8 - 8 = 619$ mm.

**(iii) Reinforcement.**

$$A_{st} = \frac{M}{\sigma_{st} \cdot jd} = \frac{102.144 \times 10^6}{230 \times 0.904 \times 619} = 794 \text{ mm}^2$$

Use 16 mm Φ bars having

$$A_\Phi = 201.06 \text{ mm}^2.$$

$\therefore$ No. of bars required = $762/201.06 \triangleq 3.9$.

Hence provide 4-16 mm Φ bars. Actual $A_{st}$ provided = $4 \times 201.06 = 804.24$ mm². Bend 2 bars up at a distance of $\sqrt{2}\, a = 1.414 \times 0.904 \times 619 = 791 \approx 800$ mm from the edge of support.

**(iv) Check for development length at the supports**

At the supports, $1.3 \dfrac{M_1}{V} + L_0 \geq L_d$ ; Available $A_{st}$ at supports = $0.5 \times 804.24 = 402.12$ mm².

$\therefore \quad M_1 = 402.12 \times 230 \times 0.904 \times 619 = 51.754 \times 10^6$ N-mm.

$$V = \frac{wl}{2} = \frac{19950 \times 6}{2} = 59850 \text{ N}.$$

$L_s = 400$ mm ; $x' = 30$ mm (say).

Taking the bars straight into supports, without providing any hook or bend,

$$\therefore \quad L_0 = \frac{L_s}{2} - x' = \frac{400}{2} - 30 = 170 \text{ mm}.$$

$$1.3 \frac{M_1}{V} + L_0 = 1.3 \times \frac{51.754 \times 10^6}{59850} + 170 = 1124 + 170 = 1294 \text{ mm}.$$

$$L_d = 45\, \varphi = 45 \times 16 = 720 \text{ mm}.$$

Hence the Code requirements are satisfied.

**(v) Check for shear.**

$$\tau_v = \frac{V}{bd} = \frac{59850}{300 \times 619} = 0.322 \text{ N/mm}^2$$

$$\frac{100\, A_{st}}{bd} = \frac{100 \times 402.12}{300 \times 619} = 0.22\%$$

$\therefore \quad \tau_c = 0.216$ N/mm² from Table 3.1, for M 20 concrete.

$\therefore \quad V_c = 0.216 \times 300 \times 619 = 40110$ N.

or $\quad V_s = V - V_c = 59850 - 40110 = 19740$ N.

The shear resistance of the bent up bars will be very much larger than this. Hence only nominal stirrups are required, the spacing of which is given by

$$s_v = \frac{2.175\, A_{sv}\, f_y}{b} = \frac{2.175 \times 415}{300} A_{sv} = 3.009\, A_{sv}$$

Using 8 mm Φ two-legged stirrups, having

$$A_{sv} = 2\, \frac{\pi}{4}\, (8)^2 = 100.5 \text{ mm}^2$$

$\therefore \quad s_v = 302$ mm. Hence provide the stirrups @ 300 mm c/c throughout. These stirrups will also act as ties to transfer the load of the slabs to the beam.

Total load transferred by the slab to the beam per metre run = 15750 N.

$\therefore \quad$ Sectional area of ties required = $(15750/230) = 68.5$ mm²

Actual area of stirrups provided per metre run = (1000/300) × 100.5 = 335 mm².
Hence O.K. Use 2-10 m Φ anchor bars at the top. The details are shown in Fig. 7.28.

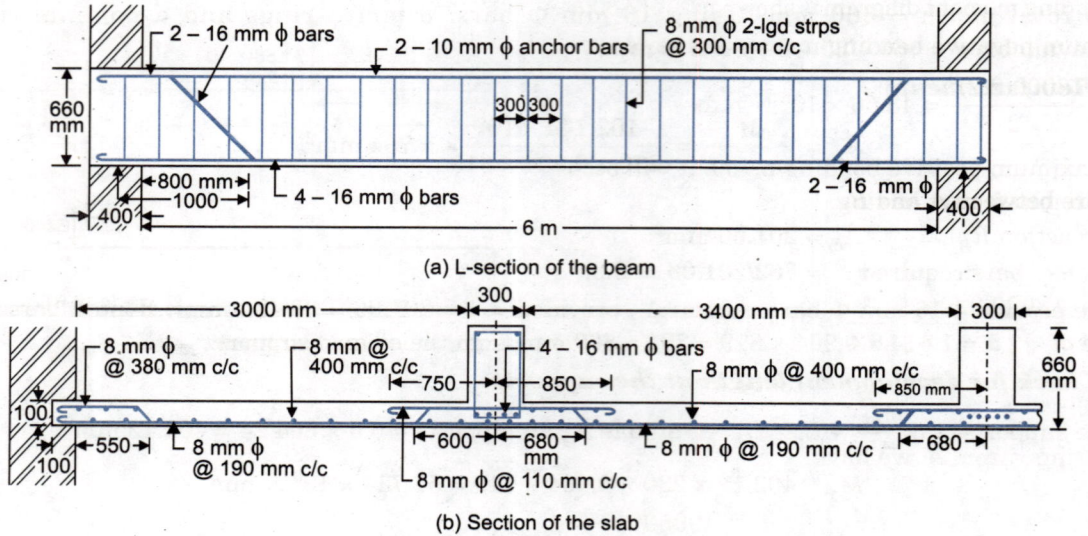

(a) L-section of the beam

(b) Section of the slab

**Fig. 7.28.** Detailing of Reinforcement

## 7.13. DESIGN OF OVERHANGING T-BEAM ROOF

**Design: Example 7.14.** *A hall measures 10 m × 6 m from inside and has walls 400 mm thick. The beams of the T-beam roof project out by 2 m on one side of the room. The slab is monolithic with the beam, and extends to the full length of the beam. Design a suitable roof system to carry a super-imposed load of 2 kN/m². Use M 20 concrete and Fe 415 steel.*

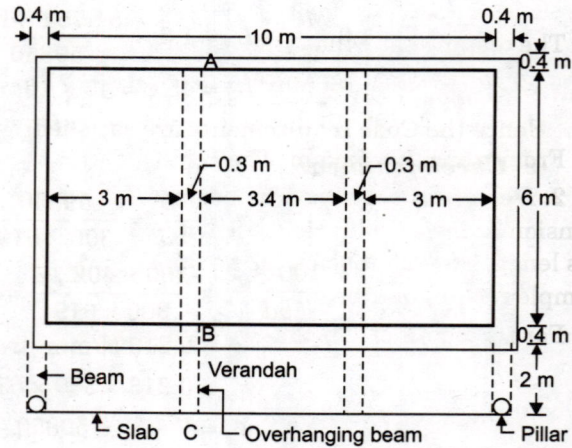

**Fig 7.29.** Positioning of Beams in the Hall

**Solution:** The problem is similar to example 7.12, except that the beams overhang to one side. Let us arrange the beams as shown in Fig. 7.29.

Since the spans of the slab are the same as in example 7.12, its design will be the same as done in example 7.12. We will use the same set of stresses. Hence thickness of the slab = 100 mm.

### Design of beam

The walls are 400 mm thick. Hence the following diagram (Fig. 7.30.) showing the spans etc. may be drawn.

*Loading per metre run*

Super-imposed load from slab = 15750 N

Let us assume the depth of rib = 300 mm and width of rib = 300 mm.

∴ Weight of rib = 0.3 × 0.3 × 25000 = 2250 N
∴ Total load/m run = 15750 + 2250 ≈ 18000 N

## Design of T-beam AB

### 1. Bending moment.

The bending moment diagram is shown in Fig. 7.30(b).
Maximum negative bending moment at support

$$B = \frac{18000(2.2)^2}{2} = 43.56 \times 10^6 \text{ N-mm.}$$

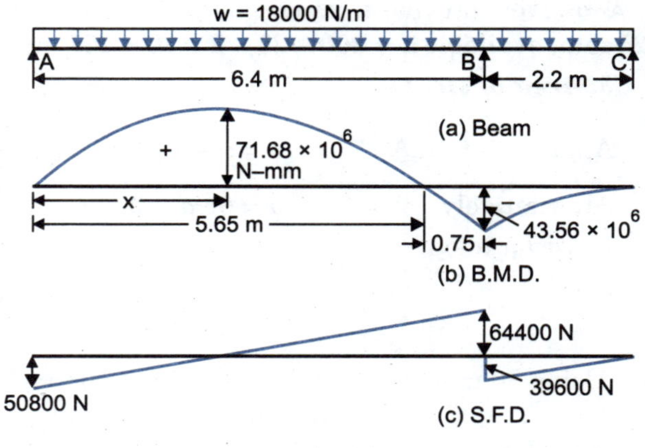

The maximum positive bending moment will occur somewhere between $A$ and $B$.

Now, reaction $R_B$

$$= \left[\frac{18000(8.6)^2}{2}\right]\frac{1}{6.4} = 104000 \text{ N}$$

Reaction $\quad R_A = \dfrac{18000 \times 8.6\,(6.4 - 4.3)}{6.4} = 50800 \text{ N}$

Measuring $x$ from $A$, we have

$$M_x = 50800\,x - 18000\,\frac{x^2}{2}$$

For maxima, $\dfrac{dM_x}{dx} = 50800 - 18000\,x = 0.$  $\therefore\ x = \dfrac{50800}{18000} = 2.82$ m.

$$\therefore\quad M_{max} = \left[(50800 \times 2.82) - \frac{18000}{2}(2.82)^2\right] \times 10^3 \text{ N-mm}$$

$$= 71.68 \times 10^6 \text{ N-mm (sagging)}$$

The distance at which B.M. is zero is given by:

$$50800\,x - \frac{18000\,x^2}{2} = 0.$$

From which $x = 5.65$ m. The B.M.D. and S.F.D. are shown in Fig. 7.30 (b) and (c) respectively.

### 2. Design of section:
The beam AB, for most of its length, has sagging bending moment, thus having tension at lower side. Hence it will work as a T-beam. However, beam BC has hogging B.M. throughout its length, having tension at the upper fibres. Thus the slab will not be helpful, and the beam will act as a simple rectangular beam.

For the depth of T-beam, we have from Eq. 6.17,

$$M = 0.45\,b_f \cdot \sigma_{cbc} \cdot D_f \left[\frac{2k_c\,d - D_f}{k_c}\right] \qquad \ldots (i)$$

The width $b_f$ of the flange will be the maximum of the following:

(i) $b_f = \dfrac{l_0}{6} + b_w + 6\,D_f$, where $l_0$ = distance between points of zero moments = 5.65 m

$\therefore\quad b_f = \dfrac{5650}{6} + 300 + 6 \times 100 = 1842$ mm

(ii) $b_f = b_w + \dfrac{1}{2}$ [sum of clear distance to this adjacent beams on either side

$$= 300 + \frac{1}{2}[3000 + 3400] = 3500 \text{ mm}$$

Hence take $b_f = 1842$ mm. Substituting the values in (i), we get

$$71.68 \times 10^6 = 0.45 \times 1842 \times 7 \times 100 \left[\frac{2 \times 0.289\,d - 100}{0.289}\right] \text{ from which}$$

$$d = 235 \text{ mm.}$$

Fig. 7.30. Beam, BM and SF Diagrams

(a) Beam — $w = 18000$ N/m, 6.4 m, 2.2 m
(b) B.M.D. — $71.68 \times 10^6$ N–mm, $43.56 \times 10^6$, 5.65 m, 0.75
(c) S.F.D. — 50800 N, 64400 N, 39600 N

However, provide total depth = 300 mm. Using 20 mm Φ bars 8 mm φ rings and a nominal cover of 25 mm, available $d = 300 - 25 - 8 - 10 = 257$ mm.

### 3. Area of steel

Approximate L.A. $\quad a = d - \dfrac{D_f}{2} = 257 - \dfrac{100}{2} = 207$ mm

Alternatively, $\quad a = 0.9d = 0.9 \times 257 = 231.3$ mm

However adopt $\quad a = 210$ mm for preliminary calculations.

$$A_{st} = \dfrac{M}{\sigma_{st} \times a} = \dfrac{71.68 \times 10^6}{230 \times 210} = 1484 \text{ mm}^2$$

Using 20 mm Φ bars having $A_\Phi = 314.16$ mm$^2$

∴ No. of bars = 1484/314.16 = 4.7.

Hence provide 5 No. of 20 mm Φ bars.

Actual $A_{st}$ provide ≅ 1571 mm$^2$. At the simply supported end, bend two bars up, one at a distance of 1.414 $a$ ≅ 290 mm and other at 580 mm form the edge of the support.

Similarly, bend two bars up near support B, one at a distance of 290 mm and the other at 580 mm from the face of the support. These bent up bars may be taken right up to end C, *to resist part of tension developed in beam BC.*

### 4. Check for shear and development length at support A

At A, $\quad \tau_v = \dfrac{V}{bd} = \dfrac{50800}{300 \times 257} = 0.66$ N/mm$^2$

$$\dfrac{100 A_s}{bd} = \dfrac{100 (3 \times 314.16)}{300 \times 257} = 1.22\%$$

Hence from Table 3.1, $\tau_c = 0.41$ N/mm$^2$. Hence shear reinforcement is necessary.

$V_c = 0.41 \times 300 \times 257 = 31610$

∴ $V_s = V - V_c = 50800 - 31610 = 19190$ N

$V_{s1}$ of one bent up bar $= 0.707 \times A_{sv} \times \sigma_{sv} = 0.707 \times 314.16 \times 230 = 51085$ N

Hence only nominal stirrups are required.

$$s_v \leq \dfrac{2.175 A_{sv} f_y}{b}$$

Using 8 mm dia. 2-lgd. stirrups,

$$A_{sv} = 2 \dfrac{\pi}{4}(8)^2 = 100.5$$

∴ $\quad s_v \leq \dfrac{2.175 \times 100.5 \times 415}{300} \leq 302.$

Hence provide these @ 300 mm c/c.

Also, $\quad \dfrac{M_1}{V} + L_0 \geq L_d$

$M_1 = (3 \times 314.16) \times 230 \times 210 = 45.52 \times 10^6$ N-mm; $V = 50800$ N

$L_s = 400$ mm; $x' = 30$ mm. Assuming no hook or bend at the ends,

$$L_0 = \dfrac{L_s}{2} - x' = \dfrac{400}{2} - 30 = 170 \text{ mm}$$

∴ $\quad \dfrac{M_1}{V} + L_0 = \dfrac{45.52 \times 10^6}{50800} + 170 = 896 + 170 = 1066$ mm

$L_d \approx 45 \Phi = 45 \times 20 = 900$ mm. Hence the Code requirements are satisfied.

## 5. Check for actual stresses in beam AB

Assuming the actual N.A. to fall below the flange, we have

$$b_f \cdot D_f \left(n - \frac{D_f}{2}\right) = m A_{st} (d - n)$$

or $\quad 1842 \times 100 \left(n - \frac{100}{2}\right) = 13.33 \times 1571 (257 - n)$.

From which $n = 71.1$ mm.

Hence the N.A. falls in the flange.

$\therefore \qquad b_f \cdot \frac{n^2}{2} = m A_{st} (d - n)$

or $\qquad 1842 \frac{n^2}{2} = 13.33 \times 1571 (257 - n)$. From which $n \approx 66$ mm

$\therefore \qquad a = d - \frac{n}{3} \approx 257 - \frac{66}{3} \approx 235$ mm

$\therefore \qquad$ Stress in steel $= \dfrac{M}{A_{st} \cdot a} = \dfrac{71.68 \times 10^6}{1571 \times 235} = 194.2$ N/mm$^2$ < 230 (safe)

Corresponding stress in concrete is given by

$$c = \frac{t}{m} \cdot \frac{n}{d-n} = \frac{194.2}{13.33} \times \frac{66}{257 - 66} = 5.034 \text{ N/mm}^2 < 7 \text{ (safe)}$$

## 6. Shear reinforcement in beam AB at end B

$$\text{S.F.} \quad V = 64000 \text{ N.}; \tau_v = \frac{64000}{300 \times 257} \triangleq 0.83 \text{ N/mm}^2.$$

$$\frac{100 A_s}{bd} = \frac{100 (3 \times 314.16)}{300 \times 257} = 1.22\%.$$

From Table 3.1, $\quad \tau_c = 0.41$ N/mm$^2$.

Hence shear reinforcement is necessary.

$$V_c = 0.41 \times 300 \times 257 = 31611 \text{ N}$$
$$V_s = V - V_c = 64000 - 31611 = 32389 \text{ N}$$

$V_{s1}$ of one bent up bar $\quad = 0.707 \times A_{sv} \cdot \sigma_{sv}$
$\qquad \qquad \qquad \qquad \quad = 0.707 \times 314.16 \times 230 = 51085$ N

Hence only nominal stirrups are required.

Provide 8 mm dia. 2 leg stirrup @ 300 mm c/c, (as provided at end $A$, step 4).

## Design of beam BC

### 1. Tensile reinforcement

Beam $BC$ will be designed as an ordinary rectangular beam with its tension side upwards. However, keep the depth and width of the beam the same as the T-beam.

| | |
|---|---|
| Thus, width of beam | = 300 mm |
| Total depth of beam | = 300 mm |
| Effective depth | = 300 – 25 – 8 – 10 = 257 mm as before |
| Moment of resistance of the beam | = $M_1 = R_c bd^2$ |

or $\qquad \qquad M_1 = 0.914 \times 300 (257)^2 = 18.11 \times 10^6$ N-mm

Actual $\qquad \qquad$ B.M. $= 43.56 \times 10^6$ N-mm.

Hence the beam will have to be designed as a doubly reinforced section.

$$A_{st1} = \frac{M_1}{\sigma_{st} \, j_c d} = \frac{18.11 \times 10^6}{230 \times 0.904 \times 257} = 339 \text{ mm}^2$$

$$M_2 = M - M_1 = (43.56 - 18.11)\,10^6 = 25.45 \times 10^6 \text{ N-mm}$$

This remaining B.M. has to be resisted by a couple provided by the tensile and compressive reinforcement. Let the centre of compressive reinforcement be placed at 40 mm below the top fiber, i.e., $d_c$ = 40 mm.

$$A_{st2} = \frac{M_2}{\sigma_{st}(d - d_c)} = \frac{25.45 \times 10^6}{230\,(257 - 40)} = 510 \text{ mm}^2$$

∴ Total  $A_{st} = A_{st1} + A_{st2} = 339 + 510 = 849 \text{ mm}^2$. Out of these, 2 bent-up bars are already available from span $AB$, having area

$$= 2 \times 314.16 \approx 628 \text{ mm}^2$$

∴ Remaining area  $= 849 - 628 = 221 \text{ mm}^2$.

Using 16 mm Φ bars, having

$$A_\Phi = 201 \text{ mm}^2, \text{ No. of bars} = 2$$

Thus the main reinforcement consists of (i) 2-20 mm Φ bars available from $AB$, and (ii) 2-16 mm Φ bars. All these are provided at the top face of the beam $BC$, with a nominal cover of 25 mm. Continue 2 – 16 φ bars throughout span $BA$, to act as anchor bars.

**2. Compressive reinforcement**

$$n = 0.289 \times 257 = 74.3 \text{ mm}$$

$$A_{sc} = \frac{m(d-n)}{(m_c - 1)(n - d_c)} A_{st2}$$

$$= \frac{13.33\,(257 - 74.3)}{(1.5 \times 13.33 - 1)\,(74.3 - 40)} \times 510 = 1906 \text{ mm}^2$$

This is much more than $A_{st}$ which is 849 mm² only. Hence redesign the beam, using steel beam theory.

**Redesign.** Let  $A_{st} = A_{sc}$.

Lever arm  $= d - d_c = 257 - 40 = 217 \text{ mm}$

$$A_{st} = A_{sc} = \frac{M}{\sigma_{st} \times a} = \frac{43.56 \times 10^6}{230 \times 217} = 873 \text{ mm}^2$$

Now two bars of 20 mm Φ, having area = 628 mm² are already available from $AB$. Remaining area to provided = 873 – 628 = 245 mm². Hence provide and 2-16 Φ bars, giving area = 2 × 201 = 402 mm². Continue 2-16 mm Φ bars throughout in span $BA$, to act as anchor bars.

Compression steel  $A_{sc} = 873 \text{ mm}^2$.

Now 3 bars of 20 mm Φ, giving area = 3 × 314.16 = 942.5 mm², are already available from span $AB$. Hence these can be continued in span $BC$ and additional reinforcement is necessary.

At a point distant $L_d$ = 45 Φ = 45 × 20 ≈ 900 mm (= 0.9 m) from end $B$, in beam $BC$,

$$M = \frac{18000\,(2.2 - 1.5)^2}{2} \times 1000 = 4.41 \times 10^6 \text{ N-mm}$$

This is much less than moment of reinforce of singly reinforced beam. Hence discontinue the 2-20 mm Φ top bars and 1-20 mm Φ bottom bars at this location, and continue the remaining bars. At this point of curtailment,

$$\text{Required } A_{st} = \frac{4.41 \times 10^6}{230 \times 0.904 \times 257} = 82.5 \text{ mm}^2$$

Area available (i.e., 2-16 mm Φ bars) = 2 × 201 = 402 mm². Hence O.K.

**3. Shear reinforcement in BC at B**

$$V = 39600 \text{ N}; \quad \tau_v = \frac{39600}{300 \times 257} = 0.51 \text{ mm}^2$$

$$\frac{100\,A_s}{bd} = \frac{100\,(628 + 402)}{300 \times 257} = 1.34\%$$

From Table 3.1, $\tau_c \triangleq 0.43$ mm². Hence shear reinforcement is necessary.
$$V_c = 0.43 \times 300 \times 257 = 33153 \text{ N};$$
$$V_s = V - V_c = 39600 - 33153 = 6447 \text{ N}.$$

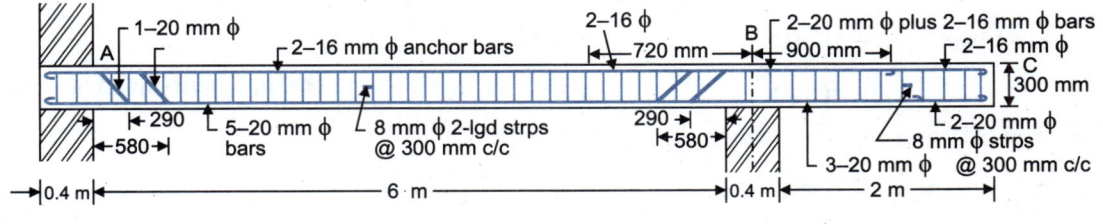

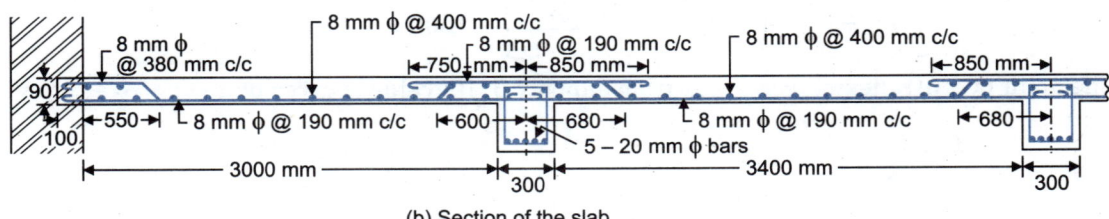

(a) L-section of the beam

(b) Section of the slab

**Fig. 7.31.** Reinforcement Detailing

Using 8 mm Φ 2-lgd stirrups,
$$A_{sv} = 2 \times \frac{\pi}{4}(8)^2 = 100.5 \text{ mm}^2$$
$$s_v = \frac{\sigma_{sv} \cdot A_{sv} \cdot d}{V_s} = \frac{230 \times 100.5 \times 257}{6447} = 921 \text{ mm}.$$

This should not be more than the spacing of nominal stirrups, given by
$$s_v \leq \frac{2.175 \, A_{sv} \, f_y}{b} \leq \frac{2.175 \times 100.5 \times 415}{300} \leq 302 \text{ mm}.$$

Hence provide 8 mm Φ 2-lgd stirrups @ 300 mm c/c throughout the length. The details of reinforcement etc. are shown in Fig. 7.31.

## 7.14. DESIGN OF CANTILEVER CANOPY

**Design: Example 7.15.** *Design a R.C. cantilever canopy having 2.5 metre effective overhang and 6 metres wide as shown in Fig. 7.32. The canopy slab is supported on two beams cantilevered out from walls and spaced at 3.5 m centre to centre, and overhangs each beam by 1.25 m on each side. The beams and slab are cast monolithic, to give a flat soffit. The super-imposed load on the slab may be taken as 1500 N/m². If the height of the wall over the anchoring length of the beam is 5 m, calculate the anchoring length of the beam required for a factor of safety of 1.5. The width of the supporting walls is 600 mm. Use M 20 concrete and Fe 415 steel.*

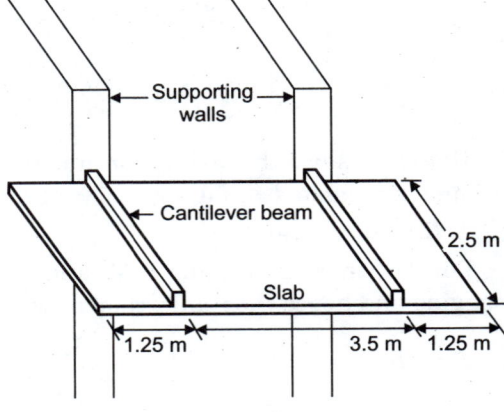

**Fig. 7.32.** Canopy with Dimensions

**Solution:** (*a*) **Design of slab**

**1.** *The design constants.* For $c = \sigma_{cbc} = 7$ N/mm²; $t = \sigma_{st} = 230$ N/mm² and $m \approx 13.33$, we have $k_c = 0.289$; $j_c = 0.904$; $R_c = 0.914$.

**2. Bending moment**

*Loads*: Let the total thickness of slab be 100 mm.

Dead load per sq. metre = $0.1 \times 1 \times 1 \times 25000$
$\quad\quad\quad\quad\quad\quad\quad\quad\quad = 2500$ N

Superimposed load $\quad = 1500$ N

Total $\quad\quad\quad\quad\quad\quad = 4000$ N

The B.M. and S.F. diagrams are shown in Fig. 7.33.

Max. hogging B.M. $= \dfrac{4000(1.25)^2}{2} \times 1000 = 3.125 \times 10^6$ N-mm

Max. sagging B.M. $= \left[ 12000 \times 1.75 - \dfrac{4000(3)^2}{2} \right] \times 1000 = 3 \times 10^6$ N-mm

Let the point of contraflexure occur at $x$ from support $B$. Its value is given by

$$12000\,x - \dfrac{4000(1.25 + x)^2}{2} = 0$$

or $\quad\quad\quad\quad\quad\quad x^2 - 3.5\,x + 1.5625 = 0.$

From which $\quad\quad\quad\quad\quad x = 0.525$ m

The slab will be designed for the greater B.M., *i.e.*, $3.125 \times 10^6$ N-mm.

**3. Design of section**

$$d = \sqrt{\dfrac{M}{R_c b}}$$

$$= \sqrt{\dfrac{3.125 \times 10^6}{0.914 \times 1000}} = 59 \text{ mm}$$

From stiffness point of view, span/eff. depth = 7 for cantilever. For a balanced section, % reinforcement

$$= \dfrac{k\,\sigma_{cbc}}{2\,\sigma_{st}} \times 100$$

$$= \dfrac{0.289 \times 7}{2 \times 230} \times 100 = 0.44,$$

for M 20 concrete. However, taking $p_t = 0.25\%$, we get from Fig. 7.1, modification factor ≈ 1.6

$\therefore \quad\quad\quad \dfrac{\text{span}}{d} = 7 \times 1.6 = 11.2$

$$d = \dfrac{\text{span}}{11.2} = \dfrac{1.25 \times 1000}{11.2} = 112 \text{ mm.}$$

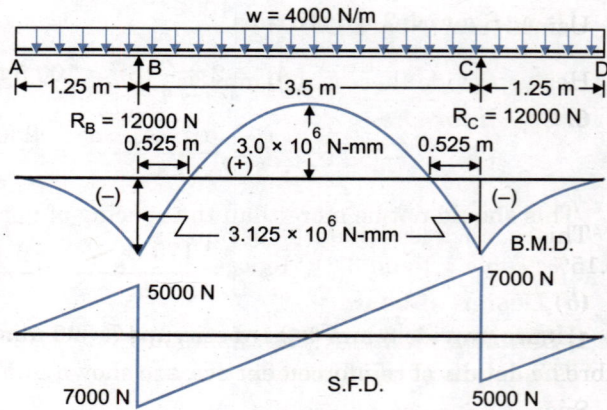

Fig. 7.33. Beam, BM and SF Diagrams

However, since the slab is continuous, keep $D = 100$ mm. Using 8 mm Φ bars and a nominal cover of 15 mm, available $d = 100 - 4 - 15 = 81$ mm.

**4. Reinforcement**

For $AB$, there is hogging B.M. and hence reinforcement will be provided at the top. For $BC$, B.M. is sagging and hence the reinforcement will be provided at the bottom.

For $AB$, $\quad\quad A_{st} = \dfrac{M}{\sigma_{st}\,j_c d} = \dfrac{3.125 \times 10^6}{0.904 \times 230 \times 81} = 185.6$ mm$^2$

For $BC$, $\quad\quad A_{st} = \dfrac{M}{\sigma_{st}\,j_c d} = \dfrac{3.0 \times 10^6}{0.904 \times 230 \times 81} = 178.1$ mm$^2$

Minimum reinforcement $= \dfrac{0.12}{100} \times b \times D = \dfrac{0.12}{100} \times 1000 \times 100 = 120$ mm$^2$

Using 8 mm Φ bars, $A_\Phi = \dfrac{\pi}{4}(8)^2 = 50.27$ mm$^2$

∴ Spacing $s = \dfrac{1000\, A_\Phi}{A_{st}} = \dfrac{1000 \times 50.27}{185.6} = 270.8$ mm

However, provide 8 mm Φ bars @ 250 mm c/c in all the three spans.

Actual $A_{st} = \dfrac{1000 \times 50.27}{250} = 201$ mm$^2$;

$$p_t = \dfrac{201}{1000 \times 101} \times 100 \approx 0.2\%$$

For AB and CD, provide reinforcement at the top and take the bars inside the span for a distance equal to 525 mm or $L_d = 45\,\Phi$ (= 45 × 8 = 360 mm), whichever is more. Hence take these bars inside by 600 mm. For BC, provide reinforcement at the bottom and take it upto the supports B and C. This provision will meet the requirements of development length.

### 5. Distribution reinforcement

$$A_{sd} = \dfrac{0.12}{100} \times 1000 \times 100 = 120 \text{ mm}^2$$

Using 8 mm Φ bars, $s = \dfrac{1000 \times 50.27}{120} = 418.9$ mm.

Hence provide these @ 400 mm c/c

### 6. Check for shear

S.F. = 7000 N.; ∴ $\tau_v = \dfrac{V}{bd} = \dfrac{7000}{1000 \times 81} = 0.0864$ N/mm$^2$.

This is much less than the permissible value of 0.18 × 1.3 = 0.234 N/mm$^2$ even for min. reinforcement of 0.15% (Table 3.1 and 3.2). Hence safe.

### (b) Design of beam

**1. Bending moment:** The beam and slab are cast monolithic. Since the compression is in the bottom fibres, flange action will be effective and the beam will act as a T-beam.

Span of cantilever = 2.5 m.

Super-imposed load from slab, per m run $= \dfrac{6}{2} \times 1 \times 4000 = 12000$ N

Self weight of web (assuming 300 mm width and 300 mm depth)
$= 0.3 \times 0.3 \times 1 \times 25000 = 2250$ N

∴ Total load per m run = 12000 + 2250 = 14250 N

∴ $M = \dfrac{wL^2}{2} = \dfrac{14250(2.5)^2}{2} \times 1000 = 44.53 \times 10^6$ N-mm

### 2. Design of section

The width of flange available will be the minimum of the following:

(i) $\quad b_f = \dfrac{l_0}{6} + b_w + 6\,D_f$

Taking $\quad l_0 = 2.5$ m = 2500 mm,

$b_f = \dfrac{2500}{6} + 300 + 6 \times 100 \approx 1316$ mm

(ii) $b_f$ = cantilever projection of slab + $\frac{1}{2}$ spacing of beam = 1250 + 1750 = 3000 mm. Hence adopt $b_f$ = 1316 mm. For effective depth of T-beam, we have

$$M = 0.45 \, b_f \cdot \sigma_{cbc} \cdot D_f \left[ \frac{2k_c \, d - D_f}{k_c} \right]$$

$$44.53 \times 10^6 = 0.45 \times 1316 \times 7 \times 100 \left[ \frac{2 \times 0.289 \, d - 100}{0.289} \right]$$

from which $d \approx 227$ mm.

Provide total $D$ = 275 mm. Using 20 mm $\Phi$ bars 8 mm $\varphi$ rings and a nominal cover of 25 mm, available $d$ = 275 − 20/2 − 8 − 25 = 232 mm.

### 3. Area of steel

Approximate L.A. $\quad a = d - \frac{D_f}{2} = 232 - \frac{100}{2} = 182$ mm;

Alternatively, $\quad a = 0.9 \, d = 0.9 \times 232 = 209$ mm

Adopt $\quad a = 195$ mm for preliminary calculation.

$$A_{st} = \frac{M}{\sigma_{st} \cdot a} = \frac{44.53 \times 10^6}{230 \times 195} = 993 \text{ mm}^2$$

Using 20 mm $\Phi$ bars, $\quad A_\Phi \approx 314.16 \text{ mm}^2$.

∴ No. of bars required $= \frac{993}{314.16} = 3.16$

Hence provide 4 Nos. 20 mm $\Phi$ bars, providing

$$A_{st} = 4 \times 314.16 = 1256.6 \text{ mm}^2.$$

### 4. Determination of actual N.A. and stresses

Assuming the N.A. to fall below the flange, we have

$$b_f \cdot D_f \left( n - \frac{D_f}{2} \right) = m \cdot A_{st} (d - n)$$

or $\quad 1316 \times 100 \left( n - \frac{100}{2} \right) = 13.33 \times 1256.6 \, (232 - n)$

From which $n$ = 71 mm. Hence the N.A. falls in the flange.

∴ $\quad b_f \cdot \frac{n^2}{2} = m A_{st} (d - n)$

or $\quad 1316 \times \frac{n^2}{2} = 13.33 \times 1256.6 \, (232 - n)$, from which $n$ = 65 mm.

∴ $\quad a = d - \frac{n}{3} = 232 - \frac{65}{3} = 210.3$ mm

∴ Stress in steel $= t = \frac{M}{A_{st} \cdot a} = \frac{44.53 \times 10^6}{1256.6 \times 210.3} = 168.5 \text{ N/mm}^2 \quad < 230$

Corresponding stress in steel is given by

$$c = \frac{t}{m} \cdot \frac{n}{d - n} = \frac{168.5}{13.33} \times \frac{65}{232 - 65} = 4.92 \text{ N/mm}^2 \quad < 7$$

### 5. Check for shear

$$V = wL = 14250 \times 2.5 = 35625 \text{ N};$$

$$\tau_c = \frac{V}{bd} = \frac{35625}{300 \times 232} = 0.51 \text{ N/mm}^2$$

$$\frac{100 A_s}{bd} = \frac{100 \times 1256.6}{300 \times 232} \approx 1.805\%.$$

Hence from Table 3.1, $\tau_c \triangleq 0.48$ N/mm²
Hence shear reinforcement is necessary.

$$V_c = 0.48 \times 300 \times 232 \triangleq 33400 \text{ N}$$

∴ $V_s = V - V_c = 35625 - 33400 = 2225$ N, which is quite small.

Using 8 mm Φ 2 lgd-stirrups,

$$A_{sv} = 2\frac{\pi}{4}(8)^2 = 100.5$$

$$s_v = \frac{\sigma_{sv} \cdot A_{sv} \cdot d}{V_s} = \frac{230 \times 100.5 \times 232}{2225} = 2410 \text{ mm}.$$

Spacing for nominal stirrups is given by

$$s_v \leq \frac{2.175 A_{sv} \cdot f_y}{V_s} \leq \frac{2.175 \times 100.5 \times 415}{300} \leq 302 \text{ mm}.$$

Hence provide 8 mm Φ 2 lgd stirrups @ 300 mm c/c. Provide 2-10 mm Φ holding bars at the bottom.

**6. Curtailment of main reinforcement:** Out of 4 main bars, let us curtail 2 bars at a distance of $x$ from cantilever end.

$$M_x = \frac{wx^2}{2} = \frac{14250 \, x^2}{1} \times 1000 = 7.125 \times 10^6 \, x^2 \text{ N-mm}.$$

This should be equal to half of the max. B.M.

∴ $7.125 \times 10^6 \, x^2 = \frac{1}{2} \times 44.53 \times 10^6$ or $x = \sqrt{\frac{44.53}{2 \times 7.125}} = 1.76$ m $= 1760$ mm

Hence curtail two bars at a distance of $(2500 - 1760) + 12\,\Phi = 740 + (12 \times 20) = 980$ mm $\approx 1000$ mm from the fixed end. This distance should be greater than $L_d = 45\,\Phi = 45 \times 20 = 900$ mm. Hence curtail the bars only at the distance of 1000 mm (say) from the fixed end.

**7. Length of embedment of the beam.** Let the length of embedment of the beam in the wall be $x$ m.

∴ Weight of wall on embeded beam $= 5 \times 0.3x \times 22500$ N

Its moment about the edge of support $= 5 \times 0.3 \times 22500 \, \frac{x^2}{2}$ N-m

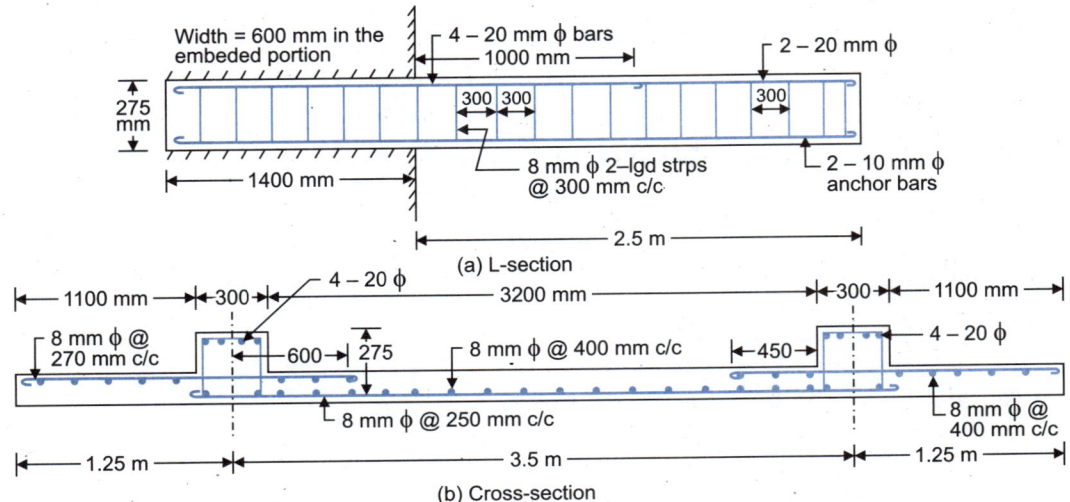

Fig. 7.34

Factor of safety against overturning = 1.5
External moment causing rotation = $44.53 \times 10^3$ N-m,

$$\therefore \quad 5 \times 0.3 \times 22500 \, \frac{x^2}{2} = 1.5 \times 44.53 \times 10^3$$

From which $x = 1.985$ m. Hence embed the beam by 2 m. If this is considered excessive, increase the width of beam to 600 mm in the embedded portion.

This will give
$$x = \sqrt{\frac{2 \times 1.5 \times 44.53 \times 10^3}{5 \times 0.6 \times 22500}} = 1.4 \text{ m.}$$

The details of reinforcement etc. are shown in Fig. 7.34.

## 7.15. DESIGN OF L-BEAM : DESIGN FOR TORSION

The design principles for L-beam are similar to those for T-beam, except that additional torsional reinforcement is provided to resist torsional moment. The width $b_f$ of the slab, acting monolithic with the beam, forming the flange of the L-beam is taken as the least of following:

(a) $b_f = \dfrac{l_0}{12} + b_w + 3 D_f$

(b) For isolated beam,

$$b_f = \frac{0.5 \, l_0}{\dfrac{l_0}{b} + 4} + b_w$$

(but in no case greater than the actual width of flange) where $l_0$ is the distance between points of zero moments in the beam. For continuous beams and frames, $l_0$ may be assumed as 0.7 times of the effective span.

The location of N.A. and calculation of the moment of resistance of L-beam are done exactly in the same way as that for T-beam, and all the equations developed for T-beam are also applicable for L-beam. The design of L-beam is done in the following steps:

1. Assume some suitable value of width of rib.
2. Determine the effective width $b_f$ of the flange, based on criteria given above.
3. Assume effective span of the beam and calculate bending moment and shear force.
4. Determine the effective depth of the beam on any one of the three considerations:
    (i) balanced section (ii) economy (iii) shear stress

For balanced section, the effective depth is given by

$$M = b_f \cdot \sigma_{cbc} \cdot D_f \left[ d - \frac{D_f(1 + k_c)}{2k_c} + \frac{D_f^2}{3k_c \, d} \right]$$

Alternatively, taking lever arm to be equal to $0.9\,d$, we have:

$$M = 0.45 \, b_f \cdot \sigma_{cbc} \cdot D_f \left[ \frac{2k_c \, d - D_f}{k_c} \right]$$

For economical depth, use Eq. 6.18,

$$d = \sqrt{\frac{r \cdot M}{\sigma_{st} \cdot j_c \cdot b_w}}$$

For shear considerations, use Eq. 6.21,

$$d_{min} = \frac{V}{b_w \, \tau_v} = \frac{V}{b_w \cdot \tau_{c\,max}}$$

where $\tau_v$ may be taken equal to $\tau_{c\,.\,max}$ given in Table 3.3.

The maximum value of $d$ obtained from the above three criteria is taken as the effective depth of the beam.

5. Take the approximate lever arm $a = d - D_f/2$, or $a = 0.9\,d$, and determine the area of tensile reinforcement: $A_{st} = \dfrac{M}{t \times a}$, where $t = \sigma_{st}$. Arrange suitably steel reinforcement.

6. Determine the actual N.A. and check the stress in concrete and steel.

Assuming the N.A. to fall below the flange, $n$ may be determined from the expression:

$$b_f \cdot D_f \left(n - \frac{D_f}{2}\right) = m\,A_{st}(d-n) \quad \text{Also,} \quad \bar{y} = \frac{3n - 2D_f}{2n - D_f} \cdot \frac{D_f}{3}$$

$\therefore \qquad a = d - \bar{y}$. If, however, N.A. falls in the flange,

$$b_f \cdot \frac{n^2}{2} = m \cdot A_{st}(d-n);$$

$\therefore \qquad a = d - \dfrac{n}{3}$

Hence $\qquad t = \dfrac{M}{A_{st} \cdot a}$ ; and $c = \dfrac{t}{m} \cdot \dfrac{n}{d-n}$

7. Calculate shear force $V$ and torsional moment $T$. Compute the *equivalent shear* $V_e$ from the expression:

$$V_e = V + 1.6\,\frac{T}{b_w}; \quad \text{and} \quad \tau_{ve} = \frac{V_e}{b_w \cdot d}$$

The value of $\tau_{ve}$ should not exceed $\tau_{c.\,max}$ given in Table 3.3. If it exceeds, the section should be redesigned by increasing the concrete area.

8. If equivalent shear stress $\tau_{ve}$ does not exceed $\tau_c$ (Table 3.1) provide nominal shear stirrups governed by the equation:

$$s_v \le \frac{2.175\,A_{sv} f_y}{b}$$

9. If $\tau_{ve}$ exceeds $\tau_c$, provide longitudinal reinforcement as well as transverse reinforcement as discussed in steps 10 and 11 respectively.

10. **Longitudinal reinforcement:** It is designed to resist an equivalent B.M. $M_{e1}$ given by

$$M_{e1} = M + M_T$$

where $M$ = B.M. at the section

and $\qquad M_T = T\dfrac{(1 + D/b_w)}{1.7}$.

If $M_T$ exceeds $M$, provide longitudinal reinforcement on compression face also, for a moment $M_{e2}$ given by:

$$M_{e2} = M_T - M$$

11. **Transverse reinforcement:** Find area of transverse reinforcement $A_{sv}$ by

$$A_{sv} = \frac{T \cdot S_v}{b_1 d_1 \sigma_{sv}} + \frac{V \cdot S_v}{2.5\,d_1 \cdot \sigma_{sv}}$$

subject to a minimum of $\dfrac{(\tau_{ve} - \tau_c) b_w \cdot S_v}{\sigma_{sv}}$

$b_1$ = centre to centre distance between corner bars in direction of width

$d_1$ = centre to distance between corner bars in direction of depth

The spacing $S_v$ should not exceed $x_1$, $(x_1 + y_1)/4$ and 300 mm, where $x_1$ and $y_1$ are respectively the short and long dimensions of the stirrup.

12. **Side face reinforcement:** Where the depth of web in a beam exceeds 750 mm, side face reinforcement shall be provided along the two faces. The total area of such reinforcement shall be not less than 0.1 percent of the web area and shall be distributed equally on two faces at spacing not exceeding 300 mm or web thickness whichever is less.

**Design: Example 7.16.** *Fig 7.35 shows the plan and section of a canopy to be provided over an entrance opening of a building. Design the beam and the slab. Use M 20 concrete and Fe 415 steel. The superimposed load on the slab may be taken as 1200 N/m².*

**Solution:** Since the beam and slab are monolithic, the beam will act as L-beam and part of the slab will resist the compression of the beam. The breadth of the beam will be kept equal to the width of the column, *i.e.,* 400 mm. Using M 20 concrete, and taking $\sigma_{st} = t = 230$ N/mm², we have $c = \sigma_{cbc} = 7$ N/mm², $m = 13.33$, $k_c = 0.289$, $j_c = 0.904$; $R_c = 0.914$.

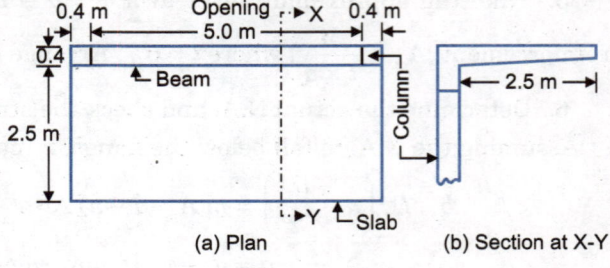

Fig. 7.35

### (A) Design of canopy slab

Let the average depth of slab = 120 mm.

**1. Loading and B.M.**

Superimposed load per sq. m = 1200 N
Dead load due to self weight = $0.12 \times 1 \times 1 \times 25000 = 3000$ N

∴ Total $\quad w = 4200$ N/m².

$$M_{max} = \frac{wL^2}{2} = \frac{4200(2.5)^2}{2} \times 1000 = 13.125 \times 10^6 \text{ N-mm}.$$

**2. Design of section**

$$d = \sqrt{\frac{M}{R_c b}} = \sqrt{\frac{13.125 \times 10^6}{0.914 \times 1000}} \approx 120 \text{ mm}.$$

As per code requirement, $L/d = 7$. For M 20 concrete, taking $p \approx 0.25\%$ modification factor = 1.6.

∴ $\quad \dfrac{L}{d} = 7 \times 1.6 = 11.2$

∴ $\quad d = \dfrac{L}{11.2} = \dfrac{2500}{11.2} = 223$ mm

Hence keep D = 250 mm at the support and reduce it to 100 mm at the edge. Average total thickness = (250 + 100)/2 = 175 mm against an assumed value of 120 mm. Hence dead load due to revised dimension = $0.175 \times 25000 = 4375$ N.

∴ Total $\quad w = 5575$ N/m²

and $\quad M = \dfrac{5575(2.5)^2}{2} \times 1000 = 17.422 \times 10^6$ N-mm

**3. Main reinforcement**

Using 8 mm Φ bars and 15 mm nominal cover available

$$d = 250 - 4 - 15 = 231 \text{ mm}$$

$$A_{st} = \frac{M}{\sigma_{st} j_c d} = \frac{17.422 \times 10^6}{230 \times 0.904 \times 231} = 363 \text{ mm}^2;$$

For 8 mm Φ bars,

$$A_\Phi = \frac{\pi}{4}(8)^2 = 50.26 \text{ mm}^2$$

∴ Spacing $\quad s = \dfrac{1000 \, A_\Phi}{A_{st}} = \dfrac{1000 \times 50.26}{363} = 138$ mm.

Hence provide 8 mm Φ bars @ 130 mm c/c.

Available  $A_{st} = \dfrac{1000 \times 50.26}{130} = 386.1$ mm$^2$;

$$p_t = \dfrac{386.1 \times 100}{1000 \times 231} = 0.167\%$$

### 4. Check for embedment length at the end

In order to develop full tensile strength at the face of the support (*i.e.*, the L-beam), each bar should be embedded in the beam by a length $L_d = 45\ \varphi = 45 \times 8 = 360$ mm. This is best achieved by providing one bend of 90° and extending it by about 100 mm vertically into the beam. The amount of anchorage length thus provided = $(400 - 40) + 8 \times 8 + 100 = 524 > L_d$. See Fig. 7.36.

### 5. Distribution reinforcement

$$A_{sd} = \dfrac{0.12}{100} \times \text{breadth} \times \text{average depth}$$

$$= \dfrac{0.12}{100} \times 1000D = 1.2D = 1.2 \times 175 = 210 \text{ mm}^2$$

Using 8 mm Φ bars,

$$A_\Phi = 50.27 \text{ mm}^2$$

$$\therefore\quad s = 1000\ \dfrac{A_\Phi}{A_{sd}} = 1000 \times \dfrac{50.27}{210} = 239 \text{ mm}.$$

However, provide 8 mm Φ @ 230 mm c/c.

## (B) Design of L-beam

Effective span = 5 + 0.4 = 5.4 m. Assume the total depth of beam = 600 mm Breadth of the beam = 400 mm.

### 1. Loading and B.M.

Dead load of beam/m = 0.6 × 0.4 × 25000 = 6000 N

Load from slab = 2.5 × 1 × 5575 ≅ 13937 N

∴ Total $w$ ≅ 19950 N/m length. Assume partial fixity at the ends,

$$M = \dfrac{wL^2}{10} = \dfrac{19950\,(5.4)^2}{10} \times 1000 = 58.17 \times 10^6 \text{ N-mm}$$

### 2. Design of beam section:
For an isolated L-beam,

$$b = 400 + 2500 = 2900 \text{ mm}$$

$$b_f = \dfrac{0.5 l_o}{\left(\dfrac{l_o}{b}\right) + 4} + b_w = \dfrac{0.5 \times 5400}{\dfrac{5400}{2900} + 4} + 400 = 860 \text{ mm}.$$

Thickness of slab at 0.86 m from the face of the beam

$$= 100 + \dfrac{250 - 100}{2.5}(2.5 - 0.86) = 198.4 \text{ mm}.$$

Average  $D_f = \dfrac{250 + 198.4}{2} \approx 224$ mm.

For a balanced section, the effective depth is given by

$$M = 0.45\, b_f\, \sigma_{cbc}\cdot D_f \left(\dfrac{2k_c d - D_f}{k_c}\right)$$

$$\therefore\quad 58.17 \times 10^6 = 0.45 \times 860 \times 7 \times 224 \left(\dfrac{2 \times 0.289 d - 224}{0.289}\right)$$

which gives $d \approx 435$ mm. keep $D = 480$ mm. Using 16 mm Φ bars, 8 mm φ rings and a nominal cover of 25 mm, available $d = 480 - 25 - 8 - 8 = 439$ mm

**3. Depth from shear point of view:** The total thickness should be such that the beam is safe in shear due to both bending as well as torsion. Torsional moment

$$T = \left\{5575 \times 2.5\left(\frac{2.5}{2} + \frac{0.4}{2}\right)\right\} \times \frac{5.4}{2} \text{ N-m} = 54565 \text{ N-m}$$

$$= 54.565 \times 10^6 \text{ N-mm}$$

Also, $\quad V = \dfrac{wl}{2} = 19700 \times \dfrac{5}{2} = 49250$ N.

Hence equivalent shear is given by

$$V_e = V + 1.6\frac{T}{b_w} = 49250 + \frac{1.6 \times 54.565 \times 10^6}{400} = 267510 \text{ N.}$$

From Table 3.3 for M 20 concrete, max. shear stress $(\tau_{c.max}) = 1.8$ N/mm².

Limiting $\tau_{ve}$ to $\tau_{c.\,max}$, we have

$$\tau_v = \tau_{c\,max,} = 1.8 = V_e/b_w d$$

$$\therefore \quad d = \frac{V_e}{b_w \tau_{c\,max}} = \frac{267510}{372 \times 1.8} = 399.5 \text{ mm.}$$

However, from bending point of views, keep $D = 480$ mm so that available $d = 439$ mm.

**4. Design of steel reinforcement**

Let $\quad a = 0.9\,d = 0.9 \times 439 = 395$ mm (say);

$$A_{st} = \frac{M}{a\,\sigma_{st}} = \frac{57.445 \times 10^6}{395 \times 230} = 633 \text{ mm}^2$$

Using 16 mm Φ bar, $\quad A_\Phi = \pi/4(16)^2 = 201$ mm².

$\therefore \quad$ No. of bars $= \dfrac{633}{201} = 3.15 \simeq 4.$

Actual $A_{st}$ provided $= 4 \times 201 = 804$ mm². Out of 4 bars, bend 2 bars upwards at a distance of $L/7 = 80$ cm from the supports.

**5. Determination of N.A. and actual stress in T-beam**

Assuming the N.A. to fall in the flange,

$$b_f \frac{n^2}{2} = m\,A_{st}\,(d - n)$$

or $\quad \dfrac{860}{2} n^2 = 13.33 \times 804\,(437 - n)$

or $\quad n^2 + 24.92\,n - 10892 = 0.$ From which $n = 92.6$ mm

$$\therefore \quad a = d - \frac{n}{3} = 439 - \frac{92.6}{3} = 408 \text{ mm}$$

$$\therefore \quad t = \frac{M}{A_{st} \times a} = \frac{57.445 \times 10^6}{804 \times 408} = 175.1 \text{ N/mm}^2 < 230 \text{ (safe)}$$

$$c = \frac{t}{m} \cdot \frac{n}{d-n} = \frac{175.1}{13.33} \times \frac{92.6}{439 - 92.6} = 3.51 \text{ N/mm}^2 < 7 \text{ (safe)}$$

**6. Design of shear reinforcement.** As found earlier, equivalent shear

$$V_e = 267510 \text{ N}$$

$$\therefore \quad \tau_v = \frac{V_e}{b_w d} = \frac{267510}{400 \times 439} = 1.52 \text{ N/mm}^2$$

which is less than $\quad \tau_{c.max} = 1.8$ N/mm².

However, shear reinforcement is required.

(a) *Longitudinal reinforcement.* Equivalent B.M. is given by

$$M_{e1} = M + M_T. \quad \text{where} \quad M = 57.445 \times 10^6 \text{ N-mm}$$

$$M_T = T\frac{(1 + D/b_w)}{1.7} = 54.565 \times 10^6 \left[\frac{1 + 480/400}{1.7}\right] = 70.61 \times 10^6 \text{ N-mm}$$

∴ $\quad M_{e1} = (57.445 + 70.61)10^6 = 128.055 \times 10^6$ N-mm

∴ $\quad A_{st} = \dfrac{M_{e1}}{\sigma_{st} \cdot j_c d} = \dfrac{128.055 \times 10^6}{230 \times 0.904 \times 439} = 1403 \text{ mm}^2$

Using 16 mm Φ bars,

$$A_\Phi = \pi/4 (16)^2 = 201 \text{ mm}^2$$

∴ No. of bars = 1403/201 ≈ 7.

Hence provide 7 Nos. 16 mm Φ bars. This reinforcement is *inclusive* of main reinforcement found in step 4. Since numerical value of $M_T$ is more than $M$, the Code specifies that longitudinal reinforcement shall be provided on the flexural compression face such that the beam can also withstand an equivalent moment given by

$$M_{e2} = M_T - M = (70.61 - 57.445)10^6 = 13.165 \times 10^6 \text{ N-mm}$$

$$A_{st} = \dfrac{M_{e2}}{\sigma_{st}(d - d_c)} = \dfrac{13.165 \times 10^6}{230 (439 - 40)} = 144 \text{ mm}^2$$

This is quite small. However, provide 3 bars of 16 mm Φ at the top, which will also work as holding bars for shear stirrups.

(b) *Transverse reinforcement*

Transverse reinforcement will be provided in the form of vertical stirrups. Let the top 16 mm corner bars be provided at a nominal cover of 25 mm.

$b_1$ = centre to centre distance between corner bars in the direction of width
= 400 – 2(25 + 10 + 8) = 314 mm

$d_1$ = centre to centre distance between corner bars in the direction of depth
= 400 – 2 (25 + 10 + 8) = 314 mm

Using 10 mm Φ 4 lgd stirrups,

$$A_{sv} = 4 \frac{\pi}{4}(10)^2 = 314 \text{ mm}^2$$

Now, $\quad A_{sv} = \dfrac{TS_v}{b_1 d_1 \sigma_{sv}} + \dfrac{Vs_v}{2.5 \, d_1 \, \sigma_{sv}}$

∴ $\quad 314 = \left[\dfrac{54.565 \times 10^6}{314 \times 314 \times 230} + \dfrac{49250}{2.5 \times 314 \times 230}\right] s_v$

from which $\quad s_v = 117.2$ mm

Hence provide 10 mm Φ four-legged stirrups @ 110 mm c/c at the supports. However, at the mid-section of the beam, where both $V$ and $T$ are zero, nominal shear stirrups may be provided. Provide 8 mm Φ 2 lgd stirrups @ 300 mm c/c at the middle 2.5 m length of the beam. The details of the reinforcement etc. are shown in Fig. 7.36.

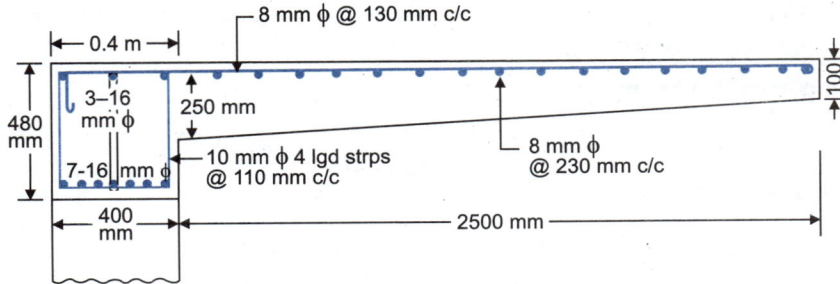

Fig. 7.36

**Design: Example 7.17.** *Solve example 7.16, if the canopy slab is flush with the bottom of the beam, so as to have a plane soffit.*

**Solution:** If the slab is attached to the bottom of the beam, it will be in the tension zone, and hence flange action will not be available. Thus, the beam will act as a simple rectangular beam, though the slab will be cast monolithic with it. *The design of the slab will remain unaltered.*

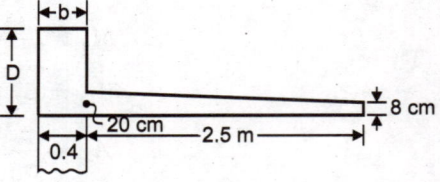

Fig. 7.37

### Design of the beam

**1. Loading and bending moment**

Width of beam = 400 mm. Let the overall depth of beam = 500 mm
Dead load of beam/m = 0.4 × 0.5 × 1 × 25000 = 5000 N
Load from slab/m = 2.5 × 1 × 5575 = 13937.5

Total = 18937.5 N/m ≅ 18940 N/m
Effective span = 5.4 m.

Assuming partial fixity at the ends,

$$M = \frac{wL^2}{10} = \frac{18940\,(5.4)^2}{10} \times 1000 = 55.23 \times 10^6 \text{ N-mm}$$

**2. Design of section.** For balanced section,

$$d = \sqrt{\frac{M}{R_c b}} = \sqrt{\frac{55.23 \times 10^6}{0.914 \times 400}} = 388.7 \text{ mm.}$$

Let us keep $D = 450$ mm.

**3. Check for shear**

Before designing reinforcement, let us check the section for shear because the beam is subjected to heavy torsional moment.

$$T = \left\{5575 \times 2.5 \left(\frac{2.5}{2} + \frac{0.4}{2}\right)\right\} \times \frac{5.4}{2} \text{ N-m} = 54565 \text{ N-mm}$$

$$= 54.565 \times 10^3 \text{ N-mm}$$

$$V = \frac{wl}{2} = \frac{18940 \times 5}{2} = 47350 \text{ N}$$

$$\therefore \quad V_e = V + 1.6 \frac{T}{b_w} = 47350 + \frac{1.6 \times 54.565 \times 10^3}{400} = 47568 \text{ N}$$

$$\tau_{ve} = \frac{V_e}{b_w \cdot d}. \text{ Limiting } \tau_{ve} \text{ to } \tau_{c\,max} \, (= 1.8 \text{ N/mm}^2, \text{ Table 3.3}),$$

$$d = \frac{V_e}{\tau_{c\,max}\, b_w} = \frac{47568}{1.8 \times 400} = 66 \text{ mm}$$

Keep depth $D = 450$ mm. Using 16 mm Φ bars 10 mm φ rings and providing 25 mm nominal cover, available $d = 450 - 25 - 10 - 8 = 407$ mm.

**4. Design of main reinforcement**

$$A_{st} = \frac{M}{\sigma_{st} \times j_c d} = \frac{55.23 \times 10^6}{230 \times 0.904 \times 407} = 653 \text{ mm}^2.$$

Using 16 mm Φ bars, having

$$A_\Phi = \frac{\pi}{4}(16)^2 = 201 \text{ mm}^2$$

∴ No. of bars = 653/201 = 3.25. Hence provide 4 bars of 16 mm Φ.

## 5. Design for torsion and shear

$$\tau_{ve} = \frac{V_e}{b_w d} = \frac{265610}{400 \times 407} = 1.63 \text{ N/mm}^2.$$

Though this is less than $\tau_{c\,max}$, it is more than $\tau_c$ and hence shear reinforcement is necessary.

### (a) Longitudinal reinforcement

$$M_{e1} = M + M_T, \quad \text{where} \quad M = 55.23 \times 10^6 \text{ N-mm}.$$

$$M_T = T\frac{(1 + D/b_w)}{1.7} = 54.565 \times 10^6 \frac{1 + (450/400)}{1.7} = 68.21 \times 10^6 \text{ N-mm}.$$

$$\therefore \quad M_{e1} = (55.23 + 68.21) \cdot 10^6 = 123.44 \times 10^6 \text{ N-mm}$$

$$\therefore \quad A_{st} = \frac{123.44 \times 10^6}{230 \times 0.904 \times 407} = 1459 \text{ mm}^2.$$

Using 16 mm φ bars,

$$A_\Phi = \frac{\pi}{4}(16)^2 = 201 \text{ mm}^2. \text{ No. of bars} = 1459/201 \approxeq 8$$

Hence provide 8 Nos. of 16 mm Φ bars. *This reinforcement is inclusive of main reinforcement found in step 4 (In fact, step 4 may be avoided altogether).* Out of these 8 bars, bend 4 bars up near supports, at a distance of $L/7 \approx 800$ mm from the support, to take negative B.M.

Since the numerical value of $M_T$ is more than $M$, the Code specifies that longitudinal reinforcement shall be provided at the flexural compression face such that the beam can also withstand an equivalent moment given by

$$M_{e2} = M_T - M = (68.21 - 55.23)10^6 = 12.98 \times 10^6 \text{ N-mm}$$

$$\therefore \quad A_{sc} = \frac{M_{e2}}{\sigma_{st}(d - d_c)} = \frac{12.98 \times 10^6}{200(407 - 45)} = 179 \text{ mm}^2.$$

This is quite small. However, provide 2 bars of 16 mm Φ at the top, which will also work as holding bars for shear stirrups.

### (b) Transverse reinforcement.

This will be provided in the form of vertical stirrups

Now $b_1$ = centre to centre distance between corner bars in the direction of width

$$= 400 - 2(25 + 10 + 8) = 314 \text{ mm}.$$

$d_1$ = centre to centre distance between corner bars in the direction of depth

$$= 450 - 2(25 + 10 + 8) = 364 \text{ mm}$$

Using 10 mm 4 lgd stirrups,

$$A_{sv} = 4\frac{\pi}{4}(10)^2 = 314 \text{ mm}^2$$

Now

$$A_{sv} = \left[\frac{T}{b_1 d_1 \sigma_{sv}} + \frac{V}{2.5 d_1 \sigma_{sv}}\right] S_v$$

or

$$314 = \left[\frac{54.565 \times 10^6}{3.4 \times 364 \times 230} + \frac{47350}{2.5 \times 364 \times 230}\right] S_v,$$

from which $S_v = 136$ mm.

Hence use 10 mm Φ two legged stirrups @ 130 mm c/c near the supports. At the middle of beam, only nominal stirrups are required. However, near midspan, provide 8 mm φ 4-lgd stirrups @ 300 mm c/c.

The details of reinforcement etc. are shown in Fig. 7.38.

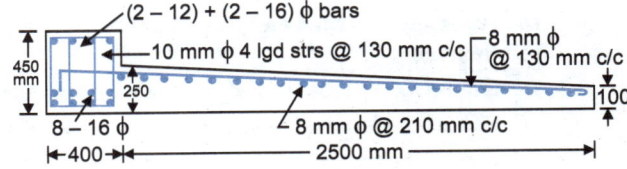

Fig. 7.38

## PROBLEMS

1. Design a reinforced concrete beam, supported on two walls 60 cm thick, over a clear opening of 5 m. The beam carries a superimposed load of 15 kN/m. Use M 20 mix and HYSD bars.

2. Design a reinforced concrete lintel over the openings of two windows, each 2 m wide separated by a wall of 50 cm. The height of wall over the lintel is 3 m, and thickness of wall is 40 cm. The lintel is to be provided with a sunshade projecting out by 80 cm and cast monolithically with the lintel. Use M 20 mix and HYSD reinforcement. The live load on the sun shade may be taken as 1500 N/m$^2$.

3. Design a simply supported reinforced concrete beam, carrying a superimposed load of 20 kN per metre run over a clear opening of 6 m. In addition to this the beam carries two concentrated loads of 30 kN each at a distance of 1.5 m from the face for each support. The width of supporting wall is 60 cm. Use M 20 concrete and Fe 415 steel.

4. Design a R.C. cantilever projecting out 3 m beyond the fixed end and carrying a superimposed load of 15 kN. The cantilever also carries a concentrated load of 5 kN at the free end. Use M 20 mix. Take $\sigma_{st}$ = 230 N/mm$^2$.

5. Design a R.C.C. floor slab for a room having inside dimensions 4 m × 10 m and supported on all side by a 40 cm thick brick wall. The superimposed load may be taken as 3 kN/m$^2$. Use M 20 mix and HYSD bars.

6. Fig. 7.39 shows the cross-section of a roof slab, the supports being 4 m part. Design the slab for a live load of 2000 N/m$^2$. The slab is to be finished with 30 mm thick granolithic flooring. Use M 20 concrete and HYSD bars.

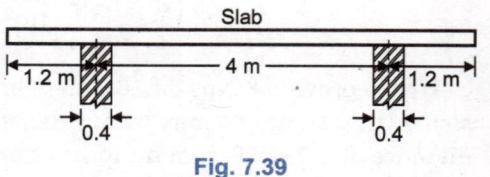

Fig. 7.39

7. Design the roof for a hall, 6 m wide and 11 m long. The slab is supported on R.C.C. beams, 30 cm wide and arranged as shown in Fig. 7.40. The live load on the roof may be taken at 3500 N/m$^2$ and floor finishing load as 700 N/m$^2$. Use M 20 mix. and HYSD bars.

8. Design a doubly reinforced concrete beam, supporting a uniformly distributed superimposed load of 40 kN/m over a clear span of 6 m. The beam carries two concentrated loads of 20 kN each at a distance of 2 m from the face of the support. The overall depth and width of the beam are restricted to 90 cm and 40 cm respectively. Use M 20 mix. and Fe 415 steel.

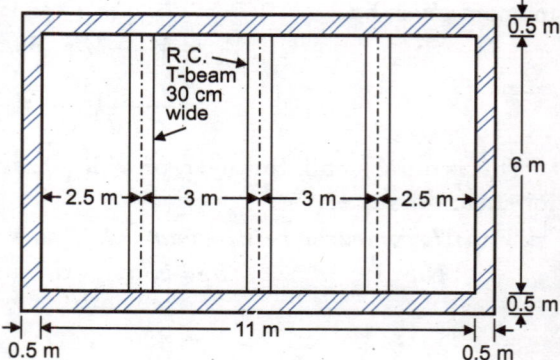

Fig. 7.40

9. Design completely a R.C.C., roof of a railway platform, consisting of a R.C. slab which cantilevers 2.5 metres on each side of a central R.C. beam (Fig. 7.41). The slab is monolithic with the beam. The beam is supported on columns 5 metres apart, and 30 cm wide. The superimposed load may be taken as 2000 N/m$^2$. Use suitable values of premisible stresses.

10. The beam of a T-beam roof, over a room 6 m × 18 m projects out by 2 m on side of the room. The roof slab which is 12 cm thick extends to the full length of the beams and its monolithic with them. The beams are spaced 3 metres centre to centre and the web which projects below the slab is 30 cm wide. Design the T-beams completely, if

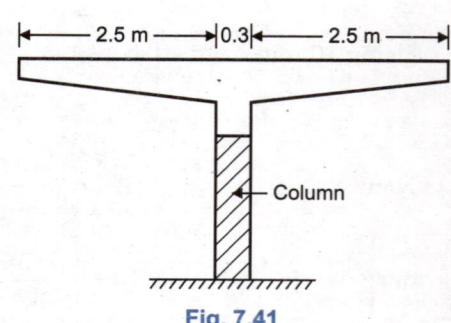

Fig. 7.41

the superimposed load on roof is 4000 N/m$^2$. The walls of the room are 40 cm thick. Use M 20 mix. and HYSD reinforcement.

11. Design a R.C.C., T-beam and a R.C.C. balcony for a lecture theatre having the following data:
    (*i*) Size of the lecture theatre ... 15 m × 9 m × 6 m high.
    (*ii*) Size of the balcony ... 1.5 m clear with 80 cm high and 1.5 cm thick R.C.C. railing.
    (*iii*) Height of balcony ... 3.5 m from floor.

(iv) Thickness of brick wall ... 40 cm.
(v) Live load for roof ... 1500 N/m².
(vi) Live load for balcony ... 4000 N/m². Use M 20 mix. and Fe 415 reinforcement.

12. Design as inverted T-beam roof for data of problem 7.
13. A R.C. cantilevered canopy 3.5 metre effective overhang and 5.6 metre wide is supported on two beams cantilevered out from wall and spaced at 3.2 meters centre to centre so that the slab overhangs each beam by 1.2 metre from the centre of each beam. The ribs of the beams are integrated with the slab so as to have flat soffit. Design the canopy completely, for a superimposed load of 1500 N/m². Use M 20 mix. and Fe 415 steel.
14. Design the porch slab and the beam shown in Fig. 7.42 for a live load of 1500 N/m², inclusive of the floor finishes, but exclusive of the self-weight of the slab. Use M 20 mix and Fe 415 steel.

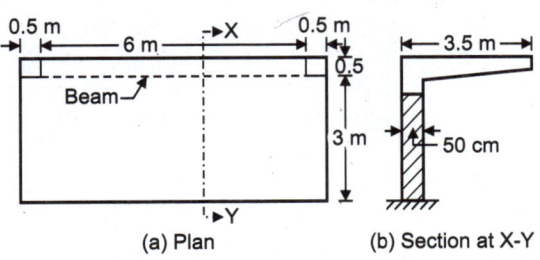
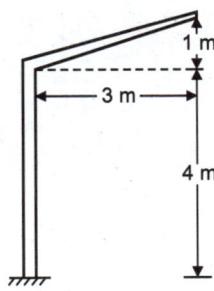

Fig. 7.42

Fig. 7.43

15. A R.C.C. bent for a city bus stand is shown in Fig. 7.43. The total length of the bus stand is 11 m. The bents are spaced at 3 m centres so that the slab overhangs by 1 m on either side beyond the centre of bent ribs. Design the slab as well as the bents, for a live load of 2000 N/m² inclusive of the finishes on the slab. The ribs are to be provided above the roof. Use M 20 mix. and HYSD bars.

# CHAPTER 8: DESIGN OF STAIR CASES

## 8.1. INTRODUCTION

Staircases are used in almost all buildings. A staircase consists of a number of steps arranged in a series, with landings at appropriate locations, for the purposes of giving access to different floors of a building. The width of a staircase may depend on the purpose for which it is provided, and may generally vary between 1 m for residential buildings to 2 m for public buildings. A *flight* is the length of the staircase situated between two landings. The number of steps in a flight may vary between 3 to 12. The *rise* of a step and the *tread* should be so proportioned that it gives comfortable access. Generally, the sum of tread plus twice the rise is kept about 500 mm and the product of tread and rise is kept about 40000 to 42000. In residential buildings, the rise may vary between 150 mm to 180 mm and tread between 200 to 250 mm. In public buildings, rise is kept between 120 to 150 mm and tread between 200 to 300 mm.

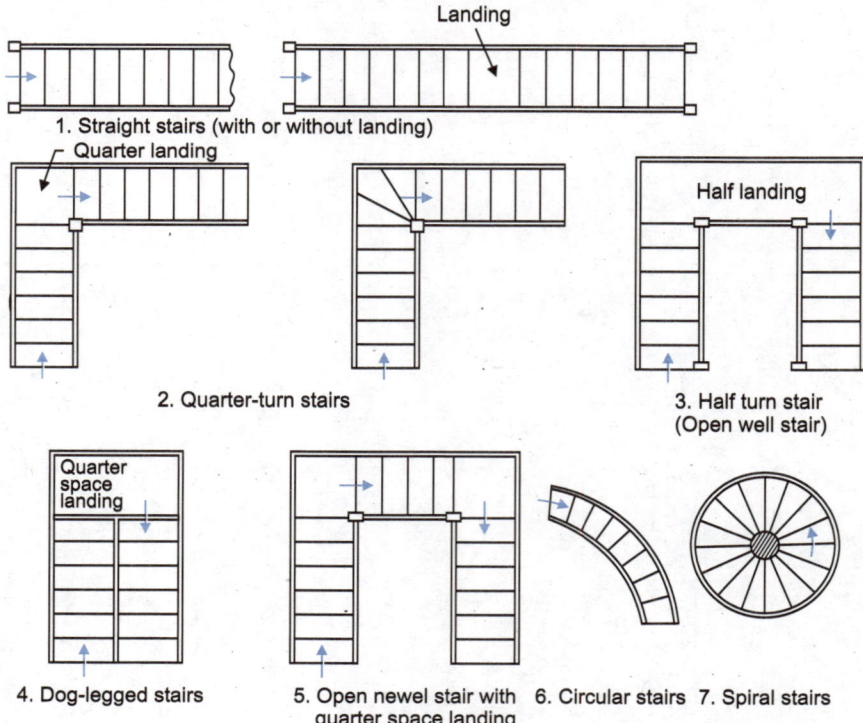

Fig. 8.1. Different Types of Stairs.

216

**Classification of stairs.** Staircases may be broadly classified in the following heads (Fig. 8.1)
1. Straight stair.
2. Quarter turn stair.
3. Half turn stair (open newel type or open well stair).
4. Dog-legged stair.
5. Open newel stair with quarter space landing.
6. Geometrical stairs such as circular stair, or spiral stair etc.

## 8.2. GENERAL NOTES ON DESIGN OF STAIRS

**1. Live load on stairs.** I.S. 875 (*Code of Practice for Structural Safety of Buildings*) gives the loads for staircases. For stairs in residential buildings, office buildings, hospital wards, hostels, etc., where there is no possibility of overcrowding, the live load may be taken as 3000 N/m², subject to a minimum of 1300 N concentrated load at the unsupported end of each step for stairs constructed out of structurally independent cantilever step. For other public buildings liable to be overcrowded, the live load may be taken to be 5000 N/m².

**2. Effective span of stairs.** Stair slab may be divided into two categories, depending upon the direction in which the stair slab spans:
  (*i*) Stair slab spanning horizontally.
  (*ii*) Stair slab spanning longitudinally.

(*i*) **Stair slab spanning horizontally.** In this category, the slab is supported on each side by side wall or stringer beam on one side and beam on the other side. Sometimes, as in the case of straight stair, the slab may also be supported on both the sides by the two side walls. The slab may also be supported horizontally by side wall on one side of each flight and the common newel on the other side between the backward and forward flights. In such a case the effective span $L$ is the horizontal distance between centre-to-centre of supports. Each step is designed as spanning horizontally with a bending moment equal to $(wL^2)/8$. Each step is considered equivalent to a rectangular beam of width $b$ (measured parallel to the slope of the stair) and an effective depth equal to $D/2$ as shown in Fig 8.2. Main reinforcement is provided in the direction of $L$, while distribution reinforcement is provided parallel to the flight direction. A waist of about 8 cm is provided.

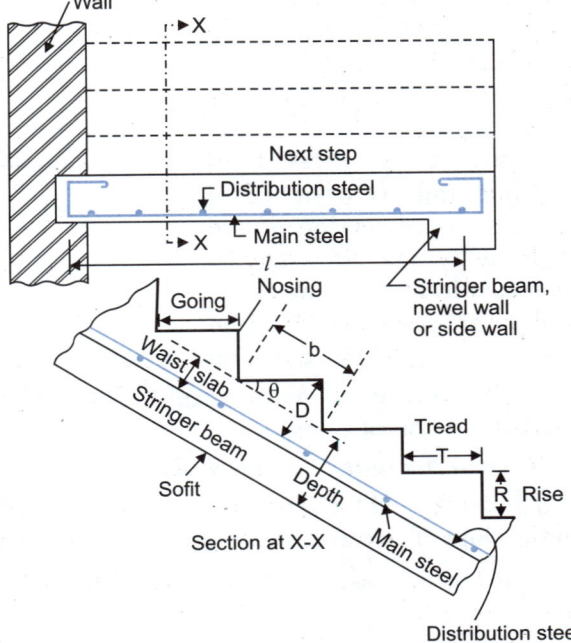

Fig. 8.2.  Stair Slab Spanning Horizontally.

(*ii*) **Stair slab spanning longitudinally.** In this category, the slab is supported at bottom and top of the flight and remain unsupported on the sides. Each flight of stairs is continuous, supported on beams at top and bottom or on landings. The effective span of such stairs, without stringer beams, should be taken as the following horizontal distances:

  (*a*) where unsupported at top and bottom risers by beams spanning parallel with the risers, the distance centre-to-centre of beam;

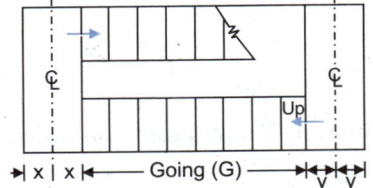

| x (m) | y (m) | Span L (m) |
|---|---|---|
| <1 | <1 | G + x + y |
| <1 | ≥1 | G + x + 1 |
| ≥1 | <1 | G + y + 1 |
| ≥1 | ≥1 | G + 1 + 1 |

Fig. 8.3.  Effective Span for Stairs Supported at each end by Landings Spanning Parallel with the Risers.

(b) where spanning on the edge of a landing slab which spans parallel to the risers (Fig. 8.3), a distance equal to the 'going' of the stairs plus at each end either half the width of the landing or one metre, whichever is smaller; and

(c) where the landing slab spans in the same direction as the stairs, they should be considered as acting together to form a single slab and the span determined at the distance centre-to-centre of the supporting beams or walls, the going being measured horizontally.

3. **Distribution of loading on stairs.** In case of stairs with open wells, where spans partly crossing at right angles occur, the load on areas common to any two such spans may be taken as one-half in each direction as shown in Fig. 8.4 (a). Where flights or landings are built into walls at a distance of not less than 110 mm and are designed to span in the direction of the flight, a 150 mm strip may be deducted from the loaded area and the effective breadth of the section increased by 75 mm for the purposes of design [Fig. 8.4 (b)].

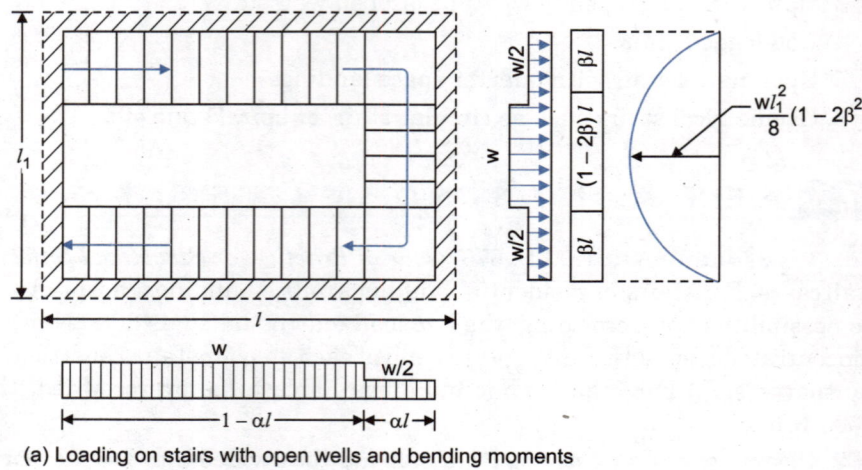

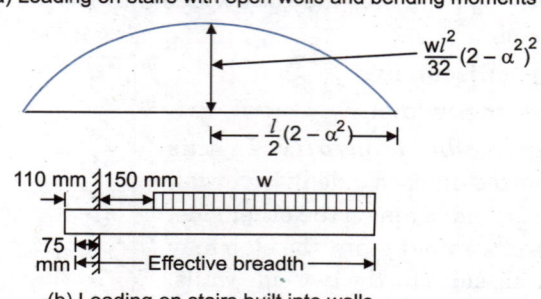

Fig. 8.4. Distribution of Loading on Stairs.

4. **Estimation of dead weight.** The dead weight of stair consists of (i) dead weight of waist slab and (ii) dead weight of steps.

**(i) Dead weight of waist slab**

The dead weight $w'$, per unit area, is first calculated at right angles to the slope. The corresponding load per unit horizontal area is then obtained by increasing $w'$ by the ratio $\sqrt{R^2 + T^2}/T$, where $R$ = rise and $T$ = tread.

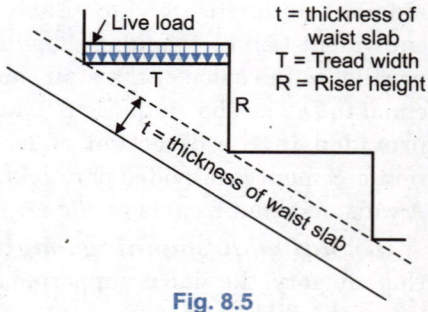

Fig. 8.5

Thus, if $t$ = thickness of waist in mm, then $w' = \dfrac{t \times 1 \times 1}{1000} \times 25000$

= $25\,t$ N/m² of inclined area. Hence dead weight $w_1$ per unit horizontal area is given by

$$w_1 = w' \times \frac{\sqrt{R^2 + T^2}}{T} = 25\,t\,\frac{\sqrt{R^2 + T^2}}{T} = 25\,t\,\sqrt{1 + (R/T)^2}$$

For example, if $R = 150$ mm, $T = 300$ mm and $t = 80$ mm.

Then $w_1 = 25\,t\,\sqrt{1 + \left(\dfrac{150}{300}\right)^2} = 27.95\,t = 2236$ N/m² of horizontal area.

### (ii) Dead weight of steps.

The dead weight of the steps is calculated by treating the step to be equivalent *horizontal* slab of thickness equal to half the rise ($R/2$). Thus, if $w_2$ is the weight of step per unit horizontal area, we have

$$w_2 = \frac{R}{2 \times 1000} \times 1 \times 1 \times 25000 = 12.5\,R\ \text{N/m}^2 \quad \text{where } R \text{ is rise in mm.}$$

Total $w = w_1 + w_2$ per unit horizontal area.

## 8.3. DESIGN OF STAIRS SPANNING HORIZONTALLY

In this type of stairs, the effective span is the horizontal distance between centre to centre of supports. Each step is designed as spanning horizontally. Each step is considered equivalent to a horizontal beam of width $b = \sqrt{R^2 + T^2}$ measured parallel to the slope of the stair, and an effective depth $D/2$, as shown in Fig. 8.2. Main reinforcement is provided in the direction of span. The distribution reinforcement, provided in the form of 6 mm diameter bars at 30 cm c/c is normally adequate.

**Design: Example 8.1.** *A straight stair in a Residential building in supported on wall on one side and stringer beam on the other side. The risers are 150 mm and treads are 250 mm, and the horizontal span of the stairs may be taken as 1.2 metres. Design the steps. Use M 20 concrete and HYSD bars.*

**Solution:** *(i) Design constants* : For M 20 concrete, $\sigma_{cbc} = 7$ N/mm$^2$ and $m = 13.33$. For HYSD bars, $\sigma_{st} = 230$ N/mm$^2$. Hence we have $k_c = 0.289$, $j_c = 0.904$ and $R = 0.914$.

*(ii) Step dimensions (Fig. 8.2)*

$R = 150$ mm ; $T = 250$ mm ; $b = \sqrt{R^2 + T^2} = \sqrt{150^2 + 250^2} = 292$ mm

Let us keep waist thickness = 80 mm. $D = 80 + \dfrac{RT}{b} = 80 + \dfrac{150 \times 250}{292} = 208$ mm. Hence the effective depth of equivalent beam = $D/2 = 104$ mm. Span $L = 1.25$ m.

*(iii) Loading and B.M.* Each step spans horizontally.

| | | | |
|---|---|---|---|
| Dead load of each step per metre | $= \left(\dfrac{1}{2} \times \dfrac{150}{1000} \times \dfrac{250}{1000} \times 25000\right)$ | | $\approx 469$ N/m |
| Dead load of waist | $= \dfrac{80 \times 292}{10^6} \times 25000$ | | $= 584$ N/m |
| Loading of finishing | | | $= 70$ N/m (say) |
| | | Total | $= 1123$ N/m |
| Live load @ 3000 N/m$^2$ | $= (250/1000 \times 3000 \times 1)$ | | $= 750$ N/m |
| | | Total | $= 1873$ N/m |

$$\therefore \quad M = \frac{wL^2}{8} = \frac{1873\,(1.25)^2}{8} = 365.8\ \text{N-m} = 36.58 \times 10^4\ \text{N-mm}$$

*(iv) Design of section*

$$d = \sqrt{\frac{M}{R_c b}} = \sqrt{\frac{36.58 \times 10^4}{0.914 \times 292}} = 37\ \text{mm. But available } d = 104\ \text{mm}$$

$$\therefore \quad A_{st} = \frac{M}{\sigma_{st}\, j_c d} = \frac{36.58 \times 10^4}{230 \times 0.904 \times 104} = 16.9\ \text{mm}^2$$

However, provide minimum steel of one bar of 8 mm diameter per step giving $A_{st} = 50$ mm$^2$. Provide distribution reinforcement in the form of 8 mm Φ bars @ 450 mm c/c. The reinforcement is arranged as shown in Fig. 8.2.

## 8.4. DESIGN OF DOG-LEGGED STAIR

**Design: Example 8.2.** *Design a dog-legged stair for a building in which the vertical distance between floors is 3.6 m. The stair hall measures 2.5 m × 5 m. The live load may be taken as 2500 N/m². Use M 20 concrete, and HYSD bars.*

**Solution:**

(*i*) **General arrangement.** Fig. 8.6 shows the plan of stair hall. Let the rise be 150 mm and tread be 250 mm.

Let us keep width of each flight = 1.2 m.
Height of each flight = 3.6/2 = 1.8 m
No. of risers required = 1.8/0.15 = 12 in each flight
No. of treads in each flight = 12 – 1 = 11.
Space occupied by treads = 11 × 25 = 275 cm.
Keep width of landing equal to 1.25 m. Hence space left for passage = 5.0 – 1.25 – 2.75 = 1 m.

(*ii*) **Design constants.** For M 20 concrete, and HYSD bars having $\sigma_{st}$ = 230 N/mm² we have $k_c$ = 0.289 ; $j_c$ = 0.904 and $R_c$ = 0.914.

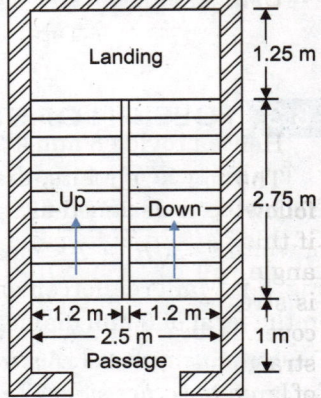

Fig. 8.6. Staircase plan and dimensions.

(*iii*) **Loading on each flight.** The landing slab is assumed to span in the same direction as the stairs, and is considered as acting together to form a single slab. Let the bearing of the landing slab in the wall be 160 mm.

Then effective span = $2.75 + 1.25 + \dfrac{0.160}{2}$ = 4.08 ≈ 4.1 m. Let the thickness of waist slab be equal to 200 mm (assumed at the rate of 40 to 50 mm per metre span).

∴ Weight of slab $w'$ on slope = 0.2 × 1 × 1 × 25000 = 5000 N/m²

Dead weight of horizontal area $w_1 = w' \dfrac{\sqrt{R^2 + T^2}}{T} = 5000 \times \dfrac{\sqrt{(150)^2 + (250)^2}}{250}$ = 5830 N/m²

Dead weight of steps is $w_2 = \dfrac{150}{2000} \times 25000$ = 1875 N/m

∴ Total dead weight per metre run = 5830 + 1875 = 7705 N
Weight of finishing etc. = 100 N (assumed)
Live load = 2500 N

Total $w$ = 10305 N/m

**Note.** The load $w$ on the landing portion will be 10305 – 1875 = 8430 N, since weight of steps will not come on it. However, a uniform value of $w$ has been adopted here.

(*iv*) **Design of waist slab**

$$M = \dfrac{wL^2}{8} = \dfrac{10305\,(4.1)^2}{8} = 21653 \text{ N-m} = 21.65 \times 10^6 \text{ N-mm}$$

∴ $$d = \sqrt{\dfrac{M}{R_c\,b}} = \sqrt{\dfrac{21.65 \times 10^6}{0.914 \times 1000}} = 154 \text{ mm}$$

Adopt 180 mm overall depth. Using 20 mm nominal cover, and 10 mm Φ bars, effective depth = 180 – 20 – 5 = 155 mm.

(*v*) **Reinforcement** $A_{st} = \dfrac{M}{\sigma_{st}\,j_c\,d} = \dfrac{21.65 \times 10^6}{230 \times 0.904 \times 155} = 671.8 \text{ mm}^2$

Using 10 mm Φ bars, No. of bars needed in 1.2 m width = (1.20 × 671.8)/78.54 ≈ 11 (say)

∴ Spacing of bars = (1200/11) = 109 mm. Hence use 10 mm Φ bars @ 109 mm c/c.

$$A_{sd} = \frac{0.12 \times 180 \times 1000}{100} = 216 \text{ mm}^2.$$

Using 8 mm Φ bars, $A_\Phi = \frac{\pi}{4}(8)^2 = 50.3 \text{ mm}^2$

∴ $s = \frac{1000 \times 50.3}{216} \approx 230 \text{ mm}.$

Hence provide 8 mm Φ bars @ 230 mm c/c.

The main reinforcement should be bent to follow the bottom profile of the stair. However, if this pattern is followed near the landing, an angle will be formed in the bar. When the bar is stressed, it will try to throw off the concrete cover. Hence near the landing, the bars are taken straight up and then bent in the compression zone of landing. In order to take tensile stresses in the landing portion, it is desirable to use separate set of bars as shown is Fig. 8.7. However, since the bending moment is very much reduced near the landing, only half the number of bars may be provided, i.e., provide 6 nos. of 10 mm Φ bars.

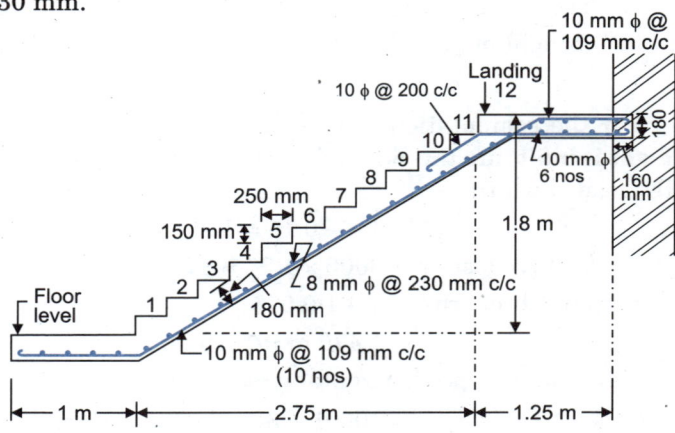

Fig. 8.7

The length of each type of bar on either side of crossing should be at least 45 Φ = 45 × 10 = 450 mm. The details of reinforcement etc. are shown in Fig. 8.7.

## 8.5. DESIGN OF STAIRS WITH QUARTER SPACE LANDING

**Design: Example 8.3.** Fig. 8.8 shows the general arrangement of a stair case for an office building. The tread is 300 mm and rise is 150 mm. The stairs is built in the side wall along the flights for a distance of 120 mm. Design the stair case for a live load of 3000 N/m² taking the span in the direction of the flight. Use M 20 concrete and Fe 415 steel.

**Solution:**

**(i) Design constants:**

For the given set of stresses, we have : $k_c = 0.289$; $j_c = 0.904$ and $R_c = 0.914$

**(ii) Effective span :** Assume 200 mm bearing of the landing in the walls.

Effective span of flight $AB$ = 3 + 1.4 + 0.10 = 4.5 m.

Effective span of flight $BC$ = 0.1 + 1.4 + 1.5 + 1.4 + 0.1 = 4.5 m

Thus, effective span of both the flights is equal. Hence any one flight (say flight $BC$) may be designed and the same reinforcement may be adopted for the other flight.

**(iii) Loading on each flight :** Let the thickness of the waist slab be 200 mm.

∴ Weight of $w'$ on slope
 = (200/1000) × 1 × 1.4 × 25000 per metre run

∴ Weight $w_1$ per horizontal metre run

$$= 0.2 \times 1.4 \times 25000 \frac{\sqrt{(150)^2 + (300)^2}}{300} = 7826 \text{ N/m run}$$

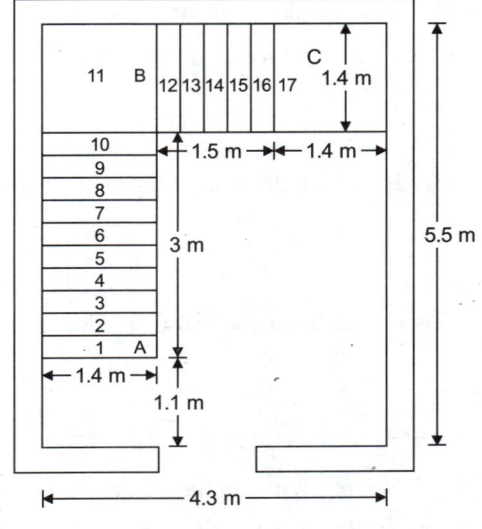

Fig. 8.8

Weight of each step

$$= 1.4 \; \frac{150}{2} \times \frac{300}{10^6} \times 25000 = 787.5 \text{ N}$$

∴ Weight $w_2$ of steps per horizontal metre run = 787.5 × (1000/300) = 2625 N

Alternatively, $w_2 = \dfrac{R}{2 \times 1000} \times 1 \times 1.4 \times 25000 = \dfrac{150}{2} \times 1.4 \times 25 = 2625$ N

∴ Total dead weight per metre run    = 7826 + 2625    = 10451 N
Weight of finishing etc.                                              = 150 N (say)

                                                      Total = 10601 N/m

For the computation of live load, consider Fig. 8.9. Since the flight is built into the side wall by a distance 120 mm (> 110 mm), the loaded width

$$= 1.4 - 0.15 = 1.25 \text{ m}.$$

∴ Live load/m = 3000 × 1.25 = 3750 N
Effective breadth $b$ = 1.4 + 0.075

$$= 1.475 \text{ m} = 1475 \text{ mm}$$

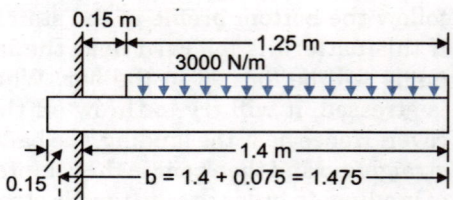

Fig. 8.9

Hence total $w$ per horizontal metre run

$$= 10601 + 3750 = 14351 \text{ N/m}$$

**(iv) Design of flight BC.** Landing B is common to both the flights. Hence $w$ for landing $B = \frac{1}{2} \times 14351 = 7175.5$ N/m, while $w$ for landing C will be taken as 14350 N/m. The loading, S.F.D. and B.M.D. are shown in Fig. 8.10.

Reaction $R_C = \dfrac{1}{4.5}\left(\dfrac{7175.5 \times 1.5 \times 1.5}{2} + 14351 \times 3 \times 3\right)$

$$= 30495.9 \text{ N}$$

$R_B = 7175 \times 1.5 + 14350 \times 3 - 30494 = 23318$ N

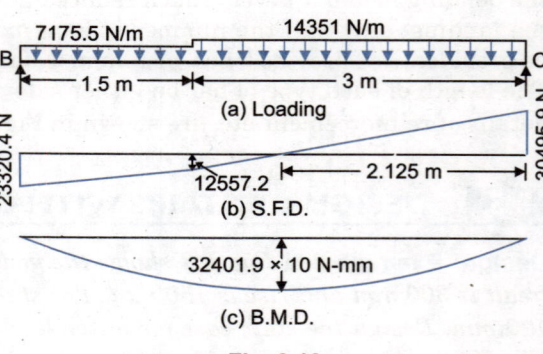

Fig. 8.10

S.F. is zero at a distance = (30494/14350) = 2.125 m from C.

B.M. will be maximum where S.F. is zero.

$$M_{max} = \left[30494 \times 2.125 - \frac{14350 (2.125)^2}{2}\right] 1000 = 32.4 \times 10^6 \text{ N-mm}$$

Breadth $b$ of slab for design = 1475 mm

$$d = \sqrt{\frac{M}{R_c b}} = \sqrt{\frac{32.4 \times 10^6}{0.914 \times 1475}} = 155 \text{ mm}$$

Keep total depth = 180 mm. Using 10 mm Φ bars and nominal cover of 20 mm,

$$d = 180 - 20 - 5 = 155 \text{ mm}.$$

$$A_{st} = \frac{32.4 \times 10^6}{230 \times 0.904 \times 155} = 1005.3 \text{ mm}^2$$

∴ No. of 10 mm φ bars required in 1475 mm width = 1005.3/78.54 = 12.8
Hence provide 13 bars of 10 mm Φ

∴ Spacing $s$ = 1475/13 ≈ 113.5 mm

Distribution reinforcement $A_{st}$ = 1.2 × 180 = 216 mm²

∴ Spacing of 8 φ bars $= \dfrac{1000 \times 50.3}{216} = 232$ mm.

Hence provide 8 Φ mm bars @ 230 mm c/c. The same reinforcement may be provided for both the flights. At the landing, provide reinforcement both at top as well as at bottom. The details of reinforcement are shown in Fig. 8.11.

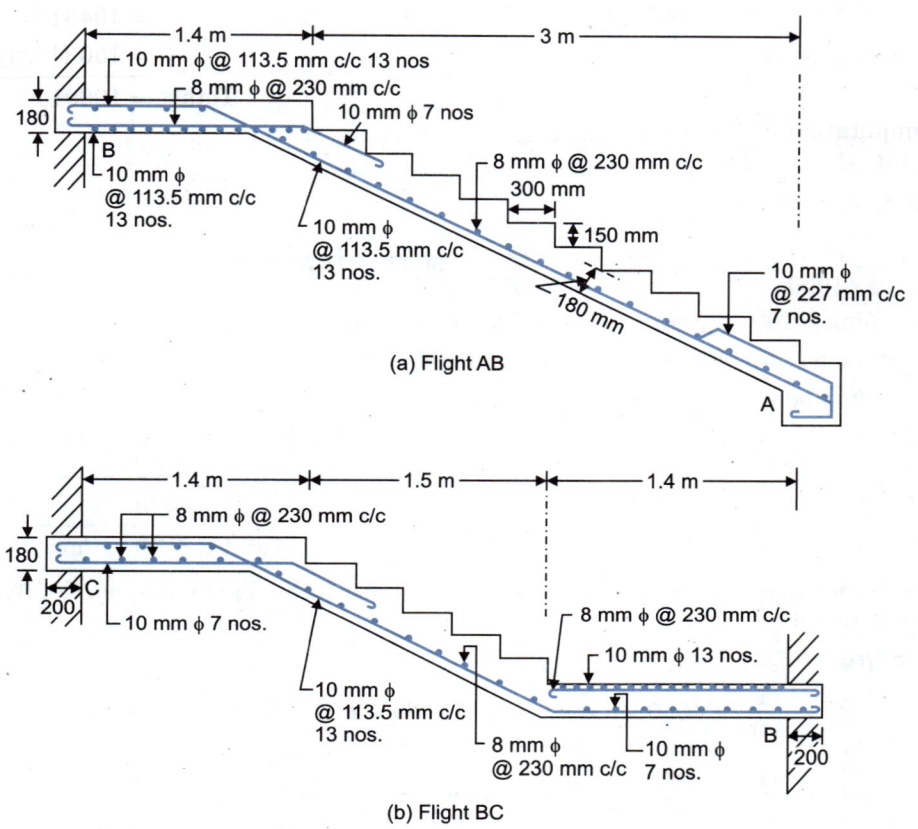

Fig. 8.11

## 8.6. DESIGN OF OPEN NEWEL STAIR WITH QUARTER SPACE LANDING

**Design: Example 8.4.** *Fig 8.12 shows the general arrangement of a staircase of a building. The risers are 150 mm and the treads are 250 mm. Design the staircase for a live load of 3000 N/m². The width of stair is 1.5 m and the width of wall is 400 mm. Use M 20 concrete and Fe 415 steel.*

**Solution:** *(i) Design constants.* For the given set of materials, we have

$k_c = 0.298 \; ; j_c = 0.904 \;$ and $R_c = 0.914.$

*(ii) Effective span* : Assume a bearing of 200 mm in the walls. For flight AB and CD.

$L = 0.1 + 2.25 + 1.5 + 0.1 = 3.95$ m

For flight BC, $L = 0.1 + 1.5 + 1.25 + 1.5 + 0.1 = 4.45$ m

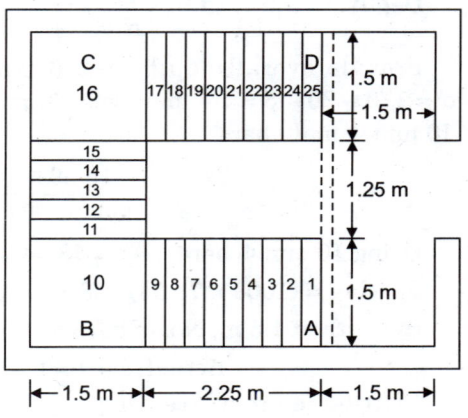

Fig. 8.12

### (iii) Loading

(a) *For the flight portion*: Let the waist slab be 200 mm thick. Also, let the landing slab be 200 mm thick. Consider a strip of slab of 1 m wide.

Weight of slab $w'$ on slope
$$= 0.2 \times 1 \times 1 \times 25000 = 5000 \text{ N/m}^2$$

Weight on horizontal area
$$= 5000 \frac{\sqrt{150^2 + 250^2}}{250} \approx 5830 \text{ N/m}^2$$

Dead weight of steps is given by
$$w_2 = \frac{R}{2 \times 1000} \times 1 \times 1 \times 25000 = \frac{150}{2000} \times 25000 \approx 1870 \text{ N}$$

Weight of finishing = 100 N (assumed). Live load = 3000 N/m².
Hence total load = 5830 + 1870 + 100 + 3000 = 10800 N/m

(b) *For the landing portion*:

| | |
|---|---|
| Dead load | $= (200/1000) \times 1 \times 1 \times 25000 = 5000$ N/m² |
| Weight of finishing | $= 100$ N/m² |
| Live load | $= 3000$ N/m² |
| Total | $= 8100$ N/m² |

However, since each quarter space landing is common to both the flights, only half of the above loading, *i.e.*, 4050 N/m will be taken.

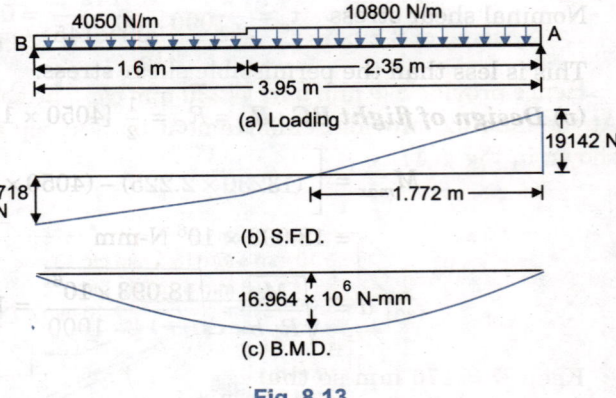

Fig. 8.13

### (iv) Design of flight AB:

$$R_A = \frac{1}{3.95}[(4050 \times 1.6 \times 0.8) + (10800 \times 2.35 \times 2.775)] \approx 19142 \text{ N}$$

$$R_B = \frac{1}{3.95}\left[\frac{10800 \times (2.35)^2}{2} + 4050 \times 1.6 \times 3.15\right] = 12718 \text{ N}$$

S.F. is zero at 19142/10800 = 1.772 m from A.

$$M_{max} = \left[(19142 \times 1.772) - \frac{10800 (1.772)^2}{2}\right] = 16.964 \times 10^6 \text{ N-mm}$$

Depth $\quad d = \sqrt{\dfrac{M}{R_c b}} = \sqrt{\dfrac{16.964 \times 10^6}{0.914 \times 1000}} = 136.2$ mm

Provide overall depth of 170 mm so that available $d = 170 - 20 - 5 = 145$ mm with 20 mm nominal cover and 10 mm Φ main bars.

$$A_{st} = \frac{M}{\sigma_{st} j_c d} = \frac{16.694 \times 10^6}{230 \times 0.904 \times 145} = 553.8 \text{ mm}^2$$

Using 10 mm Φ bars, ; $A_\Phi = 78.54$ mm²
Spacing = (1000 × 78.54)/553.8 = 141.8 mm
In width of 1.5 m, No. of bars = 1500/141.8 ≈ 11
∴ Actual, $s$ = 1500/11 ≈ 136 mm.
Distribution reinforcement
$$A_{sd} = 1.2 D = 1.2 \times 170 = 204 \text{ mm}^2.$$
Provide 8 mm Φ @ 240 mm c/c.

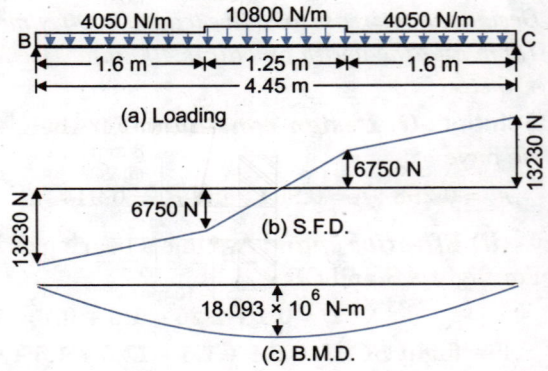

Fig. 8.14

Nominal shear stress $\tau_v = \dfrac{V}{bd} = \dfrac{19142}{1000 \times 145} = 0.132$ N/mm$^2$

This is less than the permissible shear stress.

**(v) Design of flight BC :** $R_B = R_C = \dfrac{1}{2}$ [4050 × 1.6 + 10800 × 1.25 + 4050 × 1.6] = 13230 N

$$M_{max} = \left[ (13230 \times 2.225) - (4050 \times 1.6 \times 1.425) - \dfrac{10800\,(0.625)^2}{2} \right] \times 1000$$

$$= 18.093 \times 10^6 \text{ N-mm}$$

$$d = \sqrt{\dfrac{M}{R_c\,b}} = \sqrt{\dfrac{18.093 \times 10^6}{0.914 \times 1000}} = 141 \text{ mm}$$

Keep $D = 170$ mm so that $d = 145$ mm.

$$A_{st} = \dfrac{M}{\sigma_{st}\,j_c\,d} = \dfrac{18.093 \times 10^6}{230 \times 0.904 \times 145} = 600 \text{ mm}^2$$

Using 10 mm Φ bars having $A_\Phi = 78.54$ mm$^2$.

$$\text{Spacing} = \dfrac{1000 \times 78.54}{600} = 130.9 \text{ mm.}$$

Keep, $s = 100$ mm

In width of 1.5 m, No. of bars = 1500/130.9 ≏ 12.

Actual $s = 1500/12 = 125$ mm

**Distribution reinforcement**

Provide 8 mm Φ bars @ 240 mm c/c.

The details of reinforcement for both the flights are shown in Fig. 8.15.

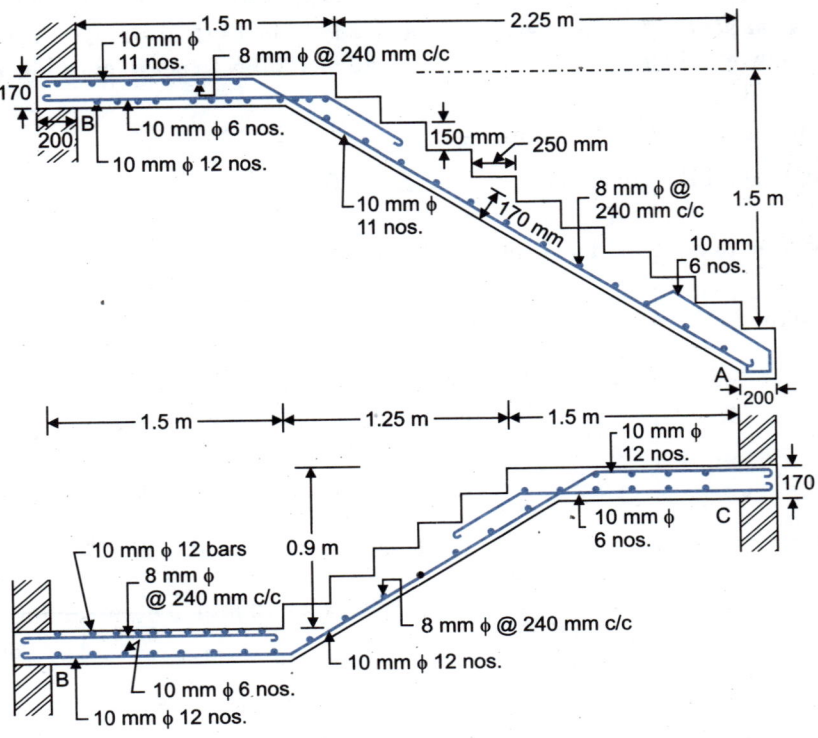

Fig. 8.15

## 8.7. DESIGN OF STAIRCASE WITH CENTRAL STRINGER BEAM

**Design: Example 8.5.** *Fig. 8.16 shows the general layout of a staircase with central stringer beam supported on columns at B, C and D. The rise and tread of the stairs are 100 mm and 250 mm respectively. The width of the steps is 1.40 m. Design the staircase for a live load of 4000 N/m². Use M 20 concrete and Fe 415 steel.*

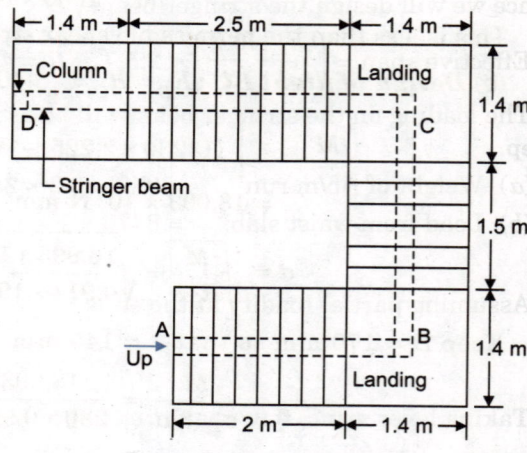

Fig. 8.16. Staircase dimensions

**Solution:** **(i) Design constants**

For the given set of materials $k_c = 0.289$; $j_c = 0.904$ and $R_c = 0.914$

**(ii) Loading on waist slab**

Let the waist slab be 80 mm thick. The weight of waist slab on the slope should be multiplied by the factor

$$\frac{\sqrt{R^2 + T^2}}{T} = \frac{\sqrt{(160)^2 + (250)^2}}{250} = 1.185$$ to get the equivalent

weight on horizontal plane. Consider 1 m width of slab.

Load per metre horizontal run will be as follows:

(a) Self weight = $(0.08 \times 1 \times 1 \times 25000) \, 1.185$ = 2370 N

(b) Wt. of steps = $\frac{1}{2} \times 0.16 \times 1 \times 1 \times 25000$ = 2000 N

(c) Finishing etc. = 100 N (say)

(d) Live load = 4000 N

Total = 8470 N

The loading on the landing will be lesser; however, for simplicity, we will take the same loading throughout.

**(iii) Design of waist slab.** The waist slab is supported on the central stringer beam. Hence the worst condition may be when we consider concentrated live load of 1300 N to act to one side only. Dead weight = $2370 + 2000 + 100 = 4470$ N

Let the width of stringer beam be 200 mm

Projection of slab beyond the rib of beam = $\frac{1.4 - 0.2}{2} = 0.6$ m

B.M. due to dead load = $\frac{1}{2}[4470 (0.6)^2] \simeq 805$ N-m

B.M. due to U.D. live load = $\frac{1}{2}[4000 (0.6)^2] = 720$ N-m

B.M. due to concentrated live load = $1300 \times 0.6 = 780$ N-m

∴ Max. B.M. $M = 805 + 780 = 1585$ N-m = $1.585 \times 10^6$ N-mm

$$d = \sqrt{\frac{M}{R_c \, b}} = \sqrt{\frac{1.585 \times 10^6}{0.914 \times 1000}} = 41.6 \text{ mm}.$$

However, keep a minimum total depth = 80 mm. Effective depth = $80 - 20 - 5 = 55$ mm

$$A_{st} = \frac{1.585 \times 10^6}{230 \times 0.904 \times 55} = 139 \text{ mm}^2.$$ Using 10 mm Φ bars, having $A_\phi = 78.5$ mm²,

$$s = \frac{1000 \times 78.5}{139} = 564 \text{ mm}.$$

However, keep spacing = 250 mm, *i.e.*, one bar per step.

Distribution reinforcement = $1.2 \times 80 = 96$ mm². Provide 8 mm Φ bars @ 300 mm c/c.

(*iv*) **Design of stringer beam.** The stringer beam will act as a T-beam. Flight *CD* is the longest, and hence we will design the stringer beam *CD*.

Effective span = $\left(1.4 - \dfrac{0.2}{2}\right) + 2.5 + \dfrac{1.4}{2} = 4.5$ m

The loading on the stringer beam will be as follows, assuming the web to be 200 mm wide and 200 mm deep.

(*a*) Weight of rib/m run  = $(0.2 \times 0.2 \times 25000) \times 1.185$ = 1185 N
(*b*) Load from waist slab = $8470 \times 1 \times 1.4$ = 11858 N

$\hspace{8cm}$ Total = 13043 ≃ 13050 N/m

Assuming partial fixidity at the ends,

$$M = \dfrac{wL^2}{10} = \dfrac{13050\,(4.5)^2}{10} \times 1000 = 26.43 \times 10^6 \text{ N-mm}$$

Taking lever arm = $0.9d$, balanced depth is given by *Eq. 6.17*

$$M = 0.45\, b_f \sigma_{cbc}\, D_f \left(\dfrac{2k_c\, d - D_f}{k_c}\right)$$

where $b_f$ = flange width of isolated T-beam given by

$$b_f = \dfrac{l_0}{\left(\dfrac{l_0}{b}\right) + 4} + b_w, \text{ where } l_0 \approx L = 4.5 \text{ m};\ b = \text{actual width} = 1.4 \text{ m};\ b_w = 0.20 \text{ m}$$

$$b_f = \dfrac{4.5}{\left(\dfrac{4.5}{1.4}\right) + 4} + 0.20 = 0.824 \text{ m} = 824 \text{ mm}.$$

Hence $26.43 \times 10^6 = 0.45 \times 824 \times 7 \times 80 \left[\dfrac{2 \times 0.289\, d - 80}{0.289}\right]$, from which $d = 202$ mm

Also, $d = \dfrac{V_e}{b_w\, \tau_{c\,max}}$, where $\tau_{c\,max} = 1.8$ N/mm² for M 20 concrete (Table 3.3)

$V_e = V + 1.6\, \dfrac{T}{b_w}$; $V = \dfrac{wL}{2} = \dfrac{13050 \times 4.5}{2} = 29363$ N

$T$ = torsional moment, which will be induced due to live load acting only to one side of step.

∴ $T = \left[\dfrac{4000\,(0.6)^2}{2}\right] \times \dfrac{4.5}{2} \times 1000$ N-mm = $1.62 \times 10^6$ N-mm

or $T = (1300 \times 0.6)\, \dfrac{4.5}{2} \times 1000$ N-mm = $1.755 \times 10^6$ N-mm whichever is more.

∴ $T = 1.755 \times 10^6$ N-mm. and $V_e = 29363 + 1.6\, \dfrac{1.755 \times 10^6}{200} = 43403$ N

Hence $d = \dfrac{43403}{200 \times 1.6} \simeq 136$ mm. However, keep total depth = 250 mm. Using 20 mm Φ bars, 8 mm Φ rings and 25 mm nominal cover, available $d = 250 - 10 - 25 - 8 = 207$ mm.

$$A_{st} = \dfrac{26.43 \times 10^6}{230 \times 0.904 \times 207} = 614 \text{ mm}^2$$

For 20 mm Φ bars, $A_\phi = 314$ mm². No. of bars = $614/314 \simeq 2$

Hence, provide 4 bars of 20 mm Φ. Actual $A_{st}$ provided = $314.16 \times 4 \simeq 1256.6$ mm².

**Note.** The above reinforcement is for bending requirements only. There will be additional longitudinal reinforcement for torsion, as computed later.

*Location of N.A.* Assuming the N.A. falls within the flange, we have

$$\frac{824}{2} n^2 = 13.33 \times 1256.6 (207 - n).$$ From which $n = 73.6$ mm.

Hence the resultant falls inside the flange. $\bar{y} = 73.6/3 \triangleq 24.5$ mm

∴  L.A. $a = d - \bar{y} = 207 - 24.5 = 182.5$ mm

$$\text{Stress in steel} = \frac{M}{A_{st} \, a} = \frac{26.43 \times 10^6}{1256.6 \times 182.5} = 115.2 \text{ N/m}^2 \text{ (safe)}$$

Corresponding stress in concrete is given by

$$c = \frac{t}{m} \cdot \frac{n}{d-n} = \frac{115.2}{13.33} \times \frac{73.6}{207 - 73.6} = 4.77 \text{ N/mm}^2 \text{ (safe)}$$

**(v) Design for torsion.** A computed earlier, $T = 1.755 \times 10^6$ N-mm

$V = 29363$ N  and  $V_e = 43403$ N.

∴ $$\tau_{ve} = \frac{V_e}{b_w \, d} = \frac{43403}{200 \times 207} = 1.048 \text{ N/mm}^2$$

$$\frac{100 A_{st}}{bd} = \frac{100 \times 1256.6}{200 \times 207} = 3.03\%.$$ Hence from Table 3.2, $\tau_c = 0.51$ N/mm$^2$.

Since $\tau_{ve} > \tau_c$, shear reinforcement is necessary.

**(a) Longitudinal reinforcement**

$$M_{e1} = M + M_T, \quad \text{where } M = 26.43 \times 10^6 \text{ N-mm}$$

$$M_T = T \frac{(1 + D/b_w)}{1.7} = 1.755 \times 10^6 \frac{(1 + 240/200)}{1.7} = 2.271 \times 10^6 \text{ N-mm}$$

∴ $M_{e1} = 26.43 \times 10^6 + 2.271 \times 10^6 \triangleq 28.71 \times 10^6$ N-mm

∴ $$A_{st} = \frac{M_{e1}}{\sigma_{st} \, jd} = \frac{28.71 \times 10^6}{(230 \times 182.5)} = 684 \text{ mm}^2$$

Hence the provision of 4 bars of 20 mm Φ, giving $A_{st} = 1256.6$ mm$^2$ is O.K.

Near the column $D$, take the bars straight up. Provide 2–20 mm Φ bars at the lower face under the landing as shown.

**(b) Transverse reinforcement** : Transverse reinforcement will be provided in the form of vertical stirrups. Let us provide 25 mm clear cover all round.

$b_1$ = centre to centre distance between corner bars in the direction of width
 = $200 - 2 \times 25 - 10 = 140$ mm

$d_1$ = centre to centre distance between corner bars in the direction of depth
 = $250 - 2 \times 25 - 10 = 190$ mm

Using 10 mm Φ 2-lgd stirrups, $A_{sv} = 2 \frac{\pi}{4} (10)^2 = 157$ mm$^2$

Now, $$A_{sv} = \frac{T \cdot s_v}{b_1 d_1 \sigma_{sv}} + \frac{V \, s_v}{2.5 d_1 \, \sigma_{sv}}$$

or $$157 = \left[\frac{1.755 \times 10^6}{140 \times 190 \times 230} + \frac{29363}{2.5 \times 190 \times 230}\right] s_v.$$

From which $s_v = 282$ mm.

However, the spacing should not exceed the least of $x_1$, $\frac{x_1 + y_1}{4}$ and 300 mm,

where  $x_1$ = short dimension of stirrup
 = $140 + 20 + 10 = 170$
and  $y_1 = 180 + 20 + 10 = 210$

DESIGN OF STAIR CASES **229**

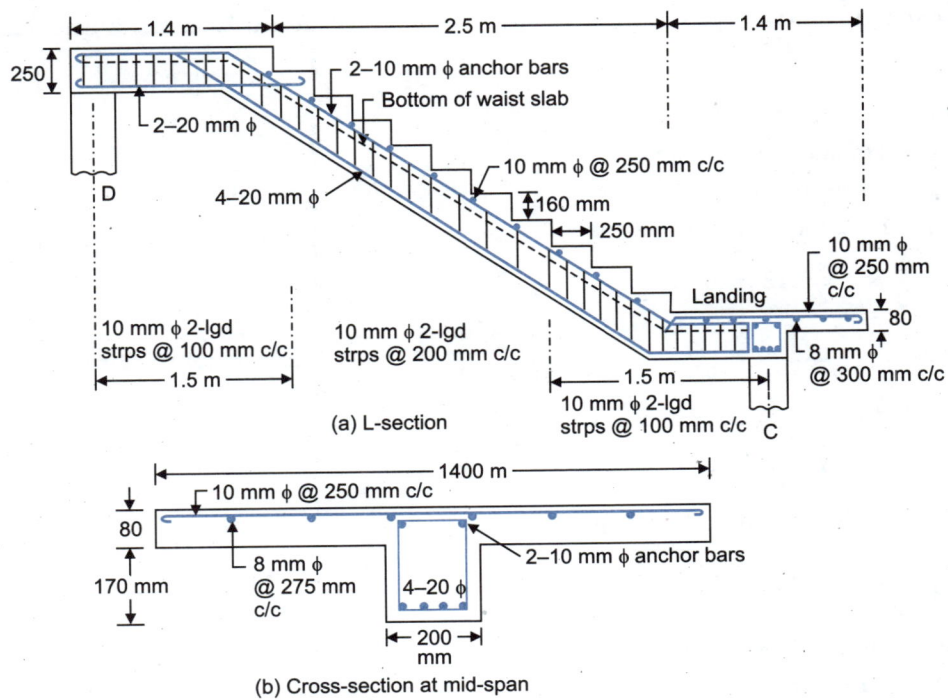

Fig. 8.17

$$\frac{x_1 + y_1}{4} = \frac{170 + 210}{4} = 95$$

Hence $s_v = 282$ mm is not permissible.

Keep $s_v = 95$ mm ≈ 100 mm c/c. Increase the spacing to 200 mm c/c in the mid span where both transverse shear as well as torsional shear are minimum. Provide 2-10 mm Φ holding bars. Keep the same section for other flights. The details of reinforcement etc. are shown in Fig. 8.17.

## 8.8. DESIGN OF CINEMA BALCONY

A cinema balcony is a huge structure designed to carry the superimposed load of furniture and audience. The furniture is arranged on a number of successive steps so that view of the audience is not obstructed. These steps along with the waist slab are supported on *rackers* which on turn are supported on wall on one side and the *fulcrum girder* on the other side. The size of the huge fulcrum girder is sometimes restricted from the point of view of obstruction in the line of sight and other head-room requirements. The treads are normally kept between 900 to 1100 mm and the risers between 100 to 125 mm. The superimposed load may vary between 4000 to 5000 N/m².

**Design: Example 8.6.** *Design a cinema balcony with the general layout shown in Fig. 8.18. The live load may be taken equal to 4500 N/m², inclusive of furniture etc. The horizontal tread of each step is 1 m and rise is 120 mm. The depth of fulcrum girder is limited to 1 m. At the end of the last step, there is a gangway 1 m wide and a R.C. balustrade 1 m high. Assume any other data not given. Use M 20 concrete and Fe 415 steel.*

**Solution:** For the given set of materials, $k_c = 0.289$ ; $j_c = 0.904$ and $R_c = 0.914$

### (A) DESIGN OF WAIST SLAB OR SLAB DECKING

The decking will be a continuous slab supported on the rackers. The slab may be designed for a maximum bending moment of $(w L^2)/10$. Since the deck slab is inclined, its weight on inclination has to be multiplied by a factor

$$\frac{\sqrt{R^2+T^2}}{T}=\frac{\sqrt{(0.12)^2+(1)^2}}{1}=1.01$$

to get the weight per unit horizontal area.

Let the thickness of slab be 100 mm

∴ Self weight per m² of slab
= (0.1 × 1 × 1 × 25000) 1.01 = 2525 N

Wt. of steps = $\frac{1}{2}$ × 1 × 0.12 × 25000 = 1500 N

Live load = 4500 N

Total = 8525 N

Effective span = 2.5 m.

∴ $$M = \frac{wL^2}{8} = \frac{8525(2.5)^2}{8} \times 1000$$

$$= 6.66 \times 10^6 \text{ N-mm.}$$

∴ $$d = \sqrt{\frac{6.66 \times 10^6}{0.914 \times 1000}} = 85.36 \text{ mm.}$$

Provide total depth = 110 mm.

Using 10 mm Φ bars, and 20 mm nominal cover,
$d = 110 - 5 - 20 = 85$ mm.

$$A_{st} = \frac{6.66 \times 10^6}{230 \times 0.904 \times 85} = 376.8 \text{ mm}^2$$

Using 10 mm Φ bars $\quad s = \frac{1000 \times 78.54}{376.8} = 208$ mm.

However, provide 10 mm Φ bars @ 200 c/c. Bend half the bars from each side up near the raker beams, to resist the hogging bending moment of equal amount.

**Check for shear**

$$V = \frac{8525 \times 2.5}{2} = 10656 \text{ N.}$$

∴ $$\tau_v = \frac{V}{bd} = \frac{10656}{1000 \times 85} = 0.125 \text{ N/mm}^2$$

At support, available $A_{st} = \frac{1000 A_\Phi}{s} = \frac{1000 \times 78.54}{200} = 392.7$ mm²

∴ $$\frac{100 A_{st}}{bd} = \frac{100 \times 392.7}{1000 \times 85} = 0.46 \%.$$

Hence from Table 3.2, $\tau_v$ = 0.29 N/mm². Hence $\tau_v < \tau_c$. (safe)

**(B) DESIGN OF RACKERS.** Let us design one of the interior rackers.

*(i) Loading.* Load due to deck slab, steps and live load = 8500 × 2.5 = 21250 N.

The racker beam will be a T-beam. Let the width of web be 300 mm and depth 300 mm.

Hence, self weight = (0.3 × 0.3 × 25000) 1.01 = 2272 N.

∴ Total load = 21250 + 2272 = 23522 N.

*Assume* the same load intensity in the gangway.

Weight of 100 mm thick balustrade, assumed as a concentrated load acting at the end of the racker
= 1 × 0.1 × 2.5 × 25000 = 6250 N.

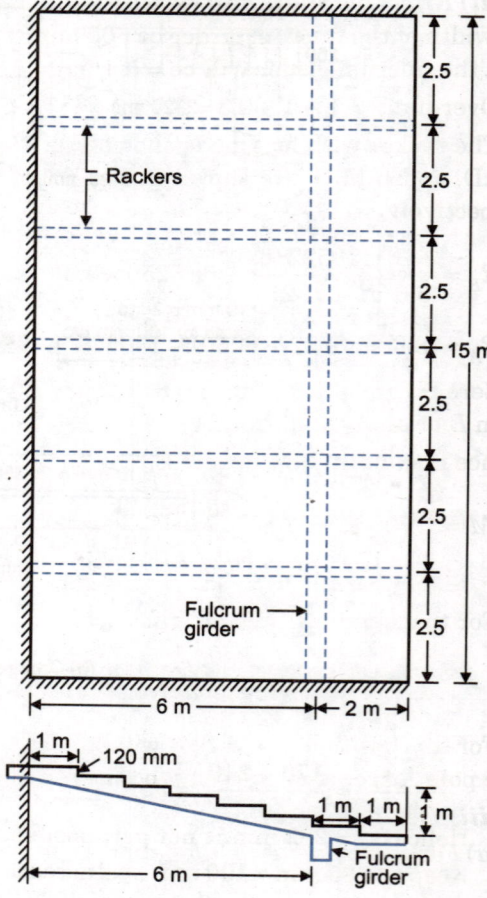

Fig. 8.18

(ii) **Effective span and bending moment.** Let the bearing of the racker in the wall be 600 mm. Also, let width of the fulcrum girder be 600 mm. Then the effective span of the racker (T-beam) between the wall and the fulcrum girder will be = 6 + 0.6 = 6.6 m.

Over-hang = 1 + 1 − 0.3 = 1.7 m.

The racker with loading is shown Fig. 8.19(a). The S.F.D. and B.M.D. are shown in Fig. 8.19 (b) and (c) respectively.

$$R_B = \frac{1}{6.6}\left\{\frac{23522(8.3)^2}{2} + 6250 \times 8.3\right\} = 130620 \text{ N}$$

$R_A = (8.3 \times 23522 + 6250) − 130620 = 70863$ N.

Zero S.F. is at a distance = 84383/28522 = 3.587 m from $B$ or at 3.013 m from $A$.

Hence maximum B.M. where S.F. is zero is given by

$$M = \left[70863 \times 3.013 - \frac{23522(3.013)^2}{2}\right] \times 1000$$

$= 1.067 \times 10^8$ N-mm.

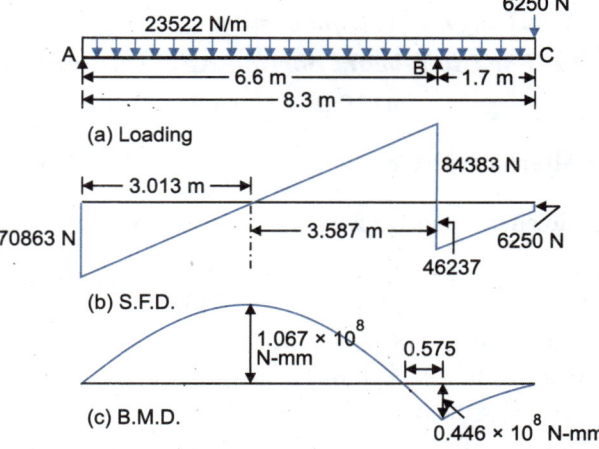

Fig. 8.19. Racker Beam.

For this bending moment, the racker behaves as a T-beam. Also, for overhang $BC$,

$$M_B = \left[\frac{23522(1.7)^2}{2} + (6250 \times 1.7)\right] \times 1000 = 0.446 \times 10^8 \text{ N-mm}.$$

For this position, the racker will behave as an ordinary rectangular beam. Fig. 8.19 (c) shows the B.M.D. The point of contraflexure will occur at a distance 0.575 m from support $B$.

(iii) **Design of section.**

(a) *For overhang portion BC.* In this portion the beam is rectangular in section having width $b$ = 300 mm

$$\therefore \quad d = \sqrt{\frac{M}{R_c\, b}} = \sqrt{\frac{0.446 \times 10^8}{0.914 \times 300}} = 404 \text{ mm} \qquad \ldots (1)$$

(b) *For portion AB.* For this portion, the beam will act as a T-beam. The width $B = b_f$ of the flange acting with the beam will be the minimum of the following:

(i)      $b_f = b_w$ + clear distance between adjacent beams

         = c/c distance between the beams = 2.5 m

(ii)      $b_f = \dfrac{l_0}{6} + b_w + 6D_f$, where $l_0$ = distance between points of zero moments

         = 6.6 − 0.57 = 6.03 m

$b_w$ = 300 mm = 0.3 m. ; $D_f$ = 110 mm = 0.11 m.

$$\therefore \quad b_f = \frac{6.03}{6} + 0.3 + 6 \times 0.11 = 1.97 \text{ m}$$

Hence    $b_f$ = 1.97 m = 1970 mm.

To find the effective depth for balanced section, we have

$$M = 0.45\, b_f c\, D_f \left[\frac{2k_c\, D - D_f}{k_c}\right].$$

Hence,     $1.067 \times 10^8 = 0.45 \times 1970 \times 7 \times 110 \left[\dfrac{2 \times 0.289d - 110}{0.289}\right]$

From which $d \approx 269$ mm.

At the end and overhang portion, the beam will behave as rectangular section. Keep total $D = 350$ mm. Using 20 mm Φ bars, 10 mm rings and a nominal cover of 25 mm and arranging them in two layers with a clear vertical distance of 20 mm, we have effective depth = $350 - 25 - 20 - 10 - 10 = 285$ mm.

### (iv) Design of reinforcement

**(a) For the T-beam portion AB.** Lever arm

$$a = 0.9d = 0.9 \times 285 \triangleq 256 \text{ mm}$$

Alternatively, $a = d - \dfrac{D_f}{2} = 285 - \dfrac{110}{2} = 230$ mm

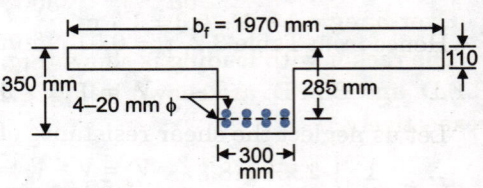

Fig. 8.20

However, use $a = 245$ mm for preliminary calculations.

$$A_{st} = \frac{M}{\sigma_{st} \times a} = \frac{1.067 \times 10^8}{230 \times 245} = 1894 \text{ mm}^2. \text{ No. of 20 mm Φ bars} = \frac{1894}{314} = 6.03$$

However, use 8 bars and arrange 4 bars in each layer as shown in Fig. 8.20. Bend two bars up at a distance equal to 1 m in given support $A$.

Actual $A_{st} = 314 \times 8 \triangleq 2512$ mm².

**(b) For the rectangular portion BC.** Effective depth = $350 - 25 - 10 - 10 = 305$ mm

$$\therefore \quad M_r = R_c bd^2 = 0.914 \times 300 (305)^2 = 0.255 \times 10^8 \text{ N-mm}$$

This is less than the actual B.M. of $0.446 \times 10^8$ N-mm.

Let us design beam $BC$ as a doubly reinforced beam, using steel beam theory. Let the compression reinforcement be placed at a cover $d_c = 35$ mm.

$$\therefore \quad A_{st} = A_{sc} = \frac{M}{\sigma_{st}(d - d_c)} = \frac{0.446 \times 10^8}{230(305 - 35)} = 718.2 \text{ mm}^2$$

Using 20 mm Φ bars, No. of bars = $718.2/314 \approx 3$. However provide 4 bars of 20 mm Φ at the top and 4 bars at the bottom. These will be arranged as follows:

*For the top.* Bend 2 bars up from the portion $AB$, at the point of contraflexure, *i.e.*, 0.575 m from the support $B$; the 2–20 mm Φ anchor bars of the T-beam portion will be continued in portion $BC$, to serve as the main reinforcement of the rectangular beam.

*For the bottom.* Out of the remaining 6 bars of the span $AB$, continue 4 bars straight into the portion $BC$. The remaining 2 bars may be curtailed at the support $B$.

### (v) Check for stresses in T-beam.
Assuming the N.A. to fall below the flange, we have

$$1970 \times 110 \left(n - \frac{110}{2}\right) = 13.33 \times 2512 (285 - n). \text{ From which } n = 87.1 \text{ mm.}$$

Hence the N.A. falls in the flange for which

$$1970 \frac{n^2}{2} = 13.33 \times 2512 (285 - n). \text{ From which } n = 84.6 \text{ mm.}$$

$$\therefore \quad a = d - \frac{n}{3} = 285 - \frac{84.6}{3} = 256.8 \text{ mm}$$

Stress in steel, $\quad t = \dfrac{M}{a \, A_{st}} = \dfrac{1.067 \times 10^8}{256.8 \times 2512} = 165.4$ N/mm²

and $\quad c = \dfrac{t}{m} \cdot \dfrac{n}{d - n} = \dfrac{165.4}{13.33} \times \dfrac{84.6}{285 - 84.6} = 5.23$ N/mm² (safe).

### (vi) Check for shear

**(a) Portion AB.** For the T-beam portion, $V_{max} = 84383$ N

$$\therefore \quad \tau_v = \frac{V}{b \times a} = \frac{84383}{300 \times 256.8} = 1.1 \text{ N/mm}^2.$$

Hence shear reinforcement is necessary.

$$\frac{100 A_s}{bd} = \frac{100 \times (8 \times 314)}{300 \times 285} = 2.95\%.$$

Hence from Table 3.2, $\tau_c = 0.51$ N/mm².

∴ $V_c = 0.51 \times 300 \times 285 = 43605$ N.

Let us neglect the shear resistance of two bent-up bars.

∴ $V_s = V - V_c = 84383 - 43605 = 40778$ N

The spacing of the stirrups is given by $s = \dfrac{\sigma_{sv} \, d \, A_{sv}}{V_s} = \dfrac{230 \times 285}{40778} A_{sv} = 1.607 A_{sv}$ mm.

Using 10 mm Φ 2-legged stirrups, $s_v = 1.607 \times 157 = 252$ mm

Maximum spacing is given by $s_v = \dfrac{2.175 \, A_{sv} \, f_y}{b} = \dfrac{2.175 \times 157 \times 415}{300} = 472$ mm

Subject to maximum of $0.75d = 0.75 \times 285 = 213.8$ mm

Hence provide 10 mm Φ 2-lgd stirrups @ 200 mm c/c throughout the length AB.

**(b) Portion BC.** $V = 46237$ N

∴ $\tau_v = \dfrac{46237}{300 \times 305} = 0.51$ N/mm² and $\dfrac{100 A_s}{bd} = \dfrac{100 (4 \times 314)}{300 \times 305} = 1.37\%$

Hence from Table 3.2, $\tau_c = 0.43$ N/mm².

∴ $V_c = 0.43 \times 300 \times 305 = 39345$ N

∴ $V_s = V - V_c = 46237 - 39345 = 6892$ N

This is too small. Hence nominal stirrups will be sufficient. Hence provide 10 mm Φ 2-lgd stirrups @ 200 mm c/c throughout the length *BC*.

### (C) DESIGN OF FULCRUM GIRDER

#### (i) Loading on the girder

The loading on the fulcrum girder will be in the form of concentrated loads transferred from the rackers. The magnitude of each concentrated load $= R_B$ (Fig. 8.19) $= 130620$ N.

Let the width of girder = 600 mm

Depth of girder = 1 m = 1000 mm

∴ Self-weight per metre run

$= 0.6 \times 1 \times 1 \times 25000 = 15000$ N

The loading diagram, S.F.D. and B.M.D. for the fulcrum girder are shown in Fig. 8.21. Assuming bearing of 800 mm on each side, the effective span of the girder will be $= 15 + 0.8 = 15.8$ m.

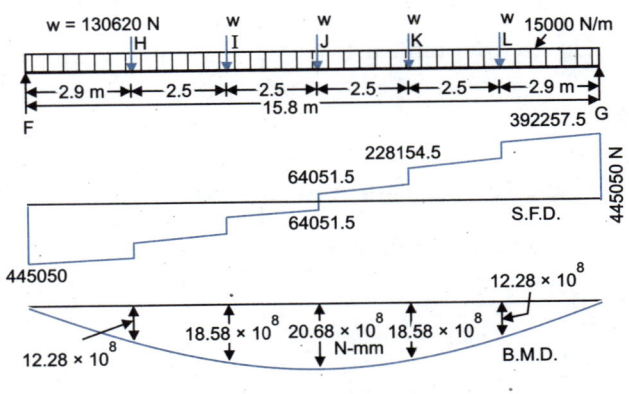

Fig. 8.21

**(ii) Bending moment and S.F.** Due to symmetrical loading, $R_F = R_G = 445050$ N

∴ $F_{max} = 445050$ N

Maximum bending moment will occur at the centre, its magnitude bending equal to

$$M = (445050 \times 7.9) - \frac{15000 (7.9)^2}{2} - 130620 (5) - 130620 (2.5)$$

$$= 20.68 \times 10^8 \text{ N-mm}$$

### (iii) Design of section at midspan.

Let us provide the main reinforcement in two layers. Using 40 mm diameter bars, 12 mm Φ rings, a nominal cover of 40 mm and vertical distance of 40 mm between the two layers, available effective depth $d = 1000 - 40 - 40 - 20 - 12 = 888$ mm

$$\therefore \quad M_r = Rbd^2 = 0.914 \times 600\,(888)^2 = 4.324 \times 10^8 \text{ N-mm}$$

This is much less than the external bending moment of $20.68 \times 10^8$ N-mm. Hence the beam will be designed as a doubly reinforced section.

Let the compression steel be also provided in two layers, so that its centre is at a distance $d_c = 85$ mm below the top fibres. Hence from the steel beam theory,

$$A_{sc} = A_{st} = \frac{M}{\sigma_{st}\,(d - d_c)} = \frac{20.68 \times 10^8}{230\,(888 - 112)} = 11587 \text{ mm}^2.$$

$\therefore$ No. of 40 mm Φ bars for each reinforcement $= 11587/1256.6 \approx 10$

Hence provide 10 bars in two tiers of 5 bars each, both near the top as well as bottom face of the beam. Provide 12 mm Φ 4-legged nominal stirrups at the spacing given by

$$s_v = \frac{2.175\,A_{sv}\,f_y}{b}, \quad \text{where} \quad A_{sv} = 4\,\frac{\pi}{4}\,(12)^2 = 452.4 \text{ cm}^2$$

$$\therefore \quad s_v = \frac{2.175 \times 452.4 \times 415}{600} = 680 \text{ mm}$$

Subject to a maximum of 450 mm. Hence provide the stirrups @ 450 mm c/c.

### (iv) Design of section at the supports.

Minimum reinforcement required at the support

$$= \frac{1}{3} \text{ of main tensile reinforcement} = 5 \text{ bars (say)}.$$

Since there is practically no bending moment at the support, the main tensile reinforcement will be provided in one layer only, and out of the compression reinforcement, only 5 bars will be retained.

**Check for shear.** $\quad V = 445050$ N ; $\quad \therefore \quad \tau_v = \frac{V}{bd} = \frac{445050}{600 \times 888} = 0.835 \text{ N/mm}^2$

$$\frac{100\,A_s}{bd} = \frac{100\,(5 \times 1256.6)}{600 \times 888} = 1.18\%.$$

Hence from table 3.1, $\tau_c \doteq 0.40$ N/mm². Hence shear reinforcement is necessary.

$$V_c = 0.40 \times 600 \times 888 = 213120 \text{ N}.$$

$$\therefore \quad V_s = V - V_c = 445050 - 213120 = 231930 \text{ N}$$

Using 12 mm Φ 4-legged stirrups, $s = \dfrac{\sigma_{sv}\,d\,A_{sv}}{V_s} = \dfrac{230 \times 888 \times 425.4}{231930} = 374.6$ mm

Maximum spacing is given by

(i) $\quad s_v = \dfrac{2.175\,A_{sv}\,f_y}{b} = \dfrac{2.175 \times 452.4 \times 415}{600} = 680$ mm

(ii) $\quad 0.75\,d = 0.75 \times 888 = 666$ mm $\quad$ or

(iii) 450 mm, whichever is less. Hence adopt $s = 350$ mm c/c at the supports. Increase this to 450 mm c/c at the mid-span.

**Check for development length.** The Code envisages that the diameter of the reinforcing bars should be such that the following equation is satisfied at the end supports:

$$1.3\,\frac{M_1}{V} + L_0 \geq L_d,$$

where $\quad M_1 = \sigma_{st}\,A_{st}\,j_c\,d = 230\,(5 \times 1256.6) \times 0.904 \times 888 = 11.6 \times 10^8$ N-mm
$\qquad V = 445050$ N

Providing a side cover of 40 mm and assuming a support width 800 mm, and taking bars straight without any hook or bend,

$$L_0 = \frac{l_s}{2} - x' = \frac{800}{2} - 40 = 360 \text{ mm}.$$

Hence, $\quad 1.3 \dfrac{M_1}{V} + L_0 = 1.3 \times \dfrac{11.6 \times 10^8}{445050} + 360 = 3388 + 360 = 3748$ mm

$$L_d = 45 \Phi = 45 \times 40 = 1800 \text{ mm}.$$

$\therefore \quad 1.3 \dfrac{M_1}{V} + L_0 > L_d$. Hence the Code requirements are satisfied.

**(v) Design of section at 2.9 m from the support**

$$M = \left[ (445050 \times 2.9) - \frac{15000 \, (2.9)^2}{2} \right] \times 1000$$

$$= 12.28 \times 10^8 \text{ N-mm}$$

$$V = 445050 - (15000 \times 2.9) = 401550 \text{ N}$$

Hence from the steel beam theory,

$$A_{st} = A_{sc} = \frac{12.28 \times 10^8}{230 \, (888 - 112)} = 6880 \text{ mm}^2$$

No. of 40 mm $\Phi$ bars required = 6880/1256.6 = 5.5

Hence provide 6 bars as follows : for tensile steel, provide 5 bars in the bottom layer and 1 bar in the second layer (*i.e.*, rest of the 4 bars of the second layer may be discontinued); for compressive steel also, provide 5 bars in the top layer and 1 bar in the second layer.

$$\tau_v = \frac{V}{bd} = \frac{401550}{600 \times 888} = 0.75 \text{ N/mm}^2$$

$$\frac{100 \, A_s}{bd} = \frac{100 \, (5 \times 1256.6)}{600 \times 888} = 1.18\%$$

Hence from Table 3.1, $\tau_c = 0.405$ N/mm$^2$. Hence shear reinforcement is necessary.
Use 12 mm $\Phi$ 4-lgd stirrups having $A_{sv} = 452.4$ mm$^2$

$$V_c = 0.405 \times 600 \times 888 = 215784 \text{ N}$$

$\therefore \quad V_s = V - V_c = 401550 - 215784 = 185766$ N

$\therefore \quad s = \dfrac{\sigma_{st} d \, A_{sv}}{V_s} = \dfrac{230 \times 888 \times 452.4}{185766} = 497$ mm, $\quad$ (subject to max. of 450 mm)

Hence provide the stirrups @ 360 mm c/c.

**(vi) Design of section at 5.4 m from the support.** B.M. at $K$ is given by

$$M = \left[ (445050 \times 5.4) - \frac{15000 \, (5.4)^2}{2} - 130620 \, (2.5) \right] \times 1000$$

$$= 18.58 \times 10^8 \text{ N-mm}$$

$$V = 445050 - (15000 \times 5.4) - 130620 = 233430 \text{ N}$$

Hence from steel beam theory:

$$A_{st} = A_{sv} = \frac{18.58 \times 10^8}{230 \, (888 - 112)} = 10410 \text{ mm}^2$$

$\therefore$ No. of 40 mm bars required = 10410/1256.6 $\approx$ 9

Hence provide 5 bars in the first layer and 4 bars in the second layer.

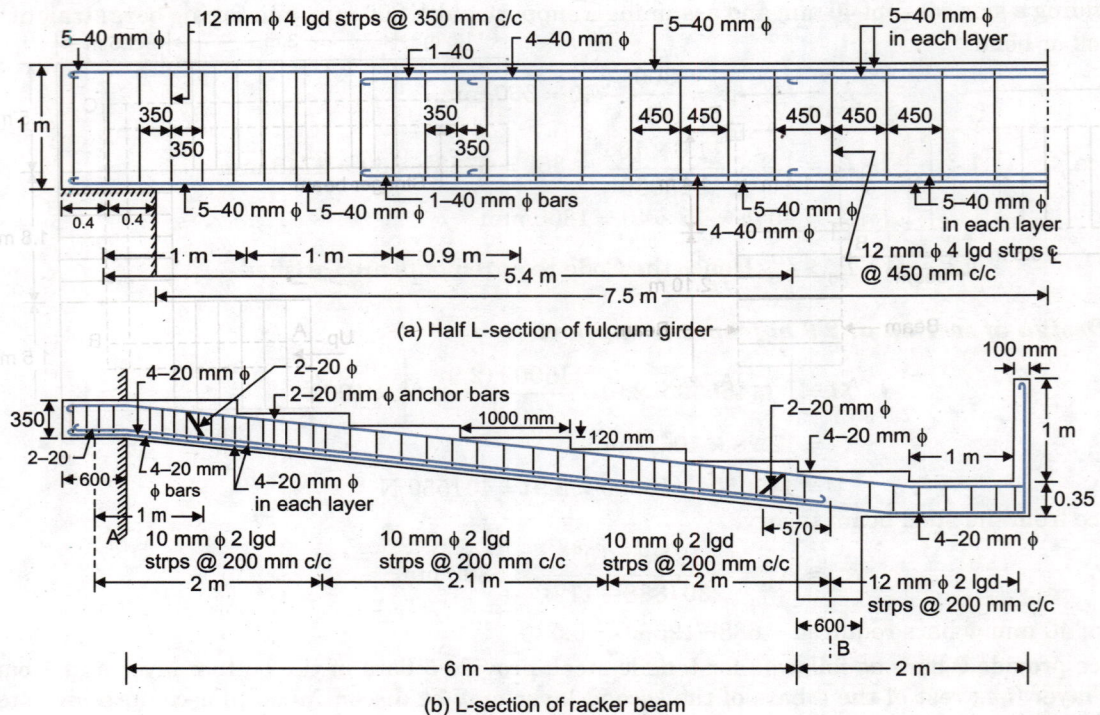

(a) Half L-section of fulcrum girder

(b) L-section of racker beam

Fig. 8.22

$$\tau_v = \frac{V}{bd} = \frac{233430}{600 \times 888} = 0.44 \text{ N/mm}^2.$$

Hence only nominal shear reinforcement is necessary. Provide 12 mm Φ 4-legged stirrups at a nominal spacing (maximum spacing) of 450 mm c/c.

The details of reinforcement etc. are shown in Fig. 8.22.

## PROBLEMS

1. Design the stairs for a public building, supported on wall on one side and stringer beam on the other side. The horizontal span of stairs if 1.4 m. The risers are 120 mm and tread are 300 mm. Use M 20 mix. and Fe 415 steel.

2. Design a suitable dog-legged stair in a public building, to be located in a staircase 6 metre long, 3.2 m wide and 3.7 m high, with a door of 1.1 m wide in each of the longitudinal walls. The doors face each other and are located with their centres at a distance of 0.9 metres from the respective corners of the staircase. Use M 20 mix. and Fe 415 steel.

3. A two storeyed building is to have a R.C. staircase from ground floor to first floor roof. The size of the staircase is 4.3 m × 4.3 m and there is one door opening in one wall and a window opening on the opposite wall. Design the staircase, giving the details of formation, R.C. slab arrangement of building, risers and treads with their top finishing with suitable sketch. The width of stair is 1.2 m and height of each storey is 3.4 m. Use M 20 mix and Fe 415 steel.

4. Fig. 8.23 shows the arrangement of a staircase for an office building. The rise of steps is 150 mm. Design completely the stair alongwith the beams, landing etc. The live load may be taken as 4000 N/m². Use M 20 mix. and Fe 415 steel.

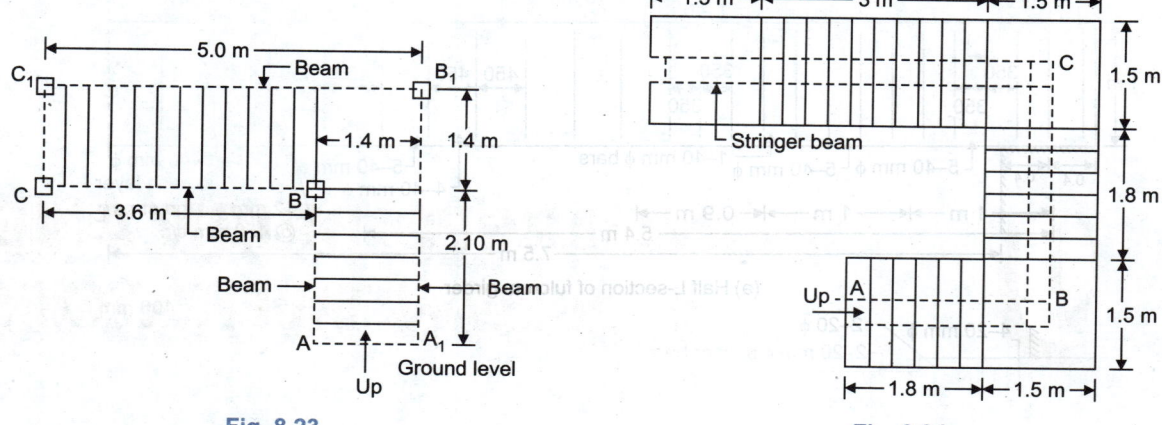

Fig. 8.23  Fig. 8.24

5. Fig. 8.24 shows the plan of stair hall for a public building. Design staircase with the central stringer beam. The tread is 300 mm and rise 150 mm. The live load may be taken as 4000 N/m². Use M 20 mix. and Fe 415 steel.

6. Explain codal recommendations for the effective span of the stair slabs with different support conditions of landings.

7. Design a flight between landing to landing of tread riser type of staircase with 10 risers each 150 mm and with tread of 270 mm the upper and lower landings are 1200 mm wide each supported on 230 mm thick masonery wall at the edges parallel to the risers. The stairs are liable to overcrowded. The materials to be used for construction are $M_{20}$ grade complete and HYSD bars of grade Fe 415.

8. Design and detail a dog-legged staircase located in stair hall measuring 2.75 × 4.50 m for a public building. The storey or floor-to-floor height is 3.30 m. The service live load on stairs is 4.2 kN/m² with the weight of finishes of 0.57 kN/m². Adopt riser and tread of 150 mm and 250 mm respectively. The grades of concrete and steel to be used are $M_{20}$ and Fe 415 respectively.

# CHAPTER 9
# REINFORCED BRICK AND HOLLOW TILE ROOFS

## 9.1. REINFORCED BRICK WORK

Reinforced brickwork is a typical type of construction in which the compressive strength of bricks is utilised to bear the compressive stress and steel bars are used to bear the tensile stresses in a slab. In other words, the usual cement concrete is replaced by the bricks. However, since the size of a bricks is limited, continuity in the slab is obtained by filling the joints between the bricks by cement mortar. The reinforcing bars are embedded in the gap between the bricks, which is filled with cement mortar. Such type of construction is quite suitable and cheap for small span floor slabs carrying comparatively lighter loads. Fig. 9.1 shows a typical section of reinforced brick slab.

The depth of reinforced brick roof is governed by the thickness of the bricks available. Normally, bricks are 100 mm thick, inclusive of plaster to be used when the brick masonry is constructed. Hence thickness of slab may be kept as 100 mm or 200 mm. If 150 mm thickness is required from design point of view, 50 mm thick tiles are used on the 100 mm thick bricks to make a total thickness of 150 mm. The joint between the two layers of tile and bricks is filled with cement mortar. Before use, the bricks should be thoroughly soaked in water. The reinforcing bars put in the joint should not come in contact with bricks.

When two layers of bricks are used, vertical joints in the bricks should be broken so that slab does not shear along the joint. The bricks near the edge should rest half on the bearing wall so that vertical joint above the edge of the wall is avoided. First class bricks should be used for such a work. Cement mortar used to fill the joints etc. should be of 1 : 3 ratio (*i.e.*, 1 part cement and 3 parts of clean coarse sand), with proper water cement ratio, to make the mortar workable. The

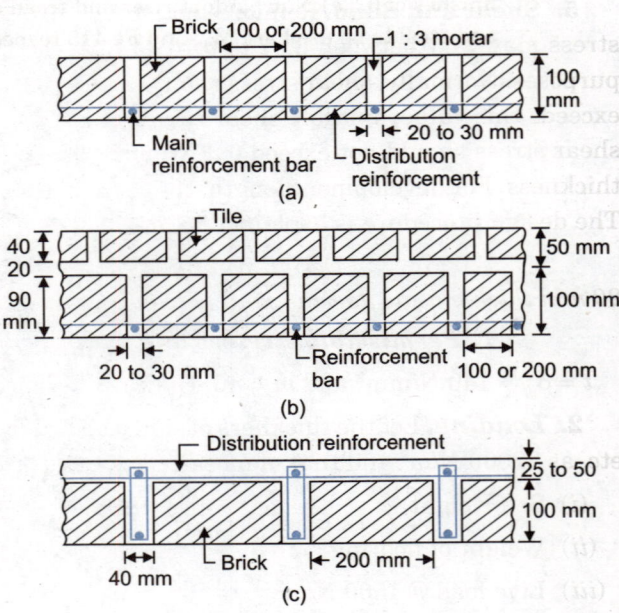

Fig. 9.1. Reinforced Brick Work.

width of joint between adjacent bricks is generally kept equal to 20 mm. The compressive strength of reinforced brickwork is sometimes increased by providing wider gap (say about 40 mm) between the bricks, and providing 25 to 50 mm thick layer of cement concrete on the top of the bricks, as shown in Fig. 9.1 (c). The construction then becomes similar to *hollow tile roof*, the design principles of which are discussed in § 9.3.

## 9.2. DESIGN OF REINFORCED BRICKS SLAB

The principles of design of reinforced brick slab are similar to those of reinforced concrete slab. The following points should be kept in mind while designing such a slab.

**1. Permissible Stresses and Design Constants.** The permissible value of compressive stress in brick depends upon the quality of brick as well as that of mortar used to fill the gap. The usual range of permissible stress may be between 2 N/mm² to 3 N/mm². In absence of suitable data, a permissible value of $c = \sigma_{cbc} = 2.5$ N/mm² may be taken. For mild steel, usual value of $t = \sigma_{st} = 140$ N/mm² may be utilised. The modular ratio $m$ is usually taken as 40. Thus adopting $c = 2.5$ N/mm², $t = 140$ N/mm² and $m = 40$, we have the following design constants:

$$k_c = \frac{mc}{mc + t} = \frac{40 \times 2.5}{40 \times 2.5 + 140} \approx 0.42 \; ; \quad j_c = 1 - k_c/3 = 1 - 0.42/3 = 0.86$$

$$R_c = \frac{1}{2} c j_c k_c = \frac{1}{2} \times 2.5 \times 0.86 \times 0.42 \approx 0.45$$

**2. Steel Reinforcement.** The bars used for tensile reinforcement should be of as small diameter as possible, and should not preferably exceed 12 mm. These should be properly embedded in the cement mortar filled in the gap between the bricks. Half the bars are usually bent upwards at 1/7 of the span from the ends.

**3. Clear Cover.** The clear cover should be equal to 15 mm or the diameter of the bar whichever is more.

**4. Distribution and Temperature Reinforcement.** The distribution reinforcement is provided @ 0.15% of the area of section. Sometimes, it is provided @ 20% of the area of the main reinforcement corresponding to maximum bending moment.

**5. Shear and Bond.** Reinforced brick work is vary weak against diagonal tension. The permissible shear stress may vary between 0.07 to 0.14 N/mm². An average value of 0.1 N/mm² may be adopted for design purposes. Normally, shear stress should not exceed this value. In exceptional cases, if the diagonal tension exceeds this value, suitable shear reinforcement should be provided. In such cases, however, the value of shear stress should not exceed 0.3 N/mm², otherwise the section should be redesigned, providing greater thickness. For development length, the same rules may be adopted as followed for reinforced concrete slab. The design procedure is explained in design example 9.1.

**Design: Example 9.1.** *Design a reinforced brick slab for a room 3 metres wide. The slab is paved with 20 mm thick mosaic flooring, and is to be designed to carry a live load 1500 N/m².*

**Solution: 1. Permissible stresses and design constants.** Let $c = \sigma_{cbc} = 2.5$ N/mm² ; $t = \sigma_{st} = 140$ N/mm² and $m = 40$. Hence $k_c = 0.42$ ; $j_c = 0.86$ and $R_c = 0.45$.

**2. Loading.** Let the thickness of slab be 150 mm. Assuming unit weight of brickwork (including) mortar etc. as 22000 N/m³ and that of mosaic as 23000 N/m³, the load on the slab per sq. metre will be as follows:

| | | |
|---|---|---|
| (i) Self weight | = 0.15 × 1 × 1 × 22000 | = 3300 N |
| (ii) Weight of flooring | = 0.02 × 1 × 23000 | = 460 N |
| (iii) Live load @ 1500 N/m² | | = 1500 N |
| | | 5260 N/m² |

**3. Effective span and bending moment.** Assuming effective depth = 130 mm, effective span = 3 + 0.13 = 3.13 m.

∴
$$M = \frac{w L^2}{8} = \frac{5260 \, (3.13)^2}{8} \times 1000 = 6.44 \times 10^6 \text{ N-mm}$$

**4. Effective depth.** $d = \sqrt{\dfrac{M}{R_c b}} = \sqrt{\dfrac{6.44 \times 10^6}{0.45 \times 1000}} = 120$ mm

However, since total depth is governed by the thickness of brick let us keep the total thickness equal to 150 mm. Using 10 mm diameter bars and 15 mm clear cover, available effective depth will be equal to 150 – 5 – 15 = 130 mm. As per IS 1077, the working size of bricks is 200 mm × 100 mm × 100 mm and that of tiles 200 mm × 100 mm × 50 mm, the modular dimensions being 10 mm less in each direction. Hence we will use one tile over one brick to get a total thickness of 150 mm. The horizontal joint between the two layers will evidently be equal to 20 mm (since the modular thickness of brick is 90 mm and that of tile is 40 mm). Leave 30 mm gap between adjacent tiles. However, both the joints should not be in same vertical line.

**5. Main reinforcement.** $A_{st} = \dfrac{M}{\sigma_{st} j_c d}$

Since the effective depth provided is slightly more than that required for the balanced design, the section will not be balanced one. However, we will use $j_c = 0.86$.

$\therefore \quad A_{st} = \dfrac{6.44 \times 10^6}{140 \times 0.86 \times 130} = 411.4$ mm². Using 8 mm Φ bars, $A_\Phi = \dfrac{\pi}{4}(8)^2 = 50.3$ mm²

$\therefore \quad s = \dfrac{1000\, A_\Phi}{A_{st}} = \dfrac{1000 \times 50.3}{411.4} = 122$ mm

Using 30 mm gap and noting that the modular width of brick is 90 mm, centre to centre spacing of bars will be 90 + 30 = 120 mm. Bend half the bars up at a distance of L/7 = 3130/7 = 450 mm from the edge of the support.

**6. Distribution reinforcement :** $A_{sd} = 0.15\%$ of area $= \dfrac{0.15}{100} \times 1000 \times 150 = 225$ mm²

Using 8 mm Φ bars. $s = \dfrac{1000\, A_\Phi}{A_{sd}} = \dfrac{1000 \times 50.3}{225} = 223$ mm

Using 3 mm gap and noting that modular length of brick is 190 mm, the centre to centre spacing of the bars will be 220 mm.

**7. Check for shear**

$V = \dfrac{wl}{2} = \dfrac{5260 \times 3}{2} = 7890$ N ; $\tau_v = \dfrac{V}{bd} = \dfrac{7890}{1000 \times 130} = 0.06$ N/mm² (safe)

**8. Check for development length.** At the supports, $1.3 \dfrac{M_1}{V} + L_0 \geq L_d$

Here, $M_1 = \sigma_{st} A_{st} j_c d = 140 \left[ \dfrac{1000 \times 50.3}{2 \times 120} \right] \times 0.86 \times 130 = 3.28 \times 10^6$ N-mm

$V = 7890$ N.; $L_0 = \dfrac{l_s}{2} - x' + 13\Phi$

Using $l_s = 200$ mm and $x' = 20$ mm.

$L_0 = \dfrac{200}{2} - 20 + 13 \times 8$

$= 184$ mm.

$\therefore \quad 1.3 \dfrac{M_1}{V} + L_0 = 1.3 \times \dfrac{3.28 \times 10^6}{7890} + 184$

$= 540 + 184 = 724$ mm.

$L_d \approx 58.3\, \Phi = 58.3 \times 8 = 466.4$ mm.

Hence Code requirements are satisfied.

The details of reinforcement etc. are shown in Fig. 9.2.

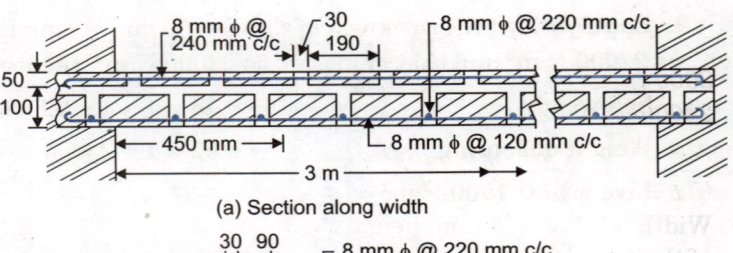
(a) Section along width

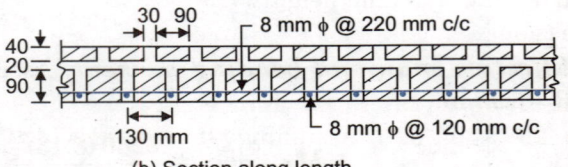
(b) Section along length

Fig. 9.2. Reinforced Brick Roof.

## 9.3. HOLLOW TILE ROOF

We have seen earlier that the area of concrete below the N.A. in a rectangular beam does not carry any appreciable moment of resistance and is neglected. From the point of view of economy, and also for reducing the self-weight, we reduce the width of the beam below the N.A., thus giving rise to T-beam. However, from architectural or hygienic considerations, sometimes plane ceiling is required. In that case, the space between the ribs of adjacent T-beams may be filled with hollow blocks or tiles of light weight concrete and ceramics of light fillers of wood, cane etc. Such a type of floor is known as *hollow tile floor (roof) or close-ribbed floor*. Some of the forms of hollow tile floors and close-ribbed floors are shown in Fig. 9.3.

Unlike T-beam floor construction, the ribs of hollow tile construction are closely spaced. The clear spacing between the ribs depends upon the size of hollow blocks available but it should normally not exceed 500 mm. The width of ribs may vary between 60 to 100 mm. The span of the ribs may be as much as 7 metres. However, when the span exceeds 5 metres, lateral ribs of the same width as the main longitudinal ones are provided at intervals between 1 to 3 metres. In that case, the main longitudinal ribs are designed as a continuous beam. Main reinforcement is provided at the bottom of the rib. To resist support moment (negative), an additional bar is placed at the top of rib section. A minimum cover of 25 mm is provided. Sometimes, ribs are reinforced with welded mats, usually one to a rib. The depth of rib is calculated on the basis of bending moment as well as on cost ratio of steel and concrete. Depth of ribs is usually taken as at least $L/20$ with free support and at least $L/25$ with fixed support.

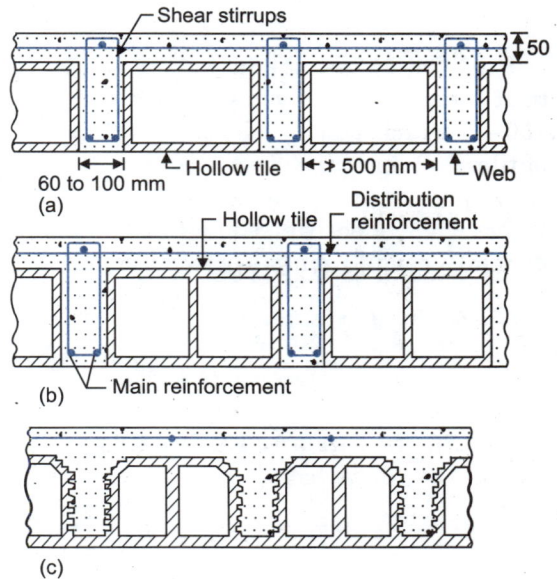

Fig. 9.3. Hollow Tile and Close-Ribbed Floors.

Due to small span, the slab is normally not analysed. Slab thickness of 40 to 50 mm is generally provided. To check its cracking and to distribute the load properly, shrinkage and temperature reinforcement is provided in the slab, in both the directions. Sometimes, a welded fabric is arranged approximately along the middle of the thickness of the slab.

Hollow tiles are available in different widths and different depths. Sometimes, to suit the requirements of the depth of rib hollow tiles of required depth may be manufactured at the site. Various forms and types of hollow tiles are available, to suit the clear distance between the ribs.

## 9.4. DESIGN OF HOLLOW TILE ROOF

**Design: Example 9.2.** *Design a hollow tile roof for a room 4 m wide. The live load may be taken as 2000 N/m². The roof supports lime concrete terracing 120 mm thick. Assume unit weight of concrete as 24000 N/m³.*

**Solution: 1. Design constants.** Let us use the following values:

$t = \sigma_{st} = 140$ N/mm² ; $c = \sigma_{cbc} = 5$ N/mm² ; $m = 18$. Unit weight of lime terracing = 19500 N/m³. Width of tile = 225 mm, height = 200 mm and thickness = 13 mm. Weight of tile = 220 N per metre length of the tile.

**2. Loading and bending moment.** Let us provided slab thickness of 50 mm and the width of rib equal to 80 mm. Hence the centre to centre spacing of the ribs will be equal to 225 + 80 = 305 mm. Assuming the effective depth of rib equal to 230 mm, the effective span of the rib = 4 + 0.23 = 4.23 m. The total depth of floor will be equal to 200 + 50 = 250 mm. The loading on the rib, pre metre run will be as follows:

1. Wt. of slab $= \dfrac{305 \times 50}{10^6} \times 24000 \qquad = 366$ N

2. Wt. of rib $= \dfrac{80 \times 200}{10^6} \times 24000 \qquad = 384$ N

3. Terracing $= \dfrac{305 \times 120}{10^6} \times 19500 \qquad = 713.7$ N

4. Live load $= \dfrac{305}{1000} \times 1 \times 2000 \qquad = 610$ N

5. Wt. of tiles @ 220 N/m $\hspace{5cm} = 220$ N

$$\text{Total} = 2293.7 \text{ N} \approx 2300 \text{ N}$$

$$\therefore \quad M = \dfrac{2300\,(4.23)^2}{8} \times 1000 = 5.14 \times 10^6 \text{ N-mm}$$

**3. Effective depth.** The beam is to be designed as a T-beam. Assuming the cost ratio of steel and concrete as 70, we have,

$$d = \sqrt{\dfrac{r\,M}{\sigma_{st}\,j_c\,b_w}} \quad (\text{Taking } j_c \approx 0.9) = \sqrt{\dfrac{70 \times 5.14 \times 10^6}{140 \times 0.9 \times 80}} = 190 \text{ mm}.$$

However, since the height of tiles = 200 mm, available total depth = 200 + 50 = 250 mm and hence available effective depth = 250 − 25 − 8 ≈ 217 mm, using 16 mm dia. bars. Thus for the T-beam section, $b_f = 305$ mm, $D_f = 50$ mm, $b_w = 80$ mm and $d = 217$ mm.

**4. Steel reinforcement.** $A_{st} = \dfrac{M}{\sigma_{st}\,jd} = \dfrac{5.14 \times 10^6}{140 \times 0.9 \times 217} = 188$ mm$^2$

Using 16 mm Φ bars, $A_\Phi = \dfrac{\pi}{4}(16)^2 = 201$ mm$^2$. Hence provide one bar per rib.

**5. Check for stresses.** To locate the N.A. we have

$$b_f D_f \left[ n - \dfrac{D_f}{2} \right] = m\,A_{st}\,(d - n) \quad \text{or} \quad 305 \times 50 \left[ n - \dfrac{50}{2} \right] = 18 \times 201\,(217 - n)$$

From which $n = 618$ mm. The stress $c_1$ at the bottom of the slab is given by

$$c_1 = \dfrac{n - D_f}{n}\,c = \dfrac{618 - 50}{618}\,c = 0.19\,c$$

The distance $\bar{y}$ of the centroid of compressive force is given by

$$\bar{y} = \left( \dfrac{c + 2c_1}{c + c_1} \right) \dfrac{D_f}{3} = \left( \dfrac{10 + 38}{10 + 19} \right) \dfrac{50}{3} = 27.5 \text{ mm}$$

$$\therefore \quad \text{L.A.} = a = d - \bar{y} = 217 - 27.5 = 189.5 \text{ mm}$$

$$\therefore \quad M_r = b_f \cdot D_f \cdot \dfrac{c + c_1}{2} \times a$$

or $\quad 5.14 \times 10^6 = 305 \times 50 \left( \dfrac{1 + 0.19}{2} \right) c \times 189.5$

From which $c = 2.99$ N/mm$^2$ (safe). The corresponding stress in steel is given by:

$$t = \dfrac{mc\,(d - n)}{2} = \dfrac{18 \times 2.99\,(217 - 61.8)}{61.8} = 135.1 \text{ N/mm}^2 \text{ (safe)}$$

## 6. Check for shear

$$V = \frac{wl}{2} = \frac{2300 \times 4}{2} = 4600 \text{ N};$$

$$\tau_v = \frac{V}{bd} = \frac{4600}{80 \times 217} = 0.26 \text{ N/mm}^2. \text{ (safe)}$$

Hence provide nominal shear stirrups, given by

$$s_v = \frac{2.175 \, A_{sv} \cdot f_y}{b}. \text{ Using 6 mm } \Phi \text{ bars,}$$

$$A_{st} = 2 \times \frac{\pi}{4} \, 6^2 = 56.5 \text{ mm}^2$$

$$\therefore \quad s_v = \frac{2.175 \times 56.5 \times 250}{80} = 384 \text{ mm}$$

subject to a max. of $0.75 \, d = 0.75 \times 217 = 163$ mm

Hence provide the nominal stirrups @ 160 mm c/c.

## 7. Temperature and distribution reinforcement

Provide nominal reinforcement in the form of 6 mm $\Phi$ bars @ 305 mm c/c, with a clear cover of 20 mm from the top of the slab, in both the directions. The details of the reinforcement etc. are shown in Fig. 9.4.

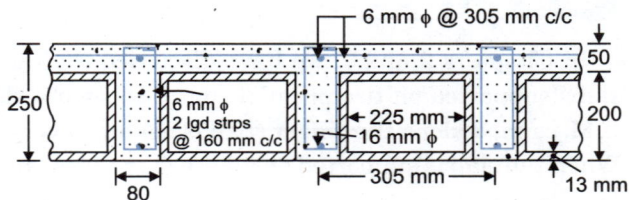

Fig. 9.4

# CHAPTER 10: TWO-WAY SLABS

## 10.1. INTRODUCTION

In Chapters 2 and 7, we have discussed the theory and design of slabs spanning in one direction, *i.e.* slabs supported on two parallel edges. In such a slab, bending takes place along the short span, and the B.M. carried along the longer span is usually a very small fraction of the B.M. carried along the short span. When the slab is supported on all the four edges, and when the ratio of long span to short span is small (say less than 2), bending takes place along both the spans. Such a slab is known as a two-way slab or a slab spanning in two directions. The maximum bending moment and deflection for such a slab is much smaller than that of a one-way slab and hence a thinner slab is required. However, reinforcement has to be provided in both the directions. When such a slab is loaded, the corners get lifted up. If the corners are held down, by fixity at the wall support etc., the bending moment and deflection are further reduced, thus requiring still thinner slab. In that case, special torsional reinforcement at the corner has to be provided to check the cracking of corners. We will divide the two-way slabs in the following three heads:

1. Slabs simply supported on the four edges, with corners not held down, and carrying uniformly distributed load.
2. Slabs simply supported on the four edges, with corners held down and carrying uniformly distributed load.
3. Slabs with edges fixed or continuous and carrying uniformly distributed load.

## 10.2. SLAB SIMPLY SUPPORTED ON THE FOUR EDGES, WITH CORNERS NOT HELD DOWN AND CARRYING U.D.L.

Such slabs are commonly used in single storey buildings, where the supporting walls do not cause any fixity. When such a slab is loaded to failure, a crack pattern at the bottom of slab, shown in Fig. 10.1 is observed. As evident from the crack pattern, the bending takes place in the direction $cd$ for the portion $EF$ of the slab, while near the corners (say $A$) bending takes place along $ab$, perpendicular to the crack direction $AE$. However, the analysis of such a slab done on the basis of directions of cracks is tedious. Grashoff-Rankine gave the following simplified method.

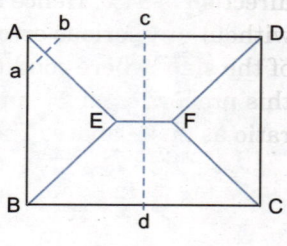

Fig. 10.1

## Grashoff-Rankine Method

Let  $L$ = Long span of a slab ; $B$ = Short span of slab

$w$ = Uniformly distributed load per unit area.

In this method, the bending moments carried in the long and short directions are determined on the basis of maximum deflection at the centre of the slab. Consider two middle strips $EF$ and $GH$ of unit width, parallel to long and short spans respectively.

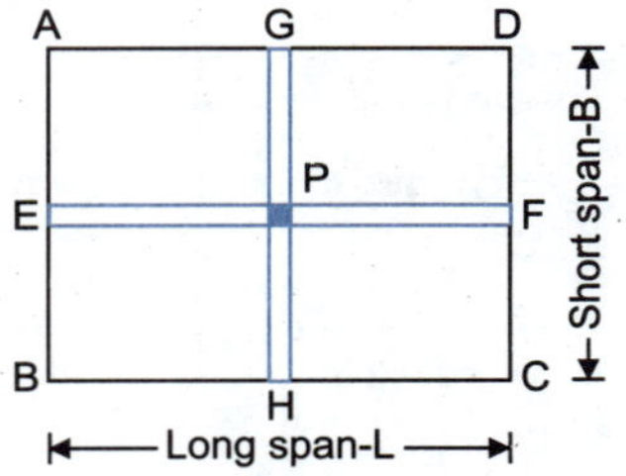

Let $w_L$ = Load carried by the slab in long direction $L$

$w_B$ = Load carried by the slab in short direction $B$.

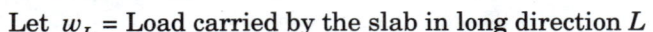

Fig. 10.2

$$\therefore \qquad w = w_L + w_B \qquad \ldots(1)$$

Consider the deflection of point $P$, at the intersection of both the strips.

For strip $EF$, $\quad \delta_P = \dfrac{5}{384} \cdot \dfrac{w_L L^4}{EI}$ ; For strip $GH$, $\delta_P = \dfrac{5}{384} \cdot \dfrac{w_B B^4}{EI}$

Equating the two, we get $\quad \dfrac{5}{384} \cdot \dfrac{w_L L^4}{EI} = \dfrac{5}{384} \cdot \dfrac{w_B B^4}{EI}$

or $\qquad\qquad\qquad\qquad w_L = w_B (B/L)^4 \qquad \ldots(2)$

Substituting in (1), we get $\quad w = w_B \left(\dfrac{B}{L}\right)^4 + w_B$

From which, $\qquad\qquad w_B = \dfrac{w}{1 + \left(\dfrac{B}{L}\right)^4} \qquad \ldots[10.1\,(a)]$

Hence, $\qquad\qquad w_L = \dfrac{w}{1 + \left(\dfrac{B}{L}\right)^4} \left(\dfrac{B}{L}\right)^4 = \dfrac{w}{1 + \left(\dfrac{L}{B}\right)^4} \qquad \ldots[10.2\,(b)]$

Since $L/B > 1$, we note that $w_B > w_L$. Putting $L/B = r$, we have

$$w_L = \dfrac{w}{1 + r^4} = r_L\, w \qquad \ldots(10.1)$$

and

$$w_B = \dfrac{w\, r^4}{1 + r^4} = r_B\, w \qquad \ldots(10.2)$$

where $r_L = \dfrac{1}{1 + r^4}$ and $r_B = \dfrac{r^4}{1 + r^4}$, and depend upon the value of $L/B$ ratio.

Table 10.1 gives the values of $r_L$ and $r_B$ for various values of $r$. From the Table it will be seen that when $L/B = r$ exceeds 2, the load carried in the longer direction is only 6% while that carried in the shorter direction is 94%. Hence for the slabs in which $L/B > 2$, one may design it as a slab supported on two edges, without any serious error. It should be clearly noted that the above relations hold good only at the middle of the slab, where conditions of maximum B.M. and maximum deflection exist. However, at other points, this proportion of $w_B$ and $w_L$ does not exist, and the design is done by approximately assuming the same ratio as given above. The maximum bending moments in the two directions are:

$$M_L = w_L \dfrac{L^2}{8} = \dfrac{w}{1 + r^4} \dfrac{L^2}{8} = w\, r_L \dfrac{L^2}{8} \qquad \ldots(10.3)$$

$$M_B = w_B \dfrac{B^2}{8} = \dfrac{w\, r^4}{1 + r^4} \dfrac{B^2}{8} = w\, r_B \dfrac{B^2}{8} \qquad \ldots(10.4)$$

**TABLE 10.1.** Values of $r_B$ and $r_L$

| $r = L/B$ | $r_B$ | $r_L$ | $r = L/B$ | $r_B$ | $r_L$ |
|---|---|---|---|---|---|
| 1.00 | 0.50 | 0.50 | 1.40 | 0.79 | 0.21 |
| 1.05 | 0.55 | 0.45 | 1.50 | 0.84 | 0.16 |
| 1.10 | 0.59 | 0.41 | 1.60 | 0.87 | 0.13 |
| 1.15 | 0.64 | 0.36 | 1.75 | 0.90 | 0.10 |
| 1.20 | 0.68 | 0.32 | 2.00 | 0.94 | 0.06 |
| 1.25 | 0.71 | 0.29 | 2.50 | 0.97 | 0.03 |
| 1.30 | 0.74 | 0.26 | 3.00 | 0.98 | 0.02 |

Normally, the bending moment in the short span will be more than the bending moment on the long span, and hence the slab thickness will be determined on the basis of $M_B$. However, reinforcement in the short span is placed *below* the reinforcement in the long span. As per IS 456 : 2000 minimum total thickness of slab should satisfy the stiffness requirements. The area of steel in both the directions are calculated as follows:

$$(A_{st})_B = \frac{M_B}{\sigma_{st} \, j_c d} \quad \text{and} \quad (A_{st})_L = \frac{M_L}{\sigma_{st} \, j_c \, (d - \Phi_B)}$$

where  $d$ = effective depth slab determined corresponding to $M_B$.

$\Phi_B$ = diameter of bars in short span.

$\sigma_{st}$ = permissible tensile stress in steel.

Alternate bars can be bent up at 1/7th span from the centre of the supports, in each direction. If both the spans are equal (for square slab), the effective depth should be taken up to the centre of the top layers of reinforcement.

**Shear force :** The relations for $w_L$ and $w_B$ developed above give reasonably good results from bending moment point of view only. The reactions at the supports obtained by the above values of $w_L$ and $w_B$ differ very much from those observed experimentally. This is because near the corners, the slab spans in a direction $ab$ at right angles at the 45° lines $AE$, $EB$ etc. It has been observed that almost the entire load of the triangle $AEB$ is transferred to the short support $AB$, while the load in trapezium $AEFD$ is transferred to the long support $AD$.

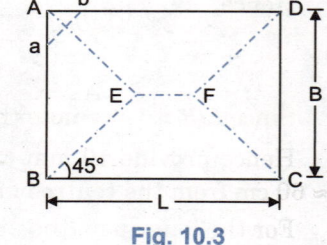

Fig. 10.3

Thus total reaction on the edge $AB$ will be equal to $\frac{1}{2} wB.B/2 = wB^2/4$. The *average* reaction per unit width, along $AB$, will therefore be $w B/4$. However, the *maximum* reactions, per unit width along $AB$ will occur near the centre of $AB$, and its value may be taken as $w B/3$ for all practical purposes.

Similarly, the S.F. along the long span $AD$ or $BC$, per unit width may be taken as $w.B. \frac{w}{2+r}$, subject to a maximum of $0.5 \, w.B.$ when $r$ exceeds 2.

**Design: Example 10.1.** *Design a R.C. slab for a room measuring 4 m × 5 m from inside. The slab carries a live load of 2000 N/m², and is finished with 20 mm thick granolithic topping. Take $c = \sigma_{cbc} = 5 \, N/mm^2$, $t = \sigma_{st} = 140 \, N/mm^2$ and $m = 19$. The slab is simply-supported at all the four edges, with corners free to lift.*

**Solution:**

1. **Design constants.** For the given set of stresses, we have

$$k_c = 0.404 \, ; \quad j_c = 0.865 \quad \text{and} \quad R_c = 0.874$$

2. **Loading and bending moments.** Let the thickness of the slab be 120 mm for the purpose of calculating the self-weight. The load per square metre of the slab will be as under:

| 1. Self weight | $= 0.12 \times 1 \times 1 \times 25000$ | $= 3000$ N |
|---|---|---|
| 2. Wt. of flooring | $= 0.02 \times 1 \times 1 \times 24000$ | $= 480$ N |
| 3. Live load @ 2000 N/m² | | $= 2000$ N |
| | Total | $= 5480$ N |

Taking effective depth $\approx 100$ mm, we have

$$L = 5 + 0.1 = 5.1 \text{ m} \quad \text{and} \quad B = 4 + 0.1 = 4.1 \text{ m}$$

$$r = \frac{L}{B} = \frac{5.1}{4.1} = 1.245 \; ; \; r^4 = 2.4$$

Hence $w_B = w \dfrac{r^4}{1+r^4} = 5480 \dfrac{2.4}{1+2.4} = 3868$ N ; $w_L = w \dfrac{1}{1+r^4} = 5480 \dfrac{1}{1+2.4} = 1612$ N

$$\therefore \quad M_B = w_B \frac{B^2}{8} = 3868 \frac{(4.1)^2}{8} \times 1000 = 8.13 \times 10^6 \text{ N-mm.}$$

$$M_L = w_L \frac{L^2}{8} = 1612 \frac{(5.1)^2}{8} \times 1000 = 5.24 \times 10^6 \text{ N-mm.}$$

**3. Design of section.** The effective depth of slab will be found on the basis of greater moment $M_B$.

$$\therefore \quad d = \sqrt{\frac{M_B}{R_c \, b}} = \sqrt{\frac{8.13 \times 10^6}{0.874 \times 1000}} = 97 \text{ mm. Adopt a total thickness} = 120 \text{ mm.}$$

Using 10 mm Φ bars and a nominal cover of 15 mm, we get $d = 120 - 15 - 5 = 100$ mm for short span. The effective depth for long span $= 100 - 10 = 90$ mm, since the reinforcement in long span is placed *above* the reinforcement in short span.

$$(A_{st})_B = \frac{M_B}{\sigma_{st} \, j_c d} = \frac{8.13 \times 10^6}{140 \times 0.865 \times 100} = 671.3 \text{ mm}^3. \text{ Using 10 mm Φ bars}$$

$$A_\Phi = \frac{\pi}{4}(10)^2 = 78.5 \text{ mm}^2. \text{ Spacing } s_B = \frac{1000 \, A_\Phi}{A_{st}} = \frac{1000 \times 78.5}{671.3} = 117 \text{ mm.}$$

Hence provide 10 mm Φ bars @ 100 mm c/c along short span. Bend-up alternate bars at $B/7 = 4.1/7 \approx 60$ cm from the centre of each support.

For the long span, the area of steel is given by

$$(A_{st})_L = \frac{5.24 \times 10^6}{140 \times 0.865 \times 90} = 481 \text{ mm}^2$$

Hence spacing of 10 mm Φ bars is given by

$$s_L = \frac{1000 \times 78.5}{481} = 163 \text{ mm.}$$

However, provide 10 mm Φ @ 150 mm c/c.

Bend up alternate bars at $L/7 = 5100/7 \approx 700$ mm from the centre of each support.

**4. Check for shear.** S.F. on the short edge per unit length is given by

$$V_B = \frac{1}{3} w \cdot B = \frac{1}{3} \times 5480 \times 4.1 = 7489 \text{ N.}$$

S.F. on the long edge, per unit length is :

$$V_L = w \, B \, \frac{r}{2+r} = 5480 \times 4.1 \, \frac{1.245}{2+1.245} = 8620 \text{ N}$$

Note carefully that the effective depth of slab at short edge is 90 mm while that at long edge is 100 mm.

$$\therefore \quad \tau_{VB} = \frac{V_B}{bd} = \frac{7489}{1000 \times 90} = 0.08 \text{ N/mm}^2$$

and
$$\tau_{VL} = \frac{8620}{1000 \times 100} = 0.086 \text{ N/mm}^2.$$

Permissible shear stress $\tau_c$ (Tables 3.1 and 3.2) is much more than these. Hence safe.

**5. Check for development length.** Let the slab have a bearing of 200 mm at the ends. Let us check for development length at the end of *short span*.

$$M_1 = \sigma_{st} A_{st} \cdot j_c d = 140 \left[ \frac{100 \times 78.5}{200} \right] \times 0.865 \times 100 = 4.75 \times 10^6 \text{ N-mm}$$

$$V = V_L = 8620 \text{ N}$$

$$L_0 = \frac{l_s}{2} - x' + 13\,\Phi$$

$$= \frac{200}{2} - 20 + 13 \times 10 = 210 \text{ mm}$$

$$1.3\,\frac{M_1}{V} + L_0 = 1.3 \times \frac{4.75 \times 10^6}{8620} + 210$$

$$= 716 + 210 = 926 \text{ mm}$$

$$L_d = 58.3\,\Phi = 58.3 \times 10 = 583 \text{ mm}$$

Thus Code requirements are satisfied. Similar check can be applied at the end of long span. The sections of the slab along the short and long spans are shown in Fig. 10.4.

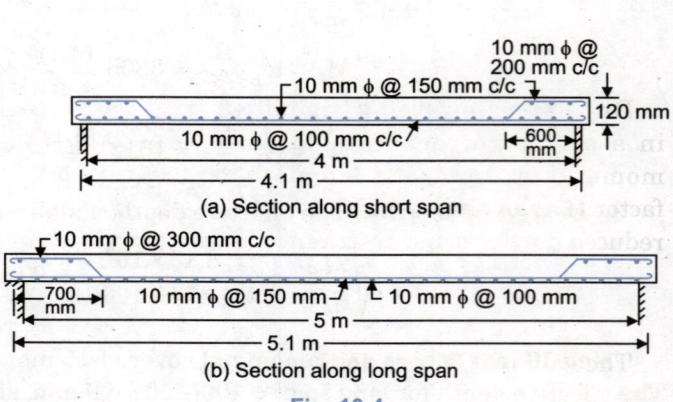

Fig. 10.4

## 10.3. SLAB SIMPLY SUPPORTED ON THE FOUR EDGES WITH CORNERS HELD DOWN AND CARRYING U.D.L.

Such a case arises when the slab is built into brick walls or when they are cast monolithically into thin beam, provided the slab is not continuous over its edges. The exact analysis of such a slab, based on theory of elasticity, is difficult and complicated. We shall discuss in brief the following three methods of analysis of such slabs:

1. Pigeaud's method.  2. Marcus's method.  3. I.S. Code method. (see §10.6)

### 1. Pigeaud's Method

When corners are held down, the maximum bending moment and maximum deflection are further reduced. However, holding down of corners super-imposes a twisting moment in the slab, for which special reinforcement is needed at top and bottom faces of the slab at each corner. The behaviour of such a slab is similar to that of a uniformly loaded thin elastic plate, and has been analysed by Pigeaud. According to this method, the bending moments $M_B$ and $M_L$ along the short and long spans, per unit width are given by the following expressions:

$$M_B = r_B' \frac{w B^2}{8} = \frac{w_B B^2}{8} \quad \ldots (10.5) \quad \text{where} \quad w_B = r_B' w \quad \ldots [10.5\,(a)]$$

$$M_L = r_L' \frac{w L^2}{8} = \frac{w_L L^2}{8} \quad \ldots (10.6) \quad \text{where} \quad w_L = r_L' w \quad \ldots [10.6\,(a)]$$

The coefficients $r_B'$ and $r_L'$ depend on $L/B$ ratio and are given in Table 10.2

TABLE 10.2. Values of $r_B'$ and $r_L'$

| $r = L/B$ | $r_B'$ | $r_L'$ | $r = L/B$ | $r_B'$ | $r_L'$ |
|---|---|---|---|---|---|
| 1 | 0.3 | 0.3 | 1.4 | 0.53 | 0.13 |
| 1.05 | 0.33 | 0.27 | 1.5 | 0.58 | 0.11 |
| 1.1 | 0.36 | 0.24 | 1.6 | 0.63 | 0.09 |
| 1.15 | 0.39 | 0.22 | 1.75 | 0.68 | 0.07 |
| 1.2 | 0.42 | 0.19 | 2.0 | 0.76 | 0.05 |
| 1.25 | 0.45 | 0.17 | 2.5 | 0.87 | 0.03 |
| 1.3 | 0.48 | 0.15 | 3.0 | 0.94 | 0.02 |

## 2. Marcus's Method

Dr. Marcus evolved a simplified but approximate method by which maximum bending moment in a slab with corners held down can be determined. According to this method, the bending moments $M_B$ and $M_L$ calculated by Rankine-Grashoff method are multiplied by a reduction factor $C$, the value of which is always less than unity since the positive bending moments in the slab are reduced due to corner restraint. The value of $C$ depends upon $L/B$ ratio, and is given by

$$C = 1 - \frac{5}{6} \cdot \frac{r^2}{1+r^4} \qquad \ldots (10.7)$$

The value of $C$ for various values of $L/B$ ratio are tabulated in Table 10.3.

TABLE 10.3. Value of factor $C$

| $r = L/B$ | $C$ | $r = L/B$ | $C$ |
|---|---|---|---|
| 1 | 0.861 | 1.4 | 0.888 |
| 1.05 | 0.862 | 1.5 | 0.898 |
| 1.1 | 0.864 | 1.6 | 0.907 |
| 1.15 | 0.866 | 1.75 | 0.919 |
| 1.2 | 0.871 | 2.0 | 0.935 |
| 1.25 | 0.874 | 2.5 | 0.957 |
| 1.3 | 0.879 | 3.0 | 0.970 |

Thus, the midspan bending moment per unit width are given by

$$M_B = \left(1 - \frac{5}{6} \cdot \frac{r^2}{1+r^4}\right) \frac{r^4}{1+r^4} \cdot \frac{wB^2}{8} \qquad \ldots [10.8\,(a)]$$

or

$$M_B = C \frac{r^4}{1+r^4} \cdot \frac{wB^2}{8} = C\, r_B \frac{wB^2}{8} \qquad \ldots (10.8)$$

$$M_L = \left(1 - \frac{5}{6} \cdot \frac{r^2}{1+r^4}\right) \frac{1}{1+r^4} \cdot \frac{wL^2}{8} \qquad \ldots [10.9\,(a)]$$

or

$$M_L = C \frac{1}{1+r^4} \cdot \frac{wL^2}{8} = C \cdot r_L \cdot \frac{wL^2}{8} \qquad \ldots (10.9)$$

The resulting bending moments calculated by Marcus's method and by the exact theory, with Poisson's ratio equal to zero, are almost identical. If, however Poisson's ratio is taken equal to 0.15, the resulting bending moments are given by the following expressions:

$$M_B = C\, r_B \left(1 + \frac{0.15}{r^2}\right) \frac{w B^2}{8} \qquad \ldots [10.10\,(a)]$$

and
$$M_L = C\, r_L \left(0.15 + \frac{1}{r^2}\right) \frac{w L^2}{8} \qquad \ldots [10.10\,(b)]$$

**Torsional reinforcement at corners.** To resist torsional moments induced at the held-down corners, it is necessary to provide a mesh reinforcement at both the faces of the slab, at each corner. The quantity and direction of such mesh reinforcement required is not definitely known either by experimental evidence or by theoretical analysis. According to common practice both top and bottom reinforcements should consist of two layers bars parallel to the sides of the slab and extending in these directions for a distance of one-fifth of the shorter span. The area of the bars in each of the four layers, per unit width of the slab should be three quarters of the area required for the maximum positive moment in the slab. To provide such mesh reinforcement effectively it is advisable to use the same diameters of bars, as used at the mid-span, and then bend these bars up through 180° in the corner strips.

**Shear Force.** The shear force, per unit width, along the long and short span is calculated by the same method explained in § 10.2.

Thus,
$$F_B = \frac{1}{3} w B \qquad \ldots [10.11\,(a)]$$

$$F_L = wB \frac{r}{2+r} \quad \text{(subject to maximum of } 0.5\, w B) \qquad \ldots [10.11\,(b)]$$

**Design: Example 10.2.** *Solve example 10.1 assuming that the corners are held down. Use Marcus's method.*

**Solution:** **1. Design constants.** $k_c = 0.404$, $j_c = 0.865$, $R_c = 0.874$

**2. Loading and bending moments.** Assuming an overall depth of 120 mm, the loading will be the same as in example 10.1. Thus $w = 5480$ N/mm²

As before, $\quad L = 5 + 0.1 = 5.1$ m; $\quad B = 4 + 0.1 = 4.1$ m

$$r = L/B = 5.1/4.1 = 1.245;\quad r^4 = 2.4\ ;\ r^2 = 1.55$$

The bending moment, according to Marcus's method, are given by Eqs. 10.8 (a) and 10.9 (a).

Thus,
$$M_B = \left(1 - \frac{5}{6}\cdot\frac{r^2}{1+r^4}\right)\frac{r^4}{1+r^4}\cdot\frac{w B^2}{8} = \left(1 - \frac{5}{6}\times\frac{1.55}{1+2.4}\right)\cdot\frac{2.4}{1+2.4}\cdot\frac{5480\,(4.1)^2}{8}\times 1000$$
$$= 0.62 \times 0.706 \times 11515 \times 1000 = 5.04 \times 10^6 \text{ N-mm}$$

$$M_L = \left(1 - \frac{5}{6}\frac{r^2}{1+r^4}\right)\cdot\frac{1}{1+r^4}\cdot\frac{w L^2}{8} = \left(1 - \frac{5}{6}\times\frac{1.55}{1+2.4}\right)\frac{1}{1+2.4}\cdot\frac{5480\,(5.1)^2}{8}\times 1000$$
$$= 0.62 \times 0.294 \times 17817 \times 1000 = 3.25 \times 10^6 \text{ N-mm}$$

**3. Design of section.** $\quad d = \sqrt{\dfrac{M_B}{R_c\, b}} = \sqrt{\dfrac{5.04 \times 10^6}{0.874 \times 1000}} = 76$ mm.

However, from I.S. code of practice, minimum total depth required from the stiffness point of view is $D = B/35 = (4.1/35) \times 1000 = 117$ mm. Hence adopt a total depth = 120 mm. Using 8 mm Φ bars and 15 mm cover, the effective depth in the short span = 120 − 15 − 4 = 101 mm and that in the long span will be = 101 − 8 = 93 mm.

**4. Reinforcement.** Area of steel in the short span is given by $(A_{st})_B = \dfrac{M_B}{\sigma_{st} j_c d} = \dfrac{5.04 \times 10^6}{140 \times 0.865 \times 101}$
$= 412$ mm². Using 8 mm Φ bars, $A_\Phi = \dfrac{\pi}{4}(8)^2 = 50.3$ mm²

$$\therefore s_B = \dfrac{1000\, A_\Phi}{A_{st}} = \dfrac{1000 \times 50.3}{412} = 122 \text{ mm. However, provide these @ 120 mm c/c.}$$

Bend alternate bars up at $B/7 = 4100/7 = 600$ mm from the centre of each support.

For the long span, $(A_{st})_L = \dfrac{M_L}{\sigma_{st} j_c d} = \dfrac{3.25 \times 10^6}{140 \times 0.865 \times 93} = 289$ mm²

Hence spacing of 8 mm Φ bars is given by $s_L = \dfrac{1000\, A_\Phi}{A_{st}} = \dfrac{1000 \times 50.3}{289} = 174$ mm.

Provide these @ 170 mm c/c. Bend alternate bars up at $L/7 = 5100/7 = 700$ mm from centre of each support.

**5. Check for shear.** As found in example 10.1,

Shear along long edge, $V_L = 8620$ N/m and shear along short edge $V_B = 7489$ N/m $\tau_{VB} = \dfrac{V_B}{bd} = \dfrac{7489}{1000 \times 93}$
$= 0.08$ N/mm² ; $\tau_{VL} = \dfrac{V_L}{bd} = \dfrac{8620}{1000 \times 101} = 0.085$ N/mm². Hence safe.

**6. Check for development length at supports.** At supports, $1.3 \dfrac{M_1}{V} + L_0 \geq L_d$.

For the ends of short span,

$$M_1 = \sigma_{st} A_{st} j_c d = 140 \left[ \dfrac{1000 \times 50.3}{2 \times 120} \right] \times 0.865 \times 101 = 2.56 \times 10^6 \text{ N-mm}$$

$$V = V_L = 8620 \text{ N.} \; ; L_0 = \dfrac{l_s}{2} - x' + 13\,\Phi = \dfrac{150}{2} - 20 + 13 \times 8 = 159 \text{ mm}$$

$$1.3 \dfrac{M_1}{V} + L_0 = 1.3 \times \dfrac{2.56 \times 10^6}{8620} + 159 = 386 + 159 = 545 \text{ mm}$$

$$L_d = 58.3\,\Phi = 58.3 \times 8 = 466 \text{ mm. Hence the Code requirements are } satisfied.$$

Similarly, at the ends of the long span,

$$M_1 = \sigma_{st} A_{st} j_c d = 140 \left[ \dfrac{1000 \times 50.3}{2 \times 170} \right] \times 0.865 \times 93 = 1.67 \times 10^6 \text{ N-mm}$$

$$V = V_B = 7489 \text{ N} \; ; L_0 = \dfrac{l_s}{2} - x' + 13\,\Phi = \dfrac{150}{2} - 20 + 13 \times 8 = 159 \text{ mm}$$

$$1.3 \dfrac{M_1}{V} + L_0 = 1.3 \times \dfrac{1.67 \times 10^6}{7489} + 159 = 290 + 159 = 449 \text{ mm}$$

$$L_d = 58.3\,\Phi = 58.3 \times 8 = 466 \text{ mm.}$$

Hence the Code requirements are *not satisfied* in this direction.

In order to satisfy Code requirements there are three alternatives: (i) to decrease the diameter of bars (ii) to decrease the spacing of 8 mm Φ bars, and (iii) to carry all the bars at the bottom of the slab right upto the edge and then bend them up through 180° so that all the bars are available at the top face of the slab, near the edges. We will adopt the third alternative, in which case full reinforcement will be available at the bottom face of the slab, at each support. Hence for the ends of long span,

$$M_1 = 140 \left[ \dfrac{1000 \times 50.3}{170} \right] \times 0.865 \times 93 = 3.34 \times 10^6 \text{ and}$$

**252** REINFORCED CONCRETE STRUCTURE

$$1.3 \frac{M_1}{V} + L_0 = 1.3 \times \frac{3.34 \times 10^6}{7489} + 159 = 579.8 + 159 = 738.8 \text{ mm, which is more than } L_d = 466 \text{ mm.}$$

The bent up bars may extend to a distance of $L/7 + 75 = 5100/7 + 75 = 800$ mm from the outer edge of the slab in the long span. For the end strip of 1000 mm, the bent up bars will be extended to a distance of 1000 mm from the outer edge of the slab so that these bars may serve as torsional reinforcement (see next step).

Use the same pattern in the short span also, *i.e.*, continue 8 mm $\Phi$ bars @ 120 mm c/c upto the support and bend these up through 180° so that these bars are available at the top face of the slab near the edges, to serve as torsional reinforcement.

**7. Torsional reinforcement at corners:**

Area of torsional reinforcement = $\frac{3}{4} \times 412 = 309$ mm². Size of mesh-reinforcement = $\frac{1}{5} \times 4100 = 820$ mm from the edge of support, or 970 mm (say) 1000 mm from the outer edge of the slab. It is always preferable to provide the same diameter of bar as that provided for the main reinforcement, and at spacing which is a convenient fraction or multiple of the spacing provided for the main reinforcement for the positive bending moments. Using 8 mm $\Phi$ bars, spacing = $\frac{1000 \times 50.3}{309} = 163$ mm.

The main reinforcement has been provided @ 120 mm c/c in the short span, and @ 170 mm c/c in the long span. Thus at the bottom face of the corner mesh-reinforcement, no additional reinforcement is necessary, if the spacing in the long span direction is reduced to 160 mm. Similarly, at the top face of the slab, bent-up bars are available in the respective spacings of 120 mm and 160 mm in both the directions, and hence no additional reinforcement is required. However, these bent up bars are extended to a distance of 1000 mm both ways. The details of reinforcement etc. are shown in Fig. 10.5.

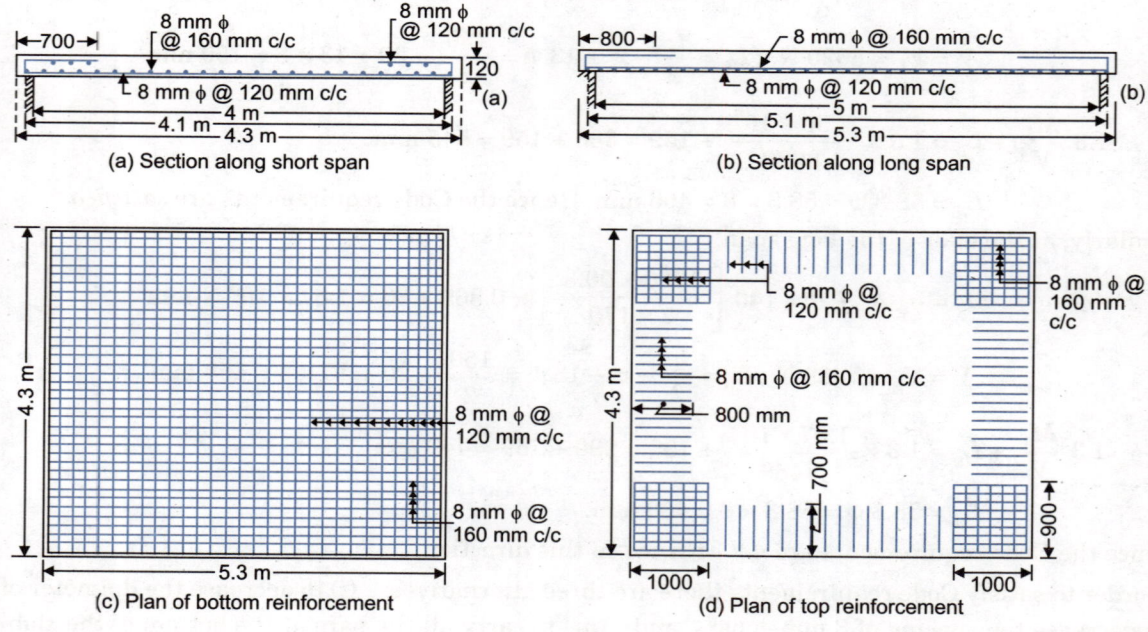

Fig. 10.5

## 10.4. SLAB WITH EDGES FIXED OR CONTINUOUS AND CARRYING U.D.L.

The following three methods may be used for the analysis of slabs with fixed edges, or continuous slabs:
1. Pigeaud's coefficients method.  2. Marcus's method.  3. I.S. code method (see § 10.6).

## 1. Pigeaud's Coefficients Method

In this method, the load $w_B$ and $w_L$ causing bending moments are calculated by *Eqs. 10.5(a)* and *10.6(a)*, using coefficients $r_B'$ and $r_L'$ from Table 10.2 for the value of $r$ based on the basis of reduced values of long and short span lengths depending on the end conditions of the slab. Table 10.4 gives the span reduction factors $f_L$ and $f_B$ by which $r$ should be multiplied for use in Table 10.2. The maximum positive and negative bending moment in the slab are also given in Table 10.4.

TABLE 10.4. Pigeaud's values for slab with fixed or free edges

| Case | Condition of support | Reduction factors for r | | Positive moments | | Negative moments | |
|---|---|---|---|---|---|---|---|
| | | $f_L$ | $f_B$ | $M_L$ | $M_B$ | $M_L$ | $M_B$ |
| 1. | (all four edges fixed) | $\frac{2}{3}$ | $\frac{2}{3}$ | $\frac{1}{10} w_L \cdot L^2$ | $\frac{1}{10} w_B \cdot B^2$ | $\frac{1}{8} w_L \cdot L^2$ | $\frac{1}{8} w_L \cdot L^2$ |
| 2. | (three edges fixed, one long edge free) | $\frac{2}{3}$ | $\frac{4}{5}$ | $\frac{1}{10} w_L \cdot L^2$ | $\frac{1}{9} w_B \cdot B^2$ | $\frac{1}{8} w_L \cdot L^2$ | $\frac{1}{8} w_L \cdot L^2$ |
| 3. | (two short edges fixed, two long edges free) | $\frac{2}{3}$ | $1$ | $\frac{1}{10} w_L \cdot L^2$ | $\frac{1}{8} w_B \cdot B^2$ | $\frac{1}{8} w_L \cdot L^2$ | Zero, but design moment is taken as $\frac{1}{8} w_B \cdot B^2$ |
| 4. | (two adjacent edges fixed, Corner C held down) | $\frac{4}{5}$ | $\frac{4}{5}$ | $\frac{1}{9} w_L \cdot L^2$ | $\frac{1}{9} w_B \cdot B^2$ | $\frac{1}{8} w_L \cdot L^2$ | $\frac{1}{8} w_L \cdot L^2$ |
| 5. | (one short edge fixed, Corners C, D held down) | $\frac{4}{5}$ | $1$ | $\frac{1}{9} w_L \cdot L^2$ | $\frac{1}{8} w_B \cdot B^2$ | $\frac{1}{8} w_L \cdot L^2$ | Zero, but design moment is taken as $\frac{1}{8} w_B \cdot B^2$ |

Table 10.4 describes 5 cases of end conditions :
(1) all the four edges fixed (or continuous)
(2) three edges fixed or continuous while one edge along long span is freely supported
(3) two edges along short span are fixed while the other two edges along long span are free
(4) two adjacent edges are fixed while the other two are free, with the common corner held down, and
(5) one edge along short span is fixed while the three edges are free, with the opposite corners held down.

For cases (1), (2) and (3), corners are always held down.

For cases (4) and (5), if the free corners are not held down, the value of $r$ may be calculated using reduction factors $f_B$ and $f_L$, but the value of $w_B$ and $w_L$ are determined using Grashoff-Rankine formulae (or Table 10.1). For case (4), the positive moments at the centre of span and negative moment at the fixed edges will be equal to $\frac{1}{10} w_B B^2$ for short span and $\frac{1}{10} w_L L^2$ for long span, when the free corner is not held down. Similarly, in case (5), if the free corners are not held down, the positive and negative moments each will be equal to $\frac{1}{10} w_L L^2$ for long span, while the positive moment for span will be equal to $\frac{1}{8} w_B B^2$.

### 2. Marcus's Method

In this method, the loads $w_B$ and $w_L$ along the two spans are calculated exactly in the same manner as done in Grashoff-Rankine method, except that proper expressions for central deflection are used. After having determined $w_B$ and $w_L$, span moments $M_B$ and $M_L$ are calculated as per support conditions. However, due to torsion of corners, the span moments are further reduced by multiplying factors given by Marcus (Table 10.5).

**TABLE 10.5.** Marcus's values for slab with fixed or free edges

| Case | Condition of support | Load coefficient | | Span moments (+) | | Multiplying factors for span moment | | Support moments | |
|---|---|---|---|---|---|---|---|---|---|
| | | $w_B/w$ | $w_L/w$ | $M_B$ | $M_L$ | For $M_B$ | For $M_L$ | $M_B'$ | $M_L'$ |
| 1. | | $\dfrac{r^4}{1+r^4}$ | $\dfrac{1}{1+r^4}$ | $\dfrac{1}{8} w_B \cdot B^2$ | $\dfrac{1}{8} w_L \cdot L^2$ | $1 - \dfrac{5}{6} \cdot \dfrac{r^2}{1+r^4}$ | $1 - \dfrac{5}{6} \cdot \dfrac{r^2}{1+r^4}$ | zero | zero |
| 2. | | $\dfrac{5r^4}{2+5r^4}$ | $\dfrac{2}{2+5r^4}$ | $\dfrac{9}{128} w_B \cdot B^2$ | $\dfrac{1}{8} w_L \cdot L^2$ | $1 - \dfrac{75}{32} \cdot \dfrac{r^2}{2+5r^4}$ | $1 - \dfrac{5}{3} \cdot \dfrac{r^2}{2+5r^4}$ | $\dfrac{1}{8} w_B \cdot B^2$ and zero | zero |
| 3. | | $\dfrac{5r^4}{1+5r^4}$ | $\dfrac{1}{1+5r^4}$ | $\dfrac{1}{24} w_B \cdot B^2$ | $\dfrac{1}{8} w_L \cdot L^2$ | $1 - \dfrac{25}{18} \cdot \dfrac{r^2}{1+5r^4}$ | $1 - \dfrac{5}{6} \cdot \dfrac{r^2}{1+5r^4}$ | $\dfrac{1}{12} w_B \cdot B^2$ each | zero |
| 4. | | $\dfrac{r^4}{1+r^4}$ | $\dfrac{1}{1+r^4}$ | $\dfrac{9}{128} w_B \cdot B^2$ | $\dfrac{9}{128} w_L \cdot L^2$ | $1 - \dfrac{15}{32} \cdot \dfrac{r^2}{1+r^4}$ | $1 - \dfrac{15}{32} \cdot \dfrac{r^2}{1+r^4}$ | $\dfrac{1}{8} w_B \cdot B^2$ and zero | $\dfrac{1}{8} w_L \cdot L^2$ and zero |
| 5. | | $\dfrac{2r^4}{1+2r^4}$ | $\dfrac{1}{1+2r^4}$ | $\dfrac{1}{24} w_B \cdot B^2$ | $\dfrac{9}{128} w_L \cdot L^2$ | $1 - \dfrac{5}{9} \cdot \dfrac{r^2}{1+2r^4}$ | $1 - \dfrac{15}{32} \cdot \dfrac{r^2}{1+2r^4}$ | $\dfrac{1}{12} w_B \cdot B^2$ each | $\dfrac{1}{8} w_L \cdot L^2$ and zero |
| 6. | | $\dfrac{r^4}{1+r^4}$ | $\dfrac{1}{1+r^4}$ | $\dfrac{1}{24} w_B \cdot B^2$ | $\dfrac{1}{24} w_L \cdot L^2$ | $1 - \dfrac{5}{18} \cdot \dfrac{r^2}{1+r^4}$ | $1 - \dfrac{5}{18} \cdot \dfrac{r^2}{1+r^4}$ | $\dfrac{1}{12} w_B \cdot B^2$ each | $\dfrac{1}{12} w_L \cdot L^2$ each |

For example, let us take the case of slab which has edges of short span fixed, while the long span is freely supported. Then the deflection at the middle of long span will be $\dfrac{5}{384} \cdot \dfrac{w_L L^4}{EI}$ while that at the middle of short span will be $\dfrac{1}{384} \cdot \dfrac{w_B B^4}{EI}$. Equating the two, we get

$$\frac{5}{384} w_L \frac{L^4}{EI} = \frac{1}{384} \cdot \frac{w_B \cdot B^4}{EI}$$

or $\qquad w_B = 5w_L(L/B)^4 = 5w_L r^4$ ...(1)

Also, $\qquad w_B + w_L = w$ ...(2)

Hence from (1) and (2), $\quad w_L = \dfrac{1}{1+5r^4} w$ ...(10.12)

and $\qquad w_B = \dfrac{5r^4}{1+5r^4} w$ ...(10.13)

Similarly, values of $w_L$ and $w_B$ for other end conditions may be calculated. Table 10.5 gives the values of $w_L/w$ and $w_B/w$ for various end conditions, in terms of factor $r$, the span ratio The values of span moments (positive moments) along with the multiplying factors for the span moments are also given. However, the support moments (fixed end moments) can be calculated. When a support is free, support moment will be $\frac{1}{12} w_B B^2$ or $\frac{1}{12} w_L L^2$ as the case may be. If one end of a span is fixed while the other end is free, the fixed end moment will be $\frac{1}{8} w_B B^2$ or $\frac{1}{8} w_L L^2$, as the case may be.

## 10.5. OTHER CASES OF SLABS

**1. Non-rectangular slabs.** Slabs with non-rectangular panels may be approximately analysed by treating the slab as an equivalent two way slab or a circular slab, depending upon the shape of the panel. If the shape of the panel is trapezoidal, the slab may be designed as an equivalent rectangular slab having one span ($L$) as the centre to centre distance between the parallel sides, and the other span ($B$) as perpendicular distance between the parallel sides, as marked in Fig. 10.6(*a*). If a panel is triangular or regular polygon in shape, it may be treated as an equivalent circular slab having a diameter equal to the mean of circumscribed and inscribed circles, as shown in Fig. 10.6 (*b*) and (*c*).

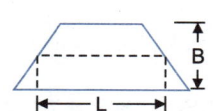

(a) Equivalent rectangular slab B × L for trapezoidal panel

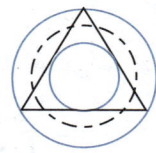

(b) Equivalent circular slab for triangular panel

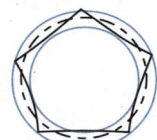

(c) Equivalent circular slab for regular polygon panel

**Fig. 10.6.** Equivalent Slabs for Non-rectangular Panels.

**2. Loads other than U.D.L.** If the slab does not carry uniformly distributed load, but carries concentrated load, spread on a small specified areas or locations of slab, it may be analysed by using curves based on Pigeaud's theory. A detailed treatment of various such cases have been dealt with separately in chapter on R.C.C. bridges.

**Example 10.3.** *Fig. 10.7 shows the plan of the roof of a building. The super-imposed load of the slab may be taken as 3000 N/m². Design all the slab panels. Also, design beam EFGH. Use Marcus's method for design. Use M 20 concrete and Fe 415 steel.*

**Solution:** The four corner slabs *ABFE, CDHG, KLPO* and *IJNM* have the same edge conditions, *i.e.* each slab is free at two adjacent edges, with the corner held down, and fixed at the other two adjacent edges. Hence design of only one slab out of these four need be done.

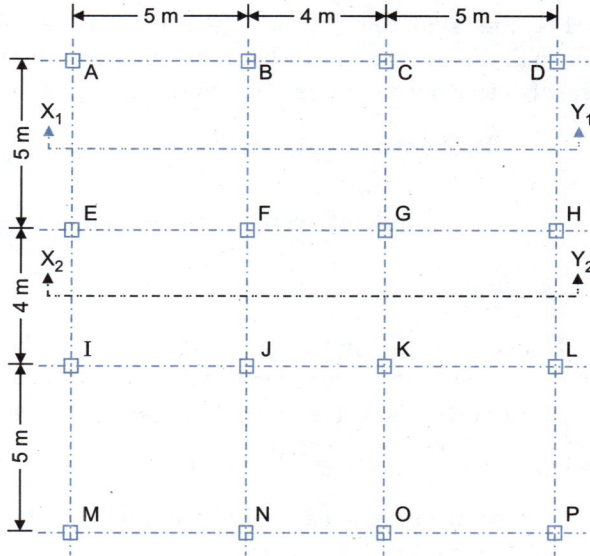

Fig. 10.7

Similarly, the four slabs *EFJI*, *BFGC*, *GHLK* and *JKON* have the same edge conditions, *i.e.* free at one edge and fixed at the other. Hence we will design only one slab out of the four. Then remains the slab *FGKJ*, which is fixed on all the four edges. For the given set of materials, the design constant will be: $k_c = 0.289$, $j_c = 0.904$ and $R_c = 0.914$.

**1. Design of slab ABFE**: This is a square panel of effective size 5 m × 5 m. Let $AB = L = 5$ m and $AE = B = 5$ m. Edges *BF* and *EF* are continuous, while *AB* and *AE* are freely supported, with corner *A* held down to the column at *A* (Fig. 10.8). Let us assume total thickness of the slab = 140 mm.

∴ Self-weight per sq. metre = 0.14 × 25000 = 3500 N/m²
Super-imposed load = 3000 N/m²
Total = 6500 N/m²

$$r = \frac{L}{B} = \frac{5}{5} = 1 \; ; r^4 = 1$$

From Table 10.5 (case 4), we have

$$w_B = \frac{r^4}{1+r^4} \times 6500 = \frac{1}{2} \times 6500 = 3250 \text{ N/m}$$

$$w_L = \frac{1}{1+r^4} \times 6500 = \frac{1}{2} \times 6500 = 3250 \text{ N/m}$$

Multiplying factor of span moment

$$= 1 - \frac{15}{32} \cdot \frac{r^2}{1+r^4} = 1 - \frac{15}{32} \times \frac{1}{2} = 0.766$$

$$M_B = M_L = \frac{9}{128} w_B B^2 \times \text{Multiplying factor}$$

$$= \frac{9}{128} \times 3250 \, (5^2) \times 0.766 \text{ N-m} = 4.376 \times 10^6 \text{ N-mm}$$

Support moments are: $M_B' = M_L' = \frac{w_B B^2}{8} = 3250 \frac{(5)^2}{8} \times 1000 = 10.156 \times 10^6$ N-mm

∴ $d = \sqrt{\frac{M_B'}{R_c \, b}} = \sqrt{\frac{10.156 \times 10^6}{0.914 \times 1000}} = 105.4$ mm.

Let us use 8 mm F bars and provide 20 mm nominal cover. Hence keeping total depth = 140 mm, available depth of the centre of the upper layer of reinforcement = 140 – 20 – 4 – 8 = 108 mm. Since the span moments in both the directions are the same, we will provide the same reinforcement, though the effective depth in one direction will be 108 mm and in other direction 116 mm.

∴ Reinforcement at centre of each span = $\frac{4.376 \times 10^6}{230 \times 0.904 \times 108} = 194.9$ mm²

Reinforcement at fixed ends = $\frac{10.156 \times 10^6}{230 \times 0.904 \times 108} = 452.3$ mm²

Spacing of 8 mm Φ bars at centre of span = $\frac{1000 \times 50.3}{194.9} = 258$ mm

Hence provide 8 mm Φ bars @ 250 mm c/c in the centre of each span. Bend alternate bars up at a distance of $L/5 = 1$ m from the fixed end and at a distance of $L/7 = 5000/7 \approx 700$ mm from the free edge.

∴ Area of steel available from half bent up bars from each adjacent span, at the fixed edges ≏ $\frac{1000 \times 50.3}{320} = 157.2$ mm².

∴ Remaining area to be provided at the fixed edges, to take up negative bending moment = 452.3 – 157.2 = 295.1 mm².

∴ Spacing of 8 mm Φ bars = $\dfrac{1000 \times 50.3}{295.1}$ = 170.5 mm

However keep spacing @ 125 mm c/c. Thus, bent up bars are available @ 250 mm c/c, while these additional bars are to be provided @ 125 mm c/c. These bars are to be provided for a length = $L/4 = 5/4 = 1.25$ m to each side of the fixed support.

Torsional reinforcement has to be provided at corner $A$, since it is held down.

$A_{st} = \dfrac{3}{4} \times 194.9 = 146.2$ mm². This is to be provided for a length $B/5 = 1$ m, in both directions and at both the faces.

Using 8 mm Φ bars, $\quad s = \dfrac{1000 \times 50.3}{146.2} = 344$ mm.

However, it is preferable to keep a spacing equal to 250 mm. It should be noted that the bent up bars @ 500 mm c/c are already available. The remaining bars need be provided @ 500 mm c/c, so that the overall spacing will be 250 mm c/c.

Torsional reinforcement will also be provided at corners $E$ and $B$, the area being half of that provided at $A$. However, as per Code requirement, maximum spacing cannot exceed 300 mm. Hence provide 8 mm Φ bars @ 250 mm c/c at these corners, for a mesh size of 1 m × 1 m. No torsional reinforcement is needed at $F$, because of symmetrical fixidity conditions there.

**Check for shear.** Since it is a square panel, the distribution of S.F. on all the four panels will be triangular, as shown in Fig. 10.8.

Total reaction on fixed edge = $\dfrac{5}{8} w_B BL = \dfrac{5}{8} \times 3250 \,(5)^2 = 50781$ N

Total reaction on free edge = $\dfrac{3}{8} w_B BL = \dfrac{3}{8} \times 3250 \,(5)^2 = 30469$ N

The distribution of each of these is triangular. Now maximum $F$ = Total reaction × $\dfrac{2}{B}$

∴ Max. S.F. at fixed edge $\quad = 50781 \times \dfrac{2}{5} = 20312$ N

Max. S.F. at free edge $\quad = \dfrac{30469 \times 2}{5} = 12188$ N

∴ Nominal shear stress at fixed edge $= \dfrac{20312}{1000 \times 108} \approx 0.188$ N/mm²

Shear stress at free edge $= \dfrac{12188}{1000 \times 108} \approx 0.113$ N/mm². Hence safe.

**Check for development length at free edge.** The Code envisages that the following equation should be satisfied at the free edge.

$$1.3 \dfrac{M_1}{V} + L_o \geq L_d$$

where $M_1 = \sigma_{st} \cdot A_{st} j_c d = 230\,[100.6] \times 0.904 \times 108 = 2.259 \times 10^6$ N-mm
$\quad V = 12188$ N; $l_s = 150$ mm (say)

Providing no hooks at the end, for the deformed bars

$$L_o = \dfrac{l_s}{2} - x' = \dfrac{150}{2} - 20 = 55 \text{ mm}$$

$$1.3 \frac{M_1}{V} + L_o = 1.3 \times \frac{2.259 \times 10^6}{12188} + 55 = 241 + 55 = 296 \text{ mm}$$

$$L_d = 45\, \Phi = 45 \times 8 = 360 \text{ mm.}$$

Hence Code requirements are not satisfied. In order to satisfy Code requirements, we will not bend half the bars up, but instead we will carry them straight upto the edge and then bend them up through 180°. In that case $M_1 = 2 \times 2.259 \times 10^6$. Hence,

$$1.3 \frac{M_1}{V} + L_o = 1.3 \times \frac{2 \times 2.259 \times 10^6}{12188} + 55 = 482 + 55 = 537 \text{ mm} > L_d$$

**2. Design of slab EFJI.** The slab in freely supported on the short edge $EI$ and is continuous on the remaining three edges. (Fig. 10.9).

$L = 5$ m and $B = 4$ m. Hence,

$$r = \frac{L}{B} = \frac{5}{4} = 1.25\, ;\, r^2 = 1.56\, ;\, r^4 = 2.44.$$

We take the same loading, *i.e.*,

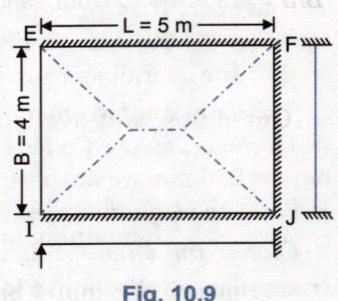

Fig. 10.9

$w = 6500$ N/m². From the Table 10.5, Case 5,

$$w_B = \frac{2r^4}{1 + 2r^4}\, w = \frac{2 \times 2.44}{1 + 2 \times 2.44} \times 6500 = 5395 \text{ N/m}^2$$

$$w_L = \frac{1}{1 + 2r^4}\, w = \frac{1}{1 + 2 \times 2.44} \times 6500 = 1105 \text{ N/m}^2$$

Multiplying factor for $M_B = 1 - \frac{5}{9} \cdot \frac{r^2}{1 + 2\, r^4}$

$$= 1 - \frac{5}{9}\left(\frac{1.56}{1 + 4.88}\right) = 0.852$$

Multiplying factor for $M_L = 1 - \frac{15}{32} \cdot \frac{r^2}{1 + 2r^4} = 1 - \frac{15}{32}\left(\frac{1.56}{1 + 4.88}\right) = 0.876$

$$M_B = \frac{1}{24}\, w_B B^2 \times \text{multiplying factor} = \frac{1}{24} \times 5395\, (4)^2 \times 0.852 \times 1000 = 3.064 \times 10^6 \text{ N-mm}$$

$$M_L = \frac{9}{128}\, w_L L^2 \times \text{multiplying factor} = \frac{9}{128} \times 1105\, (5)^2 \times 0.876 \times 1000 = 1.702 \times 10^6 \text{ N-mm}$$

Fixed end moment at each end of short span is

$$M_B' = \frac{1}{12}\, w_B B^2 = \frac{1}{12} \times 5395\, (4)^2 \times 1000 = 7.193 \times 10^6 \text{ N-mm}$$

Fixed end moment at end $FJ$ of long span is

$$M_L' = \frac{1}{8}\, w_L L^2 = \frac{1}{8} \times 1105\, (5)^2 \times 1000 = 3.453 \times 10^6 \text{ N-mm}$$

The fixed end moment at edge $EI$ will be zero. The thickness will be designed for the maximum of the above moments, *i.e.* $7.193 \times 10^6$ N-mm.

$$d = \sqrt{\frac{7.193 \times 10^6}{1000 \times 0.914}} = 89 \text{ mm.}$$

However, provide overall depth of 140 mm as provided in the panel *ABFE*. Using 8 mm Φ bars and 20 mm nominal cover, the available minimum $d = 140 - 20 - 4 - 8 = 108$ mm in long span and $140 - 20 - 4 = 116$ mm is short span. Area of steel for positive moment for short span

$$(A_{st})_B = \frac{3.064 \times 10^6}{230 \times 0.904 \times 116} = 127 \text{ mm}^2$$

Minimum reinforcement = $1.2 D = 1.2 \times 140 = 168 \text{ mm}^2$

Spacing of 8 mm Φ bars = $\frac{1000 \times 50.3}{168} = 299.4$ mm

Maximum permissible spacing = 300 mm

However, provide 8 mm Φ bars @ 280 mm c/c. Bend half the bars up near the supports at a distance = $B/5 = 4/5 = 0.8$ m from each fixed end. Reinforcement for negative moment for short span

$$= (A_{st})_B' = \frac{7.193 \times 10^6}{230 \times 0.904 \times 116} = 298.3 \text{ mm}^2$$

Out of this, area available from half bent-up bars from each adjacent span

$$= \frac{1000 \times 50.3}{280} = 179.6 \text{ mm}^2$$

∴ Remaining area = $298.3 - 179.6 = 118.7 \text{ mm}^2$

Spacing of 8 mm Φ bars = $\frac{1000 \times 50.3}{118.7} = 423$ mm

However, provide these @ 280 mm c/c, for distance of $B/4 = 1$ m from each fixed edge.

For long span, $(A_{st})_L = \frac{1.702 \times 10^6}{230 \times 0.904 \times 108} = 75.8 \text{ mm}^2$

Min. reinforcement = $1.2 D = 1.2 \times 140 = 168 \text{ mm}^2$, *i.e.* 8 mm Φ bars @ 280 mm c/c. Bend half the bars up near the support at a distance of $L/7 = 5/7 = 0.7$ m from free edge and at $L/5$ m = 1 m from fixed edge.

$$(A_{st})_L' = \frac{3.453 \times 10^6}{230 \times 0.904 \times 108} = 153.8 \text{ mm}^2$$

Min. reinforcement = $1.2 \times 140 = 168 \text{ mm}^2$ ; Area available from the bent up bars from both adjacent spans will be more than this.

**Torsional reinforcement.** Both ends $E$ and $I$ are continuous. Hence provide area of reinforcement of half of that provided at $A$. However, provide 8 mm Φ bars @ 280 mm c/c, so that some of the bent up bars may be utilised for torsional reinforcement. This is to be provided as a mesh of dimension 1 m × 1 m in each direction.

**Check for shear.** Reaction $R$ at each edge = $\frac{1}{2} w_B BL = \frac{1}{2} \times 5395 \times 4 \times 5 = 53950$ N. The distribution of these is trapezoidal. Let the maximum shear force per unit width be $V$.

Then $\quad \frac{VB}{2} + V(L - B) = R$

∴ $\quad V = \frac{2R}{2L - B} = \frac{2R}{L + (L - B)} = \frac{2 \times 53950}{5 + (5 - 4)} = 17984$ N ...(*i*)

Reaction on the fixed short edge = $\frac{5}{8} w_L BL = \frac{5}{8} \times 1105 \times 4 \times 5 = 13813$ N

Assuming this to be distributed triangularly, $V = \frac{2 \times 13813}{4} = 6906$ N ...(*ii*)

Reaction at the free short edge = $\frac{3}{8} w_L B \times L = \frac{3}{8} \times 1105 \times 5 \times 4 = 8288$ N

$$V = \frac{1}{4}(8288 \times 2) = 4144 \text{ N}.$$

∴ Max. $V = 17984$ N. Hence $\tau_v = \dfrac{17984}{1000 \times 116} = 0.155$ N/mm²

From tables 3.1 and 3.2, it will be seen that the permissible shear stress $\tau_c$ is much greater than this. Hence safe.

**Check for development length at free edge**

*Along the free short edge (i.e. Free support of long span)*

$V = 4144$ N and $A_{st} = 8$ mm @ 560 mm c/c.

∴ $M_1 = 230 \left[\dfrac{1000 \times 50.3}{556}\right] \times 0.904 \times 108 = 2.031 \times 10^6$ N-mm

$L_0 = 55$ mm (as before). Hence,

$$1.3 \frac{M_1}{V} + L_0 = 1.3 \times \frac{2.031 \times 10^6}{4144} + 55 = 692 \text{ mm}$$

$L_d = 360$ mm. Hence safe.

**3. Design of slab FGKJ** (Fig. 10.10). This slab is continuous on all the four edges

$L = B = 4$ m.  ∴  $r = 1$.

We take the same loading i.e. $w = 6500$ N/m²

From Table 10.5, case 6,

$$w_B = w_L = \frac{1}{2} w = 3250 \text{ N/m}^2$$

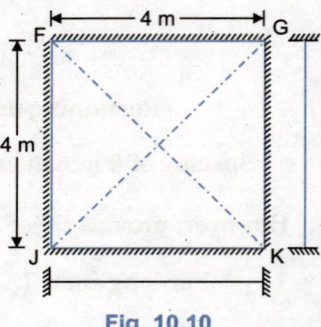

Fig. 10.10

Multiplying factor for each span moment $= 1 - \dfrac{5}{18} \cdot \dfrac{r^2}{1 + r^4} = 1 - \dfrac{5}{18} \cdot \dfrac{1}{2} = 0.861$

Span moments $\quad M_B = M_L = \dfrac{1}{24} w_B B^2 \times$ multiplying factor

$$= \frac{1}{24} \times 3250 (4)^2 \times 0.861 \times 1000 = 1.866 \times 10^6 \text{ N-mm}$$

Support moments $\quad M_B' = M_L' = \dfrac{1}{12} w_B B^2 = \dfrac{1}{12} \times 3250 (4)^2 \times 1000 = 4.333 \times 10^6$ N-mm.

Keep overall depth = 140 mm as before. Using 8 mm diameter bars and providing a nominal cover of 20 mm, the minimum available $d = 108$ mm.

$$(A_{st})_B = (A_{st})_L = \frac{1.866 \times 10^6}{230 \times 0.904 \times 108} = 83 \text{ mm}^2$$

However, minimum reinforcement = $1.2 \times 140 = 168$ mm²

Hence provide 8 mm Φ bars @ 280 mm c/c. Bend half the bars up near the support at a distance = $B/5$ = 0.8 m from the edge.

For fixing moments, $\quad (A_{st})_B' = (A_{st})_L' = \dfrac{4.333 \times 10^6}{230 \times 0.904 \times 108} = 193$ mm²

Area available from bent up bars from both the adjacent spans $= \dfrac{1000 \times 50.3}{280} = 179.6$ mm²

Hence additional area required = $193 - 179.6 = 13.4$ mm² which is quite small. Hence proved 8 mm Φ bars @ 280 mm c/c as extra bars.

No torsional reinforcement is required for this panel.

### Check for shear.

Total reaction on any edge = $\frac{1}{4} w (B.L) = \frac{1}{4} \times 6500 (4 \times 4) = 26000$ N

$\therefore \quad V_{max} = \frac{26000 \times 2}{4} = 13000$ N and $\tau_v = \frac{13000}{1000 \times 108} = 0.12$ N-mm$^2$

This is less than permissible shear stress even at the minimum reinforcement (see Table 3.1 and 3.2). Since it is an interior panel, it is not necessary to check for development length. The sections of the panels along $X_1 - Y_1$ and $X_2 - Y_2$ are shown in Fig. 10.11.

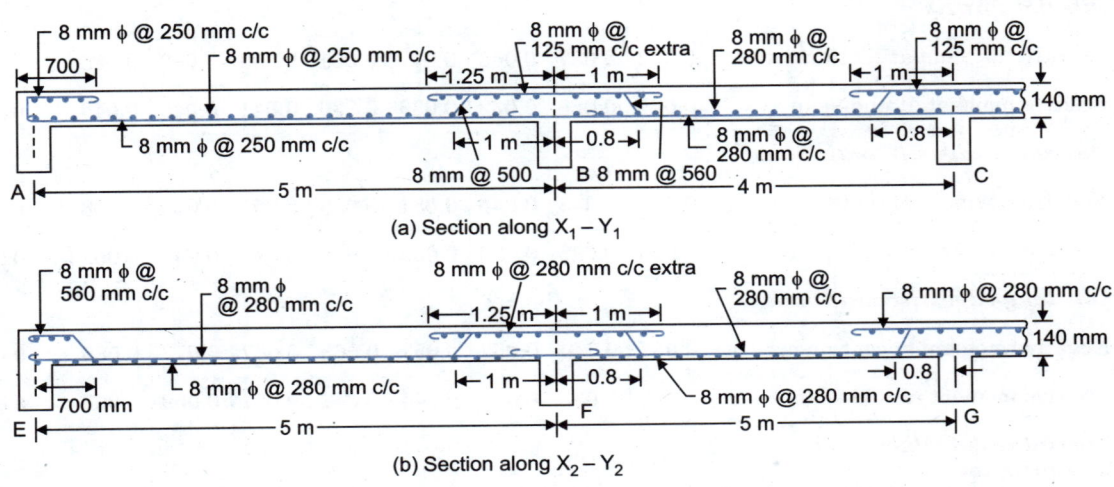

Fig 10.11

## 10.6. INDIAN STANDARD CODE METHOD (IS : 456 - 2000)

Indian Standard Code (IS : 456 - 2000) lays down design rules for two cases of slabs spanning in two directions : (*a*) restrained slabs (*b*) simply supported slabs.

**Note :** The most commonly used *elastic methods* are based on Pigeaud's or Westergaard's theory and the most commonly used *limit state of collapse method* is based on Johansen's yield line theory.

### (a) Restrained Slabs

When the corners of a slab are prevented from lifting, the slab may be designed as specified in *a*–1 to *a*–11 below.

(***a*–1**). The maximum bending moments per unit width in a slab are given by the following equations:

$$M_x = \alpha_x w l_x^2 \quad \text{and} \quad M_y = \alpha_y \cdot w \cdot l_x^2$$

where $\alpha_x$ and $\alpha_y$ as coefficients given in Table 10.6.

$w$ = total load per unit area

$M_x, M_y$ = moments on strips of unit width spanning $l_x$ and $l_y$ respectively, and

$l_x$ and $l_y$ = lengths of shorter span and longer span respectively.

(***a*–2**). Slabs are considered as divided in each direction into middle strips and edge strips as shown in Fig. 10.12, the middle strips being three quarters of the width and each edge strip one-eighth of the width.

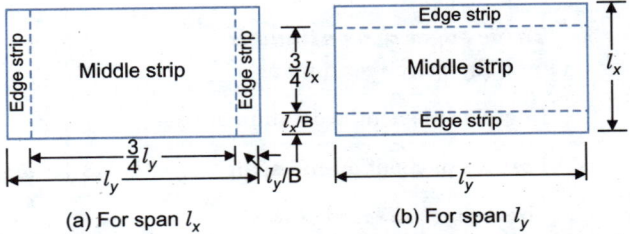

Fig. 10.12. Division of Slab into Middle and Edge Strips

**TABLE 10.6.** Bending Moment Coefficients for Rectangular Panels Supported on Four Sides with Provision for Torsion at Corners (IS : 456 - 2000)
As per Clause D - 1.1 and 24.1 of table no: 26 page no. 91 of IS: 456 – 2000

| Case No. | Type of panel and moment | Short span coefficient $\alpha_x$ for values of $l_y/l_x$ | | | | | | | | Long span coefficient $\alpha_y$ for all values of $l_y/l_x$ |
|---|---|---|---|---|---|---|---|---|---|---|
| | | 1.0 | 1.1 | 1.2 | 1.3 | 1.4 | 1.5 | 1.75 | 2.0 | |
| 1. | **Interior Panels** | | | | | | | | | |
| | Negative moment at continuous edge | 0.032 | 0.037 | 0.043 | 0.047 | 0.051 | 0.053 | 0.060 | 0.065 | 0.032 |
| | Positive moment at mid-span | 0.024 | 0.028 | 0.032 | 0.036 | 0.039 | 0.041 | 0.045 | 0.049 | 0.024 |
| 2. | **One short edge discontinuous** | | | | | | | | | |
| | Negative moment at continuous edge | 0.037 | 0.043 | 0.048 | 0.051 | 0.055 | 0.057 | 0.064 | 0.068 | 0.037 |
| | Positive moment at mid-span | 0.028 | 0.032 | 0.036 | 0.039 | 0.041 | 0.044 | 0.048 | 0.052 | 0.028 |
| 3. | **One long edge discontinuous** | | | | | | | | | |
| | Negative moment at continuous edge | 0.037 | 0.044 | 0.052 | 0.057 | 0.063 | 0.067 | 0.077 | 0.085 | 0.037 |
| | Positive moment at mid-span | 0.028 | 0.033 | 0.039 | 0.044 | 0.047 | 0.051 | 0.059 | 0.065 | 0.028 |
| 4. | **Two adjacent edges discontinuous** | | | | | | | | | |
| | Negative moment at continuous edge | 0.047 | 0.053 | 0.060 | 0.065 | 0.071 | 0.075 | 0.084 | 0.091 | 0.047 |
| | Positive moment at mid-span | 0.035 | 0.040 | 0.045 | 0.049 | 0.053 | 0.056 | 0.063 | 0.069 | 0.035 |
| 5. | **Two short edges discontinuous** | | | | | | | | | |
| | Negative moment at continuous edge | 0.045 | 0.049 | 0.052 | 0.056 | 0.059 | 0.060 | 0.065 | 0.069 | – |
| | Positive moment at mid-span | 0.035 | 0.037 | 0.040 | 0.043 | 0.044 | 0.045 | 0.049 | 0.052 | 0.035 |
| 6. | **Two long edges continuous** | | | | | | | | | |
| | Negative moment at continuous edge | – | – | – | – | – | – | – | – | 0.045 |
| | Positive moment at mid-span | 0.035 | 0.043 | 0.051 | 0.057 | 0.063 | 0.068 | 0.080 | 0.088 | 0.035 |
| 7. | **Three edges discontinuous** (one long edge continuous) | | | | | | | | | |
| | Negative moment at continuous edge | 0.057 | 0.064 | 0.071 | 0.076 | 0.080 | 0.084 | 0.091 | 0.097 | – |
| | Positive moment at mid-span | 0.043 | 0.048 | 0.053 | 0.057 | 0.060 | 0.064 | 0.069 | 0.073 | 0.043 |
| 8. | **Three edges discontinuous** (one short edge continuous) | | | | | | | | | |
| | Negative moment at continuous edge | – | – | – | – | – | – | – | – | 0.057 |
| | Positive moment at mid-span | 0.043 | 0.051 | 0.059 | 0.065 | 0.071 | 0.076 | 0.087 | 0.096 | 0.043 |
| 9. | **Four edges discontinuous** | | | | | | | | | |
| | Positive moment at mid-span | 0.056 | 0.064 | 0.072 | 0.079 | 0.085 | 0.089 | 0.100 | 0.107 | 0.056 |

(***a*–3**). The maximum moments calculated as in *a*–1 apply on to the middle strips and no redistribution shall be made.

(***a*–4**). Tension reinforcement provided at mid-span in the middle strip shall extend in the lower part of the slab to within $0.25\,l$ of a continuous edge, or $0.15\,l$ of a discontinuous edge.

(***a*–5**). Over the continuous edge of the middle strip, the tension reinforcement shall extend in the upper part of the slab a distance of $0.15\,l$ from the supports, and at least 50 percent shall extend a distance of $0.3\,l$.

(***a*–6**). At a discontinuous edge, negative moments may arise. They depend on the degree of fixidity at the edge of the slab but, in general, tension reinforcement equal to 50 percent of that provided at mid span extending $0.1\,l$ into the span will be sufficient.

(***a*–7**). Reinforcement in edge strip, parallel to that edge shall comply with minimum given in chapter 7 (*i.e.* @ 0.15%) and the requirements for torsion given in *a*–8, *a*–9 and *a*–10.

(***a*–8**). Torsion reinforcement shall be provided at any corner when the slab is simply supported on both edges meeting at that corner. It shall consist of top and bottom reinforcement, each with layers of bars placed parallel to the sides of the slab and extending from the edges to a minimum distance of one fifth of the shorter span. The area of reinforcement in each of these four layers shall be three quarters of the area required for the maximum mid-span moment in the slab.

(***a*–9**). Torsion reinforcement equal to half that described in *a*–8 shall be provided at a corner contained by edges over only one which the slab is continuous.

(***a*–10**). Torsion reinforcement need not be provided at any corner contained by edges over both of which slab is continuous.

(***a*–11**). Where $l_y/l_x$ is greater than 2, the slabs shall be designed as spanning one way.

**Restrained Slab with Unequal Conditions at Adjacent Panels**

In some cases, the support moments calculated from Table 10.6 for adjacent panels may differ significantly. The following procedure may be adopted to adjust them:

(*a*) Calculate the sum of the moments at midspan and supports (neglecting signs).

(*b*) Treat the values from Table 10.6 as fixed end moments.

(*c*) According to relative stiffness of adjacent spans, distribute the fixed end moments across the supports, giving new support moments.

(*d*) Adjust midspan moment such that, when added to the support moments from (*c*), neglecting signs, the total should be equal to that from (*a*).

If the resulting support moments are significantly greater than the value from Table 10.6, the tension steel over the supports will need to be extended further. The procedure should be as follows:

(1) Take the span moment as parabolic between the supports: its maximum value is as found from (*d*).

(2) Determine the points of contraflexure of the new support moments [from (*c*)] with the span moment [from (1)].

(3) Extend half the support tension steel at each end to at least an effective depth or 12 bar diameters beyond the nearest point of contraflexure.

(4) Extend the full area of the support tension steel at each end to half the distance from (3).

**(*b*) Simply Supported Slabs**

(***b*–1**). When simply supported slabs do not have adequate provision to resist torsion at corner and to prevent the corners from lifting, the maximum moments for unit width are given by the following equations:

$$M_x = \alpha_x\, w\, l_x^2 \quad \text{and} \quad M_y = \alpha_y\, w\, l_x^2$$

where, $M_x, M_y, l_x, l_y$ are the same as those in *a*–1, and $\alpha_x$ and $\alpha_y$ are moment coefficients given in Table 10.7.

**TABLE 10.7.** Bending Moment Coefficients for Slabs Spanning in Two Directions at Right Angles, Simply Supported on Four Sides (IS : 456 – 2000)

As per Clause D-2.1 of table no: 27 of page no.: 91 of IS : 456:2000

| $l_y/l_x$ | 1.0 | 1.1 | 1.2 | 1.3 | 1.4 | 1.5 | 1.75 | 2.0 | 2.5 | 3.0 |
|---|---|---|---|---|---|---|---|---|---|---|
| $\alpha_x$ | 0.062 | 0.074 | 0.084 | 0.093 | 0.099 | 0.104 | 0.113 | 0.118 | 0.122 | 0.124 |
| $\alpha_y$ | 0.062 | 0.061 | 0.059 | 0.055 | 0.051 | 0.046 | 0.037 | 0.029 | 0.020 | 0.014 |

**(b – 2).** At least 50 percent of the tension reinforcement provided at mid-span should extend to the supports. The remaining 50 percent should extend to within 0.1 $l_x$ or 0.1 $l_y$ of the support, as appropriate.

**Details of Reinforcement.** Fig. 10.13 shows the details of reinforcement. Half the positive moment reinforcement $(A_{st})_x$ is bent upwards, at a distance $a_x \leq 0.15\, l_x$ from the centre of the support while the remaining half is continued upto the end. The bent up bars or the negative reinforcement (top reinforcement) $(A_{st})_x'$ should be available for distance $a_x' \geq 0.1\, l_x$ from the centre of the support.

Similarly, in the span $l_y$, half the positive moment reinforcement $(A_{st})_y$ is bent up, at a distance $a_y \leq 0.15\, l_y$ from the centre of the support. The remaining half is continued upto the end. The bent bars or the negative reinforcement $(A_{st})_y'$ is available at the top of the slab for a distance $a_y' \geq 0.1\, l_y$ from the centre of the slab.

The corner reinforcement, at top and bottom of each corner is provided in the form of square mesh of size $0.2\, l_x \times 0.2\, l_x$.

**Load on supporting beams.** The loads on the beams supporting solid slabs spanning in two directions at right angles and supporting uniformly distributed loads, may be assumed as marked in Fig. 10.14.

Thus total reaction on short edge $AB$ (or $CD$) will be equal to $\frac{1}{2} w\, l_x \frac{l_x}{2} = (w\, l_x^2)/4$. The *average reaction* per unit width along short edge will therefore be $w\, l_x/4$. However the *maximum reaction* (hence max. S.F.) per unit width along $AB$ will occur near the centre of $AB$ and its value may be taken as $w\, l_x/3$ for all practical purposes.

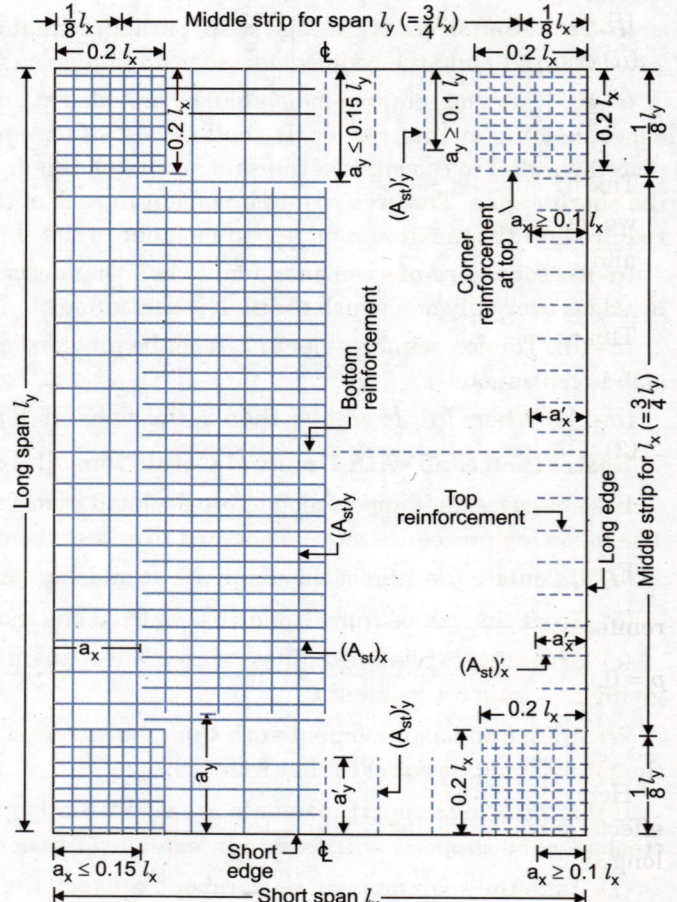

Fig. 10.13

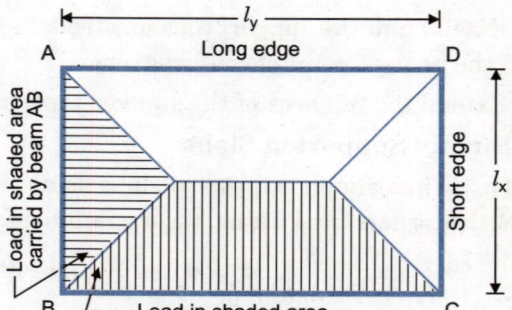

Fig. 10.14. Loads carried by supported beam.

Similarly, S.F. along the edge $AD$ or $AC$, (*i.e.* ends of short span of slab) per unit width may be taken as $w\, l_x \dfrac{r}{2+r}$, subject to a maximum of $0.5\, w\, l_x$ when $r$ exceeds 2.

**Design: Example 10.4.** *Design a R.C. slab for a room measuring 5 m × 6 m size. The slab is simply supported on all the four edges, with corners held down, and carries a superimposed load of 3000 N/m², inclusive of floor finishes etc. Use M 20 mix, Fe 415 steel and IS Code method.*

**Solution:**

(1) **Design constants.** For M 20 concrete, and Fe 415 steel we have $k_c = 0.289, j_c = 0.904$ and $R_c = 0.914$.

(2) **Loading and bending moments.** Assume an overall depth of 180 mm

(i) Weight of slab per m² $= 0.18 \times 1 \times 1 \times 25000$ $= 4500$
(ii) Super-imposed load @ 3000 N/m² $= 3000$

$$\text{Total,} \quad w = 7500 \text{ N/m}^2$$

Taking effective depth = 160 mm,
we have effective $l_y = 6 + 0.16 = 6.16$ m
and $l_x = 5 + 0.16 = 5.16$ m.
∴ $r = l_y/l_x = 6.16/5.16 \approx 1.2$

This is case 9 of Table 10.6, from which $\alpha_x = 0.072$ and $\alpha_y = 0.056$

∴ $M_x = \alpha_x\, w\, l_x^2 = 0.072 \times 7500\, (5.16)^2 = 14378$ N-m $= 14.378 \times 10^6$ N-mm
$M_y = \alpha_y\, w\, l_x^2 = 0.056 \times 7500\, (5.16)^2 = 11183$ N-m $= 11.183 \times 10^6$ N-mm

(3) **Design of section for short span**

$$d = \sqrt{\dfrac{14.378 \times 10^6}{1000 \times 0.914}} = 125.4 \text{ mm}$$

From stiffness (deflection) point of view, span/effective depth ratio = 20. For a balanced design, percentage reinforcement $\dfrac{k\, \sigma_{cbc}}{2\, \sigma_{st}} \times 100 = \dfrac{0.289 \times 7}{2 \times 230} \times 100 = 0.44\%$. However, using under-reinforced section, and taking $p = 0.25\%$ we have from Fig. 7.1, modification factor $\approx 1.6$ for HYSD bars.

∴ $\dfrac{\text{span}}{d} = 20 \times 1.6 = 32$. Hence $d = \dfrac{\text{span}}{32} = \dfrac{5160}{32} = 161.25$ mm.

Hence provide total thickness = 180 mm. Using 8 mm Φ bars and a nominal cover of 20 mm, available effective depth for short span = 180 – 20 – 4 = 156 mm for the short span, and 156 – 8 = 148 mm for the long span.

For short span, width of middle strip = $\dfrac{3}{4} l_y = \dfrac{3}{4} \times 6.16 \approx 4.62$ m

Width of edge strip = $\dfrac{1}{2} (6.16 - 4.62) = 0.77$ m

$(A_{st})_x = \dfrac{14.378 \times 10^6}{230 \times 0.904 \times 156} = 443$ mm². Using 8 mm Φ bars, $A_\Phi = 50.27$ mm²;

Spacing $s = \dfrac{1000 \times 50.27}{443} = 113.5$ mm.

However, use 8 mm Φ bars @ 100 mm c/c for the middle strip of width 4.62 m. Bend half the bars up at a distance $= 0.15 l_x = 0.15 \times 5160 \approx 770$ mm from the centre of the support, or at a distance of $770 + 80 = 850$ mm from the edge of the slab (assuming a bearing of 160 mm on the wall). Available length of bars at the top $= 770 - (160 - 20) = 630$ mm from the centre of support, assuming bending of bar at 45°. The length is more than $0.1\, l_x\, (= 0.1 \times 5160 = 516$ mm) required by the Code. Hence length of top bars from the edge of slab $= 630 + 80 = 710$ mm. Edge strip is of length 0.77 m. The reinforcement in the edge strips is given by:

$A_{st}$ (@ 0.12%) = (0.12/100) × 1000 × 180 = 216 mm².

Using 8 mm Φ bars, $s = \dfrac{1000 \times 50.27}{216} = 232$ cm c/c.

However, it is more convenient to keep the spacing as the simple fraction of the edge strip or simple multiple of spacing for the middle strip. Hence keep $s = 200$ mm.

### 4. Design of section for long span

$$(A_{st})_y = \dfrac{11.183 \times 10^6}{230 \times 0.904 \times 148} = 363 \text{ mm}^2$$

Using 8 mm Φ bars, $s = \dfrac{1000 \times 50.27}{363} = 138$ mm.

Hence provide 8 mm Φ bars @ 130 mm c/c in the middle strip of width $\dfrac{3}{4} l_x = \dfrac{3}{4} \times 5.16 = 3.87$ m. For the edge strip of width 5.16/8 = 0.645 m, provide 8 mm Φ bars @ 260 mm c/c. Bend half the bars of the middle strip, up at a distance of 0.15 $l_y$ = 0.15 × 6160 ≈ 920 mm from the centre of support or 920 + 80 = 1000 mm from the edge of the slab. Available lengths of bars of top = 920 – (150 – 20) = 790 mm from the centre of the support or 790 + 80 = 870 mm from the edge of the slab.

### (5) Check for shear and development length in short span

S.F. at long edges = $w l_x \dfrac{r}{2+r} = 7500 \times 5.16 \dfrac{1.2}{2+1.2} = 14513$ N/m.

∴ Nominal shear stress at long edges = $\dfrac{14513}{1000 \times 156} \approx 0.093$ N/mm² (safe)

At the long edges, the diameter of the bars should be so restricted that the following requirement is satisfied : $1.3 \dfrac{M_1}{V} + L_0 \geq L_d$

$A_{st}$ at supports = $\dfrac{1000 \times 50.27}{200} = 251.3$ mm².

∴ $M_1 = 251.3 \times 230 \times 0.904 \times 156 = 8.153 \times 10^6$ N-mm ; $V = 14513$ N.

Assume that support width $l_s = 160$ mm and a side cover of 20 mm. Providing no hooks,

∴ $L_0 = \dfrac{l_s}{2} - x' = \dfrac{160}{2} - 20 = 60$ mm. Hence,

$1.3 \dfrac{M_1}{V} + L_0 = 1.3 \times \dfrac{8.153 \times 10^6}{14513} + 60 = 790$ mm

$L_d \approx 45 \Phi = 45 \times 8 = 360$ mm.

∴ $1.3 \dfrac{M_1}{V} + L_0 \geq L_d$

**Note.** The Code requires that the positive reinforcement should extend into the support at least by $\dfrac{L_d}{3}$. Hence minimum support width = $\dfrac{L_d}{3} + x' = \dfrac{360}{3} + 20 \approx 140$ mm. In our case, the support width is 160 mm.

### (6) Check for shear and development length in long span

S.F. at short edges = $\dfrac{1}{3} w l_x = \dfrac{1}{3} \times 7500 \times 5.16 = 12900$ N/m

∴ Nominal shear stress = $\dfrac{12900}{1000 \times 148} \approx 0.087$ N/m² (safe).

$$A_{st} = \frac{1000 \times 50.27}{260} = 193.3 \text{ mm}^2$$

$$M_1 = 193.3 \times 230 \times 0.904 \times 148 = 5.95 \times 10^6 \text{ N-mm}$$

$$V = 12900 \text{ N/m}$$

$$L_0 = \frac{l_s}{2} - x' = \frac{250}{2} - 20 = 105 \text{ mm}$$

$$1.3 \frac{M_1}{V} + L_0 = 1.3 \times \frac{5.95 \times 10^6}{12900} + 105 = 705 \text{ mm}$$

$$L_d \approx 45 \, \Phi = 45 \times 8 = 360 \text{ mm}$$

$\therefore$ $1.3 \dfrac{M_1}{V} + L_0 > L_d$. Hence Code requirements are satisfied.

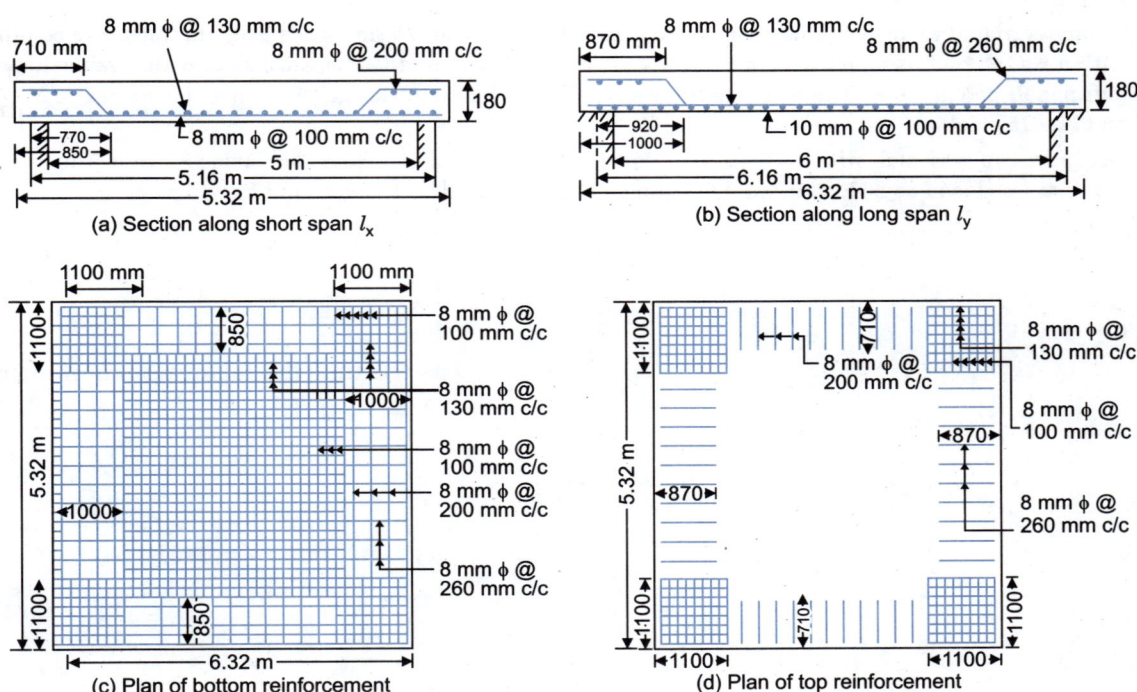

Fig. 10.15

### (7) Torsional reinforcement at corners

Size of torsional mesh = $l_x/5$ = 5.16/5 = 1.03 m or 1.03 + 0.08 ≈ 1.10 m from the edge of the slab.

Area of torsional reinforcement = $\dfrac{3}{4} (A_{st})_x = \dfrac{3}{4} \times 443 \approx 332.3 \text{ mm}^2$

Using 8 mm Φ bars, $s = \dfrac{1000 \times 50.27}{332.3} \approx 151 \text{ mm}$.

However, it is preferable to use the same spacings as provided for main reinforcement. In the short span, main reinforcement in the middle strip has been provided @ 100 mm c/c while for edge strip, it is provided @ 200 mm c/c. Hence provide 8 mm Φ bars @ 100 mm c/c in the short span direction. In the long span, main reinforcement is @ 130 @ mm c/c. Hence provide torsional reinforcement @ 130 mm c/c. This reinforcement is to be provided at both faces of the slab at each corner.

The details of the reinforcement are shown in Fig. 10.15.

## PROBLEMS

1. Design a R.C. slab for a room 4 m × 4 m measuring from inside. The thickness of wall is 400 mm. The superimposed load, exclusive of the self weight of the slab, is 2 kN/m². The slab may be assumed to be simply supported at all the four edges, with corners free to lift. Use M 20 mix. and Fe 415 steel.

2. Redesign problem 1 assuming that the corners are held down. Use Marcus's method.

3. Design a simply supported R.C. slab over a room 4.5 m × 6 m from inside, assuming that the corners are not free to lift. The thickness of all the four walls is 400 mm. The live load on the floor is 2 kN/m². The floor carries a floor finish which weighs 8.5 kN/m². Use M 20 mix and Fe 415 steel. Use Marcus's method.

4. Redesign Problem 3 using I.S. code method.

5. Redesign problem 1, assuming that the slab is fixed on all the four edges.

6. A R.C. floor is supported on R.C. columns 250 mm square at the corners of a rectangle 6 m × 7.5 m with R.C. beams connecting the columns along the perimeter of the rectangle. Design the floor as a two-way reinforced slab. Also design one of the 6 m span beams as an L-beam. Live load on floor = 8 kN/m². Use M 20 mix and Fe 415 steel.

7. An underground room of a dimensions 10 m × 12 m between the existing walls 500 mm thick is to be provided with a R.C.C. roof using one column at the centre of the room and four beams, so that the slab is divided in four panels of 5 m × 6 m. The roof is to carry a live load of 5 kN/m². Design the slab and beams. Use M 20 mix. and Fe 415 steel.

8. Design a two way slab with corners held down, for a room having clear dimensions 4 m × 5 m. A square superimposed load as 2600 N/m² and finishing load 530 N/m². Use $M_{20}$ mix and Fe 415 steel.

# CIRCULAR SLABS

## 11.1. INTRODUCTION

Circular slabs are more commonly used in the design of circular water tank containers with flat bottom and raft foundation. Circular slabs are also used for the following purposes: (*i*) roof of a room or hall circular in plan (*ii*) floor of circular water tank or tower (*iii*) roof of pump house constructed above tube well (*iv*) roof of a traffic control post at the intersection of roads, etc. The bending of such a slab is essentially different from a rectangular slab where bending takes place in distinctly two perpendicular directions along the two spans. When a circular slab simply supported at the edge is loaded with uniformly distributed load, it bends in the form of a saucer, due to which stresses are developed both in the radial as well as the in circumferential directions. The tensile radial and circumferential stresses develop towards the *convex* side of the saucer, and hence reinforcement need be provided at the convex face of the slab. Theoretically, reinforcement should be provided both in the radial as well as circumferential directions, but this arrangement would cause congestion and anchoring problem at the centre of the slab. Hence an alternative method of providing reinforcement is adopted: reinforcement is provided in the form of a mesh of bars having equal area of cross-section in both the directions, the area being equal to that required for the bigger of the radial and circumferential moments. However, if the stresses near the edge are not negligible, or if the edge is fixed, radial and circumferential reinforcement near the edge becomes essential.

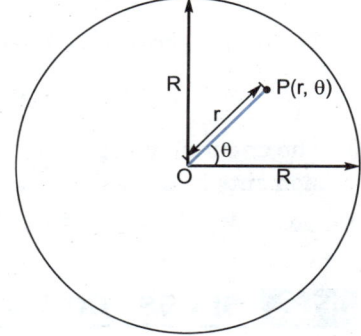

Any circular shape can be represented easily in terms of following two polar co-ordinates.

1. Radial distance ($r$) from centre.
2. Angle $\theta$ between the radius and the fixed reflexure direction.

A circular slab is essentially a two ways slab with bending in both the directions. It deflects like a saucer under the action of uniformly distributed load (udl) loads.

Fig. 11.1

The exact analysis of slab, based on theory of elasticity and assuming **Poisson's ratio** equal to zero, is beyond the scope of the present book ; only final equations are given here. Sometimes, empirical formulae are used for bending moments and shear force etc. We shall discuss in brief the following cases:

1. Slab freely supported at edges and carrying U.D.L.   2. Slab fixed at edges and carrying U.D.L.   3. Slab simply supported at the edges, with load $W$ uniformly distributed along the circumference of a concentric circle.

# 270 REINFORCED CONCRETE STRUCTURE

4. Slab simply supported at edges, with U.D.L. inside a concentric circle.  5. Slab simply supported at the edges with a central hole and carrying U.D.L.  6. Slab simply supported at the edges, with a central hole and carrying W distributed along the circumference of a concentric circle.

## 11.2. SLAB FREELY SUPPORTED AT EDGES AND CARRYING U.D.L.

Let  $w$ = uniformly distributed load, kg/m².

$a$ = radius of slab, m.

$M_r$ = radial bending moment at any point at radius $r$ from the centre of the slab.

$(M_r)_c$ & $(M_r)_e$ = radial moments at centre and edge respectively.

$M_\theta$ = circumferential bending moment at any point at radius $r$ from the centre of the slab.

$(M_\theta)_c$ & $(M_\theta)_e$ = circumferential moments at centre and edge respectively.

The circumferential moment $(M_\theta)$ distribution diagram along any diameter is shown in Fig. 11.2 (b). The moment varies parabolically with a maximum value of $\frac{3}{16} wa^2$ at the middle to a minimum value of $\frac{2}{16} wa^2$ at the edges. Similarly, the radial moment $(M_r)$ distribution along any diameter is shown in Fig. 11.2 (c). The moment varies from $\frac{3}{16} wa^2$ at the middle to zero at the edges. The values of various moments, per unit width are given below:

$$(M_\theta)_c = +\frac{3}{16} wa^2 \quad \ldots(11.1)$$

$$(M_\theta)_e = +\frac{2}{16} wa^2 \quad \ldots(11.2)$$

$$M_\theta = +\frac{w}{16}(3a^2 - r^2), \quad \ldots(11.3)$$

at any radius $r$

$$(M_r)_c = +\frac{3}{16} wa^2 \quad \ldots(11.4)$$

$$(M_r)_e = 0 \quad \ldots(11.5)$$

$$M_r = \frac{3}{16} w(a^2 - r^2) \quad \ldots(11.6)$$

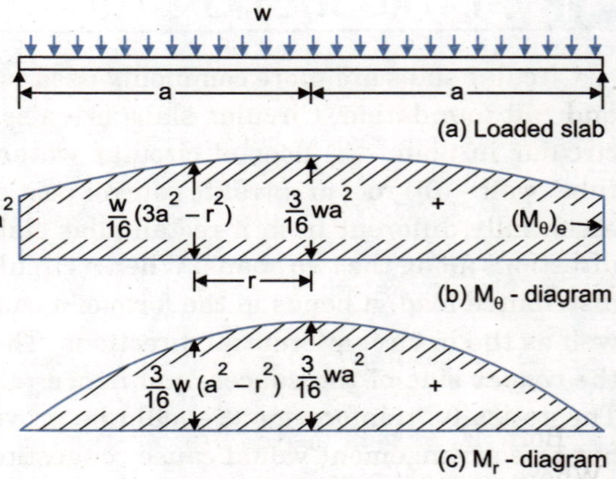

Fig. 11.2

*The radial shear force $F_r$ at any radius $r$ is given by*

$$F_r = \frac{1}{2} w \cdot r \text{ (per unit width)} \quad \ldots(11.7)$$

The circumferential shear force is zero everywhere. C.E. Reynolds assumes that there is redistribution of moments in the slab due to plasticity of materials, during failure, and therefore recommends maximum values of $M_\theta$ and $M_r$ as $\frac{1}{9} wa^2$ instead of $\frac{3}{16} wa^2$ given above.

## 11.3. SLABS FIXED AT EDGES AND CARRYING U.D.L.

Fig. 11.3 (b) shows the circumferential moment distribution diagram,

where, $M_\theta$ varies parabolically from a maximum value at the centre to zero at the edge.

The moment is *positive* throughout. Fig. 11.3(c) shows the radial moment distribution diagram. $M_r$ varies from a maximum positive value at the centre to zero at a distance $(a/\sqrt{3}) = 0.577 a$ from the centre, and then becomes negative at the edges. The various values of moments and shear per unit width are as under:

$$(M_\theta)_c = +\frac{1}{16}wa^2 \qquad \ldots(11.8)$$

$$M_\theta = +\frac{1}{16}w(a^2 - r^2) \qquad \ldots(11.9)$$

$$(M_\theta)_e = 0 \qquad \ldots(11.10)$$

$$(M_r)_c = +\frac{1}{16}wa^2 \qquad \ldots(11.11)$$

$$M_r = +\frac{1}{16}w(a^2 - 3r^2) \qquad \ldots(11.12)$$

$$(M_r)_e = -\frac{2}{16}wa^2 \qquad \ldots(11.13)$$

$$F_r = \frac{1}{2}w \cdot r \text{ (per unit width)} \qquad \ldots(11.14)$$

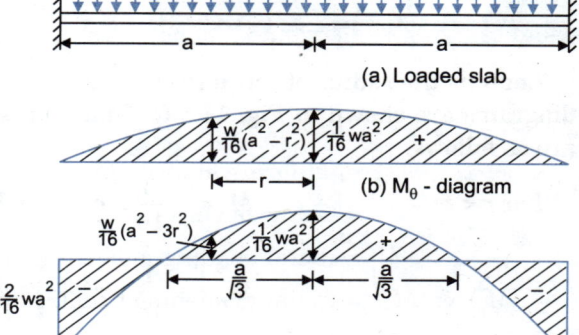

Fig. 11.3

**Slabs partially fixed at the edges.**

This case is intermediate between the case of a freely supported slab and a fixed slab. Hence the moments may be taken as the average of the corresponding moment of the two cases. The values of various moments per unit width are as follows. For radial moment the point of contraflexure occurs at a radius $r = a\sqrt{2/3}$. Thus,

$$(M_r)_c = (M_\theta)_c = +\frac{2}{16}wa^2 \qquad \ldots(11.15)$$

$$(M_r)_e = -\frac{1}{16}wa^2 \qquad \ldots(11.16)$$

$$(M_\theta)_e = +\frac{1}{16}wa^2 \qquad \ldots(11.17)$$

## 11.4. SLAB SIMPLY SUPPORTED AT THE EDGES WITH LOAD W UNIFORMLY DISTRIBUTED ALONG THE CIRCUMFERENCE OF A CONCENTRIC CIRCLE

Both $M_r$ as well as $M_\theta$ are constant from $r = 0$ to $r = b$. Where $r$ is greater than $b$ (i.e. outside the load circle), both $M_\theta$ and $M_r$ decrease parabolically, to values $(M_\theta) e$ and zero respectively at the edges. Various values of moments etc. per unit width are as follows:

For $r < b$

$$M_r = (M_r)_b = M_\theta = (M_\theta)_b$$

$$= \frac{W}{8\pi}\left[2\log_e\left(\frac{a}{b}\right) + 1 - \left(\frac{b}{a}\right)^2\right] \qquad \ldots(11.18)$$

$$F_r = 0 \qquad \ldots(11.19)$$

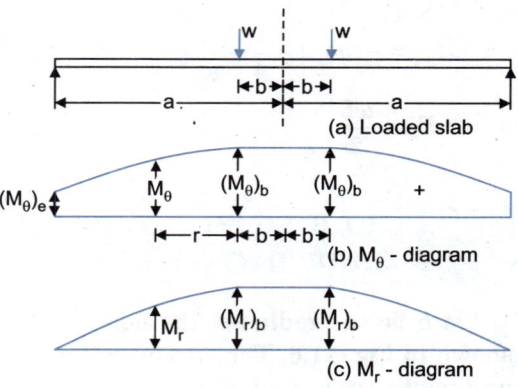

Fig. 11.4

For $r > b$:

$$M_\theta = \frac{W}{8\pi}\left[2\log_e\left(\frac{a}{r}\right) - \left(\frac{b}{r}\right)^2 + 2 - \left(\frac{b}{a}\right)^2\right] \qquad \ldots(11.20)$$

$$M_r = \frac{W}{8\pi}\left[2\log_e\left(\frac{a}{r}\right) - \left(\frac{b}{a}\right)^2 + \left(\frac{b}{r}\right)^2\right] \qquad \ldots(11.21)$$

$$F_r = \frac{W}{2\pi r} \cdot \qquad \ldots(11.22)$$

## 11.5. SLAB SIMPLY SUPPORTED AT EDGES, WITH U.D.L. INSIDE A CONCENTRIC CIRCLE

Let $b$ be the radius of concentric circle carrying U.D.L. per unit area. The bending moments $M_\theta$ and $M_r$ diagrams are shown in Fig. 11.5 (b) and (c) respectively. The various values of moments per unit width are as follows:

For $r < b$:
$$M_r = -\frac{3}{16}wr^2 + \frac{1}{4}wb^2\left[1 - \log_e\left(\frac{b}{a}\right) - \frac{b^2}{4a^2}\right] \qquad \ldots(11.23)$$

$$M_\theta = -\frac{1}{16}wr^2 + \frac{1}{4}wb^2\left[1 - \log_e\left(\frac{b}{a}\right) - \frac{b^2}{4a^2}\right] \qquad \ldots(11.24)$$

$$(M_r)_c = +\frac{1}{4}wb^2\left[1 - \log_e\left(\frac{b}{a}\right) - \frac{b^2}{4a^2}\right] \qquad \ldots[11.23\,(a)]$$

$$(M_r)_b = (M_r)_c = \frac{3}{16}wb^2 \qquad \ldots[11.23\,(b)]$$

$$(M_\theta)_c = +\frac{1}{4}wb^2\left[1 - \log_e\left(\frac{b}{a}\right) - \frac{b^2}{4a^2}\right] \qquad \ldots[11.24\,(a)]$$

$$(M_\theta)_b = (M_\theta)_c = \frac{1}{16}wb^2 \qquad \ldots[11.24\,(b)]$$

$$F_r = \frac{wr}{2} \qquad \ldots(11.25)$$

For $r > b$:
$$M_r = -wb^2\left[\frac{1}{4}\log_e\left(\frac{r}{a}\right) + \frac{b^2}{16}\left(\frac{1}{a^2} - \frac{1}{r^2}\right)\right] \qquad \ldots(11.26)$$

$$M_\theta = -wb^2\left[\frac{1}{4}\log_e\left(\frac{r}{a}\right) - \frac{1}{4} + \frac{b^2}{16}\left(\frac{1}{a^2} + \frac{1}{r^2}\right)\right] \qquad \ldots(11.27)$$

$$\therefore (M_r)_e = 0 \qquad \ldots(11.28)$$

$$(M_\theta)_e = -wb^2\left(-\frac{1}{4} + \frac{b^2}{8a^2}\right) \qquad \ldots(11.29)$$

$$F_r = \frac{wb^2}{2r} \qquad \ldots(11.30)$$

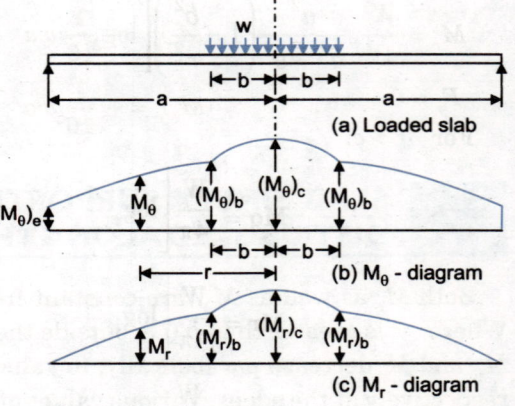

Fig. 11.5

## 11.6. SLAB SIMPLY SUPPORTED AT EDGES, WITH A CENTRAL HOLE AND CARRYING U.D.L.

Let $b$ be the radius of the hole. The B.M. diagrams are shown in Fig. 11.6. The various values of the moments per unit width are as under:

$$M_\theta = -\frac{1}{16}wr^2 + \frac{wb^2}{4}\left[\log_e\left(\frac{r}{a}\right)\right.$$
$$\left. + \frac{3}{4}\left(-\frac{1}{3} + \frac{a^2}{b^2} + \frac{a^2}{r^2}\right) - \frac{a^2 + r^2}{r^2}\cdot\frac{b^2}{a^2 - b^2}\log_e\frac{a}{b}\right]$$
$$\qquad \ldots(11.31)$$

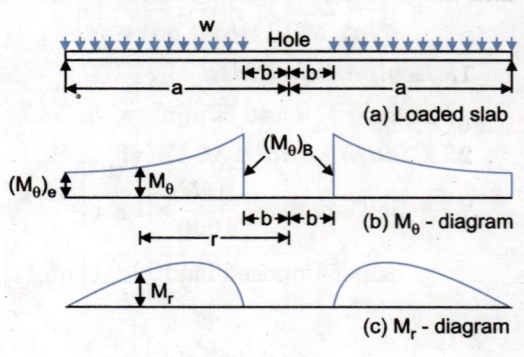

Fig. 11.6

$$M_r = -\frac{3}{16}wr^2 + \frac{wb^2}{4}\left[\log_e \frac{r}{a} + \frac{3}{4}\left(1 + \frac{a^2}{b^2} - \frac{a^2}{r^2}\right) + \frac{a^2 - r^2}{r^2} \cdot \frac{b^2}{a^2 - b^2} \log_e \frac{a}{b}\right] \quad \ldots(11.32)$$

$$F_r = \frac{wr}{2} - \frac{wb^2}{2r} \quad \ldots(11.33)$$

## 11.7. SLAB SIMPLY SUPPORTED AT THE EDGES WITH A CENTRAL HOLE AND CARRYING W DISTRIBUTED ALONG THE CIRCUMFERENCE OF A CONCENTRIC CIRCLE

Let $b$ = radius of hole and $c$ = radius of load circle. The bending moment $M_q$ and $M_r$ diagrams are shown in Fig. 11.7 (b) and (c) respectively. Various values per unit width are given below:

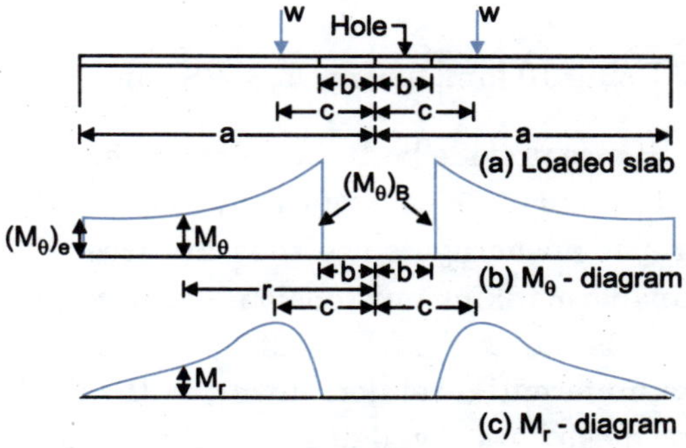

For $r < c$:

$$M_q = \frac{W}{4\pi} \cdot \frac{a^2}{a^2 - b^2}\left(1 + \frac{b^2}{r^2}\right)\left[\log_e \frac{a}{c} + \frac{1}{2} - \frac{c^2}{2a^2}\right] \quad \ldots(11.34)$$

$$M_r = \frac{W}{4\pi} \cdot \frac{a^2}{a^2 - b^2}\left(1 - \frac{b^2}{r^2}\right)\left[\log_e \frac{a}{c} + \frac{1}{2} - \frac{c^2}{2a^2}\right] \quad \ldots(11.35)$$

$$F_r = 0 \quad \ldots(11.36)$$

For $r > c$:

$$M_q = \frac{W}{4\pi}\left[\log_e \frac{c}{r} + \frac{1}{2} + \frac{a^2}{a^2 - b^2} \cdot \frac{r^2 + b^2}{r^2}\left(\log_e \frac{a}{c} + \frac{1}{2} - \frac{c^2}{2a^2}\right) - \frac{c^2}{2r^2}\right] \quad \ldots(11.37)$$

$$M_r = \frac{W}{4\pi}\left[\log_e \frac{c}{r} - \frac{1}{2} + \frac{a^2}{a^2 - b^2} \cdot \frac{r^2 - b^2}{r^2}\left(\log_e \frac{a}{c} + \frac{1}{2} - \frac{c^2}{2a^2}\right) + \frac{c^2}{2r^2}\right] \quad \ldots(11.38)$$

$$F = \frac{W}{2\pi r} \quad \ldots(11.39)$$

Fig. 11.7

**Example 11.1.** *A circular room has 5 m diameter from inside. Design a circular roof slab for room, to carry a superimposed load of 3800 N/m². Assume that the slab is simply supported at the edges. Use M 15 mix. and (a) Mild steel bars (b) HYSD bars.*

**Solution:** **(a) Mild Steel bars**

1. **Design constants**

    Taking $s_{st} = 140$ N/mm², we have $k_c = 0.404$; $j_c = 0.865$ and $R_c = 0.874$.

2. **Loading and B.M.** Let the total depth of slab be 120 mm.

    ∴ Self weight = $\frac{120}{1000} \times 1 \times 1 \times 25000$ = 3000 N/m²

    Super-imposed load = 3800 N/m²

    Total $w$ = 6800 N/m²

$$(M_r)_c = (M_\theta)_c = +\frac{3}{16}wa^2 = \frac{3}{16} \times 6800\,(2.5)^2 = 7969 \text{ N-m} = 7.969 \times 10^6 \text{ N-mm}$$

$$(M_\theta)_e = \frac{2}{16}wa^2 = \frac{2}{16} \times 6800\,(2.5)^2 = 5312 \text{ N-m} = 5.312 \times 10^6 \text{ N-mm}$$

**3. Design of section.** The slab will be designed for the greater of the two moments.

$$\therefore \quad d = \sqrt{\frac{7.969 \times 10^6}{1000 \times 0.874}} = 95.4 \text{ mm}$$

Let us keep total depth = 125 mm. Using 10 mm Φ bars and 15 mm clear cover, available $d$ = 125 − 15 − 5 = 105 mm for one layer and 105 − 10 = 95 mm for the other layer.

**4. Steel reinforcement.** Radial and circumferential reinforcement required at the centre is

$$A_{st} = \frac{7.969 \times 10^6}{140 \times 0.865 \times 95} = 693 \text{ mm}^2.$$

Using 10 mm Φ bars, $A_\Phi = 78.5$ mm², spacing $s = \dfrac{1000 \times 78.5}{693} = 113$ mm.

Hence provide both the reinforcements in the form of a mesh consisting of 10 mm diameter bars spaced @ 110 mm c/c in each of the two layers at right angles to each other. Near the edges, the bars do not have proper anchorage since they are free. There will be slipping tendency and hence the bars will not be capable of taking any tension. At the edges, radial tensile stress is zero, but circumferential tensile stresses exist because of $(M_\theta)_e = 5.312 \times 10^6$ N-mm. The tendency of slipping can be avoided by providing extra circumferential reinforcement, in the form of rings placed in a width equal to the development length of the mesh bars. Available $d$ for the rings = 95 − 10 = 85 mm.

$$\therefore \quad (A_{st})_\theta = \frac{5.312 \times 10^6}{140 \times 0.865 \times 85} = 516 \text{ mm}^2. \text{ Hence spacing } s = \frac{1000 \times 78.5}{516} \approx 150 \text{ mm}$$

$(M_\theta)_e$ is $\frac{2}{3}$ of maximum moment at the centre. The circumferential steel is to be provided for a length = $\frac{2}{3}(58.3\,\Phi) = \frac{2}{3} \times 58.3 \times 10 = 388$ mm (*i.e.* the length required to develop a tensile stress of $\frac{2}{3} \times 140 = 93.3$ N/mm² by bond). Hence provide rings of 10 mm Φ bars @ 150 mm c/c, total rings being (388/150) + 1 ≏ 3.

Since the slab is quite thin, no temperature or distribution reinforcement at the top faces of the slab is necessary. *However*, for a thicker slab (say greater than 200 mm) such reinforcement @ 0.12 % of cross-sectional area of concrete, may be provided at the top face of the slab.

**5. Check of shear.** Maximum S.F. at the edges is given by

$$V_r = \frac{1}{2}wa = \frac{1}{2} \times 6800 \times 2.5$$

= 8500 N/m width.

Hence, $\tau_r = \dfrac{V_r}{bd} = \dfrac{8500}{1000 \times 95} \approx 0.09$ N/mm² (safe)

The arrangement of the reinforcement will be as shown in Fig. 11.8

**(b) HYSD bars**

**(i) Design constants.** For HYSD bars, $\sigma_{st} = 230$ N/mm². With $\sigma_{cbc} = 5$ N/mm², we have $k_c = 0.2923$, $j_c = 0.9026$ and $R_c \approx 0.66$.

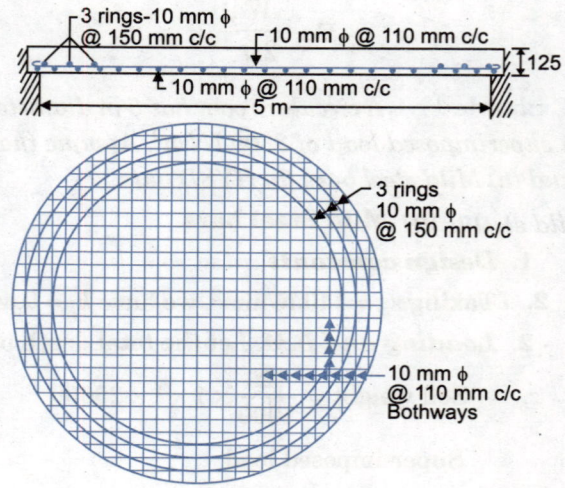

Fig. 11.8

# CIRCULAR SLABS

(ii) **Loading and B.M.** Let the total depth be 130 mm.

$$\therefore \quad \text{Self weight} = \frac{130}{1000} \times 1 \times 25000 \qquad\qquad = 3250 \text{ N/m}^2$$

$$\text{Super-imposed load} \qquad\qquad\qquad\qquad = 3800 \text{ N/m}^2$$

$$\text{Total } w \quad = 7050 \text{ N/m}^2$$

$$(M_r)_c = (M_\theta)_c = \frac{3}{16} wa^2 = \frac{3}{16} \times 7050 (2.5)^2 = 8262 \text{ N-m} = 8.262 \times 10^6 \text{ N-mm}$$

$$(M_\theta)_e = \frac{2}{16} wa^2 = \frac{2}{16} \times 7050 (2.5)^2 = 5508 \text{ N-m} = 5.508 \times 10^6 \text{ N-mm}$$

(iii) **Design of section.** Design moment = $8.262 \times 10^6$ N-mm

$$\therefore \quad d = \sqrt{\frac{8.262 \times 10^6}{1000 \times 0.66}} = 112 \text{ mm. Let us keep total depth} = 145 \text{ mm. Using 8 mm dia bars and 15 mm}$$

clear cover, $d = 145 - 15 - 4 = 126$ mm for one layer and $126 - 8 = 118$ mm for the other layer.

(iv) **Steel reinforcement.** Radial and circumferential reinforcement required at the centre is

$$A_{st} = \frac{8.262 \times 10^6}{230 \times 0.9026 \times 118} = 337.3 \text{ mm}^2. \text{ Using 8 mm dia bars having } A_\Phi = 50.27 \text{ mm}^2,$$

Spacing $s = \dfrac{1000 \times 50.27}{337.3} = 149$ mm. Hence provide a mesh 8 mm dia. bars spaced @ 140 mm c/c. To avoid slipping at the edges, provide extra circumferential reinforcement in the form of rings. Available depth = $118 - 8 = 110$ mm.

$$(A_{st})_\theta = \frac{5.508 \times 10^6}{230 \times 0.9026 \times 110} = 241.2 \text{ mm}^2. \quad \therefore \quad \text{Spacing } s = \frac{1000 \times 50.27}{241.2} = 208 \text{ mm}.$$

$(M_\theta)_e$ is $\frac{2}{3}$ of maximum moment at the centre. The circumferential steel is to be provided for a length of $\frac{2}{3}(60\,\Phi) = \frac{2}{3} \times 60 \times 8 = 320$ mm (i.e., a length required to develop a tensile stress of $\frac{2}{3} \times 230 = 153.3$ N/mm$^2$ by bond). Hence provide rings of 8 mm dia bars @ 208 mm c/c, total rings being = $(320/208) + 1 \approx 3$, i.e., spacing of rings = $(320/2) = 160$ mm.

(v) **Check for shear.** Maximum S.F. at the edges is given by

$$V_r = \frac{1}{2} wa = \frac{1}{2} \times 7050 \times 2.5 = 8812.5 \text{ N/m width}$$

$$\tau_r = \frac{V_r}{bd} = \frac{8812.5}{1000 \times 118} = 0.075 \text{ N/mm}^2 \text{ (safe)}$$

**Design: Example 11.2.** *Solve example 11.1, assuming the slab to be partially restrained at the edges. Use mild steel bars.*

**Solution: 1. Design constants:** $k_c = 0.404$; $j_c = 0.865$; $R_c = 0.874$

**2. Loading and B.M.** Let the total depth of slab = 110 mm

$$\therefore \qquad w = (110 \times 25) + 3800 = 6550 \text{ N/m}^2. \text{ Hence from } Eq.\ 11.15,$$

$$(M_r)_c = (M_\theta)_c = +\frac{2}{16} wa^2 = +\frac{2}{16} \times 6550 (2.5)^2 \times 1000 = +5.12 \times 10^6 \text{ N-mm}$$

$$(M_\theta)_e = +\frac{1}{16} wa^2 = +2.56 \times 10^6 \text{ N-mm}$$

$$(M_r)_e = -\frac{1}{16} wa^2 = -2.56 \times 10^6 \text{ N-mm}$$

The radial moment changes sign, the point of inflection being at a radius $a\sqrt{2/3} = 2.5\sqrt{2/3} = 2.04$ m = 2040 mm and or at a distance of 2500 − 2040 = 460 mm from the edge of the support.

**3. Design of section.** The thickness of the slab will be designed for a B.M. of $5.12 \times 10^6$.

$$d = \sqrt{\frac{5.12 \times 10^6}{1000 \times 0.874}} = 76.5 \text{ mm. Let us keep total thickness} = 110 \text{ mm.}$$

Using 10 mm Φ bars and a clear cover of 15 mm, available $d = 110 - 15 - 5 = 90$ mm for the bottom layer and $90 - 10 = 80$ mm for the top layer.

**4. Steel reinforcement at the centre.** The area of steel required per metre width at the centre of the slab is

$$A_{st} = \frac{5.12 \times 10^6}{140 \times 0.865 \times 80} = 528 \text{ mm}^2. \text{ Using 10 mm Φ bars, } A_\Phi = 78.5 \text{ mm}^2$$

∴ Spacing $= \dfrac{1000 \times 78.5}{528} \approx 150$ mm. Hence provide a mesh of reinforcement consisting of 10 mm Φ bars placed @ 150 mm c/c in two mutually perpendicular directions.

**5. Circumferential reinforcement at the edges.** At the edges, $(M_\theta)_e = +2.56 \times 10^6$ N-mm and area of steel required $= \frac{1}{2} \times 528 = 264$ mm². Since the mesh bars will not be in a position resist this, additional ring reinforcement will have to be provided near the edges.

Using 10 mm Φ bars for the rings, $s = \dfrac{1000 \times 78.5}{264} = 297$ mm

Since the circumferential moment at the edges is half of that at the centre, the required development length $= \frac{1}{2} \times 58.3\,\Phi = 291$ mm, i.e., the length required to develop a tensile stress $= \frac{1}{2} \times 140 = 70$ N/mm² by bond. Since $s = 297$ mm, only two rings spaced at 295 mm c/c will be required at the edges.

**6. Radial reinforcement at the edges.** Since there is negative radial moment of magnitude $= 2.56 \times 10^6$ N-mm, radial reinforcement has to be provided at the top face of the slab at the edges. The point of inflection is at a distance of 460 mm from the edge of the support. Hence the radial distance, measured from the face of the support, upto which the radial reinforcement is to be provided will be the greater of the following:

(i) 460 mm + $d$ = 460 + 90 = 550 mm.
(ii) 460 mm + 12 Φ = 460 + 12 × 10 = 580 mm.
(iii) $L_d$ = 58.3 Φ = 583 mm.

Hence provide these upto radial distance of 600 mm from the face of the support. The Code specifies that at least one third of the total negative moment reinforcement at the support shall extend beyond the point of inflection for a distance not less than one sixteenth of clear span $\left(=\frac{1}{16} \times 5000 = 312.5 \text{ mm}\right)$.

Hence provide alternate radial bars for a distance of 460 + 312.5 = 772.5 ≈ 800 mm from the face of support.

Available $d = 110 - 15 - 5 = 90$ mm.

$$A_{st} = \frac{2.56 \times 10^6}{140 \times 0.865 \times 90} = 234 \text{ mm}^2.$$

∴ Spacing of 10 mm Φ bars $= \dfrac{1000 \times 78.5}{234} \approx 330$ mm

However, maximum permissible spacing = 300 mm

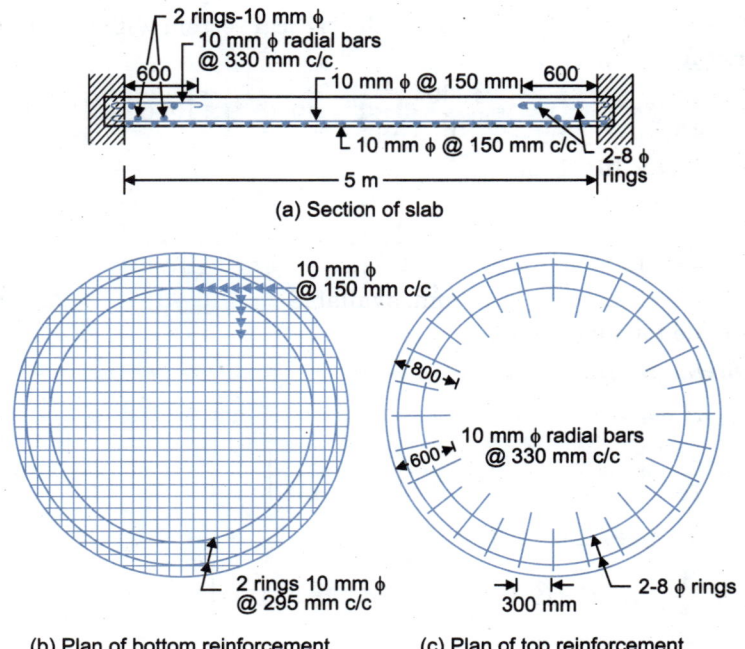

(a) Section of slab

(b) Plan of bottom reinforcement

(c) Plan of top reinforcement

Fig. 11.9

7. **Check for shear.** $F_r = \dfrac{1}{2} wa = \dfrac{1}{2} \times 6650 \times 2.5 = 8312.5$ N

$$\therefore \quad \tau_v = \dfrac{8312.5}{1000 \times 90} = 0.092 \text{ N/mm}^2$$

This is much less than permissible shear stress. (Table 3.1 and 3.2) even at the minimum reinforcement. The details of reinforcement etc. are shown in Fig. 11.9.

**Example 11.3.** *A traffic control post, 2 m in diameter is supported centrally by a reinforced concrete column, 30 cm in diameter. Design the circular slab for a super-imposed load of 1500 N/m². Use M 20 concrete and Fe 415 steel.*

**Solution:** Fig. 11.10 shows the slab supported centrally on the column and carrying U.D.L. $w$ N/m². Let the column reaction be $w_1$ N/m². Hence the slab may be considered to be simply supported at the edge and subjected to (*i*) a downward U.D.L. $w$ and (*ii*) an upward U.D.L. $w_1$ over a concentric circle of radius $b = 150$ mm.

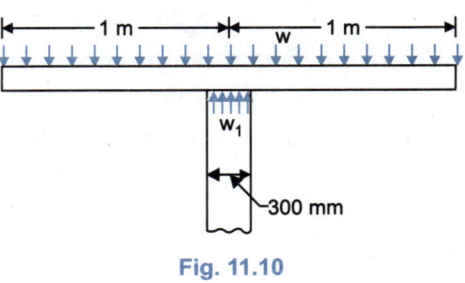

Fig. 11.10

Hence we can use § 11.2 and § 11.5 for determining net moments on the slabs.

1. **Design constants.** For the given set of materials, we have $k_c = 0.289$; $j_c = 0.904$; $R_c = 0.914$.
2. **Loading.** Let the total thickness of slab = 100 mm

$$w = (0.1 \times 25000) + 1500 = 4000 \text{ N/m}^2$$

and

$$w_1 = -\dfrac{4000 \, \pi \, (1)^2}{\pi \, (0.15)^2} = -177778 \text{ N/m}^2$$

(minus sign has been used since $w_1$ acts in an upward direction).

3. **Bending moments due to downward load $w$.** Refer § 11.2

$$(M_r)_e = 0$$

$$(M_\theta)_e = +\dfrac{2}{16} wa^2 = \dfrac{2}{16} \times 4000 \, (1)^2 \times 1000 = +5 \times 10^5 \text{ N-mm}$$

Since the slab is cast monolithic with the column, the critical section will be at the edge of the column, where $r = 0.15$ m. Hence from *Eq. 11.3*

$$M_\theta = +\frac{w}{16}(3a^2 - r^2) = +\frac{4000}{16}\{3 \times 1^2 - (0.15)^2\} \times 1000$$

$$= +7.44 \times 10^5 \text{ N-mm}$$

Also, from *Eq. 11.6*

$$M_r = +\frac{3}{16}w(a^2 - r^2) = +\frac{3 \times 4000}{16}\{1^2 - (0.15)^2\} \times 1000$$

$$= +7.33 \times 10^5 \text{ N-mm}$$

**4. Bending moments due to downward load $w_1$**

Refer § 11.5. From *Eq. 11.23*, at $r = 0.15$ m $= b$

$$M_r = -\frac{3}{16}wr^2 + \frac{1}{4}wb^2\left[1 - \log_e\left(\frac{b}{a}\right) - \frac{b^2}{4a^2}\right]$$

Putting $w = w_1 = -177778$ and $r = b = 0.15$

$$M_r = +\frac{3}{16} \times 177778\,(0.15)^2 - \frac{1}{4} \times 177778\,(0.15)^2\left[1 - \log_e\left(\frac{0.15}{1}\right) - \left(\frac{0.15}{2}\right)^2\right]$$

$$= 750 - 2891 = -2141 \text{ N-m} = -21.41 \times 10^5 \text{ N-mm}$$

Similarly from *Eq. 11.24*,

$$M_\theta = -\frac{1}{16}wr^2 + \frac{1}{4}wb^2\left[1 - \log_e\left(\frac{b}{a}\right) - \frac{b^2}{4a^2}\right]$$

Putting $w = w_1 = -177778$ and $r = b = 0.15$ m,

$$M_\theta = \frac{1}{16} \times 177778\,(0.15)^2 - \frac{1}{4} \times 177778\,(0.15)^2\left[1 - \log_e\left(\frac{0.15}{1}\right) - \left(\frac{0.15}{2}\right)^2\right]$$

$$= 250 - 2891 = -2641 \text{ N-m} = -26.41 \times 10^5 \text{ N-mm}$$

The radial moment at the edge is zero.

$$\therefore \quad (M_r)_e = 0.$$

The circumferential moment at the edge is given by *Eq. 11.29*.

$$(M_\theta)_e = -wb^2\left(-\frac{1}{4} + \frac{b^2}{8a^2}\right) = 177778\,(0.15)^2\left[-\frac{1}{4} + \frac{1}{8}\left(\frac{0.15}{1}\right)^2\right] = -989 \text{ N-m}$$

$$= -9.89 \times 10^5 \text{ N-mm}$$

**5. Net moments and thickness of slab.** The net moments are tabulated below.

| Case | Radial moments (N-mm) | | Circumferential moments (N-mm) | |
| --- | --- | --- | --- | --- |
| | $r = 0.15$ m | $r = 1$ m | $r = 0.15$ m | $r = 1$ m |
| (i) | $+7.33 \times 10^5$ | 0 | $+7.44 \times 10^2$ | $+5.00 \times 10^5$ |
| (ii) | $-21.41 \times 10^5$ | 0 | $-26.41 \times 10^5$ | $-9.89 \times 10^5$ |
| Net | $-14.08 \times 10^5$ | 0 | $-18.97 \times 10^5$ | $-4.89 \times 10^5$ |

Thus the moments throughout are negative, as expected, since the slab will bend having convexity upwards. Hence both radial and circumferential reinforcement will be placed at the top face of the slab.

The thickness will be designed for a bending moment of $18.97 \times 10^5$ N-mm.

$$d = \sqrt{\frac{18.97 \times 10^5}{1000 \times 0.904}} = 45.8 \text{ mm. Keep overall thickness} = 80 \text{ mm}.$$

Using 8 mm Φ bars and 20 mm nominal cover, available $d = 80 - 4 - 20 = 56$ mm for one layer and $56 - 8 = 48$ mm for the second layer of reinforcement.

**6. Steel reinforcement.** Let us design the reinforcement for the greater of the two moments.

$$\therefore \quad A_{st} = \frac{18.97 \times 10^5}{230 \times 0.904 \times 48} = 190 \text{ mm}^2.$$

Using 8 mm bars, $A_\Phi = 50.3$ mm²

$$\therefore \quad \text{Spacing } s = \frac{1000 \times 50.3}{190} = 264.7 \text{ mm}$$

Hence provide both the reinforcements in the form of a mesh consisting of 8 mm bars spaced at 250 mm c/c in two directions at right angles to each other. This mesh will not be in a position to resist circumferential tensile stress at the edge. The effective depth available for the rings reinforcement = $48 - 8 = 40$ mm

$$\therefore \quad A_{st} = \frac{4.89 \times 10^5}{230 \times 0.904 \times 40} = 59 \text{ mm}^2, \text{ for } 1000 \text{ mm}$$

width.

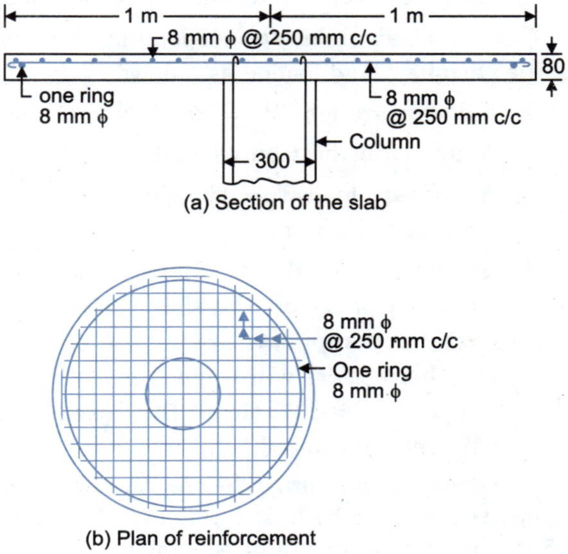

**Fig. 11.11**

(a) Section of the slab

(b) Plan of reinforcement

Since the circumferential moment at the edge is about $\frac{1}{4}$ of the moment at the centre, the circumferential reinforcement is to be provided for a width

$$= \frac{1}{4} \times 45 \, \Phi = \frac{1}{4} \times 45 \times 8 = 90 \text{ mm}.$$

Hence area of steel required for this width $= \frac{59 \times 90}{1000} \approx 6$ mm². Using 8 mm dia. bar area $A_\Phi = 50.3$ mm².

Hence one ring is sufficient.

**7. Check for shear.** Shear force at the edge of the column support is given by *Eq. 11.30* and *11.7*:

$$F = \frac{w_1 b^2}{2r} - \frac{1}{2} wr = \frac{w_1 b^2}{2b} - \frac{1}{2} wb = \frac{w_1 b}{2} - \frac{1}{2} wb = \frac{b}{2}(w_1 - w)$$

$$= \frac{0.15}{2}(177778 - 4000) = 13034 \text{ N/m}^2.$$

Hence, $\quad \tau_v = \dfrac{13034}{1000 \times 48} = 0.27$ N/mm².

The permissible shear stress, at 0.25% reinforcement = $1.3 \times 0.22 = 0.296$ N/mm² (Tables 3.1 and 3.2). Hence safe. The details of reinforcement etc. shown in Fig. 11.11.

**Design: Example 11.4.** *The base slab of an overhead circular tank is 6 m in diameter and supported centrally over a ring beam of 2 m diameter. The slab has a central hole of 1.2 m diameter. At the periphery of the central hole, the inner walls of the tank transmit a total load of 30 kN. At the outer periphery of the slab, the outer walls of the tank transmit a total load of 150 kN. In addition to this, the slab supports a uniformly distributed water load of intensity 18350 N/m², over the annular area. Design the base slab. Use M 20 concrete and Fe 415 steel.*

**Solution:** Assuming the total thickness of the slab to be 450 mm, its weight per sq. metre = $0.45 \times 25000 = 11250$ N. Hence total $w = 18350 + 11250 = 29600$ N/m². The loading on the base slab is shown in Fig. 11.12. The total load (reaction) on the support is

$$= 30000 + 150000 + \frac{\pi}{4}\{(6)^2 - (1.2)^2\} \times 29600 = 983443.5 \text{ N}.$$

For the given set of materials, we have $k_c = 0.289$, $j_c = 0.904$ and $R_c = 0.914$.

The given slab under various loadings may be looked upon a slab (with central hole), simply supported at the periphery and loaded as follows:

1. A downward load $W_1 = 30000$ N at the periphery of the hole.
2. A uniformly distributed load $w = 29600$ N/m² over the annular area.
3. An upward load $W_2 = 983443.5$ N at the outer periphery of a circle of radius 1 m.
4. A downward load $W = 150000$ N at the outer periphery of the slab. Since the imaginary supports are at the outer periphery, there will be no B.M. or S.F. due to this load $W$. Hence this case has not been considered.

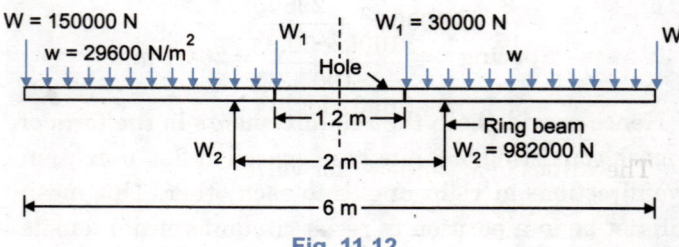

Fig. 11.12

**1. Bending moments due to $W_1$ (case 1 of loading).** Refer § 11.7. The radial and circumferential moments are given by *Eqs.* 11.38 and 11.37 respectively. Substituting $W = W_1 = 30000$ N, $a = 3$ m, $b = c = 0.6$ m, these equations reduce as follows:

$$M_\theta = \frac{W}{4\pi}\left[\log_e \frac{c}{r} + \frac{1}{2} + \frac{a^2}{a^2 - b^2} \cdot \frac{r^2 + b^2}{r^2} \times \left(\log_e \frac{a}{c} + \frac{1}{2} - \frac{c^2}{2a^2}\right) - \frac{c^2}{2r^2}\right]$$

$$= \frac{30000}{4\pi}\left[\log_e \frac{0.6}{r} + \frac{1}{2} + \frac{3^2}{3^2 - (0.6)^2} \cdot \frac{r^2 + (0.6)^2}{r^2} \times \left(\log_e \frac{3}{0.6} + \frac{1}{2} - \frac{(0.6)^2}{2(3)^2}\right) - \frac{1}{2}\left(\frac{0.6}{r}\right)^2\right]$$

$$= 2390\left[\log_e\left(\frac{0.6}{r}\right) + \frac{1}{2} + 2.18\left(\frac{r^2 + 0.36}{r^2}\right) - \frac{0.18}{r^2}\right] \quad \ldots(i)$$

$$M_r = \frac{W}{4\pi}\left[\log_e \frac{c}{r} - \frac{1}{2} + \frac{a^2}{a^2 - b^2} \cdot \frac{r^2 - b^2}{r^2} \times \left(\log_e \frac{a}{c} + \frac{1}{2} - \frac{c^2}{2a^2}\right) + \frac{c^2}{2r^2}\right]$$

$$= \frac{30000}{4\pi}\left[\log_e \frac{0.6}{2} - \frac{1}{2} + \frac{3^2}{3^2 - (0.6)^2} \cdot \frac{r^2 - (0.6)^2}{r^2} \times \left(\log_e \frac{3}{0.6} + \frac{1}{2} - \frac{(0.6)^2}{2(3)^2}\right) + \frac{1}{2}\left(\frac{0.6}{r}\right)^2\right]$$

$$= 2390\left[\log_e\left(\frac{0.6}{r}\right) - \frac{1}{2} + 2.18\left(\frac{r^2 - 0.36}{r^2}\right) + \frac{0.18}{r^2}\right] \quad \ldots(ii)$$

The values of $M_\theta$ and $M_r$ at the salient locations are tabulated below:

| $r$ (m) | 0.6 | 1 | 1.5 | 2 | 2.5 | 3 |
|---|---|---|---|---|---|---|
| $M_\theta$ (N-m) | + 10400 | + 6970 | + 4850 | + 3870 | + 3200 | + 2700 |
| $M_r$ (N-m) | 0 | + 1720 | + 1170 | + 765 | + 381 | 0 |

**2. Bending moments due to U.D.L. $w$ (case 2 of loading).** Refer § 11.6. The circumferential and radial moments are given by *Eqs.* 11.31 and 11.32 respectively.

$$M_\theta = -\frac{1}{16}wr^2 + \frac{wb^2}{4}\left[\log_e\left(\frac{r}{a}\right) + \frac{3}{4}\left(-\frac{1}{3} + \frac{a^2}{b^2} + \frac{a^2}{r^2}\right) - \frac{a^2 + r^2}{r^2} \cdot \frac{b^2}{a^2 - b^2}\log_e \frac{a}{b}\right]$$

**CIRCULAR SLABS** **281**

$$= -\frac{29600}{16}(r^2) + \frac{29600}{4}(0.6)^2 \left[\log_e\left(\frac{r}{3}\right) + \frac{3}{4}\left(-\frac{1}{3} + \left(\frac{3}{0.6}\right)^2 + \left(\frac{3}{r}\right)^2\right)\right.$$
$$\left. - \frac{3^2 + (r)^2}{r^2} \cdot \frac{(0.6)^2}{3^2 - (0.6)^2}\log_e\left(\frac{3}{3.6}\right)\right]$$

$$= -1850\,r^2 + 2660\left[\log_e\left(\frac{r}{3}\right) + \frac{3}{4}\left(24.7 + \frac{9}{r^2}\right) - \frac{9 + r^2}{r^2} \times 0.067\right] \quad ...(iii)$$

$$M_r = -\frac{3}{16}wr^2 + \frac{wb^2}{4}\left[\log_e\frac{r}{a} + \frac{3}{4}\left(1 + \frac{a^2}{b^2} - \frac{a^2}{r^2}\right) + \frac{a^2 - r^2}{r^2} \cdot \frac{b^2}{a^2 - b^2}\log_e\frac{a}{b}\right]$$

$$= -\frac{3}{16} \times 29600\,r^2 + \frac{29600(0.6)^2}{4}\left[\log_e\frac{r}{3} + \frac{3}{4}\left(1 + \left(\frac{3}{0.6}\right)^2 - \frac{9}{r^2}\right) + \frac{9 - r^2}{r^2} \cdot \frac{(0.6)^2}{9 - (0.6)^2}\log_e\left(\frac{3}{0.6}\right)\right]$$

$$= -5550\,r^2 + 2660\left[\log_e\frac{r}{3} + \frac{3}{4}\left(26 - \frac{9}{r^2}\right) + \frac{9 - r^2}{r^2} \times 0.067\right] \quad ...(iv)$$

The values of $M_\theta$ and $M_r$ for various for $r$ are tabulated below:

| $r$ (m) | 0.6 | 1 | 1.5 | 2 | 2.5 | 3 |
|---|---|---|---|---|---|---|
| $M_\theta$ (N-m) | + 89730 | + 60750 | + 50340 | + 44400 | + 39650 | + 34350 |
| $M_r$ (N-m) | 0 | + 26850 | + 30020 | + 24200 | + 13950 | 0 |

### 3. Bending moments due to upward load $W_2 = 983443.5$ N.

Refer § 11.7. The moments are given by *Eqs. 11.34, 11.35, 11.37* and *11.38*,
where $W = W_2 = -983443.5$ N, $b = 0.6$ m, $c = 1$ m and $a = 3$ m.

For $r < c$

$$M_\theta = \frac{W}{4\pi} \times \frac{a^2}{a^2 - b^2}\left(1 + \frac{b^2}{r^2}\right)\left[\log_e\left(\frac{a}{c}\right) + \frac{1}{2} - \frac{c^2}{2a^2}\right]$$

$$= -\frac{983443.5}{4\pi} \times \frac{9}{9 - 0.36}\left(1 + \frac{0.36}{r^2}\right)\left[\log_e\left(\frac{3}{1}\right) + \frac{1}{2} - \frac{1}{2 \times 9}\right] = -125791.2\left(1 + \frac{0.36}{r^2}\right) \quad ...(v)$$

$$M_r = \frac{W}{4\pi} \times \frac{a^2}{a^2 - b^2}\left(1 - \frac{b^2}{r^2}\right)\left[\log_e\frac{a}{c} + \frac{1}{2} - \frac{c^2}{2a^2}\right] = -125791.2\left(1 - \frac{0.36}{r^2}\right) \quad ...(vi)$$

For $r > c$

$$M_\theta = \frac{W}{4\pi}\left[\log_e\frac{c}{r} + \frac{1}{2} + \frac{a^2}{a^2 - b^2} \cdot \frac{r^2 + b^2}{r^2}\left(\log_e\frac{a}{c} + \frac{1}{2} - \frac{c}{2a^2}\right) - \frac{c^2}{2r^2}\right]$$

$$= -\frac{983443.5}{4\pi}\left[\log_e\frac{1}{r} + \frac{1}{2} + \frac{9}{9 - 0.36} \cdot \frac{r^2 + 0.36}{r^2} \times \left(\log_e 3 + \frac{1}{2} - \frac{1}{2 \times 9}\right) - \frac{1}{2r^2}\right]$$

$$= -78259.9\left[\log_e\frac{1}{r} + \frac{1}{2} + 1.6\left(\frac{r^2 + 0.36}{r^2}\right) - \frac{1}{2r^2}\right] \quad ...(vii)$$

$$M_r = \frac{W}{4\pi}\left[\log_e\frac{c}{r} - \frac{1}{2} + \frac{a^2}{a^2 - b^2} \cdot \frac{r^2 - b^2}{r^2} \times \left(\log_e\frac{a}{c} + \frac{1}{2} - \frac{c^2}{2a^2}\right) + \frac{c}{2r^2}\right]$$

$$= -78259.9\left[\log_e\frac{1}{r} - \frac{1}{2} + 1.6\left(\frac{r^2 - 0.36}{r^2}\right) + \frac{1}{2r^2}\right] \quad ...(viii)$$

The values of $M_\theta$ and $M_r$ are calculated for various values of $r$, from *Eqs* (v), (vi), (vii) and (viii) and tabulated below:

| $r$ (m) | 0.6 | 1 | 1.5 | 2 | 2.5 | 3 |
|---|---|---|---|---|---|---|
| $M_\theta$ (N-m) | − 251500 | − 171020 | − 136453.2 | − 110330 | − 93880 | − 78234 |
| $M_r$ (N-m) | 0 | − 80480 | − 52356.6 | − 29287 | − 13289.8 | 0 |

**4. Net moments and S.F.** The combined moments (kg – m) are tabulated on next page.

**5. Shear force.** Maximum shear will occur at the outer edge of the ring beam, its magnitude per metre width is given by *Eqs. 11.39* and *11.33*.

$$F = \frac{W_1}{2\pi r} + \left(\frac{wr}{2} - \frac{wb^2}{2r}\right) - \frac{W_2}{2\pi r} = \frac{30000}{2\pi \times 1} + \frac{29600 \times 1}{2}\left\{1 - \left(\frac{0.6}{1}\right)^2\right\} - \frac{983443.5}{2\pi \times 1}$$

$$= -142273.2 \text{ N per meter width}$$

| Moment | Loading case | \multicolumn{6}{c}{$r$ (metres)} |
|---|---|---|---|---|---|---|---|
| | | 0.6 | 1 | 1.5 | 2 | 2.5 | 3 |
| $M_\theta$ | 1. | + 10400 | + 6970 | + 4850 | + 3870 | + 3200 | +2700 |
| | 2. | + 89730 | + 60750 | + 50340 | + 44400 | + 39650 | + 34350 |
| | 3. | − 251500 | − 171020 | − 136453.2 | − 110330 | − 93880 | − 78234 |
| | Net | − 151370 | − 103300 | − 81263 | − 62060 | − 51030 | − 41184 |
| $M_r$ | 1. | 0 | + 1720 | + 1170 | + 765 | + 381 | 0 |
| | 2. | 0 | + 26850 | + 30020 | + 24200 | + 13950 | 0 |
| | 3. | 0 | − 80480 | − 52357 | − 29287 | − 13290 | 0 |
| | Net | 0 | − 51910 | − 21167 | − 4322 | + 1041 | 0 |

**6. Design of thickness.**

From B.M. point of view, $d = \sqrt{\dfrac{151370 \times 1000}{1000 \times 0.914}} = 407$ mm

From S.F. point of view, permitting $\tau_v = \tau_c = 0.3$ N/mm² corresponding to about 0.5% reinforcement (Table 3.1), we get $d = \dfrac{V}{b \cdot \tau_v} = \dfrac{142044}{1000 \times 0.3} = 473$ mm.

Let us keep total thickness = 500 mm. Using 16 mm Φ bars and a cover of 20 mm, available $d$ for circumferential reinforcement = 500 − 20 − 8 = 472 mm. Using 8 mm φ radial bars, available $d$ for radial reinforcement = 472 − 80 − 4 = 460 mm.

**7. Design of reinforcement.** *Circumferential reinforcement.* For circumferential reinforcement, to be provided in the form of rings, near the inner edge of slab:

$$(A_{st})_\theta = \frac{151370 \times 1000}{230 \times 0.904 \times 472} = 1542.4 \text{ mm}^2 \text{ per meter width.}$$

Using 16 mm Φ, having $A_\Phi = 201.06$ mm², Spacing $s = \dfrac{1000 \times 201.06}{1542.4} = 130.4$ mm

However, provide the rings @ 100 mm c/c near the inner edge. At the outer edge.

$$(A_{st})_\theta = \frac{41184 \times 1000}{230 \times 0.904 \times 472} = 419.4 \text{ mm}^2. \quad \therefore \ s = \frac{1000 \times 201.06}{419.4} = 479.4 \text{ mm}$$

But maximum permissible spacing is 300 mm as per IS code. However, the spacing of the rings can be increased gradually form 100 mm c/c at the inner edge to 300 mm c/c at the outer edge.

*Radial reinforcement*

$$(A_{st})_r = \frac{51910 \times 1000}{230 \times 0.904 \times 460} = 543 \text{ mm}^2$$

Using 8 mm $\Phi$ $A_\Phi = 50.3$ mm$^2$.

$$\therefore s = \frac{1000 \times 50.3}{543} = 92.6 \text{ mm}$$

Hence provide radial bars @ 90 mm c/c at the ring beam support. These radial bars may bent downwards near the inner edge.

**8. Check for development length.** The radial reinforcement should be checked for development length at the ring beam support, where their spacing is @ 90 mm c/c. Hence,

$$(A_{st})_r = 538 \times \frac{93}{90} = 556 \text{ mm}^2$$

$$M_1 = \sigma_{st} \cdot (A_{st})_r \cdot j_c d$$
$$= 230 \times 556 \times 0.904 \times 460$$
$$= 53.18 \times 10^6 \text{ N-mm}$$

$V = 142044$ N

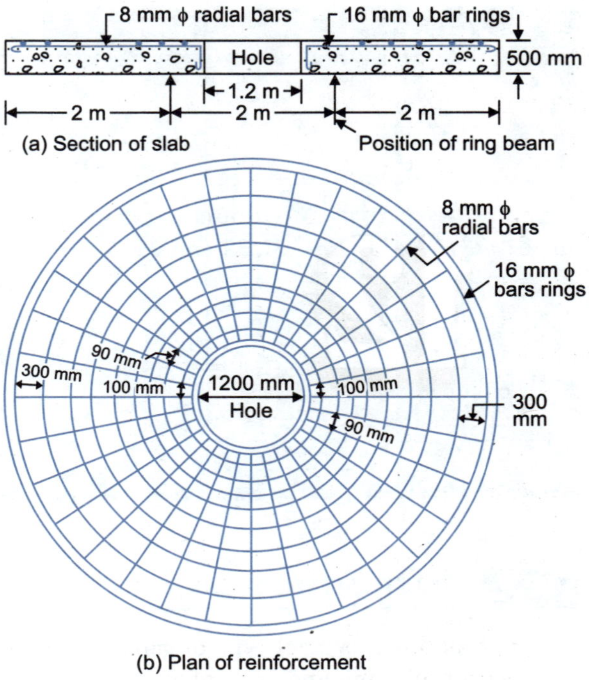

As per conditions of support (Fig. 11.13),

$$l_0 = \left(\frac{2000 - 1200}{2} - 20\right) = 380 \text{ mm}$$

$$\therefore \quad 1.3 \frac{M_1}{V} + l_0 = 1.3 \times \frac{53.18 \times 10^6}{142044} + 380 = 866.7 \text{ mm}$$

$L_d = 45 \Phi = 45 \times 10 = 450$ mm.

Hence Code requirements are satisfied. However, bend the radial bars downwards, by 90°, at the perimeter of the central hole.

The details of reinforcement etc are shown in Fig. 11.13.

**Fig. 11.13**
(a) Section of slab
(b) Plan of reinforcement

## PROBLEMS

1. Design the roof slab for a circular room 6 metre in diameter from inside and carrying a super imposed load of 5 kN/m². Assume that the slab is simply supported at the edges. Use M 20 mix and Fe 415 steel. The thickness of wall is 400 mm.

2. Redesign problem 1 assuming that the slab is restrained at the edge.

3. A circular room 8 metre in diameter from inside carries a 500 mm dia. column at its centre. Design a circular slab, simply supported at the outer edge and supported on the column at its middle. The slab carries a total uniformly distributed load of 3 kN/m² including its own weight. Use M 20 mix and Fe 415 steel.

4. Design the slab of a traffic control post, 1.5 m diameter and supported centrally on a column of 200 mm diameter. The total superimposed load, inclusive of its own weight may be taken as 1.8 kN/m². Use M 20 mix and Fe 415 steel reinforcement.

5. Design a circular slab for a room 4 m in diameter with fixed edges. The slab is subjected to a superimposed load 5.6 kN/m². Use $M_{25}$ grade and Fe 415 steel.

6. Design a simply supported circular slab a superimposed load of 4.5 kN/m². The diameter of the circular slab is 5 meters. Use $M_{20}$ grade concrete and Fe 415 steel. Assume the Poisson's Ratio for RCC as Zero.

# CHAPTER 12

# FLAT SLABS

## 12.1. INTRODUCTION

A flat slab is a typical type of construction in which a reinforced slab is built monolithically with the supporting columns and is reinforced in two or more directions, without any provision of beams. The flat slab thus transfers the load directly to the supporting columns suitably spaced below the slab. Because of exclusion of beam-system in this type of construction, a plain ceiling is obtained, thus giving attractive appearance from architectural point of view. The plain ceiling diffuses the light better and is considered less vulnerable in the case of fire than the usual beam slab construction. The flat slab is easier to construct and requires cheaper form-work. Concrete is more logically used in this type of construction, and hence in the case of large spans and heavy load, the total cost is considerably less.

## 12.2. COMPONENTS OF FLAT SLAB CONSTRUCTION

Some of the components of flat slab construction, shown in Fig. 12.1 and 12.2 are defined below.

**1. *Drop of flat slab.*** The slab in a flat slab construction may be either with drop or without drop. *Drop* is that part of the slab around the column, which is of greater thickness than the rest of the slab. The drop may have 25% to 50% more thickness, than the rest of the slab, as shown in Fig. 12.2 (*b*). Flat slab without drop is shown in Fig. 12.2 (*a*) and (*c*).

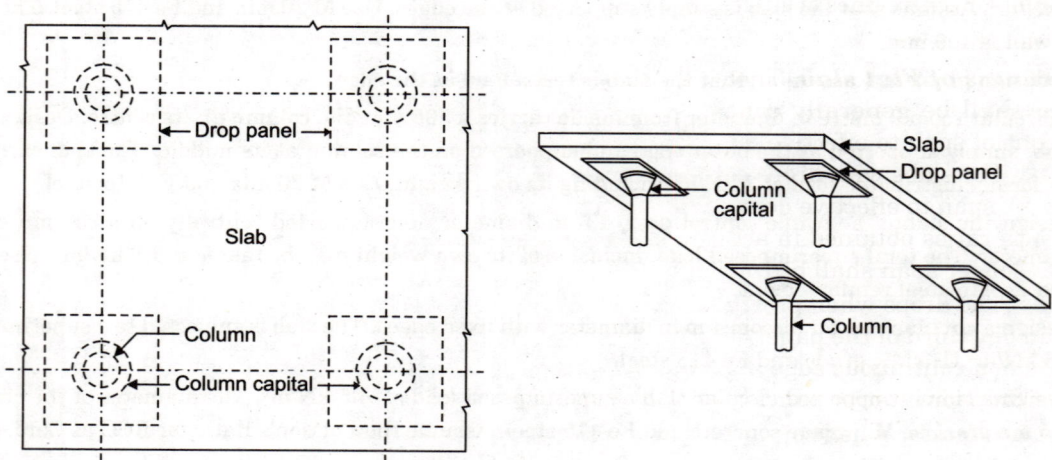

Fig. 12.1

**2. *Capital or column head.*** Sometimes, the diameter of a supporting column is increased below the slab. This part of column with increased diameter is called 'column head'. The column head increases rigidity of the slab and 'column connection' increases resistance of the slab to shear and decreases the effective span of the slab. Fig. 12.2 (c) shows a flat slab without drop but with column head.

**3. *Panel.*** A panel of a flat slab construction is the area enclosed between the centre lines connecting adjacent columns in two directions and the outline of the column heads as shown in Fig. 12.3.

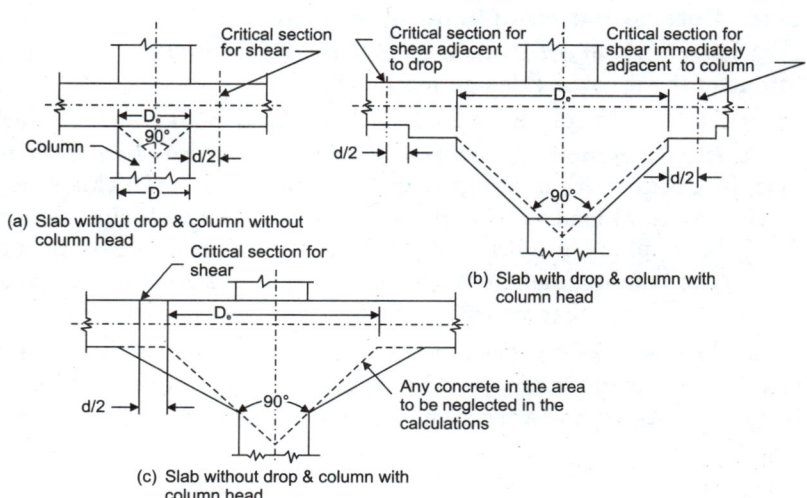

**Fig. 12.2**

## 12.3. INDIAN CODE RECOMMENDATIONS (IS : 456-2000)

### (a) General

Fig. 12.2 shows flat slab with or without drops. The following definitions shall apply:

**1. *Column strip.*** Column strip means a design strip having a width of 0.25 $l_2$, but not greater than 0.25 $l_1$ on each side of the column centre line, where $l_1$ is the span in the direction moments are being determined, measured centre to centre of supports.

**2. *Middle strip.*** Middle strip means a design strip bounded on each of its opposite sides by the column strip.

**3. *Panel.*** Panel means that part of the slab bounded on each of its four sides by the centre-line of a column or centre lines of adjacent spans.

### (b) Proportioning

**1. *Thickness of Flat slab.*** The thickness of flat slab shall be generally controlled by considerations of span to effective depth ratios given in § 7.1. For slabs with drops conforming

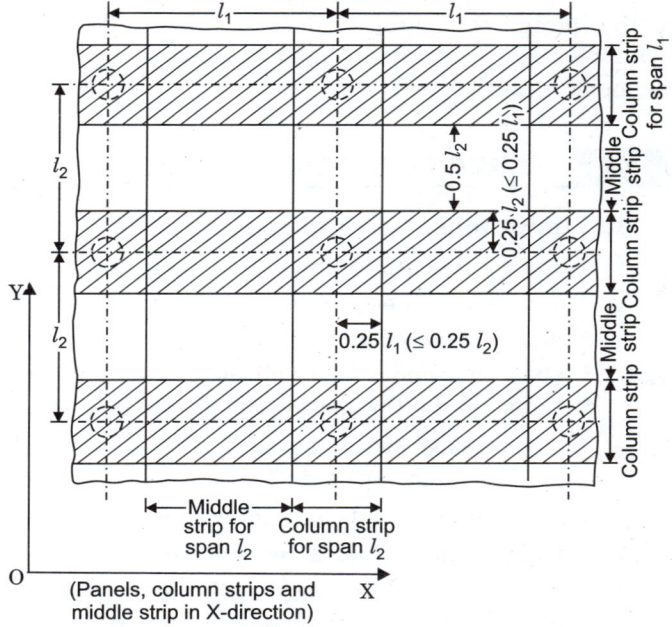

**Fig. 12.3.** Division of flat slab panel into strips.

to clause (b-2), span to effective depth ratios given in § 7.1 shall be applied directly; otherwise the span to effective depth ratios obtained in accordance with provisions of § 7.1 shall be multiplied by 0.9. For this purpose, the longer span shall be considered. The minimum thickness of slab shall be 125 mm.

**2. *Drops.*** The drops when provided shall be rectangular in plan, and have a length in each direction not less than one third of the panel length in the direction. For exterior panels, the width of drops at right angles to the non-continuous edge and measured from centre line of the columns shall be equal to one half the width of drop for interior panels.

**3. *Column heads.*** When column heads are provided, that portion of a column head which lies within the largest circular cone or pyramid that has a vertex angle of 90° and can be included entirely within the outlines of the column and the column head, shall be considered for design purposes (see Fig. 12.2).

### (c) Determination of bending moment

1. **Methods of analysis and design.** (IS: 456-2000) It shall be permissible to design the slab system by one of the following design methods.

   (i) The Direct design method. (ii) The Equivalent frame method.

2. **Bending moments in panels with marginal beams or walls.** Where the slab is supported by a marginal beam with depth greater than 1.5 times the thickness of the slab, or by a wall, then :

   (i) The total load to be carried by the beam or wall shall comprise those loads directly on the wall or beam plus a uniformly distributed load equal to one-quarter of the total load on the slab.

   (ii) The bending moments on the half column strip adjacent to the beam or wall shall be one-quarter of the bending moments for the first interior column strip.

3. **Transfer of bending moments to columns.** When unbalanced gravity load, wind, earthquake, or other lateral loads cause transfer of bending moment between slab and column, the flexural stresses shall be investigated using a fraction, $\alpha$ of the moment given by

$$\alpha = \frac{1}{1 + \frac{2}{3}\sqrt{\frac{a_1}{a_2}}}$$

where $a_1$ = overall dimension of the critical section for shear in the direction in which moment acts, and

$a_2$ = overall dimension of the critical section for shear transverse to the direction in which moment acts.

A slab width between lines that are one and one-half slab or drop panel thickness, 1.5 $D$, on each side of the column *or* capital may be considered effective, $D$ being the size of the column. Concentration of reinforcement over column head by closer spacing or additional reinforcement may be used to resist the moment on this section.

## 12.4. DIRECT DESIGN METHOD

1. **Limitations.** Slab systems designed by the direct design method shall fulfil the following conditions:

   (i) There shall be minimum of three continuous spans in each direction.

   (ii) The panel shall be rectangular, and the ratio of the longer span to the shorter span within a panel shall not be greater than 2.0.

   (iii) It shall be permissible to offset columns to a maximum of 10 percent of the span in the direction of the offset notwithstanding the provision in (ii).

   (iv) The successive span lengths in each direction shall not differ by more than one third of the longer span. The end spans may be shorter but not longer than the interior spans, and

   (v) The design live load shall not exceed three times the design dead load.

2. **Total design moment for a span**

   (i) In the direct design method, the total design moment for a span shall be determined for a strip bounded laterally by the centre line of the panel on each side of the centre line of the supports.

   (ii) The absolute sum of the positive and average negative bending moments in each direction shall be taken as :

$$M_0 = \frac{W l_n}{8},$$

where $M_0$ = Total moment ; $W$ = design load on area $l_2 \times l_n$

$l_n$ = clear span extending from face to face of columns, capitals, brackets or walls, but not less than $0.65\, l_1$

$l_1$ = length of span in the direction of $M_0$ ; and

$l_2$ = length of span transverse to $l_1$.

# FLAT SLABS

Intaking the values of $l_n$, $l_1$ and $l_2$ the following clauses are to be carefully noted:

(iii) Circular supports shall be treated as square supports having the same area.

(iv) When the transverse span of the panels on either side of the centre line of support varies, $l_2$ shall be taken as the average of the transverse spans from Fig. (12.3) it is given by $\left(\dfrac{l_{2a} + l_{2b}}{2}\right)$

(v) When the span adjacent and parallel to an edge is being considered, the distance from the edge to the centre line of the panel shall be substituted for $l_2$ in (ii) above.

**3. Distribution of bending moment into negative and positive design moments**

(i) The negative design moment shall be treated at the face of rectangular supports, circular supports being treated as square supports having the same area.

(ii) In an interior span, the total design moment $M_0$ shall be distributed in the following proportions : Negative design moment : 0.65. Positive design moment : 0.35.

(iii) In an end span, the total design moment $M_0$ shall be distributed in the following proportions :

Interior negative design moment: $\left[ 0.75 - \dfrac{0.10}{1 + \dfrac{1}{\alpha_c}} \right]$

Positive design moment: $\left[ 0.63 - \dfrac{0.28}{1 + \dfrac{1}{\alpha_c}} \right]$

Exterior negative design moment: $\left[ \dfrac{0.65}{1 + \dfrac{1}{\alpha_c}} \right]$

($\alpha_c$) is the ratio of flexural stiffness of the exterior columns to the flexural stiffness of the slab at a joint taken in the direction moments are being determined and is given by

$$\alpha_c = \dfrac{\Sigma K_c}{\Sigma K_s}$$

where $\Sigma K_c$ = Sum of the flexural stiffness of the columns meeting at the joint, and

$K_s$ = flexural stiffness of the slab, expressed as moment per unit rotation.

(iv) It shall be permissible to modify these design moments by upto 10 percent, so long as the total design moment $M_0$ for the panel in the direction considered is not less than that required by 2 (ii) above.

(v) The negative moment section shall be designed to resist the larger of the two interior negative design moments determined for the spans forming into a common support unless an analysis is made to distribute the unbalanced moment in accordance with the stiffness of the adjoining parts.

**4. Distribution of bending moments across the panel width:** Bending moments at critical cross-section shall be distributed to the column strips and middle strips as stated below:

Distribution of Moments Across the Panel Width in a Column Strip

| S. No. | Distributed Moment | Percent of total moment |
|---|---|---|
| 1. | Negative B.M. at the exterior support | 100 |
| 2. | Negative B.M. at the interior support | 75 |
| 3. | Positive bending moment | 60 |

(i) *Column strip : Negative moment at an interior support.* At an interior support, the column strip shall be designed to resist 75 percent of the total negative moment in the panel at that support.

### (ii) Column strip : Negative moment at an exterior support.

(a) At an exterior support, the column strip shall be designed to resist the total negative moment in the panel at the support.

(b) Where the exterior support consists of a column or a wall extending for a distance equal to or greater than three quarters of the value of $l_2$, the length of span transverse to the direction moments are being determined, the exterior negative moments shall be considered to be uniformly distributed across the length $l_2$.

### (iii) Column strip : Positive moment for each span :
For each span, the column strip shall be designed to resist 60 percent of the total positive moment in the panel.

### (iv) Moment in the middle strip :
The middle strip shall be designed on the following basis:

(a) The portion of the design moment not resisted by the column strip shall be assigned to the adjacent middle strips.

(b) Each middle strip shall be proportioned to resist the sum of the moment assigned to its two half middle strips.

(c) The middle strip adjacent and parallel to an edge supported by a wall shall be proportioned to resist twice the moment assigned to half the middle strip corresponding to the first row of interior columns.

## 5. Moments in columns

In this type of constructions column moments are to be modified as suggested in IS: 456-2000 (Clause No: 31.4.5) :

(i) Columns built integrally with the slab system shall be designed to resist moments arising from loads on the slab systems.

(ii) At an interior support, the supporting members above and below the slab shall be designed to resist the moment $M$ given by the following equation, in direct proportion to their stiffness unless general analysis is made:

$$M = 0.08 \frac{(w_d + 0.5 w_l) l_2 l_n^2 - w_d' l_2' l_n'^2}{1 + \frac{1}{\alpha_c}} \ ;$$

where $w_d, w_l$ = design dead and live loads respectively, per unit area

$l_2$ = length of span transverse to the direction of $M$.

$l_n$ = length of the clear span in the direction of $M$, measured face to face of supports.

$\alpha_c = \frac{\Sigma K_c}{\Sigma K_s}$ and $w_d'$, $l_2'$ and $l_n'$ refer to the shorter span.

## 6. Effect of pattern of loading

In the direct design method, when the ratio of the live load to dead load exceed 0.5 :

(i) the sum of the flexural stiffness of the columns above and below the slab, $\Sigma K_c$ shall be such that $\alpha_c$ is not less than the appropriate minimum value $\alpha_{c.min}$ specified in Table below, or

(ii) if the sum of flexural stiffnesses of the column $\Sigma K_c$ does not satisfy (i) above, the positive design moments for the panel shall be multiplied by the coefficient $\beta_s$ given by the following equation:

$$\beta_s = 1 + \left( \frac{2 - \frac{w_d}{w_l}}{4 + \frac{w_d}{w_l}} \right) \left( 1 - \frac{\alpha_c}{\alpha_{c.min}} \right)$$

$\alpha_c$ is the ratio of flexural stiffness of the columns above and below the slab to the flexural stiffness of the slabs at the joint taken in the direction moments are being determined and the given by :

$$\alpha_c = \frac{\Sigma K_c}{\Sigma K_s},$$

where $K_c$ and $K_s$ are flexural stiffness of column and slab respectively.

The method of computing flexural stiffnesses of various members has been illustrated in the design example.

**TABLE 12.1.** Minimum permissible values of $\alpha_c$ (IS : 456-2000)

| Live load/Dead load | Ratio $l_2/l_1$ | Value of $\alpha_{c\ (min.)}$ |
|---|---|---|
| 0.5 | 0.5 to 2.0 | 0 |
| 1.0 | 0.5 | 0.6 |
| 1.0 | 0.8 | 0.7 |
| 1.0 | 1.0 | 0.7 |
| 1.0 | 1.25 | 0.8 |
| 1.0 | 2.0 | 1.2 |
| 2.0 | 0.5 | 1.3 |
| 2.0 | 0.8 | 1.5 |
| 2.0 | 1.0 | 1.6 |
| 2.0 | 1.25 | 1.9 |
| 2.0 | 2.0 | 4.9 |
| 3.0 | 0.5 | 1.8 |
| 3.0 | 0.8 | 2.0 |
| 3.0 | 1.0 | 2.3 |
| 3.0 | 1.25 | 2.8 |
| 3.0 | 2.00 | 13.0 |

## 12.5. EQUIVALENT FRAME METHOD

IS: 456-2000 recommends the analysis of flat slab and column structure as a rigid frame to get design moment and shear force with the following assumptions.

**1. Assumptions.** The bending moments and shear forces may be determined by an analysis of the structure as a continuous frame and the following assumptions may be made:

(i) The structure shall be considered to be made up of equivalent frames on column lines taken longitudinally and transversely through the building. Each frame consists of a row of equivalent columns or supports, bounded laterally by the centre line of the panel on each side of the centre line of the columns or supports. Frames adjacent and parallel to an edge shall be bounded by the edge and the centre line of the adjacent panel.

(ii) Each such frame may be analysed in its entirety, or for vertical loading—each floor there-of and the roof may be analysed separately with columns being assumed fixed at their remote ends. Where slabs are thus analysed separately, it may be assumed in determining the bending moment at a given support that the slab is fixed at any support two panels distant there from provided the slab is continuous beyond that point.

(iii) For the purpose of determining relative stiffness of members, the moment of inertia of any slab or column may be assumed to be that of the gross cross-section of the concrete alone.

(iv) Variations of moment of inertia along the axis of the slab on account of provision of drops shall be taken into account. In case of recessed or coffered slab which is made solid in the region of the columns, the stiffening effect may be ignored provided the solid part of the slab does not extend more than 0.15 $l_e$ into the span measured from the centre line of the columns. The stiffening effect of flared column heads may be ignored.

**2. Loading Pattern.** (*i*) When the loading pattern is known, the structure shall be analysed for the load concerned.

(*ii*) When the live load is variable but does not exceed three quarters of the dead load, or the nature of the live load is that all panels will be loaded simultaneously, the maximum moments may be assumed to occur at all sections when full design live load is on entire slab system.

(*iii*) For other conditions of live load/dead load ratio and when all panels are not loaded simultaneously:

(*a*) maximum positive moment near mid-span of a panel may be assumed to occur when three-quarters of the full design live load is on the panel and on alternate panels; and (*b*) maximum negative moment in the slab at a support may be assumed to occur when three-quarters of the full design live load is on the adjacent panels only.

(*iv*) In no case shall design moments be taken to be less than those occuring with full design live load on all panels.

The moments determined in the beam of frame. (Flat slab) may be reduced in such proportion that the numerical sum of positive and average negative moments is not less than the value of total design moment $M_0 = \dfrac{W l_n}{8}$. The distribution of slab moment into column strip and middle strips is to be made in the same manner as specified in direct design method.

**3. Negative Design Moment**

(*i*) At interior supports, the critical section for negative moment in both the column strip and middle strip, shall be taken at the face of rectilinear supports but in no case at a distance greater than $0.175\, l_1$ from the centre of the column where $l_1$ is the length of the span in the direction moments are being determined, measured centre-to-centre of supports.

(*ii*) At exterior supports provided with brackets or capitals, the critical section for negative moment in the direction perpendicular to the edge shall be taken at a distance from the face of the supporting element not greater than one-half the projection of the bracket or capital beyond the face of the supporting element.

(*iii*) Circular or regular polygon shaped supports shall be treated as square supports having the same area.

**4. Modification of Maximum Moment.** Moments determined by means of the equivalent frame method, for slabs which fulfills the limitations of *direct design method* may be reduced in such proportion that the numerical sum of the positive and average negative moments is not less than the value of total design moment $M_0$ specified in para 2 of 12.4.

**5. Distribution of Bending Moment Across the Panel width**

(*i*) *Column strip : Negative moment at an interior support.* At an interior support, the column strip shall be designed to resist 75 percent of the total negative moment in the panel at that support.

(*ii*) *Column strip : Negative moment at an exterior support*

(*a*) As an exterior support, the column strip shall be designed to resist the total negative moment in the panel at the support.

(*b*) Where the exterior support consists of a column or a wall extending for a distance equal to or greater than three-quarters of the value of $l_2$, the length of span transvere to the direction moments are being determined, the exterior negative moment shall be considered to be uniformly distributed across the length $l_2$.

(*iii*) *Column strip : Positive moment for each span* : For each span, the column strip shall be designed to resist 60 percent of the total positive moment in the panel.

(*iv*) *Moments in the middle strip* : The middle strip shall be designed on the following bases:

(*a*) The portion of the design moment not resisted by the column strip shall be assigned to the adjacent middle strips.

(*b*) Each middle strip shall be proportioned to resist the sum of the moments assigned to its two half middle strips.

(c) The middle strip adjacent and parallel to an edge supported by a wall shall be proportioned to resist twice the moment assigned to half the middle strip corresponding to the first row of interior columns.

## 12.6. SHEAR IN FLAT SLAB

**1. The critical section :** The critical section shall be at a distance $d/2$ from the periphery of the column/capital/drop panel, perpendicular to the plane of the slab where $d$ is the effective depth of the section (See Fig. 12.2). The shape in plan is geometrically similar to the support immediately below the slab. See Fig. 12.4 (a) and 12.4 (b).

**Note :** For column sections with re-entrant angles the critical section shall be taken as indicated in Fig. 12.4 (c) and 12.4 (d).

(a) In the case of columns near the free edge of a slab, the critical section shall be taken as shown in Fig. 12.5.

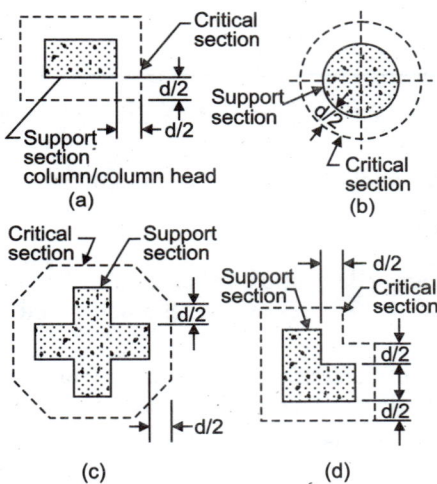

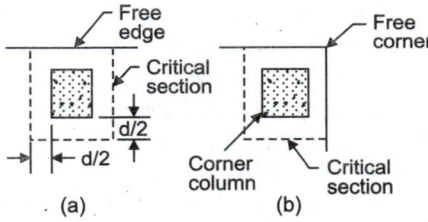

**Fig. 12.4.** Critical Sections in Plan for Shear.

**Fig. 12.5.** Effect of Free Edges on Critical Sections for Shear.

(b) When the openings in flat slabs are located at a distance less than ten times the thickness of the slab from a concentrated reaction or when the openings are located within the column strips, the critical sections specified in (a) shall be modified so that the part of the periphery of the critical section which is enclosed by radial projection of the openings to the centroid of the reaction area shall be considered ineffective (See Fig. 12.6), and openings shall not encroach upon column head.

**2. Calculation of shear stress.** The shear stress $\tau_v$ shall be the sum of the values calculated according to (i) and (ii) below :

(i) The nominal shear stress in flat slabs shall be taken as $V/b_0 d$ where $V$ is the shear force due to design load, $b_0$ is the periphery of the critical section and $d$ is the effective depth.

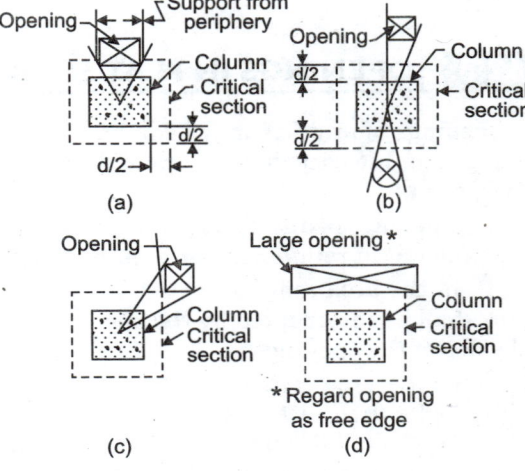

**Fig. 12.6.** Effects of Openings on Critical Section for the Shear.

(*ii*) When unbalanced gravity load, wind, earthquake or other forces cause transfer of bending moment between slab and column, a fraction $(1 - \alpha)$ of the moment shall be considered transferred by eccentricity of the shear about the centroid of the critical section. Shear stress shall be taken as varying linearly about the centroid of the critical section. The value of $\alpha$ shall be obtained from the equation given in clause C-3.

### 3. Permissible shear stress.

(*i*) When shear reinforcement is not provided, the calculated shear stress at the critical section shall not exceed $k_s \cdot \tau_c$,

where $k_s = (0.5 + \beta_c)$ but not greater than 1, $\beta_c$ being the ratio of short side to long side of the column/capital ; and

$\tau_c = 0.16 \sqrt{f_{ck}}$ in working stress method of design.

(*ii*) When the shear stress at the critical section exceeds the value given in (*i*) above, but less than $1.5 \tau_c$, shear reinforcement shall be redesigned. Shear stresses shall be investigated at successive sections more distant from the support and shear reinforcement shall be provided up to a section where the shear stress does not exceed $0.5 \tau_c$. While designing the shear reinforcement, the shear stress carried by the concrete shall be assumed to be $0.5 \tau_c$ and reinforcement shall carry the remaining shear.

## 12.7. SLAB REINFORCEMENT

**1. Spacing.** The spacing of bars in a flat slab, shall not exceed 2 times the slab thickness except where a slab is of cellular or ribbed construction.

**2. Area of reinforcement.** When drop panels are used the thickness of drop panel for determination of area of reinforcement shall be the lesser of the following:

(*i*) Thickness of drop, and

(*ii*) Thickness of slab plus one quarter the distance between edge of drop and edge of capital. The minimum percentage of the reinforcement is same as that in solid slab, *i.e.,* 0.12 percent if HYSD bars used and 0.15 percent. If mild steel is used.

**Minimum Length of Reinforcement:** At least 50 percent of bottom bars should be from support to support. The rest may be bent up. The minimum length of difficult reinforcement in flat slab should be as per Figure of IS 456-2000. If adjacent spans are not equal, the extension of the –ve reinforcement beyond each face shall be based on the longer span.

The reinforcement in flat slabs should have minimum lengths specified in Fig. 12.7.

**Note:** Bent bars at exterior supports may be used if a general analysis is made.

'D' is the diameter of the column and the dimension of the rectangular column in the direction under consideration.

## 12.8. OPENINGS IN FLAT SLAB

Openings of any size may be provided in the flat slab if it is shown by analysis that the requirement of strength and serviceability are met. However, for openings conforming to the following, no special analysis is required :

(1) Openings of any size may be placed within the middle half of the span in each direction provided the total amount of reinforcement required for the panel without the opening is maintained.

(2) In the area common to two column strips, not more than one-eighth of the width of strip in either span shall be interrupted by the openings. The equivalent of reinforcement interrupted shall be added on all sides of the openings.

(3) In the area common to one column strip and one middle strip not more than one-quarter of the reinforcement in either strip shall be interrupted by the openings. The equivalent of reinforcement interrupted shall be added on all sides of the openings.

(4) Shear requirements of § 12.6 shall be satisfied.

**Note :** *D* is the dia. of column and the dimension of rectangular column in the direction under consideration.

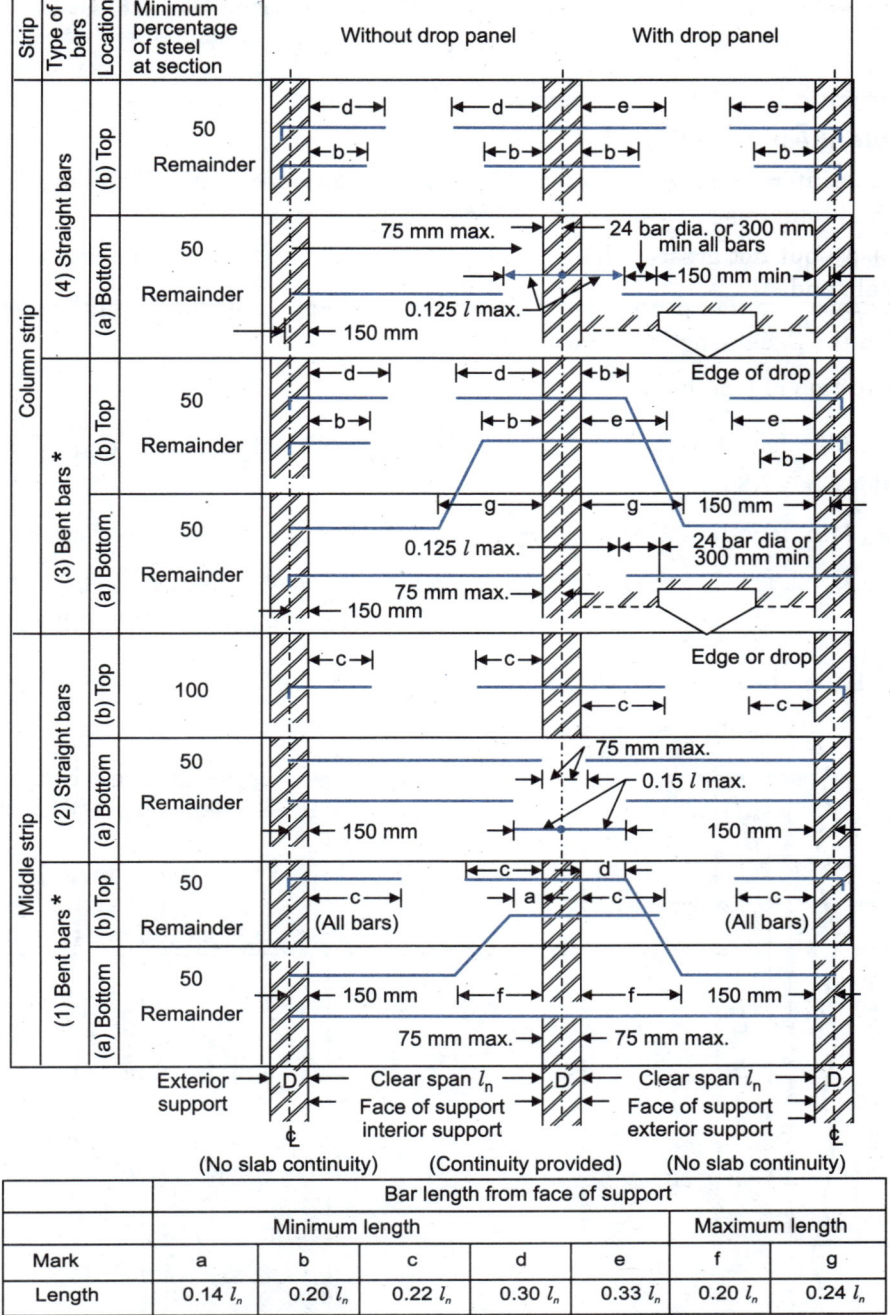

Fig. 12.7. Minimum bend joint locations and extensions for reinforcement in flat slabs.

**Design: Example :** *Design the interior panel of a flat slab 5.6 m × 6.6 m in size, for a super-imposed load of 7.75 kN/m². Provide two-way reinforcement. Use M 20 concrete and Fe 415 steel.*

**Solution:** 1. **Design constants.** For M 20 concrete and Fe 415 steel, we have $k_c = 0.289$; $j_c = 0.904$ and $R_c = 0.914$

2. **Panel dimensions.** Length of panel $L = 6.6$ m. Width of panel $B = 5.6$ m. Widths of strips will be as follows:

Along length $L : l_1 = L = 6.6$ and $l_2 = B = 5.6$ m

Width of column strip = 0.25 $l_2$ = 0.25 × 5.6 = 1.4 m, with an upper limit of 0.25 $l_1$ = 0.25 × 6.6 = 1.65 m. Hence keep width of column strip = 1.4 m on each side of column center line.

Width of middle strip = 5.6 – 2.8 m = 2.8 m

*Along width B* : $l_1$ = B = 5.6 and $l_2$ = L = 6.6 m

Width of column strip = 0.25 $l_2$ = 0.25 × 6.6 = 1.65, m with an upper limit of 0.5 $l_1$ = 0.25 × 5.6 = 1.4 m. Hence keep width of column strip = 4 m on each side of column centre line.

Width of middle strip = 6.6 – 2.8 = 3.8 m.

Let us provide drops also. The drops should be rectangular in plan, having a length in each direction not less than one third the panel length in that direction.

Thus, in the direction of length:

*Along length L* : Min. length of drop = $\frac{1}{3} L = \frac{1}{3}$ × 6.6 = 2.2 m. However, keep it equal to the total width of column strip along B = 2.8 m.

*Along width B* : Min. length of drop = $\frac{1}{3} B = \frac{1}{3}$ × 5.6 = 1.87 m. However, keep it equal to total width of column strip along L = 2.8 m. Let the column have a column head of diameter one fifth of average span.

$$l = \frac{1}{2}(L + B) = \frac{1}{2}(6.6 + 5.6) = 6.1 \text{ m. Hence, } D = l/5 \approx 1.2 \text{ m.}$$

The panel details are shown in Fig. 12.8.

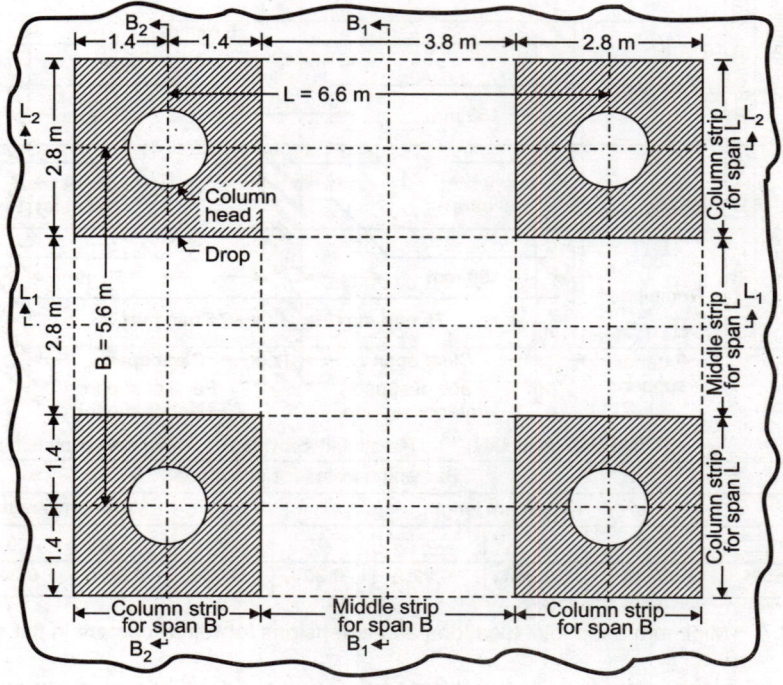

**Fig. 12.8**

**3. *Loading.*** The thickness of flat slab is generally controlled by considerations of span to effective depth ratios given in § 7.1, clause 2.

$$\frac{\text{span}}{\text{effective depth}} \not> 26$$

Assuming balanced section, percentage reinforcement for M 15 concrete = 0.72%. Hence from Fig. 7.1, modification factor for mild steel ≈ 1.6.

$$\therefore \quad \frac{\text{span}}{d} = 26 \times 1.6$$

$$\therefore \quad d = \frac{\text{span}}{26 \times 1.6} = \frac{6.6}{26 \times 1.6} = 0.159 \text{ m} \approx 160 \text{ mm}.$$

Assuming a nominal cover of 15 mm, and using 12 mm Φ bars, total thickness = 160 + 6 + 12 + 15 = 193 mm. Let us provide a total thickness equal to 200 mm.

The thickness of drops is normally kept 25% to 50% more than the thickness of the slab. Let us assume an *average* thickness of 250 mm for the calculation of dead load.

$$\therefore \quad \text{Weight of slab/m}^2 = \frac{250}{1000} \times 1 \times 1 \times 25000 = 6250 \text{ N}$$

$$\text{Super-imposed load/m}^2 = 7750 \text{ N}$$

$$\text{Total } w = 14000 \text{ N/m}^2.$$

### 4. Moments along longer span

For the longer span ($L = 6.6$ m), $l_1 = L = 6.6$ m, $l_2 = B = 5.6$ m.
The column head is circular, of diameter 1.2 m.

Hence size of equivalent square supports = $\sqrt{\frac{\pi}{4}(1.2)^2}$ = 1.06 m

$$\therefore \quad L_{nL} = l_1 - 1.06 = 6.6 - 1.06 = 5.54 \text{ m},$$

(Subject to a minimum of $0.65\, l_1 = 0.65 \times 6.6 = 4.29$ m)

$$W_L = \text{total design load on area } l_2 \times l_{nL} = (5.6 \times 5.54)\, 14000 \approx 434340 \text{ N}$$

$$\therefore \quad M_{OL} = \frac{W_L \cdot l_{nL}}{8} = \frac{434340 \times 5.54}{8} \approx 300800 \text{ N-m}.$$

### *Apportionment of moments*

Total negative moment = 0.65 $M_{OL}$ = 0.65 × 300800 = 195520 N-m.
Total positive moment = 0.35 $M_{OL}$ = 0.35 × 300800 = 105280 N-m.

**Effect of pattern of loading :** In the direct design method, the ratio of live load to dead load exceeds 0.5, the factor $\alpha_c = \frac{\Sigma K_c}{\Sigma K_s}$ should be greater than $\alpha_{c\,.\,min}$ given in Table 12.1. If not, the positive moments for panel should be multiplied by coefficient $\beta_0$. In order to compute $\alpha_c$ the flexural stiffness $K_c$ for columns and $K_s$ for the slab are needed. The flexural stiffness $K$ for any member of *uniform section* is given by

$$K = \frac{4\, E_c\, I}{l}$$

where $I$ is the moment of inertia of concrete section, neglecting reinforcement, and $l$ is the appropriate length of the member. For slab of uniform thickness $h$, span $l_1$ and transverse span $l_2$ :

$$K_s = \frac{4\, E_c\, l_2\, h^3}{12\, l_1}$$

Similarly, for uniform, circular column of size $D \times D$ and storey height $l$,

$$K_c = \frac{4\, E_c\, D^4}{12\, l}$$

If, however, the member (*i.e.* slab or column) has variable moment of inertia, the flexural stiffness of the member is computed from the expression

$$K = \frac{k E_c I}{l}$$

where $k$ = stiffness factor for the section, which is more than 4.

The values of stiffness factors ($k$) for slab without drop, slab with drop and columns are given in Tables 12.2, 12.3 and 12.4 respectively, for some idealised cases.

**TABLE 12.2.** Stiffness coefficients for slab with variable moment of inertia (*without drop*)

| Ratio $D_1/l_1$ | Stiffness factor k | Ratio $D_1/l_1$ | Stiffness factor k |
|---|---|---|---|
| 0.0 | 4.00 | 0.15 | 4.40 |
| 0.05 | 4.05 | 0.20 | 4.72 |
| 0.10 | 4.18 | 0.25 | 5.14 |

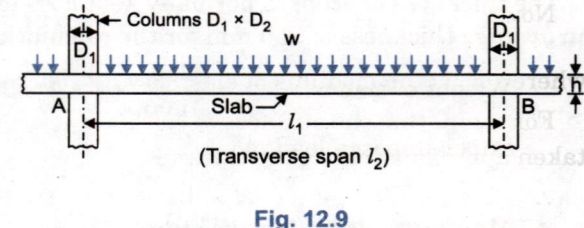

Fig. 12.9

**Note :** The above table is valid for case $D_1/l_1 = D_2/l_2$. For other cases the values are only approximate.

**TABLE 12.3.** Stiffness coefficients for slab with variable moment of inertia (*with drop*)

| Ratio $D_1/l_1$ | Stiffness factor k | Ratio $D_1/l_1$ | Stiffness factor k |
|---|---|---|---|
| 0.00 | 4.78 | 0.15 | 5.22 |
| 0.05 | 4.84 | 0.20 | 5.55 |
| 0.10 | 4.98 | 0.25 | 5.98 |

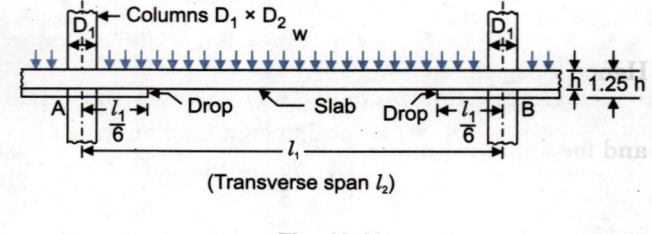

Fig. 12.10

**Note :** The above table is valid for case $D_1/l_1 = D_2/l_2$. For other cases, the values are only approximate.

**TABLE 12.4.** Stiffness coefficients for columns with variable moment of inertia

| Ratio H/l | $k_{AB}$ | $k_{BA}$ |
|---|---|---|
| 0.0 | 4.00 | 4.00 |
| 0.05 | 4.91 | 4.21 |
| 0.10 | 6.09 | 4.44 |
| 0.15 | 7.64 | 4.71 |
| 0.20 | 9.69 | 5.00 |
| 0.25 | 12.44 | 5.33 |

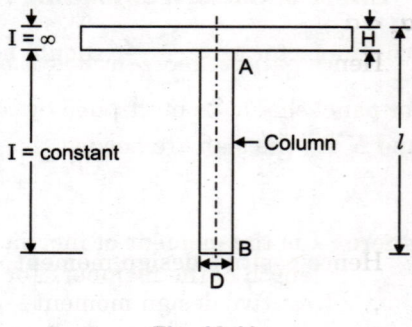

Fig. 12.11

For the present case, thickness of slab $h$ = 200 mm = 0.2 m, $l_1$ = 6.6 m and $l_2$ = 5.6 m.

Let the thickness of drops be 300 mm = 0.3 m. Also, assume that column diameter = 400 mm, height of column head = 500 mm and diameter of column head = 1200 mm.

Hence $D_1 = D_2 = D$ = Size of equivalent square column = $\sqrt{\frac{\pi}{4}(0.4)^2}$ = 0.35 m ; $D_1/l_1 = D/l_1$ = 0.35/6.6 = 0.053.

# FLAT SLABS

Table 12.2 is applicable for slabs without drops. For most of the practical cases, $D/l_1$ will be between 0.05 to 0.1, and hence $k$ will vary between 4.05 to 4.18. *In the absence of the table, therefore, the value of k may be taken as 4.1.*

Table 12.3 is applicable for slab with drops of width $l_1/6$ on either side. For $D/l_1 = 0.05$, $k \approx 4.84$, against $k = 4.05$ for $D/l_1 = 0.05$ when the slab does not have drops. For width of drops more than $l_1/6$ on either side, the value of $k$ will be more than 5. However, in absence of any table, one can adopt the value of $k = 5$ for all common cases for slabs with drops. For the present case when drops are more than $l_1/6$ on either side, assume $k = 5$.

Now, $$k_s = \frac{k E_c I}{l} = \frac{k E_c l_2 h^3}{12 l_1} = \frac{5 \times 5.6 (0.2)^3}{12 \times 6.6} E_c = 2.83 \times 10^{-3} E_c$$

where $E_c$ is the modulus of elasticity for concrete.

For computing the stiffness of columns meeting at the slab, Table 12.4 may be used, where $H$ may be taken approximately as the thickness of the drop (or the slab) plus half the height of the pedestal.

$\therefore \qquad H = 0.3 + 0.25 = 0.55$ mm.

The moment of inertia for this height $H$ is taken to be infinite. Let us assume that the height of storey is 4 m from top of one slab to top of other slab.

Hence $l = 4$ m. For $H/l = 0.55/4 = 0.1375$, we have $k_{AB} \approx 7.25$ and $k_{BA} \approx 4.64$

For column $$I = \frac{1}{12} D^4 = \frac{1}{12} (0.35)^4 = 1.25 \times 10^{-3}$$

Hence for lower column, $$K_c = \frac{k_{AB} E_c I}{l} = \frac{7.25 \times 1.25 \times 10^{-3}}{4} E_c = 2.33 \times 10^{-3} E_c$$

and for upper column, $$K_c = \frac{k_{BA} E_c I}{l} = \frac{4.64 \times 1.25 \times 10^{-3}}{4} E_c = 1.45 \times 10^{-3} E_c.$$

$\therefore \qquad \alpha_c = \frac{\Sigma K_c}{\Sigma K_s} = \frac{(2.33 + 1.45) 10^{-3}}{2 \times 2.83 \times 10^{-3}} = 0.67$

Now ratio $\frac{w_l}{w_d} = \frac{8000}{6000} = 1.33$. Also, ratio $\frac{l_2}{l_1} = \frac{5.6}{6.6} = 0.85$.

For these ratios, the value of $\alpha_{c\,min.}$ is obtained from Table 12.1, which comes out to be 0.97. Thus, $\alpha_c < \alpha_{c\,min}$.

Hence positive moment has to be multiplied by coefficient $\beta_0$, given by

$$\beta_s = 1 + \frac{\left(2 - \frac{w_d}{w_l}\right)}{\left(4 + \frac{w_d}{w_l}\right)} \left(1 - \frac{\alpha_c}{\alpha_{c.min.}}\right) = 1 + \frac{\left(2 - \frac{6}{8}\right)}{\left(4 + \frac{6}{8}\right)} \left(1 - \frac{0.67}{0.97}\right) = 1.081$$

Hence positive design moment $M_{PL} = 1.081 \times 105280 \approx 113800$ N-m

$\therefore$ Negative design moment $M_{NL} = 300800 - 113800 = 187000$ N-m.

These moments will be apportioned to the column strips and middle strip as under:

*For column strips*

    –ve moment $\qquad M_{1L} = 0.75 M_{NL} = 0.75 \times 187000 = 140250$ N-m

    +ve moment $\qquad M_{2L} = 0.60 M_{PL} = 0.6 \times 113800 = 68280$ N-m

*For middle strip*

    –ve moment $\qquad M_{3L} = M_{NL} - M_{1L} = 187000 - 140250 = 46750$ N-m

    +ve moment $\qquad M_{4L} = M_{PL} - M_{2L} = 113800 - 68280 = 45520$ N-m

## 5. Moments along shorter span

For shorter span ($B = 5.6$ m), $l_1 = B = 5.6$ m ; $l_2 = L = 6.6$ m $l_{nB} = 5.6 - 1.06 = 4.54$ m
(Subject to a minimum of $0.65\, l_1 = 0.65 \times 5.6 = 3.65$ m)

$W_B$ = total design load on area $l_2 \cdot l_{nB} = (6.6 \times 4.54) \times 14000 \approx 419500$ N

$$M_{0B} = \frac{W_B \cdot l_{nB}}{8} = \frac{419500 \times 4.54}{8} \approx 238100 \text{ N-m}$$

*Apportionment of moments*

Total negative moment = $0.65\, M_{0B} = 0.65 \times 238100 = 154765$ N-m.
Total positive moment = $0.35\, M_{0B} = 0.35 \times 238100 = 83335$ N-m.

***Effect of pattern of loading :***   $D/l_1 = 0.35/5.6 = 0.0625$.

Hence from Table 12.3, $k \approx 4.9$. However take $k = 5$ since width of drop is more than $l_1/6$ on each side of column centre.

$$\therefore \quad K_s = \frac{k\, E_c \cdot l_2\, h^3}{12\, l_1} = \frac{5 \times 6.6 (0.2)^3\, E_c}{12 \times 5.6} = 3.93 \times 10^{-3}\, E_c$$

$\Sigma K_c = (2.33 + 1.45)10^{-3}\, E_c$, as before.

$$\therefore \quad \alpha_c = \frac{\Sigma K_c}{\Sigma K_s} = \frac{(2.33 + 1.45)10^{-3}}{2 \times 3.93 \times 10^{-3}} = 0.48$$

Now ratio $w_l/w_d = 8000/6000 = 1.33$    Ratio $l_2/l_1 = 6.6/5.6 = 1.18$
Hence from Table 12.1, value of $\alpha_{c\,min}$ comes out to be 1.12

$$\therefore \quad \alpha_c < \alpha_{c\,min}:$$

$$\therefore \quad \beta_s = 1 + \frac{\left(2 - \dfrac{w_d}{w_l}\right)}{\left(4 + \dfrac{w_d}{w_l}\right)}\left(1 - \frac{\alpha_c}{\alpha_{c\,min}}\right) = 1 + \frac{\left(2 - \dfrac{6}{8}\right)}{\left(4 + \dfrac{6}{8}\right)}\left(1 - \frac{0.48}{1.12}\right) = 1.150$$

$\therefore$ Positive design moment    $M_{PB} = 1.150 \times 83335 = 95835.25$ N-m
Negative design moment    $M_{NB} = 238100 - 95835 = 142265$ N-m

These moments will be apportioned to the column strips and middle strip as follows:

*Column strips*

–ve moment    $M_{1B} = 0.75\, M_{NB} = 0.75 \times 142265 = 106699$ N-m
+ve moment    $M_{2B} = 0.6\, M_{PB} = 0.6 \times 95835 = 57501$ N-m

*Middle strips*

–ve moment    $M_{3B} = M_{NB} - M_{1B} = 142265 - 106699 = 35566$ N-m
+ve moment    $M_{4B} = M_{PB} - M_{2B} = 95835 - 57501 = 38334$ N-m.

## 6. Thickness of slab and drops

The thickness of slab will be designed on the basis of maximum positive moment $M_{2L}$ in the column strip of longer span. Width of column strip of span $L = 2.8$ m.

$$\therefore \quad d = \sqrt{\frac{M_{2L}}{R_c \cdot b}} = \sqrt{\frac{68280 \times 1000}{0.914 \times 2800}} = 164 \text{ mm}$$

Provide total thickness of 200 mm, as fixed from the deflection criterion. Using 10 mm of bars and a nominal cover of 20 mm, available $d$ for longer span = $200 - 5 - 20 = 175$ mm and for shorter span (upper layer) = $175 - 10 = 165$ mm.

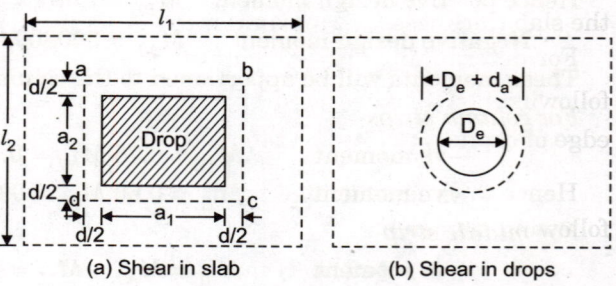

(a) Shear in slab    (b) Shear in drops

**Fig. 12.12.** Periphery of critical section.

The thickness of drops $d_d$ will be designed on the basis of maximum negative bending moment $M_{1L} = 140250$ N-m.

$$\therefore \quad d_d = \sqrt{\frac{140250 \times 1000}{0.914 \times 2800}} = 234 \text{ mm}.$$

Keep total thickness of drop = 280 mm. Using 10 mm Φ bars and 20 mm nominal cover, available $d = 280 - 20 - 5 = 255$ mm for long span and $255 - 10 = 245$ mm for shorter span.

**7. Check for shear :** The critical section for the slab will be at a distance of $d/2$ from the face of the drop. Fig. 12.12(a) shows the plan for the periphery of critical section for slab. Shear in slab is given by

$$V_s = (w_d + w_l)\,[l_1 l_2 - (a_1 + d)(a_2 + d)]$$
$$b_0 = \text{periphery of critical section in slab} = 2(a_1 + a_2 + 2b)$$

For the present case,

$$a_1 = a_2 = 2.8 \text{ m} = 2800 \text{ mm} \,;\, d = \text{effective depth of slab} = 179 \text{ mm}.$$
$$\therefore \quad V_s = [6000 + 8000]\,[6.6 \times 5.6 - (2.8 + 0.179)(2.8 + 0.179)] \approx 393197.8 \text{ N}$$
$$b_0 = 2(2800 + 2800 + 2 \times 179) = 11916 \text{ mm}$$
$$\therefore \quad \tau_v = \frac{V_s}{b_0\, d} = \frac{393200}{11916 \times 175} = 0.189 \text{ N/mm}^2$$

Again, for shear in the drops, the critical section is at a distance of $d_d/2$ from the periphery of the column head, as shown in Fig. 12.12(b), where $D_e = 1.2$ m and $d_d = 0.259$ m = 259 mm. Shear in drops is given by

$$V_d = (w_d + w_l)\left[l_1 l_2 - \frac{\pi}{4}(D_e + d_d)^2\right] = (6000 + 8000)\left[6.6 \times 5.6 - \frac{\pi}{4}(1.2 + 0.259)^2\right]$$
$$= 494034 \text{ N}$$

and $\quad b_0 =$ periphery of critical section in slab $= \pi (D_e + d_d)$
$$= \pi(1.2 + 0.259) = 4.583 \text{ m} = 4583 \text{ mm}$$
$$\therefore \quad \tau_v = \frac{V_d}{b_0\, d_d} = \frac{494034}{4583 \times 255} = 0.423 \text{ N/mm}^2$$

Thus shear stress is more critical in the drops than in slabs.

Permissible shear stress $= k_s \cdot \tau_c$, where $k_s = (0.5 + \beta_c)$ but $\not> 1$
$\beta_c =$ ratio of short side to long side of the column capital = 1, for the present case.
$$\therefore \quad k_s = (0.5 + 1) = 1.5, \text{ but } \not> 1; \text{ Hence } k_s = 1$$

For M-20 concrete, $f_{ck} = 20$ N/mm$^2$. $\quad\therefore\quad \tau_c = 0.16 \sqrt{f_{ck}} = 0.16 \sqrt{20} = 0.716$ N/mm$^2$
$\therefore$ Permissible shear stress $= k_s \cdot \tau_c = 1 \times 0.76 = 0.76$ N/mm$^2$

which is more than the actual shear stress in the drops and in the slab. Hence *safe*. Thus no shear reinforcement is required.

**8. Reinforcement along span $L$. $l_1 = L = 6.6$ m. :** The spacing of the reinforcement is limited to twice the slab thickness, *i.e.*, maximum spacing $= 2 \times 179 = 358 \approx 350$ mm.

For determining area of reinforcement in drops, the thickness of drop panel will be the lesser of the following : (*i*) thickness of drop = 255 or (*ii*) thickness of slab + $\frac{1}{4}$ the distance between edge of drop and edge of capital $= 175 + \frac{1}{4}(1400 - 600) = 175 + 200 = 375$ mm.

Hence thickness adopted for drops = 255 mm (effective). The reinforcements are to be provided for the following moments :

$$M_{1L} = -140250 \text{ N-m \& } M_{2L} = +68280 \text{ N-m for column strip of width 2800 mm.}$$
$$M_{3L} = -46750 \text{ N-m \& } M_{4L} = +45520 \text{ N-m for middle strip of width 2800 mm.}$$

**300** REINFORCED CONCRETE STRUCTURE

**(i) Middle strip.** For the *middle strip*, the effective thickness of slab = 179 mm

$$\therefore \quad A_{st4L} = \frac{M_{4L}}{\sigma_{st}\, j_c d} = \frac{45520 \times 1000}{230 \times 0.904 \times 175} = 1251 \text{ mm}^2 \text{ (+ ve reinforcement)}$$

Using 10 mm Φ bars, $A_\Phi = \frac{\pi}{4}(10)^2 = 78.54 \text{ mm}^2$.

∴ No. of bars required = 1251/78.54 ≏ 17
∴ Average spacing = 2800/17 = 165 mm c/c.

Again, $A_{st\,3L} = \dfrac{M_{3L}}{\sigma_{st}\, j_c d} = \dfrac{46750 \times 1000}{230 \times 0.904 \times 175} = 1285 \text{ mm}^2$ (– ve reinforcement)

∴ No. of 10 mm Φ bars = 1285/78.54 ≈ 17.
∴ Average spacing = 2800/17 = 165 mm c/c.

Half of the positive reinforcement is continued upto 75 mm away from the centre of the support while the remaining half may be bent up at a distance of $f = 0.20\, l_{nL} = 0.2 \times 5.54 = 1.108$ m (max.) ≈ 1100 mm from the face of the column. The bent up bars should continue upto a distance $c = 0.22\, l_n = 0.22 \times 5.54 = 1.22$ m (min.) ≈ 1250 mm to the other face of the column. Since bent up bars will be available from both the adjacent spans, reinforcement for negative moment $M_{3L}$ will be available as 10 mm Φ bars @ 165 mm c/c. It is to be ensured that such bending of bars provide a min. length $a = 0.14\, l_n = 0.14 \times 5.54 \approx 0.78$ m is available from the first face of the column. The above reinforcement has been provided according to item (1) Fig. 12.7.

**(ii) Column strips.** The column strips of span $L$ have also width equal to 2800 mm. The effective thickness for +ve reinforcement is 179 mm (slab) while the effective thickness for – ve reinforcement is 159 mm (drops).

$$A_{st2L} = \frac{M_{2L}}{\sigma_{st}\, j_c d} = \frac{68280 \times 1000}{230 \times 0.904 \times 175} = 1877 \text{ mm}^2 \text{ (+ve reinforcement)}$$

and $\quad A_{st1L} = \dfrac{M_{1L}}{\sigma_{st}\, j_c d_d} = \dfrac{140250 \times 1000}{230 \times 0.904 \times 255} = 2646 \text{ mm}^2$ (–ve reinforcement)

No. of 10 mm Φ bars for +ve reinforcement = 1877/78.54 = 24
∴ Average spacing = 2800/24 ≏ 115 mm c/c.

Continue half the positive reinforcement to a distance of $0.125L = 0.125 \times 6.6 = 0.825$ m (max.) ≈ 800 mm from the column centre. Available length of bar from the edge of drop = 1400 – 800 = 600 mm which is more than 24 Φ (or 300 mm). The remaining positive reinforcement bars will be bent up at a distance $= g = 0.24\, l_n = 0.24 \times 5.54 = 1.33$ m (max) ≈ 1300 mm from the face of the column. The bent up bars should continue upto a distance $d = 0.3\, l_n = 0.3 \times 5.54 \approx 1.662$ m (min.) ≈ 1700 mm beyond the other face of the column (see item 3 of Fig. 12.7).

Thus, available negative reinforcement is 10 mm Φ @ 115 mm c/c, giving a total reinforcement of $(2800/115) \times 78.54 = 1912 \text{ mm}^2$ against a requirement of $A_{st1L} = 2646 \text{ mm}^2$. The balance of $A_{st1L} = 2646 – 115 = 734 \text{ mm}^2$ is to be provided in the form of *straight bars* at the top. No. of straight bars required = 734/78.54 ≈ 10. Hence, provide 10 bars to have a spacing = 2800/10 = 280 mm c/c. These bars should continue to a distance $e = 0.33\, l_n = 0.33 \times 5.54 = 1.828$ m (min.) ≈ 1900 mm on either side of column face, as required in item 4(b) of Fig. 12.7.

**9. Reinforcement along span B.** $l_1 = B = 5.6$ m.

**(i) Middle strip.** Width of the middle strip = 3.8 m = 3800 mm. The middle strip carries a positive moment of $M_{4B} = 38334$ N-m at its middle and negative moment of $M_{3B} = 35566$ N-m at its ends. Available $d = 167$ mm.

$$\therefore \quad A_{st4B} = \frac{M_{4B}}{\sigma_{st}\, j_c d} = \frac{38334 \times 1000}{230 \times 0.904 \times 165} = 1118 \text{ mm}^2 \text{ (+ reinforcement)}$$

∴ No. of 10 mm Φ bars required = 1118/78.54 ≈ 15.

∴ Spacing = 3800/15 ≈ 250 mm c/c.

Also, $A_{st3B} = \dfrac{M_{3L}}{\sigma_{st} j_c d} = \dfrac{35566 \times 1000}{230 \times 0.904 \times 165} = 1036$ mm² (– ve requirement).

No. of 12 mm Φ bars required = 1036/78.54 ≈ 14. However, provide 15 bars, as for +ve reinforcement, so that bent up bars serve the purpose.

$l_n$ for span B = $l_{nB}$ = 4.54 m. Half the positive reinforcement is continued upto 75 mm away from the centre of the column, while the remaining half may be bent up at a distance of $f = 0.2\, l_n = 0.2 \times 4.54 = 0.908$ m (max.) ≈ 900 mm from the face of the column. The bent up bars should continue upto a distance $c = 0.22\, l_n = 0.22 \times 4.54 = 0.999$ m (min.) ≈ 1000 mm to the other face of the column.

(ii) **Column strips.** The width of column strips = 2800 mm. The available effective depth = 167 mm (slab) for positive reinforcement, and = 247 mm (drops) for negative reinforcement. The strip carries as positive moment of $M_{2B} = 57501$ N-m at its middle and negative moment of $M_{1B} = 106699$ N-m at the ends.

∴ $A_{st2B} = \dfrac{M_{2B}}{\sigma_{st} j_c d} = \dfrac{57501 \times 1000}{230 \times 0.904 \times 165} = 1676$ mm² (+ ve reinforcement)

and $A_{st1B} = \dfrac{M_{1B}}{\sigma_{st} j_c d} = \dfrac{106699 \times 1000}{230 \times 0.904 \times 245} = 2094$ mm² (– ve reinforcement)

No. of 10 mm Φ bars required for + ve reinforcement = 1676/78.54 ≈ 22

∴ Average spacing = 2800/22 = 127 mm c/c. Continue half the positive reinforcement to a distance of $0.125\, B = 0.125 \times 5.6 = 0.7$ m (min.) = 700 mm from the column centre. Available length of bar from the edge of drop = 1400 – 700 = 700 mm which is more than 24 Φ (or 300 mm). The remaining + ve reinforcement bars will be bent up at a distance $g = 0.24\, l_n = 0.24 \times 4.54 = 1.09$ m (min.) ≈ 1000 mm from the face of the column. The bent up bars should continue upto a distance $d = 0.3\, l_n = 0.3 \times 4.54 = 1.362$ m (min.) ≈ 1400 mm beyond the other face of the column (see item 3 of Fig. 12.7).

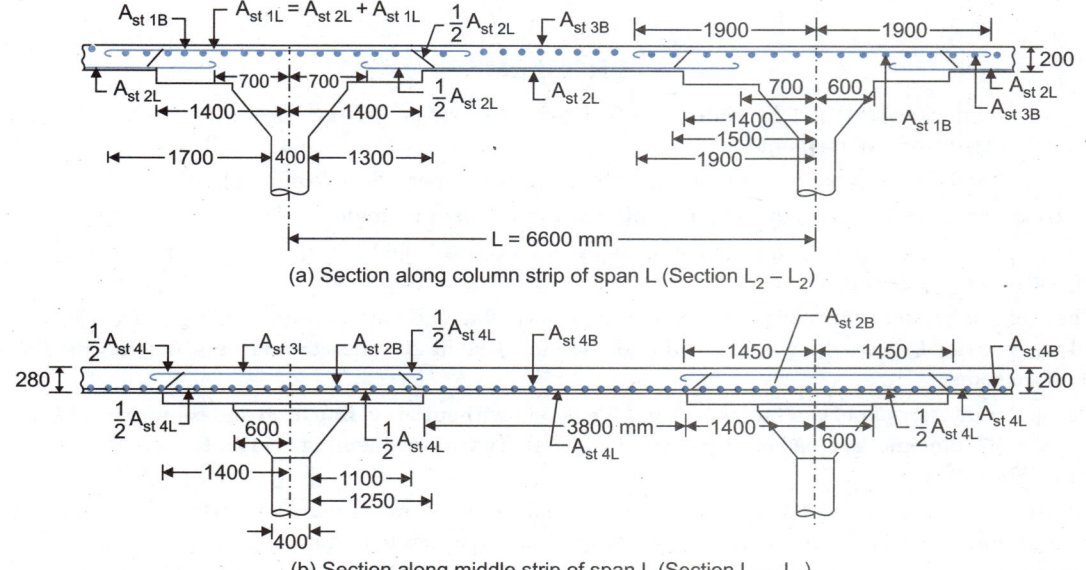

(a) Section along column strip of span L (Section $L_2 - L_2$)

(b) Section along middle strip of span L (Section $L_1 - L_1$)

Fig. 12.13

Thus, available negative reinforcement is 10 mm Φ @ 127 mm c/c, giving a total reinforcement of (2800/127 × 78.54 = 1732 mm²) against a requirement of $A_{st1B} = 2087$ mm². The balance = 2087 – 1732 = 355 mm² is to be provided in the form of *straight bars* at top.

No. of straight bars required = 355/78.54 ≈ 5. Spacing of bars = 2800/5 = 560 mm c/c. However keep spacing = 4 × 127 = 508 cm c/c for facility in construction. These bars should continue to a distance $e = 0.33\, l_n = 0.33 \times 4.54 = 1.5$ m (min.) = 1500 mm on either side of column face, as required in item 4(b) of Fig. 12.7.

### 10. Details of reinforcement

The details of reinforcement etc. are shown in Figs. 12.13 and 12.14.

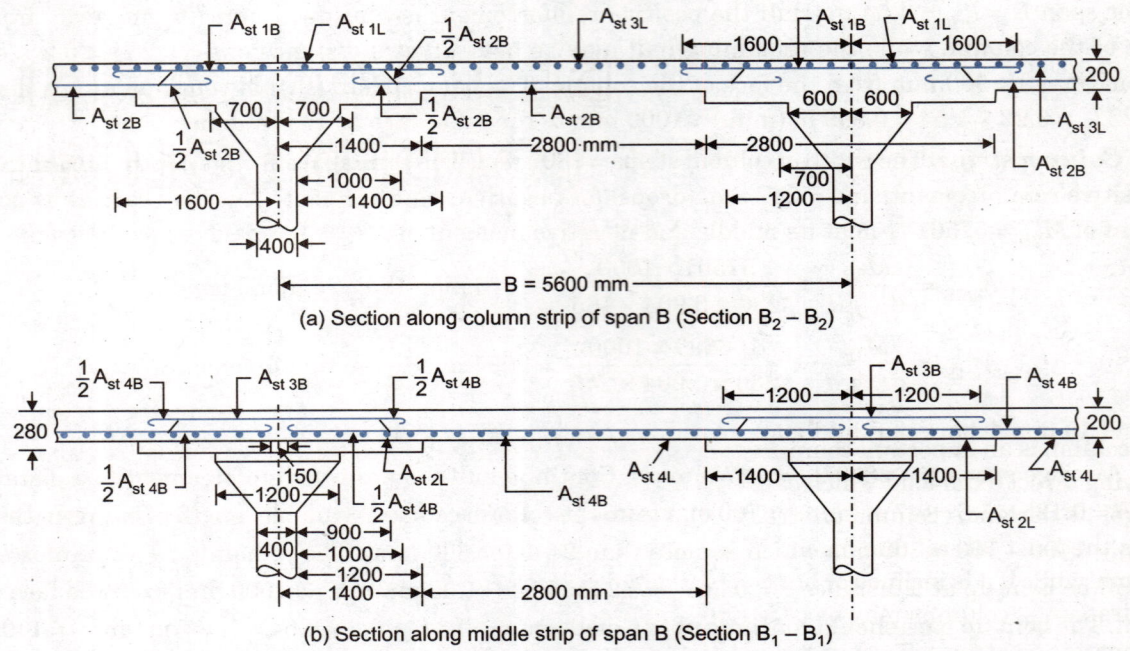

Fig. 12.14

## PROBLEMS

1. Design an interior panel of a flat slab, 6 m × 6 m, for a live load of 7 kN/m². Use M 20 concrete and Fe 415 steel. Provide two-way reinforcement.
2. Redesign problem 1 by providing four way reinforcement as per I.S. code of practice.
3. Redesign problem 1 if the slab has an opening of 1 m × 1 m at its middle.
4. Design an interior panel of a flat slab. 4 m × 4 m for a live load of 6 kN/m². Use m 20 concrete and Fe 415 steel. Provide two-way reinforcement.
5. The roof of a hall measures 26 m × 34 m, and consists of flat slab construction dividing it into 16 panels. Design all the panels of the roof for a live load of 6.5 kN/m². Use M 20 concrete and Fe 415 steel. Provide two way reinforcement.
6. Design an interior panel of a flat slab of size 5 m × 5 m without providing drop and column head size of column is 500 × 500 mm and live load on the panel is 5 kN/m². Take floor finishing load as 1.2 kN/m². Use $M_{25}$ concrete and HYSD steel.
7. Design an interior panel of a flat slab with panel size 6 m × 6 m supported by column of size 500 × 500 mm provide suitable drop. Take live load as 4 kN/m². Use $M_{20}$ concrete and Fe 415 steel.
8. Design the typical interior panel of a flat slab floor of size 5 m × 5 m with suitable drop to support a live load of 5 kN/m². The floor is supported by columns of 500 × 500 mm. Use $M_{20}$ grade concrete and Fe 415 steel. Sketch the reinforcement details by showing cross strip.

    (a) at column strip
    (b) at middle strip
    (c) column drop.

# CHAPTER 13: DESIGN OF AXIALLY LOADED COLUMNS

## 13.1. INTRODUCTION

A column is an important component of R.C. structures. A column, in general, may be defined as a member carrying direct axial load which causes compressive stresses of such magnitude that these stresses largely control its design. A column or strut is a compression member, the effective length of which exceeds three times the least lateral dimension. When a member carrying mainly axial load is vertical, it is termed as 'column' while if it is inclined or horizontal, it is termed as a 'strut'. Depending upon structural or architectural requirements, columns may be of various shapes, *i.e.,* circular, rectangular, square, hexagonal, etc.

Concrete is strong in compression. However, longitudinal steel rods are always provided to assist in carrying the direct loads. A minimum area of longitudinal steel is provided in the column, whether it is required from load point of view or not. This is done to resist tensile stresses caused by some eccentricity of the vertical loads. There is also an upper limit of amount of reinforcement in R.C. columns, because higher percentage of steel may cause difficulties in placing and compacting the concrete. Longitudinal reinforcing bars are 'tied' laterally by 'ties' or 'stirrups' at suitable interval so that the bars do not buckle.

### Functions of longitudinal and transverse reinforcements in a column

Longitudinal and transverse reinforcements are provided in a column to serve the following functions:

**(a) Longitudinal reinforcement**
1. To share the vertical compressive load, thereby reducing the overall size of the column.
2. To resist tensile stresses caused in the column due to
   (*i*) eccentric load    (*ii*) moment, or    (*iii*) transverse load.
3. To prevent sudden brittle failure of the column.
4. To impart certain ductility to the column.
5. To reduce the effect of creep and shrinkage due to sustained loading.

**(b) Transverse reinforcement**
1. To prevent longitudinal buckling of longitudinal reinforcement.
2. To resist diagonal tension caused due to transverse shear due to moment transverse load.
3. To hold the longitudinal reinforcement in position at the time of concreting.
4. To confine the concrete, thereby preventing its longitudinal splitting.
5. To impart ductility to the column.
6. To prevent sudden brittle failure of the columns.

## 13.2. SHORT AND LONG (OR SLENDER) COLUMNS

A column is a special case of a compression member that is vertical. Stability effects must be considered in the design of compression members.

Columns are divided into two types: (i) short column and (ii) long or slender column. When the ratio of *effective length* of the column to its least lateral dimension is less than 12, it is termed as a short column. A long or slender column is the one whose ratio of effective length to its least lateral dimension is not less than 12. In long columns, the permissible values of stresses in concrete and steel, used for short columns, should be multiplied by a coefficient $C_r$ given by the following formula:

$$C_r = 1.25 - \frac{l_{ef}}{48b} \qquad \ldots(13.1)$$

$C_r$ = reduction coefficient
$l_{ef}$ = effective length of column (Table 13.1)
$b$ = least lateral dimension of column.

When in a column, having helical reinforcement, the permissible load is based on the core area, the least lateral dimension should be taken as the diameter of the core.

For more exact calculations, the maximum permissible stresses in a reinforced concrete column or part thereof having a ratio of effective column length to least lateral radius of gyration above 50 should not exceed those which result from the multiplication of the appropriate maximum permissible stresses by the coefficient $C_r$ given by the following formula:

$$C_r = 1.25 - \frac{l_{ef}}{160\, i_{min}} \qquad \ldots(13.2)$$

where $i_{min}$ = least radius of gyration.

**Effective length of columns.** The effective length ($l_{ef}$) of a column of length ($l$) may be determined with the help of Table 13.1.

## 13.3. TYPES OF COLUMNS

Columns can mainly be divided into the following three types:
1. Column with longitudinal steel and lateral ties
2. Column with longitudinal steel and spirals.
3. Composite column.

Fig. 13.1(a) shows a concrete column reinforced with longitudinal bars, but with no laterals or ties. When load is applied on such a column, the concrete bulges out laterally, as shown. The bars themselves act as long slender columns and therefore tend to buckle away from the column's axis. Due to this, tension is caused in the outside shell of the concrete which opens out. The failure usually takes place suddenly. In order to check this tendency, the longitudinal reinforcement is tied transversely, at suitable intervals, with the help of ties, as shown in Fig. 13.1(b) and (c). These ties check the bars from buckling and also restrain the concrete from bulging action. When the number of bars are more than four, it is preferable to use two kinds of ties [(Fig. 13.1(b)], placing them alternatively, so that one (set *a*) holds the corner rods while the other (set *b*) holds the intermediate rods.

Each tie has to be spliced by lapping or by bending its ends around the main rod, which is quite troublesome. In order to overcome this difficulty, the longitudinal bars are tied continuously together with the help of 'spirals' shown in Fig. 13.1(d). The spiral so provided serve an additional purpose of laterally supporting the concrete inside and thus has confining effect on it.

Fig. 13.1(e) shows a typical composite column, reinforced with a centrally placed joist and with four or more longitudinal bars. Other steel sections may also be used. However, composite columns are used only for heavy loads.

## Braced and unbraced columns

In most of the cases, columns are also subjected to horizontal loads like wind, earthquake etc. If lateral supports are provided at the ends of the column, the lateral loads are borne entirely by the lateral supports. Such columns are known as *braced columns*. Other columns, where the lateral loads have to be resisted by them, in addition to axial loads and end moments are considered as *unbraced columns*. Bracings can be in one direction or in more than one direction, depending on the likely hood of the direction of external lateral loads. Thus, a column can be braced either in *x*-direction, or in *y*-direction or in both *x* and *y* directions. A braced column does not have relative lateral movements of its two ends. Thus, a braced column is not subject to *side sway*. A pin-jointed column is a simple example of a braced column. Similarly, an unbraced column is subject to *side sway* or lateral drift, *i.e.* there is significant lateral displacement between top and bottom ends of the column.

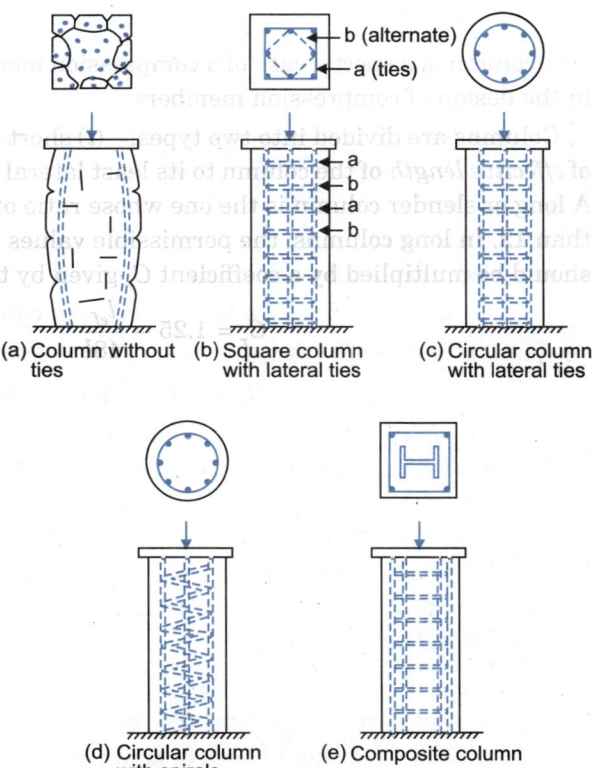

Fig. 13.1. Typical Types of Columns with Lateral Ties.

## 13.4. LOAD CARRYING CAPACITY OF SHORT COLUMNS

**(a) Short columns with lateral ties.** According to the classical *elastic theory* or the *compatible strain theory*, when a reinforced concrete column is loaded, both concrete and steel have equal strains during initial stage of loading, because they are well bonded. Hence if $P$ is the load on the column, the loads carried by steel bars ($P_s$) and that carried by surrounding concrete ($P_c$) bears the relation

$$P = P_c + P_s \qquad \ldots(i)$$

From compatibility requirements, $\varepsilon_s = \varepsilon_c$

$\therefore \qquad \dfrac{P_s}{E_s \cdot A_{sc}} = \dfrac{P_c}{E_c \cdot A_c} \qquad$ or $\qquad \dfrac{\sigma_{sc}}{E_s} = \dfrac{\sigma_{cc}}{E_c}$

or $\qquad \sigma_{sc} = \dfrac{E_s}{E_c} \cdot \sigma_{cc} = m \cdot \sigma_{cc}\ $ where $\ m$ is the modular ratio. $\qquad \ldots(ii)$

Hence from ($i$), $\qquad P = \sigma_{cc} \cdot A_c + \sigma_{sc} \cdot A_{sc}$

or $\qquad P = \sigma_{cc} \cdot A_c + m\sigma_{cc} \cdot A_{sc} \qquad \ldots(13.3)$

where $A_{sc}$ = Area of steel reinforcement; $\quad A_c$ = Area of concrete

$\sigma_{sc}$ = Compressive stress in steel; $\quad \sigma_{cc}$ = Compressive stress in concrete.

In has been found that the steel reinforcement is restrained against lateral expansion, due to concrete surrounding it, reducing the problem to be a plain strain problem rather than a plain stress problem. It is therefore essential to moderate the compatible stress in steel by restrained Poisson's effect, using modified Poisson's ratio $m_c = 1.5\ m$. Thus stress in steel is given by: $\sigma_{sc} = m_c \cdot \sigma_{cc} = 1.5\ m\ \sigma_{cc}$ and *equation 13.3* reduces to the form

$$P = \sigma_{cc} \cdot A_c + 1.5\, m\, \sigma_{cc} \cdot A_{sc} \qquad \ldots(13.4)$$

The above theoretical approach, based on strain compatibility, has been proved to be highly unrealistic due to creep and shrinkage behaviour of concrete. The above formula therefore gives very conservative results. It is therefore recommended to use *equilibrium approach* based on allowable stresses in the two materials. Such an equilibrium approach, however, violates the compatibility of strains. According to this approach, the load carrying capacity of a short column, axially loaded and provided with lateral ties, is given by

$$P = \sigma_{cc} \cdot A_c + \sigma_{sc} \cdot A_{sc} \qquad \ldots(13.5)$$

where $P$ = permissible axial load

$\sigma_{sc}$ = permissible compressive stress in column bars

$\sigma_{cc}$ = permissible compressive stress in concrete

IS : 456-2000 recommends *Eq. 13.5* for the determination of permissible axial load $P$. The minimum eccentricity mentioned in § 13.5.7 may be deemed to be incoporated in the above equation.

**(b) Short column with helical reinforcement.** For a column having longitudinal reinforcement tied with spirals (*i.e.* helical reinforcement), the load carrying capacity is taken as 1.05 times the strength of similar member with lateral ties, provided the requirements laid down in para 8*b* of § 13.5 are satisfied.

**(c) Composite columns.** The allowable load $P$ on a composite column, consisting of a structural steel or cast-iron column thoroughly encased in concrete reinforced with both longitudinal and spiral reinforcement is given by the following expression:

$$P = \sigma_{cc} \cdot A_c + \sigma_{sc} \cdot A_{sc} + \sigma_{mc} \cdot A_m \qquad \ldots(13.6)$$

where $A_c$ = net area of concrete section

$A_{sc}$ = cross-sectional area of longitudinal bar reinforcement

$\sigma_{mc}$ = allowable unit stress in metal core, not to exceed 125 N/mm² for a steel core, or 70 N/mm² for a cast iron core

$A_m$ = Cross-sectional area of steel or cast iron core.

## 13.5. INDIAN STANDARD RECOMMENDATIONS (IS : 456-2000)

### 1. Longitudinal reinforcement

(a) The cross-sectional area of longitudinal reinforcement shall be not less than 0.8 percent nor more than 6 percent of the gross cross-sectional area of the column.

**Note:** The use of 6 percent reinforcement may involve practical difficulties in placing and compaction of concrete; hence lower percentage is recommended. Where bars from the columns below have to be lapped with those in the column under consideration, the percentage of steel shall usually not exceed 4%.

(b) In any column that has a larger cross-sectional area than that required to support the load, the minimum percentage of steel shall be based upon the area of concrete required to resist the direct stress and not upon the actual area.

(c) The minimum number of longitudinal bars provided in a column shall be four in rectangular columns and six in circular columns.

(d) The bars shall not be less than 12 mm in diameter.

(e) A reinforced concrete column having helical reinforcement shall have at least six bars of longitudinal reinforcement.

(f) In a helically reinforced column, the longitudinal bars shall be in contact with the helical reinforcement and equidistant around its inner circumference.

(g) Spacing of longitudinal bars measured along the periphery of the column shall not exceed 300 mm.

(h) In case of padestals in which the longitudinal reinforcement is not taken into account in strength calculations, nominal longitudinal reinforcement not less than 0.15% of the cross-sectional area shall be provided.

**Note:** Padestal is a compression member, the effective length of which does not exceed three times the least lateral dimension.

### 2. Transverse reinforcement

**(a) General.** A reinforced concrete compression member shall have transverse or helical reinforcement so disposed that every longitudinal bar nearest to the compression face has effective lateral support against buckling subject to provisions in 2(b). The effective lateral support is given by transverse reinforcement either in the form of circular rings capable of taking up circumferential tension or by polygonal links (lateral ties) with internal angles not exceeding 135°. The ends of the transverse reinforcement shall be properly anchored.

**(b) Arrangement of transverse reinforcement**

(i) If the longitudinal bars are not spaced more than 75 mm on either side, transverse reinforcement need only to go round corner and alternate bars for the purpose of providing effective lateral supports [Fig. 13.2(a)].

(ii) If the longitudinal bars spaced at a distance of not exceeding 48 times the diameter of the tie are effectively tied in two directions, additional longitudinal bars in between these bars need to be tied in one direction by open ties. [Fig. 13.2(b)].

(iii) Where the longitudinal reinforcing bars in a compression member are placed in more than one row, effective lateral support to the longitudinal bars in the inner rows may be assumed to have been provided if (a) transverse reinforcement is provided for outermost row in accordance with (2) above, and (b) no bar of the corner row is closer to the nearest compression face than three times the diameter of the largest bar in the inner row [Fig. 13.2(c)].

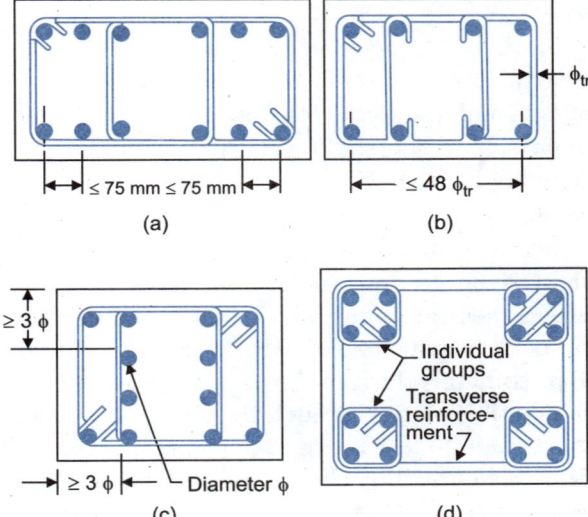

Fig. 13.2

(iv) Where the longitudinal bars in a compression member are grouped (not in contact) and each group adequately tied with transverse reinforcement in accordance with clause (2), the transverse reinforcement for the compression member as a whole may be provided on the assumption that each group is a single longitudinal bar for purpose of determining the pitch and diameter of the transverse reinforcement in accordance with clause (2). The diameter of such transverse reinforcement need not, however, exceed 20 mm [see Fig. 13.2(d)].

**(c) Pitch and diameter of lateral ties**

(i) *Pitch*. The pitch of the transverse reinforcement shall be not more than the least of the following distances: (1) the least lateral dimension of the compression member; (2) sixteen times the smallest diameter of the longitudinal reinforcement bar to be tied; (3) Fortyeight times the diameter of the transverse reinforcement.

(ii) *Diameter*. The diameter of the polygonal links or lateral ties shall not be less than one-fourth of the diameter of the largest longitudinal bar, and in no case less than 5 mm.

**(d) Helical reinforcement**

(i) *Pitch*. Helical reinforcement shall be of regular formation with the turns of the helix spaced evenly and its ends shall be anchored properly by providing one and a half extra turns of the spiral bar. Where an increased load on the column on the strength of the helical reinforcement is allowed for, the pitch of the helical turns shall be not more than 75 mm, nor more than one sixth of the case diameter of the column, nor less than 25 mm, nor less than three times the diameter of the steel bar forming the helix.

(ii) *Diameter*. The diameter of the helix reinforcement shall be in accordance with clause 2 (c ii) given above.

**3. Cover to reinforcement.** According to IS code, for a longitudinal reinforcing bar in a column, the nominal cover shall not be less than 40 mm, nor less than the diameter of such bar. In the case of columns of minimum dimension of 200 mm or under, whose reinforcing bars does not exceed 12 mm, a cover of 25 mm may be used.

**4.1. Unsupported length of column.** The unsupported length $l$ of a compression member shall be taken as the clear distance between end restraints, except that:

(a) in flat slab construction, it shall be clear distance between the floor and the lower extremity of the capital, the drop panel or slab whichever is the least.

(b) in beam and slab construction, it shall be the clear distance between the floor and the underside of the shallower beam framing into the floor and the underside of the shallower beam framing into the columns in each direction at the next higher floor level.

(c) in columns restrained laterally by struts, it shall be the clear distance between consecutive struts in each vertical plane, provided that to be an adequate support, two such struts shall meet the columns at approximately the same level, and the angle between the vertical planes through the struts shall not vary more than 30° from a right angle. Such struts shall be of adequate dimensions and shall have sufficient anchorage to restrain the member against lateral deflection.

(d) in columns restrained laterally by struts or beams with brackets used at the junction, it shall be the clear distance between the floor and the lower edge of the bracket, provided that the bracket width equals that of the beam strut and is atleast half that of the column.

**4.2. Effective length of columns.** (a) In the absence of more exact analysis, the effective length of columns in framed structures may be obtained from the ratio of effective length to unsupported length $l_{ef}/l$ given in Fig. 13.3 when relative displacement of the ends of the column is prevented and in Fig. 13.4 when relative lateral displacement is not prevented. In the latter case, it is recommended that the effective length ratio $l_{ef}/l$ may not be taken less than 1.2.

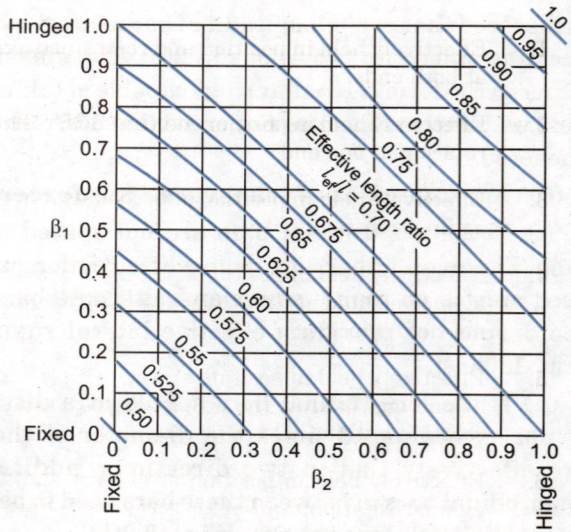

Fig. 13.3. Effective length Ratios for a Column in a frame without restraint against sway.

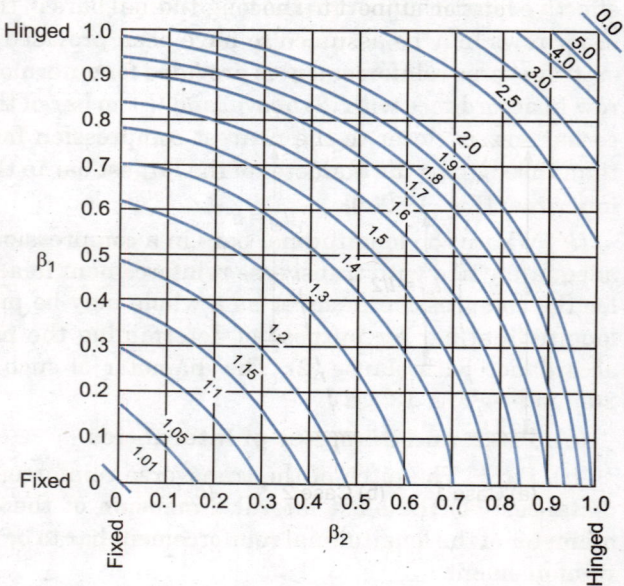

Fig. 13.4. Effective Length Ratios for a Column in a Frame with no Sway.

**Note:** In Figures 13.3 and 13.4, $\beta_1$ and $\beta_2$ are equal to $\dfrac{\Sigma k_c}{\Sigma k_c + \Sigma k_b}$ where the summation is done for the members framing into a joint at top and bottom respectively; $k_c$ and $k_b$ being the flexural stiffness for column and beam respectively.

(b) For a normal usage, assuming idealised conditions, the effective length $l_{ef}$ in a given plane may be assessed on the basis of Table 13.1 for various cases of endconditions (Fig. 13.5). In the table, $l$ is the unsupported length of the column.

## DESIGN OF AXIALLY LOADED COLUMNS

**TABLE 13.1** Effective Length of Compression Members (IS : 456-2000)

| Case No. | Degree of end restraint of compression member (Fig. 13.5) (1) | Theoretical value of effective length (2) | Recommended value of effective length (3) |
|---|---|---|---|
| 1. | Effectively held in position and restrained against rotation at both ends. | $0.5l$ | $0.65l$ |
| 2. | Effectively held in position at both ends, restrained against rotation at one end. | $0.7l$ | $0.80l$ |
| 3. | Effectively held in position at both ends, but not restrained against rotation. | $1.00l$ | $1.00l$ |
| 4. | Effectively held in position and restrained against rotation at one end, and at the other restrained against rotation but not held in position. | $1.00l$ | $1.20l$ |
| 5. | Effectively held in position and restrained against rotation at one end, and at the other partially restrained against rotation but not held in position. | — | $1.50l$ |
| 6. | Effectively held in position at one end but not restrained against rotation, and at the other end restrained against rotation but not held in position. | $2.00l$ | $2.00l$ |
| 7. | Effectively held in position and restrained against rotation at one end but not held in position nor restrained against rotation at the other end. | $2.00l$ | $2.00l$ |

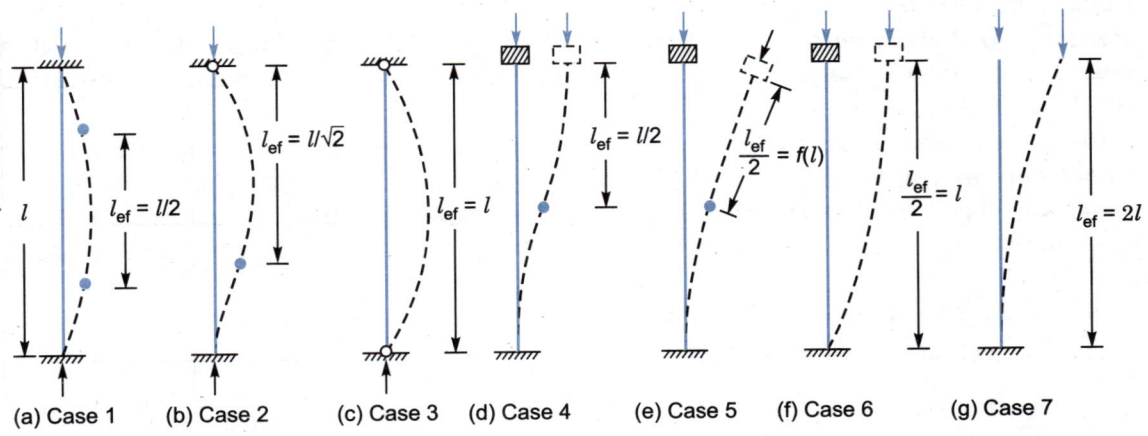

(a) Case 1  (b) Case 2  (c) Case 3  (d) Case 4  (e) Case 5  (f) Case 6  (g) Case 7

**Fig. 13.5.** End Conditions for Compression Members

**5. Short and slender compression members.** A compression member may be considered as short when both the slenderness ratios $l_{ex}/D$ and $l_{ey}/b$ are less than 12:

where $l_{ex}$ = effective length in respect of the major axis.

$D$ = Depth in respect of the major axis.

$l_{ey}$ = effective length respect of the minor axis, and $b$ = width of member.

It shall otherwise be considered as a slender compression member.

**6. Slenderness limits for columns**

(*i*) The unsupported length between end restraints shall not exceed 60 times the least lateral dimension of a column.

(*ii*) If, in a given plane, one end of a column is unrestrained, its unsupported length $l$, shall not exceed $100b^2/D$.

where: $b$ = width of that cross-section, and
$D$ = depth of the cross-section measured in the plane under consideration.

**7. Minimum eccentricity.** All columns shall be designed for minimum eccentricity equal to the unsupported length of column/500 plus lateral dimension/30, subject to a minimum of 20 mm. Where biaxial bending is considered it is sufficient to ensure that eccentricity exceeds the minimum about one axis at a time.

In actual practice, load on a column is never *axial* because of two reasons:
(*i*) Lack of homogeneity of the material, due to which the centroid may not coincode with the geometric axis of the section.
(*ii*) Load acting at the ends not truly axial.

The Code takes into account both the reasons, and gives the following expression for minimum eccentricity about either of the principal axes:

$$e_{min} = \frac{l}{500} + \frac{h}{30} \text{ but } \not< 20 \text{ mm}$$

where $l$ = unsupported length of the column
$h$ = lateral dimension of the column perpendicular to the axis of bending (*i.e.* either $b$ or $D$)

Thus, for a column, having depth $D$ and width $b$, the above expression reduces as following for bending about $x$-$x$ and $y$-$y$ axes

$$e_{min,x} = \frac{l_x}{500} + \frac{D}{30} \not< 20 \text{ mm} \quad \text{and} \quad e_{min,y} = \frac{l_y}{500} + \frac{b}{30} \not< 20 \text{ mm}$$

where $l_x$ and $l_y$ and the unsupported length of the column for bending about $x$ and $y$ axes respectively.

**8. Permissible loads in compression members**

(*a*) *Padestal and short columns with lateral ties.* The axial load $P$ permissible on a pedestal or short column reinforced with longitudinal bars and lateral ties shall not exceed that given by the following equation: $P = \sigma_{cc} A_c + \sigma_{sc} \cdot A_{sc}$

**Note:** The minimum eccentricity mentioned in clause 7 may be deemed to be incorporated in the above equation.

(*b*) *Short columns with helical reinforcement.* The permissible load for columns with helical reinforcement shall be 1.05 times the permissible load for member with lateral ties, or rings, This clause is valid only if the ratio volume of helical reinforcement to the volume of core is not less than $0.36 (A_g/A_k - 1) f_{ck}/f_{yh}$

where $A_g$ = gross area of section
$A_k$ = area of core helically reinforced column measured to the outside diameter of the helix.
$= \frac{\pi}{4} D_k^2 - A_s$
$f_{ck}$ = characteristic compressive strength of the concrete
$D_k$ = diameter of concrete core from outside of helix = $D - 2 \times$ clear cover
$f_{yh}$ = characteristic strength of helical reinforcement but not exceeding 415 N/mm$^2$.

(*c*) *Long columns.* The maximum permissible stress in a reinforced concrete column or part thereof having a ratio of effective column length to least lateral dimension above 12 shall not exceed that which results from the multiplication of the appropriate maximum permissible stresses (in concrete and steel) by the coefficient $C_r$ given by the following formula:

$$C_r = 1.25 - (l_{ef}/48b)$$

where $C_r$ = reduction coefficient
$l_{ef}$ = effective length of column, and
$b$ = least lateral dimension of column; for column with helical reinforcement, $b$ is the diameter of the core.

For more exact calculations, the maximum permissible stresses in a reinforced concrete column or part thereof having a ratio of effective column length to least lateral radius of gyration above 40 shall not exceed these which results from the multiplication of the appropriate maximum permissible stresses in concrete and steel, by the coefficient $C_r$ given by the following formula: $C_r = 1.25 - \dfrac{l_{ef}}{160\, l_{min}}$ where $i_{min}$ is the least of gyration.

### (9) Composite column

(*a*) **Allowable load.** The allowable axial load $P$ on a composite column consisting of structural steel or cast-iron column thoroughly encased in concrete reinforced with both longitudinal and spiral reinforcement shall not exceed that given by the following formula:

$$P = \sigma_{cc} \cdot A_c + \sigma_{sc} \cdot A_{sc} + \sigma_{mc} \cdot A_m$$

where $\sigma_{cc}$ = permissible stress in concrete in direct compression

$A_c$ = net area of concrete section, which is equal to the gross area of concrete section
  = $A_{sc} - A_m$

$\sigma_{sc}$ = permissible compressive stress of column bars

$A_{sc}$ = cross-sectional area of longitudinal bar reinforcement

$\sigma_{mc}$ = allowable unit stress in metal core, not to exceed 125 N/mm² for a steel core, or 70 N/mm² for a cast iron core;

$A_m$ = the cross-sectional area of the steel or cast iron core.

(*b*) **Metal core and reinforcement.** The cross-sectional area of the metal core shall not exceed 20% of the gross area of the column. If a hollow metal core is used, it shall be filled with concrete. The amount of longitudinal and spiral reinforcement and the requirements as to spacing of bars, details of splices and thickness of protective shell outside the spiral, shall conform to the requirements of clause 1. A clearance of at least 75 mm shall be maintained between the spiral and the metal core at all points, except that when the core consists of a structural steel H-column, the minimum clearance may be reduced to 50 mm.

(*c*) **Splices and connections of metal cores.** Metal cores in composite columns shall be accurately milled at splices and positive provisions shall be made for alignment of one core above another. At the column base, provisions shall be made to transfer the load to the footing at safe unit stresses.

(*d*) **Allowable load on metal core only.** The metal core of composite columns shall be designed to carry safely any constructions or other loads to be placed upon them prior to their encasement in concrete.

**Permissible stresses.** The permissible compressive stress for various grades of concrete is given in Table 1.16. The values of $\sigma_{cc}$ for M 20, M 25 and M 30 grades of concrete are 5 N/mm², 6 N/mm² and 8 N/mm² respectively. The permissible compressive stress in longitudinal bars is given in Table 1.19. For mild steel bars $\sigma_{sc}$ is 130 N/mm². For Fe 415 steel, the corresponding value is 190 N/mm².

## 13.6. DESIGN PROCEDURE

**A R.C. column may be designed in the following steps:**

1. Depending upon the grade of concrete to be used, determine the permissible stresses in concrete, longitudinal bars and ties.

2. Find the super-imposed load the column is required to carry. To this, add dead weight (assumed) of the column to get the total load the column has to carry at its base.

3. Assume some suitable value of reinforcement $A_{sc}$ say between 0.8 to 2% of gross area $A$ of column. Determine the area $A$ from the following expression.

$$P = \sigma_{cc} \cdot A_c + \sigma_{sc} \cdot A_{sc}$$

or

$$P = \sigma_{cc}(A_g - p \cdot A_g) + \sigma_{sc} \cdot pA_g$$

or

$$A_g = \dfrac{P}{\sigma_{cc}(1-p) + p \cdot \sigma_{sc}} \qquad \ldots(13.7)$$

where $A_g$ = gross area of column and $p$ = ratio of steel to total area = $A_{sc}/A_g$.

**312** REINFORCED CONCRETE STRUCTURE

4. After having known the area $A_g$, determine the dimensions of the column. If it is a square of side $b$, then $b = \sqrt{A_g}$. If a circular column of diameter $D$ is to be used, $D (= b)$ will be equal to $\sqrt{4A_g/\pi}$.

5. For the given end conditions, determine the effective length $l_{ef}$ of the column, and hence calculate $l_{ef}/b$ ratio to find whether it is short column or long column.

6. If $l_{ef}/b < 12$, it will be designed as a short column for which dimensions have already been found in steps 3 and 4. Determine the area of steel $A_{sc}$ and distribute the bars suitably around the periphery of the column, keeping suitable cover.

7. If $l_{ef}/b > 12$, it will be designed as a long column for which the reduction factor $C_r$ is determined from the expression

$$C_r = 1.25 - \frac{l_{ef}}{48\,b}$$

Calculate the design load $P'$ for an equivalent short column:

$$P' = P/C_r \qquad \qquad ...(13.8)$$

Using the revised value of $P'$ re-calculate the area $A_g$ and hence determine the size of the column. Also, calculate the area of steel $A_{sc} = pA_g$.

8. Find the diameter of bars used as ties, and determine its pitch as per rules discussed in the previous article.

**Example 13.1.** *(a) A reinforced concrete column 4 m long (effective) and 400 mm in diameter is reinforced with 8 bars of 20 mm diameter. Find safe load of columns can carry. Take M 20 concrete and Fe 415 steel reinforcement. The column carries lateral ties.*

*(b) If the effective length is increased to 8 m, what will be the safe load of column can carry?*

**Solution:** (a) $\qquad l_{ef} = 4000$ mm; $b = D = 400$ mm

Hence $\qquad \dfrac{l_{ef}}{b} = \dfrac{4000}{400} = 10 < 12$. Thus the column is short.

$$P = \sigma_{cc} \cdot A_c + \sigma_{sc} \cdot A_{sc}.$$

where $A_{sc} = 8\dfrac{\pi}{4}(20)^2 = 2513.3$ mm$^2$; $A_g = \dfrac{\pi}{4}(400)^2 = 125664$ mm$^2$

$A_c = 125664 - 2513.3 \approx 123151$ mm$^2$.

For M 20 concrete, $\sigma_{cc} = 5$ N/mm$^2$; For Fe 415 steel, $\sigma_{sc} = 190$ N/mm$^2$

$\therefore \qquad P = (5 \times 123151) + (190 \times 2513.3) = 1093282$ N $\approx$ **1093.282 kN**

(b) $l_{ef} = 8000$ mm; $D = 400$ mm

$$\dfrac{l_{ef}}{b} = \dfrac{8000}{400} = 20 > 12.$$

Hence the column is a long column for which the reduction factor

is given by $\qquad C_r = 12.5 - \dfrac{l_{ef}}{48b} = 1.25 - \dfrac{8000}{48 \times 400} = 0.8333$

$P = 0.8333 \times 1093281 \approx 911032$ N $\approx$ **911.032 kN**.

**Example 13.2.** *Solve example 13.1 (a) if the column carries spiral (helical) reinforcement of 10 mm Φ rod wound around the 20 mm Φ bars at a pitch of 60 mm. The 20 mm Φ longitudinal bars are placed with a nominal cover of 40 mm.*

**Solution:**

Outside diameter of helix $D_k = 400 - (2 \times 40) = 320$ mm. $A_k$ = Area of core of helically reinforced column measured to the outside diameter of the helix = $\dfrac{\pi}{4}(320)^2 - 2513.3 = 77911$ mm$^2$

$A_g$ = gross area of section = $\dfrac{\pi}{4}(400)^2 = 125663.7$ mm$^2$.

For M 20 concrete, $f_{ck} = 20$ N/mm$^2$. Also $f_y = 415$ N/mm$^2$ for helical reinforcement

∴ Factor $0.36 (A_g/A_k - 1) f_{ck}/f_{yh} = 0.36 (125663.7/77911 - 1) 20/415 = 0.0106$ ...(i)

Volume of helical reinforcement ($V_h$) per unit length of column

$$= \frac{\text{circumference of spiral} \times \text{its area of cross-section}}{\text{pitch of spiral}} = \left(\frac{\pi d}{s}\right)\left(\frac{\pi}{4} \Phi_s^2\right)$$

where $d$ = diameter upto centre of helix = $400 - (2 \times 40) - 10 = 310$ mm

$\Phi_s$ = dia. of spiral wire = 10 mm and $s$ = pitch = 60 mm

∴ $V_{hs}$ per 1 mm length of column = $\left(\frac{\pi \times 310}{60}\right)\frac{\pi}{4}(10)^2 = 1274.8$ mm²

Volume of core $V_k = A_k \times 1 = 77911 \times 1 = 77911$ mm³

∴ Ratio $\dfrac{V_{hs}}{V_k} = \dfrac{1274.8}{77911} = 0.0164.$ ...(ii)

Since this ratio ($V_{hs}/V_k$) is more than the factor calculated in (i), the load carried by column with helical reinforcement will be 1.05 times the permissible load for column with lateral ties or rings.

Hence for case (a) of short column (example 13.1), $P = 1.05 \times 1093.281 = $ **1147.945 kN**

**Example 13.3.** (a) *Design a short square column to carry an axial load of 1200 kN. Use M 25 concrete mix. and Fe 415 steel.*

**Solution:**

**1. Design of section.** Minimum steel = 0.8%. Let us use 1% steel *i.e.*, $p = 0.01$

$\sigma_{cc}$ for M 25 mix = 6 N/mm². For Fe 415 steel, $\sigma_{sc}$ = 190 N/mm². The load carrying capacity of a short column is given by

$P = \sigma_{cc} \cdot A_c + \sigma_{sc} \cdot A_{sc} = \sigma_{cc}(A_g - pA_g) + \sigma_{sc} pA_g$

From which $A_g = \dfrac{P}{\sigma_{cc}(1-p) + p\sigma_{sc}}$

$= \dfrac{1200 \times 1000}{6(1-0.01) + 0.01 \times 190}$

$= 153061$ mm²

∴ Size of square = $\sqrt{153061} \approx 392$ mm

However, provide square column 400 mm × 400 mm.

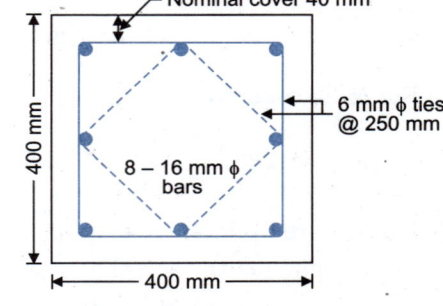

Fig. 13.6

**2. Longitudinal reinforcement.** $A_{sc} = pA = 0.01 \times 153061 \approx 1530.6$ mm²

Using 16 mm Φ bars, $A_\Phi = \dfrac{\pi}{4}(16)^2 = 201$ mm².

∴ No. of bars required = $1530.6/201 \approx 8$

Arrange the bars as shown in Fig. 13.6 at a nominal cover of 40 mm.

**3. Design of ties.** Diameter of ties should be 1/4 the diameter of longitudinal reinforcement subject to a minimum of 5 mm. However use 6 mm Φ bars of ties. The *c/c* spacing of the ties should not exceed the least of the following:

(i) least lateral dimension = 400 mm

(ii) 16 × diameter of main bar = 16 × 16 = 256 mm

(iii) 48 × dia. of ties = 48 × 6 = 288 mm.

Hence provide the ties @ 250 mm *c/c*. The ties will be square in shape in two sizes as shown in Fig. 13.6, using them alternately, so that longitudinal bars pass through the corners of the ties. Keep pitch of each set of ties at 250 mm.

**Example 13.4.** *Design a short R.C. column to take an axial load of 5000 kN. The size of the column is not to be more than 700 mm. Use spiral reinforcement. Mix of concrete to be used is M 25. Use Fe 415 steel.*

**Solution:**

**1. Steel reinforcement.** Assuming that the conditions of helical reinforcement mentioned in para 8 (b) (§ 13.5) are fulfilled, the carrying capacity of the column may be taken to be equal to 1.05 times the load carried by a column with a lateral ties.

$$\therefore \qquad P = 1.05 \, [\sigma_{cc} \cdot A_c + \sigma_{sc} \cdot A_{sc}]$$

where $A_g = \dfrac{\pi}{4} (700)^2 = 384845 \text{ mm}^2$; $A_c = A_g - A_{sc} = (384845 - A_{sc})$

For M 25 concrete, $\sigma_{cc} = 6$ N/mm$^2$. $\therefore$ $5000{,}000 = 1.05 \, [6(384845 - A_{sc}) + 190 \, A_{sc}]$

From which $A_{sc} = 13331$ mm$^2$. This is less than 6% and more than 0.8% of area of cross-section of the column, and hence satisfactory.

Using 40 mm $\Phi$ bars, No. of bars required = $13331 / \dfrac{\pi}{4}(40)^2 \simeq 11$. Keeping 40 mm nominal cover and using 10 mm $\varphi$ for spiral reinforcement, arrange the bars on a circle of diameter equal to $700 - 2(40 + 10) - 40 = 560$ mm. Hence circumferential spacing of the bars = $\pi \times 560/11 \simeq 160$ mm which is less than 300 mm.

**2. Spiral reinforcement.** Let the helix bar be of 10 mm $\Phi$.

Outside dia. of helix = $700 - (2 \times 40) = 620$ mm.

$A_k$ = area of core helically reinforced column measured to the outside diameter of helix

$$= \dfrac{\pi}{4}(620)^2 - \dfrac{\pi}{4}(40)^2 \times 11 = 288084 \text{ mm}^2$$

$$A_g = \text{gross area of section} = \dfrac{\pi}{4}(700)^2 = 384845 \text{ mm}^2$$

For M 25 concrete, $f_{ck} = 25$ N/mm$^2$; Also, $f_{yh} = 415$ N/mm$^2$ for Fe 415 steel

$$\therefore \quad \text{Factor } 0.36 \left( \dfrac{A_g}{A_k} - 1 \right) \dfrac{f_{ck}}{f_{yh}} = 0.36 \left( \dfrac{384845}{288084} - 1 \right) \dfrac{25}{415} \simeq 0.00728 \qquad \ldots(i)$$

Dia. of core upto centre of helix = $700 - (2 \times 40) - 10 = 610$ mm
Let the pitch of the spiral be $s$ mm.

$\therefore$ Volume of spiral $V_{hs}$ per 1 mm length of column

$$= \dfrac{\pi d}{s} \left( \dfrac{\pi}{4} \Phi_s^2 \right) = \dfrac{\pi \times 610}{s} \times \dfrac{\pi}{4}(10)^2 = \dfrac{150511}{s}$$

Volume of core per mm length, $V_k = A_k \times 1 = 288084 \times 1 = 288084$ mm$^3$

$$\therefore \quad \text{Ratio} \qquad \dfrac{V_{hs}}{V_k} = \dfrac{150511}{s \times 288084} = \dfrac{0.5225}{s} \qquad \ldots(ii)$$

Equating $(i)$ and $(ii)$, we get $\dfrac{0.5225}{s} = 0.00728$. From which $s = 71.8$ mm.

However the pitch should not be more than 75 mm, nor more than 1/6 core diameter (= $1/6 \times 620 \simeq 103$ mm). Also, it should not be less than $3\Phi$ (= $3 \times 10 = 30$ mm).

Hence keep the pitch equal to 70 mm. The details of reinforcement etc. are shown in Fig. 13.7.

**Example 13.5.** *Design a column 10 m long to carry an axial load of 600 kN. The column is restrained at the ends. Use M 25 concrete and Fe 415 steel reinforcement.*

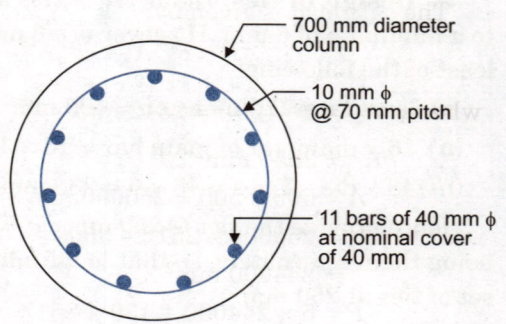

Fig. 13.7

**Solution:**

**1. Size of column.** From Table 13.1 (Case 1), $l_{ef} = 0.65l = 0.65 \times 10000 = 6500$ mm. Assume the column to be long one. Let the size of the column be $D \times D$ mm(square)

∴ Reduction factor $\quad C_r = 1.25 - \dfrac{6500}{48 \times D} = 1.25 - \dfrac{135.42}{D}$

Hence design load $\quad = \dfrac{P}{C_r} = \left( \dfrac{600000}{1.25 - \dfrac{135.42}{D}} \right)$

Providing only lateral ties, we have $\left( \dfrac{600000}{1.25 - \dfrac{135.42}{D}} \right) = \sigma_{cc} A_c + \sigma_{sc} A_{sc}$

Let us provide reinforcement @ 1% of area.
$$A_g = D^2;\ A_{sc} = 0.01\ D^2;\ A_c = A_g - A_{sc} = D^2 - 0.01\ D^2 = 0.99\ D^2.$$
For M 25 concrete, $\sigma_{sc} = 6$ N/mm². For Fe 415 steel, $\sigma_{sc} = 190$ N/mm²

∴ $\quad \dfrac{600000}{1.25 - \dfrac{135.42}{D}} = 6 \times 0.99\ D^2 + 190 \times 0.01\ D^2 = 7.84\ D^2$

or $\quad 60000 = 9.8\ D^2 - 1061.7\ D$

or $\quad D^2 - 108.336\ D - 612245 = 0$. From which $D = 307.5$ mm. Keep $D = 310$ mm.

**2. Longitudinal reinforcement.** $A_{sc} = 0.01\ (310 \times 310) = 961$ mm²
Provide 8 bars. Area of each bar = 961/8 = 120.12 mm²

∴ $\quad$ Dia. of bar $= \sqrt{\dfrac{120.12 \times 4}{\pi}} = 12.4$ mm

Provide 8 Nos. of 14 mm φ bars. Keep a nominal cover of 40 mm.

**3. Design of ties**
Use 6 mm Φ ties at a pitch not exceeding the minimum of the following:
(i) least lateral dimension = 310 mm.
(ii) 16 Φ = 16 × 14 = 224 mm.
(iii) 48 × dia. of ties = 48 × 6 = 288 mm.
However, provide the ties at 220 mm c/c.

**Example 13.6.** *Find the safe load for the composite column shown in Fig. 13.8. Take $\sigma_{cc} = 5$ N/mm², $\sigma_{sc} = 130$ N/mm² and $\sigma_{mc} = 125$ N/mm². The joist has area of cross-section of 3500 mm².*

**Solution:**
The safe load is given by
$$P = \sigma_{cc} A_c + \sigma_{sc} A_{sc} + \sigma_{mc} A_m$$
where, $A_{sc} = 8\dfrac{\pi}{4}(20)^2 = 2512$ mm²
$A_m = 3550$ mm²;
$A = 500 \times 500 = 250000$ mm
∴ $A_c = 250000 - 2512 - 3500$
$\approx 244000$ mm²
$P = 5 \times 244000 + 130 \times 2512 + 125 \times 3500$
$\approx 1984000$ N = **1984 kN.**

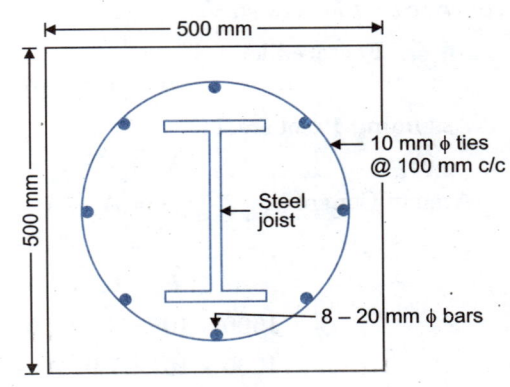

Fig. 13.8

**Example 13.7.** *Design a Short Column Square in section to carry an axial load of 1200 kN using $M_{20}$ grade concrete and Fe 415 steel.*

**Solution:** Factored load
$$P_u = 1.5 \times 1200 = 1800 \text{ kN} = 1800 \times 10^3 \text{ N}.$$

Assuming 1% of steel
$$A_{sc} = 1\% A_g = 0.01 A_g$$

Area of concrete,
$$A_c = A_g - A_{sc}$$
$$= A_g - 0.01 A_g = 0.99 A_g$$

For axially loaded short and columns
$$P_u = 0.4 f_{ck} A_c + 0.67 f_y A_{sc}.$$
$$1800 \times 10^3 = 0.4 \times 20 \times 0.99 A_g + 0.67 \times 415 \times 0.01 A_g$$
$$1800 \times 10^3 = 7.92 A_g + 2.78 A_g = 10.7 A_s.$$
$$A_g = 168224.3 \text{ mm}^2$$

Size of the Square Column = $\sqrt{168224.3}$
= 410.2 mm

Adopt 420 mm × 420 mm Square Column
$$A_{sc} = 0.01 \times A_g \text{ required}$$
$$= 0.01 \times 168224.3$$
$$= 1682.2 \text{ mm}^2$$

Provide 6 bars of 20 mm dics.
$$A_{sc} \text{ provided} = 1885 \text{ mm}^2$$

**Lateral Ties:** Diameter of lateral ties should not be less than

(a) $\dfrac{\phi_l}{4} = \dfrac{1}{4} \times 20 = 5$ mm

(b) 6 mm

Hence adopt 6 mm φ bars.

Pitch of the ties shall be minimum of

(a) Least lateral dimensions of column = 420 mm.

(b) 16 times the dia of longitudinal bar.
= 16 × 20 = 320 mm

(c) 300 mm.

Hence, provide 6 mm lateral ties at 300 mm c/c as shown in Fig. 13.9.

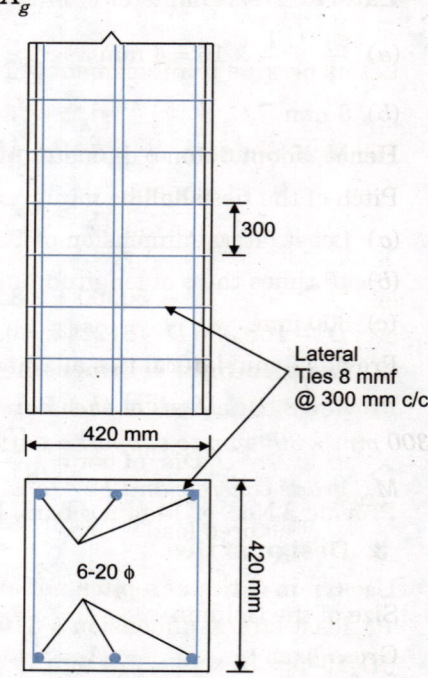

Fig. 13.9

**Example 13.8.** *Design a Circular Column to carry an axial load of 1000 kN using lateral ties. Use $M_{20}$ grade concrete and Fe 415 steel.*

**Solution:** Factored load
$$P_u = 1.5 \times 1000 = 1500 \text{ kN} = 1500 \times 10^3 \text{ N}.$$

Assuming 1% of steel
$$A_{sc} = 1\% A_g = 0.01 A_g$$

Area of Concrete,
$$A_c = A_g - A_{sc}$$
$$= A_g - 0.01 A_g = 0.99 A_g$$
$$P_u = 0.4 f_{ck} A_c + 0.67 f_y A_{sc}.$$
$$1500 \times 10^3 = 0.4 \times 20 \times 0.99 A_g + 0.67 \times 415 \times 0.01 A_g$$
$$1500 \times 10^3 = 7.92 A_g + 2.78 A_g = 10.7 A_g.$$
$$A_g = 140186.9 \text{ mm}^2.$$

Diameter of the Circular Section

$$= \frac{\pi D^2}{4} = 140186.9$$

$$D = 422.5 \text{ mm}.$$

Adopt Circular Column of diameter 430 mm

$$A_{sc} = 0.01 \times A_g \text{ required}$$
$$= 0.01 \times 140186.9$$
$$= 1401.9 \text{ mm}^2$$

Provide 8 bars of 16 mm $d$.

$A_{sc}$ provided = 1608.5 mm$^2$.

**Lateral Ties:** Diameter of lateral ties should not be less than

(a) $\dfrac{\phi_l}{4} = \dfrac{1}{4} \times 16 = 4$ mm

(b) 6 mm

Hence, adopt 6 mm $\phi$ diameter bars

Pitch of the ties shall be minimum of

(a) Least lateral dimension of Column = 430 mm.

(b) 16 times the $\phi$ of longitudinal bar = 16 × 16 = 256 mm

(c) 300 mm.

Provide 6 mm lateral ties at 250 mm c/c as shown in Fig. 13.10.

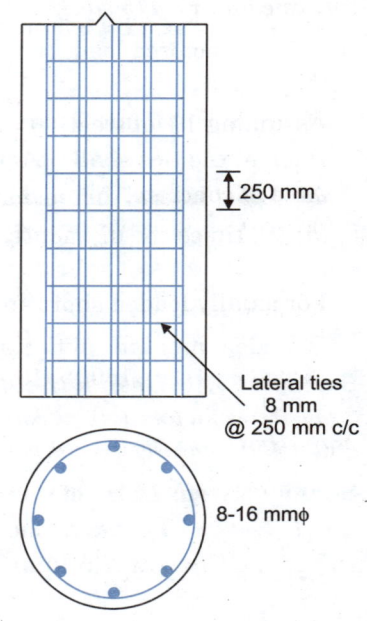

Fig. 13.10

**Example 13.9.** *Design the Reinforcement for a short axially loaded Square Column of size 300 mm × 300 mm to support a load of 750 kN. Use,*

*$M_{20}$ grade concrete and Fe 415 steel.*

**Solution:** Factored load

$$P_u = 1.5 \times 750 = 1125 \text{ kN} = 1125 \times 10^3 \text{ N}.$$

Size of the Column = 300 mm × 300 mm

Gross Area $A_g = 300 \times 300 = 90000$ mm$^2$

Area of Concrete $A_c = 90000 - A_{sc}$

For axially loaded Short Columns

$$P_u = 0.4 f_{ck} A_c + 0.67 f_y A_{sc}$$
$$1125 \times 10^3 = 0.4 \times 20 \times (900 - A_{sc}) + 0.67 \times 415 \times A_{sc}$$
$$1125 \times 10^3 = 720000 - 8 A_{sc} + 278.1 A_{sc}$$
$$405000 = 270.1 A_s$$
$$A_{sc} = 1499.5 \text{ mm}^2.$$

Minimum reinforcement = 0.8% of gross area
$$= 0.008 \times 300 \times 300 = 720 \text{ mm}^2 < A_{sc}$$

Maximum reinforcement = 6% of gross area
$$= 0.06 \times 300 \times 300 = 5400 \text{ mm}^2 > A_{sc}$$

Provide 4 bars of 22 mm $\phi$, $A_{sc}$ Provided = 1520.5 mm$^2$

**Lateral Ties:** Diameter of lateral ties should not be less than

(a) $\dfrac{\phi_1}{4} = \dfrac{1}{4} \times 22 = 5.5$ mm

(b) 6 mm

Hence, adopt 6 mm $\phi$ bars

Pitch of the ties shall be minimum of

(a) Least lateral dimension of the column = 300 mm.

(b) 16 times the diameter of longitudinal bars
$= 16 \times 22 = 352$ mm

(c) 300 mm.

Provide 6 mm lateral ties at 300 mm c/c.

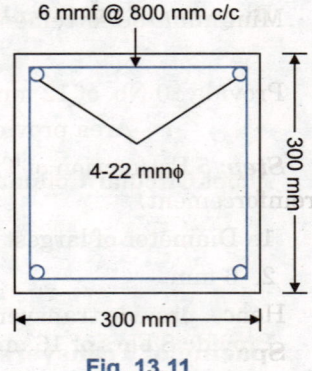

Fig. 13.11

**Example 13.10.** *Design the Rectangular Column: A short rectangular Column having effective length of 3 m carries factored axial compressive load of 200 kN. Architectural requirement dictate the column size of 600 × 450 mm. Design the column using $M_{25}$ concrete Fe 415 steel.*

**Solution:** **Step: 1** Size of Column

Factored load = 2000 kN

Self weight of column = $(0.45 \times 0.6) \times 3.0 \times 25$
$= 20.25$ kN

Assume $P_g = A_{sc}/A_g = 2\%$

$$A_c = \dfrac{P_u}{0.4(f_{ck} + 1.67 f_y P_g)}$$

$$= \dfrac{2020 \times 1000}{0.4\left(25 + 1.67 \times 415 \times \left(\dfrac{2}{100}\right)\right)} = 1,29,950 \text{ mm}^2.$$

A size of 300 × 450 mm (area = 135000 mm²) is sufficient. However for Architectural reasons, provide Column of size 600 × 450 mm.

**Step: 2** Determine Column is short (or) long column.

Assuming that the column is braced, effective length factor, 'k' may be taken equal to 1.0

Ratio = $l_x/b = 3.0/0.45 = 6.67 < 12$

$l_y/D = 3.0/0.60 = 5.0 < 12$

Hence, the column can be classified as short column.

**Step: 3** Check for minimum Eccentricity

$0.05 b = 0.05 \times 450 = 22.5$ mm

$0.05 D = 0.05 \times 600 = 30.0$ mm

$e_{x\ min} = (L_x/500) + (b/30) = (3000/500) + (450/30)$
$= 21.0$ mm $< 22.5$ mm.

$e_{y\ min} = (L_y/500) + (D/30)$
$= (3000/500) + (600/30)$
$= 26.0$ mm $< 30.0$ mm.

Hence, formula for short column capacity suggested by IS: 456; 2000 can be used.

**Step: 4** Estimation of Longitudinal Reinforcement

$2000 \times 10^3 = (0.40 \times 25 \times 450 \times 600) + (0.65 \times 415 \times A_{sc})$

$2000 \times 10^3 = 2700 \times 10^3 + 278.05 \times A_{sc}$

From the above, we see that the concrete itself is sufficient to take the load and we need to provide minimum reinforcement

Minimum reinforcement (Clause 26.5.3.1 (*b*) of IS: 456:)
$$= (0.8/100) \times 450 \times 300 = 1080 \text{ mm}^2.$$
Provide 10 No. of 12 mm dia bars, as shown
$$\text{Area provided} = 1131 \text{ mm}^2 > 1080 \text{ mm}^2.$$

**Step: 5** Estimation of Transverse reinforcement as per (Clause 26.5.3.2 (*c*)-(27) Diameter of Transverse reinforcement)

1. Diameter of largest longitudinal bars/4 = 12/4 = 3 mm
2. 6 mm

Hence, provide transverse reinforcement of diameter 8 mm.

**Spacing of Transverse reinforcement**

As per (Clause 26.5.3.2 (*c*)-(1)) of IS: 456: 2000.

1. Least lateral dimensions of column = 450 mm
2. 16 × Diameter of smallest longitudinal bar = 16 × 12 = 192 mm
3. 300 mm.

Hence, provide Transverse reinforcement of 8 mm dia bars at 180 mm c/c.

**Step: 6** Reinforcement Detailing

Provide the steel bars as shown in figure with 40 mm cover to reinforcement

Size of column = 600 × 450 mm

Steel = 10 Nos of 12 mm ϕ

ties = 8 mm ϕ bars at 180 c/c.

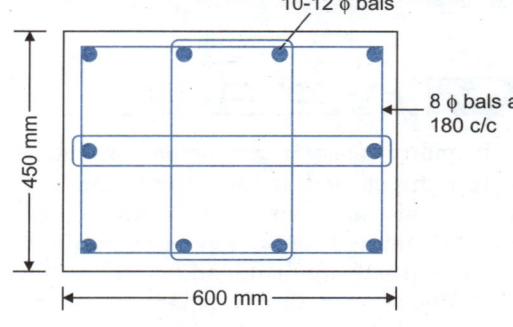

Fig. 13.12

## PROBLEMS

1. A reinforced concrete column, 3 m long (effective) and 240 mm × 240 mm in section is reinforced with 4 bars of 20 mm Φ. Find the safe load the column can carry. Use M 20 concrete and Fe 415 steel.
2. What will be the safe load for the column of problem 1 if the length of the column is 6 m?
3. A circular column of 500 mm diameter has an effective length of 5 m. The column is reinforced with 8 longitudinal bars of 20 mm Φ and spiral of 10 mm Φ rod wound around the 20 mm bars at a pitch of 60 mm. The clear cover for the main reinforcement is 40 mm. Determine the safe load for the column. Use M 20 concrete and Fe 415 reinforcement.
4. Design a short square column to carry an axial load of 1000 kN. Use M 25 mix. and Fe 415 steel.
5. Design a circular column to carry an axial load of 500 kN. The diameter of column is limited to 500 mm. Use spiral reinforcement. Provide M 20 concrete and Fe 415 steel.
6. Determine the load carrying capacity of a column of size 300 × 400 mm reinforced with six rods of 20 mm Φ *i.e.* 6 # 20. The grade of concrete $\eta_{20}$ and Fe 415 steel. Assume that the column is short.
7. Design the reinforcement in a column of six 450 × 600 mm, subject to an axial load of 2000 kN under service dead and live loads the column has an unsupported length of 3 meter and it's ends are held in position but not in direction use $\mu_{20}$ and Fe 415 steel.
8. Column of height 1.3 m, is pinned at bottom and effectively restrained against rotation but not held in position at top. It is subjected to a factored axial load of 2600 kN under the combination of Dead load and Live load. Design the column, if using $M_{30}$ Grade Concrete and Fe 415 steel.
9. Design a column subjected to an axial load of 2900 kN under Dead load and live load case. The column is braced against side sway in one direction and fixed at bottom and free at top in the other direction unsupported length of column is 2.2 m. Use $M_{25}$ Concrete and Fe 415 steel.
10. A short Rectangular column having its effective length of 3 mtrs carries factored axial compressive load of 2000 kN. The column size is 600 × 400 mm. Design the Column using $M_{25}$ Grade Concrete and Fe 415 steel.

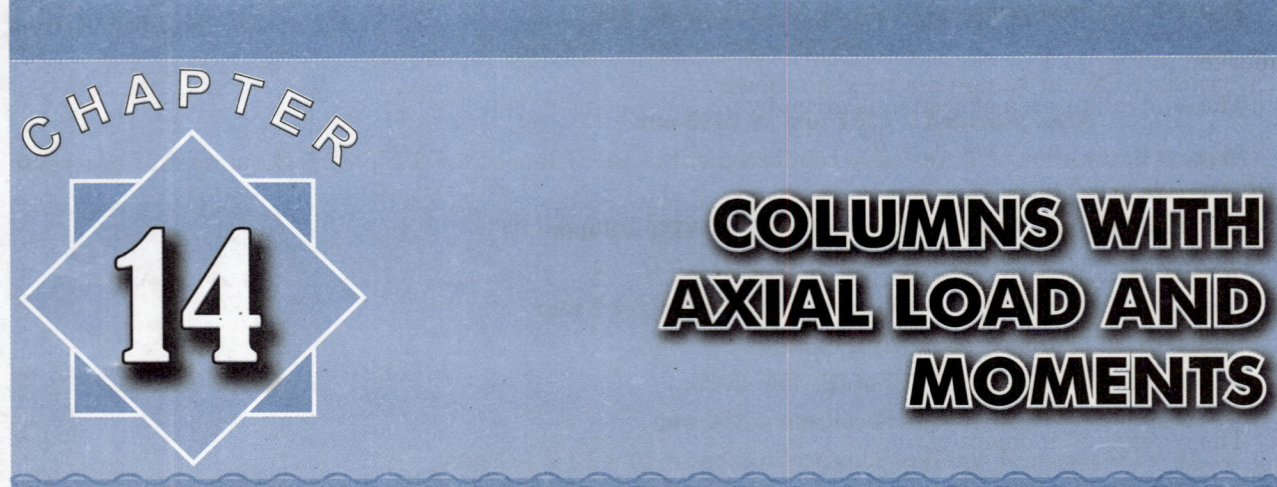

# CHAPTER 14: COLUMNS WITH AXIAL LOAD AND MOMENTS

## 14.1. INTRODUCTION

In many cases, a reinforced concrete member is subjected to both direct load and bending moment, or to a direct load acting eccentrically. Exterior columns of a framed structure, arches, chimneys, silos, bunkers etc. are some of the common examples of such combination of loading. In some cases, the forces may be compression and bending, while in other cases, the forces may be tension and bending. The effect of the centrally applied load $P$ and a moment $M$ is the same as that of a load $P$ acting at an eccentricity $e = M/P$. Hence the ratio of moment ($M$) to the direct load ($P$) is generally termed as eccentricity. The analysis and design of such members evidently depend upon the value of eccentricity. If the eccentricity is small so that section is subjected to the same kind of stress (*i.e.* either compressive or tensile), the analysis is slightly simplified. If the eccentricity is large, stress reversal may take place in the section, necessitating the determination of neutral axis. Based on the value of eccentricity, and also upon the nature of direct force (*i.e.* compressive or tensile), six cases may arise.

(*a*) **Combined compression and bending**

**Case 1.** *Eccentricity small.* When eccentricity is smaller than one-fourth the depth of the section: In such a case, no reversal of stresses takes place, or even if tension is induced, the resultant tension in concrete is not greater than 35 percent and 25 percent of the resultant compression for biaxial and uniaxial bending respectively, or exceeds three-fourths the 7-day modulus of rupture of concrete. The design of such a section is based on uncracked section.

**Case 2.** *Eccentricity large.* When the eccentricity is larger than 1.5 times the depth of the section.

**Case 3.** *Intermediate eccentricity.* When the eccentricity is between 0.25 and 1.5 times the depth of the section.

(*b*) **Combined tension and bending**

**Case 4.** *Eccentricity small.* The section is subjected to tensile stresses throughout.

**Case 5.** *Eccentricity large.*

**Case 6.** *Intermediate eccentricity.*

## 14.2. CASE I: COMPRESSIVE LOAD AT ECCENTRICITY SMALLER THAN $D/4$

Fig. 14.1 shows a R.C. section of size $b \times D$, subjected to a compressive load $P$ at a distance $e$ from the middle of the section. This eccentric load is equivalent to a direct load $P$ acting through the C.G. of the section, and a bending moment equal to $P \times e'$, where $e'$ is the eccentricity of the load measured from the centre of gravity of the section. It is assumed that eccentricity is small so that no stress reversal takes place, or even if tension is developed the whole area of concrete is effective.

Let $A_{sc1}$, $A_{sc2}$, $A_{sc3}$ be the steel areas at various levels $y_1$, $y_2$, $y_3$, etc. from the top fibre (AB) as marked in Fig. 14.1

The equivalent area of section is given by
$$A_e = bD + \Sigma (m_c - 1) A_{sc} \qquad ...(14.1)$$

where $m_c = 1.5 m$

The distance $\bar{y}$ of the centre of gravity, measured from AB, is

$$\bar{y} = \frac{(m_c - 1)(A_{sc1} y_1 + A_{sc2} y_2 + ...) + b D^2/2}{(m_c - 1)(A_{sc1} + A_{sc2} + ...) + b D}$$

$$= \frac{b D^2/2 + \Sigma (m_c - 1) A_{sc} y}{b D + \Sigma (m_c - 1) A_{sc}} \qquad ...(14.2)$$

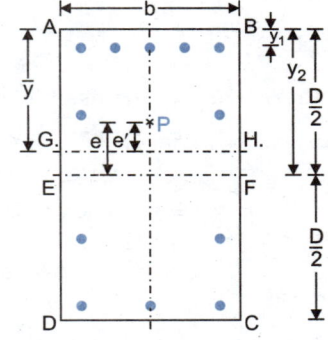

Fig. 14.1

The equivalent moment of inertia of the section is given by

$$I_e = (m_c - 1)(A_{sc1} a_1^2 + A_{sc2} \cdot a_2^2 + ...) + \frac{b D^3}{12} + BD\left(\bar{y} - \frac{D}{2}\right)^2$$

or
$$I_e = \Sigma (m_c - 1) A_{sc} \cdot a^2 + \frac{b D^3}{12} + BD\left(\bar{y} - \frac{D}{2}\right)^2 \qquad ...(14.3)$$

where $a_1, a_2, a_3, ....$ etc. are the distances of the steel areas from the centre of gravity of the section, and $m_c = 1.5 m$

Now, $M$ = load × eccentricity. $\therefore$ $M = P \times e' = P\left\{e - \left(\frac{D}{2} - \bar{y}\right)\right\}$

The stress distribution in the column may be determined from the expression:

$f = \frac{P}{A_e} \pm \frac{M \cdot y}{I_e}$. Hence the maximum and minimum compressive stresses are given by:

$$f_1 = \frac{P}{A_e} + \frac{P\left\{e - \left(\frac{D}{2} - \bar{y}\right)\right\}}{I_e} \cdot \bar{y}$$

$$f_2 = \frac{P}{A_e} - \frac{P\left\{e - \left(\frac{D}{2} - \bar{y}\right)\right\}}{I_e} \cdot (D - \bar{y})$$

Fig. 14.2 shows a symmetrically reinforced section, where $\bar{y} = D/2$.

$$A_e = b D + (m_c - 1) A_{sc} \qquad ...(14.4)$$

$$I_e = \frac{1}{12} bD^3 + (m_c - 1) A_{sc} \left(\frac{D}{2} - d_c\right)^2 \qquad ...(14.5)$$

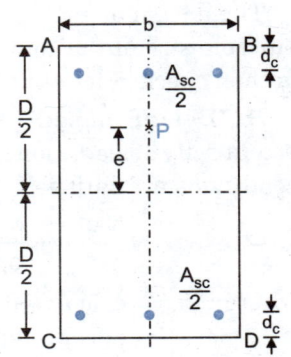

Fig. 14.2

Hence maximum compressive stress at AB is given by

$$f_1 = \frac{P}{A_e} + \frac{P \cdot e}{I_e} \cdot \frac{D}{2}.$$

Minimum stress at DC is $f_2 = \dfrac{P}{A_e} - \dfrac{P \cdot e}{I_e} \cdot \dfrac{D}{2}$

If no tension is to develop in the section, $f_2$ must be equal to zero.

Hence, $\dfrac{P}{A_e} - \dfrac{P \cdot e}{I_e} \cdot \dfrac{D}{2} = 0$, which gives $e = \dfrac{2 I_e}{A_e \cdot D}$ ...(14.6)

A member subjected to axial load and bending (due to eccentricity of load, monolithic construction, lateral forces, etc.) should be considered safe provided the following condition is satisfied:

$$\frac{\sigma_{cc'}}{\sigma_{cc}} + \frac{\sigma_{cbc'}}{\sigma_{cbc}} \leq 1 \qquad \ldots(14.7)$$

where  $\sigma_{cc}$ = permissible direct compressive stress in concrete
$\sigma_{cc'}$ = calculated direct compressive stress in concrete
$\sigma_{cbc}$ = permissible bending compressive stress in concrete
$\sigma_{cbc'}$ = calculated bending compressive stress in concrete.

$$\text{Stress in steel} = \sigma_{s'} = 1.5\, m \left[\frac{P}{A_e} + \frac{P \cdot e}{I_e}\left(\frac{D}{2} - d_c\right)\right] \qquad \ldots(14.8)$$

## 14.3. BENDING ABOUT TWO AXES

Consider a symmetrically reinforced column subjected to an axial load $P$, a bending moment $M_x$ about $x$-axis and a moment $M_y$ about $y$-axis. Let,

$I_{ex}$ = equivalent moment of inertia of the section about $x$-$x$ axis.
$I_{ey}$ = equivalent moment of inertia of the section about $y$-$y$ axis.

The stress $f$ at any point $(x, y)$ can be determined from the expression:

$$f = \frac{P}{A_e} \pm \frac{M_x \cdot y}{I_{ex}} \pm \frac{M_y \cdot x}{I_{ey}} \qquad \ldots(14.8A)$$

The maximum direct stress $\sigma_{cc'} = \dfrac{P}{A_e}$

Maximum bending stress $\sigma_{cbc'} = \dfrac{M_x}{I_{ex}} \cdot \dfrac{D}{2} + \dfrac{M_y}{I_{ey}} \cdot \dfrac{B}{2}$

The section will be safe if $\dfrac{\sigma_{cc'}}{\sigma_{cc}} + \dfrac{\sigma_{cbc'}}{\sigma_{cbc}} \leq 1.$

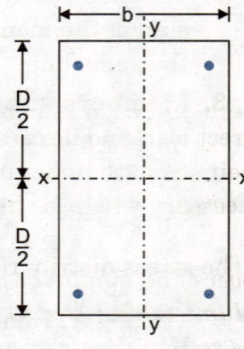

Fig. 14.3

## 14.4. DESIGN OF COLUMNS SUBJECTED TO COMBINED BENDING AND DIRECT STRESSES (IS: 456-2000)

The design of a column subjected to combined bending and direct stresses is that of trial and check. A suitable size of the column and steel reinforcement are first chosen. The preliminary section so selected is then checked for stresses in concrete and steel. The following notes refer to the design of such sections.

**1. Design based or uncracked section.** A member subjected to axial load and bending (due to eccentricity of load, monolithic construction, lateral forces etc.) shall be considered safe provided the following conditions are satisfied:

(a)  $\qquad \dfrac{\sigma_{cc'}}{\sigma_{cc}} + \dfrac{\sigma_{cbc'}}{\sigma_{cbc}} \leq 1$

where  $\sigma_{cc'}$ = calculated direct compressive stress in concrete
$\sigma_{cc}$ = permissible axial compressive stress in concrete
$\sigma_{cbc'}$ = calculated bending compressive stress in concrete
$\sigma_{cbc}$ = permissible bending compressive stress in concrete.

(b) The resultant tension in concrete is not greater than 35 percent and 25 percent of the resultant compression for biaxial and uniaxial bending respectively, or does not exceed three-fourths the 7-day modulus of rupture of concrete.

**Note:** (i) $\sigma_{cc'} = \dfrac{P}{A_c + 1.5\, m\, A_{sc}}$, for columns with ties.

(ii) $\sigma_{cbc'} = M/Z$, where $M$ equals the moment and $Z$ equals modulus of section. In the case of sections subject to moments in two directions, the stress shall be calculated separately and added algebraically.

**2. Design based on cracked section.** If the requirements specified in clause 1 above are not satisfied, the stresses in concrete and steel shall be calculated by theory of cracked section in which tensile resistance of concrete is ignored. If the calculated stresses are within the permissible stresses, the section may be assumed to be safe.

**Note:** The maximum stress in concrete and steel may be found from tables and charts based on the uncracked section theory or directly by determining the no-stress line which should satisfy the following requirements:

(a) The direct load should be equal to the algebraic sum of the forces on concrete and steel.

(b) The moment of the external loads about any reference line should be equal to the algebraic sum of the moment of the forces in concrete (ignoring the tensile force in concrete) and steel about the same line, and

(c) The moment of the external loads about any other reference lines should be equal to the algebraic sum of the moment of the forces in concrete (ignoring the tensile force in concrete) and steel about the same line.

**3. Members subjected to combined direct load and flexure.** Members subjected to combined direct load and flexure and designed by the methods based on elastic theory should be further checked for their strength under ultimate load conditions to ensure the desired margin of safety; this check is specially necessary when the bending moment is due to horizontal loads.

**Example 14.1.** *A R.C. column 400 mm × 400 mm is reinforced with 4 bars of 25 mm diameter, placed at a cover of 50 mm to the centre of steel bars. Determine the maximum and minimum stresses in concrete if the column is subjected to a load of 400 kN at an eccentricity of 50 mm about one of the axes. Also, check whether the section is safe or not. Use M 15 concrete, taking m = 19.*

**Solution:** For M 15 concrete, $\sigma_{cc} = 4$ N/mm² and $\sigma_{cbc} = 5$ N/mm². Since the eccentricity is small, the analysis of § 14.2 can be used. However while calculating the equivalent area and moment of inertia, modular ratio $m_c\,(= 1.5\, m)$ should be used with the steel area in compression.

$$A_{sc} = 4\,\dfrac{\pi}{4}\,(25)^2 = 1960 \text{ mm}^2$$

$$A_e = A_c + 1.5\, m\, A_{sc} = (A - A_{sc}) + 1.5\, m\, A_{sc}$$

$$= A + (1.5\, m - 1)\, A_{sc}$$

$$= (400 \times 400) + (1.5 \times 19 - 1)\, 1960 = 213900 \text{ mm}^2$$

$$I_e = \dfrac{1}{12}\, b\, D^3 + (m_c - 1)\, A_{sc}\left(\dfrac{D}{2} - d_c\right)^2$$

$$= \dfrac{1}{12} \times 400\,(400)^3 + (1.5 \times 19 - 1) \times 1960\,(200 - 50)^2$$

$$= 33.46 \times 10^8 \text{ mm}^4$$

$$\sigma_{cc'} = \dfrac{P}{A_e} = \dfrac{400000}{213900} = 1.87 \text{ N/mm}^2$$

$$\sigma_{cbc'} = \dfrac{P.e}{I_e} \times \dfrac{D}{2} = \dfrac{400000 \times 50}{33.46 \times 10^8} \times 200 \approx 1.20 \text{ N/mm}^2$$

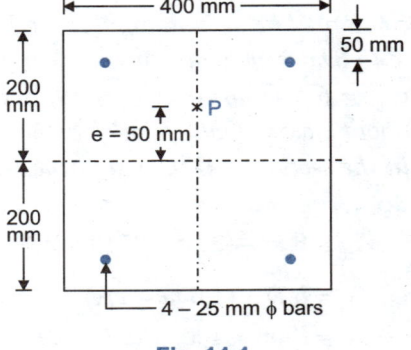

Fig. 14.4

∴ Max. stress = $\sigma_{cc'} + \sigma_{cbc'}$ = 1.87 + 1.20 = 3.07 N/mm²

Min. stress = $\sigma_{cc'} - \sigma_{cbc'}$ = 1.87 − 1.20 = 0.67 N/mm²

For the section to be safe, we have $\dfrac{\sigma_{cc'}}{\sigma_{cc}} + \dfrac{\sigma_{cbc'}}{\sigma_{cbc}} \le 1$ or $\dfrac{1.87}{4} + \dfrac{1.20}{5} \le 1$. or 0.71 ≤ 1. Hence safe.

**Example 14.2.** *Determine the maximum eccentricity of load in Example 14.1, if the column is to be safe as uncracked section.*

**Solution:** For safety, we have $\dfrac{\sigma_{cc'}}{\sigma_{cc}} + \dfrac{\sigma_{cbc'}}{\sigma_{cbc}} \le 1$

Hence, $\dfrac{\sigma_{cbc'}}{\sigma_{cbc}} = 1 - \dfrac{\sigma_{cc'}}{\sigma_{cc}} = 1 - \dfrac{1.87}{4} = 1 - 0.4675 = 0.5325$

∴ $\sigma_{cbc'} = 0.5325 \, \sigma_{cbc} = 0.5325 \times 5 = 2.66$ N/mm²

∴ $\sigma_{cbc'} = \dfrac{P \cdot e}{I_e} \times \dfrac{D}{2} = 2.66$

$e = 2.66 \times \dfrac{2 I_e}{P \cdot D} = \dfrac{2.66 \times 2 \times 33.46 \times 10^8}{400000 \times 400} = \mathbf{111\ mm^2}$.

**Example 14.3.** *If the column of Example 14.1 is subjected to an axial load of 400 kN along with a bending moment at 25 kN-m, determine the maximum and minimum stresses in concrete. Will the section be safe as an uncracked section?*

**Solution:** $A_e = 213900$ mm² and $I_e = 33.46 \times 10^8$ mm⁴, as found in Example 14.1.

∴ $\sigma_{cc'} = \dfrac{P}{A_e} = \dfrac{400000}{213900} = 1.87$ N/mm².

B.M. $M = 25$ kN-m $= 25 \times 10^6$ N-m

∴ $\sigma_{cbc'} = \dfrac{M}{I_e} \cdot \dfrac{D}{2} = \dfrac{25 \times 10^6}{33.46 \times 10^8} \times \dfrac{400}{2} = 1.49$ N/mm².

Max. Stress = $\sigma_{cc'} + \sigma_{cbc'}$ = 1.87 + 1.49 = **3.36 N/mm²**.

Min. Stress = $\sigma_{cc'} - \sigma_{cbc'}$ = 1.87 − 1.49 = **0.38 N/mm²**.

For the section to be safe as an uncracked section,

$\dfrac{\sigma_{cc'}}{\sigma_{cc}} + \dfrac{\sigma_{cbc'}}{\sigma_{cbc}} \le 1$ or $\dfrac{1.87}{4} + \dfrac{1.49}{5} \le 1$ or 0.766 ≤ 1. Hence safe.

**Example 14.4.** *A rectangular section 300 mm × 400 mm is reinforced with 8 bars of 20 mm Φ as shown in Fig. 14.5. Taking $\sigma_{cc} = 4\ N/mm^2$, $\sigma_{cbc} = 5\ N/mm^2$ and m = 19, determine (i) maximum eccentricity about x-axis at which load can be applied without developing tension in the section, (ii) the magnitude of the load.*

**Solution:**

$A_{sc} = 8 \times \dfrac{\pi}{4}(20)^2 = 2513$ mm²

$A_e = bD + (1.5 m - 1) A_{sc}$

$= (300 \times 400) + (1.5 \times 19 - 1) \times 2513$

$= 189108$ mm²

$I_e = \dfrac{1}{12} \times 300\,(400)^3 + (1.5 \times 19 - 1) \times 6 \times \dfrac{\pi}{4}(20)^2 (150)^2$

$= 27.663 \times 10^8$ mm⁴

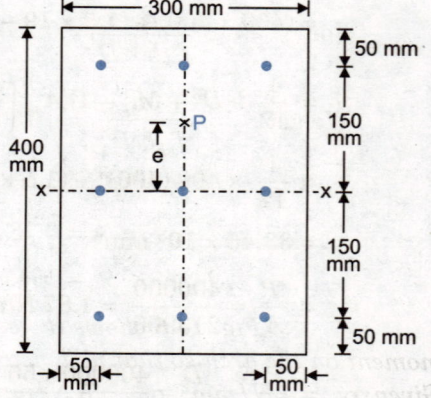

Fig. 14.5

For no tension to develop, we have

$$\frac{P}{A_e} - \frac{P \cdot e}{I_e} \cdot \frac{D}{2} = 0. \quad \text{or} \quad e = \frac{2 I_e}{A_e D} = \frac{2 \times 27.663 \times 10^8}{189108 \times 400} = 73.14 \text{ mm}$$

Again, if $P$ is the maximum load, we have

$$\sigma_{cc'} = \frac{P}{A_e} = \frac{P}{189108}$$

$$\sigma_{cbc'} = \frac{P \cdot e}{I_e} \cdot \frac{D}{2} = \frac{P}{I_e}\left[\frac{2 I_e}{A_e \cdot D}\right] \times \frac{D}{2} = \frac{P}{A_e} = \frac{P}{189108}$$

(**Note:** Both $\sigma_{cc'}$ and $\sigma_{cbc'}$ will be equal at the critical value of $e$).

Now, $\quad \dfrac{\sigma_{cc'}}{\sigma_{cc}} + \dfrac{\sigma_{cbc'}}{\sigma_{cbc}} \leq 1 \quad$ or $\quad \dfrac{P}{189108}\left[\dfrac{1}{4} + \dfrac{1}{5}\right] = 1$

or $\quad P = \dfrac{189108 \times 4 \times 5}{9} = 420240 \text{ N} = \mathbf{420.24 \text{ kN}}.$

**Example 14.5.** *Determine the bending capacity ($M_u$) for a column of size 300 × 500 mm which is reinforced with 6 bars of 20 mm diameter bars arranged on two side of the column if it is subjected to an axial load of 800 kN. Use $M_{20}$ concrete and Fe 415 steel take d = 50 mm.*

**Solution: Given:** $b = 300$ mm, $D = 500$ mm, $d' = 50$ mm

$$A_{sc} = 6 \times \frac{\pi}{4} \times (20)^2 = 1885 \text{ mm}^4$$

$$f_{ck} = 20 \text{ N/mm}^2, \quad f_y = 415 \text{ N/mm}^2, \quad P_u = 800 \text{ kN}$$

Calculating percentage of steel $p = \dfrac{A_{sc} \times 100}{b.D.} = \dfrac{1885 \times 100}{300 \times 500} = 1.257$

$$\frac{p}{f_{ck}} = \frac{1.257}{20} = 0.063,$$

$$\frac{P_u}{f_{ck} b \cdot D} = \frac{800 \times 1000}{20 \times 300 \times 500} = 0.267$$

$$\frac{d'}{D} = \frac{50}{500} = 0.1 \text{ for } \frac{d'}{D} = 0.1, \quad f_y = 415$$

Referring chart no. 32 of sp-16

For $\quad \dfrac{P_u}{f_{ck} b \times D} = 0.267, \quad \dfrac{p}{f_{ck}} = 0.063$

We get $\quad \dfrac{M_u}{f_{ck} b . D^2} = 0.13$

$\therefore \quad (M_u) = 0.13 f_{ck} b D^2$

$= 0.13 \times 20 \times 300 \times 500^2$

$= 195 \times 10^6 = 195$ kN m

**Example 14.6.** *Fig. 14.6 shows the cross-section of an arch rib. Determine the normal thrust and bending moment on the arch so that no tension is developed anywhere and the section is safe as uncracked section. Given:* $\sigma_{cc} = 4 \text{ N/mm}^2$, $\sigma_{cbc} = 5 \text{ N/mm}^2$ *and* $m = 19$.

**Solution:** Fig. (14.6)

$$A_{sc1} = \text{area of steel near face } AB = 6\frac{\pi}{4}(20)^2 = 1885 \text{ mm}^2.$$

$$A_{sc2} = \text{area of steel near face } CD = 6\frac{\pi}{4}(25)^2 = 2945 \text{ mm}^2.$$

Total $\quad A_{sc} = 1885 + 2945 = 4830 \text{ mm}^2$

$\therefore \quad A_e = (400 \times 700) + (1.5 \times 19 - 1)4830 = 412825 \text{ mm}^2$

Let $\bar{y}$ be the distance of centroidal axis from the face $AB$,

$$\bar{y} = \frac{(400 \times 700 \times 350) + (19 \times 1.5 - 1)1885 \times 50 + (19 \times 1.5 - 1) \times 2945 \times 650}{(400 \times 700) + (19 \times 1.5 - 1)1885 + (19 \times 1.5 - 1)2945} = 371.2 \text{ mm}.$$

The equivalent moment of inertia about the axis $EF$ through the centre of gravity of the section is given by,

$$I_e = \frac{1}{12} 400(700)^3 + 400 \times 700(371.2 - 350)^2$$
$$+ (19 \times 1.5 - 1)[1885(371.2 - 50)^2 + 2945(700 - 371.2 - 50)^2]$$
$$= 232.02 \times 10^8 \text{ mm}^4$$

Let $P$ be the permissible load and $M$ be the moment.
Hence, $\quad e = M/P$.

For no tension to develop, we have $\dfrac{P}{A_e} - \dfrac{M}{I_e}(D - \bar{y}) = 0$

$\therefore \quad \dfrac{P}{A_e} - \dfrac{P \cdot e}{I_e}(D - \bar{y}) = 0$

Hence, $\quad e = \dfrac{I_e}{A_e(D - \bar{y})} = \dfrac{232.02 \times 10^8}{412825(700 - 371.2)} = 170.93 \text{ mm}$

Again for an uncracked section, we have $\dfrac{\sigma_{cc'}}{\sigma_{cc}} + \dfrac{\sigma_{cbc'}}{\sigma_{cbc}} \leq 1$

where $\sigma_{cc'} = \dfrac{P}{A_e} = \dfrac{P}{412825}$; $\sigma_{cbc} = 4$; $\sigma_{cbc} = 5$

$$\sigma_{cbc'} = \frac{P \cdot e}{I_e} \cdot \bar{y} = P\left(\frac{170.93}{232.02 \times 10^8} \times 371.2\right) = \frac{P}{365678.3}$$

Hence, $\quad \dfrac{\left(\dfrac{P}{412825}\right)}{4} + \dfrac{\left(\dfrac{P}{365678.3}\right)}{5} = 1.$ or $P\left[\dfrac{1}{412825 \times 4} + \dfrac{1}{365678.3 \times 5}\right] = 1$

From which $P = 867668.99 \text{ N} = \mathbf{867.669 \text{ kN}}$

Hence $\quad M = P \cdot e = 867.669 \times 170.84 = 148232 \text{ kN-mm} \approx \mathbf{148.23 \text{ kN-m}}.$

**Example 14.7.** *Fig. 14.7 (a) Shows a reinforced concrete circular section, subjected to an axial load of 500 kN and bending moment of 25 kN-m. Taking the permissible stresses $\sigma_{cc} = 4 \text{ N}/\text{mm}^2$ and $\sigma_{cbc} = 5 \text{ N}/\text{mm}^2$, check whether the section is safe or not. Take $m = 19$.*

**Solution:** $\quad A_{sc} = 10 \times \dfrac{\pi}{4}(20)^2 \approx 3140 \text{ mm}^2.$

The steel bars can be replaced by angular ring of mean diameter $d = 350$ mm, having the same area.

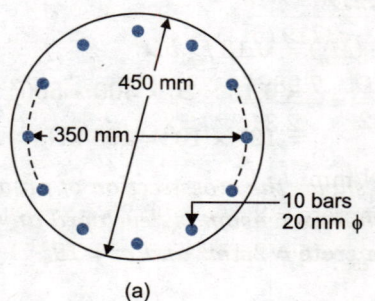
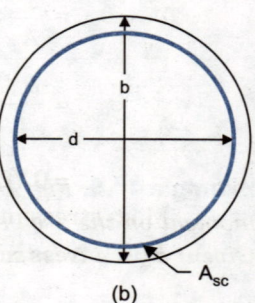

Fig. 14.7

$$A_e = \frac{\pi}{4}(450)^2 + (1.5 \times 19 - 1)\,3140 = 245393 \text{ mm}^2$$

Before finding the moment of inertia of the section, let us find the moment of inertia of the steel ring. Polar moment of inertia ($I_p$) of ring is

$$I_p = \Sigma\,\delta A_{sc} \cdot r^2 = \frac{A_{sc} \cdot d^2}{4} \qquad \therefore \quad I_{xx} = I_{yy} = \frac{I_p}{2} = \frac{A_{sc} \cdot d^2}{8}$$

$$\therefore \quad I_e = \frac{\pi}{64}(450)^4 + (19 \times 1.5 - 1)\,\frac{3140\,(350)^2}{8} = 33.35 \times 10^8 \text{ mm}^4$$

Now the section will be safe if $\dfrac{\sigma_{cc'}}{\sigma_{cc}} + \dfrac{\sigma_{cbc'}}{\sigma_{cbc}} \leq 1$

Here, 
$$\sigma_{cc'} = \frac{P}{A_e} = \frac{500 \times 1000}{245393} = 2.04 \text{ N/mm}^2; \qquad \sigma_{cc} = 4 \text{ N/mm}^2$$

$$\sigma_{cbc'} = \frac{M}{I_e} \times 225 = \frac{25 \times 10^6}{33.35 \times 10^8} \times 225 = 1.69;\ \sigma_{cbc} = 5 \text{ N/mm}^2.$$

$$\frac{\sigma_{cc'}}{\sigma_{cc}} + \frac{\sigma_{cbc'}}{\sigma_{cbc}} = \frac{2.04}{4} + \frac{1.69}{5} = 0.51 + 0.338 = 0.848 < 1. \text{ Hence safe.}$$

**Example 14.8.** *The section of a reinforced concrete two hinged arch is 300 mm wide and 700 mm deep at the crown, and is reinforced with 6 bars of 25 mm Φ each at top and bottom at an effective cover of 50 mm, as shown in Fig. 14.8. The section is subjected to a resultant thrust of 800 kN inclined to the horizontal tangent to the centre of the arch at 3° and acting on the vertical axis at a distance of 60 mm from the centre line of the arch section. Determine the stress in concrete and steel at the top and bottom edge of the crown. If $\sigma_{cc} = 4\,N/mm^2$, $\sigma_{cbc} = 5\,N/mm^2$ and m = 19, find whether the section is safe or not.*

**Solution:** $A_{sc} = 12 \dfrac{\pi}{4}(25)^2 = 5890 \text{ mm}^2$

$A_e = (300 \times 700) + (1.5 \times 19 - 1)\,5890$

$\quad = 371975 \text{ mm}^2$

$I_e = \dfrac{1}{12} \times 300\,(700)^3 + (1.5 \times 19 - 1)\,5890\,(350 - 50)^2$

$\quad = 2.315 \times 10^{10} \text{ mm}^4$

Normal thrust

$P = W \cos 3° = 800 \times 1000 \times 0.9986$

$\quad = 7.989 \times 10^5 \text{ N}$

$\sigma_{cc'} = \dfrac{P}{A_e} = \dfrac{7.989 \times 10^5}{371975} = 2.148 \text{ N/mm}^2$

$\sigma_{cbc'} = \dfrac{P \cdot e}{I_e} \cdot \dfrac{D}{2} = \dfrac{7.989 \times 10^5 \times 60}{2.315 \times 10^{10}} \times 350$

$\quad = 0.725 \text{ N/mm}^2$

$\therefore$ Max. stress in concrete = $2.148 + 0.725 = \mathbf{2.873\ N/mm^2}$ (compressive)

Min. stress in concrete = $2.148 - 0.725 = \mathbf{1.423\ N/mm^2}$ (compressive)

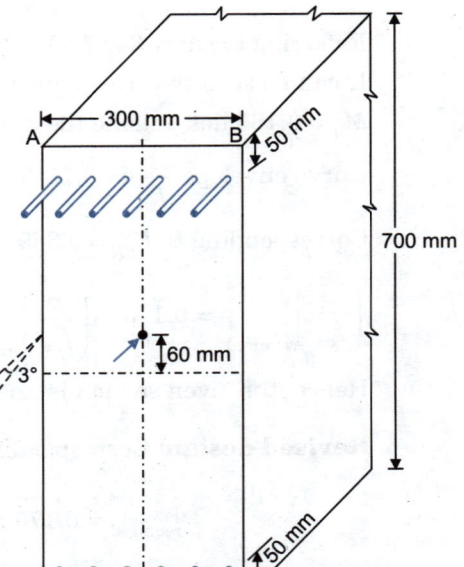

Fig. 14.8

For the section to be safe

$$\frac{\sigma_{cc'}}{\sigma_{cc}} + \frac{\sigma_{cbc'}}{\sigma_{cbc}} = \frac{2.148}{4} + \frac{0.725}{5} = 0.537 + 0.145 = 0.682 < 1. \text{ Hence safe.}$$

Stress in steel near face $AB = 1.5 \times 19 \left\{ 2.149 + \frac{7.989 \times 10^5 \times 60}{2.315 \times 10^{10}} \times 300 \right\}$

$$= 28.5 [2.149 + 0.621] = \mathbf{78.95 \ N/mm^2} \quad \text{(comp.)}$$

Stress in steel near face $CD = 1.5 \times 19 \times 2.149 - 19 \times 0.621 = \mathbf{49.45 \ N/mm^2}$ (comp.)

**Example 14.9.** *Rectangular column of size 300 mm × 600 mm. Investigate the safety of column section under uniaxial eccentric compression with respect to the minor axis considering $A_{st} = 3984 \ mm^2$, $\rho = 2.213$, effective cord $d' = 62 \ mm$ and $P_u = 1400 \ kN$ and $M_u = 200 \ kNm$, if the section is unsafe, suggest suitable modifications to reinforcement provided use $M_{20}$ and Fe 415 steel.*

**Solution: Given:** $b = 600$ mm, $\quad D = 300$ mm, $\quad f_{ck} = 20$ MPa. $f_u = 415$ MPa, $\quad P_u = 1400$ kN.

$M_u = 200$ kN m, $\quad A_{st} = 3984$ mm², $\quad \rho = 2.213$, $\quad$ effective cover $d' = 62$ mm

The arrangement of bars in this case confirms to reinforcement distributed equally on both sides.

$$\frac{d'}{D} = 62/300 = 0.207 \cong 0.2$$

$$p/f_{ck} = \frac{2.213}{20} = 0.1106 \cong 0.11$$

$$p_u = \frac{P_u}{f_{ck} bD} = \frac{1400 \times 10^3}{20 \times 600 \times 300} = 0.389$$

$$m_u = \frac{M_{ux}}{f_{acb} bD^2} = \frac{200 \times 10^6}{20 \times 600 \times 300^2} = 0.185$$

Referring to chart 34 ($d'/D = 0.2$) of sp-16,

It can be seen that the point $P_{uR} = 0.389$, $M_u = 0.185$ lies outside the design intractive

Curve envelope $\dfrac{p}{f_{ck}} = 0.11$. The value of $\dfrac{p}{f_{ck}}$.

Corresponding to $P_{uR} = 0.389$, $M_{uR} = 0.185$ given by

$$\left(\frac{p}{f_{ck}}\right)_{required} = 0.175 > \left(\frac{p}{f_{ck}}\right)_{provided} = 0.11$$

Hence the given section is unsafe.

**Revised design:** Corresponding to $\left(\dfrac{p}{f_{ck}}\right)_{required} = 0.175$

$$P_{required} = 0.175 \times 20 = 3.5$$

$$A_{st_{required}} = 3.5 \times 600 \times \frac{300}{100} = 6300 \text{ mm}^2 \text{ to be provided equally in two rows.}$$

Provide 8 – 32 mm ϕ (instead of 4 – 28ϕ + 4 – 22 ϕ)

$A_{st}$ provided = 804 × 8 = 6432 mm² > 6300 mm² ($P = 3.573$)

the detailing for the revised design section shown.

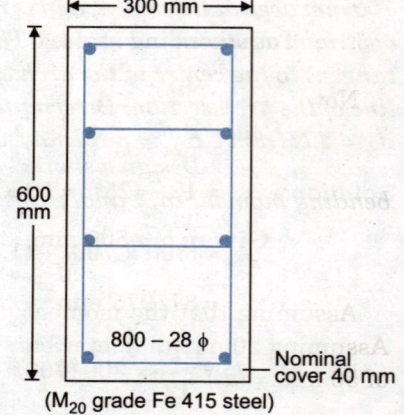

(M₂₀ grade Fe 415 steel)

Fig. 14.9

**Example 14.10.** *A R.C. section 200 mm × 300 mm is reinforced with 4 bars of 20 mm Φ arranged as shown in 14.10. The section is subjected to (i) axial load of 200 kN (ii) bending moment of 3 kN-m about x-axis, and (iii) bending moment of 2 kN-m about y-axis. Taking $\sigma_{cc} = 4\ N/mm^2$ and $\sigma_{cbc} = 5\ N/mm^2$, find whether the section is safe or not. Take m = 19.*

**Solution:**
$$A_{sc} = 4 \times \frac{\pi}{4}(20)^2 = 1256\ mm$$

$$A_e = (200 \times 300) + (1.5 \times 19 - 1)\,1256 = 94540\ mm^2$$

$$I_{ex} = \frac{1}{12}\,200\,(300)^3 + (1.5 \times 19 - 1) \times 1256\,(150 - 40)^2$$
$$= 8.679 \times 10^8\ mm^4$$

$$I_{ey} = \frac{1}{12}\,300\,(200)^3 + (1.5 \times 19 - 1) \times 1256\,(100 - 40)^2$$
$$= 3.243 \times 10^8\ mm^4$$

$$\sigma_{cc'} = \frac{P}{A_e} = \frac{200000}{94540} = 2.12\ N/mm^2$$

$$(\sigma_{cbc})_{x'} = \frac{M_x}{I_{ex}} \cdot \frac{D}{2} = \frac{3000 \times 1000}{8.679 \times 10^8} \times 150 = 0.52\ N/mm^2$$

$$(\sigma_{cbc})_{y'} = \frac{M_y}{I_{ey}} \cdot \frac{b}{2} = \frac{2000 \times 1000}{3.243 \times 10^8} \times 100 = 0.62\ N/mm^2$$

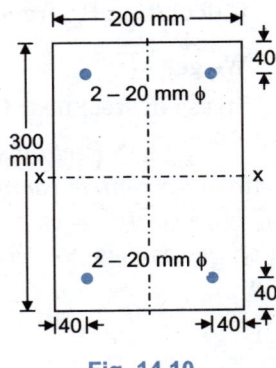

Fig. 14.10

∴ Total bending stress $\sigma_{cbc'} = (\sigma_{cbc})_{x'} + (\sigma_{cbc})_{y'} = 0.52 + 0.62 = 1.14\ N/mm^2$

Now, $\dfrac{\sigma_{cc'}}{\sigma_{cc}} + \dfrac{\sigma_{cbc'}}{\sigma_{cbc}} \leq 1$ or $\dfrac{2.12}{4} + \dfrac{1.14}{5} < 1$ or $0.758 < 1.$ Hence safe.

**Example 14.11.** *Design a short column of size 500 × 600 mm subject to an axial load $p_u$ = 2000 kN and biaxial bending moment in X and Y direction as $M_{ux}$ = 150 kNm and $M_{uy}$ = 120 kN use $M_{20}$ concrete Fe 415 steel.*

**Solution: Given:** b = 500 mm, D = 600 mm, $P_u$ = 2000 kN
$M_{ux}$ = 200 kN-m, $M_{uy}$ = 120 kN-m, $f_{ck} = M_{20}$ = 20 N/mm², $f_y$ = 415 N/mm²

Assuming that the moment due to minimum eccentricity is less than the given moment $M_{ux}$ and $M_{uy}$. Assuming 20 mm diameter bars equally distributed on all the four sides of column with an effective cover of 50 mm = (d')

$$\frac{d'}{D} = \frac{50}{600} = 0.083$$

to start with Assume ρ = 12% for trial

$$\frac{\rho}{f_{ck}} = \frac{1.2}{20} = 0.06$$

$$\frac{d'}{D} = 0.083\ \text{therefore using chart – 44 for}\ \frac{d'}{D} = 0.1\ \text{of sp-16}$$

$$\frac{P_u}{f_{ck}b.D} = \frac{2000 \times 1000}{20 \times 500 \times 600} = 0.333.$$

For $\dfrac{p}{f_{ck}} = 0.06$ and $\dfrac{P_u}{f_{ck}b.D} = 0.333.$

From chart 44, we get $\dfrac{M_u}{f_{ck}b.D^2} = 0.095$

$$M_{ux'} = 0.095 \times 20 \times 500 \times (600)^2 = 342\ kN\text{-}m$$

Calculating
$$M_{uy'} = \left(\frac{d'}{D} = \frac{50}{500} = 0.1, \text{ hence chart (44) used}\right)$$
$$M_{uy'} = 0.095 \times 20 \times 600 \times (500)^2$$
$$= 285 \text{ kN-m}.$$

Calculating $P_{uz}$ from chart (63) for $\rho = 12\%$    $f_y$ 415 N/mm² and $f_{ck}$ = 20 N/mm²

We get
$$\frac{P_{uz}}{A_g} = 12.5$$
$$P_{uz} = 12.5 \times 500 \times 600 = 3750 \text{ kN}.$$
$$\frac{P_u}{P_{uz}} = \frac{2000}{3750} = 0.53$$
$$\frac{M_{uz}}{M_{ux'}} = \frac{200}{342} = 0.585$$
$$\frac{M_{uy}}{M_{uy'}} = \frac{120}{285} = 0.421$$

For $\frac{p_u}{P_{uz}} = 0.53$ and $\frac{M_{uy}}{M_{uy'}} = 0.421$

From chart (64), we get $\left(\frac{M_{ux}}{M_{uy'}}\right)_{permissible} = 0.8.$

$$\left(\frac{M_{ux}}{M_{uy'}}\right)_{permissible} > \frac{M_{ux}}{M_{ux'}}$$

Hence safe design.

Calculating $A_{sc} = \frac{P.b.D}{100} = \frac{1.2 \times 500 \times 600}{100}$

$$A_{sc} = 3600 \text{ mm}^2.$$

Hence provide 12–20 mm diameter bars.
$$A_{sc} = 3770 \text{ mm}^2.$$

### Design of Ties:
1. Diameter of ties should not be less than
   (a) $\frac{25}{4} = 6.25$ mm
   (b) 6 mm.

Hence provide 8 mm diameter ties pitch should not be greater than
   (a) Least lateral dimension = 400 mm.
   (b) 16 dia of main bar 16 × 25 = 400 mm.
   (c) 300 mm.

Hence provide 8 mm ϕ ties @ 300 c/c.

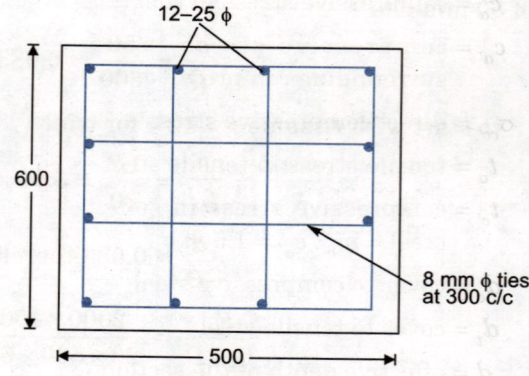

Fig. 14.11

**Example 14.12.** *Design a suitable column section to carry an axial load of 430 kN and bending moment of 8 kN-m. The allowable stresses may be taken as follows:* $\sigma_{cc} = 4 \text{ N/mm}^2$; $\sigma_{cbc} = 5 \text{ N/mm}^2$ *and m = 19.*

**Solution:** Let us assume a section of 320 mm × 320 mm, reinforced symmetrically with 4 bars of 25 mm Φ as shown in Fig. 14.12. Provide a cover of 60 mm upto the centre of steel.

$$A_s = 4 \frac{\pi}{4} (25)^2 = 1963.5 \text{ mm}^2$$

$$A_e = (320 \times 320) + (1.5 \times 19 - 1) 1963.5 = 156396 \text{ mm}^2$$

$$I_e = \frac{1}{12} \times 320 (320)^3 + (1.5 \times 19 - 1) 1963.5 \left(\frac{320}{2} - 60\right)^2$$

$$= 14.138 \times 10^8 \text{ mm}^4. \text{ Hence,}$$

$$\sigma_{cc'} = \frac{P}{A_e} = \frac{430000}{156396} = 2.75 \text{ N/mm}^2$$

$$\sigma_{cbc'} = \frac{M}{I_e} \cdot \frac{D}{2} = \frac{8000 \times 1000}{14.138 \times 10^8} \times 160 = 0.905 \text{ N/mm}^2$$

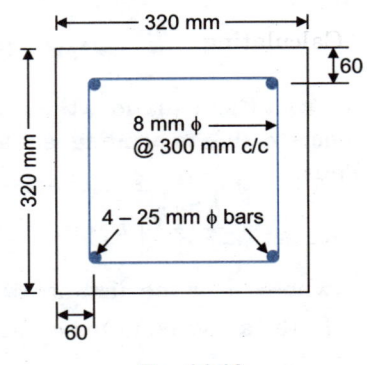

Fig. 14.12

Now, $\quad \dfrac{\sigma_{cc'}}{\sigma_{cc}} + \dfrac{\sigma_{cbc'}}{\sigma_{cbc}} = \dfrac{2.75}{4} + \dfrac{0.905}{5} = 0.869 < 1.$

Hence safe. Provide 8 mm Φ ties @ 300 mm c/c.

## 14.5. CASE 2: COMPRESSIVE LOAD AT LARGE ECCENTRICITY (e > 1.5 D)

In this case, the compressive load acts at large eccentricity, resulting in large tensile stresses. Due to this, the section below neutral axis cracks and hence the concrete below it is neglected. The compressive stresses are borne by concrete above the neutral axis and the tensile stresses below the neutral axis are entirely borne by tensile steel. In this case, it is assumed that the eccentricity is large so that the bending stresses are predominant. The super-position of the direct stresses, found by dividing the load with the effective area of the section, does not therefore materially change the position of neutral axis located by the usual method adopted in the case of pure bending. This approximation is reasonable when eccentricity $e$ is greater than 1.5 times the depth of the section.

Fig. 14.13 (a) shows a R.C. section subjected to a compressive force P acting at a large eccentricity $e$ measured from the axis passing through the mid-depth. Let,

$A_{sc}$ = area of steel in compressive zone.

$A_{st}$ = area of steel in tensile zone.

$n$ = depth of neutral axis, based on bending stresses alone

$c_a$ = compressive stress in concrete, due to bending

$c_{a'}$ = compressive stress in concrete surrounding compressive steel

$\sigma_{cc'}$ = direct compressive stress in concrete

$t_a$ = tensile stress in tensile steel

$t_{c'}$ = compressive stress in compressive steel = $m_c \cdot c_{a'} = 1.5 m c_{a'}$

$d_c$ = cover to compressive steel

$d_t$ = cover to tensile steel

$d$ = effective depth of the section

$D$ = total depth of the section

$e$ = eccentricity

$e'$ = distance of P from N.A.

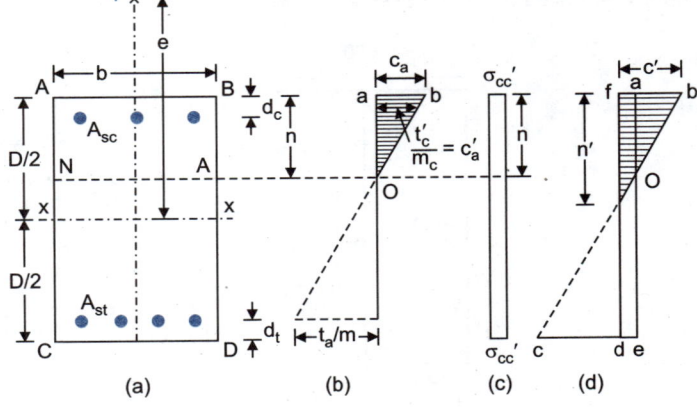

Fig. 14.13

Fig. 14.13 (b) shows the bending stress diagram. The depth of neutral axis, neglecting the direct stresses, can be found by equating the moments of effective area about the neutral axis. Thus,

**332** REINFORCED CONCRETE STRUCTURE

$$\frac{bn^2}{2} + (m_c - 1) A_{sc} (n - d_c) = m \cdot A_{st} (D - d_t - n) \qquad ...(14.9)$$

From this equation, the unknown $n$ can be found. In order to determine the compressive stress $c_a$ is concrete, due to bending, equate the moment of resistance of the section to the external bending moment. Thus:

$$\frac{b \cdot n c_a}{2} \left( D - d_t - \frac{n}{3} \right) + (m_c - 1) \cdot A_{sc} \cdot c_a' (D - d_c - d_t) = P \cdot e' \qquad ...(14.10)$$

where $e'$ is the distance of point of application of load $P$ from the N.A.

In the above equation $c_a'$ has the following value [Fig. 14.13(b)]

$$c_a' = \frac{c_a}{n} (n - d_c) \qquad ...[14.11(a)]$$

The stress in compressive steel is, $t_c' = 1.5 \, m c_{a'} = 1.5 \, m \, \frac{c_a}{n} (n - d_c) \qquad ...[14.11(b)]$

The tensile stress $t_a$ in steel is given by $t_a = m \cdot \frac{c_a}{n} (D - d_t - n) \qquad ...[14.11(c)]$

The effective area of section is given by $A_e = b \cdot n + (m_c - 1) A_{sc} + m A_{st} \qquad ...(14.12)$

Hence the direct stress in concrete is given by

$$\sigma_{cc'} = \frac{P}{A_e} = \frac{P}{b \cdot n + (m_c - 1) A_{sc} + m A_{st}} \qquad ...(14.13)$$

The final stresses can be found by superposition as under:

$$c' = \text{stress in concrete steel} = c_a + \sigma_{cc'} \qquad ...[14.14(a)]$$

$$t_c = \text{stress in compressive steel} = t_{c'} + m_c \cdot \sigma_{cc'}$$

$$= m_c \cdot c_{a'} + m_c \cdot \sigma_{cc'} = m_c (c_{a'} + \sigma_{cc'}) = 1.5 \, m (c_{a'} + \sigma_{cc'}) \qquad ...[14.14(b)]$$

$$t = \text{stress in tensile steel} = t_a - m \sigma_{cc'} \qquad ...[14.14(c)]$$

Fig. 14.13(d) shows the super position of both the stresses. It will be seen that the actual neutral axis, after superposition, shifts downwards increasing the value of $n$ and hence of the effective area. However, the error involved in negligible if $e > \frac{3}{2} D$, and is on the safer side. A more exact method of locating the neutral axis is discussed in the next article, where $e$ lies between $D/4$ and $3D/2$.

**Example 14.13.** *Fig. 14.14(a) shows a column subjected to a load of 25 kN acting at an eccentricity of 600 mm from the centre of the column. The column section is reinforced as shown in Fig. 14.14(b). Determine the stress in concrete and steel. Take $m = 19$.*

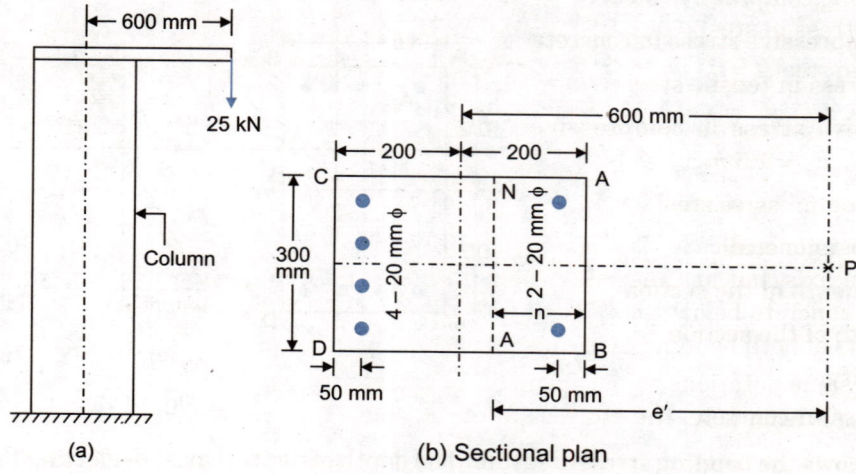

Fig. 14.14

**Solution:**

$$A_{st} = 4\frac{\pi}{4}(20)^2 \approx 1256 \text{ mm}^2$$

$$d_t = 50 \text{ mm}$$

$$A_{sc} = 2\frac{\pi}{4}(20)^2 = 628 \text{ mm}^2$$

$$d_c = 50 \text{ mm}$$

Let $n$ be the depth of N.A. from compression face AB. Equating moment of the effective areas about N.A., we have

$$\frac{300\, n^2}{2} + (1.5 \times 19 - 1)\, 628\,(n - 50) = 19 \times 1256\,(400 - 50 - n)$$

or $\quad n^2 + 274.2n - 61440 = 0.\quad$ From which $n = 146.2$ mm

$\therefore \quad e' = 600 - (200 - n) = 600 - (200 - 146.2) = 546.2$ mm.

To find the compressive stress $c_a$ in concrete, due to bending, equate the moment of resistance of the section to the external moment (*Eq. 14.10*).

$$c_{a'} = \frac{c_a}{n}(n - d_c) = \frac{c_a}{146.2}(146.2 - 50) = 0.658\, c_a$$

Substituting the values, in *Eq. 14.10*, we have

$$\frac{300 \times 146.2}{2} c_a \left[400 - 50 - \frac{146.2}{3}\right] + (1.5 \times 19 - 1)\, 628 \times 0.658\, c_a \times (400 - 50 - 50)$$

$$= 25000 \times 546.2$$

or $\quad 6606780\, c_a + 3409100\, c_a = 13655000 \quad$ or $\quad c_a = 1.36$ N/mm$^2$

Stress in compression steel, $= 1.5 \times 19 \,\dfrac{1.36}{146.2}(146.2 - 50) = 25.5$ N/mm$^2$

Stress in tensile steel $= t_a = 19 \times \dfrac{1.36}{146.2}(400 - 50 - 146.2) = 36.02$ N/mm$^2$.

Again, the effective area $A_e = (300 \times 146.2) + (1.5 \times 19 - 1)\,628 + 19 \times 1256 = 84994$ mm$^2$.

Hence the compressive stress due to direct load is $\sigma_{cc'} = 25000/84994 = 0.29$ N/mm$^2$.

The final stress will be as follows:

Compressive stress in concrete, $c' = c_a + \sigma_{cc'} = 1.36 + 0.29 = 1.65$ N/mm$^2$.

Compressive stress in steel $= t_c = t_{c'} + 1.5\, m\, \sigma_{cc'} = 25.5 + 1.5 \times 19 \times 0.29 = \mathbf{33.8}$ **N/mm$^2$**.

Tensile stress in steel, $t = t_a - m\, \sigma_{cc'} = 36.02 - 19 \times 0.29 = \mathbf{30.51}$ **N/mm$^2$**. (See example 14.12 for comparison of results)

## 14.6. CASE 3: COMPRESSIVE LOAD AT MODERATE ECCENTRICITY [D/4 < e < 3D/2]

This is the most general case. The eccentricity is such that tensile stresses are set up in the concrete below the neutral axis, but at the same time both bending stresses as well as direct stress are of equal importance. The concrete below the neutral axis is neglected and the neutral axis is located taking into account the effects of both bending moments as well as direct load.

Let us use the same notations as in the previous article. Fig. 14.15 shows the stress diagram. If $c'$ is the compressive stress in concrete, the stresses in compressive and tensile steels are given by the following expressions:

$$t_c = m_c \frac{c'}{n}(n - d_c) = 1.5\, m\, \frac{c'}{n}(n - d_c) \quad \ldots(1)$$

$$t = m\, \frac{c'}{n}(D - d_t - n) \quad \ldots(2)$$

Since the section is stable, the total force acting on the section must balance. That is, the algebraic sum of internal compressive and tensile forces must be equal to the external load $P$. Hence

$$\frac{b.n.c'}{2} + (m_c - 1)A_{sc} \cdot \frac{c'}{n}(n - d_c) - mA_{st}\frac{c'}{n}(D - d_t - n) = P \quad \ldots(14.15)$$

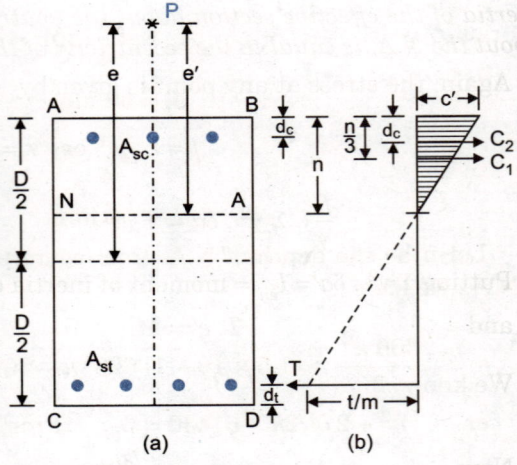

Fig. 14.15

Similarly, the moments of all forces acting on the section must balance. Taking the moments about the centre of tensile steel, we get

$$\frac{b.n.c'}{2} \times \left(D - d_t - \frac{n}{3}\right) + (m_c - 1)A_{sc}\frac{c'}{n}(n - d_c)(D - d_t - d_c) = P\left(e + \frac{D}{2} - d_t\right) \quad \ldots(14.16)$$

From *Eq. 14.15* and *14.16*, the two unknowns $c'$ and $n$ can be determined. Dividing *Eq. 14.16* by *14.15*, we get

$$\frac{\frac{b.n}{2}\left(D - d_t - \frac{n}{3}\right) + (m_c - 1)A_{sc}\frac{1}{n}(n - d_c)(D - d_t - d_c)}{\frac{b.n}{2} + (m_c - 1)A_{sc}\left(\frac{n - d_c}{n}\right) - m A_{st}\left(\frac{D - d_t - n}{n}\right)} = \left(e + \frac{D}{2} - d_t\right) \quad \ldots(14.17)$$

This is a quadratic equation in $n$ which can be solved either by trial and error, or graphically. Knowing $n$, $c'$ can be determined either from *Eq. 14.15* or from *Eq. 14.16*. The stresses in steel can then be found from (1) and (2) above.

### Alternative Method

The value of $n$ can alternatively be found as under. Let $n$ be the depth of neutral axis from the compression face. Consider as elementary strip of area $\delta a$, at distance $y$ above neutral axis. The stress at any layer will be proportional to its distance from the neutral axis.

Hence the stress at elementary strip = $r \cdot y$ where $r$ is a constant. Elementary force = $r\, y\, \delta a$.

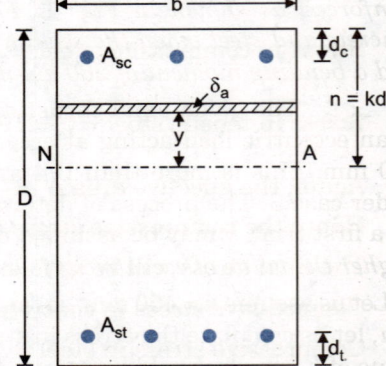

Fig. 14.16

The total force on the whole area must be equal to the external force. Hence.

$$\Sigma r \cdot y \cdot \delta a = T \quad \text{or} \quad r\, \Sigma y \cdot \delta a = T \quad \ldots(1)$$

Again moment of elementary force about the neutral axis will be equal to $r\, y^2 \cdot \delta a$. The sum of such elementary moments must be equal to the moment of the external force about the neutral axis. Hence

$$\Sigma r \cdot y^2 \cdot \delta a = T\left(e - \frac{D}{2} + n\right) = T \cdot e' \quad \text{or} \quad r\Sigma y^2 \cdot \delta a = T\left(e - \frac{D}{2} + n\right) = T \cdot e' \quad \ldots(2)$$

Dividing (2) by (1), we get

$$\frac{\Sigma y^2\, \delta a}{\Sigma y \cdot \delta a} = \left(e - \frac{D}{2} + n\right) = e' \quad \ldots(14.18)$$

*Equation 14.18* is very important conclusion. $\Sigma y^2 \cdot \delta a$ is the moment of inertia of the effective section about the neutral axis, while $\Sigma y \cdot \delta a$ is the moment of the effective area of the section about the neutral axis. The term $\left(e - \frac{D}{2} + n\right)$ or $e'$ is the distance of the load from the neutral axis. *Hence the moment of* inertia of the effective section about the neutral axis divided by the moment of area of the effective section about the N.A. is equal to the eccentricity of the load from the neutral axis.

Again, the stress at any point is given by

$$f = r \cdot y \quad \text{or} \quad r = \frac{f}{y} \quad \text{Substituting this in (2), we get}$$

$$\frac{f}{y} \cdot \Sigma y^2 \cdot \delta a = T \cdot e'$$

Putting $\Sigma y^2 \cdot \delta a = I_{en}$ = moment of inertia of equivalent section about N.A.
and $T \cdot e' = M$.

We know

$$\frac{f}{y} I_{en} = M \quad \text{or} \quad f = \frac{M}{I_{en}} y$$

Now,

$$I_{en} = \frac{bn^3}{3} + (m_c - 1) A_{sc}(n - d_c)^2 + m A_{st}(D - d_t - n)^2 \qquad \text{...(14.19)}$$

$$\Sigma y \cdot \delta a = \frac{bn^2}{2} + (m_c - 1) A_{sc}(n - d_c) - m A_{st}(D - d_t - n) \qquad \text{...(14.20)}$$

Substituting these values in *Eq. 14.18*, we get

$$\frac{\dfrac{bn^3}{3} + (m_c - 1) A_{sc}(n - d_c)^2 + m \cdot A_{st}(D - d_t - n)^2}{\dfrac{bn^2}{2} + (m_c - 1) A_{sc}(n - d_c) - m \cdot A_{st}(D - d_t - n)} = e' \qquad \text{...(14.21)}$$

which is a cubic equation in $n$ and can be solved by trial and error.

**Example 14.14.** *The section of an arch rib, 400 mm × 800 mm is reinforced as shown in Fig. 14.17. Determine maximum stress in concrete and steel when the section is subjected to a thrust of 400 kN and a bending moment of 200 kN-m. Take m = 19.*

**Solution:** The given axial thrust and bending moment is equivalent to an eccentric load acting at eccentricity $e = 200 \times 10^6/400 \times 10^3 = 500$ mm. This is more than $D/4$ and less than $3D/2$ and hence falls under case 3. The process of determining $n$ is that of trial and error. As a first trial, $k$ may be assumed to be between 0.4 to 0.7, for case 3. *Higher the value of eccentricity, lower will be the value of k.*

Let us assume $n = 450$ mm, so that $k = 450/740 \approx 0.61$. For this value of $n$, let us calculate the stress in steel and concrete, and then find the value of $k$, which should be near to the assumed value. If not, then an intermediate value of $k$ is adopted and then calculations are repeated.

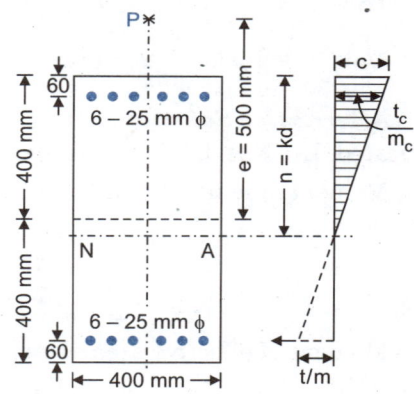

Fig. 14.17

$$A_{st} = A_{sc} = 6 \times \frac{\pi}{4}(25)^2 = 2945 \text{ mm}^2.$$

Taking the moments of internal and external forces about the centre of tensile steel, we have, from *Eq. 14.16*.

$$\frac{400 \times 450}{2} c' \left(740 - \frac{450}{3}\right) + (1.5 \times 19 - 1) 2945 \frac{c'}{450}(450 - 60) \times (740 - 60)$$

$$= 400 \times 10^3 \left(500 + \frac{800}{2} - 60\right) \qquad \text{...(1)}$$

or $53.1 \times 10^6 c' + 47.73 \times 10^6 c' = 336 \times 10^6$. From which $c' = 3.33$ N/mm$^2$.

Again, equating the sum of internal forces to the external forces, we have

$$\frac{bnc'}{2} + (m_c - 1) A_{sc} \cdot \frac{c'}{n}(n - d_c) - A_{st} \cdot t = P \qquad \text{...(2)}$$

$$\frac{400 \times 450 \times 3.33}{2} + (1.5 \times 19 - 1) 2945 \times \frac{3.33}{450} (450 - 60) - 2945 \times t = 400 \times 10^3$$

or  $3 \times 10^5 + 2.338 \times 10^5 - 2945\, t = 4 \times 10^5$. From which $t = 45.43$ N/mm².

From these values of $c$ and $t$, we have from stress diagram,

$$n = \frac{1}{1 + \dfrac{t}{m \cdot c'}} \times d = \frac{740}{1 + \dfrac{45.43}{19 \times 3.33}} = 430.8 \text{ mm.}$$

This is less than the assumed value of 450 mm. Hence adopt an intermediate value of $n = 436$ mm (say). Substituting in *Eq. (14.16)*, we get

$$\frac{400 \times 436\, c'}{2}\left(740 - \frac{436}{3}\right) + (1.5 \times 19 - 1) 2945\, \frac{c'}{436} (436 - 60) \times (740 - 60)$$

$$= 400 \times 10^3 \left(500 + \frac{800}{2} - 60\right) \text{ which gives } c' = 3.38 \text{ N/mm}^2.$$

Substituting in *Eq. (2)* above, we get

$$\frac{400 \times 436 \times 3.38}{2} + (1.5 \times 19 - 1) 2945\, \frac{3.38}{436} (436 - 60) - 2946\, t = 400 \times 10^3$$

which gives $t = 44.55$ N/mm². Hence from stress diagram,

$$n = \frac{1}{1 + \dfrac{t}{m \cdot c'}}\, d = \frac{740}{1 + \dfrac{44.55}{19 \times 3.38}} = 437 \text{ mm.}$$

This is practically the same as the assumed value, and no further trial is necessary.

Thus,  $c' = \mathbf{3.38}$ **N/mm²** (compressive) and  $t = \mathbf{44.55}$ **N/mm²**  (tensile)

and  $t_c = 1.5\, m\, \dfrac{c'}{n} (n - d_c) = 1.5 \times 19 \dfrac{3.38}{436} (436 - 60) = \mathbf{83.1}$ **N/mm²**  (compressive).

***Alternative Solution.*** Let us solve this example by alternative method in which the neutral axis is first located. Let N.A. be at a distance $n$ mm below the compression face.

Moment of inertia of effective section about N.A. is given by *Eq. 14.20*.

$$I_{en} = \frac{400 \times n^3}{3} + (1.5 \times 19 - 1) 2945 (n - 60)^2 + 19 \times 2945 (800 - 60 - n)^2$$

$$= 133.3\, n^3 + 80988 (n - 60)^2 + 55955 (740 - n)^2 \qquad \ldots(1)$$

Moment of effective area about N.A is given by *Eq. 14.21*

$$\Sigma y \cdot \delta a = \frac{400\, n^2}{2} + (1.5 \times 19 - 1) \times 2945 (n - 60) - 19 \times 2945 (800 - 60 - n)$$

$$= 200\, n^2 + 80988 (n - 60) - 55955 (740 - n) \qquad \ldots(2)$$

Distance of $P$ from N.A. is given by

$$e' = \left[e - \frac{D}{2} + n\right] = 500 - \frac{800}{2} + n = 100 + n \qquad \ldots(3)$$

Hence from *Eq. 14.18*, $\dfrac{I_{en}}{\Sigma y \cdot \delta a} = e'$

or  $\dfrac{133.3\, n^3 + 80988 (n - 60)^2 + 55955 (740 - n)^2}{200\, n^2 + 80988 (n - 60) - 55955 (740 - n)} = 100 + n$

Solving this by trial and error, trying the value of $n$ between $0.4\, d$ ($= 296$ mm) to $0.7\, d$ ($= 518$ mm), we get $n = 436.5$ mm.

Hence, $I_{en} = 133.3 \, (436.5)^2 + 80988 \, (436.5 - 60)^2 + 55955 \, (740 - 436.5)^2$

$= 1.665 \times 10^{10}$ mm$^4$

$e' = e - \dfrac{D}{2} + n = 500 - \dfrac{800}{2} + 436.5 = 536.5$

Moment of thrust about N.A. = $400000 \times 536.5$

∴ Stress in concrete, $c' = \dfrac{400000 \times 536.5}{1.665 \times 10^{10}} \times 436.5 = \mathbf{5.626 \, N/mm^2}$

Stress in tensile steel $= \dfrac{400000 \times 536.5}{1.665 \times 10^{10}} (800 - 436.5 - 60) \times 19 = \mathbf{74.34 \, N/mm^2}.$

and stress in compressive steel, $t_c = 1.5 \times 19 \, \dfrac{400000 \times 536.5}{1.665 \times 10^{10}} (436.5 - 60) = \mathbf{138.30 \, N/mm^2}.$

**Example 14.15.** *Solve Example 14.13 by the exact method.*

**Solution:** (Fig. 14.14). Since the eccentricity was more than 1.5 D, example 14.13 was solved by method of superposition. Let us now solve it by first determining the neutral axis, by method of case 3. Let the depth of neutral axis be $n$ mm below the compression face AB. The moment of inertia of effective section about N.A. is given by *Eq. 14.19*.

$A_{sc} = 628$ mm$^2$;  $A_{st} = 1256$ mm$^2$;  $d_t = d_c = 50$ mm

∴ $I_{en} = \dfrac{300 \, n^3}{3} + (1.5 \times 19 - 1) \, 628 \, (n - 50)^2 + 19 \times 1256 \, (400 - 50 - n)^2$

$= 100 \, n^3 + 17270 \, (n - 50)^2 + 23864 \, (350 - n)^2$ ...(1)

Moment of effective are about N.A. is given by *Eq. 14.21*.

$\Sigma y \cdot \delta a = \dfrac{300 n^2}{2} + (1.5 \times 19 - 1) \, 628 \, (n - 50) - 19 \times 1256 \, (400 - 50 - n)$

$= 150 \, n^2 + 17270 \, (n - 50) - 23864 \, (350 - n)$ ...(2)

Distance of P from N.A. is

$e' = \left(e - \dfrac{D}{2} + n\right) = 600 - 200 + n = 400 + n$ ...(3)

Hence from *Eq. 14.18*, $\dfrac{I_{en}}{\Sigma y \cdot \delta a} = e'$

or $\dfrac{100 \, n^3 + 17270 \, (n - 50)^2 + 23864 \, (350 - n)^2}{150 \, n^2 + 17270 \, (n - 50) - 23864 \, (350 - n)} = 400 + n$

The above relationship can be solved by trial, assuming $n$ between $0.4 \, d \, (= 140 \text{ mm})$ to $0.7 \, d \, (= 245 \text{ mm})$. By trial and error, we get $n = 176$ mm.

Hence $I_{nn} = 100 \, (176)^3 + 17270 \, (176 - 50)^2 + 23864 \, (350 - 176)^2 = 15.42 \times 10^8$ mm$^4$

$e' = e - \dfrac{D}{2} + n = 600 - 200 + 176 = 576$ mm

Moment of thrust about N.A. = $25000 \times 576 = 14.4 \times 10^6$ N/mm

∴ Stress in concrete $= c' = \dfrac{14.4 \times 10^6}{15.42 \times 10^8} \times 176 = \mathbf{1.64 \, N/mm^2}$  (compressive)

(against a value of 1.65 N/mm$^2$ calculated by approximate method in Example 14.13)

Stress in tensile steel $= \dfrac{14.4 \times 10^6}{15.42 \times 10^8} (400 - 176 - 50) \times 19 = \mathbf{30.87 \, N/mm^2}$  (tensile)

(against a value of 30.45 N/mm$^2$ in example 14.13).

**338** REINFORCED CONCRETE STRUCTURE

Stress in compression steel $= \dfrac{14.4 \times 10^6}{15.42 \times 10^8} (176 - 50) \times 1.5 \times 19 = \mathbf{33.53\ N/mm^2}$ (compressive)

(against a value of 33.8 N/mm² in Ex. 14.13)

The results show that when, $e \geq 3/2\ D$, the approximate method can be safely used without any appreciable error.

## 14.7. CIRCULAR SECTION SUBJECTED TO ECCENTRIC LOAD

The analysis of circular section subjected to load at small eccentricity has been done in example 14.7. Let us now take the case of moderate eccentricity, when tension is developed in the section.

Fig. 14.18(a) shows a column of circular section of radius $R$, having reinforcement area $= A_s$ arranged on a circle of radius $r$. The steel rods can be replaced by an equivalent steel shell of radius $r$, of equal area, as shown in Fig. 14.18(b). Let $e$ be the eccentricity of the load with respect to the axis $x$-$x$ passing through the centre of circle. Let the N.A. be at a distance $n = kd$ below the point $C$, so that angles, subtended at the centre, by its points of intersection with concrete extreme and steel shell are $\alpha_1$ and $\alpha_2$ respectively.

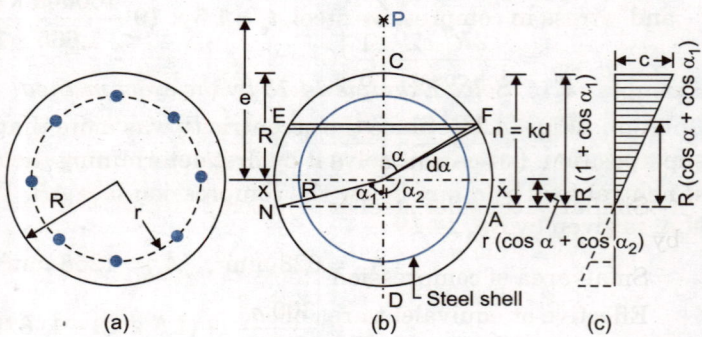

Fig. 14.18

Consider a thin layer $EF$, subtending angle $\alpha$ at the centre, with the symmetrical axis $CD$, and parallel to N.A. Length of strip $= 2R \sin \alpha$

Distance of strip from centre $= R \cos \alpha$ ; Width of strip $= \delta (R \cos \alpha) = R \sin \alpha \cdot d\alpha$

Area of strip $= 2 R^2 \sin^2 \alpha \cdot d\alpha$ ;  $n = R + R \cos \alpha_1 = R(1 + \cos \alpha_1)$

Distance of layer $EF$ above N.A. $= R \cos \alpha + R \cos \alpha_1 = R(\cos \alpha + \cos \alpha_1)$

Let $c'$ be the intensity of stress in concrete at point $C$.

$\therefore$ Intensity of stress at $EF = \dfrac{c' R(\cos \alpha + \cos \alpha_1)}{R(1 + \cos \alpha_1)} = \dfrac{c'(\cos \alpha + \cos \alpha_1)}{1 + \cos \alpha_1}$

$\therefore$ Force on strip $= 2 R^2 \sin^2 \alpha \cdot d\alpha \cdot \dfrac{c'(\cos \alpha + \cos \alpha_1)}{1 + \cos \alpha_1}$

Hence total compression $C_1$ in concrete is given by

$$C_1 = \int_0^{\pi - \alpha_1} \dfrac{2 R^2 \cdot c'}{1 + \cos \alpha_1} \cdot \sin^2 \alpha (\cos \alpha + \cos \alpha_1)\, d\alpha$$

$$= \dfrac{2 R^2 \cdot c'}{1 + \cos \alpha_1} \int_0^{\pi - \alpha_1} (\sin^2 \alpha \cdot \cos \alpha + \sin^2 \alpha \cdot \cos \alpha_1)\, d\alpha$$

$$= \dfrac{2 R^2 c'}{1 + \cos \alpha_1} \left[ \dfrac{\sin^3 \alpha}{3} + \dfrac{\cos \alpha_1}{2} \left( \alpha - \dfrac{\sin 2\alpha}{2} \right) \right]_0^{\pi - \alpha_1}$$

$$= \dfrac{2 R^2 c'}{1 + \cos \alpha_1} \left[ \dfrac{\sin^3(\pi - \alpha_1)}{3} + \dfrac{\cos \alpha_1}{2} (\pi - \alpha_1) - \dfrac{\cos \alpha_1}{4} \sin 2(\pi - \alpha_1) \right]$$

or $\qquad C_1 = \dfrac{2 R^2 c'}{1 + \cos \alpha_1} \left[ \dfrac{\sin^3 \alpha_1}{3} + \dfrac{\pi - \alpha_1}{2} \cos \alpha_1 - \dfrac{\cos \alpha_1}{4} \sin 2\alpha_1 \right] \qquad$ ...(14.22)

Moment of compression $C_1$ in concrete, about the central axis $x$-$x$ is:

$$M_1 = \int_0^{\pi - \alpha_1} 2 R^2 \sin^2 \alpha \, d\alpha \, \frac{c'(\cos \alpha + \cos \alpha_1)}{1 + \cos \alpha_1} \times R \cos \alpha$$

$$= \frac{2 R^3 c'}{1 + \cos \alpha_1} \int_0^{\pi - \alpha_1} \sin^2 \alpha \cos \alpha (\cos \alpha + \cos \alpha_1) \, d\alpha$$

$$= \frac{2 R^3 c'}{1 + \cos \alpha_1} \int_0^{\pi - \alpha_1} \sin^2 \alpha \cos^2 \alpha + \sin^2 \alpha \cos \alpha \cos \alpha_1) \, d\alpha$$

$$= \frac{2R^3 c'}{1 + \cos \alpha_1} \left[ \frac{1}{8}\left(\alpha + \frac{\sin 4\alpha}{4}\right) + \cos \alpha_1 \frac{\sin^3 \alpha}{3} \right]_0^{\pi - \alpha_1}$$

$$= \frac{2R^3 c'}{1 + \cos \alpha_1} \left[ \frac{1}{8}\left\{(\pi - \alpha_1) + \frac{\sin 4(\pi - \alpha_1)}{4}\right\} + \frac{\cos \alpha_1}{3} \sin^3 (\pi - \alpha_1) \right]$$

$$= \frac{2R^3 c'}{1 + \cos \alpha_1} \left[ \frac{\pi - \alpha_1}{8} + \frac{\sin 4\alpha_1}{32} + \frac{\cos \alpha_1}{3} \sin^3 \alpha_1 \right] \qquad \text{...(14.23)}$$

Similarly, consider steel area in compression. If $A_{st}$ is the total steel area, the thickness $z$ of steel is given by $\quad 2\pi r z = A_{st}$ \qquad\qquad ...(14.24)

Small area of compression steel at angle $\alpha = r \, \delta\alpha \cdot z$

Effective or equivalent area of compression steel element $= (m_c - 1) r \, \delta\alpha \cdot z$

Intensity of stress on elementary area $= c' \dfrac{r(\cos \alpha + \cos \alpha_2)}{R + r \cos \alpha_2}$

∴ Total compression $C_2$ in steel above neutral axis is given by

$$C_2 = 2 \int_0^{\pi - \alpha_2} (m_c - 1) r \, d\alpha \cdot z \, c' \, \frac{r(\cos \alpha + \cos \alpha_2)}{R + r \cos \alpha_2}$$

$$= \frac{2 r^2 z (m_c - 1) c'}{R + r \cos \alpha_2} \int_0^{\pi - \alpha_2} (\cos \alpha + \cos \alpha_2) \, d\alpha$$

$$= \frac{2 r^2 z (m_c - 1) c'}{R + r \cos \alpha_2} [\sin \alpha + \alpha \cos \alpha_2]_0^{\pi - \alpha_2}$$

$$= \frac{2 r^2 z (m_c - 1) c'}{R + r \cos \alpha_2} [\sin \alpha_2 + (\pi - \alpha_2) \cos \alpha_2] \qquad \text{...(14.25)}$$

Moment of total compression in steel about central axis $x$-$x$ is given by

$$M_2 = 2 \int_0^{\pi - \alpha_2} (m_c - 1) r \, d\alpha \cdot z \, c' \frac{r(\cos \alpha + \cos \alpha_2)}{R + r \cos \alpha_2} r \cos \alpha$$

$$= \frac{2r^3 z (m_c - 1) c'}{R + r \cos \alpha_2} \int_0^{\pi - \alpha_2} (\cos^2 \alpha + \cos \alpha \cos \alpha_2) \, d\alpha$$

$$= \frac{2r^3 z (m_c - 1) c'}{R + r \cos \alpha_2} \left[ \frac{\alpha}{2} + \frac{\sin 2\alpha}{4} + \cos \alpha_2 \sin \alpha \right]_0^{\pi - \alpha_2}$$

$$= \frac{2r^3 z (m_c - 1) c'}{R + r \cos \alpha_2} \left[ \frac{\pi - \alpha_2}{2} - \frac{\sin 2\alpha_2}{4} + \cos \alpha_2 \sin \alpha_2 \right]$$

$$= \frac{2r^3 z (m_c - 1) c'}{R + r \cos \alpha_2} \left[ \frac{\pi - \alpha_2}{2} + \frac{\sin 2\alpha_2}{4} \right] \qquad \text{...(14.26)}$$

where $\quad m_c = 1.5 \, m$

In the same manner total tension $T$ is tension steel below the N.A. is given by

$$T = 2 \int_0^{\alpha_2} r d\alpha . z m c' \frac{r \cos \alpha - r \cos \alpha_2}{R + r \cos \alpha_2} = \frac{2r^2 z m c'}{R + r \cos \alpha_2} \int_0^{\alpha_2} (\cos \alpha - \cos \alpha_2) d\alpha$$

$$= \frac{2 r^2 z m c'}{R + r \cos \alpha_2} [\sin \alpha_2 - \alpha_2 \cos \alpha_2] \qquad \text{...(14.27)}$$

Moment of total tension in steel, about the centre axis $x$-$x$ is given by

$$M_3 = 2 \int_0^{\alpha_2} r \, d\alpha . z m c' \frac{r(\cos \alpha - \cos \alpha_2)}{R + r \cos \alpha_2} . r \cos \alpha = \frac{2 r^3 . z.m.c'}{R + r \cos \alpha_2} \int_0^{\alpha_2} (\cos^2 \alpha - \cos \alpha \cos \alpha_2) d\alpha$$

$$= \frac{2 r^3 . z.m.c'}{R + r \cos \alpha_2} \left( \frac{\alpha_2}{2} - \frac{\sin 2\alpha_2}{4} \right) \qquad \text{...(14.28)}$$

Since the section is in equilibrium, the algebraic sum of internal forces must be equal to the external force. Hence $\quad C_1 + C_2 - T = P$ ...(14.29)

Similarly, total moment of internal force about centre line is equal to the moment of external force about the centre line.

Hence, $\qquad M_1 + M_2 + M_3 = P \cdot e$ ...(14.30)

Dividing *Eq. 14.30* by *Eq. 14.29*, we get

$$\frac{M_1 + M_2 + M_3}{C_1 + C_2 - T} = e \qquad \text{...(14.31)}$$

Substituting the values of $C_1$, $C_2$, $T$, $M_1$, $M_2$ and $M_3$, we find that there are two unknowns $\alpha_1$ and $\alpha_2$ another relationship between $\alpha_1$ and $\alpha_2$ is obtained by:

$$R \cos \alpha_1 = r \cos \alpha_2 \qquad \text{...(14.32)}$$

or $\qquad \alpha_2 = \cos^{-1} \left( \frac{R}{r} \cos \alpha_1 \right)$ ...[14.32(a)]

Thus value of $\alpha_2$ can be substituted in *Eq. 14.32*, which can then be solved by trial and error. However, it is better to assume some value of $k$ (between 0.4 to 0.7) and then find the value of $\alpha_1$ from the relation:

$$n = kd = R (1 + \cos \alpha_1) \qquad \text{...(14.33)}$$

Knowing $\alpha_1$, $\alpha_2$ is determined from *Eq. 14.32(a)*. These values of $\alpha_1$ and $\alpha_2$ can then be substituted in *Eq. 14.31* to see whether it is satisfied or not. If not satisfied, another intermediate value of $n$ is chosen and procedure repeated.

**Example 14.16.** *A circular column of 500 mm diameter is reinforced with 8 bars of 20 mm diameter arranged on a circle of 400 mm diameter as shown in Fig. 14.19. Determine the maximum stresses in concrete and steel when the column is subjected to an axial load of 200 kN and a bending moment of 40 kN-m. Take m = 19.*

**Solution:** $A_{st} = 8 \frac{\pi}{4} (20)^2 = 2513 \text{ mm}^2$

Let $z$ be the thickness of equivalent steel shell placed at a radius of 200 mm.

$2 \pi r z = A_{st} = 2513$

$\therefore \quad z = \dfrac{2513}{2 \pi r} = \dfrac{2513}{2 \pi \times 200} = 2 \text{ mm}$

$e = 40/200 = 0.2 \text{ m} = 200 \text{ mm}$

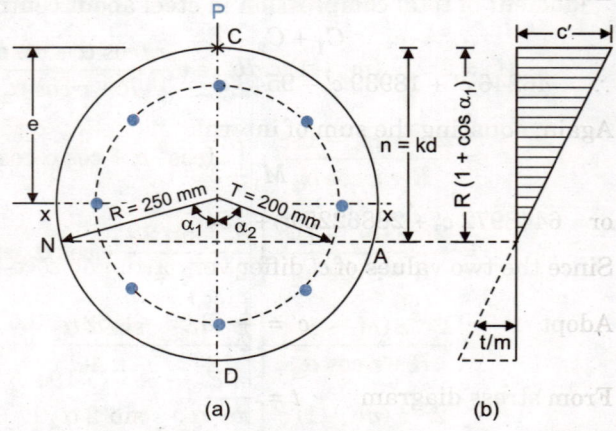

Fig. 14.19

Let $n = kd$ be the depth of N.A. The value of $k$ may be assumed between 0.4 to 0.7, smaller value for larger eccentricity.

$d = 500 - 50 = 450$ mm

Let us assume $k = 0.60$ so that $n = 450 \times 0.6 = 270$ mm

$R = 250$ mm; $r = 200$ mm    Now    $n = R(1 + \cos \alpha_1)$

$\therefore \quad \alpha_1 = \cos^{-1}\left(\dfrac{n}{R} - 1\right) = \cos^{-1}\left(\dfrac{270}{250} - 1\right) = \cos^{-1}(0.08) = 85°\,25' = 1.491$ radians

Again, from *Eq. 14.32 (a)*,

$\alpha_2 = \cos^{-1}\left(\dfrac{R}{r}\cos\alpha_1\right) = \cos^{-1}\left(\dfrac{250}{200} \times 0.08\right) = \cos^{-1}(0.1) = 84°\,16' = 1.47$ radians

The other functions useful in various equations are

$\sin\alpha_1 = 0.997$; $\cos\alpha_1 = 0.08$; $\sin 2\alpha_1 = 0.16$; $\sin 4\alpha_1 = -0.316$

$\sin\alpha_2 = 0.995$; $\cos\alpha_2 = 0.1$; $\sin 2\alpha_2 = 0.199$

Substituting these values in *Eqs. 14.22, 14.23, 14.25, 14.26, 14.27* and *14.28*, we get

Compression in concrete $'C_1' = \dfrac{2(250)^2 \, c'}{1 + 0.08}\left[\dfrac{(0.997)^3}{3} + \dfrac{\pi - 1.491}{2} \times 0.08 + \dfrac{0.08}{4} \times 0.16\right] = 46246\, c'$ ...(1)

Moment due to $C_1$ about $x$-$x$ $'M_1' = \dfrac{2(250)^3 \, c'}{1 + 0.08}\left[\dfrac{\pi - 1.491}{8} - \dfrac{0.316}{32} + \dfrac{0.08}{3}(0.997)^3\right] = 6448972\, c'$ ...(2)

Compression in steel    $'C_2' = \dfrac{2(200)^2 \times 2 \times 27.5 \, c'}{250 + 200 \times 0.1}[0.995 + (\pi - 1.47)\,0.1] = 18939\, c'$ ...(3)

Moment due to $'C_2'$ about $x$-$x$ axis    $'M_2' = \dfrac{2(200)^3 \times 2 \times 27.5 \, c'}{250 + 200 \times 0.1}\left[\dfrac{\pi - 1.47}{2} + \dfrac{0.199}{4}\right] = 2886225\, c'$ ...(4)

Tension in steel    $'T' = \dfrac{2(200)^2 \times 2 \times 19 \, c'}{250 + 200 \times 0.1}[0.995 - 1.47 \times 0.1] = 9548\, c'$ ...(5)

Moment due to $'T'$ about $x$-$x$ axis    $'M_3' = \dfrac{2(200)^3 \times 2 \times 19 \, c'}{250 + 200 \times 0.1}\left[\dfrac{1.47}{2} - \dfrac{0.199}{4}\right] = 1543082\, c'$ ...(6)

Equating the sum of internal forces to the external forces, we have

$C_1 + C_2 - T = P$

$\therefore \quad 46246\, c' + 18939\, c' - 9548\, c' = 200000$. From which $c' = 3.66$ N/mm$^2$ ...(a)

Again, equating the sum of internal moments to the external moment, we get

$M_1 + M_2 + M_3 = M$

or  $6448972\, c' + 2886225\, c' + 1543082\, c' = 40 \times 10^6$.  From which, $c' = 3.68$ N/mm$^2$.

Since the two values of $c'$ differ very little, no second trial is necessary

Adopt    $c' = \dfrac{1}{2}(3.66 + 3.68) = \mathbf{3.67\ N/mm^2}$.

From stress diagram    $t = \dfrac{m\,c'}{n}(d - n) = \dfrac{19 \times 3.67}{270}(450 - 270) = \mathbf{46.49\ N/mm^2}$.

## 14.8. CASE 4: TENSILE LOAD AT SMALL ECCENTRICITY

This case is similar to case 1, except that a tensile load acts in place of a compressive one. However, the load acts at such an eccentricity that only tensile stresses are set up in the section. This may happen when the point of application of pull happens to lie anywhere between the two areas. Because of tensile stresses, concrete becomes ineffective, and only the two steel areas bear the tensile load.

Fig. 14.20(b) shows a R.C. section reinforced with steel areas $A_{st1}$ and $A_{st2}$. Let the distances of the C.G. of steel areas from the two steel areas be $y_1$ and $y_2$. The pull $P$ acts at eccentricity $e$ from c.g. both the steel areas. Let $t_1$ and $t_2$ be the tensile stresses set up in both steel areas, as shown in Fig. 14.20(a).

Taking moments about the centre of steel area $A_{st2}$, we get

$$t_1 A_{st1} (y_1 + y_2) = P(e + y_2)$$

or $$t_1 = \frac{P(e + y_2)}{A_{st2}(y_1 + y_2)} \quad ...(14.34)$$

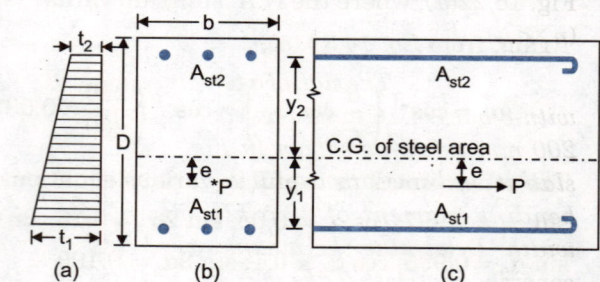

Similarly, taking moments about the centre of steel area $A_{st1}$, we get

$$t_2 A_{st2}(y_1 + y_2) = P(y_1 - e) \quad \text{or} \quad t_2 = \frac{P(y_1 - e)}{A_{st2}(y_1 + y_2)} \quad ...(14.35)$$

Fig. 14.20

**Example 14.17.** *The section of a bow-string girder suspender 200 mm × 350 mm is reinforced with 3 bars of 16 mm Φ on each side as shown in Fig. 14.21. Determine the stress in steel reinforcement, when the suspender is subjected in a pull of 80 kN and a bending moment of 5 kN-m.*

**Solution:** $$A_{st1} = A_{st2} = 3 \frac{\pi}{4}(16)^2 = 603 \text{ mm}^2$$

Since the section is symmetrically reinforced, the C.G. of steel areas will pass through the middepth of the section.

Hence, $y_1 = y_2 = 125$ mm. Also,

Eccentricity $e = \dfrac{5 \times 1000}{80} = 62.5$ mm

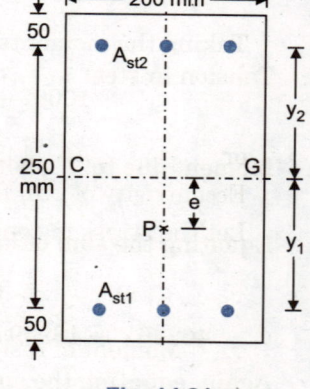

Taking moments about the centre of top steel area,

$$t_1 A_{st1}(250) = 80000(125 + 62.5)$$

∴ $$t_1 = \frac{80000 \times 187.5}{603 \times 250} = \mathbf{99.5 \text{ N/mm}^2} \quad (\text{tension})$$

Taking moments about the centre of bottom steel area

$$t_2 A_{st2}(250) = 80000(125 - 62.5)$$

∴ $$t_2 = \frac{80000 \times 62.5}{603 \times 250} = \mathbf{33.2 \text{ N/mm}^2} \quad (\text{tension}).$$

Fig. 14.21

## 14.9. CASE 5: TENSILE LOAD AT LARGE ECCENTRICITY

This case is similar to case 2. The eccentricity is so large ($e > 3/2\ D$) that bending stresses become predominant. The neutral axis is located considering the section as doubly reinforced one, subjected to bending moment. The direct tensile stresses due to axial pull are then super-imposed to get final stresses. The super-imposition will shift the neutral axis towards the compression zone, as shown in Fig. 14.22 (d), but the method gives satisfactory results when eccentricity is more than 1.5 times the depth of the section.

Fig. 14.22(b) shows the bending stress diagram, where c is the maximum compressive stress in concrete and t is the maximum tensile stress in steel due to bending. The concrete below the N.A. is neglected in the analysis. Fig. 14.22(c) shows the uniform tensile stress in the section, due to direct pull. The superimposed stress diagram is shown in Fig. 16.22(d), where the N.A. shifts upwards ($n_1 < n$).

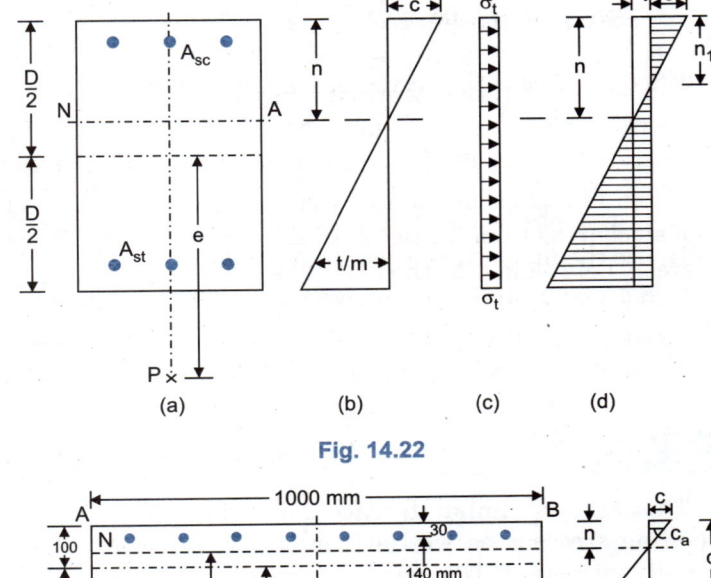

Fig. 14.22

**Example 14.18.** *The slab of a tank is reinforced with 8 bars of 16 mm Φ on each face, and is 200 mm thick, as shown in Fig. 14.23. The slab is subjected to a pull of 45 kN and a bending moment of 22.5 kN-m per metre width. Determine the maximum stresses in concrete and steel. Take m = 19.*

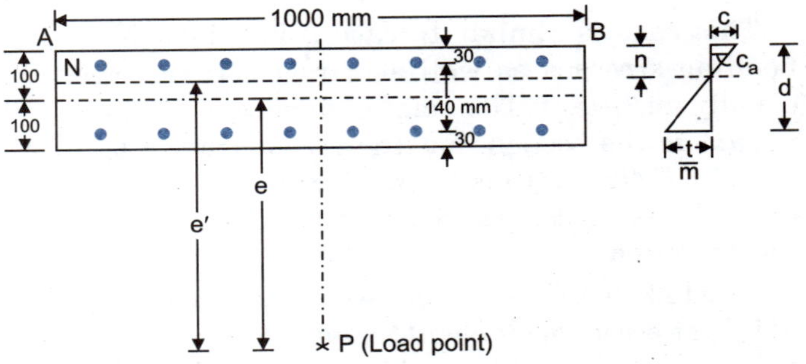

**Solution:** Eccentricity

$$e = \frac{M}{P} = \frac{22.5 \times 1000}{45} = 500 \text{ mm}$$

Fig. 14.23

Since this is more than 1.5D, case 5 will apply here, and bending stresses will be predominant.

Let the neutral axis be at a distance $n$ below the compression face AB.

$$A_{sc} = A_{st} = 8 \frac{\pi}{4}(16)^2 = 1608 \text{ mm}^2.$$

Taking the moments of effective areas about the N.A., we have

$$1000 \frac{n^2}{2} + (1.5 \times 19 - 1) 1608 (n - 30) = 19 \times 1608 (170 - n)$$

or $\quad n^2 + 149.54 n - 13041 = 0.$ From which $n = 61.7$ mm

Eccentricity of pull from N.A. = $e' = e + (100 - n) = 500 + 100 - 61.7 = 538.3$ mm

Let the stress in concrete be $c$. Let $c_a$ = stress in concrete around compressive steel

$$= \frac{c}{n}(n - 30) = \frac{c}{61.7}(61.7 - 30) = 0.514 c$$

∴ Moment of resistance of the section is

$$M_r = \frac{bnc}{2}\left(d - \frac{n}{3}\right) + (m_c - 1) A_{sc} (d - 30) c_a$$

$$= \frac{1000 \times 61.7 \times c}{2}\left(170 - \frac{61.7}{3}\right) + (1.5 \times 19 - 1) 1608 (170 - 30) 0.514 c = 7.792 \times 10^6 c.$$

Equating the moment of resistance to the moment of pull about N.A., we get

$$M_r = P e' \quad \text{or} \quad 7.792 \times 10^6 c = 45000 \times 538.3 \text{ mm}$$

∴ $$c = \frac{45000 \times 538.3}{7.792 \times 10^6} = 3.11 \text{ N/mm}^2 \text{ (comp.)}$$

Stress in compressive steel = $m_c \cdot c_a$ = $(1.5 \times 19) \times 0.514 \times 3.11 = 45.56$ N/mm² (comp.)

Stress in tensile steel $= t = m\dfrac{c}{n}(d-n) = \dfrac{19 \times 3.11}{61.7}(170 - 61.7) = 103.72 \text{ N/mm}^2$

Effective area of the section for direct tensile force is given by
$$A_e = 1000\, n + (m-1)A_{sc} + m\, A_{st}$$
$$= 1000 \times 61.7 + (19-1)1608 + 19 \times 1608 = 121196 \text{ mm}^2$$

$\therefore\ \sigma_t =$ tensile stress in concrete due to direct pull $= \dfrac{P}{A_e} = \dfrac{45000}{121196} = 0.37 \text{ N/mm}^2$ (tensile)

Hence the final stresses will be as under:

Compressive stress in concrete $= 3.11 - 0.37 = \mathbf{2.74 \text{ N/mm}^2}$
Compressive stress in top steel $= 45.56 - 19 \times 0.37 = \mathbf{38.51 \text{ N/mm}^2}$
Tensile stress in bottom steel $= 103.72 + 19 \times 0.37 = \mathbf{110.75 \text{ N/mm}^2}$.

## 14.10. CASE 6: TENSILE LOAD AT MODERATE ECCENTRICITY

This case is similar to case 3 in which both bending stresses as well as direct stresses are equally important. The point of application of pull is *outside* the section, but the eccentricity is less than $1.5\,D$. This is thus a general case, in which the neutral axis is located taking both B.M. and pull into consideration.

Fig. 14.24(a), (b) show the section subjected to a pull $P$ at eccentricity $e$ with respect to the central axis passing through the mid-depth. Let the depth of the neutral axis be $n = kd$ below the compression face $AB$. Let the stresses in concrete, compression steel and tensile steel be $c'$, $t_c$, and $t$ respectively. From the stress diagram [Fig. 14.24(c)], we have:

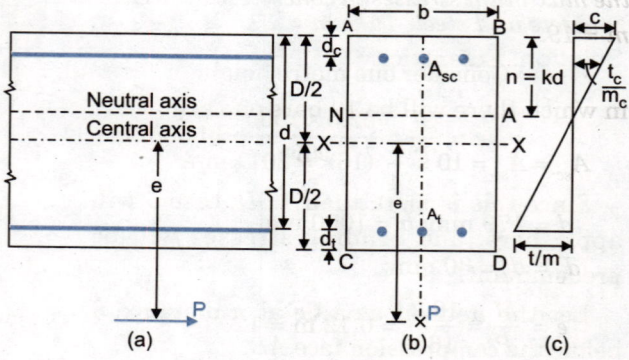

Fig. 14.24

$$t_c = m_c \dfrac{c'}{n}(n - d_c) = 1.5\, m\, \dfrac{c'}{n}(n - d_c) \qquad \ldots(1)$$

$$t = m\dfrac{c'}{n}(D - d_t - n) = m\dfrac{c'}{n}(d - n) \qquad \ldots(2)$$

Since the section is stable, the total forces acting on the section must balance. Hence

$$t\, A_{st} - \dfrac{bnc'}{2} - (m_c - 1)A_{sc}\dfrac{c'}{n}(n - d_c) = P$$

$$\dfrac{mc'}{n}(d - n)A_{st} - \dfrac{bnc'}{2} - (m_c - 1)A_{sc}\dfrac{c'}{n}(n - d_c) = P \qquad \ldots(14.36)$$

where $m_c = 1.5\, m$

Similarly, the moments of all the forces acting on the section must balance. Taking moments about the centre of tensile steel, we get

$$\dfrac{bnc'}{2}\left(D - d_t - \dfrac{n}{3}\right) + (m_c - 1)A_{sc}\dfrac{c'}{n}(n - d_c)(D - d_t - d_c) = P\left(e - \dfrac{D}{2} + d_t\right)$$

or $\qquad \dfrac{bnc'}{2}\left(d - \dfrac{n}{3}\right) + (m_c - 1)A_{sc}\dfrac{c'}{n}(n - d_c)(d - d_c) = P\left(e - \dfrac{D}{2} + d_t\right) \qquad \ldots(14.37)$

Dividing *Eq. 14.37* by *14.36*, we get

$$\frac{\frac{bnc'}{2}\left(d - \frac{n}{3}\right) + (m_c - 1) A_{sc} \frac{c'}{n} (n - d_c)(d - d_c)}{\frac{mc'}{n}(d-n) A_{st} - \frac{bnc'}{2} - (m_c - 1) A_{sc} \frac{c'}{n}(n - d_c)} = \left(e - \frac{D}{2} + d_t\right) \quad ...(14.38)$$

This is a quadratic equation in terms of *n*, which can be solved by trial and error.

The value of *k* (= *n/d*) may generally vary between 0.2 and 0.4; smaller the value of eccentricity, smaller the value of *k*. Once *n* is known, *c'* may be determined either from *Eq. 14.36* or *14.37*. The values of $t_c$ and *t* can then be determined from *Eqs* (1) and (2) above.

**Example 14.19.** *The walls of a tank is reinforced with 16 mm dia. bars spaced at 100 mm c/c on each face as shown in Fig 14.25. The total thickness of the wall is 200 mm. The wall is subjected to a pull of 160 kN and a bending moment of 19.2 kN-m per metre height. Determine the maximum stresses in concrete and steel. Take m = 19.*

**Solution:** Consider one metre height of the wall, in which there will be 10 bars on each face.

$$A_{sc} = A_{st} = 10 \times \frac{\pi}{4} (16)^2 = 2011 \text{ mm}^2$$

$d = 170$ mm; $b = 1000$ mm

$d_c = d_t = 30$ mm;

$$e = \frac{M}{P} = \frac{19.2}{160} = 0.12 \text{ m} = 120 \text{ mm}$$

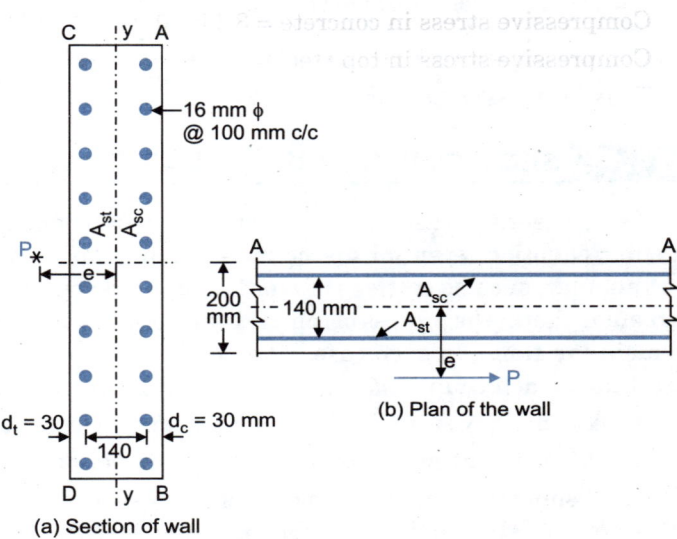

Fig. 14.25

Equating the sum of internal forces to external forces, we have, from *Eq. 14.36*:

$$m \frac{c'}{n}(d - n) A_{st} - \frac{bnc'}{2} - (m_c - 1) A_{sc} \frac{c'}{n}(n - d_c) = P$$

or

$$\frac{m}{n}(d - n) A_{st} - \frac{bn}{2} - (m_c - 1) A_{sc} \frac{n - d_c}{n} = \frac{P}{c'} \quad ...(1)$$

Similarly, equating the sum of internal moments about the centre of tensile steel to the moment of the pull, we have

$$\frac{bnc'}{2}\left(d - \frac{n}{3}\right) + (m_c - 1) A_{sc} \frac{c'}{n}(n - d_c)(d - d_c) = P\left(e - \frac{D}{2} + d_t\right)$$

or

$$\frac{bn}{2}\left(d - \frac{n}{3}\right) + (m_c - 1) A_{sc} \frac{(n - d_c)(d - d_c)}{n} = \frac{P}{c'}\left(e - \frac{D}{2} + d_t\right) \quad ...(2)$$

Dividing the two and substituting the numerical values, we get

$$\frac{\frac{1000}{2} n \left(170 - \frac{n}{3}\right) + (1.5 \times 19 - 1) 2011 \frac{(n - 30)(170 - 30)}{n}}{\frac{19}{n}(170 - n) 2011 - \frac{1000n}{2} - (1.5 \times 19 - 1) 2011 \frac{n - 30}{n}} = \left(120 - \frac{200}{2} + 30\right)$$

or

$$\frac{500 n \left(170 - \frac{n}{3}\right) + 7742350 \left(\frac{n - 30}{n}\right)}{\frac{38209}{n}(170 - n) - 500 n - 55303 \left(\frac{n - 30}{n}\right)} = 50$$

Solving this by trial and error we get $n \approx 38.9$ mm. Substituting this in *Eq. (1)*, we get

$$\frac{19}{38.9}(170-38.9)\,2011 - \frac{1000 \times 38.9}{2} - (1.5 \times 19 - 1) \times \frac{2011}{38.9}(38.9-30) = \frac{160000}{c'}$$

or
$$96668 = \frac{160000}{c'}, \text{ which gives } c' \approx 1.66 \text{ N/mm}^2 \text{(comp.)}$$

The stress in tensile steel is given by

$$t = m\,\frac{c'}{n}(d-n) = \frac{19 \times 1.66}{38.9}(170-38.9) = \mathbf{106.29\ N/mm^2}\ \text{(tensile)}$$

The stress in, compressive steel is given by

$$t_c = m_c\,\frac{c'}{n}(n-d_c) = 1.5 \times 19 \times \frac{1.66}{38.9}(38.9-30) = \mathbf{10.79\ N/mm^2}\ \text{(comp.)}$$

## 14.11. SECTIONS OF IRREGULAR SHAPE

We have seen in cases 3 and 6 that the position of neutral axis is determined by trial and error, solving a cubic equation, even for a section of rectangular cross-section. When the shape of cross-section is not rectangular, such as sections of $L$ or $T$ shapes, rectangular section bending across a diagonal or any other irregular shape, the process of computing $n$ becomes very tedius, though the basic principles of determination remain the same. In such cases, the position of neutral axis, and the maximum stresses in concrete and steel can be determined *step by step* in the following manner:

1. Assume any suitable value of $k$ and draw the neutral axis.
2. Divide the compression zone into a number of thin strips, each parallel to the neutral axis.
3. Assume the maximum compressive stress in concrete as $c'$, and determine the stresses at the middle of each strip, in terms of the maximum stress $c'$, from the stress diagram. Multiply these stresses by the corresponding areas of the strip, each assumed to be rectangular, and find compressive forces $C_1, C_2, C_3$, etc., in the strips. Also, find the compressive force in the compressive steel if any.
4. Find the moment of these compressive forces about the centre of tensile steel. Find also the moment of the external eccentric load, about the centre of tensile steel.
5. Equate both the moments, to get the maximum compressive stress $c'$.
6. Equate the algebraic sum of the compressive and tensile forces to the external load. This will give the tensile stress $t$ in tensile steel.
7. Knowing $c'$ and $t$ determine the values of $k$ from the stress relationship. If this agrees with the assumed value of $k$ within 5% then the stresses are correct. If not, asume some other suitable value of $k$ and follow steps 1 to 6, till the two values of $k$ do not differ by more than 5%.

## 14.12. DESIGN EXAMPLE

The design of eccentrically loaded sections, specially with moderate eccentricity ($D/4 < e < 3D/2$), is done by trial and error. A suitable section is first assumed and then tested for stresses in concrete and steel. Alternatively, design may be done with the help of suitable design charts which give the value of coefficient $k$, eccentricity ratio $e/d$, steel percentage $p$, cover ratio $a/d$ and compressive stress parameter. Such charts may be found in the book 'Reinforced Concrete Structures' by Peabody.

**Example 14.20.** *Design a column section to carry a direct load of 300 kN and a bending moment of 90 kN-m. The width of column is limited to 300 mm. Use M 20 concrete and Fe 415 steel.*

**Solution:** Eccentricity $e = 90 \times 1000/300 = 300$ mm. Let us keep total depth of column = 600 mm, so that eccentricity is equal to $0.5\,D$.

**(a) Design based on permissible stresses.** Let $A_{sc}$ be the area of steel in compression and $A_{st}$ be the area of steel in tension, each being provided at a cover of 60 mm to the centre of steel as shown in Fig. 14.26. Let $n_c$ be the depth of critical neutral axis, when the permissible stresses in both the materials reach simultaneously [Fig. 14.26(b)].

For M 20 concrete and Fe 415 steel combination, $\sigma_{st} = 230$ N/mm²; $c = \sigma_{cbc} = 7.0$ N/mm² and $m = 13.33$.

Hence $k_c = 0.289$. Provide 60 mm cover to the centre of steel.

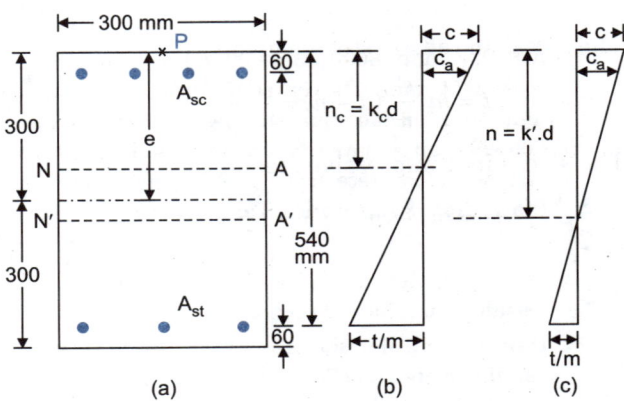

**Fig. 14.26**

∴ $d = 600 - 60 = 540$ mm

∴ $n_c = 0.289 \times 540 = 156.1$

Stress in concrete, surrounding compressive steel is given by

$$c_a = \frac{c}{n_c}(n_c - 60) = \frac{6}{156.1}(156.1 - 60)$$

$$= 3.694 \text{ N/mm}^2$$

Equating the moment of compressive forces about the centre of tensile steel, to the moment of external forces, we have:

$$(300 \times 156.1)\frac{7}{2}\left(540 - \frac{156.1}{3}\right) + (1.5 \times 13.33 - 1) A_{sc} \times 3.694(540 - 60) = 300000(300 + 300 - 60)$$

or $\quad 0.8 \times 10^8 + 33680 A_{sc} = 1.62 \times 10^8$, from which $A_{sc} = 2435$ mm²

Similarly, equating the algebraic sum of internal forces to the external forces, we get

$$300 \times 156.1 \times \frac{7}{2} + (1.5 \times 13.33 - 1) 2435 \times 3.694 - 13.33 A_{st} \times 230 = 300000.$$

From which $A_{st} = 11.34$ mm². ∴ Total area of steel $= A_{sc} + A_{st} = 2435 + 11.34 = 2446.3$ mm²

**(b) Alternative design.** It will be seen from the above design that very small steel area in tension is required. Hence if we lower down the value of $t$, keeping the value of $c = 7$ N/mm², N.A. will be shifted down, thus increasing the compressive area. It will thus be seen that a section designed on the basis of reduced value of $t$ is found to be more economical. When the value of eccentricity is about half the depth, the value of $t$ is kept between 60 to 110 N/mm². Let us keep $t = 80$ N/mm². The value of $k'$ corresponding to this is given by [Fig. 14.26(c)]:

$$k' = \frac{1}{1+\dfrac{t'}{m \cdot c}} = \frac{1}{1+\dfrac{80}{13.33 \times 7}} = 0.538; \; n = 0.538 \times 540 = 290.5 \text{ mm}$$

$$c_a = \frac{c}{n}(n - 60) = \frac{7}{290.5}(290.5 - 60) = 5.554 \text{ N/mm}^2$$

Equating the moment of compressive forces about the centre of tensile steel, to the moment of external force, we get

$$(300 \times 290.5)\frac{7}{2}\left(540 - \frac{290.5}{3}\right) + (1.5 \times 13.33 - 1) A_{sc} \times 5.554 (540 - 60) \doteq 300000 (300 + 300 - 60)$$

or $\quad 1.352 \times 10^8 + 50639 A_{sc} = 1.62 \times 10^8$, from which $A_{sc} \approx \mathbf{529 \text{ mm}^2}$

Similarly, equating the algebraic sum of internal forces to the external forces, we get

$$(300 \times 290.5)\frac{7}{2} + (1.5 \times 13.33 - 1) 529 \times 5.554 - 13.33 A_{st} \times 80 = 300000 \text{ or } A_{st} \approx \mathbf{57 \text{ mm}^2}$$

Hence total area of steel = 529 + 57 = 586 mm² against 2446.3 mm² found above. Hence it is economical to design the section on the reduced value of $t$. However, with increase in eccentricity, higher value of $t$ should be assumed. When eccentricity becomes more than 1.5 $D$, bending stress becomes predominent and hence $t$ = 230 N/mm² should be assumed.

## PROBLEMS

1. A R.C. column 600 mm × 600 mm is reinforced with 8 bars of 20 mm Φ at a cover of 50 mm to the centre of steel bars as shown in Fig. 14.27. Determine the minimum and maximum stresses in concrete if the column is subjected to a load of 800 kN at an eccentricity of 50 mm about one of the axes. Also, if the permissible stresses in concrete are: $\sigma_{cc}$ = 4 N/mm² and $c = \sigma_{cbc}$ = 5 N/mm² check whether the section is safe or not. Take $m$ = 19.
[**Ans.** 2.72 N/mm²; 1 N/mm²; safe]

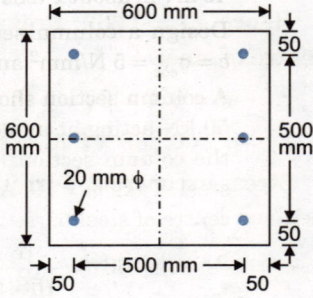

Fig. 14.27

2. Determine the maximum eccentricity of load in Problem 1 if the column is to be safe as uncracked section. [**Ans.** 156 mm]

3. If the column of problem 1 is subjected to an axial load of 800 kN along with a bending moment of 75 kN-m, determine the maximum and minimum stresses in concrete. Will the section be safe as uncracked section?
[**Ans.** 3.47 N/mm² (comp.) 0.27 N/mm² (comp.) safe]

4. Given the column section of solved Example 14.1. If it carries an axial load of 400 kN, what is the additional safe load it can carry at 50 mm eccentricity from one of the axes? Take $\sigma_{cc}$ = 4 N/mm² and $c = \sigma_{cbc}$ = 5 N/mm².
[**Ans.** 301 kN]

5. A rectangular section, 200 mm wide and 400 mm deep is reinforced with 8 bars of 20 mm Φ as shown in Fig. 14.28. Determine the maximum eccentricity about $x$-axis at which a load can be applied so that no tension is developed in the section. If $\sigma_{cc}$ = 4 N/mm², and $m$ = 19. Determine the maximum value of the load.
[**Ans.** 75 mm; 331 kN]

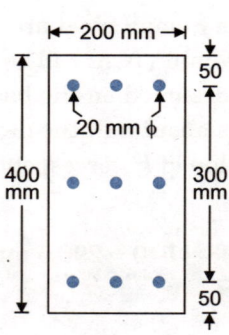

Fig. 14.28

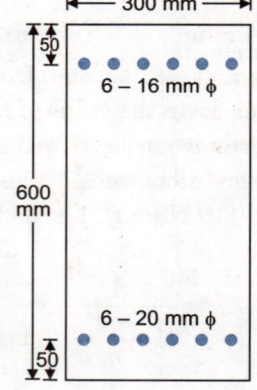

Fig. 14.29

6. Fig 14.29 shows the cross-section of an arch rib. Determine the normal thrust and bending moment on the section so that no tension is developed any where and the section is safe as uncracked section. Take $\sigma_{cc}$ = 4 N/mm²; $\sigma_{cbc}$ = 5 N/mm² and $m$ = 19. [**Ans.** 555 kN, 79.4 kN-m]

7. A reinforced concrete column of circular section, 300 mm diameter is reinforced with 8 bars of 16 mm Φ, arranged on a circle of 200 mm diameter. It is subjected to an axial load of 150 kN and B.M. of 0.5 kN-m. Determine the maximum and minimum stresses in concrete. Taking the permissible stresses as $\sigma_{cc}$ = 4 N/mm² and $c = \sigma_{cbc}$ = 5 kN/mm². Check whether the section is safe or not. Take $m$ = 19.
[**Ans.** 2.52 N/mm²(comp.); 0.09 N/mm²(comp).safe]

8. A two-hinged arch rib has a section 400 mm wide and 900 mm deep at the crown. The section is reinforced with 6 bars of 30 mm Φ at top and an equal reinforcement at the bottom. The reinforcements are placed at an effective cover of 50 mm from the respective edges. If the resultant thrust on the section is 1200 kN at 3° with

the tangent to arch centre line and acting on the vertical axis and 80 mm from the centre line of the section, determine the stresses in concrete and steel at the top and bottom of steel. Take $m = 18$. If $\sigma_{cc} = 4$ N/mm² and $\sigma_{cbc} = 5$ N/mm², determine whether the section is safe or not.

[**Ans.** Max. stress in concrete = 2.78 N/mm²; Min. Stress = 1.34 N/mm²) (comp.);

Max stress in steel = 73 N/mm²(comp.) Min. stress = 44.1 N/mm²(comp.); safe].

9. Given the section of problem 1. If the section is subjected to (*i*) axial load of 800 kN (*ii*) bending moment of 40 kN-m about *x*-axis and (*iii*) B.M. of 30 kN-m about *y*-axis, check whether the section is safe or not. [**Ans.** safe]

10. Design a column section to carry an axial load of 300 kN and B.M. of 7.5 kN-m. Take $\sigma_{cc} = 4$ N/mm², $c = \sigma_{cbc} = 5$ N/mm² and $m = 18$.

11. A column section shown in Fig. 14.30 is subjected to a load of 50 kN acting at an eccentricity of 1 m from the centre line of the column section. Determine the stresses in concrete and steel. Take $m = 19$. Assume an effective cover of 50 mm to the centre of steel bars.

[**Ans.** (*i*) Stress in concrete = 2.06 N/mm² (comp.)

(*ii*) Stress in comp steel = 48.9 N/mm² (comp.)

(*iii*) Stress in tensile steel = 29.8 N/mm² (tensile).]

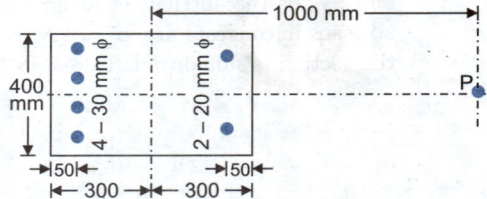

Fig. 14.30

12. The section of an arch rib is 400 mm wide and 900 mm deep, and is reinforced with 6 bars of 30 mm Φ at the top and an equal reinforcement at from the bottom. The reinforcements are placed at an effective cover of 50 mm from the respective edges. Determine the maximum stresses in concrete and steel when the section is subjected to a thrust of 500 kN and a B.M. of 350 kN-m. Take $m = 19$. [**Ans.** 0.64 N/mm²(comp.), 33 N/mm²(tensile)]

13. The suspender of a bow string girder, 200 mm × 300 mm is reinforced with 4 bars of 16 mm at each face placed at an effective cover of 40 mm. If it is subjected to a pull of 100 kN and B.M. of 6 kN-m determine the stresses in steel reinforcements. [**Ans.** 96.1 N/mm² (tensile); 28.3 N/mm²(tensile)]

14. The slab of a tank is 240 mm thick and is reinforced with 8 bars of 20 mm Φ per meter length on each face as shown in Fig. 14.31. The slab is subjected to a pull of 60 kN and a B.M. of 36 kN-m per meter width. Determine the stresses in concrete and steel. Take $m = 19$.

[**Ans.** Comp. stress in concrete = 3.06 N/mm²,

Comp. stress in steel = 41.41 N/mm².

Tensile stress in steel = 104.4 N/mm².]

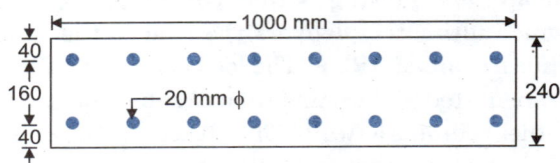

Fig. 14.31

15. Design spiral column of 400 mm diameter. Subjected to a factored compressive load of 260 kN and a factored bending moment of 78 kNm. Assess the safety of column section when subjected to a factored load.

In case of the section is found unsafe, redesign the reinforcement.

16. A corner columns of (340 × 340 mm) located in lower most story of a system of braced frames are subjected to factored loads of $P_u = 1380$ kN, $M_{ux} = 189$ kNm and $M_{uy} = 130$ kNm the unsupported length of column is 3.25 m. Design the reinforcement in column, assuming $M_{30}$ Grade concrete Fe 415 steel.

17. Determine the reinforcement required for a column with the given data and which is restrained against sway.

Size of column = 450 mm × 450 mm

Concrete grade = $M_{30}$

$F_y = 415$ N/mm² and $l_{ex} = 7.0$ mtrs

$l_{ey} = 6.2$ mtrs unsupported length = 7.0 mm.

Factored load ($P_u$) = 1630 kN, factored moment is 48 kNm at top and 32 kN at bottom, in the direction of longer dimension factored moment in direction of shorter dimension = 30 kNm at top and 28 kNm at bottom.

# CHAPTER 15
# DESIGN OF ISOLATED FOOTINGS

## 15.1. INTRODUCTION

A structure is generally considered to have two main portions (*i*) the *super-structure* and (*ii*) the *substructure*. The substructure transmits the loads of super-structure to the supporting soil and is generally termed as the *foundation*. *Footing* is that portion of the foundation which ultimately delivers the load to the soil, and is thus in contact with it. The load of the superstructure is transmitted to the foundation or sub-structure through either columns or walls. The object of providing foundation to a structure is to distribute the load to the soil in such a way that the maximum pressure on the soil does not exceed its permissible bearing value, and at the same time the settlement is within the permissible limits.

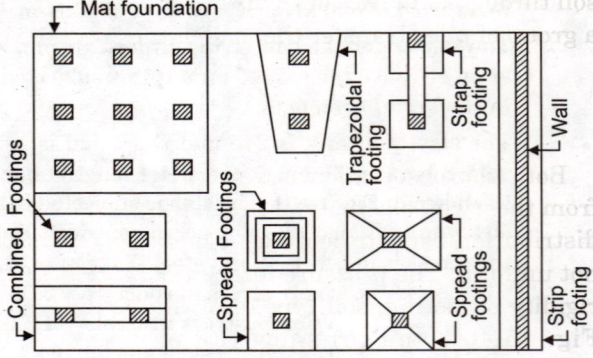

Fig. 15.1. Various Types of Shallow Footings.

Foundations may be broadly classified under two heads: *shallow foundation and deep foundation*. According to Terzaghi, a foundation is shallow if its depth is equal to or less than its width. In the case of deep foundation, the depth is greater than the width. Apart from deep strip, rectangular or square foundations, other common forms of deep foundations are: *pier foundation, pile foundation and well foundation*. The shallow foundations are of the following types: Spread footing, strap footing, combined footing and mat or raft footing. Fig. 15.1 shows the common types of shallow foundations.

**Spread footings.** A spread footing or simply footing, is a type of shallow foundation used to transmit the load of an isolated column, or that of a wall, on the subsoil. In the case of wall, the footing is *continuous* which in the case of column, it is *isolated*. Fig. 15.2 shows some common types of reinforced concrete spread footing.

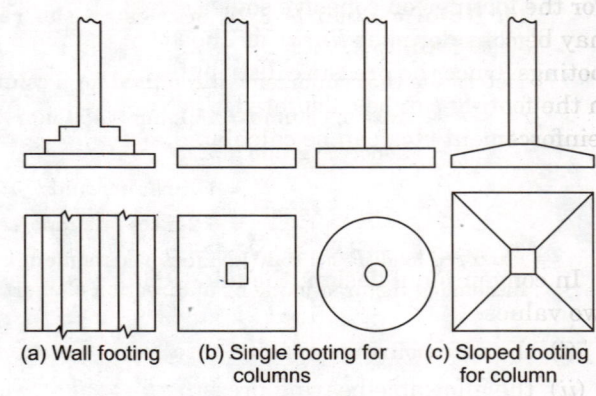

(a) Wall footing  (b) Single footing for columns  (c) Sloped footing for column

Fig. 15.2. Typical Spread Footings.

**Combined footings.** A spread footing which supports two or more columns is termed as a *combined footing*. Such a footing is provided when the individual footings are either very near to each other, or overlap. Combined footings may either be rectangular or trapezoidal (Fig. 15.3 a, b).

**Strap of cantilever footing.** A strap footing consists of spread footings of two columns connected by a strap beam. The strap beam does not remain in contact with soil, and thus does not transfer any pressure to the soil. Such a footing is generally used to combine the footing of outer column to the adjacent one so that the footing of the former does not extend in the adjoining property [Fig. 15.3 (c)].

**Mat or raft foundation.** A mat or raft is a combined footing that covers the entire area beneath a structure and supports all the walls and columns. When the available soil pressure is low or the building loads are heavy, the use of spread footings would cover more than one-half of the area and it may prove more economical to use mat or raft foundation.

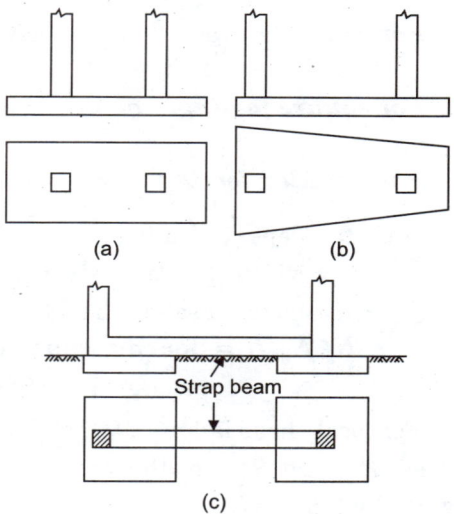

Fig. 15.3

**Pile foundation.** Pile foundation is a deep foundation used where the top soil is relatively weak. Piles transfer the load to a lower stratum of greater bearing capacity, by way of end bearing, or to the intermediate soil through skin friction. This is most common type of deep foundation generally used for buildings where a group of piles transfer the load of the super-structure to the sub-soil.

## 15.2. PRESSURE DISTRIBUTION BENEATH FOOTINGS

Both from observation as well as the analytical studies from theory of elasticity, it is known that the pressure distribution beneath footings, symmetrically loaded is not uniform. The pressure intensities depend upon the rigidity of footing, soil type, and the condition of soil. Fig. 15.4(a) and (b) show the probable pressure distribution beneath a rigid footing on a loose cohesionless soil and cohesive soil. Fig. 15.4(c) shows the usually assumed uniform pressure distribution.

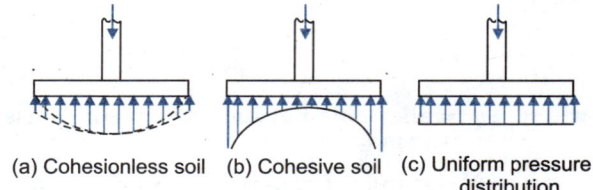

(a) Cohesionless soil  (b) Cohesive soil  (c) Uniform pressure distribution

Fig. 15.4. Pressure Distribution Beneath Footings.

When a rigid footing rests on cohesionless soil, the soil grains at the outer edge have no lateral restraint, whereas in centre the soil is relatively confined resulting in a pressure distribution shown in Fig. 15.4(a). For the footings on cohesive soils, the edge stresses may be very large. However, the pressure distribution may be considered as *linear* as shown in Fig. 15.4(c), for the purpose of the design of reinforced concrete footings. Once the pressure distribution is known, the bending moment and shear force at various locations in the footing can be calculated, and the thickness of the structural member of the footing, alongwith the reinforcement etc., can be calculated using the usual principles of reinforced concrete.

## 15.3. BEARING CAPACITY OF SOIL AND SETTLEMENT OF FOOTINGS

In conventional design, the allowable bearing capacity should be taken as the smaller of the following two values:

(*i*) the safe bearing capacity based on ultimate capacity.

(*ii*) the allowable bearing pressure based on tolerable settlement.

The safe bearing capacity may be determined from a plate load test. Alternatively, the safe bearing capacity $q_0$ may be computed from the following equations based on Terzaghi's analysis:

**For strip footing,** $\quad q_0 = \dfrac{1}{F} [cN_c + \gamma D (N_q - 1) R_{w1} + 0.5\gamma BN_\gamma R_{w2}] + \gamma D$ ...(15.1)

**For square footing,** $\quad q_0 = \dfrac{1}{F} [1.3 cN_c + \gamma D (N_q - 1) R_{w1} + 0.4\gamma BN_\gamma R_{w2}] + \gamma D$ ...(15.2)

**For circular footing,** $q_0 = \dfrac{1}{F} [1.3 cN_c + \gamma D (N_q - 1) R_{w1} + 0.3\gamma BN_\gamma R_{w2}] + \gamma D$ ...(15.3)

where  $D$ = depth of footing
$B$ = width of footing (strip or square) or diameter of circular footing
$F$ = factor of safety (2 to 3)
$N_c, N_q, N_\gamma$ = Bearing capacity factors for general shear failure.

For local shear failure, $N_c'$, $N_q'$, and $N_\gamma'$ should be used. $R_{w1}$ and $R_{w2}$ are the water reduction factors given by (Fig. 15.5).

$$R_{w1} = 0.5 \left(1 + \dfrac{Z_{w1}}{D}\right) \quad ...[15.4\,(a)]$$

$$R_{w2} = 0.5 \left(1 + \dfrac{Z_{w2}}{B}\right) \quad ...[15.4\,(b)]$$

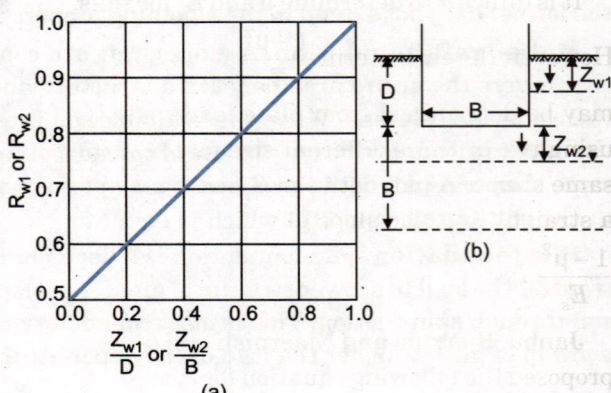

Fig. 15.5. Water Reduction Factors.

The allowable bearing pressure $q_\rho$ based on maximum settlement of individual footing to 25 mm, can be computed from the following empirical relation:

$$q_\rho = 34.3 (N - 3) \left(\dfrac{B + 0.3}{2B}\right)^2 \cdot R_{w2} R_d. \qquad ...(15.5)$$

where  $q_\rho$ = allowable net increase in soil pressure over existing soil pressure for settlement of 25 mm, kN/m$^2$
$N$ = standard penetration number with applicable corrections
$B$ = width of footings, in metres
$R_{w2}$ = water reduction factor
$R_d$ = depth factor = $\left(1 + \dfrac{0.2 D}{B}\right) \leq 1.20$

If the safe bearing capacity is taken to be smaller of the two values discussed above, the footings on granular soils will not suffer detrimental settlement. The total settlement of a footing in clay may be considered to consist of three components (Skempton and Bjerrum, 1957):

$$S = S_i + S_c + S_s \qquad ...(15.6)$$

where  $S$ = total settlement
$S_i$ = immediate elastic settlement
$S_c$ = consolidation settlement
$S_s$ = settlement due to secondary consolidation of clay.

The immediate settlement $S_i$ is the elastic settlement, and can be computed from the following expression based on the theory of elasticity:

$$S_i = q B \left(\dfrac{1-\mu^2}{E_s}\right) I_w \qquad ...(15.7)$$

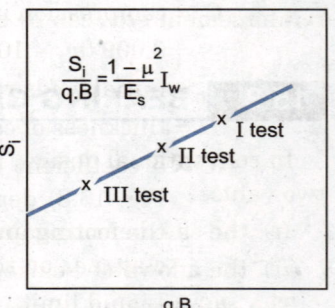

Fig. 15.6

where  $q$ = intensity of contact pressure (kN/m²)
  $B$ = least lateral dimension of footing, metres
  $E_s$ = modulus of elasticity for soil, (kN/m²)
  $I_w$ = influence factor,
   = 0.88 for rigid circular footing.
   = 0.82 for rigid square footing.
   = 1.06 for rigid rectangular footing with $L/B$ = 1.5
   = 1.70 for rigid rectangular footing width $L/B$ = 5

It is difficult to determine $\mu$ and $E_s$ for soils. However, the entire term $\dfrac{1-\mu^2}{E_s} I_w$ of Eq. 15.7 may be determined from plate load tests, by using two or three different size plates of the same shape. A plot between $S_i$ and $q \cdot B$ give a straight line, the slope of which is equal to $\dfrac{1-\mu^2}{E_s} I_w$.

Janbu, Bjerrum and Kjaernisli (1966) have proposed the following equation for computing the immediate settlement:

$$S_i = \mu_0 \cdot \mu_1 \cdot q\, B\, \frac{1-\mu^2}{E_s} \qquad \ldots(15.8)$$

where the modulus of elasticity $E_s$ is computed from the triaxial test data:

$$E_s = \frac{\sigma_1 - \sigma_3}{\Delta L/L} \qquad \ldots(15.9)$$

and $\mu_0$ and $\mu_1$ are taken from Fig. 15.7.

The *consolidation settlement* $S_c$ is computed from the following expression:

$$S_c = C\, \frac{C_c}{1+e_0} \cdot H \log_{10} \frac{\sigma_0 + \Delta \sigma}{\sigma_0} \qquad \ldots(15.10)$$

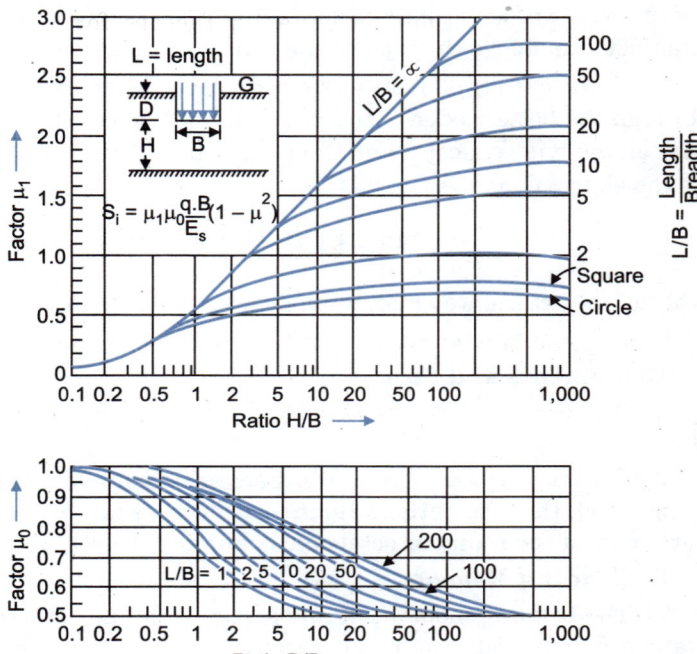

Fig. 15.7.  Value of $\mu_0$ and $\mu_1$
(After Janbu, Bjerrum and Kjaernisli, 1966)

where  $\sigma_0$ = effective overburden pressure due to soil over burden
  $\Delta\sigma$ = vertical stress on footing
  $C_c$ = compression index
   = 0.009 $(w_L - 10)$
  $e_0$ = initial voids ratio
  $H$ = thickness of compressible layer.
  $C$ = a coefficient or correction factor (given by Fig. 15.8) depending upon the geometry of the footing and the history of loading on the clay (i.e., on the pore pressure coefficient $A$).
  $w_L$ = Liquid limit.

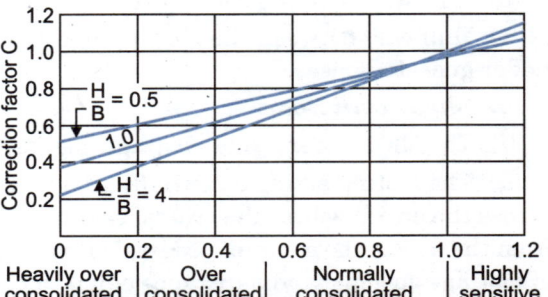

Fig. 15.8.  Value of Factor C.

## 15.4. INDIAN STANDARD CODE RECOMMENDATIONS FOR DESIGN OF FOOTINGS (IS: 456-2000)

### (A) General

1. In sloped or stepped footings, the effective cross-section in compression shall be limited by the area above the neutral plane, and the angle of slope or depth and location of steps shall be such that the design requirements are satisfied at every section. Sloped and stepped footings that are designed as a unit shall be constructed to assure action as a unit.

2. *Thickness at the edge of footing.* In reinforced and plain concrete footings, the thickness at edge shall be not less than 150 mm for footings on soils nor less than 300 mm above the tops of piles for footings on piles.

3. In the case of plain concrete pedestals, the angle $\alpha$ between plane passing through the bottom edge of the pedestal and the corresponding junction edge of the column with pedestal and the horizontal plane (Fig. 15.9) shall be governed by the expression:

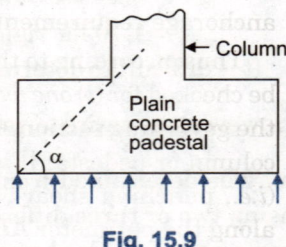

Fig. 15.9

$$\tan \alpha \not< 0.9 \sqrt{\frac{100 q_0}{f_{ck}} + 1} \qquad \ldots(15.11)$$

where  $q_0$ = calculated maximum bearing pressure at the base of the pedestal in N/mm²,
and  $f_{ck}$ = characteristic strength of concrete at 28 days in N/mm².

### (B) Moments and Forces

1. In the case of footings on piles, computation for moments and shears may be based on the assumption that the reaction from any pile is concentrated at the centre of the pile.

2. For the purposes of computing stress in footings which support a round or octagonal concrete column or pedestal, the face of the column or pedestal shall be taken as the side of a square inscribed within the perimeter of the round or octagonal column or pedestal.

3. **Bending Moment**

(*i*) The bending moment at any section shall be determined by passing through the section a vertical plane which extends completely across the footing and computing the moment of the forces acting over the entire area of the footing on one side of the said plane.

(*ii*) The greatest bending moment to be used in design of an isolated concrete footing which supports a column, pedestal or wall, shall be the moment computed in the manner prescribed in (*i*) at sections located as follows:

(*a*) At the face of the column, pedestal or wall, for footings supporting a concrete column, pedestal or wall,

(*b*) Half way between the centre line and the edge of the wall, for footings under masonry walls, and

(*c*) Half way between the face of the column or pedestal and the edge of the gussetted base, for footing under gusseted bases.

4. **Shear and Bond**

(*i*) The shear strength of footings is governed by the more severe of the following two conditions:

(*a*) The footing acting essentially as a wide beam, with a potential diagonal crack extending in a plane across the entire width, the critical section for this condition shall be assumed as a vertical section located from the face of the column, pedestal or wall at a distance equal to the effective depth of the footing in case of footings on soils, and a distance equal to half the effective depth of footing for footings on piles.

(*b*) Two-way action of the footing, with potential diagonal cracking along the surface of truncated cone or pyramid around the concentrated load: in this case, the footing shall be designed for shear in accordance with appropriate provisions discussed below (Fig 15.10)

(*ii*) In computing the external shear on any section through a footing supported on piles, the entire reaction from any pile of diameter $D_p$ whose centre is located $D_p/2$ or more outside the section shall be assumed as producing shear on the section; the reaction from any pile whose centre is located $D_p/2$ or more inside the section shall be assumed as producing no shear on the section. For intermediate positions of the pile centre, the portion of the pile reaction to be assumed as producing shear on the section shall be based on straight line interpolation between full value at $D_p/2$ outside the section and zero value at $D_p/2$ inside the section.

(iii) The critical section for checking the development length in a footing shall be assumed at the same planes as described for bending moment in B(3) and also at all other vertical planes where abrupt changes of section occur. If the reinforcement is curtailed, the anchorage requirements should be checked.

Thus, according to the above provision, shear stress is to be checked for (i) *one way action* (i.e., beam shear) for which the governing section AB is at a distance $d$ from the face of column or pedestal [Fig. 15.10(a)] and (ii) *two way action*, (i.e., punching shear), for which the governing section is along the perimeter ABCD situated at a distance $d/2$ from the face of the column or pedestal [Fig. 15.10(b)].

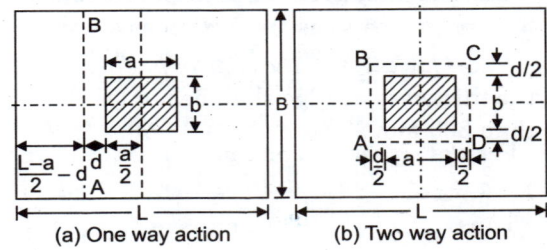

Fig. 15.10. Critical Section for Shear

For the *two way action*, the calculated shear stress $\tau_v$ should satisfy the following relation

$$\tau_v \le k_s \cdot \tau_c \qquad \text{...[15.12 (a)]}$$

where $\tau_v = \left( \dfrac{F}{2[(a+d)+(b+d)]\,d} \right) \qquad \text{...[15.12 (b)]}$

$k_s = (0.5 + \beta_c)$, but not greater than 1.0

$\beta_c = \dfrac{b}{a} = \left( \dfrac{\text{short side of column}}{\text{long side of column}} \right) \qquad \text{...[15.12 (c)]}$

$\tau_c = 0.16 \sqrt{f_{ck}}$ N/mm² in working stress method  ...[15.12 (d)]

$= 0.25 \sqrt{f_{ck}}$ in limit state design

and  $F$ = net S.F. acting on the perimeter

For the *beam shear*, the nominal shear stress across AB should satisfy the relation

$$\tau_v \le k\,\tau_c \qquad \text{...(15.13)}$$

where $\tau_c$ = the permissible shear stress for the grade of the concrete, corresponding to the reinforcement $(100\,A_s/bd)$ as given in Table 3.1.

$k$ = factor for slabs, as given in Table 3.2.

### (C) Tensile Reinforcement

The total tensile reinforcement at any section shall provide a moment of resistance at least equal to the bending moment on the section calculated in accordance with B(3). Total tensile reinforcement shall be distributed across the corresponding resisting section as given below.

(i) In one-way reinforced footing, the reinforcement shall be distributed uniformly across the full width of the footing;

(ii) In two-way reinforced square footing, the reinforcement extending in each direction shall be distributed uniformly across the full width of the footing;

(iii) In two-way reinforced rectangular footing, the reinforcement in long direction shall be distributed uniformly across the full width of the footing. For reinforcement in the short direction, a central band equal to the width of the footing shall be marked along the length of the footing and portion of the reinforcement determined in accordance with the equation given below shall be uniformly distributed across the central band:

$$\dfrac{\text{Reinforcement in central band width}}{\text{Total reinforcement in short direction}} = \dfrac{2}{\beta+1}$$

where  $\beta$ is the ratio of the long side to the short-side of the footing.

The remainder of the reinforcement shall be uniformly distributed in the outer portions of the footing.

### (D) Transfer of load at the base of column

The compressive stress in concrete at the base of a column or pedestal shall be considered as being transferred by bearing to the top of the supporting pedestal or footing. The bearing pressure ($\sigma_{cbr}$) on the loaded area shall not exceed the permissible bearing stress in direct compression multiplied by a value equal to $\sqrt{A_1/A_2}$ but not greater than 2.

where $A_1$ = supporting area for bearing, of footing, which in sloped or stepped footing may be taken as the area of the lower base of the largest frustrum of a pyramid or cone contained wholly within the footing and having for its upper base, the area actually loaded and having side slope of one vertical to two horizontal, and

$A_2$ = Loaded area at the column base.

For working stress method of design, the permissible bearing stress on full area of concrete shall be taken as $0.25 f_{ck}$; for limit state method of design the permissible bearing stress shall be $0.45 f_{ck}$.

Thus  $\sigma_{cbr} \leq 0.25 f_{ck} \sqrt{A_1/A_2}$ in working stress method  ...[15.14 (a)]

and  $\sigma_{cbr} \leq 0.45 f_{ck} \sqrt{A_1/A_2}$ in limit state method  ...[15.14 (b)]

The actual bearing pressure $\sigma_{cbr}$ = column load divided by the area of column at the base.

Thus  $\sigma_{cbr} = \dfrac{W}{a \times b}$  where $a$ and $b$ are the sides of the column.

(1) Where the permissible bearing stress on the concrete in the supporting or supported member would be exceeded, reinforcement shall be provided for developing excess force, either by extending the longitudinal bars into the supporting members or by dowels (see 3 below).

(2) Where transfer of force is accomplished by reinforcement, the development length of the reinforcement shall be sufficient to transfer the compression or tension to the supporting member.

(3) Extended longitudinal reinforcement or dowels of at least 0.5 percent of the cross-sectional area of the supported column or pedestal and a minimum of four bars shall be provided. Where the dowels are used, their diameter shall not exceed the diameter of the column bars by more than 3 mm.

(4) Column bars of diameters larger than 36 mm, in compression only can be dowelled at the footings with bars of smaller size of the necessary area. The dowel shall extend into the column, a distance equal to the development length of the column bar and into the footing, a distance equal to the development length of the dowel.

### (E) Nominal cover:
For footing, minimum nominal cover shall be 50 mm.

### (F) Nominal reinforcement:
(a) The minimum reinforcement and spacing shall be as per requirements of solid slab.

(b) The nominal reinforcement for concrete sections of thickness greater than 1 m shall be 360 mm² per meter length in each direction on each face. This provision does not supercede the requirement of minimum tensile reinforcement based on the depth of the section.

## 15.5. DESIGN OF CONTINUOUS FOOTINGS

A continuous footing is provided under walls which may be either of masonry or of concrete. For design of reinforced concrete continuous footings, it is convenient to consider the forces on unit length.

### 1. CONTINUOUS FOOTING UNDER MASONRY WALL

Let  $b$ = width of wall (metres).

$B$ = width of footing

$q_0$ = safe bearing capacity of soil (kN/m²)

$W$ = load from wall, per metre run (kN/m)

$W'$ = weight of footing in kN per metre run.

Then, the width of footing is given by

$$B = \left(\frac{W + W'}{q_0}\right)$$

Fig 15.11(b) shows the pressure diagram.

The *net* upward soil pressure $p_0$ on the footing will be

$$p_0 = \frac{W}{B \times 1} = \frac{W}{B}$$

The pressure intensity $p$ at the contact of wall and footing is given by

$$p = \frac{W}{b \times 1} = \frac{W}{b}$$

The entire footing can be considered to consist of two cantilevers, each projecting out by a length $\frac{1}{2}(B-b)$, and having a fixed support length $\frac{b}{2}$. The maximum B.M. is considered to act at a section midway between the edge of the wall and centre of the wall *i.e.*, at a distance $\frac{b}{4}$ from the edge of the wall. Its magnitude per metre length is thus given by

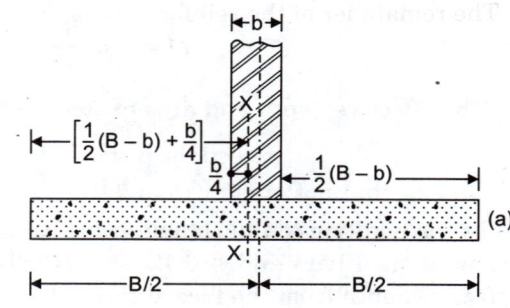

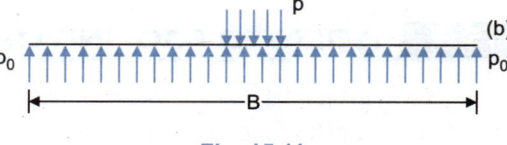

Fig. 15.11

$$M = p_0 \frac{1}{2}\left[\frac{B-b}{2} + \frac{b}{4}\right]^2 - p\frac{1}{2}\left[\frac{b}{4}\right]^2 = \frac{W}{2B}\left[\frac{2B-b}{4}\right]^2 - \frac{W}{2b} \cdot \frac{b^2}{16}$$

$$= \frac{W}{2B}\left[\frac{4B^2 + b^2 - 4bB}{16}\right] - \frac{Wb}{32} = \frac{W}{32}\left[4B - 4b + \frac{b^2}{B} - b\right] = \frac{W}{32}\left[4B - 5b + \frac{b^2}{B}\right]$$

$$= \frac{W}{32B}[4B^2 - 5Bb + b^2] = \frac{W}{32B}(4B-b)(B-b) = \frac{W}{8B}(B-b)\left[B - \frac{b}{4}\right] \text{ kN-m}$$

$$= \frac{p_0}{8}(B-b)\left[B - \frac{b}{4}\right] \times 10^6 \text{ N-mm} \qquad ...(15.15)$$

∴ Effective depth $d = \sqrt{\dfrac{M}{1000\,R_c}}$ where $R_c = \dfrac{1}{2}\sigma_{cbc} \cdot j_c\, k_c$

A minimum thickness of 150 mm should be provided at the edges.

Also $A_{st} = M/tj_c d$ where $t = \sigma_{st}$. In addition to this, longitudinal reinforcement is provided @ 0.15% of the area of section of the footings for mild steel bars and @ 0.12% for HYSD bars.

**Check for shear.** The footing slab is checked for *one way action* (*i.e.*, beam shear) for which the critical section will be at a distance of effective depth from the face of the wall. The depth of footing at that section must be sufficient to keep nominal shear stress within safe limits.

## 2. CONTINUOUS FOOTING UNDER CONCRETE WALL

A concrete footing under a concrete wall is cast monolithic with it. Hence the maximum B.M. will occur at the face of the wall.

As found earlier: $B = \dfrac{W + W'}{q_0}$;

$$p = \frac{W}{b}\,;\ p_0 = \frac{W}{B}$$

∴ Maximum B.M. per metre length is given by

$$M = p_0 \frac{1}{2}\left[\frac{B-b}{2}\right]^2 \text{ kN-m}$$

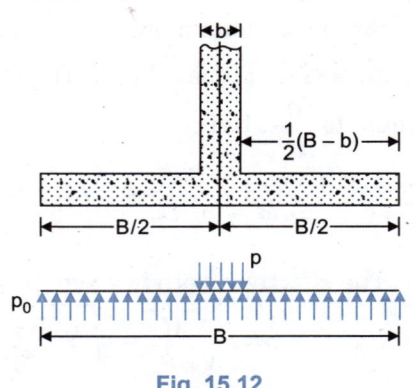

Fig. 15.12

or
$$M = p_0 \frac{(B-b)^2}{8} \times 10^6 \text{ N-mm} \qquad ...(15.16)$$

The effective depth and area of steel reinforcement are then given by
$$d = \sqrt{\frac{M}{1000 R_c}} \text{ and } A_{st} = \frac{M}{\sigma_{st} j_c d}$$

In the addition to this, longitudinal reinforcement is provided @ 0.15% of cross-sectional area of footing for mild steel bars and @ 0.12% for HYSD bars. The section is checked for beam shear at a distance of effective depth from the face of the wall.

## 15.6. ISOLATED FOOTING OF UNIFORM DEPTH

The total area at the base of such footing is determined from the point of view of the safe bearing capacity of soil. The depth of footing is determined both from the considerations of maximum bending moments as well as from punching shear.

Let     $W$ = weight transferred form the column (kN)
   $W'$ = weight of footing
   $q_0$ = safe bearing capacity of soil (kN/m²)
   $A$ = area of footing at the base.

Then $\qquad A = \dfrac{W + W'}{q_0}$.

If $p_0$ = net upward pressure at footing,

then $\qquad p_0 = \dfrac{W}{A} = \dfrac{W q_0}{W + W'}$

The weight $W'$ of the footing may be assumed to the equal to 10% $W$, for preliminary calculations.

### 1. SQUARE FOOTING

Fig. 15.13 shows a square footing of uniform thickness (depth) $D$.

The width $B$ of the footing will evidently be equal to $\sqrt{\dfrac{W+W'}{q_0}}$.

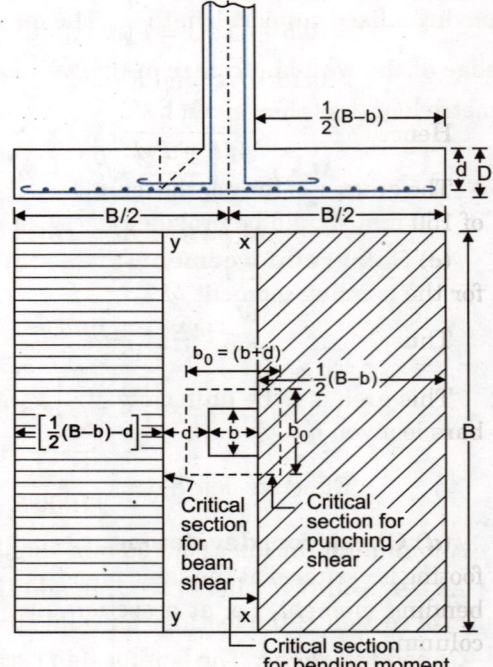

Fig. 15.13

The thickness is calculated both from the considerations of bending moment as well as punching shear.

**(a) Depth for bending moment**

Since the column and footing are cast monolithic, maximum B.M. will take place about section $X-X$ near the face of the column. The unit pressure $p_0$ causing bending of cantilever of length $\frac{1}{2}(B-b)$ will be equal to $\dfrac{W}{B^2}$. Hence,

$$M = p_0 B \left(\frac{B-b}{2}\right)\left(\frac{B-b}{4}\right) = p_0 \frac{B}{8}(B-b)^2 \text{ kN-m} = \frac{p_0 B}{8}(B-b)^2 \times 10^6 \text{ N-mm} \qquad ...(15.17)$$

The effective depth is given by
$$d = \sqrt{\frac{M}{BR_c}} = \sqrt{\frac{p_0 (B-b)^2 \, 10^6}{8 R_c}} \text{ where } R_c = \frac{1}{2}\sigma_{cbc} \cdot j_c \cdot k_c$$

### (b) Depth for shear

The footing slab is checked both for *one way action* (i.e., beam shear) and *two way action* (i.e., punching shear).

For *one way action*, the critical section Y-Y is situated at a distance $d$ from the face of the column. The nominal shear stress $\tau_v \left( = \dfrac{V}{B.d} \right)$ is equated to the permissible shear stress $k\tau_c$ found from Table 3.1 for the grade of concrete corresponding to the reinforcement $\left( 100 \dfrac{A_s}{bd} \right)$. Thus

$$\frac{V}{Bd} \leq k\tau_c \qquad \ldots [15.18\,(a)]$$

From this the effective depth $d$ is found.

For *two way action*, the calculated shear stress $\tau_v$ computed from *Eq. 15.12 (b)* should be less than $k_s \tau_c$.

Thus, $$\frac{F}{4\,[b+d]\,d} \leq k_s \cdot \tau_c$$

where $k_s = (0.5 + \beta_c) \leq 1$

$\qquad = (0.5 + 1) \leq 1 = 1$ (maximum value)

$\tau_c = 0.16 \sqrt{f_{ck}}$ and $F$ = net S.F. acting on the periphery $= p_0\,[B^2 - (b+d)^2]$

Hence, $$\frac{F}{4\,(b+d)\,d} \leq 0.16 \sqrt{f_{ck}} \qquad \ldots [15.18\,(b)]$$

The greater of the depths found from *Eqs 15.18 (a)* and *15.18 (b)* should be provided. A minimum thickness of 150 mm should be provided in case $d$ comes out to be less than 150 mm.

### (c) Steel reinforcement.
Steel reinforcement is provided in both the directions, in the form of mesh, for the bending moment $M$.

Thus, $$A_{st} = \frac{M}{tj_c\,d}, \text{ where } t = \sigma_{st}$$

This area is to be uniformly distributed over a width $B$. If $A_\phi$ is area of each bar chosen, the spacing of bars is given by

$$s = \frac{B \cdot A_\phi}{A_{st}}$$

### (d) Check for development length.
The critical section for checking the development length in the footing is assumed at the same plane as used for maximum bending moment, i.e. at a section along the face of the column.

If the reinforcement is curtailed, the anchorage requirements should be checked.

### 2. RECTANGULAR FOOTING

In the case of rectangular footing, the proportioning of the dimensions $L$ and $B$ should be done in such a way that the bending moments in each of the adjacent projections is equal to the moment of resistance of the footing. The C.G. of footing area should invariably coincide with the C.G. of column area. Fig. 15.14 shows a column of size $a \times b$. Sometimes, a rectangular footing may be provided even for a square column, where there are some restrictions on the maximum value of the width of the footing. The thickness of the footing is determined both form the B.M. as well as shear point of view.

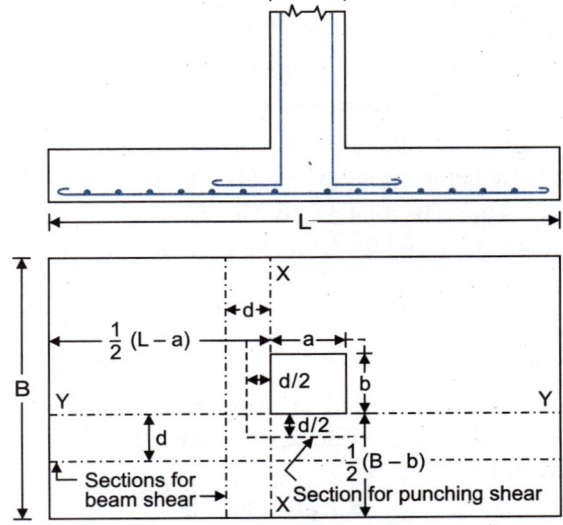

Fig. 15.14

## (a) Depth for bending moment

Bending moment $M_1$ about section XX is given by

$$M_1 = p_0 B \left(\frac{L-a}{2}\right)\left(\frac{L-a}{4}\right) = \frac{p_0 B (L-a)^2}{8} \times 10^6 \text{ N-mm} \quad \ldots(15.19)$$

$$d = \sqrt{\frac{M_1}{R_c B}} = \sqrt{\frac{p_0 (L-a)^2 \times 10^6}{8 R_c}} \quad \ldots(15.20)$$

where $R_c = \frac{1}{2} \sigma_{cbc} \cdot j_c \cdot k_c$

Bending moment $M_2$ about section Y-Y is given by

$$M_2 = p_0 L \left(\frac{B-b}{2}\right)\left(\frac{B-b}{4}\right) = \frac{p_0 L}{8} (B-b)^2 \times 10^6 \text{ N-mm} \quad \ldots(15.21)$$

$$\therefore \quad d = \sqrt{\frac{M_2}{R_c L}} = \sqrt{\frac{p_0 (B-b)^2 \times 10^6}{8 R_c}} \quad \ldots(15.22)$$

## (b) Depth for shear

From shear point of view, the depth is found for both *one way action* (*i.e.* beam section) as well as *two way action* (*i.e.* punching shear action).

For one way action, the sections are situated at a distance $d$ from each face of the column. The nominal shear stress $\tau_v$ (= $V/BD$ or $V/LD$) should be less than permissible shear $k \tau_c$ found from Table 3.1 for the grade of concrete corresponding to reinforcement ratio (100 $A_s/bd$). To begin with, the reinforcement ratio is arbitrarily selected equal to the normal value.

Thus,
$$\frac{V}{Bd} \text{ or } \frac{V}{L \cdot d} \leq k \tau_c \quad \ldots[15.23\,(a)]$$

From which effective depth is found. For *two way action*, the calculated shear stress ($\tau_v$) should satisfy the following relation:

$$\tau_v \leq k_s \cdot \tau_c$$

where $\tau_v = \dfrac{F}{2[(a+d)+(b+d)]\,d}$ ; $k_s = \left(0.5 + \dfrac{b}{a}\right) \leq 1$

$\tau_c = 0.16 \sqrt{f_{ck}}$ and $F$ = net S.F. at the periphery = $p_0 [(B \times L) - (a+d)(b+d)]$

Hence,
$$\frac{F}{2[(a+d)+(b+d)]\,d} \leq 0.16 \, k_s \sqrt{f_{ck}} \quad \ldots[15.23\,(b)]$$

The greater of the depths found from *Eq.* 15.21, 15.22, 15.23 (a) and 15.23 (b) should be provided.

## (c) Steel reinforcement

Steel reinforcement is provided in both the directions for the bending moments $M_1$ and $M_2$.

The area of steel $A_{st1}$, of *long bars* parallel to directions L, is calculated for bending moment $M_1$ as under:

$$A_{st1} = \frac{M_1}{t j_c d} \quad \ldots[15.24\,(a)]$$

where $t = \sigma_{st}$

This area should be uniformly distributed over a width B. If $A_\phi$ is the area of each bar, the spacing is given by

$$s = \frac{B A_\phi}{A_{st1}} \quad \ldots[15.24\,(b)]$$

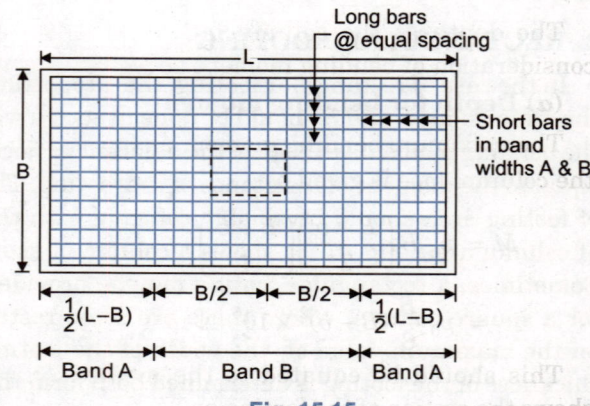

Fig. 15.15

The area of steel $A_{st2}$, of short bars parallel to direction $B$, is calculated for bending moment $M_2$ as under:

$$A_{st2} = \frac{M_2}{tj_c\, d}, \text{ where } t = \sigma_{st} \qquad \ldots(15.25)$$

This area is provided in two distinct band widths: (*i*) the central band $B$ of width $B$, and (*ii*) the end bands $A$, each of width $\frac{1}{2}(L - B)$.

The area $A_{st2\,(B)}$ in the central band $B$ is given by

$$A_{st2\,(B)} = \frac{2\,A_{st2}}{\beta + 1}, \text{ where } \beta = \frac{L}{B}. \qquad \ldots[15.25\,(a)]$$

The area $A_{st2\,(A)}$ in each of the end bands $A$ is given by

$$A_{st2\,(A)} = \frac{1}{2}(A_{st2} - A_{st2(B)}) \qquad \ldots[15.25\,(b)]$$

The reinforcement calculated above in the three bands should be distributed uniformly. (see Fig. 15.15)

**(d) Check for development length.** The section should be checked for development length at each face of the column, exactly in the same manner as for square column.

## 15.7. ISOLATED SLOPED FOOTING

In the previous article, we have seen that bending moment, beam shear and punching shear govern the thickness or depth of footing near the column face. This depth can be reduced toward the edges of the footing where the bending moment and shear decrease rapidly. If this decrease is achieved *linearly*, we get a *sloped footing* shown in Fig. 15.16 for a square footing. However, a minimum thickness of 150 mm should be kept at the edges.

The areas at the base of footing and the net upward pressure $p_0$ are found exactly in the same manner as for isolated footing of uniform depth.

### 1. SQUARE FOOTING

Fig. 15.16 shows a square footing of size $B \times B$, with variable depth.

Width $\qquad B = \sqrt{\dfrac{W + W'}{q_0}}$

and $\qquad p_0 = \dfrac{W}{B^2}$ (kN/m²)

The depth at the column face is found both from the consideration of bending moment as well as punching shear.

**(a) Depth for bending moment**

The maximum bending moment as the section $XX$ through the column face is given by

$$M = p_0\, B\left(\frac{B-b}{2}\right)\left(\frac{B-b}{4}\right)$$

$$= p_0\, \frac{B}{8}(B-b)^2 \times 10^6 \text{ N-mm} \qquad \ldots(15.26)$$

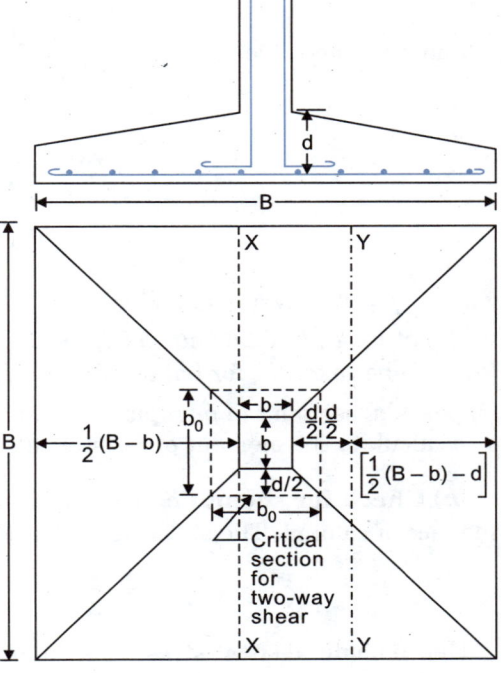

Fig. 15.16

This should be equated to the moment of resistance of the footing section at $XX$. Fig. 15.17 (*a*) shows the section of the footing at $XX$. The top width is $b$, the bottom width is $B$ and the effective depth is $d$.

The moment of resistance of the trapezoidal section consists of the moment of resistance of the central rectangular portion plus the moment of resistance of the two triangular portions.

If $\alpha$ is the inclination of the footing top face with horizontal the compressive force in the central portion '$C_1$' is given by

$$C_1 = \frac{1}{2} bkd \cdot c \cos^2 \alpha \quad \text{where } c = \sigma_{cbc}. \quad ...[15.27 (a)]$$

and the moment of resistance

$$M_{R1} = R \cdot bd^2 \cos^2 \alpha \quad ...[15.27 (b)]$$

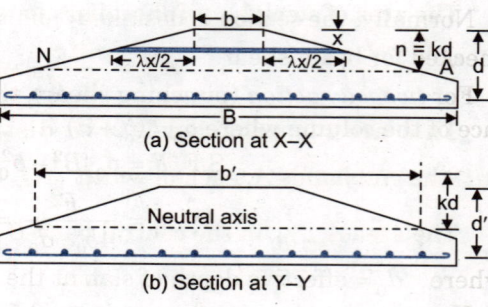

Fig. 15.17

To find the moment of resistance $M_{R2}$ of the two triangular portions consider an elementary strip at depth $x$ below the apex, and let the total width of the strip be $\lambda x$. If $c$ is the compressive stress at the top fibre, the compressive stress at the level of the strip will be $\dfrac{kd-x}{kd} c$. Hence the total compressive force in the triangular portion will be

$$C_2 = \int_0^{kd} \lambda x \cdot dx \, \frac{kd-x}{kd} c = \lambda c \frac{k^2 d^2}{6} \quad ...[15.28(a)]$$

The moment of resistance of the triangular portion is:

$$M_{R2} = \int_0^{kd} \lambda x \cdot dx \, \frac{kd-x}{kd} \cdot c(d-x) = \lambda \cdot c \frac{k^2 d^3}{12} (2-k) \quad ...[15.28(b)]$$

Hence lever arm $\quad a = \dfrac{M_{R2}}{C_2} = \dfrac{1}{2} d (2-k)$

Total $\quad M_r = M_{r1} + M_{r2} = Rbd^2 \cos^2 \alpha + \lambda \cdot c \dfrac{k^2 d^3}{12} (2-k) \quad ...(15.29)$

and combined lever arm $= \dfrac{\text{Moment}}{\text{Force}} = \dfrac{Rbd^2 \cos^2 \alpha + \lambda \cdot c \dfrac{k^2 d^3}{12}(2-k)}{\dfrac{1}{2} bkd\, c \cdot \cos^2 \alpha + \lambda \cdot c \dfrac{k^2 d^2}{6}} \quad ...(15.30)$

$$= \frac{Rbd \cos^2 \alpha + \lambda \cdot c \dfrac{k^2 d^2}{12}(2-k)}{\dfrac{1}{2} bk\, c \cdot \cos^2 \alpha + \lambda \cdot c \dfrac{k^2 d}{6}} = j'd \quad ...[15.30(a)]$$

where $j'$ = the lever arm factor

Equating *Eqs. 15.26* and *15.29*, the depth $d$ can be calculated. In the above equations, the value of $k$ and $R$ are to be taken as for balanced section. The above method is a very tedious. As an approximation, the trapezoidal section can be replaced by an equivalent rectangular section of width $b + \dfrac{1}{8}(B-b)$ and of depth $d$ to calculate the moment of resistance.

**(b) Check for shear.** For one way shear (beam action) the critical section is taken at a distance $d$ from the face of column. The shear force $V$ at the section is given by

$$V = [\frac{1}{2}(B-b) - d] B \times p_0$$

Hence nominal shear stress is given by $\tau_v = \dfrac{V}{b'd}$

where $b'$ = minimum width of section in the tensile zone

= width of the section at N.A. [Fig. 15.17 (b)]

$d'$ = effective depth of the section at the plane

Normally, the section is designed from the consideration of *two-way shear* (punching shear), and is later checked for beam shear.

For *two way action* (punching shear), the critical section located all round, is at a distance $d/2$ from the face of the column where $b_0 = (b + d)$

$$\text{S.F. } F = p_0 [B^2 - b_0^2] = p_0 [B^2 - (b+d)^2]$$

$$\tau_v = \frac{F}{4b_0 \cdot d_0} = \frac{F}{4(b+d) d_0}$$

where $d_0$ = effective depth of slab at the critical plane.

Hence
$$\tau_v \leq k_s \cdot \tau_s (0.5 + \beta_c) \cdot \tau_c$$

or
$$\frac{F}{4(b+d) d_0} \leq (0.5 + \beta_c) \times 0.16 \sqrt{f_{ck}}$$

where $k_s$ or $(0.5 + \beta_c)$ has a maximum value of 1. ...(15.31)

**(c) Steel reinforcement.** Steel reinforcement is provided in both the directions, in the form of mesh, for the bending moment $M$.

Thus,
$$A_{st} = \frac{M}{t j' d'}$$

where $t = \sigma_{st}$ and $j'$ is the lever arm factor determined from *Eq.* [15.30 (a)]

**(d) Check for development length.** The section should be checked for development length at the face of the column.

## 2. RECTANGULAR FOOTING

The procedure for design of rectangular footing is similar to that of a square sloped footing, except that each of the cantilever portion is analysed separately. The reinforcements $A_{st1}$ and $A_{st2}$ found in the two directions are provided as explained § 15.6.

## 15.8. ISOLATED STEPPED FOOTING

The construction of sloped footing is sometimes difficult and when the slope of the top face of footing is more, say more than 1 vertically to 3 horizontally, it may be difficult to finish the top without having concrete slump too much. In such cases, stepped footing shown in Fig. 15.18 is an alternative.

The design is first done at the section $X_1$-$X_1$ passing through the face of the column. The depth $d$ is determined both on the basis of bending moment and shear.

Thus,
$$M_1 = \frac{p_0 B}{8} (B - b)^2 \times 10^6 \text{ N-mm} \quad \text{(as per eq 15.15)}$$

$$\therefore \quad d_1 = \sqrt{\frac{M_1}{R_c B_1}} \quad ...(15.32)$$

where $B_1$ is the width of the footings at the top, at section $X_1$ - $X_1$.

The depth so found is checked both for one way shear (beam shear) as well as two way shear (punching shear) as explained in § 15.6.

The area of steel reinforcement is provided as usual.

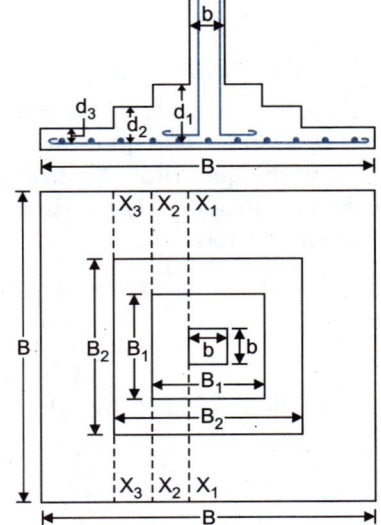

Fig. 15.18

Let the size of footings at its top, at the next step be $B_2 \times B_2$ and its depth $d_2$. The critical section for design will be at section $X_2$ - $X_2$. The curtailed depth $d_2$ should be safe in compression, tension, shear and development length. The bending moment at $X_2$ - $X_2$ will be the moment of the total pressure acting on the rectangular area to one side of the section. The resisting width will be $B_2$.

## 364 REINFORCED CONCRETE STRUCTURE

Thus, from compression point of view,

$$d_2 = \sqrt{\frac{M_2}{R_c B_2}}, \text{ where } M_2 \text{ is the bending moment at } X_2\text{-}X_2.$$

From tension point of view, providing the *same* reinforcement $A_{st}$ as at $X_1$-$X_1$,

$$A_{st2} = \frac{M_2}{t j_c A_{st}} \text{ where } t = \sigma_{st}$$

The greater of the two values of $d_2$ is selected. This value of $d_2$ is checked both for beam shear (at a section distant $d_2$ from $X_2$-$X_2$) and for punching shear (at perimeter section distant $d_2/2$ from the face of the step). Similarly, depth $d_3$ at section $X_3$-$X_3$ is determined.

## 15.9. ISOLATED FOOTING FOR CIRCULAR COLUMNS

Circular columns are usually preferred. The footing for a circular column may be either square or circular at the base.

### 1. SQUARE FOOTING

Fig. 15.19 show square sloped footing for a circular column. For the purpose of design, the equivalent square column of the same area as the circular column is assumed. The design of the square footing for the equivalent square column is done exactly in the same manner as explained in § 15.7.

The critical section (XX) for bending moment will pass through the face of the equivalent square column.

### 2. CIRCULAR FOOTING (Fig. 15.20)

Fig. 15.20 shows a column of radius $r$ with circular footing of radius $R$ given by

$$R = \sqrt{\frac{W + W'}{\pi q_0}} \quad \ldots(15.33)$$

The net upward pressure intensity of footing is given by

$$p_0 = \frac{W}{\pi R^2} \quad \ldots(15.34)$$

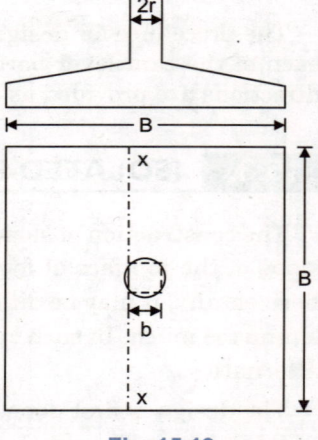

Fig. 15.19

**1. Depth for bending moment**

The maximum bending moment in the footing is calculated at the edge of the column for one quadrant such as *PQRS* shown shaded. The distance of C.G. of the quadrant of a circle of radius $R$ is at a distance of $0.6 R$ from the centre. Hence if $\bar{x}$ is the distance of C.G. of shaded area from the centre of column, we have

$$\bar{x} = \frac{\frac{\pi}{4} R^2 \times 0.6 R - \frac{\pi}{4} r^2 \times 0.6 r}{\frac{\pi}{4} R^2 - \frac{\pi}{4} r^2} = \frac{0.6 (R^3 - r^3)}{R^2 - r^2} = \frac{0.6 (R^2 + Rr + r^2)}{R + r} \quad \ldots[15.35(a)]$$

∴ Distance of C.G. of shaded area from the face of the column is

$$\bar{x}' = \frac{0.6 (R^2 + Rr + r^2)}{R + r} - r$$

$$= \frac{0.6 (R^2 + Rr + r^2) - Rr - r^2}{R + r} = \frac{0.6 R^2 - 0.4 Rr - 0.4 r^2}{R + r}$$

$$= \frac{0.2}{R + r} [3R^2 - 2Rr - 2r^2] \quad \ldots[15.35(b)]$$

Upward force on shaded area = $\frac{\pi}{4} (R^2 - r^2) p_0$

∴ Bending moment at the face of the column is

$$M = \frac{\pi}{4}(R^2 - r^2) \, p_0 \times \frac{0.2}{R+r}(3R^2 - 2Rr - 2r^2) = \frac{p_0 \, \pi}{20}(R - r)(3R^2 - 2Rr - 2r^2) \quad \ldots(15.36)$$

Breath $b$ of quadrant $= PQ = \dfrac{\pi r}{2}$

Hence the effective depth is given by

$$d = \sqrt{\frac{M}{R_c \cdot b}}$$

(where $R_c$ in the above equation is equal to $\frac{1}{2}\sigma_{cbc} \cdot j_c \cdot k_c$)

A minimum total thickness of 150 mm is provided at the outer edge of footing.

**2. Depth for two way shear** (punching shear)

Circular footing is checked for two way shear (punching shear) for which the critical circular plane exists at a radial distance of $d/2$ from the column face, as shown in Fig. 15.20. The radius $r_0$ of this critical circle $= \left(r + \dfrac{d}{2}\right)$. Net punching shear force $F = p_0 \, \pi \, (R^2 - r_0^2)$.

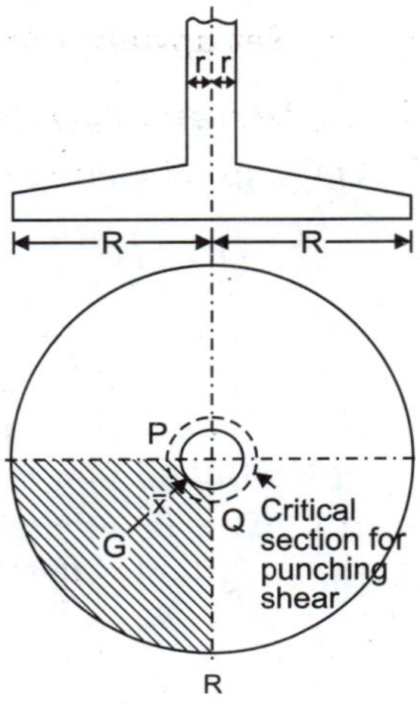

If $d_0$ is the effective depth of slab at that plane, the shear stress

$$\tau_v = \frac{F}{2\pi r_0 d_0}$$

This should be less than $k_s \cdot \tau_c$.

∴
$$\frac{F}{2 \pi r_0 d_0} \le k_s \, \tau_c. \quad \ldots(15.37)$$

where $k_s = \left(0.5 + \dfrac{b}{a}\right) \le 1 = 1$ (since $b = a = r$).

and $\tau_c = 0.16\sqrt{f_{ck}}$.

Hence, $\dfrac{F}{2 \pi r_0 d_0} \le 0.16 \sqrt{f_{ck}} \quad \ldots[15.37\,(a)]$

**Fig. 15.20**

The critical section for *one way shear* will occur at a radial distance $d$ from the face of the column, i.e., at $AB$ (Fig. 15.21). The shear force $V$ will be

$$V = p_0 \frac{\pi}{4}[R^2 - (r+d)^2] \quad \ldots(15.38)$$

The breath $b'$ of the footing at critical section $= \dfrac{\pi}{2}(r+d)$

∴
$$\tau_v = \frac{V}{b' \, d'} = \frac{V}{\dfrac{\pi}{2}(r+d)\, d'}, \quad \ldots[15.38\,(a)]$$

where $d'$ is the effective depth at the critical section.

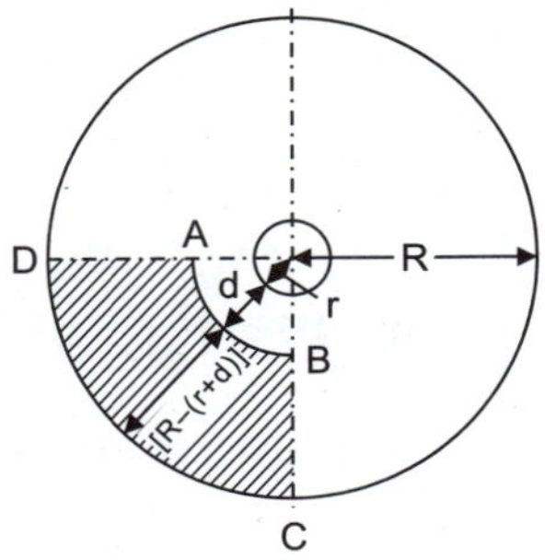

**3. Steel reinforcement.** Steel reinforcement is given by

$$A_{st} = \frac{M}{\sigma_{st} \, j_c \, d}.$$

**Fig. 15.21**

Same reinforcement ($A_{st}$) is provided in two directions at right angles, *i.e.*, the bars are arranged in form of square mesh centrally located under the column, the width of the square being equal to the side of square that can be inscribed in the circular footing of radius $R$.

**4. Check for development length.** The section is checked for development length at the face of column.

## Design: Example 15.1. *Design of footing for R.C. wall*

A reinforced concrete wall 250 mm thick carries a load of 500 kN/m inclusive of its own weight. Design a reinforced concrete footing on soil having safe bearing capacity of 160 kN/m². Use M 20 concrete and Fe 415 steel.

**Solution:**

1. **Design constants.** For M 20 concrete and Fe 415 reinforcement we have:
$$k_c = 0.289 \; ; \; j_c = 0.904 \; ; \; R_c = 0.914$$

2. **Size of footing.** $W = 500$ kN; $q_0 = 160$ kN/m²

Assume the self-weight of footing equal to 10% of the super imposed load.

$\therefore W' = 50$ kN/m ; $\therefore B = \dfrac{W + W'}{q_0} = \dfrac{500 + 50}{160} = 3.43$ m

Adopt $B = 3.5$ m $= 3500$ mm.

$\therefore$ Net upward pressure $p_0 = \dfrac{500}{3.5} = 142.86$ kN/m²/m

3. **Design of section**

Maximum bending moment at the face of the wall is given by *Eq. 15.16*.

$$M = p_0 \dfrac{(B-b)^2}{8} \times 10^6 \text{ N-mm} = \dfrac{142.86 \, (3.5 - 0.25)^2}{8} \times 10^6 = 1.89 \times 10^8 \text{ N-mm}$$

$$d = \sqrt{\dfrac{M}{BR_c}} = \sqrt{\dfrac{1.89 \times 10^8}{1000 \times 0.914}} = 455 \text{ mm}$$

Providing 50 mm nominal cover and using 20 mm φ bars, effective cover to the centre of steel will be = 50 + 20/2 = 60 mm and hence, total depth = 455 + 60 = 515 mm. However, provide total depth = 520 mm so that available effective depth will be 460 mm.

$$A_{st} = \dfrac{M}{\sigma_{st} j_c d} = \dfrac{1.89 \times 10^8}{230 \times 0.904 \times 460} = 1976 \text{ mm}^2$$

Using 20 mm Φ bars having $A_\Phi = \dfrac{\pi}{4}(20)^2 = 314.16$ mm²

$\therefore$ Spacing $s = \dfrac{1000 \, A_\Phi}{A_{st}} = \dfrac{1000 \times 314.16}{1976} \simeq 159$ mm

However, provide 20 mm Φ bars @ 150 mm c/c.

Actual $A_{st} = \dfrac{1000 \times 314.16}{150} = 2094.4$ mm² ; % reinforcement $= \dfrac{100 \times 2094.4}{1000 \times 460} \simeq 0.4\%$

Longitudinal reinforcement $= \dfrac{0.12 \times 1000 \times 520}{100} = 624$ mm²

Using 10 mm Φ bars, $A_\Phi = 78.5$ mm²  $\therefore$ Spacing $= \dfrac{1000 \times 78.5}{624} \approx 125.8$ mm

Provide 10 mm Φ bars longitudinally @ 125 mm c/c.

4. **Check for shear.** For *one way shear,* the critical section XX (Fig. 15.22) is at distance $d$ from the face of wall. Distance of section XX from the edge of footing

$= \dfrac{1}{2}(B-b) - d = \dfrac{1}{2}(3.5 - 0.25) - 0.46$

$= 1.165$ m $= 1165$ mm.

$V = p_0 \times 1.165 = 142.86 \times 1.165$

$= 166.432$ kN/m $= 166432$ N/m

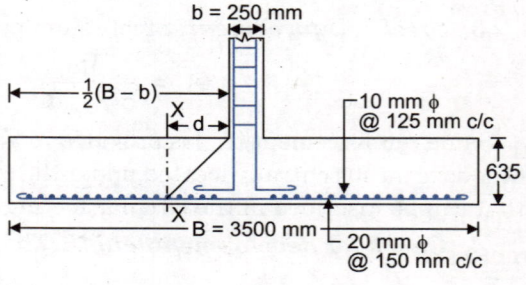

Fig. 15.22

## DESIGN OF ISOLATED FOOTINGS

Hence, $\tau_v = \dfrac{166432}{1000 \times 460} = 0.36 \text{ N/mm}^2$

Permissible shear stress $= k\,\tau_c$ where $k = 1$ for total thickness of 520 mm (Table 3.2)
Take $\tau_c \simeq 0.29$ N/mm² (Table 3.1) for reinforcement of 0.46% and M 20 concrete.
Hence $\tau_v > k.\tau_c$.

Thus, the footing is not safe from shear point of view. In order to make the footing safe in shear, increase its depth.

Taking maximum value of $\tau_v$ to equal $\tau_c \left( = 0.29 \text{ N/m}^2 \text{ for } \dfrac{100\,A_s}{bd} = 0.46\% \right)$,

we have $\tau_v = \dfrac{166432}{1000 \times d} = 0.29$ which gives $d = \dfrac{166432}{1000 \times 0.29} = 574$ mm

However, provide total depth = 635 mm so that available $d = 635 - 60 = 575$ mm.

### 5. Check for development length

$$L_d = \dfrac{\sigma_{st}}{4\,\tau_{bd}} \Phi = \dfrac{230}{4 \times 1.6 \times 0.8} \Phi \simeq 45\,\Phi = 45 \times 20 = 900 \text{ mm}$$

Providing 60 mm side cover, length of bars available $\tfrac{1}{2}[B - b] - 60 = \tfrac{1}{2}[3500 - 250] - 60 = 1565$ mm, which is greater than $L_d$. Hence O.K.

The details of reinforcement etc. are shown in Fig. 15.22

**Design: Example 15.2.** *Footing for a brickwall*

*A brick wall 300 mm thick carries a load of 180 kN/m length. Design a R.C.C. footing, if the safe bearing capacity of soil is 120 kN/m². Assume the permissible stresses as given in example 15.1.*

**Solution:**

1. **Design constants.** $k_c = 0.289;\ j_c = 0.904$ and $R_c = 0.914$.
2. **Size of footing.** $W = 180$ kN; Assume $W' = 10\%$ of $W = 18$ kN

$\therefore$ Width $B$ of footing $= \dfrac{180 + 18}{120} = 1.65$ m. Adopt $B = 1.7$ m.

Net upward pressure $p_0 = 180/1.7 = 106$ kN/m²/m

3. **Design of section.** Maximum bending moment occurs at section $X$-$X$ distant $b/4$ from the centre of the wall, and its magnitude is given by *Eq. 15.15*.

$$M = \dfrac{p_0}{8}(B - b)\left(B - \dfrac{b}{4}\right) \times 10^6 \text{ N-mm} = \dfrac{106}{8}(1.7 - 0.3)\left(1.7 - \dfrac{0.3}{4}\right) \times 10^6$$

$= 30.14 \times 10^6$ N-mm.

$\therefore \quad d = \sqrt{\dfrac{30.14 \times 10^6}{1000 \times 0.914}} = 182$ mm

Provide a total depth of 250 mm and a cover to the centre of the steel equal to 60 mm, so that available $d = 250 - 60 = 190$ mm.

### 4. Check for shear.

For balanced section, $p = 0.44\%$ for M 20 concrete and Fe 415 steel. Hence from Table 3.1, $\tau_c \approx 0.28$ N/mm². Also, from Table 3.2, $k = 1.10$ for $D = 250$ mm. Hence permissible shear stress $= k\,\tau_c = 1.10 \times 0.28 = 0.308$ N/mm².

The critical section $Y$-$Y$ lies at a distance of $d$ ($= 190$ mm) from the face of the wall. Hence distance of from the edge of $Y$-$Y$ footing

$$= \tfrac{1}{2}(B-b) - d$$
$$= \tfrac{1}{2}(1.7 - 0.3) - 0.19 = 0.51 \text{ m}.$$
$$\therefore \quad V = 106000 \times 0.51 \approx 54000 \text{ N/m}$$
$$\therefore \quad \tau_v = \frac{V}{bd} = \frac{54000}{1000 \times 190}$$
$$= 0.284 \text{ N/mm}^2.$$

This is less than the permissible shear stress. Hence safe.

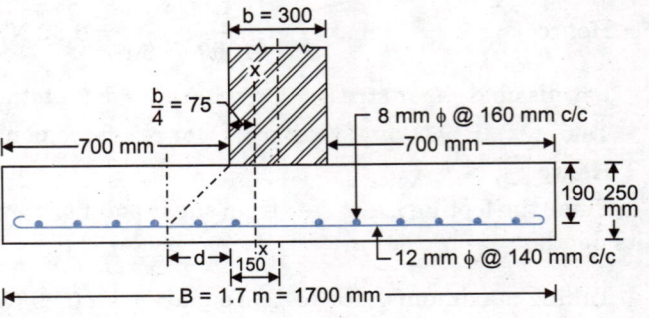

Fig. 15.23

**5. Design of reinforcement**

$$A_{st} = \frac{M}{\sigma_{st}\, j_c\, d} = \frac{30.14 \times 10^6}{230 \times 0.904 \times 190} = 763 \text{ N/mm}^2.$$

Using 12 mm Φ bars, $A_\Phi = \dfrac{\pi}{4}(12)^2 = 113 \text{ mm}^2$.

$$\therefore \quad \text{Spacing } s = \frac{1000\, A_\Phi}{A_{st}} = \frac{1000 \times 113}{763} = 148 \text{ mm}.$$

However, provide 12 mm Φ bars @ 140 mm c/c.

Area of longitudinal reinforcement = 0.12% area of cross-section.

$$= \frac{0.12}{100} \times 1000\, D = 1.2\, D = 1.2 \times 250 = 300 \text{ mm}^2$$

Use 8 mm Φ bars having $A_\Phi = \dfrac{\pi}{4}(8)^2 = 50.26 \text{ mm}^2$

$$\therefore \quad \text{Spacing } s = \frac{1000 \times 50.26}{300} = 167 \text{ mm}. \text{ However, provide 8 mm } \Phi \text{ bars @ 160 mm c/c.}$$

**6. Check for development length**

For M 20 – Fe 415 combination, $L_d = 45\, \Phi = 45 \times 12 \approx 540$ mm.

Providing 60 mm side cover, length of bars available

$$= \tfrac{1}{2}[B - b] - 60 = \tfrac{1}{2}[1700 - 300] - 60 = 640 \text{ mm}. > L_d. \text{ Hence OK.}$$

**Design: Example 15.3.** *Isolated footing of uniform thickness for a R.C. column*

*Design an isolated footing of uniform thickness of a R.C. column bearing a vertical load of 600 kN and having a base of size 500 mm × 500 mm. The safe bearing capacity of soil may be taken as 120 kN/m². Use M 20 is concrete and Fe 415 steel.*

**Solution:**

**1. Design constants.** For given set of materials, we have $k_c = 0.289$; $j_c = 0.904$ and $R_c = 0.914$.

**2. Size of footing.** $W = 600$ kN. Let $W'$ be equal to 10% $W = 60$ kN

$$\therefore \quad A = 660/120 = 5.5 \text{ m}^2$$

$\therefore$ Size $B$ of square $\sqrt{5.5} = 2.345$ m. Provide a square footing of 2.4 m × 2.4 m.

Net upward pressure $p_0 = \dfrac{600}{2.4 \times 2.4} = 104.17 \text{ kN/m}^2$

**3. Design of section.** The maximum bending moment acts at the face of column, and its magnitude is given by *Eq. 15.17*.

$$M = p_0\, \frac{B}{8}(B-b)^2 \times 10^6 \text{ N-mm} = 104.17 \times \frac{2.4}{8}(2.4 - 0.5)^2 \times 10^6 = 1.128 \times 10^8 \text{ N-mm}$$

$$\therefore \quad d = \sqrt{\frac{M}{R_c B}} = \sqrt{\frac{1.128 \times 10^8}{0.914 \times 2400}} \approx 227 \text{ mm}$$

**4. Check for shear.** The depth found above should be checked for shear. For *one way shear*, the critical section YY is located at a distance $d$ (= 227 mm) from the face of the column (Fig. 15.13), where shear force V is given by

$$V = p_0 B \left\{ \frac{1}{2}(B - b) - d \right\} = 104.17 \times 2.4 \left\{ \frac{1}{2}(2.4 - 0.5) - 0.227 \right\} = 180.755 \text{ kN}$$

$$\therefore \quad \tau_c = \frac{V}{B \cdot d} = \frac{180755}{2400 \times 227} = 0.332 \text{ N/mm}^2$$

Using 12 mm Φ bars at 60 mm clear cover, total depth will be about 227 + 6 + 60 = 293 mm ≈ 300 mm. Hence from Table 3.2, $k = 1$. Also, for a balanced section having $p = 0.44\%$ for M 20 concrete – Fe 415 steel, and hence, $\tau_c \approx 0.28$ N/mm² from Table 3.1.

Hence the permissible shear stress = $k \tau_c = 1 \times 0.28 = 0.28$ N/mm². Hence the section is unsafe. Required

$$d = \frac{180755}{2400 \times 0.28} \approx 269 \text{ mm and } D = 269 + 6 + 60 = 335 \text{ mm}$$

Also, for *two way shear* (punching shear), the section lies at $d/2$ from the column face all round. The width $b_0$ of the section = $b + d$ = 500 + 227 = 727 mm. Also, shear force around the section is

$$F = p_0 [B^2 - b_0^2] = p_0 [B^2 - (b + d)^2] = 104.17 [(2.4)^2 - (0.727)^2] = 544.962 \text{ kN}$$

$$\therefore \quad \tau_v = \frac{F}{4 b_0 d} = \frac{F}{4 [b + d] d} = \frac{544.962 \times 10^3}{4 \times 727 \times 227} = 0.826 \text{ N/mm}^2$$

Permissible shear stress = $k_s \tau_c$

where $k_s = (0.5 + \beta_c) = (0.5 + 1)$ with max. value of 1. Hence $k_s = 1$

Also, $\tau_c = 0.16 \sqrt{f_{ck}} = 0.16 \sqrt{20} = 0.715$ N/mm²

Hence permissible shear stress = $k_s \tau_c = 1 \times 0.715$ = 0.715 N/mm².

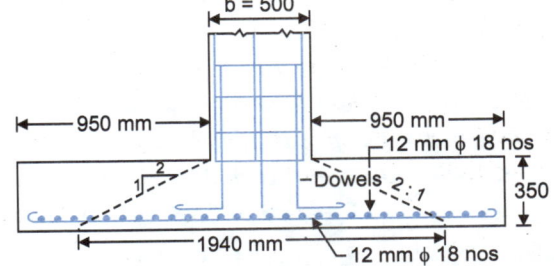

This is less that the punching shear stress. Hence the section is unsafe (*i.e.*, depth is insufficient). The required effective depth from punching shear point of view is found by equating $\tau_v$ to permissible shear stress of 0.715 N/mm².

$$\therefore \quad \tau_v = \frac{F}{4 [b + d] d} \leq k_s \tau_c$$

or $\quad \dfrac{104.17 \times 10^3 [(2400)^2 - (500 + d)^2] \times 10^{-6}}{4 [500 + d] d} = 0.715$

or $d^2 + 517.6 d - 193636 = 0$,

which gives $d = 251.7$ mm.

This is less than $d = 269$ mm required for one way shear. Hence, provide a total depth of 350 mm. Using 12 mm Φ bars and a clear cover of 60 mm, available $d = 350 - 6 - 60 = 284$ mm in one direction and $284 - 12 = 272$ mm in the other direction.

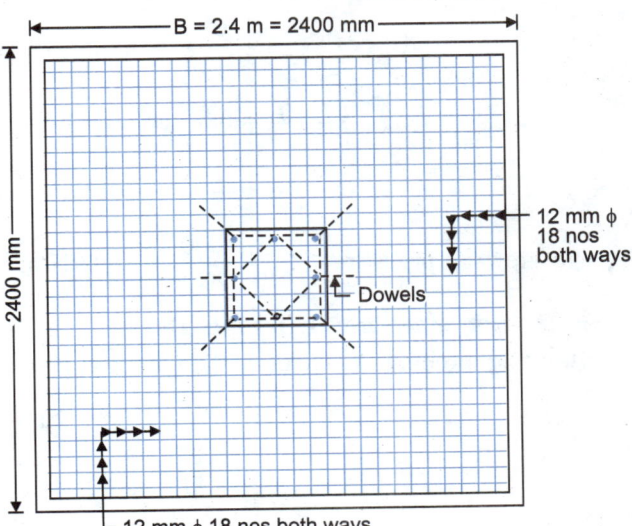

Fig. 15.24

**5. Design of steel reinforcement.** The area of reinforcement in each direction is given by

$$A_{st} = \frac{M}{\sigma_{st}\, j_c d} = \frac{1.128 \times 10^8}{230 \times 0.904 \times 272} = 1995 \text{ mm}^2$$

Using 12 mm Φ bars,
No. of bars = 1995/113.1 ≅ 18
Provide 18 bars of 12 mm Φ uniformly distributed in each direction.

**6. Check for development length**: $L_d = 45\,\Phi = 45 \times 12 \approx 540$ mm
Providing 60 mm side cover, length of bars available

$$= \tfrac{1}{2}[B - b] - 60 = \tfrac{1}{2}[2400 - 500] - 60 = 890 \text{ mm} > L_d. \text{ Hence safe.}$$

**7. Transfer of load at column base**

$A_2 = 500 \times 500 = 250000$ mm$^2$
$A_1 = [500 + 2(2 \times 350)]^2 = (1900)^2 = 3610000$ mm$^2$ for side spread rate of 2 : 1

$$\therefore \quad \sqrt{\frac{A_1}{A_2}} = \sqrt{\frac{3610000}{250000}} = 3.8$$

Adopt max. value of $\sqrt{A_1/A_2} = 2$.
Hence permissible bearing stress $= 0.25\, f_{ck}\, \sqrt{A_1/A_2} = 0.25 \times 20 \times 2 = 10$ N/mm$^2$.

Actual bearing pressure $= \dfrac{600000}{500 \times 500} = 2.4$ N/mm$^2$. Hence safe.

Thus no separate dowel bars are required for the transfer of load. However, it is advisable to continue all the bars of the column, into the foundation.

The details of reinforcement etc. are shown in Fig 15.24.

**Example 15.4.** *Design a rectangular isolated footing of uniform thickness for R.C. column bearing a vertical load of 600 kN, and having a base size of 400 × 600 mm.*

*The safe bearing capacity of soil may be taken as 120 kN/m$^2$. Use M 20 concrete and Fe 415 steel.*

**Solution: 1. Design constants.** For M 20 concrete, and Fe 415 steel,

$k_c = 0.289;\ j_c = 0.904$ and $R_c = 0.914$

**2. Size of.** $W = 600$ kN. Let $W'$ be equal to 10% $W = 60$ kN.

$$\therefore \quad A = \frac{660}{120} = 5.5 \text{ m}^2. \text{ Let ratio of } B \text{ to } L = \frac{40}{60} = \frac{2}{3}.$$

$$\therefore \quad \tfrac{2}{3} L \times L = 5.5 \quad \text{or} \quad L = 2.9 \text{ m}.$$

$$\therefore \quad B = \tfrac{2}{3} \times 2.9 = 1.93 \text{ m. However, provide a footing of size 2 m} \times 3 \text{ m}.$$

Net upward pressure $p_0 = \dfrac{600}{2 \times 3} = 100$ kN/m$^2$

**3. Design of section.** Refer Fig. 15.14.
Bending moment $M_1$ about section $X$-$X$ is given by

$$M_1 = \frac{p_0\, B}{8}(L - a)^2 \text{ kN-m} = \frac{100 \times 2}{8}(3 - 0.6)^2 \times 10^6 \text{ N-mm} = 144 \times 10^6 \text{ N-mm}$$

$$\therefore \quad d = \sqrt{\frac{M_1}{R_c B}} = \sqrt{\frac{144 \times 10^6}{0.914 \times 2000}} = 281 \text{ mm}.$$

Keep $d = 290$ mm and total depth = 350 mm.
Provide uniform thickness for the entire footing.

B.M. $M_2$ about section Y–Y is given by

$$M_2 = \frac{p_0 L}{8}(B-b)^2 \times 10^6 \text{ N-mm} = \frac{100 \times 3}{8}(2-0.4)^2 \times 10^6 = 96 \times 10^6 \text{ N-mm}.$$

Thus, $M_2 < M_1$. The effective depth found above has to be checked for shear.

**4. Check for shear.** For the *beam action* total S.F. along section AB [Fig. 15.10(a)] is

$$V = p_0 B\left(\frac{L}{2} - \frac{a}{2} - d\right) = p_0 B\left(\frac{L-a}{2} - d\right)$$

$$= 100 \times 2 \left(\frac{3-0.6}{2} - 0.29\right) = 182 \text{ kN} = 182000 \text{ N}.$$

$$\tau_v = \frac{V}{Bd} = \frac{182000}{2000 \times 290} = 0.314 \text{ N/mm}^2.$$

For M 20 concrete and Fe 415 steel, $p = 0.44\%$ and hence $\tau_c \approx 0.28$ N/mm²
Also for D > 300 mm, $k = 1$ from Table 3.2.

∴ $\tau_v > k \cdot \tau_c$ and the section is unsafe.

Required $d = \dfrac{182000}{2000 \times 0.28} = 325$ mm and $D = 325 + 60 = 385$ mm

For the *two way action or punching shear action* along ABCD [Fig. 15.10(b)],

Perimeter ABCD $= 2\{(a+d) + (b+d)\} = 2\{(0.6 + 0.29) + (0.4 + 0.29)\}$
$= 2\{0.89 + 0.69\} = 3.16$ m $= 3160$ mm.

Area ABCD $= 0.89 \times 0.69 = 0.6141$ m²

∴ Punching shear $= 100[(2 \times 3) - 0.6141] = 538.59$ kN

∴ $\tau_v = \dfrac{538.59 \times 1000}{3160 \times 290} = 0.59$ N/mm².

Allowable shear stress $\tau_c$ is given by

$$\tau_c = 0.16 \sqrt{f_{ck}} = 0.16 \sqrt{20} \approx 0.715 \text{ N/mm}^2.$$

$$k_s = (0.5 + \beta_c) = \left(0.5 + \frac{0.4}{0.6}\right) = 1.17 \text{ ; However, adopt max. } k_s = 1.$$

∴ $k_s \tau_c = 1 \times 0.715 = 0.715$ N/mm².

This is more than $\tau_v = 0.59$ N/mm². Hence safe from punching shear.

However, keep $D = 400$ mm so that $d = 400 - 60 = 340$ mm, providing an effective cover of 60 mm.

**5. Design for reinforcement.**

Area $A_{st1}$ of long bars calculated for moment $M_1$ is given by

$$A_{st1} = \frac{M_1}{\sigma_{st} \, j_c \, d} = \frac{144 \times 10^6}{230 \times 0.904 \times 340} = 2037 \text{ mm}^2$$

Using 12 mm Φ bars, $A_\phi = 113$ mm². ∴ No. of bars $= 2037/113 = 18.03 \approx 19$.

These are to be distributed uniformly in a width $B = 2$ m. Effective depth for top layer of reinforcement $= 340 - 12 = 328$ mm.

The area $A_{st2}$ of short bars calculated for $M_2$ is given by

$$A_{st2} = \frac{M_2}{\sigma_{st} \, jd} = \frac{96 \times 10^6}{230 \times 0.904 \times 328} = 1408 \text{ mm}^2.$$

This area is to be provided in two distinct band widths. Area $A_{st2(B)}$ in central band of width $B = 2$ m is given by

$$A_{st2(B)} = \frac{2A_{st2}}{\beta + 1} = \frac{2 \times 1408}{\frac{3}{2} + 1} = 1126 \text{ mm}^2.$$

∴ No. of 12 mm Φ bars = 1126/113 ≈ 10 to be provided in central bend width = 2 m.

Remaining area in each end band strip = $\frac{1}{2}$(1408 – 1126) = 141 mm².

No. of 12 mm Φ bars = 141/113 ≈ 2, to be provided in each end band of width $\frac{1}{2}(L - B)$ = $\frac{1}{2}$(3 – 2) = 0.5 m. However, provide 3 bars in each end band.

**6. Test for development length**

$$L_d = \frac{\sigma_{st}}{4\tau_{bd}}\Phi = \frac{230}{4 \times 1.6 \times 0.8}\Phi = 45\Phi = 45 \times 12 = 540 \text{ mm}.$$

Providing 60 mm side cover, length available $\frac{1}{2}[B - b] - 60$ = $\frac{1}{2}$[2000 – 400] – 60 = 740 mm, which is greater than $L_d$. Hence O.K.

**7. Check for transfer of load at the base**

$$A_2 = 600 \times 600 = 360000 \text{ mm}^2$$

At a rate of spread of 2:1,

$$A_1 = [600 + 2(2 \times 400)]^2 = (2200)^2 = 4.84 \times 10^6 \text{ mm}^2$$

∴ $\sqrt{\frac{A_1}{A_2}} = \sqrt{\frac{4.84 \times 10^6}{360000}} = 3.67 > 2$

Adopt max. value of $\sqrt{A_1/A_2}$ as 2.

∴ Permissible bearing stress
= $0.25 f_{ck} \sqrt{A_1/A_2}$
= 0.25 × 20 × 2 = 10 N/mm².

Actual bearing stress = $\frac{600000}{600 \times 600}$ = 1.67 N/mm²

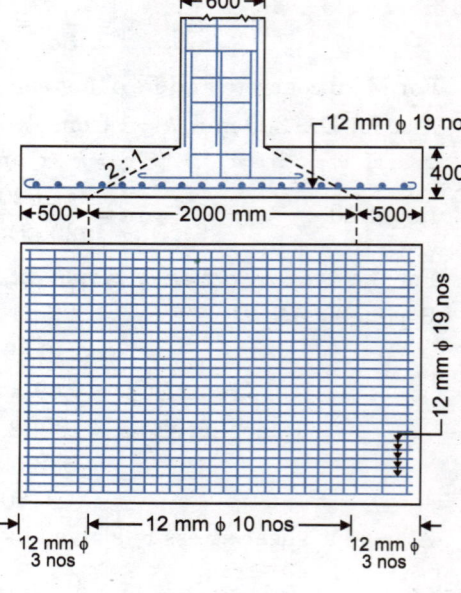

Fig. 15.25

Hence satisfactory. The details of reinforcement etc. are shown in Fig. 15.25.

**Example 15.5.** *Design an isolated footing for a column 500 mm × 500 mm, transmitting an axial load of 1200 kN. The column is reinforced with 8 bars of 20 mm diameter. The safe bearing capacity of soil is 120 tonnes/m². Use M 20 concrete and Fe 415 steel.*

**Solution:**

1. **Design constants.** For M 20 concrete and Fe 415 steel combination:
$k_c = 0.289; j_c = 0.904; R_c = 0.914$

2. **Size of footing.** W = 1200 kN. Assume W' = 10% W = 120 kN

∴ Area of footing $A = \frac{1200 + 120}{12}$ = 11 m²; Provide a footing of size 3.4 m × 3.4 m

Actual upward pressure intensity $p_0 = \frac{1200}{3.4 \times 3.4}$ = 104 kN/m².

3. **Design of footing.** Maximum bending moment occurs at the face of the column and its magnitude is given by *Eq. 15.17.*

$$M = p_0 \frac{B}{8}(B - b)^2 \text{ kN-m} = \frac{104 \times 3.4}{8}(3.4 - 0.5)^2 \times 10^6 \text{ N-mm} = 372 \times 10^6 \text{ N-mm}.$$

The section of the footing at the column face will be trapezoidal, as shown in Fig. 15.16. Let the effective depth at the column face be *d* and that at the edges be 0.2 *d*.

Slope
$$\lambda = \frac{B-b}{d-0.2d} = \frac{3400-500}{0.8\,d} = \frac{3600}{d} \qquad ...(i)$$

The moment of resistance of section is given by *Eq. 15.29*

$$M_r = Rbd^2 \cos^2 \alpha + \lambda \cdot c \, \frac{k^2 \, d^3}{12}(2-k)$$

Adopting $\cos^2 \alpha = 0.85$ for practical purposes, we have

$$M_r = 0.914 \times 500 \, d^2 \times 0.85 + \frac{3600 \times 7}{d} \frac{(0.289)^2}{12}(2-0.289)\,d^3$$

$$= 388.5 \, d^2 + 300.1 \, d^2 = 688.6 \, d^2 \qquad ...(ii)$$

Total force of compression is

$$C = \frac{1}{2} kbd\, c \cos^2 \alpha + \lambda \cdot c \, \frac{k^2 \, d^2}{6}$$

$$= \frac{1}{2} \times 500 \times 0.289 \times 7 \times 0.85\, d + \frac{3600}{d} \times \frac{7}{6}(0.289)^2\,d^2$$

$$= 429.89\,d + 350.79\,d = 780.68\,d \qquad ...(iii)$$

∴ L.A. for the section $= \dfrac{688.6\,d^2}{780.68\,d} = 0.88\,d$

∴ Lever arm factor $j = 0.88$. *Alternatively* taking the section to be an equivalent rectangle of width $= b + \dfrac{1}{8}(B-b) = 500 + \dfrac{1}{8}(3400-500) \approx 862.5$ mm.

∴ $M_r = R_c bd^2 = 0.914 \times 862.5\,d^2 = 788\,d^2$ against $688.6\,d^2$ found in *(ii)*. Since both the values are quite close, we can take the trapezoidal section to be an equivalent rectangle of width $b + \dfrac{1}{8}(B-b)$. Adopting $M_r = 688.6\,d^2$, and equating it to the B.M., we get

$$688.6\,d^2 = 372 \times 10^6$$

or $d = \sqrt{\dfrac{372 \times 10^6}{688.6}} \approx 735$ mm; Effective depth at end $= 0.2d = 140$ mm.

### 4. Check for shear

#### (a) For beam shear

The section is to be checked for beam shear is at a distance $d = 735$ mm from the column face, where shear force is given by

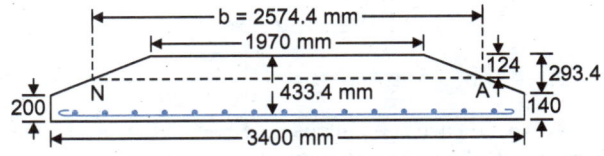

Fig. 15.26

$$V = p_0\, B\left\{\frac{1}{2}(B-b) - d\right\} = 104 \times 3.4 \left\{\frac{1}{2}(3.4-0.5) - 0.735\right\} = 252.8 \text{ kN.}$$

The section for footing is shown in Fig. 15.26.

Effective depth at that location $= 140 + \dfrac{735-140}{1450}(1450-735) = 433.4$ mm.

Top width of section $= 500 + \dfrac{3400-500}{1450} \times 735 = 1970$ mm

Depth of N.A. $= 433.4 \times 0.289 = 125.25$ mm.

Width $b'$ at N.A. $= 1970 + \dfrac{3400-1970}{293.4} \times 125.25 \approx 2580.5$ mm.

∴ $\tau_v = \dfrac{V}{b'd} = \dfrac{252.8 \times 1000}{2580.5 \times 433.4} = 0.226$ N/mm$^2$

For M 20 concrete and Fe 415 steel combination, $p = 0.44\%$ and hence $\tau_c \approx 0.28$ N/mm$^2$
Thus, $\tau_v < \tau_c$. (Hence safe).

**(b) Check for two way shear (punching shear):** [Fig. 15.10(a)]

Perimeter $ABCD = 2[(a+d)+(b+d)] = 4(a+d) = 4(500+735) = 4940$ mm.

Area $ABCD = 1.235 \times 1.235 = 1.525$ m$^2$

Punching shear $= 104[3.4 \times 3.4 - 1.525] = 1047.6$ kN.

$$\therefore \quad \tau_v = \frac{1047.6 \times 1000}{4940 \times 735} = 0.29 \text{ N/mm}^2.$$

Allowable shear stress $\tau_c$ is given by

$$\tau_c = 0.16 \sqrt{f_{ck}} = 0.16\sqrt{20} \approx 0.71 \text{ N/mm}^2$$
$$k_s = (0.5 + \beta_c) = (0.5 + 1). \text{ However adopt max. } k_s = 1.$$

$\therefore \quad k_s \tau_c = 1 \times 0.71 = 0.71$ N/mm$^2$. Hence safe.

**5. Steel reinforcement**

$$A_{st} = \frac{M}{\sigma_{st}\, jd} \text{ (where } j = 0.88) = \frac{372 \times 10^6}{230 \times 0.88 \times 735} = 2501 \text{ mm}^2$$

Using 12 mm Φ bars, $A_\Phi \triangleq 113$ mm$^2$. $\quad\therefore$ No. of bars $= 2501/113 \approx 23$.

Hence provide 23 Nos. 12 mm Φ bars, uniformly spaced in the width 3.4 m in each direction. Effective depth to the centre of bottom layer = 735 + 12 = 747 mm. Hence provide overall depth of 810 mm and reduce it to 200 mm at the edge.

**6. Check for development length** : $L_d = 45 \times 12 = 540$ mm.

Providing 60 mm side cover length, available $= \frac{1}{2}(B-b) = \frac{1}{2}(3400-500) - 60 = 1390$ mm.

Hence O.K. The details of reinforcement etc. as shown in Fig. 15.27.

**7. Check for transfer of load at the column base**

$A_2 = 500 \times 500 = 250000$ mm$^2$. At a rate of spread of 2 : 1

$A_1 = [500 + 2(2 \times 810)]^2 = (3740)^2 = 13987600$ mm$^2$. Hence,

$$\sqrt{\frac{A_1}{A_2}} = \sqrt{\frac{13987600}{250000}} = 7.48 > 2; \text{ Adopt a max. value of } \sqrt{\frac{A_1}{A_2}} = 2.$$

$\therefore$ Permissible bearing stress $= 0.25 f_{ck} \sqrt{A_1/A_2} = 0.25 \times 20 \times 2 = 10$ N/mm$^2$

Actual bearing stress $= \dfrac{1200000}{500 \times 500} = 4.8$ N/mm$^2$. Hence satisfactory.

**Note.** In case the permissible bearing stress is less than the actual bearing pressure, load transfer is done by the provision of *dowels bars*.

In the present case, the load transfer is there without exceeding the permissible bearing stress. However it is always a good practice to extend all the column bars into the foundation and anchored. If dowels bars are provided, development length requirements should be satisfied as required in § 15.4.

The details of reinforcement etc. are shown in Fig. 15.27.

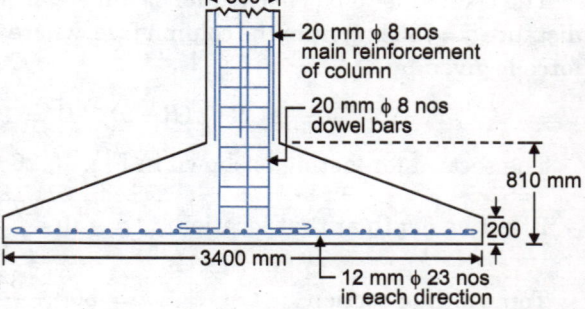

Fig. 15.27

**Design: Example 15.6.** *Design an isolated stepped foundation of column of example 15.5.*

**Solution:**

1. **Design constants.** $k_c = 0.289; j_c = 0.904$ and $R_c = 0.914$
2. **Size of footing.** As determined in example 15.5,

   $B = 3.4$ m; $p_0 = 104$ kN/m$^2$
3. **Design of footing near column face.** Refer Fig. 15.18.

Maximum bending moment at section $X_1 - X_1$ passing thorough the face of the column is given by

$$M_1 = \frac{p_0 B}{8}(B-b)^2 \text{ kN-m} = \frac{p_0 B}{8}(B-b)^2 \times 10^6 \text{ N-mm}$$

$$= \frac{104}{8} \times 3.4 \, (3.4 - 0.5)^2 \times 10^6 = 3.72 \times 10^8 \text{ N-mm}.$$

Let us give first step at 450 mm from the column face and second step at 950 mm from the column face, as shown in Fig. 15.28.

∴ $B_1 = b + (2 \times 450) = 500 + 900 = 1400$ mm
and $B_2 = b + (2 \times 950) = 500 + 1900 = 2400$ mm

The top width of footing at the column base = $B_1$ = 1400 mm. Hence the effective depth of footing near column face is given by

$$d_1 = \sqrt{\frac{M_1}{R_c B_1}} = \sqrt{\frac{3.72 \times 10^8}{0.914 \times 1400}} = 539 \cong 540 \text{ mm}.$$

Let us check this depth for shear.

For one way shear, the critical plane is located at a distance of $d = 540$ mm from the column face which lies after the step where thickness $d_2$ is yet to be found.

For two way shear, the critical plane lies at a distance of $d/2 = 540/2 \approx 270$ mm from the column face where $b_0 = (500 + 2 \times 270) = 1040$ mm. Hence perimeter or the punching shear zone = $4 \times 1040$. Also, punching shear force is given by

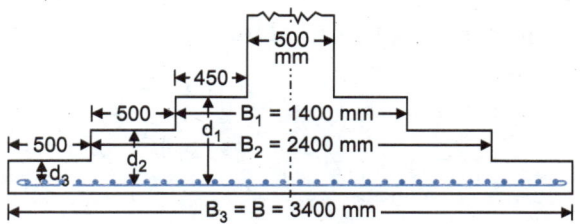

Fig. 15.28

$$F = p_0 [B_2^2 - b_0^2] = 104 \, [(3.4)^2 - (1.04)^2] = 1090 \text{ kN}$$

∴ $$\tau_v = \frac{F}{4b_0 d} = \frac{1090 \times 1000}{4 \times 1040 \times 540} = 0.485 \text{ N/mm}^2.$$

Permissible shear stress = $k_s \tau_c$ where $k_s = (0.5 + 1)$ with max. value of 1 and $\tau_c = 0.16 \sqrt{20} = 0.715$ N/mm². Hence actual shear stress is less than permissible shear stress and the depth is adequate.

$$\text{Area of steel} = \frac{M_1}{\sigma_{st} \, j_c \, d} = \frac{3.72 \times 10^8}{230 \times 0.904 \times 540} = 3314 \text{ mm}^2$$

Using 16 mm Φ bars, No. of bars = 3314/201 ≈ 17
Hence provide 17 bars in each direction. Actual $A_{st}$ = 17 × 201 = 3417 mm²
Let us check for development length of bars.
$L_d \approx 45 \, \Phi = 45 \times 16 = 720$ mm. Providing 50 mm side cover, available length of bar

$$= \tfrac{1}{2}(3400 - 500) - 50 = 1400 \text{ mm}. > L_d. \text{ Hence O.K.}$$

**4. Design of section after first step.** Let the effective depth be $d_2$ mm.
Width of footing at top = $B_2$ = 2400 mm
B.M. $M_2$ just at the face of the step is given by

$$M_2 = \frac{p_0 B}{8}(B - B_1)^2 = \frac{104 \times 3.4}{8}(3.4 - 1.4)^2 \text{ kN-m} = 1.77 \times 10^8 \text{ N-mm}.$$

Hence from compression point of view,

$$d_2 = \sqrt{\frac{M_2}{RB_2}} = \sqrt{\frac{1.77 \times 10^8}{0.914 \times 2400}} = 284.05 \cong 285 \text{ mm} \qquad \ldots(i)$$

From tension point of view, if the same reinforcement, as provided at the column face, is to be kept, we have

$$d_2 = \frac{M_2}{\sigma_{st} \, j_c \, A_{st}} = \frac{1.77 \times 10^8}{230 \times 0.904 \times 3417} = 249.1 \cong 250 \text{ mm} \qquad \ldots(ii)$$

For *One way shear,* taking permissible shear stress

$$= k \, \tau_c = \tau_c \text{ (since } k = 1 \text{ for } D > 300 \text{ mm)},$$

$$d_2 = \frac{F_2}{B_2 \, \tau_c} \quad \text{(where } \tau_c = 0.28 \text{ N/mm}^2 \text{ for } p = 0.44\%)$$

where $F_2 = p_0 \, B \left[ \frac{1}{2}(B - B_1) - d_2 \right]$; taking $d_2 \approx 285$ mm

$$F_2 = 104 \times 3.4 \left[ \frac{1}{2}(3.4 - 1.4) - 0.285 \right] = 252.8 \text{ kN}$$

$$\therefore \quad d_2 = \frac{252.8 \times 1000}{2400 \times 0.28} = 376.19 \cong 377 \text{ mm} \qquad \ldots(iii)$$

For *punching shear,* taking $d_2 = 377$ mm (max. of the above three values),

$$b_0 = B_1 + \frac{d_2}{2} = 1400 + \frac{377}{2} \approx 1588 \text{ mm}$$

$$F = p_0 \, [B^2 - b_0^2] = 104 \, [(3.4)^2 - (1.588)^2] = 940 \text{ kN}$$

$$\therefore \quad \tau_v = \frac{F}{4 b_0 \, d_2} = \frac{940 \times 1000}{4 \times 1588 \times d_2}$$

Taking $\quad \tau_v = k_c \, \tau_c = \tau_c = 0.16 \sqrt{f_{ck}} = 0.16 \sqrt{20} = 0.715$ N/mm²

We get $\quad d_2 = \dfrac{940 \times 1000}{4 \times 1588 \times 0.715} = 207$ mm. $\qquad \ldots(iv)$

Hence maximum value of $d_2 = 377$ mm.

**5. Design of section after second step.** Let the effective depth be $d_3$. This is provided upto the edge of the footing. The bending moment $M_3$ just at the face of the second steps is

$$M_3 = \frac{p_0 \, B}{8} (B - B_2)^2 = \frac{104}{8} \times 3.4 \, (3.4 - 2.4)^2 \text{ kN-m} = 44.2 \times 10^6 \text{ N-mm}$$

From compression point of view, $d_3 = \sqrt{\dfrac{M_3}{R_c \, B}} = \sqrt{\dfrac{44.2 \times 10^6}{0.914 \times 3400}} = 120$ mm $\qquad \ldots(1)$

From tension point of view, using the same steel,

$$d_3 = \frac{M_3}{\sigma_{st} \, j \, A_{st}} = \frac{44.2 \times 10^6}{230 \times 0.904 \times 3417} = 62.2 \cong 63 \text{ mm} \qquad \ldots(2)$$

From *one way shear* point of view, $d_3 = \dfrac{F}{B_3 \, \tau_c}$

where $\quad F = p_0 \, B \left[ \frac{1}{2}(B - B_2) - d_3 \right]$; taking $d_3 \approx 120$ mm

$$= 104 \times 3.4 \left[ \frac{1}{2}(3.4 - 2.4) - 0.12 \right] = 134.4 \text{ kN}$$

and $\quad \tau_c = 0.28$ N/mm² for $p = 0.44\%$ (Table 3.1)

$$\therefore \quad d_3 = \frac{134.4 \times 1000}{3400 \times 0.28} = 141.1 \cong 142 \text{ mm} \qquad \ldots(3)$$

For *punching shear,* taking $d_3 = 142$ mm (max. of the above three values),

$$b_0 = 2400 + \frac{142}{2} = 2471 \text{ mm.}$$

$\therefore \quad F = p_0 [B^2 - b_0^2] = 104 [(3.4)^2 - (2.471)^2] = 567.3$ kN

Now $\quad \tau_c = \dfrac{F}{4b_0 \, d_3}$ or $d_3 = \dfrac{F}{4b_0 \, \tau_v}$ ;

Taking $\quad \tau_v = k_s \tau_c = 1 \times 0.16 \sqrt{20} = 0.715$ N/mm$^2$,

$$d_3 = \frac{567.3 \times 1000}{4 \times 2471 \times 0.715} = 80.02 \cong 81 \text{ mm. Hence maximum value of } d_3 = 142 \text{ mm}$$

**6. Total Thickness.** At the section before first step, $d_1 = 540$ mm. Hence effective depth in the other direction $= 540 + 16 = 556$ mm. Therefore $D_1 = 556 + 60 = 616$ mm.

Provide $D_1 = 620$ mm. At the section after second step, $d_2 = 377$ mm. Hence effective depth in the other direction $= 377 + 16 = 393$ mm and $D_2 = 393 + 60 = 453$ mm. Provide $D_2 = 460$ mm. For the outer edge $d_3 = 142$ mm for one direction and $142 + 16 = 158$ mm for the other direction. Hence $D_3 = 158 + 60 = 218$ mm. Provide $D_3 = 220$ mm.

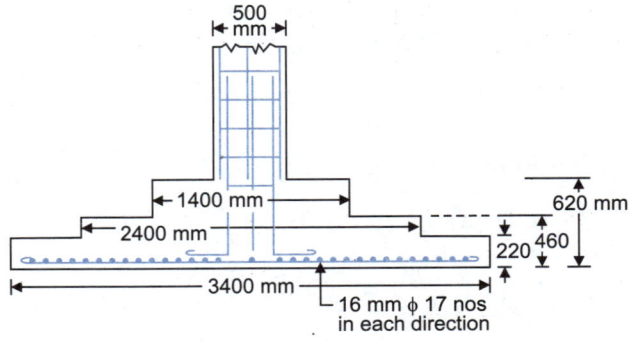

Fig. 15.29

**Design: Example 15.7.** *Design a footing for circular column 560 mm diameter, transmitting an axial load of 1200 kN. The column is reinforced with 8 bars of 10 mm diameter. The sage bearing capacity of soil is 120 kN/m$^2$. Use M 20 concrete and Fe 415 steel.*

**Solution:** (**a**) **Circular footing**

**1. Design constants.** For M 20 concrete, we have $c = \sigma_{cbc} = 7$ N/mm$^2$; For Fe 415, $t = \sigma_{st} = 230$ N/mm$^2$; Hence $k_c = 0.289$; $j_c = 0.904$; $R_c = 0.914$.

**2. Size of footing.** $W = 1200$ kN. Assume $W' = 10\% \, W = 120$ kN

$\therefore \quad$ Area of footing $A = \dfrac{1200 + 120}{120} = 11$ m$^2$.

Radius $R = \sqrt{A / \pi} = \sqrt{11 / 3.14} = 1.87$ m.
Adopt radius $R = 1.9$ m or diameter $= 3.8$ m at the base.
Radius of column $r = 280$ mm $= 0.28$ m

Net upward pressure $p_0 = -\dfrac{W}{\pi R^2} = \dfrac{1200}{\pi (1.9)^2} = 106$ kN/m$^2$

**3. Design of footing.** Refer Fig. 15.30.
Upward force on one quadrant of footing $= p_0 \dfrac{\pi}{4} (R^2 - r^2)$

Distance of C.G. of shaded area from the face of the column

$$= \frac{0.2}{R + r} [3R^2 - 2Rr - 2r^2]$$

Bending moment at the face of the column is given by

$$M = p_0 \frac{\pi}{4} (R^2 - r^2) \times \frac{0.2}{R + r} [3R^2 - 2Rr - 2r^2]$$

$$= p_0 \frac{\pi}{20} (R - r)(3R^2 - 2Rr - 2r^2)$$

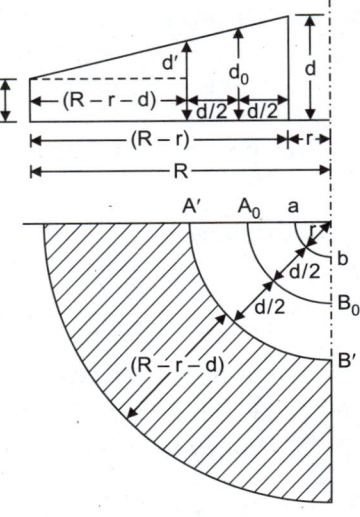

Fig. 15.30

**378** REINFORCED CONCRETE STRUCTURE

$$= \frac{106\,\pi}{20}\,(1.9 - 0.28)\,[3\,(1.9)^2 - 2 \times 1.9 \times 0.28 - 2\,(0.28)^2]\text{ kN-m}$$

$$= 2.592 \times 10^2 \text{ kN-m} = 2.592 \times 10^8 \text{ N-mm}$$

Breadth $b$ of quadrant $= \dfrac{1}{2}\pi r = \dfrac{\pi}{2}(280) = 440$ mm.

$$\therefore \quad d = \sqrt{\frac{2.592 \times 10^8}{440 \times 0.914}} = 802.8 \simeq 804 \text{ mm}.$$

**4. *Design for shear*.** The depth found above from bending compression should be adequate from shear point of view. Circular footings are checked for two way shear (punching shear) at a circular critical section distant $d/2$ ($\approx 402$ mm) away from the face of the column.

$$\therefore \quad r_0 = r + \frac{d}{2} = 280 + 402 = 682 \text{ mm}$$

Punching shear $\quad F = p_0\,\pi\,(R^2 - r_0^2) = 106\,\pi\,[(1.9)^2 - (0.682)^2] = 1047.3$ kN

Hence from *[Eq. 15.37(a)]*, $\quad \dfrac{F}{2\,\pi\,r_0\,d_0} \le 0.16\,\sqrt{f_{ck}}$

From which $d_0 = \dfrac{F}{2\,\pi\,r_0 \times 0.16\,\sqrt{f_{ck}}} = \dfrac{1047.3 \times 1000}{2\pi \times 682 \times 0.16\,\sqrt{20}} \approx 342$ mm  ...(3)

Let the effective depth $d_1$ at the outer edge be 140 mm. Then the available depth $d_0$ at the critical plane, corresponding to $d = 804$ mm (found from bending compression) is given by (Fig. 15.30)

$$d_0 = d_1 + \frac{(d - d_1)}{R - r}\left(R - r - \frac{d}{2}\right) = 140 + \frac{804 - 140}{1900 - 280}\left[1900 - 280 - \frac{804}{2}\right] = 640 \text{ mm}$$

This is more than the one found above from punching shear of view.

For *beam shear*, the critical plane $A'\,B'$ lies at a radial distance $d$ from the column face (Fig. 15.30), where the effective depth $d'$ is given by

$$d' = d_1 + \frac{(d - d_1)}{R - r}[R - r - d] = 140 + \frac{804 - 140}{1900 - 280}[1900 - 280 - 804] = 475 \text{ mm}$$

Breadth $b'$ at this section $= \dfrac{\pi}{2}(r + d) = \dfrac{\pi}{2}(280 + 804) = 1702.7$ mm

Shear force $\quad V = p_0\,\dfrac{\pi}{4}\{R^2 - (r + d)^2\} = 106 \times \dfrac{\pi}{4}[(1.9)^2 - (0.28 + 0.804)^2] = 202.7$ kN.

$$\tau_v = \frac{V}{b'd'} = \frac{202.7 \times 1000}{1702.7 \times 475} = 0.251 \text{ N/mm}^2$$

Permissible shear stress at $p = 0.44\%$ is $\tau_c = 0.28$ N/mm$^2$ (Table 3.1). Hence $d = 804$ mm found on the basis of bending compression is adequate.

**5. *Design of reinforcement*:**

$$A_{st} = \frac{M}{\sigma_{st}\,j_c\,d} = \frac{2.592 \times 10^8}{230 \times 0.904 \times 804} = 1551 \text{ mm}^2. \text{ Using 12 mm } \Phi \text{ bars, } A_\Phi = 113 \text{ mm}^2$$

$\therefore\quad$ No. of bars required in each direction $= 1551/113 \simeq 14$

Hence, provide 14 bars each, in two mutually perpendicular directions. These bars are to be accommodated in the form of square mesh, centrally located under the column. If $B$ is the size of inscribed square, its magnitude

$$= \frac{D}{\sqrt{2}} = \frac{2R}{\sqrt{2}} = R\sqrt{2}$$

$$= 1.9\sqrt{2} \approx 2.7 \text{ m.}$$

Average spacing of bars = 2700/13 = 208 mm c/c

Provide two additional bars in each sector beyond this square. Provide 60 mm cover to centre of steel.

Total depth at edge = 140 + 60 = 200 mm.
Total depth at column face = 804 + 60
= 864 ≈ 870 mm.

### 6. Check for Development Length

$$L_d \approx 45\,\Phi = 45 \times 12 = 540 \text{ mm}$$

Available length = $(R - r)$ – side cover = 1900 – 280 – 60
= 1560 mm > $L_d$

Hence safe.

The details of reinforcement etc. are shown in Fig. 15.31.

**(b) Square footing:** For square footing, the width $b$ of an equivalent square column is given by

$$b^2 = \pi r^2$$

$$\therefore \quad b = r\sqrt{\pi} = 280\sqrt{\pi} = 496 \approx 500 \text{ mm}$$

Thus the problem reduces to the design of a square footing for a square column of size 500 mm × 500 mm carrying an axial load of 1200 kN. The footing design will be practically the same as worked out in Example 15.6.

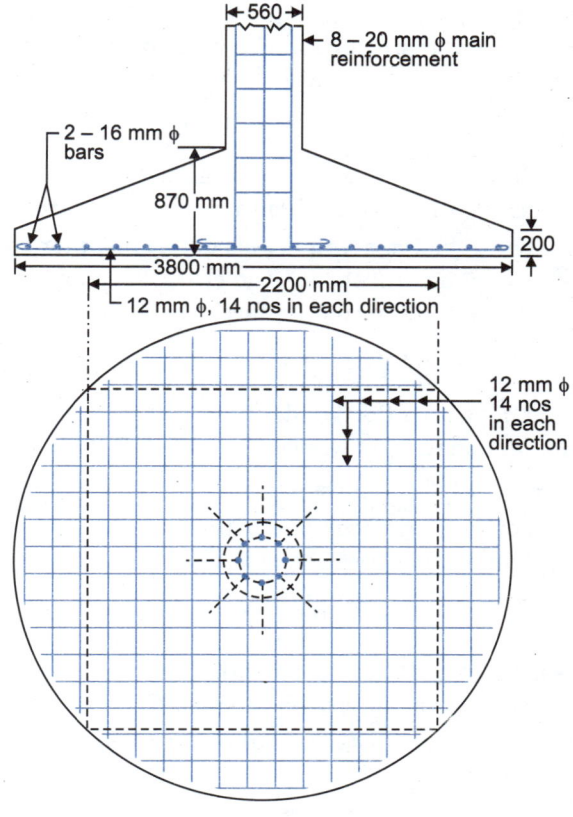

Fig. 15.31

## 15.10. ISOLATED FOOTING SUBJECTED TO ECCENTRIC LOAD

When a column is subjected to axial loading only, the C.G. of column area should coincide with the C.G. of footing area, so that the soil pressure distribution below the footing is *rectangular*. However, a column may be subjected to an axial load along with certain bending moment in one plane, or it may be subjected to an eccentric load. In both the cases, the soil pressure distribution will be trapezoidal if the C.G. of footing area coincides with the C.G. of column as shown in Fig. 15.33. If however, a rectangular pressure distribution is required, the C.G. of footing area may be placed eccentric with the C.G. of the column, by a distance $e = M/W$, or equal to eccentricity of axial load $W$, as shown in Fig. 15.32.

### (a) Design of unsymmetrical footing

Fig. 15.32 shows such a case in which the footing is arranged eccentric with respect to column, *i.e.* the axis Y-Y of footing is placed at a distance $e$ (= $M/W$) from the axis of the column. The pressure distribution will be rectangular.

As before, 
$$B^2 = \frac{W + W'}{q_0} \qquad \ldots(i)$$

and 
$$p_0 = W/B^2 \qquad \ldots(ii)$$

The bending moment will be maximum at face $AB$ of the column, since the cantilever distance in front of this face is more, its value being equal to $\dfrac{B}{2} + e - \dfrac{b}{2} = \dfrac{1}{2}(B - b) + e$.

$$\therefore \quad M = p_0\,B\left\{\frac{1}{2}(B - b) + e\right\} \times \frac{1}{2}\left\{\frac{1}{2}(B - b) + e\right\} = \frac{p_0\,B}{2}\left\{\frac{1}{2}(B - b) + e\right\}^2 \qquad \ldots(iii)$$

The section of footing at plane $AB$ will be trapezoidal and its moment of resistance can be determined by the method explained in section § 15.7. Alternatively, the trapezoidal section may be replaced by an equivalent rectangular section width $b + \frac{1}{8}(B - b)$.

Equating the moment of resistance to the external bending moment, the effective depth can be calculated.

The effective depth is also determined (or checked) from two way shear (punching shear), the critical section for which is at distance $d/2$ all round the column face. The width of the section = $b_0 = (b + d)$ and the shear stress will be $\frac{F}{4 b_0 d_0}$ where $F$ = punching shear force and $d_0$ is the effective depth at the section. The shear stress should be less than $k_s \tau_c$ where $k_s = (0.5 + \beta_c)$ but not greater than 1 and $\tau_c = 0.16 \sqrt{f_{ck}}$. The shear force $F = p_0 [B^2 - b_0^2]$.

For *one way shear*, the critical section is located at distance $d$ from column face, where shear force $V$ is given by

$$V = p_0 B \left[\frac{1}{2}(B - b) + e - d\right] \qquad \therefore \quad \tau_v = \frac{V}{b' d'}$$

where $b'$ and $d'$ are the width of the section at N.A. and the effective depth respectively, of the section $CD$ (Fig. 15.32.).

The shear stress $\tau_v$ should not exceed the permissible shear stress $\tau_c$ (Table 3.1).

The steel reinforcement, parallel to $XX$ axis is given by

$$A_{st} = \frac{M}{\sigma_{st}\, jd}$$

The same steel reinforcement is provided in the other direction, though the bending moment will be less in the other direction because of lesser cantilever projection.

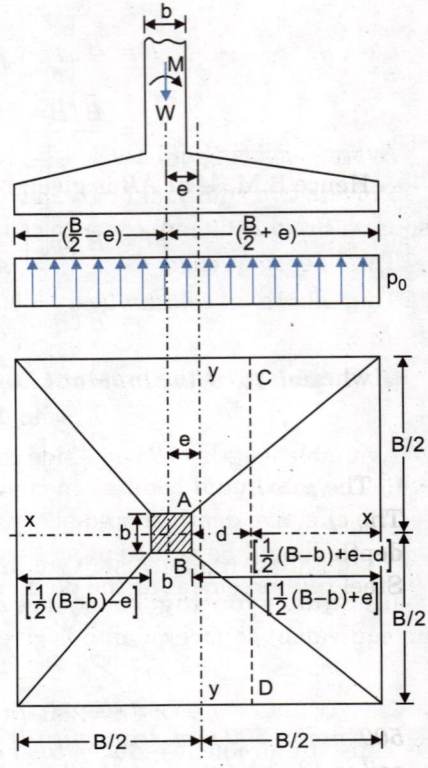

Fig. 15.32

### (b) Design of symmetrical footing

Fig. 15.33 shows a symmetrically placed footing so that C.G. of footing coincides with that of column. The pressure distribution below the footing will evidently be trapezoidal.

Let $B$ be the size of footing. The maximum stress is equated to the safe bearing capacity $q_0$ of the soil, to determine the magnitude of $B$:

$$\frac{W + W'}{B^2} + \frac{M}{B^3 / 6} = q_0$$

or $\quad 6M + (W + W') B = q_0 B^3 \qquad \ldots(15.39)$

The above equation can be solved for $B$, by trial and error.

The net upward pressure distribution is given by

$$p_0 = \frac{W}{B^2} \pm \frac{M}{B^3/6} \qquad \ldots(15.40)$$

so that pressure $p_{01}$ and $p_{02}$ at the two extremities are known. The lengths of cantilevers on all the four sides will be equal. However, the bending moment will be maximum at the face $AB$ since the pressure intensity is maximum to the right of it. The total upward force $F$ to the right of $AB$ is given by

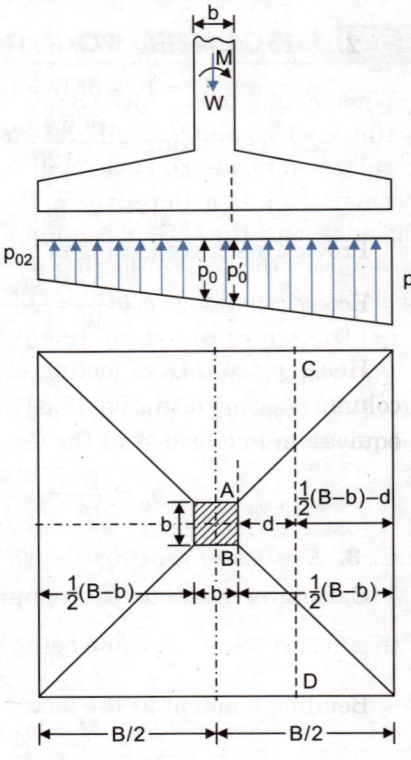

Fig. 15.33

# DESIGN OF ISOLATED FOOTINGS

$$F = B\left\{p_0'\left(\frac{B-b}{2}\right) + \frac{1}{2}(p_{01} - p_0')\left(\frac{B-b}{2}\right)\right\} = \frac{B(B-b)}{2}\left\{\frac{p_0' + p_{01}}{2}\right\}$$

$$= \frac{B(B-b)}{4}(p_0' + p_{01}) \qquad \ldots(15.41)$$

Hence B.M. $M$ at $AB$ is given by

$$M = B\left\{p_0'\left(\frac{B-b}{2}\right)\left(\frac{B-b}{4}\right) + \frac{1}{2}(p_{01} - p_0')\left(\frac{B-b}{2}\right)\frac{2}{3}\left(\frac{B-b}{2}\right)\right\}$$

$$= B(B-b)^2\left\{\frac{p_0'}{8} + \frac{p_{01} - p_0'}{12}\right\} = \frac{B(B-b)^2}{24}(2p_{01} + p_0') \qquad \ldots(15.42)$$

where $p_0'$ is the pressure intensity at $AB$, and is given by

$$p_0' = p_{02} + \frac{p_{01} - p_{02}}{B}\left\{\frac{1}{2}(B-b) + b\right\} = p_{02} + \frac{p_{01} - p_{02}}{2B}(B+b) \qquad \ldots(15.43)$$

The maximum bending moment can be equated to the moment of resistance of the section to get $d$. The effective depth should be sufficient so that punching shear stress is not exceeded. Similarly, effective depth should be sufficient so that nominal shear stress $\tau_v$ resulting from one way shear is not exceeded. Steel reinforcement at the plane of maximum B.M. is given by

$$A_{st} = \frac{M}{\sigma_{st}\, jd}.$$ The reinforcement is checked for the development length.

**Design: Example 15.8** *Design an isolated unsymmetrical square footing shown in fig. 15.34 for a column 500 mm × 500 mm, transmitting a load of 600 kN and a moment of 30 kN-m. The safe bearing capacity of soil is 120 kN/m². Use M 20 concrete and Fe 415 steel.*

**Solution:**

1. **Design constant.** For M 20 mix and Fe 415 steel,
   $k_c = 0.289; j_c = 0.904; R_c = 0.914$

2. **Size of footing.** $W = 600$ kN
   
   Assume $W' = 10\%$ of $W = 60$ kN
   
   Area of footing $= \dfrac{600 + 60}{120} = 5.5$ m²
   
   $\therefore \quad B = \sqrt{5.5} = 2.35$ m,
   
   Provide a square footing of size 2.4 m × 2.4 m.
   
   Eccentricity $\quad e = \dfrac{M}{W} = \dfrac{30 \times 1000}{600} = 50$ mm

   Hence place C.G. of footing of 50 mm away from the axis of column as shown in Fig. 15.34. Thus the load and moment is equivalent to a load $W$ at the C.G. of the footing.
   
   $$p_0 = \frac{W}{B^2} = \frac{600}{2.4 \times 2.4} = 104 \text{ kN/m}^2$$

3. **Design of footing**
   
   Cantilever length to the right of $AB$
   
   $$= \frac{1}{2}[2400 - 500] + 50 = 1000 \text{ mm} = 1 \text{ m}$$
   
   Bending moment at the face $AB$ of the column is given by
   
   $$M = p_0\, B\, \frac{(1)^2}{2} = 104 \times 2.4 \times \frac{1}{2} \text{ kN-m} = 124.8 \text{ kN-m} = 1.248 \times 10^8 \text{ N-mm}$$

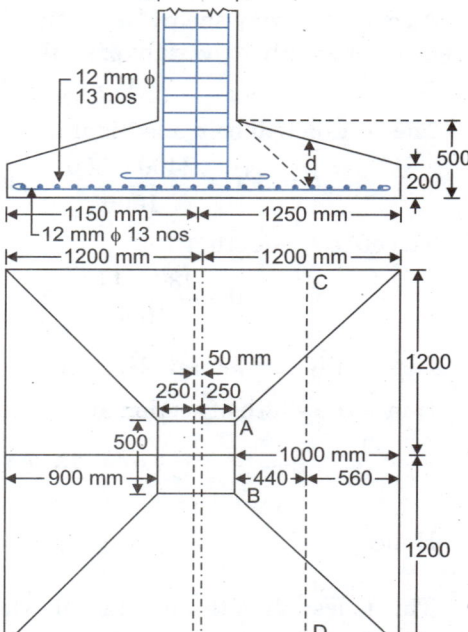

Fig. 15.34

The section of footing at AB will be trapezoidal. However its moment of resistance can be approximately determined by considering it to be a rectangle of width

$$= b + \frac{1}{8}(B - b) = 500 + \frac{1}{8}(2400 - 500) = 737.5 \approx 740 \text{ mm.}$$

$$\therefore \quad M = Rbd^2 = 0.914 \times 740\, d^2 \qquad \therefore d = \sqrt{\frac{1.248 \times 10^8}{0.914 \times 740}} = 430 \text{ mm}$$

### 4. Check for shear.

For two way shear (punching shear) the section is situated at distance $d/2$ from the column face all round. The side $b_0$ of the section is $b_0 = b + d = 500 + 430 = 930$ mm.

Shear force $F = p_0[B^2 - b_0^2] = 104\,[(2.4)^2 - (0.93)]^2 = 509.1$ kN

$$\therefore \quad \tau_v = \frac{F}{4b_0 d_0}. \text{ From which } d_0 = \frac{F}{4b_0 \tau_v}$$

Now permissible shear stress $= k_s\, \tau_c = 1 \times 0.16\sqrt{f_{sk}} = 0.16\sqrt{20} = 0.715$ N/mm²

Limiting $\tau_v$ to the above permissible shear stress,

$$d_0 = \frac{509.1 \times 1000}{4 \times 930 \times 0.715} = 191.4 \approx 192 \text{ mm.}$$

If an effective depth of 140 mm (total depth = 200 mm) is provided at the outer edge, and if effective depth $d = 430$ mm is kept at face $AB$ (found on the basis of bending compression), the available effective depth at $d/2$ from column face will be $= 140 + \dfrac{430 - 140}{1250 - 250} \times (1250 - 250 - 215) = 367.6$ mm, which is more than the required depth $d_0 = 192$ mm.

Hence the section will be safe in punching shear.

Also, for one way shear, the critical section $CD$ will be at a distance $d = 440$ mm from the column face. Cantilever length to the right of section $CD = 1000 - 430 = 570$ mm $= 0.57$ m. The shear force there is given by

$$V = p_0 B \times 0.57 = 104 \times 2.4 \times 0.57 \approx 142.3 \text{ kN}$$

The section will be trapezoidal in shape as shown in Fig. 15.35. The width $b$ at the top is given by

$$b = 500 + \frac{2400 - 500}{1000} \times 430 = 1317 \text{ mm}$$

The effective depth

$$d' = 140 + \frac{430 - 140}{1000} \times 570 = 305.3 \text{ mm}$$

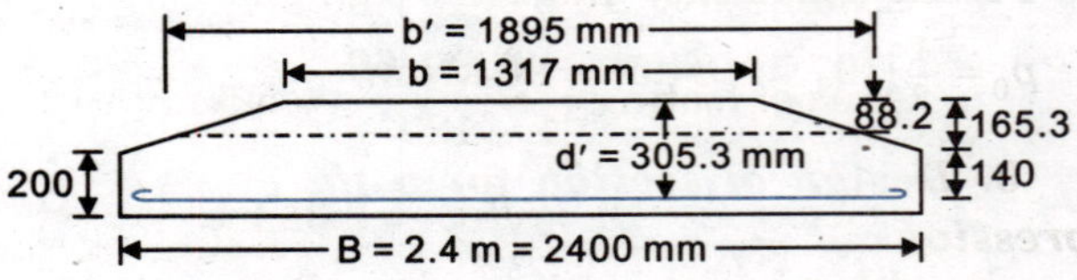

Depth of N.A. $= kd' = 0.289 \times 305.3 = 88.2$ mm

Hence the width of section at N.A. is given by

$$b' = 1317 + \frac{2400 - 1317}{165.3} \times 88.2 \approx 1895 \text{ mm}$$

Hence $$\tau_v = \frac{V}{b'd'} = \frac{142.3 \times 1000}{1895 \times 305.3} = 0.246 \text{ N/mm}^2.$$

Fig. 15.35

This is less than the permissible shear stress 0.28 N/mm² at $p = 0.44\%$ (Table 3.1) corresponding to a balanced section.

Hence $d = 430$ mm found on the basis of the bending compression is adequate.

Provide total depth = 430 + 60 = 490 mm $\approx$ 500 mm.

### 5. Steel reinforcement

$$A_{st} = \frac{M}{\sigma_{st}\, jd} = \frac{1.248 \times 10^8}{230 \times 0.904 \times 430} = 1396 \text{ N/mm}^2$$

Using 12 mm Φ bars $A_\Phi = 113$ mm².

∴ No. of bars = 1396/113 ≈ 13

Arrange them uniformly spaced in width of 2400 mm. Use the same reinforcement in the perpendicular direction.

**6. Check for the development length.** $L_d = 45\ \Phi = 45 \times 12 \approx 540$ mm

Using 60 mm side cover, available length = 1250 – 50 – 60 = 1140 mm > $L_d$.

Hence O.K. The details of reinforcement etc. are shown in Fig. 15.34.

**Design: Example 15.9.** *Redesign the isolated symmetrical square footing for the column of example 15.8.*

1. **Design constants:** As before, $k_c = 0.289$; $j_c = 0.904$ and $R_c = 0.914$
2. **Size of footing.** $W = 600$ kN.  $W' = 60$ kN; $M = 30$ kN-m

Equating the maximum soil pressure to the safe bearing capacity of soil,

$$\frac{660}{B^2} + \frac{30}{B^3/6} = 120$$

or $\quad 660\,B + 180 = 120\,B^3$

Solving this by trial and error, we get $B = 2.47$ m

However, provide the footing of size 2.5 m × 2.5 m.

The maximum and minimum soil pressure are given by

$$p_{01} = \frac{W}{B^2} + \frac{6M}{B^3} = \frac{600}{(2.5)^2} + \frac{6 \times 30}{(2.5)^3}$$

$$= 96 + 11.5 = 107.5 \text{ kN/m}^2.$$

$$p_{02} = \frac{W}{B^2} - \frac{6M}{B^3} = 96 - 11.5$$

$$= 84.5 \text{ kN/m}^2$$

Pressure intensity under the column axis is

$$p_0 = \tfrac{1}{2}[107.5 + 84.5] = 96 \text{ kN/m}^2$$

**3. Design of section for bending compression**

Arrange the footing centrally under the column as shown in Fig. 15.36.

Intensity of net soil pressure below the column face is

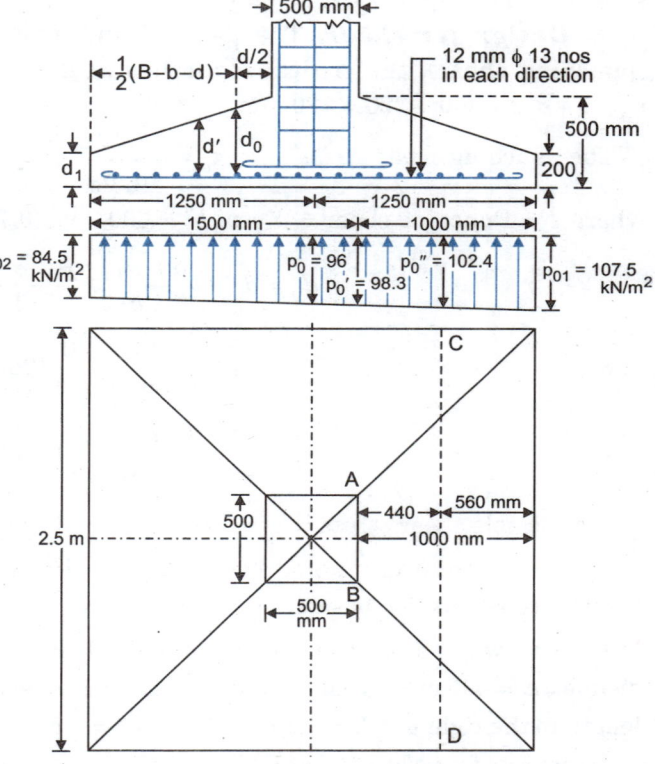

Fig. 15.36

$$p_0' = 96 + \frac{107.5 - 96}{1.25} \times 0.25 = 98.3 \text{ kN/m}^2$$

Maximum B.M. will occur at face $AB$. Cantilever length $\tfrac{1}{2}(B-b) = \tfrac{1}{2}(2.5 - 0.5) = 1$ m.

∴ Total force under cantilever = $(2.5 \times 1)\left[\dfrac{98.3 + 107.5}{2}\right] = 257$ kN

Distance of its centroid from $AB = \left(\dfrac{98.3 + 2 \times 107.5}{98.3 + 107.5}\right)\dfrac{1}{3} = 0.507$ m

∴ B.M. $M = 257 \times 0.507 = 130$ kN-m = $1.3 \times 10^8$ N-mm.

**384** REINFORCED CONCRETE STRUCTURE

Alternatively, from *Eq. 15.44*,

$$M = \frac{B(B-b)^2}{24}(2p_{01} + p_0') = \frac{2.5(2.5-0.5)^2}{24}[2 \times 107.5 + 98.3]$$

$$= 130 \text{ kN-m} = 1.3 \times 10^8 \text{ N-mm}.$$

The section will be trapezoidal in shape. Width of equivalent rectangle

$$= b + \frac{1}{8}(B-b) = 500 + \frac{1}{8}(2500-500) = 750 \text{ mm}.$$

$$\therefore \quad d = \sqrt{\frac{1.3 \times 10^8}{0.914 \times 750}} = 436 \text{ mm}.$$

Provide $d$ = 440 mm and total depth of 500 mm. Reduce the total depth at edges to 200 mm so that effective depth ($d_1$) will be 140 mm there.

**4. Design for shear.** The depth found above should be safe for shear. For two way shear (punching shear), the critical plane lies at $d/2 \approx 220$ mm from the column face for which width $b_0 = b + d/2 = 500 + 220 = 720$ mm.

The punching shear stress $\quad \tau_v = \dfrac{F}{4b_0 \, d_0}$

where $F$ = Punching shear = $W - p_0 \, b_0^2 = 600 - 96 \, (0.72)^2 \approx 550.3$ kN

$$d_0 = d_1 + \frac{d - d_1}{(B-b)/2}\left(\frac{B-b-d}{2}\right)$$

or $\quad d_0 = 140 + \dfrac{440 - 140}{2500 - 500}(2500 - 500 - 440) = 374$ mm

$$\therefore \quad \tau_v = \frac{550.3 \times 1000}{4 \times 720 \times 374} = 0.511 \text{ N/mm}^2.$$

Permissible shear stress = $k_s \times 0.16 \sqrt{f_{ck}} = 1 \times 0.16 \sqrt{20} = 0.715$ N/mm$^2 > \tau_v$

Hence the thickness found from point of view of bending consideration is safe.

For one way shear, the critical plane $CD$ lies at distance $d$ (= 440 mm) from column face. The cantilever length to the right of $CD$ = 1.25 − 0.25 − 0.44 = 0.56 m.

Intensity of pressure $p_0''$ at $CD$ is given by

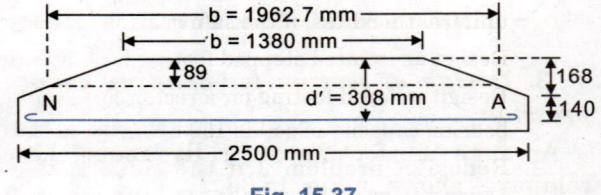

Fig. 15.37

$$p_0'' = 84.5 + \frac{107.5 - 84.5}{2.5}(2.5 - 0.56) \approx 102.4 \text{ kN/m}^2$$

The section at $CD$ will be trapezoidal as shown in Fig. 15.37. The width at top is given by

$$b = 500 + \frac{2500 - 500}{1000} \times 440 = 1380 \text{ mm}.$$

Effective depth $\quad d' = 140 + \dfrac{440 - 140}{1000} \times 560 = 308$ mm

Depth of N.A. = $kd' = 0.289 \times 308 = 89$ mm

$\therefore$ Width of the section at N.A. is given by

$$b' = 1380 + \frac{2500 - 1400}{168} \times 89 = 1962.7 \text{ mm}$$

∴ Shear force $V'$ at $CD$ is given by
$$V' = 2.5 \times 0.56 \times \frac{1}{2}[107.5 + 102.4] = 146.93 \text{ kN}$$

∴
$$\tau = \frac{V'}{b'd'} = \frac{146.93 \times 1000}{1962.7 \times 308} = 0.243 \text{ N/mm}^2$$

Permissible shear stress at $p = 0.44\%$ (balanced section) = 0.28 N/mm² (Table 3.1).
Hence thickness provided from point of view of bending compression adequate.

### 5. Design for bending tension

$$A_{st} = \frac{1.3 \times 10^8}{230 \times 0.904 \times 440} = 1421 \text{ mm}^2$$

∴ No. 12 mm Φ bars = 1450/113 ≃ 13

Provide the same reinforcement in the other direction.

### 6. Check for development length:

$$L_d = 45\ \Phi = 45 \times 12 = 540 \text{ mm}$$

Available length, providing a side a cover of 60 mm = 1000 – 60 = 940 mm.
Hence safe. The details of reinforcement etc. are shown in Fig 15.36.

## PROBLEMS

1. A steel stanchion carries 400 mm × 200 mm heavy beam section and carries a load of 1000 kN. Design a R.C.C. base for column. The safe bearing capacity of the soil may be taken as 100 kN/m² at a depth of 1 m below ground surface. Use M 20 concrete and Fe 415 steel.

2. A.R.C. column 400 mm × 400 mm in section, carries an axial load of 750 kN. Design sloping R.C. footing for the column using M 20 concrete and Fe 415 steel.
   Take safe bearing capacity of soil = 120 kN/mm²

3. Redesign problem 2, if the column is circular in section having diameter = 40 cm. Provide square footings.

4. Redesign problem 3, by providing circular base for the footing.

5. A rectangular R.C. column, 240 mm × 300 mm carries as axial load of 400 kN. Design a rectangular footing of uniform thickness, if the safe bearing capacity of the soil is 80 kN/m². Use M 20 concrete and Fe 415 steel.

6. Design as isolated stepped footing for the column of problem 2.

7. Design a sloped footing for a rectangular column 400 mm × 500 mm carrying an axial load of 800 kN. The safe bearing capacity of soil is 150 kN/m². Use M 20 concrete and Fe 415 steel.

8. Redesign problem 2 if the column also carries a moment of 150 kN-m along with axial load of 750 kN. The footing should be located centrally under the column.

9. Design a square footing of uniform thickness for reinforced concrete square column size of 500 mm transmitting an axial service load of 2700 kN. The safe bearing capacity of the soil at the site is 162 kN/m² and materials to be used all $M_{20}$ grade concrete and Fe 415 steel. Draw reinforcement details.

10. Design the reinforcement of 300 mm × 750 mm footing subjected to an axial factored load of 1380 kN and factored moments of 46 kN-m and 30 kN-m about the shorter and longer axis respectively. Assume $d' = 50$ mm and adopt $M_{30}$ grade and Fe 415 steel.

# 16
# COMBINED FOOTINGS

## 16.1. INTRODUCTION

A combined footing supports the load of two or more adjacent columns. Such a footing is provided under the following circumstances: (i) when the columns are very near to each other so that their footings overlap (ii) when the bearing capacity of soil is less, requiring more area under individual footing (iii) when the end column is near a property line, so that its footing cannot be spread in that direction. A combined footing may be rectangular or trapezoidal in plan depending on relative magnitudes of loads on the two columns which the footing supports. The aim is to get uniform pressure distribution under the footing. For this, the C.G. of footing area should coincide with the C.G. of the combined loads of the two columns. There are four types of combined footings commonly used:

1. Combined rectangular footing
2. Combined trapezoidal footing
3. Strap footing, and
4. Raft footing

Rectangular footing is used when the two columns carry equal loads. If, however, one column near the property line carries heavier load, provision of trapezoidal footing becomes essential. A strap footing consists of spread footings of two columns connected by a strap beam. A raft or mat is a combined footing that covers the entire area beneath a structure and supports all the walls and columns.

## 16.2. COMBINED RECTANGULAR FOOTING

**1. Proportioning of footing:** Fig. 16.1 (a) shows two columns, with loads $W_1$ and $W_2$ spaced at $l$ centre to centre. If $W'$ is the weight of the footing, and $q_0$ is the safe bearing capacity, the footing area is given by

$$A = \frac{W_1 + W_2 + W'}{q_0} \qquad \ldots (16.1)$$

Suitable values of length $L$ and breadth $B$ of the footing are chosen so that $B \times L = A$. The projections $a_1$ and $a_2$ should be so chosen that the C.G. of footing coincides with the C.G. of the two loads.

Let $\bar{x}$ = distance of C.G. of column loads from centre of column $A$.

Then
$$a_1 + \bar{x} = \frac{L}{2} = a_2 + l - \bar{x} \qquad \ldots (16.2)$$

From which $a_1$ and $a_2$ can be determined. The net upward pressure $p_0$ is given by

$$p_0 = \frac{W_1 + W_2}{L.B} \qquad \ldots (16.3)$$

**2. Bending pattern:** The footing will bend both in longitudinal as well as transverse directions, as shown. In the longitudinal direction, the footing will have sagging bending moment in the two cantilever portions, and hogging bending moment in some length of middle portion. In the transverse direction the footing will have sagging bending moment. Near the columns, therefore, the footing will have a tendency to bend in the form of saucer. The *transverse bending will decrease at distance away from the column, where the bending will be primarily longitudinal*. Hence sections near and around the column will be subjected to heavier bending stress.

(a) *Design for longitudinal bending*

The exact analysis of such a footing, having no transverse or longitudinal beam, is very difficult since the bending pattern is not precisely known at all points. The design is based primarily on empirical practices. One of the commonly adopted practice is to consider the footing as a longitudinal beam of width $B$. Thus, the footing will be a beam loaded with uniformly distributed load $w = p_0 B$ per unit length and supported on two columns with reactions $W_1$

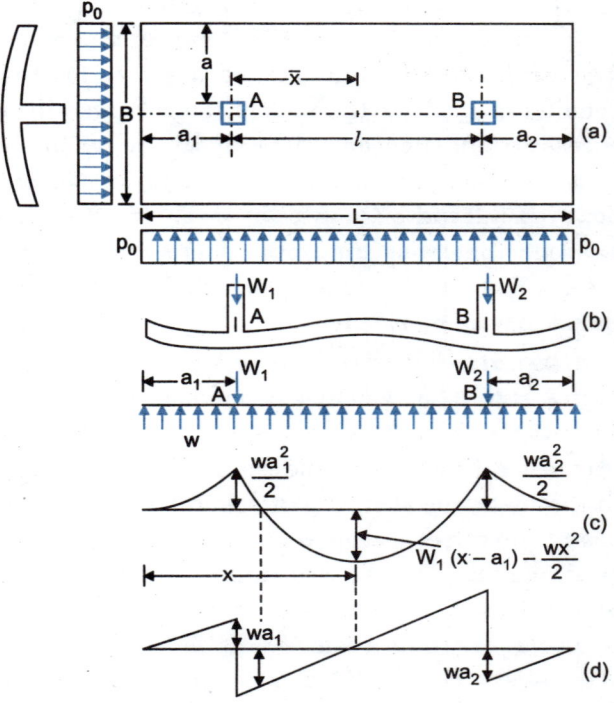

Fig. 16.1. Combined Rectangular Footing

and $W_2$ as shown in Fig 16.1 (b). The B.M. and S.F. diagrams are shown in Fig. 16.1(c) and (d) respectively. The maximum hogging bending moment will occur between $W_1$ and $W_2$ where S.F. is zero. The maximum sagging moment at the column axis will be $\frac{wx^2}{2}$, while the maximum hogging bending moment will be $W_1(x - a_1) - \frac{wx^2}{2}$, where $x$ is the distance of the section, from the edge, where shear is zero. However, the footing will be designed for sagging moment at the *outer face* of each column, and for the maximum hogging moment. The reinforcement will be placed on the bottom face for sagging bending moment and on the top face for the hogging moment.

The above analysis is based on the assumption that full width $B$ is available for longitudinal bending. However, in actual bending full width $B$ will not be available for the whole length of the footing. Fig. 16.2(a) shows a critical strip $A_1 B_1 C_1 D_1$ for transverse bending.

The total shear force on the two planes $A_1 D_1$ and $B_1 C_1$ plus the upward soil reaction on the strip $A_1 B_1 C_1 D_1$ will be equal to the column load $W_1$. Thus, the strip $A_1 B_1 B A$ will bend up like a cantilever, with a total uniformly distributed upward force equal to half the column load. If $p_0'$ is the net intensity of upward pressure on the strip, we have

$$p_0' = \frac{W_1}{\text{Area } A_1 B_1 C_1 D_1}$$

If the strip $A_1 B_1 B A$ is designed as cantilever with an upward pressure $p_0$ only, it will be over-strained and will yield, and then the effective available *width* for longitudinal bending will be $AD$ only and not the full width $B$. The available width however will increase from $AD = b$ at column face to $B$ at distance $B$ from the column face. Fig. 16.2 (b) shows the *probable* available widths shown shaded, for longitudinal bending all along the length.

There are three alternatives for the design of footing for longitudinal bending: (*i*) to design the footing for width equal to $AD = b$ at the column face, (*ii*) to provide a transverse beam of width $B_1$, of greater thickness than the rest of the slab, or (*iii*) to provide a longitudinal beam connecting the columns.

**Alternative (i):** If the footing is designed as longitudinal beam with a width $b$ only, at the column face, its thickness required for compression to resist bending moment at the face of the column will be excessive and uneconomical, specially when the cantilever length is more. For the hogging moment, however, full width $L$ may be used. This method may be adopted only when the cantilever projections $a_1$ or $a_2$ are small, so that the sagging moments are much less than the hogging moment.

**Alternative (ii): Transverse beam.** In order that strip $A\,B\,B_1\,A_1$ may not yield, another alternative is to provide a beam of width $B_1$ as shown in Fig. 16.1(c) to transfer the load $W_1$ of the column to the soil safely. In that case, the shaded portions in Fig. 16.2(c) will bend transversely, while the remaining portion will bend longitudinally as a beam utilizing full width $B$ for compression resistance. The width $B_1$ of the transverse beam may vary from a value $b$ to a value $B$. If the width of transverse beam $A_1\,B_1\,C_1\,D_1$ is kept equal to $b$ only, its thickness required will be much more than the rest of the thickness of the footing. On the other hand, if the width $B_1$ is made equal to width $B$, each portion under the two transverse beams of columns $A$ and $B$ will act as individual footings. An intermediate value of $B_1$, between $b$ and $B$, should therefore be chosen. Failure due to shear occurs approximately along $45°$ lines when footings are tested to destruction. This suggests that width of the transverse beam should be taken as $B_1 = b + 2d$, but not more than $B$. (Fig. 16.3).

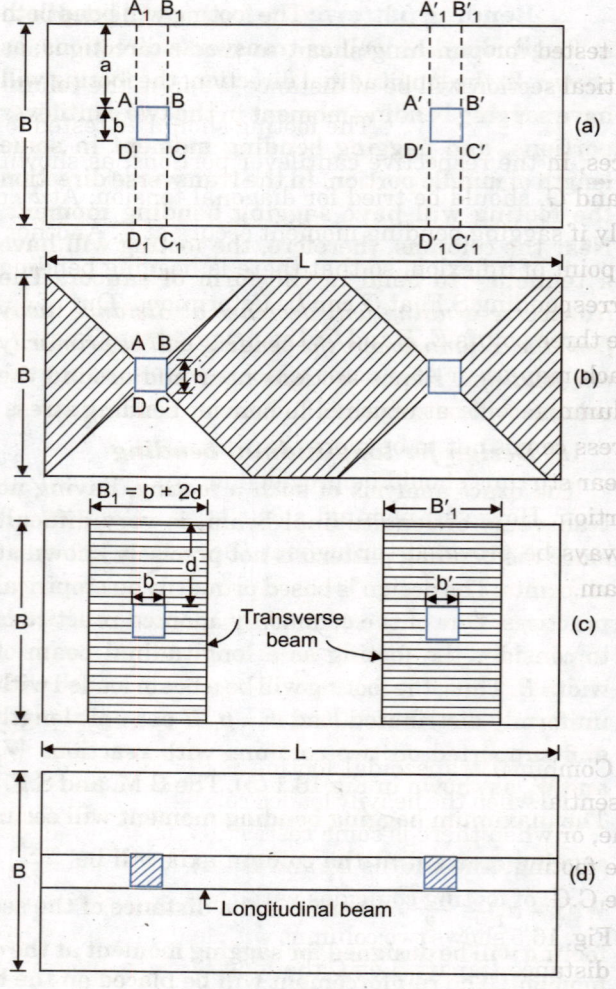

Fig. 16.2

Fig. 16.3

After having determined $B_1$, the upward soil pressure $p_0'$ for transverse bending will be $= \dfrac{W_1}{B_1 \times B}$. The thickness of the transverse beam should be found for bending moment at column face equal to $(p_0'\,B_1)\,a^2/2$. This thickness will normally be greater than the thickness $d$ for the rest of the footing. Even if both the thicknesses are equal, the reinforcement in the transverse direction should be found for the above B.M., and provided in width $B_1$. Similar reinforcement for width $B_1'$ for the other column should be provided. For the rest of the portion of footing, minimum transverse reinforcement @ 0.15% of the area of cross-section of the footing should be provided. In the longitudinal direction, the footing will be designed for B.M. and S.F. shown Fig. 16.1 (c) and (d).

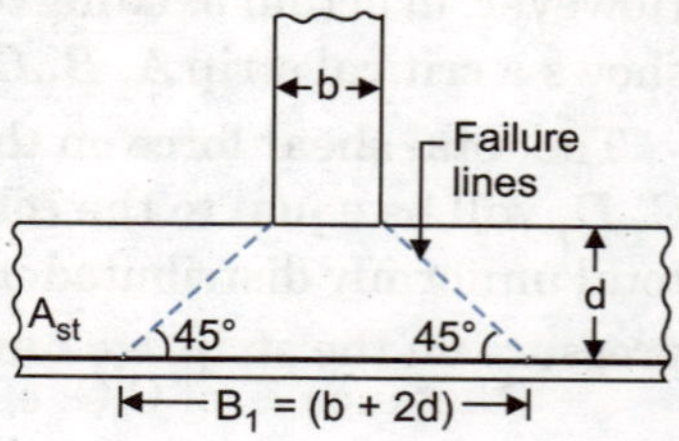

**Alternative (iii): Longitudinal beam.** A third and more common alternative is to provide a longitudinal beam, all along the length, joining the two columns as shown in Fig. 16.2(d). In this case, the beam alone will be subjected to bending moments and shear forces shown in Fig. 16.1(c), (d) in the longitudinal direction. The slab of the footing on either side of the beam is designed as cantilever, for the transverse bending. The web of the longitudinal beam should project *below* the slab if the sagging moments in the beams are more, so that T-beam action is available in the cantilever portions of the beam. If, however, the hogging bending moment is more, the web of the beam should project at the top of the slab so that T-beam action is available for central portion.

**3. Two way shear** (*punching shear*). The depth found on the basis of bending moment should always be tested for punching shear, so that punching shear stress does not exceed the permissible value. The critical section will be at distance $d/2$ from the column face.

**4. One way shear.** The footing should be tested for one way shear at a distance $d$ from the two column faces, in the respective cantilever portions, as shown in Fig. 16.4. For the central portion, two locations, $F$ and $G$, should be tried for diagonal tension. At $F$, distant $d$ from column face, tension crack will occur only if sagging bending moment occurs at $F$. Another possibility of tension crack is at $G$, just to the right of point of inflexion, so that there is hogging bending moment giving rise to a crack on the top face. The corresponding S.F. at $G$ can be determined. Out of the three points $E, F$ and $G$, the diagonal tensional crack may occur at a point which is *nearer* to the column, face, or at which S.F. is more. If the shear stress comes out to be more than permissible one, shear stirrups should be provided for the required portion. However, nominal shear stirrups should always be provided throughout the length of the beam.

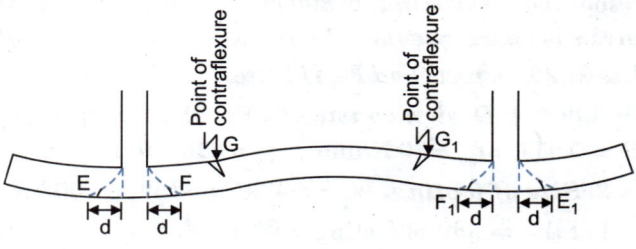

Fig. 16.4

**5. Development length.** The reinforcement should be tested for development length.

## 16.3. COMBINED TRAPEZOIDAL FOOTING

Combined trapezoidal footing for two columns becomes essential when the heavily loaded column is near the property line, or when there is some restriction on the total length of the footing. The widths $B_2$ and $B_1$ are so proportioned that the C.G. of footing coincides with the C.G. of column loads.

Fig. 16.5 shows two columns with loads $W_1$ and $W_2$, spaced at distance $l$ apart. Let $L$ the length of the footing, and $a_1$ and $a_2$ be the cantilever projections. The widths $B_1$ and $B_2$ are unknown. Hence,

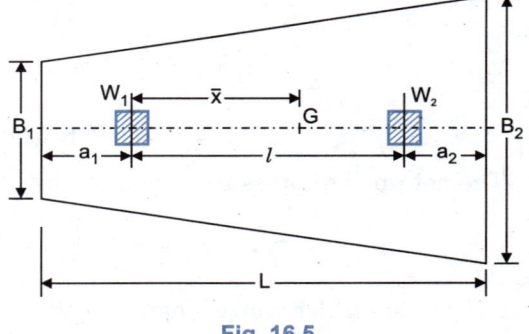

Fig. 16.5

$$\frac{(B_1 + B_2) L}{2} = \frac{W_1 + W_2 + W'}{q_0}$$

$$B_1 + B_2 = \frac{2(W_1 + W_2 + W')}{q_0 L} \qquad \text{... (1) ... (16.4)}$$

Distance $\bar{x}$ of C.G. of load $= \dfrac{W_2 \, l}{W_1 + W_2}$

∴ Distance of C.G. of load from short edge $= a_1 + \dfrac{W_2 \, l}{W_1 + W_2}$ ...(*i*)

Distance C.G. of trapezium from the short edge $= \dfrac{L}{3}\left(\dfrac{B_1 + 2B_2}{B_1 + B_2}\right)$ ...(*ii*)

Equating the two, $a_1 + \dfrac{W_2 \, l}{W_1 + W_2} = \dfrac{L}{3}\left(\dfrac{B_1 + 2B_2}{B_1 + B_2}\right)$ ...(16.5)

From *Eqs. 16.4* and *16.5*, the unknowns $B_1$ and $B_2$ are determined. The net upward soil pressure intensity $p_0$ will be uniform throughout, and its magnitude is given by

$$p_0 = \frac{W_1 + W_2}{\frac{1}{2}(B_1 + B_2)L} \qquad \ldots(16.6)$$

Knowing $p_0$, B.M. and S.F. diagrams may be drawn. The footing is designed exactly in the manner as for rectangular footing. Transverse reinforcement, for transverse bending, is provided near each column, in strip equal to width of column plus twice the effective depth of footing.

### Design: Example 16.1. *Combined rectangular footing*

*Design combined rectangular footing for two columns A and B, carrying loads of 500 and 700 kN respectively. Column A is 300 mm × 300 mm in size and column B is 400 mm × 400 mm in size. The centre to centre spacing of the columns is 3.4 metres. The safe bearing capacity of soil may be taken as 150 N/m². Use M 20 concrete and Fe 415 steel.*

**Solution: 1. *Design constants.*** For M 20 concrete and Fe 415 steel combination $k_c = 0.289$; $j_c = 0.904$ and $R_c = 0.914$, $f_{ck} = 20$ N/mm², $f_y = 415$ N/mm²

**2. *Size of footing.*** $W_1 = 500$ kN and $W_2 = 700$ kN

Let the weight of footing = $W' = 10\%$ of $(W_1 + W_2) = 120$ kN

$$\therefore \quad A = \frac{500 + 700 + 120}{150} = 8.8 \text{ m}^2. \text{ Let the size of the footing be } 1.8 \text{ m} \times 5 \text{ m}.$$

The projections $a_1$ and $a_2$ should be such that C.G. of footing coincides with the C.G. of column loads. The distance $\bar{x}$ of the C.G. of column loads from centre of column $A$ is given by

$$\bar{x} = \frac{W_2 \, l}{W_1 + W_2} = \frac{700 \times 3.4}{500 + 700} \approx 2 \text{ m}$$

$$\therefore \quad a_1 + \bar{x} = \frac{L}{2} \quad \text{or} \quad a_1 = \frac{L}{2} - \bar{x} = \frac{5}{2} - 2 = 0.5 \text{ m}$$

$$a_2 = L - (l + a_1) = 5 - (3.4 + 0.5) = 1.1 \text{ m}$$

The net upward pressure $p_0$ is given by

$$p_0 = \frac{W_1 + W_2}{B \times L} = \frac{500 + 700}{1.8 \times 5} = \frac{400}{3} \text{ kN/m}^2$$

$\therefore$ Pressure $w$ per metre length $= p_0 B = \dfrac{400}{3} \times 1.8 = \mathbf{240 \text{ kN/m}^2}$

**3. *B.M. and S.F. diagrams***

The B.M. and S.F. diagrams in the longitudinal directions are shown in Fig. 16.6.

S.F. just to the left of $A = 240 \times 0.5 = 120$ kN

S.F. just to the right of $A = 500 - 120 = 380$ kN

S.F. just to the right of $B = 240 \times 1.1 = 264$ kN

S.F. just to the left of $B = 700 - 264 = 436$ kN

S.F. will be zero at distance $x$ from $A$, its magnitude being given by $x = 380/240 = 1.585$ m.

The maximum hogging B.M. $M_1$ will therefore occur at this section, its magnitude being given by

$$M_1 = (500 \times 1.585) - \frac{240}{2}(0.5 + 1.585)^2 \approx 271 \text{ kN-m} = 271 \times 10^6 \text{ N-mm}$$

Sagging B.M. at $B = \dfrac{240 \, (1.1)^2}{2} = 145.2$ kN-m $= 145.2 \times 10^6$ N-mm

Sagging B.M. at $A = \dfrac{240}{2}(0.5)^2 = 30$ kN-m $= 30 \times 10^6$ N-mm

Sagging B.M. at the outer face of column $B$ is

$$M_2 = \frac{240}{2}(0.9)^2 = 97.2 \text{ kN-m}$$

$$= 97.2 \times 10^6 \text{ N-mm}$$

Sagging B.M. at the outer face of column $A$ is

$$M_3 = \frac{240}{2}(0.35)^2 = 14.7 \text{ kN-m}$$

$$= 14.7 \times 10^6 \text{ N-mm}$$

Let the point of contraflexure occur at $x$ from the centre of column $A$.

Then $\quad M_x = 500x - \dfrac{240}{2}(x+0.5)^2 = 0$

which gives $x = 0.081$ or $3.086$ m

Hence first point of contraflexure occurs at 0.081 m to the right centre of column $A$ and the second one occurs at 3.086 m from $A$, or $3.4 - 3.086 = 0.314$ m to the left of centre of column $B$.

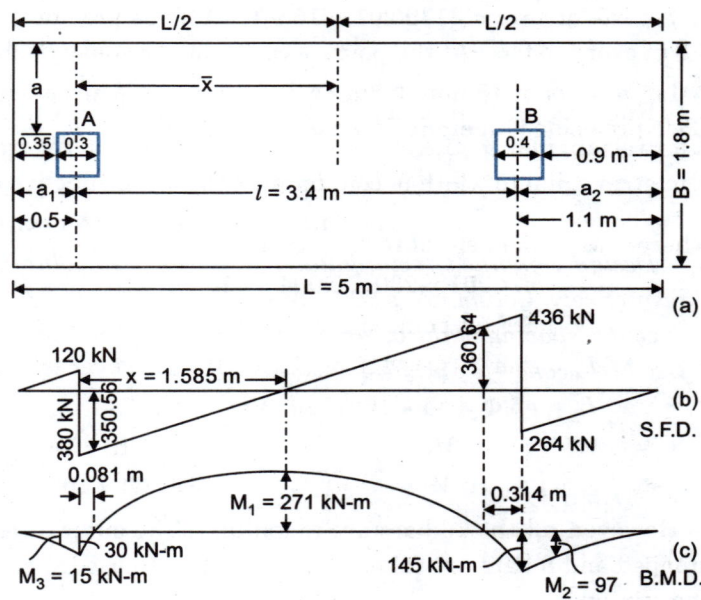

**Fig. 16.6.** Layout, Loading, Shear Force and Bending Moment Diagram of the Combined Footing

S.F. at the first point of contraflexure $= 380 - (240 \times 0.081) = 360.56$ kN

S.F. at the second point of contraflexure $= 436 - (240 \times 0.314) = 360.64$ kN

### 4. Effective depth from bending compression

The effective depth $d$ for hogging B.M. $M_1$ is given by

$$d = \sqrt{\frac{271 \times 10^6}{0.914 \times 1800}} = 406 \text{ mm}$$

### 5. Check for punching shear.

Let us check the above depth for punching shear, for which the critical plane lies at $d/2 = 203$ mm around column $B$. Width $b_0$ of critical plane $= 400 + 406 = 806$ mm.

Punching shear force at column $B$

$$= 700 - \frac{400}{3} \times (0.806)^2 \approx 613.38 \text{ kN.}$$

$\therefore \quad$ Shear stress $= \dfrac{613.38 \times 1000}{4 \times 806 \times 406} = 0.469$ N/mm$^2$.

Permissible shear stress $= 1 \times 0.16 \sqrt{f_{ck}} = 1 \times 0.16 \sqrt{20} = 0.715$ N/mm$^2$. Safe.

Using 60 mm cover to the centre of reinforcement, $D = 406 + 60 = 466$ mm.
However, provide $D = 470$ mm so that available $d = 470 - 60 = 410$ mm.

### 6. Design for bending tension in longitudinal direction

**(i) Reinforcement for hogging bending moment $M_1$**

$$A_{st} = \frac{M_1}{\sigma_{st} \, j_c d} = \frac{271 \times 10^6}{230 \times 0.904 \times 410} = 3179 \text{ mm}^2$$

Using 16 mm $\Phi$ bars, having $A_\Phi = \dfrac{\pi}{4}(16)^2 = 201$ mm$^2$

∴ No. of bars = 3179/201 ≈ 15.9 ≈ 16. Thus provide 16 bars uniformly distributed over the width of 1.8 m over top face. At the point of contraflexure near $B$, shear force $F = 360.64$ kN.

Let $n$ = no. of 16 mm Φ bars required at the point of contraflexure, from the point of requirements of Code provisions inherent in the equation

$$\frac{M_1}{V} + L_0 \geq L_d$$

where $M_1$ = moment of resistance of the section = $A_{st}\, \sigma_{st}\, j_c\, d$

$= n \times 201 \times 230 \times 0.904 \times 410 = n \times 17.13 \times 10^6$ N-mm

$V$ = S.F. at point of contraflexure = $360.64 \times 10^3$ N

$L_0 = d$ or $12\,\Phi = 410$ or $(12 \times 16)$ which ever is greater = 410 mm

$L_d \approx 45\,\Phi = 45 \times 16 = 720$ mm

∴ $\dfrac{M_1}{V} + L_0 \geq L_d$ or $\dfrac{n \times 17.13 \times 10^6}{360.64 \times 10^3} + 410 = 720$. From which $n = 6.6 \approx 7$

However, all the 16 bars are taken up to the outer faces of both the columns. Out of these, 8 bars are stopped and 8 bars are taken straight upto either edge of footing. These bars will serve as anchorage for the stirrups.

### (ii) Reinforcement for sagging B.M. $M_2$

$$A_{st2} = \frac{M_2}{\sigma_{st}\, j_c\, d} = \frac{97.2 \times 10^6}{230 \times 0.904 \times 410} = 1140.2 \text{ mm}^2$$

Using 12 mm Φ bars, No. of bars = $1140.2/113 = 10.1 \approx 11$

These no. of bars should be sufficient from development length point of view so as to satisfy the following criterion at the point of contraflexure near the inner face of column $B$.

$$\frac{M_1}{V} + L_0 \geq L_d$$

Let $n$ = number of bars required to satisfy this criterion

∴ $M_1 = A_{st}\, \sigma_{st}\, j_c\, d = (n \times 113)\, 230 \times 0.904 \times 410 = n \times 9.63 \times 10^6$ N-mm

$V$ = S.F. = 360.64 kN = $360.64 \times 10^3$ N

$L_0 = d$ or $12\,\Phi$ whichever is more = 410 mm.

$L_d = 45\,\Phi = 45 \times 12 = 540$ mm.

$$\frac{n \times 9.63 \times 10^6}{360.64 \times 10^3} + 410 = 540, \text{ which gives } n = 4.9 \approx 5.$$

Hence 11 bars of 12 mm Φ will be sufficient. These bars should be extended by $l_0$ (= 410 mm) beyond the point of contraflexure. At this point curtail 3 bars and continue 8 bars throughout the length to serve an anchor bars for stirrups.

### (iii) Reinforcement for sagging B.M. $M_3$

$$A_{st3} = \frac{15 \times 10^6}{230 \times 0.904 \times 410} \approx 176 \text{ mm}^2$$

Minimum reinforcement @ 0.12% of cross-sectional area = $\dfrac{0.12 \times 470 \times 1800}{100} = 1015.2 \text{ mm}^2$

∴ No. of 12 mm Φ bars = $1015.2/113 \approx 9$. However provide 11 bars, as for $M_2$. Let us find the reinforcement from development length point of view so that the following relation is satisfied at the point of contraflexure near the inner face of column $A$:

$$\frac{M_1}{V} + L_0 \geq L_d$$

Let    $n$ = number of bars required to satisfy the above criterion
$M_1 = A_{st}\, \sigma_{st}\, jd = (n \times 113) \times 230 \times 0.904 \times 410 = n \times 9.63 \times 10^6$ N-mm
$V$ = shear force at point of contraflexure = 360.56 kN = 360.56 × 10³ N
$L_0 = 12\,\Phi$ or $d$ whichever is more = 410 mm
$L_d = 45\,\Phi = 45 \times 12 = 540$ mm.

Substituting the values, $\dfrac{n \times 9.63 \times 10^6}{360.56 \times 10^3} + 410 = 540$

From which $n = 4.9 \approx 5$. However provide 11 bars of 12 mm φ. Continue these upto 440 mm beyond the point of contraflexure. At this point, curtail 3 bars and continue remaining 8 bars to serve as anchor bars for shear stirrups.

**7. Check for one way shear.** In the cantilever portion, test for one way shear (diagonal tension) is made at a distance $d = 0.41$ m from the column face or at $0.41 + 0.20 = 0.61$ m from the centre of column $B$. Shear $V = 264 - 240 \times 0.61 = 117.6$ kN.

∴ $\tau_v = \dfrac{117.6 \times 1000}{1800 \times 410} = 0.16$ N/mm²

Permissible shear stress = $k\,\tau_c = \tau_c = 0.28$ N/mm² for $p = 0.44\%$ (balanced section), obtained from Table 3.1. Hence safe. Similarly, in the cantilever portion near column $A$, the footing will be safe in shear.

For diagonal tension between $A$ and $B$, near column $B$, crack can occur at the bottom face of the footing (i.e. for sagging B.M.) at a distance of $d = 410$ mm, or at the top of footing (i.e., for the hogging B.M.) at the point of contraflexure distant 314 mm from the centre of the column, whichever is nearer. Shear force at the point of contraflexure = 360.64 kN which is more.

∴ $\tau_v = \dfrac{360.64 \times 1000}{1800 \times 410} = 0.49$ N/mm²

which is more than the permissible shear stress $\tau_c = 0.28$ N/mm².

Hence shear reinforcement is necessary.

Using 12 mm Φ 8-legged stirrups $A_{sv} = 8 \times 113 = 904$ mm²

Spacing    $s_v = \dfrac{\sigma_{sv}\,A_{sv}\,d}{V_s} = \dfrac{230 \times 904 \times 410}{360.64 \times 1000} \approx 236$ mm

Hence provide these @ 230 mm c/c.
The S.F. for which stress is 0.28 N/mm² is = (0.28/0.49) × 360.64 = 206.08 kN

This occurs at a distance of $\dfrac{436 - 206.08}{240} = 0.958$ m from the centre of the column $B$ or 0.758 m from the inner face of the column. Hence provide at least 4 stirrups @ 230 mm c/c, for a distance of 0.8 m from the face of the column $B$.

Similarly, near column $A$, tension crack can occur in the hogging portion at the inner face of the column since the point of contraflexure happens to fall under the column. S.F. at the inner face = 380 − 240 × 0.15 = 344 kN

∴ $\tau_v = \dfrac{344 \times 1000}{1800 \times 410} = 0.466$ N/mm²

against a permissible value of 0.28 N/mm². Hence shear reinforcement is necessary.

Hence provide 3 Nos. 12 mm Φ 8-legged stirrups @ 230 mm c/c, for a distance of 600 mm from the inner face of the column. Spacing of nominal stirrups is given by

$$s_v = \frac{2.175 A_{sv} f_y}{b} = \frac{2.175 \times 904 \times 415}{1800} = 453 \text{ mm subject to a maximum of 300 mm.}$$

Hence for the rest of the length, provide 12 mm Φ 8-lgd nominal stirrups @ 300 mm c/c.

**8. Transverse reinforcement.** The footing will bend transversely near each column face. Projection $a$ beyond the face of column $A = \frac{1}{2}(1800 - 300) = 750$ mm. Width $B_1$ of bending strip $= b + 2d = 300 + 2 \times 410 = 1120$ mm. However, width available to the left of outer face of column $A = 500 - 150 = 350$ mm only instead of 410 mm. Hence available

$$B_1 = 350 + 300 + 410 = 1060 \text{ mm} = 1.06 \text{ m}.$$

Net upward pressure $\quad p_0' = \dfrac{W_1}{B \times B_1} = \dfrac{500}{1.8 \times 1.06} = 262.05 \text{ kN/m}^2$

Maximum B.M. will occur the face of the column.

$$M = p_0' \frac{a^2}{2} = 262.05 \frac{(0.75)^2}{2} = 73.70 \text{ kN-m}$$

$$= 73.70 \times 10^6 \text{ N-mm}.$$

$$d = \sqrt{\frac{73.70 \times 10^6}{0.914 \times 1000}} = 284 \text{ mm}.$$

Hence actual $d = 410$ mm provided earlier is sufficient. The transverse beam will thus be of the same thickness as the rest of the footing. Since transverse reinforcement will be provided *above* the longitudinal one, available $d = 410 - 12 = 398$ mm.

The transverse reinforcement is given by

$$A_{st} = \frac{73.70 \times 10^6}{230 \times 0.904 \times 398} = 891 \text{ mm}^2$$

Using 12 mm Φ bars, spacing is given by

$$s = \frac{1000 A_\phi}{A_{st}} = \frac{1000 \times 113}{891} = 126.8 \approx 126 \text{ mm}$$

Provide these @ 120 mm c/c. The reinforcement provided above should have sufficient development length.

$$L_d \approx 45 \Phi = 45 \times 12 \approx 540 \text{ mm}.$$

Using 60 mm clear cover on the sides, length of bar available $= 750 - 60 = 690$ mm. Hence safe.

Similarly, for column $B$, width $B_1$ for transverse beam $= 400 + 2 \times 410 = 1220$ mm.

∴ $\quad p_0' = \dfrac{W_2}{B \times B_1} = \dfrac{700}{1.8 \times 1.22} = 318.76 \text{ kN/m}^2$

**Cantilever projection**

$$a = \frac{1}{2}[1800 - 400] = 700 \text{ mm} = 0.7 \text{ m}$$

∴ $\quad M = p_0' \dfrac{a^2}{2} = 318.76 \dfrac{(0.7)^2}{2} = 78.1 \text{ kN-m} = 78.1 \times 10^6 \text{ N-mm}$

$$A_{st} = \frac{78.1 \times 10^6}{230 \times 0.904 \times 398} = 944 \text{ mm}^2$$

Using 12 mm Φ bars, $\quad s = \dfrac{1000 \times 113}{944} \approx 120 \text{ mm}.$

Hence provide these @ 120 mm c/c in the strip of width 1220 mm. The available length of the bars will just satisfy the requirements of the development length. For the rest of the footing, provide transverse reinforcement @ 0.12% of the area of cross-section.

$$\therefore \quad A_{st} = \frac{0.12}{100}(1000 \times 470) = 564 \text{ mm}^2$$

$$\therefore \quad \text{Spacing of 12 mm } \Phi \text{ bars} = \frac{1000 \times 113}{564} \simeq 200 \text{ mm}$$

The details of reinforcement etc. are shown in Fig. 16.7.

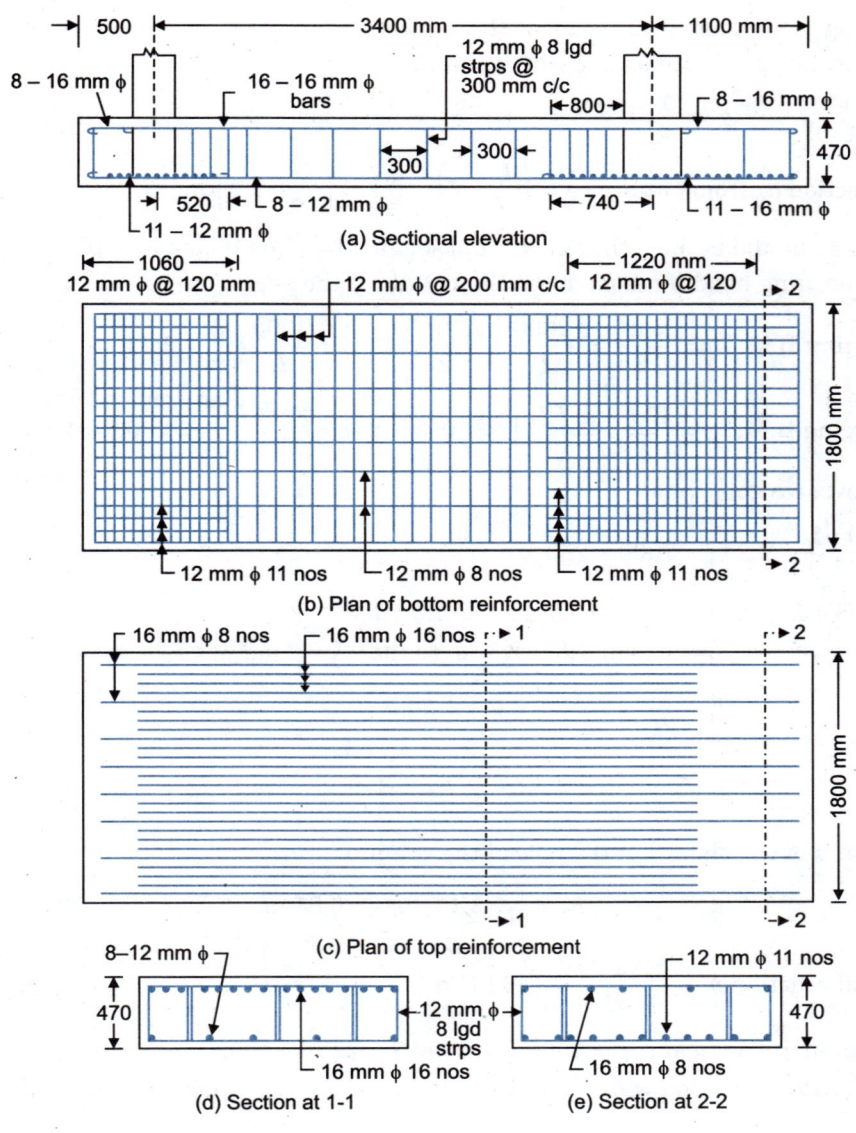

**Fig. 16.7**

## Design: Example 16.2. *Combined rectangular footing with central joining beam.*

*Two reinforced concrete columns 400 mm × 400 mm in section carry a load of 1000 kN each, inclusive of the self-weight. Design a combined footing having central beam joining the columns. The centre to centre spacing of the columns is 4 metres. The safe bearing capacity of soil is 150 kN/m². Use M 20 mix and Fe 415 steel reinforcement.*

**Solution:**

**1. Design constants.** For M 20 concrete – Fe 415 combination, we have the following:
$\sigma_{cbc} = 7$ N/mm$^2$; $\sigma_{st} = \sigma_{sv} = 230$ N/mm$^2$; $k_c = 0.289$; $j_c = 0.904$; $R_c = 0.914$

**2. Proportioning of footing.** $W_1 = W_2 = W = 1000$ kN
Let the weight of footing $W' = 0.1 (1000 + 1000) = 200$ kN

$$\therefore \quad \text{Area of footing} = \frac{1000 + 1000 + 200}{150} = 14.65 \text{ m}^2$$

For the beam, upward load $w$ per unit length $= \dfrac{1000 + 1000}{L} = \dfrac{2000}{L}$ kN/m

Since the footing is symmetrically loaded, the projection of the beam on side of either column will be equal. Projection $a_1$ measured from centre of column is

$$a_1 = \tfrac{1}{2}(L - l) = \tfrac{1}{2}(L - 4) = \frac{L}{2} - 2 \text{ metres.}$$

$\therefore$ Clear projection $a_2$ from outer face $= \left(\dfrac{L}{2} - 2\right) - 0.2 = \left(\dfrac{L}{2} - 2.2\right)$ m.

This projection $a_2$ should be such that maximum sagging bending moment at the *face* of column is equal to the maximum hogging bending moment as the middle of the beam.

Maximum sagging B.M. $= \dfrac{w\, a_2^2}{2} = \dfrac{2000}{L} \cdot \dfrac{1}{2} \left(\dfrac{L}{2} - 2.2\right)^2 = \dfrac{1000}{L}\left(\dfrac{L}{2} - 2.2\right)^2$ ...(i)

Maximum hogging B.M. $= \dfrac{W l}{2} - \dfrac{w}{2}\left(\dfrac{L}{2}\right)^2 = 1000 \times 2 - \dfrac{2000}{2L}\left(\dfrac{L}{2}\right)^2 = 2000 - 250 L$ ...(ii)

Equating the two, we get

$$\frac{1000}{L}\left(\frac{L}{2} - 2.2\right)^2 = 2000 - 250 L$$

or $\quad L^2 - 8.4 L + 9.68 = 0$. From which $L = 7.02$ m
Keep $\quad L = 7$ m and $B = 2.1$ m, so that $A = 7 \times 2.1 = 14.7$ m$^2$

$\therefore \quad a_1 = \dfrac{L}{2} - 2 = 3.5 - 2 = 1.5$ m;

and $\quad a_2 = \dfrac{L}{2} - 2.2 = 3.5 - 2.2 = 1.3$ m.

Clear projection $a$ of the slab from the face of the column is

$$a = \tfrac{1}{2}(B - b) = \tfrac{1}{2}(2.1 - 0.4) = 0.85 \text{ m.}$$

Net upward soil reaction $p_0 = \dfrac{2000}{7 \times 2.1} = 136$ kN/m$^2$

Upward load intensity on beam, $w = 2000/L = 2000/7 = 285.7$ kN/m.

**3. Design of slab.** Assuming width of beam equal to width of column, the cantilever projection of slab $= a = 0.85$ m.

B.M. $\quad M = p_0 \dfrac{a^2}{2} = \dfrac{136}{2}(0.85)^2 = 49.13$ kN-m $= 49.13 \times 10^6$ N-mm

$$d = \sqrt{\frac{M}{R_c\, b}} = \sqrt{\frac{49.13 \times 10^6}{0.914 \times 1000}} = 232 \text{ mm}$$

Provide $d = 240$ mm and total depth = 300 mm. Reduce the total depth to 200 mm at the edges so that an effective depth of 140 mm is available there.

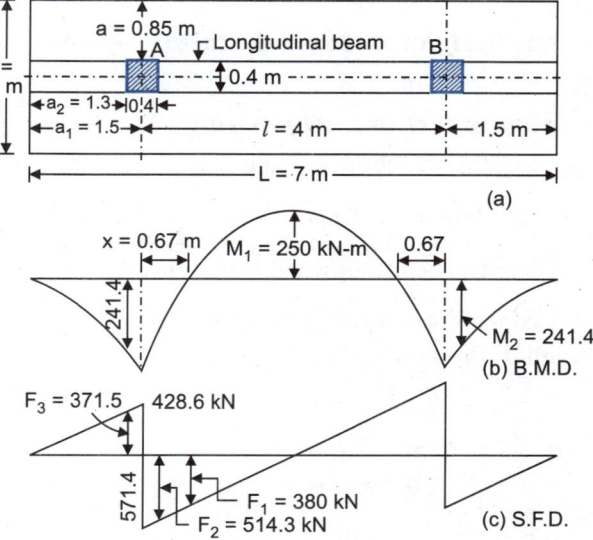

Let us check the above depth for one way shear (diagonal tension), at the critical section situated at a distance $d = 240$ mm from the column face.

The shear force $V$ at the point is
$$V = p_0 (a - 0.24) = 136 (0.85 - 0.24)$$
$$= 83 \text{ kN} = 83000 \text{ N}$$

Effective depth of slab at that point
$$= 140 + \frac{240 - 140}{0.85} (0.85 - 0.24) = 212 \text{ mm}$$

$$\therefore \quad \tau_v = \frac{V}{bd} = \frac{83000}{1000 \times 212} = 0.39 \text{ N/mm}^2$$

Permissible shear stress = $k\tau_c$, where $k = 1.05$ for $D = 212 + 60$ mm (Table 3.2)

and $\tau_c = 0.28$ N/mm² for $p = 0.44\%$ (balanced section)

**Fig. 16.8**

$\therefore$ Permissible shear stress = $1.05 \times 0.28 = 0.294$ N/mm². Hence the section is unsafe for shear. In order to make it safe against diagonal tension, the revised depth

$$d_c = \frac{V}{b(k\tau_c)} = \frac{83000}{1000(1.05 \times 0.28)} \approx 283 \text{ mm}.$$

$\therefore$ Depth at the face of the column is given by the relation

$$283 = 140 + \frac{d - 140}{0.85} \left(0.85 - \frac{d}{1000}\right)$$

or $(d - 140)\left(0.85 - \frac{d}{1000}\right) = 121.6$, which gives $d = 428.5$ mm

Hence keep $D = 490$ mm so that $d = 490 - 60 = 430$ mm. Area of steel reinforcement is given by

$$A_{st} = \frac{49.13 \times 10^6}{230 \times 0.904 \times 430} = 550 \text{ mm}^2$$

Minimum reinforcement @ 0.12% = $\frac{0.12}{100} \times 1000 \times 490 = 588$ mm²

Using 12 mm $\Phi$ bars, $A_\Phi = 113$ mm². $\therefore$ Spacing $s = \frac{1000 \times 113}{588} = 192$ mm

Hence provide 12 mm $\Phi$ bars @ 190 mm c/c at the bottom of the slab.
$$L_d = 45 \, \Phi = 45 \times 12 = 540 \text{ mm}.$$

Providing 60 mm side cover, available length of bar = $850 - 60 = 790$ mm, which is more than $L_d$. Hence safe. Area of distribution reinforcement @ 0.12% of area of cross-section will be = $\frac{0.12}{100} \times 1000 \times 490$ = 588 mm². Using 12 mm $\Phi$ bars, Spacing = $\frac{1000 \times 113}{588} = 192$ mm. However, provide 12 mm $\Phi$ bars @ 190 mm c/c.

**4. B.M. and S.F. diagrams for longitudinal beam.**

The B.M. and S.F. diagrams for the beam are shown in Fig. 16.8 (b) and (c) respectively.

**398** REINFORCED CONCRETE STRUCTURE

Sagging B.M. $M_2$ at column face $= \dfrac{w\, a_2^2}{2} = \dfrac{285.7}{2} (1.3)^2 = 241.4$ kN-m $= 241.4 \times 10^6$ N-mm

Hogging B.M. $M_1 = W\dfrac{l}{2} - \dfrac{w}{2}\left(\dfrac{L}{2}\right)^2 = 1000 \times \dfrac{4}{2} - \dfrac{285.7}{2}\left(\dfrac{7}{2}\right)^2 = 250$ kN-m $= 250 \times 10^6$ N-mm

(**Note.** Both these moments are not exactly equal since the length $L$ adopted as 7 m is slightly less than 7.02 m found on the basis of equal values of $M_1$ and $M_2$).

The point of contraflexure occurs at $x$ from the centre of column, its position is being given by :

$$\dfrac{w\,(a_1 + x)^2}{2} - W x = 0$$

or $\quad \dfrac{285.7}{2}(1.5 + x)^2 = 1000\, x$. From which $x = 0.67$ m

S.F. just to the left of $\quad A = 285.7 \times 1.5 = 428.6$ kN

S.F. just to the right of $\quad A = 1000 - 428.6 = 571.4$ kN

S.F. at the point of contraflexure is $\quad F_1 = 571.4 - (285.7 \times 0.67) = 380$ kN

S.F. at the inner face of column is $\quad F_2 = 571.4 - (285.7 \times 0.2) = 514.3$ kN

S.F. at the outer face of column is $\quad F_3 = 428.6 - (285.7 \times 0.2) = 371.5$ kN.

**5. Depth of Longitudinal beam.** Let us provide longitudinal beam projecting *below* the footing slab. In the cantilever portion, compression face is upwards and hence T-beam action will be available. In the central portion, however, T-beam action will not be available.

Let us therefore, find the depth of beam for $M_1$ where T-beam action is not available. Keep width $b$ of beam $= 400$ mm

$$d = \sqrt{\dfrac{M_2}{R_c\, b}} = \sqrt{\dfrac{250 \times 10^6}{0.914 \times 400}} \approx 830 \text{ mm}.$$

Let us check this depth for two way shear (punching shear). The critical plane occurs at a distance of half the effective depth from the column face. In this case, the effective depth $d_b$ of the beam is larger than the effective depth $d_s$ ($= 430$ mm) of the slab. The column will punch through both these depths, along its perimeter. Let us assume that the distance of critical plane $= d_b/2$ to one side and $d_s/2$ to the other side. Hence widths of the punching shear perimeter are

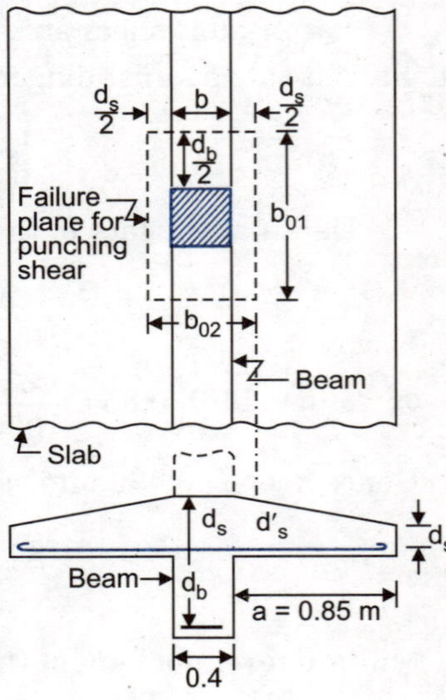

$$b_{01} = 400 + d_b$$

and $\quad b_{02} = 400 + d_s = 400 + 430 = 830$ mm

**Fig. 16.9**

∴ Punching shear area of concrete available along the perimeter for the critical plane.

$$= (2 b_{01} \times d_s') + (2 b_{02} \times d_b)$$

where $\quad d_s' =$ depth of slab at the critical plane

$$= d_{se} + \dfrac{d_s - d_{se}}{a}\left[a - \dfrac{d_s}{2}\right] = 140 + \dfrac{430 - 140}{0.85}\left[0.85 - \dfrac{0.43}{2}\right] \approx 356 \text{ mm}$$

Punching shear $\quad F = 1000 \times 10^3 - 136 \times 10^3\, [b_{01} \times b_{02}]$

$$= 1000 \times 10^3 - 136 \times 10^3\, \dfrac{[(400 + d_b) \times 830]}{(1000)^2} = 10^6 - 112.9\,(400 + d_b)$$

Punching shear stress $= \dfrac{F}{2\,[(b_{01} \times d_s{'}) + (b_{02} \times d_b)]} = \dfrac{10^6 - 112.9\,(400 + d_b)}{2\,[(400 + d_b)\,356 + 830\,d_b]}$

$= \dfrac{10^6 - 112.9\,(400 + d_b)}{2\,[142400 + 830\,d_b]}$ ...(1)

Permissible shear stress $= k_s \tau_c$ where $k_s = (0.5 + \beta_c) \not> 1 = 1$;

$\tau_c = 0.16\sqrt{f_{ck}} = 0.16\sqrt{20} = 0.715$ N/mm$^2$. Hence permissible shear stress = 0.715 N/mm$^2$

Equating this to (1), we get $\dfrac{10^6 - 112.9\,(400 + d_b)}{2\,[142400 + 830\,d_b]} = 0.715$. From which $d_b = 578$ mm.

This is less than effective depth of 830 mm found from the bending compression.

∴ Hence keep $D = 910$ mm so that providing nominal cover of 50 mm and using 20 mm φ main bars and 12 mm φ ties, $d = 910 - (50 + 20/2 + 12) = 910 - 72 = 838$ mm.

**6. Reinforcement in longitudinal beam.** For hogging B.M., the beam acts as a simple rectangular section with reinforcement at its top face. The area of steel is given by

$$A_{st1} = \dfrac{M_1}{\sigma_{st}\,j_c\,d} = \dfrac{250 \times 10^6}{230 \times 0.904 \times 838} = 1435 \text{ mm}^2$$

∴ Use bars of 20 mm Φ, having $A_\Phi = 314$ mm$^2$. ∴ No. of bars = 1435/314 = 4.6. Hence provide 5 bars of 20 mm Φ.

Let us check this reinforcement for development length at the point of contraflexure, as envisaged in the relation

$$\dfrac{M_1}{V} + L_0 \geq L_d$$

where $M_1 = \sigma_{st}\,A_{st}\,j_c d = 230\,[5 \times 314] \times 0.904 \times 838 = 2.736 \times 10^8$ N-mm

$V$ = shear force at point of contraflexure = 380 kN = $380 \times 10^3$ N

$L_0 = 12\,\Phi$ or $d$ whichever is greater = 838 mm.

$L_d = 45 \times 20 = 900$ mm

Now $\dfrac{M_1}{V} + L_0 = \dfrac{2.736 \times 10^8}{380 \times 10^3} + 838 = 1558$ mm.

which is more than $L_d$. Hence safe.

The above reinforcement (i.e. 5-20 Φ bars) may be curtailed at the face of the column. However, continue 3 bars in the cantilever portion, to serve as anchor bars for shear reinforcement.

The reinforcement in the cantilever portion is to be found for the bending moment $M_2 = 241.4$ kN-m. Here the beam will act as T-beam for which lever arm may be taken as $0.9\,d$.

∴ $A_{st2} = \dfrac{M_2}{\sigma_{st}\,jd} = \dfrac{241.4 \times 10^6}{230 \times 0.9 \times 838} = 1392$ mm$^2$

No. of 20 mm Φ bars = 1392/314 ≈ 5

This reinforcement should be safe from development length $(L_d)$ point of view, to satisfy the following equation at the point of contraflexure near the inner face of the column:

$$\dfrac{M_2}{V} + L_0 \geq L_d.$$

S.F. at this point of contraflexure is 380 kN, and it has already been established in case of $A_{st1}$ that the above criterion is satisfied. Since the point of contraflexure is the same, and since $A_{st2} = A_{st1}$, the bars of $A_{st2}$ will also satisfy the Code requirements.

It is to be noted that $A_{st2}$ bars should be extended beyond the point of contraflexure by a distance = $d$ or 12 Φ whichever is more (*i.e.* by ≃ 840 mm in this case) or to a distance of $L_d$ (= 900 mm) from the centre of column (where bending moment $M_2$ occurs), whichever is more. Hence distance upto which $A_{st2}$ bars are to be extended beyond centre of column = 0.67 + 0.84 ≈ 1.5 m. At this point all the bars except 3 Nos, may be discontinued. The remaining 3 bars may be taken throughout the length to serve as anchor bars for shear reinforcement.

### 7. Shear reinforcement

Refer Fig. 16.4. Diagonal tension crack will occur at $F$, only if the zone of sagging B.M. extends to $d$ = 840 mm beyond inner face of the column. However, point of contraflexure falls earlier. Hence there is a possibility of crack at $G$, at point of contraflexure where shear force $F_1$ = 380 kN.

$$\therefore \quad \tau_v = \frac{380 \times 1000}{838 \times 400} = 1.13 \text{ N/mm}^2.$$

This is more than the permissible value of $\tau_c$ = 0.28 N/mm² for the balanced section.

Hence shear reinforcement is required. $A_s$ at the point = 5 × 314 = 1570 mm²

$$\frac{100 A_s}{bd} = \frac{100 \times 1570}{840 \times 400} = 0.467\%$$

$\therefore \quad \tau_c$ (for 0.467%, Table 3.1) = 0.285 N/mm².

$\therefore \quad V_c = \tau_c \, bd = 0.285 \times 838 \times 400 = 95532$ N (say)

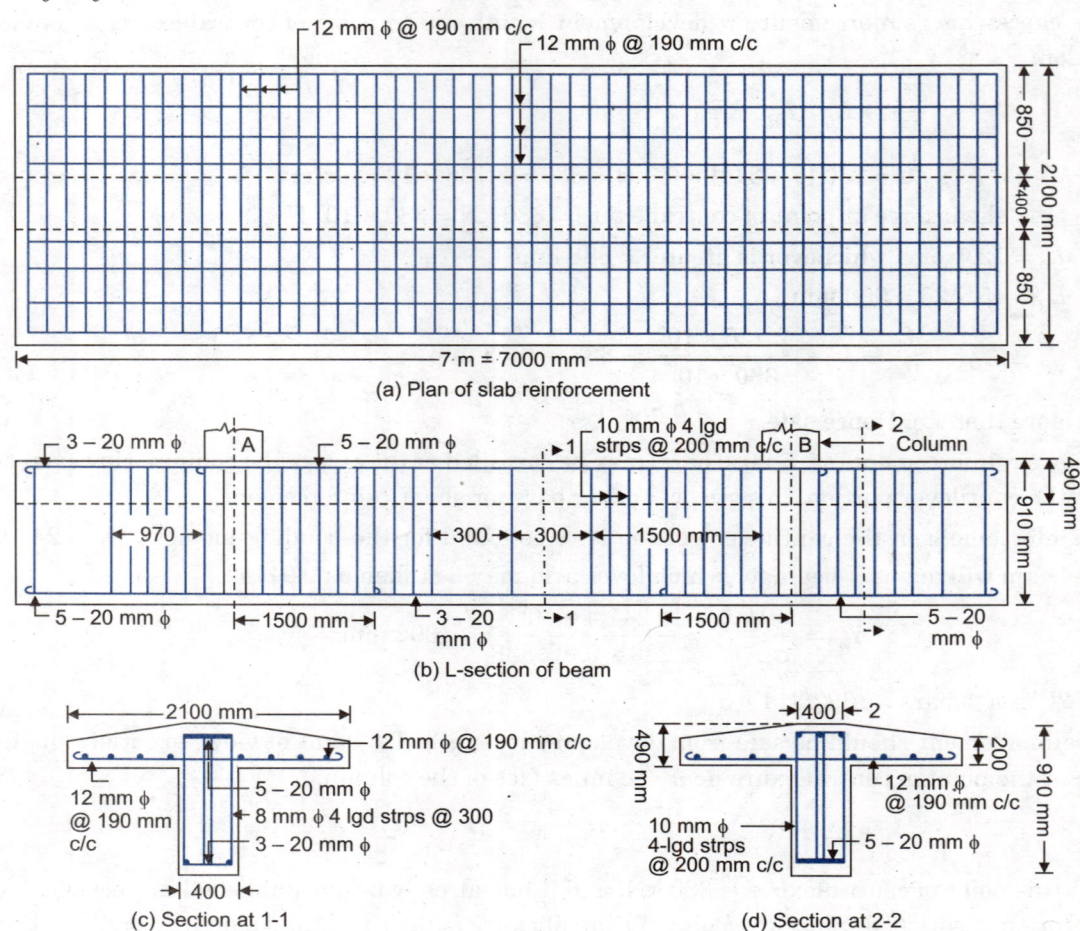

Fig. 16.10

$\therefore \quad V_s = V_1 - V_c = 380000 - 95532 = 284468$ N

Using 10 mm Φ 4-legged stirrups, $A_{sv} = 4 \times \dfrac{\pi}{4}(10)^2 = 314.16$ mm²

$\therefore \quad s_v = \dfrac{\sigma_{sv} \cdot A_{sv} \cdot d}{V_s} = \dfrac{230 \times 314.16 \times 838}{284468} = 212.8$ mm ≈ 200 mm

Distance upto which this is required (*i.e.*, at the point where S.F. = $V_c$ = 95760 N) is by

$$x_1 = \dfrac{V_2 - V_c}{285.7} = \dfrac{514.3 - 95.76}{285.7} = 1.47 \text{ m} \approx 1.5 \text{ m}.$$

Hence provide 10 mm Φ 4-legged stirrups, @ 200 mm c/c from the face of the column to a point 1.5 m, near each column.

In the cantilever portion, shear at the face of column = 371.5 kN. Distance $x_2$ of point at which shear force is equal to $V_c$ (= 95.76 kN) is given by

$$x_2 = \dfrac{V_3 - V_c}{w} = \dfrac{371.5 - 95.76}{285.7} = 0.965 \text{ m} \approx 0.97 \text{ m}$$

Hence shear stirrups are to be provided for a distance of 0.97 m from the face of the column.

$$V_s = V_3 - V_c = 371.5 - 95.76 = 275.74 \text{ kN}$$

Using 12 mm Φ 4-legged stirrups, $A_{sv} = 314.16$ mm²

$\therefore \quad s_v = \dfrac{230 \times 314.16 \times 838}{275.74 \times 1000} = 220$ mm

However provide these @ 200 mm c/c upto 0.97 m from the face of the column.

For the remaining of length of beam, provide 8 mm Φ 4-legged nominal stirrups at a spacing given by

$$s_v = \dfrac{2.175 \, A_{sv} \cdot f_y}{b} = \dfrac{2.175 \times 201.06 \times 415}{400} = 453.7 \text{ mm, subject to maximum of 300 mm.}$$

Hence provide 8 mm φ 4 lgd stirrups @ 300 mm c/c for the remaining length.

The details of reinforcement etc. are shown in Fig. 16.10.

**Design: Example 16.3.** *Combined trapezoidal footing.*

*Design a combined trapezoidal footing for two columns A and B, spaced 5 metres centre to centre. Column A is 300 mm × 300 mm in size and transmits a load of 600 kN. Column B is 400 mm × 400 mm in size and carries a load of 900 kN. The maximum length of footing is restricted to 7 metres only. The safe bearing capacity of soil may be taken as 120 kN/m². Use M 20 mix for concrete and Fe 415 steel.*

**Solution:**

**1. Design constants.** For M 20 mix and Fe 415 steel, we get:

$\sigma_{cbc} = 7$ N/mm²; $\sigma_{st} = 230$ N/mm²; $k_c = 0.289$; $j_c = 0.904$; $R_c = 0.914$.

**2. Size of footing.** Refer Fig. 16.11.

$L = 7$ m  $l = 5$ m;  Keep $a_1 = a_2 = \dfrac{1}{2}(7-5) = 1$ m.

Let the width of footings be $B_1$ and $B_2$ as shown. The values of $B_1$ and $B_2$ should be such that C.G. of footing coincides with C.G. of the column loads.

Let weight of footing $W' = 0.1(W_1 + W_2) = 0.1(600 + 900) = 150$ kN

$\therefore \quad \dfrac{(B_1 + B_2)L}{2} = \dfrac{W_1 + W_2 + W'}{q_0}$

or $\quad B_1 + B_2 = \dfrac{(600 + 900 + 150)}{120} \times \dfrac{2}{7} = 3.93$ m

Distance $\bar{x}$ of C.G. of load

$$= \frac{W_2 \, l}{W_1 + W_2} = \frac{900 \times 5}{600 + 900} = 3 \text{ m}.$$

Distance of C.G. of load from edge

$$B_1 = a_1 + \bar{x} = 1 + 3 = 4 \text{ m}. \quad \ldots(i)$$

Distance of C.G. of footing from edge

$$B_1 = \frac{L}{3}\left(\frac{B_1 + 2B_2}{B_1 + B_2}\right) = \frac{7}{3}\left(\frac{B_1 + 2B_2}{B_1 + B_2}\right) \ldots(ii)$$

Equating (i) and (ii), we get

$$\frac{7}{3}\left(\frac{B_1 + 2B_2}{B_1 + B_2}\right) = 4 \quad \ldots(2)$$

From (1) and (2), we get

$B_1 = 1.11$ m and $B_2 = 2.82$ m.

Net upward soil pressure intensity is given by

$$p_0 = \frac{W_1 + W_2}{\frac{1}{2}(B_1 + B_2)L} = \frac{(600 + 900)}{\frac{1}{2}(1.11 + 2.82)\,7}$$

$$\approx 110 \text{ kN/m}^2$$

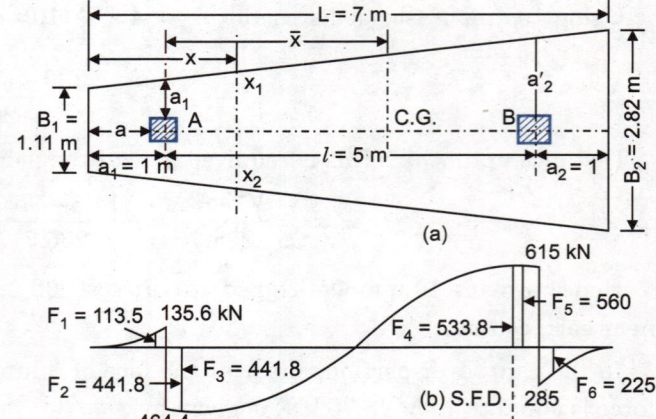

Fig. 16.11

**3. Bending moment and shear force**: Consider the bending of footing in the longitudinal direction. At any distance $x$ from the face $B_1$, the width $B_x$ of section $X_1X_2$ is given by

$$B_x = B_1 + \frac{B_2 - B_1}{L}x = 1.11 + \frac{2.82 - 1.11}{7}x = 1.11 + 0.245\,x$$

$\therefore$ Area $A_x$ to the left of section $X_1X_2 = \frac{1}{2}(B_1 + B_x)x$

$$= \frac{1}{2}\{1.11 + 1.11 + 0.245\,x\}x = \frac{x}{2}\{2.22 + 0.245\,x\} \quad \ldots(i)$$

Distance of C.G. from section $X_1X_2 = \frac{x}{3}\left\{\frac{2B_1 + B_x}{B_1 + B_x}\right\}$

$$= \frac{x}{3}\left\{\frac{2.22 + 1.11 + 0.245x}{1.11 + 1.11 + 0.245x}\right\} = \frac{x}{3}\left\{\frac{3.33 + 0.245x}{2.22 + 0.245x}\right\} \quad \ldots(ii)$$

Hence total upward force to the left of $X_1X_2$

$$= \frac{p_0 x}{2}(2.22 + 0.245x) = \frac{110}{2}x(2.22 + 0.245\,x) \quad \ldots(1)$$

$\therefore$ S.F. at outer face of column $A$, where $x = a = 0.85$ m, is

$$F_1 = \frac{110}{2} \times 0.85\,(2.22 + 0.245 \times 0.85) = 113.5 \text{ kN}.$$

$\therefore$ S.F. at $x$ to the right of $A = 600 - \frac{110}{2}x\,(2.22 + 0.245\,x) \quad \ldots(2)$

$\therefore$ S.F. at inner face of column $A$, where $x = 1.15$ m, is

$$F_2 = 600 - \frac{110}{2} \times 1.15\,(2.22 + 0.245 \times 1.15) = 441.8 \text{ kN}.$$

To find the position of section where S.F. is zero, put (2) to zero.

$$\therefore \qquad 600 - \frac{110}{2} x\,[2.22 + 0.245\,x] = 0. \text{ From which } x = 3.55 \text{ m.}$$

B.M. at section $X_1 X_2$ is given by

$$M_2 = 600\,(x - 1) - (\text{Upward force} \times \text{Distance of its C.G.})$$

$$= 600\,(x-1) - \left\{ \frac{110}{2} x\,(2.22 + 0.245\,x) \right\} \times \frac{x}{3} \left[ \frac{3.33 + 0.245\,x}{2.22 + 0.245\,x} \right]$$

$$= 600\,(x-1) - \frac{110}{6} x^2\,(3.33 + 0.245\,x) \qquad \ldots(3)$$

Maximum hogging bending moment $(M_1)$ will occur at $x = 3.55$ where S.F. is zero

$$\therefore \qquad M_1 = 600\,(3.55 - 1) - \frac{110}{6}\,(3.55)^2\,(3.33 + 0.245 \times 3.55) = 560 \text{ kN-m}$$

To find the value of $x$, where $M_x$ is zero, equate (3) to zero.

$$\therefore \qquad 600\,(x-1) - \frac{110}{6} x^2\,(3.33 + 0.245\,x) = 0$$

Solving this by trial and error, we get, $x = 1.15$ m and $5.70$ m

Thus the position of both the points of contraflexure are known.

For getting the value of S.F. $(F_3)$ at the first point of contraflexure, put $x = 1.15$ m in *Eq.* (2).

$$\therefore \qquad F_3 = 600 - \frac{110}{2}\,(1.15)\,(2.22 + 0.245 \times 1.15) = 441.8 \text{ kN}$$

Similarly S.F. $(F_4)$ at $x = 5.70$ is

$$F_4 = 600 - \frac{110}{2}\,(5.7)\,(2.22 + 0.245 \times 5.7) = 533.8 \text{ kN} \quad \text{(numerically)}$$

S.F. at inner face of column $B$, at $x = 5.8$ m is

$$F_5 = 600 - \frac{110}{2}\,(5.8)\,(2.22 + 0.245 \times 5.8) = 561.47 \approx 560 \text{ kN} \quad \text{(numerically)}$$

S.F. at outer face of column $B$, at $x = 6.2$ m is

$$F_6 = 600 - \frac{110}{2}\,(6.2)\,(2.22 + 0.245 \times 6.2) + 900 = 225 \text{ kN}$$

Sagging bending moment $(M_2)$ at the outer face of column $A$ is obtained by putting $x = 0.85$ m in the second term of Eq. (3) and neglecting the first term.

$$\therefore \qquad M_2 = \frac{110}{6}\,(0.85)^2\,(3.33 + 0.245 \times 0.85) = 47 \text{ kN-m}$$

Sagging B.M. for the cantilever portion to the right of column $B$ is given by

$$M_x = \frac{110}{6} x^2\,(3.33 + 0.245\,x) - 600\,(x-1) - 900\,(x-6)$$

Putting $x = 6.2$ m, we get B.M. $(M_3)$ at the outer face of column $B$:

$$M_3 = \frac{110}{6}\,(6.2)^2\,(3.33 + 0.245 \times 6.2) - 600(5.2) - 900(0.2) \approx 117 \text{ kN-m}$$

The B.M. and S.F. diagrams are shown in Fig. 16.11.

4. **Effective depth of footing.** The effective depth is determined on the basis of hogging moment $M_1 = 560$ kN-m which is the maximum bending moment in the footing. It occurs at a section distant $x = 3.55$ m from the edge $B_1$.

The width $B_x$ at the section is given by

$$B_x = B_1 + \frac{B_2 - B_1}{L} x = 1.11 + 0.245 x$$

$$= 1.11 + 0.245 \times 3.55 = 1.98 \text{ m} = 1980 \text{ mm}$$

$$\therefore \quad d = \sqrt{\frac{M_1}{R_c B_x}} = \sqrt{\frac{560 \times 10^6}{0.914 \times 1980}} = 556.3 \text{ mm}$$

This depth should be sufficient from punching shear point of view. Punching shear will be maximum near column B, where load $W_2 = 900$ kN.

Side $\qquad b_0 = b + d = (400 + d)$ mm

Punching shear $\qquad F = W_2 - p_0 b_0^2 = 900 - \dfrac{110(400 + d)^2}{10^3}$ kN

$$\therefore \quad \tau_v = \frac{900 \times 10^3 - 0.11(400 + d)^2}{4(400 + d)d}$$

Permissible shear stress $= k_s \tau_c = 1 \times 0.16 \sqrt{f_{ck}} = 1 \times 0.16 \sqrt{20} = 0.715$ N/mm²

$$\therefore \quad \frac{900 \times 10^3 - 0.11(400 + d)^2}{4(400 + d)d} = 0.715$$

From which $d = 376$ mm which is less than the one found from bending compression.

Hence $d = 573$ mm. Using 50 mm nominal cover, 20 mm φ bars and 10 mm φ strirups, $D = 556 + 50 + 10 + 10 = 626$ mm. However, keep $D = 650$ mm so that available $d = 650 - (50 + 10 + 10) = 580$ mm.

### 5. Reinforcement in the longitudinal direction.

(i) *Reinforcement for hogging bending moment* $M_1$ *is given by.*

$$A_{st1} = \frac{M_1}{\sigma_{st} j_c d} = \frac{560 \times 10^6}{230 \times 0.904 \times 580} = 4644 \text{ mm}^2$$

Using 20 mm Φ bars, $\quad A_\Phi = \dfrac{\pi}{4}(20)^2 = 314$ mm²

No. of bars required = 4644/314 ≈ 15. Thus provide 15 bars at the top face of the footing.

These bars should be checked for development length at the point of contraflexure, to satisfy the criteria:

$$\frac{M_1}{V} + L_0 \geq L_d.$$

Out of the two points of contraflexure in the hogging B.M. range, the one near column B, has maximum shear force

$$F_4 = 533.8 \text{ kN.}$$
$$M_1 = A_{st} \sigma_{st} j_c d = (15 \times 314) \times 230 \times 0.904 \times 580 = 5.68 \times 10^8 \text{ N-mm}$$
$$V = F_4 = 533.8 \text{ kN} = 533.8 \times 10^3 \text{ N}$$
$$L_0 = 12 \Phi \text{ or } d \text{ whichever is more} = 590 \text{ mm.}$$
$$L_d = 45 \Phi = 45 \times 20 = 900 \text{ mm.}$$

Substituting the values $\dfrac{M_1}{V} + L_0 = \dfrac{5.68 \times 10^8}{533.8 \times 10^3} + 590 = 1654$ mm

Which is more than $L_d$, provided all the 15 bars are taken upto the point of contraflexure.

These bars are to be taken beyond this point for a distance of 12 Φ (= 12 × 20 = 240 mm) or $d$ (= 590 mm), whichever is more, beyond this point. After that, curtail 7 bars and continue 8 bars straight upto either edge of footing so that these may serve as anchorage bars for shear stirrups. These 8 bars will also satisfy the requirements of minimum reinforcement @ 0.12% of the area.

(ii) *Reinforcement for sagging bending moment $M_3$ near outer face of column B*

$$A_{st3} = \frac{M_3}{\sigma_{st}\, j_c\, d} = \frac{117 \times 10^6}{230 \times 0.904 \times 580} = 970.2 \approx 971 \text{ mm}^2. \text{ (say)}$$

Width of footing near the outer face of column $B$, where $x = 6.2$ m is

$$B_x = 1.11 + 0.245\, x = 1.11 + 0.245 \times 6.2 = 2.63 \text{ m}$$

∴ Minimum reinforcement @ 0.12% = $\frac{0.12}{100} \times 2630 \times 650 = 2051.4 \approx 2052 \text{ mm}^2$

Hence provide 2052 mm² area of reinforcement. Using 12 mm Φ bars, $A_\phi = 113 \text{ mm}^2$.

∴ Spacing $\frac{B_x\, A_\phi}{A_{st}} = \frac{2630 \times 113}{2052} \approx 144 \text{ mm}.$

However, provide 12 mm φ bars @ 140 mm c/c.

Actual $A_{st} = \frac{2670 \times 113}{140} = 2155 \text{ mm}^2$ and actual $p = \frac{2155 \times 100}{2670 \times 580} \approx 0.14\%$

This reinforcement should be checked for development length at the point of contraflexure near the inner face column $B$, so as to satisfy the equation.

$$\frac{M_1}{V} + L_0 \geq L_d$$

where $M_1 = \sigma_{st}\, A_{st}\, j_c\, d = 230\,(2155) \times 0.904 \times 580 = 2.6 \times 10^8$
$V = F_4 = 533.8 \text{ kN} = 533.8 \times 10^3 \text{ N}$
$L_0 = 12\, \Phi$ or $d$, whichever is greater $= 580$ mm
$L_d = 45\, \Phi = 45 \times 12 = 540$ mm

∴ $\frac{M_1}{V} + L_0 = \frac{2.6 \times 10^8}{533.8 \times 10^3} + 580 = 1067 > L_d$. Hence safe.

Also, length of bars available, in the cantilever portion, beyond the centre of column $B = 1$ m – side cover = $1000 - 60 = 940$ mm $> L_d$. Hence safe.

(iii) *Reinforcement for sagging B.M. $M_2$ near outer face of column A.*

$$A_{st2} = \frac{M_2}{\sigma_{st}\, j_c\, d} = \frac{47 \times 10^6}{230 \times 0.904 \times 580} = 390 \text{ mm}^2.$$

Width of footing near the outer face of column $A$, where $x = 0.85$ m is

$$B_x = 1.11 + 0.245\, x = 1.11 + 0.245 \times 0.85 \approx 1.32 \text{ m}.$$

Minimum reinforcement @ 0.12% = $\frac{0.12}{100} \times 1320 \times 650 = 1030 \text{ mm}^2$

Hence provide 1030 mm² area of reinforcement. Using 12 mm φ bars having $A_\phi = 113 \text{ mm}^2$,

Spacing $s = \frac{B_x\, A_\phi}{A_{st}} = \frac{1320 \times 113}{1030} \approx 144 \text{ mm}$

However, provide 12 mm φ bars @ 140 mm c/c, as near $B$.

Since the S.F. $F_3$ at the point of contraflexure is less than S.F. $F_4$ in (ii) above, the conditions of development length will also be satisfied.

Hence provide 12 mm φ bars @ 200 mm c/c throughout the length, at the bottom face.

### 6. *Check for diagonal tension*

Test for diagonal tension should be made in the cantilever portion at a distance $d$ from the column face. In the present case, $d = 580$ mm $= 0.58$ m. At 0.58 m from the outer face of column $B$, distance

$x = 6.2 + 0.58 = 6.78$ m. At this section shear force is extremely small and available width $B_x$ is large. Hence shear stress will be small. Similarly, for the cantilever portion to the left of column $A$, shear force, and hence shear stress will be small.

For diagonal tension between $A$ and $B$, near column $B$, crack can occur at the bottom of the footing (*i.e.*, for sagging B.M.) at a distance $d = 580$ mm, or at the top of footing (*i.e.* for hogging B.M.) at the point of contraflexure distant 300 mm from the centre of column $B$. Point of contraflexure being nearer, more S.F. will be at that point. S.F. at point of contraflexure = $F_4 = 533.8$ kN. At the point of contraflexure, $x = 5.70$ m.

∴ $\qquad B_x = 1.11 + 0.245\, x = 1.11 + 0.245 \times 5.7 \approx 2.5$ m.

∴ $\qquad \tau_v = \dfrac{F_4}{bd} = \dfrac{533.8 \times 10^3}{2500 \times 580} = 0.368$ N/mm$^2$

$\qquad \dfrac{100\, A_s}{bd} = \dfrac{100\,(15 \times 314)}{2500 \times 580} = 0.325\%$. Hence from Table 3.1, $\tau_c = 0.246$ N/mm$^2$

Thus, actual shear stress is more than the permissible one, and shear reinforcement is necessary.
$$V_c = \tau_c\, bd = 0.246 \times 2500 \times 580 = 356700 \text{ N} = 356.7 \text{ kN}$$
∴ $\qquad V_s = V - V_c = 533.8 - 356.70 = 177.1$ kN

Using 10 mm Φ 8-lgd stirrups, $A_{sv} = 8 \times \dfrac{\pi}{4}\,(10)^2 = 628$ mm$^2$

∴ $\qquad s_v = \dfrac{628 \times 230 \times 580}{177.1 \times 1000} \approx 473.0$ mm.

Max. spacing of nominal stirrups is given by
$$s_v = \dfrac{2.175\, A_{sv}\, f_y}{b} = \dfrac{2.175 \times 628 \times 415}{2500} = 226.7 \text{ mm.}$$

Hence provide these nominal stirrups @ 225 mm c/c throughout the length.

**7. Transverse reinforcement.** The footing will bend transversely near column face. Projection $a_2'$ beyond face of column $B = \dfrac{1}{2}(B_x - b_2)$ where $B_x$ is the width of footing at the centre of the column $B$, where $x = 6$ m.

∴ $\qquad B_x = 1.11 + 0.245\, x = 1.11 + 0.245 \times 6 = 2.58$ m

∴ $\qquad a_2' = \dfrac{1}{2}\,(2.58 - 0.4) = 1.09$ m $= 1090$ mm.

Width of bending strip near column $B = d + b_2 + d = 580 + 400 + 580 = 1560$ mm

∴ Average area of bending strip = $2.58 \times 1.56 = 4.0248$ m$^2$

∴ Net upward pressure $p_0' = 900/4.0248 \approx 224$ kN/m$^2$

(This is approximately double the value of average pressure $p_0$).

Maximum B.M. at the face of column is
$$M = p_0'\, \dfrac{(a_2')^2}{2} = \dfrac{224}{2}\,(1.09)^2 = 133 \text{ kN-m} = 133 \times 10^6 \text{ N-mm}$$

∴ $\qquad d\text{-required} = \sqrt{\dfrac{133 \times 10^6}{0.914 \times 1000}} \approx 380$ mm.

Actual $d$ available = 580 mm. Hence the transverse beam, of width = 1.56 m will be of the same thickness as the remaining footing. Since transverse reinforcement will be provided *above* the main reinforcement, available $d = 580 - 20 = 560$ mm (approx).

The transverse reinforcement is given by

$\qquad A_{st} = \dfrac{133 \times 10^6}{230 \times 0.904 \times 560} = 1142.3$ mm$^2$. Using 16 mm Φ bars, $A_\Phi = 201$ mm$^2$.

∴ Spacing $s = \dfrac{1000 \times 201}{1142} = 176$ mm. Hence provide 16 mm Φ bars @ 175 mm c/c.

Similarly, for column A, width $B_x$ at the centre of column, where $x = 1$ m is,
$$B_x = 1.11 + 0.245\, x = 1.11 + 0.245 \times 1 = 1.36 \text{ m}.$$

Projection $a_1' = \dfrac{1}{2}(1.36 - 0.3) = 0.53$ m

Width of bending strip $= 2d + b_1 = 2 \times 580 + 300 = 1460$ mm

∴ Area of bending strip $= 1.36 \times 1.46 = 1.9856$ m²

∴ Net upward pressure $p_0' = 600/1.9856 = 302.2$ kN/m²

∴ Max. B.M. at the face of the column is

$$M = p_0 \dfrac{(a_1')^2}{2} = 302.2 \cdot \dfrac{(0.53)^2}{2} = 42.4 \text{ kN-m} = 42.4 \times 10^6 \text{ N-mm}.$$

The available depth $d = 560$ mm will be sufficient. Area of transverse reinforcement is

$$A_{st} = \dfrac{42.4 \times 10^6}{230 \times 0.904 \times 560} = 364.1 \approx 365 \text{ mm}^2$$

Minimum reinforcement $= \dfrac{0.12}{100}(1000 \times 650) = 780$ mm²

Using 16 mm Φ bars, having $A_\Phi = 201$ mm².

∴ Spacing $s = \dfrac{1000 \times 201}{780} = 257.7$ mm.

However provide 16 mm Φ bars @ 200 mm c/c in the strip of width 1.48 m.

For the remaining portion, provide 16 mm Φ bars @ 250 mm c/c, so that this reinforcement at any section is not less than 0.12% of area of cross-section. The details of reinforcement etc. are shown in Fig. 16.12.

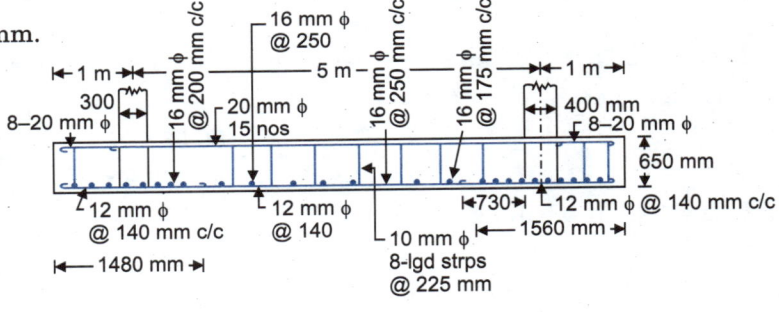

Fig. 16.12

### Design: Example 16.4. *Combined trapezoidal footing with central connecting beam*

*Redesign example 16.3 by providing a longitudinal beam connecting the two columns.*

**Solution:** For the design of trapezoidal footing with a longitudinal beam, steps 1, 2 and 3 of example 16.3 will remain the same. The bending moment and shear force diagrams drawn in Fig. 16.11 (b) and (c) will be applicable for the bending of the longitudinal beam which should be designed properly to resist these bending moments and shear forces at various sections. The slab of the footings will bend transversely.

The salient dimensions of the footing, determined in Example 16.3 are shown in Fig. 16.13. Let the beam project above the footing. Thus, for the hogging bending moment, T-beam action will be available. Let the width of beam = 400 mm.

**1. Design of slab.** The slab will bend as cantilever on either side of the beam. The maximum cantilever projection of the slab will be at the end near width $B_2$, its magnitude being given by

$$a' = \dfrac{1}{2}(2.82 - 0.4) = 1.21 \text{ m}.$$

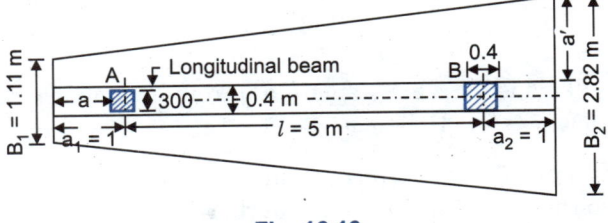

Fig. 16.13

Intensity of net upward pressure $p_0 = 110$ kN/m², determined in the previous example. Maximum bending moment at the face of column $B$ is given by

$$M = p_0 \, a'^2 = \frac{110 \, (1.21)^2}{2} = 80.53 \text{ kN-m} = 80.53 \times 10^6 \text{ N-mm}$$

$\therefore$ Thickness of slab $= \sqrt{\dfrac{80.53 \times 10^6}{0.904 \times 1000}} = 299$ mm.

However, adopt $d = 300$ mm, so that providing 60 mm cover to the centre of steel, total $D = 300 + 60 = 360$ mm. Reduce it to 200 mm at the end, where available $d = 140$ mm.

The above depth should be safe against diagonal tension crack which may occur at $d = 300$ mm from the face of the column.

S.F. $F'$ at that section $= 110 \, (1.21 - 0.30) = 100.1$ kN

Effective depth $d'$ at that section $= 140 + \dfrac{300 - 140}{1.21} \times (1.21 - 0.30) = 260$ mm

$\therefore \quad \tau_v = \dfrac{F'}{bd'} = \dfrac{100.1 \times 1000}{1000 \times 260} = 0.385$ N/mm²

Permissible $\tau_c$ (at $p = 0.44\%$) $= 0.28$ N/mm² (Table 3.1). Hence unsafe.

To make it safe, increase the effective depth in the ratio $0.385/0.28$ approximately

$\therefore \quad d$ at column face $= \dfrac{0.385}{0.28} \times 300 = 412.5$ mm. $\approx 420$ mm

Hence keep $d = 420$ mm and $D = 480$ mm at the face of the column.

The area of steel reinforcement is given by

$$A_{st} = \dfrac{M}{\sigma_{st} \, j_c d} = \dfrac{80.53 \times 10^6}{230 \times 0.904 \times 420} = 923 \text{ mm}^2.$$

Using 16 mm $\Phi$ bars, $A_\Phi = 201$ mm²

$\therefore \quad$ Spacing $s = \dfrac{1000 \times 201}{923} = 217.7$ mm.

and $\quad p = \dfrac{923 \times 100}{1000 \times 420} \approx 0.22\%$ only against assumed $p = 0.44\%$.

Hence provide 16 mm $\Phi$ bars @ 100 mm c/c at the bottom of the section. Let us check this for development length : $L_d = 45 \, \Phi = 45 \times 16 = 720$ mm.

Available length, using 60 mm side cover $= 1210 - 60 = 1150$ mm. Hence safe. Similarly, at the end strip near column $A$, the cantilever projection of slab beyond the beam $= \frac{1}{2}(1.11 - 0.4) = 0.355$ m.

$\therefore \quad$ B.M. $M = 110 \, \dfrac{(0.355)^2}{2} = 6.93$ kN-m $= 6.93 \times 10^6$ N-mm.

This is extremely small, and reinforcement found on the basis of above B.M. will be extremely small. However, for 16 mm $\Phi$ bars, $L_d = 45 \, \Phi = 720$ mm while available length of bars (after providing 60 mm side cover $= 355 - 60 = 295$ mm only). Hence maximum stress $\sigma_s$ that can be developed $= 295 \times \dfrac{230}{720} = 94.2$ N/mm², so that bending failure is not there.

$\therefore \quad A_{st} = \dfrac{M}{\sigma_s \, j_c \, d} = \dfrac{6.93 \times 10^6}{94.2 \times 0.904 \times 420} = 193.7$ mm²

Also, minimum area of steel @ 0.12% of cross-sectional area is
$$A_{st} = \frac{0.12}{100} \times 1000 \times 480 = 576 \text{ mm}^2$$

Using 16 mm Φ bars, $s = \dfrac{1000 \times 201}{576} = 348$ mm, subject to a maximum of 300 mm

Hence the spacing of 16 mm Φ bars @ 100 mm c/c at the edge near column $B$ can be gradually increased to 200 mm at the edge near column $A$.

Transverse reinforcement area @ 0.12% = 576 mm². Using 10 mm Φ bars $A_\Phi$ = 78.5 mm²

$$\therefore \quad \text{Spacing} = \frac{1000 \times 78.5}{576} = 136 \approx 130 \text{ mm c/c}.$$

**2. Depth of longitudinal beam.** Let us provide longitudinal beam of 400 mm width, projecting above the slab. In the hogging portion, therefore, T-beam action will be available. The hogging bending moment $M_1 = 560$ kN-m is much more than the sagging bending moment. Hence the depth of longitudinal beam will be designed for moment $M_1$, taking as a T-beam.

Average thickness of slab $= \dfrac{200 + 480}{2} = 340 \text{ mm} = D_f$

The maximum bending moment occurs at $x = 3.55$ m, where width $B_x$ is given by
$$B_x = 1.11 + 0.245\, x = 1.11 + 0.245 \times 3.55 = 1.98 \text{ m} = 1980 \text{ mm}.$$

The width $b_f$ of the flange will be corresponding to an isolated T-beam given by
$$b_f = \frac{l_0}{\left(\dfrac{l_0}{b} + 4\right)} + b_w$$

where $b$ = actual width of flange = 1980 mm

$l_0$ = distance between points of zero moment in the beam
$= 5.70 - 1.15 = 4.55$ m $= 4550$ mm

$b_w$ = width of web = 400 mm.

$$\therefore \quad b_f = \frac{4550}{\dfrac{4550}{1980} + 4} + 400 \approx 1122 \text{ mm}$$

Assume lever-arm = $0.9\, d$, the moment of resistance is given by
$$M = 0.45\, b_f\, \sigma_{cbc}\, D_f \left[\frac{2kd - D_f}{k}\right]$$

$$560 \times 10^6 = 0.45 \times 1122 \times 7 \times 340 \left[\frac{2 \times 0.289\, d - 340}{0.289}\right]$$

or $\quad 0.578\, d = \dfrac{560 \times 10^6 \times 0.289}{0.45 \times 1122 \times 7 \times 340} + 340$. From which $d \approx 825$ mm

Let us find the depth from punching shear point of view also. The column $B$ will have to punch through depth $d_b$ of the beam on two sides, and through depth $d_s$ of the slab on the other two sides. The critical plane lies at a distance of $d_b/2$ from column face on two sides and at $d_s/2$ on the other two sides, where $d_s$ is the depth of slab. Hence widths of the punching shear zone are:

$$b_{01} = 400 + d_b$$
and $\quad b_{02} = 400 + d_s = 400 + 420 = 820 \text{ mm}$

∴ Punching shear area of concrete available along the perimeter of the critical plane
$$= (2b_{01} \times d_s') + (2b_{02} \times d_b)$$

where $d_s'$ = depth of slab at the critical plane (Fig. 16.9)

$$= d_{se} + \frac{d_s - d_{se}}{a}\left[a - \frac{d_s}{2}\right] = 140 + \frac{420 - 140}{1000}\left[1000 - \frac{420}{2}\right] = 361 \text{ mm}$$

Punching shear $F = 900 \times 10^3 - 110 \times 10^3 \left[\dfrac{(400 + d_b) \times 820}{(1000)^2}\right] = 9 \times 10^5 - 90.2\,[(400 + d_b)]$

$\therefore$ Punching stress $= \dfrac{F}{2[(b_{01} \times d_s') + (b_{02} \times d_b)]} = \dfrac{9 \times 10^5 - 90.2\,(400 + d_b)}{2\,[(400 + d_b)\,361 + 820\,d_b]}$ ...(i)

Permissible shear stress $= k_s\,\tau_c$ where $k_s = (0.5 + \beta_c) \not> 1 = 1$

$$\tau_c = 0.16\sqrt{f_{ck}} = 0.16\sqrt{20} = 0.715 \text{ N/mm}^2$$

$\therefore \qquad \dfrac{9 \times 10^5 - 90.2\,(400 + d_b)}{2[(400 + d_b)\,361 + 820\,d_b]} = 0.715$

or $\qquad \dfrac{863920 - 90.2\,d_b}{144400 + 1181\,d_b} = 1.43$. From which $d_b = 369$ mm.

This is much less than the effective depth of 825 mm found from the bending compression.

**3. Steel reinforcement.** For hogging bending moment $M_1$, assuming lever arm $= 0.9d$, the area of steel is given by

$$A_{st} = \frac{M_1}{\sigma_{st}\,(0.9d)} = \frac{560 \times 10^6}{230 \times 0.9 \times 825} = 3280 \text{ mm}^2.$$

Using 25 mm $\Phi$ bars, $A_\Phi = 491$ mm². $\therefore$ No. of bars $= 3280/491 \approx 7$.

Provide 5 bars in upper layer and 2 bars in the next layer.

Let us check this for development length at the point of contraflexure, so that the following equation is satisfied:

$$\frac{M}{V} + L_0 \geq L_d$$

Assuming that all the 7 bars are available at the point of contraflexure near column $B$,

$M = \sigma_{st} \cdot A_{st} \cdot j_c\,d = 230\,(7 \times 491) \times 0.9 \times 825 = 5.86 \times 10^8$ N-mm.
$V = $ shear force $F_4$ at the point of contraflexure $= 533.8$ kN $= 533.8 \times 10^3$ N
$L_0 = 12\,\Phi$ or $d$ which ever is more $= 825$ mm
$L_d = 45\,\Phi = 45 \times 25 = 1125$ mm

$\therefore \qquad \dfrac{M}{V} + L_0 = \dfrac{5.86 \times 10^8}{533.8 \times 10^3} + 825 = 1922.78 \approx 1925$ mm which is more than $L_d$. Hence O.K.

Continue these 7 bars upto a point distant $L_0$ ($= 815$ mm) beyond the point of contraflexure or to a point distant $825 - 300 - 200 = 325$ mm from the outer face of the column $B$. Similarly, at the other side, continue these bars upto the outer face of column $A$. These bars are to be provided at the upper face of the beam, in two layers. Beyond the above two points, discontinue all bars except 3 bars which may be continued in the cantilever portions of the beam to serve as anchor bars. Using a nominal cover of 50 mm, and keeping vertical distance between the two layers $= 25$ mm, total depth of beam $= 825 + 12.5 + 25 + 12 + 50 = 924.5$ mm, using 12 mm $\varphi$ ties. Hence, keep total depth of beam $= 930$ mm so that available

$$d = 930 - 50 - 12 - 25 - 12.5 \approx 830 \text{ mm}.$$

The reinforcement in the cantilever portion near column $B$ is to be found for B.M. $M_3 = 117$ kN-m. Here, the beam will act as simple beam.

$$A_{st3} = \frac{M_3}{\sigma_{st}\,jd} = \frac{117 \times 10^6}{230 \times 0.904 \times 830} = 678 \text{ mm}^2$$

(Minimum tension reinforcement $= \dfrac{0.85\,bd}{f_y} = \dfrac{0.85 \times 400 \times 825}{415} = 676$ mm²)

∴ No. of 12 mm Φ bars, having $A_\phi = 113$ mm$^2$ is $n = 691/113 \approx 7$.

Hence provide 5 bars in outer layer and 2 bars in the inner layer, at the bottom face of the beam. Let us check this reinforcement for development length at the point of contraflexure near column $B$, so as to satisfy the equation

$$\frac{M}{V} + L_0 \geq L_d.$$ Assuming that all the 7 bars are available there,

$$M = \sigma_{st} A_{st} j_c d = 230 (7 \times 113) \times 0.904 \times 830 = 1.36 \times 10^8 \text{ N-mm}$$
$$V = \text{S.F. at point of contraflexure} = F_4 = 533.8 \text{ kN}$$
$$L_0 = 12 \Phi \text{ or } d \text{ whichever is more} = 830 \text{ mm}$$
$$L_d = 45 \Phi = 45 \times 12 = 540 \text{ mm}.$$

$$\frac{M}{V} + L_0 = \frac{1.36 \times 10^8}{533.8 \times 10^3} + 828 = 254 + 830 = 1084 \text{ mm, which is more than } L_d.$$

However, continue all the 7 bars upto a distance of $L_0$ (= 828 mm) from the point of contraflexure, or upto a point distant $100 + 830 = 930$ mm from the inner face of column $B$. Beyond this point, discontinue all bars, except 3 bars. These bars are to provided at the lower face of the beam in two layers.

Similarly, the steel reinforcement for the cantilever portion to the left of column $A$ can be determined for $M_2 = 47$ kN-mm. Using 12 mm φ bars and 12 mm φ ties, available $d = 930 - 50 - 12 - 6 = 862$ mm.

$$A_{st2} = \frac{M_2}{\sigma_{st} j_c d} = \frac{47 \times 10^6}{230 \times 0.904(862)} = 263 \text{ mm}^2.$$

Minimum tension reinforcement $= \dfrac{0.85 \, bd}{f_y} = \dfrac{0.85 \times 400 \times 862}{415} = 707 \text{ mm}^2$

∴ No. of 12 mm Φ bars = $707/113 = 6.25 \approx 7$

Let us check this reinforcement for development length at the point of contraflexure near column $A$, so as to satisfy the equation

$$\frac{M}{V} + L_0 \geq L_d$$

where $M = \sigma_{st} A_{st} j_c d = 230 (7 \times 113) \times 0.904 \times 862 = 141 \times 10^6$ N-mm
$V = \text{Shear force } F_2 = 441.8$ kN
$L_0 = 12 \Phi$ or $d$ whichever is greater $= 862$ mm
$L_d = 45 \Phi = 45 \times 12 = 540$ mm

∴ $$\frac{M}{V} + L_0 = \frac{141 \times 10^6}{441.8 \times 10^3} + 862 = 1181.1 \text{ m}$$

Hence the criterion is satisfied. However, continue the 7 bars of 12 mm Φ, upto a point distant $L_0$ (= 860 mm) beyond point of contraflexure, or upto a point distant $1150 + 840 \simeq 2010$ mm from the free edge of the beam. At this point, 4 bars may be discontinued, while the remaining 3 bars may be continued throughout the length, to serve an anchor bars.

**4. Shear reinforcement.** Refer Fig. 16.4. Diagonal tension crack will occur at $F$ only if the zero sagging B.M. extends to $d = 830$ mm beyond the inner face of column $A$ (or $B$). However, point of contraflxure falls earlier. Hence there is a possibility of crack at the point of contraflexure near column $B$ where S.F. $F_4 = 533.8$ kN.

∴ $$\tau_v = \frac{533.8 \times 10^3}{400 \times 830} = 1.61 \text{ N/mm}^2 \; (< \tau_{c, max} = 1.8 \text{ N/mm}^2 \text{ and hence OK)};$$

$$100 \frac{A_s}{bd} = 100 \times \frac{7 \times 491}{400 \times 830} = 1.04\%$$

Hence from Table 3.1, $\tau_c = 0.39$ N/mm². Thus $\tau_v > \tau_c$ and shear reinforcement is necessary. Hence crack will occur at the upper face of the beam if shear reinforcement is not provided.

$$V_c = \tau_c \cdot bd = 0.39 \times 400 \times 830 \times 10^{-3} \approx 129.5 \text{ kN}$$

$$V_s = 533.8 - 129.5 = 404.3 \text{ kN}$$

Using 12 mm Φ 4-lgd stirrups, $A_{sv} = 4 \times 113 = 452$ mm².

$$\therefore \quad s_v = \frac{\sigma_{sv} \cdot A_{sv} \cdot d}{V_s} = \frac{230 \times 452 \times 830}{404.3 \times 1000} = 213 \text{ mm} \approx 200 \text{ mm c/c}.$$

Distance upto which this is required (*i.e.* at a point where S.F. = $V_c$ = 129.5 kN) is obtained by putting $F_x = V_c = \pm 129.5$ in *Eq. 2* of example 16.3.

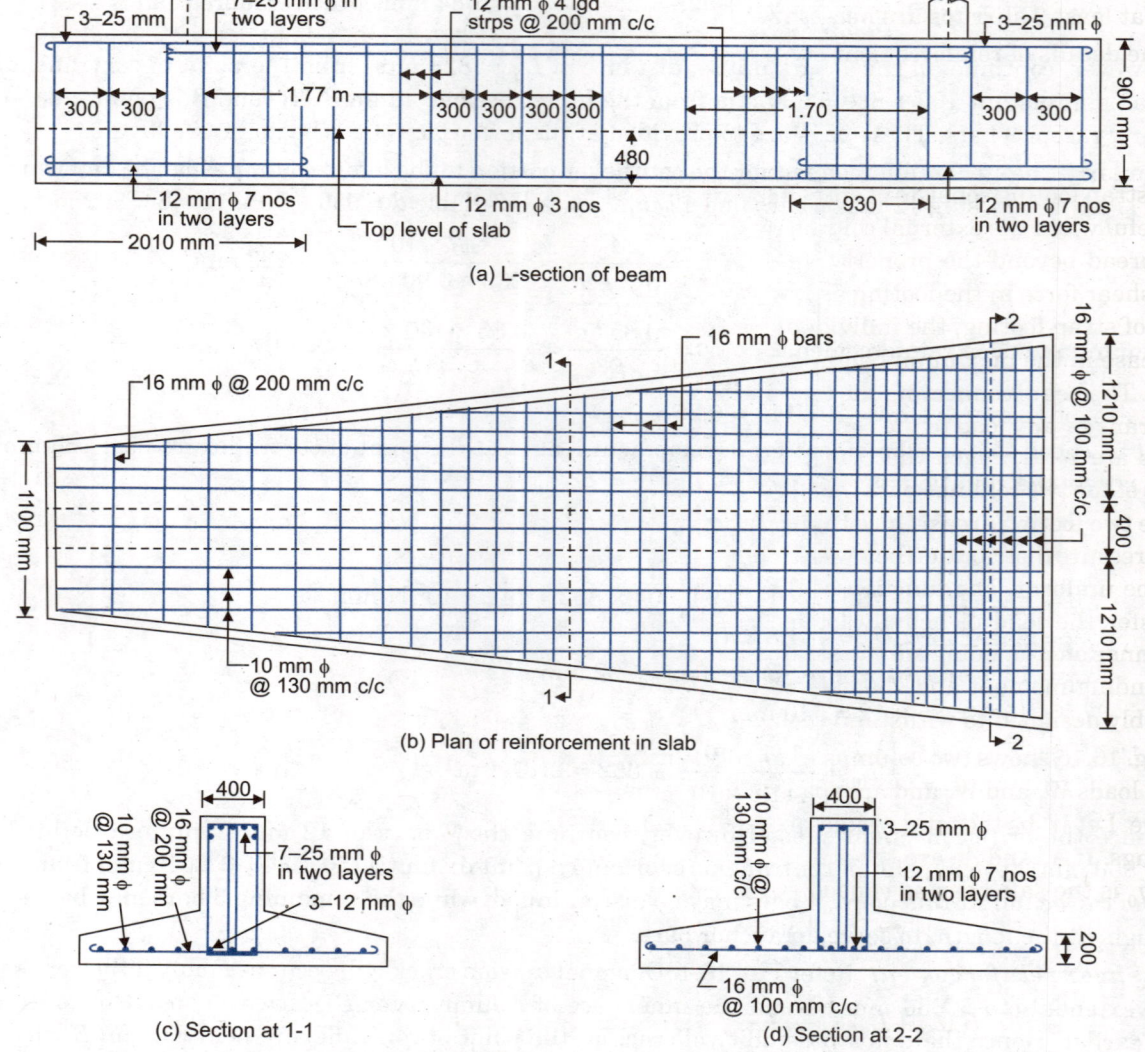

**Fig. 16.14**

Hence $F_x = \pm 129.5 = 600 - \dfrac{110}{2} x (2.22 + 0.245x)$. Solving this, we get $x = 2.92$ m and 4.14 m. These two points are correspondingly at distances of $2.92 - 1.15 = 1.77$ m from inner face of column A and $5.8 - 4.1 = 1.7$ m from the inner face of column B. Between these two values of $x_1$, provide 12 mm Φ 4-lgd stirrups at nominal spacing of

$$s_v = \frac{2.175\, A_{sv.}\, f_y}{b} = \frac{2.175 \times 201 \times 415}{400} = 450 \text{ mm, subject to a maximum of 300 mm.}$$

From the inner face of column $A$ to a distance of 1.77 m, the spacing of 12 mm Φ 4-lgd stirrups is given by

$$s_v = \frac{A_{sv.}\, \sigma_{sv.}\, d}{(F_3 - V_c)} = \frac{452 \times 230 \times 830}{(441.8 - 129.5)\, 1000} = 276.3 \text{ mm}$$

However, for the sake of symmetry, provide these @ 200 mm c/c. Thus, the shear reinforcement is summarised below:

1. From inner face of column $A$ to 1.77 m distance, provide 12 mm Φ 4-lgd stirrups @ 200 mm c/c
2. From inner face of column $B$ to a distance of 1.70 mm, provide 12 mm Φ 4-lgd stirrups @ 200 mm c/c.
3. For intermediate portion of beam, provide 8 mm Φ 4-lgd stirrups @ 300 mm c/c.
4. For the cantilever portions, where S.F. is less than $V_c$ provide 8 mm Φ 4-lgd stirrups @ 300 m c/c, but at least 2 stirrups are necessary.

The details of reinforcement etc. are shown in Fig. 16.14

## 16.4. STRAP FOOTING

A strap footing consists of spread footings of two columns connected by a *strap beam*. This type of footing is useful when the external column (wall column) is vary near to the property line so that its footing cannot be spread beyond the property line. If a combined trapezoidal footing is provided, the bending moment and shear force in the footing are very high since the footing is continuous under both the columns. In the case of strap footing, the individual spread footings have very large area under the columns, resulting in decrease in the maximum bending moment and shear force. The beam connecting the two spread footing does not transfer any load to the soil. The individual footing areas are so arranged that the C.G. of the combined loads of the two columns pass though the combined C.G. of the two footing areas. Once this criterion is achieved the pressure distribution below each individual footing will be uniform. The function of the strap beam is to transfer the load of heavily loaded outer column to the inner one. In doing so, the strap beam is subjected to bending moment and shear force, and it should be suitably designed to withstand these.

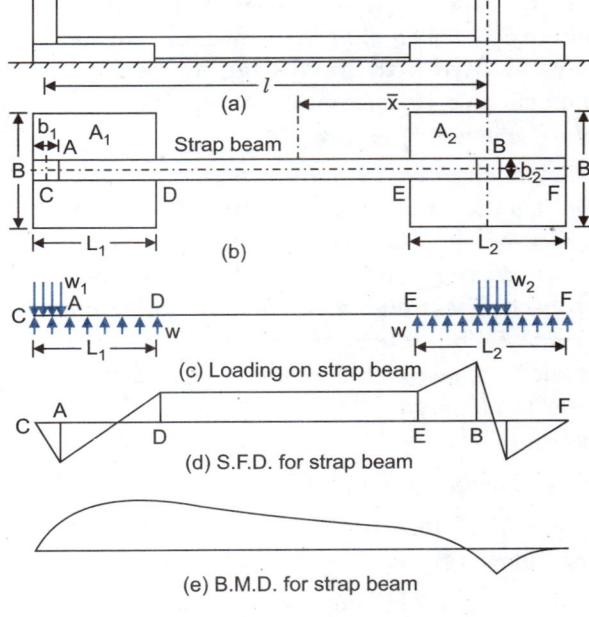

Fig. 16.15 shows two columns $A$ and $B$, transmitting axial loads $W_1$ and $W_2$ and are spaced $l$ apart, centre to centre. Let $W'$ be the total weight of both the individual footings. If $A_1$ and $A_2$ are the individual footing areas, and $q_0$ is the safe bearing capacity of soil, we have

$$A_1 + A_2 = \frac{W_1 + W_2 + W'}{q_0}$$

or $$B(L_1 + L_2) = \frac{W_1 + W_2 + W'}{q_0} \quad \ldots (1)$$

Fig. 16.15

where $B$ is the common width of each footings and $L_1$ and $L_2$ are the individual lengths of the footings.

Length $L_2$ is arranged centrally under column $B$.

The C.G. of the resultant load $W = W_1 + W_2$, falls at $\bar{x}$ from the centre of column $B$, so that

$$\bar{x} = \frac{W_1\, l}{W_1 + W_2} \quad \ldots (i)$$

Let $(b_1 \times b_1)$ and $(b_2 \times b_2)$ be the size of the columns $A$ and $B$ respectively. Taking moments of the footing areas about centre of column $B$, we get

$$\bar{x} = \frac{(B \times L_1)\left(l + \frac{1}{2}b_1 - \frac{L_1}{2}\right)}{B(L_1 + L_2)} \qquad \ldots(ii)$$

Equating (i) and (ii), we get

$$\frac{L_1\left(l + \frac{1}{2}b_1 - \frac{L_1}{2}\right)}{L_1 + L_2} = \frac{W_1 l}{W_1 + W_2} \qquad \ldots(2) \ldots(16.7)$$

From (1) and (2), the unknowns $L_1$ and $L_2$ can be known in terms of any suitable value of $B$.

Net upward soil pressure $p_0 = \dfrac{W_1 + W_2}{B(L_1 + L_2)}$

Each individual footing is designed as cantilever slab, having sagging bending moment in each of the cantilever portion.

The strap beam transfers a part of load of footing $A$ to footing $B$ in such a way that C.G. of two loads coincide with the C.G. of the footing areas. In doing so, it is subjected to bending moment and shear force all along its length, Fig. 16.15 (c) shows the loading diagram on the strap beam $CDEF$.

The upward uniform load $w$ per unit length, under $CD$ and $EF$ will be equal to $(p_0 \times B)$. The downward $w_1$ per unit length in portion under column $A = W_1/b_1$ per unit length. Similarly downward load $w_2 = W_2/b_2$ per length. The characteristic bending moment diagram and S.F. diagram are sketched in Fig. 16.15 (e) and (d) respectively. For large portion of the strap beam, it will experience hogging bending moment, the maximum being at the section where S.F. is zero. For the cantilever portion beyond $B$, it will have sagging bending moment also, its maximum value being at the place where S.F. is zero.

**Design: Example 16.5.** *Design a strap footing for two columns A and B, spaced 5 metres centre to centre. Column A, 300 mm × 300 mm in size carries a load of 600 kN and is on the property line. Column B, 400 mm × 400 mm in size, carries a load of 900 kN. The bearing capacity of soil is 120 kN/m². Use M 20 mix and Fe 415 steel.*

**Solution:**

**1. Design constants**

For M 20 mix and Fe 415 steel, we have the following:

$c = \sigma_{cbc} = 7 \text{ N/mm}^2$; $\quad t = \sigma_{st} = 230 \text{ N/mm}^2$

$k_c = 0.289$; $\quad j_c = 0.904$; $R_c = 0.914$

**2. Size of footings:** Fig. 16.16 (a) shows the general arrangement of the footing.

Let the width of the two spread footings be $B$ metres each. Length of footing under $A = L_1$ and length of footing under $B$ be $L_2$ centrally arranged under $B$.

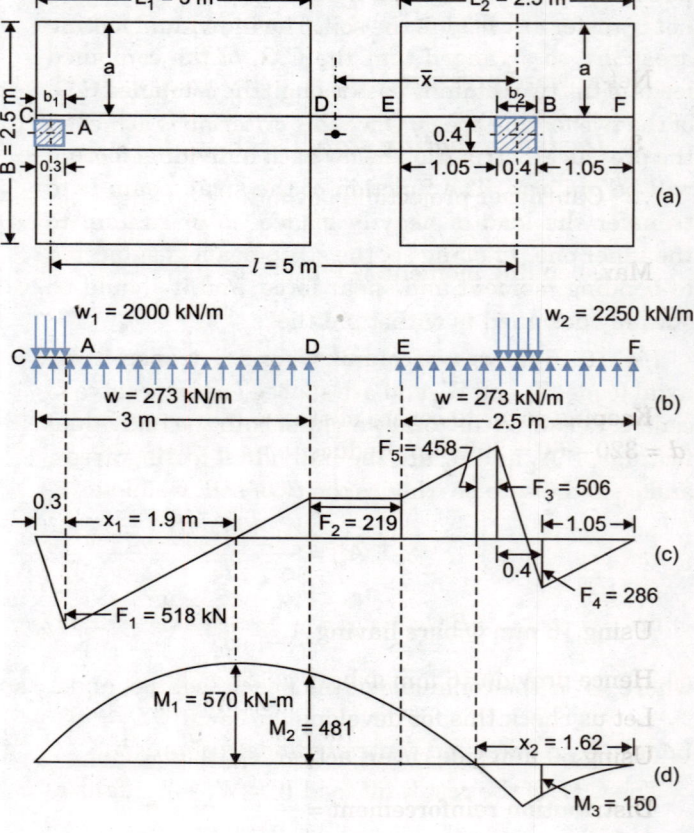

Fig. 16.16

Weight of footing, $W_1 = 600$ kN; $W_2 = 900$ kN.
$W' = 0.1 (600 + 900) = 150$ kN. Hence,

$$B(L_1 + L_2) = \frac{600 + 900 + 150}{120} = 13.75 \text{ m}^2$$

or
$$L_1 + L_2 = \frac{13.75}{B} \qquad \ldots (1)$$

Let $\bar{x}$ = distance of C.G. of loads from centre of column $B$,

$$\therefore \quad \bar{x} = \frac{W_1 l}{W_1 + W_2} = \frac{600 \times 5}{600 + 900} = 2 \text{ m}.$$

If $\bar{x}$ is also the distance of C.G. of areas, from the centre of column $B$, we have

$$\bar{x} = \frac{B \times L_1 \left( l + \dfrac{b_1}{2} - \dfrac{L_1}{2} \right)}{B(L_1 + L_2)}$$

Substituting the values of $\bar{x}$, $b_1$ and $l_1$, we get

$$2 = \frac{L_1(5 + 0.15 - 0.5 L_1)}{(L_1 + L_2)} \qquad \ldots(2)$$

Substituting the value of $(L_1 + L_2)$ from (1), and choosing $B = 2.5$ m, we get

$$\frac{L_1(5.15 - 0.5 L_1)}{13.75/2.5} = 2 \quad \text{or} \quad L_1^2 - 10.3 L_1 + 22 = 0.$$

From which $L_1 = 3.025$ m = 3 m (say).

$$\therefore \quad L_2 = \frac{13.75}{2.5} - 3.0 = 2.5 \text{ m}.$$

Net upward pressure, $p_0 = \dfrac{600 + 900}{2.5 (3 + 2.5)} = 109$ kN/m$^2$

**3. Design of footing slab.** Let the width of strap beam = 400 mm

$\therefore$ Cantilever projection, beyond the beam is $a = \dfrac{1}{2}(2.5 - 0.4) = 1.05$ m

Max. bending moment $M = p_0 \dfrac{a^2}{2} = 109 \dfrac{(1.05)^2}{2} = 60$ kN-m $= 60 \times 10^6$ N-mm.

$$\therefore \quad d = \sqrt{\frac{60 \times 10^6}{0.914 \times 1000}} = 257 \text{ mm}.$$

Keeping cover to centre of steel = 60 mm, D = 257 + 60 = 317 mm. However, keep $D$ = 320 mm so that $d$ = 320 − 60 = 260 mm. Reduce to total depth to 200 mm at the edges, so that
$d$ = 200 − 60 = 140 mm there.

$$A_{st} = \frac{M}{\sigma_{st} \, j_c \, d} = \frac{60 \times 10^6}{230 \times 0.904 \times 260} = 1110 \text{ mm}^2$$

Using 16 mm Φ bars having $A_\Phi = 201$ mm$^2$, spacing $s = \dfrac{1000 \times 201}{1110} = 181$ mm.

Hence provide 16 mm Φ bars @ 180 mm c/c.

Let us check this for development length. $L_d = 45 \Phi = 45 \times 16 = 720$ mm.

Using 60 mm side cover, actual length available = 1050 − 60 = 990 mm > $d$. Hence O.K.

Distribution reinforcement = $\dfrac{0.12}{100} \times 1000 \dfrac{(320 + 200)}{2} = 312$ mm$^2$.

∴ Spacing of 10 mm Φ bars = $\dfrac{1000 \times 78.5}{312} \approx 250$ mm c/c.

**4. Bending moment and shear force diagrams for strap beam.** Upward load $w$ per metre run, on the strap beam is given by $w = p_0 B = 109 \times 2.5 = 273$ kN/m.

Downward load intensity under column $A = w_1 = 600/0.3 = 2000$ kN/m.

Downward load intensity under of column $B = w_2 = 900/0.4 = 2250$ kN/m.

The loading diagram on the strap beam is shown in Fig. 16.16(b).

The shear force diagram for the strap beam is shown in Fig. 16.16(c).

S.F. at inner face of column $A = F_1 = (2000 \times 0.3) - (273 \times 0.3) = 518.1$ kN.

S.F. at edge $D = F_2 = 273 \times 3 - 600 = 219$ kN.

S.F. at edge $E = F_2 = 219$ kN.

S.F. at inner edge of column $B = F_3 = 219 + 273\left\{\dfrac{1}{2}(2.5 - 0.4)\right\} = 506$ kN.

S.F. at outer edge of column $B = F_4 = 900 - 506 - (273 \times 0.4) = 286$ kN.

or $= 273 \times \left\{\dfrac{1}{2}(2.5 - 0.4)\right\} = 286.7$ kN

In the range $CD$, shear force is zero at distance $x_1$ from the inner face of column $A$, its value being given by
$$x_1 = 518/273 = 1.9 \text{ m}.$$

Hogging bending moment will be maximum at this section, its values being given by
$$M_1 = (2000 \times 0.3)(1.9 + 0.15) - 273\dfrac{(0.3 + 1.9)^2}{2} = 570 \text{ kN-m}.$$

Hogging B.M. at edge $D$ is given by
$$M_2 = (2000 \times 0.3)(3 - 0.15) - \dfrac{273(3)^2}{2} = 481.5 \text{ kN-m}$$

Sagging B.M. at the outer face of column $B$ is given by
$$M_3 = \dfrac{273}{2}\left[\dfrac{1}{2}(2.5 - 0.4)\right]^2 = 150.5 \text{ kN-m}$$

Let the point of contraflexure occur at $x_2$ from point $F$.

∴ $M_x = 0 = 273\left(\dfrac{x_2^2}{2}\right) - (2250 \times 0.4)(x_2 - 1.25)$, which gives $x_2 = 1.62$ m

Shear force at the point of contraflexure is given by
$$F_5 = (2250 \times 0.4) - (273 \times 1.62) = 458 \text{ kN}$$

**5. Depth of strap beam**

Let the width of strap beam be 400 mm. Since it projects above the footings, T-beam action will be available in range $CD$. Maximum bending moment is $M_1 = 570$ kN-m. However, since T-beam action is available here, the critical section will be at $D$, where the beam acts as rectangular beam because of absence of footing slab. Bending moment at $D = M_2 = 481$ kN-m $= 481 \times 10^6$ N-mm.

∴ $d = \sqrt{\dfrac{481 \times 10^6}{0.914 \times 400}} = 1147$ mm. $\cong 1150$ mm

Let us find moment of resistance of T-beam, with $d = 1150$ mm.

Average thickness of slab $= \dfrac{320 + 200}{2} = 260$ mm $= D_f$

For an isolated T-beam, width of flange is given by

$$b_f = \frac{l_0}{\left(\frac{l_0}{b} + 4\right)} + b_w$$

where $l_0$ = distance between points of zero moments in beam
$= (5 + 0.15 + 1.25) - 1.62 = 4.78$ m = 4780 mm.
$b$ = actual width of flange = 2.5 m = 2500 mm
$b_w$ = width of web = 400 mm.

$$\therefore \quad b_f = \frac{4780}{\left(\frac{4780}{2500} + 4\right)} + 400 \approx 1208 \text{ mm.}$$

Taking lever arm = 0.9 d, the moment of resistance of T-beam is given by

$$M_r = 0.45\, b_f\, \sigma_{cbc}\, D_f \left[\frac{2kd - D_f}{k}\right] = 0.45 \times 1208 \times 7 \times 260 \left[\frac{2 \times 0.289 \times 1150 - 260}{0.289}\right]$$

$$= 1385 \times 10^6 \text{ N-mm} = 1385 \text{ kN-m}$$

which is much more than the hogging bending moment $M_1$ = 570 kN-m. Hence $d$ = 1150 mm is safe.

**6. Reinforcement in trap beam**

For hogging bending moment $M_1$, lever arm = 0.9 d and the area of steel is given by

$$A_{st1} = \frac{M_1}{\sigma_{st}\, jd} = \frac{570 \times 10^6}{230 \times 0.9 \times 1150} = 2395 \text{ mm}^2$$

For hogging B.M. $M_2$, area of steel is given by

$$A_{st2} = \frac{481 \times 10^6}{230 \times 0.904 \times 1150} = 2012 \text{ mm}^2.$$

Hence provide reinforcement equal to 2395 mm².

Using 20 mm Φ bars, $A_\Phi$ = 314 mm². ∴ No. of bars = 2395/314 = 7.63 ≈ 8.

Let us check the reinforcement for development of stresses, at point of contraflexure, so as to satisfy the criterion:

$$\frac{M}{V} + L_0 \geq L_d.$$

Assuming that the 8 bars, provided at the top face of the beam, are *available* at the point of contraflexure, we have

$M = \sigma_{st} A_{st} j_c d = 230 \times (8 \times 314) \times 0.9 \times 1150 = 598 \times 10^6$ N-mm
$V$ = S.F. at point of contraflexure = $F_5$ = 458 kN
$L_0$ = 12 Φ or $d$, whichever is more = 1150 mm.
$L_d$ = 45 Φ = 45 × 20 = 900 mm.

$$\therefore \quad \frac{M}{V} + L_0 = \frac{598 \times 10^6}{458 \times 1000} + 1150 = 1305 + 1150 = 2455 \text{ mm}$$

which is much more than $L_d$. This suggests that it is not necessary to take all the 8 bars upto the point of contraflexure. However, we will continue all the bars upto a distance of $L_0$ (= 1150 mm) from the point of contraflexure, or to a distance of 1620 − 1150 = 470 mm from the end of the beam. After that, discontinue all the bars, except 3 bars which will serve as anchor bars for shear stirrups.

Similarly at the face of support near $A$, the above reinforcement should satisfy the criterion

$$\frac{M}{V} + L_0 \geq L_d$$

where  $M = 598 \times 10^6$ N-mm, as above;  $V$ = shear force = 518 kN.

$L_0 = \frac{b}{2}$ − cover = 150 − 60 = 90 mm.;  $L_d = 45\,\Phi = 45 \times 20 = 900$ mm

$\therefore \quad \frac{M}{V} + L_0 = \frac{598 \times 10^6}{518 \times 10^3} + 90 = 1154 + 90 = 1244$ mm

which is more than $L_d$. Hence satisfactory.

Hence 8 bars of 20 mm $\Phi$ will be sufficient. These bars are arranged in two $\varphi$ stirrups, and 50 mm nominal cover, $D = d + 20/2 + 20 + 10 + 50 = d + 90 = 1150 + 90 = 1240$ mm.

Maximum sagging B.M. occurs at the outer face of column $B$, where bending moment $M_3 = 150$ kN-m. Using 12 mm $\varphi$ bars and 10 mm $\varphi$ stirrups with 50 mm nominal cover, available $d = 1240 − 12/2 − 10 − 50 = 1240 − 66 = 1174$ mm.

$\therefore \quad A_{st} = \dfrac{150 \times 10^6}{230 \times 0.904(1174)} = 615$ mm$^2$

Using 12 mm $\Phi$ bars having $A_\Phi = 113.1$ mm$^2$. No. of bars = 615/113.1 = 5.43 ≏ 6 of stress at the point of contraflexure so as to satisfy the criterion

$$\frac{M}{V} + L_0 \geq L_d$$

where  $M = \sigma_{st} A_{st} j_c d = 230\,(6 \times 113.1) \times 0.904 \times 1174 = 165 \times 10^6$ N-mm

$V$ = S.F. at the point of contraflexure = 458 kN = $458 \times 10^3$ N

$L_0 = 12\,\Phi$ or $d$ whichever is more = 1174 mm

$L_d = 45\,\Phi = 45 \times 12 = 540$ mm.

$\therefore \quad \dfrac{M}{V} + L_0 = \dfrac{165 \times 10^6}{458 \times 10^3} + 1174 = 1534$ mm

which is more than $L_d$, and hence safe.

Hence provide 6 − 12 mm $\Phi$ bars at the lower face, and continue these to a point distant $L_0 (= 1174$ mm$)$ from the point of contraflexure, or to a point distant 1174 + 1620 ≏ 2800 mm from the end of the beam. After that, discontinue all the bars, except 3 bars which will serve as anchor bars.

### 7. *Reinforcement for diagonal tension*

$100\dfrac{A_s}{bd} = \dfrac{100 \times 8 \times 314}{400 \times 1150} = 0.546\%$.  Hence from Table 3.1,  $\tau_c = 0.31$ N/mm$^2$

$\therefore$  Shear resistance of concrete, $V_c = 0.31 \times 400 \times 1150 = 142600$ N = 142.6 kN.

Hence shear reinforcement is necessary wherever shear force exceeds 142.6 kN

At the inner face of column $A$, shear force $F_1 = 518$ kN

$\therefore \quad V_s = F_1 − V_c = 518 − 142.6 = 375.4$ kN.

Using 10 mm $\Phi$ 4-lgd stirrups, $A_{sv} = 4 \times \dfrac{\pi}{4}\,(10)^2 = 314$ mm$^2$

$\therefore$  Spacing $s_v = \dfrac{A_{sv}\,\sigma_{sv}\,d}{V_s} = \dfrac{314 \times 230 \times 1150}{375.4 \times 10^3} = 221$ mm ≏ 220 mm c/c.

Distance of point where shear force is 142.6 kN, from the inner face of column = $\dfrac{518 − 142.6}{273} \approx 1.38$ m. Hence the distance from end of beam = 1.38 + 0.3 = 1.68 m.

Hence provide 10 mm $\Phi$ 4-lgd stirrups @ 220 mm c/c from end of the beam, upto a distance of 1.68 m.

At the inner face of column $B$, shear force $F_3 = 506$ kN. Hence spacing of 10 mm Φ 4-lgd stirrups is given by

$$s_v = \frac{314 \times 230 \times 1150}{(506 - 142.6)\,10^3} = 228.5 \text{ mm say 220 mm c/c.}$$

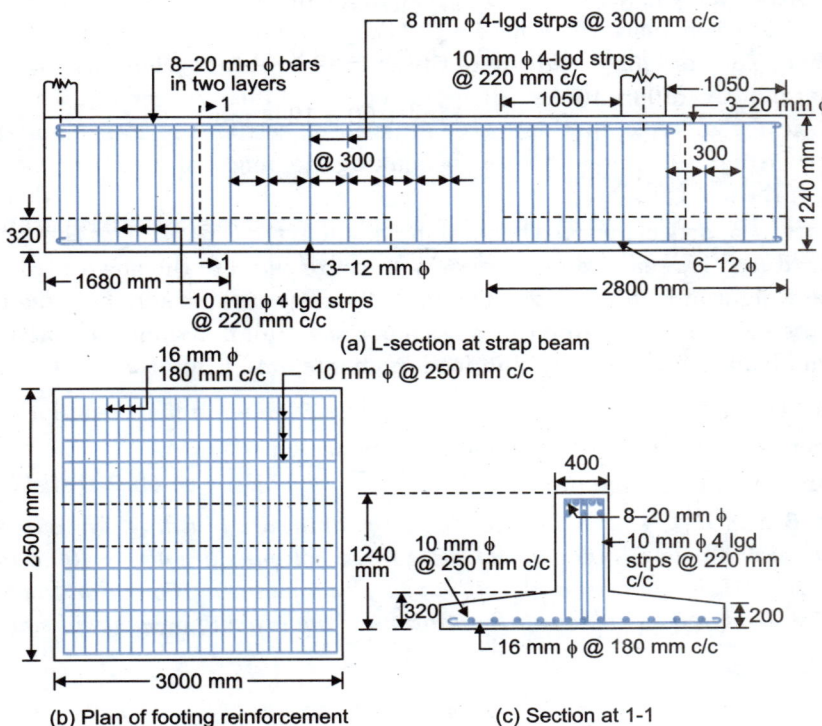

Fig. 16.17

Hence provide these stirrups @ 220 mm c/c from point $E$ to the inner face of the column. Similarly, at the outer face of column $B$, shear force $F_4 = 286$ kN.

$$\therefore \quad s = \frac{314 \times 230 \times 1150}{(286 - 142.6)\,10^3} = 579 \text{ mm.} \quad \text{This is too large.}$$

However, using 8 mm φ 4 lgd stirrups, having $A_{sv} = 4 \times \dfrac{\pi}{4}(8)^2 = 201$ mm²

$$s = \frac{201 \times 230 \times 1150}{(286 - 142.6)\,10^3} = 370 \text{ mm,} \quad \text{subject to a maximum of 300 mm}$$

Maximum spacing of stirrups is given by

$$s_v = \frac{2.175\,A_{sv}\,f_y}{b} = \frac{2.175 \times 201 \times 415}{400}$$

$$= 453 \text{ mm, subject to a maximum of 300 mm.}$$

Hence keep $s_v = 300$ mm from the outer face of the column to the end of the beam.

At the point $D$, shear force $F_2 = 219$ kN.

$$s_v = \frac{201 \times 230 \times 1150}{(219 - 142.6)\,10^3} = 695 \text{ mm, subject to a maximum of 300 mm.}$$

Hence for the remaining portion, provide 8 mm φ 4 lgd stirrups of spacing of 300 mm c/c.

**8. Check for punching shear.** The beam can be checked against punching shear at columns $A$ and $B$, as explained in Example 16.2. The section will be found to be safe at both the locations.

## 16.5. RAFT FOOTING

A raft or mat is a combined footing that covers the entire area beneath a structure and supports all the walls and columns. When the allowable soil pressure is low, or the building loads are heavy, the use of spread footings would cover more than one-half of the area and it may prove more economical to use mat or raft foundation. They are also used where the soil mass contains compressible lenses or the soil is sufficiently erratic so that the differential settlement would be difficult to control. The mat or raft tends to bridge over the erratic deposits and eliminates the differential settlement. Raft foundation is also used to reduce settlement above highly compressible soil, by making the weight of structure and raft approximately equal to the weight of the soil excavated.

Ordinarily, rafts are designed as inverted beam slab system. If the C.G. of loads coincide with the centroid of the raft, the upward load is regarded as a uniform pressure equal to the downward load divided by the area of the raft. The weight of raft is not considered in the structural design because it is assumed to be carried directly by the sub-soil. When the column loads are equal, flat slab construction can be adopted. When raft foundation is also subjected to wind load moments, and when these are considered in design, the safe stresses can be increased by $33\frac{1}{3}$%. However, when the wind load reactions are less than $\frac{1}{3}$rd reactions due to other loads, wind is not considered in the design.

Example 16.6 illustrates the design procedure.

**Design: Example 16.6.** *Fig. 16.18 shows the layout of the columns of a building. The outer columns are 300 × 300 mm in size and carry a load of 500 kN each. The inner columns are 400 × 400 mm in size and carry a load 800 kN each. In addition to this, each column carries a moment 160 kN-m due to wind load on the length of the building. Design a suitable raft foundation, if the bearing capacity of soil is 100 kN/m². Use M 20 mix and Fe 415 steel.*

### Solution:

1. **Design constants.** For M 20 mix and Fe 415 steel, we have the following values:

   $\sigma_{cbc} = c = 7$ N/mm² and $\sigma_{st} = 230$ N/mm²; $k_c = 0.289$; $j_c = 0.904$; $R_c = 0.914$

2. **Size and general arrangement of mat**

   Total weight $W = (4 \times 500) + (2 \times 800) = 3600$ kN

   Self-weight of raft

   $W' = 10\% \ W$ (say) = 360 kN

   ∴ Total load = 3600 + 360 = 3960 kN

   Total moment = 160 × 6 = 960 kN-m

   ∴ Eccentricity $e = \dfrac{960}{3960} = 0.242$ m

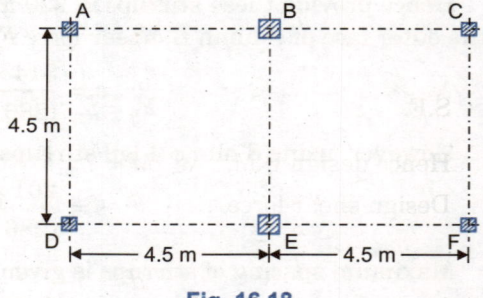

Fig. 16.18

Let the raft project 500 mm outside the outer columns all round, so that length of raft = (6 × 1.5 + 0.3 + 2 × 0.5) = 10.3 m, and width of raft = (4.5 + 0.3 + 2 × 0.5) = 5.8 m.

Hence maximum and minimum intensities of pressures would be

$$= \frac{\Sigma W}{A}\left(1 \pm \frac{6e}{B}\right) = \frac{3960}{10.3 \times 5.8}\left(1 \pm \frac{6 \times 0.242}{5.8}\right) \approx 83 \text{ N/m}^2 \text{ and } 5 \text{ N/m}^2$$

Since the maximum pressure is less than 100 kN/m², the size of the footing is satisfactory.

Net upward soil reaction $p_0 = \dfrac{3600}{10.3 \times 5.8} \approx 60$ kN/m²

Upward reaction due to wind moment is given by

$$p_w = \pm \frac{960}{\frac{1}{6} \times 10.3 \, (5.8)^2} = \pm 16.6 \text{ kN/m}^2$$

*i.e.*, + 16.6 kN/m² on leeward side and – 16.6 kN/m² on windward side.

Pressure intensity at inner face of longitudinal beams = ± 16.6 × $\frac{4.1}{5.8}$ = ± 11.7 kN/m²

Let us provide two longitudinal beams in direction AC and DF, joining three columns on each side. These two beams are joined by seven secondary beams. The general arrangement of beams etc. are shown in Fig. 16.19.

**3. *Design of slab.*** The soil pressure distribution due to load and wind moment are shown in Fig. 16.19 (*b*) and (*c*) respectively. Since the pressure $p_w$ due to wind moments is less than $\frac{1}{3}$ rd the pressure $p_0$ due to dead load, the effect of wind will not be considered in the design of slab.

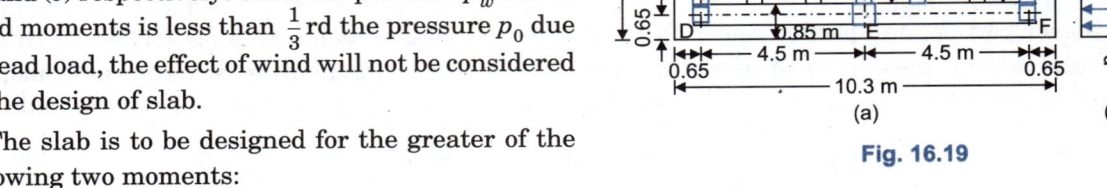

Fig. 16.19

The slab is to be designed for the greater of the following two moments:

(*a*) Bending moment in the cantilever portion of slab.

(*b*) Bending moment in the continuous slab.

Cantilever span = 0.5 m = *a*

$$M_1 = p_0 \frac{a^2}{2} = \frac{60}{2}(0.5)^2 = 7.5 \text{ kN-m} \qquad \ldots(i)$$

Shear force $\qquad F_1 = p_0 a = 60 \times 0.5 = 30 \text{ kN} \qquad \ldots(ii)$

The slab is continuous over the secondary beams. Centre to centre spacing of secondary beams = 1.5 m. Taking width of secondary beams as 0.3 m, the clear span of slab, between two beams = 1.5 – 0.3 = 1.2 m. Hence bending moment in interior spans is

$$M_2 = \frac{p_0 L^2}{12} = \frac{60(1.2)^2}{12} = 7.2 \text{ kN-m} \qquad \ldots(iii)$$

S.F. $\qquad F_2 = p_0 \frac{L}{2} = \frac{60(1.2)}{2} = 36 \text{ kN} \qquad \ldots(iv)$

Hence design B.M. $\qquad M = 7.5 \text{ kN-m} = 7.5 \times 10^6 \text{ N-mm}$

Design shear force $\qquad F = 36$ kN. The effective depth of slab is given by

$$d = \sqrt{\frac{7.5 \times 10^6}{0.914 \times 1000}} \cong 90 \text{ mm}$$

Keeping an effective cover of 60 mm and a total depth of 160 mm, available effective depth = 160 – 60 = 100 mm. Area of reinforcement is given by

$$A_{st} = \frac{7.5 \times 10^6}{230 \times 0.904 \times 100} \approx 361 \text{ mm}^2. \text{ Using 10 mm } \Phi \text{ bars, } A_\Phi = 78.5 \text{ mm}^2$$

∴ Spacing $\qquad s = \dfrac{1000 \times 78.5}{361} = 217.5$ mm. However provide 10 mm Φ bars @ 200 mm c/c.

Let us check the reinforcement for development length.

$L_d$ required = 45 Φ = 45 × 10 = 450 mm.

Providing a side cover of 50 mm, available length of each bar = 500 – 50 = 450 mm only. This is equal near to the required $L_d$. Hence O.K.

Let us check the slab for shear.

$$\tau_v = \frac{V}{bd} = \frac{F_2}{bd} = \frac{36 \times 10^3}{1000 \times 100} = 0.36 \text{ N/mm}^2$$

$$\frac{100 A_s}{bd} = 100 \left( \frac{1000 \times 78.5}{200} \right) \div (1000 \times 100) = 0.39\%$$

∴ From Table 3.1, $\tau_c = 0.265$ N/mm$^2$. Also, from Table 3.2, $k = 1.3$ for $D = 160$ mm Permissible shear stress = $k \cdot \tau_c = 1.3 \times 0.265 = 0.344$ N/mm$^2$ which is less than $\tau_v$. Hence unsafe.

To make it safe, increase $p$ from 0.39% to 0.5% so that $\tau_c$ comes out to be 0.3 N/mm$^2$ and permissible shear stress becomes = $1.3 \times 0.3 = 0.39$ N/mm$^2$. For that, decrease the spacing of reinforcement to
$$200 \times 0.265/0.3 \approx 175 \text{ mm c/c}.$$

For cantilever portion, and portion under the face of the secondary beams, this reinforcement is to be provided at the bottom face of the slab since bending moment is of sagging nature, while for intermediate portion between secondary beams, the bars are to be provided on the top face, as is done in a continuous slab.

Area of transverse reinforcement = $\frac{0.12}{100} (1000 \times 160) = 192$ mm$^2$

Using 10 mm Φ bars, spacing = $\frac{1000 \times 78.5}{192} = 408.8$ mm.

For the portion of slab projecting from the face of the main slab, cantilever projection = 0.45 m.

∴ $$M = \frac{60(0.45)^2}{2} = 6 \text{ kN-m}$$

Now, actual length of each bar in the cantilever portion, using 50 mm side cover = 450 – 50 = 400 mm. Required $L_d = 45 \times 8 = 360$ mm.

Hence actual $\sigma_{si}$ carried by each bar = 230 N/mm$^2$

Effective depth available = 100 – 10 = 90 mm

∴ $$A_{st} = \frac{6 \times 10^6}{230 \times 0.904 \times 90} = 321 \text{ mm}^2$$

and spacing of 10 mm Φ bars = $\frac{1000 \times 78.5}{321} \approx 244$ mm

For the sake of symmetry, provide these also at 175 mm c/c.

Also, the spacing of transverse reinforcement (found as 409 mm c/c) may be kept at $2 \times 175 = 350$ mm c/c, so that every alternate bar of cantilever reinforcement may be used as distribution bars.

**4. *Design of intermediate secondary beam.*** There are in all seven secondary beams spaced at 1.5 m c/c out which the two outer secondary beams and the central secondary beams, join columns. The remaining four secondary beams are intermediate beams which join the longitudinal beams. These intermediate beams are subjected to reaction from the slab beneath. Since the reaction due to wind moment is less than $\frac{1}{3}$rd the dead load reactions, it will not be considered in design.

Span of the beam = 4.5 m.

However, the load on the beam will be the upward load acting on the slab on the clear span of 4.1 m. For the remaining portion of the slab, load will be transferred directly to the longitudinal beams.

Load per metre run $w = p_0 \times 1.5$

= $60 \times 1.5 = 90$ kN/m

The loading is shown in Fig. 16.20. End reactions transferred to each longitudinal beam

= $R = \frac{1}{2} (4.1 \times 90) = 184.5$ kN

$$M = 184.5 \times \frac{4.5}{2} - \frac{90}{2}\left(\frac{4.1}{2}\right)^2 = 226 \text{ kN-m} = 226 \times 10^6 \text{ N-mm}$$

$F = R = 184.5$ kN

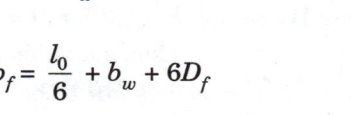

(a) Due to superimposed load

(b) Due to wind load

Fig. 16.20

Since the beam experience hogging bending moment, it will act as T-beam. Let the width of beam = $b_w$ = 300 mm. The width of flange is given by

$$b_f = \frac{l_0}{6} + b_w + 6D_f$$

where $l_0$ = 4.5 m = 4500 mm

$b_w$ = 300 mm

$D_f$ = 150 mm.

∴ $b_f$ = 4500/6 + 300 + 6 × 150 = 1950 mm

Subject to maximum of centre to centre spacing (= 1.5 m)

Hence $b_f$ = 1500 mm. Taking lever arm = 0.9 $d$, effective depth is determined from the relation:

$$M = 0.45\, b_f \cdot \sigma_{cbc} \cdot D_f \left[\frac{2kd - D_f}{k}\right]$$

∴ $226 \times 10^6 = 0.45 \times 1500 \times 7 \times 160 \left[\dfrac{2 \times 0.289\, d - 160}{0.289}\right]$

∴ $0.578\, d = \dfrac{226 \times 10^6 \times 0.289}{0.45 \times 1500 \times 7 \times 160} + 160 = 246.4$. From which $d = 427$ mm

$\tau_{cmax}$ = 1.8 N/mm² for M 20 concrete (Table 3.3)

Hence maximum depth from considerations of shear is

$$d = \frac{F}{b \cdot \tau_{c.max}} = \frac{184.5 \times 10^3}{300 \times 1.8} = 342 \text{ mm. Hence adopt } d = 430 \text{ mm.}$$

$$A_{st} = \frac{226 \times 10^6}{230 \times 0.9 \times 430} = 2539 \text{ mm}^2$$

Using 25 mm Φ bars, No. of bars = 2539/491 ≈ 6.

Provide these in two layers at the upper face of the beam. Keep a vertical distance of 25 mm between them and nominal cover of 50 mm. Hence total depth = 430 + 12.5 + 25 + 10 + 50 = 430 + 97.5 = 527.5 mm.

Keep total depth of 530 mm so that available $d$ = 530 – 97.5 = 432.5 mm

Near ends, bend 3 bars downwards to take negative bending moment induced due to fixity.

At the end, let us check for development length so as to satisfy the relation

$$\frac{M_1}{V} + L_0 \geq L_d$$

where $M_1 = \sigma_{st} \cdot A_{st} \cdot j_c\, d = 230\,(4 \times 491) \times 0.9 \times 432.5 = 1.758 \times 10^8$ N-mm

$V$ = 184.5 kN = 184.5 × 10³ N

$L_0 = \dfrac{l_s}{2} - x' + 3\,\Phi = \dfrac{400}{2} - 40 + 3 \times 25 = 235$ mm (by providing 90° bend at ends)

$L_d = 45\,\Phi = 45 \times 25 = 1125$ mm.

∴ $\dfrac{M_1}{V} + L_0 = \dfrac{1.758 \times 10^8}{184.5 \times 10^3} + 235 = 1188 > L_d$.  Hence safe.

Also,
$$\tau_v = \frac{184.5 \times 10^3}{300 \times 432.5} = 1.42 \text{ N/mm}^2 \text{ and } \frac{100 A_s}{bd} = \frac{100 \times 4 \times 491}{300 \times 432.5} = 1.51\%$$

Hence from Table 3.1, $\tau_c \approx 0.44$ N/mm². Since $\tau_v > \tau_c$, shear reinforcement is necessary.

∴
$$V_c = \tau_c \cdot bd = 0.44 \times 300 \times 432.5 = 57090 \text{ N} \approx 57.1 \text{ kN}$$
$$V_s = V - V_c = 184.5 - 57.1 = 127.4 \text{ kN}$$

Using 10 mm Φ 4-lgd stirrups, $A_{sv} = 4 \frac{\pi}{4} (10)^2 = 314$ mm²

∴
$$s_v = \frac{A_{sv} \cdot \sigma_{sv} \cdot d}{V_s} = \frac{314 \times 230 \times 432.5}{127.4 \times 1000} \approx 240 \text{ mm.}$$

The distance of the point where S.F. is equal to $V_c$ is given by $x = \frac{184.50 - 57.1}{90} \approx 1.42$ m from the face of longitudinal beam. Hence provide 10 mm Φ 4 lgd stirrups @ 240 mm c/c from the end to a point distant 1.42 m from the face of each longitudinal beam. For the remaining length, provide nominal stirrups at the maximum spacing given by

$$s_v = \frac{2.175 \cdot A_{sv} \cdot f_y}{b} = \frac{2.175 \times 314 \times 415}{300} = 945 \text{ mm.}$$

subject to a maximum of 300 mm or 0.75 $d$ (= 0.75 × 425 ≈ 318 mm). Hence provide 10 mm Φ 4-lgd nominal stirrups at a spacing of 300 mm c/c for the remaining length.

Let us calculate the reaction transferred by the secondary beam to main beam, due to wind load reaction $p_w$, since this will be useful for the design of main beam.

Load due to $p_w$, per metre run = ± 1.5 $p_w$ = ± 1.5 × 11.7 = ± 17.6 kN/m at the face of the main beam.

Couple due to this = $\left(\frac{1}{2} \times 17.6 \times 2.05\right)\left(\frac{2}{3} \times 4.1\right) = 49.3$ kN-m

∴ End reactions = 49.3/4.5 ≈ 11 kN.

∴ Net reaction at windward end = 184.5 − 11 = 173.5 kN

Net reaction at leeward end = 184.5 + 11 = 195.5 kN.

**5. Design of central secondary beam.** The central secondary beam $BE$ joins columns $B$ and $E$ and hence is subjected to end moments transmitted to it by the columns. Therefore, let us consider both $p_0$ as well as $p_w$, along with the transferred end moments, to calculate the maximum bending moment and shear force. If the bending moment is greater than the intermediate secondary beam atleast by $33\frac{1}{3}\%$, then the design will change. Otherwise it will remain the same.

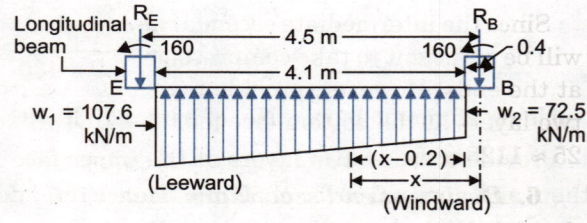

Fig. 16.21

From Fig. 16.19 (c), the pressure intensity $p_w'$ due to wind, at the face of longitudinal beam = 11.7 kN/m².

∴ Net pressure intensity at $E'$ = 60 + 11.7 = 71.7 kN/m².

Net pressure intensity at $B'$ = 60 − 11.7 = 48.3 kN/m².

Width of slab transferring load = 1.5 m.

∴ Load intensity $w_1$ at end $B'$ = 71.7 × 1.5 = 107.6 kN/m

Load intensity $w_2$ at end $B'$ = 48.3 × 1.5 = 72.5 kN/m

The effective span of the beam = 4.5 m while it will receive load only in a length of 4.1 m. The loading on the beam is shown in Fig. 16.21

To find reaction at $E$, take moments at $B$.

$$\therefore \quad R_E = \frac{1}{4.5}\left[72.5 \times 4.1\left(\frac{4.1}{2}+0.2\right) + \frac{107.6-72.5}{2} \times 4.1 \times \left\{\frac{2}{3}(4.1)+0.2\right\} - 2 \times 160\right]$$

$$= 124.5 \text{ kN}.$$

Similarly, taking moments $E_r$.

$$R_B = \frac{1}{4.5}\left[72.5 \times 4.1\left(\frac{4.1}{2}+0.2\right) + \frac{107.6-72.5}{2} \times 4.1 \times \left\{\frac{1}{3}(4.1)+0.2\right\} + 2 \times 160\right]$$

$$= 244.8 \text{ kN}$$

Bending moment will be maximum where S.F. is zero. Let us write down equation of shear force at $x$ from support $B$.

$$F_x = 244.8 - \left[72.5(x-0.2) + \frac{107.6-72.5}{4.1}(x-0.2)\frac{1}{2}(x-0.2)\right]$$

Equating this to zero, we get $x = 3.1$ m. Hence maximum bending moment is given by

$$M = (244.8 \times 3.1) - 160 - \left[\frac{72.5(2.9)^2}{2} + \left(\frac{107.6-72.5}{4.1}\right) \times 3.1 \times \frac{1}{2} \times 3.1 \times \frac{3.1}{3}\right]$$

$$= 251.5 \text{ kN-m} \quad \ldots(ii)$$

It will be seen that maximum bending moment and shear force found above by taking the wind effect into account are *less* than $1\frac{1}{3}$ times the corresponding values for the intermediate secondary beam. *Hence the same depth and same main reinforcement will be used*. If however, the B.M. and S.F. found above were more than $1\frac{1}{3}$ times the B.M. and S.F. for the intermediate secondary beam, the central beam could be designed for the above increased B.M. and S.F., taking higher permissible values of stresses.

However, due to partial fixity at the ends, the beam will be subjected to sagging B.M. = $WL/24$, in addition to wind moment of 160 kN-m.

$$\therefore \quad \text{Total sagging B.M.} = (124.5 + 244.8)\frac{4.5}{24} + 160 = 229.2 \text{ kN.m}.$$

Since the intermediate secondary beam has been designed for a B.M. of 226 kN-m, its depth $d = 432.5$ mm will be sufficient to take compression, but special reinforcement at the bottom face will have to be provided at the ends. However, provide the same reinforcement as the centre, *i.e.* 6 bars of 25 mm Φ provided in two layers at the bottom face for the portion under the columns for a distance at least equal to $L_d = 45 \times 25 \approx 1125$ mm.

6. **Design of end secondary beam.** The end secondary beam will get less load by way of reaction from the slab, because it will receive load from slab of width = $0.65 + \frac{1}{2}(1.5) = 1.4$ m only against 1.5 m width for central secondary beam. However, provide the same section and reinforcement as for the central secondary beam. The ratio of load on the outer beam to the central one will be approximately equal to $1.4/1.5 \approx 0.93$.

7. **Design of main beam.** The bending moments and shear forces in the main beam are to be found under the following three conditions:

(a) Ignoring wind effect.

(b) Taking wind effect: leeward beam.

(c) Taking wind effect; windward beam.

(a) *Ignoring wind effect.* When there is no wind acting, the reaction transferred to the main beam by all the intermediate as well as central beams will be equal, while the reaction transferred by the end beam will be 0.93 times that transferred by intermediate beam. In addition to this, the beam will get soil reaction from slab immediately below it and the slab cantilevering from its face.

Reaction transferred by each intermediate beam = 184.5 kN
Reaction transferred by end secondary beam = 0.93 × 184.5 = 171.6 kN
Reaction transferred by slab, per metre run of beam = 60 (0.4 + 0.45) = 51 kN/m.
The loading diagram is shown in Fig. 16.22.

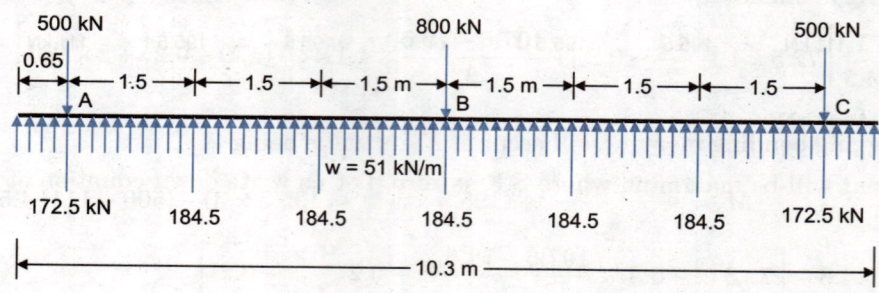

Fig. 16.22

It should be checked that total downward load of column is equal to the total upward load. Total downward load = 500 + 800 + 500 = 1800 kN

Total upward load = (184.5 × 5) + (171.6 × 2) + (10.3 × 51) = 1791 kN

Thus, both of these are approximately equal. The discrepancy is there because of rounding off the value of $p_0$ to 60 kN/m² in step 2.

Maximum sagging bending moment will occur at $B$, its value being given by

$$M_1 = 51 \frac{(5.15)^2}{2} + (184.5 \times 1.5) + (184.5 \times 3.0) - (500 - 172.5) 4.5$$

$$= 32.8 \text{ kN-m (sagging)} \quad \ldots(1)$$

Maximum hogging bending moment occurs somewhere between $A$ and $B$, where shear force is zero. This is likely to occur between the two intermediate beams at distance $x$ from $A$.

∴ $\quad F_x = 51 (x + 0.65) + 172.5 + 184.5 - 500 = 0$

This gives $x$ = 2.15 m from $A$ or 2.8 m from the edge. Hence the maximum hogging bending moment is given by

$$M_2 = (500 - 172.5) 2.15 - 184.5 (2.15 - 1.5) - \frac{51}{2} (2.8)^2$$

$$= 384.3 \text{ kN-m (hogging)} \quad \ldots(2)$$

Shear force to the right of $A$ is $F_1 = (500 - 172.5) - (51 \times 0.65) = 294.4$ kN $\quad \ldots(3)$

Shear force to the left of $B$ is $F_2 = \frac{1}{2} (800 - 184.5) = 307.8$ kN $\quad \ldots(4)$

**(b) Wind effect : leeward beam.** From step 4, reaction transferred by the intermediate secondary beam on the leeward end = 195.5 kN.

From step 5, reaction transferred by the central secondary beam on the leeward end = 124.5 kN. Reaction transferred by the end secondary beam, on the leeward end

$$= 0.93 \times 195.5 - \frac{160 + 160}{4.5} = 182 - 71 = 111 \text{ kN}$$

Total downward load = 500 + 800 + 500 = 1800 kN

∴ Upward load from slab = 1800 − (195.5 × 4 + 124.5 + 111 × 2) = 671.50 kN

Uniform upward load $w$ = 671.5/10.3 = 65.2 kN/m

The loading on the leeward beam $DF$ is shown in Fig. 16.23.

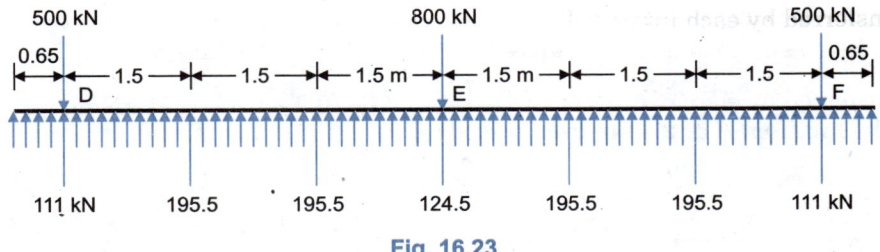

Fig. 16.23

Maximum sagging bending moment will occur at $E$, its value being

$$M_3 = \frac{65.2(5.15)^2}{2} + (195.5 \times 1.5) + (195.5 \times 3) - (500 - 111)\,4.5$$

$$= -6.1 \text{ kN-m } (i.e. \text{ hogging}) \qquad \ldots(5)$$

Maximum hogging bending moment occurs somewhere between $D$ and $E$ where shear force is zero. This is likely to occur between the two intermediate beams, at distance $x$ from $D$.

$$F_x = 65.2(0.65 + x) + 111 + 195.5 - 500 = 0$$

This gives $x = 2.32$ m from $D$, or 2.97 m from the edge. Hence maximum hogging bending moment is given by

$$M_4 = (500 - 111) \times 2.32 - 195.5\,(2.32 - 1.5) - \frac{65.2}{2}(2.97)^2$$

$$= 454.6 \text{ kN-m (hogging)} \qquad \ldots(6)$$

S.F. to the right of $D$ is $F_3 = (500 - 111) - (65.2 \times 0.65) = 346.6$ kN  ...(7)

S.F. to the left of $E$ is $F_4 = \frac{1}{2}(800 - 124.5) = 337.8$ kN  ...(8)

(c) **Wind effect: windward beam.** From step 4, reaction transferred by the intermediate secondary beam on the windward side = 173.5 kN. From step 5, reaction transferred by central secondary beam on the windward end = 244.8 kN.

Reaction transferred by the end secondary beam on the windward end

$$= (0.93 \times 173.5) + \frac{160 + 160}{4.5} = 232.5 \text{ kN}$$

Total downward load = 500 + 800 + 500 = 1800 kN

∴  Upward load from the slab = 1800 − (4 × 173.5 + 244.8 + 2 × 232.5) = 396.2 kN.

∴  Upward $w$ = 396.2/10.3 ≈ 38.5 kN/m.

The loading on the windward main beam $AC$ is shown in Fig. 16.24.

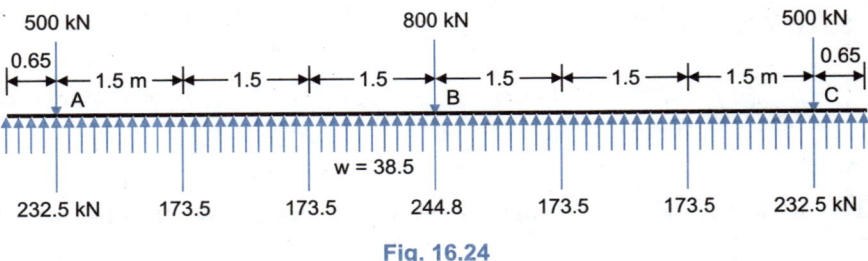

Fig. 16.24

Maximum sagging bending moment will occur at $B$:

$$M_5 = \frac{38.5\,(5.15)^2}{2} + (2 \times 173.5 \times 2.25) - (500 - 232.5) \times 4.5 = 87.6 \text{ kN-m}$$

Maximum hogging bending moment will occur where S.F. is zero, some where between $A$ and $B$. Let it occur at $x$ from $A$ between the two intermediate beams.

$$F_x = 38.5\,(x + 0.65) + 173.5 + 232.5 - 500 = 0$$

which gives $x = 1.79$ m from $A$ or 2.44 m from the edge.

Hence maximum hogging bending moment is given by

$$M_6 = (500 - 232.5) \times 1.79 - 173.5\,(1.79 - 1.5) - \frac{38.5}{2}(2.44)^2 = 314 \text{ kN-m (hogging)}$$

S.F. to the right of $A$ is given by $F_5 = (500 - 232.5) - (0.65 \times 38.5) = 242.5$ kN  ...(11)

S.F. to the left of $B$ is $F_6 = \dfrac{1}{2}(800 - 244.8) = 277.6$ kN  ...(12)

Design values. From the above cases, we get the following values for design purposes:

Without wind:  Maximum sagging moment = 32.8 kN-m
                Maximum hogging moment = 384.3 kN-m
                Maximum shear = 307.8 kN

With wind:     Maximum sagging moment = 87.6 kN-m
                Maximum hogging moment = 454.6 kN-m ($< \dfrac{4}{3} \times 384.3$)
                Maximum shear = 346.6 kN

Thus critical conditions are under 'no wind'. T-beam action will be available for maximum hogging bending moment = 384.3 kN-m. Let the width of rib = 400 mm.

The flange width of T-beam is given by

$$b_f = \frac{l_0}{6} + b_w + 6\,D_f$$

where   $l_0$ = distance between points of zero moments in the beam
        $= 10.3 - (2 \times 2.8) = 4.7$ m $= 4700$ mm
        $b_w = 400$ mm
        $D_f = 160$ mm

$\therefore \qquad b_f = 4700/6 + 400 + 6 \times 160 \approx 2143$ mm.

However, since the beam is situated nearly at the outer boundary, its half flange is restricted in its width to a value $= 0.20 + 0.45 = 0.65$ m $= 650$ mm. Hence maximum value of $b_f$ is restricted to $2 \times 650 = 1300$ mm. Hence adopt $b_f = 1300$ mm. Taking lever arm $= 0.9\,d$, the effective depth is given by the relation:

$$M = 0.45\,b_f \cdot \sigma_{cbc} \cdot D_f \left(\frac{2kd - D_f}{k}\right)$$

or $\qquad 384.3 \times 10^6 = 0.45 \times 1300 \times 7 \times 160 \left(\dfrac{2 \times 0.289\,d - 160}{0.289}\right)$

or $\qquad 0.578\,d = \dfrac{384.3 \times 10^6 \times 0.289}{0.45 \times 1300 \times 7 \times 160} + 160$

From which $d \stackrel{\sim}{=} 570$ mm.

This depth will be sufficient to take the sagging bending moment when the beam will act as a simple rectangular beam.

$$A_{st} = \frac{384.3 \times 10^6}{230 \times 0.9 \times 570} = 3257 \text{ mm}^2$$

Using 25 mm $\Phi$ bars,   No. of bars $= 3257/490 \approx 7$

Hence provide these at the top face of the beam in two layers with a gap of 25 mm between the layers. This reinforcement should be safe to satisfy the development length criterion at the point of contraflexure, so as the satisfy the relation

$$\frac{M}{V} + L_0 \geq L_d.$$

Assuming that all the bars are continued even beyond the point of contraflexure.

$$M = \sigma_{st} \cdot A_{st} j_c d = 230 \, (7 \times 490) \times 0.9 \times 570 = 4.047 \times 10^8 \text{ N-mm}$$

$V$ = Shear at the point of contraflexure

Assuming that the point of contraflexure falls just to the left of $B$, shear force $V = F_2 = 307.8$ kN

$$L_0 = 12 \, \Phi \text{ or } d \text{ whichever is more} = 570 \text{ mm}$$

$$L_d = 45 \times 25 = 1125 \text{ mm}.$$

$$\therefore \quad \frac{M}{V} + L_0 = \frac{4.047 \times 10^8}{307.8 \times 10^3} + 570 = 1884 \text{ mm}. \quad \text{Hence safe}$$

Continue all the 7 bars throughout the length, in two layers.

Using 50 mm nominal cover and 10 mm φ stirrups, minimum effective cover = 50 + 10 + 25 + 25/2 = 97.5 mm. Providing effective cover of 100 mm, $D = 570 + 100 = 670$ mm.

The beam should also have reinforcement at its bottom face to resist sagging bending moment at the centre and at the ends, for a bending moment of 87.6 kN-m under wind load conditions. The reinforcement required for bending moment will be quite small. However, this reinforcement should satisfy the criterion of development length at the point of contraflexure near the centre of beam, where S.F. is maximum.

Let us use $n$ number of 20 mm dia. bars each having $A_\Phi = 314$ mm$^2$.

$$\therefore \quad M = \sigma_{st} \cdot A_{st} j_c d = 230 \, (314 \, n) \times 0.9 \times 570 = n \times 37.05 \times 10^6 \text{ N-mm}$$

$V$ = S.F. at the point of contraflexure

$\approx 307.8$ kN (assuming this to be equal to S.F. at $B$)

$$L_0 = 12 \, \Phi \text{ or } d, \text{ whichever is more} = 570 \text{ mm}.$$

$$L_d = 45 \, \Phi = 45 \times 20 = 900 \text{ mm}.$$

$$\therefore \quad \frac{M}{V} + L_0 = \frac{n \times 37.50 \times 10^6}{307.8 \times 10^3} + 570 = (121.83 \, n + 570)$$

Equating this to $L_d$, $121.83 \, n + 570 = 900$, from which $n = 2.7 \approx 3$.

Hence provide 3 bars of 20 mm diameter at the bottom face, throughout the length of the beam. Let us now check the beam for shear force of 307.8 kN. Providing nominal cover of 50 mm, effective cover = 50 + 10 + 20/2 = 70 mm. Hence available $d = 670 - 70 = 600$.

$$\tau_v = \frac{307.8 \times 1000}{400 \times 600} = 1.28 \text{ N/mm}^2 \text{ (at } B\text{)}.$$

$$\frac{100 \, A_s}{bd} = \frac{100(7 \times 490)}{400 \times 600} = 1.43\%$$

Hence, $\tau_c = 0.44$ N/mm$^2$ (Table 3.1)

Since $\tau_v > \tau_c$, shear reinforcement is necessary.

$$V_c = 0.44 \times 400 \times 600 = 105600 \text{ N} \approx 105.6 \text{ kN}$$

$$\therefore \quad V_s = 307.8 - 106.5 = 201.3 \text{ kN}.$$

Using 10 mm Φ 4-lgd stirrups, $A_{sv} = 314$ mm$^2$.

$$\therefore \quad \text{Spacing} = \frac{\sigma_{sv} \cdot A_{sv} \cdot d}{V_s} = \frac{230 \times 314 \times 600}{201.3 \times 1000} \approx 215 \text{ mm}$$

However provide the stirups @ 200 mm c/c. for 1 m on either side of the faces of the column $B$.

Similarly, shear force at $A = 294.4$ kN. Hence provide 10 mm Φ 4-legged stirrups @ 200 mm c/c, for a distance of 0.6 m to the right of face of column $A$, and 0.6 m to left of column $C$.

For the remaining length, provide 8 mm φ 4 lgd nominal shear reinforcement at a maximum spacing of

$$s_v = \frac{2.175 \, A_{sv} \, f_y}{b} = \frac{2.175 \times 201 \times 415}{400} \approx 453.5 \text{ mm}.$$

But not more than 300 mm nor more than $0.75 \, d \, (= 0.75 \times 575 = 431 \text{ mm})$.

Hence provide 8 mm Φ 4-lgd nominal stirrups @ 300 mm c/c for the remaining length. The details of reinforcement etc. are shown in Fig. 16.25.

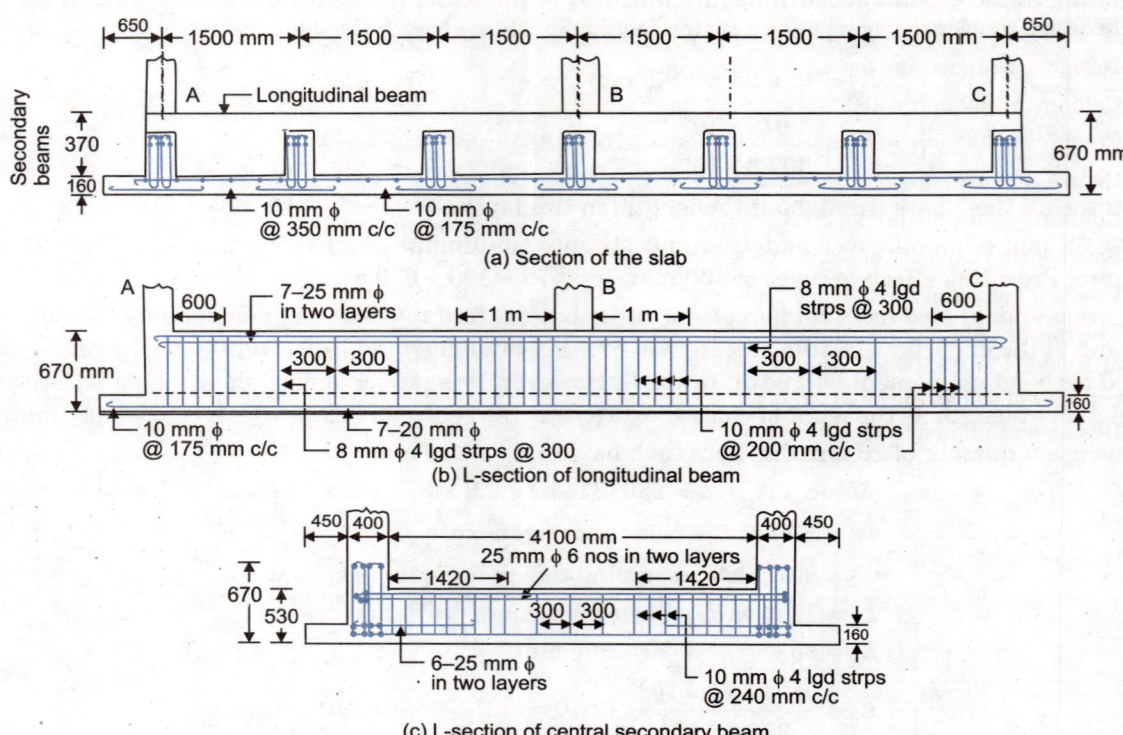

Fig. 16.25

## PROBLEMS

1. Design a combined footing for two columns, each 600 mm × 600 mm, 5 metres apart, and each carrying a load of 1800 kN. The available width is restricted to 2.4 metres. The safe bearing capacity of soil may be taken as 200 kN/m². Use M 20 concrete and Fe 415 steel.

2. Redesign problem 1 by providing a longitudinal beam joining the column. There is no restriction on the width of the footing.

3. Design a combined footing for two R.C. columns $A$ and $B$, separated by a distance of 4 m centre to centre:

   Column $A$ is 500 mm square and carries a load of 1200 kN
   
   Column $B$ is 600 mm square and carries a load of 1600 kN
   
   The safe bearing capacity of a soil is 200 kN/m². Use M 20 concrete and Fe 415 steel.

4. Redesign problem 3 if the property line is located 400 mm from centre of column $A$ measured outwards along the centre line of the columns and runs at right angles to the column line.

   No encroachment beyond the property line is permissible.

5. Design a trapezoidal combined footing for the following requirements:

   Column $A$ = 400 mm × 400 mm

   Column $B$ = 600 mm × 600 mm

   Axial load on column $A$ = 400 kN

   Axial load on column $B$ = 800 kN

   Distance between centre of columns = 2.5 m

   Safe bearing capacity of soil = 200 kN/m$^2$

   The footing is not to project more than 0.4 m beyond the outer face of the columns. Use M 20 concrete and Fe 415 steel.

6. Redesign problem 5 by providing a central longitudinal beam joining the columns.

7. Design a strap footing for the following data:

   Column $A$, 400 mm × 400 mm carries a load of 800 kN

   Column $B$, 600 mm × 600 mm carries a load of 1000 kN

   Centre to centre distance between the two column is 8 metres.

   Column $A$ is on the property line.

   Safe bearing capacity of soil = 150 kN/mm$^2$

   Use M 20 mix and Fe 415 steel.

8. The reinforced concrete column ($A_1$) and ($A_2$) of sizes 400 × 400 mm and 500 × 500 mm carry axial service loads of 1000 kN and 1500 kN respectively. These columns are placed 5 m apart centre to centre. The safe bearing capacity of soil at the site is 200 kN/m$^2$. The property lines lies at 800 mm from the outer face of column ($A_1$). Design a combined footing when the length of footing is restricted at 7 mtrs. Use $M_{20}$ mix and HYSD bars.

# CHAPTER 17: PILE FOUNDATIONS

## 17.1. TYPES OF PILES

The column loads are heavy, as in the case of multistory buildings and soil immediately below the ground surface is in capable of bearing these loads, then the pile foundations are used. The use of piles as foundation can be traced since olden times. The art of driving piles was well established in Roman times and the details of such foundations were recorded by Vitruvious in 59 A.D. Today, pile foundation is much more common than any other type of deep foundation. Based on foundation or the use. A pile foundation consists of an isolated pile (or) group of piles with an inter connecting pile cap on which the load is transferred from a column. Piles may be classified as: (1) end bearing pile (2) friction pile (3) compaction pile (4) tension pile or uplift pile (5) anchor pile (6) fender pile and dolphins (7) batter pile (8) sheet pile.

End *bearing piles* are used to transfer load through water or soft soil to a suitable bearing stratum. *Friction piles* are used to transfer loads to a depth of a friction load-carrying material by means of skin friction along the length of piles. *Compaction piles* are used to compact loose granular soils, thus increasing their bearing capacity. *Tension or uplift piles* anchor down the structures subjected to uplift due to hydrostatic pressure or due to overturning moment. *Anchor piles* provide anchorage against horizontal pull from sheet piling or other pulling forces. *Fender piles and dolphins* are used to protect water front structures against impact from ships or other floating objects. *Sheet piles* are commonly used as bulk heads or as impervious cut-off to reduce seepage and uplift under hydraulic structures. *The batter piles* are used to resist large horizontal or inclined forces.

Piles can also be classified, based on materials and composition, as follows:

1. **Concrete piles**
   (*a*) Precast.   (*b*) Cast-in-situ:
   (*i*) driven piles; cased or uncased.
   (*ii*) bored piles; pressure piles and under reamed pile.
2. **Timber piles**
3. **Steel piles**
   (*a*) H-pile.   (*b*) Pipe pile.   (*c*) Sheet pile.
4. **Composite piles**
   (*a*) Concrete and timber.   (*b*) Concrete and steel.

The *precast concrete piles* are generally used for a maximum design load of about 800 kN except for large prestressed piles. They must be reinforced to withstand handling stresses. They require space for casting and storage, more time to set and cure before installation and heavy equipment for handling and driving. They also incur large cost in cutting off extra length or adding more length. The *cast-in-situ* piles are generally used for a maximum design load of 750 kN except for compacted, pedestal piles. They are installed by pre-excavation, thus eliminating vibration due to driving and the handling stresses. Cast-in-place piles may be classified into two classes : driven piles, (cased or uncased) and bored piles (pressure piles, pedestal piles and under-reamed piles). A variety of cast-in-situ piles are in use, each bearing the name of the manufacturer. The common types are as follows:  (*i*) Raymond standard pile  (*ii*) Raymond step-taper pile  (*iii*) Union metal pile or monotub,  (*iv*) MacArthur compressed uncased pile  (*v*) Mac Arthur cased pile.  (*vi*) Franki standard pile.  (*vii*) Western button bottom pile etc.

Under-reamed pile is a special type of bored pile having an increased diameter or bulb at some point in its length, to anchor the foundation in expansive soil subjected to alternative expansion and contraction.

Concrete filled steel pipes and steel *H-piles* are used as long piles with high bearing capacity. They are rarely used unless they reach a stratum of exceptionally high supporting capacity, since their cost is very high. *Timber piles* have small bearing capacity and are not permanent unless treated. They are prone to damage by hard driving, and should not be driven through hard stratum or boulders. *Composite piles* are suitable where the upper part of the pile is to project above the water table. Such a pile consists of a lower portion of untreated timber and an upper portion of concrete. In the other type of composite pile, steel piles are attached to the lower end of cast in place concrete piles. This type is used in cases where the required length of pile is greater than that available for the cast-in-place type.

## 17.2. PILE DRIVING

Piles are commonly driven by means of a hammer supported by a crane or by a special device known as a *pile driver*. The hammer is guided between two parallel steel members known as *leads*. The leads are carried on a flange in such a way that they can be supported in a vertical position or an inclined position. Hammers are of the following types:  (*i*) drop hammer  (*ii*) single acting hammer  (*iii*) double acting hammer  (*iv*) diesel hammer and  (*v*) vibratory hammer.

A *drop hammer* (ram or monkey) is raised by which and allowed to fall by gravity on the top of a pile. A *single acting hammer* is raised by steam, compressed air or internal combustion, but is allowed to fall by gravity alone. The *double action hammer* employs steam or air for lifting the ram and for accelerating the downward stroke. It operates with succession of rapid blows. *Diesel hammer* is a small, light weight self contained and self-acting type, using gasoline for fuel. In the case of vibratory hammer, the driving unit vibrates at high frequency.

During pile driving, heads, helmets or caps are placed on the top of pile to receive the blows of hammer and to prevent damage to the head of the pile. The *cushion*, consisting of a pad of resilient material, hard wood or rope, is placed between the drive cap and the top of the pile to protect the pile head. Single acting hammers are advantageous when driving heavy piles in compact or hard soil while double acting hammers are generally used to drive piles of light or moderate weight in soils of average resistance against driving. Piles are ordinarily driven to a resistance measured by the number of blows required for the last 1 cm of penetration. Resistance of 3 to 5 blows per cm are commonly specified for concrete piles.

## 17.3. LOAD CARRYING CAPACITY OF PILES

The ultimate load carrying capacity of a pile is defined as the maximum load which can be carried by a pile, and at which the pile continues to sink without further increase of load. The *allowable load* is the load which the pile can carry safely, and is determined on the basis of:  (*i*) ultimate bearing resistance divided by suitable factor of safety  (*ii*) permissible settlement and  (*iii*) overall stability of the pile foundation. The load carrying capacity of a pile can be determined by the following methods:  (1) dynamic formulae  (2) static formulae  (3) pile load tests  (4) penetration tests.

**Dynamic Formulae.** When a pile hammer hits the pile, the total driving energy is equal to the weight of hammer times the height of drop of stroke. In addition to this, in the case of double acting hammers, some energy is also imparted by the steam pressure during the return stroke. This total downward energy is consumed by the work done in penetrating the pile and by certain losses. The various dynamic formulae are essentially based on this assumption. Based on test data several empirical expressions have been proposed for the guidance of pile driving crew for the given location for a specified type of pile to be driven by specific type of driving rig.

The formulae are proposed as per IS: 2911. Following are some commonly used dynamic formulae:

**(a) Engineering News formula.** The Engineering News formula was proposed by A.M. Wellington (1888), in the following form:

$$Q_a = \frac{WH}{F(S+C)}$$

where $Q$ = allowable load ; $W$ = weight of hammer
$H$ = height of fall ; $F$ = factor of safety = 6
$S$ = final set (penetration) per blow, usually taken as average penetration, cm per blow for the last 5 blows of drop hammer, or 20 blows of steam hammer.
$C$ = empirical constant.

Denoting $W$ in kg, $H$ in cm and $S$ in cm and $C$ = 2.5 cm for drop hammer and $C$ = 0.25 cm for single and double acting hammers the above formula reduces to the following forms:

(i) Drop hammers: $$Q_a = \frac{WH}{6(S+2.5)}$$

(ii) Single acting steam hammers: $$Q_a = \frac{WH}{6(S+0.25)}$$

(iii) Double acting steam hammers: $$Q_a = \frac{(W+ap)H}{6(S+0.25)}$$

where $a$ = effective area of piston (cm$^2$) and $p$ = mean effective steam pressure (kg/cm$^2$).

**(b) Hiley's formula.** Indian Standard IS : 2911 (Part II) 1964 gives the following formula based on original expression by Hiley:

$$Q_f = \frac{\eta_h W H \eta_b}{S + C/2}$$

where $Q_f$ = ultimate load on pile
$W$ = weight of hammer, in kg
$H$ = height of drop of hammer, in cm
$S$ = penetration or set, in cm, per blow
$C$ = total elastic compression, $C_1 + C_2 + C_3$
$C_1, C_2, C_3$ = temporary elastic compression of dolly and packing, pile and soil respectively.
$\eta_h$ = efficiency of hammer, variable from 65% for some double acting steam hammers to 100% for drop hammer released by triggers.
$\eta$ = efficiency of hammer blow (i.e. ratio of the energy after impact to striking energy of ram)
$\eta_b = \dfrac{W + e^2 P}{W + P}$ (for the case of $W > eP$)

and $\eta_b = \dfrac{W + e^2 P}{W + P} - \left\{\dfrac{W - eP}{W + P}\right\}^2$ (for the case when $W < eP$)

$P$ = weight of pile, anvil, helmet follower.
$e$ = coefficient of restitution (variable from zero for a timber pile with poor condition of head or for excess packing in the driving cap to 0.5 for double acting hammer driven steel piles without helmet but with packing on top).

The allowable load is obtained by using a factor of safety of 2 or 2.5.

**Static Formulae.** The static formulae are based on the assumption that the ultimate bearing capacity $Q_f$ of a pile is the sum of the total ultimate skin friction $R_f$ and total ultimate point or end bearing resistance $R_p$:

$$Q_f = R_f + R_p \quad \text{or} \quad Q_f = A_s r_f + A_p r_{p0}$$

where  $A_s$ = surface area of pile upon which the skin friction acts.
          $A_p$ = area of cross section of pile on which bearing resistance acts. For tapered piles, $A_b$ may be taken as the cross-sectional area of the lower one-third of the embeded length.
          $r_f$ = average skin friction.
          $r_p$ = unit point or the resistance.

A factor of safety of 3 may be adopted for finding the allowable load.

***Cohesive soil.*** For the pile in cohesive soil, the point bearing is generally neglected for individual pile action, since it is negligible as compared to friction resistance. The unit skin friction may be taken as equal to shear strength of the soil.

Hence $\quad\quad\quad\quad r_f = \tau_f = c = \dfrac{q_u}{2} \quad \text{and} \quad r_p = c N_c = 9c$

where $q_u$ = unconfined compressive strength of soil and
        $c$ = unit cohesion.

***Non-cohesive soil.*** For non-cohesive soil, the following values may be adopted:

$$r_f = K \tan \Phi (\gamma Z + q)$$

$$r_p = \frac{\gamma B}{2} N_{\gamma q} \text{ (for rectangular or square piles)}$$

or $\quad\quad\quad\quad r_p = 0.3 \gamma B N_{\gamma q}$ (for rectangular piles)

where  $K$ = coefficient or lateral earth pressure*
        $\gamma$ = density of soil
        $q$ = surcharge on the ground
        $Z$ = depth of centre of gravity of the pile below ground surface
        $B$ = least lateral dimension of pile
        $\Phi$ = angle of internal friction
        $N_{\gamma q}$ = Meyerhof's non-dimensional factor.

## 17.4. GROUP ACTION IN PILE

When several closely spaced piles are grouped, together, it is reasonable to expect that the soil pressure, developed in the soil as resistance, will overlap. The bearing capacity of a pile group may or may not be equal to the sum of the bearing capacity of individual piles constituting a group. Theory and test have shown that the bearing value $Q_f$ of a group of *friction piles*, particularly in clay, is equal to bearing capacity of individual pile multiplied by the number of piles $n$ in a group. However, no reduction due to grouping occurs in end bearing piles. For combined end bearing and friction piles, only the load carrying capacity of the friction portion is reduced. A method of estimating the bearing capacity of a group of friction piles is to multiply the quantity $nQ_f$ by a reduction factor called the *efficiency of the pile group* $\eta_g$.

$$Q_g = n Q_f \cdot \eta_g.$$

The efficiency of pile group depends upon the following factors : characteristics of pile (*i.e.* length, diameter, material, etc.), spacing of piles, total number of piles ($n$) in a row, and number of rows ($m$). Out of a number of formulae for determining efficiency of a pile group, two are given below:

---

* See Author's book 'Soil Mechanics and Foundations'.

**1. Converse-Labarre formula:** $\eta_g = 1 - \dfrac{\theta}{90}\left[\dfrac{(n-1)m + (m-1)n}{mn}\right]$

where $\theta = \tan^{-1}\dfrac{d}{s}$ (degrees)

$d$ = diameter of piles and $s$ = spacing of piles.

**2. Silver Keeney formula:** $\eta_g = \left[1 - 0.479\left(\dfrac{s}{s^2 - 0.093}\right)\left(\dfrac{m+n-2}{m+n-1}\right)\right] + \dfrac{0.3}{m+n}$

where $s$ = average spacing, centre to centre in metres.

## 17.5. STRUCTURAL DESIGN OF R.C. PILE

A R.C. pile is designed for the following: (1) total load coming on it from the structure (2) handling stresses (3) driving stresses. While designing the pile as a column, it may be considered as fixed at one end hinged at the other. The length of the pile may be taken as 2/3 rd the length embedded in firm stratum. The cross-section of pile varies with its overall length.

The main longitudinal reinforcement in a pile should not be less than the following values:

(i) 1.25 percent of the cross-sectional area of the pile, for piles having length upto 30 times their least width.

(ii) 1.5 percent of the cross-sectional area of the pile, for piles having length between 30 to 40 times its least width.

(iii) 2 percent of the gross cross-sectional area of the pile, for piles having length greater than 40 times its least width.

Lateral reinforcement should be provided in form of links, of not less than 5 mm diameter wire. The volume of such lateral reinforcement should not be less than 0.2 percent of the gross volume of the pile, and their centre to centre spacing should not exceed half the least width of the pile. Near each end of pile, upto a distance of three times the least lateral dimension, lateral reinforcement should be provided @ 0.6% of the gross volume. If the pile has to penetrate hard stratum, the lateral reinforcement at the top of the pile for a distance of three times the width, should be in the form of helix. In addition to this, the longitudinal bars should be held apart by provision of *spacers* made of 12 to 16 mm wire and provided at the interval of not exceeding 1.5 m. In order to prevent the outward displacement of the bars, 6 mm $\Phi$ links should also be provided with the spacers (Fig. 17.5). The pile should be provided with a cast iron shoes at its lower end (Fig. 17.5).

**Stresses During Handling.** Precast piles should be checked against handling stresses. When a pile is lifted by means of a derrick, it is subjected to bending moment due to its own weight. When the pile is of less than 12 metre length, it is suspended from one point at its middle. Piles longer than 12 metres are suspended at two or three points suitably spaced at its length so that handling moment is as small as possible.

**(1) Pile suspended at one point.** [*Fig. 17.1(a)*]

If $w$ is the weight of the pile per unit length, and L is the length of the pile, it will be subjected to a maximum bending moment = $\dfrac{wL}{2} \times \dfrac{L}{4} = \dfrac{wL^2}{8}$ at the point of suspension.

**(2) Pile suspended at two points.** [*Fig. 17.1(b)*].

Let the distance of each point of suspension be $x$ from the respective ends. The value of $x$ should be such that maximum bending moment anywhere in the pile should be the least. This is possible when the hogging moment is equal to maximum sagging moment.

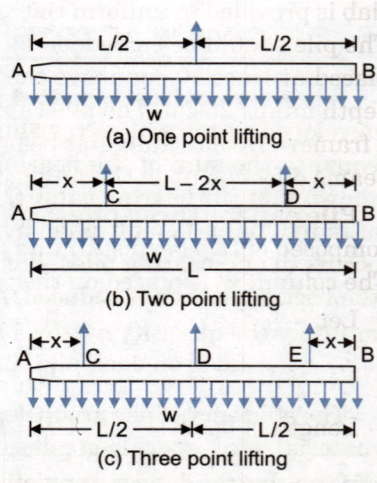

(a) One point lifting

(b) Two point lifting

(c) Three point lifting

Fig. 17.1

Reaction $R = \dfrac{wL}{2}$ at each point of suspension.

Maximum hogging moment at the end of each cantilever $= \dfrac{wx^2}{2}$ ...(i)

Maximum sagging moment at the middle of the pile $= \dfrac{wL}{2}\left(\dfrac{L}{2} - x\right) - \dfrac{wL^2}{8}$ ...(ii)

Equating the two, we get

$$\dfrac{wx^2}{2} = \dfrac{wL}{2}\left(\dfrac{L}{2} - x\right) - \dfrac{wL^2}{8} = \dfrac{wL}{2}\left(\dfrac{L}{4} - x\right)$$

or $\qquad x^2 + Lx - \dfrac{L^2}{4} = 0,\qquad$ which gives $\qquad x = 0.206\,L$

Hence maximum B.M. $= \dfrac{w}{2}(0.206\,L)^2 = \dfrac{wL^2}{47}$.

**(3) Pile suspended at three points** [*Fig. 17.1(c)*]. It can be shown that if the pile is suspended at three points, the end points should be located at $x = 0.15L$ from the corresponding ends so that B.M. is the least. The third point is located at the middle of the pile. The maximum bending moment in that case will be $wL^2/90$.

### Erecting the pile from one end

Fig. 17.2 shows a pile being erected from one point $C$, the end of the pile being resting on the ground. Let the distance of the point be $x$ from the end. In order that bending moment on the pile is the least, $x$ should be equal to $0.293\,L$. The maximum bending moment will be equal to $wL^2/23$.

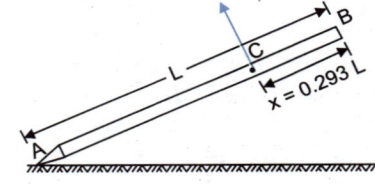

Fig. 17.2

## 17.6. DESIGN OF PILE CAP

When a column or pier is supported on the pile only, the column should rest centrally on the pile. However, when the column is supported on more than one pile, the piles should be connected through a rigid pile cap, to distribute the load to the individual piles. The pile cap consists of a rigid, deep, reinforced concrete slab which acts monolithically with the group of piles. The piles should be arranged symmetrically about the axis of the column so that the load from column is distributed uniformly to all the columns. The pile cap slab is provided in uniform thickness. The pile cap should be extended beyond exterior piles by 10 to 15 cm. The pile should be embedded by at least 15 cm in the pile cap, and the reinforcement in the cap should be placed at least 10 cm above the pile head. Thus, effective depth of the pile cap will be equal to the total depth minus 25 cm. The pile cap, provided over the entire area of the piles, is considered to be divided into a framework of rectangular beams, along which main reinforcement is provided. The arrangement of these beams depends upon the number of piles, and the width of beam is taken equal to width of pile.

**Pile cap for three piles.** Fig. 17.3 (a) shows the pile cap for three piles. The pile cap is considered to be composed of two beams $AB$ and $CD$; $A$, $B$ and $C$ being the three piles placed at distance $L$ centre to centre. The column $W$ is placed on the beam $DC$, at the centroid of the triangle $ABC$.

Let $\qquad W$ = total load of column.

∴ $\qquad$ Load on each pile = $W/3$

Length of beam $\qquad CD = l = \dfrac{L\sqrt{3}}{2}$. Distance of $W$ from $D = \dfrac{1}{3}l = \dfrac{L}{2\sqrt{3}}$.

Beam $CD$ is supported on $C$ and $D$, and is loaded with $W$, such that reaction at $C$ is $W/3$. Hence maximum bending moment is

$$M_1 = \frac{W}{3} \times \frac{2l}{3}$$

$$= \frac{W}{3} \times \frac{2}{3}\left(\frac{L\sqrt{3}}{2}\right) = \frac{WL}{3\sqrt{3}}$$

Reaction transferred to

$$D = W - \frac{W}{3} = \frac{2W}{3}$$

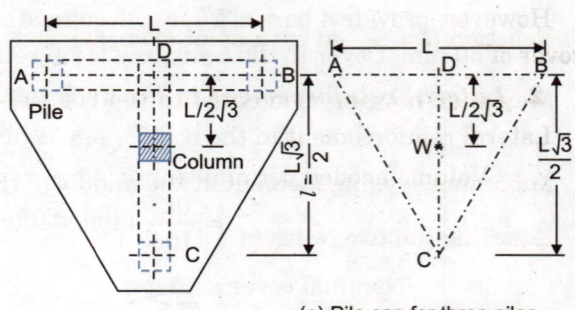

The beam $AB$ is of length $L$, and is loaded at its middle point $D$ with a load $(2W)/3$. Hence maximum bending moment is given by

$$M_2 = \frac{W}{3}\left(\frac{L}{2}\right) = \frac{WL}{6}.$$

The reinforcement for both beams can now be computed.

Pile cap for four piles [Fig. 17.3(b)]. Load on each pile = $\frac{W}{4}$.

The pile cap is considered to be composed of two beams $AB$ and $CD$.

$$\therefore \quad M_{max} = \frac{W}{4} \times \frac{L}{2} = \frac{WL}{8} \text{ for each beam.}$$

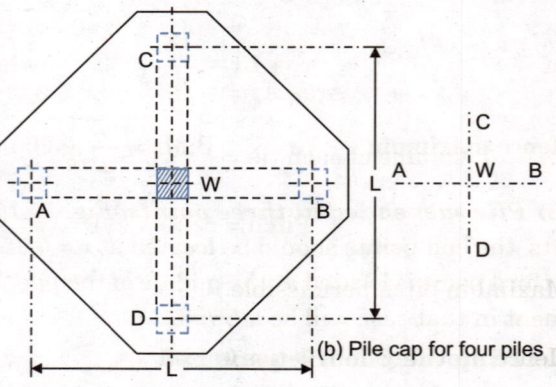

Fig. 17.3

Secondary reinforcement. In order to prevent the outward splaying tendency of piles, secondary reinforcement should always be provided. The reinforcement is provided at the bottom of the pile cap, running round the longitudinal reinforcement projecting from the piles into the cap. It should be so bent that there is change of its direction at the head of every pile. The area of secondary reinforcement changing direction at every head of the pile should not be less than 20% of the tensile reinforcement.

**Design: Example 17.1.** *Design a pile under a column transmitting an axial load of 800 kN. The pile is to be driven to a hard stratum available at a depth of 8 metres. Use M 20 concrete and Fe 415 steel.*

**Solution:**

For $M$ 20 concrete, $\sigma_{cc} = 5$ N/mm². For Fe 415 steel, $\sigma_{sc} = 190$ N/mm².

Also, $m = 13.33$ for M 20 concrete.

**1. Main reinforcement.** Let the length of the pile above ground, including pile cap, etc. = 0.6 m.

$\therefore$ Total length of pile = 8.6 m.

Let the size of the pile be 400 mm × 400 mm

$\frac{l}{D}$ ratio = $\frac{8.6}{0.4}$ = 21.5. Since this is greater than 12, the pile behaves as long column.

Hence reduction coefficient $C_r = 1.25 - \frac{l_{ef}}{48D} = 1.25 - \frac{8.6}{48 \times 0.4} \approx 0.8$

$\therefore$ Design load for a short column = 800/0.8 = 1000 kN

Load carrying capacity of column is given by $P = \sigma_{cc} A_c + \sigma_{sc} A_{sc}$

where $A_c$ = area of concrete = $(400 \times 400) - A_{sc} = 16 \times 10^4 - A_{sc}$

$\therefore \quad 1000 \times 10^3 = 5(16 \times 10^4 - A_{sc}) + 190 A_{sc}$. From which $A_{sc} = 1081$ mm².

Since the length of pile is less than 30 times the width, minimum reinforcement @ 1.25% of gross cross-sectional area = $\frac{1.25}{100}(400 \times 400) = 2000$ mm².

However, provide 4 bars of 25 mm Φ giving total area of steel = 4 × 490 = 1960 mm². Provide a nominal cover of 50 mm. Cover to the centres of main reinforcement using 8 mm φ ties = 50 + 8 + 25/2 = 70.5 mm.

**2. Lateral reinforcement in the body of the pile**

Lateral reinforcement in the body of pile is provided @ 0.2% of gross volume.

∴ Volume needed per mm length

$$= \frac{0.2}{100} (400 \times 400 \times 1) = 320 \text{ mm}^3.$$

Nominal cover = 50 mm

Using 8 mm Φ ties, length of each side of tie

$$= 400 - 2 \times 50 - 8 = 292$$

Area $A_\Phi = \frac{\pi}{4} (8)^2 = 50.3$ mm².

∴ Volume of each tie = 4 × 292 × 50.3 = 58750 mm³

∴ Pitch = 58750/320 = 183.595 ≃ 183 mm

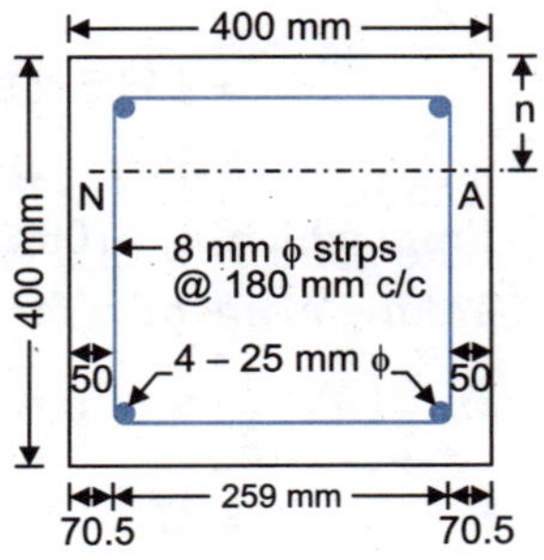

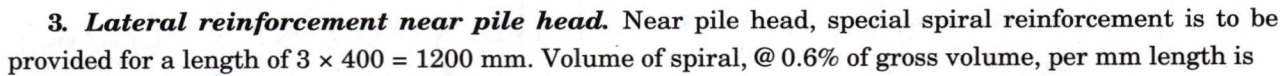

Fig. 17.4

Maximum pitch permissible = $\frac{1}{2}$ × 400 = 200 mm.

Hence provide 8 mm Φ ties @ 180 mm c/c throughout the length of the pile.

**3. Lateral reinforcement near pile head.** Near pile head, special spiral reinforcement is to be provided for a length of 3 × 400 = 1200 mm. Volume of spiral, @ 0.6% of gross volume, per mm length is

$$= \frac{0.6}{100} (400 \times 400 \times 1) = 960 \text{ mm}^3.$$

Using 8 mm Φ spiral, having $A_\Phi = 50.3$ mm², pitch is given by

$$s = \frac{\text{Circumference of spiral} \times A_\Phi}{960} = \frac{\pi \times 292 \times 50.3}{960} = 48 \text{ mm}$$

Provide the spiral at 45 mm pitch. Provide 6 additional bars of 16 mm Φ vertically within the spiral. These spirals will be in addition to the normal ties.

**4. Lateral reinforcement near pile end**

Volume of ties per mm length @ 0.6% of gross volume = 960 mm³.

∴ Volume of each tie = 58750 mm³. Hence, Pitch = 58750/960 = 61.19 ≃ 60 mm.

Provide ties @ 60 mm c/c in a bottom length of 3 × 400 = 1200 mm.

**5. Spacer forks and links.** Provide two pairs of 12 mm Φ spacer fork with 6 mm Φ links @ 1.5 m c/c along the length.

**6. Check for handling stresses.** Provide three holes in the pile as follows:

(i) One hole at $0.293L = 0.293 \times 8.6 \approx 2.5$ m from the pile for the purpose of hoisting it.

(ii) One hole each from either end, at a distance of $0.206L = 0.206 \times 8.6 \approx 1.75$ m for the purpose of stacking.

(iii) Weight of pile per meter run = 0.4 × 0.4 × 1 × 25000 = 4000 N/m

$$M = \frac{4000 (2.5)^2}{2} = 12500 \text{ N-m}$$

$$= 125 \times 10^5 \text{ N-mm}$$

Effective depth of pile section = 400 − 70.5 = 329.5 mm. Let the neutral axis be situated at $n$ below one face.

$$\frac{bn^2}{2} + (m_c - 1) A_{sc} (n - d_c) = m A_{sc} (d - n)$$

$$\frac{400}{2} n^2 + (1.5 \times 13.33 - 1) 980 (n - 70.5)$$

$$= 13.33 \times 980 (329.5 - n)$$

or $n^2 + 158.39 n - 28083 = 0$

From which $n = 106.2$ mm

Taking moment of forces about tensile steel, we get

$$bn \frac{c}{2} \left[ d - \frac{n}{3} \right] + (1.5 m - 1) A_{sc} c' (d - d_c) = M$$

where $c' = \dfrac{(n - d_c) c}{n} = \dfrac{106.2 - 70.5}{106.2} c$

$= 0.336 c$

∴ $400 \times 106.2 \dfrac{c}{2} \left[ 329.5 - \dfrac{106.2}{3} \right]$

$\qquad + 18.995 \times 980 \times 0.336 c (329.5 - 70.5)$

$= 125 \times 10^5$

which gives $c = 1.59$ N/mm².

This is much less than 5 N/mm². Hence safe. The stress in steel will also be well within safe limit. The details of reinforcement are shown in Fig. 17.5.

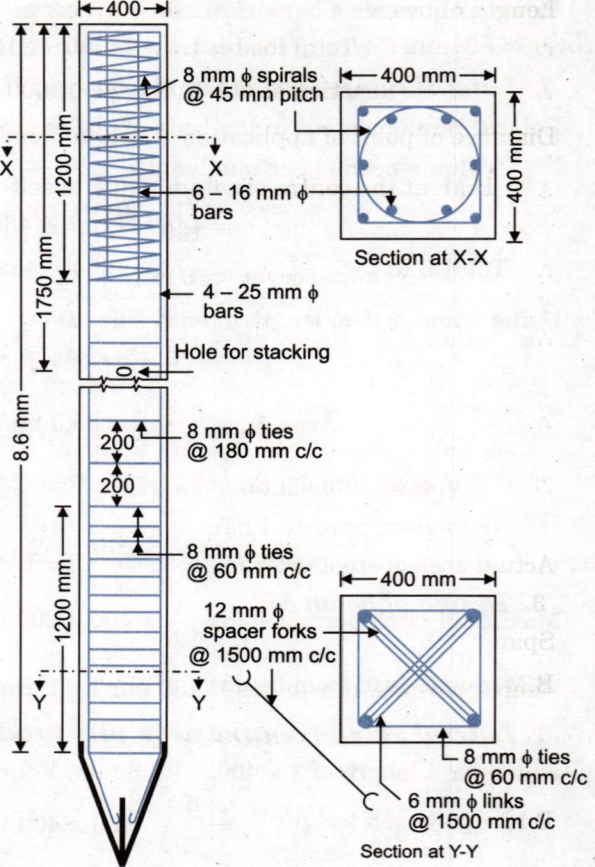

Fig. 17.5

**Design: Example 17.2.** *A R.C. column, 400 mm × 400 mm carrying a load of 600 kN is supported on three piles 400 mm × 400 mm in section. The centre to centre distance between the piles is 1.5 m. Design a suitable pile cap. Use M 20 concrete and Fe 415 steel.*

**Solution:** [Fig. 17.3 (a)]

**1. Dimensions of pile cap.** Centre to centre spacing of piles = $L$ = 1.5 m.

Keeping 200 mm clear projection of the cap beyond pile face. Overall length of the cap along the direction $AB$ = 1.5 + 0.4 + 0.4 = 2.3 m.

Length of beam $\quad CD = l = \dfrac{L \sqrt{3}}{2} = 1.5 \dfrac{\sqrt{3}}{2} = 1.3$ m.

∴ Length of cap in the direction $DC$ = 1.3 + 0.4 + 0.4 = 2.1 m.

**2. Design of beam DC**

Load on each pile = $W/3$ = 600/3 = 200 kN

Let the width of beam = width of column = 400 mm

∴ B.M. due to load = $\dfrac{WL}{3\sqrt{3}} = \dfrac{600 \times 1.5}{3\sqrt{3}} = 173$ kN-m = $173 \times 10^6$ N-mm  ...(i)

In order to calculate the bending moment due to self weight of the beam plus weight of part of slab, let us assume total thickness of slab to be 800 mm. The self-weight of the beam is calculated on the assumption that weight of slab equal to two times the width of the beam acts with the beam.

∴ $\quad w = \left[ \dfrac{3 \times 400 \times 800}{10^6} \right] \times 1 \times 25000 = 24000$ N/m

Length of beam $\quad l = L\sqrt{3}/2 = 1.3$ m.

∴ Total load $= 1.3 \times 24000 = 31200$ N

∴ Reaction at $C = 31200/2 = 15600$ N

Distance of point of application of column load $= \frac{2}{3}l = \frac{2}{3} \times 1.3 = 0.87$ m.

∴ B.M. at the centre of column, due to self-weight is $= (15600 \times 0.87) - \frac{24000}{2}(0.87)^2$

$= 4489$ N-m $= 4.489 \times 10^6$ N-mm.

∴ Total B.M. $= 173 \times 10^6 + 4.489 \times 10^6 \approx 177.5 \times 10^6$ N-mm

∴ $d = \sqrt{\dfrac{177.5 \times 10^6}{0.914 \times 400}} \approx 697$ mm. Keep $d = 700$ mm

∴ $A_{st} = \dfrac{175.5 \times 10^6}{230 \times 0.904 \times 700} \approx 1220$ mm$^2$

∴ No. of 25 mm Φ bars $= 1220/490.8 = 2.5$

However, provide 4 bars of 25 mm Φ.

Actual area of steel provided $= 4 \times 490.8 = 1963$ mm$^2$.

### 3. Design of beam AB

Span $\quad L = 1.5$ m

B.M. due to load from beam

$$CD = \frac{WL}{6} = \frac{600 \times 1.5}{6} = 150 \text{ kN-m} = 150 \times 10^6 \text{ N-mm}$$

B.M. due to self weight $= \dfrac{24000 (1.5)^2}{8} = 6750$ N-m $= 6.75 \times 10^6$ N-mm

∴ Total B.M. $= 150 \times 10^6 + 6.75 \times 10^6$

$= 156.75 \times 10^6$ N-mm

The reinforcement in direction $AB$ will be placed below the reinforcement of $CD$. Hence available $d = 700 + 25 = 725$ mm.

∴ $A_{st} = \dfrac{156.75 \times 10^6}{230 \times 0.904 \times 725} \approx 1040$ mm$^2$.

However, provide the same reinforcement, *i.e.* 4 Nos. of 25 mm Φ bars. Keep total depth = 800 mm.

### 4. Secondary reinforcement

Area of secondary reinforcement running round each pile head $= 0.2 \times 1206 = 241.2$ mm$^2$.

Using 10 mm Φ bars, $A_\Phi = 78.5$ mm$^2$.

∴ No. of bars $= 241.2/78.5 = 3.08 \approx 4$.

### 5. Check for shear

Shear is tested at a distance $d$ from the beam. The dispersion lines (at 45°) transfer the load directly to the column. Hence there is no possibility of diagonal tension cracks.

The details of reinforcement etc. are shown in Fig. 17.6.

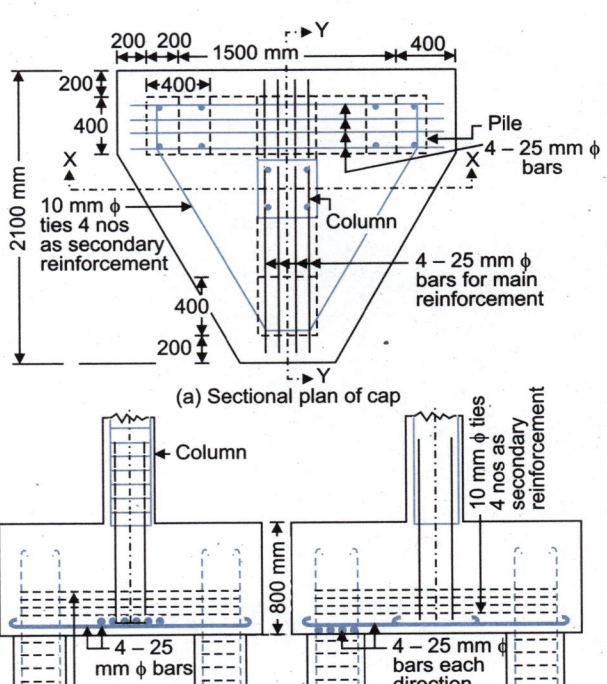

Fig. 17.6. Reinforcement details of pile cap.

## PROBLEMS

1. A column 600 mm × 600 mm carries an axial load of 1200 kN and is supported on three piles The piles are driven to hard strata available at the depth of 10 metres. Using M 20 concrete and Fe 415 steel., design the pile and the pile cap. The column is placed at the centroid of the pile group. The centre to centre distance of the piles is restricted to 2 m.

2. Design a pile cap on a group of four friction piles each of 300 mm diameter for supporting a 450 mm square reinforced concrete column an axial load of 1200 kN at service conditions. The materials to be used are $M_{20}$ concreted HYSD steel of Fe 415 grade.

3. Design a square pile for a reinforced concrete circular column of 400 mm diameter transmitting an axial service load of 1250 kN the bearing capacity of the soil at site is 150 kN/m² the material are $M_{20}$ and Fe 415. Draw the reinforcement details.

# CHAPTER 18: DESIGN OF RETAINING WALLS

## 18.1. INTRODUCTION

A *retaining wall* or retaining structure is used for maintaining the ground surfaces at different elevations on either side of it. Whenever embankments are involved in construction, retaining walls are usually necessary. In the construction of buildings having basements, retaining walls are mandatory. Similarly in bridge work, the wing walls and abutments etc. are designed as retaining walls, to resist earth pressure along with superimposed loads. The material retained or supported by a retaining wall is called *backfill* which may have its top surface horizontal or inclined. The position of the backfill lying above the horizontal plane at the elevation of the top of a wall is called the *surcharge*, and its inclination to the horizontal is called the *surcharge angle* β.

In the design of retaining walls or other retaining structures, it is necessary to compute the lateral earth pressure exerted by the retaining mass of soil. The equation of finding out the lateral earth pressure against retaining wall is one of the oldest in the Civil Engineering field. The *plastic state of stress*, when the failure is imminent, was investigated by Rankine in 1860. A lot of theoretical and experimental work has been done in this field and many theories and hypothesis have been proposed.* The common applications are in hill roads, bridges, canals, and swimming pools etc.

## 18.2. TYPES OF RETAINING WALLS

Retaining walls may be classified according to their mode of resisting the earth pressure, and according to their shape. Following are some of common types of retaining walls (Fig. 18.1).

(*i*) Gravity walls
(*ii*) Cantilever retaining walls: (*a*) T-shaped   (*b*) L-shaped
(*iii*) Counterfort retaining walls.
(*iv*) Buttressed walls.

A gravity retaining wall shown in Fig. 18.1(*a*) is the one in which the earth pressure exerted by the backfill is resisted by dead weight of the wall, which is either made of masonry or of mass concrete. The stress developed in the wall is very low. These walls are so proportioned that no tension is developed anywhere, and the resultant of forces remain within the middle third of the base.

---

* For detailed account of various earth pressure theories, the reader may refer to the Author's book 'Soil Mechanics and Foundations'.

The cantilever retaining wall resists the horizontal earth pressure as well as other vertical pressures by way of bending of various components acting as cantilevers. A common form of cantilever retaining wall is the T-shaped wall shown in Fig. 18.1(b). The wall consists of stem AB, heel slab BC and toe slab DB. Each of these bend as cantilevers, about B. They are, therefore, reinforced on the tension face. Another form of cantilever retaining walls are the L-shaped walls shown in Fig. 18.1(c) and (d). They also resist the soil pressures by bending.

A counterfort retaining wall is shown in Fig. 18.1(e). The vertical stem and the heel slab are strengthened by providing counterforts at some suitable intervals. Because of provision of counterforts, the vertical stem as well as the heel slab acts as *continuous slab*, in contrast to the cantilevers of cantilever retaining wall. The toe slab however acts as cantilever bending upwards. This type of retaining wall is used when backfill of greater height is to be retained. A buttressed wall is a modification of the counterfort retaining wall in which the counterforts, called the buttresses, are provided to the other side of the backfill. However the buttresses reduce the clearance in front of the wall, and therefore these walls are not commonly used.

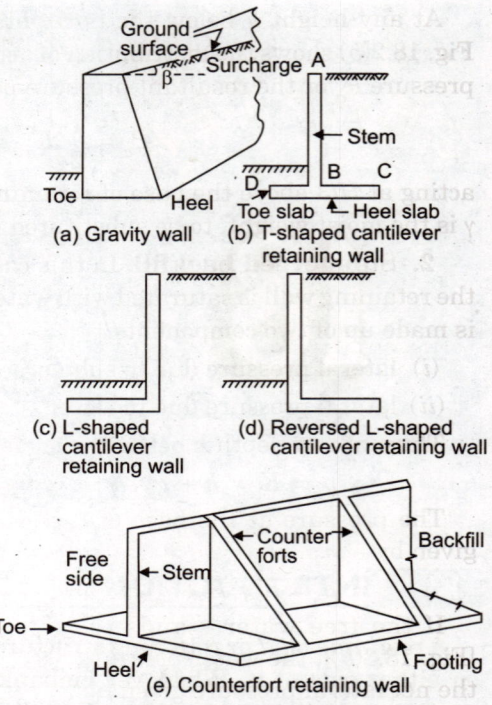

Fig. 18.1. Various Types of Retaining Walls. (commonly used)

## 18.3. ACTIVE EARTH PRESSURE: RANKINE'S THEORY

As originally proposed, Rankine's theory of lateral earth pressure is applied to uniform cohesionless soil only. Later, it was extended to include cohesive soils, by Resal and by Bell. The theory has been extended to stratified, partially immersed and submerged soils. Following are the assumptions of Rankine Theory: (1) The soil mass is semi-infinite, homogeneous, dry and cohesionless. (2) The ground surface is plane which may be horizontal or inclined. (3) The back of the wall is vertical and smooth. In other words, there are no shearing stresses between the wall and the soil, and the stress relationship for any element adjacent to the wall is the same as for any other element for away from the wall. (4) The wall yields about the base and satisfies the deformation conditions for plastic equilibrium.

However, the retaining walls are constructed of masonry or concrete, and hence the back of the wall is never smooth. Due to this, frictional forces develop. As a consequence of Rankine's assumption of non-existence of frictional forces at the wall face, the resultant pressure must be parallel to the surface of the backfill. The existence of friction makes the resultant pressure inclined to the normal to the wall at an angle that approaches the friction angle between the soil and the wall. We shall consider the following cases of cohesionless backfill:

1. Dry or moist backfill with no surcharge.
2. Submerged backfill.
3. Backfill with uniform surcharge.
4. Backfill with sloping surfaces.

**1. Dry or moist backfill with no surcharge.** Fig. 18.2 (a) shows a retaining wall of height $h$, having dry or moist backfill, with no surcharge. According to Rankine's theory, the intensity of *active earth pressure*, $p_a$, trying to move the wall away from the fill, is given by

$$p_a = K_a \gamma H \qquad \ldots(18.1)$$

where $K_a$ = co-efficient of active earth pressure,

$$= \frac{1 - \sin \Phi}{1 + \sin \Phi} \qquad \ldots(18.2)$$

$\Phi$ = angle of internal friction, for the backfill.
$\gamma$ = unit weight of soil (backfill).
$H$ = height of retaining wall.

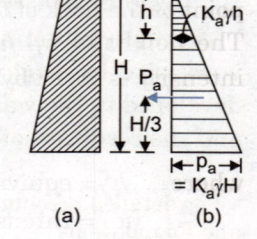

Fig. 18.2

At any height $h$ below the horizontal ground surface, the intensity of active earth pressure is $K_a\gamma h$. Fig. 18.2(b) shows the distribution of active earth pressure $p_a$ over the retaining wall. The total active earth pressure $P_a$ or the resultant pressure per unit length of the wall is given by

$$P_a = \frac{1}{2} K_a \cdot \gamma \cdot H^2 \qquad \text{...(18.3)}$$

acting at $H/3$ above the base of retaining wall. If the soil is dry, $\gamma$ is the dry weight of the soil, and if wet, $\gamma$ is the moist weight, to be substituted in Eq. 18.1 or 18.3.

**2. Submerged backfill.** In this case, the sand fill behind the retaining wall is saturated with water. The lateral pressure is made up of two components.

(i) lateral pressure due to submerged weight $\gamma'$ of soil, and
(ii) lateral pressure due to water.

Thus, at any depth $h$ below the surface,

$$p_a = K_a \gamma' h + \gamma_w \cdot h \qquad \text{...(18.4)}$$

The pressure at the base of the retaining wall ($h = H$) is given by

$$p_a = K_a \gamma' H + \gamma_w H \qquad \text{...(18.5)}$$

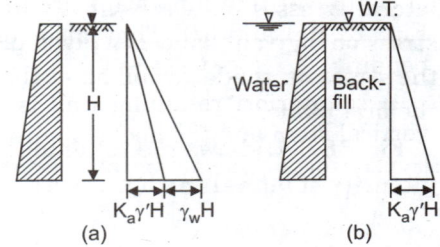

Fig. 18.3. Submerged Backfill.

If the free water stands to both the sides of the wall [Fig. 18.3(b)], the water pressure need not be considered, and the net lateral pressure given by

$$p_a = K_a \gamma' H \qquad \text{...(18.6)}$$

If the backfill is partly submerged, i.e., the backfill is moist to a depth $H_1$ below the ground level, and then it is submerged, the lateral pressure intensity at the base of the wall is given by

$$p_a = K_a \cdot \gamma H_1 + K_a \gamma' H_2 + \gamma_w \cdot H_2 \qquad \text{...(18.7)}$$

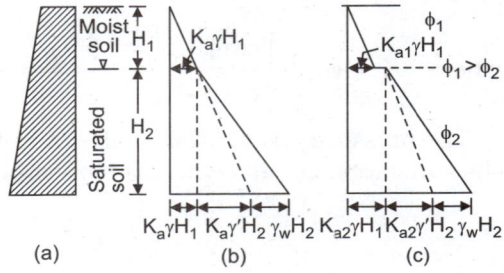

Fig. 18.4. Partly Submerged Backfill.

The above expression is based on the assumption that the value of $\Phi$ is the same for the moist as well as submerged soil. If it is different (say $\Phi_1$ and $\Phi_2$ respectively), the earth pressure coefficient $K_{a1}$ and $K_{a2}$ for both the portions will be different. As $\Phi$ decrease, $K_a$ increases. The lateral pressure intensity [Fig. 18.4(c)] at the base of wall is

$$p_a = K_{a2} \gamma H_1 + K_{a2} \gamma' H_2 + \gamma_w H_2 \qquad \text{...(18.8)}$$

**3. Backfill with uniform surcharge.** If the backfill is horizontal and carries surcharge of uniform intensity $w$ per unit area, the vertical pressure increment, at any depth $h$, will increase by $w$. The increase in the lateral pressure due to this will be $K_a w$. Hence the lateral pressure at any depth $h$ is given by

$$p_a = K_a \cdot \gamma \cdot h + K_a \cdot w \qquad \text{...(18.9)}$$

At the base of the wall, the pressure intensity is:

$$p_a = K_a \gamma H + K_a w \qquad \text{...[18.9(a)]}$$

Fig. 18.5(a) and (b) show two alternative methods of plotting the lateral pressure diagram for this case. The lateral pressure increment $K_a \cdot w$ due to the surcharge is the same at every point of the back of the wall, and does not vary with height $h$. The height of fill $h_e$ equivalent to the uniform surcharge intensity is given by the relation:

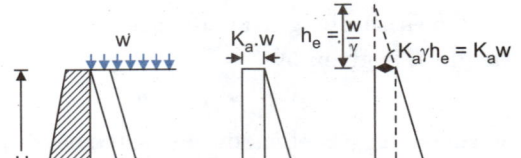

$$K_a \cdot \gamma \cdot h_e = K_a w \quad \text{or} \quad h_e = \frac{w}{\gamma} \qquad \text{...(18.10)}$$

where $h_e$ = equivalent height of soil (m)
($w$) = Intensity of surcharge (kN/m²)
($\gamma$) = Unit weight of soil (kN/m³)

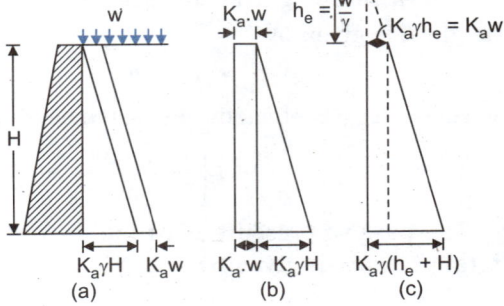

Fig. 18.5. Backfill with Uniform Surcharge

This means that the effect of the surcharge of intensity $w$ is the same as that of a fill of height $h_e$ above the ground surface [Fig. 18.5(c)].

**4. Backfill with sloping surface.** Let the sloping surface behind the wall be inclined at an angle $\beta$ with the horizontal; $\beta$ is called the surcharge angle. In finding out the active earth pressure for this case by Rankine's theory, an additional assumption that the vertical and lateral stresses are *conjugate* is made. It can be shown that if the stress on a given plane at a given point is parallel to another plane, the stress on the latter plane at the same point must be parallel to the first plane.

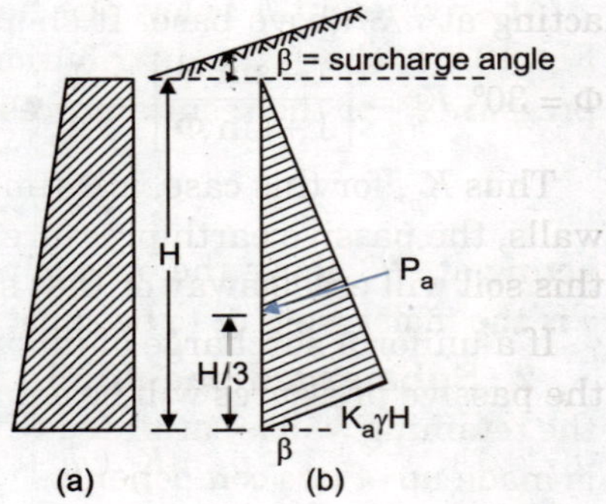

Fig. 18.6(a) shows the retaining wall with a sloping backfill. The intensity of lateral earth pressure at the base of wall is given by

**Fig. 18.6.** Lateral Pressure Distribution for Sloping Surcharge.

$$p_a = \gamma H \cdot \cos \beta \left[ \frac{\cos \beta - \sqrt{\cos^2 \beta - \cos^2 \Phi}}{\cos \beta + \sqrt{\cos^2 \beta - \cos^2 \Phi}} \right]$$

or $\quad p_a = K_a \gamma H \quad$ ...[18.11(a)]

where $\quad K_a = \cos \beta \left[ \dfrac{\cos \beta - \sqrt{\cos^2 \beta - \cos^2 \Phi}}{\cos \beta + \sqrt{\cos^2 \beta - \cos^2 \Phi}} \right] \quad$ ...[18.11(b)]

The pressure distribution is shown in Fig. 18.6(b). The pressure acts *parallel* to the sloping surface of the surcharge, i.e., at $\beta$ with the horizontal. The total active pressure $P_a$ for the wall of height $H$ is given by

$$P_a = \frac{1}{2} K_a \gamma H^2$$

The resultant acts at $H/3$ above the base, in direction parallel to the surcharge. When $\beta = 0$ (*i.e.* horizontal ground surface), Eq. 18.11(b) reduces to

$$K_a = \frac{1 - \sin \Phi}{1 + \sin \Phi} \quad \text{which is the same as Eq. 18.2.}$$

## 18.4. PASSIVE EARTH PRESSURE

Passive earth pressure is exerted on a wall when it has a tendency to move towards the backfill. Such a condition may occur when the retaining wall supports an arch, and is subjected to arch thrust, moving it towards the fill. Another condition of passive earth pressure may be when the wall supports soil of different heights on both the sides as shown in Fig. 18.7. Due to active earth pressure from the right hand side, the wall moves to the left. The soil to the left is thus compressed and in turn exert passive earth pressure, resisting such movement.

If $h$ is the height of fill, the intensity of passive pressure at height $h$ is given by

$$p_p = K_p \cdot \gamma h \quad \text{...(18.12)}$$

where $\quad K_p = $ coefficient of passive earth pressure

$$= \left[ \frac{1 + \sin \Phi}{1 - \sin \Phi} \right] = \frac{1}{K_a} \quad \text{...(18.13)}$$

The passive pressure distribution will thus be a triangle. The total pressure is given by

$$P_p = K_p \cdot \frac{\gamma h^2}{2} \quad \text{...(18.14)}$$

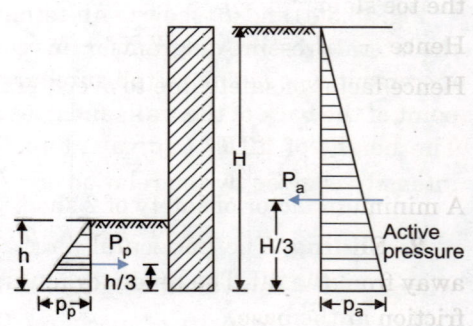

Fig. 18.7

acting at h/3 above base. It should be noted that $K_p$ is very much greater than $K_a$. For example, when $\Phi = 30°$, $K_a = \left[\dfrac{1 - \sin \Phi}{1 + \sin \Phi}\right] = \dfrac{1}{3}$ and $K_p = \left[\dfrac{1 + \sin \Phi}{1 - \sin \Phi}\right] = 3$.

Thus $K_p$, for this case, is 9 times $K_a$. In making calculations for active earth pressure behind retaining walls, the passive earth pressure due to other side of walls is generally neglected, since it is possible that this soil will erode away or that shearing will occur, thus eliminating the passive pressure.

If a uniform surcharge of intensity $w$ per unit area acts over the surface of the backfill, the increase in the passive pressures will be equal to $K_p \cdot w$. The passive pressure intensity at depth $h$ is given by

$$p_p = K_p (\gamma h + w) \quad \ldots(18.15)$$

If the backfill is having its top surface inclined at an angle $\beta$, the passive pressure is given by

$$p_p = \gamma h \cos \beta \left[\dfrac{\cos \beta + \sqrt{\cos^2 \beta - \cos^2 \Phi}}{\cos \beta - \sqrt{\cos^2 \beta - \cos^2 \Phi}}\right] \quad \text{or} \quad p_p = K_p \cdot \gamma h \quad \ldots(18.16)$$

where $K_p = \cos \beta \left[\dfrac{\cos \beta + \sqrt{\cos^2 \beta - \cos^2 \Phi}}{\cos \beta - \sqrt{\cos^2 \beta - \cos^2 \Phi}}\right] \quad \ldots[18.16(b)]$

## 18.5. STABILITY OF CANTILEVER RETAINING WALL

Fig. 18.8 shows a cantilever retaining wall subjected to the following forces:
(1) Weight $W_1$ of the stem $AB$.     (2) Weight $W_2$ of the base slab $DC$.
(3) Weight $W_3$ of the column of soil supported on heel slab $BC$.
(4) Horizontal force $P_a$, equal to active earth pressure acting at $H/3$ above the base.

The following are the modes of failure of a retaining wall:
1. Overturning about the toe. 2. Sliding. 3. Failure of soil due to excessive pressure at toe or tension at the heel. 4. Bending failure of stem or base of slab or heel slab.

**1. Overturning.** To ensure stability against overturning about the edge of the toe slab, stablizing moments due to gravity loads should be more than the overturning moments due to lateral forces so as to get factor of safety in the range of 1.5 to 2.0. According to IS: 456 code. The most hazardous mode of failure of retaining walls is due to overturning because of unbalanced moments. Thus, in Fig. 18.8, the overturning moment, due to active earth pressure, at toe is

$$M_0 = P_a \dfrac{H}{3} = K_a \dfrac{\gamma H^2}{2} \cdot \dfrac{H}{3} = K_a \cdot \dfrac{\gamma H^3}{6} \quad \ldots(18.17)$$

The resisting moment is due to the weights $W_1$, $W_2$ and $W_3$, neglecting the passive earth pressure and weight of soil above the toe slab.

Hence $M_R = W_1 \bar{x}_1 + W_2 \bar{x}_2 + W_3 \bar{x}_3 \quad \ldots[18.17(a)]$

Hence factor of safety due to overturning ($F_1$) is given by

$$F_1 = \dfrac{M_R}{M_0} \quad \ldots(18.18)$$

A minimum factor of safety of 2 should be used.

**2. Sliding.** The horizontal force $P_a$ tends to slide the wall away from the fill. The tendency to resist this is achieved by the friction at the base.

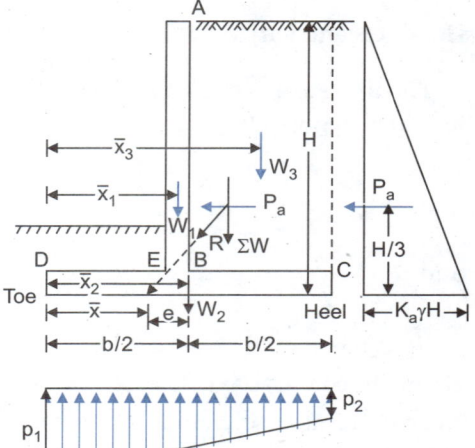
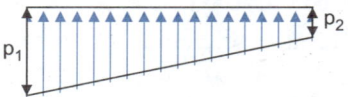

Fig. 18.8

**448** REINFORCED CONCRETE STRUCTURE

The force of resistance, $F$ is given by
$$F = \mu \Sigma W \qquad \ldots(18.19)$$
where $\mu$ is the coefficient of friction (or friction factor) between soil and concrete, and $\Sigma W$ is the sum of vertical forces. The factor of safety $F_2$ due to sliding is given by

$$F_2 = \frac{\mu \Sigma W}{H}, \qquad \text{where } H = P_a \qquad \ldots(18.20)$$

Factor of safety
where $\Sigma W$ = Total gravity load

$H$ = Horizontal component of earth pressure.

$M = (\tan \Phi)$ Friction coefficients between base and ground.

$\Phi$ = Angle of repose (or) angle of internal friction of soil.

If the wall is found to be unsafe against sliding, shear key below the base [Fig. 18.9(b)] should be provided. Such a key develops passive pressure which resists completely the sliding tendency of the wall. A factor of safety of 1.5 must be used against sliding.

In the absence of elaborate tests, the following values of $\mu$ may be adopted:

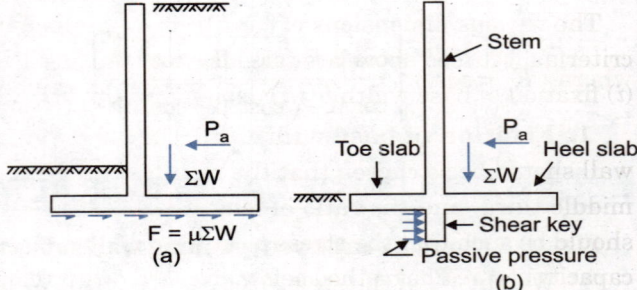

Fig. 18.9

|   |   | $\mu$ |
|---|---|------|
| 1. | Coarse grained soil without silt | 0.55 |
| 2. | Coarse grained soil with silt | 0.45 |
| 3. | Silt | 0.35 |

**3. Soil pressure distribution.** Fig. 18.8 shows the various forces acting on the wall. If $\Sigma W$ is the sum of all vertical forces, and $P_a$ is the horizontal active earth pressure, the resultant $R$ will strike the base slab at a distance $e$ (say) from the middle point of the base.

Let $\qquad \Sigma M = W_1 \bar{x}_1 + W_2 \bar{x}_2 + W_3 \bar{x}_3 - P_a \cdot \dfrac{H}{3}$ = net moment at the toe.

Then $\qquad \bar{x}$ = distance of point of application of resultant = $\dfrac{\Sigma M}{\Sigma W}$ $\qquad \ldots(18.21)$

Hence eccentricity $\qquad e = \dfrac{b}{2} - \bar{x}$ $\qquad \ldots(18.22)$

The pressure distribution below the base, is shown in Fig. 18.8. The intensity of soil pressure at the toe and heel is given by

$$p_1 = \frac{\Sigma W}{b}\left(1 + \frac{6e}{b}\right) \text{ at toe} \qquad \ldots(18.23)$$

and $$p_2 = \frac{\Sigma W}{b}\left(1 - \frac{6e}{b}\right) \text{ at heel} \qquad \ldots(18.24)$$

$p_1$ at toe should not exceed the safe bearing capacity of soil otherwise soil will fail. Similarly, $p_2$ at heel should be compressive. If $p_2$ comes to be tensile, the heel will be lifted above the soil, which is not permissible. In an extreme case, $p_2$ may be zero, where $e = b/6$. Hence in order that tension is not developed, the resultant should strike the base within the middle third.

**4. Bending failure.** There are three distinct parts of T-shaped cantilever retaining wall : the stem $AB$, the heel slab $BC$ and toe slab $DE$. The stem $AB$ will bend as cantilever, so that tensile face will be towards the backfill. The critical section will be at $B$, where cracks may occur at the inner face if it is not properly reinforced. The heel slab will have net pressure acting downwards, and will bend as a cantilever, having tensile face upwards. The critical section will be at $B$, where cracks may occur if it is not reinforced properly at the upper face as shown in Fig. 18.10(b). The net pressure on toe slab will act upwards, and hence it must be reinforced at the bottom face. The thickness of stem, heel slab and toe slab must be sufficient to withstand compressive stresses due to bending.

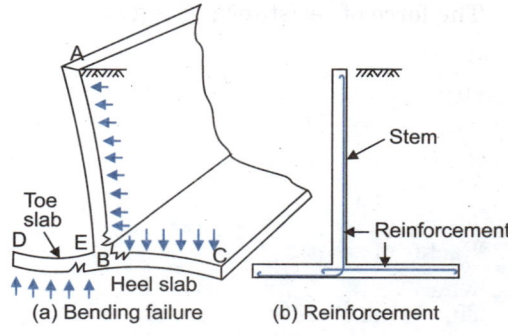

Fig. 18.10. Detailing of Flexural Behaviour.

## 18.6. DESIGN PRINCIPLES OF CANTILEVER RETAINING WALL

The various dimensions of cantilever retaining wall should be so proportioned that the various failure criteria discussed above are taken care of. The design of a cantilever retaining wall consists of the following (i) fixation of base width $b$ (ii) design of stem (iii) design of heel slab (iv) design of toe slab.

**1. Fixation of base width ($b$).** The base width $b$ of the retaining wall should be so chosen that the resultant of the forces remain within middle third, and the ratio of length of the toe slab to the base width should be such that the stress $p_1$ at toe does not exceed the safe bearing capacity of soil. The method below gives an approximate method of finding the base width $b$ and the width of toe slab $DE = \alpha b$.

Let us assume that the average unit weight of concrete and soil is equal to $1.1 \gamma$ where $\gamma$ is the unit weight of soil. Neglecting the weight of toe slab $DE$, the total weight $W$ of retaining wall plus the weight of soil in column $AFCB$ will be

$$W = (1 - \alpha)b \times H(1.1\gamma) = 1.1\gamma b H (1 - \alpha) \qquad ...(i)$$

This acts at a distance of $\frac{1}{2}(1 - \alpha)b$ from $E$ or $C$.

This horizontal earth pressure $P = K_a \dfrac{\gamma H^2}{2}$ ...(ii)

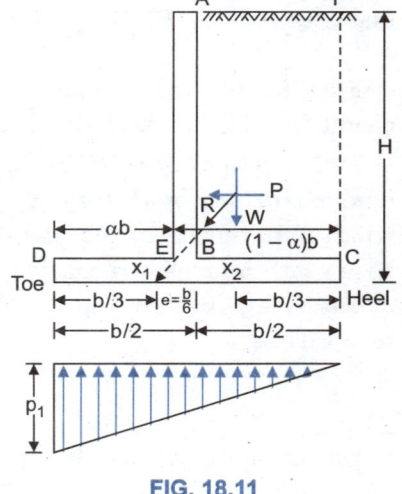

FIG. 18.11

Let $X_1$ and $X_2$ be the outer and inner third points respectively. In order that no tension is developed at heel, the resultant should pass through the outer third point $X_1$. In that case, moment of resultant $R$ will be zero, or the algebraic sum of moments of $W$ and $P$ about $X_1$ is zero. Hence, we have

$$W\left[\frac{2b}{3} - \frac{1}{2}(1-\alpha)b\right] - P \cdot \frac{H}{3} = 0 \quad \text{or} \quad 1.1\gamma b H(1-\alpha)\left\{\frac{2b}{3} - \frac{b}{2} + \frac{b\alpha}{2}\right\} - K_a \frac{\gamma H^3}{6} = 0$$

$$1.1\gamma b^2 H(1-\alpha)(1+3\alpha) = K_a \gamma H^3, \quad \text{which gives } b = 0.95 H \sqrt{\frac{K_a}{(1-\alpha)(1+3\alpha)}} \qquad ...(18.25)$$

Thus $b$ can be determined, if $\alpha$, i.e. the ratio of heel slab $DE$ to the total base width is known. The ratio $\alpha$ can be fixed on the basis of maximum pressure under the toe. When the resultant $R$ passes through the outer third point $X_1$, eccentricity $e = \dfrac{b}{2} - \dfrac{b}{3} = \dfrac{b}{6}$. Hence the maximum pressure $p_1$ under the toe is

$$p_1 = \frac{W}{b}\left(1 + \frac{6e}{b}\right) = \frac{W}{b}(1+1) = \frac{2W}{b} = \frac{2}{b}\{1.1\gamma b H(1-\alpha)\} \qquad ...(iii)$$

The value of $p_1$ should not exceed the safe bearing capacity $q_0$ of the soil.

Hence, $\quad q_0 = p_1 = 2.2 \gamma H (1 - \alpha)$

$\therefore \quad (1 - \alpha) = \dfrac{q_0}{2.2 \gamma H}$

or $\quad \alpha = 1 - \dfrac{q_0}{2.2 \gamma H} \quad$ ...(18.26)

This value of $\alpha$ can be substituted in Eq. 18.25 to get the value of $b$. Eqs. 18.25 and 18.26 are valid, though as a rough guide, only when there is no surcharge. For the case of sloping backfill with surcharge angle $\beta$, the following approximate expression may be used:

$$\alpha = 1 - \dfrac{q_0}{2.7 \gamma H} \quad ...[18.26(a)]$$

and $\quad b = H \sqrt{\dfrac{K_a \cos \beta}{(1 - \alpha)(1 + 3\alpha)}} \quad$ ...[18.26(a)]

Fig. 18.12 shows common values of $b$ for various conditions of cantilever retaining walls.

***Determination of base width from the considerations of sliding.*** Eq. 18.25. gives the base width from the stress considerations. Let us find the base width from sliding consideration also. The factor of safety against sliding should at least be equal to 1.5. Thus,

$$F.S. = \dfrac{\mu W}{P} = 1.5$$

Substituting the value of $W = 1.1 \gamma b H (1 - \alpha)$

and $\quad P = K_a \dfrac{\gamma H^2}{2}$, we get

$$\mu[1.1 \gamma b H (1 - \alpha)] = 1.5 K_a \dfrac{\gamma H^2}{2} \quad \text{or} \quad b \approx \dfrac{0.7 H}{1 - \alpha} \cdot \dfrac{K_a}{\mu} \quad ...(18.27)$$

Thus knowing $\alpha$ from Eq. 18.26, $b$ can be determined from Eq. 18.27.

**2. Design of stem.** The vertical stem $AB$ is designed as cantilever, for triangular loading. At any section $h$ below the top point $A$, the force is equal to $K_a \dfrac{\gamma h^2}{2}$ and its bending moment about the section is $K_a \gamma h^3 / 6$. The thickness at $B$ is maximum. The minimum thickness at $A$ should vary from 20 to 30 cm depending upon the height of the wall. Reinforcement is provided towards the inner face of stem, *i.e.* towards side of fill. The reinforcement towards the top of stem can be curtailed, since B.M. varies as $h^3$. Distribution reinforcement is provided @ 0.15% of the area of cross-section along the length of retaining wall at inner face. Similarly, at the outer face of the stem, temperature reinforcement is provided both in horizontal as well as in vertical direction, at the rate of 0.15% of the area of cross-section.

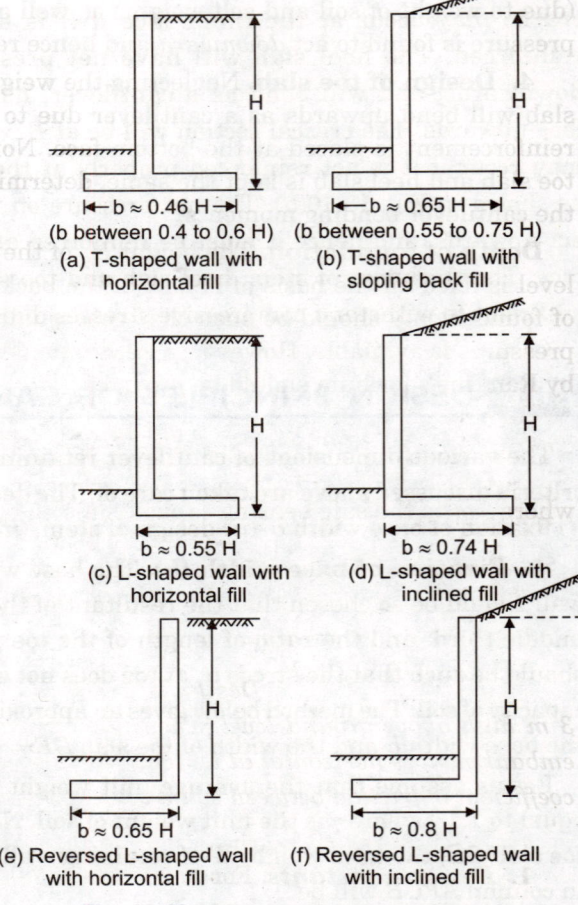

(a) T-shaped wall with horizontal fill $\quad b \approx 0.46 H$ (b between 0.4 to 0.6 H)

(b) T-shaped wall with sloping back fill $\quad b \approx 0.65 H$ (b between 0.55 to 0.75 H)

(c) L-shaped wall with horizontal fill $\quad b \approx 0.55 H$

(d) L-shaped wall with inclined fill $\quad b \approx 0.74 H$

(e) Reversed L-shaped wall with horizontal fill $\quad b \approx 0.65 H$

(f) Reversed L-shaped wall with inclined fill $\quad b \approx 0.8 H$

Fig. 18.12. Fixation of Length of Base.

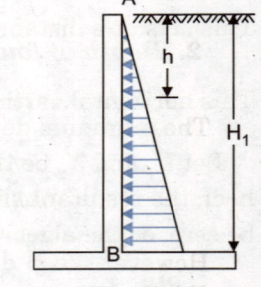

Fig. 18.13

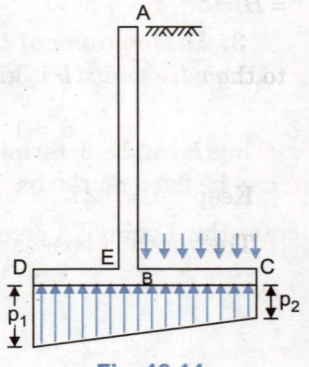

Fig. 18.14

**3. Design of heel slab.** The heel is also to be designed as a cantilever. It has both downward pressure (due to weight of soil and self-weight) as well as upward pressure due to soil reaction. However, the *net pressure* is found to act *downward* and hence reinforcement is provided at the upper face BC.

**4. Design of toe slab.** Neglecting the weight of the soil above it, the toe slab will bend upwards as a cantilever due to upward soil reaction. Hence reinforcement is placed at the bottom face. Normally, the thickness of both toe slab and heel slab is kept the same, determined on the basis of greater of the cantilever bending moments.

**Depth of foundation.** The height $H_2$ of the retaining wall, above ground level is fixed on the basis of height of the backfill to be retained. The depth of foundation $y$ should be such that good quality of soil to bear the induced pressures is available. However, a minimum depth of foundation given below by Rankine's formula should be provided:

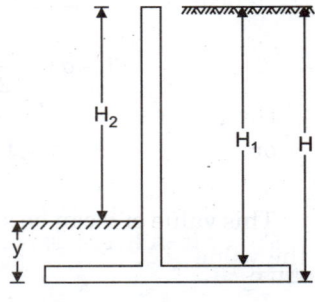

Fig. 18.15

$$y_{min} = \frac{q_0}{\gamma}\left(\frac{1-\sin\Phi}{1+\sin\Phi}\right)^2 = \frac{q_0}{\gamma} \cdot K_a^2 \qquad \ldots(18.28)$$

where $q_0$ is the safe bearing capacity of soil, or equal to the maximum pressure likely to occur on soil.

## 18.7. DESIGN OF CANTILEVER RETAINING WALL WITH HORIZONTAL BACKFILL

**Design: Example 18.1.** *Design a T-shaped cantilever retaining wall to retain earth embankment 3 m high above ground level. The unit weight of earth is 18 kN/m³ and its angle of repose is 30°. The embankment is horizontal at its top. The safe bearing capacity of soil may be taken as 100 kN/m² and the coefficient of friction between soil and concrete as 0.5. Use M 20 mix. and Fe 415 bars.*

**Solution:**

**1. Design constants.** For M 20 concrete and Fe 415 steel reinforcement we have the following values:

$c = \sigma_{cbc} = 7$ N/mm² ; $t = \sigma_{st} = 230$ N/mm² ; $m = 13.33$

$k_c = 0.289$ ; $j_c = 0.904$ and $R_c = 0.914$

**2. Depth of foundation**

$\gamma = 18$ kN/m³ = 18000 N/m³

The minimum depth of foundation is

$$y_{min} = \frac{q_0}{\gamma}\left(\frac{1-\sin\Phi}{1+\sin\Phi}\right)^2 = \frac{100}{18}\left(\frac{1-\sin 30°}{1+\sin 30°}\right)^2 = 0.62 \text{ m}.$$

However, keep depth = 1 m to accommodate thickness of base wall below the ground surface. Hence height of wall above its base = $H$ = 3 + 1 = 4 m.

**3. Dimensions of base.** The ratio of the length of toe slab (*DE*) to the base width $b$ is given by *Eq. 18.26*:

$$\alpha = 1 - \frac{q_0}{2.2\gamma H} = 1 - \frac{100}{2.2 \times 18 \times 4} = 0.369$$

Keep $\alpha = 0.37$ ...(*i*)

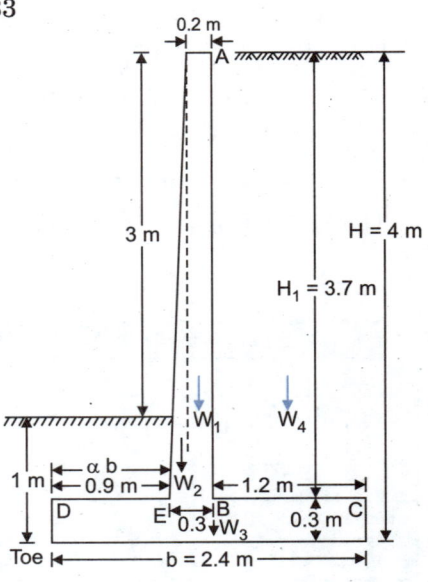

Fig. 18.16

The width of base is given by *Eq. 18.25*.

$$b = 0.95 H \sqrt{\frac{K_a}{(1-\alpha)(1+3\alpha)}}$$

$$K_a = \frac{1-\sin\Phi}{1+\sin\Phi} = \frac{1-\sin 30°}{1+\sin 30°} = \frac{1}{3}$$

$$\therefore \quad b = 0.95 \times 4 \sqrt{\frac{1}{3(1-0.37)(1+3\times 0.37)}} = 1.90 \text{ m}.$$

The base width from the considerations of sliding is given by *Eq. 18.27*.

$$b = \frac{0.7H\, K_a}{(1-\alpha)\mu} = \frac{0.7\times 4}{1-0.37} \times \frac{1}{3\times 0.5} = 2.96 \text{ m}.$$

This width is excessive. Normal practice is to provide $b$ between 0.4 to 0.6 $H$. Taking maximum value of 0.6 $H$,

$$b = 0.6\, H = 0.6 \times 4 = 2.4 \text{ m}.$$

Hence provide $b = 2.4$ m. The wall will be unsafe against sliding. This will be made safe by providing a shear key at the base.

Width of toe slab = 0.37 × 2.4 = 0.89 m. Provide toe slab 0.9 m long.

Let the thickness of base be $\frac{1}{12}\, H \approx 0.3$ m for preliminary calculations.

4. **Thickness of stem.** Height $AB = 4 - 0.3 = 3.7$ m. Consider one meter length of retaining wall.

Maximum bending moment at $B = K_a \gamma \dfrac{H_1^3}{6} = \dfrac{1}{3} \cdot \dfrac{18}{6} (3.7)^3 = 50.65$ kN-m $= 50.65 \times 10^6$ N-mm

Hence the effective depth is $d = \sqrt{\dfrac{50.65 \times 10^6}{1000 \times 0.914}} = 236$ mm

Keep $d = 240$ mm and total thickness = 300 mm so that an effective cover of 60 mm is available and nominal cover = 60 − 8 = 52 mm using 16 mm φ bars. Reduce the total thickness to 200 mm at the top so that effective depth of 140 mm is available at the top.

5. **Stability of wall.** Fully dimensioned wall is shown in Fig. 18.16.

Let     $W_1$ = weight of rectangular portion of stem

        $W_2$ = weight of triangular portion of stem

        $W_3$ = weight of base slab

        $W_4$ = weight of soil on heel slab.

The calculations are arranged in Table 18.1.

**TABLE 18.1**

| S. No. | Designation | Force (kN) | Lever arm (m) | Moment about toe (kN-m) |
|---|---|---|---|---|
| 1 | $W_1$ | 1 × 0.2 × 3.7 × 25 = 18.5 | 1.1 | 20.35 |
| 2 | $W_2$ | $\frac{1}{2}$ × 0.1 × 3.7 × 25 = 4.63 | 0.97 | 4.49 |
| 3 | $W_3$ | 1 × 2.4 × 0.3 × 25 = 18.0 | 1.2 | 21.60 |
| 4 | $W_4$ | 1 × 1.2 × 3.7 × 18 = 79.97 | 1.8 | 143.86 |
|  |  | $\Sigma W$ = 121.05 |  | $M_R$ = 190.30 |

The total resisting moment     $M_R = 190.30$ kN-m                                 ...(1)

Earth pressure              $P = K_a \gamma \dfrac{H^2}{2} = \dfrac{1}{3} \times \dfrac{18}{2} (4)^2 = 48$ kN                 ...(2)

## Overturning

Overturning moment $\quad M_0 = 48 \times \dfrac{4}{3} = 64$ kN-m $\quad\quad\quad$ ...(3)

∴ $\quad$ F.S. against overturning $= \dfrac{190.3}{64} = 2.97 > 2$. Hence safe.

## Sliding

$$\text{F.S. against sliding} = \dfrac{\mu \Sigma W}{P} = \dfrac{0.5 \times 121.05}{48} = 1.26 < 1.5. \text{ Hence unsafe.}$$

To make it safe against sliding, we will have to provide a shear key.

**Pressure distribution.** Net moment $\Sigma M = 190.3 - 64 = 126.3$ kN-m

∴ $\quad$ Distance $\bar{x}$ of the point of application of resultant, from toe is

$$\bar{x} = \dfrac{\Sigma M}{\Sigma W} = \dfrac{126.3}{121.05} = 1.04 \text{ m}$$

Eccentricity $\quad e = \dfrac{b}{2} - \bar{x} = 1.2 - 1.04 = 0.16$ m. This is less than $\dfrac{b}{6}\left(= \dfrac{2.4}{6} = 0.4 \text{ m}\right)$

Pressure $p_1$ at toe $= \dfrac{\Sigma W}{b}\left(1 + \dfrac{6e}{b}\right) = \dfrac{121.05}{2.4}\left(1 + \dfrac{6 \times 0.16}{2.4}\right) = 70.61$ kN/m² < 100. Hence safe.

Pressure $p_2$ at heel $= \dfrac{\Sigma W}{b}\left(1 - \dfrac{6e}{b}\right) = \dfrac{121.05}{2.4}\left(1 - \dfrac{6 \times 0.16}{2.4}\right) = 30.26$ kN/m²

Pressure $p$ at the junction of stem with toe slab is

$$p = 70.61 - \dfrac{70.61 - 30.26}{2.4} \times 0.9 = 55.48 \text{ kN/m}^2$$

Pressure $p'$ at the junction of stem with heel slab is

$$p' = 70.61 - \dfrac{70.61 - 30.26}{2.4} \times 1.2 = 50.44 \text{ kN/m}^2$$

**6. Design of toe slab.** The upward pressure distribution on the toe slab is shown in Fig. 18.17. The weight of the soil above the toe slab is neglected. Thus two forces are acting on it:

(i) upward soil pressure. $\quad\quad$ (ii) downward weight of slab.

Downward weight of slab per unit area $= 0.3 \times 1 \times 1 \times 25 = 7.5$ kN/m²

Hence net pressure intensities will be $= 70.61 - 7.5 = 63.11$ kN/m²
under $D$ and $55.48 - 7.5 = 47.98$ kN/m² under $E$.

Total force = S.F. at $E = \dfrac{1}{2}(63.11 + 47.98)0.9 = 50$ kN

$\bar{x}$ from $E = \left(\dfrac{47.98 + 2 \times 63.11}{47.98 + 63.11}\right)\dfrac{0.9}{3} = 0.47$ m.

∴ $\quad$ B.M. at $E = 50 \times 0.47 = 23.52$ kN-m
$\quad\quad\quad\quad\quad\quad = 23.52 \times 10^6$ N-mm

∴ $\quad d = \sqrt{\dfrac{23.52 \times 10^6}{1000 \times 0.914}} = 160.4$

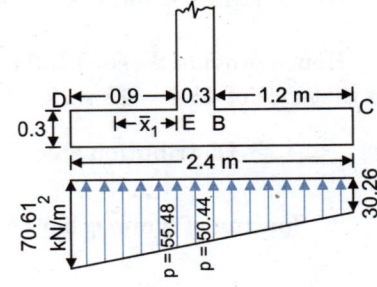

Fig. 18.17

Let us keep total depth = 260 mm and effective depth of 200 mm so that 60 mm effective cover is available. Thickness can be reduced to 200 mm at the edge.

$$A_{st} = \frac{23.52 \times 10^6}{230 \times 0.904 \times 200} = 566 \text{ mm}^2$$

This reinforcement has to be provided at the bottom face. If alternate bars of stem reinforcement are bent and continued in the toe slab, area available = $\frac{1}{2} \times 1256 = 628$ mm² (see step 8). This reinforcement will consist of 12 Φ bars @ 180 mm c/c. Let us check this reinforcement for development length.

$$L_d \approx 45\Phi = 45 \times 12 = 540 \text{ mm.}$$

Providing 50 mm clear side cover, actual length available = 900 − 50 = 850 > $L_d$ = 540 mm. Hence safe.

$$\text{Distribution reinforcement} = \frac{0.12}{100} \times 1000 \left[\frac{260 + 200}{2}\right] = 276 \text{ mm}^2$$

Using 8 mm Φ bars, $A_\Phi = 50.3$ mm². ∴ Spacing = $\frac{1000 \times 50.3}{276} = 182$ mm

However provide these @ 180 mm c/c.

**7. Design of heel slab.** Three forces act on it : (*i*) downward weight of soil 3.7 m high (*ii*) downward weight of heel slab (*iii*) upward soil pressure

Total weight of soil = 1.2 × 3.7 × 1 × 18 = 80 kN acting at 0.6 m from *B*.

Total weight of heel slab = 1.2 × 0.26 × 1 × 25 = 7.8 acting at 0.6 m from *B*.

Total upward soil reaction = $\frac{1}{2}$ (50.44 + 30.26)1.2 = 48.42 kN

acting at $\left(\frac{50.44 + 2 \times 30.26}{50.44 + 30.26}\right)\frac{1.2}{3} = 0.55$ m from *B*

∴ Total force = S.F. at *B* = 80 + 7.8 − 48.42 = 39.38 kN

∴ B.M. at *B* = (80 × 0.6) + (7.8 × 0.6) − (48.42 × 0.55)

= 26.05 kN-m = 26.05 × 10⁶ N-mm

∴ $d = \sqrt{\frac{26.05 \times 10^6}{1000 \times 0.914}} = 169$ mm.

However keep the same total depth (= 260 mm) as that of toe slab, so that available effective depth = 200 mm. The thickness is reduced to 200 mm at the edge.

$$A_{st} = \frac{26.05 \times 10^6}{230 \times 0.904 \times 200} = 626.44 \text{ mm}^2$$

Using 12 mm Φ bars, $A_\Phi$ = 113 mm². ∴ Spacing = $\frac{1000 \times 113}{626.44} \approx 180$ mm.

Hence provide these @ 180 mm c/c at the top of keep slab. Take the reinforcement into the toe slab from a distance of 45 Φ = 45 × 12 = 540 mm to the left of *B*,

$$\text{Distribution steel} = \frac{0.12}{100} \times 1000 \left[\frac{260 + 200}{2}\right] = 276 \text{ mm}^2$$

∴ Spacing of 8 mm φ bars = $\frac{1000 \times 50.3}{276}$ = 182 mm. Hence provide these @ 180 mm c/c.

$$\text{Shear stress } \tau_v = \frac{39.38 \times 1000}{1000 \times 200} = 0.20 \text{ N/mm}^2$$

This is much less than the permissible shear stress even at the minimum percentage of steel (Table 3.1).

**8. Reinforcement in the stem.** We had earlier assumed the thickness of heel slab as 0.3 m, while it has now been fixed as 0.26 m only. Hence revised $H_1$ = 4 − 0.26 = 3.74 m

$$\therefore \qquad M = K_a \gamma \frac{H_1^3}{6} = \frac{1}{3} \times \frac{18}{6} (3.74)^3 = 52.31 \text{ kN-m} = 52.31 \times 10^6 \text{ N-mm}$$

$$\therefore \qquad d = \sqrt{\frac{52.31 \times 10^6}{1000 \times 0.914}} = 239 \text{ mm}.$$

Keep $d = 250$ mm so that $D = 310$ mm. Reduce the total thickness to 200 at the top.

$$A_{st} = \frac{52.31 \times 10^6}{230 \times 0.904 \times 250} = 1007 \text{ mm}^2. \text{ Using 12 mm } \Phi \text{ bars, } A_\Phi = 113 \text{ mm}^2$$

$$\therefore \qquad s = \frac{1000 \times 113}{1007} = 112 \text{ mm. However provide 12 mm } \Phi \text{ bars @ 90 mm c/c}.$$

Actual $A_{st}$ provided $= 1000 \times \dfrac{113}{90} = 1256$ mm$^2$.

Continue alternate bars in the toe slab to serve as tensile reinforcement there. Discontinue the remaining half bars after a distance of $45 \Phi = 45 \times 12 \approx 540$ mm beyond $B$, in the toe slab.

Between $A$ and $B$, some of the bars can be curtailed. Consider a section at depth $h$ below the top of the stem. The effective depth $d'$ at that section is

$$d' = 140 + \frac{250 - 140}{3.74} h = (140 + 29.4 h) \text{ mm (where } h \text{ is in metres.)} \qquad \ldots(1)$$

Now, $\qquad A_{st} \propto \dfrac{H^3}{d}, \quad$ or $\quad H = (A_{st} d)^{1/3}$

Hence $\qquad \dfrac{h}{H_1} = \left( \dfrac{A'_{st} d'}{A_{st} d} \right)^{\frac{1}{3}}$ $\qquad \ldots(2)$

where  $A'_{st}$ = reinforcement at depth $h$
$\qquad d'$ = effective depth at depth $h$
$\qquad A_{st}$ = reinforcement at depth $H_1$
$\qquad d$ = effective depth at depth $H_1$

If $\qquad A'_{st} = \dfrac{1}{2} A_{st}, \dfrac{A'_{st}}{A_{st}} = \dfrac{1}{2}. \qquad \therefore \qquad \dfrac{h}{H_1} = \left( \dfrac{1}{2} \cdot \dfrac{d'}{d} \right)^{\frac{1}{3}}$

Substituting $d = 250$ mm and $d' = (140 + 29.4 h)$, we get

$$h = H_1 \left[ \frac{140 + 29.4 h}{2 \times 250} \right]^{\frac{1}{3}} = 3.74 \left[ \frac{140 + 29.4 h}{500} \right]^{\frac{1}{3}} = 0.471 [140 + 29.4 h]^{1/3} \qquad \ldots(3)$$

This can be solved by trial and error, noting that if the effective thickness of stem were constant, $h$ would have been equal to $\dfrac{H_1}{(2)^{1/3}} \approx 0.79 H_1 \approx 2.96$ m.

Solving (3) by trial, we get $h = 2.86$ m. Thus, half the bars can be curtailed at this point. However, the bars should be extended by a distance of $12\Phi$ ($= 12 \times 16 = 192$ mm) or $d$ ($= 250$ mm) whichever is more, beyond the point.

$\therefore \quad h = 2.86 - 0.25 = 2.61$ m. Hence curtail half the bars at a height 2.6 m below the top. If we wish to curtail half of the remaining bars so that remaining reinforcement is one-fourth of that provided at $B$, we have

$$\dfrac{A'_{st}}{A_{st}} = \dfrac{1}{4}. \quad \text{Hence from (2), } \dfrac{h}{H_1} = \left( \dfrac{1}{4} \cdot \dfrac{d'}{d} \right)^{\frac{1}{3}}$$

or $$h = H_1 \left[\frac{140 + 29.4\,h}{4 \times 250}\right]^{\frac{1}{3}} = 3.74 \left[\frac{140 + 29.4\,h}{1000}\right]^{\frac{1}{3}} = 0.374\,[143 + 29.4\,h]^{\frac{1}{3}} \qquad \ldots(4)$$

This can be solved by trial and error, noting that if the effective thickness of stem were constant, $h$ would have been equal to $\dfrac{H_1}{(4)^{\frac{1}{3}}} = 0.63\,H_1 \approx 2.36$ m.

Solving (4) by trial and error, we get $h = 2.2$ m. However the bars should be extended by 250 mm beyond this. $\therefore\ h = 2.2 - 0.25 = 1.95$ m. Hence stop half the remaining bars by 1.95 m below the top of the stem. Continue rest of the bars to the top of the step.

**Check for shear:** Shear force $= P = K_a \gamma \dfrac{H^2}{2} = \dfrac{1}{3} \times \dfrac{18}{2}(3.74)^2 = 41.96$ kN

$$\therefore\quad \tau_v = \frac{41.96 \times 1000}{1000 \times 250} = 0.17\ \text{N/mm}^2 < \tau_c. \quad \text{Hence safe.}$$

**Distribution and temperature reinforcement.** Average thickness of stem
$$= \frac{1}{2}(310 + 200) = 255\ \text{mm}$$

$\therefore\ $ Distribution reinforcement $= \dfrac{0.12}{100} \times 1000 \times 255 = 306$ mm². Using 8 mm $\Phi$ bars, $A_\Phi = 50.3$ mm²

$\therefore\ $ Spacing $= \dfrac{1000 \times 50.3}{306} = 164$ mm

However provide 8 mm $\Phi$ bars @ 150 mm c/c at the inner face of the wall, along its length.

For temperature reinforcement, provide 8 mm $\Phi$ @ 300 mm c/c both ways, in the outer face.

**9. Design of shear key.** The wall is unsafe in sliding, and hence shear key will have to be provided below the stem, as shown in Fig. 18.18.

Let the depth of key $= a$. Intensity of passive pressure $p_p$ developed in front of the key depends upon the soil pressure $p$ in front of the key.

$\therefore\qquad p_p = K_p\,p = 3 \times 55.48$
$\qquad\qquad = 166.4\ \text{kN/m}^2$

$\therefore\ $ Total passive pressure $P_p = p_p\,a = 166.4\,a$ \qquad ...(1)

Sliding force at level $D_1 C_1 = \dfrac{1}{3} \times \dfrac{18}{2}(4 + a)^2$

or $\qquad P_H = 3(4 + a)^2$ \qquad ...(2)

Weight of soil between bottom of base and
$\qquad D_1 C_1 = 2.4\,a \times 18 = 43.2\,a$

$\therefore\qquad \Sigma W = 121.05 + 43.2\,a$ (Refer Table 18.1)

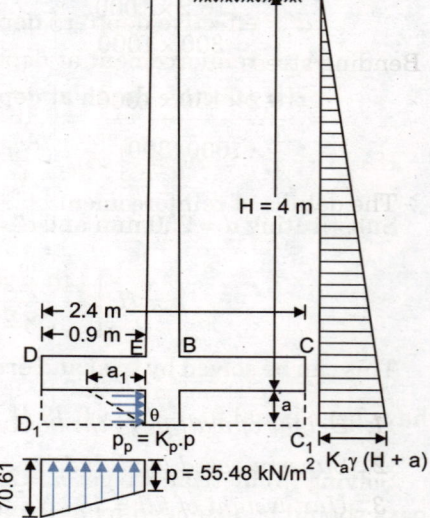

Fig. 18.18

Hence for equilibrium of the wall, permitting F.S. = 1.5 against sliding, we have

$$1.5 = \frac{\mu\,\Sigma W + P_p}{P_H} = \frac{0.5\,(121.05 + 43.2\,a) + 166.4\,a}{3\,(4 + a)^2}$$

or $\qquad a^2 - 33.8\,a + 2.54 = 0,\ $ which gives $a \approx 0.09$ m $= 90$ mm

However, provide a minimum value of $a = 0.3$ m $= 300$ mm. Keep width of key $= 300$ mm. It should be noted that passive pressure taken into account above will be developed *only when* a length $a_1$ given below is available in front of the key:

$$a_1 = a \tan \theta = a \tan\left(45° + \frac{\Phi}{2}\right) = a \sqrt{K_p},$$

where $\left(45° + \dfrac{\Phi}{2}\right) =$ shearing angle of passive resistance.

$\therefore \quad a_1 = 0.3 \sqrt{3} = 0.52$ m.

Actual length of the slab available $= DE = 0.9$ m.

Hence satisfactory.

Now size of key $= 300$ mm $\times 300$ mm

$P_H = 3(4+a)^2 = 3(4+0.3)^2 = 55.47$ kN

$\phantom{P_H} = 166.4\,a = 166.4 \times 0.3$

$\phantom{P_H} = 49.92$ kN. Hence,

$\Sigma W = 121.05 + 43.2\,a$

$\phantom{\Sigma W} = 121.05 + 43.2 \times 0.3 = 134.01$ kN

Actual force to be resisted by the key, at F.S. $= 1.5$ is

$= 1.5\,P_H - \mu\,\Sigma W$

$= 1.5 \times 55.47 - 0.5 \times 134.01$

$= 16.2$ kN

$\therefore$ Shear stress

$= \dfrac{16.2 \times 1000}{300 \times 1000} = 0.054$ N/mm$^2$ (safe)

Bending stress

$= \dfrac{16.2 \times 150 \times 1000}{\dfrac{1}{6} \times 1000\,(300)^2} = 0.16$ N/mm$^2$ (safe)

The details of reinforcement etc. are shown in Fig. 18.19.

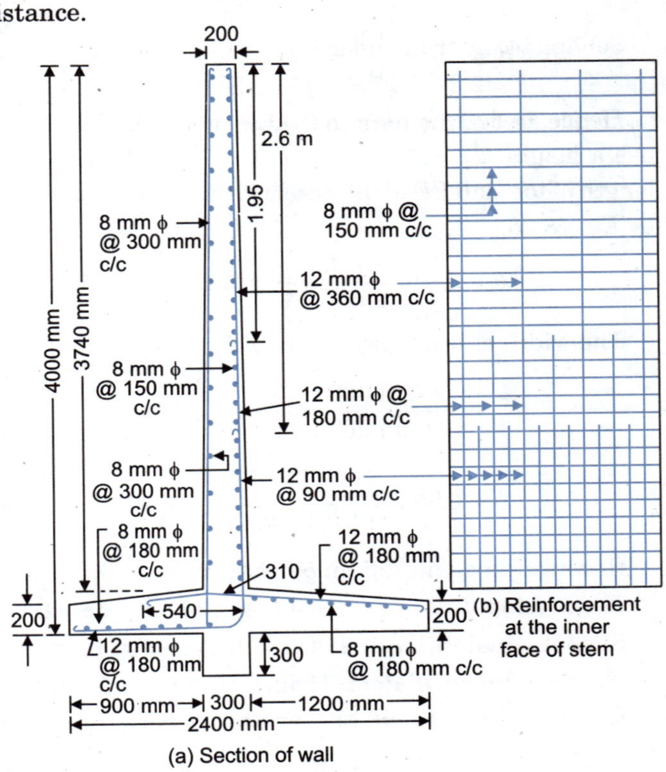

Fig. 18.19

(a) Section of wall

(b) Reinforcement at the inner face of stem

## 18.8. DESIGN OF CANTILEVER RETAINING WALL WITH HORIZONTAL BACKFILL AND TRAFFIC LOAD

**Design: Example 18.2.** *Design a cantilever retaining wall for a road for the following requirements:*

1. *Height of wall from the bottom of base to top of stem = 6 m*
2. *Superimposed load due to road traffic = 18 kN/m²*
3. *Unit weight of fill = 18 kN/m³*
4. *Angle of internal friction for fill material = 30°*
5. *Allowable bearing pressure on ground = 160 kN/m²*
6. *Coefficient of friction between concrete and ground = 0.4*

    *Also, provide a parapet wall 1 m high on the top of stem.*

    *Use M 20 concrete and Fe 415 steel.*

**Solution :** **1. Design constants.** For M 20 concrete, and Fe 415 steel reinforcement, we have the following:
$c = \sigma_{cbc} = 7$ N/m m$^2$; $t = \sigma_{st} = 230$ N/m m$^2$; $m = 13.33$; $k_c = 0.289$; $j_c = 0.904$ and $R_c = 0.914$

**2. Dimensions of base.** Assume that a horizontal force $Q = 2$ kN/m length of parapet wall will act because of person standing near the parapet.

Due to surcharge, equivalent height of fill is given by

$$h_e = \frac{w}{\gamma} = \frac{18}{18} = 1 \text{ m}$$

Hence, in determining the values of $b$ and $\alpha$ etc., we will use a height $H_2 = H + h_e = 6 + 1 = 7$ m. The ratio of the length of toe slab $EF$ to the base width $b$ may be determined by *Eq. 18.26*.

$$\alpha = 1 - \frac{160}{2.2 \times 18 \times 7} = 0.42.$$

The width of base may be found from *Eq. 18.25*.

$$K_a = \frac{1 - \sin 30°}{1 + \sin 30°} = \frac{1}{3}. \quad \text{Hence}$$

$$b = 0.95 \times 7 \sqrt{\frac{1}{3(1 - 0.42)(1 + 1.26)}} = 3.36 \text{ m}$$

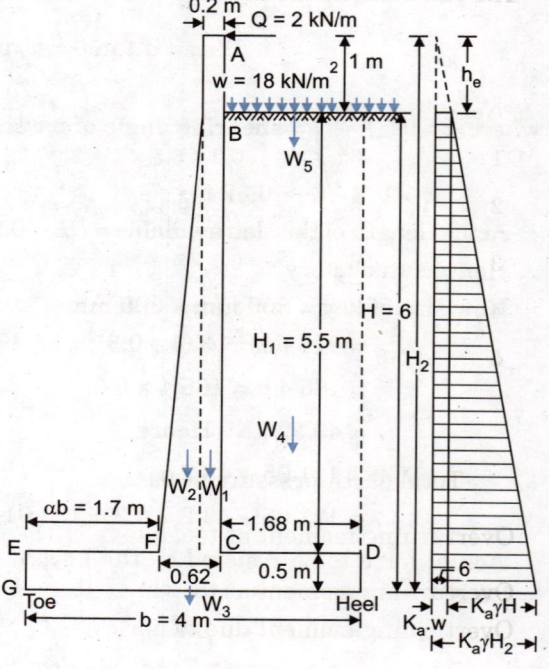

Fig. 18.20

However, keep minimum $b$ between 0.5 to 0.6 $H_2$.
Hence adopt $b = 4$ m. ∴ Length of toe slab = $\alpha b$ = 0.42 × 4 = 1.68 m.
Keep it equal to 1.7 m. Let the thickness of base = 50 cm for preliminary calculations.

**3. Thickness of stem.** Height $H_1 = 6 - 0.5 = 5.5$ m.

Due to retained soil, the earth pressure diagram will be a triangle, having an ordinate equal to $K_a \gamma h$, at $h$ below $B$. Due to surcharge $w$, there will be a uniform horizontal pressure $K_a w = \frac{1}{3} \times 18$ = 6 kN/m$^2$ throughout the height. The total bending moment at $C$ will be due to moment of horizontal force $Q = 2$ kN acting at $A$, plus moment of rectangular pressure distribution, plus moment of triangular pressure distribution.

∴ $$M = Q(H_1 + 1) + K_a w \frac{H_1^2}{2} + K_a \gamma \frac{H_1^3}{6} = 2(5.5 + 1) + \frac{1}{3} \times 18 \frac{(5.5)^2}{2} + \frac{1}{3} \times \frac{18}{6}(5.5)^3$$

$$= 13 + 90.75 + 166.35 = 270.1 \text{ kN-m} = 270.1 \times 10^6 \text{ N-mm}$$

∴ $$d = \sqrt{\frac{270.1 \times 10^6}{1000 \times 0.914}} = 544 \text{ mm}$$

Using an effective cover of 60 mm, total thickness of stem = 544 + 60 = 604 mm. However keep $D = 620$ mm so that $d = 620 - 60 = 560$ mm. Reduce the total thickness to 200 mm at $B$, so that effective depth at $B = 200 - 60 = 140$ mm. Continue uniform thickness of 200 from $B$ to $A$.

**4. Stability of wall.** Fully dimensioned wall is shown in Fig. 18.20.
Length of heel slab $CD$ = 4 − 1.7 − 0.62 = 1.68 m.
Let $W_1$ = weight of rectangular portion of stem.
$W_2$ = weight of triangular portion of stem.
$W_3$ = weight of base slab.

$W_4$ = weight of soil above heel slab.
$W_5$ = total superimposed road-traffic load, over heel slab.

The calculations are arranged in Table 18.2.

### TABLE 18.2

| S. No. | Designation | Force (kN) | Level arm (m) | Moment about toe (kN-m) |
|---|---|---|---|---|
| 1 | $W_1$ | $0.2 \times 6.5 \times 1 \times 25 = 32.5$ | 2.22 | 72.15 |
| 2 | $W_2$ | $\frac{1}{2} \times 0.42 \times 5.5 \times 1 \times 25 = 28.9$ | 1.98 | 57.17 |
| 3 | $W_3$ | $0.5 \times 4 \times 1 \times 25 = 50$ | 2.00 | 100.0 |
| 4 | $W_4$ | $1.68 \times 5.5 \times 1 \times 18 = 166.4$ | 3.16 | 525.82 |
| 5 | $W_5$ | $1.68 \times 1 \times 18 = 30.2$ | 3.16 | 95.56 |
|  |  | $\Sigma W \approx 308$ |  | $M_R = 850.7$ |

Total earth pressure at base = $K_a \dfrac{\gamma H^2}{2} = \dfrac{1}{3} \times \dfrac{18}{2} (6)^2 = 108$ kN

Overturning moment at toe, due to earth pressure = $108 \times \dfrac{6}{3} = 216$ kN-m

Overturning moment at toe, due to horizontal force $Q = 2 \times 7 = 14$ kN-m

Overturning moment due to horizontal pressure caused by live load

$$= K_a w \dfrac{H^2}{2} = \dfrac{1}{3} \times 18 \times \dfrac{(6)^2}{2} = 108 \text{ kN-m}$$

∴ Total overturning moment $\quad M_0 = 216 + 14 + 108 = 338$ kN-m

∴ F.S. against overturning $\quad = \dfrac{850.7}{338} = 2.52 > 2.$ Hence safe.

Total horizontal pressure $\quad = 2 + 108 + (6 \times 6) = 146 \text{ kN} = \Sigma P$

∴ F.S. against sliding $\quad = \dfrac{\mu \Sigma W}{\Sigma P} = \dfrac{0.4 \times 308}{146} = 0.84 < 1.5$

Hence unsafe. Special shear key will have to be designed to make the wall safe against sliding.

**Pressure distribution.** Algebraic sum of moments = $\Sigma M = M_R - M_0$

$$= 850.7 - 338 = 512.7 \text{ kN-m}$$

Algebraic sum of vertical forces $\quad = \Sigma W = 308$ kN

Hence distance $\bar{x}$ from toe, of the point of application of resultant is given by

$$\bar{x} = \dfrac{\Sigma M}{\Sigma W} = \dfrac{512.7}{308} = 1.66 \text{ m}$$

∴ Eccentricity $\quad e = \dfrac{b}{2} - \bar{x} = \dfrac{4}{2} - 1.66 = 0.34$ m

This is less than $\quad \dfrac{b}{6} = \dfrac{4}{6} = 0.67$ m. Hence no tension is developed.

Pressure $p_1$ at toe $\quad = \dfrac{\Sigma W}{b}\left(1 + \dfrac{6e}{6}\right) = \dfrac{308}{4}\left[1 + \dfrac{6 \times 0.34}{4}\right]$

$$= 116.3 \text{ kN/m}^2 < 160. \quad \text{Hence safe.}$$

Pressure $p_2$ at heel $= \dfrac{\Sigma W}{b}\left(1 - \dfrac{6e}{b}\right) = \dfrac{308}{4}\left(1 - \dfrac{6 \times 0.34}{4}\right) = 37.7$ kN/m²

Pressure $p$ at the junction of stem with toe slab is

$$p = 116.3 - \dfrac{116.3 - 37.7}{4} \times 1.7 = 82.9 \text{ kN/m}^2$$

Pressure $p'$ at the junction of stem with help slab is

$$p' = 116.3 - \dfrac{116.3 - 37.7}{4} \times 2.32 = 70.7 \text{ kN/m}^2$$

**5. Design of toe slab.** The upward pressure distribution on the toe slab is shown in Fig. 18.21. The weight of the soil above the toe slab is neglected. Thus two forces are acting on it: (i) upward soil pressure. (ii) downward weight of slab.

Downward weight of slab per unit area = $0.5 \times 1 \times 1 \times 25 = 12.5$ kN/m². Hence net pressure intensities will be $116.3 - 12.5 = 103.8$ kN/m² under $D$ and $82.9 - 12.5 = 70.4$ kN/m² under $F$.

∴  Total force = S.F. at $F = \dfrac{1}{2}(103.8 + 70.4) \times 1.7 = 148.07$ kN

C.G. of force from $F = \bar{x}_1 = \left(\dfrac{70.4 + 2 \times 103.8}{70.4 + 103.8}\right) \cdot \dfrac{1.7}{3} = 0.904$ m

∴  B.M. at $F = 148.07 \times 0.904 = 133.9$ kN-m $= 133.9 \times 10^6$ N-mm

∴  $d = \sqrt{\dfrac{133.9 \times 10^6}{1000 \times 0.914}} = 383$ mm.

Keep $d = 440$ mm and total depth equal to 500 mm so that 60 mm effective cover is available. Reduce the total depth to 200 mm at the edges.

$$A_{st} = \dfrac{133.9 \times 10^6}{230 \times 0.904 \times 440} = 1464 \text{ mm}^2$$

Bars available from stem reinforcement are 20 mm Φ @ 120 mm c/c giving $A_{st} = 2618$ mm² (step 7).

$$\tau_v = \dfrac{148.07 \times 1000}{1000 \times 440} = 0.337 \text{ N/mm}^2$$

$100\dfrac{A_s}{bd} = \dfrac{100 \times 2618}{1000 \times 440} \simeq 0.6 \%.$

Hence from Table 3.1, $\tau_c = 0.31$ N/mm²

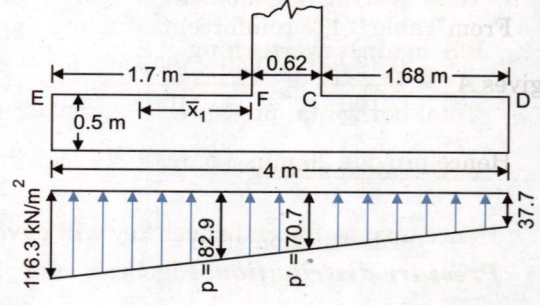

Fig. 18.21

Hence the slab is slightly unsafe. To make it safe, either increase the reinforcement or increase the depth. The reinforcement percentage required to get $\tau_c \simeq 0.34$ N/mm² is equal to 0.7%   (Table 18.1)

∴  $A_{st} = \dfrac{0.7\,bd}{100} = \dfrac{0.7 \times 1000 \times 440}{100} = 3080$ mm².  Hence spacing of 20 mm φ bars is

$$s = \dfrac{1000 \times 314.1}{3080} \simeq 100 \text{ mm c/c.}$$  Hence the slab is just safe in shear.

Distribution steel $= \dfrac{0.12}{100}\left[\dfrac{460 + 200}{2}\right] \times 1000 = 396$ mm²

Using 8 mm Φ bars, having $A_\Phi = 50.3$ mm², spacing $= \dfrac{1000 \times 50.3}{396} = 127$ mm

Hence provide these @ 120 mm c/c.

# DESIGN OF RETAINING WALLS

**6. Design of heel slab.** Four forces act on it:
(i) downward weight of soil 5.5 m high
(ii) live load of traffic
(iii) weight of heel slab
(iv) upward soil pressure.

∴ Weight of soil = $1.68 \times 1 \times 5.5 \times 18 = 166.32$ kN acting at 0.84 m from $C$.

Live load = $18 \times 1 \times 1.68 = 30.24$ kN acting at 0.84 m from $C$.

Weight of heel slab = $0.5 \times 1.68 \times 25 = 21$ kN acting at 0.84 m from $C$.

Upward soil reaction = $\frac{1}{2}(70.7 + 37.7) \times 1.68 = 91.06$ kN

acting at $\left(\dfrac{70.7 + 2 \times 37.7}{70.7 + 37.7}\right)\dfrac{1.68}{3} = 0.75$ m from $C$

∴ Total force = S.F. at $C$ = $166.32 + 30.24 + 21 - 91.06 = 126.5$ kN

B.M. at $C = (166.32 + 30.24 + 21)\,0.84 - 91.06 \times 0.75$

$= 114.46$ kN-m $= 114.46 \times 10^6$ N-mm

Keep the same depth as that of toe slab, *i.e.* $d = 440$ mm and total depth = 500 mm. Reduce total depth to 200 mm at edges.

$$A_{st} = \frac{114.46 \times 10^6}{230 \times 0.904 \times 440} = 1251.13 \text{ mm}^2$$

Using 20 mm Φ bars, $A_\Phi = 314$ mm². ∴ Spacing = $\dfrac{1000 \times 314}{1251.13} = 250.97 ≃ 250$ mm.

$$\tau_v = \frac{126.5 \times 10^3}{1000 \times 440} = 0.288 \text{ N/mm}^2$$

From Table 3.1, % reinforcement required to get $\tau_v ≃ 0.29$ N/mm² is equal to 0.49% approximately. This gives $A_{st} = \dfrac{0.49\,bd}{100} = \dfrac{0.49 \times 1000 \times 440}{100} = 2156$ mm².

Hence provide 20 mm φ bars @ 140 mm c/c. Actual $A_{st} = \dfrac{1000 \times 314}{140} = 2242.8$ mm²

Distribution steel = $\dfrac{0.12}{100}\left[\dfrac{460 + 200}{2}\right] \times 1000 = 396$ mm²

∴ Spacing of 8 mm Φ bars = $\dfrac{1000 \times 50.3}{396} \approx 127$ mm. ≃ 120 mm c/c.

**7. Reinforcement in the stem.**

$$A_{st} = \frac{270.1 \times 10^6}{230 \times 0.904 \times 560} = 2320 \text{ mm}^2$$

Using 20 mm Φ bars, $A_\Phi = 314.1$ mm². ∴ $s = \dfrac{1000 \times 314.1}{2320} = 135.38 ≃ 135$ mm

However provide 20 mm Φ bars @ 120 mm c/c

Actual $A_{st}$ provided = $\dfrac{1000 \times 314.1}{120} = 2618$ mm²

Bend all the bars in the toe slab to serve as the reinforcement there. However in order to make the toe slab safe in shear, required $A_{st} = 3080$ mm² gives $s = 100$ mm. Hence provide 20 mm φ bars @ 100 mm c/c. Due to this, sufficient bond length will be available to both the sides of point $C$ (point of maximum bending moment).

Total S.F. at $C = Q + K_a w H_1 + K_a \gamma \dfrac{H_1^2}{2} = 2 + 6 \times 5.5 + \dfrac{1}{3} \times \dfrac{18}{2} (5.5)^2 = 125.75$ kN

$$\therefore \quad \tau_v = \dfrac{125.75 \times 1000}{1000 \times 560} = 0.225 \text{ N/mm}^2$$

$$\dfrac{100 A_s}{bd} = \dfrac{100 \times 3080}{1000 \times 560} = 0.55\%. \quad \text{Hence } \tau_c = 0.31 \text{ N/mm}^2 > \tau_v. \quad \text{Hence safe.}$$

Let us curtail reinforcement between $C$ and $B$. If there were no other external force, except the earth pressure, and if the depth of stem were constant, half the bars could have been curtailed at a depth $\dfrac{H_1}{(2)^{1/3}}$ ≈ $0.79 H_1$ below the point $B$. Because of presence of other forces, let us try at depth = $0.65 H_1 = 0.65 \times 5.5 = 3.575$ m below $B$ to see whether half the bars could be curtailed there or not. Thus, depth of section below $B = h = 3.575$ m

$$\text{B.M.} = Q(h+1) + K_a w \dfrac{h^2}{2} + K_a \gamma \dfrac{h^3}{6} = 2(3.575 + 1) + \dfrac{6(3.575)^2}{2} + \dfrac{1}{3} \cdot \dfrac{18}{6}(3.575)^3$$

$$= 9.15 + 38.34 + 45.69 = 93.18 \text{ kN-m} = 93.18 \times 10^6 \text{ N-mm}$$

$$d = 140 + \dfrac{560 - 140}{5.5} \times 3.575 = 413 \text{ mm}$$

$$\therefore \quad A_{st} = \dfrac{93.18 \times 10^6}{230 \times 0.904 \times 413} = 1085.11 \approx 1086 \text{ mm}^2$$

This is less than half of that provided at $C$. Hence half the bars can be curtailed at this depth. However, the bars should be extended by a distance of $12 \Phi$ (= $12 \times 20 = 240$ mm) or $d$ (= $413$ mm) beyond this point, whichever is more. Hence $h = 3.575 - 0.413 = 3.162$ m. Hence curtail half the bars at a height of $3.1$ m below $B$. If we wish to curtail half of the remaining part, let us try it at a section at depth $h = 0.65 \times 3.575$ ≈ $2.32$ m below $B$.

$$\text{B.M.} = 2(2.32 + 1) + \dfrac{6(2.32)^2}{2} + \dfrac{1}{3} \times \dfrac{18}{6}(2.32)^3$$

$$= 6.64 + 16.15 + 12.49 = 35.28 \quad \text{kN-m} = 35.28 \times 10^6 \text{ N-mm}$$

$$d = 140 + \dfrac{560 - 140}{5.5} \times 2.32 = 317 \text{ mm}$$

$$A_{st} = \dfrac{35.28 \times 10^6}{230 \times 0.904 \times 317} = 535.27 \approx 536 \text{ mm}^2.$$

This is slightly less than one fourth of that provided at $C$. Hence we can curtail half of the remaining bars here. However, extend the bars by a distance of $d = 317$ mm beyond this, *i.e.* curtail half of the remaining bars at a distance of $3.32 - 0.317$ Ω $2$ m below $B$. Rest of the bars will be continued upto the top of the parapet.

**Distribution reinforcement.** Average thickness of stem

$$= \dfrac{200 + 620}{2} = 410 \text{ mm}$$

∴ Area distribution reinforcement

$$= (0.12/100)(1000 \times 410) = 492 \text{ mm}^2$$

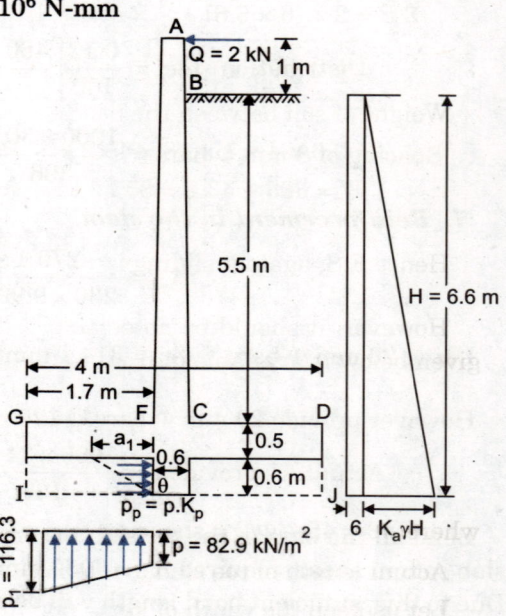

Fig. 18.22

Using 10 mm Φ bars, $A_\Phi = 78.5$ mm² ∴ Spacing = $\dfrac{1000 \times 78.5}{492} = 159.55$ mm

∴ Hence provide 10 mm bars @ 150 mm c/c at the inner face of the wall, along its length.

*Temperature reinforcement*

Total area = 492 mm². Provide 10 mm Φ bars @ 300 mm c/c both ways on the outer face.

**8. Design of shear key.** The wall is unsafe against sliding. Let us provide a shear key of depth $a$ below the stem.

Let $p_p$ be the intensity of passive pressure developed just in front of the shear key. This intensity $p_p$ depends upon the soil pressure $p$ just in front of shear key.

$$p_p = K_p \cdot p$$

where $K_p = \dfrac{1 + \sin \Phi}{1 - \sin \Phi} = \dfrac{1}{K_a} = 3$

and $p = 82.9$ kN/m², from Fig. 18.21

Hence $p_p = 3 \times 82.9 = 248.7$ kN/m²

This intensity may be considered to be constant along the depth of key, though there will be little increase in $p_p$, because of increase in $p$ with depth. We will, however, consider the constant value of $p_p = 248.7$ kN/m².

∴ Total passive force
$P_p = p_p a = 248.7\, a$ kN. Keeping $a = 600$ mm,
$P_p = 248.7 \times 0.6 = 149.22$ kN   ...(1)

Total sliding force at the bottom level of key is

$\Sigma P = 2 + (6 \times 6.6) + \dfrac{1}{3} \times \dfrac{18}{2} (6.6)^2$

$= 2 + 39.6 + 130.7 = 172.3$ kN   ...(2)

Weight of soil between the base and level $JJ$
$= 0.6 \times 4 \times 1 \times 18 = 43.2$ kN (approx.)
∴ $\Sigma W = 308 + 43.2 = 351.2$ kN   ...(3)

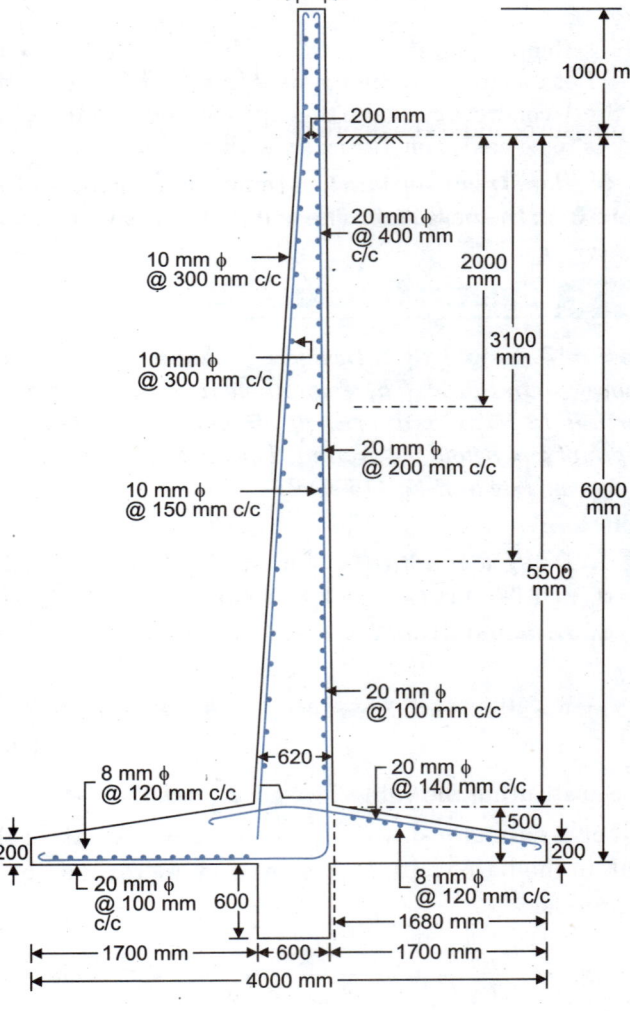

Fig. 18.23

Hence F.S. against sliding is $= \dfrac{(\Sigma \mu W) + P_p}{\Sigma P} = \dfrac{0.4 \times 351.2 + 149.22}{172.3} = 1.68$. Hence safe.

However, it should be noted that the above passive pressure will be developed *only when* a length $a_1$ given below is available in front of the key:

$$a_1 = a \tan \theta = a \tan \left(45 + \dfrac{\Phi}{2}\right)$$

$$= a \sqrt{K_p} = 0.6 \sqrt{3} = 1.04 \text{ m.}$$

where $\theta = 45 + \Phi/2 =$ shearing angle of passive resistance.

Actual length of toe slab available = $GF = 1.7$ m. Hence satisfactory.

Let us keep the width of key = 600 mm.

Actual force to be resisted by the key = $(1.5 \Sigma P) - \mu \Sigma W$
$$= 1.5 \times 172.3 - 0.4 \times 351.2$$
$$\approx 118 \text{ kN}$$

Hence shear stress $= \dfrac{118 \times 1000}{1000 \times 600} \approx 0.2$ N/mm². Safe

∴ Bending stress $= \dfrac{118 \times 1000 \times 300}{1/6 \times 1000 (600)^2} = 0.59$ N/mm²

Since concrete can take this much of tensile stress no special reinforcement is necessary for the key. The key is to be cast monolithically with the base.

**9. Construction joint.** A construction joint, in the form of a key, is to be provided at the junction of stem with the base slab. The width of key is kept equal to $d/4 = 560/4 = 140$ mm.

## 18.9. DESIGN OF CANTILEVER RETAINING WALL WITH SLOPING BACKFILL

**Design: Example 18.3.** *Design a T-shaped cantilever retaining wall to retain earth embankment 3 m high above ground level. The embankment is surcharged at angle of 16° to the horizontal. The unit weight of earth is 18 N/m³ and its angle of repose is 30°. The safe bearing capacity may be taken as 100 kN/m² at a depth of 1 m below the ground. The coefficient of friction between concrete and soil may be taken as 0.5. Use M 20 concrete and Fe 415 steel.*

**Solution:**

**1. Design constants.** For M 20 concrete, and Fe 415 steel reinforcement we have the following:
$c = \sigma_{cbc} = 7$ N/mm²; $t = \sigma_{st} = 230$ N/mm²; $m = 13.33$; $k_c = 0.289$; $j_c = 0.904$ and $R_c = 0.914$.

**2. Dimensions of base.** $\sin \beta = 0.276$; $\cos \beta = 0.961$; $\tan \beta = 0.287$;

$\sin \Phi = \sin 30° = 0.5$; $\cos \Phi = 0.866$; $K_a = \cos \beta \dfrac{\left(\cos \beta - \sqrt{\cos^2 \beta - \cos^2 \Phi}\right)}{\left(\cos \beta + \sqrt{\cos^2 \beta - \cos^2 \Phi}\right)}$

Substituting the values, we get $K_a = 0.38$

For a surcharged wall, the ratio of the length of the toe slab to the base width $b$ may be fixed approximately by the expression:

$$\alpha = 1 - \dfrac{q_0}{2.7 \gamma H} = 1 - \dfrac{100}{2.7 \times 18 \times 4} = 0.488. \text{ Adopt } \alpha = 0.49.$$

The base width can be determined from the expression

$$b = H \sqrt{\dfrac{K \cos \beta}{(1 - \alpha)(1 + 3\alpha)}}$$

$$= 4 \sqrt{\dfrac{0.38 \times 0.961}{(1 - 0.49)(1 + 3 \times 0.49)}} = 2.15 \text{ m}.$$

From sliding point of view, $b$ will be much more than this, as seen in example 18.1. However, a minimum base width $b = 0.6H = 2.4$ m should be adopted.

Hence keep $b = 2.4$ m. Length of toe slab $= 0.49 \times 2.4 \approx 1.2$ m.

Let the thickness of base slab be 300 mm for preliminary calculations.

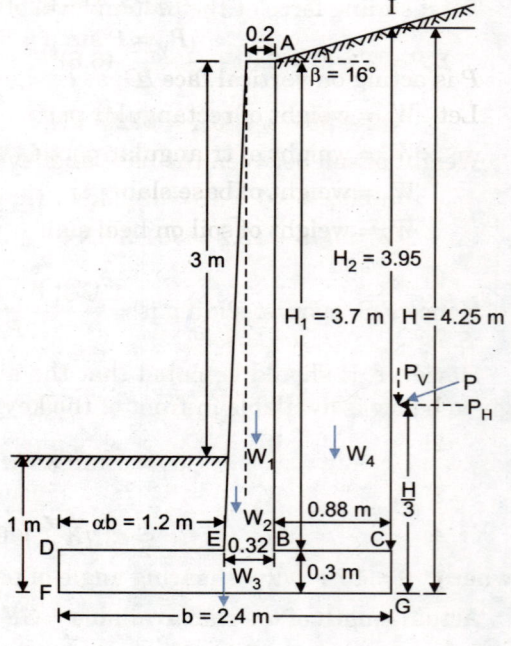

Fig. 18.24

# DESIGN OF RETAINING WALLS

### 3. Thickness of stem
Height $AB = H_1 = 4 - 0.3 = 3.7$ m.;  $K_a = 0.38$
Earth pressure on stem is

$$P = K_a \frac{\gamma H_1^2}{2}, \text{ acting at } 16° \text{ to horizontal} = \frac{0.38}{2} \times 18 (3.7)^2 = 46.82 \text{ kN}.$$

Hence the horizontal earth pressure is
$$P_H = P \cos \beta = 46.82 \cos 16° \approx 45 \text{ kN}$$

B.M. at $B = P_H \frac{H_1}{3} = 45 \times \frac{3.7}{3} = 55.5$ kN-m $= 55.5 \times 10^6$ N-mm

S.F. at $B = P_H = 45$ kN

$$d = \sqrt{\frac{55.5 \times 10^6}{0.914 \times 1000}} = 248 \text{ mm}.$$

However keep $d = 260$ mm and total thickness $= 320$ mm so that an effective cover of 60 mm is available. Reduce total thickness to 200 mm at $A$, so that effective depth is available there $= 140$ mm.

$$\tau_v = \frac{45 \times 1000}{1000 \times 260} = 0.17 \text{ N/mm}^2$$

$> \tau_c$ even at minimum steel.   Hence safe.

### 4. Stability of wall
Length of heel slab $= 2.4 - 1.2 - 0.32 = 0.88$ m
Height $H_2 = H_1 + 0.88 \tan 16° = 3.7 + 0.25 = 3.95$ m
Height $H = 3.95 + 0.3 = 4.25$ m

The fully dimensioned wall is shown in Fig. 18.24.

Earth pressure $P = k_a \frac{\gamma H^2}{2}$, acting parallel to the surcharge $= \frac{0.38}{2} \times 18 (4.25)^2 = 61.77$ kN

Its horizontal and vertical components are
$$P_H = P \cos \beta = 61.77 \times 0.961 = 59.36 \text{ kN}$$
$$P_V = P \sin \beta = 61.77 \times 0.276 = 17.05 \text{ kN}$$

$P$ is acting on vertical face $IG$, at $H/3$ and hence $P_V$ will act along the vertical line $IG$ (Fig. 18.24).

Let  $W_1$ = weight of rectangular portion of stem
  $W_2$ = weight of triangular portion of stem
  $W_3$ = weight of base slab
  $W_4$ = weight of soil on heel slab.

**TABLE 18.3**

| S. No. | Designation | Force (kN) | L.A. (m) | Moment about toe (kN-m) |
|---|---|---|---|---|
| 1 | $W_1$ | $0.2 \times 3.7 \times 25 = 18.5$ | 1.42 | 26.27 |
| 2 | $W_2$ | $\frac{1}{2} \times 0.12 \times 3.7 \times 25 = 5.55$ | 1.28 | 7.10 |
| 3 | $W_3$ | $2.4 \times 0.3 \times 25 = 18$ | 1.20 | 21.60 |
| 4 | $W_4$ | $\frac{1}{2} (3.7 + 3.95) \, 0.88 \times 18 = 60.59$ | 1.965 | 119.06 |
| 5 | $W_5$ | $P_V = 17.05$ | 2.4 | 40.92 |
|  |  | $\Sigma W = 119.69$ |  | $M_R = 214.95$ |

∴ Total resisting moment  $M_R = 214.95$ kN-m

Overturning moment $\quad M_0 = P_H \dfrac{H}{3} = 59.36 \times \dfrac{4.25}{3} = 84.1$ kN-m

$$\text{F.S. against overturning} = \dfrac{214.95}{84.1} = 2.56. \quad \text{Hence safe.}$$

$$\text{F.S. against sliding} = \dfrac{\mu \Sigma W}{P_H} = \dfrac{0.5 \times 119.69}{59.36} = 1 < 1.5. \quad \text{Hence unsafe.}$$

To make it safe against sliding, will have to provide shear key.

**Pressure distribution**

Net moment $\Sigma M = M_R - M_0 = 214.95 - 84.1 = 130.85$ kN  ...(1)

Distance $\bar{x}$ of the point of application of resultant from toe is

$$\bar{x} = \dfrac{\Sigma M}{\Sigma W} = \dfrac{130.85}{119.69} = 1.09 \text{ m.}$$

∴ Eccentricity $\quad e = \dfrac{b}{2} - \bar{x} = 1.2 - 1.09 = 0.11$ m. This is much less than $b/6$.

$$\text{Pressure } p_1 \text{ at toe} = \dfrac{\Sigma W}{b}\left(1 + \dfrac{6e}{b}\right) = \dfrac{119.69}{2.4}\left(1 + \dfrac{6 \times 0.11}{2.4}\right)$$

$$= 63.59 \text{ kN/m}^2 < 100. \quad \text{Hence safe.}$$

$$\text{Pressure } P_2 \text{ at heel} = \dfrac{119.69}{2.4}\left(1 - \dfrac{6 \times 0.11}{2.4}\right) = 36.16 \text{ kN/m}^2$$

Pressure $p$ at the junction of stem with heel slab is

$$p = 63.59 - \dfrac{63.59 - 36.16}{2.4} \times 1.2 = 49.87 \text{ kN/m}^2$$

Pressure $p'$ at the junction of stem with heel slab is

$$p' = 63.59 - \dfrac{63.59 - 36.16}{2.4} \times 1.52 = 46.22 \text{ kN/m}^2$$

**5. Design of toe slab.** The upward pressure distribution on the base is shown in Fig. 18.25. The weight of the soil above the toe slab is neglected. Thus two forces are acting on it:

(i) upward soil pressure  (ii) downward weight of the slab.

Downward weight of slab per unit area = $0.3 \times 1 \times 1 \times 25 = 7.5$ kN/m$^2$

Hence net pressure intensities will be = $63.59 - 7.5 = 56.09$ kN/m$^2$ under $D$ and $49.87 - 7.5 = 42.37$ kN/m$^2$ under $E$.

$$\text{Total force} = \text{S.F. at } E = \dfrac{1}{2}(56.09 + 42.37) \times 1.2 = 59.08 \text{ kN}$$

$$\text{C.G. of force from } E = \bar{x} = \left(\dfrac{42.37 + 2 \times 56.09}{42.37 + 56.09}\right) \cdot \dfrac{1.2}{3} = 0.63 \text{ m}$$

B.M. at $E = 59.08 \times 0.63 = 37.09$ kN-m $= 37.09 \times 10^6$ N-mm

∴ $\quad d = \sqrt{\dfrac{37.09 \times 10^6}{1000 \times 0.914}} = 201.4$ mm

Provide minimum total depth = 300 mm so that with an effective cover of 60 mm, $d = 240$ mm. Reduce the total depth to 200 mm at the edge.

$$\tau_v = \frac{59.08 \times 1000}{1000 \times 240} = 0.25 \text{ N/mm}^2.$$

This is less than the permissible shear stress of $\tau_c$ at $p = 0.35\%$ (Table 3.1)

$$A_{st} = \frac{37.09 \times 10^6}{230 \times 0.904 \times 240} = 743.27 \text{ mm}^2$$

Hence 12 mm Φ bars bent down from the stem @ 100 mm c/c will be sufficient, providing $A_{st} = 1130$ mm² (see step 7).

Actual $p = \dfrac{1130 \times 100}{1000 \times 240} \simeq 0.47$. Hence ok.

Distribution reinforcement

$$= \frac{0.12}{100} \times 1000 \left[\frac{300 + 200}{2}\right] = 300 \text{ mm}^2$$

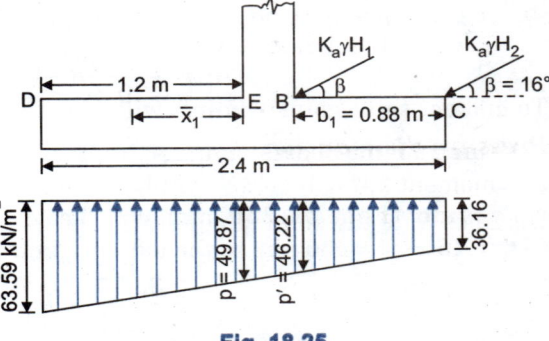

Fig. 18.25

Using 8 mm Φ bars, $A_\Phi = 50.3$ mm², ∴ $s = \dfrac{1000 \times 50.3}{300} = 167.66$ mm

However provide 8 mm Φ bars @ 150 mm c/c.

**6. Design of heel slab.** (Fig. 18.25). Let $b_1 = 0.88$ m = length of heel slab.

Four forces act on it:

    (i) weight of soil over the heel slab      (ii) weight of heel slab

    (iii) downward component of earth pressure      (iv) upward soil pressure

Weight of soil over heel slab = $\frac{1}{2}$ (3.7 + 3.95) × 0.88 × 18 = 60.59 kN

Acting at $\left(\dfrac{3.7 + 2 \times 3.95}{3.7 + 3.95}\right) \cdot \dfrac{0.88}{3} = 0.445$ m from $B$.

Weight of heel slab = 0.3 × 0.88 × 1 × 25 = 6.6 kN acting at 0.44 m from $B$.

Earth pressure intensity at $B = K_a \gamma H_1$ per unit inclined area, at β to horizontal.

∴    Earth pressure at $B$, on horizontal unit area $= K_a \gamma H_1 \cdot \tan \beta$

Vertical component of this, at $B = K_a \gamma H_1 \tan \beta \cdot \sin \beta$     ...(i)

Similarly, vertical component of earth pressure intensity at $C = K_a \gamma H_2 \tan \beta \cdot \sin \beta$     ...(ii)

Hence total force due to vertical component of earth pressure is

$$= K_a \cdot \gamma \frac{(H_1 + H_2)}{2} \times b_1 \tan \beta \cdot \sin \beta$$

$$= \frac{0.38 \times 18}{2} (3.7 + 3.95) \times 0.88 \times 0.287 \times 0.276 = 1.82 \text{ kN}$$

This acts at $\left(\dfrac{H_1 + 2H_2}{H_1 + H_2}\right) \dfrac{b_1}{3} = \dfrac{3.7 + 2 \times 3.95}{3.7 + 3.95} \times \dfrac{0.88}{3} = 0.445$ m from $B$.

Upward soil pressure = $\frac{1}{2}$ (46.22 + 36.16) × 0.88 = 36.25 kN

Acting at $\left(\dfrac{46.22 + 2 \times 36.16}{46.22 + 36.16}\right) \times \dfrac{0.88}{3} = 0.42$ m from $B$.

∴ Total force = S.F. at $B$ = 60.59 + 6.6 + 1.82 − 36.25 = 32.76 kN.

B.M. at $B$ = (60.59 × 0.445) + (6.6 × 0.44) + (1.82 × 0.445) − (36.25 × 0.42)

= 15.45 kN-m = 15.45 × 10⁶ N-mm

This is much less than the B.M. on the toe slab. However, we will keep the same depth, as that of toe slab, *i.e.* $d$ = 240 mm and $D$ = 300 mm, reducing it to 200 mm at the edges.

$$A_{st} = \frac{15.54 \times 10^6}{230 \times 0.904 \times 240} = 311.41 \approx 312 \text{ mm}^2.$$

Using 10 mm $\Phi$ bars, having $A_\Phi$ = 78.5 mm². Spacing = $\frac{1000 \times 78.4}{312}$ = 251 mm.

Provide them at 250 mm c/c. Take the reinforcement in the toe slab for a distance of 45 $\Phi$ = 45 × 10 ≈ 450 mm to the left of $B$, and its ends should be hooked.

$$\tau_v = \frac{32.76 \times 1000}{1000 \times 240} = 0.137 \text{ N/mm}^2. \quad \text{Hence safe.}$$

Distribution steel = $\frac{0.12}{100} \times 1000 \left(\frac{300 + 200}{2}\right)$ = 300 mm²

Hence provide 8 mm $\Phi$ bars @ 150 mm c/c as in the toe slab.

**7. Reinforcement in the stem.** Revised $H_1$ = 4 − 0.3 = 3.7 m

S.F. at $B$ = $\frac{0.38}{2} \times 18 \, (3.7)^2 \cos 16° = 45 \text{ kN} = P_H$

∴ B.M. at $B$ = 45 × $\frac{3.7}{3}$ = 55.5 kN-m = 55.5 × 10⁶ N-mm

$d$ = 260 mm, $A_{st} = \frac{55.5 \times 10^6}{230 \times 0.904 \times 260}$ = 1027 mm²

Using 12 mm $\Phi$ bars, $A_\Phi$ = 113 mm², ∴ Spacing = $\frac{1000 \times 113}{1027}$ = 110 mm

However provide 12 mm $\Phi$ bars @ 100 mm c/c.

Actual $A_{st}$ provided = $\frac{1000 \times 113}{100}$ Ω 1130 mm²

Bend these bars into the toe slab, to serve as reinforcement there. Sufficient development length is available.

Between $A$ and $B$, some of the bars can be curtailed. Consider a section at a depth $h$ below the top of the step. The effective depth $d'$ at the section is

$$d' = 140 + \frac{260 - 140}{3.7} h = 140 + 32.4 h \quad \ldots(1)$$

Now $A_{st} \propto \frac{H^3}{d}$ or $H \propto (A_{st} \, d)^{\frac{1}{3}}$

Hence $\frac{h}{H_1} = \left(\frac{A_{st'} \, d'}{A_{st} \, d}\right)^{\frac{1}{3}} \quad \ldots(2)$

where $A_{st'}$ = reinforcement at depth $h$

$A_{st}$ = reinforcement at depth $H_1$

$d$ = effective thickness at depth $H_1$.

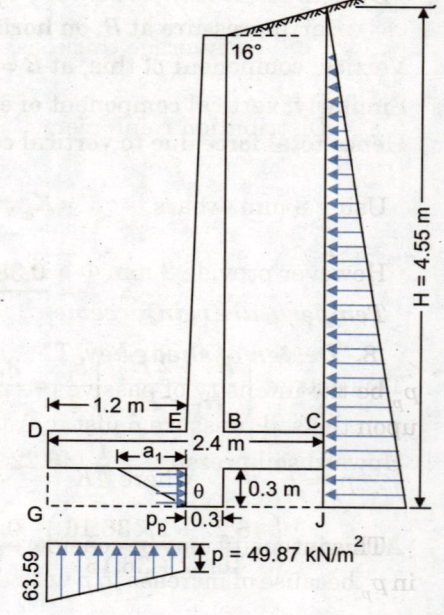

Fig. 18.26

If $A'_{st} = \frac{1}{2} A_{st}$ ; $\frac{A'_{st}}{A_{st}} = \frac{1}{2}$

$\therefore \quad \frac{h}{H_1} = \left(\frac{1}{2} \cdot \frac{d'}{d}\right)^{1/3}$. Substituting $d = 260$ mm and $d' = 140 + 32.4\, h$, we have:

$$h = H_1 \left(\frac{140 + 32.4\, h}{2 \times 260}\right)^{1/3} = 3.7 \left(\frac{140 + 32.4\, h}{520}\right)^{1/3} = 0.46\,(140 + 32.4\, h)^{1/3} \quad \ldots(3)$$

This can be solved by trial and error, noting that if the effective thickness of stem were constant, $h$ would have been equal to $H_1/(2)^{1/3} \approx 0.79\, H_1 \approx 2.92$ m. Solving (3) by trial, we get $h = 2.82$ m. Thus half the bars can be curtailed at this point. However, bars should be extended by a distance of $12\Phi$ (= 12 × 16) or $d$ (= 260) beyond this point, whichever is more.

$\therefore \quad h = 2.82 - 0.26 = 2.56$ m. However, curtail half the bars at a height 2.5 m below the top.

If we wish to curtail half of the remaining bars, so that reinforcement is one-fourth of that provided at $B$, we have

$\frac{A_{st'}}{A_{st}} = \frac{1}{4}$.  Hence from (2), $\frac{h}{H_1} = \left(\frac{1}{4} \cdot \frac{d'}{d}\right)^{1/3}$

or $\quad h = H_1 \left(\frac{140 + 32.4\, h}{4 \times 260}\right)^{1/3} = 3.7 \left(\frac{140 + 32.4}{10210}\right)^{1/3} = 0.365\,(140 + 32.4\, h)^{1/3}$

This can be solved by trial, noting that if the effective thickness of the stem were constant, $h$ would have been equal to $\frac{H}{(4)^{1/2}} \approx 0.63\, H_1 = 2.32$ m.

Hence, we get $h = 2.16$ m. However, bars should be extended by $d$ (= 260 mm) beyond this.

$\therefore \quad h = 2.16 - 0.26 = 1.90$ m

Hence stop half of the remaining bars at 1.90 m below the top of the stem. Continue rest of the bars to the top of the stem.

**Distribution reinforcement**

Average thickness = $\frac{1}{2}(320 + 200) = 260$ mm

$\therefore$ Distribution reinforcement = $\frac{0.12}{100}(1000 \times 260) = 312$ mm²

Using 8 mm $\Phi$ bars, $A_\Phi = 50.3$ mm² $\therefore$ Spacing = $\frac{1000 \times 50.3}{312} = 161$ mm

However provide 8 mm $\Phi$ bars @ 150 mm c/c at the inner face of the stem.

**Temperature reinforcement**: Provide 8 mm $\Phi$ bars @ 300 mm c/c both ways on the outer face of the stem.

**8. Design of shear key.** The wall is unsafe in shear. Let us provide a shear key 300 mm × 300 mm. Let $p_p$ be the intensity of passive pressure developed just in front of the shear key. This intensity $p_p$ depends upon the soil pressure $p$ just in front of shear key.

$p_p = K_p\, p$ where $K_p = \frac{1}{K_a} = \frac{1}{0.38} = 2.63$ $\therefore$ $p_p = 2.63 \times 49.87 = 131.2$ kN/m²

This intensity may be considered to be constant along the depth of key though there will be little increase in $p_p$ because of increase in $p$ with depth. We will, however, consider the constant value of $p_p = 131.2$ kN/m².

Total passive force $P_p = 131.2 \times 0.3 = 39.36$ kN ...(1)

Let us calculate total sliding force at level GJ.

$$P_H = \frac{0.38 \times 18}{2}(4.55)^2 \cos 16° = 68.06 \text{ kN} \qquad ...(2)$$

Weight of soil between the base and level $GJ = 0.3 \times 2.4 \times 18 = 12.96$ kN (approx) ...(3)

∴ $\Sigma W = 119.69 + 12.96 = 132.65$ kN

(There will be some more downward weight because of increasing vertical component of earth pressure).

∴ F.S. against sliding

$$= \frac{\mu \Sigma W + P_p}{P_H}$$

$$= \frac{0.5 \times 132.65 + 39.36}{68.06}$$

$$= 1.55 > 1.5 \text{ Safe.}$$

However, it should be noted that the above passive pressure will be developed *only* when a length $a_1$ given below is available in front of the key:

$$a_1 = a \tan \theta$$

$$= a \tan\left(45 + \frac{\Phi}{2}\right) = a\sqrt{K_p}$$

$$= 0.3 \sqrt{2.63} = 0.49 \text{ m.}$$

Actual length of toe slab available = 1.2 m. Hence satisfactory.

Now, force to be resisted by key, at F.S. of 1.5 is

$$= 1.5 P_H - \mu \Sigma W$$

$$= 1.5 \times 68.06 - 0.5 \times 132.65 = 35.8 \text{ kN}$$

∴ Shear stress in key = $\dfrac{35.8 \times 10^3}{1000 \times 300} = 0.12$ N/mm² (Safe)

Bending stress = $\dfrac{35.8 \times 10^3 \times 150}{\frac{1}{6} \times 1000 (300)^2} = 0.36$ N/mm²

Since concrete can take this much of tensile stress, no special reinforcement is necessary for the key. The details of reinforcement etc. are shown in Fig. 18.27.

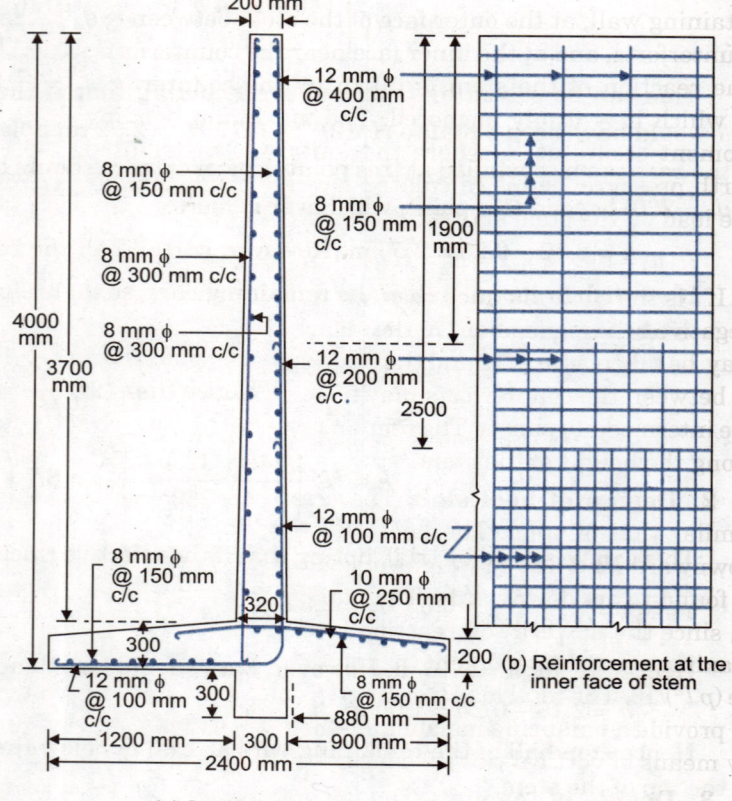

Fig. 18.27. Cantilever retaining wall reinforcement details

## 18.10 DESIGN OF COUNTERFORT RETAINING WALL

We have seen that in case of cantilever retaining wall, the bending moment in the vertical stem varies as $h^3$, where $h$ is the height of the stem. When this height exceeds 6 m, a thick section is required for the stem, and the design becomes uneconomical. In that case, counterfort retaining walls are used.

The counterforts support both the vertical stem as well as heel slab. Sometimes, counterforts are also provided over the toe slab, upto ground level, as shown in Fig. 18.28(b). Design principles for various component parts are discussed below in brief. The same criterion for fixing the length of base (b), and length of toe slab ($\alpha b$) are used as cantilever retaining wall.

**1. Design of stem.** Unlike the stem of cantilever retaining wall, the stem of a counterfort retaining wall acts as a continuous slab supported on counterforts. Due to the varying earth pressure over the height of stem, the stem slab deflects outwards, and hence main reinforcement is provided along the length of the retaining wall, at the outer face of the stem between the counterforts, and at the inner face near the counterforts. The reaction of the stem is taken by the counterforts to which it is firmly anchored. The maximum bending moment occurs at $B$, where the uniformly distributed earth pressure load is calculated for unit height. If $w$ is the load on the stem slab, at $B$, per unit length

$$w = p_a \times 1 \times 1 = K_a \gamma H.$$

If $l$ is the clear distance between the counterforts, the negative bending moment in the stem at the counterforts may be taken as $w\, l^2/12$ and the positive bending moment in between the counterforts may be taken as $w\, l^2/16$, for the intermediate panels. The reinforcement may be varied along the height of the stem.

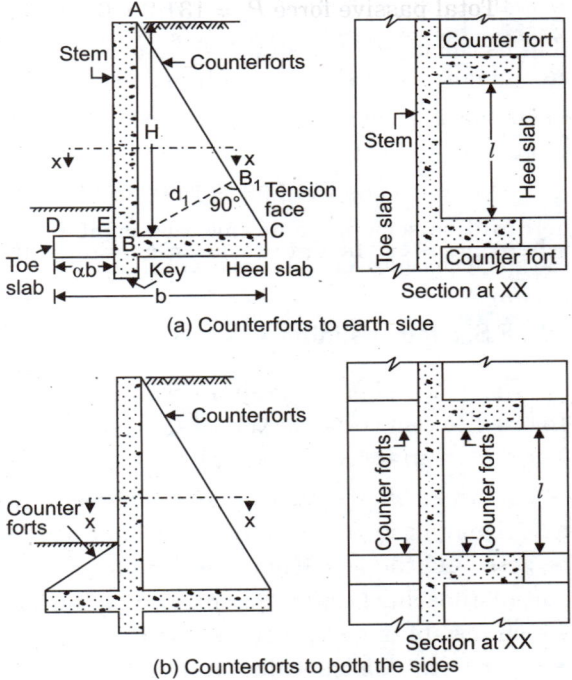

Fig. 18.28

**2. Design of heel slab.** The action of heel slab is similar that of stem. The heel slab is subjected to the downward load due to weight of soil and self-weight, and upward load due to soil reaction. The net load is found to act downwards. The maximum net pressure is found to act on a strip of unit width near edge $C$, since the upward soil reaction is minimum there. If $p$ is the net downward pressure per unit area, the maximum bending moment (negative) near counterforts will be $(p\, l^2)/12$ and positive bending moment will be $(p\, l^2)/16$. The total load (reaction) from the heel slab is transferred to the counterforts, and this load helps to provide a balancing moment against its overturning. The heel slab is firmly attached to the counterforts by means of vertical ties (two legged stirrups).

**3. Design of toe slab.** The toe slab bends as a cantilever due to upward soil reaction as in the case of cantilever retaining wall. Due to bending of the toe as cantilever, clockwise moments are induced at $E$, which are shared by both vertical stem and heel slab. The distribution of this moment between the stem and heel slab are not known. These transferred edge moments will cause bending of stem as well as heel slab, in a direction normal to their usual bending. These edge moments also cause shears in both the slabs which are not precisely known. Hence the design is done empirically, because of uncertain bending behaviour. However, if counterforts are also provided over the toe slab, upto the height of soil (*i.e.* upto ground level), their behaviour becomes certain. In such a case, the reaction is transferred directly to the counterforts, without affecting the bending behaviour of stem and heel slab.

**4. Design of counterforts.** The counterforts take reactions both from the stem as well as the heel slab. Since the active pressure on stem acts outwards, and net pressure heel slab acts downwards, the counterforts are subjected to tensile stresses along the outer face $AC$ of the counterforts. The angle $ABC$ between stem and heel slab has a tendency to increase from 90°, and this tendency is resisted by counterforts. Thus the counterfort may be considered to bend as a cantilever, fixed at $BC$. The counterfort acts as an inverted T-beam of varying rib depth. The maximum depth of this T-beam is at the junction $B$. The depth is measured perpendicular to the sloping face $AB$, *i.e.*, depth $d_1 = BB_1$ at $B$. This depth thus goes on decreasing towards $A_1$ where the bending moment also decreases. The width $b_1$ of the counterfort is kept constant throughout its height. Main reinforcement is provided parallel to $AC$. The reinforcement may be varied along $AC$, maximum being at $B_1$. The faces $AB$ and $BC$ of the counterfort remain in compression. The compressive stresses on face $AB$ are counter-balanced by the vertical upward reaction transferred by the slab. In addition to the main reinforcement, the counterforts are jointed firmly to the stem and heel slab by horizontal and vertical ties respectively, provided in the form of two legged stirrups. Normally, the spacing between the counterforts should not be kept less than 2 m. The spacing of the counterforts depends upon many factors, but usually ranges from $\frac{1}{3}$ to $\frac{1}{2}$ of the height of the wall. For an economical design the clear spacing $l$ between the counterforts may be determined from the following approximate expression:

$$l = 3.5 \left(\frac{H}{\gamma}\right)^{\frac{1}{4}} \qquad ...(18.29)$$

where $H$ is the height of wall in m and $\gamma$ is the unit weight of soil in kN/m$^3$.

## 18.11. BACK ANCHORING OF RETAINING WALL

When the height of retaining wall is much more, it becomes uneconomical to even provide counterforts. In order to reduce the section of stem etc. in the high retaining walls, the stem may be anchored at its back. The anchor practically takes all the earth pressure and B.M. and S.F. in the stem are thus greatly reduced.

Fig. 18.29 shows such a wall anchored at its back, at a height $H/3$ where the resultant earth pressure acts. Due to the provision of the anchor at this location, bending moments in the stem at $B$ and $A$ would be zero, and the whole of the earth pressure will be resisted by the anchor rod which is provided at some suitable spacing $l$ centre to centre along the length of the wall. The anchor bar is attached to a concrete slab which resists the pressure by way of development of frictional forces along both the surface $EF$ and $GH$. The concrete slab is located well beyond the plane of rupture.

Let $A$ = surface area of each face of slab, in a distance $l$.

$\mu$ = coefficient of friction between concrete and soil.

$h$ = height of embedment of slab = $\frac{2}{3}H$.

$l$ = centre to centre spacing of anchor bars.

$A_s$ = area of cross-section of each anchor bars.

Horizontal sliding force = $\frac{1}{2} K_a \gamma H^2 l$ ...(i)

Resisting friction force = $2 A \mu \cdot \gamma h$ ...(ii)

Equating the two, $2 A \mu \gamma h = \frac{1}{2} K_a \gamma H^2 l$

or $\qquad A = \dfrac{K_a H^2 l}{4 \mu h} = \dfrac{3}{8} \cdot \dfrac{K_a H l}{\mu}$ ...(18.30)

Hence width $\qquad b_1 = \dfrac{A}{l} = \dfrac{3 K_a H}{8 \mu}$ ...(18.31)

$\therefore$ Area of rod $A_s = \dfrac{\frac{1}{2} K_a \gamma H^2 l}{\sigma_{st}} = \dfrac{K_a \gamma H^2 l}{2 \sigma_{st}}$ ...(18.32)

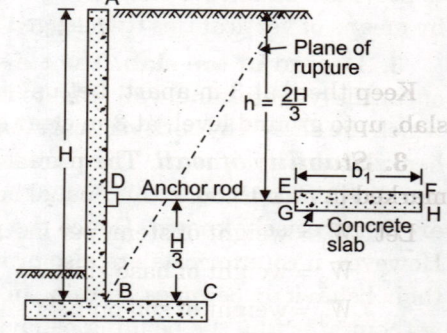

Fig. 18.29

If the height is more, two or more tie rods can used at different heights. The tie rod is fixed to the wall through horizontal or vertical beams cast monolithically with the stem. In case vertical beams are provided at interval $l$, the stem will act as a continuous slab as in the case of counterfort retaining wall. If horizontal beam is provided at height $H/3$, the portions of stem above and below it will act as cantilevers.

In some cases, such as in road or railway embankments, high retaining walls are provided to both the sides of the fill. In such a circumstances, concrete slab is not required to develop frictional force of resistance, but instead, both the walls are connected by common tie rods at suitable intervals, so that earth pressure on one is neutralised by the earth pressure on the other.

## DESIGN OF RETAINING WALLS

**Design: Example 18.4.** *Design a counterfort retaining wall to retain 7 m high embankment above ground level. The foundation is to be taken 1 m deep where the safe bearing capacity of soil may be taken as 180 kN/m². The top of earth retained is horizontal, and soil weighs 18 kN/m³ with angle of internal friction Φ = 30°. Coefficient of friction between concrete and soil may be taken as 0.5. Use M 20 concrete and Fe 415 steel.*

**Solution:**

1. **Design constants:** For M 20 concrete and Fe 415 steel reinforcement we have the following:

$c = \sigma_{cbc} = 7$ N/mm²; $t = \sigma_{st} = 230$ N/mm²; $m = 13.33$; $k_c = 0.289$; $j_c = 0.904$; $R_c = 0.914$

For the soil, $\Phi = 30°$ $\therefore$ $K_a = \dfrac{1 - \sin 30°}{1 + \sin 30°} = \dfrac{1}{3}$ and $K_p = \dfrac{1}{K_a} = 3$.

2. **Dimensions of various parts.** The ratio of length of toe slab ($DE$) to the base of width $b$ may be found by the expression:

$$\alpha = 1 - \dfrac{q_0}{2.2\gamma H} = 1 - \dfrac{180}{2.2 \times 18 \times 8} = 0.43.$$

$$b = 0.95\,H\sqrt{\dfrac{K_a}{(1-\alpha)(1+3\alpha)}} = 0.95 \times 8 \sqrt{\dfrac{1}{3(1-0.43)(1+3\times 0.43)}} = 3.64$$

Normal range of $b$ is between $0.5\,H$ to $0.6\,H$.

However, keep minimum $b = 0.5\,H = 4$ m.

$\therefore$ Length of toe slab $= 0.43 \times 4 = 1.72$ m. Keep the length of toe slab $= 1.70$ m.

Taking uniform thickness of stem = 300 mm, length of heel slab $BC$ = 4 – 1.70 – 0.3 = 2 m.

Let the thickness of base slab = 300 mm.

*Clear spacing* of counterforts is

$$l = 3.5\left(\dfrac{H}{\gamma}\right)^{\frac{1}{4}} = 3.5\left(\dfrac{8}{18}\right)^{\frac{1}{4}} = 2.86 \text{ m}.$$

Keep them at 3 m apart. Let us also provide counterforts over toe slab, upto ground level, at 3 m clear distance.

3. **Stability of wall.** The preliminary dimensions of the wall are marked in Fig. 18.30.

Let $W_1$ = weight of stem, per metre length.
$W_2$ = weight of base slab.
$W_3$ = weight of soil on heel slab.

The calculations are arranged in Table 18.4.

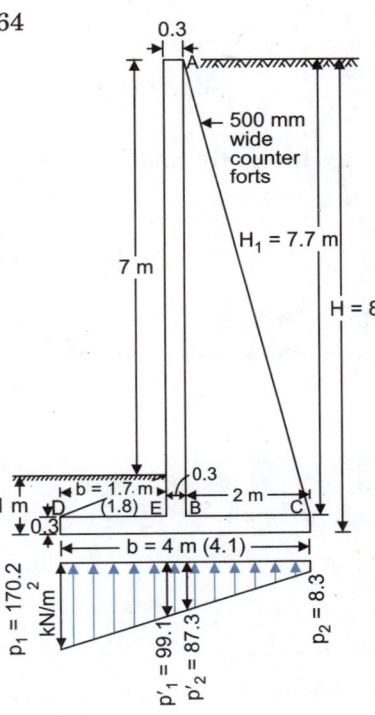

Fig. 18.30

### TABLE 18.4

| S. No. | Designation | Force (kN) | L.A. (m) | Moment about toe (kN-m) |
|---|---|---|---|---|
| 1 | $W_1$ | 0.3 × 7.7 × 1 × 25 = 57.8 | 1.85 | 106.8 |
| 2 | $W_2$ | 0.3 × 4 × 1 × 25 = 30 | 2 | 60.0 |
| 3 | $W_3$ | 2 × 7.7 × 1 × 18 = 277.2 | 3 | 831.6 |
| | | $\Sigma W = 365$ | | $M_R = 998.4$ |

**474** REINFORCED CONCRETE STRUCTURE

∴ Resisting moment $\quad M_R = 998.4$ kN-m $\hspace{6cm}$ ...(1)

Horizontal earth pressure $\quad P_H = K_a \gamma \dfrac{H^2}{2} = \dfrac{1}{3} \times \dfrac{1}{2} \times 18 \,(8)^2 = 192$ kN $\hspace{2cm}$ ...(2)

Overturning moment $\quad M_0 = P_H \cdot \dfrac{H}{3} = 192 \times \dfrac{8}{3} = 512$ kN-m $\hspace{3cm}$ ...(3)

F.S. against overturning $\quad = \dfrac{M_R}{M_0} = \dfrac{998.4}{512} = 1.95 < 2.$ Hence unsafe.

F.S. against sliding $\quad = \dfrac{\mu \Sigma W}{P_H} = \dfrac{0.5 \times 365}{192} = 0.95$

Hence the wall is unsafe against overturning and sliding. A shear key will have to be provided under the base. Also, width of the base will have to be increased.

**Pressure distribution.** Net moment $\quad \Sigma M = M_R - M_0 = 998.4 - 512 = 486.4$ kN-m.

Distance $\bar{x}$ from toe, of the point of application of the resultant is given by

$$\bar{x} = \dfrac{\Sigma M}{\Sigma W} = \dfrac{486.4}{365} = 1.333 \text{ m}$$

∴ Eccentricity $e = \dfrac{b}{2} - \bar{x} = 2 - 1.333 = 0.667$ m

This is exactly equal to $\dfrac{b}{6} = \dfrac{4}{6} = 0.667$ m. However no tension will be developed. Pressure $p_1$ under the toe is given by

$$p_1 = \dfrac{\Sigma W}{b}\left(1 + \dfrac{6e}{b}\right) = \dfrac{365}{4}\left(1 + \dfrac{6 \times 0.667}{4}\right) = 182.5 \text{ kN/m}^2 > 180. \quad \text{Hence unsafe.}$$

In order to make it safe, increase the length of toe slab DE to 1.8 m, so that total width $b = 4.1$ m. The revised computations are arranged in table 18.4 (a)

**TABLE 18.4 (a)**

| S. No. | Designation | Force (kN) | L.A. (m) | Moment about toe (kN-m) |
|---|---|---|---|---|
| 1 | $W_1$ | $0.3 \times 7.7 \times 1 \times 25 = 57.8$ | 1.95 | 112.6 |
| 2 | $W_2$ | $0.3 \times 4.1 \times 1 \times 25 = 30.8$ | 2.05 | 63.0 |
| 3 | $W_3$ | $2 \times 7.7 \times 1 \times 18 = 277.2$ | 3.1 | 859.3 |
|  |  | $\Sigma W = 365.8$ |  | $M_R = 1034.9$ |

Net moment $\quad \Sigma M = M_R - M_0 = 1034.9 - 512 = 522.9$ kN-m

$$\bar{x} = \dfrac{\Sigma M}{\Sigma W} = \dfrac{522.9}{365.8} = 1.43 \text{ m}$$

∴ $\quad e = \dfrac{b}{2} - \bar{x} = 2.05 - 1.43 = 0.62 \text{ m} < \dfrac{b}{6}$

$$p_1 = \dfrac{\Sigma W}{b}\left(1 + \dfrac{6e}{b}\right) = \dfrac{365.8}{4.1}\left(1 + \dfrac{6 \times 0.62}{4.1}\right) = 170.2 \text{ kN/m}^2 < 180.$$

Hence safe.

Pressure $p_2$ under the heel is given by $p_2 = \dfrac{\Sigma W}{b}\left(1 - \dfrac{6e}{b}\right) = \dfrac{365.8}{4.1}\left(1 - \dfrac{6 \times 0.62}{4.1}\right) = 8.3$ kN/m$^2$

The pressure intensity $p_1'$ under $E$ is $p_1' = 170.2 - \dfrac{170.2 - 8.3}{4.1} \times 1.8 = 99.1$ kN/m²

The pressure intensity $p_2'$ under $B$ is $p_2' = 170.2 - \dfrac{170.2 - 8.3}{4.1} \times 2.1 = 87.3$ kN/m²

**4. Design of heel slab.** Clear spacing between counterforts = 3 m. The pressure distribution below the heel slab is shown in Fig. 18.30. Consider a strip, 1 m wide, near the outer edge $C$. The upward pressure intensity is 8.3 kN/m², which is minimum at $C$.

Downward load due to weight of earth = 7.7 × 18 = 138.6 kN/m², and that due to selfweight of heel slab = 0.3 × 25 = 7.5 kN/m². Hence net downward intensity of load $p$ is given by

$$p = 138.6 + 7.5 - 8.3 = 137.8 \text{ kN/m}^2$$

∴ Maximum negative bending moment in heel slab, at counterforts is

$$M_1 = \frac{pl^2}{12} = \frac{137.8(3)^2}{12} = 103.4 \text{ kN-m} = 103.4 \times 10^6 \text{ N-mm}$$

∴ $d = \sqrt{\dfrac{103.4 \times 10^6}{1000 \times 0.914}} = 336.3 \cong 337$ mm;

Shear force $V = \dfrac{137.8 \times 3}{2} = 206.7 \cong 207$ kN.

For a balanced section, having $p = 0.72\%$, $\tau_c = 0.33$ N/mm². Hence depth required from shear point of view is

$$d = \frac{V}{\tau_c \, b} = \frac{207 \times 1000}{0.33 \times 1000} = 627.27 \cong 626 \text{ mm}. \text{ This is excessive.}$$

However, keep $D = 500$ mm. Providing 60 mm effective cover, $d = 500 - 60 = 440$ mm.

$$\tau_v = \frac{207 \times 1000}{1000 \times 440} = 0.47 \text{ N/mm}^2 > \tau_c. \quad \text{Hence shear reinforcement will be necessary.}$$

Area of steel at supports is given by $A_{st1} = \dfrac{103.4 \times 10^6}{230 \times 0.904 \times 440} = 1130.24$ mm²

Providing 12 mm Φ bars, $A_\Phi = 113.1$ mm². ∴ Spacing = $\dfrac{1000 \times 113.1}{1130.24} \approx 100$ mm

Provide these @ 100 mm c/c. Actual $A_{st}$ provided = $\dfrac{1000 \times 113.1}{100} = 1131$ mm²

Let us check this reinforcement for development length at the point of contraflexure, so as to satisfy the criterion: $\dfrac{M}{V} + L_0 \geq L_d$.

For a fixed beam or slab carrying U.D.L., the point of contraflexure is situated at a distance of 0.211 $L$. In our case, the slab is continuous, but we will assume the same position of point of contraflexure, i.e. at 0.211 × 3 ≈ 0.63 m from the face of counterforts. Shear force at this point is given by

$$V = \frac{pL}{2}\left(\frac{l}{2} - x\right) \div \left(\frac{L}{2}\right) = p\left(\frac{l}{2} - x\right) = 137.8\left(\frac{3}{2} - 0.63\right) = 119.9 \text{ kN}.$$

Assuming that all the bars will be available at the point of contraflexure,

$M = \sigma_{st} \cdot A_{st} jd = 230\, [1131] \times 0.904 \times 440 = 1.035 \times 10^8$ N-mm

$L_0 = 12\Phi$ or $d$, whichever is more = 440 mm

$L_d = 45\Phi = 45 \times 12 = 540$ mm.

∴ $\dfrac{M}{V} + L_0 = \dfrac{1.035 \times 10^8}{119.9 \times 10^3} + 440 = 1303$ mm $> L_d$. Hence safe.

Continue these bars by a distance $l_0$ (= $d$ = 440 mm) beyond the point of contraflexure. After that, curtail half the bars, and continue the remaining half throughout the length. At the point of curtailment, length of each bar available = 630 + 440 = 1070 mm > $L_d$. These bars will be provided at the top face of the heel slab.

$$\text{Maximum positive B.M.} = \frac{pl^2}{16} = \frac{3}{4} M_1$$

$$\therefore \text{Area of bottom steel } A_{st2} = \frac{3}{4} A_{st1} = \frac{3}{4} \times 1131 = 848.25 \approxeq 849 \text{ mm}^2$$

$$\text{Spacing of 12 mm } \Phi \text{ bars} = \frac{1000 \times 113.1}{849} \approx 133 \text{ mm} \approxeq 130 \text{ mm}$$

$$\text{Actual } A_{st} = \frac{1000 \times 113.1}{130} = 870 \text{ mm}^2$$

Let us check whether these bars satisfy the development length criterion at the point of contraflexure, inherent in the criterion : $\frac{M}{V} + L_0 \geq L_d$.

where $V$ = shear at point of contraflexure = 119.9 kN distant 0.63 m from the face of support
Assuming that all the bars are available at the point of contraflexure,

$M = 230 \times 870 \times 0.904 \times 440 = 79.6 \times 10^6$ N-mm
$L_0 = 440$ mm and $L_d = 540$ mm, as before.

$$\therefore \quad \frac{M}{V} + L_0 = \frac{79.6 \times 10^6}{119.9 \times 10^3} + 440 = 663.89 + 440 = 1103.9 \text{ mm}$$

This is more than $L_d$. Hence safe.

Thus continue all the bottom bars to a point distant $L_0$ (= 440 mm) from the point of contraflexure, *i.e.* upto a point distant 630 – 440 ≈ 190 mm from the centre of support. At this point, half the bars can be discontinued. Since this distance is quite small, it is better to continue these bars upto the centre of counterforts.

**Reinforcement near B.** The c/c spacing of reinforcement near $B$ may be increased, because $p$ decreases due to increase in upward soil reaction. Consider a strip 1 m wide near $B$. Upward soil reaction at $B$ is = 87.3 kN/m², as found earlier.

$\therefore$ Net downward load $p'$ = 138.6 + (0.5 × 25) – 87.3 = 63.8 kN/m²
This is about 63.8/137.8 = 0.463 of load intensity at $C$.
Hence spacing of steel bars = 100/0.463 ≈ 200 mm c/c. at the top face, near supports.
Spacing of steel bars at the bottom face, at midspan = 133/0.463 ≈ 280 mm c/c

**Distribution steel.** Area of steel = $\frac{0.12}{100}$ (1000 × 500) = 600 mm².

Using 12 mm $\Phi$ bars, having $A_\Phi$ = 113 mm². $\therefore$ Spacing = $\frac{1000 \times 113}{600} \approx 180$ mm c/c

**Shear reinforcement.** Shear stress at $C = \tau_v = 0.47$ N/mm².

$$\frac{100 A_s}{b\,d} = \frac{100 \times 1131}{1000 \times 440} = 0.257\%$$

Hence from Table 3.1, $\tau_c = 0.22$ N/mm² < $\tau_v$. Hence shear reinforcement is necessary.

$$V_c = \tau_c \, bd = 0.22 \times 1000 \times 440 = 96800 \text{ N} = 96.8 \text{ kN}.$$

Consider a section distant $x_1$ from the face of the counterfort, where shear force is 96.8 kN. The position is given by $\frac{96.8}{207} = \frac{1.5 - x_1}{1.5}$, which gives $x_1 = 0.8$ m.

# DESIGN OF RETAINING WALLS

Hence shear stirrups are required upto distance of 0.8 m on either side of each counterforts. The requirement is there for a strip of unit width passing through C. Let us consider a strip through $B_1$, distant $y_1$ from C, such that shear force at the counterforts is 96.8 kN.

Net downward pressure at $C = 137.8$ kN/m²

Net downward pressure at $B = 63.8$ kN/m²

Let the net downward pressure at $B_1$ be $w_1$ kN/m²

∴ S.F. at face of counterforts at $B_1 = w_1 \times \dfrac{3}{2} = 1.5 w_1$.

This should be equal to 96.8 kN.

∴ $w = \dfrac{96.8}{1.5} = 64.5$ kN/m²   ...(1)

However, at $y_1$ from C,

$w_1 = 137.8 - \dfrac{137.8 - 63.8}{2} \times y_1 = 137.8 - 37 y_1$   ...(2)

Equating the two, we get $137.8 - 37 y_1 = 64.5$

From which $y_1 = 1.98$ m ≏ 2 m

Hence shear reinforcement is required in the triangular portion on the other side of the counterforts shown hatched in Fig. 18.31. However, we will provide shear stirrups in rectangular portions $x_1 \times y_1 = 0.8$ m × 1.98 m on either side of counterforts. Let us provide 4-legged stirrups of 8 mm Φ wire.

∴ $A_{sv} = 4 \dfrac{\pi}{4}(8)^2 = 201$ mm²

and $S_v = \dfrac{A_{sv} \cdot \sigma_{sv} \cdot d}{V - V_c} = \dfrac{201 \times 230 \times 440}{(207 - 96.8)1000} \approx 184.58 \approx 150$ mm

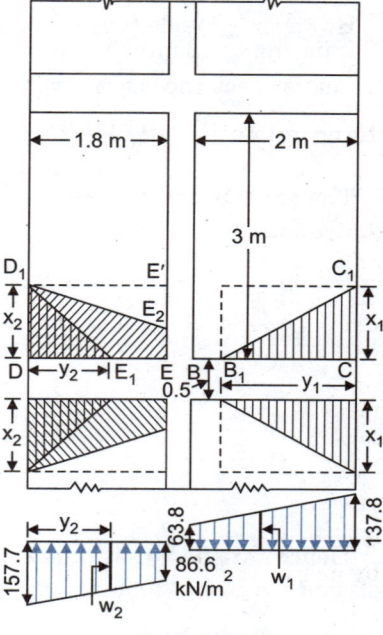

Fig. 18.31

Hence provide 8 m φ 41d stirrups @ 150 mm c/c on either side of each counterfort.

**5. Design of Toe slab.** Since the toe slab is also large, provide counterforts over it, upto ground level at 3 m clear distance face to face. The toe slab will thus bend like a continuous slab. Assume total depth of toe slab = 500 mm.

∴ Downward weight per unit area, due to self weight = $0.5 \times 1 \times 1 \times 25 = 12.5$ kN/m².

Upward pressure intensity at $D = 170.2$ kN/m²

∴ Net upward pressure intensity at $D = 170.2 - 12.5 = 157.7$ kN/m².

Similarly, net upward pressure intensity at $E = 99.1 - 12.5 = 86.6$ kN/m²

Consider strip of unit width at D. Max negative B.M. is

$$M_1 = \dfrac{wl^2}{12} = \dfrac{157.7 (3)^2}{12} = 118.3 \text{ kN-m} = 118.3 \times 10^6 \text{ N-mm}$$

$$d = \sqrt{\dfrac{118.3 \times 10^6}{0.914 \times 1000}} = 360 \text{ mm};$$

Shear force  $V = \dfrac{157.7 \times 3}{2} = 236.6$ kN.

Taking a permissible stress $\tau_c = 0.3$ N/mm² for $p = 0.5\%$ (Table 3.1) the depth of slab required from shear point of view is given by

$$d = \dfrac{V}{b \cdot \tau_c} = \dfrac{236.6 \times 1000}{1000 \times 0.3} = 789 \text{ mm}$$

**478** REINFORCED CONCRETE STRUCTURE

This is excessive. However, we will keep the same depth as that of heel, *i.e.*, $d = 440$ mm and $D = 500$ mm, and provide shear stirrups to take up excessive shearing stress.

Area of steel at supports, at bottom is $A_{st1} = \dfrac{118.3 \times 10^6}{230 \times 0.904 \times 440} = 1293.11 \cong 1294$ mm$^2$

Using 12 mm Φ bars, $A_\Phi = 113$ mm$^2$, $s = \dfrac{1000 \times 113}{1294} = 87.4$ mm

However, provide these @ 80 mm c/c. Actual $A_{st1} = \dfrac{1000 \times 113}{80} = 1412.5$ mm$^2$.

Let us check these from point of view of development length, at the point of contraflexure, so as to satisfy the criterion: $\dfrac{M}{V} + L_0 \geq L_d$.

The point of contraflexure occurs at a distance $x = 0.63$ m from the support. Hence S.F. at the point of contraflexure is

$$V = w\left(\dfrac{l}{2} - x\right) = 157.7\left(\dfrac{3}{2} - 0.63\right) = 137.2 \text{ kN}$$

$$M = 230 \times 1412.5 \times 0.904 \times 440 = 1.29 \times 10^8 \text{ N-mm}$$

$$L_0 = 12\Phi \text{ or } d, \text{ whichever is more} = 440 \text{ mm}$$

$$L_d = 45 \times 12 = 540 \text{ mm}$$

$$\therefore \quad \dfrac{M}{V} + L_0 = \dfrac{1.29 \times 10^8}{137.2 \times 10^3} + 440 = 1380 \text{ mm} > L_d$$

Hence satisfied. Continue these bars, at the bottom of toe slab, beyond the point of contraflexure by a distance of $L_0$ (= 440 mm) *i.e.*, by a distance of 630 + 440 = 1070 mm from the face of the counterforts.

Again, positive B.M. $= \dfrac{3}{4}M_1 = \dfrac{3}{4} \times 118.3 \times 10^6 = 88.72 \times 10^6$ N-mm

Area of bottom steel $= \dfrac{3}{4}A_{st1} = \dfrac{3}{4} \times 1294 = 971$ mm$^2$

Spacing of 12 mm Φ bars $= \dfrac{1000 \times 113}{971} = 116$ mm

Hence provide these @ 110 mm c/c. Actual area of steel $= \dfrac{1000 \times 113}{110} = 1027$ mm$^2$. This steel should satisfy the following criterion at the point of contraflexure

$$\dfrac{M}{V} + L_0 \geq L_d \quad \text{where } V = \text{S.F. at the point of contraflexure} = 137.2 \text{ kN}$$

Assuming that all the bars, provided at the top face, are available at the point of contraflexure.

$$M = 230 \times 1027 \times 0.904 \times 440 = 93.98 \times 10^6 \text{ N-mm}$$

$$L_0 = 12\Phi \text{ or } d, \text{ whichever is more} = 440 \text{ mm}.$$

$$L_d = 45\Phi = 45 \times 12 = 540 \text{ mm}$$

$$\therefore \quad \dfrac{M}{V} + L_0 = \dfrac{93.98 \times 10^6}{137.2 \times 10^3} + 440 = 1124.98 \text{ mm}$$

This is greater than $L_d$. Hence the criterion is satisfied. However, these bars are to be continued for a distance of $L_0$ (= 440 mm) beyond the point of contraflexure, their distance from the face of counterforts being only = 630 − 440 = 190 mm. Hence it is better to continue these bars upto counterfort, at the top face of the toe slab.

**Reinforcement at E.** At a section distant 1 m from $E$, upward soil pressure

$$= 170.2 - \frac{170.2 - 8.3}{4.1} \times 0.8 = 138.6 \text{ kN/m}^2.$$

∴ Net upward pressure = 138.6 – 12.5 = 126.1 kN/m².
This is  126.1/157.7 ≈ 0.80 of $w$ at $D$.
∴ Spacing of bottom steel = 87.4/0.80 = 109 ≈ 100 mm c/c
Spacing of top steel = 116/0.80 = 145 ≈ 140 mm c/c

**Distribution steel.** $A_{sd} = \dfrac{0.12}{100}(1000 \times 500) = 600 \text{ mm}^2$

Using 12 mm Φ bars, $A_\Phi \approx 113 \text{ mm}^2$. ∴ Spacing = $\dfrac{1000 \times 113}{600} = 188.3$ mm

Provide 12 mm ϕ @ 180 mm c/c.

**Shear reinforcement.** Shear force at $D$ = 236.6 kN

∴ $\tau_v = \dfrac{236.6 \times 1000}{1000 \times 440} = 0.54 \text{ N/mm}^2$

$\dfrac{100 A_s}{bd} = \dfrac{100 \times 1412.5}{1000 \times 440} \approx 0.32\%$. Hence from Table 3.1, $\tau_c \approx 0.24 \text{ N/mm}^2$,

Since $\tau_v > \tau_c$, shear reinforcement is necessary.
Now $V_c = \tau_c \cdot bd = 0.24 \times 1000 \times 440 = 105600$ N = 105.6 kN

Consider a section distant $x_2$ from the face of the counterfort (Fig. 18.31) where S.F. is 127.6 kN. Position $x_2$ is given by

$$\frac{105.6}{236.6} = \frac{1.5 - x_2}{1.5}, \text{ which gives } x_2 = 0.83 \text{ m}.$$

Hence shear stirrups are required upto a distance of 0.83 m on either side of counterforts. This requirement is there for a strip of unit width passing through $D$. Let us consider a strip through $E_1$, distant $y_2$ from $D$, such that shear force at the counterforts is 105.6 kN. To find the position of $y_2$, consider the net pressure distribution below the toe slab.

Self weight of toe slab = 12.5 kN/m². Hence net pressure intensity below $D$ and $E$ are respectively 170.2 – 12.5 = 157.7 kN/m² and 99.1 – 12.5 = 86.6 kN/m² respectively (Fig. 18.31). Let the net pressure intensity at $E_1$ be $w_2$ kN/m².

∴ S.F. at the counterforts at $E_1 = w_2 \times \dfrac{3}{2} = 1.5 w_2$ kN. This should be equal to 127.6 kN.

∴ $w_2 = 105.6/1.5 = 70.4 \text{ kN/m}^2$  ...(1)

However, at $y_2$ from $D$, $w_2 = 157.7 - \dfrac{157.7 - 86.6}{1.8} \cdot y_2 = 157.7 - 39.5 y_2$  ...(2)

Equating the two, we get $157.7 - 39.5 y_2 = 70.4$
From which $y_2 = 2.21$ m. This is more than $DE$ (= 1.8 m).
Hence shear force at $E$ is more than 105.6 kN.
Actual S.F. at $E$ = 1.5 × 86.8 = 130.2 kN.

Consider a section distant $z$ from the face of counterforts (Point $E$), where S.F. is 105.6 kN. The position of $z$ is given by

$$\frac{105.6}{130.2} = \frac{1.5 - z}{1.5}, \text{ which gives } z = 0.28 \text{ m}.$$

Hence shear stirrups are to be provided for a region $DEE_2D_1$, where $EE_2 = 0.28$ m only. However, we will provide stirrups for whole of rectangular area (shown dotted), for width $DD_1 = x_2 = 0.83$ m and length $DE = 1.8$ m.

Let us provide 8 legged stirrups, of 8 mm Φ wire.

$$A_{sv} = 8 \frac{\pi}{4}(8)^2 = 402 \text{ mm}^2$$

$$\therefore \quad \text{Spacing} = \frac{A_{sv} \cdot \sigma_{sv} \, d}{V - V_c} = \frac{402 \times 230 \times 440}{(236.6 - 105.6)1000} \approx 300 \text{ mm}.$$

**6. Design of stem (vertical slab).** The stem acts as a continuous slab. Consider 1 m strip at B. The intensity of earth pressure is given by

$$p_h = K_a \gamma H_1 = \frac{1}{3} \times 18 \times 7.5 = 45 \text{ kN/m}^2$$

(where revised value of $H_1 = 8 - 0.5 = 7.5$ m)

$\therefore$ Negative B.M. in slab near counterforts is

$$M_1 = \frac{p_h \cdot l^2}{12} = \frac{45(3)^2}{12} = 33.75 \text{ kN-m} = 33.75 \times 10^6 \text{ N-mm}$$

$$\therefore \quad d = \sqrt{\frac{33.75 \times 10^6}{1000 \times 0.914}} = 192.16 \cong 193 \text{ mm}.$$

Providing effective cover = 60 mm, total depth = 193 + 60 = 253 mm. However, provide total depth = 300 mm so that available $d = 300 - 60 = 240$ mm. This increased thickness will keep the shear stress within limits so that additional shear reinforcement is not required.

Shear force $V = \dfrac{45 \times 3}{2} = 67.5$ kN. $\qquad \therefore \quad \tau_v = \dfrac{67.5 \times 1000}{1000 \times 240} = 0.28$ N/mm$^2$

This is less than $\tau_c = 0.3$ N/mm$^2$ at 0.5% reinforcement (see Table 3.1)

Area of steel near counterforts is $A_{st1} = \dfrac{33.75 \times 10^6}{230 \times 0.904 \times 240} = 676.34 \cong 677$ mm$^2$

Reinforcement corresponding to $p = 0.5\%$ is $= \dfrac{pbd}{100} = \dfrac{0.5 \times 1000 \times 240}{100} = 1200$ mm$^3$

$\therefore \quad$ Spacing of 12 mm φ bars $= \dfrac{1000 \times 113}{1200} = 94$ mm

Hence provide 12 mm Φ bars @ 90 mm c/c.

Actual $A_{st} = \dfrac{1000 \times 113}{90} \cong 1255$ mm$^2$ $\qquad$ and $\qquad \dfrac{100 A_s}{bd} = \dfrac{100 \times 1255}{1000 \times 240} = 0.52\%$.

Let us check these bars for development length, near points of contraflexure, so as to satisfy the criterion: $\dfrac{M}{V} + L_0 \geq L_d$.

For a fixed beam or slab carrying U.D.L., the point of contraflexure is at a distance of $0.211 L = 0.63$ m from the face of counterforts, S.F. at this point is given by

$$V = \frac{pL}{2}\left(\frac{l}{2} - x\right) \div \left(\frac{L}{2}\right) = p\left(\frac{l}{2} - x\right) = 67.5\left(\frac{3}{2} - 0.63\right) = 58.73 \text{ kN}$$

Assuming that all the bars will be available at the point of contraflexure

$$M = \sigma_{st} \cdot A_{st} j_c d = 230 \times 1255 \times 0.904 \times 240 = 62.62 \times 10^6 \text{ N-mm}$$

$$L_0 = 12\Phi \text{ or } d, \text{ whichever is more} = 240 \text{ mm}$$

$$L_d \cong 45 \times 12 \approx 540 \text{ mm}$$

$$\therefore \quad \frac{M}{V} + L_0 = \frac{62.62 \times 10^6}{58.73 \times 1000} + 240 \approx 1306 \text{ mm}$$

Hence safe. It is thus essential to continue all the bars upto a point distant 240 mm beyond point of contraflexure, *i.e.* upto a point 630 + 240 = 870 mm ≈ 900 mm from the face of counterforts. These bars are to be provided at the inner face of the stem slab.

$$\text{Maximum positive B.M.} = \frac{3}{4} M_1 = \frac{3}{4} \times 33.75 \times 10^6 = 25.31 \times 10^6 \text{ N-mm}$$

$$\text{Area of steel} = \frac{3}{4} A_{st1} = \frac{3}{4} \times 1255 = 941.25 \triangleq 942 \text{ mm}^2$$

and spacing of 12 mm $\Phi$ bars $= \dfrac{1000 \times 113}{942} = 120$ mm. Hence, provide these @ 120 mm c/c.

Actual $A_{st} = \dfrac{1000 \times 113}{120} = 942$ mm$^2$.

Let us check these bars for development length at the point of contraflexure, so as to satisfy the criterion:
$\dfrac{M}{V} + L_0 \geq L_d$

Assuming that all the reinforcement is extended upto point of contraflexure

$$M = 230 \times 942 \times 0.904 \times 240 = 47 \times 10^6 \text{ N-mm}$$

$$L_0 = 12\Phi \quad \text{or} \quad d, \text{ whichever is more} = 240 \text{ mm}$$

$$L_d = 45 \times 12 = 540 \text{ mm}$$

$$V = 58.53 \text{ kN, as before,}$$

$$\therefore \quad \frac{M}{V} + L_0 = \frac{47 \times 10^6}{58.53 \times 10^3} + 240 = 1043 \text{ mm} > L_d. \quad \text{Hence safe}$$

The spacing of the reinforcement at $B$, found above can be increased with height. The pressure $p_h$ and hence the bending moment decreases linearly with height.

$$\therefore \quad A_{st} \propto h$$

Hence the spacing of the bars can be increased gradually to say 300 mm c/c near the top.

$$\text{Distribution steel} = \frac{0.12}{100} (1000 \times 300) = 360 \text{ mm}^2.$$

Using 10 mm $\Phi$ bars, $A_\Phi = 78.5$ mm$^2$.

$$\therefore \quad \text{Spacing} = \frac{1000 \times 78.5}{360} = 218 \text{ mm.} \quad \text{Hence provide these @ 200 mm c/c.}$$

**7. Design of main counterfort.** Let us assume thickness of counterfort = 500 mm. The counterforts will thus be spaced @ 350 cm c/c. They will thus receive earth pressure from a width of 3.5 m and downward reaction from the heel slab for a width of 3.5 m.

At any section at depth $h$ below the top $A$, the earth pressure acting on each counterfort will be

$$= \frac{1}{3} \times 18 \times h \times 3.5 = 21 h \text{ kN/m} \qquad ...(1)$$

Similarly, net downward pressure on heel at $C$ is

$$= 7.5 \times 18 + 0.5 \times 25 - 8.3 = 139.2 \text{ kN/m}^2$$

and that at $B$ is

$$= 7.5 \times 18 + 0.5 \times 25 - 87.3 = 60.2 \text{ kN/m}^2$$

Hence reaction transferred to each counterfort will be

At $C$, 139.2 × 3.5 = 487.2 kN/m. At $B$, 60.2 × 3.5 = 210.7 kN/m. The variations of horizontal and vertical forces on the counterfort are shown in Fig. 18.32.

The critical section for the counterfort will be at $F$, since below this, enormous depth will be available to resist bending.

Pressure intensity at $h = 7$ m is $= 21 \times 7 = 147$ kN/m

Shear force at $F = \dfrac{1}{2} \times 147 \times 7 = 514.5$ kN

B.M. $M = 514.5 \times \dfrac{7}{3} = 1200.5$ kN-m $= 1200.5 \times 10^6$ N-mm.

The counterfort acts as a T-beam. However, even as a rectangular beam, depth required is

$$d = \sqrt{\dfrac{1200.5 \times 10^6}{500 \times 0.914}} \approx 1621 \text{ mm}$$

∴ Total depth $= 1621 + 60 = 1681$ mm. However, keep $D = 1720$ mm so that $d = 1720 - 60 = 1660$ mm. Angle $\theta$ of face $AC$ is given by

$\tan \theta = 2/7.5 = 0.267$. ∴ $\theta = 14°\,56'$ ; $\sin \theta = 0.2577$ ; $\cos \theta = 0.9662$

Depth $F_1 C_1 = AF_1 \sin \theta = 7 \times 0.2577 = 1.8$ m $= 1800$ mm

∴ Depth $FG = 1800 + 300 = 2100$ mm. This is much more than required.

Assuming that steel reinforcement is provided in two layers with 20 mm space between them and providing a nominal cover of 50 mm and 20 mm Φ main bars, the effective depth will to be equal to $2100 - (50 + 12 + 20 + 10) = 2008$ mm.

∴ $A_{st} = \dfrac{1200.5 \times 10^6}{230 \times 0.9 \times 2008} = 2888.2 \triangleq 2889$ mm².

Using 20 mm Φ bars,

$A_\Phi = 314$ mm²

No. of bars $= 2889/314 \approx 10$. Provide these in two layers.

Effective S.F. $= Q - \dfrac{M}{d'} \tan \theta$,

where $d' = \dfrac{d}{\cos \theta} = \dfrac{2008}{0.9662} = 2078$ mm

∴ Effective S.F. $= Q - \dfrac{M}{d'} \tan \theta$

$= 514.5 \times 10^3 - \dfrac{1200.5 \times 10^6}{2078} \times 0.267$

$= 360249.04 \triangleq 360250$ N

∴ $\tau_v = \dfrac{360250}{500 \times 2078} = 0.346$ N/mm²

$100 \dfrac{A_s}{bd} = \dfrac{100\,(10 \times 314)}{500 \times 2078} = 0.3\%$.

∴ $\tau_c \approx 0.236$ N/mm².

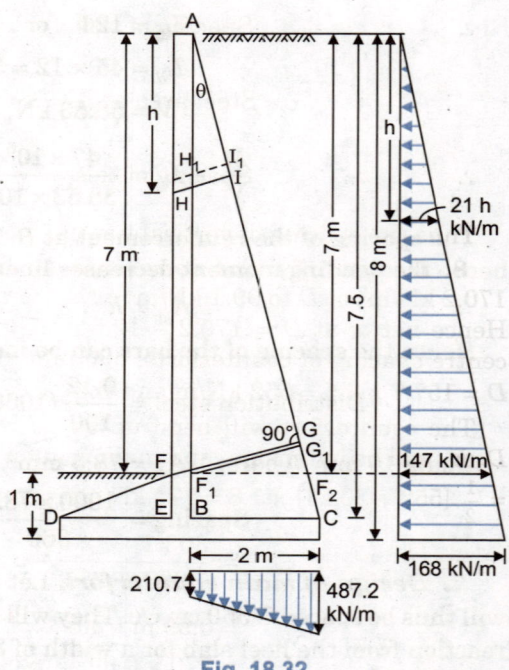

Fig. 18.32

Thus the shear stress $\tau_v$ is more than permissible shear stress $\tau_c$. However, the vertical and horizontal ties provided in the counterforts will bear the excess shear stress.

The height $h$ where half of the reinforcement can be curtailed will be equal to $\sqrt{H} = \sqrt{8} \approx 2.8$ m below $A$, i.e. at point $H$. To locate the position of point of curtailment on $AC$, draw $HI$ parallel to $FG$. Thus, half the bars can be curtailed at $I$. However, these should be extended by a distance 12 Φ $= 240$ mm beyond $I$, i.e. extended upto $I_1$. The location of $H$, corresponding to $I_1$ can be located by drawing line $I_1 H_1$ parallel to $FG$. It should be noted that $I_1 G$ should not be less than 45 Φ $= 900$ mm. Similarly, other bars can be curtailed, if desired.

**Design of horizontal ties.** The vertical stem slab has a tendency to separate out from the counterfort, and hence should be tied to it by horizontal ties. At any depth $h$ below the top, force causing separation $\dfrac{1}{3} \times 18\,h \times 3 = 18\,h$ kN/m.

At $h = 7$ m, force = $18 \times 7 = 126$ kN/m.

∴ Steel area required = $\dfrac{126 \times 1000}{230} = 548$ mm² per metre height.

Using 10 mm Φ 2 legged ties, $A_\phi = 2\dfrac{\pi}{4}(10)^2 = 157$ mm²

∴ Spacing = $\dfrac{1000 \times 157}{548} \approx 286$ mm c/c. However, provide these at 250 mm c/c.

This spacing can be increased gradually to 300 mm c/c towards top.

***Design of vertical ties.*** Similar to the stem slab, heel slab has also tendency to separate out from the counterfort, due to net downward force, unless tied properly by vertical ties. The downward force at $C$ will be $487.2 \times \dfrac{3}{3.5} = 417.6$ kN/m and $210.7 \times \dfrac{3}{3.5} = 180.6$ kN/m at $B$ (see Fig. 18.32). Near end $C$, the heel slab is tied to the counterforts with the help of main reinforcement of counterforts.

∴ Steel area at $C = \dfrac{417.6 \times 1000}{230} = 1816$ mm²

Using 12 mm Φ 2 legged ties, $A_\phi = 2 \times \dfrac{\pi}{4}(12)^2 = 226.2$ mm²

∴ Spacing of ties = $\dfrac{1000 \times 226.2}{1816} = 124.55$ mm ≈ 120 mm c/c.

Steel area at $B = \dfrac{180.6 \times 1000}{230} = 786$ mm²

∴ Spacing of ties = $\dfrac{1000 \times 226.2}{786} = 287.78 \approx 280$ mm c/c.

Thus the spacing of vertical ties can be increased gradually from 120 mm c/c at $C$ to 280 mm c/c at $B$.

**8. Design of front counterforts.** Refer Fig. 18.30. The upward pressure intensity varies from 170.2 kN/m² at $D$ to 99.1 kN/m² at $E$. Downward weight of 500 mm thick toe slab = $0.5 \times 25 = 12.5$ kN/m². Hence net $w$ at $D = 170.2 - 12.5 = 157.7$ kN/m² and at $E = 99.1 - 12.5 = 86.6$ kN/m². The centre to centre spacing of counterforts, 500 mm wide is 3.5 m. Hence upward force transmitted to counterfort at $D = 157.7 \times 3.5 = 552$ kN/m and that at $E = 86.6 \times 3.5 = 303.1$ kN/m.

The counterfort will bend up as cantilever about face $FE$. Hence $DE$ will be in compression while $D_1 E_1$ will be in tension, and main reinforcement will be provided at bottom face $D_1 E_1$. Total upward force = $\dfrac{1}{2}[552 + 303.1] \times 1.8 = 770$ kN acting at $\bar{x}$

$$= \dfrac{303.1 + 2 \times 552}{303.1 + 552} \times \dfrac{1.8}{3}$$

= 0.99 m from $E$

∴ B.M. = $770 \times 0.99 = 762.3$ kN-m

= $762.3 \times 10^6$ N-mm

∴ $d = \sqrt{\dfrac{762.3 \times 10^6}{500 \times 0.914}} = 1291.5$ mm

Hence provide a total depth of 1400 mm, so that with an effective 80 mm, available $d = 1400 - 80 = 1320$ mm.

Thus, project the counterfort 400 mm above ground level, to point $F_1$ as shown.

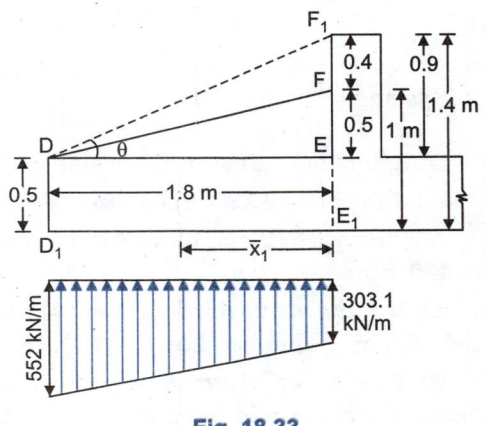

Fig. 18.33

$A_{st} = \dfrac{760 \times 10^6}{230 \times 0.904 \times 1320} = 2769.13 \approx 2770$ mm². Using 25 mm Φ bars, $A_\phi = 491$ mm².

∴ No. of bars = $2770/491 = 5.64 \approx 6$. Provide these in one layer.

These bars should be continued by a distance of 45 Φ = 45 × 25 ≈ 1125 mm beyond $E$.

$$\text{Net shear force} = F - \frac{M}{d} \tan \theta. \quad \text{From Fig. 18.33, } \tan \theta = \frac{0.9}{1.8}$$

$$\therefore \quad V = 770 - \frac{760}{1.3} \times \frac{0.9}{1.8} = 477.7 \text{ kN} \quad \text{and} \quad \tau_v = \frac{477.7 \times 1000}{500 \times 1320} = 0.724 \text{ N/mm}^2.$$

$$100 \frac{A_s}{bd} = \frac{100 (6 \times 491)}{500 \times 1320} = 0.45\%$$

Hence from Table, 3.1 $\tau_c$ = 0.285 N/mm². Since $\tau_v > \tau_c$, shear reinforcement is required.

Using 12 mm Φ 2-legged stirrups, $A_{sv} = 2 \times \frac{\pi}{4} (12)^2 = 226 \text{ mm}^2.$

$$V_c = \tau_c . bd = 0.285 \times 500 \times 1320 = 188100 \text{ N}$$

$$\therefore \quad V_s = V - V_c = 477.7 \times 10^3 - 188.1 \times 10^3 = 289.6 \times 10^3 \text{ N}$$

$$\therefore \quad s_v = \frac{\sigma_{sv} . A_{sv} . d}{V_s} = \frac{230 \times 226 \times 1320}{289.6 \times 10^3} = 236.92 \simeq 236 \text{ mm, subject to a maximum of 300 mm.}$$

However, provide these @ 230 mm c/c. Provide 2-16 mm Φ holding bars at the top.

**9. Fixing effects in stem, toe and heel slab.** At the junction of stem, toe and heel slabs fixing moments are induced, which are at right angles to their normal direction of bending. These moments are not determinate, but normal reinforcement given below may be provided:

(i) In stem @ 0.8 × 0.3 = 0.24% of cross-section, to be provided at the inner face, in vertical direction, for a length = 45 Φ.

$$\therefore \quad A_{st} = \frac{0.24}{100} \times 1000 \times 300 = 720 \text{ mm}^2$$

Using 10 mm Φ bars, $A_\Phi$ = 78.5 mm²

$$\therefore \quad \text{Spacing} = \frac{1000 \times 78.5}{720} \approx 109.02 \simeq 100 \text{ mm c/c}.$$

Length of embedment in stem, above heel slab ≈ 450 mm.

(ii) In toe slab @ 0.12% to be provided at the lower face

$$A_{st} = \frac{0.12}{100} (1000 \times 500) = 600 \text{ mm}^2$$

Spacing of 10 mm Φ bars = $\frac{1000 \times 78.5}{600} \approx 130.83 \simeq 125$ m c/c.

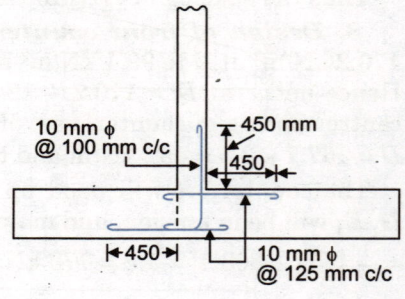

Fig. 18.34

Length of embedment in toe slab = 450 mm.

(iii) In heel, @ 0.12% to be provided at the upper face.

∴ Provide 10 mm bars @ 125 c/c to be embeded by a distance of 450 mm.

Each of the above reinforcement should be anchored properly in the adjoining slab, as shown in Fig. 18.34.

**10. Design of shear key.** The wall is unsafe in sliding, and hence shear key will have to be provided, as shown in Fig. 18.35. Let the depth of key = $a$.

Intensity of passive pressure $p_p$ developed in front of the key depends upon the soil pressure $p$ in front of key.

$$p_p = K_p \, p = 3 \times 99.1 = 297.3 \text{ kN/m}^2$$

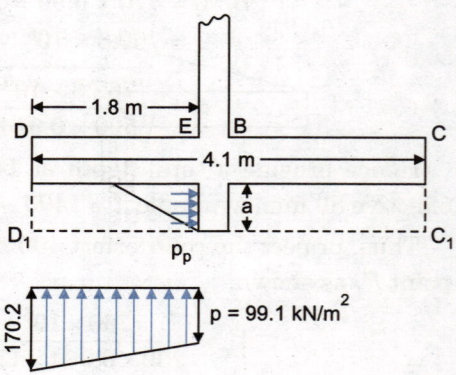

Fig. 18.35

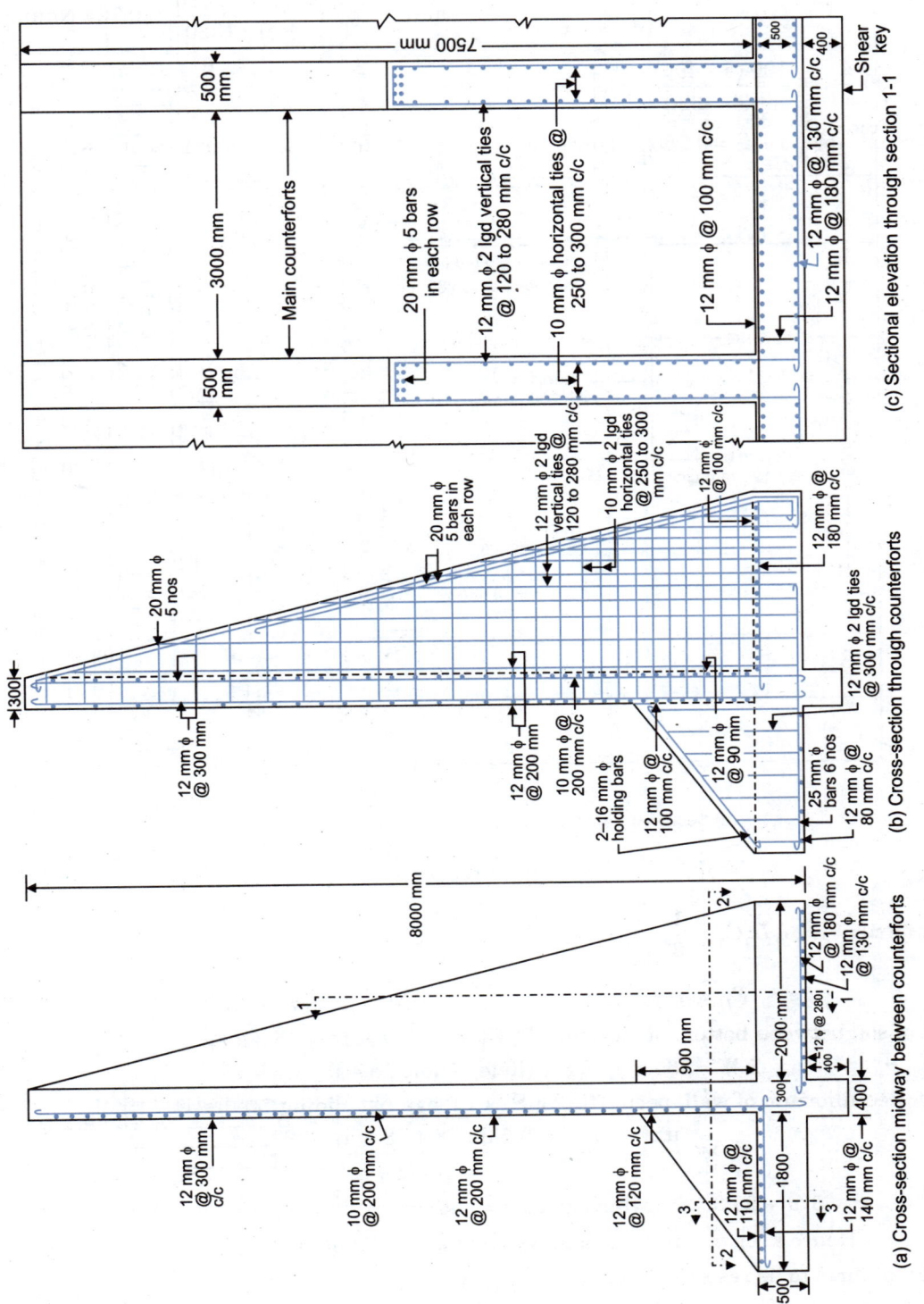

Fig. 18.36. Counterfort retaining wall reinforcement details.

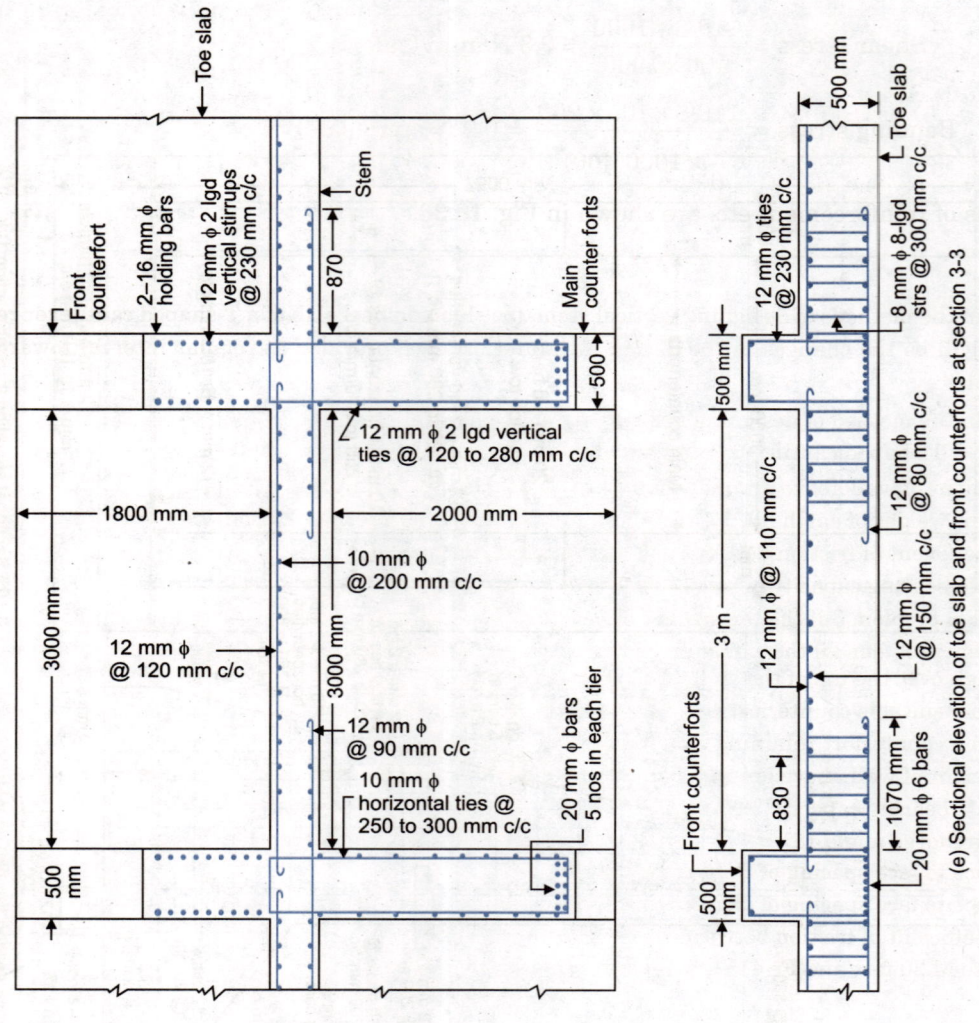

(d) Sectional plan through 2-2

**Fig. 18.36.** (Contd.)

Sliding force at level $D_1C_1 = \frac{1}{3} \times \frac{18}{2}(8+a)^2$

or $\qquad P_H = 3(8+a)^2$ ...(2)

Weight of soil between bottom of base and $D_1C_1 \approx 4.1 \times a \times 18 = 73.8\,a$ kg ...(3)

$\therefore \qquad \Sigma W = 365.8 + 73.8\,a$ (Refer Table 18.4a)

Hence for equilibrium of wall, permitting F.S. = 1.5 against sliding, we have

$$1.5 = \frac{\mu\Sigma W + P_P}{P_H} = \frac{0.5(365.8 + 73.8\,a) + 297.3\,a}{3(8+a)^2}$$

or $\qquad a^2 - 57a + 23.3 = 0 \quad$ which gives $a = 0.40$ m

Hence keep $a = 400$ mm and width of key = 400 mm.

Actual force to be resisted by key = $1.5\,P_H - \mu\Sigma W$

$= 1.5 \times 3(8+0.4)^2 - 0.5(365.8 + 73.8 \times 0.4) = 119.8$ kN

$$\therefore \quad \text{Shear stress} = \frac{119.8 \times 1000}{400 \times 1000} \approx 0.3 \text{ N/mm}^2 \text{ (safe)}$$

$$\text{Bending stress} = \frac{119.8 \times 10^3 \times 200}{\frac{1}{6} \times 1000(400)^2} = 0.9 \text{ N/mm}^2 \text{ (safe)}$$

The details of reinforcement etc. are shown in Fig. 18.36.

## PROBLEMS

1. Explain the methods of designing vertical stem, toe slab and heel slab of a T-shaped cantilever retaining wall. What will be the changes in the design if counterforts are provided at regular interval towards the side of backfill ?

2. Explain the method of designing a shear key for a retaining wall.

3. Design a T-shaped cantilever retaining wall for the following data:
   (i) Height of wall above ground, 4.5 m
   (ii) Depth of foundation, 1.5 m
   (iii) Unit weight of earth fill, 17 kN/m²
   (iv) Angle of internal friction, 20°
   (v) Coefficient of friction between soil and concrete, 0.45
   (vi) Safe bearing capacity of soil 130 kN/m². Use M 20 mix and Fe 415 steel.

4. Redesign problem 3 if the top surface of backfill is inclined at an angle of 15° with horizontal.

5. Redesign problem 3 if the embankment carries a load of 20 kN/m² over the backfill. Also provide a R.C. parapet 1 m high over the top of the stem of the retaining wall.

6. Design a suitable counterfort retaining wall for the data of problem 3.

7. Design a counterfort retaining wall for the following data:
   (i) Height of wall above ground, 8 m.
   (ii) Depth of foundation, 1.5 m
   (iii) Safe bearing capacity, 200 kN/m²
   (iv) Unit weight of earthfill, 18 kN/m²
   (v) Surcharge angle 18°
   (vi) Angle of internal friction for backfill, 30°
   (vii) Face to face spacing of front counterforts, 2 m.
   (viii) Face to face spacing of front counterforts, provided upon ground level, 2 m.
   (ix) Coefficient of friction between soil and concrete, 0.55.
   Use M 20 mix and Fe 415 steel reinforcement.

# PART – II
# WATER TANKS

19. DOMES
20. BEAMS CURVED IN PLAN
21. WATER TANKS – I : SIMPLE CASES
22. WATER TANKS – II : CIRCULAR AND INTZE TANKS
23. WATER TANKS – III : RECTANGULAR TANKS
24. WATER TANKS – IV : UNDERGROUND TANKS

# CHAPTER 19
# DOMES

## 19.1. INTRODUCTION

A dome may be defined as a thin shell generated by the revolution of a regular curve about one of its axes. The shape of the dome depends upon the type of the curve and the direction of the axis of revolution. When the segment of a circular curve revolves about its vertical diameter, a spherical dome is obtained. Similarly, conical dome is obtained by the revolution of a right angled triangle about its vertical axis, while an elliptical dome is obtained by the revolution of an elliptical curve about one of its axes. However, out of these, spherical domes are more commonly used. In the case of a spherical dome the vertical section through the axis of revolution in any direction is an arc of a circle.

Domes are used in variety of structures, such as (i) roof of circular areas (ii) circular tanks (iii) hangers, (iv) exhibition halls, auditoriums and planitoriums and (v) bottoms of tanks, bins and bunkers. Domes may be constructed of masonry, steel, timber and reinforced cement concrete. Stone and brick domes are one of the oldest architectural forms. However, reinforced concrete domes are more common now-a-days, since they can be constructed over large spans.

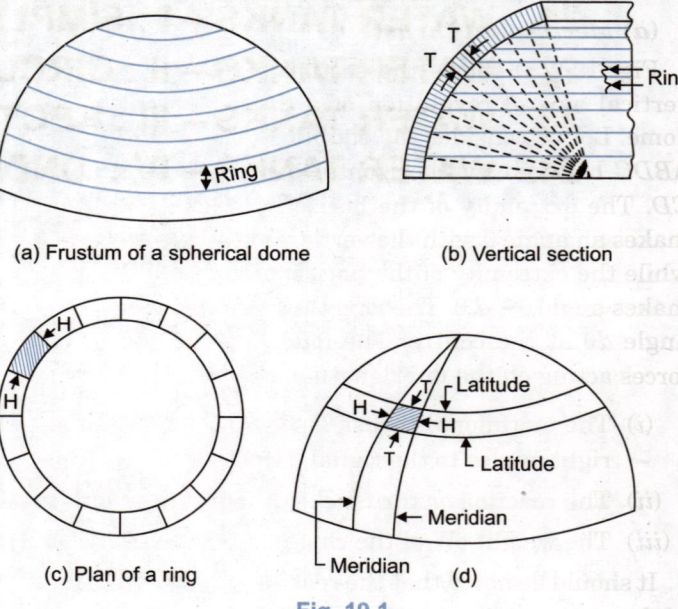

(a) Frustum of a spherical dome

(b) Vertical section

(c) Plan of a ring

(d)

Fig. 19.1

## 19.2. NATURE OF STRESSES IN SPHERICAL DOMES

A spherical dome may be imagined to consist of a number of horizontal rings placed one over the other. The diameters of the successive rings increase in the downward direction and the equilibrium is maintained independently of the rings above it. The circle of each ring is called 'latitude' while the circle drawn through two diametrically opposite points on a horizontal diameter and the crown is known as a *meridian circle*. All meridian circles converge at the crown (or top most point) of the spherical dome.

Fig. 19.1 (b) shows the vertical section of a spherical dome. The successive horizontal rings subtend equal angle at the centre of the sphere. The joints between successive horizontal rings is *radial*. Every horizontal ring supports the load of the rings above it, and transmits it to the one below it. The reaction between the rings is tangential to the curved surface, giving rise to compression along the medians. The compressive stress is called *meridional thrust or meridional compression.*

Fig. 19.1 (c) shows the plan of a horizontal ring, which may be imagined to consist of a number of *voussoirs*. The joints between adjacent voussoirs of the ring are radial. The tendency of separation of any voussoir will be prevented because of its wedge shape, and therefore, *hoop compression* will be caused in each ring.

To summarise, therefore, two types of stresses are induced in a dome [Fig. 19.1(d)].

(i) Meridional thrust ($T$) along the direction of meridian.

(ii) Hoop stress ($H$) along the latitudes.

## 19.3. ANALYSIS OF SPHERICAL DOMES

Let us now analyse stresses developed in a spherical dome of *uniform thickness*. Two cases of loading will be considered:

(i) uniformly distributed load,  (ii) concentrated load at the crown.

### 1. Uniformly distributed load

Let    $w$ = uniformly distributed load, inclusive of its own weight per unit area.

   $r$ = radius of the dome;    $t$ = thickness of dome shell

   $T$ = intensity of meridional thrust;    $H$ = intensity of hoop stress.

**(a) Meridional thrust**

Fig. 19.2(a) shows the section through the vertical axis of revolution of a thin spherical dome. Let us consider the equilibrium of a ring $ABDC$ between two horizontal planes $AB$ and $CD$. The extremity of the horizontal plane $AB$ makes an angle $\theta$ with the vertical at the centre while the extremity of the horizontal plane $CD$ makes angle $\theta + d\theta$. The ring thus subtends an angle $d\theta$ at the centre. The following are the forces acting on the unit length of the ring:

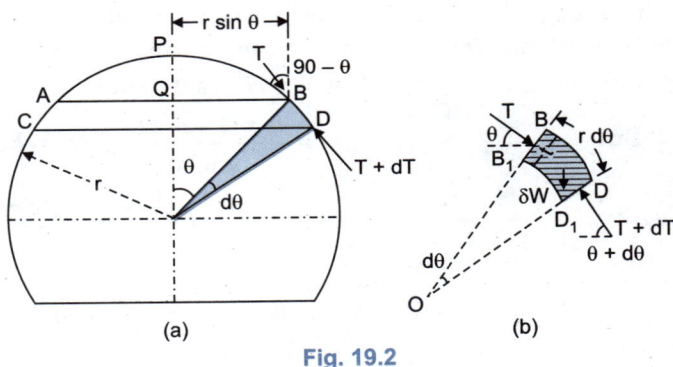

Fig. 19.2

(i) The meridional thrust $T$ per unit length of the circle of latitude $AB$, acting tangentially at $B$ (or at right angles to the radial line $OB$).

(ii) The reaction or thrust $T + dT$ per unit length of circle on latitude $CD$, acting tangentially at $D$.

(iii) The weight $\delta W$ of the ring itself, acting vertically down.

It should be noted that the reaction $T + dT$ will be greater than the thrust $T$ due to the effect of the weight of the ring and due to the change in the inclination from $\theta$ of $(\theta + d\theta)$ of the radial lines. The meridional thrust $T$ is caused due to the weight of the dome shell $APB$ above the horizontal plane $AB$.

Surface area of dome shell $APB = 2\pi r \times PQ$

But    $PQ = OP - OQ = r - r \cos \theta = r (1 - \cos \theta)$

∴ Weight of dome shell above $AB$

   $= 2\pi r \times PQ \times w = 2\pi r^2 w (1 - \cos \theta)$

Since the sum of vertical components of thrust $T$ acting along the circumference of the circle of latitude must be equal to the total weight of the dome shell $APB$, we have

$$T(2\pi \times QB) \sin \theta = 2\pi r^2 w(1 - \cos \theta)$$

or $\quad T \times 2\pi \times r \sin \theta \times \sin \theta = 2\pi r^2 w(1 - \cos \theta)$

or $\quad\quad\quad\quad T = \dfrac{w\, r(1 - \cos \theta)}{\sin^2 \theta} = \dfrac{w\, r}{1 + \cos \theta}$  ...(19.1)

This is thus the expression for the meridional thrust.

**(b) Hoop stresses.** We have seen that the meridional thrust $T$ increases to $T + dT$ at the bottom of the ring. This difference in the meridional thrust $T$ and $T + dT$ acting at $\theta$ and $\theta + d\theta$ respectively to the horizontal causes hoop stress.

Let $H$ be the hoop force per unit length of surface measured on a great circle arc.

Breadth of ring $= r d\theta$, $\quad \therefore$ Hoop force $= H \times r d\theta$ ...(i)

The horizontal component of $T$ is $T \cos \theta$, and this horizontal component cause hoop tension, tending to increase the diameter of the ring, while the horizontal component of $T + dT$ will be $(T + dT)(\cos \theta + d\theta)$ and this horizontal component will cause hoop compression.

Now magnitude of hoop tension.

$\quad\quad\quad\quad = T \cos \theta \times$ Radius of ring $AB$

$\quad\quad\quad\quad = T \cos \theta \; r \sin \theta = Tr \sin \theta \cos \theta$ ...(ii)

Magnitude of hoop compression

$\quad\quad\quad\quad = (T + dT) \cos (\theta + d\theta) \times$ radius of ring $CD$

$\quad\quad\quad\quad = (T + dT) \cos (\theta + d\theta)\, r \sin (\theta + d\theta)$ ...(iii)

The difference between (ii) and (iii) will cause the actual hoop stress. If (iii) is more than (ii), hoop stress will be compressive, while if (ii) is more than (iii), hoop stress will be tensile. Thus the hoop force, which is equal to the difference between (ii) and (iii), is due to the change in the value of $T$ when $\theta$ increases by a small amount $d\theta$. Hence in the limiting case when $d\theta$ is extremely small,

$$H r d\theta = d\, [T \cos \theta \; r \sin \theta];$$

But $\quad\quad\quad\quad T = \dfrac{w\, r\, (1 - \cos \theta)}{\sin^2 \theta} \quad$ from *Eq. 19.1*.

$\therefore \quad\quad\quad H = \dfrac{1}{r} \cdot \dfrac{d}{d\theta} \left[ r \sin \theta \cos \theta \; \dfrac{w\, r\, (1 - \cos \theta)}{\sin^2 \theta} \right]$

$\quad\quad\quad\quad = wr \dfrac{d}{d\theta} \left[ \dfrac{1 - \cos \theta}{\sin^2 \theta} \sin \theta \cos \theta \right] = wr \dfrac{d}{d\theta} \left[ \dfrac{\cos \theta}{\sin \theta} - \dfrac{\cos^2 \theta}{\sin \theta} \right]$

or $\quad\quad H = wr \left[ \dfrac{-\sin^2 \theta - \cos^2 \theta}{\sin^2 \theta} - \dfrac{-2\sin^2 \theta \cos \theta - \cos^3 \theta}{\sin^2 \theta} \right]$

$\quad\quad\quad\quad = wr \left[ \dfrac{-1 + 2\sin^2 \theta \cos \theta + \cos^3 \theta}{\sin^2 \theta} \right] = wr \left[ \dfrac{-1 + 2 \cos \theta - 2 \cos^3 \theta + \cos^3 \theta}{\sin^2 \theta} \right]$

$\quad\quad\quad\quad = wr \left[ \dfrac{-1 + 2 \cos \theta - \cos^3 \theta}{\sin^2 \theta} \right] = wr \left[ \dfrac{(1 - \cos \theta)(\cos^2 \theta + \cos \theta - 1)}{1 - \cos^2 \theta} \right]$

or $\quad\quad\quad\quad H = \dfrac{w\, r\, (\cos^2 \theta + \cos \theta - 1)}{1 + \cos \theta}.$ ...(19.2)

The above expression given the hoop stress in any horizontal ring the extremity of which subtends an angle $\theta$ with the vertical, at the centre. If the value of $H$ obtained from *Eq. 19.2* is positive, hoop force will be compressive; otherwise it will be tensile.

At the crown, $\theta = 0$. Hence from *Eq. 17.2* $\quad H = w\, r/2$

Intensity of *hoop stress at crown* $= H/t = w\, r/2\, t$ (compressive) ...(19.3)

This is the maximum value of hoop stress. The hoop stress will go on decreasing as θ increases, till $H$ becomes zero. After that $H$ becomes tensile. To find the position of the plane where hoop stress becomes zero, we have

$$H = 0 = \frac{w\,r\,(\cos^2\theta + \cos\theta - 1)}{1 + \cos\theta} \quad \text{or} \quad \cos^2\theta + \cos\theta - 1 = 0$$

or $\cos\theta = 0.618$. From which $\theta = 51°\,49'\,38''$

Hence round the circle of latitude at which the angle $\theta = 51°\,49'\,38''$, hoop stress is zero. For all portion of dome about this angle, hoop compression will be developed, while for the portion below this plane, hoop tension will be developed which will go on increasing further towards the base of the dome.

### 2. Concentrated load at the crown.

Let us now find out the stresses developed in the spherical dome due to a concentrated load $W$ situated at the crown $P$ as shown in Fig. 19.3.

As before, let us consider a ring $ABDC$ bounded by horizontal planes $AB$ and $CD$ subtending angles $\theta$ and $\theta + d\theta$ with the vertical axis. Let $T$ be the thrust per unit length of latitude $AB$ acting tangentially and $T + dT$ be thrust per unit length of the circle of latitude $CD$.

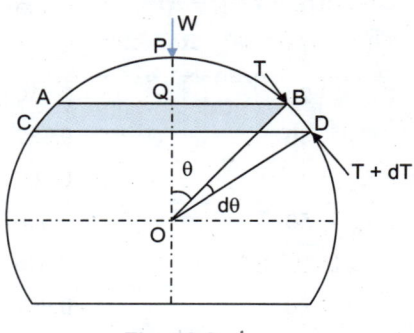

Fig. 19.3

The sum of the vertical components of thrust $T$ acting along the circumference of the circle of latitude must be equal to the load $W$; hence

$$T \times 2\pi r \sin\theta \times \sin\theta = W$$

or

$$T = \frac{W}{2\pi r \sin^2\theta} = \frac{W}{2\pi r}\operatorname{cosec}^2\theta \qquad \ldots(19.4)$$

As discussed in the previous case, the hoop stress developed in any horizontal ring is due to the difference in the meridional thrust $T$ and $T + dT$. Hence, as before,

$$H r\, d\theta = d[T \cos\theta \times r \sin\theta]$$

Substituting the value of $T$ from Fig. 19.4, we get

$$H = \frac{1}{r}\cdot\frac{d}{d\theta}\left[\frac{W}{2\pi r \sin^2\theta}\cos\theta\, r \sin\theta\right] = \frac{W}{2\pi r}\frac{d}{d\theta}\left[\frac{\cos\theta}{\sin\theta}\right]$$

Differentiating the above, we get

$$H = \frac{W}{2\pi r}\left[\frac{-\sin^2\theta - \cos^2\theta}{\sin^2\theta}\right] = -\frac{W}{2\pi r}\operatorname{cosec}^2\theta \qquad \ldots(19.5)$$

and hoop stress $= \dfrac{H}{t} = -\dfrac{W}{2\pi r\, t}\operatorname{cosec}^2\theta \qquad \ldots[19.5(a)]$

The minus signs shows that hoop stress will be tensile throughout, due to the concentrated load. At the crown, $\theta = 0$. Hence $H$ becomes infinite. Therefore, any concentrated load in the form of lantern or ornament etc., should always be distributed over sufficient area, to reduce the hoop stress at the crown. It is also desirable to thicken the dome at the top to spread the load over greater area.

**Stresses due to combined uniformly distributed load and point load.** The stresses due to point load applied at the crown should be added algebraically to the stresses developed due to uniformly distributed live load and self-weight of the dome, to get the final stresses.

Table 19.1 gives the coefficients for hoop stress and meridional stress due to uniformly distributed load and concentrated load at the crown, for different values of θ

**TABLE 19.1.** (+ For compression and − for tension)

| θ (°) | Uniformly distributed load w | | Concentrated load W at the crown | |
|---|---|---|---|---|
| | Coefficient of w.r. (for T) | Coefficient of w. r. (for H) | Coefficient of $\frac{W}{r}$ (for T) | Coefficient of $\frac{W}{r}$ (for H) |
| | $\frac{1-\cos\theta}{\sin^2\theta}$ | $\frac{\cos^2\theta+\cos\theta-1}{1+\cos\theta}$ | $\frac{\csc^2\theta}{2\pi}$ | $-\frac{\csc^2\theta}{2\pi}$ |
| (1) | (2) | (3) | (4) | (5) |
| 0 | + 0.5 | + 0.5 | ∞ | ∞ |
| 5 | + 0.500 | + 0.496 | + 21.0 | − 21.0 |
| 10 | + 0.505 | + 0.48 | + 5.3 | − 5.3 |
| 20 | + 0.516 | + 0.425 | + 1.37 | − 1.37 |
| 30 | + 0.537 | + 0.33 | + 0.64 | − 0.64 |
| 40 | + 0.566 | + 0.20 | + 0.38 | − 0.38 |
| 50 | + 0.608 | + 0.034 | + 0.27 | − 0.27 |
| 51° 49′ 38″ | + 0.618 | + 0.00 | + 0.26 | − 0.26 |
| 60 | + 0.667 | − 0.167 | + 0.21 | − 0.21 |
| 70 | + 0.747 | − 0.402 | + 0.18 | − 0.18 |
| 80 | + 0.838 | − 0.68 | + 0.16 | − 0.16 |
| 90 | + 1.00 | − 1.00 | + 0.16 | − 0.16 |
| 100 | + 1.21 | − 1.38 | + 0.16 | − 0.16 |
| 110 | + 1.52 | − 1.86 | + 0.18 | − 0.18 |
| 120 | + 2.00 | − 2.50 | + 0.21 | − 0.21 |
| 130 | + 2.79 | − 3.45 | + 0.27 | − 0.27 |
| 140 | + 4.27 | − 5.05 | + 0.38 | − 0.38 |
| 150 | + 7.48 | − 8.33 | + 0.64 | − 0.64 |
| 160 | + 16.60 | − 17.5 | + 1.37 | − 1.37 |
| 170 | + 66.60 | − 66.60 | + 5.3 | − 5.3 |
| 180 | ∞ | ∞ | ∞ | ∞ |

## 19.4. STRESSES DUE TO WIND LOAD

The analysis of stresses in domes due to wind load is very tedious since the dome has curvature in two directions. Since the thickness and reinforcement required for the vertical and uniformly distributed loads is much smaller than provided from practical considerations, exact calculation of stresses due to wind load and the effect of shrinkage and temperature variations are not necessary. Their effect is taken into account by adding and extra load of about 1000 to 1500 N per sq. m of the surface of the dome.

## 19.5. DESIGN OF R.C. DOMES

The requirements of thickness of dome and reinforcement from the point of view of induced stresses are usually very small. However, a minimum thickness of 7.5 cm is provided to protect steel. Similarly, a minimum steel provided is 0.15% for mild steel bars and 0.12% for HYSD bars, of the sectional area in each direction—meridionally as well as along the latitudes. This reinforcement will be in addition to the hoop tensile stresses. The steel reinforcement is provided in the middle of the thickness of the dome shell. Near the edges, some hogging bending moment may be developed, and hence meridional steel should be placed near the top surface.

**Provision of ring beam:** If the dome is not hemispherical, the meridional thrust at the supporting circle of latitude (*i.e.*, at the base) will not be vertical. The inclined meridional thrust at the support will have horizontal component which will cause the supporting walls to burst outwards, causing its failure. In order to bear this horizontal component of meridional thrust, a ring beam is provided at the base of the dome. The reinforcement provided in the ring beam takes this hoop tension and transfer only vertical reaction to the supporting walls. The tensile stress on the equivalent area of concrete on the ring beam section should not exceed 1.2 N/mm².

**Placement of main reinforcement in dome:** As stated earlier, a minimum reinforcement of 0.15% of area is provided both in the direction of latitude as well as of the meridians. If the reinforcement along the meridians is continued upto crown, there will be congestion of steel there. Hence from practical considerations, the meridional reinforcement is stopped at any latitude circle near crown, and a separate mesh is provided as shown in Fig. 19.4. No separate reinforcement along latitude is provided in this area at the crown.

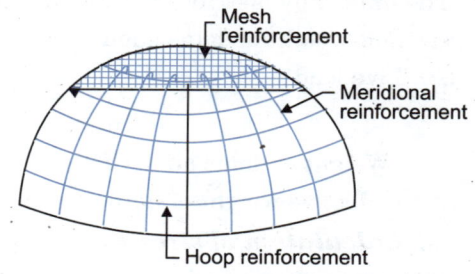

Fig. 19.4

**Provision of openings:** Openings may be provided in the dome as required from other functional *or* architectural requirements. However, sufficient trimming reinforcement should be provided all round the opening, as shown in Fig. 19.5. The meridional and hoop reinforcement reaching the opening should be well anchored to the trimming reinforcement.

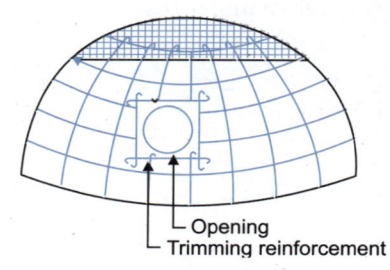

Fig. 19.5

If there is an opening at the crown of the dome, and if there is any concentrated load of lantern etc. acting there, a ring beam should be provided at the periphery of the opening, as shown in Fig. 19.7.

**Design: Example 19.1.** *Design a spherical dome over a circular room, for the following data:*

(i) *Inside diameter of room* = 12 m
(ii) *Rise of dome* = 4 m
(iii) *Live load due to wind, ice, snow etc.* = 1.5 kN/m².

*The dome has an opening of 1.6 m diameter at its crown. A lantern is provided at its top, which causes a dead load of 22 kN acting along the circumference of the opening. Use M 20 concrete and Fe 415 steel.*

**Solution:**

1. **Geometry of the dome**

Fig. 19.6 shows the dome. Let the radius of the dome be $r$. The diameter of base = $AB$ = dia. of room = 12 m.

Rise    $PS = 4$ m
        $(2r - PS) PS = 6 \times 6$
or      $(2r - 4) 4 = 36$
From which $r = AO = 6.5$ m
Again   $CD$ = diameter of opening = 1.6 m
        $PQ$ = rise at opening = $h$

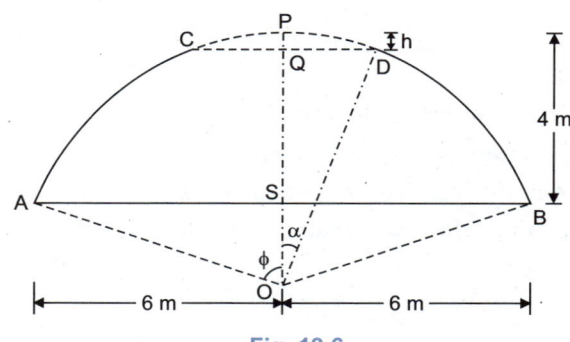

Fig. 19.6

$\therefore h(2 \times 6.5 - h) = CQ^2 = (0.8)^2$. From which $h \approx 0.05$ m

$\sin \alpha = 0.8/6.5 = 0.1231$; $\therefore \alpha = 7° 4'$; $\cos \alpha = 0.9924$

$\sin \Phi = 6/6.5 = 0.9231$; $\therefore \Phi = 67° 23'$; $\cos \Phi = 0.3840$.

**2. Loading.** The various formulae derived earlier are valid when there is no opening at the crown. In our case, there is opening of 1.6 m diameter. However, for calculation purposes, we can assume that there is no opening, and the weight of the extra portion of the dome shell *CPD* can be accounted for by reducing the load of the lantern and taking into consideration only the effective weight of lantern. That is:

*Effective wt. of lantern = Actual wt. of lantern – wt. of dome shell CPD.*

Let the thickness of the dome be 100 mm.

The uniformly distributed loads per sq. m of surface area are:

(i) Self-weight of dome shell $= 0.1 \times 25000 = 2500$ N.

(ii) Live load $= \underline{1500}$

Total $= 4000$ N/m$^2$ = 4 kN/m$^2$ = $w$

Weight of dome shell $CPD = w \times 2\pi rh = 4 \times 2\pi \times 6.5 \times 0.05 = 8.17$ kN

$\therefore$ Effective weight of lantern = $W = 22 - 8.17 = 13.83$ kN

**3. Calculation of stresses due to combined load**

The stress at any horizontal plane will be equal to the algebraic sum of stresses due to the two loading, and the dome will be designed for the maximum of these stresses.

$$\text{Total meridional stress} = \frac{wr(1-\cos\theta)}{t\sin^2\theta} + \frac{W}{2\pi rt \sin^2\theta}$$

$$= \frac{4 \times 6.5(1-\cos\theta)}{0.1 \sin^2\theta} + \frac{13.83}{2\pi \times 6.5 \times 0.1 \sin^2\theta} \text{ kN/m}^2$$

$$= \left[260 \frac{1-\cos\theta}{\sin^2\theta} + \frac{3.39}{\sin^2\theta}\right] \times 10^{-3} \text{ N/mm}^2$$

$$= 0.26 \frac{1-\cos\theta}{\sin^2\theta} + \frac{0.00339}{\sin^2\theta} \text{ N/mm}^2 \qquad \ldots(1)$$

$$\text{Hoop stress} = \frac{wr}{t}\left(\frac{\cos^2\theta + \cos\theta - 1}{1+\cos\theta}\right) - \frac{W}{2\pi rt} \cdot \frac{1}{\sin^2\theta}$$

$$= \frac{4 \times 6.5}{0.1}\left(\frac{\cos^2\theta + \cos\theta - 1}{1+\cos\theta}\right) - \frac{13.83}{2\pi \times 6.5 \times 0.1} \times \frac{1}{\sin^2\theta} \text{ kN/m}^2$$

$$= \left[260\left(\frac{\cos^2\theta + \cos\theta - 1}{1+\cos\theta}\right) - \frac{3.39}{\sin^2\theta}\right] \times 10^{-3} \text{ N/mm}^2$$

$$= 0.26 \frac{\cos^2\theta + \cos\theta - 1}{1+\cos\theta} - \frac{0.00339}{\sin^2\theta} \text{ N/mm}^2 \qquad \ldots(2)$$

The values of meridional stress and hoop stress for various values of $\theta$ are Tabulated in Table 19.2.

**4. Hoop stress in absence of live load.** Hoop stresses should also be found in absence of live load. This will increase the tensile stresses in the upper portion of the dome, specially near the periphery of the opening. However, meridional thrust will not increase by omitting the live load.

$$w = 0.1 \times 25000 = 2500 \text{ N} = 2.5 \text{ kN}$$

$\therefore$ Effective weight of lantern = $22 - 2.5 \times 2\pi \times 6.5 \times 0.05 = 16.89$ kN

# DOMES

Thus the hoop stresses due to $w$ calculated above will be *decreased* in the ratio of 250/400 (= 0.625) while the hoop stresses due to $W$ will be *increased* by a ratio of 16.89/13.84 (= 1.22). The results are tabulated in Table 19.3.

### TABLE 19.2

| θ | Meridional stress (N/mm²) | | | Hoope stress (N/mm²) | | |
|---|---|---|---|---|---|---|
| | Due to w | Due to W | Total | Due to w | Due to W | Total |
| 7° 4′ | 0.1305 | 0.224 | 0.354 | 0.127 | − 0.224 | − 0.097 |
| 10° | 0.131 | 0.112 | 0.243 | 0.125 | − 0.112 | + 0.013 |
| 20° | 0.134 | 0.029 | 0.163 | 0.102 | − 0.029 | + 0.073 |
| 30° | 0.140 | 0.014 | 0.154 | 0.086 | − 0.014 | + 0.072 |
| 40° | 0.148 | 0.008 | 0.156 | 0.052 | − 0.008 | + 0.044 |
| 50° | 0.158 | 0.006 | 0.164 | 0.009 | − 0.006 | + 0.003 |
| 60° | 0.174 | 0.004 | 0.178 | − 0.044 | − 0.004 | − 0.048 |
| 67°23′ | 0.186 | 0.004 | 0.190 | − 0.088 | − 0.004 | − 0.092 |

### TABLE. 19.3

| θ | Hoop Stress (N/mm²) | | |
|---|---|---|---|
| | Due to w | Due to W | Total |
| 7° 4′ | 0.080 | − 0.273 | − 0.193 |
| 10° | 0.078 | − 0.137 | − 0.059 |
| 20° | 0.064 | − 0.035 | + 0.029 |
| 30° | 0.054 | − 0.017 | + 0.037 |
| 40° | 0.033 | − 0.010 | + 0.023 |
| 50° | 0.006 | − 0.007 | − 0.001 |
| 60° | − 0.027 | − 0.005 | − 0.032 |
| 67° 23′ | − 0.055 | − 0.005 | − 0.060 |

Thus we see that the maximum hoop tension at the opening has been increased from 0.097 N/mm² to 0.193 N/mm².

**5. *Provision for reinforcement.*** Max. compressive stress = 0.354 N/mm² (safe)

Max. hoop tensile stress = 0.193 N/mm².

∴ Max. hoop tension per metre length of meridian
$$= 0.193 \times 100 \times 1000 = 19300 \text{ N}$$

∴ Area of steel = 19300/230 ≅ 84 mm²

Reinforcement for temp. etc. $= \dfrac{0.12}{100} \times 100 \times 1000 = 120 \text{ mm}^2/\text{m}$

∴ Total reinforcement = 84 + 120 = 204 mm²

Using 8 mm Φ bars, spacing $= \dfrac{1000 \times 50.26}{204} = 246 \text{ mm}$

However, provide 8 mm Φ bars @ 240 mm c/c where hoop tension is developed. In the portion where no hoop tension is developed, minimum area of steel @ 0.12% will be 120 mm². Hence provide 8 mm Φ bars @ 400 mm c/c.

**6. Design of lower ring beam.** Meridional thrust per metre length of the dome at its base = 0.19 × 100 × 1000 = 19000 N/m.

Horizontal component $T$ per metre length = 19000 cos 67° 23′ = 7307 N/m

∴ Hoop tension, trying to rupture the beam = $7307 \times \dfrac{12}{2}$ = 43842 N

∴ Area of steel required = 43840/230 = 191 mm²

Using 10 mm Φ bars, No. of rings = 191/78.5 ≅ 3

However provide 4 rings of 10 mm Φ bars.

Equivalent area of composite section of beam of area of cross-section $A$.
$$= A + (m-1)A_{st} = A + 12.33 \times 78.5 \times 4 = A + 3871.62$$

Allowing tensile stress of 1.2 N/mm² in the composite section, we have
$$\dfrac{43840}{A + 3871.62} = 1.2. \text{ From which } A = -32661.7 \text{ mm}^2$$

However, provide a ring beam of 200 mm × 200 mm. Provide 6 mm Φ stirrups @ 0.75 $d$ ≅ 110 mm c/c to tie the rings in the ring beam.

**7. Design of ring beam at the opening.**
Hoop compression (horizontal) in the ring beam

$$= (0.345 \times 100 \times 1000) \cos \alpha \times \dfrac{1.6}{2}$$

$$= 35400 \times 0.9924 \times 0.8 = 28105 \text{ N}$$

The horizontal component will be provided for by a beam which will from the link between the lantern and the dome.

Provide a ring beam of size 160 mm × 160 mm either above the dome or below the dome, as required from architectural requirements.

Comp. stress in ring beam = $\dfrac{28105}{160 \times 160}$

= 1.1 N/mm² (safe). Extend the 8 mm Φ rings in the ring beam also.

The details, of reinforcement are shown in Fig. 19.7.

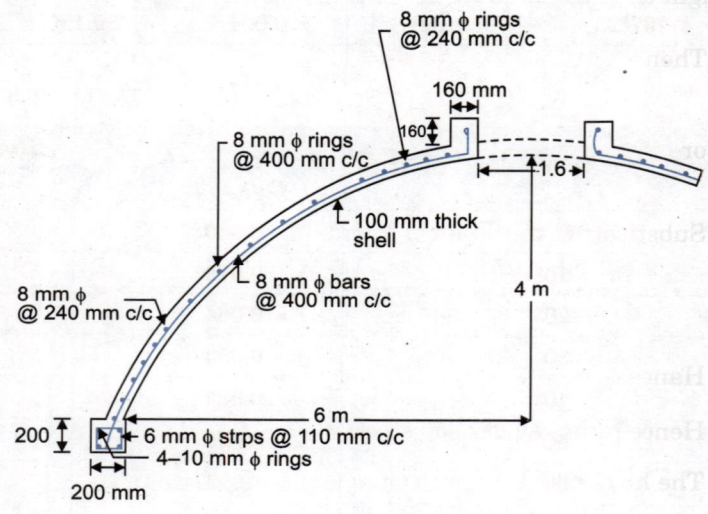

Fig. 19.7

## 19.6. CONICAL DOMES

Let us now analyse a conical dome subtending an angle $2\theta$ at the apex. Let $w$ be the intensity of uniformly distributed load inclusive of its own weight, on unit area of the dome.

Consider a horizontal ring $ABCD$ as before. Let the meridional thrust per unit length of the ring be $T$ at $B$ and $T + dT$ at $D$, acting tangentially. The vertical component of the total meridional thrust at $B$ will evidently be equal to the load on the dome shell $APB$. Let the distance of horizontal planes $AB$ and $CD$ be $y$ and $y + dy$ from apex $P$.

Diameter $AB = 2y \tan \theta$; length $AP = y/(\cos \theta)$

∴ Load on dome shell $APB$
$$= \dfrac{w}{2}(\pi \, 2y \tan \theta) \cdot \dfrac{y}{\cos \theta}$$

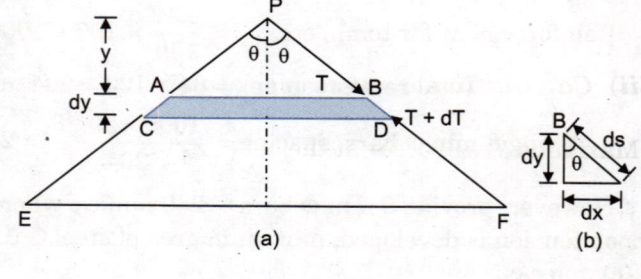

Fig. 19.8

Vertical component of total $T = (\pi\, 2y \tan \theta)\, T \cos \theta$

$\therefore \quad 2\pi y \tan \theta\, T \cos \theta = 2\pi y \dfrac{w}{2} \tan \theta \dfrac{y}{\cos \theta}$

From which $\quad T = \dfrac{w}{2} \cdot \dfrac{y}{\cos^2 \theta} \qquad \qquad \ldots(19.6)$

$\therefore$ Intensity of meridional stress $= \dfrac{w}{2t} \cdot \dfrac{y}{\cos^2 \theta} \qquad \qquad \ldots[19.6(a)]$

The horizontal component of $T$ will cause hoop tension at $B$, while the horizontal component of $T + dT$ will cause hoop compression.

Magnitude of hoop tension $= T \sin \theta \times$ radius of ring at depth $y = T \sin \theta \times y \tan \theta = Ty \dfrac{\sin^2 \theta}{\cos \theta}$

Magnitude of hoop compression $= (T + dT)(y + dy) \dfrac{\sin^2 \theta}{\cos \theta}$

The difference in these two horizontal components will give the value of hoop force.

Let $H$ be the hoop compression induced in the ring, per unit breadth. Let $ds$ be the breadth of the ring of height $dy$ [Fig. 19.8(b)].

Then $\quad ds = \dfrac{dy}{\cos \theta}$.  Hence we have $\quad H\, ds = d\left(\dfrac{T y \sin^2 \theta}{\cos \theta}\right)$

or $\quad \dfrac{H\, dy}{\cos \theta} = d\left(T y \dfrac{\sin^2 \theta}{\cos \theta}\right) \quad$ or $\quad H = \dfrac{d}{dy}(Ty \sin^2 \theta)$

Substituting the value of $T$ and differentiating, we get

$$H = \dfrac{d}{dy}\left[\dfrac{w}{2} \dfrac{y}{\cos^2 \theta} \cdot y \sin^2 \theta\right] = \dfrac{w}{2} \cdot \dfrac{\sin^2 \theta}{\cos^2 \theta} \cdot \dfrac{d}{dy}(y)^2$$

Hance $\quad H = wy \tan^2 \theta \qquad \qquad \ldots(19.7)$

Hence intensity of hoop stress $= \dfrac{w\, y}{t} \tan^2 \theta \qquad \qquad \ldots[19.7(a)]$

The hoop stress will be compressive throughout.

**Design: Example 19.2.** *Design a conical roof for a hall having a diameter of 20 m. The rise of the dome has to be 4 m. Assume the live and other loads as 1500 N/m². Use M 20 concrete and Fe 415 steel.*

**Solution:** (i) *Geometry of the dome*

Diameter $\qquad AB = 20$ m. Height $PQ = 4$ m.

$\tan \theta = 10/4 = 2.5; \quad \theta = 68° 12'$

$\sin \theta = 0.9285; \quad \cos \theta = 0.3714$

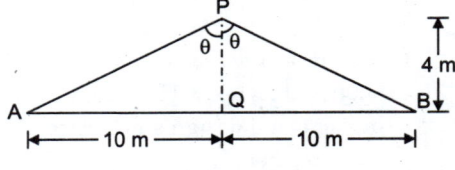

Fig. 19.9

(ii) *Loading.* Let the thickness of shell be 10 cm.

Weight of dome shell /m² $= 0.1 \times 1 \times 25000 = 2500$ N/m²

Live load $= 1500$ N/m²

Total $w = 4000$ N/m²

(iii) *Calculation of stress:* Meridional thrust $T = \dfrac{wy}{2 \cos^2 \theta}$

Maximum $T$ will occur at the base $AB$ where $y = 4$ m.

$\therefore \quad T_{max} = \dfrac{4000 \times 4}{2(0.3714)^2} = 57997.08$ N (comp.)

$\therefore \quad$ Meridional stress $= 57997/(100 \times 1000) = 5799.7$ N/mm² (Safe)

Hoop force $\quad H = wy \tan^2 \theta$. $\quad \therefore H_{max} = 4000 \times 4 (2.5)^2 = 100000$ N (comp.)

$\therefore \quad$ Hoop stress $= 100000/(100 \times 1000) = 1$ N/mm² (Safe)

**(iv) Steel reinforcement.** The stresses work out to be safe. Hence only nominal reinforcement has to be provided @ 0.12% of the area of concrete.

$$A_{st} = (0.12/100) \times 100 \times 1000 = 120 \text{ mm}^2. \quad \text{Using 8 mm } \Phi \text{ bars, } A_\Phi = 50 \text{ mm}^2$$

Spacing $= \dfrac{1000 A_\Phi}{A_{st}} = \dfrac{1000 \times 50}{120} = 416.67$ mm

However, provide 8 mm Φ bars @ 300 mm c/c both the ways.

The meridional bars may be discontinued near the apex, and a wire mesh may be provided there to avoid congestion of steel.

**(v) Design of ring beam.** Horizontal component of meridional thrust $T$ will cause an outward force on the support, causing hoop tension. Hence a ring beam is necessary.

Hoop tension $P$ in ring beam $= T \sin \theta = 58200 \times 0.9285 = 54000$ N/m.

$\therefore \quad$ Total tensile force $= P \times \dfrac{D}{2} = 54000 \times \dfrac{20}{2} = 540000$ N

$\therefore \quad$ Area of steel to resist this, $A_{st} = 540000/230 = 2348$ mm²

Using 20 mm Φ bars, this, $\quad A_\Phi = 314$ mm². $\quad \therefore$ No. of bars $= 2348/314 \approx 8$

Actual area of steel provided $= 8 \times 314 \triangleq 2512$ mm².

Tie these by 8 mm Φ 2-lgd stirrups @ 300 mm c/c

Let $A$ be the area of ring beam.

Equivalent area of composite section
$= A + (m - 1) A_{st} = A + 18 \times 3920 = A + 70560$

Assuming the allowable tensile stress in composite section to be 1.2 N/mm², we have

$$\dfrac{540000}{A + 70560} = 1.2$$

which gives $A = 379440$ mm².

Provide ring beam of size 700 mm × 700 mm.

The details of the reinforcement is shown in Fig. 19.10.

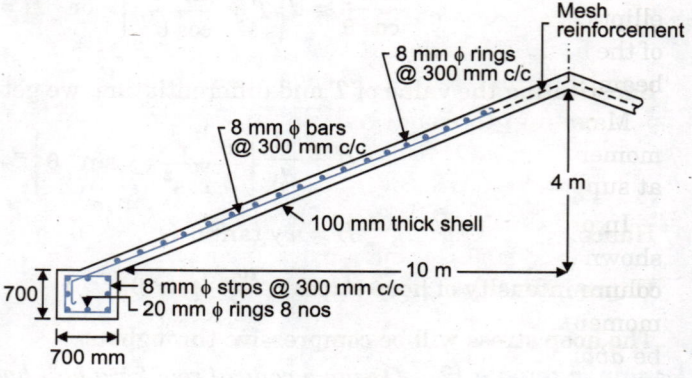

Fig. 19.10

## PROBLEMS

1. A spherical dome, span 10 m and rise 2 m, has a shell which is 120 mm thick. It carries a lantern load of 5000 N at its apex. The wind load on the dome is estimated to the equivalent to 1200 N/m². Examine the stresses in dome and design suitable reinforcement. Also, design the ring beam. Use M 20 concrete and Fe 415. steel reinforcement.

2. Design a conical dome for the above, if there is no lantern.

3. Design a spherical dome over a circular room of 20 m diameter. The rise of the dome may be taken equal to 1/5 of the diameter. The dome carries a lantern load of 30 kN attached at the circumference of an opening of 2 m diameter at the apex. Take live load due to wind etc. as 1.5 kN/m² of the surface area of the dome. Use M 20 concrete and Fe 415 steel.

4. Design a spherical dome of dia. 15 metres the rise of dome may be taken as 3.25 metres the lateral load on dome is 28 kN. Use $M_{30}$ grade conclude Fe 415 steel. Assume suitable data.

# CHAPTER 20

# BEAMS CURVED IN PLAN

## 20.1. INTRODUCTION : TORSIONAL MOMENTS IN BEAMS

Beams curved in plan are often used to support the circular water tanks (reservoirs), curved balconies, curved ramps *or* other similar structures having curved boundary. Such curved beams may be circular, elliptical *or* polygonal in plan, and the line joining the supports lies away from the curved longitudinal axis of the beam. Since the C.G. of loads and reactions to one side of any section does not lie along the axis of the beam, it is subjected to torsional moment, in addition to bending moment (flexural moment) and shear force.

Maximum torsional moments will develop at sections nearer the supports and where the bending moment is zero. The maximum torque occurs at points of contra flexure. Also shear force will be maximum at support sections.

In order to understand how torsional moments are induced in beams, let us consider a space frame shown in Fig. 20.1(*a*) having horizontal beams *AB* and *BC* mutually orthogonal to each other and a vertical column *BD*. When vertical load is applied on *AB*, both *AB* as well as *BD* will be subjected to bending moments causing joint *B* to rotate in the anticlockwise direction. The moment at joint *B* will evidently be *about* the longitudinal axis of *BC*, and will therefore, cause twisting or torsion in it. The torsional moment so induced will be constant all along the length of the beam *BC*.

Fig. 20.1 (*c*) shows a beam curved in plan, *i.e*, the longitudinal axis is curved with a radius of curvature *R*. Point *O* is the centre of curvature. When the beam is subjected to vertical loads, it bends. At any section *P*, the C.G. of loads ($C_2$) and reaction to one side of it lies away from the axis. Due to this, torsional moment is induced about the longitudinal axis at that section. Such phenomenon does not happen in the case of beam straight in plan [Fig. 20.1 (*b*)] where the C.G. of loads to one side of section *P* lies along the longitudinal axis. It should be noted that in the case of a straight beam, bending moment at any section *P* is the moment of all the forces to one side of it, *about* the transverse axis of the section. In the case of beam curved in plan, the direction of transverse axis at any point is *radial*. The bending at any point *P* of the beam is therefore the moment of all the forces to one side of it, *about the radial axis OP*. On the other hand, the torsional moment at any point is the moment, of all forces about the *longitudinal (or tangential) axis* of the beam at that section.

(a) Space frame

(b) Straight beam (plan)

(c) Curved beam (plan)

(d) Curved beam (elevation)

Fig. 20.1 Torsional moments in beams

## 20.2. CIRCULAR BEAM SUPPORTED SYMMETRICALLY

Let us first analyse a complete circular beam, supported symmetrically on columns [Fig. 20.2 (a)]. The beam is thus continuous, and forms a close loop. Let the beam be subjected to uniformly distributed load $w$ per unit length. Due to symmetry, the vertical reaction at each column will be the same. Also, the shear force and the twisting moment at the centre of each span will be zero, and the twisting moment at the supports will be zero. Let the angle subtended by two consecutive columns $A$ and $B$ be $2\theta$, and let the mean radius of the beam be $R$.

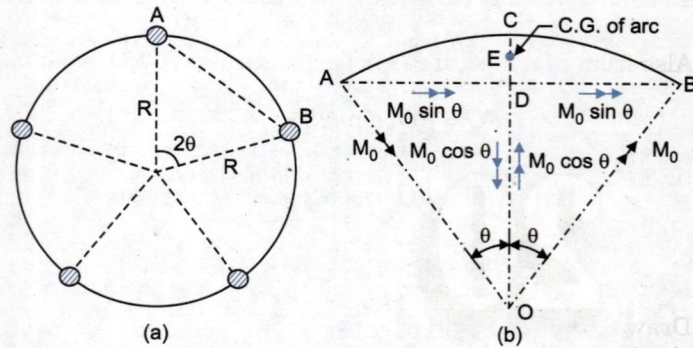

Fig. 20.2

Load $W$ of arc $AB = w \cdot R \cdot 2\theta = 2wR\theta$

The distance of C.G. of an arc, subtending an angle $2\theta$ at the centre, is given by
$$OE = R \sin\theta / \theta \qquad \ldots(20.1)$$

Thus the C.G. of the load on beam $AB$ lies at a distance of $R \sin\theta/\theta$ from the centre of curvature.

### (i) Shear Force and Moment at the Support.

Due to symmetry, the S.F. and B.M. at each support will be equal. Let $F_0$ be the S.F. and $M_0$ be the B.M. at each support.

Evidently,
$$F_0 = \frac{w \cdot R \cdot 2\theta}{2} = w \cdot R\theta \qquad \ldots(20.2)$$

The bending moment $M_0$ at each end will be hogging in nature, and it will be *about* the radial axis $AO$ and $BO$.

Representing the end moments by vectors, the direction of moments at supports, will be as shown in Fig. 20.2 (b), in which the arrows indicate the moments *about* that direction. The moment $M_0$ about $AO$ can be resolved in two components:

(i) Moment $M_0 \sin\theta$ *about* chord $AD$, and    (ii) Moment $M_0 \cos\theta$ *about* $DO$

Similarly, moment $M_0$ about $BO$ can be resolved in two components about $DB$ and $DO$ as marked.
In order to find the value of end moment $M_0$, let us balance the moment about the chord $AB$.
Components of end moments about chord $AB = 2 M_0 \sin\theta$    ...(i)
Moment of external force about the chord $AB = W \times ED$

$$= w \cdot R \cdot 2\theta \, [EO - DO] = 2wR\left[\frac{R\sin\theta}{\theta} - R\cos\theta\right] \qquad \ldots(ii)$$

[**Note:** The vertical reactions at $A$ and $B$ will not have any moments about $AB$].
Equating (i) and (ii), we get

$$2 M_0 \sin\theta = 2wR^2\theta \left[\frac{\sin\theta}{\theta} - \cos\theta\right]$$

$$\therefore \quad M_0 = wR^2 \, [1 - \theta \cot\theta] \qquad \ldots(20.3)$$

### (ii) S.F. and B.M. at any Point

After having determined end reactions and moments, the S.F. and bending moment at any point $P$ on the beam can be easily found. Let $PO$ make an angle $\phi$ with $AO$ (Fig. 20.3). Let S.F. and B.M. at $P$ be designated as $F_\phi$ and $M_\phi$ respectively.

Reaction at $A = w \cdot R \cdot \theta$.
Load on $AP = W_\phi = w \cdot R_\phi$

$$\therefore \quad F_\phi = wR\theta - wR_\phi = wR(\theta - \phi) \qquad \ldots(20.4)$$

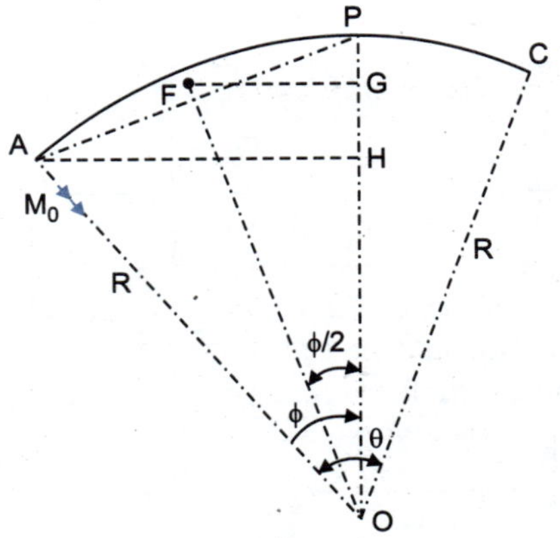

Let the C.G. of load on $AP$ be at point $F$, such that
$$FO = \frac{R \sin \phi/2}{\phi/2} \quad \text{(Ref. Eq. 20.1)}$$

Also from Fig. 20.3, if $FG$ is perpendicular to $PO$, we have
$$FG = FO \cdot \sin \phi/2 = \frac{R \sin^2 \phi/2}{\phi/2}$$
$$GO = FO \cdot \cos \phi/2 = \frac{R \sin \phi/2}{\phi/2} \cos \phi/2 = \frac{R \sin \phi}{\phi}$$
$$PG = R - R \frac{\sin \phi}{\phi} = R\left(1 - \frac{\sin \phi}{\phi}\right).$$

Draw $AH$ perpendicular to $PO$.
Then, $AH = R \sin \phi$ and $PH = R(1 - \cos \phi)$.

**Fig. 20.3**

Now B.M. at $P$ = moment about the radial axis $PO$.
Component of end moment, about the radial axis $PO = M_0 \cos \phi$ (hogging).
Hence sagging moment at $P$ is given by $M_\phi = F_0 \times AH - W_\phi \times FG - M_0 \cos \phi$.
Substituting the values of various quantities,
$$M_\phi = w \cdot R\theta \cdot R \sin \phi - wR\phi \cdot R \frac{\sin^2 \phi/2}{\phi/2} - wR^2 (1 - \theta \cot \theta) \cos \phi$$
or $$M_\phi = wR^2 [\theta \sin \phi - 2 \sin^2 \phi/2 - \cos \phi + \theta \cot \theta \cdot \cos \phi]$$
or $$M_\phi = wR^2 [\theta \sin \phi + \theta \cot \theta \cdot \cos \phi - 1] \qquad \ldots(20.5)$$

### (iii) Twisting Moment at P

Let the twisting moment at $P$ be $M_\phi^t$. Twisting moment at any point is equal to the moment of all forces on one side of it, about the tangential axis at that point. This moment may be taken as positive if the left portion twists the beam towards the centre of curvature.
$$M_\phi^t = M_0 \sin \phi - F_0 \times PH + W_\phi \times PG$$
or $$M_\phi^t = wR^2 (1 - \theta \cot \theta) \sin \phi - wR\theta \cdot R(1 - \cos \phi) + w R\phi \cdot R \left(1 - \frac{\sin \phi}{\phi}\right)$$
or $$M_\phi^t = wR^2 [\sin \phi - \theta \cot \theta \sin \phi - \theta + \theta \cos \phi + \phi - \sin \phi]$$
or $$M_\phi^t = wR^2 [\phi - \theta + \theta \cos \phi - \theta \cot \theta \cdot \sin \theta]$$
or $$M_\phi^t = wR^2 [\theta \cos \phi - \theta \cot \theta \cdot \sin \phi - (\theta - \phi)] \qquad \ldots(20.6)$$

The above equation gives the distribution of torsional moment along the beam. In order to get the position of maximum twisting moment, differentiate the above equation with respect to $\phi$ and equate it to zero. Thus, we get, $\sin \phi = \frac{1}{\theta} [\sin^2 \theta \pm \cos \theta (\theta^2 - \sin^2 \theta)^{1/2}] \qquad \ldots[20.6(a)]$

The above equation also gives the location of section of contraflexure. For the design purpose, the values of support moment ($M_0$), midspan moment ($M_C$) and the maximum twisting moment ($M_m^t$) can be represented by the following expression:
$$M_0 = C_1 \cdot wR^2 (2\theta) \qquad \ldots(20.7)$$
$$M_C = C_2 \cdot w R^2 (2\theta) \qquad \ldots(20.8)$$
and $$M_m^t = C_3 \cdot wR^2 (2\theta) \qquad \ldots(20.9)$$

The values of coefficients $C_1$, $C_2$ and $C_3$ depend upon the magnitude of $\theta$ and can be taken from Table 20.1, which also gives the value of angle $\phi_m$ where maximum twisting moment occurs.

where  $C_1$ = Negative bending moment at support.
       $C_2$ = Positive bending moment at centre of span.
       $C_3$ = Maximum twisting moment (on torque)

**TABLE 20.1.** Coefficients for B.M. and Twisting Moment in Circular Beams.

| No. of Supports | $2\theta$ | $C_1$ | $C_2$ | $C_3$ | $\phi_m$ |
|---|---|---|---|---|---|
| 4 | 90° | 0.137 | 0.070 | 0.021 | $19\frac{1°}{4}$ |
| 5 | 72° | 0.108 | 0.054 | 0.014 | $15\frac{1°}{4}$ |
| 6 | 60° | 0.089 | 0.045 | 0.009 | $12\frac{3°}{4}$ |
| 7 | $51\frac{3°}{7}$ | 0.077 | 0.037 | 0.007 | $10\frac{3°}{4}$ |
| 8 | 45° | 0.066 | 0.030 | 0.005 | $9\frac{1°}{2}$ |
| 9 | 40° | 0.060 | 0.027 | 0.004 | $8\frac{1°}{2}$ |
| 10 | 36° | 0.054 | 0.023 | 0.003 | $7\frac{1°}{4}$ |
| 12 | 30° | 0.045 | 0.017 | 0.002 | $6\frac{1°}{4}$ |

**Example 20.1.** *A curved beam is in the form of a full continuous circle in plan with a radius of 4 m and is supported continuously on six supports. The beam carries a uniformly distributed load of 2 kN/m length, inclusive of its own weight. Determine the bending moment, twisting moment and shear force at salient locations and plot B.M., T.M. and S.F. diagrams.*

**Solution:** Number of supports = 6

$$2\theta = (360°)/6 = 60° \text{ or } \theta = 30° = 0.5236 \text{ radians}$$

From Table 20.1, $\quad C_1 = 0.089 \; ; \; C_2 = 0.045 \; ; \; C_3 = 0.009$

$$\phi_m = 12.75° = 0.223 \text{ radians}$$

$$wR^2 \cdot 2\theta = 2(4)^2 \times (2 \times 0.5236) = 33.51$$

$\therefore \quad M_0 = C_1 wR^2 (2\theta) = 0.089 \times 33.51 = 2.982 \text{ kN-m (hogging)}.$

$M_C = C_2 \cdot wR^2 (2\theta) = 0.045 \times 33.51 = 1.508 \text{ kN-m}$

and $\quad M_m^t = C_3 wR^2 (2\theta) = 0.009 \times 33.51 = 0.302 \text{ kN-m}$

$F_0 = w \cdot R \cdot \theta = 2 \times 4 \times 0.5236 = 4.19 \text{ kN}.$

**(i) S.F. diagram.** The distribution of S.F. is given by *Eq. 20.4*.

$$F_\phi = wR (\theta - \phi)$$

$\therefore \quad F_\phi = 2 \times 4 (30° - \phi) \times \dfrac{\pi}{180}, \text{ where } \phi \text{ is in degrees}$

or $\quad F_\phi = 0.1396 (30° - \phi)$ ...(i)

The above equation shows that $F_\phi$ varies linearly with $\phi$. Value for $F_\phi$ at varying 10° interval is tabulated below:

| $\phi$ | $F_\phi$ (kN) | Location |
|---|---|---|
| 0° | 4.190 | Ends |
| 10° | 2.792 | |
| 20° | 1.396 | |
| 30° | 0.000 | Middle |

## (ii) B.M. diagram.

B.M. $M_\phi$ at any point is given by *Eq. 20.5*:

$$M_\phi = wR^2 [\theta \sin \phi + \theta \cot \theta \cdot \cos \phi - 1]$$

or  $M_\phi = 2(4)^2 [0.5236 \sin \phi + 0.5236 \cot 30° \cos \phi - 1]$

or  $M_\phi = 32 [0.5236 \sin \phi + 0.9069 \cos \phi - 1]$  ...(ii)

The values of $M_\phi$ at $\phi = 10°$ interval are tabulated below:

| $\phi$ | $M_\phi$ | Location |
|---|---|---|
| 0 | $-2.982$ | End of beam |
| 10° | $-0.511$ | |
| 12.75° | 0.00 | Point of max. torsion |
| 20° | $+1.00$ | |
| 30° | $+1.508$ | Centre of beam |

The distribution of bending moment is shown in Fig. 20.4 (b).

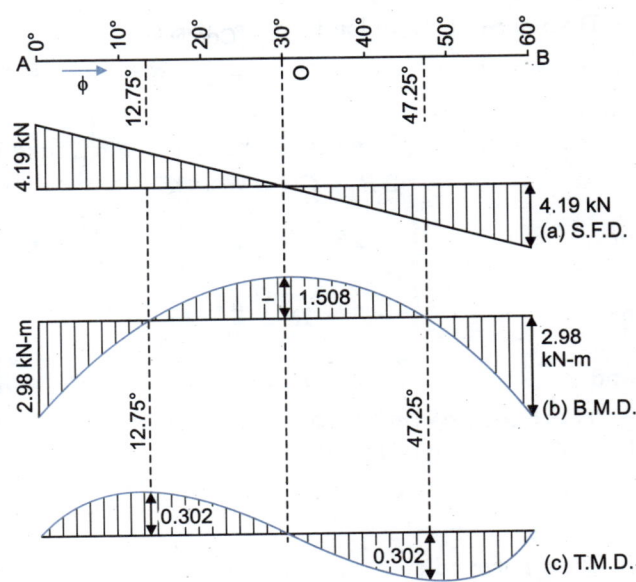

Fig. 20.4

### (iii) T.M. diagram.
The twisting moment $M_\phi^t$ at any point is given by *Eq. 20.6*:

$$M_\phi^t = wR^2 [\theta \cos \phi - \theta \cot \theta \cdot \sin \phi - (\theta - \phi)]$$

$$= 2(4)^2 \left[ 0.5236 \cos \phi - 0.5236 \cot 30° \sin \phi - (30° - \phi)\frac{\pi}{180} \right]$$

$$= 32 [0.5236 \cos \phi - 0.9069 \sin \phi - 0.01745 (30° - \phi)] \quad ...(iii)$$

where $\phi$ is in degrees.

The values of $M_\phi^t$ at 10° interval are tabulated below:

| $\phi$ | $M_\phi^t$ | Location |
|---|---|---|
| 0° | 0.00 | End of beam |
| 10° | $+0.293$ | |
| 12.75° | $+0.302$ | Point of max. torsion |
| 20° | $+0.235$ | |
| 30° | 0.00 | Middle of beam |

The twisting moment diagram is shown in Fig. 20.4(c).

## 20.3. SEMICIRCULAR BEAM SIMPLY SUPPORTED ON THREE EQUALLY SPACED COLUMNS

Let us now take the case of semi-circular beam $ABC$, simply supported over end column $A$ and $B$, and continuous over a central column at $C$ (Fig. 20.5). End moments at $A$ and $B$ will be evidently zero. Due to symmetry, reaction at $A$ and $B$ will be equal (say $R_1$) and reaction at $C$ (say $R_2$) will be different. Let us first find the reactions.

Total load on the semi-circular beam $= W = w\pi R$
(where $R$ is the radius of the semi-circle)

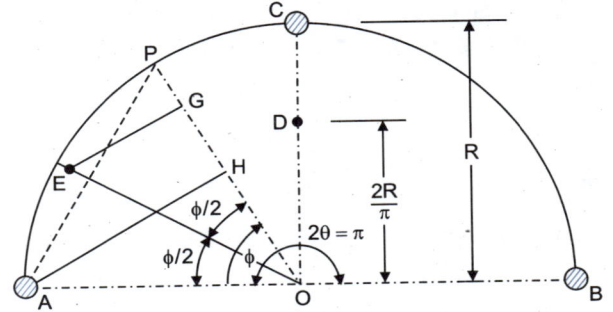

Fig. 20.5. Semicircular beam

Distance of C.G. of load from centre $O$

$$= DO = \frac{R \sin \theta}{\theta} = \frac{R \sin \pi/2}{\pi/2} = \frac{2R}{\pi}$$

In order to find the value of the reaction $R_1$, take moments about tangent at point $C$.

$\therefore \qquad 2 R_1 \times CO = W \times DC$

or $\qquad 2 R_1 \cdot R = w\pi R \left( R - \dfrac{2R}{\pi} \right)$

or $\qquad R_1 = \dfrac{wR}{2} (\pi - 2)$ ...(20.10)

and $\qquad R_2 = W - 2R_1 = w\pi R - wR(\pi - 2) \quad$ or $\quad R_2 = 2wR$ ...(20.11)

**(i) S.F. at any point:** In order to find the S.F. distribution consider any point $P$ at angle $\phi$ with $OA$. The C.G. of the load on $AP$ will be at point $E$ such that

$$EO = \frac{R \sin \phi/2}{\phi/2}$$

Draw perpendiculars $AH$ and $EG$ on the radial line $PO$.

$$AH = R \sin \phi \; ; \quad OH = R \cos \phi \; ; \quad EC = EO \sin \phi/2 = \frac{R \sin^2 \phi/2}{\phi/2}$$

$$GO = EO \cos \phi/2 = \frac{R \sin^2 \phi/2}{\phi/2} \cos \phi/2 = \frac{R \sin \phi}{\phi}$$

Load on $AP = W_\phi = wR\phi$

$\therefore \qquad F_\phi = R_1 - wR\phi = \dfrac{wR}{2}(\pi - 2) - wR\phi = wR \left[ \dfrac{\pi}{2} - 1 - \phi \right]$ ...(20.12)

The above expression is valid from $\phi = 0$ to $\phi = \pi/2$. At $C$, $\phi = \pi/2$.

$\therefore \qquad F_\phi = F_C = -wR$

**(ii) B.M. at any point:** B.M. at point $P$ is evidently equal to moment, about $OP$, of all the forces to the left of point $P$.

$\therefore \qquad M_\phi = R_1 \times AH - W_\phi \times EG = \dfrac{wR}{2}(\pi - 2) R \sin \phi - wR\phi R \dfrac{\sin^2 \phi/2}{\phi/2}$

or $\qquad M_\phi = wR^2 \left[ \dfrac{\pi - 2}{2} \sin \phi - 2 \sin^2 \dfrac{\phi}{2} \right]$ ...(20.13)

At $\phi = 0$, $\qquad M_\phi = M_A = 0$

At $\phi = \pi/2$, $\qquad M_\phi = M_C = wR^2 \left[ \dfrac{\pi - 2}{2} - 2 \times \dfrac{1}{2} \right]$

$$= wR^2 \frac{\pi - 4}{2} = -0.429 \, wR^2 \quad (i.e., \text{hogging}) \qquad ...(20.14)$$

In order to find the maximum sagging bending moment any where in $AC$ differentiate Eq. 20.13 with respect to $\phi$ and put it equal to zero.

$\therefore \qquad \dfrac{dM_\phi}{d\phi} = wR^2 \left[ \dfrac{\pi - 2}{2} \cos \phi - 2 \times 2 \sin \dfrac{\phi}{2} \times \dfrac{1}{2} \cos \phi/2 \right] = 0$

or $\qquad \dfrac{\pi - 2}{2} \cos \phi = 2 \sin \phi/2 \cos \phi/2 = \sin \phi \quad$ or $\quad \tan \phi = \dfrac{\pi - 2}{2} = 0.5708$

which gives $\qquad \phi = 29.72° = 29°44' = 0.5187$ radians.

Substituting the value of $\phi$ in Eq. 20.13, we get

$$M_{max} = wR^2 \left[ \frac{\pi - 2}{2} \sin 29.72° - 2 \sin^2 \frac{29.72°}{2} \right] = 0.1514\, w\, R^2 \quad \text{(sagging)} \quad ...(20.15)$$

*(iii) Torsional moment at any point:* Torsional moment at any point $P$ is the moment, about tangent at $P$, of all the forces to the left of $P$.

$$M_\phi^t = R_1 \times PH - W_\phi \times PG = R_1 \times (PO - HO) - W_\phi (PO - GO)$$

$$= \frac{wR}{2} (\pi - 2) \times (R - R \cos \phi) - w\, R\, \phi \left[ R - \frac{R \sin \phi}{\phi} \right]$$

or $$M_\phi^t = wR^2 \left[ \frac{\pi - 2}{2} - \frac{\pi - 2}{2} \cos \phi - \phi + \sin \phi \right] \quad ...(20.16)$$

In order to get the maximum torsional moment, differentiate the above equation with respect to $\phi$ and equate it to zero.

$$\therefore \quad \frac{dM_\phi^t}{d\phi} = wR^2 \left[ \frac{\pi - 2}{2} \sin \phi - 1 + \cos \phi \right] = 0.$$

or $\quad \frac{\pi - 2}{2} \sin \phi - 1 + \cos \phi = 0 \quad$ or $\quad 0.5708 \sin \phi - 1 = -\cos \phi$

Squaring both the sides, $\quad 0.3258 \sin^2 \phi - 1.1416 \sin \phi + 1 = \cos^2 \phi = 1 - \sin^2 \phi$

$\therefore \quad 1.3258 \sin^2 \phi = 1.1416 \sin \phi \quad$ or $\quad \sin \phi = (1.1416)/1.3258 = 0.8611$

$\therefore \quad \phi = \phi_m = 59.44° = 1.037$ radians.

Substituting the value of $\phi$ in Eq. 20.16, we get

$$M_m^t = w\, R^2 \left[ \frac{\pi - 2}{2} - \frac{\pi - 2}{2} \cos 59.44° - 1.037 + \sin 59.44° \right] = 0.1045\, wR^2 \quad ...(20.17)$$

If this value of $\phi_m$ is substitued in Eq. 20.13, we get $M_\phi = 0$. *Thus the point of maximum torsion coincides with the point of contraflexure.*

**Example 20.2.** *A semi-circular beam with radius of 4 m is simply supported at ends, and is continuous over a column at its middle. The beam carries a uniformly distributed load of 20 kN / m length of the beam, inclusive of its own weight. Determine S.F., B.M. and T.M. at salient points, and plot S.F., B.M. and T.M. diagrams.*

**Solution:** (Fig. 20.6).

$R = 4m \,;\, w = 20$ kN/m

$R_1 = \frac{wR}{2} (\pi - 2) = \frac{20 \times 4}{2} (\pi - 2)$

$\quad = 45.66$ kN

$R_2 = 2\, wR = 2 \times 20 \times 4 = 160$ kN

*(i)* S.F. $\quad F_A = R_1 = 45.66$ kN

$\quad F_C = - wR = -20 \times 4 = -80$ kN

At any other location,

$F_\phi = w\, R \left[ \frac{\pi}{2} - 1 - \phi \right] = 20 \times 4 \left[ \frac{\pi}{2} - 1 - \phi \times \frac{\pi}{180} \right]$

$\quad = 80\, [0.5708 - 0.01745\, \phi].$

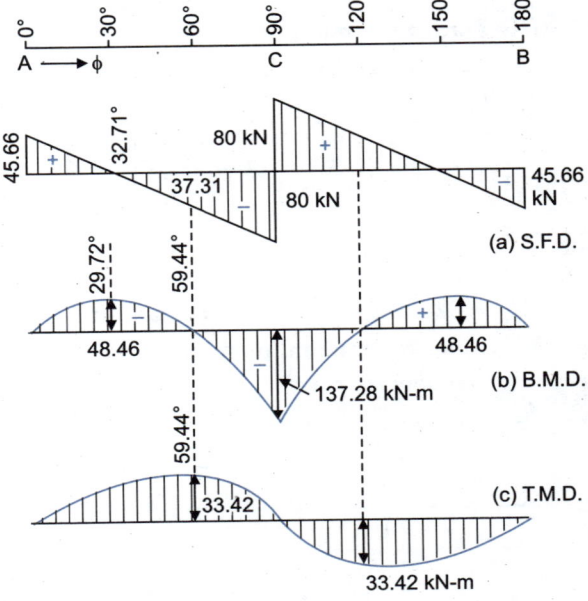

Fig. 20.6

The variation of $F_\phi$ is thus linear with $\phi$. At $\phi = 90°$,
$$F_\phi = 80\,[0.5708 - 0.01745 \times 90] = -80 \text{ kN}$$
To locate the position where $F_\phi$ is zero, we have,
$$0.5708 - 0.01745\,\phi = 0 \quad \text{or} \quad \phi = 0.5708/0.01745 = 32.71°$$

**(ii) Bending moment :**
$$M_c = -0.429\,wR^2 = -0.429 \times 20\,(4)^2 = -137.28 \text{ kN-m} \quad (i.e., \text{ hogging})$$
$$M_{max} = +0.1514\,wR^2 = +0.1514 \times 20\,(4)^2 = 48.46 \text{ kN-m} \quad (\text{sagging}).$$

The bending moment at any other location is given by *Eq. 20.13*.
$$M_\phi = wR^2\left[0.5708 \sin \phi - 2 \sin^2 \frac{\phi}{2}\right] = 320\,[0.5708 \sin \phi - 2 \sin^2 \phi/2]$$

The values of $M_\phi$ at 30° interval are tabulated below:

| $\phi$ | $M_\phi$ | Location |
|---|---|---|
| 0 | 0.00 | End |
| 29.72° | 48.46 | Max. Sagging B.M. |
| 30° | 48.45 | |
| 59.44° | 0.00 | Point of max. torsion |
| 60° | -1.82 | |
| 90° | -137.28 | Max. hogging B.M. |

In order to find the position where B.M. is zero, equate *Eq. 20.13.* to zero.
Thus, $\quad 0.5708 \sin \phi - 2 \sin^2 \phi/2 = 0 \quad$ or $\quad 0.5708 \times 2 \sin \phi/2 \times \cos \phi/2 = 2 \sin^2 \phi/2$
or $\quad \tan \phi/2 = 0.5708$
$\therefore \quad \phi/2 = 29.72° \quad$ or $\quad \phi = 59.44°$.

which is evidently the location where torsional moment is maximum. Thus, we conclude that the point of torsional moment is also the point of contra-flexure.

**(iii) Torsional moment:**

$M_m^t = 0.1045\,wR^2 = 0.1045 \times 20\,(4)^2 = 33.42$ kN-m at $\phi = 59.44°$

The distribution of torsional moment is given by *Eq. 20.16*.
$$M_\phi^t = wR^2\left[\frac{\pi - 2}{2} - \frac{\pi - 2}{2}\cos \phi - \phi + \sin \phi\right]$$
or $\quad M_\phi^t = 20\,(4)^2\left[0.5708 - 0.5708 \cos \phi + \sin \phi - \phi \frac{\pi}{180}\right]$
or $\quad M_\phi^t = 320\,[0.5708 - 0.5708 \cos \phi + \sin \phi - 0.01745\,\phi]\quad$ where $\phi$ is in degrees.

The values of torsional moment at 30° interval are tabulated below :

| $\phi$ | $M_\phi^t$ | Location |
|---|---|---|
| 0 | 00.00 | End of beam |
| 30° | 16.95 | |
| 59.44° | 33.42 | point of zero B.M. |
| 60° | 33.41 | |
| 90° | 00.00 | Middle of beam |

The S.F., B.M. and T.M. diagrams are shown in Fig. 20.6.

## 20.4. CURVED BEAM SIMPLY SUPPORTED AT ENDS AND CONTINUOUS OVER TWO EQUALLY SPACED INTERMEDIATE SUPPORTS

Fig. 20.7 shows a verandah beam continuous over four supports which are equally spaced. Hence each span subtends an angle of 90° at the centre. Due to symmetry, $R_A = R_D = R_1$ and $R_B = R_C = R_2$. In order to find the reactions, we have,

$$2(R_1 + R_2) = wR\frac{3}{2}\pi = \frac{3}{2}wR\pi \qquad ...(i)$$

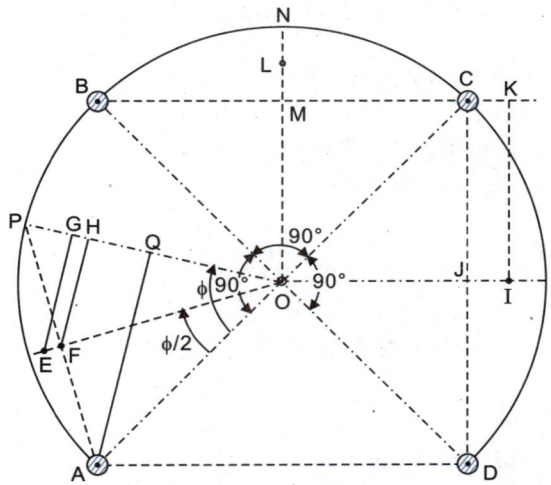

Fig. 20.7

Consider sector $CD$. Load on it $= wR\dfrac{\pi}{2}$, and its C.G. falls at $I$ whose perpendicular distance from $BC$ is equal to $IK = CJ$.

Now $\quad CD = \dfrac{AC}{\sqrt{2}} = \dfrac{2R}{\sqrt{2}} = \sqrt{2}\,R$

$\therefore \quad CJ = \dfrac{1}{2}\,CD = \dfrac{\sqrt{2}\,R}{2} = \dfrac{R}{\sqrt{2}}$

Load on sector $BC = w \cdot R \cdot \dfrac{\pi}{2}$ and its C.G. acts at point $L$, where $OL = \dfrac{R\sin \pi/4}{\pi/4}$

Distance $\quad OM = CJ = \dfrac{R}{\sqrt{2}}\,;\quad LM = OL - OM = \dfrac{R\sin\frac{\pi}{4}}{\frac{\pi}{4}} - \dfrac{R}{\sqrt{2}}$

Now taking the moments of all the forces about line BC.

$$2R_1 \times CD = 2\left(w \cdot R\frac{\pi}{2}\right)CJ - \left(w \cdot R\frac{\pi}{2}\right)(LM)$$

or $\quad 2R_1\sqrt{2}\,R = 2\left(w \cdot R\dfrac{\pi}{2}\right)\dfrac{R}{\sqrt{2}} - wR\dfrac{\pi}{2}\left(\dfrac{R\sin\frac{\pi}{4}}{\frac{\pi}{4}} - \dfrac{R}{\sqrt{2}}\right)$

$\therefore \quad R_1 = \dfrac{wR}{2\sqrt{2}}\left(\dfrac{\pi}{\sqrt{2}} - \dfrac{2}{\sqrt{2}} + \dfrac{\pi}{2\sqrt{2}}\right) = wR\left(\dfrac{3\pi}{8} - \dfrac{1}{2}\right) = 0.678\,wR \qquad ...(20.18)$

Hence from (i), $R_2 = \dfrac{3}{4}wR\pi - R_1 = \dfrac{3}{4}wR\pi - wR\left[\dfrac{3}{8}\pi - \dfrac{1}{2}\right] = wR\left[\dfrac{3}{4}\pi - \dfrac{3}{8}\pi + \dfrac{1}{2}\right]$

or $\quad R_2 = wR\left[\dfrac{3}{8}\pi + \dfrac{1}{2}\right] = 1.678\,wR \qquad ...(20.19)$

After having determined the reactions, the expression for $M_\phi$ and $M_\phi^t$ at any point can be easily written. For example, consider point $P$ at angle $\phi$, in sector $AB$.

Load on $AP = W_\phi = wR\phi$

The C.G. of load acts at point $E$, such that

$$OE = \dfrac{R\sin \phi/2}{\phi/2}\,;\quad OF = R\cos\phi/2\,;\quad EG = OE \cdot \sin\phi/2 = \dfrac{R\sin^2 \phi/2}{\phi/2}$$

$$OG = OE \cdot \cos \phi/2 = \frac{R \sin \phi/2 \cos \phi/2}{\phi/2} = \frac{R \sin \phi}{\phi}$$

$$FH = OF \cdot \sin \phi/2 = R \cos \phi/2 \sin \phi/2; \quad PG = PO - OG = R - \frac{R \sin \phi}{\phi} = R \left(1 - \frac{\sin \phi}{\phi}\right)$$

Draw $AQ$ perpendicular to $PO$.

$$AQ = R \sin \phi \,; \quad OQ = R \cos \phi \,; \quad PQ = R - R \cos \phi = R(1 - \cos \phi)$$

Now B.M. at $P$ is the moment, about $PO$, of all forces to the left of it. Hence sagging B.M. $M_\phi$ is given by

$$M_\phi = R_1 \times AQ - W_\phi \times GE = 0.678\, wR \cdot R \sin \phi - wR\phi \left(\frac{R \sin^2 \phi/2}{\phi/2}\right)$$

$$= wR^2 (0.678 \sin \phi - 1 + \cos \phi) \qquad \ldots(20.20)$$

The expression is valid for $\phi < 90°$. For $\phi > 90°$, we have

$$M_\phi = w R^2 (0.678 \sin \phi - 1 + \cos \phi) + R_2 \cdot R \sin (\phi - \pi/2)$$

or $\qquad M_\phi = wR^2 (0.678 \sin \phi - 1 + \cos \phi) + 1.678\, w R^2 \sin (\phi - \pi/2) \qquad \ldots(20.21)$

The twisting moment $M_\phi^t$ at any point is the moment, about tangent at $P$, of all the forces, to one side of it. Taking twisting moment to be positive when the left portion twists the beam towards its centre of curvature,

$$\therefore \quad M_\phi^t = -R_1 \times PQ + W_\phi \times PG = -0.678\, wR \times R(1 - \cos \phi) + w R \phi R \left(1 - \frac{\sin \phi}{\phi}\right)$$

$$M_\phi^t = wR^2 [-0.678 + 0.678 \cos \phi + \phi - \sin \phi]$$

$$\therefore \quad M_\phi^t = wR^2 [\phi + 0.678 \cos \phi - \sin \phi - 0.678] \qquad \ldots(20.22)$$

The above expression is valid when $\phi < 90°$. For $\phi > 90°$, we have,

$$M_\phi^t = w R^2 [\phi + 0.678 \cos \phi - \sin \phi - 0.678] - 1.678\, wR \left[R - R \cos \left(\phi - \frac{\pi}{2}\right)\right]$$

or $\qquad M_\phi^t = wR^2 [\phi + 0.678 \cos \phi + 0.678 \sin \phi - 2.356] \qquad \ldots(20.23)$

The S.F. at any point $P$ is given by

$$F_\phi = R_1 - W_\phi \quad (\text{for } \phi < 90°)$$

or $\qquad F_\phi = 0.678\, wR - wR\phi = wR (0.678 - \phi) \qquad \ldots(20.24)$

When $\phi > 90°$, $\qquad F_\phi = R_1 - W_\phi + R_2 = 0.678\, wR - wR\phi + 1.678\, wR = wR(2.356 - \phi) \qquad \ldots(20.25)$

## 20.5. CURVED BEAM FIXED AT ENDS

Fig. 20.8 shows a curved beam $ABC$, fixed at end $A$ and $B$, and subtending an angle $2\theta$ at the centre. The beam carries a uniformly distributed load $w$ per unit length of the beam. Because of fixidity, there will be three reaction components at each support (i) shear force $F_0$ (ii) bending moment $M_0$ and (iii) twisting moment $M_0^t$. The beam is, therefore, statically indeterminate to first degree. We shall use the method of strain energy to solve the problem.

Because of symmetry, shear force and torsional moment at the middle point $C$ will be zero. Let the bending moment at this point be $M$. If we cut the beam, at $C$, in two portions, $M_c$ will be the reaction component at $C$. This component is evidently *about* the radial axis $CO$.

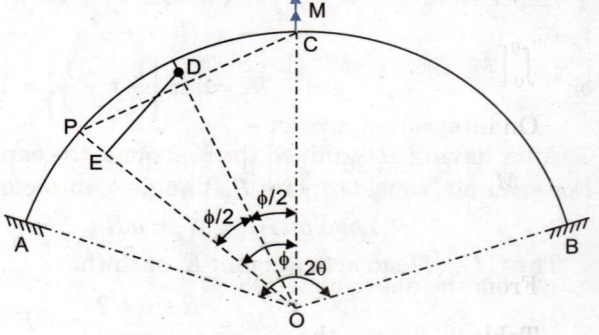

Fig. 20.8

Consider a point $P$ at angular distance $\phi$ with $OC$. Load on section $CP$ is given by $W_\phi = wR\phi$, and its $C.G.$ will be at point $D$ such that

$$OD = \frac{R \sin \phi/2}{\phi/2}$$

From $D$, drop perpendicular $DE$ on radial axis $PO$

$\therefore$
$$DE = DO \cdot \sin \phi/2 = \frac{R \sin^2 \phi/2}{\phi/2}$$

$$EO = DO \cdot \cos \phi/2 = \frac{R \sin \phi/2 \cos \phi/2}{\phi/2} = \frac{R \sin \phi}{\phi}$$

$$PE = PO \cdot EO = R - \frac{R \sin \phi}{\phi} = R\left(\frac{\phi - \sin \phi}{\phi}\right)$$

Now, $\qquad F_\phi = W_\phi = wR \cdot \phi$ ...(20.26)

The bending moment $M_c$ (sagging) can be resolved in component $M_c \cos \phi$ about $PO$ and $M_c \sin \phi$ about tangential axis at $P$. The moment $M_c \cos \phi$ about PO will evidently be the bending moment component at $P$, While $M_c \sin \phi$ about tangential axis at $P$ will be the torsional moment component.

$\therefore \qquad M_\phi = M_c \cos \phi - W_\phi ED = M_c \cos \phi - wR\phi \left(\frac{R \sin^2 \phi/2}{\phi/2}\right)$

or $\qquad M_\phi = M_c \cos \phi - wR^2 (1 - \cos \phi)$ ...(20.27)

Similarly, $\qquad M_\phi^t = M_c \sin \phi - W_\phi PE = M_c \sin \phi - wR\phi \cdot R\left(\frac{\phi - \sin \phi}{\phi}\right)$

or $\qquad M_\phi^t = M_c \sin \phi - wR^2 (\phi - \sin \phi)$ ...(20.28)

Let $U$ be the strain energy of half the portion ($CA$) of the beam consisting of strain energy due to B.M. and that due to torsional moment. The strain energy due to S.F. is extermely small and may be neglected.

Hence $\qquad U = \int_0^\theta \frac{M_\phi^2 \, ds}{2 EI} + \int_0^\theta \frac{(M_\phi^t)^2 \, ds}{2 GJ^t}$ ; where $ds = R \, d\phi$.

In order to determine $M_C$, put $\dfrac{\partial U}{\partial M_C}$ equal to zero.

$\therefore \qquad \int_0^\theta \frac{M_\phi}{EI} \frac{\partial M_\phi}{\partial M_c} R \cdot d\phi + \int_0^\theta \frac{M_\phi^t}{GJ^t} \frac{\partial M_\phi^t}{\partial M_c} R \cdot d\phi = 0$

where $G$ = modulus of rigidity = $\dfrac{E}{2(1-\mu)} \approx 0.43 \, E$ for concrete.

$J^t$ = torsion factor (rotational constant), depending upon the shape of the section.

In the expression, $\dfrac{\partial M_\phi}{\partial M_c} = \cos \phi$ and $\dfrac{\partial M_\phi^t}{\partial M_c} = \sin \phi$ ; Putting $\dfrac{EI}{GJ^t} = T_F$, we get

$$\int_0^\theta \left[M_c \cos \phi - wR^2 (1 - \cos \phi)\right] \cos \phi \, d\phi + T_F \int_0^\theta \{M_c \sin \phi - wR^2 (\phi - \sin \phi)\} \sin \phi \, d\phi = 0$$

On integration, we get

$$M_c \left\{\frac{\theta}{2} + \frac{\sin 2\theta}{4} + T_F \left(\frac{\theta}{2} - \frac{\sin 2\theta}{4}\right)\right\} + wR^2 \left\{\left(\frac{\theta}{2} - \sin \theta + \frac{\sin 2\theta}{4}\right) + T_F \left(\theta \cos \theta - \sin \theta + \frac{\theta}{2} - \frac{\sin 2\theta}{4}\right)\right\} = 0$$

...(20.29)

From the above equation, $M_c$ can be found.

Table 20.2 gives the values of factor $\dfrac{M_c}{wR^2}$ in terms of factor $T_F$ for various values of central angle $2\theta$.

Knowing the value of $M_c$, moments $M_\phi$ and $M_\phi^t$ can be computed from *Eqs. 20.27* and *20.28*.

## TABLE 20.2

| $2\theta$ | $45°$ | $60°$ | $90°$ | $120°$ | $180°$ |
|---|---|---|---|---|---|
| $\dfrac{M_c}{wR^2}$ | $\dfrac{0.0096 + 0.0004\, T_F}{0.3731 + 0.0195\, T_F}$ | $\dfrac{0.0217 + 0.0013\, T_F}{0.4783 + 0.0453\, T_F}$ | $\dfrac{0.0644 + 0.0090\, T_F}{0.6427 + 0.1427\, T_F}$ | $\dfrac{0.1259 + 0.0353\, T_F}{0.7401 + 0.3071\, T_F}$ | $\dfrac{0.2146 + 0.2146\, T_F}{0.7854 + 0.7854\, T_F} = 0.2732$ |

## SPECIAL CASE: SEMI-CIRCULAR BEAM FIXED AT THE ENDS

Special case arises when $2\theta = 180° = \pi$. Substituting $\theta = \dfrac{\pi}{2}$ in *Eq. 20.29*, we get,

$$\frac{M_c}{wR^2} = -\frac{\left(\dfrac{\theta}{2} - \sin\theta + \dfrac{\sin 2\theta}{4}\right) + \left(\theta\cos\theta - \sin\theta + \dfrac{\theta}{2} - \dfrac{\sin 2\theta}{4}\right)T_F}{\left(\dfrac{\theta}{2} + \dfrac{\sin 2\theta}{4}\right) + \left(\dfrac{\theta}{2} - \dfrac{\sin 2\theta}{4}\right)T_F}$$

$$\frac{M_c}{wR^2} = -\frac{\left(\dfrac{\pi}{4} - 1 + 0\right) + \left(0 - 1 + \dfrac{\pi}{4} - 0\right)T_F}{\left(\dfrac{\pi}{4} + 0\right) + \left(\dfrac{\pi}{4} - 0\right)T_F} = -\left(1 - \dfrac{4}{\pi}\right) = \left(\dfrac{4}{\pi} - 1\right) = 0.2732 \qquad \ldots(20.30)$$

Thus, the central moment $M_c$ is independent of the properties of the section and the material. Again from *Eqs. 20.27* and *20.28* putting $\phi = \theta = \dfrac{\pi}{2}$, we get the expressions for B.M. ($M_0$) and T.M. ($M_0^t$) at support as under :

$$M_0 = M_c \cos\frac{\pi}{2} - wR^2\left(1 - \cos\frac{\pi}{2}\right) = -wR^2 \quad (i.e.,\ \text{hogging})$$

and

$$M_0^t = M_c \sin\frac{\pi}{2} - wR^2\left(\frac{\pi}{2} - \sin\frac{\pi}{2}\right) = wR^2\left(\frac{4}{\pi} - 1\right) - wR^2\left(\frac{\pi}{2} - 1\right)$$

$$= wR^2\left(\frac{4}{\pi} - 1 - \frac{\pi}{2} + 1\right) = -wR^2\left(\frac{\pi}{2} - \frac{4}{\pi}\right) = -0.2976\, wR^2 \qquad \ldots(20.31)$$

## ALTERNATIVE SOLUTION FOR SEMI-CIRCULAR BEAM FIXED AT ENDS

At each end there will be three reaction components : (*i*) shear force $F_0$ (*ii*) bending moment $M_0$ about $OA$ axis and (*iii*) twisting moment $M_0^t$ about tangential axis.

Total load on beam $= wR\pi$ $\quad\therefore\quad F_0 = \dfrac{1}{2} wR\pi$

Load on half beam $AC = \dfrac{1}{2} wR\pi$. Its C.G. lies at $E$ at a distance $= \dfrac{R \sin\dfrac{\pi}{4}}{\pi/4}$ from the centre. Perpendicular distance of this C.G. from $OA = \dfrac{R \sin\dfrac{\pi}{4}}{\pi/4} \cdot \sin\dfrac{\pi}{4}$. Taking moments about $OA$, of all forces acting on half the beam $AC$, we get

$$M_0 = \frac{1}{2} wR\pi \left(\frac{R\sin^2\dfrac{\pi}{4}}{\pi/4}\right) = wR^2 \text{ (hogging)}$$

The twisting moment $M_0^t$ cannot be found from equation of statics alone. We shall use strain energy method. Consider any point $P$, at angle $\phi$ with $OA$ (Fig. 20.9).

Load on $AP = W_\phi = wR \cdot \phi$

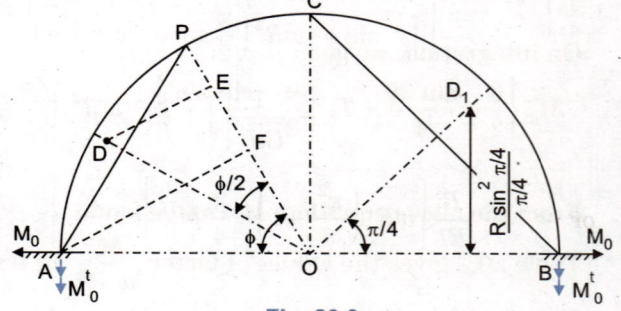

Fig. 20.9

Distance of its C.G. $= OD = \dfrac{R \sin \dfrac{\phi}{2}}{\dfrac{\phi}{2}}$. Draw $DE$ perpendicular to $OP$

Then $\quad DE = DO \sin \dfrac{\phi}{2} = \dfrac{R \sin^2 \dfrac{\phi}{2}}{\dfrac{\phi}{2}}$ ; $\quad OE = DO \cos \dfrac{\phi}{2} = \dfrac{R \sin \dfrac{\phi}{2} \cdot \cos \dfrac{\phi}{2}}{\dfrac{\phi}{2}} = \dfrac{R \sin \phi}{\phi}$

$$PE = R \left[1 - \dfrac{\sin \phi}{\phi}\right].$$

If $AF$ is drawn perpendicular to $PO$, we have $\quad AF = R \sin \phi$
$\quad OF = R \cos \phi$ and $PF = R (1 - \cos \phi)$.

Now the sagging B.M. $M_\phi$ at $P$ is given by
$$M_\phi = F_0 \times AF - M_0 \cos \phi + M_0^t \sin \phi - W_\phi \times DE$$

or $\quad M_\phi = \dfrac{1}{2} wR \pi \cdot R \sin \phi - wR^2 \cos \phi + M_0^t \sin \phi - wR \phi \dfrac{R \sin^2 \dfrac{\phi}{2}}{\dfrac{\phi}{2}}$

or $\quad M_\phi = wR^2 \left[\dfrac{\pi}{2} \sin \phi - \cos \phi - (1 - \cos \phi)\right] + M_0^t \sin \phi \qquad \ldots(i)$

Similarly, the twisting moment $M_\phi^t$ at $P$ is the moment of all forces, to its left side, about tangential axis at $P$.
$$M_\phi^t = - F_0 \times PF + M_0 \sin \phi + M_0^t \cos \phi + W_\phi \cdot PE$$

or $\quad M_\phi^t = - \dfrac{1}{2} wR \pi \cdot R (1 - \cos \phi) + wR^2 \sin \phi + M_0^t \cos \phi + wR \phi \cdot R \left(1 - \dfrac{\sin \phi}{\phi}\right)$

or $\quad M_\phi^t = wR^2 \left[- \dfrac{\pi}{2} (1 - \cos \phi) + \sin \phi + \phi - \sin \phi\right] + M_0^t \cos \phi \qquad \ldots(ii)$

Now total strain energy $U$ is given by
$$U = \int \dfrac{M_\phi^2}{2 EI} ds + \int \dfrac{(M_\phi^t)^2}{2 GJ^t} ds, \quad \text{where} \quad ds = R \cdot d\phi$$

Putting $\dfrac{\partial U}{\partial M_0^t} = 0$, we have, $\dfrac{\partial U}{M_0^t} = \int_0^{\pi/2} \dfrac{M_\phi}{EI} \cdot \dfrac{\partial M_\phi}{\partial M_0^t} R \, d\phi + \int_0^{\pi/2} \dfrac{M_\phi^t}{GJ^t} \cdot \dfrac{\partial M_\phi^t}{\partial M_0^t} R \, d\phi = 0$

where $\dfrac{\partial M_\phi}{\partial M_0^t} = \sin \phi$ and $\dfrac{\partial M_\phi^t}{\partial M_0^t} = \cos \phi$.

Substituting the values, we get

$$\dfrac{1}{EI} \left[\int_0^{\pi/2} \sin \phi \left\{wR^2 \left(\dfrac{\pi}{2} \sin \phi - \cos \phi - 1 + \cos \phi\right) + M_0^t \sin \phi\right\} R d\phi\right]$$
$$+ \dfrac{1}{GJ^t} \left[\int_0^{\pi/2} \cos \phi \left\{wR^2 \left(- \dfrac{\pi}{2} + \dfrac{\pi}{2} \cos \phi + \phi\right) + M_0^t \cos \phi\right\} R d\phi\right] = 0$$

or $\quad \dfrac{R}{EI} \left[wR^2 \left(\dfrac{\pi^2}{8} - 1\right) + \dfrac{\pi}{4} M_0^t\right] + \dfrac{R}{GJ^t} \left\{wR^2 \left(\dfrac{\pi^2}{8} - 1\right) + \dfrac{\pi}{4} M_0^t\right\} = 0$

$$\therefore \quad \left[wR^2\left(\frac{\pi^2}{8}-1\right)+\frac{\pi}{4}M_0^t\right]\left[1+\frac{EI}{GJ^t}\right]=0$$

$$\therefore \quad M_0^t = -\frac{4}{\pi}wR^2\left(\frac{\pi^2}{8}-1\right) = -wR^2\left(\frac{\pi}{2}-\frac{4}{\pi}\right) \qquad \ldots(20.31)$$

which is the same as found earlier.

Substituting the value of $M_0^t$ in *Eqs.* (*i*) and (*ii*), we can find $M_\phi$ and $M_\phi^t$ at any point. For example, at point $C$, $\phi = \frac{\pi}{2}$.

Hence, $\quad M_c = wR^2\left[\frac{\pi}{2}\sin\frac{\pi}{2}-\cos\frac{\pi}{2}-\left(1-\cos\frac{\pi}{2}\right)\right]-wR^2\left(\frac{\pi}{2}-\frac{4}{\pi}\right)\sin\frac{\pi}{2}$

$$= wR^2\left[\frac{\pi}{2}-1\right]-wR^2\left(\frac{\pi}{2}-\frac{4}{\pi}\right) = wR^2\left(\frac{4}{\pi}-1\right) = 0.2732\, wR^2. \qquad \ldots(20.30)$$

This is also the same as found earlier.

Similarly, $\quad M_c^t = wR^2\left[-\frac{\pi}{2}\left(1-\cos\frac{\pi}{2}\right)+\sin\frac{\pi}{2}+\frac{\pi}{2}-\sin\frac{\pi}{2}\right]-wR^2\left(\frac{\pi}{2}-\frac{4}{\pi}\right)\cos\frac{\pi}{2} =$ zero, as expected.

## 20.6. SEMI-CIRCULAR BEAM WITH SLAB

Fig. 20.10 shows a semi-circular beam *ABC* fixed at ends *A* and *B*, and supporting a semi-circular slab. Since the length of the slab is much greater than the width, it will behave as one way slab. Let $w$ be the uniformly distributed load per unit area of slab.

Due to symmetry, the reaction components at $A$ and $B$ will be equal. At each end, there will be three reaction components: (*i*) shear force $F_0$ (*ii*) Bending moment $M_0$ and (*iii*) Twisting moment $T_0$.

Let us consider a small element $DG$, at an angle $\phi$ with $AO$, and subtending an angle $d\phi$ at the centre. The load transferred to this element $DG$ of the beam will evidently be equal to half the load on the shaded area *DEFG*.

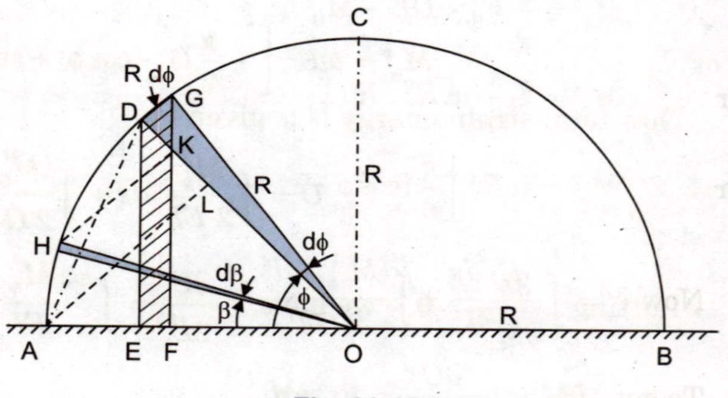

Now $\quad DE = R\sin\phi$ ; $\quad EO = R\cos\phi$

$EF = d(R\cos\phi) = R\sin\phi \cdot d\phi.$

$\therefore$ Load on shaded area $\approx w\,(DE \times EF)$

$= w\,(R\sin\phi)(R\sin\phi\, d\phi) = wR^2 \sin^2\phi\, d\phi.$

$\therefore$ Load transferred to $DG = \frac{1}{2}wR^2 \sin^2\phi\, d\phi.$

Length $\quad DG = R d\phi$

$\therefore$ Uniformly distributed load $w_\phi$ at the element $= \dfrac{\frac{1}{2}wR^2 \sin^2\phi\, d\phi}{R\, d\phi}$

or $\quad w_\phi = \frac{1}{2}wR \sin^2\phi \quad$ per unit length $\qquad \ldots(20.32)$

Fig. 20.10

By equating the vertical forces, and balancing moments about $AB$, $F_0$ and $M_0$ can be determined. But $M_0^t$ cannot be determined statically, and the problem is indeterminate to one degree. By balancing the total vertical forces, we have,

$$F_0 = \text{load on portion } AC = \int_0^{\pi/2} w_\phi\, R d\phi$$

or $\quad F_0 = \int_0^{\pi/2} (\tfrac{1}{2}wR \sin^2\phi)\, R\, d\phi = \dfrac{wR^2}{2}\left[\dfrac{1}{2}\phi - \dfrac{1}{4}\sin 2\phi\right]_0^{\pi/2} = \dfrac{\pi}{8}wR^2 \qquad \ldots(20.33)$

Now B.M at $A$ = moment of all forces on $AC$, about $AO$ axis.

$$M_0 = \int_0^{\pi/2} (w_\phi \, Rd\phi) \times DE = \int_0^{\pi/2} (\tfrac{1}{2} w R \sin^2 \phi) \, Rd\phi \, (R \sin \phi) = \frac{wR^3}{2} \int_0^{\pi/2} \sin^3 \phi \, d\phi$$

$$= \frac{wR^3}{2} \left[ -\tfrac{1}{3} \cos \phi (\sin^2 \phi + 2) \right]_0^{\pi/2} = \frac{wR^3}{2} \left[ 0 + \frac{2}{3} \right] = \frac{wR^3}{3} \qquad \ldots(20.34)$$

Thus $F_0$ and $M_0$ are known. Let the twisting moment $M_0{}^t$ at $A$ be acting clockwise. At any point $D$ at angle $\phi$, the sagging bending moment will be equal to the moment of all forces to the left side, about the radial axis $DO$.

Consider an elemental load $\delta W_\beta$ at $H$, its value being given by

$$\delta W_\beta = w_\beta \, Rd\phi = \frac{1}{2} wR \sin^2 \beta \, . R \, d\beta \qquad \ldots\text{(from Eq. 20.32)}$$

Draw $HK$ perpendicular to $DO$. Evidently $HK = R \sin (\phi - \beta)$ and $OK = R \cos (\phi - \beta)$. If a perpendicular $AL$ is drawn on $DO$, we have $AL = R \sin \phi$; $OL = R \cos \phi$.

Now $M_\phi = F_0 \times AL - M_0 \cos \phi + M_0{}^t \sin \phi - \int_0^\phi (\delta W_\beta) \times HK$

or $\quad M_\phi = \dfrac{\pi}{8} wR^2 . R \sin \phi - \dfrac{wR^3}{3} \cos \phi + M_0{}^t \sin \phi - \int_0^\phi \tfrac{1}{2} w R \sin^2 \beta \, . R \, d\beta \, . R \sin (\phi - \beta).$

or $\quad M_\phi = \dfrac{\pi}{8} wR^3 \sin \phi - \dfrac{wR^3}{3} \cos \phi + M_0{}^t \sin \phi - \dfrac{wR^3}{2} \int_0^\phi \sin^2 \beta \sin (\phi - \beta) \, d\beta$

or $\quad M_\phi = \dfrac{wR^3}{3} \left[ \dfrac{3\pi}{8} \sin \phi - \tfrac{1}{2} (1 + \cos^2 \phi) \right] + M_0{}^t \sin \phi \qquad \ldots(20.35)$

Similarly, twisting moment at $D$ is equal so moment of all the forces to the left of it, about tangent at $D$.

$\therefore \quad M_\phi{}^t = -F_0 \times DL + M_0 \sin \phi + M_0{}^t \cos \phi + \int_0^\phi (\delta W_\beta) \times DK$

or $\quad M_\phi{}^t = -\dfrac{\pi}{8} w R^2 . R (1 - \cos \phi) + \dfrac{wR^3}{3} \sin \phi + M_0{}^t \cos \phi + \int_0^\phi \tfrac{1}{2} w R \sin^2 \beta \, . R \, d\beta \, . R \{1 - \cos (\phi - \beta)\}$

or $\quad M_\phi{}^t = wR^3 \left[ \dfrac{\pi}{8} (\cos \phi - 1) + \dfrac{\phi}{4} + \dfrac{1}{24} \sin 2\phi \right] + M_0{}^t \cos \phi \qquad \ldots(20.36)$

Now $U = \int \dfrac{M_\phi^2 \, ds}{2 \, EI} + \int \dfrac{(M_\phi{}^t)^2 \, ds}{2 \, GJ^t}$ where $ds = R . d\phi$.

To get $M_0^t$, put $\dfrac{\partial U}{\partial M_0^t} = $ to zero.

$\therefore \quad \dfrac{\partial U}{\partial M_0^t} = \int_0^{\pi/2} \dfrac{M_\phi}{EI} \dfrac{\partial M_\phi}{\partial M_0^t} \times R . d\phi + \int_0^{\pi/2} \dfrac{M_\phi^t}{GJ^t} \dfrac{\partial M_\phi^t}{\partial M_0^t} Rd\phi = 0$

where $\dfrac{\partial M_\phi}{\partial M_0^t} = \sin \phi$ and $\dfrac{\partial M_\phi^t}{\partial M_0^t} = \cos \phi$

$\therefore \quad \dfrac{R}{EI} \int_0^{\pi/2} \sin \phi \left[ \dfrac{wR^3}{3} \left\{ \dfrac{3\pi}{8} \sin \phi - \tfrac{1}{2} (1 + \cos^2 \phi) \right\} + M_0^t \sin \phi \right] d\phi$

$\qquad + \dfrac{R}{GJ^t} \int_0^{\pi/2} \cos \phi \left[ w R^3 \left\{ \dfrac{\pi}{8} (\cos \theta - 1) + \dfrac{\pi}{4} + \dfrac{1}{24} \sin 2\phi \right\} + M_0^t \cos \phi \right] d\phi = 0$

or $\quad \dfrac{R}{EI}\left[\dfrac{wR^3}{3}\left\{\dfrac{3}{32}\pi^2 - \dfrac{2}{3}\right\} + \dfrac{\pi}{4} M_0^{\,t}\right] + \dfrac{R}{GJ^t}\left[wR^3\left\{\dfrac{\pi^2}{32} - \dfrac{2}{9}\right\} + \dfrac{\pi}{4} M_0^{\,t}\right] = 0$

or $\quad \left[wR^3\left\{\dfrac{1}{32}\pi^2 - \dfrac{2}{9}\right\} + \dfrac{\pi}{4} M_0^{\,t}\right]\left[1 + \dfrac{EI}{GJ^t}\right] = 0.$ From which,

$$M_0^{\,t} = -\dfrac{4}{\pi} wR^3\left[\dfrac{\pi^2}{32} - \dfrac{2}{9}\right] = -wR^3\left[\dfrac{\pi}{8} - \dfrac{8}{9\pi}\right] \qquad \ldots(20.37)$$

**Note.** The minus sign shows that the twisting moment at A will be in anticlockwise direction. Knowing the value of $M_0^{\,t}$, the values of $M_\phi$ and $M_\phi^{\,t}$ can be easily evaluated from *Eqs. 20.35* and *20.36*. For example, at the middle of the beam, $\phi = \dfrac{\pi}{2}$.

$\therefore \qquad M_c = \dfrac{wR^3}{3}\left[\dfrac{3\pi}{8}\sin\dfrac{\pi}{2} - \dfrac{1}{2}\left(1 + \cos^2\dfrac{\pi}{2}\right)\right] - wR^3\left[\dfrac{\pi}{8} - \dfrac{8}{9\pi}\right]\sin\dfrac{\pi}{2}$

$\therefore \qquad M_c = wR^3\left(\dfrac{8}{9\pi} - \dfrac{1}{9}\right) \qquad \ldots(20.38)$

and $\qquad M_c^{\,t} = wR^3\left[\dfrac{\pi}{8}\left(\cos\dfrac{\pi}{2} - 1\right) + \dfrac{\pi}{8} + \dfrac{1}{24}\sin\pi\right] - wR^3\left[\dfrac{\pi}{8} - \dfrac{8}{9\pi}\right]\cos\dfrac{\pi}{2}$

or $\qquad M_c^{\,t} = 0, \ \ \text{as expected.}$

## 20.7. TORSION FACTOR

For statically indeterminate beams, curved in plan, the torsion factor $J^t$ is used for the determination of torsional moment at any section. Torsion factor is also sometimes known as rotation constant, and its value for various shapes of the section are given below.

**1. Circular Section :** $\qquad J^t = \dfrac{\pi D^4}{32} \qquad \ldots(20.39)$

where $D$ = diameter of the circle.

**2. Elliptical section :** $\qquad J^t = \dfrac{\pi a^3 b^3}{a^2 + b^2} \qquad \ldots(20.40)$

where $a$ and $b$ are the *semi-axes* of the ellipse.

**3. Equilateral section :** $\qquad J^t = \dfrac{a^4 \sqrt{3}}{80} \qquad \ldots(20.41)$

where $a$ is the side of the triangle.

**4. Hollow circular section :** $\qquad J^t = \dfrac{4s^2 t}{p} \qquad \ldots(20.42)$

where $s$ = area enclosed ; $T$ = thickness of the tube ; $P$ = perimeter of the tube.

**5. Rectangular section :** Let $b$ = width and $D$ = depth of the section

**(i) Approximate expression**

When $\dfrac{D}{b} < 1.6,$ $\qquad J^t = \dfrac{b^3 D^3}{2.58(b^2 + D^2)} \qquad \ldots(20.43)$

When $\dfrac{D}{b} > 1.6,$ $\qquad J^t = \dfrac{b^3 D}{3}\left(1 - 0.63\dfrac{b}{D}\right) \qquad \ldots(20.44)$

**(ii) St-venant expression:** $\quad J^t = \dfrac{\beta D b^3}{16}$ ...(20.45)

where $\quad \beta = \dfrac{16}{3} - 3.36 \left(\dfrac{b}{D}\right)\left[1 - \dfrac{b^4}{12 D^4}\right]$ (approx.), ...(20.46)

or $\quad \beta$ as given by Table 20.3

### TABLE 20.3. Values of β

| D/b | β | D/b | β | D/b | β |
|---|---|---|---|---|---|
| 1.0 | 2.249 | 1.30 | 2.833 | 1.7 | 3.375 |
| 1.1 | 2.464 | 1.40 | 2.99 | 1.75 | 3.429 |
| 1.2 | 2.658 | 1.5 | 3.132 | 2.0 | 3.671 |
| 1.25 | 2.748 | 1.6 | 3.260 | 3.0 | 4.213 |

**(iii) Theoretical Solution**

Timoshenko and Goodier obtained the values of $J^t$ by using the values of $J^t$ obtained by all the three methods.

### TABLE 20.4. Values of $J^t$

| D/b | $J^t$ by approx. method (Eqs. 20.43 and 20.44) | $J^t$ by st. Venants expression | $J^t$ by Timoshenko and Goodier |
|---|---|---|---|
| 1.0 | $0.1397 D^4$ | $0.1406 D^4$ | $0.1406 D^4$ |
| 1.2 | $0.0954 D^4$ | $0.0961 D^4$ | $0.0956 D^4$ |
| 1.5 | $0.0573 D^4$ | $0.0580 D^4$ | $0.0580 D^4$ |
| 2.0 | $0.0285 D^4$ | $0.0287 D^4$ | $0.0285 D^4$ |
| 2.5 | $0.0159 D^4$ | $0.0160 D^4$ | $0.0158 D^4$ |

**Example 20.3.** *A beam is curved in plan in the form of arc of a circle with radius R = 4 m and central angle equal to 90°. The beam carries a super-imposed load of 2 kN/m, and is fixed at both the ends. If the section of the beam is rectangular, having depth = 600 mm and width = 300 mm, draw the bending moment and torsional moment diagrams for the beam. Take G = 0.4E for concrete.*

**Solution:** (Refer Fig. 20.8). The value of torsion factor is given by

$$J^t = \dfrac{b^3 D}{3}\left(1 - 0.63\dfrac{b}{D}\right) \quad \text{where} \quad \dfrac{b}{D} = \dfrac{300}{600} = \dfrac{1}{2}$$

$\therefore \quad J^t = \dfrac{D^4}{24}\left(1 - \dfrac{0.63}{2}\right) = 0.0285 D^4 = 0.0285 (0.6)^4 = 0.00369$ m$^4$ units.

$$I = \dfrac{b D^3}{12} = \dfrac{0.3(0.6)^3}{12} = 0.0054 \, ; \, T_F = \dfrac{EI}{GJ^t} = \dfrac{2.5 \times 0.0054}{0.00369} = 3.659$$

From Table 20.2, for $2\theta = 90°$

$$M_c = wR^2 \left[\dfrac{0.0644 + 0.0090 \, T_F}{0.6427 + 0.1427 \, T_F}\right] = 2(4)^2 \left[\dfrac{0.0644 + 0.0090 \times 3.659}{0.6427 + 0.1427 \times 3.659}\right] = 2.674 \text{ kN-m}$$

Hence from *Eqs. 20.27* and *20.28*, we have
$$M_\phi = 2.674 \cos\phi - 2(4)^2(1-\cos\phi)$$
$$= 34.674 \cos\phi - 32 \qquad \ldots(i)$$
and $M_\phi^t = 2.674 \sin\phi - 2(4)^2(\phi - \sin\phi)$
$$= 34.674 \sin\phi - 32\phi \qquad \ldots(ii)$$
The shear force is given by *Eq. 20.26*
$$F_\phi = wR\phi = 2 \times 4\phi = 8\phi \qquad \ldots(iii)$$
The value of $F_\phi$, $M_\phi$ and $M_\phi^t$ are tabulated below.

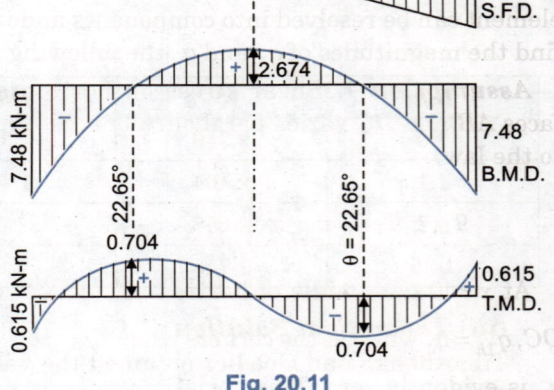

| $\phi$ | $F_\phi$ (kN) | $M_\phi$ (kN-m) | $M_\phi^t$ (kN-m) | Location |
|---|---|---|---|---|
| 0° | 0° | 2.674 | 0 | Centre line |
| 15° | 2.094 | 1.493 | + 0.597 | |
| 22.65° | 3.162 | 0.00 | + 0.704 | Point of contraflexure |
| 30° | 4.189 | 0.028 | + 0.582 | |
| 45° | 6.283 | − 7.48 | − 0.615 | Support |

For points of contraflexure, $M_\phi = 0 = 34.674 \cos\phi - 32$

or
$$\phi = \cos^{-1}\left(\frac{32}{34.674}\right) = 22.65° = 0.3953 \text{ rad.}$$

The maximum torsional moment will therefore occur at $\phi = 22.65° = 0.3953$ rad.
$$M^t_{max} = 34.674 \sin 22.65° - 32 \times 0.3953 = 0.704 \text{ kN-m}$$
The S.F., B.M. and T.M. diagrams are shown in Fig. 20.11.

## 20.8. STRESSES DUE TO TORSION IN CONCRETE BEAMS

When torsion is applied to a concrete beam, shear stresses are induced both in vertical as well as in horizontal direction. Due to this, cracks are developed on the surface of the beam ; these cracks are in the form of spiral.

### 1. Circular Section

Let $r$ = radius of the circular section ; $q'$ = maximum shear intensity due to torsion, then from simple theory of torsion of shafts
$$\frac{q'}{r} = \frac{M^t}{J}, \quad \text{or} \quad q' = \frac{M^t}{J} \times r, \quad J = \text{polar moment of inertia} = \frac{\pi}{2} r^4$$
$$q' = \frac{M^t}{\frac{\pi}{2}r^4} \times r = \frac{2 M^t}{\pi r^3} \qquad \ldots(20.47)$$

or
$$q' = \frac{16 M^t}{\pi D^3} \qquad \ldots[20.47(a)]$$

where $D$ is the diameter of the circular section.

**2. Square Section** : The stresses due to torsion in a beam of square section with side of square $a$ can be approximately determined by assuming the section to be equivalent to a circular section of diameter $a$. Thus,
$$q' = \frac{16 M^t}{\pi a^3} \qquad \ldots(20.48)$$

**3. Rectangular Section** : The rigid analysis for the determination of shear stresses due to torsion in a rectangular section is extremely difficult. An approximate analysis is given here. Fig. 20.12 (a) shows a rectangular beam $b \times D$, subjected to a torsional moment $M^t$. The torsional moment causes shear stresses of varying magnitudes at the outer faces as well as at the inner side of the section. The shear stresses at the four corners $A$, $B$, $C$ and $D$ is found to be zero, while maximum shear of values $q_1$ and $q_2$ occur at the mid-points of the horizontal and vertical faces respectively. The shear stress at the C.G. of the section will also be zero.

Consider an element of size $dx$, $dy$ at co-ordinates $(x, y)$ of the beam section. The shear stress $q$ at this element can be resolved into components and in horizontal and vertical directions respectively. In order to find the magnitudes of $q_h$ and $q_v$ the following assumptions are made:

**Assumption 1.** Shear stress on horizontal faces $AB$ and $DC$ varies parabolically according to the law:

$$q_{1h} = q_1 \left(1 - \frac{4x^2}{b^2}\right) \qquad ...(i)$$

At $x = 0$ (i.e., middle of horizontal face $AB$ or $DC$, $q_{1h} = q_1$ while at the corners where $x = \pm \frac{b}{2}$, $q_h$ is evidently zero. This variation is shown in Fig. 20.12 (b).

Similarly, the shear stress on vertical faces $AD$ and $BC$ varies parabolically according to the law [Fig. 20.12 (c)]:

$$q_{2v} = q_2 \left(1 - \frac{4y^2}{D^2}\right) \qquad ...(ii)$$

**Assumption 2.** The horizontal shear stress $q_h$ varies linearly across the depth of the beam [Fig. 20.12 (d)] with zero value as its C.G. Similarly, the variation of vertical shear stress across the width varies linearly [Fig. 20.12 (e)] with zero value at the C.G. of the section. Thus at the element with $(x, y)$ co-ordinates, the values of horizontal and vertical shear stress are:

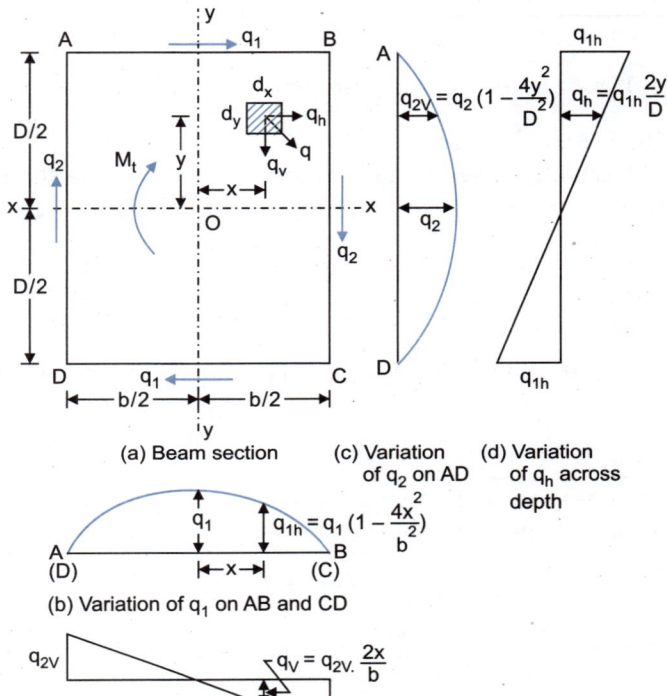

Fig. 20.12

$$q_h = q_{1h} \cdot \frac{2y}{D} = q_1 \left(1 - \frac{4x^2}{b^2}\right) \frac{2y}{D} \qquad ...(iii)$$

and

$$q_v = q_{2v} \cdot \frac{2x}{b} = q_2 \left(1 - \frac{4y^2}{D^2}\right) \frac{2x}{b} \qquad ...(iv)$$

**Assumption 3.** The shear stress $q_1$ and $q_2$ are related by the law:

$$\frac{q_1}{q_2} = \frac{b}{D} \qquad ...(v)$$

This shows that maximum shear stress ($q_2$) occurs at the centre of that side which is nearest to the centroid of the section. Generally, $b < D$, and hence $q_2 > q_1$.

Now, let us again consider the small element of size $dx$, $dy$ [Fig. 20.12 (a)]. The torsional moment of resistance developed by this element is

$$dM^t = (q_h \cdot y + q_v \cdot x) \, dx \cdot dy$$

Hence for the entire section, $M^t = 4 \int_0^{D/2} \int_0^{b/2} (q_h y + q_v x) \, dx \, dy$

Substituting the values of $q_h$ and $q_v$ from Eqs. (iii) and (iv),

$$M^t = 4 \int_0^{0.5D} \int_0^{0.5b} \left[ q_1 \left(1 - \frac{4x^2}{b^2}\right) \frac{2y^2}{D} + q_2 \left(1 - \frac{4y^2}{D^2}\right) \frac{2x^2}{b} \right] dxdy$$

Substituting $q_2 = q_1 \dfrac{D}{b}$, and performing the integration, we get

$$q_1 = \frac{4.5 \, M^t}{bD^2} \qquad \qquad ...[20.49\,(a)]$$

$$q_2 = \frac{4.5 \, M^t}{b^2 D} \qquad \qquad ...[20.49\,(b)]$$

If $b < D$, $q_2 = q'$ = maximum shear.

### Other expressions

Bach has given the following empirical formula for maximum shear stress:

$$q' = \frac{M^t}{b^2 D} \left( 3 + \frac{2.6}{4.5 + \dfrac{b}{D}} \right) \qquad \qquad ...(20.50)$$

Indian Standard, IS : 456, recommends the following expression for maximum shear stress $q'$:

$$q' = \frac{M^t}{b^2 D} \left( 3 + \frac{2b}{D} \right) \qquad \qquad ...(20.51)$$

Timoshenko and Goodier used the method of theory of elasticity for determining the shear stress due to torsion. In general, the shear stress due to torsion can be expressed by the following expression:

$$q' = C_t \frac{M^t}{b^2 D} \qquad \qquad ...(20.52)$$

where $C_t$ is the coefficient. Table 20.5 gives the comparative values of $C_t$ by various formulae:

**TABLE 20.5.** Values of $C_t$ (Eq. 20.52)

| $\dfrac{D}{b}$ | Eq. 20.49 | Bach expression (Eq. 20.50) | I.S. Formula (Eq 20.51) | Theory of elasticity method (Eq. 20.52) |
|---|---|---|---|---|
| 1.0 | 4.5 | 4.79 | 5.00 | 4.80 |
| 1.2 | 4.5 | 4.57 | 4.66 | 4.56 |
| 2.0 | 4.5 | 4.06 | 4.00 | 4.06 |
| 2.5 | 4.5 | 3.88 | 3.80 | 3.88 |

**4. Other sections :** Table 20.6 gives the expressions for maximum shear stress due to torsion, in sections of various shapes.

## 20.9. REINFORCEMENT DUE TO TORSION

Since the nature of failure of a beam subjected to torsion is similar to the failure due to diagonal tension, the reinforcement due to torsion is provided in the similar manner as for the transverse shear. It has been observed experimentally that the torsional reinforcement will be most effective if it is provided in the form of a 45° spiral having the same direction as that of torsional moment, or if it is provided in the direction of principal tensile stresses. However, due to difficulties in construction and for reasons of economy, torsional reinforcement is provided in the forms of : (i) two-legged stirrups, along with (ii) longitudinal bars.

**TABLE 20.6.** Shear Stress in Members Due to Torsion

| Description of section | Shape | Maximum shear stress ($q'$) |
|---|---|---|
| **(a) Non-tubular section** | | |
| 1. Circle | Circle of diameter $D$ | $\dfrac{16 M^t}{\pi D^3}$ |
| 2. Ellipse | Ellipse with minor $b$, major $D$ | $\dfrac{16 M^t}{\pi b^2 D}$ |
| 3. Rectangle | Rectangle $b \times D$ | $\dfrac{M^t (3 + 2b/D)}{b^2 D}$ |
| 4. Sections made up of rectangles, such as T or L | T-section with $b_1, t_1, b_2, t_2$ | $\dfrac{3 M^t\, t_i}{\sum b_1 t_1^{\,3}}$ where $t_i$ is $t_1$ or $t_2$ according to the section under consideration |
| 5. Split box | Channel / split box with $b_1, b_2, b_3, b_4, b_5, t_1, t_2, t_3, t_4, t_5$ | $\dfrac{3 M^t\, t_i}{\sum b_1 t_1^{\,3}}$ where $t_i\, t_1$ or $t_2$ or $t_3$ etc., according to the section under consideration |
| **(b) Tubular or box sections** | Box section with thicknesses $t_1, t_2$ | $\dfrac{M^t}{2 A\, t_i}$ where $A$ is the mean of the area enclosed by the inner and the outer boundaries, and $t_i$ is $t_1$ or $t_2$ etc. according to the section under consideration. |

**(i) Spiral Reinforcement : Circular Section :** In order to understand the mechanism of crack formation due to torsion, let us consider circular section of beam shown in Fig. 20.13(*a*). Let us take a plane lamina of sides $R.\, d\theta$ each, at the surface, and of $dR$ thickness. Let $q'$ be the shear stress along the sides of the lamina. Due to this, tensile stress will be induced in one diagonal plane and compressive strees along the other will be induced. Hence the lamina will try to tear-off along the diagonal *AC*. Thus cracks will be formed in directions 1 – 1, 2 – 2, 3 – 3 etc. [Fig. 20.13 (*b*)] each parallel to *AC*, forming a crack-spiral. The ideal reinforcement therefore will be at right angles to these cracks, in the form of a spiral.

Fig. 20.13(*c*) shows a circular section subjected to torsion. It is well known that shear due to torsion is maximum at the surface, and decreases toward the centre of the section. Shear reinforcement is needed only in that portion of the section (shown shaded) in which the shear stress due to combined effects is more than the permissible value. Let $R'$ correspond to that radius at which the combined shear stress is equal to the permissible value. Now consider a lamina at radial distance $r$, of thickness $dr$ and subtending an angle $d\theta$. The lamina *abcd* [Fig. 20.13 (*d*)] will thus have its sides equal to $rd\theta$ each. Let the shear stress at that plane be $q$. Due to this, diagonal tension will be induced across diagonal *ac*, of intensity $q$ *i.e.*, it will be at right angles to the principal plane *ac*. The total tensile force on *ae* will cross the side *ab*.

Total diagonal tension force crossing $ab = q \times ae \times dr = q\,(r\,d\theta \cos 45°)\,dr$. Moment of diagonal tension about the axis

$$= \int_{R'}^{R}\int_{0}^{2\pi} q\,(r\,d\theta \cos 45°)\,dr \cdot r \qquad \text{where} \quad q = \frac{q'}{R} \times r$$

This should be equal to the moment of the tensile force in the spiral reinforcement, having area of steel $A_s$ and provided at a distance $r_t$ from the axis

$$A_s \cdot \sigma_{st} \cdot r_t = \int_{R'}^{R}\int_{0}^{2\pi} \frac{q'}{R} r^3 \frac{1}{\sqrt{2}}\,dr\,d\theta = \int_{R'}^{R} \frac{2\pi q'}{R\sqrt{2}} r^3\,dr = \frac{\pi q'}{2\sqrt{2}\,R}\cdot\left[R^4 - R'^4\right]$$

For the circular section, $q' = \dfrac{2 M^t}{\pi R^3}$

$$A_s\,\sigma_{st}\cdot r_t = \frac{M_t}{\sqrt{2}}\left[1 - \left(\frac{R'}{R}\right)^4\right]$$

or

$$A_s = \frac{M_t}{\sigma_{st}\,r_t\sqrt{2}}\left[1 - \left(\frac{R'}{R}\right)^4\right] \qquad \text{...(20.53)}$$

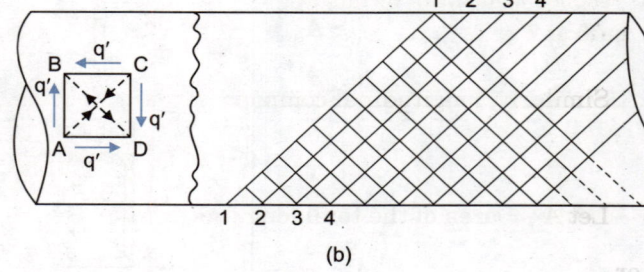

The spiral should be placed at a distance $r_t$ from the axis, such that the steel coincides with the C.G. of the tensile forces on the shaded area. From this consideration, the value of $r_t$ is given by

$$r_t = \frac{3}{4}\cdot\frac{R^4 - R'^4}{R^3 - R'^3} = \frac{3R}{4}\cdot\frac{1 - \left(\dfrac{R'}{R}\right)^4}{1 - \left(\dfrac{R'}{R}\right)^3} \qquad \text{...(20.54)}$$

Fig. 20.13

Let $s$ be the pitch of the spiral and $A_\Phi$ be the cross-sectional area of the spiral bar.

Then, $\dfrac{2\pi r_t}{s} A_\Phi = A_s \quad$ or $\quad s = 2\pi r_t \cdot \dfrac{A_\Phi}{A_s}$ ...(20.55)

Cowan (1933) has given the following expression for the torsional strength of circular section reinforced with spiral reinforcement:

$$M_t = 2\sqrt{2}\,\pi\,r_t^2 \cdot \frac{A_\Phi\,\sigma_{st}}{s} \qquad \text{...(20.56)} \qquad \text{and} \qquad A_s = 2\pi r_t \cdot \frac{A_\Phi}{s} = \frac{M^t}{\sigma_{st}\cdot r_t\sqrt{2}} \qquad \text{...(20.57)}$$

This is the same as Eq. 20.53, if $R'$ is put to zero.

**(ii) Spiral reinforcement: Rectangular section:** The analysis of rectangular section, spirally reinforced is extremely difficult. Cowan gave the following expression for the torsional strength of beam, with spiral reinforcement at 45°

$$M^t = \lambda\sqrt{2}\,A\,\frac{A_\Phi \cdot \sigma_{st}}{s} \approx 2.3\,A\,\frac{A_\Phi \cdot \sigma_{st}}{s} \qquad \text{...(20.58)}$$

where $\lambda$ depends upon $\dfrac{D}{b}$ ratio, and its average value may be taken as 1.63.

$A$ = Area of section enclosed by the spiral.

If the spiral bar is provided at an effective cover of $d_c$, we have

$$A = X_1 \times Y_1 \qquad \text{where} \quad X_1 = b - 2d_c \quad \text{and} \quad Y_1 = D - 2d_c$$

**(iii) Reinforcement in the form of hoops and longitudinal bars: Circular section**

As stated earlier, it is extremely difficult to provide spiral reinforcement. Moreover, spiral reinforcement has to suddenly change the direction at the point of contra-torsion. A more practical proposition is therefore to provide reinforcement in the form of hoops and longitudinal bars. The hoops resist the circumferential component of the diagonal tension while the longitudinal bars resist the axial component.

Referring to Fig. 20.13(c), (d), the tensile force on ae, crossing the side ab is equal to $q \cdot r \, d\theta \cdot \cos 45° \cdot dr$. The force can be resolved longitudinally along ab and tangentially along ad. The tangential component acts along the length of the beams and is resisted by hoop reinforcement provided at right angles to the axis of the beam.

Tangential component $= (q \cdot r \, d\theta \cdot dr \cdot \cos 45°) \cos 45°$

Tangential component per unit length of beam

$$= \frac{q \cdot r \, d\theta \cdot dr \cos^2 45°}{r \cdot d\theta} = \frac{q \cdot dr}{2} = \frac{q'}{2R} r \, dr$$

Its moment about the axis of beam $= \dfrac{q'}{2R} r \, dr \cdot r$

Let $A_{sv}$ = area of hoop steel required per unit length of beam, at a radius $r_1$.

$$A_{sv} \cdot \sigma_{st} \, r_1 = \int_{R'}^{R} \frac{q'}{2R} r^2 \, dr \quad \text{or} \quad A_{sv} = \frac{q'}{6R \, \sigma_{st} \cdot r_1} \left[ R^3 - R'^3 \right]. \quad \text{But } q' = \frac{2M^t}{\pi R^3}$$

$$\therefore \quad A_{sv} = \frac{M^t}{3 \pi R^4 \, \sigma_{st} \, r_1} (R^3 - R'^3) \qquad \qquad \ldots(20.59)$$

Similarly, longitudinal component of tensile force, will be

$$= q \, (r \, d\theta) \, dr \cdot \cos^2 45° = \frac{q}{2} r \, d\theta \, dr = \frac{q}{2R} r \, d\theta \cdot dr$$

Let $A_{st}$ = area of the total longitudinal steel, we have

or
$$A_{st} \cdot \sigma_{st} = \int_0^{2\pi} \int_{R'}^{R} \frac{q'}{2R} \cdot r^2 \, d\theta \, dr = \frac{\pi q}{R} \int_{R'}^{R} r^2 \, dr = \frac{\pi q'}{3R} (R^3 - R'^3)$$

or
$$A_{st} = \frac{\pi q}{3 R \, \sigma_{st}} (R^3 - R'^3). \quad \text{But } q' = \frac{2 M^t}{\pi R^3}$$

$$\therefore \quad A_{st} = \frac{2}{3} \cdot \frac{M^t}{R^4 \, \sigma_{st}} (R^3 - R'^3) \qquad \qquad \ldots(20.60)$$

**(iv) Reinforcement in the form of hoops and longitudinal bars: Rectangular section**

Cowan has given the following expression for torsional strength of beam reinforced with hoops:

$$M^t = \lambda \, A \, \frac{A_\phi \cdot \sigma_{st}}{s} \qquad \qquad \ldots(20.61)$$

where $\lambda$ depends upon $\dfrac{D}{b}$ value, and its average value may be taken usual to 1.63.

$A$ = the area enclosed by hoop ; $A_\phi$ = area of cross-section of hoop bar.

In addition to the hoops, an equal volume of steel should be placed longitudinally in the beam, well dispersed along the periphery of the hoops.

## 20.10. INDIAN STANDARD CODE FOR DESIGN FOR TORSION (IS : 456-2000)

**(a) General.** In general, where the torsional resistance or stiffness of members has not been taken into account in the analysis of structure, no specific calculation for torsion will be necessary; adequate control of any torsional cracking being provided by the required nominal shear reinforcement. Where the torsional resistance *or* stiffness of members is taken into account in the analysis, the members shall be designed for torsion.

In the Indian Standard Code (IS : 456-2000), torsional reinforcement is not calculated separately from that required for bending and shear. Instead, the total longitudinal reinforcement is determined for a fictitious (*or* equivalent) bending moment which is a function of actual bending moment and torsion. Similarly, web reinforcement is determined for a fictitious (*or* equivalent) shear which is a function of actual shear and torsion.

The design rules laid down in (*c*) and (*d*) below, shall apply to beam of solid rectangular cross-section. However, these clauses may also be applied to flanged beams by substituting $b_w$ for $b$ in which case they are generally conservative ; therefore specialist literature may be referred to.

**(b) Critical Section.** Sections located less than a distance $d$ from the face of the support may be designed for the same torsion as computed at a distance $d$ where $d$ is the effective depth.

**(c) Shear and Torsion**

**(i) *Equivalent shear* :** Equivalent shear $V_e$ shall be calculated from the formula

$$V_e = V + 1.6 \frac{T}{b} \qquad \ldots(20.62)$$

where $V_e$ = equivalent shear; $V$ = S.F.; $T$ = torsional moment ; $b$ = breadth of beam.

The equivalent nomial shear stress ($\tau_{ve}$) shall be calculated from the expression

$$\tau_{ve} = \frac{V_e}{b\,d} \quad \text{(where } b = b_w \text{ for flanged sections)}$$

The values of $\tau_{ve}$ shall not exceed the values of $\tau_{c(max)}$ given in Table 3.3 (If it exceeds, the section should be redesigned by increasing the concrete area).

(*ii*) If the equivalent nominal shear stress does not exceed $\tau_c$ given in Table 3.1, minimum shear reinforcement shall be provided in the form of stirrups, such that

$$\frac{A_{sv}}{b s_v} \geq \frac{0.4}{0.87 f_y}$$

The above equation can be rearranged in the following form to give the maximum spacing for the stirrups

$$s_v \leq \frac{2.175 A_{sv} \cdot f_y}{b} \qquad \ldots(20.63)$$

where $A_{sv}$ is the total cross-sectional area of stirrup legs effective in shear, $s_v$ is the stirrup spacing and $f_y$ is the characteristic strength of the stirrup reinforcement in N/mm², which shall not be taken greater than 415 N/mm²

(*iii*) If $\tau_{sv}$ exceeds $\tau_c$ given in Table 3.1, both longitudinal and transverse reinforcement shall be provided in accordance with clause (*d*) below :

**(d) Reinforcement in members subjected to torsion.** Reinforcement in torsion, when required, shall consist of longitudinal and transverse reinforcement.

**Longitudinal Reinforcement** : The longitudinal reinforcement shall be designed to resist an equivalent moment $M_{e1}$ given by

$$M_{e1} = M + M_T \qquad \ldots[20.64\,(a)]$$

where $M$ = B.M. at the cross-section.

and

$$M_T = T\,\frac{(1 + D/b)}{1.7} \qquad \ldots[20.64\,(b)]$$

where $T$ is the torsional moment, $D$ is the overall depth and $b$ is the breadth of the beam.

If the numerical value of $M_T$ as defined above exceeds the numerical values of $M$, longitudinal reinforcement shall be provided on the flexural compression face, such that the beam can also with– stand an equivalent moment $M_{e2}$ given by

$$M_{e2} = M_T - M$$

the moment $M_{e2}$ being taken as acting in the opposite sense to the moment $M$.

**Transverse Reinforcement :** Two legged closed hoops enclosing the corner longitudinal bars shall have an area of cross-section $A_{sv}$ given by

$$A_{sv} = \frac{T \cdot s_v}{b_1 d_1 \sigma_{sv}} + \frac{V \cdot s_v}{2.5 \, d_1 \, \sigma_{sv}} \qquad \ldots(20.65)$$

However the total transverse reinforcement shall not be less than $\dfrac{(\tau_{ve} - \tau_c) \cdot b \cdot s_v}{\sigma_{sv}}$

where  $T$ = torsional moment
  $V$ = shear force
  $s_v$ = spacing of the stirrup reinforcement
  $b_1$ = centre to centre distance between corner bars in the direction of the width
  $d_1$ = centre to centre distance between corner bars in the direction of the depth
  $b$ = breadth of member
  $\sigma_{sv}$ = permissible tensile stress in shear reinforcement
  $\tau_{ve}$ = equivalent shear stress = $V_e/(b\,d)$
  $\tau_c$ = shear strength of concrete as specified in Table 3.1

**Note:** $A_s$ is that area of longitudinal tensile reinforcement which continues atleast one effective depth beyond the section being considered, except at supports where the full area of tension reinforcement may be used.

**Note 1.** *Distribution of torsion reinforcement:* When a member is designed for torsion, torsion reinforcement shall be provided as below:

(a) The transverse reinforcement for torsion shall be rectangular closed stirrups placed perpendicular to the axis of the member. The spacing of the stirrups shall not exceed the least of $x_1$, $\dfrac{x_1 + y_1}{4}$ and 300 mm, where $x_1$ and $y_1$ are respectively the short and long dimensions of the stirrup.

(b) Longitudinal reinforcement shall be placed as close as is practicable to the corners of the cross-section and in all cases, there shall be at least one longitudinal bar in each corner of the ties. When the cross sectional dimension of the member exceeds 450 mm, additional longitudinal bars shall be provided to satisfy the requirements of minimum reinforcement and spacing given in Note 2.

**Note 2.** *Side face reinforcement:* The total area of side face reinforcement shall be not less than 0.1 per cent of the web area and shall be distributed equally on two faces at a spacing not exceeding 300 mm or web thickness whichever is less.

**Example 20.4.** *A rectangular beam 400 mm wide is subjected to the following at a section: (i) B.M. of 45 kN-m, (ii) shear force of 30 kN and (iii) torsional moment of 20 kN-m. Design the section and the torsional reinforcement. Use M 20 concrete and Fe 415 steel.*

**Solution:**

**1. Design constants :** For M 20 concrete and Fe 415 steel, we have :

$$c = \sigma_{cbc} = 7\,; \quad t = \sigma_{st} = 230 \quad \text{and} \quad m = 13.33$$
$$k_c = 0.289\,; \quad j_c = 0.904\,; \quad R_c = 0.914$$

**2. Design of section for B.M.**

$$b = 400 \text{ mm}; \, M = 45 \text{ kN-m} = 45 \times 10^6 \text{ N-mm}$$

$$d = \sqrt{\frac{M}{R_c \, b}} = \sqrt{\frac{45 \times 10^6}{0.914 \times 400}} = 351 \text{ mm. Keep total } D = 400 \text{ mm.}$$

Using 16 mm Φ bars and a nominal cover of 25 mm, with 12 mm φ ties,

$$d = 400 - 25 - 12 - \frac{16}{2} = 400 - 45 = 355 \text{ mm.}$$

## 3. Equivalent shear

$$V_e = V + 1.6 \frac{T}{b} = 30 \times 10^3 + 1.6 \frac{20 \times 10^6}{400} = 11 \times 10^4 \text{ N}$$

$$\tau_{ve} = \frac{V_e}{b\,d} = \frac{110000}{400 \times 355} = 0.77 \text{ N/mm}^2$$

For M 20 concrete, $\tau_c$ for 0.5 % reinforcement is 0.3 N/mm² and $\tau_c$ for 1% reinforcement is 0.39 N/mm² (Table 3.1). Though the percentage reinforcement is not known, it is certain that permissible shear stress $\tau_c$ will be less than $\tau_{ve}$. Hence shear reinforcement is necessary. Also from Table 3.2, $t_{c\,max}$ for M 20 concrete is 1.8 N/mm². Thus $\tau_{ve}$ is less then $\tau_{cmax}$. Hence dimensions of the beam are all right.

**4. Longitudinal Reinforcement**: Since $\tau_{ve}$ exceeds $\tau_c$, both longitudinal and transverse reinforcement are necessary. The longitudinal reinforcement is designed to resist an equivalent B.M. $M_{e1}$ given by

$$M_{e1} = M + M_T$$

where $M$ = B.M. = $45 \times 10^6$ N-mm

$$M_T = T \frac{(1 + D/b)}{1.7},$$

where $T$ = torsional moment

$$= 20 \times 10^6 \frac{(1 + 400/400)}{1.7} = 23.53 \times 10^6 \text{ N-mm}$$

$$M_{e1} = 45 \times 10^6 + 23.53 \times 10^6 = 68.53 \times 10^6 \text{ N-mm}$$

$$A_{st} = \frac{M_{e1}}{\sigma_{st}\,j_c\,d} = \frac{68.53 \times 10^6}{230 \times 0.904 \times 355} = 929 \text{ mm}^2$$

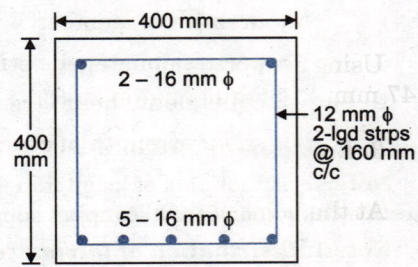

Fig. 20.14. Cross sectional details

Using 16 mm Φ bars, $A_\Phi = \frac{\pi}{4}(16)^2 = 201 \text{ mm}^2$

∴ No. of 16 mm Φ bars = 929/201 ≈ 5

Since $M_T$ is less than $M$, no longitudinal reinforcement is necessary at the top of the beam.

**5. Transverse Reinforcement**: Transverse reinforcement is provided in the form of vertical stirrups. Provide two corner bars of 16 mm Φ, at the top face, at a effective cover of 25 mm, to hold the stirrups, as shown in Fig. 20.14.

∴ $b_1$ = centre to centre distance between corner bars in the direction of width
= 400 − 2 × 45 = 310 mm

$d_1$ = centre to centre distance between corner bars in the direction of width
= 400 − 2 × 45 = 310 mm.

The area of cross-section $A_{sv}$ of the stirrups is given by: $A_{sv} = \frac{T \cdot s_v}{b_1 \cdot d_1 \cdot \sigma_{sv}} + \frac{V \cdot s_v}{2.5\,d_1\,\sigma_{sv}}$.

Using 12 mm Φ − 2 lgd stirrups, $A_{sv} = 2\frac{\pi}{4}(12)^2 = 226 \text{ mm}^2$

∴ $$226 = \left[\frac{20 \times 10^6}{310 \times 310 \times 230} + \frac{30000}{2.5 \times 310 \times 230}\right] s_v,$$ from which $s_v = 211$ mm.

However, the spacing should not exceed the least of $x_1$, $\frac{x_1 + y_1}{4}$ and 300 mm

where $x_1$ = short dimension of stirrup = 310 + 16 + 12 = 338 mm
$y_1$ = 310 + 16 + 12 = 338 mm

∴ $\frac{x_1 + y_1}{4} = \frac{338 + 338}{4} = 169$ mm.

Thus, $s_v = 211$ mm is not permissible. Hence, provide 12 mm Φ 2 lgd stirrups @ 160 mm c/c. The section of the beam is shown in Fig. 20.14.

**Example 20.5.** *Design the section of the beam analysed in example 20.2, using M 20 concrete and Fe 415 steel.*

**Solution:** From example 20.2, we have the following data:

Maximum B.M. = 137.28 kN-m hogging at the support and 48.46 kN-m sagging at the intermediate point.

S.F. at end support = 45.66 kN ;   S.F. at central support = 80 kN-m
Maximum torsional moment = 33.42 kN-m
S.F. at point of max. torsion = 37.31 kN.

**1. Design constants**

For M 20 concrete, and Fe 415 steel $k_c = 0.289$ ; $j_c = 0.904$ ; $R_c = 0.914$

**2. Design of section for B.M.** Let $b = 500$ mm

$$d = \sqrt{\frac{M}{R_c \, b}} = \sqrt{\frac{137.28 \times 10^6}{0.914 \times 500}} = 548.08 \text{ mm. Provide } D = 650 \text{ mm.}$$

Using 20 mm diameter bars, 12 mm φ ties and a nominal cover of 25 mm, effective cover = 25 + 12 + 20/2 = 47 mm. Providing 50 mm effective cover, $d = 650 - 50 = 600$ mm.

**3. Equivalent shear**   $V_e = V + 1.6 \dfrac{T}{b}$

At the point of max. torsion,

$$V = 37.31 \text{ kN}$$

$$V_e = 37.31 + 1.6 \, \frac{33.42}{0.5} = 144.25 \text{ kN}$$

$$\tau_{ve} = \frac{V_e}{bd} = \frac{144.25 \times 1000}{500 \times 600} = 0.48 \text{ N/mm}^2$$

For balanced design, percentage reinforcement ($p$) = 0.44%. From Table 3.1, $\tau_c \approx 0.28$ N/mm$^2$.

Since $\tau_{ve}$ is greater than $\tau_c$, shear reinforcement is necessary. Also, from Table 3.3, $\tau_{c(max)}$ for M 20 concrete is 1.8 N/mm$^2$. Hence $\tau_{ve}$ is less than $\tau_{c\,max}$. Therefore, the dimensions of the beam are o.k.

At the central support, S.F. = 80 kN

$$\tau_{ve} = \frac{80 \times 10^3}{500 \times 600} = 0.267 \text{ N/mm}^2$$

This is less than $\tau_c$. Hence only nomial shear reinforcement is necessary. At the end supports also,

$$\tau_{ve} = \frac{45.66 \times 10^3}{500 \times 600} = 0.15 \text{ N/mm}^2$$

This is less than half $\tau_c$. Hence even nominal shear reinforcement is not necessary at the end support.

**4. Main and Longitudinal reinforcement.**

(a) Section at point of maximum torsion :

$$T = M^t_{max} = 33.42 \text{ kN-m at } \phi = 59.44° \quad \text{where the B.M. is zero.}$$

$$M_{e1} = M + M_T, \quad \text{where} \quad M = \text{zero}$$

$$M_T = T \, \frac{(1 + D/b)}{1.7} = 33.42 \, \frac{\left(1 + \dfrac{650}{500}\right)}{1.7} = 45.22 \text{ kN-m} = 45.22 \times 10^6 \text{ N-mm}$$

$$M_{e1} = M_T = 45.22 \times 10^6$$

$$A_{st} = \frac{M_{e1}}{\sigma_{st} \, j_c \, d} = \frac{45.22 \times 10^6}{230 \times 0.904 \times 600} = 362.47 \simeq 363 \text{ mm}^2.$$

Using 20 mm Φ bars, having $A_\phi = 314$ mm². No. of bars = 363/314 ≈ 1.156 ≈ 2

Since $M$ is zero, $M_T > M$. Hence $M_{e2} = M_T - M = M_{e1}$.

Thus, the above reinforcement has also to be provided at the top of the section.

Thus provide 2 – 20 mm Φ bars both at top as well as at bottom of section, at this location.

*(b) Section at point of maximum hogging B.M.:*

Maximum hogging B.M. occurs at the central support (φ = 90°) where torsional moment is zero. $M_{max} = 137.28$ kN-m.

$$\therefore \quad A_{st} = \frac{137.28 \times 10^6}{230 \times 0.904 \times 600} = 1100 \text{ mm}^2$$

No. of 20 mm Φ bars = 1100/314 ≈ 4.

Hence use 4 Nos. 20 mm Φ bars. These bars will be provided at the top of the section since max. B.M. is hogging at the intermediate support. The bending moment is zero at a section at an angle of 59.44° with end radial line, or at an angle of (90° – 59.44°) = 30.56° from the central pillar in either direction. Distance of this section, along the axis of beam

$$= R \times \phi = 4 \times \frac{30.56}{180} \pi = 2.13 \text{ m from the support.}$$

Development length $L_d \approx 45 \Phi = 45 \times 20 = 900$ mm = 0.9 m.

Thus, 4 bars can be curtailed at a distance, say, 2 m from the mid support. It can be shown that the remaining 2 bars will easily take the hogging B.M and the torsional moment beyond 2 m $\left(i.e., \phi = \frac{2}{4} \times \frac{180}{\pi} = 28.65°\right)$ from the central support.

*(c) Section at point of max. Sagging B.M.*

At the point of max. sagging B.M. (φ = 29.72°),

$$M = 48.46 \text{ kN-m}; \quad M^t = T \approx 16.95 \text{ kN-m.} \quad \text{Shear force = zero.}$$

$$M_T = T \frac{(1 + D/b)}{1.7} = 16.95 \frac{\left(1 + \frac{650}{500}\right)}{1.7} = 22.93 \text{ kN-m}$$

$$M_{e1} = M + M_T = 48.46 + 22.93 = 71.39$$

$$\therefore \quad A_{st} = \frac{M_{e1}}{\sigma_{st} \, j_c \, d} = \frac{71.39 \times 10^6}{230 \times 0.904 \times 600} = 572.25 \approx 573 \text{ mm}^2$$

No. of 20 mm Φ bars = 573/314 ≈ 2.

Hence provide 2-20 mm Φ bars at the bottom of the section. Out of these, two bars can be curtailed at a point where they are no longer required, provided that point is at least at a distance of $L_d = 45 \Phi = 900$ mm away. The bars are to be continued to a distance of 12 Φ = 240 mm beyond that point.

Hence discontinue two bars at a distance of 900 + 240 = 1140 mm = 1.14 m on either side of the point of max. sagging B.M. (φ ≈ 30°). Thus the distance of point of curtailment from support = $4 \frac{30°}{180°} \pi \pm 1.4 = 2.10 \pm 1.4 = 3.5$ m and 0.7 m from face of support. However, these two bars are to be continued to take care of torsional B.M., and also to satisfy the bond and anchorage requirements at the end.

**5. Bond and anchorage requirements**: At end supports, there are two bars of 20 mm Φ each at top and bottom. These bars should satisfy the bond and anchorage requirements envisaged in the following equation:

$$\frac{M_1}{V} + L_0 \geq L_d$$

where $M_1$ = moment of resistance of the section assuming all reinforcement at the section to be stressed to $\sigma_{st}$

$$\sigma_{st} = A_{st} \cdot \sigma_{st} j d = (2 \times 314)(230)(0.904 \times 600) = 78.34 \times 10^6 \text{ N-mm}$$

$$V = 45.66 \text{ kN} = 45.66 \times 10^3 \text{ N}$$

$$\frac{M_1}{V} = \frac{78.34 \times 10^6}{45.66 \times 10^3} = 1715.72 \approx 1715 \text{ mm}$$

Let us assume width of support $l_s = 600$ mm. Assuming side cover $x' = 40$ mm.

$$L_0 = \frac{ls}{2} - x' = \frac{600}{2} - 40 = 260 \text{ mm} ; \qquad L_d = 45\,\Phi = 45 \times 20 = 900 \text{ mm}$$

$$\therefore \quad \frac{M_1}{V} + L_0 = 1715 + 260 = 1975 \text{ mm};$$

$$\therefore \quad \frac{M_1}{V} + L_0 > L_d. \qquad \text{Hence O.K.}$$

### 6. Transverse reinforcement

**(a) Point of max. torsional moment**

At the point of max torsion, S.F. = 37.31 kN (see Fig.20.6). As found in step 3, $\tau_{ve} = 0.49$ N/mm², which has more than $\tau_c = 0.28$ N/mm². Hence shear reinforcement is required.

The area of cross-section $A_{sv}$ of the stirrups is given by

$$A_{sv} = \frac{T \cdot s_v}{b_1 \, d_1 \, \sigma_{sv}} + \frac{V \cdot s_v}{2.5 \, d_1 \, \sigma_{sv}}$$

$b_1$ = Centre to centre distance between corner bars in the direction of width
  = $500 - 2 \times 50 = 4000$ mm

$d_1$ = Centre to centre distance between corner bars in the direction of depth
  = $650 - 2 \times 50 = 550$ mm.

$$\therefore \quad \frac{A_{sv}}{S_v} = \frac{33.42 \times 10^6}{400 \times 550 \times 230} + \frac{37.31 \times 10^3}{2.5 \times 550 \times 230} = 0.778 \qquad \ldots(i)$$

However, the minimum transverse reinforcement should be governed by

$$\frac{A_{sv}}{s_v} \geq \left(\frac{\tau_{ve} - \tau_c}{\sigma_{sv}}\right) b$$

$$\therefore \quad \frac{A_{sv}}{s_v} = \left(\frac{0.49 - 0.28}{230}\right) \times 500 = 0.46 \qquad \ldots(ii)$$

$\therefore$ Hence keep $\quad \dfrac{A_{sv}}{s_v} = 0.778$

Using 10 mm $\Phi$ 2-lgd stirrups having $A_{sv} = 2 \times 78.54 = 157$ mm²;   $s_v = 157/0.778 = 201.9$ mm.

However, the spacing should not exceed the least of $x_1$, $\dfrac{x_1 + y_1}{4}$ and 300 mm, where $x_1$ = short dimension of stirrup = $400 + 20 + 10 = 430$ mm

$$y_1 = 550 + 20 + 10 = 580 \text{ mm}.$$

$$\therefore \quad \frac{x_1 + y_1}{4} = \frac{430 + 580}{4} = 252.5 \text{ mm}.$$

Hence $s_v = 201.9$ mm is permissible. However, provide 10 mm $\Phi$ two-legged stirrups @ 200 mm c/c at the section of maximum torsional moment. This spacing may be continued for a distance of, say, 1.7 m on either side.

(b) *At point of max. S.F. (central support)* : S.F. = 80 kN

$$\tau_v = \frac{80 \times 10^3}{500 \times 600} = 0.27 \text{ N/mm}^2$$

This is less than $\tau_c$. Hence only nominal shear reinforcement is necessary. The nominal shear reinforcement is given by

$$\frac{A_{sv}}{b \cdot s_v} \geq \frac{0.4}{0.87\, fy}$$

or

$$\frac{A_{sv}}{s_v} = \frac{0.4\, b}{0.87\, fy}$$

∴

$$\frac{A_{sv}}{s_v} = \frac{0.4 \times 500}{0.87 \times 415} = 0.554$$

Choosing 10 mm Φ 2 lgd stirrups having $A_{sv}$ = 157 mm², 

$$s_v = \frac{157}{0.554} \approx 283.39 \triangleq 280 \text{ mm}$$

Maximum permissible spacing = 0.75 d = 0.75 × 600 = 450 mm, or 300 mm which ever is less. Hence provide stirrups @ 280 c/c.

(c) *At end supports* :   S.F. = 45.66 kN

∴

$$\tau_v = \frac{45.66 \times 10^3}{500 \times 600} = 0.152 \text{ N/mm}^2 < \tau_c.$$

Hence provide nominal stirrups (*i.e.*, 10 mm Φ 2 lgd stirrups 280 mm c/c) at the end supports also.

**7. Side face reinforcement** : The Code specifies that when the cross-sectional dimension exceeds 450 mm, additional longitudinal bars should be provided. Area of such side face reinforcement should not be less than 0.1 percent.

$$A_l = \frac{0.1}{100} (500 \times 650) = 325 \text{ mm}^2$$

Use two 16 mm Φ bars (one at the mid-height of each face), giving $A_l = 2 \times 201 = 402$ mm². The details of the reinforcement etc. are shown in Fig. 20.15.

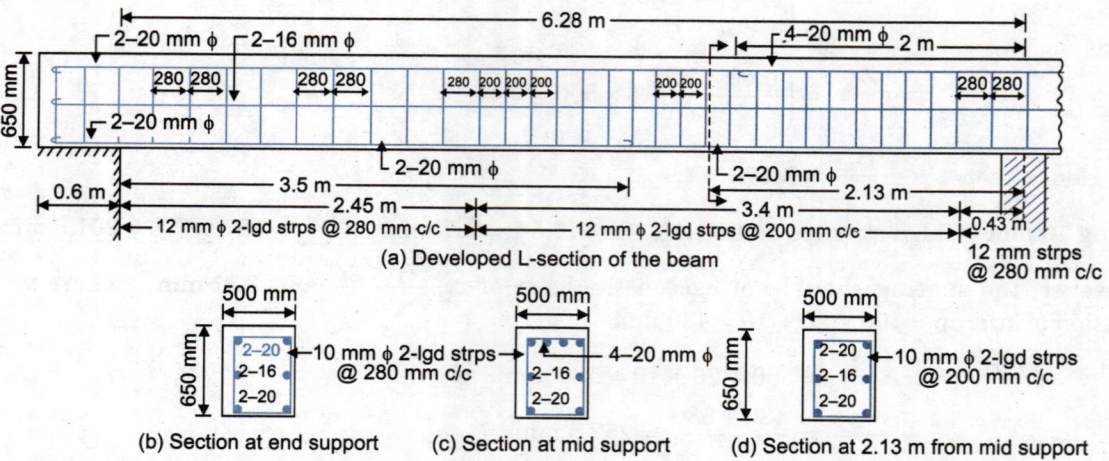

Fig. 20.15.   Typical cross section of circular RC beam.

## PROBLEMS

1. A curved beam is in the form of a full continuous circle in plan with a radius of 3 m and is supported continuously on five supports. The beam carries a uniformly distributed load of 30 kN/m length, inclusive of its own weight. Determine the bending moment, twisting moment and shear force at salient locations. Also, plot the B.M., T.M. and S.F. diagrams for one span.

2. A semi-circular beam is simply supported on three equally spaced columns. Show that the maximum bending moment and the maximum twisting moment are equal to $0.429\ wR^2$ and $0.1045\ wR^2$ respectively.

3. A semi-circular beam with radius of 5 m is simply supported at ends, and is continuous over a column at its middle. The beam carries a uniformly distributed load of 16 kN/m length of the beam, inclusive of its own weight. Determine the S.F., B.M. and T.M. at salient points and plot the S.F., B.M. and T.M. diagrams.

4. A beam is curved in plan in the form of arc of a circle with raidus of 3 m, and central angle of 120°. The beam is fixed at the ends, and carries a super-imposed load of 25 kN/m. The beam is rectanglar in cross-section, having width = 40 cm and depth = 70. Taking $G = 0.4\ E$ for concrete, draw the bending moment and torsional moment diagrams for the beam.

5. A rectangular beam, 30 cm wide is subjected to the following at a section  (*i*) bending moment = 40 kN-m, (*ii*) shear force = 20 kN and  (*iii*) torsional moment = 16 kN-m. Design the section and the torsional reinforcement in the form of  (*a*) spiral  (*b*) hoops and longitudinal reinforcement. Use M 20 concrete and Fe 415 steel.

6. Design the beam analysed in problem 1.

7. Design the beam analysed in problem 3.

8. Design a semi-circular beam supported on 3 columns equally spaced. The centre of the columns are on a curve of diameter 8.2 m. The superimposed load on the beam is 28 kN and Adopt $-M_{20}$ grade concrete and Fe 415 steel.

9. Design a semi-circular ring girder supported on three column equally spaced supports carrying a uniformly distributed load of 50 kNm. The mean diameter of the girder is 12 m. Adopt $M_{20}$ grade concrete Fe 415 steel. Design the reinforcement in the girder assuming a width of 400 mm and an overall depth 750 mm.

# CHAPTER 21

# WATER TANKS-I : SIMPLE CASES

## 21.1. INTRODUCTION

A water tank is used to store water to tide over the daily requirements. In general, water tanks can be classified under three heads : (i) tanks resting on ground  (ii) elevated tanks supported on staging, and  (iii) underground tanks. From the shape point of view, water tanks may be of several types, such as  (i) circular tanks   (ii) rectangular tanks   (iii) spherical tanks   (iv) Intze tanks and   (v) circular tanks with conical bottoms.

In the construction of concrete structures for the storage of water and other liquids, the imperviousness of concrete is most essential. The permeability of any uniform and thoroughly compacted concrete of given mix proportions is mainly dependent on the water-cement ratio. The increase in water-cement ratio results in increase in the permeability. The decrease in water-cement ratio will therefore be desirable to decrease the permeability, but very much reduced water-cement ratio may cause compaction difficulties and prove to be harmful also. For a given mix made with particular materials, there is a lower limit to the water-cement ratio which can be used economically on any job. It is essential to select a richness of mix compatible with available aggregates, whose particle shape and grading have an important bearing on workability, which must be suited to the means of compaction selected. Efficient compaction preferably by vibration is essential. All tanks are designed as crack free structure for durability and to prevent leakages and concrete should be impervious to eliminate seepage. It is desirable to specify cement content sufficiently high to ensure that thorough compaction is obtainable while maintaining a sufficiently low water-cement ratio. The quantity of cement should not be less than 330 kg/m$^3$ of concrete. It should also be less than 530 kg/m$^3$ of concrete to keep the shrinkage low. A well graded aggregate with a water cement ratio less than 0.5 is recommended for making impervious concrete. In thicker sections, where a reduction in cement content might be desirable to restrict the temperature rise due to cement hydration, a lower cement content is usually permissible. It is usual to use rich mix like M 30 grade with HYSD steel of grade Fe 415 is recommended for liquid storage/retaining structures in most of the water tanks to achieve above objectives (IS: 3370 recommendating).

Design of liquid retaining structure has to be based on the avoidance of cracking in the concrete having regard to its tensile strength. It has to be ensured in its design that concrete does not crack on its water face. Cracking may also result from the restraint to shrinkage, free expansion and contraction of concrete due to temperature and shrinkage and swelling due to moisture effects. Correct placing of reinforcement, use of small sized bars and use of deformed bars lead to a diffused distribution of cracks. The risk of cracking due to overall temperature and shrinkage effects may be minimized by limiting the changes in moisture content and temperature to which the structure as a whole is subjected. Cracks can be prevented by avoiding the use of thick timber shuttering which prevent the easy escape of heat of hydration from the concrete mass. The risk of cracking can also be minimized by reducing the restraints on the free expansion or contraction of the structure. For long walls or slabs founded at or below the ground level, restraints can be minimized by founding the structure on a flat layer of concrete with interposition of sliding layer of some material to break the bond and facilitate movement. However, it should be recognized that common and more serious causes of leakage in practice, other than cracking, are defects such as segregation and honey combing and in particular all joints are potential source of leakage.

## 21.2. GENERAL DESIGN REQUIREMENTS ACCORDING TO INDIAN STANDARD CODE OF PRACTICE (IS : 3370–PART II, 1965)

**1. Plain Concrete Structures:** Plain concrete members of reinforced concrete liquid structures may be designed against structural failure by allowing tension in plain concrete as per the permissible limits for tension in bending specified in IS : 456–2000 (*i.e.*, permissible stress in tension in bending may be taken to be the same as permissible stress in shear, $q$ measured as inclined tension). This will automatically take care of failure due to cracking. However, nominal reinforcement in accordance with the requirements of IS : 456 shall be provided for plain concrete structural members.

**2. Permissible Stresses in concrete**

**(a) For resistance to cracking:** Indian Standard Code IS: 456–2000 does not specify the permissible stresses in concrete for its resistance to cracking. However, its earlier version (IS : 456–1964) included the permissible stresses in direct tension, bending tension and shear. These values are given in Table 21.1. The permissible tensile stresses due to bending apply to the face of the member in contact with the liquid. In members with thickness less than 225 mm and in contact with the liquid on one side, these permissible stresses in bending apply also to the face remote from the liquid.

TABLE 21.1. Permissible Concrete Stresses on Water Face in Calculations Relating to Resistance to Cracking

| Grade of concrete | Permissible stresses | | Shear = $\frac{Q}{bjd}$ |
|---|---|---|---|
| | Direct Tension ($\sigma_{st}$) | Tension due to bending ($\sigma_{cbt}$) | |
| | N/mm² | N/mm² | N/mm² |
| M 15 | 1.1 | 1.5 | 1.5 |
| M 20 | 1.2 | 1.7 | 1.7 |
| M 25 | 1.3 | 1.8 | 1.9 |
| M 30 | 1.5 | 2.0 | 2.2 |
| M 35 | 1.6 | 2.2 | 2.5 |
| M 40 | 1.7 | 2.4 | 2.7 |

**(b) For strength calculations:** In strength calculations the usual permissible stresses, in accordance with IS : 456–2000 (Chapter 1), are used. Where the calculated shear stress in concrete above exceeds the permissible value, reinforcement acting in conjunction with diagonal compression in concrete shall be provided to take the whole of the shear.

**3. Permissible Stresses in Steel Reinforcement**

**(a) For resistance to cracking:** When steel and concrete are assumed to act together for checking the tensile stresses in concrete for avoidance of cracking the tensile stresses in steel will be limited by the requirement that the permissible tensile stress in concrete is not exceeded so that tensile stresses in steel shall be equal to the product of modular ratio of steel and concrete, and the corresponding allowable tensile stress in concrete.

**(b) For strength calculations:** Though the Indian Standard Code IS : 456 had its *fourth revision* in 2000, the corresponding Codes IS : 3370 (Part I, II, III and IV) for concrete structures for the storage of liquids have not been revised since 1965. The main Code on concrete–IS : 456 is in SI units. However, the *fourth reprint* (May 1982) of IS : 3370 (Part II)-1965 incorporates the *amendment* regarding the permissible stresses in steel reinforcement. The revised values of permissible stresses are given in Table 21.2 converted into SI units, using the approximation 10 kg/cm² ≈ 1 N/mm². On the liquid retaining face and also. On the exterior face for the members less than 225 mm thick.

**TABLE 21.2.** Permissible Stresses in Steel Reinforcement for Strength Calculations

| Types of stress in steel reinforcement | Permissible stresses in N/mm² | |
|---|---|---|
| | Plain round mild steel bars conforming to grade 1 of IS : 482 (Part 1) – 1966 | High yield strength deformed bars (HYSD) conforming to IS : 1789–1966 or IS : 1139–1966. |
| 1. Tensile stress in members under direct tension ($\sigma_s$) | 115 | 150 |
| 2. Tensile stress in members due to bending ($\sigma_{st}$) | | |
| (a) On liquid retaining face of members. | 115 | 150 |
| (b) On face away from liquid for members less than 225 mm. | 115 | 150 |
| (c) On face away from liquid for members 225 mm or more in thickness. | 125 | 190 |
| 3. Tensile stress in shear reinforcement ($\sigma_{sv}$) | | |
| (a) For members less than 225 mm thickness | 115 | 150 |
| (b) For members 225 or more in thickness. | 125 | 175 |
| 4. Compressive stress in columns subjected to direct load ($\sigma_{sc}$) | 125 | 175 |

**Note.** Stress limitations for liquid retaining faces shall also apply to the following:

(a) Other faces within 225 mm of the liquid retaining face.

(b) Outside or external faces of structures away from the liquid but placed in water-logged soils upto the level of highest subsoil water. In case of deformed mild steel bars, the stresses can be increased by 20 percent of the value listed in the above table.

### 4. Stresses due to drying shrinkage or temperature change

(i) Stresses due to drying shrinkage or temperature change may be ignored provided that:

(a) The permissible stresses specified in para (2) and (3), for concrete and steel respectively are not exceeded.

(b) Adequate precautions are taken to avoid cracking of concrete during the construction period and until the reservoir is put into use.

(c) The recommendations as regards the provision of joint and for suitable sliding layer (see § 21.3) are complied with, or the reservoir is to be used only for the storage of water or aqueous liquids at or near ambient temperature and the circumstances are such that the concrete will never dry out.

(ii) Shrinkage stresses may, however, be required to be calculated in special case, when a shrinkage coefficient of $300 \times 10^{-6}$ may be assumed.

(iii) When the shrinkage stresses are allowed, the permissible stresses, tensile stresses in concrete (direct and bending) as given in Table 21.1 may be increased by $33\frac{1}{3}$ percent.

(iv) Where reservoirs are protected with an internal impermeable lining, consideration should be given to the possibility of concrete eventually drying out. Unless it is established on the basis of tests or experience that the lining has adequate crack bridging properties, allowance for the increased effect of drying shrinkage should be made in the design.

### 5. Steel Reinforcement

**(a) Minimum reinforcement:**

(i) The minimum reinforcement in walls, floors and roofs in each of the two directions at right angles shall have an area of 0.3 percent of the concrete section in that direction for sections upto 100 mm thickness. For sections of thickness greater than 100 mm and less than 450 mm the minimum reinforcement in each of the two directions shall be linearly reduced from 0.3 percent for 100 mm thick section to 0.2 percent for 450 mm, minimum reinforcement in each of the two directions shall be kept at 0.2 percent. In concrete sections of thickness 225 mm or greater, two layers of reinforcing bars shall be placed one near each face of the section to make up the minimum reinforcement specified above.

1. For slab with 'D' up to 300 mm: on top D/2 only @ 0.35%
2. For slabs with 'D' 300 to 500 mm: on the top D/2 and bottom 100 mm @ 0.35% each.
3. For slabs over 500 mm thick: on the top 250 mm and bottom 100 mm @ 0.35% each.

(ii) In special circumstances, floor slabs resting directly on the ground may be constructed with percentage of reinforcement less than that specified above. In no case the percentage of reinforcement in any member be less than 0.15% of the concrete section.

**(b) Minimum cover to reinforcement:**

(i) For liquid faces of parts of members either in contact with the liquid or enclosing the space above the liquid (such as inner faces of slab), the minimum cover to all reinforcement should be 25 mm or the diameter of the main bar, whichever is greater. In the presence of sea water and soils and water of corrosive character the cover should be increased by 12 mm but this additional cover shall not be taken into account for design calculations.

(ii) For faces away from the liquid and for parts of the structure neither in contact with the liquid on any face nor enclosing the face above the liquid, the cover should be the same as provided for other reinforced concrete sections.

## 21.3. JOINTS IN WATER TANKS

The various types of joints may be categorized under three heads :
(a) Movement joints      (b) Constructions joints      (c) Temporary open joints.

**(a) Movement joints:** These require the incorporation of special materials in order to maintain water-tightness while accommodating relative movement between the side of the joints. All movement joints are essentially flexible joints. Movement joints are of three types :

(i) Contraction joint      (ii) Expansion joint      (iii) Sliding joint.

**(i) Contraction joint:** A contraction joint is a typical movement joint which accommodates the contraction of the concrete. The joint may be either a *complete contraction joint* [Fig. 21.1(a)], in which there is discontinuity of both concrete and steel, or it may be *partial contraction joint* in which there is discontinuity of concrete but the reinforcements run through the joint [Fig. 21.1(b)]. In both cases, no initial gap is kept at the joint, but only discontinuity is given during construction. In the former type, a water bar is inserted while in the later type, the mouth of the joint is filled with *joint sealing compound* and then *strip painted*. A *water bar* is a pre-formed strip of impermeable material (such as a metal, polyvinyl chloride or rubber). *Joint sealing compounds* are impermeable ductile materials which are required to provide a water-tight seal by adhesion to the concrete throughout the range of joint movement. The commonly used materials are based on asphalt, bitumen, or coal tar pitch with or without fillers such as limestone or slate dust, asbestos fibre, chopped hemp, rubber or other suitable material. These are usually applied after construction or just before the reservoir is put into service by pouring in the hot *or* cold state, by trowelling *or* gunning or as preformed strips ironed into position.

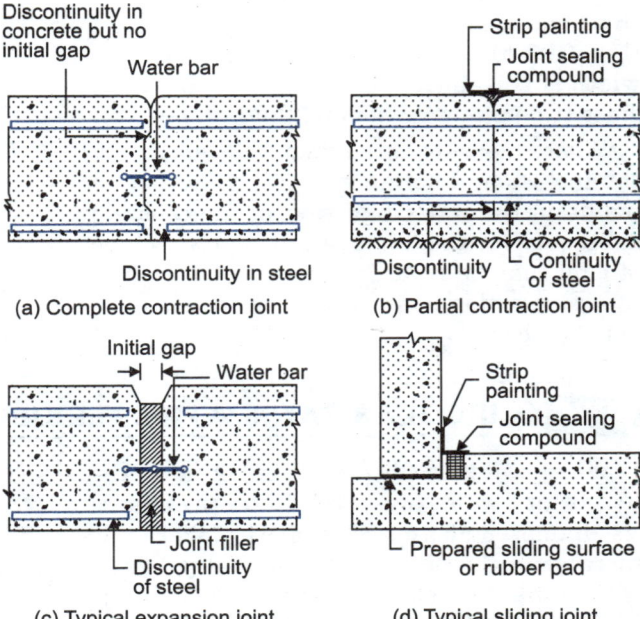

Fig. 21.1. Typical movement joints

**(ii) Expansion joint:** It is a movement joint with complete discontinuity in both reinforcement and concrete, and is intended to accommodate either expansion *or* contraction of the structure [Fig. 21.1(c)]. In general such a joint requires the provision of an initial gap between the adjoining parts of a structure which by closing *or* opening accommodates the expansion or contraction of the structure. The initial gap is filled with a joint filler. *Joint fillers* are usually compressible sheet *or* strip materials used as spacers. They are fixed to the face of the first placed concrete and against which the second placed concrete is cast. With an initial gap of 30 mm, the maximum expansion or contraction that the filler materials may allow may be of the order of 10 mm. Joint fillers, as at present available cannot by themselves function as water-tight expansion joints. But they can only be relied upon as spacers to provide the gap in an expansion joint when the gap is bridged by a water bar.

**(iii) Sliding joint:** Sliding joint is a movement joint with complete discontinuity in both reinforcement and concrete at which special provision is made to facilitate relative movement in place of the joint. A typical application of such a joint is between wall and floor in some cylindrical tank designs [Fig. 21.1(d)].

### (b) Construction joints

A construction joint is a joint in the concrete introduced for convenience in construction at which special measures are taken to achieve subsequent continuity without provision for further relative movement. It is, therefore, a rigid joint in contrast to a movement joint which is a flexible joint. Fig. 21.2 shows a typical construction joint between successive lifts in a reservoir wall. The position and arrangement of all construction joints should be predetermined by the engineer. Consideration should be given to limiting the number of such joints and to keeping them free from possibility of percolation in a manner similar to contraction joints.

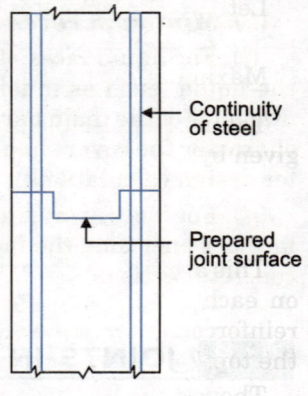

Fig. 21.2

### (c) Temporary open joints

A temporary open joint is a gap temporarily left between the concrete of adjoining parts of a structure which after a suitable interval and before the structure is put into use, is filled with mortar *or* concrete completely [Fig. 21.3(a)] *or* as provided below, with the inclusion of suitable jointing material [Fig. 21.3(b) and (c)]. In the former case the width of gap should be sufficient to allow the sides to be prepared before filling.

Where measures are taken for example, by the inclusion of suitable joining materials to maintain the water-tightness of the concrete subsequent to the filling of the joint, this type of joint may be regarded as being equivalent to a contraction joint (partial or complete) as defined above.

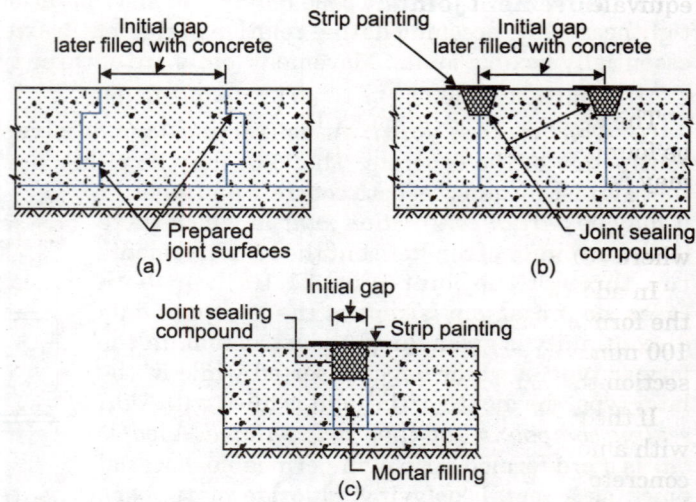

Fig. 21.3. Typical temporary open joints.

## 21.4. CIRCULAR TANK WITH FLEXIBLE JOINT BETWEEN FLOOR AND WALL

When water is filled in circular tank, the hydrostatic water pressure will try to increase its diameter at any section. However, this increase in the diameter all along the height of the tank will depend upon the nature of the joint at the junction $B$ of the wall and bottom slab. If the joint at $B$ is flexible (*i.e.,* sliding joint), it will be free to move outward to a position $B_1$. The hydrostatic pressure at $A$ is zero, and hence there will be no change in the diameter at $A$. The hydrostatic pressure at $B$ will be maximum, resulting in the maximum

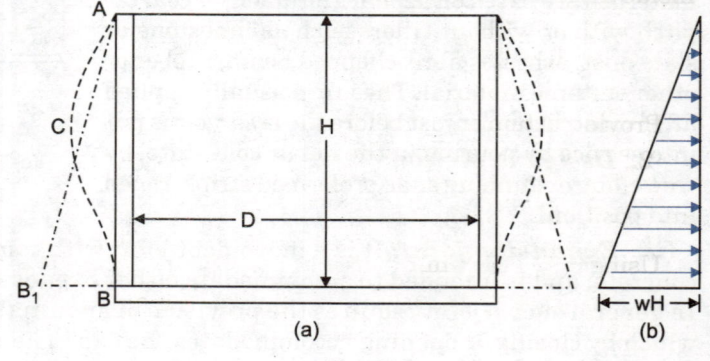

Fig. 21.4

increase in diameter there, and hence maximum movement at $B$ if the joint is flexible. Thus $AB_1$ will be the deflected position of wall $AB$. If however, the joint at $B$ is fixed no movement is possible, and a fixing moment will be induced at $B$. In that case, $ACB$ will be the deflected position.

To start with, we will consider the joint to be flexible so that the outward horizontal movement corresponding to the maximum horizontal pressure is possible. Due to this, hoop tension will be induced everywhere in the wall.

Let $D$ = diameter of the tank; $H$ = height of tank (ht. of water)

Maximum hoop tension at bottom, per unit height of wall is $= wH\dfrac{D}{2}$.

Taking permissible stress in steel in direct tension as $\sigma_s$, area of steel per metre height at the base is given by:

$$A_{sh} = \dfrac{wHD}{2\,\sigma_s} \qquad \ldots(21.1)$$

This area of steel may be provided at the centre of the wall if its thickness is small, or it may be provided on each face, keeping a minimum cover of 25 mm if the thickness is more than 225 mm. The above reinforcement is provided in the form of hoops suitably spaced, and the spacing may be increased towards the top. *However, the spacing should not be more than 3 times the thickness of the wall.*

Though the reinforcement has been provided to take the entire hoop tension, the concrete has not been prevented to take a part of this tension. The thickness of the wall should, therefore, be such that the tensile stress developed in the composite section is within safe limit. If $\sigma_{ct}$ is the permissible tensile stress in the equivalent concrete section, and $T$ is the thickness of the wall, we have

$$\sigma_{ct} = \dfrac{wHD/2}{1000\,T + (m-1)\,A_{sh}} \; (\text{N/mm}^2) \qquad \ldots(21.2)$$

The value of $\sigma_{ct}$ is taken as 1.2 N/mm$^2$ for M 20 concrete. From the above equation, the thickness $T$ can be found. Alternatively, the thickness can be fixed from the following empirical formula and then stresses may be checked from *Eq. 21.2*.

$$T = 30\,H + 50 \text{ mm} \qquad \ldots(21.3)$$

where $H$ is the height of water (liquid) retained, in metres.

In addition to the main reinforcement, temperature and distribution reinforcement may be provided in the form of vertical bars @ 0.3% of the concrete section upto 100 mm thickness. For sections thicker than 100 mm and less than 450 mm, the minimum reinforcement may be reduced from 0.3% for 100 mm thick section to 0.2% for 450 mm thick section.

If the floor slab is *resting continuously on the ground*, a minimum thickness of 150 mm may be provided, with a nomial reinforcement of 0.3% in each direction. The slab should rest on a 75 mm thick layer of lean concrete (M 10 mix). The layer of lean concrete should first be cured, and then it should be covered with a layer of tar felt to enable the floor slab to act independent of the bottom layer of lean concrete.

**Design : Example 21.1.** *Design a circular tank with flexible base for capacity of 400000 litres. The depth of water is to be 4 m, including a free board of 200 mm. Use M 20 concrete.*

**Solution: 1. Dimensions of tank:** Effective depth of water from capacity point of view = 3.8 m. If $D$ is the inside diameter of the tank, we have,

$$\dfrac{\pi}{4}D^2 \times 3.8 = \dfrac{400000 \times 10^3}{10^6}. \quad \text{From which} \quad D = \sqrt{\dfrac{400 \times 4}{\pi \times 3.8}} = 11.62 \text{ m} \approx 11.7 \text{ m}.$$

Provide a diameter of 11.7 m.

**2. Design of section:** Take $w = 9800$ N/m$^3$

$$\text{Max. hoop tension} = wH\dfrac{D}{2} = 9800 \times 4 \times \dfrac{11.7}{2} = 229320 \text{ N per metre height at base.}$$

Using $\sigma_s = 115$ N/mm$^2$, area of hoop steel at base is

$$A_{sh} = \dfrac{229320}{115} = 1994 \text{ mm}^2/\text{m height. Using 20 mm } \Phi \text{ bars,} \quad A_\Phi = 314 \text{ mm}^2$$

Spacing of hoops = $\dfrac{1000 A_\Phi}{A_{sh}} = \dfrac{1000 \times 314}{1994} = 157$ mm.

Hence provide hoops @ 150 mm c/c.

Actual $A_{sh} = \dfrac{1000 \times 314}{150} = 2093$ mm². If $T$ is thickness of wall, we have

$$\sigma_{ct} = \dfrac{wH \dfrac{D}{2}}{1000\, T + (m-1) A_{sh}}$$

where  $\sigma_{ct}$ = permissible direct tensile stress in concrete = 1.2 N/mm² for M 20 concrete

$T$ = thickness of wall, in mm

$m$ = modular ratio = 13 for M 20 concrete

∴ $1.2 = \dfrac{229320}{1000\,T + (13-1)\,2093}$  or  $1000\,T + 25116 = 191100$. From which $T = 166$ mm.

Thickness from empirical formula = $3H + 5 = 3 \times 4 + 5 = 17$ cm = 170 mm.

Hence provide a thickness of 170 mm throughout the height. Spacing of hoops can be increased near the top. Providing a minimum reinforcement of 0.3% at the top,

$A_{sh} = \dfrac{0.3}{100}(1000 \times 170) = 510$ mm². Hence spacing of hoops = $\dfrac{1000 \times 314}{510} = 615.68 \simeq 600$ mm

However, the spacing of hoops should not be more than 3 times of thickness of the wall. Hence keep a spacing of 500 mm c/c at the top. Since the thickness of wall is less than 225 mm, reinforcement will be provided at the centre of the thickness. If, however, the thickness were more than 225 mm, half the hoops would have been placed near the inner face and the other half at the outer face (*i.e.*, the spacing of the hoops would have been doubled). The spacing at the other heights can be similarly found. For example at the depth of 2 m below the top, the area of hoops is given by

$A_{sh} = \dfrac{whD/2}{115} = \dfrac{9800 \times 2 \times 11.7/2}{115} = 997$ mm².

∴ Spacing = $\dfrac{1000 \times 314}{997} = 315$ mm.

However, keep spacing = 300 mm c/c.

**3. Vertical reinforcement:** Distribution and temperature reinforcement is provided in the vertical direction, and its area is

$= 0.3 - 0.1 \left(\dfrac{170-100}{350}\right) = 0.28\%$

$A_{sd} = \dfrac{0.28}{100} \times 170 \times 1000 = 476$ mm²

Provide 10 mm Φ bars having $A_\Phi = 78.5$ mm².

Spacing = $\dfrac{1000 \times 78.5}{476} = 165$ mm.

However, provide 10 mm Φ bars at 160 mm c/c in the vertical direction. This will also serve the purpose of tieing the hoop reinforcement.

**4. Design of tank floor:** Since the tank floor is resting on ground throughout, provide a nominal thickness of 150 mm.

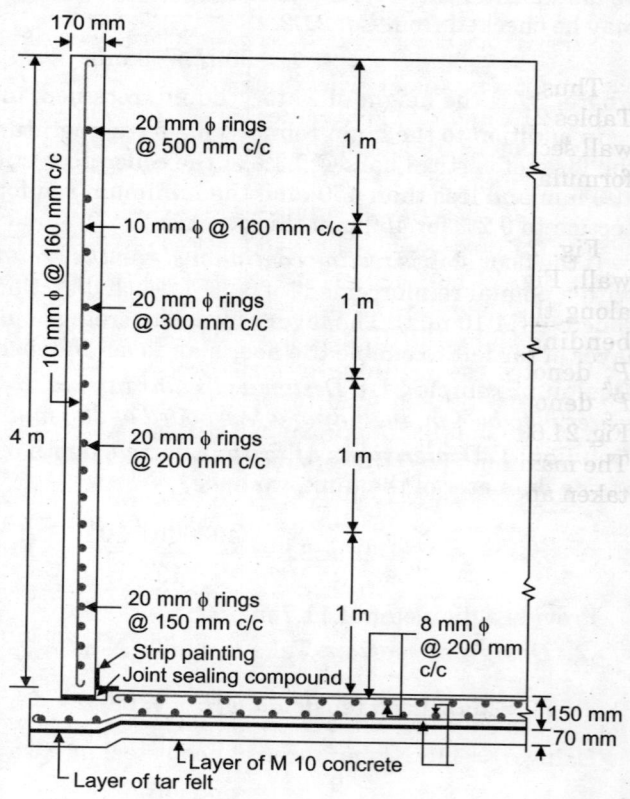

Fig. 21.5

Minimum $A_{st} = \dfrac{0.3}{100} \times 150 \times 1000 = 450$ mm² in each direction.

Providing half the reinforcement near each face, $A_{st} = 225$ mm²

Using 8 mm Φ bars, spacing $= \dfrac{1000 \times 50}{225} \approx 220$ mm c/c.

However, provide 8 mm Φ bars @ 200 mm c/c in both directions, at top and bottom of the floor slab. The floor slab will rest on 75 mm thick layer of lean concrete covered with a layer of tar felt.

**5. Details of reinforcement:** The details of reinforcement etc. are shown in Fig. 21.5.

## 21.5. CIRCULAR TANK WITH RIGID JOINT BETWEEN FLOOR AND WALL

When the joint between the wall and floor is rigid, no horizontal displacement of the wall at the joint is possible. The deflected shape of the wall will be along *ACB* [Fig. 21.4 (*a*)]. The upper part of the wall will have hoop tension, while the lower part will bend like cantilever fixed at the joint *B*. For shallow tanks with large diameter, hoop stresses are very small and the walls act more like cantilever. For deep tanks of small diameter, the cantilever action due to fixidity at the base will be small and the hoop action is predominant. The exact analysis of the tank to determine the portion of the wall upto which hoop action will be predominant and to other portion in which cantilever action will be predominant, is difficult and has been discussed in chapter 22. However, we will discuss the following methods in this chapter :

(1) Reissner's method  (2) Carpenter's simplified method
(3) Approximate method  (4) I.S. Code method.

**1. Reissner's method :** Reissner gave tables from which the resultant moment $M_f$, the maximum tension and its location from the base are given in terms of a parameter $K$ defined by

$$K = \dfrac{12 H^4}{\left(\dfrac{D}{2}\right)^2 T^2} = \dfrac{48 H^4}{D^2 T^2} \qquad \ldots(21.4)$$

Thus, for a tank of given dimensions, and of assumed thickness $T$, the parameter $K$ can be determined. Tables 21.3 and 21.4 give values of restraint moment $M_f$ and the maximum ring tension, for both rectangular wall section as well as triangular wall section. To start with, $T$ can be fixed from the following approximate formula :

$T = (30H + 50)$ mm  ...(21.5)

Fig. 21.6(*a*) shows the deformation of the wall, Fig. 21.6(*b*) shows the load distribution along the height while Fig. 21.6(*c*) shows the bending moment diagram. In Fig. 21.6 (*b*) $P_h$ denotes the load due to *hoop action* while $P_c$ denotes the load due to cantilever action. Fig. 21.6(*c*) shows the approximate B.M. diagram. The maximum positive bending moment may be taken approximately equal to $M_f/4$ to $M_f/3$.

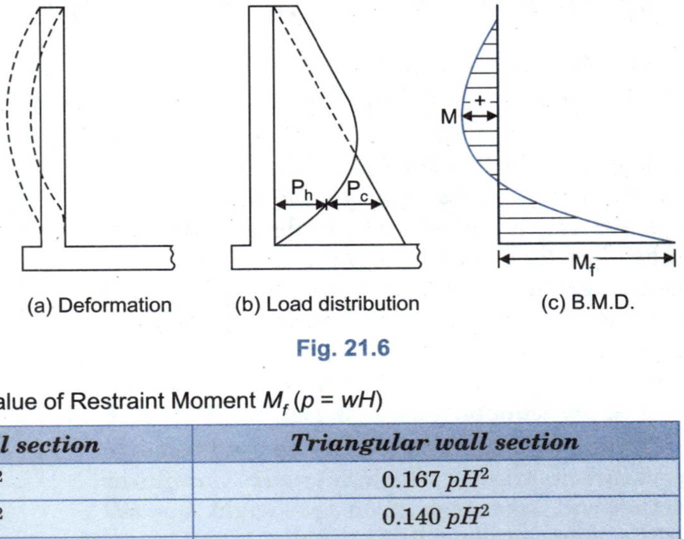

(a) Deformation  (b) Load distribution  (c) B.M.D.

**Fig. 21.6**

**TABLE 21.3.** Reissner's Value of Restraint Moment $M_f$ ($p = wH$)

| K | Rectangular wall section | Triangular wall section |
|---|---|---|
| 0 | 0.167 $pH^2$ | 0.167 $pH^2$ |
| 10 | 0.110 $pH^2$ | 0.140 $pH^2$ |
| 100 | 0.0582 $pH^2$ | 0.0707 $pH^2$ |
| 1000 | 0.024 $pH^2$ | 0.026 $pH^2$ |
| 10000 | 0.0085 $pH^2$ | 0.009 $pH^2$ |
| ∞ | 0 | 0 |

**TABLE 21.4.** Reissner's Value for Ring Tension ($p = wH$)

| K | Rectangular wall section | | Triangular wall section | |
|---|---|---|---|---|
| | Max. tension | Height from base | Max. tension | Height from base |
| 0 | 0 | — | — | — |
| 10 | $0.13\, p\left(\frac{D}{2}\right)$ | $1.0\, H$ | $0.09\, p\left(\frac{D}{2}\right)$ | $0.65\, H$ |
| 100 | $0.27\, p\left(\frac{D}{2}\right)$ | $1.0\, H$ | $0.31\, p\left(\frac{D}{2}\right)$ | $0.58\, H$ |
| 1000 | $0.47\, p\left(\frac{D}{2}\right)$ | $0.47\, H$ | $0.52\, p\left(\frac{D}{2}\right)$ | $0.44\, H$ |
| 10000 | $0.67\, p\left(\frac{D}{2}\right)$ | $0.31\, H$ | $0.70\, p\left(\frac{D}{2}\right)$ | $0.30\, H$ |
| ∞ | $1.0\, p\left(\frac{D}{2}\right)$ | 0 | $1.0\, p\left(\frac{D}{2}\right)$ | 0 |

**TABLE 21.5.** Carpenter's Values of Coefficients $F$ and $K$

| Factor | | F | | | | K | | | |
|---|---|---|---|---|---|---|---|---|---|
| $\frac{H}{T} \rightarrow$ | | 10 | 20 | 30 | 40 | 10 | 20 | 30 | 40 |
| Values of H/D | 0.2 | 0.046 | 0.028 | 0.022 | 0.015 | — | 0.50 | 0.45 | 0.40 |
| | 0.3 | 0.032 | 0.019 | 0.014 | 0.010 | 0.55 | 0.43 | 0.38 | 0.33 |
| | 0.4 | 0.024 | 0.014 | 0.010 | 0.007 | 0.50 | 0.39 | 0.35 | 0.30 |
| | 0.5 | 0.020 | 0.012 | 0.009 | 0.006 | 0.45 | 0.37 | 0.32 | 0.27 |
| | 1.0 | 0.012 | 0.006 | 0.005 | 0.003 | 0.37 | 0.30 | 0.24 | 0.21 |
| | 2.0 | 0.006 | 0.003 | 0.002 | 0.002 | 0.28 | 0.22 | 0.19 | 0.16 |
| | 4.0 | 0.004 | 0.002 | 0.002 | 0.001 | 0.27 | 0.20 | 0.17 | 0.14 |

**2. Carpenter's simplified method :** Carpenter simplified Dr. Reissner's method, and gave the values of maximum cantilever bending moment and the position and magnitude of maximum ring tension in terms of the following expressions :

(i) Position of max. hoop tension = $KH$ above base ...(21.6)

(ii) Max. hoop tension = $w\,(H - KH)\,\frac{D}{2} = w\,(H)\,\frac{D}{2}\,(1 - K)$ ...(21.7)

(iii) Maximum cantilever B.M. = $F\,w\,H^3$ ...(21.8)

The values of coefficient $K$ and $F$ depend upon $H/D$ and $H/T$ ratios, and may be taken from Table 21.5.

**3. Approximate method :** This method is followed when Reissner's table or Carpenter's tables are not available. In the approximate method, it is assumed that cantilever action will take place for a height $h = H/3$ or 1 m (whichever is greater) above base for the value of $H^2 / (DT)$ between 6 to 12 and for a height of $h = H/4$ or 1 m (whichever is more) for the value of $H^2 / (DT)$ between 12 to 30. Above this height, hoop action will be predominant.

Thus, in Fig. 21.7, $ABC$ is hydrostatic pressure diagram in which $BC = wH$. Cantilever action will take place upon the height $h = BD$ while hoop action will be predominant above this height. The maximum hoop tension is computed at level $D$, for height $(H - h)$, below $A$,

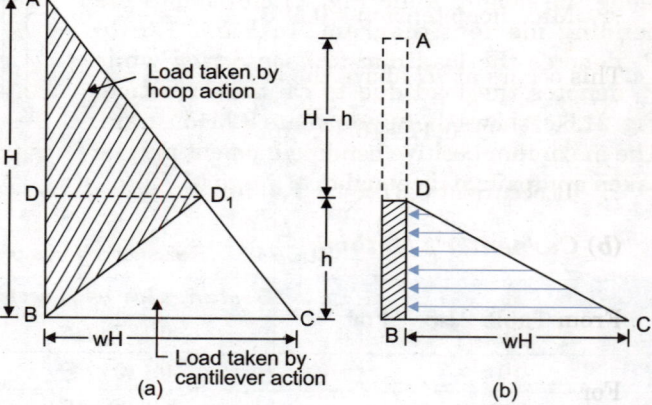

Fig. 21.7

where the water pressure will be $w(H-h)$. Fig. 21.7(b) shows the height $h$ of the wall acting as cantilever fixed at base $B$. The load on the cantilever will be the area of triangular load $DBC$, having zero ordinate at $D$ and maximum ordinate of $wH$ at $B$. Cantilever reinforcement will be provided upto height $h$, at the *inner face*. Above this height, the spacing of the cantilever reinforcement can be increased.

The maximum hoop tension at $D$, per metre height $= w(H-h)\dfrac{D}{2}$ ...(21.9)

Steel for this hoop tension is provided in form of rings provided at both the faces. The spacing of the bars for hoop tension will be kept equal upto height $h$ (*i.e.*, for portion $BD$). Above this height, the spacing of the rings can be increased.

**Example 21.2.** *A circular tank has an internal diameter of 10 m and has maximum height of water as 4 m. The walls of the tank are restrained at the base. Determine the values of maximum hoop tension and its location, and the maximum cantilever bending moment by the following methods : (a) Reissner's method. (b) Carpenter's method.*

**Solution:** **(a) Reissner's method:**

Let thickness $\quad T = 30H + 50 = (30 \times 4) + 50 = 170$ mm $= 0.17$ m

$\therefore \quad K = \dfrac{48\,H^4}{D^2\,T^2} = \dfrac{48\,(4)^4}{(10 \times 0.17)^2} = 4250.$ From Table 21.3, we have

For $K = 1000$, $M_f = 0.024\,pH^2$; For $K = 10000$, $M_f = 0.0085\,pH^2$

Hence for $K = 4250$, $M_f = \alpha\,pH^2$ where the value of $\alpha$ can be determined from the following logarithmic interpolation:

$$\alpha = 0.0085 + (0.024 - 0.0085)\,\dfrac{\log 10000 - \log 4250}{\log 10000 - \log 1000}$$

$$= 0.0085 + (0.024 - 0.0085) \times \dfrac{4 - 3.6284}{4 - 3} = 0.0143$$

$\therefore \quad M_f = 0.0143\,pH^2 = 0.0143\,wH^3 = 0.0143 \times 9800\,(4)^3 = $ **8969 N-m**

For the position and value of maximum hoop tension, we have from Table 21.4 :

For $K = 1000$, Hoop tension $= 0.47\,p\,\dfrac{D}{2}$ and For $K = 10000$, Hoop tension $= 0.67\,p\,\dfrac{D}{2}$

For $K \approx 4250$, hoop tension $= \beta\,p\,\dfrac{D}{2}$, where the value of $\beta$ may be determined from the following logarithmic interpolation:

$$\beta = 0.47 + (0.67 - 0.47)\,\dfrac{\log 4250 - \log 1000}{\log 10000 - \log 1000} = 0.47 + (0.67 - 0.47)\,\dfrac{3.6284 - 3}{4 - 3} \approx 0.596$$

$\therefore$ Max. hoop tension $= 0.596\,p\left(\dfrac{D}{2}\right) = 0.596\,w\,\dfrac{HD}{2} = 0.596 \times 9800 \times 4 \times \dfrac{10}{2} = $ **116816 N = 116.82 kN.**

This occurs at $\gamma H$ above the base, where $\gamma$ is given by

$$\gamma = 0.47 - (0.47 - 0.31) \times \dfrac{\log 4250 - \log 1000}{\log 10000 - \log 1000} = 0.47 - 0.16\,(3.6284 - 3) \approx 0.37$$

$\therefore$ It occurs at $0.37 \times 4 = $ **1.48 m** above the base.

**(b) Carpenter's Method:** $\dfrac{H}{T} = \dfrac{4}{0.17} = 23.5$; $\quad \dfrac{H}{D} = \dfrac{4}{10} = 0.4$

From Table 21.5. For $\quad \dfrac{H}{T} = 20$ and $\dfrac{H}{D} = 0.4$, $F = 0.014$ and $K = 0.39$

For $\quad \dfrac{H}{T} = 30$ and $\dfrac{H}{D} = 0.4$, $F = 0.010$ and $K = 0.35$

Hence, by linear interpolation, for $\dfrac{H}{T} = 23.5$ and $\dfrac{H}{D} = 0.4$, we have

$$F = 0.014 - \frac{(0.014 - 0.010)}{10} \times 3.5 = 0.0126 \quad \text{and} \quad K = 0.39 - \frac{0.39 - 0.35}{10} \times 3.5 = 0.376$$

Hence, (i) position of max. hoop tension = $KH$ = 0.376 × 4 ≈ **1.5 m** above base

(ii) Max. hoop tension = $w H \frac{D}{2} (1 - K)$ = 9800 × 4 × $\frac{10}{2}$ (1 − 0.376) = **122304 N = 122.3 kN**

(iii) Max. cantilever B.M. = $F w H^3$ = 0.0126 × 9800 × $(4)^3$ = **7903 N-m**

**Example 21.3.** *Redesign the circular tank for the data of example 21.1, assuming that the joint between the wall and the base is rigid. Use M 20 concrete. Follow the approximate method of analysis.*

**Solution:** 1. *Dimensions of the tank.* As found earlier, $D$ = 11.7 m and $H$ = 4 m.

2. *Design constants:* For M 20 concrete, we have $\sigma_{cbc}$ = 7 N/mm² and $m$ = 13. Taking $\sigma_{st}$ = 115 N/m², we have

$$k = \frac{m \sigma_{cbc}}{m \sigma_{cbc} + \sigma_{st}} = \frac{13 \times 7}{13 \times 7 + 115} = 0.442 \; ; \quad j = 1 - \frac{k}{3} = 0.853$$

$$R = \frac{1}{2} \sigma_{cbc} j k = \frac{1}{2} \times 7 \times 0.853 \times 0.442 = 1.32.$$

3. *Determination of bending moment and hoop tension*

Let $T$ = (30 $H$ + 50) mm = (30 × 4) + 50 = 170 mm = 0.17 m

$$\frac{H^2}{DT} = \frac{4 \times 4}{11.7 \times 0.17} = 8$$

The height $h$ above base, upto which cantilever action will be there, is given by $h = H/3$ or 1 m whichever is more. ∴ $h$ = 4/3 = 1.33 m ∴ Maximum ring tension at this level, per metre height

= $w (H - h) \frac{D}{2}$, (where $w$ = 9800 N/m³) = 9800 (4 − 1.33) $\frac{11.7}{2}$ = 153071.1 = 153.07 kN

Water pressure at bottom = $wH$ = 9800 × 4 = 39200 N/m²/m

∴ Max. cantilever B.M. = $M_f = \frac{1}{2} \times 39200 \times 1.33 \times \frac{1.33}{3}$ ≈ 11556.8 ≈ 11600 N-m/m

4. *Design of section for cantilever action:* If $d$ is the effective thickness of tank wall,

$$d = \sqrt{\frac{M_f}{1000 R}} = \sqrt{\frac{11600 \times 1000}{1000 \times 1.32}} \approx 94 \text{ mm}$$

∴ Total thickness = $d$ + cover = 94 + 35 = 129 ≈ 130 mm

However, provide a minimum thickness equal to the greater of the following :

(i) 150 mm    (ii) 30 $H$ + 50 = (30 × 4) + 50 = 170 mm.    Hence provide $T$ = 170 mm.

Providing 35 mm cover to the centre of reinforcement, available $d$ = 170 − 35 = 135 mm. Area of steel for cantilever bending is given by

$$A_{st} = \frac{M_f}{\sigma_{st} j d} = \frac{11600 \times 1000}{115 \times 0.853 \times 135} = 876 \text{ mm}^2. \text{ Using 12 mm } \Phi \text{ bars, } A_\Phi = 113 \text{ mm}^2.$$

∴ Spacing = $\frac{1000 \times 113}{876}$ = 129 mm

However, provide 12 mm Φ bars @ 120 mm c/c up to a height of 1.33 m (or say 1.4 m) from the base. Above this height, curtail half the bars and continue the other half (*i.e.* @ 240 mm c/c) upto the top. Let us test this for development length.

$$L_d = \frac{\Phi \sigma_{st}}{4 \tau_{bd}}, \text{ where } \Phi = 12 \text{ mm} ; \sigma_{st} = 115 \text{ N/mm}^2 \text{ and } \tau_{bd} = 0.8 \text{ N/mm}^2 \text{ for M 20 concrete}$$

∴ $L_d = \frac{115}{4 \times 0.8} \Phi = 36 \Phi = 36 \times 12 = 432 \approx 430 \text{ mm} = 0.43 \text{ m}.$

Hence half the bars can be curtailed at 1.4 m height above the base. These vertical bars are to be provided at inner face. Keep a clear cover of 25 mm.

## WATER TANKS-I : SIMPLE CASES

**5. Design of section for hoop action:** Maximum hoop tension = 153070 N at 1.33 m above base.
∴ Area of rings, $A_{sh}$ = 153070/115 = 1331 mm². Let us provide rings at both the faces. Hence area of rings on each face = 665 mm²

Using 12 mm Φ rings, $A_\phi$ = 113 mm². ∴ Spacing = $\dfrac{1000 \times 113}{665} \approx 170$ mm.

Hence provide 12 mm Φ rings @ 170 mm c/c on each face. This spacing is kept constant from the base to height $h$ = 1.33 m (or say 1.4 m), and above this, the spacing may be increased.

Actual $A_{sh} = 2 \times \dfrac{1000 \times 113}{170} = 1329.4$ mm².

∴ $\sigma_{ct} = \dfrac{153070}{1000 \times 170 + (13-1)1329.4} = 0.823 < 1.2$ N/mm². Hence safe.

Hoop tension at 2 m below top = $9800 \times 2 \times \dfrac{11.7}{2} = 114660$ N

∴ $A_{sh}$ = 114660/115 = 997 mm². Area at each face = 499 mm².

Spacing of rings = $\dfrac{1000 \times 113}{499} \approx 226 \approx 220$ mm

At the top, minimum $A_{sh} = \dfrac{0.3}{100} \times 170 \times 1000 = 510$ mm², giving an area of 255 mm² on each face.

∴ Spacing = $\dfrac{1000 \times 113}{255} = 443$ mm.

However, provide these @ 440 mm c/c at top.

**6. Distribution reinforcement:** Percentage area of distribution reinforcement is

$$= 0.3 - 0.1 \left( \dfrac{170 - 100}{350} \right) = 0.28\%$$

∴ $A_{sd} = \dfrac{0.28}{100} \times 170 \times 1000 = 476$ mm².

Area of steel on each face = 238 mm². However, no additional reinforcement will be provided at the inner face since the vertical steel for cantilever action will serve this purpose.

Hence provide $A_{sd}$ = 238 mm² in vertical direction at the outer face. Using 8 mm Φ bars, $A_\phi$ = 50.5 mm².

∴ Spacing = $\dfrac{1000 \times 50.5}{238} = 212$ mm.

However, provide these @ 200 mm c/c.

**7. Provision of haunches:** It is customary to provide 150 mm × 150 mm haunches at the junction of the wall and the base. A haunch reinforcement of 8 mm Φ @ 220 mm c/c may be provided.

**8. Design of base slab:** Provide a 150 mm thick base slab with the reinforcement on both faces as indicated in Example 21.1. The details of reinforcement etc. are shown in Fig. 21.8.

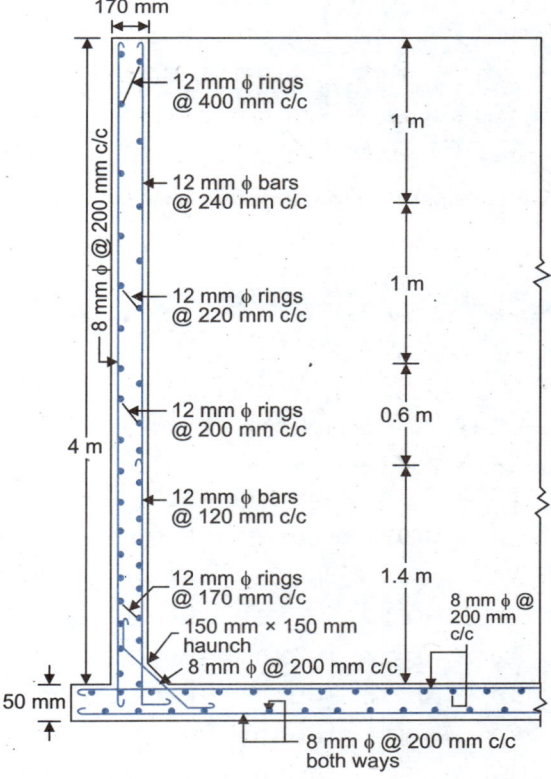

Fig. 21.8

## 21.6. I.S. CODE METHOD FOR CIRCULAR TANKS

Indian Standard Code IS : 3370 (Part IV)-1967, gives design Tables for moment and hoop tension in circular tanks for various conditions of joint and various types of loading. However, we shall describe here the case of rigid joint between the wall and the base slab, subjected to triangular distributed water pressure. Table 21.6 gives the coefficient for tension at various height in the wall for various values of $\frac{H^2}{DT}$ ratio. The tension is given by the following expression :

$$\text{Tension} = \text{coefficient} \times w\,H\,\frac{D}{2} \text{ per metre} \qquad \ldots(21.10)$$

**TABLE 21.6.** Coefficient for Tension in Cylindrical Wall Fixed to the Base

| $\frac{H^2}{DT}$ | Coefficient at point | | | | | | | | | |
|---|---|---|---|---|---|---|---|---|---|---|
| | 0.0H | 0.1H | 0.2H | 0.3H | 0.4H | 0.5H | 0.6H | 0.7H | 0.8H | 0.9H |
| 0.4 | +0.149 | +0.134 | +0.120 | +0.101 | +0.082 | +0.066 | +0.049 | +0.029 | +0.014 | +0.004 |
| 0.8 | +0.263 | +0.239 | +0.215 | +0.109 | +0.160 | +0.130 | +0.096 | +0.063 | +0.034 | +0.010 |
| 1.2 | +0.283 | +0.271 | +0.254 | +0.234 | +0.209 | +0.180 | +0.142 | +0.099 | +0.053 | +0.016 |
| 1.6 | +0.265 | +0.268 | +0.268 | +0.265 | +0.250 | +0.232 | +0.185 | +0.134 | +0.075 | +0.023 |
| 2.0 | +0.234 | +0.251 | +0.273 | +0.285 | +0.285 | +0.330 | +0.232 | +0.172 | +0.104 | +0.031 |
| 3.0 | +0.134 | +0.203 | +0.267 | +0.322 | +0.357 | +0.409 | +0.330 | +0.262 | +0.157 | +0.052 |
| 4.0 | +0.067 | +0.164 | +0.256 | +0.339 | +0.403 | +0.429 | +0.409 | +0.334 | +0.210 | +0.073 |
| 5.0 | +0.025 | +0.137 | +0.245 | +0.346 | +0.428 | +0.477 | +0.469 | +0.398 | +0.259 | +0.092 |
| 6.0 | +0.018 | +0.119 | +0.234 | +0.344 | +0.441 | +0.504 | +0.514 | +0.447 | +0.301 | +0.112 |
| 8.0 | −0.001 | +0.104 | +0.218 | +0.355 | +0.445 | +0.534 | +0.575 | +0.530 | +0.381 | +0.151 |
| 10.0 | −0.003 | +0.098 | +0.208 | +0.323 | +0.437 | +0.542 | +0.608 | +0.589 | +0.440 | +0.179 |
| 12.0 | −0.004 | +0.097 | +0.202 | +0.312 | +0.429 | +0.543 | +0.628 | +0.633 | +0.494 | +0.211 |
| 14.0 | −0.002 | +0.098 | +0.200 | +0.306 | +0.420 | +0.539 | +0.639 | +0.666 | +0.541 | +0.241 |
| 16.0 | −0.000 | +0.099 | +0.199 | +0.304 | +0.412 | +0.531 | +0.641 | +0.687 | +0.582 | +0.265 |

**Note :** Positive sign indicates tension

**TABLE 21.7.** Coefficient for Moment in Cylindrical Wall Fixed to the Base

| $\frac{H^2}{DT}$ | Coefficient at point | | | | | | | | | |
|---|---|---|---|---|---|---|---|---|---|---|
| | 0.1H | 0.2H | 0.3H | 0.4H | 0.5H | 0.6H | 0.7H | 0.8H | 0.9H | 1.0H |
| 0.4 | +0.0050 | +0.0014 | +0.0021 | +0.0007 | +0.0042 | −0.0150 | −0.0302 | −0.0529 | −0.0816 | −0.1205 |
| 0.8 | +0.0011 | +0.0037 | +0.0063 | +0.0080 | +0.0070 | +0.0023 | −0.0068 | −0.0240 | −0.0465 | −0.0795 |
| 1.2 | +0.0012 | +0.0042 | +0.0077 | +0.0103 | +0.0112 | +0.0090 | +0.0022 | −0.0108 | −0.0311 | −0.0602 |
| 1.6 | +0.0011 | +0.0041 | +0.0075 | +0.0107 | +0.0121 | +0.0111 | +0.0058 | −0.0051 | −0.0232 | −0.0505 |
| 2.0 | +0.0010 | +0.0035 | +0.0068 | +0.0099 | +0.0120 | +0.0115 | +0.0075 | −0.0021 | −0.0185 | −0.0436 |
| 3.0 | +0.0006 | +0.0024 | +0.0047 | +0.0071 | +0.0090 | +0.0097 | +0.0077 | +0.0012 | −0.0119 | −0.0333 |
| 4.0 | +0.0003 | +0.0015 | +0.0028 | +0.0047 | +0.0066 | +0.0077 | +0.0069 | +0.0023 | +0.0080 | −0.0268 |
| 5.0 | +0.0002 | +0.0008 | +0.0016 | +0.0029 | +0.0046 | +0.0059 | +0.0059 | +0.0028 | −0.0058 | −0.0222 |
| 6.0 | +0.0001 | +0.0003 | +0.0008 | +0.0019 | +0.0032 | +0.0046 | +0.0046 | +0.0029 | −0.0041 | −0.0187 |
| 8.0 | 0.0000 | +0.0001 | +0.0002 | +0.0008 | +0.0016 | +0.0028 | +0.0038 | +0.0029 | −0.0022 | −0.0146 |
| 10.0 | 0.0000 | +0.0000 | +0.0001 | +0.0004 | +0.0007 | +0.0019 | +0.0029 | +0.0028 | −0.0012 | −0.0122 |
| 12.0 | 0.0000 | −0.0001 | +0.0001 | +0.0002 | +0.0003 | +0.0013 | +0.0023 | +0.0026 | −0.0005 | −0.0104 |
| 14.0 | 0.0000 | 0.0000 | 0.0000 | 0.0000 | +0.0001 | +0.0008 | +0.0019 | +0.0022 | −0.0001 | −0.0090 |
| 16.0 | 0.0000 | 0.0000 | −0.0001 | −0.0002 | −0.0001 | +0.0004 | +0.0013 | +0.0019 | +0.0001 | −0.0079 |

**Note:** Positive sign indicates tension on outside

Table 21.7 gives the coefficients for bending moment at various heights in the wall for various values of $\frac{H^2}{DT}$ ratio. The moment is given by following expression :

$$\text{Moment} = \text{coefficient} \times wH^2 \text{ N-m/m} \qquad ...(21.11)$$

The shear at the base of cylindrical wall for the case of triangular load can be determined from the following expression:  $\text{Shear} = \text{coefficient} \times wH^2 \text{ kN}$ ...(21.12)
where the coefficient can be taken from Table 21.8.

**TABLE 21.8** Coefficients for Shear at the Base

| $\frac{H^2}{DT}$ | Coefficient | $\frac{H^2}{DT}$ | Coefficient |
|---|---|---|---|
| 0.4 | +0.436 | 5.0 | +0.213 |
| 0.8 | +0.374 | 6.0 | +0.197 |
| 1.2 | +0.339 | 8.0 | +0.174 |
| 1.6 | +0.317 | 10.0 | +0.158 |
| 2.0 | +0.299 | 12.0 | +0.145 |
| 3.0 | +0.262 | 14.0 | +0.135 |
| 4.0 | +0.236 | 16.0 | +0.127 |

**Example 21.4.** *For the data of example 21.2, determine the maximum hoop tension and its position, moment at the base and shear at the base using I.S. Code tables. Assume thickness of wall as 160 mm.*
**Solution:** Thickness $T = 160$ mm $= 0.16$ m.

Hence $\quad \dfrac{H^2}{DT} = \dfrac{(4)^2}{10 \times 0.16} = 10$

From Table 21.6, for $\dfrac{H^2}{DT} = 10$, maximum tension may be assumed to occur at $0.6 H = 2.4$ m from top, and coefficient may be taken as 0.608. Hence maximum tension $= 0.608 \times w H \dfrac{D}{2} = 0.608 \times 9800 \times 4 \times \dfrac{10}{2}$ = 119168 N. For exact location of maximum tension and its magnitude, the values of tension all along the height can be plotted by coefficients from Table 21.6. From the curve so obtained maximum value can be found.

From Table 21.7 for $\dfrac{H^2}{DT} = 10$, the moment coefficient for base $(1.0 H)$ is found to be $-0.0122$ (minus sign indicating tension at the inner face).

$\therefore \quad$ Moment $= 0.0122 \times w \, H^3 = 0.0122 \times 9800 \, (4)^3 =$ **7652 N-m/m.**

From Table 21.8, for $\dfrac{H^2}{DT} = 10$, the coefficient for shear is found to be 0.158.
Hence shear $= 0.158 \times wH^2 = 0.158 \times 9800 \times (4)^2 = 24774$ N (acting inward)

## PROBLEMS

1. Design a circular water tank of capacity 200000 litres. The depth of tank is limited to 3 m from inside. Keep the joint between the wall and base slab as flexible. The base slab rests on the ground.
2. Redesign the tank of problem 1 if the joint between wall and base is to be made rigid. Use approximate method.
3. A circular tank has 12 m diameter and 3 m water height. Determine (*i*) maximum hoop tension and its location and (*ii*) maximum bending moment. Use the following methods :
   (*a*) Reissner's method. (*b*) Carpenter's method (*c*) I.S. Code method.
4. The section of RCC wall of a rectangular tank is subjected to direct tension of 75 kN/m and moment of (*i*) 95 kNm/m and (*ii*) 5.6 kNm/m. Consider the tension steel being placed beyond 225 mm from water face. Design the section when the $M_{30}$ grade HYSD steel to be used and Class-B Exposure conditions.

# WATER TANKS-II : CIRCULAR AND INTZE TANKS

## 22.1. CIRCULAR TANK WITH RIGID JOINT BETWEEN FLOOR AND WALL

When water is filled in circular tank, the hydrostatic water pressure will try to increase in diameter at any section. However, this increase in the diameter all along the height of the tank will depend upon the nature of the joint at the junction $B$ of the wall and bottom slab (Fig. 22.1). If the joint is flexible (*i.e.*, sliding joint), it will be free to move outward to a position $B_1$. The hydrostatic pressure at $A$ is zero and hence there will be no change in the diameter at $A$. The hydrostatic pressure at $B$ will be maximum, resulting in the maximum increase in the diameter there, and maximum movement at $B$ if joint is flexible. Thus $AB_1$ will be deflected position of wall $AB$. If, however, the joint at $B$ is fixed, no movement is possible, and a fixing moment will be induced in $B$. In that case, $ACB$ will be the deflected position.

The case of flexible joint between the floor and wall has already been discussed in chapter 21. We shall analyse here the case of fixed joint. When the joint between the wall and floor is rigid, no horizontal displacement of the wall at the joints is possible. The deflected shape of the wall will be along $ACB$. The upper part will have hoop tension, while the lower part will bend like cantilever fixed at joint $B$. For shallow tanks with large diameter, hoop stresses are very small and the wall acts more like cantilever. For deep tanks of small diameter, the cantilever action due to fixity at the base will be small and the hoop action will be predominant.

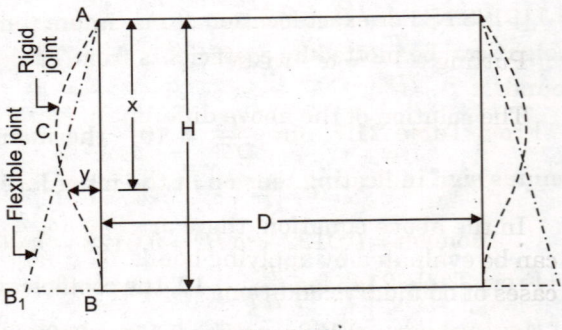

Fig. 22.1

Let $p_c$ = load transferred due to cantilever action (pressure per unit area) or horizontal shear.
$p_r$ = load carried by ring tension (pressure per unit area).
$T$ = thickness of wall.
$I$ = moment of inertia of the wall, per unit length = $T^3/12$
or $= \dfrac{T^3}{12(1-\mu^2)}$, if the effect of lateral restraint is taken into account.
$E$ = Modulus of elasticity ;   $\mu$ = Poisson's ratio = 0.2.
$y$ = Change in radius at depth $x$ ;   $y_r$ = Radial expansion due to $p_r$.
$y_c$ = Outward deflection due to cantilever action.

At any section, at depth $x$ below the top, the deformation $y_r$ due to hoop tension will be equal to displacement $y_c$ due to horizontal shear (cantilever action). The sum of loads transferred due to cantilever action, and that carried by ring tension is evidently equal to the intensity of water pressure at that section, Hence

$$p_c + p_r = w \cdot x \qquad \ldots(i)$$

For compatibility, $\qquad y_c = y_r = y$ (say) $\qquad \ldots(ii)$

Now, ring tension at depth $x = p_r \cdot \dfrac{D}{2}$ ;

Tensile stress due to ring tension $= \dfrac{p_r \cdot D}{2T}$

Tensile strain due to ring tension $= \dfrac{p_r \cdot D}{2T \cdot E}$

Change in radius $= y_r = \dfrac{p_r \, D}{2\,TE} \times \dfrac{D}{2} = \dfrac{p_r \, D^2}{4\,TE}$

or $\qquad p_r = \dfrac{4TE}{D^2} y_r = \dfrac{4TE}{D^2} \cdot y \qquad \ldots(iii)$

From simple theory of bending, we have

$$\text{B.M.} = EI\left(\dfrac{d^2 y_c}{dx^2}\right); \quad \text{shear force} = EI\left(\dfrac{d^3 y_c}{dx^3}\right)$$

and load intensity $\qquad p_c = EI\left(\dfrac{d^4 y_c}{dx^4}\right) = EI\left(\dfrac{d^4 y}{dx^4}\right) \qquad \ldots(iv)$

But $\qquad p_c = w \cdot x - p_r \qquad$ [From $(i)$]

$\therefore \qquad wx - p_r = EI\left(\dfrac{d^4 y}{dx^4}\right) \quad$ or $\quad wx - \dfrac{4\,T\,E}{D^2} y = EI \cdot \dfrac{d^4 y}{dx^4}$

or $\qquad \dfrac{d^4 y}{dx^4} + \dfrac{4T}{ID^2} y = \dfrac{wx}{EI} \qquad \ldots(22.1)$

Putting $\dfrac{T}{ID^2} = \alpha^4$ we get $\qquad \dfrac{d^4 y}{dx^4} + 4\alpha^4 y = \dfrac{wx}{EI} \qquad \ldots(22.2)$

The solution of the above differential equation is

$$y = \dfrac{wx}{4EI\alpha^4} + e^{\alpha x}(A \sin \alpha x + B \cos \alpha x) + e^{-\alpha x}(C \sin \alpha x + D \cos \alpha x). \quad \ldots(22.3)$$

In the above equation, there are four unknown constants $A$, $B$, $C$ and $D$. The values of these constants can be evaluated by applying boundary conditions at both the ends of the wall. We shall consider different cases of boundary conditions.

## CASE I : WALL WITH FIXED BASE AND FREE TOP

This case is based on the assumption that the base joint is continuous and the footing is prevented from any rotation. The constants $A$, $B$, $C$ and $D$ can be evaluated from the following conditions.

(i) At the base, deflection is zero, $\quad \therefore \quad$ At $x = H$, $\quad y = 0 \qquad \ldots(1)$

(ii) At base, slope is zero, $\quad \therefore \quad$ At $x = H$, $\quad \dfrac{dy}{dx} = 0 \qquad \ldots(2)$

(iii) At the top, B.M. is zero, $\quad$ i.e. $\quad$ At $x = 0$, $\quad \dfrac{d^2 y}{dx^2} = 0 \qquad \ldots(3)$

(iv) At the top, S.F. is zero, $\quad$ i.e. $\quad$ At $x = 0$, $\quad \dfrac{d^3 y}{dx^3} = 0 \qquad \ldots(4)$

$$\frac{dy}{dx} = \frac{w}{4EI\alpha^4} + \alpha \cdot e^{\alpha x}(A \sin \alpha x + B \cos \alpha x) + \alpha e^{\alpha x}(A \cos \alpha x - B \sin \alpha x)$$
$$- \alpha e^{-\alpha x}(C \sin \alpha x + D \cos \alpha x) + \alpha e^{-\alpha x}(C \cos \alpha x - D \sin \alpha x)$$

or $$\frac{dy}{dx} = \frac{w}{4EI\alpha^4} + \alpha e^{\alpha x}[(A-B)\sin \alpha x + (A+B)\cos \alpha x]$$
$$+ \alpha e^{-\alpha x}[-(C+D)\sin \alpha x + (C-D)\cos \alpha x] \quad \ldots[22.3(a)]$$

Differentiating it further,

$$\frac{d^2y}{dx^2} = \alpha^2 \cdot e^{\alpha x}[(A-B)\sin \alpha x + (A+B)\cos \alpha x] + \alpha^2 e^{\alpha x}[(A-B)\cos \alpha x$$
$$-(A+B)\sin \alpha x] - \alpha^2 e^{-\alpha x}[-(C+D)\sin \alpha x + (C-D)\cos \alpha x]$$
$$+ \alpha^2 e^{-\alpha x}[-(C+D)\cos \alpha x - (C-D)\sin \alpha x]$$

or $$\frac{d^2y}{dx^2} = \alpha^2 \cdot e^{\alpha x}[-2B \sin \alpha x + 2A \cos \alpha x] + \alpha^2 e^{-\alpha x}[2D \sin \alpha x - 2C \cos \alpha x] \quad \ldots[22.3(b)]$$

Differentiating it further, we get

$$\frac{d^3y}{dx^3} = \alpha^3 \cdot e^{\alpha x}[-2B \sin \alpha x + 2A \cos \alpha x] + \alpha^3 \cdot e^{\alpha x}[-2B \cos \alpha x - 2A \sin \alpha x]$$
$$- \alpha^3 \cdot e^{-\alpha x}[2D \sin \alpha x - 2C \cos \alpha x] + \alpha^3 \cdot e^{-\alpha x}[2D \cos \alpha x + 2C \sin \alpha x]$$

or $$\frac{d^3y}{dx^3} = 2\alpha^3 \cdot e^{\alpha x}[-(A+B)\sin \alpha x + (A-B)\cos \alpha x]$$
$$+ 2\alpha^3 \cdot e^{-\alpha x}[(C-D)\sin \alpha x + (C+D)\cos \alpha x] \quad \ldots[22.3(c)]$$

Now, applying condition (iii) in Eq. 22.3 (b), we get

$$\frac{d^2y}{dx^2} = 0 = \alpha^2[0 + 2A] + \alpha^2[0 - 2C]. \quad \text{From which,} \quad A = C \quad \ldots(22.4)$$

Applying, condition (iv) in Eq. 22.3 (c), we get

$$\frac{d^3y}{dx^3} = 0 = 2\alpha^3[0 + (A-B)] + 2\alpha^3[0 + (C+D)]$$

$$\therefore \quad -(A-B) = (C+D), \text{ But } A = C \quad \therefore \quad D = B - 2A \quad \ldots(22.5)$$
$$\therefore \quad C + D = A + (B - 2A) = B - A \quad \text{and} \quad C - D = 3A - B \quad \ldots(22.6)$$

Thus, constants $C$ and $D$ are known in terms of $A$ and $B$. Applying condition (i) in Eq. 22.3, we get

$$y = 0 = \frac{wH}{4EI\alpha^4} + e^{\alpha H}[A \sin \alpha H + B \cos \alpha H] + e^{-\alpha H}[A \sin \alpha H + (B - 2A)\cos \alpha H]$$

$$A[\sin \alpha H(e^{\alpha H} + e^{-\alpha H}) - 2e^{-\alpha H} \cos \alpha H] + B[\cos \alpha H(e^{\alpha H} + e^{-\alpha H})] + \frac{wH}{4EI\alpha^4} = 0 \quad \ldots(22.7)$$

Similarly, applying condition (ii) in Eq. 22.3 (a), we get

$$\frac{dy}{dx} = 0 = \frac{w}{4EI\alpha^4} + \alpha \cdot e^{\alpha H}[(A-B)\sin \alpha H + (A+B)\cos \alpha H].$$
$$+ \alpha e^{-\alpha H}[(A-B)\sin \alpha H + (3A - B)\cos \alpha H]$$

or $A[\alpha \sin \alpha H(e^{\alpha H} - e^{-\alpha H}) + \alpha \cos \alpha H(e^{\alpha H} + e^{-\alpha H}) + 2\alpha \cdot e^{-\alpha H}(\sin \alpha H + \cos \alpha H)]$
$$+ B[\alpha \cos \alpha H(e^{\alpha H} - e^{-\alpha H}) - \alpha \sin \alpha H(e^{\alpha H} + e^{-\alpha H})] + \frac{w}{4EI\alpha^4} = 0 \quad \ldots(22.8)$$

**Note:** Normally $e^{-\alpha H}$ is very small in comparison to $e^{\alpha H}$. Hence $(e^{\alpha H} - e^{-\alpha H})$ and $(e^{\alpha H} + e^{-\alpha H})$ may each be taken approximately equal to $e^{\alpha H}$. The constants $A$ and $B$ can thus be evaluated by the simultaneous solution of Eqs. 22.7 and 22.8. Substituting the values of $A$ and $B$ in Eqs. 22.4 and 22.5 constants $C$ and $D$ can be found substituting the values of $A$, $B$, $C$ and $D$, in Eqs. 22.3, 22.3 (a), 22.3 (b) and 22.3 (c), the deflection, slope, B.M. and shear force at any point on the tank wall can be found. Following are the simplified expressions for deflection and B.M.

$$y = \frac{wx}{4EI\alpha^4} + A \sin \alpha x \,[e^{\alpha x} + e^{-\alpha x}] + B \cos \alpha x \,[e^{\alpha x} + e^{-\alpha x}] - 2A \cdot e^{-\alpha x} \cos \alpha x \qquad ...(22.9)$$

Considering the B.M. *causing tension at the outer face as* positive,

$$M_x = -EI \frac{d^2y}{dx^2} = -EI\alpha^2 \cdot e^{\alpha x}[2A \cos \alpha x - 2B \sin \alpha x] - EI \cdot \alpha^2 \cdot e^{-\alpha x}[2D \sin \alpha x - 2C \cos \alpha x]$$

or $\quad M_x = -2EI\alpha^2\,[A \cos \alpha x\,(e^{\alpha x} - e^{-\alpha x}) - B \sin \alpha x\,(e^{\alpha x} - e^{-\alpha x}) - 2A \cdot e^{-\alpha x} \sin \alpha x] \qquad ...(22.10)$

## CASE II : WALL WITH HINGED BASE AND FREE TOP

The analysis given above is based on the assumption that the base joint is continuous and the footing is prevented from even smallest rotation of kind shown exaggerated in Fig. 22.2 (b). The rotation required to reduce the fixed base moment from some definite value to, say, zero is much smaller than rotations that may occur when normal settlement takes place in subgrade. It may be difficult to predict the behaviour of the subgrade and its effect upon the restraint at the base, but it is more reasonable to assume that the base is hinged than fixed, and the hinged-base assumption gives a safer design. The following are the boundary conditions.

(i) At the base, deflection is zero,
i.e. at $x = H$, $y = 0$

(ii) At the base, bending moment is zero,
i.e. at $x = H$, $\dfrac{d^2y}{dx^2} = 0$

(iii) At the top, shear force is zero,
i.e. at $x = 0$, $\dfrac{d^3y}{dx^3} = 0$

and (iv) At the top, bending moment is zero,
i.e. at $x = 0$, $\dfrac{d^2y}{dx^2} = 0$.

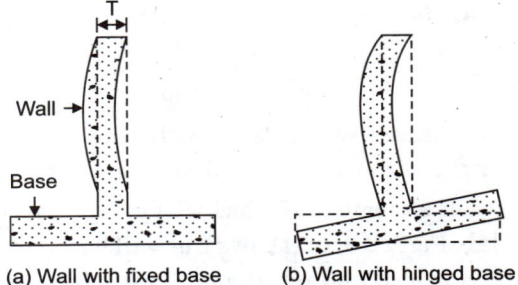

(a) Wall with fixed base  (b) Wall with hinged base

**Fig. 22.2.** Fixed and hinged base.

Applying these boundary conditions, the four constants $A$, $B$, $C$ and $D$ can be evaluated.

## CASE III : WALL MONOLITHIC WITH ELASTIC BASE AND HINGED AT THE TOP

If the base slab is thick and is supported either on the ground or on a number of continuous beams crossing each other at right angles, it can be treated as rigid and the wall may be considered to be *fixed* to the base. The angle between the tank wall and the base does not change, and the base also does not rotate. However if the base slab is supported only on a ring beam, or is supported on its edges only, the slab undergoes a rotation at its edge. The tank wall, which is monolithic with the base also undergoes the rotation. The angle between the two will either tend to increase (thereby increasing the fixing moment in the wall at its base) or will tend to decrease (thereby decreasing the moment). *Such a joint between the two may be treated as a continuous joint*. There are two methods for analysing this case.

**First method:** In the first method, the moments in the wall and slab may first be calculated by treating each one as fixed, and then the continuity of the joint may be taken into account by moment distribution at the joint. *The stiffness ($K_w$) of cylindrical wall is given by*

$$K_w = C_w \times \frac{ET^3}{H} \qquad ...(22.11)$$

in which the coefficient $C_w$ is given in Table 22.1.

*The stiffness of the circular slab is given by*

$$K_w = 0.208 \times \frac{ET^3}{D} \qquad ...(22.12)$$

**TABLE 22.1.** Stiffness coefficient for cylindrical wall

| $H^2/DT$ | $C_w$ | $H^2/DT$ | $C_w$ | $H^2/DT$ | $C_w$ |
|---|---|---|---|---|---|
| 0.4 | 0.139 | 6.0 | 0.783 | 24 | 1.566 |
| 0.8 | 0.270 | 8.0 | 0.903 | 32 | 1.810 |
| 1.2 | 0.345 | 10.0 | 1.010 | 40 | 2.025 |
| 1.6 | 0.399 | 12.0 | 1.101 | 48 | 2.220 |
| 2.0 | 0.445 | 14.0 | 1.198 | 56 | 2.400 |
| 3.0 | 0.548 | 16.0 | 1.281 | | |
| 4.0 | 0.648 | 20.0 | 1.430 | | |

**Second method:** When the tank wall is continuous with the elastic base, slope at the end of the wall and at the end of the slab will be the same. Similarly, since the tank is hinged at the top, radial deflection of the wall at top and the bottom will be the same. Thus, we have four conditions:

At $x = 0$,    $y = 0$    ...(i)

At $x = 0$,    $\dfrac{d^2y}{dx^2} = 0$ (B.M. is zero)    ...(ii)

At $x = H$,    $y = 0$    ...(iii)

At $x = H$,    $\left[\dfrac{dy}{dx}\right]_{wall} = \left[\dfrac{dy}{dx}\right]_{slab}$    ...(iv)

In order to apply the fourth condition, we may use the following formulae for deflection ($\delta$) and slope ($\psi$) for the ends of cylindrical shell and circular slab.

Let a moment $M$ (clockwise) and horizontal shear (inward) exist at the continuous joint of the wall with slab. Fig. 22.3 (ai) shows the wall with horizontal shear $Q$ at each end. Due to this the horizontal deflection $\delta_w$ (inward) and rotation $\psi_w$ (anti-clokwise) are given by

$$\delta_w = \dfrac{Q}{2\alpha^3 Z} \quad ...[22.13\,(a)] \quad \text{and} \quad \psi_w = \dfrac{Q}{2\alpha^2 Z} \quad ...[22.13\,(b)]$$

where $\alpha = \left[\dfrac{T}{ID^2}\right]^{\frac{1}{4}}$ and $Z = \dfrac{ET^3}{12(1-\mu^2)} \approx \dfrac{ET^3}{12}$    ...(22.14)

(Poisson's ratio being assumed to be zero)

Similarly, Fig. 22.3 (a ii) shows the wall with clockwise moment at each end. The horizontal deflection $\delta_w'$, (ourtward) and the rotation $\psi_w'$ (clockwise) are given by

$$\delta_w' = \dfrac{M}{2\alpha^2 Z} \quad ...[22.14\,(a)]$$

$$\psi_w' = \dfrac{M}{\alpha Z} \quad ...[22.14\,(b)]$$

Fig. 22.3 (b i) shows a circular slab of radius $a$ simply supported and loaded uniformly throughout.

The slope at each end is given by

$$\psi_s = \dfrac{3}{2} \cdot \dfrac{pa^3}{ET^3} \quad ...(22.15)$$

where $T$ is the thickness of slab.

Fig. 22.3 (b ii) shows the slab, simply supported, and loaded with a total uniformly distributed load $W$ on a concentric ring of radius $b$. The slope at each end is given by

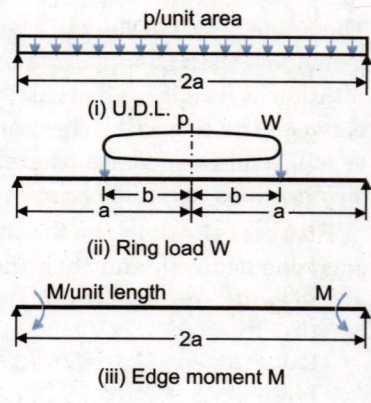

Fig. 22.3

$$\psi_s' = \frac{3W}{\pi E T^3}\left[\frac{a^2 - b^2}{a}\right] \qquad \ldots(22.16)$$

Fig. 22.3 (*b iii*) shows the same slab subjected to moment $M$ at each end. The slope at each end is given by

$$\psi_s'' = \frac{12a}{E T^3} \cdot M \qquad \ldots(22.17)$$

## CASE IV : WALL MONOLITHIC WITH DOMICAL BASE AND ROOF

Sometimes, domes are provided both at the top as well as bottom of a circular tank. The tank wall will undergo radial deflection and angular rotation at each end. These radial deflections and angular rotation at each end should be compatible with the corresponding radial deflections and angular rotations of each dome at its ends. These conditions can therefore be applied to get values of the constants $A, B, C$ and $D$.

## 22.2. I.S. CODE METHODS *AND* OTHER METHODS FOR CYLINDRICAL TANKS

The analysis given in the previous section is extremely tedious. In order to get the values of the constants $A, B, C$ and $D$, the arithmetical work is very laborious. Indian Standard Code IS : 3370 (Part IV)-1967 gives design tables for moments and hoop tension in circular tanks for various conditions of joints and various types of loadings, on the assumption of $\mu = 0.2$ for concrete. Tables 22.2, 22.3 and 22.4 have been reproduced for fixed-wall case. For other tables, reference may be made to the original Code. See example 22.2 for illustration.

**TABLE 22.2.** Tension in circular ring wall, fixed base, free top and subject to triangular load

$T = $ Coefficient $\times wHR$ N/m

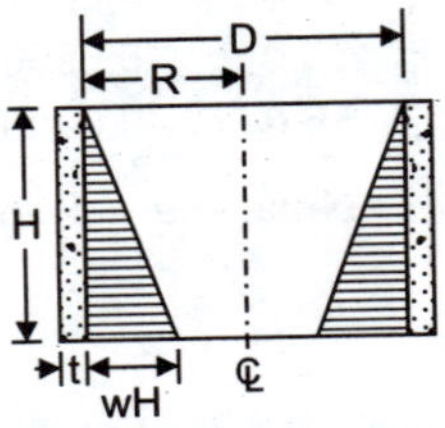

| $H^2/Dt$ | Coefficients at Point | | | | | | | | | |
|---|---|---|---|---|---|---|---|---|---|---|
| | 0.0 H | 0.1 H | 0.2 H | 0.3 H | 0.4 H | 0.5 H | 0.6 H | 0.7 H | 0.8 H | 0.9 H |
| (1) | (2) | (3) | (4) | (5) | (6) | (7) | (8) | (9) | (10) | (11) |
| 0.4 | + 0.149 | + 0.134 | + 0.120 | + 0.101 | + 0.082 | + 0.066 | + 0.049 | + 0.029 | + 0.014 | + 0.004 |
| 0.8 | + 0.263 | + 0.239 | + 0.215 | + 0.109 | + 0.160 | + 0.130 | + 0.096 | + 0.063 | + 0.034 | + 0.010 |
| 1.2 | + 0.283 | + 0.271 | + 0.254 | + 0.234 | + 0.209 | + 0.180 | + 0.142 | + 0.099 | + 0.054 | + 0.016 |
| 1.6 | + 0.265 | + 0.268 | + 0.268 | + 0.266 | + 0.250 | + 0.226 | + 0.185 | + 0.134 | + 0.075 | + 0.023 |
| 2.0 | + 0.234 | + 0.273 | + 0.273 | + 0.285 | + 0.285 | + 0.274 | + 0.232 | + 0.172 | + 0.104 | + 0.031 |
| 3.0 | + 0.134 | + 0.203 | + 0.267 | + 0.322 | + 0.357 | + 0.362 | + 0.330 | + 0.262 | + 0.157 | + 0.052 |
| 4.0 | + 0.067 | + 0.164 | + 0.256 | + 0.339 | + 0.403 | + 0.429 | + 0.409 | + 0.334 | + 0.210 | + 0.073 |
| 5.0 | + 0.025 | + 0.137 | + 0.245 | + 0.346 | + 0.428 | + 0.477 | + 0.469 | + 0.398 | + 0.259 | + 0.092 |
| 6.0 | + 0.018 | + 0.119 | + 0.234 | + 0.344 | + 0.441 | + 0.504 | + 0.514 | + 0.447 | + 0.301 | + 0.112 |
| 8.0 | − 0.001 | + 0.104 | + 0.218 | + 0.335 | + 0.443 | + 0.534 | + 0.574 | + 0.530 | + 0.381 | + 0.151 |
| 10.0 | − 0.001 | + 0.098 | + 0.208 | + 0.323 | + 0.437 | + 0.542 | + 0.608 | + 0.589 | + 0.440 | + 0.179 |
| 12.0 | − 0.005 | + 0.097 | + 0.202 | + 0.312 | + 0.429 | + 0.543 | + 0.628 | + 0.633 | + 0.494 | + 0.211 |
| 14.0 | − 0.002 | + 0.098 | + 0.200 | + 0.306 | + 0.420 | + 0.539 | + 0.639 | + 0.666 | + 0.541 | + 0.241 |
| 16.0 | + 0.000 | 0.099 | + 0.199 | + 0.304 | + 0.412 | + 0.531 | + 0.641 | + 0.687 | + 0.582 | + 0.265 |

**Note 1:** $w$ = Density of the liquid   **Note 2:** Positive sign indicates tension.

**TABLE 22.3.** Moments in cylindrical wall, fixed base, free top and subject to triangular load

$$\text{Moment} = \text{Coefficient} \times wH^3 \text{ N-m/m}$$

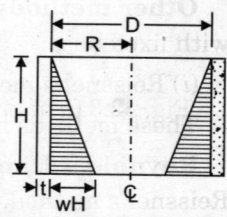

| $H^2/Dt$ | Coefficients at Point | | | | | | | | | |
|---|---|---|---|---|---|---|---|---|---|---|
| | 0.1 H | 0.2 H | 0.3 H | 0.4 H | 0.5 H | 0.6 H | 0.7 H | 0.8 H | 0.9 H | 1.0 H |
| (1) | (2) | (3) | (4) | (5) | (6) | (7) | (8) | (9) | (10) | (11) |
| 0.4 | +0.0005 | +0.0014 | +0.0021 | −0.0007 | −0.0042 | −0.0150 | −0.0302 | −0.0529 | −0.0816 | −0.1205 |
| 0.8 | +0.0011 | +0.0037 | +0.0063 | +0.0080 | +0.0070 | +0.0023 | −0.0068 | −0.0024 | −0.0465 | −0.0795 |
| 1.2 | +0.0012 | +0.0042 | +0.0077 | +0.0103 | +0.0112 | +0.0090 | +0.0022 | −0.0108 | −0.0311 | −0.0602 |
| 1.6 | +0.0011 | +0.0041 | +0.0075 | +0.0107 | +0.0121 | +0.0111 | +0.0058 | −0.0051 | −0.0232 | −0.0505 |
| 2.0 | +0.0010 | +0.0035 | +0.0068 | +0.0099 | +0.0120 | +0.1150 | +0.0075 | −0.0021 | −0.0185 | −0.0436 |
| 3.0 | +0.0006 | +0.0024 | +0.0047 | +0.0071 | +0.0090 | +0.0097 | +0.0077 | +0.0012 | −0.0119 | −0.0333 |
| 4.0 | +0.0003 | +0.0015 | +0.0028 | +0.0047 | +0.0046 | +0.0077 | +0.0069 | +0.0023 | −0.0080 | −0.0268 |
| 5.0 | +0.0002 | +0.0008 | +0.0016 | +0.0029 | +0.0046 | +0.0059 | +0.0059 | +0.0028 | −0.0058 | −0.0222 |
| 6.0 | +0.0001 | +0.0003 | +0.0008 | +0.0019 | +0.0032 | +0.0046 | +0.0051 | +0.0029 | −0.0041 | −0.0187 |
| 8.0 | 0.0000 | +0.0001 | +0.0002 | +0.0008 | +0.0016 | +0.0028 | +0.0038 | +0.0029 | −0.0022 | −0.0146 |
| 10.0 | 0.0000 | 0.0000 | +0.0001 | +0.0004 | +0.0007 | +0.0019 | +0.0029 | +0.0028 | −0.0012 | −0.0122 |
| 12.0 | 0.0000 | −0.0001 | +0.0001 | +0.0002 | +0.0003 | +0.0013 | +0.0023 | +0.0026 | −0.0005 | −0.1004 |
| 14.0 | 0.0000 | 0.0000 | 0.0000 | 0.0000 | +0.0001 | +0.0001 | +0.0019 | +0.0023 | −0.0001 | −0.0090 |
| 16.0 | 0.0000 | 0.0000 | −0.0001 | −0.0002 | −0.0001 | +0.0004 | +0.0013 | +0.0019 | +0.0001 | −0.0079 |

**Note 1:** $w$ = Density of the liquid. **Note 2:** Positive sign indicates tension on the outside.

**TABLE 22.4.** Shear at base of cylindrical wall

$$v = \text{Coefficient} \times \begin{cases} wH^2 & \text{kg} & \text{(triangular)} \\ pH & \text{kg} & \text{(rectangular)} \\ M/H & \text{kg} & \text{(moment at base)} \end{cases}$$

| $\dfrac{H^2}{Dt}$ | Triangular Load Fixed Base | Rectangular Load Fixed Base | Triangular or Rectangular Load Hinged Base | Moment at Edge |
|---|---|---|---|---|
| 0.4 | + 0.436 | + 0.755 | + 0.245 | − 1.58 |
| 0.8 | + 0.374 | + 0.552 | + 0.234 | − 1.75 |
| 1.2 | + 0.339 | + 0.460 | + 0.220 | − 2.00 |
| 1.6 | + 0.317 | + 0.407 | + 0.204 | − 2.28 |
| 2.0 | + 0.299 | + 0.370 | + 0.189 | − 2.57 |
| 3.0 | + 0.262 | + 0.310 | + 0.158 | − 3.18 |
| 4.0 | + 0.236 | + 0.271 | + 0.137 | − 3.68 |
| 5.0 | + 0.213 | + 0.243 | + 0.121 | − 4.10 |
| 6.0 | + 0.197 | + 0.222 | + 0.110 | − 4.49 |
| 8.0 | + 0.174 | + 0.193 | + 0.096 | − 5.18 |
| 10.0 | + 0.158 | + 0.172 | + 0.087 | − 5.81 |
| 12.0 | + 0.145 | + 0.158 | + 0.079 | − 6.38 |
| 14.0 | + 0.135 | + 0.147 | + 0.073 | − 6.88 |
| 16.0 | + 0.127 | + 0.137 | + 0.068 | − 7.36 |

**Note 1:** $w$ = Density of liquid. **Note 2:** Positive sign indicates shear acting inward.

**Other methods:** The following simplified methods are also available for the analysis of circular tank with fixed base:

(*i*) Reissner's method (*ii*) Carpenter's simplified method (*iii*) Approximate method.

These methods have been discussed in Chapter 21.

**Reynold's Hand Book:** Reynold's hand book gives the values of Carpenters coefficient, based on Reissner's method. These coefficients are tabulated in Table 22.5 and are applicable for tank wall with its bottom end restrained. The maximum B.M. at base, maximum circumferential tension and its location are given by the following expressions :

$$M_{max} = F \cdot w\, H^3 \text{ Per unit height} \qquad \text{...(22.18)}$$

$$T_H = \frac{1}{2} w\, H D \cdot K_2 \text{ per unit height} \qquad \text{...(22.19)}$$

$$L = \text{position of maximum circumferential tension} = K_1 H \qquad \text{...(22.20)}$$

**TABLE 22.5.** Carpenters's coefficients for cylindrical tank (*Reynold's hand book*)

| Factors | | F | | | | $K_1$ | | | | $K_2$ | | | |
|---|---|---|---|---|---|---|---|---|---|---|---|---|---|
| $H \div d_A \rightarrow$ | | 10 | 20 | 30 | 40 | 10 | 20 | 30 | 40 | 10 | 20 | 30 | 40 |
| | 0.2 | 0.046 | 0.028 | 0.022 | 0.015 | — | 0.50 | 0.45 | 0.40 | 0.32 | 0.46 | 0.53 | 0.50 |
| | 0.3 | 0.032 | 0.019 | 0.014 | 0.010 | 0.55 | 0.43 | 0.38 | 0.33 | 0.35 | 0.53 | 0.60 | 0.66 |
| Values of | 0.4 | 0.024 | 0.014 | 0.010 | 0.007 | 0.50 | 0.39 | 0.35 | 0.30 | 0.44 | 0.58 | 0.65 | 0.70 |
| $H \over D$ | 0.5 | 0.020 | 0.020 | 0.009 | 0.006 | 0.45 | 0.37 | 0.32 | 0.27 | 0.48 | 0.63 | 0.69 | 0.73 |
| | 1.0 | 0.012 | 0.006 | 0.005 | 0.003 | 0.37 | 0.28 | 0.24 | 0.21 | 0.62 | 0.73 | 0.74 | 0.83 |
| | 2.0 | 0.006 | 0.003 | 0.002 | 0.002 | 0.30 | 0.22 | 0.19 | 0.16 | 0.73 | 0.81 | 0.85 | 0.88 |
| | 4.0 | 0.004 | 0.002 | 0.002 | 0.001 | 0.27 | 0.20 | 0.17 | 0.14 | 0.80 | 0.85 | 0.87 | 0.90 |

**Note.** $d_A$ = Thickness of wall bottom.

**Example 22.1.** *Circular tank with fixed base: Analytical method.*

*Design a circular tank with fixed base for capacity of 400,000 litres. The depth of water is to be 4 m, including a free board of 0.25 m. Use M 25 concrete and tor steel reinforcement. Assume* $\mu = 0$. *The tank is tree at the top and rests on the ground.*

*Take unit weight of water as 9.8 kN/m³.*

**Solution:**

1. **Dimensions of tank**: Effective depth of water from capacity point of view = 3.75 m.

 If $D$ is the diameter of the tank, in metres, we have:

$$\frac{\pi}{4} D^2 \times 3.75 = \frac{400{,}000 \times 10^3}{10^6} \qquad \text{From which } D = \sqrt{\frac{400 \times 4}{\pi \times 3.75}} = 11.65 \text{ m.}$$

Provide diameter = 11.7 m.

2. **Determination of hoop tension and bending moment**

 We have $H = 4$ m; $D = 11.7$ m.

 $\therefore$ Specific weight $w = 9800$ N/m³ (given)

 Let the thickness of tank wall

$$= T = 3H + 5 = (3 \times 4) + 5 \text{ cm} = 17 \text{ cm} = 0.17 \text{ m.}$$

$$I = \frac{1}{12} T^3 = \frac{1}{12} (0.17)^3 = 4.094 \times 10^{-4} \text{ m}^4; \quad \alpha^4 = \frac{T}{ID^2} = \frac{0.17}{4.094 \times 10^{-4} (11.7)^2} = 3.033$$

$$\alpha^4 \cdot I = 3.033 \times 4.094 \times 10^{-4} = 12.418 \times 10^{-4}; \quad \alpha^2 = 1.7416; \quad \alpha = 1.3197; \quad \alpha^3 = 2.2984$$

$$\alpha H = 1.3197 \times 4 = 5.2788; \quad e^{\alpha H} = e^{5.2788} = 196.13; \quad e^{-\alpha H} = \frac{1}{196.13} = 5.099 \times 10^{-3}$$

$$\sin \alpha H = -0.8438; \quad \cos \alpha H = +0.5366.$$

Since $e^{-\alpha H}$ is very small in comparison to $e^{\alpha H}$

$$e^{\alpha H} \pm e^{-\alpha H} \approx e^{\alpha H} = 196.13. \text{ Hence from } Eq.\ 22.7, \text{ we have}$$

$$A \left[ -0.8438 \times 196.13 - \frac{2}{196.13} \times 0.5366 \right] + B \left[ 0.5366 \times 196.3 \right] + \frac{9800 \times 4}{4 E (12.418 \times 10^{-4})} = 0$$

or $\quad A - 0.6339 B = \dfrac{47530}{E}$  ...(i)    Similarly, from *Eq. 22.8*, we have

$$A \left[ -1.3197 \times 0.8438 \times 196.13 + 1.3197 \times 0.5366 \times 196.13 + \frac{2 \times 1.3197}{196.13}(-0.8438 + 0.5366) \right]$$

$$+ B \left[ 1.3197 \times 0.5366 \times 196.13 + 1.3197 \times 0.8438 \times 196.13 \right] + \frac{9800}{4 E \times 12.418 \times 10^{-4}} = 0$$

or $\quad A - 4.493 B = \dfrac{24814}{E}$  ...(ii)

Solving (i) and (ii), we get $B = 5887/E$ and $A = 51264/E$.

Substituting the values of $A$ and $B$ in *Eq. 22.9* and *22.10*, the values of $y$ and bending moment $M_x$ can be computed for various values of $x$.

Hoop tension $\quad P_r = \dfrac{4\,T\,E}{D^2} \cdot y = \dfrac{4 \times 0.17\,E}{(11.7)^2} y = 49.67 \times 10^{-4} E \cdot y$

and $\quad \dfrac{w\,x}{4EI\alpha^4} = \dfrac{9800\,x}{4\,E \times 12.418 \times 10^{-4}} = 197.3 \times 10^{-4} \dfrac{x}{E}$

Also, $\quad 2\,I\,\alpha^2 = 2 \times 4.094 \times 10^{-4} \times 1.7416 = 14.26 \times 10^{-4}$

Hence from *Eq. 22.9*, we have

$$E \cdot y = 197.3 \times 10^4\,x + 51264 \sin \alpha x\,[e^{\alpha x} + e^{-\alpha x}] + 5887 \cos \alpha x\,[e^{\alpha x} + e^{-\alpha x}]$$
$$- 2 \times 51264\,e^{-\alpha x} \cos \alpha x$$

Now $\quad p_r = 49.67 \times 10^{-4}\,E \cdot y$

$\quad p_r = 9800\,x + 255.4 \sin \alpha x\,[e^{\alpha x} + e^{-\alpha x}] + 29.20 \cos \alpha x\,[e^{\alpha x} + e^{-\alpha x}] - 519\,e^{-\alpha x} \cos \alpha x$  ...(a)

Similarly, from *Eq. 22.10*

$$M_x = -14.26 \times 10^{-4}\,[51264 \cos \alpha x\,(e^{\alpha x} - e^{-\alpha x}) - 5887 \sin \alpha x\,(e^{\alpha x} - e^{-\alpha x})$$
$$- 2 \times 51264\,e^{-\alpha x} \cdot \sin \alpha x]$$

or $\quad M_x = -73.1 \cos \alpha x\,(e^{\alpha x} - e^{-\alpha x}) + 8.40 \sin \alpha x\,(e^{\alpha x} - e^{-\alpha x}) + 146.2\,e^{-\alpha x} \sin \alpha x$  ...(b)

Thus values of hoop tension $p_r$ and bending moment $M_x$ at various heights can therefore be calculated with the help of *Eqs.* (a) and (b) above, and tabulated as shown in Table 22.6 and 22.7 respectively. At the base of the tank, $x = H$ and $\alpha x = \alpha H$.

$$M_H = -73.1 \times 0.5366 \times 196.13 - 8.4 \times 0.8438 \times 196.13 - \frac{146.2}{196.13} \times 0.8438$$
$$= -9083 \text{ N-m/metre width}$$

Fig. 22.4 (a) shows the variation of hoop tension with depth along with the variation of hydrostatic pressure. It is seen that hoop pressure is maximum at $x = 2.33$ m and its value at this depth is equal to 22834 N/m$^2$. At the top, there is slight hoop compression in the wall. This is due to inward radial deflection at the top. Fig. 22.4 (b) shows the variation of B.M. along the depth. The maximum B.M. occurs at the base, its magnitude being equal to $-9083$ N-m/m; the negative sign indicates that it causes tension at the inner face (water side). The maximum positive moment is $+2360$ N-m/m at a depth of 2.9 m below water surface.

**TABLE 22.6.** Values of hoop tension ($p_r$) N/m²      [α = 1.3197]

| (1) | (2) | (3) | (4) | (5) | (6) | (7) | (8) | (9) | (10) | (11) |
|---|---|---|---|---|---|---|---|---|---|---|
| $x$ m | $e^{\alpha x}$ | $e^{-\alpha x}$ | $e^{\alpha x}+e^{-\alpha x}$ | sin α$x$ | cos α$x$ | 9800$x$ | 255.4 × (5) × (4) | 29.20 × (6) × (4) | − 519 × (3) × (6) | $p_r$ = (7) + (8) + (9) + (10) |
| 0 | 1 | 1 | 2 | 0 | + 1.00 | 0 | 0 | + 58 | − 510 | − 452 |
| 1 | 3.742 | 0.267 | 4.009 | + 0.9689 | + 0.2485 | 9800 | + 992 | + 29 | − 34 | + 10787 |
| 2 | 14.004 | 0.071 | 14.075 | + 0.4813 | − 0.8765 | 19600 | + 1730 | − 360 | + 32 | + 21002 |
| 2.5 | 27.092 | 0.037 | 27.129 | − 0.1570 | − 0.9876 | 24520 | − 1088 | − 782 | + 19 | + 22669 |
| 3 | 52.410 | 0.019 | 52.429 | − 0.7294 | − 0.6840 | 29400 | − 9767 | − 1047 | + 7 | + 18593 |
| 3.5 | 101.388 | 0.010 | 101.398 | − 0.9956 | − 0.0933 | 34300 | − 25783 | − 276 | + 1 | + 8242 |
| 4 | 196.134 | 0.005 | 196.139 | − 0.8438 | + 0.5366 | 39200 | − 42270 | + 3071 | − 1 | 0 |

**TABLE 22.7.** Values of $M_x$ (N-m/m)      [α = 1.3197]

| (1) | (2) | (3) | (4) | (5) | (6) | (7) | (8) | (9) | (10) |
|---|---|---|---|---|---|---|---|---|---|
| $x$ m | $e^{\alpha x}$ | $e^{-\alpha x}$ | $e^{\alpha x}-e^{-\alpha x}$ | sin α$x$ | cos α$x$ | − 73.1 × (4) × (6) | 8.40 × (4) × (5) | 146.2 × (3) × (5) | $M_x$ = (7) + (8) + (9) |
| 0 | 1 | 1 | 0 | 0 | + 1.00 | 0 | 0 | 0 | 0 |
| 1 | 3.742 | 0.267 | 3.475 | + 0.9686 | + 0.2485 | − 63 | + 28 | + 38 | + 3 |
| 2 | 14.004 | 0.071 | 13.393 | + 0.4813 | − 0.8765 | + 859 | + 54 | + 5 | + 918 |
| 2.5 | 27.092 | 0.037 | 27.055 | − 0.1570 | − 0.9876 | + 1955 | − 36 | − 1 | + 1918 |
| 3 | 52.410 | 0.019 | 52.391 | − 0.7294 | − 0.6840 | + 2622 | − 321 | − 2 | + 2299 |
| 3.5 | 101.388 | 0.010 | 101.378 | − 0.9956 | − 0.0933 | + 692 | − 849 | − 1 | − 158 |
| 4 | 196.134 | 0.005 | 196.129 | − 0.8438 | + 0.5366 | − 7702 | − 1392 | − 1 | − 9083 |

**3. Design of side walls for hoop tension.** Maximum hoop tension = 22834 N/m² at a depth of 2.33 m below water surface. Hence maximum hoop force = 22834 × $\frac{11.7}{2}$ = 133578.9 ≈ 133580 N per metre height of wall. Permissible stress in steel = 130 N/mm² (Table 21.2).

∴ Area of rings, $A_{sh}$ = 133580/130 = 1028 mm².

Let us provide rings on both the faces. Hence area of ring reinforcement on each face = 1028/2 ≈ 514 mm². Using 12 mm Φ ring, $A_\Phi$ = 113 mm³.

∴ Spacing = $\frac{1000 \times 113}{514}$ = 219.8 mm.

However, provide the rings @ 210 mm c/c on each face. The spacing of the rings may be increased towards top and bottom section of maximum hoop load.

Actual $A_{sh}$ provided = $\frac{2 \times 1000 \times 113}{210}$ = 1076 mm².

Tensile stress in concrete wall = $\frac{133580}{(1000 \times 170) + (11-1)1076}$ = 0.74 N/mm²

This is less than the permissible value of 1.3 N/mm².

Hoop tension at 1 m below top (Table 22.6) = 10787 × $\frac{11.7}{2}$ = 63104

Area of rings = 63104/130 = 486 mm². Area of rings at each face = 243 mm².

Spacing of 12 mm of rings = $\frac{1000 \times 113}{243}$ = 465. Keep the spacing @ 450 mm c/c from top at a depth of 1 m below the top.

Similarly, at 3 m below top, hoop tension = 18593 × $\frac{11.7}{2}$ = 108769.05

∴ Area of rings on each face = $\frac{1}{2} \times \frac{108770}{130}$ = 418 mm²

∴ Spacing of 12 mm Φ rings = $\frac{1000 \times 113}{418}$ = 270 mm

However, keep the spacing @ 250 mm from depth 3 m to 4 m below the top.

**4. Design of section for cantilever action:**

For M25 concrete, we have $\sigma_{cbc}$ = 8.5 N/mm² and $m$ = 11. Taking $\sigma_{st}$ = 130 N/mm², we get

$$k = \frac{m\,\sigma_{cbc}}{m\,\sigma_{cbc} + \sigma_{st}} = \frac{11 \times 8.5}{11 \times 8.5 + 130} = 0.42$$

$$j = 1 - \frac{k}{3} = 0.86;$$

$$R = \frac{1}{2}\,\sigma_{cbc}\cdot j\cdot k$$

$$= \frac{1}{2} \times 8.5 \times 0.42 \times 0.86 = 1.535$$

Maximum bending moment = 9083 N-m at the base causing tension towards the water face.

$$d = \sqrt{\frac{9083 \times 1000}{1000 \times 1.535}} = 76.9 \text{ mm}$$

Using 12 mm Φ bars and a clear cover of 25 mm, cover to the centre of reinforcement = 31 mm.

∴ Total thickness $T$ = 76.9 + 31 = 108 mm.

However, provide a minimum thickness equal to the greater of the following:

(i) 15 cm

(ii) 3 $H$ + 5 cm = (3 × 4) + 5 = 17 cm

Hence provide $T$ = 17 cm = 170 mm
Available $d$ = 170 − 31 = 139 mm

Area of steel for cantilever bending,

$$A_{st} = \frac{9083 \times 1000}{130 \times 0.86 \times 139} = 584.5 \text{ mm}^2$$

Spacing of 12 mm Φ bars = $\frac{1000 \times 113}{584.5}$ = 193.3 mm

Hence provide 12 mm Φ bars @ 190 mm c/c at the water face. These bars are required in bottom 50 cm height only. However, provide these at least upto a minimum height equal to development length $L_d$ given by

$$L_d = \frac{\Phi\,\sigma_{st}}{4\,\tau_{bd}}$$

where Φ = 12 mm; $\sigma_{st}$ = 130 N/mm²; $\tau_{bd}$ for M 25 concrete = 0.9 N/mm²

$$L_d = \frac{130}{4 \times 0.9}\Phi = 36.11\,\Phi = 36.1 \times 12 = 433 \approx 433 \text{ mm} = 0.433 \text{ m}.$$

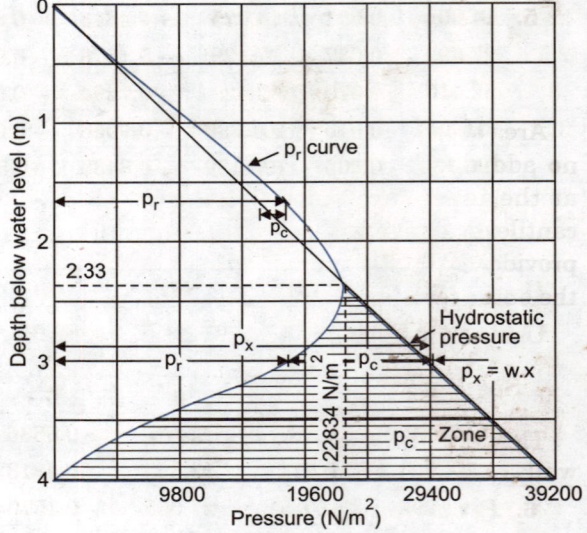

(a) Variation of ring tension

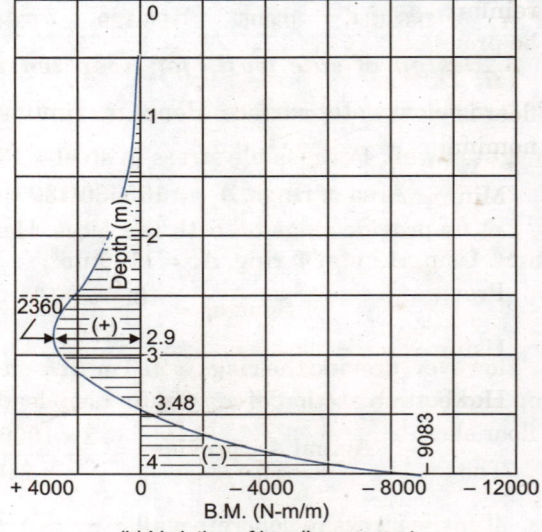

(b) Variation of bending moment

Fig. 22.4

From practical considerations, provide these bars upto 1 m height from the bottom. Above this height, curtail half the bars and continue other half (*i.e.* 320 mm c/c) upto top. Keep a clear cover of 25 mm. At the outer face, there is a positive B.M. of 2360 N-m at a depth of 2.9 m below top.

$$\text{Area of steel required} = \frac{2360 \times 1000}{130 \times 0.86 \times 139} = 152 \text{ mm}^2$$

The distribution reinforcement provided will serve this purpose.

**5. Distribution reinforcement.** Provide distribution reinforcement 0.24% of the area of concrete.

$$A_{sd} = \frac{0.24}{100} \times 170 \times 1000 = 408 \text{ mm}^2$$

Area of steel on each face = 204 mm². However, no additional reinforcement will be provided at the inner face since the vertical steel for cantilever action will serve this purpose. Hence provide $A_{sd}$ = 204 mm² in vertical direction at the outer face.

Using 8 mm Φ bars, $A_\phi$ = 50.5 mm².

$$\text{Spacing} = \frac{1000 \times 50.5}{204} = 247.5 \approx 240 \text{ mm}$$

This will also take care of positive B.M. in the wall.

**6. Provision of haunches.** It is customary to provide 150 mm × 150 mm haunches at the junction of the wall and the base. A haunch reinforcement of 8 mm Φ bars 200 mm c/c may be provided.

**7. Design of base slab.** Since the tank floor is resting on ground throughout, provide a nominal thickness of 150 mm.

$$\text{Minimum } A_{st} = \frac{0.24}{100} \times 150 \times 1000$$
$$= 360 \text{ mm}^2 \text{ in each direction.}$$

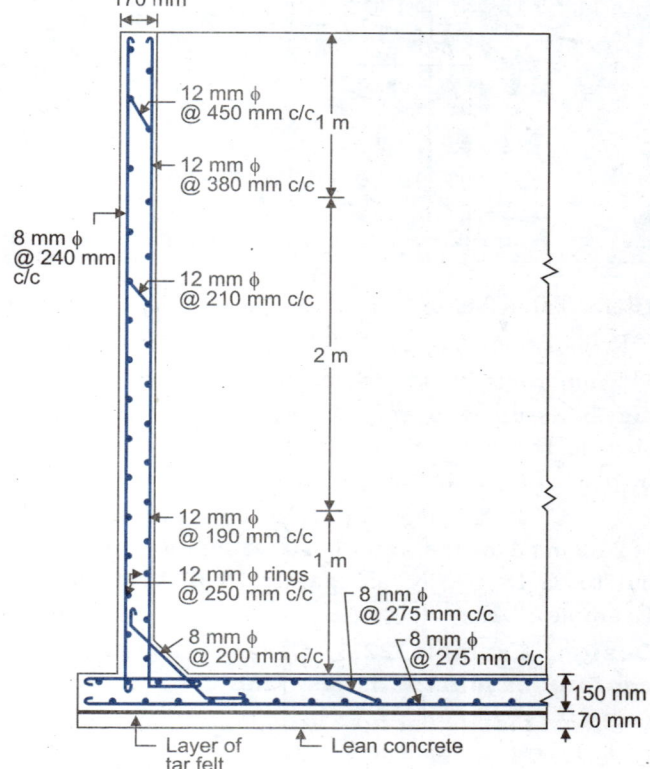

Fig. 22.5

Providing reinforcement on each face, $A_{st}$ = 180 mm²

Using 8 mm Φ bars, spacing = $\frac{1000 \times 50.5}{180}$ = 280.5 mm

However, provide 8 mm Φ bars 275 mm c/c in both directions, at top and bottom of the floor slab. The floor slab will rest on 75 mm thick layer of lean concrete covered with a layer of tar felt.

The details of reinforcement etc. are shown in Fig. 22.5.

**Example 22.2.** *Analyse the tank of Example 22.1 by I.S. Code method.*

**Solution.** Refer IS 3373 (Part IV–1967). Diameter of tank = 11.7 m

Let the thickness of the tank = 3 H + 5 cm = (3 × 4) + 5 = 17 cm = 170 mm.

$$\frac{H^2}{D \cdot T} = \frac{(4)^2}{(11.7)(0.17)} \approx 8 \; ; \; w = 9800 \text{ N/m}^3. \quad \text{The ring tension at any height}$$

$$= P_r = \text{Coefficient} \times w \, H \times \frac{D}{2} = \text{Coefficient} \times \left(9800 \times 4 \times \frac{11.7}{2}\right) = 229320 \times \text{coefficient}$$

The coefficient is given in Table 9 of the I.S. Code (See Table 22.2). Similarly, the B.M. at any height = $M$ = Coefficient × $w$ $H^3$ N-m/m = Coefficient × (9800) (4)³ = 627200 × coefficient N-m/m. The coefficient is given in Table 10 of the I.S. Code (See Table 22.3). The values of ring tension and B.M. at various heights are tabulated in Table 22.8.

**TABLE 22.8.** Distribution of ring tension and B.M. in tank wall

| Depth x below water surface (m) | Ring tension $P_r$ | | B.M. M N–m/m | |
|---|---|---|---|---|
| | Coefficient* | $P_r$ (N) | Coefficient** | M |
| 0.0 H = 0.0 | – 0.001 | – 229 | 0.0000 | 0 |
| 0.1 H = 0.4 | + 0.104 | + 23849 | + 0.0000 | 0 |
| 0.2 H = 0.8 | + 0.218 | + 49992 | + 0.0001 | + 63 |
| 0.3 H = 1.2 | + 0.335 | + 76822 | + 0.0002 | + 126 |
| 0.4 H = 1.6 | + 0.443 | + 101590 | + 0.0008 | + 502 |
| 0.5 H = 2.0 | + 0.534 | + 122457 | + 0.0016 | + 1004 |
| 0.6 H = 2.4 | + 0.574 | + 131860 | + 0.0028 | + 1756 |
| 0.7 H = 2.8 | + 0.530 | + 121540 | + 0.0038 | + 2383 |
| 0.8 H = 3.2 | + 0.381 | + 87370 | + 0.0029 | + 1819 |
| 0.9 H = 3.6 | + 0.151 | + 34627 | – 0.0022 | – 1380 |
| 1.0 H = 4.0 | 0.000 | 0 | – 0.0146 | – 9157 |

*Refer Table 9 of IS 3370 Part IV or Table 22.2 of this book. Plus sign indicates tension.
**Refer Table 10 of IS 3370 Part IV or Table 22.3 of this book. Plus sign indicates tension on the outside.

From Table 22.8, by inspection, maximum tension may be assumed to occur at 0.6 H = 2.4 m below water surface. However, for exact location, the variation can be plotted along the height, as shown in Fig. 22.6.

From Fig. 22.6, maximum tension occurs at a depth of 2.32 m below the water surface and its value comes out to be 133000 N, as against 133580 N found in Example 22.1.

### Design: Example 22.3. Circular tank with continuous joint with base slab

Design the circular tank of example 22.1 if the tank is supported as shown in Fig. 22.7, the walls have continuous joint with the base slab.

**Solution:** The bottom slab of the tank in the previous example was assumed to be resting on the ground. In the present problem, the bottom slab is supported on a circular beam (ring beam) which, in turn, is supported on a number of columns. Due to this the base slab undergoes a rotation at its ends. Since the side walls are monolithic with base, they will also undergo equal rotation, resulting in change of fixing moment at the joint.

1. **Salient data:** From example 22.1, the salient data are as follows :

$D = 11.7$ m; $H = 4$ m; $T = 0.17$ m; $I = 4.094 \times 10^{-4}$ m$^4$

$\alpha^4 = 3.033;$    $I\alpha^4 = 12.418 \times 10^{-4}$
$\alpha^2 = 1.7416;$
$\alpha = 1.3197;$    $\alpha^3 = 2.2984$
$\alpha H = 5.2788;$    $e^{\alpha H} = 196.13$

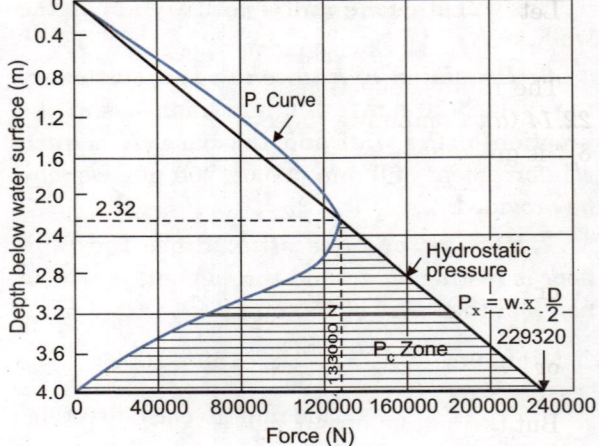

Fig. 22.6

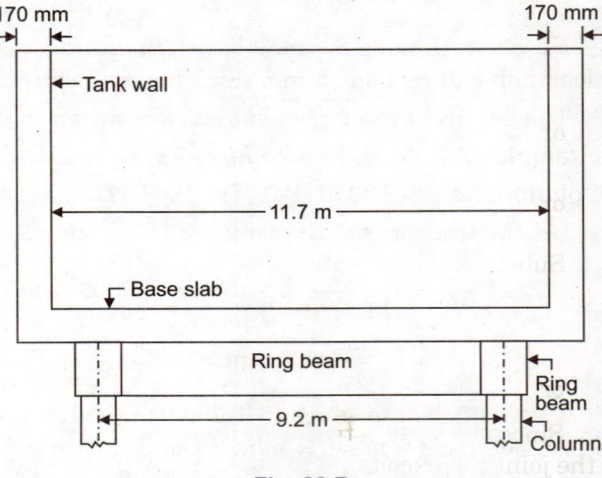

Fig. 22.7

$$e^{-\alpha H} = \frac{1}{196.13}; \quad \sin \alpha H = -0.8438; \quad \cos \alpha H = +0.5366$$

$$(e^{\alpha H} \pm e^{-\alpha H}) \approx e^{\alpha H} = 196.13$$

Diameter of supporting circle of ring beam = 9.2 m.

**2. Calculation of end slope and fixing moment:** Let us assume that the tank is hinged to the roof slab, so that we have the following conditions at the top

At $x = 0$, $\quad y = 0$ ...(1)

At $x = 0$, $\quad M = EI \dfrac{d^2y}{dx^2} = 0$ ...(2)

If the tank wall is assumed to be free at the bottom, hoop tension at bottom of wall
$$= p_r = wH = 9800 \times 4 = 39200 \text{ N/m}^2$$

$\therefore$ Radial deflection of wall $= \dfrac{p_r \cdot D^2}{4TE} = \dfrac{39200 (11.7)^2}{4 \times 0.17 \, E} = \dfrac{789.1 \times 10^4}{E}$ ...(i)

$\therefore$ Clockwise slope of wall base $= \dfrac{789.1 \times 10^4}{E \times 4} = \dfrac{197.28 \times 10^4}{E}$ ...(ii)

For the wall, $I = \dfrac{1}{12} T^3$ and $Z = \dfrac{1}{12} ET^3$ or $Z = I \cdot E = 4.094 \times 10^{-4} E$. Refer Fig. 22.3 (a).

Let $Q$ = inward radial force per unit circumferential length of wall at the bottom edge (N-m)

$M$ = clockwise moment per unit circumferential length of wall at its bottom edge (N/m/m)

The radial deflections $\delta_w$ and $\delta_w'$ of wall due to $Q$ and $M$ are given respectively by *Eqs. 22.13* and *22.14 (a)*. Combining these with the radial deflection of the wall due to hoop tension, the net deflection $\delta_w''$ is given by

$$\delta_w'' = \dfrac{p_r D^2}{4TE} - \dfrac{Q}{2\alpha^3 Z} + \dfrac{M}{2\alpha^2 Z}$$

or $\quad \delta_w'' = \dfrac{789.1 \times 10^4}{E} - \dfrac{Q}{2 \times 2.2984 \times 4.094 \times 10^{-4} \, E} + \dfrac{M}{2 \times 1.7416 \times 4.096 \times 10^{-4} \, E}$

or $\quad \delta_w'' = \dfrac{1}{E} [789.1 \times 10^4 - 531.4 \, Q + 701.25 \, M]$.

But the bottom of end wall is monolithic with base, and hence $\delta_w''$ should be equal to zero. Hence

$531.4 \, Q = 789.1 \times 10^4 + 701.25 \, M$ or $Q = 1.3196 \, M + 14849$ ...(a)

Similarly, the net clockwise slope, $\psi_w''$ is obtained by combining *Eqs. 22.13 (b)*, *22.14 (b)* and *Eq. (ii)* above.

$\therefore \quad \psi_w'' = \dfrac{p_r \cdot D^2}{4 \, T \, E \cdot H} - \dfrac{Q}{2 \alpha^2 \, Z} + \dfrac{M}{\alpha \, Z}$

or $\quad \psi_w'' = \dfrac{197.28 \times 10^4}{E} - \dfrac{Q}{2 \times 1.7416 \times 4.094 \times 10^{-4} \, E} + \dfrac{M}{1.3197 \times 4.094 \times 10^{-4} \, E}$

or $\quad \psi_w'' = \dfrac{1}{E} [197.28 \times 10^4 - 701.25 \, Q + 1850.87 \, M]$

Substituting the value of $Q$ from (a),

$\psi_w'' = \dfrac{1}{E} [197.28 \times 10^4 - 925.37 \, M - 1041.3 \times 10^4 + 1850.87 \, M]$

or $\quad \psi_w'' = \dfrac{1}{E} [925.5 \, M - 844.02 \times 10^4]$ ...(b)

Since the joint between wall and slab is continuous, the slab will have an anti-clockwise moment $M$ at the joint corresponding to a clockwise moment at the wall. Let the thickness of the base slab be 30 cm and let the unit weight of concrete be 25000 kN/m$^3$.

Weight of water per unit area of slab = $4 \times 1 \times 1 \times 9800 = 39200$ N/m$^2$

Weight of slab per unit area = $0.3 \times 25000 = 7500$ N/m$^2$

Total $p = 39200 + 7500 = 46700$ N/m$^2$.

Outer diameter of tank = $11.7 + 0.34 = 12.04$ m

Total weight of walls = $\dfrac{\pi}{4}(12.04^2 - 11.7^2) \times 4 \times 25000 = 633942$ N

Let the thickness of top cover be 20 cm. Its total weight will be

$= \dfrac{\pi}{4} \times 12.04^2 \times 0.20 \times 25000 = 569263$ N

∴ Total circumferential load on the periphery of slab = $633942 + 569263 = 1203205$ N

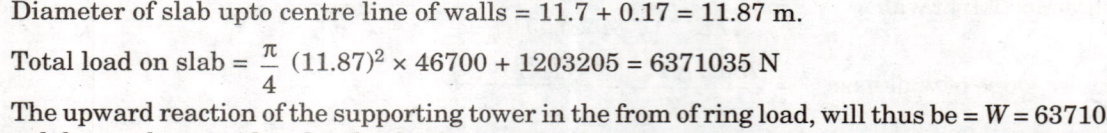

Fig. 22.8. Loading on base slab.

The loading on the base slab is shown in Fig. 22.8.

Diameter of slab upto centre line of walls = $11.7 + 0.17 = 11.87$ m.

Total load on slab = $\dfrac{\pi}{4}(11.87)^2 \times 46700 + 1203205 = 6371035$ N

The upward reaction of the supporting tower in the from of ring load, will thus be = $W = 6371035$ N. Hence the slab may be considered to be freely supported on imaginary support round its periphery of 11.87 m diameter and loaded with the following: (*i*) uniformly distributed load $p = 46700$ N/m$^2$, (*ii*) upward ring load $W = 6371035$ N distributed along the concentric circle of radius $b = 4.6$ m, and (*iii*) end moment $M$ as marked in Fig. 22.8. The clockwise slope of slab at its edges is given by the combination of *Eqs. 22.15, 22.16* and *22.17*.

$$\psi_s = \dfrac{3}{2} \cdot \dfrac{p\,a^3}{E\,T^3} - \dfrac{3\,W}{\pi\,E\,T^3}\left[\dfrac{a^2 - b^2}{a}\right] - \dfrac{12a}{ET^3}M$$

where $T$ = thickness of slab = 30 cm = 0.3 m

$$\psi_s = \dfrac{3}{2} \times \dfrac{46700\,(5.935)^3}{E\,(0.3)^3} - \dfrac{3 \times 6371035}{\pi\,E\,(0.3)^3}\left[\dfrac{5.935^2 - 4.6^2}{5.935}\right] - \dfrac{12 \times 5.935}{E\,(0.3)^3}M$$

or $\psi_s = \dfrac{1}{E}[542.38 \times 10^6 - 533.96 \times 10^6 - 2637.8\,M] = \dfrac{1}{E}[8.42 \times 10^6 - 2637.8\,M]$ ...(*c*)

This should be equal to the clockwise slope of the wall as given by *Eq.* (*b*).

∴ $925.5\,M - 844.02 \times 10^4 = 8.42 \times 10^6 - 2637.8\,M$  or  $M = 4732$ N-m/m.

This is much less than the moment of 9083 N-m/m found in example 22.1. Thus the fixing moment has been decreased. This is because the slope at edges of base slab is hogging (as obtained by substituting the value of $M$ in *Eq.* (*c*), which will rotate the wall to have the same anti-clockwise slope, thus decreasing the fixing moment which is in clockwise direction (causing tension in the water face). If however, the diameter of the supporting circle is increased (i.e. made more than 9.2 m), the fixing moment will increase. If, on the other hand, the diameter of supporting circle is decreased, the clockwise fixing moment will decrease further, and even an anti-clockwise moment (causing compression on the water face) may be created in the wall *Normally, diameter of supporting circle is kept $\dfrac{3}{4}$ times the diameter of the tank.*

**3. Boundary conditions:** After having calculated the fixing moment $M$, the boundary conditions at the upper and lower ends of wall become as follows:

At $x = 0$, $y = 0$ ...(*i*)

At $x = 0$, $\dfrac{d^2 y}{dx^2} = 0$ ...(*ii*)

At $x = H$, $y = 0$ ...(*iii*)

and $x = H$, $\dfrac{d^2 y}{dx^2} = \dfrac{4732}{EI}$ ...(*iv*)

With the help of the boundary condition equations (i), (ii), (iii) and (iv), the problem can now be solved, using *Eqs. 22.3* and *22.3 (b)*.

Applying first boundary condition (*i.e.* $x = 0$, $y = 0$) we get from *Eq. 22.3*,
$$y = 0 = 0 + 1\,(0 + B) + 1\,(0 + D) \quad \text{or} \quad D = -B \qquad \ldots(1)$$

Applying second boundary condition $\left[ i.e.\ x = 0,\ \dfrac{d^2y}{dx^2} = 0 \right]$ we get from *Eq. 22.3. (b)*,
$$0 = \alpha^2 \cdot 1\,[0 + 2A] + \alpha^2 \cdot 1\,[0 - 2C] \quad \text{or} \quad C = A \qquad \ldots(2)$$

Applying third boundary condition ( *i.e.* $x = H$, $y = 0$), we get from *Eq. 22.3*,
$$y = 0 = \frac{wH}{4EI\alpha^4} + e^{\alpha H}(A \sin \alpha H + B \cos \alpha H) + e^{-\alpha H}(A \sin \alpha H - B \cos \alpha H)$$

or $\quad A \sin \alpha H\,[e^{\alpha H} + e^{-\alpha H}] + B \cos \alpha H\,[e^{\alpha H} - e^{-\alpha H}] + \dfrac{wH}{4EI\alpha^4} = 0$

Substituting the values,
$$A(-0.8438) \times 196.13 + B \times 0.5366 \times 196.13 + \frac{9800 \times 4}{4E \times 12.418 \times 10^{-4}} = 0$$

or $\quad A - 0.6359\,B = \dfrac{47686}{E} \qquad \ldots(3)$

Finally, applying the fourth boundary condition,

*i.e.* at $x = H$, $\dfrac{d^2y}{dx^2} = \dfrac{4732}{EI}$, we get from *Eq. 22.3 (b)*

$$\frac{4732}{EI} = \alpha^2 \cdot e^{\alpha H}\,[-2B \sin \alpha H + 2A \cos \alpha H] + \alpha^2 e^{-\alpha H}\,[-2B \sin \alpha H - 2A \cos \alpha H]$$

or $\quad \dfrac{4732}{EI} = 2A\alpha^2 \cos \alpha H\,[e^{\alpha H} - e^{-\alpha H}] - 2B\alpha^2 \sin \alpha H\,[e^{\alpha H} + e^{-\alpha H}]$

$\therefore \quad (2A \times 1.7416 \times 0.5366 \times 196.13) + (2B \times 1.7416 \times 0.8438 \times 196.13) = \dfrac{4732}{E \times 4.094 \times 10^{-4}}$

or $\quad A + 1.5725\,B = \dfrac{31530}{E}$. Solving (3) and (4), we get $B = -\dfrac{7315.7}{E}$ and $A = \dfrac{43034}{E}$

Now $\quad p_r = \dfrac{4TE}{D^2} \cdot y = \dfrac{4 \times 0.17}{(11.7)^2}\,Ey = 49.67 \times 10^{-4}\,Ey$

$$\frac{wx}{4EI\alpha^4} = \frac{9800\,x}{4E \times 12.418 \times 10^{-4}} = 197.29 \times 10^4\,\frac{x}{E}$$

The equation for deflection $y$ is given by *Eq. 22.3*, which can rewritten as under:
$$y = \frac{wx}{4EI\alpha^4} + A \sin \alpha x\,[e^{\alpha x} + e^{-\alpha x}] + B \cos \alpha x\,[e^{\alpha x} - e^{-\alpha x}]$$

$\therefore \quad p_r = 49.67 \times 10^{-4}\,E \cdot y$

$\qquad = 49.67 \times 10^{-4}\,[197.29 \times 10^4\,x + 43034 \times \sin \alpha x\,(e^{\alpha x} + e^{-\alpha x}) - 7315.7 \times \cos \alpha x\,(e^{\alpha x} - e^{-\alpha x})]$

or $\quad p_r = 9799.4\,x + 213.75 \sin \alpha x\,(e^{\alpha x} + e^{-\alpha x}) - 36.34 \cos \alpha x\,(e^{\alpha x} - e^{-\alpha x}) \qquad \ldots(i)$

With the help of this equation, $p_r$ can be calculated for various values of $x$ and tabulated as shown in Table 22.9. Again, considering B.M. causing tension at the outer face as positive, we get from *Eq. 22.3. (b)*,

$$M_x = -EI\,\frac{d^2y}{dx^2}$$
$$= -EI\,\alpha^2 \cdot e^{\alpha x}\,[2A \cos \alpha x - 2B \sin \alpha x] - EI\alpha^2 \cdot e^{-\alpha x}\,[2D \sin \alpha x - 2C \cos \alpha x]$$

**562** REINFORCED CONCRETE STRUCTURE

or $\quad M_x = -2EI\alpha^2 \cdot e^{\alpha x}[A\cos\alpha x - B\sin\alpha x] + 2EI\alpha^2 \cdot e^{-\alpha x}[B\sin\alpha x + A\cos\alpha x]$

or $\quad M_x = -2EI\alpha^2 A\cos\alpha x\,[e^{\alpha x} - e^{-\alpha x}] + 2EI\alpha^2 B\sin\alpha x\,[e^{\alpha x} + e^{-\alpha x}]$

or $\quad M_x = -2I\alpha^2\,[43034\cos\alpha x\,(e^{\alpha x} - e^{-\alpha x}) + 7315.7\sin\alpha x\,(e^{\alpha x} + e^{-\alpha x})]$

But $\quad 2I\alpha^2 = 2 \times 4.094 \times 10^{-4} \times 1.7416 = 14.26 \times 10^{-4}$

$\therefore \quad M_x = -61.37\cos\alpha x\,(e^{\alpha x} - e^{-\alpha x}) - 10.43\sin\alpha x\,(e^{\alpha x} + e^{-\alpha x})$ ...(ii)

With the help of this equation, B.M. at various values of $x$ can be computed and tabulated as shown in Table 22.10. At the base of the wall, $x = H$, $\alpha x = \alpha H$.

Hence $\quad M_x = -61.37 \times 0.5366 \times 196.13 + 10.43 \times 0.8438 \times 196.13 = -4732.7$ N-m/m which is the same as the fixing moment at the base.

**TABLE 22.9.** Variation of hoop tension $p_r$ (N/m²) [$\alpha = 1.3197$]

| $x$ (m) | $e^{\alpha x}$ | $e^{-\alpha x}$ | $e^{\alpha x}+e^{-\alpha x}$ | $e^{\alpha x}-e^{-\alpha x}$ | $\sin\alpha x$ | $\cos\alpha x$ | 9799.4x | 213.75 × (4) × (6) | −36.34 × (5) × (7) | $p_r = (8) + (9) + (10)$ |
|---|---|---|---|---|---|---|---|---|---|---|
| (1) | (2) | (3) | (4) | (5) | (6) | (7) | (8) | (9) | (10) | (11) |
| 0 | 1 | 1 | 2 | 0 | 0 | 1 | 0 | 0 | 0 | 0 |
| 1 | 3.742 | 0.267 | 4.009 | 3.475 | +0.9686 | +0.2485 | 9799.4 | +830.0 | −31.4 | 10598 |
| 2 | 14.004 | 0.071 | 14.075 | 13.393 | +0.4813 | −0.8765 | 19598.8 | +1448.0 | +426.6 | 21473.4 |
| 2.5 | 27.092 | 0.037 | 27.129 | 27.055 | −0.1570 | −0.9876 | 24498.5 | −910.4 | +971.0 | 24559.1 |
| 3 | 52.410 | 0.019 | 52.429 | 52.391 | −0.7294 | −0.6840 | 29398.2 | −8174.2 | +1302.3 | 22526.3 |
| 3.5 | 101.388 | 0.010 | 101.398 | 101.378 | −0.9956 | −0.0933 | 34297.9 | −21578.5 | +343.7 | 13063.1 |
| 4 | 196.134 | 0.005 | 196.139 | 196.129 | −0.8438 | +0.5366 | 39198.6 | −35375.1 | −3823.5 | 0 |

**TABLE 22.10.** Variation of $M_x$ (kg-m/m) [$\alpha = 1.3197$]

| $x$ (m) | $e^{\alpha x}$ | $e^{-\alpha x}$ | $e^{\alpha x}+e^{-\alpha x}$ | $e^{\alpha x}-e^{-\alpha x}$ | $\sin\alpha x$ | $\cos\alpha x$ | −61.37 × (5) × (7) | −10.43 × (4) × (6) | $M_x = (8) + (9)$ |
|---|---|---|---|---|---|---|---|---|---|
| (1) | (2) | (3) | (4) | (5) | (6) | (7) | (8) | (9) | (10) |
| 0 | 1 | 1 | 2 | 0 | 0 | 1 | 0 | 0 | 0 |
| 1 | 3.742 | 0.267 | 4.009 | 3.475 | +0.9686 | +0.2485 | −53.1 | −40.5 | −93.6 |
| 2 | 14.004 | 0.071 | 14.075 | 13.393 | +0.4813 | −0.8765 | +721.7 | −70.7 | +651.0 |
| 2.5 | 27.092 | 0.037 | 27.129 | 27.055 | −0.1570 | −0.9876 | +1641.2 | +44.4 | +1685.6 |
| 3 | 52.410 | 0.019 | 52.429 | 52.391 | −0.7294 | −0.6840 | +2202.8 | +398.9 | +2601.7 |
| 3.5 | 101.388 | 0.010 | 101.398 | 101.378 | −0.9956 | −0.0933 | +581.4 | +1052.9 | +1634.3 |
| 4 | 196.134 | 0.005 | 196.139 | 196.129 | −0.8438 | +0.5366 | −6464.3 | +1732.3 | −4732 |

The maximum ring tension occurs at a depth of 2.6 m below water surface and its magnitude is found to be 24665 N/m² from the plot of $p_r$.

$\therefore \quad p_r = 24665 \times \dfrac{D}{2} = 24665 \times \dfrac{11.7}{2} = 144290$ N/m height of wall

(against 133580 N/m found in Example 22.1).

**4. Design of tank walls:** area of rings, $A_{sh} = 144290/130 = 1110$ mm²

Providing rings of 12 mm Φ bars at both the faces, spacing of rings = $\dfrac{1000 \times 113}{1110/2} = 203.6$ mm

Provide the rings @ 200 mm spacing on each face. Actual $A_{sh}$ provided = $\dfrac{2 \times 1000 \times 113}{200} = 1130$ mm².

Tensile stress in concrete wall = $\dfrac{144290}{(1000 \times 170) + (11-1)1130} = 0.796$ N/mm$^2$ < 1.2 N/mm$^2$. Hence safe.

The above spacing can be suitably varied at top and bottom.

Max. B.M. = 4732 N-m at base, causing tension at water face. For M 25 concrete (having $\sigma_{cbc}$ = 8.5 N/mm$^2$ and $m$ = 11) and $\sigma_{st}$ = 130 N/mm$^2$, we have $k$ = 0.42, $j$ = 0.86 and $R$ = 1.535.

$$d = \sqrt{\dfrac{4732 \times 1000}{1000 \times 1.535}} \approx 55.5 \text{ mm}$$

However, provide minimum total thickness = $3H + 5 = (3 \times 4) + 5 = 17$ cm. Using 12 mm Φ bars and a clear cover of 25 mm. Available $d$ = 170 − 31 = 139 mm

$$\therefore \quad A_{st} \text{ for B.M.} = \dfrac{4732 \times 1000}{130 \times 0.86 \times 139} = 304.5 \text{ mm}^2$$

Spacing of 10 mm Φ bars = $\dfrac{1000 \times 78.5}{304.5} \approx 258$ mm.

However, provide 10 mm Φ bars @ 250 mm c/c at the water face. These bars are required in the bottom 0.5 m height only. However, provide these *at least* upto a minimum height equal to development length $L_d$ given by $L_d = \dfrac{\Phi \cdot \sigma_{st}}{4 \tau_{bd}} = \dfrac{10 \times 130}{4 \times 0.9} \approx 360$ mm = 0.36 m

From practical considerations, provide these bars upto 1 m height from the bottom. Above this height, curtail half the bars and continue other half (*i.e.* @ 400 mm c/c) upto top. Keep a clear cover of 25 mm. Provide distribution reinforcement @ 0.24% of area of concrete.

$$A_{sd} = \dfrac{0.24}{100} \times 170 \times 1000 = 408 \text{ mm}^2$$

Area of steel on each face = 204 mm$^2$. However, no additional reinforcement will be provided at the inner face since the vertical steel for cantilever action will serve this purpose. At the outer face, provide 8 mm bars at spacing = $\dfrac{1000 \times 50.5}{204} = 247.5$ say 240 mm.

At the outer face, there is a positive B.M., the maximum magnitude of which is about 2650 N-m at a depth of 2.9 m below the top.

$$\text{Area of steel required} = \dfrac{2650 \times 1000}{130 \times 0.86 \times 139} = 170.5 \text{ mm}^2$$

The distribution reinforcement provided will serve this purpose. It is customary to provide 150 mm × 150 mm haunches at the junction of the wall and the base. A reinforcement of 8 mm Φ bars @ 200 mm c/c may be provided.

**5. *Design of roof slab*:** Diameter of roof slab, upto centre line of outer walls = 11.7 + 0.17 = 11.87 m. The roof slab, circular in plan, may be assumed to be simply supported at its periphery. Let the total thickness of top cover = 200 mm.

Its weight = $0.2 \times 25000 = 5000$ N/m$^2$.

Assume live load of 1000 N/m$^2$. Total U.D.L. = 5000 + 1000 = 6000 N/m$^2$.

If $(M_r)_C$ and $(M_\theta)_C$ are the radial and circumferential moments at the centre of the slab,

$$(M_r)_C = (M_\theta)_C = +\dfrac{3}{16} wa^2 = \dfrac{3}{16} \times 6000 \left(\dfrac{11.87}{2}\right)^2 \times 1000 \text{ N-mm}$$

$$= 39.63 \times 10^6 \text{ N-mm}$$

Circumferential moment at the edge is given by

$$(M_\theta)_C = +\dfrac{2}{16} wa^2 = \dfrac{2}{3}(39.63 \times 10^6) = 26.42 \times 10^6 \text{ N-mm}$$

$$d = \sqrt{\frac{39.63 \times 10^6}{1000 \times 1.535}} = 160.7 \text{ mm.}$$

Provide a total thickness of 200 mm. Using 12 mm bars and a clear cover of 15 mm, available $d = 200 - 15 - 6 = 179$ mm for the first layer and $179 - 12 = 167$ mm for the second layer.

Radial and circumferential reinforcement required at the centre:

$$A_s = \frac{39.63 \times 10^6}{130 \times 0.86 \times 167} = 2122.6 \text{ mm}^2. \quad \text{For 12 mm } \Phi \text{ bars, } A_\Phi = 113 \text{ mm}^2.$$

$$\text{Spacing} = \frac{1000 \times 113}{2122.6} = 53.2 \text{ mm.}$$

The spacing is very small. Hence use 16 mm $\Phi$ bars.

$$A_\Phi = 201 \text{ mm}^2. \quad \therefore \text{Spacing} = \frac{1000 \times 201}{2122.6} = 94.7 \text{ mm.}$$

Hence provide the reinforcement in the form of a mesh consisting of 16 mm $\Phi$ bars spaced at 90 mm c/c each of the two layers at right angles to each other. Near the edges, the bars do not have proper anchorage since they are free. There will be slipping tendency and hence bars will not be capable of taking any tension. At the edges, radial tensile stresses are zero, but circumferential tensile stresses exist because of $(M_\theta)_c = 26.42 \times 10^6$ N-mm. The tendency of slipping can be avoided by providing extra circumferential reinforcement, in the form of rings placed in a width equal to the bond length of the mesh bars. Available $d$ for rings = $200 - 15 - 2 \times 16 - 8 = 145$ mm.

$$\therefore \quad (A_s)_\theta = \frac{26.42 \times 10^6}{130 \times 0.86 \times 145} = 1630 \text{ mm}^2.$$

$$\therefore \quad \text{Spacing of ring } s = \frac{1000 \times 201}{1630} = 123.4 \text{ mm.}$$

$(M_\theta)_e$ is $\frac{2}{3}$ of maximum $M_\theta$ at the centre. The circumferential steel will be provided for a length $= 2/3 \, (45\Phi) = 2/3 \, (45 \times 16) = 480$ mm. Hence provide rings of 16 mm $\Phi$ bars @ 120 mm c/c, total rings being equal to $(480/120) + 1 \approx 5$.

*Check for shear*: Maximum S.F. at the edges is given by:

$$F_r = \frac{1}{2} wa = \frac{1}{2} \times 6000 \left(\frac{11.87}{2}\right) = 17805 \text{ N/m width.}$$

and

$$\tau_v = \frac{17805}{1000 \times 167} = 0.11 \text{ N/mm}^2. \quad \text{Hence safe.}$$

**Note.** The roof slab provided in the form of ordinary slab like this is highly uneconomical since greater thickness and heavy reinforcement is required. Normally, roof is provided in this form of a dome, the design of which has been illustrated in Example 22.5.

**6. Design of base slab:** The loading on the base slab is shown in Fig. 22.8. The design of the base slab has been done separately in Example 22.4.

**7. Design of bottom circular beam:** The tank is supported on a circular beam which in turn is supported on *six* equally spaced columns (say). The diameter of supporting circle, upto the centre of the beam = 9.2 m. The total load $W$ on the beam = 6371035 N. For the design of a circular beam, refer Chapter 20.

Super-imposed load on beam $= \dfrac{6371035}{\pi \times 9.2} = 220431$ N/m length

Assuming the beam to have a section of 800 mm × 500 mm,

Self weight $= 0.8 \times 0.5 \times 1 \times 25000 = 10000$ N/m.

$\therefore$ Total $w = 220431 + 10000 = 230431$ N/m. From Table 20.1,

$$2\theta = 60° = \frac{\pi}{3}; \quad C_1 = 0.089; \quad C_2 = 0.045; \quad C_3 = 0.009; \quad \phi_m = 12\tfrac{3°}{4}.$$

Radius of beam = 9.2/2 = 4.6 m

or
$$wR^2 \cdot 2\theta = 230431 \,(4.6)^2 \left(\frac{\pi}{3}\right) = 5.1 \times 10^6$$

$\therefore$
$$M_0 = C_1 \cdot wR^2(2\theta) = 0.089 \times 5.1 \times 10^6 = 0.4539 \times 10^6 \text{ N-m}$$
$$M_c = C_2 \cdot wR^2(2\theta) = 0.045 \times 5.1 \times 10^6 = 0.2295 \times 10^6 \text{ N-m}$$
$$M_m^t = C_3 \cdot wR^2(2\theta) = 0.009 \times 5.1 \times 10^6 = 0.0459 \times 10^6 \text{ N-m}.$$
$$F_0 = \frac{w \cdot R \cdot 2\theta}{2} = \frac{230431 \times 4.6}{2} \times \frac{\pi}{3} = 555005 \text{ N}$$

S.F. at the point of maximum torsional moment ($\phi_m = 12.75° = 0.2225$ radians) is given by:
$$F_\phi = wR\,(\theta - \phi_m) = 230431 \times 4.6 \left(\frac{\pi}{6} - 0.2225\right) = 319159.5 \text{ N} \approx 319160 \text{ N}$$

*Design of section:* Use M 25 concrete for the circular beam.
$$M_{max.} = 0.4539 \times 10^9 \text{ N-mm}.$$

$\therefore$
$$d = \sqrt{\frac{0.4539 \times 10^9}{0.86 \times 500}} = 1027.4 \text{ mm}. \quad \text{Keep total } D = 1080 \text{ mm}.$$

Using 20 mm $\Phi$ bars and a clear cover of 25 mm, $d = 1080 - 10 - 25 = 1045$ mm.

*Check of shear stress:* The equivalent shear is given by $V_e = V + 1.6\dfrac{T}{b}$

where $V$ = S.F. at the point of maximum torsional moment; $F_\phi = 319160$ N.
$T$ = maximum torsional moment = $M_m^t = 0.0459 \times 10^6$ N-m.
$b$ = breadth of the beam = 50 cm = 0.5 m

$\therefore$
$$V_e = 319160 + 1.6\,\frac{0.0459 \times 10^6}{0.5} = 466040 \text{ N}$$

$\therefore$
$$\tau_{ve} = \frac{V_e}{bd} = \frac{466040}{500 \times 1046} = 0.89 \text{ N/mm}^2$$

This is less than $\tau_{c.\,max}(= 1.6 \text{ N/mm}^2)$ for *M* 15 concrete (Table 20.8). Hence the depth is satisfactory.

*Longitudinal reinforcement:* Equivalent moment $M_{e1} = M + M_T$
where $M$ = B.M. at any section.
$$M_T = T\left(\frac{1 + D/b}{1.7}\right),$$

where $T = M^t_\phi$ = torsional moment at that section.

Let us take the section at the column support where B.M. $M (= M_0)$ is maximum.
$$M = M_0 = 0.4539 \times 10^6 \text{ N-m}.$$

The torsional moment at $\phi = 0$ is given by *Eq. 20.6*
$$M^t_\phi = wR^2\,[\theta \cos\phi - \theta \cot\theta \sin\phi - (\theta - \phi)]. \quad \text{Here } \theta = 30° = \frac{\pi}{6};\ \ \phi = 0.$$

$\therefore$
$$M^t_\phi = M^t_0 = 230421\,(4.6)^2 \left[\frac{\pi}{6}\cos 0 - \frac{\pi}{6}\cot\frac{\pi}{6}\sin 0 - \left(\frac{\pi}{6} - 0\right)\right] = 0$$

Thus, twisting moment is zero at the point of maximum B.M. Similarly at the point of maximum torsional moment $\left(\phi_m = 12\frac{3°}{4}\right)$, B.M. is zero. Hence B.M. and twisting moment are not effectively additive at these two sections.

$\therefore$
$$M_{e1} = M = 0.4539 \times 10^9 \text{ N-mm} \quad \text{and} \quad A_{st1} = \frac{0.4539 \times 10^9}{140 \times 0.87 \times 1045} = 3566 \text{ mm}^2$$

∴ Number of 20 mm Φ bars = 3566/314 = 11.3 ≈ 12. Provide 7 bars in one layer and 5 bars in the second layer. These bars will be provided at the top of the section, since the maximum B.M. is hogging at each column support. The B.M. is zero at the section of maximum torsion (*i.e.* at $\phi = 12.75°$). Distance of this section from the support, along the axis of beam = $R\phi_m$ = 4.6 × 0.2225 = 1.02 m. Hence curtail 10 bars at this section and continue the remaining two to set as anchor bars. (See revision of this arrangement, below). At the point of maximum torsional moment ($\phi_m$=12.75°) B.M. $M$ is equal to zero (see Fig. 20.6 also). Hence

$$M_{e1} = M + M_T = 0 + M_T = M_T$$

where $M_T = M_\phi^t \left(\dfrac{1+D/b}{1.7}\right) = 0.0459 \times 10^6 \left[\dfrac{1+\dfrac{1080}{500}}{1.7}\right] = 0.08532 \times 10^6$ N-m

The reinforcement has to be provided for $\pm M_T$ since $M$ is equal to zero.

∴ $\pm A_{st} = \dfrac{0.08532 \times 10^6 \times 1000}{140 \times 0.87 \times 1045} = 670$ mm².

∴ Number of 20 mm Φ bars = 670/314 = 2.13 ≈ 3.

Also, at the mid-span of the beam, maximum positive B.M. $M_c$ is 0.2295 × 10⁶ N-m. The torsional moment at this section is zero (Fig. 20.6). Hence

$$M_{e1} = M + M_T = M_c + 0 = 0.2295 \times 10^6 \text{ N-m.} \quad \therefore \quad A_{st2} = \dfrac{0.2295 \times 10^6 \times 1000}{140 \times 0.87 \times 1045} = 1803 \text{ mm}^2$$

∴ Number of 20 mm dia bars = 1803/314 ≈ 6

Hence the arrangement of 12 bars of 20 mm Φ provided for hogging B.M. (at the support) will be as under:

(*i*) Provide 12 bars of 20 mm Φ at the column support, in two layers (7 bars in top layer + 5 bars in next layer).

(*ii*) At $\phi_m$ = 12.75° (distant $R\phi_m$ = 4.6 × 0.2225 = 1.02 m from column) bend 4 bars of second layer downwards to take care of positive B.M. at the support. Provide 3 additional bars of 20 mm dia. at the bottom face of the beam running throughout.

(*iii*) Of the remaining eight bars of 20 mm Φ, discontinue 5 bars at $\phi_m$ = 12.75° location and continue 3 bars throughout at the top face.

Thus we note from the above arrangement that 3-20 mm Φ bars will be available throughout at each top and bottom face of the beam. These 3 bars will take care of torsional moment even at the location where B.M. is zero.

*Transverse reinforcement*:

$$A_{sv} = \dfrac{T \cdot s_v}{b_1 d_1 \sigma_{sv}} + \dfrac{V s_v}{2.5 d_1 \sigma_{sv}} \quad (Eq.\ 20.65)$$

Transverse reinforcement is provided in the form of vertical stirrups.

Let the alround clear cover be equal to 25 mm.

∴ $b_1$ = c/c distance between corner bars in the direction of width
= 500 − 25 × 2 − 20 = 430 mm.

$d_1$ = c/c distance between corner bars in the direction of depth
= 1080 − 25 × 2 − 20 = 1010 mm. Let us use 12 mm Φ 4 lgd stirrups.

∴ $A_{sv} = 4\left(\dfrac{\pi}{4} \times 12^2\right) = 452$ mm². Hence substituting the values,

$$452 = \left[\dfrac{0.0459 \times 10^9}{430 \times 1010 \times 140} + \dfrac{319146}{2.5 \times 1010 \times 140}\right] s_v.$$

From which $s_v$ = 272.67 mm

However, provide 12 mm Φ-4 lgd stirrups @ 250 mm c/c for a distance of 1.6 m (approximately *L*/3) from each support. For the remaining central *L*/3 portion, provide nominal 12 mm Φ 2 lgd stirrups @ 250 mm c/c.

*Side face reinforcement*: IS: 456-2000 states that when the cross-sectional dimension of beam exceeds 450 mm, additional longitudinal bars shall be provided. The total area of such longitudinal bars (side face reinforcement) should not be less than 0.1% of web area.

∴  $A_s = \dfrac{0.1}{100} \times 500 \times 1045 = 522.5 \text{ mm}^2$

Let us use 12 mm Φ bars having $A_\phi = 113 \text{ mm}^2$.  ∴  Number of bars = 522.5/113 = 4.6 However, provide 3 bars of 12 mm Φ on each vertical face.

**8. *Details of reinforcement*:** Details of reinforcement etc. are shown in Fig. 22.9.

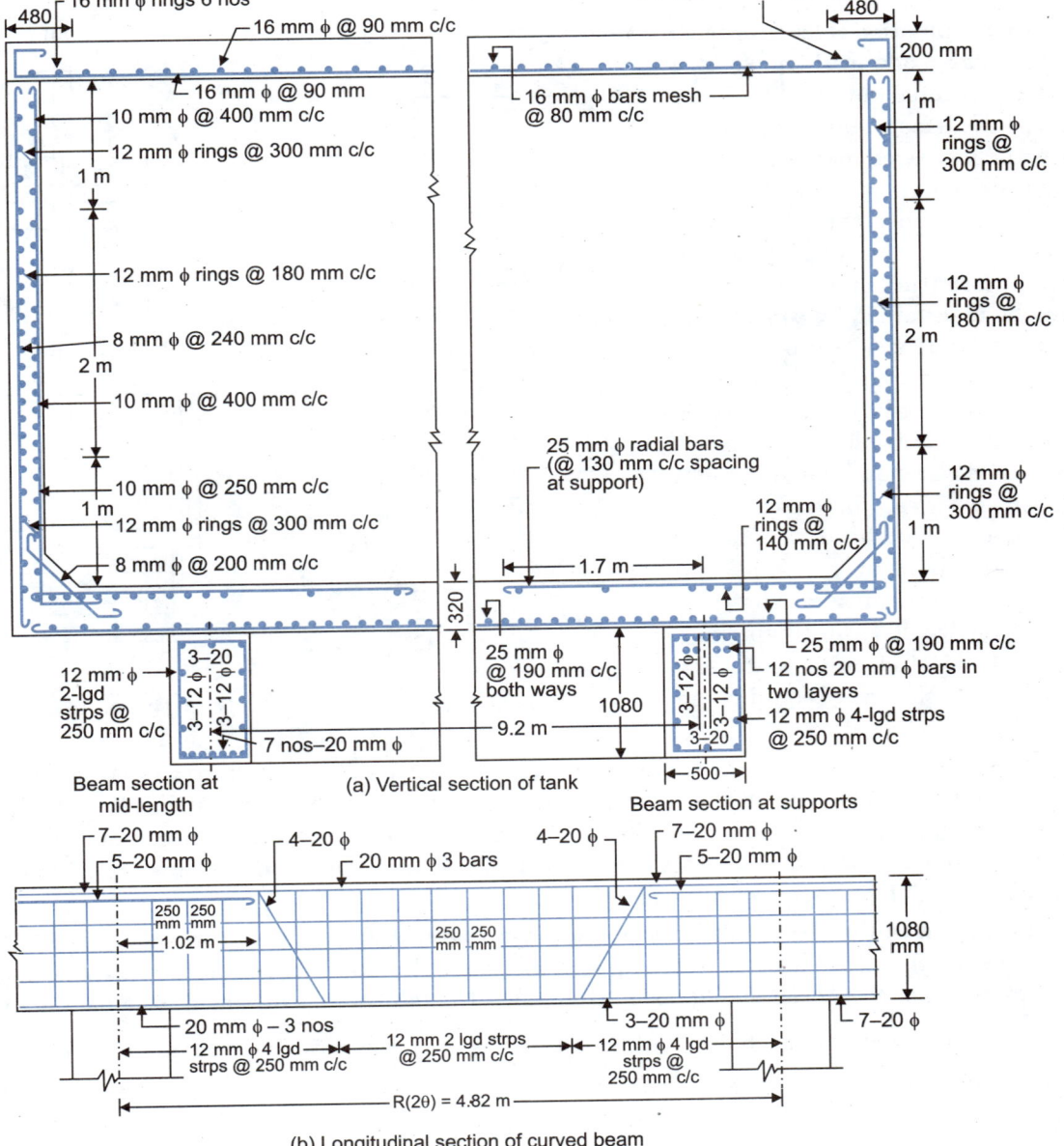

**Fig. 22.9.** Elevated circular tank with flat bottom

## 22.3. DESIGN OF FLAT BASE SLAB FOR ELEVATED CIRCULAR TANKS

When the circular tanks are elevated and supported, the analysis and design of the base slab depends on the manner in which it is supported. Fig. 22.10 shows some alternative arrangements for supporting the elevated water tanks.

**(a) Slab supported on circular beam:** Fig 22.10 (a) shows the base slab supported on a circular beam of rectangular cross-section, which in turn is supported on a number of columns, the diameter of supporting circle of the beam is normally kept $\frac{3}{4}$ of the diameter of the tank. The determination of loads on such a base slab was made in Example 22.3 where it was pointed out that the fixing moment at the base of the tank wall is decreased or increased as the diameter of the supporting circle is decreased or increased. When once the loading etc. on the base slab is known, the radial and circumferential moments in the slab may be determined by considering the slab to be freely supported on imaginary support round it's periphery at radius $a$ and loaded with the following (i) uniformly distributed load $P$ consisting of the weight of water and the self weight of slab, per unit area, (ii) upward ring load $W$ consisting of the total reaction from the circular beam at a radius $b$ and (iii) end moment $M$ at the joint with the vertical wall. Refer Chapter 11, for the analysis of circular slab. The design of such a slab has been done in Example 22.4.

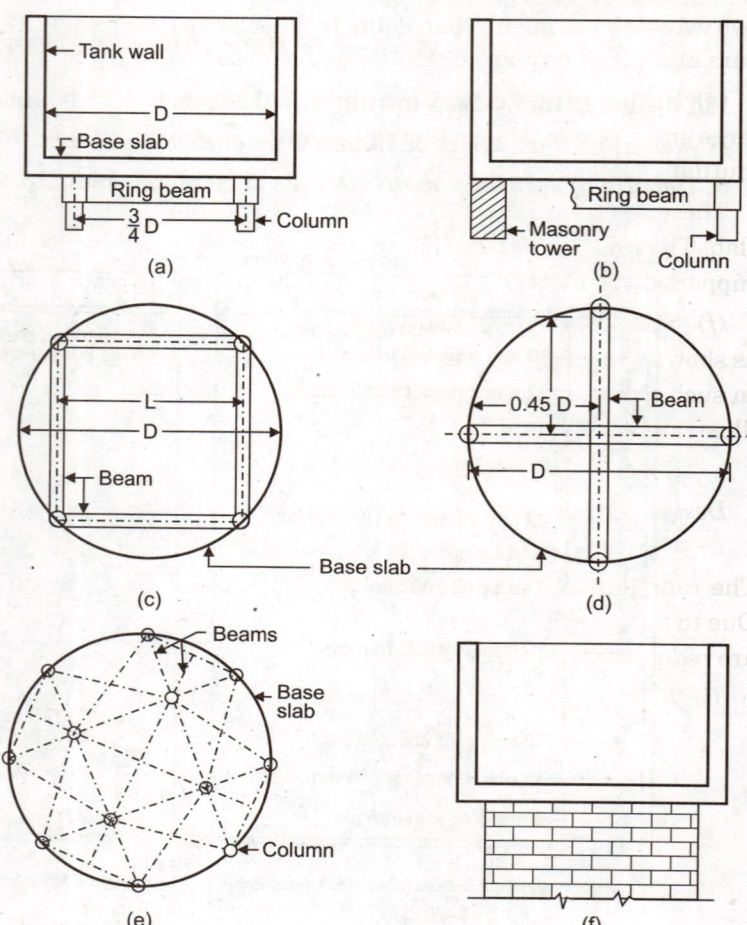

Fig. 22.10. Methods of supporting bottom slab.

**(b) Slab supported on masonry tower**: Fig. 22.10 (b) shows a slab supported on masonry tower. The slab may be assumed to be fixed at the edges and subjected to uniformly distributed load. If, however, the tank is to be supported on columns, a ring beam may be provided on the top of the columns and the tank may be supported on this. In that case also, the base slab may be assumed to be clamped at the edges. The analysis of such a slab has been made in Example 22.5.

**(c) Slab supported on four beams mutually orthogonal**: Fig. 22.10 (c) shows the slab supported on four beams which in turn are supported on four columns. The slab between the beams is designed as a two-way square slab with corners held down. Generally, the base slab is designed for a bending moment equal to $\frac{W}{32}$ per unit width in each direction, where W is the total load of water and slab. The beams are designed as T-beams, each of span L. If each beam is assumed to carry a load of $\frac{W}{4}$, distributed triangularly the maximum B.M. is taken equal to $\frac{W}{4} \times \frac{L}{6} = \frac{WL}{24}$. The tank wall overhangs the four beams. Hence the walls are reinforced with few bars at the bottom and top so that it can act as curved beam to support its own load.

(d) **Slab supported on two beams mutually orthogonal:** This system is shown in Fig. 22.10 (d). These two beams are supported on four columns. It is assumed that the tank walls act as a curved deep beam transferring half the total load, and the remaining half is assumed to be carried by the two beams. Thus, each beam is assumed to carry $\frac{W}{4}$ load, and is designed as $T$-beam of span $D$. The base slab is designed as two way slab having its span equal to 0.45 $D$ in each direction. The tank wall should be reinforced with few bars at top and bottom so that it can act as a curved beam to support it own load.

(e) **Slab supported on a number of beams:** Such a system, shown in Fig. 22.10 (e) is adopted for large size circular tanks having flat bottom. The load is transferred through a number of continuous beams mutually orthogonal to each other.

These beams are designed for triangular loadings. The slab in between the beams is designed as two-way slab. The tank wall is reinforced with a few bars at top and bottom so that it can act as a curved beam to support its own load.

(f) **Slab supported throughout:** Sometimes, elevated water tanks are supported on a masonry platform, as shown Fig. 22.10 (f). The case is similar to the one in which the base slab is supported on ground surface. In such a case, nominal thickness of slab is provided, and it is reinforced with nominal reinforcement, as illustrated in Example 22.1.

### Design: Example 22.4: Design of Base Slab

*Design the base slab of the circular tank Example 22.3.*

**Solution: 1. Calculation of B.M. and S.F.:** The loading on the base slab of the tank is shown in Fig. 22.8. The radial and circumferential bending moments are calculated with the help of formulae given below. Due to the hogging moment $M = 4732$ N-m/m, the radial and circumferential B.M. at the centre of the slab are *reduced*.

(i) *B.M. due to U.D.L.* $p = 46700$ N/m²

$$(M_r)_c = +\frac{3}{16}pa^2; \quad (M_r)_e = 0$$

$$M_r = \frac{3}{16} p(a^2 - r^2) = \frac{3}{16} \times 46700\,[(5.935)^2 - r^2] = 8756\,(35.22 - r^2)$$

$$(M_\theta)_c = +\frac{3}{16}pa^2; \quad (M_\theta)_e = +\frac{2}{16}pa^2$$

$$M_\theta = +\frac{p}{16}(3a^2 - r^2) = \frac{46700}{16}\,[3(5.935)^2 - r^2] = 2918.8\,(105.67 - r^2)\ \text{kg-m/m width.}$$

The values of $M_\theta$ and $M_r$ at various locations are tabulated below:

**TABLE 22.11**

| $r$ (m) | 0 | 2 | 4.6 | 5.935 |
|---|---|---|---|---|
| $M_r$ (kg-m) | 308418 | 273403 | 123132 | 0 |
| $M_\theta$ (kg-m) | 308418 | 296159 | 246659 | 205619 |

(ii) *B.M. due to upward load* $W = 6371035$ N

For $r < b$ (where $a = 5.935$ and $b = 4.6$ m)

$$M_r = (M_r)_b = M_\theta = (M_\theta)_o = -\frac{W}{8\pi}\left[2\log_e\left(\frac{a}{b}\right) + 1 - \left(\frac{b}{a}\right)^2\right]. \text{ For } r > b$$

$$M_r = -\frac{W}{8\pi}\left[2\log_e\left(\frac{a}{r}\right) - \left(\frac{b}{a}\right)^2 + \left(\frac{b}{r}\right)^2\right] \text{ and } M_\theta = -\frac{W}{8\pi}\left[2\log_e\left(\frac{a}{r}\right) - \left(\frac{b}{r}\right)^2 + 2 - \left(\frac{b}{a}\right)^2\right]$$

## 570 REINFORCED CONCRETE STRUCTURE

The values of $M_r$ and $M_\theta$ at various location are tabulated below:

**TABLE 22.12**

| $r$ (m) | 0 | 2 | 4.6 | 5.935 |
|---|---|---|---|---|
| $M_r$ (kg-m) | −230589 | −230589 | −230589 | 0 |
| $M_\theta$ (kg-m) | −230589 | −230589 | −230589 | −202790 |

(iii) *B.M. due to M* = − 4732 N-m. Due to hogging B.M. at the ends there will be constant B.M. of
$$M_r = -4732 \text{ and } M_\theta = -4732 \text{ throughout the slab.}$$

(iv) *Net moments*. The net moments will be the algebraic sum of the three, and are tabulated below :

**TABLE 22.13**

| $r$ (m) | 0 | 2 | 4.6 | 5.935 |
|---|---|---|---|---|
| $M_r$ (kg-m) | +73097 | +38082 | − 112189 | − 4732 |
| $M_\theta$ (kg-m) | +73097 | +60838 | +11338 | − 1903 |

From the above Table, a maximum sagging moment occurs at the centre, while maximum hogging moment occurs at the ring beam support. The circumferential moment is sagging throughout, its magnitude being maximum at the centre. Maximum shear force occurs at the outer edge of the ring beam and its magnitude is given by

$$F = \frac{pb}{2} - \frac{W}{2\pi b} = \frac{46700}{2} \times 4.6 - \frac{6371035}{2\pi \times 4.6} = -113021 \text{ N per metre width.}$$

**2. Design of slab:** The slab is to be designed for a maximum B.M. of 112189 N-m. Since this moment is hogging, tension will occur at the water face.

From B.M. point of view, $d = \sqrt{\dfrac{112189 \times 1000}{1000 \times 1.535}} = 270.3$ mm

Let us keep total thickness = 320 mm. Using 25 mm Φ bars, and a clear cover of 25 mm, available $d = 320 - 25 + 12.5 = 282.5$ mm for radial reinforcement. Since the bending moment is hogging from the outer edge of the slab to the ring beam support, it will be provided at the upper face. Area of steel for radial reinforcement

$$= \frac{112189 \times 1000}{130 \times 0.86 \times 282.5} = 3552.1 \text{ mm}^2$$

Using 25 mm Φ bars, $A_\Phi = 490.9$ mm$^2$. ∴ Spacing $= \dfrac{1000 \times 490.9}{3552} = 138$ mm

Keep the spacing of 130 mm c/c. The radial moment is found to be zero at a radius $r$ given by
$8756 (35.22 - r^2) - 230589 - 4732 = 0$

which gives $r \approx 2.9$ m. The hogging B.M. exists from $r = a = 5.935$ m to $r = 2.9$ m, with its maximum value at $r = b = 4.6$ m. Hence the above radial bars may be completely curtailed at $r = 2.9$ m, providing hooks there.

At the centre of the slab, radial moment is positive (sagging) and the circumferential moment is also positive, each being equal to 73097 N-mm. Hence reinforcement for both these may be provided in the form of mesh, at the outer face of the slab. Effective depth for second layer = 282.5 − 25 = 257.5 mm.

$$A_s = \frac{73097 \times 1000}{130 \times 0.86 \times 257.5} = 2539 \text{ mm}^2$$

Spacing of 25 mm Φ bars $= \dfrac{1000 \times 490.9}{2539} = 193$ mm. Hence provide 25 mm Φ bars @ 190 mm c/c both ways. This reinforcement may be provided for the portion of the slab inside the supporting circle, *i.e.* upto $r = b = 4.6$ m. After that, $\frac{2}{3}$rd bars can be curtailed and $\frac{1}{3}$ bars can be continued upto the end.

At the outer edge of the slab, circumferential moment is hogging. Hence special ring reinforcement will have to be provided. Using 12 mm rings, available $d = 282.5 - 12 - 6 = 264.5$ mm.

$$A_s = \frac{1903 \times 1000}{130 \times 0.86 \times 264.5} = 64.4 \text{ mm}^2$$

Minimum reinforcement @ 0.24%

$$= \frac{0.24 \times 320 \times 1000}{100} = 768 \text{ mm}^2$$

Using rings of 12 mm Φ,

$$A_\Phi = 113 \text{ mm}^2. \text{ Spacing} = \frac{1000 \times 113}{768} = 147.1 \text{ say } 140 \text{ mm}$$

Provide these rings from the outer edge ($r = 5.935$ m) to $r = 4.6$ m, at the inner face of the slab. The details of reinforcement are shown in Fig. 22.9.

**Design: Example 22.5. Design of circular tank with domical top and flat base support on masonry tower.**

*Design a circular tank, with domical top for a capacity of 400,000 litres. The depth of water is to be 4 m, including a free board of 20 cm. The tank is to be supported on masonry tower. Take unit weight of water as 9800 kN/m³.*

**Solution: 1. Dimensions of the tank.** The capacity of the tank is the same as given in Example 22.1. Hence we have $D = 11.7$ m; $H = 4$ m.

**2. Design of roof dome: Membrane Analysis:** We shall design the top dome and its ring beam on membrane analysis, considering these to be independent of the tank wall which is assumed to be free at the top. Let the rise of the dome be 2 m and its thickness be 100 mm. Let us design the dome for a live load of 1400 N/m².

The radius $r$ of the dome is given by

$$R^2 = (2r - \text{rise}) \text{ rise}$$

or $(5.85)^2 = (2r - 2) \, 2.$

From which $r = 9.56$ m

Self load of dome $= 0.1 \times 1 \times 1 \times 25000$
$= 2500$ N/m²

Live load $= 1400$ N/m²

Total $W = 3900$ N/m²

$$\sin \phi = \frac{5.85}{9.56} = 0.6119;$$

$$\cos \phi = \frac{7.56}{9.56} = 0.7908$$

$$\phi = 37.73°$$

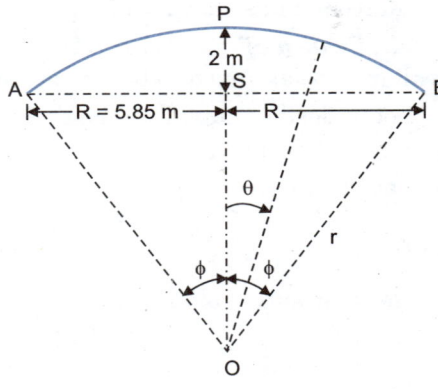

Fig. 22.11

$$\text{Hoop stress} = \frac{wr}{t}\left(\frac{\cos^2 \theta + \cos \theta - 1}{1 + \cos \theta}\right); \quad \text{Meridional stress} = \frac{wr}{t}\left(\frac{1 - \cos \theta}{\sin^2 \theta}\right)$$

Maximum hoop stress occurs at $\theta = 0$ and its magnitude

$$= \frac{3900 \times 9.56}{0.1}\left[\frac{1 + 1 - 1}{1 + 1}\right] = 186420 \text{ N/m}^2 = 0.186 \text{ N/mm}^2 \quad \text{(safe)}.$$

Maximum meridional stress will be at $\theta = \phi = 37.73°$ its magnitude being

$$= \frac{3900 \times 9.56}{0.10}\left\{\frac{1 - \cos 37.73}{\sin^2 37.73}\right\} = 208190 \text{ N/m}^2 \approx 0.21 \text{ N/mm}^2 \quad \text{(safe)}$$

The stresses are within safe limit. However provide minimum reinforcement @ 0.24 % of area in each direction.

$$\therefore \quad A_s = \frac{0.24}{100} \times 100 \times 1000 = 240 \text{ mm}^2$$

Using 8 mm Φ bars, $A_\Phi = 50 \text{ mm}^2$.

$$\therefore \quad \text{Spacing} = \frac{1000 \times 50}{240} = 208.3 \text{ mm}$$

Hence provide 8 mm Φ bars @ 200 mm c/c in both the directions.

**3. Design of ring beam:** The thickness of dome = 100 mm

Meridional thrust per metre length of the dome at its base
$$= 208190 \times 0.1 \times 1 \approx 20820 \text{ N/m}$$

Horizontal component $T$ per metre length
$$= 20820 \cos 37.73° = 20820 \times 0.7908 = 164664 \approx 16470 \text{ N/m}.$$

$$\therefore \quad \text{Hoop tension trying to rupture the beam} = 16470 \times \frac{11.7}{2} = 96349.5 \text{ say } 96350$$

Area of steel required = 96350/130 = 741 mm². Using 20 mm bars, No. of rings = 741/314 = 2.4. However provide 4 rings of 20 mm Φ bars for symmetry.

Actual $A_{sh} = 314 \times 14 = 1256 \text{ mm}^2$. Equivalent area of composite section of beam of area of cross-section $A$ is $= A + (m-1) A_{sh} = A + 10 \times 1256 = A + 12560$

Allowing a stress of 1.2 N/mm² in the composite section we have $\frac{96350}{A + 15072} = 1.2$

From which $A$ = 67733 mm². Hence provide a ring beam of size 300 mm × 230 mm.

Provide 6 mm Φ stirrups @ 200 mm c/c to tie the ring in the ring beam. These rings are lapped with dome reinforcement as shown in Fig. 22.12.

**4. Design of Tank wall:** Since the domed roof has been designed on the membrane analysis, the tank wall may be assumed to be free at the top. Let the tank be assumed to be *restrained* at the bottom.

Let us design the wall on the basis of Carpentors recommendations and coefficients (Reynold's Handbook) given in Table 22.7. $H$ = 4 m; $D$ = 11.7 m.

Let thickness of tank $= T = 3H + 5 = 3 \times 4 + 5 = 17 \text{ cm} = 0.17 \text{ m} = d_A$

$$\therefore \quad \frac{H}{d_A} = \frac{4}{0.17} = 23.5; \quad \frac{H}{D} = \frac{4}{11.7} = 0.34$$

Hence from Table 22.7, we get the following values of coefficients after linear interpolation.

$$F = 0.0154; \quad K_1 = 0.397; \quad K_2 = 0.574$$

$$\therefore \quad M_{max} = FwH^3 = 0.0154 \times 9800(4)^3 = 9659 \text{ N-m/m}$$

(against 9157 N-m/m obtained by I.S. code method in Example 22.2)

$$T_H = \frac{1}{2} wH DK_2 = \frac{1}{2} \times 9800 \times 4 \times 11.7 \times 0.574 = 131629.68 \approx 131630 \text{ N/m}$$

(against 133000 N/m obtained by I.S. code method in Example 22.2)

$L = K_1 H = 0.397 \times 4 = 1.588$ m above base (or 2.412 m below top, against 2.32 m below top as obtained by I.S. code method),

Area of rings = 131630/130 = 1013 mm². Providing 12 mm Φ rings ($A_\Phi$ = 113 mm²) on each face of wall.

$$\text{Spacing} = \frac{1000 \times 113}{1013/2} = 223 \text{ mm}.$$

Provide the rings @ 220 mm c/c on each face. The spacing of the rings may be increased towards the top and bottom from the section of maximum hoop load.

$$\text{Actual } A_{sh} \text{ provided} = \frac{2 \times 1000 \times 113}{220} = 1027 \text{ mm}^2$$

∴ Tensile stress in concrete = $\dfrac{131630}{1000 \times 170 + (11-1)1027}$ = 0.73 N/mm² (safe)

From B.M. point of view, $d = \sqrt{\dfrac{9659 \times 1000}{1000 \times 1.535}}$ = 79.3 mm

Using 12 mm Φ bars and a clear cover of 25 mm, cover to the centre of reinforcement = 25 + 6 = 31 mm. Total thickness = 79.3 + 31 = 108.3 mm

However, provide a minimum thickness equal to the greater of the following:
(i) 15 cm, (ii) $3H + 5 = (3 \times 4) + 5 = 17$ cm. Hence provide $T = 17$ cm = 170 mm.

Available $d = 170 - 31 = 139$ mm. ∴ $A_{st} = \dfrac{9659 \times 1000}{130 \times 0.86 \times 139}$ = 621.6 mm²

Spacing of 12 mm Φ bars = $\dfrac{1000 \times 113}{621.6} \approx 181.8$ mm

Hence provide 12 mm Φ vertical bars @ 180 mm c/c at the water face. These bars are required in bottom 50 cm height only. However, provide these upto bottom 1 m height. Above this height curtail half the bars and continue the other half upto top.

Provide distribution reinforcement @ 0.24 % of the area of concrete.

$$A_{sd} = \dfrac{0.24}{100} \times 170 \times 1000 = 408 \text{ mm}^2$$

Area of steel on each face = 204 mm². However, no additional reinforcement will be provided at the inner face: the vertical steel for cantilever action will serve this purpose.

Hence provide $A_{sd}$ = 222 mm² in vertical direction at the outer face.

The spacing of 8 mm bars = $\dfrac{1000 \times 50.5}{204} \approx 247.5$ mm. However, provide these @ 200 mm c/c. This will take care of positive B.M. in the wall.

**5. Design of base slab**: The bottom slab is designed as clamped at the edges.

Assume 430 mm thickness of slab.

Weight of water per m² of slab = 4 × 1 × 1 × 9800 = 39200 N

Self weight per m² of slab = 0.43 × 1 × 1 × 25000 = 10750 N

Total $p \approx 50000$ N/m²

Circumferential moment:

$$(M_\theta)_c = +\dfrac{1}{16} pa^2 = \dfrac{1}{16} \times 50000 \left(\dfrac{11.7}{2}\right)^2 = +1.07 \times 10^5 \text{ N-m}$$

$$(M_\theta)_e = 0$$

Radial moment: $(M_r)_c = +\dfrac{1}{16} pa^2 = +1.07 \times 10^5$ N-m

$$(M_r)_e = -\dfrac{2}{16} pa^2 = -2.14 \times 10^5 \text{ N-m}$$

Radial shear $F_r = \dfrac{1}{2} pa = \dfrac{1}{2} \times 50000 \left(\dfrac{11.7}{2}\right) = 146250$ N

The radial moment is zero at radius given by

$$M_r = 0 = \dfrac{1}{16} p(a^2 - 3r^2) \quad \text{or} \quad r = \dfrac{a}{2\sqrt{3}} = \dfrac{11.7}{2\sqrt{3}} = 3.38 \text{ m}.$$

(*i.e.* distance of point of contraflexure = $\dfrac{11.7}{2} - 3.38 = 2.47$ m from edge)

$$d = \sqrt{\frac{2.14 \times 10^5 \times 1000}{1000 \times 1.535}} = 373.4 \text{ mm}$$

Using 30 mm Φ bars, and a clear cover of 30 mm, total thickness

$$= 373.4 + 30 + 15 = 418.4 \text{ mm}$$

Provide total thickness

= 450 mm. Available

$d = 450 - 15 - 30 = 405$ mm $A_{st}$ for negative B.M.

$$= \frac{2.14 \times 10^5 \times 1000}{130 \times 0.86 \times 405}$$

$$= 4727 \text{ mm}^2$$

Spacing of 30 mm Φ bars having $A_\Phi = 706.9 \text{ mm}^2$

$$= \frac{1000 \times 706.9}{4727} \approx 149.5 \text{ mm}.$$

Hence provide 30 mm Φ radial bars @ 140 mm c/c from the edge to a distance of 2.5 m, at the water face of the slab. Provide two rings of 30 mm Φ wires to support these.

*Check for shear*

$$F_r = 146250 \text{ N}$$

$$\therefore \tau_v = \frac{F_r}{bd} = \frac{146250}{1000 \times 405} = 0.36 \text{ N/mm}^2$$

$$\frac{100 A_s}{bd} = \frac{100 \times 5387}{1000 \times 405} = 1.33.$$

Hence for M 25 concrete and for $\frac{100 A_s}{bd} = 1.33$, we get

$$\tau_c \approx 0.446 \text{ N/mm}^2$$

Hence the slab is safe in shear.

*Positive B.M.*: Using 20 mm Φ bars,

$d = 450 - 25 - 10 = 415$ mm for one layer and $415 - 20 = 395$ mm for the other layer.

Reinforcement for positive $M_r$ and $M_\theta$ is given by

$$A_{st} = \frac{1.07 \times 10^5 \times 1000}{130 \times 0.86 \times 393} = 2435.3 \text{ mm}^2.$$

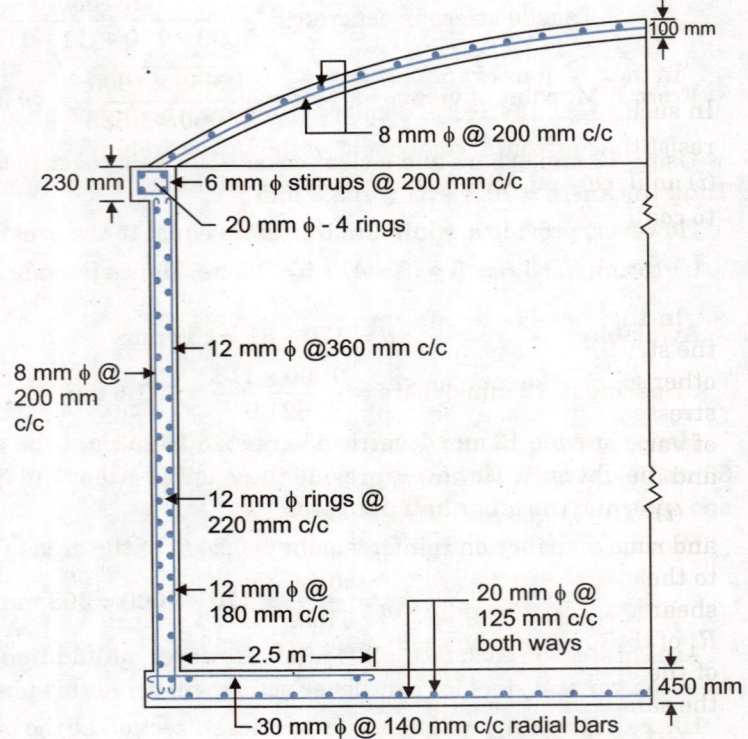

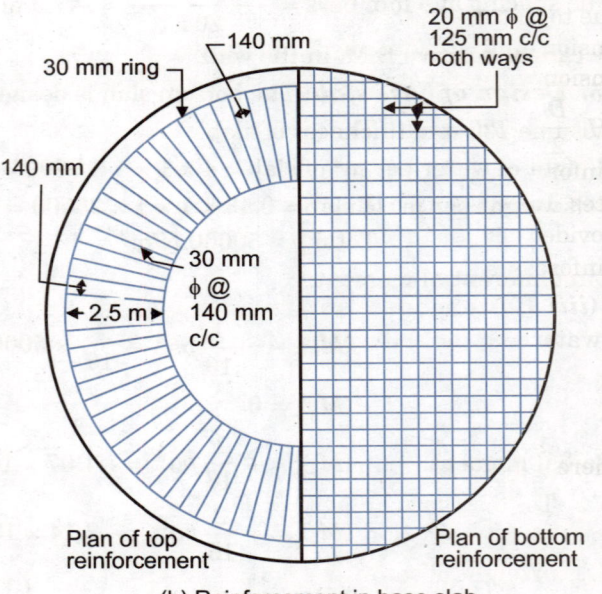

Fig. 22.12

$\therefore$ Spacing of 20 mm bars $= \dfrac{1000 \times 314}{2435.3} = 128.9$ mm. Hence provide 20 mm Φ bars @ 125 mm c/c in both the directions.

The details of reinforcement etc. are shown in Fig. 22.12.

**Note.** It will be seen that a plane slab at the bottom is highly uneconomical–requiring 45 cm thick slab and heavy reinforcement. A conical or domed bottom is preferred as used in Intze tanks.

# WATER TANKS-II : CIRCULAR AND INTZE TANKS

## 22.4. CIRCULAR TANK WITH DOMED BOTTOM AND ROOF

In the previous example, we have seen that a large circular tank with flat bottom is highly uneconomical. In such cases, therefore, domed bottom may be provided as shown in Fig. 22.13. The ring beams A and B resist the horizontal component of the thrust from the domes. The design of the tank is done in two stages: (i) analysis and design of various components on the membrane analysis (ii) effect of joint reactions due to continuity.

### 1. MEMBRANE ANALYSIS

In the membrane analysis, each component of the structure is assumed to act independent of the other so that the components carry only membrane stresses, without any bending moment. The design of various components is briefly discussed below, and illustrated in Examples 22.6 and 22.7.

(i) **Top Dome and Ring Beam A:** The top dome and ring beam are assumed to be freely connected to the cylindrical wall of the tank with the help of shear key. Corresponding to the rise $h_1$, the radius $R_1$ of the dome can be easily found, and the design of the dome and the ring beam is done exactly in the same manner as illustrated in Example 22.5.

(ii) **Cylindrical wall:** The walls of the tank are assumed to be free both at top as well as bottom. Due to this, the tank wall will be subjected to hoop tension only, without any B.M. The maximum hoop tension will occur at base, its magnitude being equal $wH . \dfrac{D}{2}$ kN/m height. The tank walls are adequately reinforced with horizontal rings provided at both faces. In addition to this, vertical reinforcement is provided on both the faces in the form of distribution reinforcement.

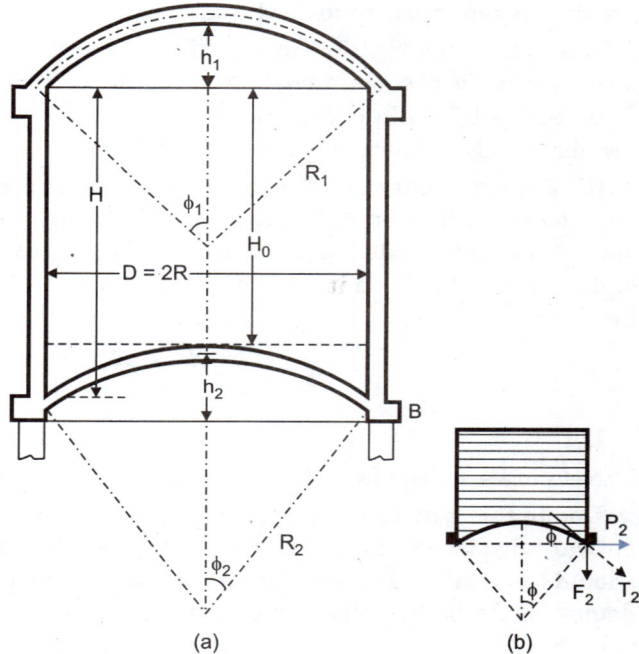

Fig. 22.13. Tank with domed top and bottom.

(iii) **Bottom dome and ring beam:** The bottom dome is subjected to vertical loads consisting of weight of water and the self weight. The weight of water over the surface of dome is given by

$$W_w = \left[ \frac{\pi}{4} D^2 H - \frac{\pi h_2^2}{3}(3 R_2 - h_2) \right] w \qquad \ldots(22.21)$$

where $R_2$ is the radius and $h_2$ is the rise of the bottom dome, and $w$ = unit weight of water ($\approx 9.80$ kN/m³).

Total surface area of bottom dome = $2\pi R_2 h_2$ ...(22.22)

∴ Weight of water per unit area $w_w = \dfrac{W w}{2 \pi R_2 h_2}$ ...(22.23)

Self weight of dome per unit area = $1 \times 1 \times t_2 \times \gamma_c$

where $\gamma_c$ = unit weight of concrete (= 25 kN/m³).

∴ Total $p_2$ per unit area of dome = $(w_w + \gamma_c . t_2)$

Total load $W_2$ = weight of water + self weight = $W_w + 2\pi R_2 h_2 t_2 \gamma_c$ ...(22.24)

Shearing force $F_2 = \dfrac{W_2}{\pi D}$ per unit length of periphery ...(22.25)

$$\text{Meridional thrust} = T_2 = \frac{W_2}{\pi D \sin \phi_2} = F_2 \operatorname{cosec} \phi_2 \qquad \ldots(22.26)$$

$$\text{Hoop stress at edge} = P_2 = T_2 \cos \phi_2 = F_2 \cot \phi_2 = \frac{W_2}{\pi D} \cot \phi_2 \qquad \ldots(22.27)$$

Knowing the hoop stress and meridional thrust, the dome can be designed.

## 2. EFFECTS OF JOINT REACTIONS DUE TO CONTINUITY

In the membrane analysis, every member is assumed to undergo deformation independent of others. Since the joints are continuous, these deformations are to be compatible. The effects of continuity of joint are thus taken into account by framing equations of continuity of deformations.

It is to be noted that continuity effect alters the hoop stresses near joints and also imposes meridional moments, while the meridional stresses are not affected and remain the same as in membrane state. The changes due to continuity are only of local nature and their effect die out soon. In the interior portion of members, only membrane stresses exist.

*(i) Junction between top dome and tank walls:* The top dome is *hinged* to the walls by means of shear key. Hence the horizontal displacement of the dome at its edge should be compatible with the displacement of the tank walls at its top. If $R_1$ is the radius of the top dome, $t_1$ its thickness, $\phi_1$ as its semicentral angle and $w_1$ is the U.D.L. on it, the horizontal *outward* displacement $\delta_{d1}$ of the freely supported dome is given by

$$\delta_{d1} = \frac{w_1 R_1^2 \sin \phi_1}{E t_1} \left[ \frac{1}{1 + \cos \phi_1} - \cos \phi_1 \right] \qquad \ldots(22.28)$$

Let $\qquad y_1$ = net outward displacement of top joint.

Net outward displacement $\delta_{nd1}$ of top dome = $y_1 - \delta_{d1}$

Due to this outward displacement $\delta_{nd1}$ of dome, the outward displacement $y_1$ of the ring beam and the outward displacement $y_1$ of the wall will exert inward horizontal reactions at the joint, the sun of which should be equal to the horizontal thrust of the dome caused by loads. Inward horizontal reaction ($H_{d1}$) of dome is given by the following equations

$$H_{d1} \cdot \frac{\alpha_1 R_1 d \sin^2 \phi_1}{E t_1} \left( K_1 + \frac{1}{K_1} \right) = \delta_{nd1} = y_1 - \delta_{d1} \qquad \ldots(22.29)$$

where $\quad \alpha_1^4 = 3 \left( \frac{R_1}{t_1} \right)^2 \quad$ and $\quad K_1 = 1 - \frac{1}{2\alpha_1} \cos \phi_1 \qquad \ldots(22.30)$

Inward horizontal reaction of ring beam is given by, $H_{b1} = \dfrac{y_1 \cdot E \cdot A_1}{(D/2)^2} = \dfrac{4 y_1 \, E A_1}{D_2} \qquad \ldots(22.31)$

where $A_1$ = cross-sectional area of beam.

Inward horizontal reaction ($H_{w1}$) of tank wall is given by

$$H_{w1} = 2\alpha^3 Z \cdot y_1 \qquad \ldots(22.32)$$

where $\quad \alpha^4 = \dfrac{12}{D^2 T^2} \quad$ and $\quad Z = \dfrac{E T^3}{12} \qquad \ldots(22.33)$

Outward horizontal thrust ($P_1$) of dome, due to external load is given by

$$P_1 = \frac{w_1 R_1}{1 + \cos \phi_1} \cdot \cos \phi_1 \qquad \ldots(22.34)$$

Hence we have $H_{d1} + H_{b1} + H_{w1} = P_1 \qquad \ldots(22.35)$

From the above equation, $y_1$ will be known. Hoop stresses in the top ring beam will be equal to $y_1 \, E/R$. The boundary condition at the top of the wall will be:

At $x = 0$, $y = y_1$ ...(i)    and    at $x = 0$, $EI \dfrac{d^2y}{dx^2}$ = B.M. = 0 ...(ii)

(ii) **Junction between bottom dome and tank wall:** The meridional thrust $T_2$ can be calculated either from Eq. 22.26 or from the following expression consisting of thrust due to (i) self weight = $\gamma_c \cdot t_2$ (ii) uniform radial pressure = $w \cdot H_0$ and (iii) radial water pressure varying from zero at the crown to $wh_2$ at the springings, where $t_2$ is the thickness of the dome in metres.

$$\therefore \quad T_2 = \underbrace{\dfrac{(\gamma_c\, t_2) R_2}{1 + \cos\phi_2}}_{(i)} + \underbrace{\dfrac{(w \cdot H_0) R_2}{2}}_{(ii)} + \underbrace{\dfrac{w \cdot R_2^2 (1 + \cos\phi_2 - 2\cos^2\phi_2)}{6(1 + \cos\phi_2)}}_{(iii)} \quad ...(22.36)$$

The horizontal thrust $(P_2)$ is given by $P_2 = T_2 \cdot \cos\phi_2$.

The outward horizontal displacement $\delta_{d2}$ of the edges of the dome is equal to the sum of the displacement due to the three loadings, and is given by

$$\delta_{d2} = \dfrac{\gamma_c \cdot t_2 R_2^2 \sin\phi_2}{t_2 \cdot E}\left[\dfrac{1}{1 + \cos\phi_2} - \cos\phi_2\right] - \dfrac{w \cdot H_0\, R_2^2 \sin\phi_2}{2 E t_2}$$

$$+ \dfrac{w \cdot R_2^3 \sin\phi}{6 E t_2 (1 + \cos\phi_2)}(2\cos 2\phi_2 + \cos\phi_2 - 3) \quad ...(22.37)$$

The *clockwise change in slope* at edges of a freely supported dome due to the above three loading is given by

$$\psi_{d2} = \dfrac{2(\gamma_c \cdot t_2) R_2 \sin\phi_2}{E t_2} + \text{zero} - \dfrac{w \cdot R_2^2 \sin\phi_2}{E t_2} \quad ...(22.38)$$

Similarly, considering the tank wall to be free at the bottom, the outward deflection, and clockwise slope of the wall as its bottom are given by

$$\delta_{w2} = \dfrac{w \cdot H \cdot R^2}{E T} \quad ...(22.39) \quad \text{and} \quad \psi_{w2} = \dfrac{\delta_{w2}}{H} \quad ...(22.40)$$

Let $y_2$ be the net outward deflection $\psi_2$ and be the net *clockwise slope* of the joint. The ring beam will thus move out by $y_2$ and rotate clockwise by $\psi_2$. The tank wall will have a net horizontal movement $\delta_{nw2} = y_2 - \delta_{w2}$ and a net clockwise rotation $\psi_{nw2} = \psi_2 - \psi_{w2}$. The dome will have a net horizontal movement $\delta_{nd2} = y_2 - \delta_{d2}$ and a net clockwise rotation $\psi_{nd2} = \psi_2 - \psi_{d2}$. Due to these movements and rotations, force and moments will be induced in the beam, tank wall and the dome, which along with the radial outward thrust of the dome should balance among themselves. Their values are as follows:

(a) *For displacement of dome*:

(i) Due to outward movement $\delta_{nd2}$ of dome, the thrust, induced so that there is no rotation of tangent, is given by

or $\quad (H_{d2})_\delta$ = Thrust stiffness × outward movement

or $\quad (H_{d2})_\delta = \dfrac{e t_2}{\alpha_2\, R_2\, K_2 \sin^2\phi_2} \cdot \delta_{nd2}$ per unit length (outward)  ...(22.41)

Corresponding moment is given by

$$(M_{d2})_\delta = \dfrac{E t_2}{2\alpha_2^2\, K_2 \sin\phi_2} \cdot \delta_{nd2} \text{ (anticlockwise)} \quad ...(22.42)$$

where $\quad \alpha_2^4 = 3\left(\dfrac{R_2}{t_2}\right)^2;\quad K_2 = 1 - \dfrac{1}{2\alpha_2}\cot\phi_2 \quad ...(22.43)$

(ii) The moment needed to cause clockwise rotation without any radial displacement of edges is given by

$$(M_{d2})_\psi = \text{moment stiffness} \times \text{rotation} = \dfrac{E R_2\, t_2}{4\alpha_2^3}\left(K_2 + \dfrac{1}{K_2}\right) \cdot \psi_{nd2} \quad ...(22.44)$$

Corresponding radial force is given by

$$(H_{d2})_\psi = \frac{Et_2}{2\alpha_2^2 K_2 \sin\phi_2} \cdot \psi_{nd2} \text{ (inward)} \qquad \ldots(22.45)$$

**(b) For displacement of wall**

(iii) The thrust needed to cause outward displacement $\delta_{nw2}$ without rotation is given by

$$(H_{w2})_\delta = \text{thrust stiffness} \times \text{displacement} = 4\alpha^3 Z \cdot \delta_{nw2} \qquad \ldots(22.46)$$

Corresponding moment is given by $(M_{w2})_\delta = 2\alpha^2 Z \cdot \delta_{nw2}$ (anticlockwise) $\qquad \ldots(22.47)$

The moment needed to cause clockwise rotation $\psi_{nw2}$ with out any radial moment is given by

$$(M_{w2})_\psi = \text{moment stiffness} \times \text{rotation} = 2\alpha Z \cdot \psi_{nw2} \qquad \ldots(22.48)$$

Corresponding radial force is given by $(H_{w2})_\psi = 2\alpha^2 Z \cdot \psi_{nw2}$ (inward) $\qquad \ldots(22.49)$

where $\alpha^4 = \dfrac{12}{D^2 T^2}$ and $Z = \dfrac{ET^3}{12}$ $\qquad \ldots(22.50)$

**(c) For displacement of ring beam:** For the outward displacement $y_2$ of the ring beam, the horizontal thrust per unit length is given by

$$(H_{b2})_\delta = \frac{y_2 \cdot E A_2}{R^2} = \frac{y_2 E b_2 d_2}{R^2} \text{ (outward)} \qquad \ldots(22.51)$$

where $b_2$ and $d_2$ are the width and depth of the ring beam.

For clockwise rotation $\psi_2$ of the ring beam, the clockwise moment is given by

$$(M_{b2})_\psi = \frac{E b_2 d_2^3}{12 R^2} \psi_2 \qquad \ldots(22.52)$$

Hence we have : $(H_{d2})_\delta - (H_{d2})_\psi + (H_{w2})_\delta - (H_{w2})_\psi + (H_{b2})_\delta = P_2 = T_2 \cos\phi_2$ $\quad\ldots(i)\quad\ldots(22.53)$

where $T_2$ meridional thrust given by Eq. 22.26

and $\qquad -(M_{d2})_\delta + (M_{d2})_\psi - (M_{w2})_\delta + (M_{w2})_\psi + (M_{b2})_\psi = 0$ $\qquad\ldots(ii)\quad\ldots(22.54)$

With the help of the above two simultaneous equations $y_2$ and $\psi_2$ will be known. Hence the moment at the bottom of the wall, $M_{w2}$ can be found from the relation

$$M_{w2} = (M_{w2})_\psi - (M_{w2})_\delta \text{ (clockwise)}$$

Hoop tension in wall at bottom $= H_{w2} = \dfrac{y_2 ET}{R}$ $\qquad \ldots(22.55)$

Similarly, hoop tension and moment in the bottom of dome and the ring beam can be calculated. For the tank wall, the boundary conditions at its bottom are as follows:

At $x = H$, $\qquad y = y_2 \quad\ldots(iii)\qquad$ At $\quad x = H$, $\quad \dfrac{dy}{dx} = \psi_2$ $\qquad\ldots(iv)$

The boundary condition equations (i), (ii), (iii) and (iv) can now be substituted in *Eq. 22.3* to get the values of the four constants A, B, C and D, and the wall can be completely analysed. However, in practice, it is sufficient to determine the hoop tension and B.M. at top and bottom of the wall, in which case it is not necessary to apply *Eq. 22.3*. See Example 22.7 for the analysis and design of such a tank.

**Design: Example 22.6.** **Design of circular tank with domed bottom and top: Membrane Analysis**

*Redesign the tank of example 22.5 if the bottom of the tank consists of a dome having a central rise of 2.2 m. The tank is supported on masonry tower.*

**Solution:** (Fig. 22.13).

1. **Dimensions of tank:** We have $D = 11.7$ m; $R = D/2 = 5.85$ m; $H = 4$ m.

For the top dome, $R_1 = 9.56$ m; $H_1 = 2$ m; $\phi_1 = 37.73°$; $\sin\phi_1 = 0.6119$; $\cos\phi_1 = 0.7908$.

For the bottom dome, the radius $R_2$ is given by

$$R^2 = (2R_2 - h_2) h_2, \text{ where } h_2 = 2.2 \text{ m or } (5.85)^2 = (2R_2 - 2.2)2.2$$

$\therefore \qquad R_2 = 8.88$ m; $\sin\phi_2 = \dfrac{5.85}{8.88} = 0.6588$; $\phi_2 = 41.21°$; $\cos\phi_2 = 0.7523$.

**2. Design of top dome and the top ring beam**: Since the design is being done by membrane analysis, the dimensions and reinforcement etc. of top dome and top ring beam will be the same as computed in example 22.5. The salient data are summarised below:

Maximum hoop stress = 186420 N/m² = 0.1864 N/mm²

Maximum meridional stress = 208190 N/m² = 0.2082 N/mm²

Thickness of top dome = $t_1$ = 10 cm = 0.1 m.

Reinforcement in top dome consists of 8 mm Φ bars @ 160 mm c/c in both the directions. For the top ring beam.

Horizontal component of $T$ = 16470 N/m.

$$\text{Hoop tension in beam} = 16470 \times \frac{11.7}{2} = 96349.5 \text{ N}$$

Width of beam = $b_1$ = 300 mm = 0.3.; Depth of beam = $d_1$ = 230 mm = 0.23 m.

Reinforcement consists of 4 rings of 20 mm Φ bars tied by 6 mm Φ wire stirrups @ 200 mm c/c.

**3. Design of tank walls**: Since the design is to be done on membrane analysis, the tank walls will be assumed to be free at top and bottom, and the tank walls will be subjected to purely hoop stresses.

$$\text{Maximum hoop-tension at base} = wH\frac{D}{2} = 9800 \times 4 \times \frac{11.7}{2} = 229320 \text{ N/m}$$

(against 131630 N/m obtained in Example 22.5).

$$\text{Area of rings} = 229320/130 = 1764 \text{ mm}^2$$

Area required at each face = 882 mm²

$$\text{Using 12 mm Φ bars, spacing} = \frac{1000 \times 113}{882} = 128 \text{ mm.}$$

Hence provide 12 mm Φ rings @ 120 mm c/c on each face.

$$\text{Actual area provided} = \frac{1000 \times 113}{120} \times 2 = 1883.3 \text{ mm}^2.$$

The above spacing can be increased at the top, since pressure varies linearly.

Using a tensile stress of 1.2 N/mm² for the combined section, thickness $T$ is given by

$$\frac{229320}{1000\,T + (11-1)1883.3} = 1.2 \text{ from which } T = 172 \text{ mm.}$$

Minimum thickness = $3H + 5 = (3 \times 4) + 5 = 17$ cm = 170 mm.

Hence provide 170 mm thickness throughout the height, though the thickness at the top can be reduced. Distribution reinforcement 0.24%

$$= \frac{0.24}{100} \times 170 \times 1000 = 408 \text{ mm}^2$$

Area at each face = 204 mm². Spacing of 8 mm Φ bars = $\frac{1000 \times 50.5}{204}$ = 247.5 mm.

However, provide 8 mm Φ vertical bars @ 240 mm c/c on both faces. Keep a clear cover of 25 mm.

**4. Design of bottom dome**: For the bottom dome,

$h_2 = 2.2$ m; $R_2 = 8.88$ m; $\sin\phi_2 = 0.6588$; $\cos\phi_2 = 0.7523$; $\phi_2 = 41.21°$. Let the thickness $t_2 = 20$ cm = 0.2 m.

Weight of water over the surface of dome is given by *Eq. 22.21*

$$W_w = \left[\frac{\pi}{4}D^2H - \frac{\pi h_2^2}{3}(3R_2 - h_2)\right] \times w = 9800\pi \left[\frac{(11.7)^2 \times 4}{4} - \frac{(2.2)^2}{3}(3 \times 8.88 - 2.2)\right] = 3000564 \text{ N}$$

Total surface area of dome = $2\pi R_2 h_2 = 2\pi \times 8.88 \times 2.2$ = 122.75 m².

Self weight = 122.75 × 0.2 × 25000 = 613750 N

∴ Total $W_2$ = 3000564 + 613750 = 3614314 N

Load $P_2$ per unit area = 3614314/122.75 = 29445 N/m²

Maximum hoop stress at centre $= \dfrac{p_2 R_2}{2t_2} = \dfrac{29445 \times 8.88}{2 \times 0.2} = 653679 \text{ N/m}^2 = 0.654 \text{ N/mm}^2$ (Safe)

Maximum meridional stress

$$= \dfrac{p_2 R_2}{t_2}\left(\dfrac{1 - \cos\phi_2}{\sin^2 \phi_2}\right) = \dfrac{29445 \times 8.88}{0.2}\left(\dfrac{1 - 0.7523}{0.6588 \times 0.6588}\right)$$

$$= 746129 \text{ N/m}^2 = 0.746 \text{ N/mm}^2 \text{ (safe)}$$

*Alternatively* shear force $F_2 = \dfrac{W_2}{\pi D} = \dfrac{3614314}{\pi \times 11.7} = 98331$ N/m

Meridional thrust $T_2 = \dfrac{F_2}{\sin \phi_2} = \dfrac{98331}{0.6588} = 149258$ N/m

$\therefore$ Meridional stress $= \dfrac{149258}{1000 \times 200} = 0.746$ N/mm$^2$

Stresses are within safe limit. However, provide minimum reinforcement 0.24%

$$= 2.4 \times 200 = 480 \text{ mm}^2 \text{ in each direction.}$$

Using 8 mm $\Phi$ bars, $A_\Phi = 50$ mm$^2$

$\therefore$ Spacing $= \dfrac{1000 \times 50}{480} = 104$ mm

Hence provide 8 mm $\Phi$ rods @ 100 mm c/c in both directions.

### 5. Design of bottom ring beam

Meridional thrust

$$= 746128 \times 0.2 \times 1$$
$$= 149225.6 \approx 149226 \text{ N/m.}$$

Horizontal component

$$= 149226 \cos \phi_2$$
$$= 149226 \times 0.7523$$
$$= 112262.7 \approx 112263 \text{ N}$$

(Alternatively,

$$P_2 = F_2 \cot \phi_2 = 98331 \cot 14.21°$$
$$= 112262)$$

Hoop tension, trying to rupture the beam

$$= 112263 \times \dfrac{11.7}{2} = 656738.6 \text{ N}$$

Area of steel required

$$= 656739/130 = 5051.8 \text{ mm}^2$$

Using 30 mm $\Phi$ bars $A_\Phi = 706.86$ mm$^2$

Number of rings required

$$= 5052/706.86 = 7.15.$$

However, provide 8 rings of 30 mm $\Phi$ rod.

Actual $A_{sh} = 706.86 \times 8 = 5654.88$ mm$^2$

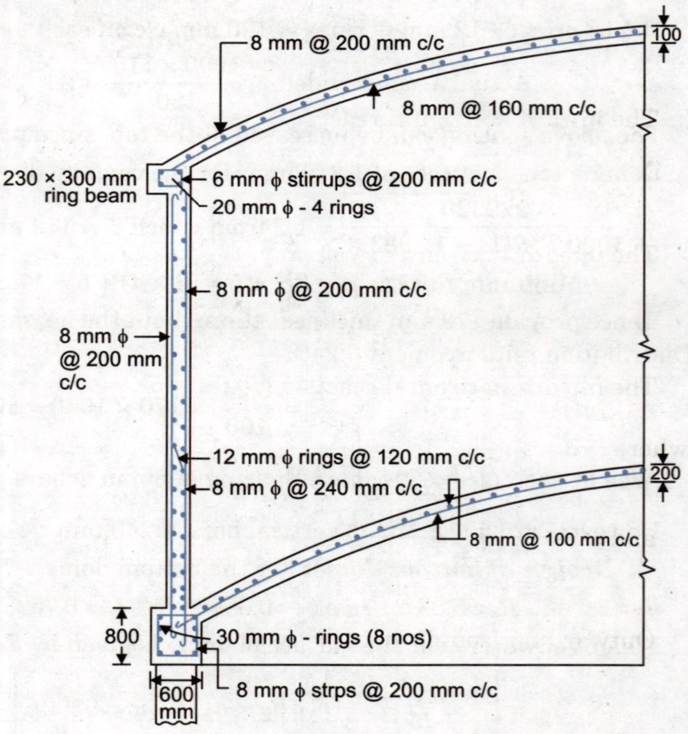

Fig. 22.14

Equivalent area of composite section of beam of area of cross-section

$$A_{sh} = A_2 + (m-1) A_{sh} = A_2 + (11-1) 5654.88 = A_2 + 56548$$

Allowing a stress of 1.2 N/mm² in the composite section, we have

$$\frac{656739}{A_2 + 56548} = 1.2 \qquad \text{From which } A_2 = 490734 \text{ mm}^2$$

Hence provide $d_2 = 850$ mm and $b_2 = 600$ mm, providing $A_2 = 510000$ mm².

Provide 8 mm Φ stirrups @ 200 mm c/c to tie the rings. Alternatively, the 8 mm Φ vertical bars provided @ 200 mm c/c on the inner face of the tank wall may be taken around the rings. The details of reinforcement etc. are shown in Fig. 22.14.

**Design: Example 22.7. Design of circular tank with domed bottom and top : Effect of continuity.**

*Modify the design of previous example, taking into account the effect of continuity of joints.*

**Solution:**

**1. *Top joint*:** The outward displacement of top dome is given by *Eq. 22.28* :

$$\delta_{d1} = \frac{3900(9.56)^2 \times 0.6119}{E \times 0.1}\left[\frac{1}{1+0.7908} - 0.7908\right] = -\frac{506850}{E}$$

Let $y_1$ be the outward movement of the top joint. The net outward displacement $\delta_{nd1}$ of top dome will be

$$= y_1 - \delta_{d1} = y_1 + \frac{506850}{E}$$

Now

$$\alpha_1^4 = 3\left(\frac{R_1}{t_1}\right)^2 = 3\left(\frac{9.56}{0.1}\right)^2 = 27418; \quad \alpha_1 = 12.87;$$

$$K_1 = 1 - \frac{1}{2\alpha_1}\cot\phi_1 = 1 - \frac{1}{1 \times 12.87} \times \cot 37.73° = 0.949.$$

$$K_1 + \frac{1}{K_1} = 0.949 + \frac{1}{0.949} = 2.003$$

The inward horizontal reaction $H_{d1}$ of dome is given by *Eq. 22.29*.

$$H_{d1}\frac{12.87 \times 9.56(0.6119)^2}{E(0.1)} \times 2.003 = y_1 + \frac{506850}{E} \quad \text{or} \quad H_{d1} = \left(\frac{y_1 E}{922.7} + 549.3\right)\text{N/m} \quad ...(i)$$

The *inward* horizontal reaction of ring beam is given by *Eq. 22.31*.

$$H_{b1} = \frac{4y_1 E(0.3 \times 0.23)}{(11.7)^2} = 20.16 \times 10^{-4} y_1 E \quad ...(ii)$$

The *inward* horizontal reaction $H_{w1}$ of tank wall is given by *Eq. 22.32*,

where $\alpha = \left(\frac{12}{D^2 T^2}\right)^{1/4} = \left[\frac{12}{(11.7)^2(0.17)^2}\right]^{1/4} = 1.3197$

and $Z = \frac{ET^3}{12} = \frac{E(0.17)^3}{12} = 4.094 \times 10^{-4} E$

∴ $H_{w1} = 2(1.3197)^3 \times 4.094 \times 10^{-4} E y_1 = 18.8 \times 10^{-1} y_1 E \quad ...(iii)$

Outward horizontal thrust $P_1$ of dome due to external load is given by *Eq. 22.34*:

$$P_1 = \frac{3900 \times 9.56}{1 + 0.7908} \times 0.7908 = 16464 \text{ N/m} \quad ...(iv)$$

Hence from *Eq. 22.35* we have : $H_{d1} + H_{b1} + H_{w1} = P_1$

or $\left(\frac{y_1 E}{922.7} + 549.3\right) + (20.16 \times 10^{-4} y_1 E) + (18.8 \times 10^{-4} y_1 E) = 16464.$

From which $y_1 = \dfrac{319.7 \times 10^4}{E}$

Hoop stress in ring beam = $\dfrac{y_1 E}{R} = \dfrac{319.7 \times 10^4}{(11.7/2)} = 54.65 \times 10^4$ N/m$^2$ = 0.5465 N/mm$^2$ = $54.65 \times 10^4$ N/m$^2$

Hoop tension in ring beam = $54.65 \times 10^4 \times 0.3 \times 0.23 = 37708.5 \approx 37710$ N

(Against 96330 N found by membrane analysis).

Hoop tension in dome = $54.65 \times 10^4 \times 0.1 \times 1 = 54650$ N/m

The boundary conditions at the top of the wall will be

At $x = 0$, $\quad y_1 = \dfrac{319.7 \times 10^4}{E}$  ...(a) $\qquad$ At $x = 0$, $\quad \dfrac{d^2 y}{dx^2} = 0$  ...(b)

**2. Bottom Junction:** From previous example, meridional thrust $T_2 = 149258$ N/m. Alternatively, $T_2$ can be computed from *Eq. 22.36*.

$$T_2 = \dfrac{25000 \times 0.2 \times 8.88}{1 + 0.7523} + \dfrac{9800(4 - 2.2)8.88}{2} + \dfrac{9800(8.88)^2 [1 + 0.7523 - 2(0.7523)^2]}{6(1 + 0.7523)}$$

$= 25338 + 78322 + 45599 = 149259$ N/m

The horizontal thrust $P_2$ is given by $P_2 = 149259 \times 0.7523 = 112287$ N

The outward horizontal displacement of dome $\delta_{d2}$ is given by *Eq. 22.37*:

$$\delta_{d2} = \dfrac{25000(8.88)^2 \times 0.6588}{E}\left[\dfrac{1}{1+0.7523} - 0.7523\right] - \dfrac{9800(4-2.2)(8.88)^2 \times 0.6588}{2E(0.2)}$$

$$+ \dfrac{9800(8.88)^3 \times 0.6588}{6E(0.2)(1+0.7523)}\{2\cos 82.42° + 0.7523 - 3\}$$

$= -\dfrac{23.59 \times 10^4}{E} - \dfrac{229.1 \times 10^4}{E} - \dfrac{426.5 \times 10^4}{E} = -\dfrac{679.2 \times 10^4}{E}$ metres.

The clockwise change in slope of dome edges is given by *Eq. 22.38*:

$$\psi_{d2} = \dfrac{2 \times 25000 \times 0.2 \times 8.88 \times 0.6588}{E(0.2)} - \dfrac{9800(8.88)^2 \times 0.6588}{E(0.2)} = -\dfrac{225.3 \times 10^4}{E}$$ radians.

The outward deflection and clockwise slope of tank wall are given by *Eqs. 22.39* and *22.40*

$\delta_{w2} = \dfrac{9800 \times 4(5.85)^2}{E(0.17)} = \dfrac{789.1 \times 10^4}{E}$ metres; $\quad \psi_{w2} = \dfrac{789.1 \times 10^4}{4E} = \dfrac{197.3 \times 10^4}{E}$ radians.

Hence the net displacements and rotations of the dome and wall are:

$\delta_{nd2} = y_2 - \delta_{d2} = y_2 + \dfrac{679.2 \times 10^4}{E}$; $\quad \psi_{nd2} = \psi_2 - \psi_{d2} = \psi_2 + \dfrac{225.3 \times 10^4}{E}$

$\delta_{nw2} = y_2 - \delta_{w2} = y_2 - \dfrac{789.1 \times 10^4}{E}$; $\quad \psi_{nw2} = \psi_2 - \psi_{w2} = \psi_2 - \dfrac{197.3 \times 10^4}{E}$

The forces and moments induced in the tank wall, dome and bottom ring beam are given by *Eqs. 22.41* to *22.52*. The salient parameters required to compute these are as follows:

$$\alpha_2^4 = 3\left(\dfrac{R_2}{t_2}\right)^2 = 3\left(\dfrac{8.88}{0.20}\right)^2 = 5914 ; \quad \alpha_2 = 8.77$$

$$K_2 = 1 - \dfrac{1}{2 \times 8.77}\cot 41.21° = 0.935 ; \quad K_2 + \dfrac{1}{K_2} = 2.005$$

$$\alpha^4 = \dfrac{12}{D^2 T^2} = \dfrac{12}{(11.7 \times 0.17)^2} = 3.033 ; \quad \alpha = 1.3197$$

$$Z = \frac{ET^3}{12} = \frac{(0.17)^3}{12} E = 4.094 \times 10^{-4} E$$

$\therefore \quad (H_{d2})_\delta = \dfrac{E(0.2)}{8.77 \times 8.88 \times 0.935 \sin^2 41.21°}\left[y_2 + \dfrac{679.2 \times 10^4}{E}\right]$     From (22.41)

or $\quad (H_{d2})_\delta = 63.28 \times 10^{-4} y_2 E + 42963$    (outward)

$\quad (M_{d2})_\delta = \dfrac{E(0.2)}{2(8.77)^2 \times 0.935 \sin 41.21°}\left[y_2 + \dfrac{679.2 \times 10^4}{E}\right]$     From (22.42)

$\quad\quad\quad\quad = 21.11 \times 10^{-4} y_2 E + 14336$   (Anticlockwise)      ...(ii)

$\quad (M_{d2})_\psi = \dfrac{E(8.88 \times 0.2)}{4(8.77)^3}(2.005) \times \left\{\psi_2 + \dfrac{225.3 \times 10^4}{E}\right\}$     From (22.44)

or $\quad (M_{d2})_\psi = 13.2 \times 10^{-4} \psi_2 E + 2973$   (clockwise)      ...(iii)

$\quad (H_{d2})_\psi = \dfrac{E(0.20)}{2(8.77)^2 \times 0.935 \sin 41.21°}\left\{\psi_2 + \dfrac{225.3 \times 10^{-4}}{E}\right\}$     From (22.45)

or $\quad (H_{d2})_\psi = 21.11 \times 10^{-4} \psi_2 E + 4755$   (Inward)      ...(iv)

$\quad (H_{w2})_\delta = 4(1.3197)^3 \times 4.094 \times 10^{-4} E \left\{y_2 - \dfrac{789.1 \times 10^4}{E}\right\}$     From (22.46)

or $\quad (H_{w2})_\delta = 37.64 \times 10^{-4} y_2 E - 29700$   (outward)      ...(v)

$\quad (M_{w2})_\delta = 2(1.3197)^2 \times 4.094 \times 10^{-4}\left\{y_2 - \dfrac{789.1 \times 10^4}{E}\right\}$     From (22.47)

or $\quad (M_{w2})_\delta = 14.26 \times 10^{-4} y_2 E - 11253$   (anticlockwise)      ...(vi)

$\quad (M_{w2})_\psi = 2 \times 1.3197 \times 4.094 \times 10^{-4} E\left\{\psi_2 - \dfrac{197.3 \times 10^2}{E}\right\}$     From (22.48)

or $\quad (M_{w2})_\psi = 10.81 \times 10^{-4} \psi_2 E - 2132$   (clockwise)      ...(vii)

$\quad (H_{w2})_\psi = 2(1.3197)^2 (4.094 \times 10^{-4} E)\left\{\psi_2 - \dfrac{197.3 \times 10^4}{E}\right\}$     From (22.49)

$\quad (H_{w2})_\psi = 14.26 \times 10^{-4} \psi_2 E - 2814$   (Inward)      ...(vi)

$\quad (H_{b2})_\delta = \dfrac{y_2 E(0.6 \times 0.8)}{(5.85)^2}$     From (22.51)

or $\quad (H_{b2})_\delta = 140.26 \times 10^{-4} y_2 E$   (outward)      ...(vii)

$\quad (M_{b2})_\psi = E\psi_2 \dfrac{0.6(0.8)^3}{12(5.85)^2}$     From (22.52)

or $\quad (M_{b2})_\psi = 7.48 \times 10^{-4} \psi_2 E$   (clockwise).

For the balance of horizontal forces, we have from *Eq. 22.53.*

$(H_{d2})_\delta - (H_{d2})_\psi + (H_{w2})_\delta - (H_{w2})_\psi + (H_{b2})_\delta = P_2.$

$\therefore \quad [63.28 \times 10^{-4} y_2 E + 42963] - [21.11 \times 10^{-4} \psi_2 E + 4755] + [37.64 \times 10^{-4} y_2 E - 29700]$
$\quad\quad\quad\quad\quad\quad\quad\quad\quad\quad\quad\quad - [14.26 \times 10^{-4} \psi_2 E - 2814] + [140.26 \times 10^{-4} y_2 E] = 112287$

or $\quad 241.18 y_2 E - 35.37 \psi_2 E = 100965 \times 10^4$

or $\quad y_2 E - 0.1467 \psi_2 E = 418.6 \times 10^4$      ...(A)

Similarly, for balance of moments, we have from *Eq. 22.54*

$-(M_{d2})_\delta + (M_{d2})_\psi - (M_{w2})_\delta + (M_{w2})_\psi + (M_{b2})_\psi = 0$

or $\quad -[21.11 \times 10^{-4} y_2 E + 14336] + [13.2 \times 10^{-4} \psi_2 E + 2973] - [14.26 \times 10^{-4} y_2 E - 11253]$
$$+ [10.81 \times 10^{-4} \psi_2 E - 2132] + [7.48 \times 10^{-4} \psi_2 E] = 0$$
or $\quad -35.37 y_2 E + 31.49 \psi_2 E - 2242 \times 10^4 = 0$
or $\quad y_2 E - 0.8903 \psi_2 E = -63.4 \times 10^4$ ...(B)

Solving A and B, we get $\quad y_2 = \dfrac{513.7 \times 10^4}{E}$ and $\psi_2 = \dfrac{648.2 \times 10^4}{E}$

Hence the boundary condition equations at the bottom of wall are :

At $x = H$, $\quad y = y_2 = \dfrac{513.7 \times 10^4}{E}$. and $x = H$, $\dfrac{dy}{dx} = \psi_2 = \dfrac{648.2}{E} \times 10^4$

Applying equations (a), (b), (c), (d), in *Eq. 22.3*, the four constants A, B, C, and D can be evaluated, and then B.M. and hoop tension at any location in the wall can be easily determined.

### 3. Hoop tension and B.M. at joints

Hoop tension in the wall at top $= \dfrac{y_1 E \cdot t}{R} = \dfrac{319.7 \times 10^4 (0.17)}{5.85} = 92904$ N/m. ...(i)

Hoop tension in the wall at bottom $= \dfrac{y_2 Et}{R} = \dfrac{513.7 \times 10^4 (0.17)}{5.85} = 149280$ N/m ...(ii)

(Against 229320 in Example 22.6)

Clockwise moment in the wall at base $= M_w = (M_{w2})_\psi - (M_{w2})_\delta$
$= \{10.81 \times 10^{-4} \psi_2 E - 2132\} - \{14.26 \times 10^{-4} y_2 E - 11253\}$
$= 8803$ N-m/m (clockwise)

Hoop tension in bottom dome at edges $= \dfrac{513.7 \times 10^4}{5.85} (0.20) = 175624$ N/m.

Clockwise moment in the dome at edge $= M_d = (M_{d2})_\psi - (M_{d2})_\delta$
$= \{13.2 \times 10^{-4} \psi_2 E + 2973\} - \{21.11 \times 10^{-4} y_2 E + 14336\}$
$= -13651$ N-m/m (*i.e.* clockwise)

Hoop tension in bottom ring beam $= \dfrac{y_2 E}{R} \cdot b_2 d_2 = \dfrac{513.7 \times 10^4}{5.85} (0.6 \times 0.8) = 421497$ N.

(As against 656733) N found in the previous example)

Clockwise radial moment in bottom ring beam $= (M_{b2})_\psi = 7.48 \times 10^{-4} \psi_2 E = 4849$ N-m/m.

B.M. in ring beam $= M_{b2} \left(\dfrac{D}{2}\right) = 4849 \times 5.85 = 28366.5$ N-m.

Thus, the bottom edge of the wall and the outer edges of the bottom dome are subjected to considerable bending moment. Since the B.M. induced due to joint reactions is of localised nature, it dies out quickly, and provision for the joint reactions is made only for a distance of $0.76\sqrt{Rt}$ in the wall and $0.76\sqrt{R_2 t_2}$ in the dome.

The distribution of B.M. and hoop tension can be found by the solution of Eq. 22.3. However, it is found that maximum values are generally obtained at the bottom of the wall. Hence it may not always be necessary to solve *Eq. 22.3*.

### 4. Design of components (a) Top dome

Meridional thrust at edges = 16464 N/m.

Meridional stress $= \dfrac{16464}{1000 \times 100} = 0.16$ N/mm$^2$ (safe)

Hoop tension at edge = 54650 N/m.; Hoop stress $= \dfrac{54650}{1000 \times 100} = 0.55$ N/mm$^2$ (safe)

The entire hoop tension is to be resisted by reinforcement.

$$A_s = \frac{54650}{130} = 420.4 \text{ mm}^2 \text{ (Minimum } A_s = 3 \times 100 = 300 \text{ mm}^2 \text{ in each direction)}$$

Using 8 mm Φ bars, spacing = $\frac{1000 \times 50.3}{420.4}$ = 119.6 mm.

Hence provide 8 mm Φ bars @ 100 mm c/c both ways

(b) **Top Ring beam**: Hoop tension = 37710 N;

∴ Hoop steel = 37710/115 = 328 mm². Hence provide 4 rings of 12 mm Φ bars and tie them with 6 mm Φ wire stirrups @ 200 mm c/c.

(c) **Tank wall**

Hoop tension at bottom = 149280 N/m.

B.M. at bottom = 8803 N-m/m (clockwise).

Area of hoop reinforcement = 149280/130 = 1148.3 mm²

Area of ring on each face = 574.2 mm².

Providing 12 mm Φ ring, spacing = $\frac{1000 \times 113}{574.2}$ = 196.8 mm.

Hence provide 12 mm Φ rings @ 190 mm c/c on each face.

Hoop tension in the wall at top = 92904

∴ $A_{sh}$ = 92904/130 = 714.7 mm².

Area of rings on each face = 357.3 mm².

Spacing of 12 mm Φ rings = $\frac{1000 \times 113}{357.3}$ = 316.2 mm *say* 300 mm c/c.

From B.M. point of view, $d = \sqrt{\frac{8803 \times 1000}{1000 \times 1.535}}$ = 75.72 mm.

Minimum thickness = 3H + 5 = 17 cm = 170 mm. Hence provide T = 17 cm = 170 mm.

Using a clear cover of 25 mm and using 12 mm Φ vertical bars, available

$$d = 170 - 25 - 6 = 139 \text{ mm}; \quad A_{st} = \frac{8803 \times 1000}{130 \times 0.86 \times 139} = 566.5 \text{ mm}^2.$$

Spacing of 12 mm Φ bars = $\frac{1000 \times 113}{566.5}$ = 199.5 mm

However, provide 12 mm Φ vertical bars @ 190 mm c/c at the water face. These bars are required for a distance = $0.76\sqrt{5.85 \times 0.17}$ = 0.76 m only. However, provide these for 1 m height. Above this height, curtail half the bars and continue the remaining half upto the top to act as distribution reinforcement.

At the outer face provide distribution reinforcement, in the form of vertical bars, of 8 mm bars @ 200 mm c/c. This will take care of positive B.M. in the wall.

(d) **Bottom dome**

Meridional thrust in the dome = 149258 N/m.

Hoop tension in bottom dome at edges = 175624 N-m/m

Meridional moment in the dome at edges = – 13651 N-m/m

(*i.e.* anticlockwise, causing tension on water side)

Meridional stress = $\frac{149258}{1000 \times 200}$ = 0.75 N/mm² (safe)

$$\text{Hoop stress} = \frac{175624}{1000 \times 200} = 0.88 \text{ N/mm}^2 \text{ (safe)}$$

$$\text{Area of Hoop steel} = 175624/130 = 1351 \text{ mm}^2.$$

$$\text{Using 16 mm } \Phi \text{ bar hoops, spacing} = \frac{1000 \times 201}{1351} = 148.8 \text{ say } 140 \text{ mm}.$$

The effect of hoop tension will be only upto a length $0.76\sqrt{R_2 t_2} = 0.76\sqrt{8.88 \times 0.2} \approx 1$ mm. Hence provide total 8 hoops in one metre length.

For the B.M., $\quad d = \sqrt{\dfrac{13651 \times 1000}{1000 \times 1.535}} = 94.3$ mm.

Actual total $D = 200$ mm. Using 12 mm $\Phi$ bars and a clear cover of 25 mm, available
$$d = 200 - 25 - 6 = 169 \text{ mm}.$$

$$\therefore \quad A_{st} = \frac{13651 \times 1000}{130 \times 0.86 \times 169} = 722.5 \text{ mm}^2.$$

Using 16 mm $\Phi$ bars, spacing $= \dfrac{1000 \times 201}{722.5} = 278.2$ mm.

Hence provide 16 mm $\Phi$ bars @ 260 mm c/c in the direction of meridion. These bars may be curtailed at a distance of 1 m from the edge.

Though the meridional stress (= 0.75 N/mm²) is within safe limit, and the hoop stress at the centre (= 0.654 N/mm², Example 22.6) is also within safe limits. Provide a minimum reinforcement @ 0.24% in both the direction.

$$\therefore \quad A_s = \frac{0.24}{100} \times 1000 \times 200 = 480 \text{ mm}^2 \text{ in each direction}.$$

$$\therefore \quad \text{Spacing of 8 } \Phi \text{ mm bar} = \frac{1000 \times 50}{480} = 104.2 \text{ mm}.$$

∴ Provide 8 mm $\Phi$ rods @ 100 mm c/c in both directions.

### (e) Bottom Ring beam

Hoop tension = 421500 N.

∴ Hoop steel

$A_{sh} = 421500/130 = 3242.3$ mm²

Using 30 mm $\Phi$ bars,

$A_\Phi = 706.9$ mm²

∴ Number of rings required 3242.3/706.9

$= 4.6 \approx 5.$

Actual $A_{sh}$ provided

$= 706.9 \times 5 = 3534.5$ mm².

Stress in composite section

$$= \frac{421500}{(800 \times 600) + (10 \times 3534.5)}$$

$= 0.82$ N/mm² (safe)

Continue the reinforcement of wall (*i.e.* 12 mm $\Phi$ @ 150 mm c/c) in the ring beam, to act as stirrups. The details of reinforcement etc. are shown in Fig. 22.15.

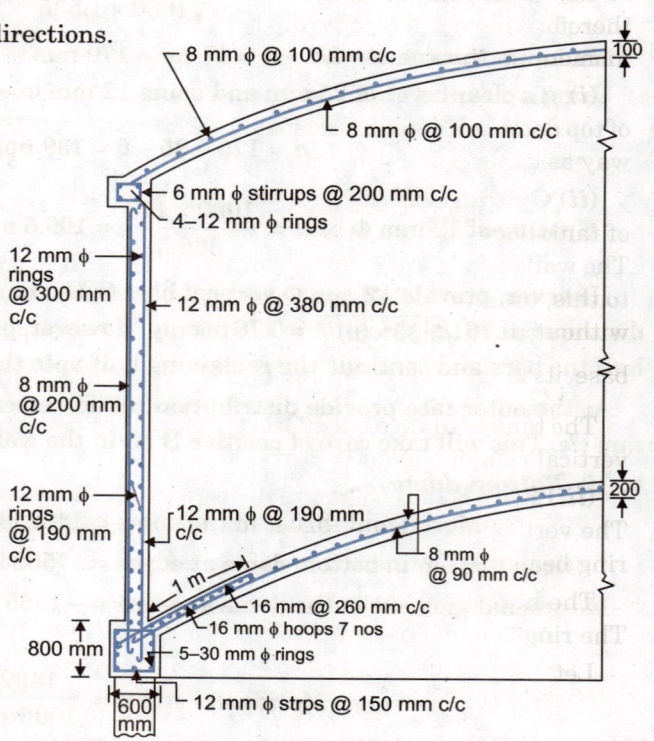

Fig. 22.15

## 22.5. INTZE TANK

In the previous articles, we have analysed circular tanks with flat bottom as well as with domical bottom. In the flat bottom, the thickness and reinforcement is found to be heavy. In the domed bottom, though the thickness and reinforcement in the dome is normal, the reinforcement in the ring beam is excessive. In cases of large diameter tanks, an economical alternative would be to reduce its diameter at its bottom by conical dome, as shown in Fig. 22.16. Such a tank is known as Intze tank and is very commonly used. The main advantage of such a tank is that the outward thrust from the top of the conical part is resisted by the ring beam $B_3$ while the difference between the inward thrust from the bottom of conical dome and the outward thrust from the bottom dome are resisted by ring beam $B_2$. The proportions of the conical dome and the bottom dome are so arranged that the outward thrust from bottom dome balances the inward thrust due to the conical dome.

The most economical dimensions of Intze tank have been obtained by using 'systems approach'. The overall economy of a design depends on the economy reliable both in the superstructures and foundation. Since the design of tank superstructures and foundation are different. It is convenient to deal with them separately.

Fig. 22.16 shows suitable proportions for Intze tank with internal diameter $D$. The volume of water stored in the tank with the above proportions is $0.585 D^3$. In general, the volume of water stored is given by

$$V = \frac{\pi}{4} D^2 h + \frac{\pi h_0}{12} (D^2 + D_0^2 + DD_0) - \frac{\pi h_2^2}{3} (3 R_2 - h_2) \qquad ...(22.56)$$

From economical considerations, the inclination of the conical dome should be 50° to 55° with the horizontal.

The design of such a tank is done in two stages:
(i) Membrane analysis  (ii) Analysis taking into account continuity effect at joints.

### MEMBRANE ANALYSIS

In the membrane analysis, the members are assumed to act independent of the others. The members are therefore subjected to only direct stresses and no bending moment is introduced.

(i) **Top dome and Top Ring Beam $B_1$:** The design of top dome and top ring beam is done exactly in the same way as explained in the previous article.

(ii) **Cylindrical portion of tank:** Let the diameter of tank be $D$ and the height of cylindrical portion be $H$. The walls are assumed to be free at top and bottom. Due to this, tank walls will be subjected to hoop tension only without any B.M. Maximum hoop tension will occur at base, its magnitude being equal to $wh \dfrac{D}{2}$ per unit height.

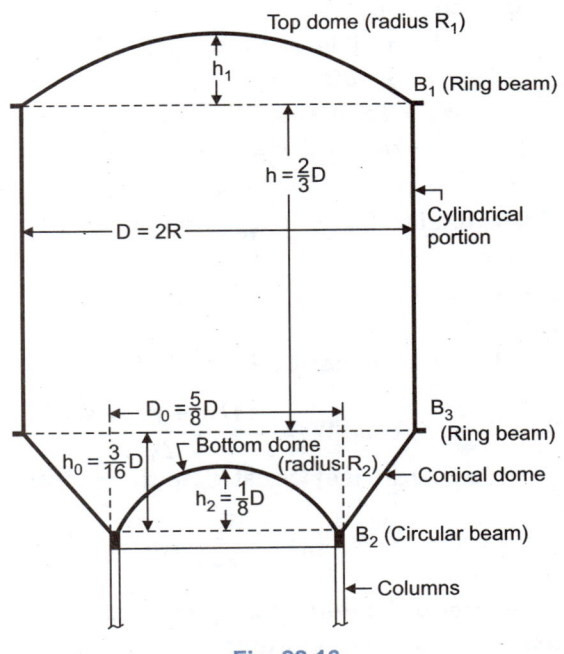

Fig. 22.16

The tank walls are adequately reinforced with horizontal rings provided at both faces. In addition to this, vertical reinforcement is provided on both the faces in the form of distribution reinforcement.

(iii) **Ring Beam $B_3$ at the junction of Cylindrical wall and Conical dome:** The vertical load at the junction of wall with conical dome is transferred to the ring beam ($B_3$) by meridional thrust in the conical dome.

The horizontal component of this thrust causes hoop tension at the junction. The ring beam is provided to take up this hoop tension.

Let  $W$ = load transmitted through tank wall, at the top of conical dome, per unit length
 $\phi_0$ = Inclination of conical dome with vertical.
 $T$ = meridional thrust in conical dome, at the junction.

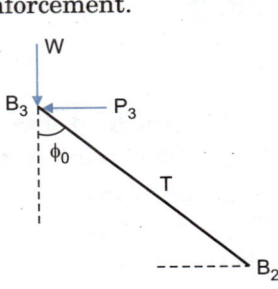

Fig. 22.17

Horizontal component of this is given by
$$P_W = W \tan \phi_0 \qquad \ldots[22.57(a)]$$
Horizontal force $P_w$ caused due to water pressure at top of conical dome is given by
$$P_w = w \cdot h \cdot d_3 \qquad \ldots[22.57(b)]$$
where $h$ = depth of water upto centre of ring beam and $d_3$ is the depth of the ring beam.

Hence hoop tension trying to burst the beam $B_3$ is given by
$$P_3 = (P_W + P_w)\frac{D}{2} \qquad \ldots(22.58)$$

It is more economical to keep lesser depth ($d_3$) and more width for the ring beam. Greater width may serve as a walk way around the tank. Reinforcement in the form of ring is provided to take up full hoop tension $P_3$. The tensile stress in the composite section should be within permissible limit of 1.2 N/mm².

*(iv) Conical dome:* The conical dome is subjected to both meridional thrust as well as hoop tension.

*(a) Meridional thrust:* The meridional thrust in the conical dome is due to vertical forces (weights) transferred to it at its base. The total load consists of

(*i*) Weight of top dome, cylindrical wall etc.
(*ii*) Weight of water
(*iii*) Self weight.

Let $W$ = weight of top dome, cylindrical wall etc. per metre length.

$W_w$ = Total weight of water on the conical dome.

Let $D$ and $D_0$ be the diameters at the top and bottom of the conical dome and $h_0$ be the height of conical dome.

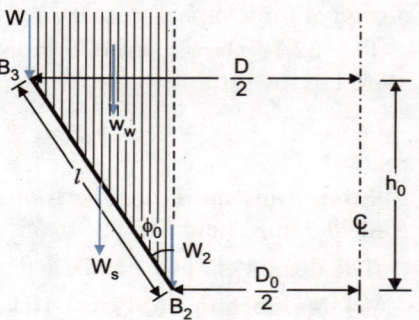

Fig. 22.18

The total self weight $W_s$ of the conical dome of slanting length $l$ and thickness $t_0$ given by
$$W_s = \left[\pi\left(\frac{D + D_0}{2}\right) \cdot l \cdot t_0\right] \gamma_c \qquad \ldots(22.59)$$

Hence total vertical load = $[\pi DW + W_w + W_s]$

$\therefore$ Load $W_2 = \dfrac{\pi DW + W_w + W_s}{\pi D_0}$ per unit length $\qquad \ldots(22.60)$

Hence meridional thrust $T_0$ in the conical dome is given by
$$T_0 = \frac{W_2}{\cos \phi_0} \text{ per unit length.} \qquad \ldots(22.61)$$

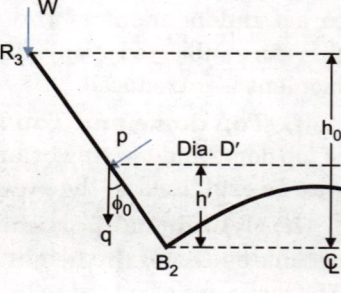

Fig. 22.19

*(b) Hoop tension:* Due to water pressure and self weight, the conical dome will be subjected to hoop tension. Let $P$ be the water pressure at any height $h'$ above the base of conical dome, and let $D'$ be the diameter of the conical dome at that depth. The water pressure $P$ will act normal to the inclined slab surface. Let $q$ be the weight of conical slab per square metre of the surface area. Then the hoop tension $P_0'$ at height $h'$ above base is given by
$$P_0' = \left(\frac{p}{\cos \phi_0} + q \tan \phi_0\right)\frac{D'}{2} \qquad \ldots(22.62)$$

With the help of the above formula, hoop tension at the top, middle and base of the conical dome can be found. Steel reinforcement is provided to bear the whole of hoop tension.

*(v) Bottom dome:* Bottom dome develops compressive stresses both meridionally as well as along hoops, due to weight of water supported by it and also due to its own weight. Let $H_0$ be the total depth of water above the edges of the dome, (Fig. 22.20). The weight of water above the surface of the dome is given by

$$W_0 = \left[\frac{\pi}{4} D_0^2 H_o - \frac{\pi h_2^2}{3}(3R_2 - h_2)\right] \times w \qquad \ldots(22.63)$$

where $R_2$ is the radius and $h_2$ is the rise of the bottom dome.
Total surface area of dome = $2\pi R_2 h_2$.
Self weight of dome = $2\pi R_2 h_2 t_2 \times \gamma_c$ where $t_2$ is the thickness of bottom dome.

∴ Total load $W_T = W_0 + 2\pi R_2 h_2 t_2 \times \gamma_c$

Meridional thrust $\qquad T_2 = \dfrac{W_T}{\pi D_0 \sin \phi_2} \qquad \ldots(22.64)$

Intensity of road $\qquad p_2 = \dfrac{W_T}{2\pi R_2 h_2}$

Max. hoop stress at centre $= \dfrac{p_2 R_2}{2 t_2} \qquad \ldots(22.65)$

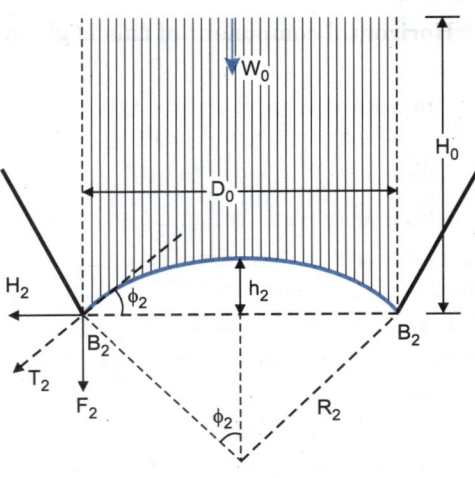

Fig. 22.20

Knowing the meridional thrust and hoop stress, the dome can be designed.

(*vi*) **Bottom Ring Beam** $B_2$. The ring beam receives an inward inclined thrust $T_0$ from the conical dome and an outward thrust $T_2$ from the bottom dome. The horizontal components of both of these oppose each other. Net horizontal force $P$ is given by

$$P = T_0 \sin \phi_0 - T_2 \cos \phi_2 \qquad \ldots(22.66)$$

If $T_0 \sin \phi_0 > T_2 \cos \phi_2$, the beam will be subjected to hoop compression. If however $T_0 \sin \phi_0 < T_2 \cos \phi_2$, it will be subjected to hoop tension. The dimensions of the tank should be so adjusted that either $P$ is zero or $P$ is compressive.

The hoop force is given by

$$P_H = P \cdot \frac{D_0}{2} \qquad \ldots(22.67)$$

If $b_2$ is the width and $d_2$ is the depth of the ring beam, the stress is given by

$$= p_H = \frac{P D_0}{2} \times \frac{1}{b\, d} \qquad \ldots(22.68)$$

The vertical load per unit length is given by

$$P_v = T_0 \cos \phi_0 + T_2 \sin \phi_2 \text{ per unit length} \qquad \ldots(22.69)$$

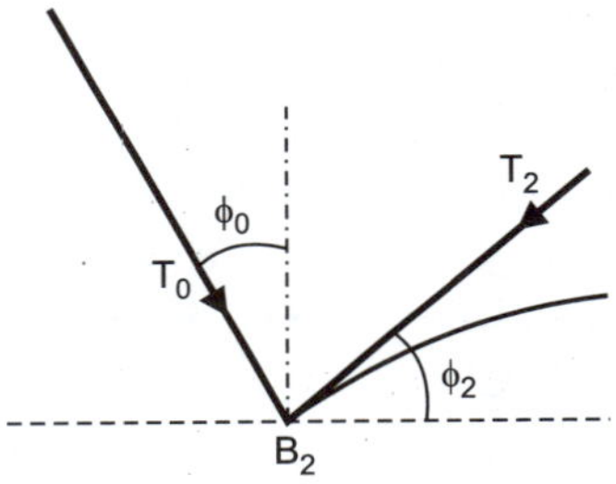

Fig. 22.21

The circular ring beam can now be designed for the above super-imposed load. The design of the Intze tank on the basis of membrane analysis is illustrated in Example 22.8.

## 22.6. EFFECTS OF CONTINUITY

In the membrane analysis, it was assumed that each member is independent of the other and therefore subjected to direct stresses only. However, due to continuity of joints between the members, joint reactions are introduced, due to which secondary stresses in the form of edge moment and hoop stresses are induced in the members. Stresses due to continuity can be obtained by applying the principle of consistent deformations. At each joint, the horizontal deformation and angular displacement between the shells should be consistent.

The analysis of Intze tank is therefore done in two stages : (*i*) *membrane analysis*, in which membrane stresses in each member are calculated and the members are designed, (*ii*) *analysis of effects due to continuity*, in which the deformations due to membrane stresses are first calculated and equations of consistent deformations are formulated to know the secondary stresses. The final stresses are then found by adding the stresses due to the above two cases. The analysis of effects due to continuity has been illustrated in Example 22.9. The various equations for calculating horizontal and angular deformations due to membrane stresses, and the equations for calculating moment and thrust stiffnesses of each shell at the joint are given in that example itself.

## 22.7. DESIGN OF TANK SUPPORTING TOWERS

Water tanks are generally elevated above the ground, to obtain the desired head of water. This is accomplished either by supporting the tank on masonry walls raised upto the desired height, or by supporting it on a number of columns suitably braced at various heights. In the latter case, the columns are subjected to (*i*) dead load of tank, water and other connected structures, (*ii*) wind loads and seismic forces. Generally, all the columns are made of equal dimensions and are placed symmetrically. Therefore, the dead load may be assumed to be equally distributed amongst the columns. The forces due to wind and other horizontal loads will depend upon the arrangement of the columns and their support conditions. We shall consider here several cases, and analyse them by approximate methods only.

Design requirements for the liquid contains of proper staging/tower components and footing slab are the same as for the containers resting on the ground.

**Case 1.** Two equal columns (*or rows of columns*) **with rigid top and Fixed at the Footings.**

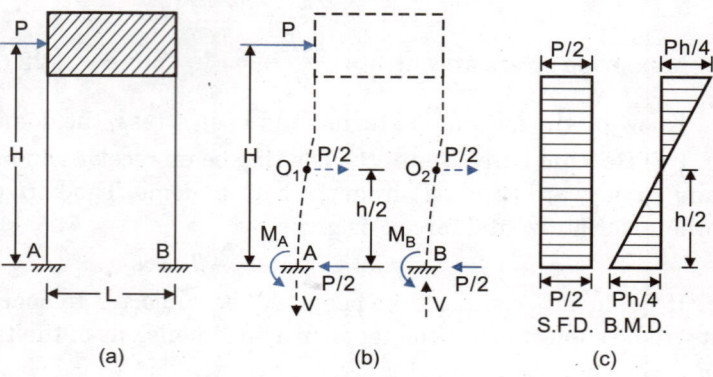

Fig. 22.22 (*a*) shows a tank supported on two equal columns. Let $P$ be the total wind load on the tank surface. The columns are fixed at the base, and are rigidly connected to the tank. Fig. 22.22 (*b*) shows the deflected shape. The analysis is based on the assumption that the point of contraflexure ($O_1$ and $O_2$) occurs at the mid height of each column. At the point of contraflexure, there is no B.M. and the column is subjected to only horizontal shear ($Q$) and axial force ($V$). In general, however, there are three effects of wind and other horizontal forces :
(*i*) bending moment $M$, (*ii*) horizontal shear $Q$ and (*iii*) axial force $V$. At the base of each column, the B.M. is $M_A$, horizontal shear is $P/2$ and axial force is $V$, which is tensile in column $A$ and compressive in column $B$.

Taking moments of external forces about $B$, we get : $P \cdot H - V \cdot L - M_A - M_B = 0$ ...(1)

However, considering the equilibrium of $O_1 A$,

$$M_A = \frac{P}{2} \times \frac{h}{2} = \frac{Ph}{4}$$

∴ Similarly, $M_B = \dfrac{Ph}{4}$. Hence from (1), $V = \dfrac{P \cdot H - \dfrac{Ph}{2}}{L} = \dfrac{P}{L}\left[H - \dfrac{h}{2}\right]$

If there are *n* columns in each row, we have

$$M_A = M_B = \frac{Ph}{4n} \quad \text{...[22.70 (a)]} \qquad V = \frac{P}{nL}\left(H - \frac{h}{2}\right) \quad \text{...[22.70 (b)]}$$

and shear in each column $= Q = \dfrac{P}{2n}$ ...[22.70 (c)]

The total stress in each column is that due to
(*i*) Dead load of the structure and the contents.
(*ii*) Axial force $\pm V$.
(*iii*) Flexural stress due to $M$ and
(*iv*) Shear stress due to shear force $Q$, which is considered to be negligibly small. In the above case the wind load on the column faces has not been considered.

**Case 2.** **Two rows of Columns with Horizontal Braces.**

Fig. 22.23 shows a tank supported on two columns (or two rows of columns), subjected to a horizontal wind load $P$ on the exposed tank surface. Here again, the wind load on exposed column faces has been neglected for simplicity. The two rows of columns have been connected with horizontal braces. It is assumed that the braces are so stiff that the columns are constrained to maintain their axes vertical at their junctions with braces.

It is also assumed that the columns develop points of contraflexure, there will be only horizontal shear ($=P/2$ in the present case), and axial force, the B.M. being zero. The bending moment at the junction of the column with the brace, such as point $C$, will be given by

$$M_{CE} = \frac{P}{2} \cdot \frac{h_2}{4} = \frac{P h_2}{4},$$

above the brace and

$$M_{CA} = \frac{P}{2} \cdot \frac{h_3}{2} = \frac{P h_3}{4},$$

below the brace

Moment in the brace will be the sum of the two:

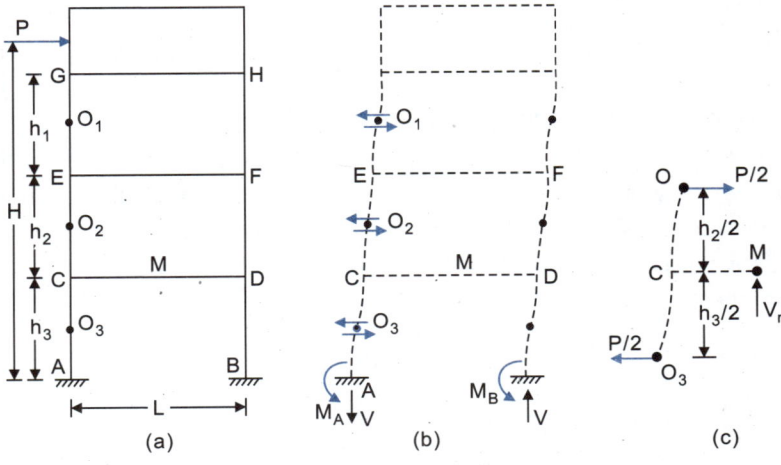

Fig. 22.23

$$M_{CD} = \frac{P h_2}{4} + \frac{P h_3}{4} = V_m \cdot \frac{L}{2} \qquad \ldots[22.71\,(a)]$$

Hence

$$V_m = P \frac{(h_2 + h_3)}{2L} \qquad \ldots[22.71\,(b)]$$

The moments at $A$ and $B$ will evidently be:

$$M_A = M_B = \frac{P}{2} \cdot \frac{h_3}{2} = \frac{P h_3}{4} \qquad \ldots[22.71\,(c)]$$

To find axial force $V$, take moments about $B$: $P \cdot H - V \cdot L - M_A - M_B = 0$

$$V = \frac{P \cdot H - \dfrac{2 P \cdot h_3}{4}}{L} = \frac{P}{L}\left[H - \frac{h_3}{2}\right] \qquad \ldots[22.71\,(d)]$$

If there are $n$ columns in each row, the above expressions are modified as follows :

$$M_A = M_B = \frac{P h_3}{4n} \quad \ldots[22.72\,(a)] \qquad V = \frac{P}{nL}\left[H - \frac{h_3}{2}\right] \quad \ldots[22.72\,(b)]$$

**Case 3. Frame work with three or more rows of columns**: Fig. 22.24 shows the frame work with three rows of columns, having $n$ columns in each row, and stiffened with braces. Let $P$ be the total load due to wind on the exposed surface of the tank. Since the interior columns ($C$) are braced on both sides and are held more stiffly than the exterior ones they are assumed to *take double horizontal shear than the exterior ones.*

Thus, the horizontal shear at the points of contraflexure in each external column will be $\dfrac{P}{4n}$, while that in each middle column will be $P/(2n)$. If $h_0$ is the height of lower panel,

$$M_A = M_B = \frac{P}{4n} \cdot \frac{h_0}{2} = \frac{P h_0}{8n} \qquad \ldots[22.73\,(a)]$$

and

$$M_C = \frac{P}{2n} \cdot \frac{h_0}{2} = \frac{P h_0}{4n} \qquad \ldots[22.73\,(b)]$$

The whole frame work will rotate about the horizontal axis passing through $C$. Hence the vertical (axial) force in $A$ and $B$ will be equal, while the force in $C$ will be zero.

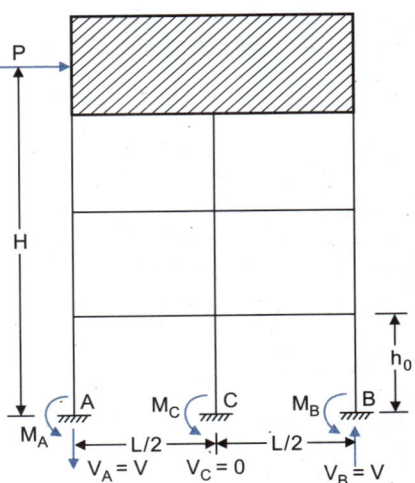

Fig. 22.24

$$\therefore \qquad V_A = V_B = \frac{P}{nL}\left[H - \frac{h_0}{2}\right] \text{ as before} \qquad \ldots[22.73\ (c)]$$

### Case 4. Circular group of columns

Fig. 22.25 (a) shows the tower, subjected to a wind force $P$ on the water tank. Let there be *n columns arranged symmetrically on a circle of radius r*. Fig. 22.25 (b) shows the plan. The whole framework will have a tendency to rotate about the axis of bending perpendicular to the direction of wind. Let $V_0$, $V_a$, $V_b$, $V_r$ etc. be the axial forces in the columns situated at distances $o, a, b, r$ from the bending axis. Due to wind moment $P.H$, at the column base the axial loads are related as follows:

$$V_b = V_r \cdot \frac{b}{r}; \quad V_a = V_r \cdot \frac{a}{r}; \quad V_0 = 0$$

*If the columns are assumed to be hinged at the bottom*, the external moment $M_w = P.H$ will be equal to the moment of resistance $M_R$:

$$M_w = P \cdot H = M_R = 2 V_r \cdot r + 4 V_b \cdot b + 4 V_a \cdot a$$

or 
$$M_w = M_R = 2 V_r \cdot r + 4 V_r \cdot \frac{b^2}{r} + 4 V_r \cdot \frac{a^2}{r} = \frac{V_r}{r}(2r^2 + 4b^2 + 4a^2)$$

Hence in general terms, $M_w = M_R = \dfrac{V_r}{r}\Sigma a^2$.

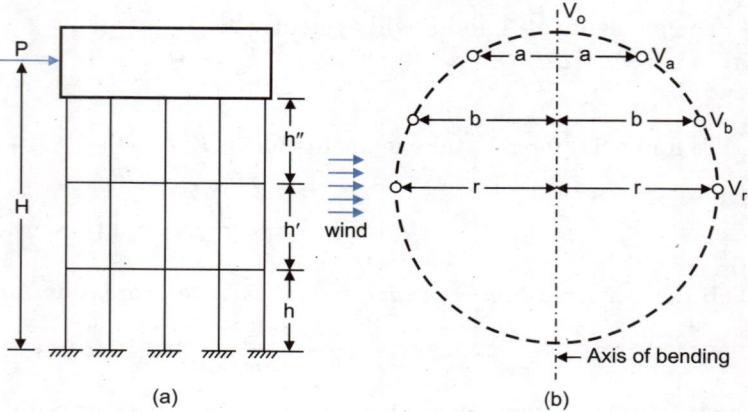

However, the columns are generally fixed to the base and hence the above term for resisting moment should include the sum of fixing moments at the bases of separate columns. If the lowest panel has height $h$, the horizontal shear force at the point of contraflexure at $\dfrac{h}{2}$ height above base will be $= \dfrac{P}{n}$ in each column, and fixing moment will be $\dfrac{P}{n} \cdot \dfrac{h}{2} = \dfrac{Ph}{2n}$ in each column.

(a)

Fig. 22.25

The sum of fixing moment in $n$ symmetrical columns will therefore be $n \cdot \dfrac{Ph}{2n} = \dfrac{Ph}{2}$.

$$\therefore \qquad M_w = M_R = \frac{P \cdot h}{2} + \frac{V_r}{r}\Sigma a^2 \qquad \ldots(22.74)$$

Let us consider the wind load on column faces also, as shown in Fig. 22.26. Let $P$ be the wind load on the exposed surface of tank and $P'$ be the total wind load on exposed faces of column and braces. If two sets of braces are provided, dividing the total height of columns in three equal parts, the wind load $P'$ can be assumed to act as shown.

In that case, total external moment at the base will be:

$$M_w = P \cdot H + \frac{P'}{6}(3h) + \frac{P'}{3}(2h) + \frac{P'}{3}(h) \qquad \ldots[22.75\ (a)]$$

The horizontal shear at horizontal plane through point of contraflexure $O_3$ (assumed at mid-height) will be:

$$Q_3 = P + \frac{P'}{6} \qquad \ldots[22.75\ (b)]$$

Similarly, horizontal shears at the points of contraflexure $O_2$ and $O_1$ (each assumed to be at mid-height) will be:

$$Q_2 = P + \frac{P'}{6} + \frac{P'}{3} \quad \ldots[22.75\,(c)]$$

$$Q_1 = P + \frac{P'}{6} + \frac{P'}{3} + \frac{P'}{3} \quad \ldots[22.75\,(d)]$$

Total fixing moment at base $= Q_1 \cdot \frac{h}{2}$. Hence equating the external moment to moment of resistance we have, as before (Eq. 22.74):

$$M_w = M_R = Q_1 \cdot \frac{h}{2} + \frac{V_r}{r} \Sigma a^2 \quad \ldots(22.76)$$

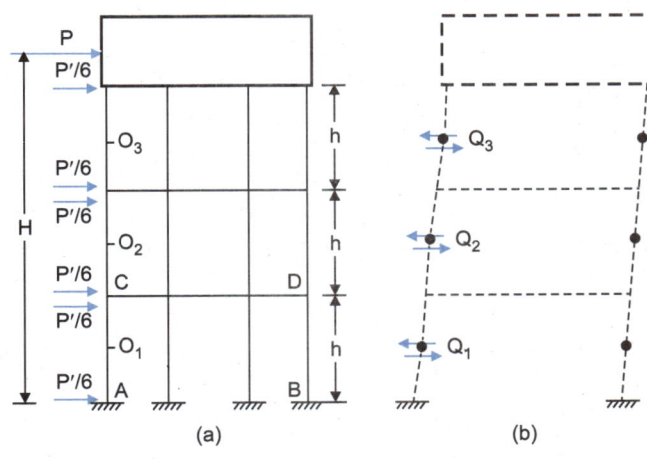

Fig. 22.26

**Alternative analysis for case 4:** In the alternative analysis the magnitudes of horizontal shear ($S$) the vertical force ($V$) at any point of contraflexure is approximately calculated by considering the tower as a whole with a single cantilever beam, with its cross-sectional area equal to the total cross-sectional area of the columns. Fig. 22.27 shows the plan of such an equivalent cantilever of hollow circular cross-section, with diameter $D_0$ and thickness $t$. If there are $n$ columns, each of area of cross-section $a$, the thickness $t$ of the equivalent cantilever beam is given by

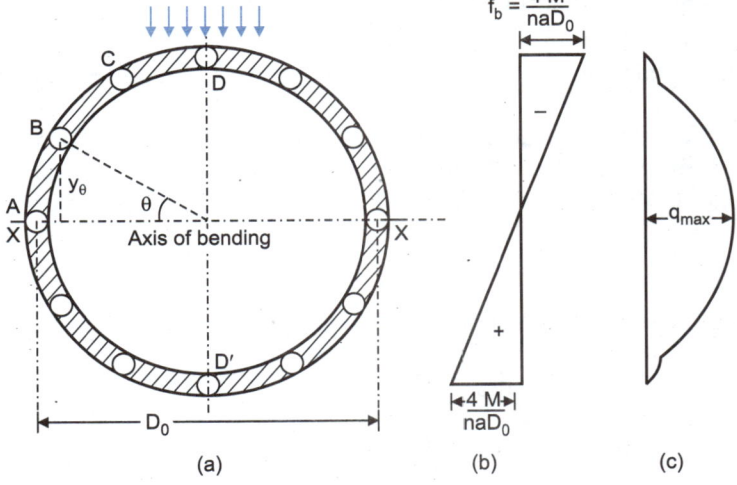

Fig. 22.27

$$t = \frac{na}{\pi D_0} \quad \ldots[22.77(a)]$$

The moment of inertia of the equivalent section, about the axis of bending $x$-$x$ is given by

$$I_e = \frac{\pi D_0^3 t}{8} = \frac{na D_0^2}{8} \quad \ldots[22.77(b)]$$

The horizontal shear force ($Q_w$) and bending moment ($M_w$) on the equivalent vertical cantilever can be calculated at any horizontal section passing through the points of contraflexure $O_1$, $O_2$, or $O_3$ (Fig. 22.26). The vertical force in any column will then be given by bending stress in that column multiplied by its area of cross-section. Similarly, horizontal force will be given by the shearing stress multiplied by the area of cross-section of the column. Let the bending moment at any horizontal section be $M_w$.

Then the bending stress is given by $f_b = \dfrac{M_w}{I_e} \cdot y$

For the farthest column $D'$ (Fig. 22.27), maximum bending stress occurs,

$$(f_b)_{max.} = \frac{M_w}{n a D_0^2 / 8} \times \frac{D_0}{2} = \frac{4 M_w}{n a D_0}.$$

Hence vertical force $V_{max.}$ in the farthest column $D'$ on the leeward side is

$$V_{max.} = (f_b)_{max.} \cdot a = \frac{4 M_w}{n D_0} \quad \ldots(22.78)$$

This will be compressive. Similarly vertical force on the column $D$ situated in the windward side will be $\dfrac{4 M_w}{n D_0}$ (tensile). Vertical force in any other column making an angle $\theta$ is given by:

$$V_\theta = \dfrac{M_w}{I_e} \times y_\theta \times a = \dfrac{M_w}{n a D_0^2 / 8} \times \dfrac{D_0}{2} \sin \theta \times a = \dfrac{4 M_w}{n D_0} \sin \theta \qquad \text{...[22.78 (a)]}$$

Vertical force in column lying on bending axis will be zero. Again, if $Q_w$ is shear force at that horizontal section, shear stress in any column making an angle $\theta$ with the bending axis is given by

$$q = \dfrac{Q_w}{I_e \cdot z}(A \bar{y}) = \dfrac{Q_w}{\dfrac{n a D_0^2}{8} \times 2 t \sec \theta} \times 2 (D_0/2)^2 \, t \cos \theta = \dfrac{2 Q_w \cos^2 \theta}{n \cdot a}.$$

This is evidently maximum at $\theta = 0$.

$$\therefore \quad q_{max} = \dfrac{2 Q_w}{n a}, \quad \text{for columns lying on bending axis.}$$

Shear force $S$ in any column $B$ is given by $S = q \cdot a = \dfrac{2 Q_w \cos^2 \theta}{n}$ \qquad ...[22.79 (a)]

Maximum shear force in column on bending axis is given by $S_{max} = \dfrac{2 Q_w}{n}$ \qquad ...(22.79)

The bending moment $M_w$ and shear force $Q_w$ in the equivalent cantilever, at any horizontal plane passing through the point of contraflexure of columns can be computed from Eq. 22.75. For example, shear force $Q_w$ at point of contraflexure $O_1$ (Fig. 22.26) of the bottom plane of height $h$ is given by

$$M = P + \dfrac{5 P'}{6}$$

and the bending moment in the equivalent cantilever, at section through $O_1$ is given by

$$M_w = P\left(H - \dfrac{h}{2}\right) + \dfrac{P'}{6}\left(\dfrac{5}{2}h\right) + \dfrac{P'}{3}\left(\dfrac{3}{2}h\right) + \dfrac{P'}{3}\left(\dfrac{1}{2}h\right) = P\left(H - \dfrac{h}{2}\right) + \dfrac{13}{12} P' \cdot h$$

Knowing $M_w$ and $Q_w$, the vertical force $V$ and shear force $S$ in any column can be computed with the help of Eqs. 22.78 (a) and 22.79 (a). Knowing $S_1$ at $O_1$ the bending moment at $C = M_1 = S_1 \times \dfrac{h}{2}$. Similarly, if $S_2$ is is the shear force at point of contraflexure $O_2$, the B.M. in the column at its junction $C$ with brace will be equal to $M_2 = \dfrac{S_2 h}{2}$. The total joint moment $M$ at the junction ($C$) of the columns with the two horizontal braces will be $= M_1 + M_2$.

The column is therefore to be designed for the following:
1. Axial thrust due to weight of tank, water and self weight.
2. Axial force in the columns due to wind load. This is critical for columns situated farthest from the bending axis, and will be a thrust for lee-ward columns and tension for wind ward columns.
3. Bending moment due to wind pressure. Maximum bending moment will be at its base, and will be equal to $S \times \dfrac{h}{2}$, where $S$ is the shear force at the next point of contraflexure and $h$ is the height of lowest panel. Bending moment will be maximum for columns lying on the bending axis.
4. Shear force ($S$) due to wind pressure. This is generally small.

## ANALYSIS OF BRACES

Fig. 22.28 shows a column joint $B$ where two braces $BA$ and $BC$ are meeting. If $S_1$ and $S_2$ are the internal shear forces at the points of contraflexure $O_1$ and $O_2$, the bending moments at the two ends of column meeting at $B$ will be $M_1 = S_1 \times \dfrac{h_1}{2}$ and $M_2 = S_2 \times \dfrac{h_2}{2}$. The total *joint moment* at $B$ will be $M = M_1 + M_2$. Due to this joint moments at $B$, moments $m_1$ and $m_2$ will be induced in brace $BA$ and $BC$. The magnitudes of $m_1$ and $m_2$ will depend upon the direction of wind with respect to the direction of the braces.

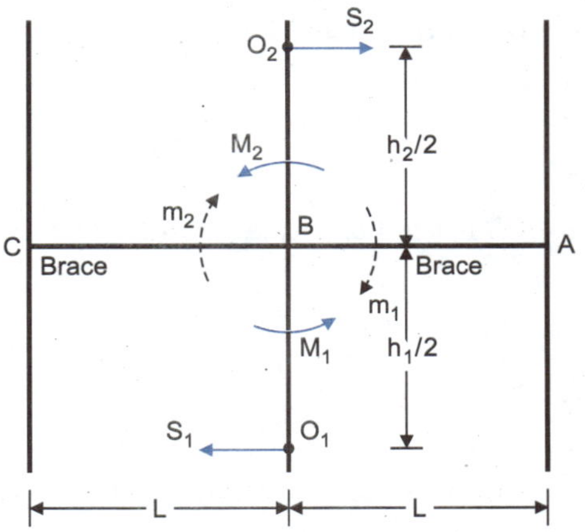

Fig. 22.29 shows the direction of wind. Let column $B$ make an angle $\theta$ with the bending axis. The joint moment at $B$ is $M$ while the induced moment in $AB$ and $BC$ are $m_1$ and $m_2$. Prolong $BC$ and $AB$ to meet the wind direction in $C_1$ and $B_1$ respectively. Triangle $BC_1B_1$ represents the triangle of moment form which we obtain

$$\frac{m_1}{\sin\left(\theta + \dfrac{\pi}{n}\right)} = \frac{m_2}{\sin\left(\theta - \dfrac{\pi}{n}\right)} = \frac{M}{\sin\left(\dfrac{2\pi}{n}\right)}$$

Hence $m_1 = \dfrac{M}{\sin\dfrac{2\pi}{n}} \sin\left(\theta + \dfrac{\pi}{n}\right)$

and $m_2 = \dfrac{M}{\sin\dfrac{2\pi}{n}} \sin\left(\theta - \dfrac{\pi}{n}\right)$

But $M = S_1 \times \dfrac{h_1}{2} + S_2 \times \dfrac{h_2}{2}$,

where $S$ in general, is given by *Eq.* 22.79(a).

$\therefore \quad M = \dfrac{Q_{w1} \cdot h_1 + Q_{w2} \cdot h_2}{n} \cos^2\theta \quad$ ...(22.80)

where $Q_{w1}$ and $Q_{w2}$ are the shear at the equivalent cylinder, at the points of contraflexure.

Substituting the value of $M$ in $m_1$ and $m_2$, we get

$$m_1 = \frac{Q_{w1} \cdot h_1 + Q_{w2} \cdot h_2}{n \sin\dfrac{2\pi}{n}} \cos^2\theta \cdot \sin\left(\theta - \dfrac{\pi}{n}\right) \quad ...(22.81)$$

and $m_2 = \dfrac{Q_{w1} \cdot h_1 + Q_{w2} \cdot h_2}{n \sin\dfrac{2\pi}{n}} \cos^2\theta \cdot \sin\left(\theta - \dfrac{\pi}{n}\right)$

...[22.81(a)]

**Fig. 22.28**

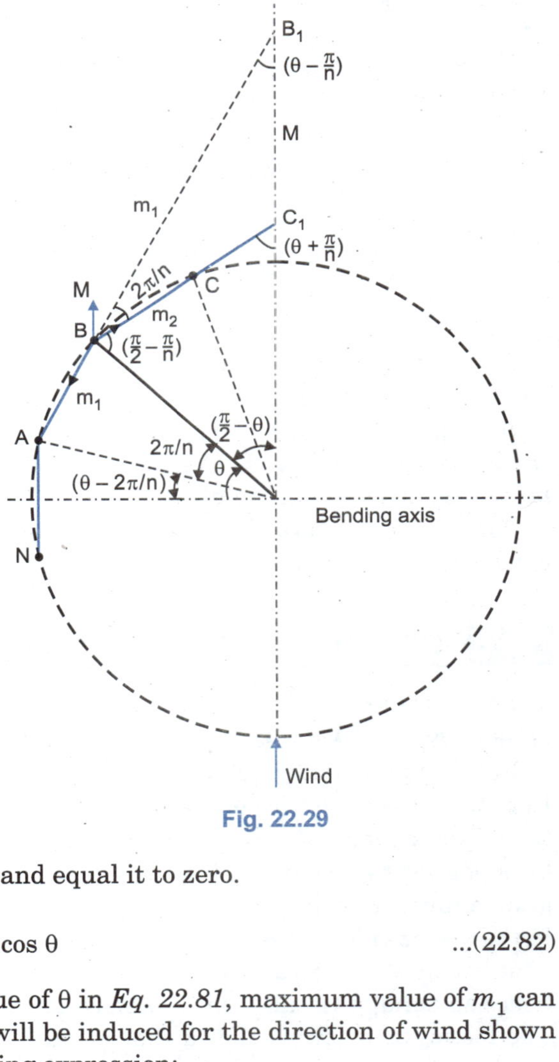

**Fig. 22.29**

For $m_1$ to be maximum, differentiate it with respect to $\theta$ and equal it to zero.

$\therefore \quad \dfrac{d}{d\theta}\left[\cos^2\theta \cdot \sin\left(\theta + \dfrac{\pi}{n}\right)\right] = 0 \quad$ or $\quad \tan\left(\theta + \dfrac{\pi}{n}\right) = \dfrac{1}{2}\cos\theta \quad$ ...(22.82)

Solving the above, $\theta$ can be found. Substituting that value of $\theta$ in *Eq.* 22.81, maximum value of $m_1$ can be computed. At the end $A$ of the brace $AB$, a moment $m_1'$ will be induced for the direction of wind shown in Fig. 22.29. The value of $m_1'$ can be found from the following expression:

$$m_1' = \frac{M'}{\sin\left(\dfrac{2\pi}{n}\right)} \sin\left(\theta - \dfrac{3\pi}{n}\right)$$

where $M'$ = Joint moment at joint $A$ = $\dfrac{Q_{w1} \cdot h_1 + Q_{w2} \cdot h_2}{n} \cos^2\left(\theta - \dfrac{2\pi}{n}\right)$

$$m_1' = \dfrac{Q_{w1} \cdot h_1 + Q_{w2} \cdot h_2}{n \sin \dfrac{2\pi}{n}} \cos^2\left(\theta - \dfrac{2\pi}{n}\right) \cdot \sin\left(\theta - \dfrac{3\pi}{n}\right)$$

If $L$ is the horizontal length of brace $AB$, shear force in it is given by

$$S_b = \dfrac{m_1 - m_1'}{L}$$

or  $$S_b = \dfrac{1}{L} \cdot \dfrac{Q_{w1} \cdot h_1 + Q_{w2} \cdot h_2}{n \sin \dfrac{2\pi}{n}} s \left[\cos^2 \theta \cdot \sin\left(\theta + \dfrac{\pi}{n}\right) - \cos^2\left(\theta - \dfrac{2\pi}{n}\right) \sin\left(\theta - \dfrac{3\pi}{n}\right)\right] \qquad ...(22.83)$$

Differentiating the above for maximum value, we get $\theta = \pi/n$. The angle at $B_1$ (Fig. 22.29) will then be = $\theta - \dfrac{\pi}{n} = \dfrac{\pi}{n} - \dfrac{\pi}{n}$ = zero.

Hence maximum shear force in a brace occurs when the wind blows parallel to it.

$$\therefore \quad (S_b)_{max} = \dfrac{Q_{w1} \cdot h_1 + Q_{w2} \cdot h_2}{L \cdot n \sin \dfrac{2\pi}{n}}$$

$$\times \left[\cos^2 \dfrac{\pi}{n} \sin \dfrac{2\pi}{n} + \cos^2 \dfrac{\pi}{n} \sin \dfrac{2\pi}{n}\right]$$

$$= \dfrac{Q_{w1} \cdot h_1 + Q_{w2} \cdot h_2}{L \cdot n \sin \dfrac{2\pi}{n}} \left(2 \cos^2 \dfrac{\pi}{n} \sin \dfrac{2\pi}{n}\right)$$

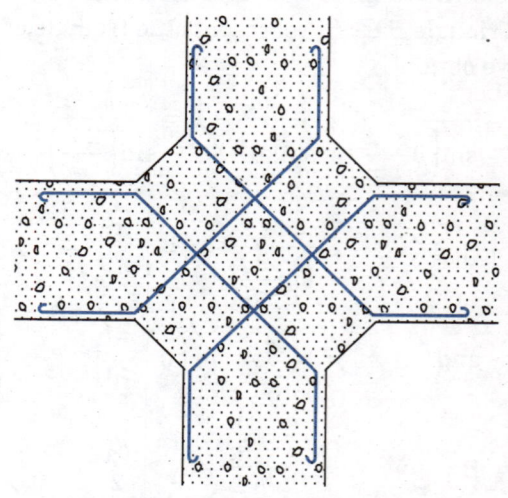

...(22.84)   **Fig. 22.30.** Junction of columns and braces

In addition to bending moment ($m_1$) and shear force $S_b$, the brace is also subjected to twisting moment, the value of which may be taken equal to 5% of $m_1$. It should be clearly noted that maximum values of both $m_1$ and $S_b$ do not occur simultaneously.

**Joint with Column:** The braces are assumed to be rigid. Hence its joint with column should be stiff. This can be achieved by providing haunches at the junction and reinforcing it by bars as shown in Fig. 22.30.

## 22.8. DESIGN OF FOUNDATIONS

The type of foundation depends upon the nature of soil and type of tower. At the sites with low lying areas or low safe bearing capacity and with high ground water table, pile foundations are provided.

In order to obtain rigidity at the column base, combined footing is generally provided for Intze tank. Depending upon the allowable soil pressure the combined footing may be either in the form of a solid circular raft [Fig. 22.31 (a)] or an annular circular raft [Fig. 22.31 (b)]. In the case of solid circular raft, the bending moment at the centre is very high, and hence an annular circular raft is preferred. The footing is so proportioned that equal overhang is obtained to both the sides of the circle jointing the column centres.

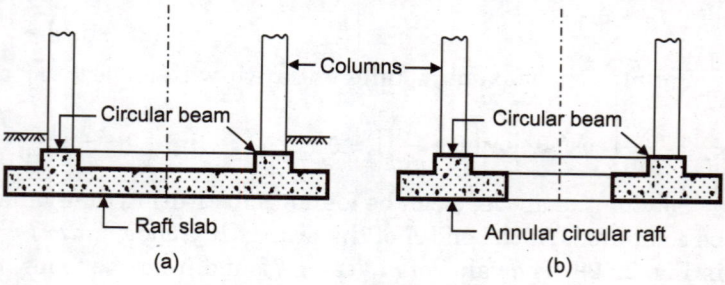

**Fig. 22.31.** Raft foundation

WATER TANKS-II : CIRCULAR AND INTZE TANKS **597**

In order to distribute the column load to the raft uniformly, the columns are connected by a circular beam at its base. The annular area of the raft depends upon the bearing capacity of soil, which can be increased by $33\frac{1}{3}\%$ while considering the wind load along with the superimposed load. If the *additional* soil pressure due to wind load is *less than* $33\frac{1}{3}\%$ of the uniform soil pressure due to superimposed load, the raft slab is designed only for pressure due to superimposed load. Also, when the tank is empty and wind is blowing, the pressure at the windward edge of the raft should be in the downward direction.

**Design: Example 22.8. Design of Intze Tank: Membrane Analysis**

*Design an Intze tank of 900,000 litres capacity. The height of staging is 16 m upto the bottom of tank. The bearing capacity of soil may be assumed to be 150 k N/m². Assume the intensity of wind pressure as 1500 N/m². Use M 20 concrete and HYSD bars.*

**Solution: 1. Dimensions of the tank.** (Fig. 22.32)

Let the diameter of cylindrical portion = $D$ = 14 m; $R$ = 7 m.

Let diameter of ring beam $B_2 = D_0$ = 10 m. Height $h_0$ of conical dome = 2 m.

Rise $h_1$ = 1.8 m; Rise $h_2$ = 1.6 m.

The radius $R_2$ of bottom dome is given by
$1.6(2R_2 - 1.6) = 5^2$ or $R_2 = 8.61$ m.

$$\sin\phi_2 = \frac{5}{8.61} = 0.5807 \; ; \; \phi_2 = 35.50°\;;$$

$\cos\phi_2 = 0.8141\;;\quad \tan\phi_2 = 0.7133$

$\cot\phi_2 = 1.4019.$

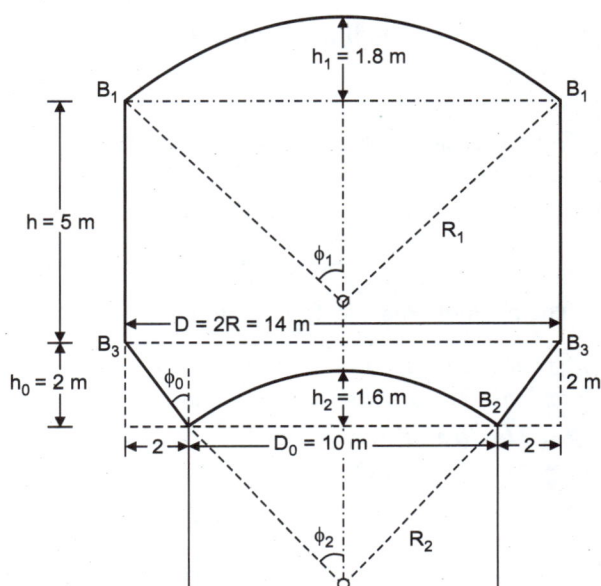

Fig. 22.32

Let $h$ be the height of cylindrical portion.
Capacity of tank is given by (Eq. 22.56)

$$V = \frac{\pi}{4}D^2 h + \frac{\pi h_0}{12}(D^2 + D_0^2 + DD_0) - \frac{\pi h_2^2}{3}(3R_2 - h_2)$$

Required volume = 900,000 litres = 900 m³.

$$\therefore \quad 900 = \frac{\pi}{4}(14)^2 h + \frac{\pi \times 2}{12}(14^2 + 10^2 + 14 \times 10) - \frac{\pi(1.6)^2}{3}(3 \times 8.61 - 1.6)$$

From which $h$ = 4.78 m. Allowing for free board, keep $h$ = 5 m.

For the top dome, the radius $R_1$ is given by: $1.8(2R_1 - 1.8) = 7^2$

From which $\quad R_1 = 14.51$ m ; $\sin\phi_1 = \dfrac{7}{14.51} = 0.4824$ ; $\phi_1 = 28.84°$ ; $\cos\phi_1 = 0.8760$.

**2. Design of top dome:**

$$R_1 = 14.51 \text{ m}\;;\; \sin\phi_1 = 0.4824\;;\; \cos\phi_1 = 0.8760$$

Let thickness $t_1$ = 100 mm = 0.1 m.

Taking a live load of 1500 N/m², total $p$ per sq. m of dome = $0.1 \times 25000 + 1500 = 4000$ N/m².

Meridional thrust at edges

$$= T_1 = \frac{pR_1}{1 + \cos\phi_1} = \frac{4000 \times 14.51}{1 + 0.8760} = 30938 \text{ N/m}.$$

$$\therefore \quad \text{Meridional stress} = \frac{30938}{100 \times 1000} = 0.31 \text{ N/mm}^2 \text{ (safe)}$$

Maximum hoop stress occurs at the centre and its magnitude is

$$= \frac{pR_1}{t_1} \times \frac{1}{2} = \frac{4000 \times 14.51}{2 \times 0.1} = 290200 \text{ N/m}^2 = 0.29 \text{ N/mm}^2 \text{ (safe).}$$

**598** REINFORCED CONCRETE STRUCTURE

Since the stresses are within the safe limits, provide nominal reinforcement @ 0.24%.
$$A_s = \frac{0.24}{100} \times 100 \times 1000 = 240 \text{ mm}^2.$$
Using 8 mm Φ bars, $A_\Phi = 50 \text{ mm}^2$. Spacing $= \frac{1000 \times 50}{240} = 208.3$ mm,

Hence provide 8 mm Φ bars @ 200 mm c/c in both the directions.

**3. Design of top ring beam $B_1$:** Horizontal component of $T_1$ is given by
$$P_1 = T_1 \cos \phi_1 = 30938 \times 0.8760 = 27102 \text{ N/m}$$
Total tension tending to rupture the beam
$$= P_1 \times \frac{D}{2} = 27102 \times \frac{14}{2} = 189714 \text{ N}.$$
Permissible stress in high yield strength deformed bars (*HYSD* bars) = 150 N/mm².

∴ $A_{sh} = 189714/150 = 1265 \text{ mm}^2$
∴ Number of 20 mm Φ bars = 1265/314.16 ≈ 4.
Actual $A_{sh}$ provided = 314.16 × 4 = 1257 mm².

The area of cross-section of ring beam is given by: $\frac{189712}{A + 12 \times 1257} = 1.2$

From which $A = 143014$ mm². Provide ring beam of 360 mm depth and 400 mm width. Tie the 20 mm Φ rings by 6 mm dia. nominal stirrups @ 200 mm c/c.

**4. Design of cylindrical wall**: In the membrane analysis, the tank wall is assumed to be free at top and bottom. Maximum hoop tension occurs at the base of the wall, its magnitude being given by
$$P = wh.\frac{D}{2} = 9800 \times 5 \times \frac{14}{2} = 343000 \text{ N/m height}$$
Area of steel, $A_{sh} = 343000/150 = 2286 \text{ mm}^2$ per metre height.
Providing rings on both the faces, $A_{sh}$ on each face = 1143 mm²

Spacing of 12 mm Φ rings = $\frac{1000 \times 113}{1143} = 98.9$ mm. Provide 12 mm Φ rings @ 95 mm c/c at bottom. This spacing can be increased at the top.

Actual $A_{sh}$ provided = $\frac{1000 \times 113}{95} \approx 1190$ mm² on each face.

Permitting 1.2 N/mm² stress on composite section, $\frac{343000}{1000\, t + 12 \times 1190 \times 2} = 1.2$.

From which $t = 257.3$ mm. Minimum thickness = $3H + 5 = (3 \times 5) + 5 = 20$ cm
However provide $t = 300$ mm at bottom and taper it to 200 mm at top.

Average $t = \frac{300 + 200}{2} = 250$ mm ; % of distribution steel = 0.24

∴ $A_{sh} = \frac{0.24 \times 250 \times 1000}{100} = 600$ mm². Area of steel on each face = 300 mm²

Spacing of 8 mm Φ bars = $\frac{1000 \times 50.3}{300} = 167.6$ mm. Hence provide 8 mm Φ bars @ 150 mm c/c on both faces. Keep a clear cover of 25 mm. Extend the vertical bars of outer face into the dome to take care of the continuity effects.

To resist the hoop tension at 2 m below top, $A_{sh} = \frac{2}{5} \times 2286 = 914.4$ mm²

∴ Spacing of 12 mm Φ rings = $\frac{1000 \times 113}{914.4/2} = 247$ mm.

Hence provide the rings @ 240 mm c/c in the top 2 m height.

At 3 m below the top, $\quad A_{sh} = \frac{3}{5} \times 2286 = 1372 \text{ mm}^2$

$$\text{Spacing of 12 mm } \Phi \text{ rings} = \frac{1000 \times 113}{1372/2} = 164.7 \text{ mm.}$$

Hence provide rings @ 160 mm c/c in the next 1 m height. At 4 m below the top,

$$A_{sh} = \frac{4}{5} \times 2286 = 1829 \text{ mm}^2$$

$$\text{Spacing of 12 mm } \Phi \text{ rings} = \frac{1000 \times 113}{1829/2} = 123.6 \text{ mm}$$

Hence provide rings @ 120 mm c/c for the next 1 m height. In the last 1 m height (4 m to 5 m) provide rings 95 mm c/c, as found earlier.

### 5. Design of ring beam $B_3$

This ring beam connects the tank wall with conical dome. The vertical load at the junction of the wall with conical dome is transferred to ring them $B_3$ by meridional thrust in the conical dome. The horizontal component of the thrust causes hoop tension at the junction. The ring beam is provided to take up this hoop tension. Refer Fig. 22.17. The load W transmitted through tank wall, at the top of conical dome consists of the following:

(i) Load of top dome = $T_1 \sin \phi_1 = 30938 \times 0.4824 = 14924$ N/m.
(ii) Load due to the ring beam $B_1 = 0.36 \times (0.4 - 0.2) \times 1 \times 25000 = 1800$ N/m.
(iii) Load due to tank wall = $5 \left( \frac{0.2 + 0.3}{2} \right) \times 1 \times 25000 = 31250$ N/m.
(iv) Self load of beam $B_3$ (1 m × 0.6 m, say) = $(1 - 0.3) \times 0.6 \times 25000 = 10500$ N/m.

$\therefore \qquad$ Total $W = 58474$ N/m.

Inclination of conical dome wall with vertical = $\phi_0 = 45°$

$\therefore \quad \sin \phi_0 = \cos \phi_0 = 0.7071 = \frac{1}{\sqrt{2}}$ ; $\tan \phi_0 = 1$; $\quad \therefore P_W = W \tan \phi_0 = 58474 \times 1 = 58474$ N/m

$$P_w = whd_3 = 9800 \times 5 \times 0.6 = 29400 \text{ N/m.}$$

Hence hoop tension in the ring beam is given by

$$P_3 = (P_W + P_w) \frac{D}{2} = (58474 + 29400) \frac{14}{2} = 615118 \text{ N}$$

This to be resisted entirely by steel hoops, the area of which is

$$A_{sh} = 615118/150 = 4100.78 \approx 4100 \text{ mm}^2. \text{ No. of 30 mm } \Phi \text{ bars}$$
$$= 4100/706.9 = 5.8.$$

Hence provide 6 rings of 30 mm $\Phi$ bars. Actual $A_{sh} = 4241.4$

Stress in equivalent section

$$= \frac{615118}{(1000 \times 600) + 12 \times 4241.4} = 0.95 \text{ N/mm}^2 < 1.2$$

Hence safe. The 8 mm $\Phi$ distribution bars (vertical bars) provided in the wall @ 150 mm c/c should be taken round the above rings to act as stirrups.

### 6. Design of conical dome

(a) *Meridional thrust*: The weight of water (Fig. 22.18 and 22.33) is given by *Eq. 22.56*:

$$W_w = \frac{\pi}{4}(14^2 - 10^2) \times 5 \times 9800 + \left\{ \left( \frac{\pi \times 2 \times 9800}{12} \right) \times \left[ 14^2 + 10^2 + 14 \times 10 \right] \right\}$$
$$- \frac{\pi}{4} \times 10^2 \times 2 \times 9800 = 4392368 \text{ N.}$$

Let the thickness of conical slab be 400 mm.

∴ Total self weight $W_s$ is given by Eq. 22.59.

$$W_s = 25000\,\pi\left(\frac{14+10}{2}\right) \times 2\sqrt{2} \times 0.4 = 1066292 \text{ N}$$

Weight $W$ at $B_3 = 58474$ N/m

Hence vertical load $W_2$ per metre run is given by Eq. 22.60

$$W_2 = \frac{(\pi \times 14 \times 58474) + 4392368 + 1066292}{\pi \times 10} = 255618 \text{ N/m}$$

Meridional thrust $T_0$ in the conical dome is

$$T_0 = \frac{W_2}{\cos\phi_0} = 255618\sqrt{2} = 361500 \text{ N/m}$$

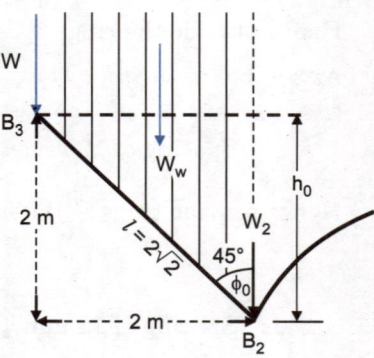

Fig. 22.33

∴ Meridional stress = $\dfrac{361500}{1000 \times 400} = 0.90 \text{ N/mm}^2$ (safe)

**(b) Hoop tension**: Refer Fig. 22.19. Diameter of conical dome at any height $h'$ above base is

$$D' = 10 + \left(\frac{14-10}{2}\right)h' = 10 + 2h'$$

Intensity of water pressure $p = (5 + 2 - h') \times 9800 = (7 - h')\,9800 \text{ N/m}^2$

Self weight $q = 0.4 \times 1 \times 1 \times 25000 = 10000 \text{ N/m}^2$

Hence hoop tension $P_0'$ is given by Eq. 22.62

$$P_0' = \left(\frac{p}{\cos\phi_0} + q\tan\phi_0\right)\frac{D'}{2} = \{(7-h')\,9800\sqrt{2} + (10000 \times 1)\}\left\{\frac{10+2h'}{2}\right\}$$

$$= [13859(7-h') + 10000](5+h') = (535075 + 37720\,h' - 13859\,h'^2)$$

The values of $P_0'$ at $h' = 0$, $h' = 1$ and $h' = 2$ are tabulated below:

| $h'$ | hoop tension |
|---|---|
| 0 | 535075 N |
| 1 | 558936 N |
| 2 | 555079 N |

For maxima, $\dfrac{dP_0'}{dh'} = 0 = 37720 - 2 \times 13859\,h'$. From which $h' = 1.361$ m

Max. $P_0' = 535075 + 37720(1.361) - 13859(1.361)^2 = 560702$ N

**(c) Design of walls.** Meridional stress = 0.9 N/mm². (safe). Max. hoop stress = 560739, whole of which is to be resisted by steel. $A_s = 560702/150 = 3738 \text{ mm}^2$

∴ Area of each face = 1869 mm². Spacing of 16 mm Φ bars = $\dfrac{1000 \times 201}{1869} = 107.5$ mm.

Hence provide 16 mm Φ hoops @ 100 mm c/c on each face.

Actual $A_s = \dfrac{1000 \times 201}{100} = 2010 \text{ mm}^2$.

Max. tensile stress in composite section

$$= \frac{560702}{(400 \times 1000) + (10 \times 2010 \times 2)} = 1.274 \text{ N/mm}^2.$$

This is more than the permissible value of 1.2 N/mm². Hence increase the thickness to 430 mm. This will reduce the tensile stress to 1.193 N/mm².

In the meridional direction, provide reinforcement @ 0.24%.

∴ $\quad A_{sd} = 0.24 \times 4200 = 1008$ mm², or 504 mm² on each face.

Spacing of 10 mm Φ bars $= \dfrac{1000 \times 78.5}{504} = 155.7$ mm.

Hence provide 10 mm bars Φ @ 150 mm c/c on each face. Provide a clear cover of 25 mm.

**7. Design of bottom dome.** $R_2 = 8.61$ m ; $\sin \phi_2 = 0.5807$ ; $\cos \phi_2 = 0.8141$

Weight of water $W_0$ on the dome is given by *Eq. 22.63*.

$$W_0 = \left[\dfrac{\pi}{4}(10)^2 \times 7 - \dfrac{\pi(1.6)^2}{3} \times (3 \times 8.61 - 1.6)\right] \times 9800 \text{ N} = 4751259 \text{ N}$$

Let the thickness of bottom dome be 250 mm.

Self weight $= 2\pi R_2 h_2 t_2 \times 25000 = 2\pi \times 8.61 \times 1.6 \times 0.25 \times 25000 = 540982$ N

Total weight $W_T = 4751259 + 540982 = 5292241$ N

Meridional thrust $= T_2 = \dfrac{5292241}{\pi \times 10 \times 0.5807} = 290093$ N/m

∴ Meridional stress $= \dfrac{290093}{250 \times 1000} = 1.16$ N/mm² (safe)

Intensity of load per unit area

$$= p_2 = \dfrac{5292241}{2\pi \times 8.61 \times 1.6} = 61142 \text{ N/m}^2$$

∴ Max. hoop stress at centre of dome

$$= \dfrac{p_2 R_2}{2 t_2} = \dfrac{61142 \times 8.61}{2 \times 0.25} = 1052865.24 \approx 1052860 \text{ N/m}^2 = 1.05 \text{ N/mm}^2 \text{ (safe)}$$

Area of minimum steel = 0.24%

∴ $\quad A_s = 0.24 \times 2500 = 600$ mm² in each direction.

Spacing of 10 mm Φ bars $= \dfrac{1000 \times 78.5}{600} = 130.8$ mm.

Hence provide 10 mm Φ bars @ 120 mm c/c in both the directions. Also, provide 16 mm Φ meridional bars @ 100 mm c/c near water face, for 1 m length, to take care of the continuity effect. The thickness of the dome may be increased from 250 mm to 280 mm gradually in 1 m length (see example 22.9).

**8. Design of bottom circular beam $B_2$**

Inward thrust from conical dome

$$= T_0 \sin \phi_0 = 361437 \times \dfrac{1}{\sqrt{2}} = 255613 \text{ N/m}$$

Outward thrust from bottom dome

$$= T_2 \cos \phi_2 = 290093 \times 0.8141 = 236165 \text{ N/m}$$

∴ Net inward thrust = 255613 − 236165 = 19448 N/m

Hoop compression in beam $= 19448 \times \dfrac{10}{2} = 97240$ N

Assuming the size of the beam to be 600 mm × 1200 mm

Hoop stress $= \dfrac{97240}{600 \times 1200} = 0.135$ N/mm². This is extremely small.

Vertical load on beam, per metre run = $T_0 \cos \phi_0 + T_2 \sin \phi_2$

$$= 361437 \times \frac{1}{\sqrt{2}} + 290093 \times 0.5807 = 424070 \text{ N/m}$$

$$\left[ \text{Alternatively, vertical load} = W_2 + \frac{W_T}{\pi D_0} = 255613 + \frac{5292241}{\pi \times 10} = 424070 \text{ N/m} \right]$$

Self weight = $0.6 \times 1.20 \times 1 \times 25000 = 18000$ N/m.

∴ The load on beam = $w = 424070 + 18000 = 442070$ N/m.

Let us support the beam on 8 equally spaced columns at a mean diameter of 10 m. Mean radius of curved beam is $R = 5$ m. Refer Table 20.1, chapter 20.

$$2\theta = 45° = \frac{\pi}{4}; \quad \theta = 22.5° = \frac{\pi}{8} \text{ radians.}; \quad C_1 = 0.066; C_2 = 0.030; C_3 = 0.005; \phi_m$$

$$= 9\frac{1}{2}° \quad w R^2 (2\theta) = 442070 (5)^2 \left(\frac{\pi}{4}\right) = 8680024 \text{ N-m}$$

Maximum – ve B.M. at support = $M_0 = C_1 . w R^2 . 2\theta = 0.066 \times 8680024 = 572882$ N-m

Maximum + ve B.M. at support = $M_c = C_2 . w R^2 . 2\theta = 0.030 \times 8680024 = 260401$ N-m

Maximum torsional moment = $M_m^t = C_3 . w R^2 . 2\theta = 0.005 \times 8680024 = 43400$ N-m

For M-25 concrete ($\sigma_{cbc} = 8.5$ N/mm²) and HYSD bars ($\sigma_{st} = 150$ N/mm²) we have $k = 0.378$; $j = 0.874$ and $R = 1.423$.

∴ Required effective depth = $\sqrt{\dfrac{572882 \times 1000}{600 \times 1.423}} = 819.1$ mm

However, keep total depth = 1200 mm from shear point of view. Let $d = 1140$ mm.

Max. shear force at supports, $F_0 = w R \theta = 442070 \times 5 \left(\dfrac{\pi}{8}\right) = 868002$ N

S.F. at any point is given by $F = w R (\theta - \phi)$

At $\phi = \phi_m$, $\quad F = 442070 \times 5 (22.5° - 9.5°) \dfrac{\pi}{180} = 501512$ N

B.M. at the point of maximum torsional moment ($\phi = \phi_m = 9\frac{1}{2}°$) is given by *Eq. 20.5*.

$$M_\phi = w R^2 (\theta \sin \phi + \theta \cot \theta \cos \phi - 1) \text{ (sagging)}$$

$$= 442070 (5)^2 \left[ \frac{\pi}{8} \sin 9.5° + \frac{\pi}{8} \cot 22.5° \cos 9.5° - 1 \right]$$

$$= -1421 \text{ N-m (sagging)} = 1421 \text{ N-m (hogging)}$$

The torsional moment at any point is given by *Eq. 20.6*,

$$M_\theta^t = w R^2 [\theta \cos \phi - \theta \cot \theta \sin \phi - (\theta - \phi)]$$

At the supports, $\phi = 0$; $M_0^t = w R^2 [\theta - \theta]$ = zero.

At the mid-span, $\phi = \theta = 22.5° = \dfrac{\pi}{8}$ rad.

$$M_\phi^t = w R^2 \left[ \theta \cos \theta - \theta \frac{\cos \phi}{\sin \phi} \sin \phi \right] = \text{zero}.$$

Hence we have the following combinations of B.M. and torsional moment:

(a) *At the supports,*

$\quad M_0 = 572882$ N-m (hogging or negative); $M_0^t$ = zero.

(b) *At mid-span*:

$\quad M_c = 260401$ N-m (sagging or positive); $M_0^t$ = zero.

(c) At the point of max. torsion $\left(\phi = \phi_m = 9\frac{1°}{2}\right)$

$M_\phi = 1421$ N-m (hogging or negative); $M_m^t = 43400$ N-m

## Main and Longitudinal Reinforcement

(a) Section at point of maximum torsion

$$T = M_{max}^t = 43400 \text{ N-m}; \quad M_\phi = M = 1421; \quad M_{e1} = M + M_T$$

where $M_T = T\left[\dfrac{1+D/b}{1.7}\right] = 43400\left[\dfrac{1+\dfrac{1.2}{0.6}}{1.7}\right] = 76588$ N-m

$\therefore \quad M_{e1} = 1421 + 76588 = 78009$ N-m or,

$$A_{st1} = \dfrac{M_{e1}}{\sigma_{st}\, j\, d} = \dfrac{78009 \times 1000}{150 \times 0.872 \times 1160} = 514.1 \text{ mm}^2$$

Number of 25 mm Φ bars = 514/491 = 1.05. Let us provide a minimum of 2 bars.

Since $M_T > M$, $\quad M_{e2} = M_T - M = 76588 - 1421 = 75167$ N-m

$$A_{st2} = \dfrac{75167 \times 1000}{150 \times 0.872 \times 1160} = 495.4 \text{ mm}^2$$

∴ Number of 25 mm Φ bars ≈ 1. However, provide a minimum of 2 bars. Thus, at the point of maximum torsion, provide 2-25 mm Φ bars each at top and bottom.

(b) Section at max. hogging B.M. (support)

$$M_0 = 572882 \text{ N-m} = M_{max}; M_0^t = 0$$

$$A_{st} = \dfrac{572882 \times 1000}{150 \times 0.872 \times 1160} = 3775.7 \text{ mm}^2 .$$

∴ Numbers of 25 mm Φ bars = 3776/491 = 7.7 ≈ 8. Hence provide 6 Numbers of 25 mm Φ bars in one layer and 2 bars in the second layer. These will be provided at the top of the section, near supports.

(c) Section at max. sagging B.M. (mid-span)

$$M_c = 260401 \text{ N-m}; M_c^t = 0$$

∴ For positive B.M., steel will be to the other face, where stress in steel ($\sigma_{st}$) can be taken as 190 N/mm². The constants for M 25 concrete having $c = 8.5$ N/mm² and $m = 11$ will be $k = 0.33$; $j = 0.89$ and $R = 1.248$;

$$A_{st} = \dfrac{260401 \times 1000}{190 \times 0.89 \times 1160} = 1328 \text{ mm}^2$$

Number of 25 mm Φ bars = 1328/491 = 2.7.

Hence the *scheme of reinforcement* will be as follows : At the supports, provide 6 – 25 mm Φ bars at top layer and 2 – 25 mm Φ bars in the second layer. Continue these upto the section of maximum torsion (*i.e.* at $\phi_m = 9.5° = 0.166$ rad.) at a distance $= R\,\phi_m = 5 \times 0.166 = 0.83$ m or equal to $L_d = 52\,\Phi = 1300$ mm from supports.

At this point, discontinue four bars while continue the remaining four bars. Similarly, provide 4 bars of 25 mm Φ at the bottom, throughout the length. These bars will take care of both the max. positive B.M. as well as maximum torsional moment.

## Transverse Reinforcement

(a) *At point of max. torsional moment*: At the point of max. torsion, $V = 501512$ N

$$V_e = V + 1.6\dfrac{T}{b} \quad \text{where } T = M_m^t = 43400 \text{ N-m}; b = 600 \text{ mm} = 0.6.$$

$\therefore \quad V_e = 501512 + 1.6 \times \dfrac{43400}{0.6} = 617245$ N

∴ $\tau_{ve} = \dfrac{V_e}{b\,d} = \dfrac{617245}{600 \times 1160} = 0.887$ N/mm²

This is less than $\tau_{c\,max} = 1.8$ N/mm² for M 25 concrete (Table 20.8); hence O.K.

$$\dfrac{100\,A_s}{bd} = \dfrac{100(4 \times 491)}{600 \times 1160} = 0.282.$$

Hence from Table 20.7; $\tau_c \approx 0.24$ N/mm²

Since $\tau_{ve} > \tau_c$, shear reinforcement is necessary. The area of cross-section $A_{sv}$ of the stirrups is given by

$$A_{sv} = \dfrac{T \cdot s_v}{b_1\,d_1 \cdot \sigma_{sv}} + \dfrac{V \cdot s_v}{2.5\,d_1\,\sigma_{sv}}$$

where $b_1 = 600 - (40 \times 2) - 25 = 495$ mm; $d_1 = 1200 - (40 \times 2) - 25 = 1095$ mm

∴ $\dfrac{A_{sv}}{s_v} = \dfrac{43400 \times 1000}{495 \times 1095 \times 150} + \dfrac{501512}{2.5 \times 1095 \times 150} = 1.755$

Minimum transverse reinforcement is governed by $\dfrac{A_{sv}}{s_v} \geq \left(\dfrac{\tau_{ve} - \tau_c}{\sigma_{sv}}\right) b$

∴ $\dfrac{A_{sv}}{s_v} = \dfrac{0.887 - 0.23}{150} \times 600 = 2.628.$ Hence depth $\dfrac{A_{sv}}{s_v} = 2.628$

Using 12 mm Φ 4 lgd stirrups,

$$A_{sv} = 4 \times 113 = 452 \text{ mm}^2 \text{ or } s_v = 452/2.628 = 172 \text{ mm}$$

However, the spacing should not exceed the least of $x_1$, $\dfrac{x_1 + y_1}{4}$ and 300 mm,

where $x_1$ = short dimension of stirrup = 495 + 25 + 12 = 532 mm

$y_1$ = long dimension of stirrup = 1095 + 25 + 12 = 1132 mm

$$\dfrac{x_1 + y_1}{4} = \dfrac{532 + 1132}{4} = 416 \text{ mm}$$

Hence provide 12 mm Φ 4 lgd stirrups @ 170 mm c/c.

(b) *At the point of max. shear (supports).*

At supports, $F_0 = 868002$ N; $\tau_v = \dfrac{868002}{600 \times 1160} = 1.25$ N/mm²

At supports, $\dfrac{100\,A_s}{b\,d} = \dfrac{100\,(8 \times 491)}{600 \times 1160} = 0.564$

$\tau_c \approx 0.323$ N/mm². Hence shear reinforcement is necessary.

$V_c = 0.323 \times 600 \times 1160 = 224808$

∴ $V_s = F_0 - V_c = 868002 - 224808 = 643194$ N

The spacing of 10 mm Φ 4 lgd stirrups having

$A_{sv} = 314$ mm² is given by

$$s_v = \dfrac{\sigma_{sv} \cdot A_{sv} \cdot d}{V_s} = \dfrac{150 \times 314 \times 1160}{643194} = 84.9 \text{ mm}$$

This is small. Hence use 12 mm Φ 4 lgd stirrups, having:

$$A_{sv} = 4 \times \dfrac{\pi}{4}(12)^2 = 452.39 \text{ mm}^2 \text{ at spacing}$$

$$s_v = \dfrac{150 \times 452.39 \times 1160}{643194} \approx 122.4 \text{ mm} \quad say \quad 120 \text{ mm}$$

(c) *At mid-span*: At the mid-span, S.F. is zero. Hence provide minimum/nominal shear reinforcement, given by $\dfrac{A_{sv}}{b \cdot s_v} \geq \dfrac{0.4}{f_y}$

or $\qquad \dfrac{A_{sv}}{s_v} = \dfrac{0.4\,b}{f_y}$.  For HYSD bars, $f_y = 415$ N/mm²

$\therefore \qquad \dfrac{A_{sv}}{s_v} = \dfrac{0.4 \times 600}{415} = 0.578$

Choosing 10 mm Φ 4 lgd stirrups, $A_{sv} = 314$ mm².

$$s_v = \dfrac{314}{0.578} = 543 \text{ mm}$$

Max. permissible spacing = $0.75\,d = 0.75 \times 1160 = 870$ or 300 mm, whichever is less.
Hence provide 10 mm Φ 4 lgd stirrups @ 300 mm c/c.

**Side Face Reinforcement**: Since the depth is more than 450 mm, provide side face reinforcement @ 0.1%.

$\therefore \qquad A_l = \dfrac{0.1}{100}(600 \times 1200) = 720 \text{ mm}^2.$

Provide 3-16 mm Φ bars on each face, having total $A_l = 6 \times 201 = 1206$ mm².

**9. Design of Columns**: The tank is supported on 8 columns, symmetrically placed on a circle of 10 m mean diameter. Height of staging above ground level is 16 m. Let us divide this height into four panels, each of 4 m height. Let the column be connected to raft foundation by means of a ring beam, the top of which is provided at 1 m below the ground level, so that the actual height of bottom panel is 5 m.

(a) *Vertical loads on columns*

1. Weight of water = $W_w + W_0$ = 4392368 + 4751259 = 9143627 N
2. Weight of tank :
  (i) Weight of top dome + cylindrical walls etc. = $W = 58474 \times \pi \times 14 = 2571821$ N
  (ii) Weight of conical dome = $W_s = 1066131$ N
  (iii) Weight of bottom dome = 540982 N
  (iv) Weight of bottom ring beam = $18000 \times \pi \times 10 = 565487$ N

$\therefore \quad$ Total weight of tank = 4744421 N

Total superimposed load = 9143627 + 4744421 = 13888048 N

Check: Total load = load on bottom beam per metre × π × 10 = $442070 \times \pi \times 10 = 13888038$

$\therefore \quad$ Load per column = 13888038/8 ≈ 1736004 N. Let the column be of 700 mm diameter.

Weight of column per metre height = $\dfrac{\pi}{4}(0.7)^2 \times 1 \times 25000 = 9621$ N

Let the brace be of 300 mm × 600 mm size.

Length of each brace = $L = R\,\dfrac{\sin\frac{2\pi}{n}}{\cos\frac{\pi}{n}} = 5 \times \dfrac{\sin\frac{\pi}{4}}{\cos\frac{\pi}{8}} = 3.83$ m  $\quad \left(\text{Alternatively}, L \approx \dfrac{\pi \times 10}{8} = 3.93 \text{ m}\right)$

Clear length of each brace = 3.83 − 0.7 = 3.13 m.

$\therefore \quad$ Weight of each brace = $0.3 \times 0.6 \times 3.13 \times 25000 = 14085$ N

Hence total weight of column *just above* each brace is tabulated below:

Brace *GH*: $\qquad W = 1736000 + 4 \times 9620 = 1774480$ N
Brace *EF*: $\qquad W = 1736000 + 8 \times 9620 + 14085 = 1827045$ N
Brace *CD*: $\qquad W = 1736000 + 12 \times 9620 + 2 \times 14085 = 1879610$ N

Bottom of column:

$W = 1736004 + 17 \times 9621 + 3 \times 14085$

$\approx 1941800$ N

(b) *Wind loads*: Intensity of wind pressure = 1500 N/m². Let us take a shape factor of 0.7 for sections circular in plan.

Wind load on tank, domes and ring beam
$= [(5 \times 14.4) + (14.2 \times \frac{2}{3} \times 1.9) + (2 \times 12.8)$
$\quad + (10.6 \times 1.2)] \times 1500 \times 0.7$
$= 134721.3 \approx 134720$ N

This may be assumed to act at about 5.7 m above the bottom of ring beam. Wind load on each panel of 4 m height of columns
$= (4 \times 0.7 \times 8) 1500 \times 0.7 + (0.6 \times 10.6) 1500$
$= 23520 + 9540 = 33060$ N

Wind load at the top end of top panel
$= \frac{1}{2} \times 23520 = 11760$ N. Wind loads are shown marked in Fig. 22.34. The points of contraflexure $O_1, O_2, O_3$ and $O_4$ are assumed to be at the mid-height of each panel. The shear forces $Q_w$ and moments $M_w$ due to wind at these planes are given below :

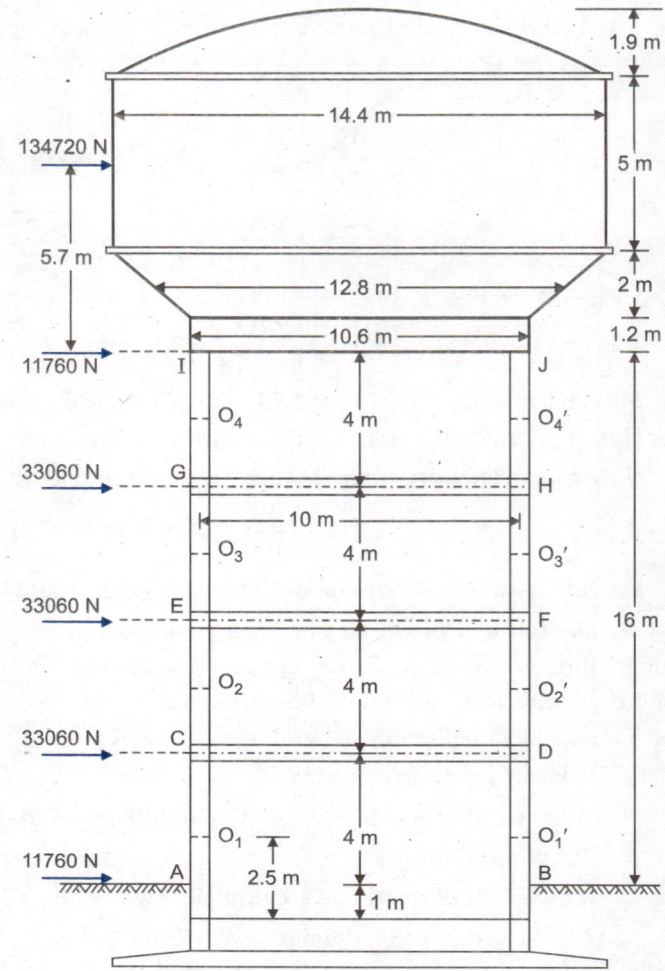

Fig. 22.34

| Level | $Q_w$ (N) | $M_w$ (N-m) |
|---|---|---|
| $O_4$ | 146480 | $134720 \times 7.7 + 11760 \times 2 = 1060864 \approx 1060860$ |
| $O_3$ | 179540 | $134720 \times 11.7 + 11760 \times 6 + 33060 \times 2 = 1712904 \approx 1712900$ |
| $O_2$ | 212600 | $134720 \times 15.7 + 11760 \times 10 + 33060 (6 + 2) = 2497184 \approx 2497180$ |
| $O_1$ | 245660 | $134720 \times 20.2 + 11760 \times 14.5 + 33060 (10.5 + 6.5 + 2.5) = 3536534$ |

The axial thrust $V_{max} = \dfrac{4M_w}{nD_0} = \dfrac{4M_w}{8 \times 10} = 0.05\, M_w$ in the farthest leeward column, the shear force $S_{max} = 2Q_w/n = 0.25 Q_w$ in the column on the bending axis at each of the above levels and the bending moment $M = S_{max} \times \dfrac{h}{2}$ in the columns are tabulated below:

**TABLE 22.14**

| Level | $V_{max}$ | $S_{max}$ (N) | M (N-m) |
|---|---|---|---|
| $O_4$ | 53040 | 36620 | 73240 |
| $O_3$ | 85650 | 44890 | 89780 |
| $O_2$ | 124860 | 53150 | 106300 |
| $O_1$ | 176830 | 61420 | 153550 |

The farthest leeward column will be subjected to the superimposed axial load plus $V_{max}$ given above. The column on the bending axis, on the other hand, will be subjected to super-imposed axial load plus a bending moment $M$ given above. These critical combinations for various panels of these columns are tabulated below:

**TABLE 22.15**

| Panel | Farthest Leeward column | | Column on Bending axis | |
|---|---|---|---|---|
| | Axial load (N) | $V_{max}$ (N) | Axial load (N) | M (N-m) |
| $O_4 O_4'$ | 1774480 | 53040 | 1774480 | 73240 |
| $O_3 O_3'$ | 1827045 | 85650 | 1827045 | 89780 |
| $O_2 O_2'$ | 1879610 | 124860 | 1879610 | 106300 |
| $O_1 O_1'$ | 1941795 | 176830 | 1941795 | 153550 |

According to I.S. Code, when effect of wind load is to be considered, the permissible stresses in the materials may be increased by $33\frac{1}{3}\%$. For the farthest leeward column the axial thrust $V_{max}$ due to wind load is less than even 10% of the super-imposed axial load. Hence the effect of wind is not critical for the farthest leeward column. However, column situated on the bending axis need be considered to see the effect of maximum B.M. of 153550 N-m due to wind, along with the super imposed axial load of 1941795 N at the lowest panel. Use M20 concrete, for which and $\sigma_{cbc}$ = 7 N/mm² and $\sigma_{cc}$ = 5 N/mm². For steel, $\sigma_{st}$ = 230 N/mm². All the three can be increased by $33\frac{1}{3}\%$ when taking into account wind action.

Diameter of column = 700 mm. Use 12 bars of 30 mm dia. at an effective cover of 40 mm.

$$A_{sc} = \frac{\pi}{4}(30)^2 \times 12 = 8482 \text{ mm}^2$$

Equivalent area of column = $\frac{\pi}{4}(700)^2 + (13-1)\,8482 = 486629 \text{ mm}^2$

Equivalent moment of inertia = $\frac{\pi}{64}d^4 + (n-1)\frac{A_{sc}\,d'^2}{8}$,

where $d$ = 700 mm;
$d'$ = 700 − 2 × 40 = 620 mm

$$I_c = \frac{\pi}{64}(700)^4 + (13-1)\frac{8482 \times (620)^2}{8} = 1.66766 \times 10^{10} \text{ mm}^4$$

∴ Direct stress in column = $\sigma_{cc}'$ = 1941795/486629 = 3.99 N/mm²

Bending stress in column = $\sigma_{cbc}' = \frac{153550 \times 1000}{1.6676 \times 10^{10}} \times 350 = 3.22$ N/mm²

For the safety of the column, we have the condition,

$\frac{\sigma_{cc}'}{\sigma_{cc}} + \frac{\sigma_{cbc}'}{\sigma_{cbc}} \leq 1$  ∴  $\frac{3.99}{1.33 \times 6} + \frac{3.22}{1.33 \times 8.5} < 1$

or 0.5 + 0.28 < 1 or 0.78 < 1. Hence safe.

Use 10 mm Φ wire rings of 250 mm c/c to tie up the main reinforcement. Since the columns are of 700 mm diameter, increase the width of curved beam $B_2$ from 600 mm to 700 mm.

**10. Design of braces**: The bending moment $m_1$ in a brace is given by *Eq. 22.81*, its maximum value being governed by *Eq. 22.82*:

$$\tan\left(\theta + \frac{\pi}{8}\right) = \frac{1}{2}\cot\theta. \quad \text{Solving this graphically, we get } \theta = 24.8°$$

∴ $$(m_1)_{max.} = \frac{Q_{w1} \cdot h_1 + Q_{w2}\,h_2}{n \sin\frac{2\pi}{n}} \cos^2\theta \sin\left(\theta + \frac{\pi}{n}\right)$$

**608** REINFORCED CONCRETE STRUCTURE

For the lowest junction $C$ (Fig. 22.34),

$$h_1 = 5 \text{ m and } h_2 = 4 \text{ m}$$

$$(m_1)_{max.} = \frac{(245660 \times 5) + (212600 \times 4)}{8 \sin \frac{2\pi}{8}} \cos^2 24.8° \sin\left(24.8° + \frac{180°}{8}\right)$$

$$= 222540 \text{ N-m}$$

The maximum shear force $(S_b)_{max}$ in a brace is given by *Eq. 22.84*, for $\theta = \pi/8$:

$$(S_b)_{max.} = \frac{(245660 \times 5) + (212600 \times 4)}{3.93 \times 8 \sin \frac{2\pi}{8}}\left(2\cos^2\frac{\pi}{8} \times \sin\frac{2\pi}{8}\right) = 112870 \text{ N}$$

For $\theta = \frac{\pi}{8}$, the value of $m_1$ is given by *Eq. 22.81*:

$$[(m_1)]_{\theta = \frac{\pi}{8}} = \frac{(245650 \times 5) + (212600 \times 4)}{8 \sin \frac{2\pi}{8}} \cos^2\frac{\pi}{8} \sin\left(\frac{\pi}{8} + \frac{\pi}{8}\right) = 221786 \text{ N-m}$$

Twisting moment at $\theta = \frac{\pi}{8}$ is $M^t = 0.05 \, m_1 = 0.05 \times 221786 = 11090$ N-m

Thus the brace will be subjected to a critical combination of max. shear force $(S_b)_{max}$ and a twisting moment $(M^t)$ when the wind blows parallel to it (*i.e.* when $\theta = \pi/8$).

The brace is reinforced equally at top and bottom since the sign of moment $(m_1)$ will depend upon the direction of wind.

For M 25 concrete, $c = \sigma_{cbc} = 8.5$ N/mm², $m = 11$, also, $\sigma_{st} = t = 230$ N/mm²; $k = 0.289$, $j = 0.904$ and $R = 1.11$. Depth of N.A. $= 0.289 \, d$. Let $A_{sc} = A_{st} = pbd$ and $d_c = 0.1 \, d$

Equating the moment of equivalent area about N.A.

$$\frac{1}{2}b(0.289 \, d)^2 + (11 - 1)pbd(0.289 \, d - 0.1 \, d) = 11 \, pbd(d - 0.289 \, d).$$

From which $p = 0.007$

Since the brace is subjected to both the B.M. as well as twisting moment, we have

$$M_{e1} = M + M_T \quad \text{where} \quad M = \text{B.M.} = (m_1)_{max.} = 222540 \text{ N-m}$$

$$M_T = T\left(\frac{1 + D/b}{1.7}\right), \quad \text{where} \quad T = M^t = 11090 \text{ N-m}$$

Let $D = 700$ mm. $\therefore M_T = 11090 \left[\dfrac{1 + \dfrac{700}{300}}{1.7}\right] = 21745$ N-m

$$M_{e1} = 222540 + 21745 = 244285$$

In order to find the depth of the section, equate the moment of resistance of the section to the external moment.

$$b.n.\frac{c}{2}\left[d - \frac{n}{3}\right] + (m_c - 1)A_{sc} \cdot c'(d - d_c) = M_{e1}$$

Here $c = 1.33 \times 8.5 = 11.305$ N/mm²; $m_c = 1.5 \, m = 1.5 \times 11 = 16.5$;

$$c' = \frac{11.305 \, (0.283 - 0.1)}{0.283} = 7.39 \text{ N/mm}^2. \text{ Hence,}$$

$$300 \times 0.289 \, d \times \frac{11.305}{2}\left[1 - \frac{0.289}{3}\right]d + (16.5 - 1)(0.007 \times 300 \, d)\, 7.39 \, (1 - 0.1)\, d = 244285 \times 10^3$$

or $442.86 \, d^2 + 216.5 \, d^2 = 244285 \times 10^3$. From which $d = 609$ mm

Adopt $D = 650$ mm so that $d = 650 - 25 - 10 = 615$ mm

$$A_{sc} = A_{st} = pbd = 0.007 \times 300 \times 650 = 1365 \text{ mm}^2$$

Number of 20 mm Φ bars = 1365/314 = 4.35

Hence provide 5 Nos. of 20 mm Φ bars each at top and bottom.

$$\frac{100 A_s}{bd} = \frac{100 \times 5 \times 491}{300 \times 650} = 1.26\%. \text{ Maximum shear} = 112870 \text{ N}$$

$$V_e = V + 1.6 \, T/b = 112870 + 1.6 \times \frac{11090}{0.3} = 172017 \text{ N}$$

$$\therefore \quad \tau_{ve} = \frac{172017}{300 \times 650} = 0.88 \text{ N/mm}^2$$

This is less then $\tau_{c.max}$ (Table 20.8) but more than $\tau_c$ = 0.441 N/mm². Hence transverse reinforcement is necessary.

$$A_{sv} = \frac{T \cdot s_v}{b_1 \, d_1 \, \sigma_{sv}} + \frac{V \cdot s_v}{2.5 \, d_1 \, \sigma_{sv}}$$

where $b_1$ = 300 – (25 × 2) – 20 = 230 mm; $d_1$ = 650 – (25 × 2) – 20 = 580 mm.

Using 12 mm Φ 2 lgd stirrups,

$$A_{sv} = 2 \frac{\pi}{4}(12)^2 = 226 \text{ mm}^2.$$

$$\therefore \quad \frac{A_{sv}}{s_v} = \left[\frac{11090 \times 1000}{230 \times 580 \times 230} + \frac{112870}{2.5 \times 580 \times 230}\right] = 0.361 + 0.338 = 0.699$$

Minimum transverse reinforcement is given by

$$\frac{A_{sv}}{s_v} \geq \left(\frac{\tau_{ve} - \tau_c}{\sigma_{sv}}\right) b \quad \therefore \quad \frac{A_{sv}}{s_v} = \frac{0.88 - 0.441}{230} \times 300 = 0.573$$

$$\therefore \quad s_v = \frac{A_{sv}}{0.699} = \frac{226}{0.699} = 323 \text{ mm}$$

This spacing should not exceed least of $x_1$, $\frac{x_1 + y_1}{4}$ and 300 mm

where  $x_1$ = short dimension of stirrup = 230 + 20 + 12 = 262 mm
 $y_1$ = long dimension of stirrup = 630 + 20 + 12 = 662 mm.

$$\therefore \quad \frac{x_1 + y_1}{4} = \frac{262 + 662}{4} = 231 \text{ mm}.$$

Hence provide 12 mm Φ 2-lgd stirrups at 230 mm c/c throughout. Since the depth of section exceeds 450 mm, provide side face reinforcement @ 0.1 percent.

$$A_l = \frac{0.1}{100} (300 \times 650) = 195 \text{ mm}^2.$$

Provide 2-10 mm Φ bars at each face, giving total $A_l$ = 4 × 78.5 = 314 mm².

Provide 300 mm × 300 mm haunches at the junction of braces with columns and reinforce it with 10 mm Φ bars.

## 11. Design of raft foundation

Vertical load from filled tank and columns = 1941795 × 8 = 15534360 N

Weight of water = 9143627. Vertical load of empty tank and columns = 6390733 N

$V_{max}$ due of wind load = 170950 × 8, which is less than $33\frac{1}{3}$% of the superimposed load. Assume self weight etc. as 10% = 1553436 N

$\therefore$ Total load = 15534360 + 1553436 = 17087796 N

$\therefore$ Area of foundation required = 17087796/150000 = 113.9 m²

Circumference of column circle = π × 10 = 31.42 m

∴ Width of foundation = $\frac{113.9}{31.32}$ = 3.64 m. Hence inner diameter = 10 – 3.64 = 6.36 m.

Outer diameter = 10 + 3.64 = 13.64 m. Area of annular raft

$$= \frac{\pi}{4}(13.64^2 - 6.36^2) = 114.35 \text{ m}^2$$

Moment of inertia of slab about a diametrical axis

$$= \frac{\pi}{64}[13.64^4 - 6.36^4] = 1618.8 \text{ m}^4$$

Total load, tank empty = 6390733 + 1553436 = 7944169 N

∴ Stabilising moment = 7944169 × $\frac{13.64}{2}$ = 54179233 N-m

Let the base of the raft be 2 m below ground level.

∴ $M_w$ at base = (134720 × 23.7) + (11760 × 18) + 33060(14 + 10 + 6)

= 4396344 N-m

Hence the soil pressures at the edges along a diameter are

(a) tank full: $\frac{17087796}{114.35} \pm \frac{4396344}{1618.8} \times \frac{13.64}{2}$

= 167956 N/m² or 130912 N/m²

(b) tank empty: $\frac{7944169}{114.35} \pm \frac{4396344}{1618.8} \times \frac{13.64}{2}$

= 87994 N/m² or 509551 N/m²

Under the wind load, the allowable bearing capacity is increased to 150 × 1.333 = 200 kN/m², which is greater than the maximum soil pressure of 167.96 kN/m². Hence the foundation raft will be designed only for super-imposed load.

The layout of the foundation is shown in Fig. 22.35. A ring beam of 700 mm width may be provided. The foundation will be designed for an average pressure $P$:

$$p = \frac{15534360}{114.35} \approx 135849 \text{ N/m}^2$$

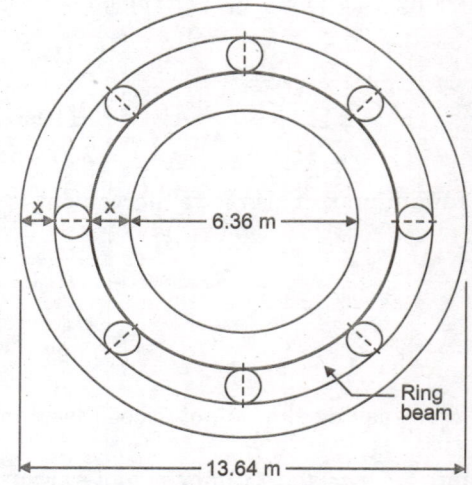

Ring beam

Fig. 22.35

The overhang $x$ of raft slab = $\frac{1}{2}\left[\frac{1}{2}(13.64 - 6.36) - 0.7\right]$ = 1.47 m

B.M. = 135849 $\frac{(1.47)^2}{2}$ = 146778 N-m;

S.F. = 135849 × 1.47 = 199698 N

$$d = \sqrt{\frac{146778 \times 1000}{1000 \times 1.11}} = 364 \text{ mm}$$

Provide 450 mm thick slab with effective depth of 410 mm. Decrease the total depth of 250 mm at the edges.

$$A_{st} = \frac{146778 \times 1000}{230 \times 0.904 \times 410} = 1722 \text{ mm}^2$$

Spacing of 16 mm Φ bars = $\frac{1000 \times 201}{1722}$ = 117 mm. Hence provide 16 mm Φ radial bars @ 110 mm c/c at the bottom of slab.

Area of distribution steel = $\frac{0.15}{100} \times 1000 \times 450 = 675 \text{ mm}^2$

Spacing of 10 mm Φ bars = $\dfrac{1000 \times 78.5}{675}$ = 116 mm. Hence provide 10 mm Φ bars @ 110 mm c/c at the support. Increase this spacing to 200 mm at the edge.

**Design of circular beam of raft**: The design of circular beam of raft will be practically similar to the circular beam $B_2$ provided at the top of the columns.

$$\text{Design load} = \dfrac{15534360}{\pi \times 10} = 494474 \text{ N/m}$$

The circular beam $B_2$ was designed for $w = 442070$ N/m. Hence the B.M. etc. will be increased in this ratio of 494474/442070 = 1.11854

∴      Max. (–) B.M. at support = $M_0$ = 572882 × 1.11854 = 640790 N-m
     Max. (+) B.M. at midspan = $M_c$ = 260401 × 1.11854 = 291268 N-m
     Max. torsional moment $M_m^t$ = 43400 × 1.11854 = 48545 N-m
     B.M. at the point of max. torsion = 1421 × 1.11854 = 1589 N-m

At $\phi = \phi_m = 9\dfrac{1°}{2}$,      $F = 501512 \times 1.11854 = 560960$ N

Max. shear force at supports = 868002 × 1.11854 = 970893 N

Use      $b = 700$ mm = diameter of columns. Use M 20 concrete.

$\sigma_{st} = 230$ N/mm².      $d = \sqrt{\dfrac{640790 \times 1000}{700 \times 1.11}} = 908.1$ mm

However, keep total depth of 1200 mm from shear point of view. Using an effective cover of 60 mm, $d = 1140$ mm

**Main or longitudinal reinforcement**: (a) *Section at point of maximum torsion*

$$T = M^t_{max} = 48545 \text{ N-m}; \quad M_\phi = M = 1589 \text{ N-m}; \quad M_{e_1} = M + M_T$$

where      $M_T = T\left[1 + \dfrac{D/b}{1.7}\right] = 48545\left[1 + \dfrac{1200/700}{1.7}\right] = 97498$ N-m

∴      $M_{e_1} = 1589 + 97498 = 99087$ N/m

$$A_{st} = \dfrac{M_{e1}}{\sigma_{st}\, j\, d} = \dfrac{99087 \times 1000}{230 \times 0.904 \times 1140} = 418 \text{ mm}^2;$$

No. of 25 Φ bars = 418/491 ≈ 1

Since $M_T > M$,      $M_{e_2} = M_T - M = 77509 - 1589 = 75920$

$$A_{st_2} = \dfrac{75920 \times 1000}{230 \times 0.904 \times 1140} = 320.3 \text{ mm}^2$$

∴      Number of 25 mm Φ bars = 320.3/491 ≈ 1

However provide minimum of 2 bars each at top and bottom.

(b) *Section at max. hogging B.M. (support)*

$$M_0 = 640760 \text{ N-m} = M_{max}; M_0^t = 0$$

$$A_{st} = \dfrac{640790 \times 1000}{230 \times 0.904 \times 1140} = 2703 \text{ mm}^2;$$

Number of 25 mm Φ bars = 2703/491 = 5.51

However, provide 6 bars of 25 mm Φ at the bottom of the section, near supports.

(c) *Section at max. sagging B.M. (mid-span)*

$$M_c = 291268 \text{ N-m}; M_c^t = 0$$

∴      $A_{st} = \dfrac{291268 \times 1000}{230 \times 0.904 \times 1140} = 1229$ mm²

∴      Number of 25 mm Φ bars = $\dfrac{1229}{491} = 2.5$.

Hence the *scheme of reinforcement* along the span will be as follows:

At supports, provide 6-25 mm Φ bars at bottom of section. Continue these upto the section of maximum torsion (*i.e.* at $\phi_m = 9.5° = 0.116$ rad.), at a distance $= R\,\phi_m = 5 \times 0.166 = 0.83$ or equal to $L_d = \dfrac{\Phi \cdot \sigma_{st}}{4\,\tau_{bd}}$
$= \dfrac{\Phi \times 230}{4 \times 1.12} = 52\,\Phi = 52 \times 25 = 1300$ mm whichever is more.

Beyond this, discontinue 2 bars, while the remaining 4 bars may be continued throughout the length.

Similarly, provide 4-25 mm Φ bars at top, throughout the length. These bars will take care of both the maximum positive B.M. as well as maximum torsional moment.

***Transverse reinforcement***: (*a*) *At the point of maximum torsional moment*
$$V = 560960 \text{ N-m}$$
$$\therefore \quad V_e = V + 1.6\,\dfrac{T}{b} = 560960 + 1.6 \times \dfrac{48545}{0.7} = 671920 \text{ N}$$
$$\tau_{ve} = \dfrac{671920}{700 \times 1140} = 0.84 \text{ N/mm}^2.$$

This is less than $\tau_{c\,max} = 1.8$ N/mm². Hence O.K.
$$\dfrac{100\,A_s}{b\,d} = \dfrac{100\,(3 \times 491)}{700 \times 1140} = 0.185$$
$$\therefore \quad \tau_c = 0.20 \text{ N/mm}^2. \quad \text{Hence shear reinforcement is necessary.}$$
$$A_{sv} = \dfrac{T \cdot s_v}{b_1\,d_1 \sigma_{sv}} + \dfrac{V \cdot s_v}{2.5\,d_1\,s_v} \quad \text{where } b_1 = 700 - (40 \times 2) - 25 = 595 \text{ mm}$$
$$d_1 = 1200 - (40 \times 2) - 25 = 1095 \text{ mm}$$
$$\therefore \quad \dfrac{A_{sv}}{s_v} = \left[\dfrac{48545 \times 1000}{595 \times 1095 \times 230} + \dfrac{560960}{2.5 \times 1095 \times 230}\right] = 1.215$$

Minimum transverse reinforcement is governed by
$$\therefore \quad \dfrac{A_{sv}}{s_v} \geq \left(\dfrac{\tau_{ve} - \tau_c}{\sigma_{sv}}\right) b \quad \therefore \quad \dfrac{A_{sv}}{s_v} = \dfrac{0.84 - 0.20}{230} \times 700 = 1.936$$

Hence adopt $\dfrac{A_{sv}}{s_v} = 1.887$. Using 12 mm Φ 4 lgd stirrups,
$$A_{sv} = 4 \times 113 = 452 \text{ mm}^2.$$
$$\therefore \quad s_v = \dfrac{452}{1.936} \approx 233.5 \text{ mm.}$$

However, spacing should not exceed least of $x_1$, $\dfrac{x_1 + y_1}{4}$ and 300 mm,

where $x_1 =$ short dimension of stirrup $= 595 + 25 + 12 = 632$ mm

$y_1 =$ long dimension of stirrup $= 1095 + 25 + 12 = 1132$ mm
$$\therefore \quad \dfrac{x_1 + y_1}{4} = \dfrac{632 + 1132}{4} = 441 \text{ mm.}$$

Hence provide 12 mm Φ 4 lgd stirrups @ 230 mm c/c.

(*b*) *At the point of max. shear (supports)*

At supports, $F_0 = 970893$ N
$$\therefore \quad \tau_v = \dfrac{970893}{700 \times 1140} = 1.22 \text{ N/mm}^2.$$

At supports, $\dfrac{100\,A_s}{b\,d} = \dfrac{100\,(6 \times 491)}{700 \times 1140} \approx 0.37\%$

Hence $\tau_c \approx 0.268$ N/mm². Hence shear reinforcement is necessary.

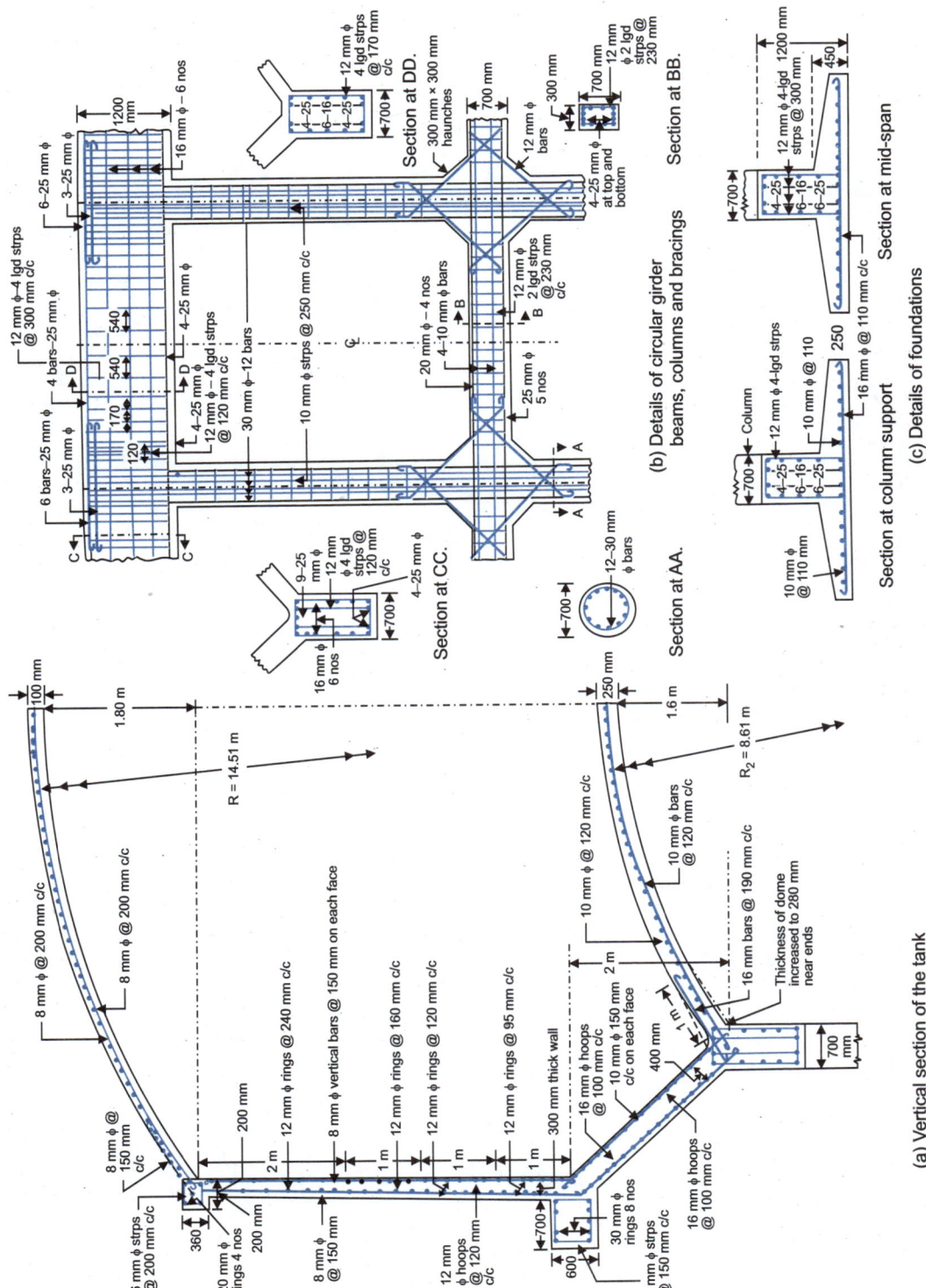

Fig. 22.36

$$V_c = 0.268 \times 700 \times 1140 = 213864 \text{ N}$$
$$\therefore \quad V_s = F_0 - V_c = 970893 - 213864 = 757029 \text{ N}$$

The spacing of 12 mm Φ 4-lgd stirrups having

$$A_{sv} = 4 \times \frac{\pi}{4}(12)^2 = 452.4 \text{ mm}^2 \text{ is given by}$$

$$s_v = \frac{\sigma_{sv} \cdot A_{sv} \cdot d}{V} = \frac{230 \times 452.4 \times 1140}{757029} = 156.7 \text{ mm}.$$

Hence provide 12 mm Φ 4-ldg stirrups @ 150 mm c/c.

(c) *At mid-span*: At the mid-span, S.F. is zero. Hence provide minimum/nominal shear reinforcement given by

$$\frac{A_{sv}}{b \cdot s_v} \geq \frac{0.4}{f_y} \quad \text{or} \quad \frac{A_{sv}}{s_v} = \frac{0.4\, b}{f_y} = \frac{0.4 \times 700}{415} = 0.675$$

Choosing 10 mm Φ 4 lgd stirrups, $A_{sv} = 314 \text{ mm}^2$. $\therefore s_v = 314/0.675 = 465 \text{ mm}$
Max. permissible spacing = $0.75\, d = 0.75 \times 1140 = 855$ or 300 mm, whichever is less.
Hence provide 10 mm Φ 4 lgd stirrubs @ 300 mm.

**Side Face Reinforcement**: Since depth is more than 450 mm, provide side face reinforcement @ 0.1 %

$$A_l = \frac{0.1}{100}(700 \times 1200) = 840 \text{ mm}^2$$

Provide 3-16 mm Φ bars on each face, having total $A_l = 6 \times 201 = 1206 \text{ mm}^2$.

**12. Details of reinforcement**: Shown in Fig. 22.36.

**Design: Example 22.9. Intze Tank: Effects of continuity**

Redesign the Intze tank of Example 22.8, taking into account the effects of continuity of joints.

**Solution:**

**1. Effects of continuity at the junction of top-dome, ring beam ($B_1$) and cylindrical portion:**

(a) **Membrane deformations and stiffnesses**: In order to determine the effects of continuity of joints, let us first calculate the membrane deformations and stiffness at edges.

(i) **Top dome**: The slope at the left edge is given by

$$\psi_d = \frac{2pR_1 \sin \phi_1}{E\, t_1} \qquad \qquad ...(22.85)$$

where $p$ = U.D.L. on dome = 4000 N/m²
$R_1$ = Radius = 14.51 m; $t_1 = 0.1$ m; $\sin \phi_1 = 0.4824$; $\cos \phi_1 = 0.8760$

$$\psi_d = \frac{2 \times 4000 \times 14.51 \times 0.4824}{E \times 0.1} = \frac{559970}{E} \text{ radians (clockwise)}$$

The horizontal outward deflection $\delta_d$ is given by

$$\delta_d = \frac{p\, R_1^2 \sin \phi_1}{E \cdot t_1}\left[\frac{1}{1+\cos \phi_1} - \cos \phi_1\right] \qquad \qquad ...(22.86)$$

$$= \frac{4000(14.51)^2 \times 0.4824}{E \times 0.1}\left[\frac{1}{1+0.8760} - 0.8760\right]$$

$$= -\frac{1393266}{E} \text{ m } (i.e. \text{ inwards})$$

**Moment stiffness**: The moment per unit length of periphery at edges to cause unit clockwise rotation, without any displacement of edges is called *moment stiffness* and its value is given by

$$M = \frac{ER_1 t_1}{4\alpha_1^{\,3}}\left(K_1 + \frac{1}{K_1}\right) \qquad \qquad ...(22.87)$$

and the corresponding radial force, applied at edges towards the centre, is given by

$$H/\text{unit length} = \frac{E\, t_1}{2\, \alpha_1^2\, K_1 \sin \phi_1} \qquad \ldots(22.88)$$

where $\alpha_1^4 = 3\left(\dfrac{R_1}{t_1}\right)^2 = 3\left(\dfrac{14.51}{0.1}\right)^2 = 63162.$ ∴ $\alpha_1 = 15.85$

and $K_1 = 1 - \dfrac{1}{2\alpha_1} \cot \phi_1 = 1 - \dfrac{1}{2 \times 15.85}\left(\dfrac{0.8760}{0.4824}\right) = 0.9427$

∴ $M = \dfrac{E \times 14.51 \times 0.1}{4\,(15.85)^3}\left(0.9427 + \dfrac{1}{0.9427}\right)$

$= 182.5 \times 10^{-6}\, E$ N-m/m/radian

and $H/\text{unit length} = \dfrac{E \times 0.1}{2(15.85)^2 \times 0.9427 \times 0.4824} = 437.6 \times 10^{-6}\, E$ N-m/radian.

**Thrust stiffness**: The thrust needed to cause a unit outward movement of edges without any rotation of tangent is called the thrust stiffness, and its value is given by

$$H = \frac{E\, t_1}{\alpha_1 R_1 K_1 \sin^2 \phi_1} \qquad \ldots(22.89)$$

$= \dfrac{E \times 0.1}{15.85 \times 14.51 \times 0.9427\,(0.4824)^2} = 1982 \times 10^{-6}\, E$ N-m/unit movement.

The corresponding moment per unit length at edges is given by

$$M/\text{unit length (clockwise)} = \frac{E\, t_1}{2\, \alpha_1^2 K_1 \sin \phi_1} \qquad \ldots(22.90)$$

$= \dfrac{E \times 0.1}{2 \times (15.85)^2 \times 0.9427 \times 0.4824} = 437.6 \times 10^{-6}\, E$ N-m/unit displacement.

**(ii) Ring Beam $B_1$**: Depth $d = 360$ mm and width $b = 400$ mm.
Radial thrust to cause unit outside deflection is given by

$$H = \frac{Ebd}{R^2} = \frac{E\,(0.40 \times 0.36)}{(7)^2} = 2939 \times 10^{-6}\, E \text{ N/m} \qquad \ldots(22.91)$$

Moment per unit circumference to cause unit rotation of the beam is given by

$$M = \frac{Ebd^3}{12\, R^2} = \frac{E \times 0.4\,(0.36)^3}{12\,(7)^2} = 31.74 \times 10^{-6}\, E \text{ N-m/m} \qquad \ldots(22.92)$$

**(iii) Cylindrical wall**: Average thickness = 250 mm

$$\alpha = \left(\frac{12}{D^2\, T^2}\right)^{1/4} = \left[\frac{12}{(14)^2\,(0.25)^2}\right]^{1/4} = 0.9948$$

$$Z = \frac{E\, T^3}{12} = \frac{E\,(0.25)^3}{12} = 13 \times 10^{-4}\, E$$

The *moment stiffness* (clockwise) is given by $M = 2\alpha Z$ $\qquad \ldots(22.93)$
$= 2 \times 0.9948 \times 13 \times 10^{-4} E = 2586 \times 10^{-6}\, E$ N-m/m

The corresponding thrust (outward) is given by $H = 2\alpha^2 Z$ $\qquad \ldots(22.94)$
$= 2\,(0.9948)^2 \times 13 \times 10^{-4}\, E = 2573 \times 10^{-6}\, E$ N/m/radian

The *thrust stiffness* (inward) is given by $H = 4\alpha^3 Z$ $\qquad \ldots(22.95)$
$= 2\,(0.9948)^3 \times 13 \times 10^{-4}\, E = 2560 \times 10^{-6}\, E$ N/m

**616** REINFORCED CONCRETE STRUCTURE

The corresponding moment (anti-clockwise) is given by $M = 2\alpha^2 Z$ ...(22.96)

$$= 2(0.9948)^2 \times 13 \times 10^{-4} E = 2573 \times 10^{-6} E \text{ N-m/m}$$

Membrane displacement (outward) of the tank at bottom

$$= \frac{PR}{TE} = \frac{343000 \times 7}{0.25 E} = \frac{9604000}{E}$$

Clockwise slope of wall $= \dfrac{9604000}{5E} = \dfrac{1920800}{E}$ radians.

**(b) Reaction due to continuity at joint:** Let the net rotation $\psi$ (clockwise) and net displacement be $\delta$ (inward). The changes in the slope and displacement of various members, from the membrane state are as follows (Table 22.16)

### TABLE 22.16

| Member | Clockwise slope | Inward Displacement |
|---|---|---|
| Top dome | $\psi - 559970/E$ | $\delta - 1392820/E$ |
| Beam $B_1$ | $\psi$ | $\delta$ |
| Tank wall | $\psi - 1920800/E$ | $\delta$ |

The reactions imposed by the joint on each member, will therefore be equal to the product of the above changes in slopes and deflections and the corresponding stiffnesses. They are tabulated below. (Table 22.17). Since the dome imposes an outward thrust of 27102 N/m, the joint must react with an inward thrust of 27102 N/m.

### TABLE 22.17. Joint $B_1$

| Member | Clockwise moment M | Inward thrust H |
|---|---|---|
| 1. Dome | $(\psi - 559970/E)\, 182.5 \times 10^{-6}$<br>$= 182.5 \times 10^{-6}\, \psi E - 102.2$<br>$437.6 \times 10^{-6} E\, (\delta - 1392820/E)$<br>$= 437.6 \times 10^{-6}\, \delta E - 609.5$ | $437.6 \times 10^{-6} E\, (\psi - 559970/E)$<br>$= 437.6 \times 10^{-6}\, \psi E - 245$<br>$(\delta - 1392820\, E) \times 1982 \times 10^{-6} E$<br>$= 1982 \times 10^{-6}\, \delta E - 2761$ |
| 2. Beam | $31.74 \times 10^{-6}\, \psi E$ | $2939 \times 10^{-6} \times \delta E$ |
| 3. Wall | $(\psi - 1920800/E) \times 2586 \times 10^{-6} E$<br>$= 2586 \times 10^{-6}\, \psi E - 4986 - 2573$<br>$\times 10^{-6}\, \delta E$ | $-2573 \times 10^{-6} E\, (\psi - 1920800/E)$<br>$= -2573 \times 10^{-6}\, \psi E + 4961$<br>$5119 \times 10^{-6}\, \delta E$ |
| 4. Reaction due to thrust from dome | | 27102 |
| 5. Total | $2800.24 \times 10^{-6}\, \psi E - 2135.4 \times 10^{-6}\, \delta E$<br>$- 5697.7$ | $-2135.4 \times 10^{-6}\, \psi E + 10040 \times 10^{-6}\, \delta E$<br>$+ 29057$ |

For the equilibrium of the joint, $\Sigma M = 0$ and $\Sigma H = 0$.

Hence from Table 22.17. $2800.24 \times 10^{-6}\, \psi E - 2135.4 \times 10^{-6}\, \delta E - 5697.7 = 0$

and $-2135.4 \times 10^{-6}\, \psi E + 10040 \times 10^{-6}\, \delta E + 29057 = 0$.

Solving the above two equations, we get

$$\psi = -0.2056 \times 10^6/E \text{ and } \delta = -2.9379 \times 10^6/E$$

Substituting the values of $\psi$ and $\delta$ in Table 22.17, the values of $M$ and $H$ will be as under (Table 22.18).

The net hoop tension in any member will be $= -\dfrac{\delta E}{D/2} \times A$ where $A$ is sectional area of member in 1.0 m length.

## TABLE 22.18

| Member | M (clockwise) N-m | H (inward) N | Net hoop tension = $-\frac{\delta E}{7} \times A$ (N/m) |
|---|---|---|---|
| Dome | − 2034.95 | − 8918.7 | 41970 |
| Beam | − 6.60 | − 8634.3 | 60437 |
| Wall | +2041.55 | − 9549.0 | 104925 |
| Sum | 0.0 | − 27102.0 | |

**(c) Design of members:** From Table 22.18, it clear that the beam $B_1$ has been considerably relieved from hoop tension, while hoop tension has been increased in the dome and the wall.

**(i) Top dome:** The compressive stresses in the dome were within safe limits. Due to joint effect a hogging moment of 2034.95 N-m/m has been imposed.

Bending stress at edges due to this moment = $\frac{M}{Z} = \frac{2034.95 \times 1000}{\frac{1}{6} \times 1000 (100)^2} = 1.22$ N/mm²

This is more than the compressive stress of 0.31 N/mm² caused due to meridional thrust. Hence reinforcement will be required to resist the tensile stresses

$$A_{st} = \frac{2034.95 \times 1000}{150 \times 0.874 \times 82} = 189 \text{ mm}^2$$

The reinforcement of the tank wall from one face (8 mm Φ @ 150 mm c/c having $A_{st}$ = 325 mm²) may be continued in the dome for some distance, since the effect of the joint reaction is just localised. The dome section may be tested for the combined effect of thrust = 30938 N/m and B.M. = 2034.95 N-m/m having

$$e = \frac{2034.95 \times 1000}{30938} = 65.8 \text{ mm}.$$

The stresses in concrete and steel would be found to be within safe limits.

In addition to this, the dome is subjected to hoop tension of 41970 N/m near edges.

$$\therefore \quad A_s = \frac{41970}{150} = 280 \text{ mm}^2$$

Using 8 mm Φ bars, $A_\Phi$ = 50 mm². ∴ Spacing = $\frac{1000 \times 50}{280}$ = 178.6 mm

Hence reinforcement provided earlier (*i.e.* 8 mm Φ @ 160 mm c/c) is sufficient.

**(ii) Ring Beam $B_1$:** The total radial force has been reduced from 27102 N/m to 8634.3 N/m. The reduced reinforcement will be

$$A_{sh} = \frac{8634.3 \times 7}{150} = 403 \text{ mm}^2$$

Hence number of 20 mm Φ rings ≈ 2. However, retain 4 rings of 20 mm Φ provided earlier. The bending moment induced in the beam will be negligible.

**(iii) Tank wall:** The tank wall is subjected to a hoop tension of 104925 N/m. If the tank wall were free at top and bottom, it would be subjected to this hoop tension at a depth of 104925/ (9800 × 7) = 1.53 m. Hence it will be assumed that hoop tension remains constant from top to a depth of 1.53 m.

Hoop steel at top = 104925/150 = 699.5 ≈ 700 mm²/m

Providing rings on both faces, $A_{sh}$ on each face = 350 mm².

Spacing of 12 mm Φ rings = $\frac{1000 \times 113}{350}$ = 323 mm.

Hence the 12 mm Φ rings provided @ 240 mm c/c earlier will serve the purpose.

The B.M. of 2041.55 N-m (clockwise) at the top will cause tension at the outside face.

$$A_{st} = \frac{2041.55 \times 1000}{150 \times 0.874 \times 180} = 86.5 \text{ mm}^2$$

The distribution reinforcement of 325 mm² will therefore be sufficient to resist the B.M.

**618** REINFORCED CONCRETE STRUCTURE

### 2. Effects of Continuity at the Junction of Cylindrical Portion, Beam $B_3$ and Conical Dome

**(a) Membrane Deformations and Stiffnesses**

**(i) Cylindrical wall:** The moment and thrust stiffnesses will be the same, as determined earlier.

**(ii) Ring Beam $B_3$:** Width = 1000 mm; depth = 600 mm.

Radial thrust to cause unit outward deflection is given by

$$H = \frac{Ebd}{R^2} = \frac{E \times 1 (0.6)}{7^2} = 12200 \times 10^{-6} \text{ E N/m}$$

Radial moment to cause unit rotation of the beam is given by

$$M = \frac{Ebd^3}{12 R^2} = \frac{E \times 1 (0.6)^3}{12 (7)^2} = 367.3 \times 10^{-3} \text{ E N-m/m}.$$

**(iii) Conical dome:** Hoop tension at top = 555079 N

Hoop tension at bottom = 535075 N

The outward deflection $\delta$ at top $= \dfrac{555079 \times 7}{E \times 0.4} = \dfrac{9713883}{E}$ m

The outward deflection $\delta_0$ at bottom $= \dfrac{535075 \times 5}{E \times 0.4} = \dfrac{6688438}{E}$ m

Slope at top edges is given by $\psi = -\dfrac{\tan \phi_0}{E t}(2 T_2 + T_1)$ ...(22.97)

where $T_2$ = hoop tension at top = 555079 N.

$T_1$ = meridional thrust
$= W \sec \phi_0 = 58474 \sec 45° = 82695$

$$\psi = -\frac{\tan 45°}{E \times 0.4}(2 \times 555079 + 82695)$$

$$= -\frac{2982132}{E} \text{ radians } (i.e \text{ anticlockwise})$$

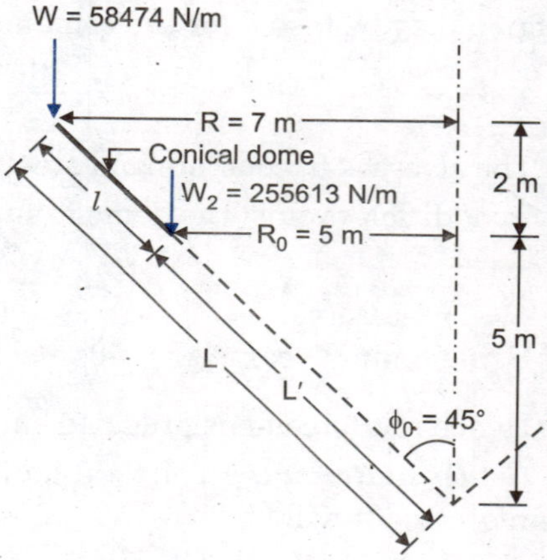

Slope at bottom edge, $\psi_0 = -\dfrac{\tan \phi_0}{Et}(2 T_2 + T_1)$,

where $T_2$ = hoop tension at bottom = 535075

$T_1$ = meridional thrust
$= W_2 \sec \phi_0 = 255613 \sec 45° = 361491$

$$\therefore \psi = -\frac{\tan 45°}{E \times 0.4}(2 \times 535075 + 361491) = -\frac{3579103}{E} \text{ radians } (i.e. \text{ anticlockwise})$$

Moment Stiffness at top $M = \dfrac{E t K_4 L}{(K_1 K_4 - K_2 K_3) \tan^2 \phi_0}$ ...(22.98)

and corresponding thrust $H$ is $H = \dfrac{E t K_2}{(K_1 K_4 - K_2 K_3) \sin \phi_0 \tan \phi_\theta}$ ...(22.99)

The values of coefficients $K_1, K_2, K_3$ and $K_4$ can be found from Table 22.19,

where $Z = 2\Delta \sqrt{L}$ ...(22.100)

and $\Delta = \left(\dfrac{12 \cot^2 \phi_0}{t^2}\right)^{1/4}$ ...(22.101)

Fig. 22.37

For our case (Fig. 22.37), $l = 2\sqrt{2} = 2.83$ m

$\therefore$ $\qquad L' = R_0 \csc \phi_0 = 5\sqrt{2} = 7.07$ m; $L = l + L' = 2.83 + 7.07 = 9.9$ m

and $\qquad \Delta = \left[\dfrac{12 \cot^2 45°}{(0.4)^2}\right]^{1/4} = 2.94;\quad Z = 2 \times 2.94 \sqrt{9.9} = 18.5$

**TABLE 22.19.** Coefficients for conical dome

| Z | $K_1$ | $K_2 = K_3$ | $K_4$ | Z' | $K_1'$ | $K_2' = K_3'$ | $K_4'$ |
|---|-------|-------------|-------|----|--------|---------------|--------|
| 10 | 183.0 | 25.20 | 6.79 | 10 | 169.0 | 23.55 | 7.30 |
| 11 | 220.5 | 27.25 | 7.21 | 11 | 236.5 | 29.35 | 7.90 |
| 12 | 288.0 | 32.60 | 7.91 | 12 | 312.0 | 36.0 | 8.70 |
| 13 | 366.5 | 38.40 | 8.60 | 13 | 396.0 | 42.20 | 9.40 |
| 14 | 463.0 | 46.20 | 9.36 | 14 | 490.0 | 48.85 | 10.09 |
| 15 | 568.5 | 52.50 | 10.10 | 15 | 596.0 | 56.30 | 10.82 |
| 16 | 696.0 | 60.30 | 10.82 | 16 | 728.0 | 63.20 | 11.42 |
| 17 | 834.0 | 68.40 | 11.52 | 17 | 880.0 | 72.30 | 12.13 |
| 18 | 990.0 | 76.65 | 12.24 | 18 | 1038 | 81.20 | 12.97 |
| 19 | 1172 | 86.00 | 12.98 | 19 | 1222 | 89.70 | 13.61 |
| 20 | 1370 | 95.90 | 13.69 | 20 | 1430 | 100.0 | 14.30 |
| 21 | 1574 | 105.6 | 14.39 | 21 | 1658 | 111.0 | 15.10 |
| 22 | 1830 | 116.3 | 15.10 | 22 | 1900 | 121.1 | 15.70 |
| 23 | 2100 | 127.8 | 15.88 | 23 | 2190 | 134.0 | 16.50 |
| 24 | 2400 | 139.9 | 16.45 | 24 | 2470 | 145.0 | 17.16 |
| 25 | 2695 | 175.0 | 17.24 | 25 | 2800 | 156.9 | 17.80 |

Hence from Table 22.19, by interpolation

$\therefore \qquad K_1 = 1081;\quad K_2 = K_3 = 81.3;\quad K_4 = 12.61$

$\therefore \qquad M = \dfrac{E \times 0.4 \times 12.61 \times 9.9}{(1081 \times 12.61 - 81.3 \times 81.3) \times 1 \times 1} = 7111 \times 10^{-6} E$ (clockwise)

$\qquad H = \dfrac{E \times 0.4 \times 81.3}{(1081 \times 12.61 - 81.3 \times 81.3)\, 0.707 \times 1} = 6551 \times 10^{-6} E$ (outward)

Thrust stiffness at top: $\quad H = \dfrac{E t K_1}{(K_1 K_4 - K_2 K_3) L \sin^2 \phi_0}$ ...(22.102)

$\qquad = \dfrac{E \times 0.4 \times 1081}{(1081 \times 12.61 - 81.3 \times 81.3) \times 9.9\,(0.707)^2} = 12444 \times 10^{-6} E$ (inward)

Corresponding moment is given by

$\qquad M = \dfrac{E t K_3}{(K_1 K_4 - K_2 K_3) \sin \phi_0 \tan \phi_0}$ ...(22.103)

$\qquad = \dfrac{E \times 0.4 \times 81.3}{(1081 \times 12.61 - 81.3 \times 81.3)\, 0.707 \times 1} = 6551 \times 10^{-6} E$ (anticlokwise).

*Moment and thrust stiffness at bottom*: These are given by the same expressions (Eq. 22.98 to 22.103) as used for the top, except that $K_1'$, $K_2'$, $K_3'$, and $K_4'$, and $L'$ should be used in place of $K_1$, $K_2$, $K_3$, $K_4$, and $L$ respectively. The values of these coefficients are given in table 22.19, where $Z' = 2\Delta\sqrt{L'}$ $= 2 \times 2.94 \sqrt{7.07} = 10.63$.

Hence from Table 22.19 by interpolation,

$\qquad K_1' = 679.2\,;\quad K_2' = K_3' = 60.6\,;\quad K_4' = 11.2.$ Hence we have,

**620   REINFORCED CONCRETE STRUCTURE**

Moment stiffness $\quad M = \dfrac{E \times 0.4 \times 11.2 \times 7.07}{(679.2 \times 11.2 - 60.6 \times 60.6) \times 1 \times 1} = 8050 \times 10^{-6} E \quad \text{(clockwise)}$

Corresponding $\quad H = \dfrac{E \times 0.4 \times 60.6}{(679.2 \times 11.2 - 60.6 \times 60.6) \times 0.707 \times 1} = 8714 \times 10^{-6} E \quad \text{(inward)}$

Thrust stiffness $\quad H = \dfrac{E \times 0.4 \times 679.2}{(679.2 \times 11.2 - 60.6 \times 60.6) \times 0.707 (0.707)^2} = 19358 \times 10^{-6} E \quad \text{(inward)}$

Corresponding $\quad M = \dfrac{E \times 0.4 \times 60.6}{(679.2 \times 11.2 - 60.6 \times 60.6) \, 0.707 \times 1} = 8714 \times 10^{-6} E \quad \text{(clockwise)}$

**(b) Reaction due to continuity:** Let the net rotation be $\psi$ (clockwise) and net displacement be $\delta$ (inward). The changes in the slope and displacement of various members from the membrane state will be as tabulated in Table 22.20.

**TABLE 22.20**

| Member | Clockwise slope | Inward displacement |
|---|---|---|
| Wall | $\psi - 1920800/E$ | $\delta + 9620800/E$ |
| Beam | $\psi$ | $\delta$ |
| Conical dome | $\psi + 2982132/E$ | $\delta + 9713883/E$ |

The reactions imposed by the joint on each member will be equal to the product of the above changes in slopes and deflections, and the corresponding stiffness, and they are tabulated in Table 22.21. In addition to these, there will be following three reactions:

(i) *Balcony moment.* An anticlockwise moment of about 700 N-m per m may be assumed. The reactive moment will be 700 N-m/m in clockwise direction.

(ii) Water pressure on beam height : Height of beam = 0.6 m.

$\therefore \quad$ Water pressure = $(9800 \times 5) \, 0.6 = 29400$ N/m($\leftarrow$)

Hence reactive thrust = 29400 N/m (inward)

(iii) Thrust from conical dome = $W \tan \phi_0$ (Fig. 22.37) = $58474 \tan 45° = 58474$ N

$\therefore \quad$ Reactive inward thrust = 58474 N

All the above three reactions have been included in Table 22.21. For the equilibrium of the joint, $\Sigma M = 0$ and $\Sigma H = 0$

**TABLE 22.21.   Joint $B_3$**

| Member | Clockwise Moment M | Inward Thrust H |
|---|---|---|
| 1. Wall | $(\psi - 1920800/E) \times 2586 \times 10^{-6} E$ | $2573 \times 10^{-6} E \, (\psi - 1920800/E)$ |
| | $= 2586 \times 10^{-6} \psi E - 4967$ | $= 2573 \times 10^{-6} \psi E - 4942$ |
| | $2573 \times 10^{-6} E \, (\delta + 9620800/E)$ | $(\delta + 9620800/E) \, 5119 \times 10^{-6} E$ |
| | $= 2573 \times 10^{-6} \delta E + 24754$ | $= 5119 \times 10^{-6} \delta E + 49249$ |
| 2. Beam $B_3$ | $367.3 \times 10^{-6} E \, \psi$ | $12200 \times 10^{-6} E \, \delta$ |
| 3. Conical dome | $(\psi + 2982132/E) \, 7111 \times 10^{-6} E$ | $-6551 \times 10^{-6} E \, (\psi + 2982132/E)$ |
| | $= 7111 \times 10^{-6} \psi E + 21206$ | $= -6551 \times 10^{-6} \psi E - 19536$ |
| | $-6551 \times 10^{-6} E \, (\delta + 9713883)$ | $(\delta + 9713883/E) \, 12444 \times 10^{-6} E$ |
| | $= -6551 \times 10^{-6} \delta E - 63636$ | $= 12444 \times 10^{-6} \delta E + 120880$ |
| 4. Balcony moment | + 700 | |
| 5. Water pressure on beam $B_3$ | | + 29400 |
| 6. Thrust from conical dome | | + 58474 |
| | $\Sigma M = 10064.3 \times 10^{-6} \psi E$ $- 3978 \times 10^{-6} \delta E - 21943$ | $\Sigma H = -3978 \times 10^{-6} \psi E + 29763$ $\times 10^{-6} \delta E + 233525$ |

# WATER TANKS-II : CIRCULAR AND INTZE TANKS

Hence from Table 22.21, $10064.3 \times 10^{-6} \psi E - 3978 \times 10^{-6} \delta E - 21943 = 0$
and $-3978 \times 10^{-6} \psi E + 29763 \times 10^{-6} \delta E + 233525 = 0$

Solving these, we get

$$\psi = -0.9723 \times 10^6 / E \text{ and } \delta = -7.976 \times 10^6/E$$

Substituting the values of $\psi$ and $\delta$, the moment and thrust in the members will be as tabulated in Table 22.22.

The net hoop tension $= -\dfrac{\delta E}{D/2} A$ (N/m), where $A$ is the area of cross-section of the member.

**TABLE 22.22**

| Member | M (N-m) | H (N) | Net hoop tension = $-\dfrac{\delta E}{7} \times A$ (N/m) |
|---|---|---|---|
| Wall | – 3250 | + 976 | 284857 |
| Beam $B_3$ | – 357 | – 97307 | 683657 |
| Conical done | + 2907 | + 8460 | 455771 |

**(c) Design of members**

**(i) Tank wall:** Hoop tension at bottom edge = 284857 N

Moment at bottom edge = 3250 N-m (anticlockwise)

Hoop steel = 284857/150 = 1899 mm². $A_s$ on each face = 950 mm².

∴ Spacing of 12 mm Φ rings = $\dfrac{1000 \times 113}{950}$ = 118.9 mm

Hence provide 12 mm Φ rings on each face @ 110 mm c/c at bottom; increase this spacing gradually to 240 mm c/c upto 1.48 m from top. After that, keep spacing constant @ 240 mm c/c. The B.M will cause tension at the outer face.

$$A_{st} = \dfrac{3250 \times 1000}{150 \times 0.874 \times 280} = 88.5 \text{ mm}^2.$$

Distribution reinforcement already provided = 325 mm²; hence it will take care of the bending moment.

**(ii) Ring Beam $B_3$:** The ring beam $B_3$ is subjected to a hoop tension of 683657 N against 615118 N found by membrane analysis. Area of rings is given by

$$A_s = 683657/150 = 4558 \text{ mm}^2. \quad \text{Number of 30 mm } \Phi \text{ bars}$$
$$= 4558/706.8 = 6.5$$

However provide 8 rings of 30 mm Φ bars. Actual $A_s$ = 5654 mm²

∴ Hoop stress = $\dfrac{683657}{1000 \times 600 + 11 \times 5654}$ = 1.03 N/mm². (safe)

The beam is also subjected to moment of 357 N-m which is negligibly small.

**(iii) Conical dome:** Hoop tension at top edge = 455771 N against 555079 N found earlier by membrane analysis. Hence no change in the hoop reinforcement is necessary. The conical dome is also subjected to a bending moment of 2907 N-m.

$$A_{st} = \dfrac{2907 \times 1000}{150 \times 0.874 \times 380} = 58 \text{ mm}^2.$$

Distribution reinforcement of 430 mm² provided earlier at the outer face will take care of this.

**3. Effects of continuity at the junction of conical dome, beam $B_2$ and bottom dome**

**(a) Member deformations and stiffnesses**

**(i) Conical dome:** Calculations for this have already been made earlier.

**(ii) Ring beam $B_2$:** The axis of the ring beam does not coincide with the intersection of the axes of bottom dome and conical dome. Due to this the beam will be subjected to a radial moment also, in addition to the moment and bending and radial thrust at the top edge. The radial moment will be equal to the radial thrust multiplied by half the height of the beam, since it is assumed that the bending moment and radial thrust act at the top edge of the beam, though they strictly act at the point of intersection of the axes of the conical dome and the bottom dome. In order to derive the expression for *moment stiffness*, let us apply a clockwise moment $M$/unit circumference at the top edge of the beam. For no radial movement of the top edge, a corresponding outward radial force $H$ should be applied at the top.

$$\text{Net clockwise moment} = \left(M - H \cdot \frac{d}{2}\right); \quad \text{Angular rotation } \phi = \frac{12 R_0^2}{bd^3 E}\left(M - H\frac{d}{2}\right) \quad \ldots(i)$$

Inward movement of the top edge due to angular rotation $= \phi \cdot \dfrac{d}{2}$.

Outward movement of the top edge due to hoop force $H = \dfrac{HR_0^2}{bdE}$

$$\therefore \quad \phi \frac{d}{2} = \frac{HR_0^2}{bd\,E} \quad \ldots(ii)$$

Eliminating $H$ from ($i$) and ($ii$), we get

$$\phi = \frac{3 R_0^2}{bd^3 E} M$$

or  Moment stiffness $M = \dfrac{bd^3 E}{3 R_0^2}$ per radian  $\ldots(22.104)$

Corresponding outward thrust $H = \dfrac{bd^2 E}{2R_0^2}$ per radian  $\ldots(22.105)$

Again, to derive an expression for thrust stiffness, apply an inward radial thrust $H$, and a corresponding anti-clockwise moment $M$ at the top edge.

For no angular rotation, $H \cdot \dfrac{d}{2} = M$. Total inward radial movement $= \dfrac{HR_0^2}{bdE}$

$\therefore$ Thrust stiffness $H = (bdE)/(R_0^2)$ per unit movement  $\ldots(22.106)$

Corresponding anti-clockwise edge moment

$$M = \frac{bd^2 E}{2 R_0^2} \text{ per unit movement} \quad \ldots(22.107)$$

For our case, $b = 600$ mm, $d = 1200$ mm, $R_0 = 5$ m

$\therefore$ Moment stiffness $M = \dfrac{0.6(1.2)^3 E}{3(5)^2}$ (clockwise) $= 13824 \times 10^{-6} E$ N-m per radian

Corresponding $H = \dfrac{0.6(1.2)^2 E}{2(5)^2}$ (outward) $= 17280 \times 10^{-6}$ N per radian.

Thrust stiffness $H = \dfrac{0.6(1.2) E}{(5)^2}$ per unit movement $= 28800 \times 10^{-6} E$ N (inward)

Corresponding moment $= \dfrac{0.6(1.2)^2 E}{2(5)^2}$ (anticlockwise) $= 17280 \times 10^{-6} E$ N-m per unit movement.

*It should be noted since the beam is monolithically joined to the columns, the stiffness calculated above will increase.*

(iii) **Bottom dome:** The bottom dome is subjected to three types of loads:

(a) U.D.L. $p$ due to self weight:
   $p = 0.25 \times 25000 = 6250$ N/m$^2$

(b) U.D.L. $p'$ due to weight of water contained above the crown level
   $p' = 9800(5 + 2 - 1.6) = 52920$ N/m$^2$

(c) Radial water pressure acting radially from zero at the crown to maximum near edges.

The clockwise slopes and horizontal outward movement of the left edge, due to the above three forces are given by the following expressions:

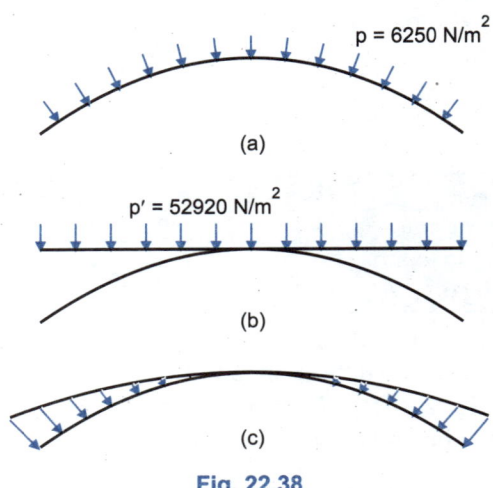

Fig. 22.38

$$\psi = \frac{2 p R_2 \sin \phi_2}{Et_2} + 0 - \frac{wR_2^2 \sin \phi_2}{Et_2} \qquad \ldots(22.108)$$

$$= \frac{2 \times 6250 \times 8.61 \times 0.5807}{E \times 0.25} - \frac{9800(8.61)^2 \times 0.5807}{E \times 0.25} = -\frac{1437510}{E} \quad (i.e.\ \text{anticlockwise})$$

$$\delta = \frac{pR_2^2 \sin \phi_2}{Et_2}\left[\frac{1}{1+\cos\phi_2} - \cos\phi_2\right] - \frac{p' R_2^2 \sin \phi_2}{2Et_2} + \frac{wR^3 \sin \phi_2}{6Et_2(1+\cos\phi_2)}[2\cos 2\phi_2 + \cos\phi_2 - 3] \ldots(22.109)$$

$$= \frac{6250(8.61)^2 \times 0.5807}{E \times 0.25}\left[\frac{1}{1+0.8141} - 0.8141\right] - \frac{52929(8.61)^2 \times 0.5807}{2E(0.25)}$$

$$+ \frac{9800(8.61)^3 \times 0.5807}{6E(0.25)(1+0.8141)}[2 \times 0.3256 + 0.8141 - 3]$$

$$= -\frac{282896}{E} - \frac{4557029}{E} - \frac{2048606}{E} = -\frac{6888531}{E} \quad (i.e.\ \text{inward})$$

For the dome $\quad \alpha_2^4 = 3(R_2/t_2)^2 = 3(8.61/0.25)^2 = 3558;\quad \alpha_2 = 7.72$

$\alpha_2^3 = 460.5;\quad \alpha_2^2 = 59.65;\quad K_2 = 1 - \dfrac{1}{2\alpha_2}\cot\phi_2 = 1 - \dfrac{1}{2 \times 7.72} \times 1.4019 = 0.9092$

The moment and thrust stiffness are given by Eqs. 22.89 to 22.90.

Moment stiffness $\quad M = \dfrac{ER_2 t_2}{4\alpha_2^3}\left(K_2 + \dfrac{1}{K_2}\right)$ (clockwise)

$$= \frac{E(8.61)(0.25)}{4 \times 460.5}\left[0.9092 + \frac{1}{0.9092}\right] = 2348 \times 10^{-6}\ E\ \text{N-m/m/ radian}$$

Corresponding $\quad H = \dfrac{Et_2}{2\alpha_2^2 K_2 \sin \phi_2} = \dfrac{E \times 0.25}{2 \times 59.65 \times 0.9092 \times 0.5807}$

$= 3969 \times 10^{-6}\ E$ N/m/radian ($\rightarrow$)

Thrust stiffness $\quad H = \dfrac{Et_2}{\alpha_2 R_2 K_2 \sin^2 \phi_2} = \dfrac{E \times 0.25}{7.72 \times 8.61 \times 0.9092(0.5807)^2}$

$= 12267 \times 10^{-6}\ E$ per unit displacement ($\rightarrow$)

Corresponding moment $M = \dfrac{Et_2}{2\alpha_2^2 K_2 \sin \phi_2} = \dfrac{E(0.25)}{2 \times 59.65 \times 0.9092 \times 0.5807}$

$= 3969 \times 10^{-6}\ E$ N-m/m unit displacement (clockwise)

**(b) Reactions due to continuity:** Let the reaction be $\psi$ (clockwise) and net displacement be $\delta$(inward). The changes in the slope and displacement of various members from the membrane state will be as tabulated in Table 22.23.

**TABLE 22.23**

| Member | Clockwise slope | Inward displacement |
|---|---|---|
| Conical dome | $\psi + 3579103/E$ | $\delta + 6688438/E$ |
| Beam $B_2$ | $\psi$ | $\delta$ |
| Bottom dome | $\psi + 1437510/E$ | $\delta - 6888531/E$ |

The reactions imposed by the joint on each member will be equal to the product of the above changes in slopes and deflection, and the corresponding stiffness, and they are tabulated in Table 22.24. In addition to this, the beam $B_2$ is subjected to a net inward thrust of 19448 N/m (Example 22.8). Hence the joint reaction will be an outward force of 19448 N/m.

**TABLE 22.24.** Joint $B_2$

| Member | Clockwise Moment M | Inward Thrust |
|---|---|---|
| 1. Conical dome | $(\psi + 3579103/E) \times 8050 \times 10^{-6} E$ <br> $= 8050 \times 10^{-6} \psi E + 28812$ <br> $8714 \times 10^{-6} E (\delta + 6688438/E)$ <br> $= 8714 \times 10^{-6} E \delta + 58283$ | $8714 \times 10^{-6} E (\psi + 3579103/E)$ <br> $= 8714 \times 10^{-6} E \psi + 31188$ <br> $(\delta + 6688438/E) \times 19358 \times 10^{-6} E$ <br> $= 19358 \times 10^{-6} \delta + 129475$ |
| 2. Beam $B_2$ | $13824 \times 10^{-6} E \psi$ <br> $-17280 \times 10^{-6} E \delta$ | $-17280 \times 10^{-6} E \psi$ <br> $28800 \times 10^{-6} E \delta$ |
| 3. Bottom dome | $(\psi + 1437510/E) \times 2348 \times 10^{-6} E$ <br> $= 2348 \times 10^{-6} \psi E + 3375$ <br> $3969 \times 10^{-6} E (\delta - 6888531/E)$ <br> $= 3969 \times 10^{-6} \delta E - 27340$ | $3969 \times 10^{-6} E(\psi + 1437510/E)$ <br> $= 3969 \times 10^{-6} E \psi + 5705$ <br> $(\delta - 6888531/E) \times 12267 \times 10^{-6} E$ <br> $= 12267 \times 10^{-6} E \delta - 84502$ |
| Net thrust from both domes | | $-19448$ |
| | $\Sigma M = 24222 \times 10^{-6} E \psi$ <br> $-4597 \times 10^{-6} E \delta + 63130$ | $\Sigma H = -4597 \times 10^{-6} E \psi + 60425$ <br> $\times 10^{-6} E \delta + 62418$ |

For the equilibrium of the joint, $\Sigma M = 0$ and $\Sigma H = 0$. Hence from Table 22.24

$$24222 \times 10^{-6} E\psi - 4597 \times 10^{-6} E \delta + 63130 = 0$$

and $\quad -4597 \times 10^{-6} E\psi + 60425 \times 10^{-6} E\delta + 62418 = 0$

Solving these, we get $\quad \psi = -2.843 \times 10^6/E$ and $\delta = -1.249 \times 10^6/E$

Substituting these value of $\psi$ and $\delta$ the moment and thrust in the members will be as tabulated in Table 22.25. The net hoop tension $= -\dfrac{\delta E}{D_0/2} \times A$ (N/m), where $A$ is the area of cross-section of the member.

**TABLE 22.25**

| Member | M (N-m) | H (N) | Net hoop tension $= -\dfrac{\delta \cdot E}{5} \times A$ (N/m) |
|---|---|---|---|
| Conical dome | +53325 | +111711 | 99920 |
| Beam $B_2$ | -17719 | +13156 | – |
| Bottom dome | -35598 | -105402 | 62450 |

**(e) Design of Members**

**(i) Conical dome:** B.M. at the bottom edge = 53325 (clockwise)

Hoop tension at bottom edge = 99920 N, which is much less that the maximum hoop tension of 560739 N for which the dome has been designed earlier.

Bending stress at bottom = $\dfrac{53325 \times 6}{(400)^2}$ = 2.0 N/mm². Since there is already a meridional compressive stress of 0.9 N/mm², the resulting tensile stress will be within the reasonable limits of 1.7 N/mm².

$$A_{st} = \dfrac{53325 \times 1000}{150 \times 0.874 \times 380} = 1070 \text{ mm}^2.$$

Spacing 16 mm Φ bars = $\dfrac{1000 \times 201}{1070}$ = 187.85 mm.

Hence provide them @ 185 mm c/c on water face. Two third of this reinforcement may be curtailed at a distance of 42 Φ ≈ 700 mm from the joint.

(ii) **Bottom dome:** The bottom dome is subjected to the following :
Hoop tension = 62450 N/m. Meridional thrust = 290093 N/m ≈ 290 N/mm
Bending moment = 35598 N-m/m (anti-clockwise thickness from tensile stress point of view is given by:

$$\dfrac{35598 \times 6}{t^2} - \dfrac{290}{t} = 1.7, \quad \text{which gives } t \approx 280 \text{ mm}.$$

Thickness provided earlier = 250 mm. Hence increase the thickness from 250 to 280 mm near the junction, gradually in a length of $0.76 \sqrt{R_0\, t} \approx 1$ m

$$A_{st} = \dfrac{35598 \times 1000}{150 \times 0.874 \times 260} = 1044 \text{ mm}^2.$$

Spacing of 16 mm Φ bars = $\dfrac{1000 \times 201}{1044}$ = 192.52 mm.

Hence provide 16 mm Φ bars @ 190 mm c/c near water face meridionally to take up the bending stresses. These bars may be curtailed at 1 m from the edges.

(iii) **Ring beam $B_2$:** The ring beam is subjected to a radial inward force of 13156 N/m and an anticlockwise moment of 17719 N-m/m both acting at the top edge. The inward radial force will give rise to an clockwise moment of 13156 × 0.6 = 7894 N-m/m.

Hence the beam is subjected to a net clockwise moment of 17719 − 7894 = 9825 N-m/m along the circumference. This gives rise to a hogging bending moment = $9825 \times \dfrac{D_0}{2}$ = 9825 × 5 = 49125 N-m at every section. The radial force will give rise to a hoop tension = 13156 × 5 = 65780 N. These reactions should be considered along with the bending moment and shear force for which the beam has been designed in Example 22.8. The hoop tension caused due to reactions is negligibly small. In Example 22.8, maximum hogging moment at supports = 572882 N-m. Hence it has now to be designed for a total hogging moment of 572882 + 65780 = 638662 N-m. Effective depth = 1140 mm.

$$A_{st} = \dfrac{638662 \times 1000}{150 \times 0.874 \times 1140} = 4273 \text{ mm}^2.$$

No. of 25 mm bars = $\dfrac{4273}{491}$ = 8.70

Hence provide 6 number of 25 mm Φ bars in one layer and 3 bars in the second layer, at the top of the section near the supports, as provided in Example 22.8.

**Note.** It will be seen from this example that the effect of the continuity of joints is to alter the hoop stresses near the joints and also to impose the meridional moments. *The meridional stresses are not affected.* Also these changes are only of local nature, and die out soon with the result that only membrane stresses exist in the interior of the members.

## PROBLEMS

1. Design a circular water tank of capacity 200,000 litres. The depth of the tank is limited to 3 m from inside. Keep the joint between the wall and base slab as flexible. The base slab rests on the ground. Use M 20 concrete.

2. Redesign the tank of problem 1 if the joint between the wall and the base is to made rigid. Use approximate method.

3. A circular tank has 12 m diameter and 3 m water height. Determine (i) maximum hoop tension and its location, and (ii) maximum bending moment. Use the following methods: (a) Reissner's method, (b) Carpenter's method, (c) I.S. Code method.

4. Using the analytical method, design a circular tank with fixed base, for a capacity of 200,000 litres. The depth of water is to be 3 m, including a free board of 25 cm. Use M 20 concrete. Assume $\eta = 0$. The tank is free at the top, and rests on the ground.

5. Redesign the circular tank of problem 4, if the tank is supported on a number of columns, through a ring beam. The diameter to the centre of the width of the ring beam is to be 75% of the internal diameter of the tank. The walls have continuous joint with the base slab. Design the roof slab and base slab also.

6. Redesign the circular tank of problem 4, if it has dome bottom and top. The top dome has a central rise of 1.6 m while the bottom dome has a central rise of 1.2 m. Use membrane analysis.

7. Redesign the circular tank of problem 6, taking into account the joint reactions.

8. Design an Intze tank of 60,000 litres capacity. The height of staging is 12 m upto bottom of tank. Use membrane analysis. Assume the wind pressure to be 1.5 kN/m² and bearing capacity of soil as 120 kN/m². Use M 20 concrete.

9. Redesign the Intze tank of problem 8, taking into account the continuity effects of various joints.

10. Design a circular water tank of capacity 450 m³ (450 kL) resting on the ground and having a fixed base condition due to a rigid joint between the wall and the base slab. The materials to be used are $M_{30}$ grade Fe 415 steel. Use the method recommended in IS: 3370 (Part-IV).

# CHAPTER 23

# WATER TANKS – III : RECTANGULAR TANKS

## 23.1. INTRODUCTION

For tanks of smaller capacity, the cost of shuttering for circular tanks becomes high. Rectangular tanks are, therefore, used in such circumstances. However, rectangular tanks are normally not used for large capacities since they are uneconomical and also its exact analysis is difficult. For a given capacity, perimeter is least for circular tank than for a rectangular tank. From structural and aesthetic considerations, the longer side is normally less than the twice the smaller side. The rectangular tanks should preferably be close to square in plan from point of view of economy.

The walls of a rectangular tank are subjected to bending moments both in the horizontal as well as in vertical direction. The analysis of moment in the wall is difficult since water pressure results in a triangular load on them. The magnitude of the moment will depend upon several factors such as length, breadth and height of tank, and the conditions of support of the wall at top and bottom edges. If the length of the wall is more in comparison to its height, the moments will be mainly in the vertical direction *i.e.,* the panel will bend as a cantilever. If, however, height is large in comparison to length, the moments will be in the horizontal direction, and the panel will bend as a thin slab supported on the edges. For intermediate condition bending will take place both in horizontal as well as vertical directions. In addition to the moments, the walls are also subjected to direct pull exerted by water pressure on some portion of side walls. The wall of the tank will thus be subjected to both bending moment as well as direct tension. The design of the walls is done on the premise that no cracks are developed in it. Though reinforcement is provided both for moments as well as direct tension, the maximum permissible value of tensile stresses for M 20 concrete may be taken as 1.2 N/mm$^2$ and 1.7 N/mm$^2$ respectively for direct tension and due to bending. There are two methods of analysis of rectangular tanks:

1. Approximate analysis
2. Exact analysis based on elastic theory.

## 23.2. APPROXIMATE METHOD

For the design by approximate method, rectangular tanks may be divided into two categories:
(*a*) tanks in which ratio of length to breadth is less than 2;
(*b*) tanks in which ratio of length to breadth is more than 2.

Consider the center line dimensions of tank in plan to be L × B where 'L' and 'B' are the longer and shorter spans, respectively, when the plan dimensions L × B are very large as compared to water height 'H' the wall will bend mainly as cantilever from the base and practically no load is carried horizontally.

**(a) Ratio L/B less than 2.** Let $L$ be the length of the tank and $B$ be the width of the tank from inside, and $H$ be its height. If the ratio $L/B$ is less than 2, the tank walls are designed as continuous frame subjected to a triangular load. Such a bending is assumed to take place from top to a height $h = H/4$ or 1 m above base (whichever is more). For the bottom height $h$, bending is assumed to act in vertical plane as cantilever, subjected to triangular load $D_1 BC$, having zero intensity at $D_1$ and $wH$ at the base, as shown in Fig. 23.1(a).

For horizontal bending, the maximum horizontal force per unit height, at $D$ is taken as equal to $p = w(H - h)$ per metre run. The panels are first assumed to be fixed at the supports and fixed end moments are calculated. The final moments are then computed by moment distribution. Thus, in Fig. 23.1 (b), if $p$ is water pressure at level $D$ [Fig. 23.1 (a)] the fixing moments for long slab $AB$ will be $(pL^2)/12$ while fixing moments for short slab $AD$ will be $(pB^2)/12$. Moment distribution can be carried out for the quarter frame $EAF$, with fixed corner at $A$. After having found the moment $M_f$ at $A$, the moment at the centre of $AB$ may be taken equal to $\left(\dfrac{pL^2}{8} - M_f\right)$.

Similarly, moment at the centre of the short span may be taken equal to $[(pB^2)/8] - M_f$. In the absence moment distribution, the bending moments may be computed by the following approximate expressions:

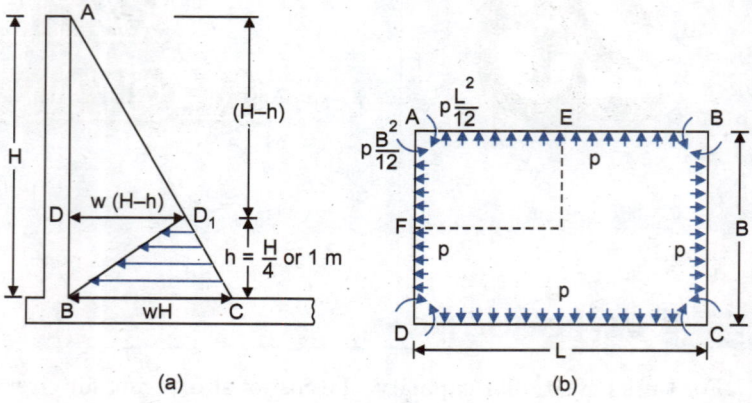

**Fig. 23.1.** Moment and forces for analysis of rectangular tank

B.M. at centre of span = $(pB^2)/16$ (producing tension on outer face) ...(23.1)

B.M. at ends of span = $(pB^2)/12$ (producing tension on water face) ...(23.2)

The bottom portion of height $h$ of each panel is designed as cantilever, in a manner similar to circular tank fixed at base. In addition to the bending moments, the walls are subjected to direct tension of following values:

Direct tension on long walls = $P_L = w(H - h)\dfrac{B}{2}$ ...(23.3)

Direct tension on short walls = $P_B = w(H - h)\dfrac{L}{2}$ ...(23.4)

The bending moments at the ends (corners) produce tension on the water face and mid span moment produces tension away from the water face. In addition to bending moments the walls are subjected to direct tension of magnitudes $\left(w(H-h)\dfrac{B}{2}\right)$ and $\left(w(H-h)\dfrac{L}{2}\right)$ on the long and short walls, respectively.

**(b) Ratio L/B greater than 2.** If the ratio $L/B$ is greater than 2, the long walls are *assumed* to bend vertically as cantilever fixed at base and subjected to triangularly distributed load. The short wall is assumed to bend horizontally, supported on long walls, for the portion from top to the point $D$ [Fig. 23.1 (a)]. The load intensity for such a bending is taken as $p = w(H - h)$. The bottom portion of height $h = H/4$ or 1 m (whichever is more) of the short wall is designed as cantilever subject to triangular load $D_1 BC$ of Fig. 23.1 (a). Thus, for long wall, maximum B.M. at the base, per unit length of the wall $(wH^3)/6$. For short walls, the maximum bending moment at level $D$, at ends and centre may be taken equal to $\dfrac{w(H-h)B^2}{16}$. Similarly, the maximum cantilever B.M. for short wall is equal to $wH\dfrac{h}{2}\cdot\dfrac{h}{3} = \dfrac{wHh^2}{6}$. In addition to above bending moments, the long and short walls are subjected to direct pulls. Since the short walls are assumed to be supported on long walls, at its ends, the water pressure acting on the short walls, at the level $D$, is transferred to long walls thus causing pull in it. The direct tension, so developed, on long walls is given by

$$P_L = w(H-h) \cdot \frac{B}{2} \qquad \ldots(23.5)$$

The long walls behave as cantilever, and hence they do not transfer any water load to short wall in the form of pull. However, it is observed that some portion of long walls near the corners transfer its water load to short walls. It is assumed that the end one metre width of long wall contributes to pull or direct tension in the short walls, its magnitude being given by

$$P_B = w(H-h) \times 1 \qquad \ldots(23.6)$$

Thus we see that long wall is subjected to cantilever bending moment as wall as direct pull. The reinforcement due to direct pull is provided in the horizontal direction while the reinforcement due to cantilever action is provided in the vertical direction on the water face. However, short walls bend horizontally for major portion of its height and hence reinforcement provided for B.M. is in the horizontal direction. The reinforcement provided for direct tension is also in the horizontal direction. Hence the section is to be designed for combined bending and direct pull effect. The approximate method of design is given below.

## Design of section subjected to combined effects of bending and pull

We have seen that walls of rectangular tanks of $L/B$ ratio less than 2, and the short walls of rectangular tanks having $L/B$ greater than 2, have reinforcements for B.M. and pull in horizontal direction. Fig. 23.2 shows such a section of wall subjected to axial pull $P$ and bending moment $M$. The nature of the B.M. is such that tension is induced to the right side of the axis. Let the distance of reinforcement from the central axis be $x$. Assume equal and opposite force $P$ (pulls) acting along the reinforcement. The section may then be considered to be subjected to a net B.M. equal to $(M - Px)$ and a pull $P$ in the reinforcement. Areas of steel are separately calculated for B.M. and pull by the following expression:

For B.M., $\qquad A_{st1} = \dfrac{M - Px}{t_1 \, j \, d} = \dfrac{M - Px}{\sigma_{st} \, j \, d} \qquad \ldots(23.7)$

For pull, $\qquad A_{st2} = \dfrac{P}{t_2} = \dfrac{P}{\sigma_{sh}} \qquad \ldots(23.8)$

where $t_1 (= \sigma_{st})$ is the safe stress in steel in bending while $t_2 = (\sigma_{sh})$ is the safe stress in steel in direct pull.

The value of $t_2$ is taken as 115 N/mm² as usual. The value of tensile stress in bending $(t_1)$ is taken equal to 115 N/mm² if it is placed on the water face and 125 N/mm² if it is used more than 225 mm away from water.

Total steel $\qquad A_{st} = A_{st1} + A_{st2}.$

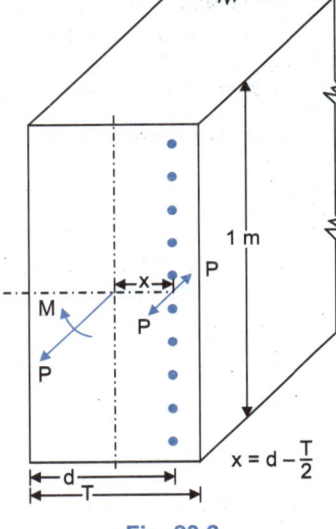

Fig. 23.2

**Example 23.1.** *An open rectangular tank 4m × 6m × 3m deep rests on firm ground. Design the tank. Use M 20 mix.*

**Solution: 1. Design constants:** For M 25 concrete, we have $\sigma_{cbc} = 8.5$ N/mm² and $m = 11$. Taking $\sigma_{st} = 130$ N/mm², we have, $k = 0.42$, $j = 0.86$, $R = 1.535$.

**2. Determination of B.M. for horizontal bending:** $L/B = 6/4 = 1.5 < 2$

Hence both long and short walls will bend horizontally for the upper portion, upto point $D$ [Fig. 23.1 (a)], where horizontal water pressure is $p = w(H - h)$.

Here $h = H/4$ or 1 m whichever is greater. $\therefore h = 1$ m.

Thus top 2 m height of walls will bend horizontally while the bottom 1 m will bend as vertical cantilever. The bending moments for horizontal bending may be determined by moment distribution by considering tank as continuous frame of unit height at level of $D$.

Water pressure $p$ at point $D$ is given by $p = w(H - h) = 9800 (3 - 1) = 19600$ N-m

The fixed end moments for long wall $= \dfrac{pL^2}{12} = \dfrac{p(6)^2}{12} = 3p$ N-m

Fixed end moments for short wall $= \dfrac{pB^2}{12} = \dfrac{p(4)^2}{12} = \dfrac{4}{3} p$ N-m.

Refer Fig. 23.1 (b). Consider quarter frame *FAE* with joint *A* as rigid. Taking clockwise moment as positive and anticlockwise moment as negative, the fixed end moment $M_{AF}$ for long wall will be $+3p$ while the fixed end moment $M_{AF}$ for short wall will be $-\dfrac{4}{3}p$.

Considering area *A* and moment of inertia *I* for both the walls to be the same, the stiffness of walls will be inversely proportional to these lengths. Thus we have following table:

| Member | Stiffness | Relative stiffness | Sum | Distribution factor |
|---|---|---|---|---|
| AE | $\frac{1}{3}$ | $\frac{1}{3} \times 6 = 2$ | 5 | $\frac{2}{5} = 0.4$ |
| AF | $\frac{1}{2}$ | $\frac{1}{2} \times 6 = 3$ |   | $\frac{3}{5} = 0.6$ |

The moment distribution is carried out in the following Table:

| Joint | A | |
|---|---|---|
| Member | AE | AF |
| Distribution Factors | 0.4 | 0.6 |
| Fixed end moments | $+3p$ | $-\frac{4}{3}p$ |
| Balancing moments | $-\frac{2}{3}p$ | $-p$ |
| Final moments ($M_f$) | $+\frac{7}{3}p$ | $-\frac{7}{3}p$ |

Hence moment at supports,

$$M_f = \frac{7}{3}p = \frac{7}{3} \times 19600 \approx 45740 \text{ N-m/m}.$$

This support moment will cause tension at the water force.

B.M. at the centre long span $= \dfrac{pL^2}{8} - M_f = \dfrac{19600\,(6)^2}{8} - 45740 = 42460$ N-m/m.

This B.M. will cause tension at the outer face.

B.M. at the centre of short span $= \dfrac{pB^2}{8} - M_f = \dfrac{19600\,(4)^2}{8} - 45740 = -6540$ N-m/m.

This will cause tension at the water face. ∴ Max. design B.M. = 45740 N-m/m.

**3. Design of section.** Considering bending effect alone, effective depth

$$d = \sqrt{\dfrac{45740 \times 1000}{1.535 \times 1000}} = 172.6 \text{ mm}$$

Provide an overal depth $T = 220$ mm so that effective depth $= 220 - 35 = 185$ mm.

**4. Determination of pull.**

Direct tension on long wall $= P_L = p \times \dfrac{B}{2} = 19600 \times \dfrac{4}{2} = 39200$ N.

Direct tension on short wall $= P_B = p \times \dfrac{L}{2} = 19600 \times \dfrac{6}{2} = 58800$ N.

**5. Cantilever moment.** Cantilever moment at the base, per unit height

$= wH \dfrac{h^2}{6} = 9800 \times 4 \dfrac{(1)^2}{6} \approx 6535$ N-m. This will cause tension on the water face.

**6. Reinforcement at corners of long walls:** The upper portion of long walls is subjected to both bending in horizontal direction as well as pull. The reinforcement for both will be in horizontal direction.

Hence reinforcement has to be provided for a net moment $(M_f - Px)$, where $M_f$ is the moment at ends (causing tension on the water face). Similarly reinforcement has to be provided for pull $P_L$ in the horizontal direction. Fig. 23.3 shows vertical section of unit height (1 m) of long wall, at its end, at the level of 1 m above the base, where reinforcement is provided at the water face.

$$x = d - \frac{T}{2} = 185 - \frac{220}{2} = 75 \text{ mm}$$

$$A_{st1} \text{ for B.M.} = \frac{M_f - P_L x}{\sigma_{st} \, jd} = \frac{(45740 \times 1000) - 39200 \times 75}{130 \times 0.86 \times 185} = 2069.3 \text{ mm}^2$$

$$A_{st} \text{ for pull} = \frac{P_L}{\sigma_s} = 39200/130 = 301.6 \text{ mm}^2$$

Total $A_{st} = A_{st1} + A_{st2} = 2069.3 + 301.6 = 2371 \text{ mm}^2$ per meter height.

Using 20 mm Φ bars, $A_\Phi = 314 \text{ mm}^2$. ∴ spacing $= \dfrac{1000 \times 314}{2371} = 132.4$ mm.

Hence provide 20 mm Φ bars @ 130 mm c/c. The above reinforcement is to be provided at the inner face, near the corners, and at a height 1 m above the base. For other heights the above spacing may be varied, since bending moment will be reduced.

**7. Reinforcement at the middle of long walls.** Tension occurs at the outer face. However, since the distance of centre of steel from water face will be less than 225 mm, permissible stress in steel will be 130 N/mm² only. Hence the same design constants, found in step 1 will be used. Design B.M. = 42460 N-m per metre height. Also

$$P_L = 39200 \text{ N}$$

$$A_{st1} = \frac{M - P_L x}{\sigma_{st} \, jd} = \frac{(42460 \times 1000) - 39200 \times 75}{130 \times 0.86 \times 185} = 1910.7 \text{ mm}^2$$

$$A_{st2} = P_L / \sigma_{sh} = 39200/130 = 301.5 \text{ mm}^2$$

Total $A_{st} = 1910.7 + 301.5 = 2212 \text{ mm}^2$ per meter height.

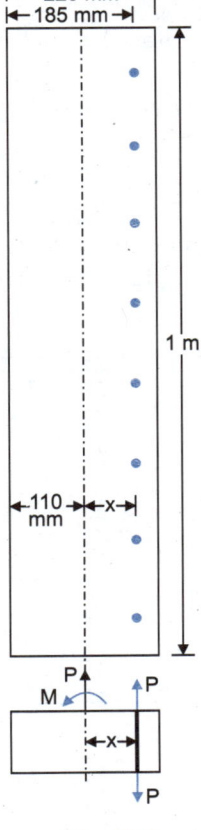

Fig. 23.3

This is very near to the reinforcement provided at the ends. Hence provide 20 mm Φ bars @ 110 mm c/c. Bend half the bars provided at ends, outwards, at distance of $L/4$ = 1.5 m from ends. Provide additional 20 mm Φ bars @ 220 mm c/c. This reinforcement is to be provided at the outer face. The additional 20 mm Φ bars provided @ 220 mm c/c are to continued upto the end.

**8. Reinforcement for short walls**

B.M. at ends = $M_f$ = 45740 N-m;     Direct pull $P_B$ = 58800 N

$$A_{st1} \text{ (for moment)} = \frac{M_f - P_B \, x}{\sigma_{st} \, jd} = \frac{45740 \times 1000 - 58800 \times 75}{130 \times 0.86 \times 185} = 1998 \text{ mm}^2.$$

$A_{st2}$ (for pull) = 58800/130 = 452 mm². Total $A_{st} = A_{st1} + A_{st2}$ = 1998 + 452 = 2450 mm²

Spacing of 20 mm Φ bars $= \dfrac{1000 \times 314}{2450} = 128$ mm

Hence provide 20 mm Φ bars @ 120 mm c/c at the inner face near the ends of short span. The B.M. at the centre of short walls causes tension at the water face (unlike that in the centre of long walls where tension is produced at the outer face). Since this B.M. is small, only nominal reinforcement is required. Similarly, we have to provide nominal reinforcement at the outer face. Hence bend half the bars outward at distance of $B/4$ = 1 m from each end, and continue remaining half throughout. Thus, at the centre of span, the reinforcement on each face will consist of 20 mm Φ bars @ 220 mm c/c.

### 9. Reinforcement for cantilever moment and distribution reinforcement

Max. cantilever moment = 6535 N-m; $A_{st} = \dfrac{6535 \times 1000}{130 \times 0.86 \times 185} = 316 \text{ mm}^2$.

But minimum reinforcement in vertical direction = $\dfrac{0.24}{100} \times (220 \times 1000) = 528 \text{ mm}^2$.

Since half of this area of steel can resist cantilever moment, we will provide 330 mm² steel area vertically on the inner face and the remaining area *i.e.* 330 mm² vertically at the outer face to serve as distribution reinforcement. ∴ Area of steel on each face = 330 mm².

Using 10 mm Φ bars, spacing = $\dfrac{1000 \times 78.5}{330} = 237.87$ *say* 230 mm c/c.

**10. Design of base slab.** Since the tank rests on ground, provide a 100 mm thick base slab. Provide nominal reinforcement of 8 mm Φ bars @ 220 mm c/c in both directions, at top and bottom of base slab. The details of reinforcement etc. are shown in Fig. 23.4.

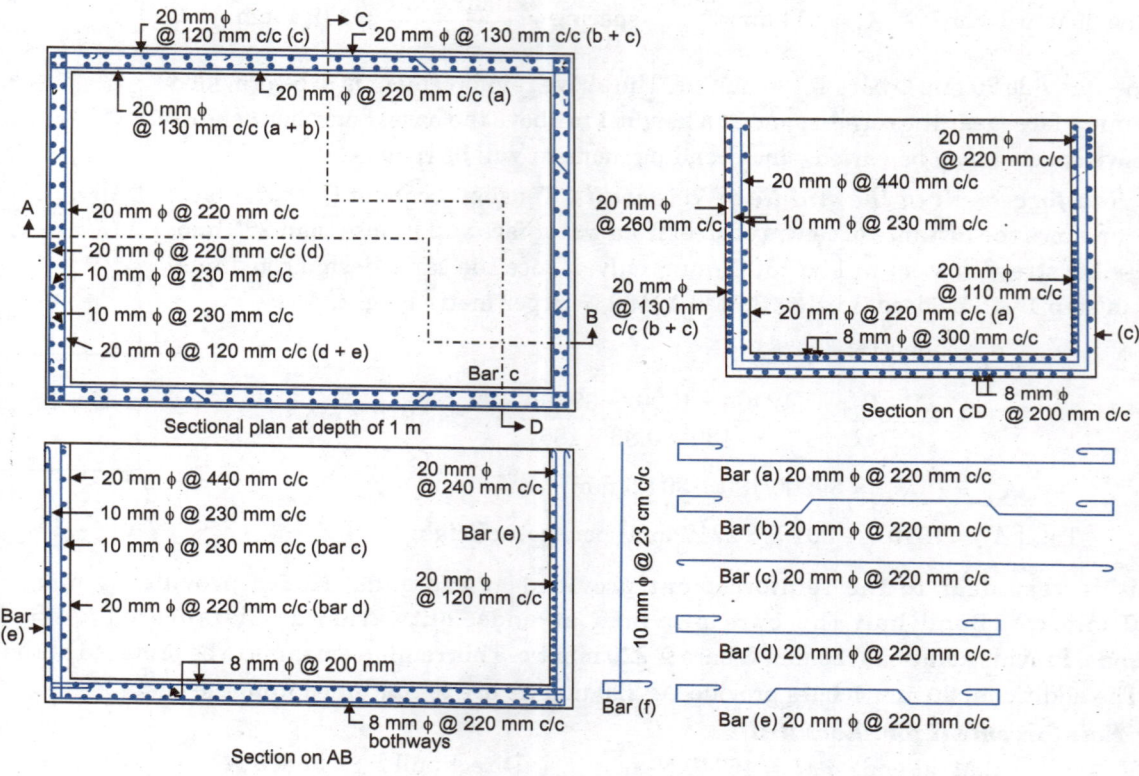

**Fig. 23.4**

## 23.3. EXACT METHOD

For the tanks where L/H and B/H ratios are neither too large nor too small, the walls are to be designed for horizontal and vertical lands. The exact analysis is based on elastic theory. The resulting differential equations are very difficult to be solved directly. IS : 3370 (Part IV) – 1967 gives the Tables for moments and shear forces in walls for certain edge conditions. In most practical cases, the corner joints between two adjacent wall can be treated rigid the top edge of the wall which supports relatively light roof slab can be treated free *(or)* hinged. The bottom edge of wall is normally built monolithic with base slab.

Moment coefficients for individual panels considered fixed along vertical edges, but having different edge conditions at top and bottom are given in Tables 23.1 to 23.3.

**TABLE 23.1.** Moment Coefficients for Individual wall Panel, Top and Bottom Hinged, Vertical Edge Fixed

| L/H | x/H | y = 0 (i.e., line CD) | | y = L/2 (i.e., edge AB) | |
|---|---|---|---|---|---|
| | | $m_x$ | $m_y$ | $m_x$ | $m_y$ |
| 3.00 | 1/4 | + 0.035 | + 0.010 | − 0.008 | − 0.039 |
| | 1/2 | + 0.057 | + 0.016 | − 0.013 | − 0.063 |
| | 3/4 | + 0.051 | + 0.013 | − 0.011 | − 0.055 |
| 2.50 | 1/4 | + 0.031 | + 0.011 | − 0.008 | − 0.038 |
| | 1/2 | + 0.052 | + 0.017 | − 0.012 | − 0.062 |
| | 3/4 | + 0.047 | + 0.015 | − 0.011 | − 0.055 |
| 2.00 | 1/4 | + 0.025 | + 0.013 | − 0.007 | − 0.037 |
| | 1/2 | + 0.042 | + 0.020 | − 0.012 | − 0.059 |
| | 3/4 | + 0.041 | + 0.016 | − 0.011 | − 0.053 |
| 1.50 | 1/4 | + 0.015 | + 0.013 | − 0.006 | − 0.032 |
| | 1/2 | + 0.028 | + 0.021 | − 0.010 | − 0.052 |
| | 3/4 | + 0.030 | + 0.017 | − 0.010 | − 0.048 |
| 1.00 | 1/4 | + 0.005 | + 0.009 | − 0.004 | − 0.020 |
| | 1/2 | + 0.011 | + 0.016 | − 0.007 | − 0.035 |
| | 3/4 | + 0.016 | + 0.014 | − 0.007 | − 0.035 |
| 0.5 | 1/4 | + 0.000 | + 0.003 | − 0.001 | − 0.005 |
| | 1/2 | + 0.001 | + 0.005 | − 0.002 | − 0.010 |
| | 3/4 | + 0.004 | + 0.007 | − 0.003 | − 0.014 |

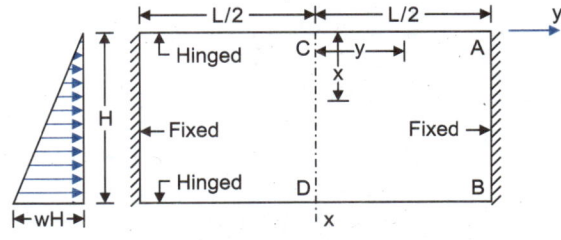

Fig. 23.5

The coefficients for individual panels with fixed side edges apply without modifications to continuous walls, provided there is no rotation about the vertical edges. In a square tank, therefore, moment coefficients may be taken direct from Tables 23.1 to 23.3. In a rectangular tank, however, an adjustment has to be made in a manner similar to the modification of fixed end moments in frame analysed by the method of moment distribution.

Table 23.4 gives the moment coefficients for rectangular tank with continuous walls, free at the top and hinged at bottom. Table 23.5 gives the moment coefficients, when the walls are hinged both at the top and the bottom. In all these tables, the values of the moments are given by the following expressions.

$$\text{Horizontal moment} \quad M_H = m_y \cdot wH^3 \quad \ldots(23.9)$$

and
$$\text{Vertical moment} \quad M_v = m_x \, wH^3 \quad \ldots(23.10)$$

**TABLE 23.2.** Moment Coefficients for Individual Wall Panel, Top Free and Bottom Hinged, Vertical Edges Fixed

| L/H | x/H | y = 0 (i.e., line CD) | | y = L/2 (i.e., edge AB) | |
|---|---|---|---|---|---|
| | | $m_x$ | $m_y$ | $m_x$ | $m_y$ |
| 3.00 | 0 | 0 | + 0.070 | 0 | − 0.196 |
| | 1/4 | + 0.028 | + 0.061 | − 0.034 | − 0.170 |
| | 1/2 | + 0.049 | + 0.049 | − 0.027 | − 0.137 |
| | 3/4 | + 0.046 | + 0.030 | − 0.017 | − 0.087 |
| 2.50 | 0 | 0 | + 0.061 | 0 | − 0.138 |
| | 1/4 | + 0.024 | + 0.053 | − 0.026 | − 0.132 |
| | 1/2 | + 0.042 | + 0.044 | − 0.023 | − 0.115 |
| | 3/4 | + 0.041 | + 0.027 | − 0.016 | − 0.078 |
| 2.00 | 0 | 0 | + 0.045 | 0 | − 0.091 |
| | 1/4 | + 0.016 | + 0.042 | − 0.019 | − 0.040 |
| | 1/2 | + 0.033 | + 0.036 | − 0.018 | − 0.089 |
| | 3/4 | + 0.035 | + 0.024 | − 0.013 | − 0.065 |
| 1.50 | 0 | 0 | + 0.027 | 0 | − 0.052 |
| | 1/4 | + 0.009 | + 0.028 | − 0.012 | − 0.059 |
| | 1/2 | + 0.022 | + 0.027 | − 0.013 | − 0.063 |
| | 3/4 | + 0.027 | + 0.020 | − 0.010 | − 0.052 |
| 1.00 | 0 | 0 | + 0.010 | 0 | − 0.019 |
| | 1/4 | + 0.002 | + 0.013 | − 0.005 | − 0.025 |
| | 1/2 | + 0.010 | + 0.017 | − 0.007 | − 0.036 |
| | 3/4 | + 0.015 | + 0.015 | − 0.007 | − 0.036 |
| 0.5 | 0 | 0 | + 0.002 | 0 | − 0.003 |
| | 1/4 | + 0.000 | + 0.004 | − 0.001 | − 0.005 |
| | 1/2 | + 0.002 | + 0.006 | − 0.002 | − 0.010 |
| | 3/4 | + 0.007 | + 0.008 | − 0.003 | − 0.014 |

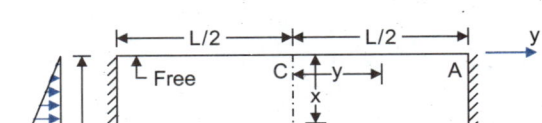

Fig. 23.6

Table 23.6 gives the shear at wall panel hinged at top and bottom, while Table 23.7 gives shear at edges of all panel free at top and hinged at bottom. The design of the walls is done on the premise that no cracks are developed in it. Though reinforcement is provided both for moments as well as direct tension the maximum permissible value of tensile stresses for M 20 concrete may be taken as 1.2 N/mm² and 1.7 N/mm² respectively for direct tension and tension due to bending.

When the tank wall is subjected to both bending moments as well as direct tension the criterion for safe design of wall thickness is governed by the following expression:

$$\frac{\sigma_{cbt'}}{\sigma_{cbt}} + \frac{\sigma_{ct'}}{\sigma_{ct}} \leq 1 \qquad \ldots(23.11)$$

where  $\sigma_{cbt}'$ = calculated bending tension stress in concrete
$\sigma_{cbt}$ = permissible bending tensile stress in concrete
$\sigma_{ct}'$ = calculated direct tensile stress in concrete
$\sigma_{ct}$ = permissible direct tensile stress in concrete.

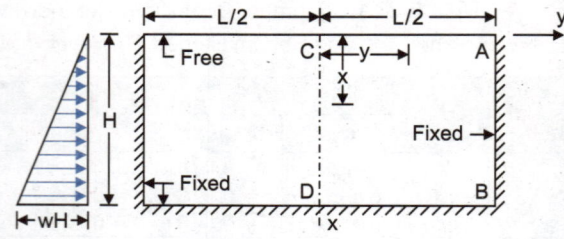

Fig. 23.7

**TABLE 23.3.** Moment Coefficient for Individual Wall Panel, Top Free Bottom and Vertical Edge Fixed

| L/H | x/H | y = 0 (i.e. line CD) | | y = L/2 (i.e. edge AB) | |
|---|---|---|---|---|---|
| | | $m_x$ | $m_y$ | $m_x$ | $m_y$ |
| 3.00 | 0 | 0 | + 0.025 | 0 | − 0.082 |
| | 1/4 | + 0.010 | + 0.019 | − 0.014 | − 0.071 |
| | 1/2 | + 0.005 | + 0.010 | − 0.011 | − 0.055 |
| | 3/4 | − 0.033 | − 0.004 | − 0.006 | − 0.028 |
| | 1 | − 0.126 | − 0.025 | 0 | 0 |
| 2.50 | 0 | 0 | + 0.027 | 0 | − 0.074 |
| | 1/4 | + 0.012 | + 0.022 | − 0.013 | − 0.066 |
| | 1/2 | + 0.011 | + 0.014 | − 0.011 | − 0.053 |
| | 3/4 | − 0.021 | − 0.001 | − 0.005 | − 0.027 |
| | 1 | − 1.108 | − 0.022 | 0 | 0 |
| 2.00 | 0 | 0 | + 0.027 | 0 | − 0.060 |
| | 1/4 | + 0.013 | + 0.023 | − 0.012 | − 0.059 |
| | 1/2 | + 0.015 | + 0.016 | − 0.100 | − 0.049 |
| | 3/4 | − 0.008 | + 0.003 | − 0.005 | − 0.027 |
| | 1 | − 0.086 | − 0.017 | 0 | 0 |
| 1.50 | 0 | 0 | + 0.021 | 0 | − 0.040 |
| | 1/4 | + 0.008 | + 0.020 | − 0.009 | − 0.044 |
| | 1/2 | + 0.016 | + 0.016 | − 0.008 | − 0.042 |
| | 3/4 | − 0.003 | − 0.006 | − 0.005 | − 0.026 |
| | 1 | − 0.060 | − 0.042 | 0 | 0 |
| 1.00 | 0 | 0 | + 0.009 | 0 | − 0.018 |
| | 1/4 | + 0.002 | + 0.011 | − 0.005 | − 0.023 |
| | 1/2 | + 0.009 | + 0.013 | − 0.006 | − 0.029 |
| | 3/4 | + 0.008 | − 0.008 | − 0.004 | − 0.020 |
| | 1 | − 0.035 | − 0.007 | 0 | 0 |
| 0.5 | 0 | 0 | + 0.001 | 0 | − 0.002 |
| | 1/4 | + 0.000 | + 0.005 | − 0.001 | − 0.004 |
| | 1/2 | + 0.002 | + 0.006 | − 0.002 | − 0.009 |
| | 3/4 | + 0.004 | + 0.006 | − 0.001 | − 0.007 |
| | 1 | − 0.015 | − 0.003 | 0 | 0 |

Similarly, when the tank wall is subjected to both bending moment as well as direct thrust, the criterion for safe design of wall thickness is governed by the following expression:

$$\frac{\sigma_{cbc'}}{\sigma_{cbt}} + \frac{\sigma_{cc'}}{\sigma_{cc}} \leq 1 \qquad \ldots(23.12)$$

where  $\sigma_{cbc'}$ = calculated bending compressive stress in concrete
$\sigma_{cbc}$ = permissible bending compressive stress in concrete
$\sigma_{cc'}$ = calculated direct compressive stress in concrete
$\sigma_{cc}$ = permissible direct compressive stress in concrete.

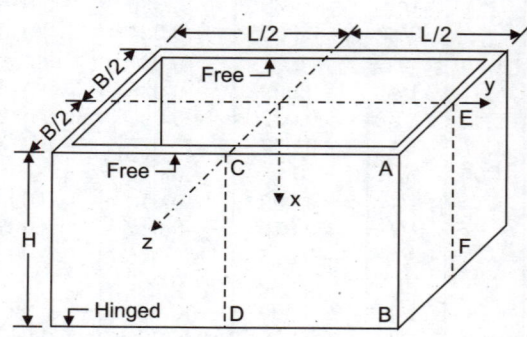

Fig. 23.8

**TABLE 23.4.** Moment Coefficients for Tanks with Walls Free at Top and Hinged at Bottom: L/H = 3

| B/H | x/H | y = 0 (line CD) | | y = L/2 (edge AB) | | z = 0 (line EF) | |
|---|---|---|---|---|---|---|---|
| | | $m_x$ | $m_y$ | $m_x$ | $m_y$ | $m_x$ | $m_y$ |
| 3.00 | 0   | 0       | + 0.070 | 0       | − 0.196 | 0       | + 0.070 |
|      | 1/4 | + 0.028 | + 0.061 | − 0.034 | − 0.170 | + 0.028 | + 0.061 |
|      | 1/2 | + 0.049 | + 0.049 | − 0.027 | − 0.137 | + 0.049 | + 0.049 |
|      | 3/4 | + 0.046 | + 0.030 | − 0.017 | − 0.087 | + 0.046 | + 0.030 |
| 2.50 | 0   | 0       | + 0.073 | 0       | − 0.169 | 0       | + 0.057 |
|      | 1/4 | + 0.028 | + 0.063 | − 0.030 | − 0.151 | + 0.022 | + 0.050 |
|      | 1/2 | + 0.049 | + 0.050 | − 0.025 | − 0.126 | + 0.041 | + 0.043 |
|      | 3/4 | + 0.046 | + 0.030 | − 0.017 | − 0.084 | + 0.040 | + 0.027 |
| 2.00 | 0   | 0       | + 0.075 | 0       | − 0.146 | 0       | + 0.031 |
|      | 1/4 | + 0.029 | + 0.065 | − 0.027 | − 0.133 | + 0.013 | + 0.032 |
|      | 1/2 | + 0.050 | + 0.051 | − 0.023 | − 0.113 | + 0.030 | + 0.029 |
|      | 3/4 | + 0.046 | + 0.031 | − 0.016 | − 0.078 | + 0.034 | + 0.020 |
| 1.50 | 0   | 0       | + 0.077 | 0       | − 0.129 | 0       | − 0.006 |
|      | 1/4 | + 0.035 | + 0.010 | − 0.007 | − 0.035 | + 0.014 | + 0.013 |
|      | 1/2 | + 0.057 | + 0.015 | − 0.011 | − 0.057 | + 0.027 | + 0.020 |
|      | 3/4 | + 0.051 | + 0.013 | − 0.010 | − 0.051 | + 0.029 | + 0.017 |
| 1.00 | 0   | 0       | + 0.079 | 0       | − 0.118 | 0       | − 0.060 |
|      | 1/4 | + 0.035 | + 0.010 | − 0.006 | − 0.029 | + 0.002 | + 0.008 |
|      | 1/2 | + 0.057 | + 0.015 | − 0.010 | − 0.048 | + 0.007 | + 0.014 |
|      | 3/4 | + 0.051 | + 0.013 | − 0.009 | − 0.044 | + 0.013 | + 0.013 |
| 0.5  | 0   | 0       | + 0.078 | 0       | − 0.130 | 0       | − 0.123 |
|      | 1/4 | + 0.036 | + 0.010 | − 0.004 | − 0.021 | − 0.005 | − 0.008 |
|      | 1/2 | + 0.057 | + 0.015 | − 0.007 | − 0.035 | − 0.006 | − 0.010 |
|      | 3/4 | + 0.052 | + 0.013 | − 0.007 | − 0.033 | − 0.001 | − 0.004 |

**TABLE 23.4.** Contd. (Fig. 23.8). Moment Coefficients for Tanks with Walls Free at Top and Hinged at Bottom: L/H = 2.5

| B/H | x/H | y = 0 (line CD) | | y = L/2 (edge AB) | | z = 0 (line EF) | |
|---|---|---|---|---|---|---|---|
| | | $m_x$ | $m_y$ | $m_x$ | $m_y$ | $m_x$ | $m_y$ |
| 2.5 | 0   | 0       | + 0.061 | 0       | − 0.138 | 0       | + 0.061 |
|     | 1/4 | + 0.024 | + 0.053 | − 0.026 | − 0.132 | + 0.024 | + 0.053 |
|     | 1/2 | + 0.042 | + 0.044 | − 0.023 | − 0.115 | + 0.042 | + 0.044 |
|     | 3/4 | + 0.041 | + 0.027 | − 0.016 | − 0.078 | + 0.041 | + 0.027 |
| 2   | 0   | 0       | + 0.065 | 0       | − 0.118 | 0       | + 0.038 |
|     | 1/4 | + 0.025 | + 0.055 | − 0.023 | − 0.113 | + 0.015 | + 0.037 |
|     | 1/2 | + 0.043 | + 0.046 | − 0.020 | − 0.102 | + 0.032 | + 0.033 |
|     | 3/4 | + 0.042 | + 0.028 | − 0.014 | − 0.070 | + 0.034 | + 0.022 |
| 1.5 | 0   | 0       | + 0.068 | 0       | − 0.100 | 0       | + 0.008 |
|     | 1/4 | + 0.026 | + 0.058 | − 0.019 | − 0.097 | + 0.004 | + 0.013 |
|     | 1/2 | + 0.045 | + 0.047 | − 0.018 | − 0.089 | + 0.017 | + 0.017 |
|     | 3/4 | + 0.043 | + 0.029 | − 0.013 | − 0.063 | + 0.024 | + 0.015 |
| 1.0 | 0   | 0       | + 0.070 | 0       | − 0.087 | 0       | − 0.032 |
|     | 1/4 | + 0.026 | + 0.060 | − 0.017 | − 0.083 | − 0.008 | − 0.021 |
|     | 1/2 | + 0.046 | + 0.048 | − 0.015 | − 0.077 | − 0.001 | − 0.008 |
|     | 3/4 | + 0.044 | + 0.029 | − 0.011 | − 0.056 | + 0.011 | − 0.000 |
| 0.5 | 0   | 0       | + 0.069 | 0       | − 0.080 | 0       | − 0.080 |
|     | 1/4 | + 0.025 | + 0.059 | − 0.015 | − 0.075 | − 0.019 | − 0.068 |
|     | 1/2 | + 0.044 | + 0.046 | − 0.014 | − 0.068 | − 0.017 | − 0.048 |
|     | 3/4 | + 0.042 | + 0.028 | − 0.010 | − 0.052 | − 0.002 | − 0.026 |

## L/H = 2.0

| B/H | x/H | $m_x$ | $m_y$ | $m_x$ | $m_y$ | $m_x$ | $m_y$ |
|---|---|---|---|---|---|---|---|
| 2.0 | 0   | 0       | + 0.045 | 0       | − 0.091 | 0       | + 0.045 |
|     | 1/4 | + 0.016 | + 0.042 | − 0.019 | − 0.094 | + 0.016 | + 0.042 |
|     | 1/2 | + 0.033 | + 0.036 | − 0.018 | − 0.089 | + 0.033 | + 0.036 |
|     | 3/4 | + 0.036 | + 0.024 | − 0.013 | − 0.065 | + 0.036 | + 0.024 |
| 1.5 | 0   | 0       | + 0.050 | 0       | − 0.072 | 0       | + 0.018 |
|     | 1/4 | + 0.018 | + 0.046 | − 0.015 | − 0.077 | + 0.007 | + 0.020 |
|     | 1/2 | + 0.035 | + 0.039 | − 0.015 | − 0.076 | + 0.020 | + 0.022 |
|     | 3/4 | + 0.036 | + 0.025 | − 0.012 | − 0.058 | + 0.025 | + 0.017 |
| 1.0 | 0   | 0       | + 0.054 | 0       | − 0.058 | 0       | − 0.023 |
|     | 1/4 | + 0.019 | + 0.050 | − 0.012 | − 0.062 | − 0.005 | − 0.013 |
|     | 1/2 | + 0.037 | + 0.042 | − 0.013 | − 0.064 | + 0.001 | + 0.000 |
|     | 3/4 | + 0.037 | + 0.026 | − 0.010 | − 0.051 | + 0.008 | + 0.004 |
| 0.5 | 0   | 0       | + 0.054 | 0       | − 0.065 | 0       | − 0.061 |
|     | 1/4 | + 0.018 | + 0.052 | − 0.014 | − 0.068 | − 0.014 | − 0.051 |
|     | 1/2 | + 0.038 | + 0.044 | − 0.013 | − 0.064 | − 0.012 | − 0.034 |
|     | 3/4 | + 0.037 | + 0.026 | − 0.010 | − 0.050 | − 0.004 | − 0.018 |

**TABLE 23.4.** Contd. (Fig. 23.8). L/H = 1.5

| B/H | x/H | y = 0 (line CD) | | y = L/2 (edge AB) | | z = 0 (line EF) | |
|---|---|---|---|---|---|---|---|
|     |     | $m_x$ | $m_y$ | $m_x$ | $m_y$ | $m_x$ | $m_y$ |
| 1.5 | 0   | 0       | + 0.027 | 0       | − 0.052 | 0       | + 0.027 |
|     | 1/4 | + 0.009 | + 0.028 | − 0.012 | − 0.059 | + 0.009 | + 0.028 |
|     | 1/2 | + 0.022 | + 0.027 | − 0.013 | − 0.063 | + 0.022 | + 0.027 |
|     | 3/4 | + 0.027 | + 0.020 | − 0.010 | − 0.052 | + 0.027 | + 0.020 |
| 1.0 | 0   | 0       | + 0.035 | 0       | − 0.038 | 0       | − 0.006 |
|     | 1/4 | + 0.011 | + 0.034 | − 0.008 | − 0.042 | − 0.001 | + 0.001 |
|     | 1/2 | + 0.025 | + 0.032 | − 0.010 | − 0.049 | + 0.006 | + 0.010 |
|     | 3/4 | + 0.028 | + 0.022 | − 0.009 | − 0.045 | + 0.009 | + 0.010 |
| 0.50| 0   | 0       | + 0.040 | 0       | − 0.036 | 0       | − 0.028 |
|     | 1/4 | + 0.010 | + 0.037 | − 0.008 | − 0.040 | − 0.007 | − 0.027 |
|     | 1/2 | + 0.024 | + 0.034 | − 0.009 | − 0.044 | − 0.006 | − 0.020 |
|     | 3/4 | + 0.028 | + 0.022 | − 0.008 | − 0.040 | − 0.004 | − 0.010 |

**TABLE 23.4.** Contd. (Fig. 23.8). L/H = 1.00

| B/H | x/H | y = 0 (line CD) | | y = L/2 (edge AB) | | z = 0 (line EF) | |
|---|---|---|---|---|---|---|---|
|     |     | $m_x$ | $m_y$ | $m_x$ | $m_y$ | $m_x$ | $m_y$ |
| 1.0 | 0   | 0       | + 0.010 | 0       | − 0.019 | 0       | + 0.010 |
|     | 1/4 | + 0.002 | + 0.013 | − 0.005 | − 0.025 | + 0.002 | + 0.013 |
|     | 1/2 | + 0.010 | + 0.017 | − 0.007 | − 0.036 | + 0.010 | + 0.017 |
|     | 3/4 | + 0.015 | + 0.015 | − 0.007 | − 0.036 | + 0.015 | + 0.015 |
| 0.50| 0   | 0       | + 0.020 | 0       | − 0.011 | 0       | − 0.005 |
|     | 1/4 | + 0.003 | + 0.018 | − 0.004 | − 0.018 | − 0.003 | − 0.007 |
|     | 1/2 | + 0.012 | + 0.016 | − 0.006 | − 0.032 | + 0.002 | − 0.005 |
|     | 3/4 | + 0.017 | + 0.013 | − 0.006 | − 0.031 | + 0.006 | + 0.001 |

WATER TANKS – III : RECTANGULAR TANKS  637

**TABLE 23.5.** Moment Coefficients for Tanks with Walls Hinged at Top and Bottom: L/H = 1

| B/H | x/H | y = 0 (line CD) | | y = L/2 (edge AB) | | z = 0 (line EF) | |
|---|---|---|---|---|---|---|---|
| | | $m_x$ | $m_y$ | $m_x$ | $m_y$ | $m_x$ | $m_y$ |
| 1.0 | 1/4 | + 0.005 | + 0.009 | – 0.004 | – 0.020 | + 0.005 | + 0.009 |
| | 1/2 | + 0.011 | + 0.016 | – 0.007 | – 0.035 | + 0.011 | + 0.016 |
| | 3/4 | + 0.016 | + 0.015 | – 0.007 | – 0.035 | + 0.016 | + 0.015 |
| 0.50 | 1/4 | + 0.007 | + 0.011 | – 0.002 | – 0.010 | – 0.003 | – 0.002 |
| | 1/2 | + 0.015 | + 0.018 | – 0.004 | – 0.021 | – 0.003 | – 0.002 |
| | 3/4 | + 0.018 | + 0.016 | – 0.005 | – 0.026 | – 0.000 | + 0.001 |

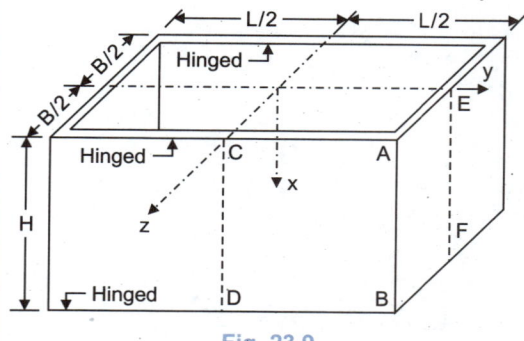

Fig. 23.9

**TABLE 23.5.** Contd. (Fig. 23.9) L/H = 1.5

| B/H | x/H | y = 0 (line CD) | | y = L/2 (edge AB) | | z = 0 (line EF) | |
|---|---|---|---|---|---|---|---|
| | | $m_x$ | $m_y$ | $m_x$ | $m_y$ | $m_x$ | $m_y$ |
| 1.50 | 1/4 | + 0.015 | + 0.013 | – 0.006 | – 0.032 | + 0.015 | + 0.013 |
| | 1/2 | + 0.028 | + 0.021 | – 0.010 | – 0.052 | + 0.028 | + 0.021 |
| | 3/4 | + 0.030 | + 0.017 | – 0.010 | – 0.048 | + 0.030 | + 0.017 |
| 1.0 | 1/4 | + 0.016 | + 0.013 | – 0.005 | – 0.025 | + 0.003 | + 0.008 |
| | 1/2 | + 0.030 | + 0.021 | – 0.009 | – 0.043 | + 0.008 | + 0.014 |
| | 3/4 | + 0.031 | + 0.017 | – 0.008 | – 0.041 | + 0.014 | + 0.014 |
| 0.50 | 1/4 | + 0.020 | + 0.016 | – 0.003 | – 0.017 | – 0.004 | – 0.006 |
| | 1/2 | + 0.035 | + 0.024 | – 0.006 | – 0.031 | – 0.005 | – 0.007 |
| | 3/4 | + 0.034 | + 0.020 | – 0.007 | – 0.033 | – 0.001 | – 0.001 |

**TABLE 23.5.** Contd. (Fig. 23.9): L/H = 2.0

| B/H | x/H | y = 0 (line CD) | | y = L/2 (edge AB) | | z = 0 (line EF) | |
|---|---|---|---|---|---|---|---|
| | | $m_x$ | $m_y$ | $m_x$ | $m_y$ | $m_x$ | $m_y$ |
| 2.00 | 1/4 | + 0.025 | + 0.013 | – 0.007 | – 0.037 | + 0.025 | + 0.013 |
| | 1/2 | + 0.042 | + 0.020 | – 0.012 | – 0.059 | + 0.042 | + 0.020 |
| | 3/4 | + 0.040 | + 0.016 | – 0.011 | – 0.053 | + 0.040 | + 0.016 |
| 1.50 | 1/4 | + 0.025 | + 0.013 | – 0.007 | – 0.034 | + 0.014 | + 0.013 |
| | 1/2 | + 0.043 | + 0.020 | – 0.011 | – 0.056 | + 0.027 | + 0.021 |
| | 3/4 | + 0.041 | + 0.016 | – 0.010 | – 0.050 | + 0.029 | + 0.017 |
| 1.0 | 1/4 | + 0.026 | + 0.013 | – 0.006 | – 0.028 | + 0.002 | + 0.008 |
| | 1/2 | + 0.044 | + 0.020 | – 0.009 | – 0.046 | + 0.007 | + 0.014 |
| | 3/4 | + 0.041 | + 0.016 | – 0.009 | – 0.044 | + 0.013 | + 0.013 |
| 0.50 | 1/4 | + 0.027 | + 0.013 | – 0.004 | – 0.021 | – 0.004 | – 0.007 |
| | 1/2 | + 0.046 | + 0.020 | – 0.007 | – 0.034 | – 0.006 | – 0.009 |
| | 3/4 | + 0.042 | + 0.016 | – 0.007 | – 0.037 | – 0.002 | – 0.003 |

### TABLE 23.5. Contd. (Fig. 23.9). L/H = 2.5

| B/H | x/H | y = 0 (line CD) | | y = L/2 (edge AB) | | z = 0 (line EF) | |
|---|---|---|---|---|---|---|---|
| | | $m_x$ | $m_y$ | $m_x$ | $m_y$ | $m_x$ | $m_y$ |
| 2.50 | 1/4 | + 0.031 | + 0.011 | − 0.008 | − 0.038 | + 0.031 | + 0.011 |
| | 1/2 | + 0.052 | + 0.017 | − 0.012 | − 0.062 | + 0.052 | + 0.017 |
| | 3/4 | + 0.047 | + 0.015 | − 0.011 | − 0.055 | + 0.047 | + 0.015 |
| 2.00 | 1/4 | + 0.031 | + 0.011 | − 0.008 | − 0.038 | + 0.025 | + 0.012 |
| | 1/2 | + 0.052 | + 0.017 | − 0.012 | − 0.061 | + 0.042 | + 0.020 |
| | 3/4 | + 0.047 | + 0.015 | − 0.011 | − 0.054 | + 0.041 | + 0.016 |
| 1.50 | 1/4 | + 0.032 | + 0.011 | − 0.007 | − 0.035 | + 0.014 | + 0.013 |
| | 1/2 | + 0.052 | + 0.018 | − 0.011 | − 0.057 | + 0.027 | + 0.021 |
| | 3/4 | + 0.047 | + 0.015 | − 0.010 | − 0.051 | + 0.029 | + 0.017 |
| 1.0 | 1/4 | + 0.032 | + 0.011 | − 0.006 | − 0.028 | + 0.002 | + 0.008 |
| | 1/2 | + 0.053 | + 0.018 | − 0.010 | − 0.048 | + 0.007 | + 0.014 |
| | 3/4 | + 0.048 | + 0.015 | − 0.009 | − 0.044 | + 0.013 | + 0.013 |
| 0.50 | 1/4 | + 0.033 | + 0.012 | − 0.004 | − 0.021 | − 0.005 | − 0.008 |
| | 1/2 | + 0.054 | + 0.018 | − 0.007 | − 0.035 | − 0.006 | − 0.010 |
| | 3/4 | + 0.049 | + 0.015 | − 0.007 | − 0.034 | − 0.001 | − 0.004 |

### TABLE 23.5. Contd. (Fig. 23.9). L/H = 3.0

| B/H | x/H | y = 0 (line CD) | | y = L/2 (edge AB) | | z = 0 (line EF) | |
|---|---|---|---|---|---|---|---|
| | | $m_x$ | $m_y$ | $m_x$ | $m_y$ | $m_x$ | $m_y$ |
| 3.00 | 1/4 | + 0.035 | + 0.010 | − 0.008 | − 0.039 | + 0.035 | + 0.010 |
| | 1/2 | + 0.057 | + 0.016 | − 0.013 | − 0.063 | + 0.057 | + 0.016 |
| | 3/4 | + 0.051 | + 0.013 | − 0.011 | − 0.055 | + 0.051 | + 0.013 |
| 2.50 | 1/4 | + 0.035 | + 0.010 | − 0.008 | − 0.039 | + 0.031 | + 0.011 |
| | 1/2 | + 0.057 | + 0.016 | − 0.012 | − 0.062 | + 0.052 | + 0.017 |
| | 3/4 | + 0.051 | + 0.013 | − 0.011 | − 0.055 | + 0.047 | + 0.014 |
| 2.00 | 1/4 | + 0.035 | + 0.010 | − 0.008 | − 0.038 | + 0.025 | + 0.013 |
| | 1/2 | + 0.057 | + 0.016 | − 0.012 | − 0.062 | + 0.043 | + 0.020 |
| | 3/4 | + 0.051 | + 0.013 | − 0.011 | − 0.054 | + 0.041 | + 0.016 |
| 1.50 | 1/4 | + 0.035 | + 0.010 | − 0.017 | − 0.035 | + 0.014 | + 0.013 |
| | 1/2 | + 0.057 | + 0.015 | − 0.011 | − 0.057 | + 0.027 | + 0.020 |
| | 3/4 | + 0.051 | + 0.013 | − 0.010 | − 0.051 | + 0.029 | + 0.017 |
| 1.0 | 1/4 | + 0.035 | + 0.010 | − 0.006 | − 0.029 | + 0.002 | + 0.008 |
| | 1/2 | + 0.057 | + 0.015 | − 0.010 | − 0.048 | + 0.007 | + 0.014 |
| | 3/4 | + 0.051 | + 0.013 | − 0.009 | − 0.044 | + 0.013 | + 0.013 |
| 0.50 | 1/4 | + 0.036 | + 0.010 | − 0.004 | − 0.021 | − 0.005 | − 0.008 |
| | 1/2 | + 0.057 | + 0.015 | − 0.007 | − 0.035 | − 0.006 | − 0.010 |
| | 3/4 | + 0.052 | + 0.013 | − 0.007 | − 0.033 | − 0.001 | − 0.004 |

**TABLE 23.6.** Shear at Edges of Wall Panel Hinged at Top and Bottom (Fig. 4.5)
(Note: Negative Sign Indicates that Reaction Acts in Direction of Load)

| Location | L/H | | | | |
|---|---|---|---|---|---|
| | 1/2 | 1 | 2 | 3 | Infinity |
| 1. Mid-point of bottom edge | $+0.141\,wH^2$ | $+0.242\,wH^2$ | $+0.329\,wH^2$ | — | $-0.333\,wH^2$ |
| 2. Corner at bottom edge | $-0.258\,wH^2$ | $-0.444\,wH^2$ | $-0.583\,wH^2$ | — | $-0.600\,wH^2$ |
| 3. Mid-point of fixed side edge | $+0.128\,wH^2$ | $+0.258\,wH^2$ | $+0.360\,wH^2$ | — | $+0.391\,wH^2$ |
| 4. Lower third-point of side edge | $+0.174\,wH^2$ | $+0.311\,wH^2$ | $+0.402\,wH^2$ | — | $+0.412\,wH^2$ |
| 5. Lower third-point of side edge | $+0.192\,wH^2$ | $+0.315\,wH^2$ | $+0.390\,wH^2$ | — | $+0.398\,wH^2$ |
| 6. Total at bottom edge | $0.00\,wH^2L$ | $0.005\,wH^2L$ | $0.054\,wH^2L$ | $0.120\,wH^2L$ | $0.167\,wH^2L$ |
| 7. Total at bottom edge | $0.048\,wH^2L$ | $0.096\,wH^2L$ | $0.182\,wH^2L$ | $0.272\,wH^2L$ | $0.333\,wH^2L$ |
| 8. Total at one fixed side edge | $0.226\,wH^2L$ | $0.199\,wH^2L$ | $0.132\,wH^2L$ | $0.054\,wH^2L$ | $0.275\,wH^2L$ |
| 9. Total at four edges | $0.500\,wH^2L$ | $0.500\,wH^2L$ | $0.500\,wH^2L$ | $0.500\,wH^2L$ | $0.500\,wH^2L$ |

**TABLE 23.7** Shear at Edges of Wall Panel free at Top and Hinged at Bottom (Fig. 4.6)
(Note: Negative Sign Indicates that the Reaction Acts in Direction of the Load)

| Location | L/H | | | | |
|---|---|---|---|---|---|
| | 1/2 | 1 | 2 | 3 |
| 1. Mid-point of bottom edge | $+0.141\,wH^2$ | $+0.242\,wH^2$ | $+0.38\,wH^2$ | $+0.45\,wH^2$ |
| 2. Corner at bottom edge | $-0.258\,wH^2$ | $-0.440\,wH^2$ | $-0.583\,wH^2$ | $-0.590\,wH^2$ |
| 3. Top of bottom edge | $+wH^2$ | $+0.010\,wH^2$ | $+0.100\,wH^2$ | $+0.165\,wH^2$ |
| 4. Mid-point of fixed side edge | $+0.128\,wH^2$ | $+0.258\,wH^2$ | $+0.375\,wH^2$ | $+0.406\,wH^2$ |
| 5. Lower third-point of side edge | $+0.174\,wH^2$ | $+0.311\,wH^2$ | $+0.406\,wH^2$ | $+0.416\,wH^2$ |
| 6. Lower-quarter-point of side edge | $+0.192\,wH^2$ | $+0.315\,wH^2$ | $+0.390\,wH^2$ | $+0.398\,wH^2$ |
| 7. Total at bottom edge | $0.048\,wH^2L$ | $0.096\,wH^2L$ | $0.204\,wH^2L$ | $0.286\,wH^2L$ |
| 8. Total at one fixed side edge | $0.226\,wH^2L$ | $0.202\,wH^2L$ | $0.148\,wH^2L$ | $0.107\,wH^2L$ |
| 9. Total at four sides | $0.500\,wH^2L$ | $0.500\,wH^2L$ | $0.500\,wH^2L$ | $0.500\,wH^2L$ |

**Example 23.2.** *A rectangular water tank 4.5 m long, 2.25 m wide and 2.25 m high has its walls rigidly jointed at the vertical edges and pin jointed at their horizontal edges. Design the tank if it is supported on all sides under the wall. Use M 25 concrete and HYSD reinforcement.*

**Solution: 1. *Design Constants***

For M 25 concrete, $\sigma_{cbc} = 8.5$ N/mm² and $m = 11$. Taking $\sigma_{st} = 130$ N/mm², we have

$$k = \frac{m\,\sigma_{cbc}}{m\,\sigma_{cbc} + \sigma_{st}} = \frac{11 \times 8.5}{11 \times 8.5 + 130} = 0.42$$

$$j = 1 - k/3 = 0.86$$

$$R = \frac{1}{2}\sigma_{cbc}\,j.k = \frac{1}{2} \times 8.5 \times 0.86 \times 0.42 = 1.535$$

**2. *Moment and Shear***

$$L = 4.5 \text{ m}; \quad B = 2.25 \text{ m}; \quad H = 2.25 \text{ m}$$

$$\frac{L}{H} = \frac{4.5}{2.25} = 2 \text{ m}\,; \quad \frac{B}{H} = \frac{2.25}{2.25} = 1$$

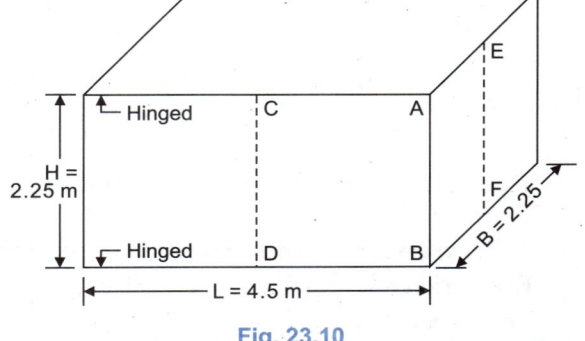

Fig. 23.10

From Table 23.2 we get the following values of moment coefficients:

| x/H | Line CD | | Line AB | | Line EF | |
|---|---|---|---|---|---|---|
| | $m_x$ | $m_y$ | $m_x$ | $m_y$ | $m_x$ | $m_y$ |
| 1/4 | + 0.026 | + 0.013 | − 0.006 | − 0.028 | + .0002 | + 0.008 |
| 1/2 | **+ 0.044** | + 0.020 | − 0.009 | **− 0.046** | + 0.007 | + 0.014 |
| 3/4 | + 0.041 | + 0.016 | − 0.009 | − 0.044 | + 0.013 | + 0.013 |

Horizontal moment $\quad M_H = m_y \cdot wH^3$
Vertical moment $\quad M_V = M_x wH^3$

Maximum value of $m_y = -0.046$. Hence maximum moment in horizontal direction occurs at the mid-point of edge $AB$.

Maximum value of $m_x = +0.044$. Hence maximum moment on vertical direction occurs at the mid-point of edge $CD$. Specific weight $w = 9800$ N/m$^3$

Maximum moment $M_H$ at edge $AB$ $\quad = -0.046 \times 9800 \times (2.25)^3$
$\quad = -5135$ N-m

Maximum moment $M_H$ at $EF$ $\quad = +0.014 \times 9800 (2.25)^3$
$\quad = +1562$ N-m

Maximum moment $M_H$ at $CD$ (at mid point) $\quad = 0.020 \times 9800 (2.25)^3$
$\quad = +2233$ N-m

Max. moment $M_V$ at $CD$ (at mid point) $\quad = +0.044 \times 9800 (2.25)^3$
$\quad = +4912$ N-m

Max. moment $M_V$ at $EF$ (at lower quarter point) $\quad = +0.013 \times 9800 (2.25)^3$
$\quad = +1452$ N-m

The horizontal moment $M_H$ in the wall will be combined with the direct tension due to shear force on adjacent wall. Similarly, vertical moment $M_V$ in the wall will be combined with the direct thrust due to weight of roof slab and wall itself, though this effect will be of minor importance. From Table 23.3, shear force at mid-point of fixed side edge $AB$ of long wall ($L/H = 2$) = $0.36\ wH^2 = 0.36 \times 9800 \times (2.25)^2 = 17860$ N. Similarly, shear force at mid-point of fixed side edge $AB$ of short wall ($B/H = 1$) = $0.258\ wH^2 = 0.258 \times 9800 (2.25)^2 = 12800$ N.

**3. Design of thickness of tank walls:** The thickness of wall is governed by moment $M_H$ and S.F. at mid-point of $AB$. $\quad M_H = 5135$ N-m; $\quad$ S.F. $S = 17860$ N.

Let the thickness of the wall be 150 mm.

The criterion for the safe design is:

$$\frac{\sigma_{cbt}'}{\sigma_{cbt}} + \frac{\sigma_{ct}'}{\sigma_{ct}} \leq 1$$

where $\sigma_{cbt}'$ = calculated bending tensile stress in concrete $\quad = \dfrac{5135 \times 1000 \times 6}{1000 (150)^2} = 1.37$ N/mm$^2$

$\sigma_{cbt}$ = permissible bending tensile stress in concrete $\quad = 1.7$ N/mm$^2$

$\sigma_{ct}'$ = calculated direct tensile stress in concrete $\quad = \dfrac{17860}{1000 \times 150} = 0.12$ N/mm$^2$

$\sigma_{ct}$ = permissible direct tensile stress in concrete $\quad = 1.2$ N/mm$^2$

$\therefore \quad \dfrac{1.37}{1.7} + \dfrac{0.12}{1.2} \leq 1 \quad$ or $\quad 0.81 + 0.1 = 0.91 < 1$. Hence safe.

Thus keep total thickness $T = 150$ mm, and an effective thickness of 120 mm. Depth of N.A $120 \times 0.442 \approx 53$ mm

**4. Reinforcement in horizontal direction**

Eccentricity of tensile force = 5135/17860 = 0.287 m = 287 mm

This eccentricity is from the centre of the thickness of the section. Eccentricity of the tensile force, measured from the centre of the steel will be = 287 − 75 + 30 = 242 mm (Fig. 23.11).

Let the area of tensile steel = $A_{st}$. To find its value, take moment about the C.G. of the compression zone. Distance of reinforcement from the C.G. of the compression zone

$$= 120 - \frac{53}{3} \approx 102 \text{ mm}.$$

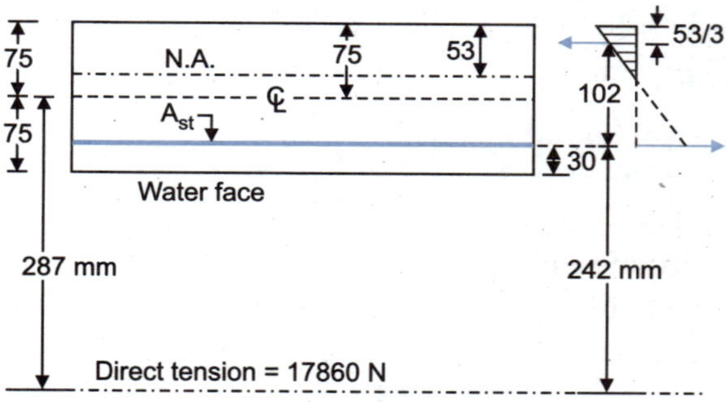

∴ $17860 (242 + 102) = A_{st} \times 130 \times 102$.

From which $A_{st} = 463.3 \text{ mm}^2$

∴ Spacing of 10 mm Φ bars $= \dfrac{1000 \times 78.5}{463.3} \approx 169.4 \text{ mm}$

Hence provide these @ 150 mm c/c on the water face.

*Maximum sagging moment* in horizontal direction occurs at mid-point of CD, and its value is 2233 N-m. The direct tension at this point = S.F. at the mid-point of vertical edge AB of short wall = 12800 N.

Eccentricity of tension = 2233/12800 = 0.174 m
= 174 mm

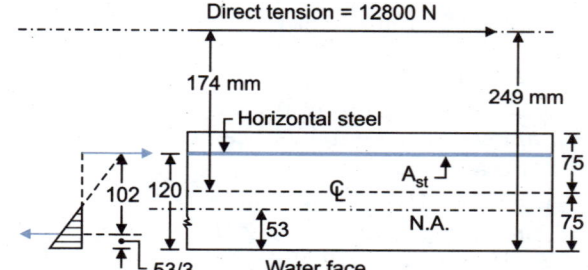

Fig. 23.11

∴ Distance of line of action of tension from water face = 174 + 75 = 249 mm
Taking moments about the centre of compression zone (Fig. 23.12),

$$A_{st} \times 130 \times 102 = 12800\left(249 - \frac{53}{3}\right).$$

From which $A_{st} = 223.3 \text{ mm}^2$

Minimum $A_{st} = \dfrac{0.24}{100}(150 \times 1000) = 360 \text{ mm}^2$

∴ Spacing of 8 mm Φ bars $= \dfrac{1000 \times 50.2}{360} = 139.4 \text{ mm}$

Hence provide 8 mm Φ bars @ 110 mm c/c.

The horizontal bending moment (sagging) at the line EF of the short wall is 1562 N-m only, but the direct tension is 17860 N. Hence provide the same reinforcement, *i.e.*, 8 mm Φ bars @ 110 mm c/c at the outer face.

Maximum shear stress

$$= \frac{17860}{1000 \times 0.86 \times 120}$$

$$= 0.173 \text{ N/mm}^2.$$

Since this is less then $\tau_c \approx 0.23$ N/mm² for 0.24% reinforcement for M 25 concrete, it is safe.

**5. Reinforcement in vertical direction:** The B.M. $M_V$ in vertical direction combines with the direct compressions due to weight of wall and weight of roof. Let the thickness of roof slab be 10 cm = 0.1 m. Assuming unit weight of concrete as 24000 N/m³,

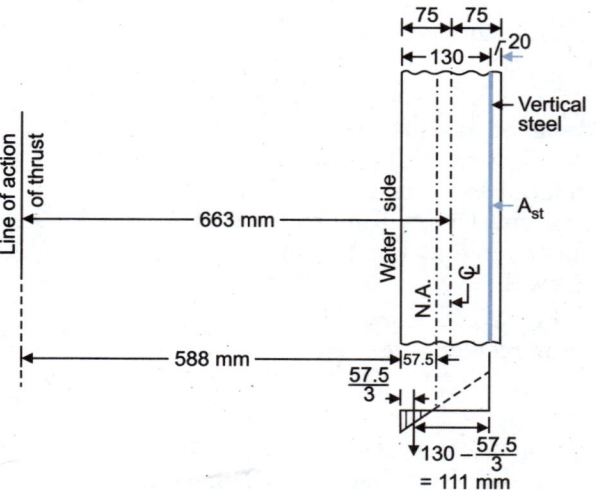

Fig. 23.12

Fig. 23.13

Dead load transferred to walls $= \left(\dfrac{2.25}{2} + 0.15\right) \times 1 \times 0.1 \, (25000) = 3188 \text{ N/m}$

Weight of wall, upto its mid-height $= \dfrac{2.25}{2} \times 0.15 \times 1 \, (25000) = 4219 \text{ N/m}$

∴ Vertical thrust in wall at its mid-height = 3188 + 4219 = 7407 N/m

B.M. in vertical direction, at mid-point of CD = 4912 N-m   (sagging)

∴ Eccentricity of thrust in vertical direction in long wall = $\dfrac{4912}{7407}$ = 0.663 m = 663 mm

Its distance from water face = 663 – 75 = 588 mm.

Let the cover to the centre of steel be 20 mm.

Effective depth = 150 – 20
= 130 mm.

Depth of N.A. = 130 × 0.442
= 57.5 mm.

Taking moments about the C.G. of compression area,

$A_{st} \times 130 \times 111$

$= 7407 \left(615 - \dfrac{57.5}{3}\right)$

From which

$A_{st}$ = 305.8 mm$^2$

Minimum area of steel @ 0.24%

$= \dfrac{0.24}{100} (1000 \times 150)$

= 360 mm$^2$

Hence provide 8 mm Φ bars @ 110 mm c/c at the outside face, in the vertical direction.

The B.M. $M_V$ in the vertical direction in short wall is 1452 N-m (sagging), against 4912 N-m in the long wall. Hence provide minimum area of steel @ 0.3% = 450 mm$^2$.

Therefore provide 8 mm Φ @ 110 mm c/c.

The top and bottom slabs can be designed as usual. While designing the bottom slab, direct tension induced in it should be taken into account. This will be equal to the shear force at the bottom edge of the wall.

Fig. 23.14 shows the details of reinforcement etc.

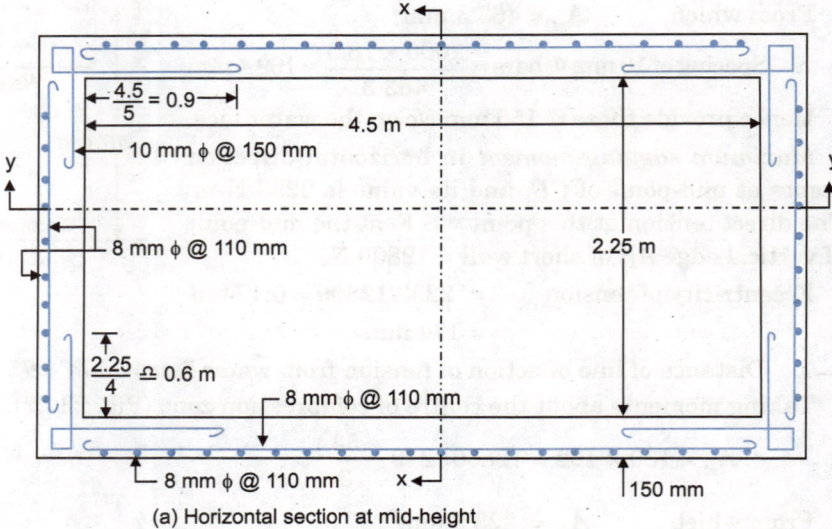

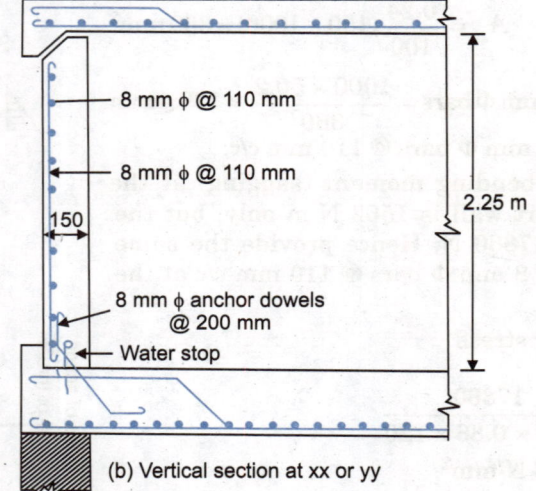

Fig. 23.14

## PROBLEMS

1. Design a square tank 5 m × 5 m × 3 m high. The tank is open at top and the walls are rigidly fixed to the base which rests on firm ground. Use approximate analysis.

2. Design a rectangular tank 5 m × 8 m × 3 m deep. The tank is open at top and the walls are rigidly fixed to the base which rests on firm ground.

3. A rectangular tank 5 m × 10 m × 4 m deep has its walls rigidly jointed at the vertical edges and pin jointed at their horizontal edges. The tank is supported on all sides under the wall using I.S. Code table, design the tank. Use M 30 mix.

4. Design a rectangular water tank resting on the ground and having a capacity of 110 kl. Overall height of tank is restricted to 4 m with a free board of 300 mm. The bearing capacity of soil at site is 180 kN/m$^2$. The materials concrete $M_{30}$ and Fe 415 steel.

# CHAPTER 24

# WATER TANKS – IV : UNDERGROUND TANKS

## 24.1. INTRODUCTION

Underground water tanks are quite common, as they are used for storage of water received from water supply mains operating at low pressures, or received from other source. Underground water tanks are usually of two shapes: (i) circular shape, and (ii) rectangular shape. Generally, circular tanks are used for large capacity. For tanks of smaller capacity, the cost of shuttering for circular tanks becomes high. Rectangular tanks are, therefore, used in such circumstances. However rectangular tanks are normally not used for large capacities since they are uneconomical and also, its exact analysis is difficult. *For a given capacity, perimeter is least for circular tank.*

When circular and rectangular tanks are situated underground, the walls of the tank are to be designed for earth pressure, as well as water pressure acting separately, and also acting simultaneously. Similarly the floors of the tanks are to be designed for hydrostatic water pressure (if water table is higher) acting upwards.

## 24.2. EARTH PRESSURE ON TANK WALLS

The walls of underground tanks are designed for earth pressure, specially under the condition when the tank is empty. Such a condition is very frequently experienced as and when the tank is emptied for cleaning purposes. The active earth pressure, which varies triangularly along the depth of tank wall, depends upon the conditions of side fill. If the water table rises upto ground, or even upto the level above the bottom of the tank, additional hydrostatic pressure due to sub-soil water will be experienced.

Fig. 24.1. shows a tank wall supporting a dry or moist backfill of cohesionless soil. The intensity of active earth pressure at the base of wall of height $H$ is given by

$$p_a = K_a \gamma H \qquad \ldots[24.1\,(a)]$$

where $K_a$ = coefficient of active earth pressure

$= \dfrac{1 - \sin \phi}{1 + \sin \phi}$ and $\gamma$ = unit weight of the soil.

The total active earth pressure is given by

$$P_a = \tfrac{1}{2} K_a \gamma H^2 \qquad \ldots[24.1\,(b)]$$

acting at $H/3$ above the base.

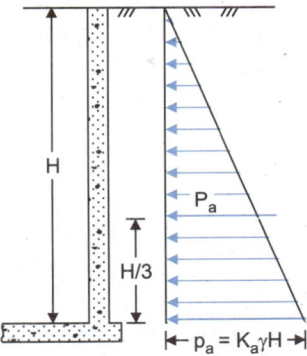

Fig. 24.1

Fig. 24.2 shows a submerged backfill. The sand fill is thus saturated with water. The lateral pressure is made up of two components:

(i) lateral pressure due to submerged weight of the soil, and

(ii) lateral pressure due to water. Thus, at any depth $z$ below the surface

$$p_a = K_a \gamma' z + \gamma_w \cdot z \qquad \ldots[24.2(a)]$$

The pressure at the base of wall $(z = H)$ is given by

$$p_a = K_a \gamma' H + \gamma_w H \qquad \ldots[24.2(b)]$$

If the backfill is partly submerged, i.e. the backfill is moist to a depth $H_1$, below the ground level, and then it is submerged, the lateral pressure intensity at the base of the wall is [Fig. 24.3(a)]

$$p_a = K_a \gamma H_1 + K_a \gamma' H_2 + \gamma_w H_2 \qquad \ldots[24.3(a)]$$

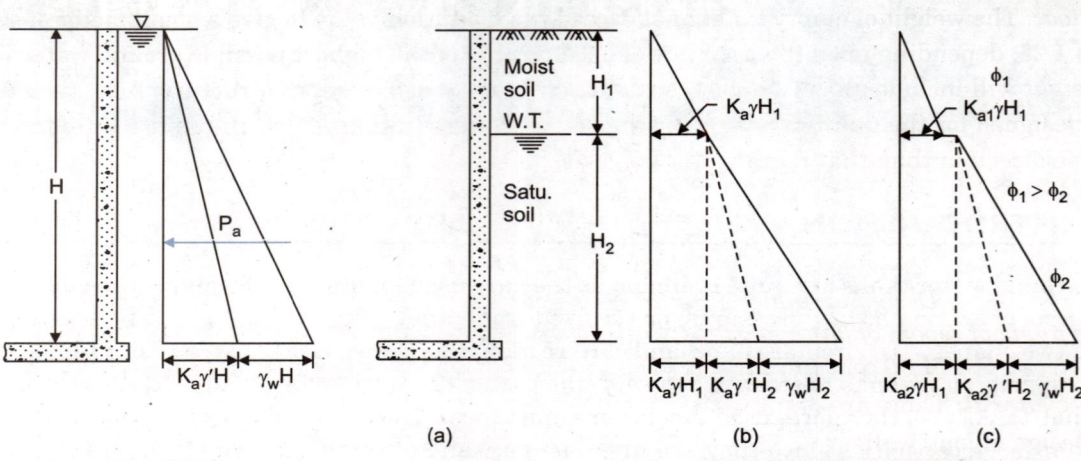

Fig. 24.2

Fig. 24.3. Partly submerged backfill

The above expression is on the assumption that the value of $\phi$ is the same for the moist as well as submerged soil. If it is different (say, $\phi_1$ and $\phi_2$ respectively), the earth pressure coefficient $K_{a1}$ and $K_{a2}$ for both the portions will be different. As $\phi$ decreases, $K_a$ increases. The lateral earth pressure at the base of the wall is given by

$$p_a = K_{a2} \gamma H_1 + K_{a2} \gamma' H_2 + \gamma_w H_2 \qquad \ldots[24.3(b)]$$

If the backfill however consists of a $c$-$\phi$ soil, the intensity of earth pressure at any depth $z$ is given by Bell's equation:

$$p_a = \gamma \cdot z \cot^2 \alpha - 2 c \cot \alpha \qquad \ldots[24.4(a)]$$

where $\alpha = 45° + \phi/2$

From Fig. 24.4 (a), at $z = 0$ (ground level), we get

$$p_a = -2 c \cot \alpha$$

Suggesting that there will be negative pressure (tension) at top level. This tension decreases to zero at a depth $z_0 = \dfrac{2c}{\gamma} \tan \alpha$

At $z = H (> z_0)$, $p_a = \gamma H \cot^2 \alpha - 2 c \cot \alpha$ ...[24.4(b)]

Fig. 24.4 shows the variation of lateral pressure $p_a$ along the height of the tank wall. It will be seen that if $c$ were equal to zero, the pressure at the base would be equal to $\gamma H \cot^2 \alpha$.

*Thus the effect of cohesion in soil is to be reduce the pressure intensity everywhere by $2 c \cot \alpha$.*

The *total net pressure* is given by

$$P_a = \tfrac{1}{2} \gamma H^2 \cot^2 \alpha - 2 c H \cot \alpha \qquad \ldots[24.4(c)]$$

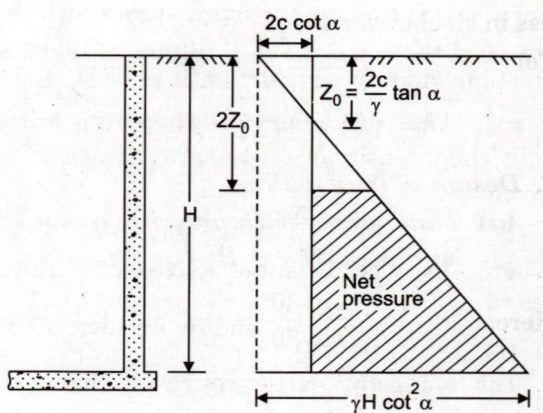

Fig. 24.4. Active earth pressure of cohesive backfill.

If however, the water table exists at a depth $H_1$ below the top of the wall, the lateral pressure at depth $z$ ($z > H_1$), is given by

$$p_a = [\gamma H_1 + \gamma'(z - H_1)] \cot^2 \alpha - 2c \cot \alpha + \gamma_w (z - H_1) \qquad \ldots(24.5)$$

## 24.3. UPLIFT PRESSURE ON THE FLOOR OF THE TANK

If the water table is below the floor level, the floor of the tank is designed for the load of tank wall, tank roof etc, assumed to be distributed evenly; the weight of water standing on the floor and the self weight of the floor are assumed to pass directly to the foundation. If however, the sub-soil water level (or ground water level) is above the floor level of the tank, uplift pressure, will be induced. *When the tank is empty, it should not float.* The weight of empty tank must exceed the floatation value to give a small factor of safety of say 1.1 to 1.25, depending upon the accuracy of local records of the highest possible ground water level. The *total weight* will include all permanent construction such as roof or super-structure. Also, the floor of the tank is designed for the uplift pressure, for the empty tank condition. This will require reinforcement in the reverse direction than that required for downward loading.

## 24.4. DESIGN OF RECTANGULAR TANK

**Design: Example 24.1.** *Design an underground water tank 4 m × 10 m × 3 m deep. The sub soil consists of sand having angle of repose of 30°, and saturated unit weight of 17 kN/m³. The water table is likely to rise upto ground level. Use M 20 concrete and HYSD bars. Take unit weight of water as 9.81 kN/m³.*

**Solution:** **1. General:** There are four components of design:
  (i) Design of long walls        (ii) Design of short walls
  (iii) Design of roof slab       (iv) Design of base slab.

The design of the walls will be done under two conditions:
  (a) Tank full with water, with no earthfill outside.
  (b) Tank empty, with full earth pressure due to saturated earth fill.

The base slab will be designed for uplift pressure and the whole tank is to be tested against floatation. As the $L/B$ ratio is greater than 2 the long walls will be designed as cantilevers. The bottom one metre (> $H/4$) of short walls will be designed as cantilever while the top portion will designed as slab supported by long walls.

**2. Design constants:** For M 20 concrete, we have $\sigma_{cbc} = 7$ N/mm² and $m \approx 13$. Since the face of wall will be in contact with water for each condition, $\sigma_{st} = 150$ N/mm² for HYSD bars. Also, permissible compressive stress in steel under direct compression is $\sigma_{sc} = 175$ N/mm².

For $\sigma_{cbc} = 7$ N/mm², $\sigma_{st} = 150$ N/mm² and $m = 13$, we have

$$k = \frac{13 \times 7}{13 \times 7 + 150} = 0.378; \quad j = 1 - 0.378/3 = 0.874; \quad R = \frac{1}{2} \times 7 \times 0.874 \times 0.378 = 1.156$$

**3. Design of long walls.**
  (a) *Tank empty with pressure of saturated soil from outside*

$$p_a = K_a \gamma' H + \gamma_w H \qquad \ldots[24.2(b)]$$

Here, $K_a = \dfrac{1 - \sin 30°}{1 + \sin 30°} = \dfrac{1}{3}$; $\gamma' = 17 - 9.81 = 7.19$ kN/m³ $= 7190$ N/m³

$\gamma_w = 9.81$ kN/m³ $= 9810$ N/m³ $\quad \therefore \quad p_a = \dfrac{1}{3} \times 7190 \times 3 + 9810 \times 3 = 36620$ N/m²

Max. B.M. at the base of wall $= 36620 \times \dfrac{3}{2} \times \dfrac{3}{3} = 54930$ N-m

$$= 54.93 \times 10^6 \text{ N-mm}$$

$$\therefore \quad d = \sqrt{\frac{54.93 \times 10^6}{1.156 \times 1000}} \approx 218 \text{ mm}$$

Provide total depth $T = 260$ mm so that $d = 260 - 35 = 225$ mm

$$A_{st} = \frac{54.93 \times 10^6}{150 \times 0.874 \times 225} = 1862.2 \text{ mm}^2$$

Using 16 mm $\Phi$ bars, spacing $= \dfrac{1000 \times 201.1}{1862.2} \approx 108$ mm. However, provide 16 mm $\Phi$ bars @ 100 mm c/c on the outside face, at the bottom of long wall.

*Curtailment of reinforcement.* Since the B.M. is proportional to $h^3$, we have:

$$\frac{A_{sth}}{A_{st}} = \frac{h^3}{H^3}. \quad \text{From which} \quad h = H \left(\frac{A_{sth}}{A_{st}}\right)^{1/3}$$

If $A_{sth} = \dfrac{1}{2} A_{st}$ (*i.e.* half the bars being curtailed), $h = H\left(\dfrac{1}{2}\right)^{\frac{1}{3}} = 3 \left(\dfrac{1}{2}\right)^{\frac{1}{3}} = 2.38$ m.

$\therefore$ Height from base $= 3 - 2.38 = 0.62$ m. However, as per Code requirements, the bars are to be continued further for a distance of $12\Phi$ ($= 12 \times 16 = 192$ mm) or $d(= 225$ mm), whichever is more, beyond this point. Hence curtail half the bars at $0.62 + 0.225 \approx 0.85$ m from base. Similarly, depth where only 1/4 th reinforcement is required is

$$h = H\left(\frac{1}{4}\right)^{\frac{1}{3}} = 3\left(\frac{1}{4}\right)^{\frac{1}{3}} = 1.89 \text{ m}. \quad \therefore \text{ Height from base} = 3 - 1.89 = 1.11 \text{ m. However, as per Code}$$

requirement, the bars are to be continued further for a distance of $12\Phi$ ($= 192$ mm) or $d(= 225$ mm), whichever is more, beyond this point. Hence curtail 3/4 of the bars at $1.11 + 0.225 \approx 1.35$ m above base.

$$\text{Minimum, \% reinforcement} = 0.3 - 0.1 \times \frac{(260 - 100)}{(450 - 100)} = 0.254 \%$$

$$\text{Min. } A_{st} = 0.254 \times \frac{260 \times 1000}{100} = 660.4 \text{ mm}^2$$

This is more than 1/4 of $A_{st}$ at the bottom. Hence the above curtailment is *not permissible*. Hence the reinforcement will be provided as under:

(i) At base: 16 $\Phi$ bars @ 100 mm c/c

(ii) At 0.85 m above base, up to top: 16 $\Phi$ bars @ 200 mm c/c.

*Distribution steel:*

$$\% \text{ distribution steel} = 0.3 - 0.1 \frac{(260 - 100)}{(450 - 100)} = 0.254 \%$$

$$\therefore \quad A_{std} = 0.254 \times \frac{260 \times 1000}{100} = 660.4 \text{ mm}^2$$

Area to be provided on each face = 330.2 mm$^2$

$$\therefore \quad \text{Spacing of 8 mm } \Phi \text{ bars} = \frac{1000 \times 50.26}{330.2} = 152.2 \text{ mm}$$

Hence provide 8 mm $\Phi$ bars @ 150 mm c/c on each face.

Actual $\quad A_{sd} = 2 \times \dfrac{1000 \times 50.26}{150} = 670 \text{ mm}^2$

**Direct compression in long walls**: The earth pressure acting on short walls will cause compression in long walls, because top portion of short walls act as slab supported on long walls.

At $h = 1$ m $\left(> \dfrac{H}{4}\right)$ above the base

$$p_a = K_a \gamma' (H - h) + \gamma_w (H - h) = \tfrac{1}{3} \times 7190 (3 - 1) + 9810 (3 - 1)$$

$$= 4793 + 19620 = 24413 \text{ N/m}^2$$

This direct compression developed on long walls is given by

$$P_{LC} = p_a \cdot \dfrac{B}{2} = 24413 \times \dfrac{4}{2} = 48826 \text{ N}$$

This will be well taken by the distribution steel and wall section.

**(b) Tank full with water, and no earth fill outside**

$$p = \gamma_w H = 9810 \times 3 = 29430 \text{ N/m}^2$$

$$M = p \cdot \dfrac{H}{2} \cdot \dfrac{H}{3} = 29430 \times \dfrac{3}{2} \times \dfrac{3}{3} = 44145 \text{ N-m} = 44.145 \times 10^6 \text{ N-mm}$$

$$A_{st} = \dfrac{44.145 \times 10^6}{150 \times 0.874 \times 225} = 1496.6 \text{ mm}^2$$

Using 16 mm Φ bars, spacing = $\dfrac{1000 \times 201.1}{1496.6} = 134.4$ mm

However, provide 16 mm Φ bars @ 130 mm c/c at the inside face.

*Curtailment of reinforcement.*

$$\dfrac{A_{sth}}{A_{st}} = \left(\dfrac{h}{H}\right)^3 \quad \therefore \quad h = H \left(\dfrac{A_{sth}}{A_{st}}\right)^{\tfrac{1}{3}}$$

If $A_{sth} = \tfrac{1}{2} A_{st}$ (*i.e.* half the bars being curtailed), $h = H \left(\dfrac{1}{2}\right)^{\tfrac{1}{3}} = 3 \left(\dfrac{1}{2}\right)^{\tfrac{1}{3}} = 2.38$ m.

As decided earlier, curtail half the bars at 0.85 m from the base.

Min. reinforcement @ 0.254 % = 660.4 mm². Hence further curtailment is not possible.

Thus, the reinforcement at the inner face will be provided as follows:

(i) At base: 16 mm Φ bars @ 130 mm c/c

(ii) At 0.85 m above base, upto top. : 16 m Φ bars @ 260 mm c/c

**Direct tension in long walls:** Since the top portion of short walls act as slab supported on long walls, the water pressure acting on short walls will cause tension in long walls.

$$P_L = p \dfrac{B}{2}, \text{ where } p = 9810 \times 2 = 19620 \text{ N/m}^2 \text{ at 1 m above base}$$

$$\therefore \quad P_L = 19620 \times \dfrac{4}{2} = 39240 \text{ N}; \quad A_s \text{ required} = \dfrac{39240}{150} = 261.6 \text{ mm}^2$$

Area of distribution steel provided in horizontal direction = 670 mm².

Hence distribution steel will take direct tension.

**4. Design of short walls**

**(a) Tank empty, with pressure of saturated soil from outside**

(i) *Top Portion* The bottom 1 m (> H/4) acts as cantilever, while the remaining 2 m acts as slab supported on long walls.

At $h = 1$ m $\left(> \dfrac{H}{4}\right)$ above the base of short wall,

$$p_a = K_a \gamma' (H - h) + \gamma_w (H - h) = \tfrac{1}{3} \times 7190 \times 2 + 9810 \times 2 = 4793 + 19620 = 24413 \text{ N/m}^2$$

$$M_f \text{ (at supports)} = \frac{p_a L^2}{12} = \frac{24413(4)^2}{12} = 32551 \text{ N-m (causing tension outside)}$$

$$M \text{ (at centre)} = \frac{p_a L^2}{8} - M_f = \frac{p_a L^2}{24} = 16276 \text{ N-m (causing tension inside)}$$

$$d = 260 - (25 + 16 + 8) = 211 \text{ mm}.$$

At supports, $A_{st} = \dfrac{32551 \times 1000}{150 \times 0.874 \times 211} = 1177 \text{ mm}^2.$

Using 12 mm Φ bars, $s = \dfrac{1000 \times 113.1}{1177} = 96 \text{ mm}$

Hence provide 12 mm Φ bars @ 95 m c/c at the outer face, at 2 mm below the top.

At midspan, $A_{st} = \tfrac{1}{2} \times 1177 = 588.5 \text{ mm}^2$

Min. $A_{st} = 660.4 \text{ mm}^2$ (as found earlier)

∴ Spacing of 12 mm Φ bars $= \dfrac{1000 \times 113.1}{660.4} = 171.3$

Hence provide 12 mm Φ bars at 170 mm c/c at the inner face.

*(ii) Bottom portion*: The bottom 1 m will bend as cantilever.

Intensity of earth pressure at bottom = 36620 N/m² (step 3)

∴ $M = \left(\tfrac{1}{2} \times 36620 \times 1\right) \times \tfrac{1}{3} = 6103 \text{ N-m}$ (with tension at outside face)

∴ $A_{st} = \dfrac{6103 \times 1000}{150 \times 0.874 \times 225} = 207 \text{ mm}^2.$

Minimum steel @ 0.254 % = 660.4 mm² (found earlier)

∴ Spacing of 12 mm Φ bars $= \dfrac{1000 \times 113.1}{660.4} = 171.3 \text{ mm}$

Hence provide 12 mm Φ bars @ 170 mm c/c at the outside face, in the vertical direction for bottom 1 m height. The spacing can be doubled for the upper portion.

*(iii) Direct compression in short walls*: Though the long walls bend as cantilever, it is observed that end one meter width of long wall contributes to push in short walls, due to earth pressure, and its magnitude is given by

$$P_{BC} = p_a \times 1 = 24413 \times 1 = 24413 \text{ N}$$

This is quite small, and hence its effect has not been considered.

**(b) Tank full with water, and no earthfill outside**

*(i) Top Portion*

The bottom portion $h = 1$ m (> H/4) acts as a cantilever, while the remaining 2 m acts as slab supported on long walls.

At $h = 1$ m (> H/4) above the base of short wall,

$$p = w(H-h) = 9810 \times 2 = 19620 \text{ N/m}^2$$

$$M_f \text{ (at supports)} = \frac{pB^2}{12} = \frac{19620(4)^2}{12} = 26160 \text{ N-m}$$

(causing tension at the inside)

$$M_c \text{ (at centre)} = \frac{pB^2}{24} = \tfrac{1}{2} \times 26160 = 13080 \text{ N-m}$$

(causing tension at the outside)

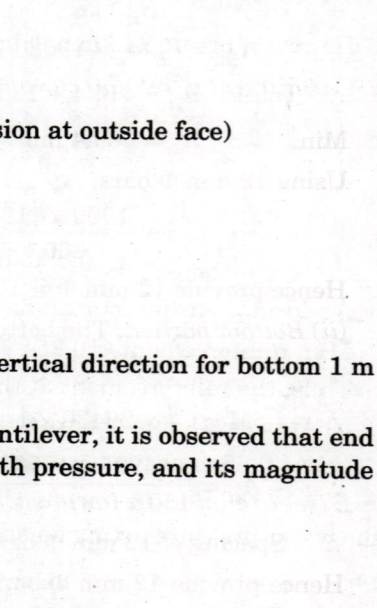

Fig. 24.5

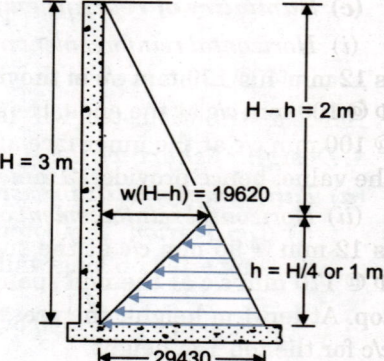

Fig. 24.6

Direct tension in short wall, due to water pressure on the end one metre width of long walls is
$$P_B = w(H - h) \times 1 = 19620 \times 1 = 19620 \text{ N}$$
Effective depth $d$, for horizontal steel = 211 mm

∴ Distance $x = d - \dfrac{T}{2} = 211 - 130 = 81$ mm. Hence Net B.M. $= M - P_B \cdot x$

∴ $A_{st1} = \dfrac{M - P_B \cdot x}{\sigma_{st} \, j \, d}$ and $A_{st2} = \dfrac{P_B}{\sigma_s}$

*At the inside face (end of short walls)*

$A_{st1} = \dfrac{M_f - P_B \cdot x}{\sigma_{st} \, j \, d} = \dfrac{26160 \times 1000 - 19620 \times 81}{150 \times 0.874 \times 211} = 888.2 \text{ mm}^2$

$A_{st2} = P_b/\sigma_{sh} = 19620/150 = 130.8 \text{ mm}^2$.   ∴ Total $A_{st} = A_{st1} + A_{st2} = 888.2 + 130.8 = 1019 \text{ mm}^2$

Using 12 mm Φ bars, $s = \dfrac{1000 \times 113.1}{1019} = 111$ mm

Hence provide 12 mm Φ bars @ 110 mm *at the inner face.*

*At the outside face (midle of short walls)*

$A_{st1} = \dfrac{M_c - P_B \cdot x}{\sigma_{st} \cdot j \, d} = \dfrac{13080 \times 1000 - 19620 \times 81}{150 \times 0.874 \times 211} = 415.4 \text{ mm}^2$

$A_{st2} = P_B/\sigma_s = 19620/150 = 130.8 \text{ mm}^2$

∴ Total $A_{st} = 415.4 + 130.8 = 546.2 \text{ mm}^2$

Min. $A_{st} = 660.4 \text{ mm}^2$ (as found earlier)

Using 12 mm Φ bars,

$s = \dfrac{1000 \times 113.1}{660.4} = 171$ mm

Hence provide 12 mm Φ @ 170 mm c/c *at the outside face*

(ii) *Bottom portion*: The bottom 1 m will bend as cantilever.

$p$ (at bottom) = 29430 N/m² (step 3)

∴ $M = (\tfrac{1}{2} \times 29430 \times 1) \times \tfrac{1}{3} = \dfrac{29430}{6} = 4905$ N-m (with tension at inside face)

∴ $A_{st} = \dfrac{4905 \times 1000}{150 \times 0.874 \times 225} = 166.3 \text{ mm}^2$. Min. steel @ 0.254% = 660.4 mm² (found earlier)

∴ Spacing of 12 mm Φ bars = $(1000 \times 113.1)/660.4 = 171.3$ mm

Hence provide 12 mm Φ bars @ 170 mm c/c at the inside face, in the vertical direction for bottom 1 m height. The spacing can be doubled for the upper portion.

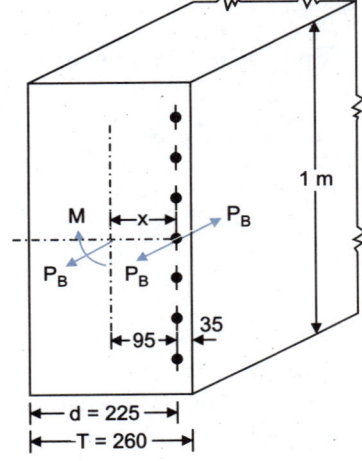

Fig. 24.7

**(c) Summary of reinforcement in short walls**

(i) *Horizontal reinforcement at inner face*: For lateral earth pressure, horizontal reinforcement required is 12 mm Φ @ 170 mm c/c in the mid-span. For water pressure, horizontal reinforcement required is 12 mm Φ @ 100 mm c/c at the ends. Hence provide greater one of the two reinforcements. Thus, provide 12 mm Φ @ 100 mm c/c at the inner face at 2 m from the top. At top 1 m height, the pressure will be reduced to half the value; hence provide 12 mm Φ @ 170 mm c/c for the top 1 m height.

(ii) *Horizontal reinforcement at outer face*: For lateral earth pressure, horizontal reinforcement required is 12 mm @ 95 mm c/c at the supports. For water pressure, horizontal reinforcement required is 12 mm Φ @ 170 mm c/c at the mid span. Hence provide 12 mm Φ @ 95 mm c/c at the outer face, at 2 m from the top. At top 1 m height, the pressure will be reduced to half the value, hence provide 12 mm Φ @ 170 mm c/c for the top 1 m height.

(iii) *Vertical reinforcement at inner face*: Provide 12 mm Φ bars @ 170 mm c/c at the inner face, for bottom 1 m height; the spacing can be doubled for the upper portion.

(iv) *Vertical reinforcement at outer face*: Provide 12 mm Φ bars @ 170 mm c/c at the outer face, for bottom 1 m height. The spacing can be doubled for the upper portion.

**Step 5. *Design of top slab*:** $L/B = 10/4 = 2.5$. Hence the top slab will be designed as one way slab. Let the live load on top slab = 2000 N/m²

Assuming a tickness of 20 cm including finishes etc.,

Self weight = $0.20 \times 1 \times 1 \times 25000 = 5000$ N/m²

∴ Total $w = 2000 + 5000 = 7000$ N/m²

$$M = \frac{wB^2}{8} = \frac{7000(4+0.26)^2}{8} = 15879.15 \text{ N-m} \quad \therefore d = \sqrt{\frac{15879.15 \times 1000}{1.156 \times 1000}} = 117.2 \text{ mm}$$

Provide total thickness = 150 mm. Keeping a clear cover of 25 mm and using 12 mm Φ bars,

$$d = 150 - 25 - 6 = 119 \text{ mm}.$$

$$A_{st} = \frac{15879.15 \times 1000}{150 \times 0.874 \times 119} = 1018 \text{ mm}^2$$

Spacing of 12 mm Φ bars $= \frac{1000 \times 113.1}{1018} = 111.1 \text{ mm} \approx 110 \text{ mm}$

Distribution reinforcement $= 0.3 - 0.1 \left[\frac{150-100}{450-100}\right] = 0.286\%$

∴ $A_{sd} = \frac{0.286 \times 150 \times 1000}{100} = 429 \text{ mm}^2$

∴ Spacing of 10 mm Φ bars $= \frac{1000 \times 78.54}{429} = 183.07 \text{ mm}$

Hence provide 10 mm Φ bars @ 180 mm c/c the other direction.

**Step 6. *Design of bottom slab*:** If there were no sub-soil water, only nominal reinforcement would be required. However, because of saturated subsoil, there will be uplift pressure on the bottom slab, of the magnitude given by

$p_u = wH_1 = 9810 \times 3.3 = 32373$ N/m²  (Assuming thickness of base slab to be 300 mm)

(a) *Check against floatation*: The whole tank must be checkced against floation when the tank is empty

Total upward floatation force = $P_u = p_u \times B \times L = 32373 \times 4 \times 10 = 1294920$ N.

Total downward force consists of weight of the tank. Let us assume thickness of bottom slab = 300 mm.

Weight of walls = $0.26(4 + 4 + 10 + 10) \times 3 \times 25000 = 546000$ N.

Weight of roof slab and finishes = $0.2 \times 4 \times 10 \times 25000 = 200000$ N

Weight of base slab = $4 \times 10 \times 0.3 \times 25000 = 300000$ N

∴ Total $W = 546000 + 200000 + 300000 = 1046000$ N

This is much less than the floatation force. Hence provide projections of base slab, beyond the face of vertical walls, by an amount $x$ m alround, so that weight of soil column supported by the projections will provide additional downward force. It is assumed that if the tank is floated, the earth would rupture on the vertical planes shown by dotted lines (Fig. 24.8). Most soils would tend to rupture on an inclined plane, thus tending to increase the effective downward load from the earth but this increase might be small in waterlogged non-cohesive ground.

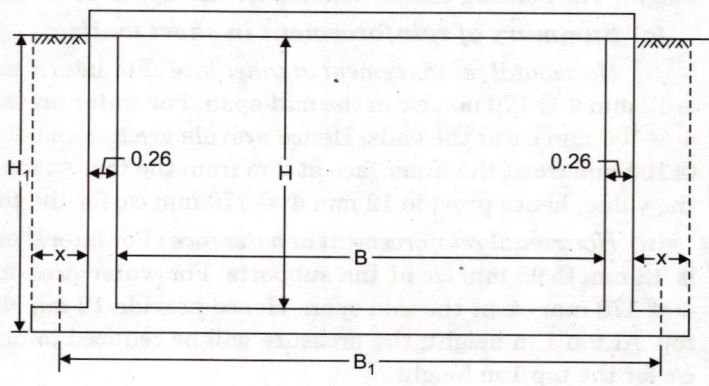

Fig.24.8

Weight of soil supported by projection
$$x \approx 2(L+B) \times H \gamma_{sat} = 2(4+10)x(3 \times 17000) = 1428000\,x\text{ N}$$
Weight of roof slab = 200000 N   (as before)
Weight of walls = 546000 N   (as before)
Weight of base slab = $(4 + 2 \times 0.26 + 2x)(10 + 2 \times 0.26 + 2x) \times 0.3 \times 25000$
$$= (4.52 + 2x)(10.52 + 2x) \times 7500$$
Total uplift force = $29430\,(4.52 + 2x)(10.52 + 2x)$

Equating total upward force to the total downward forces, we get

$29430\,(4.52 + 2x)(10.52 + 2x) = 1428000\,x + 200000 + 546000 + (4.52 + 2x)(10.52 + 2x)7500$

or $\quad 21930\,(4.52 + 2x)(10.52 + 2x) = 1428000\,x + 746000$

or $\quad (4.52 + 2x)(10.52 + 2x) = 65.116\,x + 34.017$

which gives $4x^2 - 35.036\,x + 13.533 = 0$. From which $x \approx 0.4$ m.

Check: $\quad$ Width $B_1 = 4 + (2 \times 0.26) + 0.4 = 4.92$ m

$\quad\quad\quad$ Length $L_1 = 10 + (2 \times 0.26) + 0.4 = 10.92$ m

∴ Weight of soil supported on projection $x = 2(L_1 + B_1) \times H \gamma_{sat}$
$$= 2(4.92 + 10.92)\,0.4 \times 3 \times 17000 = 646272\text{ N} \quad\ldots(i)$$
Weight of walls = 546000 N $\quad\ldots(ii)$
Weight of roof slab = 200000 N $\quad\ldots(iii)$
Weight of base slab = $(4.52 + 2 \times 0.4)(10.52 + 2 \times 0.4) \times 0.3 \times 25000 = 451668$ $\quad\ldots(iv)$

∴ Total downward weight = $646272 + 546000 + 200000 + 451668 = 1843940$ N

Total uplift force = $29430\,(4.52 + 0.8)(10.52 + 0.8) = 1772345$ N

∴ F.S. against floation = $1843940/1772345 = 1.04$

A factor of safety of about 1.1 is needed because (i) concrete may weigh less than 25000 N/m³ (ii) earth may weigh less than 17000 N/m³, or (iii) ground water may turn saline, and may weigh more than 9810 N/m³.

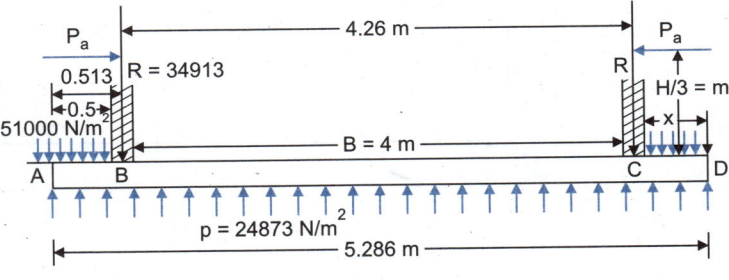

Fig. 24.9

*Hence keep $x = 0.5$ m. The base slab will be designed as one way slab. Consider one metre length of the slab.*

Upward water pressure $(p_u) = 32373$ N/m²

Self weight of slab = $1 \times 1 \times 0.3 \times 25000 = 7500$ N/m²

∴ Net upward pressure, $p = 32373 - 7500 = 24873$ N/m²

Weight of wall per m run = $0.26 \times 3 \times 1 \times 25000 = 19500$ N/m

Weight of roof slab, transferred to each wall, per m run
$$= 0.2 \times (2 + 0.26) \times 1 \times 25000 = 11300\text{ N}$$
Weight of earth of projection = $17000 \times 3 \times 1 \times 0.5 = (51000) \times 1 \times 0.5 = 25500$ N/m

∴ Net unbalanced force/m run = $32373\,(5.286 \times 1) - 2\,(19500 + 25500 + 11300) \approx 58524$ N

∴ Reaction on each wall = $58524/2 = 29262$ N

$p_a = K_a\,\gamma'\,H + \acute{w}\,H = 36620$ N/m² $\quad$ ∴ $P_a = 36620 \times \dfrac{3}{2} \times 1 = 54930$ N/m

Acting at $(\frac{3}{3} + 0.3) = 1.3$ m above the bottom of base slab.

B.M. at the edge of cantilever portion

$$= \frac{24873\,(0.5)^2}{2} + 54930 \times 1.3 - \frac{51000\,(0.5)^2}{2} = +68143 \text{ N-m (causing tension at the bottom face)}$$

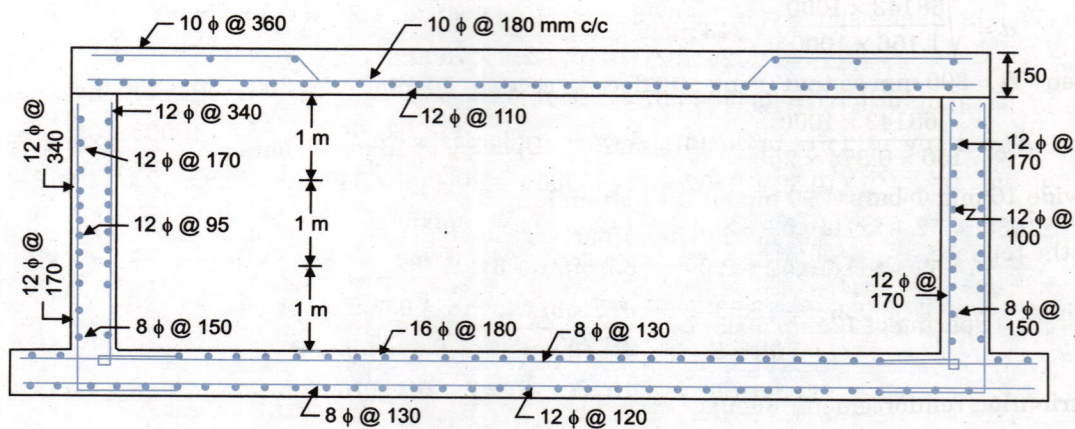

Section X-X

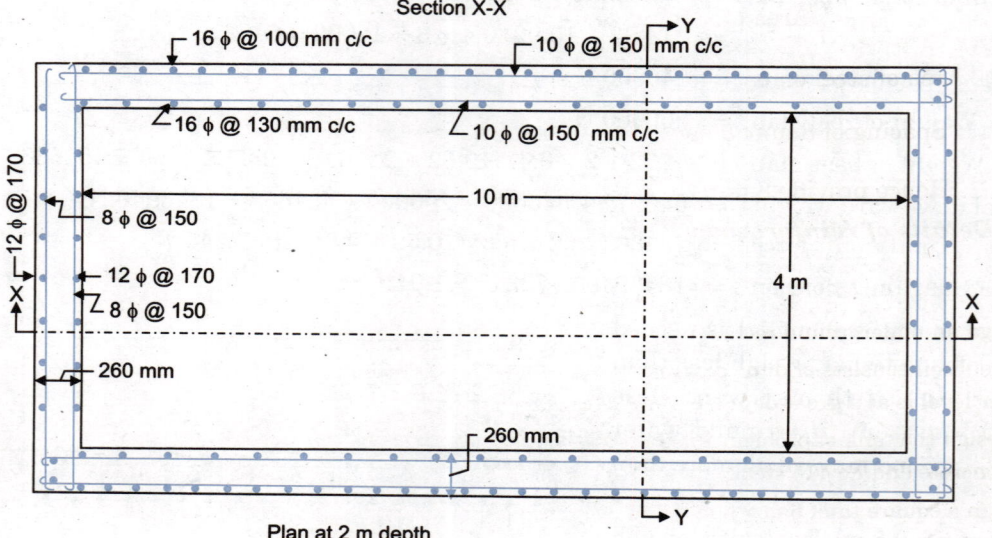

Plan at 2 m depth

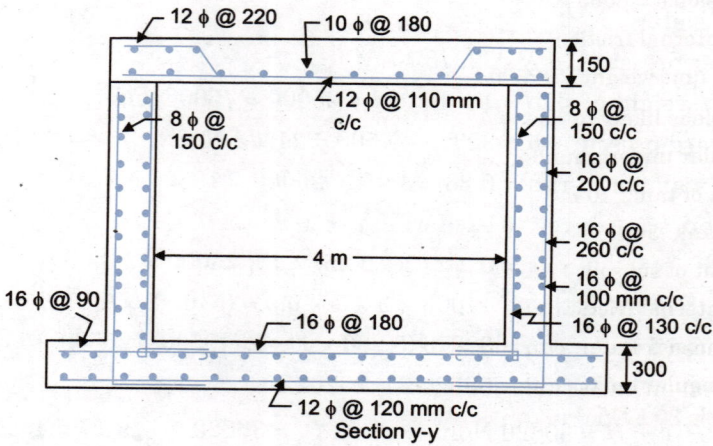

Section y-y

Fig. 24.10

B.M. at the centre of span

$$= \frac{24873}{2}\left(\frac{5.286}{2}\right)^2 + 54930 \times 1.3 - (19500 + 11300 + 29262)\frac{4.26}{2} - 51000 \times 0.5 \left(\frac{5.286}{2} - 0.25\right)$$

$= 86874.5 + 71409 - 127932 - 61021.5 \approx -30670$ N-m (causing tension at the top face)

$$d = \sqrt{\frac{68143 \times 1000}{1.156 \times 1000}} \approx 243 \text{ mm.}$$

Keep $D = 300$ mm so that using an effective cover of 50 mm, $d = 300 - 50 = 250$ mm.

$$A_{st} = \frac{68143 \times 1000}{150 \times 0.874 \times 250} = 2079 \text{ mm}^2 \; ; \quad \text{Spacing of 16 mm } \Phi \text{ bars} = \frac{1000 \times 201}{2079} = 97 \text{ mm.}$$

Provide 16 mm Φ bars @ 90 mm at the bottom face.

For the top face, $\quad A_{st} = \dfrac{30670 \times 1000}{150 \times 0.874 \times 250} = 935.8 \text{ mm}^2$

∴ Spacing of 12 mm Φ bars $= \dfrac{1000 \times 113.1}{935.8} \approx 120$ mm c/c

Distribution reinforcement in longitudinal direction $= 0.3 - 0.1\left[\dfrac{300-100}{450-100}\right] = 0.243\%$

∴ Area of steel $= \dfrac{0.243 \times 1000 \times 300}{100} = 729 \text{ mm}^2$

∴ Area of steel on each face $= 729/2 = 364.5 \text{ mm}^2$

∴ Spacing of 8 mm Φ bars $= \dfrac{1000 \times 50.3}{364.6} = 138$ mm

Hence provide 8 mm @ 130 m c/c on each face.

**Step 7. *Details of reinforcement*:** Shown in Fig. 24.10.

## PROBLEMS

1. Design an underground rectangular tank 4 m × 6 m × 2 m deep.

   The subsoil consists of dune sand having unit weight of 16000 N/m³ and angle of friction of 34°. The subsoil water level is at a great depth. Use M 20 concrete and Fe 415 steel.

2. Redesign the tank of problem 1 if the water table is likely to rise to a level 1 m below the ground level. Use M 20 concrete and Fe 415 steel.

3. Design a square tank 6 m × 6 m × 4 m deep in a c-φ soil having the following properties.
   (i) Unit weight = 16500 N/m³
   (ii) Unit cohesion c = 8000 N/m²
   (iii) Angle of internal friction φ = 12°
   (iv) Saturated unit weight = 17600 N/m³

   The water table is likely to rise to a level 2 m below ground level. Use M 20 concrete and Fe 415 steel.

4. Design a circular underground tank, with domical top with the following data:
   (i) Inside dia. of tank: 10 m
   (ii) Depth of tank: 3 m
   (iii) Unit weight of subsoil: 17200 N/m³
   (iv) Angle of internal friction: 30°
   (v) Depth of subsoil water: 8 m below G.L. Use M 30 concrete and Fe 415 steel.

5. Design a rectangular tank of capacity 75,000 gallons capacity and height 6 mtrs. The bearing capacity of medium soil at the site is 80 KN/m² respectively use $M_{30}$ grade and Fe 415 steel.

# PART – III
# MISCELLANEOUS STRUCTURES

25. REINFORCED CONCRETE PIPES
26. BUNKERS AND SILOS
27. CHIMNEYS
28. PORTAL FRAMES
29. BUILDING FRAMES

# CHAPTER 25
# REINFORCED CONCRETE PIPES

## 25.1. LOADS ON PIPES

Reinforced cement concrete pipes are widely used for carrying drainage and irrigation water, and culverts over national highways carrying very heavy traffic.

Reinforced concrete pipes are generally used for three purposes: (i) Carrying water in water supply system, (ii) Carrying waste water or storm water and (iii) Conducting small streams or drains, under embankments. If the pipe is resting on the ground surface (specially for the first purpose), it is subjected to hydrostatic pressure of water. For conveying liquids under high pressure, prestressed concrete pipes are widely used. If, however, the pipe is embedded under the ground surfaces, it is subjected to the following loads.

1. Hydrostatic pressure ($p_w$) due to fluid pressure.
2. Self weight.
3. Weight of water inside the pipe.
4. Earth fill over haunches *AEC* and *AFD*.
5. Uniformly distributed load ($q$) on its top, due to weight of earth fill and the traffic load etc.
6. Uniform earth pressure ($p_H$) from sides (Fig. 25.1).
7. Varying earth pressure ($p_H'$) from sides.
8. Point load ($W$) on its top.

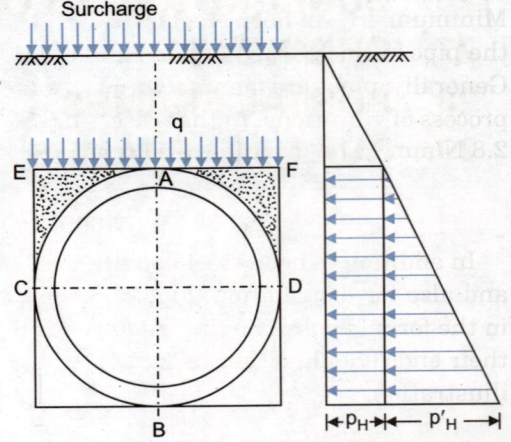

Fig. 25.1

The hydrostatic pressure induces only hoop stresses while the other forces induce both the hoop stresses as well as bending moments. The problem of determination of internal stresses and moments in the pipe is essentially statically indeterminate. However strain energy method can be used to solve the problem, provided the manner in which pipe is supported is known. Generally, pipe is supported at its base upto its horizontal diameter *CD* as shown in Fig. 25.2.

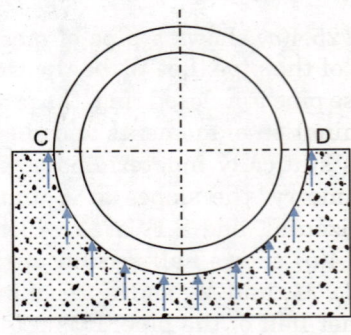

Fig. 25.2

We shall consider all the above loading conditions separately. According to IS: 458–1971, RCC pipes are classified as non-pressure and pressure pipes with their applications.

The longitudinal reinforcement is designed to support the RCC pipe as a circular beam loaded with twice the self weight of the pipe and twice the weight of water to fill the pipe across a span equal to the length of the pipe. Under these loading condition, the stress in the reinforcement should not exceed the permissible stresses.

## 25.2. STRESSES DUE TO HYDROSTATIC PRESSURE

Let the pipe carry water under a head of $h$ metres. The hydrostatic water pressure $p$ from inside will evidently be equal to $wh$. If we consider unit length of the pipe, hoop tension is given by

$$P_T = \frac{p \cdot D}{2} \qquad \ldots(25.1)$$

where $D$ is the internal diameter of the pipe.

In order to protect the pipe against water hammer action due to sudden closure of valve etc., the effective pressure $p$ inside the pipe is taken four times the normal water pressure. After having determined the hoop pressure $P_T$, the pipe can be designed. The whole of hoop tension is taken by steel, to be provided in the form of rings. The area of steel is given by

$$A_{sh} = \frac{P_T}{\sigma_{st}} \qquad \ldots(25.2)$$

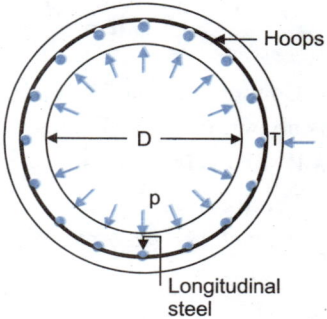

Fig. 25.3

where $\sigma_{st}$ is the permissible tensile stress in steel, the value of which may be taken from Table 3.2. Minimum area of hoop steel is 0.3%. If the pipe is thin, the rings are provided in the centre of thickness. If the pipe is thick, it is provided on both the faces. The thickness of pipe is designed on the no crack basis. Generally, pipes are manufactured in rich concrete and is thoroughly compacted by centrifugal or some other process of vibrations. In that case, limiting tensile strength in the composite section may be permitted upto 2.8 N/mm². The modular ratio is generally taken as 12. The thickness $T$ of the pipe is then given by

$$\frac{P_T}{1000\,T + (12+1)A_{sh}} = 2.8 \qquad \ldots(25.3)$$

In addition to hoop steel, longitudinal steel is provided @ 0.3%, for temperature and shrinkage stresses, and also for distribution. Fig. 25.3 shows the reinforcement in the pipe. The hoop steel may be provided in the form of spiral having pitch equal to the spacing of hoops. If however, individual hoops are provided, their ends should either be lapped or welded, so that they may develop full tension. See example 25.1 for illustration.

## 25.3. STRESSES DUE TO SELF WEIGHT

Fig. 25.4(a) shows a pipe of mean radius $r$ supported on its horizontal diameter. Consider one metre length of the pipe. Let $W_c$ be the weight of pipe per metre length and per metre run of its circumference. Because pipe is a closed ring, the problem of determination of moments and shear force in it is statically indeterminate. Because of symmetry, the slopes at $A$ and $B$ will be horizontal, and S.F. at $A$ will be zero. Fig 25.4(b) shows half section of the pipe with $R_1$, $M_1$ and $M_2$ as the reactions from the other half of the pipe. It is required to determine these reaction components by strain energy method, and then to develop an expression for bending moment at any point $P$, subtending an angle $\theta$ with the vertical diameter $AB$.

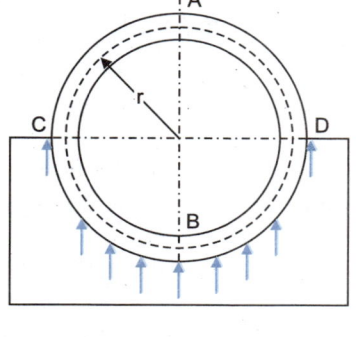
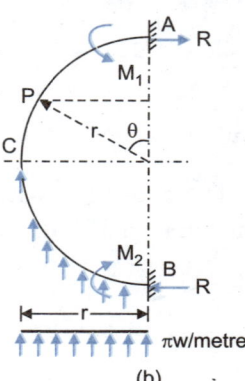

Fig. 25.4

Total reaction = total weight = $2\pi r\, W_c$

Reaction per metre run of horizontal diameter

$$= \frac{2\pi r\, W_c}{2r} = \pi W_c$$

Total weight of half pipe from $A$ to $B = \pi r\, W_c$ and it acts at a distance of $2r/\pi$ from vertical diameter $AB$. To find $R$, take moments about $B$.

$$R \cdot 2r - M_1 - \pi r\, W_c \cdot \frac{2r}{\pi} + M_2 + \pi W_c \cdot r\frac{r}{2} = 0$$

$$R = \frac{M_1 - M_2}{2r} + W_c \cdot r\left(1 - \frac{\pi}{4}\right) \qquad \ldots(1)$$

Consider an element of pipe, at angle $\alpha$, and subtending an angle $d\alpha$ as shown in Fig. 25.5. The length of the element is $r\,d\alpha$, and its weight is $W_c \cdot r\,d\alpha$. Its moment about point $P$ will be $W_c \cdot r\,d\alpha \cdot PF$.

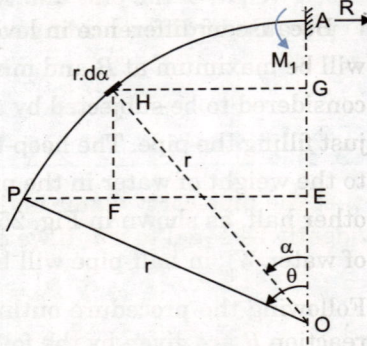

Fig. 25.5

From Fig. 25.5, $\quad PF = PE - HG = r\sin\theta - r\sin\alpha = r(\sin\theta - \sin\alpha)$

Now *clockwise moment* $M_\theta$ at point $P$ is given by

$$M_\theta = R \cdot (AE) - M_1 + \int_0^\theta (W_c\, r \cdot d\alpha)\, PF + \left|\frac{\pi W_c}{2}\left[r(1 - \sin\theta)\right]^2\right.$$

The last term after Macaulay's bar is valid for $\theta > 90°$.

$$\therefore \quad M_\theta = R \cdot r(1 - \cos\theta) - M_1 + \int_0^\theta (W_c\, r \cdot d\alpha)\, r(\sin\theta - \sin\alpha) + \left|\frac{\pi W_c}{2}\left[r(1 - \sin\theta)\right]^2\right.$$

or $\quad M_\theta = R\, r(1 - \cos\theta) - M_1 + W_c\, r^2(\theta\sin\theta + \cos\theta - 1) + \left|\frac{1}{2}\pi W_c\, r^2(1 - \sin\theta)^2\right.$

Substituting the value of $R$,

$$M_\theta = -\frac{1}{2}[M_1(1 + \cos\theta) + M_2(1 - \cos\theta)] + W_c\, r^2\left[\theta\sin\theta - \frac{\pi}{4}(1 - \cos\theta)\right] + \left|\frac{\pi}{2} W_c \cdot r^2(1 - \sin\theta)^2\right. \qquad \ldots(2)$$

The points $A$ and $B$ do not move horizontally, and the slopes at these points are also zero. Hence we have

$$\int_A^B \frac{M \cdot y\,ds}{EI} = 0 \quad \text{and} \quad \int_A^B \frac{M \cdot ds}{EI} = 0$$

where $ds = r\,d\theta$, $y = r(1 - \cos\theta)$ and $M = M_\theta$.

Hence, $\quad \int_A^B \frac{M \cdot y\,ds}{EI} = \frac{1}{EI}\int_0^\pi M_\theta\, r(1 - \cos\theta)\, r\,d\theta = 0$

or $\quad \int_0^\pi M_\theta \cdot r^2(1 - \cos\theta)\, d\theta = 0 \qquad \ldots(3) \quad \text{and} \quad \int_0^\pi M\,ds = \int_0^\pi M_\theta \cdot r\,d\theta = 0 \qquad \ldots(4)$

Substituting the values of $M_\theta$ from (2), integrating and solving the above two, we get:

$$M_1 = \frac{W_c \cdot r^2}{2}\left[\frac{3\pi}{4} - \frac{5}{3}\right] \qquad \ldots[25.4(a)] \quad \text{and} \quad M_2 = \frac{W_c \cdot r^2}{2}\left[\frac{5}{3} - \frac{\pi}{4}\right] \qquad \ldots[25.4(b)]$$

and $\quad R = \frac{W_c\, r}{6} \qquad \ldots[25.4(c)]$

The plus signs with $M_1$, $M_2$ and $R$ show that the assumed directions of these are correct. Substituting the value of $M_1$, $M_2$ in (2), the following expression for $M_\theta$ is obtained:

$$M_\theta = W_c\, r^2\left[\theta\sin\theta + \frac{5}{6}\cos\theta - \frac{3\pi}{8}\right] + \left|\frac{\pi}{2} W_c\, r^2(1 - \sin\theta)^2\right. \qquad \ldots[25.4(d)]$$

# REINFORCED CONCRETE PIPES

## 25.4. STRESSES DUE TO WEIGHT OF WATER INSIDE

Because of difference in levels at point $A$ and $B$, the water pressure in the pipe will not be uniform; it will be maximum at $B$ and minimum at $A$. If $p_1$ is the minimum water pressure at top $A$, the pipe can be considered to be subjected by an internal uniform pressure $p_1$ along with stresses due to weight of water just filling the pipe. The hoop-tension due to $p_1$ will be uniform and can be determined from *Eq. 25.1*. Due to the weight of water in the pipe, let $M_1, M_2$ and $R$ be the reaction components on half the pipe, from the other half, as shown in Fig. 25.6. Let $\gamma_w$ be the unit weight of water. The weight of water $W_w$ in half-pipe will be $\frac{1}{2}\pi r^2 \cdot \gamma_w$, acting at a distance of $\frac{4r}{3\pi}$ from $AB$. Following the procedure outlined in § 25.3, the values of moments $M_1, M_2$ and reaction $R$ are given by the following expressions:

$$M_1 = \frac{r^3}{4}\left[\frac{3\pi}{4} - \frac{5}{3}\right]\gamma_w \quad ...(25.5) ; \quad M_2 = \frac{r^3}{4}\left(\frac{5}{3} - \frac{\pi}{4}\right)\gamma_w \quad ...(25.6)$$

and 
$$R = \frac{7}{12} r^2 \cdot \gamma_w \quad ...(25.7)$$

The clockwise (or hogging) B.M. at $P$ is given by

$$M_\theta = r^3 \gamma_w \left(\frac{5}{12}\cos\theta + \frac{1}{2}\theta\sin\theta - \frac{3\pi}{16}\right) + \left|\frac{\pi r^2}{4}\gamma_w(1-\sin\theta)^2\right. \quad ...(25.8)$$

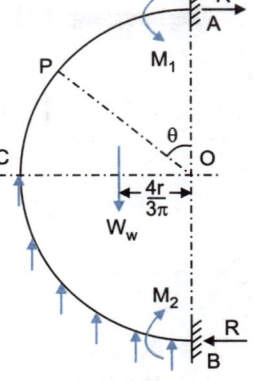

Fig. 25.6

## 25.5. STRESSES DUE TO EARTHFILL OVER HAUNCHES

Pipes are generally embedded in the ground, due to which they are subjected to stresses due to weight of earthfull over it. This weight can be divided into two parts: (*a*) uniformly distributed load over horizontal plane through the crown $A$, and (*b*) load of soil over haunches, such as in $AEC$. The effect of (*a*) has been discussed in the next article. If $\gamma$ is the unit weight of the earthfill, the weight of soil in the haunch $AEC$ will be

$$W = \gamma \left(r^2 - \frac{\pi}{4}r^2\right) \text{ with its C.G. acting at distance } \bar{x} = \frac{r\left(\frac{5}{6} - \frac{\pi}{4}\right)}{1 - \frac{\pi}{4}} \cdot \text{ as shown.}$$

Following the same procedure as outlined on § 25.3, we get

$$M_1 = 0.024\, r^3 \gamma \quad ...(25.9)$$
$$M_2 = 0.047\, r^3 \cdot \gamma \quad ...(25.10)$$
and 
$$R = 0.018\, r^2 \cdot \gamma \quad ...(25.11)$$

The hogging B.M. from $\theta = 0°$ to $90°$ is given by

$$M_\theta = -r^3 \gamma (0.006 + 0.018 \cos\theta)$$
$$+ r^3 \gamma \left[\frac{1}{2}\sin^2\theta - \frac{1}{6}\sin^2\theta \cos\theta\right.$$
$$\left. + \frac{1}{3}(1-\cos\theta) - \frac{1}{2}\theta\sin\theta\right] \quad ...[25.12\,(a)]$$

The hogging B.M. from $\theta = 90°$ to $180°$ is given by

$$M_\theta = -r^3 \gamma (0.006 + 0.018 \cos\theta]$$
$$+ r^3 \gamma [0.1075 \sin\theta - 0.059]$$

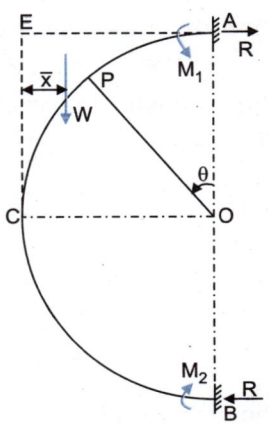

Fig. 25.7

## 25.6. STRESSES DUE TO UNIFORMLY DISTRIBUTED LOAD ON TOP

Let $q$ be the U.D.L. per metre horizontal run. The reaction components are given by the following expressions:

$$M_1 = \frac{1}{4} q \cdot r^2 = M_2 \qquad \ldots(25.13)$$

$$R = 0 \qquad \ldots(25.14)$$

The hogging B.M. at $P$ is given by

$$M_\theta = \frac{1}{2} q r^2 \sin^2 \theta - \frac{1}{4} q r^2 \qquad \ldots(25.15)$$

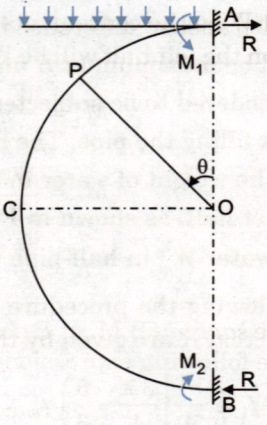

Fig. 25.8

## 25.7. STRESSES DUE TO UNIFORM PRESSURE FROM SIDES

Let the intensity of horizontal pressure be $p_H$, uniform throughout. Due to this, the pipe will be subjected to sagging B.M. throughout. This case is similar to the case of § 25.6, except that the U.D.L. is vertical.

Hence $\quad M_c = \frac{1}{4} p_H \cdot r^2 \quad$ (sagging) $\qquad \ldots[25.16\,(a)]$

$$R = -p_H \cdot r \quad (i.e.\ \text{thrust}) \qquad \ldots[25.16\,(b)]$$

The Sagging B.M. at $P$ is given by,

$$M_\theta = \frac{1}{4} p_H \cdot r^2 (1 - 2 \cos^2 \theta) \qquad \ldots(25.17)$$

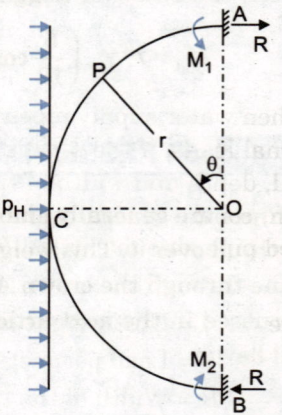

Fig. 25.9

## 25.8. STRESSES DUE TO TRIANGULARLY DISTRIBUTED LOAD

Let the pressure intensity at a depth $x$ below the crown is equal to $p'_x = k \cdot x$.

The stress intensity at the bottom level of the pipe is

$$p'_H = k \cdot 2r = 2k \cdot r \qquad \ldots(25.18)$$

The values of the reaction components are given by the following expressions:

$$M_1 = -\frac{5}{24} k \cdot r^3 \quad (i.e.\ \text{hogging}) \qquad \ldots(25.19)$$

$$M_2 = -\frac{7}{24} k r^3 \quad (i.e.\ \text{hogging}) \qquad \ldots(25.20)$$

$$R = -\frac{15}{24} k r^2 \quad (i.e.\ \text{thrust}) \qquad \ldots(25.21)$$

The *sagging* B.M. at $P$ is given by

$$M_\theta = \frac{5}{12} k r^3 (1 - 1.5 \cos \theta) - \frac{1}{6} k \cdot r^3 (1 - \cos \theta)^3 \qquad \ldots(25.22)$$

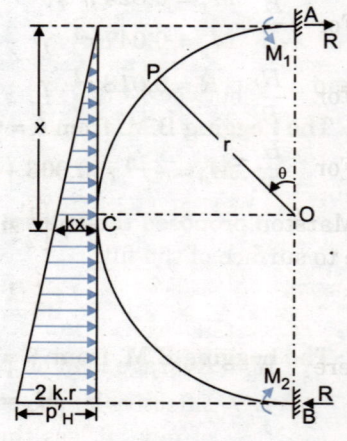

Fig. 25.10

# REINFORCED CONCRETE PIPES

## 25.9. STRESSES DUE TO POINT LOAD ON CROWN

Let $W$ be the concentrated load at crown, just to the left of point $A$. The S.F. due to the other half of the pipe on the left half will be $W/2$ as marked. The reaction components are given by the following expressions:

$$M_1 = \frac{W \cdot r}{2\pi}\left[\frac{2}{3} + \frac{3\pi}{8}\right] \quad \ldots(25.23)$$

$$M_2 = \frac{W \cdot r}{2\pi}\left[\frac{4}{5} - \frac{\pi}{8}\right] \quad \ldots(25.24)$$

$$R = \frac{W}{6\pi} \quad \ldots(25.25)$$

The *sagging* B.M. at $P$, for $\theta < 90°$ and for $\theta > 90°$ are respectively given by the following expressions:

$$M_\theta = W \cdot r\left[\frac{3}{16} + \frac{1}{2\pi} - \frac{\cos\theta}{6\pi} - \frac{\sin\theta}{2}\right] \quad \ldots[25.26(a)]$$

$$M_\theta = W \cdot r\left[\frac{3}{16} + \frac{1}{2\pi} - \frac{\cos\theta}{6\pi} - \frac{\sin\theta}{2}\right] - \frac{W \cdot r}{4}(1 - \sin\theta)^2 \quad \ldots[25.26(b)]$$

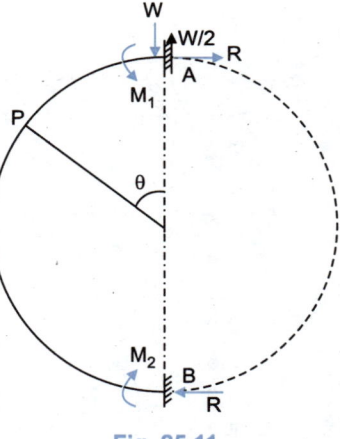

Fig. 25.11

## 25.10. STRESSES DUE TO OVER-BURDEN AND EXTERNAL LOADS

When water supply pipes are buried underground, they are subjected to weight of the backfill, and the external loads, if the pipe is buried under a load. The load transferred to the pipe depends upon the type of soil, depth and width of trench, diameter of the pipe and the support condition of pipe. Due to arching action, complete load of the backfill is not transferred to the pipe. According to Marston, the load $W$ on buried pipe, due to earth filling is given by

$$W = C \cdot \gamma \cdot B^2 \quad \ldots(25.27)$$

where  $\gamma$ = Unit weight of backfill material (N/m³)
 $W$ = Load per meter of pipe
 $B$ = Width of the trench
 $C$ = Coefficient, which depends upon the type of soil and ratio of the depth to the width of the trench.

Table 25.1 gives the values of $C$ for various values of $H/B$ ratio. In absence of any other data, the following values of $C$ may be adopted:

For $\dfrac{H}{B} \leq 4$, $C \approx \dfrac{2}{3}\dfrac{H}{B}$

For $\dfrac{H}{B}$ between 4 to 7, $C \approx \dfrac{1}{2} \cdot \dfrac{H}{B}$

For $\dfrac{H}{B} > 7$, $C \approx 3.5$

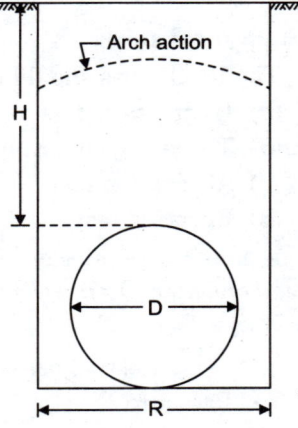

Fig. 25.12

Marston proposed the following formula for computing additional load on the pipes due to loads applied due to surface of the fill:

$$W_t = \frac{1}{L_e} I_c \cdot C_t \cdot P \quad \ldots(25.28)$$

where  $W_t$ = Average load on pipe due to wheel load (N/m)
 $L_e$ = Effective length of the pipe. It is taken equal to the actual length of a precast segment section of pipe, if less than 1 m. If the segment length is more than 1 m, the maximum value of $L_e$ is recommended as 1 m.

$I_e$ = Impact factor. ; $C_t$ = Load coefficient.
P = Concentrated wheel load on the embankment.

The load coefficient $C_t$ is dependent on the length and width of the conduit cross-section and the depth of cover over it. It may be evaluated in accordance with the Boussinesq formula of stress distribution.

**TABLE 25.1.** Values of C

| Ratio of depth to trench width | Sand and damp top soil | Saturated top soil | Wet clay | Saturated clay |
|---|---|---|---|---|
| 0.5 | 0.46 | 0.46 | 0.47 | 0.47 |
| 1.0 | 0.85 | 0.86 | 0.88 | 0.90 |
| 1.5 | 1.18 | 1.21 | 1.24 | 1.28 |
| 2.0 | 1.46 | 1.50 | 1.56 | 1.62 |
| 2.5 | 1.70 | 1.76 | 1.85 | 1.92 |
| 3.0 | 1.90 | 1.98 | 2.08 | 2.20 |
| 3.5 | 2.08 | 2.17 | 2.30 | 2.44 |
| 4.0 | 2.22 | 2.33 | 2.49 | 2.66 |
| 5.0 | 2.45 | 2.59 | 2.80 | 3.03 |
| 6.0 | 2.61 | 2.78 | 3.04 | 3.33 |
| 7.0 | 2.73 | 2.93 | 3.22 | 3.57 |
| 8.0 | 2.81 | 3.03 | 3.37 | 3.76 |
| 9.0 | 2.90 | 3.14 | 3.52 | 3.98 |
| 10.0 | 2.92 | 3.17 | 3.56 | 4.04 |
| 12.0 | 2.97 | 3.24 | 3.68 | 4.22 |
| 15.0 | 3.01 | 3.30 | 3.77 | 4.38 |
| Very Great | 3.03 | 3.33 | 3.85 | 4.55 |

## Tests on Pipes

The RCC pipes should confirm to the following tests specified in Indian Standard Code IS: 3597–1966.
(a) Hydrostatic test
(b) Three edge bearing test (or) sand bearing test
(c) Absorption test
(d) Bursting test.

**Example 25.1.** *A pipe of 1500 mm internal diameter is subjected to a normal water pressure of 12 metres of head of water. Design the pipe, if it is supported at ground level, at its horizontal diameter. Use HYSD bars.*

**Solution:**

Effective pressure = 4 times working pressure = 4 × 12 = 48 m head of water.

∴ Pressure $P = wh = 9800 \times 48 = 470400$ N/m²

Hoop tension $P_T = \dfrac{p \cdot D}{2} = 470400 \left(\dfrac{1.5}{2}\right) = 352800$ N/m

Let us use rich concrete for which permissible tensile stress is 2.8 N/mm² and m = 12. Permissible stress in steel will be taken as 150 N/mm² for HYSD bars.

Area of hoops, $A_{sh} = \dfrac{P_T}{\sigma_{st}} = \dfrac{352800}{150} = 2352$ mm²

Spacing of 20 mm Φ bars hoops = $\dfrac{1000 \times 314}{2352} = 133.5$ mm. Hence use 20 mm Φ bar hoops at a spacing of 130 mm c/c. Actual $A_{sh} = 2415$ mm².

The thickness $T$ of the pipe is given by

$$\frac{P_T}{1000\,T + (m-1)\,A_{sh}} = 2.8 \quad \text{or} \quad \frac{352800}{1000\,T + (12-1)\,2415} = 2.8$$

For which $\quad T = 99.4$ mm
Provide $\quad T = 100$ mm.

Area of longitudinal reinforcement per metre of the circumference

$$= 1000 \times T \left(\frac{0.3}{100}\right) = 100 \times 3 = 300 \text{ mm}^2$$

∴ Spacing of 6 mm Φ bars

$$= \frac{1000 \times 28.27}{300} = 94 \text{ mm}$$

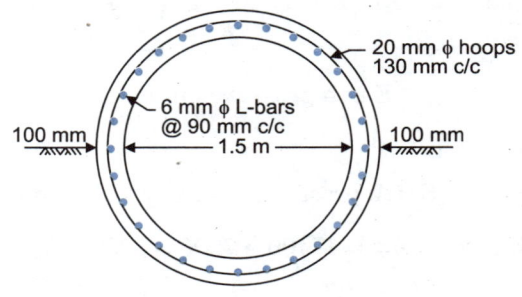

Fig. 25.13

Hence, provide 6 mm Φ longitudinal bars @ 90 mm c/c, evenly distributed along the circumference as shown Fig. 25.13.

**Example 25.2.** *Redesign the pipe of example 25.1, if it passes under a road, and the depth of filling over its crown is 1.2 m. The maximum live load expected on the road is equivalent to a concentrated load of 40 kN/m length of the pipe. The unit weight of soil around the pipe is 18 kN/m³ and its angle of repose is 30°. The pipe is supported upto its horizontal diameter.*

**Solution:**

**(i) Preliminary dimensions and hoop tension due to water pressure.**

The pipe will be subjected to stresses due to the following (a) water pressure, (b) self weight of pipe, (c) uniformly distributed load due to earthfill over its crown, (d) weight of water in the pipe, (e) weight of soil cover its haunches, and (f) earth pressure. All these forces are shown in Fig. 25.14. Let the thickness of the pipe be = 200 mm = 0.2 m. (Thickness may be assumed to be about 12 to 15% of its diameter)

∴ Mean diameter of pipe = 1500 + 200 = 1700 mm = 1.7 m

Let us assume that the concentrated load on the pipe is distributed uniformly over the width of the pipe, with a load intensity

$$q = \frac{40000}{1.7} = 23529.4 \approx 23530 \text{ N/m}^2$$

Effective water pressure = 12 × 4 = 48 m head of water.

$$p = 48 \times 9800 = 470400 \text{ N/mm}^2$$

Hoop tension $P_T = 470400 \times \dfrac{1.5}{2} = 352800$ N/m

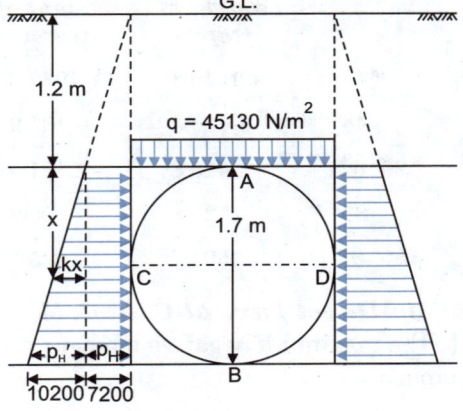

Fig. 25.14

**(ii) Total R at A**

Self weight $W_C$ of pipe = 1 × T × 25000 = 1 × 0.2 × 25000 = 5000 N/m

∴ R due to self weight = $\dfrac{W_c \cdot r}{6} = \dfrac{5000\,(0.85)}{6} = 708$ N (tension)

R due to weight of water = $\dfrac{7}{12} r^2 \cdot \gamma_w = \dfrac{7}{12}(0.85)^2 \times 9800 = 4130$ N (tension)

R due to earthfill over haunches = $0.018\, r^2 \cdot \gamma = 0.018\,(0.85)^2 \times 18000 = 234$ N (tension)
R due to uniformly distributed load, due to live load and earthfill on the top = zero.

∴ Total tension R at   A = 708 + 4130 + 234 = 5072 N

Total hoop tension at $A = T_A = P_T + R = 352800 + 5072 = 357872$ N
Total hoop tension at $B = T_B = P_T - R = 352800 - 5072 = 347728$ N

### (iii) Bending moment $M_\theta$ at point P

Uniform load due to earth over top = $\gamma H = 18000 \times 1.2 = 21600$ N/m²

∴ Total $q$ due to overburden and live load = $23530 + 21600 = 45130$ N/m²

$$K_a = \frac{1 - \sin \phi}{1 + \sin \phi} = \frac{1 - \sin 30°}{1 + \sin 30°} = \frac{1}{3}$$

∴ Earth pressure intensity at top = $\frac{1}{3} \times 1.2 \times 18000 = 7200$ N/m²

∴ $p_H = 7200$ N/m²

Earth pressure intensity at bottom = $\frac{1}{3} \times 2.9 \times 18000 = 17400$

Hence, $p_H' = 17400 - 7200 = 10200$ N/m². Referring to Fig. 25.10, we have
$k \cdot x = p_H'$ or $k(1.7) = 10200$ Hence, $k = 10200/1.7 = 6000$

∴ Triangular load intensity at any depth $x$ below the top of the pipes is $k \cdot x = 6000 x$.

The moments $M_\theta$ at various value of $\theta$ due to various loadings are tabulated in Table 25.2. It will be seen that the moments due to earth pressure, given by the sum of *Eq. 25.17* and *Eq. 25.22*, are of opposite sign than other moments. Hence the design moment entered in the last column of Table 25.2 are *exclusive* of those due to earth pressure. Due to this reason, earth pressure has been neglected while calculating $R$ in step 2 above, so as to have critical combination of tensile forces.

From Table 25.2, it is clear that maximum B.M. occurs at the bottom of pipe, and is of sagging nature, while B.M. at C is hogging.

**TABLE 25.2.** Moments $M_\theta$ (Kg-m) [+ ve for sagging, – ve for hogging]

| $\theta$ | $M_\theta$ due to | | | | | |
|---|---|---|---|---|---|---|
| | Weight of pipe | Weight of water | Earth over haunches | Uniform live load and overburden (q) | Earth pressure | Design moment |
| 0° (A) | + 1245.8 | + 1037.8 | + 265 | + 8152 | – 2068 | + 10700.6 |
| 45° | + 120.8 | + 100.9 | +85 | 0.0 | – 102 | + 306.7 |
| 90° (C) | – 1418.8 | – 1181.9 | – 460 | – 8152 | + 2222 | – 11212.7 |
| 135° | – 120.8 | – 100.9 | – 262 | 0 | + 108 | – 483.7 |
| 180° (B) | + 1591.7 | + 1325.9 | + 520 | + 8152 | – 2375 | + 11589.6 |

**(iv) Direct force at C.** At C, $M_c = -11212.7$ N-m. The maximum direct force $R_C$ at C, corresponding to the maximum negative moment conditions will naturally be equal to half the total vertical force on the pipe.

$R_c$ due to weight of pipe $= \frac{1}{2} W_c \pi r = \frac{1}{2} \times 5000 \times \pi (0.85) = 6675.9$ N (thrust)

$R_c$ due to upward pressure of water

$$= -\gamma_w r^3 \left(1 - \frac{\pi}{4}\right) = -9800 (0.85)^3 \left(1 - \frac{\pi}{4}\right) = -1291.6 \text{ N (i.e. tension)}$$

$R_C$ due to earth filling over haunches

$$= \frac{\gamma}{2}\left(2r^2 - \frac{\pi r^2}{2}\right) = \frac{18000}{2}\left(2 \times 0.85^2 - \frac{\pi \times (0.85)^2}{2}\right) = 2790 \text{ N (thrust)}$$

$R_c$ due to earth fill and live load

$$= \frac{q(2r)}{2} = 45130 \times 0.85 = 38360 \text{ N (thrust)}$$

∴ Total $R_c$ = 6675.9 − 1291.6 + 2790 + 38360 = 46534.3 N

∴ Net hoop tension $T_c$ = 352800 − 46534.3 = 306265.7 N

**(v) Eccentricities at A, B and C**

$M_A$ = 10700.6 N-m; $\quad T_A$ = 357872 N $\quad\quad \therefore \quad e_a = \dfrac{10700.6}{357872} \times 1000 \approx 30$ mm

$M_B$ = 11589.6 N-m; $\quad T_B$ = 347728 N $\quad\quad \therefore \quad e_b = \dfrac{11589.6}{347728} \times 1000 \approx 33$ mm

$M_C$ = 11212.7 N-m; $\quad T_C$ = 306265.7 N $\quad \therefore \quad e_c = \dfrac{11212.7}{306265.7} \times 1000 \approx 37$ mm

**(vi) Design of pipe:** Let $T$ be the thickness of the pipe. Let us provide 20 mm Φ bar hoops at $T/4$ effective cover, at a spacing of 100 mm c/c on each face of the pipe shell. Area of each bar

$$= A_\Phi = \dfrac{\pi}{4}(20)^2 = 314 \text{ mm}^2.$$

$$A_{sh} \text{ for 1 m length} = (2A_\Phi) \times \dfrac{1000}{s} = \dfrac{2(314) \times 1000}{100} = 6280 \text{ mm}^2$$

$$A_e = 1000\,T + (m-1)A_{sh} = 1000\,T + (12-1) \times 6280 = (1000\,T + 69080) \text{ mm}^2$$

$$I_e = \dfrac{1000\,T^3}{12} + (m-1)A_{sh}\left(\dfrac{T}{4}\right)^2 = (83.33\,T^3 + 4318\,T^2) \text{ mm}^4$$

The pipe has biggest moment at $B$. Assuming a permissible tensile stress of 2.8 N/mm² in concrete, at $B$, we get

$$\dfrac{T_B}{A_e} + \dfrac{M_B}{I_e} \times \dfrac{T}{2} = 2.8. \text{ Hence,}$$

$$\dfrac{347728}{1000\,T + 69080} + \dfrac{11589.6 \times 1000}{83.33\,T^3 + 4318\,T^2} \cdot \dfrac{T}{2} = 2.8$$

or $\quad \dfrac{124.2}{T + 69.1} + \dfrac{24836}{T^2 + 51.82\,T} = 1$

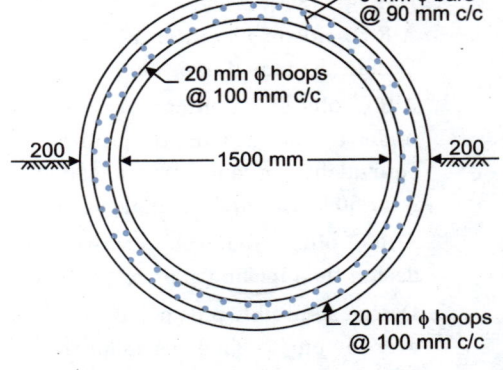

Fig. 25.15

Solving this by trial, we get $T$ = 193 mm.

Keep $\quad T$ = 200 mm.

Provide 20 mm Φ hoop bars @ 100 mm c/c on both faces, at a cover of 35 mm to the centre of steel.

Area of longitudinal (distribution) reinforcement @ 0.3% = $1000 \times 200 \left(\dfrac{0.3}{100}\right) = 600$ mm² per metre of its circumference.

Area on each face = 300 mm².

Hence spacing of 6 mm Φ bars = $\dfrac{1000 \times 28.27}{300} = 94.23$ mm (say 90 mm c/c).

The reinforcement is shown in Fig. 25.15.

**(vii) Check of tensile stresses in steel**

The previous step ensures that concrete will not crack. In case the pipe cracks, the reinforcement alone should be sufficient to take up tensile stresses.

Refer Fig. 25.16. The eccentricity $e_b$ = 33 mm.

The c/c distance between two layers of hoop = $a$ = 200 − 2(35) = 130 mm.

∴ Distance of line of action of pull $T_B$

$$= \frac{a}{2} + e_b = \frac{130}{2} + 33$$

= 98 mm from the layer XX

and $$= \frac{a}{2} - e_b = \frac{130}{2} - 33$$

= 32 mm from the layer YY

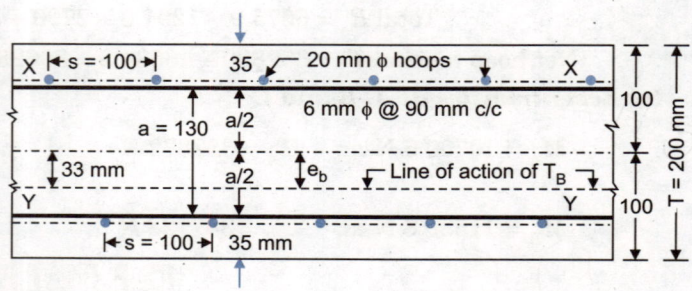

Fig. 25.16

The layer YY will be stressed more since it is near to the line of action to $T_B$.

In order to find the stress in layer YY, take moments about layer XX. If $A_\phi$ is the area of hoop bar, Area of steel in one metre length $= A_\phi \times \dfrac{1000}{s}$

∴ $\left(A_\phi \times \dfrac{1000}{s}\right) t \times a = T_B \left(\dfrac{a}{2} + e_0\right)$

∴ $\left(\dfrac{314 \times 1000}{100}\right) \times t \times 130 = 347728 \left(\dfrac{130}{2} + 3.3\right)$

From which $t = 58.18$ N/mm² < 150 (safe)

## PROBLEMS

1. Discuss in brief the various forces that contribute to the stresses and bending moment in reinforced concrete pipe laid below ground level.

2. A R.C. pipe of 1 m internal diameter is subjected to a normal water pressure of 10 metres of head of water. Design the pipe if its is supported at ground level, at its horizontal diameter.

3. A R.C. pipe of 1 m internal diameter, and carrying water under a normal pressure of 15 metres of head passes under a road. The depth of filling over its crown is 1.6 m. The maximum live load expected on the road is equivalent to a concentrated load of 30 kN per metre length of the pipe. The unit weight of soil around the pipe is 17.50 kN/m³ and its angle of repose is 24°. Design the pipe if it is supported on its horizontal diameter.

4. A RCC pipe is required to withstand 15 m head of water using $M_{20}$ grade concrete and hard drawn steel wire, design the pipe and sketch the details of reinforcement.

5. Design a suitable pipe culvert to carry a discharge of 1 m³/s with a velocity of 2 m/s. The depth of earth filling over the pipe is 2 metres adopt IRC class 'AA' loading with $M_{20}$ grade concrete and steel confirming to IS: 432. Sketch the details of reinforcement and bedding for the pipe.

# CHAPTER 26
# BUNKERS AND SILOS

## 26.1. INTRODUCTION

Bunkers and silos are the structures used for the storage of materials like food grain, cereals, coal, cement and other granular materials. Both bunkers and silos are commonly called as *bins*. If the depth and breadth of a bin are such that the *plane of rupture* meets the surface of the material, before it strikes the opposite side of the bin, it is called a *shallow bin* or a *bunker*. However, when the plane of rupture drawn from the bottom edge of the bin does not intersect the surface level of the material, it is called a *deep bin* or a *silo*. Ordinarily, a bin may be said to be a silo, if its depth is greater than twice the breadth. Hoppers are rectangular bins with the bottom floor consisting of four sloping slabs. Reinforced Concrete bunkers and silos have almost replaced the steel storage structures because of their ease of maintance and superior architectural qualities.

Silos are generally circular in cross-section. For self-cleansing and for emptying, it is supported on a number of columns, through a ring beam. Its bottom height is fixed in such a way that a truck can pass its underneath. It is covered with shallow spherical or conical dome, or with a beam and slab type flat roof with suitable man-hole.

The stored material exerts pressure on the side of a bin. This pressure varies during the filling and emptying processess, and also with the location of the discharging hole. The exact analysis of pressure is extremely difficult because of many variable factors. Therefore, approximate methods suggested by Janssen and Airy are commonly followed.

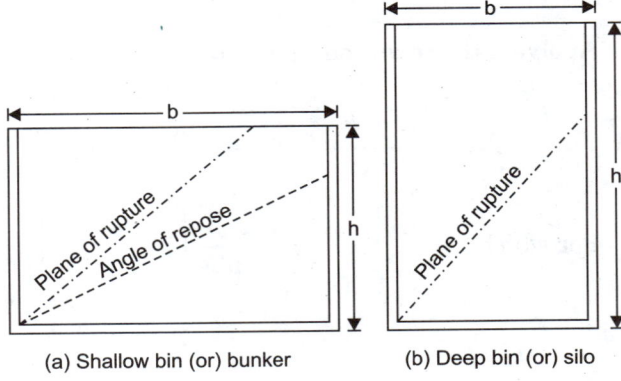

(a) Shallow bin (or) bunker    (b) Deep bin (or) silo

Fig. 26.1

## 26.2. JANSSEN'S THEORY

In the analysis that follows, it is assumed that a large portion of the weight of the contained material is supported by friction between material and the wall, and only a small portion of the weight is transferred to the hopper bottom. Due to this, Rankine's or Coulomb's lateral pressure theories can not be directly applied. The vertical walls of the silo are subjected to direct compression as well as lateral pressure. Let,

$w$ = density of material contained

$A$ = area of horizontal section

**668** REINFORCED CONCRETE STRUCTURE

$P$ = perimeter of horizontal section
$r$ = $A/P$ = hydraulic mean depth of the section
$\phi$ = angle of repose of the filling
$\phi'$ = angle of friction of the filling on the walls of the bin
$\mu'$ = $\tan \phi'$.
'$R$' = Hydraulic mean radius.

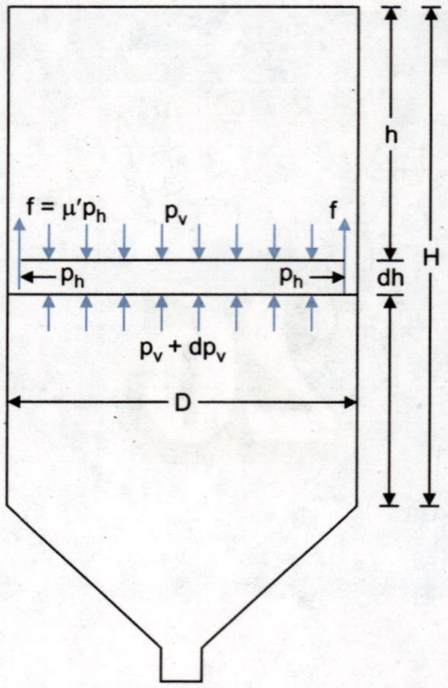

Consider a lamina of small thickness $dh$ at a depth $h$ below the top. Let $p_v$ be the vertical pressure intensity above the slice and $p_v + dp_v$ be the pressure below the slice. Let $p_h$ be the horizontal stress intensity, and $f$ be the frictional stress. Equating the vertical forces of the strip, we have

$$p_v \cdot A + w \cdot A \cdot dh = (p_v + dp_v) A + f \cdot P \cdot dh$$

But $\qquad f = \mu' \cdot ph$

$\therefore \qquad p_v \cdot A + w \cdot A \cdot dh = (p_v + dp_v) A + \mu' \cdot p_h \cdot P \cdot dh$

From which $\qquad w \cdot dh = dp_v + \mu' \cdot \dfrac{p_h}{r} \cdot dh$

or $\qquad dp_v = \left(w - \mu' \dfrac{p_h}{r}\right) dh$

Now $\qquad p_h = K \cdot p_v \qquad \therefore \qquad dp_v = \left(w - \mu' p_v \dfrac{K}{r}\right) dh$

or $\qquad \displaystyle\int \dfrac{dp_v}{w - \mu' p_v \dfrac{K}{r}} = \int dh \qquad \therefore \qquad \dfrac{\log\left(w - \mu' p_v \dfrac{K}{r}\right)}{-\mu' \dfrac{K}{r}} = h + \text{constant}.$

or $\qquad \log\left(w - \mu' p_v \dfrac{K}{r}\right) = -\mu' \dfrac{K}{r} h + C$

Applying the condition that $p_v = 0$ when $h = 0$, we get $C = \log w$

$\therefore \qquad \log\left(\dfrac{w - \mu' p_v \dfrac{K}{r}}{w}\right) = -\mu' \dfrac{K}{r} h \quad \text{or} \quad 1 - \dfrac{\mu'}{w} \cdot \dfrac{K}{r} \cdot p_v = e^{-\mu' \dfrac{K}{r} h}$

For which $\qquad p_v = \dfrac{wr}{\mu' K}\left[1 - e^{-\dfrac{\mu' K h}{r}}\right]$ ...(26.1)

and $\qquad p_h = \dfrac{wr}{\mu'}\left[1 - e^{-\dfrac{\mu' K h}{r}}\right]$ ...(26.2)

Generally, silos are circular in cross-section. The hoop tension on the wall will then be = $p_h \cdot D/2$, where $D$ is the diameter. In addition to the hoop tension, it will also be subjected to vertical pressure $P_w$ per metre run, where total vertical load taken by wall for a depth '$h$' is given by Expression:

$$P_w = \int_0^h \mu' \, p_h \cdot dh = \int_0^h wr \left[1 - e^{-\dfrac{\mu' K h}{r}}\right] dh$$

or $\qquad P_w = wr\left[h - \dfrac{r}{\mu' K}\left\{1 - e^{-\dfrac{\mu' K h}{r}}\right\}\right]$ ...(26.3)

Total vertical pressure is

$$P_{WT} = P_w \cdot P = wA\left[h - \dfrac{r}{\mu' K}\left\{1 - e^{-\dfrac{\mu' K h}{r}}\right\}\right] = w \cdot Ah - Ap_v \qquad ...(26.4)$$

Fig. 26.2

The value of $K$ lies between $\dfrac{1-\sin\phi}{1+\sin\phi}$ and $\dfrac{1+\sin\phi}{1-\sin\phi}$, its exact value can be found where 'A' = Cross sectional area of silo only experimentally. Some Codes specify $K = K_f = 0.5$ during filling and $K = K_e = 1.0$ during emptying. Similarly, during filling, $\mu' = \mu_f' \approx \tan(0.75\phi)$ while during emptying $\mu' \approx \mu_e' \approx \tan(0.6\Phi)$. Table 26.1 gives the values of the function $(1 - e^{-x})$, where $x = \dfrac{\mu' K h}{r}$ in Janssen's formula.

**TABLE 26.1.** Value of $(1-e^{-x})$

where $x = \dfrac{\mu' K h}{r}$ in Janssen's formula and $x = \dfrac{h}{z_0}$ in IS code formula (Eq. 26.24)

| $x$ | $1-e^{-x}$ | $x$ | $1-e^{-x}$ | $x$ | $1-e^{-x}$ | $x$ | $1-e^{-x}$ | $x$ | $1-e^{-x}$ |
|---|---|---|---|---|---|---|---|---|---|
| 0.01 | 0.010 | 0.42 | 0.343 | 0.92 | 0.599 | 1.60 | 0.798 | 2.85 | 0.941 |
| 0.02 | 0.020 | 0.44 | 0.346 | 0.94 | 0.609 | 1.65 | 0.808 | 2.90 | 0.945 |
| 0.03 | 0.030 | 0.46 | 0.369 | 0.96 | 0.617 | 1.70 | 0.817 | 2.95 | 0.948 |
| 0.04 | 0.040 | 0.48 | 0.381 | 0.98 | 0.625 | 1.75 | 0.826 | 3.00 | 0.950 |
| 0.05 | 0.049 | 0.50 | 0.393 | 1.00 | 0.632 | 1.80 | 0.835 | 3.05 | 0.953 |
| 0.06 | 0.058 | 0.52 | 0.405 | 1.02 | 0.639 | 1.85 | 0.843 | 3.10 | 0.955 |
| 0.07 | 0.068 | 0.54 | 0.417 | 1.04 | 0.646 | 1.90 | 0.850 | 3.15 | 0.957 |
| 0.08 | 0.077 | 0.56 | 0.429 | 1.06 | 0.653 | 1.95 | 0.858 | 3.20 | 0.959 |
| 0.09 | 0.086 | 0.58 | 0.440 | 1.08 | 0.660 | 2.00 | 0.865 | 3.25 | 0.961 |
| 0.10 | 0.095 | 0.60 | 0.451 | 1.10 | 0.667 | 2.05 | 0.871 | 3.30 | 0.963 |
| 0.12 | 0.113 | 0.62 | 0.462 | 1.12 | 0.674 | 2.10 | 0.878 | 3.35 | 0.965 |
| 0.14 | 0.131 | 0.64 | 0.473 | 1.14 | 0.680 | 2.15 | 0.884 | 3.40 | 0.967 |
| 0.16 | 0.148 | 0.66 | 0.483 | 1.16 | 0.687 | 2.20 | 0.889 | 3.45 | 0.968 |
| 0.18 | 0.165 | 0.68 | 0.493 | 1.18 | 0.693 | 2.25 | 0.895 | 3.50 | 0.970 |
| 0.20 | 0.181 | 0.70 | 0.503 | 1.20 | 0.699 | 2.30 | 0.900 | 3.60 | 0.973 |
| 0.22 | 0.198 | 0.72 | 0.512 | 1.22 | 0.705 | 2.35 | 0.905 | 3.70 | 0.975 |
| 0.24 | 0.213 | 0.74 | 0.523 | 1.24 | 0.711 | 2.40 | 0.909 | 3.80 | 0.978 |
| 0.26 | 0.229 | 0.76 | 0.532 | 1.26 | 0.716 | 2.45 | 0.914 | 3.90 | 0.980 |
| 0.28 | 0.244 | 0.78 | 0.542 | 1.28 | 0.722 | 2.50 | 0.918 | 4.00 | 0.982 |
| 0.30 | 0.259 | 0.80 | 0.551 | 1.30 | 0.727 | 2.55 | 0.922 | 5.00 | 0.993 |
| 0.32 | 0.274 | 0.82 | 0.560 | 1.35 | 0.741 | 2.60 | 0.926 | 6.00 | 0.997 |
| 0.34 | 0.288 | 0.84 | 0.568 | 1.40 | 0.753 | 2.65 | 0.929 | 8.00 | 0.999 |
| 0.36 | 0.302 | 0.86 | 0.577 | 1.45 | 0.765 | 2.70 | 0.933 | – | – |
| 0.38 | 0.316 | 0.88 | 0.585 | 1.50 | 0.777 | 2.75 | 0.936 | – | – |
| 0.40 | 0.330 | 0.90 | 0.593 | 1.55 | 0.788 | 2.80 | 0.939 | $\infty$ | 1.000 |

## 26.3. AIRY'S THEORY

Airy's analysis is based on Coulomb's wedge theory of earth pressure. By this theory, it is possible to calculate the horizontal pressure per unit length of the periphery and the position of the plane of rupture. We shall consider two cases (a) shallow bin and (b) deep bin.

(a) **Case I: Shallow bin.** Fig. 26.3 shows the case of a shallow bin (bunker) in which the rupture plane, inclined at an angle cuts the top horizontal surface. Consider the wedge $ABC$ of unit thickness, and let $W$ be its weight. Let $R_1$ and $R_2$ be the two reactions to $BC$ and $BA$ respectively. Due to these, $R$ and $P_h$ are the normal reactions, while $\mu R$ and $\mu' P_h$ are the frictional forces from the material and the wall, as shown in Fig. 26.3. If $h$ is the height of the bin, we have $AC = h \cot\theta$

where $AE$ = plane of rupture

'$b$' = Diameter of the silo/bin

$$W = \tfrac{1}{2} h \cot \theta . h . w = \frac{wh^2}{2} \cot \theta$$

Resolving the forces on the wedge in vertical direction, we have

$$W = \mu' P_h + \mu R \sin \theta + R \cos \theta$$

$$R = \frac{W - \mu' P_h}{\mu \sin \theta + \cos \theta} \qquad \ldots(a)$$

Resolving the forces on the wedge in horizontal direction,

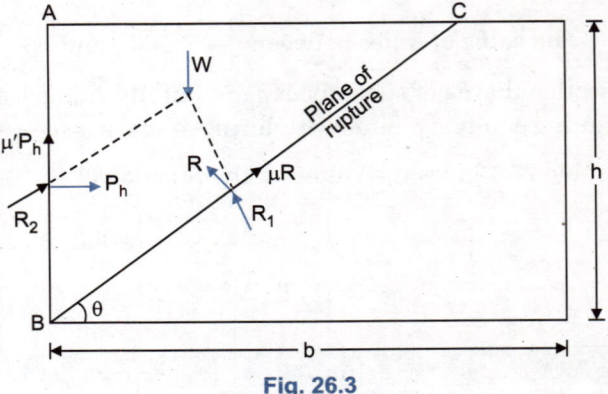

Fig. 26.3

$$P_h + \mu R \cos \theta = R \sin \theta$$

$$R = \frac{P_h}{\sin \theta - \mu \cos \theta} \qquad \ldots(b)$$

Equating (a) and (b)

$$\frac{W - \mu' P_h}{\mu \sin \theta + \cos \theta} = \frac{P_h}{\sin \theta - \mu \cos \theta}$$

or $P_h [\mu \sin \theta + \cos \theta + \mu' (\sin \theta - \mu \cos \theta)] = W (\sin \theta - \mu \cos \theta)$

$$P_h = \frac{W (\sin \theta - \mu \cos \theta)}{\sin \theta (\mu + \mu') + \cos \theta (1 - \mu \mu')} \qquad \ldots(c)$$

Substituting the value of $W$ and simplifying, we get

$$P_h = \frac{wh^2}{2} \times \frac{\tan \theta - \mu}{(\mu + \mu') \tan^2 \theta + (1 - \mu \mu') \tan \theta} \qquad \ldots(26.5)$$

or $$P_h = \frac{wh^2}{2} \left( \frac{u}{v} \right)$$

where $u = \tan \theta - \mu$ and $v = (\mu + \mu') \tan^2 \theta + (1 - \mu \mu') \tan \theta$

For $(P_h)_{max}$, $\dfrac{dP_h}{d\theta} = 0$ $\therefore$ $\dfrac{dP_h}{d\theta} = \dfrac{wh^2}{2} \left[ \dfrac{u dv - v du}{v^2} \right] = 0$

$\therefore$ $$\frac{u}{v} = \frac{du}{dv}$$

Therefore, $$\frac{\tan \theta - \mu}{(\mu + \mu') \tan^2 \theta + (1 - \mu \mu') \tan \theta} = \frac{\sec^2 \theta}{2 (\mu + \mu') \tan \theta \sec^2 \theta + (1 - \mu \mu') \sec^2 \theta}$$

Rearranging and simplifying, we get $\tan^2 \theta - 2 \mu \tan \theta - \dfrac{\mu (1 - \mu \mu')}{\mu + \mu'} = 0$

From which $$\tan \theta = \mu + \sqrt{\frac{\mu (1 + \mu^2)}{\mu + \mu'}} \qquad \ldots(26.6)$$

Substituting this in Eq. 26.5

$$P_h = \frac{wh^2}{2} \times \frac{\mu + \sqrt{\dfrac{\mu (1 + \mu^2)}{\mu + \mu'}} - \mu}{(\mu + \mu') \left[ \mu + \sqrt{\dfrac{\mu(1+\mu^2)}{\mu + \mu'}} \right]^2 + (1 - \mu \mu') \left[ \mu + \sqrt{\dfrac{\mu(1+\mu^2)}{\mu + \mu'}} \right]}$$

or $$P_h = \frac{wh^2}{2} \left[ \frac{1}{\sqrt{1 + \mu^2} + \sqrt{\mu (\mu + \mu')}} \right]^2 \qquad \ldots(26.7)$$

Since $P_h$ denotes the total horizontal force per unit length of the wall, at depth $h$, the pressure per unit area is

$$p_h = \frac{dP_h}{dh} = wh \left[ \frac{1}{\sqrt{1+\mu^2} + \sqrt{\mu(\mu+\mu')}} \right]^2 \qquad \ldots(26.8)$$

Substituting $\mu = \tan \phi$ and $\mu' = \tan \phi'$, an alternative expression for $p_h$ is:

$$p_h = wh \left[ \frac{\cos \phi}{1 + \sqrt{\sin \phi \sec \phi' \sin(\phi + \phi')}} \right]^2 \qquad \ldots[26.8(a)]$$

The vertical load taken by the walls is

$$P_w = \mu' \cdot P_h \qquad \ldots(26.9) \qquad \text{and} \qquad p_w = \mu' p_h \qquad \ldots[26.9(a)]$$

Total vertical load carried by wall is

$$P_{WT} = \pi b(\mu' P_h) = \frac{wh^2 \pi b}{2} \cdot \frac{\mu'}{\sqrt{1+\mu^2} + \sqrt{\mu(\mu+\mu')^2}} \qquad \ldots(26.10)$$

The depth upto which the bin will act as a shallow one is given by

$$\tan \theta = \frac{h}{b} = \mu + \sqrt{\frac{\mu(1+\mu^2)}{\mu+\mu'}} \qquad \text{or} \qquad h = b \left[ \mu + \sqrt{\frac{\mu(1+\mu^2)}{\mu+\mu'}} \right] \qquad \ldots(26.11)$$

This is the maximum depth upto which equation 26.8 is applicable.

**(b) Case 2: Deep bin (silo)** In case of deep bin or a silo, the plane of rupture $BC$ cuts the opposite side (Fig. 26.4).

where '$w$' = weight of wedge $ABCD$

$\therefore \qquad CD = (h - b \tan \theta)$.

$\therefore \qquad W = wbh - \dfrac{wb}{2} \cdot b \tan \theta$

or $\qquad W = wb \left[ h - \dfrac{b \tan \theta}{2} \right]$

or $\qquad W = \dfrac{wb}{2}(2h - b \tan \theta) \qquad \ldots(d)$

where $w$ = wt of wedge $ABCD$

As obtained for shallow bin,

$$P_h = \frac{W(\sin \theta - \mu \cos \theta)}{\sin \theta(\mu + \mu') + \cos \theta(1 - \mu \mu')}$$

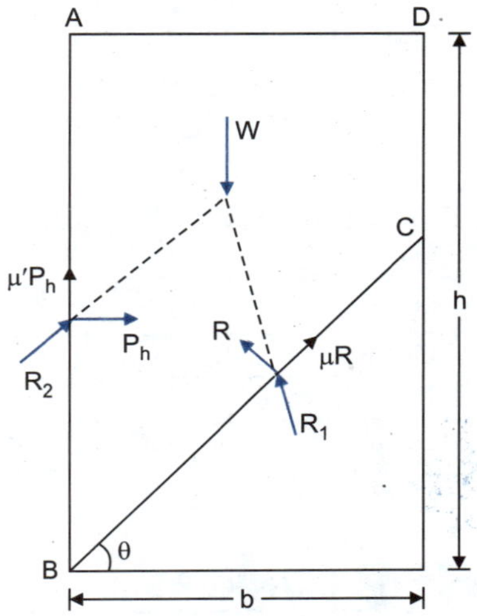

Substituting the value of $W$ and simplifying

$$P_h = \frac{wb \left[ h - \dfrac{b \tan \theta}{2} \right][\tan \theta - \mu]}{\tan \theta(\mu + \mu') + (1 - \mu \mu')} = wb \frac{\left[ \left( h + \dfrac{b\mu}{2} \right) \tan \theta - \dfrac{b}{2} \tan^2 \theta - h\mu \right]}{\tan \theta(\mu + \mu') + (1 - \mu \mu')}$$

For maxima, $\dfrac{dP_h}{d\theta} = 0$. Hence,

$$\frac{\left( h + \dfrac{b\mu}{2} \right) \tan \theta - \dfrac{b}{2} \tan^2 \theta - h\mu}{\tan \theta(\mu + \mu') + (1 - \mu \mu')} = \frac{\left( h + \dfrac{b\mu}{2} \right) \sec^2 \theta - b \tan \theta \sec^2 \theta}{(\mu + \mu') \sec^2 \theta}$$

Simplifying and rearranging, we get

$$\tan^2 \theta + \frac{2(1 - \mu \mu')}{\mu + \mu'} \tan \theta - \frac{2h(\mu^2 + 1) + b\mu(1 - \mu \mu')}{b(\mu + \mu')} = 0. \quad \text{From which}$$

$$\tan \theta = \frac{1}{\mu + \mu'} \left[ -(1 - \mu \mu') + \sqrt{(1 - \mu \mu')(1 + \mu^2) + \frac{2h}{b}(1 + \mu^2)(\mu + \mu')} \right] \qquad \ldots(26.12)$$

Fig. 26.4

Substituting this value of tan θ, we get

$$P_h = \frac{wb^2}{2(\mu+\mu')^2}\left[\sqrt{\frac{2h}{b}(\mu+\mu')+(1-\mu\mu')} - \sqrt{1+\mu^2}\right] \qquad \ldots(26.13)$$

The pressure $\quad p_h = \dfrac{dP_h}{dh} = \dfrac{wb}{(\mu+\mu')}\left[1 - \dfrac{\sqrt{1+\mu^2}}{\sqrt{\dfrac{2h}{b}(\mu+\mu')+(1-\mu\mu')}}\right] \qquad \ldots(26.14)$

**Note:** For rectangular silos, $b$ may be taken as the length of the side adjacent to the one on which pressure is being calculated. The vertical load taken by the wall is

$$P_w = \mu' P_h \quad \ldots[26.15\,(a)] \quad \text{and} \quad P_{WT} = \pi\,b\,\mu'\,P_h \qquad \ldots(26.15)$$

## Surcharge on conical hopper

Let $\quad H =$ depth of cylindrical portion (Fig. 26.2)

$P_H =$ horizontal pressure at depth $H$, per unit length

∴ $\quad P_{WT} = \pi\,b\,\mu'\,P_H$

∴ $\quad$ Surcharge pressure $= \dfrac{\dfrac{\pi b^2}{4}wH - \pi\,b\,P_H\cdot\mu'}{\pi b^2/4} = wH - \dfrac{4\mu'}{b}P_H. \qquad \ldots[26.15(a)]$

The following are the values of $w$, $\mu$ and $\mu'$ for some common materials:

where $\quad w =$ Density of material

$(\mu) =$ Filling on filling

$(\mu') =$ Filling on concrete.

| *Materials* | $w$ (N/m³) | $\mu$ | $\mu'$ |
|---|---|---|---|
| 1. Wheat | 7850 | 0.466 | 0.444 |
| 2. Maize | 6870 | 0.521 | 0.432 |
| 3. Cement | 14150 | 0.316 | 0.554 |
| 4. Bituminous coal | 7850 | 0.700 | 0.700 |
| 5. Sand | 16000 | 0.674 | 0.577 |
| 6. Coke | 4500 | 0.839 | 0.839 |

## 26.4. BUNKERS

In the bunkers, because of shallow depth, it is assumed that the friction between the wall and the fill in negligible. Fig. 26.5 shows a bunker with other structural components such as top rib, junction beam, hopper bottom with central opening, and supporting columns.

**Pressure and moments on walls**

Let $\alpha$ be the angle of surcharge of the fill. The pressure against the vertical wall is given by Rankine's formula:

$$p = wh\cos\alpha\,\frac{\cos\alpha - \sqrt{\cos^2\alpha - \cos^2\phi}}{\cos\alpha + \sqrt{\cos^2\alpha - \cos^2\phi}} \qquad \ldots[26.16\,(a)]$$

The pressure acts in a direction parallel to the top surface of the retained material. The horizontal component is

$$p_h = p\cos\alpha = wh\cos^2\alpha\,\frac{\cos\alpha - \sqrt{\cos^2\alpha - \cos^2\phi}}{\cos\alpha + \sqrt{\cos^2\alpha - \cos^2\phi}} \qquad \ldots[26.16\,(b)]$$

where $\quad\alpha =$ Angle of Surcharge

$\phi =$ Angle of repose

$w =$ density of material stored

The pressure acts in a direction parallel to the surface of the retained material Horizontal Component = $p = p_\alpha \cos \alpha$

∴ Service load pressure = $p = wh \cos^2 \theta$

Design ultimate pressure = $(p_u) = 1.5\, wh \cos^2 \phi$

If $\alpha = \phi$,

$$p_h = wh \cos^2 \alpha \qquad \text{...[26.16 (b)]}$$

If $\alpha = 0$, (i.e. top surface is horizontal),

$$p_h = wh\, \frac{1 - \sin \phi}{1 + \sin \phi} \qquad \text{...[26.16 (c)]}$$

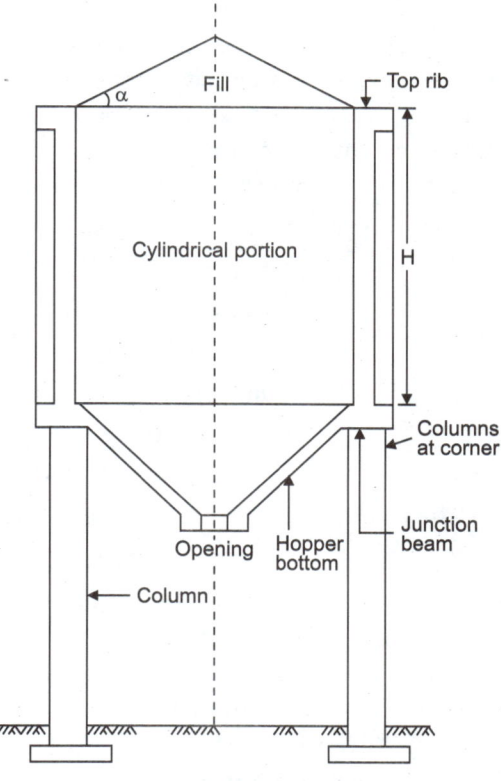

The edge conditions are such that the top is free and the bottom as well as the vertical edges are restrained due to continuity. For this case, the moment due to triangularly distributed lateral pressure may be computed with the help of Table 3 of IS : 3370- Part IV. A simpler approach is to assume that the lateral pressure is taken entirely in the horizontal direction. This is reasonably true if the height of the wall is large as compared to its length. In that case, for a *square bunker*, the bending moment at any depth will be $-p_h \cdot \dfrac{L^2}{12}$ at the corners and $p_h \cdot \dfrac{L^2}{24}$ at mid span, where $L$ is the span between the centre lines of the supporting walls. Generally, haunches are provided at the corners, due to which negative moments are much lesser than $p_h \cdot L^2/12$, and the design is made for a uniform value of $p_h \cdot L^2/16$.

Fig. 26.5

For a *rectangular bunker* (Fig. 26.6), the negative moments at the corners are:

$$M_A = M_B = M_D = M_C = -\frac{p_h}{12}(B^2 - BL + L^2) \qquad \text{...(26.17)}$$

The positive moment at centre of span $B$ is

$$M_{CB} = \frac{p_h B^2}{8} - \frac{p_h}{12}(B^2 - BL + L^2)$$

$$= \frac{p_h}{24}(B^2 + 2BL - 2L^2) \qquad \text{...[26.18 (a)]}$$

and the positive moment at the centre of span is

$$M_{CB} = \frac{p_h}{24}(L^2 + 2BL - 2B^2) \qquad \text{...[26.18 (b)]}$$

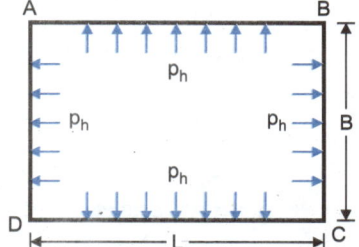

Fig. 26.6. Pressure Intensity on Walls of a Bunker.

In addition to the bending moment, the walls are also subjected to direct tension due to pressure on the adjoining walls. Thus, direct tension on wall $L$ will be $0.5\, p_h \cdot B$ and that on wall $B$ will be $0.5\, p_h \cdot L$. Apart from the bending moment and direct tension, the walls are also subjected to vertical weight of material transferred to it by hopper bottom. The vertical wall is therefore designed as a deep beam supported between the columns.

**Circular bunkers.** The vertical wall of the circular bunker of diameter $b$ is designed for a hoop tension of $p_h \cdot b/2$.

## 26.5. HOOPER BOTTOM

**(a) Conical hopper.** Conical hoppers are subjected essentially to meridional and hoop tensions. The total meridional tension at any horizontal plane passing through the hopper is such that its vertical component is equal to total vertical pressure on the plane plus the weight of the hopper and contents below the plane. Consider any horizontal plane $AD$ at depth $h$ below the top surface of the material. Let $W_g$ be the weight of the grain and $W_c$ be the weight of the cone below this plane. The meridional tension $N$ is given by

$$N \sin \alpha \cdot 2 \pi b = p_v \pi b^2 + W_g + W_c$$

$$\therefore \quad N = \frac{p_v \pi b^2 + W_g + W_c}{2 \pi b \sin \alpha} \quad ...(26.19)$$

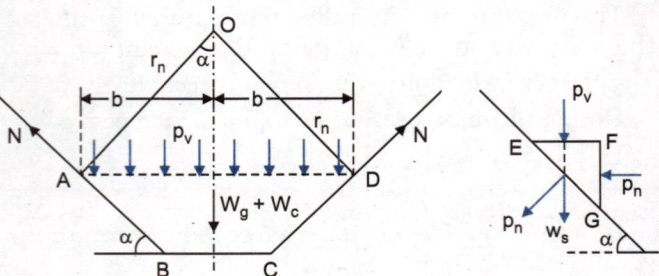

Fig. 26.7. Conical Hoppers.

Let $p_h$ be the normal pressure. If $w_s$ is the self weight of the cone per unit area, we have, from Fig. 26.7 (b)

$$p_n \cdot EG = p_v \cdot EF \cos \alpha + p_h \cdot FG \sin \alpha + w_s \, EG \cos \alpha$$

or 
$$p_n = p_v \cos^2 \alpha + p_h \sin^2 \alpha + w_s \cos \alpha \qquad ...(26.20)$$

The ring tension $T$ at any plane is given by

$$T = p_n \cdot r_n \qquad ...[26.21\,(a)]$$

where $r_n$ = radius of curvature [Fig. 26.7 (a)] = $b \, \text{cosec} \, \alpha$

For deep bins, $p_v$ and $p_h$ are obtained from the expressions developed earlier. However, the value of $\mu'$ will be taken equal to $\mu$ in the equation for horizontal pressure. For shallow bins, $p_v$ on any horizontal cross-section will be equal to the weight of the material above the section and $p_h$ will be given by *Eq. 26.23*, given by IS Code (§ 26.6) substituting $\phi'$ for $\phi$.

(*b*) **Pyramidal hoppers:** Pyramidal hoppers are subjected to bending moments and direct tensions besides meridional tension along the slope.

*Meridional tension.* The weight of retained material, weight of sloping bottom and the weight of rib at the opening etc. will give rise to a vertical pull in the downward direction, at the bottom of the side wall. If $W$ is the weight mentioned above, per unit length, the maximum meridional tension at $A$, per unit length will be given by

$$N = W \, \text{cosec} \, \alpha \qquad ...(26.22)$$

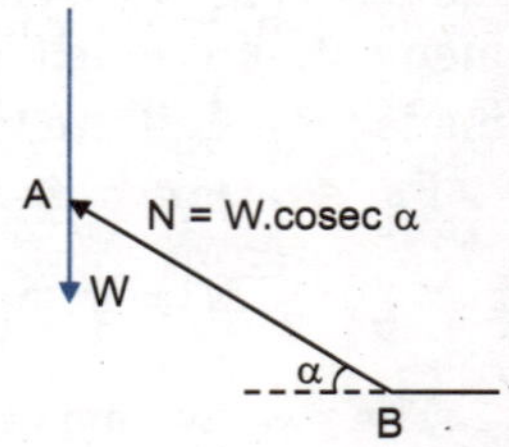

This direct tension will gradually decrease to zero at the opening $B$.

Fig. 26.8

The hopper bottom is also subjected to bending moment due to bending of the slab spanning horizontally between the intersection of adjacent sloping faces. For example, if we consider unit length of the slab between levels $XX$ and $YY$ [Fig. 26.9 (b)], the span of the slab $AB$ will be $l$ and it will be subjected to a normal load intensity $p_n$, where $p_n$ is given by *Eq. 26.20*. Since the strip is continuous on all the four sides of the hopper, it forms a closed frame, developing negative moments at joints and positive moment at the centre of each span. The magnitude of negative moment may be taken as $(p_n \cdot l^2)/12$ and the positive moment equal to $(p_n \cdot l^2)/12$. The reinforcement will be placed at the outer face for +ve moments and inside for negative moments. The horizontal span $l$ of the slab will increase from minimum at the opening to maximum at the junction with vertical wall. The pressure $p_n$ will, however, decrease from maximum at the opening to minimum at junction with vertical wall. The section of the slab should be designed at the centre. In addition to this, the strip will also be subjected to a direct pull equal to $p_n \, l/2$.

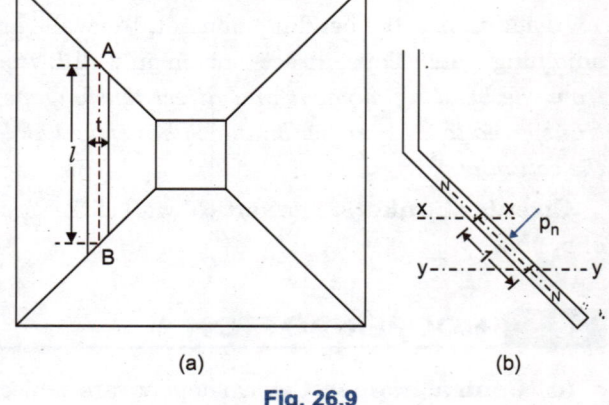

Fig. 26.9

## 26.6. INDIAN STANDARD ON DESIGN OF BINS (IS: 4995-1968)

### 1. Permissible stresses

**(a) In concrete**

*(i) Resistance to cracking.* The permissible stresses in tension (direct and due to bending) and shear shall conform to the values specified in Table 26.2. The values given in IS: 4995-1968 are in kg/cm² units. These have been converted in N/mm² unit by the approximate relation: 10 kg/cm² ≈ 1 N/mm². The permissible tensile stress due to bending apply to the outside face of the bin. In members less than 225 mm thick and in contact with the grain on one side, the permissible stresses in bending apply also on the side in contact with the grain.

*(ii) Resistance to buckling.* The maximum compressive stress on the net wall section deducting all openings, recesses, etc. shall not exceed $0.15 f_c$ where $f_c$ is the compressive strength of concrete at the age of 28 days. Where wind or earth quake forces are taken into account, the stress on net wall section shall not exceed $0.2 f_c$.

**TABLE 26.2.** Permissible stresses in concrete

| S. N. | Grade of concrete | Permissible stresses (N/mm²) | | Shear (N/mm²) |
|---|---|---|---|---|
| | | Direct tension | Tension due to bending | |
| 1. | M 20 | 1.2 | 1.7 | 1.7 |
| 2. | M 25 | 1.3 | 1.8 | 1.9 |
| 3. | M 30 | 1.5 | 2.0 | 2.2 |
| 4. | M 35 | 1.6 | 2.2 | 2.5 |
| 5. | M 40 | 1.7 | 2.4 | 2.7 |

**(b) In steel**: In strength calculations, the stresses in plain mild steel reinforcement, and in high yield strength bars (*HYSD* bars), the values given in Table 21.2 are adopted.

### 2. Design parameters

*(i) Unit weight and angle of internal friction.* Table 26.3 gives the values of unit weight and angle of internal friction for same of the common types of grains.

**TABLE 26.3.** Unit weight and angles of internal friction

| Materials | Unit weight (N/m³) | Angle of internal friction $\phi$ |
|---|---|---|
| Wheat | 8340 | 28° |
| Paddy | 5640 | 36° |
| Rice | 8830 | 33° |
| Maize | 7850 | 30° |
| Barley | 6770 | 27° |
| Corn | 7850 | 27° |

*(ii) Wall friction.* In the absence of test results, the angle of wall friction may be assumed to be 0.75 $\phi$ during filling and 0.6 $\phi$ during emptying, where $\phi$ is the angle of internal friction of the material.

*(iii) Pressure ratio.* The ratio of horizontal pressure to vertical pressure shall be assumed to be 0.5 during filling and 1.0 during emptying, i.e. $K_f = 0.5$ and $K_e = 1.0$.

### 3. Design of walls

*(i) Pressure on walls of a shallow bin:* The horizontal pressure $p_h$ on a vertical wall of a shallow bin shall be calculated by the following formula

$$p_h = \frac{wh \cos^2 \phi}{\left[1 + \sqrt{\frac{\sin(\phi + \phi') \sin \phi}{\cos \phi'}}\right]^2} \qquad \ldots(26.23)$$

where $\phi'$ = angle of wall friction.

**(ii) Pressure in a deep bin:** In deep bins, the maximum values of the horizontal pressure on the wall ($p_h$), the vertical pressure on the horizontal cross-section of the stored material ($p_v$) and the vertical load transferred to the wall per unit area due to friction ($p_w$) may be calculated using formulae given below:

| Nature of pressure | During filling | During emptying |
|---|---|---|
| Maximum $p_v$ | $\dfrac{wr}{K_f \mu_f'}$ | $\dfrac{wr}{K_e \mu_e'}$ |
| Maximum $p_h$ | $\dfrac{wr}{\mu_f'}$ | $\dfrac{wr}{\mu_e'}$ |
| Maximum $p_w$ | $wr$ | $wr$ |

where  $w$ = unit weight of stored material;  $R$ = area/perimeter ratio
  $K_f$ = pressure ratio during filling condition = 0.5
  $K_c$ = pressure ratio during emptying condition = 1.0
  $\mu_f'$ = coefficient of wall friction during filling conditions.
  $\mu_e'$ = coefficient of wall friction during emptying conditions.

The variation of $p_v$, $p_h$ and $p_w$ with depth (Fig. 26.10) may be obtained from the following expression.

$$(p_i)_h = (p_i)_{max.} (1 - e^{-h/z_0}) \qquad ...(26.24)$$

where  $p$ stands for pressure and suffix $i$ stands for $v$, $h$ or $w$, and  $z_0$ assumes following values:

During filling,  $z_0 = \dfrac{r}{K_f \mu_f'}$

During emptying,  $z_0 = \dfrac{r}{K_e \mu_e'}$   ...(26.25)

The value of $(1 - e^{-h/z_0})$ may be taken from Table 26.1.

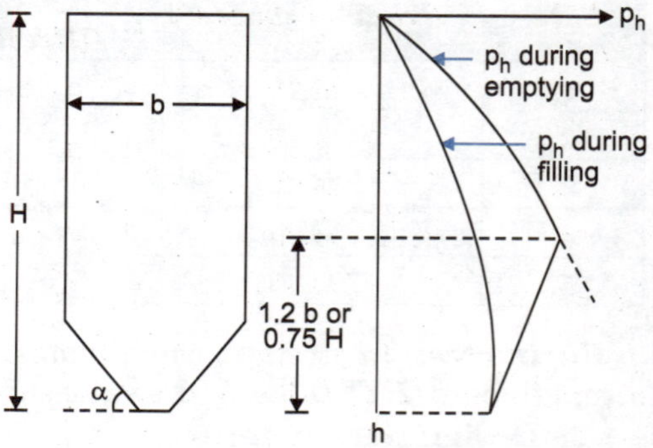

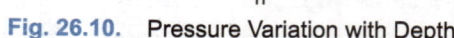

Fig. 26.10.  Pressure Variation with Depth.

To take advantage of the decrease in lateral pressure caused by the bottom of a bin, the horizontal pressure due to emptying, over a height of 1.2 $b$ or 0.75 $H$ whichever is smaller, may be taken to be varying linearly from the emptying pressure at this height to the filling pressure at bottom as shown in Fig. 26.10.

The wall of a storage bin shall be examined for strength as well as stability under the following two cases of loading separately:

(a) *When the bin is empty*. All possible vertical loads (including the weight of the material transferred to the wall due to friction) and lateral loads due to wind or quake (whichever is critical) shall be considered in this case.

(b) *When the bin is empty*. In this case, the self weight of the bin and other permanent loads on the bin shall be considered to act vertically in combination with the maximum lateral load caused by wind or earthquake.

**(iii) Thickness of walls:** The thickness of walls shall be computed on no crack basis. The minimum thickness of the wall of a concrete bin having an internal diameter of 6 m or less shall be 15 cm. When the internal diameter exceeds 6 m, the minimum thickness shall be determined from the equation:

$$T = 15 + \dfrac{D - 600}{120} \text{ cm} \quad \text{where} \quad D \text{ is the internal diameter of the bin in cm.}$$

Walls of circular bins shall be designed essentially for hoop stresses. For walls of rectangular bins, the lateral pressure at any depth may be obtained from *Eq. 26.24*, taking approximate value of $r$ given below:

For square bin,  $r = 0.283 a$,  where  $a$ is the side of the square.

For rectangular bin, with short side $a$ and long side $b$,

$r = 0.283\ a$, for obtaining pressure on short side

$$r = 0.283 \left(\frac{2\ ab - a^2}{b}\right), \quad \text{for obtaining pressure on the long side.}$$

The walls of rectangular bins are subjected to bending moment, besides direct tension. The wall shall, therefore, be designed for bending moments and tensile force caused by the horizontal pressure exerted by the stored material.

**4. Ring Girder:** When the conical hopper is supported on a ring girder provided at the junction of the wall hopper, the girder is to be designed to resist compression, bending as well as tension. The compressive force subjected to the girder is equal to the horizontal component of the downward pull of the loaded hopper minus the outward lateral thrust exerted by the grain. The girder is however designed for the net horizontal component obtained when (a) the bin is assumed to be full, and (b) when the surface of the grain is assumed at the ring level whichever gives critical values.

The bending and twisting moments in the ring girders having 4, 6, 8 and 12 columns are calculated from the following formulae:

(a) Maximum negative bending moment at supports = $C_1 wD/2$
(b) Maximum positive bending moment midway between the supports = $C_2 wD/2$
(c) Maximum torsional moment = $C_3 wD/2$

where $w$ = unit weight of stored material; $D$ = Diameter of circular bin
$C_1, C_2, C_3$ = coefficients of bending and torsional moment as given in Table 26.4.

**TABLE 26.4.** Coefficients of bending and torsional moments of ring girder in bins

| No. of columns | Load on columns | Max. shear | Coeff. of negative bending moment at columns $C_1$ | Coeff. of positive bending moment midway between columns $C_2$ | Angular distance from columns points of max. torsion | Coeff. of max. torsional moment $C_3$ |
|---|---|---|---|---|---|---|
| 1 | 2 | 3 | 4 | 5 | 6 | 7 |
| 4 | W/4 | W/8 | 0.03415 | 0.0176 | 19° 12′ 0″ | 0.0053 |
| 6 | W/6 | W/12 | 0.01482 | 0.0075 | 12° 44′ 0″ | 0.00151 |
| 8 | W/8 | W/16 | 0.00827 | 0.00416 | 6° 33′ 0″ | 0.00063 |
| 12 | W/12 | W/24 | 0.00365 | 0.00190 | 6° 21′ 0″ | 0.000185 |

**5. Bin bottom:** *(i) Level bottom.* The vertical load for the design of level bottom of a deep bin shall be (a) the gross weight of the material stored to the full capacity of the bin minus the total vertical load carried by walls due to friction during filling, or (b) the weight of the material in the bin standing upto a height corresponding to that of a shallow bin, whichever is more.

*(ii) Hopper bottom.* The intensity of normal pressure, $p_n$, at any point on the hopper bottom shall be given by

$$p_n = p_v \cos^2 \alpha + p_h \sin^2 \alpha + w_s \cos \alpha.$$

**6. Reinforcement:** *(i) Circumferential reinforcement.* The minimum circumferential reinforcement shall be 0.3 per cent of the cross-sectional area. Splices in bars shall be well saggered. The bars shall be at least 10 mm in diameter and spacing not more than the thickness of the shell or in any case not more than 200 mm centres.

*(ii) Vertical Reinforcement.* Vertical reinforcement shall be at least 0.3% of the cross-sectional area. Half the number of bars on the inside and half on the outer side may be provided to take care of temperature and shrinkage stress. Where the base of the wall is fixed to the bottom, vertical reinforcement duly calculated shall be provided on the tension face.

A minimum cover of 5 cm shall be provided for the reinforcement.

**Example 26.1.** *Design a bunker to store 300 kN of coal, for the following data:*

*Unit weight of coal = 8340 N/m³; Angle of repose = 30°. The stored coal is to be surcharged at its angle of repose. Take permissible stress in steel as 140 N/mm².*

**Solution:**

**1. *Capacity and dimensions*:** Let us provide a square bunker of size 3 m × 3 m. Let the hopper portion have a height of 1.25 m with a central hole of size 0.5 m × 0.5 m.

Height of surcharge = 1.5 tan 30° = 0.87 m.

Volume required = $\dfrac{300 \times 1000}{8340}$ = 35.97 m

Volume provided by top surcharge

$$= \tfrac{1}{3}(3 \times 3)\,0.87 = 2.61 \text{ m}^3$$

Volume provided by conical bottom

$$= \tfrac{1}{3}\left[3 \times 3 + 0.5 \times 0.5 + \sqrt{3^2 \times 0.5^2}\right] \times 1.25 = 4.48 \text{ m}.$$

Alernatively, volume

$$= \tfrac{1}{3}[3 \times 3 \times 1.5 - 0.50 \times 0.5 \times 0.25] = 4.48 \text{ m}$$

∴  Remaining volume to be provided by the chamber

= 35.97 − (2.61 + 4.48) = 28.88 m³

∴  Height ($h$) = $\dfrac{28.88}{3 \times 3}$ = 3.21 m.

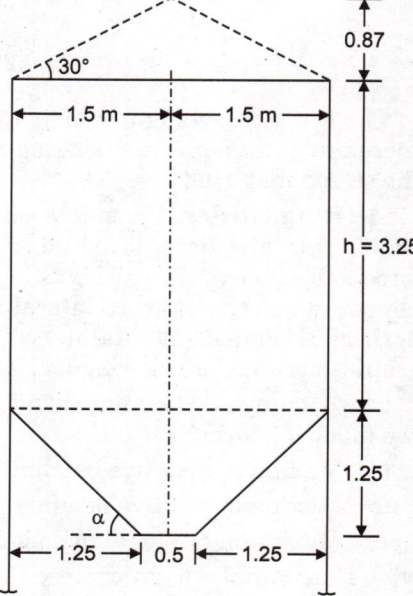

Fig. 26.11

Let the height $h$ be 3.25 m, so that its volume will be = 3.25 × 9 = 29.25 m³, making a total capacity of 2.61 + 4.48 + 29.25 = 36.34 m³ and storing 36.34 × 8.34 ≈ 303 kN of coal.

**2. *Design of side walls*:** The side walls will be designed as continuous slab. Since the angle of surcharge is equal to the angle of repose ϕ, the horizontal pressure at any level is

$$p_h = w\,h\,\cos^2 \phi \qquad \text{At 3.25 m depth,}$$
$$p_h = 8340 \times 3.25\,(\cos 30°)^2 = 20329 \text{ N/m}^2$$

Using M 15 concrete mix, $m = 19$, $k = 0.404$, $j = 0.865$ and $R = 0.875$.
Let the thickness of the wall be 180 mm.

Effective span of slab = $l = 3 + 0.18 = 3.18$ m.

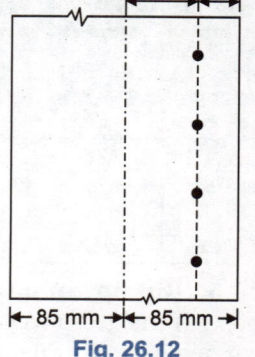

Fig. 26.12

∴ B.M. at the corners of the square frame = $\dfrac{p_h \cdot l^2}{12} = \dfrac{20329\,(3.18)^2}{12} = 17131$ N-m

Direct tension in the wall = $\dfrac{20329 \times 3.18}{2} = 32323$ N.

Let the cover to the centre of steel be 30 mm, so that effective depth will be 170 − 30 = 140 mm. The distance of centre of steel from the centre of the slab will be = 85 − 30 = 55 mm as marked in Fig. 26.12.

∴ Net B.M. = 17131 − 32323 × 0.055 = 15353 N-m at the corners.

B.M. at the centre of span = $(p_h \cdot l^2)/24$ = 8566 N-m

∴ Net B.M. at the centre of span = 8566 − 32323 × 0.055 = 6788 N-m

$$\text{Effective depth} = \sqrt{\dfrac{15353 \times 1000}{1000 \times 0.874}} = 132.53 \approx 133 \text{ mm}.$$

Provide a total depth of 170 mm, so that effective depth $d = 170 − 30 = 140$ mm.
The total thickness may be varied from 120 mm at top to 170 mm at the bottom.

$$\text{Area of steel at corners} = \dfrac{15353 \times 1000}{0.865 \times 140 \times 140} + \dfrac{32323}{140} = 1136 \text{ mm}^2$$

Using 12 mm Φ bars, spacing = $\dfrac{1000 \times 113}{1136}$ = 99.5 mm

Hence provide 12 mm Φ bars 95 mm c/c at the corners, at the inner face. Half of these bars may be curtailed at the centre of the span.

Steel required at centre = $\dfrac{6788 \times 1000}{0.865 \times 140 \times 140} + \dfrac{32323}{140}$ = 631 mm²

∴ Spacing of the 12 mm Φ bars = $\dfrac{1000 \times 113}{631}$ = 179 mm c/c.

Hence provide 12 mm Φ bars 175 mm c/c at the outer face, at the centre of the span. Half of the bars can be curtailed at the corners.

Vertical reinforcement = $0.3 \left( \dfrac{12 + 17}{2} \right) 100$ = 435 mm²

Area on each face = 218 mm². Using 10 mm Φ bars, spacing = $\dfrac{1000 \times 78.5}{218}$ = 360 mm.

*Reinforcement at 2 m below top*

$p_h = 20329 \times \dfrac{2}{3.25}$ = 12510 N-m ; B.M. at supports = $17131 \times \dfrac{2}{3.25}$ = 10542 N-m

B.M. at centre = $8566 \times \dfrac{2}{3.25}$ = 5271 N-m ; Direct tension = $32323 \times \dfrac{2}{3.25}$ = 19891 N

Thickness of wall = $120 + \dfrac{2}{3.25} \times 50$ = 150.8 mm

∴ Net B.M. at supports = $10542 - 19891 \left[ \dfrac{\dfrac{150.8}{2} - 30}{1000} \right]$ = 9639 N-m

Net B.M. at centre = $5271 - 19891 \left[ \dfrac{\dfrac{150.8}{2} - 30}{1000} \right]$ = 4368 N-m

$A_{st}$ at support = $\dfrac{9639 \times 1000}{0.865 \times 140 \times 120.8} + \dfrac{19891}{140}$ = 801 mm²

$A_{st}$ at centre = $\dfrac{4368 \times 1000}{0.865 \times 140 \times 120.8} + \dfrac{19891}{140}$ = 441 mm².

Spacing of 12 mm Φ bars at centre = $\dfrac{1000 \times 113}{441}$ = 256 mm.

Spacing of 12 mm Φ bars at ends = $\dfrac{1000 \times 113}{801}$ = 141 mm.

**3. *Design of hopper bottom*:** The hopper bottom is subjected to meridional tension, as well as to bending moment due to normal pressure acting on it. The meridional tension along the slope is due to weight of coal above the section, weight of coal in the hopper below the section, weight of the hopper below the section and the weight of the gate at the opening.

$\tfrac{1}{4}$ of weight of coal in bunker = $\tfrac{1}{4} \times 303000$ = 75750 N

Let the thickness of hopper slab be 160 mm, with 20 mm lining.

∴ $\frac{1}{4}$ of weight of hopper slab = $\left(\frac{3+0.5}{2}\right) \times 1.25 \sqrt{2} \times \frac{180}{1000} \times 25000 = 13919$ N

$\frac{1}{4}$ weight gate etc = 200 N (Say)

Total $W = 75750 + 13919 + 200 = 89869$ N

Direct tension = $W \csc \alpha = 89869 \csc 45° = 127094$ N

$A_s = 127094/140 = 908$ mm$^2$

No. of 10 mm Φ bars = 908/78.5 ≈ 12.  Provide half bars on each face.

Since the meridional tension decreases towards the opening, these bars may be curtailed to half at the opening. The slab is also subjected to B.M. due to bending of the slab spanning horizontally between the intersection of adjacent sloping faces. The horizontal span of the slab will increase from minimum at the opening to maximum at the junction, while the pressure will decrease from maximum at the opening to minimum at junction.

The section of the slab is normally designed at the centre of the hopper,   where depth upto coal surface

$= 3.25 + \frac{1.25}{2} + \frac{0.87}{1.5} \times 0.625 \approx 4.24$ m.

Span of the strip = $\frac{3+0.5}{2} = 1.75$ m.    Lateral pressure $p = wh \cos \phi$

∴ $p_v = wh + p \sin \phi = wh + wh \sin \phi \cos \phi$

$p_h = p \cos \phi = wh \cos^2 \phi$

Self weight of hopper slab lining = $0.18 \times 25000 = 4500$ N/m$^2$.

Inclination $\alpha = 45°$ (Fig. 26.7).   Hence from *Eq. 26.20*.

$p_n = p_v \cos^2 \alpha + p_h \sin^2 \alpha + w_s \cos \alpha = wh [\cos^2 \alpha + \cos^2\alpha.\sin \phi \cos \phi + \cos^2\phi \sin^2 \alpha] + w_s \cos \alpha$

$= 8340 \times 4.24 [\cos^2 45° + \cos^2 45° \sin 30° \cos 30° + \cos^2 30° \sin^2 45°] + 4500 \cos 45°$

$= 41779$ N/m$^2$

∴ B.M. = $\frac{41779 (1.75)^2}{12} = 10662$ N-m ;  Pull = $\frac{p_n \cdot l}{2} = \frac{41779 \times (1.75)}{2} = 36557$ N

Overall depth = 160 mm, effective depth = 130 mm

Net B.M. = $10662 - \frac{36557}{1000}\left(\frac{160}{2} - 30\right) = 8834$ N-m

Effective $d = \sqrt{\frac{8834 \times 1000}{0.874 \times 1000}} = 100.5$ mm  ;  Actual $d$ provided = 130 mm.

$A_{st} = \frac{8834 \times 1000}{140 \times 0.865 \times 130} + \frac{36557}{140} = 822$ mm$^2$ ; Spacing of 12 mm Φ bars = $\frac{1000 \times 113}{822} = 137$ mm

Hence provide 12 mm Φ bars @ 130 mm c/c at the inner face, at the corners. At the middle of the span, the B.M. will be half of the above value, *i.e.* $M = 5331$ N-m, but pull will be the same, i.e. 36557 N.  Hence Net   B.M. = $5331 - 36557 \times 0.05 = 3503$ N-m. Hence,

$A_{st} = \frac{3503 \times 1000}{140 \times 0.865 \times 130} + \frac{36557}{140} = 484$ mm$^2$

∴ Spacing = $1000 \times 113/484 \approx 230$ mm.

These bars are to be provided at the outer face.

**4. Top and bottom ribs:** Provide a top rib of size 300 × 300 mm with a nominal reinforcement of 4-12 mm Φ bars, and 6 mm Φ stirrups @ 200 mm c/c. Similarly, provide a bottom rib (junction beam) of size 400 × 400 mm at the junction of the wall with hopper, with the same reinforcement as the top rib. The top rib allows for the attachement of stanchion bases, conveyor supports and other super-structure. The junction beam is provided to accommodate the reinforcement in the wall behaving as a deep beam. The details of reinforcement etc. are shown in Fig. 26.13.

**Example 26.2.** *Design a silo for storing wheat, with the overall dimensions as shown in Fig. 26.14. The conical dome has central opening of 50 cm diameter. Use Airy's theory, and the concrete mix of M 20 grade and mild steel bars. For wheat, take* $w = 7850$ $N/m^3$, $\mu = 0.466$ *and* $\mu' = 0.444$.

**Solution: 1. Horizontal pressure in shallow portion.**

The height upto which it behaves as a shallow bin is given by *Eq. 26.11*

$$h = b \left[ \mu + \sqrt{\frac{\mu(1 + \mu^2)}{\mu + \mu'}} \right]$$

$$= 5 \left[ 0.466 + \sqrt{\frac{0.466[1 + (0.466)^2]}{0.466 + 0.444}} \right] = 6.28 \text{ m}.$$

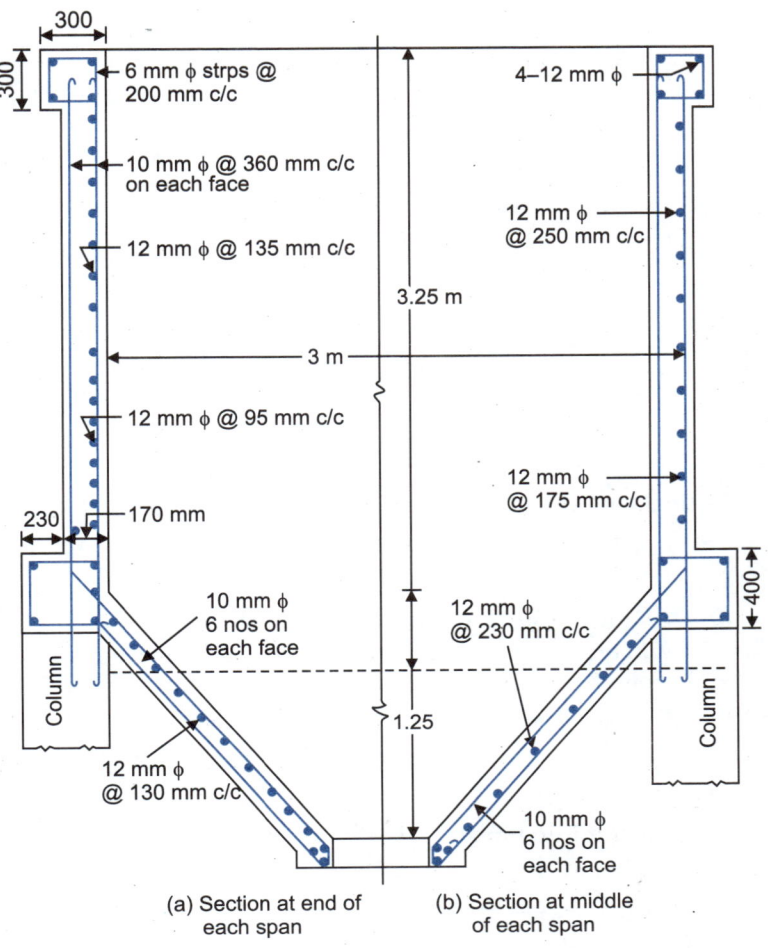

Fig. 26.13

(a) Section at end of each span

(b) Section at middle of each span

The horizontal pressure upto the height is given by *Eq. 26.8*

$$p_h = w\,h \left[ \frac{1}{\sqrt{1 + \mu^2} + \sqrt{\mu(\mu + \mu')}} \right]^2$$

$$= 7850\,h \left[ \frac{1}{\sqrt{1 + (0.466)^2} + \sqrt{0.466\,(0.466 + 0.444)}} \right]^2 = 2550\,h \quad \ldots(i)$$

$$\therefore\quad P_h = p_h \times \frac{h}{2} = 1275\,h^2$$

Vertical load taken upto this height, per meter length of perimeter is

$$P_w = \mu' P_h = 0.444 \times 1275\,(6.28)^2 = 22326 \text{ N/m}$$

**2. Horizontal pressure in deep portion.**

The horizontal pressure given by *Eq. 26.14*

$$P_h = \frac{w\,b}{(\mu + \mu')} \left[ 1 - \frac{\sqrt{1 + \mu^2}}{\sqrt{\frac{2h}{b}(\mu + \mu') + (1 - \mu\mu')}} \right]$$

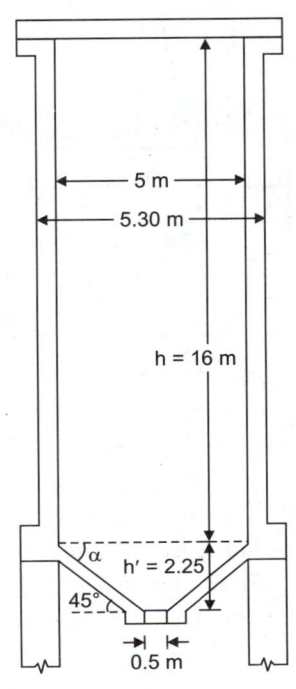

Fig. 26.14

$$= \frac{7850 \times 5}{0.466 + 0.444}\left(1 - \frac{\sqrt{1+(0.466)^2}}{\sqrt{\frac{2h}{5}(0.466+0.444)+(1-0.466\times 0.444)}}\right)$$

$$= 43132 - 47585\,[0.364\,h + 0.793]^{-1/2} \qquad \ldots(ii)$$

The horizontal force is given by *Eq. 26.13*

$$P_h = \frac{wb^2}{2(\mu+\mu')^2}\left[\sqrt{\frac{2h}{b}(\mu+\mu')+(1-\mu\mu')} - \sqrt{1+\mu^2}\right]^2$$

$$= \frac{7850\times(5)^2}{2(0.466+0.444)^2}\left[\sqrt{\frac{2h}{5}(0.466+0.444)+(1-0.466\times 0.444)} - \sqrt{1+(0.466)^2}\right]^2$$

$$= 118494\,[(0.364\,h + 0.793)^{1/2} - 1.103]^2 \qquad \ldots(iii)$$

Vertical load taken by the cylinderical portion, at $h = 16$ m, per metre length of perimeter is

$$P_w = \mu' P_h = 0.444 \times 118494\,[(0.364\,h + 0.793)^{1/2} - 1.103]^2$$
$$= 52611\,[(0.364 \times 16 + 0.793)^{1/2} - 1.103]^2 = 113587 \text{ N/m}$$

**3. Design of wall portion:** If $b$ is the diameter of the cylinder, the hoop tension is equal to $T = p_h \times b/2$, at any section. The reinforcement will be equal to $T/150$. However, a minimum reinforcement 0.3% of concrete area is to be provided. The thickness of wall determined on no crack basis is too small to be adopted in practice. IS : 4995-1968 recommends of following minimum thickness.

$$T = 15 + \frac{b - 600}{120} \text{ cm. For the present case } T = 15 \text{ cm} = 150 \text{ mm.}$$

∴ Area of minimum steel $= \dfrac{0.3}{100} \times 150 \times 1000 = 450$ mm$^2$

Spacing of 10 mm Φ bars $= \dfrac{1000 \times 78.5}{450} = 174.4 \approx 170$ mm.

The horizontal pressure $(p_h)$, hoop tension $T$, hoop steel area $A_{sh}$ and the reinforcement at various heights are tabulated in Table 26.5.

**TABLE 26.5**

| Depth (m) | $p_h$ (N/m$^2$) | $T = p_h \cdot \frac{b}{2} = 2.5\,p_h$ | $A_{sh} = \frac{T}{115}$ (mm$^2$) | Reinforcement |
|---|---|---|---|---|
| 2 | 4548 | 11378 | 99 | 10 mm Φ @ 170 mm c/c |
| 4 | 11402 | 28505 | 248 | 10 mm Φ @ 170 mm c/c |
| 6.28 | 16013 | 40043 | 348 | 10 mm Φ @ 170 mm c/c |
| 8 | 18410 | 46026 | 400 | 10 mm Φ @ 170 mm c/c |
| 10 | 20531 | 51328 | 446 | 10 mm Φ @ 170 mm c/c |
| 12 | 22186 | 55465 | 482 | 10 mm Φ @ 160 mm c/c |
| 14 | 23523 | 48808 | 511 | 10 mm Φ @ 150 mm c/c |
| 16 | 24633 | 61583 | 536 | 10 mm Φ @ 140 mm c/c |

Hence provide the reinforcement as follows:

For the first 10 m, 10 mm Φ @ 170 mm c/c.
For the next 2 m, 10 mm Φ @ 160 mm c/c.
For the next 2 m, 10 mm Φ @ 150 mm c/c.
For the last 2 m, 10 mm Φ @ 140 mm c/c.

Vertical steel @ 0.3% = 0.3 × 1500 = 450 mm$^2$.

Hence provide 10 mm Φ bars @ 170 mm c/c, in the vertical direction.

**4. Test of wall portion as column:** The wall portion is subjected to compressive stresses due to the following : (i) vertical load $P_w$ of the grains, transferred to it due to friction, (ii) self load, (iii) load of the top cover, if any and (iv) wind load. Let us provide a flat roof of 12 cm thickness (say). Allowing for a live load of 1500 N/m², load per 1 metre of perimeter of the wall is

$$P_{w1} = \frac{(0.12 \times 25000 + 1500)\frac{\pi}{4}(5)^2}{\pi \times 5} = 5625 \text{ N/m}.$$

The vertical load due to self weight of the wall is
$$P_{w2} = (0.15 \times 1) \times 16 \times 25000 = 60000 \text{ N/m}.$$
Grain load carried by the wall is $P_W = 113587$ N/m

Total $P = 5625 + 60000 + 113587 = 179212$ N/m $\quad \therefore \quad f_c = \dfrac{179212}{150 \times 1000} = 1.19$ N/mm²

To calculate the wind stress let us assume wind pressure @ 1500 N/m² with a shape factor of 0.7. B.M. at the bottom of the cylindrical portion is

$$M = (1500 \times 5.30 \times 16 \times 0.7) \times \frac{16}{2} = 712320 \text{ N-m}; \quad I = \frac{\pi}{64}[(5.30)^4 - (5)^4] = 8.053 \text{ m}^4$$

$Z = 8.053/2.65 = 3.039 \quad \therefore \quad f_{max} = M/Z = 712320/3.039 = 234393$ N/m² $= 0.234$ N/mm²

∴ Total compressive stress = $1.19 + 0.234 = 1.424$ N/mm² (Hence safe).

**5. Conical bottom:** (a) *Meridional tension*. Grain load intensity on conical portion:

$$= \frac{\frac{\pi}{4}b^2 h \cdot w - \pi b\, P_W}{\frac{\pi}{4}b^2} = \frac{\frac{\pi}{4}(5)^2 \times 16 \times 7850 - \pi(5) \times 113587}{\frac{\pi}{4}(5)^2} = 34730 \text{ N/m}^2$$

Weight of grain in conical portion

$$= \frac{w\,h'}{3}\left[A_1 + A_2 + \sqrt{A_1 \cdot A_2}\right] = \frac{7850 \times 2.25}{3}\left[\frac{\pi}{4} \times 5^2 + \frac{\pi}{4}(0.5)^2 + \sqrt{\frac{\pi}{4}(5)^2 \times \frac{\pi}{4}(0.5)^2}\right] = 128315 \text{ N}$$

Assuming a thickness of 15 cm (min.), the weight of concrete in the conical portion is

$$= 25000 \times \frac{2.25}{3}\left[\frac{\pi}{4}(5.3^2 - 5^2) + \frac{\pi}{4}(0.8^2 - 0.5^2) + \frac{\pi}{4}\sqrt{(5.3^2 - 5^2)(0.8^2 - 0.5^2)}\right] = 67448 \text{ N}$$

Weight of gate etc = 1000 N (say). Hence the load $W$ per 1 m perimeter, at the junction of the cone with the cylinder is given by

$$W = \frac{34730\,\frac{\pi}{4}(5)^2 + 128315 + 67448 + 1000}{\pi(5)} = 55939 \text{ N}$$

Inclination $\alpha = 45°$. Meridional tension $T_m = W \operatorname{cosec} \alpha = 55939 \operatorname{cosec} 45° = 79109$ N/m

Meridional reinforcement = $79109/115 = 688$ mm²

∴ Spacing of 10 mm Φ bars $= \dfrac{1000 \times 78.5}{688} = 114$ mm.

However, provide 10 mm Φ @ 100 mm c/c, converging to the lower end of the cone. Half the bars may be stopped half way.

(b) *Hoop tension*: The hoop tension depends upon the height or level selected. The intensity of normal pressure $(p_n)$ is greatest at the bottom level (BB), while the diameter is greatest at the top level CC. The hoop tension is usually found at mid-height of the cone, corresponding to level AA. The diameter of cone at level AA

$= 0.5 + (5 - 0.5)/2 = 2.75$ m $\quad \therefore \quad b = 1.375$ m

$r_n = 1.375 \operatorname{cosec} \alpha = 1.375 \operatorname{cosec} 45° = 1.945$ m

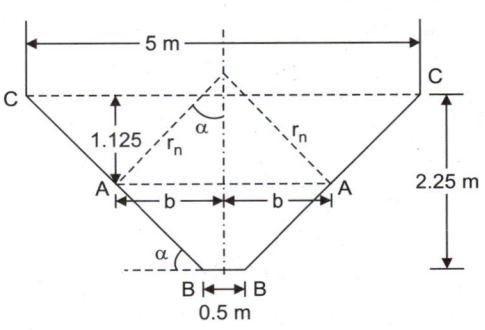

Fig. 26.15

There is no accurate method of determining vertical pressure at any level in the hopper. Approximately, it can be found by adding to the vertical pressure at the junction of the wall and hopper, the weight of the grain contained between the junction and that level. Vertical pressure, per square metre of the horizontal area = 34730 + 7850 × 1.125 = 43561 N/m²

Weight of concrete, per sq. m of horizontal area = $0.15 \times (\sqrt{2} \times 1) \times 25000 = 5303.3$ N

Total $p_1 = 43561 + 5303.3 = 48864$ N/m²

Grain pressure $p_2 = 43561 \left(\dfrac{1-\sin\phi}{1+\sin\phi}\right) = 43561 \left[\dfrac{1-\sin(\tan^{-1} 0.466)}{1+\sin(\tan^{-1} 0.466)}\right]$

$= 17689$ N/m².

∴ Normal pressure $p_n = \dfrac{p_1 \cos\alpha}{\sqrt{2}} + \dfrac{p_2 \sin\alpha}{\sqrt{2}} = \dfrac{48863}{2} + \dfrac{17689}{2} = 33276$ N/m².

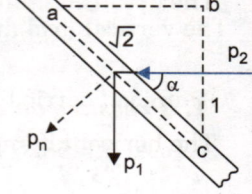

Fig. 26.16

Alternatively, $p_n$ can be found from Eq. 26.20:

$p_n = p_v \cos^2\alpha + p_h \sin^2\alpha + w_s \cos\alpha$ where $w_s = 0.15 \times 25000 = 3750$ N/m² ; $\alpha = 45°$

$p_h = \dfrac{wb}{(\mu+\mu')}\left[1 - \dfrac{\sqrt{1+\mu^2}}{\sqrt{\dfrac{2h}{b}(\mu+\mu') + (1-\mu\mu')}}\right]$ in which $\mu'$ is to be taken as equal to $\mu$ (= 0.466)

$p_h = \dfrac{7850 \times 2.75}{0.466 + 0.466}\left[1 - \dfrac{\sqrt{1+(0.466)^2}}{\sqrt{\dfrac{2 \times 17.125}{2.75}(0.466+0.466) + (1-0.466^2)}}\right] = 15903$ N/m²

(against a value of $p_2 = 17689$ found earlier)

$p_v = p_h \div \left[\dfrac{1-\sin(\tan^{-1} 0.466)}{1+\sin(\tan^{-1} 0.466)}\right] = 39169$ N/m².

∴ $p_n = 39169 \cos^2 45° + 15906 \sin^2 45° + 3750 \cos 45°$

$= 30189$ N/m²

(against a value of 33276 N/m² found earlier)

Now hoop tension $T = p_n \cdot r_n$ (Eq. 26.20)

$T = 33276 \times 1.945 = 64722$ N

∴ Hoop steel = 64722/115 = 563 mm²

Spacing of 10 mm Φ bars = $\dfrac{1000 \times 78.5}{563} \approx 139$ mm

Hence provide 10 mm Φ bar rings @ 130 mm c/c throughout the height of the cone.

**6. Ring beam** (*junction beam*): The ring beam at the junction of the cylindrical portion with the domical bottom is subject to hoop compression which is equal to the horizontal component of the downward pull of the loaded hopper minus the outward lateral thrust exerted by the grain.

Ring compression = $T_m \cos\alpha - T = 2268 \cos 45° - 61583$

$= -3411$ (*i.e.* tension).

Also when the surface of the grain is at the level of the ring

$W$ = (wt. of grain in hopper + wt. of hopper and gate)/m

$= \dfrac{163399 + 67448 + 1000}{\pi(5)} = 14760$

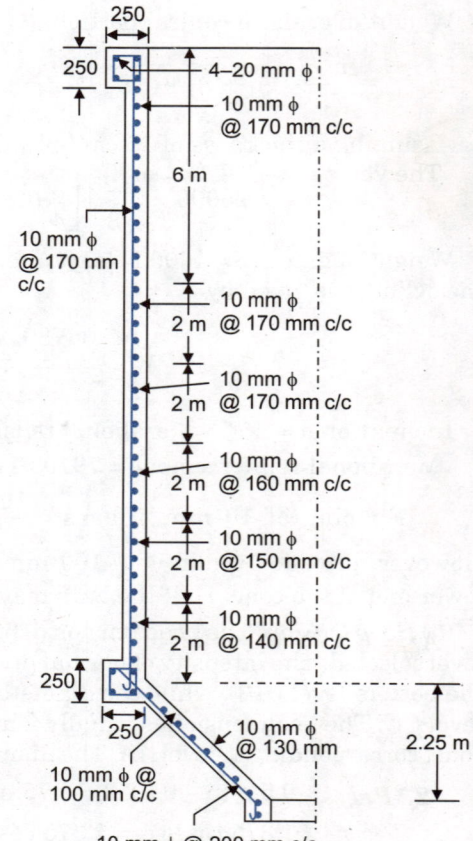

Fig. 26.17

∴ $T_m$ = W cosec α = 14760 cosec 45° = 20874 N/m

$T$ = zero.   ∴ Ring compression = 20874 N/m

The ring should be designed for this value. Provide a 250 mm × 250 mm section with 4 bars of 20 mm Φ stirrups and 20 mm φ stirrups @ 250 mm c/c. The above section should be tested for bending and twisting moment depending upon support conditions (*i.e.* number of columns). Provide a top rib of size 250 mm × 250 mm with a nominal reinforcement of 4-12 mm Φ bars and 6 mm Φ stirrups @ 200 mm c/c.

The details of reinforcement etc. are shown in Fig. 26.17

**Example 26.3.** *Analyse Example 26.2, using Janssen's theory.*
**Solution:**

1. *Pressure on cylindrical portion during filling.*

The horizontal and vertical pressure are given by *Eq. 26.2* and *26.1* :

$$p_h = \frac{w\,r}{\mu_f'}\left[1 - e^{-\frac{\mu_f' \cdot k_f \cdot h}{r}}\right] \qquad \ldots(26.2)$$

$p_v = \dfrac{p_h}{K_f}$   Here $K_f = 0.5$ ;   φ = tan$^{-1}$(0.466) = 24.99°

$\mu_f'$ = tan (0.75 φ) = tan (0.75 × 24.99) = 0.339

$w$ = 7850 N/m³ ;   $b$ = diameter = 5 m

$$r = \frac{A}{P} = \frac{\frac{\pi}{4}b^2}{\pi b} = \frac{b}{4} = \frac{5}{4} = 1.25 \text{ m}$$

∴ $$p_h = \frac{7850 \times 1.25}{0.339}\left[1 - e^{-\frac{0.399 \times 0.5 \times h}{1.25}}\right] = 28945\,[1 - e^{-0.1356\,h}]$$

$p_v = \dfrac{p_h}{0.50} = 2\,p_h$.   Hoop tension $T = p_h \cdot \dfrac{b}{2} = 2.5\,p_h$

$A_{sh} = T/115$

The values of $p_h$, $T$, $A_{sh}$ and reinforcement are tabulated in Table 26.6.

**TABLE 26.6**

| Depth (m) | $p_h$ | $T = 2.5\,p_h$ | $A_{sh} = \frac{T}{115}$ (mm²) | Reinforcement |
|---|---|---|---|---|
| 2 | 6875 | 17189 | 149.5 | 10 mm Φ @ 170 mm c/c |
| 4 | 12118 | 30296 | 163 | 10 mm Φ @ 170 mm c/c |
| 6 | 16112 | 40280 | 350 | 10 mm Φ @ 170 mm c/c |
| 8 | 19164 | 47910 | 417 | 10 mm Φ @ 170 mm c/c |
| 10 | 21489 | 53723 | 467 | 10 mm Φ @ 160 mm c/c |
| 12 | 23256 | 58129 | 505 | 10 mm Φ @ 150 mm c/c |
| 14 | 24610 | 61524 | 535 | 10 mm Φ @ 140 mm c/c |
| 16 | 25640 | 64100 | 557 | 10 mm Φ @ 140 mm c/c |

$A_{sd}$ = 0.3 × 1500 = 450 mm².   Hence provide 10 mm Φ bars @ 170 mm c/c in the vertical direction.

$[(p_v)_{h=16}] = 2 \times 25640 = 51280$ N/m²

$$P_{WT} = \frac{\pi}{4} \times (5)^2\,[7850 \times 16 - 51280] = 1459270 \text{ N} ; \quad P_W = \frac{1459270}{\pi \times 5} = 92900 \text{ N/m}$$

2. *Pressure on cyclindrical portion during emptying*

$K_e = 1$

$\mu_e'$ = tan (0.6 φ) = tan (0.6 × 24.99) = 0.268

$$p_h = \frac{wr}{\mu_e'}\left[1 - e^{-\frac{\mu' K_{e.h}}{r}}\right] = \frac{7850 \times 1.25}{0.268}\left[1 - e^{-\frac{0.268 \times 1}{1.25}h}\right] = 36614\,[1 - e^{-0.2144\,h}]$$

The value of $p_h$, $T$, $A_{sh}$ and reinforcement are tabulated in Table 26.7.

### TABLE 26.7

| Depth (m) | $p_h$ | $T = p_h \cdot \frac{b}{2} = 2.5\,p_h$ | $A_{sh} = \frac{T}{115}$ (mm²) | Reinforcement |
|---|---|---|---|---|
| 2 | 12815 | 32038 | 279 | 10 mm Φ @ 170 mm c/c |
| 4 | 21077 | 52693 | 458 | 10 mm Φ @ 170 mm c/c |
| 6 | 26494 | 66234 | 576 | 10 mm Φ @ 130 mm c/c |
| 8 | 30026 | 75066 | 653 | 10 mm Φ @ 120 mm c/c |
| 10 | 32322 | 80805 | 703 | 10 mm Φ @ 110 mm c/c |
| 12 | 33814 | 84535 | 735 | 10 mm Φ @ 100 mm c/c |
| 14 | 34795 | 86988 | 756 | 10 mm Φ @ 100 mm c/c |
| 16 | 35423 | 88558 | 770 | 10 mm Φ @ 100 mm c/c |

$$[(p_v)_{h=16}] = \frac{p_h}{K_e} = p_h = 35423\,;\quad P_w = \frac{\frac{\pi}{4}(5)^2\,(7850 \times 16 - 35423)}{\pi \times 5} = 112721\text{ N/m}$$

From Table 26.6 and 26.7 it is clear that both $p_h$ and $P_W$ are maximum during emptying, while the pressure $p_v$ is maximum during filling.

**3. Conical hopper : filling condition**

$$p_v = 51280\text{ N/m}^2\,;\quad P_v = 51280 \times \frac{\pi}{4}(5)^2 \approx 1006880\text{ N}$$

Weight of grain in conical portion, as obtained in Ex. 26.2 = 128315 N
Self weight of cone = 67448 ;  Weight of gate etc. = 1000

$$W \text{ per metre perimeter} = \frac{1006880 + 128315 + 67448 + 1000}{\pi(5)} \approx 76626\text{ N/m}$$

∴ $\quad T_m = W\,\text{cosec}\,\alpha = 76626\,\text{cosec}\,45° = 108366\text{ N/mm}$

$p_n$ is obtained from Eq. 26.20

$$p_n = p_v \cos^2\alpha + p_h \sin^2\alpha + w_s \cos\alpha \quad \text{where}\quad w_s = 0.15 \times 1 \times 25000 = 3750\text{ N/m}^2$$

$$p_h = \frac{wr}{\mu_f'}[1 - e^{-\mu_f' K_f h/r}]$$

Let us find $p_n$ at the mid-height of cone, where $h = 17.125$, $b = 1.375$; $r = \text{dia.}/4$
$\quad = 2 \times 1.375/4 = 0.6875\,;\ K_f = 0.5\,;\ \mu_f' = 0.339.$

∴ $\quad p_h = \frac{7850 \times 0.6875}{0.339}\left[1 - e^{-0.339 \times 0.5 \times 17.125/0.6875}\right] = 15690\text{ N/m}^2$

$$p_v = \frac{p_h}{K_f} = \frac{15690}{0.5} = 31380$$

∴ $\quad p_n = 31380 \cos^2 45° + 15690 \sin^2 45° + 3750 \cos 45° = 26187$

∴ Hoop tension $T = p_n \cdot r_n = 26187 \times 1.945 = 50933.7 \approx 50933$ N

Also, at $h = 16$ m, $p_h = 25640$ (Table 26.6), and $p_v = 51280$

∴ $\quad p_n = 51280 \cos^2 45° + 25640 \sin^2 45° + 3750 \cos 45° = 41112$

∴ Hoop tension $T = p_n \cdot r_n = 41112\left[\dfrac{5}{2}\,\text{cosec}\,45°\right] = 145353$ N

Thus the hoop tension is maximum at the junction.

**4. Conical hopper : emptying condition**

$$p_v = 35423 \text{ (at } h = 16 \text{ m)}$$

$$\therefore \quad p_v = 35423 \times \frac{\pi}{4}(5)^2 = 695529 \text{ N}$$

$$\therefore \quad W = \frac{695529 + 128315 + 67448 + 1000}{\pi(5)} = 56805 \text{ N/m}$$

$$T_m = 56805 \text{ cosec } 45° = 80334 \text{ N/m}.$$

Let us calculate hoop tension at $h = 16$ m.

$$p_h = 35423 \text{ (Table 26.7), and } p_v = 35423$$
$$p_n = 35423 \cos^2 45° + 35423 \sin^2 45 + 3750 \cos 45° = 38075 \text{ N}$$

This is less than the one obtained for the 'filling' condition.
Thus both $T_m$ and $T$ are more for filling condition.

**5. Design of hopper bottom**

$$T_m = 108366 \quad \therefore \quad A_s = 108366 / 115 = 942 \text{ mm}^2.$$

Spacing of 12 mm Φ bars = $\dfrac{1000 \times 113}{942}$ = 119 mm.

Hence provide 12 mm Φ bars @ 110 mm c/c. Half the bars may be stopped half way.

Hoop tension $T = 145353$ N (at $h = 16$ m)

$$\text{Area of hoops} = \frac{145353}{115} = 1264 \text{ mm}^2$$

Spacing of 12 mm Φ hoops = $\dfrac{1000 \times 113}{1264}$ = 89 mm

Hence provide these @ 80 mm c/c
Hoop tension at mid-height of cone = 50933N

Spacing of 12 mm Φ hoops at mid height of cone = $\dfrac{1000 \times 113}{50933/115} \approx 250$ mm c/c

## PROBLEMS

1. Distinguish clearly between a bunker and a silo. Using Airy's theory, show that the height upto which a bin behaves as a shallow one is given by

$$h = b\left[\mu + \sqrt{\frac{\mu(1+\mu^2)}{\mu + \mu'}}\right]$$

2. Using Janssen's theory, derive an expression for horizontal pressure at any depth $h$ below the top, in a silo. Also derive an expression for total vertical load of the grain transferred to the walls.
3. Discuss the Indian Standard method of determining the vertical and horizontal pressure a bin.
4. Discuss the procedure for designing the hoppers of
   (a) Rectangular bin            (b) Cylindrical bin.
5. Design a bunker to store 500 tonnes of coal. The unit weight and angle of repose of coal may be taken as 8100 N/m³ and 30° respectively. The stored coal is surcharged at its angle of repose.
6. Design a silo for storing maize, having unit weight of 6870 N/m³. The silo has 6 m internal diameter and the height of the cylindrical portion is 15 m. The conical dome has a slope of 40° with horizontal, and has an opening of 60 cm diameter. Use Airy's theory. Take μ = 0.521 and μ' = 0.432. Use M20 grade concrete.
7. Design the silo of problem 6, using Janssen's analysis.
8. A rectangular shallow bin 3 m by 2.5 m in plan is required to store 230 kN of foamed slag aggregates. The unit weight of Aggregates being 9.42 kN/m³. The bin is supported on four R.C. columns at the corners. Using M-20 grade concrete and Fe - 415 steel, design the side wall, hopper bottom and column and sketch the structural details. Angle of repose = 30 degrees, Angle of surcharge = 25°.

# CHAPTER 27: CHIMNEYS

## 27.1. INTRODUCTION

During the last few decades the use of reinforced concrete chimneys in place of brick masonry and steel chimneys have become very popular due to their low cost and durability. Chimneys are relatively tall structures subjected to three types of stresses: (i) stresses due to self weight (ii) stresses due to wind moment, and (iii) stresses due to temperature variation between the inside and outside of the chimney. Brick chimneys are suitable only for short height as they become bulky with the increase in height, and require heavy foundations. Also, due to large temperature gradient, brick chimney frequently cracks and becomes unstable. In contrast to this, concrete chimneys are lighter and stronger, and are less vulnerable to cracks due to the temperature difference. If the temperature of the flue gases does not exceed 400°C, concrete chimneys can be used without any special fire brick lining. For higher temperatures, fire brick lining is provided with an air gap between the inner face of the chimney and the lining as shown in Fig. 27.5. Concrete chimneys are generally constructed vertical, without batter, so that sliding shuttering may be used. The thickness may vary from a minimum of 12 cm at the top to 45 cm at the bottom, depending upon the height of the chimney. Sufficient amount of vertical steel is provided to resist the bending moments due to wind. In addition to this, horizontal steel (hoop steel) is provided to resist horizontal shear, and also reduce the effects of temperature gradient on concrete. Concrete stacks with lesser maintenance costs are architecturally superior to masonry and steel chimneys.

## 27.2. WIND PRESSURE

When wind meets an obstruction, it exerts pressure on the obstruction. The intensity of pressure depends on the wind velocity, which in turn depends upon the elevation above ground. Hence intensity of wind pressure depends upon the height of the chimney. IS 875 specifies the wind pressure at various heights, as given in Table 27.1. The shape factors for various shapes of the chimney are given in Table 27.2.

TABLE 27.1. Wind Pressure at Various Heights

| Height above ground (m) | Wind pressure (kg/m$^2$) | Height above ground (m) | Wind pressure (kg/m$^2$) | Height above ground (m) | Wind pressure (kg/m$^2$) |
|---|---|---|---|---|---|
| 0 | 39 | 18 | 112.0 | 46 | 169.0 |
| 3 | 58.6 | 21 | 121.0 | 54 | 177.0 |
| 6 | 72.0 | 24 | 127.0 | 62 | 186.5 |
| 9 | 86.0 | 27 | 135.0 | 77 | 200.0 |
| 12 | 97.0 | 30 | 141.0 | 92 | 212.0 |
| 15 | 107.0 | 38 | 151.5 | 107 | 225.0 |
| | | | | 122 | 235.0 |

688

CHIMNEYS **689**

TABLE 27.2. Shape Factors for Chimneys

| Ratio of height to base width | 0 to 4 | 4 to 8 | 8 and over |
|---|---|---|---|
| Cross-sectional Shape of Chimney | Shape factors | | |
| 1. Circular | 0.7 | 0.7 | 0.7 |
| 2. Octagonal | 0.8 | 0.9 | 1.0 |
| 3. Square (wind perpendicular to diagonal) | 0.8 | 0.9 | 1.0 |
| 4. Square (wind perpendicular to face) | 1.0 | 1.15 | 1.3 |

## 27.3. STRESSES IN CHIMNEY SHAFT DUE TO SELF-WEIGHT AND WIND LOADS

For the purposes of analysis of stresses due to self weight and wind, it is assumed that the steel reinforcement is replaced by a thin steel cylinder, and that it is located at the centre of its thickness. Also, it is assumed that the variation of stresses in the thickness of the shell is small and the total compressive or tensile stresses may be taken acting on the centre line of the shell thickness. Fig. 27.1(a) shows the concrete shell with the actual reinforcement. Fig. 27.1(b) shows the equivalent cylindrical (ring) reinforcement placed at the centre of the thickness. Let the radius to the centre of the thickness of shell be $R$, and the thickness of concrete be $T_c$. If $T_s$ is the thickness of the steel ring, and $A_s$ is its area of cross-section, then we have

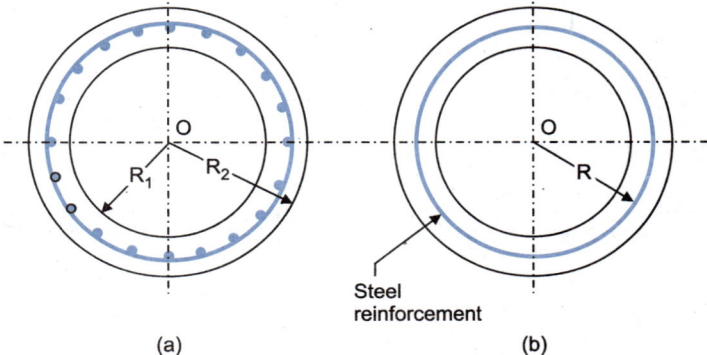

Fig. 27.1

$$2\pi R T_s = A_s \quad \text{or} \quad T_s = \frac{A_s}{2\pi R}$$

Fig. 27.2 shows the section of the chimney at any height. Let $W$ be the weight of the chimney above that section, and $P$ be the resultant wind force acting at height $h$ above the section. Due to wind, the section will be subjected at bending moment $M = Ph$, and a shear force $= P$. Due to this, tensile stresses will be set up in the windward side and compressive stresses in the leeward side. Let $c_1$ and $t_1$ be the extreme compressive and tensile stresses, as shown in Fig. 27.2. The neutral axis will be situated towards the windward side, subtending and angle $2\phi$ at the centre $O$.

Consider an element of ring, at an angle $\theta$ with the wind direction and subtending an angle $\delta\theta$ at the centre. The stress $c_1'$ at this level is given by

$$c_1' = c_1 \frac{R \cos\theta + R \cos\phi}{R + R \cos\phi} = c_1 \frac{\cos\theta + \cos\phi}{1 + \cos\phi}$$

Area of concrete in strip $= R\, \delta\theta \cdot T_c$ ;
Area of steel in the strip $= R\, \delta\theta \cdot T_s$

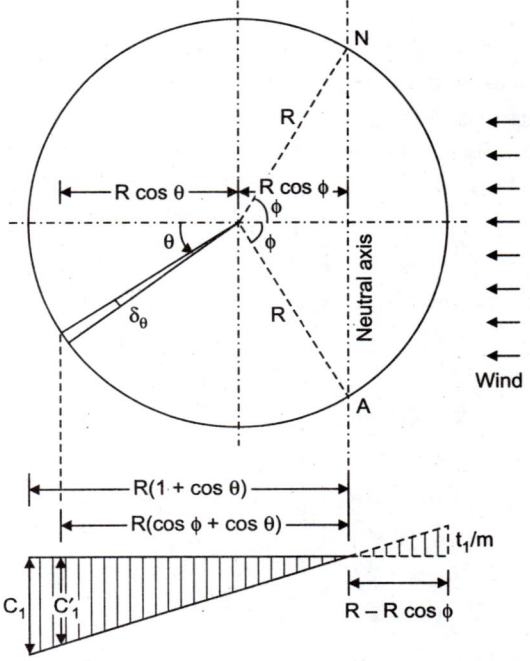

Fig. 27.2

$\therefore$ Total compression in concrete and steel $= 2\int_0^{\pi-\phi} R\, \delta\theta\, [T_c + (m-1)\, T_s] \cdot c_1 \dfrac{\cos\theta + \cos\phi}{1 + \cos\phi}$

$$= \frac{2R \cdot c_1[T_c + (m-1)T_s]}{1 + \cos \phi} \int_0^{\pi - \phi} (\cos \theta + \cos \phi) \, d\theta$$

$$= \frac{2R \, c_1[T_c + (m-1)T_s]}{1 + \cos \phi} [\sin \theta + \theta \cos \phi]_0^{\pi - \phi}$$

$$= \frac{2R \, c_1[T_c + (m-1)T_s]}{1 + \cos \phi} [\sin (\pi - \phi) + (\pi - \phi) \cos \phi]$$

$$= \frac{2R \, c_1[T_c + (m-1)T_s]}{1 + \cos \phi} [\sin \phi + (\pi - \phi) \cos \phi] \qquad \ldots(i)$$

The moment of compressive force in concrete and steel about the *centre line* of the section is

$$= 2\int_0^{\pi - \phi} R\delta \theta \cdot c_1 [T_c + (m-1)T_s] \frac{\cos \theta + \cos \phi}{1 + \cos \phi} \times R \cos \theta$$

$$= \frac{2R^2 \, c_1[T_c + (m-1)T_s]}{1 + \cos \phi} \int_0^{\pi - \phi} (\cos^2 \theta + \cos \theta \cos \phi) \, d\theta$$

$$= \frac{2R^2 \, c_1[T_c + (m-1)T_s]}{1 + \cos \phi} \left[ \frac{\sin 2\theta}{4} + \frac{\theta}{2} + \sin \theta \cos \phi \right]_0^{\pi - \phi}$$

$$= \frac{2R^2 \, c_1[T_c + (m-1)T_s]}{1 + \cos \phi} \left[ \frac{\sin 2(\pi - \phi)}{4} + \frac{\pi - \phi}{2} + \sin(\pi - \phi) \cos \phi \right]$$

$$= \frac{2R^2 \, c_1[T_c + (m-1)T_s]}{1 + \cos \phi} \left[ -\frac{\sin 2\phi}{4} + \frac{\pi - \phi}{2} + \frac{\sin 2\phi}{2} \right]$$

$$= \frac{2R^2 \, c_1 [T_c + (m-1)T_s]}{1 + \cos \phi} \left[ \frac{\sin 2\phi}{4} + \frac{\pi - \phi}{2} \right] \qquad \ldots(ii)$$

Similarly, to find the tensile force in steel on the windward side consider in Fig. 27.3, and element of the ring at an angle θ. It is assumed that tensile force is taken up by steel only. Tensile stress in steel, at the level of the element ring

$$= mc_1 \frac{R \cos \theta - R \cos \phi}{R + R \cos \phi} = mc_1 \frac{\cos \theta - \cos \phi}{1 + \cos \phi} \qquad \ldots(27.1)$$

∴ Total tension in steel

$$= 2\int_0^\phi R \, d\theta \cdot T_s \, mc_1 \frac{\cos \theta - \cos \phi}{1 + \cos \phi}$$

$$= \frac{2RT_s \cdot mc_1}{1 + \cos \phi} \int_0^\phi (\cos \theta - \cos \phi) \, d\theta$$

$$= \frac{2RT_s \cdot mc_1}{1 + \cos \phi} [\sin \theta - \theta \cos \phi]_0^\phi$$

$$= \frac{2RT_s \cdot mc_1}{1 + \cos \phi} [\sin \phi - \phi \cos \phi] \qquad \ldots(iii)$$

Moment of tensile force in steel, about the centre line of the section is

$$= 2\int_0^\phi R d\theta \cdot T_s \cdot mc_1 \frac{\cos \theta - \cos \phi}{1 + \cos \phi} R \cos \theta$$

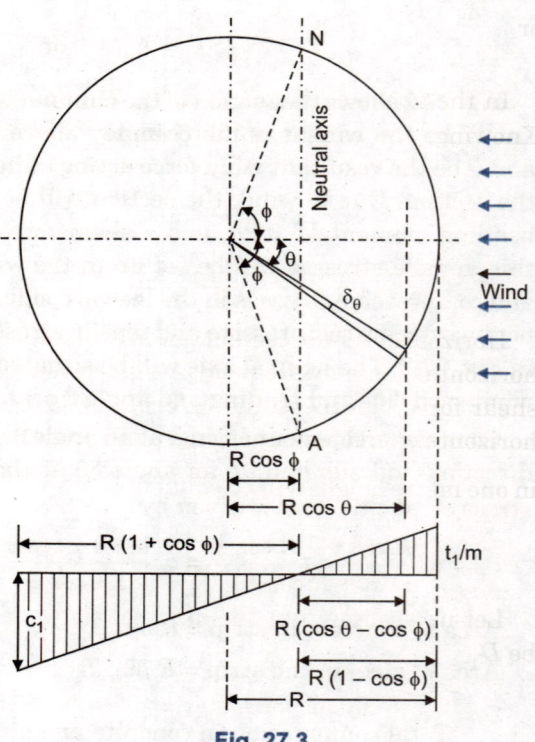

Fig. 27.3

$$= \frac{2R^2T_s\,mc_1}{1+\cos\phi}\int_0^\phi(\cos^2\theta-\cos\theta\cos\phi)d\theta = \frac{2R^2T_s\,mc_1}{1+\cos\phi}\left[\frac{\sin 2\theta}{4}+\frac{\phi}{2}-\sin\theta\cos\phi\right]_0^\phi$$

$$= \frac{2R^2T_s\,mc_1}{1+\cos\phi}\left[\frac{\sin 2\phi}{4}+\frac{\phi}{2}-\frac{\sin 2\phi}{2}\right] = \frac{2R^2T_s\,mc_1}{1+\cos\phi}\left[\frac{\phi}{2}-\frac{\sin 2\phi}{4}\right] \qquad ...(iv)$$

Equating the algebraic sum of internal forces to the external force, we get

$$W = (i) + (iii)$$

$$W = \frac{2Rc_1[T_c + (m-1)T_s]}{1+\cos\phi}\,[\sin\phi + (\pi-\phi)\cos\phi] - \frac{2RT_s\,mc_1}{1+\cos\phi}\,[\sin\phi - \phi\cos\phi]$$

$$W = \frac{2Rc_1}{1+\cos\phi}\times[(T_c - T_s)\{\sin\phi + (\pi-\phi)\cos\phi\} + \pi\,mT_s\cos\phi] \qquad ...[27.2(a)]$$

Similarly equating the sum of the moments of internal forces about the centre of the section to the external moment, we get

$$M = (ii) + (iv) = \frac{2R^2c_1[T_c + (m-1)T_s]}{1+\cos\phi}\left[\frac{\sin 2\phi}{4}+\frac{\pi-\phi}{2}\right] + \frac{2R^2T_s\,mc_1}{1+\cos\phi}\left[\frac{\phi}{2}-\frac{\sin 2\phi}{4}\right]$$

or $$M = \frac{2R^2c_1}{1+\cos\phi}\left[(T_c - T_s)\left(\frac{\sin 2\phi}{4}+\frac{\pi-\phi}{2}\right)+\frac{m\pi}{2}T_s\right] \qquad ...[27.2(b)]$$

Hence eccentricity $= e = \dfrac{M}{W} = \dfrac{\dfrac{2R^2c_1}{1+\cos\phi}\left[(T_c-T_s)\left(\dfrac{\sin 2\phi}{4}+\dfrac{\pi-\phi}{2}\right)+\dfrac{m\pi}{2}T_s\right]}{\dfrac{2Rc_1}{1+\cos\phi}[(T_c-T_s)\{\sin\phi+(\pi-\phi)\cos\phi\}+m\pi T_s\cos\phi]}$

or $$e = R\frac{\left[(T_c-T_s)\left\{\dfrac{\sin 2\phi}{4}+\dfrac{\pi-\phi}{2}\right\}+\dfrac{m\pi}{2}T_s\right]}{[(T_c-T_s)\{\sin\phi+(\pi-\phi)\cos\phi\}+m\pi T_s\cos\phi]} \qquad ...(27.3)$$

In the above equation, $e$, $R$, $T_c$, $T_s$ and $m$ are known, hence $\phi$ can be found by trial the error solution. Knowing, $\phi$, $c_1$ can be found from *Eq. 27.1*. Max. tensile stress in steel is given by

$$t_1 = mc_1\frac{R(1-\cos\phi)}{R(1+\cos\phi)} = mc_1\frac{1-\cos\phi}{1+\cos\phi} \qquad ...[27.1(a)]$$

## 27.4. STRESSES IN HORIZONTAL REINFORCEMENT DUE TO FORCE SHEAR

Horizontal reinforcement, mainly provided in the form of hoops, resists the horizontal shear due to wind. At any horizontal section, let $p$ be the horizontal shear force, and $D$ be the external diameter of the chimney. Let $A_\phi$ be the horizontal reinforcement in the pitch $s$ mm. Hence area of steel resisting shear in one metre height will be equal to $\left(2A_\phi \times \dfrac{1000}{s}\right)$. If $t$ is the stress in steel, the

$$\text{S.F. resisted} = \left(2A_\phi \cdot \frac{1000}{s}\times t\right) \qquad ...(i)$$

Let us assume that the distance between the reinforcement on both side be $D_1$

$$\therefore \quad \text{Shear per meter} = \left(\frac{P}{bjd}\times b\times 1000\right) = \frac{1000\,P}{jd}$$

Let us assume that lever arm $jd \approx D_1$.

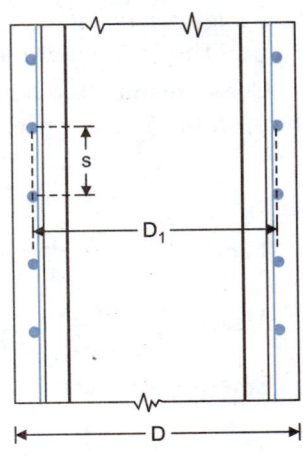

Fig. 27.4

$$\therefore \quad \text{Shear} = \frac{1000\, P}{D_1} \quad \ldots(ii)$$

Equating (i) and (ii), we get: $2A_\phi \cdot \dfrac{1000}{s} \cdot t = \dfrac{1000\, P}{D_1}$, $\quad t = \dfrac{P \cdot s}{2 A_\phi D_1}$  ...(27.4)

where '$d$' and '$s$' are expressed in '$mm$' units and $A_\phi$ is in $mm^2$ units.

## 27.5. STRESSES DUE TO TEMPERATURE DIFFERENCE

Because of high temperature of flue gases, there is a large temperature gradient. The temperature outside the chimney shell is only slightly higher than the outside atmospheric temperature. The fall of temperature through the concrete shell causes severe stresses. To reduce temperature effects, 10 cm thick fire brick lining is provided, spaced slightly away from the concrete shell so that an air gap 8 to 10 cm wide is available between the two. The lining is supported at every 5 to 7 metres, as shown in Fig. 27.5(b).

As shown in Fig. 27.5(a), let $T_1°$ be the internal temperature of the flue gases, while $T_3°$ is the temperature at the external face of the chimney. The temperature $T_2°$ in the air gap is practically constant. It is estimated that $(T_1° - T_2°)$ is equal to 5 to 8 times $(T_2° - T_3°)$. Thus, the fire bricks play the major role in the temperature drop.

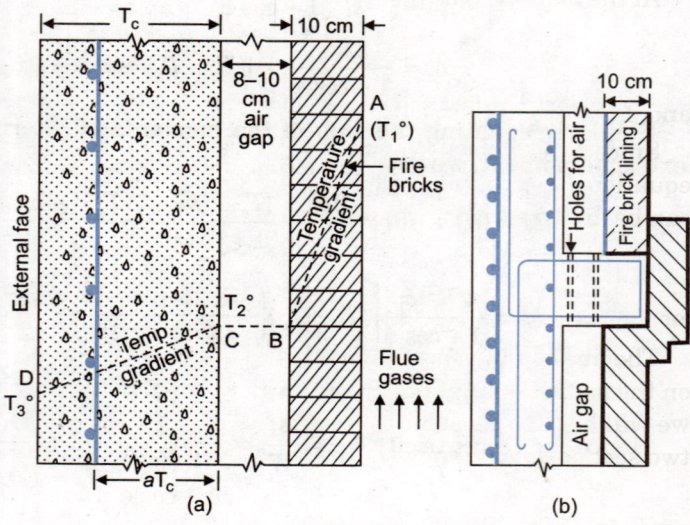

Fig. 27.5. A Temperature Stresses in Chimney Walls.

Let $T_2° - T_3° = T°$ be temperature drop in the chimney shell. Due to the higher temperature at the inner face, it tries to expand, while the external face will not expand by the same amount. Due to this, compressive stresses will be set up at the inner face of the shell and tensile stresses will be set up at the external face, as shown in Fig. 27.6(b).

Let the thickness of concrete shell be $T_c$, and let the vertical reinforcement be placed at a distance of $a.T_c$ from the inner face. Let the compressive stress in concrete be $c_2$ and tensile stress in steel be $t_2$, and the N.A. be at a distance at $k_2 T_c$ from the inner face.

Considering unit length of the circumference and equating total tension to total compression, we get :

$$\tfrac{1}{2} c_2 \cdot k_2 T_c = T_s \cdot t_2. \quad \text{Let} \quad T_s = p T_c$$

Then $\quad \tfrac{1}{2} c_2 k_2 T_c = p T_c \cdot t_2.\quad$ or $\quad t_2 = \dfrac{c_2 k_2}{2p}\quad$ ...(i)

Also, $\quad \dfrac{t_2/m}{aT_c - k_2 T_c} = \dfrac{c_2}{k_2 T_c} \quad \therefore \quad t_2 = \dfrac{a - k_2}{k_2} \cdot m c_2 \quad$ ...(ii)

Equating (i) and (ii), we get

$$\dfrac{c_2 k_2}{2p} = \dfrac{a - k_2}{k_2} \cdot m c_2 \quad \text{or} \quad k_2^2 = 2 p m (a - k_2)$$

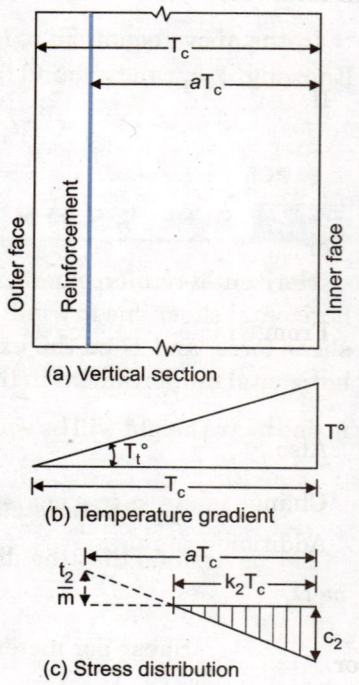

(a) Vertical section

(b) Temperature gradient

(c) Stress distribution

Fig. 27.6

From which $\quad k_2 = -mp + \sqrt{2\,mpa + m^2 p^2}\quad$ ...(27.5)

From the above equation, $k_2$ can be found.

Now, from Fig. 27.6(b) rise in temperature at steel level is $T_t^\circ = (1-a)\,T^\circ$

Let the coefficient of thermal expansion of both steel and concrete be $\alpha$. Let $\varepsilon$ be the final strain in concrete and steel. Free expansion of steel = $\alpha(1-a)T^\circ$ while free expansion of concrete = $\alpha T^\circ$ per unit width.

$\therefore\quad$ Stress in steel = $t_2 = E_s\,[\varepsilon - (1-a)\alpha\,T^\circ]$ ...(iii)

Max. stress in concrete = $c_2 = E_c\,[T^\circ \cdot \alpha - \varepsilon]$ ...(iv)

At the neutral axis, the free expansion will be equal to strain $\varepsilon$ since there is no temperature stress there.

$\therefore\quad \varepsilon = (1-k_2)\,\alpha T^\circ$. Substituting this value of $\varepsilon$ in (iii) and (iv), we get

$$t_2 = E_s\,[(1-k_2)\alpha T^\circ - (1-a)\alpha T^\circ] = E_s\,\alpha T^\circ\,(a - k_2) \quad ...(27.6)$$

and $\quad c_2 = E_c\,[\alpha T^\circ - (1-k_2)\alpha T^\circ] = E_c \cdot \alpha T^\circ\,k_2 \quad$ ...(27.7)

Since $k_2$ is known from Eq. 27.5, the stress in steel, and concrete can be found from the above two equations. If '$k_2$' is more than unity, the whole thickness of concrete '$t_c$' will be in compression and stress can be analysed using the same procedure.

## 27.6. COMBINED EFFECT OF SELF LOAD, WIND AND TEMPERATURE

The above analysis of temperature stress is applicable on the vertical section of the chimney shell lying on the neutral axis of Fig. 27.2, where there will be no stresses due to external forces. At the other sections, we will have to consider the combined effect of self load, wind and temperature stresses. We will consider two sections of chimney shell (a) section on the leeward side and (b) section on the windward side.

(a) **Leeward side compression zone:** The leeward side of the chimney shell is the compression zone, in which the compressive stress due to self load and wind has uniform value $c_1$ throughout the shell thickness, as shown in Fig. 27.7(b). Fig. 27.7(c) shows the stress diagram due to combined effect, in which the new neutral axis is located at a distance of $kT_c$ from the inside face. Since there is no change in total compression due to self load and wind, we have:

$$c_1 T_c + (m-1)\,T_s \cdot c_1 = \tfrac{1}{2}\,ckT_c - T_s \cdot t$$

If $\quad T_s = pT_c$, we have

$$c_1 T_c + (m-1)pT_c \cdot c_1 = \tfrac{1}{2}\,ckT_c - pT_c\,mc\left(\frac{a-k}{k}\right)$$

or $\quad c_1 T_c\,[1 + (m-1)\,p] = c \cdot T_c\left[\dfrac{k}{2} - mp\left(\dfrac{a-k}{k}\right)\right]$

From which $\quad c = \dfrac{c_1[1+(m-1)p]}{\dfrac{k}{2} - mp\left(\dfrac{a-k}{k}\right)}\quad$ ...(27.8)

Also $\quad t = \dfrac{mc\,(a-k)}{k}\quad$ ...(27.9)

Change in stress in concrete at inside face = $c - c_1$

Additional compressive strain on inside face = $(\alpha T^\circ - \varepsilon)$

$\therefore\quad \dfrac{c - c_1}{E_c} = \alpha T^\circ - \varepsilon$

or $\quad \varepsilon = \alpha T^\circ - \dfrac{c - c_1}{E_c}\quad$ ...(i)

Similarly, change in stress in steel = $t + mc_1$

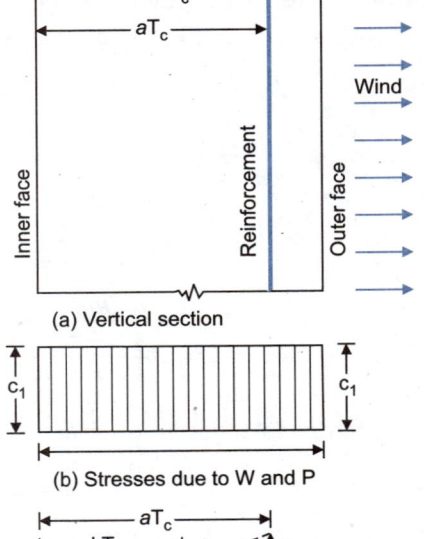

(a) Vertical section

(b) Stresses due to W and P

(c) Combined effect

Fig. 27.7

Additional tensile strain in steel $= \varepsilon - (1-a)\alpha T°$

$\therefore \quad \dfrac{t + mc_1}{E_s} = \varepsilon - (1-a)\alpha T°$

or $\quad \varepsilon = \dfrac{t + mc_1}{E_s} + (1-a)\alpha T°$ ...(ii)

Equating (i) and (ii), we get

$$\alpha T° - \dfrac{c - c_1}{E_c} = \dfrac{\dfrac{mc(a-k)}{k} + mc_1}{E_s} + (1-a)\alpha T°$$

or $\quad a\alpha T° E_c = c - c_1 + \dfrac{c(a-k)}{k} + c_1,\quad$ from which

$$c = \dfrac{a\alpha T° E_c}{1 + \dfrac{a-k}{k}} = k\alpha T° E_c$$

Equating this to *Equation 27.8*, we get

$$\dfrac{\dfrac{c_1[1+(m-1)p]}{\dfrac{k}{2} - mp\left(\dfrac{a-k}{k}\right)}}{} = \dfrac{a\alpha T° E_c}{1 + \dfrac{a-k}{k}} \quad ...[27.10\,(a)]$$

or $\quad \dfrac{c_1[1+(m-1)p]}{0.5k^2 - mp(a-k)} = \alpha T° E_c \quad ...(27.10)$

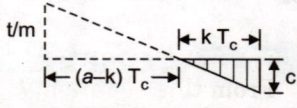

(a) Vertical section

(b) Stresses due to W and P

(c) Combined stresses

**Fig. 27.8**

From the above equation, $k$ can be determined. Knowing $k$, the values of final stresses $c$ and $t$ can be determined from *Eqs. 27.8* and *27.9*. It should be noted that the above analysis is valid only if $k$ is less than $a$. If $k$ is more than 1, the whole thickness of the concrete will be in compression and stresses should be re-analysed following the same procedure.

**(b) Windward side: Tension zone:** Due to self load and wind, the whole concrete shell on the windward side is in tension, which is borne entirely by steel having stress $t_1$. Fig. 27.8(c) shows the stress diagram due to the combined effect. Since there is no change in the total force due to temperature difference, we have:

$$T_s \cdot t_1 = T_s \cdot t - \tfrac{1}{2}c \cdot kT_c$$

But $\quad t = mc\dfrac{a-k}{k}\quad$ and $\quad T_s = pT_c$

$\therefore \quad pT_c \cdot t_1 = pT_c \cdot mc\dfrac{a-k}{k} - \tfrac{1}{2}ckT_c \quad \therefore \quad p \cdot t_1 = c\left[mp\left(\dfrac{a-k}{k}\right) - \dfrac{k}{2}\right]$. From which

$$c = \dfrac{pt_1}{mp\left(\dfrac{a-k}{k}\right) - \dfrac{k}{2}} \quad ...[27.11\,(a)]$$

Change in strain in concrete at inside face $= \dfrac{t_1}{E_s} + \dfrac{c}{E_c} \quad \therefore \quad \dfrac{t_1}{E_s} + \dfrac{c}{E_c} = \alpha T° - \varepsilon$

$$\varepsilon = \alpha T - \dfrac{t_1}{E_s} - \dfrac{c}{E_c} \quad ...(i)$$

Similarly, change in strain in steel $= \dfrac{t - t_1}{E_s} \quad \therefore \quad \dfrac{t - t_1}{E_s} = [\varepsilon - (1-a)\alpha T°]$

or $\quad \varepsilon = \dfrac{t - t_1}{E_s} + (1-a)\alpha T° \quad ...(ii)$

Equating (i) and (ii), we get

$$\alpha T° - \frac{t_1}{E_s} - \frac{c}{E_c} = \frac{t-t_1}{E_s} + (1-a)\alpha T° \quad \text{or} \quad a.\alpha T° - \frac{c}{E_c} = \frac{1}{E_s}$$

or

$$c = \left(a.\alpha T° - \frac{t}{E_s}\right) E_c = a.\alpha T° E_c - \frac{t}{m}$$

Substituting the value of $t$, $\quad c = a\alpha T° E_c - \dfrac{c(a-k)}{k}$

From which $\quad c = \alpha T° E_c . k \quad$ ...[27.11 (b)]

Equating this to *Eq. 27.11 (a)*, we get

$$\frac{pt_1}{mp\left(\dfrac{a-k}{k}\right) - \dfrac{k}{2}} = \alpha T° E_c . k \qquad \text{...[27.12 (a)]}$$

or

$$\frac{pt_1}{mp(a-k) - 0.5k^2} = \alpha T° E_c \qquad \text{...(27.12)}$$

From this equation, $k$ can be found. Substituting the value of $k$ in *Eq. 27.11*, $c$ can be determined, and hence stress in steel can be calculated.

## 27.7. TEMPERATURE STRESSES IN HORIZONTAL REINFORCEMENT

At high temperatures, the inner surface try to expand more. But, as stated earlier, it is prevented from expansion and is therefore subjected to the compressive stresses due to temperature gradient, while the outer surface expands more than its natural expansion and is subjected to tensile stresses. Since the horizontal reinforcement (hoops) are located near the outer face, it is subjected to tensile stresses.

Let horizontal reinforcement be placed at $a' T_c$ from inner face, and let the N.A. be at $k' T_c$, as shown in Fig. 27.9.

Let $A_\phi$ be the horizontal reinforcement at pitch $s$. If we consider 1 mm height of shell, the area of horizontal reinforcement per unit height is

$$As' = \frac{A_\phi}{s} p'T_c \quad \text{(say)}.$$

Now, the compressive force in concrete on the inner side will be equal to the tensile force in horizontal steel.

$\therefore \quad \dfrac{1}{2} c' . k' T_c = A'_s t' = p'T_c mc' \left(\dfrac{a'-k'}{k'}\right)$

or $\quad \dfrac{k'}{2} = mp'\left(\dfrac{a'-k'}{k'}\right)$

From which, $\quad k' = -mp' + \sqrt{2m p'a' + p'^2 m^2} \quad$ ...(27.13)

From this, $k'$ can be found. Now, strain in concrete is

$$\frac{c'}{E_c} = \alpha T° - \varepsilon$$

or $\quad \varepsilon = \alpha T° - \dfrac{c'}{E_c} \qquad$ ...(i)

Strain in steel is $\dfrac{t'}{E_s} = \varepsilon - \alpha T° (1-a')$

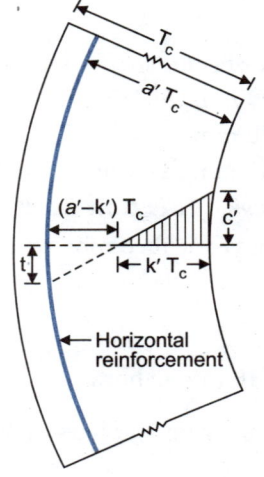

Fig. 27.9. Stress in Horizontal Reinforcement Due to Temperature Difference.

or
$$\varepsilon = \frac{t'}{E_s} + \alpha T° (1 - a') \qquad ...(ii)$$

Equating (i) and (ii),

$$\alpha T° - \frac{c'}{E_c} = \frac{t'}{E_s} + \alpha T° (1 - a') \quad \text{or} \quad \frac{t'}{E_s} + \frac{c'}{E_c} = \alpha T° a'$$

From which
$$\frac{t'}{m} + c' = E_c \cdot \alpha T° a' \qquad ...(iii)$$

But
$$\frac{t'}{m} = c'\left(\frac{a' - k'}{k'}\right) \qquad ...(iv)$$

$$\therefore \quad c'\left(\frac{a' - k'}{k'}\right) + c' = E_c \cdot \alpha T° a'$$

$$\therefore \quad c' = \frac{E_c \cdot \alpha T° a'}{1 + \frac{a' - k'}{k'}} = k' E_c \alpha T° \qquad ...(27.14)$$

From this, $c'$ can be found, and $t'$ can then be calculated from (iv).

## 27.8. DESIGN OF R.C. CHIMNEY

**Design: Example.** *Design a chimney of 66 m height, having external diameter of 4 m throughout the height. the chimney has fire brick lining of 100 mm thickness, provided upto a height of 42 m above ground level, with an air gap of 100 mm. The temperature of gases above surrounding air is 200°C. Take the coefficient of expansion of concrete and steel = $11 \times 10^{-6}$ per degree C, and $E_s = 2.05 \times 10^5$ N/mm². Use M 25 grade concrete mix.*

**Solution: 1. Dimensions of chimney and forces**: Fig. 27.10 shows the vertical section, along with wind pressure intensity at various heights. Let the thickness of chimney shell at the top be 200 mm, and let it be increased to 300 mm and 400 mm at 24 m and 48 m below top. Let the lining be supported at every 6 m.

Let us assume a constant wind pressure intensity of 1800 N/m² for the top 24 m portion, 1600 N/m² for the central 24 m portion and 1400 N/m² for the bottom 18 m portion. The shape factor will be 0.7. Actually, the section should be tested at every 6 m height. However, for illustration purposes, the section has been tested at 24 m, 48 m and 66 m below the top. Using 100 mm thick fire brick lining and taking its unit weight 19000 N/m³, its weight per metre height

$$= \pi[4 - 2(0.4 + 0.10 + 0.05)] \times 0.1 \times 1 \times 19000 \approx 17310 \text{ N}$$

*Weight of concrete per metre height*:
For 200 mm thick shell, $w = \pi[4 - 0.2] \times 0.2 \times 1 \times 25000 \approx 59690$ N/m
For 300 mm thick shell, $w = \pi[4 - 0.30] \times 0.30 \times 1 \times 25000 \approx 87179$ N/m
For 400 mm thick shell, $w = \pi[4 - 0.40] \times 0.40 \times 1 \times 25000 \approx 113097$ N/m

**2. Stress at Section 24 m below top**

Let the vertical reinforcement be 1% of the concrete area placed at a cover of 50 mm.

$$A_s = \frac{1}{100} \times \frac{\pi}{4}(4^2 - 3.6^2) \times 10^6 \approx 23876 \text{ mm}^2$$

No. of 16 mm Φ bars $= \frac{23876}{201} = 118.8$

Hence provide 120 bars of 16 mm Φ, suitably spaced along the circumference.
Actual $A_s = 24127$ mm²

Equivalent thickness of steel ring placed at the centre of the shell thickness ($R = 2 - 0.1 = 1.9$ m) is

$$T_s = \frac{24127}{2\pi R} = \frac{24127}{2\pi \times 1900} = 2.02 \text{ mm}.$$

Horizontal steel (hoops) may be provided @ 0.2% of sectional area. Area of steel per metre height of chimney

$$= \frac{0.2}{100} \times 200 \times 1000 = 400 \text{ mm}^2.$$

Hence pitch s of 12 mm Φ bar hoops

$$= \frac{1000 \times 113}{400} = 282.5 \text{ mm}.$$

Provide these at 250 mm centre to centre.

$$W = 24 \times 59690 = 1432560 \text{ N}$$
$$P = 0.7 \times 1800 (4 \times 24) = 120960 \text{ N};$$

acting at 12 m below the top.

∴ $M = 120960 \times 12 = 1451520$ N/m.

∴ Eccentricity $e = \dfrac{M}{W} = \dfrac{1451520}{1432560}$

$$= 1.013 \text{ m} = 1013 \text{ mm}.$$

For M 25 concrete, $m = 11$.

∴ Equivalent area $= A = \dfrac{\pi}{4}(4^2 - 3.6^2) \times 10^6$
$\qquad\qquad + (11 - 1) \times 24127$

$$= 2628880 \text{ mm}^2$$

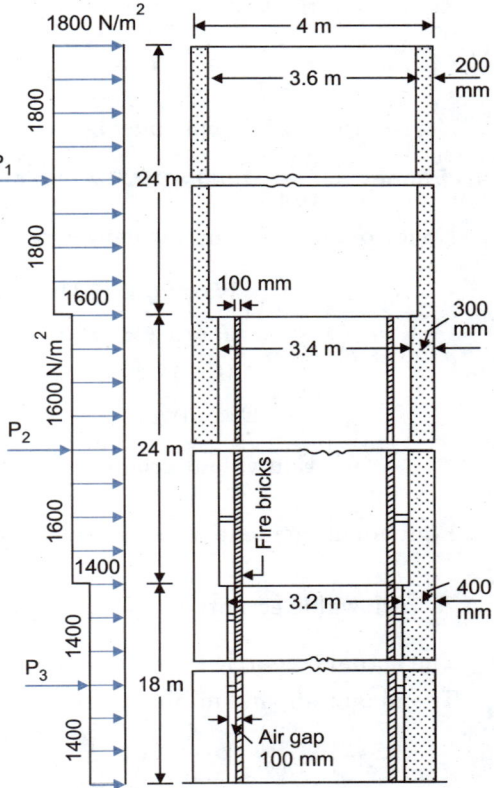

Fig. 27.10

Equivalent moment of inertia $= I = \dfrac{\pi}{64}(D^4 - d^4) + (m - 1)\pi R T_s (R)^2$

$$= \frac{\pi}{65}[4^4 - 3.6^4] \times (1000)^4 + (11 - 1)\pi (1900)^3 (2.02) = 4.7568 \times 10^{12} \text{ mm}^4$$

For no tension to develop, allowable eccentricity

$$= \frac{2I}{AD} = \frac{2 \times 4.7568 \times 10^{12}}{2628880 \times 4000} = 904.7 \text{ mm}.$$

The actual eccentricity is 1055 mm. Hence some tension will be developed in the leeward side. The maximun and minimum stresses are given by

$$\sigma = \frac{W}{A} \pm \frac{MD}{2I} = \frac{1432560}{2628880} \pm \frac{1451520 \times 10^3 \times 4000}{2 \times 4.7568 \times 10^{12}} = 0.545 \pm 0.610.$$

Compressive stress = 1.115 N/mm² (safe); Tensile stress = 0.065 N/mm² (safe)

**3. Stresses at 48 m below top**

Thickness of shell = 300 mm². Area of vertical steel @ 1% of area is

$$A_s = \frac{1}{100} \frac{\pi}{4}(4^2 - 3.4^2)10^6 \approx 34872 \text{ mm}^2$$

If 120 bars are to be provided, area of each bar $= A_\Phi = \dfrac{34872}{120} = 290.6$ mm²

∴ Diameter of bar $= \left(\dfrac{290.6 \times 4}{\pi}\right)^{1/2} = 19.2$ mm. Provide 20 mm Φ bars.

Actual area $= 120 \times \dfrac{\pi}{4}(20)^2 \approx 37699$ mm²

Equivalent thickness of steel ring placed at the centre of the shell thickness ($R = 2 - 0.15 = 1.85$ m) is

$$T_s = \frac{37699}{2\pi R} = \frac{37699}{2\pi (1850)} = 3.24 \text{ mm.}$$

Horizontal steel (hoops) may be provided equal to 0.2% of sectional area. Area of steel per metre height of chimney = $\frac{0.2}{100} \times 300 \times 1000 = 600$ mm$^2$.

Hence pitch $s$ of 12 mm Φ bar hoops = $\frac{1000 \times 113}{600} = 188.33 \approx 180$ mm.

$$W = (24 \times 17310) + 24 \times 59690 + 24 \times 87179 = 3940296 \text{ N}$$
$$P = 0.7 \times 1800 (4 \times 24) + 0.7 \times 1600 (4 \times 24) = 120960 + 107520 = 228480 \text{ N}$$
$$M = 120960 (12 + 24) + 107520 \times 12 = 5644800 \text{ N-m}$$

Eccentricity $e = M/W = 5644800/3940296 = 1.433$ m = 1433 mm.

Equivalent area $A = \frac{\pi}{4}(4000^2 - 3400^2) + (11-1)37699 = 3864158$ mm$^4$.

Equivalent moment of inertia = $\frac{\pi}{64}[4^4 - 3.4^4](1000)^4 + (11-1)\pi(1850)^3 \times 3.24 = 6.651 \times 10^{12}$ mm$^4$.

∴ Allowable eccentricity for no tension to develop = $\frac{2I}{AD} = \frac{2 \times 6.651 \times 10^{12}}{3864158 \times 4000} = 860$ mm.

The actual eccentricity is 1433 mm. Hence some tension will be developed.

The maximum and minimum stresses are given by

$$\sigma = \frac{W}{A} \pm \frac{MD}{2I} = \frac{3940296}{3864158} \pm \frac{5644800 \times 1000 \times 4000}{2 \times 6.651 \times 10^{12}} = 1.020 \pm 1.697.$$

∴ Max. compressive stress = 2.717 N/mm$^2$ (safe).

Max. tensile stress = 0.677 N/mm$^2$. This is less than the allowable tension of 0.8 N/mm$^2$ in M 25 grade concrete. Hence safe.

**4. Stress at base of chimney:**

**(a) Stress due to W and M**

Thickness of shell = 400 mm = 0.40 m. Area of vertical steel @ 1% of area is

$$A_s = \frac{1}{100} \cdot \frac{\pi}{4}(4^2 - 3.2^2) 10^6 \approx 45239 \text{ mm}^2$$

Provide 120 Nos. of 25 mm Φ bars getting total area = 58905 mm$^2$. Equivalent thickness of steel ring placed at the centre of the shell (thickness $R = 2 - 0.20 = 1.80$ m) is

$$T_s = \frac{58905}{2\pi R} = \frac{58905}{2\pi(1800)} = 5.2 \text{ mm}$$

$$W = (24 \times 59690) + (24 \times 87179) + (18 \times 113097) + (42 \times 17310) = 6287622 \text{ N}$$
$$P = 0.7 \times 1800 (4 \times 24) + 0.7 \times 1600 (4 \times 24) + 0.7 \times 1400 (4 \times 18)$$
$$= 120960 + 107520 + 70560 = 299040 \text{ N}$$
$$M = 120960 (12 + 42) + 107520 (12 + 18) + 70560 \times 9 = 10392480 \text{ N-m}$$

Eccentricity $e = \frac{10392480}{6287622} = 1.653$ m.

The eccentricity is quite high. Due to this, tensile stresses in the windward side are expected to be greater than 0.8 N/mm$^2$ resulting in cracking of concrete. Hence it is assumed that only steel will take the tensile stresses and concrete in the tensile zone will be ignored. Thus, the method of analysis used at 24 m and 48 m will not be applicable. We shall analyse the section for stresses by method discussed in § 27.3.

$T_c = 400$ mm; $T_s = 5.2$ mm; $R = 2 - 0.2 = 1.80$ m.

In order to find the position of N.A., use equation 27.3:

$$e = R\frac{\left[(T_c - T_s)\left\{\frac{\sin 2\phi}{4} + \frac{\pi - \phi}{2}\right\} + \frac{m\pi}{2}T_s\right]}{[(T_c - T_s)\{\sin\phi + (\pi - \phi)\cos\phi\} + m\pi T_s \cos\phi]}$$

$\therefore$ 
$$e = 1800\frac{\left[(400 - 5.2)\left\{\frac{\sin 2\phi}{4} + \frac{\pi - \phi}{2}\right\} + \frac{11\pi}{2} \times 5.2\right]}{[(400 - 5.2)\{\sin\phi + (\pi - \phi)\cos\phi\} + 11 \times \pi \times 5.2 \cos\phi]}$$

or
$$e = \frac{710640\left\{\frac{\sin 2\phi}{4} + \frac{\pi - \phi}{2}\right\} + 161729}{394.8\{\sin\phi + (\pi - \phi)\cos\phi\} + 179.7 \cos\phi}$$

Assuming $\phi = 90° = \frac{\pi}{2}$ radians; $\sin\phi = 1$; $\cos\phi = 0$; $\sin 2\phi = 0$

$\therefore$
$$e = \frac{710640\left[0 + \frac{\pi}{4}\right] + 161729}{394.8(1 + 0) + 0} = 1823 \text{ mm}$$

This is slightly more than actual value of $e = 1653$ mm.
Adopt next value of $\phi = 85° = 1.4835$ radians $\sin\phi = 0.9962$; $\cos\phi = 0.0872$; $\sin 2\phi = 0.1736$

$\therefore$
$$e = \frac{710640\left\{\frac{0.1736}{4} + \frac{\pi - 1.4835}{2}\right\} + 161729}{394.8\{0.9962 + (\pi - 1.4835)\,0.0872\} + 179.7 \times 0.0872}$$

$= 1677$ mm, which is practically the same as $e = 1653$.

Hence $\phi = 85°$. The maximum stress $c_1$ in concrete is found from *Eq. 27.2 (a)*.

$$W = \frac{2Rc_1}{1 + \cos\phi}[(T_c - T_s)\{\sin\phi + (\pi - \phi)\cos\phi\} + \pi m T_s \cos\phi]0$$

$$6287622 = \frac{2 \times 1800\, c_1}{1 + 0.0872}[(400 - 5.2)\{0.9962 + (\pi - 1.4835) \times 0.0872\} + \pi \times 11 \times 5.2 \times 0.0872]$$

or $6287622 = 1543218\, c_1.$    From which $c_1 = 4.07$ N/mm$^2$

$\therefore$ $t_1 = mc_1\dfrac{1 - \cos\phi}{1 + \cos\phi} = 11 \times 4.07\left[\dfrac{1 - 0.0872}{1 + 0.0872}\right] = 37.63$ N/mm$^2$

**(b) Stress in horizontal reinforcement:** Horizontal steel (hoops) may be provided @ 0.2% of sectional area. Area of steel per metre height of chimney $= \dfrac{0.2}{100} \times 400 \times 1000 = 800$ mm$^2$.

Hence pitch $s$ of 12 mm $\Phi$ bars hoop $= \dfrac{1000 \times 113}{800} = 141.3$ mm.

$A_\Phi = 113$ mm$^2$ in pitch $s = 140$ mm. If the cover is 40 mm.
$$D_1 = 4000 - 80 = 3920 \text{ mm}.$$

$\therefore$
$$t = \frac{P \cdot s}{2\, A_\Phi\, D_1} = \frac{299040 \times 140}{2 \times 113 \times 3920} = 47.26 \text{ N/mm}^2$$

**(c) Stress on leeward side due to temperature gradient**

$T_c = 400$ mm;   $T_s = 5.2$ mm.
Cover to vertical steel $= 50$ mm ;   $\therefore$ $a\,T_c = 350$ mm

$$a = 350/400 = 0.875; \quad c_1 = 4.07 \text{ N/mm}^2$$

$$E_s = 2.05 \times 10^5; \quad E_c = \frac{2.05 \times 10^5}{11} = 0.186 \times 10^3 \text{ N/m}^2; \quad p = \frac{T_s}{T_c} = \frac{5.2}{400} = 0.013$$

$$\alpha = 11 \times 10^{-6} \text{ per °C}. \quad \text{Temperature difference} = 200°C$$

Let us assume that 80% of temperature drops through the lining and shell. Drop in temperature = 200 × 0.8 =160°C. Assuming that drop in lining is 5 times more than that in shell, per unit thickness, the drop of temperature through concrete is given by

$$T° = \frac{160}{400 + 5 \times 100} \times 400 \approx 71°C$$

To locate neutral axis in the shell thickness, use *Eq. 27.10* $\quad \dfrac{c_1 [1 + (m-1)\, p]}{0.5k^2 - mp(a-k)} = \alpha T° E_c$

$$\therefore \quad \frac{4.07 [1 + (11-1)\, 0.013]}{0.5\, k^2 - 11 \times 0.013\, (0.875 - k)} = 11 \times 10^{-6} \times 71 \times 0.186 \times 10^5$$

or $\quad k^2 + 0.286\, k - 0.8836 = 0 \quad \therefore \quad k = 0.81$

$$\therefore \quad c = \frac{a\, \alpha\, T°\, E_c}{1 + \dfrac{a-k}{k}} = k\, \alpha\, T°\, E_c = 0.81 \times 11 \times 10^{-6} \times 71 \times 0.186 \times 10^5 = 11.77 \text{ N/mm}^2$$

Since wind stresses are taken into account, permissible $c = \dfrac{4}{3} \times 8.5 = 11.33 \text{ N/mm}^2$

Thus, the compressive stress induced is slightly more than the permissible. The above analysis is based on the assumption that the tension caused by temperature variation cannot be taken by concrete, and it is taken entirely by steel.

Stress in steel $\quad = t = mc \dfrac{a-k}{k} = 11 \times 11.77 \dfrac{0.875 - 0.81}{0.81} = 10.39 \text{ N/mm}^2$

### (d) Stresses on windward side, due to temperature gradient

$$\frac{pt_1}{mp(a-k) - 0.5k^2} = \alpha\, T°\, E_c \quad \text{where} \quad t_1 = 37.63 \text{ N/mm}^2$$

$$\therefore \quad \frac{0.013 \times 37.63}{11 \times 0.013(0.875 - k) - 0.5\, k^2} = 11 \times 10^{-6} \times 71 \times 0.186 \times 10^5$$

or $\quad k^2 + 0.286\, k - 0.1829 = 0. \quad$ From which $k = 0.308$

$$\therefore \quad c = \alpha\, T\, E_c\, k = 11 \times 10^{-6} \times 71 \times 0.186 \times 10^5 \times 0.308 = 4.47 \text{ N/mm}^2$$

and $\quad t = mc \dfrac{a-k}{k} = 11 \times 4.47 \dfrac{0.875 - 0.308}{0.308} = 90.52 \text{ N/mm}^2 \quad$ (safe)

### (e) Stresses on the N.A. (i.e. temperature effect alone)

$$k_2 = -mp + \sqrt{2mpa + m^2 p^2} \qquad \qquad \ldots(27.5)$$

or $\quad k_2 = -11 \times 0.013 + \sqrt{(2 \times 11 \times 0.013 \times 0.875) + (11 \times 0.013)^2} = 0.377$

$$c_2 = E_c \cdot \alpha\, T°\, k_2 = 0.186 \times 10^5 \times 11 \times 10^{-6} \times 71 \times 0.377$$
$$= 5.48 \text{ N/mm}^2 \qquad \qquad \ldots(27.7)$$

$$t_2 = mc_2 \frac{a - k_2}{k_2} = 11 \times 5.48 \left( \frac{0.875 - 0.377}{0.377} \right) = 79.63 \text{ N/mm}^2$$

## (f) Stresses in horizontal reinforcement due to temperature

$$p' = \frac{A_\Phi}{S.T_c} = \frac{113}{140 \times 400} = 0.002; \quad a' = \frac{360}{400} = 0.9$$

From Eq. 27.13.

$$k' = -mp' + \sqrt{2mp'a' + p'^2 m^2}$$

$$= -11 \times 0.002 + \sqrt{(2 \times 11 \times 0.002 \times 0.9) + (0.002 \times 11)^2} = 0.178$$

$$c' = k' E_c . T° = 0.178 \times 0.186 \times 10^5 \times 11 \times 10^{-6} \times 71 = 2.59 \text{ N/mm}^2 \quad ...(27.14)$$

$$t' = mc' \frac{a' - k'}{k'}$$

$$= 11 \times 2.59 \left(\frac{0.9 - 0.178}{0.178}\right) = 115.5 \text{ N/mm}^2.$$

These stresses are due to temperature effect alone. To this, we must add the stresses due to wind. Hence total stress in steel

$$= 115.5 + 47.26 = 162.76 \text{ N/mm}^2.$$

Since wind is also acting, permissible $t = \frac{4}{3} \times 140 = 186.7 \text{ N/mm}^2$.

Hence O.K.

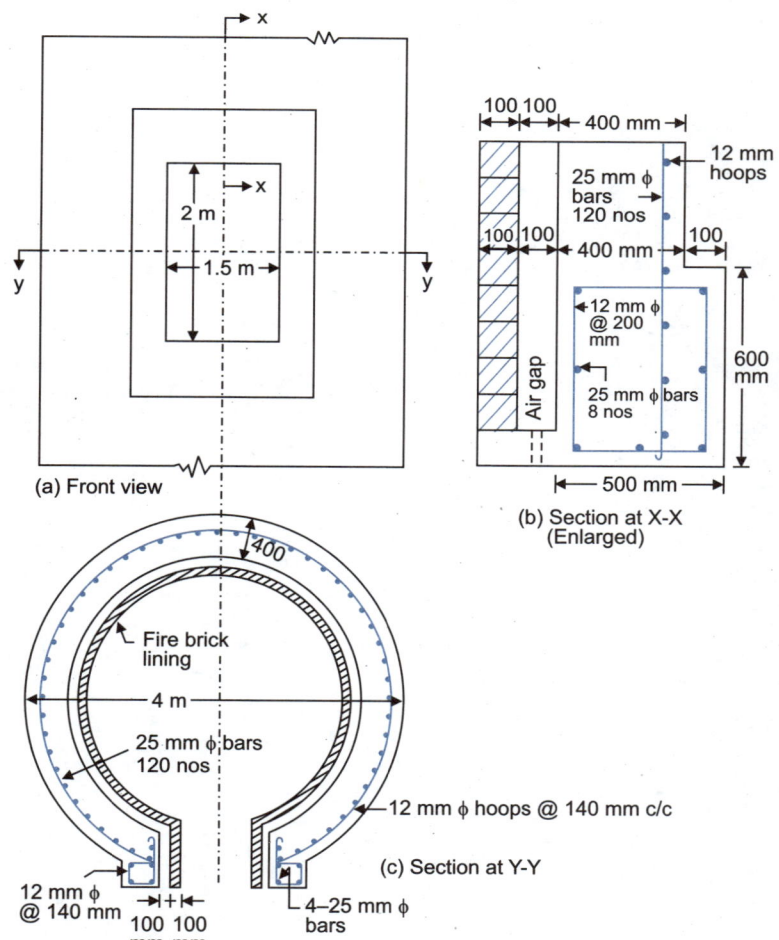

(a) Front view

(b) Section at X-X (Enlarged)

(c) Section at Y-Y

Fig. 27.11

**5. Flue Opening:** Provide a flue opening 1.5 m wide and 2 m high at the bottom. The boundary of the opening is thickened and reinforced as shown in Fig. 27.11. The vertical steel bars are bent on either side of the opening as shown.

## PROBLEMS

1. Design a chimney of 30 m height, having external diameter of 2.6 m throughout the height. The chimney has a fire brick lining of 100 mm thickness, provided upto a height of 24 m above base, with an air gap of 100 mm. Assume the temperature of gases above the surrounding air as 240°C. Take $\alpha$ for R.C.C. as $11 \times 10^{-6}$ per °C and $E_s = 2.05 \times 10^5$ N/mm². Use M 20 grade concrete.
2. Derive an expression for temperature stresses in horizontal reinforcement of a R.C. chimney.
3. A reinforced concrete chimney 50 m high above ground has an outside diameter of 4 m. The thickness of the shell 260 mm at the top and it is increased to 250 mm and 300 mm at 18 m and 30 m from the top vertical steel bars = 1% of cross-sectional area throughout. The total wind load above the section at 18 metres from top may be taken as 96 kN. Find stresses developed due to wind and dead loads at the section 18 m from the top of the Chimney. (Assume modular ratio m = 13).

# CHAPTER 28
# PORTAL FRAMES

## 28.1. INTRODUCTION

Portal frames are widely used in the construction of large sheds for industrial buildings.

They are also used in stiffening large span bridge girders or as viaducts. A portal frame essentially consists of vertical members (called columns) and top member which may be horizontal, curved or pitched. The vertical and top members are rigidly jointed. The frames may be fixed or hinged at the base. The portal frames are spaced at suitable distance, and support the roof which may consist of either a continuous slab, or a beam-slab system. Fig. 28.1 shows various forms of portal frames. For workshop and storage sheds, portal frames with sloping roof is generally preferred. For highways and buildings, portal frames, with flat roof is used.

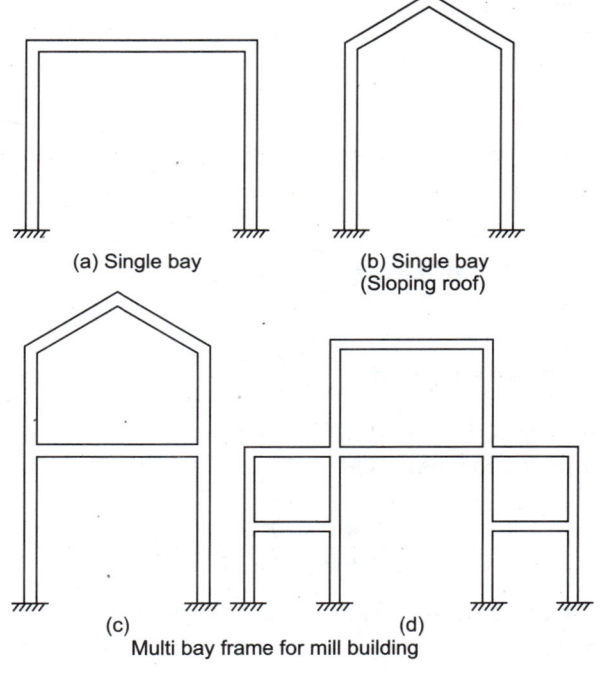

Fig. 28.1. Portal Frames

## 28.2. ANALYSIS OF PORTAL FRAMES

In case of buildings, the portal frames are generally spaced at intervals of 3 to 4 m with a reinforced concrete slab cast monolithically between the frames. A portal frame is subjected to vertical loads from its roof etc. as well as horizontal loads due to wind. The loads coming on the columns as well as roof can be easily found. The analysis of the portal frames for determining the bending moment and shear force may be done by the following methods:

 1. Slope-deflection method
 2. Moment distribution method
 3. Strain energy method
 4. Kani's method.

For details about these methods, the reader may refer to Author's book 'Strength of Materials and Theory of Structures' Vol.II. The members of the portal frame are designed for B.M., S.F. and axial forces.

## 28.3. DESIGN OF RECTANGULAR PORTAL FRAME WITH VERTICAL LOADS

A simple portal frame consists of a horizontal beam resting on two columns. The junction of the beam with the columns consist of rigid joints. If the loading is symmetrical, and geometry of the frame is also symmetrical, there will be no joint translation or sway. Example 28.1 illustrates the method of analysis and design.

**Design: Example 28.1.** *For a hall 10 m wide and 20 m long, portal frames are to be provided at 4 m centre to centre, dividing the hall into five equal parts. The portals are fixed at the base, and its height upto centre of horizontal member is 6.5 m. Design the roof and portal frame, if it carries a live load of 1.5 kN/m². Assume the safe bearing capacity of soil as 120 kN/m². Use M 15 concrete and take its unit weight as 24000 N/m³.*

**Solution: 1. Design constants**

Let $\sigma_{st} = 140$ N/mm². Using M 15 concrete, $\sigma_{cbc} = 5$ N/mm² and $m = 19$

$$\therefore \quad k = \frac{m\,\sigma_{cbc}}{m\,\sigma_{cbc} + \sigma_{st}} = \frac{19 \times 5}{19 \times 5 + 140} = 0.405 \; ; \quad j = 1 - \frac{k}{3} = 1 - \frac{0.405}{3} = 0.865$$

$$R = \tfrac{1}{2}\sigma_{cbc} \cdot jk = \tfrac{1}{2} \times 5 \times 0.865 \times 0.405 = 0.875$$

**2. Design of roof slab :** The roof slab will be continuous over portals, having a centre to centre span of 4 m. Assume the thickness of slab as 100 mm. Dead load of slab, $w_d = 1 \times 1 \times 0.1 \times 24000 = 2400$ N/m². Superimposed live load $w_s = 1500$ N/m². As per IS Code, the maximum negative B.M. occurs at the first interior support, its value being given by

$$M = \frac{w_d L^2}{10} + \frac{w_s L^2}{9} = \frac{2400\,(4)^2}{10} + \frac{1500\,(4)^2}{9} = 6507 \text{ N-m}$$

The maximum positive B.M. occurs at the middle of the end spans, its value being given by

$$M = \frac{w_d\,L^2}{12} + \frac{w_s\,L^2}{10} = \frac{2400\,(4)^2}{12} + \frac{1500\,(4)^2}{10} = 5600 \text{ N-m}$$

The S.F. at the exterior face of the support, next to the end support is

$$F = 0.6\,w_d L + 0.5\,w_s L = 0.6 \times 2400 \times 4 + 0.5 \times 1500 \times 4 = 8760 \text{ N}.$$

Effective depth $d = \sqrt{\dfrac{6507 \times 1000}{0.875 \times 1000}} = 86.2$ mm. Keep overall depth = 105 mm, and a clear cover equal to 13 mm. Using 10 mm Φ bars, available effective depth = 105 − 13 − 5 = 87 mm. Reinforcement will be provided for max. B.M.

$$A_{st} = \frac{6507 \times 1000}{140 \times 0.865 \times 87} = 618 \text{ mm}^2. \quad \text{Spacing of 10 mm Φ bars} = \frac{1000 \times 78.5}{618} = 127 \text{ mm}.$$

Keep pitch = 120 mm. Alternate bars from the bottom can be bent up near supports to bear the negative B.M.

$$A_{sd} = \frac{0.15}{100} \times 1000\,D = 1.5\,D = 1.5 \times 105 = 157.5 \approx 158 \text{ mm}^2$$

$$\therefore \text{ Spacing of 6 mm Φ bars} = \frac{1000 \times 28.3}{158} = 179.11 \approx 180 \text{ mm}$$

$$\text{Nominal shear stress} = \frac{8760}{1000 \times 87} = 0.1 \text{ N/mm}^2 \quad (\text{safe})$$

Let us check the bar diameter to meet the following requirement of anchorage and bond at the ends:

$$\frac{M_1}{V} + L_0 \geq L_d$$

where $M_1$ = moment of resistance of the section, assuming all reinforcement at the section to be stressed to $\sigma_{st}$

$$= A_{st} \cdot \sigma_{st} \cdot jd = \left[\frac{1}{2}\frac{1000 \times 78.5}{120}\right] \times 140 \times 0.865 \times 87 = 3.45 \times 10^6 \text{ N-mm}$$

$$V = \text{S.F.} = 8760 \text{N}; \quad L_0 = \left(\frac{l_s{'}}{2} - x' - 3\,\Phi\right) + 16\,\Phi = \frac{l_s}{2} - x' + 13\,\Phi.$$

where $l_s$ = length of support = 300 mm (say); $x'$ = Side cover = 30 mm (say)

$$\therefore L_0 = \frac{300}{2} - 30 + (13 \times 10) = 250 \text{ mm}$$

$$L_d = \text{development length} = \frac{\Phi\,\sigma_{st}}{4\,\tau_{bd}} = \frac{\Phi\,(140)}{4 \times 0.6} \approx 58\,\Phi = 58 \times 10 = 580$$

$$\therefore \frac{M_1}{V} + L_0 \geq L_d. \quad \text{or} \quad \frac{3.45 \times 10^6}{8760} + 250 \geq 580$$

or $393.8 + 250 \geq 580$. Hence the requirement is just satisfied.

**Note:** If the requirements were not satisfied, it would be necessary to reduce the bar diameter. By reducing the bar diameter (say from 10 mm to 8 mm), $M_1$ will remain practically the same (since $A_{st}$ will be the same), while $L_0$ will decrease slightly and $L_d$ will decrease greatly.

Revised $w_d = 1 \times 1 \times 0.105 \times 24000 = 2520 \text{ N/m}^2$.

### 3. Analysis of Portal Frame

Load on each portal, from slab = $4\,(2520 + 1500) = 16080$ N/m

Let the depth of the beam $\approx \frac{1}{15}$ span $\approx 700$ mm

Let the width be 300 mm.

Let the vertical members (columns) be of size 500 mm × 300 mm.
Self load of beam

$$= \left(\frac{700 - 105}{1000}\right)\left(\frac{300}{1000}\right) \times 24000 = 4284 \text{ N/m}$$

$\therefore$ Total $w = 16080 + 4284 = 20364$ N/m.

The moment of inertia and the columns are as follows:

$$I_{BC} = \frac{300\,(700)^3}{12} \quad \therefore I_{BA} = I_{CD} = \frac{300\,(500)^3}{12}$$

$$\therefore \frac{I_{BC}}{I_{BA}} = \left(\frac{700}{500}\right)^3 = 2.74$$

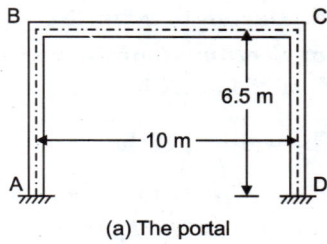

(a) The portal

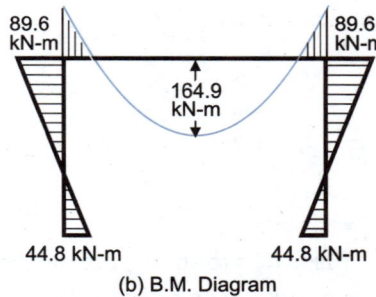

(b) B.M. Diagram

Fig. 28.2

At the joint B, distribution factor for the beam is

$$= \frac{I_{BC}/10}{(I_{BC}/10) + (I_{BA}/6.5)} = \frac{2.74/10}{(2.74/10) + (1/6.5)} = 0.64$$

$\therefore$ Distribution factor for column = 0.36.

$$M_{FBC} = -\frac{20364(10)^2}{12} = -169700 \text{ N-m} \approx -170 \text{ kN-m}$$

$$M_{FCB} = +\frac{20364\,(10)^2}{12} = +169700 \text{ N-m} \approx +170 \text{ kN-m}$$

The moment distribution is done as shown in Table 28.1 below:

# 706 REINFORCED CONCRETE STRUCTURE

## TABLE 28.1

| A | B | | C | | D | |
|---|---|---|---|---|---|---|
|   | 0.36 | 0.64 | 0.64 | 0.36 | | D.F. |
| — | — | – 170 | + 170 | — | — | F.E.M. |
| — | + 61.2 | + 108.8 | – 108.8 | – 61.2 | — | Balance |
| + 30.6 | — | – 54.4 | + 54.4 | — | – 30.6 | Carry over |
| — | + 19.6 | + 34.8 | – 34.8 | – 19.6 | — | Balance |
| + 9.8 | — | – 17.4 | + 17.4 | — | – 9.8 | Carry over |
| — | + 6.2 | + 11.2 | – 11.2 | – 6.2 | — | Balance |
| + 3.1 | — | – 5.6 | + 5.6 | — | – 3.1 | Carry over |
| — | +2.0 | + 3.6 | – 3.6 | – 2.0 | — | Balance |
| + 1.0 | — | – 1.8 | + 1.8 | — | – 1.0 | Carry over |
| + 0.3 | ←+ 0.6 | + 1.2 | – 1.2 | – 0.6 → | – 0.3 | Balance & Carry over |
| + 44.8 | +89.6 | – 89.6 | + 89.6 | – 89.6 | – 44.8 | Final moments |

B.M. at the centre of simply supported beam = $\dfrac{20360\,(10)^2}{8}$ = 254500 N-m ≈ 254.5 kN-m

∴ Net B.M. at the centre of $BC$ = 254.5 – 89.6 = 164.9 kN-m = 164.9 × $10^6$ N-mm.

Fig. 28.2 (b) shows the B.M. diagram.

**4. Design of beam BC**

**(a) Section at mid-span:** The beam $BC$ behaves as $T$-beam at the mid-span, where it carries a maximum B.M. of 254.5 KN-m. Let the width of the web, $b_w$ = 300 mm.

The width $b_f$ of the flange is given by $b_f = \dfrac{l_0}{6} + b_w + 6D_f$

where $l_0$ = distance between points of zero moments in the beam.

In order to find $l_0$, let $x$ be the distance of point of contraflexure in the beam, measured from $A$. End reaction for beam $BC$ = 20.36 × $\dfrac{10}{2}$ = 101.8 kN.

∴ $89.6 + \dfrac{20.36\,x^2}{2} - 101.8\,x = 0$ or $x^2 - 10x + 8.8 = 0$, which gives $x$ = 0.975 m

Hence $l_0$ = 10 – 2 × 0.975 = 8.05 m ; $D_f$ = 105 mm

∴ $b_f = \dfrac{8.05 \times 1000}{6} + 300 + 6 \times 105 \approx 2270$ mm.

This is less than $b_w$ plus half the sum of clear distances to the adjacent beam on either side.

∴ $b_f$ = 2270 mm

Assuming lever arm $a$ equal to $0.9\,d$, the depth of balanced section of $T$-beam is given by

$$M_f = 0.45\,b_f \cdot c \cdot D_f \left(\dfrac{2\,k\,d - D_f}{k}\right)$$

or $164.9 \times 10^6 = 0.45 \times 2270 \times 5 \times 105 \left(\dfrac{2 \times 0.404\,d - 105}{0.404}\right)$.

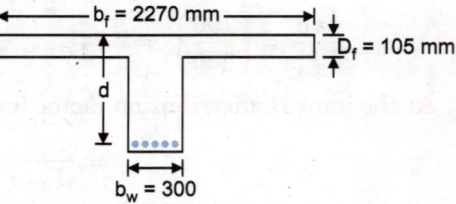

Fig. 28.3

From which $d$ = 284 mm

However, the beam has to behave as rectangular beam at the ends, where the effective depth is given by $d = \sqrt{\dfrac{89.6 \times 10^6}{300 \times 0.875}}$ = 584 mm.

Keep total depth $D = 630$ mm. Using a clear cover of 25 mm, available $d = 630 - 25 - \frac{25}{2} = 592.5$ mm. Approximate lever arm at the middle of span

$$= d - \frac{D_f}{2} = 592.5 - \frac{105}{2} = 540 \text{ mm}$$

$$\therefore \quad A_{st} = \frac{164.9 \times 10^6}{140 \times 540} = 2181.2 \text{ mm}^2$$

Using 25 mm Φ bars, $A_\Phi = 490.9$ mm². ∴ No. of 25 mm Φ bars $= 2181.2/490.9 = 4.44$

Provide 5-25 mm Φ bars, having actual $A_{st} = 2454$ mm². Out of these, we can curtail 3 bars at 2 m from the support, provided the following equation is satisfied at the point of contraflexure:

$$\frac{M_1}{V} + L_0 \geq L_d$$

where $M_1$ = moment of resistance of the section, assuming all reinforcement at the section to be stressed to $\sigma_{st} = 2 \times 490.9 \times 140 \times 0.865 \times 592.5 = 70.45 \times 10^6$ N-mm

$V$ = S.F. at the point of contraflexure = $101.8 - 20.36 \times 0.975 \approx 81.95$ kN = 81950 N
$L_0 = d$ or 12 Φ whichever is greater = 592.5 or $12 \times 25 = 300$ mm.

Therefore $L_0 = 592.5$ mm
$L_d = 58\,\Phi = 58 \times 25 = 1450$ mm.

$$\frac{M_1}{V} + L_0 = \frac{70.45 \times 10^6}{81950} + 592.5 = 859.67 + 592.5 = 1452.2 \text{ mm}$$

This is greater than $L_d = 1450$ mm. Hence three bars can be curtailed without violating bond and anchorage requirements. Let $n$ = depth of N.A.

Assuming that the actual N.A. falls below the flange, we have

$$b_f \cdot D_f \left( n - \frac{D_f}{2} \right) = m\, A_{st}\, (d - n) \quad \therefore \quad 2270 \times 105 \left( n - \frac{105}{2} \right) = 19 \times 2454\, (592.5 - n).$$

From which $n = 140.85$ mm. The L.A. is given by.

$$a = d - \frac{D_f}{3} + \frac{D_f^2}{6(2n - D_f)} = 592.5 - \frac{105}{3} + \frac{(105)^2}{6(2 \times 140.85 - 105)} = 567.9 \approx 568 \text{ mm}$$

Stress in steel $= \dfrac{M}{A_{st} \times a} = \dfrac{164.9 \times 10^6}{2454 \times 568} = 118.3$ N/mm² (safe)

Stress in concrete $= \dfrac{t}{m} \cdot \dfrac{n}{d - n} = \dfrac{118.3}{19} \times \dfrac{140.85}{592.5 - 140.85} = 1.94$ N/mm² (safe)

**(b) Section at support:** Because of negative B.M., the beam will behave as a rectangular beam. Effective depth $d = 590$ mm;

$j = 0.865$. $A_{st} = \dfrac{89.6 \times 10^6}{140 \times 0.865 \times 592.5} = 1248.8$ mm².

Provide 4 bars of 20 mm Φ, having total $A_{st} = 4 \times 314 = 1256$ mm². Two of these bars may be curtailed at 2 m from the supports, since this distance is greater than development length ($= 58\,\Phi = 58 \times 20 = 1160$ mm), and is beyond the point of contraflexure.

S.F. at support $= V = 101.8$ kN = 101800 N

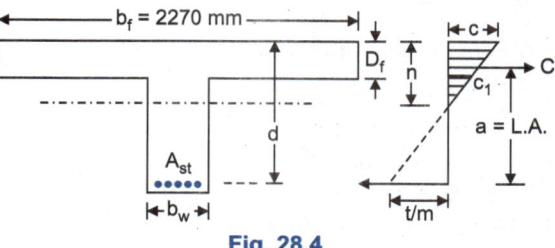

Fig. 28.4

Nominal shear stress $\tau_v = \dfrac{101.8 \times 1000}{300 \times 592.5} = 0.57$ N/mm²

Percentage reinforcement $= \dfrac{A_s}{b\,d} \times 100 = \dfrac{1256}{300 \times 592.5} \times 100 = 0.71$

Hence permissible shear stress $\tau_c = 0.32$ N/mm². Since $\tau_v > \tau_c$ shear reinforcement is necessary. Also, $\tau_{c\,max}$ for M 15 concrete = 1.6 N/mm²; thus $\tau_v$ is less than $\tau_{c\,max}$.

Shear resistance of concrete = $V_c = \tau_c\,b\,d = 0.32 \times 300 \times 592.5 = 56880$ N

Shear resistance to be provided by shear reinforcement
$$V_s = V - V_c = 101800 - 56880 = 44920 \text{ N}.$$

Spacing of vertical stirrups is given by
$$s_v = \frac{\sigma_{sv} \cdot A_{sv} \cdot d}{V_s} = \frac{140 \times A_{sv} \times 592.5}{44920} = 1.847\, A_{sv}$$

Using 10 mm Φ 2-lgd stirrups, $A_{sv} = 2\dfrac{\pi}{4}(10)^2 = 157$ mm²

∴ $s_v = 1.847 \times 157 = 289.9$ mm.

The maximum spacing of stirrups is given by
$$s_v = \frac{2.5 A_{sv}\, f_y}{b} = \frac{2.5 \times 157 \times 250}{300} = 327 \text{ mm}.$$

However, provide 10 mm Φ 2 lgd stirrups @ 250 mm c/c at the supports. This spacing may be increased to 300 mm c/c at 2 m from the supports to the middle of the beam.

**5. Design of vertical members**

(a) *Section at the top:* $P$ = Reaction from the beam = 101800 N

$M$ = B.M. at head = $89.6 \times 10^6$ N-mm.

$$\text{Eccentricity} = e = \frac{89.6 \times 10^6}{101800} \approx 880 \text{ mm}.$$

Let the size of the member be 500 × 300 mm. The eccentricity is thus more than 1.5 D. Hence this corresponds to case 2 (*i.e.*, large eccentricity) of chapter 13. Let the section be reinforced with 5 bars of 20 mm Φ on each side, with a cover of 40 mm to the centre of steel.

$$A_{st} = A_{sc} = 5\frac{\pi}{4}(20)^2 \approx 1571 \text{ mm}^2$$

The depth of the neutral axis, neglecting the direct stresses can be found by equating the moments of effective area about the N.A.

Thus, 
$$\frac{bn^2}{2} + (1.5\,m - 1)\,A_{sc}\,(n - d_c) = m\,A_{st}\,(D - d_t - n)$$

$$\frac{300}{2} n^2 + (1.5 \times 19 - 1)\,1571\,(n - 40)$$

$$= 19 \times 1571\,(500 - 40 - n)$$

$$n^2 + 288\,(n - 40) = 199\,(460 - n)$$

or $\qquad n^2 + 477n - 103060 = 0$

From which $n = 161$ mm.

Eccentricity $e'$ of the load from N.A.

$$= e - \left(\frac{D}{2} - n\right)$$

$$= 880 - (250 - 161) = 791 \text{ mm}$$

In order to determine the com-pressive stress $c$ in concrete, due to bending, equate the moment of resistance of the section to the external bending moment. Thus,

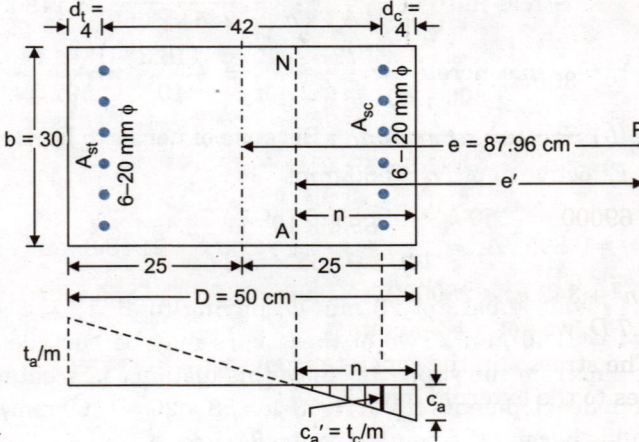

Fig. 28.5

$$\frac{bn \cdot c_a}{2}\left(D - d_t - \frac{n}{3}\right) + (1.5\,m - 1)\,A_{sc} \cdot c_a'(D - d_c - d_t) = P \cdot e'$$

where $\qquad c_a' = \dfrac{c_a}{n}(n - d_c) = \dfrac{c_a}{161}(161 - 40) = 0.752\,c_a$

Substituting the values, we get

$$\frac{300 \times 161 \times c_a}{2}\left(500 - 40 - \frac{161}{3}\right) + (1.5 \times 19 - 1)1571 \times 0.752\, c_a \times (500 - 40 - 40) = 101800 \times 791$$

From which $c_a = 3.43$ N/mm². The tensile stress $t_a$ in steel due to B.M. is given by

$$t_a = m\frac{c_a}{n}(D - d_t - n) = \frac{19 \times 3.43}{161}(500 - 40 - 161) = 121 \text{ N/mm}^2$$

The effective area of the section is given by

$$A_e = bn + (1.5\, m - 1)A_{sc} + m\, A_{st} = 300 \times 161 + (1.5 \times 19 - 1)1571 + 19 \times 1571 \approx 121352 \text{ mm}^2$$

∴ Direct compressive stress in concrete is given by

$$\sigma_c' = \frac{P}{A_e} = \frac{101800}{121352} = 0.84 \text{ N/mm}^2.$$

Final stresses are as follows: $\quad c_1 = c_a + \sigma_c' = 3.43 + 0.84 = 4.27$ N/mm² $< 5$

$$t' = t_a - 1.5\, m\, \sigma_c' = 121 - 1.5 \times 19 \times 0.84 = 97 \text{ N/mm}^2 \text{ (safe)}$$

Thus the section and the reinforcement provided are all right. Diameter of ties = $\frac{1}{4}$ dia. of main reinforcement. However, provide 6 mm Φ ties, with the spacing of the least of the following:

(i) Least lateral dimension       = 300 mm
(ii) 16 times dia. of main bars   = 16 × 20 = 320 mm
(iii) 48 times dia. of ties           = 48 × 6 = 288 mm.

However, provide 6 mm Φ ties @ 250 mm c/c.

**(b) Section at the bottom**

Direct load = 101800 + (0.5 × 0.3 × 6.5 × 24000) = 125200 N; B.M. = 44.8 × 10⁶ N-mm.

Provide the same section (i.e. 300 × 500 mm) of the column. However, since the B.M. has reduced to half, provide only 3 bars of 20 mm Φ on each face, with a cover of 40 mm to the centre of steel.

$$A_{st} = A_{sc} = 3 \times \frac{\pi}{4}(20)^2 \approx 942 \text{ mm}^2. \qquad \text{Eccentricity, } e = \frac{44.8 \times 10^6}{125200} = 358 \text{ mm}$$

This is less than $1.5\, D$ but greater than $D/4$. Hence case 3 will apply.
The N.A. is located by the following equation:

$$\frac{\dfrac{bn}{2}\left(D - d_t - \dfrac{n}{3}\right) + (1.5\, m - 1)A_{sc}\dfrac{1}{n}(n - d_c)(D - d_t - d_c)}{\dfrac{bn}{2} + (1.5\, m - 1)A_{sc}\left(\dfrac{n-d}{n}\right) - m\, A_{st}\dfrac{1}{n}(D - d_t - n)} = \left(e + \dfrac{D}{2} - d_t\right)$$

or

$$\frac{\dfrac{300\, n}{2}\left(500 - 40 - \dfrac{n}{3}\right) + (1.5 \times 19 - 1)\dfrac{942}{n}(n - 40)(500 - 40 - 40)}{\dfrac{300\, n}{2} + (1.5 \times 19 - 1)\dfrac{942}{n}(n - 40) - \dfrac{19 \times 942}{n}(500 - 40 - n)} = (358 + 250 - 40)$$

or $\quad 69000\, n^2 - 50\, n^3 + 10880100\, n - 4.352 \times 10^8$

$$= 85200\, n^2 + 14714040\, n - 5.886 \times 10^8 - 46.764 \times 10^8 - 10166064\, n$$

or $\quad n^3 + 342\, n^2 + 280000\, n - 0.966 \times 10^8 = 0$. Solving this by trial and error with value of $n$ between $0.4\, D$ to $0.7\, D$, we get $n = 235$ mm.

The stress $c$ in the concrete is given by equating the algebraic sum of internal compressive and tensile forces to the external force $P$.

∴ $$c\left\{\frac{bn}{2} + (1.5\, m - 1)A_{sc}\left(\frac{n - d_c}{n}\right) - m\, A_{st}\frac{1}{n}(D - d_t - n)\right\} = P$$

∴ $$c\left\{\frac{300}{2} \times 235 + (1.5 \times 19 - 1)942\left(\frac{235 - 40}{235}\right) - \frac{19 \times 942}{235} \times (500 - 40 - 235)\right\} = 125200$$

From which $c = 3.16$ N/mm². The stress in tensile steel will be

$$t = m \frac{c}{n}(D - d_c - n)$$

$$= \frac{19 \times 3.16}{235}(500 - 40 - 235) = 57.5 \text{ N/mm}^2.$$

The stress in compressive steel is given by

$$\frac{\frac{t_c}{1.5\, m}}{n - d_c} = \frac{c}{n}$$

$$\therefore \quad t_c = 1.5\, m\, (n - d_c)\frac{c}{n}$$

$$= 1.5 \times 19\,(235 - 40) \times \frac{3.16}{235} = 74.7 \text{ N/mm}^2.$$

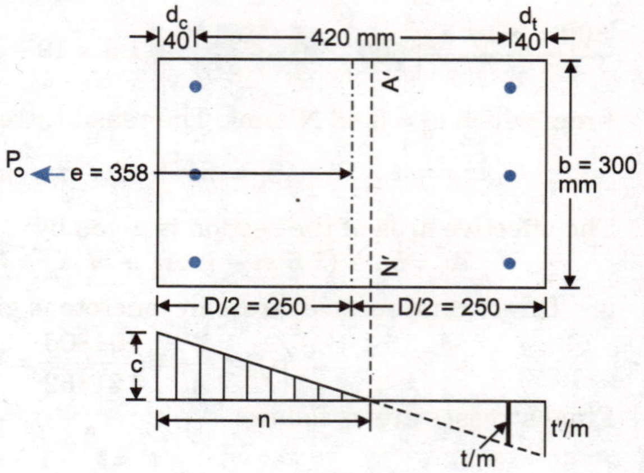

Provide 6 mm Φ ties @ 250 mm c/c.

**6. Design of Foundation**

Load on foundation = $P = 125200$ N. Self wight of foundation @ 10% = 12520 N.

$$\therefore \quad \text{Total } W = 125200 + 12520 = 137720 \text{ N}.$$

B.M. $M = 44.8 \times 10^6$ N-mm. $\therefore$ Eccentricity, $e = \dfrac{44.8 \times 10^6}{137720} \approx 325$ mm

The foundation should be so proportioned that the C.G. of the loads coincides with the C.G. of the footing. Let the width of R.C. footing be 500 mm.

Length required $= \dfrac{137.720}{0.5 \times 120} \approx 2.3$ m.

Hence provide footing 500 mm wide, and 2300 mm long. Also provide a padestal 500 mm wide, 700 mm long and 500 mm high. The footing is shown in Fig. 28.7.

Net upward soil pressure

$$= \frac{125200}{2.3 \times 0.5} = 108869.5 \approx 108870 \text{ N/m}^2.$$

$$\therefore M_{max} = 108870 \times 0.5\,\frac{(1.125)^2}{2}$$

$$= 34450 \text{ N-m}$$

$$= 34.45 \times 10^6 \text{ N-mm}$$

$$\therefore d = \sqrt{\frac{34.45 \times 10^6}{500 \times 0.875}} = 280.6 \approx 280 \text{ mm}$$

Depth required for punching, taking $\sigma_{sp} = 1$ N/mm² is given by

$$2 \times 500\, d \times 1 = 125200.$$

$$\therefore \quad d = 125.2 \approx 125 \text{ mm}.$$

Hence provide an overall depth of 340 mm, with an effective cover of 60 mm, so that available $d = 280$ mm.

$$A_{st} = \frac{34.45 \times 10^6}{140 \times 0.865 \times 280} = 1016 \text{ mm}^2$$

No. of 16 mm Φ bars = 1016 / 201 ≈ 5.
Hence provide 5 bars of 16 mm Φ.

Area of distribution reinforcement @ 0.15% $\dfrac{0.15}{100}(340 \times 1000) = 510$ mm².

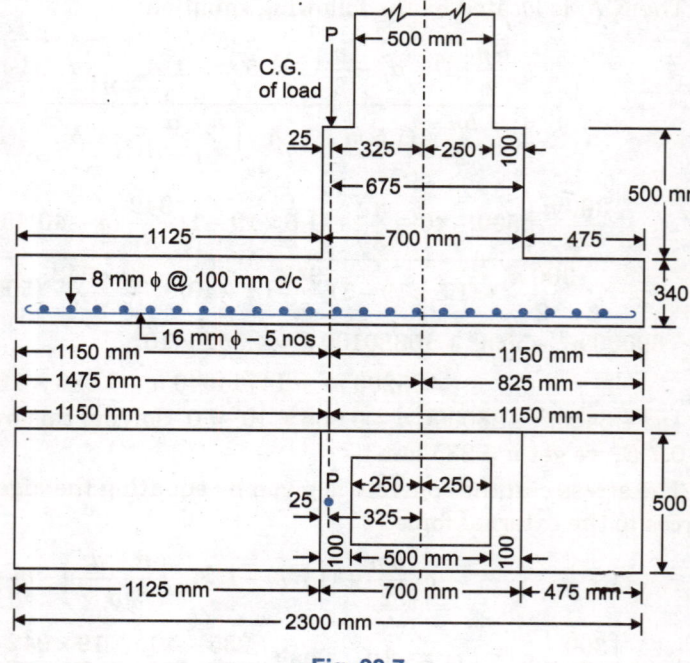

Fig. 28.6

Fig. 28.7

Spacing of 8 mm bars = $\dfrac{1000 \times 50.3}{510} \approx 100$ mm.

The details of reinforcement etc. are shown in Fig. 28.8.

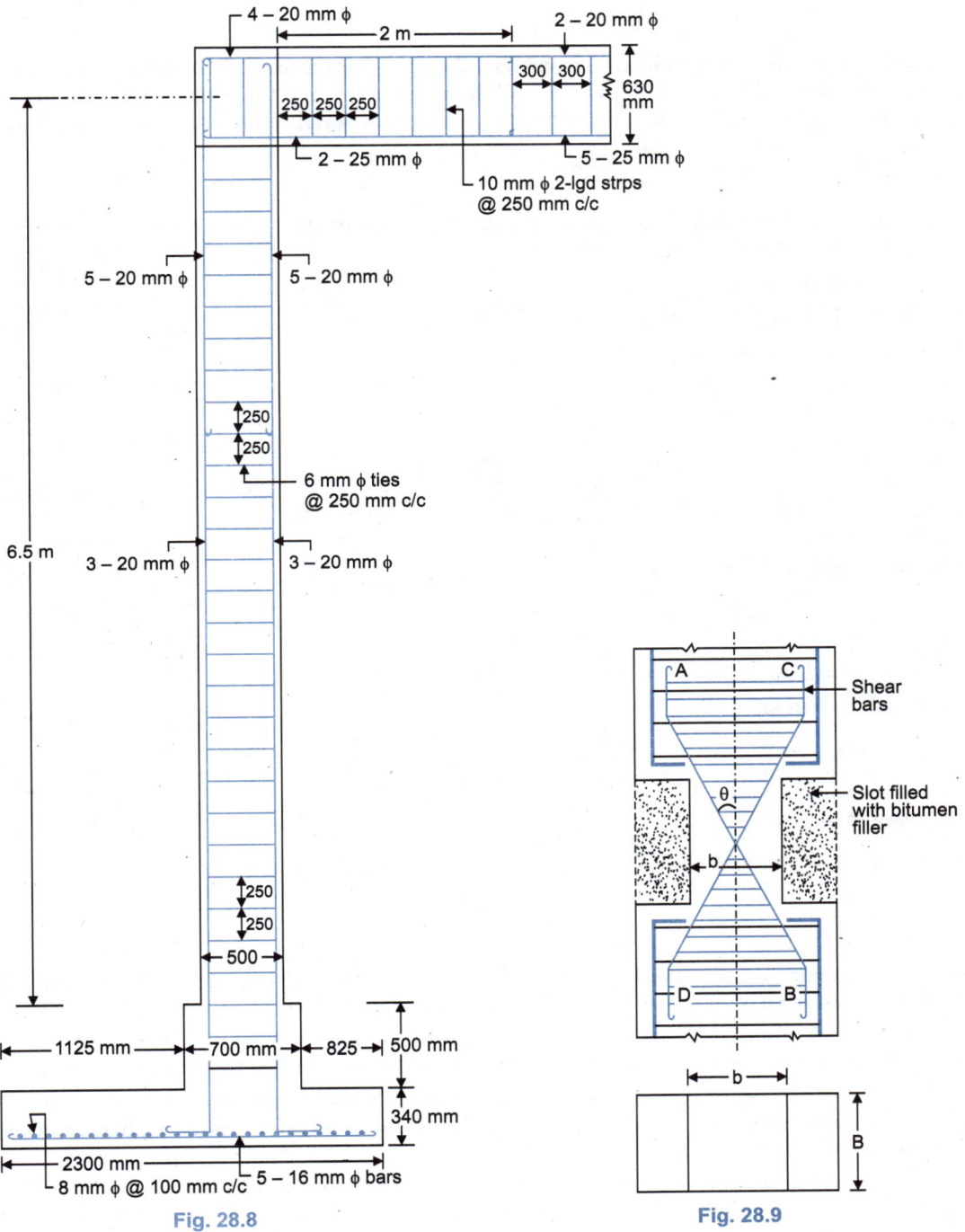

Fig. 28.8

Fig. 28.9

## 28.4. DESIGN OF HINGE AT THE BASE

If the columns of the portal are hinged at the base, the hinges are to be properly designed so that they can transmit the thrust and shearing force, and permit rotation without restraint. Fig. 28.9 shows the details of a hinge at the column base. The width $b$ of concrete at the hinge is found by allowing a high compressive stress equal to half of ultimate compressive stress ($\sigma_{cu}$) so that concrete becomes plastic enough to permit the required angular movements.

Thus, $$b = \frac{P}{B \times \sigma_{cu}/2}$$

Hinge bars or *shear bars* are provided to resist the whole of shear force $F$ at the base by diagonal tension. Out of the two sets of bars $AB$ and $CD$, one set will be in tension while the other set will compression. If $\theta$ is the inclination of the shear bars with vertical, the area of section $A_s$ of one set of bars is given by,

$$A_s = \frac{F}{\sigma_{st} \sin \theta}$$

**Design: Example 28.2.** *The columns of a portal frame are hinged at the base. Design the hinge for the following data :*

(i) *Axial load in the column = 200 kN*
(ii) *Horizontal shear at the base = 60 kN*
(iii) *Size of the column = 500 × 300 mm.*

**Solution:**

Taking $\sigma_{cu} = 15$ N/mm²,

we have $b = \dfrac{P}{B \cdot \sigma_{cu}/2} = \dfrac{200000 \times 2}{300 \times 15} = 88.9$ mm.

Hence provide a hinge of size 300 mm × 100 mm. Provide the height of slot as 150 mm.

Let 20 m be the effective cover to the shear bars

∴ $\tan \theta = \dfrac{50 - 20}{75} = 0.4$

$\theta = 21.8°$   $\sin \theta = 0.3714$

Hence area of steel in shear bar (diagonal bars) is

$A_s = \dfrac{F}{\sigma_{sc} \sin \theta} = \dfrac{60000}{130 \times 0.3714} = 1242.6 \approx 1243$ mm².

No. of 20 mm Φ bars = $\dfrac{1243}{314} \approx 4$.

Hence provide 4 bars on each side. Provide 10 mm Φ closely spaced ties.

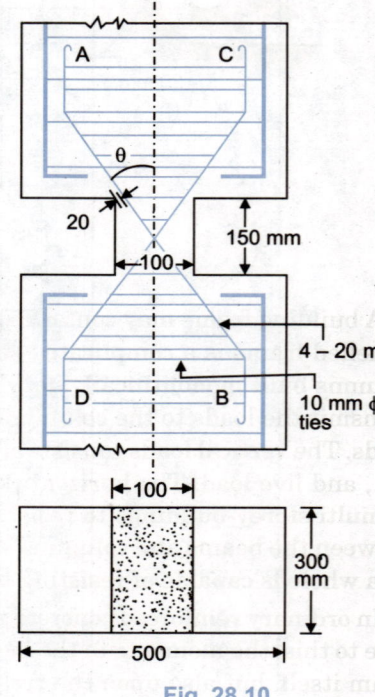

Fig. 28.10

## PROBLEMS

1. A portal has a clear span of 8 m and clear height of 5 m above the base. It carries a uniformly distributed load of 20 kN/m and a central point load of 20 kN. Design the portal if it is fixed at the base. The safe bearing capacity of soil may be taken as 150 kN/m².

2. Redesign the portal of problem 1 if the portal is hinged at the base. Design the hinge also.

3. For a hall 7 m wide and 15 m long, portal frames are provided at 5 m c/c. The height of the portal upto the centre of horizontal member is 5 m. Design the roof and portal frame if it carries a live load of 1.6 kN/m². Assume the safe bearing capacity of soil as 150 kN/m².

4. A portal frame ABCD has fixed supports at 'A' and 'D'. The columns AB and CD are 5 m in height while the transom BC is 10 m in length. The frames are spaced at 3.5 m intervals. The live load on the root slab which 100 mm thick may be taken 1.55 kN/m². Design the transom 'BC' and sketch the details of reinforcement. Adopt $M_{20}$ grade concrete Fe 415 steel.

# BUILDING FRAMES

## 29.1. INTRODUCTION

A building frame may contain a number of bays, and may have several storeys. A multi-storeyed, multi-panelled frame is a complicated statically indeterminate structure. It consists of a number of beams and columns built monolithically, forming a net work. The floors and the walls are supported on beams which transmit the loads to the columns. A building frame is subjected to both the vertical as well as horizontal loads. The vertical loads consists of the dead weight of structural components such as beams, slabs, columns etc., and live load. The horizontal loads consist of the wind forces and the earthquake forces. The ability of multi-storey-buildings to resist wind and other lateral forces depends upon the rigidity of connections between the beams and columns. When the connections of beams and columns are fully rigid, the structure as a whole is capable of resisting the lateral forces acting on the structure.

In ordinary reinforced concrete skeleton buildings, a continuous beam is rigidly connected with columns. Due to this, the moments in the beam depend not only upon the number and length of spans composing the beam itself, but also upon the rigidity of columns with which it is connected. The bending moment at the end of any one span of the continuous beam cannot be transferred to the beam in the next span without subjecting the columns to bending. Instead of transmitting the bending moment in full to the beam in the next span, part of the moment is transferred to the columns above and below the beam, and the balance to the beam. Due to this, the effect of loading on one span upon the other spans is much lower than in ordinary continuous beams which are not connected to the columns.

There are several methods of analysis such as moment distribution, Kani's rotation contribution, Takabeya's method and matrix methods.

Matrix methods with use of computer are ideally suited for the analysis of multistorey frames with large number of redundants.

## 29.2. SUBSTITUTE FRAMES

The analysis of a multi-storeyed multi-panelled building frame is very cumbersome, since the frame contains a number of continuous beams and columns. As stated in the previous article, the effect of loading on the span upon other spans is much smaller. The moments in any beam or column are mainly due to the loads on spans very close to it. Loads on distant spans do not have appreciable effect. Due to this, a simple method of analysis, accurate enough for practical purpose, is used by analysing a small portion of the frame, called '*substitute frame*' rather than analysis of the whole frame.

**714** REINFORCED CONCRETE STRUCTURE

It has been found by exact analysis that the moments carried from floor to floor, through columns, are very small in comparison to the beam moments. In other words, the moments in one floor have negligible effect of the moments of the floors above and below it. Therefore, a substitute frame consists of one floor, connected above and below with their far end either hinged or fixed or restrained. Fig. 29.1(a) shows a building frame consisting of five storeys and three bays. Fig. 29.1(b) shows the substitute frame for determining bending moment in the second floor. Generally, it is sufficient to consider two adjacent spans on each side of joint considered. The substitute frame gives the results which are safe for all practical purposes.

**Types of Substitute Frames:** Under ordinary conditions, the following four types of substitute structures are considered sufficient:

(a) three-span structure with two storey columns
(b) substitute frame for wall columns
(c) substitute frame for two panel wide building
(d) substitute frame for one panel wide building.

Fig. 29.2(a) shows the most general substitute frame consisting of *three span, two-storey substitute structure* with irregular spacing of columns. Fig. 29.2(b) shows the substitute frame for finding the bending moments in wall columns. This consists of three spans and three two-storey columns, one of which is the wall column. Fig. 29.2(c) shows the substitute frame for structures two panels wide. Fig. 29.2(d) shows the substitute frame for one span multi-storey frame.

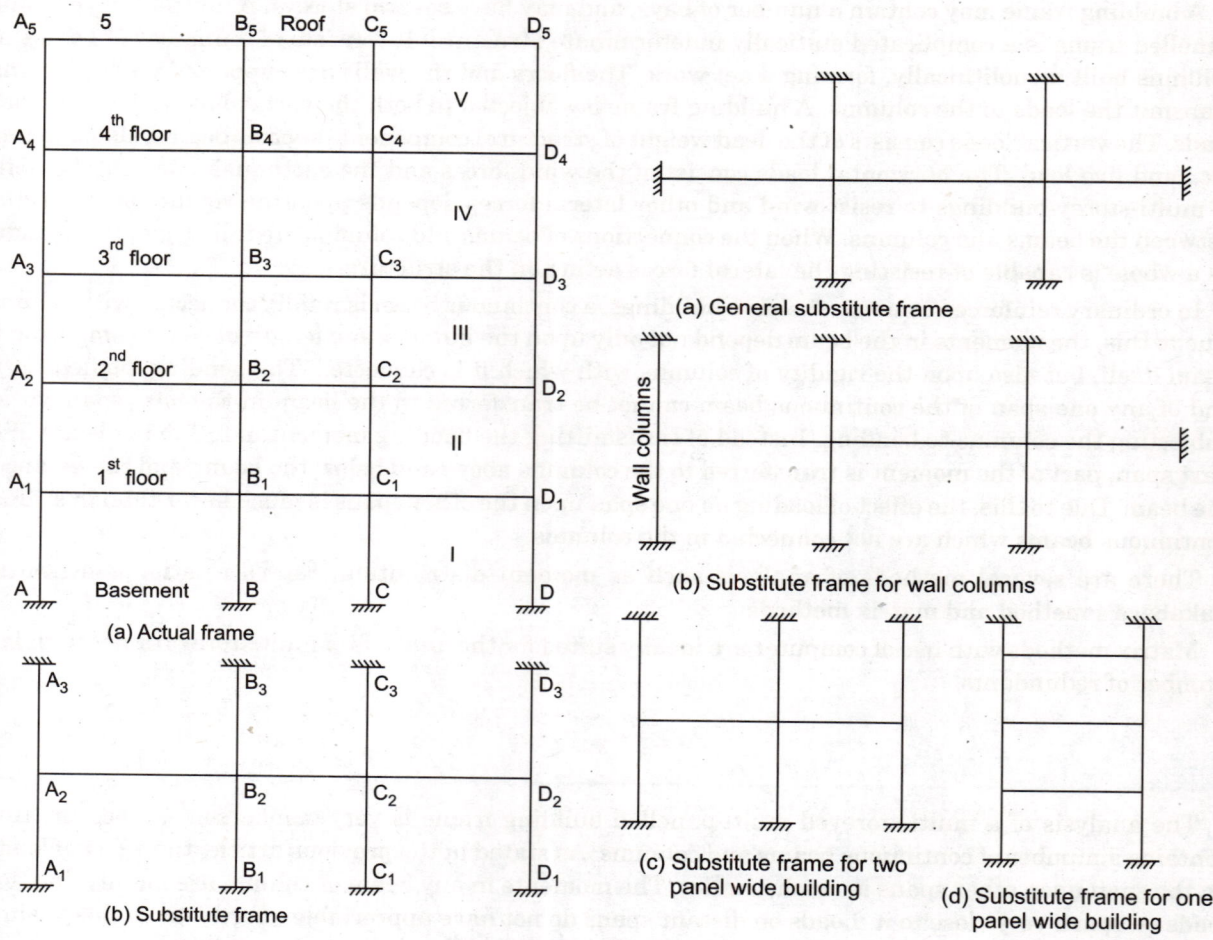

Fig. 29.1

Fig. 29.2. Various types of substitute frames.

### End conditions for substitute frames

The restraining effect of any one member, upon other members forming a joint depends also upon the condition existing on the other end of the restraining member. The other end may have three conditions: (*i*) free to turn (*i.e.*, hinged), (*ii*) partially restrained, or (*iii*) rigidly fixed. The restraining effect is largest for the rigidly fixed conditions of the end and smallest for free end. It should be noted that the restraining effect of a fixed member is one third larger than the restraining effect if it were free to turn. The rigidity of any member is expressed by the ratio $I/l$ where $I$ is its moment of inertial and $l$ is its length (for beam) or height (for column). If the loaded member has rigidity $I/l$, and the restraining member has rigidity $I_1/l_1$ then this restraining member is considered as fixed at the other end if $\dfrac{I_1}{l_1} \div \dfrac{I}{l}$ is equal to or greater than 10. The end of a member is considered as partly restrained when it runs into another joint composed of several members, a condition which is often found in concrete skeleton structure. No restraint exists if $\dfrac{I_1}{l_1} \div \dfrac{I}{l} = 0$. The outer ends of the member of the substitute frame are sometimes taken as hinged (except for columns fixed in the ground). This gives severset condition for a particular reaction under investigation. The moments obtained by assuming the ends hinged gives the moments nearest to the value obtained from full frame analysis and compensates to some extent for the error caused due to neglecting loads on members of distant span.

## 29.3. ANALYSIS FOR VERTICAL LOADS

A building frame is a three dimensional structure consisting of a number of bays in two directions at right angles to each other. A building structure may be assumed to be consisting of two sets of plane frames crossing each other at right angles. The vertical members (*i.e.*, columns) are common to both these sets of frames. Each set of frames are analysed separately. Since moments in the vertical members occur in two planes, the stresses in columns should be found for moments acting in two planes simultaneously and the corresponding vertical loads.

### (a) Maximum bending moments in beams

The magnitude of bending moments in beams and columns respectively depend upon their relative rigidity *Generally*, the beams are made of the same dimensions in all floors, while the dimensions of column vary from storey to storey. Columns have smallest dimensions at the top and largest dimensions at the bottom. Due to this reason, the ratio of the rigidity of the beam to that of the column is larger in the upper floors than in the lower floors. The positive bending moments in the beams increase with decrease of the rigidity of the columns, while the negative B.M. in them increase with the increase in the rigidity of the columns. Due to this, the positive B.M. are largest in the upper storeys where the columns are least rigid and the negative bending moments are maximum in the lower storeys where the columns are rigid.

In order to find the maximum moment in a given span of the beam, the substitute frame is so selected that span under investigation forms the centre span. This substitute frame may be moved from floor to floor. However, since the beams in all floors are made of the same dimensions and provided with same amount of steel, only one

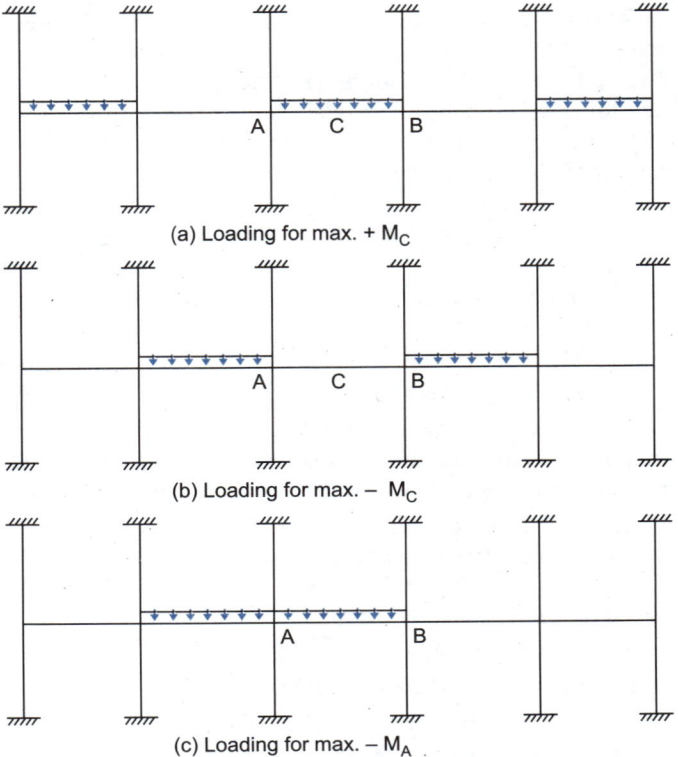

Fig. 29.3. Loading for bending moments in beams.

substitute frame may be sufficient when placed in a position in the structure for which the bending moments are the largest. The beam should be loaded with live loads as follows for maximum effects:

*(i) For maximum positive B.M.* At the mid-point C of a span AB, the loads should be placed on the span and on alternative spans, as shown in Fig. 29.3(a).

*(ii) For maximum negative B.M.* At the mid-point C of a span AB, the span AB should be unloaded while load should be placed on spans adjacent to the span under consideration, as shown in Fig. 29.3 (b).

*(iii) For maximum negative B.M.* At the support A, loads should be placed on the two spans adjacent to the support, as shown in Fig. 29.3(c).

When the spans of beams are not equal, substitute frames should be selected in which the largest span forms the centre span and also frames in which the smallest span forms the centre span. Several trial computations may be necessary to get the frame for which the bending moments are maximum. To get bending moment in the wall columns and wall beams, substitute frame shown in Fig. 29.2(b) should be used.

The bending moments due to dead loads are found separately. The bending moments for dead and live loads are then added, and the beam is designed.

### (b) Maximum bending moment in columns

The bending moments in columns increase with increase in their rigidity. Hence they are largest in the lower storeys, and smallest in the upper storeys. The maximum compressive stresses in columns is found by combining maximum vertical loads with maximum bending moments. The maximum tensile stresses in columns is found by combining the maximum bending moment with minimum vertical loads. Though the bending moment is smallest in the upper floors, its effect is much larger since the dimensions of the columns are the smallest there and also the vertical loads are much smaller than in lower storeys. Also the possibility of tensile stresses in columns is much larger in upper storeys than in lower storeys.

The maximum moments is columns occur when alternative spans are loaded as shown in Fig. 29.4(a),(b). The corresponding axial loads are found. The column is designed to resist the stresses provided by every combination of axial load and the corresponding moment.

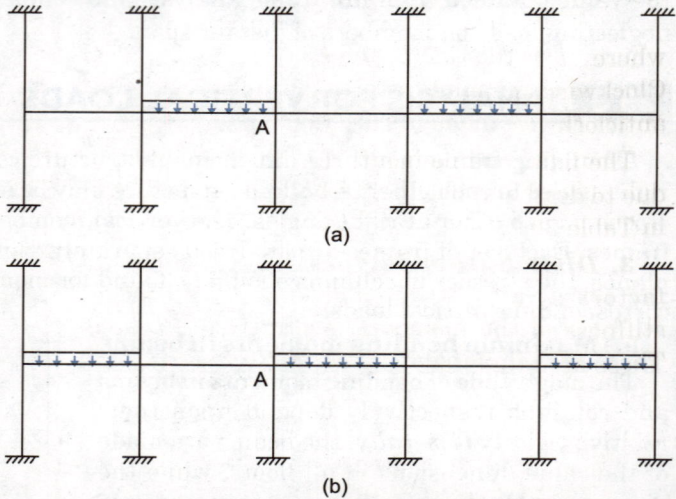

Fig. 29.4. Loading for max. B.M. at column A.

## 29.4. METHODS OF COMPUTING B.M.

The bending moments in the beams and columns of a substitute frame may be computed by the following methods:

1. Slope-deflection method.
2. Moment distribution method.
3. Building frame formulae.
4. Kani's method.

The slope deflection method results in too many equations to be solved simultaneously. Hence moment distribution method, using two cycles is used. Taylor, Thomson and Smulski recommend the use of building frame formulae which they have developed using slope deflection equations.

**Example 29.1.** *Fig. 29.5 shows an intermediate frame of a multistoreyed building. The frames are spaced at 4 metres centre to centre. Analyse the frame taking live load of 4000 N/m² and dead load as 3000 N/m², 3250 N/m² and 2750 N/m² for the panels AB, BC and CD respectively. The self weight of the beams may be taken as under:*

   Beams of  7 m span     : 5000 N/m
   Beams of  5 m span     : 3500 N/m
   Beams of  3.5 m span  : 2500 N/m.

*The relative stiffness of the members are marked on the figure itself.*

# BUILDING FRAMES

**Solution:**

**1. Substitute frame:** Let us analyse the second floor *ABCD*. The substitute frame is shown in Fig. 29.6, assuming the far ends of the columns fixed. Other floors can be analysed in a similar manner.

**2. Loading and fixed end moments:** Since the frames are spaced @ 4 m c/c, the live loads transferred from the floors will be 4000 × 4 = 16000 N/m.

The total dead load on a beam will be equal to the dead load from the floors plus the dead load due to the self-weight of the beam. Thus, the total dead load on beam *AB*, per metre run = (3000 × 4) + 5000 = 17000 N/m. Dead loads for other beams are tabulated in Table 29.1. The fixed end moment is calculated from the following expressions:

$$M_F = \pm \frac{wL^2}{12}$$

where $w$ is the U.D.L. and $L$ is span of the beam. Clockwise moments are taken as positive and anticlockwise moments are taken as negative.

The fixed end moments due to dead load, and due to dead and live load combined are tabulated in Table 29.1.

**3. Distribution factors:** The distribution factors at a joint depends upon the relative stiffness of the members meeting at the joints. These are tabulated in Table 29.2.

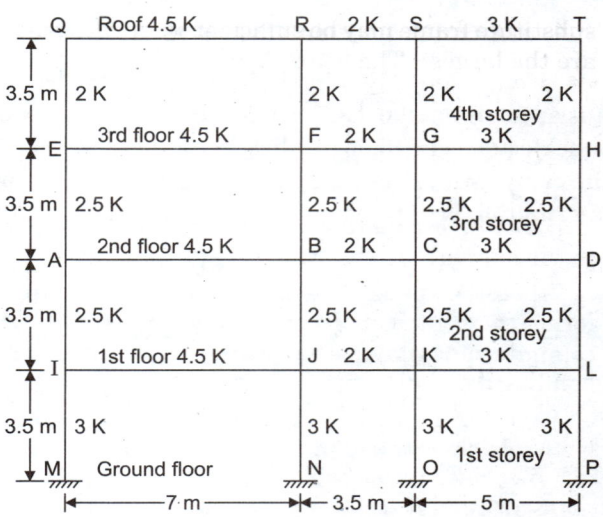

Fig. 29.5

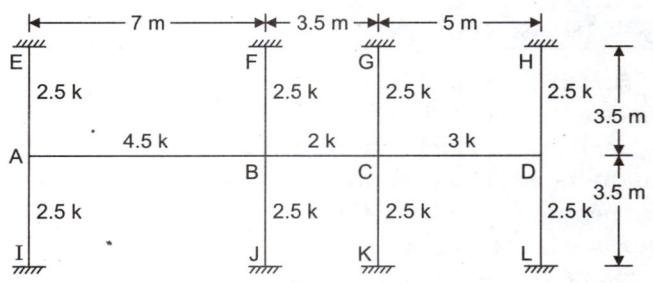

Fig. 29.6

**TABLE 29.1.** Loading and Fixed End Moments

| Member | Dead load (N/m) | Live load (N/m) | F.E.M. due to dead load (N-m) | F.E.M. due to dead and live load combined (N-m) |
|---|---|---|---|---|
| AB | 17000 | 16000 | 69420 | 134750 |
| BC | 15500 | 16000 | 15820 | 32160 |
| CD | 14500 | 16000 | 30210 | 63540 |

**TABLE 29.2.** Distribution Factors

| Joint | Members | Relative stiffness | Sum | Distribution factors |
|---|---|---|---|---|
| A | AE | 2.5 K | | 0.263 |
|   | AI | 2.5 K | 9.5 K | 0.263 |
|   | AB | 4.5 K | | 0.474 |
| B | BA | 4.5 K | | 0.392 |
|   | BF | 2.5 K | | 0.217 |
|   | BC | 2.0 K | 11.5 K | 0.174 |
|   | BJ | 2.5 K | | 0.217 |
| C | CB | 2.0 K | | 0.200 |
|   | CG | 2.5 K | | 0.250 |
|   | CD | 3.0 K | 10 K | 0.300 |
|   | CK | 2.5 K | | 0.250 |
| D | DC | 3.0 K | | 0.375 |
|   | DH | 2.5 K | 8 K | 0.3125 |
|   | DL | 2.5 K | | 0.3125 |

## (A) MAXIMUM NEGATIVE B.M. AT SUPPORTS

**4. Maximum negative B.M. at joint A:** The condition of loading to obtain max. negative B.M. at joint A is as follows: Live load on AB only, while the dead load is on AB and CD. The effect of load on other spans is neglected. The moment distribution is carried out in Table 29.3. The distribution is done at joint B and the carry over effect (*i.e.*, half the moment) is transferred to joint A. After adding the total moments at A, distribution is done at A. The distribution at joint B has not been recorded in the Table 29.3.

**TABLE 29.3.** Moment Distribution for – ve B.M. at A

| Joint | A | B | | C | | D |
|---|---|---|---|---|---|---|
| Member | AB | BA | BC | CB | CD | DC |
| D.F. | 0.474 | 0.392 | 0.174 | 0.20 | 0.30 | 0.375 |
| 1. F.E.M. due to D.L. | | | – 15820 | + 15820 | | |
| 2. F.E.M. due to total load | – 134750 | + 134750 | | | | |
| 3. Distribution at B and carry over to A | – 23310 | | | | | |
| 4. Add (2) and (3) | – 158060 | | | | | |
| 5. Distribution | + 74920 | | | | | |
| 6. Total sum of (4) and (5) | – 83140 | | | | | |

**5. Maximum negative B.M. at joint B:** The loading conditions are : Live load on AB and BC, while dead load on whole of ABCD. The Moment distribution is carried out in Table 29.4. In the first cycle, joints A and C are balanced, and half the moments are carried over to joint B for the beams BA and BC respectively. In the second cycle, joint B is balanced and final moments are found. Thus, in the first cycle, unbalanced moment at C is + 1950, the distributed moment to CB will be 1950 × 0.2 ≈ – 400, the carry over moment at B = – 400/2 = – 200.

Similarly, the unbalanced moment at A is – 134750, the distributed moment of AB = + 134750 × 0.474 = + 63880, the carried over moment to B = + 31940. The total moments at BA and BD will be + 166690 and – 32260, leaving an unbalanced moment of + 134330. The distributed moments to BA and BC will be – 134330 × 0.392 = – 52660 and – 134330 × 0.174 = – 23370 respectively.

**TABLE 29.4.** Moment Distribution for – ve B.M. at B

| Joint | A | B | | C | | D |
|---|---|---|---|---|---|---|
| Member | AB | BA | BC | CB | CD | DC |
| D.F. | 0.474 | 0.392 | 0.174 | 0.20 | 0.30 | 0.375 |
| 1. F.E.M. due to D.L. | | | | | +30210 | |
| 2. F.E.M. due to total load | – 134750 | +134750 | – 32160 | +32160 | | |
| 3. Distribution at A and C and carry over to B | | +31940 | – 200 | | | |
| 4. Add (2) and (3) | | +166690 | – 32360 | | | |
| 5. Distribution | | – 52660 | – 23370 | | | |
| 6. Total sum of (4) and (5) | | +114030 | – 55730 | | | |

**6. Maximum negative B.M. at C:** The conditions of loadings are : Live load on BC and CD, and dead load on ABCD. In the first cycle, joints B and D are balanced and effects are carried over to C. In the second cycle, joint C is balanced, as shown in Table 29.5.

**7. Maximum negative B.M. at D:** The conditions for loadings are : Live load on CD and dead load on ABCD. In the first cycle, joint C is balanced and its effect is carried over to D. In the second cycle, joint D is balanced as shown in Table 29.6.

**TABLE 29.5.** Moment Distribution for – ve B.M. at C

| Joint | A | B | | C | | D |
|---|---|---|---|---|---|---|
| Member | AB | BA | BC | CB | CD | DC |
| D.F. | 0.474 | 0.392 | 0.174 | 0.20 | 0.30 | 0.375 |
| 1. F.E.M. due to D.L. | – 69420 | +69420 | | | | |
| 2. F.E.M. due to total load | | | – 32160 | +32160 | – 63540 | +63540 |
| 3. Distribution at B and D carry over to C | | | | – 3240 | – 11910 | |
| 4. Add (2) and (3) | | | | +28920 | – 75450 | |
| 5. Distribution | | | | +9310 | +13960 | |
| 6. Total sum of (4) and (5) | | | | +38230 | – 61490 | |

**TABLE 29.6.** Moment Distribution for – ve B.M. at D

| Joint | A | B | | C | | D |
|---|---|---|---|---|---|---|
| Member | AB | BA | BC | CB | CD | DC |
| D.F. | 0.474 | 0.392 | 0.174 | 0.20 | 0.30 | 0.375 |
| 1. F.E.M. due to D.L. | | | – 15820 | +15820 | | |
| 2. F.E.M. due to total load | | | | | – 63540 | +63540 |
| 3. Distribution at C and carry over to D | | | | | | +7160 |
| 4. Add (2) and (3) | | | | | | + 70700 |
| 5. Distribution | | | | | | – 26510 |
| 6. Total sum of (4) and (5) | | | | | | +44190 |

### (B) MAXIMUM POSITIVE B.M. AT MID-SPANS

**8. Maximum +ve B.M. in mid-span of AB:** The conditions of loadings are :

Live load on AB and CD, and Dead load on ABCD. In the first cycle, distribution is performed at joints A, B and C. Half of these distributed moments are carried over to the opposite ends, *i.e.* from A to B and B to A, and from C to B. In the second cycle, distribution is performed at A and B as illustrated in Table 29.7. Thus, the end moments at A and B for beam AB are – 83140 and + 105680 respectively. The free B.M. at mid-span of $AB = \dfrac{wL^2}{8} = \dfrac{(17000 + 16000)(7)^2}{8} \approx 202120$ N-m

$\therefore$ Net B.M. at centre of $AB = 202120 - \dfrac{(83140 + 105680)}{2} = 107710$ N-m

**9. Maximum +ve B.M. in mid-span of BC:** The conditions for loadings are : Live load on BC and dead load on ABCD. In the first cycle, moments are distributed at A, B, C and D. These distributed moments are carried over from A to B, from B to C, from C to D and from D to C. Finally, the moment are distributed at joints B and C, as shown in Table 29.8.

**TABLE 29.7.** Moment Distribution for + ve B.M. at AB

| Joint | A | B | | C | | D |
|---|---|---|---|---|---|---|
| Member | AB | BA | BC | CB | CD | DC |
| D.F. | 0.474 | 0.392 | 0.174 | 0.20 | 0.30 | 0.375 |
| 1. F.E.M. due to D.L. | | | – 15820 | +15820 | | |
| 2. F.E.M. due to total load | – 134750 | +134750 | | | – 63540 | +63540 |
| 3. Distribution at A, B and C | +63870 | – 46620 | +4770 | | | |
| 4. Carry over | – 23310 | +31940 | | +9540 | | |
| 5. Distribution at A and B | +11050 | – 14390 | | | | |
| 6. Total moments (sum of 2, 3, 4, 5) | – 83140 | +105680 | | | | |

**TABLE 29.8.** Moment Distribution for – ve B.M. at BC

| Joint | A | B | | C | | D |
|---|---|---|---|---|---|---|
| Member | AB | BA | BC | CB | CD | DC |
| D.F. | 0.474 | 0.392 | 0.174 | 0.20 | 0.30 | 0.375 |
| 1. F.E.M. due to D.L. | – 69420 | +69420 | | | – 30210 | +30210 |
| 2. F.E.M. due to total load | | | – 32160 | +32160 | | |
| 3. Distribution at A, B, C and D | +32900 | | – 6480 | – 390 | | – 11330 |
| 4. Carry over | | +16450 | – 200 | – 3240 | – 5660 | |
| 5. Distribution at B and C | | | – 2830 | +1780 | | |
| 6. Total moments | | | – 41670 | +30310 | | |

Thus, the end moments at B and C are – 41670 and + 34190 respectively. The free B.M. at the centre of span BC is $= \dfrac{wL^2}{8} = \dfrac{(15500 + 16000)}{8}(3.5)^2 \approx 48230$

Net B.M. at centre of BC $= 48230 - \dfrac{41670 + 30310}{2} = 12240$ N-m

**10. Maximum +ve B.M. in mid-sapn of CD:** Conditions of loadings are : Live load on CD and AB and dead load on ABCD. In the first cycle, moment distribution is done at joints B, C and D, and half the distributed moments are carried over to the opposite ends, i.e., from D to C and C to D, and from B to C. In the second cycle, distribution is performed at C and D, as illustrated in Table 29.9.

Thus the end moments at C and D are – 54450 and +44200 respectively. The free B.M. at the centre of span CD

$$= \dfrac{wL^2}{8} = \dfrac{(14500 + 16000) \times 5^2}{8} \approx 95310 \text{ N-m}$$

∴ Net B.M. at the centre of CD $= 95310 - \dfrac{54450 + 44200}{2} \approx 45980$ N-m

**TABLE 29.9.** Moment Distribution for – ve B.M. at CD

| Joint | A | B | | C | | D |
|---|---|---|---|---|---|---|
| Member | AB | BA | BC | CB | CD | DC |
| D.F. | 0.474 | 0.392 | 0.174 | 0.20 | 0.30 | 0.375 |
| 1. F.E.M. due to D.L. | | | – 15820 | +15820 | | |
| 2. F.E.M. due to total load | – 134750 | +134750 | | | – 63540 | +63540 |
| 3. Distribution at B, C and D | | | – 20690 | | +14320 | – 23820 |
| 4. Carry over | | | | – 10350 | – 11910 | +7160 |
| 5. Distribution at C and D | | | | | + 6680 | – 2680 |
| 6. Total moments | | | | | – 54450 | +44200 |

**(C) MAXIMUM NEGATIVE B.M. AT CENTRE OF SPANS**

**11. Maximum Negative B.M. at the centre of span BC**

The condition for loadings are : Live loads on AB and CD, and dead load on ABCD.

In the first cycle, moment distribution is carried out at all the four joints A, B, C and D. These moments are then carried over to joints B and C from joints A and B, as well as between themselves. The second distribution is carried out at joints B and C, as shown in Table 29.10.

BUILDING FRAMES 721

TABLE 29.10. Moment Distribution for – ve B.M. at Centre of Span BC

| Joint | A | B | | C | | D |
|---|---|---|---|---|---|---|
| Member | AB | BA | BC | CB | CD | DC |
| D.F. | 0.474 | 0.392 | 0.174 | 0.20 | 0.30 | 0.375 |
| 1. F.E.M. due to D.L. | | | – 15820 | +15820 | | |
| 2. F.E.M. due to total load | – 134750 | + 134750 | | | – 63540 | +63540 |
| 3. Distribution at A, B, C and D | +63880 | – 46620 | – 20700 | +9540 | + 14320 | – 23820 |
| 4. Carry over to B and C | | +31940 | +4770 | – 10350 | – 11910 | |
| 5. Distribution at B and D | | | – 6380 | +4450 | | |
| 6. Final moments | | | – 38130 | +19460 | | |

Thus, the end moment at B and C are – 38130 and +19460 respectively. Free B.M. at the centre of span

$BC$ = Dead load intensity × $\dfrac{L^2}{8}$ = $\dfrac{15500\,(3.5)^2}{8}$ ≈ 23730

Net B.M. at centre of $BC$ = $23730 - \dfrac{38130 + 19460}{2}$ = $- 5065$.

**12. Maximum Negative B.M. at centre of spans AB and CD:** Since spans AB and CD are large, free B.M. at their mid-span will be large. It will be seen that the net B.M. at the centre of these spans will either be positive, or will be negative but of negligibly small magnitude. Due to this reason, these spans are not being investigated for max. negative B.M. However, conditions for max. negative B.M. at the centre of span AB will be when live load is on BC and dead load is on ABCD. Similarly, the loading condition for max. negative B.M. at centre of CD will be when span BC is loaded with live load, and dead load is on ABCD.

**(D) BENDING MOMENTS IN COLUMNS:** For maximum B.M. in columns, alternate spans should be loaded with live load, while the whole floor is loaded with dead load.

The two possible load conditions are shown in Fig. 29.4. For the present case, the loadings will be as shown in para (13) and (14) below. See Tables 29.11 and 29.12.

**13. Max. B.M. in columns:** The loading conditions are : live load on AB and CD and dead load on ABCD. The moment distribution is carried out as illustrated in Table 29.11. In the first cycle, distribution is done at all the four joints A, B, C and D. The carry over moments are then transferred to the appropriate points. These carried-over moments are added to the original F.E.M. to get new moments at each joint. These new moments are distributed to the columns meeting at the joints.

TABLE 29.11. Bending Moment in Columns

| Joint | A | B | | C | | D |
|---|---|---|---|---|---|---|
| Column D.F. | | | | | | |
| (a) Just above floor | 0.263 | 0.217 | | 0.25 | | 0.3125 |
| (b) Just below floor | 0.263 | 0.217 | | 0.25 | | 0.3125 |
| Horizontal members | AB | BA | BC | CB | CD | DC |
| D.F. | 0.474 | 0.392 | 0.174 | 0.20 | 0.30 | 0.375 |
| 1. F.E.M. due to D.L. | | | – 15820 | + 15820 | | |
| 2. F.E.M. due to total load | – 134750 | +134750 | | | – 63540 | +63540 |
| 3. Distribution | +63880 | – 46620 | – 20700 | +9540 | +14320 | – 23820 |
| 4. Carry over | – 23310 | +31940 | +4770 | – 10350 | – 11910 | +7160 |
| 5. New moments (total 1, 2, 4) | – 158060 | +166690 | –11050 | +5470 | – 75450 | +70700 |
| 6. Distribution to Columns | | | | | | |
| (i) just above floor | +41570 | – 33770 | | +17500 | | – 22160 |
| (ii) just below floor | +41570 | – 33770 | | +17500 | | – 22160 |

**TABLE 29.12.** Bending Moment in Columns

| Joint | A | B | | C | | D |
|---|---|---|---|---|---|---|
| Column D.F. | | | | | | |
| (a) Just above floor | 0.263 | 0.217 | | 0.25 | | 0.3125 |
| (b) Just below floor | 0.263 | 0.217 | | 0.25 | | 0.3125 |
| Horizontal members | AB | BA | BC | CB | CD | DC |
| D.F. | 0.474 | 0.392 | 0.174 | 0.20 | 0.30 | 0.375 |
| 1. F.E.M. due to D.L. | −69420 | +69420 | | | −30210 | +30210 |
| 2. F.E.M. due to total load | | | −32160 | +32160 | | |
| 3. Distribution | +32900 | −14600 | −6480 | −400 | −580 | −11320 |
| 4. Carry over | −7300 | +16450 | −200 | −3240 | −5660 | −290 |
| 5. New moments (total 1, 2, 4) | −76720 | +85870 | −32360 | +28920 | −35870 | +29920 |
| 6. Distribution to Columns | | | | | | |
| (i) just above floor | +20180 | −11610 | | +1740 | | −9350 |
| (ii) just below floor | +20180 | −11610 | | +1740 | | −9350 |

**14. Max. B.M. in columns: Alternative loading :** The loading conditions are: live load on BC and dead load on ABCD. The moment distribution is carried out as illustrated in Table 29.12. In the first cycle, distribution is done at all the four joints. The carry over moments are then transferred to appropriate points. These carried over moments are added to the original F.E.M. to get new moments at each joint. These new moments are distributed to the columns meeting at the joints.

## 29.5. ANALYSIS OF FRAMES SUBJECTED TO HORIZONTAL FORCES

A building frame is subjected to horizontal forces due to wind pressure and seismic effects. These horizontal forces cause axial forces in columns and bending moment in all the members of the frame. As stated earlier, a building frame is a highly indeterminate structure. The degree of indeterminacy of a building bent (Fig. 29.7) is found by providing a cut near mid-span of each beam. Each cut beam will thus have three unknown reaction components: moment ($M$), shear ($F$) and axial thrust ($H$). Each column with its cut beams will act as a cantilever, which is in a statically determinate structure. Thus, if $n$ is the number of beams in a bent, the degree of indeterminancy will be $3n$. For the building bent shown in Fig. 29.7, there are eight beams and hence the bent is statically indeterminate upto 24th degree. An ordinary 20 storey building with 20 storeys and 5 stacks of columns has 80 beams, thus having the degree of indeterminacy of 240.

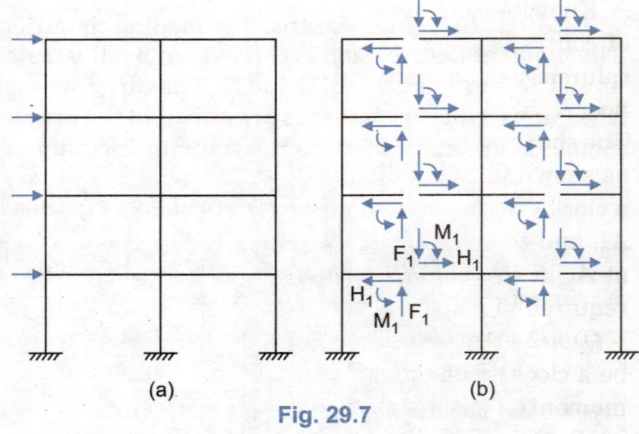

Fig. 29.7

Due to this reason, suitable assumptions are made so that the frame subjected to horizontal forces can be analysed by using simple principles of mechanics. Following approximate methods are commonly used for the analysis of building frames subjected to lateral forces:

1. Portal method.  2. Cantilever method.  3. Factor method.

## 29.6. PORTAL METHOD

For the purposes of analysis it is assumed that the horizontal forces are acting on the joints. The portal method is based on the following two important assumptions:

(i) the points of contraflexure in all the members lie at their mid-points and

(ii) horizontal shear taken by each interior column is double the horizontal shear taken by each of exterior column.

Fig. 29.8 shows a three storey building frame with three spans. Let $P_1, P_2, P_3$ be the external horizontal forces acting at the joints of the wall columns. Under the action of the horizontal forces, the frame will deflect. The point of contraflexure will lie at the middle of each member. Only horizontal shears will act at these points of contraflexure, since B.M. will be zero at these points.

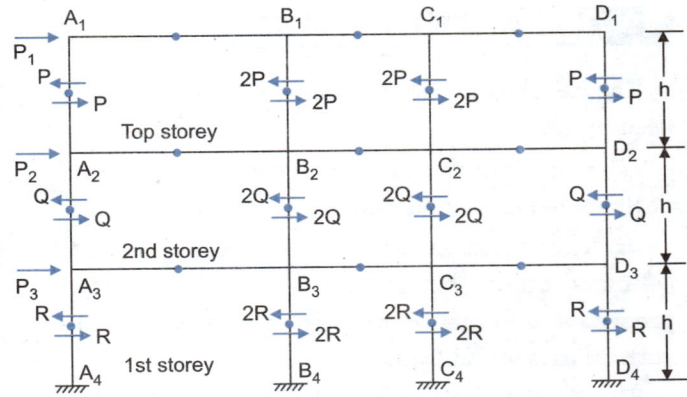

Fig. 29.8. Portal method of analysis for wind loads

Consider the top storey having vertical members $A_1A_2, B_1B_2, C_1C_2$ and $D_1D_2$. The horizontal shear for the outer columns $A_1A_2$ and $D_1D_2$ will be $P$ each while that for the inner columns $B_1B_2$ and $C_1C_2$ will be $2P$ each, as marked.

The value of $P$ is given by

$$P_1 = P + 2P + 2P + P$$

∴ $$P = \frac{1}{6} P_1$$

Similarly, consider the second storey, where the exterior columns $A_2A_3$ and $D_2D_3$ have shear $Q$. The value of shear $Q$ is found by :

$$P_1 + P_2 = Q + 2Q + 2Q + Q \qquad ∴ \qquad Q = \frac{1}{6}(P_1 + P_2)$$

Similarly, for the bottom storey, the shear $R$ is given by

$$P_1 + P_2 + P_3 = R + 2R + 2R + R \qquad ∴ \qquad R = \frac{1}{6}(P_1 + P_2 + P_3)$$

Knowing the horizontal shears at the points of contraflexure, the bending moments in the columns can be easily found. Let us consider the floor $A_2 B_2 C_2 D_2$ between third and second storey. The shear acting at the point of contraflexure are as shown in Fig. 29.9. The joint $A_2$ is subjected a clockwise moment of $P\,h/2$ at $A_2$ in column $A_1 A_2$, and to a clockwise moment equal to $Q\,h/2$ at $A_2$ in column $A_2A_3$. The beam $A_2 B_2$ is thus required to resist a clockwise moment of $m = (P + Q)\,h/2$ at $A_2$. Similarly, at joint $B$, there will

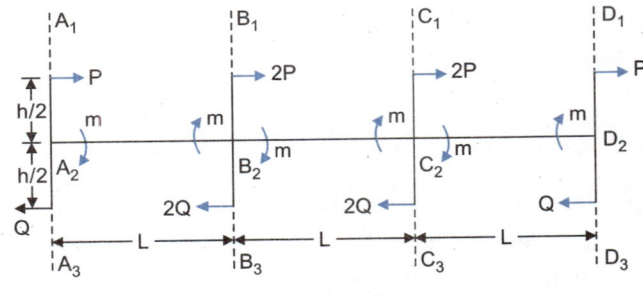

Fig. 29.9

be a clockwise moment equal to $(2P + 2Q)\,h/2$. But there are two beams to resist this. Hence clockwise moment in each beam will be $(P + Q)\,h/2$. Thus, the ends of each beam receive the same clokwise moment of $(P + Q)\,h/2$, with the result that points of contraflexure will lie in the middle of the beams.

The moment $m$ acting at each end of the beams $A_2B_2, B_2C_2, C_2D_2$, give rise to vertical reactions in columns. If $L$ is the span of these beams, each beam will impose an upward pull of $2m/L$ on windward column and a push of $2m/L$ on leeward column connected to the beam, for each span. The vertical reactions will neutralize for any intermediate column, provided span of beams on either side are equal. Only the end columns will experience vertical reactions. The windward column will have an upward pull of $2m/L$ and the leeward column will have a downward push of $2m/L$. The method of analysis is illustrated in Example 29.2.

## 29.7. CANTILEVER METHOD

The cantilever method is based on the following assumptions:

(i) Points of contraflexure in each member lies at its mid-span or mid-height.

(ii) The direct stresses (axial stress) in the columns, due to horizontal forces, are directly proportional to their distance from the centroidal vertical axis of the frame.

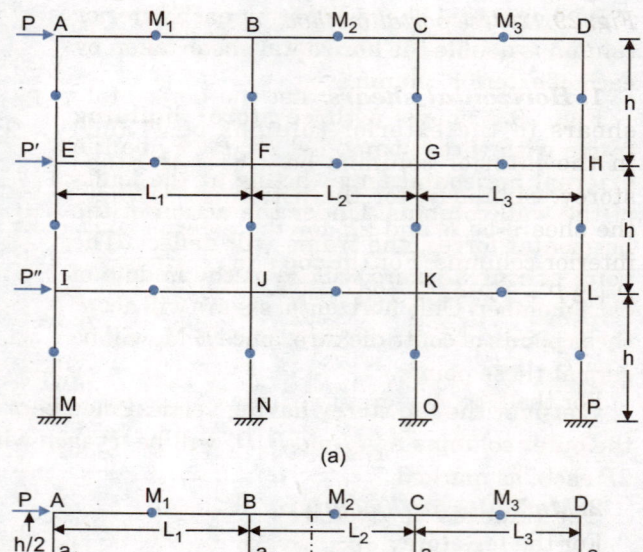

Fig. 29.10. (a) shows a building frame subjected to horizontal forces. Fig 29.10(b) shows the top storey, up to the points of contraflexure of columns. The reactions at the points of contraflexure will be direct and shear forces only. Let $V_1$, $V_2$, $V_3$ and $V_4$ be the axial forces in the columns $AE$, $BF$, $CG$ and $DH$, having areas of cross-sections $a_1$, $a_2$, $a_3$ and $a_4$ respectively.

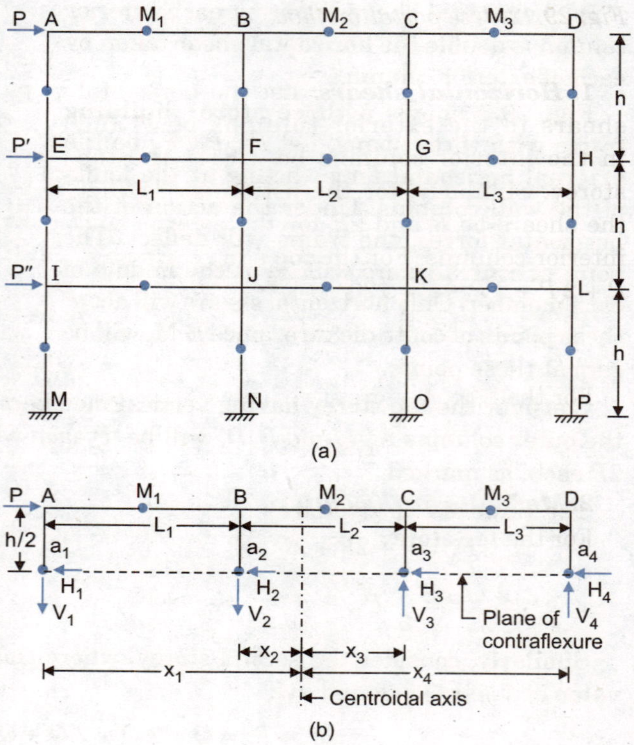

From statics, we have
$$P = H_1 + H_2 + H_3 + H_4 \qquad \ldots(i)$$

From assumption (2), we have
$$\frac{V_1/a_1}{x_1} = \frac{V_2/a_2}{x_2} = \frac{V_3/a_3}{x_3} = \frac{V_4/a_4}{x_4} \qquad \ldots(ii)$$

where $x_1, x_2, x_3$ and $x_4$ are the centroidal distances of the columns from the vertical centroidal axis of the frame.

By taking moment about the point of intersection of the vertical centroidal axis and the top beam, we get
$$(H_1 + H_2 + H_3 + H_4)\frac{h}{2} = V_1 x_1 + V_2 x_2 + V_3 x_3 + V_4 x_4$$

or
$$V_1 x_1 + V_2 x_2 + V_3 x_3 + V_4 x_4 = \frac{Ph}{2} \qquad \ldots(iii)$$

From (ii) and (iii), axial forces $V_1$, $V_2$, $V_3$ and $V_4$ can be determined.

In order to determine $H_1$, take moments about the point of contraflexure $M_1$ in beam $AB$ [Fig. 29.11(a)].
$$H_1 \frac{h}{2} = V_1 \cdot \frac{L_1}{2}$$

$$\therefore \quad H_1 = \frac{V_1 L_1}{h} \qquad \ldots(a)$$

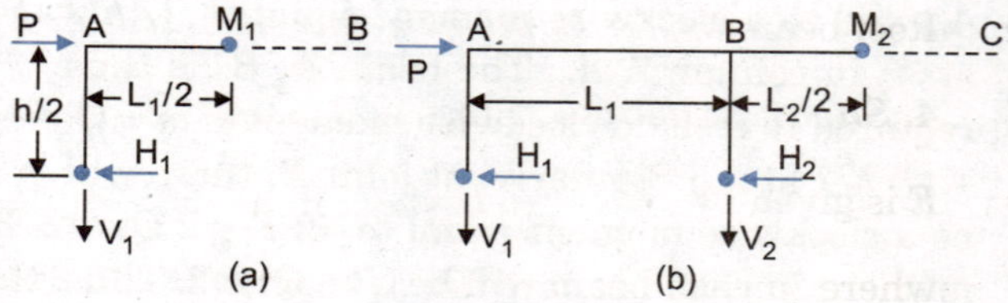

Fig. 29.11

Similarly, taking moments about point of contraflexure $M_2$ in beam $BC$,
$$H_1 \cdot \frac{h}{2} + H_2 \cdot \frac{h}{2} = V_1\left(L_1 + \frac{L_2}{2}\right) + V_2 \frac{L_2}{2}$$

$$\therefore \quad (H_1 + H_2) = \frac{2\left[V_1 L_1 + (V_1 + V_2)\frac{L_2}{2}\right]}{h} \qquad \ldots(b)$$

Since $H_1$ is known from (a), $H_2$ can be determined. In a similar manner, $H_3$ and $H_4$ can be determined.

**Example 29.2.** *Analyse the building frame, subjected to horizontal forces, as shown in Fig. 29.12. Use portal method.*

**Solution:**

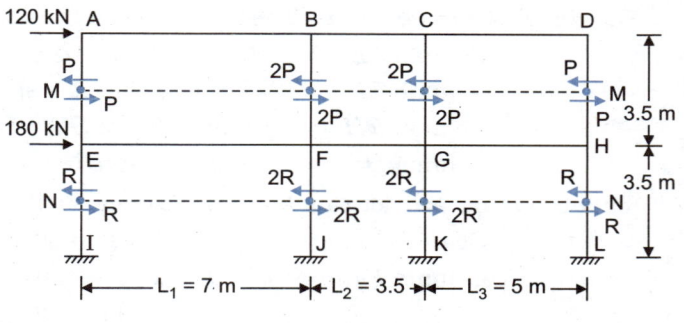

Fig. 29.12

**1. Horizontal shears:** Let the horizontal shears in the exterior columns be $P$ and in the interior columns be $2P$ for the top storey. Similarly, for the bottom storey, let the shears be $R$ and $2R$ for the exterior and interior columns. For the top storey, we have
$P + 2P + 2P + P = 120$

$$\therefore \quad P = \frac{120}{6} = 20 \text{ kN}$$

For the bottom storey, we have $R + 2R + 2R + R = 120 + 180$

$$\therefore \quad R = \frac{300}{6} = 50 \text{ kN}$$

**2. Moments at the ends of columns**

For the top storey,
$$M_{EA} = M_{AE} = M_{HD} = M_{HD}$$
$$= P \times \frac{h}{2} = 20 \times \frac{3.5}{2} = 35 \text{ kN-m}$$
$$M_{FB} = M_{BF} = M_{GC} = M_{CG} = 2P \times \frac{h}{2} = 20 \times 3.5 = 70 \text{ kN-m}$$

For the bottom storey. $M_{IE} = M_{EI} = M_{LH} = M_{HL} = R \times \frac{h}{2} = 50 \times \frac{3.5}{2} = 87.5 \text{ kN-m}$

$$M_{JF} = M_{FJ} = M_{KG} = M_{GK} = 2R \times \frac{h}{2} = 50 \times 3.5 = 175 \text{ kN-m}$$

**3. Moments at the ends of beams**

First floor beams: $m_{EF} = M_{EA} + M_{EI} = 35 + 87.5 = 122.5$ kN-m

Similarly, $m_{FE} = m_{FG} = m_{GF} = m_{GH} = m_{HG} = 122.5$ kN-m, since the point of contraflexure lies at the middle of each span.

In general, $m = (P + R)\frac{h}{2} = (20 + 50) \times \frac{3.5}{2} = 122.5$

Roof beams: $m_{AB} = m_{BA} = m_{BC} = m_{CB} = m_{CD} = m_{DC} = P \cdot \frac{h}{2} = 20 \times \frac{3.5}{2} = 35$ kN-m

**4. Shear in beams:** Since no external vertical force is acting on the beam, shear $F$ is given by
$$F = \frac{m_1 + m_2}{L}$$

where $m_1$ and $m_2$ are the moments at ends of the beams of span $L$.

Thus, $F_{EF} = \dfrac{122.5 + 122.5}{7} = 35$ kN ↑ $\qquad F_{FE} = 35$ ↓

$F_{FG} = F_{GF} = \dfrac{122.5 + 122.5}{3.5} = 70$ kN $\qquad F_{GH} = F_{HG} = \dfrac{122.5 + 122.5}{5} = 49$ kN

$F_{AB} = F_{BA} = \dfrac{35 + 35}{7} = 10$ kN $\qquad F_{BC} = F_{CB} = \dfrac{35 + 35}{3.5} = 20$ kN

$F_{CD} = F_{DC} = \dfrac{35 + 35}{5} = 14$ kN

**5. Axial force in columns.** The axial force in the columns will be as under :

Column $AE$ = shear in beam $AB = 10 \uparrow$
Column $EI$ = axial force in $AE$ + shear in $EF = 10 + 35 = 45 \uparrow$
Column $DH$ = shear in beam $DC = 14$ kN $\downarrow$
Column $HL$ = axial force in $DH$ + shear in $HG = 14 + 49 = 63 \downarrow$

Since the spans are not equal, interior columns will also have axial forces.

Column $BF$ = $F_{BA} - F_{BC} = 10 - 20 = -10$ (i.e., $\uparrow$)
Column $FJ$ = $(-10) + (F_{FE} - F_{FG}) = (-10) + (35 - 70) = -45$ (i.e., $\uparrow$)

Alternatively, axial force in $BF = \dfrac{2m'}{L_1} - \dfrac{2m'}{L_2} = \dfrac{2 \times 35}{7} - \dfrac{2 \times 35}{3.5} = -10$ kN

and axial force in column $FJ = (-10) + \left(\dfrac{2m}{L_1} - \dfrac{2m}{L_2}\right) = (-10) + \left(\dfrac{2 \times 122.5}{7} - \dfrac{2 \times 122.5}{3.5}\right)$

$= -45$ ($\uparrow$)

Axial force in column $CG = \dfrac{2m'}{L_2} - \dfrac{2m'}{L_3} = \dfrac{2 \times 35}{3.5} - \dfrac{2 \times 35}{5} = 6$ ($\downarrow$)

Axial force in column $GK = 6 + \left(\dfrac{2m}{L_2} - \dfrac{2m}{L_3}\right) = 6 + \left(\dfrac{2 \times 122.5}{3.5} - \dfrac{2 \times 122.5}{5}\right) = 27 \downarrow$

**Check.** Total axial force at the base = $-45(\uparrow) - 45(\uparrow) + 27(\downarrow) + 63(\downarrow) =$ zero.

**Example 29.3.** *Re-analyse the frame of example 29.2 by cantilever method, assuming that all the columns have the same area of cross-section.*

**Solution: 1. *Location of centroidal axis of the columns***

Let the centroidal axis be at a distance of $\bar{x}$ from the wind-ward columns $AEI$. Taking the moment of areas of the columns about $AEI$, we get

$$\bar{x} = \dfrac{(2 \times 0) + (2 \times 7) + (2 \times 10.5) + (2 \times 15.5)}{8} = 8.25 \text{ m. Hence,}$$

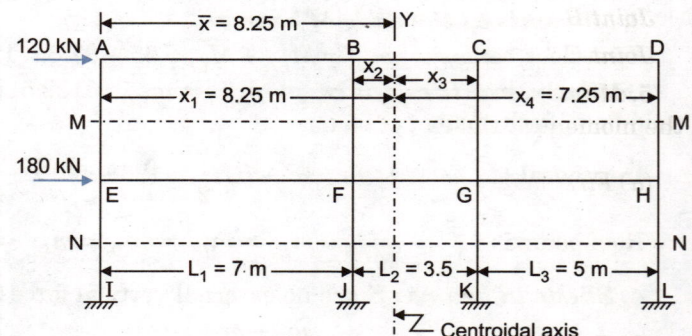

$x_1 = 8.25 \,(= \bar{x})$ m
$x_2 = 8.25 - 7 = 1.25$ m
$x_3 = 3.5 - 1.25 = 2.25$ m
$x_4 = (7 + 3.5 + 5) - 8.25 = 7.25$ m.

**2. *Axial forces in columns of first storey***

Let the axial force in column $EI = V_1 = V$.

Since the areas are equal, the axial forces in other columns will be in proportion to their distances from the centroidal axis.

Fig. 29.13

$\dfrac{V_1}{x_1} = \dfrac{V_2}{x_2} = \dfrac{V_3}{x_3} = \dfrac{V_4}{x_4}$

$V_2 = V\dfrac{x_2}{x_1} = \dfrac{1.25}{8.25}V = 0.1515\,V\,(\downarrow)$

$V_3 = V\dfrac{x_3}{x_1} = \dfrac{2.25}{8.25}V = 0.2727\,V\,(\uparrow)$

$V_4 = V\dfrac{x_4}{x_1} = \dfrac{7.25}{8.25}V = 0.8788\,V\,(\uparrow)$

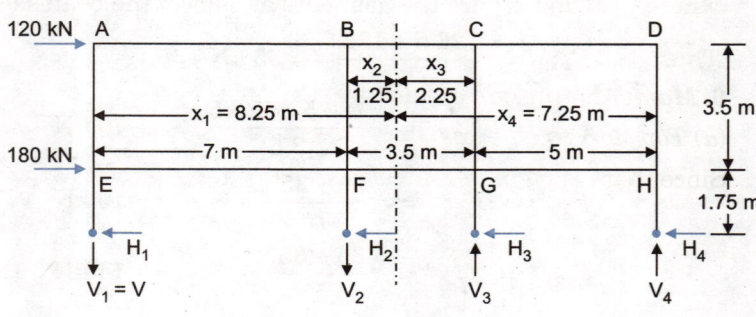

Fig. 29.14

Taking moments of all forces about the point of contraflexure $N$ of the leeward column,
we get,  $(120 \times 5.25) + (180 \times 1.75) - (V \times 15.5) - (0.1515 \, V \times 8.5) + (0.2727 \, V \times 5) = 0$
which gives  $V = V_1 = 61.267$ kN ($\downarrow$)    $\therefore$  $V_2 = 0.1515 \times 61.267 = 9.282$ ($\downarrow$)
$V_3 = 0.2727 \times 61.267 = 16.707$ ($\uparrow$)    $V_4 = 0.8788 \times 61.267 = 53.842$ ($\uparrow$)
Check    $\Sigma V = 61.267 + 9.282 - 16.707 - 53.842 = 0$

### 3. Axial forces in the columns of second storey
Let    $V_1' = V' =$ axial force in column $AE$
$\therefore$    $V_2' = 0.1515 \, V'$;    $V_3' = 0.2727 \, V'$ and $V_4' = 0.8788 \, V'$
Taking moments about point of contraflexure $M$, we get
$(120 \times 1.75) - (V' \times 15.5) - (0.1515 \, V' \times 8.5) + (0.2727 \, V' \times 5) = 0$
From which  $V' = V_1' = 13.615$ ($\downarrow$)
$V_2' = 0.1515 \times 13.615$
    $= 2.063$ ($\downarrow$)
$V_3' = 0.2727 \times 13.615$
    $= 3.713$ ($\uparrow$)
$V_4' = 0.8788 \times 13.615$
    $= 11.965$ ($\uparrow$)

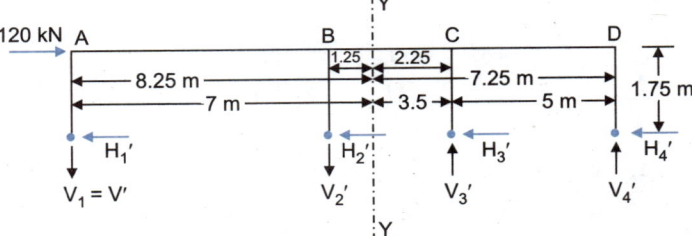

Fig. 29.15

Check    $\Sigma V = 13.615 + 2.603 - 3.713 - 11.965 = 0$   (Rounding off error)

### 4. Shears at ends of beams
The shears at the ends of beams can be determined from the axial forces in the columns at various joints. Let us assume downward force as negative.

Joint $E$:    $F_{EF} = V_1' - V_1 = 13.615 - 61.267 = -47.652$
Joint $F$:    $F_{FG} = -47.652 + 2.063 - 9.282 = -54.871$
Joint $G$:    $F_{GH} = -54.871 - 3.713 + 16.707 = -41.877$
Joint $A$:    $F_{AB} = -13.615$
Joint $B$:    $F_{BC} = -13.615 - 2.063 = -15.678$
Joint $C$:    $F_{CD} = -15.678 + 3.713 = -11.965$

### 5. Moments at the ends of beams
Since there is a point of contraflexure at the middle of each beam, the moment at the end of beams is found by multiplying shear at the end by half the length of the beam.

(a) *First floor*:    $M_{EF} = M_{FE} = F_{FE} \times \dfrac{L_1}{2} = 47.652 \times \dfrac{7}{2} = 166.8$

$M_{FG} = M_{GF} = 54.871 \times \dfrac{3.5}{2} = 96.0$;    $M_{GH} = M_{HG} = 41.877 \times \dfrac{5}{2} = 104.7$

(b) *Second floor*:    $M_{AB} = M_{BA} = 13.615 \times \dfrac{7}{2} = 47.6$

$M_{BC} = M_{CB} = 15.678 \times \dfrac{3.5}{2} = 27.4$;    $M_{CD} = M_{DC} = 11.965 \times \dfrac{5}{2} = 29.9$.

### 6. Moments at ends of column
(a) *Top storey*:    $M_{AE} = M_{AB} = 47.6$ kN-m
Since there is point of contraflexure at the middle of column $AE$, $M_{EA} = 47.6$
$M_{BF} = M_{BA} + M_{BC} = 47.6 + 27.4 = 75$ kN-m    $\therefore$  $M_{FB} = 75$ kN-m
$M_{CG} = M_{CB} + M_{CD} = 27.4 + 29.9 = 57.3$    $\therefore$  $M_{GC} = 57.5$
$M_{DH} = M_{DC} = 29.9$ kN-m    $\therefore$  $M_{HD} = 29.9$ kN-m

**(b) Bottom storey**

$$M_{EI} + M_{EA} = M_{EF}$$
$$\therefore \quad M_{EI} = M_{EF} - M_{EA} = 166.8 - 47.6 = 119.2$$
$$\therefore \quad M_{IE} = 119.2 \text{ kN-m}$$
$$M_{FJ} + M_{FB} = M_{FE} + M_{FG}$$
$$\therefore \quad M_{FJ} = 166.8 + 96 - 75 = 187.8 \text{ kN-m}$$
Hence $\quad M_{JF} = 187.8 \text{ kN-m}$
$$M_{GK} = M_{GF} + M_{GH} + M_{GC}$$
$$\therefore \quad M_{GK} = 96 + 104.7 - 57.5 = 143.2$$
$$\therefore \quad M_{KG} = 143.2 \text{ kN-m}$$
$$M_{HL} + M_{HD} = M_{GH}$$
$$\therefore \quad M_{HL} = 104.7 - 29.9 = 74.8$$
$$\therefore \quad M_{LH} = 74.8.$$

*Alternatively*, the moment at the column ends can be found by first determining horizontal shears ($H$) at the point of contraflexure, and multiplying these by half the height of the column. Thus,

$$M_{AE} = H_1' \times \frac{h}{2} \; ; \quad M_{BF} = H_2' \times \frac{h}{2} \text{ etc.}$$

Similarly, $\quad M_{EI} = H_1 \times \dfrac{h}{2} \; ; \quad M_{FJ} = H_2 \times \dfrac{h}{2}$ etc.

The method of determining horizontal shears have been explained in § 29.7.

For example, $\quad H_1' = \dfrac{V_1' L_1}{h} = 13.615 \times \dfrac{7}{3.5} = 27.23$

$$H_2' = \frac{2\left[V_1' L_1 + (V_1' + V_2')\dfrac{L_2}{2}\right]}{h} - H_1' = \frac{2\left[13.615 \times 7 + (13.615 + 2.063) \times \dfrac{3.5}{2}\right]}{3.5} - 27.23 = 42.908$$

$$\therefore \quad M_{AE} = H_1' \times \frac{h}{2} = 27.23 \times \frac{3.5}{2} = 47.65.$$

$$M_{BF} = H_2' \times \frac{h}{2} = 42.908 \times \frac{3.5}{2} = 75, \text{ which is the same as found earlier.}$$

## 29.8. FACTOR METHOD

This method is more accurate than either the portal method or cantilever method, and is more useful when the moments of inertia of various members [*i.e.*, of columns and beams] are different. Both cantilever method as well as portal method assume uniform moments of inertia of members. These methods therefore depend on some stress assumptions, thus limiting the analysis to be based on equations of statics only. The factor method is based on assumptions regarding the elastic action of the structure. For analysis by factor method, the relative stiffness $K (= I/L)$ for each member of the frame or structure should be known. The procedure consists of the following steps :

**Step 1.** Calculate the *girder factor g* for each joint from the following expression

$$g = \frac{\Sigma K_c}{\Sigma K} \qquad \qquad ...(29.1)$$

where $\quad \Sigma K_c$ = Sum of relative stiffnesses of all column members at the joint considered
$\quad \Sigma K$ = Sum of relative stiffnesses of all the members at the joint considered.

These values of girder factor $g$ are entered in a tabular from as shown in Table 29.13. The values are entered at the end girder meeting at that joint.

**Step 2.** Calculate the *column factor c* for all joints from the following expression:

$$c = 1 - g \qquad ...(29.2)$$

where $g$ = girder factor of the joint.

The values of column factors are entered in Table 29.13 at the end of each column at the joint.

For columns which are fixed at the base, the column factor $c$ is taken as 1.00.

Fig. 29.16 shows a simple frame with two storeys and two bays, used for illustration purpose.

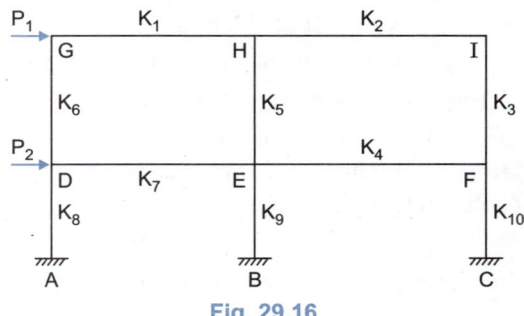

Fig. 29.16

The relative stiffnesses $k_1, k_2, k_3, ...., k_{10}$ of all the ten members are entered on/near each member. Table 29.13 is used for computation of *column factor (c), girder factor(g), column moment factor (C)* and *girder moment factor (G)*.

**TABLE 29.13**

| Joint | Member | Column/girder factors $c$ or $g$ | Half value of factor at opposite end of member | (3) + (4) | $K = I/L$ | Column/girder moment factor $C$ or $G$ = (5) × (6) |
|---|---|---|---|---|---|---|
| (1) | (2) | (3) | (4) | (5) | (6) | (7) |
| D | DG | $c_1$ | $c_2/2$ | $c_1 + c_2/2$ | $K_6$ | |
|   | DA | $c_1$ | $c/2$ from $AD$ | $c_1 + (c/2)_{AD}$ | $K_8$ | |
|   | DE | $g_1$ | $g/2$ from $ED$ | $g_1 + (g/2)_{ED}$ | $K_7$ | |
| G | GD | $c_2$ | $c_1/2$ | $c_2 + c_1/2$ | $K_6$ | |
|   | GH | $g_2$ | $g/2$ from $HG$ | $g_2 + (g/2)_{HG}$ | $K_1$ | |

**Note:** Refer Table 29.14 for further illustration.

**Step 3.** As shown in Table 29.13, in the first column of the table, the name of all the joints are entered. The 2nd column contains all the members at each joint. In 3rd column, the corresponding girder or column factors are entered against each girder or column. In column 4 of the table, half the values of the column factor/girder factor of opposite end of the members are entered. For example, if $c_1$ = column factor of member $DG$, it is entered in column 3 opposite $DG$, while column factor $c_2$ of member $GD$ is entered in column 3 opposite member $GD$. Hence in column 4, half the column factor of opposite end i.e. $c_2/2$ is entered opposite member $DG$. Similarly, for member $GD$, a value of $c_1/2$ is entered opposite it for the same reason. So in this way column no. 4 is entered. The values in column no. (3) and (4) of Table 29.13 for each member are added and entered in column no. 5. In column no. 6, the relative stiffness values $K = I/L$ for each member is entered.

**Step 4.** The sum of columns (3) and (4), which is entered in column no. (5) is multiplied by relative stiffness of respective members (which are entered in column no. 6). This product is termed as *column moment factor* $C$ for columns and *girder moment factor* $G$ for girders. This is entered in column No. (7) of Table 29.13.

The column moment factor $C$, gives the relative values of moments at the ends of columns for each storey in which the column occurs. The sum of column end moments is equal to the horizontal shear on that storey multiplied by the storey height. Hence the column moment factors ($C$) are converted into end moments for columns by direct proportion for each storey. Similarly, the girder moment factor $G$, gives the relative values of moment at ends of each girder for the joint. The sum of girder end moment at each joint is equal to the sum of end moments in the columns at the joint. Hence the girder moment factors are converted into end moments for girders by direct proportion for each storey.

**Step 5. Calculation of column moment:**

(a) Total column moments (A) for each storey is found by the relation

$$A = \frac{H \times h}{\Sigma C} \qquad ...(29.3)$$

**730** REINFORCED CONCRETE STRUCTURE

where  $A$ = Total column moment for each storey
  $H$ = Total horizontal force above the storey considered
  $h$ = height of the storey considered.
  $\Sigma C$ = Sum of column end moment factors of that storey

Thus, for each storey, different column moments $A_1, A_2, ...$ etc. are calculated.

(b) The column moment factor $C$ of each member is multiplied by the total column moment ($A$) of that storey in which the column occurs. For example, in Fig. 29.16 if we want to find the column moment $M_{GD}$ of column $GD$, we have

$$M_{GD} = A_1 \times C_{GD} \qquad \qquad ...(29.4)$$

where  $A_1$ = Total column moments of first storey (Eq. 29.3) and
  $C_{GD}$ = Column moment factor for column $GD$.

Similarly, moment $M_{HE}$ in column $HE$ of first storey is $M_{HE} = A_1 \times C_{HE}$

Also for column $DA$ of ground storey, $M_{DA} = A_0 \times C_{DA}$

where  $A_0$ = total column moments of ground storey.

**Step 6. Calculations of Girder/beam moments:**

(a) For calculation of girder/beam moments, a constant $B$ is found for each joint.

$$B = \frac{\text{Sum of column moments at the joint}}{\text{Sum of the girder moment factors at that joint}}$$

(b) This constant $B$ is multiplied by the girder moment factor ($G$) to obtain the girder moments. For example (Fig. 29.16),

$$M_{DE} = B_D \times G_{DE}; \qquad M_{HI} = B_H \times G_{HI}$$
$$M_{IH} = B_I \times G_{IH} \text{ and so on.}$$

Here, $B_D, B_H, B_I$ are the constants for joints $D$, $H$ and $I$ respectively.

The factor method of analysing the building frame has been illustrated in Example 29.4

**Example 29.4.** *Analyse the frame as shown in Fig. 29.17 by factor method. Sketch the B.M.D.*

*The relative k values are written on the members.*

**Solution: Step 1.** Calculate the girder factor '$g$' at all joint by Eq. 29.1.

$$g = \frac{\Sigma K_C}{\Sigma K}$$

where  $\Sigma K_C$ = Sum of relative Stiffness of columns at that joint
  $\Sigma K$ = Sum of relative Stiffness of all the members at that join.

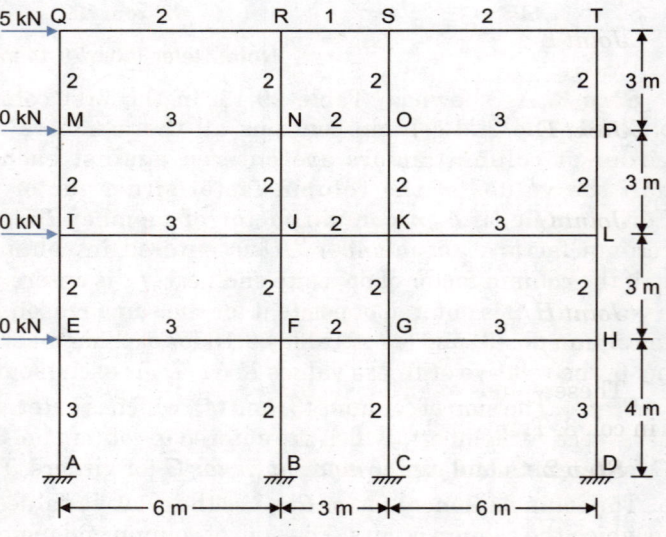

Fig. 29.17

Joint Q  $\quad g_Q = \dfrac{K_{QM}}{K_{QM} + K_{QR}} = \dfrac{2}{2+2} = 0.5$

Joint R  $\quad g_R = \dfrac{K_{RN}}{K_{RN} + K_{RQ} + K_{RS}} = \dfrac{2}{2+2+1} = 0.4$

Joint S  $\quad g_S = \dfrac{K_{SO}}{K_{SO} + K_{SR} + K_{ST}} = \dfrac{2}{2+1+2} = 0.4$

Joint T  $\quad g_T = \dfrac{K_{TP}}{K_{TP} + K_{TS}} = \dfrac{2}{2+2} = 0.5$

Joint M  $\quad g_M = \dfrac{K_{MI} + K_{MQ}}{K_{MI} + K_{MQ} + K_{MN}} = \dfrac{2+2}{2+2+3} = 0.57$

Joint N  $\quad g_N = \dfrac{K_{NJ} + K_{NR}}{K_{NJ} + K_{NR} + K_{NM} + K_{NO}} = \dfrac{2+2}{2+2+3+2} = 0.44$

Joint O  $\quad g_O = \dfrac{K_{OK} + K_{OS}}{K_{OK} + K_{OS} + K_{ON} + K_{OP}} = \dfrac{2+2}{2+2+2+3} = 0.44$

Joint P  $\quad g_P = \dfrac{K_{PL} + K_{PT}}{K_{PL} + K_{PT}\ K_{PO}} = \dfrac{2+2}{2+2+3} = 0.57$

Joint I  $\quad g_I = \dfrac{K_{IE} + K_{IM}}{K_{IE} + K_{IM} + K_{IJ}} = \dfrac{2+2}{2+2+3} = 0.57$

Joint J  $\quad g_J = \dfrac{K_{JF} + K_{JN}}{K_{JF} + K_{JN} + K_{JI} + K_{JK}} = \dfrac{2+2}{2+2+3+2} = 0.44$

Joint K  $\quad g_K = \dfrac{K_{KO} + K_{KG}}{K_{KO} + K_{KG} + K_{KJ} + K_{KL}} = \dfrac{2+2}{2+2+2+3} = 0.44$

Joint L  $\quad g_L = \dfrac{K_{LH} + K_{LP}}{K_{LH} + K_{LP} + K_{LK}} = \dfrac{2+2}{2+2+3} = 0.57$

Joint E  $\quad g_E = \dfrac{K_{EA} + K_{EI}}{K_{EA} + K_{EI} + K_{EF}} = \dfrac{3+2}{3+2+3} = 0.63$

Joint F  $\quad g_F = \dfrac{K_{FB} + K_{FJ}}{K_{FB} + F_{FJ} + K_{FE} + K_{FG}} = \dfrac{2+2}{2+2+3+2} = 0.44$

Joint G  $\quad g_G = \dfrac{K_{GC} + K_{GK}}{K_{GC} + K_{GK} + K_{GF} + K_{GH}} = \dfrac{2+2}{2+2+2+3} = 0.44$

Joint H  $\quad g_H = \dfrac{K_{HD} + K_{HL}}{K_{HD} + K_{HL} + K_{HG}} = \dfrac{3+2}{3+2+3} = 0.63.$

These values of girder factors are written at the ends of girders, beams meeting at each joint as shown in col. 3, Table 29.14.

**Step 2.** Calculate the column factor $c$ at the joints by the relation

$$c = 1 - g \quad \text{where} \quad g = \text{girder factor at that joint} \qquad \ldots(29.2)$$

Joint Q  $\quad c_Q = 1 - g_Q = 1 - 0.5 = 0.5$
Joint R  $\quad c_R = 1 - g_R = 1 - 0.4 = 0.6$
Joint S  $\quad c_S = 1 - g_S = 1 - 0.4 = 0.6$
Joint T  $\quad c_T = 1 - g_T = 1 - 0.5 = 0.5$
Joint M  $\quad c_M = 1 - g_M = 1 - 0.57 = 0.43$
Joint N  $\quad c_N = 1 - g_N = 1 - 0.44 = 0.56$
Joint O  $\quad c_O = 1 - g_O = 1 - 0.44 = 0.56$
Joint P  $\quad c_P = 1 - g_P = 1 - 0.57 = 0.43$

| Joint I | $c_I = 1 - g_I = 1 - 0.57 = 0.43$ |
| Joint J | $c_J = 1 - g_J = 1 - 0.44 = 0.56$ |
| Joint K | $c_K = 1 - g_K = 1 - 0.44 = 0.56$ |
| Joint L | $c_L = 1 - g_L = 1 - 0.57 = 0.43$ |
| Joint E | $c_E = 1 - g_E = 1 - 0.63 = 0.37$ |
| Joint F | $c_F = 1 - g_F = 1 - 0.44 = 0.56$ |
| Joint G | $c_G = 1 - g_G = 1 - 0.44 = 0.56$ |
| Joint H | $c_H = 1 - g_H = 1 - 0.63 = 0.37$ |
| Joint A | $c_A = 1.00$ |
| Joint B | $c_B = 1.00$ |
| Joint C | $c_C = 1.00$ |
| Joint D | $c_D = 1.00$ |

For columns fixed at the base, column factor is taken as 1.00

These values of column factors are written at the end of columns meeting at the joint, and have been entered in column 3 of Table 29.14.

**Step 3.** Half the values of column/girder factors of the opposite ends are entered in column 4. The values in col. 3 and col. 4 are added and entered in col. 5 of Table 29.14.

**Step 4.** Enter the values of relative stiffnesses of all members in col.6. These values of stiffnesses are multiplied by the values of column 5, to get the values of girder/column moment factors ($G$ or $C$), and are entered in col. 7 of Table 29.14.

Thus Table 29.14 is completed.

**Step 5. Calculation column moments:** Total column moments for each storey

$$A = \frac{H \times h}{\Sigma C}$$

where  $H$ = Total horizontal force above the storey considered
 $h$ = height of the storey considered
 $\Sigma C$ = sum of *column end moment factors* of that storey

Let  $A_3$ = Total column moments of third/top storey
 $A_2$ = Total column moments of second storey
 $A_1$ = Total column moments of first storey
 $A_0$ = Total column moments of ground storey.

$$\therefore A_3 = \frac{5 \times 3}{C_{QM} + C_{MQ} + C_{RN} + C_{NR} + C_{SO} + C_{OS} + C_{TP} + C_{PT}}$$

$$= \frac{15}{1.42 + 1.36 + 1.76 + 1.72 + 1.76 + 1.72 + 1.42 + 1.36} = 1.2 \text{ kN-m}$$

$$A_2 = \frac{(10 + 5) \times 3}{C_{MI} + C_{IM} + C_{NJ} + C_{JN} + C_{OK} + C_{KO} + C_{PL} + C_{LP}}$$

$$= \frac{45}{1.28 + 1.28 + 1.68 + 1.68 + 1.68 + 1.68 + 1.28 + 1.28} = 3.80 \text{ kN-m}$$

$$A_1 = \frac{(10 + 10 + 5) \times 3}{C_{IE} + C_{EI} + C_{JF} + C_{FJ} + C_{KG} + C_{GK} + C_{LH} + C_{HL}}$$

$$= \frac{75}{1.22 + 1.16 + 1.68 + 1.68 + 1.68 + 1.68 + 1.22 + 1.16} = 6.533 \text{ kN-m}$$

## TABLE 29.14

| Joint | Member | Girder/Column factor (c or g) | Half values of the factors from opposite end | (3)+(4) | Relative stiffness ($K = I/L$) | Girder/column Moment factor (C or G) = (5) × (6) |
|---|---|---|---|---|---|---|
| 1 | 2 | 3 | 4 | 5 | 6 | 7 |
| Q | QR | 0.5 | 0.2 | 0.7 | 2 | 1.4 |
|   | QM | 0.5 | 0.21 | 0.71 | 2 | 1.42 |
| R | RQ | 0.4 | 0.25 | 0.65 | 2 | 1.30 |
|   | RS | 0.4 | 0.2 | 0.60 | 1 | 0.60 |
|   | RN | 0.6 | 0.28 | 0.88 | 2 | 1.76 |
| S | SR | 0.4 | 0.2 | 0.6 | 1 | 0.6 |
|   | ST | 0.4 | 0.25 | 0.65 | 2 | 1.3 |
|   | SO | 0.6 | 0.28 | 0.88 | 2 | 1.76 |
| T | TS | 0.5 | 0.2 | 0.7 | 2 | 1.4 |
|   | TP | 0.5 | 0.21 | 0.71 | 2 | 1.42 |
| M | MN | 0.57 | 0.22 | 0.79 | 3 | 2.37 |
|   | MI | 0.43 | 0.21 | 0.64 | 2 | 1.28 |
|   | MQ | 0.43 | 0.25 | 0.68 | 2 | 1.36 |
| N | NM | 0.44 | 0.28 | 0.72 | 3 | 2.16 |
|   | NO | 0.44 | 0.22 | 0.66 | 2 | 1.32 |
|   | NR | 0.56 | 0.3 | 0.86 | 2 | 1.72 |
|   | NJ | 0.56 | 0.28 | 0.84 | 2 | 1.68 |
| O | ON | 0.44 | 0.22 | 0.66 | 2 | 1.32 |
|   | OP | 0.44 | 0.28 | 0.72 | 3 | 2.16 |
|   | OS | 0.56 | 0.30 | 0.86 | 2 | 1.72 |
|   | OK | 0.56 | 0.28 | 0.84 | 2 | 1.68 |
| P | PO | 0.57 | 0.22 | 0.79 | 3 | 2.37 |
|   | PL | 0.43 | 0.21 | 0.64 | 2 | 1.28 |
|   | PT | 0.43 | 0.25 | 0.68 | 2 | 1.36 |
| I | IJ | 0.57 | 0.22 | 0.79 | 3 | 2.37 |
|   | IM | 0.43 | 0.21 | 0.64 | 2 | 1.28 |
|   | IE | 0.43 | 0.18 | 0.61 | 2 | 1.22 |
| J | JI | 0.44 | 0.28 | 0.72 | 3 | 2.16 |
|   | JK | 0.44 | 0.22 | 0.66 | 2 | 1.32 |
|   | JN | 0.56 | 0.28 | 0.84 | 2 | 1.68 |
|   | JF | 0.56 | 0.28 | 0.84 | 2 | 1.68 |
| K | KJ | 0.44 | 0.22 | 0.66 | 2 | 1.32 |
|   | KL | 0.44 | 0.28 | 0.72 | 3 | 2.16 |
|   | KO | 0.56 | 0.28 | 0.84 | 2 | 1.68 |
|   | KG | 0.56 | 0.28 | 0.84 | 2 | 1.68 |
| L | LK | 0.57 | 0.22 | 0.79 | 3 | 2.37 |
|   | LP | 0.43 | 0.21 | 0.64 | 2 | 1.28 |
|   | LH | 0.43 | 0.18 | 0.61 | 2 | 1.22 |
| E | EF | 0.63 | 0.22 | 0.85 | 3 | 2.55 |
|   | EI | 0.37 | 0.21 | 0.58 | 2 | 1.16 |
|   | EA | 0.37 | 0.5 | 0.87 | 3 | 2.61 |
| F | FE | 0.44 | 0.31 | 0.75 | 3 | 2.25 |
|   | FG | 0.44 | 0.22 | 0.66 | 2 | 1.32 |
|   | FB | 0.56 | 0.5 | 1.06 | 2 | 2.12 |
|   | FJ | 0.56 | 0.28 | 0.84 | 2 | 1.68 |

**734 REINFORCED CONCRETE STRUCTURE**

| 1 | 2 | 3 | 4 | 5 | 6 | 7 |
|---|---|---|---|---|---|---|
| G | GF | 0.44 | 0.22 | 0.66 | 2 | 1.32 |
|   | GH | 0.44 | 0.31 | 0.75 | 3 | 2.25 |
|   | GK | 0.56 | 0.28 | 0.84 | 2 | 1.68 |
|   | GC | 0.56 | 0.50 | 1.06 | 2 | 2.12 |
| H | HG | 0.63 | 0.22 | 0.85 | 3 | 2.55 |
|   | HL | 0.37 | 0.21 | 0.58 | 2 | 1.16 |
|   | HD | 0.37 | 0.50 | 0.87 | 3 | 2.61 |
| A | AE | 1.00 |      | 0.18 | 1.18 | 3 | 3.54 |
| B | BF | 1.00 |      | 0.28 | 1.28 | 2 | 2.56 |
| C | CG | 1.00 |      | 0.28 | 1.28 | 2 | 2.56 |
| D | DH | 1.00 |      | 0.18 | 1.18 | 3 | 3.54 |

$$A_0 = \frac{(10 + 10 + 10 + 5) \times 4}{C_{EA} + C_{AE} + C_{FB} + C_{BF} + C_{GC} + C_{CG} + C_{HD} + C_{DH}}$$

$$= \frac{140}{2.61 + 3.54 + 2.12 + 2.56 + 2.12 + 2.56 + 2.61 + 3.54} = 6.464 \text{ kN-m}.$$

## Column Moments:

*Top Storey:* $A_3 = 1.2$ kN-m

$M_{QM} = A_3 \times C_{QM} = 1.2 \times 1.42 = 1.704$ kN-m
$M_{MQ} = A_3 \times C_{MQ} = 1.2 \times 1.36 = 1.632$ kN-m
$M_{RN} = A_3 \times C_{RN} = 1.2 \times 1.76 = 2.112$ kN-m
$M_{NR} = A_3 \times C_{NR} = 1.2 \times 1.72 = 2.064$ kN-m
$M_{SO} = A_3 \times C_{SO} = 1.2 \times 1.76 = 2.112$ kN-m
$M_{OS} = A_3 \times C_{OS} = 1.2 \times 1.72 = 2.064$ kN-m
$M_{TP} = A_3 \times C_{TP} = 1.2 \times 1.42 = 1.704$ kN-m
$M_{PT} = A_3 \times C_{PT} = 1.2 \times 1.36 = 1.632$ kN-m

*Second Storey:* $A_2 = 3.80$ kN-m

$M_{MI} = 3.80 \times 1.28 = 4.864$ kN-m
$M_{IM} = 3.80 \times 1.28 = 4.864$ kN-m
$M_{NJ} = 3.80 \times 1.68 = 6.384$ kN-m
$M_{JN} = 3.80 \times 1.68 = 6.384$ kN-m
$M_{OK} = 3.80 \times 1.68 = 6.384$ kN-m
$M_{KO} = 3.80 \times 1.68 = 6.384$ kN-m
$M_{PL} = 3.80 \times 1.28 = 4.864$ kN-m
$M_{LP} = 3.80 \times 1.28 = 4.864$ kN-m

*First Storey:* $A_1 = 6.533$ kN-m

$M_{IE} = 6.533 \times 1.22 = 7.97$ kN-m
$M_{EI} = 6.533 \times 1.16 = 7.58$ kN-m
$M_{JF} = 6.533 \times 1.68 = 10.98$ kN-m
$M_{FJ} = 6.533 \times 1.68 = 10.98$ kN-m
$M_{KG} = 6.533 \times 1.68 = 10.98$ kN-m
$M_{GK} = 6.533 \times 1.68 = 10.98$ kN-m
$M_{LH} = 6.533 \times 1.22 = 7.97$ kN-m
$M_{HL} = 6.533 \times 1.16 = 7.58$ kN-m

*Ground Storey:* $A_0 = 6.464$ kN-m

$M_{EA} = 6.464 \times 2.61 = 16.871$ kN-m
$M_{AE} = 6.464 \times 3.54 = 22.882$ kN-m
$M_{FB} = 6.464 \times 2.12 = 13.703$ kN-m
$M_{BF} = 6.464 \times 2.56 = 16.547$ kN-m
$M_{GC} = 6.464 \times 2.12 = 13.703$ kN-m
$M_{CG} = 6.464 \times 2.56 = 16.547$ kN-m
$M_{HD} = 6.464 \times 2.61 = 16.871$ kN-m
$M_{DH} = 6.464 \times 3.54 = 22.882$ kN-m

### Step 6. Calculation of beam moments

(a) $\quad \text{Constant } B = \dfrac{\text{Sum of column moments at the joint}}{\text{Sum of girder moment factors at that joint}}$

Joint Q $\qquad B_Q = \dfrac{M_{QM}}{G_{QR}} = \dfrac{1.704}{1.4} = 1.217$

Joint R $\qquad B_R = \dfrac{M_{RN}}{G_{RQ} + G_{RS}} = \dfrac{2.112}{1.30 + 0.60} = 1.111$

Joint S $\qquad B_S = \dfrac{M_{SO}}{G_{SR} + G_{ST}} = \dfrac{2.112}{0.6 + 1.3} = 1.111$

Joint T $\quad B_T = \dfrac{M_{TP}}{G_{TS}} = \dfrac{1.704}{1.4} = 1.217$

Joint M $\quad B_M = \dfrac{M_{MQ} + M_{MI}}{G_{MN}} = \dfrac{1.632 + 4.864}{2.37} = 2.741$

Joint N $\quad B_N = \dfrac{M_{NR} + M_{NJ}}{G_{NM} + G_{NO}} = \dfrac{2.064 + 6.384}{2.16 + 1.32} = 2.427$

Joint O $\quad B_O = \dfrac{M_{OS} + M_{OK}}{G_{ON} + G_{OP}} = \dfrac{2.064 + 6.384}{1.32 + 2.16} = 2.427$

Joint P $\quad B_P = \dfrac{M_{PT} + M_{PL}}{G_{PO}} = \dfrac{1.632 + 4.864}{2.37} = 2.741$

Joint I $\quad B_I = \dfrac{M_{IE} + M_{IM}}{G_{IJ}} = \dfrac{7.97 + 4.864}{2.37} = 5.415$

Joint J $\quad B_J = \dfrac{M_{JN} + M_{JF}}{G_{JI} + G_{JK}} = \dfrac{6.384 + 10.98}{2.16 + 1.32} = 4.99$

Joint K $\quad B_K = \dfrac{M_{KO} + M_{KG}}{K_{KJ} + G_{KL}} = \dfrac{6.384 + 10.98}{1.32 + 2.16} = 4.99$

Joint L $\quad B_L = \dfrac{M_{LP} + M_{LH}}{G_{LK}} = \dfrac{4.864 + 7.97}{2.37} = 5.42$

Joint E $\quad B_E = \dfrac{M_{EI} + M_{EA}}{G_{EF}} = \dfrac{7.58 + 16.871}{2.55} = 9.59$

Joint F $\quad B_F = \dfrac{M_{FJ} + M_{FB}}{G_{FE} + G_{FG}} = \dfrac{10.98 + 13.703}{2.25 + 1.32} = 6.91$

Joint G $\quad B_G = \dfrac{M_{GK} + M_{GC}}{G_{GF} + G_{GH}} = \dfrac{10.98 + 13.703}{1.32 + 2.25} = 6.91$

Joint H $\quad B_H = \dfrac{M_{HD} + M_{HL}}{G_{HG}} = \dfrac{16.871 + 7.58}{2.55} = 9.59$

## (b) Beam Moments

Beam moments = B × Girder Moment Factor.

$M_{QR} = B_Q \times G_{QR} = 1.217 \times 1.40 = 1.704$ kN-m
$M_{RQ} = B_R \times G_{RQ} = 1.111 \times 1.30 = 1.444$ kN-m
$M_{RS} = B_R \times G_{RS} = 1.111 \times 0.60 = 0.667$ kN-m
$M_{SR} = B_S \times G_{SR} = 1.111 \times 0.60 = 0.667$ kN-m
$M_{ST} = B_S \times G_{ST} = 1.111 \times 1.30 = 1.444$ kN-m
$M_{TS} = B_T \times G_{TS} = 1.217 \times 1.40 = 1.704$ kN-m
$M_{MN} = B_M \times G_{MN} = 2.741 \times 2.37 = 6.496$ kN-m
$M_{NM} = B_N \times G_{NM} = 2.427 \times 2.16 = 5.242$ kN-m
$M_{NO} = B_N \times G_{NO} = 2.427 \times 1.32 = 3.203$ kN-m
$M_{ON} = B_O \times G_{ON} = 2.427 \times 1.32 = 3.203$ kN-m
$M_{OP} = B_O \times G_{OP} = 2.427 \times 2.16 = 5.242$ kN-m
$M_{PO} = B_P \times G_{PO} = 2.741 \times 2.37 = 6.496$ kN-m
$M_{IJ} = B_I \times G_{IJ} = 5.415 \times 2.37 = 12.83$ kN-m

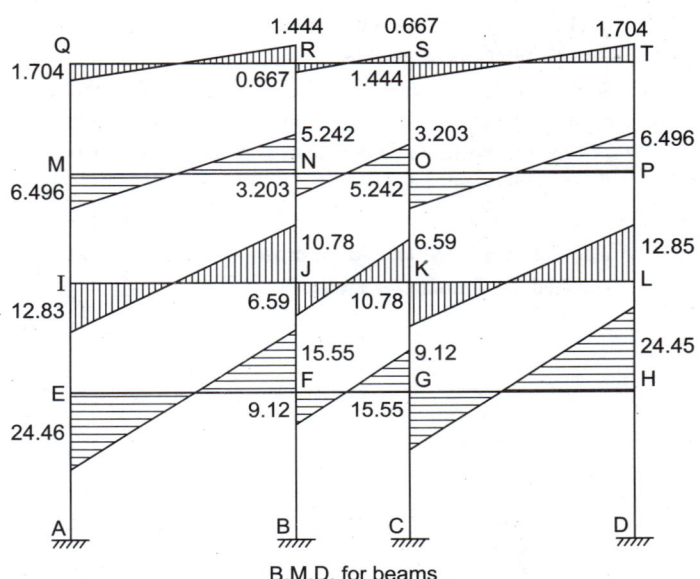

B.M.D. for beams

Fig. 29.18

$M_{JI} = B_J \times G_{JI} = 4.99 \times 2.16 = 10.78$ kN-m

$M_{JK} = B_J \times G_{JK} = 4.99 \times 1.32 = 6.59$ kN-m

$M_{KJ} = B_K \times G_{KJ} = 4.99 \times 1.32 = 6.59$ kN-m

$M_{KL} = B_K \times G_{KL} = 4.99 \times 2.16 = 10.78$ kN-m

$M_{LK} = B_L \times G_{LK} = 5.42 \times 2.37 = 12.85$ kN-m

$M_{EF} = B_E \times G_{EF} = 9.59 \times 2.55 = 24.45$ kN-m

$M_{FE} = B_F \times G_{FE} = 6.91 \times 2.25 = 15.55$ kN-m

$M_{FG} = B_F \times G_{FG} = 6.91 \times 1.32 = 9.12$ kN-m

$M_{GF} = B_G \times G_{GF} = 6.91 \times 1.32 = 9.12$ kN-m

$M_{GH} = B_G \times G_{GH} = 6.91 \times 2.25 = 15.55$ kN-m

$M_{HG} = B_H \times G_{HG} = 9.59 \times 2.55 = 24.45$ kN-m

Thus, the moments in all the columns and girder are found. Fig. 29.18 shown the B.M. diagram for girder/beams while Fig. 29.19 shows the B.M. diagram for columns.

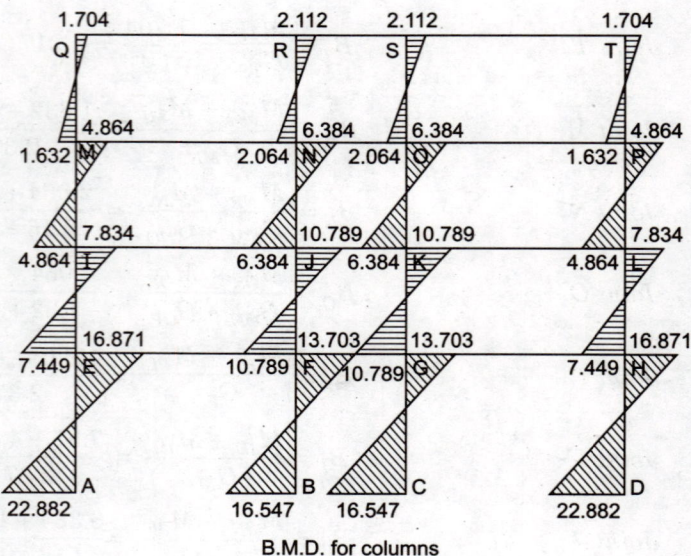

B.M.D. for columns

Fig. 29.19

## PROBLEMS

1. What do you understand by a substitute frame? How do you select it? Discuss in brief the method of analysis.
2. Explain the portal method for analysing a building frame subjected to horizontal forces.
3. Explain the cantilever method for analysing a building frame subjected to horizontal forces.
4. A two-span intermediate frame of a multi-storeyed building is shown in Fig. 29.20. The frames are spaced at 5 m intervals. The dead load and live load per metre run of the beam may be taken as 15 kN/m and 20 kN/m respectively. Analyse the frame using two cycle method of moment distribution.
5. If wind loads of 15 kN, and 30 kN are acting at joint A, B and C respectively, analyse the frame (Fig. 29.20) by (a) portal method, (b) cantilever method. Assume that all the columns have equal area of cross-section for the purpose of analysis.
6. A four storeyed multistoreyed building frame has four equal bays of 4 m each and the height between floors is 4 m. The wind loads acting at roof level and various floor levels are:

$H_1 : 5$ kN, $H_2 = 10$ kN, $H_3 = 10$ kN, and $H_4 = 10$ kN

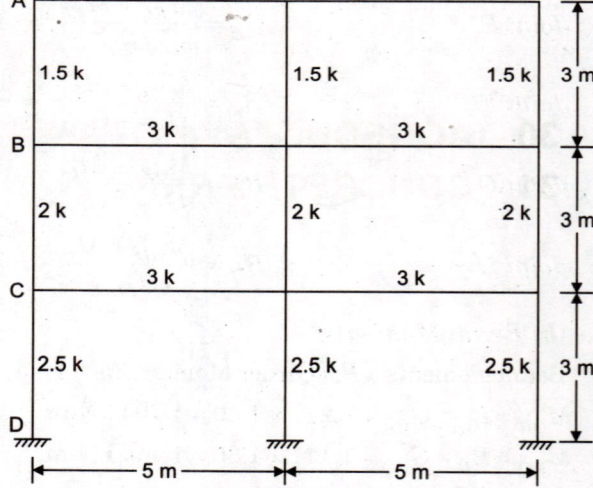

Fig. 29.20

The column have the same cross-section. Estimate the moments in the columns and beams using (a) portal method (b) Cantilever method.

# PART – IV
# CONCRETE BRIDGES

**30. AQUEDUCTS AND BOX CULVERTS**
**31. CONCRETE BRIDGES**

# CHAPTER 30

# AQUEDUCTS AND BOX CULVERTS

## 30.1. AQUEDUCTS AND SYPHON AQUEDUCTS

An aqueduct is a rectangular channel meant for carrying water of a channel when it crosses a natural stream. It is generally made of reinforced concrete. The design of an aqueduct is similar to that of a rectangular tank, except that its length is infinite. Therefore, the design of aqueduct is done for its one metre length.

Fig. 30.1 (a) shows an aqueduct in which the bottom of canal trough (aqueduct) lies above the expected high flood level of the drain. The drain water, in this case, does not exert any pressure on the aqueduct structure. Fig. 30.1 (b), on the other hand, shows a syphon aqueduct in which the high flood level of the drain is higher than the bottom of the canal trough. Due to syphonic action, some head loss takes place through the syphon barrel, resulting in the upward

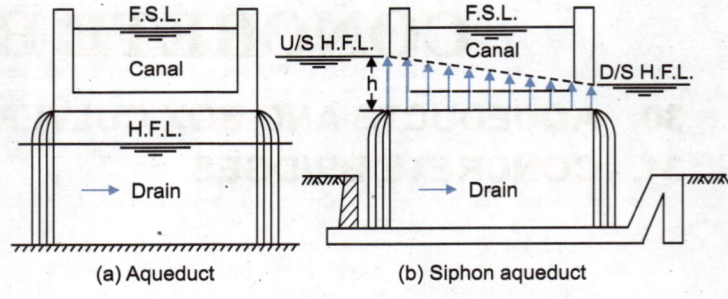

Fig. 30.1. Types of Aqueducts

(uplift) pressure on the bottom of the aqueduct trough. The aqueduct (Fig. 30.1a) is subjected to water loads only when the canal is running full. However, the syphon aqueduct (Fig. 30.1b) is subjected to two sets of forces : (i) water pressure when canal is running full, (ii) upward water pressure when the drain is running full. The worst cases for the design of a syphon aqueduct will be (i) canal running full, but the drain is dry, and (ii) drain running full, but there is no water in the canal.

## 30.2. DESIGN OF AN AQUEDUCT

Consider an aqueduct trough of height $H$ and span $L$ from inside. The maximum water pressure at the bottom of the vertical walls will be $wH$, and the B.M. there will be $(wH^3)/6$. Since the vertical walls are rigidly fixed to the bottom slab, the negative B.M. transferred to each end of the slab will be $(wH^3)/6$.

The bottom slab will be subjected to a uniform water load $p = wH$.

Simply supported B.M. at centre of slab = $\dfrac{pL^2}{8} = \dfrac{wHL^2}{8}$

Hence net B.M. at centre of slab, $M = \dfrac{wHL^2}{8} - \dfrac{wH^3}{6}$.

For this positive B.M. to be maximum, we have $\frac{dM}{dH} = 0 = \frac{wL^2}{8} - \frac{wH^2}{2}$, which gives $H = \frac{L}{2}$.

Thus maximum positive B.M. in the slab occurs when depth of water is equal to half the span.

$$M_{max} = \frac{wL^3}{16} - \frac{wL^3}{48} = \frac{wL^3}{24}$$

However, if $H$ is greater than $L/2$ (i.e. if the depth of water is more), the net B.M. at the centre may be negative. The depth of water, corresponding to this condition is given by:

$$\frac{wH^3}{6} > \frac{wHL^2}{8}$$

or $\quad H > L\sqrt{3/4} > 0.866\,L$.

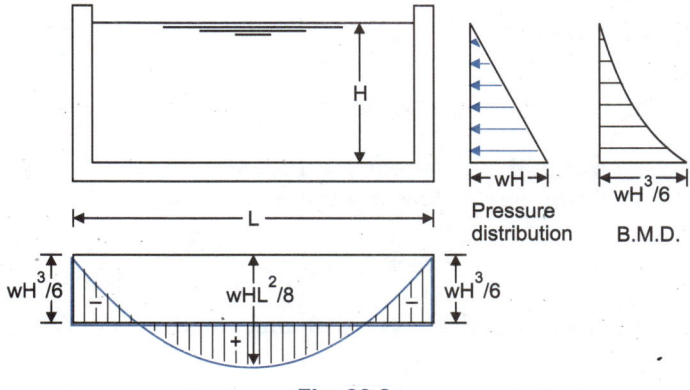

Fig. 30.2

Hence if the depth of water is equal to $0.866\,L$, there will be zero B.M. at the centre of slab.

If the length of the aqueduct is finite, the vertical walls may be designed as beams.

**Example 30.1.** *Design an aqueduct 5 m wide and 2.5 m deep. The aqueduct is expected to carry water with a maximum depth of flow of 1.3 m. Use M20 concrete and HYSD bars. The aqueduct has a span of 8 m.*

**Solution:**

**1. Design constants:** For M 20 concrete, $\sigma_{cbc} = 7$ N/mm² and $m = 13$. For HYSD bars on waterside, $\sigma_{st} = 150$ N/mm² and hence $k = 0.375$, $j = 0.874$ and $R = 1.155$. For steel to other side, $\sigma_{st} = 190$ N/mm² and hence $k = 0.324$; $j = 0.892$ and $R = 1.011$.

**2. Design of vertical walls:** $\quad$ B.M $= \frac{wH^3}{6} = \frac{9800\,(2.3)^3}{6} \approx 19873$ N-m

Since tension occurs at water side,

$\sigma_{st} = 150$ N/mm² and $R = 1.155$. $\quad \therefore \quad d = \sqrt{\frac{19873 \times 1000}{1.155 \times 1000}} = 131$ mm.

The walls are also to be designed as girder having a span of 8 m between the supports. Hence the minimum thickness should be equal to span / 30 = 800 / 30 ≈ 27 cm = 270 mm. Keeping a cover of 50 mm, $d = 270 - 50 = 220$ mm

$\therefore \quad A_{st} = \frac{19873 \times 1000}{150 \times 0.874 \times 220} = 689$ mm²

Using 10 mm Φ bars, $A_\phi = 78.5$ mm². Hence, Spacing $= \frac{1000 \times 78.5}{689} = 114$ mm.

However provide 10 mm Φ bars @ 100 mm c/c. Half of these bars may be curtailed at 1 m from the base.

Area of distribution steel $= 0.3 - \frac{0.1\,(27-10)}{35} = 0.25\%$

$\therefore \quad A_{sd} = \frac{0.25}{100} \times 270 \times 1000 = 675$ mm². Area on each face $= 338$ mm².

Using 8 mm Φ bars, spacing $= \frac{1000 \times 50.2}{338} \approx 140$ mm.

**3. Design of horizontal slab**

Let the thickness of slab be 300 mm. Effective span of slab $= 5 + 0.27 = 5.27$ m.
Load of water per sq. m of slab $= 2.3 \times 9800 = 22540$ N.
Self weight per sq. m of slab $= 1 \times 1 \times 0.3 \times 25000 = 7500$ N
$\therefore \quad$ Total load $p = 22540 + 7500 = 30040$ N.

Total water pressure on the vertical wall = $\dfrac{9800(2.3)^2}{2} = 25921$

∴ Fixing moment at the end of slab = $25921\left(\dfrac{2.3}{3} + 0.15\right) = 23761$ N-m.

Simply supported B.M. at the centre of span of slab = $\dfrac{pL^2}{8} = \dfrac{30040\,(5.27)^2}{8} = 104287$ N-m.

∴ Net B.M. at the centre = $104287 - 23761 = 80526$ N-m. The slab is to be designed for this B.M. Since tension face is outside, $\sigma_{st} = 190$ and $R = 1.011$

$$\text{Effective depth} = \sqrt{\dfrac{80526 \times 1000}{1.011 \times 1000}} = 282 \text{ mm}$$

Provide total thickness = 340 mm, with cover of 50 mm

$$A_{st} = \dfrac{80526 \times 1000}{190 \times 0.892 \times 290} = 1638 \text{ mm}^2. \quad \text{Using 16 mm } \Phi \text{ bars, } A_\Phi = 201 \text{ mm}^2.$$

∴ Spacing = $\dfrac{1000 \times 201}{1638} = 122.7$ mm. However, provide 16 mm Φ bars @ 120 mm c/c.

Area of steel required at the end (near supports) = $\dfrac{23761 \times 1000}{150 \times 0.874 \times 290} = 625$ mm².

This is less than half the steel provided at the centre of span. However, half the bars from the centre of the span may be bent up at $L/2 \approx 1$ m form the supports. Let us check whether this bending of half bars satisfies the anchorage and development length requirements envisaged in the equation $M_1/V + L_0 \geq L_d$

where $M_1 = A_{st} \cdot \sigma_{st} j d = \left[\dfrac{1}{2} \cdot \dfrac{1000 \times 201}{120}\right] \times 190 \times 0.892 \times 290 = 41.16 \times 10^6$ N-mm

$V$ = shear force at the end = $\dfrac{30040 \times 5.27}{2} = 79155$ N

$L_0 = \left(\dfrac{l_s}{2} - x' - 3\,\Phi\right) + 16\,\Phi = \dfrac{l_s}{2} - x' + 13\,\Phi$

where $I_s$ = length of support = 270 mm and $x'$ = side cover = 40 mm

$\dfrac{M_1}{V} + L_0 = \dfrac{41.16 \times 10^6}{79155} + \left(\dfrac{270}{2} - 40 + 13 \times 16\right) = 823$ mm

$L_d = \dfrac{\Phi \sigma_{st}}{4\,\tau_{bd}} = \dfrac{\Phi\,(150)}{4 \times 0.8} = 46.88\,\Phi = 46.88 \times 16 = 750$ mm

Thus the requirement is satisfied.

Area of distribution steel = $0.3 - \dfrac{0.1\,(340 - 100)}{(450 - 100)} = 0.23\%$

∴ $A_{sd} = \dfrac{0.23}{100} \times 340 \times 1000 = 782$ mm². Area of each face = 391 mm²

∴ Spacing of 8 mm Φ bar = $\dfrac{1000 \times 50.2}{391} = 128$ mm

Hence provide 8 mm Φ distribution bars @ 125 mm c/c on each face.

### 4. Design of side walls as beam

Load from slab = $30040 \times 5/2 = 75100$ N/m
Self load = $2.5 \times 0.27 \times 25000 = 16875$ N/m
∴ Total load = $75100 + 16875 = 91975$ N/m

∴ Max. B.M. = $\frac{91975\,(8)^2}{8}$ = 735800 N/m

Using $\sigma_{st}$ = 190 N/mm², we have $k$ = 0.324, $j$ = 0.892 and $R$ = 1.011

∴ Effective depth = $\sqrt{\frac{735800 \times 1000}{1.011 \times 270}}$ = 1642 mm

Actual total depth = 2.5 + 0.3 = 2.8 m = 2800 mm, with an effective depth of say 2720 mm.

$$A_{st} = \frac{735800 \times 1000}{190 \times 0.892 \times 2720} = 1596 \text{ mm}^2.$$

Using 25 mm Φ bars,

No. of bars = $\frac{1596}{491} \approx 3.26$.

However, provide 4 bars of 25 mm Φ; these bars may be provided in two rows, 2 bars in each row. Actual $A_{st}$ = 1963 mm²   Shear force = 91975 × 8/2 = 367900 N

∴ $\tau_v = \frac{367900}{270 \times 2720} = 0.5$ N/mm²

Permissible shear stress for $\frac{A_{st}}{bd} \times 100 = \frac{1963}{270 \times 2720} \times 100 = 0.27$ is $\tau_c$ = 0.23 N/mm²

Hence shear reinforcement is necessary.

$V_c$ = shear resistance of concrete = $\tau_c\, b\, d$ = 0.23 × 270 × 2720 = 168912 N
$V_s = V - V_c$ = 367900 – 168912
= 198988 N.

Spacing of stirrups is given by :

$s_v = \frac{\sigma_{sv} \cdot d \cdot A_{sv}}{V_s} = \frac{190 \times 2720}{198988} A_{sv}$

= 2.597 $A_{sv}$

$(s_v)_{max} \leq \frac{2.175\, A_{sv}\, f_y}{b} \leq \frac{2.175 \times 415}{270} A_{sv}$

$\leq 3.343\, A_{sv}$

Hence $s_v$ = 3.343 $A_{sv}$. Using 8 mm dia. 2 lgd stirrups, $A_{sv}$ = 100.5 mm²

$s_v$ = 3.343 × 100.5 ≈ 336 mm, subject to a max. of 300 mm.

Hence provide stirrups @ 300 c/c. Provide 2-12 mm dia holding bars at the top.

The details of reinforcement etc. are shown in Fig. 30.3.

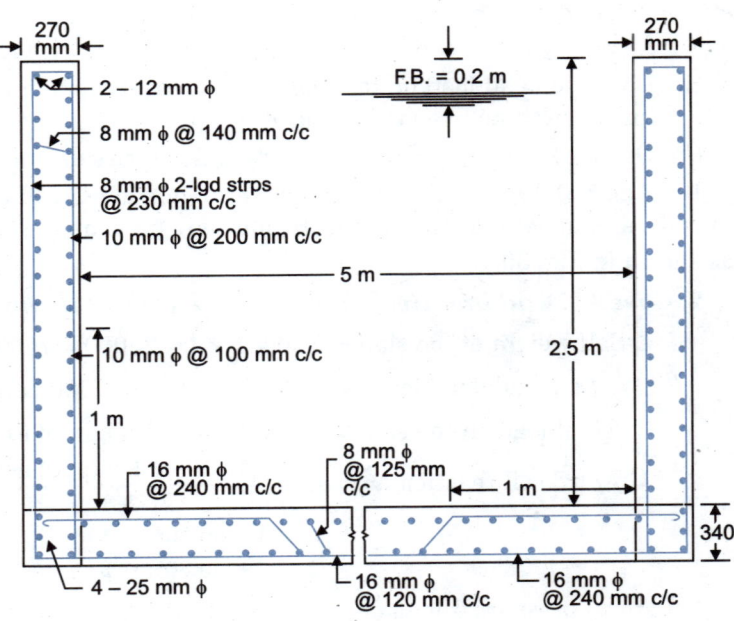

Fig. 30.3

## 30.3. BOX CULVERT

Box culverts consisting of two horizontal and two vertical slabs built Monolithically. A box culvert is used where a small drain crosses a high embankment of a road or a railway or a canal—specially when the bearing capacity of soil is low. A box culvert is continuous rigid frame of rectangular section in which the abutment and the top and bottom slabs are cast monolithic. In the case of high embankments, an

**742** REINFORCED CONCRETE STRUCTURE

ordinary culvert will require very heavy abutments which will be expensive and will transfer heavy loads to the foundations, while a R.C. box culvert will be cheaper. Due to light section of the abutments, load transferred to the foundations will also be lighter. A box culvert is subjected to soil load from outside and water load from inside. The vertical walls are subjected to earth pressure from outside and water pressure from inside. Similarly the bottom slab will be subjected to soil pressure from outside and water pressure from inside. The top slab will, however, be subjected to embankment weight and traffic loads, if any. A box culvert is therefore designed for two conditions.

(i) No water flowing in the drain. The box culvert will be dry from inside, and the sidewalls will be subjected to earth pressure from outside.

(ii) Water in box, which will be subjected to earth pressure from outside and water pressure from inside. Box culverts ideally suited for a road (or) railway bridge crossing with high embankment crossing a stream with a limited flow. The analysis is usually done using moment distribution method, as illustrated in example 30.2.

## 30.4. DESIGN OF BOX CULVERT

**Design: Example 30.2.** *Design a box culvert having inside dimensions 3.5 m × 3.5 m. The box culvert is subjected to a superimposed dead load of 12000 N/m² and a live load of 45000 N/m² from the top. Assume unit weight of soil as 18000 N/m³ and angle of repose of 30°. Use M 20 concrete and Fe 415 steel.*

**Solution:**

**1. General.** For the purpose of design, one metre length of the box is considered. The analysis is done for the following cases:

(i) Live load, dead load and earth pressure acting, with no water pressure from inside.

(ii) Live and dead load on top and earth pressure acting from outside, and water pressure acting from inside, with no live load on sides.

(iii) Dead load and earth pressure acting from outside and water pressure acting from inside.

Let the thickness of the vertical and horizontal slabs be 300 mm throughout. For analysis purposes, the centre line of the frame will be considered, having dimensions of 3.80 m × 3.80 m as shown in Fig. 30.4.

**2. Case I : Dead and live loads from outside while no water pressure from inside.**

Self weight of top slab = 0.30 × 1 × 1 × 25000 = 7500 N/m²

Live and dead loads = 45000 + 12000 = 57000 N/m²

∴ Total load on top slab = 64500 N/m². Weight of each side wall = 3.8 × 0.3 × 25000 = 28500 N/m

∴ Upward soil reaction at base = $\dfrac{(64500 \times 3.8) + (2 \times 28500)}{3.8}$ = 79500 N/m²

$$K_a = \dfrac{1 - \sin 30°}{1 + \sin 30°} = \dfrac{1}{3}$$

∴ Lateral pressure due to dead and live load = $p_v \cdot K_a$ = (12000 + 45000) $\dfrac{1}{3}$ = 19000 N/m².

Lateral pressure due to soil = $K_a \cdot wh = \dfrac{1}{3} \times 18000\, h = 6000\, h$

Hence total lateral pressure at depth $h$ is = 19000 + 6000 $h$

Lateral pressure intensity at top = 19000 N/m²

Lateral pressure intensity at bottom = 19000 + (6000 × 3.8) = 41800 N/m².

Fig. 30.4 shows the box culvert frame *ABCD*, along with the external loads. Due to symmetry, half of the frame (*i.e. AEFD*) of the box culvert is considered for moment distribution. Since all the members have uniform thickness, and of uniform dimensions, the relative stiffness $K$ for *AD* will be equal to 1 while

the relative stiffness for AE and DF will be 1/2.
Distribution factors for AD and DA

$$= \frac{1}{1+1/2} = \frac{2}{3}$$

Distribution factors for AB and DC

$$= \frac{1/2}{1+1/2} = \frac{1}{3}$$

The fixed end moments will be as under:

$$M_{FAB} = -\frac{wL^2}{12} = -\frac{64500\,(3.8)^2}{12}$$
$$= -77615 \text{ N-m}$$

$$M_{FDC} = +\frac{wL^2}{12} = +\frac{79500\,(3.8)^2}{12}$$
$$= +95665 \text{ N-m}$$

$$M_{FAD} = +\frac{pL^2}{12} + \frac{WL}{15} \text{ ; where } W \text{ is the total triangular earth pressure.}$$

$$M_{FAD} = \frac{19000\,(3.8)^2}{12} + (\frac{1}{2} \times 22800 \times 3.8) \times \frac{3.8}{15} \approx 22860 + 10970 = 33830 \text{ N-m}$$

$$M_{FDA} = -\frac{pL^2}{12} - \frac{WL}{10} = -\frac{19000\,(3.8)^2}{12} - (\frac{1}{2} \times 22800 \times 3.8) \times \frac{3.8}{10}$$
$$\approx -22860 - 16460 = -39320 \text{ N-m}$$

The moment distribution is carried out as illustrated in Table 30.1.

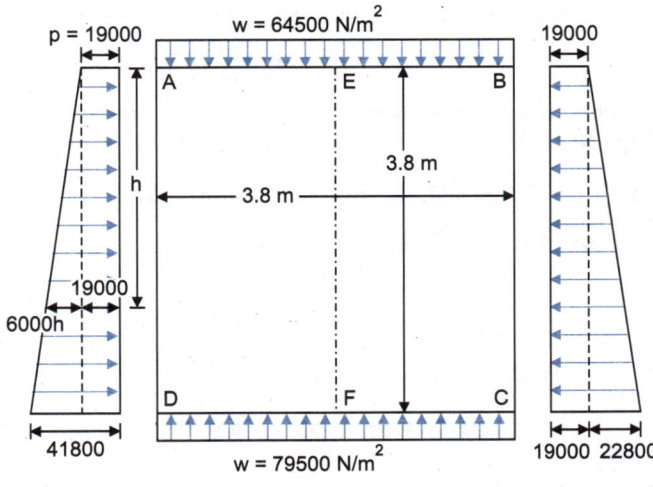

Fig. 30.4

**TABLE 30.1.** Moment Distribution for Case 1

| Joint | D | | A | |
|---|---|---|---|---|
| Member | DC | DA | AD | AB |
| Distribution factors | 1/3 | 2/3 | 2/3 | 1/3 |
| F.E.M. | +95665 | −39320 | +33830 | −77615 |
| Balance | −18782 | −37563 | +29190 | +14595 |
| Carry over | | +14595 | −18782 | |
| Balance | −4865 | −9730 | +12521 | +6261 |
| Carry over | | +6261 | −4865 | |
| Balance | −2087 | −4174 | +3243 | +1622 |
| Carry over | | +1622 | −2087 | |
| Balance | −541 | −1081 | +1391 | +696 |
| Carry over | | +696 | −541 | |
| Balance | −232 | −464 | +361 | +180 |
| Carry over | | +180 | −232 | |
| Balance | −60 | −120 | +155 | +77 |
| Carry over | | +77 | −60 | |
| Balance | −26 | −51 | +40 | +20 |
| Final moments | +69072 | −69072 | +54164 | −54164 |

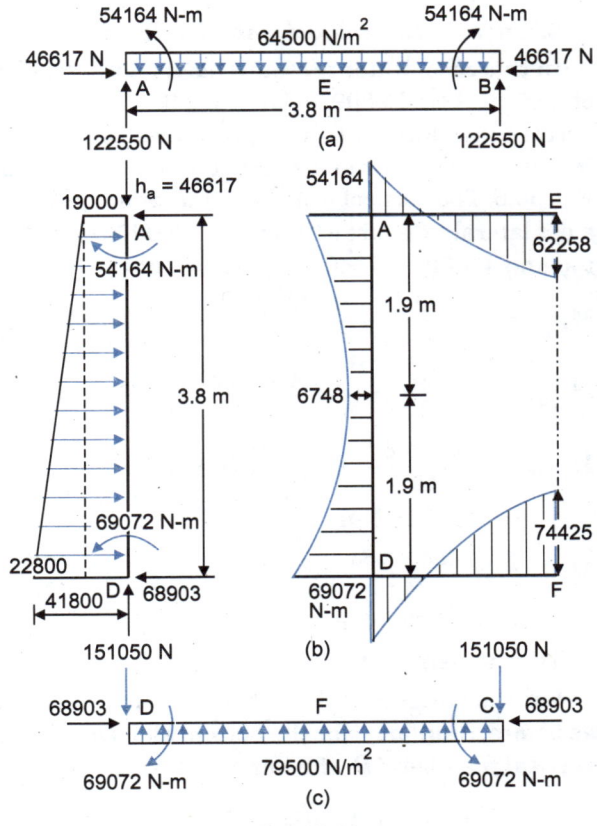

Fig. 30.5

For the horizontal slab AB, carrying U.D.L. @ 64500 N/m², vertical reactions at A and B

$$= \frac{1}{2} \times 64500 \times 3.8 = 122550 \text{ N}.$$

Similarly for the bottom slab $DC$ carrying U.D.L. @ 79500 N/m² vertical reactions at $D$ and $C$

$$= \frac{1}{2} \times 79500 \times 3.8 = 151050 \text{ N}.$$

The free body diagrams for various members, indicating loading, B.M. and reactions are shown in Fig. 30.5. For the vertical member $AD$, the horizontal reaction at $A$ is found by taking moments at $D$. Thus,

$$(-h_a \times 3.8) + 54164 - 69072 + 19000 \times 3.8 \times \frac{3.8}{2} + \frac{1}{2} \times 22800 \times 3.8 \times \frac{3.8}{3} = 0$$

From which $h_a = 46617$ N. Hence,

$$h_d = \left(\frac{19000 + 41800}{2}\right) \times 3.8 - 46617 = 68903 \text{ N}.$$

Free B.M. at mid-point $E = \dfrac{64500 \times 3.8^2}{8} \approx 116422$ N-m

∴ Net B.M. at $E = 116422 - 54164 = 62258$ N-m

Similarly, free B.M. at $F = \dfrac{79500 \times 3.8^2}{8} \approx 143497$ N-m

∴ Net B.M. at $F = 143497 - 69072 = 74425$ N-m

For the vertical member $AD$, simply supported B.M. at the mid-span

$$= \frac{19000 \times 3.8^2}{8} + \frac{1}{16}(22800)(3.8)^2 \approx 54870 \text{ N-m}$$

∴ Net B.M. $= \dfrac{69072 + 54164}{2} - 54870 \approx 6748$ N-m

### 3. Case 2 : Dead load and live load from outside and water pressure from inside.

In this case, water pressure having an intensity of zero at $A$ and $9800 \times 3.8 = 37240$ N/m² at $D$ is acting, in addition to the pressures considered in case 1. The various pressures are marked in Fig. 30.6. The vertical walls will thus be subjected to a net lateral pressure of intensity 19000 N/m² at the top and $41800 - 37240 = 4560$ N/m² at the bottom.

$M_{FAB} = -\dfrac{wL^2}{12} = -\dfrac{64500(3.8)^2}{12} = -77615$ N-m

$M_{FDC} = -\dfrac{w'L^2}{12} = +\dfrac{79500(3.8)^2}{12} = +95665$ N-m

$M_{FAD} = \dfrac{4560(3.8)^2}{12} + (\dfrac{1}{2} \times 14440 \times 3.8) \times \dfrac{3.8}{10}$

$= +15913$ N-m

$M_{FDA} = -\dfrac{4560(3.8)^2}{12} - (\dfrac{1}{2} \times 14440 \times 3.8) \times \dfrac{3.8}{15}$

$= -12438$ N-m

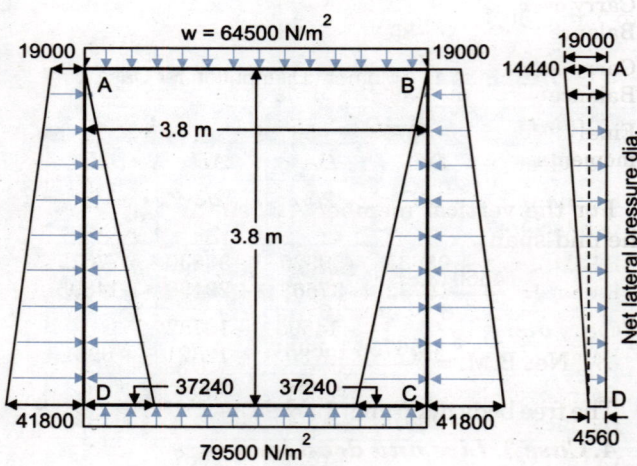

Fig. 30.6

The moment distribution is carried out as illustrated in Table 30.2.

For the horizontal slab $AB$, carrying U.D.L. @ 64500 N/m², vertical reactions at $A$ and $B = 122550$ N-m as before. Similarly, for the bottom slab $DC$, the vertical reactions at $D$ and $C$ will be 151050 N each. For vertical member $AD$, the horizontal reaction at $A$ is found by taking moments at $D$. Thus,

$$(-h_a \times 3.8) + 44093 - 56758 + 4560 \times \frac{3.8^2}{2} + \left(\frac{1}{2} \times 14440 \times 3.8\right)\left(\frac{2}{3} \times 3.8\right) = 0.$$

From which $h_a \approx 23622$ ;   ∴ $h_d = \left(\dfrac{19000 + 4560}{2}\right) \times 3.8 - 23622 = 21142$

Free B.M. at mid-point $E = \dfrac{64500\,(3.8)^2}{8} \approx 116423$ N-m

$\therefore$ Net B.M. at $E = 116423 - 44093 = 72330$ N-m

Similarly, free B.M. at $F = \dfrac{79500\,(3.8)^2}{8} \approx 143498$ N-m

Net B.M. at $F = 143498 - 56758 = 86740$ N-m

**TABLE 30.2.** Moment Distribution for Case 2

| Joint | D | | A | |
|---|---|---|---|---|
| Member | DC | DA | AD | AB |
| Distribution factors | 1/3 | 2/3 | 2/3 | 1/3 |
| F.E.M. | +95665 | −12438 | +15913 | −77615 |
| Balance | −27742 | −55485 | +41135 | +20567 |
| Carry over | | +20567 | −27742 | |
| Balance | −6856 | −13711 | +18495 | +9247 |
| Carry over | | +9247 | −6856 | |
| Balance | −3082 | −6165 | +4571 | +2285 |
| Carry over | | +2285 | −3082 | |
| Balance | −762 | −1523 | +2055 | +1027 |
| Carry over | | +1027 | −762 | |
| Balance | −342 | −685 | +508 | +254 |
| Carry over | | +254 | −342 | |
| Balance | −85 | −169 | +228 | +114 |
| Carry over | | +114 | −85 | |
| Balance | −38 | −76 | +57 | +28 |
| Final moments | +56758 | −56758 | +44093 | −44093 |

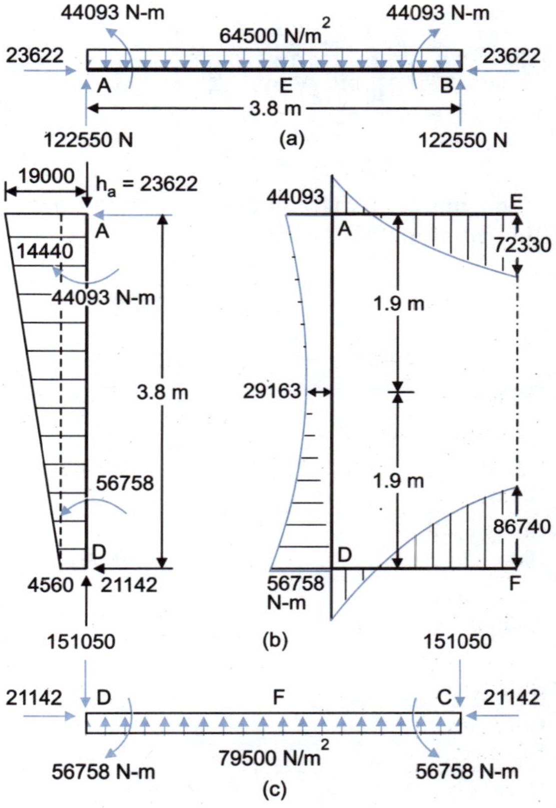

For the vertical member $AD$, simply support B.M. at the mid-span

$$= \dfrac{4560\,(3.8)^2}{8} + \dfrac{1}{16}(14440)(3.8)^2 = 21263 \text{ N-m}$$

$\therefore$ Net B.M. $= \dfrac{56758 + 44093}{2} - 21263 \approx 29163$ N-m.

The free body diagrams for various members, indicating loading, B.M. and reactions are shown in Fig. 30.7.

**4. Case 3. Live and dead load on top and water pressure from inside: no live load on the sides.**

In this case, it is assumed that there is no lateral pressure due to live load. As before, the top slab is subjected to a load of 64500 N/m² and the bottom slab is subjected to load intensity of 79500 N/m².

Lateral pressure due to dead load = (1/3) × 12000 = 4000 N/m²
Lateral pressure due to soil = (1/3) × 18000 = 6000 $h$
Hence earth pressure at depth $h$ is = 4000 + 6000 $h$
Earth pressure intensity at top = 4000 N/m².
Earth pressure intensity at bottom = 4000 + 6000 (3.8) = 4000 + 22800 = 26800 N/m²

In addition to these, the vertical walls will be subjected to water pressure of intensity zero at top and 37240 N/m² at bottom, acting from inside. The lateral pressure on vertical walls is shown in Fig. 30.8(*b*).

The fixed end moments will be as under :

$$M_{FAB} = -\dfrac{wL^2}{12} = -77615 \text{ kg-m} \quad \text{(as before)}$$

Fig. 30.7

$$M_{FDC} = -\frac{w'L^2}{12}$$
$$= +95665 \text{ N-m (as before)}$$

$$M_{FAD} = \frac{pL^2}{12} - \frac{WL}{15}$$
$$= \frac{4000(3.8)^2}{12} - \left(\frac{1}{2} \times 14440 \times 3.8\right)\frac{3.8}{15}$$
$$= 2137 \text{ N-m.}$$

$$M_{FDA} = -\frac{pL^2}{12} + \frac{WL}{10}$$
$$= -\frac{4000(3.8)^2}{12} + \left(\frac{1}{2} \times 14440 \times 3.8\right)\frac{3.8}{10}$$
$$= +5612 \text{ N-m}$$

The moment distribution is carried out as illustrated in Table 30.3.

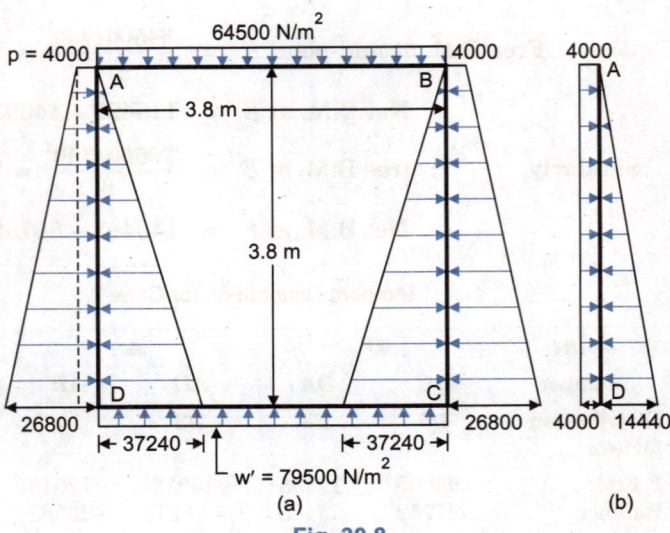

Fig. 30.8

**TABLE 30.3.** Moment Distribution for Case 3

| Joint | D | | A | |
|---|---|---|---|---|
| Member | DC | DA | AD | AB |
| Distribution factors | 1/3 | 2/3 | 2/3 | 1/3 |
| F.E.M. | +95665 | +5612 | −2137 | −77615 |
| Balance | −33759 | −67518 | +53168 | +26584 |
| Carry over | | +26584 | −33759 | |
| Balance | −8861 | −17723 | +22506 | +11253 |
| Carry over | | +11253 | −8861 | |
| Balance | −3751 | −7502 | +5907 | +2954 |
| Carry over | | +2954 | −3751 | |
| Balance | −985 | −1969 | +2501 | +1250 |
| Carry over | | +1250 | −985 | |
| Balance | −417 | −833 | +657 | +328 |
| Carry over | | +328 | −417 | |
| Balance | −109 | −219 | +278 | +139 |
| Carry over | | +139 | −109 | |
| Balance | −46 | −93 | +73 | +36 |
| Final moments | +47737 | −47737 | +35071 | +35071 |

For the horizontal slab. $AB$, carrying U.D.L. @ 64500 N/m², vertical reactions at $A$ and $B$ = 122550 N, as before. Similarly, for bottom slab $DC$, the vertical reactions at $D$ and $C$ will be 151050 N each.

For vertical member $AD$, the horizontal reaction at $A$ is found by taking moments at $D$. Thus,

$$(h_a \times 3.8) + 35071 - 47737 + \frac{4000(3.8)^2}{2} - \left(\frac{1}{2} \times 14440 \times 3.8\right)\frac{3.8}{3} = 0$$

From which $h_a \approx 4878$ N;

$$\therefore \quad h_d = (\frac{1}{2} \times 14440 \times 3.8) - (4000 \times 3.8) - 4878 = 7358 \text{ N}$$

Free B.M. at midpoint $E = \dfrac{64500(3.8)^2}{8} \approx 116423$ N-m

$\therefore \quad$ Net B.M. at $E$ = 116423 − 35071 = 81352 N-m

Similarly, free B.M. at $F = \dfrac{79500(3.8)^2}{8} \approx 143498$ N-m

∴ Net B.M. at $F$
$$= 143498 - 47737 = 95761 \text{ N-m}$$

For the vertical member $AD$, simply supported, B.M. at mid span

$$= -\frac{4000 \,(3.8)^2}{8} + \frac{1}{16}(14440)(3.8)^2$$
$$= 5812 \text{ N-m}$$

∴ Net B.M. $= \dfrac{47737 + 35071}{2} + 5812$

$$= 47216 \text{ N-m}$$

Let the maximum hogging B.M. occur at a distance $x$ below $A$.

$$M_x = 35071 + 4878\,x + \frac{4000\,x^2}{2}$$
$$-\left(\frac{14440}{3.8}x\right)\left(\frac{1}{2}x\right)\left(\frac{x}{3}\right)$$

or $M_x = 35071 + 4878\,x + 2000\,x^2 - 633.3\,x^3$

For maximum,

$$\frac{dM_x}{dx} = 4878 + 4000\,x - 633.3 \times 3\,x^2 = 0,$$

which gives $x \approx 2.966$ m

∴ $M_{max} = 35071 + 4878\,(2.966) + 2000\,(2.966)^2$
$- 633.3\,(2.966)^3 \approx 50609.2$ say $50610$ N-m

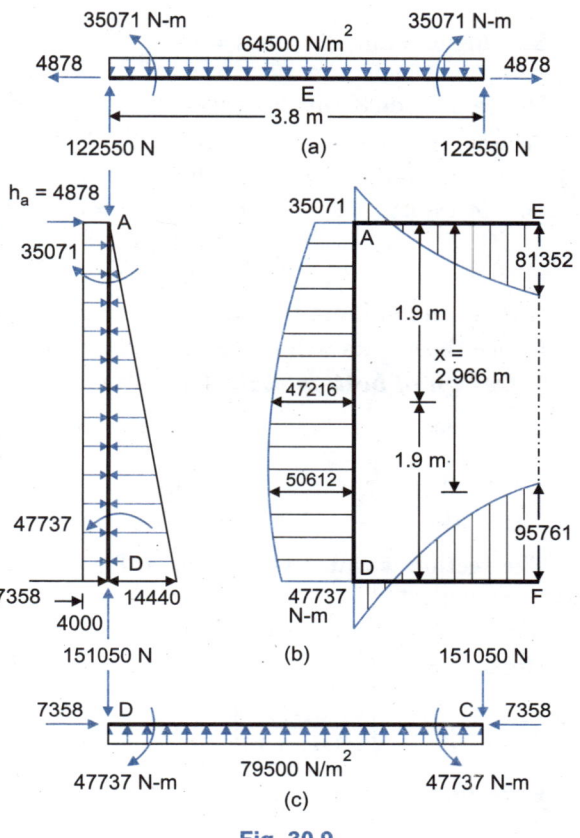

Fig. 30.9

**5. Design of top slab:** The top slab is subjected to the following values of B.M. and direct forces :

| Case | B.M. at centre (N-m) | B.M. at ends (N-m) | Direct force (N) |
| --- | --- | --- | --- |
| I | 62258 | 54164 | 46617 (comp.) |
| II | 72330 | 44093 | 23622 (comp.) |
| III | 81352 | 35071 | 4878 (tensile) |

The section will be designed for a maximum B.M. of 81352 N-m, induced at mid-span. The direct force of 4878 N(tensile) is quite small and may be neglected. Overall depth = 300 mm ; using a cover of 50 mm to the centre of steel, effective depth = 250 mm.

Use M 20 concrete. Since tension occurs at the inner face (*i.e.* towards water side),

$\sigma_{st} = 150$ N/mm$^2$ for HYSD bars. Hence $k = 0.375$ ; $j = 0.874$ and $R = 1.155$.

∴ $M_r = Rbd^2 = 1.155 \times 1000\,(250)^2 = 72187 \times 10^3$ N-mm against a required value of $81352 \times 10^3$

Hence increase the thickness to 320 mm, so that $d = 270$ mm

∴ $M_r = 1.155 \times 1000\,(270)^2 = 84200 \times 10^3$ N-mm

$$A_{st} = \frac{81352 \times 10^3}{150 \times 0.874 \times 270} = 2298 \text{ mm}^2$$

Using 20 mm Φ bars having $A_\Phi = 314$ mm$^2$, spacing $= \dfrac{1000 \times 314}{2298} \approx 137$ mm

However, provide 20 mm dia. bars @ 130 mm c/c. Actual $A_{st}$ provided $= \dfrac{1000 \times 314}{130} = 2415$ mm$^2$.

Bend half the bars up near supports at a distance of $\dfrac{L}{5} = \dfrac{3.5}{5} \approx 0.8$ m from edges.

Area of distribution steel $= 0.3 - \dfrac{0.1\,(320 - 100)}{(450 - 100)} \approx 0.24\,\%$

∴ $A_{st} = 0.24 \times 320 \times 10 = 768$ mm$^2$. Area on each face = 384 mm$^2$

Spacing of 8 mm dia. bars = $\dfrac{1000 \times 50.2}{384}$ = 130.7 mm

Hence provide 8 mm Φ bars @ 130 mm c/c on each face.

*Section at supports* : Max. B.M. = 54164 N-m. There is direct compression of 46617 N also, but its effect is not considered because the slab is actually reinforced both at the top and bottom. Since steel is at top, $\sigma_{st}$ = 1900 N/m². Hence $k$ = 0.324 ; $j$ = 0.892 and $R$ = 1.011

$$\therefore \quad A_{st} = \dfrac{54164 \times 1000}{190 \times 0.892 \times 270} \approx 1184 \text{ mm}^2.$$

Area available from the bars bent up from the middle section = 2415/2 ≈ 1207 mm²

Hence these bars will serve the purpose. However, provide 8 mm dia. additional bars @ 200 mm c/c.

**6. *Design of bottom slab:*** The bottom slab has the following values of B.M. and direct forces :

| Case | B.M. at centre (N-m) | B.M. at ends (N-m) | Direct force (N) |
|------|----------------------|--------------------|--------------------|
| I    | 74425                | 69072              | 68903 (comp.)    |
| II   | 86740                | 56758              | 21142 (comp.)    |
| III  | 95761                | 47737              | 7358 (tension)   |

The section is thus to be designed for a B.M. of 95761 N-m. Since the direct tension is small, its effect may be neglected.

Since tension is towards water side, we have $\sigma_{st}$ = 150 N/mm², for HYSD bars.

Hence $k$ = 0.375, $j$ = 0.874 and $R$ = 1.155.

$$\therefore \quad d = \sqrt{\dfrac{95761 \times 1000}{1000 \times 1.155}} = 288 \text{ mm, and } D = 288 + 50 = 338 \text{ mm}$$

Provide $D$ = 340 mm so that $d$ = 290 mm $\quad \therefore \quad A_{st} = \dfrac{95761 \times 1000}{150 \times 0.874 \times 290}$ = 2519 cm²

$\therefore$ Spacing of 20 mm Φ bars = $\dfrac{1000 \times 314}{2519} \approx 124$ cm

However, provide these @ 120 mm c/c. Actual area = 2616 mm². Bend half the bars at the outer face near supports.

Provide distribution reinforcement of 8 mm dia. bars @ 130 mm c/c on each face.

At the ends, the slab is subjected to a B.M. of 69072 N-m and direct force of 68903N. Since steel is at the outer face, $\sigma_{st}$ = 190 N/mm² for HYSD bars. Hence $k$ = 0.324, $j$ = 0.892 and $R$ = 1.011.

$$A_{st} = \dfrac{69072 \times 1000}{190 \times 0.892 \times 290} = 1405 \text{ mm}^2$$

Area available from the bars bent down from the middle section = 2616/2 = 1308 mm². additional area required = 1405 – 1308 = 97 mm² only.

However, provide 8 mm dia. additional bars @ 200 mm c/c throughout the slab, at its bottom.

**7. *Design of side walls*** : The side walls are subjected to the following values of B.M. and direct forces:

| Case | B.M. at centre or intermediate point (N-m) | Max. B.M. at ends (N-m) | Direct force (N) |
|------|--------------------------------------------|--------------------------|------------------|
| I    | 6748                                       | 69072                    | 151050           |
| II   | 29163                                      | 56758                    | 151050           |
| III  | 50610                                      | 47737                    | 151050           |

The side walls are thus subjected to a maximum B.M. of 69072 N-m and a direct compression of 151050 N.

$$\text{Eccentricity} = e = \dfrac{69072 \times 1000}{151050} = 457.3 \text{ mm}$$

Let the thickness of side walls be 320 mm.

$$\therefore \quad \dfrac{e}{D} = \dfrac{457.3}{320} = 1.43 < 1.5$$

Hence the eccentricity corresponds to case 3 (See Ch. 13).

Let us reinforce the section with 20 mm Φ bars @ 300 mm c/c provided on both the faces, as shown in Fig. 30.10, with a cover of 50 mm.

$$A_{sc} = A_{st} = \frac{1000}{300}\left(\frac{\pi}{4} \times 20^2\right) = 1047 \text{ m}^2.$$

The depth of N.A. is computed from the following expression: (*Eq. 13.17*) :

$$\frac{\dfrac{bn}{2}\left(D - d_t - \dfrac{n}{3}\right) + (m-1)\,A_{sc}\,\dfrac{1}{n}(n-d_c)(D-d_t-d_c)}{\dfrac{bn}{2} + (m-1)\,A_{sc}\left(\dfrac{n-d_c}{n}\right) - m\,A_{st}\left(\dfrac{D-d_t-n}{n}\right)} = \left(e + \dfrac{D}{2} - d_t\right)$$

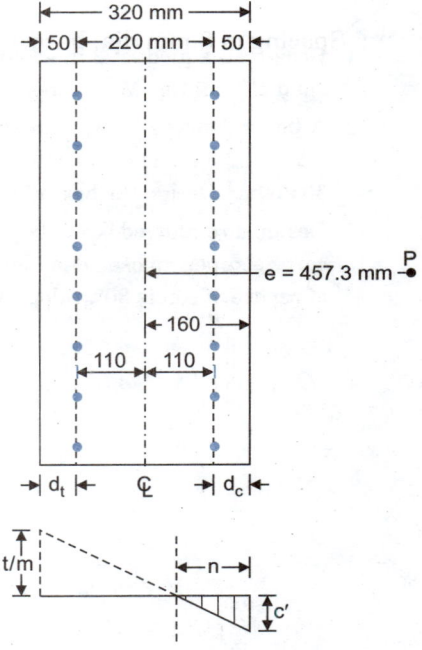

Fig. 30.10

$$\therefore \quad \frac{500\,n\left(320 - 50 - \dfrac{n}{3}\right) + 12 \times \dfrac{1047}{n}(n-50)(320-50-50)}{500\,n + 12 \times \dfrac{1047}{n}(n-50) - 13 \times \dfrac{1047}{n}(320-50-n)}$$

$$= (457.3 + 160 - 50)$$

or

$$\frac{500\,n\left(270 - \dfrac{n}{3}\right) + \dfrac{2764080}{n}(n-50)}{500\,n + \dfrac{12564}{n}(n-50) - \dfrac{13611}{n}(270-n)} = 567.3$$

Simplifying and rearranging the terms, we get $n^3 = 892\,n^2 + 72495\,n - 13815143 = 0$

Solving this by trial and error, we get $n = 87.5$ mm

$$\therefore \quad c'\left[(500 \times 87.5) + \frac{12 \times 1047}{87.5}(87.5 - 50) - \frac{13 \times 1047}{87.5}(320 - 50 - 87.5)\right] = 151050$$

$$\therefore \quad c' = \frac{151050}{20746} = 7.28 \text{ N/mm}^2$$

This is more than permissible stress of 7 N/mm².

Also, stress in steel,

$$t = \frac{m\,c'}{n}(D - d_t - n)$$

$$= \frac{13 \times 7.28}{87.5}(320 - 50 - 87.5)$$

$$= 197.4 \text{ N/mm}^2 > 190.$$

Thus we see that stress in steel is also more than permissible. Hence revision of section is necessary.

Adopt $D = 340$ mm.

Provide 20 mm Φ bars @ 250 mm c/c.

It will be seen that stresses in both the materials come within safe limits. (The readers are advised to work out stresses in both concrete and steel.)

Provide distribution reinforcement of 8 mm dia. bars @ 130 mm c/c on both faces. The details of reinforcement etc. are shown in Fig. 30.11.

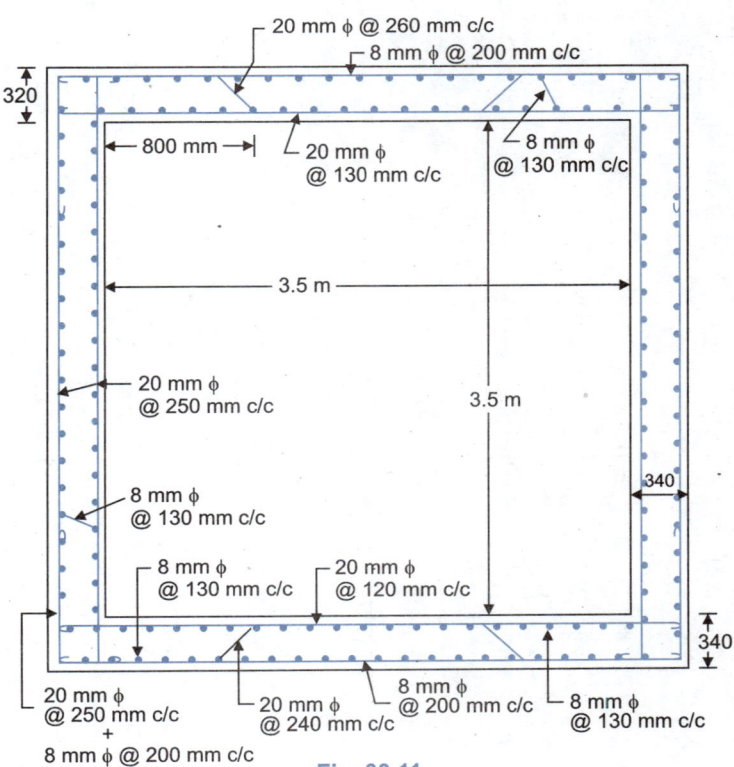

Fig. 30.11

## PROBLEMS

1. Design an aqueduct 3 m wide and 1.5 deep, to carry a maximum depth of 1.3 m of water. The aqueduct has a span of 6 m. Use M 20 concrete and HYSD bars.
2. A box culvert is to be provided for a drain under a highway. The inside dimensions of the box culvert is 2.5 m × 2.5 m. The box culvert is expected to carry a super-imposed dead load of 15 kN/m² and a live load of 50 kN/m². Design the box culvert. Take $\gamma = 18.3$ kN/m³ and $\phi = 30°$ for the soil. Use M 20 concrete and HYSD bars.
3. Design a reinforced box culvert with inside dimensions of 3.5 m height and 4.25 m width. The box culvert has to carry a superimposed dead load of 10 kN/m² and live load of 52 kN/m² the density of earth is 18 kN/m³. Angle of repose of soil is 30°. Adopt $M_{30}$ grade and HYSD bars and also sketch reinforcement details.

# 31 CONCRETE BRIDGES

## 31.1. INTRODUCTION: VARIOUS TYPES OF BRIDGES

Reinforced concrete is increasingly used for highway and railway bridge construction due to its durability, rigidity, economy, ease of construction and ease with which pleasing appearance can be made in it. Reinforced concrete bridges may be of following types and most commonly used type of reinforced concrete bridge decks are:

1. Solid slab bridge *or* deck slab bridge. 2. Deck girder bridge (T-beam bridge). 3. Balanced cantilever bridge. 4. Rigid frame culvert *or* bridge (single span as well as multispan) 5. Arch bridge. 6. Bowstring grider bridge. 7. Continuous girder *or* arch bridge.

The reinforced concrete slab type deck is generally used for small spans.

A deck slab bridge [Fig. 31. 1 (*a*)] *or* solid slab bridge is the simplest type of construction, used mostly for culverts *or* small bridges with a span not exceeding 8 m. Though the thickness of deck slab is considerable, its construction is much simpler and the cost of form work is also minimum.

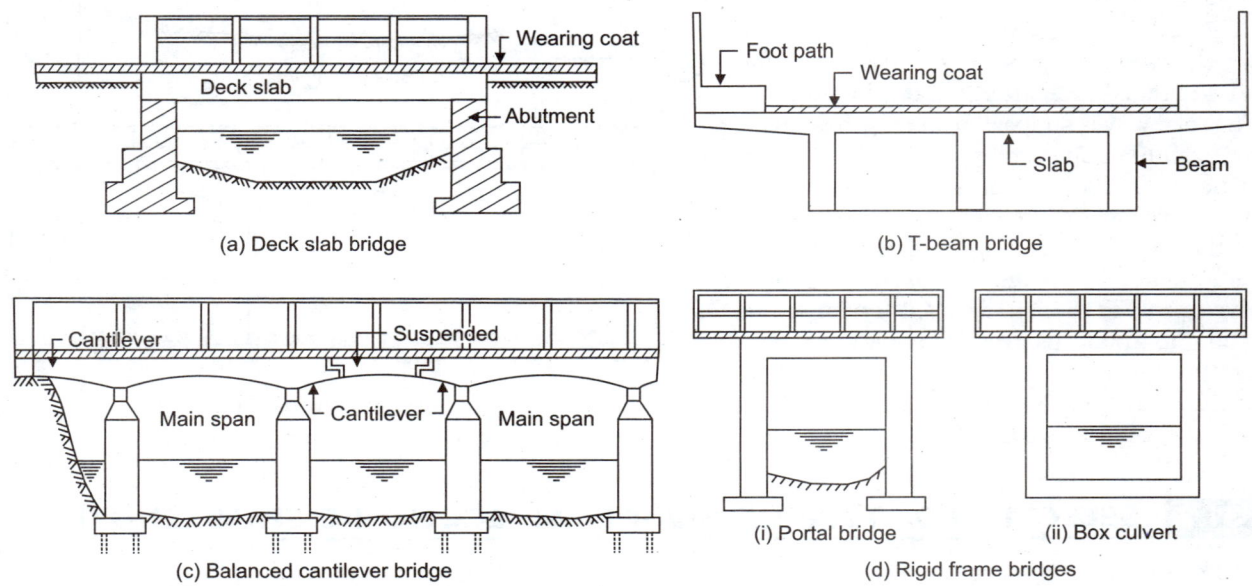

Fig. 31.1. Beam-slab Bridges.

Deck girder bridge [Fig. 31.1 (b)] *or* T-beam bridge is another type of a simple R.C. bridge used for spans between 10 to 20 metres. The number of longitudinal girders depends upon the road width. The slab is generally built monolithic with girders so that T-beam effect is achieved.

A balanced cantilever bridge is a statically determinate structure, used for spans between 25 to 50 metres. It consists of alternate spans with projecting cantilevers, the ends of which are used to support a suspended span [Fig. 31.1 (c)]. The joint between suspended span and edge of cantilever is known as *articulation*. A balanced cantilever bridge is used where the width of river is large, but where due to the possibility of unequal settlement, a continuous girder bridge cannot be used. Generally, a parabolic profile for the main girder is used.

A rigid frame bridge is used for only small drains, in which the vertical abutments are cast monolithic with deck-slab. If the foundation conditions are good, the rigid frame may be of *portal type* [Fig. 31.1(d-i)]. However, if the bearing capacity is poor, a *box culvert* [Fig. 31.1(d-ii)] may be used.

Arch bridges are used for long spans where the use of beams becomes uneconomical. If properly designed, the dead load moment in an arch are almost absent. Due to this, arch bridges offers a large head room for navigation. Arch bridges also offer pleasing appearance. The arches may be in the form of *arch slab* or *rib*. In arch slab or arch barrel, the deck is generally supported on earth filling placed on arch slab; such an arch is known as open spandrel arch [Fig. 31.2(a)]. The earth is retained by spandrel walls. In arch rib, called *open spandrel arch*, [Fig. 31.2(b)] the deck is supported on columns. The columns are supported on the arch rib. Both the filled spandrel arch as well as open spandrel arch employ the fixed arch as the main supporting member.

Fig. 31.2 (c) shows a *bow string girder bridge*, which is a special type of arch bridge employing a two hinged arch standing on the abutments. The road way is supported on cross beams which in turn are supported by vertical suspenders at either end. The ends of the arch are connected by a tie rod called the *string*. The arch is termed as the *bow*. The arches are supported on roller and rocker bearings. A *continuous bridge*, shown in Fig. 31.2(d) is used for large spans, where foundations are of unyielding type. The continuous spans may have either girders, or arches. Generally, continuous girders with parabolic shape are used.

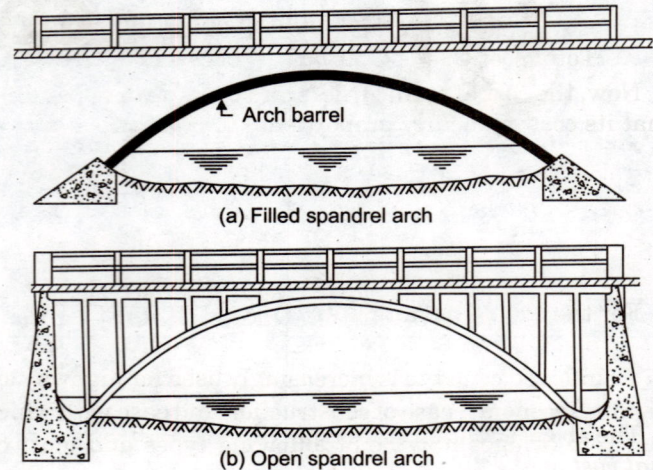
(a) Filled spandrel arch

(b) Open spandrel arch

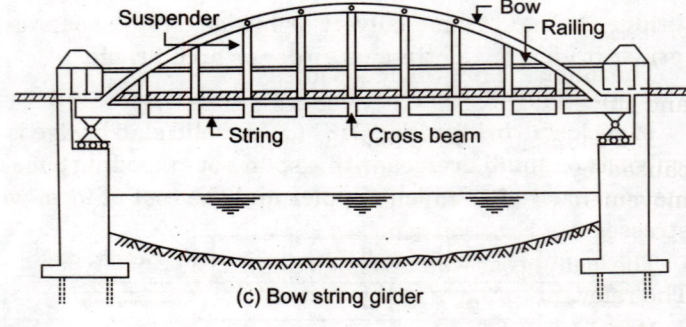
(c) Bow string girder

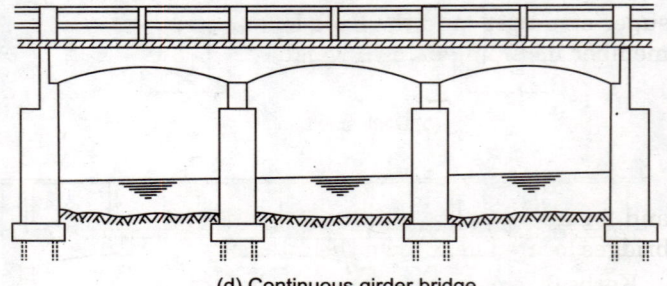
(d) Continuous girder bridge

**Fig. 31.2.** Types of Girder Bridges.

## 31.2. SELECTION OF TYPE OF BRIDGE AND ECONOMIC SPAN LENGTH

The selection of a type of bridge at a certain location depends upon two factors: (i) cost consideration and (ii) natural conditions such as nature of foundation, water way to be provided, width of bridge, type of loading and construction methods. The factors involved are so large and their effect so complicated that it is difficult to formulate the selection of the type of bridge. Generally, experience is invaluable in deciding the type of bridge. The total cost of bridge consists of  (i) cost of super-structure and  (ii) cost of substructure, such as piers, foundation etc. The cost in the pier and its foundation will not be affected much for small variation in the length of span. Let $L$ be the total length of bridge and $n$ be the number of piers. Then the number of spans $= n + 1$, and the length of each spans $l = \dfrac{L}{n+1}$.

Let $\quad C_A$ = Cost of abutments ; $\quad C_P$ = Cost of one pier
$\quad C_G$ = Cost of main girder and cross girders per unit length
$\quad C_F$ = Cost of flooring per unit length.

∴ Total cost $C = C_A + n\, C_P + LC_G + LC_F$ ...(1)

Now, the cost of main girders and cross girders per unit length will depend upon its span $l$. Let us assume that its cost is linearly proportional to its span:

$$C_G \propto l = kl. \qquad \text{Also} \qquad n = \dfrac{L}{l} - 1$$

∴ $$C = C_A + C_P\left(\dfrac{L}{l} - 1\right) + Lkl + LC_F$$

For the cost to be minimum, $\dfrac{dC}{dl} = 0.$ ∴ $-C_P \dfrac{L}{l^2} + Lk = 0$

or $\qquad \dfrac{C_P}{l} = kl = C_G \qquad$ or $\qquad C_P = lC_G$ ...(31.1)

The above equation states that *the total cost of the bridge will be minimum when the spans are so arranged that cost of one pier is equal to the cost of girders of one span.*

## 31.3. TYPES OF LOADS, FORCES AND STRESSES

The following are the loads forces and stresses to be considered in designing super structures of bridges and culverts:

1. Dead load  2. Live load  3. Impact or dynamic effect of live load  4. Wind load  5. Longitudinal forces caused by the tractive effort of vehicles *or* by braking of vehicles and/*or* those caused by restraint to movement of free bearings.  6. Centrifugal forces  7. Seismic forces  8. Temperature stresses  9. Secondary stresses  10. Erection stresses.

The loadings as well as design criteria are laid down by the Indian Roads Congress (IRC) Codes of practice. The relevant extracts of the Code are reproduced below.

**Dead Load:** The dead load carried by a girder *or* member shall consist of the portion of the weight of the super-structure (and the fixed loads carried there on) which is supported wholly *or* in part by the girder *or* member including its own weight.

## 31.4. LIVE LOAD

**Footway and Kerb Loading:** For all parts of bridge floors accessible only to pedestrians and animals and for all footways, the loading shall be 4000 N/m². Where crowd loads are likely to occur, such as on bridges located near town the intensity of footway loading shall be increased from 4000 N/m² to 5000 N/m².

Kerbs 60 cm *or* more in width shall be designed for the above loads. If the kerb width is less than 60 cm, no live load shall be applied in addition to the lateral load.

In calculating stresses in members of structure with cantilevered footways, the footways shall be considered as located on one side *or* on both sides, *or* unloaded, whichever condition gives the maximum stresses. The main girders, trusses, arches *or* other members sporting the footways shall be designed for the following live loads per square metre of the footway area, the loaded length of the footway taken in each case being such as to produce the worst effects on the member under consideration:

(a) For effective span of 7.5. m or less : 4000 N/m² (*or* 5000 N/m² for crowded area.).

(b) For effective spans of over 7.5 m but not exceeding 30 m, the intensity of load shall be determined according to the equation.

$$P = P' - \left(\frac{400L - 3000}{9}\right) \qquad \ldots(31.2)$$

(c) For effective spans of over 30 m, the intensity of load shall be determined according to the equation:

$$P = \left(P' - 2600 + \frac{48000}{L}\right)\left(\frac{16.5 - W}{15}\right) \qquad \ldots(31.3)$$

where $P' = 4000$ N/m$^2$ or $5000$ N/m$^2$, as the case may be ; $P$ = live load in N/m$^2$
$L$ = effective span of the main girder, truss or arch, in metres.
and $W$ = width of roadway, in metres.

Each part of footway shall be capable of carrying a wheel load of 40 kN, which shall be deemed to include impact, distributed over a contact area of 300 mm in diameter; the working stresses shall be increased by 25% to meet this provision. This provision need not be made where vehicles cannot mount the footway as in the case of a footway separated from the roadway by means of an insurmountable obstacle, such as truss or a main girder. A footway kerb shall be considered mountable by vehicles.

(d) **For Road Bridges:** Road bridges are designed for the following classes of loadings. These loadings consist of trains of vehicles with known axle loads.

Though SI unit are widely used in India, the IRC loadings have yet not been converted into SI units as yet. However, since the present book is entirely in SI units, the IRC loadings have been converted into SI units, using the following approximate relations:

1 kg ≈ 10 N        1 tonne ≈ 10 kN

**IRC Class AA Loading.** This loading is to be adopted within certain municiple limits, in certain existing or contemplated industrial areas, in other specified areas and along certain specified highways. Bridges designed for class AA loading should be checked for class A loading also, since under certain conditions, heavier stresses may be obtained under class A loading.

**IRC Class A Loading.** This is the most common type of loading adopted for prominent bridges and culverts.

**IRC Class B Loading.** This loading is adopted for temporary structures, and for bridges in specified areas. The structures with timber spans are to be regarded as temporary structures.

The standard wheeled or tracked vehicles or trains of vehicles, for the above types of loadings, are illustrated in Figs. 31.3, 31.4 and 31.5. The trailers attached to the driving unit are not be considered as detachable.

Within the kerb to kerb width of the roadway, the standard vehicle or train shall be assumed to travel parallel to the length of bridge, and to occupy any position which will produce maximum stresses provided that the minimum clearance between a vehicle and the roadway face of a kerb and between two passing or crossing vehicles are not encroached upon.

For each standard vehicle or train, all the axles of a unit of vehicles shall be considered as

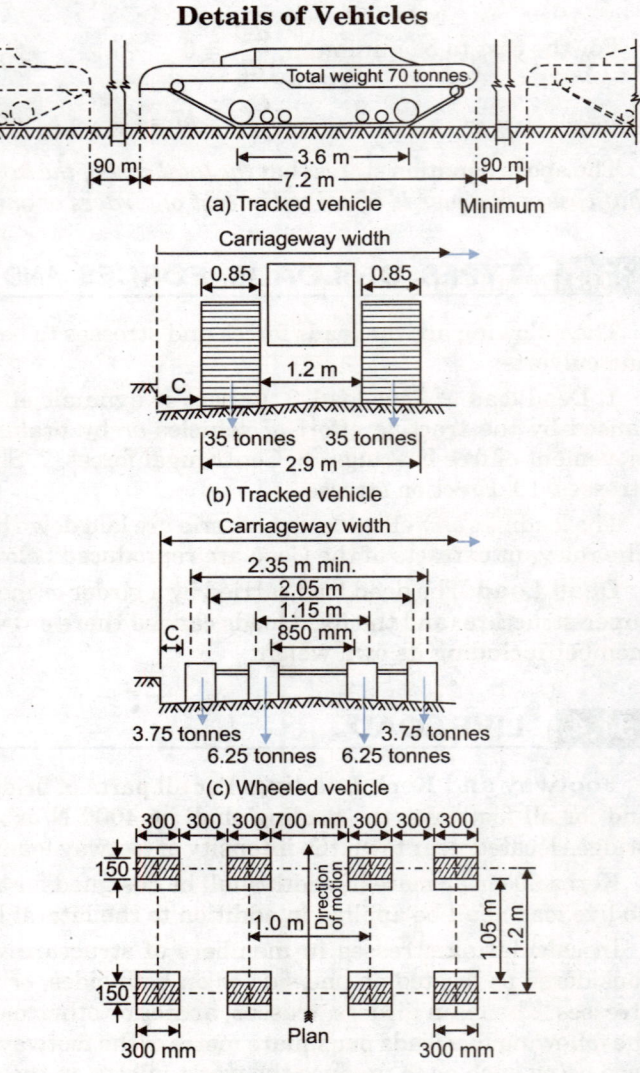

Fig. 31.3. IRC class AA loading. (Note: 1 Tonne ≈ 10 kN)

acting simultaneoulsy in a position causing maximum stresses. Vehicles in adjacent lanes shall be taken as headed in the direction producing maximum stresses.

The spaces on the carriageway left uncovered by the standard train *or* vehicle shall not be assumed as subject to any additional live load.

## IRC CLASS AA LOADING

### General notes

1. The nose to tail spacing between two successive vehicles shall not be less than 90 m.

2. For multi-lane bridges and culverts, one train of class AA tracked *or* wheeled vehicles whichever creates severer conditions shall be considered for every two traffic lane width.

No other live load shall be considered on any part of the said 2-lane width carriageway of the bridge when the above mentioned train of vehicle is crossing the bridge.

3. The maximum loads for the wheeled vehicle shall be 20 tonnes ($\approx$ 200 kN) for a single axle *or* 40 tonnes ($\approx$ 400 kN) for a bogie of two axles spaced not more than 1.2 m centres.

4. The minimum clearance between the road face of the kerb and the outer edge of the wheel *or* track, $C$, shall be as under:

**(a) Single Lane Bridges**

| Carriage way width | Minimum value of C |
|---|---|
| 3.8 m and above | 0.3 m |

**(b) Multi Lane Bridges**

| | |
|---|---|
| Less than 5.5 m | 0.6 m |
| 5.5 m or above | 1.2 m |

## IRC CLASS A VEHICLES

(*a*) **Details of vehicles**

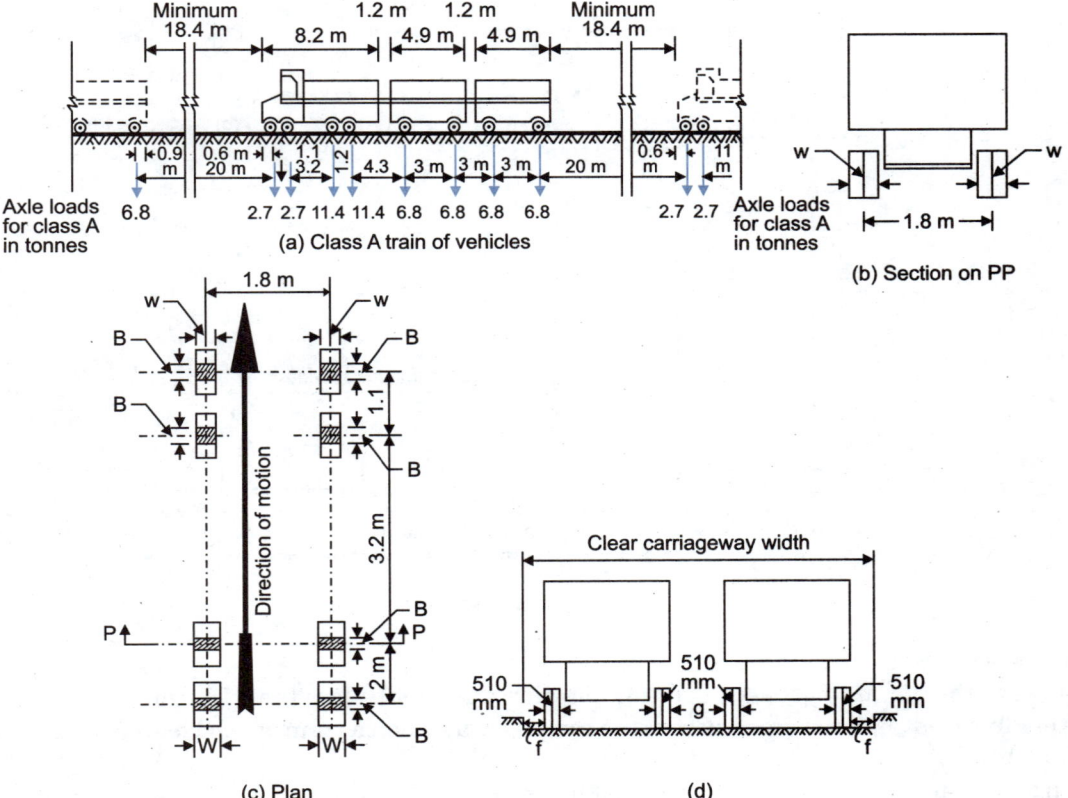

Fig. 31.4. Class A loading. (Note : 1 Tonne $\approx$ 10 kN)

## 756 REINFORCED CONCRETE STRUCTURE

(b) **General Notes**

1. The nose to tail distance between successive trains shall not be less than 18.4 m.
2. No other live load shall cover any part of the carriage way when a train of vehicles (or trains of vehicles in multi-lane bridge) is crossing the bridge.
3. The ground contact area of the wheels shall be as under (Fig. 31.4 b and c).

| Axle load | | Ground contact area | |
|---|---|---|---|
| *Tonnes* | *kN* | *(B) (mm)* | *(W) (mm)* |
| 11.4 | 114 | 250 | 500 |
| 6.8 | 68 | 200 | 380 |
| 2.7 | 27 | 150 | 200 |

4. The minimum clearance $f$, between outer edge of the wheel and the roadway face of the kerb, and the minimum clearance $g$, between the outer edges of passing or crossing vehicles on multilane bridges shall be as given below (refer Fig. 31.4 d).

| *Clear carriage width* | *g* | *f* |
|---|---|---|
| 5.5 m to 7.5 m | Uniformly increasing from 0.4 m to 1.2 m | 150 mm for all carriageway vehicles |
| above 7.5 m | 1.2 m | |

## IRC CLASS B LOADING

(a) **Details of vehicles**

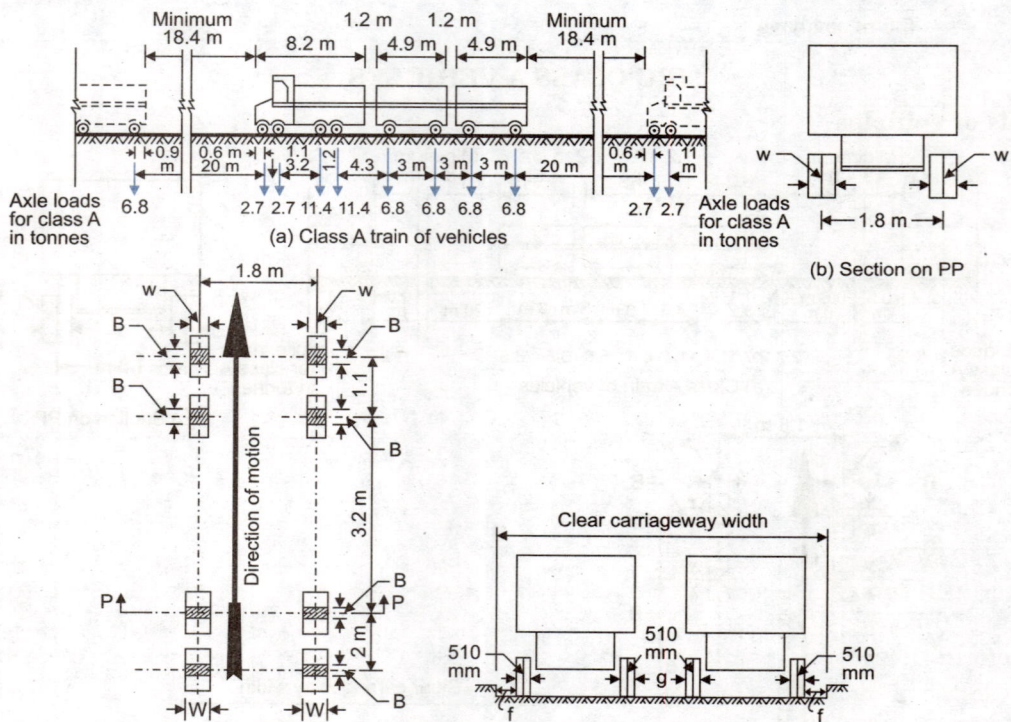

Fig 31.5. IRC Class B loading.

(b) **General Notes**

1. The nose to tail distance between successive trains shall not be less than 18.4 m.
2. No other live load shall cover any part of the carriage way when a train of vehicles (or trains of vehicles in multi-lane bridge) in crossing the bridge.
3. The ground contact area of the wheels shall be as under [Fig. 31.5 (b, c)].

| Axle load | | Ground contact area | |
|---|---|---|---|
| Tonnes | kN | B (mm) | W (mm) |
| 6.8 | 68 | 200 | 380 |
| 4.1 | 41 | 150 | 300 |
| 1.6 | 16 | 125 | 175 |

4. The minimum clearance $f$, between the outer edge of the wheel and the roadway face of the kerb, and the minimum clearance, $g$, between the outer edges of passing of crossing vehicles on multi-lane bridges shall be as given below [Fig. 31.5 (d)].

| Clear carriage way width | g | f |
|---|---|---|
| 5.5 m to 7.5 m | Uniformly increasing from 0.4 to 1.2 m | 150 mm for all carriageway widths |
| above 7.5 m | 1.2 m | |

## 31.5. IMPACT EFFECT

Provision for impact *or* dynamic action shall be made by an increment of live load by an impact allowance expressed as a fraction *or* a percentage of applied live load.

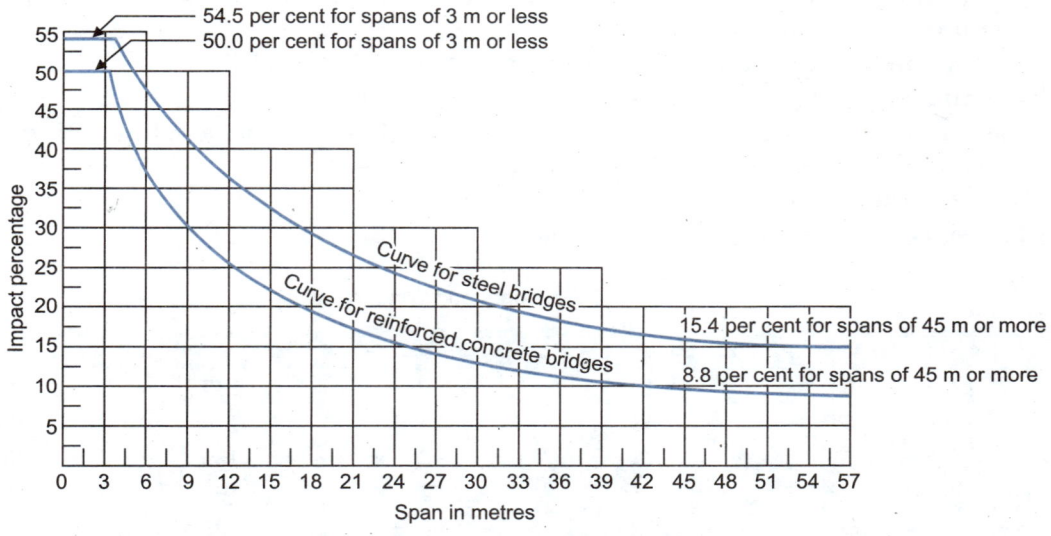

Fig. 31.6. Impact Percentage Curves.

**For Class A *or* Class B loading.** In the members of any bridge designed either for class A or class B, loading, the impact percentage shall be determined from the curves indicated in Fig. 31.6. The impact factor shall be determined from the following equations which are applicable for spans between 3 m and 45 m.

Impact factor for R.C. bridges, $I = 4.5/(6 + L)$ ...(31.4)

where $L$ is the length of the span in metres.

**For Class AA loading.** The value of the impact percentage shall be taken as follows:

**(a) For spans less than 9 m:**
(*i*) For tracked vehicles : 25% for spans upto 5 m, linearly reducing to 10% for spans of 9 m.
(*ii*) For wheeled vehicles : 25%

**(b) For span of 9 m or more:**
(*i*) Tracked vehicles : 10% upto a span of 40 m and in accordance with the curve indicated in Fig. 31.6 for spans in excess of 40 m.

(*ii*) Wheeled vehicles : 25% for spans upto 12 m and in accordance with the curve indicated in Fig. 31.6 for spans in excess of 23 m.

In any bridge structure, where there is a filling of not less than 0.6 m including road crust, the impact percentage to be allowed in the design should be assumed to be one-half of what is prescribed above.

## 31.6. WIND LOAD

All structures shall be designed for the lateral wind forces given below. These forces shall be considered to act horizontally and in such a direction that the resultant stresses in the member under consideration are maximum.

The wind force on structure shall be assumed as a horizontal force of the intensity specified in Table 31.1. and acting on an area calculated as follows:

(*a*) *For a deck structure*

The area of the structure as seen in elevation including the floor system and railings.

(*b*) *For a through or half through structure*

The area of the elevation of the windward truss as specified in (*a*) above plus half the area of elevation above the deck level of all other trusses or girders.

The pressure given in Table 31.1 shall be doubled for bridges situated in areas such as the Kathiawar, Peninsula and Odisha coasts.

where   $H$ = The average height in metres of the exposed surface above the mean retarding surface (ground *or* bed level *or* water level)

$V$ = Horizontal velocity of wind in kilometres per hour at height $H$.

$P$ = horizontal wind pressure in kg per sq. m. at height $H$.

The lateral wind force against any exposed moving load shall be considered as acting at 1.5 m above the roadway and shall be assumed to have the following values:

Highway bridges, ordinary: 300 kg ($\approx$ 3000 N) per linear metre.

Highway bridges carrying tramway: 450 kg ($\approx$ 4500 N) per linear metre.

**TABLE 31.1.** Wind Pressures and Wind Velocities

| $H$ | $V$ | $P$ | $H$ | $V$ | $P$ |
|---|---|---|---|---|---|
| 0 | 80 | 40 | 30 | 147 | 141 |
| 3 | 98 | 58 | 38 | 154 | 151 |
| 6 | 107 | 73 | 46 | 159 | 166 |
| 9 | 115 | 88 | 54 | 163 | 176 |
| 12 | 123 | 98 | 62 | 168 | 185 |
| 15 | 128 | 107 | 76 | 175 | 200 |
| 18 | 133 | 112 | 91 | 181 | 210 |
| 21 | 138 | 122 | 106 | 186 | 224 |
| 24 | 143 | 127 | 122 | 190 | 234 |
| 27 | 145 | 136 | | | |

**Note :** 1 kg/m$^2$ $\approx$ 10 N/m$^2$

The bridges shall not be considered to be carrying any live load when the wind velocity at deck level exceeds 130 km per hour. The total assumed wind force shall not be less than 450 kg (4500 N) per linear metre in the plane of the loaded chord and 225 kg (2250 N) per linear metre in the plane of the unloaded chord on through *or* half through truss, latticed *or* other similar spans and not less than 450 kg (4500 N) per linear metre on deck spans.

A wind pressure of 240 kg (2400 N) per square metre of the unloaded structure shall be used if it produces greater stresses than those produced by the combined wind forces as stated above.

## 31.7. LONGITUDINAL FORCES

In all road bridges, provision shall be made for longitudinal forces arising from any one or more of the following causes:
 (a) Tractive effect caused through acceleration of the driving wheels;
 (b) Breaking effect resulting from the application of brakes to braked vehicles; and
 (c) Frictional resistance offered to the movement of free bearings due to change of temperature or any other cause.

The braking effect on a simply supported span or a continuous units of spans or any other type of bridge unit shall be assumed to have the following values:

 (a) In the case of single lane or two lane bridge: twenty per cent of the loads of the succeeding trains or part therefore, the train loads in one lane only being considered for this purpose.

The force due to braking effect shall be assumed to act along a line parallel to the roadway and 1.2 m above it. While transferring the force to the bearings, the change in the vertical reaction at the bearings should be accounted for. The longitudinal force at any free bearing shall be limited to the sum of dead and live load reactions at the bearing multiplied by the appropriate coefficient of friction. The coefficient of friction at the bearings shall be assumed to have the following values:

| | |
|---|---|
| For roller bearings ... | 0.03 |
| For sliding bearings of hard copper alloy ... | 0.15 |
| For sliding bearings of steel on cast iron or steel on steel ... | 0.25 |
| For sliding bearings of steel on ferro-asbestos ... | 0.20 |

The longitudinal force at the fixed bearings shall be taken as the algebraic sum of the longitudinal forces at the free bearings in the bridge unit under consideration and the forces due to braking effect on the wheels as mentioned above. The effect of braking force on bridge structures without bearings such as arches, rigid frames etc. shall be calculated in accordance with approved methods of analysis of indeterminate structures.

The effect of the longitudinal forces and all other horizontal forces should be calculated upto a level where the resultant passive earth resultance of the soil below the deepest scour level (floor level in the case of a bridge having pucca floor) balances these forces.

## 31.8. LATERAL LOADS

**Forces on parapet.** Railings or parapets shall have a minimum height above the adjacent roadway or footway surface of 4 m less one half the horizontal width of the top rail or top of the parapet. They shall be designed to resist a lateral longitudinal force and a vertical force each of 150 kg/m (1500 N/m) applied simultaneously at the top of the railing or parapet. These forces shall also be considered in the design of main structural members of the bridges provided with foot paths. Where, however, foot paths are not provided these forces need not be considered in the design of the main structural members.

**Kerbs.** Kerbs shall be designed for lateral loading of 750 kg/m (7500 N/m) run of the kerb applied horizontally at the top of the kerb. This load need not be taken for the design of supporting structure.

## 31.9. CENTRIFUGAL FORCE

Where a road bridge is situated on a curve, all portions of the structure affected by the centrifugal action of moving vehicles are to be proportioned to carry safely the stress induced by this action in addition to all other stresses to which they may be subjected. The centrifugal force shall be determined from the following equation:

$$C = \frac{WV^2}{127\,R} \qquad \ldots(31.5)$$

where  $C$ = centrifugal force in kN acting normally to the traffic (1) at the point of action of the wheel loads or (2) uniformaly distributed over every metre length on which a uniformly distributed load acts.

$W$ = live load (1) in kN in case of wheel loads, each wheel load being considered as acting over the ground contact length specified in § 31.4. and (2) in kN per linear meter in case of a uniformly distributed live load.

$V$ = the design speed of the vehicles using the bridge, in km per hour, and

$R$ = the radius of curvature in metres.

The centrifugal force shall be considered to act at a height of 1.2 m above the level of the carrige way.

## 31.10. WIDTH OF ROADWAY AND FOOTWAY

(a) For high level bridges constructed for the use of road traffic only, the width of roadway shall not be less than 3.8 m for a single lane bridge and shall be increased by a minimum of 3 m for every additional lane of traffic on multiple lane bridges. Road bridges shall be either one lane, two lanes *or* four lanes. Three lane bridges shall not be constructed. In the case of four lanes *or* multiples of two-lanes, a central verge of at least 1.2 m width shall be provided.

(b) For bridges constructed for the use of combined road and tramway traffic, the roadway width given in (a) above shall be increased by 4 m for a single track tramway and by 7.6 m for a double track tramway.

(c) When a footway is provided its width shall not be less than 1.5 m.

## 31.11. GENERAL DESIGN REQUIREMENTS

**1. Size of bars.** (i) The maximum size of reinforcement shall be 45 mm diameter *or* a section of equivalent area, unless a bigger size is permitted by the competent authority.

(ii) Excepting of wire mesh and similar reinforcement, the diameter of any reinforcing bar including transverse ties, spirals, stirrups and all secondary reinforcements shall generally be not less than 6 mm.

(iii) The diameter of longitudinal reinforcing bars in columns shall not be less than 12 mm.

(iv) The diameter of wire under tensile stress in connected mesh and similar reinforcement in slabs shall not be less than 3 mm.

**2. Distance between bars.** The horizontal distance between two parallel reinforcing bars shall not be less than the greatest of the following dimensions:

(a) Diameter of the bar if the diameters are equal.

(b) The diameter of the largest bar – if the diameters are unequal.

(c) 6 mm more than the nominal maximum size of coarse aggregate used in concrete.

When immersion vibrators are intended to be used, sufficient space shall be left between groups of bars to enable the vibrator to be inserted.

The minimum vertical distance between two horizontal main reinforcing bars shall be 12 mm *or* the maximum diameter of the coarse aggregate *or* the maximum size of bar, whichever is greater.

When contact of bars along the lap length cannot be avoided, such bars shall preferably be grouped in the vertical plane. In no case, however, shall there be more than three bars in contact. The pitch of bars *or* wires of main tensile reinforcement in slab shall not exceed 300 mm *or* twice the effective depth of the slab whichever is smaller.

All mesh reinforcement shall be of such dimensions as will enable the coarsest material in the concrete to pass easily through meshes of such reinforcement.

**3. Curtailment of bars.** To prevent large changes in the moment of resistance, the points at which the bars are curtailed shall be suitably spread.

In simply supported spans, at least 20 percent of the steel required to resist the maximum bending moment shall be carried over to the supports along the tension side of the beam *or* slab. For a span continuous beyond a support, at least 10 percent of the steel required to resist the maximum positive bending moment shall be similarly carried over the support.

**4. Cover.** The thickness of concrete cover (exclusive of plaster *or* other decorative finish) shall be as follows:

(*i*) At each end of a reinforcing bar, neither less than 25 mm nor less than twice the diameter of such bar.

(*ii*) For longitudinal reinforcing bar in a column *or* beam, neither less than 40 mm, nor less than the diameter of such bar.

(*iii*) In no case, the cover to any reinforcement including stirrups *or* binders be less than 25 mm nor less than the diameter of such bar.

(*iv*) The cover of concrete in direct contact with earth face shall be 25 mm more than specified above.

(*v*) For members totally immersed in sea water, the cover shall be increased by 40 mm and for members subject to sea spray, the cover shall be increased by 50 mm over the values given above.

**5. Curved or sloped reinforcement.** Where the alignment of the reinforcement deviated from the normal to the plane of bending as in the case of a beam with curved *or* sloped soffit, only, the area of reinforcement effective in the direction normal to the plane of bending shall be considered.

**6. Distribution reinforcement.** (*i*) For solid slabs spanning in one direction, distributing bars shall be provided at right angles to the main tensile bars to provide for the lateral distribution of the loads. The distribution reinforcement shall be such as to produce a resisting moment in the direction perpendicular to the span equal to 0.3 times the moment due to concentrated live load plus 0.2 times the moment due to other loads such as dead load shrinkage, temperature etc.

(*ii*) In cantilever slabs, the distribution steel, to resist a moment equal to that of 0.3 times live load moment plus 0.2 times dead load moment shall be provided half at top and half at bottom of the slab.

(*iii*) The pitch of the distributing bars shall not be greater than thrice the effective depth of the slab *or* 450 mm whichever is less.

**7. Bond, anchorage, lap.** The local bond shall not exceed 1.5 times the permissible average bond stress. The permissible average bond stress is $0.04\ \sigma_{cu}$ for concrete whose works cube strength $\sigma_{cu} < 20$ N/mm$^2$ and is equal to $[0.02\ \sigma_{cu} + 0.4]$ N/mm$^2$ subject to a maximum of 1.1 N/mm$^2$ for concrete whose cube strength $\sigma_{cu} > 20$ N/mm$^2$. For deformed bars the bond strength may be increased by 25%. All bent up bars acting as shear reinforcement should be fully anchored in both flanges of the beam, the anchorage length being measured from the end of the sloping portion of the bar.

**8. Shear stress.** In no case shall the shear stress exceed (*i*) 4 times the permissible shear stress when it is not associated with high bending moment; (*ii*) $2\frac{1}{2}$ times the permissible shear stress when it is associated with high bending moment.

When web reinforcement is not required to carry shear, minimum reinforcement of area not less than 0.15% of the area of web in plan shall be provided.

**9. Allowable pressure on the loaded area**

Allowable bearing pressure $\sigma_{c1}$ on loaded area shall not exceed 0.6 times $\sigma_{cu}$ nor the value given by $\sigma_{c1} = \sigma_c (A/A_1)^{1/3}$, where $\sigma_c$ is the allowable value in direct compression. $A_1$ is the loaded area and $A$ is the area of concrete at the bottom of footing *or* bed plate contained in a frustrum of a pyramid *or* cone having for its upper base, the loaded area $A_1$ and having side slopes for one horizontal to two vertical for plain concrete and two horizontal to one vertical for reinforced concrete. For concrete in beams, value of $A/A_1$ may be taken as 3.

For resisting spalling tension, continuous mesh of small diameter. *i.e.*, 8 mm Φ @ 60 mm centres both ways shall be provided at 20 mm from the surface.

**10. Permissible stresses.** The IRC Code allows the following permissible stress:

*Concrete :*

| | |
|---|---|
| Direct compression | $0.25\ \sigma_{cu}$ |
| Bending compression | $0.33\ \sigma_{cu}$ |
| Shear stress and tensile stress in flexure | $0.033\ \sigma_{cu}$ |
| Average bond stress | $0.04\ \sigma_{cu}$ |

Tensile stress in flexure for members cast with construction joints     0.0264 $\sigma_{cu}$

Modular ratio     2800/$\sigma_{cu}$

**Stresses in steel** : Table 31.2

**TABLE 31.2.** Stresses in Steel (N/mm²)

| | *Mild steel as per I.S. 432 grade 1* | *Medium tensile steel as per I.S. 432 guaranted yield stress of 315 to 360 N/mm²* |
|---|---|---|
| 1. Tension other than in spiral in columns | 125 | 140 |
| 2. Tension in spiral in columns | 95 | 125 |
| 3. Compression in columns | 115 | 140 |
| 4. Compression in steel in beams and slabs when resistance of concrete is taken into account | Modular ratio times the stress in surrounding concrete | |
| 5. Compression in steel in beam and slabs when resistance of concrete is not taken into account | 115 | 140 |

**11. Thermal coefficient and shrinkage.** The values of co-efficients of thermal expansion and shrinkage of concrete shall be taken as follows:

Thermal expansion coefficient    :    $117 \times 10^{-7}/°C$

Shrinkage coefficient    :    $2 \times 10^{-4}$.

**12. Increase in permissible stresses.** (*i*) When the effect of temperature, shrinkage and creep is taken into consideration the permissible stresses should not be exceeded in the case of determinate structures but may be exceeded by 15% in the case of indeterminate structures.

(*ii*) When the effect of wind forces is taken into consideration, permissible stresses may be exceeded by 25%.

(*iii*) When the worst combination of loads meet with during erection is considered, the permissible working stresses may be exceeded by 25%.

(*iv*) When the effect of seismic forces is also considered, the permissible working stresses may be exceeded by 50%.

## 31.12. SOLID SLAB BRIDGES

A solid slab bridge *or* deck slab bridge is the simplest type of construction used mostly for culverts *or* small bridges with a span not exceeding 8 m. Though the thickness of deck slab is considerable its construction is much simpler and the cost of form work is also minimum.

### WHEEL LOAD ON SLAB

A load from a wheel *or* similar concentrated load bearing on a small but definite area of supporting surface (called the contact area) is considered as being further dispersed over the area dependent upon the thickness of the road *or* other surfacing material. A mathematical treatment for the analysis of moments and shears for flat plates, under concentrated load was developed by Navier, and other methods have been evolved from time to time. These methods are all some what complicated and their full applicability to concrete slab is doubtful. The analysis is therefore done in semiempirical manner, based on the modification of results available from elastic analysis. There are three methods used: (*i*) Effective width method, (*ii*) Pigeaud's coefficient method and (*iii*) Westergaard's method. The first method is applicable for those slabs which are supported on two opposite edges only, *i.e.*, for the present case of solid slab bridges. The second method has been discussed in § 31.14.

## EFFECTIVE WIDTH METHOD: IRC FORMULA

When a concrete slab, supported on two opposite edges (*or* when a very long slab supported on all the four edges) is subjected to a point bearing a certain definite contact area, it deflects forming a saucer. Due to this bending moments in the slab are created in the plane of the span as well as normal to it. The load is borne not only by the strip of slab immediately below the load that bears it but also by strips on either side of the load. It is therefore assumed that the load is supported by a certain width of slab, known as the *effective width*. IRC Code recommends formulae for the effective width for two cases: (*a*) slabs supported on opposite edges, (*b*) cantilever slabs. Refer Fig. 31.7.

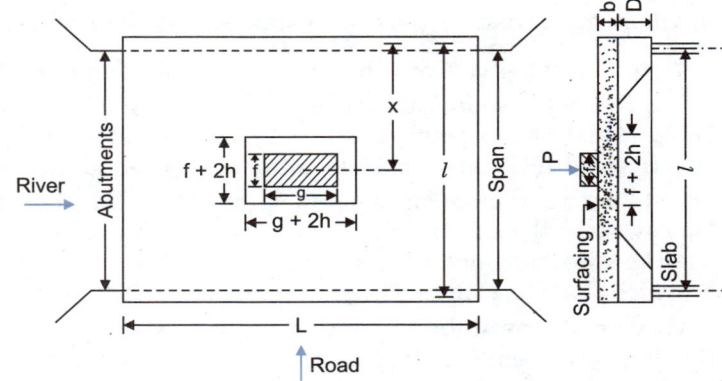

Fig. 31.7. Dispersion of Live Load Through Deck Slab.

(*a*) **Slab supported on opposite edges**

If a solid slab is supported on two opposite edges and carries concentrated loads, it should be designed to risist the maximum bending moment caused by the loading system. Such bending moment shall be assumed to be resisted by an effective width of the slab (measured parallel to the supported edges), the effective width being calculated as follows:

(*i*) For a single concentrated load, the effective width shall be calculated by the expression

$$e = Kx\left(1 - \frac{x}{l}\right) + W \qquad \ldots(31.6)$$

where
- $e$ = the effective width of slab on which the load acts
- $l$ = the effective span in the case of simply supported slab and equal to the clear span in the case of continuous slabs.
- $x$ = the distance of the centre of gravity of the concentrated load from the nearest support.
- $W$ = the breadth of concentration area of load = $(g + 2h)$ where $g$ is the length of the area of contact of the tyre on the road surface at right angles to the span and $h$ is the thickness of the wearing coat *or* surface finish above the structural slab.
- $K$ = a constant having the following values (Table 31.3). depending upon $L/l$ where $L$ is the width of the slab.

**TABLE 31.3**

| $\frac{L}{l}$ | K for simply supported slab | K for continuous slab | $\frac{L}{l}$ | K for simply supported slab | K for continuous slab |
|---|---|---|---|---|---|
| 0.1 | 0.40 | 0.40 | 1.1 | 2.60 | 2.28 |
| 0.2 | 0.80 | 0.80 | 1.2 | 2.64 | 2.36 |
| 0.3 | 1.16 | 1.16 | 1.3 | 2.72 | 2.40 |
| 0.4 | 1.48 | 1.44 | 1.4 | 2.80 | 2.44 |
| 0.5 | 1.72 | 1.68 | 1.5 | 2.84 | 2.48 |
| 0.6 | 1.96 | 1.84 | 1.6 | 2.88 | 2.52 |
| 0.7 | 2.12 | 1.96 | 1.7 | 2.92 | 2.56 |
| 0.8 | 2.24 | 2.08 | 1.8 | 2.96 | 2.60 |
| 0.9 | 2.36 | 2.16 | 1.9 | 3.00 | 2.60 |
| 1.0 | 2.48 | 2.24 | 2 & more | 3.00 | 2.60 |

Provided that the effective width shall not exceed the actual width of slab, and provided further that in case of a load near the unsupported edge of a slab, the effective width shall not exceed the above value nor it will be more than $\frac{1}{2}e$ plus the distance of the C.G. of load from the unsupported edge.

**Note:** The values of $K$ given by IS : 456–2000 are given in Table 31.3 (*a*).

(*ii*) For two *or* more concentrated loads in a line in the direction of span, the bending moment per unit width of slab at any point of span shall be calculated separately for each load according to its appropriate effective width of slab as calculated in (*c*) above, and then added.

(*iii*) For two *or* more loads not in a line in the direction of span, if the effective width of the slab for one load overlaps the effective width of slab for an adjacent load, the resultant effective width for the two loads shall be taken as equal to the sum of the respective effective widths for each load minus the width of overlap, provided that the slab so designed is tested for the two loads acting separately.

(*b*) **Cantilever slab:** If a cantilever solid slab carries concentrated loads, it should be designed to resist the maximum bending moment caused by the loading system. Such bending moment shall be assumed to be resisted by an effective width of slab (measured parallel to the supported edge) as follows:

(*i*) For a single concentrated load, the effective width shall be calculated by the equation

$$e = 1.2\,x + W \qquad \ldots(31.7)$$

where    $e$ = effective width

$x$ = the distance of the centre of gravity of the concentrated load from the face of the cantilever support, and

$W$ = the breadth of concentration area of the load = $(g + 2h)$, as before.

Provide that the effective width of the cantilever slab shall not exceed the length of the cantilever slab measured parallel to the support. And provided further that when the concentrated load is placed near one of the two extreme ends of the length of cantilever slab in the direction parallel to the support, the effective width shall not exceed the above values, nor shall it exceed half the above values plus distance of the concentrated load from the nearer extreme end measured in the direction parallel to the fixed edge.

(*ii*) For two *or* more concentrated loads, if the effective width of slab for one load overlaps the effective width of slab for an adjacent load, the resultant effective width for the two loads shall be taken equal to the sum of the respective effective widths for each load minus the width of overlap, provided that the slab so designed is tested for two loads acting separately.

(*c*) **Dispersion of loads along the span:** The effective length of slab on which a wheel load *or* track load acts shall be taken equal to the dimension of the tyre contact area over the wearing surface of slab in the direction of the span plus twice the overall depth of the slab inclusive of the thickness of the wearing surface, *i.e.*, $e = f + 2\,(h + D)$.

The general design requirements have already been discussed in § 31.11. Other requirement relevant to slab bridges are given below:

**1. Stiffening Unsupported Edges in Slabs:**  (*i*) Each unsupported edge of the slab parallel to traffic and beyond the clear road width, shall be stiffened as to give a resisting moment for any type of flexure equal to *or* in excess of that of 300 mm strip of main roadway slab adjoining the edge. In case of a roadway slab of uniform depth, whether reinforced in one way *or* two way, the maximum resisting moment of the roadway slab adjoining the edge and given by a 300 mm strip in any direction shall be taken as the criterion for the resisting moment of the stiffened edge. When the roadway slab is of varying depth in direction parallel to the edge concerned, the stiffening at any particular point along the length of the edge shall be adjusted according to the resisting moment of the 300 mm adjacent strip at that particular point.

Stiffening of edge may consist of a reinforced kerb section *or* an edge stiffening beam.

(*ii*) Unsupported edge along a line across the traffic of a road way slab (as at the cantilever end of a solid slab cantilever bridge) shall, for a strip of at least 300 mm width (in addition to the articulation if any) be suitably stiffened by providing top and bottom reinforcement across the direction of traffic. The sectional area of such reinforcement of a 300 mm width of this strip, both at the top and bottom, shall each be not less than the average area of the and longitudial reinforcement for 300 mm width at the end of cantilever.

(*iii*) Unsupported edges of cantilever foot path slabs *or* similar cantilevers shall be suitably stiffened on line similar to those indicated in (*i*) above.

**2. Distribution Reinforcement:** (i) Distribution reinforcement in solid slabs spanning in one direction shall be provided at right angles to the main tensile bars so to produce a resisting moment in the direction perpendicular to span equal to 0.3 times the moment due to concentrated live load plus 0.2 times the moment due to other loads such as dead load, shrinkage, etc.

(ii) In cantilever slabs, the distribution steel to resist a moment equal to that of 0.3 times the live load moment plus 0.2 times the dead load moment shall be provided half at top and half at the bottom of the slab.

(iii) The pitch of distribution bars shall not be greater than thrice the effective depth of the slab or 450 mm whichever is less.

**Example 31.1.** *Design a solid slab bridge for class A loading for the following data:*

*Clear span*           $= 4.5$ m
*Clear width of roadways*  $= 7$ m
*Average thickness of wearing coat* $= 80$ mm
*Use M 20 mix. Take unit weight of concrete as 24000 N/m³.*

**Solution:**

**1. General :** Let the width of the kerb be 500 mm = 0.5 m.

∴ Over all width of bridge,  $L = 7 + 2 \times 0.5 = 8$ m.

Let the overall thickness of the slab be 330 mm, with effective depth equal to 300 mm.

Effective span, $l = 4.5 + 0.3 = 4.80$ m

For M 20 concrete, $m = 13$ and $\sigma_{cbc} = 7$ N/mm²

Taking $\sigma_{st} = 125$ N/mm², we have the following design constants:

$$k = \frac{13 \times 7}{13 \times 7 + 125} = 0.421 \; ; \; j = 1 - \tfrac{1}{3} \times 0.421 \approx 0.86 \; ; \; R = \tfrac{1}{2} \times 7 \times 0.86 \times 0.421 = 1.27$$

**2. Dead Load B.M.** Dead load of slab = $0.33 \times 1 \times 1 \times 24000 = 7920$ N/m²

Dead load of wearing coat = $0.08 \times 22000 = 1760$ N/m²

Total dead load = 9680 N/m²

Dead load B.M. = $\dfrac{9680 \,(4.8)^2}{8}$

$\approx 27880$ N-m

**3. Live Load B.M.** Since the width of carriage way is 7 m, two trains of load can pass or cross each other, with a clear distance

$$g = 0.4 + \frac{1.2 - 0.4}{7.5 - 5.5}(7 - 5.5) = 1 \text{ m}.$$

For axle load of 114 kN, $W = 500$ mm = 0.5 m

∴ Centre to centre distance between wheels of two trains = $1 + 0.5 = 1.5$ m

Distance of face of wheel from kerb = $f$ = 150 mm = 0.15 m

∴ Distance of centre of wheel from edge of kerb = $0.15 + \dfrac{0.50}{2} = 0.4$ m

Dispersion width $b$ of load along span = $B + 2(80 + 330) = 250 + 820 = 1070$ mm

As there are two axles, moment could be maximum under one of them. Let the axles, be arranged as shown in Fig. 31.9 for maximum B.M.

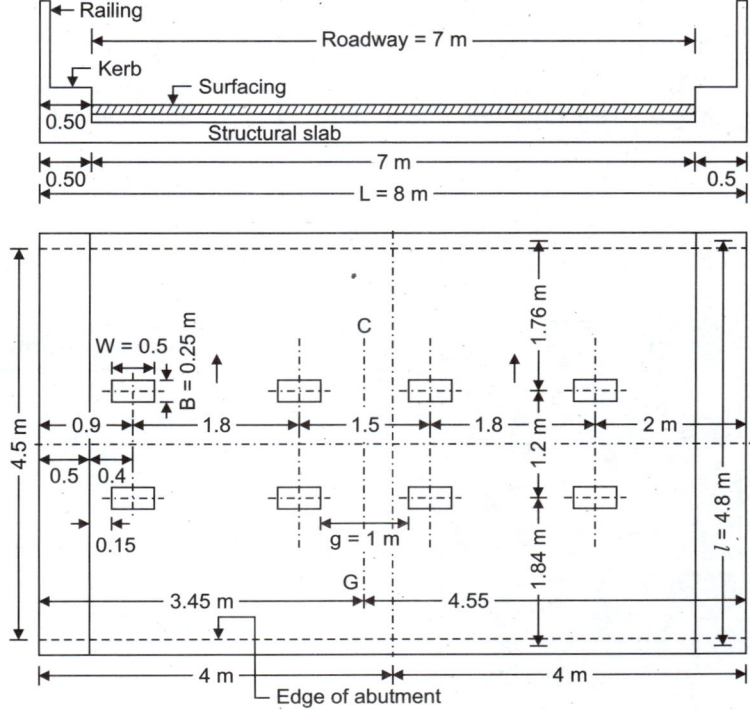

Fig. 31.8

Let the distance of the C.G. of loads from support $B$ be $y$. The distance $a = 1.2$ m for the present case, while $b = 1.07$ m. Let the maximum B.M. occur at $C$, the point of zero shear.

$$R_A = \frac{2P \cdot y}{l}$$

Hence the value of $z$, for the location of point $C$ of zero shear is given by:

$$R_A - \frac{P}{b} z = 0 \quad \text{or} \quad \frac{2P \cdot y}{l} - \frac{P}{b} z = 0$$

or

$$z = \frac{2by}{l} \qquad \qquad \ldots(31.8)$$

$$\therefore \qquad M_C = \frac{2P \cdot y}{l}\left[l - \left(y + \frac{a}{2} + \frac{b}{2}\right) + z\right] - \frac{P}{b} \cdot \frac{z^2}{2} \qquad \ldots(31.9)$$

For maxima, $\dfrac{dM}{dy} = 0$, which gives $y = \dfrac{2l - (a+b)}{4\left(1 - \dfrac{b}{l}\right)}$ $\qquad \ldots(31.10)$

For the present case, $y = \dfrac{(2 \times 4.8) - (1.2 + 1.07)}{4\left(1 - \dfrac{1.07}{4.8}\right)} = 2.36$ m

This is very near to $\dfrac{l}{2}$. $\quad \therefore \quad z = \dfrac{2 \times 1.07 \times 2.36}{4.8} = 1.05$ m.

$\therefore$ Distance of $C$ from $A = l - \left(y + \dfrac{a}{2} + \dfrac{b}{2}\right) + z = 4.8 - \left(2.36 + \dfrac{1.2}{2} + \dfrac{1.07}{2}\right) + 1.05 = 2.35$ m.

$\therefore$ Distance of axle II from support $B = \left(y - \dfrac{a}{2}\right) = 2.36 - 0.6 = 1.76$ m

Distance of axle I from support $A = l - \left(y + \dfrac{a}{2}\right) = 4.8 - (2.36 + 0.6) = 1.84$ m

These positions have been marked in Fig. 31.8 and 31.9.

When the two trains are running parallel, the effective width for various wheels will overlap. Hence all the four wheels are taken together for calculation of effective width and the quantity $W$ in *Eq. 31.6* will be

$$= \left(1.8 + 1.5 + 1.8 + 2 \times \frac{0.5}{2}\right) + 2 \times 0.08 = 5.76 \text{ m}$$

$L/l = 8/4.8 = 1.67$

**TABLE 31.3 (a)** Values of $K$ for Simply Supported and Continuous Slab

| $\dfrac{L}{l}$ | *K for Simply Supported Slabs* | *K for Continuous Slabs* |
|---|---|---|
| 0.1 | 0.4 | 0.4 |
| 0.2 | 0.8 | 0.8 |
| 0.3 | 1.16 | 1.16 |
| 0.4 | 1.48 | 1.44 |
| 0.5 | 1.72 | 1.68 |
| 0.6 | 1.96 | 1.84 |
| 0.7 | 2.12 | 1.96 |
| 0.8 | 2.24 | 2.08 |
| 0.9 | 2.36 | 2.16 |
| 1.0 and above | 2.48 | 2.24 |

The effective width of slab on which the load acts of given by Eq. 31.6

$$e = K \cdot x \left(1 - \frac{x}{l}\right) + W$$

where the values of $K$ as per I.S. : 456–2000 are given Table 31.3(a).

Thus the maximum value of $K$ is for $\frac{L}{l} \geq 1$. For the present case, $\frac{L}{l} = 1.67$, for which $K = 2.48$. Hence from Eq. 31.6,

Effective width for axle I = $2.48 \times 1.84 \left(1 - \frac{1.84}{4.8}\right) + 5.76 = 8.57$ m

Effective width for axle II = $2.48 \times 1.76 \left(1 - \frac{1.76}{4.8}\right) + 5.76 = 8.52$ m

Since the effective width extend beyond the free edges of slabs, the modified effective widths will be as under:

For axle I,    $e = (8.57/2)$ + distance of C.G. of axle from free edge = $4.285 + 3.45 \approx 7.74$ m

For axle II,    $e = (8.52/2) + 3.45 = 7.71$ m.

It should be noted that axle load $P$, per unit width of slab, marked in Fig. 31.9 will be slightly different because of different values of $e$ for each load. Thus, for 1st axle load, $P = \dfrac{2 \times 114}{7.74} = 29.46$

and for 2nd axle load, $P = \dfrac{2 \times 114}{7.71} = 29.57$. In the analysis for determining $y$, however, it was assumed that $P$ is the same for both axles.

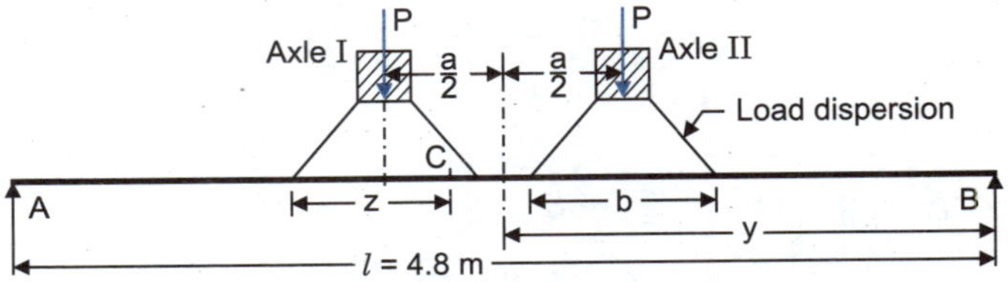

Impact factor = $\dfrac{4.5}{6 + l} = \dfrac{4.5}{6 + 4.8} = 0.42$

∴    Effective axle load
         = $2 \times 114 \times 1.42 = 323.76$ kN

For the load position marked in Fig. 31.10,

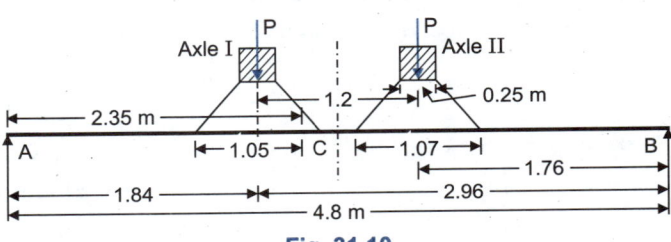

Fig. 31.9

Fig. 31.10

$$R_A = \left[\left(\frac{323.76}{7.71} \times 1.76\right) + \left(\frac{323.76}{7.74} \times 2.96\right)\right] \frac{1}{4.8} = 41.19 \text{ kN}$$

∴    $M_C = R_A \times 2.35 - \dfrac{323.76}{7.74 \times 1.07} \cdot \dfrac{(1.07)^2}{2} = 74.42$ kN-m $= 74.42 \times 10^3$ N-m.

The maximum B.M. due to live load occurs at a section 2.35 m from the support and not at the middle of the slab. However, maximum dead load B.M. occurs at the mid span.

Dead load B.M. at $C = \dfrac{wl}{2} x - \dfrac{wx^2}{2} = 9680 \left[\dfrac{4.8}{2} \times 2.35 - \dfrac{(2.35)^2}{2}\right] \approx 27870$ N-m

∴    Total B.M. = $74420 + 27870 = 102290$ N-m.

**Alternative Method for Live Load B.M.** In the alternative method, it is assumed that maximum B.M. occurs when the loads are symmetrically placed about centre of span. B.M. at the centre will be taken for design purpose. This is reasonably correct since we have seen that the actual location of point $C$ of maximum live load B.M. is away from the centre only by 0.05 m.

For this arrangement of loads (Fig. 31.11),

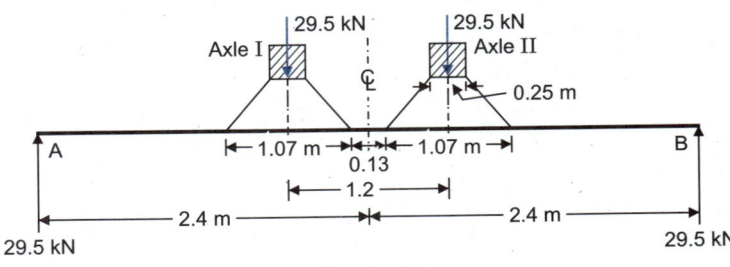

Fig. 31.11

$x$ = distance of axle load from the nearest support = $\frac{1}{2}$ (4.8 − 1.2) = 1.8 m.

Since the effective width for individual loads will overlap, the quantity $W$ in *Eq. 31.6* will be

$$= 1.8 + 1.5 + 1.8 + 2 \times \frac{0.5}{2} + 2 \times 0.08 = 5.76 \text{ m as before.}$$

$$L/l = 8/4.8 = 1.67\ ;\ K = 2.48$$

$$e = 2.48 \times 1.8 \left(1 - \frac{1.8}{4.8}\right) + 5.76 = 8.55$$

Since the effective width extends beyond the free edge of slabs, the modified effective width

$$= \frac{8.55}{2} + 3.45 \approx 7.72 \text{ m.}$$

∴ Load per meter width $= \dfrac{2 \times 114}{7.72} = 29.5$ kN

Dispersion along span = 250 + 2 (80 + 330) = 1070 mm

$$\text{Impact factor} = \frac{4.5}{6+l} = \frac{4.5}{6+4.8} = 0.42$$

∴  $$M_{max} = 1.42 \left[ 29.5 \times \frac{4.8}{2} - 29.5 \times \frac{1.2}{2} \right] = 75.4 \text{ kN-m} = 75400 \text{ N-m}$$

Dead load B.M. = 27880 N-m.

∴ Total B.M. = 75400 + 27880 = 103280 N-m, as against 102290 N-m found earlier.

Hence it is better to follow the alternative method, which is much simpler.

**4. *Design of Section*:** Design B.M. = 103280 N-m = 103280 × 10³ N-mm

∴ $$d = \sqrt{\frac{103280 \times 10^3}{1000 \times 1.27}} \approx 285 \text{ mm}$$

Provide an overall depth of 330 mm. Using 20 mm Φ bars and a clear cover of 25 mm, available

$$d = 330 − 25 − 10 = 295 \text{ mm.}$$

∴ $$A_{st} = \frac{103280 \times 1000}{125 \times 0.86 \times 295} \approx 3257 \text{ mm}^2$$

Spacing of 20 mm Φ bars $= \dfrac{1000 \times 314}{3257} = 96.4$ mm

Hence provide these @ 95 mm c/c. Bend half the bars up near the support.

**5. *Distribution Reinforcement*:** Distribution steel is provided to resist 0.3 times the live load B.M. and 0.2 times the dead load moment.

∴  $M = 0.3 \times 75400 + 0.2 \times 27880 = 28196$ N-m = 28196 × 1000 N-mm

Using 12 mm Φ bars, Effective depth = 295 − 10 − 6 = 279 mm

$$A_{sd} = \frac{28196 \times 1000}{125 \times 0.86 \times 279} = 940 \text{ mm}^2. \quad \therefore \text{ Spacing of 12 mm } \Phi \text{ bars} = \frac{1000 \times 113}{940} \approx 120 \text{ mm}$$

**6. *Live load S.F.*:** The maximum S.F. occurs when the load is near the support. Since the dispersed width in the direction of span = 1070 mm = 1.07 m, the wheel loads will be placed as shown in Fig. 31.12 for maximum S.F.

Distance of 1st axle from support

$$A = \frac{1.07}{2} + 0.15 = 0.685 \text{ m.}$$

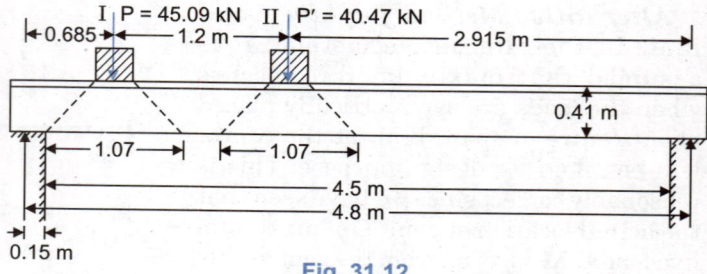

Fig. 31.12

For $\dfrac{L}{l} = \dfrac{8}{4.8} = 1.67$, $K = 2.48$ (Table 31.3 a).

W for individual load = 0.5 + 2 × 0.08 = 0.66 m

∴ Effective width for first axle = $2.48 \times 0.685 \left(1 - \dfrac{0.685}{4.8}\right) + 0.66 \approx 2.12$ m.

The effective widths will overlap. Hence let us find the combined effective width of all the four wheels of first axle of the two trains.

$$W = 1.8 + 1.5 + 1.8 + 2 \times \dfrac{0.5}{2} + 2 \times 0.08 = 5.76 \text{ m. as before.}$$

∴ $$e = 2.48 \times 0.685 \left(1 - \dfrac{0.685}{4.8}\right) + 5.76 = 7.22 \text{ m}$$

Since the effective width may extend beyond the free edge of slab (specially when the wheel load is 0.4 m away from the kerb as shown in Fig. 31.8) modified effective width = $\dfrac{e}{2}$ + dist. of C.G. of loads from free edge = $\dfrac{7.22}{2} + 3.45 = 7.06$ m. Since this is less than $e$, it will be adopted. Similarly, distance of 2nd axle from the nearest support = 0.685 + 1.2 = 1.885 m (Fig. 31.12).The effective width of individual wheels will overlap, and hence combined W for all the four wheels = 5.76 m as before.

∴ $e = 2.48 \times 1.885 \left(1 - \dfrac{1.885}{4.8}\right) + 5.76 \approx 8.6$ m. Since the effective width extends beyond the free edge of slabs modified effective width = 8.6/2 + 3.45 = 7.75 m.

Thus the effective width for first axle of loads is 7.06 m and that for the second axle of loads is 7.75 m. Impact factor = 0.42.

∴ Effective axle load = 2 × 114 × 1.42 = 323.8 kN
∴ P for first axle = 323.8/7.06 = 45.86 kN;
P' for second axle = 323.8/7.75 = 41.78 kN

Now, $$R_A = \dfrac{(41.78 \times 2.915) + 45.86 (2.915 + 1.20)}{4.8} = 64.69 \text{ kN}$$

∴ Live load S.F. = 65.69 kN ; Dead load S.F. ≈ (9.68 × 4.8)/2 = 23.23 kN
∴ Total V = 64.69 + 23.23 = 87.92 kN.

**7. Check for Shear:** Nominal shear stress on the beam is

$$\tau_v = \dfrac{V}{bd} = \dfrac{87.92 \times 1000}{1000 \times 295} = 0.298 \text{ N/mm}^2$$

At the support $A_s = \dfrac{1}{2}\left[\dfrac{1000 \times 314}{95}\right] \approx 1653$ mm$^2$ ∴ $\dfrac{100 A_s}{bd} = \dfrac{1653}{1000 \times 295} \times 100 \approx 0.56\%$

Hence for M 20 concrete, having 0.56% reinforcement

$\tau_c = 0.32$ N/mm$^2$. Thus $\tau_v$ is less than $\tau_c$. Hence safe.

**8. Check for Development Length at Supports:** The Code stipulates that at simple supports, the diameter of reinforcement should be such that $\dfrac{M_1}{V} + L_0 \geq L_d$.

In the present case, only half the bars are available.

∴ $$A_{st} = \dfrac{1}{2}\left[\dfrac{1000 \times 314}{95}\right] \approx 1653 \text{ mm}^2$$

Now, $M_1 = 125 \times 1653 \times 0.86 \times 295 = 52420 \times 10^3$ N-mm
$V$ = S.F. at support = $87.92 \times 10^3$ N
$L_0$ = Sum of anchorage beyond centre line of the support and equivalent anchorage value of hooks.

Let us assume a support width $l_s$ = 300 mm. Assuming a clear side cover $x'$ = 40 mm, and providing U-hook for which the anchorage value is 16Φ we get

$$L_0 = \left(\frac{l_s}{2} - x' - 3\Phi + 16\Phi\right) = \left(\frac{l_s}{2} - x' + 13\Phi\right) = 150 - 40 + 13\Phi = 110 + 13 \times 20 = 370 \text{ mm}.$$

$$\therefore \qquad L_d = \frac{\Phi \times 125}{4 \times 0.8} = 39.06\,\Phi = 39.06 \times 20 \approx 781 \text{ mm}.$$

Now, $\qquad \dfrac{M_1}{V} + L_0 = \dfrac{52420 \times 10^3}{87.92 \times 10^3} + 370 = 966 \text{ mm} > L_d.\quad$ Hence safe.

### 9. Design of Kerb

The kerb is designed for a live load of 4 kN/m² and for a horizontal load of 7.5 kN/m length. Width of kerb = 500 mm. Let the weight of railings etc. be 0.5 kN/m run.

The minimum height of kerb is to be 225 mm above the road surface. Keeping this height as 240 mm, the total depth of kerb = 330 + 80 + 240 = 650 mm.

Live load per metre run of kerb = 0.5 × 1 × 4000 = 2000 N
Dead load of kerb = 0.65 × 0.5 × 24000 = 7800 N
Weight of railings etc. = 500 N
Total = 10300 N = 10.3 kN

$$\text{B.M.} = \frac{10.3\,(4.8)^2}{8} = 29.66 \text{ kN-m} = 29.66 \times 10^6 \text{ N-mm}.$$

While determining the live load B.M. in slab, the width of kerb was also taken into account. Hence traffic live load B.M. for a width of 500 mm should also be considered.

Live load B.M. for class A loading = 0.5 × 75.4 = 37.7 kN-m = 37.7 × 10⁶ N-mm

$\therefore \quad$ Total B.M. = (29.66 + 37.7) 10⁶ = 67.36 × 10⁶ N-mm

$$\therefore \quad d = \sqrt{\frac{67.36 \times 10^6}{500 \times 1.27}} = 325.69 \text{ mm. Total depth provided} = 650 \text{ mm}.$$

Using 20 mm Φ bars, available $d$ = 650 − 25 − 10 = 615 mm

$$\therefore \quad A_{st} = \frac{67.36 \times 10^6}{125 \times 0.86 \times 615} \approx 1018 \text{ mm}^2 \quad \therefore \text{ No. of 20 mm Φ bars} = 1018/314 = 3.24.$$

Hence provide 4 bars of 20 mm Φ. However, the available bars of main reinforcement of slab (i.e., 20 mm @ 95 mm c/c) = 500/95 which are more than 5. Hence these bars will be sufficient.

#### Design of kerb for horizontal force

Horizontal force = 7500 N; Height of kerb above road surface = 240 mm

Height of kerb above level of slab
= 240 + 80 = 320 mm.

Cantilever B.M. due to horizontal load
= 7500 × 320 = 2400000 N-mm

The width of kerb is 500 mm, which is sufficient to take the compressive stresses.

$$A_{st} = \frac{2400000}{125 \times 0.86\,(500 - 25 - 5)} = 47.5 \text{ mm}^2$$

This will be provided by the vertical legs of stirrups of 8 mm Φ @ 200 mm c/c. Provide two bars of 12 mm Φ at the corners for supporting the stirrups.

The details of reinforcement etc. are shown in Fig. 31.13.

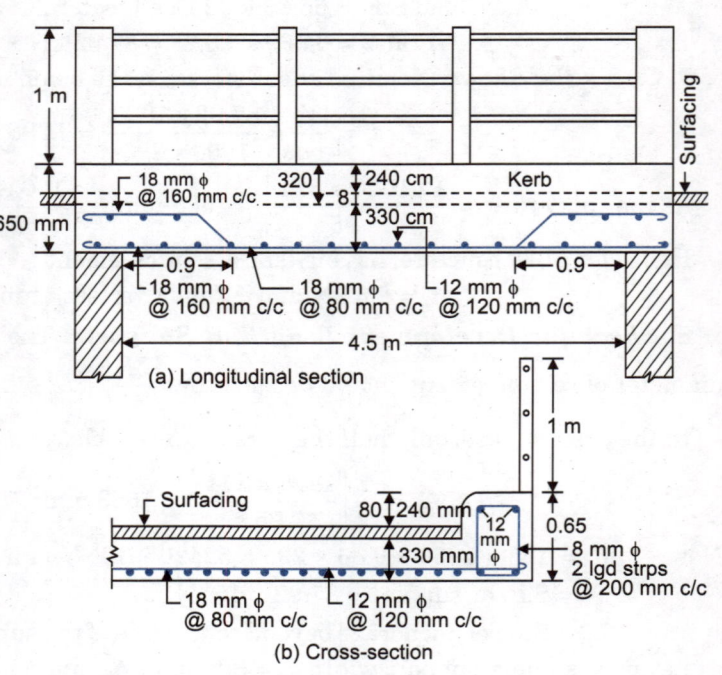

Fig. 31.13

## 31.13. DECK GIRDER BRIDGES

A deck girder bridge *or* T-beam bridge is another type of a simple R.C. bridge used for spans between 10 to 20 metres. The number of longitudinal girders depends upon the road width. The slab is generally built monolithic with girders so that T-beam effect is achieved. For wider bridges, the roadway is supported on a number of longitudinal girders with transverse beams *or* diaphragms.

The general requirements prescribed by IRC have already been given in § 31.11. Some additional design requirements relevant to beams are given below:

**1. Distribution of live loads on longitudinal beams**

(*a*) When the longitudinal beams are connected together by transverse members like deck slab, cross-girders, diaphragms, soffit slab etc. the distribution of bending moments between the longitudinals shall be calculated by one of the following methods:

(*i*) Finding the reaction on the longitudinals assuming the support of deck slab as unyielding. This method is applicable where there are two longitudinals;

(*ii*) Distributing the loads between longitudinals by Courbon's method (See § 31.17) strictly within its limitations, *i.e.*, when the effective width of deck is less than half the span and when the stiffness of cross-girders is very much greater than that of the longitudinals, and

(*iii*) Distributing the loads between logitudinals by any rational method of grid analysis, *i.e.* the method of harmonic analysis as given by Hendry and Jaeger *or* Morrice and Little's version of the isotropic plate theory of Guyon and Massonet etc. (see § 31.16).

(*b*) In calculating the shear force on sections of longitudinal beams, wheel loads *or* track loads shall be allocated to respective longitudinal beams by any rational method. Alternatively, the following method shall be followed. For loads at *or* within 5.5 metres from either supports : The reaction on the longitudinal beams shall be the greatest of the results obtained by (*a*) assuming the deck slab simply supported *or* continuous as the case may be, the supports being assumed unyielding, and (*b*) following one of the three methods as used for distribution of bending moments. Distribution of the loads between the longitudinals for the purpose of finding shearing forces shall be assumed to be the same as for bending moments.

The dispersion of the live load along the span length through the wearing coat, deck slab and filling shall not be considered.

**2. Distribution of live loads on transverse beams**

Dispersion of live loads along the span length through the wearing coat, deck slab and filling shall not be considered.

**3. Distribution of live loads on intermediate transverse floor beams**

Distribution of loads between longitudinal beams for the purpose of finding bending moments and stress in intermediate transverse floor beams shall be made by one of the methods as used for distribution of bending moments.

**4. T and L-beams**

For T and L-beams the slab shall be considered as an integral part of the beam if adequate bond and shear resistance is provided at the junction of the slab and the web of the beam.

For the purpose of calculation of stress at any section of a T-beam *or* L-beam *or* in the calculation of its moment of inertia the effective width of slab, to function as the compression flange of the beam, shall be least of the following:

**(*I*) *In the case of T-beams***

(*i*) one fourth the effective span of the beam

(*ii*) the distance between the centres of the ribs of the beam

(*iii*) the breadth of the rib plus twelve times the thickness of the slab.

**(*II*) *In case of L-beams***

(*i*) one-tenth the effective span of the beam

(*ii*) the breadth of the rib plus one-half the clear distance between the ribs

(*iii*) the breadth of the rib plus four times the thickness of the slab.

If T-form *or* L-form is used only for the purpose of providing additional compressive area, such as the continuous beam over supports, the flange thickness shall not be less than one half of the width of the web. For effective stress transfer, it is desirable to splay the junction of the web and flange so as to form an angle of not less than 110°.

Where the principal reinforcement in a slab which is considered as the flange if a T-beam *or* L-beam is parallel to the beam, transverse reinforcement shall be provided at the top of the flange. This reinforcement shall be equal to sixty percent of the main reinforcement of the slab at its mid-span unless it is specially calculated.

For T-beams *or* L-beams with spans over 10 m in length, properly designed diaphragms shall be placed at suitable points at the discretion of the designer. The spacings of such diaphragms shall not be more than thirty times thickness of web. Cross girders monolithic with deck slab shall be provided at the bearings and also at the ends of cantilevers.

### B.M. in slabs and Longitudinal Beams

The method of determining B.M. in the deck slab has been discussed in § 31.14 while the distribution of live loads on longitudinal beams has been discussed in § 31.15.

## 31.14. B.M. IN SLAB SUPPORTED ON FOUR EDGES

In deck girder bridge, if transverse beams are used, each panel of slab may be considered to be freely supported at its edges with corners not free to lift. Alternatively, the slab may be considered to be continuous over supporting beams. Two method are available for analysis. (*i*) Pigeaud's theory, and (*ii*) Westergaard's theory. Only Pigeaud's theory is given here, since it is widely followed in the bridge design.

### PIGEAUD'S THEORY

Pigeaud curves are used for computing bending moments in a panel freely supported along four edges with restrained corners and carrying symmetrically placed load distributed over some well defined area. Pigeaud derived these curves for thin plates, using elastic theory of flexure, and assuming Possion's ratio of 0.15.

**Centrally placed load:** Fig. 31.14 (*a*) shows a wheel load placed centrally over a panel of R.C. slab of length $L$ and width $B$. Let $a$ be the assumed contact length and $b$ be the tyre width. If $h$ is the thickness of the wearing coat, the load may be assumed to be dispersed at 45°, through this height, such that the dispersed loaded dimensions in Fig. 31.14.(*c*) are:

$v = a + 2h$ and $u = b + 2h$

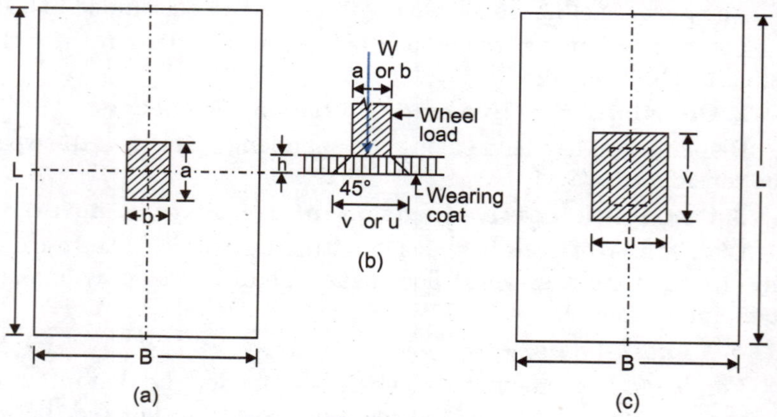

**Fig. 31.14**

The bending moments in the slab are given by the following expressions:

Short span B.M. $\quad M_B = W(M_1 + 0.15 M_2)$  ...(31.11)
Long span B.M. $\quad M_L = W(0.15 M_1 + M_2)$

The values of the moment coefficients $M_1$ and $M_2$ are given in the Pigeaud's curves for various values of span ratio $k = \dfrac{B}{L}$ and ratios $\dfrac{u}{B}$ and $\dfrac{v}{L}$.

Fig. 31.15, 31.16, 31.17, 31.18, 31.19, 31.20, 31.21, 31.22 and 31.23 give the coefficients $M_1$ and $M_2$ for span ratio $B/L = 0.9$, 0.8, 0.707, 0.6, 0.5, 0.4, 0.3, 0.2 and zero respectively. When the slab is supported on two opposite supports only (*i.e.*, when there are no transverse beams), its length $L$ will be very large. Hence $B/L$ may be treated as zero, and the moment coefficients can be determined from Fig. 31.23. For a square panel ($B/L = 1$) the moment coefficients $M_1 = M_2$ can be determined from Fig. 31.24.

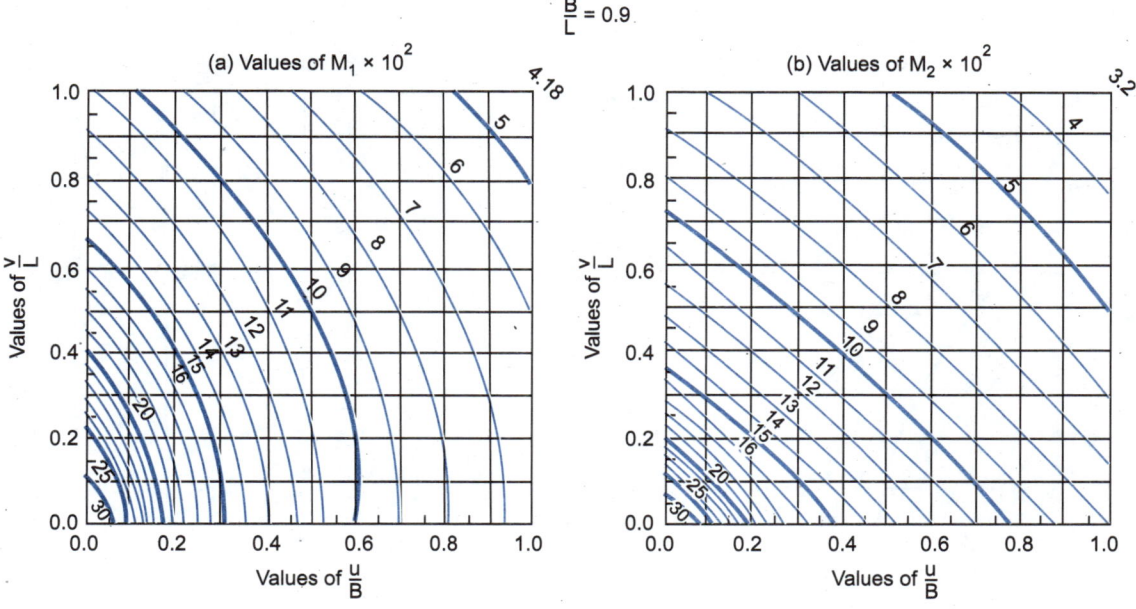

Fig. 31.15. Values for $M_1$ and $M_2$ for $B/L = 0.9$

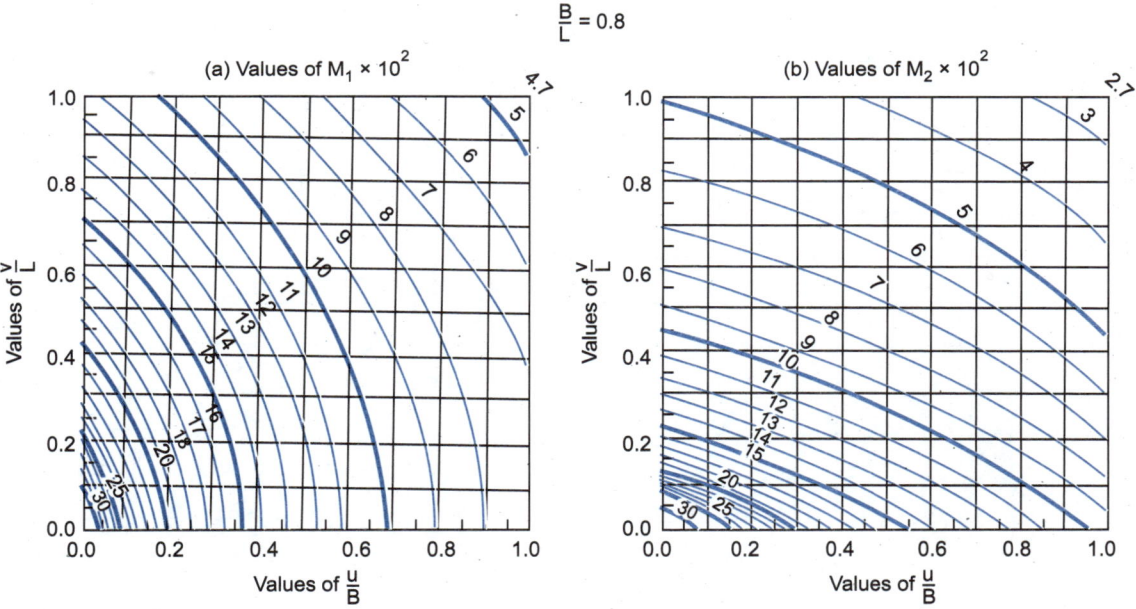

Fig. 31.16. Values of $M_1$ and $M_2$ for $B/L = 0.8$

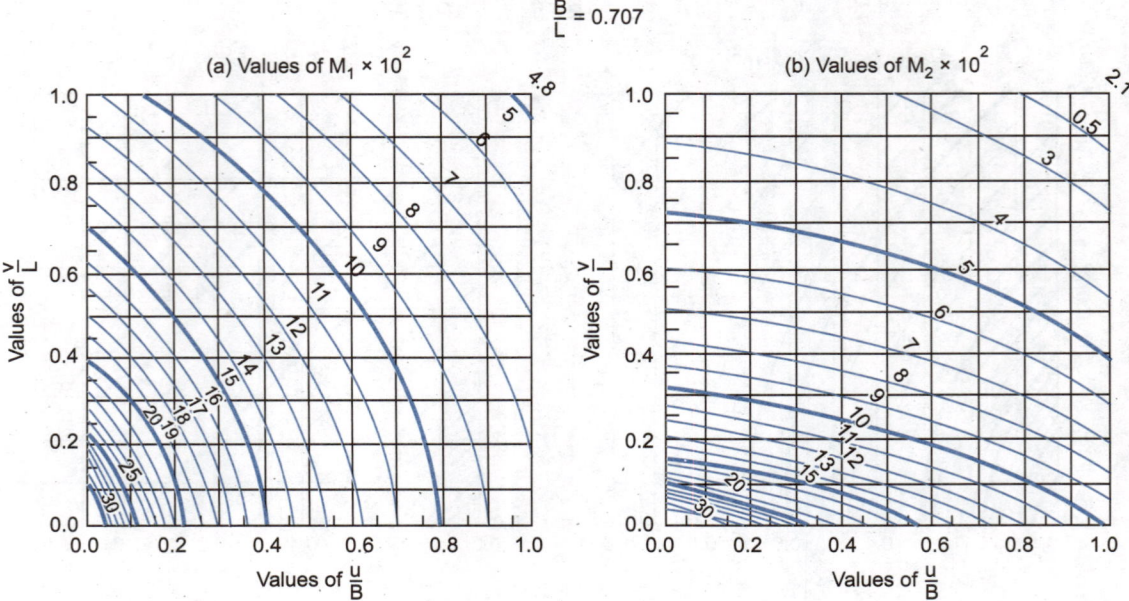

Fig. 31.17. Values of $M_1$ and $M_2$ for $B/L = 0.707$

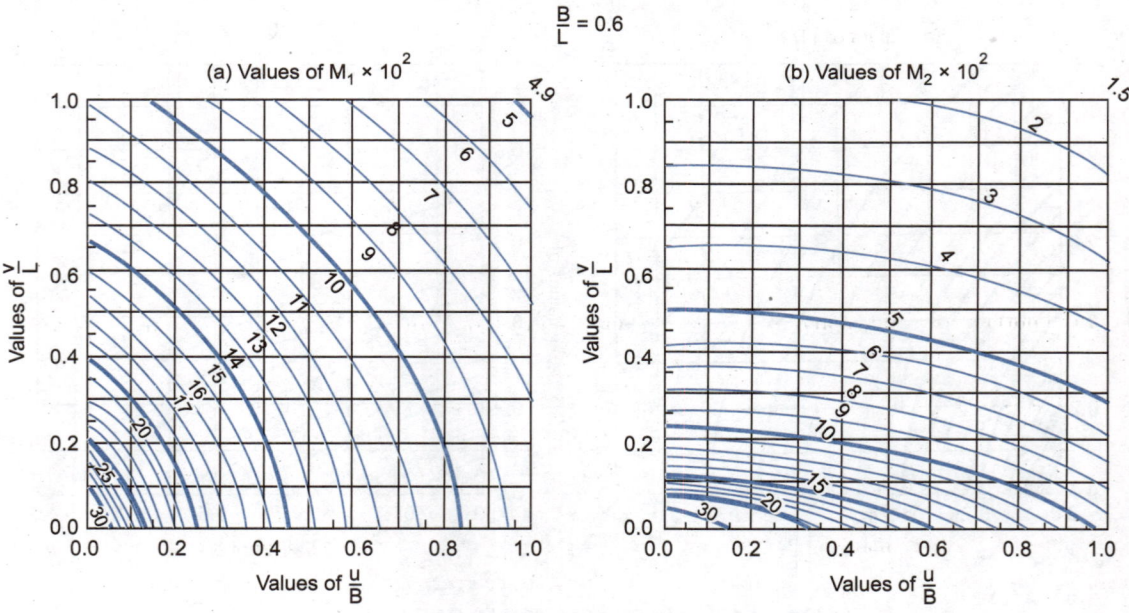

Fig. 31.18. Values of $M_1$ and $M_2$ for $B/L = 0.6$.

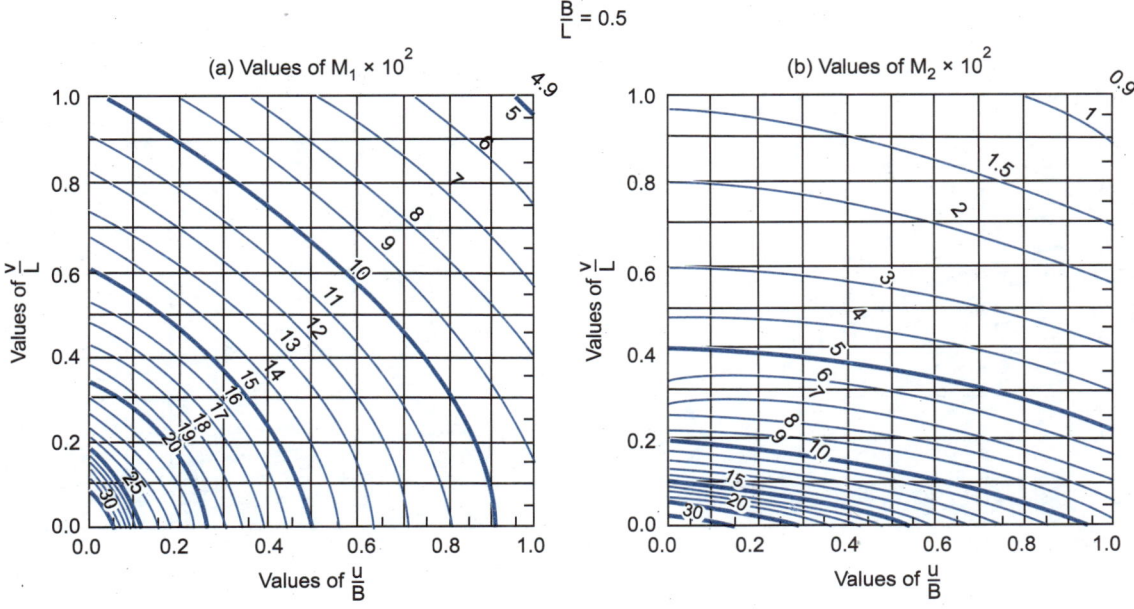

**Fig. 31.19.** Values of $M_1$ and $M_2$ for $B/L = 0.5$.

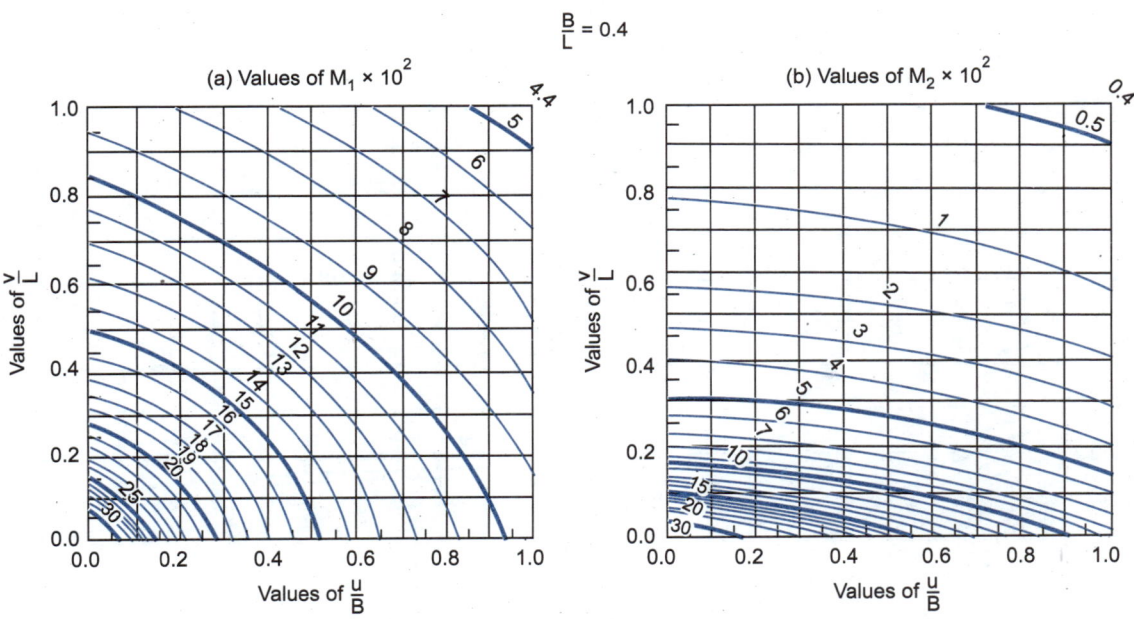

**Fig. 31.20.** Values of $M_1$ and $M_2$ for $B/L = 0.4$.

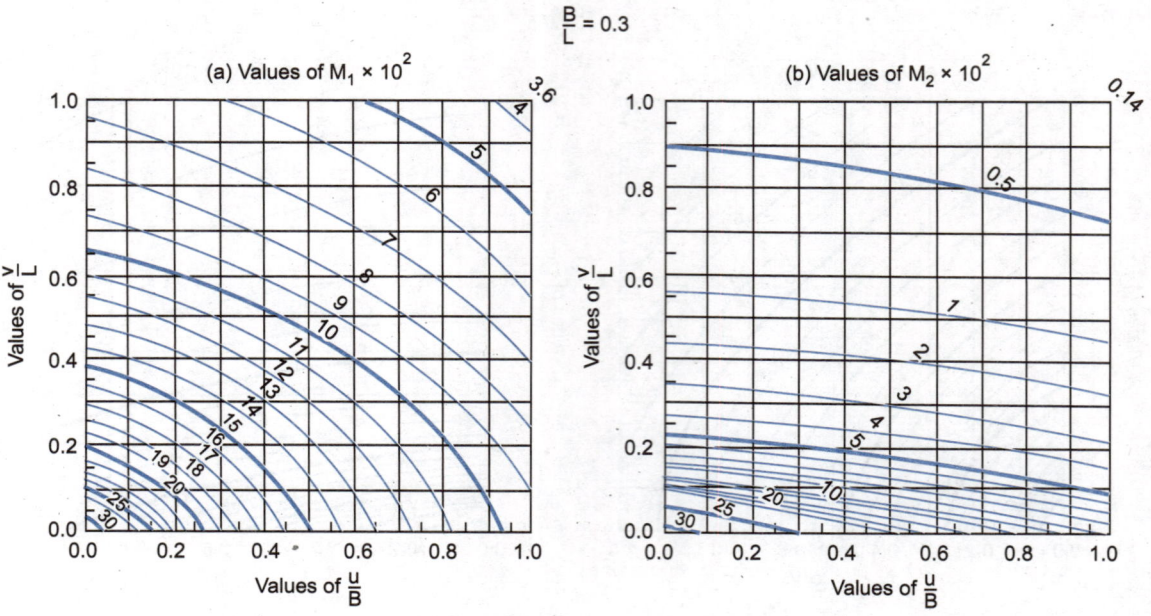

**Fig. 31.21.** Values of $M_1$ and $M_2$ for $B/L = 0.3$.

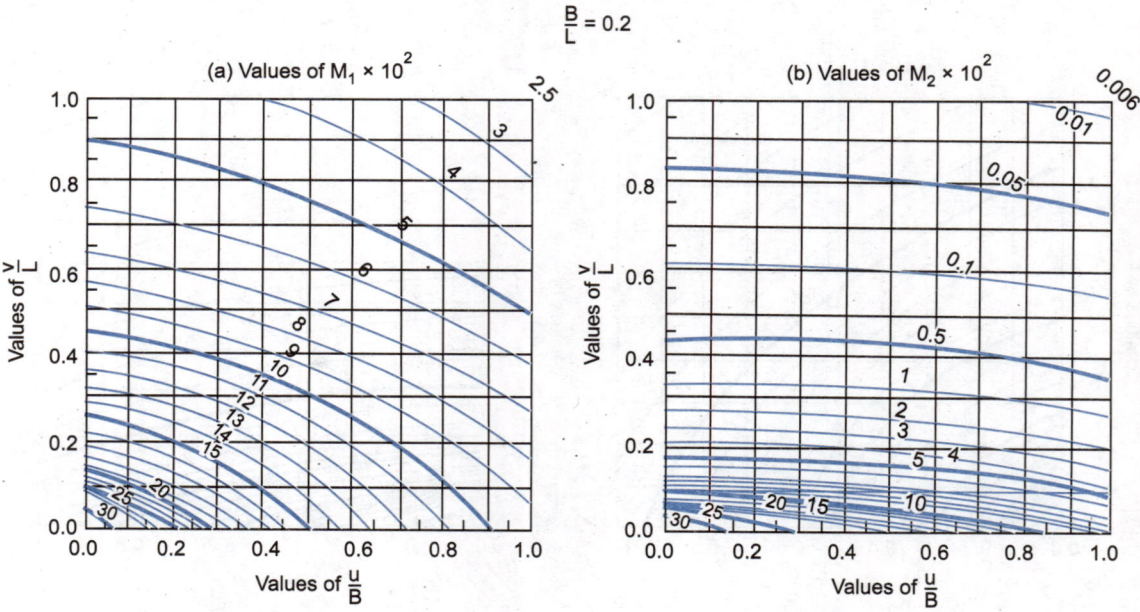

**Fig. 31.22.** Values of $M_1$ and $M_2$ for $B/L = 0.2$.

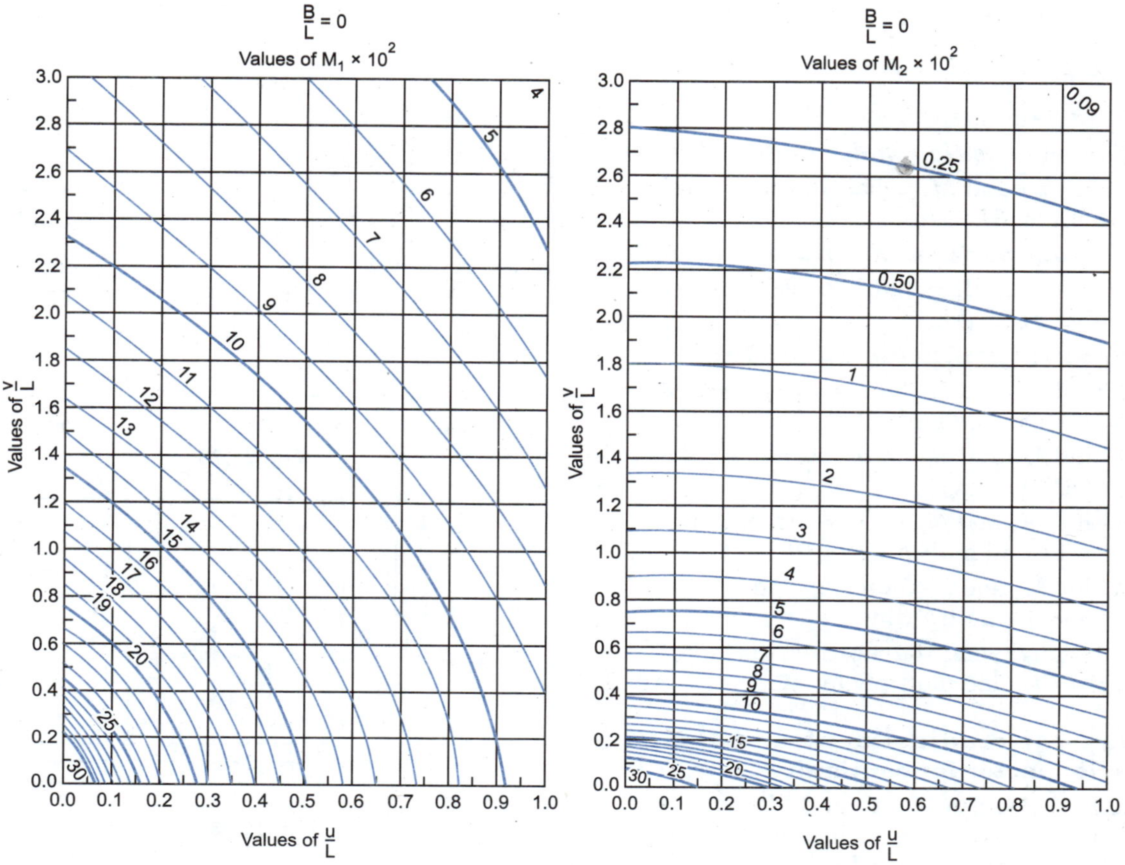

(a) Values of $M_1$ for $B/L = 0$

(b) Values of $M_2$ for $B/L = 0$

Fig. 31.23

## ECCENTRIC AND MULTIPLE CONCENTRIC LOADS

Pigeaud's curves given in Fig. 31.15 to 31.24 are applicable only for a single wheel load placed centrally. For multiple concentric loads *or* an eccentric load, the B.M. coefficients $M_1$ and $M_2$ can be found as discussed below:

### (a) Two Concentric Loads

**1. *For load position shown in Fig. 31.25 (a):***
Follow the following procedure:

(i) Find $M_1$ and $M_2$ for $u = 2(u_1 + x)$ and $v = v$ and multiply by $(u_1 + x)$.

(ii) Find $M_1$ and $M_2$ for $u = 2x$ and $v = v$ and multiply by $x$.

(iii) Deduct (ii) from (i) Design $M_1$ and $M_2$ = (iii) multiplied by $\dfrac{2W}{u_1}$.

**2. *For load position shown in Fig. 31.25 (b)***

(i) Find $M_1$ and $M_2$ for $u = u$ and $v = 2(v_1 + y)$ and multiply by $(v_1 + y)$.

(ii) Find $M_1$ and $M_2$ for $u = u$ and $v = 2y$ and multiply for $y$.

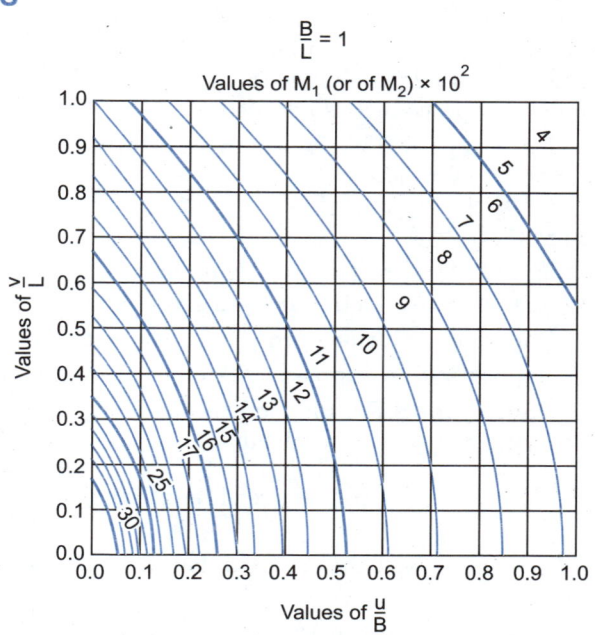

Fig. 31.24. Values of $M_1 = M_2$ for $B/L = 1$.

(*iii*) Deduct (*ii*) from (*i*).

Design $M_1$ and $M_2$ = (*iii*) multipled by $\dfrac{2W}{v_1}$.

### (b) Single Eccentric Load

**1. *For load position shown in Fig. 31.26(a)*.**

Design $M_1$ and $M_2$ = half of those obtained for case of Fig. 31.25 (*a*).

**2. *For load position shown in Fig. 31.26(b)***

Design $M_1$ and $M_2$ = half of those obtained for case of Fig. 31.25 (*b*).

**3. *For load position in Fig. 31.26 (c)***

(*i*) Find $M_1$ and $M_2$ for $u = 2(u_1 + x)$ and $v = 2(v_1 + y)$ and multiply by $(u_1 + x)(v_1 + y)$.

(*ii*) Find $M_1$ and $M_2$ for $u = 2x$ and $v = 2y$ and multiply by $xy$.

(*iii*) Find $M_1$ and $M_2$ for $u = 2(u_1 + x)$ and $v = 2y$ and multiply by $y(u_1 + x)$.

(*iv*) Find $M_1$ and $M_2$ for $u = 2x$ and $v = 2(v_1 + y)$ and multiply by $x(v_1 + y)$.

For design $M_1$ and $M_2$, substract $[(iii) + (iv)]$ from $[(i) + (ii)]$ and multiply by $\dfrac{W}{u_1 v_1}$.

**Span ratio adjustment and B.M. reduction factor for continuity**

The B.M. coefficients $M_1$ and $M_2$ given in Figs. 31.15 to 31.24 are applicable for mid-span of panels that are freely supported with restrained corners. If the panel is either fixed *or* continuous along all the four edges, Pigeaud recommends that mid-span B.M. be reduced by 20%. The negative moments at the support may be taken equal to the reduced moment at the centre of the panel. For one *or* more continuous edges shown in Fig. 31.27, the following procedure is adopted.

Find $M_1$ and $M_2$ for modified value of $k$, i.e., $k_1 = \dfrac{B}{f_1 L}$ where $f_1$ is taken from Fig. 31.27.

The B.M. coefficient thus found are multiplied by reduction factors given below:

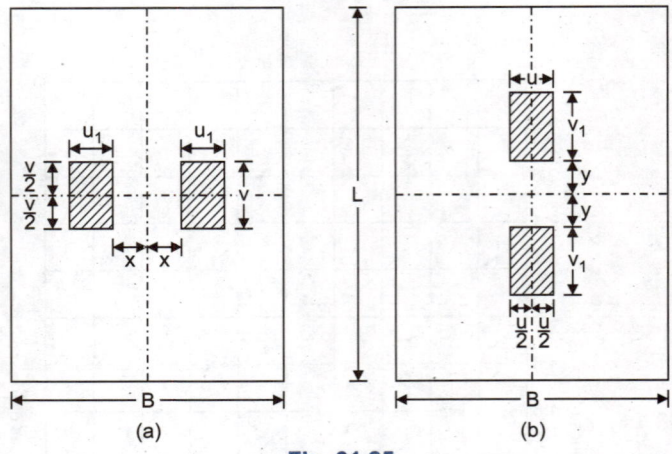

Fig. 31.25

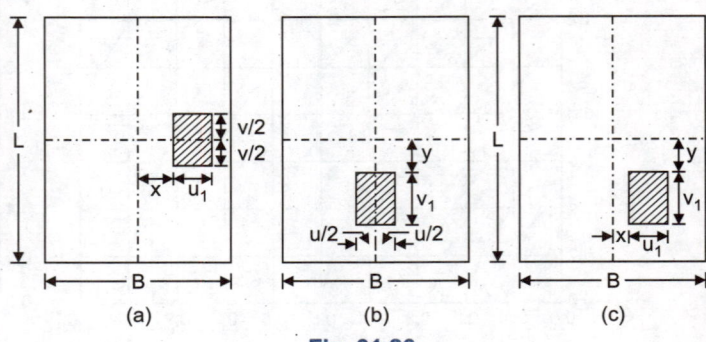

Fig. 31.26

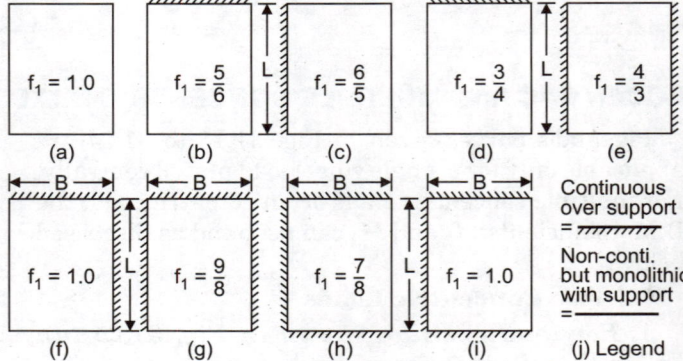

Fig. 31.27

Mid-span: Interior span = 0.7 ; End span = 0.85

Supports: End support = 0.25; penultimate = 0.95 ; Interior (except penultimate) = 0.90.

**Maximum S.F. :** Pigeaud recommends the following values of maximum S.F. per unit length of panel carrying a concentrated load:

$u \geq v$ :     At the centre of length $u$, $Q = \dfrac{W}{2u + v}$.

At the centre of length $v$, $Q = \dfrac{W}{3u}$.

$v \geq u$ :     At the centre of length $u$, $Q = \dfrac{W}{3v}$. At the centre of length $v$, $Q = \dfrac{W}{2v + u}$.

However S.F. is mostly found by the *effective width method*.

## 31.15. DISTRIBUTION OF LIVE LOADS ON LONGITUDINAL BEAMS

In the case of deck girder bridge, the bridge structure consists of a number of longitudinal beams (main beams), cross beams and the deck slab which rests on the beams and is monolithic with them. Fig. 31.28(a) shows the plan of such a structure. When a point load (wheel load) acts on one of the longitudinal beams, it undergoes deflection. But since it is monolithic with the deck slab and the cross-beams, this point load is partly transferred to longitudinal beams. In other words, the point load acting on any one longitudinal beam is distributed to other beams because of the transverse strength of the deck, thereby relieving the loaded beam to some extent.

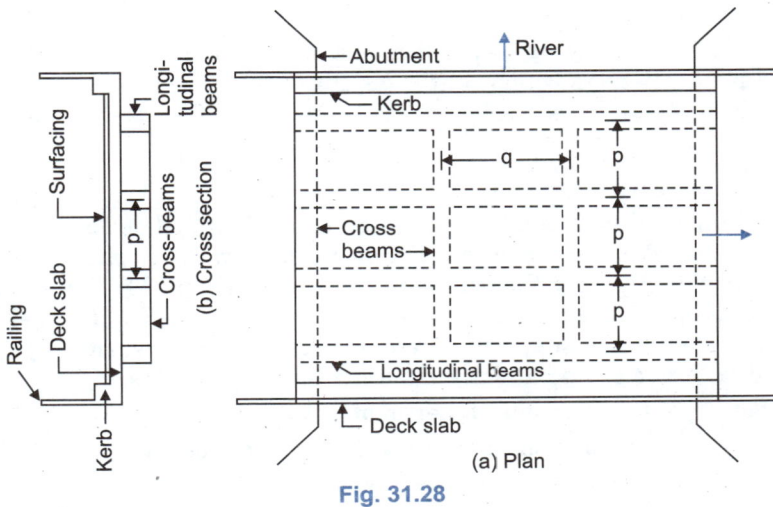

Fig. 31.28

Basically, the various methods of analysis of the grid and slab structures fall under three categories. In the *first category or* approach used by Lazarides, the structure is divided into individual longitudinal and transverse members each possessing the appropriate flexural and torsional stiffness. For each point of intersection of members, compatibility equations for deflection and slope can be set up, and these equations can be solved. This approach is extremaly cumbersome involving great deal of arithmatic work, and cannot be generalised.

In the *second category* the analysis separate the longitudinal *or* primary members of the structure and consider some form of secondary cross-connection which represents the behaviour of transverse members. Henry and Jaeger have developed this approach, based on the simplifying assumption that the transverse members can be replaced by a uniform continuous transverse medium of equivalent stiffness. Harmonic analysis is used to derive the amplitudes of the deflection and bending moment for each longitudinal member.

The *third category* of analysis are based on anisotropic *or* orthotropic plate theory. These analysis replace the actual bridge structure by an *equivalent orthotropic plate* which is then treated according to the classical theory. Guyon (1946) first developed this approach, which was generalized by Massonnet (1950) to include the effects of torsion. Extensions and developments of Guyon's and Massonnet's work have been produced by Rowe (1955), and Morrice and Little (1954, 55, 56), which generalized the use of this method and from which a design procedure has been formulated. This particular approach has the merit that a single set of *distribution of coefficients* for two extreme cases of non-torsion grillage and a full torsion slab, enable the distribution behaviour of any type of bridge structure to be found. The method, commonly known as the *method of distribution coefficients* is given here.

*Courbon* has given an empirical approach according to which the load distribution between the various longitudinal beams of a bridge can be determined. The method is applicable only when the effective width of deck is less than half the span, and when stiffness of cross-girders is very much greater than that of the longitudinals. However, because of simplicity of approach, the method is given here.

## 31.16. METHOD OF DISTRIBUTION COEFFICIENTS

This method is based upon isotropic and orthotropic plate analysis, in which the actual bridge structure is replaced by an equivalent orthotropic plate which possess the same longitudinal and transverse stiffness as the actual deck.

**Equivalent orthotropic plate:** If $p$ is the centre to centre spacing of longitudinal beams, and their number is $n$, the effective width $(2b)$ of the equivalent orthotropic plate will be given by $2b = np$.

Similarly, if there are $m$ number of transverse beams spaced at $q$, the span $2a$ of the equivalent plate is given by
$$2a = mq.$$
In a T-beam bridge, the effective width and the actual width will be identical provided the edge members have cantilevers which cantilever out for half the beam spacing. Where this is not the case, the effective width is simply reduced from the ratio of the $I$ for edge member and the $I$ for the internal member, i.e.,
$$2b = (n-2)p + 2\,\frac{I(external)}{I(internal)}p.$$

It should be noted that since the equivalent orthotropic plate is being derived, no restriction on flange width applies. Also with regard to the transvers members, it should always be assumed that transverse members *or* diaphragms are provided at the supports. These are essential in ensuring distribution of load and in sustaining bending stresses.

Let $I_L$ = moment of inertia of longitudinal beams
$I_T$ = moment of inertia of transverse beams
$q$ = centre to centre spacing of cross-beams.
Moment of inertia of plate per unit width, $i_L = I_L/p$
Moment of inertia of plate per units length $i_T = I_T/q$

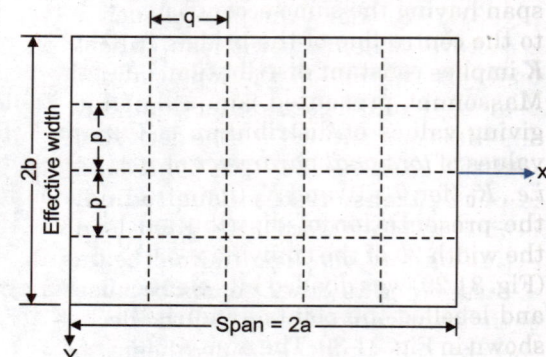

Fig. 31.29. Equivalent Plate.

For simple slab of thickness $D$, supported on opposite edges : $i_L = i_T = D^3/12$

**Flexural Parameter** $\theta$ : The relationship between longitudinal and transverse stiffnesses is defined by *flexural parameter* $\theta$ :

$$\theta = \frac{b}{2a}\left(\frac{i_L}{i_T}\right)^{1/4} = \frac{b}{2a}\left(\frac{I_L}{I_T}\cdot\frac{q}{p}\right)^{1/4} \qquad \ldots(31.12)$$

For slabs
$$\theta = \frac{b}{2a} \qquad \ldots[31.12\,(a)]$$

For most practical structure, $\theta$ lies between 0.3 to 1.0.

**Torsional Parameter**

Let $J_L$ = Torsional inertia of longitudinal beams
$J_T$ = Torsional inertia of transverse beams.
Torsional inertia of plate, per unit width, $j_L = J_L/p$
Torsional inertia of plate, per unit length $j_T = J_T/q$.

The relationship between the longitudinal and transverse torsional stiffnesses is defined by *torsional parameter* $\alpha$ :

$$\alpha = \frac{G(j_L + j_T)}{2E\sqrt{j_L \cdot i_T}} = \frac{G}{2E}\frac{\left(\dfrac{J_L}{p} + \dfrac{J_T}{q}\right)}{\sqrt{\dfrac{I_L}{p}\cdot\dfrac{I_T}{q}}} \qquad \ldots(31.13)$$

where $G$ is the modulus of rigidity and $E$ is the Young's modulus of elasticity. For R.C.C., $G/E = 0.435$. For deck slab $\alpha = 1$. For T-beam bridges, the value of $\alpha$ is always less than 1.

## DISTRIBUTION COEFFICIENT FOR LONGITUDINAL MOMENT

The distribution coefficient ($K$) for longitudinal bending moment ($M_x$) is defined by the following equation:
$$M_x = K\cdot M_{x,\,mean} \qquad \ldots(31.14)$$
where
$M_{x\,mean}$ = mean *or* average longitudinal moment found by considering the load to be equally divided in all the beams
$M_x$ = actual moment in the longitudial beam.

$$K' = f(k) = \frac{\Sigma K W}{\Sigma W}$$

$K$ = moment distribution coefficient ; $W$ = load transferred at standard positions.

The moment distribution coefficient is affected only by the transverse location of the load and is the same for all positions of the load along the span having the same eccentricity with respect to the centre line of the bridge. In other words $K$ implies constant distribution along the span. Massonnet presented comprehensive tables giving values of distribution coefficient $K$ for values of *torsional parameter* of zero and unity, i.e., $K_0$ (for $\alpha = 0$) and $K_1$ (for $\alpha = 1$). To simplify the presentation of distribution coefficients, the width $2b$ of the equivalent orthotropic plate (Fig. 31.29) was divided into eight equal sections and lebelled the points dividing the section as shown in Fig. 31.30. The nine points are referred to as the *standard positions* or occasionally as the *reference stations*.

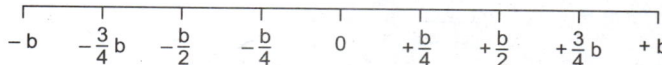

**Fig. 31.30.** Standard Positions or Reference Stations.

Fig. 31.31 to 31.41 (Eleven plates) gives the graphs for distribution coefficients $K_0$ and $K_1$ and various reference stations. Let us consider Fig. 31.31; the ordinates give the value of $K_0$ (value of distribution coefficient for $\alpha = 0$) as a function of three factors:

(i) the value of flexural parameter $\theta$.
(ii) the position of concentrated load.
(iii) the standard position or reference station considered.

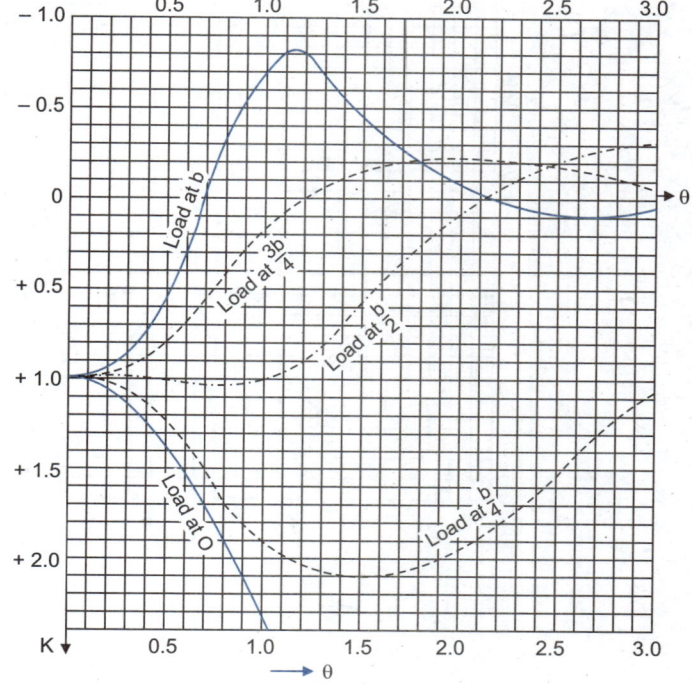

**Fig. 31.31.** $K_0$ For Reference Station 0 (Plate 1).

The curves in Fig. 31.31 are for reference station 0, i.e., the longitudinal centre line of the bridge. Because of symmetry, only five load positions have been considered, i.e., 0, $\frac{b}{4}$, $\frac{b}{2}$, $\frac{3b}{4}$ and $b$. In Fig. 31.32 reference station is $\frac{b}{4}$. Since there is no symmetry, curves are given for a load at each of the nine standard positions, i.e., $0, \pm \frac{b}{4}, \pm \frac{b}{2}, \pm \frac{3b}{4}$ and $\pm b$.

**Distribution coefficient for intermediate values of $\alpha$ :** As stated earlier, the curves are available only for two values of $\alpha$ i.e., for $\alpha = 0$ ($K_0$) and $\alpha = 1$ ($K_1$). The distribution coefficient for a specific value of torsional parameter may be obtained by using the interpolation formula:

$$K_\alpha = K_0 + (K_1 - K_0) \sqrt{\alpha} \qquad \text{...(31.15)}$$

**Procedure for finding longitudinal moment:** To start with, the maximum average longitudinal moments are found in the bridge, considering the load to be equally divided in all the beams. While computing the maximum average longitudinal B.M. ($M_{av.}$) for various load positions, the actual loads should be changed to equivalent loads acting at the standard positions on the width $2b$. This is done by simple statics by considering the equivalent load as the reaction from a simple beam of span $b/4$.

The load distribution coefficients $K_0$ (for $\alpha = 0$) and $K_1$ (for $\alpha = 1$) are tabutated for given value of $\theta$. The value of $K_\alpha$ is then computed from Eq. 31.15. From these tables, the values of $K_\alpha$ for reference stations corresponding to the actual location of longitudinal girders are then computed for various load position $-b, -\frac{3b}{4}, -\frac{b}{2}, -\frac{b}{4}, 0, \frac{b}{4}, +\frac{b}{2}, +\frac{3b}{4}$ and $+b$.

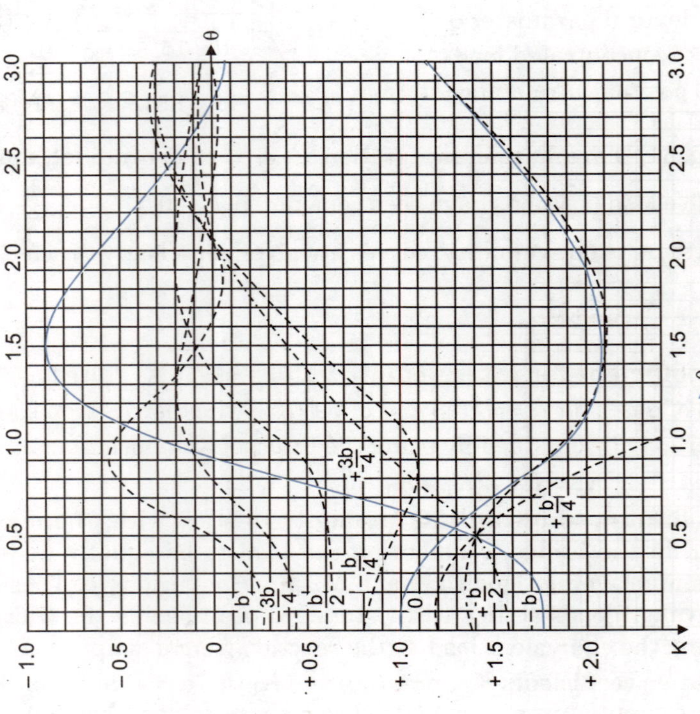

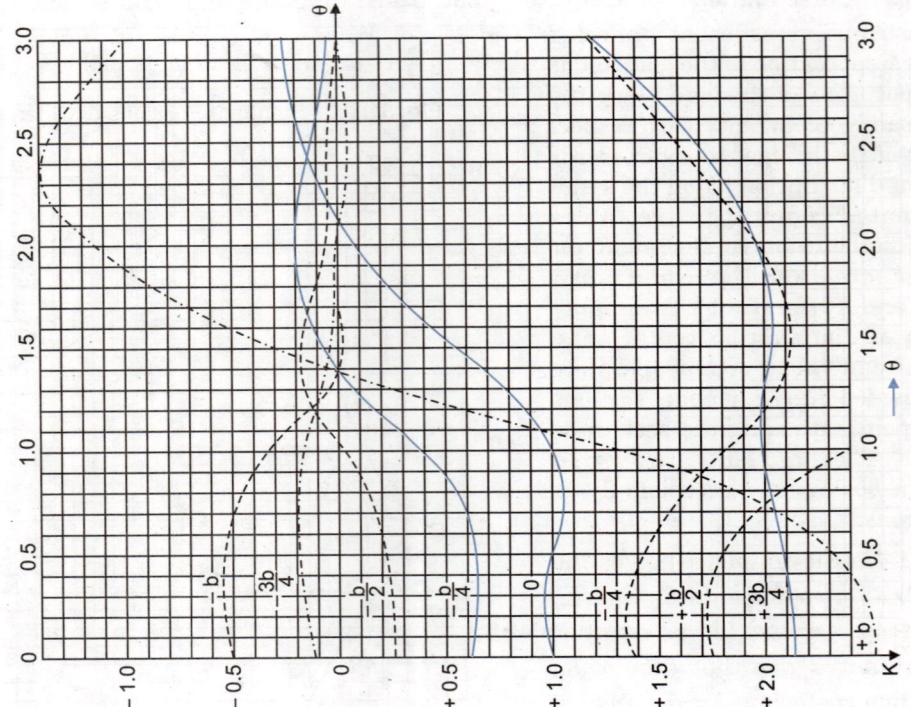

Fig. 31.33. $K_0$ For Reference Station $b/2$ (Plate 3).

Fig. 31.32. $K_0$ For Reference Station $b/4$ (Plate 2).

# CONCRETE BRIDGES

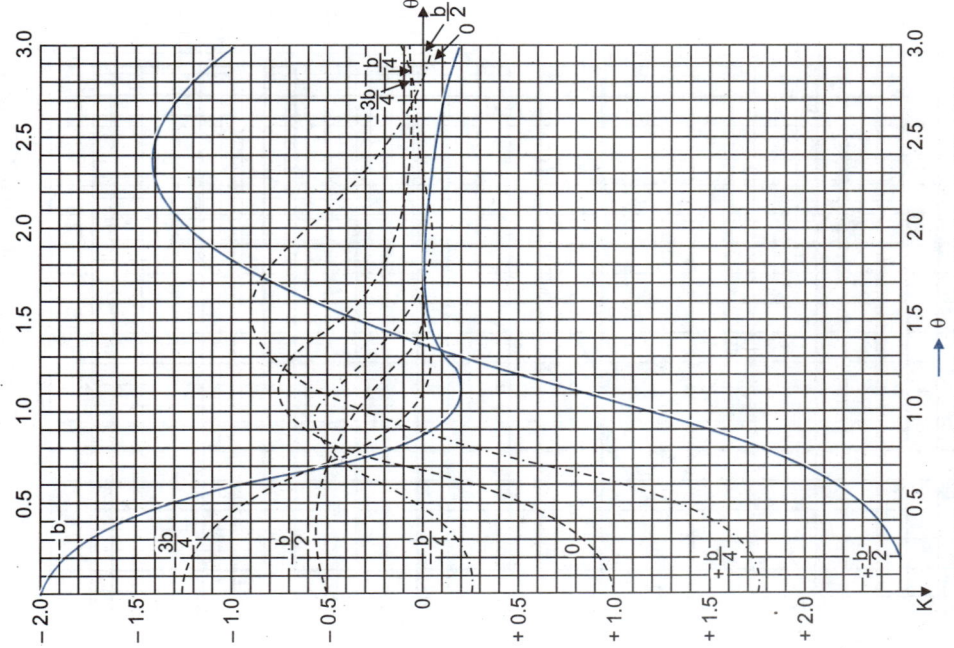

Fig. 31.35. $K_0$ For Reference Station b (Plate 5).

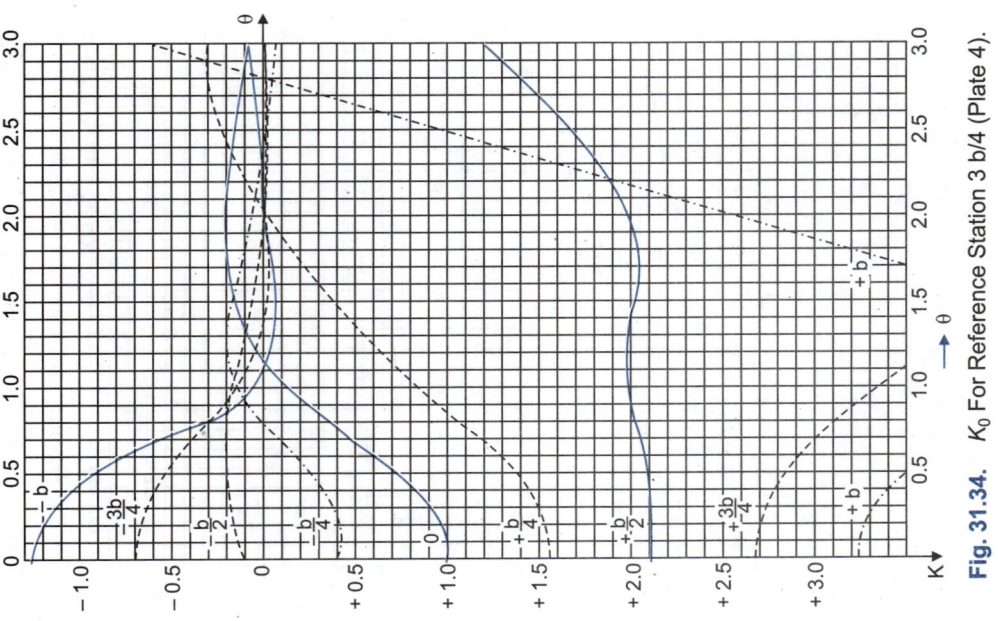

Fig. 31.34. $K_0$ For Reference Station 3 b/4 (Plate 4).

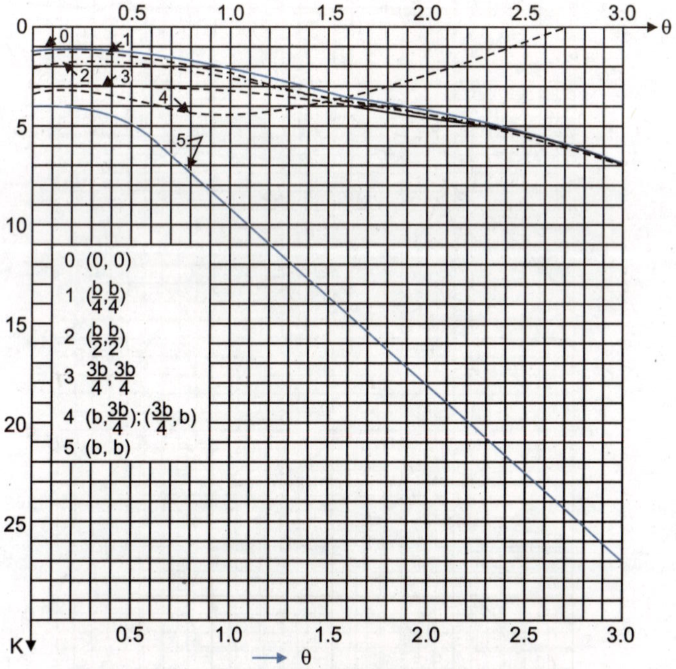

# 784 REINFORCED CONCRETE STRUCTURE

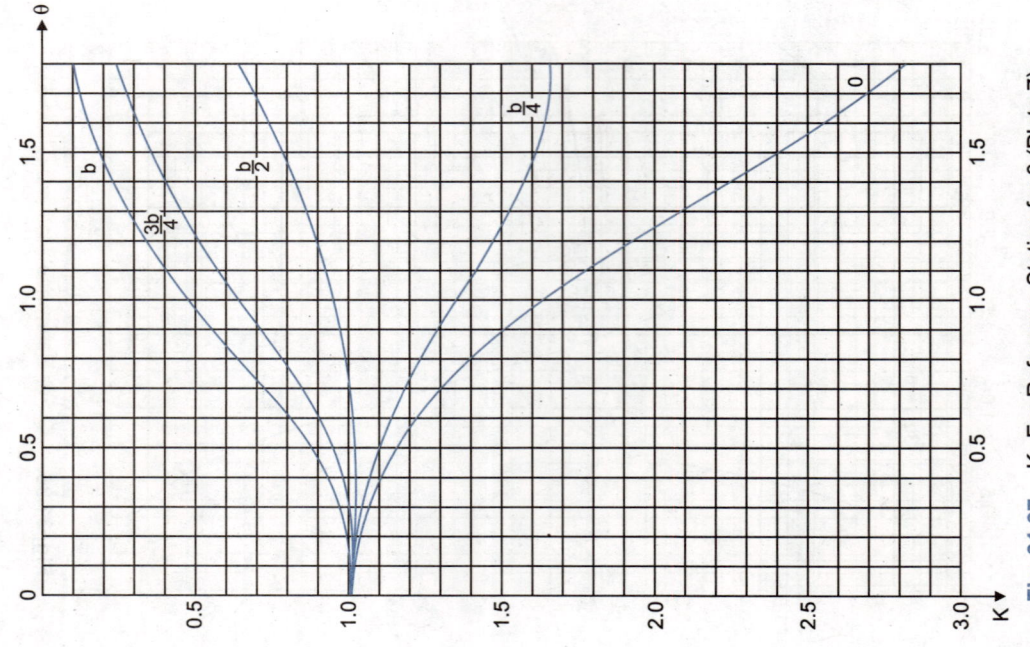

**Fig. 31.37.** $K_1$ For Reference Station for 0 (Plate 7).

**Fig. 31.36.** Large Range Coefficients $K_0$ (Plate 6).

**Note:** For each curve, the first figure shows reference station while the second figure shows load eccentricity. Thus $(b, 3b/4)$ means reference station $b$, load position $3b/4$.

0  (0, 0)
1  $(\frac{b}{4}, \frac{b}{4})$
2  $(\frac{b}{2}, \frac{b}{2})$
3  $\frac{3b}{4}, \frac{3b}{4}$
4  $(b, \frac{3b}{4}); (\frac{3b}{4}, b)$
5  $(b, b)$

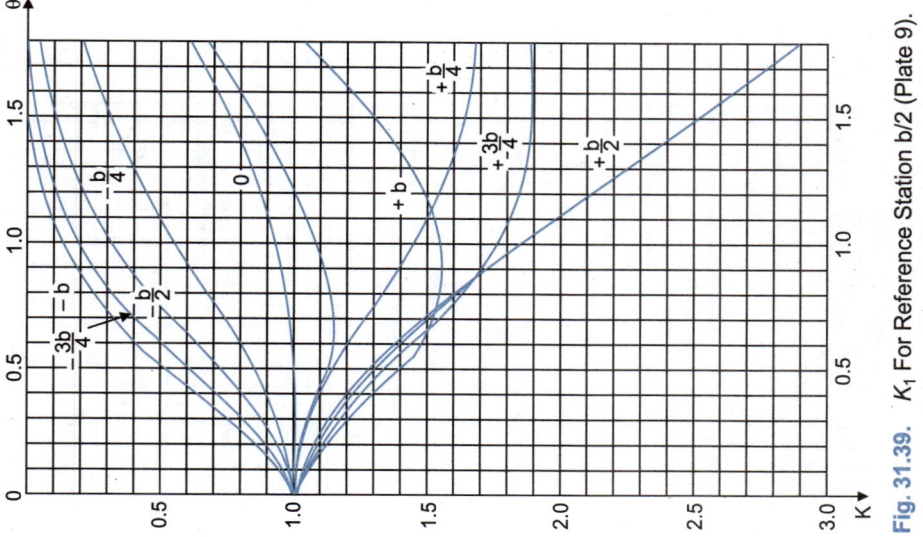

Fig. 31.39. $K_1$ For Reference Station $b/2$ (Plate 9).

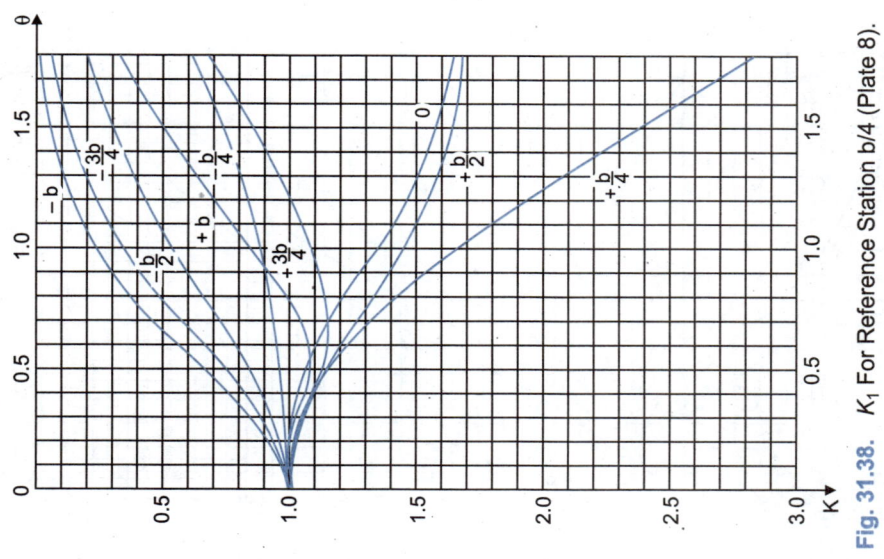

Fig. 31.38. $K_1$ For Reference Station $b/4$ (Plate 8).

**786** REINFORCED CONCRETE STRUCTURE

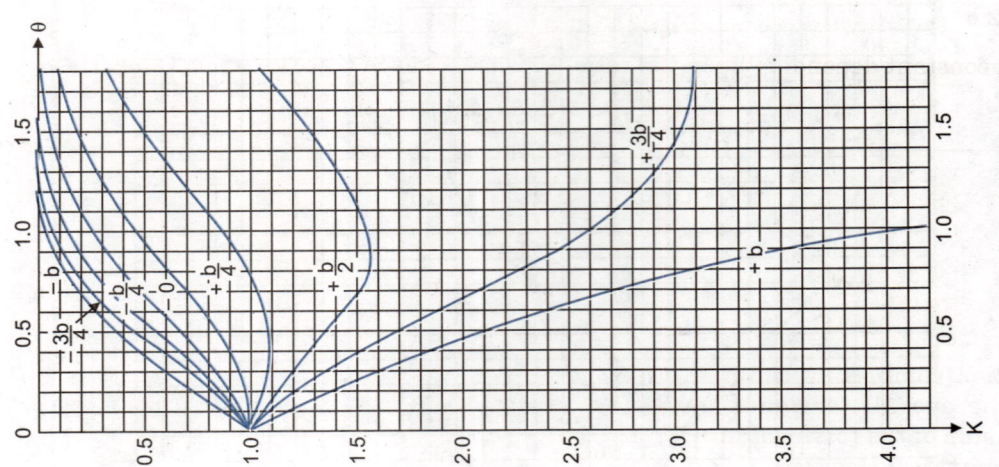

Fig. 31.41. $K_1$ For Reference Station $b$ (Plate 11).

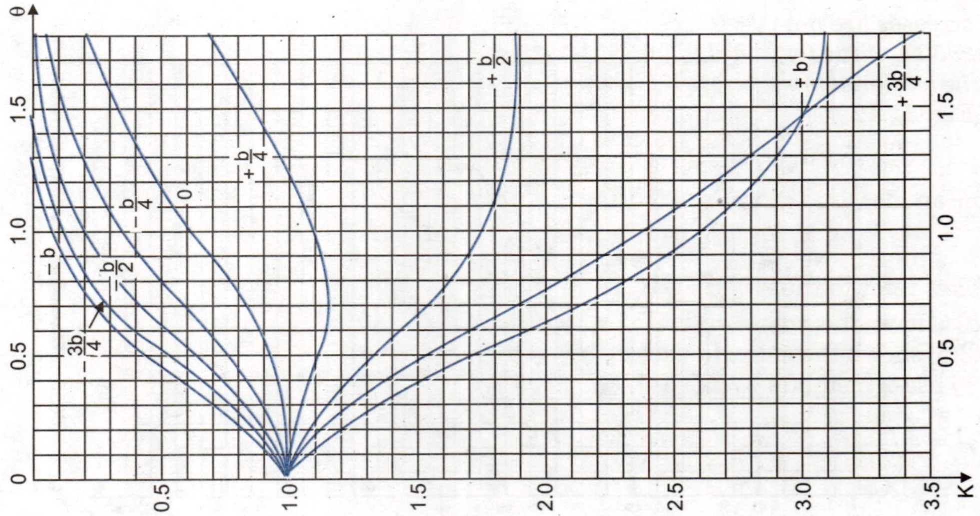

Fig. 31.40. $K_1$ For Reference Station $3b/4$ (Plate 10).

These coefficient $K_\alpha$ are then multiplied by loads $W$ transferred at the standard load position to get $Q$ (i.e., $Q = K_\alpha \cdot W$) at each load position. The summation of $Q$ (i.e. $\Sigma Q$) and of $W$ (i.e., $\Sigma W$) is then found. A new coefficient $K'$ is then found from the relation $K' = \Sigma Q/W$, for each girder. The final moment in each girder are then computed from the relation $M_x = M_{av} \times K'$. The procedure is illustrated in Example 31.3.

Since the coefficients $K_0$ and $K_1$ are calculated by taking only the first term of the Fourier series, Morice recommended that these should be increased by 10%. That is Eq. 31.14. should be modified as under:

$$M_x = 1.1 \, K \, M_{x,\,mean} \text{ (mean)} \quad \ldots[31.14(a)]$$

**Torsional Inertia (J):** The torsional inertia of a rectangular area of width $2a$ and length $2b$ is given by
$$J = \beta \, (2a)^3 \cdot 2b \quad \ldots(31.16)$$
where $\beta$ is a constant depending upon the ratio $\dfrac{b}{a}$. The values of $\beta$ are given in Table 31.4.

**TABLE 31.4.** Values of $\beta$

| b/a | β | b/a | β | b/a | β |
|---|---|---|---|---|---|
| 1.0 | 0.141 | 2.0 | 0.229 | 4.0 | 0.281 |
| 1.2 | 0.166 | 2.25 | 0.240 | 5.0 | 0.291 |
| 1.5 | 0.196 | 2.5 | 0.249 | 10.0 | 0.312 |
| 1.75 | 0.213 | 3.0 | 0.263 | ∞ | 0.333 |

For a section comprising a number of rectangles the overall torsional inertia would be the sum of the torsional interia of individual rectangles. However, this is not correct in the case of a T-beam in which the top flange is the part of a continuous slab. In each individual T-beam, only the shear stress parallel to the top surface can exist and in an individual T-beam isolated *or* split, for convenience in determining $J$, the vertical shear stresses are not present. *This means that only 50% of the torsional inertia contributes to the torsional parameter.* Thus, for T-beam of Fig. 31.42,

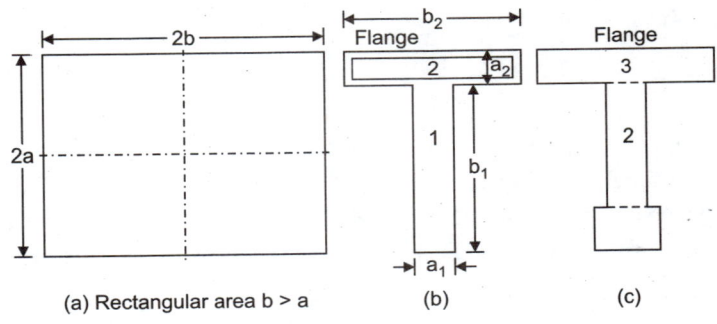

Fig. 31.42

$$J = \beta_1 \, (2a_1)^3 \cdot 2b_1 + \tfrac{1}{2} \beta_2 \, (2a_2)^3 \cdot 2b_2 \quad \ldots[31.17(a)]$$

Similarly, for an idealised girder section shown in Fig. 31.42 (c),

$$J = \beta_1 \, (2a_1)^3 \cdot 2b_1 + \beta_2 \, (2a_2)^2 \cdot 2b_2 + \tfrac{1}{2} \beta_3 \, (2a_3)^3 \cdot 2b_3 \quad \ldots[31.17(b)]$$

where the suffixes refer to the rectangles 1, 2, 3.

Fig. 31.43(a) shows the cross-section of a bridge while Fig. 31.43(b) shown the individual cell, for which the value of $J$, by use of membrane analogy, is given by

$$J = \dfrac{4A^2}{\oint \dfrac{ds}{t}} \quad \ldots[31.18(a)]$$

If the perimeter of the hole consists of a series of straight lines as shown in Fig. 31.43(c), the above equation reduced to

$$J = \dfrac{4A^2}{\left[\dfrac{p - 2t_3}{t_1} + \dfrac{p - 2t_3}{t_2} + \dfrac{2(d - t_1 - t_2)}{t_3}\right]},$$

where $A$ = area of hole in section $\quad \ldots[31.18\,(b)]$

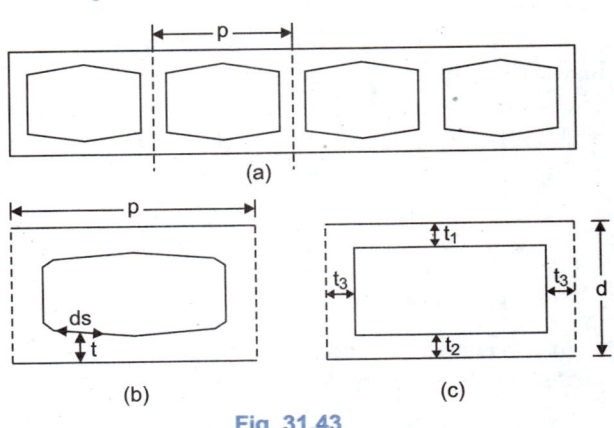

Fig. 31.43

# DISTRIBUTION COEFFICIENT FOR TRANSVERSE MOMENT

When a point load acts on the bridge, it also produces moment $M_y$ in the transverse direction. Mossonnet showed that this moment is given by

$$M_y = \sum_{m=1}^{\infty} \mu_m \frac{P \cdot b}{a} \sin \frac{m\pi u}{2a} \sin \frac{m\pi x}{2a} \qquad \ldots(31.19)$$

where   $P$ = point load acting at distance $u$ from abutment
   $x$ = distance from abutment of the section where transverse B.M. is required.
   $\mu_m$ = distribution coefficient similar to $K$.

It has been found that maximum value of $M_y$ occurs at the midspan at the centre of transverse section when the eccentricity is the least and when the load is also acting at the mid-span. Hence putting $x = a$

$$M_y = \frac{P \cdot b}{a} \sum \mu_m \sin \frac{m\pi u}{2a} \cdot \sin \frac{m\pi}{2} \qquad \ldots[31.19(a)]$$

or

$$M_y = \frac{P \cdot b}{a} \left[ \mu_1 \frac{\sin \pi u}{2a} - \mu_3 \sin \frac{3\pi u}{2a} + \mu_5 \sin \frac{5\pi u}{2a} - \ldots \right] \qquad \ldots[31.19(b)]$$

For distributed load $w$ per unit length, $M_y = \frac{4wb}{\pi} \left[ \mu_1 - \frac{\mu_3}{3} + \frac{\mu_5}{5} - \ldots \right] \qquad \ldots[31.19(c)]$

If distributed load $p$ per unit length is in the central length $2c$ of the span,

$$M_y = \frac{4wb}{\pi} \left[ \mu_1 \sin \frac{\pi c}{2a} + \frac{\mu_3}{3} \sin \frac{3\pi c}{2a} + \frac{\mu_5}{5} \sin \frac{5\pi c}{2a} + \ldots \right] \qquad \ldots[31.19(d)]$$

Generally only first three terms are considered sufficient to evaluate $M_y$. The terms $\mu_m$ are equal to value of $\mu$ for flexural parameter $m$. Thus $\mu_1 = \mu$ for $\theta$, $\mu_3 = \mu$ for $3\theta$, $\mu_5 = \mu$ for $5\theta$ etc. Massonnet computed the value of $\mu$ for two values of $\alpha$ i.e., for $\mu_0$ for $\alpha = 0$ and $\mu_1$ for $\alpha = 1$. Massonnet showed that the coefficient for any value $\alpha$ $(0 < \alpha < 1)$ could be obtained from the interpolation formula

$$\mu_\alpha = \mu_0 + (\mu_1 - \mu_0) \sqrt{\alpha} \qquad \ldots(31.20)$$

Design curves for $\mu_0$ and $\mu_1$ have been prepared for the standard positions (reference stations) $0$, $\frac{b}{4}$, $\frac{b}{2}$ and $\frac{3b}{4}$ with load position $0, \pm \frac{b}{4}, \pm \frac{b}{2}, \pm \frac{3b}{4}$ and $\pm b$. However for most design purposes, only the curves relevant to the standard position (reference station) 0 are required. These are given in plate 12 for $\mu_0$ (Fig. 31.44) and plate 13 for $\mu_1$ (Fig. 31.45). Since both these plates are applicable for refernce station 0 (i.e., the longitudinal centre line of the bridge), symmetry applies and the curves are necessary for load position $0, \frac{b}{4}, \frac{b}{2}, \frac{3b}{4}$ and $b$ only. For any particular value of $\theta$ the design curves enable the influence lines for $\mu_0$ and $\mu_1$ at the centre of the bridge, to be derived.

**Derivation of maximum transverse moment:** The maximum transverse B.M. caused by the abnormal load will occur at the centre of the bridge in the vast majority of cases. Fig. 31.46 shows the wheel positions of the abnormal load to cause maximum transverse moment at the centre of bridge, with the line of one of the internal wheels coincident with the longitudinal centre line of the bridge. The generalised equation giving the transverse moment is given by

$$M_y = \sum_{m=1}^{\infty} \mu_{m\theta} \cdot b \cdot H_n \sin \frac{n\pi x}{2a} \qquad \ldots(31.21)$$

where   $\mu_{m\theta}$ is value of $\mu$ (i.e., $\mu_0$ or $\mu_1$) for flexural parameter $m\theta$ (i.e., for $\theta, 2\theta, 3\theta$ etc).

At the centre of bridge, $x = a$ and hence $\sin \frac{n\pi x}{2a}$ is alternatively plus and minus units for odd terms of series and zero for the even terms.

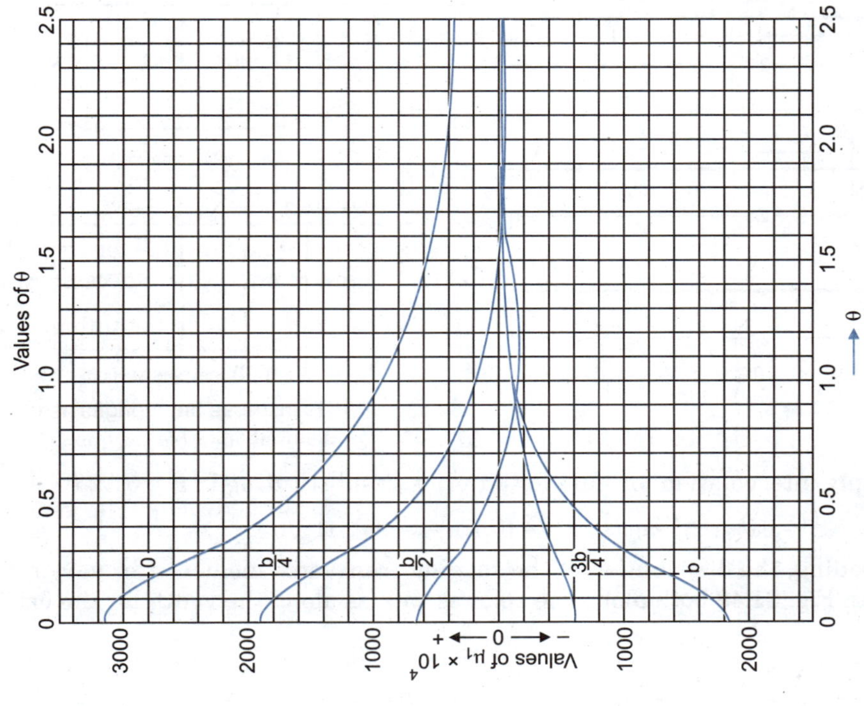

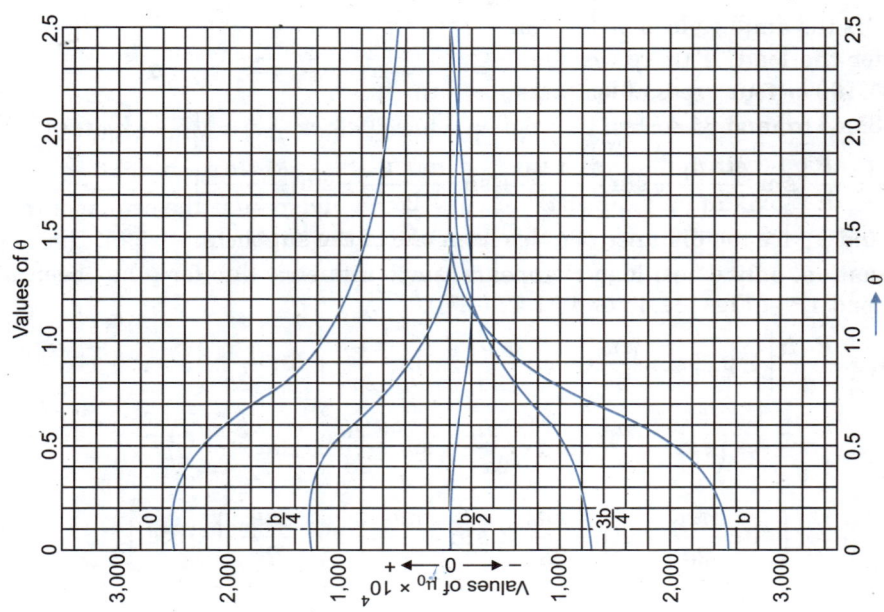

**Fig. 31.45.** Coefficient $\mu_1$ for Reference Station 0 (Plate 13).

**Fig. 31.44.** Coefficient $\mu_0$ for Reference Station 0 (Plate 12).

The value of $\mu_{m\theta}$ in the above equation can be determined from the influence lines for $\mu_0$ and $\mu_1$ for flexural parameters of $\theta, 3\theta, 5\theta$ etc. For each influence line, and wheels are placed in the specified positions and the ordinates of influence lines are summed; let these be $\Sigma\mu_\theta, \Sigma\mu_{3\theta}, \Sigma\mu_{5\theta}$ etc.

In *Eq. 31.21*, $H_m$ is the amplitude of $m$th term in Fourier series for the load. If there are four point loads (Fig. 31.46.) *or* four types of loadings as shown in Figs. 31.47 (*a*) and 31.47 (*b*).

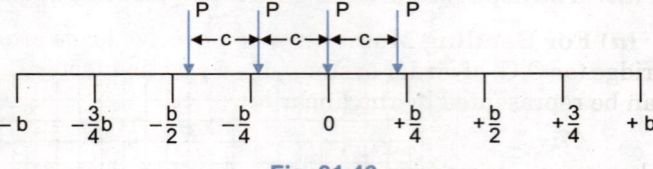

Fig. 31.46

$$H_m = \frac{P}{a}\left[\sin\frac{m\pi u_1}{2a} + \sin\frac{m\pi u_2}{2a} + \sin\frac{m\pi u_3}{2a} + \sin\frac{m\pi u_4}{2a}\right] \qquad \ldots(31.22)$$

where $u_1, u_2, u_3$ and $u_4$ are the distance of axles from one of the supports.

In the above equation, only wheel load $P$ appears, since influence line for $\mu$ has been used to find the effect of the axles load $4P$ of Figs. 31.46 and 31.47 (*a, b*).

$$\therefore M_{y\,max.} = \frac{P \cdot b}{a}\left[\Sigma\mu_\theta\left(\sin\frac{\pi u_1}{2a} + \sin\frac{\pi u_2}{2a} + \sin\frac{\pi u_3}{2a} + \sin\frac{\pi u_4}{2a}\right)\right.$$
$$-\Sigma\mu_{3\theta}\left(\sin\frac{3\pi u_1}{2a} + \sin\frac{3\pi u_2}{2a} + \sin\frac{3\pi u_3}{2a} + \sin\frac{3\pi u_4}{2a}\right)$$
$$\left.+\Sigma\mu_{5\theta}\left(\sin\frac{5\pi u_1}{2a} + \sin\frac{5\pi u_2}{2a} + \sin\frac{5\pi u_3}{2a} + \sin\frac{5\pi u_4}{2a}\right) - \text{etc.}\right] \qquad \ldots(31.23)$$

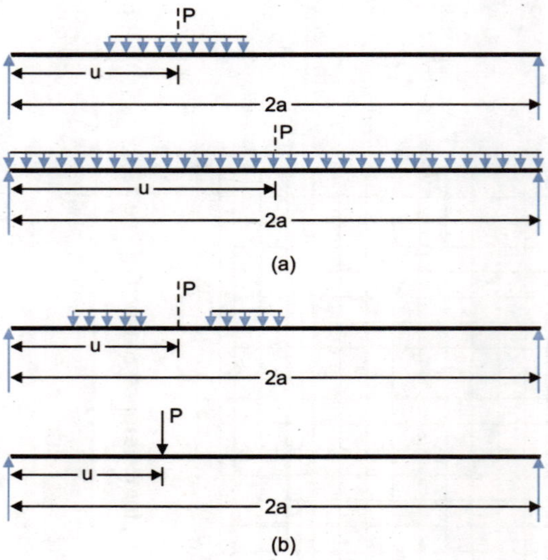

Fig. 31.47

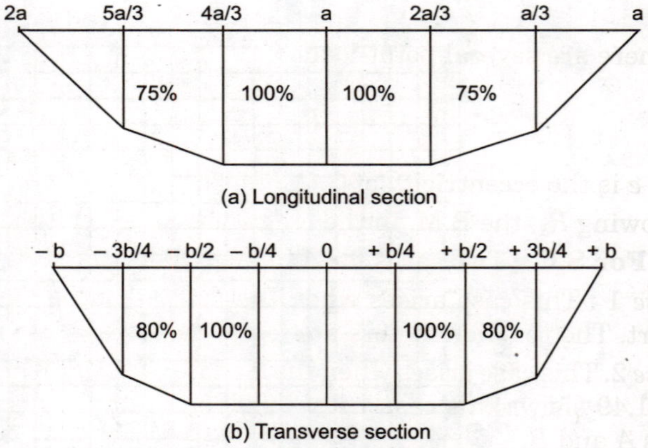

Fig. 31.48. Transverse and Longitudinal Distribution of $M_y$.

It is better to apply interpolation formula for $\mu\alpha$ prior to substituting in *Eq. 31.23*:

$$(\Sigma\mu_\alpha)_{m\theta} = (\Sigma\mu_0)_{m\theta} + [(\Sigma\mu_1)_{m\theta} - (\Sigma\mu_0)_{m\theta}]\sqrt{\alpha} \qquad \ldots(31.24)$$

For abnormal loading, the distribution of the sagging transverse moments for varying positions of the vehicle are shown in Fig. 31.48 both along the span as well as along the width for the bridge.

## 31.17. COURBON'S METHOD

Courbon's theory of distribution of live load on longitudinal beams is applicable when the following conditions are satisfied:

(*i*) The span-with ratio is greater than 2 and less than 4.

(*ii*) There are at least five symmetrical cross-girders *or* diaphragms connecting the longitudinal girders.

(*iii*) The depth of the cross-girders *or* diaphragms is atleast $\frac{3}{4}$th of the depth of longitudinal girders.

**(a) For Bending Moment:** When the live loads are eccentrically placed with respect to the axis of the bridge (*or* C.G. of girder system), then reaction factor $R_x$ for any given girder distant $x$ from the bridge axis can be represented by the linear law.

$$R_x = c + d \cdot x \qquad \ldots(i)$$

where $c$ and $d$ are constants.

The above law is justified if the cross beams are so rigid that the deflections vary linearly in the transverse direction, so that the load supported by each beams is proportional to its deflection.

$$\Sigma(c + d \cdot x) = P \quad \ldots(ii) \qquad \text{and} \qquad \Sigma(c + d \cdot x)x = P \cdot e \qquad \ldots(iii)$$

where $e$ is the eccentricity of load $P$ with respect to the axis of the bridge.

Measuring $x$ to be positive towards the load $P$ and –ve in the other direction, we have $\Sigma x = 0$

Also, $\qquad \Sigma c = n \cdot c \qquad$ where $n$ is the number of longitudinal girders.

Hence, from (*i*), $\quad \Sigma c + d \Sigma x = P \qquad$ or $\qquad nc + 0 = P \qquad$ or $\quad c = \dfrac{P}{n} \qquad \ldots(iv)$

Also, from (*iii*), $\quad c \Sigma x + d \Sigma x^2 = P \cdot e \qquad$ or $\qquad 0 + d \Sigma x^2 = P \cdot e$

$\therefore \qquad\qquad\qquad d = \dfrac{P \cdot e}{\Sigma x^2} \qquad \ldots(v)$

Substituting the values of $c$ and $d$ in (*i*)

$$R_x = \frac{P}{n} + \frac{Pe}{\Sigma x^2} \cdot x \qquad \ldots[31.25\,(a)]$$

or $\qquad\qquad\qquad R_x = \dfrac{P}{n}\left[1 + \dfrac{ne \cdot x}{\Sigma x^2}\right] \qquad \ldots[31.25\,(b)]$

If there are several point-loads

$$R_x = \frac{\Sigma P}{n}\left[1 + \frac{nex}{x^2}\right] \qquad \ldots(31.25)$$

where $e$ is the eccentricity of C.G. of loads.

Knowing $R_x$, the B.M. in the longitudinal girder can be computed.

**(b) For S.F. :** Two cases for S.F. may arise.

*Case* 1 : This case arises when the load is placed beyond the diaphragm (*or* cross beam) closest to the support. The reaction factors are found as for B.M.

*Case* 2. This case arises when the load is placed between the support and the first intermediate diaphragm. Fig. 31.49 shows this case with four longitudinal girders $A$, $B$, $C$ and $D$, with the load $P$ placed between beams $A$ and $B$.

The load $P$ is distributed to $A$ and $B$, as $P_A$ and $P_B$ by considering the slab to be simply supported. The loads $P_A$ and $P_B$ are distributed to the support $AB$ and diaphragm $A' B'$, considering the beam as simply supported. The portion of the load going to the support is the direct S.F. in the beam, while that going to the diaphragm is redistributed among all the beams by normal Courbon's method.

Hence S.F. $Q_A$ due to $P_A$ is given by

$$Q_A = P_A\left[\frac{y}{q} + \frac{z}{nq}\left(1 + \frac{n \cdot e x}{x^2}\right)\frac{l-q}{l}\right] \qquad \ldots(31.26)$$

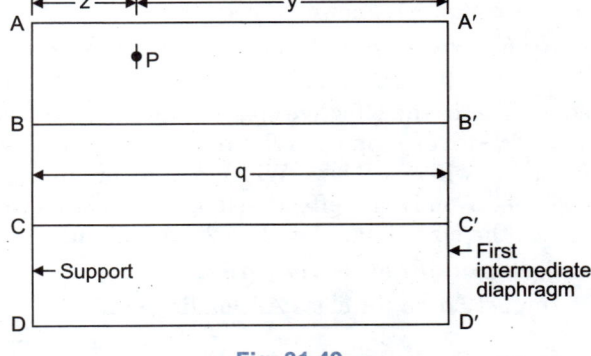

Fig. 31.49

where $\quad l$ = span of the brige

$\qquad q$ = distance between support and first intermediate diaphragm.

# 31.18. DESIGN OF A T-BEAM BRIDGE

**Design: Example 31.2.** *Design a T-beam bridge for the following data:*
*Clear width of roadway = 7 m*
*Span centre to centre of bearings = 16 m*
*Live load = one lane of class AA or two lanes of class A loading.*
*Average thickness of wearing coat = 8 cm.*
*Use M 20 concrete for deck slab and M 15 concrete for beams. Take unit weight of concrete as 24000 N/m³.*

**Solution:**

**1. Preliminary Dimensions:** Let us provide three longitudinal beams at centre to centre spacing of 2.6 m. Let the rib width be 30 cm. Keeping 40 cm wide kerbs, the arrangement will be as shown in Fig. 31.50.

Let us assume a slab thickness of 20 cm for interior panels. Let the cantilever slab be 22 cm thick at its junction with rib and 12 cm thick at the end. Let thickness of the kerb be 24 cm above the wearing coat or 32 cm above the slab. Let the cross-girders be

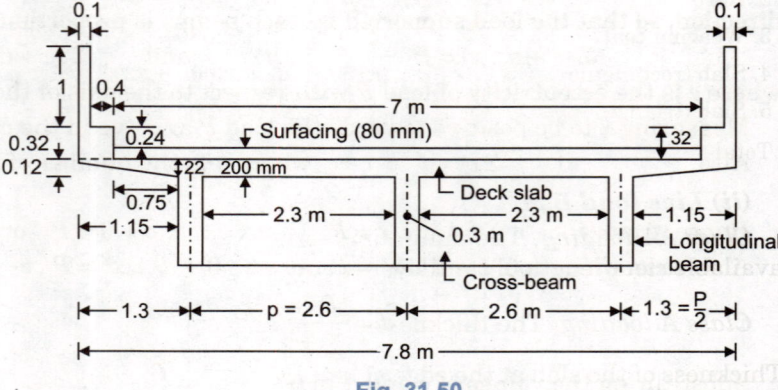

**Fig. 31.50**

provided at every 4 m c/c. There will be five cross girders. Let the breadth of the cross-girders be 30 cm and depth the same as that of the longitudinal girders.

**2. Effective Size of Panel:** The inner dimensions of the interior panel of the slab are: width = 2.3 m, length = 4 − 0.3 = 3.7 m.

**(a) Effective size of panel for live load moments.**

Live load moments will be calculated by Pigeaud's curves. These curves are based on assuming the slab to be simply supported at the edges, with corners held down. Hence the effective span of the slab in the width and length directions will be decided on the basis of normal rules followed for the effective span of a simply supported slab. Let us assume the effective depth of slab = 17 cm. Hence:

(i) Effective width $B$ = distance of c/c of supports = 2.6 m.
or = clear span + effective thickness. = 2.3 + 0.17 = 2.47 m
whichever is smaller. Hence adopt $B$ = 2.47 m.

(ii) Effective length $L$ = c/c of supports = 4 m
or = clear span + 0.17 = 3.7 + 0.17 = 3.87. Hence adopt $L$ = 3.87 m.

Thus, the effective size of the panel will be 2.47 × 3.87 m.

**Note:** Some Authors (for example see Concrete Bridge Design by Rowe) use the effective size of the panel as equal to the dimensions c/c of the ribs. This may be adopted for convenience since neither the thickness of the rib nor the thickness of slab is known in beginning.

**(b) Effective size of panel for dead load moments**

Dead load moments are found by I.S. Code method, assuming the slab to be continuous on all the four edges. Hence the effective span in the two direction will be fixed on the basis of rules followed for continuous slabs. Width of support in $B$ direction = 30 cm which is more than 1/12 of clear span of 2.3 m. Hence effective $B$ = clear width = 2.3 m. Width of support in $L$ direction = 30 cm, which is slightly less than 1/12 of clear span of 3.7 m. Hence effective length of panel = clear span + effective thickness of slab = 3.7 + 0.17 = 3.87 m. Hence the size of panel will be 2.3 × 3.87 m.

**3. Impact Factors for panels**

Impact factor for class AA loading = 25%

Impact factor for class A loading = $\dfrac{4.5}{6 + 2.47}$ = 0.53.

However, maximum value of impact factor for class A loading is 0.5 only.

**4. Design of Cantilever Slab:** The live load moments and S.F. in cantilever slab are determined by the effective width method.

**(i) Dead load B.M**

TABLE 31.5

| Component | Dead load per m run (N) | Dist. of C.G. from edge of cantilever (m) | Moment (kN-m) |
|---|---|---|---|
| 1. Parapet | 500 | 1.10 | 550 |
| 2. Kerb | 0.32 × 0.40 × 1 × 24000 = 3072 | 0.95 | 2918 |
| 3. Wearing coat | 0.75 × 0.08 × 22000 = 1320 | 0.375 | 495 |
| 4. Slab (rectangular) | 0.12 × 1.15 × 24000 = 3312 | 0.575 | 1904 |
| 5. Slab (triangular) | $\frac{1}{2}$ × 1.15 × 0.1 × 24000 = 1380 | 0.383 | 528 |
| Total | 9584 | | 6395 |

**(ii) Live load B.M.**

*Class AA loading.* The minimum distance of class AA loading from the kerb is to be 1.2 m. Since the available clear length of cantilever is only 0.75 m, class AA loading will not come on the cantilever portion.

*Class A loading.* The thickness of the cantilever slab is 12 cm at one end and 22 cm on the other end. Thickness of the slab at the edge of kerb = $12 + \frac{22-12}{1.15} \times 0.4 = 15.5$ cm. Hence the calculation for the load spread along the span can be done by taking an average thickness of slab = $\frac{15.5 + 22}{2} \approx 19$ cm. The total thickness, including the wearing coat = 19 + 8 = 27 cm.

Distance of C.G. of wheel from the edge of cantilever support = 75 − (15 + 25) = 35 cm.

Dispersed width of load along span = 50 + 2 × 27 = 104 cm, out of which only $\left(\frac{104}{2} + 35\right)$ = 87 cm length will be on the cantilever as shown in Fig. 31.51. Hence load coming on cantilever = $\frac{114000}{2} \times \frac{87}{104} \approx 47680$ N.

Now effective width of cantilever is :
$$e = 1.2x + W$$

where  $x$ = distance of C.G. of load from the edge. = 87/2.

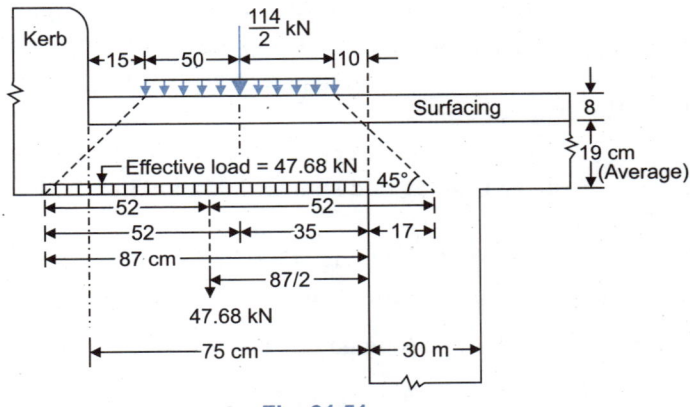

Fig. 31.51

$W$ = breadth of concentration area of load perpendicular to span = 25 + 2 × 8 = 41 cm.

∴ $e = 1.2\left(\frac{87}{2}\right) + 41 = 93.2$ cm = 0.932 m.  Impact factor = 1.5.

∴ B.M. due to live load = $1.5 \times \frac{47680}{0.932} \times \frac{0.87}{2} \approx 33381.1$ N-m.

Live load shear force = $1.5 \times \frac{47680}{0.932} \approx 76738$ N.

∴ Design moment = 6395 + 33381 = 39776 N-m. Design S.F. = 9584 + 76738 = 86322 N.

**(iii) Design of cantilever slab.** Use M 20 concrete for which:
$$m = 13\,;\quad \sigma_{cbc} = 7\,;\quad k = 0.394\,;\quad j = 0.869\,;\quad R = 1.2$$

$$d = \sqrt{\frac{39776 \times 1000}{1.2 \times 1000}} \approx 182 \text{ mm}.$$

Provide 220 mm as overall thickness using 25 mm as clear cover and 16 mm Φ bars,
effective $d = 220 - 25 - 8 = 187$ mm.

$$A_{st} = \frac{39776 \times 1000}{125 \times 0.869 \times 187} \approx 1958.2 \text{ mm}^2.$$

∴ Spacing of 16 mm bars = $\frac{1000 \times 201.1}{1958.2} \approx 102.7$ mm.

Hence provide these @ 100 mm c/c

*Distribution steel* is to be provided for 0.3 times live load moment plus 0.2 times dead load moment.

∴ $M = (0.3 \times 33380) + (0.2 \times 6395) = 11383$ N-m.

Effective depth available, using 8 mm Φ bars = $187 - 8 - 4 = 175$ cm.

∴ $$A_{sd} = \frac{11293 \times 1000}{125 \times 0.869 \times 175} \approx 594 \text{ mm}^2$$

Providing these both at top bottom, area on each face of cantilever = 300 mm².

∴ Spacing of 8 mm Φ bars = $\frac{1000 \times 50.3}{300} \approx 170$ mm.

Shear stress = $\frac{86322}{1000 \times 0.869 \times 187} = 0.53$ N/mm² < 0.7.  Hence safe.

### 5. Analysis of inner panels

**(i) Dead load moments:** Effective size of panel = 2.3 × 3.87 m.

For the purpose of calculating the dead load moments, the slab may be considered to be continuous at all the four edges. Hence as per IS 456, the dead load moments will be as follows:

Span ratio = 3.87 / 2.3 = 1.68.

∴ $Z_B$ for negative B.M. = 0.070 ; $Z_L$ = 0.033

$Z_B$ for positive B.M. = 0.052 ; $Z_L$ = 0.025.

Dead load = $(0.2 \times 1 \times 1 \times 24000) + (0.08 \times 1 \times 1 \times 22000) = 6560$ N/m².

∴ *For short sapan:*

Negative B.M. at edges = $0.070 \times 6560 (2.3)^2 \approx 2430$ N-m/m

Positive B.M. at mid span = $0.052 \times 6560 (2.3)^2 \approx 1805$ N-m/m

*For long span:*

Negative B.M. at edges = $0.033 \times 6560 (2.3)^2 \approx 1145$ N-m/m

Positive B.M. at mid span = $0.025 \times 6560 (2.3)^2 \approx 868$ N-m/m.

S.F. on long edge = $6560 \times \frac{2.3}{2} = 7544$ N.

**(ii) Class AA (tracked) vehicle:** Live load moments are calculated by Pigeaud's theory, assuming that the slab is simply supported at the edges with corners held down. Hence the effective size of the panel will be 2.47 m × 3.87 m. Effect of continuity will be accounted for by multiplying the midspan moments by 0.8.

For maximum B.M. one of the tracks will be placed centrally on the panel. The track load is 350 kN spread over area of 3.6 m × 0.8 m.

After dispersion through the wearing coat of 8 cm thickness, the dispersed dimensions (Fig. 31.52) are:

$u = 0.85 + 2 \times 0.08 = 1.01$ m.
$v = 3.6 + 2 \times 0.08 = 3.76$ m.

∴ $\frac{u}{B} = \frac{1.01}{2.47} \approx 0.41$ ; $\frac{v}{L} = \frac{3.76}{3.87} \approx 0.97$ ;

$\frac{B}{L} = \frac{2.47}{3.87} \approx 0.64$.

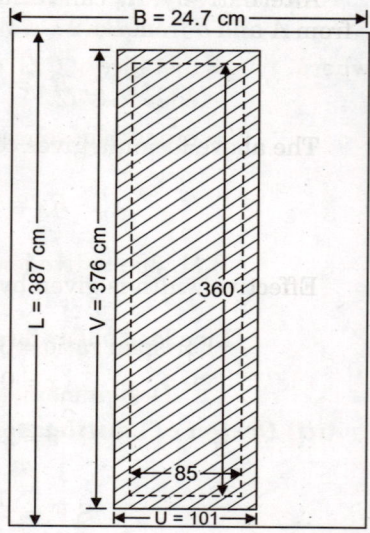

Fig. 31.52

Hence from Pigeaud's curves, by interpolation between Fig. 31.17 and 31.18, we get
$M_1 = 8.4 \times 10^{-2}$ and $M_2 = 2.7 \times 10^{-2}$

Taking impact factor of 0.25,

Moment $M_B = W(M_1 + 0.15 M_2) = \dfrac{350}{100} (8.4 + 0.15 \times 2.7)$

$= 30.82$ kN-m

Taking a continuity factor of 0.8 and impact factor of 1.25,
$M_B = 1.25 \times 0.8 \times 30.82 = 30.82$ kN-m $= 30820$ N-m

Similarly, $M_L = W(M_2 + 0.15 M_1) = \dfrac{350}{100}(2.7 + 0.15 \times 8.4) = 13.86$ kN-m

Taking a continuity factor of 0.8 and impact factor of 1.25,
$M_L = 1.25 \times 0.8 \times 13.86 = 13.86$ kN-m

### Shear force

Shear force is usually calculated by the effective width method, considering the panel to be fixed on all the four edges. Hence the effective size of the panel will be 2.3 m × 3.7 m (*i.e.*, the interior dimensions). For maximum S.F. the load will be so placed that its spread upto slab bottom reaches upto the face of the rib, as shown in Fig. 31.53. Width of track = 85 cm. Dispersed width = 85 + 2 (8 + 20) = 141 cm.

∴ P will be at 70.5 cm away from the rib face. The equivalent loading on a fixed beam is shown in Fig. 31.53(*b*), in which the total uniformly distributed load P is spread over a distance α B. The reactions $R_A$ and $R_B$ are given by

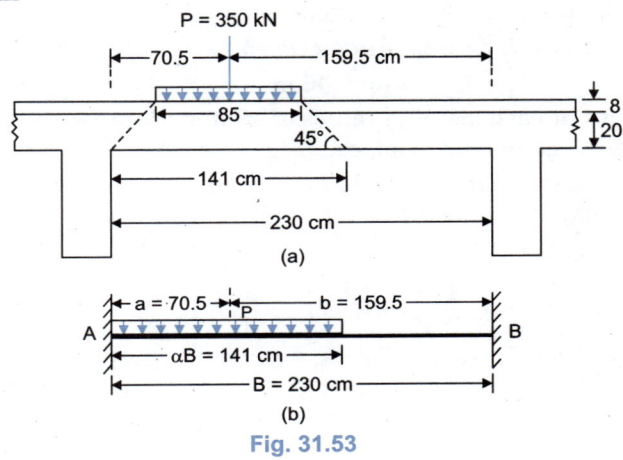

Fig. 31.53

$R_A = \dfrac{P}{2} [\alpha^3 - 2\alpha^2 + 2]$ ...[31.27 (a)]    and    $R_B = \dfrac{P}{2} [2\alpha^2 - \alpha^3]$ ...[31.27 (b)]

For the present case, $\alpha B = 141$ cm, $B = 230$ cm. ;  ∴  $\alpha = 141/230 \approx 0.613$

$R_A = \dfrac{350}{2} [(0.613)^3 - 2(0.613)^2 + 2] = 258.8$ kN.

Alternatively, $R_A$ can be found from the following expression treating P as a knife-edge load, distant *a* from A and *b* from B:

$R_A = \dfrac{Pb^2}{L^3}(3a + b).$

The above formula gives conservative result. For the present case

$R_A = \dfrac{350 (1.595)^2}{(2.30)^3}(3 \times 0.705 + 1.595) = 271.5$ kN.

Effective width is given by *Eq. 31.6*.,   $e = Kx\left(1 - \dfrac{x}{l}\right) + W.$

For span ratio = 3.7 / 2.3 = 1.61,  $K \approx 2.52$ ;  $l = 2.30$ m.
$x = 0.705$ ;  $W = 3.6 + 2 \times 0.08 = 3.76$ m.

∴ $e = 2.52 \times 0.705 \left(1 - \dfrac{0.705}{2.3}\right) + 3.76 = 4.99$ m

∴ S.F. $= \dfrac{1.25 \times 258.8}{4.99} = 64.83$ kN $= 64830$ N.

(iii) **Class AA** (Wheeled) **Vehicle : Bending Moment**

Since the effective width of panel is limited to 2.47 m only three wheels of the axle can be accommodated on the panel. Three possibilities should be considered for finding maximum B.M. in the panel. In the first possibility two loads of 37.5 kN each and two loads of 62.5 kN each are placed symmetrically as shown in Fig. 31.54. In the second possibility, two loads of 37.5 kN each and 4 loads of 62.5 kN each are tried with the central two loads of 62.5 kN placed centrally as shown in Fig. 31.55. A third possibility should also be tried in which three wheel loads (37.5 kN, 62.5 kN and 62.5 kN) of the first axle are so placed that the middle 62.5 kN wheel load is placed centrally, with the three wheel loads of second axle following it as shown in Fig. 31.56.

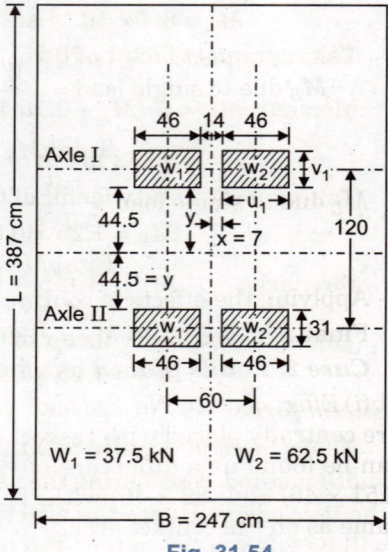

Fig. 31.54

The wheel dimensions are 30 cm × 15 cm ; after dispersion through the wearing coat, the dimensions will be 46 cm × 31 cm.

**Case 1: Loads placed as shown in Fig. 31.54**

$$u_1 = 0.46 \text{ m} ; \quad v_1 = 0.31 \text{ m} ; \quad y = 0.445 ; \quad x = 0.07 \text{ m}.$$

For each load the values of $M_1$ and $M_2$ are found as under, treating it to be eccentrically placed [Fig. 31.26 (c)].

*Step 1.* Find $M_1$ and $M_2$

For $\quad u = 2(u_1 + x) = 2(0.46 + 0.07) = 1.06$
and $\quad v = 2(v_1 + y) = 2(0.31 + 0.445) = 1.51 ;$

$\therefore \quad \dfrac{u}{B} = \dfrac{1.06}{2.47} \approx 0.43 ; \quad \dfrac{v}{L} = \dfrac{1.51}{3.87} = 0.39 ; \quad \dfrac{B}{L} = \dfrac{2.47}{3.87} \approx 0.64.$

From Pigeaud's curves, we get, by interpolation, $M_1 = 13.2 \times 10^{-2}$ ; and $M_2 = 6.6 \times 10^{-2}$

Multiply these by $(u_1 + x)(v_1 + y) = (0.46 + 0.07)(0.31 + 0.445) = 0.4$

$\therefore \quad M_1 = 0.4 \times 13.2 \times 10^{-2} = 5.28 \times 10^{-2}$
and $\quad M_2 = 0.4 \times 6.6 \times 10^{-2} = 2.64 \times 10^{-2}.$

*Step 2.* Find $M_1$ and $M_2$ for $u = 2x = 0.14$ and $v = 2y = 0.89$

$\therefore \quad \dfrac{u}{B} = \dfrac{0.14}{2.47} \approx 0.06 \quad \text{and} \quad \dfrac{v}{L} = \dfrac{0.89}{3.87} \approx 0.23$

$\therefore \quad M_1 = 23.2 \times 10^{-2} \quad \text{and} \quad M_2 = 10.7 \times 10^{-2}$

Multiply these by $xy$ where $xy = 0.07 \times 0.445 = 0.031$

$\therefore \quad M_1 = 0.72 \times 10^{-2} \quad \text{and} \quad M_2 = 0.33 \times 10^{-2}$

*Step 3.* Find $M_1$ and $M_2$ for $u = 2(u_1 + x) = 2(0.46 + 0.07) = 1.06$ and $v = 2y = 0.89$

$\dfrac{u}{B} = \dfrac{1.06}{2.47} \approx 0.43 ; \quad \dfrac{v}{L} = \dfrac{0.89}{3.87} \approx 0.23.$

$\therefore \quad M_1 = 14.2 \times 10^{-2} \quad \text{and} \quad M_2 = 9.8 \times 10^{-2}$

Multiply these by $y(u_1 + x) = 0.445(0.46 + 0.07) = 0.236$

$\therefore \quad M_1 = 3.35 \times 10^{-2} \quad \text{and} \quad M_2 = 2.31 \times 10^{-2}$

*Step 4.* Find $M_1$ and $M_2$ for $u = 2x = 2 \times 0.07 = 0.14$
and $\quad v = 2(v_1 + y) = 2(0.31 + 0.445) = 1.51$

$\therefore \quad \dfrac{u}{B} = \dfrac{0.14}{2.47} \approx 0.06 ; \quad \dfrac{v}{L} = \dfrac{1.51}{3.87} \approx 0.39$

$\therefore \quad M_1 = 19 \times 10^{-2} \quad \text{and} \quad M_2 = 7.2 \times 10^{-2}$

Multiply these by $x(v_1 + y) = 0.07(0.31 + 0.444) \approx 0.053$

$\therefore \quad M_1 = 1.01 \times 10^{-2} \quad \text{and} \quad M_2 = 0.38 \times 10^{-2}$

Design $M_1 = M_1$ of (1) + $M_1$ of (2) − $M_1$ of (3) − $M_1$ of (4)
$= 10^{-2}(5.28 + 0.72 - 3.35 - 1.01) = 1.64 \times 10^{-2}$

$$M_2 = M_2 \text{ of } (1) + M_2 \text{ of } (2) - M_2 \text{ of } (3) - M_2 \text{ of } (4)$$
$$= 10^{-2} (2.64 + 0.33 - 2.31 - 0.38) = 0.28 \times 10^{-2}$$

$\therefore \quad M_B$ due to single load $= \dfrac{W}{u_1 \, v_1} (M_1 + 0.15 \, M_2) = \dfrac{W}{0.46 \times 0.31} (1.64 + 0.15 \times 0.28) \, 10^{-2} = W \times 11.8 \times 10^{-2}$

$\therefore \qquad$ Total $M_B = 11.8 \, (2 \times 62500 + 2 \times 37500) \, 10^{-2} = 23600$ N-m.

$M_L$ due to single load $= \dfrac{W}{u_1 \, v_1} (M_2 + 0.15 \, M_1) = \dfrac{W}{0.46 \times 0.31} (0.28 + 0.15 \times 1.64) \times 10^{-2} = W \times 3.69 \times 10^{-2}$

$\therefore \qquad$ Total $M_L = 3.69 \times 10^{-2} \, (2 \times 62500 + 2 \times 37500) = 7380$ N-m.

Applying the effects of continuity and impact

Final $\qquad M_B = 1.25 \times 0.8 \times 23600 = 23600$ N-m $\quad$ and $\quad M_L = 1.25 \times 0.8 \times 7380$ N-m.

**Case 2: Loads placed as shown in Fig. 31.55.**

(*i*) *Effect of wheel No. 2 of both the axles* : Wheels No. 2 of both axles are centrally placed with respect to $y$ axis. The effect of these loads can be found as a difference of two centrally placed loads on area (151 × 46) and (89 × 46) with the intensity of loading in each case same as on the contact area.

For the larger load, $\dfrac{u}{B} = \dfrac{46}{247} \approx 0.19$

$\dfrac{v}{L} = \dfrac{151}{387} \approx 0.39 \, ; \quad \dfrac{B}{L} = \dfrac{247}{387} \approx 0.64.$

$M_1 = 16.5 \times 10^{-2} \, ; \quad M_2 = 7.1 \times 10^{-2}$

$M_B = 62500 \left[ \dfrac{151 \times 46}{31 \times 46} \right] [16.5 + 0.15 \times 7.1] \times 10^{-2}$

$\approx 53470$ N-m

$M_L = 62500 \left[ \dfrac{151 \times 46}{31 \times 46} \right] [7.1 + 0.15 \times 16.5] \times 10^{-2}$

$\approx 29150$ N-m.

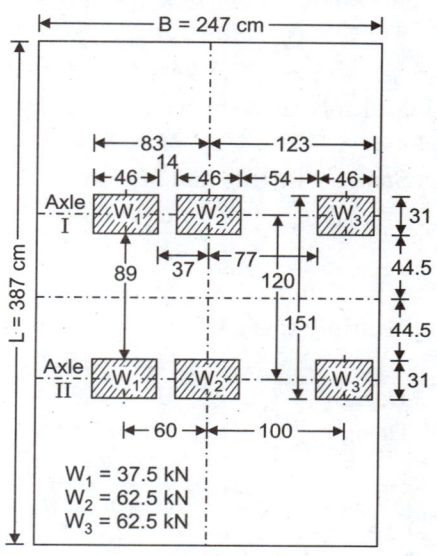

Fig. 31.55

For the smaller load, $\dfrac{u}{B} = \dfrac{46}{247} \approx 0.19. \, ; \quad \dfrac{v}{L} = \dfrac{89}{387} \approx 0.23$

$M_1 = 18.9 \times 10^{-2} \, ; M_2 = 10.7 \times 10^{-2}$

$M_B = 62500 \left[ \dfrac{89 \times 46}{31 \times 46} \right] [18.9 + 0.15 \times 10.7] \times 10^{-2} \approx 36790$ N-m

$M_L = 62500 \left[ \dfrac{89 \times 46}{31 \times 46} \right] [10.7 + 0.15 \times 18.9] \times 10^{-2} \approx 24290$ kg-m

$\therefore \qquad$ Net moment $(M_B)_1 = 53470 - 36790 = 16680$ N-m; $\quad (M_L)_1 = 29150 - 24290 = 4860$ N-m.

(*ii*) *Effects of wheel No. 1 of both the axles*: They are not centrally placed on any of the axes. Their effect will be analysed by treating each wheel load as eccentrically placed [Fig. 31.26(*c*)]. $u_1 = 0.46 \, ; v_1 = 0.31 \, ;$ $x = 0.37 \, ; y = 0.445$

**Step 1.** Find $M_1$ and $M_2$ for $u = 2 \, (u_1 + x) = 2 \, (0.46 + 0.37) = 1.66$

and $\qquad v = 2 \, (v_1 + y) = 2 \, (0.31 + 0.445) = 1.51$

$\therefore \qquad \dfrac{u}{B} = \dfrac{1.66}{2.47} = 0.67 \, ; \quad \dfrac{v}{L} = \dfrac{1.51}{3.87} = 0.39$

$\therefore \qquad M_1 = 10.4 \times 10^{-2} \quad$ and $\quad M_2 = 5.8 \times 10^{-2}$

Multiply these by $(u_1 + x)(v_1 + y)$, i.e., by $0.83 \times 0.775$

$\therefore \quad M_1 = 0.83 \times 0.775 \times 10.4 \times 10^{-2} = 6.69 \times 10^{-2}$

$\quad M_2 = 0.83 \times 0.775 \times 5.8 \times 10^{-2} = 3.73 \times 10^{-2}$

*Step 2.* Find $M_1$ and $M_2$ for $u = 2x = 0.74$ and $v = 2y = 0.89$

$$\frac{u}{B} = \frac{0.74}{2.47} = 0.3 \; ; \quad \frac{v}{L} = \frac{0.89}{3.87} = 0.23$$

$\therefore \quad M_1 = 16.5 \times 10^{-2}$ and $M_2 = 10.3 \times 10^{-2}$

Multiply these by $xy$ or by $0.37 \times 0.445$

$\therefore \quad M_1 = 2.72 \times 10^{-2}$ and $M_2 = 1.70 \times 10^{-2}$

*Step 3.* Find $M_1$ and $M_2$ for $u = 2(u_1 + x) = 2(0.46 + 0.37) = 1.66$ and $v = 2y = 0.89$

$$\frac{u}{B} = \frac{1.66}{2.47} = 0.67 \; ; \quad \frac{v}{L} = \frac{0.89}{3.87} = 0.23$$

$\therefore \quad M_1 = 11.3 \times 10^{-2}$ and $M_2 = 8.3 \times 10^{-2}$

Multiply these by $y(u_1 + x)$ or by $0.455 \times 0.83$

$\therefore \quad M_1 = 4.17 \times 10^{-2}$ and $M_2 = 3.07 \times 10^{-2}$

*Step 4.* Find $M_1$ and $M_2$ for $u = 2x = 0.74$ and $v = 2(v_1 + y) = 1.51$

$$\frac{u}{B} = \frac{0.74}{2.47} = 0.3 \; ; \quad \frac{v}{L} = \frac{1.51}{3.87} = 0.39$$

$\therefore \quad M_1 = 14.9 \times 10^{-2}$ and $M_2 = 7 \times 10^{-2}$

Multiply these by $x(v_1 + y)$ or by $0.37 \times 0.755$

$\therefore \quad M_1 = 4.16 \times 10^{-2}$ and $M_2 = 1.96 \times 10^{-2}$

Design $\quad M_1 = (6.69 + 2.72 - 4.17 - 4.16)10^{-2} = 1.08 \times 10^{-2}$

Design $\quad M_2 = (3.73 + 1.70 - 3.07 - 1.96)10^{-2} = 0.4 \times 10^{-2}$

$$(M_B)_2 = \frac{2W}{u_1 v_1}(M_1 + 0.15 M_2) = \frac{2 \times 37500}{0.46 \times 0.31}(1.08 + 0.15 \times 0.4)10^{-2} \approx 6000 \text{ N-m}$$

$$(M_L)_2 = \frac{2W}{u_1 v_1}(M_2 + 0.15 M_1) = \frac{2 \times 37500}{0.46 \times 0.31}(0.4 + 0.15 \times 1.08)10^{-2} \approx 2960 \text{ N-m}$$

*(iii) Effect of wheel No. 3 of both axles*: These are also not centrally placed. Hence follow the same procedure as above.

$$u_1 = 0.46 \; ; \quad v_1 = 0.31 \; ; \quad x = 0.77 \; ; \quad y = 0.445$$

*Step 1.* Find $M_1$ and $M_2$ for $u = 2(u_1 + x) = 2(0.46 + 0.77) = 2.46$
and $\quad v = 2(v_1 + y) = 2(0.31 + 0.445) = 1.51$

$\therefore \quad \dfrac{u}{B} = \dfrac{2.46}{2.47} \approx 1$ and $\dfrac{v}{L} = \dfrac{1.51}{3.87} = 0.39$

$\therefore \quad M_1 = 7.5 \times 10^{-2}$ and $M_2 = 4.5 \times 10^{-2}$

Multiply these by $(u_1 + x)(v_1 + y)$ or by $1.23 \times 0.755$

$\therefore \quad M_1 = 6.96 \times 10^{-2}$ and $M_2 = 4.18 \times 10^{-2}$

*Step 2.* Find $M_1$ and $M_2$ for $u = 2x = 1.54$ and $v = 2y = 0.89$

$$\frac{u}{B} = \frac{1.54}{2.47} = 0.62 \; ; \quad \frac{v}{L} = \frac{0.89}{3.87} = 0.23$$

$\therefore \quad M_1 = 11.6 \times 10^{-2}$ and $M_2 = 8.6 \times 10^{-2}$

Multiplying these by $xy$ or by $0.77 \times 0.445$

$\therefore \quad M_1 = 3.97 \times 10^{-2}$ and $M_2 = 2.95 \times 10^{-2}$

*Step 3.* Find $M_1$ and $M_2$ for $u = 2(u_1 + x) = 2(0.46 + 0.77) = 2.46$
and $\quad v = 2y = 2 \times 0.445 = 0.89$

$$\therefore \quad \frac{u}{B} = \frac{2.46}{2.47} = 1 \; ; \; \frac{v}{L} = \frac{0.89}{3.87} = 0.23$$

$$M_1 = 8.1 \times 10^{-2} \quad \text{and} \quad M_2 = 5.9 \times 10^{-2}$$

Multiply these by $y(u_1 + x)$ or by $0.445 \times 1.23$

$$\therefore \quad M_1 = 4.43 \times 10^{-3} \quad \text{and} \quad M_2 = 3.23 \times 10^{-3}$$

**Step 4.** Find $M_1$ and $M_2$ for $u = 2x = 1.54$ and $y = 2(v_1 + y) = 2(0.31 + 0.445) = 1.51$

$$\therefore \quad \frac{u}{B} = \frac{1.54}{2.47} = 0.62 \; ; \; \frac{v}{L} = \frac{1.51}{3.87} = 0.39 \; ; \; \therefore M_1 = 10.8 \times 10^{-2} \quad \text{and} \quad M_2 = 6.0 \times 10^{-2}$$

Multiply these by $x(v_1 + y)$ or by $0.77 \times 0.755$

$$\therefore \quad M_1 = 6.28 \times 10^{-2} \quad \text{and} \quad M_2 = 3.49 \times 10^{-2}$$

$$\therefore \quad \text{Design } M_1 = (6.96 + 3.97 - 4.43 - 6.28) \, 10^{-2} \approx 0.22 \times 10^{-2}$$

$$M_2 = (4.18 + 2.95 - 3.23 - 3.49) \, 10^{-2} = 0.41 \times 10^{-2}$$

$$\therefore \quad (M_B)_3 = \frac{2W}{u_1 v_1}(M_1 + 0.15 M_2) = \frac{2 \times 62500}{0.46 \times 0.31}(0.22 + 0.15 \times 0.41) \times 10^{-2} \approx 2470 \text{ N-m.}$$

$$(M_L)_3 = \frac{2W}{u_1 v_1}(M_2 + 0.15 M_1) = \frac{2 \times 62500}{0.46 \times 0.31}(0.41 + 0.15 \times 0.22) \, 10^{-2} \approx 3880 \text{ N-m}$$

Final bending moments, without taking into account the effects of impact and continuity are

$$M_B = M_{B1} + M_{B2} + M_{B3} = 16680 + 6000 + 2470 = 25150 \text{ N-m}$$

$$M_L = M_{L1} + M_{L2} + M_{L3} = 4860 + 2960 + 3880 = 11700 \text{ N-m}$$

Applying the effects of continuity and impact

$$M_B = 1.25 \times 0.8 \times 25150 = 25150 \text{ N-m}$$

$$M_L = 1.25 \times 0.8 \times 11700 = 11700 \text{ N-m.}$$

**Case 3. Loads placed as shown in Fig. 31.56.**

(i) *Effect of wheel 2 of axle 1* : The wheel is centrally placed.

$$u = 0.46 \; ; \; v = 0.31 \; ; \; \frac{u}{B} = \frac{0.46}{2.47} \approx 0.19 \; ; \; \frac{v}{L} = \frac{0.31}{3.87} = 0.08$$

$$\therefore \quad M_1 = 21.2 \times 10^{-2} \quad \text{and} \quad M_2 = 16.3 \times 10^{-2}$$

$$(M_B)_1 = 62500 \, (21.2 + 0.15 \times 16.3) \times 10^{-2} \approx 14780 \text{ kg-m.}$$

$$(M_L)_1 = 62500 \, (16.3 + 0.15 \times 21.2) \times 10^{-2} \approx 12180 \text{ N-m.}$$

(ii) *Effect of wheel 1 of axle 1*

The wheel is placed centrally with respect to $x$-axis but eccentric with respect to $y$-axis. To analyse the case, place an imaginary load of the same dimensions to the other side and find the moments due to two loads having dimensions $(166 \times 31)$ cm and $(74 \times 31)$ cm. The central moment will be half the difference of moments due to these two loads.

*For larger load,*

$$\frac{u}{B} = \frac{1.66}{2.47} = 0.67 \; ; \; \frac{v}{L} = \frac{0.31}{3.87} = 0.08.$$

$$M_1 = 11.5 \times 10^{-2} \quad \text{and} \quad M_2 = 11.3 \times 10^{-2}$$

$$M_B = 37500 \left(\frac{166 \times 31}{46 \times 31}\right)(11.5 + 0.15 \times 11.3) \, 10^{-2} \approx 17860.$$

$$M_L = 37500 \left(\frac{166 \times 31}{46 \times 31}\right)(11.3 + 0.15 \times 11.5) \, 10^{-2} \approx 17630.$$

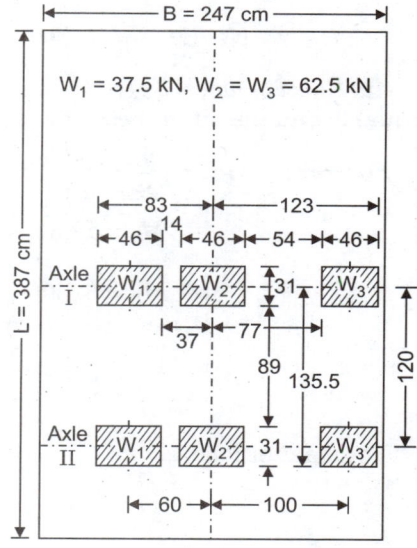

Fig. 31.56

*For smaller load,*

$$\frac{u}{B} = \frac{0.74}{2.47} = 0.3 \; ; \; \frac{v}{L} = \frac{0.31}{3.87} = 0.08. \quad \text{Hence, } M_1 = 17.8 \times 10^{-2} \quad \text{and} \quad M_2 = 15.3 \times 10^{-2}$$

$$M_B = 37500 \left(\frac{74 \times 31}{46 \times 31}\right)(17.8 + 0.15 \times 15.3)\, 10^{-2} \approx 12120$$

$$M_L = 37500 \left(\frac{74 \times 31}{46 \times 31}\right)(15.3 + 0.15 \times 17.8)\, 10^{-2} \approx 10840. \quad \therefore \text{ Net moments}$$

$(M_B)_2 = \frac{1}{2}(17860 - 12120) = 2870$ N-m. and $(M_L)_2 \approx \frac{1}{2}(17630 - 10840) = 3400$ N-m.

*(iii) Effect of wheel 3 of axle 1.*

This case is similar to *(ii)* above. Actual moments will be equal to half the difference of moments due to two loads having dimensions $(246 \times 31)$ cm and $(154 \times 31)$ cm.

For larger load, $\quad \dfrac{u}{B} = \dfrac{2.46}{2.47} \approx 1 \,; \quad \dfrac{v}{L} = \dfrac{0.31}{3.87} = 0.08$

$\therefore \qquad M_1 = 8.3 \times 10^{-2} \quad \text{and} \quad M_2 = 8.3 \times 10^{-2}$

$$M_B = M_L = 62500 \left(\frac{246 \times 31}{46 \times 31}\right)[8.3 + 0.15 \times 8.3] \times 10^{-2} \approx 31900.$$

For smaller load, $\quad \dfrac{u}{B} = \dfrac{1.54}{2.47} = 0.62 \,; \quad \dfrac{v}{L} = \dfrac{0.31}{3.17} = 0.098.$

$\qquad M_1 = 12.2 \times 10^{-2} \quad \text{and} \quad M_2 = 12.1 \times 10^{-2}$

$$M_B = 62500 \left(\frac{154 \times 31}{46 \times 31}\right)[12.2 + 0.15 \times 12.1] \times 10^{-2} \approx 29320.$$

$$M_L = 62500 \left(\frac{154 \times 31}{46 \times 31}\right)[12.1 + 0.15 \times 12.2] \times 10^{-2} \approx 29150.$$

Net $(M_B)_3 = \frac{1}{2}(31900 - 29320) = 1290 \quad \text{and} \quad (M_L)_3 = \frac{1}{2}(31900 - 29150) \approx 1380.$

*(iv) Effect of wheel 2 of axle 2* : This load is centrally placed with respect to $y$ axis. Actual moments will be equal to half the difference of moments due to two loads having dimensions $(46 \times 271)$ cm and $(46 \times 206)$ cm.

For larger load, $\quad \dfrac{u}{B} = \dfrac{0.46}{2.47} = 0.19 \,; \quad \dfrac{v}{L} = \dfrac{2.71}{3.87} = 0.7$

$\qquad M_1 = 12.6 \times 10^{-2} \quad \text{and} \quad M_2 = 4.2 \times 10^{-2}$

$$M_B = 62500 \left(\frac{46 \times 271}{46 \times 31}\right)[12.6 + 0.15 \times 4.2] \times 10^{-2} \approx 72280$$

$$M_L = 62500 \left(\frac{46 \times 271}{46 \times 31}\right)[4.2 + 0.15 \times 12.6] \times 10^{-2} \approx 33270$$

For smaller load, $\quad \dfrac{u}{B} = \dfrac{0.46}{2.47} = 0.19 \,; \quad \dfrac{v}{L} = \dfrac{2.09}{3.87} = 0.54$

$\qquad M_1 = 14.5 \times 10^{-2} \quad \text{and} \quad M_2 = 5.3 \times 10^{-2}$

$$M_B = 62500 \left(\frac{46 \times 209}{46 \times 31}\right)[14.5 + 0.15 \times 5.3] \times 10^{-2} \approx 64450$$

$$M_L = 62500 \left(\frac{46 \times 209}{46 \times 31}\right)[5.3 + 0.15 \times 14.5] \times 10^{-2} \approx 31500$$

$\therefore \quad$ Net $(M_B)_4 = \frac{1}{2}(72280 - 64450) = 3915$ N-m. $\quad$ and $\quad (M_L)_4 = \frac{1}{2}(33270 - 31500) = 885$ N-m.

It will be seen that these moments are 53% and 14.5% respectively of $(M_B)_1$ and $(M_L)_1$ obtained for the centrally placed load.

## CONCRETE BRIDGES

**(v) *Effect of wheel load 1 of axle 2***

This load is eccentrically placed with respect to both axes. Its effect at the centre will be small, through it can be found by the lengthy procedure laid down for case of Fig. 31.26 (c) of an eccentric load. Alternatively, its effect can be approximately proportioned out from moments of load 1 of axle 1.

$$(M_B)_5 = 0.53 \times 2870 \approx 1520$$
$$(M_L)_5 = 0.145 \times 3400 \approx 490.$$

However, the exact procedure is as follows: $u_1 = 0.46$; $v_1 = 0.31$; $x = 0.37$; $y = 1.045$

**Step 1.** Find $M_1$ and $M_2$ for $u = 2(u_1 + x) = 2(0.46 + 0.37) = 1.66$
$$v = 2(v_1 + y) = 2(0.31 + 1.045) = 2.71$$
$$\frac{u}{B} = \frac{1.66}{2.47} = 0.67 \ ; \quad \frac{V}{L} = \frac{2.71}{3.87} = 0.7$$
$$\therefore \quad M_1 = 8.5 \times 10^{-2} \quad \text{and} \quad M_2 = 3.7 \times 10^{-2}$$

Multiply these by $(u_1 + x)(v_1 + y)$ or by $0.83 \times 1.355$
$$\therefore \quad M_1 = 9.6 \times 10^{-2} \quad \text{and} \quad M_2 = 4.2 \times 10^{-2}$$

**Step 2.** Find $M_1$ and $M_2$ for $u = 2x = 0.74$ and $v = 2y = 2.9$
$$\frac{u}{B} = \frac{0.74}{2.47} = 0.3 \ ; \quad \frac{v}{L} = \frac{2.09}{3.87} = 0.54.$$
$$M_1 = 13.2 \times 10^{-2} \quad \text{and} \quad M_2 = 5.0 \times 10^{-2}$$

Multiply these by $xy$ or by $(0.37 \times 1.045)$
$$M_1 = 5.1 \times 10^{-2} \quad \text{and} \quad M_2 = 1.9 \times 10^{-2}$$

**Step 3.** Find $M_1$ and $M_2$ for $u = 2(u_1 + x) = 2(0.46 + 0.37) = 1.66$ and $v = 2y = 2.09$
$$\therefore \quad \frac{u}{B} = \frac{1.66}{2.47} = 0.67 \ ; \quad \frac{v}{L} = \frac{2.09}{3.87} = 0.54$$
$$\therefore \quad M_1 = 9.4 \times 10^{-2} \quad \text{and} \quad M_2 = 4.6 \times 10^{-2}$$

Multiply these by $y(u_1 + x)$, i.e., by $1.045 \times 0.83$
$$\therefore \quad M_1 = 8.2 \times 10^{-2} \quad \text{and} \quad M_2 = 4 \times 10^{-2}$$

**Step 4.** Find $M_1$ and $M_2$ for $u = 2x = 0.74$ and $v = 2(v_1 + y) = 2(0.31 + 1.045) = 2.71$
$$\therefore \quad \frac{u}{B} = \frac{0.74}{2.47} = 0.3 \ ; \quad \frac{v}{L} = \frac{2.71}{3.87} = 0.7$$
$$\therefore \quad M_1 = 11.6 \times 10^{-2} \quad \text{and} \quad M_2 = 4.2 \times 10^{-2}$$

Multiply these by $x(v_1 + y)$, i.e. by $0.37 \times 1.355$
$$\therefore \quad M_1 = 5.8 \times 10^{-2} \quad \text{and} \quad M_2 = 2.1 \times 10^{-2}$$
$$\therefore \quad \text{Design } M_1 = (9.6 + 5.1 - 8.2 - 5.8) \, 10^{-2} = 0.7 \times 10^{-2}$$
$$M_2 = (4.2 + 1.9 - 4 - 2.1) \, 10^{-2} = \text{zero.}$$
$$\therefore \quad (M_B)_5 = \frac{37500}{0.46 \times 0.31} (0.7 + 0.15 \times 0) \, 10^{-2} \approx 1840$$
$$(M_L)_5 = \frac{37500}{0.46 \times 0.31} (0 + 0.15 \times 0.7) \, 10^{-2} \approx 280.$$

**(vi) *Effect of wheel load 3 of axle 2***

This load is also eccentrically placed. Its effect can be found either approximately by proportioning or by exact method given in the following steps:

$u_1 = 0.46$; $v_1 = 0.31$; $x = 0.77$; $y = 1.045$.

**Step 1.** Find $M_1$ and $M_2$ for $u = 2(u_1 + x) = 2(0.46 + 0.77) = 2.46$
and
$$v = 2(v_1 + y) = 2(0.31 + 1.045) = 2.71$$
$$\frac{u}{B} = \frac{2.46}{2.47} = 1 \ ; \quad \frac{v}{L} = \frac{2.71}{3.87} = 0.7$$
$$M_1 = 6.2 \times 10^{-2} \quad \text{and} \quad M_2 = 2.8 \times 10^{-2}$$

Multiply these by $(u_1 + x)(v_1 + y)$, i.e. by $1.23 \times 1.355$
$$M_1 = 10.3 \times 10^{-2} \quad \text{and} \quad M_2 = 4.7 \times 10^{-2}.$$

**Step 2.** Find $M_1$ and $M_2$ for $u = 2x = 1.54$ and $v = 2y = 2.09$

$$\frac{u}{B} = \frac{1.54}{2.47} = 0.62 \text{ and } \frac{v}{L} = \frac{2.09}{3.87} = 0.54$$

$\therefore \qquad M_1 = 9.8 \times 10^{-2}$ and $M_2 = 4.8 \times 10^{-2}$. Multiply these by $xy$, i.e., by $0.77 \times 1.045$
$$M_1 = 7.9 \times 10^{-2} \text{ and } M_2 = 3.9 \times 10^{-2}.$$

**Step 3.** Find $M_1$ and $M_2$ for $u = 2(u_1 + x) = 2(0.46 + 0.77) = 2.46$
and $\qquad v = 2y = 2.09.$

$$\frac{u}{B} = \frac{2.46}{2.47} = 1 \; ; \quad \frac{v}{L} = \frac{2.09}{3.87} = 0.54$$

$\therefore \qquad M_1 = 6.9 \times 10^{-2}$ and $M_2 = 3.6 \times 10^{-2}$

Multiply these by $y(u_1 + x)$, i.e. by $1.045 \times 1.23$
$$M_1 = 8.9 \times 10^{-2} \quad \text{and} \quad M_2 = 4.6 \times 10^{-2}.$$

**Step 4.** Find $M_1$ and $M_2$ for $u = 2x = 1.54$ and $v = 2(v_1 + y) = 2(0.31 + 1.055) = 2.71$

$$\frac{u}{B} = \frac{1.54}{2.47} = 0.62 \; ; \quad \frac{v}{L} = \frac{2.71}{3.87} = 0.7.$$

$$M_1 = 8.8 \times 10^{-2} \quad \text{and} \quad M_2 = 3.8 \times 10^{-2}$$

Multiply these by $x(v_1 + y)$, i.e. by $0.77 \times 1.355$
$$M_1 = 9.2 \times 10^{-2} \text{ and } M_2 = 4 \times 10^{-2}$$

Design $M_1 = (10.3 + 7.9 - 8.9 - 9.2) \, 10^{-2} = 0.1 \times 10^{-2}$

$M_2 = (4.7 + 3.9 - 4.6 - 4) \, 10^{-2} = \text{zero}; (M_B)_6 = \dfrac{62500}{0.46 \times 0.31} (0.1 + 0.15 \times 0) \, 10^{-2} \approx 440$

$(M_L)_6 = \dfrac{62500}{0.46 \times 0.31} (0 + 0.15 \times 0.1) \, 10^{-2} \approx 60.$ By approximate method of proportioning,

$(M_B)_6 = 0.53 \times (M_B)_3 = 0.53 \times 1290 \approx 680$

$(M_L)_6 = 0.145 \, (M_L)_3 = 0.145 \, (1380) \approx 200.$

*Final moments for case 3*

$M_B = (M_B)_1 + (M_B)_2 + \ldots (M_B)_6 = 14780 + 2870 + 1290 + 7830 + 1840 + 440 = 29050$

$M_L = (M_L)_1 + (M_L)_2 + \ldots (M_L)_6 = 12180 + 3400 + 1380 + 1770 + 280 + 60 = 19070.$

Applying the effect of continuity and impact.

Final $\qquad M_B = 1.25 \times 0.8 \times 29050 = 29050$ N-m

$\qquad\qquad M_L = 1.25 \times 0.8 \times 19070 = 19070$ N-m

Thus, out of the three cases of loading considered, case: 3 gives the highest moments.

Thus for class AA (wheeled) loading, $M_B = 29050$ kg-m and $M_L = 19070$ N-m.

**(iv) Class AA (wheeled) Vehicle : Shear force**

S.F. is computed by effective width method by considering the panel to be fixed. The effective size of the panel will be $2.3 \times 3.7$ m (*i.e.*, the interior dimensions). For maximum S.F., the loading is to be arranged by trial and error, keeping in mind the following two points: (*i*) wheel 1 is 1.2 m from the kerb, (*ii*) the outer spread line of wheel 3 should be as near to the face of the right hand support as is possible. It should be noted that wheel 3 is heavier than wheel 1. The load positions are shown in Fig. 31.57(*a*). The total length of each load, after dispersion $= 30 + 2(8 + 20) = 86$ cm. Hence effective $W_1'$

$$= 37500 \times \frac{70}{860} = 3052.32 \approx 30520 \text{ N},$$

with its C.G. distant 0.35 m from $A$ as shown in (*b*).

The reaction at $A$ and $B$ can be calculated by *Eq. 31.37*.

Fig. 31.57 (c) shows load $W_2 = 62500$ N with its C.G. distant 0.87 m from $A$.

The reaction $R_A$ and $R_B$ can be determined from the following Eqs.

$$R_A = W\left[2\beta^3 - 3\beta^2 + \frac{\alpha^2 \beta}{2} - \frac{\alpha^2}{4} + 1\right] \quad …(31.28)$$

and

$$R_B = W\left[-2\beta^3 + 3\beta^2 - \frac{\alpha^2 \beta}{2} + \frac{\alpha^2}{4}\right] = W - R_A$$

Fig. 31.57 (d) shows the load $W_3 = 62500$ N with its C.G. 0.43 from $B$. The reactions can be determined by $Eq.\ 31.27$ by transposing $A$ and $B$. The effective width of each load is determined from $Eq.\ 31.3$ in which $W = 15 + 2 \times 8 = 31$ cm $= 0.31$ m

For $W_1$, $x_1 = 0.35$; for $W_2$, $x_2 = 0.87$ and for $W_3$, $x_3 = 0.43$ m.

$l = B = 2.30$ m; $\dfrac{L}{l} = \dfrac{3.7}{2.3} = 1.61$.
Hence $k = 2.52$.

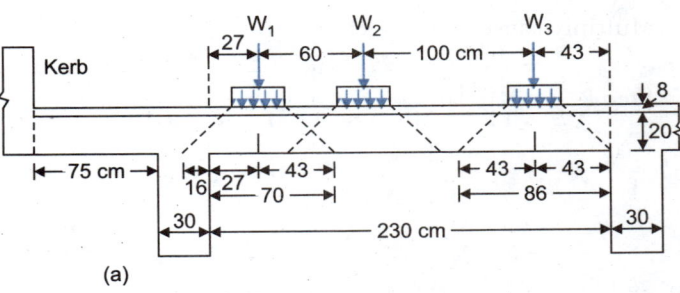

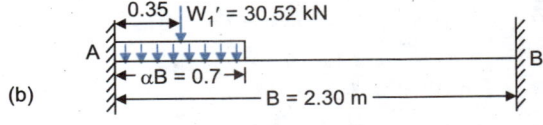

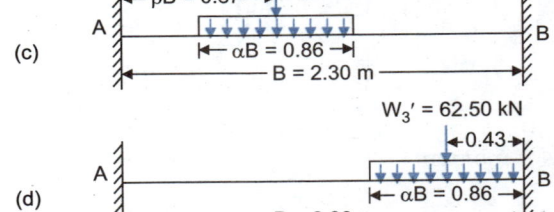

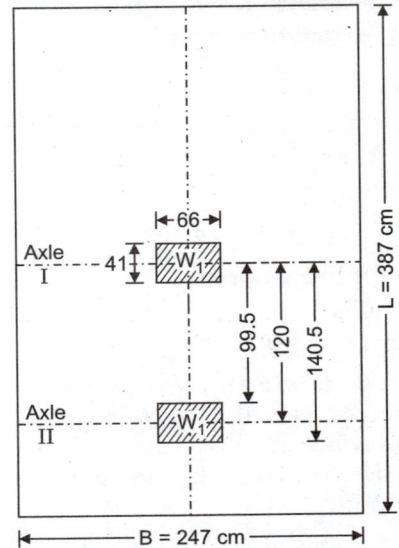

Fig. 31.57

The values of effective width, reactions and shear force ($F$) at both the ends, due at all the three loads are tabulated below. The effective width for load $W_2$ comes out to be 1.673. However, since c/c distance between wheels in the longitudinal direction is 1.2. net effective $e = \dfrac{1.673 + 1.2}{2} \approx 1.436$ m for each wheel.

| Load | $x$ | Effective width $e$ | $R_A$ | $R_B$ | $F_A$ | $F_B$ |
|---|---|---|---|---|---|---|
| $W_1 = 30520$ | 0.35 | 1.058 | 28130 | 2390 | 26590 | 2260 |
| $W_2 = 62500$ | 0.87 | 1.436* (net) | 41900 | 20600 | 29180 | 14350 |
| $W_3 = 62500$ | 0.43 | 1.191 | 7100 | 55400 | 5960 | 46520 |
| | | | | Total | 61730 | 63130 |

Taking into account impact effect, Design shear $= 1.25 \times 63130 \approx 78910$ N.

(*v*) **Class A loading : B.M.** Fig. 31.58 shows the placement of loading for the maximum B.M. in which wheel 1 of axle 1 is placed centrally with wheel 1 of axle 2 behind it. Each $W = 114/2 = 57$ kN. Width of tyre after dispersion through surfacing $= 50 + 2 \times 8 = 66$ cm while $v = 25 + 2 \times 8 = 41$ cm. Since the c/c distance between the wheels of axle is 1.8 m, other set of wheels will be out of span.

(*i*) **Effect of load 1 of axle 1**

$u = 0.66$; $v = 0.41$

$\dfrac{u}{B} = \dfrac{0.66}{2.47} = 0.27$; $\dfrac{v}{L} = \dfrac{0.41}{3.87} = 0.11$

$M_1 = 18.3 \times 10^{-2}$ and $M_2 = 14.8 \times 10^{-2}$

$(M_B)_1 = 57000 (18.3 + 0.15 \times 14.8) 10^{-2} \approx 11700$ N-m.

$(M_L)_1 = 57000 (14.8 + 0.15 \times 18.3) 10^{-2} \approx 10000$ N-m.

(*i*) **Effect of load 1 of axle 2**

Imagine a similar load to the other side of the axis. The moment then will be equal to half the difference of moments obtained for area $(281 \times 66)$ cm and $(199 \times 66)$ cm.

Fig. 31.58

*For the bigger load,* $u = 0.66$ and $v = 2.81$ m.

$$\frac{u}{B} = \frac{0.66}{2.47} = 0.27 \quad \text{and} \quad \frac{v}{L} = \frac{2.81}{3.87} = 0.73$$

$\therefore \quad M_1 = 11.6 \times 10^{-2}$ and $M_2 = 4.0 \times 10^{-2}$

$\therefore \quad M_B = 57000 \left(\dfrac{281 \times 66}{41 \times 66}\right) [11.6 + 0.15 \times 4.0] \times 10^{-2} \approx 47660$

$\quad M_L = 57000 \left(\dfrac{281 \times 66}{41 \times 66}\right) [4.0 + 0.15 \times 11.6] \times 10^{-2} \approx 22420$

*For the smaller load* $u = 0.66$ and $v = 1.91$ m

$$\frac{u}{B} = \frac{0.66}{2.47} = 0.27 \quad \text{and} \quad \frac{v}{L} = \frac{1.99}{3.87} = 0.514$$

$\therefore \quad M_1 = 13.8 \times 10^{-2}$ and $M_2 = 5.4 \times 10^{-2}$

$\therefore \quad M_B = 57000 \left(\dfrac{199 \times 66}{41 \times 66}\right) [13.8 + 0.15 \times 5.4] \times 10^{-2} \approx 40420$

$\quad M_L = 57000 \left(\dfrac{199 \times 66}{41 \times 66}\right) [5.4 + 0.15 \times 13.8] \times 10^{-2} \approx 20670$

Net moment $\quad (M_B)_2 = \dfrac{1}{2}(47660 - 40220) = 3720$

$\quad (M_L)_2 = \dfrac{1}{2}(22420 - 20670) = 875$

Thus total moment due to class. A loading, taking into account impact and continuity factors are

$M_B = 1.5 \times 0.8\,(11700 + 3720) = 18504$ N-m

$M_L = 1.5 \times 0.8\,(10000 + 875) = 13050$ N-m

(*vi*) **Class A loading: S.F.** S.F. will be maximum when the dispersed edge of the load touches the face of the support as shown in Fig. 31.59. In that position, the other load will be on the other support.

Dispersed width of load in the direction of span = $50 + 2 \times 28 = 106$. For effective width, dispersed width of load perpendicular to the span = $25 + 2 \times 8 = 41$ cm. In the effective width formula, $k = 2.52$ (as before), $x = 0.53$.

$\therefore \quad e = 2.52 \times 0.53 \left(1 - \dfrac{0.53}{2.30}\right) + 0.41 = 1.44$ m

Since this is more than the c/c spacing of 12 m between the axles, effective widths will overlap. Net effective width per wheel = $\dfrac{1}{2}(1.44 + 1.20) = 1.32$ m. Reaction $R_A$ is calculated from Fig. 31.27, treating the span to be fixed.

$$R_A = \frac{W}{2}(\alpha^3 - 2\alpha^2 + 2) \quad \text{where} \quad \alpha = \frac{106}{230} = 0.461$$

$\therefore \quad R_A = \dfrac{57000}{2}\,[(0.461)^3 - 2(0.461)^2 + 2] = 47678.5$ (say 47680)

$\therefore$ S.F. taking into account effect of impact, is $F = 1.5 \left(\dfrac{47680}{1.32}\right) \approx 54180$ N

(*vii*) **Summary**

The maximum live load B.M. and shear are :
$M_B = 30820$ N-m, $M_L = 19070$ N-m, $F = 78910$

**6. Design of Inner Panel:** The final design moments will be greatest of moments found in the previous step (5). The negative live load B.M. at the edges may be taken to be the same as the mid-span B.M.

$\therefore \quad (+M_B) = 1805 + 30820 = 32625$ N-m;
$\quad (-M_B) = 2430 + 30820 = 33250$ N-m

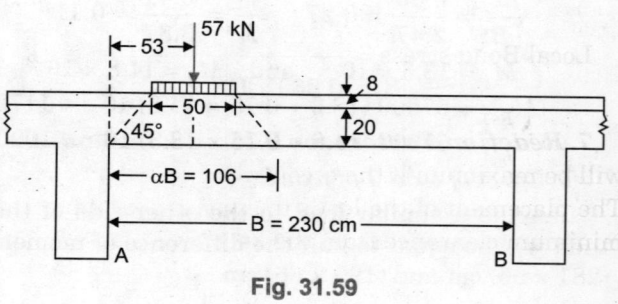

Fig. 31.59

$(+ M_L) = 1145 + 19070 = 20215$ N-m;  $(- M_L) = 868 + 19070 = 19938$ N-m

S.F. $= 7544 + 78910 = 86454$ N.

Effective depth from shear point of view $= \dfrac{86454}{1000 \times 0.869 \times 0.7} \approx 142$ mm.

From B.M. point of view, $d = \sqrt{\dfrac{33250 \times 1000}{1000 \times 1.2}} \approx 166$ mm

Provide 200 cm overall thickness. Using 25 mm as clear cover and 16 mm Φ bars, $d$ along width = 200 − 25 − 8 = 167 mm and $d$ along length = 168 − 16 = 152 mm.

$$\therefore \quad (A_{st})_B = \dfrac{33250 \times 1000}{125 \times 0.869 \times 167} = 1833 \text{ mm}^2 \ ; \ (A_{st})_L = \dfrac{20215 \times 1000}{125 \times 0.869 \times 152} \approx 1224 \text{ mm}^2$$

Spacing, $S_B = \dfrac{1000 \times 201}{1833} \approx 109.7$ mm. However, provide these as 100 mm c/c i.e., same spacing that is provided for cantilever portion.  Spacing $S_L = \dfrac{1000 \times 201}{1224} \approx 164$ mm.

The negative reinforcement of 16 mm Φ @ 1000 c/c may be curtailed by 50% at 1/5 span plus 12 Φ or 230 mm, whichever is more with a minimum of 52 Φ = 52 × 16 ≈ 832 mm. The remaining half of the top reinforcement should continue. The bottom reinforcement should continue throughout. In the cantilever portion however, 2/3 bars may be curtailed. Near stiffners, equal reinforcement should be placed at the top of the slab extending 52 Φ = 832 mm on either side. The details of the reinforcement in the slab, in both the direction is shown in Fig. 31.60.

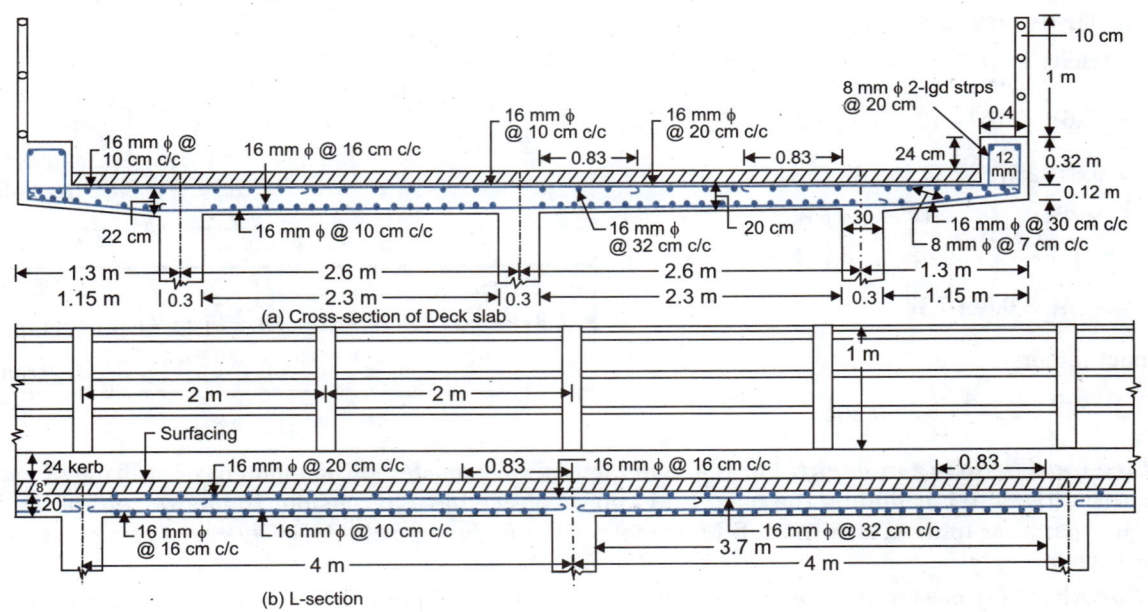

Fig. 31.60

Local Bond stress $= \dfrac{86455}{0.869 \times 167 \times \dfrac{1000}{100}(16\pi)} = 1.19$ N/mm$^2$  (safe).  (As per IRC Code)

**7. Reaction Factors for B.M. in Longitudinal Girders by Courbon's Method:** The reaction factors will be maximum if the eccentricity of the C.G. of loads with respect to the axis of the bridge is maximum. The placement of the loads to given maximum eccentricity will be subject to the provision of the specified minimum clearance from the kerb.

(a) **Class AA (tracked) vehicle**: Min. clearance = $1.2 + \frac{0.85}{2} = 1.625$ m, upto centre of track. The load positions are shown in Fig. 31.61, in which $e = 0.85$ m.

If $W$ is the axle load, $P = \frac{W}{2}$.

Let us assume that all the girders have the same moment of inertia.

$\Sigma x^2 = (2.6)^2 + (0)^2 + (2.6)^2 = 2(2.6)^2$

For the outer girder,
$$x = 2.6$$
For the inner girder,
$$x = 0.$$

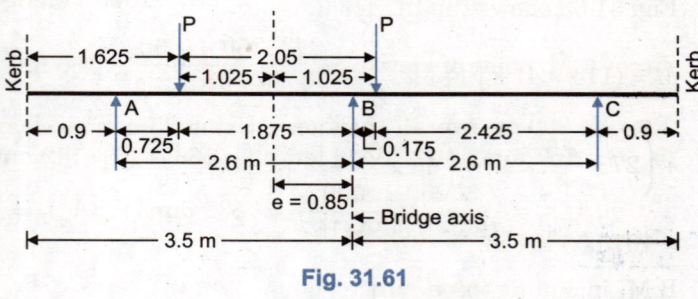

Fig. 31.61

Hence from Eq. 31.25,

$$R_A = \frac{\Sigma P}{n}\left[1 + \frac{n e x}{\Sigma x^2}\right] = \frac{2P}{3}\left[1 + \frac{3 \times 0.85 \times 2.6}{2(2.6)^2}\right] = 0.9936\,P = 0.9936\,\frac{W}{2} = 0.4968\,W$$

$$R_B = \frac{2P}{3}[1+0] = \frac{2P}{3} = \frac{2}{3}\left(\frac{W}{2}\right) = \frac{W}{3}$$

Class AA wheeled vehicles will give lesser B.M. and S.F., and will not be considered here. Impact factor for class AA loading = 10%.

(b) **Class A loading**: Permissible clearance upto centre of first wheel = $0.15 + 0.25 = 0.4$ m. The load positions for two trains is shown in Fig. 31.62.

Eccentricity $e = 0.55$ m. If $W$ is the axle load, $P = W/2$.

$\Sigma x^2 = (2.6)^2 + (0)^2 + (2.6)^2 = 2(2.6)^2$ ; $R_A = \frac{4P}{3}\left[1 + \frac{3 \times 0.55 \times 2.6}{2(2.6)^2}\right] = 1.7564\,P = 1.7564\left(\frac{W}{2}\right)$

$= 0.8782\,W.$

$R_B = \frac{4P}{3}(1+0) = \frac{4P}{3} = \frac{4}{3}\left(\frac{W}{2}\right)$

$= \frac{2}{3}\,W = 0.6667\,W.$

Impact factor

$= \frac{4.5}{6+L} = \frac{4.5}{6+16} = 0.204.$

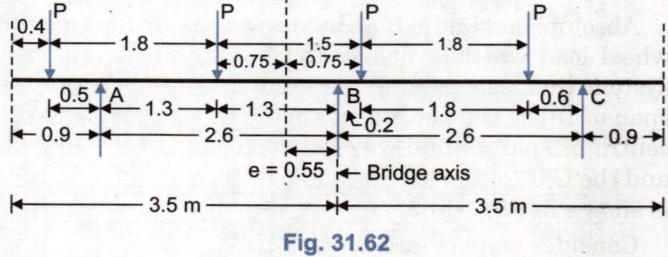

Fig. 31.62

**8. Live load bending moment**: For reaction factors, we considered the moment or shifting of loads in transverse direction. For finding the max. B.M., however, we have to consider the movement of the loads along the span. For finding maximum B.M. we follow the following rules (see Theory of Structures Vol. II by the Authors).

(a) **For absolute maximum bending moment**: (i) The maximum B.M. always occurs under a wheel load and not any where between wheel loads.

(ii) The maximum B.M. under any wheel load occurs when the centre of the span is midway between the C.G. of load system and the wheel load under consideration.

(iii) Absolute maximum B.M. always occurs at a section near the centre of span (it never occurs at the centre unless the C.G. of the resultant load coincides with the centre line of some heavy wheel load).

(iv) The absolute max. B.M. generally occurs under the heavier wheel load specially that which is very near to the C.G. of the load system.

(b) **For maximum B.M. at a given section.** The maximum B.M. at any section of a simply supported beam, due to a given system of point loads crossing the beam occurs when the average loading on the portion to the left is equal to the average loading to the right of it, i.e., when the section divides the load in the same ratio as it divides the span. To get the maximum B.M. at a given section, one of the wheel loads should be placed at the section. We shall try these rules for both class A loading as well as class AA (tracked) loadings.

(a) *Class A loading:* (i) *Maximum B.M. at the middle of the span*

Fig. 31.63 shows the I.L. for B.M. at C.

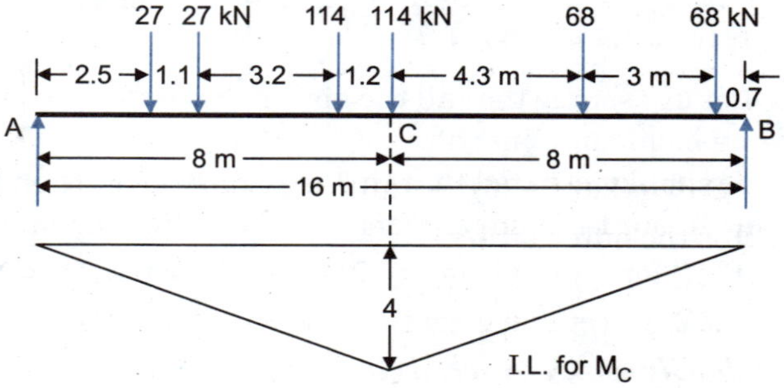

$$M = (114 \times 4) + \left(114 \times \frac{6.8}{8} \times 4\right) + \left(27 \times \frac{3.6}{8} \times 4\right)$$

$$+ \left(27 \times \frac{2.5}{8} \times 4\right) + \left(68 \times \frac{3.7}{8} \times 4\right) + \left(68 \times \frac{0.7}{8} \times 4\right)$$

$$= 1075.55 \text{ kN-m}.$$

B.M. including impact factor and reaction factor for the outer girder = 1.204 × 0.8782 × 1075.55
= 1137.2 kN-m. B.M. including impact and reaction factor for the middle girder = 1.204 × 0.6667 × 1075.55
= 863.4 kN-m.

(ii) *Maximum B.M. at the quarter span*

$$M = (114 \times 3) + \left(27 \times \frac{0.8}{4} \times 3\right) + \left(68 \times \frac{5.0}{12} \times 3\right) + \left(68 \times \frac{3.5}{12} \times 2\right)$$

$$+ \left(68 \times \frac{6.5}{12} \times 3\right) + \left(114 \times \frac{10.8}{12} \times 3\right) = 901.17 \text{ kN-m}.$$

B.M. for the outer girder, taking into account impact factor and reaction factor = 1.204 × 0.8782 × 901.17 = 952.85 say 953 kN-m.

B.M. for inner girder
= 1.204 × 0.6667 × 844.5 ≈ 678 kN-m.

(ii) *Absolute maximum B.M.*

Absolute maximum B.M. occurs under that heavier wheel load which is nearer to the C.G. of the load system that can possibly be accommodated on the span of 16 m. The placement should be such that the centre of span is mid-way between the wheel load and the C.G. of the load system. This position is shown in Fig. 31.65.

Consider train of load consisting of 6 wheel loads, having a total magnitude of 418 kN, and total length of 12.8 m. Taking moment about the outer 27 kN load.

$$\bar{x} = \frac{1}{41.8} [(27 \times 1.1) + (114 \times 4.3) + (114 \times 5.5)$$

$$+ (68 \times 9.8) + (68 \times 12.8)]$$

$$= 6.42 \text{ m}.$$

Thus the C.G. of load lies at a distance of 6.42 − (1.1 + 3.2 + 1.2) = 0.92 m from the fourth load of 114 kN. This load should, therefore, be so placed that the centre of span is midway between this load and the C.G., *i.e.*, the centre of the span will be 0.46 m away from the fourth load as shown. Distances of this load (point E) from supports A and B work out be 7.54 m and 8.46 m respectively.

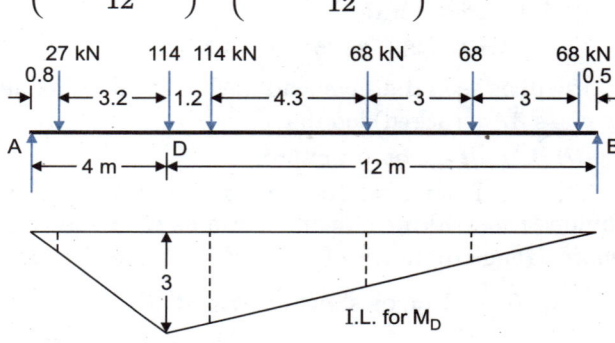

Fig. 31.63

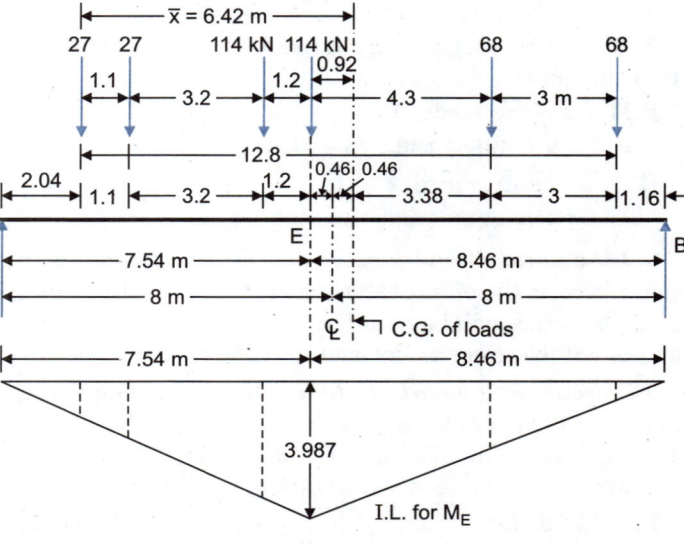

Fig. 31.64

Fig. 31.65

Hence ordinate of influence line diagram = $\dfrac{7.54 \times 8.46}{16} = 3.987$.

$\therefore M_{max} = (114 \times 3.987) + \left(27 \times \dfrac{2.04}{7.54} \times 3.987\right) + \left(27 \times \dfrac{3.14}{7.54} \times 3.987\right) + \left(114 \times \dfrac{6.34}{7.54} \times 3.987\right)$

$\qquad\qquad + \left(68 \times \dfrac{1.16}{8.46} \times 3.987\right) + \left(68 \times \dfrac{4.16}{8.46} \times 3.987\right) = 1081$ kN-m

It should be noted that this is slightly more than the central B.M. of 1075.55 kN-m found earlier. Taking into account the impact factor and reaction factor,

B.M. for outer girder = 1081 × 1.204 × 0.8782 = 1143 kN-m.

B.M. for central girder = 1081 × 1.204 × 0.6667 = 867.7 kN-m.

(b) **Class AA (tracked) loading**: This is a uniformly distributed loading of 700 kN covering a length of 3.6 m.

(i) *B.M. at the centre of the span : Absolute Max. B.M.*

Absolute max. B.M. for U.D.L. always occurs at the centre of span. From Fig. 31.66, $M = 700 \left(\dfrac{3.1 + 4}{2}\right)$ = 2485 kN-m

Taking into account the impact factor and the reaction factor,

B.M. for outer girder
 = (1.1 × 0.4968) 2485 = 1358 kN-m,

B.M. for inner girder
 = (1.1 × 0.3333) 2485 = 911 kN-m,

It will be seen that live load B.M. is maximum due to class AA (tracked) loading.

(ii) *B.M. at the quarter span*

The U.D.L. should be so arranged that the section divides the load in the same ratio as it divides the span. Hence distance of edge of load from section = $\dfrac{1}{4} \times 3.6 = 0.9$ m, as shown in Fig. 31.67.

Hence,

$\qquad$ B.M. = $\dfrac{3 + 2.325}{2} \times 700 = 1863.75$

Taking into account the impact factor and the reaction factor,

B.M. for outer girder
 = (1.1 × 0.4968) 1863.75 = 1018.5 kN-m.

B.M. for central girder
 = (1.1 × 0.3333) 1863.75 = 683.3 kN-m.

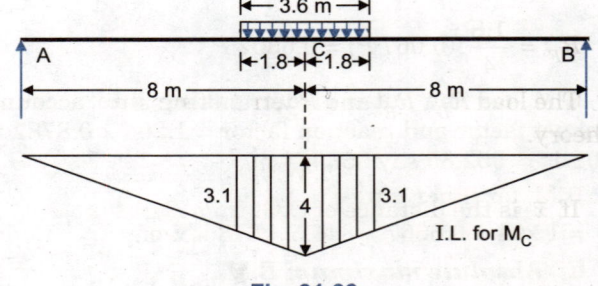

Fig. 31.66

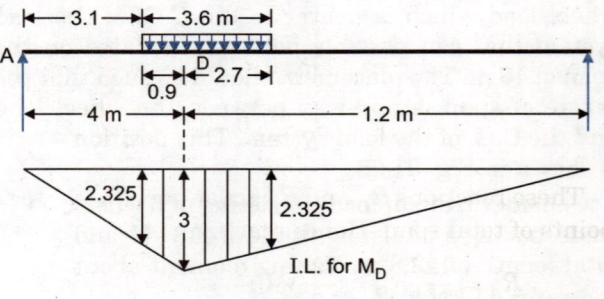

Fig. 31.67

**9. Live Load Shear force.** Shear force will be maximum due to class AA tracked vehicle. For maximum shear force at the ends of the girder, the load will be placed between the support and the first intermediate girder, and S.F. will be found by the reaction factors derived below. For intermediate section, same reaction factors will be used, as derived for B.M.

(a) **Shear at the end of girder.** Since the length of the track is 3.6 m, maximum shear will occur when the C.G. of load is at 1.8 m away from support A of the girder. The load will be confined between the end and the first stiffner. Along the width of the bridge, the track will be so placed that it maintains a maximum clearance of 1.2 m. Hence distance of C.G. of load from kerb = 1.2 + 0.425 = 1.625 m.

Fig. 31.68 shows the position of the two wheel loads P each of 350 tonnes. If $P_1, P_2$ and $P_3$ are the loads coming on the three girders, considering the slab to be simply supported, we have

$$P_1 = \frac{1.875}{2.6} P = 0.721 P; \qquad P_2 = \frac{0.752}{2.6} P + \frac{2.425}{2.6} P = 1.221 P$$

$$P_3 = \frac{0.175}{2.6} P = 0.067 P$$

The reactions at end of each longitudinal girders due to transfer of these loads at 1.8 m from left supports are:

$$R_A' = \frac{2.2}{4} (0.721 P) = 0.3966 P$$

$$R_D' = \frac{1.8}{4} (0.721 P) = 0.3245 P$$

$$R_B' = \frac{2.2}{4} (1.212 P) = 0.6666 P$$

$$R_E' = \frac{1.8}{4} (1.212 P) = 0.5454 P$$

$$R_C' = \frac{2.2}{4} (0.067 P) = 0.0369 P$$

$$R_F' = \frac{1.8}{4} (0.067 P) = 0.0302 P$$

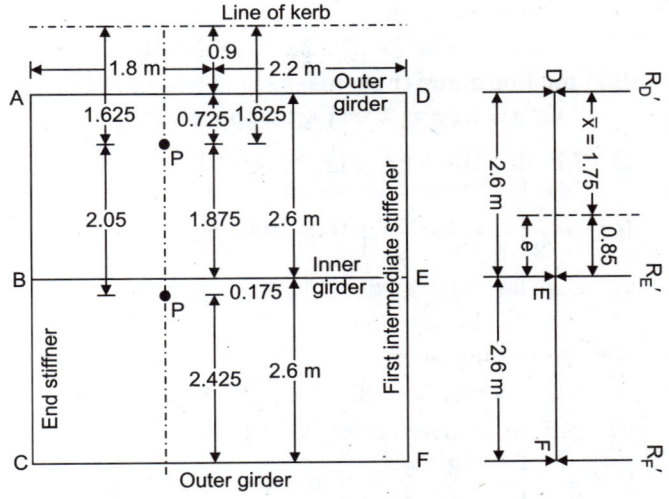

Fig. 31.68

The load $R_D'$, $R_E'$ and $R_F'$ transferred at the cross-girders should be distributed according to Courbons's theory.

$$\Sigma W = 0.3245 P + 0.5454 P + 0.0302 P = 0.9001 P.$$

If $\bar{x}$ is the distance of C.G. from D, we have

$$\bar{x} = \frac{1}{0.9001 P} [0.5454 P \times 2.6 + 0.0302 \times 5.2 P] = 1.75 \text{ m}.$$

$$\therefore \quad e = 2.6 - 1.75 = 0.85; \qquad \Sigma x^2 = 2(2.6)^2$$

$$\therefore \quad R_D = \frac{0.9001 P}{4} \left[ 1 + \frac{3 \times 0.85 \times 2.6}{2.(2.6)^2} \right] = 0.3354 P$$

$$R_E = \frac{0.9001 P}{3} (1 + 0) = 0.3 P$$

These reactions $R_D$ and $R_E$ act as point loads on the outer and inner longitudinal girders at their quarter points of total span. Hence reactions at A and B due these will be.

$$R_A = \frac{12}{16} R_D = \frac{12}{16} \times 0.4472 P = 0.3354 P; \qquad R_B = \frac{12}{16} R_E = \frac{12}{16} \times 0.3 P = 0.2250 P$$

Hence shear at $A = R_A' + R_A = 0.3966 P + 0.3354 P = 0.7320 P$

Shear at $B = R_B' + R_B = 0.6666 P + 0.2250 P = 0.8916 P$.

$P = 350$ kN

Taking into account the impact,

Max. S.F. at support of outer girder = $1.1 \times 0.7320 \times 350 = 281.82$ kN.

Max. S.F. at support of inner girder = $1.1 \times 0.8916 \times 350 = 343.26$ kN.

(b) **Shear at intermediate points:** Shear at other points will be found on the basis of the same reaction coefficients as found for B.M. Thus, for class AA (tracked) loading, reaction coefficient for outer girder = $0.4968 W$ and that for inner girder = $0.3333 W$, where W is axle load. Fig. 31.69 shows the I.L. for shear at various points, along with load position for maximum shear. Length of load = 3.6 m.

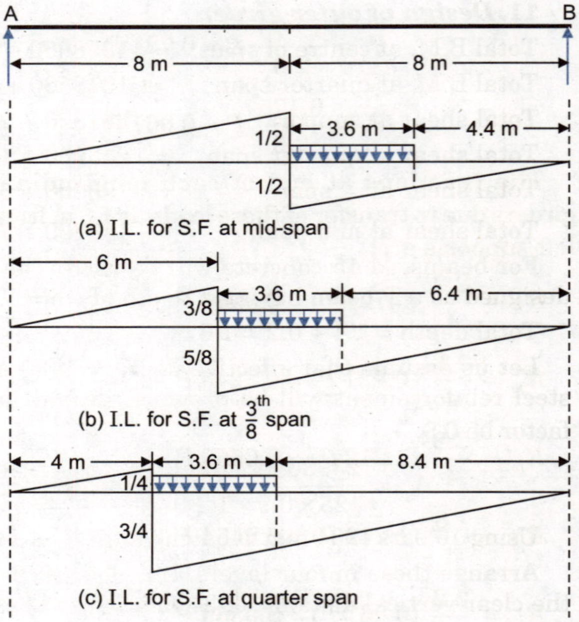

(i) S.F. at mid-span [Fig. 31.69 (a)].

S.F. at mid-span
$$= \frac{1}{2}\left[\frac{1}{2} + \frac{1}{2} \times \frac{4.4}{8}\right] \times 700 = 271.25 \text{ kN}$$

Taking into account impact factor and reaction factor,

S.F. for outer girder
$$= 1.1 \times 0.4968 \times 271.25 \approx 148.20 \text{ kN}.$$

S.F. for inner girder
$$= 1.1 \times 0.3333 \times 271.25 \approx 99.5 \text{ kN}.$$

(ii) S.F. at $\frac{3}{8}$ th span [Fig. 31.69 (b)]

$$\text{S.F.} = \frac{1}{2}\left(\frac{5}{8} + \frac{5}{8} \times \frac{6.4}{10}\right) 700 = 358.75.$$

∴ S.F. for outer girder
$$= 1.1 \times 0.4968 \times 358.75 = 196 \text{ kN}.$$

S.F. for inner girder
$$= 1.1. \times 0.3333 \times 358.75 = 131.5 \text{ kN}$$

(iii) S.F. at quarter span [Fig. 31.69 (c)].

$$\text{S.F.} = \frac{1}{2}\left(\frac{3}{4} + \frac{3}{4} \times \frac{8.4}{12}\right) 700 = 446.25 \text{ kN}$$

∴ S.F. for outer girder $= 1.1 \times 0.4968 \times 446.25 = 243.9$ kN.

S.F. for inner girder $= 1.1 \times 0.3333 \times 446.25 = 163.6$ kN.

Fig. 31.69

## 10. Dead Load B.M. and S.F. in girders.

Dead load from each cantilever portion = 9584 N/m.

Dead load of slab and wearing coat = $(0.08 \times 22000) + (0.2 \times 24000) = 6560$ N/m².

∴ Total load from deck = $(2 \times 9584) + (2 \times 2.6 + 0.3) \times 6560 = 55248$ N/m.

Assuming this load to be equally shared by all the three girders, dead load per girder = 55248/3 ≈ 18420 N/m. Let the depth of rib of girder be 1.4 m.

∴ Weight of rib = $1 \times 0.3 \times 1.4 \times 24000 = 10080$

∴ Total U.D.L. on girder = 18420 + 10080 = 28500 N/m.

Let the depth of rib of the cross girder also be 1.4 m.

Weight of rib of cross-girder = $1 \times 0.3 \times 1.4 \times 24000 = 10080$ N/m.

Let the length of each cross-grider = $2 \times 2.3 = 4.6$ m.

Assuming the weight of cross girders to be equally shared by all the three longitudinal girders, point load on each girder

$$= \tfrac{1}{3} (4.6 \times 10080) \approx 15460 \text{ N}.$$

The loading on the main girder is shown in Fig. 31.70. The point loads are due to cross-girder.

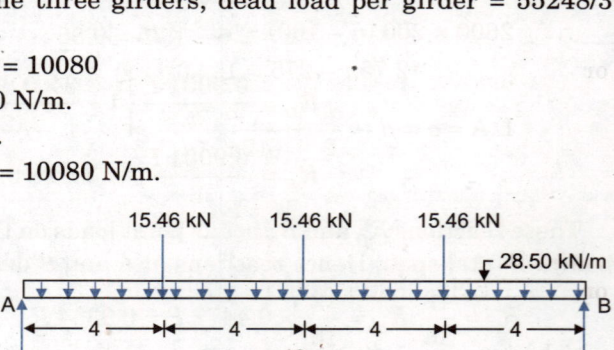

Fig. 31.70

$$R_A = \tfrac{1}{2} (3 \times 15460 + 16 \times 28500) = 251190$$

B.M. at mid span $= (251190 \times 8) - (15460 \times 4) - \dfrac{28500(8)^2}{2} = 1035680$ N-m.

B.M. at quarter span $= 251190 \times 4 - \dfrac{28500(4)^2}{2} = 776760$ N-m.

S.F. at support $= R_C = 251190$ N.

S.F. at quarter span $= 251190 - 28500 \times 4 = 137190$ N.

S.F. at $\tfrac{3}{8}$ span, i.e., at 6 m from A $= 251190 - (28500 \times 6) - 15460 = 64730$ N.

S.F. at mid span $= 15460/2 = 7730$ N.

### 11. *Design of outer girder*

| | | | |
|---|---|---|---|
| Total B.M. at centre of span | = | 1358000 + 1035680 | = 2393680 N-m |
| Total B.M. at quarter span | = | 1018500 + 776760 | = 1795260 N-m. |
| Total shear at support | = | 281820 + 251190 | = 533010 N. |
| Total shear at quarter span | = | 243900 + 137190 | = 381090 N. |
| Total shear at $\frac{3}{8}$ span | = | 196000 + 64760 | = 260760 N. |
| Total shear at mid-span | = | 148200 + 7730 | = 155930 N. |

For beams, M 15 concrete will be used. The outer girder will be designed as a T-beam having a depth of rib = 1.4 m.

Total depth = 1.4 + 0.2 = 1.6 m.

Let us assume that effective depth = 1480 mm, since the heavy steel reinforcement will be provided in four layers. Let lever arm factor be 0.9.

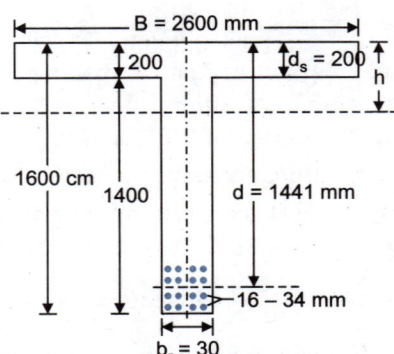

$$\therefore \quad A_{st} = \frac{2393680 \times 1000}{125 \times 0.9 \times 1480} \approx 14370 \text{ mm}^2.$$

Using 16 bars of 34 mm Φ having total $A_{st}$ = 14530 mm².

Arrange these in four layers. If a clear cover of 4 cm is used and the clear vertical distance of 34 mm is used between layers, height of centre of steel area, above bottom = 40 + 3 × 34 + 17 = 159 mm.

Hence $d$ = 1600 − 159 = 1441 mm.

Fig. 31.71

The concrete used for slab is of M 20 grade while that for girder is of M 15 grade. Since the whole of the compression is being resisted by flange (slab), $m$ = 13 for M 20 concrete will be used in calculations.

Flange width will be the least of the following:

(i)             $12 d_s + b_r = 12 \times 200 + 300 = 2700$ mm

(ii)            c/c of spacing = 2600 mm.

(iii)           Span / 3 = 16 / 3 = 5.333 m = 5333 mm

$\therefore$    $B$ = 2600 mm. Let the depth of N.A. be $n$, lying in the web.

$\therefore \quad 2600 \times 200 (n - 100) = 13 \times 14530 (1441 - n)$

or $\quad 2.75n - 275 = 1441 - n. \quad \therefore \quad n = 457$ mm.

$$L.A = a = d - \frac{d_s}{2} + \frac{d_s^2}{6(2n - d_s)} = 1441 - \frac{200}{2} + \frac{(200)^2}{6(2 \times 457 - 200)} = 1350 \text{ mm}.$$

$$M = Bd_s \cdot c_a \left[1 + \frac{n - d_s}{n}\right] a$$

or $\quad (239,3680 \times 1000) = 2600 \times 200\, c_a \left[1 + \frac{457 - 200}{200}\right] \times 1350$

which gives $c_a$ = 1.49 N/mm². Hence safe.

*Area of steel at quarter span*

$$A_{st} = \frac{1795260 \times 1000}{125 \times 0.9 \times 1441} \approx 11080 \text{ mm}^2. \quad \therefore \text{ No. of 34 mm Φ bars} = \frac{11080}{908} \approx 12$$

*Check for local bond stress as per IRC code.*

Assume effective depth at support = 1540 mm, and a lever arm of 0.9 $d$

$$\therefore \quad \Sigma \text{ required} = \frac{533010}{1 \times 0.9 \times 1540} \approx 385. \quad \therefore \text{ No. of 34 mm Φ bars required} = \frac{385}{\pi \times 34} = 3.6$$

Hence at least 4 bars are to be taken straight.

### Check for shear

Nominal Shear stress at support = $\dfrac{533010}{300 \times 1540}$ = 1.15 N/mm². Hence shear reinforcement is required.

Nominal Shear stress at mid-span = $\dfrac{155930}{300 \times 1350}$ = 0.39 N/mm². Hence no shear reinforcement is required.

Norminal Shear stress at $\dfrac{3}{8}$ span = $\dfrac{260760}{300 \times 1350}$ = 0.64 N/mm². Hence shear reinforcement is required.

Approximate distance from support, at which shear stress is 0.5 N/mm² is = $\dfrac{1}{2}$ (6 + 8) = 7 m. Let us bend up 2 bars at a time, at spacing of 0.707 × $a$ = 0.707 (0.9 × 1440) = 916.3 mm. If five such sets are bent up, bent up bars will be effective at distance of 5 × 0.95 = 4.75 m from support. Since the sets of bars are bent up at interval of 0.707 $a$, 4 bars will be effective at each section.

∴ Shear taken by 4 bent-up 34 mm Φ bars = 4 × 908 × 125 × 0.707 ≈ 320980 N.

∴ Net remaining shear at support, for which shear reinforcement is to be provided = 533010 – 320980 = 212030 N. Using 10 mm Φ 2 legged stirrups,

$$\text{Spacing} = \dfrac{2 \times 78.5 \times 125 \times 1540}{212030} = 142.5 \text{ mm}.$$

Hence provide 10 mm Φ 2 legged stirrups @ 140 mm c/c upto 4 × 0.95 = 3.8 m from support. After 3.8 m from support, only 2 bars will be effective.

At quarter span, remaining shear = 381090 – $\dfrac{320980}{2}$ = 220600 N.

∴ Spacing of 10 mm Φ 2 lgd stirrup = $\dfrac{2 \times 78.5 \times 125 \times 1440}{220600}$ = 128.

Hence provide 10 mm Φ stirrups @ 120 mm c/c from 3.8 m to 4.75 m from support. Beyond 4.75 m, no ben-tup bar will be available. Shear at 3/8 span (*i.e.*, 6 m) = 260760 N.

∴ Spacing of 12 mm Φ stirrup = $\dfrac{2 \times 113 \times 125 \times 1440}{260760}$ ≈ 156 mm.

At 4.75 m from support approximate value of shear

$$= 260760 + (381090 - 260760)\left(\dfrac{1.25}{2}\right) \approx 335970 \text{ N}.$$

∴ Spacing of 12 mm Φ stirrups = $\dfrac{2 \times 113 \times 125 \times 1440}{335970}$ ≈ 120 mm.

At the mid-span no shear reinforcement is required. The summary of the stirrups is as follows :

(*i*) From support to 3.8 m : 10 mm Φ 2 lgd stirrups @ 140 mm c/c.
(*ii*) From 3.8 m to 4.75 m : 10 mm Φ 2 lgd stirrups @ 120 mm c/c.
(*iii*) From 4.75 m to 6 m : 12 mm Φ 2 lgd stirrups @ 120 mm c/c.
(*iv*) From 6m to 7m : 12 mm Φ 2 lgd stirrups @ 150 mm c/c.
(*v*) For remaining distance : 12 mm Φ 2 lgd stirrups @ 300 mm c/c.

### 12. Design of Inner Girder

| | |
|---|---|
| Total B.M. at centre of span | = 911000 + 1035680 = 1946680 N-m |
| Total B.M. at quarter span | = 683300 + 776760 = 1460060 N-m. |
| Total shear at support | = 343230 + 251190 = 594420 N. |
| Total shear at quarter span | = 163600 + 137190 = 300790 N. |
| Total shear at 3/8 span | = 131500 + 64760 = 196260 N. |
| Total shear at mid-span | = 99500 + 7730 = 107230 N. |

Adopt the same section, as for main girder. Thus $B$ = 2600 mm; $b$ = 300 mm; $D$ = 1600 mm. Let the effective depth be 1440 mm. The section will be safe in compression.

$$M = 1946680 \text{ N-m.}$$

$$\therefore \quad A_{st} = \frac{1946680 \times 1000}{125 \times 0.9 \times 1440} \approx 12017 \text{ mm}^2.$$

Use 16 bars of 32 mm Φ, having $A_{st}$ = 8040 mm²

Hence total $A_{st}$ = 16 × 8040 = 128640 mm².

$$A_{st} \text{ at quarter span} = \frac{1460000 \times 1000}{125 \times 0.9 \times 1440} \approx 9010 \text{ mm}^2. \text{ Hence 12 bars of 32 mm } \Phi \text{ are required.}$$

### Check for local bond as per IRC code.

Shear at support = 594420 N. Let the effective depth be 1500 mm at the support.

$$\Sigma 0 \text{ required for bond} = \frac{594420}{1 \times 0.9 \times 1500} = 440.$$

$$\therefore \quad \text{No. of 32 mm } \Phi \text{ bars required} = \frac{440}{\pi \times 32}$$

= 4.4. However, take 6 bars straight upto the support.

### Check for shear

$\tau_v$ at support = $\frac{594420}{300 \times 1500}$ = 1.32 N/mm².

Hence shear reinforcement is required.

$\tau_v$ at centre = $\frac{107230}{300 \times 1440}$ = 0.25 N/mm².

Hence no shear reinforcement is required.

$\tau_v$ at $\frac{3}{8}$ span = $\frac{196260}{330 \times 1440}$ = 0.41 N/mm².

Thus, shear reinforcement is required from the support upto 3/8th span (*i.e.*, upto 6 m).

Let us bend the sets of bars at spacing of 0.707 a ≈ 950 mm with two bars in one set. Five such sets of bent bars are provided, and these bents will be effective upto 5 × 0.95 = 4.75 m from the support. At every section two sets of bars, *i.e.*, 4 bars will be effective. Hence shear taken up by 4 bent up 32 mm Φ bars = 4 × 804 × 125 × 0.707 ≈ 284200 N.

∴ Net shear remaining at support
= 594420 – 284200 = 310220 N.

Spacing of 12 mm Φ 2 lgd stirrups

$$= \frac{2 \times 113 \times 125 \times 1500}{310220} \approx 136.6 \text{ mm.}$$

Hence provide this spacing from the support upto 4 × 0.95 = 3.8 m from the support. Beyond this point, only 2 bent-up bars will be available upto 4.75 m. Hence shear taken up by 2 bent-up bars = 142100 N. At quarter span. S.F. = 300790 N. Hence remaining shear = 300790 – 142100 = 158690.

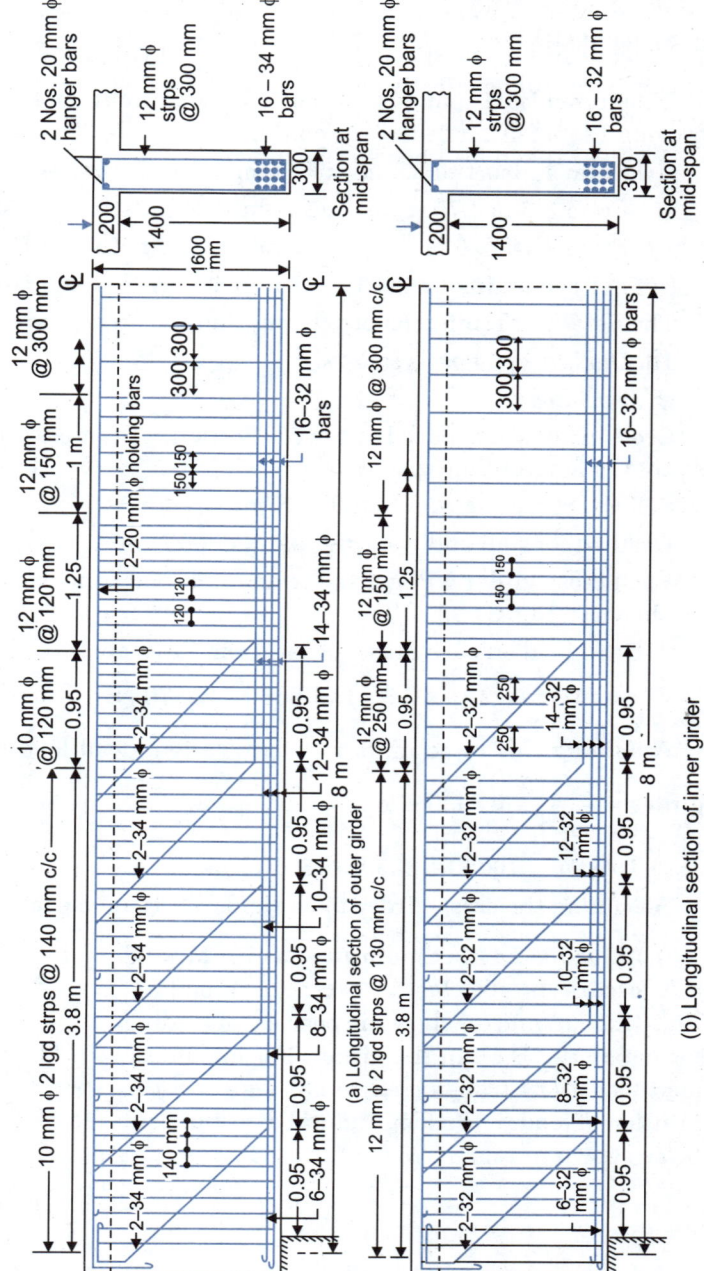

Fig. 31.72

∴ Spacing of 12 mm Φ stirrups = $\dfrac{2 \times 113 \times 125 \times 1440}{158690} \approx 256.35$ mm.

However, provide 12 mm Φ stirrups @ 250 mm c/c from 3.8 to 4.75 m away from the support. At 6 m away from the support, no shear reinforcement is required. Between 4.65 m to 6 m, no bent up bars will be available. Shear at 4.75 m away from support is approximately

$$= 196260 + \dfrac{300790 - 196260}{2} \times 1.25 \approx 261600 \text{ N}.$$

Spacing of 12 mm Φ stirrups $= \dfrac{2 \times 113 \times 125 \times 1440}{261600} = 155.5$ mm.

Thus provide 12 mm Φ stirrups @ 150 mm c/c from 4.75 to 6 m away from support. Beyond this point provide 12 mm Φ 2 *lgd* stirrups nominal spacing of 300 mm c/c. To summarise, the spacing of stirrups are:

(*i*) From support to 3.8 m    : 12 mm Φ 2 *lgd* stirrups @ 130 mm c/c.
(*ii*) From 3.8 to 4.75 m        : 12 mm Φ 2 *lgd* stirrups @ 250 mm c/c.
(*iii*) From 4.75 to 6 m          : 12 mm Φ 2 *lgd* stirrups @ 150 mm c/c.
(*iv*) For remaining length    : 12 mm Φ 2 *lgd* stirrups @ 300 mm c/c.

The details of reinforcement in longitudinal girders are shown in Fig. 31.72.

### 13. Design of Cross-girders.

(*i*) *Dead Load*

Cross girders are spaced at 4 m c/c. Section of the cross girders is the same as that of longitudinal girder.

Self wt. of cross-girder = 10080 N/m.

Dead wt. of slab and wearing coat = 6560 N/m².

Each cross-girder will get the triangular load from each side of the slab, as shown in Fig. 31.73.

∴ Dead load on each cross girder, from the slab,

$$= 2 \left[\tfrac{1}{2} \times 2.6 \times 1.3\right] 6560 \approx 22170 \text{ N}.$$

Assuming this to be uniformly distributed, dead load per metre run of girder = $\dfrac{22170}{2.6} \approx 8530$ N/m.

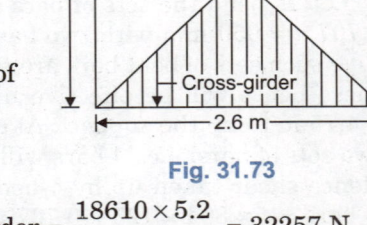

Fig. 31.73

∴ Total $w$ = 10080 + 8530 = 18610 N/m

Assuming the cross-girder to be rigid, reaction on each longitudinal girder = $\dfrac{18610 \times 5.2}{3} = 32257$ N.

(*ii*) *Live load* : Maximum live load B.M. and S.F. occurs due to class *AA* (tracked) loading. Fig. 31.75 (*a*) and (*b*) show the position of loading for maximum B.M. in the girder. For maximum load (reaction) transferred to cross-girder, the position of load in the longitudinal direction will be as shown in Fig. 31.75 (*a*).

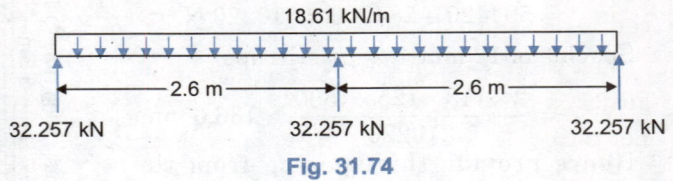

Fig. 31.74

$$R = \dfrac{700000 \times 3.1}{4} = 542500 \text{ N}.$$

Fig. 31.75 shows the loading transferred on the cross girder. Assuming the cross-girder to be rigid, reaction on each longitudinal girder – $\dfrac{542500}{3}$ N.

∴ Max. live load B.M. [Fig. 31.75 (c)] occur under wheel load:

$$M = \frac{542500}{3} \times 1.575 \approx 284810 \text{ N-m.}$$

Taking into account the impact factor,

$M = 1.1 \times 284810 \approx 313290$ N-m.

Dead load B.M. from 1.575 m support (Fig. 31.74)

$= 32257 \times 1.575 - \dfrac{18610 (1.575)^2}{2} \approx 27720$ N-m.

∴ Total B.M. = 284810 + 27720 = 312530 N-m.

Live load shear, including impact

$= 1.1 \times \dfrac{542500}{3} \approx 198917$ N.

∴ Dead load shear (Fig. 31.74) = 32257 N.

∴ Total $F$ = 198917 + 32257 = 231174 kg.

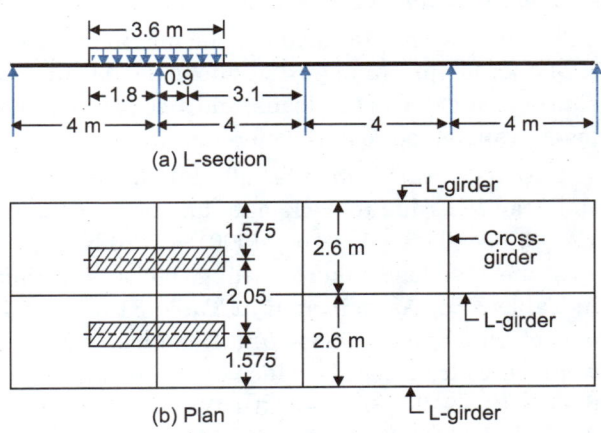

*(iii) Design of section*

Total depth = 1600 mm. Let the effective depth be 1540 mm.

$$A_{st} = \frac{312530 \times 1000}{125 \times 0.9 \times 1540} = 1804 \text{ mm}^2$$

Provide 4 bars of 25 mm Φ having area = 4 × 490.9 ≈ 1964 mm².

Bend two of these bars at $L/5$ = 0.80 m from the support shear

$$\text{stress} = \frac{231174}{300 \times 0.9 \times 1540} = 0.5 \text{ N/mm}^2$$

Fig. 31.75

However, bent up bars will take care of this shear. Provide nominal 2 *lgd* stirrups of 12 mm Φ @ 300 mm c/c throughout the length. The cross section of the cross-girder is shown in Fig. 31.76.

**Example 31.3.** **Use of Distribution Coefficients**

*Find longitudinal and transverse moments in the bridge deck of example 31.2, using distribution coefficients, if only three cross-beams are provided.*

**Solution:**

**1. Moments due to weight of rib:** Centre to centre spacing of cross-beams = 8 m.

Let the width of web of both the longitudinal girders as well as that of cross-girder be 300 mm. Let the depth of rib be 140 cm. Weight of rib

$= 1.4 \times 0.3 \times 1 \times 24000 = 10080$ N-m.

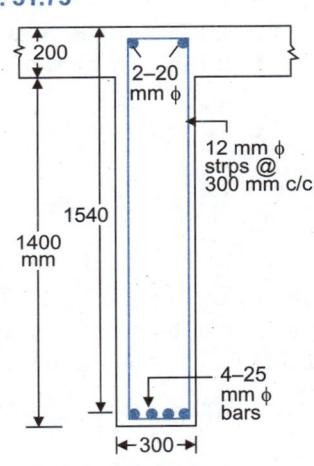

Fig. 31.76

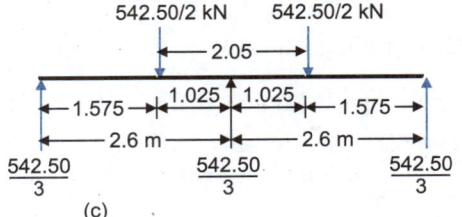

Fig. 31.77

Total weight of rib of each cross-girder = 10080 (2 × 2.3) = 46368 N. Assuming this to be equally divided to the three longitudinal girders, dead load at the centre of each longitudinal girder = $\dfrac{1}{3} \times 46368 \approx 15460$ N.

The longitudinal girder will thus be loaded as shown in Fig. 31.77.

## 2. Maximum average B.M. per girder

The next step is to calculate the maximum average B.M., per longitudinal girder, due to dead load of deck slab etc., and due to live loads, and also to calculate the loads transferred to each reference station.

*Case (i) : Dead loads of deck* : Fig. 31.78 (a) shows the dead load due to deck slab, having $w = (0.20 \times 1 \times 24000) + (0.08 \times 1 \times 1 \times 24000) = 6720$ N/m², along with the dead load of kerb, railings etc, as determined in Table 31.5. As indicated in Table 31.5, dead load of each cantilever = 9584 N/m. Dead load of slab and wearing coat of deck, exclusive of the cantilevers = $6560 \times (0.3 + 2.3 + 0.3 + 2.3 + 0.3) = 36080$ N. Total dead load = $2 \times 9584 + 36080 = 55248$ N/m length of

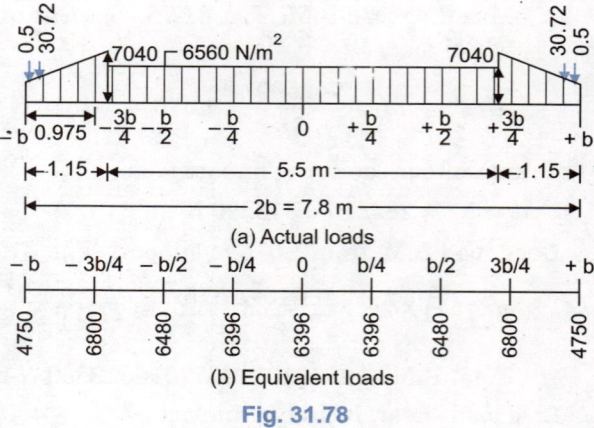

Fig. 31.78

the longitudinal girders. Max. B.M. at the centre of the longitudinal girders = $\dfrac{55248 \,(16)^2}{8} = 1767936$ N-m.

Max average B.M. per girder = $\dfrac{1}{3}(1767936) \approx 589310$ N-m. Fig. 31.78 (b) shows the equivalent loads transferred at the reference station, assuming the span between the reference stations as simply supported. For the end stations, the transferred loads are approximate. It should be noted here that $2b = 7.8$ m and hence $b/4 = 0.975$ m.

*Case (ii) : Class AA (tracked) load centrally located*

Fig. 31.79 (a) shows the centrally placed class AA loading. Equivalent load at $\pm b/2$ station

$= \dfrac{0.013\, b}{0.25\, b} \times P = 0.052\, P$ and at that $\pm \dfrac{b}{4}$ station

$= \dfrac{0.237\, b}{0.25\, b} \times P = 0.948\, P.$

Equivalent load at other stations is zero, as shown in Fig. 31.79 (b). Fig. 31.79 (c) shows the position of load in longitudinal direction for Max. B.M. which is equal to

$(350 \times 8) - \dfrac{700}{3.6} \cdot \dfrac{(1.8)^2}{2} = 2485$ kN-m.

Hence maximum average B.M. per

beam = $\dfrac{1}{3} \times 2485 \times 1000 \approx 828330$ N-m.

*Case (iii) : Class AA (tracked) load at maximum eccentricity*

Fig. 31.61 shows the placement of loading, with centre of wheel load at a distance of 1.625 m from the kerb, or at $1.625 + 0.4 = 2.025$ m from end station. This is equal to $2.025/3.9\, b = 0.519\, b$ from the end station, as shown in Fig. 31.80 (a).

Fig 31.80 (b) shows equivalent loading. At 0 station, this load = $\dfrac{0.205}{0.25} P = 0.82\, P$ while that at $+(b/4)$ it is $= (0.045)/(0.25)\, P = 0.18\, P.$

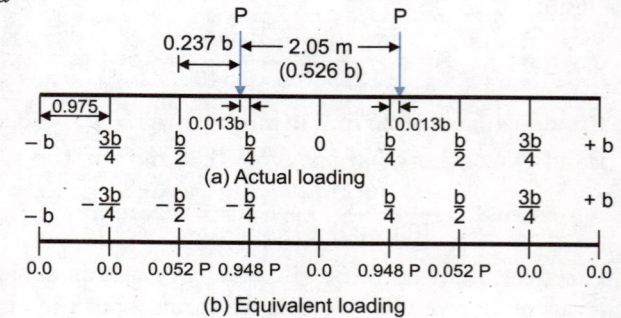

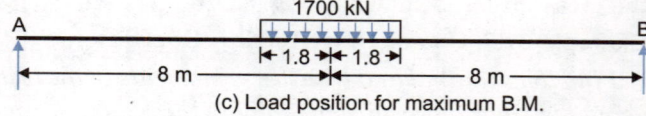

Fig. 31.79

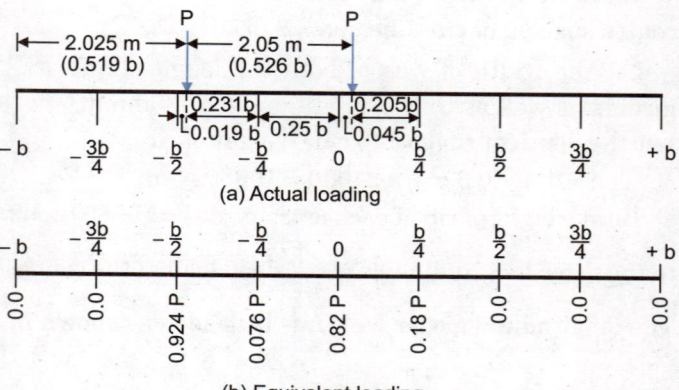

Fig. 31.80

The placement of loading on the longitudinal girder for maximum B.M will be the same as shown in Fig. 31.79(c). Hence maximum average B.M. per beam = 828330 N-m as found earlier.

*Case (iv) : Class A loading, centrally placed* : Fig. 31.81 (a) shows the actual loads, placed centrally along the width, while Fig. 31.81 (b) shows equivalent loading.

The position of loads for absolute max. B.M. in the longitudinal girder is shown in Fig. 31.65, in which the fourth wheel load (*i.e.*, 114 kN) is placed at a distance of 0.46 m from the centre of span. In that figure, the loads marked are for one train of loads. Hence absolute maximum B.M in the girder = 2 × 1081 = 2162 kN-m.

Average max. B.M per girder = $\frac{1}{3}$ × 2162000 ≈ 720667 N-m.

*Case (v): Class A loading at maximum eccentricity*

Permissible clearance upto centre of first wheel = 0.15 + 0.25 = 0.4 m from kerb.

∴ Distance from edge = 0.4 + 0.4

= 0.8 m = $\frac{0.8}{3.9}$ b = 0.2051 b.

The placement of loading is shown in Fig. 31.82 (a). Fig. 31.82 (b) shows the equivalent loading.

The max. average B.M. per girder = 720667 N-m, as before. It will be seen that max. average B.M. for class AA (tracked) vehicle is more than that for class A loading. Hence only class AA loading will be considered for further calculations.

**3. Calculations for θ and α**

Let total depth of main beams and cross-beams be = 160 cm.

Width of rib = 30 cm. ; Total width of deck = 7.8 m

∴ Effective width of flange assigned to each main beam = 7.8/3 = 2.6 m. (Alternatively, this is equal to $p = 2b/n = 7.8/3$). Effective width of flange for central cross-girder

= (16 + 0.3) / 2 = 8.15 m.

The section of the two girders are shown in Fig. 31.83. While calculating the moment of inertia, the section is assumed to be of plain concrete only.

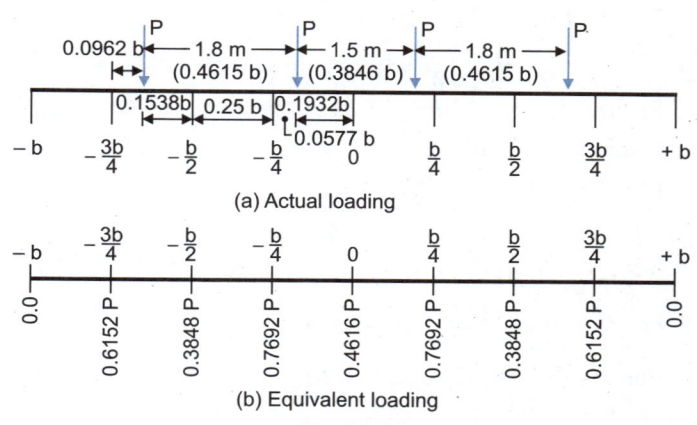

Fig. 31.81

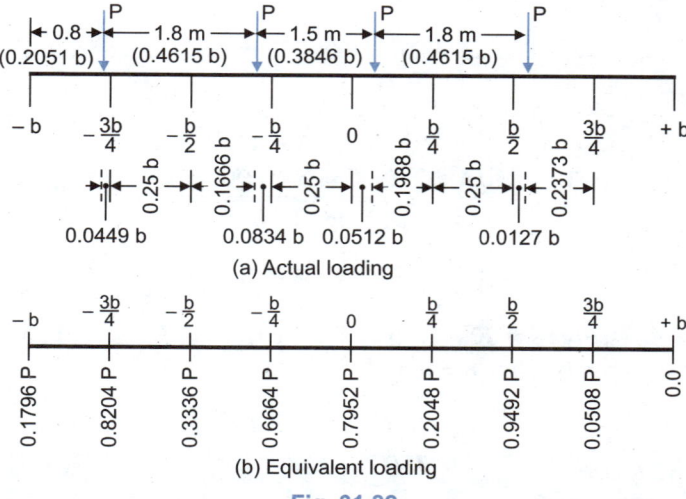

Fig. 31.82

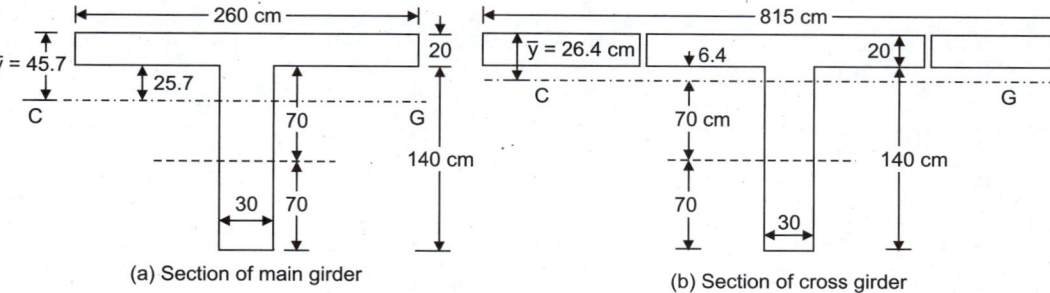

Fig. 31.83

For the main girder, depth $\bar{y}$ of centroidal plane is given by

$$\bar{y} = \frac{(260 \times 20 \times 10) + (140 \times 30 \times 90)}{(260 \times 20) + (140 \times 30)} = 45.7 \text{ cm.}$$

$$I_L = \left[\tfrac{1}{12} 260 (20)^3 + (260 \times 20)(45.7 - 10)^2\right] + \left[\tfrac{1}{12} \times 30 (140)^3 + (30 \times 140)(70 - 25.7)^2\right] \text{cm}^4 = 0.219 \text{ m}^4$$

$\therefore \quad i_L = 0.219/2.6 = 0.0842 \text{ m}^3/\text{m}$. Similarly, for the cross-girders,

$$\bar{y} = \frac{(815 \times 20 \times 10) + (140 \times 30 \times 90)}{(815 \times 20) + (140 \times 30)} = 26.4$$

$$I_T = \left[\tfrac{1}{12} \times 815 (20)^3 + (815 \times 20)(26.4 - 10)^2\right] + \left[\tfrac{1}{12} \times 30 (140)^3 + (30 \times 140)(70 - 6.4)^2\right] \text{cm}^4 = 0.2878 \text{ m}^4.$$

$\therefore \quad i_T = 0.2878/8.15 = 0.0353 \text{ m}^4/\text{m}$

*For the longitudinal beam*

$$J_L = \tfrac{1}{2}(J)_{flange} + (J)_{web}. \quad \text{In general,} \quad J = \beta (2a)^3 \cdot 2b \; ...(Eq.\, 31.16), \quad \text{where } b \geq a$$

For the flange, $2b = 2.6$ m ; $2a = 0.20$ ; $b/a = 1.3/0.1 = 13$ ;
Hence from Table 31.4, $\beta = 0.32$.

For the web, $2b = 1.4$ ; $2a = 0.30$ ; $b/a = 0.7/0.15 = 4.67$. $\therefore \beta = 0.285$

$\therefore \quad J_L = \tfrac{1}{2}[0.32(0.2)^3(2.6)] + [0.285(0.3)^3(1.4)] = \tfrac{1}{2} \times 0.00666 + 0.01077 = 0.0141 \text{ m}^4$

$\therefore \quad J_T = 0.0141 / 2.6 = 0.00542 \text{ m}^3/\text{m}$

*For the cross-beam* $\quad J_T = \tfrac{1}{2} \cdot (J)_{flange} + (J)_{web}.$

For the flange, $2b = 8.15$ ; $2a = 0.20$ ; $b/a = 8.15/0.2 \approx 40$ ; $\therefore \beta = 0.333$.
For the web, $2b = 1.4$ ; $2a = 0.30$ ; $b/a = 1.4/0.3 = 4.67$. $\therefore \beta = 0.285$.

$\therefore \quad J_T = \tfrac{1}{2}[0.333(0.2)^3(8.15)] + [0.285(0.3)^3(1.4)] = \tfrac{1}{2} \times 0.02171 + 0.01077 = 0.02162 \text{ m}^4$

$\therefore \quad j_T = 0.02162 / 8.15 = 0.00265 \text{ m}^3/\text{m}$. Hence from *Eq. 31.12,*

$$\theta = \frac{b}{2a}\left[\frac{i_L}{i_T}\right]^{1/4} = \frac{(7.8/2)}{16.3}\left[\frac{0.0842}{0.0353}\right]^{1/4} = 0.3 \quad \text{and} \quad \alpha = \frac{G}{2E} \frac{j_L + j_T}{\sqrt{i_L \cdot i_T}} \; ; \quad \text{for concrete, } G/E = 0.435.$$

$$\therefore \quad \alpha = \frac{0.435}{2} \cdot \frac{(0.00542 + 0.00265)}{\sqrt{0.0842 \times 0.0353}} = 0.0322. \text{ The beam positions in terms of effective width}$$

$2b (= 7.8)$ of the bridge are : $\dfrac{2.6\, b}{3.9}\left(= +\dfrac{2}{3} b\right)$, $0$ and $= \dfrac{2}{3} b$.

**4. Tables for $K_0$, $K_1$ and $K_a$:** Values of $K_0$ ($\alpha = 0$) and $K_1$ ($\alpha = 1$) are tabulated in Tables 31.6. and 31.7 respectively, for $\theta = 0.3$.

**TABLE 31.6.** Values of $K_0$ for $\theta = 0.3$

| Load position | Reference station | | | | | | | | |
|---|---|---|---|---|---|---|---|---|---|
| | $-b$ | $-3b/4$ | $-b/2$ | $-b/4$ | $0$ | $+b/4$ | $+b/2$ | $+3b/4$ | $+b$ |
| $-b$ | 4.20 | 3.31 | 2.48 | 1.68 | 0.86 | 0.20 | $-0.54$ | $-1.14$ | $-1.78$ |
| $-3b/4$ | 3.40 | 2.73 | 2.10 | 1.53 | 0.94 | 0.40 | $-0.16$ | $-0.62$ | $-1.15$ |
| $-b/2$ | 2.46 | 2.10 | 1.73 | 1.34 | 0.98 | 0.63 | 0.24 | $-0.15$ | $-0.54$ |
| $-b/4$ | 1.67 | 1.52 | 1.37 | 1.22 | 1.08 | 0.86 | 0.54 | 0.40 | 0.20 |
| $0$ | 0.86 | 0.96 | 0.98 | 1.05 | 1.11 | 1.05 | 0.98 | 0.96 | 0.86 |
| $+b/4$ | 0.20 | 0.40 | 0.54 | 0.86 | 1.08 | 1.22 | 1.37 | 1.52 | 1.67 |
| $+b/2$ | $-0.54$ | $-0.15$ | 0.24 | 0.63 | 0.98 | 1.34 | 1.73 | 2.10 | 2.46 |
| $+3b/4$ | $-1.15$ | $-0.62$ | $-0.16$ | 0.40 | 0.94 | 1.49 | 2.10 | 2.73 | 3.40 |
| $+b$ | $-1.78$ | $-1.14$ | $-0.54$ | 0.20 | 0.86 | 1.68 | 2.48 | 3.31 | 4.20 |

**TABLE 31.7.** Values of $K_1$ for $\theta = 0.3$

| Load position | Reference station | | | | | | | | |
|---|---|---|---|---|---|---|---|---|---|
| | $-b$ | $-3b/4$ | $-b/2$ | $-b/4$ | $0$ | $+b/4$ | $+b/2$ | $+3b/4$ | $+b$ |
| $-b$ | 1.60 | 1.37 | 1.20 | 1.02 | 0.94 | 0.84 | 0.78 | 0.69 | 0.63 |
| $-3b/4$ | 1.39 | 1.28 | 1.16 | 1.04 | 0.97 | 0.88 | 0.82 | 0.78 | 0.71 |
| $-b/2$ | 1.22 | 1.18 | 1.12 | 1.06 | 1.00 | 0.94 | 0.88 | 0.82 | 0.76 |
| $-b/4$ | 1.06 | 1.06 | 1.06 | 1.05 | 1.02 | 0.97 | 0.94 | 0.88 | 0.86 |
| $0$ | 0.94 | 0.97 | 1.00 | 1.02 | 1.04 | 1.02 | 1.00 | 0.97 | 0.95 |
| $+b/4$ | 0.86 | 0.88 | 0.94 | 0.97 | 1.00 | 1.05 | 1.06 | 1.06 | 1.06 |
| $+b/2$ | 0.76 | 0.82 | 0.88 | 0.94 | 1.00 | 1.06 | 1.12 | 1.18 | 1.22 |
| $+3b/4$ | 0.71 | 0.74 | 0.82 | 0.88 | 0.97 | 1.04 | 1.16 | 1.28 | 1.39 |
| $+b$ | 0.63 | 0.69 | 0.78 | 0.84 | 0.94 | 1.02 | 1.20 | 1.37 | 1.60 |

The values of $K_\alpha$ for $\alpha = 0.0322$ can be computed by $K_\alpha = K_0 + (K_1 - K_0)\sqrt{\alpha}$
The values of $K_\alpha$ are tabulated in Table 31.8.

**TABLE 31.8.** Values of $K_\alpha$ for $\theta = 0.3$

| Load position | Reference station | | | | | | | | |
|---|---|---|---|---|---|---|---|---|---|
| | $-b$ | $-3b/4$ | $-b/2$ | $-b/4$ | $0$ | $+b/4$ | $+b/2$ | $+3b/4$ | $+b$ |
| $-b$ | 3.733 | 2.962 | 2.250 | 1.562 | 0.874 | 0.315 | $-0.303$ | $-0.812$ | $-1.348$ |
| $-3b/4$ | 3.039 | 2.470 | 1.931 | 1.442 | 0.945 | 0.486 | 0.016 | $-0.376$ | $-0.816$ |
| $-b/2$ | 2.237 | 1.935 | 1.621 | 1.290 | 0.984 | 0.686 | 0.355 | 0.024 | $-0.307$ |
| $-b/4$ | 1.561 | 1.437 | 1.314 | 1.189 | 1.069 | 0.880 | 0.612 | 0.486 | 0.318 |
| $0$ | 0.874 | 0.962 | 0.984 | 1.045 | 1.097 | 1.045 | 0.984 | 0.962 | 0.874 |
| $+b/4$ | 0.318 | 0.486 | 0.612 | 0.880 | 1.069 | 1.189 | 1.314 | 1.437 | 1.561 |
| $+b/2$ | $-0.307$ | 0.024 | 0.355 | 0.686 | 0.984 | 1.290 | 1.621 | 1.935 | 2.237 |
| $+3b/4$ | $-0.816$ | $-0.376$ | 0.016 | 0.486 | 0.945 | 1.442 | 1.931 | 2.470 | 3.039 |
| $+b$ | $-1.348$ | $-0.812$ | $-0.303$ | 0.315 | 0.874 | 1.562 | 2.250 | 2.962 | 3.733 |

In Table 31.8, the value of $K_\alpha$ have been tabulated at various reference stations spaced at $b/4$. The values of $K_\alpha$ for $\pm 2/3\, b$ can be obtained by interpolation. These are shown in Table 31.9.

**TABLE 31.9.** Values of $K_\alpha$ for Reference Station $\pm 2/3\, b$ and $0$

| Ref. station | Load position | | | | | | | | |
|---|---|---|---|---|---|---|---|---|---|
| | $-b$ | $-3b/4$ | $-b/2$ | $-b/4$ | $0$ | $+b/4$ | $+b/2$ | $+3b/4$ | $+b$ |
| $-2b/3$ | 2.726 | 2.291 | 1.831 | 0.396 | 0.969 | 0.528 | 0.134 | $-0.246$ | $-0.643$ |
| $0$ | 0.874 | 0.945 | 0.984 | 1.069 | 1.097 | 1.069 | 0.984 | 0.945 | 0.874 |
| $+2b/3$ | $-0.643$ | $-0.246$ | 0.134 | 0.528 | 0.969 | 1.396 | 1.831 | 2.291 | 2.726 |

**5. Computation of coefficient $K'$:** The new coefficient $K'$ at a reference station is such that when this is multiplied by average moment it gives the final moment for that beam at that reference station. Let us now compute $K'$ for reference stations $\mp 2/3\, b$ and $0$. Calculations for $K'$ are done in the following steps.

(i) Multiply $K_\alpha$ for each station with appropriate load $(W)$ at standard positions in each of the five sets of loading considered in step 2, to get $Q$. Thus, $Q = K_\alpha \cdot W$.

(ii) Take the summation of the products of step (i) to get $\Sigma Q$.
Also find the total load $\Sigma W$ on the beam of each case.

(iii) Divide $\Sigma Q$ by $\Sigma W$ to get $K'$. The calculations are tabulated in Table 31.10.

**6. Computation of final longitudinal moments** $(M_x)$: Final moments are now found by multiplying the average maximum moments by the appropriate distribution coefficient $K'$.

**820** REINFORCED CONCRETE STRUCTURE

Thus $M_x = M_{av} \times K'$.

The computations are arranged in Table 31.11. Since class $AA$ loading gives much higher moments than class $A$ loading, the later has not been included in the Table. To account for the approximation the computed values of $K_0$ and $K_1$, the average live load moments have been increased by 10%. Also, to account for impact (= 10%) for class $AA$ loading, the moments have further been multiplied by the factor 1.1

**TABLE 31.10.** Computation of Final Moment $M_X$ ($M_x = M_{av} \times K'$)

| | Loading | Average moment per beam | Outer beam | | Inner beam | |
|---|---|---|---|---|---|---|
| | | | $K'$ | $M_x$ | $K'$ | $M_x$ |
| 1. | Dead load of decking Case (*i*) | 589310 | 0.9962 | 587070 | 0.9882 | 582360 |
| 2. | Class $AA$ : central Case (*ii*) | 828330 × 1.1 × 1.1 ≈ 1002280 | (0.9631) | (965300) | 1.0646 | 1067030 |
| 3. | Class $AA$ : eccentric Case (*iii*) | 828330 × 1.1 × 1.1 ≈ 1002280 | 1.3438 | 1346860 | 1.0412 | (1043570) |
| 4. | Dead load of rib | | | 384240 | | 384240 |
| | Total | | | 2318170 (N-m) | | 2033630 (N-m) |

**Note.** Total = (1) + [(2) or (3) whichever is greater] + (4).

It should be noted that maximum B.M. in the outer longitudinal girder was found to be 2393680 N-m in Example 31.2, calculated on the basis of Courbon's method. However, the max. B.M. at the inner girder was found to be 1946680 N-m only as against 2033630 N-m found above. In the detailed design of girders, moments are required at other point also, such as at $\frac{1}{4}$ and $\frac{3}{8}$ span points. Average max. moments at these points can found by influence lines, keeping in mind the guide lines dissused in Example 31.2. The average moments so found can be multiplied by $K$ to get final moments at these points.

**TABLE 31.11.** Computation of $K' = \dfrac{\Sigma Q}{\Sigma W}$

| Load position | | | $-b$ | $-3b/4$ | $-b/2$ | $-b/4$ | 0 | $+b/4$ | $+b/2$ | $+3b/4$ | $+b$ | $\Sigma W$ | $\Sigma Q$ | $K'$ |
|---|---|---|---|---|---|---|---|---|---|---|---|---|---|---|
| Case (*i*) Loading | Load W | | 475 | 680 | 648 | 639.6 | 639.6 | 639.6 | 648 | 680 | 475 | 5524.8 | | |
| | Ref. station | $-2b/3$ | 1294.9 | 1557.9 | 1186.5 | 892.9 | 619.8 | 337.7 | 86.8 | $-167.3$ | $-305.5$ | | 5505.8 | 0.9962 |
| | | 0 | 415.1 | 642.6 | 637.6 | 683.7 | 701.0 | 683.7 | 637.6 | 642.6 | 415.2 | | 5459.7 | 0.9882 |
| Case (*ii*) Loading | Load W | | 0 | 0 | 0.052 P | 0.948 P | 0 | 0.948 P | 0.052 P | 0 | 0 | 2 P | | |
| | Ref. station | $-2b/3$ | 0 | 0 | 0.0952P | 1.3234 P | 0 | 0.5005 P | 0.007 P | 0 | 0 | | 1.926 P | 0.9631 |
| | | 0 | 0 | 0 | 0.0512 P | 1.0134 P | 0 | 1.0134 P | 0.0512 P | 0 | 0 | | 2.1292 P | 1.0646 |
| Case (*iii*) Loading | Load W | | 0 | 0 | 0.924 P | 0.076 P | 0.82 P | 0.18 P | 0 | 0 | 0 | 2 P | | |
| | Ref. station | $-2b/3$ | 0 | 0 | 1.6918P | 0.1061P | 0.7946P | 0.0950P | 0 | 0 | 0 | | 2.6875P | 1.3438 |
| | | 0 | 0 | 0 | 0.9092P | 0.0812P | 0.8995P | 0.1924P | 0 | 0 | 0 | | 2.0824P | 1.0412 |
| Case (*iv*) Loading | Load W | | 0 | 0.6162P | 0.3848P | 0.7692P | 0.4616P | 0.7692P | 0.3848P | 0.6152P | 0 | 4 P | | |
| | Ref. station | $-2b/3$ | 0 | 1.4094P | 0.7046P | 1.0739P | 0.4517P | 0.4061P | 0.0516P | 0.1513P | 0 | | 3.950P | 0.989 |
| | | 0 | 0 | 0.5814P | 0.3786P | 0.8223P | 0.5113P | 0.8223P | 0.3786P | 0.5814P | 0 | | 4.0750P | 1.019 |
| Case (*v*) Loading | Load W | | 0.1796P | 0.8204P | 0.3336P | 0.6664P | 0.7952P | 0.2048P | 0.9492P | 0.0508P | 0 | 4 P | | |
| | Ref. station | $-2b/3$ | 0.4896P | 1.8795P | 0.6108P | 0.9303P | 0.7705P | 0.1081P | 0.1272P | 0.0124P | 0 | | 4.9037P | 1.2259P |
| | | 0 | 0.1546P | 0.7353P | 0.3283P | 0.7124P | 0.8723P | 0.2189P | 0.9340P | 0.0480P | 0 | | 4.0438P | 1.0110 |

**7. Computation of transverse moment ($M_y$).** Transverse moments ($M_y$) at the centre of bridge can be computed by using *Eq. 31.23*, for any combination of loads. For this it is necessary to compute influence line ordinates for $\mu_0$ and $\mu_1$ (and hence for $\mu_\alpha$) for station 0, for the placement of load in transverse direction and for values of $\theta$, $3\theta$ and $5\theta$. From these, $\Sigma \mu_\theta$, $\Sigma \mu_{3\theta}$, and $\Sigma \mu_{5\theta}$ can be computed.

For the present case, $\theta = 0.3$; $3\theta = 0.9$ and $5\theta = 0.5$; Also $b/a = 3.9/8.15 = 0.478$

The values of $\mu$ for reference station are tabulated in Table 31.12, in which only half the width has been considered since $\mu$ values for station 0 are symmetrical.

**TABLE 31.12.** Values of $\mu_0$ and $\mu_1$ for station 0

| $\theta$ | $\mu_0$ | | | | | $\mu_1$ | | | | |
|---|---|---|---|---|---|---|---|---|---|---|
| | Load position | | | | | Load position | | | | |
| | 0 | b/4 | b/2 | 3b/4 | b | 0 | b/4 | b/2 | 3b/4 | b |
| 0.3 | 0.247 | 0.122 | 0 | − 0.122 | − 0.241 | 0.227 | 0.118 | 0.032 | − 0.022 | − 0.11 |
| 0.9 | 0.132 | 0.034 | − 0.017 | − 0.042 | − 0.06 | 0.100 | 0.022 | − 0.012 | − 0.013 | − 0.017 |
| 1.5 | 0.075 | − 0.002 | − 0.018 | − 0.012 | + 0.002 | 0.060 | 0.003 | − 0.004 | − 0.003 | − 0.003 |

Coefficient $\mu_\alpha$ can be computed from the relation:

$\mu_\alpha = \mu_0 + (u_1 - u_0)\sqrt{\alpha}$. The values of $\mu_\alpha$ have been tabulated in Table 31.13.

**TABLE 31.13.** Values of $\mu_\alpha$ for station 0

| $\theta$ | Load Position | | | | |
|---|---|---|---|---|---|
| | 0 | b/4 | b/2 | 3b/4 | b |
| 0.3 | 0.243 | 0.121 | 0.006 | − 0.104 | − 0.217 |
| 0.9 | 0.126 | 0.032 | − 0.016 | − 0.037 | 0.052 |
| 1.5 | 0.072 | − 0.02 | − 0.015 | − 0.009 | 0.001 |

Table 31.13 shows that $\mu_\alpha$ is maximum when the load is nearest to the central point 0.

*(i)* $M_y$ *due to dead loads.* If the dead load were uniformly distributed across the width of the bridge there would be no transverse B.M. However, due to cantilever portions, the load is slightly more at the end $b/4$ span (see Fig. 31.78). Due to this non-uniformity some $M_y$ may be induced. In most of the cases the magnitude of $M_y$ due to this may vary between 2 to 4% of $M_y$ due to live load. Due to this reason, it may not be necessary to calculate $M_y$ due to dead loads.

*(ii)* $M_y$ *due to class AA (tracked) loading.* Fig. 31.80 (a) shows the class AA loading. For maximum moment one track of wheel is at a distance of $0.345\ b$ from centre while the other track is at $-0.481\ b$ away from centre. $\Sigma \mu_\alpha$ is found by summing $\mu_\alpha$ values for these two load positions. The values of $\mu_\alpha$ for these load positions can either be found by linear interpolation from Table 31.13 or by drawing influence lines for $\mu_\alpha$ as shown in Fig. 31.84.

The values of $\mu_\alpha$ at these load positions, derived either from Table 31.13 or Fig. 31.84 are given below.

**TABLE 31.14**

| $\theta$ | $\mu_\alpha$ | | $\Sigma \mu_\alpha$ |
|---|---|---|---|
| | Load Position | | |
| | 0.045 b | − 0.481 b | |
| 0.3 | 0.205 | 0.010 | 0.225 |
| 0.9 | 0.105 | − 0.017 | 0.088 |
| 1.5 | 0.050 | − 0.017 | 0.033 |

For single U.D.L. $w$, $M_y$ is given by Eq. 31.19 (d). For two such loads (i.e., two tracks of class AA loading), the equation is modified as follows:

$$M_y = \frac{4\,wb}{\pi}\left[\sum \mu_1\,\frac{\sin \pi c}{2a} + \frac{\sum \mu_3}{3}\sin\left(\frac{3\pi c}{2a}\right) + \frac{\sum \mu_5}{5}\sin\frac{5\pi c}{2a}\right]$$

where $\sum \mu_1 = (\sum \mu_\alpha$ for $\theta) = 0.225$
$\sum \mu_3 = (\sum \mu_\alpha$ for $3\theta) = 0.088$   and   $\sum \mu_5 = \sum \mu_\alpha$ for $5\theta) = 0.033$ (Table 31.14)

$w = \dfrac{P}{c} \times 1.1$ (including impact)

$= \dfrac{350000}{3.6} \times 1.1 \approx 107000$ N/m.

$c = \dfrac{3.6}{2} = 1.8$ m;  $b = \dfrac{7.8}{2} = 3.9$ m

$2a = 16.3$ m

$\dfrac{\pi c}{2a} = \dfrac{\pi \times 1.8}{16.3} = 0.3469$ radians.

$\therefore \sin \dfrac{\pi c}{2a} = 0.340$ ;

$\sin \dfrac{3\pi c}{2a} = 0.8628$ ;  $\sin \dfrac{5\pi c}{2a} = 0.9866$

$\therefore M_y = \dfrac{4 \times 107000 \times 3.9}{\pi} \times \left[(0.225 \times 0.340)\right.$

$\left. + \dfrac{0.088}{3} \times 0.8628 + \dfrac{0.033}{5} \times 0.9866\right]$

$= 57553$ N-m/m

(ii) $M_y$ *due to class A loading*  Fig. 31.85 (a) shows the placement of two trains of class A loading, in which wheel load 3 has been placed nearest to the centre section 0, subject to the permissible clearance. The distance of the four wheels from the centre of the span are shown. Table 31.15 gives values for $\mu_\alpha$ for each load and hence $\sum \mu_\alpha$, derived from Fig. 31.84.

The distance of the four wheels are  $-0.7949\,b$, $-0.3334\,b + 0.0512\,b$ and $+0.5127\,b$ respectively.

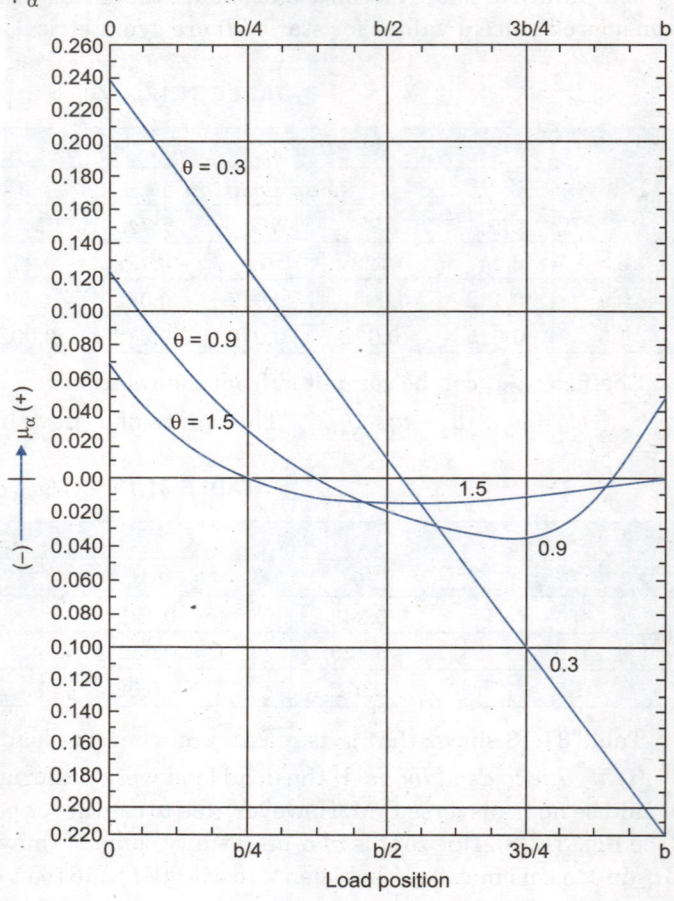

Fig. 31.84

**TABLE 31.15.**  $\sum \mu_\alpha$ For Two Trains Placed as Shown In Fig. 31.85 (a)

| $\theta$ | Position of wheel loads | | | | $\sum \mu_\alpha$ for four wheels | $\sum \mu_\alpha$ for wheels 3 and 4 |
| | Wheel 1 $-0.7949\,b$ | Wheel 2 $-0.3334\,b$ | Wheel 3 $+0.0512\,b$ | Wheel 4 $+0.5127\,b$ | | |
|---|---|---|---|---|---|---|
| 0.3 | $-0.122$ | $+0.08$ | $+0.212$ | $0.000$ | $+0.170$ | $+0.212$ |
| 0.9 | $-0.032$ | $+0.01$ | $+0.102$ | $-0.02$ | $+0.062$ | $+0.082$ |
| 1.5 | $-0.008$ | $-0.01$ | $+0.046$ | $-0.015$ | $+0.013$ | $+0.031$ |

Table 31.15 also shows $\sum \mu_\alpha$ for wheels 3 and 4, i.e., $\theta$ when only one vehicle is there. It will be seen that $\sum \mu_\alpha$ for wheels 3 and 4 is greater than $\sum \mu_\alpha$ for all the four wheels. This shows that $M_y$ will not be maximum when both the vehicles are passing ; it will be more when only one vehicles is passing.

Fig. 31.85 (b) shows only one vehicle passing with wheel 2 exactly on station 0, and wheel 1 at –0.4615 b from 0. Table 31.16 shows $\mu_\alpha$ for both the wheel positions, as well as $\Sigma \mu_\alpha$.

TABLE 31.16. $\Sigma \mu_\alpha$ for One Train

| $\theta$ | Position of wheel loads | | $\Sigma \mu_\alpha$ |
|---|---|---|---|
| | Wheel 1<br>0.4615 b | Wheel 2<br>(0) | |
| 0.3 | + 0.020 | + 0.243 | + 0.263 |
| 0.9 | – 0.012 | + 0.126 | + 0.114 |
| 1.5 | – 0.014 | + 0.072 | + 0.068 |

Comparison of Tables 31.15 and 31.16 will reveal that $\Sigma \mu_\alpha$ is maximum when only one train of loads is crossing the bridge with lateral placement as shown in Fig. 31.85 (a).

$\therefore (\Sigma \mu_\alpha)_1 = \Sigma \mu_1 = 0.263 ; (\Sigma \mu_\alpha)_3$
$= \Sigma \mu_3 = 0.114 ; (\Sigma \mu_\alpha)_5$
$= \Sigma \mu_5 = 0.068.$

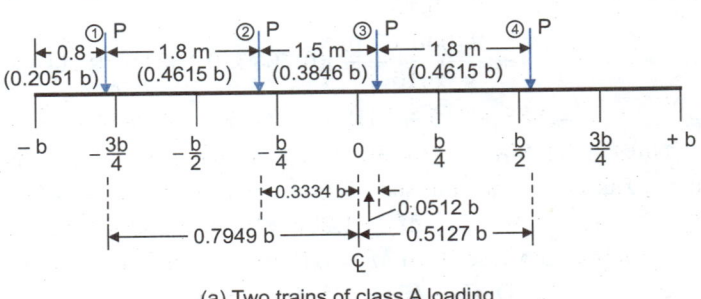

Fig. 31.86 shows the position of the axles in the longitudinal direction, with fourth axle at the middle of the span where $M_y$ is being sought. Only six axles will cover the span $2a$ = 16.3 m. It should be noted that $P$ = single wheel load of the axle, since $\Sigma \mu_\alpha$ covers the effect of two loads on an axle. $M_y$ due to one axles of loads is computed by the following equation :

$$M_y = \frac{Pb}{a}\left[\Sigma \mu_1 \cdot \frac{\sin \pi u}{2a} - \Sigma \mu_3 \sin \frac{3\pi u}{2a} + \Sigma \mu_5 \cdot \frac{5\pi u}{2a}\right]$$

where $\Sigma \mu_1, \Sigma \mu_3, \Sigma \mu_5$ take into account the effects of multiple wheels on the same axle of one or more vehicles, $u$ is the distance of the axle from the nearest abutment and $P$ is the load on each wheel of the axle. For the present case, there are six axles ; the values of $u$ and $P$ for each axle are marked in Fig. 31.86. The computations of $\sin \frac{\pi u}{2a}$ etc. for each axle are shown in Table 31.17. For the present case, $a = 8.15$ m and $b = 3.9$ m.

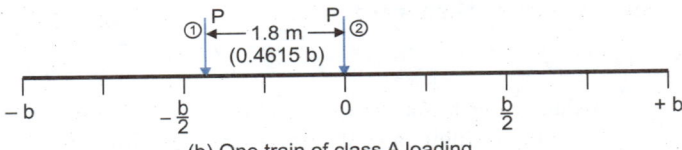

Fig. 31.86

TABLE 31.17. Computation of Sine Functions

| Axle | $u$ | $\frac{\pi u}{2a}$ (radians) | $\sin \frac{\pi u}{2a}$ | $\sin \frac{3\pi u}{2a}$ | $\sin \frac{5\pi u}{2a}$ |
|---|---|---|---|---|---|
| 1 | 2.65 | 0.5107 | 0.4888 | 0.9993 | 0.5564 |
| 2 | 3.75 | 0.7228 | 0.6615 | 0.8268 | – 0.4548 |
| 3 | 6.95 | 1.3395 | 0.9734 | – 0.7688 | 0.4026 |
| 4 | 8.15 | 1.5708 | 1.0000 | – 1.0000 | + 1.000 |
| 5 | 3.85 | 0.7420 | 0.6758 | 0.7979 | – 0.5384 |
| 6 | 0.85 | 0.1638 | 0.1631 | 0.4719 | 0.7305 |

$$M_y = \frac{13500 \times 3.9}{8.15} [(0.263 \times 0.4888) - (0.114 \times 0.9993) + (0.068 \times 0.5546)]$$

$$+ \frac{13500 \times 3.9}{8.15} [(0.263 \times 0.6615) - (0.114 \times 0.8268) + (0.068)(-0.4548)]$$

$$+ \frac{57000 \times 3.9}{8.15} [(0.263 \times 0.9734) - (0.114)(-0.7688) + (0.068 \times 0.4026)]$$

$$+ \frac{57000 \times 3.9}{8.15} [(0.263 \times 1.0) - (0.114)(-1.00) + (0.068 \times 1.00)]$$

$$+ \frac{34000 \times 3.9}{8.15} [(0.263 \times 0.6758) - (0.114 \times 0.7929) + (0.068)(-0.5384)]$$

$$+ \frac{34000 \times 3.9}{8.15} [(0.263 \times 0.1631) - (0.114 \times 0.4719) + (0.068 \times 0.7305)]$$

$$= 338 + 315 + 10120 + 12138 + 825 + 631 = 24367 \text{ N-m/m}.$$

**Note.** It is interesting to note that the effect of axles 1, 2, 5 and 6, which are away from the central station 0, is very little. Taking into account the effect of impact, using impact factor of 1.204,

$$M_y = 1.204 \times 24367 = 29338 \text{ N-m/m}.$$

This is much less than $M_y = 57540$ N-m/m found for class $AA$ loading.

∴ Design $M_y = 57540$ N-m/m.

Since the cross-beams are spaced at 8 m c/c, sagging B.M. in central cross beam = $57540 \times 8 = 460320$ N-m.

## PROBLEMS

1. A slab bridge has a clear span of 3.5 m and clear width of roadway as 4.5 m. Compute maximum B.M. and max. S.F. due to single lane of class $A$ loading, using the effective width method.

2. Design a solid slab bridge for class $A$ loading for the following data : Clear span : 5 m; Clear width of roadway : 7.5 m ; Thickness of wearing coat : 7.5 cm. Use M 20 mix.

3. For the data of problem 2, compute the maximum B.M. and maximum S.F. due to class $AA$ loading. Use effective width method.

4. (a) Write a note on 'Impact Factor' for bridges.

   (b) Explain in brief Pigeaud's method of determining B.M. in slabs, due to a wheel load.

5. A roof slab, freely supported over the four edges, has inner dimensions of 4 m × 5 m. The slab carries a load of 50 kN at its centre distributed over an area of 1 m × 1 m as shown in Fig. 31.87 (a). Compute the bending moments in the two directions using Pigeaud's curves.

6. Compute the bending moments $M_1$ and $M_2$ in the two directions, for a slab simply supported over spans of 4 m × 5 m, and loaded by two loads of 25 kN each, as shown in Fig. 31.87 (b). Each load is distributed over an area of 0.5 m × 0.5 m. Use Pigeaud's curves.

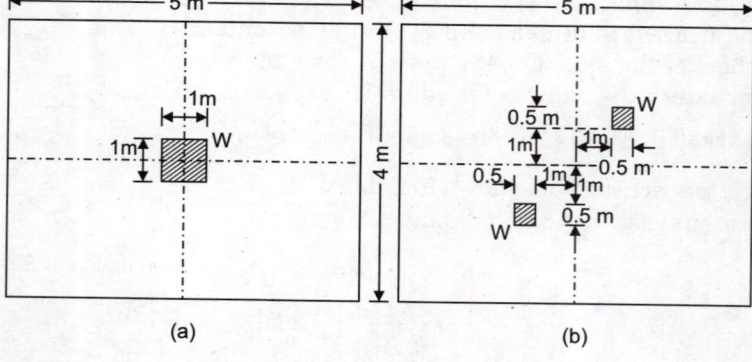

Fig. 31.87

7. Explain the method of finding reaction factors for B.M. in longitudinal girders, using Courbon's method.

8. Design a T-beam bridge for the following data : Clear width of roadway : 7.5 m. ; Span c/c of bearings : 12 m. Live load : One lane of class $AA$ or two lanes of class $A$ loading. Average thickness of wearing coat : 7.5 cm. Use M-20 mix.

9. A Road bridge deck consist of a reinforced concrete slab continuous over the beam spaced at 2 m centres and cross girders spaced at 6.2 metres thickness of wearing coat is 100 mm type of loading as per IRC class 'AA' or A whichever gives worst effect. Adopt $M_{20}$ grade Fe 415 steel Design Deck slab and sketch the details of reinforcement.

# PART – V
# LIMIT STATE DESIGN

32. DESIGN CONCEPTS
33. SINGLY REINFORCED SECTIONS
34. DOUBLY REINFORCED SECTIONS
35. T AND L-BEAMS
36. SHEAR, BOND AND TORSION IN BEAMS
37. DESIGN OF RCC BEAMS AND SLABS
38. AXIALLY LOADED COLUMNS
39. COLUMNS WITH UNIAXIAL AND BIAXIAL BENDING
40. DESIGN OF STAIR CASES
41. TWO WAY SLABS
42. CIRCULAR SLABS
43. YIELD LINE THEORY AND DESIGN OF SLABS
44. FOUNDATIONS

# CHAPTER 32: DESIGN CONCEPTS

## 32.1. METHODS OF DESIGN

The following design methods are used for the design of reinforced concrete prestressed, as well as steel structures:

(a) The working stress method   (b) Ultimate load method, and   (c) Limit state method

**(a) Working stress method of design:** The working stress method was the principal method used from early 90's. In this method, the structures are analysed by the classical *elastic theory*. The stresses in the members are considered for normal working load condition, and no attention is given to the conditions that arise at the time of structural collapse. The working loads are fixed by *limiting the stresses* in concrete and steel to a fraction of the stresses at which the material fails when tested as cubes and cylinders of concrete and bars of steel. The ratio which the yield stress of the steel or the cube strength of the concrete bears to the corresponding *permissible* or *working stress* is usually called *factor of safety*, which is actually a *stress factor of safety*. As per clause 18.2.2 of IS 456-2000 requires that the working stress method may be used only if it is not possible to adopt the limit state design method.

**(b) Ultimate load method of design:** An alternative method of design that was developed was the *ultimate load method* or the *load factor method*. It is more rational approach in which the ultimate load is some known *multiple* of the maximum working load which the structure is likely to carry. The ratio of the collapse load to the working load is known as *load factor*. The load factor gives exact margin of safety against collapse.

Since the method utilizes a large reserve of strength in plastic region (inelastic region) and of ultimate strength of member, the resulting section is very slender or thin. This gives rise to excessive deformations and cracking. Also, the method does not take into consideration the effects of creep and shrinkage. The method thus does not take into consideration *serviceability requirements* of avoidance of excessive deflection and cracking.

**(c) Limit state method of design:** Limit state design has originated from ultimate or plastic design. We have seen that while the *working stress method* gives satisfactory performance of the structure at working loads, it is unrealistic at ultimate state of collapse. Similarly, while the *ultimate load method* provides realistic assessment of safety, it does not guarantee the satisfactory serviceability requirements at service loads. An ideal method is the one which takes into account not only the ultimate strength of the structure but also the *serviceability* and *durability requirements*. The newly emerging 'limit state method'

**DESIGN CONCEPTS** 827

of design is oriented towards the simultaneous satisfaction of all these requirements. This new method makes a judicious combination of the working stress philosophy as well as the ultimate load philosophy, thus avoiding the demerits of both. *The acceptable limit of safety and serviceability requirements, before failure occurs is called a **limit state**.* IS 456-2000 code for the design of reinforced concrete structures and IS : 1343-1980 code for the design of prestressed concrete are based on the limit state design philosophy. Two prominent types of *limit states* are considered in the design:

    1. Limit state of collapse (strength limit state)

and  2. Limit state of serviceability

In India, IS Code has completely replaced the ultimate load method by the *limit state method*.

## 32.2. SAFETY AND SERVICEABILITY REQUIREMENTS (IS : 456-2000)

In the *Limit State Method*, the structure is designed to withstand *safely* all the loads liable to act on it throughout its life and also to satisfy the serviceability requirements, such as limitations to deflection and cracking. *The acceptable limit of safety and serviceability requirements before failure occurs is called a limit state*. The structure will not reach the limit state in it's life time; All the relevant limit states have to considered in the design. A structure to be designed on the basis of *limit state* should satisfy the following limit states :

**1. Limit state of collapse (Safety requirements) :** It is the limit state at which the structure as likely to collapse. The limit state of collapse of the structure or part of the structure could be assessed from rupture of one or more critical sections and from buckling due to elastic or plastic instability or overturning. The resistance to bending, shear, torsion and axial loads at every section shall not be less than the appropriate value at that section produced by the probable most unfavourable combination of loads on the structure using the appropriate partial factors.

The following *limit states of collapse* are considered in design :

    (*i*) Limit state of collapse in flexure    (*ii*) Limit state of collapse in compression

    (*iii*) Limit state of collapse in shear    (*iv*) Limit state of collapse in torsion

**2. Limit state of serviceability :** Limit state of serviceability relate to the performance of the structure at working loads. The limit states of serviceability consists of

    (*i*) Excessive deflection    (*ii*) Premature or excessive cracking.

    (*iii*) Other limit states (like Vibration, Durability, Fire Resistance).

Out of these, the important limit states of serviceability are excessive deflection and cracking.

The deflection of a structure or part thereof shall not adversely affect the appearance or efficiency of the structure or finishes or partitions. Cracking of concrete should not adversely affect the appearance or durability of the structure. The limit state of excessive deflection and crackwidth is applicable at *service loads* and is estimated on the basis of elastic analysis (working stress method). The limit state of collapse (or failure), however, depends upon *ultimate strength*.

## 32.3. CHARACTERISTIC AND DESIGN VALUES AND PARTIAL SAFETY FACTORS

All relevant limit states should be considered in design to ensure adequate degree of *safety* and *serviceability*. The structure should be designed on the basis of the most critical limit state and then checked for other limit states. For this to be achieved, the design should be based on characteristic values of applied loads, which take into account the variations in the loads to be supported. *The design values are derived from the characteristic values through the use of partial safety factors.*

## (a) MATERIAL STRENGTH

(i) **Characteristic strength of materials**: The term *characteristic strength* means that value of strength of the material below which not more than 5 percent of the results are expected to fall. The characteristic strength of concrete is denoted by $f_{ck}$ (N/mm$^2$) and its values for different grades of concrete are given in the Chapter 1. Concrete grades are specified based on this strength.

For steel the minimum yield strength (or) 0.2% proof strength is taken as the characteristic strength of steel. Until the relevant Indian Standard Specifications for reinforcing steel are modified to include the concept of characteristic strength, the characteristic value shall be assumed as minimum yield / 0.2 percent proof stress specified in the relevant Indian Standard Specifications.

(ii) **Design Values**: The design strength of the material, $F_d$ is given by

$$F_d = f / \gamma_m \qquad \qquad ...(32.1)$$

$f$ = characteristic strength of material (i.e. $f_{ck}$ for concrete and $f_y$ for steel) and

$\gamma_m$ = partial safety factor appropriate to material and limit state being considered.

(iii) **Partial safety factor $\gamma_m$**: The value of partial safety factor for material strength should account for the following parametres (i) Possibility of deviation of the strength of materials (ii) Deviation of the sectional Dimensions (iii) Accuracy of the calculation proceeding and (iv) Risk and life and economic consequences. When assessing the strength of a structure or structural member for limit state of collapse, the value of partial safety factor $\gamma_m$ should be taken as 1.5 for concrete and 1.15 for steel. Thus $\gamma_{mc}$ = 1.5 and $\gamma_{ms}$ = 1.15. It suggests that a greater variation is excepted in the strength of concrete than that in steel reinforcement.

It should be noted that $\gamma_m$ values are already incorporated in the equations and tables given in IS : 456-2000 for the limit state design. A higher value of partial safety factor ($\gamma_{mc}$) for concrete has been adopted because there are greater chances of variation of strength of concrete due to improper compaction, inadequate curing, improper batching and mixing and variations in the properties of ingradients. The chances of variations in the strength of reinforcement are known to be small, and hence a lower value ($\gamma_{ms}$ = 1.15) has been adopted. Thus, in the limit state method, the design stress ($f_{yd}$) for steel reinforcement

$$= \frac{f_y}{\gamma_{ms}} = \frac{f_y}{1.15} \simeq 0.87 f_y.$$

According to the IS Code (IS : 456-2000) for design purposes, the compressive strength of concrete in the structure shall be assumed to be 0.67 times the characteristic strength.

The partial safety factor $\gamma_{mc}$ = 1.5 shall be applied in addition to this.

## (b) IMPOSED LOADS

(i) **Characteristic loads**: The term *characteristic load* means that value of load which has a 95 percent probability of not being exceeded during the life of the structure. Since sufficient data are not available to express loads in statistical terms, the loads given in various relevant Codes shall be assumed as the characteristic loads.

(ii) **Design loads**: The design load $F_d$ is given by : $F_d = F(\gamma_f)$ ...(32.2)

where $F$ = characteristic load and $\gamma_f$ = *partial safety factor* appropriate to the nature of loading and the limit state being considered.

(iii) **Partial safety factor ($\gamma_f$)**: The partial safety factors ($\gamma_f$) for loads shall be as given in Table 32.1 in accordance with clause 36.4 of the code.

IS : 875–1987 Design loads other than earthquake loads.
IS : 875–1987 Part-I – Dead loads (DL)
IS : 875–1987 Part-II – Imposed loads (IL)
IS : 875–1987 Part-III – Wind loads (WL)
IS : 875–1987 Part-IV – Snow loads (SL)
IS : 875–1987 Part-V – Special load and load combinations

**TABLE 32.1.** Partial Safety Factor $\gamma_f$ For Loads

| Load Combinations | Limit state of collapse | | | Limit state of serviceability | | |
|---|---|---|---|---|---|---|
| | DL | LL | WL | DL | LL | WL |
| (1) | (2) | (3) | (4) | (5) | (6) | (7) |
| DL + LL | 1.5 | 1.5 | – | 1.0 | 1.0 | – |
| DL + WL | 1.5 or 0.9* | – | 1.5 | 1.0 | – | 1.0 |
| DL + LL + WL | 1.2 | 1.2 | 1.2 | 1.0 | 0.8 | 0.8 |

**Note:** While considering earthquake effects, substitute EL for WL. (*This value is to be considered when stability against overturning or stress reversal is critical).

## Comparison with Working Stress Method:

For the *working stress method*, we have

$$\gamma_f = 1.0 \text{ for all loads} \quad \text{and} \quad \gamma_{ms} = \frac{f_y}{\sigma_{st}} = \frac{250}{140} = 1.79 \text{ for mild steel bars}$$

and

$$\gamma_{ms} = \frac{f_y}{\sigma_{st}} = \frac{415}{230} = 1.80 \quad \text{for HYSD bars of grade Fe 415.}$$

Thus average $\gamma_{ms} = \frac{1}{2}(1.79 + 1.80) = 1.795$.

Also, $\gamma_{mc} = \frac{f_{cu}}{\sigma_{cbc}} \approx 3.0$ for all grades of concrete.

Hence $\gamma_f \times \gamma_{ms} = 1.0 \times 1.795 = 1.795$ for reinforcement
and $\gamma_f \times \gamma_{mc} = 1.0 \times 3.0 = 3.0$ for concrete.

For the *limit state design*, taking maximum compressive stress in concrete = $0.67 f_{cu}$, and $\gamma_f = 1.5$; $\gamma_{ms} = 1.15$ and $\gamma_{mc} = 1.5$, we have, $\gamma_f \times \gamma_{ms} = 1.5 \times 1.15 = 1.73$ for reinforcement which is slightly less than 1.795 (for working stress design), and $\gamma_f \times \gamma_{mc} = 1.5 \times \frac{1.5}{0.67} = 3.36$ for concrete, which is more than 3.0 (for working stress design).

## Shrinkage creep and temperature (high (or) low)

May produce stresses and cause deformations like other loads and forces. Hence these are also considered as loads which are time dependent. The safety and serviceability of structures are to be checked following the stipulations of clause 6.2.4, 5 and 6 of IS : 456-2000 and Part (5) of IS : 875-1987.

# 33. SINGLY REINFORCED SECTIONS

## 33.1. LIMIT STATE OF COLLAPSE IN FLEXURE

As stated in the previous chapter, the following *limit states of collapse* are considered in design: (*i*) Limit state of collapse in flexure. (*ii*) Limit state of collapse in compression. (*iii*) Limit state of collapse in shear. (*iv*) Limit State of collapse in torsion.

Design for the limit state of collapse in flexure is based on the following assumptions. These assumptions are valid only for shallow beams where the span to depth ratio is greater than 2.5.

### Assumptions in limit state of collapse in flexure

(*a*) Plane sections normal to the axis remain plane after bending.

(*b*) Maximum strain in concrete at the outermost compression fibre is taken as 0.0035 in bending.

(*c*) The relationship between the compressive stress distribution in concrete and the strain in concrete may be assumed to be rectangular, trapezoidal, parabolic or of any other shape which results in the prediction of strength in substantial agreement with the results of tests. An acceptable stress-strain curve is given in Fig. 33.1. For design purposes, the *compressive strength of concrete in structure* shall be assumed to be 0.67 times the characteristic strength. The partial safety factor $\gamma_{mc} = 1.5$ shall be applied in addition to this.

(*d*) The tensile strength of the concrete is ignored.

(*e*) The stresses in the reinforcement are derived from representative stress-strain curve for the type of steel used. For design purposes, the partial safety factor $\gamma_{ms} = 1.15$ shall be applied.

(*f*) The maximum strain in the tension reinforcement in the section at failure shall not be less than

$$\frac{f_y}{1.15 \, E_s} + 0.002$$

where $f_y$ = characteristic strength of steel, and $E_s$ = Modulus of Elasticity of steel.

**Comments on the above assumptions:** Assumption (*a*) implies linear variation of strain across the depth of the section (See Fig. 33.5). This has been verified by numerous tests on reinforced concrete beams. In assumption (*b*), the maximum strain in concrete is the strain at which the section reaches its maximum moment capacity. The value of such a strain is liable to large variations depending upon rate of deformation, loading time, grade of concrete, shape of cross-section, percentage of reinforcement etc. The Code adopts a value of 0.0035 which will give conservative results in most cases of pure bending. Assumption (*c*) giving acceptable stress-strain curve in concrete (Fig. 33.1) gives the shape of compressive stress distribution in concrete. Such a stress distribution curve across the depth, popularly known as the *stress block* is a combination of a parabola and a rectangle [See Fig. 33.5 *b* (*iii*)]. Assumption (*d*) is similar to the one made for the working stress method, in which the tensile strength of concrete is neglected. Assumption (*e*) is with reference to stress strain curve for steel reinforcement. For cold twisted bars which do not have definite yield point, the characteristic strength $f_y$ is taken corresponding to 0.2% proof strain. Assumption (f) is intended to ensure ductile failure, that is, the tensile reinforcement at the critical section has to undergo a certain degree of inelastic deformation before the concrete fails in compression. In the expression for minimum value of maximum strain in steel, laid down by the Code, the first term, $f_y/1.15\,E_s$, corresponds to the yield stress $f_y$ for the case of mild steel and other types which have well defined yield point. However, the cold twisted steel (HYSD bars) do not have a well defined yield point and the yield stress $f_y$ is taken as the conventional value of 0.2 percent proof stress. Hence a strain of 0.002 is added so that there is sufficient yielding of steel before failure at a constant stress. It is to be noted that assumption (b) and (f) govern the maximum depth of neutral axis in flexural members.

## 33.2. STRESS STRAIN RELATIONSHIP FOR CONCRETE

The Code permits the use of any appropriate curve for relationship between the compressive stress and strain distribution in concrete, subject to the condition that it results in the prediction of strength in substantial agreement with test results. An acceptable stress-strain curve (Fig. 33.1) given in Fig. 21 of the Code forms the basis of the design in this book. The compressive strength of concrete in the structure is assumed to be $0.67\,f_{ck}$ (curve 2 of Fig. 33.1). The 0.67 factor is introduced to account for difference in strength indicated by a *cube test* and the strength of concrete *in structure*. The initial portion of the curve is parabolic. At a strain of 0.002 (*i.e.* 0.2% strain) the stress remains constant with increasing load, until a strain of 0.35% is reached when the concrete is said to have failed. The design curve (curve 3 of

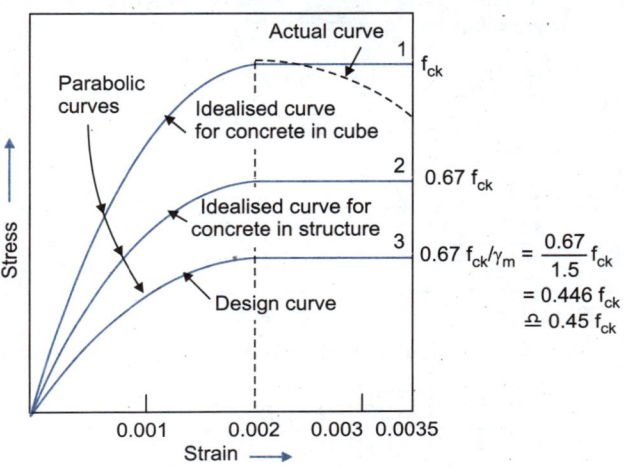

Fig. 33.1. Stress Strain Curve for Concrete

Fig. 33.1) is obtained by using a partial safety factor $\gamma_m = 1.5$. Thus the maximum compressive stress in concrete for design purpose is equal to

$$\frac{0.67}{\gamma_m} f_{ck} = \frac{0.67}{1.5} f_{ck} = 0.446\, f_{ck} \approx 0.45\, f_{ck}.$$

## 33.3. STRESS-STRAIN RELATIONSHIP FOR STEEL

The stress-strain curve for mild steel is shown in Fig. 33.2 while the stress strain curve for Fe 415 and Fe 500 steel are shown in Fig. 33.3 and Fig. 33.4 respectively. The modulus of elasticity of all types of reinforcing steel, $E_s$, is taken as $2 \times 10^5$ N/mm². Also, *the stress strain relationship for steel in tension and compression is assumed to be the same*.

For mild steel, the stress is proportional to strain upto yield point and thereafter the strain increases at constant stress. The design yield stress of steel ($f_{yd}$) is equal to $f_y/\gamma_{ms}$. With a value of 1.15 for $\gamma_{ms}$, the design yield stress $f_{yd}$ becomes $0.87\,f_y$. For cold worked bars (Fe 415 and Fe 500 HYSD bars), there is no definite yield point; hence yield stress is taken as 0.2 percent proof stress. The stress-strain curves (Fig. 33.3 and Fig. 33.4) for these two types of steels are linear upto a stress of $0.8\,f_y$ and the strains are elastic. Thereafter, the stress-strain curves are defined as given below:

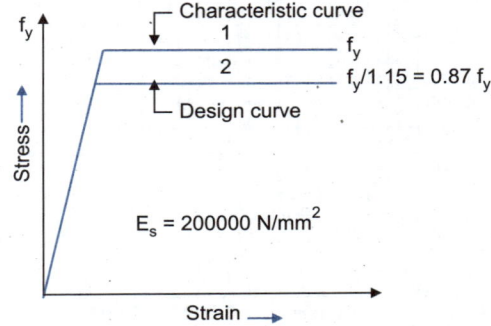

Fig. 33.2. Stress Strain Curve for Mild steel

| Stress | Inelastic strain | Stress | Inelastic strain |
|---|---|---|---|
| $0.80\, f_y$ | Nil | $0.95\, f_y$ | 0.0007 |
| $0.85\, f_y$ | 0.0001 | $0.975\, f_y$ | 0.001 |
| $0.90\, f_y$ | 0.0003 | $1.0\, f_y$ | 0.002 |

The stress-strain curve for design purposes is obtained by substituting $f_{yd}$ for $f_y$ in the above, where $f_{yd} = f_y/\gamma_{ms} = f_y/1.15 \cong 0.87\, f_y$. The salient points on the design stress-strain curve for two types of cold-worked steel are given in Table 33.1.

**TABLE 33.1.** Salient Points on the Design Stress-Strain Curve for Cold Worked Bars
($f_{yd}$ = Design yield stress; $f_y$ = Characteristic stress)

| Stress level | Fe 415 steel : $f_y$ = 415 N/mm² | | Fe 500 steel : $f_y$ = 500 (N/mm²) | |
|---|---|---|---|---|
| | Strain | Stress (N/mm²) | Strain | Stress (N/mm²) |
| $0.8\, f_{yd}$ | 0.00144 | 288.7 | 0.00174 | 347.8 |
| $0.85\, f_{yd}$ | 0.00163 | 306.7 | 0.00195 | 369.6 |
| $0.90\, f_{yd}$ | 0.00192 | 324.8 | 0.00226 | 391.3 |
| $0.95\, f_{yd}$ | 0.00241 | 342.8 | 0.00277 | 413.0 |
| $0.975\, f_{yd}$ | 0.00276 | 351.8 | 0.00312 | 423.9 |
| $1.0\, f_{yd}$ | 0.00380 | 360.9 | 0.00417 | 434.8 |

**Note 1.** $f_{yd} = f_y/\gamma_{ms} = f_y/1.15 \cong 0.87\, f_y$.

**Note 2.** Linear interpolation may be made for intermediate value.

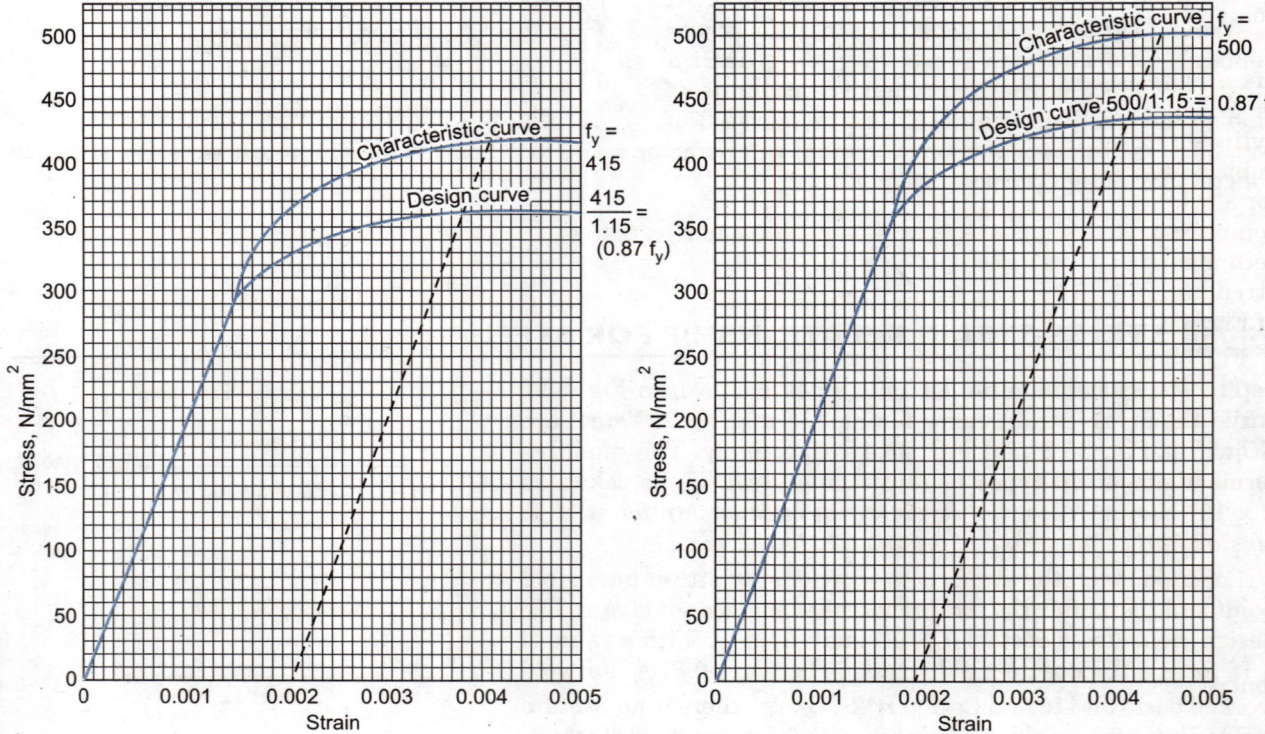

Fig. 33.3. Stress Strain Curve for Fe 415 Steel.

Fig. 33.4. Stress Strain Curve for Fe 500 Steel.

## 33.4. STRESS BLOCK PARAMETERS

The various theories advanced for estimating the ultimate flexural resistance of reinforced concrete sections can be broadly classified into two categories. In the first category of theories (such as those of Suenson, Menaeh, Kempton, Dyson and Whitney) no assumptions are made regarding the distribution of strain across the section, and neutral axis is located empirically, for analysing compression failures. These theories are based on equilibrium only.

The second category of theories (such as those by Stussi or Indian Standard) *include the strain compatibility*. A limiting strain in concrete is used as a criterion for the crushing of concrete in flexural compression.

The stress-strain behaviour of concrete under compression is generally obtained from cylinder or cube of concrete

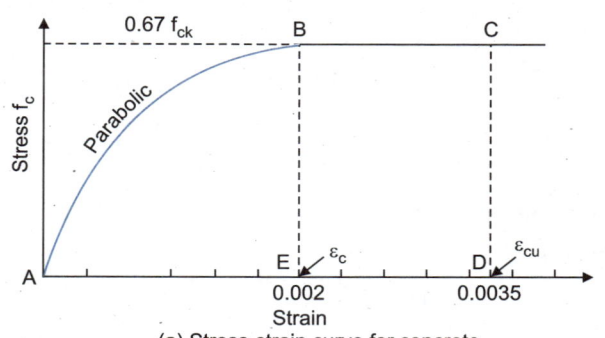

(a) Stress-strain curve for concrete

(i) Section  (ii) Strain diagram  (iii) Stress block
(b) Strain and stress variations across section

Fig. 33.5

subjected to longitudinal compressive loading. Whereas the stress and strains are uniform for a cube (or for a cylinder), they vary across the depth of a flexural member. The ratio between the maximum flexural compression in a beam and the crushing strength of a cube is likely to be different from unity. The I.S. Code recommends the compressive strength of concrete in the structure equal to 0.67 times the characteristic strength. The corresponding stress-strain diagram is shown in Fig. 33.5(a), which is the same as curve 2 of Fig. 33.1.

Fig. 33.5(*bii*) shows the strain diagram, while Fig. 33.5(*biii*) shows the variation of stresses across the depth. The stress diagram *ABCDEA* [Fig. 33.5 (*biii*)] has parabolic shape from *A* to *B* (corresponding to the curve *AB* of Fig. 33.5(a), and then linear from *B* to *C* at a constant stress of $0.67 f_{ck}$ corresponding to line *BC* of Fig. 33.5(a). The total compressive force $C_u$ and its location below the top fibre can be expressed in terms of stress block factors $k_1$, $k_2$ and $k_3$ as follows:

$k_1$ = shape factor, defined as the ratio between the area of stress block *ABCD* and area of rectangle *AFCD*

$$k_1 = \frac{\text{Area } ABCD}{\text{Area } AFCD} \quad \text{or} \quad k_1 = \frac{\text{Area } ABCD}{\text{Area } (x \times DC)} \quad \text{[Fig. 33.5(biii)]}$$

Now ultimate strain in concrete $\varepsilon_{cu} = AD = 0.0035$ [Fig. 33.5(a)] while the strain $\varepsilon_c = AE$, after which concrete yields at constant stress of $0.67 f_{ck}$, is found to be equal to 0.002 for all grades concrete. Ratio

$$\frac{AE}{AD} = \frac{\varepsilon_c}{\varepsilon_{cu}} = \frac{0.002}{0.0035} = \frac{4}{7}.$$

Also,

$$\frac{ED}{AD} = \frac{0.0035 - 0.002}{0.0035} = \frac{0.0015}{0.0035} = \frac{3}{7}$$

∴

$$AE = \frac{4}{7} AD. \quad \text{and} \quad ED = \frac{3}{7} AD.$$

Now  Area $ABCD$ = Area $ABE$ + Area $BCDE$

$$= \left(\frac{2}{3} \times AE \times BE\right) + (ED \times CD) = \left(\frac{2}{3} \times \frac{4}{7} AD \times CD\right) + \left(\frac{3}{7} AD \times CD\right)$$

$$= \frac{8}{21}(AD \times CD) + \frac{9}{21}(AD \times CD) = \frac{17}{21} \times \text{Area } AFCD$$

$\therefore \quad k_1 = \dfrac{17}{21} = 0.8095 \approx 0.81$ ...(33.1)

The resultant compressive force is located at a depth $k_2 \cdot x$ below the top fibre, where $x$ is the depth of the neutral axis.

Now  $k_2 \cdot AD = \dfrac{(\text{Area } ABE) \times \bar{x}_1 + (\text{Area } BCDE) \times \bar{x}_2}{\text{Area } ABCD}$

$$k_2 \cdot AD = \frac{\dfrac{8}{21}\left(ED + \dfrac{3}{8} AE\right) + \dfrac{9}{21}\left(\dfrac{1}{2} DE\right)}{\dfrac{17}{21}} = \frac{\dfrac{8}{21}\left(\dfrac{3}{7} + \dfrac{3}{8} \times \dfrac{4}{7}\right) AD + \dfrac{9}{21}\left(\dfrac{1}{2} \times \dfrac{3}{7}\right) AD}{\dfrac{17}{21}}$$

or  $k_2 = \dfrac{\dfrac{36}{7} + \dfrac{27}{14}}{17} = \dfrac{99}{238} = 0.416 \approx 0.42.$ ...(33.2)

Hence resultant compressive force is located at a depth of $k_2 x$ (= 0.416 $x$) below the top fibre. Let the maximum ordinate ($DC$) of the stress block be $k_3 f_{ck}$. This is the maximum compressive stress in flexure on the structure, in relation to characteristic strength $f_{ck}$. IS : 456–2000 recommends $k_3 = 0.67$. The compressive force $C_u$ in concrete as given by

$C_u = b \times \text{area } ABCD$ [Fig. 33.5($biii$)] $= k_1 \times x \times DC \times b = k_1 k_3 f_{ck} \cdot bx$ ...(33.3)

The moment of resistance is given by

$$M_u = C_u(d - k_2 x) = k_1 k_3 f_{ck} \cdot bd^2 \left(\frac{x}{d}\right)\left(1 - k_2 \frac{x}{d}\right)$$ ...(33.4)

## 33.5. DESIGN STRESS BLOCK PARAMETERS (IS : 456 – 2000)

The general expressions for $C_u$ and $M_u$ derived above can be easily modified by incorporating partial safety factors $\gamma_{mc}$ and $\gamma_{ms}$ for concrete and steel. The stress strain curve for concrete is corresponding to curve 3 of Fig. 33.1 while the stress strain curve for steel are corresponding to curves 2 of Fig. 33.2, 33.3 and 33.4. Fig. 33.6(b) shows the strain diagram while Fig. 33.6(c) shows the stress block *for design* as recommended by IS : 456 – 2000.

The maximum stress  $DC = \dfrac{k_3 f_{ck}}{\gamma_{mc}} = \dfrac{0.67 f_{ck}}{1.5}$

$= 0.446 f_{ck} \approx 0.45 f_{ck}.$

Factor $k_2 = 0.416 \triangleq 0.42$ while depth of neutral axis is $x_u$. The strain $\varepsilon_{cu}$ in concrete = 0.0035.

The maximum stress in steel is limited to $\dfrac{f_y}{\gamma_{ms}} = \dfrac{f_y}{1.15} \approx 0.87 f_y$

Area $c_u$ of design stress block is given by

$$c_u = \text{area } ABCD = k_1 \times x_u \times DC = k_1 \times x_u \times \frac{k_3 f_{ck}}{\gamma_{mc}}$$

Substituting  $k_1 = 0.8095$, $k_3 = 0.67$ and $\gamma_{mc} = 1.5$,
we get  $c_u = 0.36 f_{ck} \cdot x_u$ ...[33.5($a$)]

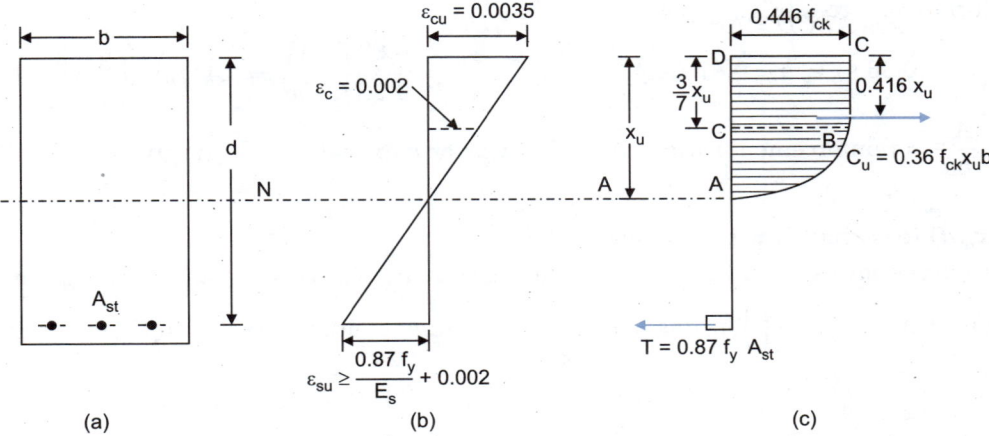

Fig. 33.6. Strain and Stress Distribution for Design (IS: 456—2000)

Total compressive force for a rectangular beam of width $b$ is given by
$$C_u = 0.36 f_{ck} \cdot x_u \cdot b \qquad \ldots(33.5)$$
The maximum strain $\varepsilon_{cu}$ in concrete is taken equal to 0.0035.

The Code requires that at failure the strain in reinforcement ($\varepsilon_{su}$) should not be less than $\left(\dfrac{0.87 f_y}{E_s} + 0.002\right)$, thus ensuring a stress of $\dfrac{f_y}{1.15}$ ($\approx 0.87 f_y$). This limits the depth $x_u$ of N.A. to its maximum or limiting value. The depth $x_u$ of the N.A. is given by
$$x_u = \dfrac{\varepsilon_{cu}}{\varepsilon_{cu} + \varepsilon_{su}} d \qquad \ldots(33.6)$$

Substituting minimum value of $\varepsilon_{su} = 0.002 + \dfrac{0.87 f_y}{E_s}$, we get

$$\dfrac{x_{u(max.)}}{d} = \dfrac{0.0035}{0.0055 + \dfrac{0.87 f_y}{E_s}} = \dfrac{0.0035 E_s}{0.0055 E_s + 0.87 f_y} \qquad \ldots[33.7(a)]$$

Substituting $E_s = 2 \times 10^5$ N/mm², we get $\dfrac{x_{u(max)}}{d} = \dfrac{700}{1100 + 0.87 f_y} \qquad \ldots(33.7)$

The limiting values of the depth of the neutral axis, for different grades of steel, based on the above expression are as follows

| $f_y$ | 250 | 415 | 500 |
|---|---|---|---|
| $\dfrac{x_{u(max.)}}{d}$ | 0.53 (0.531) | 0.48 (0.479) | 0.46 (0.456) |

**Note:** The values of $x_{u\cdot max}/d$ in the bracket are the exact values.
The above limiting values are corresponding to assumptions (para $f$) of the IS Code.

## 33.6. SINGLY REINFORCED RECTANGULAR BEAMS

The total compressive force $C_u$ in a rectangular beam, singly reinforced, is given by *Eq. 33.5*, while the total tensile force $T_u$ in reinforcement will be equal to $0.87 f_y \cdot A_{st}$.

## 836 REINFORCED CONCRETE STRUCTURE

For equilibrium of forces [Fig. 33.6(c)], $C_u = T_u$

$$\therefore \quad 0.36 f_{ck} x_u b = 0.87 f_y \cdot A_{st} \quad \text{or} \quad \frac{x_u}{d} = \frac{0.87 f_y A_{st}}{0.36 f_{ck} bd} = 2.417 p_t \frac{f_y}{f_{ck}} \quad ...(33.8)$$

where $p_t = \dfrac{A_{st}}{bd}$ = reinforcement ratio and the limiting or max. value of $\dfrac{x_u}{d}$ is given by Eq. 33.7.

### Three cases may arise.

**Case 1: ($x_u/d$) less than limiting value**

The case corresponds to under-reinforced section in working stress method, in which the reinforcement ratio $p_t \left( = \dfrac{A_{st}}{bd} \right)$ is less than $p_{t\ lim.}$ corresponding to the limiting value $\dfrac{(x_{u,max.})}{d}$,

where $\quad p_{t\ lim.} = \dfrac{1}{2.417} \dfrac{f_{ck}}{f_y} \cdot \dfrac{x_{u,max.}}{d} \quad$ or $\quad p_{t\ lim.} = 0.414 \dfrac{f_{ck}}{f_y} \cdot \dfrac{x_{u,max.}}{d} \quad ...(33.9)$

The values of $p_{t\ lim.}$ for various values of $f_{ck}$ and $f_y$ are given in Table 33.2.

**TABLE 33.2.** Value of $p_{t,\ lim.}$ (%)

| Grade of concrete | $f_{ck}$ (N/mm²) | Values of 100 $p_{t,\ lim}$ for | | |
|---|---|---|---|---|
| | | $f_y$ = 250 N/mm² | $f_y$ = 415 N/mm² | $f_y$ = 500 N/mm² |
| M 15 | 15 | 1.319 | 0.716 | 0.566 |
| M 20 | 20 | 1.758 | 0.955 | 0.754 |
| M 25 | 25 | 2.198 | 1.194 | 0.943 |
| M 30 | 30 | 2.638 | 1.433 | 1.132 |
| M 35 | 35 | 3.077 | 1.672 | 1.320 |
| M 40 | 40 | 3.517 | 1.910 | 1.509 |

**Note:** According to ACI and Australian Codes, normal structures should have $p_{t,\ max.} \not> 0.75\ p_{t,\ lim.}$. The moment of resistance for such a case is conveniently expressed as

$$M_u = T_u (d - 0.416 x_u)$$

where $T_u = 0.87 f_y A_{st} = 0.87 f_y \cdot p_t\ bd$

$$\therefore \quad M_u = 0.87 f_y A_{st} (d - 0.416 x_u) = 0.87 f_y A_{st} \cdot d \left( 1 - 0.416 \frac{x_u}{d} \right) \quad ...[33.10(a)]$$

Substituting the value of $\dfrac{x_u}{d}$ from Eq. 33.8, we get

$$M_u = 0.87 f_y A_{st} \cdot d \left( 1 - 0.416 \frac{0.87 f_y}{0.36 f_{ck}} \cdot \frac{A_{st}}{bd} \right) \approx 0.87 f_y A_{st} \cdot d \left( 1 - \frac{f_y}{f_{ck}} \cdot \frac{A_{st}}{bd} \right) \quad ...(33.10)$$

**Case 2: ($x_u/d$) equal to limiting value:** This case corresponds to *balanced design* in which the steel reinforcement reaches its yield stress at the same instant when ultimate strain is reached in concrete. The stress block is fully developed as shown in Fig. 33.6(c). The moment of resistance $M_u$ reaches its limiting value $M_{u,\ limit}$, and given by

$$M_{u,\ lim.} = 0.36 f_{ck} \cdot (x_u)_{max.}\ b (d - 0.416\ x_{u(max.)})$$

or $\quad M_{u,\ lim.} = 0.36 f_{ck} \cdot \dfrac{x_{u(max.)}}{d} \left( 1 - 0.416 \dfrac{x_{u,max.}}{d} \right) bd^2 \quad ...(33.11)$

**Case 3: ($x_u/d$) greater than the limiting value:** The case corresponds to the beam in which the percentage of steel is sufficient to ensure that steel yield does not take place. Failure occurs when the strain in extreme fibres in concrete reaches its ultimate value. Thus the failure takes place due to crushing of concrete while strain $\varepsilon_s$ in steel still remains below the yield strain $\varepsilon_{su}$. It should be noted that compression

# SINGLY REINFORCED SECTIONS

failure is *sudden,* and therefore not desirable. The Code recommends that if $(x_u/d)$ is found greater than the limiting value, *the beam should be redesigned.*

## 33.7. PROCEDURE FOR FINDING MOMENT OF RESISTANCE

The moment of resistance of rectangular section without compression reinforcement should be obtained as follows (IS : 456–2000):

(a) Determine the depth of neutral axis from the following equation:

$$\frac{x_u}{d} = \frac{0.87 f_y \cdot A_{st}}{0.36 f_{ck} \, bd} = 2.417 \frac{f_y}{f_{ck}} \cdot \frac{A_{st}}{bd} \qquad \ldots(33.8)$$

(b) If the value of $x_u/d$ is less than the limiting (*i.e.* max.) value given by Eq. 33.7, calculate the moment of resistance by the following equation.

$$M_u = 0.87 f_y \cdot A_{st} \cdot d \left(1 - \frac{A_{st}}{bd} \cdot \frac{f_y}{f_{ck}}\right) \qquad \ldots(33.10)$$

(c) If the value of $x_u/d$ is equal to the limiting value, the moment of resistance of the section is given by the following expression:

$$M_{u \, . \, lim.} = 0.36 f_{ck} \frac{x_{u(max.)}}{d}\left(1 - 0.416 \frac{x_{u(max.)}}{a}\right) bd^2 \qquad \ldots(33.11)$$

(d) If $x_u/d$ is greater than the limiting value, the section is *over-reinforced*. The moment of resistance of such a section is limited to $M_{u(lim.)}$ given by Eq. 33.11. It should be noted that in the case of an over-reinforced section, not only that there is wastage of excess steel, the concrete reaches its ultimate capacity before steel yields resulting in a sudden failure. *The Code therefore recommends that such a section should be redesigned.*

## VARIATION OF MOMENT OF RESISTANCE WITH $p_t$ :

The variation of moment of resistance $M_u$ with reinforcement ratio $p_t$ is given by

$$\frac{M_u}{bd^2} = 0.87 f_y p_t \left[1 - \frac{f_y}{f_{ck}} p_t\right] \qquad \ldots(33.12)$$

Thus, $M_u$ increases with the increase in the value of $p_t$. When $p_t$ reaches limiting value $p_{t,\,lim.}$ given by Eq. 33.9, $M_u$ also reaches the limiting or maximum value $M_{u,lim}$ given by Eq. 33.11, rearranged as follows:

$$\frac{M_{u, lim.}}{bd^2} = R_{u, lim.}$$

$$= 0.36 f_{ck} \frac{x_{u, max.}}{d}\left(1 - 0.416 \frac{x_{u, max.}}{d}\right)$$

For Fe 415 steel having $f_y = 415$ N/mm², we have

$$\frac{x_{u, max.}}{d} = 0.479.$$

Hence, $p_{t, lim.} = \dfrac{0.414}{f_y} \cdot \dfrac{x_{u, max.}}{d} \cdot f_{ck}$

$= \dfrac{0.414}{415} \times 0.479 f_{ck} = 4.778 \times 10^{-4} f_{ck}$

For M 20 concrete

$$p_{t, lim.} = 4.778 \times 10^{-4} \times 20 = 9.56 \times 10^{-3}$$

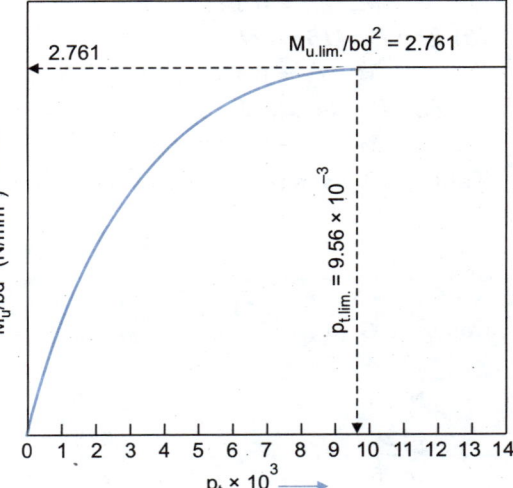

Fig. 33.7. Variation of $M_u/bd^2$ with $p_t$ (M 20 Concrete; Fe 415 Steel)

Also, $\dfrac{M_{u, lim.}}{bd^2} = 0.36 \times 20 \times 0.479 (1 - 0.416 \times 0.479) = 2.761 = R_{u, lim.}$

For other values of $p_t$, $M_u/bd^2$ can be computed from Eq. 33.12.

Thus, $\dfrac{M_u}{bd^2} = 0.87 f_y p_t \left(1 - \dfrac{f_y}{f_{ck}} p_t\right) = 0.87 \times 415 p_t \left(1 - \dfrac{415}{20} p_t\right)$

or $\dfrac{M_u}{bd^2} = 361.05 \, p_t (1 - 20.75 \, p_t)$

The values of $M_u/bd^2$ can thus be computed for different values of $p_t$, ranging from $p_t = 0$ to $p_t = p_{t, lim.} = 9.56 \times 10^{-3}$. Fig. 33.7 shows the variation of $M_u/bd^2$ with $p_t$ for Fe 415 steel and M 20 concrete.

## LIMITING MOMENT OF RESISTANCE

In a balanced section, the steel reinforcement reaches its yield stress at the same instant when the ultimate strain is reached in the concrete. The stress block is fully developed as shown in Fig. 33.6(c). The moment of resistance $M_u$ reaches its limiting value $M_{u, lim.}$ which can be determined by taking the moment of compressive force about the centre of tensile reinforcement.

Now $\qquad C_u = (b \cdot x_{u, max.}) \times (0.36 f_{ck}) = 0.36 f_{ck} b \cdot x_{u, max.}$

Lever arm $= d - 0.416 \, x_{u, max.}$

$\therefore \qquad M_{u, lim.} = 0.36 f_{ck} b \cdot x_{u, max.} (d - 0.416 \, x_{u, max.})$

or $\qquad M_{u, lim.} = 0.36 f_{ck} \dfrac{x_{u, max.}}{d} \left(1 - 0.416 \dfrac{x_{u, max.}}{d}\right) bd^2$

or $\qquad \dfrac{M_{u, lim.}}{bd^2} = R_{u, lim.} = 0.36 f_{ck} \cdot \dfrac{x_{u, max.}}{d} \left(1 - 0.416 \dfrac{x_{u, max.}}{d}\right) \qquad \ldots(33.11)$

In the above expression, the values of $x_{u, max.}/d$ for mild steel, Fe 415 steel and Fe 500 steel are 0.531, 0.479 and 0.456 respectively. Substituting these values, we get the following simplified expressions for the three types of steel.

**(a) For mild steel (Fe 250):** $x_{u, max.}/d = 0.531$

$M_{u, lim.} = 0.1489 f_{ck} bd^2 = R_{u, lim.} bd^2 \qquad$ (where $R_{u, lim.} = 0.1489 f_{ck}$) $\qquad \ldots[33.11(a)]$

**(b) For Fe 415 steel:** $x_{u, max.}/d = 0.479$

$M_{u, lim.} = 0.1381 f_{ck} bd^2 = R_{u, lim.} bd^2 \qquad$ (where $R_{u, lim.} = 0.1381 f_{ck}$) $\qquad \ldots[33.11(b)]$

**(c) For Fe 500 steel:** $x_{u, max.}/d = 0.456$

$M_{u, lim.} = 0.1330 f_{ck} bd^2 = R_{u, lim.} bd^2 \qquad$ (where $R_{u, lim.} = 0.1330 f_{ck}$) $\qquad \ldots[33.11(c)]$

Table 33.3 gives the values of $R_{u, lim.}$ for various grades of concrete.

**TABLE 33.3.** Values of $R_{u, lim.}$ for Balanced Design

| Concrete Grade | Reinforcement | | |
|---|---|---|---|
| | Fe 250 ($R_{u, lim.} = 0.1489 f_{ck}$) | Fe 415 ($R_{u, lim.} = 0.1381 f_{ck}$) | Fe 500 ($R_{u, lim.} = 0.1330 f_{ck}$) |
| M 15 | 2.233 | 2.071 | 1.995 |
| M 20 | 2.978 | 2.761 | 2.660 |
| M 25 | 3.722 | 3.452 | 3.325 |
| M 30 | 4.467 | 4.142 | 3.990 |
| M 35 | 5.211 | 4.833 | 4.655 |
| M 40 | 5.956 | 5.523 | 5.320 |

The readers are advised to compare these values of $R_{u,\,lim.}$ with $1.5\,R$ values of working stress method. It will be seen that $R_{u,\,lim.}$ is greater than $1.5\,R$, resulting in smaller concrete section. *It will be seen from Table 33.2 for $p_{t,\,lim.}$ that the balanced design gives smallest concrete section and maximum area of reinforcement*. Since the cost of steel is quite high, an *under-reinforced section*, which gives larger concrete area and smaller value of $p_t$, is always desirable in the limit state design. This can be achieved by (i) providing $b$ or $d$ or both greater than those required by balanced design, or  (ii) using lesser values of $x_u/d$ than the limiting value, or  (iii) assuming lesser value of $R_{u,\,lim}$ than the limiting value given in Table 33.3.

## 33.8. DESIGN OF RECTANGULAR BEAM SECTION

The design of a section consists of determination of  (i) cross-sectional dimensions $b$ and $d$, and  (ii) area of steel, so as to develop a given moment of resistance. Though the objective of a designer would be to design a *balanced section* so that the ultimate stresses in both the materials are developed simultaneously, such a design may not be the most economical. It has been stated in the previous para that a balanced design gives smallest concrete section and maximum area of reinforcement. Since the cost of steel is very high in comparison to that of concrete, a balanced design may not be economical. Also, for practical considerations, sometimes, it may be necessary to fix some uniform cross-sectional dimensions. In such a case, the design may result in a singly reinforced balanced section, under-reinforced section or doubly reinforced section. However, if the section has to be singly reinforced, an under-reinforced section is always more desirable in the limit state design.

**1. Design to determine cross-sectional dimensions and reinforcement.** This is the most usual case of design in which the ultimate moment of resistance ($M_u$) to be developed by the section is given, and it is required to determine $b$, $d$ and $A_{st}$. The design is done in the following steps.

**Step 1.** Determine the limiting depth of N.A. by *Eq. 33.7*

$$x_{u\,(max.)} = \frac{0.0035}{0.0055 + 0.87\,f_y/E_s}\,d = \frac{700}{1100 + 0.87 f_y} \qquad ...(33.7)$$

**Step 2.** Choose some suitable *ratio* of $d$ and $b$. The value of $d/b$ in the range of 1.5 to 3 is usually taken.

**Step 3.** Find $d$ from the relation

$$M_u = 0.36\,f_{ck}\,\frac{x_{u\,(max.)}}{d}\left(1 - 0.416\,\frac{x_{u\,(max.)}}{d}\right) bd^2 = R_u\,bd^2 \quad ...(33.11)$$

**Step 4.** Knowing $b$ and $d$, determine area of reinforcement from the relation

$$M_u = 0.87\,f_y\,A_{st}\,(d - 0.416\,x_{u\,(max.)}) \qquad ...\,[33.10(a)]$$

Alternatively, find $p_{t\,lim.}$ from *Eq. 33.9*:

$$p_{t\,lim.} = 0.414\,\frac{f_{ck}}{f_y}\,\frac{x_{u\,(max.)}}{d} \qquad ...(33.9)$$

and
$$A_{st} = p_{t\,lim.} \times bd.$$ The values of $p_{t\,lim.}$ are given in Table 33.2.

**2. Design to determine area of reinforcement if $b$ and $d$ are known.** In this case $M_u$, $b$ and $d$ are given. To start with, we have to ascertain whether it will result in a singly reinforced section or a doubly reinforced section. The design is done in the following steps:

**Step 1.** Determine the limiting moment of resistance from *Eq. 33.11*

$$M_{u,\,lim.} = 0.36\,f_{ck}\,\frac{x_{u\,(max.)}}{d}\left(1 - 0.416\,\frac{x_{u\,(max.)}}{d}\right) bd^2 = R_u\,bd^2$$

**Step 2.** (a) If $M_u = M_{u,\,lim.}$, it will result in a balanced section, and area of steel (or $p_{t,\,lim.}$) is found either from *Eq. 33.9* or from Table 33.2.

(b) If $M_u > M_{u,\,lim.}$, it will result in a doubly reinforced section (see chapter 34).

(c) If $M_u < M_{u,\,lim.}$, it will result in an *under-reinforced* section, and the design is done in the following steps:

**Step 3.** If $M_u < M_{u,\,lim.}$, determine actual $x_u$ from relationship:

$$M_u = 0.36\,f_{ck}\,b\,x_u\,(d - 0.416\,x_u)$$

This will result in a quadratic equation in terms of $x_u$ which can be easily solved.

**Step 4.** Determine area of steel from the relation

$$M_u = 0.87 f_y A_{st} (d - 0.416 x_u) \qquad ...[33.10(a)]$$

or

$$A_{st} = \frac{M_u}{0.87 f_y (d - 0.416 x_u)}$$

Alternatively, find $p_t$ from *Eq. 33.8*:

$$p_t = 0.414 \frac{f_{ck}}{f_y} \cdot \frac{x_u}{d} \qquad ...[33.8(a)]$$

and

$$A_{st} = p_t \, bd$$

*Alternatively*, steps 3 and 4 can be avoided and $A_{st}$ can be found from *Eq. 33.10* applicable for under-reinforced section.

$$M_u = 0.87 f_y A_{st} d \left(1 - \frac{A_{st}}{bd} \cdot \frac{f_y}{f_{ck}}\right) \qquad ... (33.10)$$

This gives a quadratic equation in terms of $A_{st}$, the solution of which works out as under:

$$A_{st} = \frac{0.5 f_{ck}}{f_y} \left[1 - \sqrt{1 - \frac{4.6 M_u}{f_{ck} bd^2}}\right] bd \qquad ... (33.13)$$

**Example 33.1.** **M.R. of singly reinforced section:** *Find the M.R. of a singly reinforced concrete beam of 200 mm width and 400 mm effective depth, reinforced with 4 bars of 16 mm dia. of Fe 415 steel. Take M 20 concrete. Use I.S. Code method. Redesign the beam if necessary.*

**Solution:** $A_{st} = 4 \, \frac{\pi}{4} (16)^2 = 804.25 \text{ mm}^2$; $p_t = \frac{A_{st}}{bd} = \frac{804.25}{200 \times 400} = 0.0101$

For Fe 415 steel bars, $f_y = 415 \text{ N/mm}^2$. For M 20 concrete, $f_{ck} = 20 \text{ N/mm}^2$.

Hence from *Eq. 33.8*, $\frac{x_u}{d} = 2.417 \, p_t \, \frac{f_y}{f_{ck}} = 2.417 \times 0.0101 \times \frac{415}{20} = 0.5065$

The limiting value of $x_u/d$ is given by *Eq. 33.7*

$$\frac{x_{u,max.}}{d} = \frac{700}{1100 + 0.87 \times 415} = 0.479$$

Thus, the actual N.A. depth is more than the limiting one. Such a beam is over-reinforced and hence is *undesirable*, since the failure is sudden and without warning. The Code recommends that such a beam should be *redesigned*. The limiting moment of resistance for such a beam is found from *Eq. 33.11*.

$$M_{u,lim.} = 0.36 \times 20 \times 0.479 (1 - 0.416 \times 0.479) \, 200 \, (400)^2$$
$$= 88.37 \times 10^6 \text{ N-mm} = 88.37 \text{ kN-m}$$

Also, $\frac{x_{u,max.}}{d} = 2.417 \, p_t \, \frac{f_y}{f_{ck}}$

or

$$p_t = \left(\frac{x_{u,max.}}{d}\right) \cdot \frac{f_{ck}}{f_y} \times \frac{1}{2.417} = \frac{0.479 \times 20}{415 \times 2.417} = 9.551 \times 10^{-3}$$

∴ $A_{st} = p_t \, bd = 9.551 \times 10^{-3} \times 200 \, (400)$
$= \mathbf{764.1 \text{ mm}^2}$ (as against 804.25 mm² at present).

**Example 33.2.** **M.R. of singly reinforced section.** *For the beam of Example 33.1, find the M.R. if the reinforcement consists of 3–16 mm dia. bars of Fe 415 steel.*

**Solution:** $A_{st} = 3 \times \frac{\pi}{4} (16)^2 = 603.19 \text{ mm}^2$; $p_t = \frac{603.19}{200 \times 400} = 7.54 \times 10^{-3}$

∴ $\frac{x_u}{d} = 2.417 \, p_t \, \frac{f_y}{f_{ck}} = 2.417 \times 7.54 \times 10^{-3} \times \frac{415}{20} = 0.378;$

$$\frac{x_{u,max.}}{d} = 0.479, \text{ as before.}$$

Hence the beam is *under-reinforced*. The M.R. is given by *Eq. 33.10*:

$$M_u = 0.87 f_y A_{st} d \left(1 - \frac{f_y}{f_{ck}} \frac{A_{st}}{bd}\right) = 0.87 \times 415 \times 603.19 \times 400 \left(1 - \frac{415}{20} \times \frac{603.19}{200 \times 400}\right)$$

$$= 73.48 \times 10^6 \text{ N-mm} = \textbf{73.48 N-mm}$$

(This is less than $M_{u,lim.}$ = 88.37 kN-m found in the previous example)

**Example 33.3.** **Determination of actual stresses.** *For the beam section of example 33.1, determine the actual stresses when the section is subjected to the limiting moment of resistance.*

**Solution:** Given: $A_{st}$ = 804.25 mm² and $M_u = M_{u,lim.}$ 88.37 × 10⁶ N-mm

As found earlier, $\dfrac{x_{u,max.}}{d} = 0.479$ and $\dfrac{x_u}{d} = 0.504$

It is very clear that if $x_u/d$ is taken = 0.504, the stress in concrete will exceed 0.45 $f_{ck}$. However, at $M_u = M_{u,lim}$, $x_u$ will be equal to $x_{u,max.}$ and the stress in concrete will be equal to 0.45 $f_{ck}$ = 0.45 × 20 = 9 N/mm². In such a circumstance, the stress $f_s$ in steel will be less than $f_y$. From equilibrium of forces, $T = C$

$$\therefore \quad A_{st} \cdot f_s = 0.36 f_{ck} \cdot b \cdot x_u \quad \text{or} \quad f_s = \frac{0.36 f_{ck} \cdot b \cdot x_u}{A_{st}}, \quad \text{where } x_u = x_u, \text{max.}$$

$$\therefore \quad f_s = \frac{0.36 f_{ck} bd (x_{u,max.}/d)}{A_{st}} = \frac{0.36 \times 20 \times 200 \times 400 (0.479)}{804.25}$$

$$= \textbf{343.1 N/mm}^2 \text{ (against a value of } f_y = 415 \text{ N/mm}^2)$$

The corresponding strain in steel is

$$\varepsilon_s = \frac{0.0035 (d - x_{u,max.})}{x_{u,max.}} = \frac{0.0035 (1 - x_{u,max.}/d)}{x_{u,max.}/d} = \frac{0.0035 (1 - 0.479)}{0.479} = 0.0038$$

**Check:** Lever arm $= d - 0.416 x_{u,max.} = d \left(1 - 0.416 \dfrac{x_{u,max.}}{d}\right)$

$$= 400 (1 - 0.416 \times 0.479) = 320.29 \text{ mm}$$

$$\therefore \quad M_u = C \times \text{lever arm} = (0.36 f_{ck} b x_{u,max.}) \times 320.29$$

$$= 0.36 \times 20 \times 200 (0.479 \times 400) \times 320.29 = 88.37 \times 10^6 \text{ N-mm}$$

Also, $M_u = T \times \text{lever arm} = A_{st} f_s \text{ (Lever arm)}$

$$= 804.25 \times 343.1 \times 320.29 = 88.38 \times 10^6 \text{ N-mm}$$

Each one of these is equal to $M_{u,lim}$.

**Example 33.4.** **Design of a singly reinforced beam.** *Design a balanced singly reinforced concrete beam section for an applied moment of 60 kN-m. The width of the beam is limited to 175 mm. Use M 20 concrete and Fe 415 steel bars.*

**Solution:** $M_D$ = 60 kN-m = 60 × 10⁶ N-mm

Since $\gamma_f$ = 1.5 for both dead and live loads, $M_{uD}$ = 1.5 × 60 × 10⁶ = 90 × 10⁶ N-mm

For a balanced design, we have from *Eq. 33.7*

$$\frac{x_{u,max.}}{d} = \frac{700}{1100 + 0.87 f_y} = \frac{700}{1100 + 0.87 \times 415} = 0.479$$

The moment of resistance of a balanced beam section is given by

$$M_{u,\,lim.} = 0.36\, f_{ck} \cdot \frac{x_{u,\,max.}}{d}\left(1 - 0.416 \times \frac{x_{u,\,max.}}{d}\right) bd^2$$

Equating $M_{u,\,lim.}$ and $M_{u\,D}$, we get

$$90 \times 10^6 = 0.36 \times 20 \times 0.479\,(1 - 0.416 \times 0.479)\,175\,d^2$$

From which, we get $d = \mathbf{432\ mm}$

The reinforcement is given by Eq. 33.8:

$$\therefore \quad p_{t,\,lim.} = \frac{x_{u,\,max.}}{d} \cdot \frac{f_{ck}}{f_y} \times \frac{1}{2.417} = \frac{0.479 \times 20}{415 \times 2.417} = 9.55 \times 10^{-3}$$

This is the same as given in Table 33.2 for $f_{ck} = 20$ and $f_y = 415$.

$$\therefore \quad A_{st} = 9.55 \times 10^{-3} \times 175 \times 432 = \mathbf{722\ mm^2}$$

**Example 33.5.** **Design of singly reinforced beam:** *Find the reinforcement for the beam of Example 33.4, if the effective depth of the beam is kept equal to 500 mm.*

**Solution:** Since the depth provided by the beam is more, if will be under-reinforced beam. The limiting ultimate moment of resistance of this beam is given by Eq. 33.11.

$$M_{u,\,lim.} = 0.36\, f_{ck}\, \frac{x_{u,\,max.}}{d}\left(1 - 0.416\,\frac{x_{u,\,max.}}{d}\right) bd^2 = 0.36 \times 20 \times 0.479\,(1 - 0.416 \times 0.479)\, bd^2$$

$$= 2.761\, bd^2 \text{ (same as the value given in Table 33.3 for } f_{ck} = 20 \text{ and } f_y = 415)$$

$$= 2.761 \times 175\,(500)^2 = 120.8 \times 10^6 \text{ N-mm}$$

$M_{uD} = 90 \times 10^6$ N-mm only, and the reinforcement has to be found for this value.

From Eq. 33.10, $\quad M_u = M_{uD} = 0.87\, f_y \cdot A_{st} \cdot d\left(1 - \frac{f_y}{f_{ck}}\frac{A_{st}}{bd}\right)$

or $\quad 90 \times 10^6 = 0.87 \times 415\, A_{st} \times 500\left(1 - \frac{415}{20} \times \frac{A_{st}}{175 \times 500}\right)$

or $\quad A_{st}\,(1 - 2.371 \times 10^{-4}\, A_{st}) = 498.54$

or $\quad A_{st}^2 - 4217\, A_{st} + 2102302 = 0$

From which $A_{st} \triangleq \mathbf{578\ mm^2}$

Alternatively from Eq. 33.13,

$$A_{st} = \frac{0.5\, f_{ck}}{f_y}\left[1 - \sqrt{1 - \frac{4.6\, M_u}{f_{ck}\, bd^2}}\right] bd = \frac{0.5 \times 20}{415}\left[1 - \sqrt{1 - \frac{4.6 \times 90 \times 10^6}{20 \times 175\,(500)^2}}\right] 175 \times 500$$

$$= \mathbf{578\ mm^2} \text{ against } 722\ mm^2 \text{ found in the previous example.}$$

**Example 33.6.** **Design of singly reinforced beam** *A beam, simply supported over an effective span of 7 m carries a live load of 20 kN/m. Design the beam, using M 20 concrete and HYSD bars of grade Fe 415. Keep the width equal to half the effective depth. Assume unit weight of concrete as 25 kN/m³.*

**Solution:** For the computation of self weight, let us assume that the beam has a size of 300 mm × 600 mm.

Self weight $= 0.3 \times 0.6 \times 1 \times 25 = 4.5$ kN/m; Live load $= 20$ kN/m

∴ Total load $\quad w = 4.5 + 20 = 24.5$ kN/m

∴ Ultimate design load, $w_{uD} = 1.5 \times 24.5 = 36.75$ kN/m

$$\therefore \quad M_{uD} = \frac{36.75\,(7)^2}{8} = 225.09 \text{ kN-m} = 225.09 \times 10^6 \text{ N-mm}$$

For M 20 concrete and Fe 415 steel, $f_{ck} = 20$ N/mm² and $f_y = 415$ N/mm², respectively.

$$\therefore \qquad \frac{x_{u,max.}}{d} = \frac{700}{1100 + 0.87 f_y} = \frac{700}{1100 + 0.87 \times 415} = 0.479$$

Let $d$ = effective depth, so that $b = 0.5\,d$. Also, from *Eq. 33.11*.

$$M_{u,lim.} = 0.36 \times 20 \times 0.479\,(1 - 0.416 \times 0.479) \times 0.5\,d \times d^2 = 1.381\,d^3$$

Equating this to $M_{uD}$, we get $1.381\,d^3 = 225.09 \times 10^6$

From which we get $d = 546$ mm and $b = d/2 = 273$ mm

From *Eq. 33.9*, $\quad p_{t,lim.} = 0.414\,\dfrac{f_{ck}}{f_y} \cdot \dfrac{x_{u,max.}}{d} = 0.414 \times \dfrac{20}{415} \times 0.479 = 9.55 \times 10^{-3}$

$\therefore \qquad A_{st} = 9.55 \times 10^{-3} \times 273 \times 546 = \mathbf{1423.50 \approx 1424\ mm^2}$

**Example 33.7.** *Design of singly reinforced beam:* *A reinforced concrete beam has width equal to 300 mm and total depth equal to 700 mm, with a cover of 40 mm to the centre of the reinforcement. Design the beam if it is subjected to a total bending moment of 150 kN-m. Use M 20 concrete and HYSD bars of grade 415.*

**Solution:** $\qquad b = 300$ mm; $d = 700 - 40 = 660$ mm

Taking $\qquad \gamma_f = 1.5$, $M_{uD} = 150 \times 1.5 = 225$ kN-m $= 225 \times 10^6$ N-mm

$$\frac{x_{u,max.}}{d} = \frac{700}{1100 + 0.87 f_y} = \frac{700}{1100 + 0.87 \times 415} = 0.479$$

$$M_{u,lim.} = 0.36 \times 20 \times 0.479\,(1 - 0.416 \times 0.479)\,bd^2$$
$$= 2.761\,bd^2 = 2.761 \times 300\,(660)^2 = 360.8 \times 10^6 > M_{uD}$$

Hence compression reinforcement is not required. Also, the beam is under-reinforced.

Now, $\qquad M_{uD} = 0.87\,f_y\,A_{st}\,d\left(1 - \dfrac{f_y}{f_{ck}} \cdot \dfrac{A_{st}}{bd}\right)$

$\therefore \qquad 225 \times 10^6 = 0.87 \times 415\,A_{st}\,(660)\left[1 - \dfrac{415}{20} \times \dfrac{A_{st}}{300 \times 660}\right]$

or $\quad A_{st}\,[1 - 1.048 \times 10^{-4}\,A_{st}] = 944.2 \quad$ or $\quad A_{st}^2 - 9542\,A_{st} + 9.0097 \times 10^6 = 0$

From which $\qquad A_{st} = \mathbf{1062.5\ mm^2}$. Alternatively from *Eq. 33.13*, we get

$$A_{st} = \frac{0.5\,f_{ck}}{f_y}\left[1 - \sqrt{1 - \frac{4.6\,M_u}{f_{ck}\,bd^2}}\right]bd = \frac{0.5 \times 20}{415}\left[1 - \sqrt{1 - \frac{4.6 \times 225 \times 10^6}{20 \times 300\,(660)^2}}\right]300 \times 660$$

$$= \mathbf{1063.2\ mm^2}$$

## PROBLEMS

1. A beam section 300 mm wide and 500 mm deep is reinforced with a tension reinforcement of 3000 mm² at an effective cover of 300 mm. Determine the ultimate moment of resistance of beam section. Use M 20 concrete and Fe 250 grade steel.
2. Solve problem 1 if the grade of concrete is M 20 and grade of steel Fe 415.
3. Design a balanced singly reinforced concrete beam with a span of 5 m to carry a dead load of 25 kN/m and a working live load of 20 kN/m. Use M 20 concrete and Fe 415 grade steel.
4. Redesign the beam of problem 2 if the cross-section of the beam is limited to 300 mm width and 600 mm effective depth. Use M 20 mix and Fe 415 grade steel.

# CHAPTER 34
# DOUBLY REINFORCED SECTIONS

## 34.1. NECESSITY

Beams which are reinforced in both compression and tension sides are called doubly reinforced beam. We have seen in the previous chapter that a singly reinforced section has a limiting value of moment of resistance, corresponding to limiting value of steel reinforcement. However, if the applied moment $M_{ua}$ is larger than $M_{u\,lim}$, two alternatives will be available: (i) to increase the depth of the section or (ii) to provide compression reinforcement. In many cases, the maximum value of depth of the section may be limited or restricted from architectural or other considerations. In that case, the only alternative that will be left will be to provide reinforcement in the compression zone, giving rise to a *doubly reinforced section*.

## 34.2. STRESS BLOCK AND NEUTRAL AXIS (NA)

Fig. 34.1(a) shows a doubly reinforced section having compression reinforcement at a depth of $d'$ below the outermost compression fibre. Fig. 34.1(b) shows the strain diagram while Fig. 34.1(c) the stress block. Let,

$A_{st}$ = Total reinforcement at tension face ;    $A_{sc}$ = Reinforcement in compression side.
$x_u$ = Depth of N.A. ;    $C_{cu}$ = Compressive force in concrete = $0.36 f_{ck} x_u b$
$C_{su}$ = Compressive force in compression steel = $f_{sc} A_{sc}$
$f_{sc}$ = *Design stress* in compression reinforcement read off from the stress strain curve corresponding to the strain $\varepsilon_{sc}$ in compression reinforcement.

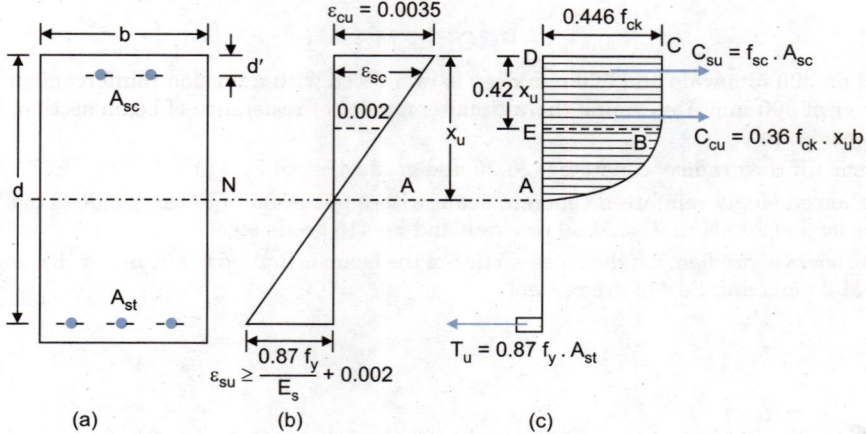

Fig. 34.1. Doubly Reinforced Section.

$\varepsilon_{sc}$ = Strain in compression reinforcement = $\dfrac{0.0035 (x_u - d')}{x_u}$ ...(34.1)

Total compressive force is given by
$$C_u = C_{cu} + C_{su}$$
or
$$C_u = 0.36 f_{ck} x_u b + f_{sc} A_{sc}$$ ...(34.2)
(neglecting the loss of concrete area occupied by compressive steel).

Total tensile force is given by
$$T_u = 0.87 f_y A_{st}$$ ...(34.3)

In order to locate the N.A., equate the total compressive force to the total tensile force:

### Depth of Neutral Axis:
The depth of neutral axis can be calculated by equating total force of compression to total force of tension.
$$C_{cu} + C_{su} = T_u$$
$$\therefore \quad 0.36 f_{ck} x_u b + f_{sc} A_{sc} = 0.87 f_y A_{st}$$ ...(34.4)
$$x_u = \dfrac{0.87 f_y A_{st} - f_{sc} A_{sc}}{0.36 f_{ck} \cdot b}.$$

From the above relation, $x_u$ can be found. However, for the solution of the above equation, an iterative procedure will have to be adopted, since $f_{sc}$ depends upon $\varepsilon_{sc}$, which in turn depends upon $x_u$. If the loss of compressive area, occupied by the compressive steel is taken into account, *Eqs. 34.2* and *34.4* are modified as under:

$$C_u = 0.36 f_{ck} x_u b + f_{sc} A_{sc} - 0.446 f_{ck} A_{sc} = 0.36 f_{ck} x_u b + (f_{sc} - 0.446 f_{ck}) A_{sc} \quad ...[34.2(a)]$$
and
$$0.36 f_{ck} x_u b + (f_{sc} - 0.446 f_{ck}) A_{sc} = 0.87 f_y A_{st} \quad ...[34.4(a)]$$

**Note:** Normally, the term $0.446 f_{ck} A_{sc}$ is very small, and can be neglected without causing any appreciable error.

### Iterative procedure for computation of $x_u$

**Step 1.** Compute the depth of N.A. of a balanced section given by strain compatibility [Fig. 34.1(b)]:

$$x_{u.max} = \dfrac{0.0035}{0.0055 + 0.87 f_y/E_s} \; ; \text{Taking } E_s = 2 \times 10^5 \text{ N/mm}^2$$

$$x_{u.max} = \dfrac{700}{1100 + 0.87 f_y}$$

It is to be noted that this expression is the same as *Eq. 33.7* used for a singly reinforced section.

**Step 2.** For the given doubly reinforced section, *assume* $x_u$ equal to $x_{u.max}$.

**Step 3.** Compute the value of $\varepsilon_{sc}$ from *Eq. 34.1*.

**Step 4.** Compute value of $f_{sc}$ from the stress strain curve of steel or Table 33.1, corresponding to this value of $\varepsilon_{sc}$.

**Step 5.** Substitute the value of $f_{sc}$ in *Eq. 34.4* or *34.4(a)* and compute the modified value of $\varepsilon_{sc}$.

**Step 6.** Repeat steps 3 to 5 till convergence for the value of $x_u$ is achieved.

**Note 1.** A slightly different form of iterative procedure has been illustrated in Example 34.1.

**Note 2.** *For mild steel bars,* $f_{sc} = E_s \times \varepsilon_{sc} \leq 0.87 f_y$.

Taking $E = 2 \times 10^5$ N/mm$^2$, we get $f_{sc} = 2 \times 10^5 \times 0.0035 \left(1 - \dfrac{d'}{x_u}\right) \leq 0.87 f_y$

or $\quad 700 \left(1 - \dfrac{d'}{x_u}\right) \leq 0.87 \times 250 \leq 217.5 \quad$ or $\quad d' \not> 0.689 x_u$

In most of the cases, doubly reinforced sections are provided when $M_u > M_{u.lim}$ and therefore, $x_u$ is normally equal to $x_{u, lim} = 0.53 d$ mild steel bars. Assuming the lowest value of $x_{u, lim} = 0.3 d$ corresponding to maximum 30% redistribution of moments,

$$d' \not> 0.689 \times 0.3 d \not> 0.21 d.$$

Since in actual practice $d'$ hardly exceeds $0.2d$, $f_{sc}$ in most of the cases can be taken equal to $0.87 f_y$.

**Note 3.** *For HYSD bars*, the value of $f_{sc}$ is obtained corresponding to $\varepsilon_{sc}$ using Table 33.1.

**Note 4.** From Table 33.1, we find that a stress of $0.87 f_y$ (= $1.0 f_{yD}$) reaches at a strain of 0.00380 for Fe 415 and 0.00417 for Fe 500 steel. Hence the stress in steel never reaches a value of $0.87 f_y$ prior to the crushing of concrete, because concrete crushes/fails at a strain of 0.0035. *In other words, the strength of compression steel is never fully utilised, specially for HYSD bars.*

## 34.3. TYPES OF PROBLEMS

As in the case of singly reinforced section, there may be three types of problems in doubly reinforced sections:

**Type I:** *Analysis problem:* Determination of moment of resistance of the section, given the material properties ($f_{ck}$ and $f_y$) and the sectional properties (i.e. $b$, $d$, $d'$, $A_{st}$ and $A_{sc}$)
   (a) Using mild steel bars           (b) Using HYSD bars

**Type II:** *Analysis problem:* Determination of actual stress in the given section, subjected to a given B.M.

**Type III:** Design of a given section to resist given ultimate design moment:
Given:     $b$, $d$, $d'$ and $M_u$
To find (a) Both $A_{st}$ and $A_{sc}$ and (b) $A_{st}$ if $A_{sc}$ is also given.

## 34.4. DETERMINATION OF MOMENT OF RESISTANCE

For a given section, the depth of neutral axis, $x_u$ can be determined as illustrated in the previous article. If this value of $x_u$ is compared to the limiting depth ($x_{u.max}$) of the N.A. axis, three cases may arise:

*Case 1.* $x_u/d$ less than limiting value
*Case 2.* $x_u/d$ equal to the limiting value.
*Case 3.* $x_u/d$ greater than limiting value.

**Case 1.** $x_u/d$ **less than limiting value**

The case corresponds to the under reinforced-section in working stress method. The moment of resistance is given by

$$M_u = 0.36 f_{ck} x_u b (d - 0.42 x_u) + A_{sc} f_{sc} (d - d') \qquad ...[34.5(a)]$$

or
$$M_u = 0.36 f_{ck} \frac{x_u}{d} \left(1 - 0.42 \frac{x_u}{d}\right) bd^2 + A_{sc} f_{sc} (d - d') \qquad ...(34.5)$$

The centroid $\bar{x}$ of the compressive forces, from the top fibre is given by

$$\bar{x} = \frac{0.36 f_{ck} x_u b \times 0.42 x_u + f_{sc} A_{sc} d'}{0.36 f_{ck} x_u b + f_{sc} A_{sc}} \qquad ...[34.5(b)]$$

$\therefore$ Lever arm    $a = d - \bar{x} = d - \dfrac{0.36 f_{ck} x_u b \times 0.42 x_u + f_{sc} A_{sc} d'}{0.36 f_{ck} x_u b + f_{sc} A_{sc}} \qquad ...(34.6)$

Hence, with respect to tensile force,
$$M_u = 0.87 f_y A_{st} \times a \qquad ...(34.7)$$

The moment of resistance found by *Eqs 34.5* and *34.7* should be equal.

The above equations have been obtained by *neglecting* the loss of concrete occupied by compressive steel. The equations are modified as under if this loss is taken into account:

$$M_u = 0.36 f_{ck} x_u b (d - 0.416 x_u) + (A_{sc} f_{sc} - 0.446 f_{ck} A_{sc}) (d - d') \qquad ...(34.8)$$

where   $x_u$ is determined on the basis of *Eq. 34.4(a)*

$$\bar{x} = \frac{0.36 f_{ck} x_u b (0.42 x_u) + (f_{sc} - 0.446 f_{ck}) A_{sc} d'}{0.36 f_{ck} x_u b + (f_{sc} - 0.446 f_{ck}) A_{sc}} \qquad ...(34.9)$$

$$a = d - \bar{x} \quad \text{and} \quad M_u = 0.87 f_y A_{st} \times a$$

### Case 2. $x_u/d$ equal to limiting value

This case corresponds to a *balanced design* in which the steel reinforcement reaches its yield stress at the same instant when ultimate strain is reached in concrete. The moment of resistance $M_u$ reaches its limiting value $M_{u.lim.}$ and is given by

$$M_{u.lim.} = 0.36 f_{ck} x_{u.lim.} b (d - 0.42 x_{u.lim.}) + A_{sc} f_{sc} (d - d') \qquad ...(34.10)$$

where $f_{sc}$ = stress in compressive steel, found from stress strain curve (or Table 33.1) corresponding to $\varepsilon_{sc}$ given by *Eq. 34.1* in which $x_u = x_{u.lim.}$

If, however, loss of concrete area occupied by compressive steel is taken into account,

$$M_{u.lim} = 0.36 f_{ck} x_{u.lim.} b (d - 0.42 x_{u.lim.}) + (f_{sc} - 0.446 f_{ck}) A_{sc} (d - d') \qquad ...[34.10(a)]$$

### Case 3. $x_u/d$ greater than the limiting value

This corresponds to an over-reinforced section. Failure occurs when strain in extreme fibres in concrete reaches its limiting value, while strain in tensile steel remain below the yield strain $\varepsilon_{su}$. The moment of resistance of such a section is limited to $M_{u.lim.}$ given by *Eq. 34.10*. It should be noted that compression failure is *sudden*, and therefore not desirable. The Code recommends that if $x_u/d$ is found to be greater than the limiting value, the *section should be redesigned*.

## 34.5. DESIGN OF A DOUBLY REINFORCED SECTION

We have seen earlier that a doubly reinforced section is required when the applied moment $M_{ua}$ is larger than $M_{u.lim.}$, given by *Eq. 34.10* for a singly reinforced rectangular section. The difference between the applied moment $M_{ua}$ and the limiting moment $M_{u.lim.}$ is carried by additional tensile reinforcement $A_{st2}$ and compression reinforcement $A_{sc}$. Thus, a doubly reinforced section is equivalent to a singly reinforced balanced section and a section with additional tension and compression reinforcement, as illustrated in Fig. 34.2.

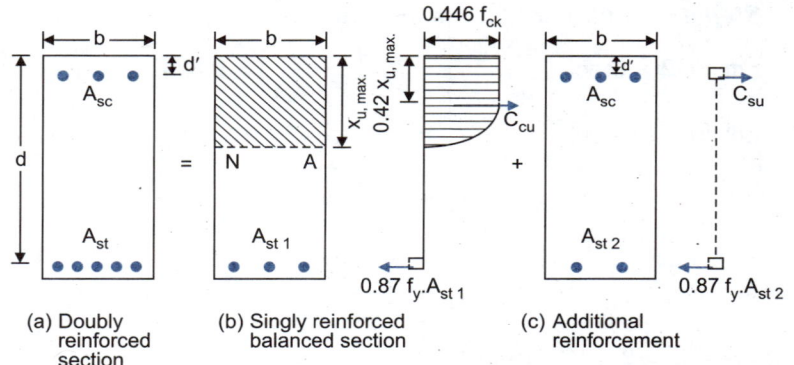

Fig. 34.2. Doubly Reinforced Section

Let, $A_{st1}$ = area of tensile reinforcement for a singly reinforced section, for $M_{u.lim.} = p_{lim} \cdot bd$.

$A_{st2}$ = additional tensile reinforcement, given by

$$A_{st2} = \frac{M_{ua} - M_{u.lim.}}{0.87 f_y (d - d')} = \frac{A_{sc} f_{sc}}{0.87 f_y} \qquad ...(34.11)$$

The compression reinforcement $A_{sc}$ is given by

$$A_{sc} = \frac{M_{ua} - M_{u.lim.}}{f_{sc}(d - d')} \qquad ...(34.12)$$

If however, loss of concrete area, occupied by compressive steel is taken into account, we have

$$A_{sc} = \frac{M_{ua} - M_{u.lim.}}{(f_{sc} - 0.446 f_{ck})(d - d')} \qquad \left(\text{where } d' < \frac{3}{7} x_u\right) \qquad ...[34.12(a)]$$

where, $f_{sc}$ = *design stress* in compression reinforcement read off from the stress-strain curve corresponding to the strain $\varepsilon_{sc}$ in the compression reinforcement.

$$\varepsilon_{sc} = \text{Strain in compression reinforcement} = \frac{0.0035 (x_{u.max} - d')}{x_{u.max}} \qquad ...(34.13)$$

and $x_{u.max}$ = limiting value of $x_u$.

## Design of the beam when $A_{sc}$ is given

In many circumstances, specially in case of a continuous beam, some of the bottom bars of mid-span tensile reinforcement are continued at bottom over the intermediate support. This may result in an excess of the requirement as compression reinforcement. The section therefore, no longer, remains balanced but becomes *under reinforced* and $A_{st1}$ is no more equal to $A_{st,\,lim}$ for a singly reinforced section

**Procedure:** Given $b, d, d', M_{uD}$ and $A_{sc}$. It is required to find $A_{st}$.

**Step 1.** Assuming $f_{sc}$, compute $M_{u2} = f_{sc} A_{sc} (d - d')$

$$\therefore \qquad M_{u1} = M_u - M_{u2}$$

**Step 2.** Determine $x_u$ from relation : $M_{u1} = 0.36 f_{ck}\, b\, x_u\, (d - 0.42\, x_u)$

This gives rise to a quadratic equation for $x_u$, the solution for which is

$$x_u = 1.2\left[1 - \sqrt{1 - \frac{4.62\, M_{u1}}{f_{ck}\, bd^2}}\right]d \qquad \ldots(34.14)$$

**Step 3.** Determine $x_{u,max}$ for given values of $f_{ck}$, and $f_y$.

Check that $\qquad x_u < x_{u,\,max}$

Also, find $\qquad M_{u,\,lim} = 0.36\, f_{ck}\, (x_{u,\,max})\, b\, (d - 0.42\, x_{u,\,max})$

Check that $\qquad M_1 < M_{u,\,lim}$.

If both the conditions are satisfied, go to steps 4(*a*) and 4(*b*). If not, go to steps 5.

**Step 4** (*a*) Check for $f_{sc}$, assumed in step 1.

For Fe 250 steel, $\qquad f_{sc} = 700\left(1 - \dfrac{d'}{x_u}\right) \not> 0.87\, f_y$

For Fe 415 steel, $\qquad f_{sc} = \varepsilon_{sc} \times \varepsilon_s$

$$\varepsilon_{sc} = \left\{\frac{0.0035\,(x_{u,\,max} - d')}{x_{u,\,max}}\right\}$$

$$f_{sc} = \left\{\frac{0.0035\,(x_{u,\,max} - d')}{x_{u,\,max}}\right\}.\varepsilon_s$$

where $\varepsilon_s = 2 \times 10^5$ N/mm$^2$

$$\varepsilon_{sc} = 0.0035\left(1 - \frac{d'}{x_u}\right) \text{ and } f_{sc} \text{ is obtained corresponding to } \varepsilon_{sc}$$

**Step 4** (*b*) Compute $\quad A_{st1} = \dfrac{M_{u1}}{0.87 f_y (d - 0.42 x_u)} \quad$ and $\quad A_{st2} = \dfrac{f_{sc} A_{sc}}{0.87 f_y}$

Total $\qquad A_{st} = A_{st1} + A_{st2}$. see Example 34.5 for illustration.

**Step 5.** If $M_1 > M_{u,\,lim}$ design a doubly reinforced sections the original area $A_{sc}$ falls short of the actual area $A_{sc}$ for a doubly reinforced beam for that design moment $M_{uD}$. In that case, redesign the beam by the process outlined in § 34.5. The additional area $A_{sc}$ required to be provided will be equal to the total $A_{sc}$ required for the doubly reinforced beam minus area $A_{sc}$ already available. See Example, 34.6 for illustration.

**Example 34.1. M.R. of doubly reinforced beam.** *Find the M.R. of an existing beam of M 15 concrete, 200 mm wide and 400 mm effective depth, reinforced with 4 bars (mild steel) of 20 mm $\Phi$ for tension and 2 bars of 20 mm dia. for compression. The cover to compressive reinforcement is 50 mm.*

**Solution.** Given

$$f_{ck} = 15 \text{ N/mm}^2; \quad f_y = 250 \text{ N/mm}^2; \quad b = 200 \text{ mm}; \quad d = 400 \text{ mm}; \quad d' = 50 \text{ mm};$$

$$A_{st} = 4\frac{\pi}{4}(20)^2 = 1256.6 \text{ mm}^2; \quad A_{sc} = 2\frac{\pi}{4}(20)^2 = 628.3 \text{ mm}^2$$

Let us assume that $d' < \dfrac{3}{7} x_u$.

$$T_u = 0.87 f_y A_{st} = 0.87 \times 250 \times 1256.6 = 273310.5 \approx 273310 \text{ N} \qquad ...(i)$$

## DOUBLY REINFORCED SECTIONS

$$C_u = 0.36 f_{ck} x_u b + f_{sc} A_{sc} - 0.446 f_{ck} A_{sc}$$
$$= 0.36 \times 15 \times 200 x_u + f_{sc} \times 628.3 - 0.446 \times 15 \times 628.3$$
$$= 1080 x_u + 628.3 f_{sc} - 4203 \text{ N} \qquad ...(ii)$$

where $f_{sc}$ = stress in the compressive reinforcement to be read off from the stress strain curve (or Table 33.1) for strain $\varepsilon_{sc}$,

where $\varepsilon_{sc} = \dfrac{0.0035(x_u - d')}{x_u}$

The depth $x_u$ is best found by trial solution. We know that when $f_y = 250$ N/mm², the value of $(x_{u.max})/d = 0.531$, for singly reinforced sections. However, for most of the cases the value of $x_u/d$ is less than this, and for doubly reinforced sections, $x_u/d$ is still much less since the N.A. will be shifted up due to the presence of compressive steel.

The limiting depth of Newton was for Fe-415 steel $x_{u.max} = 0.48 d$. As a first trial, let us assume $x_u = 0.48 d = 0.48 \times 400 = 160$ mm

$$\therefore \quad \varepsilon_{sc} = \dfrac{0.0035(160 - 50)}{160} = 0.0024$$

The stress-strain curve for mild steel bars is linear, with the yield point reaching at 0.2 percent strain. Hence $f_{sc} = 0.87 f_y = 0.87 \times 250 = 217.5$ N/mm². Substituting in (ii), we get

$$C_u = (1080 \times 160) + (628.3 \times 217.5) - 4203$$
$$= 172800 + 136655 - 4203 = 309455 - 4203 = 305252 \text{ N}.$$

(**Note:** It will be seen that $0.446 f_{ck} A_{sc} = 4203$ N only in comparison to total $C_u$, and hence can be neglected).

We find that $C_u > T_u$. In the second trial, therefore, let us reduce $x_u$ from the 160 mm to 135 mm. (**Note:** $f_{sc}$ is constant $= 0.87 f_y$ when strain $\varepsilon_{sc}$ is greater than 0.002).

$$\varepsilon_{sc} = \dfrac{0.0035(135 - 50)}{135} = 0.0022$$

$\therefore \qquad f_{sc} = 0.87 f_y = 0.87 \times 250 = 217.5$

$\therefore \qquad C_u = 1080 \times 135 + 628.3 \times 217.5 - 4203 = 145800 + 136655 - 4203 = 278252$ N

This is quite near to $T_u = 273310$ N. However, adopt $x_u = 130$ mm.

$\therefore \qquad C_u = 1080 \times 130 + 136655 - 4203 = 140400 + 132452 = (C_{cu}) + (C_{su})$

$\therefore \qquad M_u = C_{cu}(d - 0.416 x_u) + C_{su}(d - d') = 140400(400 - 0.416 \times 130) + 132452(400 - 50)$
$$= 48.58 \times 10^6 + 46.35 \times 10^6 = 94.94 \times 10^6 \text{ N-mm} = \mathbf{94.94 \text{ kN-m}} = M_{uD}$$

$\therefore$ Design moment $M_D = 94.94/1.5 = 63.29$ kN-m

**Note 1.** The moment of resistance of the same beam without compression reinforcement was found to be 71.4 kN-m only, in Example 33.1.

**Note 2.** The moment of resistance of the doubly reinforced beam, by the working stress method will be as under:

For M 15 concrete, with mild steel bars, $k = 0.404$; $j = 0.865$; $R = 0.874$.

For the N.A. location, $\dfrac{bn^2}{2} + (m_c - 1) A_{sc}(n_a - d') = m A_{st}(d - n_a)$

or $\quad \dfrac{200}{2} n^2 + (1.5 \times 19 - 1) 628.3 (n - 50) = 19 \times 1256.6 \times (400 - n_a)$

or $\quad 100 n_a^2 + 17278 n_a - 863912 = 9550160 - 23875 n$

or $\quad n_a^2 + 411.54 n_a - 104141 = 0$. From which $n_a = 177$ mm.

Critical depth $n_a = 0.404 \times 400 = 161.6$ mm. The actual N.A. is below critical; hence the stress in concrete will reach its maximum value first.

$\therefore \qquad c = \sigma_{cbc} = 5$ N/mm².

$$c' = c\frac{b-d'}{n_a} = \frac{5}{177}(177-50) = 3.59 \text{ N/mm}^2$$

$$M_r = bn\frac{c}{2}\left(d - \frac{n_a}{3}\right) + (m_c - 1)A_{sc} \cdot c'(d-d')$$

$$= 200 \times 177 \times \frac{5}{2}\left(400 - \frac{177}{3}\right) + (1.5 \times 19 - 1)\,628.3 \times 3.59 \times (400-50)$$

$$= 30.179 \times 10^6 + 21.71 \times 10^6$$

$$= 51.89 \times 10^6 \text{ N-mm} \approx 51.9 \text{ kN-m}$$

This is less than $M_D = 63.29$ kN-m found from the limit analysis. The ratio of the two moments is 1.23.

### Example 34.2. M.R. of doubly reinforced beam.

*Determine the moment of resistance of beam having the following data: b = 350 mm; d = 900 mm; d' = 50 mm; Tension reinforcement: 5-20 mm HYSD bars (Fe 415); compression reinforcement 2-20 mm HYSD bars (Fe 415). Grade of concrete: M 15.*

**Solution:**

$$A_{st} = 5 \times \frac{\pi}{4}(20)^2 = 1570.8 \text{ mm}^2; \quad A_{sc} = 2 \times \frac{\pi}{4}(20)^2 = 628.3 \text{ mm}^2.$$

$$T_u = 0.87\,f_y\,A_{st} = 0.87 \times 415 \times 1570.8 = 567137.3 \text{ N} \qquad \ldots(i)$$

$$C_u = 0.36\,f_{ck}\,x_u\,b + f_{sc}\,A_{sc} - 0.446\,f_{ck}\,A_{sc}$$

$$= 0.36 \times 15 \times 350\,x_u + 628.3\,f_{sc} - 0.446 \times 15 \times 628.3$$

$$= 1890\,x_u + 628.3\,f_{sc} - 4203 \qquad \ldots(ii)$$

Let us assume $x_u = 230$ mm.

$$\frac{3}{7}x_u = \frac{3}{7} \times 230 = 98.6 > d'.$$

$$\varepsilon_{cu} = \frac{0.0035\,(x_u - d')}{x_u} = \frac{0.0035\,(230-50)}{230} = 0.00274.$$

Hence from stress-strain curve or Table 33.1, we get $f_{sc} = 351$ N/mm$^2$

$$\therefore \quad C_u = 1890 \times 230 + (628.3 \times 351) - 4203 = 434700 + 220533 - 4203$$

$$= 651030 \text{ N}$$

This is much more than $T_u = 567137.3$ N. Hence take $x_u = 190$ mm.

$$\varepsilon_{cu} = \frac{0.0035\,(190-50)}{190} = 0.002580$$

$\therefore \quad f_{sc} \approx 347$ N/mm$^2$ (Table 33.1)

$\therefore \quad C_u = 1890 \times 190 + (628.3 \times 347) - 4203$

$$= 359100 + 218020 - 4203 = 359100 + 213817 = 572917 \approx T_u$$

$$\qquad\qquad\qquad (C_{cu}) \qquad (C_{su})$$

$\therefore \quad M_u = C_{cu}(d - 0.416\,x_u) + C_{su}(d - d')$

$$= 359100\,(900 - 0.416 \times 190) + 213817\,(900 - 50)$$

$$= 294.8 \times 10^6 + 181.7 \times 10^6 = 476.5 \times 10^6 \text{ N-mm} = \mathbf{476.5 \text{ kN-m}.}$$

### Example 34.3. Design of beam.
*A concrete beam has 350 mm breadth and 700 mm effective depth. Design the beam if it is subjected to a super-imposed bending moment of 400 kN-m. Use HYSD bars of Fe 415 grade and concrete of M 20 grade. Take d' = 50 mm.*

**Solution.** Using $\gamma_f = 1.5$, $M_{uD} = 400 \times 1.5 = 600$ kN-m $= 600 \times 10^6$ N-mm.

For Fe 415 steel,

$$\frac{x_{u,max}}{d} = \frac{700}{1100 + 0.87 \times 415} = 0.479$$

$$x_{u,max} = 0.479 \times 700 = 335.3 \text{ mm}$$

# DOUBLY REINFORCED SECTIONS 851

From *Eq. 33.9*, $\quad p_{t,lim.} = 0.414 \times \dfrac{20}{415} \times 0.479 = 0.00956 = 0.956\%$

From *Eq. 33.11*, $\quad M_{u,lim.} = 0.36 f_{ck} \dfrac{x_{u,max}}{d}\left(1 - 0.42\dfrac{x_{u,max}}{d}\right) bd^2$

$\quad\quad\quad\quad\quad\quad = 0.36 \times 20 \times 0.479(1 - 0.42 \times 0.479)\, 350\,(700)^2 = 473.6 \times 10^6$ N-mm

$\quad\quad\quad\quad A_{st1} = p_{t,lim.}\, bd = 0.00956 \times 350 \times 700 = 2342.2$ mm$^2$

$\quad\quad\quad\quad A_{st2} = \dfrac{M_{uD} - M_{u,lim.}}{0.87 f_y\,(d - d')} = \dfrac{(600 - 473.6)10^6}{0.87 \times 415\,(700 - 50)} = 538.6$ mm$^2$

Total $\quad\quad\quad A_{st} = A_{st1} + A_{st2} = 2342.2 + 538.6 = \mathbf{2880.8\ mm^2}$

$\quad\quad\quad\quad A_{sc} = \dfrac{M_{uD} - M_{u,lim.}}{f_{sc}\,(d - d')} = \dfrac{(600 - 473.6)10^6}{f_{sc}(700 - 50)} = \dfrac{0.1945 \times 10^6}{f_{sc}}$

$\quad\quad\quad\quad \varepsilon_{sc} = \dfrac{0.0035\,(x_{u\,max} - d')}{x_{u,max}} = \dfrac{0.0035\,(335.3 - 50)}{335.3} = 0.00298$

Hence from Table 33.1, for Fe 415 bars, $f_{sc} \triangleq 353.7$ N/mm$^2$ for $\varepsilon_{sc} = 0.00298$

$\therefore \quad\quad\quad A_{sc} = \dfrac{0.1945 \times 10^6}{353.7} = \mathbf{549.9\ mm^2}$

If the loss of concrete area, occupied by compressive steel is taken into account,

$\quad\quad\quad\quad A_{sc} = \dfrac{M_{uD} - M_{u,lim.}}{(f_{sc} - 0.446 f_{ck})(d - d')} = \dfrac{(600 - 473.6)10^6}{(353.7 - 0.446 \times 20)(700 - 50)}$

$\quad\quad\quad\quad\quad = \mathbf{564\ mm^2}$

**Example 34.4. Design of beam:** *Design the reinforcement for a reinforced concrete beam 300 mm wide and 400 mm deep of grade M 20, to resist an ultimate moment of 150 kN-m, using mild steel bars of grade Fe 250.*

**Solution:** Given $\quad\quad b = 300$ mm; $\quad D = 400$ mm; $\quad M_{uD} = 150$ kN-m;

$\quad\quad\quad\quad f_{ck} = 20$ N/mm$^2$; $\quad f_y = 250$ N/mm$^2$.

Let us assume cover to tensile reinforcement as 50 mm and cover to compressive reinforcement as 35 mm, each to the centre of reinforcement.

$\therefore \quad\quad\quad d = 400 - 50 = 350$ mm; $\quad d' = 35$ mm

For Fe 250 steel, $\quad \dfrac{x_{u,max}}{d} = \dfrac{700}{1100 + 0.87 \times 250} = 0.531$

$\therefore \quad\quad\quad x_{u,max} = 0.531 \times 350 \triangleq 186$ mm

From *Eq. 33.9*, $\quad p_{t,lim} = 0.414 \times \dfrac{20}{250} \times 0.531 = 0.01759 = 1.759\%$

From *Eq. 33.11*, $\quad M_{u,lim} = 0.36 f_{ck} \dfrac{x_{u,max}}{d}\left(1 - 0.42\dfrac{x_{u,max}}{d}\right) bd^2$

$\quad\quad\quad\quad\quad = 0.36 \times 20 \times 0.531\,(1 - 0.42 \times 0.531)\, 300\,(350)^2$

$\quad\quad\quad\quad\quad = 109.4 \times 10^6$ N-mm

Since $\quad\quad\quad M_{u,lim.} < M_{uD}$, the section will be doubly reinforced.

$\therefore \quad\quad\quad M_{u2} = M_{uD} - M_{u,lim.} = (150 - 109.4)10^6 = 40.6 \times 10^6$ N-mm.

$\quad\quad\quad\quad A_{st1} = p_{t,lim}\, bd = 0.01759 \times 300 \times 350 \approx 1847$ mm$^2$

$\quad\quad\quad\quad A_{st2} = \dfrac{M_{u2}}{0.87 f_y\,(d - d')} = \dfrac{40.6 \times 10^6}{0.87 \times 250\,(350 - 35)} = 592.6$ mm$^2$

Total $\quad\quad\quad A_{st} = A_{st1} + A_{st2} = 1847 + 592.6 = \mathbf{2439.6\ mm^2}$

$$A_{sc} = \frac{M_{u2}}{(f_{sc} - 0.446 f_{ck})(d - d')} = \frac{40.6 \times 10^6}{(f_{sc} - 0.446 \times 20)(350 - 35)} = \frac{0.1289 \times 10^6}{f_{sc} - 8.92}$$

$$\varepsilon_{sc} = \frac{0.0035(x_{u,max} - d')}{x_{u\,max}} = \frac{0.0035(186 - 35)}{186} = 0.00284$$

Since this is more than 0.002, $f_{sc} = 0.87 f_y = 0.87 \times 250 = 217.5$ N/mm²

$$\therefore \quad A_{sc} = \frac{0.1289 \times 10^6}{217.5 - 8.92} = \mathbf{618 \ mm^2}.$$

**Example 34.5. Design of beam:** *A reinforced concrete beam of M 20 grade concrete, 300 mm wide and 500 mm deep is required to resist a super-imposed moment of 152 kN-m at an intermediate support of a continuous beam. Using mild steel bars, calculate $A_{st}$ at top, if 4 Nos. 16 mm dia. bars are required to be continued at bottom from one span to the other. Assume effective cover to compression steel as 45 mm and that to the tension steel as 50 mm. (b) Redesign the beam as doubly reinforced beam, if no $A_{sc}$ is available from adjacent spans.*

**Solution:** Given $\quad f_{ck} = 20$ N/mm²; $\quad f_y = 250$ N/mm²; $\quad b = 300$ mm; $\quad D = 500$ mm
$\quad d' = 45$ mm; $\quad d = 500 - 50 = 450$ mm; $\quad A_{sc} = 4$ bars of 16 mm φ;

$M_{uD} = 1.5 \times 152 = 228$ kN-m; $\quad A_{sc} = 4 \frac{\pi}{4}(16)^2 = 804.25$ mm²

**Step 1.** Assume $\quad f_{sc} = 0.87 f_y = 0.87 \times 250 = 217.5$ N/mm²
$\quad M_{u2} = f_{sc} \cdot A_{sc}(d - d') = 217.5 \times 804.25(450 - 45) = 70.84 \times 10^6$ N-mm

and $\quad M_{u1} = M_{uD} - M_{u2} = (228 - 70.84)10^6 = 157.16 \times 10^6$ N-mm

**Step 2.** But $\quad M_{u1} = 0.36 f_{ck} b x_u (d - 0.416 x_u)$, the solution of which is

$$\therefore \quad x_u = 1.2 \left[1 - \sqrt{1 - \frac{4.62 \times 157.16 \times 10^6}{20 \times 300 (450)^2}}\right] 450 = 197.4 \text{ mm}$$

**Step 3.** $\quad x_{u,max} = \frac{700}{1100 + 0.87 \times 250} \times 450 = 239$ mm. $\quad$ Hence $x_u < x_{u,max}$

Also, $\quad M_{u,lim} = 0.36 \times 20 \times 239(450 - 0.42 \times 239) \times 300 = 180.48 \approx 181 \times 10^6$ N-mm.
Hence $\quad M_{u1} < M_{u,lim}$.

**Step 4.** Check for $f_{sc}$.

$$f_{sc} = 700\left(1 - \frac{d'}{x_u}\right) = 700\left(1 - \frac{45}{197.4}\right) = 540.4 \text{ N/mm}^2 \not> 0.87 f_y \not> 0.87 \times 250$$

Hence $\quad f_{sc} = 0.87 \times 250 = 217.5$ N/mm², and assumption 1 is correct.

**Step 5.** $\quad A_{st2} = \frac{f_{sc} \cdot A_{sc}}{0.87 f_y} = \frac{0.87 f_y \cdot A_{sc}}{0.87 f_y} = A_{sc} = 804.25$ mm²

$$A_{st1} = \frac{M_{u1}}{0.87 f_y (d - 0.416 x_u)} = \frac{157.16 \times 10^6}{0.87 \times 250 (450 - 0.416 \times 197.4)} = 1964.15 \text{ mm}^2$$

∴ Total $\quad A_{st} = A_{st1} + A_{st2} = 1964.15 + 804.25 = \mathbf{2768.4 \ mm^2}$

**Design as a doubly reinforced beam if no $A_{sc}$ is available from adjacent spans**

$$\frac{x_{u,max}}{d} = \frac{700}{1100 + 0.87 \times 250} = 0.531$$

$x_{u,max} = 0.531 \times 450 = 239$ mm

$p_{t,lim} = 0.414 \times \frac{20}{250} \times 0.531 = 0.01759$

$M_{u,lim} = 0.36 \times 20 \times 0.531(1 - 0.416 \times 0.531) \times 300 (450)^2 = 181 \times 10^6$ N-mm

$A_{st1}$ for $M_{u,\,lim} = p_{t,\,lim}\,bd = 0.01759 \times 300 \times 450 = 2374.2$ mm²

$$A_{st2} = \frac{(228 - 181)10^6}{0.87 \times 250\,(450 - 45)} = 533.6 \text{ mm}^2$$

Total $\quad A_{st} = A_{st1} + A_{st2} = 2374.2 + 533.6 = \mathbf{2907.8 \text{ mm}^2}$

and $\quad A_{sc} = \dfrac{M_{uD} - M_{u,lim}}{(f_{sc} - 0.446\,f_{ck})\,(d - d')} = \dfrac{(228 - 181)10^6}{(f_{sc} - 0.446 \times 20)(450 - 45)} = \dfrac{0.116 \times 10^6}{f_{sc} - 8.92}$

$$\varepsilon_{sc} = \frac{0.0035\,(x_{u,\,max} - d')}{x_{u,\,max}} = \frac{0.0035\,(0.531 \times 450 - 45)}{0.531 \times 450} = 0.00284$$

Since this is more than 0.002, $f_{sc} = 0.87\,f_y = 0.87 \times 250 = 217.5$ N/mm²

$\therefore \quad\quad\quad\quad A_{sc} = \dfrac{0.116 \times 10^6}{217.5 - 8.92} = \mathbf{556 \text{ mm}^2.}$

**Example 34.6.** **Design of beam:** $A_{sc}$ **given but small.** *Redesign the beam of example 34.5 if only 2 Nos. 16 mm dia. bars are available at bottom in the form of compression reinforcement.*

**Solution.** Given $\quad A_{sc} = 2\dfrac{\pi}{4}(16)^2 = 402.1$ mm²

**Step 1.** Assume $\quad f_{sc} = 0.87\,f_y = 0.87 \times 250 = 217.5$ N/mm²

$\therefore \quad\quad\quad M_{u2} = f_{sc}\,A_{sc}\,(d - d') = 217.5 \times 402.1\,(450 - 45) = 35.42 \times 10^6$ N-mm

and $\quad\quad M_{u1} = M_{uD} - M_{u2} = (228 - 35.42)\,10^6 = 192.58 \times 10^6$ N-mm

**Step 2.** But $\quad M_{u1} = 0.36\,f_{ck}\,b\,x_u\,(d - 0.416\,x_u)$, from which

$$x_u = 1.2\left[1 - \sqrt{1 - \frac{4.62 \times 192.58 \times 10^6}{20 \times 300\,(450)^2}}\,\right] \times 450 = 260.6 \text{ mm}$$

**Step 3.** $\quad x_{u,\,max} = 239$ mm as before, $\quad\quad\quad \therefore \quad x_u > x_{u,\,max}$

Also, $\quad M_{u,\,lim.} = 181 \times 10^6$ N/mm², as before. $\quad\quad \therefore \quad M_{u1} > M_{u,\,lim.}$

Hence the beam is to be redesigned as an ordinary doubly reinforced beam, neglecting the existing $A_{sc}$.

**Step 4.** The design of beam is given in part (*b*) of example 34.5 in which we have

$$A_{st} = \mathbf{2907.8 \text{ mm}^2} \text{ and } A_{sc} = 556 \text{ mm}^2$$

Since already available $A_{sc} = 402.1$ mm²

Additional $A_{sc}$ required $= 556 - 402.1 = \mathbf{153.9 \text{ mm}^2.}$

## PROBLEMS

1. A rectangular beam 300 mm wide and 600 mm effective depth is reinforced with a tensile reinforcement of 9000 mm² and compressive reinforcement of 3000 mm². The compressive reinforcement has an effective cover of 50 mm. Determine its ultimate moment of resistance. Use M 15 grade concrete and Fe 250 grade steel.
2. A concrete beam has 300 mm breadth and 500 mm effective depth. Design the beam if it is subjected to a superimposed bending moment of 200 kN-m. Use M 20 grade concrete and Fe 415 steel.
3. Design a reinforced concrete beam of rectangular sections using following data: effective span = 5 m, width of beam = 250 mm, overall depth 550 mm, service load (DL + LL) = 40 kN/m, effective cover = 50 mm materials used $M_{20}$ grade concrete and Fe 415 steel.

# CHAPTER 35

# T AND L-BEAMS

## 35.1. INTRODUCTION

Generally most of the construction process of concrete structures, beams and slabs are always cast monolithically. Form works are errected for beams and slabs together and concrete is poured in beam to the top of the slab. When the slab occurs on both the sides of the beams the beam is known as T-beam and one side slabs (edge beam) is known as L-beam. We have seen that the concrete below the N.A. does not resist any bending moment but simply serves to embed the tensile steel. Also, the portion of the concrete just above the N.A. carries only very little compressive force since the intensity of compressive stress there is of a very small magnitude. This suggests that the section of the beam should be such that it has greater width at the top (compression side) in comparison to the width below the neutral axis. Such a section is known as a T-section as shown in Fig. 35.1 $a$ $(ii)$.

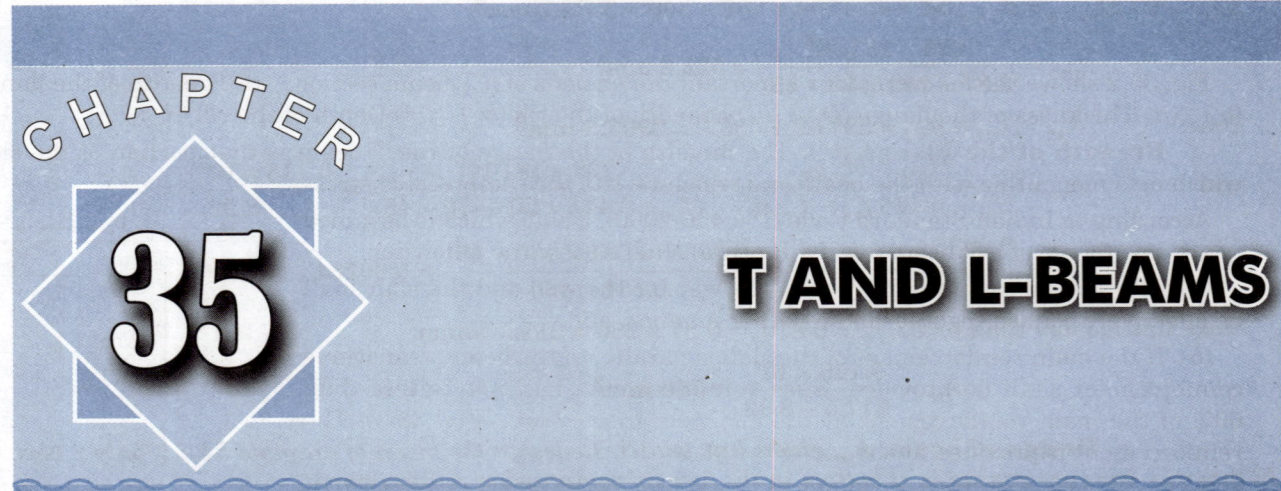

In actual practice, a beam of T-section, known as T-beam is more common than the rectangular section. Fig. 35.1 $(b)$ shows the plan of a big size room. The roof system of the room may be of two types: $(i)$ a number of beams of rolled steel section with a concrete roof slab, both acting independently, as shown in Fig. 35.1 $(c)$ or $(ii)$ a number of R.C. beams with a roof slab acting monolithically [Fig. 35.1 $(d)$]. The second system is more common because in this system a portion of the slab of width $B$ acts monolithically with the rectangular section of the beam, thus giving rise to T-beam. In fact, the slab forms the compression flange of the T-beam, while the comparatively narrow rectangular section forms the web of the T-beam.

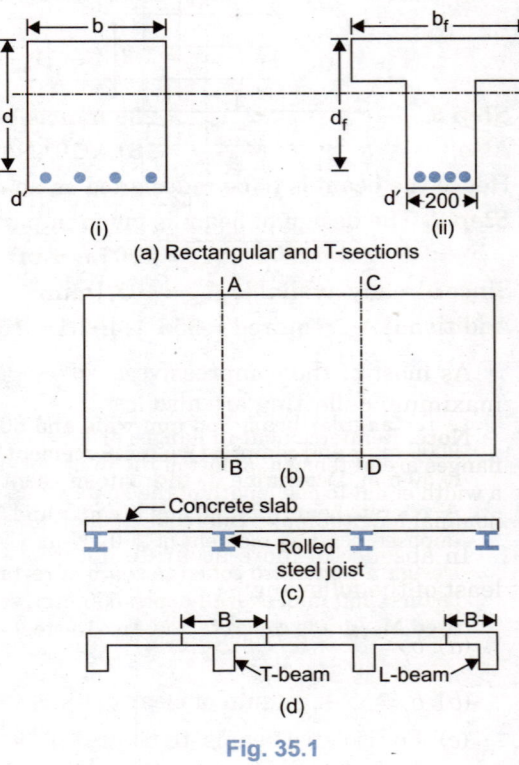

Fig. 35.1

In such a system, the roof slab is designed to bend over the span $BD$ and the reinforcement for positive and negative bending moment is provided in the direction at right angles to the direction $AB$ of the beam. The beam, on the other hand, takes uniformly distributed load from the slab, and bends over the span $AB$. The slab, which forms the upper part of the T-beam, is stressed laterally in compression, but this does not reduce its capacity to carry longitudinal compression as a part of the beam.

## 35.2. T-BEAM

Fig. 35.2 shows the following four important dimensions of a T-beam section: (i) Breadth of the flange ($b_f$) (ii) Thickness of the flange ($D_f$) (iii) Breadth of the rib ($b_w$) (iv) Depth of the rib ($d_w$)

**(1) Breadth of the Flange ($b_f$):** The breadth of the flange of the T-beam is that portion of the slab which acts monolithic with the beam and which resists the compressive stresses.

According to Indian Standard Code (IS : 456-2000), a slab which is assumed to act as a flange of a T-beam (or of a L-beam) shall satisfy the following:

(a) The slab be cast integrally with the web, or the web and the slab shall be effectively bonded together in any other manner, and

(b) If the main reinforcement of the slab is parallel to the beam, transverse reinforcement shall be provided. Such reinforcement shall not be less than 60% of the main reinforcement at the mid-span of the slab. (Fig. 35.3). The reinforcement is provided at the top of the slab.

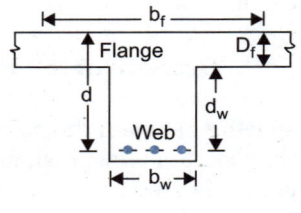

Fig. 35.2

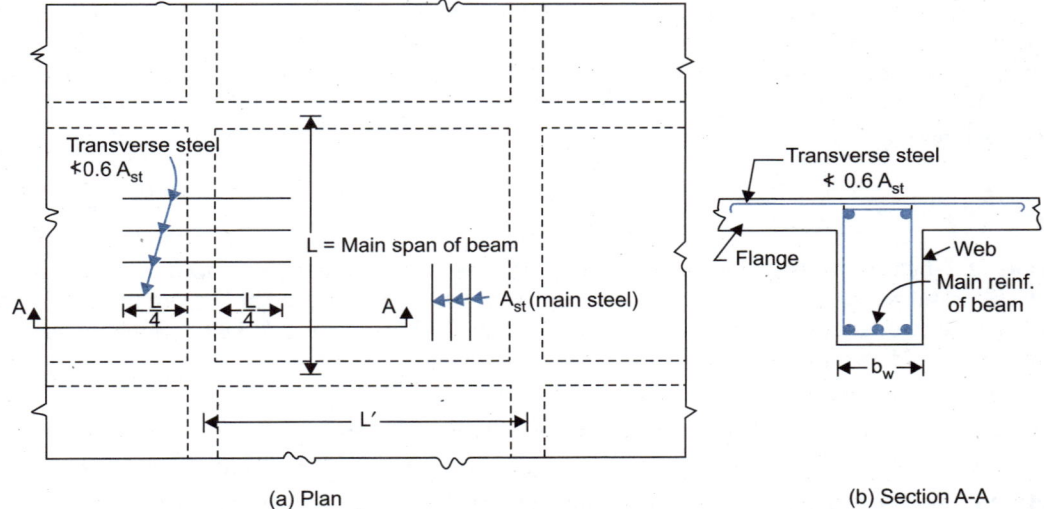

(a) Plan  (b) Section A-A

**Fig. 35.3.** Provision of transverse steel for flanged section.

As most of the compressive force is shared by flange the beam depth required is less and hence the maximum deflecting are also less.

**Note.** Reinforcement in flanges of T-beams (and also of L-beams) shall satisfy the above requirement 1 (b). Where flanges are in tension, a part of the main tension reinforcement shall be distributed over the effective flange width or a width equal to one length of the span, whichever is smaller. If the effective flange width exceeds one length of span, nominal longitudinal reinforcement shall be provided in the outer portion of flange.

In absence of more accurate determinations, the effective width $b_f$ of the flange may be taken as the least of the following:

(a) $b_f = \dfrac{l_0}{6} + 6 D_f + b_w$ ...(35.1)

(b) $b_f = b_w + \dfrac{1}{2}$ (sum of clear distances to the adjacent beams on either side) ...(35.2)

(c) For isolated beams, the flange shall be obtained as below but in no case greater than actual width (b);

$$b_f = \dfrac{l_0}{\left(\dfrac{l_0}{b}\right)+4} + b_w \qquad ...(35.3)$$

where   $b_f$ = effective width of flange
    $l_0$ = distance between points of zero moments in the beam
    $b_w$ = breadth of web

$D_f$ = thickness of flange; and

$b$ = actual width of flange.

**Note:** For continuous beams and frames, $l_0$ may be assumed as 0.7 times the effective span.

2. **Thickness of the Flange ($D_f$):** The thickness of the flange is taken equal to the total thickness of the slab, including cover. The slab spans in a direction transverse to the span of the beam, and, therefore, the thickness of the slab is determined on the basis of its bending in transverse direction. As far as design of the T-beam is concerned, $D_f$ is the fixed dimension known from the considerations of bending etc. of the slab.

3. **Breadth of the Rib or Web ($b_w$):** The breadth of the rib is equal to width of the portion of the beam in the tensile zone. This width should be sufficient to accommodate the tensile reinforcement with proper spacing between the bars. The width of the beam should also be sufficient to provide lateral stability; from this point of view, it should atleast be equal to one-third of the height (depth) of the rib. From architectural point of view, the breadth of the rib should be equal to the width of the column supporting the beam.

4. **Depth of Rib or Web ($d_w$):** The depth of rib is the vertical distance between the bottom of the flange and the centre of the tensile reinforcement, and is dependent on the effective depth ($d$) of the beam. The effective depth ($d$) of the T-beam is the distance between the top of the flange and the centre of the tensile reinforcement. Thus $d = d_w + D_f$. The assumed overall depth of the T-beam may be taken as 1/12 to 1/15 of the span when it is simply supported at the ends. When it is continuous, the assumed overall depth may be taken as 1/15 to 1/20 of span for light loads, 1/12 to 1/15 of span for medium loads and 1/10 to 1/12 of span for heavy loads.

## 35.3. L-BEAM

In the case of T-beam, the rectangular beam has flanges on both the sides and hence load is transferred to it in vertical plane passing through the middle of its width. However, in the case of monolithic beam-slab construction, the end beams have flanges to one side only, as shown in Fig. 35.4(a). Such a beam having the shape of letter 'L' is known as L-beam.

1. It is subjected to torsion and bending moment.
2. If beam is resisting on another beam it can be called as L-beam.
3. If beam is resisting on column it cannot be called as L-beam. It becomes (–ve) beam.

Fig. 35.4 (b) shows the loading on the beam received from the slab. Evidently, the beam is not symmetrically loaded. Due to this, an eccentric load $W$ is transferred to the beam. This eccentric load may be considered as central load $W$ plus a moment $M_t$, as shown in Fig. 35.4(c). The central load $W$ causes bending moment in the beam, which is jointly resisted by the rectangular portion of the beam as well as the flange (of width $B$), similar to that of a T-beam. The moment $M_t$ causes torsion in the beam, and is known as the torsional moment which is resisted by the rectangular portion alone, and the flange does not contribute to any torsional moment of resistance. Separate torsional resistance has to be provided. According to Indian Standard Code (IS : 456-2000), the width $B$ (or $b_f$) of the slab, acting monolithic with the beam, forming the flange of the L-beam is taken as the least of the following:

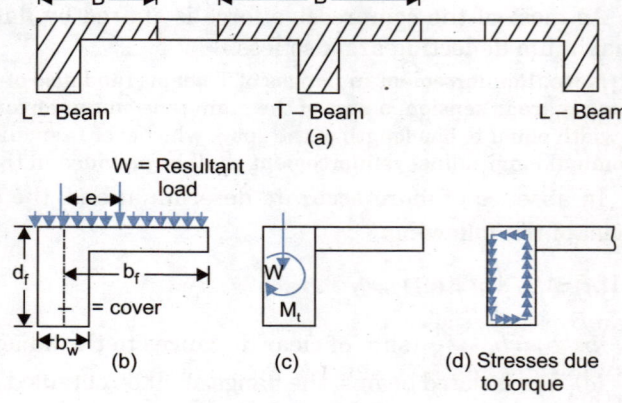

Fig. 35.4.  L-Beam

(a) $\quad b_f = \dfrac{l_0}{12} + b_w + 3 D_f$ ...(35.4)

(b) For isolated beam, $\quad b_f = \dfrac{0.5 \, l_0}{\left[\dfrac{l_0}{b} + 4\right]} + b_w$ ...(35.5) (but in no case greater than the actual width of flange) where $l_0$ is the distance between points of zero moments in the beam. For continuous beams and frames, $l_0$ may be assumed as 0.7 times the effective span.

The location of the N.A. and the calculations of the moment of resistance of L-beam are done exactly in the same way as that for T-beam, and all the equations developed for T-beam are also applicable for L-beam. The design for bending moment is also done exactly in the way as that for T-beam. However, additional reinforcement, has to be provided to resist shear induced by torsional moment.

## 35.4. STRESS BLOCK AND NEUTRAL AXIS

Consider a T-beam having flange width $b_f$, web width $b_w$ and flange thickness $D_f$, as shown in Fig. 35.5(a). The beam is singly reinforced. The stress block, shown in Fig. 35.5(c) is similar to the one for the rectangular section, with maximum compressive stress equal to $0.446 f_{ck}$. The strain diagram, shown in Fig. 35.5(b) is triangular, as for the rectangular section, with a maximum strain in concrete equal to 0.0035. The stress block shown in Fig. 35.5(c) has a rectangular portion DCBE and parabolic portion EBA with depth $DE = \frac{3}{7} x_u$ and level EB is corresponding to a strain of 0.002 in concrete. The remaining portion $EA = \frac{4}{7} x_u$. Assuming that the N.A. falls in the flange, its position is given by equating the total compressive force to the total tensile force:

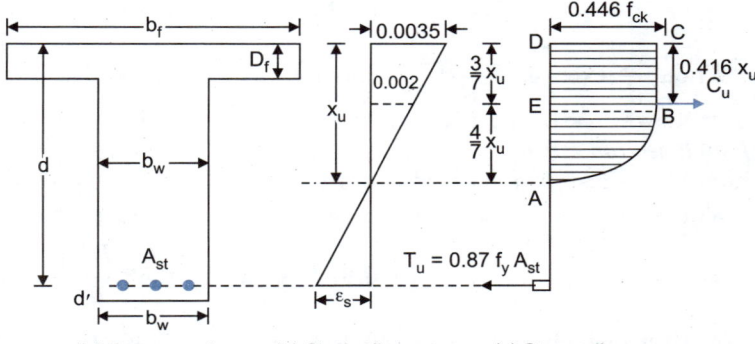

(a) Cross-section    (b) Strain diagram    (c) Stress diagram

Fig. 35.5. Flanged Beam

$$\therefore \qquad 0.36 f_{ck} b_f x_u = 0.87 f_y A_{st}$$

or
$$\frac{x_u}{d} = \frac{0.87 f_y A_{st}}{0.36 f_{ck} b_f d} = 2.417 p_t \frac{f_y}{f_{ck}} \qquad \ldots(35.6)$$

If this value of $x_u$ comes out to be less than $D_f$, it will be treated as a rectangular section of width $b_f$. The section will then be under-reinforced, balanced or over-reinforced, depending upon whether $\frac{x_u}{d}$ is less than, equal to or more than $\frac{x_{u.max}}{d}$. The depth of critical N.A. is given by Eq. 33.7:

$$\frac{x_{u.max}}{d} = \frac{0.0035}{0.0055 + 0.87 f_y / E_s} = \frac{700}{1100 + 0.87 f_y}$$

If $x_u$ found by Eq. 35.6 comes out to be more than $D_f$, the N.A. will fall outside the flange, and compression of the flange will have to be taken into account. The method of locating the depth of N.A. for such a case has been discussed in § 35.6.

## 35.5. MOMENT OF RESISTANCE WHEN $x_u \leq D_f$ : (AS PER IS: 456–2000 ANNEXURE-G)

If $x_u \leq D_f$, the section will behave as a rectangular section with width equal to $b_f$. However, if $x_u$ is greater than $D_f$, the analysis will depend upon whether $x_u$ is greater than $\frac{7}{3} D_f$ or lesser than $\frac{7}{3} D_f$, i.e., whether the N.A. falls outside the rectangular portion of the stress block [Fig. 35.5 (c)] or it falls inside the rectangular portion DCBE. Thus, in all, three cases may arise in the analysis of a T-beam.

Case 1.          $x_u \leq D_f$

Case 2.          $D_f \leq \frac{3}{7} x_u \left( \text{or } \frac{D_f}{x_u} \leq 0.43 \right)$

Case 3.          $D_f > \frac{3}{7} x_u \left( \text{or } \frac{D_f}{x_u} > 0.43 \right)$

We will be considering the first case in this article and the other two cases in the next article.

**Case 1.** $x_u \leq D_f$: In this case, the neutral axis is entirely within the flange. Hence the moment of resistance can be found from the equations applicable for rectangular sections substituting $b_f$ for $b$.

Here again three cases may arise depending upon whether $x_u$ is less than, equal to or greater than the limiting value $x_{u.max}$. When $x_u < x_{u.gmax}$, we have the case of under-reinforced section and the moment of resistance is given by *Eq. 33.10* of rectangular section.

$$M_u = 0.87 f_y A_{st} d \left(1 - \frac{f_y}{f_{ck}} \cdot \frac{A_{st}}{b_f d}\right) \qquad ...(35.7)$$

When $\frac{x_u}{d}$ is equal to $\frac{x_{u.max}}{d}$, we have the *balanced* rectangular section, and the moment of resistance $M_u$, which reaches its $M_{u.lim}$ is given by *Eq. 33.11*.

$$M_{u.lim} = 0.36 f_{ck} \frac{x_{u.max}}{d} \left(1 - 0.42 \frac{x_{u.max}}{d}\right) b_f d^2 \qquad ...(35.8)$$

When $\frac{x_u}{d}$ exceeds $\frac{x_{u.max}}{d}$, we have the over-reinforced section. The failure takes place due to crushing of concrete while strain $\varepsilon_s$ in steel still remains below the yield strain $\varepsilon_{su}$. Therefore, *the moment of resistance of such section is limited to $M_{u.lim}$* given by *Eq. 35.8*. However, the Code recommends that such a section should be *redesigned* because compression failure is *sudden* and therefore not desirable.

When $x_u = D_f$, the moment of resistance is given by

$$M_{u1} = 0.87 f_y A_{st} d \left(1 - 0.42 \frac{D_f}{d}\right) = 0.87 f_y A_{st}(d - 0.42 D_f) \qquad ...(35.9)$$

or Alternatively, $\qquad M_{u1} = 0.36 f_{ck} b_f D_f (d - 0.42 D_f) \qquad ...[35.9(a)]$

## 35.6. MOMENT OF RESISTANCE WHEN N.A. FALLS IN THE WEB

When the value of $x_u$, found from *Eq. 35.6* exceeds $D_f$, the N.A. falls in the web, and *a new value of $x_u$ will have to be found by equating the total compressive force to the total tensile force*. Fig. 35.6 (a) shows the flanged section in which the N.A. falls outside the flange. The total compressive force is equal to $C_u$ which can be thought to be equal to the sum of compressive force $C_{uw}$ developed by web contribution [Fig. 35.6(b)] and the compressive force $C_{uf}$ developed by contribution of remaining flange width $(b_f - b_w)$ [Fig. 35.6(c)].

The compressive force $C_{uw}$ from the web contribution is given by

$$C_{uw} = 0.36 f_{ck} x_u b_w \qquad ...(35.10)$$

If out of the total tensile reinforcement $A_{st}$, a reinforcement $A_{sw}$ is the component required to balance the compressive force $C_{uw}$ and the remaining reinforcement $A_{sf}$ is required to balance the compressive force $C_{uf}$ in the flange, we have $A_{st} = A_{sw} + A_{sf}$, and the tensile force $T_{uw}$ to balance the compressive force $C_{uw}$ will be equal to $0.87 f_y A_{sw}$.

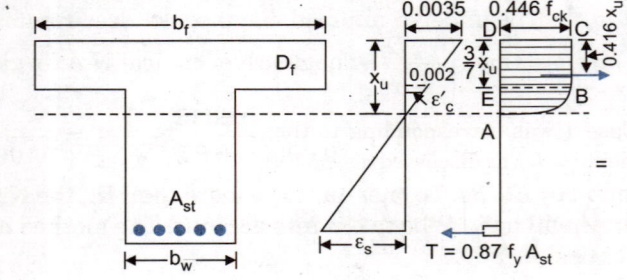

(a) Flanged section

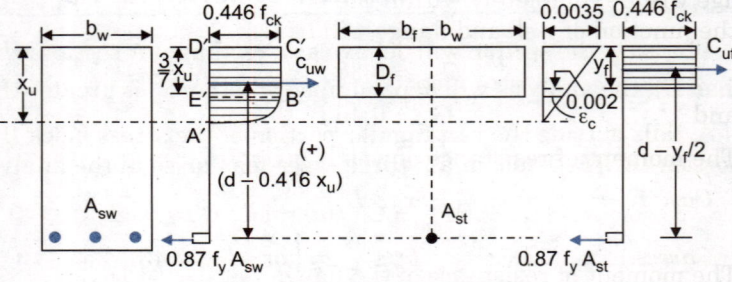

(b) Web contribution  (c) Flange contribution

**Fig. 35.6.** Singly Reinforced Flanged Section

## T AND L-BEAMS

Hence, $\quad 0.36 f_{ck} x_u b_w = 0.87 f_y A_{sw}$ ...[35.10 (a)]

The flange contribution $C_{uf}$ is shown in Fig. 35.6 (c). The compressive force $C_{uf}$ can be computed by approximating the actual stress block (rectangular-cum-parabolic) by a rectangular stress block of depth $y_f$, having the same peak stress value of $0.446 f_{ck}$.

The depth of equivalent rectangular stress block is given by
$$y_f = 0.15 x_u + 0.65 D_f \quad \text{(but not greater than } D_f\text{)} \quad ...(35.11)$$

The compressive force contribution of the flange is given by
$$C_{uf} = 0.446 f_{ck} (b_f - b_w) y_f \quad ...(35.12)$$

The tensile force $T_{uw}$ to balance the compressive force $C_{uf}$ will be equal to $0.87 f_y A_{sf}$.

The total compressive force $C_u$ is then given by
$$C_u = 0.36 f_{ck} \cdot x_u b_w + 0.446 f_{ck} (b_f - b_w) y_f \quad ...(35.13)$$

The total tensile force $\quad T_u = 0.87 f_y A_{st}$ ...(35.14)

Hence equating the two, we get, $\quad 0.36 f_{ck} x_u b_w + 0.446 f_{ck} (b_f - b_w) y_f = 0.87 f_y A_{st}$ ...(35.15)

It should be clearly noted that the value of $y_f$ in Eq. 35.15 in terms of $x_u$ is given by Eq. 35.11. Hence the value of $x_u$ found from Eq. 35.15 is valid only if the value of $y_f$, calculated from Eq. 35.11 by using this value of $x_u$, is equal to or less than $D_f$. If however, the value of $y_f$, comes out to be greater than $D_f$, the value of $x_u$ should be *recalculated* from Eq. 35.15, taking the value of $y_f = D_f$. The modified expression is that case will be
$$0.36 f_{ck} x_u b_w + 0.446 f_{ck} (b_f - b_w) D_f = 0.87 f_y A_{st} \quad ...[35.15 (a)]$$

Thus, the value of $x_u$ is known. This value of $x_u$ should now be compared with the limiting value $x_{u.max}$. If $x_u < x_{u.max}$, we have *under reinforced section*. If $x_u = x_{u.max}$, we have the *balanced section* while if $x_u > x_{u.max}$, we have *over-reinforced section*. We will take each one of these cases separately.

### (a) UNDER-REINFORCED SECTION ($x_u < x_{u.max}$)

Here again, two cases may arise, depending upon the comparison of $x_u$ with $D_f$.

Case 2. $\quad D_f \leq \dfrac{3}{7} x_u \left( \text{or } \dfrac{D_f}{x_u} \leq 0.43 \right)$

Case 3. $\quad D_f > \dfrac{3}{7} x_u \left( \text{or } \dfrac{D_f}{x_u} > 0.43 \right)$

(**Note.** Case 1 was corresponding to the situation when $x_u < D_f$, as discussed in § 35.5).

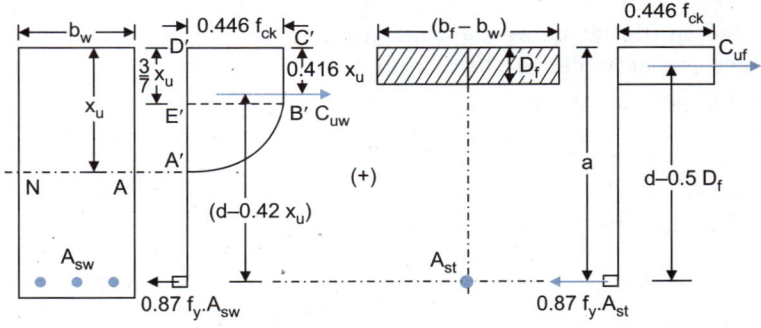

(a) Web contribution    (b) Flange contribution

Fig. 35.7.  $M_u$ For $\dfrac{D_f}{x_u} \leq 0.43$.

Case 2. $\quad D_f \leq \dfrac{3}{7} x_u \left( \text{or } \dfrac{D_f}{x_u} \leq 0.43 \right)$

If $D_f$ is equal to or less than $\dfrac{3}{7} x_u$, the depth $DE$ of the stress block will be greater than $D_f$. Hence the flange will lie completely within the rectangular portion of the stress block, and the strain $\varepsilon_c'$ in concrete, at the junction of web and flange will be less than 0.002. The compressive forces $C_{uw}$ and $C_{uf}$ are given by

$$C_{uw} = 0.36 f_{ck} x_u b_w = 0.87 f_y A_{sw} \quad ...[35.10(a)]$$
and $\quad C_{uf} = 0.45 f_{ck} (b_f - b_w) D_f = 0.87 f_y A_{sf}$ ...[35.12(a)]

The moment of resistance of web is given by
$$M_{uw} = C_{uw} (d - 0.42 x_u) = 0.36 f_{ck} \dfrac{x_u}{d} b_w \left(1 - 0.42 \dfrac{x_u}{d}\right) d^2 \quad ...[35.16(a)]$$

The moment of resistance of the flange is given by
$$M_{uf} = C_{uf} (d - 0.5 D_f) = 0.45 f_{ck} (b_f - b_w) D_f (d - 0.5 D_f) \quad ...[35.16(b)]$$

Hence
$$M_u = 0.36 \frac{x_u}{d}\left(1 - 0.42 \frac{x_u}{d}\right) f_{ck} b_w d^2 + 0.45 f_{ck} (b_f - b_w) D_f (d - 0.5 D_f) \qquad ...(35.16)$$

When $D_f = \frac{3}{7} x_u$ $\left(i.e., \text{ when } x_u = \frac{7}{3} D_f\right)$ the moment of resistance is given by

$$M_{u2} = 0.36 f_{ck} b_w \left(\frac{7}{3} D_f\right)\left(d - 0.42 \times \frac{7}{3} D_f\right) + 0.45 f_{ck} (b_f - b_w) \times D_f (d - 0.5 D_f) \qquad ...[35.16(c)]$$

**Case 3.** $D_f > \frac{3}{7} x_u$ $\left(\text{or } \frac{D_f}{x_u} > 0.43\right)$.

Since $D_f$ is greater than $\frac{3}{7} x_u$ in this case, the depth $DE$ of the stress block will be smaller than the depth of the flange. Due to this, the distribution of compressive stress in the flange will not be rectangular. However, the distribution of stress in web will be the same as an Case 2. The total moment of resistance will be the sum of the contribution of the web and the flange. The actual distribution is shown in Fig. 35.8(b). This actual stress block can be replaced by a rectangular stress

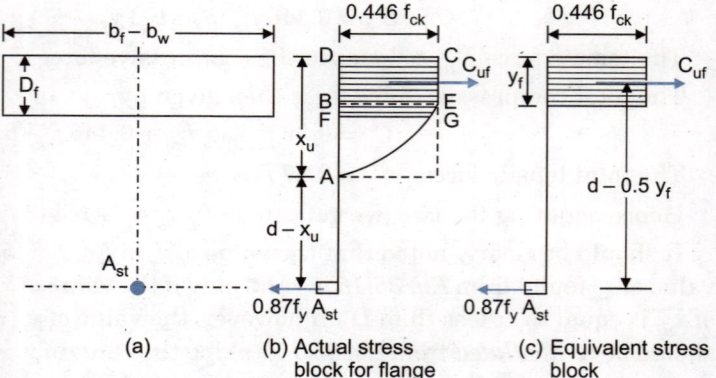

Fig. 35.8. Flange Contribution when $D_f > \frac{3}{7} x_u$

block of depth $y_f$, having the same peak stress value of $0.446 f_{ck}$. The depth of equivalent rectangular stress block is given by

$$y_f = 0.15 x_u + 0.65 D_f \text{ (but not greater than } D_f). \qquad ...(35.11)$$

The idealisation of actual stress block for the flange into a simple rectangular stress block of height $y_f$ is based on the consideration that the compressive force and its centre of gravity of the idealised rectangular block and actual stress block compare within the acceptable limits of accuracy. The compressive forces $C_{uw}$ and $C_{uf}$ are given by

$$C_{uw} = 0.36 f_{ck} x_u b_w = 0.87 f_y A_{sw} \text{ (as before)} \qquad ...[35.10(a)]$$

and
$$C_{uf} = 0.446 f_{ck} (b_f - b_w) y_f = 0.87 f_y A_{sf} \qquad ...(35.12)$$

The moment of resistance of the web is given by

$$M_{uw} = C_{uw} (d - 0.42 x_u) = 0.36 f_{ck} \frac{x_u}{d} b_w \left(1 - 0.42 \frac{x_u}{d}\right) d^2 \qquad ...[35.17(a)]$$

The moment of resistance of the flange is given by

$$M_{uf} = C_{uf} (d - 0.5 y_f) = 0.45 f_{ck} (b_f - b_w) y_f (d - 0.5 y_f) \qquad ...[35.17(b)]$$

Hence,
$$M_u = 0.36 \frac{x_u}{d}\left(1 - 0.42 \frac{x_u}{d}\right) f_{ck} b_w d^2 + 0.45 f_{ck} (b_f - b_w) y_f (d - 0.5 y_f) \qquad ...(35.17)$$

**(b) BALANCED SECTION ($x_u = x_{u.max}$)**

When $x_u = x_{u.max}$ the moment of resistance is given by Eq. 35.17, by replacing $x_u$ by $x_{u.max}$. Thus,

$$M_{u.lim} = 0.36 \frac{x_{u.max}}{d}\left(1 - 0.42 \frac{x_{u.max}}{d}\right) f_{ck} b_w d^2 + 0.446 f_{ck} (b_f - b_w) y_f (d - 0.5 y_f) \qquad ...(35.18)$$

where $y_f = 0.15 x_{u.max} + 0.65 D_f$ (subject to a maximum of $D_f$). $\qquad ...[35.11(b)]$

**(c) OVER-REINFORCED SECTION ($x_u > x_{u.max}$):** When $x_u$ is greater than $x_{u.max}$ the section will be *over-reinforced*. The failure takes place due to crushing of concrete while strain $\varepsilon_s$ in steel still remains below the yield strain $\varepsilon_{su}$. The moment of resistance of such a section is limited to $M_{u.lim}$ given by *Eq. 35.18*. The over-reinforced-section results in wastage of excess steel provided, and the concrete reaches its ultimate capacity before steel yields, resulting in a sudden failure. Such a section should therefore be redesigned.

## 35.7. IS CODE PROCEDURE FOR FINDING MOMENT OF RESISTANCE (IS : 456 – 2000; ANNEXURE-G)

IS Code procedure for finding moment of resistance of the flanged section depends upon the following three cases.

Case 1. When $x_u < D_f$
Case 2. When $(x_{u.max} = x_u) > D_f$
Case 3. When $x_{u.max} > x_u > D_f$

**Case 1. When $x_u < D_f$:** If $x_u < D_f$ the N.A. falls entirely in the flange. Hence the moment of resistance is computed, treating the section as a rectangular section, taking $b = b_f$

(a) The value of $x_u$ is found from:

$$\frac{x_u}{d} = \frac{0.87 f_y A_{st}}{0.36 f_{ck} b_f d} \quad \text{or} \quad x_4 = \frac{0.87 f_y A_{st}}{0.36 f_{ck} b_f} \quad \ldots(35.6)$$

(b) If the value of $\frac{x_u}{d}$ is less than $\frac{x_{u.max}}{d}$, we have

$$M_u = 0.87 f_y A_{st} d \left(1 - \frac{A_{st} f_y}{b_f d f_{ck}}\right) \quad \ldots(35.19)$$

(c) If the value of $\frac{x_u}{d}$ is equal to $\frac{x_{u.max}}{d}$, the moment of resistance of the section is given by:

$$M_{u.lim.} = 0.36 \frac{x_{u.max}}{d}\left(1 - 0.42 \frac{x_{u.max}}{d}\right) b d^2 f_{ck} \quad \ldots(35.20)$$

(b) If $\frac{x_u}{d}$ is greater than $\frac{x_{u.max}}{d}$, the section should be redesigned.

All these equations and recommendations are as per para G-2.1 of the I.S. Code.

**Case 2. $(x_{u, max} = x_u) > D_f$ : Balanced section**

This is the case when $x_u$ is greater than $D_f$ but at the same time it is equal to $x_{u.max}$. It is thus the case of a *balanced section*. In such a case, the moment of resistance of the section, known as *limiting moment of resistance* will depend upon the *value* of $D_f/d$ ratio either lesser than or greater than a certain *value* discussed below.

Taking $\quad x_u = x_{u, max}, \quad \frac{x_{u, max}}{D_f} = \frac{7}{3}$

or $\quad \frac{x_{u, max}/d}{D_f/d} = \frac{7}{3} \quad$ or $\quad \frac{D_f}{d} = \frac{3}{7}\frac{x_{u.max}}{d}$

The values of $D_f/d$, dividing *case 2* into *case 2 (a)* and *case 2 (b)*, for different grades of steel are as under:

| Grade of steel | $x_{u, max}/d$ | $D_f/d = \frac{3}{7} x_{u, max}/d$ |
|---|---|---|
| Fe 250 | 0.531 | 0.228 |
| Fe 415 | 0.479 | 0.205 |
| Fe 500 | 0.456 | 0.195 |

Indian Standard Code IS: 456-2000 recommends only a single value of $D_f/d = 0.2$ for all the grades of steel. Thus, we have two dividing cases:

**Case 2 (a):** $(x_{u.max} = x_u) > D_f$ and $D_f/d \leq 0.2$
and **Case 2 (b):** $(x_{u.max} = x_u) > D_f$ and $D_f/d > 0.2$

**Case 2. (a) $(x_{u.max} = x_u) > D_f$ and $D_f/d \leq 0.2$:** The moment of resistance is given by

$$M_{u.lim.} = 0.36 \frac{x_{u.max}}{d}\left(1 - 0.42 \frac{x_{u.max}}{d}\right) f_{ck} b_w d^2 + 0.446 f_{ck}(b_f - b_w) D_f \left(d - \frac{D_f}{2}\right) \quad \ldots(35.21)$$

Thus, in this case, the I.S. Code has taken the depth $y_f$ of the equivalent rectangular stress block equal to its maximum value of $D_f$. This equation is the same as given in para G-2.2 of the Code. Note that, this equation is same as *Eq. 35.18*, where $y_f$ has been taken equal to $D_f$.

**Case 2. (b) $(x_{u,max} = x_u) > D_f$ and $D_f/d > 0.2$:** The moment of resistance is given by

$$M_{u.lim.} = 0.36 \frac{x_{u.max}}{d}\left(1 - 0.42 \frac{x_{u.max}}{d}\right) f_{ck} b_w d^2 + 0.46 f_{ck} (b_f - b_w) y_f \left(d - \frac{y_f}{2}\right) \quad ...(35.22)$$

where $y_f = (0.15 x_{u.max} + 0.65 D_f)$ (But not greater than $D_f$).

The above equation is the same as given in para G-2.2.1 of the Code. Also, this equation is same as *Eq. 35.18*.

**Case 3. $x_{u,max} > x_u > D_f$:** This is a case of under-reinforced section in which the N.A. falls outside the flange. Here again, the moment of resistance will depend upon whether the value of $D_f/x_u$ is equal to or less than 0.43 (or 3/7) or the value of $D_f/x_u$ is greater than 0.43.

**Case 3. (a) $x_{u,max} > x_u > D_f$ and $D_f/x_u < 0.43$:** When $D_f$ is equal to or less than $\frac{3}{7} x_u$ (or 0.43 $x_u$) the depth $DE$ of the stress block will be greater than $D_f$. Thus the flange will fall entirely within the rectangular portion of the stress block. Hence the moment of resistance is given by

$$M_u = 0.36 \frac{x_u}{d}\left(1 - 0.42 \frac{x_u}{d}\right) f_{ck} b_w d^2 + 0.446 f_{ck}(b_f - b_w) D_f \left(d - \frac{D_f}{2}\right) \quad ...(35.23)$$

This is same as *Eq. 35.16*. Also, this equation is the same as given in para G-2.2 of the code except that $x_{u.max}$ has been replaced by $x_u$.

**Case 3. (b) $x_{u.max} > x_u > D_f$ and $D_f/x_u < 0.43$:** When $D_f$ is equal to or greater than $\frac{3}{7} x_u$ (or 0.43 $x_u$) the depth $DE$ of the stress block will be smaller than the depth of the flange. Due to this, the distribution of compressive stress in the flange will not be rectangular. The moment of resistance is given by:

$$M_u = 0.36 \frac{x_u}{d}\left(1 - 0.42 \frac{x_u}{d}\right) f_{ck} b_w d^2 + 0.446 f_{ck} (b_f - b_w) y_f \left(d - \frac{y_f}{2}\right) \quad ...(35.24)$$

where $y_f = (0.15 x_u + 0.65 D_f)$ (But not greater than $D_f$)

The above equation is the same as *Eq. 35.17*. Also, it is the same as given in para G-2.2.1 of the Code, except that $x_{u.max}$ has been replaced by $x_u$.

**Note 1:** When the N.A. falls outside the flange, as concluded from the operation of *Eq. 35.6*, a new value of $x_u$ should be found, by taking the compressive force of the web into account, as discussed in § 35.6. Though the Code has not specifically mentioned this, it is implied that this new value of $x_u$ should be used while comparing it with $x_{u.max}$ or $D_f$. The value of $x_u$ for such a case can be found either by *Eq. 35.15* or by *Eq. 35.15 (a)*, as the case may be.

**Note 2.** In para G-2.2.1, while computing the value of $y_f$, the value of $x_u$ has been erroneously printed in the place of $x_{u.max}$. Hence the value of $y_f$ should be taken as equal to, $(0.15 x_{u.max} + 0.65 D_f)$, subject to a maximum of $D_f$.

**Note 3.** The Code has not specifically dealt with the case when $x_u$ is greater than $D_f$ and is also greater than $x_{u.max}$. Hence, when $x_u > x_{u.max} > D_f$, the section will be *over-reinforced*. The failure will take place due to the crushing of concrete while strain $\varepsilon_s$ in steel still remains below the yield strain $\varepsilon_{su}$. Therefore, the moment of resistance of such a section is limited to $M_{u.lim}$, and is given by *Eq. 35.21* (if $D_f/d \le 0.2$) or by *Eq. 35.22* if $D_f/d > 0.2$, as the case may be. Such a section should be redesigned because compression failure is *sudden* and therefore not desirable.

## Minimum and Maximum Reinforcement

(a) The minimum percentage of reinforcement to be provided in flange beams as per IS: 456-2000. Clause–26.5.1.1 to be calculated only on the width of the web of the T-beam.

$$\frac{A_{st}}{b_w . d} = \frac{0.85}{f_y}$$

(b) The maximum area of tension reinforcement should not exceed 4% of the gross surface area (based on web width)

$$A_{st\,max} < 0.04\, b_w . D.$$

(c) Transverse reinforcement as per clause 23.1.1 (b) of IS: 456. When the main reinforcement of slabs is parallel to the beam, Transverse reinforcement should be provided at the top of slab. It should not be less 60% of main steel at mid span.

## 35.8. TYPES OF PROBLEMS

There may be following types of problems in flanged beams:

(a) **Analysis problems**

1. **Type I problem:** Given (i) Material properties (i.e. $f_{ck}$ and $f_y$)
   and (ii) Section properties (i.e. $b_f, D_f, b_w, D, A_{st}$)
   To find : Moment of resistance ($M_u$)
   (See examples 35.1, 35.2, and 35.3)

2. **Type II problem:** Given (i) Material properties (i.e. $f_{ck}$ and $f_y$)
   (ii) Concrete section properties (i.e. $b_f, D_f, b_w, D$)
   To find : (i) Limiting moment $M_{u,\,lim}$.
   (ii) Limiting area of steel ($A_{st,\,lim}$ or $A_{st,\,max}$)
   (See example 35.4)

(b) **Design problem**

3. **Type III problem:** Given (i) Material properties (i.e. $f_{ck}$ and $f_y$)
   and (ii) Ultimate design moment ($M_{uD}$)
   To find area of tension steel $A_{st}$
   (See example 35.5, 35.6, and 35.7)

4. **Type IV problem:** Given (i) Material properties (i.e. $f_{ck}$ and $f_y$)
   and (ii) Ultimate design moment ($M_{uD}$)
   To find: $A_{st}$ and $A_{sc}$ : See § 35.9 and 35.10 on doubly reinforced T-beams. Also see example 35.11 for design of doubly reinforced T-beam.

## 35.9. DESIGN OF T-BEAM

The design of T-beam of given dimensions (i.e. $b_w, b_f, d$ known) and given design moment, is done in the following steps:

**Step 1.** Assuming $x_u = D_f$, compute the moment of resistance $M_{u1}$ from Eq. 35.9(a)
$$M_{u1} = 0.36 f_{ck} b_f D_f (d - 0.416 D_f) \qquad ...[35.9(a)]$$

**Step 2.** If $M_{u1} > M_{uD}$, the N.A. will fall inside the flange (i.e. $x_u \leq D_f$). In that case the reinforcement is given by Eq. 35.7:
$$M_u = M_{uD} = 0.87 f_y A_{st} d \left(1 - \frac{f_y}{f_{ck}} \frac{A_{st}}{b_f d}\right) \qquad ...(35.7)$$

The solution of the above quadratic equation gives:
$$A_{st} = \frac{0.5 f_{ck}}{f_y}\left[1 - \sqrt{1 - \frac{4.6 M_{uD}}{f_{ck} b_f d^2}}\right] b_f d \qquad ...[35.7\,(a)]$$

**Step 3.** If $M_{u1} < M_{uD}$, the N.A. falls outside the flange. Assuming $x_u = \frac{7}{3} D_f$, compute $M_{u2}$ from Eq. 35.16 (c):
$$M_{u2} = 0.36 f_{ck} b_w \left(\frac{7}{3} D_f\right)\left(d - 0.42 \times \frac{7}{3} D_f\right) + 0.45 f_{ck} (b_f - b_w) \times D_f (d - 0.5 D_f) \qquad ...[35.16(c)]$$

**Step 4.** If $M_{u2} < M_{uD}, x_u > \frac{7}{3} D_f$. Compute $A_{sw}$ from Eq. 35.10(a) as under:
$$C_{uw} = 0.36 f_{ck} x_u b_w = 0.87 f_y A_{sw}$$

or $$A_{sw} = \frac{0.36 f_{ck} x_u b_w}{0.87 f_y}$$ ...(35.25)

Similarly, compute $A_{sf}$ from Eq. 35.12(a)

$$C_{uf} = 0.45 f_{ck} (b_f - b_w) D_f = 0.87 f_y A_{sf}$$

or $$A_{sf} = \frac{0.45 f_{ck} (b_f - b_w) D_f}{0.87 f_y}$$ ...(35.26)

Total $$A_{st} = A_{sw} + A_{sf}$$

When $M_{uD} \gg M_{u2}$, it is always advisable to compute $M_{u.lim}$, since it is likely that $x_u > x_{u.max}$. If $M_{uD} > M_{u.lim}$, the section will be doubly reinforced.

**Step 5.** If $M_{u2} > M_{uD}$, $x_u < \frac{7}{3} D_f$. The design procedure in this case will be that of trial and error. The following steps are recommended:

(i) Assume $x_u < \frac{7}{3} D_f$

(ii) Compute $y_f = 0.15 x_u + 0.65 D_f$ (subject to a max. of $D_f$)

(iii) Compute $M_u$ from Eq. 35.17:

$$M_u = 0.36 \frac{x_u}{d}\left(1 - 0.42 \frac{x_u}{d}\right) f_{ck} b_w d^2 + 0.45 f_{ck} (b_f - b_w) y_f (d - 0.5 y_f)$$ ...(35.17)

(iv) Compare $M_u$ with $M_{uD}$.

If $M_u = M_{uD}$, assumed value of $x_u$ is correct.

If $M_{uD} > M_u$, increase $x_u$ in next trial.

If $M_{uD} < M_u$, decrease $x_u$ in next trial. Repeat till $M_u = M_{uD}$

(v) Knowing $x_u$, compute: $C_{uw} = 0.36 f_{ck} x_u b_w$ ...[35.11(a)]

$C_{uf} = 0.45 f_{ck} (b_f - b_w) y_f$ ...(35.12)

∴ $C_u = C_{uw} + C_{uf}$.

(vi) Compute $$A_{st} = \frac{C_u}{0.87 f_y}$$ ...(35.27)

Alternatively, $x_u$ can be found by substituting the value of $y_f$ in terms of $x_u$ in Eq. 35.17. and then solving the quadratic equation in terms of $x_u$. This value of $x_u$ will be acceptable only if the value of $y_f$ obtained from this is less than (or equal to) $D_f$.

**Example 35.1. M.R. of T-beam.** *Find the moment of resistance of an existing T-beam having the following data:* $b_f = 740$ mm; $d = 400$ mm; $b_w = 240$ mm

$A_{st} = 5\text{-}20$ mm $\Phi$ HYSD steel bars

$D_f = 100$ mm. M 20 concrete.

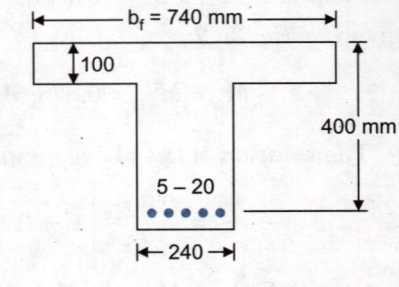

Fig. 35.9

**Solution:** $A_{st} = 5 \frac{\pi}{4} (20)^2 \approx 1570$ mm$^2$

Assuming the N.A. to fall in the flange, we get, from Eq. 35.6

$$\frac{x_u}{d} = \frac{0.87 f_y A_{st}}{0.36 f_{ck} b_f d} = \frac{0.87 \times 415 \times 1570}{0.36 \times 20 \times 740 \times 400} = 0.2659$$

or $x_u = 0.2659 \times 400 = 106.36$ mm

∴ $x_u < D_f$. Hence the N.A. lies in the flange.

Also, $x_{u.max} = 0.48 \times d$ for Fe 415 steel $= 0.48 \times 400 = 192$ mm $> x_u$.

Hence, the moment of resistance is given by

$$M_u = 0.87 f_y \cdot A_{st} \cdot d \left(1 - \frac{f_y}{f_{ck}} \cdot \frac{A_{st}}{b_f \times d}\right)$$

$$= 0.87 \times 415 \times 1570 \times 400 \left(1 - \frac{415}{20} \times \frac{1570}{740 \times 400}\right) = 201.78 \times 10^6 \text{ N-mm} = \mathbf{201.78 \text{ kN-m}}.$$

**Example 35.2. M.R. of T-beam**

*Determine the M.R. of T-beam of example 35.1 if $D_f = 80$ mm.*

**Solution:** Assuming the N.A. to fall in the flange, we get $x_u = 85.5$ mm, as obtained in example 35.1. Since $x_u > D_f (= 80$ mm), the N.A. falls outside the flange. Let us obtain a new value of $x_u$ by equating the total compressive force to the total tensile force.

Total tensile force $\quad T_u = 0.87 f_y A_{st} = 0.87 \times 415 \times 1570 = 566848.5$ N

$\quad\quad\quad\quad\quad y_f = 0.15 x_u + 0.65 D_f = 0.15 x_u + 0.65 \times 80 = 0.15 x_u + 52$

Hence $\quad\quad C_u = 0.36 f_{ck} x_u b_w + 0.45 f_{ck} (b_f - b_w) y_f$

$\quad\quad\quad\quad = (0.36 \times 20 x_u \times 240) + 0.45 \times 20 (740 - 240)(0.15 x_u + 52)$

$\quad\quad\quad\quad = 1728 x_u + 675 x_u + 234000 = 2403 x_u + 234000$

Equating $\quad C_u = T_u$, we get, $2403 x_u + 234000 = 566848.5$

From which $\quad x_u = 138.5$ mm. This value will be acceptable only if $y_f \leq D_f$

Now $\quad\quad y_f = 0.15 x_u + 0.65 D_f = 0.15 \times 138.5 + 52 \approx 72.77$ mm $< D_f$

Hence $\quad\quad x_u = 138.5$ mm is acceptable.

$\quad\quad\quad x_{u.max} = 0.48 d$ (for Fe 415 steel) $= 0.48 \times 400 = 192$ mm $> x_u$

Now $\quad \frac{3}{7} x_u = \frac{3}{7} \times 138.5 = 59.35$ mm. $\therefore D_f > \frac{3}{7} x_u =$ or $\frac{D_f}{x_u} > 0.43$ which corresponds to case 3 (b) of § 35.7 of IS Code procedure. Hence, the moment of resistance is given by Eq. 35.24.

$$M_u = 0.36 \frac{x_u}{d}\left(1 - 0.42 \frac{x_u}{d}\right) f_{ck} b_w d^2 + 0.45 f_{ck} (b_f - b_w) y_f \left(d - \frac{y_f}{2}\right)$$

where $\quad y_f = 0.15 x_u + 0.65 D_f = 66$ mm (computed above)

$$M_u = \frac{0.36 \times 138.50}{400}\left(1 - 0.42 \frac{138.50}{400}\right) \times 20 \times 240 (400)^2 + 0.45 \times 20 (740 - 240) \times 66 (400 - 0.5 \times 66)$$

$\quad\quad = 81.80 \times 10^6 + 108.99 \times 10^6 = 190.790 \times 10^6$ N-mm $= \mathbf{190.79 \text{ kN-m}}.$

**Example 35.3. M.R. of T-beam.** *Determine the M.R. of T-beam shown in Fig. 35.10. Use M 15 concrete and Fe 415 steel.*

**Solution:** $\quad A_{st} = 5 \times \frac{\pi}{4} (25)^2 \approx 2454.4$ mm$^2$

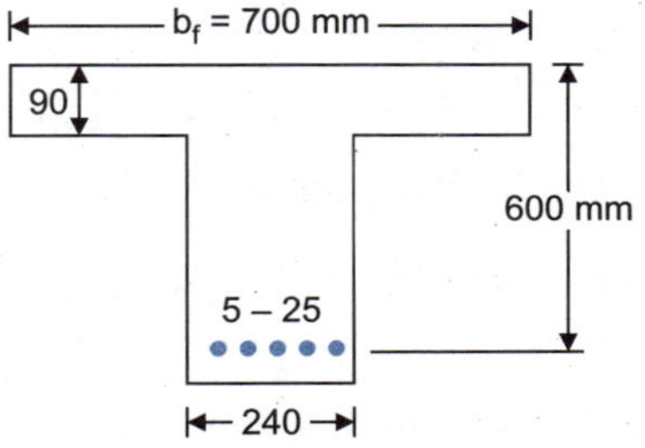

Assuming the N.A. to fall in the flange, we have,

$$\frac{x_u}{d} = \frac{0.87 f_y A_{st}}{0.36 f_{ck} b_f d} = \frac{0.87 \times 415 \times 2454.4}{0.36 \times 15 \times 700 \times 600} = 0.3907$$

$\therefore \quad x_u = 0.3907 \times 600 = 234.4$ mm $> D_f.$

Hence N.A. falls outside flange. To obtain a new value of $x_u$, equate total compressive force to the total tensile force.

$\quad\quad\quad T_u = 0.87 f_y A_{st} = 0.87 \times 415 \times 2454.4 \approx 886161$ N

$\quad\quad\quad y_f = 0.15 x_u + 0.65 D_f = 0.15 x_u + 0.65 \times 90 = 0.15 x_u + 58.5$

with a maximum equal to $D_f$. Since approximate value of $x_u$ found above is 234.4, $y_f \approx 0.15 \times 234.4 + 58.5 \approx 93.7 > D_f.$

Fig. 35.10

Hence take $\quad y_f = D_f = 90$ mm

$$C_u = 0.36 f_{ck} x_u b_w + 0.45 f_{ck} (b_f - b_w) D_f$$
$$= 0.36 \times 15 \times x_u (240) + 0.45 \times 15 (700 - 240) 90 = 1296 x_u + 279450.$$

Equating $C_u$ and $T_u$, we get, $1296 x_u + 279450 = 886161 \quad \therefore \quad x_u = 468.1$ mm.

$x_{u.max} \triangleq 0.48 d$ (for Fe 415 steel) $= 0.48 \times 600 = 288$ mm.

Hence $x_u > x_{u.max}$. Thus the beam is *over-reinforced*. Though the Code recommends that such a beam should be redesigned, its moment of resistance is limited to $M_{u.lim}$. Corresponding to the balanced section.

$\therefore$ Take $\quad x_u = x_{u.max} = 288$ mm.

Again, $\quad \dfrac{D_f}{d} = \dfrac{90}{500} = 0.18 < 0.2$. Hence moment of resistance is given by *Eq. 35.21* (corresponding to para G-2.2 of code):

$$M_u = M_{u.lim} = 0.36 \frac{x_{u.max}}{d} \left(1 - 0.42 \frac{x_{u.max}}{d}\right) f_{ck} b_w d^2 + 0.45 f_{ck} (b_f - b_w) D_f \left(d - \frac{D_f}{2}\right)$$

$$= 0.36 \times \frac{288}{600}\left(1 - 0.416 \frac{288}{600}\right) 15 \times 240 \,(600)^2 + 0.45 \times 15 \,(700 - 240) \times 90 \left(600 - \frac{90}{2}\right)$$

$$= 179.2 \times 10^6 + 155.09 \times 10^6 = 334.3 \times 10^6 \text{ N-mm} = \mathbf{334.3 \text{ kN-m}}.$$

**Example 35.4.** **Limiting M.R. and Limiting area of steel**

*A reinforced concrete T-beam has the following data:*
*Width of flange = 1600 mm;  Effective depth = 350 mm*
*Thickness of flange = 105 mm;  Width of web = 250 mm*
*Grade of concrete : M 20 ;  Grade of steel Fe 500*
*Determine (a) Limiting moment of resistance and (b) Limiting area of tensile steel.*

**Solution:** Given $b_f = 1600$ mm; $d = 350$ mm; $D_f = 105$ mm; $b_w = 250$ mm

For Fe 500 steel, $\quad \dfrac{x_{u,max}}{d} = \dfrac{700}{1100 + 0.87 f_y} = \dfrac{700}{1100 + 0.87 \times 500} = 0.456$

$\therefore \quad x_{u,max} = 0.456 d = 0.456 \times 350 = 159.6$ mm $> D_f (= 105$ mm$)$

Also, $\quad \dfrac{D_f}{d} = \dfrac{105}{350} = 0.3$. Since $\dfrac{D_f}{d} > 0.2$, it is case (b) of § 35.7.

$$y_f = (0.15 x_{u,max} + 0.65 D_f), \text{ but not greater than } D_f (= 105 \text{ mm})$$
$$= 0.15 \times 159.6 + 0.65 \times 105 = 92.19 \text{ mm, which is less than } D_f.$$

The limiting moment of resistance is given by *Eq. 35.22.*

$$M_{u, lim} = 0.36 \frac{x_{u,max}}{d}\left(1 - 0.42 \frac{x_{u,max}}{d}\right) f_{ck} b_w d^2 + 0.45 f_{ck} (b_f - b_w) y_f \left(d - \frac{y_f}{2}\right)$$

$$= 0.36 \times 0.456 \,(1 - 0.42 \times 0.456) \times 20 \times 250\,(350)^2 + 0.45 \times 20 \,(1600 - 250) \times 92.19 \left(350 - \frac{92.19}{2}\right)$$
$$= 421.69 \times 10^6 \text{ N-mm} = \mathbf{421.69 \text{ kN-m}}$$

Also, $\quad A_{st, lim.} = \dfrac{0.36 f_{ck} b_w \cdot x_{u, max} + 0.45 f_{ck} (b_f - b_w) y_f}{0.87 f_y}$

$$= \frac{0.36 \times 20 \times 250 \times 159.6 + 0.45 \times 20 \,(1600 - 250)\, 92.19}{0.87 \times 500}$$

$$= \mathbf{3235.4 \text{ mm}^2}$$

# T AND L-BEAMS

## Example 35.5. Design of T-beam

A T-beam has the following data : Width of flange = 750 mm ; Breadth of beam = 250 mm. Effective depth = 500 mm ; Thickness of flange = 90 mm ; Applied moment = 130 kN-m

Design the beam. Use M 20 concrete and Fe 415 steel.

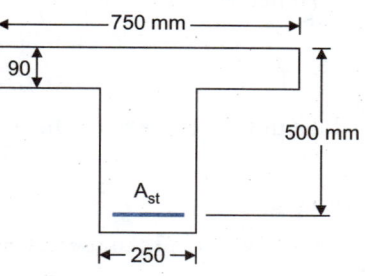

Fig. 35.11

**Solution:** Given: $b_f$ = 750 mm; $b_w$ = 250 mm; $d$ = 500 mm; $D_f$ = 90 mm; $M_D$ = 130 kN-m. Hence $M_{uD}$ = 1.5 × 130 = 195 kN-m = 195 × $10^6$ N-mm

Assuming $x_u = D_f$, we get

$$M_{u1} = 0.36\, f_{ck}\, b_f\, D_f\, (d - 0.42\, D_f)$$
$$= 0.36 \times 20 \times 750 \times 90\, (500 - 0.42 \times 90)$$
$$= 224.6 \times 10^6 \text{ N-mm} > M_{uD}$$

∴ $x_u < D_f$. The reinforcement is found from $M_{uD} = 0.87\, f_y\, A_{st}\, d \left(1 - \dfrac{f_y}{f_{ck}} \cdot \dfrac{A_{st}}{b_f \cdot d}\right)$

or $\quad 195 \times 10^6 = 0.87 \times 415 \times 500\, A_{st} \left(1 - \dfrac{415}{20} \times \dfrac{A_{st}}{750 \times 500}\right) \quad$ or $\quad A_{st}(1 - 5.533 \times 10^{-5} A_{st}) = 1080.2$

or $\quad A_{st}^2 - 18072\, A_{st} + 19.52 \times 10^6$, from which $A_{st} \triangleq$ **1154 mm²**

Alternatively, $\quad A_{st} = \dfrac{0.5\, f_{ck}}{f_y} \left[1 - \sqrt{1 - \dfrac{4.6\, M_{uD}}{f_{ck}\, b_f\, d^2}}\right] b_f\, d \quad$ ...[35.7(a)]

$$= \dfrac{0.5 \times 20}{415} \left[1 - \sqrt{1 - \dfrac{4.6 \times 195 \times 10^6}{20 \times 750 \times 500^2}}\right] \times 750 \times 500 \approx \mathbf{1154 \text{ mm}^2}$$

## Example 35.6. Design of T-beam

Redesign the T-beam of example 35.5 if the applied moment is 230 kN-m.

**Solution:** $M_{uD}$ = 1.5 × 230 = 345 kN-m = 345 × $10^6$ N-mm

Since $M_{uD}$ is more than $M_{u_1}$ (= 224.8 × $10^6$) determined in the previous example, $x_u > D_f$.

Let us calculate compressive force for $x_u = \dfrac{7}{3} D_f = \dfrac{7}{3} \times 90 = 210$ mm

∴ $\quad C_{uw} = 0.36\, f_{ck}\, x_u\, b_w = 0.36 \times 20 \times 210 \times 250 = 378000$ N
$\quad C_{uf} = 0.45\, f_{ck}\, (b_f - b_w)\, D_f = 0.45 \times 20\, (750 - 250)\, 90 = 405000$ N
$\quad M_{uw} = 378000\, (500 - 0.416 \times 210) = 155.98 \times 10^6$ N-mm
$\quad M_{uf} = 401400\, (500 - 0.5 \times 90) = 182.64 \times 10^6$ N-mm

∴ $\quad M_{u2} = M_{uw} + M_{uf} = (155.98 + 182.64)\, 10^6 = 338.62 \times 10^6$ N-mm

Since $\quad M_{u2} < M_{uD}, \quad x_u > \dfrac{7}{3} D_f$

Now, $M_{uf}$ does not depend on $x_u$, and thus remains constant.

∴ $\quad M_{uw} = M_{uD} - M_{uf} = (345 - 182.64)\, 10^6 = 162.36 \times 10^6$ N-mm.

Now $\quad M_{uw} = 0.87\, f_y\, A_{sw}\, d \left(1 - \dfrac{f_y}{f_{ck}} \dfrac{A_{sw}}{b_w\, d}\right) \quad$ ...(35.28)

∴ $\quad 162.36 \times 10^6 = 0.87 \times 415 \times A_{sw} \times 500 \left(1 - \dfrac{415}{20} \times \dfrac{A_{sw}}{250 \times 500}\right)$

or $\quad A_{sw}(1 - 1.66 \times 10^{-4} A_{sw}) = 899$
or $\quad A_{sw}^2 - 6024\, A_{sw} + 5.418 \times 10^6 = 0;\quad$ From which $A_{sw} \triangleq$ **1100.5 mm²**

Alternatively, $\quad A_{sw} = \dfrac{0.5 \times 20}{415} \left[1 - \sqrt{1 - \dfrac{4.6 \times 162.36 \times 10^6}{20 \times 250\, (500)^2}}\right] 250 \times 500 \triangleq$ **1101.1 mm²**

Also, $A_{sf} = \dfrac{C_{uf}}{0.87 f_y} = \dfrac{401400}{0.87 \times 415} \approx 1111.8 \text{ mm}^2$

∴ Total $A_{st} = A_{sw} + A_{sf} = 1101.1 + 1111.8 = 2212.9 \text{ mm}^2$

It is now necessary to check that actual $x_u$ is less than the critical value.

$$x_u = \dfrac{0.87 f_y A_{sw}}{0.36 f_{ck} b_w} = \dfrac{0.87 \times 415 \times 1101.1}{0.36 \times 20 \times 250} = 220.86 \text{ mm}$$

$x_{u,\,max} \approx 0.48 \times 500 = 240 \text{ mm} > x_u$.    Hence OK.

### Example 35.7. Design of T-beam

*Redesign the T-beam of Example 35.5 if the applied moment is 196 kN-m.*

**Solution:**   $M_{uD} = 1.5 \times 196 = 294 \text{ kN-m}$

In example 35.5,   $M_{u1} = 224.8 \text{ kN-m}$; In example 35.6. $M_{u2} = 338.62 \text{ kN-m}$

∴   $M_{u1} < M_{uD} < M_{u2}$. Hence for this case $x_u < \dfrac{7}{3} D_f < 210 \text{ mm}$

The design procedure will be essentially that of trial and error.

**Trial 1:** Let us assume $x_u = 150$ mm. Using I.S. Code method,

$y_f = 0.15 x_u + 0.65 D_f = 0.15 \times 150 + 0.65 \times 90 = 81 \text{ mm} \le D_f$.    Hence OK.

$M_u = 0.36 \dfrac{x_u}{d}\left(1 - 0.42 \dfrac{x_u}{d}\right) f_{ck} b_w d^2 + 0.45 f_{ck} (b_f - b_w) y_f \times (d - 0.5 y_f)$  ...(35.17)

$= 0.36 \times \dfrac{150}{500}\left(1 - 0.42 \times \dfrac{150}{500}\right) \times 20 \times 250 (500)^2 + 0.45 \times 20 (750 - 250) 81 (500 - 0.5 \times 81)$

$= 117.99 \times 10^6 + 167.49 \times 10^6 = 285.5 \times 10^6 \text{ N-mm} < M_{uD}$

**Trial 2.** Let $x_u = 160$ mm

$y_f = 0.15 \times 160 + 0.65 \times 90 = 82.5 \text{ mm} \le D_f$

$M_u = 0.36 \times \dfrac{160}{500}\left(1 - 0.416 \times \dfrac{160}{500}\right) \times 20 \times 250 (500)^2 + 0.446 \times 20 (750 - 250) 82.5 (500 - 0.5 \times 82.5)$

$= 124.83 \times 10^6 + 168.80 \times 10^6 = 293.63 \approx M_{uD}$

Hence adopt    $x_u = 160$ mm

Now   $C_{uw} = 0.36 f_{ck} x_u b_w = 0.36 \times 20 \times 160 \times 250 = 288000 \text{ N}$ (*Eq. 35.10 a*)

$C_{uf} = 0.45 f_{ck} (b_f - b_w) y_f = 0.45 \times 20 (750 - 250) 82.5 = 371250 \text{ N}$

∴ Total compression  $C_u = 288000 + 371250 = 659250 \text{ N}$

∴   $A_{st} = \dfrac{C_u}{0.87 f_y} = \dfrac{659250}{0.87 \times 415} = \mathbf{1825.9 \text{ mm}^2}$

**Note:** $x_u$ can be alternatively obtained by substituting the value of $y_f$ in terms of $x_u$ in *Eq. 35.17* and then solving the quadratic equation for $x_u$.

Thus,    $y_f = 0.15 x_u + 0.65 \times 90 = 0.15 x_u + 58.5$

Hence from *Eq. 35.17*, putting $M_u = M_{uD} = 294 \times 10^6$ N-mm, we get

$$294 \times 10^6 = 0.36 \dfrac{x_u}{500}\left(1 - \dfrac{0.416 x_u}{500}\right) 20 \times 250 (500)^2 + 0.446 \times 20 (750 - 250)$$
$$\times (0.15 x_u + 58.5) \times [500 - 0.5 (0.15 x_u + 58.5)]$$

Simplifying and arranging, we get $x_u^2 - 1496 x_u + 214295 = 0$

From which $x_u = 160.4$ mm. This value of $x_u$ will be acceptable only if the value of $y_f$ obtained from this is less than $D_f$.

Thus,    $y_f = 0.15 \times 160.4 + 0.65 \times 90 = 82.56 \text{ mm} < D_f$

Hence the exact value of   $x_u = 160.4$ mm

$$C_{uw} = 0.3 \times 20 \times 160.4 \times 250 = 240600 \text{ N}$$
$$C_{uf} = 0.45 \times 20 (750 - 250) 82.56 = 371520 \text{ N}$$

Total $\quad C_u = 240600 + 371520 = 612120 \text{ N} \quad \therefore \quad A_{st} = \dfrac{612120}{0.87 \times 415} = 1695.4 \text{ mm}^2$

## Example 35.8. Design of T-beam

Design the reinforcement for a T-beam for the following data:

| | | |
|---|---|---|
| Effective span | : | 8 m; Ends simply supported. |
| Spacing of beams | : | 3.3 m centre to centre. |
| Thickness of slab | : | 130 mm |
| Width of web | : | 300 mm |
| Total Depth | : | 450 mm |
| Live load on the floor | : | 10 kN/m² |
| Floor finish load | : | 0.75 kN/m². |

The beam also supports a partition wall which transmits a load of 12 kN/m run.
Use M 20 concrete and Fe 415 steel

**Solution:** Weight of slab = $25 \times 0.13$ = 3.25 kN/m²
Weight of finishing = 0.75 kN/m²
Live load = 10.0 kN/m²
Total 14.0 kN/m².

Load on beam, transferred from slab = $14 \times 3.3$ = 46.2 kN/m
Load transferred from wall = 12 kN/m
Weight of rib = $0.3 (0.45 - 0.13) \times 25$ = 2.4 kN/m
Total working load $w$ = 60.6 kN/m.
Load factor = 1.5.
∴ Ultimate load
$\quad w_u = 1.5 \times 60.6$ = 90.9 kN/m
∴ Ultimate design moment
$$M_{uD} = \dfrac{w_u L^2}{8} = \dfrac{90.9 (8)^2}{8}$$

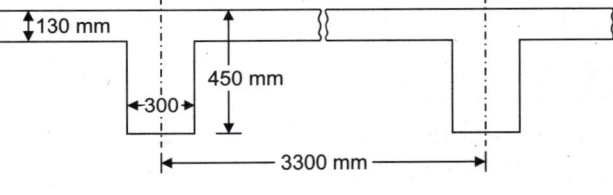

Fig. 35.12

$= 727.2$ kN-m. $= 727.2 \times 10^6$ N-mm.

The flange width $b_f$ given by,
$$b_f = \dfrac{l_0}{6} + b_w + 6 D_f = \dfrac{8}{6} + 0.3 + 6 (0.13) = 2.413 \text{ m} = 2413 \text{ mm}$$
$$d = D - \text{cover} = 450 - 80 \text{ (say)} = 370 \text{ mm}$$

**Step 1.** Assuming $x_u = D_f$. we get
$$M_{u1} = 0.36 f_{ck} b_f D_f (d - 0.42 D_f)$$
$$= 0.36 \times 20 \times 2413 \times 130 (370 - 0.42 \times 130) = 712.35 \times 10^{-6} \text{ N-m.}$$

**Step 2.** Since $M_{uD} > M_{u1}$, N.A. falls outside the flange

Assuming $\quad x_u = \dfrac{7}{3} D_f \approx 303.3 \text{ mm}$, we get
$$C_{uw} = 0.36 f_{ck} x_u b_w = 0.36 \times 20 \times 303.3 \times 300 = 655128 \text{ N.}$$
$$y_f = 0.15 \times 303.3 + 0.65 \times 130 = 45.5 + 84.5 = 130 \text{ mm, but} \leq D_f$$
$$C_{uf} = 0.45 f_{ck} (b_f - b_w) y_f = 0.45 \times 20 (2413 - 300) 130 = 2472210 \text{ N}$$

Hence $\quad M_{uw} = C_{uw} (d - 0.416 x_u) = 655128 (370 - 0.416 \times 303.3) = 159.7 \times 10^6$ N-mm
and $\quad M_{uf} = C_{uf} (d - 0.5 D_f) = 2450235 (370 - 0.5 \times 130) = 747.3 \times 10^6$ N-mm
$\quad M_{u2} = M_{uw} + M_{uf} = (159.7 + 747.3) 10^6 = 907 \times 10^6$ N-mm

Since $M_{uD} < M_{u2}$, $x_u < \frac{7}{3} D_f$. Hence the design procedure is done by trial and error.

**Step 3.** *Let us follow I.S. Code method:*

(i) Assume $x_u < \frac{7}{3} D_f < 303.3$ mm. Let us assume $x_u = 230$ mm.

Then $y_f = 0.15 x_u + 0.65 D_f = 0.15 \times 230 + 0.65 \times 130 = 119$ mm

and $M_u = 0.36 \dfrac{x_u}{d}\left(1 - 0.42 \dfrac{x_u}{d}\right) f_{ck} b_w d^2 + 0.45 f_{ck} (b_f - b_w) \times y_f (d - 0.5 y_f)$

$= 0.36 \times \dfrac{230}{370}\left(1 - 0.42 \dfrac{230}{370}\right) \times 20 \times 300 \,(370)^2 + 0.45 \times 20 \,(2413 - 300)\, 119\, (370 - 0.5 \times 119)$

$= 135.8 \times 10^6 + 702.7 \times 10^6 = 838.5 \times 10^6$ N-mm

This is quite close to $M_{uD} = 727.2 \times 10^6$ N-mm. However, in the second trial, let $x_u = 150$ mm.

$y_f = (0.15 \times 150) + 0.65 \times 130 = 107$

$M_u = 0.36 \times \dfrac{150}{370}\left(1 - 0.42 \times \dfrac{150}{370}\right) \times 20 \times 300\, (370)^2 + 0.45 \times 20\, (2413 - 300)\, 107\, (370 - 0.5 \times 107)$

$= 99.5 \times 10^6 + 644 \times 10^6 = 743.5 \times 10^6 \approx M_{uD}$

Hence $x_u = 150$ mm

∴ $C_{uw} = 0.36 f_{ck} x_u b_w = 0.36 \times 20 \times 150 \times 300 = 324000$ N

$C_{uf} = 0.45 f_{ck} (b_f - b_w) y_f = 0.45 \times 20\, (2413 - 300)\, 107 = 2034819$ N

∴ $C_u = C_{uw} + C_{uf} = 324000 + 2034819 = 2358819$ N

∴ $A_{st} = \dfrac{C_u}{0.87 f_y} = \dfrac{2358819}{0.87 \times 415} \approx \mathbf{6533.2 \text{ mm}^2}$

## 35.10. ANALYSIS OF DOUBLY REINFORCED T-BEAMS

We have seen in the previous articles that a singly reinforced-section has limiting value of moment of resistance. However, if the applied moment is more than $M_{u.lim.}$ two alternatives will be available: (i) to increase the depth of the section, or (ii) to provide compressive reinforcement. In many cases the maximum value of depth of section may be limited or restricted from architectural or other considerations. In that case, the only alternative that will be left will be to provide reinforcement in the compression zone, giving rise to *doubly reinforced-section*.

Fig. 35.13(a) shows a doubly reinforced T-beam with compression reinforcement $A_{sc}$ provided at a cover of $d'$. The strain in the compressive reinforcement is $\varepsilon_{sc}$, as shown in Fig. 35.13(b). The compressive force contribution of $A_{sc}$ will be $C_{su} = f_{sc} A_{sc}$ [Fig. 35.13(c)] where $f_{sc}$ is the compression stress in steel at a strain of $\varepsilon_{sc}$, and is found from the stress strain curve.

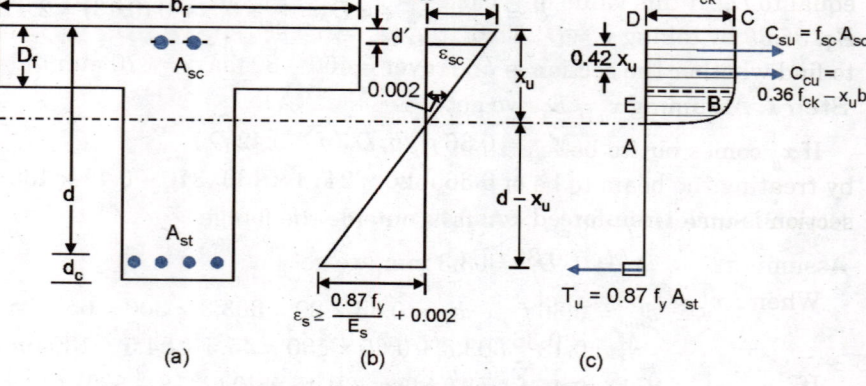

Fig. 35.13. Doubly Reinforced T-beam.

Now $\varepsilon_{sc} = \dfrac{0.0035\,(x_u - d')}{x_u}$ ...(35.29)

**DETERMINATION OF $x_u$:** For a given section, $x_u$ can be determined by equating the total tensile force to the total compressive force. Assuming that the N.A. lies in the flange,
$$0.36 f_{ck} b_f x_u + f_{sc} A_{sc} = 0.87 f_y A_{st} \qquad \ldots(35.30)$$
If the loss of concrete area occupied by compressive steel is taken into account, we have
$$0.36 f_{ck} b_f x_u + (f_{sc} - 0.446 f_{ck}) A_{sc} = 0.87 f_y A_{st} \qquad \ldots[35.30(a)]$$
(This is subject to the condition that $d' < \frac{3}{7} x_u$ so that stress in concrete at the level of compressive steel is equal to $0.446 f_{ck}$.)

Stress $f_{sc}$ is found from stress-strain curve (or Table 33.1), corresponding to the value of $\varepsilon_{sc}$ given by Eq. 35.29. Thus, $f_{sc}$ indirectly depends upon $x_u$. Hence the following *iterative procedure* is adopted:

**Step 1.** Find $x_{u.max}$ for a balanced section from strain compatibility:
$$x_{u.max} = \frac{0.0035}{0.0055 + 0.87 f_y / E_s} = \frac{700}{1100 + 0.87 f_y}$$

**Step 2.** Assume $x_u = x_{u.max}$ or $D_f$ (whichever is less) and compute $\varepsilon_{sc}$ from Eq. 35.29.

**Step 3.** Find $f_{sc}$ from stress-strain curve, corresponding to the above value of $\varepsilon_{sc}$. (Alternatively, in the first trial, take $f_{sc}$ slightly less than $0.87 f_y$).

**Step 4.** Substitute this value of $f_{sc}$ in Eq. 35.30 (or Eq. 35.30 a) and compute the modified value of $x_u$.

**Step 5.** Repeat step 2 to 4, till convergence of the value of $x_u$ is achieved.

If $x_u < D_f$, the N.A. lies in the flange and hence Eq. 35.30 [or 35.30(a)] is valid. If however, $x_u$ comes out to be greater than $D_f$, the N.A. will be outside the flange, *and a new value of $x_u$ will have to be found* by equating the total compressive force to the total tensile force. The determination of $x_u$ will be done exactly in the same manner as done for the singly reinforced T-beam discussed in § 35.6. The total compressive force $C_u$ will consists of (i) web contribution, $C_{uw}$ and (ii) flange contribution $C_{uf}$, where
$$C_{uw} = 0.36 f_{ck} x_u b_w + (f_{sc} - 0.446 f_{ck}) A_{sc} = 0.87 f_y A_{sw} \qquad \ldots(35.31)$$
The compressive force $C_{uf}$ can be computed by approximating the actual stress block into an equivalent rectangular block of height $y_f$ given by
$$y_f = 0.15 x_u + 0.65 D_f \text{ (subject to a max. of } D_f)$$
$$\therefore \quad C_{uf} = 0.446 f_{ck} (b_f - b_w) y_f = 0.87 f_y A_{sf} \qquad \ldots(35.32)$$
Equating the total compressive force to the total tensile force, we get
$$\therefore \quad 0.36 f_{ck} \cdot x_u b_w + (f_{sc} - 0.446 f_{ck}) A_{sc} + 0.446 f_{ck} (b_f - b_w) y_f = 0.87 f_y A_{st} \qquad \ldots(35.33)$$
Since $f_{sc}$ depends upon $x_u$, an iterative procedure will have to be adopted. Also, $y_f$ depends upon $x_u$, and hence the final value of $x_u$ is valid only if the value of $y_f$ calculated by using this value of $x_u$, is less than or equal to $D_f$. If this value of $y_f$ comes out to be greater than $D_f$, the value of $x_u$ should be *recalculated* from Eq. 35.33 by taking $y_f = D_f$. Thus, the value of $x_u$ is known. This value of $x_u$ should be compared with $x_{u.max}$ to find whether the section is under reinforced, balanced or over-reinforced.

### 1. MOMENT OF RESISTANCE WHEN $x_u < D_f$

If $x_u$ comes out to be less than $D_f$, from Eq. 35.30 or 35.30(a), the moment of resistance can be found by treating the beam to be of a doubly reinforced rectangular section of width $b_f$. Again, if $x_u < x_{u.max}$, the section is under-reinforced, and the moment of resistance is given by
$$M_u = 0.36 f_{ck} x_u b_f (d - 0.42 x_u) + (f_{sc} - 0.45 f_{ck}) A_{sc} (d - d') \qquad \ldots(35.34)$$
When $x_u = D_f$, we get
$$M_{u1} = 0.36 f_{ck} b_f D_f (d - 0.42 D_f) + (f_{sc} - 0.45 f_{ck}) A_{sc} (d - d') \qquad \ldots[35.34(a)]$$
If $x_u = x_{u.max}$, we have the balanced section, for which,
$$M_{u.lim} = 0.36 f_{ck} x_{u.max} b_f (d - 0.42 x_{u.max}) + (f_{sc} - 0.45 f_{ck}) A_{sc} (d - d') \qquad \ldots(35.35)$$
If $x_u > x_{u.max}$, the section is *over-reinforced*, and its moment resistance $M_u$ will be limited to $M_{u.lim}$ given by Eq. 35.35. However, the Code recommends that such section should be redesigned.

## 2. MOMENT OF RESISTANCE WHEN $x_u > D_f$

The value of $x_u$ obtained from Eq. 35.33 should be compared with $x_{u.max}$ to find whether the section is under-reinforced, balanced or over-reinforced.

The procedure for finding $M_u$ will be the same as that discussed for singly reinforced section except that an additional moment of resistance contribution by compressive steel = $(f_{sc} - 0.446 f_{ck}) A_{sc} (d - d')$ will have to be taken into account.

**(a) Thus, for Under-reinforced Section** $(x_u < x_{u.max})$,

If $D_f / x_u \leq 0.43$ or 3/7, we have Case 2, for which

$$M_{uw} = 0.36 f_{ck} \frac{x_u}{d} b_w \left(1 - 0.42 \frac{x_u}{d}\right) d^2 + (f_{sc} - 0.45 f_{ck}) A_{sc}(d - d') \quad \text{...[35.36(a)]}$$

$$M_{uf} = 0.446 f_{ck} (b_f - b_w) D_f (d - 0.5 D_f) \quad \text{...[35.36(b)]}$$

and

$$M_u = 0.36 f_{ck} \frac{x_u}{d} b_w \left(1 - 0.42 \frac{x_u}{d}\right) d^2 + (f_{sc} - 0.45 f_{ck}) A_{sc}(d - d')$$
$$+ 0.446 f_{ck} (b_f - b_w) D_f (d - 0.5 D_f) \quad \text{...(35.36)}$$

When $D_f = \frac{3}{7} x_u$ (or $x_u = \frac{7}{3} D_f$), we have

$$M_{u2} = 0.36 f_{ck} b_w \left(\frac{7}{3} D_f\right)\left(d - 0.42 \times \frac{7}{3} D_f\right) + (f_{sc} - 0.45 f_{ck}) A_{sc} (d - d')$$
$$+ 0.446 f_{ck} (b_f - b_w) D_f (d - 0.5 D_f) \quad \text{...[35.36(c)]}$$

**(b) Similarly, when** $x_u < x_{u.max}$ **and** $D_f / x_u > 0.43$ or 3/7 we have Case 3, for which

$$M_{uw} = 0.36 f_{ck} \frac{x_u}{d} b_w \left(1 - 0.42 \frac{x_u}{d}\right) d^2 + (f_{sc} - 0.446 f_{ck}) A_{sc}(d - d') \quad \text{...(35.37a)]}$$

and

$$M_{uf} = 0.45 f_{ck} (b_f - b_w) y_f (d - 0.5 y_f)$$

and

$$M_u = 0.36 \frac{x_u}{d} \left(1 - 0.42 \frac{x_u}{d}\right) f_{ck} b_w d^2 + (f_{sc} - 0.45 f_{ck}) A_{sc} (d - d')$$
$$+ 0.45 f_{ck} (b_f - b_w) y_f (d - 0.5 y_f) \quad \text{...(35.37)}$$

## 3. MOMENT OF RESISTANCE WHEN $(x_u = x_{u.max}) > D_f$

The limiting moment of resistance can be computed by modifying the equations (Eq. 35.21 and 35.22) suggested by IS Code, by including the moment of resistance term of the compression steel. Thus, we have the following modified equations.

**(a) When** $\frac{D_f}{d} \leq 0.2$.

$$M_{u,lim} = 0.36 \frac{x_{u.max}}{d} \left(1 - 0.42 \frac{x_{u.max}}{d}\right) f_{ck} b_w d^2 + (f_{sc} - 0.446 f_{ck}) A_{sc} (d - d')$$
$$+ 0.45 f_{ck} (b_f - b_w) D_f \left(d - \frac{D_f}{2}\right) \quad \text{...(35.38)}$$

where $f_{sc}$ is found from stress-strain curve corresponding to $\varepsilon_{sc}$ given by Eq. 35.29 in which $x_u = x_{u,max}$.

**(b) When** $\frac{D_f}{d} > 0.2$.

$$M_{u,lim} = 0.36 \frac{x_{u.max}}{d} \left(1 - 0.42 \frac{x_{u.max}}{d}\right) f_{ck} b_w d^2 + (f_{sc} - 0.45 f_{ck}) A_{sc} (d - d')$$
$$+ 0.446 f_{ck} (b_f - b_w) y_f \left(d - \frac{y_f}{2}\right)$$

where $y_f = (0.15 x_{u.max} + 0.65 D_f) \leq D_f$ ...(35.39)

## 4. MOMENT OF RESISTANCE WHEN $(x_u > x_{u.max}) > D_f$

If $x_u > x_{u.max}$ the section is over-reinforced, and its moment of resistance $M_u$ will be limited to $M_{u.lim}$ given by Eq. 35.38 (when $D_f/d \le 0.2$) or by Eq. 35.39 (when $D_f/d > 0.2$). However, the Code recommends that such a section should be redesigned.

### 35.11. DESIGN OF DOUBLY REINFORCED T-BEAM

A doubly reinforced section will be required when $M_{uD} > M_{u.lim}$. For a T-beam of given cross-sectional dimensions, (i.e. given $b_w$, $b_f$, $D_f$ and $d$), the design is done in the following steps.

**Step 1.** Assuming $x_u = D_f$, and assuming singly reinforced section, compute $M_{u1}$ from Eq. 35.9(a).

**Step 2.** If $M_{u1} < M_{uD}$, the N.A. will fall outside the flange.

Taking $x_u = \frac{7}{3} D_f$ and taking singly reinforced section, compute $M_{u2}$ from Eq. 35.16(c).

**Step 3.** If $M_{u2} < M_{uD}$, $x_u > \frac{7}{3} D_f$. Compute $M_{u,lim}$ by taking $x_u = x_{u,max}$, either from Eq. 35.21 (if $D_f/d \le 0.2$) or from Eq. 35.22 (if $D_f/d \ge 0.2$).

If $M_{u,lim} < M_{uD}$, a doubly reinforced section will be required.

**Step 4.** (a) Compute $A_{st1}$ for $M_{u.lim}$ as under:

$$A_{sw} = \frac{0.36 f_{ck} x_{u.max} b_w}{0.87 f_y} \quad \text{and} \quad A_{sf} = \frac{0.446 f_{ck} (b_f - b_w) D_f}{0.87 f_y}$$

$$\therefore A_{st1} = [0.36 f_{ck} x_{u.max} b_w + 0.446 f_{ck} (b_f - b_w) D_f] \div (0.87 f_y) \quad \ldots(35.40)$$

(b) Compute the additional tensile area $A_{st2}$ from the relation

$$A_{st2} = \frac{M_{uD} - M_{u.lim.}}{0.87 f_y (d - d')} \quad \ldots(35.41)$$

(c) Compute the compressive area $A_{sc}$ corresponding to $A_{st2}$, by the relation.

$$A_{sc} = \frac{M_u - M_{u.lim.}}{(f_{sc} - 0.45 f_{ck})(d - d')} \quad \ldots(35.42)$$

**Example 35.9. M.R. of doubly reinforced T-beam** *Find the moment of resistance of T-beam of example 35.3, if the beam also carries a compression reinforcement in the form of 3 bars of 25 mm dia. provided with a cover of 40 mm from the centre of bars.*

**Solution:** $A_{st} = 2454.4 \text{ mm}^2$ (as before);

$A_{sc} = 3 \times \frac{\pi}{4} (25)^2 = 1472.6 \text{ mm}^2$.

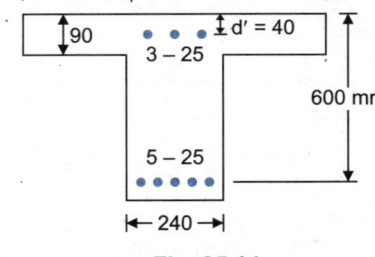

Fig. 35.14

Assuming the N.A. to fall in the flange, we get,

$$0.36 f_{ck} b_f x_u + (f_{sc} - 0.45 f_{ck}) A_{sc} = 0.87 f_y A_{st} \quad \ldots[35.30(a)]$$

Take $f_{sc} < 0.87 f_y < 0.87 \times 415 < 361$, in the first trial.

Hence take $f_{sc} = 350 \text{ N/mm}^2$ (say)

$\therefore \quad 0.36 \times 15 \times 700 x_u + (350 - 0.45 \times 15) \times 1472.6 = 0.87 \times 415 \times 2454.4$

From which $x_u = 100.7 \text{ mm} > D_f$

(Note that $d'$ is less than $3/7 x_u$ so that stress in concrete at $A_{sc}$ level is $0.446 f_{ck}$)

Hence the N.A. lies outside the flange. To obtain a new value of $x_u$, equate total compressive force to the total tensile force.

$$T_u = 0.87 f_y A_{st} = 0.87 \times 415 \times 2454.4 = 886161 \text{ N}$$

$$y_f = 0.15 x_u + 0.65 D_f = 0.15 x_u + 0.65 \times 90 = 0.15 x_u + 58.5$$

with a maximum equal to $D_f$. Since to approximate value of $x_u$ found above is 100.7 mm,

$$y_f = 0.15 \times 100.7 + 58.5 = 73.6 \text{ mm} < D_f.$$

$\therefore \quad C_u = 0.36 f_{ck} x_u b_w + (f_{sc} - 0.45 f_{ck}) A_{sc} + 0.45 f_{ck} (b_f - b_w) y_f$

or $\quad C_u = 0.36 \times 15\, x_u \times 240 + (f_{sc} - 0.45 \times 15)\, 1472.6 + 0.45 \times 15\, (700 - 240) \times (0.15\, x_u + 58.5)$

or $\quad C_u = 1296\, x_u + 1472.6\, f_{sc} - 9852 + 461.6\, x_u + 180028 = 1757.6\, x_u + 170176 + 1472.6\, f_{sc}$

Equating $C_u$ and $T_u$, we get $1757.6\, x_u + 170176 + 1472.6\, f_{sc} = 886161$

or $\quad x_u = 407.4 - 0.8378\, f_{sc}$ ...(i)

In the above equation, $f_{sc}$ can be obtained from stress-strain curve (or Table 33.1), corresponding to $\varepsilon_{sc}$ which itself depends upon $x_u$, as is evident from *Eq. 35.29*:

$$\varepsilon_{sc} = \frac{0.0035\,(x_u - d')}{x_u} \qquad ...(ii)$$

In the first approximation, taking $x_u = 100.7$ mm, $\varepsilon_{sc} = \dfrac{0.0035\,(100.7 - 40)}{100.7} = 0.0021$

From stress-strain curve for Fe 415 or from Table 33.1, we get $f_{sc} \approx 328$ N/mm².

Hence from (i), $\quad x_u = 407.4 - 0.8378 \times 328 = 132.6$ mm

For this new value of $x_u$, $y_f = 0.15 \times 132.6 + 58.5 = 78.4$ mm $< D_f$. Hence O.K.

In the next trial, recalculate $\varepsilon_{sc}$ by taking $x_u = 132.6$ mm, from (ii)

$\therefore \quad \varepsilon_{sc} = \dfrac{0.0035\,(132.6 - 40)}{132.6} = 0.0024$

From Table 33.1, we get $f_{sc} = 342$ N/mm² for $\varepsilon_{sc} = 0.0024$

Hence from (i), $\quad x_u = 407.4 - 0.8378 \times 342 = 120.8$ mm.

In the next trial, $\quad \varepsilon_{sc} = \dfrac{0.0035\,(120.8 - 40)}{120.8} = 0.00234$

From Table 33.1, $\quad f_{sc} = 340$ N/mm². for $\varepsilon_{sc} = 0.00234$

Hence from (i), $\quad x_u = 407.4 - 0.8378 \times 340 = 122.5$ mm

In the next trial, the solution will converge to a value of $x_u = 122$ mm, for which $\varepsilon_{sc} = 0.00235$ and $f_{sc} \approx 340.5$ N/mm²

$\quad y_f = 0.15\, x_u + 58.5 = 0.15 \times 122 + 58.5 = 76.8$ mm

$x_{u.max} \approx 0.48\, d$ (for Fe 415 steel) $\approx 0.48 \times 600 = 288$ mm

$\therefore \quad x_u < x_{u.max}$ and the section is under-reinforced.

Again, $\dfrac{3}{7} x_u = \dfrac{3}{7} \times 122 = 52.3$ mm $< D_f$. Hence $D_f > \dfrac{3}{7} x_u$

or $\quad \dfrac{D_f}{x_u} > 0.43$. Hence $M_u$ is given by *Eq. 35.37*

$$M_u = 0.36 \frac{x_u}{d}\left(1 - 0.42 \frac{x_u}{d}\right) f_{ck}\, b_w\, d^2 + (f_{sc} - 0.45 f_{ck}) A_{sc}(d - d') + 0.45 f_{ck}(b_f - b_w) y_f (d - 0.5 y_f) \quad ...(35.37)$$

$M_u = 0.36 \dfrac{122}{600}\left(1 - 0.42 \dfrac{122}{600}\right) 15 \times 240\, (600)^2 + (340.5 - 0.45 \times 15) \times 1472.6 \times (600 - 40) + 0.45 \times 15\, (700 - 240) \times 76.8\, (600 - 0.5 \times 76.8)$

$\quad = 86.76 \times 10^6 + 275.2 \times 10^6 + 133.9 \times 10^6 = 495.86 \times 10^6$ N-mm = **495.86 kN-m.**

### Example 35.10. M.R. of doubly reinforced T-beam

*Find the moment of resistance of the T-beam of Example 35.3 if the beam carries a compression reinforcement in the form of 2 bars of 20 mm dia. with $d' = 40$ mm and tensile reinforcement consists of 6 bars of 25 mm dia.*

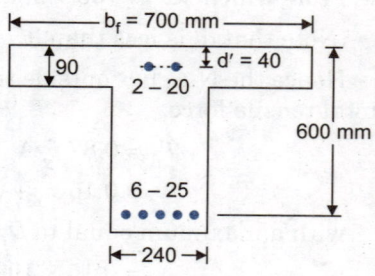

Fig. 35.15

**Solution:**

$A_{st} = 6 \times \dfrac{\pi}{4}(25)^2 = 2945.2$ mm²

$A_{sc} = 2 \dfrac{\pi}{4}(20)^2 = 628.3$ mm²

Assuming the N.A. to fall in flange, we have,

$0.36 f_{ck}\, b_f\, x_u + (f_{sc} - 0.45 f_{ck}) A_{sc} = 0.87 f_y\, A_{st}$

Taking $f_{sc} < 0.87 f_y < 0.87 \times 415 < 361$ or equal to 350 N/mm² in the first trial, we get
$$0.36 \times 15 \times 700 \, x_u + (350 - 0.45 \times 15) \, 628.3 = 0.87 \times 415 \times 2945.2$$
From which $x_u = 224.2$ mm $> D_f$

(Note that $d'$ is less than $\frac{3}{7} x_u$ so that stress in concrete at the level of compressive steel is equal to $0.446 f_{ck}$).

Hence the N.A. lies outside the flange. To obtain a new value of $x_u$, equate $T_u$ and $C_u$.
$$T_u = 0.87 f_y A_{st} = 0.87 \times 415 \times 2945.2 \approx 1063365 \text{ N}$$
$$y_f = 0.15 \, x_u + 0.65 \, D_f = 0.15 \, x_u + 0.65 \times 90 = 0.15 \, x_u + 58.5$$
with a maximum equal to $D_f$. Since the approximate value of $x_u$ found above is 224.2 mm,
$$y_f = 224.2 \times 0.15 + 58.5 = 92.1 > D_f. \text{ Hence take } y_f = D_f$$
$$\therefore \quad C_u = 0.36 f_{ck} x_u b_w + (f_{sc} - 0.45 f_{ck}) A_{sc} + 0.45 f_{ck} (b_f - b_w) D_f$$
or $\quad C_u = 0.36 \times 15 \, x_u \times 240 + (f_{sc} - 0.45 \times 15) \, 628.3 + 0.45 \times 15 \, (700 - 240) \, 90$

or $\quad C_u = 1296 \, x_u + 628.3 f_{sc} - 4203 + 276966 = 1296 \, x_u + 272763 + 628.3 f_{sc}$

Equating $C_u$ and $T_u$, $1296 \, x_u + 272763 + 628.3 f_{sc} = 1063365 \quad x_u = 610 - 0.4848 f_{sc}$  ...(i)

In the above equation, $f_{sc}$ can be obtained from stress-strain curve, corresponding to $\varepsilon_{sc}$ which itself depends upon $x_u$, as is evident from *Eq. 35.29*.
$$\varepsilon_{sc} = \frac{0.0035 \, (x_u - d')}{x_u} \qquad ...(ii)$$

In the first approximation, taking $x_u \approx 224.2$ mm $\varepsilon_{sc} = \dfrac{0.0035 \, (224.2 - 40)}{224.2} = 0.002875$

From Table 33.1, we get : $f_{sc} \approx 353$ N/mm². Hence from (i), $x_u = 610 - 0.4848 \times 353 \approx 439$ mm

Substituting this value of $x_u$ in (ii), we get $\varepsilon_{sc} = \dfrac{0.0035 \, (439 - 40)}{439} = 0.0032$

$\therefore \quad f_{sc} = 356$ N/mm² from Table 33.1 $\quad \therefore \quad$ Next value of $x_u = 610 - 0.4848 \times 356 \approx 437$ mm.

Since the value has converged, take $x_u = 437$ mm for which $f_{sc} = 356$ N/mm²

Now, $\quad x_{u.max} = 0.48 \, d = 0.48 \times 600 = 288$ mm. Hence $x_u > x_{u.max}$.

Thus, the beam is over-reinforced. Though the Code recommends that such a beam should be *redesigned*, its moment of resistance is limited to $M_{u.lim}$ corresponding to the balanced section.

$\therefore$ Take $\quad x_u = x_{u.max} = 288$ mm.

For this value of $x_u$, $\quad \varepsilon_{sc} = \dfrac{0.0035 \, (288 - 40)}{288} = 0.003.$ $\qquad \therefore \quad f_{sc} \approx 354$ N/mm².

Again $\quad \dfrac{D_f}{d} = \dfrac{90}{600} = 0.15 < 0.2.$ Hence moment of resistance is given by

$$M_u = M_{u.lim.} = 0.36 \, \frac{x_{u.max}}{d} \left(1 - 0.42 \frac{x_{u.max}}{d}\right) f_{ck} b_w d^2 + (f_{sc} - 0.446 f_{ck})$$
$$A_{sc} (d - d') + 0.45 f_{ck} (b_f - b_w) D_f \left(d - \frac{D_f}{2}\right)$$

$$= 0.36 \times \frac{288}{600} \left(1 - 0.42 \times \frac{288}{600}\right) 15 \times 240 \, (600)^2 + (354 - 0.45 \times 15)$$
$$\times 628.3 \, (600 - 40) + 0.45 \times 15 \, (700 - 240) \times 90 \times (600 - 0.5 \times 90)$$
$$= 178.8 \times 10^6 + 122.2 \times 10^6 + 155.1 \times 10^6 = 456.1 \times 10^6 \text{ N-mm} = \mathbf{456.1 \text{ kN-m}}.$$

**Example 35.11. Design of doubly reinforced T-beam**

*Redesign the T-beam of example 35.5 if the applied moment is 300 kN-m.*

**Solution:** $M_{uD} = 1.5 \times 300 = 450$ kN-m

**Step 1:** Assuming $x_u = D_f$, and *assuming* singly reinforced section, compute $M_{u_1}$.

$$M_{u_1} = 0.36 f_{ck} b_f D_f (d - 0.416 D_f)$$
$$= 0.36 \times 20 \times 750 \times 90 (500 - 0.416 \times 90)$$
$$= 224.80 \times 10^6 \text{ N-mm} < M_{uD}$$

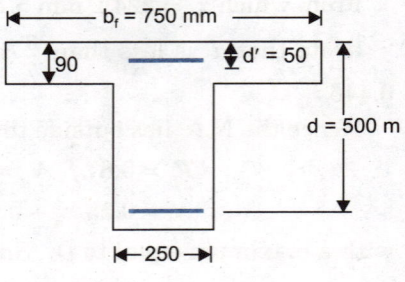

Fig. 35.16

**Step 2:** Since $M_{uD} > M_{u_2}$, N.A. will fall out side the flange.

Taking $x_u = \dfrac{7}{3} D_f = \dfrac{7}{3} \times 90 = 210$ mm,

compute $M_{u_2}$ treating the section as singly reinforced,

$$M_{u_2} = 0.36 f_{ck} b_w \left(\dfrac{7}{3} D_f\right)\left(d - 0.42 \times \dfrac{7}{3} D_f\right) + 0.45 f_{ck}(b_f - b_w) D_f (d - 0.5 D_f)$$
$$= 0.36 \times 20 \times 250 (210)(500 - 0.42 \times 210) + 0.45 \times 20(750 - 250)90(500 - 0.5 \times 90)$$
$$= 155.66 \times 10^6 + 184.27 \times 10^6 = 339.93 \times 10^6 \text{ N-mm} < M_{uD}$$

**Step 3:** Since $M_{uD} > M_{u_2}$, $x_u > \dfrac{7}{3} D_f > 210$ mm. Compute $M_{u, lim}$ by taking $x_u = x_{u, max}$

$\dfrac{x_{u, max}}{d} = 0.479$ (for Fe 415); $x_{u, max} = 0.479 \times 500 = 239.5$ mm

Also, $\dfrac{D_f}{d} = \dfrac{90}{500} = 0.18 < 0.2$

Hence compute $M_{u, lim}$ from *Eq. 35.21*, treating the section as singly reinforced.

$$M_{u, lim} = 0.36 \dfrac{x_{u, max}}{d}\left(1 - 0.42 \dfrac{x_{u, max}}{d}\right) f_{ck} b_w d^2 + 0.45 f_{ck}(b_f - b_w) D_f \left(d - \dfrac{D_f}{2}\right)$$
$$= 0.36 \times 0.479 (1 - 0.42 \times 0.479) 20 \times 250 (500)^2 + 0.45 \times 20 (750 - 250) 90 \left(500 - \dfrac{90}{2}\right)$$
$$= 172.18 \times 10^6 + 184.27 \times 10^6 = 356.45 \times 10^6 < M_{uD}$$

Since $M_{uD} > M_{u, lim}$ a doubly reinforced section will be required.

**Step 4. Design of doubly reinforced section**

(a) *Computation of $A_{st1}$ for $M_{u.lim}$* as under:

$$A_{sw} = \dfrac{0.36 f_{ck} x_{u, max} b_w}{0.87 f_y} = \dfrac{0.36 \times 20 \times 239.5 \times 250}{0.87 \times 415} = 1194 \text{ mm}^2$$

$$A_{sf} = \dfrac{0.45 f_{ck}(b_f - b_w) D_f}{0.87 f_y} = \dfrac{0.45 \times 20 (750 - 250) 90}{0.87 \times 415} = 1121.7 \text{ mm}^2$$

∴ $A_{st1} = 1194 + 1121.7 = 2315.7$ mm²

(b) *Computation of additional tensile area $A_{st_2}$*: Taking $d' = 50$ mm (say)

$$A_{st2} = \dfrac{M_{uD} - M_{u, lim}}{0.87 f_y (d - d')} = \dfrac{(450 - 355.24) 10^6}{0.87 \times 415 (500 - 50)} \approx 583 \text{ mm}^2$$

(c) *Computation of compressive steel area $A_{sc}$ corresponding to $A_{st_2}$*

$$A_{sc} = \dfrac{M_u - M_{u, lim}}{(f_{sc} - 0.45 f_{ck})(d - d')} = \dfrac{(450 - 355.24) 10^6}{(f_{sc} - 0.45 \times 20)(500 - 50)} = \dfrac{210577.7}{f_{sc} - 9}$$

$$\varepsilon_{sc} = \dfrac{0.0035(x_{u, max.} - d')}{x_{u, max}} = \dfrac{0.0035(239.5 - 50)}{239.5} = 0.00276$$

Hence from Table 33.1 for HYSD bars

$f_{sc} = 351.8$ N/mm² for $\varepsilon_{sc} = 0.00276$

$$\therefore \quad A_{sc} = \frac{210577.7}{351.8 - 9} = \mathbf{614.3 \text{ N/mm}^2}$$

(d) Total tensile area: $A_{st} = A_{st_1} + A_{st_2} = 2306 + 583 = \mathbf{2889 \text{ mm}^2}$

## ANALYSIS PROBLEMS

**Problem 1.** *Find the effective flange width of the following simply supported Tee-beam. Effective span is 5.0 m, centre to centre distance of adjacent panels = 4.0 m., Breadth of the web = 300 mm, Thickness of slab = 110 mm.*

**Solution:** Given data  length $(l) = 5$ m,   $b_w = 300$ m,   $D_f = 110$ mm

Since the beam is simply supported, the distance between the points of zero moment
$$l_0 = l = 5 \text{ m}$$

Clear span of the slab to the left or right of the beam = c/c distance of adjacent panels – $b_w$
$$= 4000 - 300 = 3700 \text{ mm}.$$

Effective width of the flange is the least of the following

1. $b_f = \dfrac{l_0}{6} + b_w + 6D_f = \dfrac{5000}{6} + 300 \text{ m} + 6 \times 110 = 1793.3$ mm

2. $b_f = b_w$ + half of the clear distance to the adjacent beams on either side
$$= b_w + \frac{3700}{2} + \frac{3700}{2} = 4000 \text{ mm}$$

Hence $b_f = 1793.3$ mm

**Problem 2.** *Find the effective flange width of the following simply supported isolated Tee-beam, Effective span = 5.0 m, breadth of the web = 230 mm, thickness of slab = 110 mm, width of the support = 230 mm, Actual width of the flange = 750 mm.*

**Solution:**   $l = 5$ m,   $b_w = 230$ mm,   $D_f = 110$ mm,   $b = 750$ mm

Since the beam is simply supported, the distance between the points of zero moments
$$l_0 = l = 5 \text{ m}$$

For isolated T-beam, effective width of the flange is the least of the following.

1. $b_f = \dfrac{l_0}{(l_0/b + u)} + b_w = \dfrac{5000}{\left(\dfrac{5000}{750} + 4\right)} + 230 = 698.8$ mm

2. $b_f$ = actual width of the flange = 750 mm.   Hence $b_f = 698.8$.

**Problem 3.** *A T-beam of effective flange width of 1200 mm, thickness of slab 100 mm, width of rib 300 mm and effective depth 460 mm is reinforced with 4 numbers of 12 mm $\phi$ bars. Calculate the moment of resistance of the section. The materials are $M_{20}$ grade* **$M_{20}$ grade concrete Fe 415 steel**.

**Solution:** Breadth of flange,
$b_f = 1200$ mm

Depth of flange,   $D_f = 100$ mm
Breadth of web,   $b_w = 300$ mm
Effective depth,   $d = 460$ mm
$f_{ck} = 20$ N/mm²
$f_y = 415$ N/mm².

Area of steel,   $A_{st} = 4 \times \dfrac{\pi}{4} \times 12^2 = 452.4$ mm².

**1. Assuming Neutral axis is within the flange:**

Equating compressive force in concrete to tensile force in steel
$$0.36 f_{ck} b_f x_u = 0.87 f_y A_{st}$$

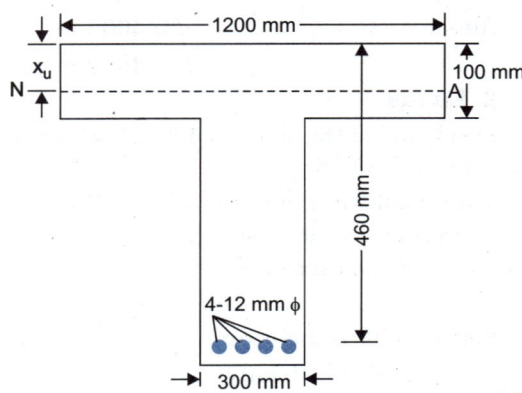

Fig. 35.17

$$x_u = \frac{0.87 f_y A_{st}}{0.36 f_{ck} b_f}$$

$$x_u = \frac{0.87 \times 415 \times 452.4}{0.36 \times 20 \times 1200} = 18.9 \text{ mm} < D_f.$$

Therefore, our assumption is correct. So, the analysis is similar to that of rectangular beam of width $b_f$ and effective depth $d$.

2. $x_{u,\,max} = 0.48\, d$ for Fe 415 steel = $0.48 \times 460 = 220.8$ mm
$x_u < x_{u,\,max}$.
Therefore, the section is under reinforced section.

3. **Moment of Resistance:**

$$M_u = 0.36 f_{ck} b_f x_u (d - 0.42 x_u)$$
$$= 0.36 \times 20 \times 1200 \times 18.9 (460 - 0.42 \times 18.9)$$
$$= 73.82 \times 10^6 \text{ N-mm} = 73.82 \text{ kN-m}$$

(This can also be calculated from)

$$M_u = 0.87 f_y A_{st} (d - 0.42 x_u)$$
$$= 0.87 \times 415 \times 452.4 (460 - 0.42 \times 18.9)$$
$$= 73.82 \times 10^6 \text{ N-mm} = 73.82 \text{ kN-m}$$

## DESIGN PROBLEMS

**Problem 1.** *A Tee-beam of floor consists of 150 mm thick R.C.C. slab monolithic with 300 mm wide beams. The beams are spaced at 3.5 m centre to centre and their effective span is 6 m if the superimposed load on the slab is kN/m³, design an intermediate Tee-beam. Use* **$M_{20}$ mix and Fe 415** *grade steel.*

**Solution:**  
Effective span ($l$) = 6 m  
Spacing of beams = 3.5 m c/c  
Depth of flange ($D_f$) = 150 mm  
Breadth of web = 300 mm  
$f_{ck}$ = 20 N/mm²  
$f_y$ = 415 N/mm².

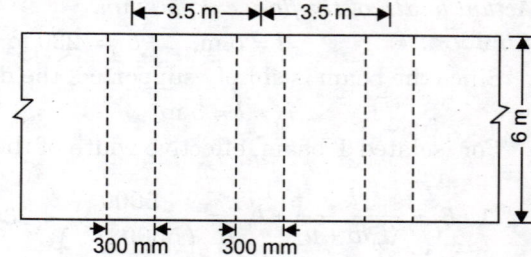

Fig. 35.18

### 1. Depth of the Beam
Selecting the depth in range of $\frac{l}{12}$ to $\frac{l}{15}$ based on stiffness

$$d = \frac{6000}{15} = 400 \text{ mm}$$

Adopt $d = 400$ mm.  
$D = 450$ mm.

### 2. Loads
Dead load of the slab = $0.15 \times 25 = 3.75$ kN/m² live load from the slab = 5 kN/m²

Total load from the slab = 8.75 kN/m².

Load from the slab per metre run of the beam = load on slab per m² × c/c distance between beams
$$= 8.75 \times 3.5 = 30.63 \text{ kN/m.}$$

Self weight of the beam = $0.3 \times 0.3 \times 1 \times 25 = 2.25$ kN/m  
(weight of rib only)  

Total load = 32.88 kN/m  
Factored load $w_u = 1.5 \times 32.88 = 49.32$ kN/m

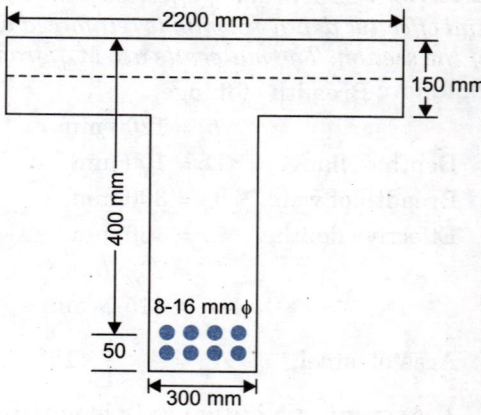

Fig. 35.19

Factored bending moment $M_u = \dfrac{w_u l^2}{8} = \dfrac{49.32 \times 6^2}{8} = 221.94$ kN-m.

**3. Effective width of flange.**

$$bf = \dfrac{l_0}{6} + b_w + 6D_f = \dfrac{6000}{6} + 300 + 6 \times 150$$
$$= 2200 \text{ mm, limited to 3500 mm. Hence } b_f = 2200 \text{ mm.}$$

**4. Assuming $x_u$ is within the flange.**

Equating compressive force in concrete to tensile force in steel

$$0.36 f_{ck} b_f x_u = 0.87 f_y A_{st}$$
$$x_u = \dfrac{0.87 \times 415 \times A_{st}}{0.36 \times 20 \times 2200} = 0.0228 A_{st}$$

**5. Reinforcement**

$$M_u = 0.87 f_y A_{st} (d - 0.42 x_u)$$
$$221.94 \times 10^6 = 0.87 \times 415 \times A_{st} (400 - 0.42 \times 0.228 A_{st})$$
$$221.94 \times 10^6 = 144420 A_{st} + 3.454 A_{st}^2$$
$$A_{st}^2 - 41776 A_{st} + 64.2 \times 10^6 = 0$$

Solving the above quadratic equation

$$A_{st} = 1597.9 \text{ mm}^2$$
$$x_u = 0.0228 A_{st} = 0.0228 \times 1597.9$$
$$= 36.4 \text{ mm} < D_f$$

Hence, our assumption is correct

Minimum area of tension steel $A_{st\,min} = \dfrac{0.85 b_w d}{f_y} = \dfrac{0.85 \times 300 \times 400}{415}$
$$= 245.8 < A_{st} \text{ provided}$$

Maximum area of tension steel

$$A_{st, max} = 0.04 b_w D$$
$$= 0.04 \times 300 \times 450 = 5400 \text{ mm}^2 > A_{st} \text{ provided}$$

Provide 8–16 mm $\phi$ bars, $A_{st}$, provided = 1608.5 mm².

## By Using SP16

$$\dfrac{M_u}{bd^2} = \dfrac{221.94 \times 10^6}{2200 \times 400^2} = 0.63$$

Refer, Table no. 2 of SP-16 and read out the value of Percentage of reinforcement corresponding to $f_y = 415$ N/mm² and $f_{ck} = 20$ N/mm²

For = $\dfrac{M_u}{bd^2} = 0.63, P_t = 0.181\%$

$$A_{st} = \dfrac{P_t \times bd}{100} = \dfrac{0.181 \times 2200 \times 400}{100} = 1592.8 \text{ mm}^2$$

Provide 8–16 mm $\phi$ bars, $A_{st}$, provided = 1608.5 mm².

**Problem 2.** *Calculate the area of reinforcement request red for a T-beam with the following data span = 8 m, ends simply supported, spacing of beams = 3 m, Superimposed load on slab = 3 kN/m², Thickness of slab = 130 mm, weight of wall on beam = 15 kN/m Assume width of web 230 mm, total depth 680 mm, concrete–$M_{20}$ and steel-Fe 250.*

**Solution:** Effective span = $(l) = 8$ m. Depth of flange $D_f = 130$ mm.

Gross depth $D = 680$ mm. Breadth of web $b_w = 230$ mm

$$f_{ck} = 20 \text{ N/mm}^2$$
$$f_y = 250 \text{ N/mm}^2$$

### 1. Loads:

Dead load of the slab = 0.13 × 25 = 3.25 kN/m²
Live load from the slab = 3 kN/m²
Total load from the slab = 6.25 kN/m².

Load from the slab per metre run of the beams load on slab per m² × c/c distance between beams
$$= 6.25 \times 3 = 18.75 \text{ kN/m}.$$

Self weight of the beam = 0.55 × 0.23 × 1 × 25
(weight of rib only) = 3.16 kN/m

Weight of wall on the beam = 15 kN/m
Total load = 36.91 kN/m

Factored load $w_u$ = 1.5 × 36.91 = 55.37 kN/m.

Factored bending moment

$$M_u = \frac{w_u l^2}{8} = \frac{55.37 \times 8^2}{8} = 442.96 \text{ kN-m}.$$

### 2. Effective width of flange:

$$b_f = \frac{l_0}{6} + b_w + 6D_f = \frac{8000}{6} + 230 + 6 \times 130 = 2343.33 \approx 2343 \text{ mm, limited to 3000 mm}.$$

Hence $b_f$ = 2343 mm.

### 3. Assuming $x_u$ is within the flange:

Equating compressive force in concrete to tensile force in steel

$$0.36 f_{ck} b_f x_u = 0.87 f_y A_{st}$$

$$x_u = \frac{0.87 f_y A_{st}}{0.36 f_{ck} b_f}$$

$$x_u = \frac{0.87 \times 250 \times A_{st}}{0.36 \times 20 \times 2343} = 0.0129 A_{st}$$

### 4. Reinforcement:

$$M_u = 0.87 f_y A_{st} (d - 0.42 x_u)$$
$$442.96 \times 10^6 = 0.87 \times 250 \times A_{st} (650 - 0.42 \times 0.0129 A_{st})$$
$$442.96 \times 10^6 = 14.375 A_{st} - 1.178 A_{st}^2$$
$$A_{st}^2 - 120012.7 A_{st} + 376 \times 10^6 = 0$$

Solving the above quadratic equation $A_{st}$ = 3219.4 mm²

$$x_u = 0.0129 A_{st} = 0.0129 \times 3219.4 = 41.5 \text{ mm} < D_f$$

Hence, our assumption is correct

Minimum area of tension steel

$$A_{st, min} = \frac{0.85 b_w d}{f_y} = \frac{0.85 \times 230 \times 650}{250} = 508.3 \text{ mm}^2 < A_{st}$$

Maximum area of tension steel

$$A_{st, max} = 0.04 b_{co} D = 0.04 \times 230 \times 680 = 6256 \text{ mm}^2 > A_{st}$$

Provide 11–20 mm ϕ bars, $A_{st}$, provided = 3455.8 mm².

**Problem 3.** *Design the reinforcement for a T-beam for the following data span 8 m, end simply supported, spacing of beams = 3 m, superimposed load = 5 kN/m², floor finish = 0.7 kN/m², Thickness of slab = 130 mm, weight of wall on beam = 15 kN/m, Assume width of web 230 mm, Total depth 680 mm, concrete-$M_{20}$ and steel-Fe 250.*

**Solution:** Given:

Simply supported Beam,

$L$ = 8 m;   $w_s$ = 5 kN/m²;   c/s spacing = 3 m;   floor finish (ff) = 0.7 kN/m²;   wall load = 15 kN/m
$b_w$ = 230 mm;   $D_p$ = 680 mm;   $D_f$ = 130 mm;   $f_{ck}$ = 20 N/mm²;   $f_y$ = 250 N/mm²

Let $d' = 80$ mm
$$d = D - d' = 680 - 80 = 600 \text{ mm.}$$

Calculation for $M_u$:

Total load/m² = weight of slab + $ff$ + $w_s$ = $25 \times 0.13 + 0.7 + 5 = 8.95$ kN/m².

Loads on beam from floor = load intensity × beam spacing = $8.95 \times 3 = 26.85$ kN/m.

Load from wall = 15.00 kN/m.

Weight of rib = $25(0.68 - 0.13) \times 0.23 = 3.16$ kN/m

Total working = $26.85 + 15 + 3.16 = 45.01$ kN/m

Load ($w$)

Ultimate load ($w_u$) = $1.5 \, w = 1.5 \times 45.01 = 67.51$ kN-m.

Ultimate design moment $M_u = \dfrac{w_u l^2}{8}$

$$= \dfrac{67.51 \times 8^2}{8} = 540.08 \approx 540 \text{ kN-m}$$

Effective depth of flange

$$b_f = \left(\dfrac{l_0}{6} + 6D_f\right) + b_w \text{ for T-beam}$$

$$b_f = \left(\dfrac{8000}{6} + 6 \times 130\right) + 230 = 2343.33 \approx 2343 \text{ mm} < 3000 \text{ mm}$$

Calculations for $A_{st}$

For $x_u = D_f$, $\quad M_{url} = 0.36 \, f_{ck} \, b_f \, D_f \, (d - 0.42 \, D_f)$

$$= 0.36 \times 20 \times 2343 \times 130 \, (600 - 0.42 \times 130) \, 10^{-6} = 1196.1 \text{ kNm.}$$

Given $M_u < M_{url}$ and hence $x_u < D_f$ [from Eqn. 35.7(a)]

$$A_{st} = \dfrac{0.5 \times 20}{250} \left[1 - \sqrt{\dfrac{4.6 \times 540 \times 10^6}{20 \times 2343 \times 600^2}}\right] \times 2342 \times 600 = 34639.4 \text{ mm}^2.$$

## PROBLEMS

1. Find the moment of resistance of a T-beam having the following data
   Width of flange: 800 mm; Thickness of slab: 120 mm
   Width of rib: 200 mm; Effective depth: 400 mm
   Tensile steel area: 3500 mm². Use M 20 concrete and Fe 250 steel.

2. Solve problem 1 with M 20 – Fe 415 combination.

3. Find the moment of resistance of a T-beam section having $b_w$ = 300 mm, $b_f$ = 1600 mm, $D_f$ = 100 mm and $d$ = 510 mm. The reinforcement consists of 4 bars of 25 mm dia. Use M 20 concrete and Fe 415 Steel.

4. Determine area of steel required and moment of resistance corresponding to obtained section of a T-beam with the following data $b_f$ = 1000, $D_f$ = 100 mm then $b_w$ = 300 mm effective concrete 50 mm, $d$ = 450 mm, Use $M_{20}$ concrete and Fe 415 steel.

5. Solve problem 3 if the reinforcement consists of 6 bars of 25 mm dia.

6. Solve problem 3 if the reinforcement consists of 8 bars of 25 mm dia.

7. Solve problem 3 if the beam also carries compressive reinforcement in the form of 2 bars of 16 mm dia.

8. Solve problem 4, if the beam also carries compressive reinforcement in the form of 4 bars of 16 mm dia.

9. For the T-beam of the dimensions of problem 1, find the reinforcement for an applied B.M. of 100 kN-m.

10. A Tee beam floor consists of 140 mm thick RCC slab monolithic with 250 mm wide beams. The beam span at 4.0 mm centre to centre and their effective span is 5 m. If the superimposed load on the slab is 4 kN/m², design an intermediate Tee beam: Use materials $M_{20}$ mix and Fe 415 grade steel.

# SHEAR, BOND AND TORSION IN BEAMS

## 36.1. LIMIT STATE OF COLLAPSE : SHEAR

**1. *Nominal Shear Stress*:** The nominal shear stress $\tau_v$ in beams of uniform depth is determined from the expression:

$$\tau_v = \frac{V_u}{bd} \qquad \ldots(36.1)$$

where  $V_u$ = shear force due to design load.
   $b$ = width of beam (= $b_w$ for flanged sections) and $d$ = effective depth

In the case of beams of varying depth, the equation is modified as under:

$$\tau_v = \frac{V_u \pm \dfrac{M_u}{d} \tan \beta}{bd} \qquad \ldots(36.2)$$

where  $M_u$ = bending moment at the section
and    $\beta$ = angle between the top and bottom edges of the beam.

The negative sign in the formula applies when the B.M. $M_u$ increases as the effective depth $d$ increases, and the positive sign when the moment decreases.

**2. *Design Shear Strength of Concrete*:** The design shear strength ($\tau_c$) of concrete in beams without shear reinforcement is given in Table 36.1.

**Minimum Shear Reinforcement:** When $\tau_v$ is less than $\tau_c$ given above minimum shear reinforcement shall be provided in the form of stirrups such that

$$\frac{A_{sv}}{b\, s_v} \geq \frac{0.4}{0.87\, f_y} \qquad \ldots(36.3)$$

where  $A_{sv}$ = cross sectional area of stirrup legs effective in shear,
   $s_v$ = stirrup spacing along the length of member.
   $b$ = breadth of the beam or breadth of web of flanged beam, and
   $f_y$ = characteristic strength of the stirrup reinforcement in N/mm² which shall not be taken greater than 415 N/mm².

Hence spacing based on minimum shear reinforcement is given by

$$s_v \leq \frac{0.87\, f_y\, A_{sv}}{0.4\, b} \leq \frac{2.175\, A_{sv}\, f_y}{b} \qquad \ldots[36.3(a)]$$

However, in members with minor structural importance, such as lintels or where the minimum shear stress calculated in less than half the permissible value, this provision need not be complied with.

**TABLE 36.1.** Design shear strength ($\tau_c$) of concrete (N/mm²)
(As per Table No: 19 in IS: 456–2000)

| $\dfrac{100 A_s}{bd}$ | Concrete grade | | | | | |
|---|---|---|---|---|---|---|
| | M 15 | M 20 | M 25 | M 30 | M 35 | M 40 and above |
| ≤ 0.15 | 0.28 | 0.28 | 0.29 | 0.29 | 0.29 | 0.30 |
| 0.25 | 0.35 | 0.36 | 0.36 | 0.37 | 0.37 | 0.38 |
| 0.50 | 0.46 | 0.48 | 0.49 | 0.50 | 0.50 | 0.51 |
| 0.75 | 0.54 | 0.56 | 0.57 | 0.59 | 0.59 | 0.60 |
| 1.00 | 0.60 | 0.62 | 0.64 | 0.66 | 0.67 | 0.68 |
| 1.25 | 0.64 | 0.67 | 0.70 | 0.71 | 0.73 | 0.74 |
| 1.50 | 0.68 | 0.72 | 0.74 | 0.76 | 0.78 | 0.79 |
| 1.75 | 0.71 | 0.75 | 0.78 | 0.80 | 0.82 | 0.84 |
| 2.00 | 0.71 | 0.79 | 0.82 | 0.84 | 0.86 | 0.88 |
| 2.25 | 0.71 | 0.81 | 0.85 | 0.88 | 0.90 | 0.92 |
| 2.50 | 0.71 | 0.82 | 0.88 | 0.91 | 0.93 | 0.95 |
| 2.75 | 0.71 | 0.82 | 0.90 | 0.94 | 0.96 | 0.98 |
| 3.00 and above | 0.71 | 0.82 | 0.92 | 0.96 | 0.99 | 1.01 |

$\tau_{c,max}$ = max shear stress allowable even with shear reinforcement, is as given in Table 36.2.

**TABLE 36.2.** Maximum shear stress $\tau_{c,max}$ (N/mm²)
(As per Table No: 20 of (IS : 456–2000)

| Grade | M 15 | M 20 | M 25 | M 30 | M 35 | M 40 and above |
|---|---|---|---|---|---|---|
| $\tau_{c,max}$ (N/mm²) | 2.5 | 2.8 | 3.1 | 3.5 | 3.7 | 4.0 |

For solid slabs (not that slab) the *design shear strength* shall be $k\, \tau_c$, where $k$ has the following values:

Values of $k$

| Overall Depth of slab (mm) | 300 or more | 275 | 250 | 225 | 200 | 175 | 150 or less |
|---|---|---|---|---|---|---|---|
| k | 1.00 | 1.05 | 1.10 | 1.15 | 1.20 | 1.25 | 1.30 |

**3. Design of Shear Reinforcement:** When $\tau_v$ exceeds $\tau_c$ given above shear reinforcement is provided in one of the following forms:
(a) Vertical stirrups   (b) Bent up bars and stirrups, and   (c) Inclined stirrups.

Where bent up bars are available, their contribution towards shear resistance shall not be more than half that of the total shear reinforcement.

Shear reinforcement shall be provided to carry a shear $V_{us}$ equal to $V_u - \tau_c\, bd$. The strength of shear reinforcement $V_{us}$ shall be worked out as follows:

(a) Vertical stirrups: $V_{us} = \dfrac{0.87 f_y A_{sv} d}{s_v}$   ...(36.4)

The spacing of vertical stirrups is given by   $s_v = \dfrac{0.87 f_y A_{sv}}{(\tau_v - \tau_c) b}$   ...[36.4(a)]

(b) For inclined stirrups or a series of bars bent up at different cross-sections,

$V_{us} = \dfrac{0.87 f_y A_{sv} d}{s_v} (\sin \alpha + \cos \alpha)$   ...(36.5)

(c) For single bar or group of parallel bars, all bent-up at the same cross-section.

$V_{us} = 0.87 f_y A_{sv} \sin \alpha$   ...(36.6)

where   $\alpha$ = angle between the inclined stirrup or bent-up bar and the axis of the member, not less than 45°

**1.** Where more than one type of shear reinforcement is used, the total shear resistance shall be computed as the sum of the resistances for various types respectively.

2. The area of stirrups shall not be less than the minimum specified in by *Eq.* below;

3. The shear resistance of minimum shear reinforcement is found by substituting the value $\dfrac{0.87 f_y A_{sv}}{s_v}$ = 0.4 b (*Eq. 36.3*) into *Eq. 36.4*

Thus, $\quad V_{us.min.} = \left(0.87 \dfrac{f_y A_{sv}}{s_v}\right) d = (0.4 \, b) \, d = 0.4 \, bd$

Thus, the shear carried by concrete and that carried by stirrups is given by

$$V_{uc.min} = \tau_c \, bd + 0.4 \, bd$$

4. The maximum spacing of shear reinforcement shall not exceed 0.75 d for vertical stirrups and d for inclined stirrups, where d is the effective depth of the section. In no case shall the spacing exceed 300 mm.

### Example 36.1. Design of shear stirrups of Fe 250 grade steel.

*A reinforced concrete beam 250 mm wide and 400 mm effective depth is subjected to ultimate design shear force of 150 kN at the critical section near supports. The tensile reinforcement at the section near supports is 0.5 percent. Design the shear stirrups near the supports. Also, design the minimum shear reinforcement at the mid span. Assume concrete of grade M 20 and mild steel bars of Fe 250 grade.*

**Solution:** Given : $b = 250$ mm; $\quad d = 400$ mm; $\quad A_{st}/bd = 0.5\% = 0.005$

$$\tau_v = \dfrac{V_u}{bd} = \dfrac{150 \times 10^3}{250 \times 400} = 1.5 \text{ N/mm}^2$$

From Table 36.1, $\quad \tau_c = 0.48$ N/mm² for M 20 concrete and $100 \, A_{st}/bd = 0.5$
Also, from Table 36.2, $\quad \tau_{c, max} = 2.8$ N/mm² for M 20 concrete.
Thus, $\tau_v$ is less than $\tau_{c, max}$, but greater than $\tau_c$. Hence shear reinforcement is necessary.

$$V_{uc} = \tau_c \, bd = 0.48 \times 250 \times 400 = 48000 \text{ N}$$

Hence $\quad V_{us} = V_u - V_{uc} = 150000 - 48000 = 102000$ N

The shear resistance of nominal stirrups is given by

$$V_{us, min} = 0.4 \, bd = 0.4 \times 250 \times 400 = 40000 \text{ N} < V_{us}$$

Hence nominal stirrups are *not* sufficient at the section near supports.

From *Eq. 36.4*, $\quad s_v = \dfrac{0.87 f_y A_{sv}}{V_{us}} \cdot d$

Using two legged stirrups of 10 mm dia. bars, $A_{sv} = 2 \dfrac{\pi}{4} (10)^2 = 157.08$ mm²

$$\therefore \quad s_v = \dfrac{0.87 \times 250 \times 157.08}{102000} \times 400 \approx 134 \text{ mm}$$

Alternatively, from 36.4 (a), $s_v = \dfrac{0.87 f_y A_{sv}}{(\tau_v - \tau_c) \, b} = \dfrac{0.87 \times 250 \times 157.08}{(1.5 - 0.48) \, 250} \approx 134$ mm

Maximum spacing = 0.75 d or 300 mm, which ever is less

Hence provide 10 mm dia. two legged stirrups @ 130 mm c/c at the section near supports.

At mid-span, the spacing of minimum shear reinforcement for 10 mm φ – 2 lgd stirrups is given by (*Eq. 36.3 a*)

$$s_v = 0.87 \dfrac{f_y A_{sv}}{0.4 \, b} = \dfrac{0.87 \times 250 \times 157.08}{0.4 \times 250} = 341.6 \text{ mm}$$

However, maximum spacing is limited to 0.75 d or 300 mm which ever is less. Hence $s_v = 300$ mm.
Hence provide 10 mm dia. two legged stirrups @ 300 mm c/c at the mid-span.

### Example 36.2. Design of shear stirrups of Fe 415 grade steel.

*Redesign shear stirrups for example 36.1, using Fe 415 steel for shear stirrups.*

**Solution:** As found in example 36.1, $\tau_v = 1.5$ N/mm²; $\quad \tau_c = 0.48$ N/mm²; $\quad V_{us} = 102000$ N.

Using two legged stirrups of 8 mm dia. bar, $A_{sv} = 2 \dfrac{\pi}{4} (8)^2 = 100.53$ mm²

$$\therefore \quad s_v = \frac{0.87 \times 415 \times 100.53}{102000} \times 400 = 142.3 \text{ mm}$$

Maximum spacing = 300 mm.
Hence provide 8 mm φ 2 *lgd* stirrups @ 140 mm c/c near supports.
At mid-span, using minimum shear reinforcement,

$$s_v = \frac{0.87 \times 415 \times 100.53}{0.4 \times 250} \approx 363 \text{ mm}$$

However, maximum spacing is limited to 0.75 $d$ or 300 mm.
Hence provide 8 mm dia. two legged stirrups @ 300 mm c/c at the mid-span.

**Example 36.3. Design of shear reinforcement. Bent up bars with vertical stirrups**

*A simply supported beam, 300 mm wide and 600 mm effective depth carries a uniformly distributed load of 74 kN/m including its own weight over an effective span of 6 m. The reinforcement consists of 5 bars of 25 mm diameter. Out of these, two bars can be safely bent up at 1 m distance from the support. Design the shear reinforcement for the beam.*

*Given: Grade of concrete: M 20; Grade of steel: Fe 415*
*Assume width of supports = 400 mm.*

**Solution:** $w_u = 1.5 w = 1.5 \times 74 = 111$ kN/m $\quad \therefore$ Max. shear $V_{u.\ max} = \dfrac{w_u L}{2} = \dfrac{111 \times 6}{2} = 333$ kN

According to IS Code, the critical section for shear occurs at a distance of $d$ (= 600 mm) from the face of the support, *i.e.* at a distance of 400/2 + 600 = 800 mm from the centre of support.

$\therefore$ Design shear $V_{uD} = 333 - 111 \times 0.8 = 244.2$ kN; $\quad \therefore \quad \tau_v = \dfrac{244.2 \times 1000}{300 \times 600} \approx 1.36$ N/mm²

$A_{st}$ at support $= (5-2)\dfrac{\pi}{4}(25)^2 = 1472.6$ mm²; $\quad \therefore \quad \dfrac{100 A_{st}}{bd} = \dfrac{100 \times 1472.6}{300 \times 600} = 0.818\%$

Hence from Table 36.1 $\quad \tau_c = 0.56 + \dfrac{0.62 - 0.56}{0.25}(0.818 - 0.75) = 0.576$ N/mm²

Also, from Table 36.2. $\tau_{c,\ max} = 2.8$ N/mm² $> \tau_v$
Since $\tau_v > \tau_c$, shear reinforcement is necessary.
Shear resistance of concrete $V_{uc} = 0.576 \times 300 \times 600 = 103680$ N = 103.68 kN
$\therefore \quad V_{us} = V_{uD} - V_{uc} = 244.2 - 103.68 = 140.52$ kN

Area of cross-section of two bent-up bass $= A_{sv} = 2\dfrac{\pi}{4}(25)^2 = 981.7$ mm²

Shear resistance of two bent-up bars $= 0.87 f_y \cdot A_{sv} \sin \alpha$
$= 0.87 \times 415 \times 981.7 \sin 45° = 250630$ N = 250.63 kN

However, the maximum shear resistance, to be assigned to the bent up bars should not exceed half the resistance to be provided by the shear reinforcement, i.e. it should not exceed $\dfrac{1}{2} \times 140.52 = 70.26$ kN.
Hence the remaining resistance to be provided by the vertical stirrups = 140.52 − 70.26 = 70.26 kN. Thus, $V_{us} = 70.26$ kN for vertical stirrups.

From *Eq. 36.4*, $\quad s_v = \dfrac{0.87 f_y A_{sv} \cdot d}{V_{us}}$

Using 8 mm dia 2-lgd stirrups, $A_{sv} = 2 \times \dfrac{\pi}{4}(8)^2 = 100.54$ mm²

$$\therefore \quad s_v = \frac{0.87 \times 415 \times 100.54 \times 600}{70260} \approx 310 \text{ mm}$$

Also, spacing corresponding to minimum steel is given by

$$s_v = \frac{0.87 f_y A_{sv}}{0.4 b} = \frac{0.87 \times 415 \times 100.54}{0.4 \times 300} = 302.49 \approx 302.5 \text{ mm}$$

## 886 REINFORCED CONCRETE STRUCTURE

However, max. $s_v = 0.75 \, d$ or 300 mm (whichever is less) = 300 mm

Hence provide 8 mm dia. 2-lgd stirrups @ 300 mm c/c from the face of the support to the point at which the bars are bent. These two bars are bent up at a distance of $1.414 \times a \approx 1.414 \times 0.9 \times 600 \approx 760$ mm from the face of the support (*i.e.* at a distance of 760 + 200 = 960 mm from the centre of the support). Beyond this point, bent up bars are not available.

At 0.96 m away from the support, $V_u = 333 - 111 \times 0.96 = 226.44$ kN

$$A_{st} = 5 \, \frac{\pi}{4} (25)^2 = 2454.4 \text{ mm}^2; \quad \frac{100 \, A_{st}}{bd} = \frac{100 \times 2454.4}{300 \times 600} = 1.36\%$$

Hence from Table 36.1, $\tau_c = 0.67 + \dfrac{0.72 - 0.67}{0.25} (1.36 - 1.25) = 0.692$ N/mm$^2$

$$V_{uc} = 0.692 \times 300 \times 600 = 124560 \text{ N} = 124.56 \text{ kN}$$
$$\therefore \quad V_{us} = V_u - V_{uc} = 226.44 - 124.56 = 101.88 \text{ kN}$$

The spacing of 8 mm dia. 2-lgd stirrups is given by

$$s_v = \frac{0.87 \times 415 \times 100.54 \times 600}{101880} = 213.78 \approx 213.8 \text{ mm}$$

Hence provide these @ 210 mm c/c.

To locate the section where maximum spacing of 300 mm, corresponding to minimum shear reinforcement, is provided, proceed as follows:

Let the distance of the section be $x$ m from the mid-span.

$$V_u = \frac{333}{3} x = 111 \, x \text{ kN} = 111000 \, x \text{ N}$$

$V_{uc} = 124560$ N (as before); $\therefore V_{us} = 111000 \, x - 124560$

Now $\quad s_v = \dfrac{0.87 f_y \cdot A_{sv} \cdot d}{V_{us}} \quad$ or $\quad 300 = \dfrac{0.87 \times 415 \times 100.54 \times 600}{111000 \, x - 124560}$

This gives $x \approx 1.776$ m from mid-span (or 1.224 m from the centre of the support).

### Summary

| | Distance from centre of support | Stirrups | Spacing |
|---|---|---|---|
| 1. | 0 to 0.76 m | 8 φ 2 lgd | 300 mm c/c |
| 2. | 0.76 to 1.224 m | 8 φ 2 lgd | 210 mm c/c |
| 3. | 1.224 m to 3 m | 8 φ 2 lgd | 300 mm c/c |

**Example 36.4. Design of shear reinforcement: Vertical Stirrups**

*A simply supported beam, 300 mm wide and 500 mm effective depth carries a uniformly distributed load of 50 kN/m, including its own weight over an effective span of 6 m. Design the shear reinforcement in the form of vertical stirrups. Assume that the beam contains 0.75% reinforcement throughout the length. The concrete is of M 20 grade and steel for stirrups is of Fe 250 grade. Take width of supports as 400 mm.*

**Solution:** $\quad w_u = 1.5 \times 50 = 75$ kN/m; $\quad V_{u, max} = \dfrac{w_u L}{2} = \dfrac{75 \times 6}{2} = 225$ kN

The critical section lies at a distance of $d = 500$ mm from the face of the support or at a distance of 500 + 400/2 = 700 mm from the centre of the support.

$\therefore \quad V_{uD} = 225 - 75 \times 0.7 = 172.5$ kN $\quad$ and $\quad \tau_v = \dfrac{172.5 \times 10^3}{300 \times 500} = 1.15$ N/mm$^2$

From Table 36.1, for $100 \, A_s/bd = 0.75\%$, we get $\tau_c = 0.56$ N/mm$^2$ for M 20 concrete.

$\therefore \quad V_{uc} = 0.56 \times 300 \times 500 = 84000$ N = 84 kN

Also, $\tau_{v, max} = 2.8$ N/mm$^2$ for M 20 concrete. Since $\tau_v < \tau_{v, max}$ it is OK.

However, $\tau_v > \tau_c$; hence shear reinforcement is necessary.

$$V_{us} = V_{uD} - V_{uc} = 172500 - 84000 = 88500 \text{ N}$$

Using 10 mm φ 2-lgd vertical stirrups, $A_{sv} = 2 \cdot \frac{\pi}{4}(10)^2 = 157.1 \text{ mm}^2$

∴ Spacing $s_v = \dfrac{0.87 f_y \cdot A_{sv} \cdot d}{V_{us}} = \dfrac{0.87 \times 250 \times 157.1 \times 500}{88500} = 193 \text{ mm} \approx 190 \text{ mm (say)}$

Spacing corresponding to minimum shear reinforcements is

$$s_v = \dfrac{0.87 f_y A_{sv}}{0.4\, b} = \dfrac{0.87 \times 250 \times 157.1}{0.4 \times 300} = 284.7 \text{ m} \approx 280 \text{ mm (say)}$$

However in no case should the spacing exceed $0.75\, d = 0.75 \times 500 = 375$ mm, or 300 mm whichever is less. Hence the spacing is to vary from 190 mm at the end section to 280 mm at a section distance $x$ m (say) from the mid-span. Let us locate this section where the S.F. is $V_{ux}$.

∴ $V_{ux} = \dfrac{V_{u,max}}{3} x = \dfrac{225000}{3} x = 75000\, x$

∴ $V_{us} = V_{ux} - V_{uc} = 75000\, x - 84000; \quad s_v = 280 = \dfrac{0.87 \times 250 \times 157.1 \times 500}{75000\, x - 84000}$

from which, we get, $x = 1.93$ m from mid-span or 1.07 m from supports. Hence provide 8 mm φ 2 lgd stirrups at a spacing of 190 mm c/c from supports to a section distant 1.07 m from the centre of either supports.

For the remaining length, provide the stirrups @ 280 mm c/c.

## 36.2. DEVELOPMENT LENGTH

The development length in tension is given by

$$L_d = \dfrac{0.87 f_y}{4\, \tau_{bd}} \Phi = k_d\, \Phi \qquad \ldots(36.7)$$

where $k_d$ = development length factor = $0.87 f_y / 4 \tau_{bd}$

For bars in compression, the values of $\tau_{bd}$ given in Table 36.3 are to be increased by 25%. The development length $(L_{dc})$ for bar in compression is given by

$$L_{dc} = \dfrac{0.87 f_y}{5\, \tau_{bd}} \varphi = k_d\, \varphi \qquad \ldots[36.7(a)]$$

Eq. 36.7 is the expression used for working stress method except that $\sigma_s$ has been replaced by $0.87 f_y$. Values of the design bond stress $(\tau_{bd})$ for plain bars in tension are given in Table 36.3. For deformed bars conforming to IS 1786, these values shall be increased by 60%. For bars in compression, the values of bond stress in tension shall be increased by 25%. Table 36.4 gives the values of development length factor $(k_d)$ for various grades of concrete and steel, both in tension and compression.

**TABLE 36.3.** Bond stress for plain bars in tension (IS 456 : 2000)

| Grade of concrete | M 20 | M 25 | M 30 | M 35 | M 40 and above |
|---|---|---|---|---|---|
| Design bond stress $\tau_{bd}$ (N/mm²) | 1.2 | 1.4 | 1.5 | 1.7 | 1.9 |

**TABLE 36.4.** Development length factor $(k_d)$

| 1. Grade of concrete | M 20 | | | M 25 | | | M 30 | | |
|---|---|---|---|---|---|---|---|---|---|
| 2. Grade of steel | Fe 250 | Fe 415 | Fe 500 | Fe 250 | Fe 415 | Fe 500 | Fe 250 | Fe 415 | Fe 500 |
| 3. Bars in tension | 46 | 47 | 57 | 39 | 41 | 49 | 37 | 38 | 46 |
| 4. Bars in comp. | 37 | 38 | 46 | 31 | 33 | 39 | 29 | 31 | 37 |

| 1. Grade of concrete | M 35 | | | M 40 and above | | | | | |
|---|---|---|---|---|---|---|---|---|---|
| 2. Grade of steel | Fe 250 | Fe 415 | Fe 500 | Fe 250 | Fe 415 | Fe 500 | | | |
| 3. Bars in tension | 32 | 34 | 40 | 29 | 30 | 36 | | | |
| 4. Bars in comp. | 26 | 27 | 32 | 23 | 24 | 29 | | | |

## 36.3. DEVELOPMENT LENGTH REQUIREMENTS AT SIMPLE SUPPORTS

The B.I.S code stipulates that at the simple supports (and at the point of inflexion), the positive moment tension reinforcement shall be *limited* to a diameter such that:

$$L_d \leq \frac{M_1}{V} + L_0 \qquad \ldots [36.8(a)]$$

where $L_d$ = development length, computed from *Eq. 36.7*.

$M_1$ = moment of resistance of the section at the support, assuming all the reinforcement at the section to be stressed to $0.87 f_y$

$V$ = shear force at the section, due to ultimate load.

$L_0$ = sum of anchorages beyond the centre of support and the equivalent anchorage value of any hook or mechanical anchorage at the simple support (at the point of inflection, $L_0$ is limited to $d$ or $12\,\Phi$ whichever is greater).

The method of computing the value of $L_0$ is same as discussed for the working stress method of design.

The Code further recommends that the value of $M_1/V$ in the above expression may be increased by 30 percent when the ends of the reinforcement are confined by a compressive reaction. This condition of 'confinement' of reinforcing bar may not be available at all the types of simple supports.

Four types of *simple supports* are shown in Fig. 36.1(a), a beam is simply supported on a wall which offers a compressive reaction which confines the ends of the reinforcement. Hence a factor 1.3 will be applicable with the term $M_1/V$. Similarly in Fig. 36.1(b), a slab is simply supported on a wall and hence factor 1.3 will be applicable. However, in Fig. 36.1(c) and 36.1(d) though a simple support is available, the reaction does not confine the ends of the reinforcement and hence the factor 1.3 will not be applicable with $M_1/V$ term.

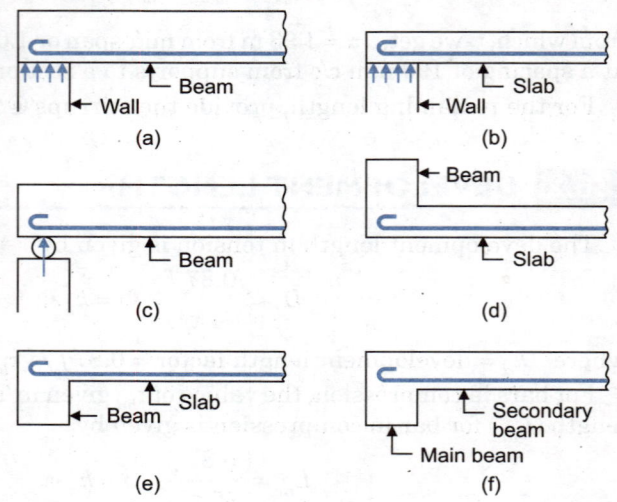

Fig. 36.1

Similarly, for the case of a slab connected to a beam Fig. 36.1(e) or for the case of a secondary beam connected to main beam [Fig. 36.1(f)] tensile reaction is induced and hence factor 1.3 will not be available. Thus, at simple supports, where the compressive reaction confines the ends of reinforcing bars, we have

$$L_d \leq 1.3 \frac{M_1}{V} + L_0 \qquad \ldots [36.8(b)]$$

In *Eq. 36.8(a)* or *36.8(b)*, $M_1$ is the moment of resistance of the bars available at the support.

∴ $M_1 = 0.87 f_y A_{st1} z_u$. where $z_u$ is the lever arm, which may be assumed to be constant.

Also, at the mid-span, $M = 0.87 f_y A_{st} z_u$

∴ $$\frac{M_1}{M} = \frac{A_{st1}}{A_{st}} = \frac{N_1}{N}$$

or $$M_1 = M \frac{N_1}{N} \qquad \ldots [36.8(c)]$$

where $N_1$ = No of bars at support; $N$ = No. of bars at mid span

$M$ = moment of resistance available

It should be clearly noted that $M_1$ is approximate, since it assumes a constant value of lever arm $z_u$. Its actual value is $(d - 0.42\, x_u)$ where $x_u$ is depth of N.A. of the section at support.

Hence $$M_1 = 0.87 f_y A_{st1} (d - 0.42\, x_u) \qquad \ldots [36.8(d)]$$

where $$x_u = \frac{0.87 f_y A_{st1}}{0.36 f_{ck} b} \qquad \ldots [36.8(e)]$$

## 36.4. TORSION: LIMIT STATE OF COLLAPSE

Torsional reinforcement is not calculated separately from that required for bending and shear. Instead, the total longitudinal reinforcement is determined for an equivalent bending moment which is a function of actual bending moment and torsion. Similarly, shear reinforcement (web reinforcement) is determined for an equivalent shear which is a function of actual shear and torsion.

**1. Shear and Torsion:** Equivalent shear $V_e$ is calculated from the expression

$$V_e = V_u + 1.6 \frac{T_u}{b} \qquad \ldots(36.9)$$

where  $V_e$ = equivalent shear;   $V_u$ = shear
   $T_u$ = torsional moment;   $b$ = breadth of the beam.

Equivalent nominal shear stress to be given by

$$\tau_{ve} = \frac{V_e}{bd} \qquad \ldots(36.10)$$

The values of $\tau_{ve}$ shall not exceed the values of $\tau_{c.max}$ given in Table 36.2.

**2. Reinforcement in Members Subjected to Torsion:** If the equivalent nominal shear stress $\tau_{ve}$ does not exceed $\tau_c$ given in Table 36.1, minimum shear reinforcement shall be provided as per *Eq. 36.3*. If $\tau_{ve}$ exceeds $\tau_c$, reinforcement is required. Reinforcement consists of (*a*) longitudinal reinforcement and (*b*) transverse reinforcement:

(*a*) **Longitudinal Reinforcement:** The longitudinal reinforcement shall be designed to resist an equivalent bending moment, $M_{e1}$ given by

$$M_{e1} = M_u + M_t \qquad \ldots(36.11)$$

where  $M_u$ = bending moment at the cross-section, and

$$M_t = T_u \left( \frac{1 + D/b}{1.7} \right) \qquad \ldots(36.12)$$

$T_u$ = torsional moment;   $D$ = overall depth of beam.

If the value of $M_t$ given by *Eq. 36.12* exceeds the value of $M_u$, longitudinal reinforcement shall be provided on the flexural compression face, such that beam can also withstand an equivalent B.M. $M_{e2}$ given by

$$M_{e2} = M_t - M_u \qquad \ldots(36.13)$$

In which $M_{e2}$ is taken as acting opposite to the moment $M_2$.

(*b*) **Transverse Reinforcement:** Two-legged closed hoops enclosing the corner longitudinal bars shall have an area of cross-section $A_{sv}$ given by

$$A_{sv} = \frac{T_u s_v}{b_1 d_1 (0.87 f_y)} + \frac{V_u s_v}{2.5 d_1 (0.87 f_y)} \qquad \ldots(36.14)$$

However, the total reinforcement shall not be less than $\frac{(\tau_{vs} - \tau_c) b s_v}{0.87 f_y}$.

where  $s_v$ = spacing of stirrups.
   $b_1$ = centre to centre distance between corner bars in the direction of width.
   $d_1$ = centre to centre distance between corner bars in the direction of depth.

**Example 36.5. Design for Torsion.**

*Determine the reinforcement required for a rectangular beam section with the following data:*
Width of section:   300 mm;   Depth of section:   500 mm
Factored B.M:   80 kN-m;   Factored torsional moment:   40 kN-m
Factored shear force: 70 kN. Use M 15 grade concrete and Fe 415 grade steel.

**Solution: 1. Equivalent Shear**

Total depth $D$ = 500 mm. Assume 20 mm dia. bars and 25 mm nominal cover and 10 mm dia. striups.

∴    $d = 500 - 25 - 10 - 20/2 = 455$ mm.

$$V_e = V_u + 1.6 \frac{T_u}{b} = 70 + 1.6 \frac{40}{0.3} = 283.333 \text{ kN.}$$

$$\therefore \quad \tau_{ve} = \frac{283.333 \times 1000}{300 \times 455} = 2.07 \text{ N/mm}^2$$

From Table 36.2, $\tau_{c.max} = 2.5$ N/mm$^2$. Hence the section does not require revision.

**2. Longitudinal Reinforcement.** Equivalent B.M. is given by

$$M_{e1} = M_u + M_t$$

where $\quad M_t = T_u \left(\dfrac{1 + D/b}{1.7}\right) = 40 \left[\dfrac{1 + 0.5/0.3}{1.7}\right] = 62.745$ kN-m

$$\therefore \quad M_{e1} = 80 + 62.745 = 142.745 \text{ kN-m}$$

The moment of resistance of a balanced section is given by

$$M_{u.lim.} = 0.36 f_{ck} \frac{x_{u.max}}{d} \left(1 - 0.42 \frac{x_{u.max}}{d}\right) bd^2$$

where $\quad \dfrac{x_{u.max}}{d} = 0.479$ for Fe 415 steel

$$\therefore \quad M_{u.lim.} = 0.36 f_{ck} (0.479)(1 - 0.42 \times 0.479) bd^2 = 0.138 f_{ck} bd^2$$

or $\quad = 0.138 \times 15 (300)(455)^2 = 128.56 \times 10^6$ N-mm.

Since $M_{e1} < M_{u.lim}$ the section is under-reinforced. Hence

$$M_u = M_{e1} = 0.87 f_y A_{st} d \left(1 - \frac{f_y}{f_{ck}} \frac{A_{st}}{bd}\right)$$

or $\quad 142.745 \times 10^6 = 0.87 \times 415 A_{st} \times 455 \left(1 - \dfrac{415}{15} \times \dfrac{A_{st}}{300 \times 455}\right)$

or $\quad A_{st} (1 - 202.7 \times 10^{-6} A_{st}) = 868.92$

or $\quad A_{st}^2 - 4934 A_{st} + 4.287 \times 10^6 = 0$. From which $A_{st} = 1126$ mm$^2$.

Hence provide 4 bars of 20 mm dia., having total $A_{st} = 4 \times 314 = 1256$ mm$^2$.

Again, as $M_t$ is less than $M_u$, we need not consider $M_{u2}$. Therefore provide only bars of 12 mm dia. on the compression face to act as holding bars for transverse reinforcement.

**3. Transverse Reinforcement**

$$\frac{100 A_s}{bd} = \frac{100 \times 1256}{300 \times 455} \approx 0.92\%$$

Hence from Table 36.1, $\tau_c = 0.54 + \dfrac{0.60 - 0.54}{0.25} \times 0.17 = 0.58$ N/mm$^2$

Since $\tau_{ve} > \tau_c$, transverse shear reinforcement is essential. The area of two legs of the stirrups should satisfy the following:

$$A_{sv} = \frac{T_u s_v}{b_1 d_1 (0.87 f_y)} + \frac{V_u s_v}{2.5 d_1 (0.87 f_y)}$$

Assuming dia. of stirrups as 10 mm,

$$d_1 = 500 - \left(25 + 10 + \frac{20}{2}\right) - \left(25 + 10 + \frac{12}{2}\right) = 414 \text{ mm}$$

$$b_1 = 300 - 2\left(25 + 10 + \frac{20}{2}\right) = 210 \text{ mm.}$$

$$\therefore \quad \frac{A_{sv}(0.87 f_y)}{s_v} = \frac{40 \times 10^6}{210 \times 414} + \frac{70 \times 1000}{2.5 \times 414} = 527.7 \text{ N/mm}^2.$$

Also, area of all the legs of the stirrups should satisfy the following condition

$$A_{sv} = \frac{(\tau_{ve} - \tau_c) b\, s_v}{0.87 f_y}$$

or $\quad \dfrac{A_{sv}(0.87 f_y)}{s_v} = (\tau_{ve} - \tau_c) b = (2.07 - 0.58)\,300 = 447$ N/mm. ...(ii)

Choosing the greater one of the two, we have

$$\frac{A_{sv}}{s_v} = \frac{527.7}{0.87 f_y} = \frac{527.7}{0.87 \times 415} = 1.462.$$

Hence spacing of 10 mm dia. 2-legged stirrups is

$$s_v = \frac{A_{sv}}{1.462} = \frac{2 \times 75.5}{1.462} \approx 103 \text{ mm}.$$

Hence provide the stirrups @ 100 mm c/c.

As per Code requirements, this spacing should not exceed $x_1$, $(x_1 + y_1)/4$ and 300 mm.

where $\quad x_1 =$ short dimension of stirrup $= 300 - 2(25 + 5) = 240$ mm

$y_1 =$ long dimension of stirrup $= 500 - 2(25 + 5) = 440$ mm

$$\frac{x_1 + y_1}{4} = \frac{240 + 440}{4} = 170 \text{ mm}.$$

Hence 10 mm dia. two legged stirrups @ 100 mm c/c satisfy the above Code requirements.

## PROBLEMS

1. A rectangular beam, 240 mm wide and 450 mm effective depth is reinforced with 3 bars of 20 mm dia. at supports. Design the shear reinforcement if it carries a shear of 60 kN at service state. Use M 20 concrete and Fe 415 steel. Also, determine the nominal shear stirrup at mid-span.

2. Redesign beam of problem 1 if the S.F. is 200 kN at service state.

3. Check whether the beam of problem 1 is safe against the requirements of bond at the simple support. Assume that the width of support is 400 mm.

4. An RCC beam 230 mm wide and 450 mm deep is reinforced with 5 bars of 16 mm $\phi$ and of grade Fe 415 on tension side. If shear force is 60 kN, design the shear reinforcement consisting only of vertical stirrups and $M_{20}$ grade used for beam.

# CHAPTER 37: DESIGN OF R.C.C. BEAMS AND SLABS

## 37.1. DESIGN OF RCC BEAMS

(a) **Basic Rules for Design** (*IS : 456-2000*): The basic design rules in the limit state method are the same as given in chapter 7 for working stress method.

(b) **Design procedure:** The design of a singly reinforced beam is done in the following steps:

**Useful IS Code Clauses**

(a) Selection of width and depth of the beam
The width of the beam selected shall satisfy the slenderness limits specified in IS 456–2000 Clause 23.3 to ensure lateral stability.

(b) Calculation of effective span (*le*) (As per Clause 22.2) of IS : 456–2000.

**Step 1. Determination of limiting depth of N.A.**

Determine $\frac{x_{u,max}}{d}$ from Eq. 33.7. $x_{u,max} = \frac{700}{1100 + 0.87 f_y}$

The values of $\frac{x_{u,max}}{d}$ for $f_y$:

| $f_y$ | $\frac{x_{u,max}}{d}$ |
|---|---|
| 250 | 0.531 |
| 415 | 0.479 |
| 500 | 0.456 |

**Step 2. Computation of design B.M.** Assume suitable values of overall depth (*D*) and *b* of the beam, determine *d* and then the effective span. This depth *d* should satisfy the requirements of *limit state of serviceability*.

Calculate self weight, and add to it the given super-imposed load, to get total U.D.L. *w* per metre run. Apply partial safety factors $\gamma_f$ for loads, and compute factored load $w_u$ per unit length of the beam. Compute max. B.M. $M_u$ in the beam.

**Step 3. Computation of effective depth d.** Find *d* from the relation using the equation G.1.1 (*c*) Annexure (G) of IS: 456–2000

$$M_{u,lim} = 0.36 \frac{x_{u,max}}{d}\left(1 - 0.42 \frac{x_{u,max}}{d}\right) bd^2 f_{ck}$$

or $\quad M_u = R_u bd^2$ (if *b* has already been chosen)

892

**Step 4. Steel Reinforcement**: Knowing $b$ and $d$, compute $D$ by adding effective cover, and round off the value of $D$ to slightly higher side. Find $d$ by subtracting effective cover from $D$. Determine the area of reinforcement from

$$M_u = 0.87 f_y \cdot A_{st}(d - 0.416 x_{u,\,max}) \qquad \ldots[33.10(a)]$$

or Alternatively, find $p_{t,\,lim}$ from Eq. 33.9

$$p_{t,\,lim} = 0.414 \frac{f_{ck}}{f_y} \cdot \frac{x_{u,\,max}}{d} \quad \text{and} \quad A_{st} = p_{t,\,lim} \times bd.$$

The reinforcement so computed should be equal to or more than minimum reinforcement given by

$$\frac{A_s}{bd} = \frac{0.85}{f_y}$$

**Curtailment of reinforcement**: Curtail the reinforcement at other locations where it is no longer required from the B.M. point of view, using the relation.

$$x = \frac{L}{2}\sqrt{\frac{N_x}{N}}$$

where $x$ is the distance from the mid-span, where, $N_x$ bars can be curtailed out of total bars $N$.

However, the bars have to develop full tension at the centre of span, and hence $x$ should be equal to or greater than the development length $L_d$ of the bar which is to be curtailed. The Code also specifies that the reinforcement shall extend the point at which it is no longer required to resist flexure for a distance equal to effective depth $d$ or $12\,\Phi$ whichever is greater. Hence the bar should be curtailed at a distance $x + d$ or $(12\Phi)$ from the centre, and this distance should be greater than $L_d$ and where $\phi$ is diameter of bar.

**Note**: If the depth $d$ is fixed from other considerations and its value is more than the one found by bending compression (step 3), we have the *under-reinforced section*, for which the reinforcement is given by *Eq. 33.13*:

$$A_{st} = \frac{0.5 f_{ck}}{f_y}\left[1 - \sqrt{1 - \frac{4.6 M_u}{f_{ck}\, bd^2}}\right]bd \qquad \ldots(33.13)$$

**Step 5. Shear reinforcement**: Compute max. shear force $V_u$ at the critical section. The position of critical section is shown in Fig. 7.4. Compute nominal shear stress

$$\tau_v = \frac{V_u}{bd}; \quad \text{Also, compute } \tau_c \text{ from Table 36.1.}$$

(i) If $\tau \le \frac{1}{2}\tau_c$ or $V_u = \frac{1}{2}V_{uc}$ where $(V_{uc} = \tau_c\, bd)$, no shear reinforcement is really necessary.

(ii) If $\frac{1}{2}V_{uc} \le V_u \le V_{uc}$. Provide nominal shear reinforcement given by:

$$\frac{A_{sv}}{b \cdot s_v} \ge \frac{0.4}{0.87 f_y} \quad \text{or} \quad s_v = \frac{0.87 f_y\, A_{sv}}{0.4\, b} = \frac{2.175\, A_{sv}\, f_y}{b}$$

Choosing bar dia., $A_{sv}$ is known, and hence $s_v$ can be computed from the above expression. However, max. spacing is restricted to $0.75\, d$ or 300 mm, whichever is less.

(iii) If $V_{uc} \le V_u \le V_{uc,\,max} =$ (where $V_{uc,\,max} = \tau_{c\,max} \cdot bd$)

Design the transverse reinforcement for the net S.F. $V_{us} = V_u - V_{uc}$

For vertical stirrups, $\quad V_{us} = \dfrac{0.87 f_y \cdot A_{sv} \cdot d}{s_v}$

(iv) If $V_u \ge V_{uc,\,max}$, redesign by increasing the depth such that $V_{cu,\,max}$ becomes equal to or more than the external shear $V_u$.

**Step 6. Anchorage of bars (or) check for development length at end**: At simple supports and at points of inflection position moment tension reinforcement shall be limited to a diameter such that $l_d$ computed for $(f_d)$ by 26.2.1 of IS : 456–2000 does not exceed $\dfrac{M_1}{V} + L_0$ where $f_d = 0.87 f_y$

$$L_d \le 1.3\, \frac{M_1}{V} + L_0$$

The factor 1.3 in the above equation should be chosen for only those simple supports which offer compressive reaction that confines the ends of reinforcing bars.

In accordance with clause 26.2 of IS : 456–2000

$$L_d = \frac{\phi \sigma_{st}}{4 \tau_{bd}}$$

where  $\phi$ = Nominal diameter of the bar

$\sigma_{st}$ = Stress in bar at the section considered at design load

$\tau_{bd}$ = Design bond stress given in table 26.2.1.1 of IS : 456–2000

### Example 37.1. Design of singly reinforced beam

A reinforced concrete beam is supported on two walls 750 mm thick, spaced at a clear distance of 6 m. The beam carries a super-imposed load of 9.8 kN/m. Design the beam using M 20 concrete mix and HYSD bars of Fe 415 grade.

**Solution:**

**Step 1. Limiting depth of N.A.**

For Fe 415 steel, $f_y$ = 415 N/mm². For M 20 concrete, $f_{ck}$ = 20 N/mm²

$\frac{x_{u\,max}}{d}$ depends only a grade of steel and is independent on grade of concrete. For Fe 415 steel,

$$\frac{x_{u\,max}}{d} = \frac{700}{1100 + 0.87 \times 415} = 0.479.$$

**Step 2. Computation of design B.M.:** The minimum depth of beam is based on limit state of serviceability requirements. For simply supported beam, $L/d$ = 20

For a balanced section, $p_{t,\,lim}$ = 0.96% (Table 33.2). Also for Fe 415 steel, $f_s$ = 0.58 × 415 = 240.7 ≈ 240. Hence modification factor ≈ 1.

$$\therefore \quad d = L/20 \approx 6000/20 = 300 \text{ mm}.$$

This is the minimum value of $d$. Actual value of $d$, based on bending may be more than this. In working stress design method, $d$ is kept equal to $l/10$ for preliminary computations. However, the limit state design method results in smaller depth. Hence take

$$d = \frac{l}{15} = \frac{6000}{15} = 400 \text{ mm.};$$

$D$ = 400 + 30 = 430 mm (say); Keep $b$ = 300 mm.

$\therefore$ Self weight of beam = 0.43 × 0.3 × 1 × 25000 = 3225 N/m

External load = 9800 N/m.

$\therefore$ Total $w$ = 9800 + 3225 = 13025 N/m

$w_u$ = 1.5 $w$ = 1.5 × 13025 = 19537.5 ≈ 19538 N/m

Effective span $L = l + d$ = 6 + 0.4 = 6.4 m, which is smaller than centre to centre distance of 6.75 m between the supports.

$$M_u = \frac{w_u \cdot L^2}{8} = \frac{19538(6.4)^2}{8} = 100034.56 \approx 100035 \quad \text{N-m} \approx 100.04 \times 10^6 \text{ N-mm}$$

**Step 3. Computation of effective depth d**

$$M_u = 0.36 f_{ck} \frac{x_{u,\,max}}{d}\left(1 - 0.416 \frac{x_{u,\,max}}{d}\right)bd^2 = R_u \cdot bd^2$$

where  $R_u$ = 0.36 × 20 × 0.479 (1 – 0.416 × 0.479) = 2.761

(Compare this with value given in Table 33.3)

$$\therefore \quad d = \sqrt{\frac{M_u}{R_u\, b}} = \sqrt{\frac{100.04 \times 10^6}{2.761 \times 300}} = 347.5 \approx 347 \text{ mm}.$$

Providing 25 mm nominal cover and using 20 mm bars with 8 mm dia strrups
$$D = 347 + 25 + 8 + 20/2 = 390 \text{ mm.}$$

Keep $D = 430$ mm. Since this value is the one assumed earlier, no revised computations for self load are recommended.

$d = 430 - 25 - 8 - 20/2 = 387$ mm. Since available $d$ (= 387 mm) is more than the one required for balanced design, we will have an under-reinforced section.

**Step 4. Steel reinforcement**: The reinforcement, for an under-reinforced section, is given by *Eq. 33.13*:

$$A_{st} = \frac{0.5 f_{ck}}{f_y}\left[1 - \sqrt{1 - \frac{4.6 M_u}{f_{ck} bd^2}}\right] bd$$

$$= \frac{0.5 \times 20}{415}\left[1 - \sqrt{1 - \frac{4.6 \times 100.04 \times 10^6}{20 \times 300 \times (387)^2}}\right] 300 \times 387 = 843.48 \approx 843.5 \text{ mm}^2$$

No. of 20 mm dia. bars = 843.5/314.15 = 2.68 Nos. Provide 3 No. of 20 mm dia. bars at mid span.

Actual $A_{st} = 3 \times 314.15 = 942.45 \approx 942.5$ mm$^2$

Hence actual moment of resistance available at mid span is

$$M_{ur} = 0.87 f_y \cdot A_{st} \cdot d \left(1 - \frac{f_y}{f_{ck}} \cdot \frac{A_{st}}{bd}\right) \qquad \ldots(33.10)$$

$$= 0.87 \times 415 \times 942.5 \times 387 \left(1 - \frac{415}{20} \times \frac{942.5}{300 \times 387}\right) \approx 110 \times 10^6 \text{ N/mm}$$

$$M_{u.lim} = 2.761 bd^2 = 2.761 \times 300(387)^2 = 124.05 \approx 124.1 \times 10^6 \text{ N/mm}$$

Since $M_{ur} < M_{u,lim}$, the design is OK.

The Code recommends that at least one third the positive moment reinforcement in simple members shall extend along the same face of the member into the support. However bend one bar upwards, at a distance $x_1$ from the support.

B.M. at $x_1 = \dfrac{w_u \cdot L}{2} x_1 - \dfrac{w_u x_1^2}{2}$. This should be 2/3 of the maximum B.M.

$\therefore \quad \dfrac{w_u \cdot L}{2} x_1 - \dfrac{w_u \cdot x_1^2}{2} = \dfrac{2}{3} \dfrac{w_u L^2}{8}$, which gives $x_1 = 0.211 L = 0.211 \times 6.4 = 1.35$ m

However, the bars are to be taken further by a distance $d$ or 12Φ. Hence distance of point from the centre of support = 1.35 – 0.387 = 0.963 m, or at 0.963 – 0.75/2 ≅ 0.59 m. Hence bend one bar up, at a distance of 0.6 m from the face of the support, and continue the remaining two bars into the support.

**Step 5. Shear reinforcement**

The critical section for shear is at a distance of $d$ (= 0.365 m) from the face of the support.

$$\therefore \quad V_{uD} = \frac{w_u L}{2} - w_u \left(\frac{0.75}{2} + 0.365\right)$$

$$= \frac{19538 \times 6.4}{2} - 19538 [0.375 + 0.365] = 48063.48 \approx 48063 \text{ N}$$

$$\therefore \quad \tau_v = \frac{V_{uD}}{bd} = \frac{48063}{300 \times 365} \approx 0.439 \text{ N/mm}^2$$

$$\frac{100 A_{st}}{bd} \text{ (at support)} = \frac{100}{300 \times 365} (314 \times 2) = 0.57\%.$$

Hence from Table 36.1, $\tau_c \approx 0.5$ N/mm$^2$.

Since $\tau_v < \tau_c$, no shear reinforcement is necessary. However, as per Code recommendations, provide nominal shear reinforcement, given by the expression

$$\frac{A_{sv}}{b \cdot s_v} > \frac{0.4}{0.87 f_y} \quad \text{or} \quad s_v = \frac{2.175 A_{sv} \cdot f_y}{b}.$$

Choosing 8 mm Φ bars, $A_{sv} = 2 \times \frac{\pi}{4}(8)^2 = 100.5$ mm²

$$\therefore \quad s_v = \frac{2.175 \times 100.5 \times 415}{300} = 302.37 \approx 302 \text{ mm}.$$

Max. $s_v$ = least of (0.75 $d$ or 300 mm) = 0.75 $d$ = 0.75 × 387 = 290.25 ≈ 290 mm.

Hence provide 8 Φ 2 logged stirrups @ 290 mm c/c throughout the length of the beam. Provide 2-10 Φ holding bars at the top.

**Step 6. Check for development length at the ends:** Since the beam is supported on wall, the compressive reaction will confine the reinforcement. Hence

$$L_d \leq 1.3 \frac{M_1}{V} + L_0$$

Only two bars are available at the supports.

$$x_u = \frac{0.87 f_y \cdot A_{st1}}{0.36 f_{ck} \cdot b} = \frac{0.87 \times 415 (2 \times 314.15)}{0.36 \times 20 \times 300} = 105 \text{ mm}$$

$$\therefore \quad M_1 = 0.87 f_y A_{st1}(d - 0.416 x_u)$$
$$= 0.87 \times 415 (2 \times 314.15)(387 - 0.416 \times 105)$$
$$\approx 77.88 \times 10^6 \text{ N-mm}$$

$$V = \text{S.F.} = \frac{w_u \cdot L}{2} \text{ N} = \frac{19538 \times 6.4}{2} = 62521.6 \approx 62522$$

$$L_d = \frac{0.87 f_y}{4 \tau_{bd}} \times \Phi = \frac{0.87 \times 415}{4 \times 1.2 \times 1.6} \times \Phi = 47\Phi = 47 \times 20 = 940 \text{ mm}$$

$L_0$ = sum of anchorage beyond centre line of support and anchorage value of hook.

**Bends and hooks:** Shall confirm to IS : 2502.

1. **Bends:** The anchorage value of bend shall be taken as 4 times the diameter of the bars for each 45° bend subject to a maximum of 16 times the diameter of bar.
2. **Hooks:** The anchorage value of a standard U-type hook shall be equal to 16 times the diameter of bar.
3. **Anchorage bars in compression:** As per clause 26.2.1 of IS: 456–2000

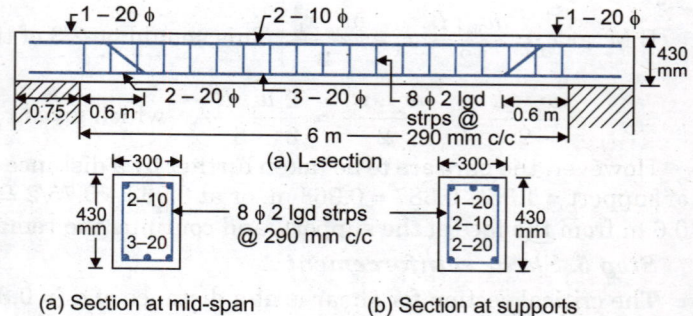

(a) Section at mid-span   (b) Section at supports

Fig. 37.1

If no hook is provided, $L_0 = \frac{L_s}{2} - x' = \frac{750}{2} - 40 = 335$ mm,

where $x'$ = side cover assumed to be 40 mm. The Code further recommends that each bar should be extended by a distance of $L_d/3 = 940/3 = 313$ mm beyond the face of the support. Available distance, beyond the face of support = $L_0 + \frac{L_s}{2} = 335 + \frac{750}{2} = 710$ mm.

which is more than $L_d/3$. Hence no hook is required.   Hence $L_0$ = 335 mm.

$$\therefore \quad 1.3 \frac{M_1}{V} + L_0 = \left(1.3 \times \frac{77.88 \times 10^6}{62522}\right) + 335 = 1954.33 \approx 1954 \text{ mm} > L_d. \quad \text{Hence O.K.}$$

**Step 7. Details of Reinforcement etc.**

The details of reinforcement etc. are shown in Fig. 37.1.

## DESIGN OF R.C.C. BEAMS AND SLABS

### 37.2. DESIGN OF CANTILEVER

When a cantilever beam or slab is subjected to downward loads, it bends downwards with its convexity upwards. Due to this, the upper fibres of the section are subjected to tensile stress and the lower fibres are subjected to compressive stress. Hence the main reinforcement is provided at the *upper face*, and not at the lower face. The B.M. varies from a zero value at the free section to maximum value at the fixed end(support). Hence the depth of the section and the steel reinforcement is varied as per requirements of bending moment. In order that the whole cantilever does not overturn, it should be properly fixed in the wall. The reinforcement should be embedded in the support for a distance of $L_d$ from the edge of the support.

When curtailment of bar is done, it should be kept in mind that the embedment of the bar is not less than $L_d$ from the point it is needed the most, nor less than 12 dia. (or $d$) from the point it can be spared from bending moment point of view.

**Effective length:** The effective length of a cantilever shall be taken as its length to the face of the support plus half the effective depth except where it forms the end of a continuous beam where the length to the centre of the support shall be taken.

**Example 37.2.** *Design of a cantilever beam.*

*A cantilever beam projects 2.5 m beyond the fixed end, and carries a super-imposed load of 12 kN per metre run. Design the cantilever, using M 20 concrete and Fe 415 steel. The width of support is 350 mm.*

**Solution:** **Step 1. Design constants and limiting depths of N.A.**

For Fe 415 steel, $f_y = 415$ N/mm². For M 20 concrete, $f_{ck} = 20$ N/mm².

$$\frac{x_{u,max}}{d} \text{ (for Fe 415 steel)} = \frac{700}{1100 + 0.87 \times 415} = 0.479$$

$$R_u = 0.36 f_{ck} \cdot \frac{x_{u,max}}{d}\left(1 - 0.42 \frac{x_{u,max}}{d}\right)$$

$$= 0.36 \times 20 \times 0.479 (1 - 0.42 \times 0.479) = 2.761$$

**Step 2. Computation of design B.M. and S.F.**

The minimum depth of beam is based on limit state of serviceability requirement.

For a cantilever, $L/d = 7$. ∴ $d = L/7 = 2500/7 \approx 375$ mm.

Keep $D = 400$ mm at fixed end and 200 mm at free end. Keep width $b = 250$ mm.

Self weight of beam = $\frac{1}{2}(0.4 + 0.2) \times 0.25 \times 2.5 \times 25000 = 4687.5 \approx 4688$ N acting at

$$\frac{0.4 + 2 \times 0.2}{0.4 + 0.2} \times \frac{2.5}{3} = 1.11 \text{ m from fixed end.}$$

∴ $$M = (4688 \times 1.1) + \frac{12000(2.5)^2}{2} = 42656.8 \approx 42657 \text{ N-m}$$

∴ $M_u = 1.5M = 1.5 \times 42657 = 63985$ N-m $\approx 63.99 \times 10^6$ N-mm

S.F. $V$ at the edge of support = $4688 + (12000 \times 2.5) = 34688$ N

∴ $V_u = 1.5 V = 52032$ N

**Step 3. Computation of effective depth:** For a balanced section, $M_u = R_u bd^2 = 2.761 bd^2$

∴ $$d = \sqrt{\frac{M_u}{2.761 b}} = \sqrt{\frac{63.99 \times 10^6}{2.761 \times 250}} = 304.47 \approx 305 \text{ mm.}$$

However, keep $D = 400$ mm, so that using 16 mm dia. bars, and 8 mm dia stirrups with 25 mm nominal cover, $d = 400 - 25 - 8 - 8 = 359$ mm. This is more than the one required for the balanced section. Hence, we will have an under-reinforced section. Keep $D = 200$ mm at free end.

**Step 4. Steel Reinforcement**

The reinforcement for an under-reinforced section is given by:

$$A_{st} = \frac{0.5 f_{ck}}{f_y}\left[1 - \sqrt{1 - \frac{4.6 M_u}{f_{ck} bd^2}}\right]bd$$

$$= \frac{0.5 \times 20}{415}\left[1 - \sqrt{1 - \frac{4.6 \times 63.99 \times 10^6}{20 \times 250(359)^2}}\right] \times 250 \times 359 = 568.7 \approx 569 \text{ mm}^2$$

∴ No. of 16 mm dia. bars = 569/201.06 = 2.83. Hence provide 3 bars of 16 mm dia.

Actual $A_{st}$ = 3 × 201.06 = 603.2 mm². Hence actual moment of resistance available is using equation G. 1.1(b) Annexure – G of IS : 456–2000

$$M_{ur} = 0.87 f_y \cdot A_{st} d \left(1 - \frac{f_y}{f_{ck}} \cdot \frac{A_{st}}{bd}\right)$$

$$= 0.87 \times 415 \times 603.2 \times 359 \left(1 - \frac{415}{20} \times \frac{603.2}{250 \times 359}\right)$$

$$= 67.28 \times 10^6 \text{ N-mm} > M_u.$$

Also, $M_{u,lim} = 2.761\,bd^2 = 2.761 \times 250\,(359)^2 = 88.96 \times 10^6$ N-mm.

Since $M_{ur} < M_{u,lim}$, the design is OK.

**Curtailment of reinforcement**

Since the B.M. decreases to zero at the end, let us curtail one bar at a distance $x$ mm from free end.

Effective depth at free end ≈ 200 – 25 – 8 – 8 = 159 mm

Effective depth $d_x$ from $x$ free end = $159 + \left(\dfrac{359 - 159}{2500}\right) x = (159 + 0.08x)$ mm.

Assuming B.M.D. to be parabolic, the B.M. at this section may be approximately taken to be equal to

$$\left(\frac{x}{2500}\right)^2 \times 63.99 \times 10^6 = 10.24\,x^2 \text{ N-mm}.$$

Area of remaining 2 bars = 2 × 201.06 ≈ 402.1 mm²

$$M_{ur} = 0.87 f_y \cdot A_{st} \left[d_x - \frac{f_y}{f_{ck}} \cdot \frac{A_{st}}{b}\right]$$

$$= 0.87 \times 415 \times 402.1 \left[(159 + 0.08x) - \frac{415}{20} \times \frac{402.1}{250}\right]$$

$$= 145178\,[125.6 + 0.08x] = 18.234 \times 10^6 + 11614\,x$$

∴ $10.24\,x^2 = 18.234 \times 10^6 + 11614\,x$

or $x^2 - 1134\,x - 1.780 \times 10^6 \approx 0$. From which $x \approx 2016$ mm

$d_x = 159 + 0.08 \times 2016 \approx 320$ mm.

Hence one bar can be curtailed at a distance = 2500 – 2016 + 320 ≈ 804 mm away from the face of fixed end. However, minimum distance at which any bar can be curtailed from the fixed end is $L_d$ so that all the bars can develop their full tensile strength at the fixed end.

Now $L_d = \dfrac{0.87 f_y}{4 \tau_{bd}} \cdot \Phi = \dfrac{0.87 \times 415}{4 \times 1.6 \times 1.2} \Phi$

$= 47\,\Phi = 47 \times 16 = 752$ mm < 804 mm.

Hence one bar can be curtailed at a distance of 804 mm from the fixed end. However, curtail one bar at 810 mm from the fixed end. Continue the remaining two bars upto the end. At the bottom, provide 2 – 10 Φ bar as holding bars at the bottom face.

**Step 5. Shear reinforcement:** $V_u = 52032$ N

Nominal shear stress $\tau_v = \dfrac{V_u - \dfrac{M_u}{d} \tan\beta}{bd}$, where $\tan\beta = \dfrac{400 - 200}{2500} = 0.08$. Hence,

$$\tau_v = \frac{52032 - \dfrac{63.99 \times 10^6}{365} \times 0.08}{250 \times 365} = 0.416 \approx 0.42 \text{ N/mm}^2.$$

$$p_t = \frac{100 A_s}{bd} = \frac{100 \times 603.2}{250 \times 365} = 0.66\%$$

$\tau_c \approx 0.5$ N/mm² $> \tau_v$.

Hence provide only nominal shear reinforcement, given by *Eq. 36.3*, rearranged below.

$$s_v = \frac{2.175 \, A_{sv} \cdot f_y}{b}$$

Using 8 mm Φ 2 lgd. stirrups,

$$A_{sv} = 2 \times 50.3 = 100.6.$$

Hence, $\quad s_v = \dfrac{2.175 \times 100.6 \times 415}{250} = 363$ mm.

However, max. $s_v = 0.75 \, d = 0.75 \times 359 \triangleq 270$ mm. Hence provide 8 Φ, 2-lgd stirrups @ 270 mm c/c, at the supports. Reduce this spacing at the ends, to a value $= 0.75 \times 165 = 123.75 \approx 120$ mm

### Step 6. Embedment of reinforcement in the support.

In order to develop full tensile strength at the face of the support, each of the three bars must be embedded into the support by a length equal to $L_d = 752$ mm. This could be best achieved by providing two bends of 90°, as shown in Fig. 37.2, where the anchorage value of each 90° bend = 8 Φ = 8 × 16 = 128 mm.

### Step 7. Details of Reinforcement: Shown in Fig. 37.2.

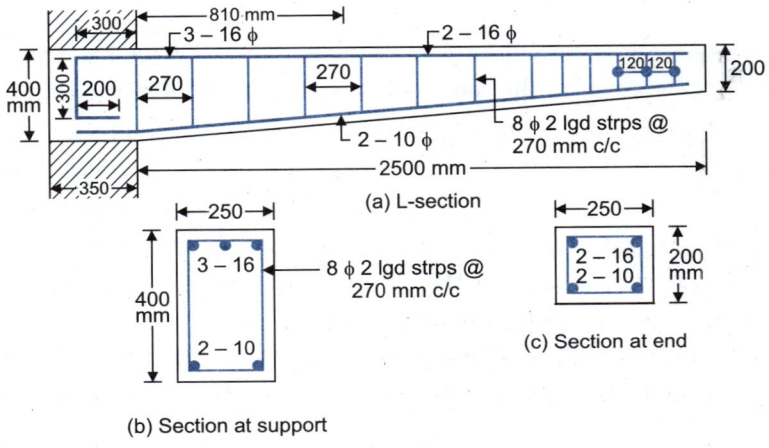

**Fig. 37.2**

## 37.3. DESIGN OF DOUBLY REINFORCED BEAM

**Design: Example 37.3.** *Design a reinforced concrete beam supported on two walls 500 mm thick, spaced at a clear distance of 6 m. The beam carries a super-imposed load of 30 kN/m. The size of the beam is restricted to 300 mm × 500 mm. Use M 20 concrete and Fe 415 steel.*

**Solution:**

### Step 1. Computation of design constants and limiting depth of N.A.

For Fe 415 steel, $\quad f_y = 415$ N/mm² and $\dfrac{x_{u,max}}{d} = \dfrac{700}{1100 + 0.87 \times 415} = 0.479$

For M 20 concrete, $\quad f_{ck} = 20$ N/mm².

$$R_u = 0.36 \, f_{ck} \cdot \frac{x_{u,max}}{d} \left(1 - 0.42 \frac{x_{u,max}}{d}\right)$$

$\qquad\qquad = 0.36 \times 20 \times 0.479 \, (1 - 0.42 \times 0.479) = 2.761$

Here $\qquad d \approx 500 - 25 - 10 - 8 = 457$ mm, using 20 mm Φ bars and 8 mm Φ rings

∴ $\qquad M_{u,lim} = R_u \, bd^2 = 2.761 \times 300 \, (457)^2 = 173.005 \times 10^6$ N-mm,

$\qquad x_{u,max} = 0.479 \, d = 0.479 \times 457 = 218.9$ mm.

### Step 2. Computation of design B.M.: $L \approx 6000 + 457 = 6457$ mm < 6500 mm

∴ Available $L/d = 6457/457 = 14.13$

Max. permissible value of $L/d$, even without modification factor = 20.

Hence deflection requirements are satisfied.

Dead load of beam = 0.3 × 0.5 × 25000 = 3750 N/m
Total $w$ = 3750 + 30,000 = 33750 N/m.
$w_u$ = 1.5 × 33750 = 50625 N

$$M_{uD} \text{ (at midspan)} = \frac{w_u L^2}{8} = \frac{50625 (6.457)^2}{8} = 263837.5 \approx 263838 \text{ N-m} \approx 263.84 \times 10^6 \text{ N-mm.}$$

Since $M_{uD} > M_{u,lim}$, a doubly reinforced section will be required.

### Step 3. Determination of tensile reinforcement.

$$p_{t,\,lim} = 0.414 \frac{f_{ck}}{f_y} \cdot \frac{x_{u,max}}{d} \qquad \ldots(33.9)$$

$$= 0.414 \times \frac{20}{415} \times 0.479 = 0.0095569$$

∴ $\qquad A_{st1} = p_{t,\,lim} \times bd = 0.0095569 \times 300 \times 457 = 1310 \text{ mm}^2$

Let the cover to the centre of compressive reinforcement $(d')$ = 25 + 10 + 8 = 43 mm

∴ $\qquad A_{st2} = \dfrac{M_{uD} - M_{u,lim}}{0.87 f_y (d - d')} = \dfrac{(263.84 - 173.05) 10^6}{0.87 \times 415 (457 - 43)} = 607.4 \text{ mm}^2$

∴ Total $\quad A_{st} = A_{st1} + A_{st2} = 1310 + 607.4 = 1917.4 \text{ mm}^2$

(Max. reinforcement = $0.04 \, bD = 0.04 \times 300 \times 500 = 6000 \text{ mm}^2 > A_{st}$). Hence O.K.

N – 20 mm Φ bars = 1917.4/314.2 = 6.1 ≈ 7    and    N – 25 Φ bars = 1917.4 / 490.9 = 3.905 ≈ 3.91 ≈ 4.

Hence provide 4 bars of 25 mm dia.

### Step 4. Determination of compressive reinforcement

$$A_{sc} = \frac{M_{uD} - M_{u,lim}}{(f_{sc} - 0.446 f_{ck})(d - d')} = \frac{(263.84 - 173.05)10^6}{(f_{sc} - 0.446 \times 20)(457 - 43)} = \frac{219300}{(f_{sc} - 8.92)}$$

$$\varepsilon_{sc} = \frac{0.0035 (x_{u,max} - d')}{x_{u,max}} = \frac{0.0035(218.9 - 43)}{218.9} = 0.00281$$

Hence from Table 33.1, for Fe 415 steel, $f_{sc} \approx 352 \text{ N/mm}^2$ for $\varepsilon_{sc} = 0.00281$

∴ $\qquad A_{sc} = \dfrac{219300}{352 - 8.92} = 639.2 \text{ mm}^2$.

Provide 2 – 16 mm Φ and 1 – 20 mm Φ
bars having total $\quad A_{sc} = 2 \times 201.06 + 1 \times 314.2 = 716.32 \text{ mm}^2$.

### Step 5. Curtailment of tensile and compressive reinforcement

Development length $(L_d)$ of 25 dia. bars in tension.

$$= \frac{0.87 f_y}{4 \tau_{bd}} \Phi = \frac{0.87 \times 415}{4 \times 1.2 \times 1.6} \cdot \Phi = 47 \Phi = 47 \times 25 = 1175 \text{ mm}$$

Development length $(L_d)$ of 20 mm dia. bars in compression.

$$= \frac{0.87 f_y}{4(1.25 \tau_{bd})} \Phi = \frac{0.87 \times 415}{5 \times 1.92} \Phi = 37.6 \Phi \approx 750 \text{ mm}$$

Hence the tension steel cannot be curtailed at less than 1.18 m from the centre of span, while the compression cannot be curtailed at less than 0.75 m from the centre.

Let us curtail 1 compression bar and two tension bars at a theoretical point of cut off distant 1.20 m from the centre of span.

∴    Actual point of cut-off from the centre of the span will be larger of the two values:
 (i) 1.2 + d = 1.2 + 0.457 = 1.657 m.    (ii) 1.2 + 12 Φ = 1.2 + 12 × 0.025 = 1.50 m

Keep the *actual* point of cut-off at 1.70 m from centre of span. Distance of theoretical point of cut off from centre of span = 1.70 – 0.457 = 1.243 m.    ∴    Distance of theoretical point of cut-off from centre of support = 6.465/2 – 1.243 = 1.98 ≈ 2.2 m.

**DESIGN OF R.C.C. BEAMS AND SLABS** | **901**

It can be shown that the moment of resistance of the section at this point will be larger than the actual B.M. at this location.

**Step 6. Shear reinforcement:** The critical section for shear is at a distance of $d$ (= 457 mm) from the face of the support, or at a distance of $457 + \dfrac{457}{2} = 685.5$ mm ≈ 0.69 m from the centre of support.

$$V_u = \left(50625 \times \dfrac{6.457}{2}\right) - (50625 \times 0.69) = 128511.56 \approx 128511 \text{ N}$$

$$\tau_v = \dfrac{V_u}{bd} = \dfrac{128511}{300 \times 457} \approx 0.94 \text{ N/mm}^2.$$

$$\dfrac{100 A_s}{bd} = \dfrac{100 (2 \times 490.9)}{300 \times 457} = 0.716 \approx 0.72\%$$

Hence from Table 36.1. $\tau_c = 0.34$ N/mm². Hence shear reinforcement is necessary.
Let us provide 8 mm Φ 2 lgd stirrups, having $A_{sv} = 100.5$ mm²

$$\therefore \quad s_v = \dfrac{0.87 f_y \cdot A_{sv}}{(\tau_v - \tau_c) b} = \dfrac{0.87 \times 415 \times 100.5}{(0.94 - 0.34) \times 300} = 201.58 \approx 201 \text{ mm}$$

Spacing based on minimum shear reinforcement

$$s_v = \dfrac{2.175 A_{sv} f_y}{b} = \dfrac{2.175 \times 100.5 \times 415}{300} = 302.27 \approx 302 \text{ mm}$$

Spacing based on effective depth of beam

$$s_v = 0.75 d = 0.75 \times 457 = 342.75 \approx 342 \text{ mm} \quad \text{or} \quad 300 \text{ mm which ever is less.}$$

Hence provide 8 mm Φ 2 lgd strip @ 200 mm c/c near support and increase this gradually to 300 mm c/c towards the centre.

**Step 7. Check for development length at support**

At the support, the tensile reinforcement should satisfy the following relation

$$1.3 \dfrac{M_1}{V} + L_0 \geq L_d$$

There are two tensile bars available at supports, having

$$A_{st} = 2 \times 490.9 = 981.7 \text{ mm}^2$$

$$x_u = \dfrac{0.87 f_y A_{st1}}{0.36 f_{ck} \cdot b} = \dfrac{0.87 \times 415 \times 981.7}{0.36 \times 20 \times 300} = 164 \text{ mm}$$

$$\therefore \quad M_1 = 0.87 f_y A_{st}(d - 0.416 x_u) = 0.87 \times 415 \times 981.7 (457 - 0.416 \times 164)$$

$$= 137.8 \times 10^6 \text{ N-mm}$$

$$V = V_u = 50625 \times \dfrac{6.457}{2} = 163442.8 \approx 163443 \text{ N}$$

$$L_d = 47 \Phi = 47 \times 25$$
$$= 1175 \text{ mm (as computed earlier)}$$

Minimum embedment required from the face of the support = $L_d/3 = 1175/3 = 391.67 \approx 392$ mm.
Width of support available ($b_0$) = 500 mm.
Providing end cover $x'$ of 40 mm, and providing 90° bend to the tensile bars,

$$L_0 = (500 - 40 - 228.6 + 3 \Phi)$$
$$= 231.4 + 3 \times 25 = 306.4 \text{ mm}$$

$$1.3 \dfrac{M_1}{V} + L_0 = 1.3 \times \dfrac{137.8 \times 10^6}{163443} + 306.4$$

$$= 1402 \text{ mm} > L_d. \quad \text{Hence O.K.}$$

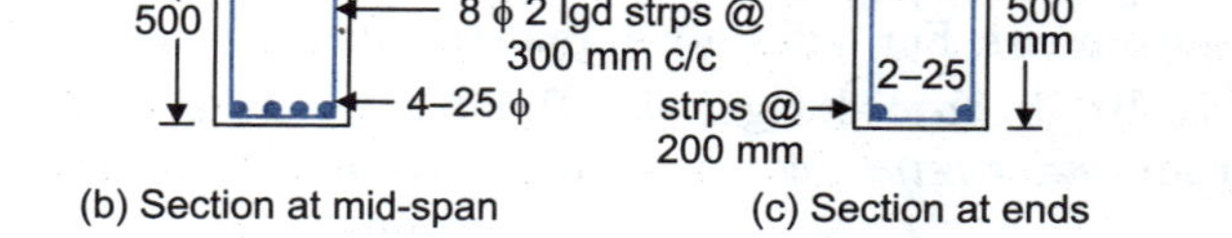

**Fig. 37.3.** Doubly Reinforced Beam.

(a) L-section
(b) Section at mid-span
(c) Section at ends

**902** REINFORCED CONCRETE STRUCTURE

The compression reinforcement at the top has to be embedded into the support by a length = $b_0$ – end cover = 500 – 40 = 460 mm, which is more than $L_d/3$ (= 750/3 = 250 mm).

**Step 8. Details of reinforcement etc.** As shown in Fig. 37.3.

## 37.4. DESIGN OF ONE WAY SLAB

If the ratio of longer span ($l_y$) to the shorter span ($l_x$) i.e., $\left(\dfrac{l_y}{l_x}\right)$ is greater than 2, then the slab is called one way slab.

### (A) LONGITUDINAL AND TRANSVERSE REINFORCEMENTS

The analysis of slab spanning in one direction is done by assuming it to be a beam of 1 m width. The reinforcement etc. are calculated for 1 m width and the bars are distributed accordingly.

In addition to the main reinforcement (called *longitudinal reinforcement*), *transverse reinforcement* (also called distribution reinforcement) is provided in a direction at right angles to the span of the slab. The transverse reinforcement is provided to serve the following purposes:

(*i*) It distributes the effects of point load on the slab more evenly and uniformly.

(*ii*) It distributes the shrinkage and temperature cracks more evenly.

(*iii*) It keeps main reinforcement in position.

The reinforcement in either direction in slab shall not be 0.15% of the total cross-sectional area of mild steel bars. However, this value can be reduced to 0.12% when high strength deformed bars or welded wire fabric are used. The maximum diameter of the reinforcing bars shall not exceed one eighth of total thickness of slab.

### (B) BASIC RULES FOR DESIGN (*IS 456 : 2000*):

For basic rules for design, refer to § 7.4 (*b*).

### (C) BAR BENDING SCHEME

For bar bending scheme, refer § 7.4 (*c*) and Fig. 7.12.

For solid slabs, the design shear strength of concrete is taken as $k\,\tau_c$, where $k$ may be taken from Table 36.3. The other requirements and basic rules are the same as discussed in § 7.4 for design of one way slab by working stress method.

**Example 37.4.** *Design of one way slab*

*Design a R.C. slab for a room having inside dimensions 3 m × 7 m. The thickness of supporting wall is 300 mm. The slab carries 75 mm thick lime concrete at its top, the unit weight of which may be taken as 20 kN/m³. The live load on the slab may be taken as 2 kN/m². Assume the slab to be simply supported at the ends. Use M 20 concrete and Fe 415 steel.*

**Solution:**

**Step 1. Design constants and limiting depth of N.A.:** For Fe 415 steel, $f_y = 415$ N/mm².

For M 20 concrete, $f_{ck} = 20$ N/mm². $\dfrac{x_{u,max}}{d}$ (for Fe 415 steel) = $\dfrac{700}{1100 + 0.87 \times 415}$ = 0.479

$$R_u = 0.36 f_{ck} \dfrac{x_{u,max}}{d}\left(1 - 0.42 \dfrac{x_{u,max}}{d}\right) = 0.36 \times 20 \times 0.479\,(1 - 0.42 \times 0.479) = 2.755$$

**Step 2. Computation of design B.M. and S.F.** The minimum depth of slab is based on serviceability requirement. For a simply supported slab $L/d$ = 20

In limit state design, $p$ = 0.955% for M 20 – Fe 415 combination. However, assuming $p_t$ = 0.35% for an *under-reinforced* section, we get modification factor ≈ 1.5 from Fig. 7.1.

Taking $L \approx 3.2$ m, we have $d = \dfrac{3200}{20 \times 1.5} \simeq 107$ mm

Keep $D$ = 125 mm. Using 15 mm nominal cover and 8 mm dia. bars, $d$ = 125 – 15 – 4 = 106 mm

Effective span $L = (l + d) = 3 + 0.106 \approx 3.1$ m (or equal to c/c distance between supports, whichever is less)

| Self weight of slab | $= 0.125 \times 25000$ | $= 3125 \text{ N/m}^2$ |
|---|---|---|
| Live load | $= 2 \text{ kN/m}^2$ | $= 2000 \text{ N/m}^2$ |
| Weight of lime terracing | $= 0.075 \times 20000$ | $= 1500 \text{ N/m}^2$ |
| | Total | $w = 6625 \text{ N/m}^2$ |

$\therefore \quad w_u = 1.5 w = 1.5 \times 6625 = 9937.5 \approx 9938 \text{ N/m}^2$

$$M_u = \frac{w_u \, l^2}{8} = \frac{9938 \, (3.1)^2}{8} = 11938.02 \text{ N-m} \approx 11.94 \times 10^6 \text{ N-mm}$$

$$V_u = \frac{w_u \cdot L}{2} = 9938 \, \frac{(3.1)}{2} = 15403.9 \approx 15404 \text{ N}$$

**Step 3. Computation of effective depth d:** For a balanced section,

$$\therefore \quad d = \sqrt{\frac{M_u}{R_u \cdot b}} = \sqrt{\frac{11.94 \times 10^6}{2.755 \times 1000}} = 65.83 \text{ mm}$$

Actual $d$ available = 106 mm (based on the limit state of serviceability requirements). Hence the section will be *under-reinforced*.

**Step 4. Steel Reinforcement.** For an under-reinforced section,

$$A_{st} = \frac{0.5 f_{ck}}{f_y} \left[ 1 - \sqrt{1 - \frac{4.6 M_u}{f_{ck} bd^2}} \right] bd = \frac{0.5 \times 20}{415} \left[ 1 - \sqrt{1 - \frac{4.6 \times 11.94 \times 10^6}{20 \times 1000 \, (106)^2}} \right] 1000 \times 106$$

$$= 334 \text{ mm}^2$$

$$(A_{st})_{min} = \frac{0.12}{100} \times 1000 \times 125 = 150 \text{ mm}^2 = A_{sd}$$

Hence spacing of 8 Φ mm bars as main reinforcement is

$$s = \frac{1000 \, A_\Phi}{A_{st}} = \frac{1000 \times 50.3}{334} = 150.6 \text{ mm c/c}$$

However, provide 8 mm Φ bars @ 150 mm c/c. This spacing is less than lesser of $3d$ or 300 mm.

Actual $\quad A_{st} = \frac{1000 \times 50.3}{150} = 335.3 \text{ mm}^2 \quad$ and $\quad p_t = \frac{335.3}{1000 \times 106} \times 100 = 0.32\%$.

Spacing of 8 Φ mm bars as distribution reinforcement is

$$s = \frac{1000 \, A_\Phi}{A_{sd}} = \frac{1000 \times 50.3}{150} = 335.33 \approx 335 \text{ mm} \approx 300 \text{ mm}$$

(which is less than max. spacing $5d = 5 \times 106$ mm = 530 or 450 mm whichever is smaller).

Bend alternate bars of main reinforcement up, at a distance of $l/7 = 3000/7 = 428.57 \approx 430$ mm from the face of the support. Continue the remaining half bars into the support.

**Step 5. Check for limit state of serviceability of deflection:** Basic $L/d$ = 20. Actual $p_t = 0.32\%$. Hence for Fe 415 steel, $f_s = 0.58 f_y \times \frac{A_{st} \text{ required}}{A_{st} \text{ provided}} = 0.58 \times 415 \times \frac{334}{335.3} = 239.77 \text{ N/mm}^2$.

Hence from Fig. 7.1, for $p_t = 0.32$ and $f_s = 239.77$ N/mm$^2$, we get $m_{f1} \approx 1.6$. Hence required $d = 3100/(20 \times 1.6) \triangleq 97$ mm, against available $d$ = 106 mm. Hence safe.

**Step 6. Check for shear:** The critical section for shear is at a distance of $d$ ( = 106 mm) from the face of the support or at a distance of $d + \frac{d}{2} = 106 + \frac{106}{2} \approx 159$ mm from the centre of support. $V_{uD} = V_u - w_u(0.159)$ = 15404 − 9938 (0.159) = 13824 N.

$$\tau_v = \frac{V_{uD}}{bd} = \frac{13824}{1000 \times 106} = 0.13 \text{ N/mm}^2;$$

$\dfrac{100 A_s}{bd}$ (at supports) = $\dfrac{1}{2} \times 0.32 = 0.16$

Hence $\tau_c \approx 0.285$ N/mm² from Table 36.1. Also $k = 1.30$ (from Table 36.3)

Since $\tau_v \ll k\tau_c$, the section is safe in shear.

**Note:** Slabs are normally found safe in shear. Hence it will be reasonably correct if we take face of the support as the critical section for shear.

Thus, at the face of support, $V_u = \dfrac{1}{2} \times 9938 \times 3 = 14907$ N and $\tau_v = 14907/(1000 \times 106) = 0.14$ N/mm², against 0.13 found earlier.

### Step 7. Check for development length and anchorage at ends

$$L_d = \dfrac{0.87 f_y}{4 \tau_{bd}} \Phi = \dfrac{0.87 \times 415}{4 \times 1.2 \times 1.6} \Phi = 47 \Phi = 47 \times 8 = 376 \text{ mm}.$$

The Code requires that the reinforcement should extend by a length equal to $L_d/3 = 376/3 \approx 125$ mm beyond the face of support. This suggests that the width of support should not be less than $125 + 25 = 150$ mm. In the present case, the support width is 300 mm and hence the anchorage condition is satisfied. Also, at the simple support like this, where compressive reaction confines the ends of bars,

$$1.3 \dfrac{M_1}{V} + L_0 \gg L_d$$

Here $V = V_u = 15404$ N, at the centre of support.

$M_1$ = moment of resistance of the remaining bars, which is computed as under:

$$x_u = \dfrac{0.87 f_y A_{st1}}{0.36 f_{ck} \cdot b} = \dfrac{0.87 \times 415 \times (335.3/2)}{0.36 \times 20 \times 1000} = 8.4 \text{ mm}$$

$\therefore \quad M_1 = 0.87 f_y A_{st1}(d - 0.416 x_u)$

$= 0.87 \times 415 \times \dfrac{335.3}{2} (106 - 0.416 \times 8.4)$

$= 6.2 \times 10^6$ N-mm

Providing 90° bend and a clear cover = 25 mm at the side end,

$$L_0 = \dfrac{l_s}{2} - x' + 3\Phi = \dfrac{300}{2} - 25 + 3 \times 8 = 149 \text{ mm}.$$

Hence, $\quad 1.3 \dfrac{M_1}{V} + L_0 = 1.3 \times \dfrac{6.2 \times 10^6}{15404} + 149 = 672.24 \approx 672 \text{ mm} > L_d$

Hence the Code requirements are satisfied.

**Note:** In limit state design, effective depth required from the point of view of bending will be very much less than the one required from deflection point of view. This will result in an under-reinforced section. Hence while taking the modification factor from Fig. 7.1, the value of $p_t$ should be assumed equal to 40 to 50% of $p_{t, lim}$ for mild steel bars, and equal to about 30% of $p_{t, lim}$ for HYSD bars.

### Step. 8. Details of reinforcement.
Shown in Fig. 37.5.

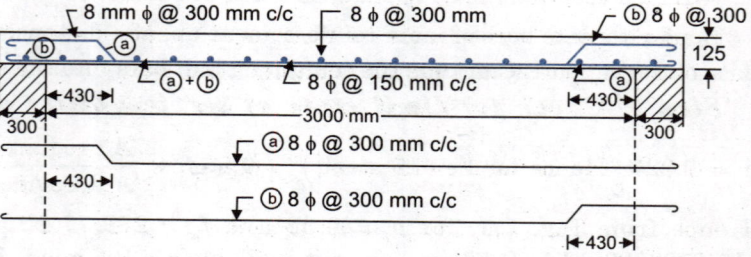

Fig. 37.4

## 37.5. DESIGN OF ONE WAY CONTINUOUS SLAB

When a slab is continuous over several spans, negative (*i.e.* hogging) bending moment is induced over intermediate supports and hence reinforcement has to be provided at the top of the slab portion over the intermediate supports. These intermediate supports may either be monolithic with the slab, or may be of some different material such as rolled steel joist etc. The *basic rules* for design, as per IS : 456-2000 are given in § 7.1.

**Example 37.5. Design of one way continuous slab**

*Design a continuous R.C. slab for a hall 6.5 m wide and 13.5 m long. The slab is supported on R.C.C. beams, each 240 mm wide which are monolithic. The ends of the slab are supported on walls, 300 mm wide. Design the slab for a live load of 2 kN/m². Assume the weight of roof finishing equal to 1.5 kN/m². Use M 20 concrete and Fe 415 steel.*

**Solution:**

**Step 1. Arrangement of spans:** Let us arrange the spans as shown in Fig. 37.5. Assuming $d = 100$ mm. The width of intermediate supports is less than 1/12 of the clear span. Hence the effective span will be as under:

(i) *For end spans*: $L$ = centre to centre of supports = $3.04 + \dfrac{0.3 + 0.24}{2} = 3.31$ m  or  $L$ = clear span + $d$
= $3.04 + 0.10 = 3.14$ m, whichever is less.  Hence $L = 3.14$ m.

(ii) *For intermediate spans*:

$L$ = centre to centre of supports = $3.35 + 0.24 = 3.59$ m

or  $L$ = clear span + $d$ = $3.35 + 0.10 = 3.45$ whichever is less.  Hence $L = 3.45$ m

Thus, in this case, the length of effective spans is not equal. However, the difference is very much less than 0.15% of the longer spans; hence they may be treated as being equal for the purposes of using the B.M. coefficients given in the I.S. Code.

**Step 2. Fixation of d and D:** From the limit state of serviceability, the effective depth is given by $L/d = 26$ for continuous spans and $L/d = 20$ for simply supported spans.

Hence for end span,  $L/d \approx \dfrac{1}{2}(26 + 20) = 23$.

Assume a modification factor (corresponding to $p_t \approx 0.25\%$) equal to 1.6 for Fe 415 steel. Hence, on the basis of end span.

$$d = \dfrac{L}{23 \times 1.6} = \dfrac{3140}{23 \times 1.6} = 85.32 \approx 86 \text{ mm}$$

On the basis of intermediate spans,

$$d = \dfrac{L}{26 \times 1.6} = \dfrac{3450}{26 \times 1.6} = 82.9 \approx 83 \text{ mm}.$$

Keep $D = 110$ mm,  so that $d = 110 - 15 - 4$
= 91 mm.

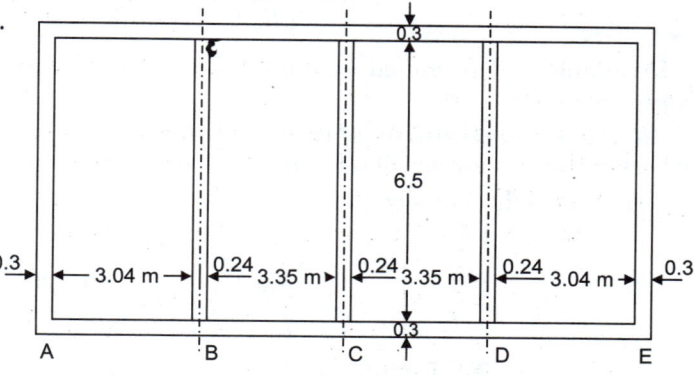

Fig. 37.5.  Arrangement of spans.

**Step 3. Design constants and limiting depth of N.A.**

For Fe 415 steel, $f_y = 415$ N/mm². For M 20 concrete, $f_{ck} = 20$ N/mm².

$$\dfrac{x_{u,max}}{d} \text{ (For Fe 415 steel)} = \dfrac{700}{1100 + 0.87 \times 415} = 0.479$$

$$\therefore R_u = 0.36 f_{ck} \dfrac{x_{u,max}}{d}\left(1 - 0.42 \dfrac{x_{u,max}}{d}\right)$$

$$= 0.36 \times 20 \times 0.479 (1 - 0.42 \times 0.479) = 2.755$$

**Step 4. Computation of design B.M. and effective depth.**

| | | |
|---|---|---|
| Self weight of slab | = 0.11 × 25000 | = 2750 N/m² |
| Weight of roof finishing | = 1.5 kN/m² | = 1500 N/m² |
| | Total | $w_d$ = 4250 N/m² |
| Live load | $w_s$ = 2 kN/m² | = 2000 N/m² |

*For end spans*

$$M \text{ (near the centre)} = \frac{w_d L^2}{12} + \frac{w_s L^2}{10} = \frac{4250(3.14)^2}{12} + \frac{2000(3.14)^2}{10} = 5463.86 \approx 5464 \text{ N-m}$$

$$= 5.464 \times 10^6 \text{ N-mm}$$

$M$ (at support next to end support)

$$= -\left[\frac{w_d L^2}{10} + \frac{w_s L^2}{9}\right] = -\left[\frac{4250(3.14)^2}{10} + \frac{2000(3.14)^2}{9}\right]$$

$$= 6381.35 \approx 6382 \text{ N-m} = 6.382 \times 10^6 \text{ N-mm}$$

*For Intermediate spans*

$$M \text{ (at the middle)} = +\frac{w_d L^2}{24} + \frac{w_s L^2}{12} = \frac{4250(3.45)^2}{24} + \frac{2000(3.45)^2}{12} = 4091.48 \approx 4092 \text{ N-m}$$

$$= 4.092 \times 10^6 \text{ N-mm}$$

$$M \text{ (at interior supports)} = -\left[\frac{w_d L^2}{12} + \frac{w_s L^2}{9}\right] = -\left[\frac{4250(3.45)^2}{12} + \frac{2000(3.45)^2}{9}\right]$$

$$= 6860.46 \approx 6861 \text{ N-m} = 6.861 \times 10^6 \text{ N-mm}$$

Out of the four values of moments, the effective depth will be determined for max. value.

∴ Design moment $= 6.861 \times 10^6$ N-mm.

Hence $M_{uD} = 1.5 \times 6.861 \times 10^6 = 10.291 \times 10^6$ N-mm

$$\therefore \quad d = \sqrt{\frac{M_{uD}}{R_u \cdot b}} = \sqrt{\frac{10.291 \times 10^6}{2.755 \times 1000}} = 61.12 \text{ mm}.$$

Available $d$, determined from limit state of serviceability, is equal to 91 mm. Hence the section will be *under-reinforced*.

**Step 5. Determination of reinforcement**: It is proposed to provide type II detailing shown in Fig. 7.19, wherein there is a provision for different reinforcements at various critical locations.

(a) *At Middle of end span*

$M_{uD1} = 1.5 \times 5.464 \times 10^6 = 8.196 \times 10^6$ N-mm

$$A_{st1} = \frac{0.5 f_{ck}}{f_y}\left[1 - \sqrt{1 - \frac{4.6 M_u}{f_{ck} bd^2}}\right]bd = \frac{0.5 \times 20}{415}\left[1 - \sqrt{1 - \frac{4.6 \times 8.196 \times 10^6}{20 \times 1000(91)^2}}\right]1000 \times 91 \quad ...(33.13)$$

$$= 265.7 \text{ mm}^2$$

∴ $s$-8 mm Φ bars $= \dfrac{1000 \times 50.3}{265.7} = 189.3$ mm.

Max. spacing $= 3d = 3 \times 91 = 273$ mm. Hence provide 8 Φ bars @ 180 mm c/c.

Actual $A_{st1}$ provided $= \dfrac{1000 \times 50.3}{180} = 279.4$ mm$^2$

$$p_t = \frac{100 \times 279.4}{1000 \times 91} = 0.31\% \quad \left(A_{st,min} = \frac{0.12}{100} \times 1000 \times 110 = 132 \text{ mm}^2\right)$$

Bend half bars up at a distance of $0.15 L = 0.15 \times 3140 = 471$ mm from the centre of support, or from $471 - d/2 = 471 - 101/2 \approx 420$ mm from the face of the support. Continue remaining half bars into the support for a distance $L_d/3$. At the other end, bend half bars up at a distance of $0.25 L = 0.25 \times 3140 = 785$ from the centre of the support, and continue the remaining half into the support, for a distance or $L/10 \approx 320$ mm from the face of the support.

(b) *At middle of intermediate span*
$$M_{uD2} = 1.5 \times 4.092 \times 10^6 = 6.138 \times 10^6 \text{ N-mm}$$
$$\therefore A_{st2} = \frac{0.5 \times 20}{415}\left[1 - \sqrt{1 - \frac{4.6 \times 6.138 \times 10^6}{20 \times 1000 (91)^2}}\right] 1000 \times 91$$
$$= 195.6 \text{ mm}^2 \, (A_{st,min} = 132 \text{ mm}^2).$$

However keep $A_{st2} = A_{st1}$ because of practical considerations, and also because of the reason that the section is already very much under-reinforced.

(c) *At support next to end support*
$$M_{uD1}' = 1.5 \times 6.382 \times 10^6 = 9.573 \times 10^6 \text{ mm}$$
$$\therefore A_{st1}' = \frac{0.5 \times 20}{415}\left[1 - \sqrt{1 - \frac{4.6 \times 9.573 \times 10^6}{20 \times 1000 (91)^2}}\right] 1000 \times 91 = 314 \text{ mm}^2.$$

Area of steel available in the form of bent up bars from the adjacent spans
$$= \frac{1}{2}(A_{st1} + A_{st2}) = 279.4 \text{ mm}^2.$$

Additional steel required = 314 − 279.4 = 34.6 mm².

and $\quad s$-8 mm $\Phi$ bars $= \dfrac{1000 \times 50.3}{34.6} = 1453.75 \approx 1450$ mm.

However, this has to be in multiple of earlier spacing of 180 mm. Hence provide extra 8 mm $\Phi$ bars @ 900 mm c/c for length equal to 0.3 L = 0.3 × 3140 ≈ 950 mm to one side and 0.3 × 3450 = 1050 mm 1050 mm to the other side of the support.

(d) *At interior supports*
$$M_{uD2}' = 1.5 \times 6861 \times 10^6 = 10.292 \times 10^6 \text{ N-mm}.$$
$$\therefore A_{st2}' = \frac{0.5 \times 20}{415}\left[1 - \sqrt{1 - \frac{4.6 \times 10.292 \times 10^6}{20 \times 1000 (91)^2}}\right] \times 1000 \times 91 = 339.72 \approx 340 \text{ mm}^2.$$

Steel available in the form of bent-up bars from adjacent spans = 279.4 mm².

∴ Additional steel required = 340 − 279.4 = 60.6 mm²

$\therefore \quad s$-8 mm $\Phi = \dfrac{1000 \times 50.3}{60.6} = 830$ mm.

However, this has to be in multiple of earlier spacing of 180 mm c/c. Hence provide 8 $\Phi$ extra bars @ 720 mm c/c, for a length = 0.3L = 0.3 × 3450 ≈ 1050 m to the either side of the support.

*Distribution reinforcement*
$$A_{sd} = \frac{0.12}{100} \times 1000 \times 110 = 132 \text{ mm}^2.$$
$$\therefore \quad s\text{-8 mm } \Phi = \frac{1000 \times 50.3}{132} \approx 381 \text{ mm}$$

However provide 8 $\Phi$ bars @ 375 mm c/c.

**Step 5. Check for anchorage and development length at end support.**
The end support is a free support: $L_d = 47\Phi = 47 \times 8 \approx 376$ mm
$$L_d/3 = 376/3 \approx 126 \text{ mm}.$$

Width of support available = 300 mm. Length by which half the bars can be taken inside the support = 300 − 25 = 275 mm > 126 mm. Hence safe.

Again, at the end support, the following relation should satisfied
$$1.3 \frac{M_1}{V} + L_0 \geq L_d$$

where $\quad V = V_u = 1.5[0.4 \, w_d L + 0.45 w_s \cdot L]$
$$= 1.5[0.4 \times 4250 \times 3.14 + 0.45 \times 2000 \times 3.14] = 12246 \text{ N}.$$

Area of steel available at supports = $A_{st1}/2 = 279.4/2 \approx 139.7$ mm²

$$x_u = \frac{0.87 f_y \cdot A_{st}}{0.36 f_{ck} \cdot b} = \frac{0.87 \times 415 \times 139.7}{0.36 \times 20 \times 1000} = 7 \text{ mm}$$

∴ $M_1 = 0.87 f_y A_{st}(d - 0.416 x_u) = 0.87 \times 415(139.7)(91 - 0.416 \times 7)$
$= 4.443 \times 10^6$ N-mm

∴ $L_0 = \dfrac{b_s}{2} - x' = \dfrac{300}{2} - 25 = 125$ mm.

$1.3 \dfrac{M_1}{V} + L_0 = 1.3 \times \dfrac{4.443 \times 10^6}{12246} + 125 = 596.65 \approx 596$ mm.

$L_d = 376$ mm.   Hence Safe.

**Step 6. Check for shear.** Normally, slabs are found to be safe in shear. However, let us check for shear at the support next to the end support, where S.F. is maximum.

$$V_u = 1.5 [0.6 w_d L + 0.6 w_s L] = 1.5 \times 0.6 \times 3.14 [4250 + 2000]$$
$= 17662.5 \approx 17662$ N.

Though the critical section will be at a distance of $d$ from the support, where S.F. will be slightly less, we will adopt the same value of $V_u$ as that the centre of the support.

$$\tau_v = \frac{17662}{1000 \times 91} = 0.194 \text{ N/mm}^2; \quad \frac{100 A_s}{bd} = 0.31\%$$

Hence     $\tau_c \triangleq 0.24$ N/mm²

∴     $\tau_v < 1.3 \tau_c$.  Hence safe.

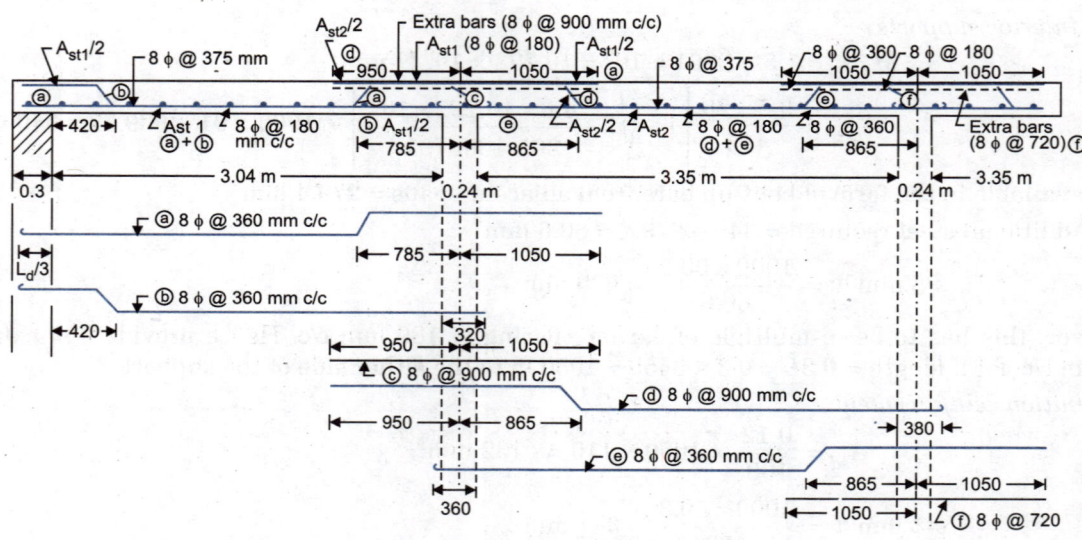

**Fig. 37.6**

**Step 7. Check for deflection**

For the end span,     $p_t = 0.31\%$.   ∴  Modification factor ≈ 1.5

∴     $d = \dfrac{L}{\frac{1}{2}(20 + 26) \times 1.5} = \dfrac{3140}{23 \times 1.5} = 91$ mm

Available $d = 91$ mm. Hence safe.

**Step 8. Details of reinforcement etc.:** Shown in Fig. 37.6.

## 37.6. DESIGN OF T-BEAM ROOF

A T-beam roof consists of two components:
(i) Continuous slab cast monolithic with beams.  (ii) T-beams.

The design of continuous slab has been given in the previous article. The design of T-beam has been thoroughly discussed and illustrated in Chapter 35.

## PROBLEMS

1. Design a reinforced concrete beam, supported on two walls, 450 mm thick, over a clear span of 8 m. The beam carries a super-imposed load of 15 kN/m. Use M 20 mix. and Fe 415 reinforcement.
2. Design a reinforced concrete lintel over the openings of two windows, each 2 m wide, separated by a wall of 50 cm. The height of wall over the lintel is 3 m and the thickness of wall is 30 cm. The lintel is to be provided with a sunshade projecting out by 80 cm and cast monolithically with the lintel. Use M 20 mix and mild steel reinforcement. The live load on sunshade may be taken as 1 kN/m².
3. Design a R.C. cantilever beam projecting out 3 m beyond the fixed end and carrying a super-imposed load of 15 kN. The cantilever also carries a concentrated load of 5 kN at the free end. Use M 20 mix and Fe 415 steel.
4. Design a R.C.C. floor slab for a room having inside dimensions 4 m × 10 m and supported on all sides by a 40 cm thick brick wall. The super-imposed load may be taken as 3 kN/m². Use M 20 concrete and Fe 415 reinforcement.
5. A simply supported RCC slab has to be provided for two roof of a room of clear dimensions 3 m × 8 m used for school building. The weight of weathering course over slab is 1 kN/m². Taken the live as per usage. Design the slab using $M_{20}$ grade of concrete and HYSD bars. Check the design for structures.

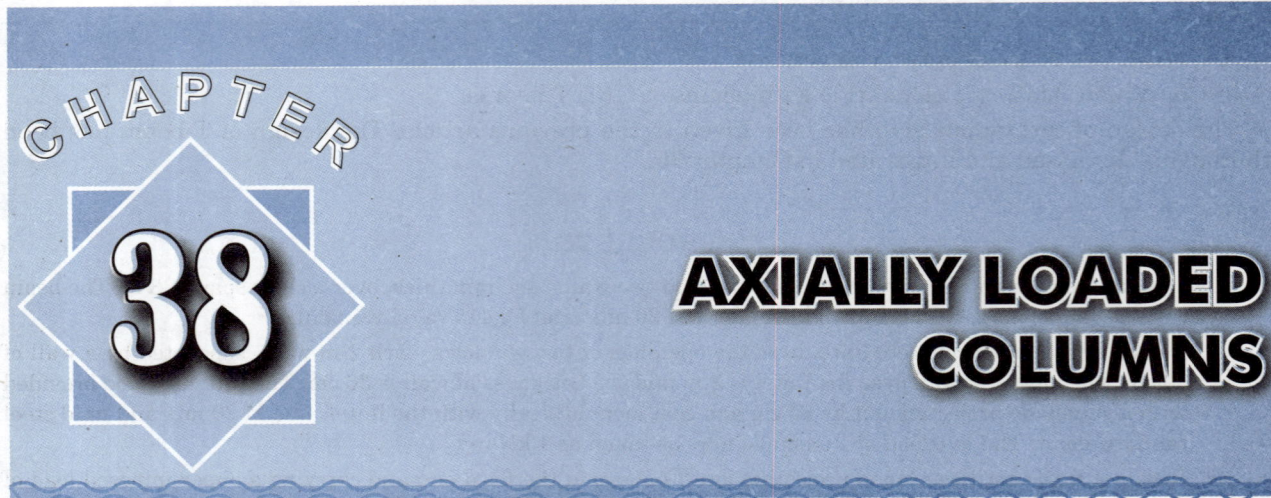

# CHAPTER 38: AXIALLY LOADED COLUMNS

## 38.1. INTRODUCTION

A column is defined as a vertical compression member which is mainly subjected to axial loads and the effective length of which exceeds three times, its least lateral dimensions. Column transfer the loads from the beams (*or*) slabs to the footings (*or*) foundations.

**Classification of Columns:** Columns are classified based on different criteria such as;

1. Shapes of cross section
2. Material of construction
3. Type of loading
4. Slenderness ratio
5. Type of lateral reinforcement

## 38.2. LIMIT STATE OF COLLAPSE : COMPRESSION (*As per clause 39 of IS : 456-2000*)

**Assumptions:** The limit state method of design of compression member is based on the following assumptions.

1. The plane section normal to the axis of column before deformation remain plane after deformation. This means that the strain at any point is proportional to its distance from the N.A.

2. The relationship between compressive stress distribution in concrete and strain in concrete may be assumed to be rectangle, trapezoid, parabola or any other shape which results in prediction of strength in substantial agreement with the results of tests. An acceptable stress-strain curve is given in Fig. 33.1. For design purposes, the *compressive strength of concrete in structure* is assumed to be 0.67 times the characteristic strength. The partial safety factor $\gamma_{mc} = 1.5$ is applied in addition to this. Thus the design strength of concrete is taken as $0.67 f_{ck}/1.5 = 0.45 f_{ck}$.

3. The tensile strength of concrete is ignored.

4. The stresses in the reinforcement are derived from representative stress-strain curve for the type of steel used. Typical curves are shown in Fig. 33.2, 33.3 and 33.4. For design purposes, partial safety factor $\gamma_{ms} = 1.15$ shall be applied.

5. The maximum compressive strain in concrete in axial compression is taken as 0.002.

6. The maximum strain in concrete, at the outer most compression fibre is taken as 0.0035 in bending when N.A. lies within the section.

7. The maximum compressive strain at the highly compressed extreme fibre in concrete subjected to axial compression and bending, and where there is no tension on the section shall be 0.0035 minus 0.75 times the strain at the least compressed extreme fibre.

According to assumption 5, for purely axial compression, the strain is assumed to be uniform, equal to 0.002, across the section [Fig. 38.1 (b)]. Also, for the case of axial compression along with bending, the maximum strain in concrete at the outer most fibre is 0.0035 [Fig. 38.1 (c)] and this assumption is applicable when the N.A. lies within the section and in the limiting case when the N.A. lies along the edge of the section. In the later case, the strain varies from 0.0035 at the highly compressed edge to zero at the opposite edge. The strain distribution for these two cases (of assumptions 5 and 6) intersect each other at a depth of $\frac{3}{7}D$ from the highly compressed edge. This point is assumed to act as *fulcrum* for the strain distribution line when the N.A. lies out side the section [Fig. 38.1 (c)]. This leads to the assumption (No. 7) that the strain at highly compressed edge is 0.0035 minus 0.75 times the strain at least compressed edge.

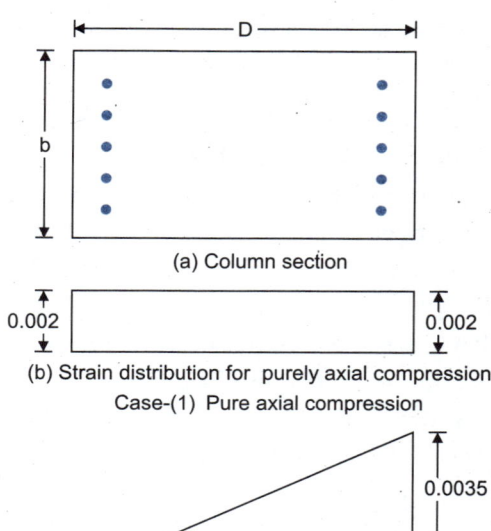

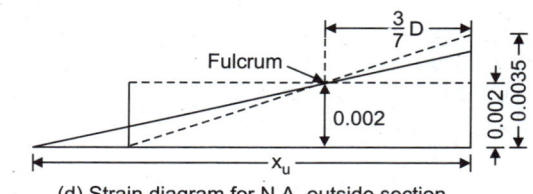

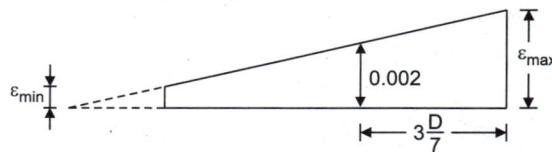

Fig. 38.1. Strain Distribution Across the Section.

## 38.3. SHORT COLUMNS

A compression member may be considered as short when both the slenderness ratios $(l_{ex})/D$ and $(l_{ey})/b$ are less than 12. A short column fails by crushing (Pure compression failure) where,

$l_{ex}$ = effective length in respect of the major axis
$D$ = Depth in respect of the major axis
$l_{ey}$ = effective length in respect of the minor axis and $b$ = width of the member.

## 38.4. SHORT AXIALLY LOADED MEMBERS IN AXIAL COMPRESSION

Experiments on column show that load carrying capacity $(P_u)$ of an axially loaded R.C. member at collapse is made up of ultimate strength of concrete member at $(P_{uc})$ at collapse plus the ultimate strength of steel $(P_{us})$ in compression. Thus

$$P_u = P_{uc} + P_{us} = \alpha_c f_{ck} A_c + \alpha_s f_y A_s \qquad \text{...(1)}$$

where $\alpha_c f_{ck} = f_c$ = stress in concrete at failure at uniform strain of 0.002
$\alpha_s f_y = f_s$ = stress in steel at failure at strain of 0.002

When a short column is axially loaded, the strain distribution across the section will be rectangular. At failure, the strain in concrete will be uniform at a value of 0.002, as stated in assumption No. 4 of § 38.1. The stress in concrete will be = $0.67 f_{ck}/\gamma_{mc} = 0.67 f_{ck}/1.5 \approx 0.45 f_{ck}$ (assumption 2). Thus $\alpha_c$ is take as equal to 0.446.

When the concrete attains a limiting strain of 0.002, the stress in mild steel reinforcement may develop full design stress ($f_{yd} = 0.87 f_y$). However with cold twisted reinforcement or HYSD bars ($f_y = 415$ N/mm² or $f_y = 500$ N/mm²), full design stress will not develop at a strain of 0.002. In general therefore, stress in steel reinforcement at strain of 0.002, can be taken equal to $\alpha_s f_y A_s$, where the value of $\alpha_s$ will depend upon the type of reinforcement, as given below :

| Type of reinforcement | Value of $\alpha_s$ | Stress in steel ($f_s$) |
|---|---|---|
| (i) Mild steel | 0.87 | $0.87 f_y$ |
| (ii) Fe 415 | 0.79 | $0.79 f_y$ |
| (iii) Fe 500 | 0.75 | $0.75 f_y$ |

Hence the load carrying capacity of a member, subjected to axial load only, is given by

$$P_u = 0.45 f_{ck} A_c + \alpha_s f_y A_{sc} \qquad ...(2)$$

For *simplicity*, I.S. Code (IS : 456-2000) adopts only the lowest value of $\alpha_s$ (= 0.75), which is for steel of Fe 500 grade. Also the Code has redesignated $P_u$ as $P_{uz}$ in section 39.6 of the Code, and has given the following expression in *design aids* (SP : 16-1980).

$$P_{uz} = 0.45 f_{ck} A_c + 0.75 f_y A_s = 0.45 f_{ck} A_g + (0.75 f_y - 0.45 f_{ck}) A_s \qquad ...(38.1)$$

where $A_g$ = gross area of concrete and $A_c$ = net area of concrete = $A_g - A_s$

## 38.5. SHORT AXIALLY LOADED COLUMN WITH MINIMUM ECCENTRICITY

The *ideal* condition of axial loading, indicated in the previous article hardly ever exists. There is always certain inherent minimum eccentricity in the columns. According to IS : 456-2000, all compression members are to be designed for a *minimum eccentricity* of the load in two principal directions. Clause 25.4 of the Code specifies the following minimum eccentricity $e_{min}$ for the design of columns :

$$e_{min} = \frac{l}{500} + \frac{D}{30} \text{ , subject to a minimum of 20 mm}$$

where $l$ = unsupported length of the column in the direction under consideration
$D$ = lateral dimension of the column in the direction under consideration

If $x$-axis is the major axis and $y$-axis is the minor axis of bending, we have

$$e_{x, min} = \frac{l_x}{500} + \frac{D}{30} \text{ and } e_{y, min} = \frac{l_y}{500} + \frac{b}{30}, \text{ each not less than 20 mm}$$

where $e_{x, min}$ and $e_{y, min}$ are minimum eccentricities for bending about $x$ and $y$ axes respectively.
and $l_x$ and $l_y$ = unsupported lengths of the column for bending in the two directions respectively.

If the value of minimum eccentricity is less than or equal to $0.05 D$, clause 39.3 of the Code permits the design of short axially loaded compression member by the following equation:

Factored axial load on column

$$P_u = 0.4 f_{ck} A_c + 0.67 f_y A_s = 0.4 f_{ck} A_g + (0.67 f_y - 0.4 f_{ck}) A_s \qquad ...(38.2)$$

It should be noted that *Eq. 38.2* is obtained by reducing the capacity of column, given by *Eq. 38.1*, by approximately 10% thereby allowing for the minimum eccentricity of $0.05 D$.

If, however, the minimum eccentricity is greater than $0.5 D$, the section is designed for combined axial load and bending, as discussed in Chapter 39.

**Note :** The minimum eccentricity to be adopted in the design, as specified by the Code, is 20 mm. (As per Clause 25.4 of IS: 456 – 2000).

$\therefore \qquad 0.05 D \geq 20$ mm which gives $D \geq 20/0.05 \geq 400$ mm $\qquad ...[38.3(a)]$

Also, $\qquad e_{min} = \dfrac{l}{500} + \dfrac{D}{30} \ngtr 0.05 D \ngtr \dfrac{D}{20}$

or $\qquad \dfrac{l}{500} \ngtr \left( \dfrac{D}{20} - \dfrac{D}{30} \right) \ngtr \dfrac{D}{60}$

Hence $\qquad D \ngtr 0.12 \, l \qquad ...[38.3(b)]$

Hence in all columns, which have $D$ equal to or less than 400 mm or 0.12 $l$, the minimum eccentricity will be greater than 0.05 $D$, and hence *Eq. 38.2* will not be applicable. Such sections are therefore, to be designed for the combined axial load and bending, as discussed in chapter 39.

## 38.6. DESIGN CHARTS (Sp 16 design charts 24 to 26)

Design of Axially loaded short column as per clause 39.3 of IS : 456–2000.
Short column with Rectangular ties.
(i) Minimum eccentricity does not exceed 0.05 times the lateral dimension *i.e.*

$$P_u = 0.4 f_{ck} \cdot A_c + 0.67 f_y \cdot A_{sc}$$

were $P_u$ = factored axial load
$A_c$ = area of concrete
$A_{sc}$ = area of longitudinal reinforced of column
$f_{ck}$ = characteristic strength of the compression longitudinal reinforcement.

*Eq. 38.2* can be rearranged as follows

$$P_u = 0.4 f_{ck} \left( A_g - \frac{pA_g}{100} \right) + 0.67 f_y \frac{p A_g}{100}, \text{ where}$$

$A_g$ = Gross area of cross section = $b \times D$ for rectangular section or = $\frac{\pi}{4} D^2$ for a circular section.

$p$ = percentage of reinforcement = $\frac{A_s}{A_g} \times 100$. Dividing both sides by $A_g$

$$\frac{P_u}{A_g} = 0.4 f_{ck} \left( 1 - \frac{p}{100} \right) + 0.67 f_y \frac{p}{100} = 0.4 f_{ck} + \frac{p}{100} (0.67 f_y - 0.4 f_{ck}) \qquad ...(38.4)$$

*Eq. 38.4* suggests that design charts can be prepared with $P_u/A_g$, $f_{ck}$ and $p$ are variables, for different types of steels (*i.e.* for different values of $f_y$). Such design charts are given in Design Aids for Reinforced Concrete to IS : 456 prepared by ISI under its publication No. SP: 16-1980. Two such charts are reproduced here in Fig. 38.2 and 38.3 for $f_y$ = 250 N/mm² (mild steel bars) and $f_y$ = 415 N/mm² (Fe 415 bars) respectively.
The working of these charts are explained on the chart itself.

## 38.7. COMPRESSION MEMBERS WITH HELICAL REINFORCEMENT

The Code permits larger load in compression members with helical reinforcement because columns with helical reinforcement have greater ductility or toughness when they are loaded concentrically or with small eccentricity. As per the Code, the strength of compression members with helical reinforcement satisfying the requirement given below shall be taken as 1.05 times the strength of similar members with lateral ties.

**Requirement.** The ratio of volume of helical reinforcement to the volume of core shall not be less than

$$0.36 \left( \frac{A_g}{A_k} - 1 \right) \frac{f_{ck}}{f_{yn}}$$

where $A_g$ = gross area of the section (say $\frac{\pi}{4} D^2$ for a circular column)

$A_k$ = area of core of the helically reinforced column measured to the outside diameter of the helix.

$D_k$ = diameter of concrete core, measured from outside of helix
= $D - 2 \times$ clear cover

$f_y$ (or) $f_{yh}$ = characteristic strength of helical reinforcement but not exceeding 415 N/mm²

$\phi_s$ = diameter of spiral,

$P$ = pitch of spiral

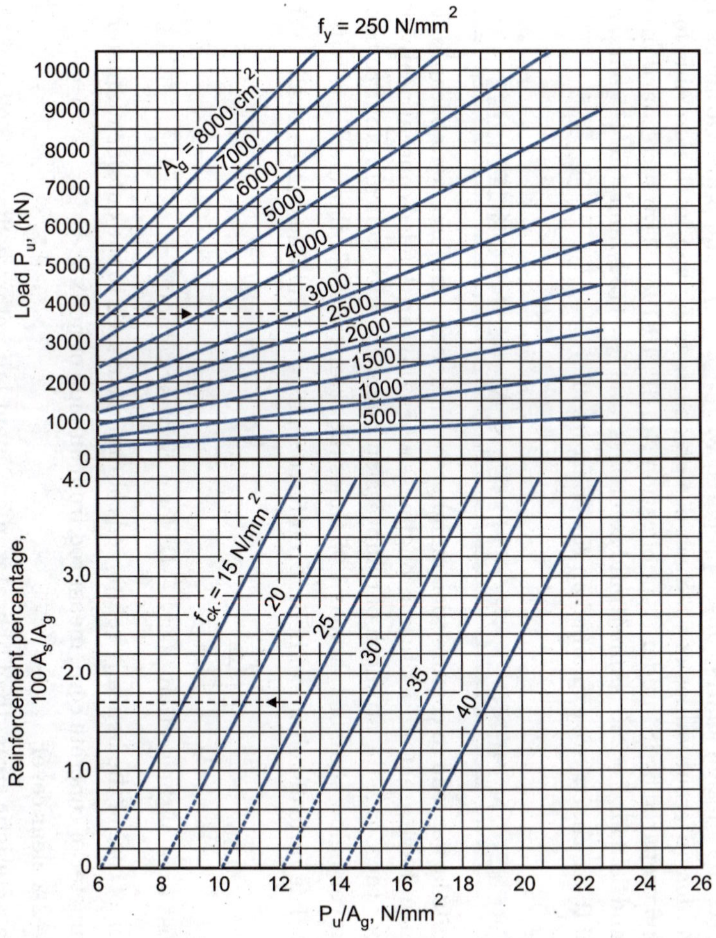

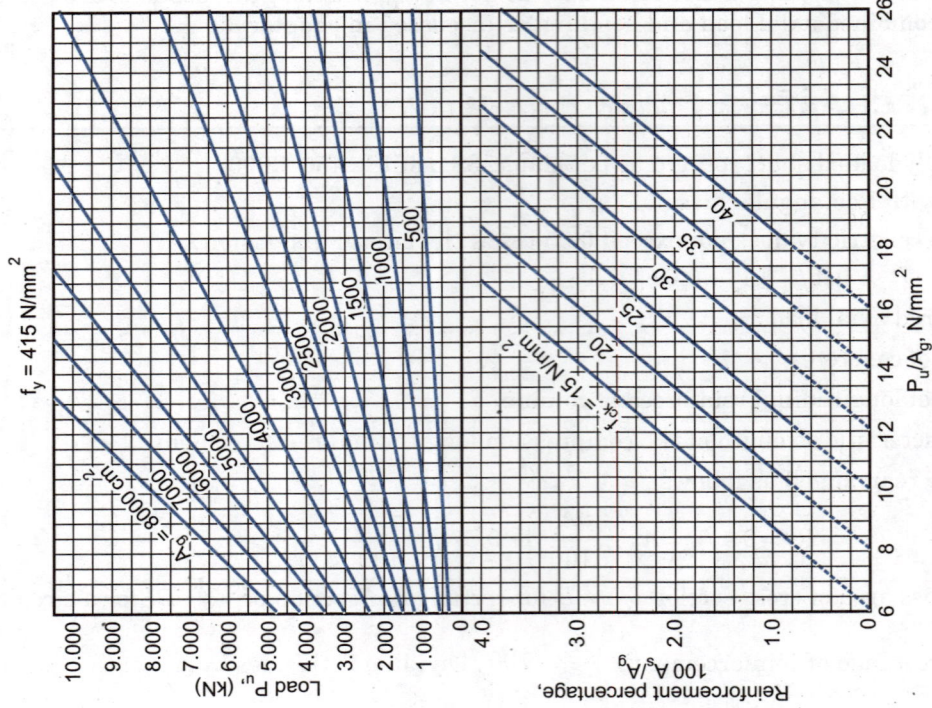

**Fig. 38.3.** Design Charts for Axial Compression (HYSD bars : Fe 415)

**Fig. 38.2.** Design Charts for Axial Compression (mild steel bars)

$$A_k = \frac{\pi}{4} D^2$$

$d = D - 2 \times$ clear cover $-$ dia. of helical reinforcement $(= \phi_s)$

## 38.8. DESIGN SPECIFICATIONS (IS : 456-2000)

The design specifications, as per IS : 456-2000, have been given in § 13.5.4 of Chapter 13. These specifications are also applicable for limit state design of compression members.

**Example 38.1.** *Load capacity of a short column*

A concrete column is reinforced with 4 bars of 20 mm dia. Determine the ultimate load capacity of the column, using M 20 grade concrete and Fe 415 grade steel, if the size of the column is (a) 450 mm × 450 mm (b) 300 mm × 300 mm. What will be the allowable service load in each case ?

**Solution:** Adopting a min. eccentricity of 20 mm, the minimum value of $D$, for which *Eq. 38.2* will be applicable, comes out to be 400 mm. Hence the column of 300 mm × 300 mm of case (b) will have to be solved for a case of combined axial load and bending (see Chapter 39).

For column of size 450 mm × 450 mm, $D$ = 450 mm.

Hence 0.05 $D$ = 0.05 × 450 mm = 22.5 mm. Thus, the given minimum eccentricity of 20 mm is less than 0.05 $D$ (= 22.5 mm), and *Eq. 38.2* will thus be applicable

$$A_g = 450 \times 450 = 202500 \text{ mm}^2 \; ; \quad A_s = 4 \times \frac{\pi}{4}(20)^2 = 1256.64 \text{ mm}^2$$

$$A_c = A_g - A_s = 202500 - 1256.64 = 201243.36 \text{ mm}^2$$

For M 20 concrete, $f_{ck} = 20$ N/mm$^2$, For Fe 415 steel, $f_y = 415$ N/mm$^2$

For Axially loaded short column, factored load is given by

$$P_u = 0.4 f_{ck} A_c + 0.67 f_y A_s = 0.4 \times 20 \times 201243.36 + 0.67 \times 415 \times 1256.64$$
$$\approx 1609947 + 349408 = 1959355 \text{ N} = 1959.355 \text{ kN}$$

Allowable service load $= \dfrac{P_u}{1.5} = \dfrac{1959.355}{1.5} \approx \mathbf{1306.2 \text{ kN}}$ (or) $\mathbf{1306.2 \times 10^3 \text{ N}}$.

**Example 38.2.** *Design of short column*

Design a short axially loaded square column, 500 mm × 500 mm for a service load of 2000 kN. Use M 20 concrete and Fe 415 grade steel.

**Solution:** Service load = 2000 kN, Load factor = 1.5

$P_u = 2000 \times 1.5 = 3000$ kN $= 3000 \times 10^3$ N

Min. eccentricity when the effective length of the column is not given, is equal to 20 mm. But 0.05 $D$ = 0.05 × 500 = 25 mm. Hence min. eccentricity is less than 0.05 $D$, and therefore *Eq. 38.2* is applicable.

$\therefore \qquad P_u = 0.4 f_{ck} A_c + 0.67 f_y A_s$

$3000 \times 10^3 = 0.4 \times 20 (500 \times 500 - A_{sc}) + 0.67 \times 415 A_{sc}$

or $\qquad 278.05 A_s - 8 A_s = 1000000$

From which $\qquad A_s = 3703.02$ mm$^2$

Min. area of longitudinal reinforcement = 0.8 %

$$= \frac{0.8}{100} \times 500 \times 500 = 2000 \text{ mm}^2$$

$A_s = 3703.02$ mm$^2$

$\therefore$ No. of 20 mm dia. bars = 3703.02/314.16 = 11.79

This number is large

$\therefore$ No. of 25 mm dia. bars = 3703.02/490.87 = 7.54

Hence provide 8 Nos. of 25 mm dia. bars.

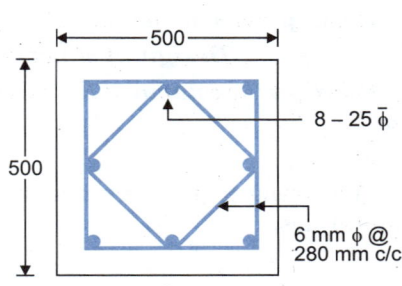

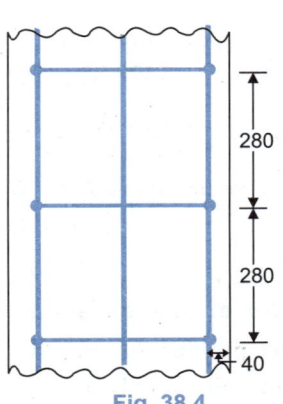

Fig. 38.4

Actual $A_s = 8 \times 490.87 \approx 3927$ mm$^2$

Using 6 mm dia. ties, spacing is the least of the following:

Pitch of ties shall be minimum of

(i) Least lateral dimension = 500 mm

(ii) 16 times of the dia. of longitudinal bar $16 \times 25 = 400$ mm

(iii) $48 \times 6 = 288$ mm

(iv) 300 mm

Hence provide 6 mm dia. ties @ 280 mm c/c, as shown in Fig. 38.4.

**Example 38.3.** *Design of short column:* Design a circular column to carry an axial load of 1000 kN. Use M 20 concrete and Fe 415 steel.

**Solution:** $P_u = 1000 \times 1.5 = 1500$ kN $= 15 \times 10^3$

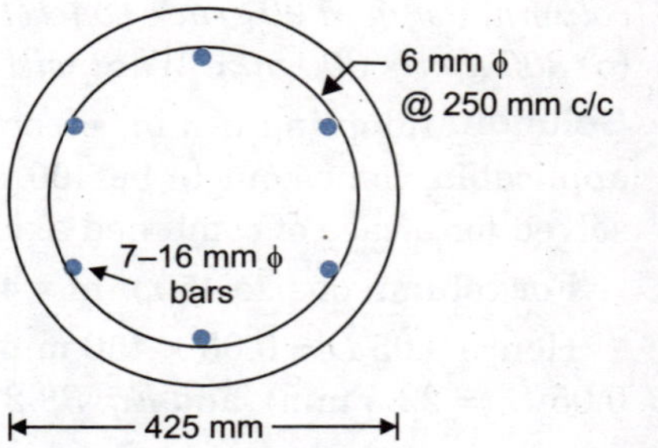

For M 20 concrete
$$f_{ck} = 20 \text{ N/mm}^2.$$

For Fe 415 steel, $f_y = 415$ N/mm$^2$

Assuming that $e_{min} \leq 0.05 D$,
$$P_u = 0.4 f_{ck} A_c + 0.67 f_y A_s.$$

Assuming 1% steel :

$\therefore \qquad 1500 \times 10^3 = 0.4 \times 20 (A_g - 0.01 A_g) + 0.67 \times 415 \times 0.01 A_g$

From which $A_g = 140180 = \dfrac{\pi}{4} D^2$

$\therefore \qquad D = 422.5$ mm. Adopt $D = 425$ mm.

Area of longitudinal steel $= 0.01 \times 140180 = 1401.8$ mm$^2$

No. of 16 mm dia. bars $= 1401.8/201.06 = 6.97$

Hence provide 7 bars of 16 mm dia.

Using 6 mm dia. lateral ties, spacing is the least of the following :

(i) Least lateral dimension = 425 mm   (ii) $16 \times 16 = 256$ mm

(iii) $48 \times 6 = 288$ mm.   (iv) 300 mm whichever is less

Hence provide 6 mm dia. lateral ties @ 250 mm c/c.

**Fig. 38.5**

**Example 38.4.** *Design of short column using helical reinforcement*

Redesign the column of Example 16.3 using helical reinforcement, M 20 grade concrete and Fe 415 grade steel.

**Solution:**

The strength of a column with helical reinforcement is 1.05 times the strength of similar member with lateral ties.

$$P_u = \dfrac{1000 \times 1.5}{1.05} = 1428.57 \text{ kN}$$

Assuming that $e_{min} \leq 0.05 D$, and providing 1% steel.
$$1428.57 \times 10^3 = 0.4 \times 20 (A_g - 0.01 A_g) + 0.67 \times 415 \times 0.01 A_g$$

From which $A_g = 133505$ mm$^2 = \dfrac{\pi}{4} D^2$   and   $D = 412.3$ mm

Provide overall dia. $D = 420$ mm.

$0.05 D = 0.05 \times 420 = 21$ mm   $\therefore$   $e_{min} < 0.05 D$

$A_s = 0.01 \times 133505 \approx 1335.1$ mm$^2$.

No. of 16 mm dia. bars $= 1335.1 / 201.06 = 6.64$

Hence provide 7 bars of 16 mm dia. giving total area $= 1407.4$ mm$^2$, at a nominal cover of 40 mm. Let us use 8 mm dia. HYSD bars for helical reinforcement.

Outside dia. of helix = $D_k$ = 420 – 2 × 40 = 340 mm

$A_k$ = area of core of helically reinforced column measured to the outside of the helix

$$= \frac{\pi}{4}(340)^2 - 1407 = 89385 \text{ mm}^2$$

$A_g$ = gross area of section = $\frac{\pi}{4}(420)^2 = 138544 \text{ mm}^2$

Factor $0.36\left(\frac{A_g}{A_k} - 1\right)\frac{f_{ck}}{f_{yh}} = 0.36\left(\frac{138544}{89385} - 1\right)\frac{20}{415} = 0.00954$ ...(i)

Dia. $d$ of core upto centre of helix = 420 – (2 × 40) – 8 = 332 mm

Let the pitch of the spiral be $s$ mm. Volume of spiral $V_{hs}$ per 1 mm length of column

$$= \frac{\pi d}{s}\left(\frac{\pi}{4}\Phi_s^2\right) = \frac{\pi \times 332}{s}\left(\frac{\pi}{4}\right)(8)^2 = \frac{52427.3}{s}$$

Volume of core per mm length = $V_k = A_k \times 1 = 89385 \times 1 = 89385$

Ratio $\qquad \dfrac{V_{hs}}{V_k} = \dfrac{52427.3}{s \times 89385} = \dfrac{0.5865}{s}$ ...(ii)

Equating (i) and (ii), we get $\dfrac{0.5865}{s} = 0.00954$ or $s \approx 61.5$ mm

However, the pitch should not be more than 75 mm, nor more than 1/6 core dia. (= 1/6 × 340 = 56.7 mm). Also, it should not be less 3 $\varphi_s$ (= 3 × 8 = 24 mm). Hence keep the pitch equal to 55 mm.

It should be noted that since the helically reinforced columns are very ductile as compared with columns with lateral ties, they are more desirable in highly seismic zones.

**Example 38.5.** *Design of circular column: A circular column, 4.6 m high is effectively held in position at both the ends and restrained against rotation at one end. Design the column, to carry an axial load of 1200 kN, if its dia. is restricted to 450 mm. Use M 20 mix and Fe 415 steel.*

**Solution:** For a column with the above end conditions, we get effective length = 0.80 × 4600 = 3680 mm. Using 450 mm dia. column, slenderness ratio = 3680/450 = 8.18 < 12. Hence it is a short column.

Also, $\qquad e_{min} = \dfrac{l}{500} + \dfrac{D}{30} = \dfrac{3680}{500} + \dfrac{450}{30} = 22.36$ mm

Also, $\qquad 0.05\, D = 0.05 \times 450 = 22.5$

∴ $\qquad e_{min} < 0.05\, D$. Hence strength of the column is given by

$$P_u = 0.4 f_{ck} A_c + 0.67 f_y A_s$$

where $P_u = 1.5 \times 1200 = 1800$ kN $= 1800 \times 10^3$ N

$$A_c = \frac{\pi}{4}(450)^2 - A_s = (159043 - A_s) \text{ mm}^2$$

∴ $\qquad 1800 \times 10^3 = 0.4 \times 20 (159043 - A_s) + 0.67 \times 415 A_s$.

From which, $\qquad A_s = 1954$ mm²

Min. steel @ 0.8% $\qquad = \dfrac{0.8}{100} \times \dfrac{\pi}{4}(450)^2 = 1272$ mm². Hence $A_s = 1954$ mm²

∴ No. of 20 mm Φ bars = 1954/314.56 = 6.22. Hence provide 7 No. 20 mm Φ longitudinal bars. Using 6 mm dia. ties its spacing should not exceed the least of the following : Pitch of the ties shall be minimum of

(i) Least lateral dimension = 450 mm

(ii) 16 × dia. of main bar = 16 × 20 = 320 mm

(iii) 48 × dia. of ties mild steel ties = 48 × 6 = 288 mm.

(iv) 300 mm

Hence provide 6 mm dia. mild steel ties @ 280 mm c/c.

**Example 38.6. Design of rectangular column:** Design a rectangular column of 4.5 m unsupported length, restrained in position and direction at both the ends, to carry an axial load of 1200 kN. Use M 20 concrete and Fe 415 steel.

**Solution:** For a column with the above end conditions,
effective length = 0.65 times the actual length = 0.65 × 4500 = 2925 mm.

Let us use 1 % steel. Also let width $b = \frac{1}{2} D$.

Assuming that $e_{min} \leq 0.05 D$ in appropriate direction, Eq. 38.2 will be applicable.

∴ Since the column is axially loaded short column, its load carrying capacity is given by
$$P_u = 0.4 f_{ck} A_c + 0.67 f_y A_{sc}. \quad \text{Hence,}$$
$$1.5 \times 1200 \times 10^3 = 0.4 \times 20 \, [A_g - 0.01 A_g] + 0.67 \times 415 \times 0.01 A_g.$$

From which $A_g = 168216 \text{ mm}^2$

∴ $b . D = \frac{1}{2} D.D = 168216.$ From which $D = 580$ mm and $b = \frac{1}{2} D = 290$ mm

However, provide 300 mm × 600 mm column, from practical considerations.
$$A_s = 0.01 \times 168216 \approx 1682.2 \text{ mm}^2$$

For column of rectangular shape, let us provide 8 bars.
Area of each bar = 1682.2/8 = 210.27 mm². ∴ Dia. of bar = 16.36 mm

Since the actual area of concrete provided is more than required, let us provide 8 bars of 16 mm dia. Actual $A_s = 8 \times 201.06 = 1608.5$

$$\% \text{ steel} = \frac{1608.5}{300 \times 600} \times 100 = 0.89\% > 0.8 \%.$$

Before checking for the load capacity of the column, let us check for the slenderness and eccentricity requirements.

$l_e = 2925$ mm ; $l_e/b = 2925/300 = 9.75 < 12$ ; $l_e/D = 2925/600 = 4.875 < 12$

Hence the column is a short column, in both the directions

*In one direction* (i.e. x-direction)
$$e_{min.} = \frac{l}{500} + \frac{D}{30} = \frac{4500}{500} + \frac{600}{30} = 29 \text{ mm}$$
$0.05 D = 0.05 \times 600 = 30$ mm.

∴ $e_{min} < 0.05 D$.

*In the other direction* (i.e. y-direction)
$$e_{min} = \frac{l}{500} + \frac{D}{30} = \frac{4500}{500} + \frac{300}{30} = 19 \text{ mm} \quad \text{(subject to a min. of 20 mm)}$$

∴ $e_{min} = 20$ mm ; $0.05 b = 0.05 \times 300 = 15$ mm ; ∴ $e_{min} > 0.05 b$

Hence Eq. 38.2 is not strictly applicable, and the column is to be designed as an *eccentrically loaded column.* Alternatively, let us keep $b = 400$ mm, so that $0.05 b = 0.05 \times 400 = 20$ mm $= e_{min.}$, and redesign the column.

**Redesign:** $b = 400$ mm.

∴ $400 \times D = 168216$

∴ $D = 168216/400 = 420.5$ mm.

Hence adopt a column of size 400 × 450 mm, and provide 8 bars of 16 mm dia, arranged as shown in Fig. 38.6.

Using 6 mm dia. ties, spacing will we the least of the following
  (i) Least lateral dimension = 400 mm
  (ii) 16 × Φ = 16 × 16 = 256 mm
  (iii) 48 × $\Phi_s$ = 48 × 6 = 288 mm. Hence provide 6 mm dia. ties @ 250 mm c/c, including two open ties as shown in Fig. 38.6.
  (iv) 300 mm

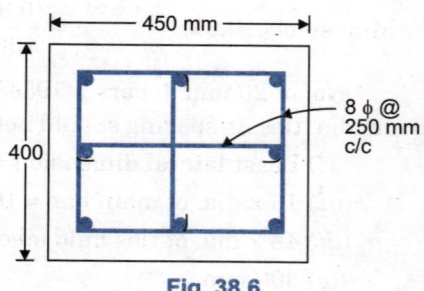

Fig. 38.6

## PROBLEMS

1. Discuss various assumptions used in the limit state method of design of compression members.
2. A reinforced concrete column, 500 mm × 500 mm is reinforced with 8 bars of 20 mm dia, symmetrically arranged. Determine the serice load which the column can carry, assuming it to be a short one.
   Use (a) M 20 grade concrete and Fe 250 steel and (b) M 20 grade concrete and Fe 415 steel.
3. A short R.C. column has a diameter of 450 mm, and is reinforced with 8 bars of 16 mm dia. placed with a clear cover of 40 mm. Determine load capacity of the column if the transverse reinforcement consists of (a) 8 mm dia. mild steel ties @ 250 mm c/c (b) 8 mm spirals of Fe 415 steel, arranged at a pitch of 50 mm. Use M 20 concrete and Fe 415 steel for longitudinal bars.
4. Design a short square column to carry a service load of 1600 kN. Use M 20 concrete Fe 415 steel combination.
5. Redesign a circular column for problem 4.
6. Redesign a rectangular column for problem 4, taking width of column equal to 400 mm.
7. Design a short reinforced concrete circular column with lateral ties to carry an axial load of 1500 kN. Use $M_{25}$ concrete and Fe 415 steel.
8. Design a circular column of diameter of 360 mm, 2.8 m. long subjected to a working load of 1380 kN. The column is effectively held in position and direction at both the ends $M_{20}$ grade concrete and Fe 415 steel is to be used.
9. Design a short column square in section to carry an axial load of 2000 kN using $M_{25}$ grade concrete and Fe 415 steel, take load factor as 1.5.

# CHAPTER 39
# COLUMNS WITH UNIAXIAL AND BIAXIAL BENDING

## 39.1. INTRODUCTION

As discussed in chapter 38, load on a column is rarely axial. Due to nonhomogeneity of material, inperceptible crookedneess of the member, inaccuracies in loading, inaccuracies in construction, etc, there is always some eccentricity inherent in the compression member. Due to this, IS : 456-2000 recommends that every compression member should be designed for a certain minimum eccentricity $e_{min}$. If this $e_{min}$ is equal to or less than $0.05\,D$, the column may be designed as an axially loaded column, using Eq. 38.2. If, however, $e_{min}$ is greater than $0.05\,D$, the axially loaded column has to be designed for axial load ($= P_u$) and moment $M_u = e_{min} P_u$. Besides this, there are many occassions when a column is subjected to (i) eccentric load, or (ii) end moments, on account of monolithic connection of beams and columns. Such a column may be subjected to either uniaxial moment or biaxial moments, in addition to the axial load.

**Note:** The design of members subjected to combined axial load and uniaxial or biaxial bending will involve. Lengthy calculation by trial and error. In order to overcome these difficulties, interaction curves given in **SP: 16**. (Design aids for reinforced concrete to IS: 456-2000 published by **BIS**) may be used.

## 39.2. COMBINED AXIAL LOAD AND UNIAXIAL BENDING

**(A) BASIC ASSUMPTIONS:** The basic assumptions for limit state of collapse due to compression are given in § 38.1. Also, refer to Fig. 38.1. According to assumption 5, for purely axial compression, the strain is assumed to be uniform, equal to 0.002, across the section [Fig. 38.1 (b)]. Also, for the case of axial compression along with bending, the maximum strain in concrete at the outer most fibre is 0.0035 [Fig. 38.1 (c)] and this assumption is applicable when the N.A. lies within the section, and in the limiting case when the N.A. lies at the edge of the section. In the later case, the strain varies from 0.0035 at the highly compressed edge to zero at the opposite edge. The strain distribution for these two cases (of assumption 5 and 6) intersect each other at a depth of $\frac{3}{7} D$ from the highly compressed edge. This point is assumed as *fulcrum* for the strain distribution line when the N.A. lies out side the section [Fig. 38.1 (d)]. This leads to the assumption (No. 7) that the strain at the highly compressed edge is 0.0035 minus 0.75 times the strain at the least compressed edge.

**(B) MODES OF FAILURE:** Let us consider a symmetrically reinforced short column subject to axial load $P_u$ and moment $M_u$. This moment can be considered to be equivalent to a load $P_u$ acting at an eccentricity $e = M_u/P_u$. Depending upon the relative magnitudes of $P_u$ and $M_u$, there may be three modes of failure as under:

### Compression Failure
(a) Failure by crushing of concrete without tension on the other face.
$$(\epsilon_{c1} = 0.0035 - 0.75\,\epsilon_{c2})$$
(b) Failure by crushing of the concrete with some tension in the other face.
$$(\epsilon_{c1} = 0.0035)$$

### Tension Failure
(c) The tensile stress in the reinforcement reaches the yield limit before the strain in the compression concrete reaches the crushing strain.
$$(\epsilon_{st} = \epsilon_y\,;\;\epsilon_c < 0.0035)$$

### Code requirements on reinforcement and detailing
(a) Longitudinal reinforcement.   (refer clause 26.5.3.1 of IS: 456-2000)
(b) Transverse reinforcement    (refer clause 26.5.3.2 of IS: 456-2000)

**(C) INTERACTION CURVE:** The design conditions of the columns for each of the above modes are different, and these modes can be best represented in the form of interaction curve between $P_u$ and $M_u$, shown in Fig. 39.1. When the relative magnitudes of $P_u$ and $M_u$ are such that the entire column is under compression, we have *the compression control region*, in which case the N.A. lies outside the section. When the relative values of $P_u$ and $M_u$ are such that a part of the section is in tension such that the strain in tension steel is greater than the yield point strain when the compressive strain in the concrete reaches 0.0035, we have the *tension control region*, in which case the N.A. lies inside the section.

The *compression control region*, where the entire column section is under compression, can further be subdivided in two regions: (i) *region* I in which the eccentricity $e$ is less than $e_{min}$ specified in the Code and (ii) region II in which eccentricity $e$ exceeds $e_{min}$. The first region is indicated by line $EA$ of the interaction curve (Fig. 39.1). Point $E$ indicates the failure load $P_{uz}$ when a column section is subjected to a perfectly axial load, with zero eccentricity. The value of $P_{uz}$ is given by *Eq. 38.1*. Point $A$ indicates the failure load of the column subjected to axial load with nominal eccentricity equal to $e_{min}$. The moment corresponding the this eccentricity is equal to $P_u \cdot e_{min}$.

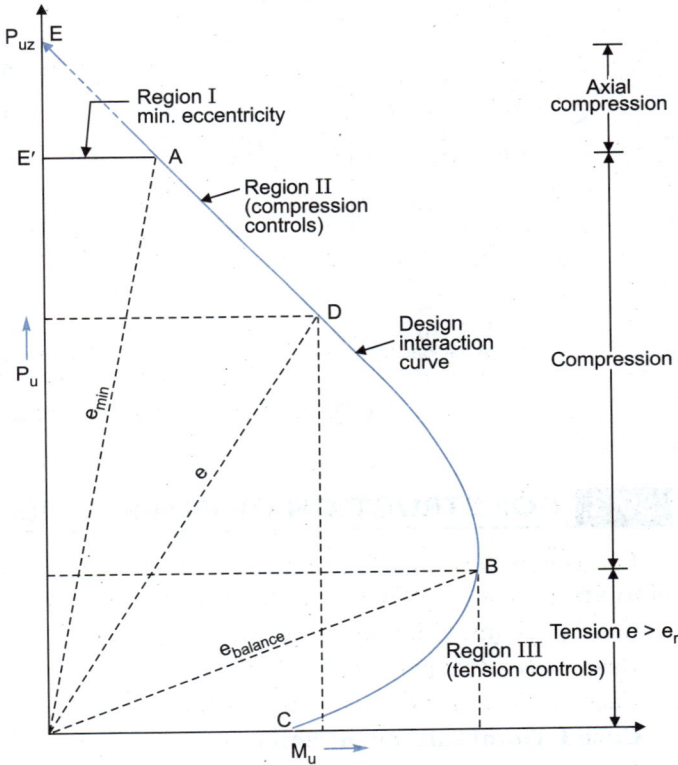

Fig. 39.1. Typical $P_u - M_u$ Interaction Curve Diagram.

The region II where compression controls, indicated by $AB$, is controlled by crushing strain in concrete, indicated by increasing moment on the column. Point $B$ on the interaction curve is called the *balanced load point*, (or) "Balanced failure condition" where in the column will fail by simultaneous occurrence of limiting strains in concrete and reinforcement.

With further increase in moment, we enter into the region III (curve $BC$), where tension failure occurs. In region III, tension controls. The axial load carrying capacity of the section decreases rapidly in the zone $BC$, as the moment increases, which is because of higher tensile force of reinforcement. When the compressive axial load is zero, corresponding to point $C$, the column section behaves as a doubly reinforced beam and its *moment capacity* (or pure bending moment capacity) $M_0$ is equal to $R_u bd^2$.

# 922 REINFORCED CONCRETE STRUCTURE

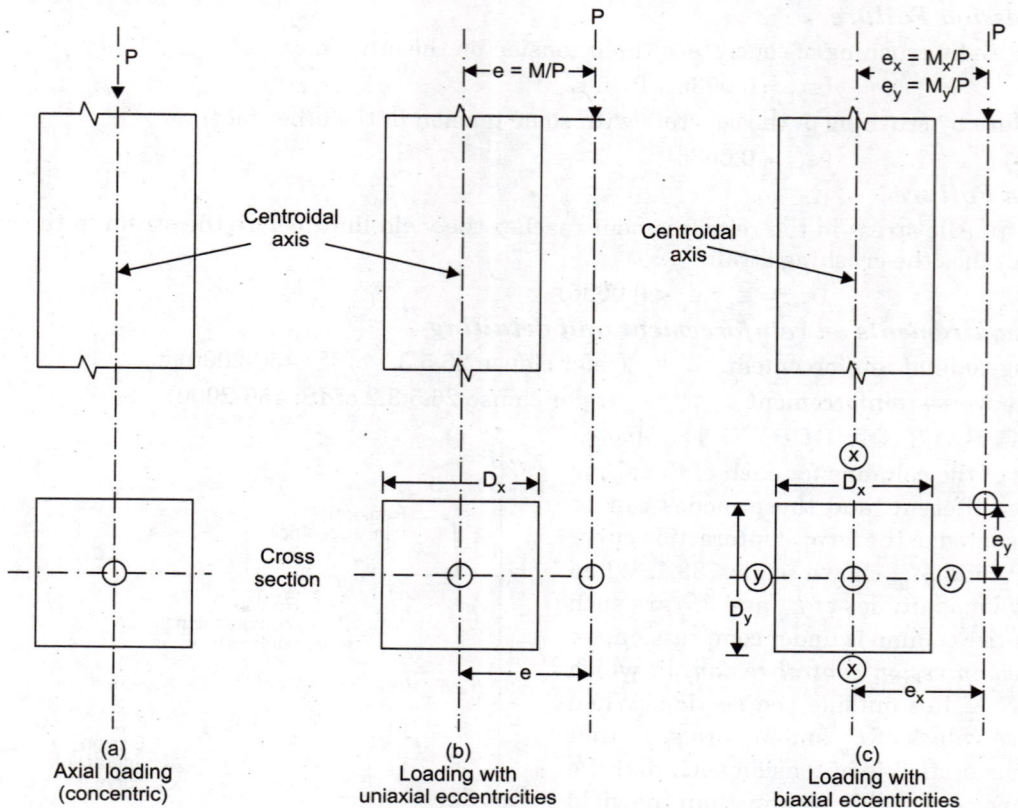

Fig. 39.2. Different Loading Situations of Columns Comp. Members

## 39.3. CONSTRUCTION OF INTERACTION CURVES FOR COLUMN DESIGN

Let us now derive basic equations for the construction of interaction curve. We will consider three cases.

Case 1 : Nominal Eccentricity ($e \leq e_{min}$)
Case 2 : Neutral axis outside the section.
Case 3 : Neutral axis inside the section.

### Case I. Nominal Eccentricity ($e \leq e_{min}$)

Fig. 39.3 (a) shows a symmetrically reinforced column section axially loaded, with *nominal eccentricity* $e < e_{min}$. Fig. 39.3. (c) shows the *stress diagram*, with uniform stress of $0.446 f_{ck}$.

Balancing the axial forces on the section, we have

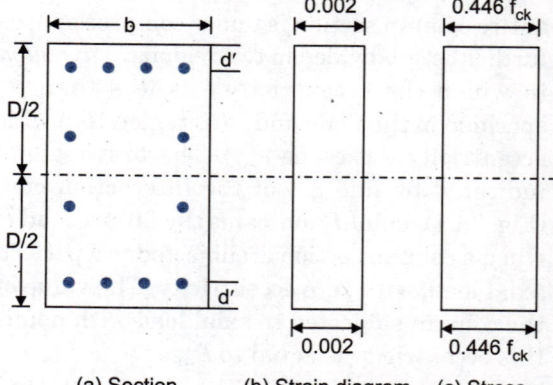

(a) Section    (b) Strain diagram    (c) Stress diagram

Fig. 39.3. Nominal Eccentricity ($e < e_{min}$.)

$$P_u = 0.446 f_{ck} A_c + f_s A_s$$
$$P_u = 0.446 f_{ck} (bD - A_s) + f_s A_s$$
$$P_u = 0.446 f_{ck} bD + A_s (f_s - 0.446 f_{ck})$$
$$\frac{P_u}{f_{ck} bD} = 0.446 + \frac{A_s}{f_{ck} bD} (f_s - 0.446 f_{ck})$$
$$\frac{P_u}{f_{ck} bD} = 0.446 + \frac{P}{100 f_{ck}} (f_s - 0.446 f_{ck}) \qquad \ldots(39.1)$$

where  $p$ = percent of *total steel* in section = $\dfrac{A_s}{bD} \times 100$

$f_s$ = stress in steel corresponding to a strain of 0.002
= $0.87 f_y$ for Fe 250 (mild steel)
= $0.79 f_y$ for Fe 415 (HYSD Bars)    and = $0.75 f_y$ for Fe 500 (HYSD Bars)

However, as per **clause 38.3 of IS : 456-2000**, a short column axially loaded, should be designed by *Eq. 38.2*, when the minimum eccentricity, $e_{min}$ does not exceed 0.05 D :

$$P_u = 0.4 f_{ck} A_c + 0.67 f_y A_s = 0.4 f_{ck} bD + (0.67 f_y - 0.4 f_{ck}) A_s,$$

which reduces to the form

$$\dfrac{P_u}{f_{ck} bD} = 0.4 + \dfrac{P}{100 f_{ck}} (0.67 f_y - 0.4 f_{ck}) \qquad \text{...(39.2)}$$

In other words, when the moment $M_u$ is equal to or less than $0.05 P_u.D$, axial strength of the short column is given by *Eq. 39.2*. Thus, we have two equations for $P_u$ : *Eq. 39.1* and *Eq. 39.2*, and the ISI design hand book (SP : 16 – 1980) uses both these equations. When the min. eccentricity ($e_{min}$) does not exceed 0.05 D, *Eq. 39.2* has been used, while for eccentricities less than $e_{min.}$, *Eq. 39.1* has been used. At this stage, it is essential to know the difference between *Eqs. 39.1* and *39.2*. In *Eq. 39.1*, a factor 0.446 has been used for concrete and $f_s$ corresponds to the actual stress in steel corresponding to a strain of 0.002. In *Eq. 39.2*, a factor 0.4 has been used for concrete and $f_s$ has been taken equal to $0.67 f_y$ for all types of steel.

In *Eq. 39.1*, the second term within perenthesis represent the deduction of concrete replaced by reinforcement bars. This term is usually neglected for convenience. However, as a better approximation a constant value corresponding to M 20 grade concrete has been used in the design charts (interaction curves) in the ISI hand book, so that the error is negligibly small over the range of concrete mixes normally used. An accurate consideration of this term will necessitate the preparation of separate charts for each grade of concrete which is not considered worth while.

## Case II : Neutral axis outside the section

This case corresponds to the situation of a column, subjected to both the axial load and moment, when the entire column section is in compression, and the strain varies across the section as shown is Fig. 39.4 (a). Obviously, *the N.A. will lies outside the section*. For such a case, the points for plotting the charts (interaction curves) are obtained by *assuming* different positions across the section and the stress block parameters are determined, explained below. The stresses in the reinforcement are also calculated from the known strains. Therefore, the resultant axial force and the moment about the centroid of the section are calculated and the equilibrium equations are applied, so that (i) axial forces on the section balance, and (ii) moment on the section also balance.

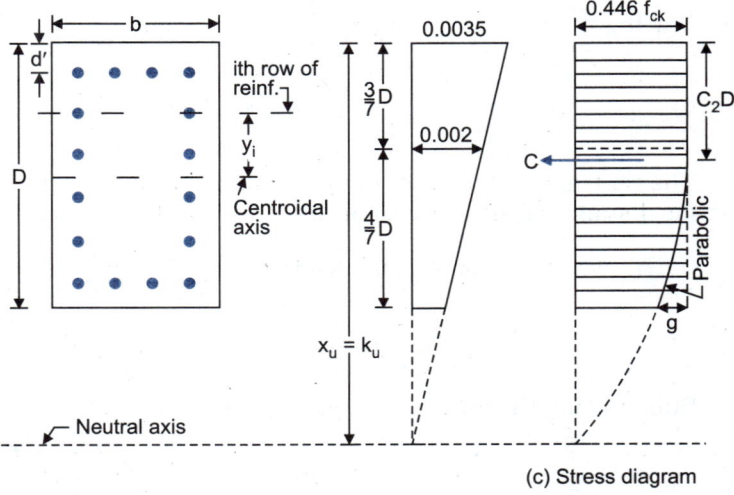

Fig. 39.4. N.A. Outside the Section.

Fig. 39.4 (a) shows a symmetrically reinforced section, with (i th) row of the reinforcing bar situated at a distance $y_i$ from the centroidal axis. Fig. 39.4 (b) shows the strain diagram, with a strain of 0.002 at the fulcrum point, $\dfrac{3}{7} D$ from the most compressed edge. Fig. 39.4 (c) shows the stress diagram, which is partly rectangular (up to depth $\dfrac{3}{7} D$) and partly parabolic (for the remaining depth), with a max. value of $0.446 f_{ck}$ at the highly compressed edge. Before applying the equilibrium equation for forces and moment, let us first determine the stress block parameters.

### Stress block parameters

Let
$x_u = k_u D$ = depth of N.A.
$C_1$ = coefficient for area of the stress block
$C_2 D$ = distance of centroid from highly compressed edge
$g$ = difference between the stress at the highly compressed edge and the stress at the least compressed edge.

From the equation of the parabola $y = x^2$, we get

$$g = \left(\frac{4}{7}D\right)^2 \qquad \text{...(i)}$$

Also,
$$0.446 f_{ck} = \left[k_u D - \frac{3}{7}D\right]^2 \qquad \text{...(ii)}$$

Dividing Eq. (i) by Eq. (ii), we get

$$g = 0.446 f_{ck} \left[\frac{\frac{4}{7}D}{k_u D - \frac{3}{7}D}\right]^2 = 0.446 f_{ck}\left[\frac{4}{7k_u - 3}\right]^2 \qquad \text{...[39.3 (a)]}$$

Area of stress block is

$$A = 0.446 f_{ck} D - \frac{g}{3}\left(\frac{4}{7}D\right) = 0.446 f_{ck} D - \frac{4}{21} gD$$

or
$$A = 0.446 f_{ck} D \left[1 - \frac{4}{21}\left(\frac{4}{7k_u - 3}\right)^2\right] = C_1 f_{ck} D \qquad \text{...[39.3 (b)]}$$

where
$$C_1 = 0.446 \left[1 - \frac{4}{21}\left(\frac{4}{7k_u - 3}\right)^2\right] \qquad \text{...[39.3 (c)]}$$

Putting
$$C_0 = \frac{8}{7}\left[\frac{4}{7 k_u - 3}\right]^2 \qquad \text{...[39.3 (d)]}$$

We get
$$C_1 = 0.446\left(1 - \frac{C_0}{6}\right) \qquad \text{...[39.3 (e)]}$$

The centroid of the stress block is found by taking moments about the highly compressed edge. It $C_2 D$ is the distance of the centroid of the stress block from highly compressed edge, we have

$$A(C_2 D) = 0.446 f_{ck} D \left(\frac{D}{2}\right) - \frac{4}{21} gD \left[\frac{3}{7}D + \frac{3}{4}\left(\frac{4}{7}D\right)\right]$$

$$A \cdot C_2 D = 0.446 f_{ck} \frac{D^2}{2} - \frac{8}{49} gD^2 \qquad \text{...[39.3 (f)]}$$

Substituting the value of $g$ from Eq. 39.3 (a), we get

$$A \cdot C_2 D = 0.446 f_{ck} D^2 \left[0.5 - \frac{8}{49}\left(\frac{4}{7 k_u - 3}\right)^2\right] = 0.446 f_{ck} D^2 \left[0.5 - \frac{C_0}{7}\right]$$

$$\therefore \quad C_2 = \frac{0.446 f_{ck} D^2 \left[0.5 - \frac{C_0}{7}\right]}{C_1 f_{ck} D^2} = \frac{0.446 f_{ck} D^2 \left(0.5 - \frac{C_0}{7}\right)}{0.446\left(1 - \frac{C_0}{6}\right) f_{ck} D^2}$$

or
$$C_2 = \frac{0.5 - C_0/7}{1 - C_0/6} \qquad \text{...(39.4)}$$

For different values of $k_u$, the area of stress block ($= C_1 f_{ck} D$) and the position of its centroid ($= C_2 D$) are given in Table 39.1. In this table, the values of stress block parameters have been tabulated for values of $k_u$ upto 4.00 for information only. For construction of interaction diagrams (design charts), it is generally adequate to consider values of $k_u$ upto about 1.2.

**TABLE 39.1.** Stress block parameters when N.A. falls outside the section

| $k_u = \dfrac{x_u}{D}$ | $C_0$ | Area of stress block $(C_1 f_{ck} \cdot D)$ | Distance of centroid from highly compressed edge $(C_2 D)$ |
|---|---|---|---|
| 1.00 | 1.143 | $0.361 f_{ck} D$ | $0.416 D$ |
| 1.05 | 0.966 | $0.374 f_{ck} D$ | $0.431 D$ |
| 1.10 | 0.828 | $0.384 f_{ck} D$ | $0.443 D$ |
| 1.20 | 0.627 | $0.399 f_{ck} D$ | $0.458 D$ |
| 1.30 | 0.419 | $0.409 f_{ck} D$ | $0.468 D$ |
| 1.40 | 0.395 | $0.417 f_{ck} D$ | $0.475 D$ |
| 1.50 | 0.325 | $0.422 f_{ck} D$ | $0.480 D$ |
| 2.00 | 0.151 | $0.435 f_{ck} D$ | $0.491 D$ |
| 3.00 | 0.056 | $0.442 f_{ck} D$ | $0.497 D$ |
| 4.00 | 0.029 | $0.444 f_{ck} D$ | $0.499 D$ |

**Equations of equilibrium:** Balancing the axial forces on the section, we get from *Eq. 39.1*

$$P_u = P_{uc} + P_{us} = C_1 f_{ck} bD + \sum_{i=1}^{n} \frac{p_i \, bD}{100} (f_{si} - f_{ci}) \qquad \ldots[39.5\,(a)]$$

where $p_i = \dfrac{A_{si}}{bD}$, where $A_{si}$ is the area of reinforcement in $i^{th}$ row

$f_{si}$ = stress in the $i^{th}$ row of reinforcement, compression being positive and tension being negative

$f_{ci}$ = stress in concrete at the level of $i^{th}$ row of reinforcement

and $n$ = number of rows of reinforcement.

Rewriting *Eq. 39.5 (a)*

$$\frac{P_u}{f_{ck} bD} = C_1 + \sum_{i=1}^{n} \frac{p_i}{100 f_{ck}} (f_{si} - f_{ci}) \qquad \ldots(39.5)$$

Similarly, taking moments of the forces about the centroid of the section, we get

$$M_u = M_{uc} + M_{us} = M_{uc} + \sum_{i=1}^{n} M_{usi}$$

$$= C_1 f_{ck} bD \left(\frac{D}{2} - C_2 D\right) + \sum_{i=1}^{n} \frac{p_i \, bD}{100} (f_{si} - f_{ci}) y_i \qquad \ldots[39.6\,(a)]$$

Dividing both sides by $f_{ck} bD^2$, we get

$$\frac{M_u}{f_{ck} bD^2} = C_1 (0.5 - C_2) + \sum_{i=1}^{n} \frac{p_i}{100 f_{ck}} (f_{si} - f_{ci}) \left(\frac{y_i}{D}\right) \qquad \ldots(39.6)$$

**Variation of strain across the section:** The values of stress in steel ($f_{si}$) and that in concrete ($f_{ci}$) are obtained corresponding to $\varepsilon_i$ at the $i^{th}$ row of reinforcement, from the respective stress-strain curve of steel and concrete.

The value $\varepsilon_i$, is obtained from the following linear expression, based on the premise that the reference strain is taken equal to 0.002 at a depth of $3D/7$ from the highly compressed edge, Thus,

$$\varepsilon_i = \frac{0.002 \, a_i}{x_u - \dfrac{3}{7} D} = \frac{0.002 \, (x_u - 0.5 D + y_i)}{x_u - \dfrac{3}{7} D} \qquad \ldots(39.7)$$

where $a_i$ = distance of $i^{th}$ row of reinforcing bars from the N.A. = $x_u - D/2 + y_i$

$y_i$ = distance of $i^{th}$ row of reinforcing bars from the centroidal axis, with proper algebraic sign.

**Determination of value of $f_{si}$**

**(a) For mild steel bars of Fe 250 grade**

(i) For $\varepsilon_i > \dfrac{0.87 f_y}{E_s} = \dfrac{0.87 \times 250}{2 \times 10^5} = 0.0010875$, $f_{si} = 0.87 f_y = 0.87 \times 250 = 217.5$ N/mm$^2$

(In other words, $\varepsilon_{su} = 0.0010875$ for Fe 250)

(ii) For $\varepsilon_i < \dfrac{0.87 f_y}{E_s} = 0.0010875$, $f_{si} = \varepsilon_i \cdot E_s$

**(b) For Fe 415 bars**

(i) For $\varepsilon_i > \dfrac{0.87 f_{yd}}{E_s} = \dfrac{0.8 \times 0.87 f_y}{E_s} = \dfrac{0.8 \times 0.87 \times 415}{2 \times 10^5} = 0.00144$

$f_{si}$ is obtained either from stress-strain curve or from Table 33.1

(ii) For $\varepsilon_i < \dfrac{0.8 f_{yd}}{E_s} = 0.00144$, $f_{si} = \varepsilon_i E_s$

**(c) For Fe 500 bars**

(i) For $\varepsilon_i > \dfrac{0.87 f_{yd}}{E_s} = \dfrac{0.8 \times 0.87 f_y}{E_s} = \dfrac{0.8 \times 0.87 \times 500}{2 \times 10^5} = 0.00174$

$f_{si}$ is obtained either from stress-strain curve or from Table 33.1

(ii) For $\varepsilon_i < 0.8 f_{yd}/E_s = 0.00174$, $f_{si} = \varepsilon_i E_s$

**Determination of value of $f_{ci}$**

The value of $f_{ci}$ corresponding to $\varepsilon_i$ can be found from equation of the parabolic part of idealised stress-strain curve for concrete, expressed as under:

$$f_{ci} = 0.446 f_{ck} \left[ \dfrac{2 \varepsilon_i}{0.002} - \left( \dfrac{\varepsilon_i}{0.002} \right)^2 \right] \text{ for } \varepsilon_c < 0.002 \qquad \ldots[39.8\,(a)]$$

or
$$f_{ci} = 446 f_{ck} \varepsilon_i (1 - 250 \varepsilon_i) \text{ for } \varepsilon_i < 0.002 \qquad \ldots(39.8)$$

However, for $\varepsilon_i \geq 0.002$, $f_{ci} = 446 f_{ck}$

Alternatively, $f_{ci}$ can be directly found from the following expression obtained by substituting the value of $\varepsilon_i$ (given by *Eq.* 39.7 in *Eq.* 39.8).

$$f_{ci} = 0.446 f_{ck} \left[ \dfrac{2 a_i}{\left(x_u - \dfrac{3}{7} D\right)} - \dfrac{a_i^2}{\left(x_u - \dfrac{3}{7} D\right)^2} \right] \qquad \ldots[39.8\,(b)]$$

where  $a_i$ = distance of $i^{th}$ row from the N.A.

**Case III : Neutral axis inside the section :** This case arises when the eccentricity is *moderate* or *large*, due to which some part of the section is under tension. In this case, the *stress block parameters* are simpler and they can be directly incorporated into the expressions which are otherwise same as for the earlier case. Fig. 39.5 (a) shows the section, in which N.A. is within the section. Fig. 39.5 (b) shows the strain diagram, in which there is a max. strain of 0.0035 at the highly compressed edge, and there is tension below the N.A. Fig. 39.5 (c) shows the stress diagram. Evidently, the stress block parameters are as under:

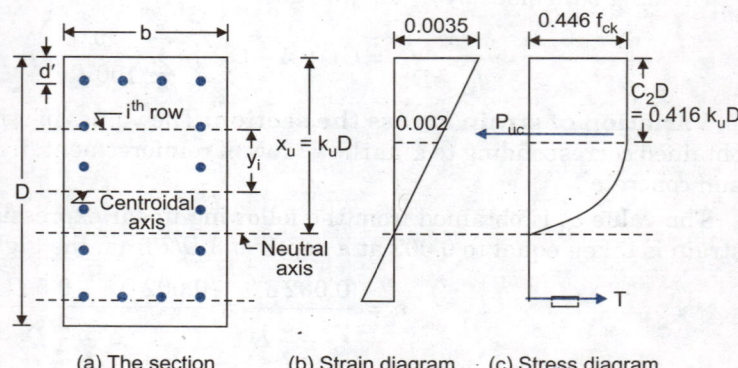

(a) The section  (b) Strain diagram  (c) Stress diagram

**Fig. 39.5.**  N.A. Inside the Section.

(i) Area of stress block = $0.36 f_{ck} k_u D$

(ii) Distance of c.g. of compressive force from highly compressed edge = $C_2 D = 0.416 k_u D$

It should be noted that unlike flexural members, there is no limitation on the depth of N.A. ($k_u D$) in compression members, and in extreme case, $k_u$ can be equal to one. Balancing the axial forces on the section, we have, from Eq. 39.1

$$P_u = P_{uc} + \sum_{i=1}^{n} P_{usi} = 0.36 f_{ck} k_u bD + \sum_{i=1}^{n} \frac{p_i \, bD}{100}(f_{si} - f_{ci})$$

or

$$\frac{P_u}{f_{ck} bD} = 0.36 k_u + \sum_{i=1}^{n} \frac{p_i}{100 f_{ck}}(f_{si} - f_{ci}) = C_1 + \sum_{i=1}^{n} \frac{p_i}{100 \, f_{ck}}(f_{si} - f_{ck}) \qquad ...(39.9)$$

Similarly, taking moments of forces, about the centroid of the section

$$M_u = M_{uc} + \sum_{i=1}^{n} M_{usi} = 0.36 f_{ck} k_u bD (0.5 D - 0.416 k_u D) + \sum_{i=1}^{n} \frac{p_i \, bD}{100}(f_{si} - f_{ci}) y_i$$

or

$$\frac{M_u}{f_{ck} bD^2} = 0.36 k_u (0.5 - 0.416 k_u) + \sum_{i=1}^{n} \frac{p_i}{100 \, f_{ck}}(f_{si} - f_{ci})\frac{y_i}{D} \qquad ...(39.10)$$

Design charts (for uniaxial and centric compressor) in SP–16:

The design charts (non dimensional interaction curves) given in SP: 16. Design handbook cover the following three cases of symmetrically arranged reinforcement.

An approximation is made for the value of $f_{ci}$ for M 20, as in the case of case I. For circular sections, the procedure is the same as above, except that the stress block parameters given earlier are not applicable ; hence the section is divided into strips and summation is done for determining the forces and moments due to the stresses in concrete. Fig. 39.7 gives specimen design charts for $f_y = 250$ N/mm² and $d'/D = 0.1$ while Fig. 39.8 give the charts for $f_y = 415$ N/mm² and $d'/D = 0.1$, for rectangular section where reinforcement is distributed equally on *two sides*. Fig. 39.9 and 39.10 give the corresponding charts for rectangular section where reinforcement is distributed equally on *four sides*. Fig. 39.11 gives the design Chart for circular section when $f_y = 250$ N/mm² and $d'/D = 0.1$, while Fig. 39.12 gives the design charts for circular section when $f_y = 415$ N/mm² and $d'D = 0.1$.

All the above design charts, given in ISI design handbook (SP; 16-1980), are in the form of interaction diagrams in which curves for $P_u/f_{ck} bD$ versus $M_u/f_{ck} bD^2$ have been plotted for different values of $p/f_{ck}$ where $p$ is the reinforcement percentage.

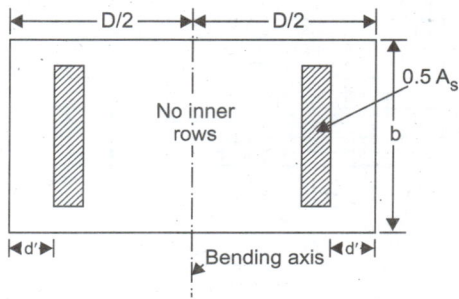

(a) Rectangular Section with reinforcement distributed equally on two sides. (Charts 27–38)

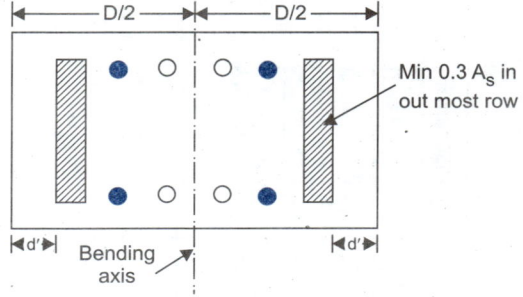

(b) Rectangular Section with reinforcement distributed equally on four sides (Charts 39–50)

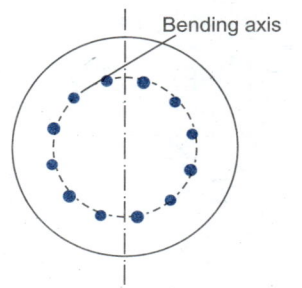

(c) Circular Column Section (Charts 51–62)

**Fig. 39.6.** At least 6 Bars (equal dia)

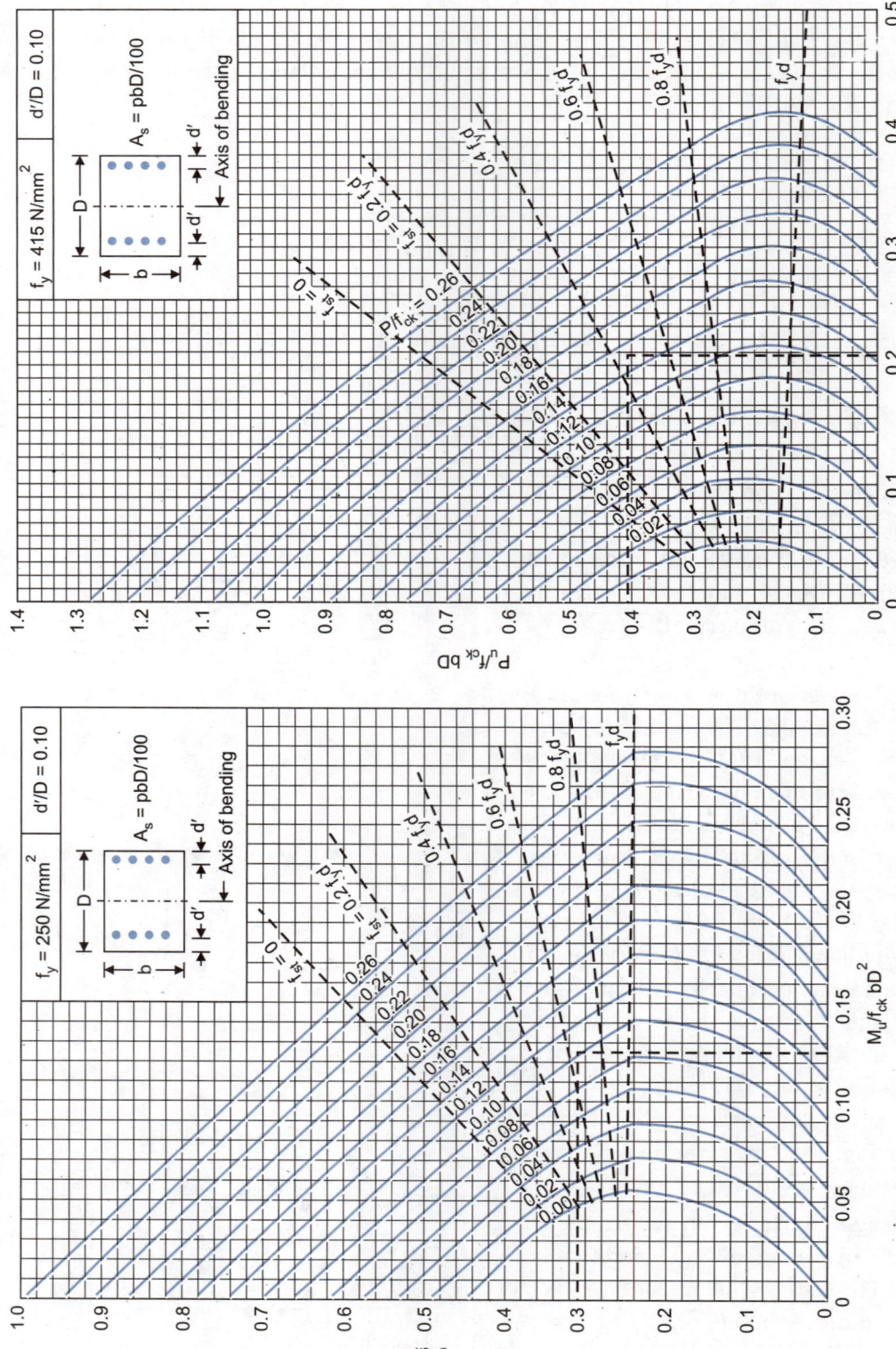

**Fig. 39.8.** Design Chart for Rectangular Section: Reinforcement Distributed Equally on Two Sides ($f_y = 415$ N/mm$^2$; $d'/D = 0.1$)

**Fig. 39.7.** Design Chart for Rectangular Section: Reinforcement Distributed Equally on Two Sides ($f_y = 250$ N/mm$^2$; $d'/D = 0.1$)

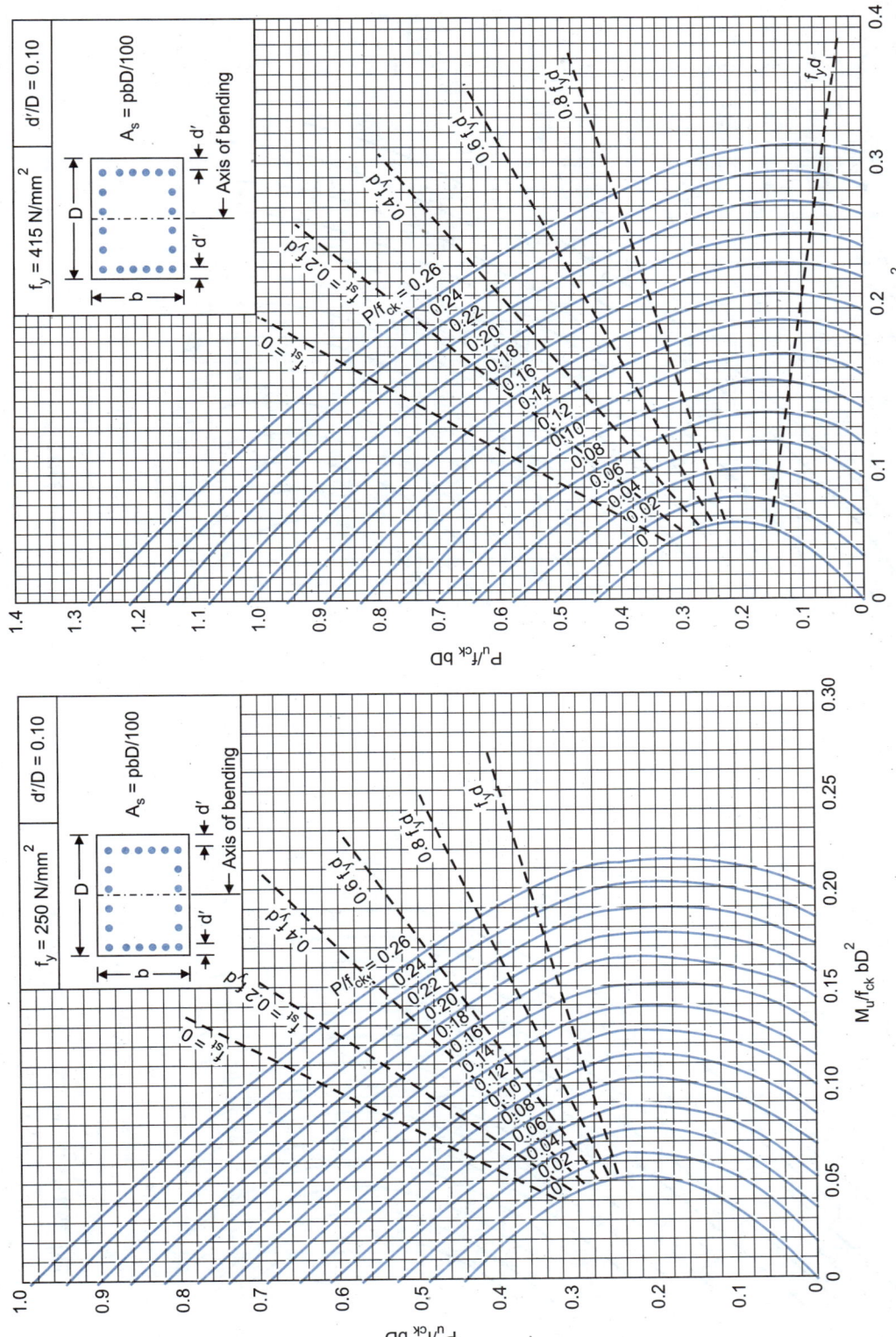

Fig. 39.10. Design Chart for Rectangular Section: Reinforcement Distributed Equally on Four Sides ($f_y = 415$ N/mm$^2$; $d'/D = 0.1$)

Fig. 39.9. Design Chart for Rectangular Section: Reinforcement Distributed Equally on Four Sides ($f_y = 250$ N/mm$^2$; $d'/D = 0.1$)

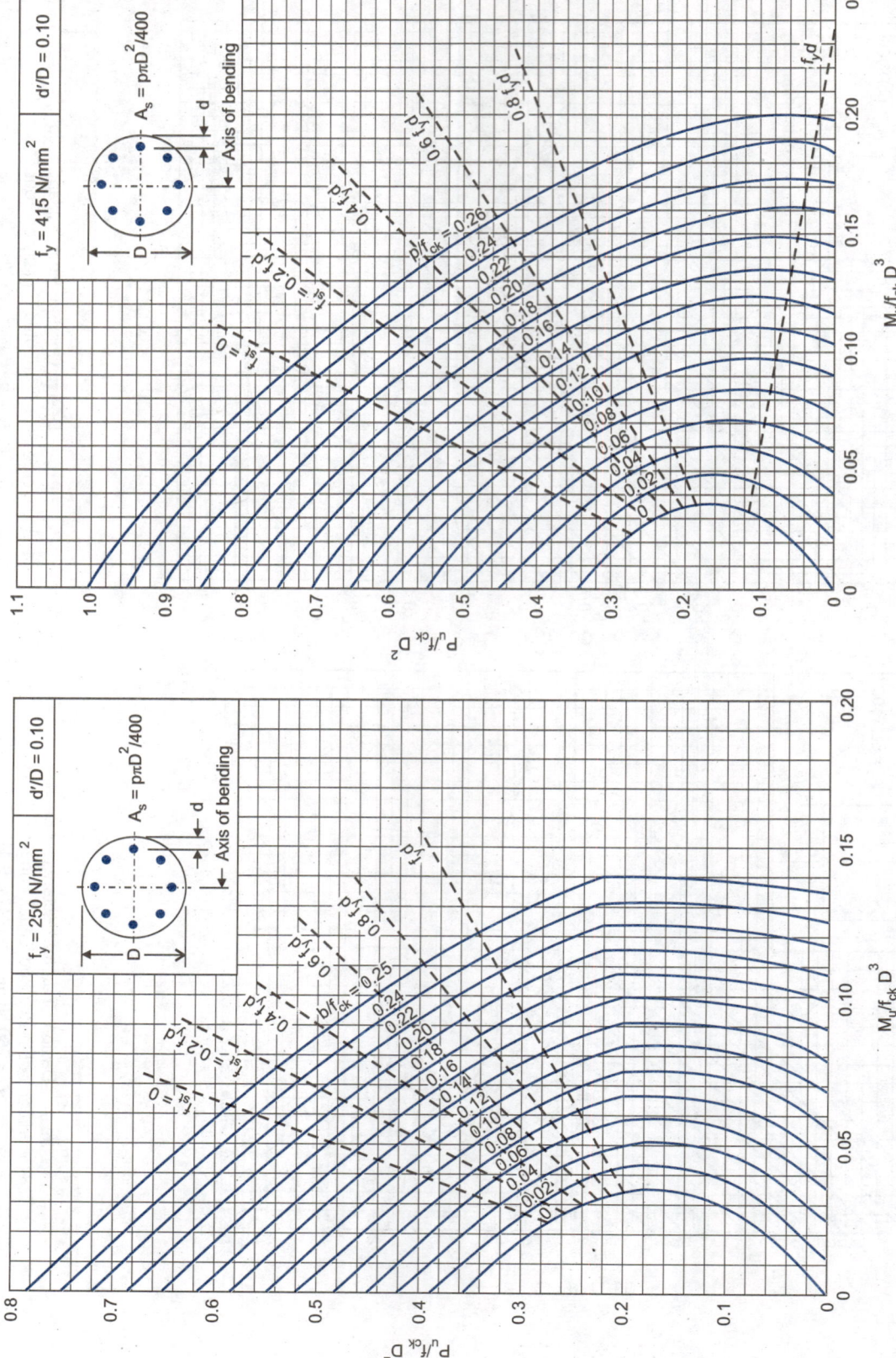

**Fig. 39.12.** Design Chart for Circular Section
($f_y$ = 415 N/mm²; $d'/D$ = 0.1)

**Fig. 39.11.** Design Chart for Circular Section
($f_y$ = 250 N/mm²; $d'/D$ = 0.1)

# COLUMNS WITH UNIAXIAL AND BIAXIAL BENDING

**Example 39.1.** *Determination of $M_u$ for given $P_u$:* A short R.C. column 300 mm wide and 500 mm deep is reinforced with 6 bars of 20 mm diameter, arranged as shown in Fig. 39.13. Determine the bending moment $M_u$ about an axis bisecting the depth, when it is also subjected to $P_u$ = 800 kN. Assume M 20 grade concrete and Fe 415 grade steel.

**Solution:** Here $d'$ = 50 mm;  $D$ = 500 mm. $d'/D$ = 50/500 = 0.1.

Also, $f_y$ = 415 N/mm². Hence use Chart 32 of ISI Hand book.

$$f_{ck} = 20 \text{ N/mm}^2$$

$$\frac{P_u}{f_{ck} \, bD} = \frac{800 \times 1000}{20 \times 300 \times 500} = 0.267$$

$$\therefore \quad A_s = 6 \times \frac{\pi}{4}(20)^2 \approx 1885 \text{ mm}^2$$

$$p = \frac{A_s}{bD} \times 100 = \frac{1885}{300 \times 500} \times 100 \approx 1.257$$

$$\frac{p}{f_{ck}} = \frac{1.257}{20} = 0.063$$

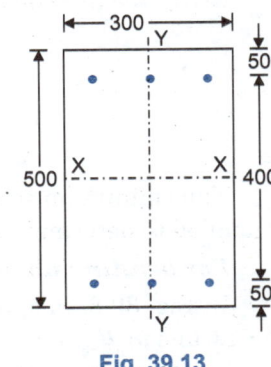

Fig. 39.13

Hence from Chart 32 of ISI Hand book, for $\frac{P_u}{f_{ck} bD}$ = 0.267 and $\frac{p}{f_{ck}}$ = 0.063, we get $\frac{M_u}{f_{ck} bD^2} \approx 0.13$

$M_u$ = 0.13 $f_{ck}$ $bD^2$ = 0.13 × 20 × 300 (500)² = 195 × 10⁶ N-mm = **195 kN-m.**

**Example 39.2.** *Determination of $P_u$ for given $M_u$:*

For the data of example 39.1. determine the value of $P_u$ if $M_u$ = 100 kN-m

**Solution:** $P_u = \frac{M_u}{f_{ck} \, bD^2} = \frac{100 \times 10^6}{20 \times 300 \, (500)^2} = 0.067$

Hence from Fig. 39.8 (Chart 32 of SP 16 of B.I.S.), for $M_u/f_{ck} bD^2$ = 0.067 and $p/f_{ck}$ = 0.063, we get

$$P_u/f_{ck} \, bD = 0.49.$$

$$P_u = 0.49 \times 20 \times 300 \, (500) = 1470 \times 10^3 \text{ N} = \mathbf{1470 \text{ kN}.}$$

**Example 39.3.** *Determination of $P_u$ for given eccentricity:* For the data of example 39.1, determine the value of $P_u$, if it is applied at an eccentricity of (a) 150 mm (b) 80 mm.

**Solution:** (a) Given : $e$ = 150 mm. $\therefore$ $e/D$ = 150/500 = 0.3

In Fig. 39.8, imagine a line, by holding a straight edge, joining the origin and a point having $P_u/f_{ck} \, bD$ = 1.0 and $M_u/f_{ck} \, bD^2 = (P_u/f_{ck} \, bD)(e/D)$ = 0.3. The point of intersection of this line with visually interpolated curve corresponding to $p/f_{ck}$ = 0.063 gives the required point. The value of $P_u/f_{ck} \, bD$ corresponding to this point comes out to be 0.36.

$$\therefore \quad P_u = 0.36 \times 20 \times 300 \times 500 = 1080 \times 10^3 \text{ N} = \mathbf{1080 \text{ kN}}$$

(b) Given : $e$ = 80 mm.   $\therefore$   $e/D$ = 80/500 = 0.16

In Fig. 39.8, imagine a line, by holding a straight edge, joining the origin and a point having $P_u/f_{ck} \, bD$ = 1.0 and $M_u/f_{ck} \, bD^2 = (P_u/f_{ck} \, bD)(e/D)$ = 0.16. The point of intersection of this line with visually interpolated curve corresponding to $p/f_{ck}$ = 0.063 gives the required point. The value of $P_u/f_{ck} \, bD$ corresponding to this point comes out to be 0.47

$$P_u = 0.47 \times 20 \times 300 \times 500 = 1410 \times 10^3 \text{ N} = \mathbf{1410 \text{ kN}}$$

**Example 39.4.** *Determination of $P_u$ at minimum eccentricity for given length of column:* For the data of example 39.1, determine the value of $P_u$ that can be permitted at the minimum eccentricity specified by the Code, if the effective length of the column is 5 m.

**Solution:** Let us first find the maximum eccentricities in the two directions

(i) In the direction of longer dimension,

$$e_{min} = \frac{l}{500} + \frac{D}{30} = \frac{5000}{500} + \frac{500}{30} = 26.67 \text{ mm}.$$

$$\frac{e_{min}}{D} = \frac{26.67}{500} = 0.0533 > 0.05$$

(ii) *In the direction of shorter dimension,*

$$e_{min} = \frac{l}{500} + \frac{b}{30} = \frac{5000}{500} + \frac{3000}{300} = 20 \text{ mm.}$$

$$\frac{e_{min}}{b} = \frac{20}{300} = 0.067 > 0.05$$

Thus minimum eccentricity ratio is more than 0.05 in both the directions, and hence *Eq. 38.2* cannot be applied to determine $P_u$.

**For bending about x-axis (Fig. 39.13)** : $e/D = 0.0533$

In Fig. 39.8, imagine a line by holding a straight edge joining the origin and a point having $P_u/f_{ck} bD$ = 1.0 and $M_u/f_{ck}bD^2 = (P_u/f_{ck} bD)(e/D) = 0.0533$. The point of intersection of this line with visually interpolated curve corresponding to $p/f_{ck} = 0.063$ gives the required point. The value of $P_u/f_{ck} bD$ corresponding to this point comes out to be 0.58.

∴ $\quad P_u = 0.58 \times 20 \times 300 \times 500 = 1740 \times 10^3 \text{ N} = 1740 \text{ kN}$

**For bending about y-axis (Fig. 39.13)** : For the purpose of using the Chart, width of the section (b) = 500 mm and depth of the section (D)= 300 mm.

$$e/D = 20/300 = 0.067$$

In Fig. 39.8, imagine a line having $P_u/f_{ck} bD = 1.0$ and $M_u/f_{ck} bD^2 = (P_u f_{ck} bD) (e/D) = 0.067$. The point of intersection of this line, with visually interpolated curve corresponding to $p/f_{ck} = 0.063$ gives the required point. The value of $P_u/f_{ck} bD$ corresponding to this point comes out to be 0.565.

∴ $\quad P_u = 0.565 \times 20 \times 500 \times 300 = 1695 \times 10^3 \text{ N} = 1695 \text{ kN.}$

Permissible $P_u$ will be the lesser of the two values. Hence $P_u$ = **1695 kN**

**Example 39.5.** *Design of square column with eccentric load*

*Design a reinforced concrete column, 400 mm square, to carry an ultimate load of 1000 kN at an eccentricity of 160 mm. Use M 20 grade concrete and Fe 250 grade steel.*

**Solution:** Let cover to the centre of steel = $d'$ = 40 mm

Here $b = D = 400$ mm. $\quad d'/D = 40/400 = 0.1$

$P_u = 1000 \text{ kN} = 1000 \times 10^3 \text{ N}$; $\quad M_u = e \cdot P_u = 1000 \times 10^3 \times 160$

The non-dimensional parametres are :

$$\frac{P_u}{f_{ck} \cdot bD} = \frac{1000 \times 10^3}{20 \times 400 \times 400} = 0.3125; \quad \frac{M_u}{f_{ck} \cdot bD^2} = \frac{1000 \times 10^3 \times 160}{20 \times 400 (400)^2} = 0.125$$

Referring to the interaction diagram for $f_y$ = 250 N/mm² and $d'/D$ = 0.1, *i.e.* Chart 28 of ISI Handbook (*i.e.* Fig. 39.7), the corresponding value of $p/f_{ck}$ is obtained as shown in dotted line.

∴ $\quad p/f_{ck} = 0.105$

∴ $\quad p = 0.105 \times f_{ck} = 0.105 \times 20 = 2.1.$ or $\quad p = \dfrac{100 A_s}{bD} = 2.1$

∴ $\quad A_s = \dfrac{2.1 \, bD}{100} = \dfrac{2.1 \times 400 \times 400}{100} = 3360 \text{ mm}^2$

Provide 2-25 mm dia. bars and 1-30 mm dia. bar on each face.

$$\text{Total } A_s = 4 \times \frac{\pi}{4}(25)^2 + 2 \times \frac{\pi}{4}(30)^2 = 1963.5 + 1413.7 = 3377.2 \text{ mm}^2.$$

Use 6 mm dia. lateral ties, at a spacing of the least of the following :

(i) Least lateral dimension = 400 mm,  (ii) 16 × 25 = 400 mm,  (iii) 48 × 6 = 288 mm.

Hence provide 6 mm dia. lateral ties at a spacing of 250 mm/cc.

# COLUMNS WITH UNIAXIAL AND BIAXIAL BENDING

**Example 39.6.** *Design of circular column : Hoop reinforcement*
Design a short circular column of 500 mm dia. with the following data :
Factored load : 800 kN.   Factored moment : 162.5 kN-m
Provide hoop reinforcement. Take M 20 concrete mix and use Fe 415 steel.

**Solution:** Let us provide 20 mm dia. bars at a clear cover of 40 mm.

$d' = 40 + 10 = 50$ mm

$d'/D = 50/500 = 0.1$. Hence use Chart 56 of ISI hand book.

Let us assume that moment due to minimum eccentricity will be less than the given moment.

$$\frac{P_u}{f_{ck} D^2} = \frac{800 \times 10^3}{20 (500)^2} = 0.16 \ ; \quad \frac{M_u}{f_{ck} \cdot D^3} = \frac{162.5 \times 10^6}{20 (500)^3} = 0.065$$

Referring to Chart 56 (or Fig. 39.11), we get $p/f_{ck} \approx 0.05$, for the above pair of two values.

$p = 0.05 f_{ck} = 0.05 \times 20 = 1\%$

$\therefore \quad \dfrac{100 A_s}{\dfrac{\pi}{4}(500)^2} = p = 1.0 \quad$ or $\quad A_s = \dfrac{1.0 \times \dfrac{\pi}{4}(500)^2}{100} \approx 1963.5$ mm$^2$

No. of 20 mm dia. bars = 1963.5/314.16 = 6.25.   Hence provide 7 bars of 20 mm dia.
Using 6 mm dia lateral ties, spacing will be the least of the following:
(i) Least lateral dimension = 500 mm   (ii) 16 Φ = 16 × 20 = 320 mm and
(iii) 48 Φ$_s$ = 48 × 6 = 288 mm.   Hence provide 6 mm dia. lateral ties @ 280 mm c/c.

**Example 39.7.** *Design of circular column : Helical reinforcement*
Redesign the column of example 39.6, using helical reinforcement.

**Solution:** According to clause 39.4 of the Code, the strength of a compression member with helical reinforcement is 1.05 times the strength of a similar member with lateral ties.
Therefore, the given load and moment should be divided by 1.05 before using the Chart.

$\therefore \quad \dfrac{P_u}{f_{ck} D^2} = \dfrac{0.16}{1.05} = 0.152 \ ; \quad \dfrac{M_u}{f_{ck} D^2} = \dfrac{0.065}{1.05} = 0.062$

Hence from Fig. 39.11, we get $p/f_{ck} = 0.045$.

$\therefore \quad p = 0.045 \times 20 = 0.9\%$

$\therefore \quad A_s = \dfrac{p \pi D^2}{400} = \dfrac{0.9 \times \pi (500)^2}{400} = 1767.1$ mm$^2$

$\therefore$ No. of 20 mm dia. bars = 1767.1/314.16 = 5.62
Hence provide 6 nos. of 20 mm dia. bars.

Now $\quad A_g = \dfrac{\pi}{4}(500)^2 = 196349.5$ mm$^2$

Let us use 8 mm dia. mild steel bars for helical reinforcement.
Core diameter = $D_c = 500 - 2 \times 40 + 2 \times 8 = 436$ mm

Area of core = $A_c = \dfrac{\pi}{4}(436)^2 = 149301$ mm$^2$

$\therefore \quad \dfrac{A_g}{A_c} = \dfrac{196349.5}{149301} = 1.315$

Factor $0.36 \left( \dfrac{A_g}{A_c} - 1 \right) \dfrac{f_{ck}}{f_{yh}} = 0.36 (1.315 - 1) \times \dfrac{20}{250} = 0.00907$ ...(i)

Dia. of core upto centre of helix = $500 - (2 \times 40) + 8 = 428$ mm

Let the pitch of the spiral be $s$ mm

$$V_h = \frac{\pi d}{s}\left(\frac{\pi}{4}\Phi_s^2\right) = \frac{\pi \times 428}{s} \times \frac{\pi}{4}(8)^2 = \frac{67587}{s} \; ; \; V_c = A_c \times 1 = 149301$$

$$\therefore \frac{V_h}{V_c} = \frac{67587}{s \times 149301} = \frac{0.4527}{s} \qquad \ldots(ii)$$

Equating (i) and (ii), we get $\frac{0.4527}{s} = 0.00907.$ or $s \approx 50$ mm.

However, the pitch should not be more than 75 mm, nor more than $\frac{1}{6}$ core dia. $\left(\frac{1}{6} \times 436 = 72.6 \text{ mm}\right)$. Also, it should not be less than $3\Phi_s$ ($= 3 \times 8 = 24$ mm). Hence keep the pitch equal to 50 mm.

## 39.4. SHORT COLUMNS SUBJECTED TO AXIAL LOAD AND BIAXIAL BENDING

Let a R.C. column be subjected to an axial force $P_u$, and biaxial moments $M_{ux}$ and $M_{uy}$ about $x$ and $y$ axes respectively. Such a section may be considered to be acted upon by an axial load $P_u$ at eccentricities $e_x = M_{ux}/P_u$ and $e_y = M_{uy}/P_u$, as shown in Fig. 39.14. In other words, if $P_u$, $e_x$ and $e_y$ are given, we have $M_{ux} = e_x P_u$ and $M_{uy} = e_y P_u$. The exact analysis or design of members subjected to axial load biaxial bending is extremely difficult, and is beyond the scope of the present book. The method set out in clause 39.6 is based on Bresler's formation for load contour of the Code (IS : 456–2000) is based on an assumed failure surface that extends the axial load moment ($P_u - M_u$) diagrams for single axis bending in three dimensions. Such an approach is also known as *Bresler's Load Contour method*. Based on this, the Code permits the design of such members by the following equation:

$$\left(\frac{M_{ux}}{M_{ux1}}\right)^{\alpha_n} + \left(\frac{M_{uy}}{M_{uy1}}\right)^{\alpha_n} \leq 1.0 \qquad \ldots(39.11)$$

where $M_{ux}$ and $M_{uy}$ the factored biaxial moments about $x$ and $y$ axes respectively due to design loads. and $M_{ux1}$ and $M_{uy1}$ are the maximum uniaxial moment capacities with an axial load $P_u$, bending about $x$ and $y$ axes respectively.

$\alpha_n$ is an exponent, whose value depends on the ratio $P_u/P_{uz}$, as given below, where $P_{uz}$ is given by
$$P_{uz} = 0.45 f_{ck} A_c + 0.75 f_y A_{sc} \;\Rightarrow\; P_{uz} = 0.45 f_{ck} A_g + (0.75 f_y - 0.45 f_{ck}) A_{sc} \qquad \ldots(38.1)$$

| $P_u/P_{uz}$ | $\alpha_n$ |
|---|---|
| $\leq 0.2$ | 1.0 |
| $\geq 0.8$ | 2.0 |

For intermediate values, linear interpolation may be done. Alternatively, the value of $\alpha_n$ can be determined by the following equation applicable for $P_u/P_{uz}$ ratio between 0.2 and 0.8 :

$$\alpha_n = \frac{2}{3}\left[1 + \frac{5}{2}\frac{P_u}{P_{uz}}\right]. \qquad \ldots(39.12)$$

Load $P_{uz}$ may be evaluated from Chart 62 of ISI handbook, reproduced here in Fig. 39.15. Fig. 39.16 shows graphical representation of *Eq. 39.11* for appropriate values of $\alpha_n$ corresponding to different values of $P_u/P_{uz}$.

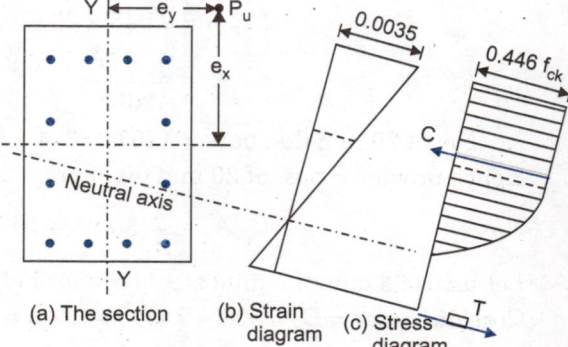

Fig. 39.14. Section Subjected to $P_u$, $M_{ux}$ and $M_{uy}$

(a) The section  (b) Strain diagram  (c) Stress diagram

**Design Procedure:** The design of a section for the given axial load $P_u$ and biaxial moments $M_{ux}$ and $M_{uy}$ is made by assuming the section and testing its adequacy. The design is done in the following steps.

**Step 1 :** Assume the cross-section of the column, and area of reinforcement along with its distribution.

**Step 2 :** Compute $P_{uz}$ either from *Eq. 38.1* or from Fig. 39.15. Find ratio $P_u/P_{uz}$.

**Step 3 :** Determine *uniaxial moment capacity* $M_{ux1}$ and $M_{uy1}$ combined with axial load $P_u$, using appropriate interaction curves (design Chart) for the case of column subjected to axial load ($P_u$) and uniaxial moment.

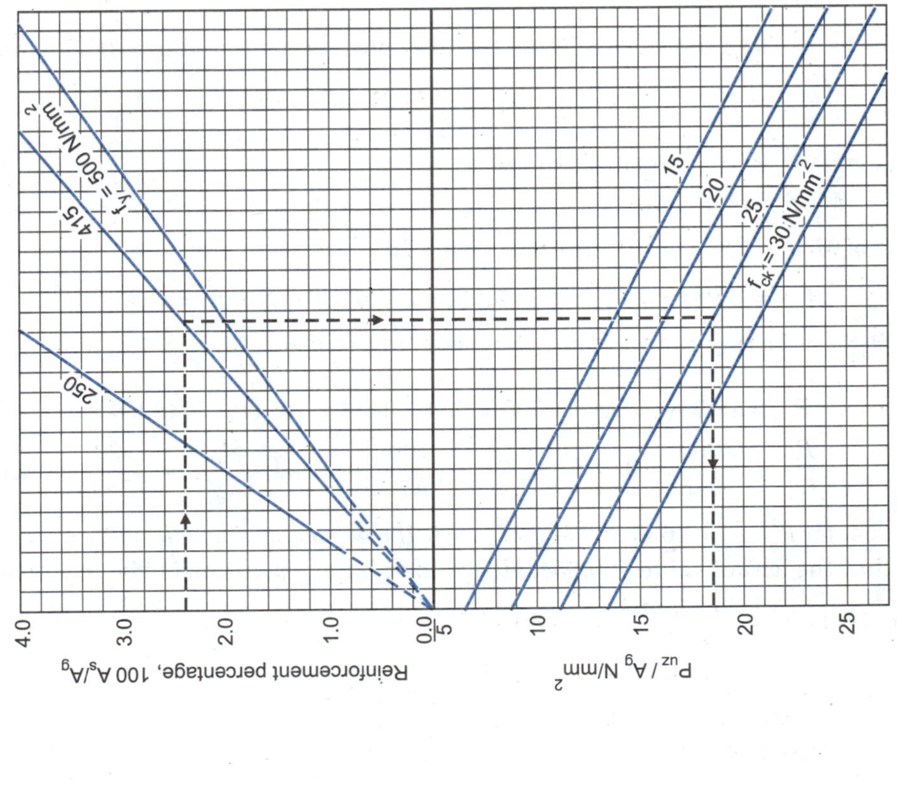

# COLUMNS WITH UNIAXIAL AND BIAXIAL BENDING

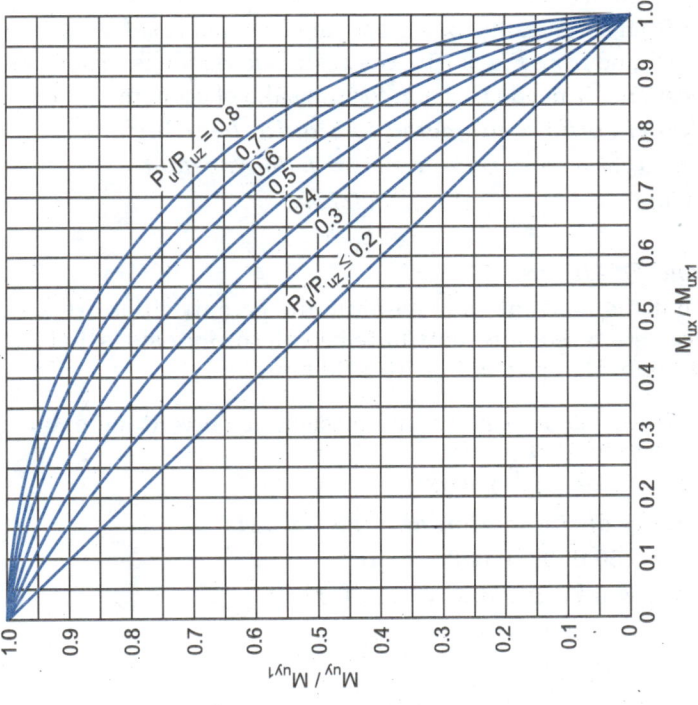

**Fig. 39.16.** Interaction Curves for Biaxial Bending in Compression in Members

**Fig. 39.15.** Values of $P_{uz}$ for Compression Members ($f_y = 250$ N/mm²; $d'/D = 0.1$)

**Step 4 :** Compute the values $M_{ux}/M_{ux1}$ and $M_{uy}/M_{uy1}$. From Fig. 39.16, find, the permissible value $M_{ux}/M_{ux1}$ corresponding to the above values of $M_{uy}/M_{uy1}$ and $P_u/P_{uz}$. If the *actual value* of $M_{ux}/M_{ux1}$ is less than the above value found from the Fig. 39.16, the assumed section is safe. If the actual value of $M_{ux}/M_{ux1}$ is *more* than the above value found from Fig. 39.16, the assumed section is unsafe, and needs revision. Even if the assumed section is *oversafe*, it needs revision for the sake of economy.

**Example 39.8.** *Design of short column subjected to biaxial bending*

*Determine the reinforcement for a short column for the following data:*

Column size : 400 × 600 mm ;        $P_u$ : 2000 kN
    $M_{ux}$ : 160 kN ;                    $M_{uy}$ : 120 kN

Use M 20 concrete mix and Fe 415 steel.

**Solution:** Let us assume that the moments due to minimum eccentricities are less than the values given above. Also, let the reinforcement be equally distributed on all the four sides. Let the dia. of bars be 20 mm, placed at a clear cover of 40 mm. Hence $d' = 40 + 10 = 50$ mm

***First trial :*** As a first trial, let us assume reinforcement percentage equal to $1\frac{1}{2}$ times the min. reinforcement, i.e. $p = 0.8 \times 1.5 = 1.25$ %

$$\therefore \quad p/f_{ck} = 1.25/20 = 0.0625$$

***Uniaxial moment capacity of the section about x – x axis***

$$d'/D = 50/600 = 0.083$$

Hence Charts for $d'/D = 0.1$ will be used (i.e. Chart 44)

$$\frac{P_u}{f_{ck}\, bD} = \frac{2000 \times 10^3}{20 \times 400\,(600)} = 0.417$$

Hence from Chart 44 of ISI Hand book, for $P_u/f_{ck}\, bD = 0.417$ and $p/f_{ck} = 0.0625$, we get $M_u/f_{ck}\, bD^2 = 0.085$

$$M_{ux1} = 0.085 \times 20 \times 400\,(600)^2 = 244.8 \times 10^6 \text{ N-mm} = 244.8 \text{ kN-m}$$

***Uniaxial moment capacity of the section about y–y axis***

$$\frac{d'}{D} = \frac{50}{400} = 0.125$$

Hence we will have to use both the charts for $d'/D = 0.1$ (Chart 44) and for $d'/D = 0.15$ (Chart 45)

$$\frac{P_u}{f_{ck}\, bD} = 0.417 \quad \text{and} \quad p/f_{ck} = 0.0625 \text{ as before.}$$

Hence from Chart 44,   $M_u/f_{ck}\, bD^2 = 0.085$ (as before)
and from Chart 45,     $M_u/f_{ck}\, bD^2 = 0.08$

$\therefore$   Average value of $M_u/f_{ck}\, bD^2 = \frac{1}{2}(0.085 + 0.08) = 0.0825$

$$\therefore \quad M_{uy1} = 0.0825 \times 20 \times 600\,(400)^2 = 158.4 \times 10^6 \text{ N-mm} = 158.4 \text{ kN-m}$$

***Calculation of*** $P_{uz}$ : Referring to Chart 63, corresponding to $p = 1.25$, $f_y = 415$ and $f_{ck} = 20$, we get $P_{uz}/A_g \approx 12.7$ N/mm²

$$\therefore \quad P_{uz} = 12.7\, A_g = 12.7 \times 400 \times 600 = 3048 \times 10^3 \text{ N} = 3048 \text{ kN}$$

$$\frac{P_u}{P_{uz}} = \frac{2000}{3048} = 0.656 \;;\quad \frac{M_{ux}}{M_{ux1}} = \frac{160}{244.8} = 0.654;\quad \frac{M_{uy}}{M_{uy1}} = \frac{120}{158.4} = 0.758$$

Hence from Chart 64, for $P_u/P_{uz} = 0.656$ and $M_{uy}/M_{uy1} = 0.758$ we get *permissible value* of $M_{ux}/M_{ux1} = 0.58$. This value is lower that the actual value of 0.654. Hence the reinforcement has to be increased.

**Second trial.** Revised $p = \dfrac{1.25 \times 0.654}{0.58} = 1.409$

$$A_s = \frac{p\,bD}{100} = \frac{1.409 \times 400 \times 600}{100} \approx 3380 \text{ mm}^2$$

No. of 20 mm Φ bars = 3380/314.16 ≈ 10.8

However, for symmetrical arrangement, provide 12 bars of 20 mm dia. as shown in Fig. 39.17.
Actual $A_s = 12 \times 314.16 = 3769.9$ mm$^2$

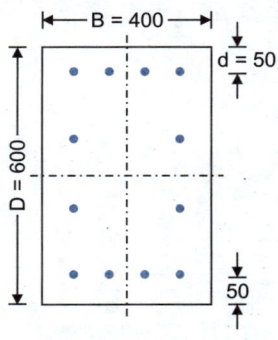

∴ Actual $p = \dfrac{100 A_s}{bD} = \dfrac{100 \times 3769.9}{400 \times 600} \approx 1.571$

With this percentage, the section may be re-checked as follows :
$$p/f_{ck} = 1.571/20 = 0.079.\ \text{Hence from Chart 44,}$$
$$M_u/f_{ck}\,bD^2 = 0.098$$
$$M_{ux1} = 0.098 \times 20 \times 400\,(600)^2$$
$$= 282.2 \times 10^6 \text{ N-mm} = 282.2 \text{ kN-m}$$

Similarly, from Chart 45 (for $d'/D = 0.15$) we get $M_u/f_{ck}\,bD^2 = 0.09$.

∴ Av. value of $M_u/f_{ck}\,bD^2 = (0.098 + 0.09)\,\dfrac{1}{2} = 0.094$

Fig. 39.17

∴ $M_{uy1} = 0.094 \times 20 \times 600\,(400)^2 \approx 180.5 \times 10^6$ N-mm $= 180.5$ kN-m

Again, from Chart 63, corresponding to $p = 1.571$, $f_y = 415$ and $f_{ck} = 20$, we get $p_{uz}/A_g = 13.7$ N/mm$^2$

∴ $P_{uz} = 13.7 \times 400 \times 600 = 3288 \times 10^3$ N $= 3288$ kN
$P_u/P_{uz} = 2000/3288 = 0.608$
$M_{ux}/M_{ux1} = 160/282.2 = 0.567$
$M_{uy}/M_{uy1} = 120/180.5 = 0.665$

Hence from Chart 64, for $M_u/M_{uy1} = 0.665$ and $P_u/P_{uz} = 0.608$, we get $M_u/M_{ux1} = 0.67$.
The actual value of $M_{ux}/M_{ux1}$ is 0.567. Hence the section is safe.

## PROBLEMS

1. A column 300 mm × 300 mm, has an effective length of 3.5 m. It is reinforced with 4 bars of 20 mm dia. at a clear cover of 40 mm. Determine the safe axial load the column can carry. Assume concrete grade M 20 and steel grade Fe 250.
2. A short column 300 mm × 450 mm is subjected to an axial load $P_u$ of 800 kN. The column is reinforced with 8 bars of 16 mm dia. distributed equally on all the faces. Assume concrete grade M 20 and steel grade Fe 415. Clear cover is 40 mm. Determine the bending moment $M_u$ about an axis bisecting the depth.
3. For the data of problem 2, determine the eccentricity $e$ at which the load $P_u = 800$ kN can be placed.
4. For the data of problem 2, determine the value of $P_u$ when the eccentricity from the $x$-axis is 100 mm.
5. Design a short column to carry a working load of 1000 kN and an uniaxial moment of 250 kN. Use M 20 concrete and Fe 415 steel.
6. Design a short column for the following data :
   Column size         : 350 mm × 450 mm
   Factored load $P_u$ : 1200 kN
   Factored moment acting parallel to the larger dimension, $M_{ux}$ : 100 kN
   Factored moment acting parallel to the shorter dimension, $M_{uy}$ : 80 kN
   Use M 20 concrete and Fe 415 steel.
7. For column section 500 × 300 determine the design strength compare corresponding to the condition of 'balanced failure': Assume M$_{25}$ concrete and Fe 415 steel. Considering loading eccentricity with respect to the major axis alone. Assume 8 $\phi$ ties and 40 mm clear cover.
8. For a column section 300 × 500 using M$_{25}$ and Fe 415 materials $A_{sc}$ 2946 mm$^2$, constructed the Design Interaction curve for axial compression consigned with uniaxial bending about the major axis. Hence investigate the safety of column section under the following factored load effects.
   (i) $P_u = 2348$ kN    $M_{ux} = 46.5$ kNm (max. axial compression)
   (ii) $P_u = 1195$ kN   $M_{ux} = 165$ kNm (max. acceleration).
9. A corner column 350 × 350 mm located in the lower most storey of a system of braced frame is subjected to factored loads; $P_u = 1800$ kN; $M_{ux} = 180$ kNm and $M_{uy} = 140$ kNm. The unsupported length of column is 3.5 m; Design the reinforcement in column, assuming M$_{25}$ grade of concrete and Fe 415 steel.

# CHAPTER 40: DESIGN OF STAIR CASES

**Staircase:** The stair consists of series of steps with landing at appropriate intervals. Stairs provide access for the various floors of building.

**Type of stair cases:** Depending upon the Geometry/shape; they are classified into the following.

(i) Single flight stair case.
(ii) Quater term stair case.
(iii) Doglegged stair case.
(iv) Open well stair case.
(v) Geometrical stair case.
(vi) Spiral stair case.

## 40.1. GENERAL NOTES ON DESIGN OF STAIRS

**1. Live load on stairs:** I.S. 875-1987 (*Code of practice for Structural Safety of Buildings*) gives the loads for staircases. For stairs in residential buildings, office buildings, hospital wards, hostels, etc., where there is no possibility of overcrowding, the live load may be taken to be 3000 N/m$^2$, or 3 kN/m$^2$ (if not crowded) subject to a minimum of 1300 N concentrated load at the unsupported end of each step for stairs constructed out of structurally independent cantilever step. For other public buildings liable to be overcrowded, the live load may be taken to be 5000 N/m$^2$ or 5 kN/m$^2$ (if crowded).

**Based on structural Behaviour (support conditions)**

The stairs are classified into the following categories depending upon the structural behaviour.

**2. Effective span of stairs:** Stair slab may be divided into two categories, depending upon the direction in which the stair slab spans:

(i) Stair slab spanning horizontally.
(ii) Stair slab spanning longitudinally.

(i) *Stair slab spanning horizontally* (*with side supports*). In this category, the slab is supported on each side by side wall or stringer beam on one side and beam on the other side. Sometimes, as in the case of straight stair, the slab may also be supported on both the sides by the two side walls. The slab may also be supported horizontally by side wall on one side of each flight and the common newel on the other side between the backward and forward flights. In such a case the

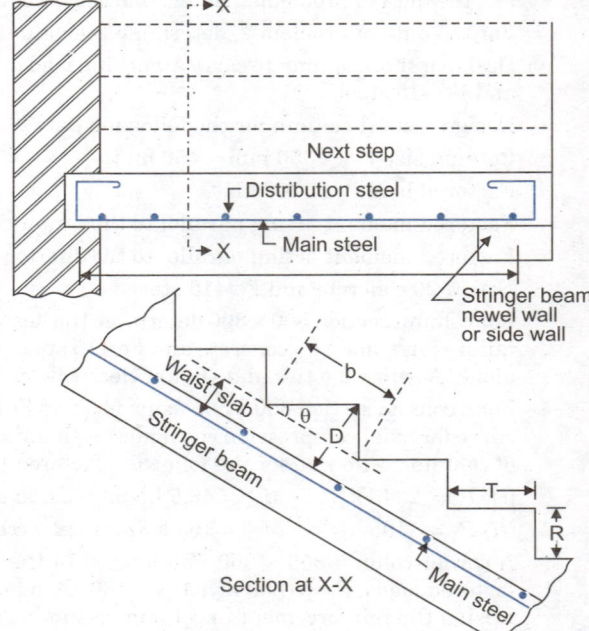

**Fig. 40.1.** Stair Slab Spanning Horizontally.

effective span $L$ is the horizontal distance between centre-to-centre of supports. Each step is designed as spanning horizontally with a bending moment equal to $(w L^2)/8$. Each step is considered equivalent to a rectangular beam of width $b$ (measured parallel to the slope of the stair) and an effective depth equal to $D/2$ (i.e. $d = D/2$) as shown in Fig. 40.1. Main reinforcement is provided in the direction of $L$, while distribution reinforcement is provided parallel to the flight direction. A waist of about 8 cm is provided.

(ii) *Stair slab spanning longitudinally.* In this category, the slab is supported at bottom and top of the flight and remain unsupported on the sides. Each flight of stairs is continuous, supported on beams at top and bottom or on landings. The effective span of such stairs, without stringer beams, should be taken as the following horizontal distances:

(a) where unsupported at top and bottom risers by beams spanning parallel with the risers, the distance centre-to-centre of beam;

(b) where spanning on the edge of a landing slab which spans parallel to the risers (Fig. 40.2), a distance equal to the 'going' of the stairs plus at each end either half the width of the landing or one metre, whichever is smaller; and

(c) where the landing slab spans in the same direction as the stairs, they should be considered as acting together to from a single slab and the span determined at the distance centre-to-centre of the supporting beams or walls, the going being measured horizontally.

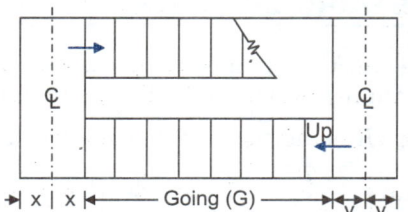

| x (m) | y (m) | Span L (M) |
|---|---|---|
| <1 | <1 | G + x + y |
| <1 | ≥1 | G + x + 1 |
| ≥1 | <1 | G + y + 1 |
| ≥1 | ≥1 | G + 1 + 1 |

**Fig. 40.2.** Effective Span for Stairs Supported at each by Landings Spanning Parallel with the Risers.

**3. Distribution of loading on the stairs :** In case of stairs with open wells, where spans partly crossing at right angles occur, the load on areas common to any two such spans may be taken as one-half in each direction as shown in Fig. 40.3 (a). Where flights or landings are built into walls at a distance of not less than 110 mm and are designed to span in the direction of the flight, a 150 mm strip may be deducted from the loaded area and the effective breadth of the section increased by 75 mm for the purposes of design [Fig. 40.3 (b)].

**4. Estimation of dead weight:** The dead weight of stair consists of (i) dead weight of waist slab and (ii) dead weight of steps.

(i) *Dead weight of waist slab*: The dead weight $w'$, per unit area, is first calculated at right angles to the slope. The corresponding load per unit horizontal area is then obtained by increasing $w'$ by the ratio $\sqrt{R^2 + T^2}/T$ where $R$ = rise and $T$ = tread. Thus, if $t$ = thickness of waist in mm, then $w' = \dfrac{t \times 1 \times 1}{1000} \times 25000 = 25\, t$ N/m² of inclined area.

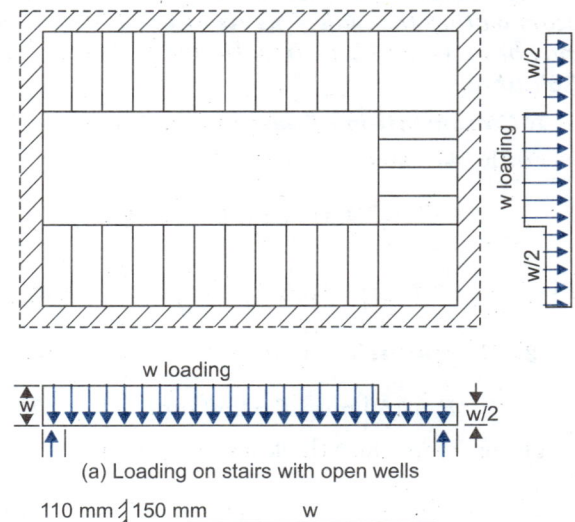

**Fig. 40.3.** Distribution of Loading on Stairs.

Hence dead weight $w_1$ per unit horizontal area is given by

$$w_1 = w' \times \dfrac{\sqrt{R^2 + T^2}}{T} = 25\, t\, \dfrac{\sqrt{R^2 + T^2}}{T} = 25\, t\, \sqrt{1 + (R/T)^2}$$

For example, if $R = 150$ mm, $T = 300$ mm and $t = 80$ mm

Then $w_1 = 25\, t\, \sqrt{1 + \left(\dfrac{150}{300}\right)^2} = 27.95\, t = 2236$ N/m² of horizontal area.

**(ii) Dead weight of steps:** The dead weight of the steps is calculated by treating the step to be equivalent *horizontal* slab of thickness equal to half the rise (R/2). Thus, if $w_2$ is the weight of step per unit horizontal area, we have

$$w_2 = \frac{R}{2 \times 1000} \times 1 \times 1 \times 25000$$

$$= 12.5\ R\ \text{N/m}^2 \quad \text{where } R \text{ is rise in mm.}$$

Total $w = w_1 + w_2$ per unit horizontal area.

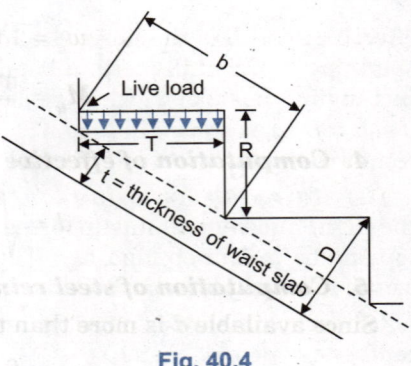

Fig. 40.4

## 40.2. DESIGN OF STAIRS SPANNING HORIZONTALLY

In this type of stairs, the effective span is the horizontal distance between centre to centre of supports. Each step is designed as spanning horizontally. Each step is considered equivalent to a horizontal beam of width $b = \sqrt{R^2 + T^2}$ where $R$ being rise and $T$ being tread.

$D$ = thickness of waist slab + $\frac{R.T}{b}$, effective depth $(d) = \frac{D}{2}$ measured parallel to the slope of the stair, and an effective depth $D/2$, as shown in Fig. 40.1. Main reinforcement is provided in the direction of span. The distribution reinforcement, provided in the form of 6 mm diameter bars at 30 mm c/c is normally adequate.

**Design: Example 40.1.** *A straight stair in a residential building is supported on wall on one side and stringer beam on the other side. The risers are 150 mm and treads are 250 mm and the horizontal span of the stairs may be taken as 1.2 metres. Design the steps. Use M 20 concrete and Fe 415 steel bars.*

**Solution:**

### 1. Computation of design constants

For Fe 415 steel, $f_y = 415\ \text{N/mm}^2$, For M 20 concrete, $f_{ck} = 20\ \text{N/mm}^2$

$$\frac{x_{u,\ max}}{d} \text{ (For Fe 415 steel)} = \frac{700}{1100 + 0.87 \times 415} = 0.479$$

$$R_u = 0.36\ f_{ck}\ \frac{x_{u,\ max}}{d}\left(1 - 0.42\ \frac{x_{u,\ max}}{d}\right) = 0.36 \times 20 \times 0.479\ (1 - 0.42 \times 0.479) = 2.755$$

### 2. Computation of step dimensions (Fig. 40.2)

$R = 150\ \text{mm}; T = 250\ \text{mm}\ ;\ b = \sqrt{R^2 + T^2} = \sqrt{150^2 + 250^2} \approx 292\ \text{mm}$

Let us keep waist thickness = 80 mm

$$D = 80 + \frac{RT}{b} = 80 + \frac{150 \times 250}{292} \approx 208\ \text{mm}$$

Hence the effective depth of equivalent beam = $D/2$ = 104 mm

Width $b$ = 292 mm ; span $L$ = 1.25 m

### 3. Computation of loading and B.M. Each step spans horizontally.

Dead load of each step per meter = $\frac{1}{2} \times \frac{150}{1000} \times \frac{250}{1000} \times 25000 \approx 469$ N/m

Dead load of waist slab = $\frac{80 \times 292}{10^6} \times 25000$ = 584 N/m

Load of finishing = 70 N/m (say)

Total = 1123 N/m

Live load @ 3000 N/m² = $\frac{250 \times 3000 \times 1}{1000}$ = 750 N/m

Total $w$ = 1873 N/m

$$\therefore \quad w_u = 1.5\, w = 1.5 \times 1873 \approx 2810 \text{ N/m}$$

$$M_u = \frac{w_u L^2}{8} = \frac{2810\,(1.25)^2}{8} = 548.8 \text{ N-m} = 54.88 \times 10^4 \text{ N-mm}$$

### 4. Computation of effective depth

$$d = \sqrt{\frac{M_u}{R_u\, b}} = \sqrt{\frac{54.88 \times 10^4}{2.755 \times 292}} \approx 26.1 \text{ mm. But available } d = 104 \text{ mm}$$

### 5. Computation of steel reinforcement

Since available $d$ is more than that required from B.M., we have an under-reinforced section, for which

$$A_{st} = \frac{0.5\, f_{ck}}{f_y}\left[1 - \sqrt{1 - \frac{4.6\, M_u}{f_{ck}\, bd^2}}\right] bd$$

$$= \frac{0.5 \times 20}{415}\left[1 - \sqrt{1 - \frac{4.6 \times 54.88 \times 10^4}{20 \times 292\,(104)^2}}\right] 292 \times 104 \approx 14.8 \text{ mm}^2$$

However, provide minimum steel in the form of one bar of 8 mm diameter per step giving $A_{st} = 50.3$ mm². Provide distribution reinforcement in the form of 8 mm φ bars @ 450 mm c/c. The reinforcement is arranged as shown in Fig. 40.1.

## 40.3. DESIGN OF DOG-LEGGED STAIR

**Design: Example 40.2.** *Design a dog-legged stair for a building in which the vertical distance between floors is 3.6 m. The stair hall measures 2.5 m × 5 m. The live load may be taken as 2500 N/m². Use M 20 concrete and Fe 415 steel bars.*

**Solution:**

### 1. General arrangement of stair

Fig. 40.5 shows the plan of stair hall. Let the rise be 150 mm and tread be 250 mm. Let us keep width of each flight = 1.2 m.

| | |
|---|---|
| Height of each flight | = 3.6/2 = 1.8 m |
| No. of risers required | = 1.8/0.15 = 12 in each flight. |
| No. of treads in each flight | = 12 – 1 = 11 |
| Space occupied by treads | = 11 × 25 = 275 cm. |

Keep width of landing equal to 1.25 m. Hence space left for passage = 5.0 – 1.25 – 2.75 = 1 m.

### 2. Computation of design constants

For Fe 415 steel, $f_y = 415$ N/mm². For M 20 concrete, $f_{ck} = 20$ N/mm²

$$\frac{x_{u,\max}}{d} \text{ (for Fe 415 steel)} = \frac{700}{1100 + 0.87 \times 415} = 0.479$$

$$\therefore \quad R_u = 0.36\, f_{ck}\, \frac{x_{u,\max}}{d}\left(1 - 0.42\, \frac{x_{u,\max}}{d}\right)$$

$$= 0.36 \times 20 \times 0.479\,(1 - 0.42 \times 0.479) = 2.755$$

### 3. Computation of loading and B.M.:

The landing slab is assumed to span in the same direction as the stairs, and is considered as acting together to form a single slab. Let the bearing of the landing slab in the wall be 160 mm. The effective span = 2.75 + 1.25 + 0.160/2 = 4.08 ≈ 4.1 m.

Let the thickness of waist slab be equal to 200 mm (assumed at the rate of 40 to 50 mm per metre span).

∴ Weight of slab $w'$ on slope

$$= \frac{200}{1000} \times 1 \times 1 \times 25000 = 5000 \text{ N/m}^2$$

Dead weight on horizontal area,

$$w_1 = w' \frac{\sqrt{R^2 + T^2}}{T}$$

$$= 5000 \times \frac{\sqrt{(150)^2 + (250)^2}}{250} \approx 5830 \text{ N/m}^2$$

Dead weight of steps is given by

$$w_2 = \frac{R}{2 \times 1000} \times 1 \times 1 \times 25000 = \frac{150}{2000} \times 25000 = 1875 \text{ N/m}$$

∴ Total dead weight per m run = 5830 + 1875 = 7705 N
Weight of finishing etc. = 100 N (assumed)
Live load = 2500 N
_____
Total $w$ = 10305 N/m

∴ $w_u = 1.5 \times 10305 \approx 15460$ N/m.

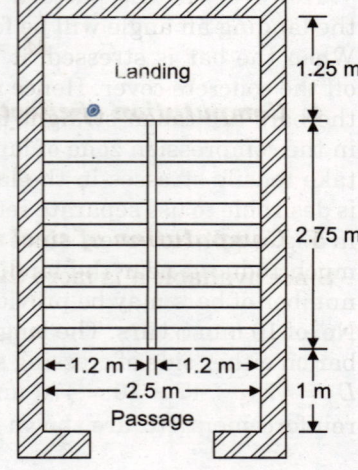

Fig. 40.5. General Arrangement

**Note :** The loading $w$ on the landing portion will be 10305 – 1875 = 8430 N, since weight of steps will not come on it. However, a uniform value of $w$ has been adopted here.

$$M_u = \frac{w_u L^2}{8} = \frac{15460 (4.1)^2}{8} = 32485 \text{ N-m} = 32.485 \times 10^6 \text{ N-mm}.$$

### 4. Design of waist slab

$$d = \sqrt{\frac{M_u}{R_u b}} = \sqrt{\frac{32.485 \times 10^6}{2.755 \times 1000}} = 108.5 \text{ mm}$$

Adopt 150 mm overall depth. Using 20 mm nominal cover and 10 mm φ bars, effective depth
= 150 – 20 – 5 = 125 mm.

### 5. Computation of reinforcement

Since actual $d$ provided is more than that required for B.M., we have an under-reinforced section for which

$$A_{st} = \frac{0.5 \times 20}{415} \left[ 1 - \sqrt{1 - \frac{4.6 \times 32.485 \times 10^6}{20 \times 1000 (125)^2}} \right] \times 1000 \times 125$$

$$= 836.3 \text{ mm}^2.$$

Using 10 mm φ bars having $A_\varphi$ = 78.54 mm², No. of bars required in 1.2 m width

$$= \frac{1.2 \times 836.3}{78.54} \approx 12.8 \approx 13 \text{ (say)}$$

∴ Spacing of bars = 1200/13 = 92.3 mm

Distribution reinforcement, $A_{sd} = \frac{0.12 \times 150 \times 1000}{100} = 180 \text{ mm}^2$

Hence spacing of 8 mm φ bars $= \frac{1000 \times 50.3}{180} \approx 279$ mm

Hence provide 8 mm φ bars @ 250 mm c/c, i.e. one bar per step.

The main reinforcement should be bent to follow the bottom profile of the stair.

However, if this pattern is followed near the landing an angle will be formed in the bar. When the bar is stressed, it will try to throw off the concrete cover. Hence near the landing, the bars are taken straight up and then bent in the compression zone of landing. In order to take tensile stresses in the landing portion, it is desirable to use separate set of bars as shown in Fig. 40.6. However, since the B.M. is very much reduced near the landing, only half the number of bars may be provided, i.e. provide 7 No. of 10 mm φ bars. The length of each type of bar on either side of crossing should be at least $L_d = 47\ \varphi = 47 \times 10 = 470$ mm. The details of reinforcement etc. are shown in Fig. 40.6.

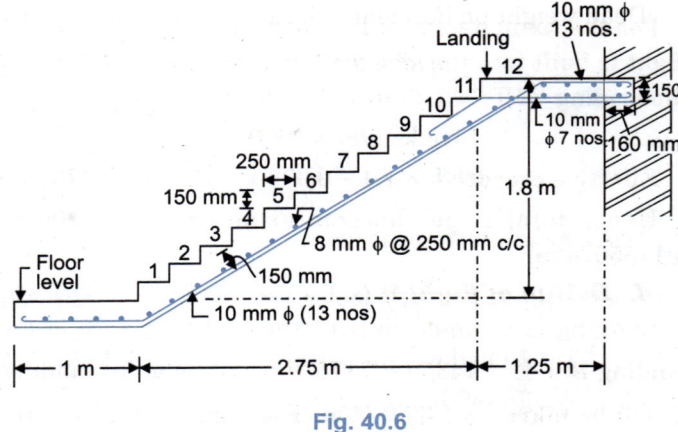

Fig. 40.6

## 40.4. DESIGN OF STAIR WITH QUARTER SPACE LANDING

**Design: Example 40.3.** *Fig. 40.7 shows the general arrangement of stair case for an office building. The tread is 300 mm and rise is 150 mm. The stairs is built-in the side wall along the flights for a distance of 120 mm. Design the stair case for a live load of 3000 N/mm², taking the span in the direction of the flight. Use M 20 concrete and F 415 steel.*

**Solution:**

1. **Design constants:** For Fe 415–M 20 combination,

$$\frac{x_{u,\max}}{d} = 0.479 \text{ and } R_u = 2.761 \text{ as in example 40.1.}$$

2. **Effective span:** Assume 200 mm bearing of the landing in the wall. Effective span of flight $AB = 3 + 1.4 + 0.1 = 4.5$ m.

Effective span of flight
$$BC = 0.1 + 1.4 + 1.5 + 1.4 + 0.1 = 4.5 \text{ m.}$$

Thus, effective span of both the flights is equal. Hence any one flight (say flight $BC$) may be designed and the same design may be adopted for the other flight.

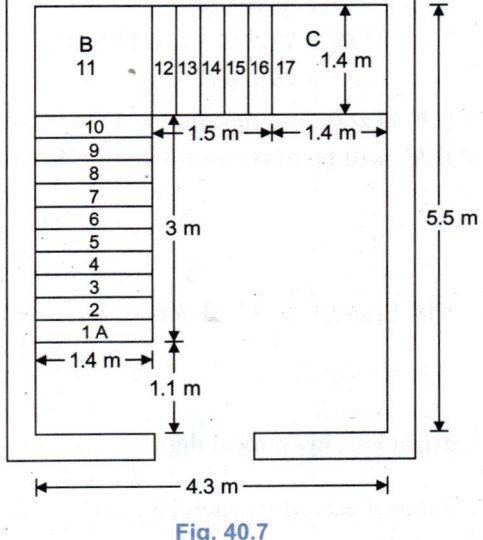

Fig. 40.7

3. **Loading on each flight:** Let the thickness of the waist slab be 200 mm

∴   Weight $w'$ on slope $= 0.2 \times 1 \times 1.4 \times 25000$ N/m

Weight $w_1$ per horizontal metre run $= 0.2 \times 1 \times 1.4 \times 25000 \dfrac{\sqrt{(150)^2 + (300)^2}}{300} \approx 7825$ N/m

Weight of each step $= 1.4 \times \dfrac{0.15}{2} \times 0.3 \times 25000 \approx 788$ N

∴   Weight $w_2$ of steps per horizontal metre run $= 788\,(1000/300) \approx 2625$ N

Alternatively, $w_2 = \dfrac{0.15}{2} \times 1 \times 1.4 \times 25000 = 2625$ N

∴   Total dead weight/m run $= 7825 + 2625\qquad = 10450$ N
Weight of finishing etc.$\qquad\qquad\qquad\quad =\ \ \ 150$ N

$\qquad\qquad\qquad$ Total $w\quad = 10600$ N/m

For the computation of live load, consider Fig. 40.8. Since the flight is built into the side wall by a distance 120 mm > 110 mm, the loading width = 1.4 − 0.15 = 1.25 m.

∴ Live load/m = 3000 × 1.25 = 3750 N

Effective breadth $b$ = 1.4 + 0.075 = 1.475 m = 1475 mm

Hence total $w$ per horizontal metre run = 10600 + 3750 = 14350 N/m

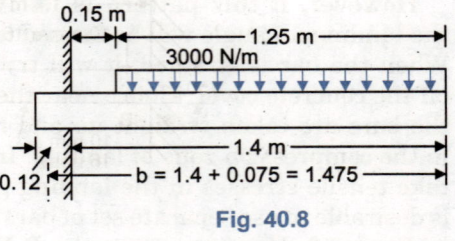

Fig. 40.8

### 4. Design of flight B.C.

Landing is common to both the flights. Hence $w$ for landing B = $\frac{1}{2}$ × 14350 = 7175 N/m, while $w$ for landing C will be taken as 14350 N/m. The loading, B.M.D. and S.F.D. are shown in Fig 40.9.

Reaction

$$R_c = \frac{1}{4.5}\left(\frac{7175 \times 1.5 \times 1.5}{2} + 14350 \times 3 \times 3\right)$$

$$\approx 30494 \text{ N}$$

$$R_B = 7175 \times 1.5 + 14350 \times 3 - 30494$$

$$\approx 23318 \text{ N}$$

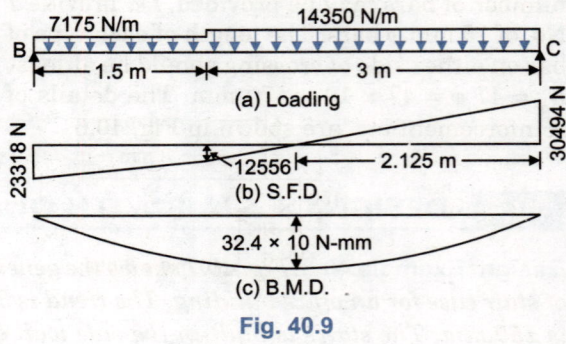

Fig. 40.9

S.F. is zero at a distance = 30494/14350 = 2.125 m from C

B.M. will be maximum where S.F. in zero

∴ $$M_{max} = \left[30494 \times 2.125 - \frac{14350 (2.125)^2}{2}\right] 1000 = 32.4 \times 10^6 \text{ N-mm}$$

∴ $$M_{u, max} = 1.5 \times 32.4 \times 10^6 = 48.6 \times 10^6 \text{ N-mm}$$

The breadth $b$ of slab for design = 1475 mm

∴ $$d = \sqrt{\frac{M_u}{R_u b}} = \sqrt{\frac{48.6 \times 10^6}{2.761 \times 1475}} = 109.3 \text{ mm}.$$

However, keep total depth = 150 mm. Using 10 mm φ bars and a nominal cover of 20 mm,

$$d = 150 - 20 - 5 = 125 \text{ mm}.$$

Since $d$ actually provided is more than that required from bending, we have an under-reinforced section for which

$$A_{st} = \frac{0.5 \times 20}{415}\left[1 - \sqrt{1 - \frac{4.6 \times 48.6 \times 10^6}{20 \times 1475 (125)^2}}\right] \times 1475 \times 125 \approx 1254.6 \text{ mm}^2$$

∴ Number of 10 mm φ bars required in a width of 1475 mm = 1254.6/78.54 = 15.97.

Hence provide 16 bars of 10 mm φ. spacing $s$ = 1475/16 ≃ 92 mm.

Distribution reinforcement $A_{sd}$ = 1.2 × 150 = 180 mm²

Hence spacing of 8 mm φ bars = $\frac{1000 \times 50.3}{180} \approx 279$ mm.

Hence provide 8 mm φ bars @ 250 mm c/c. The same reinforcement may be provided for both the flights. At the landing, provide reinforcement both at top as well as at bottom. The details of reinforcement etc. are shown in Fig. 40.10.

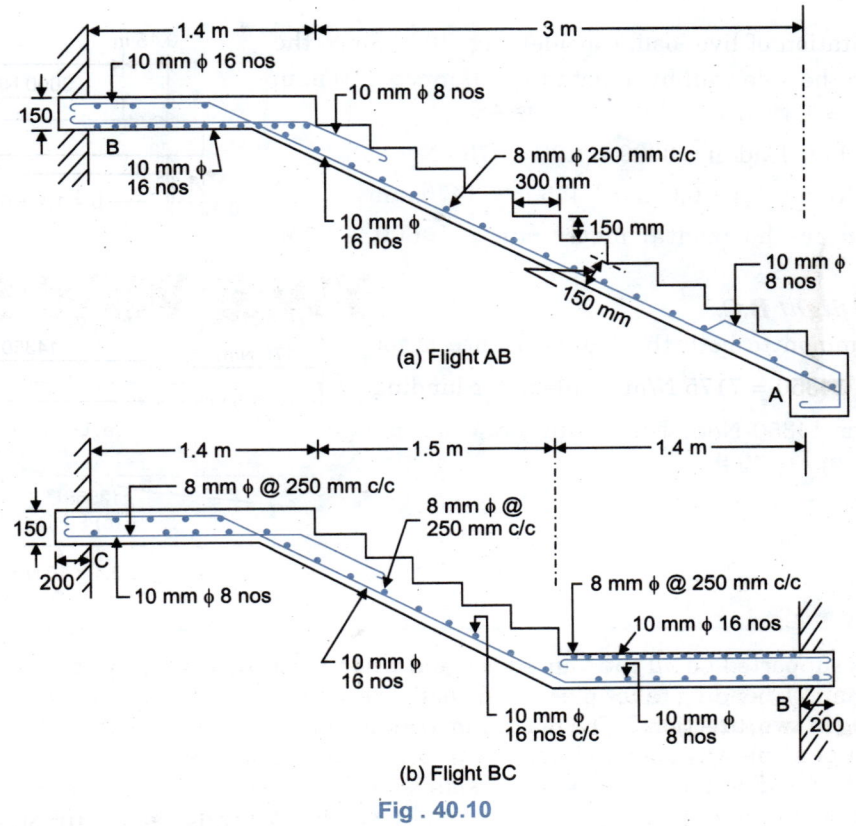

Fig. 40.10

## PROBLEMS

1. Design the stairs for a public building, supported on wall on one side and stringer beam on the other side. The horizontal span of stairs is 1.4 m. The risers are 120 mm and tread are 300 mm. Use M 20 mix and Fe 415 steel.

2. Design a suitable dog-legged stair in a public building, to be located in a staircase 6 metre long, 3.2 m wide and 3.7 m high, with a door of 1.1 m wide in each of the longitudinal walls. The doors face each other and are located with their centres at a distance of 0.9 metres from the respective corners of the staircase. Use M 20 mix and Fe 415 steel.

3. A two storeyed building is to have a R.C. staircase from ground floor to first floor roof. The size of the staircase is 4.3 m × 4.3 m and there is one door opening in one wall and a window opening on the opposite wall. Design the staircase, giving the details of formation, R.C. slab arrangement of building, risers and treads with their top finishing with suitable sketch. The width of stair is 1.2 m and height of each storey is 3.4 m. Use M 20 mix and Fe 415 steel.

# TWO WAY SLABS

## 41.1. INTRODUCTION

When a slab is supported on all the four edges, and when the ratio of long span ($l_y$) to short span ($l_x$) is small, (say less than 2), bending takes place along both the spans. Such a slab is known as a *'two way slab'* or a 'slab spanning in two directions'. The maximum bending moment and deflection for such a slab is much smaller than that of a one-way slab and hence a thinner slab is required. However, reinforcement has to be provided in both the directions. When such a slab is loaded, the corners get lifted up. If the corners are held down, by fixidity at the wall support etc., the bending moment and deflection are further reduced, thus requiring still thinner slab. In that case, special torsional reinforcement at the corners has to be provided to check the cracking of corners.

Two way slabs can be divided into the following categories depending on support conditions.
1. Slabs simply supported on the four edges, with corners not held down, and carrying uniformly distributed load.
2. Slabs simply supported on the four edges, with corners held down and carrying uniformly distributed load.
3. Slabs with edges fixed or continuous and carrying uniformly distributed load.

The analysis of a two-way slab may be done by the following methods.
1. Grashoff-Rankine method.   2. Pigeaud's method.
3. Marcus's method,   and   4. I.S. Code method.

The two way slab, simply supported on the four edges with corners not held down can be analysed either by the first method or by the fourth method. The other cases of two-way slabs can be analysed by the second, third or fourth method. For a detailed discussion, reference may be made to chapter 10. We will discuss here only the fourth method (*i.e.* I.S. Code Method).

## 41.2. SIMPLY SUPPORTED SLAB WITH CORNERS FREE TO LIFT *(I.S. CODE METHOD)*

When simply supported slabs do not have adequate provision to resist torsion at corner and to prevent the corners from lifting, the maximum moments for unit width are given by the following equation : (As per clause D.2.1 of IS: 456-2000)

$$M_x = \alpha_x w l_x^2 \qquad \qquad \text{...[41.1 (a)]}$$

$$M_y = \alpha_y w l_x^2 \qquad \qquad \text{...[41.1 (b)]}$$

where, $M_x$, $M_y$, $l_x$, $l_y$ are the same as those in § 41.3, and $\alpha_x$ and $\alpha_y$ are moment coefficients given in Table 41.1. As per Table no: 27 of IS: 456-2000.

**TABLE 41.1.** Bending Moment Coefficients for Slabs Spanning in Two Directions at Right Angles, Simply Supported on Four Sides (IS : 456-2000)

| $l_y/l_x$ | 1.0 | 1.1 | 1.2 | 1.3 | 1.4 | 1.5 | 1.75 | 2.0 | 2.5 | 3.0 |
|---|---|---|---|---|---|---|---|---|---|---|
| $\alpha_x$ | 0.062 | 0.074 | 0.084 | 0.093 | 0.099 | 0.104 | 0.113 | 0.118 | 0.122 | 0.124 |
| $\alpha_y$ | 0.062 | 0.061 | 0.059 | 0.055 | 0.051 | 0.046 | 0.037 | 0.029 | 0.020 | 0.014 |

At least 50 percent of the tension reinforcement provided at mid-span should extend to the supports. The remaining 50 percent should extend to within $0.1\,l_x$ or $0.1\,l_y$ of the support, as appropriate.

### Details of Reinforcement

Fig. 41.1 shows the details of reinforcement. Half the positive moment reinforcement $(A_{st})_x$ is bent upwards, at a distance $a_x \leq 0.15\,l_x$ from the centre of the support while the remaining half is continued upto the end. The bent up bars or the negative reinforcement (top reinforcement) $(A_{st})_x'$ should be available for distance $a_x' \geq 0.1\,l_x$ from the centre of the support.

Similarly, in the span $l_y$, half the positive moment reinforcement $(A_{st})_y$ is bent up, at a distance $a_y \leq 0.15\,l_y$ from the centre of the support. The remaining half is continued upto the end. The bent bars or the negative reinforcement $(A_{st})_y'$ is available at the top of the slab for a distance $a_y' \geq 0.1\,l_y$ from the centre of the slab. The corner reinforcement, at top and bottom of each corner is provided in the form of square mesh of size $0.2\,l_x \times 0.2\,l_x$.

### Load on supporting beams

The loads on the beams supporting solid slabs spanning in two directions at right angles and supporting uniformly distributed loads, may be assumed as marked in Fig. 41.2.

Thus total reaction on short edge AB (or CD) will be equal to $\frac{1}{2}\,w\,l_x\,\frac{l_x}{2} = \frac{w\,l_x^2}{4}$. The average reaction per unit width along short edge will therefore be $w\,l_x/4$. However the maximum reaction (hence max. S.F.) per unit width along AB will occur near the centre of AB and its value may be taken as $w\,l_x/3$ for all practical purposes.

Similarly, S.F. along the edge AD or AC, (i.e. ends of short span of slab) per unit width may be taken as $w\,l_x\,\frac{r}{2+r}$, subject to a maximum of $0.5\,w\,l_x$ when $r$ exceeds 2.

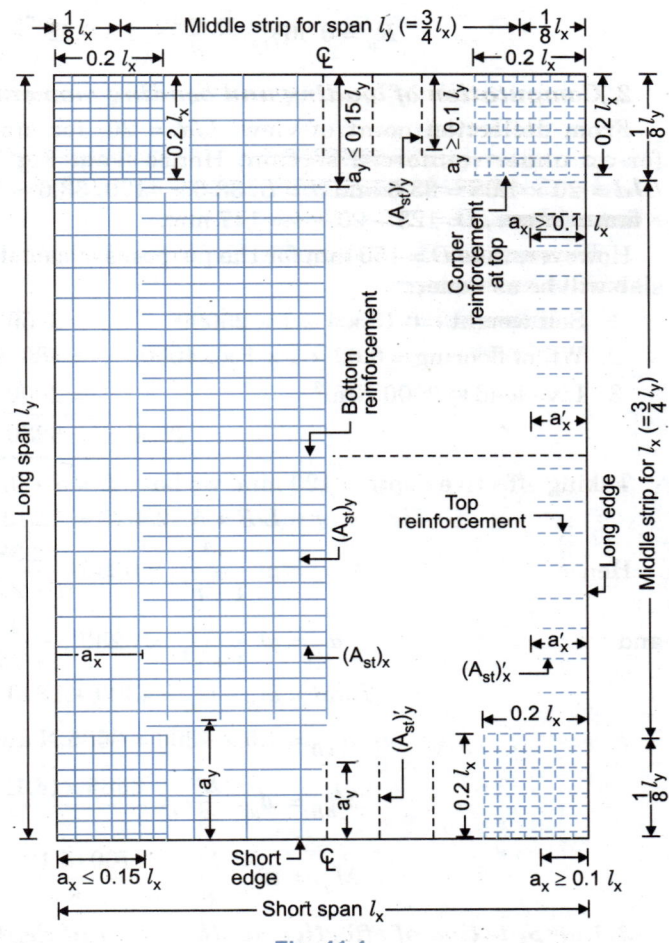

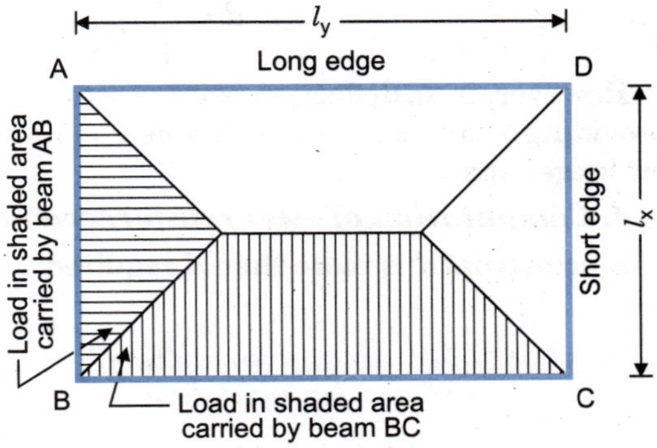

Fig. 41.1

Fig. 41.2. Loads carried by supporting beams.

**Design: Example 41.1.** *Design a R.C. slab for room measuring 4 m × 5 m from inside. The slab carries a live load of 2000 N/m² and is finished with 20 mm thick granolithic topping. Use M 20 concrete and Fe 415 steel. The slab is simply supported at all the four edges, with corners free to lift.*

**Solution:**

### 1. Design constants and limiting depth of N.A.

Given : $f_{ck}$ = 20 N/mm² for M 20 concrete and $f_y$ = 415 N/mm² for Fe 415 steel.

For Fe 415, $\dfrac{x_{u,\max}}{d} = \dfrac{700}{1100 + 0.87 \times 415} = 0.479$

$$R_u = 0.36 f_{ck} \dfrac{x_{u,\max}}{d}\left(1 - 0.42\dfrac{x_{u,\max}}{d}\right) = 0.36 \times 20 \times 0.479\,(1 - 0.42 \times 0.479) = 2.755$$

### 2. Computation of loading and bending moment

From deflection point of view, $L/d$ = 20, for simply supported slab. Let us assume $p_t$ = 0.2% for an under-reinforced section. Hence from Fig 7.1, we get modification factor ≈ 1.68. Hence $L/d = 20 \times 1.68 = 33.6$ and $d = L/33.6 ≈ 4120/33.6 ≈ 123$ mm. Keeping 20 mm nominal cover and using 8 mm φ bars, $D = 123 + 20 + 4 = 147$ mm.

However, take $D = 150$ mm for the purposes of calculating self weight. The load per square metre of the slab will be as under:

1. Self weight = 0.15 × 1 × 1 × 25000    = 3750 N
2. Wt. of flooring = 0.02 × 1 × 1 × 24000   = 480 N
3. Live load @ 2000 N/m²             = 2000 N
                           Total  = 6230 N

Taking effective depth ≈ 120 mm, we have $L = 5 + 0.12 = 5.12$ m and $B = 4 + 0.12 = 4.12$ m

$$r = L/B = 5.12/4.12 = 1.2427; \quad r^4 = 2.385$$

Hence $w_B = w\,\dfrac{r^4}{1+r^4} = 6230\,\dfrac{2.385}{1 + 2.385} ≈ 4390$ N

and $w_L = w\,\dfrac{1}{1+r^4} = 6230\,\dfrac{1}{1 + 2.385} ≈ 1840$ N

Total = $w_B + w_L$ = 4390 + 1840 = 6230 = w

∴ $w_{uB} = 1.5 \times 4390 = 6585$ N and $w_{uL} = 1.5 \times 1840 = 2760$

$$M_{uB} = w_{uB}\dfrac{B^2}{8} = \dfrac{6585 \times (4.12)^2}{8} = 13972 \text{ N-m} = 13.972 \times 10^6 \text{ N-mm}$$

$$M_{uL} = w_{uL}\dfrac{L^2}{8} = \dfrac{2760\,(5.12)^2}{8} = 9044 \text{ N-mm} = 9.044 \times 10^6 \text{ N-mm}$$

### 3. Computation of effective depth and total depth

$$d = \sqrt{\dfrac{M_{u,\max}}{R_u\,b}} = \sqrt{\dfrac{13.972 \times 10^6}{2.762 \times 1000}} = 71.2 \text{ mm}$$

However, from deflection point of view, provide $D = 150$ mm so that keeping nominal cover = 20 mm and providing 8 mm dia. bars available $d = 150 - 20 - 4 = 126$ mm for shorter span and $d = 126 - 8 = 118$ mm for longer span.

### 4. Computation of steel reinforcement

Since actual $d$ is more than $d$ required from B.M., we have an under-reinforced section for which

$$A_{st,B} = \dfrac{0.5 \times 20}{415}\left[1 - \sqrt{1 - \dfrac{4.6 \times 13.972 \times 10^6}{20 \times 1000\,(126)^2}}\right] 1000 \times (126) = 324.6 \text{ mm}^2$$

∴ Spacing of 8 mm dia. bars, $s_B = \dfrac{1000\,A_\varphi}{A_{st,B}} = \dfrac{1000 \times 50.3}{324.6} = 154.9$ mm

Hence provide 8 mm φ bars @ 150 mm c/c along the short span. Bend-up alternate bars at $B/7 = 4.12/7 = 0.59$ m from the centre of each support.

For long span, $A_{st\,L} = \dfrac{0.5 \times 20}{415}\left[1 - \sqrt{1 - \dfrac{4.6 \times 9.044 \times 10^6}{20 \times 1000\,(118)^2}}\right] 1000 \times 118 = 221\ \text{mm}^2$

Spacing $s_L = \dfrac{1000\,A_\varphi}{A_{st,L}} = \dfrac{1000 \times 50.3}{221} = 227.6\ \text{mm}$

However, provide 8 mmφ bars @ 220 mm c/c. Bend-up alternate bars at $L/7 = 5.12/7 \approx 0.73$ m from the centre of each supports.

### 5. Check for shear

S.F. on the short edge per unit length is given by

$$V_{uB} = \dfrac{1}{3}\,w_u\,B = \dfrac{1}{3} \times 1.5 \times 6230 \times 4.12 \approx 12834\ \text{N}$$

Similarly, $V_{uL} = w_u\,B\,\dfrac{r}{2+r} = 1.5 \times 6230 \times 4.12 \times \dfrac{1.2427}{2 + 1.2427} \approx 14755\ \text{N}$

∴ $\tau_{vB} = \dfrac{V_{uB}}{b d_L} = \dfrac{12834}{1000 \times 118} = 0.109\ \text{N/mm}^2$  and  $\tau_{vL} = \dfrac{V_{uL}}{b d_B} = \dfrac{14755}{1000 \times 126} = 0.117\ \text{N/mm}^2$

Note carefully that the effective depth of slab at short edge is 118 mm while that at long edge is 126 mm. Permissible shear stress $\tau_c$ (Table 36.1 and 36.3) is much more than these values. Hence safe.

### 6. Check for development length

Let the slab have a bearing of 200 mm at the ends. Let us check for development length at the end of short span. Available $A_{st1} = \dfrac{1000 \times 50.3}{300} = 167.7\ \text{mm}^2$

$x_u = \dfrac{0.87\,f_y\,A_{st1}}{0.36\,f_{ck}\,b} = \dfrac{0.87 \times 415\,(167.7)}{0.36 \times 20 \times 1000} \approx 8.41\ \text{mm}$

$M_{1u} = 0.87\,f_y\,A_{st1}\,(d - 0.416\,x_u) = 0.87 \times 415 \times 167.7\,(126 - 0.416 \times 8.41) = 7.417 \times 10^6\ \text{N-m}$

$V = V_{uL} = 14755\ \text{N}$

$L_0 = \dfrac{l_s}{2} - x' = \dfrac{200}{2} - 20 = 80\ \text{mm}$ (with no end hooks)

$L_d = 47\,\varphi = 47 \times 8 = 376\ \text{mm}$

development length

$L_d \le \dfrac{M_1}{V} + l_0$

$1.3\,\dfrac{M_{1u}}{V_u} + L_0$

$= 1.3 \times \dfrac{7.417 \times 10^6}{14755} + 80 \approx 733\ \text{mm} > L_d$

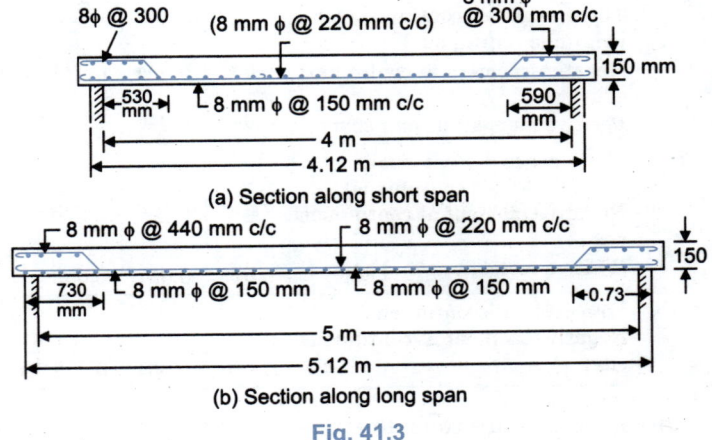

(a) Section along short span

(b) Section along long span

Fig. 41.3

Hence the Code requirements are satisfied. Similar check can be applied at the end of the long span. The sections of the slab along the short and long spans are shown in Fig. 41.3.

## 41.3. RESTRAINED SLABS (I.S. CODE METHOD)

When the corners of a slab are prevented from lifting, the slab may be designed as specified in $a - 1$ to $a - 11$ below.

**(a–1).** The maximum bending moments per unit width in a slab are given by the following equation:

$M_x = \alpha_x\,w\,l_x^2$

$M_y = \alpha_y \cdot w \cdot l_x^2$, where $\alpha_x$ and $\alpha_y$ are coefficients given in Table 41.2. $w$ = total load per unit area

$M_x, M_y$ = moments on strips of unit width spanning $l_x$ and $l_y$ respectively, and

$l_x$ and $l_y$ = lengths of shorter span and long span respectively. As per table no: 26 of IS: 456-2000.

**TABLE 41.2.** Bending moment coefficients for rectangular panels supported on four sides with provision for torsion at corners (IS: 456-2000) As per table no: 26 of IS: 456-2000.

| Case No. | Type of panel and moment considered | Short span coefficient $\alpha_x$ for values of $l_y/l_x$ | | | | | | | | Long span coefficient $\alpha_y$ for all values of $l_y/l_x$ |
|---|---|---|---|---|---|---|---|---|---|---|
| | | 1.0 | 1.1 | 1.2 | 1.3 | 1.4 | 1.5 | 1.75 | 2.0 | |
| 1. | Interior Panels | | | | | | | | | |
| | Negative moment at continuous edge | 0.032 | 0.037 | 0.043 | 0.047 | 0.051 | 0.053 | 0.060 | 0.065 | 0.032 |
| | Positive moment at mid-span | 0.024 | 0.028 | 0.032 | 0.036 | 0.039 | 0.041 | 0.045 | 0.049 | 0.024 |
| 2. | One short edge discontinuous | | | | | | | | | |
| | Negative moment at continuous edge | 0.037 | 0.043 | 0.048 | 0.051 | 0.055 | 0.057 | 0.064 | 0.068 | 0.037 |
| | Positive moment at mid-span | 0.028 | 0.032 | 0.036 | 0.039 | 0.041 | 0.044 | 0.048 | 0.052 | 0.028 |
| 3. | One long edge discontinuous | | | | | | | | | |
| | Negative moment at continuous edge | 0.037 | 0.044 | 0.052 | 0.057 | 0.063 | 0.067 | 0.077 | 0.085 | 0.037 |
| | Positive moment at mid-span | 0.028 | 0.033 | 0.039 | 0.044 | 0.047 | 0.051 | 0.059 | 0.065 | 0.028 |
| 4. | Two adjacent edges discontinuous | | | | | | | | | |
| | Negative moment at continuous edge | 0.047 | 0.053 | 0.060 | 0.065 | 0.071 | 0.075 | 0.084 | 0.091 | 0.047 |
| | Positive moment at mid-span | 0.035 | 0.040 | 0.049 | 0.049 | 0.053 | 0.056 | 0.064 | 0.069 | 0.035 |
| 5. | Two short edges discontinuous | | | | | | | | | |
| | Negative moment at continuous edge | 0.045 | 0.049 | 0.052 | 0.056 | 0.059 | 0.060 | 0.065 | 0.069 | – |
| | Positive moment at mid-span | 0.035 | 0.037 | 0.040 | 0.043 | 0.044 | 0.045 | 0.049 | 0.052 | 0.035 |
| 6. | Two long edges continuous | | | | | | | | | |
| | Negative moment at continuous edge | – | – | – | – | – | – | – | – | 0.045 |
| | Positive moment at mid-span | 0.035 | 0.043 | 0.051 | 0.075 | 0.063 | 0.068 | 0.080 | 0.088 | 0.035 |
| 7. | Three edges discontinuous (one long edge continuous) | | | | | | | | | |
| | Negative moment at continuous edge | 0.057 | 0.064 | 0.071 | 0.076 | 0.080 | 0.084 | 0.091 | 0.097 | – |
| | Positive moment at mid-span | 0.043 | 0.048 | 0.053 | 0.057 | 0.060 | 0.064 | 0.069 | 0.073 | 0.043 |
| 8. | Three edges discontinuous (one short edge continuous) | | | | | | | | | |
| | Negative moment at continuous edge | – | – | – | – | – | – | – | – | 0.057 |
| | Positive moment at mid-span | 0.043 | 0.051 | 0.059 | 0.065 | 0.071 | 0.076 | 0.087 | 0.096 | 0.043 |
| 9. | Four edges discontinuous | | | | | | | | | |
| | Negative moment at continuous edge | 0.056 | 0.064 | 0.072 | 0.079 | 0.085 | 0.089 | 0.100 | 0.107 | 0.056 |

**(a–2).** Slabs are considered as divided in each direction into middle strips and edge strips as shown Fig. 41.4, the middle strips being three quarters (3/4) of the width and each edge strip one-eighth (1/8) of the width of slab.

**(a–3).** The maximum moments calculated as in (a–1) apply on to the middle strips only and no redistribution shall be made.

**(a–4).** Tension reinforcement provided at mid-span in the middle strip shall extend in the lower part of the slab to within $0.25\,l$ of a continuous edge, or $0.15\,l$ of a discontinuous edge.

**(a–5).** Over the continuous edge of the middle strip, the tension reinforcement shall extend in the upper part of the slab a distance of $0.15\,l$ from the supports, and at least 50 percent shall extend a distance of $0.3\,l$.

**(a–6).** At a discontinuous edge, negative moments may arise. They depend on the degree of fixidity at the edge of the slab but, in general, tension reinforcement equal to 50 percent of that provided at mid span extending 0.1 $l$ into the span will be sufficient.

**(a–7).** Reinforcement in edge strip, parallel to that edge shall comply with minimum and the requirements for torsion given in (a – 8), (a – 9) and (a – 10).

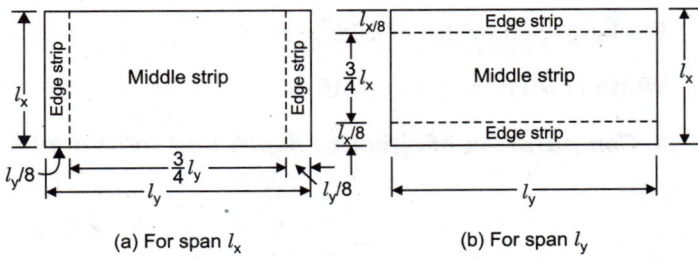

Fig. 41.4. Division of slab into middle and edge strips.

**(a–8).** Torsion reinforcement shall be provided at any corner when the slab is simply supported on both edges meeting at that corner. It shall consist of top and bottom reinforcement, each with layers of bars placed parallel to the sides of the slab and extending from the edges to a minimum distance of one fifth of the shorter span. The area of reinforcement in each of these four layers shall be three quarters of the area required for the maximum mid-span moment in the slab.

**(a–9).** Torsion reinforcement equal to half that described in (a – 8) shall be provided at a corner contained by edges over only one which the slab is continuous.

**(a–10).** Torsion reinforcement need not be provided at any corner contained by edges over both of which slab is continuous.

**(a–11).** Where $l_y/l_x$ is greater than 2, the slabs shall be designed as one way slab.

**Example 41.2.** Design a R.C. slab for a room measuring 5 m × 6 m size. The slab is simply supported on all the four edges, with corners held down and carries a superimposed load of 3000 N/m², inclusive of floor finishes etc. Use M 20 mix, Fe 415 steel and IS Code method.

**Solution:**

### 1. Design constants and limiting depth of N.A.

Give : $f_{ck} = 20$ N/mm² for M 20 concrete and $f_y = 415$ N/mm² for Fe 415 steel.

For Fe 415, $\dfrac{x_{u,\max}}{d} = \dfrac{700}{1100 + 0.87 \times 415} = 0.479$

$R_u = 0.36 f_{ck} \dfrac{x_{u,\max}}{d}\left(1 - 0.416 \dfrac{x_{u,\max}}{d}\right) = 0.36 \times 20 \times 0.479 (1 - 0.416 \times 0.479) = 2.761$

### 2. Computation of loading and bending moment

From deflection point of view, $L/d = 20$ for simply supported slab. Let us assume $p_t = 0.2\%$ for an under-reinforced section. Hence from Fig. 7.1, we get modification factor (from Fig 4 of IS: 456-2000.) ≈ 1.68. Hence $L/d = 20 \times 1.68 = 33.6$ and $d = L/33.6 = 5000/33.6 = 148.8$ mm. Providing 20 mm nominal cover and 8 mm φ bars, $D = 148.8 + 20 + 4 = 172.8$ mm. Hence assume an overall depth of 180 mm for the purposes of computing dead weight.

(i) Weight of slab per m² = 0.18 × 1 × 1 × 25000     = 4500
(ii) Super-imposed load @ 3000 N/m²     = 3000
                                  Total $w$  = 7500 N/m²

Hence $w_u = 1.5 w = 1.5 \times 7500 = 11250$ N/m²

Taking an effective depth of 150 mm, we have effective $l_y = 6 + 0.15 = 6.15$ m and effective $l_x = 5 + 0.15 = 5.15$ m. Hence $r = l_y/l_x = 6.15/5.15 \approx 1.2$

This is case 9 of Table 41.2, from which $\alpha_x = 0.072$ and $\alpha_y = 0.056$ from table no: 26 of IS: 456-2000.

∴  $M_{ux} = \alpha_x w_u l_x^2 = 0.072 \times 11250 (5.15)^2 = 21483$ N-m = $21.483 \times 10^6$ N-mm
   $M_{uy} = \alpha_y w_u l_x^2 = 0.056 \times 11250 (5.15)^2 = 16709$ N-m = $16.709 \times 10^6$ N-mm.

For short span width of middle strip = $\dfrac{3}{4} l_y = \dfrac{3}{4} \times 6.15 \approx 4.61$ m

Width of edge strip = $\dfrac{1}{2}(6.15 - 4.61) = 0.77$ m.

For long span, width of middle strip = $\frac{3}{4} l_x = \frac{3}{4} \times 5.15 \approx 3.87$

Width of edge strip = $\frac{1}{2}(5.15 - 3.87) = 0.64$ m.

### 3. Computation of effective depth and total depth

$$d = \sqrt{\frac{M_{ux}}{R_u b}} = \sqrt{\frac{21.483 \times 10^6}{2.761 \times 1000}} = 88.2 \text{ mm}.$$

However, from the requirements of stiffness (deflection), keep $D = 180$. Keeping nominal cover of 20 mm and providing 8 mm φ bars, available $d$ from short span = $180 - 20 - 4 = 156$ mm and that for long span, $d = 156 - 8 = 148$ mm.

### 4. Computation of steel reinforcement for short span

Since actually provided $d$ is more than that required from bending we have an under reinforced section. Hence

$$A_{stx} = \frac{0.5 f_{ck}}{f_y}\left[1 - \sqrt{1 - \frac{4.6 M_{ux}}{f_{ck} bd^2}}\right] bd = \frac{0.5 \times 20}{415}\left[1 - \sqrt{1 - \frac{4.6 \times 21.483 \times 10^6}{20 \times 1000 (156)^2}}\right] 1000 \times 156 = 403.2 \text{ mm}^2$$

Spacing of 8 mm φ bars, $s_x = \dfrac{1000 \times 50.3}{403.2} = 124.7$ mm.

However, use 8 mm φ bars @ 120 mm c/c for the middle strip of width 4.61 m. Bend half the bars up at a distance = $0.15 \, l_x = 0.15 \times 5150 \approx 770$ mm from the centre of the support, or at a distance of $770 + 80 = 850$ mm from the edge of the slab (assuming a bearing of 160 mm on the wall). Available length of bars at the top = $770 - (160 - 20) = 630$ mm from the centre of the support, assuming bending of bar at 45°. The length is more than $0.1 \, l_x$ ($0.1 \times 5150 = 515$ mm) required by the Code. Hence length of top bars from the edge of slab = $630 + 80 = 710$ mm.

Edge strip is of length 0.77 m. The reinforcement in the edge strips is = $1.2 D = 1.2 \times 180 = 216$ mm². Hence spacing of 8 mm φ bars, $s = 1000 \times 50.3/216 \approx 232$ mm c/c. However, it is more convenient to keep the spacing as the simple fraction of the edge strip or simple multiple of spacing for the middle strip. Hence keep $s = 770/4 = 192.5$ mm c/c.

### 5. Computation of reinforcement for long span

$$A_{sty} = \frac{0.5 \times 20}{415}\left[1 - \sqrt{1 - \frac{4.6 \times 16.709 \times 10^6}{20 \times 1000 (148)^2}}\right] 1000 \times 148 = 327.9 \text{ mm}^2$$

using 8 mm φ bars, $s = \dfrac{1000 \times 50.3}{327.9} = 153.4$ mm

Hence provide 8 mm φ bars @ 150 mm c/c in the middle strip of width 3.87 m. Bend half the bars of the middle strip up at a distance of $0.15 \, l_y = 0.15 \times 6150 \approx 920$ mm from the centre of the support or $920 + 80 = 1000$ mm from the edge of the slab. Available length of bars of top = $920 - (150 - 20) = 790$ mm from the centre of the support or $790 + 80 = 870$ mm from the edge of the slab.

For the edge strip of width 0.64 m, $A_{st} = 1.2 D = 1.2 \times 180 = 216$ mm² giving rise to a spacing $s = 1000 \times 50.3/216 \approx 232$ mm c/c. Hence provide 8 mm φ bar @ $640/3 \approx 213$ mm c/c.

### 6. Check for shear and development length in short span.

$$\text{S.F. at long edges} = w_u l_x \frac{r}{2 + r} = 11250 \times 5.15 \times \frac{1.2}{2 + 1.2} \approx 21727 \text{ N}$$

$$\therefore \text{ Nominal shear stress at long edges} = \frac{21727}{1000 \times 156} = 0.139 \text{ N/mm}^2$$

At the long edges, the diameter of bars should be so restricted that the following requirement is satisfied:

$$1.3 \frac{M_{u1}}{V_u} + L_0 \geq L_d$$

$A_{st}$ at supports of short span = $\dfrac{1000 \times 50.3}{240} = 209.6$ mm²

$$x_u = \frac{0.87 f_y A_{st1}}{0.36 f_{ck} b} = \frac{0.87 \times 415 \times 209.6}{0.36 \times 20 \times 1000} = 10.5 \text{ mm}$$

$$M_{1u} = 0.87 f_y A_{st1} (d - 0.416 x_u) = 0.87 \times 415 \times 209.6 (156 - 0.416 \times 10.5) = 11.475 \times 10^6 \text{ N-mm}$$

$V_u = 21727 \text{ N}$ ; $L_d = 47 \varphi = 47 \times 8 = 376 \text{ mm}$

Assume that support width $l_s = 160$ mm and a side cover of 20 mm.

Providing no hooks, $L_0 = \dfrac{l_s}{2} - x' = \dfrac{160}{2} - 20 = 60$ mm. Hence

$$1.3 \frac{M_{1u}}{V_u} + L_0 = 1.3 \times \frac{11.475 \times 10^6}{21727} + 60 \approx 747 \text{ mm} > L_d.$$

Hence Code requirements are satisfied. Also the Code requires that the positive reinforcement should extend into the support atleast by $L_d/3$. Hence minimum support width $= \dfrac{L_d}{3} + x' = \dfrac{360}{3} + 20 = 140$ mm. In our case, the support width is 160 mm.

### 7. Check for shear and development length in long span

S.F. at short edges $= \dfrac{1}{3} w_u l_x = \dfrac{1}{3} \times 11250 \times 5.15 \approx 19313$ N/m

$\therefore$ Nominal shear stress $= \dfrac{19313}{1000 \times 148} = 0.13$ N/mm² (safe)

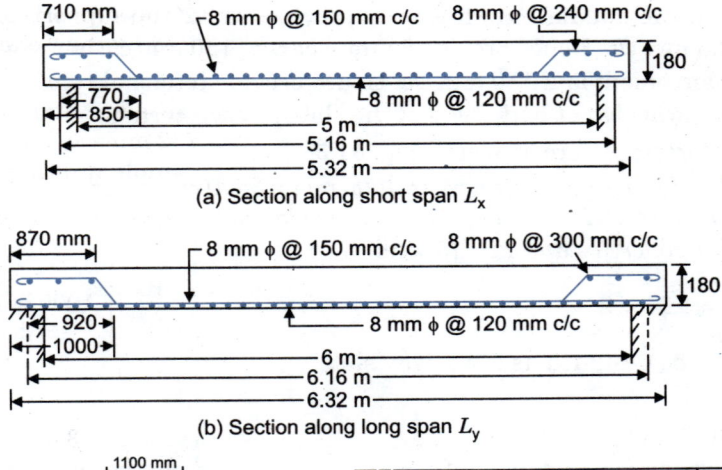

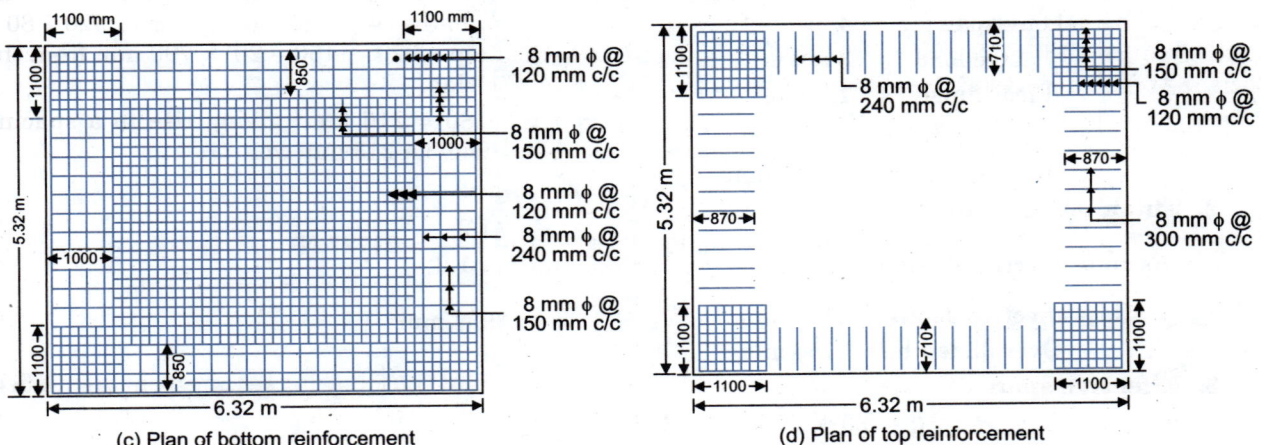

Fig. 41.5

$$A_{st} = \frac{1000 \times 50.3}{300} = 167.7 \text{ mm}^2$$

$$x_u = \frac{0.87 \times 415 \times 167.7}{0.36 \times 20 \times 1000} = 8.4 \text{ mm}$$

$$M_{1u} = 0.87 \times 415 \times 167.7 \,(148 - 0.416 \times 8.4)$$
$$= 8.749 \times 10^6 \text{ N-mm}$$

$$V_u = 19313 \text{ N}$$

$$L_0 = \frac{160}{2} - 20 = 60 \text{ mm}$$

$$L_d = 47 \times 8 = 376 \text{ mm}$$

$$\therefore \quad 1.3 \frac{M_{1u}}{V_u} + L_0 = 1.3 \frac{8.749 \times 10^6}{19313} + 60 \approx 649 \text{ mm} > L_d$$

Hence Code requirements are satisfied.

### 8. Torsional reinforcement at corners

Size of torsional mesh = $l_x/5 = 5.15/5 = 1.03$ m from the centre of support or $1.03 + 0.08 \approx 1.10$ m from the edge of the slab. Area of torsional reinforcement = $\frac{3}{4} A_{stx} = \frac{3}{4} \times 403.2 = 302.4$ mm². Spacing of 8 mm φ bar is

$$s = \frac{1000 \times 50.3}{302.4} \approx 166 \text{ mm. (say) } 160 \text{ mm.}$$

However, it is preferable to use the same spacing as provided for main reinforcement. In the short span, main reinforcement in the middle strip has been provided @ 120 mm c/c while for the edge strip, it has been provided @ 192.5 mm c/c. Hence provide 8 mm φ bars @ 120 mm c/c in the short span direction. In the long span, main reinforcement is @ 150 mm c/c. Hence provide torsional reinforcement @ 150 mm c/c. This reinforcement is to be provided at both faces of the slab, at each corner.

The details of reinforcement are shown in Fig. 41.5.

**Example 41.3.** *Design a two way slab for a room 4000 mm × 3500 mm clear in size if the super imposed load is 3 kN/m² and floor finish of 1 kN/m². The edges of the slab are simply sumported and corners are not held down. Use $M_{20}$ grade concretes and Fe 415 steel?*

**Solution:** $\dfrac{l_y}{l_x} = \dfrac{4}{3.5} = 1.14 < 2.$

Hence the slab is to be designed as a two way slab.

### 1. Data

$$\begin{aligned}
\text{Short span } l_x &= 3.5 \text{ m} \\
\text{Long span } l_y &= 4 \text{ m} \\
\text{Live load} &= 3 \text{ kN/m}^2 \\
\text{Floor finish} &= 1 \text{ kN/m}^2 \\
f_{ck} &= 20 \text{ N/mm}^2 \\
f_y &= 415 \text{ N/mm}^2.
\end{aligned}$$

### 2. Thickness of slab

Assume effective depth $d = \dfrac{\text{Span}}{28} = \dfrac{3500}{28} = 125$ mm

Adopt effective depth $d = 125$ mm

Overall depth $D = 150$ mm

### 3. Effective span

$$l_x = 3.5 + 0.125 = 3.625 \text{ m}$$
$$l_y = 4.0 + 0.125 = 4.125 \text{ m}$$
$$\frac{l_y}{l_x} = \frac{4.125}{3.625} = 1.14.$$

## 4. Loads

Per unit area of slab

$$\text{Self weight of the slab} = 0.15 \times 25 \times 3.75 \text{ kN/m}^2$$
$$\text{Live load} = 3 \text{ kN/m}^2$$
$$\text{Floor finish} = 1 \text{ kN/m}^2$$
$$\text{Total load} = 7.75 \text{ kN/m}^2$$
$$\text{Factored load } W_u = 1.5 \times 7.75 = 11.625 \text{ kN/m}^2.$$

## 5. Design moments and shear forces

The slab is simply supported on all the four sides. The corners are not held down. Hence moment coefficients are obtained from Table-27 of IS-456

$$\alpha_x = 0.074 + (0.084 - 0.074) \times \frac{4}{10} = 0.078$$

$$\alpha_y = 0.061 - (0.061 - 0.059) \times \frac{4}{10} = 0.06$$

$$M_{ux} = w_x w l_x^2 = 0.078 \times 11.625 \times 3.625^2 = 11.92 \text{ kN-m}$$
$$M_{uy} = w_y w l_x^2 = 0.06 \times 11.625 \times 3.625^2 = 9.17 \text{ kN-m}$$

$$Y_u = \frac{w_x l}{2} = \frac{11.625 \times 3.625}{2} = 21.07 \text{ kN}$$

## 6. Minimum depth required

The minimum depth required to resist bending moment

$$M_u = 0.138 f_{ck} b d^2$$
$$11.92 \times 10^6 = 0.138 \times 20 \times 1000 \times d^2$$

$$d = \sqrt{\frac{11.92 \times 10^6}{0.138 \times 20 \times 1000}} = 65.7 \text{ mm} < 125 \text{ mm}. \quad \text{Provided depth.}$$

Hence provided depth is adequate.

(Provided depth is not reduced to satisfy the stiffness requirements)

## 7. Reinforcement: Along x-direction

$$M_{ux} = 0.87 f_y A_{st} d \left(1 - \frac{f_y A_{st}}{f_{ck} b d}\right)$$

$$11.92 \times 10^6 = 0.87 \times 415 \times A_{st} \times 125 \left(1 - \frac{415 \times A_{st}}{20 \times 1000 \times 125}\right)$$

$$264.1 = A_{st} \left(1 - \frac{A_{st}}{6024.1}\right)$$

$$A_{st}^2 - 6024.1 A_{st} + 6024.1 \times 264.1 = 0 \qquad A_{st} = 278.8 \text{ mm}^2.$$

Using 8 mm φ bars, spacing of bars

$$S = \frac{a_{st}}{A_{st}} \times 1000 = \frac{\frac{\pi}{4} \times 8^2}{276.8} \times 1000 = 181.6 \text{ mm}.$$

Maximum spacing is (i) $3d = 3 \times 125 = 375$ mm (ii) 300 mm whichever is less.

Hence provide 8 mm bars at 180 mm c/c

Along y-direction

These bars will be placed above the bars in x-direction.

Hence, $\qquad d = 125 - 8 = 117$ mm

$$M_{uy} = 0.87 f_y A_{st} d \left(1 - \frac{f_y A_{st}}{f_{ck} b d}\right)$$

$$9.17 \times 10^6 = 0.87 \times 415 \times A_{st} \times 117 \left(1 - \frac{415 \times A_{st}}{20 \times 1000 \times 117}\right)$$

$$217.1 = A_{st}\left(1 - \frac{A_{st}}{5638.6}\right)$$

$$A_{st}^2 - 5638.6\,A_{st} + 5638.6 \times 217.1 = 0$$

$$A_{st} = \frac{5638.6 - \sqrt{5638.6^2 - 4 \times 5638.6 \times 217.1}}{2} = 226.2 \text{ mm}^2.$$

Using 8 mm φ bars, spacing of bars

$$S = \frac{a_{st}}{A_{st}} \times 1000 = \frac{\frac{\pi}{4} \times 8^2}{226.2} \times 1000 = 222.2 \text{ mm}.$$

Maximum spacing is  (i) $3 \times d = 3 \times 117 = 351$ mm  (ii) 300 mm whichever is less

Hence provide 8 mm bars at 220 mm c/c.

### 8. Reinforcement in Edge strip

$A_{st}$ = 0.12% of gross area

$= \dfrac{0.12}{100} \times 1000 \times 150 = 180 \text{ mm}^2.$

Using 8 mm φ bars, spacing of bars.

$$S = \frac{\frac{\pi}{4} \times 8^2}{180} \times 1000 = 279.3 \text{ mm}$$

Maximum spacing is

(i) $5d = 5 \times 125 = 625$

(ii) 450 mm whichever is less

Hence provide 8 mm bars at 275 mm c/c edge strip in both directions.

### 9. Check for Deflection

For simply supported slabs basic value of $\dfrac{l}{d}$ ratio is 20.

$$P_t = \frac{\frac{\pi}{4} \times 8^2}{180 \times 125} \times 100 = 0.223\%$$

$f_x = 0.58 \times f_y = 0.58 \times 415 = 240 \text{ N/mm}^2$

From Figure 4 of IS: 456,

modification factor = 1.6

Maximum permitted $\dfrac{l}{d}$ ratio = $1.6 \times 20 = 32$

$\dfrac{l}{d}$ provided = $\dfrac{3625}{125} = 29 < 32.$

Hence deflection control is safe, Details of Reinforcement are shown in Fig 41.6.

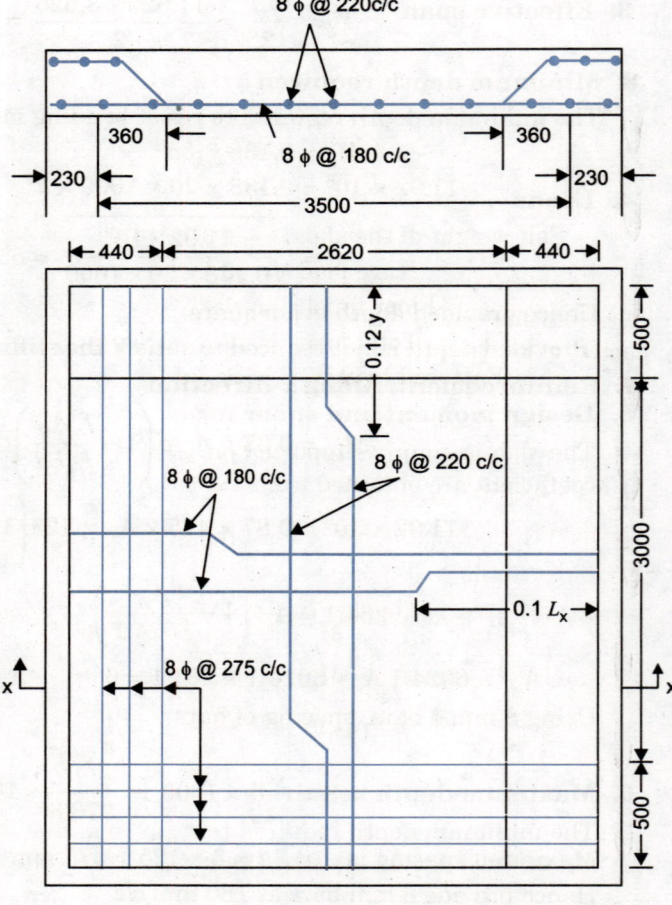

Fig. 41.6

**Example 41.4.** *The floor slab of a class room of 3 m × 5 m is discontinuous on all four sides. The concrete of the slabs are prevented from lifting 50 mm thick floor. Finish of unit weight 20 kN/m² is to be provided over the slab. Live load on the slab is 3 kN/m² width of the support is 250 mm. Design the slab using $M_{20}$ grade concrete and Fe 415 steel. Design the torsion reinforcement also.*

## TWO WAY SLABS

**Solution:** $\dfrac{l_y}{l_x} = \dfrac{5}{3} = 1.67 < 2$

Hence, the slab is to be designed as a two way slab.

### 1. Data
Short span $l_x$ = 3 m
Long span $l_y$ = 5 m
Live span = 3 kN/m²
Floor finish = 0.05 × 20 = 1 kN/m²
$f_{ck}$ = 20 N/mm²
$f_y$ = 415 N/mm².

### 2. Thickness of slab
Assume effective depth $d = \dfrac{\text{Span}}{28} = \dfrac{3000}{28} = 107$ mm

Adopt effective depth $d = 110$ mm
Overall depth $D = 135$ mm

### 3. Effective span
$l_x = 3.0 + 0.11 = 3.11$ m
$l_y = 5.0 + 0.11 = 5.11$ m.
$\dfrac{l_y}{l_x} = \dfrac{5.11}{3.11} = 1.64$

### 4. Loads
Self weight of the slab = 0.135 × 25 = 3.375 kN/m²
Live load = 3 kN/m²
Floor finish = 1 kN/m²
Total load = 7.375 kN/m²
Factored load $W_u$ = 1.5 × 7.375 = 11.06 kN/m²

### 5. Design moment and shear force
The slab is simply supported on all the four sides. The corners are not held down. Hence moment coefficient are obtained from Table: 27 of IS: 456

$\alpha_x$ = $0.089 + (0.10 - 0.089) \times \dfrac{14}{25}$ = 0.095
$\alpha_y$ = 0.056
$M_{uy}$ = $\alpha_x w l_x^2$ = 0.095 × 11.06 × 3.11² = 10.16 kN-m
$M_{uy}$ = $\alpha_y w l_x^2$ = 0.056 × 11.06 × 3.11² = 5.99 kN-m.
$V_u$ = $\dfrac{w_u l}{2} = \dfrac{11.06 \times 3.11}{2}$ = 17.2 kN.

### 6. Maximum depth required
The minimum depth required to resis bending moment
$M_u = 0.138 f_{ck} b d^2$
$10.16 \times 10^6 = 0.136 \times 20 \times 1000 \times d^2$

$d = \sqrt{\dfrac{10.16 \times 10^6}{0.138 \times 20 \times 1000}}$ = 60.67 mm < 110 mm

Hence provided depth is adequate.

**958** REINFORCED CONCRETE STRUCTURE

7. **Reinforcement**

Along x-direction (Short span) $M_{ux} = 0.87 f_y A_{st} \left(1 - \dfrac{f_y A_{st}}{f_{ck} bd}\right)$

$$10.16 \times 10^6 = 0.87 \times 415 \times A_{st} \times 110 \left(1 - \dfrac{415 A_{st}}{26 \times 1000 \times 110}\right)$$

$$255.82 = A_{st}\left(1 - \dfrac{A_{st}}{5301.2}\right)$$

$$A_{st}^2 - 5301.2 A_{st} + 5301.2 \times 255.82 = 0$$

$$A_{st} = \dfrac{5301.2 - \sqrt{5301.2^2 - 4 \times 5301.2 \times 255.82}}{2}$$

$$A_{st} = 269.5 \text{ mm}^2.$$

Using 8 mm φ bars, φ spacing of bars

$$S = \dfrac{\sigma_{st}}{A_{st}} \times 1000 = \dfrac{\frac{\pi}{4} \times 8^2}{269.5} \times 1000 = 186.5 \text{ mm}.$$

Maximum spacing   (i) $3d = 3 \times 110 = 330$ mm   (ii) 300 mm whichever is less.
Hence provide 8 mm bars at 180 mm c/c.
Along y-direction. These bars will be placed above the bar in x-direction.
Hence, $d = 110 - 8 = 102$ mm

$$M_u = 0.87 f_y A_{st} d \left(1 - \dfrac{f_y A_{st}}{f_{ck} bd}\right)$$

$$5.99 \times 10^6 = 0.87 \times 415 \times A_{st} \times 102 \left(1 - \dfrac{415 A_{st}}{20 \times 1000 \times 102}\right) \quad 162.65 = A_{st}\left(1 - \dfrac{A_{st}}{4915.66}\right)$$

$$A_{st}^2 - 4915.66 A_{st} + 4915.66 \times 162.65 = 0$$

$$A_{st} = \dfrac{4915.66 - \sqrt{4915.66^2 - 4 \times 4915.66 \times 162.65}}{2}$$

$$A_{st} = 168.4 \text{ mm}^2$$

Using 8 mm φ bars, spacing bar.   $S = \dfrac{\alpha_{st}}{A_{st}} \times 1000 = \dfrac{\frac{\pi}{4} \times 8^2}{168.4} \times 1000 = 298.5$ mm.

Maximum spacing is   (i) $3d = 3 \times 102 = 306$ mm   (ii) 300 mm whichever is less.
Hence provide 8 mm bars at 290 mm c/c.

8. **Reinforcement in edge strip**

$$A_{st} = 0.12\% \text{ of gross area} = \dfrac{0.12}{100} \times 1000 \times 135 = 162 \text{ mm}^2.$$

Using 8 mm φ bars, spacing of bars

$$S = \dfrac{\frac{\pi}{4} \times 8^2}{162} \times 1000 \approx 310 \text{ mm}$$

Maximum spacing  (i) $5d = 5 \times 110 = 550$   (ii) 450 mm whichever less.
Hence provide 8 mm bars at 300 mm c/c in edge strips in both directions.

9. **Torsion reinforcement**

Area of reinforcement in each layer

$$A_{st} = \dfrac{3}{4} A_{st} x = \dfrac{3}{4} \times 269.5 = 202 \text{ mm}^2.$$

Distance over which torsion reinforcement is to be provided = $\frac{1}{5}$ Short span

$$= \frac{1}{5}l_x = \frac{3110}{5} = 622 \text{ mm}$$

Using 6 mm bars, spacing

$$S = \frac{\frac{\pi}{4} \times 6^2}{202} \times 1000 = 139.9 \text{ mm.}$$

Hence, provide 6 mm bars at 130 mm c/c at all the four corners in four layers.

## 10. Check for deflection

For simply supported slabs basic value of $\frac{l}{d}$ ratio = 20

Modification factor for tension steel $f_y$

% of steel at mid span

$$P_t = \frac{\frac{\pi}{4} \times 8^2}{180 \times 110} \times 100 = 0.253\%$$

$$f_x = 0.58 \times f_y = 0.58 \times 415 \approx 240 \text{ N/mm}^2.$$

From Fig. 4 of IS: 456, modification factor = 1.5

Maximum permitted $\frac{l}{d}$ ratio = 1.5 × 20 = 30

$$\frac{l}{d} \text{ provided} = \frac{3110}{110} = 28.27 < 30$$

Hence deflection control is safe.

Details of reinforcements are shown in Fig. 41.7.

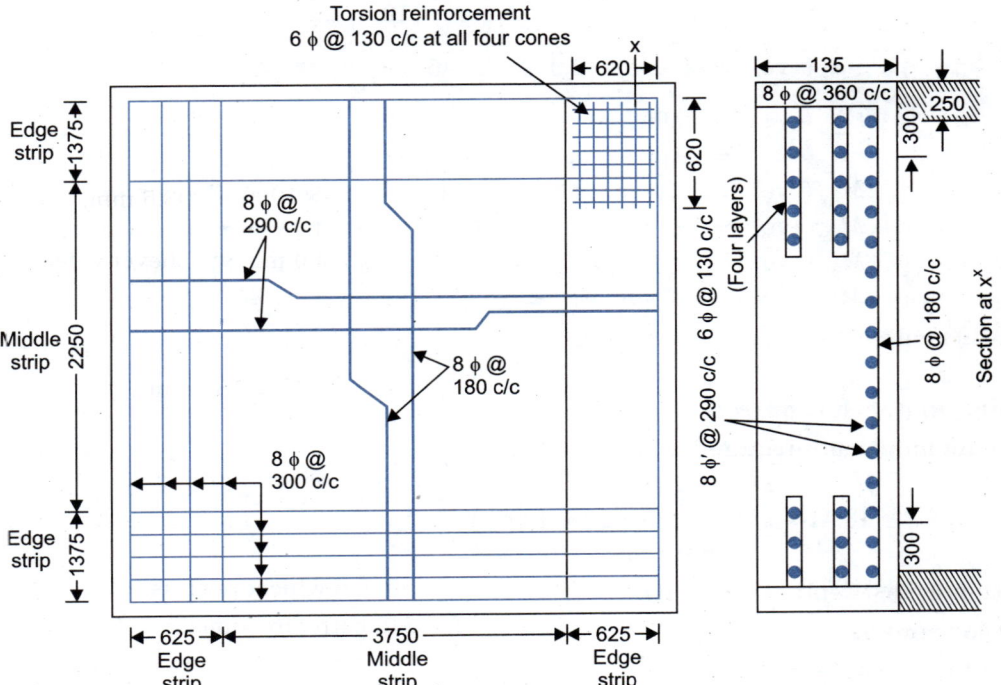

Fig. 41.7

**Example 41.5.** *Design the floor slab for a hall 4 m × 5 m to carry a live load of 3 kN/m² and floor finish of 1 kN/m². The slab is continuous over two adjacent walls of the hall. Walls are 300 mm wide. Use concrete of grade $M_{20}$ & steel of Fe 415.*

**Solution:** $\dfrac{l_y}{l_x} = \dfrac{5}{4} = 1.25 < 2.$

Hence, the slab is to be designed as a two way slab.

### 1. Thickness of slab

Assume effective depth $d = \dfrac{\text{Span}}{28} = \dfrac{4000}{28} = 142.9$ mm

Adopt effective depth $d = 150$ mm

Overall depth $D = 175$ mm.

### 2. Effective span

$l_x = 4.0 + 0.15 = 4.15$ m
$l_y = 5.0 + 0.15 = 5.15$ m
$\dfrac{l_y}{l_x} = \dfrac{5.15}{4.15} = 1.24.$

### 3. Loads

Self weight of the slab = $0.175 \times 25 = 4.375$ kN/m²
Live load = 3 kN/m²
Floor finish = 1 kN/m²
Total load = 8.375 kN/m²
Factored load $W_u$ = $1.5 \times 8.375 = 12.56$ kN/m².

### 4. Design moments and shear forces

This slab corresponds case-4 of Table-26 of IS: 456 as two adjacent edges are continuous.

$\alpha_x(-ve) = 0.06 + (0.065 - 0.06) \times \dfrac{4}{10} = 0.062$

$\alpha_x(+ve) = 0.045 + (0.049 - 0.045) \times \dfrac{4}{10} = 0.047$

$\alpha_y(-ve) = 0.047$
$\alpha_y(+ve) = 0.035$
$M_{ux}(-ve) = \alpha_x(-ve)\, wl_x^2 = 0.062 \times 12.56 \times 4.15^2 = 13.41$ kN-m.
$M_{ux}(+ve) = \alpha_x(+ve)\, wl_x^2 = 0.047 \times 12.56 \times 4.15^2 = 10.17$ kN-m.
$M_{uy}(-ve) = \alpha_y(-ve)\, wl_x^2 = 0.047 \times 12.56 \times 4.15^2 \times 10.17$ kN-m.
$M_{uy}(+ve) = \alpha_y(+ve)\, wl_x^2 = 0.035 \times 12.56 \times 4.15^2 = 7.57$ kN-m.
$Y_u = \dfrac{w_u l}{2} = \dfrac{12.56 \times 4.15}{2} = 26.06$ kN.

### 5. Minimum depth required

The minimum depth required to resist B.M.

$M_u = 0.138\, f_{ck} bd^2$
$13.41 \times 10^6 = 0.138 \times 20 \times 1000 \times d^2$
$d = 69.7$ mm $< 150$ mm.

Hence provided depth is adequate.

### 6. Reinforcement

Along short span (–ve) @ support

$M_{ux} = 0.87\, f_y A_{st}\, d \left(1 - \dfrac{f_y A_{st}}{f_{ck} bd}\right)$

$$13.41 \times 10^6 = 0.87 \times 415 \times A_{st} \times 150 \left(1 - \frac{415 A_{st}}{20 \times 1000 \times 150}\right)$$

$$247.6 = A_{st}\left(1 - \frac{A_{st}}{7228.9}\right)$$

$$A_{st}^2 - 7228.9\, A_{st} + 7228.9 \times 247.6 = 0$$

$$A_{st} = \frac{7228.9 - \sqrt{7228.9^2 - 4 \times 7228.9 \times 247.6}}{2}$$

$$A_{st} = 256.7 \text{ mm}^2.$$

Using 8 mm φ bars, spacing of bars

$$S = \frac{\sigma_{st}}{A_{st}} \times 1000 = \frac{\frac{\pi}{4} \times 8^2}{256.7} \times 1000 = 195.8 \text{ mm}$$

Maximum spacing is  (i) $3d = 3 \times 150 = 450$ mm  (ii) 300 mm whichever is less

Hence, provide 8 mm bars at 190 mm c/c.   Along shorter span (+ve) @ mid span

$$M_u = 0.87\, f_y A_{st}\, d\left(1 - \frac{f_y A_{st}}{f_{ck} bd}\right)$$

$$10.17 \times 10^6 = 0.87 \times 415 \times A_{st} \times 150 \left(1 - \frac{415 \times A_{st}}{20 \times 1000 \times 150}\right)$$

$$187.8 = A_{st}\left(1 - \frac{A_{st}}{7228.9}\right)$$

$$A_{st}^2 - 7228.9\, A_{st} + 7228.9 \times 187.8 = 0$$

$$A_{st} = 193.4 \text{ mm}^2.$$

Minimum reinforcement $= A_{st}$ min. $= 0.12\%$ gross area $= \frac{0.12}{100} \times 1000 \times 175 = 210$ mm$^2$.

Adopt $A_{st} = 210$ mm$^2$.

Using 8 mm φ bars, spacing of bars $S = \dfrac{\sigma_{st}}{A_{st}} \times 1000 = \dfrac{\frac{\pi}{4} \times 8^2}{210} \times 1000 = 239.4$ mm.

Maximum spacing is  (i) $3d = 3 \times 150 = 450$ mm   (ii) 300 mm whichever is less.

Hence, provide 8 mm bars at 230 mm c/c. Along y-direction (–ve BM).

As the moment is same (+ 10.17 kN-m) provide 8 mm bars at 230 mm c/c.

As the +ve BM is still less, provide minimum reinforcement as calculated above i.e. 8 mm bars at 230 mm c/c.

7. **Reinforcement in edge strip**

$$A_{st} = 0.12\% \text{ of gross area} = \frac{0.12}{100} \times 1000 \times 175 = 210 \text{ mm}^2.$$

Using 8 mm bars, spacing $S = \dfrac{\frac{\pi}{4} \times 8^2}{210} \times 1000 = 239.4$ mm.

Maximum spacing is  (i) $5d = 5 \times 150 = 750$   (ii) 450 mm whichever is less

Hence, provide 8 mm bars of 230 mm c/c in edge strip in both directions.

8. **Torsion reinforcement**

At the corner,  where both edges are discontinuous. Area of reinforcement in each layer.

$$A_t = \%\, A_{st} x \times 210 = 157.5 \text{ mm}^2.$$

Distance over which torsion reinforcement is to provided = $\frac{1}{5}$ short span

$$= \frac{1}{5}l_x = \frac{4150}{5} = 830 \text{ mm}$$

Using 6 mm bars, spacing $S = \dfrac{\frac{\pi}{4} \times 6^2}{157.5} \times 1000 = 179.5 \text{ mm}$

Hence, provide 6 mm bars at 170 mm c/c in four layers at corner A where both edges are discontinuous. At the corner where one edge is discontinuous and one edge is continuous, Area of reinforcement in each layer

$$A_t = \frac{3}{8} A_{st} x = \frac{3}{8} \times 210 = 78.75 \text{ mm}^2.$$

Using 6 mm bars, space is

$$S = \dfrac{\frac{\pi}{4} \times 6^2}{78.75} \times 1000 = 359 \text{ mm}.$$

Hence provide 6 mm bars at 300 mm c/c @ corner where one edge is discontinuous and one edge is continuous.

At the corner, where both edges are continuous torsion reinforcement is not required.

9. **Check for deflection**

For simply supported slabs basic value of $\dfrac{l}{d}$ ratio = 20

Modification factor for tension steel $f$, % of steel = 0.12
$$f_s = 0.58 \times f_y = 0.58 \times 415 \approx 240 \text{ N/mm}^2$$

From Fig. 4 of IS: 456, modification factor = 1.6

Maximum permitted $\dfrac{l}{d}$ ratio
$$= 1.6 \times 20 = 32$$

$\dfrac{l}{d}$ provided $= \dfrac{4150}{150} = 27.67 < 32$. Hence deflection control is safe.

## PROBLEMS

1. Design a R.C. slab for a room 4 m × 4 m measuring from inside. The thickness of wall is 400 mm. The superimposed load, exclusive of the self weight of the slab, is 2 kN/m². The slab may be assumed to be simply supported at all the four edges, with corners free to lift. Use M 20 mix. and Fe 415 steel.

2. Redesign problem 1 assuming that the corners are held down. Use Marcus's method.

3. Design a simply supported R.C. slab over a room 4.5 m × 6 m from inside, assuming that the corners are not free to lift. The thickness of all the four walls is 400 mm. The live load on the floor is 2 kN/m². The floor carries a floor finish which weighs 8.5 kN/m². Use M 20 mix. and Fe 415 steel. Use Marcus's method.

4. Redesign Problem 3 using I.S. Code method.

5. Redesign problem 1, assuming that the slab is fixed on all the four edges.

6. A R.C. floor is supported on R.C. columns 250 mm square at the corners of a rectangle 6 m × 7.5 m with R.C. beams connecting the columns along the perimeter of the rectangle. Design the floor as a two-way reinforced slab. Also design one of the 6 m span beams as an L-beam. Live load on floor = 8 kN/m². Use M 20 mix. and Fe 415 steel.

# CIRCULAR SLABS

## 42.1. INTRODUCTION

Circular slabs are used for the following purposes ; (i) roof of a room or hall circular in plan (ii) floor of circular water tanks or towers (iii) roof of pump houses constructed above tube wells (iv) roof of a traffic control post at the intersection of roads, (v) circular water tank containers with flat bottom and raft foundations etc. The bending of such a slab is essentially different from a rectangular slab where bending takes place in distinctly two perpendicular directions along the two spans. When a circular slab simply supported at the edge is loaded with uniformly distributed load, it bends in the form of a saucer, due to which stresses are developed both in the radial as well as the in circumferential directions. The tensile radial and circumferential stresses develop towards the *convex* side of the saucer, and hence reinforcement need be provided at the convex face of the slab. Theoretically, reinforcement should be provided both in the radial as well as circumferential directions, but this arrangement would cause congestion and anchoring problem at the centre of the slab. Hence an alternative method of providing reinforcement is adopted : reinforcement is provided in the form of a mesh of bars having equal area of cross-section in both the directions, the area being equal to that required for the bigger of the radial and circumferential moments. However, if the stresses near the edge are not negligible, or if the edge is fixed, radial and circumferential reinforcement near the edge becomes essential.

The exact analysis of slab, based on theory of elasticity and assuming Poisson's ratio equal to zero, is beyond the scope of the present book ; only final equations are given here. Sometimes, empirical formulae are used for bending moments and shear force etc. We shall discuss in brief the following cases :

1. Slab freely supported at edges and carrying U.D.L.
2. Slab fixed at edges and carrying U.D.L.
3. Slab simply supported at the edges, with load $W$ uniformly distributed along the circumference of a concentric circle.
4. Slab simply supported at edges, with U.D.L. inside a concentric circle.

## 42.2. SLAB FREELY SUPPORTED AT EDGES AND CARRYING U.D.L.

For complete treatment (i.e. equations for bending moments, shear etc. ), please refer Chapter 11 (§ 11.2)

$$M_r = \frac{w}{16}\left[(3+\mu)(a^2-r^2)\right]$$

$$M_\theta = \left\{\frac{w}{16}(a^2(3+\mu)-r^2(1-3\mu))\right\}; \quad V = 0.5\, wr$$

where  $M_r$ = radial moment
$M_\theta$ = circumferential moment
$V$ = radial shear force
$a$ = radius of slab fully restraint slab with uniformly distributed load
$\mu$ = Poisson's ratio,
$r$ = any section at distance from the origin
$w$ = Intensity of uniformly distributed load in kN/m²

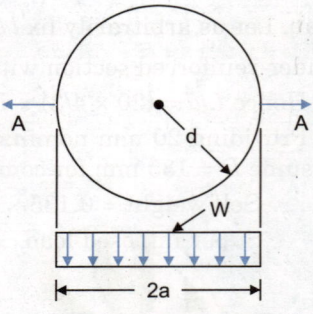

Fig. 42.1. Section at 'A-A'

## 42.3. SLABS FIXED AT EDGES AND CARRYING U.D.L.

For complete treatment (i.e. equations for bending moments, shear etc.) please refer Chapter 11 (§ 11.3)

## 42.4. SLAB SIMPLY SUPPORTED AT THE EDGES WITH LOAD W UNIFORMLY DISTRIBUTED ALONG THE CIRCUMFERENCE OF A CONCENTRIC CIRCLE

For complete treatment (i.e. equations for B.M. and S.F. etc.) please refer Chapter 11 (§ 11.4)

Simply supported slabs into concentric load. Let radius of slab be '$a$' and radius of the concentric load circle be '$b$' then for

$$M_r = M_\theta = \frac{P}{8\pi}\left[2\log_e\left(\frac{a}{b}\right) + 1 - \left(\frac{b}{a}\right)^2\right]$$

$$V = 0.$$

$$M_r = \frac{P}{8\pi}\left[2\log_e\left(\frac{a}{r}\right) + \left(\frac{b}{r}\right)^2 - \left(\frac{b}{a}\right)^2\right]$$

$$M_\theta = \frac{P}{8\pi}\left[2\log_e\left(\frac{a}{r}\right) - \left(\frac{b}{r}\right)^2 - \left(\frac{b}{a}\right)^2\right]$$

where  $P$ = total load.

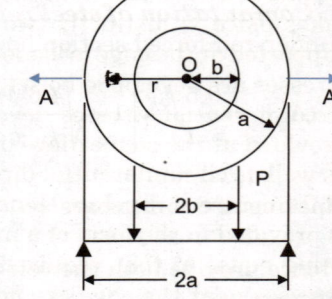

Fig. 42.2. Section at 'A-A'

## 42.5. SLAB SIMPLY SUPPORTED AT EDGES, WITH U.D.L. INSIDE A CONCENTRIC CIRCLE

For complete treatment (i.e. equations for B.M. and S.F. etc.) please refer Chapter 11 (§ 11.5)

**Example 42.1.** *A circular room has 5 m diameter from inside. Design a circular roof slab for room, to carry a superimposed load of 3750 N/m². Assume that the slab is simply supported at the edges. Use M 20 concrete and HYSD bars of Fe 415 grade.*

**Solution:** **1. Design constants and limiting depth of N.A.**

Given: $f_{ck}$ = 20 N/mm² for M 20 concrete and $f_y$ = 415 N/mm² for Fe 415 steel.

For Fe 415, $\dfrac{x_{u,max}}{d} = \dfrac{700}{1100 + 0.87 \times 415} = 0.479$

$$R_u = 0.36 f_{ck} \frac{x_{u,max}}{d}\left(1 - 0.42 \frac{x_{u,max}}{d}\right) = 0.36 \times 20 \times 0.479 (1 - 0.42 \times 0.479) = 2.755$$

**2. Computation of loading and bending moment**: From deflection point of view, $L/d$ = 20 for simply supported on way slab. Unfortunately, Code has not given any recommendation for circular slab. Let us arbitrarily fix $l/d$ ratio for circular slab equal to $1\frac{1}{3}$ times that for a one way slab. Assuming under-reinforced section with $p_t = 0.2\%$, we get a modification factor $\approx 1.68$.

Hence $L/d = (20 \times 4/3) \times 1.68 \approx 45$. Hence $d = L/45 = 5000/45 = 111$ mm.

Providing 20 mm nominal cover and using 8 mm φ bars, we get $D = 111 + 20 + 4 = 135$ mm. Hence assume $D = 135$ mm for computation of dead load.

∴ Self weight = $0.135 \times 1 \times 1 \times 25000 = 3375$ N/m²
Super-imposed load = $3750$ N/m²
Total $w$ = $7125$ N/m²

∴ $w_u = 1.5 w = 1.5 \times 7125 \approx 10688$ N/m²

∴ $(M_{ur})_c = (M_{u\theta})_c = +\dfrac{3}{16} w_u a^2 = +\dfrac{3}{16} \times 10688 (2.5)^2 \times 1000 = 12.525 \times 10^6$ N-mm.

$(M_{u\theta})_e = \dfrac{2}{16} w_u a^2 = \dfrac{2}{16} \times 10688 (2.5)^2 \times 1000 = 8.35 \times 10^6$ N-mm

### 3. Computation of effective depth and total depth

$$d = \sqrt{\dfrac{M_{u,\max}}{R_u \cdot b}} = \sqrt{\dfrac{12.525 \times 10^6}{2.761 \times 1000}} = 67.4 \text{ mm}$$

However from deflection point of view, provide $D = 135$ mm, so that keeping nominal cover = 20 mm and providing 8 mm dia. bars, available $d = 135 - 20 - 4 = 111$ mm for one layer and $111 - 8 = 103$ mm for the other layer.

**4. Computation of steel reinforcement**: Since actual $d$ is more than that required from B.M., we have an under-reinforced section for which radial and circumferential reinforcement required at the centre is

$$(A_{st})_c = \dfrac{0.5 f_{ck}}{f_y}\left[1 - \sqrt{\dfrac{4.6 M_{uc}}{f_{ck} bd^2}}\right] bd = \dfrac{0.5 \times 20}{415}\left[1 - \sqrt{1 - \dfrac{4.6 \times 12.525 \times 10^6}{20 \times 1000 (103)^2}}\right] \times 1000 (103) = 363.6 \text{ mm}^2$$

Spacing of 8 mm φ bars = $\dfrac{1000 \times 50.3}{363.6} \approx 138$ mm.

Hence provide both the reinforcements in the form of a mesh consisting of 8 mm φ bars spaced at 130 c/c in each of the two layers at right angles to each other.

Near the edges, the bars do not have proper anchorage since they are free. There will be slipping tendency and hence the bars will not be capable of taking any tension. At the edges, radial tensile stress is zero, but circumferential tensile stresses exist because $(M_{u\theta})_e = 8.35 \times 10^6$ N-mm. The tendency of slipping can be avoided by providing extra circumferential reinforcement in the from of rings placed in a width equal to the development length of the mesh bars. Available $d$ for rings = $103 - 8 = 95$ mm.

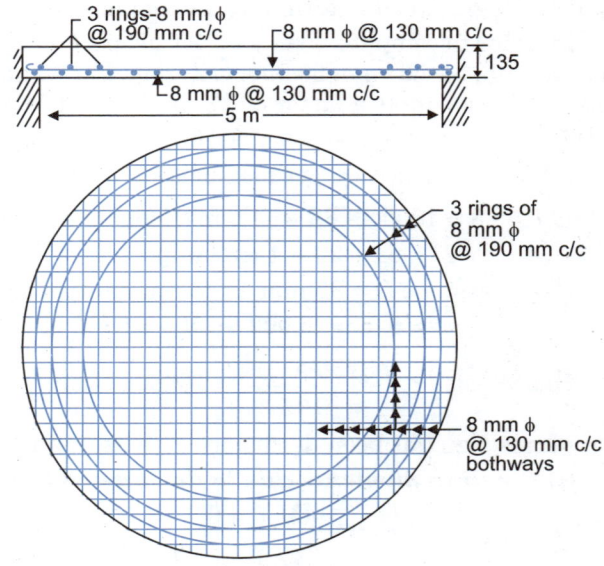

Fig. 42.3

∴ $(A_{st\theta})_e = \dfrac{0.5 \times 20}{415}\left[1 - \sqrt{1 - \dfrac{4.6 \times 8.35 \times 10^6}{20 \times 1000 (95)^2}}\right] 1000 \times 95 = 258$ mm²

and spacing $s$ of 8 mm dia. bars = $\dfrac{1000 \times 50.3}{258} = 194.9$ mm $\approx 190$ mm

$(M_\theta)_e$ is $\frac{2}{3}$ of the maximum moment at the centre. The circumferential steel is to be provided for a length $= \frac{2}{3} L_d = \frac{2}{3} \times 47 \times 8 \approx 251$ mm. (*i.e.* the length required to develop a tensile stress of $\frac{2}{3} \sigma_s = \frac{2}{3} (0.87 \times 415) = 240.7$ N/mm² by bond). Hence provide rings of 8 mm φ bars @ 190 mm c/c, total number of rings being $= (251/190) + 1 \approx 3$.

Since the slab is quite thin, no temperature or distribution reinforcement at the top face of the slab is necessary. *However*, for a thicker slab (say greater than 200 mm), such reinforcement @ 0.12% of cross-sectional area of concrete may be provided at the top face of the slab.

The details of reinforcement etc. are shown in Fig. 42.3.

**Design: Example 42.2.** *A traffic control post, 2 m in diameter is supported centrally by a reinforced concrete column 300 mm in diameter. Design the circular slab for a super-imposed load of 1500 N/m². Use M 20 concrete and Fe 415 steel.*

**Solution:**

### 1. Design constants and limiting depth of N.A.

Given:      $f_{ck} = 20$ N/mm² for M 20 concrete and $f_y = 415$ N/mm² for Fe 415 steel,

For Fe 415,    $\dfrac{x_{u,\max}}{d} = \dfrac{700}{1100 + 0.87 \times 415} = 0.479$

$$R_u = 0.36 f_{ck} \dfrac{x_{u,\max}}{d}\left(1 - 0.42 \dfrac{x_{u,\max}}{d}\right) = 0.36 \times 20 \times 0.479 (1 - 0.42 \times 0.479) = 2.755$$

### 2. Computation of loading and B.M.

Fig. 42.4. shows the slab supported centrally on the column and carrying U.D.L. $w$ N/m². Let the column reaction be $w_1$ N/m². Hence the slab may be considered to be simply supported at the edge and subjected to (*i*) a downward U.D.L. $w$ and (*ii*) an upward U.D.L. $w_1$ over a concentric circle of radius $b = 150$ mm. Hence we can use § 11.2 and § 11.5 for determining net moments on the slab.

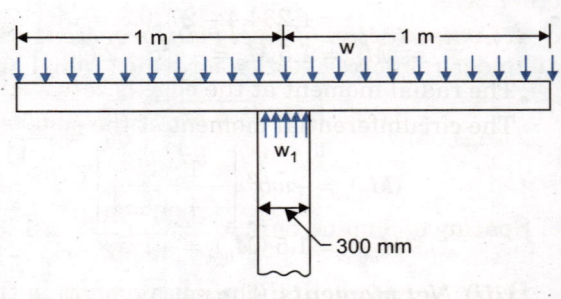

Fig. 42.4

In order to compute $w$ and $w_1$, let us assume thickness of the slab, based on deflection criterion. For a cantilever, $L/d = 7$. Also, apply a factor of 4/3 for the cantilever being circular. Assuming an under-reinforced section with $p_t \approx 0.2\%$, for which we get modification factor $\approx 1.68$ from Fig. 7.1. Hence $L/d = \frac{4}{3} \times 7 \times 1.68 = 15.68$ and $d = L/15.68 = 1000/15.68 \approx 64$ mm. Keeping a nominal cover of 20 mm and using 8 mm dia. bars,

$D = 64 + 20 + 4 = 88$ mm.

Hence assume $D = 90$ mm for the purpose of computing self weight.

∴      $w = 0.09 \times 25000 + 1500 = 3750$ N/m²

Hence    $w_1 = -\dfrac{3750\,\pi\,(1)^2}{\pi\,(0.15)^2} \approx 166667$ N/m²

Minus sign has been used since $w_1$ acts in an upward direction.

(*i*) **Bending moments due to downward load $w$** (Refer § 11.2)

$$(M_r)_e = 0$$

$$(M_\theta)_e = +\dfrac{2}{16} wa^2 = +\dfrac{2}{16} \times 3750 (1)^2 \times 1000 = +4.688 \times 10^5 \text{ N-mm}$$

∴      $(M_{u\theta})_e = +1.5 \times 4.688 \times 10^5 = 7.031 \times 10^5$ N-mm

Since the slab is cast monolithic with the column, the critical section will be at the edge of the column, where $r = 0.15$ m. Hence from *Eq. 11.3*:

$$M_\theta = +\dfrac{w}{16}(3a^2 - r^2) = +\dfrac{3750}{16}\left[3 \times 1^2 - (0.15)^2\right] \times 1000 = +6.979 \times 10^5 \text{ N-mm}$$

From (*Eq. 11.6*) $\quad M_r = +\dfrac{3}{16} w(a^2 - r^2) = +\dfrac{3 \times 3750}{16}\left[1^2 - (0.15)^2\right] \times 1000 = +6.873 \times 10^5$ N-mm

$\therefore \qquad M_{u\theta} = +1.5 \times 6.979 \times 10^5 = +10.468 \times 10^5$ N-mm

and $\qquad M_{ur} = +1.5 \times 6.873 \times 10^5 = +10.31 \times 10^5$ N-mm

### (ii) Bending moments due to downward load $w_1$

Refer § 11.5. From *Eq. 11.23* at $r = b = 0.15$ m, $M_r = -\dfrac{3}{16} wr^2 + \dfrac{1}{4} wb^2 \left[1 - \log_e\left(\dfrac{b}{a}\right) - \dfrac{b^2}{4a^2}\right]$

Putting $\quad w = w_1 = -166667 \quad$ and $\quad r = b = 0.15$ m

$$M_r = +\dfrac{3}{16} \times 166667 (0.15)^2 - \dfrac{1}{4}(166667)(0.15)^2 \left[1 - \log_e\left(\dfrac{0.15}{1}\right) - \left(\dfrac{0.15}{2 \times 1}\right)^2\right]$$

$\qquad = 703.1 - 2710.9 = -2007.8$ N-m $= -20.078 \times 10^5$ N-mm

$\therefore \qquad M_{ur} = 1.5 M_r = -1.5 \times 20.078 \times 10^5 = -30.117 \times 10^5$ N-mm

Also, From *Eq. 11.24*, $\quad M_\theta = -\dfrac{1}{16} wr^2 + \dfrac{1}{4} wb^2\left[1 - \log_e\left(\dfrac{b}{a}\right) - \dfrac{b^2}{4a^2}\right]$

Putting $\quad w = w_1 = -166667$ N $\quad$ and $\quad r = b = 0.15$ m

$\therefore \qquad M_\theta = \dfrac{1}{16} \times 166667 (0.15)^2 - \dfrac{1}{4} \times 166667 (0.15)^2 \left[1 - \log_e\left(\dfrac{0.15}{1}\right) - \left(\dfrac{0.15}{2 \times 1}\right)^2\right]$

$\qquad = +234.4 - 2710.8 = -2476.4$ N-m $= 24.764 \times 10^5$ N-mm

$\therefore \qquad M_{u\theta} = 1.5 M_\theta = -1.5 \times 24.764 \times 10^5 = -37.146 \times 10^5$ N-mm

The radial moment at the edge is zero, *i.e.* $(M_r)_e = 0$. Hence $(M_{ur})_e = 0$.

The circumferential moment at the edge is given by *Eq. 11.29*, where $w = w_1 = -166669$

$(M_\theta)_e = -wb^2\left[-\dfrac{1}{4} + \dfrac{b^2}{8a^2}\right] = 166667 (0.15)^2\left[-\dfrac{1}{4} + \dfrac{1}{8}\left(\dfrac{0.15}{1}\right)^2\right] = -927$ N-m $= -9.27 \times 10^5$ N-mm

$\therefore \qquad (M_{u\theta})_e = 1.5 (M_\theta)_e = -1.5 \times 9.27 \times 10^5 = -13.904 \times 10^5$ N-mm

### (iii) Net moments: The net moments are tabulated below

| Case | Radial moment (N-mm) | | Circumferential moments (N-mm) | |
|---|---|---|---|---|
| | $r = 0.15$ m | $r = 1$ m | $r = 0.15$ m | $r = 1$ m |
| (i) | $10.31 \times 10^5$ | 0 | $+10.468 \times 10^5$ | $+7.031 \times 10^5$ |
| (ii) | $-30.117 \times 10^5$ | 0 | $-37.146 \times 10^5$ | $-13.904 \times 10^5$ |
| Net | $-19.807 \times 10^5$ | 0 | $-26.678 \times 10^5$ | $-6.873 \times 10^5$ |

Thus the moments are throughout negative, as expected, since the slab will bend having convexity upwards. Hence both radial and circumferential reinforcements will be placed at the top face of the slab.

### 3. Computation of effective depth and total depth

The thickness will be designed for a bending moment of $26.678 \times 10^5$ N-mm.

$$d = \sqrt{\dfrac{M_{u,\max}}{R_u \cdot b}} = \sqrt{\dfrac{26.678 \times 10^5}{2.755 \times 1000}} = 31.1 \text{ mm}$$

However, keep $D = 90$ mm, required from deflection point of view. Using 8 mm φ bars and providing 20 mm nominal cover, available $d = 90 - 20 - 4 = 66$ mm for one layer and $66 - 8 = 58$ mm for the second layer of reinforcement.

### 4. Computation of steel reinforcement

Let us design the reinforcement for the greater of the two moments. Since actual $d$ is more than that required for bending, we have an under reinforced section for which

$$A_{st} = \frac{0.5 f_{ck}}{f_y}\left[1 - \sqrt{1 - \frac{4.6 M_{u,\text{max}}}{f_{ck} bd^2}}\right] bd = \frac{0.5 \times 20}{415}\left[1 - \sqrt{1 - \frac{4.6 \times 26.678 \times 10^5}{20 \times 1000 (58)^2}}\right] 1000 \times 58 = 133.9 \text{ mm}^2$$

Minimum reinforcement = $1.2 D = 1.2 \times 90 = 108 \text{ mm}^2$

Using 8 mm φ bars, spacing $s = \dfrac{1000 \times 50.3}{133.9} \approx 375$ mm.

Maximum permissible spacing = 300 mm.

Hence provide both the reinforcements in the form of mesh consisting of 8 mm φ bars spaced at 300 mm c/c in two directions at right angles to each other. This mesh will not be in a position to resist circumferential tensile stress at the edge. The effective depth available for rings reinforcement = 58 − 8 = 50 mm

$$A_{st} = \frac{0.5 \times 20}{415} \times \left[1 - \sqrt{1 - \frac{4.6 \times 6.873 \times 10^5}{20 \times 1000 (50)^2}}\right] \times 1000 \times 50 = 38.7 \text{ mm}^2 \text{ for 1000 mm width.}$$

Minimum reinforcement = $1.2 D = 1.2 \times 90 = 108 \text{ mm}^2$ for 1000 mm width. Since the circumferential moment at the edge is about $\frac{1}{4}$ of the moment at the centre, the circumferential moment is to be provided for a width = $\frac{1}{4} L_d = \frac{1}{4} \times 47 \times 8 = 94$ mm. Hence area of steel required for this width = $\dfrac{108 \times 94}{1000} = 10.2 \text{ mm}^2$. Using 8 mm φ bars, area $A_\varphi = 50.3 \text{ mm}^2$. Hence one ring is *sufficient*

### 5. Check for shear

Shear force at the edge of the column support is given by *Eqs 11.30* and *11.7*

$$F = \frac{w_1 b^2}{2r} - \frac{1}{2} wr = \frac{w_1 b^2}{2b} - \frac{1}{2} wb$$

$$= \frac{w_1 b}{2} - \frac{1}{2} wb = \frac{b}{2}(w_1 - w)$$

$$= \frac{0.15}{2}(166667 - 3750) \approx 12219 \text{ N}$$

$$\therefore F_u = 1.5 F = 1.5 \times 12219 \approx 18328 \text{ N}$$

$$\therefore \tau_v = \frac{18328}{1000 \times 66} = 0.278 \text{ N/mm}^2$$

The permissible shear stress even at minimum reinforcement = $1.3 \times 0.28 = 0.364 \text{ N/mm}^2$

Hence safe. The details of reinforcement etc are shown in Fig. 42.5.

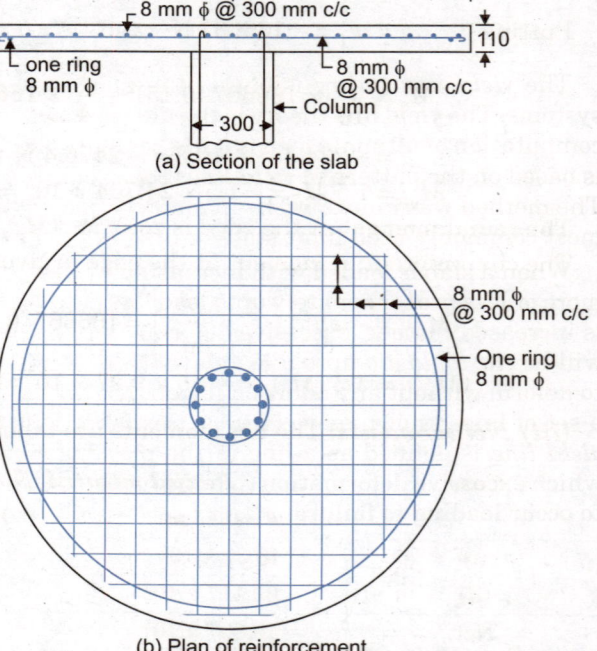

Fig. 42.5

## PROBLEMS

1. Design the roof slab for a circular room 6 metre in diameter from inside and carrying a super imposed load of 5 kN/m². Assume that the slab is simply supported at the edges. Use M 20 mix. and Fe 415 steel. The thickness of wall is 400 mm.
2. Redesign problem 1 assuming that the slab is restrained at the edge.
3. A circular room 8 metre in diameter from inside carries a 500 mm dia. column at its centre. Design a circular slab, simply supported at the outer edge and supported on the column at its middle. The slab carries a total uniformly distributed load of 3 kN/m² including its own weight. Use M 20 mix. and Fe 415 steel.
4. Design the slab of a traffic control post, 1.5 m diameter and supported centrally on a column of 200 mm diameter. The total superimposed load, inclusive of its own weight may be taken as 1.8 kN/m². Use M 20 mix. and Fe 415 steel.
5. Design circular slab of dia 8.5 m with supported uniformly distributed load of 4.5 kN/m² use $M_{25}$ and Fe 415 steel.

# 43 YIELD LINE THEORY AND DESIGN OF SLABS

## 43.1. INTRODUCTION

The *yield line theory* is one of the most important developments in the analysis and design of slab systems. The yield line theory is the ultimate load theory for the design of R.C. slabs. In the case of slabs, the computation of ultimate load is quite complicated. In the yield line method, the computation of ultimate load is based on the pattern of *yield lines* that are developed in the slabs under conditions approaching collapse. The method was innovated by Ingerslav (1923) and was greatly extended and advanced by Johanssen. The most commonly used limit state or collapse method is based on Johanssen's yield line theory.

When a slab is loaded with increasing loads, the stresses in the reinforcement and the concrete increase more or less proportionately upto load level corresponding to the yield stress in the reinforcement. If the load is increased further, excessive deformations and rapid increase in strains will result. These deformations will be *elastoplastic* upto a load level called *limit load*. When limit load is reached, the slab will continue to deform without any additional load, leading to total collapse. At this stage, a pattern of cracks will form a set of lines known as *yield lines*, resulting into a mechanism leading to the total collapse of the slab. A *yield line* is defined as a line in the plane of slab across which reinforcing bars have yielded and about which excessive deformation (plastic rotations) under constant limit moment (ultimate moment) continues to occur leading to failure.

## 43.2. YIELD LINE PATTERNS

(*a*) **One Way Slabs:** The behaviour of one way slab regarding its ultimate moment capacity is the same as that of beam of unit width. Consider a one way slab of unit width, simply supported at the ends, and loaded with uniformly distributed load [Fig. 43.1 (*a*)]. The maximum positive B.M. will be at the mid-span of the slab, its value being equal to $+ wL^2/8$. Similarly, Fig. 43.1 (*b*) shows a one way slab of unit width fixed at the ends. The maximum negative B.M. will be $- wL^2/12$ at the ends and maximum mid-span positive B.M. will be $+ wL^2/24$.

At service loads, the slab will be uncracked, but as the load is increased, cracks will be developed at the top of slab

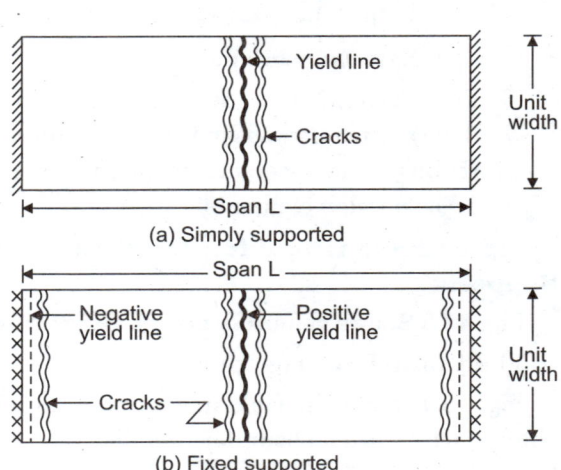

**Fig. 43.1.** Yield lines in one way slabs.

in negative moment regions, and at bottom in the positive moment regions. As the load is further increased, the negative and/or positive steel will start yielding. At the yield load $w_y$, the entire width of the slab will develop cracks, and *yield lines* will be formed. In the case of simply supported slab, such an yield line will be formed at mid-span, while for slab with fixed supports, *negative yield line* will be formed at the supports and *positive yield line* will be formed at the mid-span. At the yield lines, plastification takes place, creating '*plastic hinges*'. These plastic hinges withstand plastic moments. It is to be noted that the direction of yield line is perpendicular to the direction of reinforcement.

(*b*) **Two Way Slabs:** In one way slab, the direction of yield line is perpendicular to the direction of steel. But in two way slab (square or rectangular slabs) the direction of yield lines are not perpendicular to the two directions of reinforcement.

Fig. 43.2 (*a*) shows the initiation, growth and development of yield lines in square slabs. The cracks first develop along with small patches of yield of diagonals at mid-span. These cracks spread towards the corners, till yield lines are fully developed. Similarly, in a rectangular slabs, [Fig. 43.2 (*b*)], the initial crack is developed in a direction perpendicular to the short-span, at the mid-region of the slabs. Further increase in the load results in continuous growth of cracks as shown in Fig. 43.2 (*b*) (*ii*) and (*iii*). It is seen that except for the length AB, the cracks are not transverse to the directions of reinforcements.

**Characteristic Features of Yield Lines:** The following characteristics are useful in selecting the yield line patterns for slab with various boundary conditions.

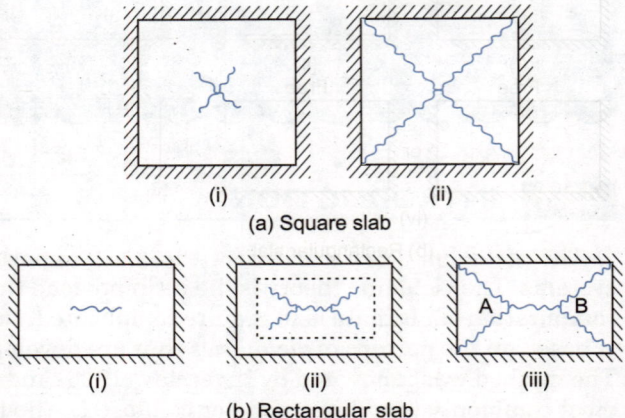

Fig. 43.2. Initiation, growth and completion of yield lines in simply supported two way slabs.

1. Yield lines are *straight lines* so that they may act as plastic hinges of a collapse mechanism.
2. Yield lines terminate at a slab boundary or at the intersection of other yield lines.
3. Yield lines act as axes of rotation for the movements of adjoining segments.
4. Each of the segments of the slab will tend to rotate in a rigid body motion. The axes of rotation generally lie along the lines of supports and pass over the columns.
5. If an edge is fixed or continuous, a yield line may form along the support.

$$\text{Initial work done} = \text{moment} \times \text{rotation}$$
$$\text{External work done} = \text{load} \times \text{deformation}$$

6. For mechanism to develop (*i.e.* for compatibility of deformations) yield lines or yield lines produced, pass through the intersection of the axes of rotation of adjacent slab elements.

### Sign Conventions for Yield Line Patterns and Supports

Fig. 43.3 Shows generally accepted sign conventions.

### Yield Line Patterns for Slabs

Fig. 43.4 shows the yield line patterns for slabs of various shapes, with various boundary conditions.

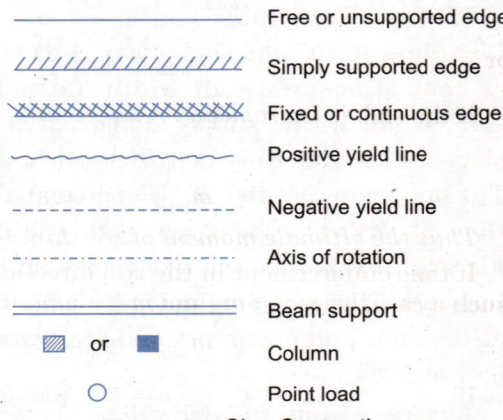

Fig. 43.3. Sign Conventions.

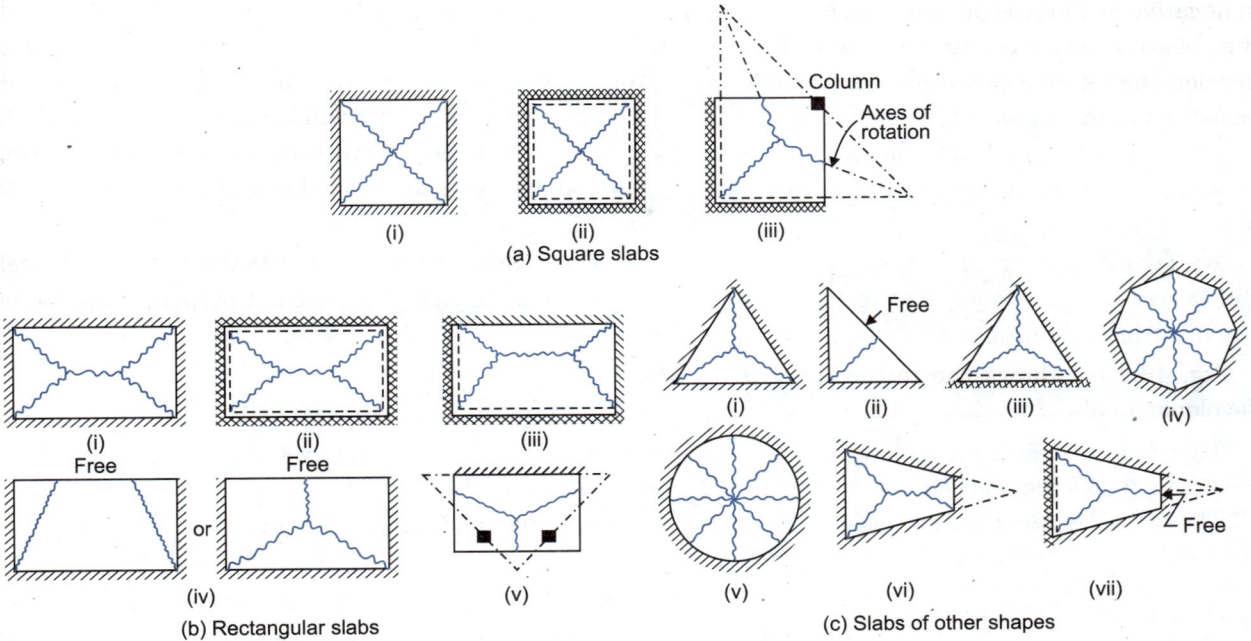

Fig. 43.4. Typical Yield Line Patterns in R.C. Slabs.

## 43.3. MOMENT CAPACITY ALONG AN YIELD LINE

In the case of one way slab, the yield line is perpendicular to the direction of reinforcement (Fig. 43.1). However, in the case of rectangular slabs having two-way reinforcement, the direction of the yield line may be inclined to the principal directions of reinforcement. Fig. 43.5 shows an inclined yield line $AB$ in orthogonally reinforced slab. The normal to the yield line makes an angle $\alpha$ with the $x$-direction, *i.e.* the yield line makes an angle $\alpha$ with the $y$-direction.

Let $m_{ux}$ = ultimate moment of resistance per unit length across a plane whose normal is in $x$-direction.
$m_{uy}$ = ultimate moment of resistance per unit length, across a plane whose normal is in $y$-direction.
$m_{u\alpha n}$ = ultimate moment of resistance per unit length, across a plane whose normal makes an angle $\alpha$ with $x$-direction.
$m_{u\alpha t}$ = ultimate torsional moment of resistance per unit length of yield line.

Fig. 43.5 (*c*) shows the moment vectors. Taking moments about the side $ce$ (parallel to the yield line),

$$m_{u\alpha n}(ce) = m_{ux}(cd)\cos\alpha + m_{uy}(de)\sin\alpha$$

or
$$m_{u\alpha n} = m_{ux}\left(\frac{cd}{ce}\right)\cos\alpha + m_{uy}\left(\frac{de}{ce}\right)\sin\alpha = m_{ux}\cos^2\alpha + m_{uy}\sin^2\alpha \qquad \text{...(43.1)}$$

If the slab is *isotropically reinforced* (*i.e.* equal reinforcement in the two directions), $m_{ux} = m_{uy} = m_u$. Hence,

$$m_{u\alpha n} = m_u(\cos^2\alpha + \sin^2\alpha) = m_u \qquad \text{...(43.2)}$$

*Thus the ultimate moment of resistance in an isotropically reinforced slab, in any direction, is the same.*

If the reinforcement in the two directions is not the same, it is said to be *orthotropically reinforced*. In such a case, let $m_{ux} = m_u$ and $m_{uy} = \mu m_u$. Then from *Eq 43.2*, we get

$$m_{u\alpha n} = m_u \cos^2\alpha + \mu m_u \sin^2\alpha, \quad \text{or} \quad m_{u\alpha n} = m_u(\cos^2\alpha + \mu \sin^2\alpha) \qquad \text{...(43.3)}$$

If $\alpha = 45°$, $\cos^2\alpha = \sin^2\alpha = \dfrac{1}{2}$

$$\therefore \quad m_{u\alpha n} = \frac{m_u(1+\mu)}{2} \qquad \text{...(43.4)}$$

Again, considering the moment vectors of Fig. 43.5 (c) and taking moments about an axis perpendicular to ce, to get torsional moment per unit length of yield line as :

$$m_{u\alpha t} = (m_{ux} - m_{uy}) \sin \alpha \cos \alpha \qquad ...(43.5)$$

For an isotropically reinforced slab, $m_{ux} = m_{uy}$, and hence torsional moment is zero.

For an orthotropically reinforced slab,

$$m_{u\alpha t} = m_u (1 - \mu) \sin \alpha \cos \alpha \qquad ...(43.6)$$

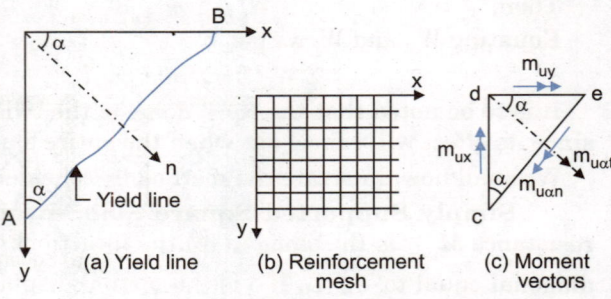

Fig. 43.5. Inclined yield line in orthotropically reinforced slab.

## 43.4. ULTIMATE LOAD ON SLABS

There are two methods of determining the ultimate load capacity of reinforced concrete slabs.
1. Virtual work method.
2. Equilibrium method.

The *virtual work method* is based on the principle that the work done by the external forces in undergoing a small virtual displacement is equal to the internal work done (or energy dissipated) in rotation along the yield lines. The elastic deformations are however ignored in comparison to the plastic deformations that take place along the yield lines. In the equilibrium method, the collapse load is calculated from the equilibrium of the individual segments of a mechanism.

Both the virtual work method as well as the equilibrium method give the upper bound solution for the collapse load, *i.e.* the computed collapse load on the basis of an assumed yield line patterns is bound to be larger than the actual collapse load. Hence it is essential to examine several yield line patterns and a minimum value of collapse load has to be obtained. Test results have shown that the actual failure load of slab is greater than the one predicted by the upper bound solution of the yield line analysis, because of membrane action. Hence the results of yield line analysis can be used in design with reasonable degree of safety.

## 43.5. ANALYSIS BY VIRTUAL WORK METHOD

The principle of *virtual work* states that :

(i) 'If a deformable structure in equilibrium under the action of a system of external forces is subjected to a virtual deformation compatible with its condition of support, the work done by these forces on the displacements associated with the virtual deformation is equal to the work done by the internal stresses on the strains associated with this deformation'.

(ii) Thus, the work done during small motion of collapse mechanism is equal to work absorbed by the plastic hinges formed along the yield lines.

(iii) The segments of the slab within the yield lines go through rigid body displacements with the collapse load acting on the structure.

Thus work done by external forces = work done by internal forces (or energy absorbed by hinges)

or $$W_E = W_I \qquad ...(43.7)$$

where $W_E$ = work done by external forces
$W_I$ = Energy absorbed by plastic hinges in undergoing corresponding rotations.

If $w_u$ is the uniformly distributed external load, we have

$$W_E = \iint w_u \, \delta_{x,y} \, dx \, dy = \Sigma W_u \Delta \qquad ...(43.8)$$

where $\delta_{x,y}$ = virtual displacement at any point.
$W_u$ = resultant load on each segment.
$\Delta$ = corresponding displacement at the centroid of load in each segment.

If $m_{u\alpha n}$ = ultimate moment across and yield line.
$l_0$ = length of yield line.

and $\theta_n$ = relative rotation of two adjacent plates, perpendicular to the yield line.

Then, $\quad\quad\quad\quad\quad\quad W_I = \Sigma\, m_{uan}\, \theta_n\, l_0$ ...(43.9)

Equating $W_E$ and $W_I$ we get

$$\Sigma\, W_u\, \Delta = \Sigma\, m_{uan}\, \theta_n\, l_0 \quad\quad\quad ...(43.10)$$

It is to be noted that the work done by the twisting moment $(m_{uat})$ has not been considered in *Eq. 43.9* since its effect will cancel out when the entire slab is eventually considered.

We shall now illustrate the method by considering several cases.

**1. Simply Supported Square Slab:** A square slab is isotropically reinforced. Hence moment of resistance $M_{uan}$ is the same in all the directions. Fig. 43.6 shows such a slab of span $L$, having length of diagonal equal to $\sqrt{2}\,L$. If $\delta$ is the virtual displacement at the middle, the total rotation of the diagonal segments $= 2\,\theta_n = 2\,\dfrac{\delta}{\dfrac{\sqrt{2}\,L}{2}} = \dfrac{2\sqrt{2}\,\delta}{L}$

Length of each diagonal yield line $= \sqrt{2}\,L$

Hence for the two diagonal yield lines,

$$W_I = \Sigma\, m_{uan}\, \theta_n\, l_0 = 2\left(m_u\, \dfrac{2\sqrt{2}\,\delta}{L}\, \sqrt{2}\,L\right) = 8\, m_u\, \delta \quad\quad ...(1)$$

There are four segments, each segment carrying equal total load $W_u = w_u\, L\, \dfrac{L}{2} = w_u\, \dfrac{L^2}{4}$ acting through its centroid which undergoes a virtual displacement $\Delta = \dfrac{\delta}{3}$. Hence

$$W_E = \Sigma\, W_u\, \Delta = 4\left(\dfrac{w_u\, L^2}{4}\cdot\dfrac{\delta}{3}\right) = w_u\, L^2\, \dfrac{\delta}{3} \quad\quad ...(2)$$

Equating $W_E$ and $W_I$, we have

$$w_u\, L^2\, \dfrac{\delta}{3} = 8\, m_u\, \delta \quad\text{or}\quad w_u = \dfrac{24\, m_u}{L^2} \quad\quad ...(43.11)$$

or $\quad\quad\quad\quad\quad\quad m_u = \dfrac{w_u\, L^2}{24}$ ...[43.11 (a)]

**Alternative Solution**

The problem can be solved alternatively by combination of components of moments and rotations in the co-ordinate directions $x$ and $y$. Refer Fig. 43.6 (b).

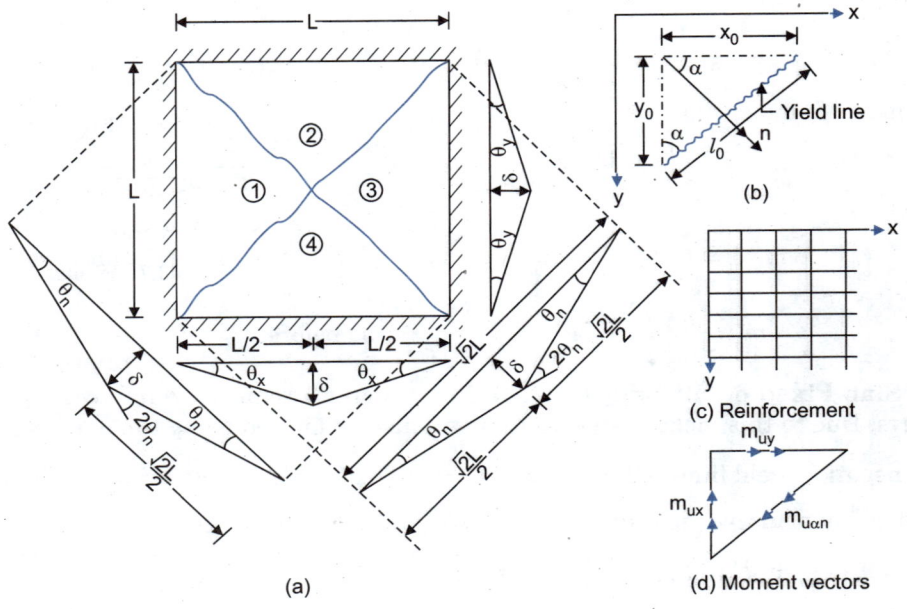

Fig. 43.6. Square slab : Freely supported.

From *Eq. 43.1*, we have
$$m_{uan} = m_{ux}\cos^2\alpha + m_{uy}\sin^2\alpha.$$
Hence
$$\Sigma m_{uan}\theta_n l_0 = \Sigma(m_{ux}\cos^2\alpha + m_{uy}\sin^2\alpha)\theta_n l_0$$
or
$$\Sigma m_{uan}\theta_n l_0 = \Sigma m_{ux} y_0 \cos\alpha . \theta_n + \Sigma m_{uy} x_0 \cos\alpha . \theta_n$$
(since $\cos\alpha = y_0/l_0$ and $\sin\alpha = x_0/l_0$)

Hence
$$\Sigma m_{uan}\theta_n l_0 = \Sigma m_{ux} y_0 \theta_x + \Sigma m_{uy} x_0 \theta_y \qquad \text{...(43.12)}$$

where  $\theta_x$ = Component of $\theta_n$ in $x$ direction.
$\theta_y$ = Component of $\theta_n$ in $y$ direction.
$x_0$ = Projected length of yield line in $x$ direction.
$y_0$ = Projected length of yield line in $y$ direction.

It is to be noted that $\theta_x$ and $\theta_y$ are the sum of the rotations of segments adjacent to the yield line under consideration.

Substituting the value of $\Sigma m_{uan}\theta_n l_0$ in *Eq 43.10*, we get
$$\Sigma m_{ux} y_0 \theta_x + \Sigma m_{uy} x_0 \theta_y = \Sigma W_u \Delta \qquad \text{...(43.13)}$$

Now, for the square slab of Fig. 43.6 (*a*), we have four segments. Segment 1 and 3 rotate about *y*-axis and there is no rotation about *x*-axis. Similarly, segments 2 and 4 rotate about *x*-axis and there is no rotation about *y*-axis.

Rotation  $\theta_x = \theta_y = \dfrac{\delta}{L/2} = \dfrac{2\delta}{L}$

Also,  $x_0 = y_0 = L$, and $m_{ux} = m_{uy} = m_u$

∴ For segment 1,  $m_{ux} y_0 \theta_x + m_{uy} x_0 \theta_y = \left(m_u L \dfrac{2\delta}{L}\right) + 0$

For segment 2,  $m_{ux} y_0 \theta_x + m_{uy} x_0 \theta_y = 0 + m_u L \dfrac{2\delta}{L}$

For segment 3,  $m_{ux} y_0 \theta_x + m_{uy} x_0 \theta_y = m_u L \dfrac{2\delta}{L} + 0$

and for segment 4,  $m_{ux} y_0 \theta_x + \Sigma m_{uy} \theta_y x_0 = 0 + m_u L \dfrac{2\delta}{L}$

∴ $\Sigma m_{ux}\theta_x y_0 + \Sigma m_{uy}\theta_y x_0 = 4 m_u L \dfrac{2\delta}{L} = 8 m_u \delta,$

which is the same as found earlier.

$$\Sigma W_u \Delta = 4\left(\dfrac{w_u L^2}{4}\right)\dfrac{\delta}{3} = w_u L^2 \dfrac{\delta}{3}$$

∴  $8 m_u \delta = w_u L^2 \dfrac{\delta}{3}$

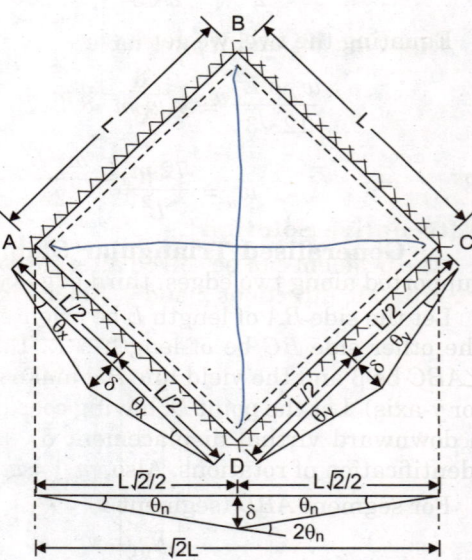

**Fig. 43.7.** Square slab : Fixed edges.

or  $m_u = \dfrac{w_u L^2}{24}$  or  $w_u = \dfrac{24 m_u}{L^2}$, as before.

**2. Square Slab Fixed on all Edges**: Fig. 43.7 shows an isotropically reinforced square slab fixed at the four supports. Due to this, negative yield lines will also be formed along the fixed edges. The rotation $\theta_x = \theta_y$ of these negative yield lines will be equal to $\dfrac{\delta}{L/2}$. The length of each negative yield line = $L$. Hence internal work done by negative yield lines *AB*, *BC*, *CD* and *DA* is

$$\Sigma m_{uan}\theta_n l_0 = 4 m_u L \dfrac{\delta}{L/2} = 8 m_u \delta \qquad \text{...(i)}$$

Internal work done along positive yield lines along the diagonals $AC$ and $BD$, as found in case 1, is

$$\Sigma\, m_{uan}\, \theta_n\, l_0 = 8\, m_u\, \delta \qquad \ldots(ii)$$

Hence $\quad W_I = 8\, m_u\, \delta + 8\, m_u\, \delta = 16\, m_u\, \delta \qquad \ldots(a)$

It is to be noted that the energy absorbed by both negative yield line as well as positive yield line has positive sign.

Again, as found in the previous case,

$$W_E = \Sigma\, W_u\, \Delta = w_u\, L^2\, \frac{\delta}{3} \qquad \ldots(b) \qquad\qquad 16\, m_u\, \delta = w_u\, L^2\, \frac{\delta}{3}$$

or $\quad m_u = \dfrac{w_u L^2}{48} \qquad \ldots[43.14\,(a)] \qquad$ or $\quad w_u = \dfrac{48\, m_u}{L^2} \qquad \ldots(43.14)$

**3. Equilateral Triangular Slab Isotropically Reinforced** Fig. 43.8 shows isotropically reinforced triangular slab. The ultimate moment of resistance $m_u$ will be same in all directions. When it is simply supported along all the three edges, the three yield lines will divide the slab in three symmetrically placed sectors. Hence we need consider only one sector.

The distance of point $O$ from edge $AB$ will be $\dfrac{L}{2\sqrt{3}}$.

For sector 1, $W_I = m_u\, \theta\, l_0 = m_u\, \dfrac{\delta}{L/2\sqrt{3}}\, L = 2\sqrt{3}\, m_u \delta \qquad \ldots(1)$

Also, $\quad W_E = \dfrac{1}{2} L \times \dfrac{L}{2\sqrt{3}} \times w_u \times \dfrac{\delta}{3} = w_u\, \dfrac{\delta L^2}{12\sqrt{3}} \qquad \ldots(2)$

Equating the two, we get

$$\dfrac{w_u\, \delta\, L^2}{12\sqrt{3}} = 2\sqrt{3}\, m_u\, \delta$$

$\therefore \quad w_u = \dfrac{72\, m_u}{L^2} \qquad \ldots[43.15\,(a)] \qquad$ or $\quad m_u = \dfrac{w_u L^2}{72} \qquad \ldots(43.15)$

**4. Generalised Triangular Slab:** Let us now take the general case of a triangular slab, simply supported along two edges, third edge being free, and isotropically reinforced.

Let the side $BA$ of length $L$ be oriented along $y$-axis, and the other side $BC$ be of length $\alpha L$. Edge $AC$ is free. Let $\angle ABC$ be $\beta$ and the yield line $BD$ make an angle $\phi$ with $BA$ (or $y$-axis). Let the point $D$, having coordinates $(x_d, y_d)$ have a downward virtual displacement $\delta$. Let us have proper identification of rotations. Also, $m_{ux} = m_{uy} = m_u$.

For segment $ABD$ (segment 1),

$$y_{01} = y_0$$
$$x_{01} = x_d = y_0\, \tan\phi$$
$$\theta_{x1} = \dfrac{\delta}{x_d} = \dfrac{\delta}{y_0\, \tan\phi}$$
$$\theta_{y1} = 0$$

For segment $BCD$ (segment 2)

$$\theta_{x2} = \dfrac{\delta}{DE} = \dfrac{\delta}{y_0\, \tan\beta - y_0\, \tan\phi}$$

$$\theta_{y2} = \dfrac{\delta}{FD} = \dfrac{\delta}{y_0 - x_d\, \cot\beta} = \dfrac{\delta}{y_0 - y_0\, \tan\phi\, \cot\beta}$$

$$y_{02} = y_0$$
$$x_{02} = x_d = y_0\, \tan\phi$$

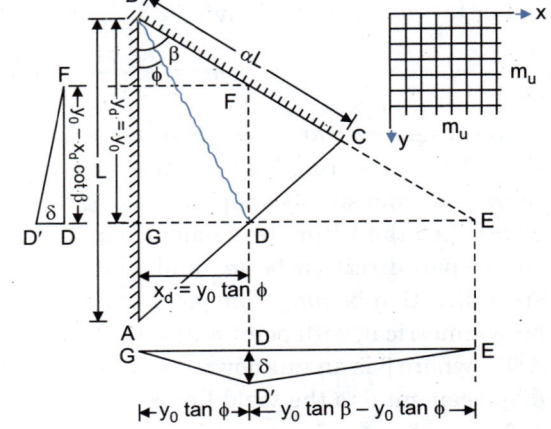

Fig. 43.9

**976** REINFORCED CONCRETE STRUCTURE

Now from *Eq. 43.12*
$$W_I = \Sigma\, m_{ux}\, y_0\, \theta_x + \Sigma\, m_{uy}\, x_0\, \theta_y$$
$$W_I = [m_u\, y_0\, \theta_{x1} + m_u\, y_0\, \theta_{x2}] + [m_u\, x_{01}\, \theta_{y1} + m_u\, x_{02}\, \theta_{y2}]$$
$$= m_u \left[ y_0\, \frac{\delta}{y_0 \tan \phi} + y_0\, \frac{\delta}{y_0 (\tan \beta - \tan \phi)} \right] + m_u \left[ y_0 \tan \phi \times \text{zero} + y_0 \tan \phi\, \frac{\delta}{y_0 (1 - \tan \phi \cot \beta)} \right]$$
$$= m_u \left[ \frac{\delta}{\tan \phi} + \frac{\delta}{\tan \beta - \tan \phi} + 0 + \frac{\delta \tan \Phi}{1 - \tan \phi \cot \beta} \right] = m_u\, \delta \left[ \cos \phi + \frac{1}{\tan \beta - \tan \phi} + \frac{\tan \phi \tan \beta}{\tan \beta - \tan \phi} \right]$$

or
$$W_I = m_u\, \delta \left[ \cos \phi + \frac{1 + \tan \beta \tan \phi}{\tan \beta - \tan \phi} \right] = m_u \delta\, [\cot \phi + \cot (\beta - \phi)] \qquad \ldots(1)$$

Also,
$$W_E = \Sigma\, W_u\, \Delta = w_u \times \text{area of segments} \times \frac{1}{3}\, \delta$$

or
$$W_E = w_u\, \frac{1}{2} L\, (\alpha L \sin \beta)\, \frac{\delta}{3} = w_u\, \delta\, \alpha L^2\, \frac{\sin \beta}{6} \qquad \ldots(2)$$

Now $W_I = W_E$. $\therefore\ m_u\, \delta\, [\cot \phi + \cot (\beta - \phi)] = w_u\, \delta\, \alpha L^2\, \dfrac{\sin \beta}{6}$

or
$$w_u = \frac{6\, m_u}{\alpha L^2} \cdot \frac{\cot \phi + \cot (\beta - \phi)}{\sin \beta} = \frac{6\, m_u}{\alpha L^2}\, \frac{1}{\sin \phi\, \sin (\beta - \phi)} \qquad \ldots(43.16)$$

For $w_u$ to be minimum, $\dfrac{\partial w_u}{\partial \phi} = 0$

This gives $\cos \phi \sin (\beta - \phi) - \sin \phi \cos (\beta - \phi) = 0$ or $\tan \phi = \tan (\beta - \phi)$. $\therefore \phi = \beta/2$
Hence for minimum yield load, the yield line must bisect the angle $\beta$.
Substituting $\phi = \beta/2$ in *Eq. 43.16*, we get

$$w_u = \frac{6\, m_u}{\alpha L^2}\, \frac{1}{\sin \beta/2\, \sin \beta/2} = \frac{6\, m_u}{\alpha L^2\, \sin^2 \beta/2} \qquad \ldots[43.17\,(a)]$$

or
$$m_u = \frac{w_u\, \alpha L^2}{6}\, \sin^2 \beta/2 \qquad \ldots(43.17)$$

For an equilateral triangle, $\alpha = 1$ and $\beta = 60°$.

$$m_u = \frac{w_u\, L^2}{6}\, \sin^2 30° = \frac{w_u\, L^2}{24} \qquad \ldots(43.18)$$

**5. Rectangular Slab, Simply Supported:**
Fig. 43.10 shows rectangular slab, orthotropically reinforced, and simply supported along all the four edges. Let the ultimate moment capacity in the short span direction be $m_u$ and that in the long span direction be $\mu m_u$. The yield line pattern will be symmetrical, with point $E$ at a distance $\beta L$ from $AD$, where $\beta$ is an unknown. Let us give a virtual displacement $\delta$ to the yield line $EF$.

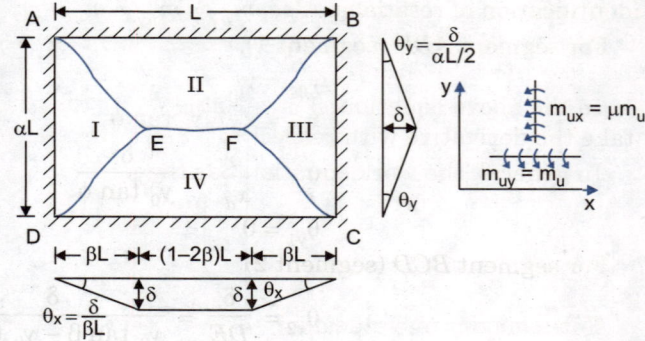

Fig. 43.10

Let $L$ = Long span
$\alpha L$ = Short span

$\theta_x$ = Rotation in the $x$-direction = $\dfrac{\delta}{\beta L}$;

$\theta_y$ = Rotation in the $y$-direction = $\dfrac{\delta}{\alpha L/2} = \dfrac{2\delta}{\alpha L}$; $m_{ux} = \mu\, m_u$; $m_{uy} = m_u$

For segment $AED$, $\quad m_{ux} y_0 \theta_x + m_{uy} x_0 \theta_y = (\mu m_u)(\alpha L)\left[\dfrac{\delta}{\beta L}\right] + 0$

For segment $ABFE$, $\quad m_{ux} y_0 \theta_x + m_{uy} x_0 \theta_y = 0 + m_u L \dfrac{2\delta}{\alpha L}$

For segment $BCF$, $\quad m_{ux} y_0 \theta_x + m_{uy} x_0 \theta_y = (\mu m_u)(\alpha L)\left(\dfrac{\delta}{\beta L}\right) + 0$

For segment $CDEF$, $\quad m_{ux} y_0 \theta_x + m_{uy} x_0 \theta_y = 0 + m_u L \dfrac{2\delta}{\alpha L}$

$\therefore\ \Sigma m_{ux} y_0 \theta_x + \Sigma m_{uy} x_0 \theta_y = \dfrac{2\mu m_u \alpha L \delta}{\beta L} + 2 m_u L \dfrac{2\delta}{\alpha L} = 2\delta\left[\dfrac{2 m_u}{\alpha} + \dfrac{\alpha \mu m_u}{\beta}\right]$ ...(1)

*Work done by external load*

For $AED$, $\quad W_u \Delta = \dfrac{1}{2} \alpha L \beta L w_u \dfrac{\delta}{3} = \alpha \beta L^2 \dfrac{\delta}{6} w_u$

For $ABEF$, $\quad W_u \Delta = 2\left\{\dfrac{1}{2}\beta L \dfrac{\alpha L}{2} w_u \dfrac{\delta}{3}\right\} + (1-2\beta) L \dfrac{\alpha L}{2} w_u \dfrac{\delta}{2}$

$\quad = \alpha \beta L^2 \dfrac{\delta}{6} w_u + (1-2\beta) \alpha L^2 \dfrac{\delta}{4} w_u$

For $BCF$, $\quad W_u \Delta = \alpha \beta L^2 \dfrac{\delta}{6} w_u$.

and for $CDEF$, $\quad W_u \Delta = \alpha \beta L^2 \dfrac{\delta}{6} w_u + (1-2\beta)\alpha L^2 \dfrac{\delta}{4} w_u$

$\therefore\ \Sigma W_u \Delta = 2\left[\alpha\beta L^2 \dfrac{\delta}{6} w_u\right] + 2\left[\alpha\beta L^2 \dfrac{\delta}{6} w_u + (1-2\beta)\alpha L^2 \dfrac{\delta}{4} w_u\right]$

$\quad = \delta w_u L^2 \left[\dfrac{\alpha\beta}{3} + \dfrac{\alpha\beta}{3} + \dfrac{\alpha}{2} - \alpha\beta\right] = \delta w_u L^2 \left[\dfrac{\alpha}{2} - \dfrac{\alpha\beta}{3}\right] = \delta w_u L^2 \alpha\left(\dfrac{3-2\beta}{6}\right)$ ...(2)

Equating internal and external work, we get

$\alpha \delta w_u L^2 \left(\dfrac{3-2\beta}{6}\right) = 2\delta\left[\dfrac{2m_u}{\alpha} + \dfrac{\alpha\mu m_u}{\beta}\right]$

or $\quad \alpha w_u L^2 \left(\dfrac{3-2\beta}{6}\right) = 2 m_u \left(\dfrac{2}{\alpha} + \dfrac{\alpha\mu}{\beta}\right) = 2 m_u \left(\dfrac{2\beta + \alpha^2 \mu}{\alpha\beta}\right)$

$\therefore\ m_u = \dfrac{\alpha L^2 (3-2\beta)}{6} \times \dfrac{\alpha\beta}{2(2\beta+\alpha^2\mu)} \times w_u = \dfrac{w_u \alpha^2 L^2}{12}\left[\dfrac{3\beta-2\beta^2}{2\beta+\mu\alpha^2}\right]$ ...(43.19)

In the above equation, $\beta$ is variable. For a maximum value of $m_u$ (or minimum value of $w_u$), we should take the derivative with respect to $\beta$ and equate it to zero.

In general, the work equation has the form

$$m_u = w_u \left[\dfrac{f_1(x_1, x_2)}{f_2(x_1, x_2)}\right]$$

For a maximum value of $m_u$, $\dfrac{\partial m_u}{\partial x_1} = 0$. By differential process,

$$\dfrac{\partial m_u}{\partial x_1} = \dfrac{\partial}{\partial x_1}\left[\dfrac{f_1(x_1, x_2)}{f_2(x_1, x_2)}\right] = \dfrac{f_2(x_1, x_2)\dfrac{\partial}{\partial x_1}[f_1(x_1, x_2)] - f_1(x_1, x_2)\dfrac{\partial}{\partial x_1}[f_2(x_1, x_2)]}{[f_2(x_1, x_2)]^2}$$

**978　REINFORCED CONCRETE STRUCTURE**

Equating this to zero, we get

$$f_2(x_1, x_2) \frac{\partial}{\partial x_1}[f_1(x_1, x_2)] - f_1(x_1, x_2) \frac{\partial}{\partial x_1} f_2[(x_1, x_2)] = 0$$

or
$$\frac{f_1(x_1, x_2)}{f_2(x_1, x_2)} = \frac{\dfrac{\partial}{\partial x_1}[f_1(x_1, x_2)]}{\dfrac{\partial}{\partial x_1}[f_2(x_1, x_2)]} \qquad \ldots(43.20)$$

Eq. 43.20 is a very important governing equation for maximum value of $m_u$.

Applying this to our present case, we get

$$\frac{3\beta - 2\beta^2}{2\beta + \mu\alpha^2} = \frac{3 - 4\beta}{2} \qquad \text{or} \qquad 6\beta - 4\beta^2 = 6\beta + 3\mu\alpha^2 - 8\beta^2 - 4\mu\beta\alpha^2$$

or $\quad 4\beta^2 + 4\mu\alpha^2\beta - 3\mu\alpha^2 = 0.$ Solution of the above quadratic equation, we get,

$$\beta = \frac{1}{2}\left[\sqrt{\mu^2\alpha^4 + 3\mu\alpha^2} - \mu\alpha^2\right] \qquad \ldots(43.21)$$

Substituting this value of $\beta$ in Eq 43.19, we get

$$m_u = \frac{1}{12} w_u \alpha^2 L^2 \frac{\frac{3}{2}\left\{\sqrt{\mu^2\alpha^4 + 3\mu\alpha^2} - \mu\alpha^2\right\} - \frac{2}{4}(\mu^2\alpha^4 + 3\mu\alpha^2 + \mu^2\alpha^4 - 2\mu\alpha^2\sqrt{\mu^2\alpha^4 + 3\mu\alpha^2})}{\frac{2}{2}\left\{\sqrt{\mu^2\alpha^4 + 3\mu\alpha^2} - \mu\alpha^2\right\} + \mu\alpha^2}$$

Simplifying the above, we get

$$m_u = \frac{w_u \alpha^2 L^2}{24}\left[\sqrt{3 + \mu\alpha^2} - \alpha\sqrt{\mu}\right]^2 \qquad \ldots(43.22)$$

*Special cases*:
(i) *For an isotropically reinforced* square slab,
$\alpha = 1$ and $\mu = 1$

$\therefore \quad m_u = \dfrac{w_u \alpha^2 L^2}{24}\left[\sqrt{3+1} - 1\sqrt{1}\right]^2 = \dfrac{w_u L^2}{24} \qquad \ldots(43.23)$

which is the same as *Eq 43.11 (a)*

(ii) *For an isotropically reinforced rectangular slab*,
$\mu = 1$

$$m_u = \frac{w_u \alpha^2 L^2}{24}\left[\sqrt{3 + \alpha^2} - \alpha\right]^2 \qquad \ldots(43.24)$$

**6. Polygonal and Circular Slabs:** Let us first consider polygonal slab, fixed along its edges, with isotropic positive moments $m_u$ and isotropic negative moment $m_u'$. Let the length of each side be $L$, and let the perpendicular distance of each side from centre be $r$, which is also the radius of the inscribed circle. The positive yield lines will be along the radial lines as shown in Fig. 43.11 (a). The negative yield lines will be along the outer edges.

Let us give a virtual displacement $\delta$ at the apex O.
For any sector ABO,

$$M_I = (m_u + m_u') L \frac{\delta}{r} \qquad \ldots(1)$$

$$M_E = \frac{1}{2} L r \frac{\delta}{3} w_u \qquad \ldots(2)$$

Equating the two, we get

$$w_u L r \frac{\delta}{6} = (m_u + m_u') L \frac{\delta}{r}$$

or $\quad w_u = \dfrac{6(m_u + m_u')}{r^2} \qquad \ldots(43.25)$

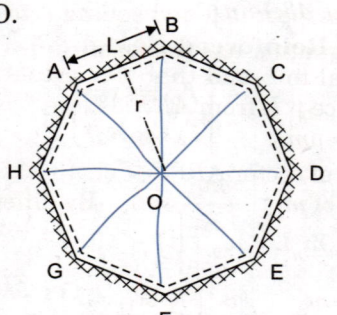
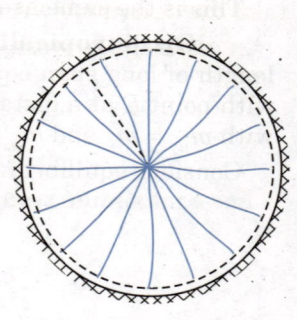

(a) Polygonal slab　　　(b) Circular slab

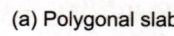

For a circular slab, there will be innumerable number of positive yield lines, such that number of sides $n$ will tend to be infinity and $L$ will tend to be zero. However, $r$ will be the radius of the circle, and for any sector $ABO$,

$$w_u = \frac{6(m_u + m_u')}{r^2}$$

Thus, *Equation 43.25* is applicable to both polygonal as well as circular slabs.

For circular slabs simply supported along the edge, $m_u' = 0$

$$\therefore \quad w_u = \frac{6 m_u}{r^2} \quad ...[43.26] \quad \text{or} \quad m_u = \frac{w_u r^2}{6} \quad ...[43.26(a)]$$

For hexagonal slab, $\quad r = \sqrt{3}\, L/2$

Hence from Eq 43.25, $\quad w_u = \dfrac{6(m_u + m_u')}{(\sqrt{3}\, L/2)^2} = \dfrac{8(m_u + m_u')}{L^2}$

If the hexagonal slab is simply supported.

$$w_u = \frac{8 m_u}{L^2} \quad ...(43.27)$$

or $\quad m_u = \dfrac{w_u L^2}{8} \quad ...[43.27(a)]$

## 43.6. ANALYSIS BY EQUILIBRIUM METHOD

**1. Isotropically Reinforced Square Slab:** Fig. 43.12 (a) shows a square slab, simply supported along the edges. $m_u$ will be the same along all the directions. The yield lines will be along the two diagonals. Considering the equilibrium of the triangular sector $ABO$, and taking moments about edge $AB$, we get

$$m_u L = \frac{1}{2} \cdot L \cdot \frac{L}{2} \cdot w_u \cdot \frac{L}{6}$$

or $\quad m_u = \dfrac{w_u L^2}{24} \quad ...[43.11(a)]$

This is the same as *Eq. 43.11 (a)*.

Again, for a fixed slab of Fig. 43.12 (b), positive yield lines will be along the diagonals, while the negative yield lines will be developed along the fixed edges. Considering the equilibrium of sector $ABO$, and taking moments about edge, $AB$, we get

$$m_u L + m_u L = \frac{1}{2} L \cdot \frac{L}{2} \cdot w_u \cdot \frac{L}{6}$$

or $\quad m_u = \dfrac{w_u L^2}{48} \quad ...[43.14(a)]$

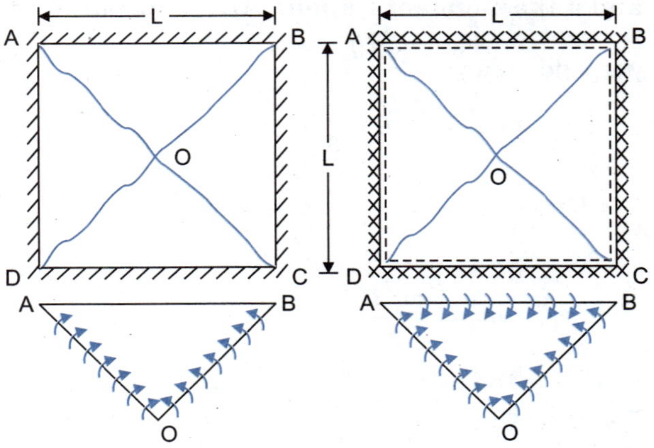

(a) Simply supported   (b) Fixed

**Fig. 43.12.** Square slabs.

This is the same as *Eq. 43.14(a)*.

**2. Orthotropically Reinforced Rectangular Slab:** Fig. 43.13 (a) shows a rectangular slab, with length of long span equal to $L$ and that of short edge as $\alpha L$. The yield line pattern will be symmetrical with point $E$ at a distance $\beta L$ from $AD$, where $\beta$ is unknown. Let the slab be orthotropically reinforced with $m_{uy} = m_u$ and $m_{ux} = \mu m_u$.

Consider equilibrium of sector $ABFE$ [Fig. 43.13 (b)]. Taking moments about edge $AB$, we get.

$$m_u L = w_u \left[ \left\{ (1 - 2\beta) L \frac{\alpha L}{2} \frac{\alpha L}{4} \right\} + 2 \left\{ \frac{1}{2} \beta L \frac{\alpha L}{2} \frac{\alpha L}{6} \right\} \right]$$

or $\quad m_u = w_u\, \alpha^2 L^2 \left[ \dfrac{(1-2\beta)}{8} + \dfrac{\beta}{12} \right] = \dfrac{w_u\, \alpha^2 L^2}{24}(3 - 4\beta) \quad ...(1)$

Similarly, consider equilibrium of sector *FBC*. Taking moments about edge *BC*, we get

$$\mu m_u \alpha L = \frac{1}{2} \alpha L \beta L \frac{\beta L}{3} w_u = \frac{\alpha \beta^2 L^3}{6} w_u$$

or $\quad m_u = \dfrac{\beta^2 L^2}{6\mu} w_u \quad$ ...(2)

Equating the two equilibrium equations, we get

$$\frac{w_u \alpha^2 L^2}{24} (3 - 4\beta) = \frac{\beta^2 L^2}{6\mu} w_u$$

or $\quad \mu \alpha^2 (3 - 4\beta) = 4\beta^2$

or $\quad 4\beta^2 + 4\mu \alpha^2 \beta - 3\mu \alpha^2 = 0$

$$\therefore \quad \beta = \frac{1}{2}\left[\sqrt{\mu^2 \alpha^4 + 3\mu \alpha^2} - \mu \alpha^2\right]$$

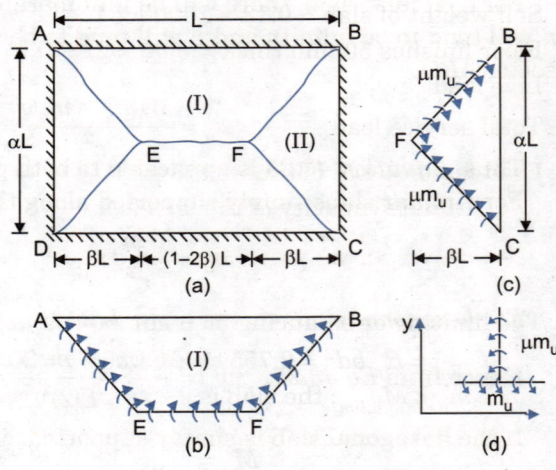

Fig. 43.13. Rectangular slab.

This is the same as *Eq. 43.21*. Substituting this value of β in *Eq. (2)*, we get

$$m_u = \frac{w_u L^2}{6\mu} \times \frac{1}{4}\left[\sqrt{\mu^2 \alpha^4 + 3\mu\alpha^2} - \mu\alpha^2\right]^2$$

or $\quad m_u = \dfrac{w_u L^2}{24\mu} \times \mu\alpha^2 \left[\sqrt{3 + \mu\alpha^2} - \alpha\sqrt{\mu}\right]^2 = \dfrac{w_u \alpha^2 L^2}{24}\left[\sqrt{3 + \mu\alpha^2} - \alpha\sqrt{\mu}\right]^2$

This is the same as *Eq. 43.22*.

**3. Isotropically Reinforced Hexagonal Slab:** Consider the equilibrium of sector *ABO*, and take moments about *AB*. The value of perpendicular $r = \dfrac{\sqrt{3} L}{2}$.

$\therefore \quad m_u L = \dfrac{1}{2} L r \dfrac{r}{3} w_u$

or $\quad m_u = \dfrac{w_u}{6} r^2 = \dfrac{w_u}{6}\left(\dfrac{\sqrt{3} L}{2}\right)^2$

$\therefore \quad m_u = \dfrac{w_u L^2}{8}$

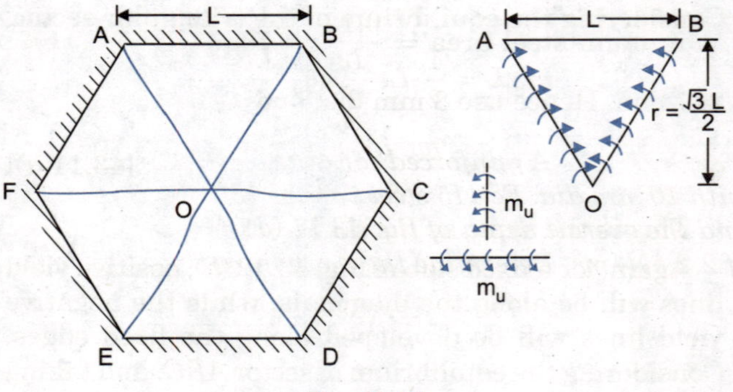

Fig. 43.14

This is the same as *Eq. 43.27 (a)*

**Example 43.1.** *A square slab of side length 4 m is simply supported at the ends and carries a service live load of 3 kN/m². Design the slab. Use M 20 concrete and Fe 415 steel bars.*

**Solution:** Given : $L = 4$ m ; $f_{ck} = 20$ N/mm² ; $f_y = 415$ N/mm²., Service load = 3 kN/m²

For Fe 415 steel, $\quad \dfrac{x_{u,\,max}}{d} = \dfrac{700}{1100 + 0.87 \times 415} = 0.479$

$\therefore \quad R_u = 0.36 f_{ck} \dfrac{x_{u,\,max}}{d}\left(1 - 0.42 \dfrac{x_{u,\,max}}{d}\right)$

$= 0.36 \times 20 \times 0.479 (1 - 0.42 \times 0.479) = 2.755$

Minimum effective depth $= \dfrac{\text{span}}{35} = \dfrac{4000}{35} \approx 114$ mm

Let us provide an overall depth = 130 mm.

Using 15 mm clear cover and 10 mm dia. bars, average $d = 130 - 15 - 10 = 105$ mm.

Self weight of slab = 0.13 × 25000 × 1 × 1 = 3250 N/m²
Floor finishes 50 mm thick = 0.05 × 22000 = 1100 N/m² (say)
Live load = 3000 N/m²
Total service load = 7350 N/m².

Load factor = 1.5. Ultimate design load $w_u$ = 1.5 × 7350 = 11025 N/m²
∴ Ultimate capacity of the slab is given by *Eq. 43.11 (a)*

$$m_u = \frac{w_u L^2}{24} = \frac{11025 (4)^2}{24} = 7350 \text{ N-m} = 7.35 \times 10^6 \text{ N-mm}$$

The limiting or balancing moment capacity of the slab is given by
$M_{u.lim} = R_u bd^2 = 2.755 (1000) (105)^2 = 30.37 \times 10^6$ N-mm
Since $m_u < M_{u.lim}$, the slab is under-reinforced. The reinforcement in the slab is given by

$$M_u = m_u = 0.87 f_y A_{st} d \left[1 - \frac{A_{st} f_y}{bd f_{ck}}\right]$$

∴ $7.35 \times 10^6 = 0.87 \times 415 A_{st} \times 105 \left[1 - \frac{A_{st} \times 415}{1000 \times 105 \times 20}\right]$

$A_{st}^2 - 5060 A_{st} + 0.981 \times 10^6 = 0$, which gives $A_{st} = 202$ mm²
Alternatively, from *Eq. 33.13* applicable for under-reinforced sections,

$$A_{st} = \frac{0.5 f_{ck}}{f_y}\left[1 - \sqrt{1 - \frac{4.6 M_u}{f_{ck} bd^2}}\right] bd = \frac{0.5 \times 20}{415}\left[1 - \sqrt{1 - \frac{4.6 \times 7.35 \times 10^6}{20 \times 1000 (105)^2}}\right] 1000 \times 105 = 202 \text{ mm}^2$$

Minimum steel area = $\frac{0.12}{100} \times 1000 \times 130 = 156$ mm². Using 8 mm dia. bars, spacing = $\frac{1000 \times 50}{202}$
≈ 247 mm. Hence use **8 mm dia. bars @ 240 c/c both ways.**

**Example 43.2.** *A reinforced concrete slab 5 m × 5 m is simply supported along the four edges and is reinforced with 10 mm dia. Fe 415 steel bars at 150 mm c/c both ways. The average effective depth of the slab is 100 mm and the overall depth of the slab is 130 mm. The slab carries a flooring of 50 mm thick having unit weight of 2.2 kN/m². Determine the maximum permissible service load, if M 20 concrete is used.*

**Solution:** $A_{st} = \frac{1000 \times 78.54}{150} = 523.6$ mm²/m

Now, $M_u = 0.87 f_y A_{st} d \left[1 - \frac{A_{st} f_y}{bd f_{ck}}\right] = 0.87 \times 415 \times 523.6 \times 100 \left[1 - \frac{523.6 \times 415}{1000 \times 100 \times 20}\right]$

$= 16.85 \times 10^6$ N-mm
$M_{u.lim} = R_u bd^2 = 2.761 (1000) (100)^2 = 27.61 \times 10^6$ N-mm

Since $M_u > M_{u.lim}$ the slab is under-reinforced.
Again, $m_u = M_u = 16.85 \times 10^6$ N-mm = 16.85 kN-m

But $m_u = \frac{w_u L^2}{24}$ [Eq. 43.11 (a)].

∴ $w_u = \frac{24}{L^2} m_u = \frac{24}{(5)^2} \times 16.85 = 16.176$ kN/m².

∴ Service load = 16.176/1.5 ≈ 10.78 kN/m²
Dead load of slab = 0.130 × 25 = 3.25 kN/m²
Dead load of finishing = 0.05 × 22 = 1.10 kN/m²
Total dead load = 3.25 + 1.10 = 4.35 kN/m²
∴ Permissible service load = 10.78 − 4.35 = **6.43 kN/m².**

**Example 43.3.** *A reinforced concrete slab, 4 m × 6 m is reinforced with 10 mm diameter bars at 150 mm spacing in the short direction and 200 mm spacing in the long direction. The slab is 100 mm thick with a average effective depth of 80 mm. If the yield lines are inclined at 45° to either direction of reinforcement, find the ultimate moments $m_{uan}$ and $m_{uat}$ for unit length along yield line. Use M 20 concrete and Fe 415 steel.*

**Solution:** Refer Fig. 43.10 in which the yield lines *AE, DE, BF* and *CF* are inclined at 45° to the principal *x* and *y* directions. The reinforcement in *y*-direction is spaced at 150 mm c/c while that in *x*-direction is spaced at 200 mm c/c.

Now $M_{uy} = m_{uy}$ = moment across plane whose normal is in *y*-direction.

$$= 0.87 f_y A_{st} d \left(1 - \frac{A_{st} f_y}{bd f_{ck}}\right)$$

where $A_{st} = \dfrac{1000}{150} \times 78.6 = 524$ mm²

∴ $m_{uy} = 0.87 \times 415 \times 524 \times 80 \left(1 - \dfrac{524 \times 415}{1000 \times 80 \times 20}\right) = 13.078 \times 10^6$ N-mm = 13.078 kN-m

Also, $M_{ux} = m_{ux}$ = moment across plane whose normal is in *x*-direction.

$$= 0.87 f_y A_{st} d \left(1 - \frac{A_{st} f_y}{bd f_{ck}}\right)$$

where $A_{st} = \dfrac{1000}{200} \times 78.6 = 393$ mm²

∴ $m_{ux} = 0.87 \times 415 \times 393 \times 80 \left(1 - \dfrac{393 \times 415}{1000 \times 80 \times 20}\right) = 10.194 \times 10^6$ N-mm = 10.194 kN-m.

Now, from *Eqs. 43.1*, and *43.6*

$m_{uan} = m_{ux} \cos^2 \alpha + m_{uy} \sin^2 \alpha$; and $m_{uat} = (m_{ux} - m_{uy}) \sin \alpha \cos \alpha$.

Here $\alpha = 45°$ ; $\cos \alpha = \sin \alpha = 0.707$

$m_{uan} = 10.194 (0.707)^2 + 13.078 (0.707)^2 = $ **11.636 kN-m**

$m_{uat} = (10.194 - 13.078) 0.707 \times 0.707 = $ **− 1.442 kN-m**

(Negative sign shows that the direction of torsional moment is opposite to the one marked in Fig. 43.5 (c)).

**Example 43.4.** *A rectangular slab 3.5 × 4.5 m is isotropically reinforced with 8 mm dia. bars spaced at 150 mm bothways. The average effective depth may be taken as 80 mm and the total depth of the slab is 100 mm. If Fe 415 steel and concrete of grade M 20 are used, determine the safe service live load. The dead load of floor finishing may be assumed as 1.5 kN/m².*

**Solution:** Given $L = 4.5$ m ; $\alpha L = 3.5$ m ; $\alpha = 3.5/4.5 = 0.7778$.

Since the slab is isotropically reinforced, $m_{ux} = m_{uy} = m_u$. Hence $\mu = 1$

$$A_{st} = \frac{1000 \times 50.27}{150} \approx 335 \text{ mm}^2/\text{m}.$$

The ultimate moment is given by

$$M_u = m_u = 0.87 f_y A_{st} d \left[1 - \frac{A_{st} f_y}{bd f_{ck}}\right] = 0.87 \times 415 \times 335 \times 80 \left[1 - \frac{335 \times 415}{1000 \times 80 \times 20}\right]$$

$$= 8.835 \times 10^6 \text{ N-mm} = 8.835 \text{ kN-m}$$

The yield moment of the slab is given by *Eq. 43.24*

$$m_u = \frac{w_u \alpha^2 L^2}{24} \left[\sqrt{3 + \alpha^2} - \alpha\right]^2$$

or $w_u = \dfrac{24 m_u}{\alpha^2 L^2 \left[\sqrt{3 + \alpha^2} - \alpha\right]^2} = \dfrac{24 \times 8.835}{(0.7778)^2 \times (4.5)^2 [\sqrt{3 + (0.7778)^2} - 0.7778]^2}$

$= 13.76$ kN/m².

∴ Total serviced load = 13.76/1.5 ≈ **9.17 kN/m²**

Dead load of slab = 0.1 × 25 = 2.5 kN/m²
Dead load of floor finishing = 1.5 kN/m²
∴    Service live load = 9.17 – (2.5 + 1.5) = **5.17 kN/m².**

**Example 43.5.** *For the slab of example 43.3 determine the safe permissible service live load. The dead load of flooring may be assumed as 1 kN/m².*

**Solution:** Refer Fig. 43.10

Given : $L = 6$ m. ;   $\alpha L = 4$ m.

∴    $\alpha = 4/L = 4/6 = 0.667$ ;   $D = 100$ mm ;   $d = 80$ mm

As found in example 43.3.

$$m_{uy} = 13.078 = m_u; \quad m_{ux} = 10.194 = \mu m_u.$$
$$\mu = 10.194/13.078 = 0.779$$

Now from *Eq. 43.22*

$$m_u = \frac{w_u \alpha^2 L^2}{24}\left[\sqrt{3+\mu\alpha^2} - \alpha\sqrt{\mu}\right]^2$$

$$w_u = \frac{24 m_u}{[\alpha^2 L^2 \sqrt{3+\mu\alpha^2} - \alpha\sqrt{\mu}]^2}$$

$$= \frac{24 \times 13.078}{(0.667)^2 (6)^2 [\sqrt{3 + 0.779(0.667)^2} - 0.667\sqrt{0.779}]^2}$$

$$= \frac{24 \times 13.078}{0.445 \times 36 \times 1.539} = 12.729 \text{ kN/m}^2$$

∴    Service load = 12.729/1.5 = 8.486 kN/m²
Dead load of slab = 0.1 × 25 = 2.5 kN/m²
Dead load of flooring = 1.0 kN/m²
∴    Service live load = 8.486 – 3.5 = **4.986 kN/m².**

**Example 43.6.** *A rectangular slab 3.5 × 5 m in size, simply supported at the edges. The slab is expected to carry a service live load of 3 kN/m² and a floor finishing load of 1 kN/m². Use M 20 concrete and Fe 415 steel. Design the slab if (a) it is isotropically reinforced (b). If it is orthotropically reinforced with $\mu = 0.75$.*

**Solution:** Given : $L = 5$ m ;   $\alpha L = 3.5$ ;   ∴   $\alpha = 3.5/5 = 0.7$

$$\left(\frac{\text{span}}{\text{depth}}\right) = 35. \quad ∴ \quad \text{Depth of slab} = 350/35 = 10 \text{ cm} = 100 \text{ mm}$$

Use 10 mm dia. bars both ways. Using 15 mm clear cover, assume average effective depth
= 100 – 15 – 10 = 75 mm

Dead weight of slab = 0.1 × 25 = 2.5 kN/m²
Dead weight of flooring    = 1.0 kN/m²
Live load    = 3.0 kN/m²
∴    Total service load    = 6.5 kN/m²
∴    Ultimate design load = $w_u$ = 1.5 × 6.5 = 9.75 kN/m²

**(a) Isotropically reinforced slab :**   $\mu = 1$

Now, ultimate moment on slab is given by *Eq 43.22* :

$$m_u = \frac{w_u \alpha^2 L^2}{24}\left[\sqrt{3+\mu\alpha^2} - \alpha\sqrt{\mu}\right]^2 = \frac{9.75 (0.7)^2 (5)^2}{24}\left[\sqrt{3+1(0.7)^2} - 0.7\sqrt{1}\right]^2$$

= 6.791 kN-m/m = 6.791 × 10⁶ N-mm/m

The reinforcement is given by

$$M_u = m_u = 0.87 f_y A_{st} d \left[1 - \frac{A_{st} f_y}{bd f_{ck}}\right]$$

or $\qquad 6.791 \times 10^6 = 0.87 \times 415 A_{st} \times 75 \left[1 - \dfrac{A_{st} \times 415}{1000 \times 75 \times 20}\right]$, which simplifies to

$A_{st}^2 - 3614 A_{st} + 906459 = 0$.  Hence $A_{st} = 272$ mm$^2$

Using 8 mm dia. bars,   spacing $= \dfrac{1000 \times 50.3}{272} \approx 185$ mm.

However provide 8 mm dia. bars @ 175 mm c/c both ways.

**(b) Orthotropically reinforced slab**

$\mu = 0.75$

$\therefore \qquad m_u = \dfrac{w_u\, \alpha^2\, L^2}{24} \left[\sqrt{3 + \mu\, \alpha^2} - \alpha\sqrt{\mu}\right]^2$

$\qquad = \dfrac{9.75\,(0.7)^2\,(5)^2}{24} \left[\sqrt{3 + 0.75\,(0.7)^2} - 0.7\sqrt{0.75}\right]^2$

$\qquad = 8.541$ kN-m/m $= 8.541 \times 10^6$ N-mm/m

But $\qquad m_u = 0.87 f_y A_{st}\, d \left[1 - \dfrac{A_{st}\, f_y}{bd\, f_{ck}}\right]$

or $\qquad 8.541 \times 10^6 = 0.87 \times 415 A_{st}\,(75) \left[1 - \dfrac{A_{st} \times 415}{1000 \times 75 \times 20}\right]$

or  $A_{st}^2 - 3614.4 A_{st} + 1140000 = 0$, which gives $A_{st} = 350$ mm$^2$.

$\therefore$ Spacing of 8 mm dia. bars in short direction $= \dfrac{1000 \times 50.3}{350} \approx 144$ mm and spacing of 8 mm dia. in long direction $= 144/0.75 = 192$ mm.

Hence provide 8 mm dia. bars @ 140 mm c/c in short direction and @ 190 mm c/c in long direction.

**Example 43.7.** *A right angled rectangular slab is simply supported along the two edges. The length of the supported edges are 3 m and 4 m respectively. The slab is isotropically reinforced with 8 mm dia. bars spaced at 100 mm centres both ways. The average effective depth of the slab is 95 mm and total depth is 120 mm. Using M 20 concrete and HYSD bars (Fe 415), determine the safe permissible service live load on the slab. Assume dead load of floor finishing at 1.5 kN/m$^2$.*

**Solution:**

Given : $\qquad L = 4$ m ; $\alpha L = 3$ m   $\therefore \quad \alpha = \dfrac{3}{4} = 0.75$

$A_{st} = \dfrac{1000 \times 50.3}{100} = 503$ mm$^2$.  The ultimate moment is given by

$m_u = M_u = 0.87 f_y A_{st}\, d \left[1 - \dfrac{A_{st}\, f_y}{bd\, f_{ck}}\right] = 0.87 \times 415 \times 503 \times 95 \left[1 - \dfrac{503 \times 415}{1000 \times 95 \times 20}\right]$

$\qquad = 15.357 \times 10^6$ N-mm $= 15.357$ kN-m.  Now, from *Eq. 43.17*,

$\qquad m_u = \dfrac{w_u\, \alpha\, L^2}{6} \sin^2 \beta/2 \qquad$ or $\qquad w_u = \dfrac{6\, m_u}{\alpha\, L^2 \sin^2 \beta/2}$

Here, $\qquad \beta = 90°$.

$\therefore \qquad w_u = \dfrac{6 \times 15.357}{0.75\,(4)^2\,(\sin 45°)^2} = 15.357$ kN/m$^2$

$\therefore$ Total service load $= 15.357/1.5 \qquad\qquad = 10.238$ kN/m$^2$

Dead load of slab $\qquad = 0.12 \times 25 \qquad\qquad = 3$ kN/m$^2$

Dead load of flooring $\qquad\qquad\qquad\qquad = 1.5$ kN/m$^2$

Total dead load $\qquad\qquad\qquad\qquad\qquad = 4.5$ kN/m$^2$

$\therefore$ Permissible service live load $= 10.238 - 4.5 = 5.738$ kN/m$^2$.

# YIELD LINE THEORY AND DESIGN OF SLABS

**Example 43.8.** *A hexagonal slab is simply supported along the edges with a side length of 3.5 m. It is isotropically reinforced with 10 mm dia. bars at 125 mm centres both ways, at an average effective depth of 120 mm. The overall depth of the slab is 150 mm. Calculate (i) ultimate load capacity of slab, (ii) safe permissible service live load. Use M 20 concrete and Fe 415 steel. Take the load of floor finishing as 1.25 kN/m².*

**Solution:**
$$L = 3.5 \text{ m.} \; ; \quad A_{st} = \frac{1000 \times 78.54}{125} = 628.3 \text{ mm}^2$$

For M 20 – Fe 415 combination, $R_u = 2.761$, as in Example 43.1

$$m_u = M_u = 0.87 f_y A_{st} d \left[1 - \frac{A_{st} f_y}{bd f_{ck}}\right] = 0.87 \times 415 \times 628.3 \times 120 \left[1 - \frac{628.3 \times 415}{1000 \times 120 \times 20}\right]$$

$$= 24.264 \times 10^6 \text{ N-mm} = \mathbf{24.264 \text{ kN-m.}}$$

Now, from Eq. 43.28

$$w_u = \frac{8 m_u}{L^2} = \frac{8 \times 24.264}{(3.5)^2} = 15.845 \text{ kN/m}^2$$

∴ Total service load = 15.845/1.5 = 10.564 kN/m²
Dead load of slab = 0.15 × 25 = 3.75 kN/m²
Load of floor finishing = 1.25 kN/m²

Total dead load = 5 kN/m²

∴ Permissible service live load = 10.564 – 5 = **5.564 kN/m²**.

**Example 43.9.** *Design a reinforced circular slab for the following data :*
(i) *Diameter of slab :* 5.5 m
(ii) *Service live load :* 4 kN/m²
(iii) *Floor finishing load :* 1 kN/m²
(iv) *Grade of concrete :* M 20
(v) *Grade of steel :* Fe 415. *The slab is simply supported along the edge.*

**Solution:** For a simply supported two way slab,

$$\frac{\text{Span}}{\text{Depth}} = 35 \quad \therefore \quad \text{Depth} = \frac{5500}{35} \approx 157 \text{ mm.}$$

Let us provide overall depth = 160 mm. Using 10 mm dia. bars and a clear cover of 15 mm, average effective depth = 160 – 15 – 10 = 135 mm.

Self load of slab = 0.16 × 25 = 4 kN/m²
Dead load of flooring = 1 kN/m²
Live load = 4 kN/m²

Total service load = 9 kN/m²

$$w_u = 9 \times 1.5 = 13.5 \text{ kN/m}^2$$

Now, from Eq. 43.26 (a),

$$m_u = \frac{w_u r^2}{6} = \frac{13.5 (5.5/2)^2}{6} = 17.016 \text{ kN-m/m} = 17.016 \times 10^6 \text{ N-mm/m.}$$

Limiting moment capacity:
$$M_{u.lim} = R_u bd^2 = 2.755 (1000)(135)^2 = 50.21 \times 10^6 \text{ N-mm.}$$

Since $m_u < M_{u.lim}$ the slab is under reinforced.

$$\therefore \quad m_u = M_u = 0.87 f_y A_{st} d \left(1 - \frac{A_{st} f_y}{bd f_{ck}}\right)$$

or $\quad 17.016 \times 10^6 = 0.87 \times 415 \, A_{st} \times 135 \left(1 - \dfrac{A_{st} \times 415}{1000 \times 135 \times 20}\right)$

or $\quad A_{st}^2 - 6506 \, A_{st} + 2.271 \times 10^6 = 0$, which gives $A_{st} = 370.4 \text{ mm}^2$

Alternatively,

$$A_{st} = \frac{0.5 f_{ck}}{f_y}\left[1 - \sqrt{1 - \frac{4.6 M_u}{f_{ck} bd^2}}\right] bd = \frac{0.5 \times 20}{415}\left[1 - \sqrt{1 - \frac{4.6 \times 17.016 \times 10^6}{20 \times 1000 (135)^2}}\right] \times 1000 \times 135$$

$$= 370.4 \text{ mm}^2$$

Spacing of 10 mm dia. bars = $\dfrac{1000 \times 78.5}{370.4}$ = 212 mm

However, use 10 mm dia. bars @ 200 mm centres both ways.

## PROBLEMS

1. A reinforced concrete square slab, 3.5 m × 3.5 m is simply supported at the ends and is reinforced with 8 mm diameter bars spaced at 150 mm centres both ways. Determine the safe service live load if the average effective depth of slab is 100 mm and the total thickness of slab, inclusive of flooring, is 160 mm. Use M 20 concrete and HYSD (Fe 415) bars.

2. Redesign the slab of problem 1 to carry a service load of 4 kN/m², inclusive of floor finishes, but exclusive of self load.

3. A rectangular slab 4 m × 5 m is simply supported at the ends. Design the slab to carry super-imposed service load of 5 kN/m², if the slab is to be isotropically reinforced. Use M 20 concrete and Fe 415 steel.

4. Redesign the slab of problem 3 if the slab is to be orthotropically reinforced, with μ = 0.8.

5. Find the safe superimposed service load on the slab of problem 3, if it is reinforced with 8 mm diameter bars spaced at 125 mm centres both ways. The total thickness of the slab may be assumed as 125 mm.

6. A triangular slab has equal side lengths of 4.5 m is supported on two edges and is isotropically reinforced with 8 mm dia. bars of Fe 415 grade, spaced at 125 mm centres both ways. Determine (*i*) ultimate moment capacity, (*ii*) ultimate collapse load. The total thickness of slab is 125 mm. in M 20 Grade concrete.

7. A hexagonal slab, simply supported on all the sides has a side lenth of 3 m. Find the uniformly distributed load which would cause failure of the isotropically reinforced slab, if the ultimate moment of resistance of slab is 2.7 kN-m/m. Work from the first principles.

8. A rectangular slab 6 m × 4.5 m, simply supported at its edge is to be designed as an isotropically reinforced slab to support an uniformly distributed working load of 18 kN/m² which includes the self weight of the slab. Calculate the ultimate moment of resistance required for the slab section from first principles.

9. Design a circular slab of 5 m diameter, simply supported along the edge, to carry a service live load of 5 kN/m². Use M 20 concrete and Fe 415 steel.

10. Design a rectangular slab of size 5 m × 4 m, is simply supported at edges to support a service load of 4 kN/m², adopt $M_{25}$ and Fe 415 grade materials (take orthotropically reinforced with μ = 0.7).

11. Design a simply supported slab of 3.75 metre side length to supports a live load of 3.5 kN/m² adopt $M_{25}$ and Fe 415 grade materials.

# 44 FOUNDATIONS

## 44.1. INDIAN STANDARD CODE RECOMMENDATIONS FOR DESIGN OF FOOTINGS (IS : 456–2000)

### (A) General

Foundation is an important part of the structure which transfers the load of super structure the sub soil. Foundation increase the stability of structures.

1. In sloped or stepped footings, the effective cross-section in compression shall be limited by the area above the neutral plane, and the angle of slope or depth and location of steps shall be such that the design requirements are satisfied at every section. Sloped and stepped footings that are designed as a unit shall be constructed to assure action as a unit.

2. *Thickness at the edge of footing.* In reinforced and plain concrete footings, the thickness at edge shall be not less than 15 cm (or) 150 mm for footings on soils nor less than 30 cm (or) 300 mm above the tops of piles for footings on piles.

3. In the case of plain concrete pedestals, the angle $\alpha$ between plane passing through the bottom edge of the pedestal and the corresponding junction edge of the column with pedestal and the horizontal plane (Fig. 44.1) shall be governed by the expression:

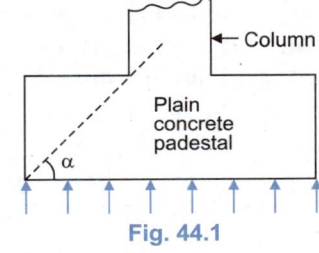

Fig. 44.1

$$\tan \alpha \not< 0.9 \sqrt{\frac{100 q_0}{f_{ck}} + 1} \qquad ...(44.1)$$

where $q_0$ = calculated maximum bearing pressure at the base of the pedestal in N/mm², and
$f_{ck}$ = characteristic strength of concrete at 28 days in N/mm².

### (B) Moments and Forces

1. In the case of footings on piles, computation for moments and shears may be based on the assumption that the reaction from any pile is concentrated at the centre of the pile.

2. For the purposes of computing stress in footings which support a round or octagonal concrete column or pedestal, the face of the column or pedestal shall be taken as the side of a square inscribed within the perimeter of the round or octagonal column or pedestal.

3. *Bending Moment*
   (i) The bending moment at any section shall be determined by passing through the section a vertical plane which extends completely across the footing and computing the moment of the forces acting over the entire area of the footing on one side of the said plane.

(*ii*) The greatest bending moment to be used in design of an isolated concrete footing which supports a column, pedestal or wall, shall be the moment computed in the manner prescribed in (*i*) at sections located as follows:

(*a*) At the face of the column, pedestal or wall, for footings supporting a concrete column, pedestal or wall,

(*b*) Half way between the centre line and the edge of the wall, for footings under masonry walls, and

(*c*) Half way between the face of the column or pedestal and the edge of the gusseted base, for footing under gusseted bases.

**4. Shear and bond**

(*i*) The shear strength of footings is governed by the more severe of the following two conditions:

(*a*) The footing acting essentially as a wide beam, with a potential diagonal crack extending in a plane across the entire width; the critical section for this condition shall be assumed as a vertical section located from the face of the column, pedestal or wall at a distance equal to the effective depth of the footing in case of footings on soils, and a distance equal to half the effective depth of footing for footings on piles.

(*b*) Two-way action of the footing, with potential diagonal cracking along the surface of truncated cone or pyramid around the concentrated load; in this case, the footing shall be designed for shear in accordance with appropriate provisions discussed below (Fig. 44.2).

(*ii*) In computing the external shear on any section through a footing supported on piles, the entire reaction from any pile of diameter $D_p$ whose centre is located $D_p/2$ or more outside the section shall be assumed as producing shear on the section; the reaction from any pile whose centre is located $D_p/2$ or more inside the section shall be assumed as producing no shear on the section. For intermediate positions of the pile centre, the portion of the pile reaction to be assumed as producing shear on the section shall be based on straight line interpolation between full value at $D_p/2$ outside the section and zero value at $D_p/2$ inside the section.

(*iii*) The critical section for checking the development length in a footing shall be assumed at the same planes as described for bending moment in B(3) and also at all other vertical planes where abrupt changes of section occur. If the reinforcement is curtailed, the anchorage requirements should be checked.

Thus, according to the above provision, shear stress is to be checked for (*i*) *one way action* (*i.e.,* beam shear) for which the governing section *AB* is at a distance *d* from the face of column or pedestal [Fig. 44.2(*a*)] and (*ii*) *two way action,* (*i.e.,* punching shear), for which the governing section is along the perimeter *ABCD* situated at a distance *d*/2 from the face of the column or pedestal [Fig. 44.2(*b*)].

For the *two way action,* the calculated shear stress $\tau_v$ should satisfy the following relation

$$\tau_v \le k_s \cdot \tau_c \qquad \ldots[44.2(a)]$$

where $\tau_v = \dfrac{F}{2\,[(a+d)+(b+d)]\,d} \qquad \ldots[44.2(b)]$

$k_s = (0.5 + \beta_c)$, but not greater than 1.0

$\beta_c = \dfrac{b}{a} = \dfrac{\text{short side of column}}{\text{long side of column}} \qquad \ldots[44.2(c)]$

$\tau_c = 0.25\,\sqrt{f_{ck}}$ N/mm² in limit stress method
$\qquad \ldots[44.2(d)]$

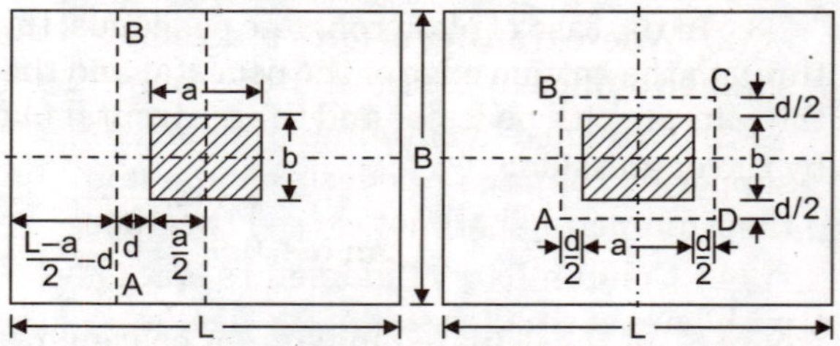

and $F$ = net S.F. acting on the perimeter

**Fig. 44.2.** Critical section for shear.

For the *beam shear*, the nominal shear stress across *AB* should satisfy the relation

$$\tau_v \le k\,\tau_c \qquad \ldots(44.3)$$

where $\tau_c$ = the permissible shear stress for the grade of the concrete, corresponding to the reinforcement

$= \left(100\,\dfrac{A_s}{bd}\right)$ as given in Table 36.1.

$k$ = factor for slabs, as given in Table 36.3.

### (C) Tensile Reinforcement

The total tensile reinforcement at any section shall provide a moment of resistance at least equal to the bending moment on the section calculated in accordance with B(3). Total tensile reinforcement shall be distributed across the corresponding resisting section as given below.

(i) In one-way reinforced footing, the reinforcement shall be distributed uniformly across the full width of footing;

(ii) In two-way reinforced square footing, the reinforcement extending in each direction shall be distributed uniformly across the full width of the footing;

(iii) In two-way reinforced rectangular footing, the reinforcement in long direction shall be distributed uniformly across the full width of the footing. For reinforcement in the short direction, a central band equal to the width of the footing shall be marked along the length of the footing and portion of the reinforcement determined in accordance with the equation given below shall be uniformly distributed across the central band:

$$\frac{\text{Reinforcement in central band width}}{\text{Total reinforcement in short direction}} = \frac{2}{\beta + 1}$$

where $\beta$ is the ratio of the long side to the short-side of the footing.

The remainder of the reinforcement shall be uniformly distributed in the outer portions of the footing.

### (D) Transfer of load at the base of column

The compressive stress in concrete at the base of a column or pedestal shall be considered as being transferred by bearing to the top of the supporting pedestal or footing. The bearing pressure ($\sigma_{cbr}$) on the loaded area shall not exceed the permissible bearing stress in direct compression multiplied by a value equal to $\sqrt{A_1/A_2}$ but not greater than 2.

where  $A_1$ = supporting area for bearing, of footing, which in sloped or stepped footing may be taken as the area of the lower base of the largest frustum of a pyramid or cone contained wholly within the footing and having for its upper base, the area actually loaded and having side slope of one vertical to two horizontal, and

$A_2$ = Loaded area at the column base.

For limit state method of design the permissible bearing stress shall be $0.45 f_{ck}$.

Thus  $\sigma_{cbr} \leq 0.45 f_{ck} \sqrt{A_1/A_2}$ in limit state method ...[44.4]

The actual bearing pressure $\sigma_{cbr}$ = column load divided by the area of column at the base.

Thus  $\sigma_{cbr} = \dfrac{W}{a + b}$, where $a$ and $b$ are the sides of the column.

(1) Where the permissible bearing stress on the concrete in the supporting or supported member would be exceeded, reinforcement shall be provided for developing excess force, either by extending the longitudinal bars into the supporting members or by dowels (see 3 below).

(2) Where transfer of force is accomplished by reinforcement, the development length of the reinforcement shall be sufficient to transfer the compression or tension to the supporting member.

(3) Extended longitudinal reinforcement or dowels of at lest 0.5 percent of the cross-sectional area of the supported column or pedestal and a minimum of four bars shall be provided. Where the dowels are used, their diameter shall not exceed the diameter of the column bars by more than 3 mm.

(4) Column bars of diameters larger than 36 mm, in compression only can be dowelled at the footings with bars of smaller size of the necessary area. The dowel shall extend into the column, a distance equal to the development length of the column bar and into the footing, a distance equal to the development length of the dowel.

## 44.2. ISOLATED FOOTING OF UNIFORM DEPTH

The analysis of isolated footing of uniform depth has been thoroughly discussed in § 15.6, Chapter 15, and a reference may be made to that article for various formulae and expressions. The design procedure is explained in Example 44.1.

**Example 44.1.** *Isolated rectangular footing of uniform thickness for R.C. column:*

*Design a rectangular isolated footing of uniform thickness for R.C. column bearing a vertical load of 600 kN, and having a base size of 400 × 600 mm. The safe bearing capacity of the soil may be taken as 120 kN/m². Use M 20 concrete and Fe 415 steel.*

**990** REINFORCED CONCRETE STRUCTURE

**Solution:**

1. **Design constants.** For M 20 – Fe 415 combination, we have

$$\frac{x_{u,\max}}{d} = 0.479 \quad \text{and} \quad R_u = 2.761.$$

2. **Size of footing** $W = 600$ kN. Let $W'$ be equal to 10% $W = 60$ kN.

$$\therefore \quad A = \frac{660}{120} = 5.5 \text{ m}^2. \text{ Let ratio of } B \text{ to } L = \frac{400}{600} = \frac{2}{3}.$$

$$\therefore \quad \frac{2}{3} L \times L = 5.5 \quad \text{or} \quad L = 2.87 \approx 2.9 \text{ m}.$$

$$\therefore \quad B = \frac{2}{3} \times 2.9 = 1.93 \text{ m}. \text{ However, provide a footing of size } 2 \text{ m} \times 3 \text{ m}.$$

Net upward pressure $p_0 = \dfrac{600}{2 \times 3} = 100$ kN/m$^2$. $= 0.1$ N/mm$^2$

3. **Design of section.** Refer Fig. 15.14.

**(a) Design on the basis of bending compression**

Bending moment $M_1$ about section X-X is given by

$$M_1 = \frac{p_0 B}{8}(L-a)^2 \text{ kN-m} = \frac{100 \times 2}{8}(3-0.6)^2 \times 10^6 \text{ N-mm} = 144 \times 10^6 \text{ N-mm}$$

$$\therefore \quad M_{1u} = 1.5 \, M_1 = 1.5 \times 144 \times 10^6 = 216 \times 10^6 \text{ N-mm}$$

B.M. $M_2$ about section Y-Y is given by

$$M_2 = \frac{p_0 L}{8}(B-b)^2 \times 10^6 \text{ N-mm} = \frac{100 \times 3}{8}(2-0.4)^2 \times 10^6 = 96 \times 10^6 \text{ N-mm}.$$

$$\therefore \quad M_{2u} = 1.5 \, M_2 = 1.5 \times 96 \times 10^6 = 144 \times 10^6 \text{ N-mm}; \quad \text{Thus, } M_{2u} < M_{1u}.$$

$$\therefore \quad d = \sqrt{\frac{M_{1u}}{R_u B}} = \sqrt{\frac{216 \times 10^6}{2.761 \times 2000}} = 197.778 \approx 198 \text{ mm} \approx 200 \text{ mm}$$

and $\quad D = 200 + 60 = 260$ mm, providing 60 mm effective cover.

Provide uniform thickness for the entire footing.

**(b) Depth on the basis of one way shear.** For the *beam action*, total S.F. along section AB [Fig. 15.10(a)] is

$$V = p_0 B\left(\frac{L}{2} - \frac{a}{2} - d\right) = p_0 B\left(\frac{L-a}{2} - d\right)$$

$$= 100 \times 2 \left(\frac{3-0.6}{2} - 0.001 d\right) 10^3 = 200000 (1.2 - 0.001 d) \text{ N/m}.$$

$$V_u = 1.5 \, V = 300000 (1.2 - 0.001 d) \text{ N/m}$$

$$\tau_v = \frac{V_u}{Bd} = \frac{300000}{2000 \, d}(1.2 - 0.001 d) = \frac{150}{d}(1.2 - 0.001 d). \quad \ldots(i)$$

Assuming under-reinforced section, with $p = 0.3\%$, we get $\tau_c = 0.384$ N/mm$^2$, from Table 36.1 for M 20 concrete. Also, $k = 1$ (from Table 36.3) for $D \geq 300$ mm.

$\therefore$ Permissible shear stress $= 1 \times 0.384 = 0.384$ N/mm$^2$

Equating this to (i), we get

$$0.384 = \frac{150}{d}(1.2 - 0.001 d), \text{ from which } d = 337 \text{ mm} \approx 340 \text{ mm (say)}$$

### (c) Check for two way shear action

For the *two way shear action or punching shear action* along $ABCD$ [Fig. 15.10(b)],

Perimeter $ABCD = 2\{(a + d) + (b + d)\} = 2\{(0.6 + 0.34) + (0.4 + 0.34)\}$
$= 2\{0.94 + 0.74\} = 3.36$ m $= 3360$ mm.

Area $ABCD = 0.94 \times 0.74 = 0.6956$ m²

∴ Punching shear $= 1.5 \times 100 [(2 \times 3) - 0.6956] = 795.66$ kN

∴ $\tau_v = \dfrac{795.66 \times 1000}{3360 \times 340} = 0.696$ N/mm².

Allowable shear stress $\tau_c$ is given by

$$\tau_c = 0.25\sqrt{f_{ck}} = 0.25\sqrt{20} \approx 1.118 \text{ N/mm}^2.$$

$$k_s = (0.5 + \beta_c) = \left(0.5 + \dfrac{0.4}{0.6}\right) = 1.17; \quad \text{However, adopt max. } k_s = 1.$$

∴ $k_s \tau_c = 1 \times 1.118 = 1.118$ N/mm².

This is more than $\tau_v = 0.696$ N/mm². Hence safe.

Thus the effective depth $d = 340$ mm is alright. Keep $D = 420$ mm, so that using effective cover of 60 mm, $d = 420 - 60 = 360$ mm in one direction and $d = 360 - 12 = 348$ mm in the other direction, with 12 mm φ bars.

### 4. Design of reinforcement.

Since actual $d$ provided is more than that required for bending compression, we have an under-reinforced section. Hence area $A_{st1}$ of long bars calculate for B.M. $M_{1u}$ is given by

$$A_{st1} = \dfrac{0.5 f_{ck}}{f_y}\left[1 - \sqrt{1 - \dfrac{4.6 M_{1u}}{f_{ck} B d_1^2}}\right] B d_1$$

$$= \dfrac{0.5 \times 20}{415}\left[1 - \sqrt{1 - \dfrac{4.6 \times 216 \times 10^6}{20 \times 2000 \times (360)^2}}\right] 2000 \times 360 = 1751.33 \approx 1752 \text{ mm}^2$$

Hence number of 12 mm φ bars $= 1752/113 \triangleq 16$

These are to be distributed uniformly in a width $B = 2$ m

Effective depth for top layer of reinforcement $= d_2 = 348$ mm

The area $A_{st2}$ of short bars, calculated for $M_{2u}$ is given by

$$A_{st2} = \dfrac{0.5 f_{ck}}{f_y}\left[1 - \sqrt{1 - \dfrac{4.6 M_{2u}}{f_{ck} L d_2^2}}\right] L d_2 = \dfrac{0.5 \times 20}{415}\left[1 - \sqrt{1 - \dfrac{4.6 \times 144 \times 10^6}{20 \times 3000 (348)^2}}\right] 3000 \times 348$$

$\approx 1175$ mm²

This area is to be provided in two distinct band widths. Area $A_{st2(B)}$ in central band of width $B = 2$ m is given by

$$A_{st2(B)} = \dfrac{2 A_{st2}}{\beta + 1} = \dfrac{2 \times 1175}{\dfrac{3}{2} + 1} = 940 \text{ mm}^2.$$

∴ No. of 12 mm Φ bars $= 940/113 \approx 9$ to be provided in central band width $= 2$ m.

Remaining area in each end band strip $= \dfrac{1}{2}(1175 - 940) = 117.5$ mm².

No. of 12 mm Φ bars $= 117.5/113 = 1.04$. However provide minimum 3 bars in each end band of width $\dfrac{1}{2}(L - B) = \dfrac{1}{2}(3 - 2) = 0.5$ m.

**5. Test for development length**

$L_d = 47 \Phi = 47 \times 12 = 564$ mm.

Providing 60 mm side cover, length available
$\frac{1}{2}[B-b] - 60 = \frac{1}{2}[2000 - 400] - 60 = 740$ mm, which is greater than $L_d$. Hence O.K.

**6. Check for transfer of load at the base**

$A_2 = 600 \times 600 = 360000$ mm²

At a rate of spread of 2:1,
$$A_1 = [600 + 2(2 \times 420)]^2$$
$$= (2280)^2$$
$$= 5.198 \times 10^6 \text{ mm}^2$$

$\therefore \sqrt{\frac{A_1}{A_2}} = \sqrt{\frac{5.198 \times 10^6}{360000}} = 3.8 > 2$

Adopt max. value of $\sqrt{A_1/A_2}$ as 2.

$\therefore$ Permissible bearing stress
$$= 0.45 f_{ck} \sqrt{A_1/A_2}$$
$$= 0.45 \times 20 \times 2 = 18 \text{ N/mm}^2.$$

Actual bearing stress = $\frac{1.5 \times 600000}{600 \times 600} = 2.5$ N/mm²

Hence satisfactory.

The details of reinforcement etc. are shown in Fig. 44.3.

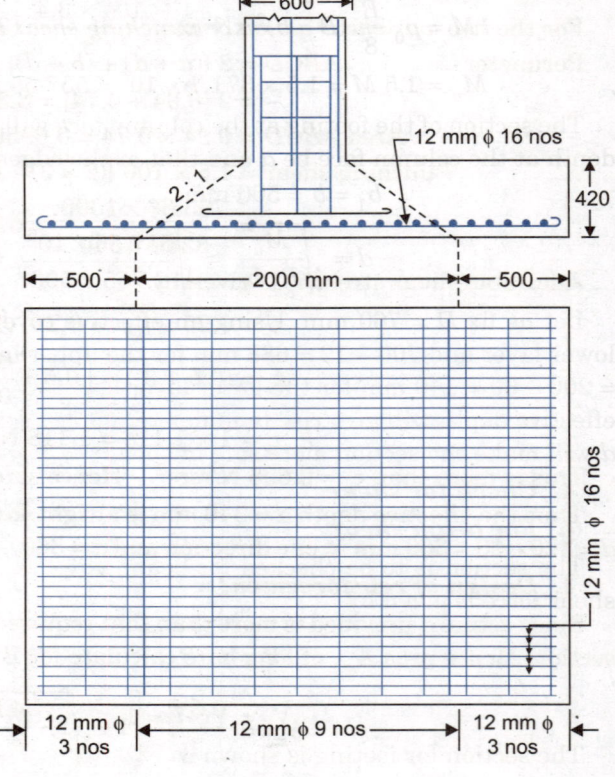

Fig. 44.3

## 44.3. ISOLATED SLOPED FOOTING

The bending moment, beam shear and punching shear govern the thickness or depth of footing near the column face. This depth can be reduced towards the edges of the footing where the bending moments and shear decrease rapidly. If this decrease is achieved linearly, we get a *sloped footing*. The analysis of isolated sloped footing has been thoroughly discussed in § 15.7, Chapter 15, and reference may be made to that article for various formulae and expressions. The design procedure is explained in Example 44.2 given below.

**Example 44.2.** *Isolated square sloped footing: Design an isolated square sloped footing for a column 500 mm × 500 mm, transmitting an axial load of 1200 kN. The column is reinforced with 8 bars of 20 mm diameter. The safe bearing capacity of soil is 120 kN/m². Use M 20 concrete and Fe 415 steel reinforcement.*

**Solution:**

**1. Design constants.**

For M 20 – Fe 415 combination, we have:
$$\frac{x_{u,\max}}{d} = 0.479 \quad \text{and} \quad R_u = 2.755$$

**2. Size of footing.** $W = 1200$ kN. Assume $W' = 10\% W = 120$ kN

$\therefore$ Area of footing $A = \frac{1200 + 120}{120} = 11$ m² ; Provide a footing of size 3.4 m × 3.4 m.

Actual upward pressure intensity $p_0 = \frac{1200}{3.4 \times 3.4} = 103.8 \approx 104$ kN/m².

**3. Design of footing.** Maximum bending moment occurs at the face of the column and its magnitude is given by:

$$M = p_0 \frac{B}{8}(B-b)^2 \text{ kN-m} = \frac{104 \times 3.4}{8}(3.4-0.5)^2 \times 10^6 \text{ N-mm} \approx 371.7 \times 10^6 \text{ N-mm}.$$

$$\therefore \quad M_u = 1.5 M = 1.5 \times 371.7 \times 10^6 = 557.55 \times 10^6 \text{ N-mm}$$

The section of the footing at the columm face will be trapezoidal, as shown in Fig. 15.16. Let the effective depth at the column face be $d$ and that at the edges be $0.2\, d$.

$$b_1 = b = 500 \text{ mm}$$

$$d = \sqrt{\frac{M_u}{R_u\, b_1}} = \sqrt{\frac{557.55 \times 10^6}{2.755 \times 500}} = 636.2 \text{ mm}.$$

Let us fix $D = 760$ mm. Using an effective cover of 60 mm, available $d = 760 - 60 = 700$ mm for the lower layer and $700 - 12 = 688$ mm for the upper layer. At the end, provide $D = 200$ mm, so that available $= 200 - 60 = 140$ mm for the lower layer and $140 - 12 = 128$ mm for the upper layer. Provision of greater effective depth will give rise to under-reinforced section, as desired by the Code. Also, provision of greater $d$ will make the section more safe in shear (next step).

**4. Check for shear**

**(a) For beam shear**

The section is to be checked for beam shear at a distance $d = 700$ mm from the column face, where shear force is given by

$$V = p_0 B\left\{\frac{1}{2}(B-b) - d\right\} = 104 \times 3.4 \left\{\frac{1}{2}(3.4-0.5) - 0.7\right\} = 265.2 \approx 266 \text{ kN}.$$

$$\therefore \quad V_u = 1.5\, V = 1.5 \times 266 = 399 \text{ kN}$$

The section for footing is shown in Fig. 44.4.

Effective depth $d'$ at that location $= 140 + \dfrac{700-140}{1450}(1450 - 700) = 429.65 \approx 430$ mm.

Top width of section $= 500 + \dfrac{3400-500}{1450} \times 700 = 1900$ mm

Factor $x_{u.\max}/d = 0.479$ for balanced section. However, for an under-reinforced section, adopt $x_u/d \approx 0.4$, so that $x_u = 0.4\, d' = 0.4 \times 430 = 172$ mm.

Width $b_n$ at N.A. $= 1900 + \dfrac{3400-1900}{290} \times 172 \approx 2790$ mm.

$$\therefore \quad \tau_v = \frac{V_u}{b_n\, d'} = \frac{399 \times 1000}{2790 \times 430} = 0.333 \text{ N/mm}^2$$

Assuming $p = 0.3\%$ for an under-reinforced section $\tau_c = 0.384$ N/mm$^2$, for M 20 concrete.

Also, $k = 1$ for $D > 300$ mm. Thus, $\tau_v < \tau_c$. (Hence safe).

**(b) Check for two way shear (punching shear) :**
[Fig. 15.10(a)]

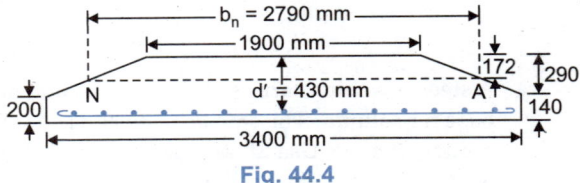

Fig. 44.4

Perimeter $ABCD = 2[(a+d)+(b+d)] = 4(a+d) = 4(500+700) = 4800$ mm.

Area $ABCD = 1.2 \times 1.2 = 1.44$ m$^2$

Punching shear $= 104\,[3.4 \times 3.4 - 1.44] = 1052.5$ kN.

$$\therefore \quad \tau_v = \frac{1.5 \times 1052.5 \times 1000}{4800 \times 700} = 0.47 \text{ N/mm}^2.$$

Allowable shear stress $\tau_c$ is given by

$$\tau_c = 0.25\sqrt{f_{ck}} = 0.25\sqrt{20} \approx 1.118 \text{ N/mm}^2$$

$$k_s = (0.5 + \beta_c) = (0.5 + 1). \quad \text{However adopt max. } k_s = 1.$$

$$\therefore \quad k_s \tau_c = 1 \times 1.118 = 1.118 \text{ N/mm}^2. \quad \text{Hence safe.}$$

5. **Steel reinforcement**: For an under-reinforced section,

$$A_{st} = \frac{0.5 f_{ck}}{f_y}\left[1 - \sqrt{1 - \frac{4.6 M_u}{f_{ck} b_1 d^2}}\right] b_1 d = \frac{0.5 \times 20}{415}\left[1 - \sqrt{1 - \frac{4.6 \times 557.55 \times 10^6}{20 \times 500 (700)^2}}\right] 500 \times 700$$

$$\approx 2611.5 \text{ mm}^2$$

Number of 12 mm Φ bars = 2611.5/113.1 = 23.1 ≅ 24

Hence provide 12 mm Φ bars, 24 Nos. uniformly spaced in the width 3.4 m in each direction.

6. **Check for development length**: $L_d = 47 \times 12 = 564$ mm.

Providing 60 mm side cover, length available = $\frac{1}{2}(B - b) = \frac{1}{2}(3400 - 500) - 60 = 1390$ mm.

Hence O.K. The details of reinforcement etc. as shown in Fig. 44.5.

7. **Check for transfer of load at the column base**

$A_2 = 500 \times 500 = 250000$ mm$^2$.

At a rate of spread of 2 : 1,

$A_1 = [500 + 2(2 \times 760)]^2 = (3540)^2 = 12531600$ mm$^2$.  Hence,

$$\sqrt{\frac{A_1}{A_2}} = \sqrt{\frac{12531600}{250000}} = 7.08 > 2 \quad \text{Adopt a max. value of } \sqrt{\frac{A_1}{A_2}} = 2.$$

∴ Permissible bearing stress = $0.45 f_{ck} \sqrt{A_1/A_2} = 0.45 \times 20 \times 2 = 18$ N/mm$^2$

Actual bearing stress = $\frac{1.5 \times 1200000}{500 \times 500} = 7.2$ N/mm$^2$. Hence satisfactory.

In the present case, the load transfer is there without exceeding the permissible bearing stress. However it is always a good practice to extend all the column bars into the foundation and anchored. If dowels bars are provided, development length requirements should be satisfied as required in § 15.4.

The details of reinforcement etc. are shown in Fig. 44.5.

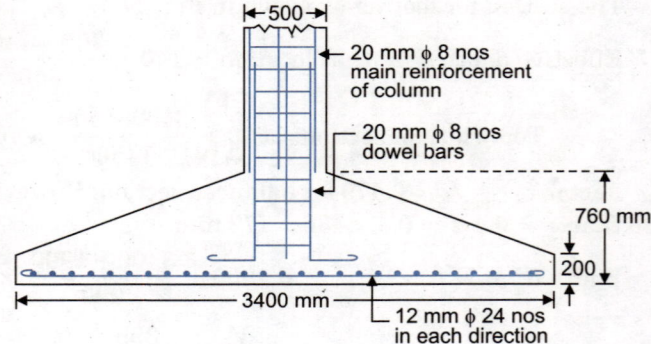

Fig. 44.5

## PROBLEMS

1. A steel stanchion carries 400 mm × 200 mm heavy beam section and carries a load of 1000 kN. Design a R.C.C. base for column. The safe bearing capacity of the the soil may be taken as 100 kN/m$^2$ at a depth of 1 m below ground surface. Use M 20 concrete and Fe 415 steel.

2. A. R.C. column 400 mm × 400 mm in section, carries an axial load of 750 kN. Design sloping R.C. footing using M 20 concrete and Fe 415 steel.

   Take safe bearing capacity of soil = 120 kN/m$^2$.

3. Redesign problem 2, if the column is circular in section having diameter = 40 cm. Provide square footings.

4. Redesign problem 3, by providing circular base for the footing.

5. A rectangular R.C. column, 240 mm × 300 mm carries an axial load of 400 kN. Design a rectangular footing of uniform thickness, if the safe bearing capacity of the soil is 80 kN/m$^2$. Use M 20 concrete and Fe 415 steel.

6. Design a reinforced concrete sloping footing for a RCC column of 600 × 600 mm size carry of an axial load 1600 kN using $M_{25}$ concrete and Fe 415 steel and SBC of soil-250 kN/m$^2$.

7. Design an RCC footing of uniform thickness to carry an axial load of 1200 kN from a square column of size 350 × 350 mm. The safe bearing capacity of soil is 180 kN/m$^2$. Use $M_{25}$ grade concrete and Fe 415 steel.

# PART – VI
# PRESTRESSED CONCRETE AND MISCELLANEOUS TOPICS

45. PRESTRESSED CONCRETE
46. SHRINKAGE AND CREEP
47. FORMWORK
48. TESTS ON CEMENT AND CEMENT CONCRETE

# CHAPTER 45: PRESTRESSED CONCRETE

## 45.1. INTRODUCTION

In ordinary reinforced concrete, consisting of concrete and mild steel as basic components, the compressive stresses are borne by concrete while tensile stresses are borne entirely by steel. The concrete surrounding steel reinforcement does not take part in resisting the external forces/moments since concrete is considered weak in tension. It simply acts as a bonding material. Thus only that portion of concrete, which lies above the neutral axis, is considered to be useful in resisting the external forces. This results in heavy sections. In the case of *prestressed concrete*, comprising of concrete and high tensile steel as basic components, both steel and concrete are stressed prior to the application of external loads. If such induced *pre-stress* in concrete is of compressive nature, it will balance the tensile stress produced in concrete surrounding steel due to external loads. Due to this, the whole of the concrete can participate in resisting the external forces.

The load carrying capacity of such concrete sections can be increased if steel and concrete both are stressed before the application of external loads.

According to ACI Committee on Pre-stressed Concrete, *pre-stressed concrete is the one in which there have been introduced internal stresses of such magnitude and distribution that the stresses resulting from given external loadings are counteracted to a desired degree.* In reinforced concrete member, the prestress is commonly introduced by tensioning the steel reinforcement. Thus, prestressing is the intentional creation of permanent stresses in a structure or assembly, for the purpose of improving its behaviour and strength under various service conditions (Lin, 55).

In prestressed concrete, compression is induced prior to loading in the zones where external loads would normally cause tensile stresses. Simple reinforced concrete has two major disadvantages.

(1) It has weak crack-resistance in consequence of the low tensile strength—all the reinforced concrete beams and other bending members have cracks at working loads. Due to this, corrosion of both the reinforcement and concrete takes place.

(2) It is impossible to use high tensile steel as reinforcement for simple reinforced concrete. If high tensile steel is used, the high stresses in the tensile reinforcement will result in such wide cracks in the tensile zone of concrete that the load carrying capacity of the members will practically be lost. A prestressed concrete construction on the other hand, has no cracks at working loads and has offered the possibility of employing high-tensile steel as reinforcement.

In 1904, Freyssinet attempted to introduce permanently acting forces in concrete to resist the elastic forces developed under loads and this idea was later developed under the name of "prestressing".

The earlier attempts of prestressing were made immediately after the development of reinforced concrete. P. Jackson (1886) of Sanfrancisco U.S.A. obtained patents for "pretensioning steel tie rods in artificial stones and concrete arches to serve as floor slabs". K. Doring (1888) of Germany suggested pretensioning of wires in reinforced concrete floor structures. However, in all the earlier attempts, low tensile steel was used as prestressing material with the result that low pretension was lost in *shrinkage* and *creep* in concrete. If an ordinary mild steel bar is prestressed to a working stress of 140 N/mm$^2$, the resulting strain in it will be

$$\varepsilon = \frac{140}{2.1 \times 10^5} = 0.00066.$$

The permanent negative strain due to both shrinkage and creep is of order of the 0.0008. Since this permanent strain is greater than the initial strain in the mild steel caused by tensioning it, the prestress induced in mild steel would soon disappear, leaving the member simply reinforced.

To avoid this trouble, C.R. Steiner (1908) of U.S.A. recommended the tightening of reinforcing rods after some shrinkage and creep of concrete had taken place. R.E. Dill (1925) of Nebraska used high strength steel bars coated to prevent bond with concrete. The bars were tensioned and anchored to concrete by means of nuts, after the concrete had set. For economical reasons, these methods could not be applied to an appreciable extent. It was E. Freyssinet (1928 and 1933) of France who made the first specific contribution of prestressing as it is today. He used high strength steel wires for prestressing. These wires has an ultimate strength in tension of 1750 N/mm$^2$ and yield point of about 1260 N/mm$^2$. If these are prestressed to 1050 N/mm$^2$, the resulting strain will be

$$\varepsilon = \frac{1050}{2.1 \times 10^5} = 0.005.$$

Allowing for a total loss of 0.0008 due to shrinkage and creep a net strain of 0.005 – 0.0008 = 0.0042 will remain, giving rise to an effective or final prestress in steel equal to

$$f_s = 2.1 \times 10^5 \times 0.0042 = 882 \text{ N/mm}^2$$

According to Freyssinet, to prestress a structure is artificially to create either prior to or simultaneously with the application of external loads such permanent stresses that in combination with the stress due to external load, total stresses will every where and for all states of loads envisaged, be within limits of stress that the material can support indefinitely. In his earlier attempts, he tried pretensioning in which steel was bonded to concrete without end anchorages. However, in 1939, he developed conical wedges for end anchorages and designed double acting jacks which tensioned the wires and then thrust the male cones into the female cones for anchoring them. After this successful work, prestressed concrete was very widely developed in many countries and many famous workers like Guyon, Kani, Leonhardt, Magnel and others, started working on this. G. Magnel (1940) of Belgium developed the Magnel system, wherein two wires were stretched at a time and anchored with a simple metal wedge at each end. In the U.S.S.R., investigations on prestressed concrete were initiated by V.V. Michailov and then by A.A. Gvosdev, S.A. Dmitriev and others.

## 45.2. BASIC CONCEPTS

As stated earlier, *prestressing is the initial application of stresses, controlled in magnitude and direction, to a structural member to counteract the undesirable stresses due to working loads*. Prestressing is commonly introduced by tensioning the steel reinforcement. According to Lin, three different concepts may be applied to explain and analyse the basic behaviour of prestressed concrete. These concepts are as follows: (1) Stress concept, (2) Strength concept and (3) Balanced load concept.

The application of permanent compressive stress to a material like concrete, which is strong in compression but weak in tension, increasing the apparent tensile strength of that material because the subsequent application of tensile stress must first nullify the compressive prestress.

1. **Stress concept: Prestressing to transform concrete into an elastic material.**

This concept is credited to Eugene Freyssinet who visualised prestressed concrete as essentially concrete which is transformed from a brittle material into an elastic one by precompression given to it. If an ordinary concrete, whether plain or reinforced, is subjected to only compressive stresses, it behaves as a perfect elastic material because no tension cracks are there. But if it is subjected to flexural stresses, some portion of it will be in tension resulting in tension cracks; the material under such circumstance no longer remains elastic. In prestressed concrete, on the other hand, concrete is visualised as being subjected to two system of forces: internal prestress, which is compressive and external load causing tensile stresses. The tensile stresses caused due to external load are counterbalanced by the compressive stress due to prestress, with the result that final stress in the extreme fibre is either compressive or zero. Due to absence of final tensile stress, no tension cracks would be there in concrete, and it will thus be transformed from brittle to elasic material. To elaborate this point, let us consider two cases: (a) concentric tendon (steel reinforcement) and (b) eccentric tendon.

Fig. 45.1(a) shows a concentrically prestressed concrete beam. Due to prestress force $T$ in tendon, a uniform compressive stress = $T/A$ will be induced in concrete. If the beam is subjected to a moment $M$ due to external load, inclusive of its own weight, the stress at any point will be $\dfrac{M}{I}y$ where $y$ is the distance of the point from the centriodal axis and $I$ is the moment of inertia of the section. The final stress at any point is given by

$$f = \frac{T}{A} \pm \frac{M y}{I} \qquad ...(45.1)$$

Fig. 45.2(a) shows an eccentrically prestressed beam with external loading. Due to prestress force $T$ in tendon, applied eccentrically, the moment produced due to prestress will be $T.e$. Hence the stress $f'$ due to prestress at any point will be

$$f' = \frac{T}{A} \mp \frac{T.e.y}{I} \qquad ...[45.2(a)]$$

If $M$ is the external moment, the final stress at any point is given by

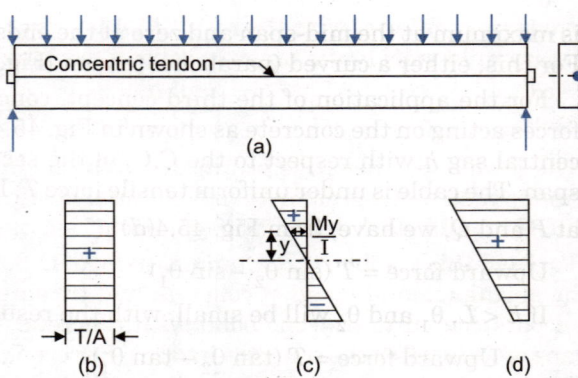

Fig. 45.1. Concentrically Prestressed.

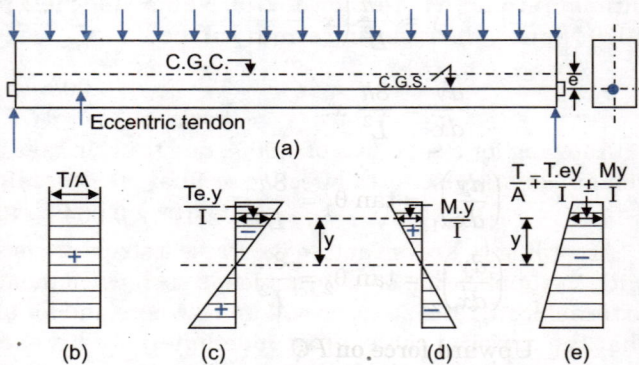

Fig. 45.2. Eccentrically Prestressed Section.

$$f = \frac{T}{A} \mp \frac{T.e.y}{I} \pm \frac{M.y}{I} \qquad ...(45.2)$$

The first concept is further illustrated through Example 45.1

### 2. Strength concept: Prestressing for combination of high strength steel and concrete

In reinforced concrete, steel takes tension while concrete takes compression and the couple formed by the resultant compressive force $C$ and tensile force $T$ (where $C = T$) resists the external moment [Fig. 45.3(a)].

The same concept can be applied for a prestressed concrete section. The prestressed concrete is considered as a combination of steel and concrete, with steel taking $T$ (passing through the tendon) and concrete taking compression $C$ [passing through the C.G. of the stress distribution shown shaded in Fig. 45.3(a)], so that the two materials form a resisting couple to resist the external moment. This concept has been well utilised to determine the ultimate strength of prestressed concrete beams. This concept has been illustrated in Example 45.2.

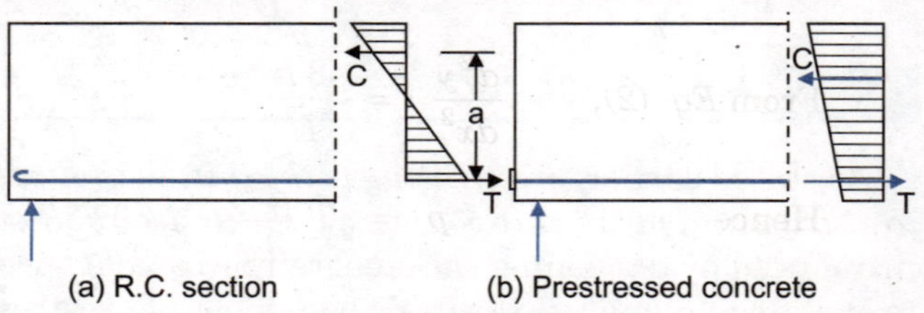

(a) R.C. section        (b) Prestressed concrete

Fig. 45.3

The minimum 28 day cube compressive strength prescribed in IS:1343–1980 is 40 N/mm² for pre-tensioned members and 30 N/mm² for post tensioned members.

### 3. Load Balancing concept: Prestressing to achieve load balancing.
This concept was developed by Lin, in which a flexural member is transformed into a member under direct stress. In the overall design of a prestressed concrete structure, the effect of prestressing is essentially viewed as the balancing of gravity loads (both self-weight as well as external loads) so that members under bending will not be subjected to flexural stresses under a given load conditions. It will be clear in Example 45.1 that by placing the tendon eccentrically, moment can be resisted. Flexural members, such as beam, slabs etc. are normally subjected to bending moments which are not uniform along their length. In the case of simply supported beam, moment

is maximum at the mid-span and zero at the ends. Hence the eccentricity $e$ should also vary along the span. For this, either a curved (parabolic) tendon [Fig. 45.4(a)] or inclined tendon [Fig. 45.5(a)] is used.

For the application of the third concept, concrete is taken as free body and the tendon is replaced by forces acting on the concrete as shown in Fig. 45.4(b). Fig. 45.4(a) shows the parabolic profile of tendon with central sag $h$ with respect to the C.G. of the section. Consider two points $P$ and $Q$ at $x_1$ and $x_2$ from mid-span. The cable is under uniform tensile force $T$. If $\theta_1$ and $\theta_2$ are the inclinations of the cable with horizontal, at $P$ and $Q$, we have, from Fig. 45.4(d)

Upward force $= T(\sin\theta_2 - \sin\theta_1)$

If $h < L$, $\theta_1$ and $\theta_2$ will be small, with the result that $\sin\theta \approx \tan\theta$

∴ Upward force $= T(\tan\theta_2 - \tan\theta_1)$ ...(1)

The equation of the parabola [Fig. 45.4(c)] is

$$y = \frac{4h}{L^2}\left(x^2 - \frac{L^2}{4}\right) \quad ...(2)$$

∴ $\dfrac{dy}{dx} = \dfrac{8h}{L^2} x.$

∴ $\left(\dfrac{dy}{dx}\right)_1 = \tan\theta_1 = \dfrac{8h}{L^2} x_1$

$\left(\dfrac{dy}{dx}\right)_2 = \tan\theta_2 = \dfrac{8h}{L^2} x_2$

∴ Upward force on $PQ$

$= T \dfrac{8h}{L^2} (x_2 - x_1).$

If $(x_2 - x_1)$ is taken as unity, upward force $p$ per unit length is

$$p = T \frac{8h}{L^2} \quad ...(45.3)$$

From Eq. (2), $\dfrac{d^2y}{dx^2} = \dfrac{8h}{L^2}$

Hence $p = T \dfrac{d^2y}{dx^2}$ ...[45.3(a)]

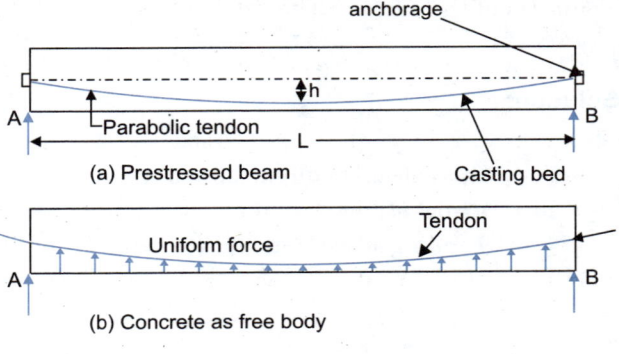

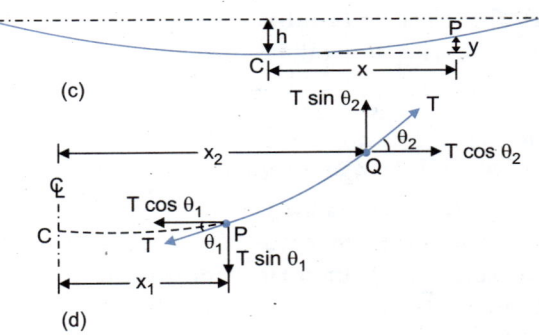

Fig. 45.4. Prestressed Beam with Parabolic Tendon.

Thus, the parabolic cable supports a uniformly distributed load $p$ given by the above equations. If the external U.D.L. $w$ is equal to $p$, transverse load on the beam is balanced and the beam is subjected to only the axial force $T$, which produces uniform stress in concrete $= f = T/A$. If the external load $w$ is greater than $p$, the mid-span moment $M$ for the remaining force $(w - p)$ can be computed and fibre stresses due to that moment would be $My/I$. The resulting stresses will then be equal to $\dfrac{T}{A} \pm \dfrac{My}{I}$.

Fig. 45.5(a) shows a beam with bent tendon while Fig. 45.5(b) shows the freebody of concrete with tendon replaced by forces. In the figure, C.G.C. stands for

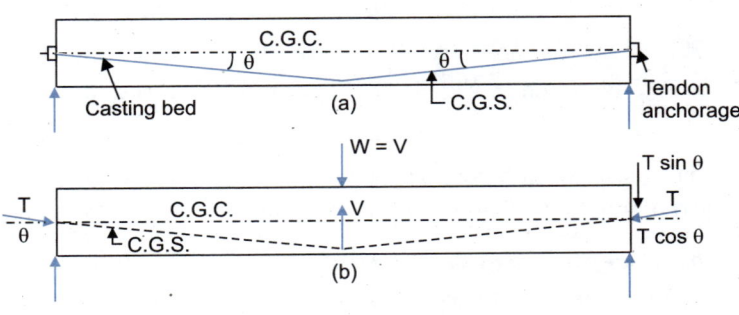

Fig. 45.5

centre of gravity line for concrete while C.G.S. stands for C.G. line of steel cable. The balancing force $V$ is given by
$$V = 2T \sin \theta \qquad \ldots[45.4(a)]$$
If this $V$ exactly counter-balances a concentrated load $W$ applied at midspan, the beam is not subjected to any transverse load (neglecting the weight of the beam).

The uniform compressive stress in beam is given by
$$f = \frac{T \cos \theta}{A_c} \approx \frac{T}{A_c} \text{ (if } \theta \text{ is small)} \qquad \ldots[45.4(b)]$$

While this 'load balancing concept' often represents the simplest approach to prestressed design and analysis, its advantage over the other two concepts is not significant for statically determinate structures. However, the method offers tremendous advantages when dealing with statically indeterminate structures.

**Example 45.1. Stress concept.** *A simply supported prestressed concrete beam of rectangular cross-section 400 mm × 600 mm, is loaded with a total uniformly distributed load of 256 kN over a span of 6 m. Sketch the distribution of stresses at mid-span and end sections if the prestressing force is 1920 kN and the tendon is (a) concentric, (b) eccentric, located at 200 mm above the bottom fibre.*

**Solution:**
Let
- $f$ = final stresses at extreme fibre
- $(f_1)$ = direct stress due to prestressing force
- $(f_b)$ = bending stress and mid span due to applied loads
- $(f_2)$ = Bending stress due to eccentricity of tendons.
- $f'$ = stress due to prestressing force in tendons and eccentricity.

**(a) Concentric tendon**
$$A = 400 \times 600 = 240000 \text{ mm}^2$$
$$I = \frac{1}{12}(400)(600)^3 = 72 \times 10^8 \text{ mm}^4$$

$f_1$ due to prestressing force
$$= \frac{1920 \times 1000}{240000} = 8 \text{ N/mm}^2.$$

At the end section, B.M. is zero, and hence the section will be subjected to uniform compressive stress of 8 N/mm² throughout its depth.

Midspan moment $= \frac{WL}{8} = \frac{256 \times 6}{8} = 192$ kN-m
$$= 192 \times 10^6 \text{ N-mm}$$

∴ Extreme fibre stresses $= f_b = \pm \dfrac{192 \times 10^6}{72 \times 10^8}(300) = \pm 8$ N/mm²

Hence the final stresses at extreme fibres
$$f = f_1 \pm f_b = 8 \pm 8 = 16 \text{ or } 0.$$
The stress distributions at end and mid-span are shown in Fig. 45.6(a).
It is to be noted in Fig. 45.6(b) that for final stress diagram,
$$C = \tfrac{1}{2} \times 16 \times 240{,}000 = 1920{,}000 \text{ N} = 1920 \text{ kN}.$$

The compressive force $C$ acts at a distance of 100 mm from $T$, giving rise to a couple $= 100 \times C = 100\,T$ $= 100 \times 1920 = 192000$ kN-mm $= 192$ kN-m = applied moment.

**(b) Eccentric tendon.** Eccentricity; $e = 300 - 200 = 100$ mm.
The stresses due to eccentric prestress force is
$$f' = \frac{T}{A} \mp \frac{T \cdot e}{I} y \,; f' = f_1 \mp f_2,$$

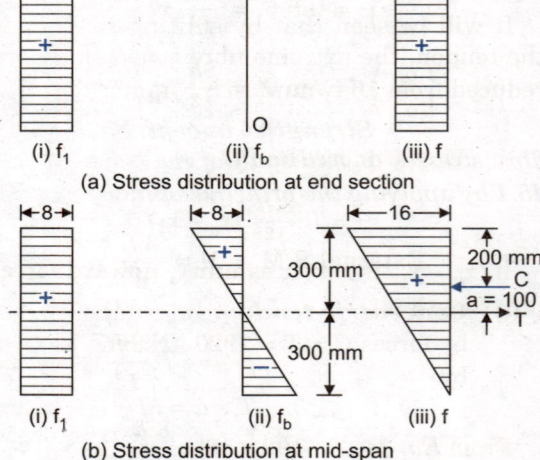

Fig. 45.6. Stress Distributions Due to Concentric Tendon.

where $f_1 = \dfrac{1920 \times 1000}{240000} = 8$ N/mm²

$f_2 = \mp \dfrac{1920 \times 1000 \times 100}{72 \times 10^8} \times 300$

$= \mp 8$ N/mm²

$f_2$ will be tensile at the top and compressive at the bottom fibres.

At the end section, where B.M. is zero, the final stresses $f$ will be $8 \mp 8$ *i.e.* zero at top fibre and 16 N/mm² at the bottom fibre, as shown in Fig. 45.7 (*b*). In the final stress distribution, line of action of $C$ and $T$ coincide, giving rise to zero moment.

At the mid-span, the bending stress $f_b$ is equal to $\pm 8$ N/mm², as found earlier. Hence the final stress $f$ is given by

$f = f_1 \mp f_2 \pm f_b = 8 \mp 8 \pm 8$.

Hence $f$ at top fibre = 8 N/mm²; $f$ at bottom fibre = 8 N/mm². The stress distribution is shown in Fig. 45.6(*b*).

It will be seen that by shifting the position of the tendon, the extreme fibre stress in concrete is reduced from 16 N/mm² to 8 N/mm².

**Example 45.2.** *Strength Concept. Find the extreme fibre stresses at mid-span of the beam of example 45.1 by applying the principle of internal resisting couple.*

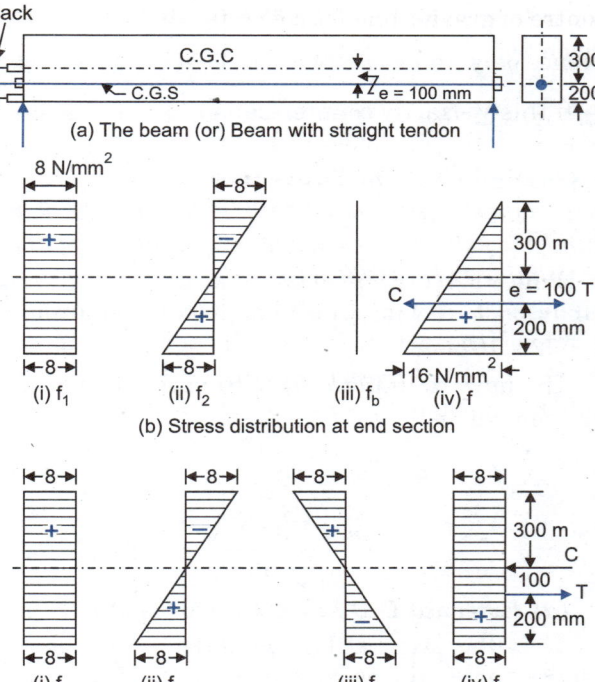

Fig. 45.7. Stress Distribution Due to Eccentric Tendon.

**Solution:** External B.M. = $M = \dfrac{WL}{8} = \dfrac{256 \times 6}{8} = 192$ kN-m $= 192 \times 10^6$ N-mm

(*a*) *Concentric tendon:* Internal couple is furnished by forces $C = T = 1920$ kN. The lever arm $a$ given by

$T \times a = M$

or $a = \dfrac{M}{T} = \dfrac{192 \times 10^6}{1920 \times 1000} = 100$ mm

[**Note:** See Fig. 45.6(*b*) (*iii*) also.]

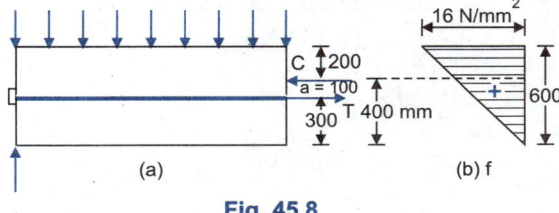

Fig. 45.8

Since $T$ acts at 300 mm above bottom fibre, distance of $C$ from bottom fibre = 300 + 100 = 400 mm. Thus $C$ is located. It acts at a distance of $e' = 400 - 300 = 100$ mm from the c.g.c., as shown in Fig. 45.8(*a*).

Hence extreme fibre stresses are given by

$f = \dfrac{C}{A} \pm \dfrac{C \cdot e'}{I} \cdot y = \dfrac{1920 \times 1000}{240000} \pm \dfrac{(1920 \times 1000)\,100}{72 \times 10^8} (300)$

= 16 N/mm² at top fibre and 0 N/mm² at bottom fibre.

The stress distribution is sketched in Fig. 45.8(*b*), which is the same as that shown in Fig. 45.6(*b*) (*iii*).

(*b*) *Eccentric tendon*

$C = T = 1920$ kN

Also, $T \times a = M$

$a = \dfrac{M}{T} = \dfrac{192 \times 10^6}{1920 \times 1000} = 100$ mm

[See Fig. 45.7 (*c*) (*iii*) also.]

Since $T$ acts at 200 mm above bottom fibre, distance of $C$ from bottom fibre = 200 + $a$ = 200 + 100 = 300 mm.

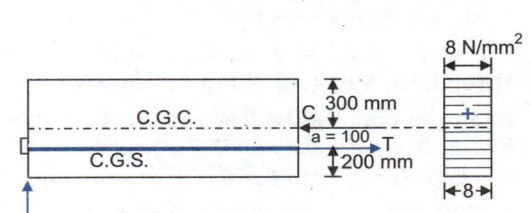

Fig. 45.9

Thus C is located. It acts at a distance of $e' = 300 - 300 =$ zero cm from c.g.c., as shown in Fig. 45.9(a). Hence only uniform stress will be induced, having magnitude $f = \dfrac{C}{A} = \dfrac{1920 \times 1000}{240000} = 8$ N/mm$^2$, as shown in Fig. 45.9 (b).

**Example 45.3. Load Balancing Concept.** *The beam of Example 45.1 is pre-stressed by a parabolic tendon with a prestressing force of 1920 kN. The tendon has a sag of 100 mm at mid-span as shown in Fig. 45.10(a). Find the extreme fibre stresses by load balancing concept, if it is subjected to (a) total U.D.L. of 256 kN, (b) total U.D.L. of 360 kN.*

**Solution:** $h = 0.1$ m; $A = 400 \times 600 = 240000$ mm$^2$; $I = 72 \times 10^8$ mm$^4$

The upward uniform force from the tendon on the concrete is

$$p = T \cdot \dfrac{8h}{L^2} \qquad \ldots(45.3)$$

$$= \dfrac{1920 \times 8 \times 0.1}{6 \times 6} = 42.67 \text{ kN/m.}$$

**(a) External U.D.L.** $= w = 256/6 = 42.67$ kN/m.

Thus the external load is perfectly balanced. Hence there will be no moment, and the beam will be subjected to a uniform compressive stress.

$$= f = \dfrac{1920 \times 1000}{400 \times 600} = 8 \text{ N/mm}^2$$

**(b) External U.D.L.** $= w = \dfrac{360}{6}$

$= 60$ kN/m.

$w' = w - p = 60 - 42.67$

$= 17.33$ kN/m.

$$M = \dfrac{w' L^2}{8}$$

$$= \dfrac{17.33 (6)^2}{8} \approx 78 \text{ kN-m.}$$

$= 78 \times 10^6$ N-mm.

$\therefore \quad f_b = \pm \dfrac{78 \times 10^6 \times 300}{72 \times 10^8}$

$= \pm 3.25$ N/mm$^2$. ;

$f_1 = \dfrac{T}{A} = \dfrac{1920 \times 1000}{240000}$

$= 8$ N/mm$^2$

Hence final stress are

$f = f_1 \pm f_b = 8 \pm 3.25 = 11.25$ N/mm$^2$ at top fibre and 4.75 N/mm$^2$ at bottom fibre.

**Example 45.4. Load Balancing Concept.** *Solve Example 45.3 if the parabolic profile of the tendon has an eccentricity of 100 mm at ends and 200 mm at mid-span, as shown in Fig. 45.11. The beam carries a U.D.L. of 60 kN/m.*

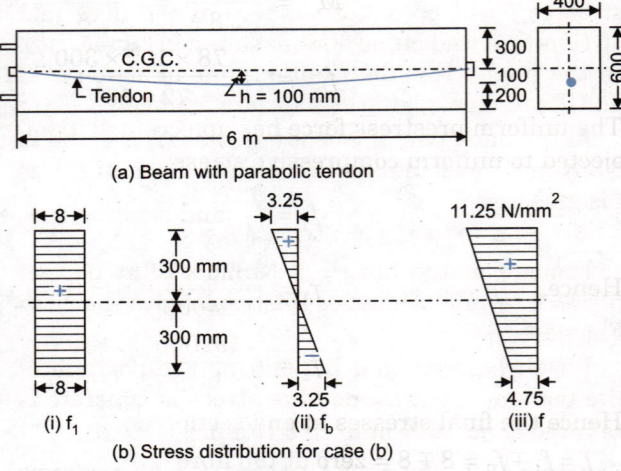

Fig. 45.10

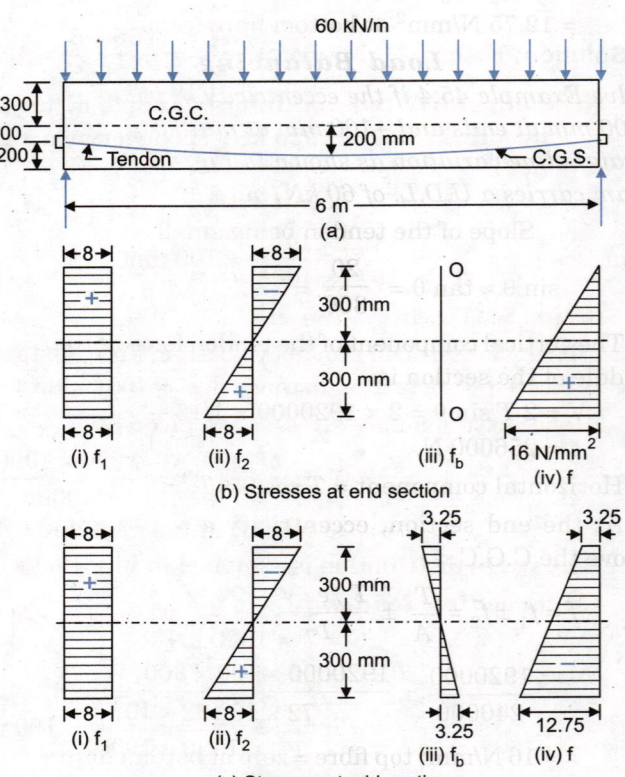

Fig. 45.11

**Solution:** The upward uniform force from tendon on the concrete is

$$p = T \cdot \frac{8h}{L^2},$$

where $h$ = sag of centre of tendon with respect to ends = 200 – 100 = 100 mm = 0.1 m. Hence,

$$p = \frac{1920 \times 8 \times 0.1}{6 \times 6} = 42.67 \text{ kN/m}.$$

Effective vertical load = $w' = w - p$ = 60 – 42.67 = 17.33 kN/m.

$$M = \frac{w'L^2}{8} = \frac{17.33(6)^2}{8} = 78 \text{ kN-m} = 78 \times 10^6 \text{ N-mm}.$$

$$\therefore \quad f_b = \pm \frac{78 \times 10^6 \times 300}{72 \times 10^8} = \pm 3.25 \text{ N/mm}^2.$$

The uniform prestress force has an eccentricity of 100 mm at ends. Due to this, the entire beam will be subjected to uniform compressive stress

$$f_1 = \frac{T}{A} \text{ and bending stress } f_2 = \mp \frac{T \cdot e}{I} \cdot y.$$

Hence, 
$$f_1 = \frac{1920 \times 1000}{240000} = 8 \text{ N/mm}^2 \text{ (comp.)}$$

$$f_2 = \mp \frac{1920000 \times 100 \times 300}{72 \times 10^8} = \mp 8 \text{ N/mm}^2$$

Hence the final stresses at end section are

$f = f_1 \mp f_2 = 8 \mp 8$ = zero at top fibre
   = 16 N/mm² at bottom fibre.   Final stresses at mid-span are

$f = f_1 \mp f_2 \pm f_b = 8 \mp 8 \pm 3.25$ = 3.25 N/mm² at top fibre (comp.)
   = 12.75 N/mm² at bottom fibre (comp.)

### Example 45.5. Load Balancing Concept.

Solve Example 45.4 if the eccentricity of tendon is –100 mm at ends and +100 mm at mid-span, with straight line variation as shown in Fig. 45.12. The beam carries a U.D.L. of 60 kN/m.

**Solution:** Slope of the tendon being small,

$$\sin\theta \approx \tan\theta = \frac{20}{300} = \frac{1}{15}$$

The vertical component of the tendon force at the middle of the section is

$V = 2T \sin\theta = 2 \times 1920000 \times 1/15$
  = 256000 N.

Horizontal component = $T \cos\theta \approx T$.

At the end section, eccentricity $e$ = 100 mm, above the C.G.C.

$$f = f_1 \pm f_2 = \frac{T}{A} \pm \frac{T \cdot e \cdot y}{I}$$

$$= \frac{1920000}{240000} \pm \frac{1920000 \times 100 \times 300}{72 \times 10^8} = 8 \pm 8$$

= 16 N/mm² top fibre = zero at bottom fibre.

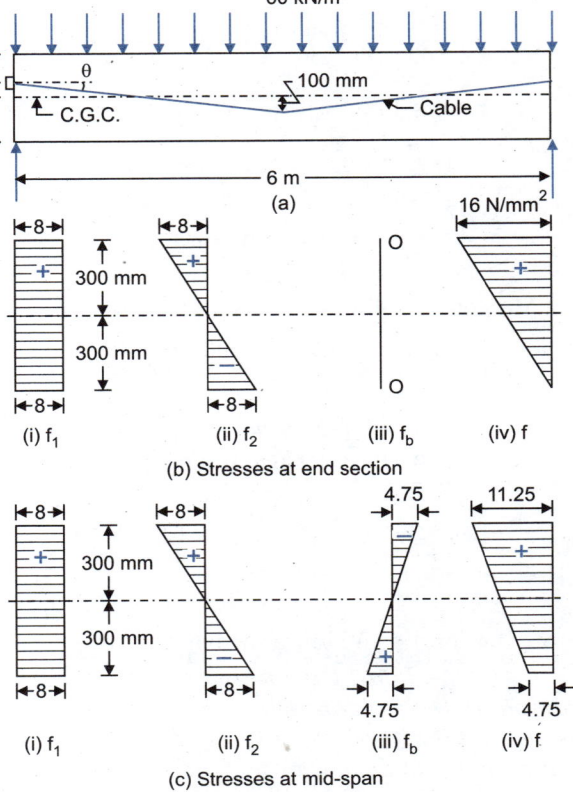

Fig. 45.12

At the mid-section, B.M. due to live load is

$$M_l = \frac{wL^2}{8} = \frac{60(6)^2}{8} = 270 \text{ kN-m} = 270 \times 10^6 \text{ N-mm}.$$

B.M. due to V is $\quad M_T = -\dfrac{VL}{4} = -\dfrac{256000 \times 6000}{4} = -384 \times 10^6$ N-mm.

∴ Net B.M. $M = (270 - 384) 10^6 = -114 \times 10^6$ N-mm (*i.e.* hogging)

$$f_b = \mp \frac{My}{I} = \mp \frac{114 \times 10^6 \times 300}{72 \times 10^8} = \mp 4.75 \text{ N/mm}^2.$$

Hence final stresses at mid-span section are

$$f = f_1 \pm f_2 \mp f_b = 8 \pm 8 \mp 4.75$$
$$= 11.25 \text{ N/mm}^2 \text{ (comp.) at top fibre and}$$
$$= 4.75 \text{ N/mm}^2 \text{ (comp.) at bottom fibre.}$$

**Example 45.6.** *Load Balancing method.*

*Solve Example 12.4 if the eccentricity of tendon is +100 mm at ends and +200 mm at mid-span, with straight line variation as shown in Fig. 45.13. The beam carries U.D.L. of 60 kN/m.*

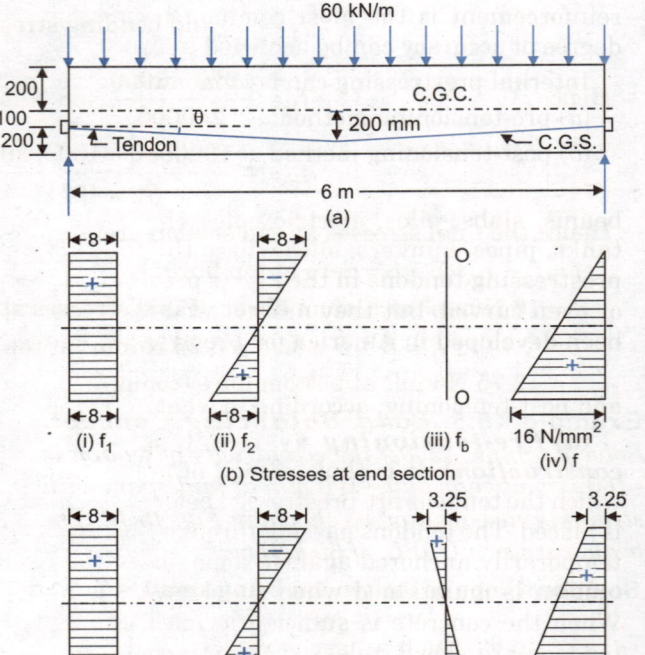

Fig. 45.13

**Solution:** $\sin \theta \approx \tan \theta = \dfrac{10}{300} = \dfrac{1}{30}.$

$V = 2 T \sin \theta$

$= 2 \times 1920000 \times \dfrac{1}{30}$

$= 128000$ N

Horizontal component $= T \cos \theta \approx T$

At the end section,

$e = 100$ mm below C.G.C.

∴ $f = f_1 \mp f_2 = \dfrac{T}{A} \mp \dfrac{T \cdot e \cdot y}{I}$

$= \dfrac{1920000}{240000} \mp \dfrac{1920000 \times 100 \times 300}{72 \times 10^8}$

$= 8 \mp 8 =$ zero at top fibre

$= 16$ N/mm² at bottom fibre.

Moment at mid-span, due to live load is

$M_l = \dfrac{wL^2}{8} = \dfrac{60(6)^2}{8} = 270$ kN-m

$= 270 \times 10^6$ N-mm

$M_T = -\dfrac{VL}{4} = -\dfrac{128000 \times 6000}{4} = -192 \times 10^6$ N-mm

∴ Net $\quad M = (270 - 192) 10^6 = 78 \times 10^6$ N-mm (sagging)

$$f_b = \pm \frac{78 \times 10^6 \times 300}{72 \times 10^8} = \pm 3.25 \text{ N/mm}^2$$

∴ Final stresses at mid-span section are

$$f = f_1 \mp f_2 \pm f_b = 8 \mp 8 \pm 3.25$$
$$= 3.25 \text{ N/mm}^2 \text{ (comp.) at top fibre}$$
$$= 12.75 \text{ N/mm}^2 \text{ (comp.) at bottom fibre.}$$

Compare the stress distribution with that of example 45.4.

## 45.3. CLASSIFICATION AND TYPES OF PRESTRESSING

Prestressed concrete may be classified into a number of ways depending upon design, construction, method of applying prestressing and purpose of structure. Some of these are given below.

**1. External or Internal Prestressing System:** The prestress in a concrete structure is maintained either by adjusting its external reactions, or by a built-up internal thrust using high tensile steel. The *external prestressing*, though not common, is achieved by external reactions introduced through different support conditions. Fig. 45.14(a) shows a simply supported beam, externally prestressed by jacking against abutments. Fig. 45.14(b) shows a continuous beam prestressed externally by jacking the appropriate supports. External prestressing system is not common because it requires greater degree of accuracy in planning, maintenance and execution. *Internal prestressing* obtained by tensioning the steel reinforcement is the most common because greater degree of accuracy can be achieved in execution.

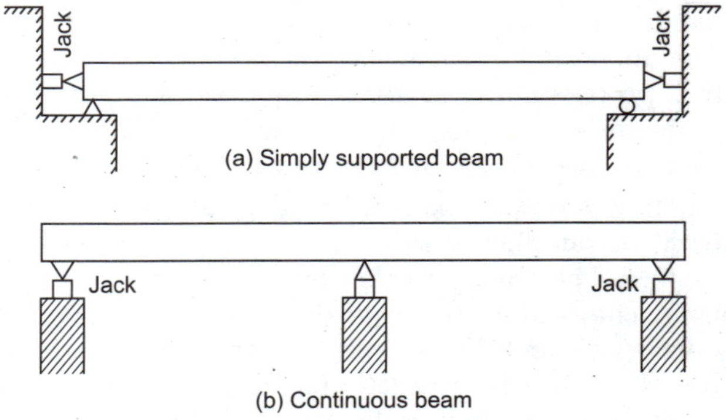

Internal prestressing can be done by two methods
(a) pre-tensioning method
(b) post-tensioning method

**Fig. 45.14**

**2. Linear or Circular Prestressing:** Linear prestressing is applied in straight members such as beams, slabs, piles, electric poles etc. Circular prestressing is applied to circular structure such as tanks, pipes, bunkers, silos where the prestressing tendons are wound around in circles. However, the prestressing tendons in the linear prestressing need not be necessarily straight. They can be either bent or even curved, but they are not wound in round circles around the structure. The *Preload system* has been developed in America for prestressing circular structures.

**3. Pretensioning and Post Tensioning System:** Internal prestressing has two systems: pretensioning and post-tensioning, according to whether the steel is tensioned before or after casting the concrete.

(a) *Pre-tensioning system (fully bonded construction):* It is that method of prestressing in which the tendons are prestressed before the concrete is placed. The tendons passing through the mould are temporarily anchored against some abutments. The tension is maintained when the concrete is placed. When the concrete is sufficiently hard, the ends of tendons are slowly released from anchorages, thereby transferring the prestress from steel to concrete

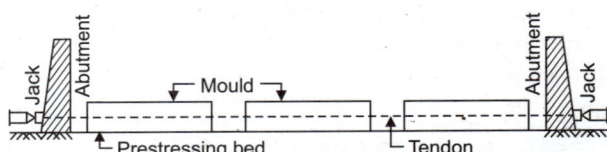

**Fig. 45.15.** Prestressing bed.

*through bond.* Pretensioning is applied to relatively simple units precast in factories. In mass production, the wires may be stretched the full length of the casting shop (say 100 to 150 m) between anchorages rigidly held in the ground; the wires pass through a whole series of 'moulds' as shown in Fig. 45.15. Each mould is mounted on rollers and spaced with small end gaps between adjacent moulds to allow the wires to be cut later and the individual units freed. When the concrete in the mould has set, the wires are released from end anchorages, thereby transferring the prestress to concrete through bond. Since tendon force is transferred from steel to concrete through bond, it is desirable to use small diameter wires. The diameter of wires often used is 2 to 3 mm. They are sometimes twisted or notched to improve bond. When wires are released, their ends loose their stress (within bond length) and recover to their original diameter. This increase in diameter provides mechanical anchorage. The *bond length or transmission length* varies from 84 Φ (for 2 mm wires) to 120 Φ (for 5 mm wire), within which the stress in the wire increase from zero at free end to the maximum value at the end of bond length. If the profile of the tendon is curved, necessary profile fixing arrangements are needed.

This method is commonly used for small sized members like, beams, slabs, piles, sleepers, and electric poles, etc, which can be casted easily in factories.

(b) *Post-tensioning system (end anchored construction):* In the post-tensioning system is done in factories, so it is more reliable and durable techniques, but it is used for small sections, and the member is concreted and a duct is formed in the member either with a tube or with a metal sheath. When the concrete is sufficiently hard, tendon consisting of either a bar, cable or strand is threaded through the duct, tensioned and anchored to the concrete member at its ends. Thus the tendon force is transferred to the member through necessary anchorage wedges or similar blocks at the end of the member. In some cases the wires or cables are assembled and lie freely in tubes or ducts carefully located in the form work before the concrete is cast.

The cable ducts can be cast in the members to any desired curved paths, and by sweeping the cables up at the ends of the beams the eccentricity can be reduced as required to suit the tailing off of the bending moments; the upward component of the cable force acts also to assist in carrying the end shear force. The anchoring of the wires to the concrete itself needs more care than the anchoring to bulk heads done in a pretensioning system. Many popular post-tensioning systems have been developed under patented propritories, notable among these being the Freyssinet system, Magnel system, Hobling system, Leonhardts system, Lee McCall system and the Ryerson BBRV system.

After applying the prestressing force to the tendons through jack and end anchorages, the ducts may be filled completely with cement grout injected under pressure. Such a system is known as fully bonded system. The wires in this system are protected against rusting. They are also doubly secured, primarily by mechanically anchorage devices and secondly can by bond by the grouting. It has been claimed that the anchorage can be safely released and re-used after grout has set; however, invariably they are retained and concreted over for protection against corrosion. If the duct is not grouted, an unbonded system is obtained, in which the protection for the tendons from corrosion is provided by galvanising, greasing or some other means. In some cases, the bonded tendon may be purposely unbonded along certain portions of their length.

In post-tensioned structure, complications arise from friction of wires in the ducts, at the anchorage and in the jack itself at the time of prestressing. In pretensioned structure, a major loss is the elastic deformation of the concrete when prestress is first applied. Pretensioned work also suffers more severely from effects of shrinkage, about half of which takes place in the first three weeks from the date of casting. The total losses in post-tensioned work may amount to about 15% while those in pre-tensioned work may be as high as 25%. Post tensioning method of prestressing is used for both precast and cast in situ construction. It is used for large span structures, like bridges.

    **4. Full Prestressing** *or* **Partial Prestressing:** The concrete is said to be fully prestressed if there are no tensile stresses in it under working load. However, if some tensile stresses are produced in the concrete under working load conditions, it is said to be partially prestressed. In partially prestressed concrete, additional mild steel bars are frequently provided to reinforce the portion under tension.

## 45.4. PRESTRESSING SYSTEMS : END ANCHORAGES

A *prestressing system* comprises essentially a method of stressing the steel combined with a method of anchoring it to the concrete. A number of different systems and their patents for tensioning and anchoring the tendons exist. The practicing engineer who simply wants to design some structures of prestressed concrete, is free to specify and design for any system without studying the intrigues of patent rights. In fact, the royalty of the patent owner is indirectly included in the bid price for the supplying of prestressing steel and anchorages, which sometimes also includes the furnishing of equipment for prestressing and some technical supervision for jacking. Table 45.1 gives various types of linear prestressing systems. Some prominent and more common systems are discussed here.

    **1. Pre-Tensioning Systems and End Anchorages:** The simplest pre-stressing system consists of two bulkheads anchored against the ends of a stressing bed. The tendons are pulled between the two bulk heads. In some cases, the bulk heads may be independent. In either case the bulk head and the bed must be strong enough to resist the force of pre-stress and its eccentricity. A prestressing bed is generally costly, but its use has two advantages:

    (*i*) intermediate bulkheads can be inserted in the bed to use and tension shorter wires for casting shorter units and

    (*ii*) it can support vertical reactions, due to which prestressing of bent cable can be done.

    (*iii*) it is cheaper because the cost of sheating is not involved.

    (*iv*) it is more reliable and durable.

    (*v*) small sections are to be constructed.

**PRESTRESSED CONCRETE** 1007

**TABLE 45.1.** Linear Prestressing Systems

| Type (1) | Classification (2) | Description (3) | | Name of system (4) | Country of origin (5) |
|---|---|---|---|---|---|
| Pre-tensioning | Method of stressing | (i) Against buttresses or stressing beds | | Hoyer | Germany |
| | | (ii) Against central steel tube | | Shorer Chalos | U.S. France |
| | | (iii) Continuous stressing against moulds | | Continuous wire winding | U.S.S.R. |
| | | (iv) Electric current to heat steel | | Electro-thermal | U.S.S.R. |
| | Method of anchoring | During Pre-stressing | (i) Wire | Various wedges | |
| | | | (ii) Strands | Strandwise | U.S. |
| | | For transfer of prestress | (i) Bond, for strands and small wires | | Europe, U.S. |
| | | | (ii) Corrugated clips, for big wires | Dorland | U.S. |
| Post-tensioning | Methods of stressing | (i) Steel against concrete | | Most systems | Germany/U.S. |
| | | (ii) Concrete against concrete | | Leonhardt/Billner | |
| | | (iii) Expanding cement | | | France |
| | | (iv) Electrical Prestressing | | Lossier | U.S. |
| | | (v) Bending steel beams | | Biliner | Belgium |
| | | | | Preflex | |
| | Methods of anchoring | (i) *Wires*, by frictional grips | | Freyssinet | France |
| | | | | Magnel | Belgium |
| | | | | Morandi | Italy |
| | | | | Holzmann | Germany |
| | | | | Preload | U.S. |
| | | (ii) *Wires*, by bearing | | B.B.R.V. | Switzerland |
| | | | | Prescon | U.S. |
| | | (iii) *Wires*, by loops and combination of methods | | Texas, P.I. | U.S. |
| | | | | Biller | U.S. |
| | | | | Monier bau | Germany |
| | | (iv) *Bars*, by bearing and by grips | | Leoba | Germany |
| | | | | Leonhardt | Germany |
| | | | | Lee Mc Call | England |
| | | | | Stress-steel | U.S. |
| | | (v) *Strands*, by bearing | | Stress rods | U.S. |
| | | | | Dywidag | Germany |
| | | | | Wet's | Belgium |
| | | (vi) *Strands*, by friction grips | | Bakker | Holland |
| | | | | Roebling | U.S. |
| | | | | Wayss and Freytag | Germany |
| | | | | CCL | England |
| | | | | Freyssinet | U.S. |
| | | | | Anderson | U.S. |
| | | | | Atlas | U.S. |

For mass production of pre-tensioned members, Hoyer system is used, which is the extension of the above method. In this system, the wires are stretched the full length of the casting stop (say 100 to 150 m) between anchorages rigidly held in the ground. The wires pass through a whole series of moulds, as shown in Fig. 45.15. Each mould is mounted on rollers and spaced with small ends gaps between adjacent moulds to allow the wires to be cut later and the individual units freed. When the concrete in the mould has set, the wires are released from end anchorages, thereby transferring the prestress to concrete through bond.

The anchoring devices for holding pretensioning wires to the bulkheads are made on the *wedge and friction* principle. One common device consists of a split cone wedge, which is made from a tapered conical pin. The tapered conical pin is drilled axially and tapped, and then cut in half longitudinally to form pair of wedges [Fig. 45.16(a)]. The anchoring block also has a conical hole, in which the tapered conical pin holds the wires. These grips can be used for single wires as well as for twisted wire strands.

In some cases, the pin is not drilled, but is cut in half longitudinally and the flat surface is machined and serrated, as shown in Fig. 45.16(a). In some cases, quick release grips, which are more complicated and costly, are used–specially when wires are to be held in tension only for short periods.

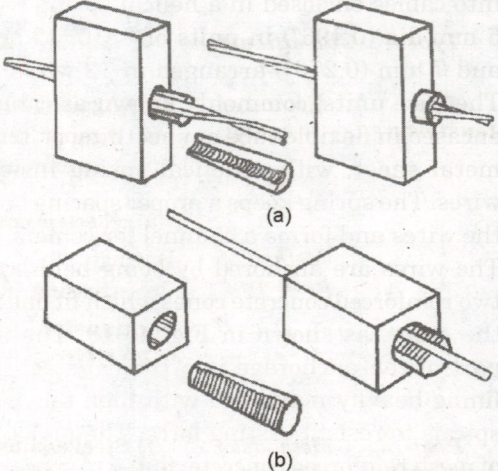

Fig. 45.16. Wire Gripping Devices

As stated earlier, the dependence on bond to transmit prestress between steel and concrete necessitates the use of small wires to ensure good anchorage. When the wires are relased, their ends lose their stress and recover to their original diameter, thus providing mechanical anchorage. If the *bond length or transfer length* (Fig. 45.17) is insufficient, specially when cracks occur near the end of a beam, the bond may be broken and the wires may slip. A more reliable method is to add mechanical end anchorages to the pre-tensioned wires.

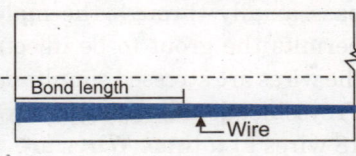

Fig. 45.17

For example, *Dorland anchorage*, consisting of clips, can be gripped to the tendon under high pressure, and the edges of the clips can then be welded together at several points. If such mechanical anchorages are used, tendons of greater diameter can be permitted.

**2. Post-Tensioning Systems:** In this method, the prestress is induced for tendons are tensioned only after the concrete has hardened and various post-tensioning systems can be classified under four groups: (a) mechanical prestressing, using jacks, (b) electrical prestressing by application of heat, (c) chemical prestressing by use of expanding cement, and (d) miscellaneous methods.

**(A) MECHANICAL PRESTRESSING:** This is the most common method, in which the stretching of wire is done by means of hydraulic jacks, screw jacks or by a system of levers. For large prestress, hydraulic jacks are used having capacity between 30 to 1000 kN. Systems of jacking vary from pulling one or two wires upto several hundred wires at a time. For example, the Magnel system employs a hydraulic jack which pulls two wires at a time while the Freyssinet double acting jack can pull up to 18 wires or 12 strands at a time. Pressure gauges for jacks are calibrated either to read the pressure on the piston or to read directly the amount of tension applied to the tendon. Usually, the elongation of steel is measured to check against the gauge readings. When several tendons in a member are to be tensioned in succession, care should be taken to pull them in the proper order so that no serious eccentric loading will result during the tensioning process.

Before the pull in the jack is released, the wires are anchored in their extended position against the end of concrete, by some wedging device. These are three basic principles or methods by which steel wires are anchored to concrete after stretching:

 (i) By method of wedge action, producing a friction grip on the wires ;
 (ii) By method of direct bearing from rivet or bolt heads formed at the end of wires; and
 (iii) By method of looping the wires around concrete.

Variations on the wire arrangement and jacking and anchorage devices constitute the difference between various systems. A number of stressing systems are now available, each varying detail and protected by patent rights. Some of the common systems are described below, in brief.

**(i) Freyssinet system.** This system was the first method of post-tensioning to be developed. The method, developed by French engineer Freyssinet is one of the most widely used method of prestressing. It uses high tensile steel tendons are grouped together into cables enclosed in a helical spring wires of 5 mm dia (0.196″) in units of 8, 10, 12 and 18, and 7 mm (0.276″) arranged in 12 wires units. The wire units, commonly known as cables are encased in flexible tube or sheething of 32 gauge metal sheet, with a helical spring inside the wires. The spring keeps a proper spacing between the wires and forms a channel for cement grout. The wires are anchored by being held between two reinforced concrete cones which fit one inside the other, as shown in Fig. 45.18. The female part of the anchorage is a conical steel wound lining heavily reinforced with high tensile steel spirals to resist bursting force. The male cone is of mesh reinforced concrete, fluted to space evenly the requisite number of wires. A central tube pass axially through the male cone. This tube permits the grout to be injected through it. All the wires are stressed simultaneously by means of Freyssinet double acting jack, which can pull upto 18 wires at a time. Wires are wedge around the jack casting and are stretched by means of main ram that reacts against the embeded anchorage [Fig. 45.19(*a*)]. When the required tension is reached the inner piston pushes the plug into the enchorage to secure the wires. The pressure on the main ram and that on the inner piston are then released gradually and the jack is removed. Since the anchoring unit is buried flush with the face of the concrete, it helps to transmit the reaction of the jack as well as the prestress of the concrete. After the completion of prestressing, grout is injected through the hole at the centre of male cone. Fig. 45.19(*b*) shows the concreted face.

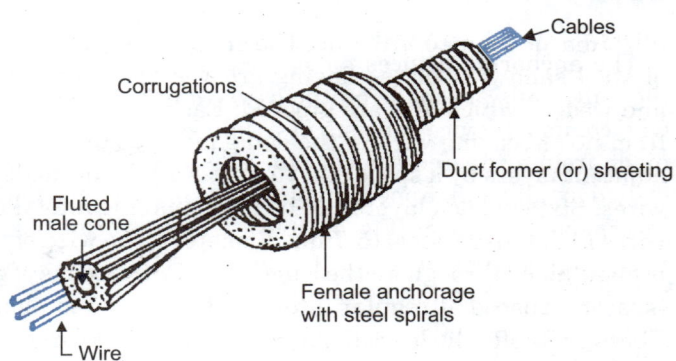

Fig. 45.18. Freyssinet Anchorage System.

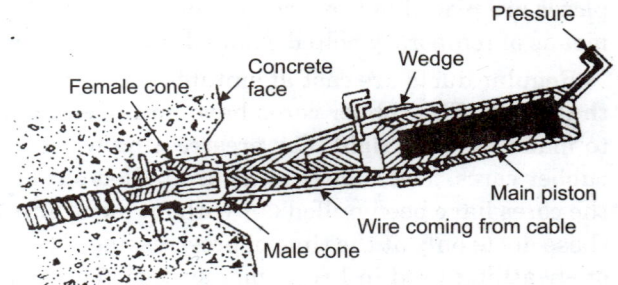

Fig. 45.19(*a*). Prestressing Jack.

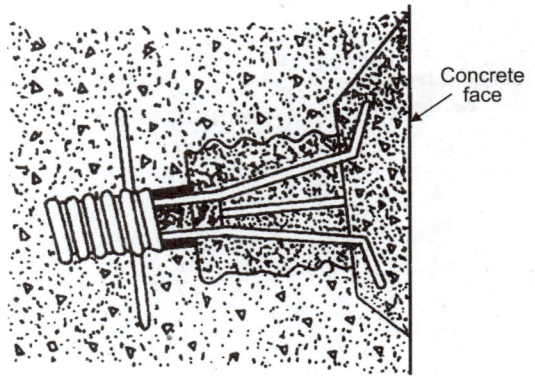

Fig. 45.19(*b*). Final Concreted Face.

This method is not very costly as the wires can be secured easily and pluge can be left in the concrete as they do not project out side.

Table 45.2 gives the details about standard size wires.

**TABLE 45.2**

| Types of cables | 8 × 5 mm | 10 × 5 mm | 12 × 5 mm | 18 × 5 mm |
| --- | --- | --- | --- | --- |
| Steel area | 1.57 cm² | 1.96 cm² | 2.35 cm² | 3.53 cm² |
| Weight | 1.2 kg/m | 1.5 kg/m | 1.8 kg/m | 2.7 kg/m |
| O.D. metal hose | 28.6 mm | 28.6 mm | 31.5 mm | 41.3 mm |
| I.D. metal hose | 25.4 mm | 25.4 mm | 28.6 mm | 38.1 mm |

(*ii*) **Magnel Blaton system.** This system was developed by Prof. Magnel and contractors Blaton, both of Belgium. The unit of stretching is two wires at a time, as against 8 to 18 wires in the Freyssinet system. The cable consists of any even number of wires upto 64, using high tensile steels wires of 5 mm

(0.2″) dia. or 7 mm (0.276″) dia. The anchorage device consists of steel sandwich plates having grooves to hold the wires and wedges which also are grooved. Each wire is separated from neighbouring wire in the same layer and those in the adjacent layers by a space of 0.75 mm. Each layer holds four wires. Sixteen such layers can be used, making a total of 64 wires high tensile steel (5-7mm diameter). The wire pattern is maintained throughout the length of cable by means of grills (spacers) spaced at regular interval of 1.5 to 2.5 m along it. These grills offer little frictional resistance to the wires which are free to move relative to each other during tensioning. The steel plate wedges are about 25.4 mm thick, with two wedge shaped grooves in their upper and lower faces. The sandwich plates are placed in the form of a bank, one above the other against a distribution plate and held thus by means of temporary bolted clamp during tensioning operations.

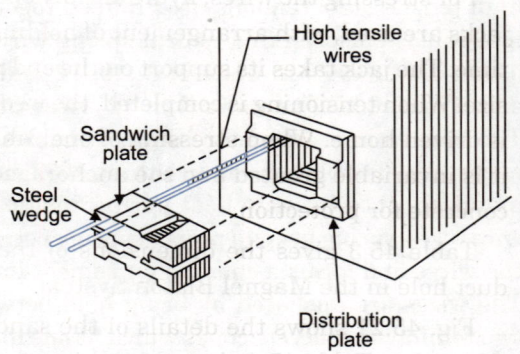

**Fig. 45.20.** Magnel Blaton system.

Regular ducts are cast at suitable places along the length of the beam by introducing rubber cores in the mould. The rubber cores have an initial diameter of 3 cm. Steel tubes are introduced in the holes, to make the core rigid. For prestressing wires, duct are embedded into member by the moulds having rubber cores. The rubber cores are pulled out after 8 to 10 hours of concreting. Even after 1 or 2 days, the cores have been pulled out without any difficulty. Ducts are thus formed; the wires are introduced in these ducts only at the time of prestressing. Thus, the Magnel Blaton System results in the saving of cost of sheathing used in Freyssinet system. The distribution plates are either cast into the member at the extremity of the duct or placed on thin mortar beds after the concrete has hardened, prior to assembling the anchorage.

**TABLE 45.3(a).** Details for 5 mm wire system

| No. of wires in cable | End plate | | Duct hole | |
|---|---|---|---|---|
| | Length (cm) | Width (cm) | Width (cm) | Depth (cm) |
| 16 | 13 | 13 | 5.5 | 5.5 |
| 24 | 16 | 16 | 5.5 | 7.5 |
| 32 | 20 | 17 | 5.5 | 10.0 |
| 40 | 23 | 18 | 5.5 | 12.5 |
| 48 | 26 | 19 | 5.5 | 15.0 |
| 56 | 29 | 20 | 5.5 | 17.5 |
| 64 | 32 | 21 | 5.5 | 20.0 |

**TABLE 45.3(b).** Details for 7 mm wire system

| No. of wires in cable | End plate | | Duct hole | |
|---|---|---|---|---|
| | Length (cm) | Width (cm) | Width (cm) | Depth (cm) |
| 16 | 15 | 20 | 6.5 | 6 |
| 24 | 18.5 | 26 | 6.5 | 9 |
| 32 | 24 | 26 | 6.5 | 13 |
| 40 | 29 | 27 | 6.5 | 15 |
| 48 | 34 | 28 | 6.5 | 18 |
| 56 | 37 | 29 | 6.5 | 19 |
| 64 | 39 | 30 | 6.5 | 22 |

For stressing the wires, hydraulically operated stressing jacks are used, with arrangement of holding two wires at a time. The jack takes its support on the end plate of suitable size. When tensioning is completed, the wedge for two wires is driven home. When stressing of one cable is completed, it is invariable grouted and the anchorage is covered with concrete for protection.

Table 45.3 gives the dimensions of the end plate and duct hole in the Magnel Blaton System.

Fig. 45.21 shows the details of the sandwich plate and the wedge. Table 45.4 gives the values of width $a$, length $e$ and other dimensions of the plate and the wedge.

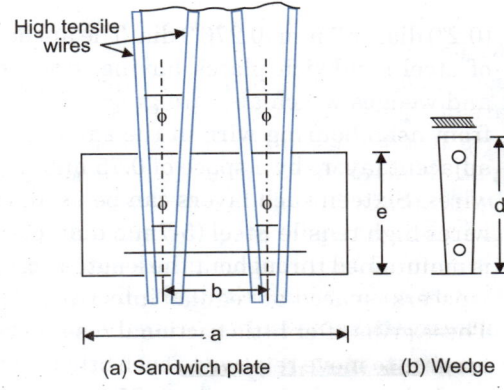

(a) Sandwich plate  (b) Wedge

Fig. 45.21. Sandwich Plate and Wedge.

**TABLE 45.4**

| Dia. of wire | Dimensions (cm) | | | | |
|---|---|---|---|---|---|
| | a | b | c | d | e |
| 5 mm | 11.4 | 4.0 | 2.1 | 6.0 | 5.4 |
| 7 mm | 16.0 | 5.0 | 2.5 | 10.4 | 8.9 |

*(iii) Gifford Udall system.* Gifford Udall system is one of the most widely used systems. This system developed in U.K. using 7 mm wires. The cable consists of parallel wires which are separated by means of circular spacers placed at one metre c/c. The diameter of spacer is 39 mm for cable upto 8 wires and of 51 mm for cable upto 12 wires. The system uses two types of anchorage systems:

(a) plate anchorage  (b) tube anchorage.

*(a) Plate anchorage.* Fig. 45.22 shows the plate anchor system originally used by Gifford Udall system. It essentially consists of a bearing plate, a thrust ring anchor grips and a steel helix. The wires are stressed and anchored individually by small wedge type Udall grips seating against a bearing plate. The thrust ring is cast into the concrete. The bearing plate locates against the thrust ring. The end of the dust is encircled by a helix. The anchor grips consist of an outer barrel and two semi circular wedges which fit in the tapered hole of the barrel. The wire is gripped between the serratted parallel faces of the wedge.

*(b) Tube anchorage.* This is the most recent anchorage system in which the wedges fit directly into the tappered recesses in the bearing plate which is thick. Thus the need for separate grips is eliminated (Fig. 45.23). The bearing plate includes a threaded hole for grouting purposes. The bearing plate seats against a *tube unit* incorporating the thrust ring and helix in a single element which bolts into the form work and is cast into the concrete. The tube unit provides the necessary spreading out of the wire in the cable so as to pass through the hole of the bearing plate. This type of anchorage is available for 8 wire and 12 wire cables. The jack pulls one wire at a time.

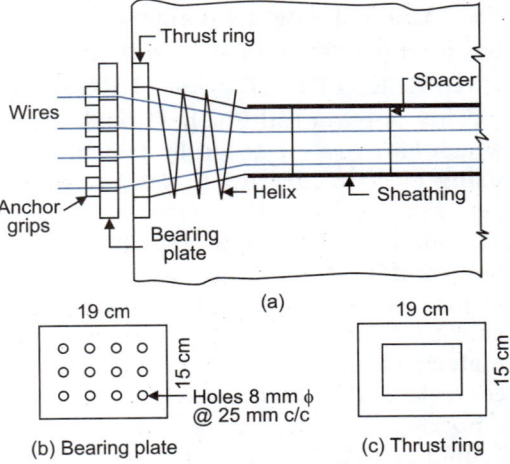

Fig. 45.22. Plate Anchorage with Grips.

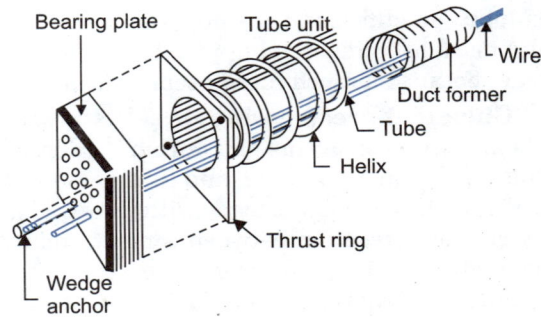

Fig. 45.23. Gifford Udall Tube Anchorage System.

(*iv*) **P.S.C. Mono-wire system:** In this system, wires are tensioned individually. The system uses *collect sleeves* wedging in conical holes. These sleeves are of single piece serrated internally to give positive grip to wire passing through them. A steel trunked guide leads each wire from its cable position to its point of anchorage through a gentle curvature (Fig. 45.24). The mono-wire anchorage are available for 1-, 2-, 4-, 8-, and 12 wire cables of size upto 7 mm. For 8- and 12-wire cable, a high strength plastic spacer is used to make up the cable and separate the wires for stressing.

(*v*) **Lee McCall system**

Lee-McCall system uses high-tensile alloy steel bars as the prestressing tendons, in the place of high-tensile steel wires used in other systems. *Macalloy steel* used in the system is an open-hearth Silico-Manganese steel, hot rolled into bars and subsequently processed to give the required physical properties. The diameter of the rods may vary from 12 to 40 mm. The rods are threaded at the ends. Holes are made in the member by means of rubber core. When the concrete has set, bars are introduced in the hole, after removing the core. After the desired stretching, a nut is tightened at its end to prevent its return to original length. Fig. 45.25 shows the details of the system. The bars can be either bonded or unbonded to the concrete.

### (B) ELECTRIC (*Thermal*) PRESTRESSING

This is essentially a post-tensioning method in which bars are stretched by means of heating these using electric current. In the high tensile wires is generally reffered as 'thermo-electric prestressing'. This method, widely used since 1958 for pre-tensioning bar reinforcement of pre-cast units. Thus the use of jack is eliminated altogether. The method uses smooth reinforcing bars coated with thermoplastic material such as sulphur, before embeding them in concrete. After the concrete has set, a low voltage high amperage current is passed through the bar to heat it upto 170°C. The sulphur melts and allows steels to expand. The nuts on the protruding ends of the bars are then tightened against heavy washers. On cooling, the cooling period is reckoned to be 12-15 min., the sulphur solidifies and restores the bond. The method did not find commercial application with high tensile steel because a much higher temperature would be required for its prestressing.

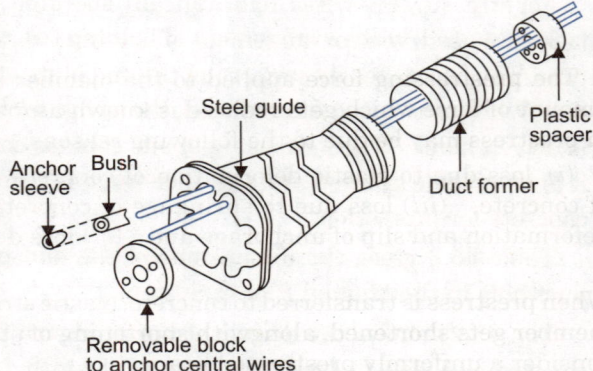

Fig. 45.24. P.S.C. Mono-wire system.

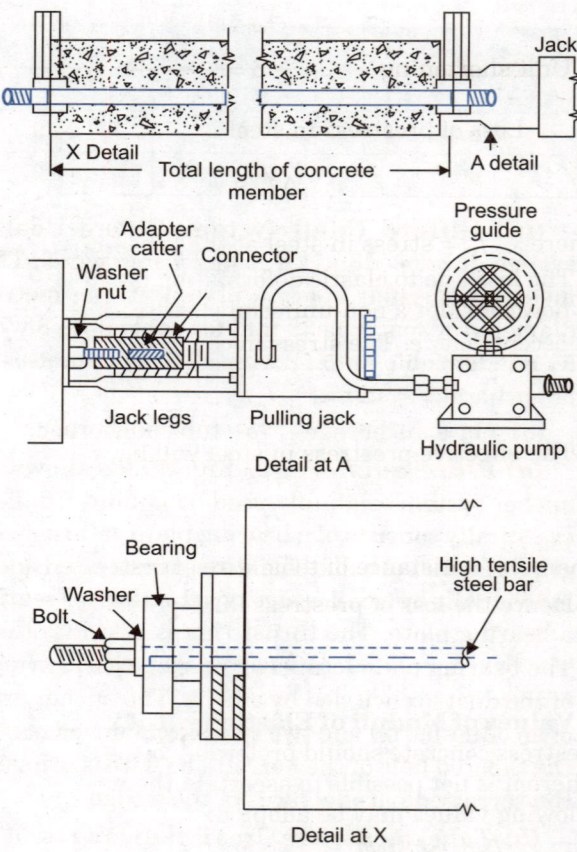

Fig. 45.25. Lee-Mccall System.

**Chemical prestressing or self-stressing:** In this method, *self stressing cement* is used that expand chemically after setting and during hardening. When steel is embedded in such a concrete, it is elongated by the expansion of concrete, and thus gets prestressed. The steel, in turn, produces compressive prestress in concrete. If the concrete made with the expanding cement is un-restrained, the amount of expansion produced by chemical reaction between cement and water could be between 3 to 5%, resulting in disintegration of concrete by itself. However, if the concrete is restrained internally (with steel) or externally (by other means), the expansion would be controlled, resulting in compressive prestress of the concrete. Such cement has been successfully applied for many interesting projects in France. Extensive application of expansive cement may not be too far in future.

## 45.5. LOSSES OF PRESTRESS

The prestressing force applied to the member does not remain constant, but gradually reduces. The amount of force which gets reduced is known as *loss of prestress* and may vary from 15 to 20%. The losses in prestress may be due to the following reasons:

(i) loss due to elastic deformation of concrete when prestress is applied to it. (ii) loss due to creep of concrete, (iii) loss due to shrinkage of concrete, (iv) loss due to relaxation of steel, (v) loss due to deformation and slip of anchorage and (vi) loss due to friction between tendon and concrete.

**1. Loss due to elastic deformation of concrete:** This loss takes place only in pretensioned beams. When prestress is transferred to concrete, elastic stress and strains are induced in it. Due to this the concrete member gets shortened, alongwith shortening of steel, thereby reducing the prestress in steel. Let us first consider a uniformly prestressed beam.

Let
$P_i$ = initial prestress force.
$A_e$ = area of transformed section = $A_c + m A_{st}$
$A_{st}$ = area of tensioned steel.

Unit shortening $\delta = \dfrac{P_i}{A_c E_c + A_{st} E_s}$

∴ Loss of prestress in steel is

$$\Delta f_{el} = E_s \delta = \frac{E_s P_i}{A_c E_c + A_{st} E_s} = \frac{m P_i}{A_c + m A_{st}} = \frac{m P_i}{A_e} = m \frac{f_{si} A_{st}}{A_e} \quad ...(45.5)$$

where $f_{si}$ = stress in steel at the time of initial tension.

The loss due to elastic deformation due to concrete may range from 3 to 6% in pre-tensioned beams. Let us now consider a non-uniformly prestressed beam in which the prestressing force ($P_i$) has been applied at an eccentricity $e$. The stress in concrete at any height $y$ from the centroidal axis is

$$f_{ci} = \frac{P_i}{A_e} \pm \frac{P_i \cdot e}{I_e} y$$

The effective prestress in steel will be

$$f_{se} = f_{si} - m f_{ci} = f_{si} - m \left( \frac{P_i}{A_e} + \frac{P_i \cdot e}{I_e} y_t \right)$$

where $y_t$ = distance of tensile steel from centriodal axis.

Hence the loss of prestress in steel is

$$\Delta f_{el} = f_{si} - f_{se} = m \left[ \frac{P_i}{A_e} + \frac{P_i \cdot e}{I_e} y_t \right] = m \left[ \frac{f_{si} A_{st}}{A_e} + \frac{f_{si} A_{st}}{I_e} \cdot e \cdot y_t \right] \quad ...(45.6)$$

**Values of Moduli of Elasticity:** The value of the modulus of elasticity of steel ($E_s$) used for the design of prestress concrete should preferably by determined by tests on samples of steel to be used for construction. Where it is not possible to ascertain the modulus of elasticity by test or from the manufacturer of steel, the following values may be adopted:

| Type of Steel | $E_s$ (N/mm$^2$) |
|---|---|
| Plain-drawn wires (cold drawn or cold worked) | $2.1 \times 10^5$ |
| High tensile steel bars (rolled or heat treated) | $2.0 \times 10^5$ |

The modulus of elasticity of concrete may be assumed to have the values given by the following expression:

$$E_c = 5700 \sqrt{f_{ck}} \ (N/mm^2) \quad ...(45.7)$$

where $f_{ck}$ is the characteristic cube strength of concrete.

**Post-Tensioned Beams:** In the case of post tensioned beam with a single concentric tendon, as the tendon is stretched and anchored against concrete, all the elastic strain has occurred in concrete when the jacking force is reached in tendon. Since the force in the cable is measured after the elastic shortening of the concrete has taken place, no loss in prestress due to this need be accounted for. However, in actual practice there are more than one tendon. If the tendons are stressed in succession, the prestress is applied gradually. Due to this, the tendon that is stressed first will suffer the maximum loss while the one that is tensioned last will not suffer any loss. Accurate calculation for losses is complicated, but for all practical purposes, it may be assumed that average loss of each cable is equal to half the loss in first cable.

In order to counteract elastic loss all the tendons are anchord at an *overstress* equal to the average elastic loss in practice. The average loss in post-tensioned beam may be of the order 40%.

**2. Loss due to creep of concrete:** Creep in concrete is defined as its time-dependent deformation resulting from the presence of stress. It is the plastic flow of concrete under compression. Creep strain varies with the intensity of stress, and is about two to three times the elastic strain. The rate of creep is high initially, and then decreases as time increases. The creep increases with higher water-cement ratio and with a lower aggregate cement ratio. It varies inversely with strength of concrete. It also depends upon the humidity of surrounding atmosphere and the strength at the time of loading. It is known that the failure of early efforts at prestressing was attributed largely to the lack of knowledge concerning creep in concrete. It is still one of the main source of loss, and a serious one, if the prestress in the steel is low and the compression in concrete is high. Pre-tensioned members have more loss than post-tensioned ones, because transfer of prestress usually takes place earlier in pre-tensioned members. Creep coefficient can be taken from IS:456–2000 as per clause 6.2.5.1 of code.

The ratio of the final creep strain to the elastic strain in concrete is defined as *creep coefficient* ($C_c$) the values of which are given in Table 45.5 for different atmospheric conditions. Creep strain ($\varepsilon_{cr}$) is given by

$$\varepsilon_{cr} = (C_c - 1)\varepsilon_{el} \qquad ...(45.8)$$

where $\varepsilon_{el}$ = elastic strain in concrete.

**TABLE 45.5** Creep Coefficient $C_c$

| Condition of exposure | Final creep coefficient ($C_c$) |
|---|---|
| In water | $0.5 K_c$ to $K_c$ |
| In very humid atmosphere | $1.5 K_c$ to $2.0 K_c$ |
| In average humidity | $2.0 K_c$ to $3.0 K_c$ |
| In dry atmosphere (*i.e.* in dry internal rooms) | $2.5 K_c$ to $4.0 K_c$ |

The value of $K_c$ depends upon the ratio of cube strength ($f_{cr}$) of concrete at transfer of prestress on 15 cm cube and cube strength ($f_{cu}$) of concrete at 28 days where Portland cement is used and 7 days where rapid hardening cement is used. Its values may be taken from Table 45.6

**TABLE 45.6** Values of $K_c$

| $f_{cr}/f_{cu}$ at the time of stressing | $K_c$ |
|---|---|
| 0.50 | 2.20 |
| 0.84 | 1.50 |
| 1.00 | 1.00 |
| 1.08 | 0.75 |
| 1.30 | 0.50 |

The loss of prestress due to creep of concrete under load should be determined for all the permanently applied loads including the prestress. The creep loss due to live load stresses, erection stresses and other stresses of short duration may be ignored. The loss of prestress due to creep of concrete is obtained as the product of modulus of elasticity of prestress steel ($E_s$) and the integrated ultimate creep strain of concrete fibre along the line of centre of gravity of the prestressing steel over its entire length. The total creep strain during any specific period should be assumed, for all practical purpose, to be the creep strain due to a sustained stress equal to the average of the stresses at the beginning and end of the period.

As per IS : 1343-1980, the ultimate creep strain may be estimated from the following values of creep-coefficient (that is, ultimate creep strain/elastic strain at the age of loading).

| Age of Loading | Creep coefficient |
|---|---|
| 7 days | 2.2 |
| 28 days | 1.6 |
| 1 year | 1.0 |

The above ultimate creep strain does not include the elastic strain. For the calculation of deformation at some stage before the total creep is reached, it may be assumed that about half the total creep takes place in the first month after the loading and that about three quarters of the total creep takes place in the first six months after loading.

The loss in prestress due to creep may range from 5 to 10%.

**3. Loss due to shrinkage of concrete:** Shrinkage in concrete is its contraction due to drying and chemical changes. It depends upon the quantity of water, type of aggregates used in the mix and surrounding atmospheric conditions. If minimum shrinkage is desired, the water cement ratio and the proportion of cement paste should be kept to a minimum. Aggregates of larger size, well graded for minimum void, need a smaller amount of cement paste, and shrinkage will be smaller. Harder and denser aggregates of low absorptions and high modulus of elasticity will exhibit small shrinkage. Shrinkage is relatively small for cements high in tricalcium silicate and low in the alkalies and the oxides of sodium and potassium. Shrinkage is also time dependent, though bulk of the shrinkage takes place in the early days of the hardening of concrete. In pre-tensioned work, wires are subjected to some compression due to creep of concrete, before the release of wires. The concrete around the wires is subjected to tension as the anchored wires are not free to shorten its length. The concrete thus tends to slip in the early stages and develop fine cracks. When the wires are released, these cracks close, causing a loss of prestress. After that, the time dependent shrinkage continue to shorten the length of member, though at a much reduced rate. In post-tensioning work, shrinkage takes place unhampered and hence loss of shrinkage is relatively small. As per IS : 1343-1980, the approximate value of shrinkage strain for design is assumed as follows:

$$\text{For pre-tensioning} = 0.0003$$
$$\text{For post-tensioning} = \frac{0.0002}{\log_{10}(t+2)} \qquad \ldots(45.9)$$

where $t$ = age of concrete at transfer, in days.

The value of shrinkage strain for design of post-tensioned concrete may be increased by 50 percent in dry atmospheric conditions, subject to a maximum value of 0.0003. Thus if $\varepsilon_{sh}$ is the shrinkage strain, the loss of prestress $\Delta F_{sh}$ due to shrinkage is given by

$$\Delta f_{sh} = \varepsilon_{sh} \cdot E_s \qquad \ldots(45.10)$$

The loss of prestress due to shrinkage may be of the order of 4 to 6% for pre-tensioned work and 3 to 4% for post-tensioned work.

**4. Loss due to relaxation of steel.** When the stress in steel is more than half the yield stress, there is creep of steel also. It is sometimes measured by the amount of lengthening when maintained under a constant stress for a period of time. Relaxation varies with steel of different compositions and treatments. The percentage of creep increases with increasing stress, and when a steel is under low stress, the creep is negligible. While creep in steel is a function of time, there is evidence to show that under the ordinary working stress for high-tensile steel, creep takes place mostly during of the first few days.

The magnitude of the loss of prestress due to relaxation of steel may be taken to vary from 2 to 8 percent of the average initial stress depending on the quantity of the steel and the order of the initial prestress under constant strain. It is time dependent and also depends upon the magnitude of stress. Temporary overstressing for the purpose of the reducing the loss of prestress due to relaxation of steel may be permitted provided such overstressing does not exceed 85% of the ultimate tensile strength ($f_{su}$) of the prestressing steel.

**5. Loss due to deformation and slip of anchorage.** This loss is applicable in post-tensioning system in which, after tensioning a tendon, the jack is released to transfer prestress to concrete by bearing of the anchorage. During the transfer, the anchorage fixures deform under stress, and the friction wedges used in several system of prestressing slip slightly before wires are firmly gripped. The amount of slippage ($\delta_s$) depends on the type of wedge and stress in the wires, and an average value of 2 to 3 mm may be taken. For heavy strands, the slip may be 5 mm. Generally, manufacturers supply the value of such slip. In the case of Freyssinet cones, the slip is 6 mm for 5 mm of wires and 9 mm for 7 mm of wire cables. If the release of strain is assumed to be uniform throughout the length of wire, the loss of prestress is given by

$$\Delta f_{slip} = E_s \frac{\delta_s}{L} \qquad \ldots(45.11)$$

where $L$ is the length of wire.

It is necessary to note that this loss of prestress depends upon the length of the tendons, being high when a tendon is short and low when a tendon is long.

Due to friction forces between the piston and cylinder etc. of the jack, there may be some loss of stress in jack itself. Due to this, the actual force in the wire will be less than the one registered by the gauge of the jack. The value of this loss depends upon the type of jack. For Freyssinet jack, it is about 36 kN.

Any loss of prestress which may occur due to slip of wires during anchoring or due to the strain of anchorage may be allowed for in the design. Loss due to slip in anchorage may be allowed for in the design. Loss due to slip in anchorage is of special importance with short members and the necessary additional elongation should be provided for at the time of tensioning to compensate for this loss.

**6. Loss due to friction.** During tensioning operations in post-tensioned members, there will be considerable movement or sliding of tendon relative to the surrounding. Since the tendon is in direct contact with the duct or with spacers provided, the friction will cause a reduction in the prestressing force as the distance from the jack increases. The frictional loss can be conveniently considered in two parts:

(*a*) the curvature effect and (*b*) length or wobble effect.

**(*a*) Curvature effect.** The loss of prestress due to curvature effect results from the intended curvature of the tendons. The loss depends upon:

(*i*) coefficient of friction between the contact materials.

(*ii*) Stress in the tendon and (*iii*) total change in the angle.

Fig. 45.26 shows a small length $dx$ of a prestressing tendon whose centroid follows the arc of a circle of radius $R$. The change $d\alpha$, as the tendon goes around the length $dx$ is

$$d\alpha = \frac{dx}{R} \qquad ...(i)$$

The normal component of pressure produced by stress $P$ bending around an angle $d\alpha$ is given from the vector diagram:

$$N \approx P \cdot d\alpha = P \cdot \frac{dx}{R} \qquad ...(ii)$$

Hence frictional loss $dP$ around the length $dx$ is

$$dP = -\mu N = -\mu P d\alpha$$

where $\mu$ is the coefficient of friction.

(The minus sign indicates that $P$ decreases as $\alpha$ increases.)

$$\therefore \qquad \frac{dP}{P} = -\mu \, d\alpha \qquad ...(iii)$$

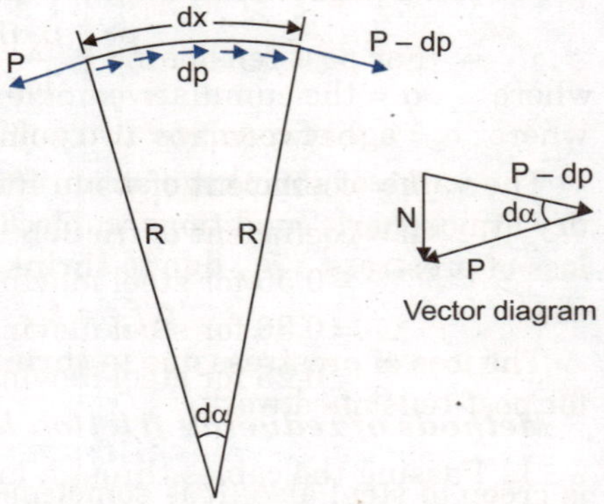

**Fig. 45.26**

If $P_1$ is the force at one end, and $P_2$ is force at the other end of the curved cable, subtending an angle $\alpha$ we get, by integrating *Eq.* (*iii*)

$$\left[\log_e P\right]_{P_1}^{P_2} = -\left[\mu \alpha\right]_0^{\alpha} = -\mu \alpha$$

or $$P_2 = P_1 \cdot e^{-\mu \alpha} \qquad ...(45.12)$$

The loss of force $(P_1 - P_2)$ is called *loss due to curvature*.

*Eq. 45.12* may also be written in the alternative form:

$$P_2 = P_1 \cdot e^{-\mu L/R} \qquad ...[45.12(a)]$$

since $\alpha = L/R$ for a length $L$ of constant radius $R$.

For tendons with a succession of curves of varying radii, it is necessary to apply this formula to the different sections in order to obtain the total loss.

**(*b*) Wobble or length effect.** In post-tensioning there is another loss known as *length, wobble* or *wave loss*. This is due to the cable touching the duct or surrounding concrete. This is mainly caused by the deviation of the duct from its intended depth. The length effect is the amount of friction that would be encountered if the tendon is a straight one, that is one that is not purposely bent or curved. This loss is dependent on the length and stress of tendon and coefficient of friction between the contact materials, and the workmanship and method used in aligning and obtaining the duct. The prestressing force $P_2$ at the other end can be computed by substituting a factor $K_f L$ in the place of $\mu \alpha$ in *Eq. 45.12*:

$$P_2 = P_1 e^{-K_f L} \qquad \ldots(45.13)$$

Combining both the effects, we have

$$P_2 = P_1 e^{-(\mu \alpha + K_f L)} = P_1 e^{-\left[\frac{\mu L}{R} + K_f L\right]} \qquad \ldots(45.14)$$

If the total difference in tension in the tendon at the start and that at end of the curve is not excessive (say not more than 15 to 20%) the following approximate formula may be used:

$$\left(\frac{P_2 - P_1}{P_1}\right) = -\left(K_f + \frac{\mu}{R}\right) L \qquad \ldots(45.15)$$

or

$$\frac{P_1 - P_2}{P_1} = (K_f L + \mu \alpha) \qquad \ldots[45.15\,(a)]$$

Strain loss,

$$\varepsilon_f = \frac{P_1 - P_2}{E_s \cdot A_{st}} \qquad \ldots[45.15\,(b)]$$

As per IS : 1343-1980, for straight or moderately curved structures with curved or straight cables, the values of prestressing force $P_x$ at a distance $x$ from the tensioning end may be determined by the formula,

$$P_x = P_0 e^{-(\mu \alpha + k \cdot x)} \qquad \ldots[45.14\,(a)]$$

where $P_0$ = Prestressing force in the prestressed steel at the tensioning end.

For small values of $(\mu \alpha + kx)$, the above equation may be reduced to the form

$$P_x = P_0 (1 - \mu \cdot \alpha - k \cdot x)$$

where
  $\alpha$ = the cumulative angle in radians through which the tangent to the cable profile has turned between any two point under consideration.
  $k$ = coefficient of wave effect, varying from $15 \times 10^{-4}$ to $50 \times 10^{-4}$ per metre.
  $\mu$ = coefficient of friction in curve, to be taken as follows:
    = 0.55 for steel moving on smooth curve,
    = 0.30 for steel moving fixed to duct,
    = 0.25 for steel moving on lead.

**Methods of reducing friction loss:** The friction loss can be reduced by the following methods:

1. Passing the cables through metal tubes at bends
2. Making the bends through as small an angle as possible.
3. Making the radius of curvature at bends as large as possible, but not lesser than 800 times the diameter of the wire.
4. Avoiding double curvatures and using catenary curves.
5. Supporting the cables at closer intervals.
6. Using lubricants of unbonded concrete. For bonded concrete, water-soluble oils have been successfully employed to reduce the friction while tensioning. The lubricant is flushed off with water afterwards.
7. Prestressing the wires from both the ends. This involves more work in the field, but is often resorted to when the tendons are long or when the angles of bending are large.
8. Over-tensioning the wire. When friction loss in not excessive the amount of overtension may be equal to maximum frictional loss. The expected extra elongation can be computed to serve as a check. This extra pull required for overcoming the loss due to slip in anchorage or that required for minimising the creep in steel. The extra pull should be the greatest of the three required values. At this pull, the desired prestress will be created at the centre of the cable and the tension will gradually increase towards the pulling ends. If the wire is released now, part of stress will be lost at the end due to slip in anchorage. This will result in lessening the strain throughout the length of wire, and the resulting movement of wire toward the centre will take place. Due to this, the friction force will act in the opposite direction, without letting the stress at the centre be reduced. Thus, the effect of overtensioning with a subsequent release-back is to put the frictional difference in the reverse direction. However, when the frictional loss is very high, it cannot be totally overcome by over tensioning, since maximum amount of tensioning is limited by strength or yield point on the tendon. In such a case, overtensioning may be limited to 10% only and the remaining loss must be allowed for in the design.

**Total Amount of Losses:** The amount of losses to be deducted will differ depending upon the definition of the term *initial prestress* and other terms defined below:

*Apparent jacking stress*. It is the force shown on the gauge of the jack.

*Jacking stress*. It is the actual pull in the wire near the jack before the anchorage. This stress is always less than the apparent jacking stress.

*Initial prestress*. The jacking stress minus the anchorage loss will be the stress at anchorage, after release, and is frequently called the initial prestress.

*Effective prestress or design prestress*. It is equal to the initial prestress minus the losses.

If the jacking stress minus the anchorage loss is taken as the initial prestress as defined above, the losses to be deducted will consist of elastic shortening, creep and shrinkage in concrete and creep in steel. For points away from the jacking end, the effect of friction must be considered in addition.

Magnitude of losses are generally expressed in the following four ways:

(i) *In unit strains*. Losses like creep, shrinkage and elastic shortening of concrete are more conveniently expressed by unit strains.

(ii) *In total strains*. Generally, anchorage losses are expressed in terms of total strain.

(iii) *In unit stresses*. If $E_s$ is known, all losses expressed in strains can be converted into unit stresses in steel.

(iv) *In per cent of prestress*. This is more common form, specially for losses due to creep in steel and friction. Other losses expressed in unit stresses can be converted into percentage of initial prestress.

For average steel and concrete properties, cured under average air conditions, the following values may be taken as representative of average losses:

**TABLE 45.7**

| Loss due to | Pre-tensioning % | Post-tensioning % |
|---|---|---|
| 1. Elastic shortening and bending of concrete | 3 | 1 |
| 2. Creep of concrete | 6 | 5 |
| 3. Shrinkage of concrete | 7 | 6 |
| 4. Creep in steel | 2 | 3 |
| Total loss | 18 | 15 |

The above table is based on the assumption that overtensioning has been applied to reduce creep in steel and to overcome friction and anchorage losses.

If the frictional loss is excessive, that part which has not been balanced by overtensioning should be accounted for separately, in the design.

When the average prestress ($P_0/A_c$) is high (say about 7 N/mm²) the above losses should be increased to 25% for pretensioning and 20% for post-tensioning.

When the average prestress is low (say about 1.75 N/mm²) the above total losses may be reduced to 14% for pre-tensioning and 12% for post-tensioning.

Accurate computation of various losses is possible only after a preliminary design of a member is made. However, the preliminary design must take into account the losses. Due to this, some losses are assumed, and for this guidance may be taken from the above table.

## 45.6. COMPUTATION OF ELONGATION OF TENDONS

Pre-tensioning results in the elongation of tendons. The computation of this elongation is important, so that it can be compared with the measured elongation, thus serving as a check on the accuracy of the gauge readings or on the magnitude of frictional loss along the length of the tendon. We shall consider two cases.

**Case 1. Friction along tendon neglected:** If the friction along the tendon is neglected, it will be subjected to uniform tension $P$ along its entire length, resulting in uniform stress.

Thus, 
$$\Delta_s = \frac{PL}{A_{st} \cdot E_s}$$ 
...(45.16)

This formula is valid only if the prestress does not exceed the proportional limit. If, however, it exceeds the proportional limit, it is necessary to refer to the stress-section diagram for the corresponding strain.

**Case 2. Considering friction along tendon:** Let $L$ be the total length of the cable and $\alpha$ be the total change in the angle. At a distance $x$ from the pulling end, the change in angle

$$\alpha_x = \frac{x \cdot \alpha}{L}$$

The pull at distance $x$ from the support is given by *Eq. 45.14(b)*.

$$P_x = P_1 \cdot e^{-(\mu \cdot \alpha_x + K_f \cdot x)} = P_1 \cdot e^{-x\left(\frac{\alpha}{L}\mu + K_f\right)}$$

∴ Extension in length $\quad dx = \dfrac{P_1 \cdot e^{-x\left(\frac{\alpha}{L}\mu + K_f\right)}}{A_{st} \cdot E_s} \cdot dx$

Hence total extension of steel is $\Delta s = \dfrac{P_1}{A_{st.} \cdot E_s} \int_0^L e^{-x\left(\frac{\alpha}{L}\mu + K_f\right)} \cdot dx$

$$\Delta s = \frac{P_1 L}{-A_{st} E_s (\alpha\mu + K_f L)} \left[ e^{-(\alpha\mu + K_f L)} - 1 \right] = \frac{P_1 L}{A_{st} E_s (\alpha\mu + K_f L)} \left[ 1 - \frac{P_2}{P_1} \right] \quad \ldots(45.17)$$

$$\frac{\text{Actual extension}}{\text{Extension of frictionless cable}} = \frac{1 - \dfrac{P_2}{P_1}}{(\mu\alpha + K_f L)}$$

**Example 45.7.** *A pretensioned prestress concrete beam of 9 m span has a cross-section of 400 mm × 800 mm, and is prestressed with 2400 kN at transfer. The cable has cross-sectional area of 2000 mm² of steel and has a parabolic profile with maximum eccentricity of 120 mm at the middle of span. Determine the loss of prestress, given that $E_s = 2.1 \times 10^5$ N/mm². Use M 30 concrete. Assume minimum ultimate tensile strength of prestressing steel as 1500 N/mm².*

**Solution:**

Prestress at transfer,

$$f_s = \frac{2400 \times 1000}{2000} = 1200 \text{ N/mm}^2$$

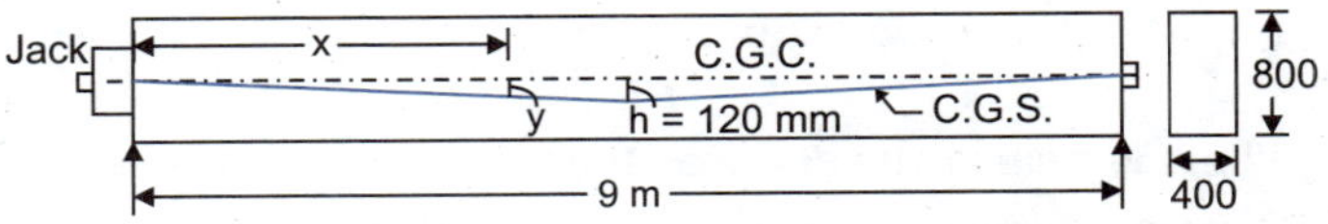

$$E_c = 5700 \sqrt{f_{ck}} = 5700 \sqrt{30}$$

$$\approx 31220 \text{ N/mm}^2$$

Strain in steel $\quad \epsilon_s = \dfrac{f_s}{E_s} = \dfrac{1200}{2.1 \times 10^5} = 57.1 \times 10^{-4}$

**Fig. 45.27**

*(i) Loss due to elastic shortening of concrete.* If the cable were straight and parallel to the centroidal axes, at a constant eccentricity $y$, Eq. 45.6 could be used for computing the elastic loss. However, in the present case, the cable is parabolic. Let us derive an expression for strain in concrete ($E_c$) at the level of steel.

The equation for parabola is $y = \dfrac{4\,hx\,(L-x)}{L^2}$

where $y$ is the eccentricity of cable at distance $x$ from the end of the span, and $h$ is the maximum eccentricity of the mid-span.

Hence compressive deformation in concrete, at steel level, due to prestressing force is

$$\delta_{cx} = -\frac{1}{E_c}\left(\frac{P_i}{A_e} + \frac{P_i \cdot y^2}{I_e}\right)$$

If $\epsilon_{el}$ is the mean elastic strain, we have

$$\epsilon_{el} = \frac{1}{LE_c}\int_0^L \left(\frac{P_i}{A_e} + \frac{P_i \cdot y^2}{I_e}\right) dx$$

Substituting the value of $y$, we get

$$\epsilon_{el} = \frac{P_i}{L E_c} \int_0^L \left[ \frac{1}{A_e} + \frac{16 h^2 x^2 (L-x)^2}{I_e \cdot L^4} \right] dx = \frac{P_i}{L E_c} \int_0^L \left[ \frac{1}{A_e} + \frac{16 h^2}{I_e \cdot L^4} \{x^2 L^2 + x^4 - 2Lx^3\} \right] dx$$

$$= \frac{P_i}{L E_c} \left[ \frac{x}{A_e} + \frac{16 h^2}{I_e L^4} \left\{ \frac{x^3 L^2}{3} + \frac{x^5}{5} - \frac{2 L x^4}{4} \right\} \right]_0^L$$

or $\qquad \epsilon_{el} = \dfrac{P_i}{E_c} \left[ \dfrac{1}{A_e} + \dfrac{8}{15} \dfrac{h^2}{I_e} \right] \qquad\qquad$ ...(45.18)

Since the losses are of small nature, $A_e$ can be approximately taken equal to $A_c$ and $I_e$ can be taken equal to $I_c$ without any appreciable error.

$\therefore \quad A_e \approx A_c = 400 \times 800 = 32 \times 10^4 \text{ mm}^2; I_e \approx I_c = \dfrac{1}{12}(400)(800)^3 = 1.7 \times 10^{10}$.

Substituting the values in *Eq. 45.18*

$$\epsilon_{el} = \frac{2400 \times 1000}{E_c} \left[ \frac{1}{32 \times 10^4} + \frac{8}{15} \frac{120 \times 120}{1.71 \times 10^{10}} \right] \approx \frac{8.58}{E_c}$$

An approximate value of strain can be obtained by taking the average of strains at maximum and minimum strains points. Minimum compressive strain occures at the ends while maximum compressive strain occurs at mid span.

$$\text{Strain at end section} = \frac{P_i}{A E_c} = \frac{2400 \times 1000}{32 \times 10^4 E_c} = \frac{7.5}{E_c}$$

$$\text{Strain at mid span} = \frac{1}{E_c} \left[ \frac{P_i}{A} + \frac{P h^2}{I} \right] = \frac{2400 \times 1000}{E_c} \left[ \frac{1}{32 \times 10^4} + \frac{120 \times 120}{1.71 \times 10^{10}} \right] \approx \frac{9.53}{E_c}$$

$\therefore \quad$ Approximate mean strain $\epsilon_{el} = \dfrac{(7.5 + 9.53)}{2 E_c} = \dfrac{8.515}{E_c}$

This is nearly the same as found by exact expression. Hence the approximate method can be used. For the present case,

$$\epsilon_{el} = \frac{8.58}{E_c} = \frac{8.58}{31220} = 2.75 \times 10^{-4}$$

$$\text{Percentage loss} = \frac{2.75 \times 10^{-4}}{60 \times 10^{-4}} \times 100 = \mathbf{4.58\%}$$

$\therefore \quad$ Loss of prestress $= 0.0458 \times 1200 \approx 55 \text{ N/mm}^2$

(*ii*) **Loss due to shrinkage of concrete:** As per I.S. Code, shrinkage strain = 0.0003 for pretensioned members.

$$\% \text{ Loss} = \frac{3 \times 10^{-4}}{60 \times 10^{-4}} \times 100 = \mathbf{5\%}; \text{ Loss of prestress} = 0.05 \times 1200 = 60 \text{ N/mm}^2$$

(*iii*) **Loss due to creep of concrete:** As per IS : 1343-1980 the creep coefficient at transfer of prestress at 28 days of curing of concrete may be taken as 1.6.

$\therefore \qquad\qquad$ Creep strain $= C \epsilon_{el} = 1.6 \times 2.75 \times 10^{-4} = 4.4 \times 10^{-4}$

$\text{Percentage loss} = \dfrac{4.4 \times 10^{-4}}{60 \times 10^{-4}} \times 100 = 7.33\%; \text{ Loss of prestress} = 0.0733 \times 1200 \approx 88 \text{ N/mm}^2$

**(iv) Loss due to relaxation of steel:** As per IS : 1343-1980, the characteristic strength or prestressing steel ($f_p$) shall be assumed as the minimum ultimate tensile stress.

∴ $f_p = 1500$ N/mm²

Ratio of initial stress in steel to $f_p = \dfrac{1200}{1500} = 0.8$

Thus, initial stress in steel = $0.8 f_p$. Hence from Table 45.13, the relaxation loss = 90 N/mm²

Per cent loss = $\dfrac{90}{1200} \times 100 = 6.5\%$; Loss of strain = $60 \times 10^{-4} \times 0.075 = 4.5 \times 10^{-4}$

**(v) Anchorage loss:**

Let us assume loss of prestress due to anchorage take-up and friction of spacers and end block as equivalent to 2% of strain loss,

Loss of strain = $0.02 \times 60 \times 10^{-4} = 1.2 \times 10^{-4}$; Loss of prestress = $0.02 \times 1200 = 24$ N/mm²

The various losses are tabulated below:

**TABLE 45.8(a). Summary of losses**

| Loss due to | Loss of strain | Percentage Loss | Loss of prestress (N/mm²) |
|---|---|---|---|
| 1. Elastic shortening | $2.75 \times 10^{-4}$ | 4.58 | 55 |
| 2. Shrinkage | $3.00 \times 10^{-4}$ | 5.00 | 60 |
| 3. Creep in concrete | $4.4 \times 10^{-4}$ | 7.33 | 88 |
| 4. Creep in steel | $4.50 \times 10^{-4}$ | 7.50 | 90 |
| 5. Anchorage etc. | $1.20 \times 10^{-4}$ | 2.00 | 24 |
| Total | $15.85 \times 10^{-4}$ | 26.41 | 317 |

Thus total loss is 26.41 %

$$\eta = \dfrac{\epsilon_s - \text{loss}}{\epsilon_s} = \dfrac{P_e}{P_i} = (1 - 0.2641) = 0.7359 \quad \text{and} \quad P = \eta P_i = 0.7359 \times 1200 = 883 \text{ N/mm}^2$$

Total loss of prestress = $1200 - 883 = $ **317 N/mm²**

**Example 45.8.** *A post-tensioned prestress concrete beam of 30 m span is subjected to a transfer prestress force of 2500 KN at 28 day's strength. The profile of the cable is parabolic with maximum eccentricity of 200 mm at midspan. Determine the loss of prestress, and the jacking force required if jacking is done from both ends of the beam. The beam has a cross-section of 500 mm × 800 mm, and is prestressed with 9 cables, each cable consisting of 12 wires of 5 mm diameter. Take $E_s = 2.1 \times 10^5$ N/mm² and $E_c = 3.5 \times 10^4$ N/mm². One cable is tensioned at a time.*

**Solution:**

$A_{st} = A_p = 9 \times 12 \times \dfrac{\pi}{4}(5)^2 \approx 2121$ mm²;

$f_s = \dfrac{2500 \times 1000}{2121} = 1178.7$ N/mm²

Elastic strain in steel at transfer $\epsilon_s = \dfrac{f_s}{E_s} = \dfrac{1178.7}{2.1 \times 10^5} = 56.12 \times 10^{-4}$

**(i) Loss due to elastic shortening:** We have seen in Example 45.7 that this loss can be found by taking the average of elastic strains at the ends and the middle of the beam.

Area of concrete $A_c = 500 \times 800 = 40 \times 10^4$ mm²; $I = \dfrac{1}{12} \times 500(800)^3 = 213.33 \times 10^8$ mm⁴

Elastic strain at the end section = $\dfrac{P_i}{A_c E_c} = \dfrac{2500 \times 1000}{40 \times 10^4 \times 3.5 \times 10^4} \approx 1.79 \times 10^{-4}$ N

Elastic strain at midspan section.

$= \dfrac{P_i}{A E_c} + \dfrac{P_i \cdot h^2}{I E_c} = \dfrac{P_i}{E_c}\left[\dfrac{1}{A} + \dfrac{h^2}{I}\right] = \dfrac{2500 \times 1000}{3.5 \times 10^4}\left[\dfrac{1}{40 \times 10^4} + \dfrac{(200)^2}{213.33 \times 10^8}\right] \approx 3.13 \times 10^{-4}$

∴ Mean elastic strain = $\frac{1}{2}(1.79 + 3.13) \times 10^{-4} = 2.46 \times 10^{-4}$

Since tendons are tensioned one after the other, the first tendon will loose stress due to elastic shortening of subsequent eight tendons. Hence loss of strain in first tendon

$$= \frac{8}{9} \times 2.46 \times 10^{-4} = 2.18 \times 10^{-4}$$

Average elastic loss for all tendons is gives by

$$\epsilon_{el} = \frac{1}{2}(2.18 \times 10^{-4}) = 1.09 \times 10^{-4}$$

∴ Percentage loss = $\frac{1.09 \times 10^{-4}}{56.12 \times 10^{-4}} \times 100 = \mathbf{1.94\%}$

**(ii) Loss due to shrinkage:** As per IS : 1343-1980, the shrinkage strain is given by

$$\epsilon_{sh} = \frac{0.0002}{\log_{10}(t+2)} = \frac{0.0002}{\log_{10}(28+2)} = 1.35 \times 10^{-4}$$

Percentage loss = $\frac{1.35 \times 10^{-4}}{56.12 \times 10^{-4}} \times 100 \approx \mathbf{2.41\%}$

**(iii) Loss due to creep in concrete:** As per IS : 1343-1980, the creep coefficient at transfer of prestress at 28 days of curing of concrete may be taken as 1.6

∴ $\epsilon_{cr} = (C \cdot \epsilon_{el}) = (1.6) \times 2.46 \times 10^{-4} = 3.936 \times 10^{-4}$

Percentage loss = $\frac{3.936 \times 10^{-4}}{56.12 \times 10^{-4}} \times 100 = \mathbf{7.01\%}$

**(iv) Loss due to relaxation of steel:** As per IS : 1343-1980, the characterstic strength ($f_p$) of prestressing steel shall be assumed as the minimum ultimate tensile strain.

From Table 45.11, ultimate tensile strain for 5 mm wires = 1600 N/mm²

∴ $f_p = 1600$ N/mm²

Ratio of initial stress to characteristic strength = 1178.7/1600 ≈ 0.74

∴ Initial stress in steel = $0.74 f_p$. Hence from Table 45.73,

The relaxation loss = 78 N/mm²

Percent loss = $\frac{78}{1178.7} \times 100 \approx \mathbf{6.62\%}$;  Loss of strain = $56.12 \times 10^{-4} \times 0.0662 = 3.7 \times 10^{-4}$

**(v) Loss due to anchorage take up:** Let 2.5 mm be the anchorage take-up at each jacking end.

Loss of strain = $\frac{2.5}{L/2} = \frac{2.5}{15 \times 1000} = 1.67 \times 10^{-4}$

Percentage loss = $\frac{1.67 \times 10^{-4}}{56.12 \times 10^{-4}} \times 100 \approx \mathbf{2.98\%}$

**(vi) Friction loss:** Let us assume μ = 0.3 and $k = 15 \times 10^{-4}$ per metre.

Since jacking is done from both ends, effective length of cable

$L_e = L/2 = 15$ m.  The cable profile is given by

$$y = \frac{4hx(L-x)}{L^2}, \text{ where } x \text{ is measured from the free end.}$$

∴ $y = \frac{4 \times 0.2 \, x(30-x)}{900} = \frac{8}{9000}(30x - x^2)$

∴ $\left(\frac{dy}{dx}\right)_{x=0} = \frac{8}{9000}(30) = 0.0267$ radians = α

$$P_2 = P_1 \cdot e^{-(\mu\alpha + k \cdot L_e)}. \quad \text{Hence} \quad \frac{P_2}{P_1} = e^{-(\mu \cdot \alpha + k \cdot L_e)}$$

or
$$\frac{P_1 - P_2}{P_1} = 1 - e^{-(\mu\alpha + k \cdot L_e)} = 1 - e^{-(0.3 \times 0.0267 + 0.0015 \times 15)} = 0.03.$$

Alternatively, from approximate method, $P_2 = P_1(1 - \mu\alpha + k L_e)$

or
$$\frac{P_1 - P_2}{P_1} = \mu\alpha + k L_e = 0.3 \times 0.0267 + 0.0015 \times 15 = 0.0305$$

Loss of prestress = $0.03 P_1 = 3\%$

∴ Strain loss $\in_f = \dfrac{P_1 - P_2}{E_s \cdot A_{st}} = \dfrac{0.03 P_1}{E_s \cdot A_{st}} = \dfrac{0.03 \times 2500 \times 1000}{2.1 \times 10^5 \times 2131} \approx 1.68 \times 10^{-4}$

It is assumed that the friction loss will be neutralised by applying additional force at the jack. It will therefore not be counted in the total loss.

Jacking force = Transfer force + Friction loss = $P_1 + 0.03 P_1 = 1.03 P_1$

The loss of prestress due to various causes are tabulated below.

**TABLE 45.8(b)** Summary of Losses

| Loss due to | Loss of strain | Percent loss |
|---|---|---|
| (i) Elastic shortening | $1.09 \times 10^{-4}$ | 1.94 |
| (ii) Shrinkage | $1.35 \times 10^{-4}$ | 2.41 |
| (iii) Creep in concrete | $3.936 \times 10^{-4}$ | 7.01 |
| (iv) Creep in steel | $3.70 \times 10^{-4}$ | 6.62 |
| (v) Anchorage take up | $1.67 \times 10^{-4}$ | 2.98 |
| Total | $11.746 \times 10^{-4}$ | 20.96 |

Percent loss in prestress = 20.96; Loss of prestress = $0.2096 \times 1178.7 \approx 248$ N/mm².

Effective prestress force $P_e = (1 - 0.2096) P_1 = 0.7904 \times 2500 =$ **1976 kN.**

**Example 45.9.** *If the beam of Example 45.8 carries a superimposed load of 20 kN/m, find the total losses.*

**Solution:** B.M. $\quad M = \dfrac{w L^2}{8} = \dfrac{20000(30)^2}{8} \times 1000 = 2.25 \times 10^9$ N-mm.

Strain in concrete at the steel level at midspan, due to B.M.

$$= -\frac{M \cdot h}{I E_c}, \quad \text{where } h \text{ is the eccentricity of steel at midspan.}$$

$$= -\frac{2.25 \times 10^9 \times 200}{213.33 \times 10^8 \times 3.5 \times 10^4} = -6.03 \times 10^{-4}$$

This strain is negative, i.e. there is a gain in strain instead of loss.

Percentage gain = $\dfrac{6.03 \times 10^{-4}}{56.12 \times 10^{-4}} \times 100 = 10.74\%$

Hence final loss of strain = $(11.746 - 6.03) 10^{-4} = 5.716 \times 10^{-4}$

Percentage loss = $20.96 - 10.74 =$ **10.22%**

**Example 45.10.** *Fig. 45.28 shows a prestressed concrete beam, continuous over two spans. The curved tendon is to be tensioned from both ends. Compute the percentage loss of prestress due to friction, from one end to the centre of beam (i.e from A to E). Take the coefficient of friction between the cable and duct as 0.45, and the coefficient of wobble effect as 0.0015 per metre.*

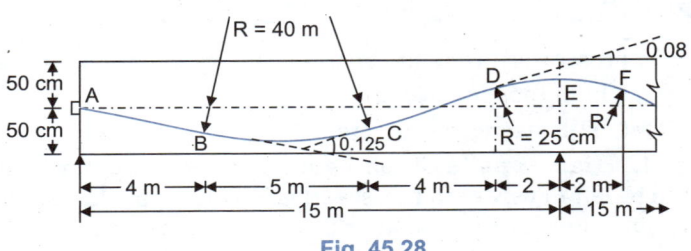

Fig. 45.28

## 1024 REINFORCED CONCRETE STRUCTURE

**Solution:**

(a) *Approximate solution:*

Change in angle for curve $BC = L'/R = 5/40 = 0.125$.

Change in angle for curve $DE = L'/R \approx 2/25 = 0.08$

$$\left(\frac{P_1 - P_2}{P_1}\right) = kL + \mu\alpha = 0.0015 \times 15 + 0.45 (0.125 + 0.08) \approx 0.1148$$

∴ Loss due to friction = $P_1 - P_2 = 0.1148 P_1 = $ **11.48%**

(b) *Exact solution:* In this method, the cable is divided into segments $AB$ (straight) $BC$ (curved), $CD$ (straight) and $DE$ (curved), and stress at the end of each segment is computed, using *Eq. 45.14(b)*. The calculation are tabulated below.

**TABLE 45.9**

| Segment | $x$ | $k \cdot x$ | $\alpha$ | $\mu\alpha$ | $kx + \mu\alpha$ | $e^{-(k \cdot x + \mu\alpha)}$ | Stress at end of segments |
|---|---|---|---|---|---|---|---|
| AB | 4 | 0.006 | 0 | 0 | 0.006 | 0.9940 | $0.9940 P_1$ |
| BC | 5 | 0.0075 | 0.125 | 0.0563 | 0.0638 | 0.9382 | $0.9326 P_1$ |
| CD | 4 | 0.006 | 0 | 0 | 0.006 | 0.9940 | $0.9270 P_1$ |
| DE | 2 | 0.003 | 0.08 | 0.036 | 0.039 | 0.9618 | $0.8915 P_1$ |

∴ Total frictional loss = $(1 - 0.8915) P_1 = 0.1085 P_1 = $ **10.85%**

**Example 45.11.** *Fig 45.29 shows a prestressed concrete beam, in which the cable is pulled from both the ends, with a pulling stress of 1200 N/mm² at each end. Assuming $\mu = 0.40$ and $k = 0.0015$ per metre, compute the extension at each end. Take $E_s = 2.1 \times 10^5$ N/mm²*

**Solution:** The mid-point $C$ will have least tension.

*For portion AB*

$\mu = L'/R \approx 10/50 = 0.20$ radians; $k \cdot x + \mu\alpha = (0.0015 \times 10) + (0.4 \times 0.20) = 0.095$

Stress at $B = 1200 \, e^{-0.095} = 1091$ N/mm²

∴ $$\Delta_{ab} = \frac{f_1 x}{E_s (kx + \mu\alpha)}\left[1 - \frac{f_2}{f_1}\right] = \frac{1200 (10 \times 1000)}{2.1 \times 10^5 \times 0.095}\left[1 - \frac{1091}{1200}\right] = 54.6 \text{ cm}$$

*For portion BC*

$\alpha = 0$. Hence,

$k \cdot x + \mu\alpha = 0.0015 \times 10 = 0.015$

Stress at $B = f_1 = 1091$ N/mm²

∴ Stress at $C = f_2 = 1091 \, e^{-0.015}$

$= 1075$ N/mm²

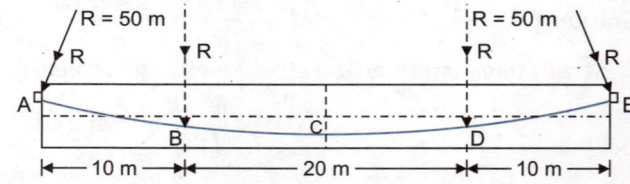

Fig. 45.29

∴ $$\Delta_{bc} = \frac{1091 (10 \times 1000)}{2.1 \times 10^5 \times 0.015}\left[1 - \frac{1075}{1091}\right] = 50.8 \text{ mm}$$

∴ Total extension = $54.6 + 50.8 = $ **105.4 mm.**

### 45.7. PROPERTIES OF MATERIALS

In prestressed cement concrete, the principal materials used are:

(i) high tensile steel and

(ii) high strength concrete.

**1. High Tensile Steel:** As stated earlier, ordinary mild steel wires are not useful because no prestress will be left in wires after losses (specially the shrinkage and creep losses) have taken place. The initial tensile strain in steel is reduced after tensioning by as much as 15 to 20 % (or even more) due to various losses. The reduction of strain must be small compared to initial strain otherwise most of the prestress will be lost.

Hence the first principal requirement of a steel for use in prestressing is high *ultimate tensile strength*. The second principal requirement is high ultimate elongation. High tensile steel have ultimate strength capacity as high as 2100 N/mm² and the use of such steel will provide considerable amount of effective prestressing force even after losses in prestress. Mild and hard steel used in reinforced concrete construction have a yield limit of 200 to 300 N/mm² only.

High tensile steel is produced by *alloying*. Carbon is extremely economical element for alloying, since it is cheap and easy to handle. Other alloys include manganese and silicon. High tensile steel contains about 3/4 per cent of carbon against only about 0.2% in mild steel. Approximate chemical composition of the high tensile steel wires is as follows:

| | | |
|---|---|---|
| Carbon : 0.6 to 0.85% | Sulphur : 0.055% | Manganese : 0.7 to 1.0 % |
| Silicon : 0.1 to 0.35 % | Phosphorus : 0.05% | |

High tensile steel is hot rolled, and then given heat treatment or cold treatment by rolling or drawing. The most common method for increasing the tensile strength of steel for prestressing is by cold drawing high tensile bars through a series of dyes. The process of cold drawing tends to realign the crystals and the strength of the wires is increased by each drawing so that the strength of wire increases as its diameter decreases. The ductility of wires is some what decreased as a result of cold drawing, which is a disadvantage.

Another method of increasing the tensile strength is to obtain the wire as *as-drawn* wires. These wires have low proportional limit, and in order to increase the proportional limit, the wires are subjected to some type of stress relieving processes. Fig. 45.30 shows the stress-strain curves for mild steel, high tensile rod and high tensile wires. Unlike mild steel, high tensile steel does not have any well defined yield point. While the ultimate strength of high tensile steel can be easily determined by testing, its elastic limit or its yield point cannot be so simply ascertained. The yield point of high tensile steel is generally defined at 0.2% set, *i.e.* at 0.2% of permanent inelastic deformation. Thus in Fig. 45.30, $OA$ is the straight portion of the curve for high tensile wire, while $P$ is the so selected yield point on the curve portion. If the specimen is loaded upto $P$ and then unloaded, the unloading curve will follow line $PQ$ which is practically parallel to the straight portion $OA$ of the loading curve. $OQ$ is then the permanent strain which is equal to 0.2%. The yield stress, defined as the stress which produces a residual strain equal to 0.2% on unloading, is also termed as 0.2 *percent proof stress*. The approximate values of yield point and proportional limit for various types of high steel wires are given in Table 45.10. On an average, 0.2% proof stress of wires is about 80% of their ultimate strength. The working stress in high tensile steel is taken equal to 80% of its 0.2% proof stress. Also, 0.2% proof stress should not be less than 80% of the minimum specified ultimate strength.

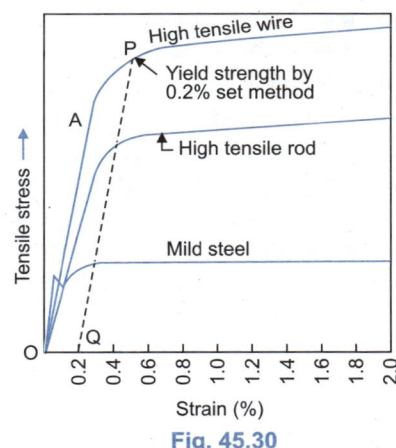

Fig. 45.30

TABLE 45.10. Approximate Yield Stress and Proportional Limits

| Wire | Yield point | Proportional limits |
|---|---|---|
| 1. Wires as drawn | $0.75 f_{su}$ | $0.35 f_{su}$ |
| 2. Prestretched | $0.85 f_{su}$ | $0.55 f_{su}$ |
| 3. Temperature treated wires | $0.87 f_{su}$ | $0.70 f_{su}$ |
| 4. Strands-stress relieved | $0.90 f_{su}$ | $0.75 f_{su}$ |

The modulus of elasticity for cold drawn wires and heat treated wires may be taken as $2.1 \times 10^5$ N/mm² and $2.0 \times 10^5$ N/mm² respectively. In most of the high tensile wires the creep is negligible upto $0.45 f_{su}$ and it is about 3% at $0.5 f_{su}$ and $0.55 f_{su}$ stress level.

Wires are supplied in drums. In order to have least permanent set due to winding on drums, the drum diameter are made as big as possible, with a minimum of 1.5 to 2 m. Most wires have slight permanent set when unwound, and require some straightening. Small diameter wires possess higher unit strength and furnish better bond, which is often vital in pre-tensioning. However, large diameter are preferred for pre-tensioning to save labour and anchorage costs. The prestressing steel is also available in the form of strands which are obtained by twisting wires together. Mostly, seven-wire strands are used, having a centre wire slightly larger then the outer six wires which enclose it tightly in a helix with a uniform pitch between 12 to 16 times the nominal diameter of the strand. After stranding, all strands are subjected to a stress-relieving continuous heat treatment to produce the prescribed mechanical properties.

The high tensile steel wire used for prestressed concrete should conform to Table 45.11 regarding minimum ultimate tensile strength.

The proof stress shall not be less than 80% of the minimum ultimate tensile strength. The elongation at rupture shall not be less than 2% over a gauge of 200 mm. The wire shall be free from rust, scale and other similar deleterious matter liable to affect adversely proper tensioning or its bond with concrete. Slight rust may be permitted provided it is not loose. High tensile steel wires shall be supplied in coils having a sufficiently large diameter for the wire to layout straight. Cables used as prestressing steel shall consists of a group of single wires arranged in parallel or stranded formation. For the purposes of tests, the strength of a cable shall be taken as the multiple of the strength of a single or stranded wires forming the cable.

**TABLE 45.11.** Minimum Ultimate Tensile Strength of Tensile Steel Wires

| Nominal dia. of wires (mm) | Min. ultimate tensile strength $(f_{su})$ (N/mm²) |
|---|---|
| 8.0 | 1500 |
| 7.0 | 1500 |
| 5.00 | 1600 |
| 4.00 | 1750 |
| 3.00 | 1900 |
| 2.50 | 2050 |
| 2.00 | 2200 |
| 1.50 | 2350 |

High tensile alloy steel bars shall have a minimum ultimate tensile strength $(f_{su})$ of 950 N/mm². Proof stress $(f_{su})$ of the bars shall not be less than 80% of its ultimate tensile strength. The bars as supplied shall be free from end jagged edges. Coupling units and similar fixtures used in conjunction with the wires or bars shall have an ultimate tensile strength of not less than the individual strength of the wires or bars being joined.

**2. High Strength Concrete:** Because of the use of high tensile steel in pre-stressed concrete construction, the concrete has to be of good quality and of high strength. A good, well compacted dense concrete has less elastic strain, and has less shrinkage plastic flow, thus reducing the loss of prestress considerably. It is cheaper comparatively to prepare a high strength concrete as its cost does not increase in the same proportion as its strength does. High quality concrete is also essential to bear the high concentration of stresses under the end anchorages. Ordinary concrete would require bigger size of anchorages. In some cases, high grade concrete can be very useful from the point of view of crack resistance of the members and also resistance against corrosion. Concrete of high compressive strength offers high resistance in tension and shear, as well as in bond and bearing, and is desirable for prestressed concrete structures whose various portions are under higher stresses. Indian Standard Code recommends a minimum cube strength of 40 N/mm² for pre-tensioned system.

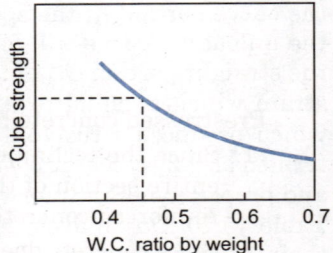

Fig. 45.31

The strength of concrete decreases with increase in water-cement ratio as is evident from Fig. 45.31. To attain a strength of 36.5 N/mm², the water cement ratio should not be higher than 0.45 by weight. However, if water cement ratio is less than 0.4, the workability will decrease and consequently compact and high density concrete would not be obtained. High workability with less water-cement ratio would require a higher percentage of cement and well graded aggregate. Since excessive cement tends to increased shrinkage, a lower cement factor is desirable. To this end, good vibration is advised. Slumps of 12 to 25 cm are sometimes used with controlled vibration. Workability can also be increased by proper admixtures. Air entrainment of 3 to 5% improves workability and reduces bleeding.

The high compressive strength mentioned above is of 15 cm cube at 28 days. Actually, the strength of concrete increase with age, as is evident from Fig. 45.32.

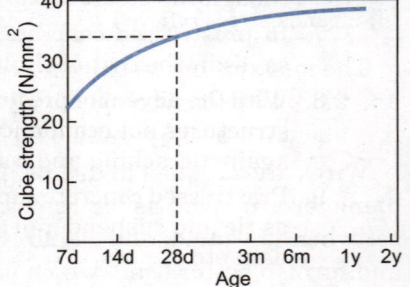

Fig. 45.32

Fig. 45.33 shows the stress strain curve for concrete. The stress-strain relation can be expressed by the following relations:

$$f_c = E_c \cdot \varepsilon \left(1 - \frac{\varepsilon}{2\varepsilon_u}\right) \qquad \ldots(45.19)$$

where $\varepsilon_u$ = ultimate stain = 0.003.

The modulus of elasticity $E_s$ used in the above equation is an average one. It is well known that the modulus of elasticity of concrete decrease with increase in stress. Since the compression on the section of beam varies from top to bottom, each fibre of concrete will have different $E_c$, though the difference will not be much. Also, modulus of elasticity of concrete during a compression operation is lower than during decompression and there is a residual strain after every decompression. If these cycles of compression and decompression are repeated, $E_c$ tends to equalise and behaviour of concrete becomes almost elastic. An average modulus of elasticity can conveniently be calculated from the deflection measurements of a prestressed concrete beam.

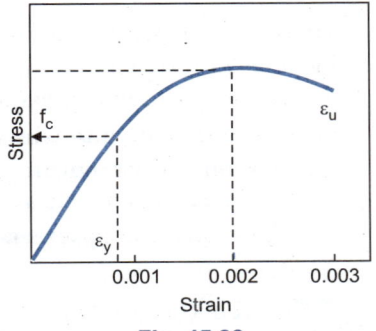

Fig. 45.33

**3. Cement Grout:** In the case of bonded post-tensioning, cement grouting is done which also serves to protect steel against corrosion. Conduits can be made of aluminium, steel, tin or other metal sheathing or tubes. It is also possible to form the duct by withdrawing steel tubing or rod before concrete hardens. Entry of the grout into the cable way is provided by means of holes in the anchorage heads and cones. Where the space between the wires is large, such as in Magnel system, a 1:1 cement-sand mix is used with a water cement ratio of 0.5 by volume. In other systems, such as in freyssinet or Strescon cable, neat cement paste with 0.5 water-cement ratio is injected. To save cement, from economical considerations, fine sand of about 0.4 mm grain size may be used, with water : cement : sand proportion of about 1:1.3:0.7 by volume. The grout pressure may be between 0.5 to 0.7 N/mm². The injection can be applied at one end of the member until it is forced out at the other end ; the end is then plugged and pressure is again applied at the injecting end to compact the grout.

## 45.8. MERITS AND DEMERITS OF PRESTRESSED CONCRETE

Prestressed concrete has the following merits:
1. Since the technique of prestressing eliminates cracking of concrete under all stage of loading, the entire section of the structures takes part in resisting the external load. In contrast to this, in the reinforced concrete, only portion of concrete above the neutral axis is effective.
2. Since concrete does not crack, the possibility of steel of rust and concrete to deteriorate is minimised.
3. Absence of cracks results in higher capacity of the structure to bear reversal of stresses, impact, vibration and shock.
4. In prestressed concrete beams, dead loads are practically neutralised. The reactions required are therefore, much smaller than required in reinforced concrete. The reduced dead weight of structure results in saving in the cost of foundation.

The neutralisation of dead weight is of importance in large bridges.

5. The use of curved tendons and the precompression of concrete helps to resist shear.
6. The quantity of steel required for prestressing about 1/3 of that required for reinforced concrete, though the steel for the former should have high tensile strength.
7. In prestressed concrete, precast blocks and elements can be assumed and used as one unit. This saves in the cost of shuttering and centring for large structures.
8. With the advent of prestressed concrete, it has been possible now to construct large size liquid retaining structures not economical to build otherwise. Such structures have low cost and are preferably safe against cracking and consequent leakage.
9. Prestressed concrete can be used with advantage in all those structures where tension develops, such as tie and suspender of a bow string girder, railway sleepers, electric poles, upstream face of gravity dam etc.
10. Prestressed concrete beams have usually low deflection.

## 1028 REINFORCED CONCRETE STRUCTURE

However, prestressed concrete construction has the following demerits:
1. It requires high quality dense concrete of high strength. Perfect quality control in production, placement and compaction is required.
2. It requires high tensile steel, which is 2.5 to 3.5 times costlier than mild steel.
3. It requires complicated tensioning equipment and anchoring devices, which are usually covered under patented rights.
4. Construction requires perfect supervision at all stages of construction.

### 45.9. BASIC ASSUMPTIONS

The design of prestressed concrete members is based on the following assumptions:
1. A plane section before bending remains plane after bending.
2. Within the range of working stresses both concrete and steel behave elastically notwithstanding the small amount of creep which occurs in both materials under sustained loading.
3. Up to the limit of the modulus of rupture of concrete, any change in the loading produces a change of stress in the concrete only, the sole function of the prestressing tendon being to impart and maintain the prestress in the concrete notwithstanding the comparatively small changes in the stress of steel due to changes in strain of concrete surrounding it.

### 45.10. ANALYSIS OF BEAMS FOR FLEXURE

**1. Concentrically Prestressed Section:** If a horizontal force $P$ is applied at the centroid of the concrete section, the stress in concrete is uniform across the section and is given by

$$f = \frac{P}{A}$$

The release of resistance from bulkheads is equivalent to the application of an opposite force $P_i$ to the member. Using transformed section method,

$$f_c = \frac{P_i}{A_c + m A_{st}} = \frac{P_i}{A_e} \qquad \ldots(45.20)$$

where $A_e$ is the area of equivalent or transformed section of concrete.

The stress induced in steel is

$$\Delta f_s = m f_c = \frac{m P_i}{A_c + m A_{st}} = \frac{m P_i}{A_e} \qquad \ldots(45.21)$$

where $\Delta f_s$ represents immediate reduction of the prestress in the steel as a result of the transfer.

In practice, however, the loss in prestress is computed by the following approximate expression:

$$\Delta f_s = \frac{m P_i}{A_c} \qquad \ldots[45.21(a)]$$

**2. Eccentrically Prestressed Section:** Let the prestressed force be applied at an eccentricity $e$. Using elastic theory, the fibre stress at any point is given by

$$f = \frac{P}{A} \pm \frac{P \cdot e \cdot y}{I}$$

Using transformed section method, and taking $P = P_i$ = initial prestressing force,

$$f_c = \frac{P_i}{A_e} \pm \frac{P_i \cdot e \cdot y}{I_e} \qquad \ldots(45.22)$$

where $I_e$ is the equivalent moment of inertia of transformed section.

However, in practice, the gross or net concrete section is used, and either inital or reduced prestress ($P_e$) is used:

$$f_c = \frac{P}{A_e} \pm \frac{P\,e\cdot y}{I_e} \qquad \ldots[45.22\,(a)]$$

where $P$ is either $P_i$ or $P_e$.

**3. Stresses due to Live Load:** Let the beam be subjected to a B.M. $M$, either due to weight of beams, or due to externally applied loads, or due to both. The resulting stress in concrete is given by

$$f_c = \frac{P}{A} \pm \frac{P\,.\,e\,y}{I} \pm \frac{M\,.\,y}{I} = \frac{P}{A}\left(1 \pm \frac{e\,.\,y}{r^2}\right) \pm \frac{M\,y}{I} \qquad \ldots(45.23)$$

where $r$ = radius of gyration = $\sqrt{I/A}$.

In the above expression, $P = P_i$ if $M$ is due to dead load only, while $P = P$ if $M$ is both due to dead and live loads.

For a pretensioned beam, steel is always bonded to concrete before $M$ is applied. Hence $y$ and $I$ are computed on the basis of transformed section. As an approximation, either gross or net section of concrete may be used without any appreciable error. In the post-tensioned bonded construction, transformed section should be used. If however, weight of the beam is applied before bonding takes place, net section should be used. For post-tensioned unbonded beam, net section is to be used.

**Example 45.12.** *Fig. 45.34 shows the midspan cross-section of a prestressed post-tensioned beam of 24 m span. There are 6 cables each of 12 wires of 5 mm diameter, and are stressed with initial prestress of 1150 N/mm². Assuming an effective prestress of 1000 N/mm² after all losses have taken place, compute the extreme fibre stresses in concrete. The profile of cable is parabolic with zero ecccentricity at the ends. The beam carries a live load of 9 kN/m in addition to its own weight. Take unit weight of concrete as 24 kN/m³.*

**Solution:** Area of prestressing steel

$$= 6 \times 12 \left(\frac{\pi}{4} \times 25\right) = 1413.716 \text{ mm}^2$$

$P_i = 1150 \times 1413.716 = 1625774$ N
$P_e = 1000 \times 1413.716 = 1413716$ N
$A_e = (1000 \times 150) + (150 \times 750) + (400 \times 400)$
$= 150000 + 112500 + 160000 = 422500 \text{ mm}^2$

$$y_1 = \frac{(150000 \times 75) + (112500 \times 525) + (160000 \times 1100)}{422500}$$

$= 583$ mm.

$$y_2 = \frac{(150000 \times 1225) + (112500 \times 775) + (160000 \times 200)}{422500}$$

$= 717$ mm.

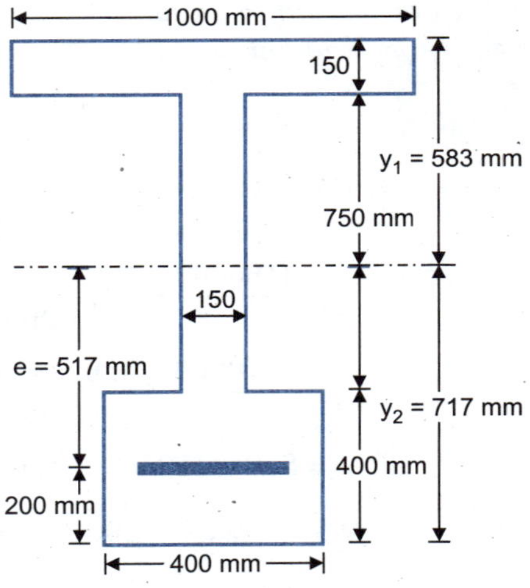

**Check:** $y_1 + y_2 = 583 + 717 = 1300$ mm.

Eccentricity $e = y_2 - 200 = 717 - 200 = 517$ mm

$$I = \left[\frac{1}{12}1000(150)^3 + 150000(583-75)^2\right] + \left[\frac{1}{12} \times 150(750)^3 + 112500(583-150-375)^2\right]$$

$$+ \left[\frac{1}{12} \times 400(400)^3 + 160000(717-200)^2\right] = 8.954 \times 10^{10} \text{ mm}^4$$

Dead load of beam = $(422500 \times 10^{-6}) \times 1 \times 24000 = 10140$ N/m.

Dead load B.M. $M_d = \dfrac{10140 \times 24^2}{8} \times 1000 = 730 \times 10^6$ N-mm.

Live load B.M. $M_l = \dfrac{9000 \times 24^2}{8} \times 1000 = 648 \times 10^6$ N-mm.

$\therefore$ Total $M = M_d + M_l = 1378 \times 10^6$ N-mm.

Fig. 45.34

**Initial Condition:** Due to moment caused by eccentrically applied pre-stress, the top fibres will be subjected to tensile stress while the bottom fibres will be subjected to compressive stresses. However, due to dead weight of beam, top fibres will be subjected to compressive stresses. Using plus sign for compressive stress and minus sign for tensile stresses, we get

$$f = \frac{P_i}{A_c} \mp \frac{P_i \cdot e \cdot y}{I} \pm \frac{M_d \cdot y}{I}$$

$P_i/A_c = 1625774/422500 \approx 3.848$; $\quad \dfrac{P_i \cdot e}{I} = \dfrac{1625774 \times 517}{8.954 \times 10^{10}} = 93.87 \times 10^{-4}$

$\dfrac{M_d}{I} = \dfrac{730 \times 10^6}{8.954 \times 10^{10}} \approx 81.53 \times 10^{-4}$

$f_1$ (top fibre) $= 3.848 - 93.87 \times 10^{-4} \times 583 + 81.53 \times 10^{-4} \times 583 \approx$ **3.13 N/mm²**.
$f_2$ (bottom fibre) $= 3.848 + 93.87 \times 10^{-4} \times 717 - 81.53 \times 10^{-4} \times 717 =$ **4.73 N/mm²**.

**Final Condition** $f = \dfrac{P_e}{A_c} \mp \dfrac{P_e \cdot e \cdot y}{I} \pm \dfrac{M \cdot y}{I}$; $\quad P_e/A_c = 1413716/422500 = 3.346$

$\dfrac{P_e \cdot e}{I} = \dfrac{1413716 \times 517}{8.954 \times 10^{10}} \approx 81.63 \times 10^{-4}$; $\quad \dfrac{M}{I} = \dfrac{1378 \times 10^6}{8.954 \times 10^{10}} = 153.9 \times 10^{-4}$

$\therefore \quad f_1$ (top fibre) $= 3.346 - 81.63 \times 10^{-4} \times 583 + 153.9 \times 10^{-4} \times 583 =$ **7.56 N/mm²**
$f_2$ (bottom fibre) $= 3.346 + 81.63 \times 10^{-4} \times 717 - 153.9 \times 10^{-4} \times 717 \approx$ **−1.84 N/mm²**.

**Example 45.13.** *A prestressed concrete beam of inverted T-section has dimensions as shown in Fig. 45.35 and is simply supported over a span of 16 m. The beam is post-tensioned with 3 Freyssinet cables, each containing 12 wires of 7 mm dia. placed as shown, at the midspan. If the initial prestress is 1000 N/mm², calculate maximum uniformly distributed load if the maximum compressive stress in concrete is limited to 14 N/mm² and tensile stress is limited to 1 N/mm². Assume loss of prestress = 15%.*

**Solution:** $A_{st} = A_p = 3 \times 12 \times \dfrac{\pi}{4}(7)^2 = 1385.4$ mm²

$P_i = 1385.4 \times 1000 = 1385400$ N; $\quad P_e = 0.85 \times 1385400 \approx 1177600$ N
$A_c = (300 \times 900) + (300 \times 600) = 270000 + 180000 = 450000$ mm²

$y_2 = \dfrac{(180000 \times 150) + (270000 \times 750)}{450000} = 510$ mm.

$y_1 = \dfrac{(270000 \times 450) + (180000 \times 1050)}{450000} = 690$ mm.

$e = 510 - 150 = 360$ mm.

$I = \left[\dfrac{1}{12} \times 600 (300)^3 + 180000 \times (510 - 150)^2\right]$

$\quad + \left[\dfrac{1}{12} \times 300 (900)^3 + 270000 \times (690 - 450)^2\right]$

$\quad = 5.8455 \times 10^{10}$ mm⁴.

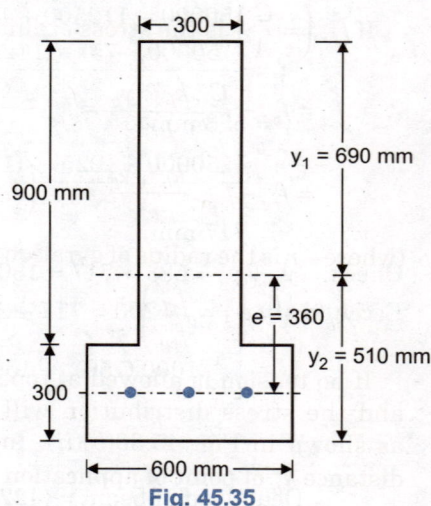

Fig. 45.35

Dead load $w_d = 450000 \times 10^{-6} \times 24000 = 10800$ N/m.
Compressive stress due to $P_e = 1177600/450000$
$\approx 2.62$ N/mm²

Stress due to eccentricity of prestress $= \mp \dfrac{P_e \cdot e}{I} y$

$f_1$ (top fibre) $= -\dfrac{1177600 \times 360}{5.8455 \times 10^{10}} \times 690 = -5$ N/mm²

$$f_2 \text{ (bottom fibre)} = + \frac{1177600 \times 360}{5.8455 \times 10^{10}} \times 510 = + 3.7 \text{ N/mm}^2$$

Total stress at top = 2.62 − 5 = − 2.38 N/mm².

Total stress at bottom = 2.62 + 3.7 = + 6.32 N/mm².

Due to super-imposed load (inclusive of self load), there will be compressive stress at top and tensile stress at bottom. In order to limit the compressive stress to 14 N/mm², the stress due to super-imposed load = 14 − (− 2.38) = 16.38 N/mm² at top. Similarly, in order to limit the tensile stress to 1 N/mm², the stress due to supper-imposed load = 1 + 6.32 = 7.32 N/mm². Let $M_1$ and $M_2$ be the respective moments to cause these.

$$\therefore \quad M_1 = \frac{16.38 \times 5.8455 \times 10^{10}}{690} = 13.38 \times 10^8 \text{ N-mm}$$

and

$$M_2 = \frac{7.32 \times 5.8455 \times 10^{10}}{510} = 8.39 \times 10^8 \text{ N-mm}$$

∴ Minimum B.M = $8.39 \times 10^8$ N-mm.

$$M = \frac{wL^2}{8} = \frac{w \times 256 \times 1000}{8} = 32000 \, w \text{ N-mm}$$

$$w = \frac{8.39 \times 10^8}{32000} = 2.62 \times 10^4 \text{ N/m} = 26.2 \text{ kN/m}; \quad \text{Dead load} = 10.8 \text{ kN/m}.$$

∴ External load = 26.2 − 10.8 = **15.4 kN/m**.

## 45.11. KERN DISTANCES AND EFFICIENCY OF SECTION

Let us consider an unsymmetrical beam section subjected to a prestressing force $P_e$ acting at an eccentricity $e$, with respect to the C.G. of concrete section as shown in Fig. 45.36(a). Fig. 45.36(b) shows the stress distribution due to combined effect of eccentric prestress and dead load moment $M_d$. Let the resultant compressive force in $C (= P_e)$ act at distance of $k_b'$ from the C.G. of concrete area.

If $f_{cg} \left( = \dfrac{C}{A} \right)$ is the stress ordinate at C.G. of the section, the stress at the top $(f_{ct})$ is given by

$$f_{ct} = f_{cg} - \frac{C \cdot k_b' \cdot y_t}{I} = f_{cg} - \frac{C \cdot k_b' \cdot y_t}{A r^2}$$

$$= f_{cg} - f_{cg} \cdot \frac{k_b' \cdot y_t}{r^2} = f_{cg} \left( 1 - \frac{k_b' y_t}{r^2} \right)$$

(where $r$ is the radius of gyration for the section).

$$k_b' = \frac{r^2}{y_t} \left( 1 - \frac{f_{ct}}{f_{cg}} \right) \qquad \ldots [45.24\,(a)]$$

If no tension is allowed at top, $f_{ct}$ will be zero, and the stress distribution will be triangular, as shown in Fig. 45.36(b)(ii). In that case, the distance $k_b$ of point of application of $C$ is given by

$$k_b = \frac{r^2}{y_t} = \frac{I}{A\,y_t} \qquad \ldots (45.24)$$

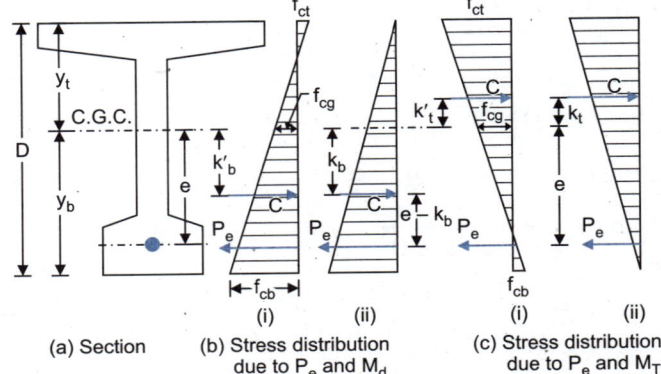

(a) Section  (b) Stress distribution due to $P_e$ and $M_d$  (c) Stress distribution due to $P_e$ and $M_T$

Fig. 45.36

The distance $k_b$ is known as *bottom kern distance*. It is the distance from the C.G.C. where the resultant compressive force acts, resulting in zero stress at the top.

Let us now consider the effect of total moment $M$ consisting of dead load moment $M_a$ and live load moment $M_L$. Fig. 45.36(c)(i) shows the final stress distribution. Let the resultant compressive force $C$ act at a distance $k_t'$ from C.G. of concrete. If $f_{cg}$ is the stress in concrete at C.G., the stress at bottom ($f_{cb}$) is given by

$$f_{cb} = f_{cg} - \frac{C \cdot k_t' \, y_b}{I} = f_{cg} - \frac{C \, k_t' \cdot y_b}{A \, r^2} = f_{cg} - \frac{f_{cg} \, k_t' \cdot y_b}{r^2} = f_{cg}\left(1 - \frac{k_t' \, y_b}{r^2}\right)$$

$$k_t' = \frac{r^2}{y_b}\left(1 - \frac{f_{cb}}{f_{cg}}\right) \qquad \ldots[45.25(a)]$$

If no tension is allowed at bottom, $f_{cb}$ will be zero, and the stress distribution will be triangular, as shown in Fig. 45.36(c)(ii). In that case, the distance $k_t$ of point of application of $C$ is given by

$$k_t = \frac{r^2}{y_b} = \frac{I}{A \, y_b} \qquad \ldots(45.25)$$

The distance $k_t$ is known as *top kern distance*. Let us consider Fig. 45.36(b)(ii). The lever arm between $C$ and $P_e$ is $e - k_b$. Equating the internal moment to the external moment, we get

$$M_d = P_e (e - k_b) \qquad \ldots(45.26)$$

Similarly, consider Fig. 45.36(c)(ii), where lever arm between $C$ and $P_e$ is $(k_t + e)$. Equating the external moment to the internal moment, we get

$$M_T = (M_d + M_l) = P_e(k_t + e) \qquad \ldots(45.27)$$

or $M_l = P_e(k_t + e) - M_d = P_e(k_t + e) - P_e(e - k_b)$

$$\therefore M_l = P_e(k_t + k_b) \qquad \ldots(45.28)$$

*Equation 45.28* is very important, suggesting that greater the sum of the two kern distances $(k_t + k_b)$, greater will be the capacity of the beam to carry the external moment. If $D$ is the depth of the beam the *efficiency factor* $\rho$ on crack resistance may be expressed as

$$\rho = \frac{k_t + k_b}{D} = \frac{\dfrac{r^2}{y_b} + \dfrac{r^2}{y_t}}{y_b + y_t} \quad \text{or} \quad \rho = \frac{r^2}{y_t \cdot y_b} = \frac{I}{A \, y_t \, y_b} \qquad \ldots(45.29)$$

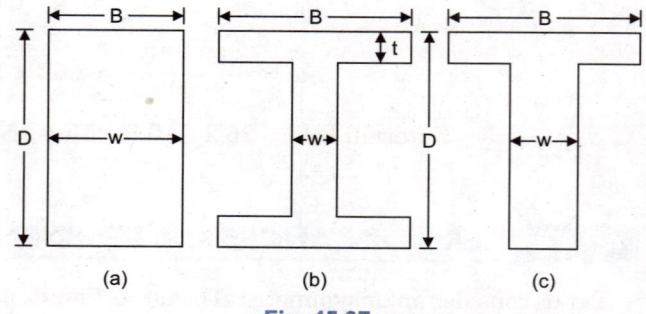

Fig. 45.37

For a rectangular section of width $B$ and depth $D$,

$$\rho = \frac{\frac{1}{12} BD^3}{BD\left(\dfrac{D}{2} \times \dfrac{D}{2}\right)} = \frac{1}{3}$$

For a symmetrical I-section,

$$\rho = \frac{I}{A \, y_t \, y_b} = \frac{4 I}{A D^2}$$

$$= \frac{1}{3} \frac{BD^3 - (B-W)(D - 2t)^3}{D^2 \, [BD - (B - W)(D - 2t)]}$$

$$= \frac{1}{3}\left[\frac{1 - (1 - W/B)(1 - 2t/D)^3}{1 - (1 - W/B)(1 - 2t/D)}\right]$$

$$\ldots(45.30)$$

Similar expressions can be derived for T and other sections. Fig. 45.38 shows the efficiency curves for I, T

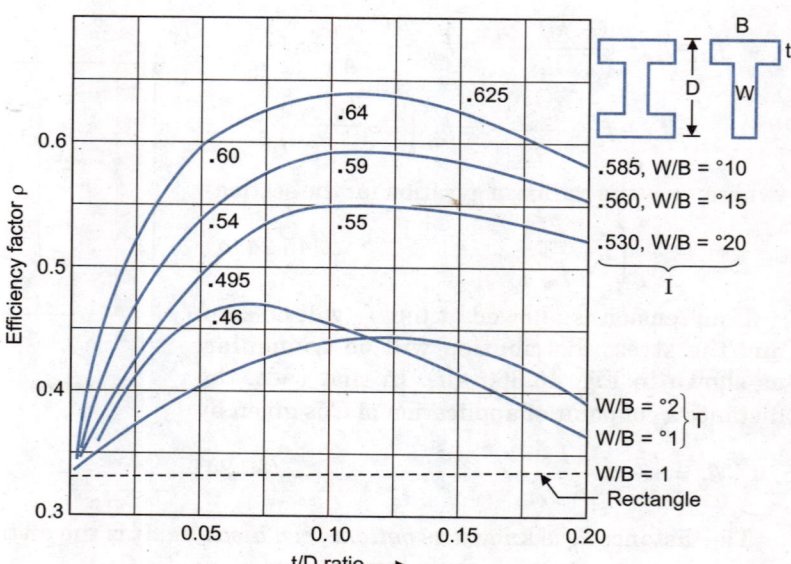

Fig. 45.38. Efficiency of Sections.

and rectangular sections for various *W/B* and *t/D* ratios. It is found that the rectangular section (*B/W* = 1) has a constant ρ = 0.333. Efficiency of I-section is maximum for $t/D \approx 0.1$. Also, smaller the *W/B* ratio, greater the efficiency. A rectangular section gives minimum efficiency on crack resistance.

**Example 45.14.** *Fig. 45.39 shows four sections each having equal area of cross section. Determine the kern distance and efficiency of each section.*

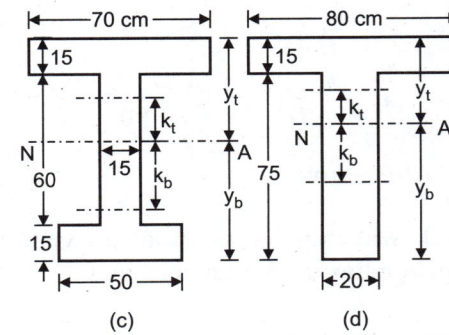

Fig. 45.39

**Solution:** **(a) Rectangular section**

$A = 30 \times 90 = 2700$ cm²;

$y_t = y_b = 45$ cm $= y$

$I = \frac{1}{12} \times 30 (90)^3 = 1822500$ cm⁴

$\therefore k_t = k_b = \dfrac{I}{A\,y} = \dfrac{1822500}{2700 \times 45}$

= 15 cm and

$\rho = \dfrac{I}{A\,y_t \cdot y_b} = \dfrac{k_t}{y} = \dfrac{15}{45} = \dfrac{1}{3} = \mathbf{0.3333}$

**(b) Symmetrical I-section**

$A = 2(60 \times 15) + (60 \times 15) = 2700$ cm²

$I = \frac{1}{12} \times 60 (90)^3 - \frac{1}{12} (60-15)(90-30)^3 = 2835000$ cm⁴

$y_t = y_b = 45$ cm $= y$; $k_t = k_b = \dfrac{I}{A\,y} = \dfrac{2835000}{2700 \times 45} = 23.33$ cm

$\rho = \dfrac{I}{A\,y_t\,y_b} = \dfrac{I}{A\,y^2} = \dfrac{k_t}{y} = \dfrac{23.33}{45} = \mathbf{0.5185.}$

**(c) Unsymmetrical I-section [Fig. 45.39(c)]**

$A = (70 \times 15) + (50 \times 15) + (60 \times 15) = 2700$ cm²

$y_t = \dfrac{(70 \times 15 \times 7.5) + (60 \times 15 \times 45) + (50 \times 15 \times 82.5)}{2700} = 40.83$ cm

$y_b = 90 - 40.83 = 49.17$ cm

$I = \left[\frac{1}{12} \times 70 \times 15^3 + 1050 (40.83-7.5)^2\right] + \left[\frac{1}{12} \times 15 \times 60^3 + 60 \times 15 (45-40.83)^2\right]$

$+ \left[\frac{1}{12} \times 50 \times 15^2 + 750 (49.17-7.5)^2\right] = 2788124$ cm⁴

$k_t = \dfrac{I}{A\,y_b} = \dfrac{2788214}{2700 \times 49.17} = \mathbf{21.00}$ **cm.**; $k_b = \dfrac{I}{A\,y_t} = \dfrac{2788214}{2700 \times 40.83} = \mathbf{25.29}$ **cm.**

$\rho = \dfrac{k_t + k_b}{D} = \dfrac{21.00 + 25.29}{90} = \mathbf{0.514}$

Alternatively, $\rho = \dfrac{I}{A\,y_t\,y_b} = \dfrac{2788124}{2700 \times 40.83 \times 49.17} = 0.514$

**(d) T-section [Fig. 45.39(d)]**

$A = (80 \times 15) + (75 \times 20) = 2700$ cm²

$y_t = \dfrac{(80 \times 15 \times 7.5) + (75 \times 20 \times 52.5)}{2700} = 32.5$

$y_b = 90 - 32.5 = 57.5$ cm.

$$I = \left[\tfrac{1}{12} \times 80 \times 15^3 + 80 \times 15 \,(32.5 - 7.5)^2\right] + \left[\tfrac{1}{12} \times 20 \times 75^3 + 75 \times 20 \,(57.5 - 37.5)^2\right] = 2075625 \text{ cm}^4$$

$$k_t = \frac{I}{A\, y_b} = \frac{2075625}{2700 \times 57.5} = \mathbf{13.37 \text{ cm}}; \qquad k_b = \frac{I}{A\, y_t} = \frac{2075625}{2700 \times 32.5} = \mathbf{23.65 \text{ cm}}$$

$$\rho = \frac{k_t + k_b}{D} = \frac{13.37 + 23.65}{90} = 0.41$$

Alternatively, $\qquad \rho = \dfrac{I}{A\, y_t\, y_b} = \dfrac{2075625}{2700 \times 32.5 \times 57.5} = 0.41$

It will thus be seen that unsymmetrical I-section has the maximum efficiency while the rectangular section has minimum efficiency.

## 45.12. DESIGN OF SECTIONS FOR FLEXURE : MAGNEL'S METHOD

**Notations.** In the notations given below, suffix $t$ stands for top fibre, while suffix $b$ stands for the bottom fibre of the concrete section. Suffix $i$ stands for initial stress condition, suffix $d$ corresponds to dead loads while $l$ corresponds for live load conditions.

- $P_i$ = Initial prestressing force.
- $P_e$ = Effective prestressing force = $\lambda P_i$.
- $\lambda$ = Loss factor or proportion of prestressing that remains permanently.
- $A$ = Sectional area.    $I$ = Moment of inertia of the section.
- $r$ = Radius of gyration.    $e$ = Eccentricity of pre-stressing force.
- $M_d$ = Moment due to dead weight.
- $M_l$ = Moment due to external or superimposed load.
- $M_T$ = Moment due to total load = $M_d + M_l$.
- $f_{yp}$ = Stress at any level $y$ due to prestressing force.
- $f_{td}$ = Tensile stress in concrete at top, due to initial dead load only.
- $f_{bd}$ = Compressive stress in concrete at bottom, due to initial dead load only.
- $f_{ti}$ = Total stress at top, under initial stress condition.
- $f_{bi}$ = Total stress at bottom under initial stress condition.
- $f_{tl}$ = Compressive stress at top due to external load.
- $f_{bl}$ = Tensile stress at bottom due to external load.
- $f_{tT}$ = Total stress at top (compressive) under final stress conditions.
- $f_{bT}$ = Total stress at bottom (tensile) under final stress condition.
- $p_{ci}$ = Permissible compressive stress in concrete, under initial stress condition.
- $p_{ti}$ = Permissible tensile stress in concrete, under initial stress condition.
- $p_c$ = Permissible compressive stress in concrete under final stress condition.
- $p_t$ = Permissible tensile stress in concrete under final stress condition.
- $y_t$ = Distance of top fibre from C.G.C.
- $y_b$ = Distance of bottom fibre from C.G.C.
- $k_t$ = Distance of top kern level.    $k_b$ = Distance of bottom kern level.

**Initial stress conditions:** Fig. 45.40(*a*) shows a concrete section, prestressed with a force $p_i$ acting at eccentricity $e$. Fig. 45.40(*b*) shows the stresses induced by the eccentric prestress force. The stress at any distance $y$ is given by

$$f_{yp} = \frac{P_i}{A} \mp \frac{P_i \cdot e \cdot y}{I} = \frac{P_i}{A} \mp \frac{P_i \cdot e \cdot y}{A r^2} = \frac{P_i}{A}\left(1 \mp \frac{e y}{r^2}\right),$$

with minus sign for upper fibers and plus sign for lower fibres.

Fig. 45.40(c) shows the stress distribution due to dead load moment $M_d$, which will cause compressive stress ($f_{td}$) at top and tensile stress ($f_{bd}$) at bottom. The stress at any level is given by

$$f_{bi} = \pm \frac{M_d \cdot y}{I}$$

Fig. 45.40(d) shows the total stress due to eccentric prestress force and the dead load moment. The stresses for top and bottom fibres, due to this stress conditions are as follows: Tensile stress at top fibre:

$$f_{ti} = \frac{P_i}{A}\left(-1 + \frac{e\, y_t}{r^2}\right) - \frac{M_d\, y_t}{I} \quad ...(a)$$

Compressive stress at bottom fibre:

$$f_{bi} = \frac{P_i}{A}\left(1 + \frac{e\, y_b}{r^2}\right) - \frac{M_d\, y_b}{I} \quad ...(b)$$

By inspection of Fig. 45.40(d), there is possibility of tensile stress at the top. *The stress should be lesser than the permissible tensile stress* ($p_{ti}$). Similarly, maximum compressive stress at bottom ($f_{bi}$) should be less than the permissible compressive stress ($p_c$) for concrete. Hence we have two condition equations for initial stress conditions

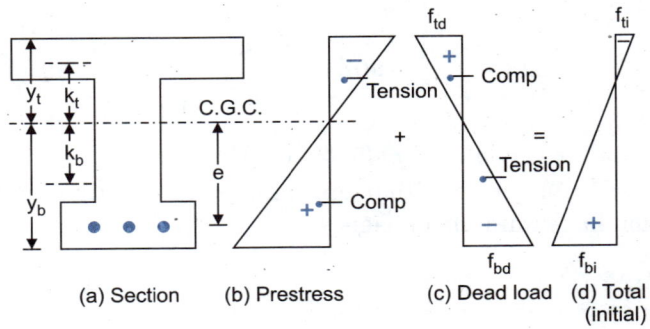
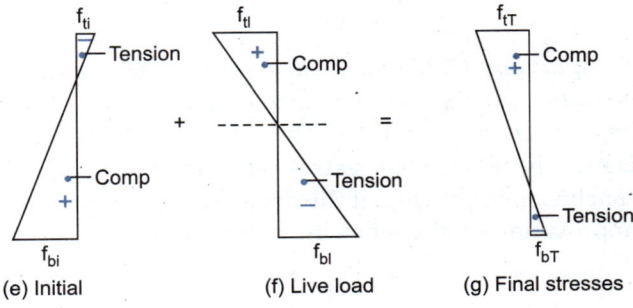

Fig. 45.40. Initial and Final Stress Conditions.

$$\frac{P_i}{A}\left(-1 + \frac{ey_t}{r^2}\right) - \frac{M_d\, y_t}{I} \leq p_{ti} \quad ...(45.31)$$

$$\frac{P_i}{A}\left(1 + \frac{e\, y_b}{r^2}\right) - \frac{M_d\, y_b}{I} \leq p_{ci} \quad ...(45.32)$$

**Final stress conditions:** Final stress conditions are included when external moment ($M_l$) is applied to the section. Fig. 45.40(f) shows the stress distribution due to $M_l$ the stress being compressive at top ($f_{tl}$) and tensile at bottom ($f_{bl}$). Fig. 45.40(g) shows the final stress distribution due to eccentric prestress, dead load moment $M_d$ and live load moment $M_l$. The stress at top fibre ($f_{tT}$) will be compressive while stress at bottom fibre ($f_{bT}$) is likely to be tensile. Both these stresses should be larger than the permissible values. If the effective prestress $P_e = \lambda P_i$, we get the following two condition equations:

Compressive stress at top, $f_{tT} \leq p_c$

or

$$-\frac{\lambda P_i}{A}\left(-1 + \frac{e\, y_t}{r^2}\right) + \frac{M_d\, y_t}{I} + \frac{M_l \cdot y_t}{I} \leq p_c \quad ...(45.33)$$

Similarly tensile stress at bottom, $f_{bT} \leq p_t$

or

$$-\frac{\lambda P_i}{A}\left(1 + \frac{e\, y_b}{r^2}\right) + \frac{M_d\, y_b}{I} + \frac{M_l\, y_b}{I} \leq p_t \quad ...(45.34)$$

In the above four equations, there are four unknowns: $A, I, P_i$ and $e$. The values of these unknowns should be such that these satisfy all the above equations simultaneously. Magnel suggested a graphical solution. But before that, let us find the section modulus.

Adding *Eq. 45.31* and *45.33*, we get

$$\frac{P_i}{A}(1-\lambda)\left(-1+\frac{e\,y_t}{r^2}\right)+\frac{M_l\,y_t}{I} \le p_c + p_{ti} \quad \text{or} \quad \frac{I}{y_t} \ge \frac{M_l}{(p_c + p_{ti}) - \frac{P_i}{A}(1-\lambda)\left(-1+\frac{e\,y_t}{r^2}\right)}$$

Let $P_i/A$ = average compressive stress = $f_c$

$$\frac{I}{y_t}\,(\text{or } Z_t) \ge \frac{M_l}{(p_c + p_{ti}) - f_c(1-\lambda)\left(-1+\frac{e\,y_t}{r^2}\right)} \qquad \text{...(45.35)}$$

Similarly, adding *Eq. 45.32* and *45.34*, we get

$$\frac{P_i}{A}(1-\lambda)\left(1+\frac{e\,y_b}{r^2}\right)+\frac{M_l\,y_b}{I} \le p_{ci} + p_t$$

or
$$\frac{I}{y_b}\,(\text{or } Z_b) \ge \frac{M_l}{(p_{ci} + p_t) - f_c(1-\lambda)\left(1+\frac{e\,y_b}{r^2}\right)} \qquad \text{...(45.36)}$$

Equations 45.35 and 45.36 given the section moduli. It is clear from the above equations that *the section modulus is independent of dead load moments*. The prestressed beam is known as 'carry itself' beam, since its section does not have to be increased to carry its weight, as is required for reinforced concrete beams. Hence the section of prestressed beam is much smaller than the ordinary reinforced concrete beam. This conclusion is not valid if the dead load moment is very large in comparison to $M_l$. For obtaining dimensions approximately, the following values may be adopted, for use in *Eq. 45.35* and *45.36*.

$$\lambda = 0.85 \ (\textit{i.e. loss equal to 15\%})$$

$$\frac{P_i}{A} = f_c = \frac{1}{2}\,p_c \quad \text{or} \quad \frac{1}{2}\,p_{ci} \text{ as the case be.}$$

$$\frac{e\,y_t}{r^2} = \frac{e\,y_b}{r^2} = 2 \text{ in beams which have more or less uniform section.}$$

For sections with heavy top flange, $\frac{e\,y_t}{r^2}$ may be taken as 2 and $\frac{e\,y_b}{r^2}$ equal to 3.

For a section to be most economical, the sectional area ($A$) should be the least for a required value $\frac{I}{y_b}$. In other words $\frac{I}{A\,y_b}\left(\text{or } \frac{r^2}{y_b}\right)$ should be as large as possible.

But $r^2/y_b = k_t$ = distance of upper kern point from the centroid, and its maximum value can be only $y_t$. For that extreme condition,

$$\frac{r^2}{y_b} = y_t \quad \text{or} \quad \frac{r^2}{y_b\,y_t} = 1.$$

When $r^2/y_b$ is less that $y_t$, for a general case, the factor $r^2/y_b y_t$ ($=\rho$) is a measure of the efficiency of the section (see *Eq.* § 45.29 also). For rectangular section, $r^2/y_b\,y_t = 1/3$ (see § 45.11). For I-section, $r^2/y_b\,y_t \ge 0.5$ shows an economical section, while $r^2/y_b\,y_t \le 0.45$ gives a heavy section. However, it is usually not possible to raise $\rho$ higher than 0.55 without making the web thinner than desirable. After having determined the section moduli ($I/y_t$ and $I/y_b$) the other quantities to be found are $P_i$ and $e$. In order that these values satisfy *Eqs. 45.31* to *45.34*, Magnel suggested the following graphical procedure:

**(a) Equation 45.31** (*Line $T_1$*)

$$\frac{P_i}{A}\left(\frac{e\,y_t}{r^2}-1\right) \le \frac{M_d}{Z_t} + p_{ti}$$

or $\quad\dfrac{A}{P_i} \geq \dfrac{\dfrac{e\, y_t}{r^2} - 1}{\dfrac{M_d}{Z_t} + p_{ti}}$  ...[45.31(a)]

If a graph is plotted between $A/P_i$ and $e$, a straight line will be obtained as shown in Fig. 45.41. For $A/P_i = 0$, $e \leq r^2/y_t$.

Similarly, for $\quad e = 0,\ \dfrac{A}{P_i} \geq \dfrac{1}{\dfrac{M_d}{Z_t} + p_{ti}}$

These give the values of co-ordinates at points of crossing the line with the axis. Any value of $A/P_i$ and $e$ lying in the shaded area will satisfy this equation.

**(b) Equation 45.32 (Line $B_1$)**

$$\dfrac{P_i}{A}\left(1 + \dfrac{e\, y_b}{r^2}\right) \leq \dfrac{M_d}{Z_b} + p_{ci}$$

or $\quad\dfrac{A}{P_i} \geq \dfrac{\left(1 + \dfrac{e\, y_b}{r^2}\right)}{\dfrac{M_d}{Z_b} + p_{ci}}$  ...[45.32(a)]

Fig. 45.42 shows the plot of Eq. 45.32(a). Any value of $A/P_i$ and $e$ lying in the shaded area will satisfy Eq. 45.32(a).

**(c) Equation 45.33 (Line $T_2$)**

$$\dfrac{\lambda P_i}{A}\left(1 - \dfrac{e\, y_t}{r^2}\right) \leq p_c - \dfrac{M_T}{Z_t} \quad \text{or} \quad \dfrac{A}{P_i} \geq \dfrac{\lambda\left(1 - \dfrac{e\, y_t}{r^2}\right)}{\left[p_c - \dfrac{M_T}{Z_t}\right]} \quad ...[45.33(a)]$$

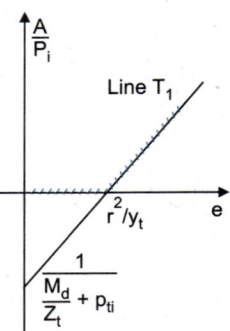

Fig. 45.41

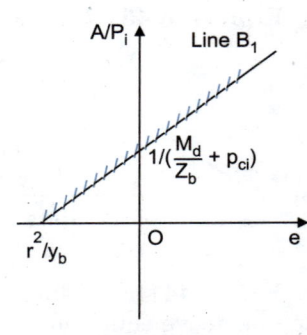

Fig. 45.42

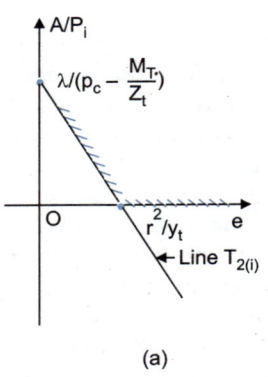
(a)

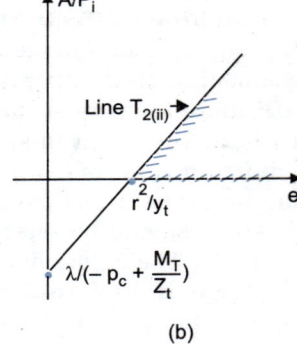
(b)

Fig. 45.43

Two conditions, may arise depending upon the relative magnitudes of $p_c$ and $M_T/Z_t$

*Case (i)* $p_c \geq M_T/Z_t$ : If $e$ more than $r^2/y_t$, equation 45.33(a) will be satisfied for all values of $P_i$. If $e$ is less than $r^2/y_t$, the prestress force will be at point higher than the lower kern, causing compressive stress at top, instead of the usual tensile stress.

or $\quad\dfrac{A}{P_i} \geq \dfrac{\lambda\left(1 - \dfrac{e\, y_t}{r^2}\right)}{\left[p_c - \dfrac{M_T}{Z_t}\right]}$  ...[45.33(a)(i)]

This is represented by line $T_2(i)$ in Fig. 45.43(a)

*Case (ii)* $p_c \leq M_T/Z_t$ : If $e$ is less than $r^2/y_t$ Eq. 45.33(a) will never be satisfied. For $e > r^2/y_t$.

$\quad\dfrac{A}{P_i} \leq \dfrac{\lambda\left(-1 + \dfrac{e\, y_t}{r^2}\right)}{\left[-p_c + \dfrac{M_T}{Z_t}\right]}$  ...[45.33(a)(ii)]

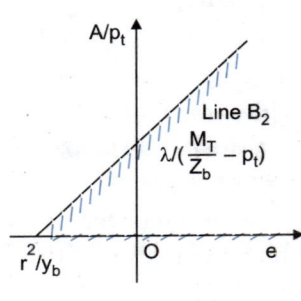

Fig. 45.44

This is represented by line $T_2(ii)$ in Fig. 45.43(b)

**Equation 45.34** (Line $B_2$)

$$\frac{\lambda P_i}{A}\left(1+\frac{ey_b}{r^2}\right) \leq \frac{M_T}{Z_b} - p_t$$

or $$\frac{A}{P_i} \geq \frac{\lambda\left(\frac{ey_b}{r^2}+1\right)}{\left[\frac{M_T}{Z_b}-p_t\right]} \qquad ...[45.34(a)]$$

Fig. 45.44 shows the straight line $B_2$ represented by the above equation.

Values of $A/P_i$ and $e$ lying in the shaded area will satisfy Eq. 45.34(c).

**Combined effect:** Figs. 45.45(a) and (b) show the four equations plotted together. The shaded area common to all the four lines give the magnitude of $A/P_i$ and $e$ that would satisfy all the four equations. In both the diagrams, point $R$ and $R'$ indicate minimum value of $e$, while points $S$ and $S'$ indicate minimum values of $P_i$. Hence for design purposes $P_i$ and $e$ should be selected with respect to points $S$ and $S'$. The values of $e$ corresponding to $S$ and $S'$ are large. If these values of $e$ are not available, the maximum values of $e$ available should be used.

For a simply supported beam, both $M_d$ and $M_l$ go on decreasing towards the support, and become zero at the supports. Since $P_i$ remains constant, $e$ should be changed from section to section, giving rise to a curved cable profile.

**Cable zone:** It is clear from the above that the C.G. of the cable should remain within some limits so that the stresses are within the allowable limits.

Consider Eq. 45.31 corresponding to initial prestress and dead load moment $M_d$. If the minimum stress is equal to zero, we have

$$\frac{P_i}{A}\left(1-\frac{ey_t}{r^2}\right) + \frac{M_d y_t}{I} = 0$$

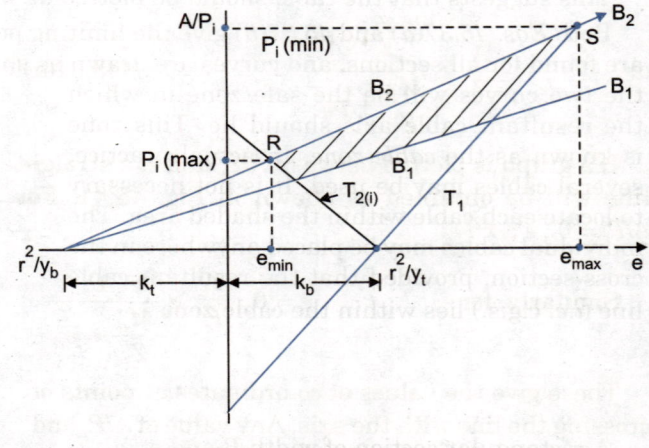

(a) Lines $B_1$, $B_2$, $T_1$ and $T_{2(i)}$

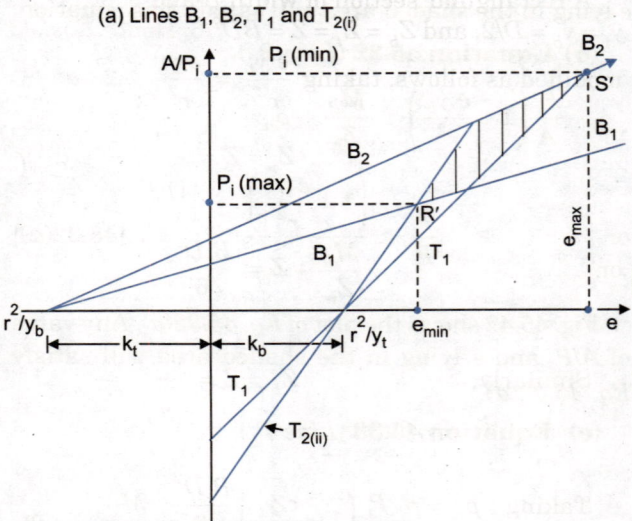

(b) Lines $B_1$, $B_2$, $T_1$ and $T_{2(ii)}$

**Fig. 45.45**

Putting $I = Ar^2$, $P_i\left(-1+\frac{ey_t}{r^2}\right) = \frac{M_d y_t}{r^2}$ or $e = \frac{r^2}{y_t} + \frac{M_d}{P_i}$

or $$e = k_b + \frac{M_d}{P_i} \qquad ...[45.37(a)]$$

Eq. 45.37(a) suggests that for no tension to be developed, the cable should be placed at $M_d/P_i$ below the lower kern point.

Again, consider Eq. 45.35 corresponding to eccentric prestress $M_d$ and $M_l$. Minimum stress will occur at the bottom fibre. If this minimum stress is to be zero, we get

$$\frac{\lambda P_i}{A}\left(1+\frac{ey_b}{r^2}\right) - \frac{M_d y_b}{Ar^2} - \frac{M_l y_b}{Ar^2} = 0 \quad \text{or} \quad \lambda P_i\left(1+\frac{ey_b}{r^2}\right) = \frac{M_T y_b}{r^2}$$

$$e = \frac{M_T}{\lambda P_i} - \frac{r^2}{y_b} = \frac{M_T}{\lambda P_i} - k_t \qquad ...(45.37)$$

This suggests that the cable should be plotted at $M_T/\lambda P_i$ below the upper kern point.

Both *Eqs. 45.37(a)* and *45.37(b)* give the limiting positions of the cable for the section. If these positions are found for all sections, and curves are drawn as shown in Fig. 45.46, the zone (shown shaded) between the two curves will be the safe zone in which the resultant cable axis should lie. This zone is known as the *cable zone*. In actual practice, several cables may be used. It is not necessary to locate each cable within the shaded area. The individual cables may be placed anywhere in the cross-section, provided that the resultant cable line (*i.e.* c.g.s.) lies within the cable zone.

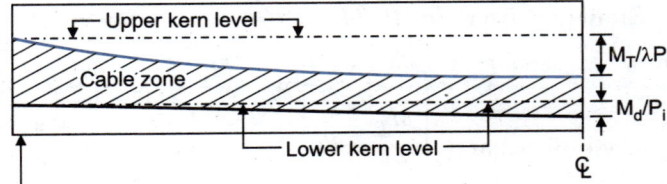

Fig. 45.46. Cable Zone.

## 45.13. RECTANGULAR SECTION

A rectangular section of width $B$ and depth $D$ is symmetrical about the C.G. of the section, thus having $y_t = y_b = D/2$, and $Z_b = Z_t = Z = BD^2/6$. Hence *Eqs. 45.35* and *45.36*, giving the values of section modulus are modified as follows, taking $\dfrac{e\, y_t}{r^2} = \dfrac{e\, y_b}{r^2} = 2$ and $f_c = \dfrac{1}{2} P_c$.

$$Z_t = Z = \frac{M_l}{(p_c + p_{ti}) - \frac{1}{2} P_c (1-\lambda)(-1+2)} \qquad \ldots[45.35(a)]$$

or

$$Z = \frac{BD^2}{6} = \frac{M_l}{\dfrac{p_c}{2}(\lambda+1) + p_{ti}} \qquad \ldots[45.35(b)]$$

Similarly,

$$Z_b = Z = \frac{M_l}{(p_{ci} + p_t) - \frac{1}{2} P_c (1-\lambda)(1+2)} \qquad \ldots[45.36(a)]$$

Taking $p_{ci} = p_c$,

$$Z = \frac{BD^2}{6} = \frac{M_l}{\dfrac{p_c}{2}(3\lambda-1) + p_t} \qquad \ldots[45.36(b)]$$

It will be seen that *Eq. 45.36(b)* gives larger section modulus, should be adopted for design. If $\lambda$ is assumed to be 0.85, *Eq. 45.36(b)* simplifies to:

$$Z = \frac{M_l}{0.775\, p_c + p_t} \qquad \ldots[45.36(c)]$$

Similarly, *Eq. 45.35(b)* simplifies to:

$$Z = \frac{M_l}{0.925\, p_c + p_{ti}} \qquad \ldots[45.37(b)]$$

For a rectangular section, the depth of the beam should be between 1/15 to 1/20 th of span. From the bucking point of view, the width of the beam should not be less than 1/30 th of the length between effective restrains or 1/4 th of the depth of the beam.

After having obtained the section modulus, $B$ and $D$, the other unknowns to be determined are $P_i$ and $e$.

Consider *Eqs. 45.31* (Line $T_1$) and *45.34* (Line $B_2$), to get the minimum value of prestressing force (max. value of $1/P_i$). From *Eq. 45.31*, we get

$$\frac{P_i}{A}\left(e - \frac{r^2}{y_t}\right)\frac{y_t}{r^2} = M_d \frac{y_t}{A\cdot r^2} + p_{ti} \quad \text{or} \quad P_i\left(e - \frac{r^2}{y_t}\right) = M_d + p_{ti}\frac{A\, r^2}{y_t}$$

But

$$\frac{r^2}{y_t} = \frac{r^2}{y_b} = \frac{r^2}{y} = \frac{I}{A\cdot y} = \frac{Z}{A} = \frac{D}{6} \quad \text{and} \quad \frac{A\, r^2}{y} = \frac{I}{y} = Z$$

$$\therefore \quad P_i \left(e - \frac{D}{6}\right) = M_d + p_{ti} Z$$

Similarly, from Eq. 45.34

$$-\frac{\lambda P_i}{A}\left(e + \frac{r^2}{y_b}\right)\frac{y_b}{r^2} = p_t - \frac{M_d y_b}{A r^2} - \frac{M_l y_b}{A r^2} \quad \text{or} \quad \lambda P_i\left(e + \frac{D}{6}\right) = (M_d + M_l) - p_t \cdot Z \quad \text{...(b)}$$

Dividing (a) and (b), we get

$$\frac{\left(e - \dfrac{D}{6}\right)}{\left(e + \dfrac{D}{6}\right)} = \frac{\lambda (M_d + p_{ti} Z)}{(M_d + M_l) - p_t Z} \quad \text{...(45.38)}$$

From this equation, $e$ can be found. If $p_t$ and $p_{ti}$ are taken as zero, this equation further simplifies to:

$$\frac{e - \dfrac{D}{6}}{e + \dfrac{D}{6}} = \frac{\lambda M_d}{M_T} \quad \text{...[45.38(a)]}$$

Knowing $e$, value of $P_i$ can be obtained from Eq. 45.34 (line $B_2$) simplified as follows:

$$-\frac{\lambda P_i}{A}\left(e + \frac{r^2}{y_b}\right)\frac{y_b}{r^2} + \frac{M_d y_b}{I} + \frac{M_l y_b}{I} = p_t$$

$$\lambda P_i\left(e + \frac{D}{6}\right) = (M_d + M_l) - p_t Z \quad \text{or} \quad P_i = \frac{(M_d + M_l) - p_t Z}{\lambda\left(e + \dfrac{D}{6}\right)} \quad \text{...(45.39)}$$

If, however, available eccentricity is less than the one found from Eq. 45.38(a), the corresponding value of $P_i$ is obtained by putting the available value of $e$ in Eq. 45.34 (Eq. 45.39) or Eq. 45.33 whichever gives higher value. Eq. 45.33 simplifies as follows:

$$P_i = \frac{(M_d + M_l) - p_c Z}{\lambda\left(e - \dfrac{D}{6}\right)} \quad \text{...[45.39(a)]}$$

**Critical span for slabs and rectangular beams:** Critical span ($L_c$) is that span of the beam upto which the dead load does not affect the design of the section, and only live load need be considered as is evident from Eq. 45.36(c). Let us derive the expression for the same.

Fig. 45.47(a) shows the rectangular section with $k_t = k_b = D/6$ while Fig. 45.47(b) shows the stress distribution due to dead load. For stress to be zero at the top fibre (i.e. for triangular stress distribution, the resultant compressive force $C$ must pass through the lower kern point, so that

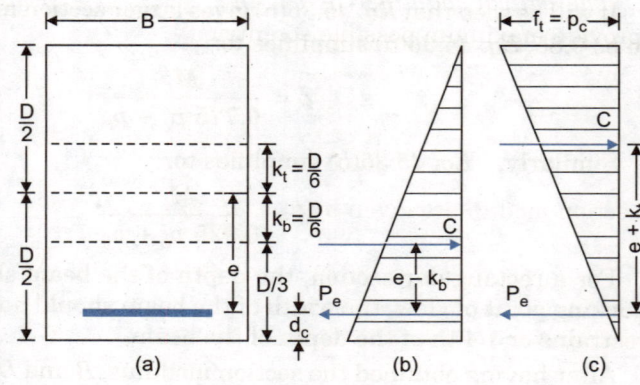

Fig. 45.47

$$M_d = C(e - k_b) = P_i(e - k_b) \quad \text{(See Eq. 45.26 also)} \quad \text{...(a)}$$

Now let $\Delta M$ be increase in moment, due to external load, and $\Delta f_t$ be the increase in compressive stress at top fibre due to this.

Then $\quad \Delta M = \Delta f_t Z$

If $\Delta M = M_l$ and $\Delta f_t = p_c$ = allowable compressive stress at top [Fig. 45.47(c)], we have $M_l = p_c \cdot Z \quad \text{...(b)}$

Now for a live load of $w_l$ and critical span $L_c$,

$$M_l = \frac{w_l L_c^2}{8}, \text{ and } Z = \frac{BD^2}{6} \quad \therefore \quad \frac{w_l L_c^2}{8} = p_c \frac{BD^2}{6}$$

$$\therefore \quad \frac{D}{L_c} = \sqrt{\frac{3}{4}\frac{w_l}{p_c}} \quad \ldots(45.40)$$

Fig. 45.47(c) shows the final stress distribution due to $P_e$, $M_d$ and $M_l$.

For the triangular stress distribution, the resultant compressive force must act at the upper kern, thus getting a lever arm of $(e + k_b)$ for the internal moment.

$$\therefore \quad M_d + M_l = P_e (e + k_t) \quad \ldots(\text{see } Eq. \text{ 45.27 also})$$

Subtracting Eq. (a) from this, we get

$$M_l = P_e (k_b + k_t) \quad \ldots(\text{see } Eq. \text{ 45.27 also})$$

$$\frac{M_d}{M_l} = \frac{w_d}{w_l} = \frac{e - k_b}{k_b + k_t} \quad \ldots(45.41)$$

But $k_b = k_t = D/6$. If the distance of centre of steel from bottom fibre is $d_c$, we have $e = 0.5 D - d_c$. Also, $w_d = 0.025\, BD$ N/m$^2$ where $B$ and $D$ are in mm.

$$\frac{0.025\, B D}{w_l} = \frac{0.5 D - d_c - D/6}{D/3} \quad \text{or} \quad D = \frac{w_l}{0.025\, B}\left(1 - \frac{3 d_c}{D}\right)$$

Equating this to Eq. 45.40, we get $\dfrac{L_c}{2}\sqrt{3\dfrac{w_l}{p_c}} = \dfrac{w_l}{0.025\, B}\left(1 - 3\dfrac{d_c}{D}\right)$

or

$$L_c = \frac{1}{0.0125\, B}\sqrt{\frac{w_l \cdot p_c}{3}}\left(1 - 3\frac{d_c}{D}\right) \quad \ldots(45.42)$$

In a slab, $B = 1000$ mm

$$\therefore \quad L_c = \frac{1}{1.25}\sqrt{\frac{w_l\, p_c}{3}}\left(1 - \frac{3 d_c}{D}\right) \quad \ldots[45.42(a)]$$

If the actual span ($L$) is less than $L_c$ given above, the dead load is fully carried.

The critical span increases with eccentricity $e$ of the prestressing force, and attempts should be made to provide maximum possible eccentricity.

## 45.14. I-SECTION

**(a) Determination of section modulus:** Let us start with the case of an unsymmetrical I-section. The section moduli are given by Eqs. 45.35 and 45.36.

$$Z_t = \frac{M_l}{(p_c + p_{ti}) - f_c (1 - \lambda)\left(-1 + \dfrac{e\, y_t}{r^2}\right)} \quad \ldots(45.35)$$

$$Z_b = \frac{M_l}{(p_{ci} + p_t) - f_c (1 - \lambda)\left(1 + \dfrac{e\, y_b}{r^2}\right)} \quad \ldots(45.36)$$

Taking $f_c = \dfrac{p_c}{2}$, $\dfrac{e\, y_c}{r^2} = \dfrac{e\, y_b}{r^2} = 2$ (to start with), $p_{ci} = p_c$ and $p_{ti} = p_t$, we get

$$Z_t = \frac{M_l}{\dfrac{p_c}{2}(\lambda + 1) + p_t} \quad \ldots[45.35(A)]$$

$$Z_b = \frac{M_l}{\frac{p_c}{2}(3\lambda - 1) + p_t} \qquad ...[45.36(A)]$$

By the inspection of *Eqs. 45.35(A)* and *45.36(A)*, it is clear that $Z_b$ is more that $Z_t$, *i.e.* the bottom flange should be heavier (or wider) from the top flange. However, from buckling point of view, top flange is kept wider than the bottom flange. Bottom flange is kept thicker for accommodating prestressing cables. For the prestressing force to be minimum, eccentricity should be maximum. Eccentricity can be increased by providing heavier top flange so that C.G. is shifted up. All these factors are to be kept in mind while selecting the section. Taking, $\lambda = 0.85$,

$$Z_t = \frac{M_l}{0.925\, p_c + p_t} \qquad ...[45.35(B)]$$

and
$$Z_b = \frac{M_l}{0.775\, p_c + p_t} \qquad ...[45.36(B)]$$

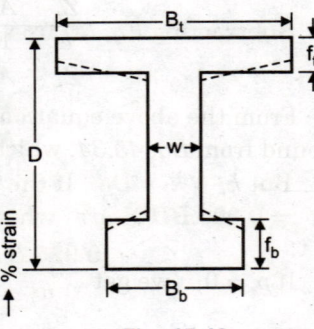

Fig. 45.48

When the dead load is greater, the depth of the section has to be increased beyond the value given by *Eq. 45.36(B)*.

Fig. 45.48 shows an unsymmetrical I-section, with wider top flange, and a thicker bottom flange. Based on experience and practical considerations, Guyon has suggested the following proportions. However, the flanges are splayed for easiness of pouring concrete and to avoid sharp corners, as shown by dotted lines.

(*i*) **Depth D:** If the dead moment ($M_d$) is 20% or more of total moment ($M_T$),

$$D = k\sqrt{M_T} \qquad ...[45.43\,(a)]$$

where $k$ varies from 10 to 13 and $M_T$ is in tonne-cm.

If $M_d$ less than 20% of $M_T$, $D = k\sqrt{M_l}$ ...[45.43 (b)]

where $M_l$ is the live load B.M.

According to Guyon:
$$D = \frac{L}{25} + 10 \text{ cm (for spans 18 to 36 m)} \qquad ...[45.43\,(c)]$$

and
$$D = \frac{L}{20} \text{ (for spans 36 to 45 m)} \qquad ...[45.43\,(d)]$$

(*ii*) **Web thickness (w):** The thickness should be adequate to accommodate the cables with sufficient cover. For this the details about the size of duct for various system should be known. A minimum cover of 30 mm or the size of cable whichever is more is provided. For example, if Freyssinet cable consisting of 12 wires of 5 mm wires are provided, the external diameter of sheathing required is 34 mm. Hence minimum $w = 3 \times 3.34 = 10.2$ cm for a single cable.

From other considerations,

For $D < 75$ cm, $\quad w = \dfrac{D}{7}$ to $\dfrac{D}{8}$ ...[45.44 (a)]

For $D > 75$ cm, $\quad w = \dfrac{D}{40} + 10$ cm ...[45.44 (b)]

(*iii*) **Top flange**

$$B_t = 0.6\, D \text{ to } 0.8\, D \qquad ...(45.45)$$

Minimum width of flange may be taken equal to $L/30$.

The depth of flange may be taken equal to 0.1 to 0.2 times overall depth.

(*iv*) **Bottom flange:** The thickness and width of the bottom flange are decided on the basis of accommodating all the cables at the section of maximum moment. The minimum clear distance between cable should not be less than 40 mm or the minimum dimension of the cable or 6.5 mm in excess of the larger size of aggregate used, whichever is greater.

(b) **Determination of $e$ and $P_i$:** After having determined section modulus and other dimensions of the section, the next step is to determine the eccentricity $e$ of the prestressing force. For minimum prestressing force (or maximum value of $A/P_i$) consider *Eq. 45.31* (Line $T_1$) and *Eq. 45.34* (Line $B_2$).

$$P_i\left[-\frac{1}{A} + \frac{e}{Z_t}\right] \le \frac{M_d}{Z_t} + p_{ti} \qquad \text{...[45.31 (b)]};$$

and $$\lambda P_i\left[\frac{1}{A} + \frac{e}{Z_b}\right] \le \frac{M_T}{Z_b} - p_t \qquad \text{...[45.34 (b)]}$$

$$\therefore \quad \frac{\dfrac{e}{Z_t} - \dfrac{1}{A}}{\dfrac{e}{Z_b} + \dfrac{1}{A}} \le \frac{\lambda\left(\dfrac{M_d}{Z_t} + p_{ti}\right)}{\left(\dfrac{M_T}{Z_b} - p_t\right)} \qquad \text{...(45.46)}$$

From the above equation, maximum allowable $e$ can be found. If this eccentricity is available, $P_i$ can be found from *Eq. 45.34*, which simplifies as follows:

$$P_i = \frac{M_T - p_t Z_b}{\lambda(e + k_t)} \qquad \text{...(45.47)}$$

If $p_t = 0$, we get $P_i = \dfrac{M_T}{\lambda(e + k_t)}$, which is the same as *Eq. 45.27*.

If, however, eccentricity $e$ available is lesser than that given by *Eq. 45.46*, the corresponding value of $P_i$ is obtained by putting the available value of $e$ in *Eq. 45.34* or *45.33* whichever gives higher value. *Eq. 45.33* simplifies as follows:

$$P_i = \frac{M_T - p_c Z_t}{\lambda(e - k_b)} \qquad \text{...(45.48)}$$

Alternatively, $e$ and $P_i$ can be obtained by plotting lines $T_1$ (*Eq. 45.31*), $B_1$ (*Eq. 45.32*), $T_2$ (*Eq. 45.33*) and $B_2$ (*Eq. 45.34*), as explained earlier.

**Design: Example 45.15.** *Design of prestressed concrete slab.;* Design a prestressed concrete slab over a span of 16 metres to carry a superimposed load of 15 kN/m². Assume permissible stresses in concrete of 12 N/mm² and no tension. Also, assume that the initial losses in prestress in steel will amount to 15 per cent. The live load will be brought to bear after the ducts are grouted. The initial prestress in the wires will be 1100 N/mm².

**Solution:**

1. **Determination of the Section**

Consider one metre width of slab. $p_c = p_{ci} = 12$ N/mm². $p_t = p_{ti} = 0$.
The section modulus is given by *Eq. 45.37*,

$$Z = \frac{BD^2}{6} = \frac{M_l}{\dfrac{p_c}{2}(3\lambda - 1) + p_t}$$

Taking $\lambda = 0.85$, this reduces to

$$\frac{BD^2}{6} = \frac{M_l}{0.775\, p_c + p_t} \qquad \text{...[45.36 (c)]}$$

$$M_l = \frac{15000\,(16)^2}{8} \times 1000 = 480 \times 10^6 \text{ N-mm}$$

$$p_c = 12 \text{ N/mm}^2; \quad p_t = 0; \quad B = 1000 \text{ mm}$$

$$\therefore \quad \frac{1000\, D^2}{6} = \frac{480 \times 10^6}{0.775 \times 12}$$

$$D = \sqrt{\frac{480 \times 10^6 \times 6}{0.775 \times 12 \times 1000}} \approx 556 \text{ mm}; \quad \text{Adopt } D = 560 \text{ mm}$$

Dead load = $0.56 \times 1 \times 1 \times 25000 = 14000$ N/m

$$\therefore \quad M_d = \frac{14000 (16)^2}{8} \times 1000 = 448 \times 10^6 \text{ N-mm}$$

Total moment $\quad M_T = M_d + M_l = (480 + 448) 10^6 = 928 \times 10^6$ N-mm

Section modulus $\quad Z = \frac{1}{6} B D^2 = \frac{1}{6} \times 1000 (560)^2 = 52.267 \times 10^6$ mm$^3$

**2. Determination of Eccentricity, $e$ :** The eccentricity $e$ is given by *Eq. 45.38*

$$\frac{e - \dfrac{D}{6}}{e + \dfrac{D}{6}} = \frac{\lambda (M_d + p_{ti} Z)}{(M_d + M_l) - p_t Z}$$

Taking $p_t = p_{ti} = 0$, we get

$$\frac{e - \dfrac{D}{6}}{e + \dfrac{D}{6}} = \frac{\lambda M_d}{M_T} \quad \text{or} \quad \frac{e - \dfrac{560}{6}}{e + \dfrac{560}{6}} = \frac{0.85 \times 448 \times 10^6}{928 \times 10^6} = 0.4103$$

$$e - \frac{560}{6} = 0.4103 \left(e + \frac{560}{6}\right); \quad \text{From which } e = 223.3 \text{ mm}.$$

This is the maximum permissible value of $e$ to get the minimum value of $P_i$. Let us assume that the size of the cable duct is such that this much eccentricity will be available.

**3. Determination of Prestressing Force $P_i$:** The prestressing force is given by *Eq. 45.39*

$$P_i = \frac{(M_d + M_l) - p_t . Z}{\lambda \left(e + \dfrac{D}{6}\right)}$$

Taking $\quad p_t = 0, \quad P_i = \dfrac{M_T}{\lambda \left(e + \dfrac{D}{6}\right)} = \dfrac{928 \times 10^6}{0.85 \left(223.3 + \dfrac{560}{6}\right)} = 3.44 \times 10^6$ N.

**4. Design of Cables:** Permissible prestress in cable = 1100 N/mm$^2$.

$$\therefore \quad \text{Area of steel} = \frac{3.44 \times 10^6}{1100} \approx 3127 \text{ mm}^2.$$

Using 7 mm Φ wires, number of wires required per metre width of slab = $\dfrac{3127}{\dfrac{\pi}{4} (7)^2} \approx 81$ wires.

Adopting Magnel system, and providing 16 wires in a cable, the width of duct hole will be 65 mm and depth will be 60 mm. The duct spacing = $\dfrac{16}{81} \times 1000 \approx 200$ mm c/c.

Hence use ducts @ 200 mm c/c. Area of steel in each duct = $16 \times \dfrac{\pi}{4} (7)^2 \approx 615.8$ mm$^2$

**5. Check for Stresses before Grouting Ducts:** Take a unit of slab 200 mm wide.
There will be one duct of size 65 × 60 mm in the unit. Let the N.A. be at $\bar{y}$ from the top fibre.
Net area = $(200 \times 560) - (65 \times 60) = 112000 - 3900 = 108100$ mm$^2$

$$\bar{y} = \frac{(112000 \times 280) - (3900 \times 503.3)}{108100} = 271.9 \text{ mm}$$

$$\therefore \quad \bar{x} = 280 - 271.9 = 8.1 \text{ mm}$$

$$I_{NA} = \left[\frac{1}{12} \times 200 \, (560)^3 + 112000(8.1)^2\right]$$
$$- \left[\frac{1}{12} \, 65 \, (60)^3 + 3900(8.1 + 223.3)^2\right]$$
$$= 2934.28 \times 10^6 - 210.0 \times 10^6$$
$$= 2724.28 \times 10^6 \text{ mm}^4$$

$$Z_t = \frac{2724 \times 10^6}{280 - 8.1} \approx 10.019 \times 10^6 \text{ mm}^3$$

$$Z_b = \frac{2724.28 \times 10^6}{280 + 8.1} = 9.456 \times 10^6 \text{ mm}^3$$

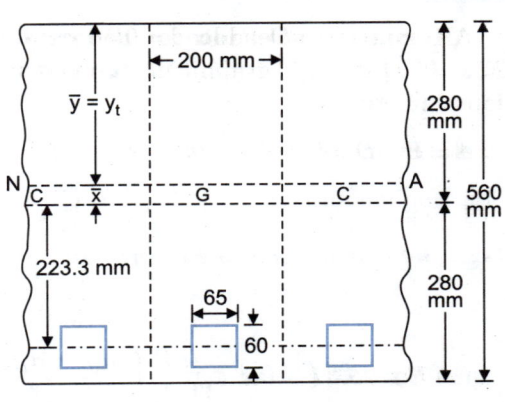

Fig. 45.49

∴ Stresses due to dead load are

$$f_{td} = + \frac{M_d{}'}{Z_t} = \frac{448 \times 10^6}{10.019 \times 10^6} \left(\frac{200}{1000}\right) = + 8.94 \text{ N/mm}^2 \quad (i.e. \text{ compression})$$

$$f_{bd} = - \frac{M_d{}'}{Z_b} = - \frac{448 \times 10^6}{9.456 \times 10^6} \left(\frac{200}{1000}\right) \approx - 9.48 \text{ N/mm}^2 \quad (i.e. \text{ tensile})$$

Stresses due to prestressing force are

**Top fibre:** $\quad \dfrac{P_i}{A} - \dfrac{P_i \cdot e}{Z_t} = 3.44 \times 10^6 \left[\dfrac{1}{108100} - \dfrac{223.3}{10.019 \times 10^6}\right] \times \dfrac{200}{1000} \approx - 8.97 \text{ N/mm}^2$

**Bottom fibre:** $\quad \dfrac{P_i}{A} + \dfrac{P_i \cdot e}{Z_b} = 3.44 \times 10^6 \left[\dfrac{1}{108100} + \dfrac{223.3}{9.456 \times 10^6}\right] \times \dfrac{200}{1000} \approx + 22.60 \text{ N/mm}^2$

Hence the total stresses for the initial condition are

$$f_{ti} = 8.94 - 8.97 = - 0.03 \text{ N/mm}^2; \qquad f_{bi} = - 9.48 + 22.60 = 13.12 \text{ N/mm}^2$$

It is seen that maximum stresses are slightly exceeded but accepted since it diminishes as soon as any losses in cable pull materializes.

**6. Check for Stresses after Grouting:** Since live load is applied after grouting, section moduli with steel are used. Let us use $m = 7$.

Equivalent area $= (200 \times 560) + (7 - 1) \, 615.8 \approx 112000 + 3695 = 115695 \text{ mm}^2$

$$\bar{y} = \frac{(112000 \times 280) + 3695 \times 503.3}{115695} = 287 \text{ mm}; \quad \bar{x} = 287 - 280 = 7 \text{ mm}.$$

$$I_e = \left[\frac{1}{12} \times 200(560)^3 + 112000(7)^2\right] + [0 + 3695(223.3 - 7)^2]$$
$$= 2932.42 \times 10^6 + 172.87 \times 10^6 \text{ mm}^4 = 3105.29 \times 10^6 \text{ mm}^4$$

$$Z_t = \frac{3105.29}{280 + 7} = 10.82 \times 10^6 \text{ mm}^3; \quad Z_b = \frac{3105.29}{280 - 7} = 11.375 \times 10^6 \text{ mm}^3$$

$$\frac{\lambda P_i}{A_e} = \frac{0.85 \times 3.44 \times 10^6}{115695} = 25.27 \text{ N/mm}^2$$

$$M_T{}' = 928 \times 10^6 \left(\frac{200}{1000}\right) = 185.6 \times 10^6 \text{ N-mm per 200 mm width}$$

Hence the final stresses are as follows:

$$f_t = \left[25.27 - \frac{0.85 \times 3.44 \times 10^6 \times 223.3}{10.82 \times 10^6}\right] \frac{200}{1000} + \frac{185.6 \times 10^6}{10.82 \times 10^6} \approx 10.14 \text{ N/mm}^2$$

$$f_b = \left[25.27 + \frac{0.85 \times 3.44 \times 10^6 \times 223.3}{11.375 \times 10^6}\right] \frac{200}{1000} - \frac{185.6 \times 10^6}{11.375 \times 10^6} = 0.2175 \text{ N/mm}^2$$

The stresses are within safe limit.

**Alternative solution for determining $P_i$ and $e$:** $P_i$ and $e$ can be alternatively found by solving Eqs. 45.31 to 45.34 graphically. Consider one metre width of slab. Thus, $B = 1000$ mm and $D = 560$ mm as found in step 1.

$$A = B \times D = 560000 \text{ mm}^2; \quad Z = \tfrac{1}{6} BD^2 = \tfrac{1}{6} \times 1000 (560)^2 \approx 52267000 \text{ mm}^2$$

$$\frac{y_t}{r^2} = \frac{y_b}{r^2} = \frac{Ay}{I} = \frac{A}{Z} = \frac{560000}{52267000} = 0.01071;$$

$$\frac{r^2}{y} = \frac{1}{0.01071} \approx 93.33 \text{ mm}$$

**(a) Eq. 45.31 (Line $T_1$):** $\dfrac{P_i}{A}\left(-1 + \dfrac{e\, y_t}{r^2}\right) - \dfrac{M_d}{Z_t} \le p_{ti}$

or $\quad \dfrac{P_i}{A}(-1 + 0.01071\, e) \le \dfrac{448 \times 10^6}{52267000} \le 8.571 \quad$ or $\quad \dfrac{A}{P_i} \ge \dfrac{-1 + 0.01071\, e}{8.571} \quad$ ...($T_1$)

Hence for $e = 0$, $\quad \dfrac{A}{P_i} = -\dfrac{1}{8.571} = -11.67 \times 10^{-2}$

For $\dfrac{A}{P_i} = 0, \quad e = +\dfrac{1}{0.01071} \approx +93.33$

**(b) Eq. 45.32 (Line $B_1$):** $\dfrac{P_i}{A}\left(1 + \dfrac{e\, y_b}{r^2}\right) \le p_{ci} + \dfrac{M_d}{Z_b}$

or $\quad \dfrac{P_i}{A}(1 + 0.01071\, e) \le 12 + \dfrac{448 \times 10^6}{52267000} \le 20.571 \quad$ or $\quad \dfrac{A}{P_i} \ge \dfrac{1 + 0.01071\, e}{20.571} \quad$ ...($B_1$)

For $e = 0$; $\quad \dfrac{A}{P_i} = \dfrac{1}{20.571} = 4.86 \times 10^{-2}$

For $\dfrac{A}{P_i} = 0, \quad e = -\dfrac{1}{0.01071} \approx -93.33$

**(c) Eq. 45.33 (Line $T_2$):** $-\lambda \dfrac{P_i}{A}\left(-1 + \dfrac{e\, y_t}{r^2}\right) + \dfrac{M_T}{Z_t} \le p_c$

$$= -\frac{0.85\, P_i}{A}(-1 + 0.01071\, e) \le 12 - \frac{928 \times 10^6}{5226700} \le -5.75$$

or $\quad \dfrac{A}{P_i} \ge \dfrac{-0.85 + 0.009104\, e}{5.75} \quad$ ...($T_2$)

For $e = 0$, $\quad \dfrac{A}{P_i} = -\dfrac{0.85}{5.75} = -14.78 \times 10^{-2}$

For $\dfrac{A}{P_i} = 0, \quad e = \dfrac{0.85}{0.009104} \approx 93.33$ mm

**(d) Eq. 45.34 (Line $B_2$):**

$-\dfrac{\lambda P_i}{A}\left(1 + \dfrac{e\, y_b}{r^2}\right) + \dfrac{M_T}{Z_b} \le p_t \quad$ or $\quad -0.85\, \dfrac{P_i}{A}(1 + 0.01071\, e) \le -\dfrac{928 \times 10^6}{52267000} \le -17.75$

or $\quad \dfrac{A}{P_i} \ge \dfrac{0.85(1 + 0.01071\, e)}{17.75} \ge \dfrac{0.85 + 0.009104\, e}{17.75} \quad$ ...($B_2$)

For $e = 0$, $\quad \dfrac{A}{P_i} = +\dfrac{0.85}{17.75} = 4.789 \times 10^{-2}$

For $\dfrac{A}{P_i} = 0, \quad e = -\dfrac{0.85}{0.009104} = -93.33.$

Fig. 45.50 shows the plot of all the four lines. The eccentricity $e$ is found corresponding to the intersection of lines $T_1$ and $B_2$.

From Fig. 45.50

$$e = 224 \text{ mm} \quad \text{and} \quad \frac{A}{P_i} \times 10^2 = 16.4$$

$$\therefore \quad P_i = \frac{A \times 10^2}{16.4} = \frac{560000 \times 10^2}{16.4}$$

$$= 3.415 \times 10^6 \text{ N.}$$

Both these values are very near to the one found by analytical solution. However, graphical solution is cumbersome and does not give very accurate results. Hence analytical solution is always preferred.

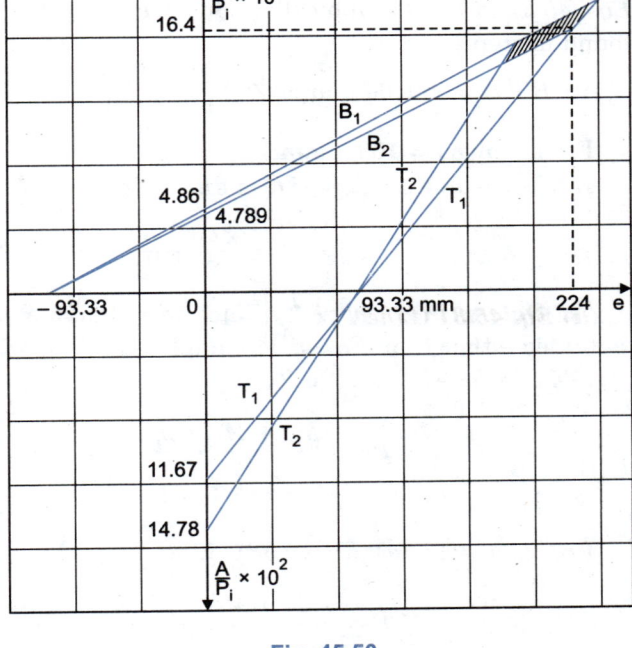

Fig. 45.50

### Design: Example 45.16. *Rectangular beam*

*A rectangular beam of prestressed concrete is required to support a dead load moment of $15 \times 10^6$ N-mm (inclusive of its own weight) and a live load moment of $40 \times 10^6$ N-mm at its midsection. Determine the initial prestressing force and its eccentricity at the mid-span section. Take the following values:*

*Allowable initial compressive stress = 17 N/mm²*

*Allowable final compressive stress = 14 N/mm²*

*Allowable initial or final tensile stress = 1 N/mm².*

*Ultimate stress in steel = 1500 N/mm². Assume losses 15%.*

**Solution:**

**1. Section Modulus:** Given $p_{ci} = 17$ N/mm² ; $p_c = 14$ N/mm² ; $p_{ti} = p_t = 1$ N/mm²

As the dead load moment is less than live load moment, the section modulus will be governed by Eqs. 45.35(a) and 45.36(a):

$$Z_t = Z = \frac{40 \times 10^6}{(14+1) - \frac{14}{2}(1 - 0.85) \times 1} = 2.8674 \times 10^6 \text{ mm}^3$$

and

$$Z_b = Z = \frac{40 \times 10^6}{(17+1) - \frac{14}{2}(1 - 0.85) \times 3} = 2.6936 \times 10^6 \text{ mm}^3$$

Hence adopt $Z = \frac{1}{6} B D^2 = 2.8674 \times 10^6 \text{ mm}^3$

Taking $B = 150$ mm, $D = \sqrt{\frac{2.8674 \times 10^6 \times 6}{150}} \approx 338.7$ mm;

Adopt $D = 340$ mm

Actual $Z = \frac{1}{6} \times 150(340)^2 = 2.89 \times 10^6 \text{ mm}^3$; $A = 150 \times 340 = 51000 \text{ mm}^2$

$y_t = y_b = y = 170$ mm

$$\frac{r^2}{y} = \frac{I}{Ay} = \frac{Z}{A} = \frac{D}{6} = 56.7 \text{ mm.} \quad ; \quad \frac{y}{r^2} = \frac{1}{56.7} \approx 0.01765.$$

**2. Determination of $e$:** The maximum value of $e$ is given by solving (*i.e.* dividing) Eqs. 45.31 (Line $T_1$) and 45.34 (Line $B_2$). The solution is given by Eq. 45.38.

$$\frac{e - D/6}{e + D/6} = \frac{\lambda(M_d + p_{ti} \cdot Z)}{(M_d + M_l) - p_t Z} \qquad \ldots(45.38)$$

$$\frac{e - 56.7}{e + 56.7} = \frac{0.85\,(15 \times 10^6 + 1 \times 2.89 \times 10^6)}{(15 + 40)\,10^6 - 1 \times 2.89 \times 10^6} = 0.2918.$$

From which $e = 103.4$ mm.

Let us assume average size of duct = 60 mm. Providing 40 mm clear cover to the duct, available

$$e = \frac{340}{2} - \frac{60}{2} - 40 = 100 \text{ mm. Hence adopt } e = \mathbf{100 \text{ mm}}.$$

**3. Determination of $P_i$:** Since the available value of $e$ is lesser than that given by *Eq. 45.38*, $P_i$ is obtained either from *Eq. 45.34* (simplified to *Eq. 45.39*) or by *Eq. 45.33* [simplified to *Eq. 45.39(a)*]. From *Eq. 45.39 (Eq. 45.34)*: Line $B_2$

$$P_i = \frac{(M_d + M_l) - p_t \cdot Z}{\lambda \left( e + \dfrac{D}{6} \right)} = \frac{(15 + 40)\,10^6 - 1 \times 2.89 \times 10^6}{0.85\,(100 + 56.7)} \approx 391230 \text{ N}$$

From *Eq. 45.39(a) (Eq. 45.33)* : Line $T_2$

$$P_i = \frac{(M_d + M_l) - p_c \cdot Z}{\lambda \left( e - \dfrac{D}{6} \right)} = \frac{(15 + 40)\,10^6 - 14 \times 2.89 \times 10^6}{0.85\,(100 - 56.7)} \approx 395050 \text{ N}$$

Hence adopt $P_i = \mathbf{395050 \text{ N}}$.

**4. Alternative solution for $e$ and $P_i$:** Alternatively, $e$ and $P_i$ can be obtained graphically from plotting *Eqs. 45.31* to *45.34*.

**(a) Line $T_1$** [*Eq. 45.31(a)*]

$$\frac{A}{P_i} \geq \frac{\dfrac{ey}{r^2} - 1}{\dfrac{M_d}{Z} + p_{ti}} \quad \text{or} \quad \frac{A}{P_i} \geq \frac{0.01765\,e - 1}{\dfrac{15 \times 10^6}{2.89 \times 10^6} + 1} \geq \frac{0.01765\,e - 1}{6.19} \qquad \ldots(\text{Line } T_1)$$

For $e = 0$, $\quad \dfrac{A}{P_i} = -\dfrac{1}{6.19} = -16.15 \times 10^{-2}$;

For $\dfrac{A}{P_i} = 0$, $\quad e = \dfrac{1}{0.01765} = 56.7$ mm

**(b) Line $B_1$** [*Eq. 45.32(a)*]

$$\frac{A}{P_i} \geq \frac{1 + \dfrac{ey}{r^2}}{\dfrac{M_d}{Z} + p_{ci}} \quad \text{or} \quad \frac{A}{P_i} \geq \frac{1 + 0.01765\,e}{\dfrac{15 \times 10^6}{2.89 \times 10^6} + 17} \geq \frac{1 + 0.01765\,e}{22.19} \qquad \ldots(\text{Line } B_1)$$

For $e = 0$, $\quad \dfrac{A}{P_i} = -\dfrac{1}{22.19} = -4.51 \times 10^{-2}$; For $\dfrac{A}{P_i} = 0$, $e \approx \dfrac{1}{0.01765} \approx -56.7$ mm

**(c) Line [$T_2$** [*Eq. 45.33(a)*]

$$\frac{A}{P_i} \geq \frac{\lambda \left(1 - \dfrac{ey}{r^2}\right)}{p_c - \dfrac{M_T}{Z}} \quad \text{or} \quad \frac{A}{P_i} \geq \frac{0.85(1 - 0.01765\,e)}{14 - \dfrac{(15+40)10^6}{2.89 \times 10^6}} \geq \frac{0.015\,e - 0.85}{5.03} \qquad \ldots(\text{Line } T_2)$$

For $e = 0$, $\quad \dfrac{A}{P_i} = -\dfrac{0.85}{5.03} = -16.89 \times 10^{-2}$;

For $\dfrac{A}{P_i} = 0$, $\quad e = \dfrac{0.85}{0.015} = +56.7$ mm

**(d) Line $B_2$** [Eq. 45.34(a)]

$$\dfrac{A}{P_i} \geq \dfrac{\lambda\left(\dfrac{ey}{r^2}+1\right)}{\dfrac{M_T}{Z}-p_t} \quad \text{or} \quad \dfrac{A}{P_i} \geq \dfrac{0.85\,(0.01765\,e+1)}{\dfrac{(15+40)\,10^6}{2.89\times 10^6}-1} \geq \dfrac{0.015\,e+0.85}{18.031} \quad \text{...(Line } B_2\text{)}$$

For $e \approx 0$, $\quad \dfrac{A}{P_i} = 0.85/18.031 \approx 4.71 \times 10^{-2}$.

For $\dfrac{A}{P_i} = 0$, $\quad e = -0.85/0.015 \approx -56.7$ mm

Fig. 45.51 shows the plotted positions of all the four lines. Lines $B_1$ and $B_2$ nearly coincide, though $B_2$ is slightly higher than $B_1$. The maximum value of $e$, at the intersection of $B_2$ and $T_1$ comes out to be 104 mm. However, available $e = 100$ mm, corresponding to which value of $A/P_i$ comes out to be $12.90 \times 10^{-2}$. Hence,

$$P_i = \dfrac{A}{12.90 \times 10^{-2}}$$

$$= \dfrac{51000}{12.90 \times 10^{-2}} \approx 395350 \text{ N}.$$

**5. Design of reinforcement**

Allowable stress in steel after losses
$\quad = 0.6 \times 1500 = 900$ N/mm²

Allowable stress in steel before losses
$\quad = 900/0.85 \approx 1059$ N/mm²

Initial prestressing force $= P_i = 395050$

Use 7 mm wires. Area of steel
$\quad = 395050/1059 \approx 373$ mm²

No. of 5 mm Φ bars
$$= \dfrac{373}{\dfrac{\pi}{4}(5)^2} \approx 19$$

Hence provide two cables of 10 wires each and use the Freyssinet system.

Actual $A'_{st} = 20 \times \dfrac{\pi}{4}(5)^2 \approx 392.7$ mm²

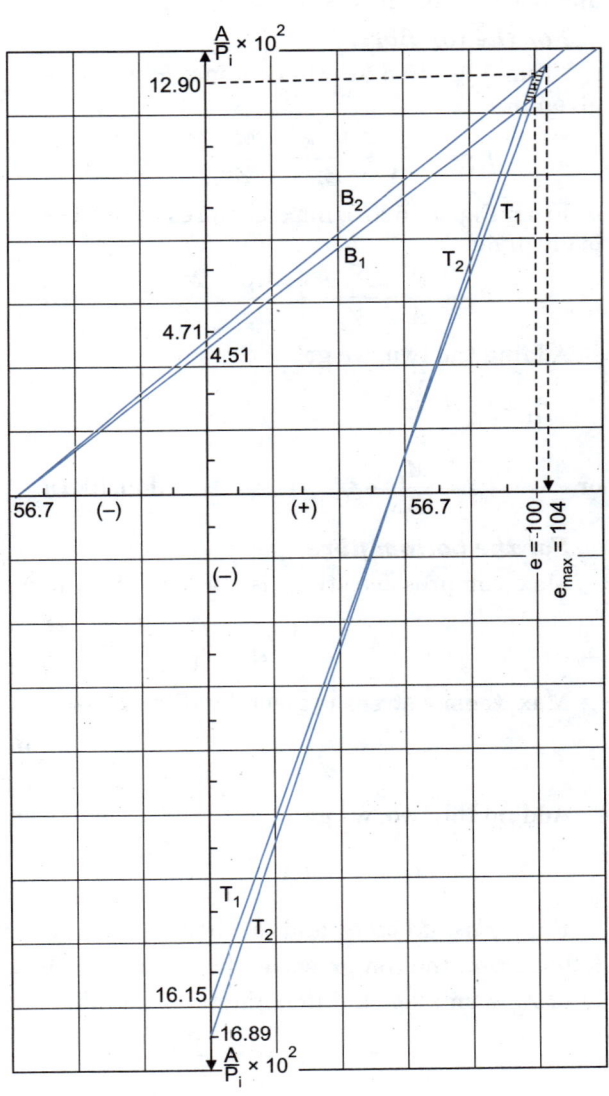

Fig. 45.51

## 45.15. ALTERNATIVE DESIGN PROCEDURE

The design procedure discussed in § 45.12 is cumbersome, since it involves simultaneous solution of four condition equations, as suggested by Magnel. A simplified design procedure is given below. The procedure takes into account two stress conditions, after the losses have taken place:

(1) Stresses due to $P$ and $M_d$ (2) Stresses due to $P$, $M_d$ and $M_l$

It is justified to neglect the initial stress conditions due to $P_i$ and $M_d$ since higher compressive stress ($p_{ci}$) is normally permitted.

(a) **Determination of section modulus:**
Fig. 45.52(b), (c) and (d) show the stress distribution due to $P$, $P.e$ and $M_d$, while Fig. 45.52(e) shows the combined effect due to these (first condition). Let the top fibre have tensile stress $f_{t1}$ and bottom fibre has compressive stress $f_{c1}$. Fig 45.52(h) shows the final stress distribution due to $P$, $M_d$ and $M_l$ in which the top fibre has compressive stress $f_{c2}$ and bottom fibre has tensile stress $f_{t2}$.

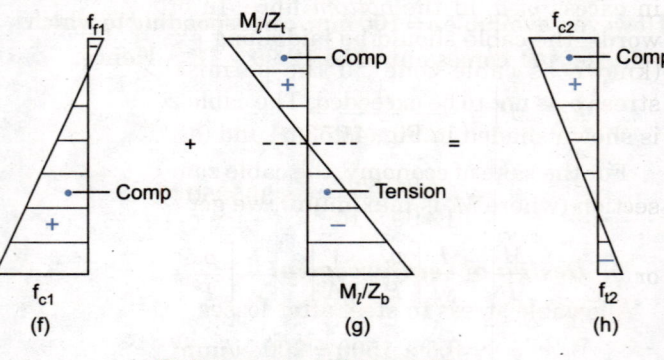

**For the top fibre:**
From Fig. 45.52(e), max. tensile stress is given by

$$f_{t1} = -\frac{P}{A} + \frac{P.e}{Z_t} - \frac{M_d}{Z_t} \qquad ...(i)$$

From Fig. 45.52(h), max. compressive stress is given by

$$f_{c2} = \frac{P}{A} - \frac{P.e}{Z_t} + \frac{M_d}{Z_t} + \frac{M_l}{Z_t} \qquad ...(ii)$$

Adding the two, we get

$$f_{t1} + f_{c2} = \frac{M_l}{Z_t}$$

or $\qquad f_{c2} = \dfrac{M_l}{Z_t} - f_{t1} \qquad ...[45.49(a)]$

**For the bottom fibre**
Max. compressive stress is given by [Fig. 45.52(e)]

$$f_{c1} = \frac{P}{A} + \frac{P.e}{Z_t} - \frac{M_d}{Z_b} \qquad ...(iii)$$

Max. tensile stress is given by [Fig. 45.52(h)]

$$f_{t2} = -\frac{P}{A} - \frac{P.e}{Z_b} + \frac{M_d}{Z_b} + \frac{M_l}{Z_b} \qquad ...(iv)$$

Adding the two, we get

$$f_{c1} + f_{t2} = \frac{M_l}{Z_b} \quad \text{or} \quad f_{c1} = \frac{M_l}{Z_b} - f_{t2} \qquad ...[45.49(b)]$$

From Eqs. 45.49(a) and (b), it is evident that if $f_{t1}$ and $f_{t2}$ are limited to equal permissible value of $p_t$ (often zero), the compressive fibre stress will be maximum at the top or bottom, depending upon whether $Z_t$ or $Z_b$ is smaller. Let maximum permissible compressive stress be $p_c$. Then,

$$p_c = \frac{M_l}{Z_{min}} - p_t \qquad ...(45.49)$$

From the above equation, the section modulus can be found.

**(b) Determination of Prestress Force (P):** For a given prestress force, the cable eccentricities must be properly determined so as to limit tensile stress to permissible value $p_t$. Let $e_1$ and $e_2$ be the cable eccentricities to limit the tensile fibre stresses under condition 1 [Fig. 45.52(e)] and condition 2 [Fig. 45.52(h)], to permissible value $p_t$.

$$\therefore \quad f_{t1} = -\frac{P}{A} + \frac{P \cdot e_1}{Z_t} - \frac{M_d}{Z_t} = p_t \qquad \ldots(i)$$

Re-arranging the terms, we get

$$e_1 = \frac{M_d}{P} + \frac{I}{y_t}\left[\frac{p_t}{P} + \frac{1}{A}\right] \qquad \ldots[45.50(a)]$$

and

$$f_{t2} = \frac{M_d}{Z_b} + \frac{M_l}{Z_b} - \frac{P}{A} - \frac{P \cdot e_2}{Z_b} = p_t \qquad \ldots(iv)$$

Re-arranging the terms we get

$$e_2 = \frac{M_d + M_l}{P} - \frac{I}{y_b}\left[\frac{p_t}{P} + \frac{1}{A}\right] \qquad \ldots[45.50(b)]$$

It will be seen from *Eq. 45.50(a)* that an increase of eccentricity above $e_1$ will give a tensile stress in excess of $p_t$ in the *top* fibre. Similarly, from *Eq. 45.50(b)*, a decrease in eccentricity below $e_2$ will give a tensile stress in excess of $p_t$ in the *bottom* fibre. In order words, the cable should be laid along a zone (known as cable zone), if the permissible stress $p_t$ is not to be exceeded. The cable zone is shown shaded in Fig. 45.53(a) and (b).

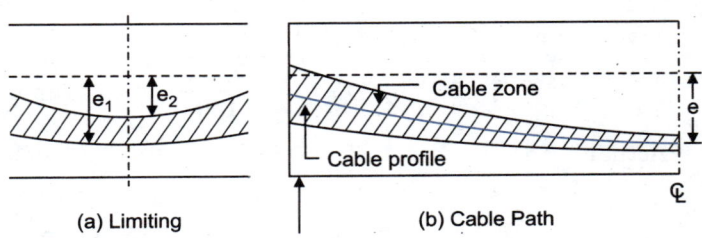

(a) Limiting     (b) Cable Path

**Fig. 45.53.** Cable Path.

For the sake of economy, the cable zone should be limited to a minimum. Equating $e_1$ and $e_2$ at the critical section (where $M_l$ is maximum), we get

or $$\frac{M_l}{P} - \frac{I}{y_b}\left[\frac{p_t}{P} + \frac{1}{A}\right] = \frac{I}{y_t}\left[\frac{p_t}{P} + \frac{1}{A}\right] \quad \text{or} \quad \frac{M_l}{P} \cdot \frac{y_b \cdot y_t}{I\left(\frac{p_t}{P} + \frac{1}{A}\right)} = y_b + y_t = D$$

or $$\left(p_t + \frac{P}{A}\right) = \frac{y_b \cdot y_t \, M_l}{ID}$$

From which $$P = A\left[\frac{y_b \, y_t}{ID} M_l - p_t\right] \qquad \ldots(45.51)$$

**(c) Determination of eccentricity:** Re-arranging *Eq. 45.51*,

$$\frac{I}{y_t} = \frac{M_l \cdot y_b}{D\left(\frac{P}{A} + p_t\right)} = \frac{M_l \cdot y_b}{PD\left(\frac{1}{A} + \frac{p_t}{P}\right)}$$

Substituting this in *equation 45.50(a)* and noting that $e_1 = e_2 = e$ at the critical section, we get

$$e = \frac{M_d}{P} + \frac{M_l \cdot y_b}{P \cdot D} = \frac{1}{P}\left[M_d + \frac{M_l \cdot y_b}{D}\right] \qquad \ldots(45.52)$$

If the available eccentricity is less than the one given by *Eq. 45.52*, the available value of $e$ should be re-substituted in *Eq. 45.52* to get the revised prestressing force. Alternatively, a large section may be provided and both $P$ and $e$ may be determined.

Thus the position of the cable at the critical section is determined. It will not, in general, be possible to achieve zero cable-zone (*i.e.* $e_1 = e_2 = e$) elsewhere. Thus the designer can fix the path of the cable at his discretion, keeping in mind the friction losses and the end-shear requirements.

Knowing $P$ and estimating the losses (or assuming the losses), the initial prestressing force ($P_i$) can be determined.

Thus, $$P_i = P + \text{Losses} = \frac{P}{\lambda}$$

The above method is extremely useful in designing unsymmetrical sections, specially T-section and unsymmetrical I-section.

### Design: Example 45.17. *Rectangular Beam*

Redesign the rectangular beam of example 45.16. Check the stresses in the section.

**Solution:** Given: $p_c = 14$ N/mm$^2$; $p_t = 1$ N/mm$^2$

**1. Determination of section modulus:** The minimum section modulus is given *Eq. 45.49*:

$$p_c = \frac{M_l}{Z_{min}} - p_t \quad \text{or} \quad Z_{min} = \frac{M_l}{p_c + p_t}$$

or $$\frac{BD^2}{6} = \frac{40 \times 10^6}{14 + 1} \approx 2666667 \text{ mm}^3$$

Let $B = 150$ mm.

∴ $$D = \sqrt{\frac{6 \times 2666667}{150}} \approx 327 \text{ mm. Adopt } D = 330 \text{ mm}$$

Actual $Z = \frac{1}{6} \times 150 (330)^2 = 2722500$ mm$^3 \approx 2.722 \times 10^6$

$y_t = y_b = y = 165$ mm; $I = 2722500 \times 165 = 449.21 \times 10^6$
$A = 150 \times 330 = 49500$ mm$^2$

**2. Determination of tendon force:** From *Eq. 45.51*

$$P = A\left[\frac{y_b \cdot y_t}{ID} M_l - p_t\right] = A\left[\frac{M_l}{2Z} - p_t\right] = 49500\left[\frac{40 \times 10^6}{2 \times 2.722 \times 10^6} - 1\right] = 314203 \text{ N}.$$

**3. Determination of eccentricity e:** From *Eq. 45.52*

$$e = \frac{1}{P}\left[M_d + \frac{M_l \cdot y_b}{D}\right] = \frac{1}{314203}\left[15 \times 10^6 + \frac{40 \times 10^6 \times 165}{330}\right] = 111.4 \text{ mm}$$

Let us assume average size of duct = 60 mm. Providing 40 mm clear cover to the duct, available $e = \frac{330}{2} - \frac{60}{2} - 40 = 95$ mm. This is much less than the required one. If we adopt this value, corresponding $P$ will be high, giving rise to high initial compressive stress.

Hence let us increase the depth from 330 to 350 mm.

Available $$e = \frac{350}{2} - \frac{60}{2} - 40 = 105 \text{ mm}$$

∴ $y_t = y_b = \frac{D}{2} = \frac{350}{2} = 175$ mm; $A = 150 \times 350 = 52500$ mm$^2$

$Z = \frac{1}{6} \times 150 (350)^2 = 3.0625 \times 10^6$ mm$^3$; $I = 3.0625 \times 10^6 \times 175 = 535.94 \times 10^6$ mm$^4$

Revised $P = \frac{1}{105}\left[15 \times 10^6 + \frac{40 \times 10^6 \times 175}{350}\right] = 333333$ N

Assuming losses @ 15%, we get: $P_i = \frac{333333}{0.85} \approx \mathbf{392157 \text{ N}}$

**4. Check for stresses:** *Initial condition*

$$f = \frac{P_i}{A} \mp \frac{P_i \cdot e}{Z} \pm \frac{M_d}{Z} = \frac{392157}{52500} \mp \frac{392157 \times 105}{3.0625 \times 10^6} \pm \frac{15 \times 10^6}{3.0625 \times 10^6} = 7.47 \mp 13.45 \pm 4.90$$

∴ $f_{(top)} = -1.08$ N/mm² (tension) > 1; $f_{(bottom)} = 16.02$ N/mm² < 17

Hence the stresses are within safe limits, except that the tensile stress is slightly higher.

**Final condition**

$$f = \frac{P}{A} \mp \frac{P \cdot e}{Z} \pm \frac{M_d}{Z} \pm \frac{M_l}{Z} = \frac{333333}{52500} \mp \frac{333333 \times 105}{3.0625 \times 10^6} \pm \frac{(15+40)10^6}{3.0625 \times 10^6}$$

$$= 6.35 \mp 11.43 \pm 17.96$$

∴ $f_{(top)} = 12.88$ N/mm² < 14 (safe); $f_{(bottom)} = -0.18$ N/mm² < 1 (safe)

## 45.16. SHEAR AND DIAGONAL TENSION

In the case of prestressed concrete, the whole section of the beam is effective since the section remains uncracked. Hence the calculation of shear resistance of a prestressed member at safe working condition is more straight forward than for ordinary R.C. sections. The shear stress at any plane of a pre-stressed concrete section is given by the usual formula:

$$q = \frac{Q}{I\,b} A_a \, \overline{y}_a \qquad ...(45.53)$$

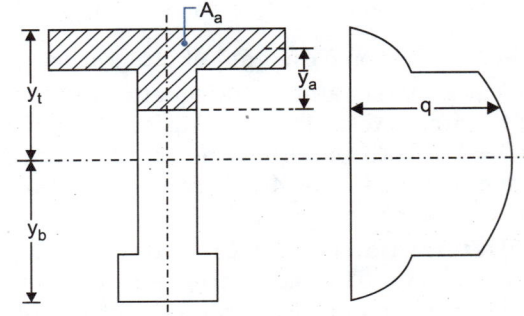

Fig. 45.54

where  $Q$ = net shear force.
$I$ = moment of inertia of the section.
$b$ = width of the section

$A_a$ and $\overline{y}_a$ are as marked in Fig. 45.54.

The shear force at any section is decreased by the vertical component ($V$) of the cable tension, and the net shear force $Q$ used in *Eq.* 45.53 will be equal to shear force due to dead and live loads minus $V$. The up-swept cable has thus the balancing effect, and this effect can be determined if we first decide the cable slope. It is also necessary to check for shear stresses by combining dead load S.F. with $V$.

**Determination of cable slope** (*shear cable*): The available latitude of cable eccentricity were determined in *Eq.* 45.50 (*a*) and (*b*) from considerations of bending. At the end section, where B.M. is zero, the cable eccentricity can be $\pm \dfrac{D}{6}$ without producing tensile stress.

Let us now determine the cable path from considerations of shear. Let,

$Q_d$ = S.F. due to dead load;   $Q_l$ = S.F. due to live load.
$Q$ = Net shear force;
$V$ = Vertical component of prestressing force = $P \tan \theta$.

When live load and dead load act, net S.F. = $Q_d + Q_l - V$

When only dead load acts, net S.F. = $V - Q_d$
(*i.e.* there is reversal of direction of S.F.)

From design considerations, the cable slope should be such that at the ends, where the S.F. is maximum, the net shear when dead load is acting is equal to the net shear when both dead and live loads are acting.

$$Q_d + Q_l - V = V - Q_d$$

or $\qquad V = Q_d + \dfrac{Q_l}{2} \qquad ...(45.54)$

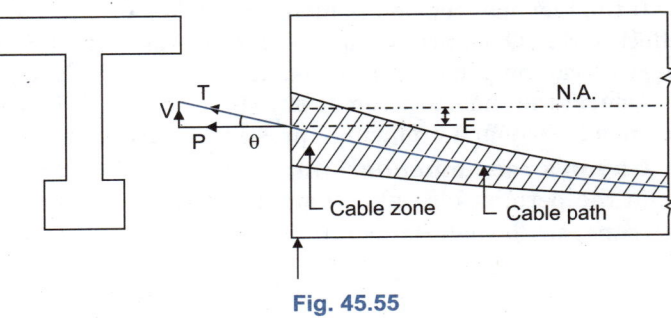

Fig. 45.55

Substituting $V = P \tan \theta = T \sin \theta$ we get

$$\tan \theta = \frac{1}{P}\left(Q_d + \frac{Q_l}{2}\right) \qquad ...[45.55(a)]$$

or
$$\sin\theta = \frac{1}{T}\left(Q_d + \frac{Q_l}{2}\right) \qquad ...[45.55(b)]$$

Let us assume that the conditions for worst shear occur with the same loading that cause worst moment conditions. If $e$ is the cable eccentricity, and $x$ is the distance along the member.

$$\tan\theta = \frac{de}{dx} = \frac{V}{P}; \quad V = P\frac{de}{dx} \qquad ...(45.56)$$

For simple beam theory, $Q = \dfrac{dM}{dx}$

Hence from Eq. 45.54,
$$V = \frac{d}{dx}\left(M_d + \frac{M_l}{2}\right) \qquad ...(45.57)$$

Equating Eqs. 45.56 and 45.57 and intergrating

$$e = \frac{1}{P}\left(M_d + \frac{M_l}{2}\right) + E \qquad ...(45.58)$$

where $E$ is some constant of integration.

The above equation is the equation for shear cables, which is similar to Eq. 45.52 found from considerations of bending, except for the addition of constant $E$. For rectangular and other symmetrical section, where $y/D = 1/2$, both the equations are identical with value of $E = 0$. For non-symmetrical section, the value of $E$, as marked in Fig. 45.55, can be determined by substituting the value of $e$, $M_d$ and $M_l$ as calculated for bending at the critical section.

**Determination of principal tensile stresses:** Thus the cable path (*i.e.* inclination $\theta$), $V$ and net shear $Q$ are known. The shear stress $q$ can then be computed from Eq. 45.53 and it may then be combined with the bending stress to compute the principal tension, using the following expression;

$$\sigma_t = \frac{1}{2}\sqrt{f_c^2 + 4q^2} - f_c \qquad ...(45.59)$$

where  $\sigma_t$ = principle tensile stress
$f_c$ = bending stress at the point considered.
$q$ = shear stress at the point considered.

From study of Eq. 45.59 as well Fig. 45.56, it is clear that the principal tension (diagonaly tension) will be less than the shear stress of $q$. This is one of the great advantages of prestressing techniques. The elimination of principal tension, or its reduction to a small value enables the adoption of a thin web. The safe value of diagonal tension in prestressed concrete may be taken as equal to $0.4\sqrt{f_{su}}$. Under ultimate load conditions, it should not exceed $1.9\sqrt{f_{cu}}$.

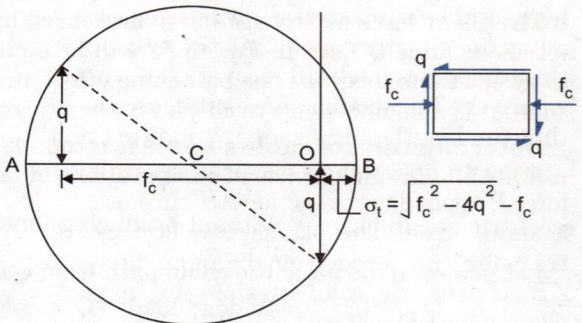

Fig. 45.56. Mohr circle.

If the diagonal tension is more than the permissible one, shear stirrups are provided to bear the total effective S.F. $Q_r$ allowing usual working stress in steel. Some times, *vertical prestressing* is done to neutralise the vertical component of principal tension. Generally, the I-section of a beam is thickened into a rectangular section, at the ends, to accommodate the anchorage. This further reduces the shear stresses. Hence diagonal tension is seldom a problem in prestressed concrete beams.

**Guyon's recommendations:** Guyon recommends that even though the principal tension may be less than the permissible value, nominal shear reinforcement should be provided. The *pitch* of the vertical stirrups may be determined from the following expression:

$$p = \frac{Z\cot\phi\, A_w\, f_w}{Q_d + Q_l - T\sin\theta} \qquad ...(45.60)$$

where  $p$ = pitch of the stirrups.

$$\phi \approx \tan^{-1}\left(\frac{2q}{f}\right)$$

$q$ = shear stress in concrete
$f_c$ = compression stress in concrete
$A_w$ = area of stirrup
$f_w$ = yield stress in shear reinforcement
$Z$ = lever arm = $I/A_a \cdot \bar{y}_a$
$A_a \cdot \bar{y}_a$ = first moment of area, above the level of the point at which shear stress is to be calculated, about the N.A.
$I$ = moment of inertia of the section.

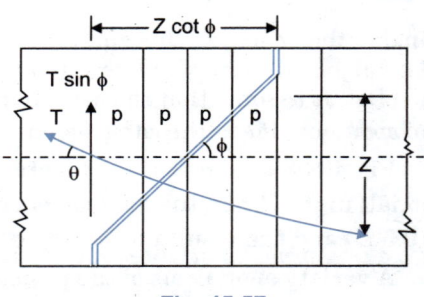

Fig. 45.57

The above expression is based on the assumption of the formation of a full crack, as shown in Fig. 45.57. Guyon recommended the following values of tan φ and $f_w$

| $\sigma_t$ or $p$ | $tan\ \phi$ | $f_w$ |
|---|---|---|
| $\sigma_t < p_c/20$ | $2\ q/f_c$ | 210 N/mm² |
| $\sigma_t > p_c/20$ | $2\ q/f_c$ | 140 N/mm² |
| $q > f_c/2$ | 1 | 140 N/mm² |

where  $\sigma_t$ = principal tensile stress in concrete
$p_c$ = permissible compressive stress in concrete
$f_c$ = actual compressive stress in concrete.

## 45.17. STRESSES AT ANCHORAGE

The anchorage devices in post-tensioned beams bear against the ends of the concrete, resulting in very high local bearing stresses. The stresses in the immediate vicinity of the anchorage are different from those at sections farther away. At the ends, the prestressing force is transmitted over a small area ; the force then disperses itself till it covers the whole section. The length of the beam between the end and the section where the full section shares the force is known as *lead-in-zone* or *anchorage zone*. It is generally believed that the length of the zone is equal to the depth of the beam.

Fig. 45.58(*a*) shows a prestressing force $P$ applied to the section $AB$, through a small area. The stress trajectories transfer the force from bearing face $AB$ across the lead-in-zone to face $CD$, where the stress $p = P/BD$. These lines of forces (or stress trajectories) can be considered to be acting as individual struts which are first convex and then concave towards the centroidal axis. Because of curvatures of the stress trajectories, transverse stresses are induced as shown in Fig. 45.58(*b*) and (*c*). These stresses are compressive near the face $AB$, and tensile at a greater distance from it. Fig. 45.58(*d*) shows the isobars of transverse stresses, known from photoelastic studies. The distribution of stresses near anchorage is discontinuous, and besides longitudinal compressive stresses, transverse tensile stresses, normal to the beam axis also occur. The transverse tensile stresses as high as 0.3 to 0.5 $p$ may occur deep

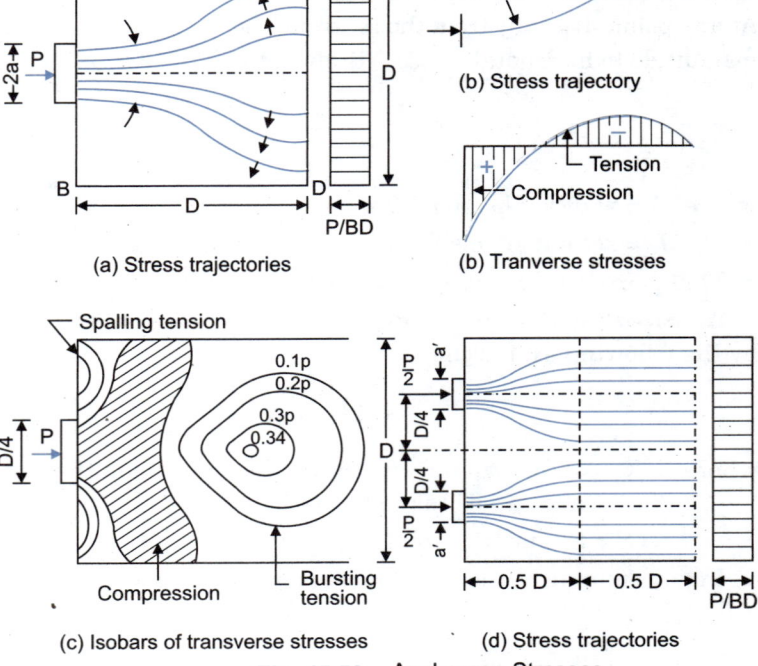

Fig. 45.58. Anchorage Stresses.

inside the concrete; this zone is known as *bursting zone*. Also, the corners of the end section are found to be subjected to high spalling tensile stresses the tensile stress may be as high as 0.68 p. It is, therefore necessary to provide one layer of reinforcement very near to the end (just behind the anchorage), to take care of *spalling zone* and another layer of reinforcement away from the anchorage to take care of bursting zone.

Fig 45.58(d) shows the advantage of distributing the total prestressing force P over a large number of small units. When the force on each bearing surface is P/2, the tensile stresses act only on half the length (0.5 D) and the resultant forces are halved.

A variety of analysis of the problem of stresses at anchorage have been developed. However, the method suggested by Prof. Magnel is given here.

**Magnel's Method:** Magnel assumes that the local bearing stresses are dissipated over a length of the beam equal to its depth. Within the anchorage zone, any elemental unit is acted on by three stresses (i) horizontal stress $f_x$, (ii) vertical stress $f_y$ and (iii) shear q, shown in Fig. 45.59(a).

**1. Horizontal stress $f_x$:** Consider Fig. 45.59(b) in which it is assumed that the compression from the anchorage device is assumed to be spread out at 45°, cutting the top and bottom faces at H and G respectively. Upto vertical section FG, the stressed zone is symmetrical about the cable, while beyond this the stressed zone has greater depth above the cable and small depth below the cable. Thus, line EKLM is the *line of centroid* of the stressed zone, and can be easily located. At any vertical section RS, of height h (i.e. RS = h), the cable eccentricity $e_v$ is thus known. At any point, distant y from the centroid, the magnitude to horizontal stress $f_x$ is given by

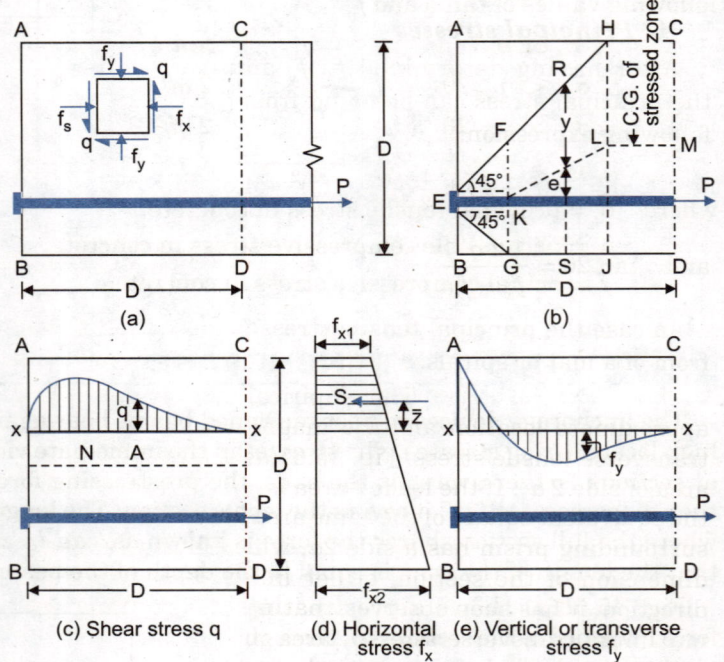

Fig. 45.59. Magnel's Method.

$$f_x = \frac{P}{b.h} \pm \frac{P.e_v.y}{I_v} \qquad ...(45.61)$$

where  b = width of beam section;

$I_v$ = second moment of area of zone under stress.

Thus horizontal stress distribution across any vertical plane RS can be found.

**2. Shear stress (q):** The distribution of shear stress, at any horizontal plane XX [Fig. 45.59(c)] is given by the following expression:

$$q = K_q \frac{S}{bD} \qquad ...(45.62)$$

where   $K_q$ = *shear stress factor*, which depends upon the distance of the point from the end section AB, and can be taken from Fig. 45.6.

S = Summation of all horizontal stresses ($\Sigma f_x$) above plane XX [Fig. 45.59(d)].

In determining S, the amount of opposing cable force should be deduced if it occurs above the plane XX.

### 3. Vertical stress (or transverse stress) $f_y$

Fig. 45.59(e) shows the distribution of vertical stress on any horizontal plane $XX$. Magnel gives the following expression for $f_y$:

$$f_y = K_y \cdot \frac{M}{bD^2} \qquad ...(45.63)$$

where  $K_y$ = vertical stress factor, to be taken from Fig. 45.60

$M = S \times \bar{z}$

$\bar{z}$ = distance of centroid $S$ above $XX$.

### 4. Principal stresses

After having determined $f_x$, $f_y$ and $q$, the principal stress can be found from the following expression:

$$\sigma = \frac{1}{2}(f_x + f_y) \pm \frac{1}{2}\sqrt{(f_x - f_y)^2 + 4q^2}$$

and  $\tan 2\theta = \dfrac{2q}{f_x - f_y}$

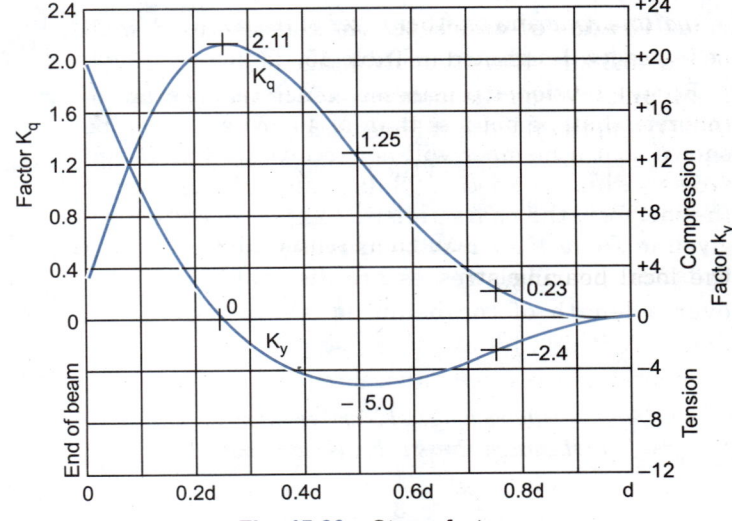

Fig. 45.60. Stress factors.

In case the principal tensile stress is more than the allowable, mild steel reinforcement is provided in from of a mat or spirals.

**Empirical Method:** The empirical method gives the approximate location and the magnitude of maximum transverse tensile stress. Fig. 45.61(a) shows the loaded area of side $2a_1$. If the loaded area is circular, the side of the equivalent square of the same area may be found. The surrounding prism has a side $2a$, which is the smallest dimension of the section, either in width or in depth direction. it has been observed that positions of zero and maximum transverse stresses area independent on the ratio $a_1/a$, through its magnitude is dependent on this ratio.

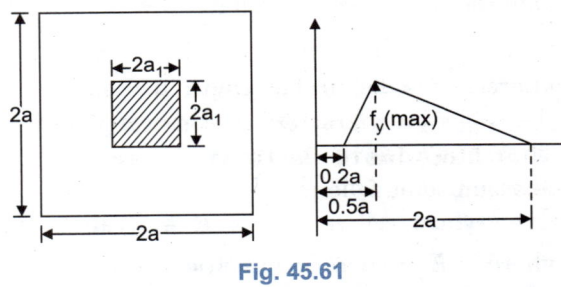

Fig. 45.61

Fig. 45.61(b) shows the distribution of the transverse tensile stress. It increases from *zero* at $0.2a$ to a maximum at $0.5\,a$, and then again decreases to zero at $2a$ from the end section. The maximum transverse tensile stress is given by

$$f_{y(max)} = f_c\left[0.98 - 0.825\,\frac{a_1}{a}\right]; \quad \text{where } f_c = \frac{P}{(2a_1)^2} \qquad ...(45.64)$$

Total tensile force,  $F_t = P\left[0.48 - 0.4\left(\dfrac{a_1}{a}\right)\right] \qquad ...(45.65)$

Reinforcement is provided in the form of mats, helices or links, in the region $0.2\,a$ to $1.0\,a$. The area of steel is equal to $F_t/1400$ cm$^2$, where $F_t$ is the total tensile force in kg. If the concrete can take some tensile stress $p_t$,

$$\text{Corrected } F_t = F_t\left[1 - \left(\frac{p_t}{f_{y(max)}}\right)^2\right] \qquad ...(45.66)$$

Generally, a rectangular mesh of bars of 6 to 10 mm Φ is provided. In addition to this, a light spiral of 6 mm Φ at a pitch of 5 cm is provided round individual anchorages.

## 45.18. INDIAN STANDARD CODE RECOMMENDATIONS (IS : 1343-1980)

### 1. Concrete

(a) **Grades of concrete:** The concrete used shall be in grades designated in Table 45.12.

For pre-tensioned prestressed concrete, the grade of concrete shall be not less than M 40. Where it can be shown that a member will not receive its full design stress within a period of 28 days after the casting of the member, the characteristic compressive strength given in Table 45.12 may be increased by multiplying by factors given below:

**TABLE 45.12.** Grades of Concrete

| Grade Designation | Specified characteristic compressive strength at 28 days $f_{ck}$ (N/mm²) |
|---|---|
| M 30 | 30 |
| M 35 | 35 |
| M 40 | 40 |
| M 45 | 45 |
| M 50 | 50 |
| M 55 | 55 |
| M 60 | 60 |

| Minimum Age of Member when Full Design Stress is Expected (months) | Age Factor |
|---|---|
| 1 | 1.00 |
| 3 | 1.10 |
| 6 | 1.15 |
| 12 | 1.20 |

(b) **Tensile strength of concrete:** When the designer wishes to use an estimate of flexural strength from the compressive strength, the following formula may be used:

$$f_{cr} = 0.7\sqrt{f_{ck}} \text{ N/mm}^2 \qquad \ldots(45.67)$$

where   $f_{cr}$ = flexural strength in N/mm²
$f_{ck}$ = characteristic compressive strength of concrete (N/mm²)

(c) **Modulus of elasticity:** In absence of test data, the modulus of elasticity for structural concrete may be assumed as follows:

$$E_c = 5000\sqrt{f_{ck}} \qquad \ldots(45.68)$$

where   $E_c$ = short-term static modulus of elasticity, in N/mm².

(d) **Shrinkage strain:** The shrinkage of concrete depends upon the constituents of concrete, size of the member and environmental conditions. For a given environment, the shrinkage of concrete is most influenced by the total amount of water present in the concrete at the time of mixing and, to a lesser extent by the cement content. In the absence of test date, the approximate value of shrinkage strain for design shall be assumed as follows:

For pre-tensioning = 0.0003;   For post-tensioning = $\dfrac{0.0002}{\log_{10}(t+2)}$

where   $t$ = age of concrete at transfer, in days.

The value of shrinkage strain for design of post-tensioned concrete may be increased by 50 per cent in dry atmospheric conditions, subjected to a maximum value of 0.0003. For the calculation of deformation of concrete at some stage before the maximum shrinkage is reached, it may be assumed that half of the shrinkage taken place during the first month and that about three-quarters of the shrinkage takes place in the first six months after commencement of drying.

(e) **Creep of concrete:** Creep of concrete depends, in addition to the factors listed in (d) above, on the stress in concrete, age at loading and the duration of loading. As long as the stress in concrete does not exceed one third of the characteristic compressive strength, creep may be assumed to be proportional to the stress.

In the absence of experimental data and detailed information of the effect of variables, the ultimate creep strain may be estimated from the following values of crep coefficient (that is, ultimate creep strain/elastic strain at the age of loading).

| Age of loading | Creep coefficient |
|---|---|
| 7 days | 2.2 |
| 28 days | 1.6 |
| 1 year | 1.1 |

The ultimate creep strain estimated as per above clause does not include the elastic strain.

For the calculation of deformation at some stage before the total creep is reached, it may be assumed that about half the total creep takes place in the first month after loading and that about three quarters of the total creep takes place in the first six months after loading.

### 2. Prestressing steel

(a) *Type of steel and modulus of elasticity*

The prestressing steel shall be any one of the following types:

| Type of Steel | Modulus of elasticity $E$ (kN/mm²) |
|---|---|
| (i) Plain hard-drawn wire conforming to IS : 1785 (Part I)-1966 and IS 1785 (Part II)-1967 | 210 |
| (ii) Cold drawn indented wire conforming to IS : 6003-1970 | 210 |
| (iii) High tensile steel bar conforming to IS : 2090-1962 | 200 |
| (iv) Uncoated stress relieved strand conforming to IS : 6006-1900 | 195 |

(b) *Untensioned steel:* Reinforcement used as untensioned steel shall be any of the following:

(i) Mild steel and medium tensile steel bars conforming to IS : 432 (Part I)-1966.

(ii) Hot rolled deformed bars conforming to IS : 1139-1966.

(iii) Cold twisted bars conforming to IS : 1786-1979.

(iv) Hard drawn steel wire fabric conforming to IS : 1566-1967

(c) *Characteristic strength ($f_p$):* For prestressing steel, the characteristic strength shall be assumed as the minimum ultimate tensile stress. For steel bars, $f_p$ = 1400 N/mm². For steel wires, $f_p$ = 1500, 1600, 1650 and 1700 N/mm² depending upon the diameter of wires (higher value for thinner wire). Table 45.11 gives the minimum ultimate tensile strength of tensile steel wires.

(d) *Cover:* In pre-tensioned work, the cover of concrete measured from the outside of the pretressing tendon shall be atleast 20 mm. In post-tensioned work, where cables and large-sized bars are used, the maximum clear cover from sheathing/duct shall be at least 30 mm or the size of the cable or bar whichever is bigger. Where prestressed concrete members are located in aggressive environment, the cover specified above shall be increased by 10 mm.

(e) *Spacing:* In the case of single wires used in pre-tension system, the minimum clear spacing shall not be less than greater of the following:

(i) 3 times diameter of wire, and  (ii) $1\frac{1}{3}$ times the maximum size of aggregate.

In the case of cables or large bars, the minimum clear spacing (measured between sheathings/ducts, wherever used) shall not be less than greater of the following:

(i) 40 mm

(ii) Maximum size of cable or bar

(iii) 5 mm plus maximum size of aggregate.

(f) *Grouped cables:* Cables or ducts may by grouped together in group of not more than four, as shown in Fig. 45.62. The minimum clear spacing between groups of cables or ducts of grouped cables shall be greater of the following:

(i) 40 mm and  (ii) 5 mm plus maximum size of aggregates.

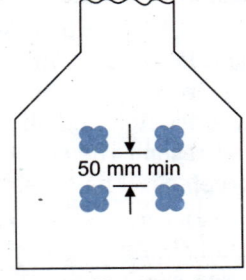

**Fig. 45.62.** Spacing of groups of cables.

The vertical distance between groups shall not be less than 50 mm (See Fig. 45.62).

(g) *Minimum longitudinal reinforcement:* (i) A minimum longitudinal reinforcement of 0.2 per cent of the total concrete area shall be provided in all cases except in the case of pretensioned units of small sections. This reinforcement may be reduced to 0.15 per cent in the case of high yield strength deformed reinforcement. The percentage of steel provided, both tensioned and untensioned taken together, should be sufficient so that when the concrete in the precompressed tensile zone cracks, the steel is in a position to take up the additional tensile stress transferred on to it by the cracking of the adjacent fibres of concrete and a sudden failure is avoided.

(*ii*) When the depth of web exceeds 50 cm, longitudinal distribution reinforcement not less than 0.5 per cent of the area of the web shall be provided on each face. The spacing of the individual bars of such reinforcement shall not exceed 20 cm.

(*h*) **Transverse reinforcement:** (*i*) The amount and spacing of transverse reinforcement shall be governed by shear and torsion requirements. It is however, desirable to provide transverse reinforcement in the web when the web is thin and cables are located in the web.

(*ii*) In case of all members subjected to dynamic loading, webs shall be provided with transverse reinforcement not less than 0.3 per cent of the sectional area of the web in plan. This percentage of reinforcement may be reduced to 0.2 percent in members where the depth of the web is not more than four times the thickness of the web. These values may be reduced to 0.2 and 0.15 percent respectively when high strength reinforcement is used.

(*iii*) In case of members not subjected to dynamic loading, reinforcement shall be provided when the depth of the web is more than 4 times the thickness. Such reinforcement shall not be less than 0.1 percent of the sectional area of the web in plan. The reinforcement shall be spaced at a distance not greater than the clear depth of web and the size of such reinforcement shall be as small as possible.

(*iv*) Reinforcement in the form of links or helix shall be provided perpendicular to the line of heavy compression or shock loading to resist the induced tensile stresses.

(*i*) **Deductions for Prestressing Tendons**

In calculating areas, centroid and moment of inertia of cross-section, deduction for prestressing tendons shall be made as follows:

(*i*) In the case of pre-tensioned members, where the prestressing tendons are single wires distributed on the cross-section or strands of wires of relatively small cross-sectional area, allowances for the prestressing tendons need not be made. Where allowance is made, it shall be on the basis of $(m - 1)$ times the area of the prestressing tendons, $m$ being the modular ratio.

(*ii*) In the case of post-tensioned members, deductions shall invariably be made for prestressing tendons, cable ducts or sheaths and such other openings whether they are formed longitudinally or transversely. These deductions need not however, be made for determining the effect of loads applied after the ducts, sheaths or openings have been grouped or filled with concrete. Where such deductions are not made, a transformed area equivalent to $(m - 1)$ times the area of the prestressing tendon shall be taken in calculation.

**Note:** Modular ratio $m$ shall be calculated as $E_s/E_c$. Wherever necessary, creep effects shall also be taken into consideration.

**3. Minimum Initial Prestress:** At the time of initial tensioning, the maximum tensile stress $f_{pi}$ ($f_{si}$) immediately behind the anchorages shall not exceed 80 per cent of the ultimate tensile strength of the wire or bar or strand.

**4. Losses in Prestress:** While assessing the stresses in concrete and steel during tensioning operations and later in service, due regard shall be paid to all losses and variations in stress resulting form creep of concrete, shrinkage of concrete, relaxation of steel, shortening (elastic deformation) of concrete at transfer, and friction and slip of anchorages. In computing the losses in prestress when untensioned reinforcement is present, the effect of tensile stresses developed by the untensioned reinforcement due to shrinkage and creep shall be considered.

(*a*) **Loss of prestress due to creep of concrete:** The loss of prestress due to creep of concrete under load shall be determined for all the permanently applied loads including the prestress.

The creep loss due to live load stresses, erection stress and other stresses or short duration may be ignored. The loss of prestress due to creep of concrete is obtained as the product of modulus of elasticity of the prestressing steel and the ultimate creep strain of the concrete fibre integrated along the line of centre of gravity of the prestressing steel over its entire length. The total creep strain during any specific period shall be assumed for all practical purposes, to be creep strain due to sustained stress equal to the average of the stresses at the beginning and end of the period.

(b) **Loss of prestress due to shrinkage of concrete:** The loss of prestress due to shrinkage of concrete shall be the product of the modulus of elasticity of steel and the shrinkage strain of concrete.

(c) **Loss of prestress due to relaxation of steel:** The relaxation losses in prestressing steels vary with type of steel, initial prestress, age and temperature, and therefore shall be determined from experiments. When experimental values are not available, the relaxation losses may be assumed as given in Table 45.13.

**TABLE 45.13.** Relaxation Losses for Prestressing Steel (At 1000 H at 27° C)

| Initial stress | Relaxation loss ($N/mm^2$) |
|---|---|
| $0.5 f_p$ | 0 |
| $0.6 f_p$ | 35 |
| $0.7 f_p$ | 70 |
| $0.8 f_p$ | 90 |

**Note:** $f_p$ is the characteristic strength of prestressing steel. For tendons at higher temperatures or subjected to large lateral loads, greater relaxation losses as specified by engineer-in-charge shall be allowed for. No reduction in the value of relaxation losses should be made for a tendon with a load equal to or greater than the relevant jacking force that has been applied for a short time prior to the anchoring of the tendon.

(d) **Loss of prestress due to shortening of concrete:** This loss is proportional to the modular ratio and initial prestress in the concrete and shall be calculated as below, assuming that the tendons are located at their centroid:

(i) For pretensioing, the loss of prestress in the tendons at transfer shall be calculated on a modular ratio basis using the stress in the adjacent concrete.

(ii) For members with post-tensioned tendons which are not stressed simultaneously, there is progressive loss of prestress during transfer, due to the gradual application of the prestressing forces. This loss of prestress should be calculated on the basis of half the product of the stress in the concrete adjacent to the tendons averaged along their lengths and the modular ratio. Alternatively, the loss of prestress may be exactly computed on the sequence of tensioning.

(e) **Loss of prestress due to slip anchorage:** Any loss of prestress which may occur due to slip of wires during anchoring or due to the strain of anchorage shall be allowed for in the design. Loss due to slip in anchorage is of special importance with short member and the necessary additional elongation should be provided for at the time of tensioning to compensate for this loss.

(f) **Loss of prestress due to friction:** The design shall take into consideration all losses in prestress that may occur during tensioning due to friction between the prestressing tendons and the surrounding concrete or any flexure attached to the steel or concrete.

For straight or moderately curved structures with curved or straight cables, the value of prestressing force $P_x$ at a distance $x$ metres from tensioning end and acting in the direction of the tangent to the curve of the cable shall be calculated as below:

$$P_x = P_0 \cdot e^{-(\mu\alpha + kx)} \qquad \ldots(45.69)$$

where  $P_0$ = Prestressing force in the prestressed steel at the tensioning end acting in the direction of the tangent to the curve of the cable.

$\alpha$ = cumulative angle in radians through which the tangent to the cable profile has turned between any two points under consideration.

$\mu$ = coefficient of friction in curve; unless otherwise proved by tests, $\mu$ may be taken as:

0.55 for steel moving on smooth concrete.

0.30 for steel moving on steel fixed to duct, and

0.25 for steel moving on lead.

$k$ = coefficient of wave effect varying from $15 \times 10^{-4}$ to $50 \times 10^{-4}$ per metre.

**Note 1:** Expansion of equation for $P_x$ for small values of $(\mu\alpha + kx)$ may be
$$P_x = P_0 (1 - \mu\alpha - k \cdot x)$$

**Note 2:** In circular constructions, where circumferential tendons are tensioned by jacks, values of $\mu$ for calculating friction may be taken as :

0.45 for steel moving in smooth concrete,

0.25 for steel moving on steel bearers fixed to the concrete, and

0.10 for steel moving on steel rollers.

**Note 3:** The effect of reverse friction shall be taken into consideration in such cases where the initial tension applied to a prestressing tendon is partially released and action of friction in the reverse direction causes an alteration in the direction of stress along the length of the tendon.

5. **Bearing Stress:** (*i*) On the areas immediately behind external anchorages the permissible unit bearing stress on the concrete, after accounting for all losses due to relaxation of steel, elastic shortening, creep of concrete, slip and/or seating of anchorages etc., should not exceed $0.48 f_{ci} \sqrt{\dfrac{A_{br}}{A_{pun}}}$ or $0.8 f_{ci}$ whichever is smaller,

where, $f_{ci}$ is the cube strength at transfer,

$A_{br}$ is the bearing area and $A_{pun}$ is the punching area.

(*ii*) During tensioning, the allowable bearing stress specified in (*i*) may be increased by 25 per cent, provided that this temporary value does not exceed $f_{ci}$.

(*iii*) The bearing stress specified in (*i*) and (*ii*) for permanent and temporary bearing stress may be increased suitabily if adequate hoop reinforcement is provided at anchorages.

(*iv*) When the anchorages are embedded in concrete, the bearing stress shall be investigated after accounting for the surface friction between the anchorage and the concrete.

(*v*) The effective punching area shall generally be the contact area of the anchorage devices which, if circular in shape, shall be replaced by a square of equivalent area. The bearing area shall be the maximum area of that portion of the member which is geometrically similar and concentric to the effective punching area.

(*vi*) Where a number of anchorages are used, the bearing area $A_{br}$ shall not overlap. Where there is already a compressive stress prevailing over the bearing area, as in the case of anchorage placed in the body of a structure, the total stress shall not exceed the limiting values specified in (*i*), (*ii*) and (*iii*). For stage stressing of cables, the adjacent unstressed anchorages shall be neglected while determining the bearing area.

### 6. Structural Design: Limit State Method

(*a*) **Limit state design:** The structural design shall be based on limit state concept. In this method of design, the structure shall be designed to withstand safely all loads liable to act on it throughout the life; it shall also satisfy the serviceability requirements, such as limitation on serviceability requirements before failure occurs is called as '*limit state*'. The aim of the design is to achieve acceptable probabilities that the structure will not become unfit for use for which it is intended, that is, it will not reach a limit state.

The design should be based on *characteristic values* for material strength and applied loads. The *design values* are derived from the characteristic values through the use of partial safety factors, one for the material strengths and the other for loads. The values of partial safety factors for material strength ($\gamma_m$) should be taken as 1.5 for concrete and 1.15 for steel. It should be noted that $\gamma_m$ values are already incorporated in the equation and tables given in IS : 1343-1980. The partial safety factors ($\gamma_f$) for loads shall be the same as given in the previous chapter on limit state design.

(*b*) **Limit state of collapse:** The limit state of collapse of the structure or part of the structure could be assessed from rupture of one or more critical sections and from buckling due to elastic or plastic instability (including the effects of sway where appropriate) or overturning. The resistance to bending, shear, torsion and axial load at every section shall not be less than appropriate value at that section produced by the probable most unfavourable combination of loads on the structure using the appropriate partial safety factor.

(*c*) **Limit state of serviceability: Deflection.**

The deflection shall be limited to the following:

(*i*) The final deflection, due to all loads including the effects of temperature, creep and shrinkage and measured from the as-cast level of the supports of floors, roofs and all other horizontal members, should not normally exceed span / 250.

(*ii*) The deflection including the effects of temperature creep and shrinkage occurring after erection of partitions and the application of finishes should not normally exceed span / 350 or 20 mm whichever is less.

(iii) If finishes are to be applied to prestressed concrete members, the total upward deflection should not exceed span/300, unless uniformity of camber between adjacent units can be ensured.

### (d) Limit state of serviceability: Cracking

Cracking of concrete shall not affect the appearance of durability of structure. The criteria of limit state of cracking for the *three types* of prestressed concrete members shall be as follows :

(i) **For type 1**: No tensile stress.

(ii) **For type 2**: Tensile stresses are allowed but no visible cracking.

(iii) **For type 3**: Cracking is allowed, but should not affect the appearance or durability of the structure; the acceptable limits of cracking would vary with the type of structure and environment and will vary between wide limits and the prediction of absolute maximum width is not possible.

**Note:** For design of type 3 members, as a guide, the following paragraph may be regarded as reasonable limits: The flexural tensile stress at any section of the structure, both at transfer and under the most unfavourable combination of design loads, shall satisfy the criteria for the corresponding type of structure.

(e) **Limit state of serviceability: Maximum compression:** The compressive stresses both at transfer and under design loads shall be limited to the values given in paragraph follows, for all types of structures.

### 7. Limit State of Collapse: Flexure

(a) **Assumption:** Design for the limit-state of collapse in flexure shall be based on the following assumptions:

(i) Plane sections normal to the axis remain plane after bending.

(ii) The maximum strain in concrete at the outermost compression fibre is taken as 0.0035 in bending.

(iii) The relationship between the compressive stress distribution in concrete and the strain in concrete may be assumed to be rectangle, trapezoid, parabola or any other shape which results in prediction of strength in substantial agreement with the result of tests. An acceptable stress-strain curve is given in Fig. 32.2 (Chapter 32). For design purposes the compressive strength of concrete on the structure shall be assumed to be 0.67 times the characteristic strength. The partial safety factor $\gamma_m = 1.5$ shall be applied in addition to this.

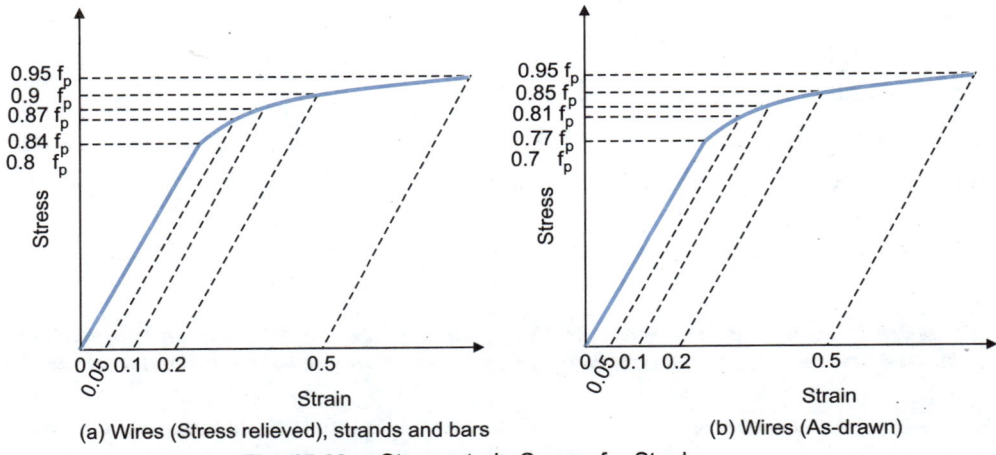

Fig. 45.63. Stress-strain Curves for Steel.

For the stress block is shown in Fig. 33.5(a) for which the parameters may be adopted as follows:

Area of stress block = $0.36 f_{ck} \cdot x_u$

Depth of centre of compressive force from the extreme fibre in compression = $0.42 x_u$.

(iv) The tensile strength of the concrete is ignored.

(v) The stresses in bonded prestressing tendons, whether initially tensioned or untensioned, and in additional reinforcement are derived from the representative stress-strain curve for the type of steel used given the manufactures of typical curves given in Fig. 45.63 for prestressing tendons and in Fig. 33.3 and 33.4 for reinforcement of Fe415 and Fe500 grades. For design purposes, the partial safety factors $\gamma_m$ equal to 1.15 shall be applied.

(vi) If tendons are unbonded in post-tensioned members, the stress in tendons may be obtained from a rigorous analysis or from tests.

### (b) Design Formulae

(i) The moment of resistance of rectangular section or T-section in which neutral axis lies within the flanges be obtained as follows:

$$M = f_{pu} \cdot A_p (d - 0.42 \, x_u) \qquad \qquad ...(45.70)$$

where  $M$ = moment of resistance of the section,
$f_{pu}$ = ultimate tensile stress in the tendons,
$f_p$ = characteristic strength of prestressing steel,
$A_p$ = area of pretensioning tendons,
$d$ = effective depth, and
$x_u$ = neutral axis depth.

For pretensioned members and for post-tensioned members with effective bond between the concrete and tendons values of $f_{pu}$ and $x_u$ are given in Table 45.14. The effective prestress after all losses should not be less than $0.45 \, f_p$ where $f_p$ is the characteristic strength of prestressing steel.

**TABLE 45.14.** Conditions at the Ultimate Limit State for Rectangular Beams with Pre-Tensioned Tendons or with Post-Tensioned Tendons Having Effective Bond

| $\dfrac{A_s \, f_p}{bd \, f_{ck}}$ | Stress in tendon as a proportion of the design strength $f_{pu}/0.87 f_p$ | | Ratio of the depth of N.A. to that of the centroid of the tendon in the tension zone $x_u/d$ | |
|---|---|---|---|---|
| | Pre-tensioning | Post-tensioning with effective bond | Pre-tensioning | Post-tensioning with effective bond |
| 0.025 | 1.0 | 1.0 | 0.054 | 0.054 |
| 0.05 | 1.0 | 1.0 | 0.109 | 0.109 |
| 0.10 | 1.0 | 1.0 | 0.217 | 0.217 |
| 0.15 | 1.0 | 1.0 | 0.326 | 0.316 |
| 0.20 | 1.0 | 0.95 | 0.435 | 0.414 |
| 0.25 | 1.0 | 0.90 | 0.542 | 0.488 |
| 0.30 | 1.0 | 0.85 | 0.655 | 0.558 |
| 0.40 | 0.9 | 0.75 | 0.783 | 0.653 |

The Post-tensioned members with unbonded tendons, the values of $f_{pu}$ and $x_u$ are given in Table 45.15.

**TABLE 45.15.** Conditions at the Ultimate Limit State for Post-Tensioned Rectangular Beams Having Unbonded Tendons

| $\dfrac{A_p \, f_p}{bd \cdot f_{ck}}$ | Stress in tendon as a proportion to the effective pressure $f_{pu}/f_p$ for values of $l/d$ $\left(\dfrac{\text{Effective span}}{\text{Effective depth}}\right)$ | | | Ratio of depth of N.A. to that of the centroid of the tendon in the tension zone, $x_u/d$ for values of $l/d$ $\left(\dfrac{\text{Effective span}}{\text{Effective depth}}\right)$ | | |
|---|---|---|---|---|---|---|
| | 30 | 20 | 10 | 30 | 20 | 10 |
| 0.025 | 1.23 | 1.34 | 1.45 | 0.10 | 0.10 | 0.10 |
| 0.05 | 1.21 | 1.32 | 1.45 | 0.16 | 0.16 | 0.18 |
| 0.10 | 1.18 | 1.26 | 1.45 | 0.30 | 0.32 | 0.36 |
| 0.15 | 1.14 | 1.20 | 1.36 | 0.44 | 0.46 | 0.52 |
| 0.20 | 1.11 | 1.16 | 1.47 | 0.56 | 0.58 | 0.64 |

(ii) For flanged section in which the N.A. lies outside the flange, the moment of resistance shall be determined using assumption in 7(a) above.

8. **Limit State of Collapse: Compression:** Prestressed concrete compression members in framed structures, where the mean stress in concrete section imposed by tendons is less than 2.5 N/mm², may be analysed as reinforced concrete compression members in accordance with IS : 456 ; in other cases, special literature may be referred to.

9. **Limit State of Collapse: Tension:** Tensile strength of tension member shall be based on the design strength (0.87 times characteristic strength of prestressing tendons) and the strength developed by any additional reinforcement. The additional reinforcement may usually be assumed to be acting at its design stress (0.87 times characteristic strength of reinforcement); in special cases it may be necessary to check the stress in the reinforcement using strain compatibility.

10. **Limit State of Collapse: Shear:** The ultimate shear resistance of concrete along $V_c$ should be considered at both sections uncracked and cracked in flexure, and lesser value taken, and if necessary, shear reinforcement provided.

(a) **Section uncracked in flexure:** The ultimate shear resistance of a section uncracked in flexure, $V_c = V_{co}$ is given by

$$V_{co} = 0.67\, b\, D\, \sqrt{f_t^2 + 0.8\, f_{cp}\, f_t} \qquad \ldots(45.71)$$

where  $b$ = breadth of the members which for $T$, $I$ and $L$ beams should be replaced by breadth of rib $b_w$.

$D$ = overall depth of the member.

$f_t$ = maximum principal stress given by $0.24\sqrt{f_{ck}}$ taken as positive, where is the characteristic compressive strength of concrete.

$f_{cp}$ = compressive stress at centroidal axis due to prestress taken as positive.

In flanged members where the centroidal axis occurs in the flange, the principal tensile stress should be limited to $0.24\sqrt{f_{ck}}$ at the intersection of the flanged web; in this calculation, 0.8 of the stress due to prestress at this intersection may be used, in calculating $V_{co}$. For a section uncracked in flexure and with inclined tendons or vertical prestress, the component of prestressing force normal to the longitudinal axis of the member may be added to $V_{co}$.

(b) **Section cracked in flexure:** The ultimate shear resistance of a section cracked in flexure $V_c = V_{cr}$ is given by:

$$V_{cr} = \left(1 - 0.55\,\frac{f_{pe}}{f_p}\right) \xi_c \cdot bd + M_0\,\frac{V}{M} \qquad \ldots(45.72)$$

where  $f_{pe}$ = effective prestress after losses have occurred, which shall not be put greater than $0.6 f_p$.

$f_p$ = characteristic strength of prestressing steel.

$\xi_c$ = ultimate shear stress capacity of concrete, obtained from Table 45.16.

$b$ = breadth of members, which, for flanged sections shall be taken as the breadth of the web, $b_w$.

$d$ = distance from extreme compression-fibre to the centroid of the tendons at the section considered.

$M_0$ = moment necessary to produce zero stress in concrete at the depth given by

$$M_0 = 0.8\, f_{pt}\, I/y \qquad \ldots(45.73)$$

where  $f_{pt}$ = is the stress due to prestress only at depth $d$ and distance $y$ from the centroid of the concrete section which has second moment of area $I$, and

$V$ and $M$ = shear force and bending moment respectively at the section considered due to ultimate loads.;

$V_{cr}$ should be taken not less than $0.1\, bd\, \sqrt{f_{ck}}$

The value of $V_{cr}$ calculated at a particular section may be assumed to be constant for a distance equal to $d/2$, measured in the direction of increasing moment, from that particular section.

For a section cracked in flexure and with inclined tendons, the component of prestressing force normal to the longitudinal axis of the member should be ignored.

**TABLE 45.16.** Design Shear Strength of Concrete $\xi_c$ N/mm²

| $100 \dfrac{A_p}{bd}$ | Concrete grade | | |
|---|---|---|---|
| | M 30 | M 35 | M 40 and above |
| 0.20 | 0.37 | 0.37 | 0.38 |
| 0.50 | 0.50 | 0.50 | 0.51 |
| 0.75 | 0.59 | 0.59 | 0.60 |
| 1.00 | 0.66 | 0.67 | 0.68 |
| 1.25 | 0.71 | 0.73 | 0.74 |
| 1.50 | 0.76 | 0.78 | 0.79 |
| 1.75 | 0.80 | 0.82 | 0.84 |
| 2.00 | 0.84 | 0.86 | 0.88 |
| 2.25 | 0.88 | 0.90 | 0.92 |
| 2.50 | 0.91 | 0.93 | 0.95 |
| 2.75 | 0.94 | 0.96 | 0.98 |
| 3.00 | 0.96 | 0.99 | 1.01 |

**Note:** $A_p$ is the area of prestressing tendon.

(c) **Shear reinforcement**

(i) When V, the shear force due to the ultimate loads, is less than $V_c$ the shear force which can be carried by the concrete, minimum shear reinforcement should be provided in the form of stirrups such that

$$\frac{A_{sv}}{b \cdot s_v} = \frac{0.4}{0.87 f_y} \qquad \ldots(45.74)$$

where  $A_{sv}$ = total cross-sectional area of stirrups legs effective in shear.

$b$ = breadth of the members which for T, I and L beams should be taken as the breadth of the rib, $b_w$.

$s_v$ = stirrup spacing along the length of the member.

$f_y$ = characteristic strength of stirrup reinforcement which shall not be taken greater than 415 N/mm².

However, shear reinforcement need not be provided in the following cases: (i) where V is less than 0.5 $V_c$, and (ii) in members of minor importance.

(ii) When V exceeds $V_c$ shear reinforcement shall be provided such that

$$\frac{A_{sv}}{s_v} = \frac{V - V_c}{0.87 f_y \cdot d_t} \qquad \ldots(45.75)$$

In rectangular beams, at both corners in tensile zone, a stirrup should pass around a longitudinal bar, a tendon or a group of tendons t having a diameter not less than the diameter of the stirrup. The depth $d_t$ is then taken as the depth from extreme compression fibre either to the longitudinal bars or to the centriod of the tendons which ever is greater.

The spacing of stirrpus along a member should not exceed 0.75 $d_t$ nor 4 times the web thickness for flanged members. When V exceeds 1.8 $V_c$ the maximum spacing should be reduced to 0.5 $d_t$. The lateral spacing of the individual legs of the stirrups provided at a cross-section should not exceed 0.75 $d_t$.

(iii) In no circumstances should the shear force V, due to ultimate loads, exceed the appropriate values given in Table 45.17 multiplied by b.d.

**TABLE 45.17.** Maximum Shear Stress

| Grade of Concrete | M 30 | M 35 | M 40 | M45 | M50 | M 55 and over |
|---|---|---|---|---|---|---|
| Maximum shear stress (N/mm²) | 3.5 | 3.7 | 4.0 | 4.3 | 4.6 | 4.8 |

# PRESTRESSED CONCRETE

## 11. Limit State of Serviceability: Deflection

### (a) Type 1 and type 2 members.

The *instantaneous deflection* (short-term deflection) due to design loads may be calculated using elastic analysis based on uncracked section and modulus of elasticity of concrete.

The total *long term deflection* due to the prestressing force, dead load and any sustained imposed load may be calculated using elastic analysis taking into account the effects of cracking and of creep and shrinkage. Due allowance shall be made for the loss of prestress after the period considered. The deflection should comply with limits given in clause 6(c) above.

### (b) Type 3 members.
Where the permanent load is less than or equal to 25 per cent of the design imposed load, the deflection may be calculated as in (i) above. When the permanent load is more than 25 per cent of the design imposed load, the vertical deflection limits of beams and the design imposed load, the vertical deflection limits of beam and slabs may generally be assumed to be satisfied provided that the span to effective depth ratio are not greater than values obtained as below:

For values of span to effective depth ratios for spans upto 10 m:

Cantilever: 7;     Simply supported: 20;     Continuous: 26

For spans above 10 m, the above values may be multiplied by 10/span in metres except for cantilever in which case deflection calculation should be made.

## 12. Limit State of Serviceability: Cracking

For the members made up of precast units, no tension shall be allowed at any stage at mortar or concrete joints. For a member which is free of joints, the tensile stress shall not exceed the values specified below for the 3 types of members:

(a) **Type 1.** No tensile stress.

(b) **Type 2.** The tensile stress shall not exceed 3 N/mm². However, where part of service loads is temporary in nature, this value may be exceeded by 1.5 N/mm² provided under the permancent component of the service load the stress remains compressive.

(c) **Type 3.** For type 3 members in which cracking is permitted, it may be assumed that the concrete section is uncracked, and that hypothetical tensile stresses exist at the maximum size of cracks. The hypothetical tensile stresses for use in these calculations for member with either pre-tensioned or post tensioned tendons are given in Table 45.18. The values in Table 45.18 shall be multiplied by the depth factors obtained from Fig. 45.64.

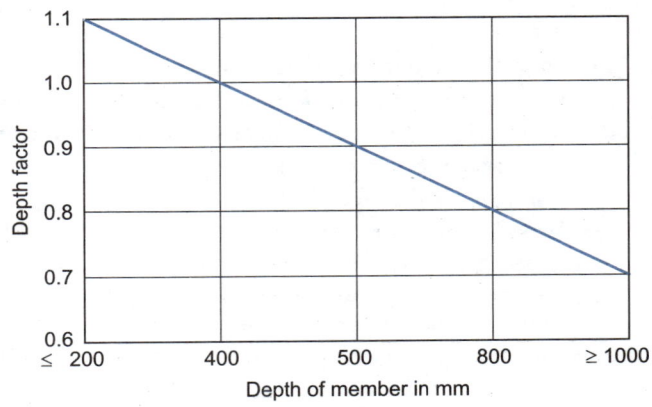

Fig. 45.64. Depth factors for tensile stresses for Types 3 Members.

**TABLE 45.18.** Hypothetical Flexural Tensile Stresses for Type 3 Members

| Type of Tendons | Limiting crack width (mm) | Stress of Concrete of Grade | | | | |
|---|---|---|---|---|---|---|
| | | M 30 | M 35 | M 40 | M 45 | M 50 & above |
| (i) Pre tensioned tendons | 0.1 | – | – | 4.1 | 4.4 | 4.8 |
| | 0.2 | – | – | 5.0 | 5.4 | 5.8 |
| (ii) Grouted post tensioned tendons | 0.1 | 3.2 | 3.6 | 4.1 | 4.4 | 4.8 |
| | 0.2 | 3.8 | 4.4 | 5.0 | 5.4 | 5.8 |
| (iii) Pre tensioned tendons distributed in the tensile zone and positioned close to the tension face of concrete. | 0.1 | – | – | 5.3 | 5.8 | 6.3 |
| | 0.2 | – | – | 6.3 | 6.8 | 7.3 |

**Note:** When additional reinforcement is distributed within the tension zone and positioned close to the tension face of concrete, the hypothetical tensile stresses may be increased by an amount which is proportioned to the cross-sectional areas of the additional reinforcement expressed as a percentage of the cross-sectional area of the concrete. For 1 per cent of additional reinforcement, the stress may be increased by 4 N/mm² for members with pre-tensioned and grouted post-tensioned tendons and by 3 N/mm² for other members. For other percentages of additional reinforcement the stresses may be increased in proportion excepting that total hypothetical tensile stress shall not exceed 0.25 times the characteristic compressive strength of concrete.

### 13. Limit State of Serviceability: Maximum Compression

**(a) Maximum Stress Under Service Conditions**

*(i) Compressive stress in flexure.* The maximum permissible compressive stress, prestress and service loads after deduction of the full losses in the specified prestress shall be determined by a straight line relation as shown in Fig. 45.65 but different stress limits shall apply to the concrete of the structure depending on whether if falls in a part of the section where the compressive stresses are not likely to increase in service (zone I) or in part of the section where the compressive stresses are likely to increase in service (zone II).

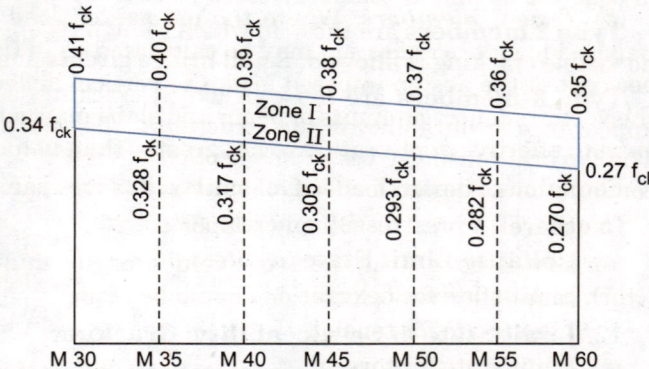

**Fig. 45.65.** Computation of Maximum Permissible Compressive Stresses on Flexure due to Final Prestress.

For zone I, the straight line relation of permissible stress shall be determined by straight line joining a point given by a permissible stress of $0.41 f_{ck}$ for concrete of grade M 30 to another point given by a permissible stress of $0.35 f_{ck}$ of concrete of grade M 60.

For zone II, the determining points of the graph shall be reduced to $0.34 f_{ck}$ and $0.27 f_{ck}$ respectively.

*(ii) Stress in direct compression.* Except in part immediately behind the anchorage, the maximum stress in direct compression shall be limited to 0.8 times the permissible stress obtained in *(i)* above.

**(b) Maximum Stress at Transfer**

*(i) Compressive stress in flexure.* The maximum permissible compressive stress due to bending and direct force at the time of transfer of prestress shall be determined from a graph in which a straight line joins a point given by $0.54 f_{ci}$ for a concrete of grade M 30 to a second point giving a permissible stress of $0.37 f_{ci}$ for concrete of grade M 60, [Fig. 45.66(a)]; $f_{ci}$ being cube strength of concrete at transfer which in no case shall be less than half the corresponding characteristic compressive strength of concrete. These values apply to post-tensioned work; for pre-tensioned work the variation represented by Fig. 45.66(b) will apply.

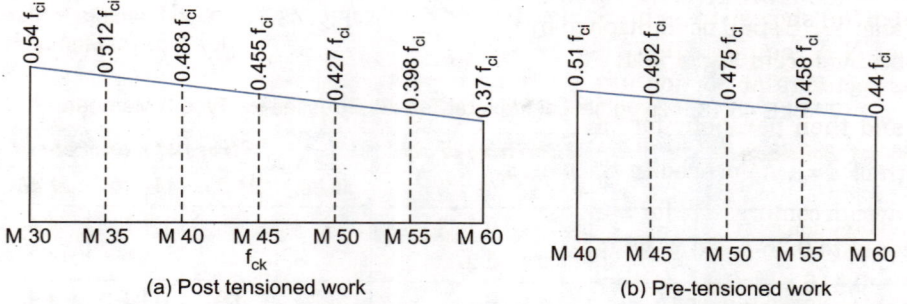

**Fig. 45.66.** Computation of Maximum Permissible Compressive Stress in Flexure at Transfer

**Note.** The strength of concrete at the time of transfer, $f_{ci}$ shall be established by tests carried out on cubes at the age of the concrete at transfer for bridges and such other major structures and in other cases, if more convenient, from straight line graph, joining the characteristic compressive strength of concrete and cube strength at 7 days. The transfer of prestress shall be made only after the concrete has attained a strength of atleast half the characteristic compressive strength of concrete.

(*ii*) *Stress in direct compression.* Except in the parts immediately behind the anchorages, the maximum stress in direct compression shall be limited to 0.8 times the permissible stress obtained in (*i*) above.

## 45.19. PROCEDURE FOR LIMIT STATE DESIGN

As mentioned in § 45.18, there are three types of prestressed concrete members: Type 1, Type 2 and Type 3.

**Type 1 members** are those in which no tensile stresses are permitted. The common examples of such members are the water retaining structures, nuclear structures, bridge girders etc. The compressive stress in concrete as well as tensile stresses in steel are also limited to specified values.

**Type 2 members** are those in which though tensile stress are allowed to certain permissible value but no visible cracking is allowed. Small bridge girders and certain building elements fall under this category.

**Type 3 members** are those in which cracking is allowed; however such cracking should not effect appearance or durability of the structure. Office and residential buildings etc. fall under these category. The flexural tensile stress at any section of the structure both at transfer and under most unfavourable combination of design loads shall satisfy the criteria for corresponding type of structure.

In general, a prestressed concrete structure is designed for the following limiting states:

1. **Collapse Limit State** (*or Strength limit state*)
    (*a*) Limit state for flexure      (*b*) Limit state for compression
    (*c*) Limit state for tension      (*d*) Limit state for shear
    (*e*) Limit state for torsion      (*f*) Limit state for bursting and bond.
2. **Serviceability Limit States**
    (*a*) Limit state for allowable stresses both for service conditions as well as transfer condition.
    (*b*) Limit state for deflection.      (*c*) Limit state for cracking.
3. **Limit State of Durability:**

Out of these, the *collapse limit state* (or strength limit state) are *common* to all the types of structures. The *serviceability limit state* are same for type 1 and type 2 structures but different for type 3 structures.

The partial safety factors for material strength are 1.5 for concrete and 1.15 for steel. The partial safety factor $\gamma_f$ for loads shall be the same as discussed in the chapter for limit state design. The design is carried out in the following steps.

**Step 1. Determination of design bending moment at collapse:** The superimposed dead load and live loads are generally given. The self weight of the beams is estimated on the basis of assumed dimensions of the beam. For this purpose, the width $b$ and depth $d$ of rectangular beams, simply supported at the ends, may be taken as $L/30$ and $L/15$ respectively. The design bending moment at collapse ($M_c$) is then computed, after applying partial safety factor for loads. $\gamma_f = 1.5$.

**Step 2. Design for limit state of flexure:** Fig. 45.67(*c*) shows stress block, having depth of N.A. equal to $x_u$. The stress block has rectangular portion upto depth of $\frac{3}{7} x_u$ and then parabolic for the remaining depth of $\frac{4}{7} x_u$. As proved in the chapter 33, the design compressive force $C_u$ in concrete is equal to $0.36 f_{ck} \cdot b \, x_u$ which acts at a depth $= 0.416 \, x_u \approx 0.42 \, x_u$ below the top fiber.

The neutral axis can be located from the strain compatibility [Fig. 45.67(*b*)].

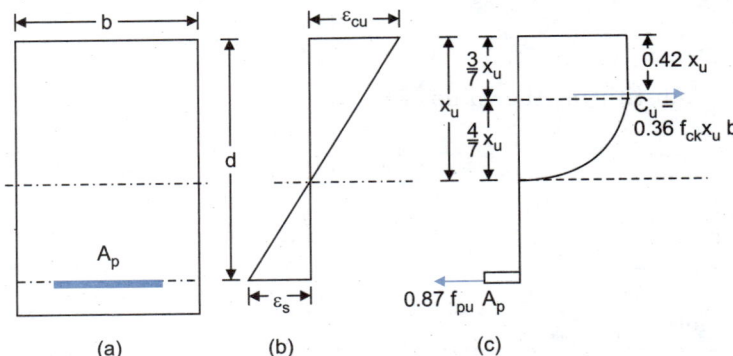

Fig. 45.67. Computation of Maximum Permissible Compressive Stresses on Flexure Due to Final Prestress.

$$\frac{\varepsilon_{cu}}{\varepsilon_{cu} + \varepsilon_s} = \frac{x_u}{d} \quad \text{or} \quad x_u = \frac{\varepsilon_{cu}}{\varepsilon_{cu} + \varepsilon_s} d$$

where  $\varepsilon_{cu}$ = strain in concrete at limit state = 0.0035
$\varepsilon_s$ = compatible strain in steel = $\varepsilon_y - \varepsilon_{sc}$
$\varepsilon_{sc}$ = effective prestrain in steel
$\varepsilon_y$ = total strain in steel

$$= 0.0020 + \frac{0.87 f_p}{E_s} \quad \text{(IS Code)} \quad \ldots(45.76)$$

Normally accepted value of $\varepsilon_{sc}$ = 0.004. Taking $E_s = 2 \times 10^5$ N/mm² and substituting the values in *Eq. 45.76*, we get

$$x_u = \frac{0.0035\, d}{0.0035 + \left(0.0020 + \frac{0.87 f_p}{\varepsilon_s}\right) - 0.004}$$

$$= \frac{0.0035\, d}{0.0015 + \frac{0.87 f_p}{2 \times 10^5}} = \frac{35\, d}{15 + 0.0435 f_p} \quad \ldots(45.77)$$

Now the design moment caused by the compressive force about the centre of steel is given by
$$M_u = C_u (d - 0.42\, x_u)$$

Substituting the values of $C_u$ and $x_u$, we get

$$M_u = 0.36 f_{ck} \cdot b\, \frac{35\, d}{15 + 0.0435 f_p} \left(d - 0.42 \times \frac{35\, d}{15 + 0.0435 f_p}\right)$$

or $$M_u = f_{ck}\, b\, d^2\, \frac{12.6}{15 + 0.0435 f_p}\left(1 - \frac{14.6}{15 + 0.0435 f_p}\right)$$

or $$M_u = R_u\, b\, d^2 \cdot f_{ck} \quad \ldots(45.78)$$

where  $$R_u = \frac{12.6}{15 + 0.0435 f_p}\left(1 - \frac{14.6}{15 + 0.0435 f_p}\right) \quad \ldots[45.78(a)]$$

The area of pretensioned steel can be obtained by equating the compressive and tensile forces:
$$C_u = T_u \quad \text{or} \quad 0.36 f_{ck} \cdot b\, x_u = 0.87 f_p \cdot A_p$$

where  $A_p$ = area of pretensioned steel;  $f_p$ = characteristic strength of pretensioned steel.

$$\therefore \quad A_p = \frac{0.36}{0.87}\, \frac{f_{ck}}{f_p} \cdot b\, x_u$$

Substituting the value of $x_u$, we get

$$A_p = \frac{0.36}{0.87}\, \frac{f_{ck}}{f_p} \cdot b\, \frac{35\, d}{15 + 0.0435 f_p} = \frac{14.49\, b\, d}{15 + 0.0435 f_p} \cdot \frac{f_{ck}}{f_p}$$

or $$\frac{A_p}{b\, d} = p_0\, \frac{f_{ck}}{f_p} \quad \ldots(45.79)$$

where  $$p_0 = \frac{14.49}{15 + 0.0435 f_p} = \frac{A_p\, f_p}{b\, d\, f_{ck}} \quad \ldots[45.79(a)]$$

Lever arm factor $j = 1 - 0.42\, \dfrac{x_u}{d} = 1 - 0.42 \times \dfrac{35\, d}{(15 + 0.0435 f_p)\, d} = 1 - \dfrac{14.6}{15 + 0.0435 f_p} \quad \ldots(45.80)$

The design coefficients $\dfrac{x_u}{d}$, $R_u$, $j$ and $p_0$ depend upon the characteristic strength $f_p$ of pretensioned steel. Their values for various values of $f_p$ are given in Table 45.19.

**TABLE 45.19. Design Coefficients**

| $f_p$ (N/mm²) | $\dfrac{x_u}{d}$ | $j$ | $R_u$ | $p_0$ |
|---|---|---|---|---|
| 1400 | 0.461 | 0.808 | 0.134 | 0.191 |
| 1500 | 0.436 | 0.818 | 0.128 | 0.181 |
| 1600 | 0.414 | 0.827 | 0.123 | 0.171 |
| 1700 | 0.393 | 0.836 | 0.118 | 0.163 |
| 1800 | 0.384 | 0.840 | 0.116 | 0.159 |
| 1900 | 0.358 | 0.850 | 0.110 | 0.148 |

*Eq. 45.79* is applicable for balanced sections. Type 1 and Type 2 members are likely to be over-reinforced, while Type 3 members are often under-reinforced in which the steel reaches the yield stress before the crushing in concrete is initiated. The condition is represented by

$$\frac{A_p}{b\,d} < p_0 \frac{f_{ck}}{f_p} \qquad \ldots[45.79(b)]$$

For over-reinforced section, we have

$$\frac{A_p}{b\,d} \geq p_0 \frac{f_{ck}}{f_p} \qquad \ldots[45.79(c)]$$

The amount of prestressing steel in a section is limited by the following equation:

$$\frac{A_p}{b\,d} \leq 0.24 \frac{f_{ck}}{f_p} \qquad \ldots(45.81)$$

For under-reinforced section, the moment of resistance is given by

$$M_u = f_{pu} \cdot A_p (d - 0.42\, x_u) \qquad \ldots(45.82)$$

where $f_{pu}$ = ultimate tensile stress in tendons = $\dfrac{\alpha \cdot f_p}{1.15} = \alpha\,(0.87) f_p$

$\alpha = \dfrac{f_{pu}}{0.87\, f_p}$ = coefficient in the range of 1 to 0.9 depending upon steel ratio

$\left(\dfrac{A_p}{b\,d} \cdot \dfrac{f_p}{f_{ck}}\right)$ and is given in Table 45.14.

For over-reinforced section, ultimate moment of resistance is limited to that one for balanced section. Thus, the area of steel in excess of the balanced proportion should be neglected.

For flanged section, if the N.A. lies in the flange (i.e. if $x_u \leq D_f$),

$$M_u = R_u\, b_f \cdot d^2 f_{ck} \qquad \ldots(45.83)$$

where $b_f$ = width of flange and $D_f$ = thickness of flange.

If $x_u \geq D_f$, the ultimate moment of resistance is given by

$$M_u = R_u\, b_w\, d^2 f_{ck} + 0.45\,(b_f - b_w)\, D_f\,(d - 0.5\, D_f) f_{ck} \qquad \ldots(45.84)$$

The area of steel is given by balancing the compressive and tensile forces:

$$0.87\, A_p \cdot f_{pu} = 0.36\, b_w\, x_u\, f_{ck} + 0.45\,(b_f - b_w)\, D_f \cdot f_{ck}$$

or

$$A_p = 1.15 \left[ 0.36\, b_w\, x_u + 0.45\,(b_f - b_w) D_f \right] \frac{f_{ck}}{f_{pu}} \qquad \ldots(45.85)$$

where $x_u = \dfrac{35\, d}{15 + 0.0435\, f_p}$ $\qquad \ldots(45.77)$

In *Eq. 45.85*, $b_f$ and $D_f$ are considered to be known, while $b_w$ is suitably assumed. Hence $d$ and $A_p$ can be found.

Thus, in the *second step*, the depth $d$ and area of steel $A_p$ can be determined from the relation $M_u \geq M_c$. where $M_u$ is given by *Eq. 45.78* for rectangular section and *Eq. 45.84* for flanged section.

Determine ultimate prestressing force from the relation
$$P_u = \gamma_{fp} \cdot A_p \cdot f_p \qquad \ldots(45.86)$$
where $\gamma_{fp}$ can be taken between 1.0 to 1.20.

**Step 3. Design for limit state strength at transfer condition:** At the transfer state, the prestressing force may be considered as an external load, along with the dead load.

The ultimate tendon force $P_u$ is given by *Eq. 45.86*, in which is $\gamma_{fp}$ is the partial safety factor which may be taken equal to 1.0 to 1.20.

If $M_{dy}$ is the dead load moment due to self weight of beam, the tensile force produced in the tendons is given by
$$F_{dg} = \frac{\gamma_{fd} \cdot M_{dg}}{d - 0.42\, x_u} = \frac{\gamma_{fd} M_{dg}}{j\, d} \qquad \ldots(45.87)$$
where $\gamma_{fd}$ = partial safety factor for dead load which may be taken equal to 0.9 since the dead load moment is of opposite nature than the moment due to prestressing force.

Net compressive force to be resisted by concrete section is
$$F_u = P_u - F_{dg} \qquad \ldots(45.88)$$

The area of concrete has to resist this force, with the centre of compression assumed at the centroid of prestressing steel, its value being given by
$$A_{ct} \geq \frac{\gamma_m\, F_u}{k_p\, f_{ci}} \qquad \ldots(45.89)$$

In the above expression, $\gamma_m$ = partial safety factor for concrete strength = 1.5; $f_{ci}$ = concrete cube strength at the time of transfer of prestress and $k_p$ = ratio of the compressive strength of concrete in flexure to that of the cube and taken equal to $\frac{2}{3}$ and $A_{ct}$ is the area of concrete with its centroid at the prestressing steel level.
$$A_{ct} \geq \frac{1.5\, F_u}{\frac{2}{3} f_{ci}} \geq 2.25\, \frac{F_u}{f_{ci}} \qquad \ldots(45.90)$$

For rectangular beams, the area of concrete centred at the steel level is
$$A_{ct} = 2(D-d)b; \qquad 2(D-d)b \geq 2.25\, \frac{F_u}{f_{ci}}$$
or
$$(D-d)b \geq 1.125\, \frac{F_u}{f_{ci}} \qquad \ldots(45.91)$$

The above equation is applicable both for rectangular as well as T-beams.

For I-beams (Flanged beam) such as shown in Fig. 45.68,
$$A_{ct} = A_{ctw} + A_{ctb},$$
where $A_{ctw}$ = area of web which contributes to the concrete action = $2(D-d)b_w$
$A_{ctb}$ = area of bottom flange overhangs = $(b_b - b_w) D_b$
$$A_{ct} = 2(D-d)b_w + (b_b - b_w) D_b \qquad \ldots(45.92)$$

Substituting in *Eq. 45.90*, we get
$$2(D-d)b_w + (b_b - b_w) D_b \geq 2.25\, \frac{F_u}{f_{ci}}$$
or
$$(D-d)b_w + \frac{1}{2}(b_b - b_w) D_b \geq 1.125\, \frac{F_u}{f_{ci}} \qquad \ldots(45.93)$$

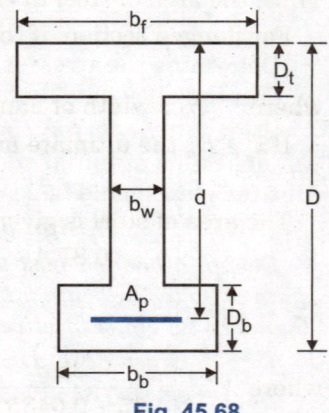

Fig. 45.68

From the above relation, the width of bottom flange ($b_b$) can be found for assumed value ($D_b$) of its thickness.

**Step 4. Design for limit state of collapse in shear:** The section for critical shear failure is at a distance $d$ from the effective support. The shear force $V$ at this section is computed after applying partial safety factor $\gamma_f$.

The ultimate shear resistance of section uncracked in flexure is given by *Eq. 45.94*

$$V_c = 0.67\, b\, D\, \sqrt{f_t^2 + 0.8\, f_{cp} \cdot f_t} \qquad ...(45.94)$$

If the tendon has inclination $\theta$ at the section, the component $P_c \cdot \theta$ of cable force in transverse direction should be *added* to $V_c$, where

$$P_c = \frac{A_p \cdot f_{pu}}{\gamma_{fp}} \qquad ...(45.95)$$

where $\gamma_{fp}$ = partial safety factor for prestressing force, the value of which is normally taken as 1.67.

(a) In no case should the S.F. $V$ due to ultimate load exceed the appropriate values given in Table 45.17 multiplied by $b \cdot d$.

(b) If $V < V_c$, minimum reinforcement given by *Eq. 45.74* should be provided.

(c) If $V > V_c$, shear reinforcement should be provided in accordance with *Eq. 45.75*:

$$\frac{A_{sv}}{s_v} = \frac{V - V_c}{0.87\, f_y\, d_t} \qquad ...(45.75)$$

(d) No shear reinforcement is necessary if $V$ is less than $0.5\, V_c$ or if the member is of minor importance.

**Step 5. Determination of prestressing steel area**

For the case of *rectangular section*, we have, from *Eq. 45.82*

$$A_p = \frac{M_u}{(d - 0.42\, x_u)\, f_{pu}} = \frac{M_u}{j\, d\, f_{pu}} \qquad ...[45.82(a)]$$

For balanced flanged sections $A_p$ is given by *Eq. 45.85*

$$A_p = 1.15\, [0.36\, b_w x_u + 0.45\, (b_f - b_w)\, D_f]\, \frac{f_{ck}}{f_{pu}}$$

**Step 6. Design for limit state of serviceability: allowable stresses**

The initial prestressing force is computed from the relation

$$P_i = A_p \cdot f_{pi}$$

where $f_{pi}$ = maximum tensile stress immediately behind the anchorages. As per IS : 1343-1980, this should not exceed 0.8 times ultimate tensile strength of the wire = $0.8\, f_p$.

$$\therefore \qquad P_i = 0.8\, A_p f_p \qquad ...(45.96)$$

**Note:** $f_{pi\,(max)}$ is equal to $0.8\, f_p$. Generally, $f_{pi}$ is choosen between $0.7\, f_p$ to $0.8\, f_p$.

The losses of prestress are estimated, generally to a value equal to $0.15\, f_p \cdot A_p$. Subtracting these losses from $P_i$, we get effective prestressing force given by

$$P_e = P_i - 0.15\, f_p A_p = 0.8\, A_p f_p - 0.15\, A_p f_p = 0.65\, A_p f_p \qquad ...(45.97)$$

Determine the stresses in concrete at transfer and working load condition and check whether these are within the allowable limits as specified by IS : 1343-1980.

**Step 7. *Check for limit state of serviceability: deflection:*** Compute (*i*) deflection due to creep, (*ii*) deflection at initial stage, (*iii*) deflection at working load condition. The total deflection including that due to creep should be less than the permissible value.

**Example 45.18.** *Design of Rectangular Beam.*

*Design a bonded post-tensioned prestressed concrete beam of type 2 simply supported over an effective span of 9 m. The beam has to be of rectangular cross-section. It carries superimposed dead and live loads of 6 and 12 kN/m respectively. Use grade M 35 concrete. Take unit weight of concrete as 24 kN/m³.*

**Solution:** *Given data:* $L = 9$ m

Superimposed dead load $w_d = 6$ kN/m. Superimposed live load $w_l = 12$ kN/m

$$f_{ck} = 35 \text{ N/mm}^2;\quad f_{ci} = 30 \text{ N/mm}^2$$

The allowable stresses at the time of transfer are:

$$\sigma_{cci} = \left[0.54 - \frac{0.54 - 0.37}{6} \times 1\right] f_{ci} = 0.512 \times 30 \approx 15.35 \text{ N/mm}^2 \qquad \text{[Fig. 45.66 (a)]}$$

$$\sigma_{cti} = 3 \text{ N/mm}^2.$$

The allowable stresses under service condition are

$$\sigma_{cce} = \left[0.34 - \frac{0.34 - 0.27}{6} \times 1\right] f_{ck} = 0.328 \times 35 \approx 11.49 \text{ N/mm}^2$$

and $\sigma_{cte} = 3$ N/mm². (Fig. 45.65 zone II)

**Step 1.** *Determination of design bending moment*
For the purpose of computing self weight of the beam, let

$$b = \frac{L}{30} = \frac{9000}{30} = 300 \text{ mm}; \quad D = \frac{L}{15} = \frac{9000}{15} = 600 \text{ mm}$$

∴ Self weight $w_{dg} = (300 \times 600 \times 10^{-6}) \times 1 \times 24000 \approx 4300$ N/m = 4.3 kN/m
∴ Characteristic load $w_k = w_{dg} + w_d + w_l = 4.3 + 6 + 12 = 22.3$ kN/m
$\gamma_f$ for load = 1.5. ∴ Design bending moment at collapse is

$$M_c = \frac{\gamma_f w_k L^2}{8} = \frac{1.5 \times 22.3 \times 1000}{8} (9)^2 \times 1000 \text{ N-mm}$$

$$= 338.68 \times 10^6 \text{ N-mm}$$

**Step 2.** *Design for limit state of flexure:* Let the characteristic strength of steel wires (7 mm Φ) be 1500 N/mm² (See Table 45.11). From *Eq. 45.77*,

$$\frac{x_u}{d} = \frac{35}{15 + 0.0435 \times 1500} = 0.436$$

From *Eq. 45.80*, $\quad j = 1 - \dfrac{14.6}{15 + 0.0435 \times 1500} = 0.818$

From *Eq. 45.78 (a)* $\quad R_u = \dfrac{12.6}{15 + 0.0435 \times 1500}\left(1 - \dfrac{14.6}{15 + 0.0435 \times 1500}\right) = 0.128$ N/mm²

From *Eq. 45.79 (a)* $\quad p_0 = \dfrac{14.49}{15 + 0.0435 \times 1500} \approx 0.181$

(Alternatively, all the above four values could be taken from Table 45.19).
The ultimate moment of resistance of the beam is given by $M_u = R_u \cdot b d^2 f_{ck}$
Hence the design criterion is
∴ $\quad M_u \geq M_c$.
∴ $\quad R_u bd^2 f_{ck} \geq M_c$

or $\quad d = \sqrt{\dfrac{M_c}{R_u \cdot b \cdot f_{ck}}}$

Choosing $\quad b = L/30 = 300$ mm,

$$d = \sqrt{\frac{338.68 \times 10^6}{0.128 \times 300 \times 35}} \approx 502 \text{ mm}$$

The area of pre-tensioned steel is estimated from *Eq. 45.79*

$$A_p = p_0 b d \cdot \frac{f_{ck}}{f_p} = 0.181 \times 300 \times 502 \times \frac{35}{1500} = 636 \text{ mm}^2$$

Limit of ultimate prestressing force is given by *Eq. 45.86*:

$$P_u = \gamma_{fp} \cdot A_p \cdot f_p$$

Choosing $\gamma_{fp} = 1.2$, $P_u = 1.2 \times 636 \times 1500 = 1.14 \times 10^6$ N

**Step 3.** *Design for limit state of strength at transfer condition*
Self weight $\quad W_{dg} = 4300$ N/m

$$\therefore \qquad M_{dg} = \frac{4300 \times (9)^2}{8} \times 1000 \approx 43.54 \times 10^6 \text{ N-mm}$$

Axial tension caused by the above moment is given by *Eq. 45.87*:

$$F_{dg} = \frac{\gamma_{fd} \, M_{dg}}{d - 0.42 \, x_u} = \frac{\gamma_{fd} \cdot M_{dg}}{j \cdot d}$$

Taking $\gamma_{fd} = 0.9$, and $j = 0.818$,

$$F_{dg} = \frac{0.9 \times 43.54 \times 10^6}{0.818 \times 500} \approx 95800 \text{ N}$$

Hence from *Eq. 45.88*,

$$F_u = P_u - F_{dg} = 1.14 \times 10^6 - 0.096 \times 10^6 = 1.044 \times 10^6 \text{ N}$$

For rectangular beams, the design conditions is given by *Eq. 45.91*:

$$(D - d) \, b \geq 1.125 \, \frac{F_u}{f_{ci}}$$

Taking $f_{ci} = 30$ N/mm², we get

$$D = \frac{1.125}{b} \frac{F_u}{f_{ci}} + d = \frac{1.125}{300} \times \frac{1.044 \times 10^6}{30} + 500$$

$$= 130.5 + 500 = 630.5 \text{ mm}$$

The assumed value of $D = 600$ mm.

The increase in the self weight due to increased value of $D$ will be only marginal. Hence effective depth can be increased marginally.

Thus, choose the following dimensions:

$$D = 650 \text{ mm}; \quad b = 300 \text{ mm}; \quad d = 510 \text{ mm}.$$

$$w_{dg} = 0.3 \times 0.65 \times 24000 = 4680 \text{ N} = 4.68 \text{ kN}$$

$$w = w_{dg} + w_d + w_L = 4.68 + 6 + 12 = 22.68 \text{ kN/m}.$$

$$M_c = \gamma_f \frac{w L^2}{8} = \frac{1.5 \times 22.68 \, (9)^2}{8} \times 10^6 \text{ N-mm} = 334.5 \times 10^6 \text{ N-mm}.$$

The ultimate moment of resistance of the beam is

$$M_u = R_u \, b \, d^2 \, f_{ck} = 0.128 \, (300) \, (510)^2 \times 35 = 349.6 \times 10^6 \text{ N-mm}.$$

Since, $M_u > M_c$, the section is safe against its limit state of collapse in flexure.

**Step 4.** *Design for limit state of collapse in shear*

S.F. at distance $d$ from the support is

$$V = \gamma_f \left( \frac{w L}{2} - w \, d \right) = 1.5 \left( \frac{22.68 \times 9}{2} - 22.68 \times 0.51 \right)$$

$$= 135.74 \text{ kN} = 135740 \text{ N}$$

Let us provide the prestressing tendons parabolic in shape, with say $h = 0$ at end section, and maximum sag at the centre $= h = d - 0.5 \, D = 510 - 0.5 \times 650 = 185$ mm.

The slope $\theta$ of the cable at distance $d$ from the support is

$$\theta = \frac{4h}{L^2}(L - 2 \, d) = \frac{4 \times 0.185}{(9)^2} (9 - 2 \times 0.51) = 0.073 \text{ radians}$$

$$P_c = \frac{A_p \cdot f_{pu}}{\gamma_{fp}} \qquad (Eq. \; 45.95)$$

Taking $f_{pu} = f_p = 1500$ N/mm² and $\gamma_{fp} = 1.67$, $\quad P_c = \dfrac{633.5 \times 1500}{1.67} \approx 569000$ N

$$P_c \cdot \theta = 569000 \times 0.073 \approx 41500 \text{ N}$$

The ultimate shear resistance of uncracked section is given by *Eq. 45.71*.

$$V_{co} = 0.76 \, b \, D \, \sqrt{f_t^2 + 0.8 \, f_{cp} \cdot f_t}$$

where $f_t = 0.24 \sqrt{f_{ck}} = 0.24 \sqrt{35} = 1.42$ N/mm², $f_{cp} = \dfrac{P_c}{A} = \dfrac{569000}{300 \times 650} = 2.92$ N/mm²

$\therefore \qquad V_{co} = 0.67 \times 300 \times 650 \sqrt{(1.42)^2 + 0.8 \, (1.42) \, (2.92)} = 301730$ N

$\therefore \qquad V_c = V_{co} + P_c \cdot \theta = 301730 + 41500 = 343230$ N

Since the shear resistance $V_c$ is more than twice the actual ultimate S.F., *no shear reinforcement is necessary*.

**Step 5.** *Determination of prestressing steel area*

The area of steel is given by *Eq. 45.82*, in which $f_{pu} = f_p$

$\therefore \qquad A_p = \dfrac{M_u}{j \, d \, f_p} = \dfrac{349.6 \times 10^6}{0.818 \times 510 \times 1500} = 558.7$ mm²

Using 7 mm dia. wires, having $A_\Phi = \dfrac{\pi}{4} (7)^2 = 38.48$ mm². No of wires = $\dfrac{558.7}{38.48} = 14.5$

Provide 15 wires of 7 mm $\Phi$. Actual $A_p = 15 \times 38.48 \approx 577.3$ mm²

**Step 6.** *Design for limit state of serviceability: allowable stresses*

The initial prestressing force is given by $= P_i = A_p \cdot f_{pi}$

Choosing $\qquad f_{pi} = 0.75 \, f_p$, we have

$P_i = A_p \, (0.75 \, f_p) = 577.3 \times 0.75 \times 1500 \approx 649500$ N

Assuming losses $\quad = 0.15 \, f_p \, A_p = 0.15 \times 1500 \times 577.3 \approx 130000$ N

$P_e = 649500 - 130000 = 519500$ N

The properties of the section, neglecting steel area, are as follows:

$$A = 300 \times 650 = 195000 \text{ mm}^2; \quad Z = \dfrac{300 \, (650)^2}{6} = 21.125 \times 10^6 \text{ mm}^3.$$

Eccentricity $\qquad e = 185$ mm at midspan.

*(a) At the transfer condition*

$$f = \dfrac{P_i}{A} \mp \dfrac{P_i \cdot e}{Z} \pm \dfrac{M_{dg}}{Z}$$

where $M_{dg} = \dfrac{w_{dg} \cdot L^2}{8} = \dfrac{4680 \, (9)^2}{8} \times 1000 = 47.38 \times 10^6$ N-mm

$\dfrac{P_i}{A} = \dfrac{649500}{195000} = 3.33$ N/mm²; $\quad \dfrac{P_i \cdot e}{Z} = \dfrac{649500 \times 185}{21.125 \times 10^6} = 5.69$ N/mm²

$\dfrac{M_{dg}}{Z} = \dfrac{47.38 \times 10^6}{21.125 \times 10^6} = 2.24$ N/mm²

$\therefore \qquad f_1$ (top fibre) $= 3.33 - 5.69 + 2.24 = -0.12$ N/mm² (i.e. tensile)

This is less than permissible tensile stress of 3 N/mm²

$f_2$ (bottom fibre) $= 3.33 + 5.69 - 2.24 = 6.78$ N/mm²

This is less than permissible value of 15.35 N/mm²

*(b) At the service load condition*

$$f = \dfrac{P_e}{A} \mp \dfrac{P_e \cdot e}{Z} \pm \dfrac{M}{Z}$$

where $M = \dfrac{w \, L^2}{8} = \dfrac{22.68 \, (9)^2}{8} \times 10^6 = 229.6 \times 10^6$ N-mm; $\quad P_e = 519500$ N

$$\therefore \quad \frac{P_e}{A} = \frac{519500}{195000} = 2.66 \text{ N/mm}^2; \quad \frac{P_e \cdot e}{Z} = \frac{519500 \times 185}{21.125 \times 10^6} = 4.55 \text{ N/mm}^2$$

$$\frac{M}{Z} = \frac{229.6 \times 10^6}{21.125 \times 10^6} = 10.87 \text{ N/mm}^2$$

$\therefore \quad f_1$ (top fibre) = $2.66 - 4.55 + 10.87 = 8.98$ N/mm² (Safe)

$f_2$ (bottom fibre) = $2.66 + 4.55 - 10.87 = -3.66$ N/mm² (*i.e.*, tensile)

This is slightly more than the permissible tensile stress of 3 N/mm². In order to bring it back to the permissible value, let us increase the prestressing force. The net stress due to prestressing force = $2.66 + 4.55 = 7.21$ against requirement of $10.87 - 3 = 7.87$.

Hence increase the prestressing force by ratio $7.87/7.21 = 1.09$.

$\therefore \quad$ No. of wires = $15 \times 1.09 = 16.4$

Let us provide 17 wires of 7 mm $\Phi$. $\quad \therefore \quad A_p = 17 \times 38.48 = 654.16$ mm²

$$P_i = A_p (0.75 f_p) = 654.16 \times 0.75 \times 1500 = 735930 \text{ N}$$

$$\text{Losses} = 0.15 f_p A_p = 0.15 \times 1500 \times 654.16 = 147186 \text{ N}$$

$\therefore \quad P_e = 735930 - 147186 = 588744$ N.

The raised stresses will be as under:

(*a*) *At transfer condition*

$$f = \frac{P_i}{A} \mp \frac{P_i \cdot e}{Z} + \frac{M_{dg}}{Z}; \quad \frac{P_i}{A} = \frac{735930}{195000} = 3.77 \text{ N/mm}^2$$

$$\frac{P_i \cdot e}{Z} = \frac{735930 \times 185}{21.125 \times 10^6} = 6.44 \text{ N/mm}^2; \quad \frac{M_{dg}}{Z} = \frac{47.38 \times 10^6}{21.125 \times 10^6} = 2.24 \text{ N/mm}^2$$

$\therefore \quad f_1$ (*top fibre*) = $3.77 - 6.44 + 2.24 = -0.43$ N/mm² (safe)

$f_2$ (bottom fibre) = $3.77 + 6.44 - 2.24 = 7.97$ N/mm² (safe)

(*b*) *At service load condition*

$$f = \frac{P_e}{A} \mp \frac{P_e \cdot e}{Z} \pm \frac{M}{Z}; \quad \frac{P_e}{A} = \frac{588744}{195000} = 3.02 \text{ N/mm}^2$$

$$\frac{P_e \cdot e}{Z} = \frac{588744 \times 185}{21.125 \times 10^6} = 5.16 \text{ N/mm}^2; \quad \frac{M}{Z} = \frac{229.6 \times 10}{21.125 \times 10^6} = 10.87 \text{ N/mm}^2$$

$\therefore \quad f_1$ (top fibre) = $3.02 - 5.16 + 10.87 = 8.73$ N/mm² (safe)

$f_2$ (bottom fibre) = $3.02 + 5.16 - 10.87 = -2.69$ N/mm² (safe)

Since $A_p$ has been increased, the section should be re-checked for limit state of strength at transfer condition (step 3).

Revised $P_u = \gamma_{fp} \cdot A_p \cdot f_p = 1.2 \times 654.16 \times 1500 = 1.178 \times 10^6$ N.

$$F_{dg} = \frac{\gamma_{fd} \cdot M_{dg}}{j d} = \frac{0.9 \times 47.38 \times 10^6}{0.818 \times 510} = 102050 \text{ N} = 0.102 \times 10^6 \text{ N}$$

$\therefore \quad F_u = P_u - F_{dg} = 1.178 \times 10^6 - 0.102 \times 10^6 = 1.076 \times 10^6$

The criterion for safety is

$$(D - d) b \geq 1.125 \frac{F_u}{f_{ci}} \quad \text{...(Eq. 45.91)}$$

$\therefore \quad (650 - 510) \, 300 \geq 1.125 \, \dfrac{1.076 \times 10^6}{30}$; $42000 \geq 40350$. Hence satisfied.

**Step 7.** *Check for limit state of serviceability: deflection*

$$I = \frac{b D^3}{12} = \frac{300 \, (650)^3}{12} = 6866 \times 10^6 \text{ mm}^4;$$

$$E_c = 5700 \sqrt{f_{ck}} = 5700 \sqrt{35} = 33722 \text{ N/mm}^2$$

The equivalent distributed load of cables at the initial and transfer stage are as under.

$$w_{pi} = \frac{8 P_i \cdot e}{L^2} = \frac{8 \times 735930 \times 0.185}{(9)^2} = 13450 \text{ N/m} = 13.45 \text{ kN/m}.$$

$$w_{pe} = \frac{8 P_e \cdot e}{L^2} = \frac{8 \times 588744 \times 0.185}{(9)^2} = 10760 \text{ N/m} = 10.76 \text{ kN/m}.$$

The effective distributed load on the beam at the transfer and service load conditions are:

$$w_i = w_{dg} - w_{pi} = 4.68 - 13.45 = -8.77 \text{ kN/m} = -8770$$

$$w_e = w - w_{pe} = 22.68 - 10.76 = 11.92 \text{ kN/m} = 11920 \text{ N/m}$$

The effective permanent load is given by

$$w_{pd} = w_{dg} + w_d - w_{pe} = 4.68 + 6 - 10.76 = -0.08 \text{ kN/mm}^2 = -80 \text{ N/mm}^2$$

The deflections at various stages will be as follows:

**(i) Deflection at initial stage**

$$\delta_i = \frac{5 w_i L^4}{384 E_c I} = \frac{5 (8770)(9)^4 (1000)^3}{384 \times 32722 \times 6866 \times 10^6}$$

$$= -3.24 \text{ mm} \quad (i.e. \text{ upwards})$$

Allowable upward deflection $= \dfrac{L}{300} = \dfrac{9000}{300} = 30$ mm.  Hence safe.

**(ii) Deflection at working load condition**

$$\delta_e = \frac{5 w_e L^4}{384 E_c I} = \frac{5 (11920)(9)^4 (1000)^3}{384 \times 32722 \times 6866 \times 10^6} = 4.4 \text{ mm}.$$

**(iii) Deflection due to effective permanent load**

$$\delta_{pd} = \frac{5 w_{pd} L^4}{384 E_c I} = \frac{5 (-80)(9)^4 (1000)^3}{384 \times 3372 \times 6866 \times 10^6} = -0.03 \text{ mm}$$

**(iv) Total deflection, including effect of creep**

Net creep coefficient $= 1.6$

$$\delta_t = \delta_e + (\delta_{pd} \times \text{creep coefficient}) = 4.4 + (-0.03)(1.6) = 4.35 \text{ mm}$$

Allowable downward deflection $= \dfrac{L}{250} = \dfrac{9000}{250} = 36$ mm.  Hence safe.

## PROBLEMS

1. A simply supported prestressed concrete beam of rectangular cross-section 300 mm × 500 mm is loaded with a total uniformly distributed load of 200 kN over a span of 5 m. Sketch the distribution of stresses at the mid-span and end sections if the prestressing force is 1500 kN and the tendon is eccentric, located at 150 mm above the bottom fibre.

2. Find the extreme fibre stresses at mid-span of the beam of problem 1 by applying the principle of internal resisting couple.

3. The beam of problem 1 is prestressed by a parabolic tendon with a prestressing force of 1500 kN. The tendon has a sag of 100 mm at the mid-span. Find the extreme fibre stress by load balancing concept if it is subjected to (a) total U.D.L. of 200 kN, (b) total U.D.L. of 300 kN.

4. A pretensioned prestress concrete beam of 7 m span has a cross-section of 300 mm × 500 mm, and is prestressed with 1500 kN force at transfer. The cable has cross-sectional area of 1500 mm² of steel, and has a parabolic profile with a maximum eccentricity of 1500 mm at the mid-span. Determine the loss of prestress, given that $E_s = 2.1 \times 10^5$ N/mm² and $E_c = 3 \times 10^4$ N/mm².

5. A post-tensioned prestressed concrete beam of 24 m span is subjected to a transfer prestress force of 2000 kN at 25 days strength. The profile of the cable is parabolic with maximum eccentricity of 160 mm at the mid-span. Determine the losses of prestress and the jacking force required if jacking is done from both the ends of the beam. The beam has a cross-sectional area of 450 mm × 700 mm and is prestressed with 8 cables, each cable consisting of 12 wire of 5 mm dia. One cable is tensioned at a time. Take $E_s = 2.1 \times 10^5$ N/mm² and $E_c = 3.5 \times 10^4$ N/mm².

6. (a) Write notes on (i) kern distance (ii) efficiency factor (b). Find the kern distance, and compare the efficiency factors of (i) rectangular section, (ii) symmetrical I-section, (iii), T-section, if all the three have the same cross-sectional area of 2800 cm², and same depth of 70 cm. Assume suitable thickness of web and flange.

7. Design a prestressed concrete slab over a span of 12 metres to carry a super-imposed load of 15 kN/m². Assume permissible stresses in concrete of 12 N/mm² and no tension. Also assume that the initial losses in prestress in steel will amount to 15%. The live load will be brought to bear after the ducts are grouted. The initial prestress in the wires is to be 1100 N/mm².

8. A rectangular beam of prestressed concrete is required to support a dead load moment of $12 \times 10^6$ N-mm (inclusive of its own weight) and a live load moment of $30 \times 10^6$ N-mm at its mid-section. Determine the initial prestressing force and its eccentricity at the mid-span section. Adopt the following values:

Allowable initial compressive stress = 16 N/mm²
Allowable final compressive stress = 13 N/mm²
Allowable initial or final tension = 1 N/mm²
Ultimate stress in steel = 1500 N/mm²
Assume losses as 15%.

# CHAPTER 46: SHRINKAGE AND CREEP

## 46.1. INTRODUCTION

If concrete is subjected to sustained loads, it continuous to deform with time. This phenomenon, discovered in 1907 by "Halt" is commonly referred to as "Creep".

The deformation of a loaded concrete specimen may be made up of the following three factors:

(*a*) The immediate elastic deformation.

(*b*) The shrinkage and/or moisture movement. and (*c*) The creep.

The *elastic deformation* of concrete is proportional to the stress applied. Since time has its effect through gain in maturity, the elastic deformation under a given stress will also vary with time, specially during the maturity period.

*Shrinkage* may be defined as the shortening in the length of the concrete member due to reduction in the moisture content. It is a complex physico-chemical phenomenon. Unrestrained concrete members exhibit progressive shortening over a long period while they are hardening.

*Creep* is the slow deformation, additional to the elastic deformation exhibited by concrete under sustained stress, and proceeds at a decreasing rate over many years. At ages upto one year, creep is proportional to the stress, provided the stress is upto $\frac{1}{4}$ of the ultimate strength of concrete. For higher stresses, creep increases rapidly, and is not in proportion to stress.

## 46.2. SHRINKAGE OF CONCRETE

Concrete also exhibits stress independent deformations which in addition to thermal dilation include shrinkage (or swelling), *i.e.* volumetric deformation due to change in water content and long time chemical process.

Types of shrinkage, caused due to loss of water.

(*a*) **Plastic shrinkage:** Occurs due to loss of water by evaporation from freshly placed concrete while the cement paste is plastic.

(*b*) **Drying shrinkage:** Occurs due to loss of water by evaporations from freshly hardened concrete exposed to air.

Shrinkage is the shortening of the concrete member due to drying and other complex phenomenon. The *drying shrinkage* is partly recoverable on wetting. In general, shrinkage may be partly reversible and partly permanent. The reversible part of shrinkage may be due to colloidal swelling or contraction of cement gel, depending upon water saturation. It is a physical action in which the particles of cement are pulled together due to capillary and cohesive forces as moisture evaporates. The irreversible part or permanent part of shrinkage is due to chemical changes that take place as cement sets and hardens. The gel formed by cement with water crystalises in due course, giving out free water, resulting in shrinkage. The permanent shrinkage can not be reversed.

For concrete that can dry completely and where shrinkage is unrestrained, the shrinkage strain may be approximately 0.00025 at 28 days and 0.00035 at three months, after which period the increase is less rapid until at the end of 12 months, it may approach a maximum of 0.0005. However where the concrete does not completely become dry, such as in reservoir and other structures, a maximum value of 0.0002 is reasonable. When concrete is stored in water or very wet conditions, shrinkage may be zero or concrete may even expand.

## Factors affecting shrinkage

The shrinkage of a concrete member depends upon the following factors :

(i) volume of cement paste
(ii) water cement ratio
(iii) chemical properties of cement
(iv) size and grading of aggregates
(v) mineral and physical character of aggregates
(vi) admixtures, and entrained air
(vii) size and shape of concrete member
(viii) presence of fine material.
(ix) effect of time.
(x) effect of relative humidity

### 1. Volume of Cement Paste (*mix proportion*)

A concrete containing large volume of cement paste has more shrinkage. This is because a richer mix absorbs more water in the beginning and releases later. Fig. 46.1 shows the effect of mix proportion and water/cement ratio on drying shrinkage. It will be seen that at constant water/cement ratio, the shrinkage increases considerably with the richness of the mix.

### 2. Water Cement Ratio :
The shrinkage of concrete increases with the water/cement ratio, as is evident from Fig. 46.1. A wet mix which shrinks more purely for physical reasons as the volume of concrete decreases by the evaporation of water. It should be noted that the cement content of mix has only a small effect on shrinkage, provided the amount of water per unit volume of concrete is maintained constant. Shrinkage can be greatly reduced by reducing the amount of mixing water. An increase in 1% in the amount of mixing water may increase the shrinkage by 2%.

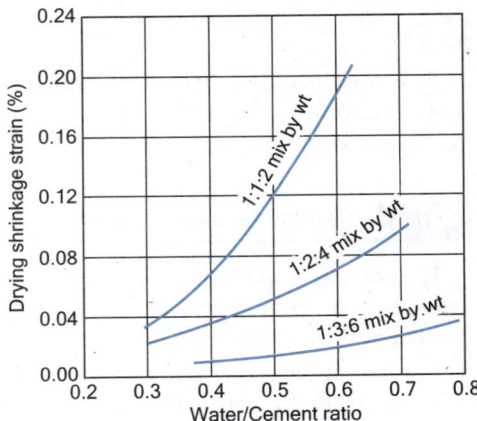

Fig. 46.1. Effect of Mix Proportion and Water/Cement Ratio on Drying Shrinkage

To keep the shrinkage low, the w/c ratio and quantity of cement should be kept a minimum. To get a dense concrete, however proper grading of aggregates, and use of vibrators will be essential.

### 3. Chemical Properties of Cement :
Shrinkage of concrete also depends upon the chemical composition of cement. The amount and condition of the magnesia, as well as the amounts of uncombined lime, calcium sulphate, tricalcium aluminate, the tetracalcium aluminoferrite and the so-called free alkali constituents of the cement are predominent factors influencing permanent volume change. Carlson, Woods and Bogue studied the relative effects of the principal compounds of cement in producing shrinkage. According to them, tricalcium silicates ($C_3S$) expands slightly under moist curing and then when cured in air contracts to about 70% of the volume of Portland cement. Dicalcium silicate ($C_2S$) does not change in volume under moist curing and under air curing shrinkage at about double the rate of normal cement. Tricalcium aluminate ($C_3A$) showed great expansion under moist curing and rapid contraction in air. Shrinkage is low for these cements which have large quantity of tricalcium silicate and have lower amounts of alkalies (*i.e.*, sodium or potassium oxides). Higher the loss on ignition of cement, the greater will be the shrinkage.

### 4. Size and Grading of Aggregates :
The shrinkage of concrete is also dependent on the size of coarse and fine aggregates and on their grading. According to Carlson, the measured percent shrinkage with 10 to 20 mm size aggregate was 0.07 while that for 1.20 to 2.0 mm size, it was as high as 0.122. The measured shrinkage for neat cement paste was as high as 0.27%. Thus, lower the percentage of voids in the mixed aggregates and larger the maximum size of particles, lesser will be the shrinkage. It has been observed that shrinkage of sand motars is 2 to 3 times as great as that for concrete with 20 mm size aggregate.

5. **Mineral and Physical Characteristics of Aggregates :** The shrinkage of concrete is largely governed by the compressibility of aggregate and its own shrinkage properties on drying. Aggregate in concrete is surrounded by cement paste which on shrinkage places the aggregate in compression. Aggregates which are hard and dense with low absorption and high modulus of elasticity give a concrete with less shrinkage. Thus quartz, feldsper, lime stone and dolomite produce low shrinkages while pyroxene and hornblende produce high shrinkages. Granite gives intermediate values. On the other hand sandstone and slate not only have low rigidity but shrink themselves on drying and concrete made of these aggregates has high shrinkage. It should, however, be noted that the effect of the aggregate on shrinkage is greater on leaner mix. Aggregate shape appears to have little effect on shrinkage except in so far as it affects the amount of mixing water required to maintain workability.

6. **Admixtures and Entrained Air :** Admixtures like $CaCl_2$ increases shrinkage. The amount of gypsum within normal range has little effect on shrinkage. Similarly, most commercial wetting and dispersing agents also have very little effect on shrinkage. Air entrainment within limits does not increase shrinkage and may be employed to reduce the water content.

7. **Size and Shape of Concrete Member :** The rate and magnitude of shrinkage is affected by the size of the specimen or mass of concrete being observed. Large size members have less ultimate shrinkage, and take more time to achieve this. According to Lea, a specimen of 36 in. × 4 in. × 4 in. gave a shrinkage strain of 0.06% at one year and increased very little at four years while a specimen 108 in. × 30 in. × 20 in. containing 0.4 percent steel gave a shrinkage strain of only 0.025% at one year and of slightly less than 0.05 % at 7 years.

8. **Presence of Fine Material :** The presence of material finer than 0.1 mm increases shrinkage. It has been found that shrinkage is proportional to the ratio of volume of (cement + dust + water) to the total volume of concrete.

## 46.3. SHRINKAGE STRESSES IN SYMMETRICALLY REINFORCED SECTIONS

If a concrete member is restrained so that reduction in length due to shrinkage cannot take place, tensile stresses are caused in concrete. A shrinkage coefficient of 0.0002 (*i.e.*, shrinkage strain of 0.02%) may correspond to a stress of 3.5 N/mm² when restrained.

Let us now consider a symmetrically reinforced concrete section of cross-section area $A$ and length $AB = L$ (Fig. 46.2). If the concrete member were un-reinforced, it will shrink to a length $AB_1 = L_1$, the shrinkage being equal to $\Delta_{sh}$. However, the presence of reinforcement will restrict the member to shrink to $AB'$, with the result that concrete will be subjected to tensile stress ($f_c$) while steel will be subjected to compressive stress ($f_s$). Fig. 46.2 (c) shows the final state of the bar, with a length $AB' = L'$, the final shrinkage being restricted to $\Delta'$ only.

Let,
$E_c$ = modulus of elasticity of concrete
$E_s$ = modulus of elasticity of steel
$f_c$ = stress induced in concrete (tensile)
$f_s$ = stress induced in steel (compressive)
$p'$ = steel ratio = $\dfrac{A'}{A}$
$\varepsilon_{sh}$ = shrinkage coefficient = $\Delta_{sh}/L$
Shortening of unreinforced concrete = $\Delta_{sh}$
Shrinkage strain of unreinforced concrete = $\Delta_{sh}/L = \varepsilon_{sh}$
Net shortening = $\Delta' = \Delta_{sh} - \Delta$

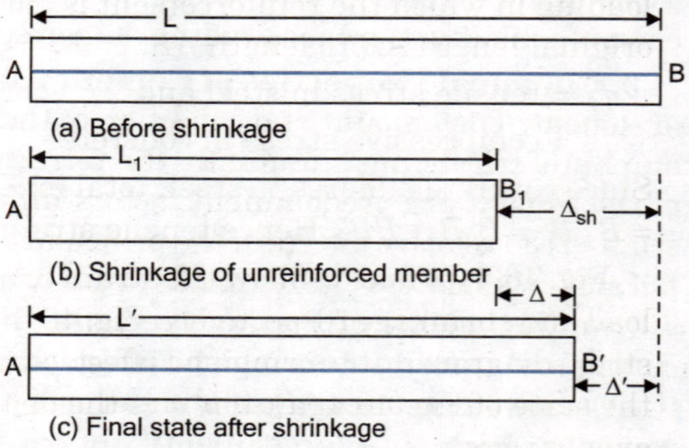

**Fig. 46.2.** Symmetrically Reinforced Section.

(a) Before shrinkage
(b) Shrinkage of unreinforced member
(c) Final state after shrinkage

∴   Net shortening strain in concrete = $\varepsilon_{sh} - f_c/E_c$ ...(*i*)
Shortening of steel = $\Delta'$
∴   Shortening strain in steel = $f_s/E_s$ ...(*ii*)

Since there is no slip between concrete and steel, both the strains will be equal.

∴   $$\varepsilon_{sh} - (f_c / E_c) = (f_s / E_s)$$ ...(1)

For the equilibrium, the total compressive force in steel should be equal to the total tensile force in concrete.

$$\therefore \quad p' \cdot A \cdot f_s = (A - p'A) f_c \quad \text{or} \quad p' \cdot f_s = (1 - p') f_c \qquad \ldots(2)$$

Substituting the value of $f_c$ from (2) to (1), we get

$$\text{or} \quad \varepsilon_{sh} - \frac{p'}{1-p'} \cdot \frac{f_s}{E_c} = \frac{f_s}{E_s} \quad \text{or} \quad f_s \left(1 + \frac{mp'}{1-p'}\right) = E_s \cdot \varepsilon_{sh}$$

From which
$$f_s = \frac{1-p'}{1+(m-1)p'} E_s \cdot \varepsilon_{sh} \qquad \ldots(46.1)$$

and
$$f_c = \frac{p'}{1-p'} \cdot f_s = \frac{p'}{1+(m-1)p'} E_s \cdot \varepsilon_{sh} \qquad \ldots(46.2)$$

In the above expression, $m$ is equal to the ratio of the moduli of elasticity of steel ($E_s$) and concrete ($E_c$). Since the stress induced due to shrinkage is of permanent nature, it gives rise to *creep strains* also. As indicated in § 46.8, $E_c$ decrease due to creep, and hence $m$ increases from its normal value (i.e. 19 for 1 : 2 : 4 concrete) to as high as 40 to 60. This increased value of $m$ should be used in the above equations.

## 46.4. SHRINKAGE STRESSES IN SINGLY REINFORCED BEAMS

In the previous section, we have considered symmetrically reinforced member. A singly reinforced beam is unsymmetrically reinforced with respect to the C.G. of the section. The top fibres of such a beam shrink more freely than the bottom ones. Due to this, the beam warps and deflects downwards.

The shrinkage of concrete member exerts compressive force in reinforcement, which exerts back tensile force on concrete surrounding it. This tensile force in concrete around the steel acts eccentrically on the concrete section, causing compressive stress at top and tensile stresses in the bottom. Due to this, the cracked area increase and the neutral axis is shifted up.

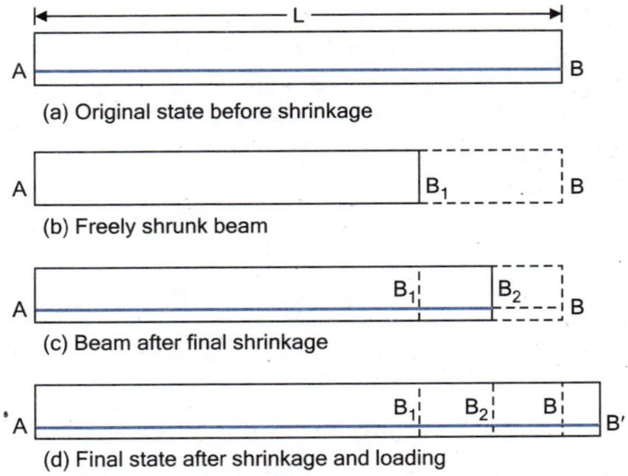

Fig. 46.3 (a) shows the original state of the beam, without shrinkage, with reinforcement length $AB$. Fig. 46.3 (b) shows a freely shrunk unreinforced beam, the shrinkage strain being equal to $\varepsilon_{sh}$ (= $BB_1/L$). Fig. 46.3 (c) shows the beam after final shrinkage. Fig. 46.3 (d) shows the final state after shrinkage and loading in which the reinforcement is elongated from its original length $AB$ to length $AB'$.

$f_s'$ = tensile stress in steel and
$f_c'$ = compressive stress in concrete

(a) Original state before shrinkage
(b) Freely shrunk beam
(c) Beam after final shrinkage
(d) Final state after shrinkage and loading

Fig. 46.3. Singly Reinforced Beam.

Since concrete adheres to steel, total extension of concrete, with respect to its free shrunk position, will be = $B_1 B' = B_1 B + BB'$. Hence tensile strain in concrete = $\varepsilon_{sh} + f_s'/E_s$.

Fig. 46.4 (b) and (c) show the strain diagrams due to load and shrinkage respectively. Fig. 46.4 (d) shows the strain diagram due to combined effect, corresponding to the state of Fig. 46.4 (d). If $k'd$ is the depth of neutral axis, we have

$$\frac{f_c'/E_c}{k'} = \frac{(\varepsilon_{sh} + f_s'/E_s)}{1-k'} \qquad \ldots(I)$$

Also, total tension is equal to total compression

or
$$f_s' \cdot p' bd = \frac{1}{2} f_c' b \cdot k'd \qquad \ldots(II)$$

(neglecting tension taken by concrete below N.A.)

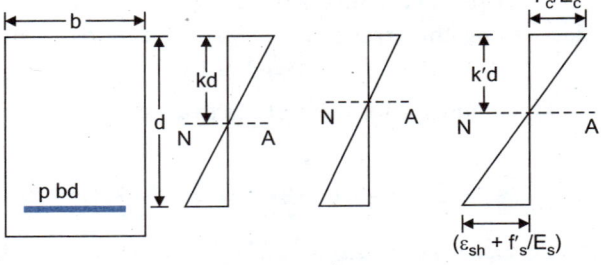

(a) Section  (b) Strain dia due to load  (c) Strain dia due to shrinkage  (d) Combined strain dia

Fig. 46.4. Strain Diagrams.

Substituting the value of $f_c'$ from I to II, we get

$$f_s' \, p' \, bd = \frac{1}{2} \cdot bk'd \, \frac{E_c \, k'}{1-k'} [\varepsilon_{sh} + f_s'/E_s] \quad \text{or} \quad \frac{f_s'}{E_s \cdot \varepsilon_{sh} + f_s'} = \frac{k'^2}{2 \, mp'(1-k')}$$

Putting $\quad \dfrac{f_s'}{E_s \cdot \varepsilon_{sh} + f_s'} = \delta \quad$ ...(46.3)

we get $\quad \delta = \dfrac{k'^2}{2 \, mp'(1-k')} \quad$ or $\quad k'^2 - 2mp'\delta + 2mp'\delta k' = 0$

From which

$$k' = \sqrt{(mp' \, \delta)^2 + 2mp'\delta} - mp' \, \delta \qquad ...(46.4)$$

From the above equation, the N.A. can be located. The stresses $f_c'$ and $f_s'$ are found by equating the moment of resistance of the beam to the external moment. The procedure is essentially that of trial and error, as illustrated in Example 46.1.

**Example 46.1.** *The balanced section of a singly reinforced beam is designed for the following permissible stresses : $c = \sigma_{cb}$ 5 $N/mm^2$ ; $t = \sigma_{st} = 140 \, N/mm^2$; $m = 18$ ; $E_s = 2.1 \times 10^5 \, N/mm^2$. Determine the final stresses in concrete and steel, taking into account shrinkage of concrete having a shrinkage coefficient of 0.0004.*

**Solution:** (a) **Properties of original section** : For a balanced section,

$$k = \frac{18 \times 5}{18 \times 5 + 140} = 0.3913 \; ; \quad j = 1 - k/3 = 0.8696 \; ; \quad R = \frac{1}{2} \times 5 \times 0.8696 \times 0.3913 = 0.851$$

$$\therefore \quad M_r = Rbd^2 = 0.851 \, bd^2 \; ; \quad p' = \frac{k \cdot c}{2t} = \frac{0.3913 \times 5}{2 \times 140} \approx 0.007$$

(b) **Effect of shrinkage** : The N.A., after shrinkage, is located with the help of *Eq. 46.4*. As a first approximation, let $f_s' \approx t = 140 \, N/mm^2$. Them, from *Eq. 46.3*,

$$\delta = \frac{f_s'}{E_s \cdot \varepsilon_{sh} + f_s'} = \frac{140}{2.1 \times 10^5 \times 0.0004 + 140} = 0.625$$

$$k' = \sqrt{(18 \times 0.007 \times 0.625)^2 + (2 \times 18 \times 0.007 \times 0.625)} - 18 \times 0.007 \times 0.625$$

$$= 0.3258 \; (\text{as against } k = 0.391)$$

$$j' = 1 - k'/3 = 0.891$$

The above values of $k'$ and $j'$ have been found on the assumption that $f_s' = t$. Restricting $f_s' = 140 \, N/mm^2$, the stress is given by equating total tension to total compression

$$p' \, bd \cdot f_s' = \frac{1}{2} f_c' \cdot bk' \, d \quad \text{or} \quad f_c' = \frac{2 \, p' f_s'}{k'} = \frac{2 \times 0.007 \times 140}{0.3258} \approx \mathbf{6.02 \, N/mm^2}$$

Now corresponding to $f_s' = t = 140 \, N/mm^2 \quad$ and $\quad k' = 0.3258$

$$M_r' = p' \, bd \cdot f_s' \cdot j' \, d = 0.007 \times 140 \times 0.891 \, bd^2 = 0.873 \, bd^2$$

However, beam has been designed for a moment of resistance of $0.851 \, bd^2$. Hence, as an approximation, decreasing the stress in *steel proportionately*, we get

$$f_s' \text{ (revised)} = (0.851/0.873) \times 140 \approx 136.5 \, N/mm^2$$

This shows, that concrete is over-stressed while steel is relieved of part of its tensile stress.

## 46.5. INSTANTANEOUS AND REPEATED LOADING ON CONCRETE

(a) **Instantaneous Loading** : A concrete member is said to be loaded instantaneously if the load is applied for a very short period, say for 1 minute. If a curve is plotted between stress and strain, it will be of the shape shown in Fig. 46.5 (a). The curve will be straight in the initial portion, and then it becomes curvilinear indicating that strain increases more rapidly than stress. If the stress is removed, the whole of the strain is not reversible. At small stress, strain may decrease to zero on the removal of load but for higher stress, some residual strain remains on the removal of the applied stress. Fig. 46.5 (b) shows the curve between stress and residual strain. The magnitude of residual strain may be about 1/6 of the instantaneous strain.

**(b) Repeated Loading :** If a concrete specimen is loaded to a certain stress $f_c$, it attains a certain strain $OA$ instantaneously. This strain is known as

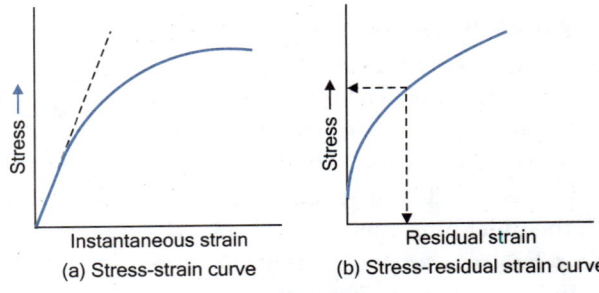

Fig. 46.5

initial strain, while $OA$ is the loading curve. If the specimen is unloaded it follows an unloading curve $AA_1$, which is more or less linear, leaving behind certain residual strain $OA_1$ ($= e_r$), as shown in Fig. 46.6 (a). If the specimen is loaded for the second time, to the *same stress f*, the loading curve will be $AB$, giving an initial strain $FB$, which is more than the earlier initial strain $FA$ of the first cycle. On unloading, it will follow the straight path $BB_1$ leaving behind a total residual strain $OB_1$. On repeated loading and the unloading, it is seen that the total initial strain increases and residual strain also increases, though at slower rate. The unloading curves have the same slopes, suggesting the unloading to be elastic.

Fig. 46.6 (b) shows the curve between total strain and No. of loadings. It is observed that after certain loadings, strain does not increase further, *i.e.* max. strain is reached. The concrete at this stage is said to be *work-hardened*. At this stage, the loading and unloading curve terminate at the same points $N$ and $N_1$ as shown in Fig. 46.6 (a).

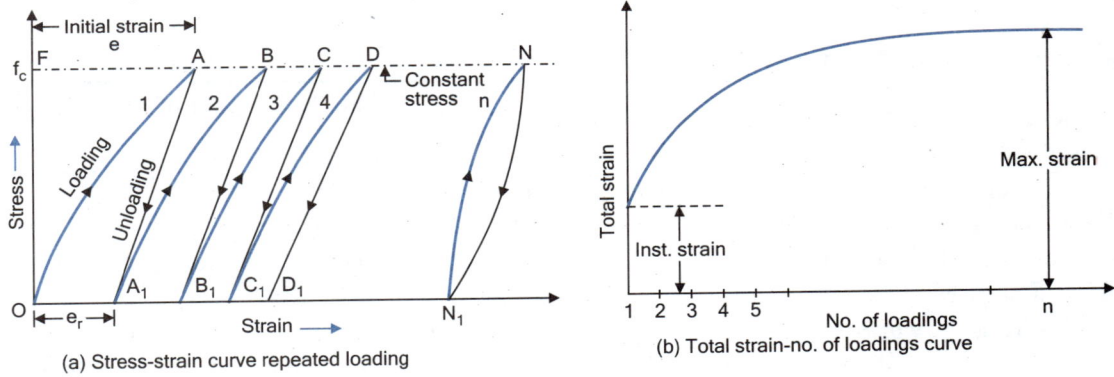

Fig. 46.6

## 46.6. SUSTAINED LOADING : CREEP

In the previous article, creep is the increase in strain under a sustained constant stress after taking into account other time dependent deformations not associated with stress (*viz*, shrinkage, swelling, and thermal deformations) we have considered only instantaneous loading, giving rise to instantaneous strain only.

However, the strain in concrete, at a given constant stress, is a function of time. The strain which increases with time at a given constant stress, is called *deferred strain or creep*. The *ultimate strain* may be 2 to 3 times the initial or instantaneous strain (elastic strain) and it may take 2 to 3 years to achieve this. Characteristic values for creep, expressed as deformation per unit length (or expressed as strain), for 1 : 2 : 4 concrete loaded at 28 days with a sustained stress of 4 N/mm² are 0.0003 at 28 days after loading and 0.0006 at one year. Creep may be represented either by a *creep coefficient*, which is the ratio of total strain to initial strain, or by value of strain per unit of concrete stress.

The creep is counted from the initial elastic strain $\dfrac{\sigma_0}{E}$ (where $\sigma_0$ is compressive stress applied to the concrete after curing for time ($t_0$) and (E) is the secant modulus of elasticity to concrete)

Fig. 46.7 shows the curve between strain and time elapsed since loading, when the stress is maintained at constant level. The initial portion OA represents *instantaneous strain*. The creep strain increases with time till ultimate strain corresponding to point C is reached. If the specimen is unloaded at point B, the unloading curve $B B_1 B_2 B'$ will have straight drop $BB_1$ representing instantaneous strain recovery which may be upto 70% of the instantaneous strain OA. After that, the unloading curve is curvilinear upto point $B_2$. Point $B'$ corresponds to the maximum strain recovery. The rate of strain, or creep can be expressed by the following relation:

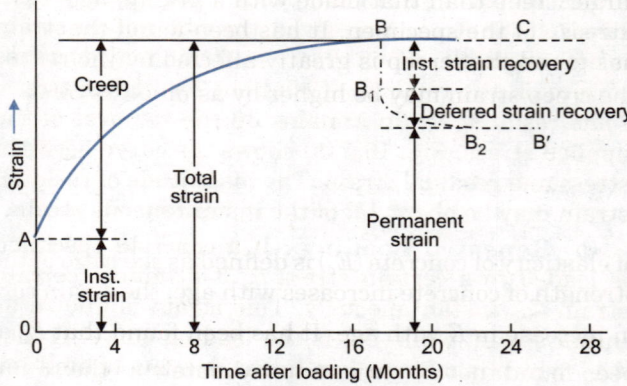

Fig. 46.7. Strain-time Curve.

$$\varepsilon_t = \varepsilon_u \left[ 1 - 10^{-\frac{1}{4}\sqrt{t}} \right] \qquad \ldots(46.5)$$

where  $\varepsilon_t$ = creep after time $t$ months

$\varepsilon_u$ = ultimate strain

If ultimate strain is reached after $2\frac{1}{2}$ years, the creep after 1 month, 3 month and 1 year will be approximately 44%, 63% and 86% of the ultimate strain.

## 46.7. FACTORS AFFECTING CREEP

Creep depends upon the following factors:
(1) Magnitude of stress and rate of its application
(2) Duration of stress
(3) Age of concrete at commencement of loading
(4) Type of cement
(5) Method of mixing and water cement ratio
(6) Fineness of cement
(7) Characteristics of aggregate
(8) Volume/space ratio of member
(9) Atmospheric condition, *i.e.* temperature and humidity.

Within the range of working stress, creep is proportional to stress. For higher stresses it increases rapidly and not in proportion to the increase in stress. Also creep increases with magnitude of stress. The amount of creep depends upon the rate of application of the load—it being greater if load is applied slowly. As is clear from *Eq. 46.5*, creep increases with time, though the rate of creep decreases. Glanville found that creep may continue even upto 7 years. It has been found that the creep for about first month after applying the load depends on the age of concrete on loading, its value decreases with increasing age. However, the later rates of creep depend on the actual age of the concrete at the time of observation of creep and *not* on the age when loaded.

Creep also depends upon type of cements. For Portland cements, the creep increases very rapidly over the first year and then only very slow. In high alumina cement, the creep increases almost in proportion with time for ages between 1 to 5 years, being almost double at 5 years than what it is at 1 year. Creep for Portland blast furnace cement is much greater than that of ordinary Portland cement. Davis found that creep for low heat cement is greater than that of ordinary Portland cement.

Creep increases with increase in water cement ratio. The creep for hand mixed concrete having higher water/cement ratio is found to be much higher than that obtained with vibration compacted concrete of much low water/cement ratio. However, the increase in creep at higher water/cement ratio helps to counteract the increased tendency for cracking caused through the reduction in the strength of concrete and the increased drying shrinkage.

Fineness of grinding of cement affects the strength of concrete. Creep is less for stronger concrete. Lyse found that creep is also proportional to the amount of cement paste in the concrete.

According to Lorman, increase on fineness modulus of the aggregate increases the creep. Also concrete made up of porous aggregate has a greater creep. Concrete made up of sandstone aggregate has a very much larger creep than that made with a granite aggregate. Davis found that creep decreases with increase in the size of the specimen. It has been found that creep also increases with increase in temperature. Lyse has found that creep is greatly affected by the relative humidity of the atmosphere. In dry atmospheres, the creep strain may be higher by as much as 50%.

## 46.8. EFFECT OF CREEP ON $E_c$ AND $m$

(a) **Variation of instantaneous modulus of elasticity with age.** The value of instantaneous modulus of elasticity of concrete ($E_c$) is defined as the ratio of the stress applied to the instantaneous strain. Since the strength of concrete increases with age, the strain corresponding to a given stress also decreases, resulting in increase in $E$ with age. It has been found that $E_c$ attains a limiting value at an age of $4\frac{1}{2}$ years. It has been found that $E_c$ varies approximately as the square root of strength. Thus

$$E_c \propto \sqrt{\sigma_c} \qquad \ldots(46.6)$$

If the strength of concrete and the value of $E_c$ are known at 28 days, the value of $E_c$ at any other period can be known provided the strength at that time is known. For example, if a concrete has attained 64% of its days strength, its $E_c$ will be $100\sqrt{0.64}$ = 80 % of its value at 28 days. The % values of $E_c$ at various ages are as follows :

| Age | 7 days | 28 days | 90 days | 1 year | $4\frac{1}{2}$ years |
|---|---|---|---|---|---|
| $E_c$ | 66% | 80% | 90% | 94% | 100%. |

(b) **Variation of $E_c$ and $m$ with creep :** The above discussions are with respect to the instantaneous strain or elastic strain. Since the total strain at any time after loading is the sum of instantaneous strain and creep strains, the Youngs modulus of elasticity computed on the basis of measured strains will decrease with time. Ordinary M 15 concrete has $E_c$ = 0.11 × 10⁵ N/mm², on the basis of instantaneous strain, giving a value of $m$ = 19 if $E_s$ is taken equal to 2.1 × 10⁵ N/mm². If the creep strain is taken as twice the elastic strain, the *total strain* will decrease to of its original value, and hence $m$ will increase from 19 to 57.

Ross observed that if creep strains are also taken into account, the modular ratio $m$ for Portland cement varies as follows with times :

| Times after loading | 14 days | 1 month | 2 months | 3 months | 5 months |
|---|---|---|---|---|---|
| Modular ratio | 18 | 23 | 32 | 40 | 62 |

## 46.9. EFFECT OF SHRINKAGE AND CREEP IN COLUMNS

We have seen earlier that shrinkage causes tensile stress in concrete and compressive stress in steel. Hence a loaded column will have less compressive stress in concrete and more compressive stress in steel, when the effect of shrinkage is considered.

Due to creep, the modular ratio increases. In a loaded column, the share of load between concrete and steel depends upon the modular ratio. When modular ratio increases (say from 19 to 57 for M 15 concrete), steel shares more load and concrete shares less load. When a column is tested to failure, $E_c$ decreases as stress in concrete increase, with the result that $m$ increases. Hence at higher stress, steel shares more load, till it reaches its yield stress at which stage $E_s$ becomes zero and hence $m$ also becomes zero. When once steel has yielded, concrete starts taking the additional load till it is crushed at its yield stress. Hence a reinforced concrete ultimately fails when steel has reached its yield stress and concrete has crushed.

The method of calculating the stresses in reinforced concrete column, taking into account the effects of shrinkage and creep, is illustrated in example 46.2.

**Example 46.2.** *A reinforced concrete column, 200 mm × 200 mm is reinforced with 1.25 % vertical reinforcement. It is subjected to a total load of 200 kN out of which 50% is the dead load. Compute the stresses in concrete and steel, with and without live load, taking into account the effect of shrinkage and creep. Given : coefficient of creep = 8.1 × 10⁻⁵ per unit stress; shrinkage strain = 0.045% $E_c$ = 0.14×10⁵ N/mm² and $E_s$ = 2.1 × 10⁵ N/mm².*

**Solution:** Given : $p' = 0.0125$ and $\varepsilon_{sh} = 0.00045$ ; $A_{st} = 0.0125 \times 40000 = 500$ mm²

### (a) Stresses under dead load

$$\text{Elastic strain per unit stress} = \frac{1}{E_c} = \frac{1}{0.14 \times 10^5} = 7.142 \approx 7.15 \times 10^{-5}$$

$$\text{Creep strain per unit stress} = 8.1 \times 10^{-5}$$

$$\text{Total stain per unit stress} = (7.15 + 8.1) \times 10^{-5} = 15.25 \times 10^{-5}$$

$$\therefore \quad \text{Modified } E_c = \frac{1}{15.25 \times 10^{-5}} = 0.06557 \times 10^5 \text{ N/mm}^2$$

$$\text{Modified } m = \frac{E_s}{E_c} = \frac{2.1 \times 10^5}{0.06557 \times 10^5} \approx 32$$

This modified value of $m$ should be used for calculating stresses in concrete and steel, due to shrinkage, given by *Eqs. 46.1* and *46.2*. Thus, from *Eq. 46.1*,

$$f_s = \frac{1-p'}{1+(m-1)p'} \cdot E_s \cdot \varepsilon_{sh} = \frac{1 - 0.0125}{1 + (32-1)0.0125} \times 2.1 \times 10^5 \times 0.00045$$

$$= 67.26 \text{ N/mm}^2 \text{ (compressive)}$$

$$f_c = \frac{p'}{1-p'} \cdot f_s = \frac{0.0125}{1-0.0125} \times 67.26 = 0.85 \text{ N/mm}^2 \quad \text{(tensile)}$$

Compressive stress in concrete, due to dead load

$$= \frac{100000}{(200 \times 200) + (32-1) \times 500} = 1.802 \text{ N/mm}^2$$

Compressive stress in steel, due to dead load = $1.802 \times 32 = 57.66$ N/mm²

$$\therefore \quad \text{Net stress in concrete} = 1.802 - 0.85 = \mathbf{0.952 \text{ N/mm}^2} \quad \text{(Comp.)}$$

$$\text{Net stress in steel} = 67.26 + 57.66 = \mathbf{124.92 \text{ mm}^2} \quad \text{(Comp.)}$$

### (b) Stress due to live load :
The effect of live load has to be considered immediately after its application. Since they are of short duration, creep strains are caused only by dead loads.

Hence

$$m = \frac{2.1 \times 10^5}{0.14 \times 10^5} = 15$$

$$\therefore \quad \text{Compressive stress in concrete} = \frac{100000}{(200 \times 200) + (15-1)\,500} = 2.128 \text{ N/mm}^2$$

Comp. stress in steel = $2.128 \times 15 \approx 31.915$ N/mm²

$$\therefore \quad \text{Total stress in concrete} = 0.952 + 2.128 = \mathbf{3.08 \text{ N/mm}^2}$$

$$\text{Total stress in steel} = 124.92 + 31.915 = \mathbf{156.835 \text{ N/mm}^2}$$

## 46.10. EFFECT OF SHRINKAGE *AND* CREEP IN BEAMS

Out of the dead load and live load, only dead load causes creep since it is of a permanent nature while live load is only for a short duration. Due to creep caused by dead load, the value of modular ratio changes. Hence the calculation for shrinkage stresses are to be done with respect to the changed value of modular ratio. In the case of column, we had calculated the effects of dead load and live load separately and superimpose them. However, in the case of R.C. beam, the stresses are dependent on the position of N.A., which in turn, is dependent on modular ratio. Modular ratio is different for dead and live load effects. Hence their combined effects are to be computed on the basis of the position of actual neutral axis for the combined loads. The method of computing the stresses is illustrated in Example 46.3. *It will be seen there that the effects of shrinkage and creep practically neutralise each other.*

**Example 46.3.** *Compute the stresses in beam of Example 46.1 taking into account creep also, if dead loads are 1/3rd of the total load. Take the changed value of m due to creep as 54.*

**Solution:** Given : $t = \sigma_{st} = 140$ N/mm$^2$ ; $c = \sigma_{cbc} = 5$ N/mm$^2$ ; $m = 18$ ; $\varepsilon_{sh} = 0.0004$ ; $p' = 0.007$

### (a) Effects of dead load alone

Since dead load is equal to $\frac{1}{3}$ rd the total load, the net tensile stress in steel due to dead load

$$= \frac{1}{3} \times 140 \approx 46.67 \text{ N/mm}^2 = f_s'. \quad \text{Hence from Eq. 46.3,}$$

$$\delta = \frac{f_s'}{f_s' + E_s \cdot \varepsilon_{sh}} = \frac{46.67}{46.67 + 2.1 \times 10^6 \times 0.0004} = 0.053$$

Hence from Eq. 46.4, taking $m = 54$, we get

$$k' = \sqrt{(54 \times 0.007 \times 0.053)^2 + (2 \times 54 \times 0.007 \times 0.053)} - 54 \times 0.007 \times 0.053 = 0.181$$

$$f_c' = \frac{2 \, p' \, f_s'}{k'} = \frac{2 \times 0.007 \times 46.67}{0.501} = 1.304 \text{ N/mm}^2 \text{ (tensile)}$$

Thus stress in concrete, taking into account both shrinkage and creep, is 1.304 N/mm$^2$.

**Note 1.** If *creep is neglected* and effect of shrinkage is taken into account, we have $m = 18$. Hence from *Eq. 46.4* taking $\delta = 0.053$, we get

$$k' = \sqrt{(18 \times 0.007 \times 0.053)^2 + (2 \times 18 \times 0.007 \times 0.053)} - (18 \times 0.007 \times 0.053) = 0.109$$

$$f_c' = \frac{2p' f_s'}{k} = \frac{2 \times 0.007 \times 46.67}{0.109} = 5.99 \text{ N/mm}^2$$

Thus, the effect of creep is to *decrease* the stress in concrete by a value 5.99 – 3.61 = 2.38 N/mm$^2$.

**Note 2.** Again, if shrinkage and creep are neglected, we have

$$k = 0.3913 \quad \text{(Example 46.1)}$$

$$f_c = \frac{2 \, p' \, f_s}{k} = \frac{2 \times 0.007 \times 46.67}{0.3913} = 1.67 \text{ N/mm}^2$$

This is practically the same as that found by taking shrinkage and creep into account. *Hence it is concluded that the effects of shrinkage and creep neutralise each other.* Due to this reason, shrinkage and creep are not taken into account in the usual design of beams.

### (b) Effect of combined loads

When beam is subjected to full loads, $t = 140$ N/mm$^2$ and $c = 6.02$ N/mm$^2$, considering shrinkage, as founded in Example 46.1. As found in step (a) above, effect of creep is to cause a stress of 0.9 N/mm$^2$. Hence superimposing the effect of creep (due to dead load), the stress in concrete = 6.02 – 0.9 = 5.12 N/mm$^2$.

If both creep and shrinkage are neglected, $k = 0.3913$. and $j = 0.869$, as for normal design (See Example 46.1).

If creep is neglected but shrinkage is taken into account $k' = 0.3258$. (See Example 46.1), corresponding to $t = f_s' = 140$ N/mm$^2$.

If creep and shrinkage both are taken into account for $t = f_s' = 46.67$ N/mm$^2$, we get $k' = 0.4018$.

Therefore, taking into account the combined load, and the effect of creep and shrinkage, $k'$ lies between 0.3258 and 0.4018, which will be a value certainly less than 0.3913. Hence lever arm $j$ will be greater than 0.869 used for normal design. Hence the values of $f_s'$ and $f_c'$ will be less than 140 N/mm$^2$ and 5.12 N/mm$^2$, while in the usual design procedure (neglecting shrinkage and creep), the values of stress are 140 N/mm$^2$ and 5 N/mm$^2$. *Hence shrinkage and creep do not change the design materially.*

# CHAPTER 47: FORMWORK

## 47.1. INTRODUCTION

The *formwork* or *shuttering* is a temporary ancillary construction used as a mould for the structure, in which concrete is placed and in which it hardens and matures. The construction of formwork involves considerable expenditure of time and material. The cost of formwork may be upto 15 to 25% of the cost of structure in building work and even higher in bridges. In order to reduce this expenditure, it is necessary to design economical types of formwork and to mechanize its construction. When the concrete has reached a certain required strength, the formwork is no longer needed and is removed. Formwork must be adequately strong and stiff to carry the loads produced by concrete, the workers placing and finishing the concrete and any equipment (*or*) material supported by the forms. The operation of removing the formwork is commonly known as *stripping*. When stripping takes place, the components of formwork are removed and then reused for the forms of another part of the structure. Such forms, whose components can be reused several times are known as *panel forms*. In contrast to this are *stationary forms* which are made for individual non standard member and structures, which have no repeatable elements, and also for structural members the formwork of which cannot be stripped.

Forms are classified as wooden, plywood, steel, combined wood steel, reinforced concrete and plain concrete. Timber is the most common material used for formwork. The disadvantage of wooden formwork is the possibility of warping, swelling and shrinkage of the timber. However, these defects can be overcome by applying to the shuttering, water-impermeable coatings. This coating also prevents the shuttering from adhering to concrete and hence makes stripping easier. Steel shuttering is used for major work where every thing is mechanised. Steel formwork has many advantages, such as follows (*i*) it can be put to high number of uses (*ii*) it provides ease of stripping (*iii*) it ensures an even and smooth concrete surface (*iv*) it posseses greater rigidity (*v*) it is not liable too shrinkage or distortion. However, steel formwork is comparatively more costly.

### A good formwork should satisfy the following requirements :

(*i*) The material of the formwork should be cheap and it should be suitable for re-use several times.

(*ii*) It should be practically waterproof so that it does not absorb water from concrete. Also, its shrinkage and swelling should be minimum.

(*iii*) It should be strong enough to withstand all loads coming, on it, such as dead load of concrete and live load during its pouring, compaction and curing.

(*iv*) It should be stiff enough so that deflection is minimum.

(*v*) It should be as light as possible.

(*vi*) The surface of the formwork should be smooth, and it should afford easy stripping.

(*vii*) All joints of the formwork should be stiff so that lateral deformation under loads is minimum. Also, these joints should be leakproof.

(*viii*) The formwork should rest on non-yielding supports.

## 47.2. INDIAN STANDARD ON FORMWORK (IS : 456-2000)

**1. *General*.** The formwork shall conform to the shape, lines and dimensions as shown on the plans and be so constructed as to remain sufficiently rigid during the placing and compacting of the concrete, and shall be sufficiently tight to prevent loss of liquid from the concrete.

**2. *Cleaning and Treatment of forms*.** All rubbish, particularly chippings, shavings and sawdust, shall be removed form the interior of the forms before the concrete is placed and the formwork in contact with the concrete shall be cleaned and thoroughly wetted or treated with an approved composition. Care shall be taken that such approved composition is kept out of contact with the reinforcement.

**3. *Stripping time*.** In no circumstances shall forms be struck until the concrete reaches a strength of atleast twice the stress to which the concrete may be subjected at the time of striking.

The stripping time of concrete formwork depends on strength of structures member. The strength development of concrete member depends on

1. Grade of concrete
2. Grade of cement
3. Type of cement
4. Temperatures
5. Six of the concrete member
6. Curing method/accelerated covering

The strength referred to shall be that of concrete using the same cement and aggregate, with the same proportions ; and cured under condition of temperature and moisture similar to those existing on the work. Where possible, the formwork should be left longer, as it would assist the curing.

In normal circumstances (generally where temperatures are above 20°C), and where ordinary cement is used, forms may be struck after expiry of the following periods:

| | |
|---|---|
| (*a*) Walls, columns and vertical sides of beams | 24 to 48 hours as may be decided by the engineer-in-charge. |
| (*b*) Slab soffits (props left under) | 3 days. |
| (*c*) Beam soffits (props left under) | 7 days. |
| (*d*) Removal of props to slab: | |
|    (*i*) Spanning upto 4.5 m | 7 days. |
|    (*ii*) Spanning over 4.5 m | 14 days. |
| (*e*) Removal of props to beams and arches: | |
|    (*i*) Spanning upto 6 m | 14 days. |
|    (*ii*) Spanning over 6 m | 21 days. |

**Note:** The number of props, their sizes and disposition, shall be such as to be able to safely carry the full load of the slab, beam or arch as the case may be.

**4. *Procedure when Removing the Formwork*.** All formwork shall be removed without such shock or vibration as would damage the reinforced concrete. Before the soffit and struts are removed, the concrete surface shall be exposed, where necessary in order to ascertain that the concrete has sufficiently hardened. Proper precautions shall be taken to allow for the decrease in the rate of hardening that occurs with all cements in the cold water.

**5. *Camber*.** It is generally desirable to give forms an upward camber to ensure that the beams do not have a sag when they have taken up their deflection, but this should not be done unless allowed for in design calculations of the beams.

**6. *Tolerances*.** Formwork shall be so constructed that the internal dimensions are within the permissible tolerance specified by the designer.

Generally the table of concrete strength is considered for removal at formwork for various type of concrete structural member. Strength of concrete Vs. structural member type and span for formwork removal.

| Structural Member Type and Span | Concrete Strength |
|---|---|
| Lateral parts of the formwork for all structural members can be removed. | 2.5 N/mm² |
| Interior parts of formwork of slabs and beams with a span of up to 6 m can be removed. | 70% of design strength |
| Interior parts of formwork of slabs and beams with a span of more than 6 m can be removed. | 85% of design strength |

## 47.3. LOADS ON FORMWORK

The formwork has to bear mainly the following loads apart from its own weight: (i) live load due to labour etc. (ii) dead weight of wet concrete (iii) hydrostatic pressures of the fluid concrete acting against the vertical or inclined faces of form, and (iv) impact due to pouring concrete. The temporary live loads of workmen and equipment, including the impact, may be taken equal to 3700 N/mm², for the design of planks and joists in bending and shear. The hydrostatic pressure due to fluidity of concrete in the initial stages of pouring depends upon several factors such as quantity of water in concrete, size of aggregates, rate of pouring and temperature. The hydrostatic pressure is maximum during pouring, and then decreases as concrete sets. Therefore, the *main* factor influencing this pressure is the depth of concrete poured before the concrete sets. The setting time may be taken between $\frac{3}{4}$ to 1 hour. Hence while computing the pressure, only the height of concrete poured in $\frac{3}{4}$ to 1 hour need only be taken into account. For heights of concrete upto 1.5 m, the equivalent fluid weight of concrete may be taken as 23000 N/m³. For higher heights, the equivalent fluid weight is reduced. When the height of concrete in one pour is 6 m, the equivalent fluid weight may be taken as only 12000 N/m³. For intermediate heights between 1.5 to 6 m poured within the setting time of $\frac{3}{4}$ to 1 hour, linear interpolation of unit weight between 23000 to 12000 N/m³ may be done.

Table 47.1 gives the safe values of stresses etc. for some common types of soft wood used for formwork. The maximum permissible deflection of sheathing and joists etc. should not exceed 2.5 mm.

TABLE 47.1. Permissible Stresses in Timber

| Property | Types of timber | | | |
|---|---|---|---|---|
| | Fir | Deodar | Kail | Chir |
| Density (N/m³) | 4500 | 5450 | 5150 | 5750 |
| Modulus of elasticity $E$ (N/mm²) | 9400 | 9500 | 6800 | 9800 |
| Permissible stresses in bending and tension (N/mm²) | | | | |
|    (i) Inside | 7.8 | 10.2 | 6.6 | 8.4 |
|    (ii) Outside | 6.6 | 8.8 | 5.6 | 7.0 |
|    (iii) Wet | 5.6 | 7.0 | 5.0 | 6.0 |
| Permissible stresses in shear (N/mm²) | | | | |
|    (i) Horizontal | 0.6 | 0.7 | 0.6 | 0.4 |
|    (ii) Along grain | 0.8 | 1.0 | 0.8 | 0.9 |
| Permissible compressive stresses [Parallel] (N/mm²) | | | | |
|    (i) Inside | 6 | 7.8 | 5.2 | 6.4 |
|    (ii) Outside | 5.2 | 7.0 | 4.6 | 5.6 |
|    (iii) Wet | 4.2 | 5.6 | 3.8 | 4.5 |
| Permissible compressive stresses [perpendicular] (N/mm²) | | | | |
|    (i) Inside | 1.6 | 2.6 | 1.7 | 2.2 |
|    (ii) Outside | 1.2 | 2.1 | 1.3 | 1.7 |
|    (iii) Wet | 1.0 | 1.7 | 1.0 | 1.4 |

## Standards o Formwork

### (i) International standards
- ACI 347r – 94 – Guide to formwork for concrete
- ACI SP – 4 – Formwork for concrete
- OHSAS – Occupational healths and safety act standards
- BS5975 – British standards
- CAN/CSA – 6269.3 Canadian standard
- SAA 1509 – 1974 Australian standards
- DIN 4420 – German standards

The above standards cover all topics in details, including lateral pressure of concrete superimposed loads like D and M loads, impact, loads and environmental loads like wind loads etc.

### (ii) Indian standards
IS: 14687-1999 – Indian standard – Formwork for concrete structures guidelines

IS: 456-2006 (Plain and reinforced concrete)

The occur code available for reference IRC–87–1984 Guidelines on design and errection of formwork for road bridges is derived form British and American codes.

## 47.4. SHUTTERING FOR COLUMNS

Shuttering for a column is probably the simplest. It consists of the following main components: (i) sheathing all round the column periphery (ii) side yokes and end yokes (iii) wedges and (iv) bolts with washers. Fig. 47.1 shows the formwork for a square column. The side yokes and end yokes consists of two number each, and are suitably spaced along the height of the column. The two-side yokes are comparatively of heavier section, and are connected together by two long bolts of 16 mm dia. Four wedges, one at each corner, are inserted between the bolts and the end yokes. The sheathing is nailed to the yokes.

**Spacing of yokes.** The main design element for the formwork of a column is the spacing of yokes along the height of the columns. The sheathing may be considered to be a continuous slab over the yokes with its span(s) equal to the spacing of the yokes. Let $b$ the width of the column. The fluid pressure of concrete over the width will be transferred to the yokes.

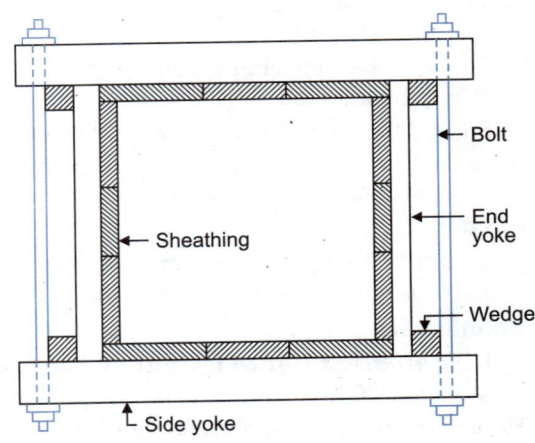

Fig. 47.1. Formwork for Square or Rectangular Column.

Let $w$ = equivalent fluid weight of concrete (N/m$^3$)

$h$ = depth of concrete poured within the setting time of $\frac{3}{4}$ to 1 hours (m).

$W$ = total load on each yoke, uniformly distributed over the length $b$ (N).

$l$ = span of the side yoke, centre to centre between the bolts (mm).

$s$ = centre to centre spacing of yokes (mm).

$$\therefore \quad W = \frac{wh \cdot s \cdot b}{1000 \times 1000} = wh \cdot sb \times 10^{-6} \text{ N} \quad \ldots(i)$$

The bending moment in sheathing = $\frac{W \times s}{10}$ (N-mm)

If $f$ is the permissible stress in sheathing (N/mm$^2$) and $d$ is the thickness of sheathing (in mm) we have

$$f \cdot \frac{b d^2}{6} = \frac{W \cdot s}{10} \quad \ldots(ii)$$

Substituting the value of W from (i), we get

$$f \frac{bd^2}{6} = whbs^2 \times 10^{-7}$$

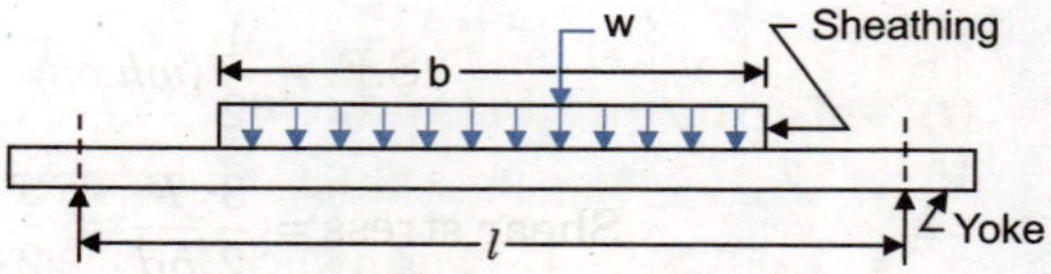

or $\qquad s = 1291 \, d \sqrt{\dfrac{f}{wh}}$ mm  ...(47.1)

The spacing should also be checked for shear and deflection in the sheathing. The above expression for '$s$' is based on permissible stress in sheathing. Let us also find the value of $s$ based on permissible stress in the yoke. The span of the yoke is '$l$' mm centre to centre between the bolts, while it receives a total uniformly distributed load of $W$ over length $b$.

Hence B.M. $= \dfrac{W}{4}\left(l - \dfrac{b}{2}\right) = \dfrac{W}{8}(2l - b)$ N-mm

If '$Z$' is the section modulus of yoke, and $f'$ is the permissible stress in bending (N/mm²) for the timber of yoke, we have

$$\frac{W}{8}(2l - b) = f' Z \qquad ...(iii)$$

Substituting the value of W from (i), we get

$$\frac{whsb \times 10^{-6}}{8}(2l - b) = f' Z$$

$$s = \frac{8 \times 10^6 \, f' Z}{whb(2l - b)} \qquad ...(47.2)$$

The lesser of spacing given by Eqs. 47.1 and 47.2 may be adopted.

Fig. 47.3 shows shutterings for octagonal and round columns. Example 47.1 illustrates the method of designing the shuttering of a square column.

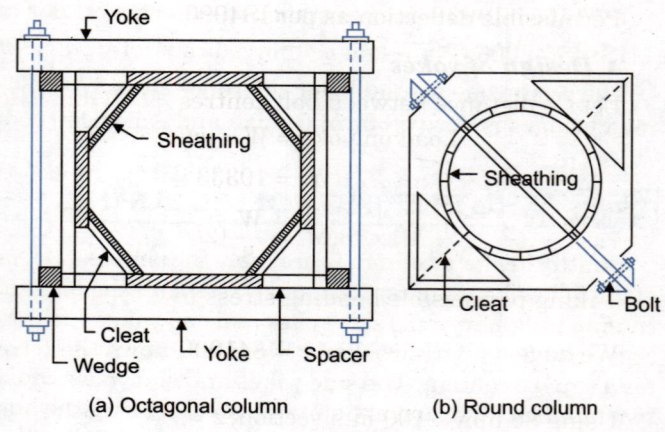

Fig. 47.2

(a) Octagonal column  (b) Round column

Fig. 47.3

**Example 47.1.** *Design the formwork for a column 350 mm × 350 mm, having a height of 3 m. It is proposed to pour the entire concrete in one stage.*

**Solution:**

**1. Equivalent liquid weight :** When the height of pour is 1.5 m, the equivalent fluid weight of concrete is taken as 23000 N/m³, while for height of pour of 6 m, equivalent fluid weight may be taken as 12000 N/m³. Hence by linear interpolation when the height of pour is 3 m the unit fluid weight may be taken as

$$= 23000 - \left(\frac{23000 - 12000}{6 - 1.5}\right)(3 - 1.5) = 23000 - 3667 = 19333 \text{ N/mm}^3.$$

**2. Bending stress in sheathing :** Let the spacing of yokes be $s$ mm c/c.

Also, let the distance between the bolt centres = 600 mm.

From bending stress point of view, the spacing of yokes is given by Eq. 47.1. Taking the sheathing of chir wood, the permissible bending stress may be taken equal to 8.4 N/mm². Hence using 25 mm (= $d$) thick sheathing, we have

$$s = 1291 \, d \sqrt{\frac{f}{wh}} = 1291 \times 25 \sqrt{\frac{8.4}{19333 \times 3}} \approx 388 \text{ mm}$$

Keep spacing as 390 mm.

**3. Check for shear and deflection in sheathing**

Let us check this spacing for shear and deflection.

Permissible shear stress = 0.9 N/mm²

$$\text{S.F.} = \frac{1}{2}(wh.b) \times \frac{390}{1000} = \frac{1}{2}(19333 \times 3 \times 0.35)\frac{390}{1000} \approx 3958 \text{ N}$$

∴ $\qquad$ Shear stress $= \frac{3}{2}\frac{F}{bd} = \frac{3}{2} \times \frac{3958}{350 \times 25} = 0.68 \text{ N/mm}^2 < 0.9 \quad$ Hence safe.

Again $\qquad \delta = \frac{5}{384} \cdot \frac{(whb)s^4}{EI} \quad$ where $\quad EI = (9800)\frac{350(25)^3}{12} = 4.47 \times 10^9$ N-mm units

$whb$ = load per mm run of sheathing $= (19333 \times 3 \times 0.35)\frac{1}{1000} = 20.3$ N/mm

Substituting the values, we get

deflection $\qquad (\delta) = \frac{5}{384} \times \frac{20.3 \times (390)^4}{4.47 \times 10^9} \approx 1.368$ mm.

Permissible deflection as per IS4990 – 2011, $\quad S/270 = \frac{390}{270} = 1.44$ mm $> 1.368$ mm (safe).

**4. Design of yokes**

Let $l$ = distance between bolt centres = 600 mm. (Refer Fig. 47.2.)

$\qquad$ Load on yoke $= W = wh\,bs$ (where $b$ and $s$ are in metres).

$\qquad (W) = 19333 \times 3 \times 0.35 \times 0.39 \approx 7917$ N

∴ $\qquad$ Max. B.M. $= \frac{W}{8}(2l - b) = \frac{7917}{8}(2 \times 0.6 - 0.35)\,1000 \approx 841000$ N-mm

Taking permissible bending stress in yoke as 8.4 N/mm²

We have $\qquad 8.4\,z = 841000 \quad$ (or) $\quad z = \frac{841000}{8.4} = 1 \times 10^5$ mm³

Using 80 mm × 100 mm section, $z = \frac{1}{6} \times 80 \times (100)^2 \approx 13400$ mm²

Let us check this section for deflection.

$$\text{Deflection }(\delta) = \frac{5}{384} \times \frac{Wl^3}{EI} = \frac{5}{384} \cdot \frac{7917(600)^3}{9800 \times \frac{1}{12} 80(100)^3} = 0.34 \text{ mm} < 2.5 \text{ mm.} \quad \text{Hence safe.}$$

**5. Design of bolts**: Bolts are subjected to both direct pull as well as bending since the load is being transferred through wedges.

$\qquad W = 7917$ N ; $\quad$ Load in each bolt $= 7917/2 \approx 3959$ N.

Assuming that the distance of load transmitted through wedges is 40 mm,

$\qquad$ B.M. in bolt = 3959 × 40 = 158360 N-mm.

Using 25 mm dia. bolts, area of bolt $(A)$

$$= \frac{\pi}{4}(25)^2 \approx 491 \text{ mm}^2$$

∴ $\qquad$ Tensile stress in bolt = 3959/491 ≈ 8.07 N/mm²

$$\text{Bending stress} = \frac{158360}{\frac{\pi}{64}(25)^4}\left(\frac{25}{2}\right) \approx 103 \text{ N/mm}^2$$

∴ $\qquad$ Total stress = 103 + 8.07 = 111.07 N/mm².

## 47.5. SHUTTERING FOR BEAM AND SLAB FLOOR

Fig. 47.4 shows the formwork for beam and slab floor. The slab is continuous over a number of beams. The slab is supported on 25 mm thick sheathing laid parallel to the main beams. The sheathing is supported on wooden battens which are laid between the beams, at some suitable spacing. In order to reduce the deflection, the battens may be propped at middle of the span through joists. The side forms of the beam consists of 30 mm thick sheathing. The bottom sheathing of the beam form may be 50 to 70 mm thick. The ends of the battens are supported on the ledger which is fixed to the cleats throughout the length. Cleats 100 mm × 20 mm to 30 mm are fixed to the side forms at the same spacing as that of battens, so that battens may be fixed to them. The beam form is supported on a head tree. The shore or post is connected to head tree through cleats. At the bottom of shore, two wedges of hard wood are provided over a sole piece.

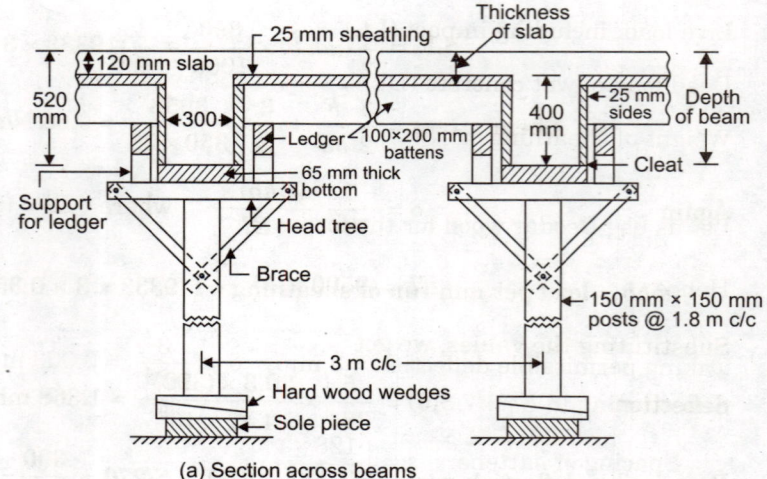

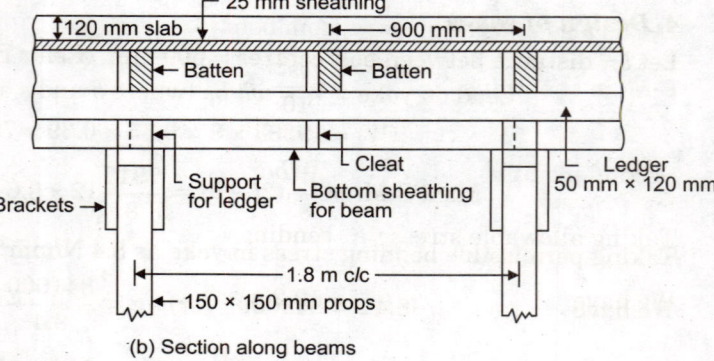

Fig. 47.4

### Spacing of battens

Let $s$ be the centre to centre spacing, in metres, of the battens which support the sheathing of the deck slab. The sheathing is continuous over battens.

Let  $'\delta'$ = permissible deflection in metres

$'w'$ = load on decking, N/mm$^2$

$'E'$ = modulus of elasticity in N/m$^2$ and $EI$ in N-m$^2$ units.

Taking the deflection to be equal to $\dfrac{3}{384} \dfrac{ws^4}{EI}$.

we have deflection $\quad '\delta' = \dfrac{3}{384} \dfrac{ws^4}{EI}$

From which $\quad s = \left(\dfrac{128 \, E \, I \, \delta}{w}\right)^{1/4}$ metres ...(47.3)

The above spacing is based on deflection criterion. However, the sheathing should be thick enough to take up the bending stresses caused due to loading.

**Example 47.2.** *Design the formwork for the beam and slab floor, for the following data:*

 1. *Thickness of floor : 120 mm*
 2. *Centre to centre spacing of beams = 3 m*
 3. *Width of beam = 300 mm and depth 400 mm below slab.*
 4. *Height of ceiling of the roof = 4 m above the floor.*

*Take live load on sheathing equal to 4000 N/m$^2$ and dead weight of wet concrete as 26500 N/m$^3$.*

**Solution:**

**1. Sheathing for the slab** : Let the sheathing be 25 mm thick, laid parallel to the beam. Let the spacing of battens be $s$ metre centre to centre. Load per square metre of the decking will be as under :

Live load, including impact (LL) $\qquad = 4000$ N/m²

Dead load of wet concrete (DL) $= \left[ 26500 \times \dfrac{120}{1000} \right] = 3180$ N/m²

Weight of sheathing $\qquad = 200$ N/m² (say)

Total ($w$) $\quad = 7380 \approx 7400$ N/m²

Let us use Deodar wood for the formwork.

Hence $\qquad EI = (9500 \times 10^6) \dfrac{1}{12} \times \left( \dfrac{25}{1000} \right)^3$ N-m² $= 12369.79 \approx 12370$ N-m².

Taking permissible deflection = 3 mm, $\delta = \dfrac{3}{1000} = 0.003$ m

Substituting in *Eq. 47.3*,

$\therefore$ Spacing of battens $\quad s = \left( \dfrac{128 \times 12370 \times 0.003}{7400} \right)^{1/4} \approx 0.9$ m.

Let us also find the spacing from bending stress point of view.

B.M. in sheathing $\quad = \dfrac{ws^2}{10}$ N-m $= \dfrac{7400\, s^2}{10} = 740\, s^2$ N-m

Section modulus $\quad = \dfrac{1}{6} bd^2 = \dfrac{1}{6} \times \left( \dfrac{25}{1000} \right)^2$ m²

Taking allowable stress $f$ in bending = 10.2 N/mm² = $10.2 \times 10^6$ N/m²

We have $\quad = 10.2 \times 10^6 \times \dfrac{1}{6} \left( \dfrac{25}{1000} \right)^2 = 740\, s^2$

From which centre to centre spacing ($s$) = $\sqrt{\dfrac{10.2 \times 10^6 \times 625}{6 \times 10^6 \times 740}} = 1.198 \approx 1.2$ m

Keep the spacing $s$ equal to lesser of the two values.

Hence provide battens at 900 mm centre to centre.

**2. Design of battens :** Centre to centre spacing of battens = 900 mm = 0.9 m.

Span of battens = $l = 3 - 0.3 = 2.7$ m.

Load on each batten = $7400 \times 0.9 = 6660$ N/m.

$$\text{B.M.} = \dfrac{6660(2.7)^2}{8} \times 1000 = 6.07 \times 10^6 \text{ N-mm}$$

$6.07 \times 10^6 = f \cdot z = 10.2\, z$ ; or $z = \dfrac{6.07 \times 10^6}{10.2} = 5.95 \times 10^5$ mm³

$\therefore \qquad bd^2 = 6 \times 5.95 \times 10^5 = 35.7 \times 10^5$

Taking 100 mm wide battens, $d = \sqrt{\dfrac{35.7 \times 10^5}{100}} = 189$ mm

However, use 100 mm × 200 mm section.

Max. S.F. = $\dfrac{5}{8} wl = \dfrac{5}{8} \times 6660 \times 2.7$ N

$\therefore \qquad$ Shear stress = $1.5 \dfrac{F}{bd} = \dfrac{1.5}{100 \times 200} \times \dfrac{5}{8} \times 6660 \times 2.7 = 0.84$ N/mm² < 1 N/mm².

**3. Design of side forms of the beam**

The side forms of beam will be subjected to hydrostatic pressure of fluid concrete. Let the equivalent fluid weight of concrete be 23000 N/m³.

Total height of beam = 120 + 400 = 520 mm
Height of side form sheathing above bottom of beam = 520 − (120 + 25) = 375 mm
Pressure at top level of side form

$$= 23000 \left(\frac{120+25}{1000}\right) = 23000 \times \frac{145}{1000}$$

Pressure at bottom level of side form $= 23000 \times \dfrac{520}{1000}$

Hence average pressure intensity on side form

$$= 23000 \left(\frac{145+520}{2}\right) \times \frac{1}{1000} \approx 7650 \text{ N/m}^2$$

The side form sheathing will span between battens.
Hence span of sheathing of side form may be taken as 0.9 m.

Height of side form = 375 mm = 0.375 m
Load $w = 7650 \times 0.375 \times 1$ N/m

$$\text{B.M.} = (7650 \times 0.375) \times \frac{(0.9)^2}{10} \times 1000 = 232368.7 \text{ N-mm}$$

Let the thickness of side = 25 mm.

∴ $z = \dfrac{1}{6} \times 375 \,(25)^2 \approx 39100 \text{ N/mm}^3$

∴ Bending stress = 232000/39100 = 5.95 N/mm² < 10.2.   Hence safe.

**4. Design of bottom of beams**: For the bottom of beams, the dead load of the full height of beam, including the thickness of slab, is to be taken into account. Similarly, the live load coming on the width of the beam alone is to be taken. Load per metre run of the bottom will be as follows :

| | | |
|---|---|---|
| Dead weight of concrete | = 26500 × 0.3 × 0.52 | ≈ 4140 N |
| Live load | = 4000 × 0.3 | = 1200 N |
| Weight of boarding/m | | = 400 N(say) |
| | Total $w$ | = 5740 N/m. |

The bottom of the beam is supported on props. Let the spacing of the props be double the spacing of the battens, i.e. 2 × 900 = 1800 mm = 1.8 m c/c.

$$\text{B.M. in bottom sheathing} = \frac{5740(1.8)^2}{8} \times 1000 = 2324700 \text{ N-mm}$$

Width of bottom = 300 + 2 × 25 = 350 mm

Let the thickness be $t$ cm.

∴ $z = \dfrac{1}{6} \times 350 \, t^2$

Allowable $f = 10.2$ N/mm²

∴ $\left(\dfrac{1}{6} \times 350 \, t^2\right) \times 10.2 = 2324700$

From which $t = \sqrt{\dfrac{2324700 \times 6}{350 \times 10.2}} \approx 62.5$ mm

Hence provide 65 mm thick sheathing at the bottom.

**5. Design of props**

| | | |
|---|---|---|
| Load from beam | = 5740 × 1.8 | ≈ 10350 N |
| Load from slab | = 6660 × 2.7 × 1.8 | = 32367.6 N |
| | Total load | = 42717.6 N |

Taking allowable compressive stress perpendicular to grains as 2.1 N/mm².

$$\text{Area of section} = \frac{42717.6}{2.1} = 20342 \text{ mm}^2.$$

Use 150 mm × 150 mm section. Let us test the prop against bucking,

$$\text{Safe stress} = f\left(1 - \frac{h}{50\,d}\right)$$

where  $f$ = safe compressive stress as short strut, parallel to grains = 7 N/mm²
$h$ = height of prop ≈ 4 − 0.4 = 3.6 m
$d$ = least dimension = 0.15 m

∴  $\quad\text{Safe stress} = 7\left(1 - \dfrac{3.6}{50 \times 0.15}\right) = 3.64$ N/mm²

∴  $\quad\text{Allowable load} = 3.64 \times 150 \times 150 \approx 82000$ N > 42717.6

Hence safe.

The details of formwork are shown in Fig. 47.4.

# CHAPTER 48

# TESTS ON CEMENT AND CEMENTS CONCRETE

## 48.1. INTRODUCTION

The quality of cement, and that of concrete manufactured with it, is judged with the tests conducted in the laboratory. The tests described in this chapter are conducted in all projects. There are many tests for cement, but we shall discuss the following important tests : (*i*) fineness test  (*ii*) consistency test  (*iii*) soundness test  (*iv*) test for setting time  (*v*) compressive strength test and  (*vi*) tensile strength test.

Following are the tests conducted on cement concrete :

(*i*) Workability test  
(*ii*) Test for compressive strength  
(*iii*) Test for flexural strength  
(*iv*) Test for modulus of elasticity  
(*v*) Test for analysis of fresh concrete  
(*vi*) Test for cement content of hardened cement concrete.

### (A) TESTS ON CEMENT

## 48.2. TEST ON FINENESS OF CEMENT

The rate of hydration and hydrolysis and the setting of cement depends upon fineness of its particles. The gain of strength is rapid for finer cement, though the final strength is not affected by fineness. The shrinkage and cracking of cement is greater for fine cement. The fineness of cement is measured in terms of its *specific surface area*. There are several methods of measuring specific surface area of cement, such as  (*i*) particle size distribution method  (*ii*) air permeability method  (*iii*) Wagner's turbidimeter method etc. Particle size distribution can be determined by hydrometer test using kerosene as sedimentation medium and oleic acid as dispersing agent. The fineness can also be determined by sieve analysis but this is an unsatisfactory test and is not used much. Indian Standard specifies the last two methods. However, we shall describe Blaine's air permeability method specified in IS : 4031–1968.

**Blaine's air permeability method** : The Blaine's air permeability apparatus consists essentially of a means of drawing a definite quantity of air through a prepared bed of cement of definite porosity. The number and size of the pores in the prepared bed of cement of definite porosity is a function of the size of the particles and determines the rate of air flow through it.

Fig. 48.1 shows the Blaine's air permeability apparatus. The apparatus consists of three parts : (*i*) the permeability cell  (*ii*) the plunger and  (*iii*) the manometer tube assembly. The permeability cell consists of a rigid cylinder of glass or non-corroding metal of 12.7 + 1 mm in diameter. The cell is fitted to the manometer tube assembly. The bottom of the cell forms an airtight connection with the top of the manometer. A perforated disc of 0.9 ± 0.1 mm thickness, perforated with 30 to 40 holes of 1 mm dia. equally distributed over its area, fits the inside of the cell snugly. The plunger is used to compact the bed of cement over the perforated disc. The height of the plunger is such that when the head of the plunger touches the top of the cell, the difference in height between the plunger bottom and the perforated disc is 15 mm. The top of one arm of the manometer forms an air tight connection with the permeability cell. A side outlet is provided at 250 to 305 mm above the bottom of the manometer used in the evacuation of the manometer arm connected to the permeability cell. A positive airtight valve or clamp is provided on the side out let. The manometer is filled to the mid-point with a non-volatile, non-hygroscopic liquid of low viscosity and density, such as dibutylphthalate or a light grade of mineral oil.

The test is conducted on two samples : (i) test on *standard sample* and (ii) test on sample of cement under investigation. For standard sample, *National Bureau of Standard's standard sample No. 114* is used.

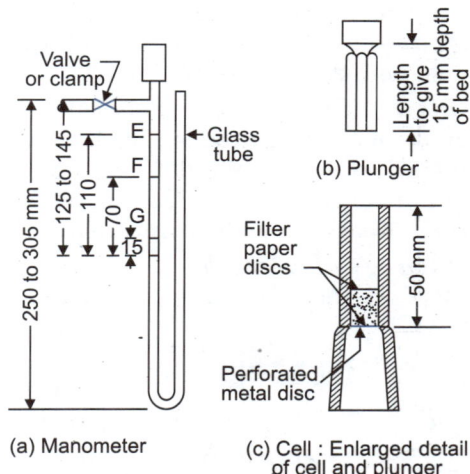

Fig. 48.1. Blaine's air permeability apparatus.

**Note :** In the absence of NBS standard sample No. 114, samples of Indian cement having specific surface equivalent to NBS standard sample No. 114 may be used. Such samples may be obtained from the ACC Central Research Station, Bombay.

The test on the standard sample is commonly known as 'Calibration of Apparatus'. The complete test is done in the following steps:

### Step 1. Determination of bulk volume of compacted bed of Cement.

A thin bed of cement, of height 15 mm is compacted in the cylinder. This bed of cement is actually the test specimen. In order to know exactly the bulk volume of cement compacted in the cell, mercury is used. The perforated metal disc is first placed in position in the cell and two filter paper discs are placed over it. The filter paper should be corresponding to No. 40 Whatman. The filter papers placed over the disc inside the cell are pressed with the help of a rod slightly smaller than the cell diameter until the filter discs are flat on the perforated disc. Mercury is then filled in the cell, to its top. Use a glass plate to level the mercury with the top of the cell. The mercury is removed from the cell and weighed ($W_a$ gm). One of the filter discs is removed from the cell. Use a trial quantity of 2.80 g of cement and place over the filter paper in the cell. Tap lightly the side of the cell in order to level the bed of cement. Put the other filter paper disc on the top of the cement and compress it with the plunger until the plunger collar is in contact with the top of the cell. The plunger is then removed. Fill the space remaining in the top of the cell with mercury and level off the top as before. Remove the mercury from the cell and weigh it ($W_b$ gm).

The bulk volume occupied by the cement is calculated by the following expression:

$$V = \frac{W_a - W_b}{D} \text{ cm}^3 \qquad \qquad ...(48.1)$$

where $D$ = Density of mercury at temperature of test in g/cm$^3$, and its value may be taken from Table 48.1.

### Step 2. Test on standard sample :
The air permeability is first conducted on the *standard sample*. The standard sample is generally available in the vial. The contents of the vial are enclosed in a 125 g jar and shaken vigorously. The weight ($W$) of standard sample used for the calibration test should be that required to produce a bed of cement having a porosity of $0.500 \pm 0.005$, and is calculated from the following expression :

$$W = \rho V (1 - e) \qquad \qquad ...(48.2)$$

Where $\rho$ = specific gravity of test sample (for Portland cement, a value of 3.15 is used).

$V$ = bulk volume of bed of cement, determined from *Eq. 48.1*

$e$ = desired porosity of cement ($0.500 \pm 0.005$).

Put a filter paper disc above the perforated disc and place the quantity of standard cement determined above. Place a filter paper disc over the cement bed and compress it with the plunger till the plunger collar is in contact with the top of the cell. Remove the plunger.

Attach the cell to the manometer tube, making certain that an airtight connection is obtained and taking care not to jar or disturb the prepared bed of cement. The air in one arm of the manometer U-tube is slowly evacuated until the liquid reaches the top mark (E) and the valve is then closed tightly. The stopwatch is started as the bottom of the meniscus of the manometer liquid reaches the second mark (F) and is stoped as the bottom of meniscus of liquid reaches the third mark (G). Measure the time interval between F and G and note the temperature of test.

**Step 3. Test on the test sample** : Repeat completely step 2 on the cement to be tested. The weight of sample to be compacted is calculated from *Eq. 48.2*. It is customary to use the same porosity (*i.e.*, 0.500 ± 0.005). However, while determining the fineness of high-early-strength cements, a porosity of 0.530 ± 0.005 should be adopted. Determine the time taken by the liquid to drop from point F to G as described in step 2. Measure the temperature during the test.

**Step 4. Calculations**

The specific surface of the cement is calculated from the following expression :

$$S = S_s \sqrt{\frac{T}{T_s}} \qquad \ldots(48.3)$$

where $S$ = specific surface in sq. cm per gram of the test sample.

$S_s$ = specific surface of the standard sample.

$T$ and $T_s$ are the time intervals, in seconds, of manometer drop for the test sample and standard sample respectively.

The above expression in valid *only if* the temperature during both the tests is constant. If the temperature varies, the following expression is used:

$$S = S_s \sqrt{\frac{T}{T_s}} \sqrt{\frac{n_s}{n}} \qquad \ldots(48.4)$$

where $n_s$ = viscosity of air in poise at the temperature of test of the standard sample (Table 48.1).

$n$ = viscosity of air in poise at temperature of test of the test sample (Table 48.1).

Both *Eqs. 48.3* and *48.4* are valid only if the test sample is compacted at the same porosity as the standard sample. If the porosity is different, use the following expression:

$$S = S_s \sqrt{\frac{T}{T_s}} \sqrt{\frac{n_s}{n}} \sqrt{\frac{e^3}{e_s^3} \cdot \frac{1-e_s}{1-e}} \qquad \ldots(48.5)$$

where $e_s$ = porosity of bed of standard sample

$e$ = porosity of bed of test sample.

**TABLE 48.1.** Density of Mercury, Viscosity of Air (n) and $\sqrt{n}$

| Room Temperature °C | Density of mercury g/cm³ | Viscosity of air (n) poise | $\sqrt{n}$ |
|---|---|---|---|
| 16 | 13.56 | 0.0001788 | 0.01337 |
| 18 | 13.55 | 0.0001798 | 0.01341 |
| 20 | 13.55 | 0.0001808 | 0.01344 |
| 22 | 13.54 | 0.0001818 | 0.01348 |
| 24 | 13.54 | 0.0001828 | 0.01352 |
| 26 | 13.53 | 0.0001837 | 0.01355 |
| 28 | 13.53 | 0.0001847 | 0.01359 |
| 30 | 13.52 | 0.0001857 | 0.01362 |
| 32 | 13.52 | 0.0001867 | 0.01366 |
| 34 | 13.51 | 0.0001876 | 0.01369 |

## 48.3. TEST FOR CONSISTENCY OF CEMENT PASTE

The Tests on cement, such as the determination of setting time, soundness, compressive strength and tensile strength etc. require the preliminary determination of the amount of water required to produce a cement paste of *standard consistency*. The test for determination of water content to form the cement paste of standard consistency is conducted on Vicat's apparatus. The standard consistency of a cement paste is defined as that which will permit the Vicat plunger (Fig. 48.2) to penetrate to a point 5 to 7 mm from the bottom of the Vicat mould.

*Vicat's apparatus* : Fig. 48.2 shows a Vicat's apparatus. It consists of a stand $D$ in which a sliding rod $B$ weighing 300 g. At the end of the rod. $B$, either a needle ($C$) of 1 mm sq., 5 cm long is fitted (for determination of setting time) or a plunger ($G$) of 10 mm $\Phi$ can be fitted. A movable indicator is attached to rod $B$. The indicator travels over scale graduated from 0 to 40 mm to measure the travel of the rod. The cement paste is filled in the Vicat mould $E$ restive on a non-porous plate.

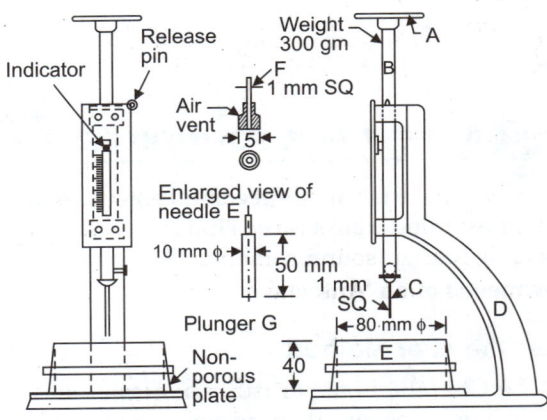

Fig. 48.2. Vicat's apparatus.

*Test procedure* : The test consists of making cement pastes with different amounts of water content. Each paste is placed in the mould under the Vicat plunger which is lowered into contact with surface of the paste and quickly released. To prepare the paste, take weighed quantity (say about 300 g) of cement and place it in a crucible. Mix a weighed quantity of water to it and mix it with spatula. The time of mixing or gauging should not be less than 3 minutes, nor more than 5 min, and gauging time should be counted from the time of adding water to the dry cement until commencing to fill the mould. The Vicat mould is filled with the paste, and levelled off at its top. The mould is placed under the rod with a plunger ($G$). The plunger is lowered gently to touch the surface of the test block and then quickly released, allowing it to sink in the paste. The reading in the indicator is noted. The test is repeated on a number of pastes prepared with different amounts of water. The paste of standard consistency is that which allows the Vicat plunger to settle to a point within 5 to 7 mm from the bottom of the Vicat mould. The amount of water required to prepare the paste of standard consistency is expressed as a percentage by weight of the dry cement. The test is performed at a temperature of 27° ± 2° C.

## 48.4. TEST FOR DETERMINATION OF INITIAL/FINAL SETTING TIMES

The setting time of cement may be divided into the *initial setting time* and the *final setting time*. It is difficult to say what exactly each time means as they are arbitrary points in relationship connecting the strength of cement with the time from adding water. The initial set is a stage in the process of hardening of cement after which any cracks that may occur or appear will not re-unite. The two setting times are defined in the Vicat's test, as described below.

**Preparation of Test Block.** Take about 300 g of cement and prepare a cement paste by mixing it with 0.85 times the water required to give a paste of standard consistency. The paste is mixed in the manner described in the consistency test. Stop watch is started at the instant when water is added to the cement. The Vicat mould is filled with cement paste mixed above, with the mould resting on a non-porous plate. The mould is filled completely and the surface of the paste is made level with the top of the mould. The cement block thus prepared in the mould is known as the *test block*.

**Determination of Initial Setting Time.** The mould along with the plate is placed under the rod bearing the needle ($C$). The needle is lowered slowly to bring it in contact with the surface of the test block. The needle is then released quickly, allowing it to penetrate into the test block. In the beginning, the needle will completely pierce the test block. The above procedure is repeated until the needle, when brought in contact with the test block and released, fails to pierce the block for 5 ± 0.5 mm measured from the bottom of the mould. *The period elasping between the time when water is added to the cement and the time at which the needle fails to pierce the test block by 5 ± 0.5 mm is taken as initial setting time.*

**Determination of Final Setting Time.** To determine the Final Setting Time, needle $C$ is replaced by needle $F$ having an annular attachment. This needle also has a diameter equal to 1 mm, but the needle projects out of annular attachment by 0.5 mm. The cement is considered as finally set when upon applying the needle gently to the surface of the test block, the needle makes an impression thereon, and the attachment fails to do so. *The period elasping between the time when water is added to the cement and* *the time at which the needle makes an impression on the surface of test block and the attachment fails to do so, is taken as the final setting time.*

## 48.5. TEST FOR SOUNDNESS OF CEMENT

*Soundness* is the ability of a cement to maintain a constant volume. The cement having free lime and magnesia undergoes large changes of volume as the time elapses, tending to cause cracks. Such a cement is called an 'unsound cement'. The soundness of cement is determined either by 'Le Chatelier Method' or by means of an 'Autoclave Test'.

### Le Chatelier Method

Le Chatelier's apparatus consists of small split cylinder of spring brass, or other suitable metal of 0.5 mm thickness, forming a mould of 30 mm internal diameter and 30 mm high. On either side of the split are attached two indicators with pointed ends $AA$, the distance from these ends to the centre of the cylinder being 165 mm.

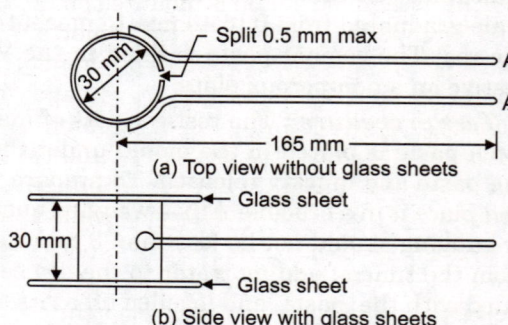

**Fig. 48.3.** Apparatus for Le Chatelier test.

Take about 100 g of cement. Mix it with 0.78 times the water required to give a paste of standard consistency. The mould is placed on a glass plate and filled with the cement paste. The mould is covered with another piece of glass sheet and a small weight is placed over the glass sheet. The whole assembly is immediately placed in water having a temperature of 27° ± 0° C, and kept there for 24 hours. The assembly is taken out after 24 hours and the distance separating the indicator points is measured. The mould is then immersed in a water. The water is brought to boiling point, with the mould submerged for 25 to 30 minutes and kept boiling for 3 hours. The mould is taken out and allowed to cool. The distance between the points is then measured. The difference between these *two measurements* represents the expansion of cement. Ordinary Portland cement should not have an expansion of more than 10 mm.

## 48.6. TEST FOR DETERMINATION OF COMPRESSIVE STRENGTH OF CEMENT

The compressive strength of concrete depends upon the strength of cement used for the manufacture of the concrete. The test for the strength of cement should reflect the ability of the cement to make concrete of a given strength. The compressive strength test of cement is conducted on mortar cube, formed by mixing one part of cement with 3 parts of standard sand, compacted by means of a vibration machine as outlined below :

**1. Mix proportion and mixing** : Standard re-graded Ennore sand conforming to IS : 650–1966 is used for preparing the cement mortar. The following proportions are used:

| | |
|---|---|
| Cement | 200 g |
| Standard Sand | 600 g |
| Water | $\left(\dfrac{P}{4} + 3.0\right)$ per cent of combined weight of cement and sand, |

where $P$ is the percentage of water required to produce a paste of standard consistency.

However, if standard sand is not available, the selected sand should pass through 850 micron IS sieve and not more than 10 percent by weight should pass through 600 micron IS sieve. In that case mix proportions may be taken as under:

| | |
|---|---|
| Cement | 185 g |
| Standard | sand 555 g |
| Water | $\left(\dfrac{P}{4} + 3.5\right)$ percent of combined weight. |

Cement and sand are first mixed dry on a non porous plate, with a trowel for one minute and then with water until the mixture is of uniform colour. The time of mixing should not be less than 3 minutes and should the time taken to obtain a uniform colour exceed 4 min., the mixture should be rejected and the operation repeated with a fresh cement, sand and water.

**2. Moulding specimens :** Compressive strength test is done on mortar cubes having area of face equal to 50 cm². The moulds for the cube specimens of 50 cm² face area should be of metal. Normally, split moulds are used so that the moulded specimens could be removed easily. The parts of the mould when assembled together should be held rigidly together. In assembling the mould, the joints between the two halves of the mould are covered with a thin film of petroleum jelly. A coating of petroleum jelly between the contact surfaces of the bottom of the mould and its base plate is applied. The interior faces of the mould are treated with a thin coating of mould oil.

The assembled mould is mounted on the table of the vibrating machine and a hopper is attached to the top of it. Immediately after mixing the mortar, place the mortar in the cube mould and prod it with the standard packing rod made of non-absorptive, nonabrasive, non-brittle material, such as a rubber compound. The mortar should be proded 20 times in about 8 seconds to ensure the elimination of entrained air and honey combing. The remaining quantity of mortar is placed in the hopper of the cube, and it is proded again. The vibrating machine is then started and the mould is vibrated for 2 minutes. The vibrating machine should have a speed of $12000 \pm 400$ vibrations per minute. At the end of vibration, the mould assembly is removed from the machine, and its top is levelled with a trowel.

**3. Curing specimens :** Several such cubes are made in separate moulds. The mixture is made separately for each mould. The moulds are kept at a temperature of $27° \pm 2°$ C and in an atmosphere of at least 90% relative humidity for 24 hours after the completion of vibration. After that, the cube specimens are removed from the moulds and immediately submerged in water and kept there until taken out for testing.

**4. Compression testing :** Three such cubes are tested at the end of specified period, and the compressive strength should be the average of the strength of the three cubes for each period of curing. Normally 3 cubes are tested at the end of 30 days curing and 3 cubes at the end of 7 days curing. The testing can be done on any standard compression testing machine. The cubes should be tested on their sides without any packing between the cube and the steel plattens of the testing machine. One of the plattens should be carried on a base and should be self adjusting, and the load should be steadily and uniformly applied, starting form zero at a rate of 350 kg/cm²/min. The compressive strength is calculated from the crushing load and the average area over which the load is applied. The minimum desirable values of compressive strengths at various curing periods, for different cements, are given in Chapter 1.

## 48.7. TEST FOR TENSILE STRENGTH OF CEMENT

The tensile strength of cement is determined by testing of specimens of cement sand mortar in direct tension.

**1. Mix proportion and mixing :** Standard sand is used for preparing the cement mortar. The following proportions are used :

| | |
|---|---|
| Cement | 250 g |
| Standard sand | 750 g |
| Water | $\dfrac{P}{5}$ + 2.5 percent of total weight of sand and cement, |

where  $P$ is the percent of water required for standard consistency.

The mortar is prepared exactly in the same way as for the compression test.

**2. Moulding specimens and curing :** Specimens are prepared in standard briquette moulds placed on non-porous plate. The moulds are of split type so that specimen can be easily taken out later. These moulds are open on both the faces. Fig. 48.4 (a) shows the dimensions of standard briquette. The thickness of the briquette is 25.4 mm (1 inch) and it has a minimum sectional area of 645 sq. mm

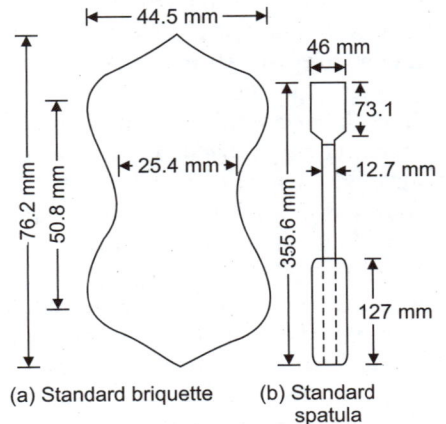

(a) Standard briquette  (b) Standard spatula

Fig. 48.4

(1 sq. inch) at the central section. Before filling the mould, the two halves of the mould are lightly greased, and the inner surface of the mould is coated with oil. Mortar is then filled in the mould. To ensure complete filling of the mould the mortar should be heaped on the mould and the excess mortar is beaten down by means of standard spatula shown in Fig. 48.4(b), until the mortar is in level with the top of the mould and water appears on the surface. This process of heaping the motar and beating it down on the other-open face of the briquette mould is repeated. Finally, the faces are smoothened with the help of a trowel. Twelve specimens are prepared, six of which are tested after 3 days' curing, and the other six after 7 days. The briquette moulds are kept for 24 hours at $27° \pm 2° C$ and 90% humidity. The specimens are then taken out of mould and submerged in water for curing.

**3. Test for tensile strength :** The briquettes so prepared are tested on a special *tension testing machine* shown in Fig. 48.5. The machine utilises the principle of compound lever for applying tensile force. The briquette is held in position in specially shaped jaws. A pan is attached to the free end of the lever. The loading is done by allowing lead shots to fall from a hopper to the pan. This causes tensile force in the briquette which ultimately breaks. When fracture occurs, the supply of lead shots is automatically cutoff. The lead shots collected in the pan are weighed by moving the counterpoise until balance is obtained. The reading on the yard gives the tensile load during fracture. The load divided by area of neck gives the tensile strength of the briquette.

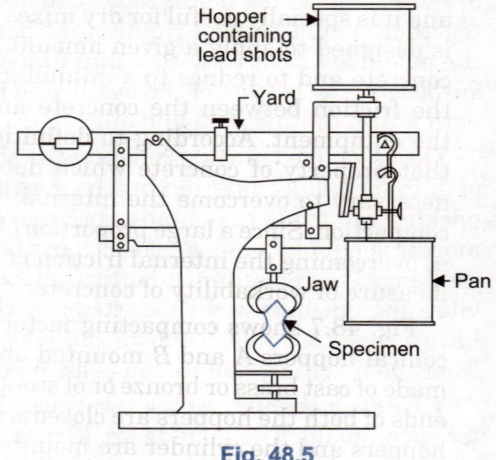

Fig. 48.5

## (B) TESTS ON CEMENT CONCRETE

## 48.8. TEST FOR WORKABILITY

Following are some of the tests carried for determining the workability of cement concrete: (*i*) slump test (*ii*) compacting factor test (*iii*) flow table test (*iv*) Vee-Bee consistometer method.

**1. SLUMP TEST :** It can be done either in the laboratory or in the field where the maximum size of aggregate does not exceed 38 mm. The mould used for the test specimen is in the form of frustum of a cone having the following internal dimensions: bottom dia. = 20 cm ;   top diameter = 10 cm and height = 30 cm. It is provided with suitable foot pieces and also handle to facilitate lifting it from the moulded concrete test specimen in vertical direction as required by the test. A mould provided with suitable guide attachment may be used, such as shown in Fig. 48.6. In addition to the mould a tamping rod of steel or other suitable material 16 mm in diameter, 0.6 m long and rounded at one end is used for tamping the concrete in the mould. Steel measure for measuring the slump is usually incorporated in the stand. To prepare the test specimen, the cleaned mould is placed on a smooth horizontal, rigid, non-absorbant surface (base). The mould is filled with freshly mixed concrete in *four* layers, each

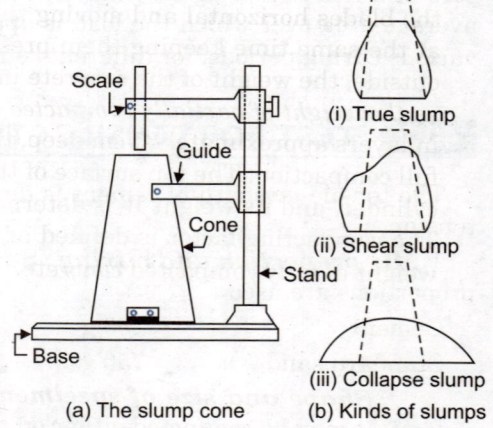

Fig. 48.6.  The Slump Cone.

approximately one quarter of the height of the mould. Each layer is tamped with twenty five strokes of the rounded end of the tamping rod, the strokes being distributed in a uniform manner over the area. The strokes for the second and subsequent layers should penetrate into the underlying layer. The bottom layer should be tamped through out its depth. After the top layer has been rodded, the concrete is struck off level with a trowel or the tamping rod. The mould is then removed from the concrete immediately by raising it slowly and carefully in the vertical direction. This allows the concrete to subside and the slump is measured immediately by determining the difference between the highest point of the mould and that of the highest point of the slumped specimen. The slump measured is recorded in terms of millimeters of subsidence of the specimens during the test. Fig. 48.6 (*b*) shows three forms of slumps that may occur. The first is a *true slump*, the second is known as a *shear slump* and the third a *collapse slump* which is obtained with lean harsh or very wet mixes. Any slump specimen which collapses or shears off laterally gives incorrect results and if this occurs, the test should be repeated with another sample. If, in the repeat test also, the specimen should shear, the slump shall be measured and the fact that the specimen sheared, shall be recorded. Generally, if shear and collapse slumps are obtained, the concrete will be unsatisfactory for placing.

**2. THE COMPACTING FACTOR TEST :** The compacting factor test gives more consistent and sensitive results than the slump test, and it is specially useful for dry mixes which give no slump. The test is designed to apply a given amount of work to a given amount of concrete and to reduce to a minimum the work lost in overcoming the friction between the concrete and the containing surfaces of the equipment. According to definition, workability of concrete is that property of concrete which determines the amount of work necessary to overcome the internal friction in order to obtain full compaction. Since a large proportion of the work is expanded usefully in overcoming the internal friction of concrete, the test gives a good measure of workability of concrete.

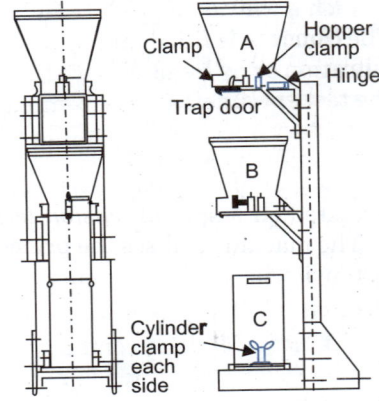

Fig. 48.7. Compacting factor apparatus.

Fig. 48.7 shows compacting factor apparatus. It consists of two conical hoppers $A$ and $B$ mounted above a cylindrical mould $C$, all made of cast brass or bronze or of stout sheet brass or steel. The lower ends of both the hoppers are closed with tightly fitting hinged trap-doors having quick release catches. The hoppers and the cylinder are mounted on a rigid frame. The hoppers and cylinder can be easily detached from the frame. In addition to the main apparatus, two ordinary bricklayers trowels, one hand scoop, a rod of steel and a balance to weigh upto 30 kg, to the nearest 10 g, are required.

**Test procedure.** The freshly mixed concrete sample is placed gently in upper hopper. The hopper is filled level with its brim and the trap door of the first, hopper is opened so that concrete falls into the lower hopper. If concrete has a tendency to stick in hopper, the concrete may be helped through by pushing the rod gently into the concrete from the top. Immediately after the concrete has come to rest, the trap-door of the lower hopper is opened, so that the concrete falls into the lower hopper. The excess of concrete remaining above the lever of the top of the cylinder is cut off by holding a trowel in each hand, with the plane of the blades horizontal and moving them simultaneously one from each side across the top of the cylinder, at the same time keeping them pressed on the top edge of the cylinder. After cleaning the cylinder from outside, the weight of the concrete in the cylinder is determined to nearest 10 g. This weight ($w$) is known as *the weight of partially compacted concrete*. The cylinder is refilled with concrete from the same sample in layers approximately 5 cm deep, the layers being heavily rammed or preferably vibrated so as to obtain full compaction. The top surface of the fully compacted concrete is carefully struck off level with the top of cylinder, and its weight $W$ is determined. This weight is known as *the weight of fully compacted concrete*. The compacting factor is defined as the ratio ($w/W$) of the weight of partially compacted concrete, to the weight of fully compacted concrete.

## 48.9. TEST FOR COMPRESSIVE STRENGTH

**1. Shape and size of specimen :** The most important strength test for concrete is the compression test. It may be conducted either on specimens prepared in the laboratory or on specimens prepared in the field during the progress of the work. The test conducted on laboratory prepared specimen is called the *preliminary test*, while the test conducted on specimen made of concrete taken at the mixer is called the *works test*. The compression test can be conducted either on cubical specimens or on cylindrical specimens. If the maximum size of aggregate does not exceeds 30 mm, the size of the cube should be 15 × 15 × 15 cm. If the largest nominal size of the aggregate does not exceed 20 mm, 10 cm cubes may be used as an alternative. Cylindrical test specimens should have a length equal to twice the diameter. They should be 15 cm diameter and 30 cm long. The cube moulds, made of metal, should be such that moulded specimen can be removed easily. The moulds are of split type, and are provided with a metal base plate having a plane surface. The cylindrical mould should be capable of being opened longitudinally to facilitate removal of the specimen. Each cylindrical mould is provided with a metal base plate, and with a capping plate of glass or other suitable material. In addition to the moulds, tamping bar of steel, 16 mm in dia., 60 cm long and bullet pointed at the lower end is used for hand compaction of specimen.

**2. Preparation of specimen :** Before placing the concrete in the mould, its interior surface and the base plate should be lightly oiled. The test specimens should be made as soon as practicable after mixing, and in such a way as to produce full compaction of the concrete with neither segregation nor excessive liatance. The concrete is filled into the mould in layers of approximately 5 cm deep. Each layer is compacted either by vibration or by hand. Where the concrete at the actual work site is to be compacted by mechanical vibrator, the test specimen may be compacted with a mechanical vibrator. The mode and quantum of vibration of the laboratory specimen should be as nearly the same as those adopted in actual concreting operation. When compacting by hand, the standard tamping bar should be used and the number of strokes should be evenly distributed. For cubical specimens, the number of strokes should not be less than 35 strokes per layer for 10 cm cubes. For cylindrical specimens, the number of strokes should not be less than 30 per layer. The strokes should penetrate into the underlying layer and the bottom layer should be rodded throughout its depth. In the case of cylindrical samples, it becomes very necessary to have a uniform and smooth surface. The ends of all cylindrical test specimens should be capped. The cylinders may be capped with a thin layer of stiff, neat Portland cement paste after the concrete has ceased setting in the moulds generally for two to four hours or more after moulding. The cement for capping should be mixed to a stiff paste for about two to four hours before it is to be used in order to avoid the tendency of the cap to shrink. The stiff cement paste is placed at the top of the mould and the glass capping plate is worked on the cement paste. No gap is allowed to remain between the plate and the top surface of the mould. Adhesion of paste to the capping plate is avoided by coating the plate with a thin coat of oil or grease.

**3. Curing of test specimens :** As soon as moulding is complete, the mould is stored, in moist air of at least 90% relative humidity and at a temperature of $27° \pm 2°C$ for 24 hours. After this period, the specimens are marked, removed from the moulds and submerged in water, and kept there till testing.

**4. Test for compressive strength :** The specimens are tested for compressive strength on *compression test machine* provided with two steel bearing platens with hardened faces. The test should be made at recognized ages of test specimen, the most usual being 7 and 28 days. At least three specimens, preferably from different batches, should be made for testing at each selected age. Specimens stored in water should be tested immediately on removal from the water and while they are in the wet condition. In the case of cubes, the specimen should be placed in the machine in such a manner that the load is applied to the opposite sides of the cubes as cast, that is, not to the top and bottom. The load should be applied without shock and increased continuously at a rate of approx. 140 $kg/cm^2$/min. until the resistance of the specimen to the increasing load breaks down and no greater load can be sustained. The measured compressive strength of the specimen is calculated by dividing the maximum load applied to the specimen during the test by the cross-sectional area, calculated from the mean dimensions of the section. Average of three values should be taken as the representative of the batch, provided the individual variation is not more than ± 15 percent of the average. Otherwise repeat test should be made.

## 48.10. TEST FOR FLEXURAL STRENGTH

For flexural strength, specimens of rectangular beams with square cross-sections are used. Following are the recommended sizes of the concrete beams, depending upon the maximum size of the aggregate :

| Max. size of aggregate | Cross-section | Length |
|---|---|---|
| Less than 20 mm | 10 cm × 10 cm | 50 cm |
| Between 22 to 38 mm | 15 cm × 15 cm | 70 cm |
| Exceeding 38 mm | 4 times max. size of aggregate | 4 times the cross-sectional dimension |

Each mould is provided with a metal base plate and two loose top plates of 4.0 × 0.6 cm cross-section and 5 cm longer than the width of the mould. The mould should be constructed in such a manner as to facilitate the removal of the moulded specimen without damage. For compacting concrete in the mould a *tamping bar* of steel, weighing 2 kg, 40 cm long and having a ramming face 25 mm square as used. The procedure of compaction is the same as adopted for preparing cubical specimens for compression test. The procedure for curing is also the same.

Fig. 48.8 shows arrangement for loading of flexure test specimens. The specimen is placed in the machine in such a manner that load is applied to the uppermost surface as cast in the mould, along two lines spaced 48.0 cm for 70 cm long beam or 13.3 cm for the 50 cm long beam. The axis of the specimen is carefully aligned with axis of the loading device. No packing should be used between the bearing surfaces of the specimen and the rollers. The load should be applied without shock and increasing continuously at a rate such that the extreme fibre stress increases at approximately 7 kg/cm²/min, that is, at rate of loading of 400 kg/min. for 15.0 cm specimens and at a rate of 180 kg/min. for the 10.0 cm specimen. The load should be increased until the specimen fails, and the maximum load applied to the specimen during the test should be recorded. The appearance of the fractured faces of concrete and any unusual features in the type of failure should be noted.

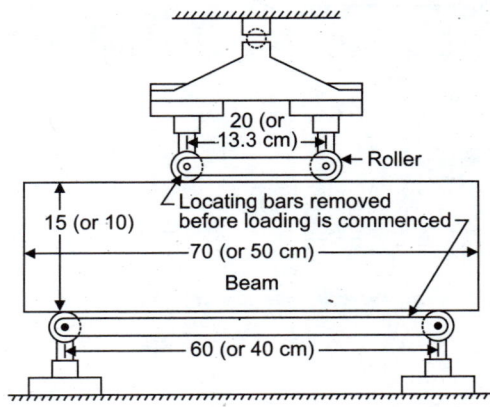

Fig. 48.8. Arrangement for Loading of Flexure Test Specimen.

Let     'a' = distance between the line of fracture and the nearer support, measured on the centre line of the tensile side of the specimen (cm).

$f_b$' = modulus of rupture (kg/cm²)

'P' = maximum load applied to the specimen (kg)

'l' = length of the span on which the specimen is supported (cm)

'b' = measured width of specimen (cm) or (mm)

'd' = measured depth of specimen at the point of failure (cm).

If $a$ is *greater* than 48.0 cm for 15.0 cm specimen, or greater than 13.3 cm for 10.0 cm specimen, the modulus of rupture is calculated from the following expression:

$$f_b = \frac{P \times l}{bd^2} \qquad \text{...(48.6)}$$

If $a$ is *less* than 20 cm but greater than 17.0 cm for 15 cm specimen, or less than 13.3 cm but greater than 11.0 cm for a 10.0 cm specimen, the modulus of rupture is calculated from the following :

$$f_b = 3\,P\,a/bd^2 \qquad \text{...(48.7)}$$

However, if $a$ is less than 17.0 cm for a 15.0 cm specimen or less than 11.0 cm for a 10.0 cm specimen, the results of the test should be discarded.

# APPENDIX-A
# B.M. AND S.F. COEFFICIENTS

**TABLE A-1.** Continuous Beams

**(a) B.M. Coefficients**    B.M. = (Coefficient) × (Total load on one Span) × Span

| Load | Dead Load (All spans loaded) | Live Load (Sequence of loaded spans to give max. B.M.) |
|---|---|---|
| **Uniformly distributed** | 0.125<br>0.071   0.071 | 0.125<br>0.096   0.096 |
| | 0.100   0.100<br>0.080   0.025   0.080 | 0.117   0.117<br>0.101   0.075   0.101 |
| | 0.107   0.072   0.107<br>0.077   0.056   0.056   0.077 | (0.116) (0.107) (0.116)<br>0.121   0.107   0.121<br>0.099   0.081   0.081   0.099 |
| | 0.105   0.08   0.08   0.105<br>0.078   0.053   0.046   0.053   0.078 | (0.116) (0.107) (0.107) (0.116)<br>0.120   0.111   0.111   0.120<br>0.100   0.080   0.086   0.080   0.100 |
| **Concentrated at mid-span** | 0.188<br>0.156   0.156 | 0.188<br>0.203   0.203 |
| | 0.150   0.150<br>0.175   0.100   0.175 | 0.175   0.175<br>0.213   0.175   0.213 |
| | 0.161   0.107   0.161<br>0.164   0.116   0.116   0.164 | (0.174) (0.160) (0.174)<br>0.181   0.160   0.181<br>0.210   0.183   0.183   0.210 |
| | 0.158   0.119   0.119   0.158<br>0.171   0.110   0.130   0.110   0.171 | (0.174) (0.160) (0.160) (0.174)<br>0.174   0.167   0.167   0.174<br>0.211   0.181   0.141   0.81   0.211 |
| **Concentrated at third points** | 0.167<br>0.111   0.111 | 0.167<br>0.139   0.139 |
| | 0.133   0.133<br>0.123   0.054   0.123 | 0.157   0.157<br>0.145   0.100   0.145 |
| | 0.143   0.075   0.143<br>0.119   0.056   0.056   0.119 | (0.155) (0.143) (0.155)<br>0.160   0.144   0.160<br>0.143   0.111   0.111   0.143 |
| | 0.141   0.106   0.106   0.141<br>0.120   0.050   0.061   0.050   0.120 | (0.155) (0.142) (0.142) (0.155)<br>0.159   0.148   0.148   0.159<br>0.144   0.108   0.115   0.108   0.144 |

(For notes, see next page)

## TABLE A-1. (contd.)

### (b) S.F. coefficients

S.F. = Coefficient × (Total load on one span)

| Load | Dead Load (All spans loaded) | Live Load (Sequence of loaded, spans to give max. B.M.) |
|---|---|---|
| Uniformly distributed | 0.38  0.62 <br> 0.62  0.38 | 0.44  0.62 <br> 0.62  0.44 |
| | 0.40  0.50  0.60 <br> 0.60  0.50  0.40 | 0.45  0.58  0.62 <br> 0.62  0.58  0.45 |
| | 0.39  0.54  0.46  0.61 <br> 0.61  0.46  0.54  0.39 | 0.45  0.60  0.57  0.62 <br> 0.62  0.57  0.60  0.45 |
| | 0.40  0.53  0.50  0.47  0.60 <br> 0.60  0.47  0.50  0.53  0.40 | 0.45  0.60  0.59  0.58  0.62 <br> 0.62  0.58  0.59  0.60  0.45 |

**Notes**

1. B.M. coefficients above line apply to negative B.M. at supports.
   B.M. coefficients below line apply to positive B.M. in span

2. S.F. coefficients above line apply to S.F. at right hand side of support.
   S.F. coefficients below line apply to S.F. at left hand side of support.

3. Coefficients apply when all spans are equal.
   Loads on each loaded span are equal.
   Moment of inertia are same throughout.

4. B.M. coefficients (live load) in brackets apply if two spans only are loaded.

**TABLE A-2.** Simply Supported Beams (Reaction = $rW$, Max. B.M. = $K_C \cdot WL$)

| Loading | Total load $W$ | Reaction coefficients $r_A$ | Reaction coefficients $r_B$ | Coefficient $K_C$ for max. B.M. |
|---|---|---|---|---|
| Point load at $\alpha L$ from A | $W$ | $1-\alpha$ | $\alpha$ | $\alpha(1-\alpha)$ |
| Central load, $\alpha = 1/2$ | $W$ | $1/2$ | $1/2$ | $1/4$ |
| Partial UDL over $\beta L$ starting at $\alpha L$ | $w\beta L$ | $1-\left(\alpha+\dfrac{\beta}{2}\right)$ | $\alpha+\dfrac{\beta}{2}$ | $\dfrac{\left(\alpha+\dfrac{\beta}{2}\right)\left(1-\alpha-\dfrac{\beta}{2}\right)(2-\beta)}{2}$ |
| Span fully loaded, ($\alpha=0$; $\beta=1$) | $wL$ | $1/2$ | $1/2$ | $1/8$ |
| Triangular load (max at A) | $\dfrac{wL}{2}$ | $2/3$ | $1/3$ | $1/7.81$ |
| Triangular load (max at center) | $\dfrac{wL}{2}$ | $1/2$ | $1/2$ | $1/6$ |
| Trapezoidal load | $wL(1-\alpha)$ | $1/2$ | $1/2$ | $\dfrac{3-4\alpha^2}{24(1-\alpha)}$ |
| Two point loads $W/2$ at $L/3$ | $W$ | $1/2$ | $1/2$ | $1/6$ |

**TABLE A-3.** Cantilevers (Max. B.M. = $K_A \cdot WL$)

| Loading | Total load $W$ | B.M. Coeff. $K_A$ | Loading | Total load $W$ | B.M. Coeff. $K_A$ |
|---|---|---|---|---|---|
| Point load $W$ at $\alpha L$ | $W$ | $\alpha$ | Triangular load (max at B) | $\dfrac{wL}{2}$ | $2/3$ |
| Partial UDL over $\beta L$ at $\alpha L$ | $w\beta L$ | $\alpha+\dfrac{\beta}{2}$ | Triangular load over $\alpha L$ at B | $\dfrac{w\alpha L}{2}$ | $\left(1-\dfrac{\alpha}{3}\right)$ |
| Triangular load (max at A) | $\dfrac{wL}{2}$ | $1/3$ | Parabolic load | $\dfrac{2}{3}wL$ | $1/2$ |
| Triangular load over $\alpha L$ at A | $\dfrac{w\alpha L}{2}$ | $\dfrac{\alpha}{3}$ | Triangular load (max at center) | $\dfrac{wL}{2}$ | $1/2$ |

## TABLE A-4. Fixed Beams

**Reaction at $B = r_B \cdot W$; Reaction at $A = r_A \cdot W$**
$M_A = -k_A \cdot WL \qquad M_B = -k_B \cdot WL \qquad M_C = +k_C \cdot WL$
Position of plane of max. B.M. $= \; C \, L \; = \;$ distance $CB$

| Loading | Total load W | Reaction Coeff. $r_B$ | $k_A$ | $k_B$ | $k_C$ | $C$ |
|---|---|---|---|---|---|---|
| Point load $W$ at $\alpha L$ from A | $W$ | $\alpha^2(3-2\alpha)$ | $\alpha(1-\alpha)^2$ | $\alpha^2(1-\alpha)$ | $2\alpha^2(1-\alpha)^2$ | $(1-\alpha)$ |
| Point load $W$ at centre | $W$ | $1/2$ | $1/8$ | $1/8$ | $1/8$ | $1/2$ |
| UDL $w$ over span | $wL$ | $1/2$ | $1/12$ | $1/12$ | $1/24$ | $1/2$ |
| Triangular load (peak at centre) | $\dfrac{wL}{2}$ | $1/2$ | $5/48$ | $5/48$ | $1/16$ | $1/2$ |
| Trapezoidal load | $wL(1-\alpha)$ | $1/2$ | \multicolumn{2}{c}{$k_A = k_B = \dfrac{1-2\alpha^2+\alpha^3}{12(1-\alpha)}$} | $\dfrac{1-2\alpha^3}{24(1-\alpha)}$ | $1/2$ |
| Triangular load over $\alpha L$ from A | $\dfrac{w\alpha L}{2}$ | $\dfrac{\alpha^2}{10}(5-2\alpha)$ | $\dfrac{\alpha}{3}-\dfrac{\alpha^2}{3}+\dfrac{\alpha^3}{10}$ | $\dfrac{\alpha^2}{6}-\dfrac{\alpha^3}{10}$ | $r_B \cdot C - k_B - \dfrac{1}{3\alpha^2}(\alpha+C-1)^3$ | $\alpha^2\sqrt{\dfrac{5-2\alpha}{10}}\Big/(1-\alpha+\ldots)$ |

UDL $w$ over $\alpha L$ from A: Total load $w\alpha L$
$$r_B = \tfrac{1}{2}(2\alpha^2 - \alpha^3)$$
$$r_A = \tfrac{1}{2}(\alpha^3 - 2\alpha^2 + 2)$$
$$k_A = \dfrac{\alpha}{12}(3\alpha^2 - 8\alpha + 6)$$
$$k_B = \dfrac{\alpha^2}{12}(4 - 3\alpha)$$

UDL $w$ over $\alpha L$ centred at mid-span (length $\beta L$ offset): Total load $w\alpha L$
$$r_B = \left(-2\beta^3 + 3\beta^2 - \dfrac{\alpha^2\beta}{2} + \dfrac{\alpha^2}{4}\right)$$
$$r_A = \left(2\beta^3 - 3\beta^2 + \dfrac{\alpha^2\beta}{2} - \dfrac{\alpha^4}{4} + 1\right)$$
$$k_A = \beta(1-\beta)^2 - \alpha^2\left(\dfrac{1}{6} - \dfrac{\beta}{4}\right)$$
$$k_B = \beta^2(1-\beta) - \alpha^2\left(\dfrac{\beta}{4} - \dfrac{1}{12}\right)$$

# APPENDIX-B
# PROPERTIES OF MATERIALS AND CONCRETE

**TABLE B-1.** Grades of Concrete (IS : 456 – 2000)

| Group | Grade designation | Specified characteristic compressive strength ($f_{ck}$) of 150 mm cube at 28 days (N/mm$^2$) |
|---|---|---|
| *Ordinary concrete* | M 10 | 10 |
| | M 15 | 15 |
| | M 20 | 20 |
| *Standard concrete* | M 25 | 25 |
| | M 30 | 30 |
| | M 35 | 35 |
| | M 40 | 40 |
| | M 45 | 45 |
| | M 50 | 50 |
| | M 55 | 55 |
| *High strength concrete* | M 60 | 60 |
| | M 65 | 65 |
| | M 70 | 70 |
| | M 75 | 75 |
| | M 80 | 80 |

**Note :** Grades of concrete lower than M 20 shall not be used in reinforced concrete.

**TABLE B-2.** Permissible stresses in concrete (IS : 456 – 2000)

| Grade of concrete | Permissible stress in compression ($N/mm^2$) | | Permissible stress in Bond (Average) for plain bars in tension ($N/mm^2$) ($\tau_{bd}$) |
|---|---|---|---|
| | Bending ($\sigma_{cc}$) | Direct ($\sigma_{cc}$) | |
| (1) | (2) | (3) | (4) |
| M 10 | 3.0 | 2.5 | — |
| M 15 | 5.0 | 4.0 | 0.6 |
| M 20 | 7.0 | 5.0 | 0.8 |
| M 25 | 8.5 | 6.0 | 0.9 |
| M 30 | 10.0 | 8.0 | 1.0 |
| M 35 | 11.5 | 9.0 | 1.1 |
| M 40 | 13.0 | 10.0 | 1.2 |
| M 45 | 14.5 | 11.0 | 1.3 |
| M 50 | 16.0 | 12.0 | 1.4 |

**Notes :**
1. The bond stress given in col. 4 shall be increased by 25 percent for bars in compression.
2. In the case of deformed bars conforming to IS 1786, the bond stresses given in col. 4 may be increased by 60 percent.

**TABLE B-3.** Tolerable Concentrations of Some Impurities in Mixing Water

| | Impurity | Tolerable concentration |
|---|---|---|
| 1. | Sodium and potassium carbonates and bicarbonate | 1000 ppm (total). If this is exceeded, it is advisable to make tests both for setting time and 28 days strength. |
| 2. | Chlorides | 10,000 ppm. |
| 3. | Sulphuric anhydride | 3000 ppm. |
| 4. | Calcium chloride | 2% by wt. of cement in non prestressed concrete. |
| 5. | Sodium iodate, Sodium phosphate, Sodium arsonate, Sodium borate | very low |
| 6. | Sodium sulphate | Even 100 ppm warrants testing. |
| 7. | Sodium hydroxide | 0.5% by wt. of cement, provided quick set is not induced. |
| 8. | Silt and suspended particles | 2000 ppm. Mixing water with a high content of suspended solids should be allowed to stand in a setting basin before use. |
| 9. | Total dissolved salts | 1500 ppm. |
| 10. | Organic material | 3000 ppm. Water containing humic acid or such organic acids may adversely affect the hardening of concrete ; 780 ppm of humic acid are reported to have seriously impaired the strength of concrete. In the case of such waters, therefore, further testing in necessary. |
| 11. | pH | 4.5 to 8.5 |

**Note :** When a water of unknown performance contains impurities of possible adverse effect, it is certainly good practice to make for setting time strength in which known, acceptable water is used for compression. For the indications, test should be made using the particular cement involved. The procedure for the tests is set down in BS 3148 : 1959 'Tests for water for making concrete.' The water can be considered suitable for making concrete if :

(i) initial setting times of the cement do not differ by more than 30 minutes.
(ii) the average compressive strength of concrete test cubes using the suspected water is not less than 90% of he average strength of the control concrete test cubes using known good water or distilled water.

# APPENDIX-C
# REINFORCEMENT

**TABLE C-1.** Area, Perimeter and Mass of Round Bars

| Diameter of bar, mm | Area ($cm^2$) | Perimeter (cm) | Mass ($kg/m^3$) |
|---|---|---|---|
| 5 | 0.20 | 1.57 | 0.15 |
| 6 | 0.28 | 1.88 | 0.22 |
| 8 | 0.50 | 2.51 | 0.39 |
| 10. | 0.79 | 3.14 | 0.62 |
| 12 | 1.13 | 3.77 | 0.89 |
| 16 | 2.01 | 3.03 | 1.58 |
| 20 | 3.14 | 6.28 | 2.47 |
| 22 | 3.80 | 6.91 | 2.98 |
| 25 | 4.91 | 7.85 | 3.85 |
| 28 | 6.16 | 8.80 | 4.80 |
| 32 | 8.04 | 10.05 | 6.31 |
| 36 | 16.18 | 11.31 | 7.99 |
| 40 | 12.57 | 12.57 | 9.86 |
| 45 | 15.90 | 14.14 | 12.49 |
| 50 | 19.64 | 15.71 | 15.41 |

# APPENDIX-D
# LOADINGS

**TABLE D-1.** Imposed Floor Loads for Different Occupancies (IS : 875-1987)

| S.N. (1) | Occupancy Classification (2) | Uniformly Distributed Load UDL (kN/m²) (3) | Concentrated Load kN (4) |
|---|---|---|---|
| 1. | **Residential Buildings** | | |
| | (i) Dwelling Houses | 2.0 – 3.0 | 1.8 – 4.5 |
| | (ii) Hotels, hostels, boarding houses, lodging houses, domitories, residential clubs | 2.0 – 4.0 | 1.8 – 4.5 |
| | (iii) Boiler rooms and plant rooms | 5.0 | 6.7 |
| | (iv) Store rooms | 5.0 | 4.5 |
| | (v) Garrages | 2.5 – 5 | 9.0 |
| | (vi) Balconies | 3 | 1.5 kN/m at the outer edge |
| 2. | **Educational Buildings** | | |
| | (i) Class rooms, restaurants, offices, staff rooms, kitchens, toilets | 2 – 3.0 | 2.7 |
| | (ii) Store rooms etc. | 5 | 4.5 |
| | (iii) Libraries and archives | 6.0 kN/m² for a min. height of 2.2 m + 2.0 kN/m² per m additional height | 4.5 |
| | (iv) Reading rooms | 3.0 – 4.0 | 4.5 |
| | (v) Corridors, lobbies, staircases | 4.0 | 4.5 |
| | (vi) Boiler rooms and plant rooms | 4.0 | 4.5 |
| | (vii) Balconies | Same as for rooms with a min. of 4 | 1.5 kN/m at the outer edge. |
| 3. | **Institutional buildings** | | |
| | (i) Bed rooms, wards, dormitories, lounges | 2.0 | 1.8 |
| | (ii) Kitches, laundaries, laboratores, dining rooms, cafeteria, toilets, | 2.0 – 3.0 | 2.7 – 4.5 |
| | (iii) Corridors, passages, lobbies, staircases | 4.0 | 4.5 |
| | (iv) Office rooms and OPD rooms | 2.5 | 2.7 |
| | (v) Boiler rooms and plant rooms | 5.0 | 4.5 |
| | (vi) Balconies | same as for (2 vii) | same as for (2 vii) |

**TABLE D-1.** Imposed Floor Loads for Different Occupancies (IS : 875-1987) (Contd.)

| S.N. (1) | Occupancy Classification (2) | Uniformly distributed Load UDL ($kN/m^2$) (3) | Concentrated Load kN (4) |
|---|---|---|---|
| 4. | **Business and office buildings** | | |
| | (i) Rooms with separate store | 2.5 | 2.7 |
| | (ii) Banking halls | 3.0 | 2.7 |
| | (iii) Vaults and strong rooms | 5.0 | 4.5 |
| | (iv) Record rooms/store rooms | 5.0 | 4.5 |
| 5. | **Mercantile Buildings** | | |
| | (i) Retail shops | 4.0 | 3.6 |
| | (ii) Wholesale shops | 6.0 (min.) | 4.5 (min.) |
| | (iii) Dining rooms, restaurants, cafeteria | 3.0 | 2.7 |
| | (iv) Corridors, passages, staircases | 4.0 | 4.5 |
| | (v) Office rooms | 2.5 | 2.7 |
| 6. | **Industrial buildings** | | |
| | (i) Work areas without machinery/equipment | 2.5 | 4.5 |
| | (ii) Work area with machinery/equipment | 5.0–10.0 | 4.5 |
| | (iii) Cafeteria, dining rooms | 3.0 | 4.5 |
| | (iv) Corridors, passages, staircases | 4.0 | 4.5 |
| 7. | **Storage Buildings** | | |
| | (i) Storage rooms (other than cold storage) | 2.4 $kN/m^2$ per each metre of storage height with a min. of 7.5 $kN/m^2$ | 7.0 |
| | (ii) Cold storage | 5.0 $kN/m^2$ per each metre of storage height with a min. of 15 $kN/m^2$ | 9.0 |
| | (iii) Corridors, passages etc. | 5.0 | 4.5 |
| | (iv) Boiler rooms and plant rooms | 7.5 | 4.5 |

APPENDIX 1119

**TABLE D-2. SI Units and Metric/Imperical Equivalents**
(Imposed Loads on Floors and Roofs)

| Concentrated Load | | | | Loading per unit length | | | | Intensity of Load | | | |
|---|---|---|---|---|---|---|---|---|---|---|---|
| lb | kg | Newtons N | kN | lb per foot | kg per metre | Newtons per metre | lb per metre | lb per sq. ft. | kg per sq. ft. | Newtons per sq. m. | kN per sq. m |
| 202 | 91.8 | 900 | 0.9 | 15.0 | 22.4 | 220 | 0.22 | 15.7 | 76.5 | 750 | 0.75 |
| 225 | 102 | 1000 | 1.0 | 24.6 | 36.7 | 360 | 0.36 | 20.9 | 102 | 1000 | 1.0 |
| 415 | 143 | 1400 | 1.4 | 50.7 | 75.5 | 740 | 0.74 | 31.3 | 253 | 1500 | 1.5 |
| 405 | 184 | 1800 | 1.8 | 103 | 153 | 1500 | 1.5 | 41.8 | 204 | 2000 | 2.0 |
| 603 | 275 | 2700 | 2.7 | 168 | 251 | 2460 | 2.46 | 52.6 | 255 | 2500 | 2.5 |
| 809 | 367 | 3000 | 3.6 | 206 | 306 | 3000 | 3.0 | 62.7 | 306 | 3000 | 3.0 |
| 1012 | 459 | 4500 | 4.5 | 280 | 418 | 4100 | 4.1 | 83.5 | 408 | 4000 | 4.0 |
| 2023 | 918 | 9000 | 9.0 | 308 | 459 | 4500 | 4.5 | 104 | 510 | 5000 | 5.0 |
| | | | | 560 | 836 | 8200 | 8.9 | 136 | 663 | 6000 | 6.5 |
| | | | | 672 | 1018 | 10000 | 10.0 | 157 | 765 | 7500 | 7.5 |
| | | | | | | | | 209 | 1020 | 10000 | 10.0 |
| | | | | | | | | 261 | 1275 | 12500 | 12.5 |
| | | | | | | | | 313 | 1570 | 15000 | 15.0 |
| | | | | | | | | 418 | 2040 | 20000 | 20.0 |

Conversion factors expressed to four significant figures but are calculated on more exact values

1 ft = 0.3048 m
1 sq. ft. = 0.0929 sq. m

1 m = 3.281 ft.
1 sq. m = 10.76 sq. ft.

## TABLE D-2. (contd.)
### SI : Units and Metric/Imperical Equivalents

| Concentrated Load | | | | Loading per unit length | | | | Intensity of Load | | | |
|---|---|---|---|---|---|---|---|---|---|---|---|
| lb | kg | Newtons N | kN | lb per foot | kg per metre | Newtons per metre | lb per metre | lb per sq. ft. | kg per sq. ft. | Newtons per sq. m. | kN per sq. m |
| 1 lb = | 0.4536 kg | = 4.448 N | | | | | | | | | |
| | | = 0.00448 kN | | lb/ft²/ft | | kg/m²/m | | N/m²/m | | kN/m²/m | |
| 1 kg = | 2.2046 | = 9.807 N | | 15.3 | | 245 | | 2400 | | 2.4 | |
| | | = 0.009807 kN | | | | | | | | | |
| 1 N = | 22243 lb | = 0.1020 kg | | 25.3 | | 408 | | 4000 | | 4.0 | |
| | | = 0.001020 kN | | 30.1 | | 510 | | 5000 | | 5.0 | |
| 1 kN = | 224.8 lb | = 102.0 kg | = 1000 N | | | | | | | | |
| | lb per foot | | | kg per metre | | Newtons per metre | | Kg per metre | | | |
| | 1 | | | 1.488 | | 14.59 | | 0.01452 | | | |
| | 0.6720 | | | 1 | | 9.807 | | 0.09807 | | | |
| | 0.06542 | | | 0.1020 | | 1 | | 0.001000 | | | |
| | 68.52 | | | 102.2 | | 1000 | | 1 | | | |
| | lb per sq. ft. | | | kg per sq. m | | Newtons per sq. m | | kN per sq. m | | | |
| | 1 | | | 4.882 | | 47.88 | | 0.04781 | | | |
| | 0.2048 | | | 1 | | 9.807 | | 0.009807 | | | |
| | 0.02089 | | | 0.1020 | | 1 | | 0.001000 | | | |
| | 20.89 | | | 102.0 | | 1000 | | 1 | | | |
| | lb per sq. ft. per ft. | | | kg. per sq. m per m | | N per sq. m per m | | kN per sq. m per m | | | |
| | 1 | | | 1.600 | | 15.69 | | 0.01556 | | | |
| | 0.06250 | | | 1 | | 9.807 | | 0.009827 | | | |
| | 0.09375 | | | 0.1020 | | 1 | | 0.001000 | | | |
| | 63.75 | | | 102.0 | | 1000 | | 1 | | | |

# INDEX

## A

| | |
|---|---|
| Airy's Theory | 669 |
| Alternative Design Procedure | 1050 |
| Analysis by Equilibrium Method | 979 |
| Analysis for Vertical Loads | 715 |
| Active Earth Pressure: Rankine's Theory | 444 |
| Aggregates | 7 |
| Analysis by Virtual Work Method | 972 |
| Analysis of Beams for Flexure | 1028 |
| Analysis of Doubly Reinforced T-beams | 870 |
| Analysis of Frames Subjected to Horizontal Forces | 722 |
| Analysis of Portal Frames | 703 |
| Analysis of Spherical Domes | 491 |
| Anchorage Bond Stress: Development Length | 91 |
| Approximate Method | 627 |
| Aqueducts and Box Culverts | 735–750 |
| Aqueducts and Syphon Aqueducts | 738 |
| Axially Loaded Columns | 303–319, 910–919 |

## B

| | |
|---|---|
| B.M. in Slab Supported on Four Edges | 772 |
| Balanced, Under-reinforced and Over-reinforced Section | 37 |
| Basic Assumptions | 1028 |
| Beam of Trapezoidal Section | 56 |
| Beam of Triangular Section | 54 |
| Bunkers | 672 |
| Back Anchoring of Retaining Wall | 472 |
| Basic Concepts | 997 |
| Beams Curved in Plan | 501–531 |
| Bearing Capacity of Soil and Settlement of Footings | 351 |
| Bending about Two Axes | 322 |
| Bond, Anchorage and Development Length | 89 |
| Box Culvert | 741 |
| Building Frames | 713–736 |
| Bunkers and Silos | 667 |

## C

| | |
|---|---|
| Cantilever Method | 724 |
| Case 1: Compressive Load at Eccentricity Smaller than D/4 | 320 |
| Circular Tank with Rigid Joint between Floor and Wall | 539 |
| Combined Direct and Bending Stresses | 320–349 |
| Combined Footings | 386–431 |
| Concrete Mix Proportioning (IS : 456–2000) | 18 |
| Case 2: Compressive Load at Large Eccentricity ($e > 1.5 D$) | 331 |
| Case 3: Compressive Load at Moderate Eccentricity [$D/4 < e < 3 D/2$] | 333 |
| Case 4: Tensile Load at Small Eccentricity | 342 |
| Case 5: Tensile Load at Large Eccentricity | 342 |
| Case 6: Tensile Load at Moderate Eccentricity | 344 |
| Cement Concrete | 1 |
| Centrifugal Force | 759 |
| Characteristic and Design Values and Partial Safety Factors | 827 |
| Checking Development Lengths of Tension Bars | 94 |

1121

| | |
|---|---|
| Chimneys | 688–702 |
| Circular Beam Supported Symmetrically | 502 |
| Circular Section Subjected to Eccentric Load | 338 |
| Circular Slabs | 269–283, 963–968 |
| Circular Tank with Domed Bottom and Roof | 575 |
| Circular Tank with Flexible Joint between Floor and Wall | 536 |
| Circular Tank with Rigid Joint between Floor and Wall | 546 |
| Classification and Composition of Cement | 1 |
| Classification and Types of Prestressing | 1005 |
| Columns with Uniaxial and Biaxial Bending | 920–937 |
| Combined Axial Load and Uniaxial Bending | 920 |
| Combined Effect of Self Load, Wind and Temperature | 693 |
| Combined Rectangular Footing | 386 |
| Combined Trapezoidal Footing | 389 |
| Components of Flat Slab Construction | 284 |
| Compression Members with Helical Reinforcement | 913 |
| Computation of Elongation of Tendons | 1018 |
| Concrete Bridges | 751–823 |
| Conditions for Curtailment of Reinforcement | 97 |
| Conical Domes | 498 |
| Construction of Interaction Curves for Column Design | 922 |
| Continuous and Isolated Footings | 350–385 |
| Courbon's Method | 790 |
| Critical Section for Design Shear : IS : 456–2000 | 82 |
| Curved Beam Fixed at Ends | 510 |
| Curved Beam Simply Supported at Ends and Continuous over Two Equally Spaced Intermediate Supports | 509 |

### D

| | |
|---|---|
| Deck Girder Bridges | 771 |
| Depth of Balanced Section of T-Beam | 126 |
| Design Charts (*Sp* 16 Design Charts 24 to 26) | 913 |
| Design of Foundations | 596 |
| Design of Pile Cap | 437 |
| Design of R.C. Chimney | 696 |
| Doubly Reinforced T-Beams | 137 |
| Design Concepts | 825–829 |
| Design Example of Combined Direct and Bending Stresses | 346 |
| Design of a Doubly Reinforced Section | 847 |
| Design of a T-Beam Bridge | 792 |
| Design of an Aqueduct | 738 |
| Design of Beams and Slabs | 147–215, 892–909 |
| Design of Beams | 147 |
| Design of Box Culvert | 742 |
| Design of Cantilever | 165 |
| Design of Cantilever Canopy | 201 |
| Design of Cantilever Chajja | 170 |
| Design of Cantilever Retaining Wall with Horizontal Backfill and Traffic Load | 457 |
| Design of Cantilever Retaining Wall with Horizontal Backfill | 451 |
| Design of Cantilever Retaining Wall with Sloping Backfill | 464 |
| Design of Cantilever | 897 |
| Design of Cinema Balcony | 229 |
| Design of Columns Subjected to Combined Bending and Direct Stresses (IS: 456–2000) | 322 |
| Design of Continuous Footings | 356 |
| Design of Continuous Slab | 174 |
| Design of Counterfort Retaining Wall | 470 |
| Design of Cycle Stand Shade | 186 |
| Design of Dog-legged Stair | 220, 941 |
| Design of Doubly Reinforced Beam | 179, 899 |
| Design of Doubly Reinforced T-Beam | 873 |
| Design of Flat Base Slab for Elevated Circular Tanks | 568 |
| Design of Hinge at the Base | 711 |
| Design of Hollow Tile Roof | 241 |
| Design of Inverted T-Beam Roof | 194 |
| Design of L-Beam : Design for Torsion | 206 |
| Design of Lintel Beams | 159 |
| Design of Lintel with Sunshade | 172 |
| Design of One-way Continuous Slab | 904 |

# INDEX

| | |
|---|---|
| Design of One-way Slab | 167, 902 |
| Design of Open Newel Stair with Quarter Space Landing | 223 |
| Design of Overhanging T-Beam Roof | 196 |
| Design of R.C. Domes | 494 |
| Design of RCC Beams | 892 |
| Design of Rectangular Beam Section | 839 |
| Design of Rectangular Portal Frame with Vertical Loads | 704 |
| Design of Rectangular Tank | 645 |
| Design of Reinforced Bricks Slab | 239 |
| Design of Sections for Flexure : Magnel's Method | 1034 |
| Design of Stair Cases | 216–237, 938–945 |
| Design of Stair with Quarter Space Landing | 943 |
| Design of Staircase with Central Stringer Beam | 226 |
| Design of Stairs Spanning Horizontally | 219, 940 |
| Design of Stairs with Quarter Space Landing | 221 |
| Design of Tank Supporting Towers | 590 |
| Design of T-Beam Roof | 190, 909 |
| Design of T-Beam | 182, 863 |
| Design Principles of Cantilever Retaining Wall | 449 |
| Design Procedure | 311 |
| Design Specifications (IS : 456–2000) | 915 |
| Design Stress Block Parameters (IS : 456–2000) | 834 |
| Determination of Moment of Resistance | 846 |
| Development Length at Point of Inflexion | 96 |
| Development Length Requirements at Simple Supports | 94, 888 |
| Development Length | 887 |
| Dimensions of a T-Beam | 122 |
| Direct Design Method | 286 |
| Distribution of Live Loads on Longitudinal Beams | 779 |
| Domes | 490–500 |
| Doubly Reinforced Beams | 107–120 |
| Doubly Reinforced Sections | 844–853 |
| Durability of Concrete | 12 |

### E

| | |
|---|---|
| Earth Pressure on Tank Walls | 643 |
| Effect of Shrinkage and Creep in Columns | 1087 |
| Effects of Continuity | 589 |
| Exact Method | 632 |
| Economical Depth of T-Beam | 127 |
| Effect of Creep on $E_c$ and $m$ | 1087 |
| Effect of Shrinkage and Creep in Beams | 1088 |
| Effects of Shear : Diagonal Tension | 69 |
| Enhanced Shear Strength of Sections Close to Supports (IS : 456–2000) | 88 |
| Equivalent Frame Method | 289 |

### F

| | |
|---|---|
| Factor Method | 728 |
| Factors Affecting Creep | 1086 |
| Factors Affecting Shear Resistance of a R.C. Member | 73 |
| Flat Slabs | 284–302 |
| Flexural Bond Stress | 90 |
| Formwork | 1089–1099 |
| Foundations | 987–994 |
| Formwork (IS : 456–2000) | 20 |

### G

| | |
|---|---|
| Grades of Concrete and Characteristic Strength (IS : 456–2000). | 17 |
| General Design Requirements According to Indian Standard Code of Practice (IS : 3370 Part II, 1965) | 533 |
| General Design Requirements | 760 |
| General Notes on Design of Stairs | 217, 938 |
| Group Action in Pile | 435 |

### H

| | |
|---|---|
| Hollow Tile Roof | 241 |
| Hooks and Bends | 91 |
| Hooper Bottom | 673 |

### I

| | |
|---|---|
| I.S. Code Method for Circular Tanks | 543 |

| | |
|---|---|
| Indian Standard Code Recommendations (IS : 1343–1980) | 1058 |
| Indian Standard Recommendations (IS : 456–2000) | 306 |
| Introduction of Axially Loaded Columns | 303 |
| Introduction of Bunker and Silos | 667 |
| Introduction of Columns with Uniaxial and Biaxial Bending | 920 |
| Introduction of Torsion | 101 |
| Introduction of Two Way Slabs | 244 |
| Introduction: Various Types of Bridges | 751 |
| Intze Tank | 587 |
| IS Code Procedure for Finding Moment of Resistance (IS: 456–2000; Annexure-G) | 861 |
| I-Section | 1041 |
| Isolated Footing of Uniform Depth | 358 |
| Isolated Sloped Footing | 361, 992 |
| I.S. Code Methods and Other Methods for Cylindrical Tanks | 551 |
| Impact Effect | 757 |
| Inclined Bars | 76 |
| Indian Code Recommendations (IS : 456–2000) | 285 |
| Indian Standard Code for Design for Torsion (IS : 456–2000) | 523 |
| Indian Standard Code Method (IS : 456–2000) | 261 |
| Indian Standard Code Recommendations (IS : 456–2000) | 78 |
| Indian Standard Code Recommendations for Design of Footings (IS : 456–2000) | 987 |
| Indian Standard Code Recommendations for Design of Footings (IS: 456–2000) | 354 |
| Indian Standard on Design of Bins (IS: 4995–1968) | 675 |
| Indian Standard on Formwork (IS : 456–2000) | 1091 |
| Indian Standard Recommendations on Design for Torsion (IS : 456–2000) | 102 |
| Instantaneous and Repeated Loading on Concrete | 1084 |
| Introduction : Torsional Moments in Beams | 501 |
| Introduction of Combined Footings | 386 |
| Introduction of T and L-Beams | 121 |
| Introduction of Axially Loaded Columns | 910 |
| Introduction of Beams and Slabs | 30 |
| Introduction of Building Frames | 713 |
| Introduction of Chimney's | 688 |
| Introduction of Circular Slabs | 269, 963 |
| Introduction of Combined Direct and Bending Stresses | 320 |
| Introduction of Continuous and Isolated Footings | 350 |
| Introduction of Domes | 490 |
| Introduction of Doubly Reinforced Beams | 107 |
| Introduction of Flat Slabs | 284 |
| Introduction of Formwork | 1090 |
| Introduction of Portal Frames | 703 |
| Introduction of Rectangular Tanks | 627 |
| Introduction of Retaining Walls | 443 |
| Introduction of Shrinkage and Creep | 1080 |
| Introduction of Stair Cases | 216 |
| Introduction of T-Beams and L-Beams | 854 |
| Introduction of Test for Cement and Concrete | 1100 |
| Introduction of Two Way Slabs | 946 |
| Introduction of Underground Tanks | 643 |
| Introduction of Water Tanks | 532 |
| Introduction of Yield Line Theory and Design of Slabs | 969 |
| IS Code on Bond and Anchorage Requirements (IS : 456–2000) | 92 |
| Isolated Footing for Circular Columns | 364 |
| Isolated Footing of Uniform Depth | 989 |
| Isolated Footing Subjected to Eccentric Load | 379 |
| Isolated Stepped Footing | 363 |

**J**

| | |
|---|---|
| Janssen's Theory | 667 |
| Joints in Water Tanks | 535 |

**K**

| | |
|---|---|
| Kern Distances and Efficiency of Section | 1031 |

# INDEX

## L

| | |
|---|---|
| Lateral Loads | 759 |
| Lattice Girder Effect | 77 |
| L-Beam | 142 |
| Lever Arm and Moment of Resistance | 124 |
| Limit State Design | 826 |
| Limit State of Collapse : Compression (As per Clause 39 of IS : 456–2000) | 910 |
| Limit State of Collapse : Shear | 882 |
| Limit State of Collapse in Flexure | 830 |
| Live Load | 753 |
| Load Carrying Capacity of Piles | 433 |
| Loads on Formwork | 1092 |
| Loads on Pipes | 656 |
| Location of Neutral Axis | 108 |
| Longitudinal Forces | 759 |
| Losses of Prestress | 1013 |
| L-Beam | 856 |
| Load Carrying Capacity of Short Columns | 305 |

## M

| | |
|---|---|
| Measurement of Materials | 10 |
| Methods of Computing B.M. | 716 |
| Methods of Proportioning Concrete Mixes | 13 |
| Moment of Resistance | 35 |
| Mechanisms of Shear Transfer in R.C. Beam without Shear Reinforcement | 70 |
| Merits and Demerits of Prestressed Concrete | 1027 |
| Method of Distribution Coefficients | 779 |
| Methods of Design | 826 |
| Modes of Shear Failure | 71 |
| Modular Ratio | 30 |
| Moment Capacity Along an Yield Line | 971 |
| Moment of Resistance Taking Compression in Rib into Account | 125 |
| Moment of Resistance when N.A. Falls in the web | 858 |
| Moment of Resistance when $x_u \leq D_f$ : (As per IS: 456–2000 Annexure-G) | 857 |
| Moment of Resistance | 109 |

## N

| | |
|---|---|
| Nature of Stresses in Spherical Domes | 490 |
| Neutral Axis of Beam Section | 33 |
| Necessity of Doubly Reinforced Sections | 844 |

## O

| | |
|---|---|
| Openings in Flat Slab | 292 |
| Other Cases of Slabs | 255 |

## P

| | |
|---|---|
| Passive Earth Pressure | 446 |
| Pile Driving | 433 |
| Pile Foundations | 432–442 |
| Portal Method | 722 |
| Position of Neutral Axis | 123 |
| Pressure Distribution Beneath Footings | 351 |
| Prestressed Concrete and Miscellaneous Topics-Prestressed Concrete | 996–1079 |
| Prestressing Systems : End Anchorages | 1006 |
| Procedure for Finding Moment of Resistance | 837 |
| Properties and Tests on Concrete | 12 |
| Properties of Materials | 1024 |
| Permissible Stresses in Concrete (IS : 456–2000) | 23 |
| Portal Frames | 703–712 |
| Procedure for Limit State Design | 1069 |

## R

| | |
|---|---|
| Raft Footing | 420 |
| Rectangular Section | 1039 |
| Regions of Cracks in Beams | 69 |
| Reinforced Brick and Hollow Tile Roofs | 238–243 |
| Reinforced Brick Work | 238 |
| Reinforcement Due to Torsion | 520 |
| Reinforcement for Diagonal Tension | 73 |
| Reinforcement Splicing | 99 |
| Restrained Slabs (I.S. Code Method) | 949 |
| Retaining Walls | 443–488 |
| Reinforced Concrete Pipes | 656–666 |

## S

| | |
|---|---|
| Semicircular Beam Simply Supported on Three Equally Spaced Columns | 505 |

| Entry | Page |
|---|---|
| Short and Long (*or* Slender) Columns | 304 |
| Shrinkage and Creep | 1080–1088 |
| Shrinkage of Concrete | 1080 |
| Singly Reinforced Beam | 31 |
| Slab Simply Supported at the Edges with a Central Hole and Carrying W Distributed Along the Circumference of a Concentric Circle | 273 |
| Slab Spanning in One Direction | 63 |
| Specifications for Portland Cement | 6 |
| Steel Reinforcement | 25 |
| Stresses at Anchorage | 1055 |
| Stresses Due to Hydrostatic Pressure | 657 |
| Stresses Due to Self Weight | 657 |
| Stresses Due to Temperature Difference | 692 |
| Stresses Due to Uniform Pressure from Sides | 660 |
| Substitute Frames | 713 |
| Safety and Serviceability Requirements (IS : 456–2000) | 827 |
| Sections of Irregular Shape | 346 |
| Selection of Type of Bridge and Economic Span Length | 752 |
| Semi-circular Beam with Slab | 514 |
| Shear and Bond | 71–100 |
| Shear and Diagonal Tension | 1053 |
| Shear in Flat Slab | 291 |
| Shear Span | 71 |
| Shear Stress in R.C. Beams | 67 |
| Shear Stress, Bond Stress and Development Length | 117 |
| Shear, Bond and Development Length | 128 |
| Shear, Bond and Torsion | 882–891 |
| Short Axially Loaded Column with Minimum Eccentricity | 912 |
| Short Axially Loaded Members in Axial Compression | 911 |
| Short Columns Subjected to Axial Load and Biaxial Bending | 934 |
| Short Columns | 911 |
| Shrinkage Stresses in Singly Reinforced Beams | 1083 |
| Shrinkage Stresses in Symmetrically Reinforced Sections | 1082 |
| Shuttering for Beam and Slab Floor | 1095 |
| Shuttering for Columns | 1093 |
| Simply Supported Slab with Corners Free to Lift (I.S. Code Method) | 946 |
| Singly Reinforced Rectangular Beams | 835 |
| Singly Reinforced Sections | 830–843 |
| Slab Freely Supported at Edges and Carrying U.D.L. | 270, 963 |
| Slab Reinforcement | 292 |
| Slab Simply Supported at Edges, with a Central Hole and Carrying U.D.L. | 272 |
| Slab Simply Supported at Edges, with U.D.L. Inside a Concentric Circle | 272, 964 |
| Slab Simply Supported at the Edges with Load W Uniformly Distributed Along the Circumference of a Concentric Circle | 271, 964 |
| Slab Simply Supported on the Four Edges with Corners Held Down and Carrying U.D.L. | 248 |
| Slab Simply Supported on the Four Edges, with Corners not Held Down and Carrying U.D.L. | 244 |
| Slab with Edges Fixed *or* Continuous and Carrying U.D.L. | 252 |
| Slabs Fixed at Edges and Carrying U.D.L. | 270, 964 |
| Solid Slab Bridges | 762 |
| Stability of Cantilever Retaining Wall | 447 |
| Steel Beam Theory | 110 |
| Strap Footing | 413 |
| Stress Block and Neutral Axis (NA) | 844 |
| Stress Block and Neutral Axis | 857 |
| Stress Block Parameters | 833 |
| Stress Strain Relationship for Concrete | 831 |
| Stresses Due to Earthfill over Haunches | 659 |
| Stresses Due to over-burden and External Loads | 661 |

# INDEX

| | |
|---|---|
| Stresses Due to Point Load on Crown | 661 |
| Stresses Due to Torsion in Concrete Beams | 518 |
| Stresses Due to Triangularly Distributed Load | 660 |
| Stresses Due to Uniformly Distributed Load on Top | 660 |
| Stresses Due to Weight of Water Inside | 659 |
| Stresses Due to Wind Load | 494 |
| Stresses in Chimney Shaft Due to Self-weight and Wind Loads | 689 |
| Stresses in Horizontal Reinforcement Due to Force Shear | 691 |
| Stress-strain Relationship for Steel | 831 |
| Structural Design of R.C. Pile | 436 |
| Sustained Loading : Creep | 1085 |

## T

| | |
|---|---|
| T-Beam | 855 |
| T-Beams and L-Beams | 121–146, 854–881 |
| Temperature Stresses in Horizontal Reinforcement | 695 |
| Test for Soundness of Cement | 1104 |
| Transporting, Placing, Compaction and Curing (IS : 456–2000) | 21 |
| Types of Problems in Singly-Reinforced Beams | 38 |
| Test for Compressive Strength | 1107 |
| Test for Consistency of Cement Paste | 1102 |
| Test for Determination of Compressive Strength of Cement | 1104 |
| Test for Determination of Initial/Final Setting Times | 1103 |
| Test for Flexural Strength | 1108 |
| Test for Tensile Strength of Cement | 1105 |
| Test for Workability | 1106 |
| Test on Fineness of Cement | 1100 |
| Tests for Cement and Concrete | 1100–1109 |
| Theory of Reinforced Beams and Slabs | 30–66 |
| Torsion Factor | 516 |
| Torsion | 101–106 |
| Torsion: Limit State of Collapse | 889 |
| Torsional Resistance : Elastic Behaviour | 101 |
| Two Way Slabs | 244–268, 946–962 |
| Types of Columns | 304 |
| Types of Loads, Forces and Stresses | 753 |
| Types of Piles | 432 |
| Types of Problems in Doubly Reinforced Beams | 110 |
| Types of Problems in T-Beam | 129 |
| Types of Problems | 846, 863 |
| Types of Retaining Walls | 443 |
| Types of Shear Reinforcement | 74 |

## U

| | |
|---|---|
| Ultimate Load on Slabs | 972 |
| Uplift Pressure on the Floor of the Tank | 645 |

## V

| | |
|---|---|
| Variation of $M_r$ with $p$ | 49 |
| Vertical Stirrups | 75 |

## W

| | |
|---|---|
| Water Tanks-II : Circular and Intze Tanks | 547–626 |
| Workability of Concrete | 12 |
| Water | 10 |
| Water Cement Ratio | 11 |
| Water Tanks-III : Rectangular Tanks | 627–642 |
| Water Tanks-IV : Underground Tanks | 643–654 |
| Water Tanks-I : Simple Cases | 532–545 |
| Width of Roadway and Footway | 760 |
| Wind Load | 758 |
| Wind Pressure | 688 |

## Y

| | |
|---|---|
| Yield Line Patterns | 969 |
| Yield Line Theory and Design of Slabs | 969–986 |